Organic Chemistry

Organic Chemistry

SIXTH EDITION

Marc Loudon
PURDUE UNIVERSITY

Jim Parise
UNIVERSITY OF NOTRE DAME

ROBERTS AND COMPANY PUBLISHERS
Greenwood Village, Colorado

Roberts and Company Publishers, Inc.
4950 South Yosemite Street, F2 #197
Greenwood Village, CO 80111 USA
Telephone: (303) 221-3325
Fax: (303) 221-3326
Email: info@roberts-publishers.com
Internet: www.roberts-publishers.com

Publisher: Ben Roberts; Production Editor: Julianna Scott Fein; Manuscript Editor: John Murdzek;
Creative Director: Emiko-Rose Paul; Text and Cover Designer: Jeanne Schreiber; Cover Art: Quade Paul;
Illustrators: Marc Loudon and Quade Paul; Photo Researcher: Sharon Donahue; Proofreader: Kate St.Clair.
The text was set in 10/12 Times LT Std by TECHarts of Colorado and printed on #45 Utopia Silk Thinbook
by Transcontinental Printing.

About the cover: **The Aldol Condensation**

The reaction on the cover is a stylized representation of the aldol condensation, one of the most important reactions in organic chemistry and a very important reaction in biology. The cover reaction is shown in the text as Eq. 22.45 on p. 1122. Research on improvements to the aldol condensation has continued to the present time. Discovery of the aldol condensation was announced independently in 1872 by Charles-Adolphe Wurtz, a widely recognized French chemist, and by Alexandr Borodin, holder of a professorial chair at the Medical-Surgical Academy in St. Petersburg. Borodin was not only a chemist and a surgeon, but was also an accomplished amateur musician and composer whose compositions are widely played to this day. Borodin was an advocate of women's rights, and he also championed medical education for women. (For more on Borodin, see p. 1122.) The cover reaction symbolizes the continuity of organic chemistry from its medical origins to the important position it holds today as both a basic chemical discipline and a foundation of modern biology.

Library of Congress Cataloging-in-Publication Data

Loudon, G. Marc.
  Organic chemistry / Marc Loudon, Purdue University, Jim Parise, University of Notre Dame. -- Sixth edition.
    pages cm
  Includes bibliographical references and index.
  ISBN 978-1-936221-34-9
1. Chemistry, Organic--Textbooks. I. Parise, Jim, 1978- II. Title.

  QD251.3.L68 2016
  547--dc23

                            015014647

Manufactured in Canada

10 9 8 7 6 5 4 3 2 1

To Tarra and Shala . . .
Wives of the sons have become daughters of the father.

and to Julian . . .
His effervescence foretells a prized vintage.
—ML

To Kat, and my family.
—JP

Brief Contents

Contents

9   # The Chemistry of Alkyl Halides   382

22   The Chemistry of Enolate Ions, Enols, and α,β-Unsaturated Carbonyl Compounds   1103

23   The Chemistry of Amines   1183

Preface

A PREVIEW OF THE SIXTH EDITION

We believe, and research in chemical education shows, that students who make the effort to learn but still have trouble in organic chemistry are in many cases trying to memorize their way through the subject. One of the keys to students' success, then, is to provide them with help in relating one part of the subject to the next—to help them see how various reactions that seem very different are tied together by certain fundamentals. An overarching goal of our text is to help students achieve a *relational understanding of organic chemistry*. Here are some of the ways that we have tried to help students meet this goal.

Use of an Acid–Base Framework Is a Key to Understanding Mechanisms

Although we have organized *Organic Chemistry*, Sixth Edition, by functional group, we have used mechanistic reasoning to help students understand the "why" of reactions. Mechanisms alone, however, do not provide the relational understanding that students need. Left to their own devices, many students view mechanisms as something else to memorize, and they are baffled by the "curved-arrow" notation. We believe that an understanding of acid–base chemistry is the key that can unlock the door to a mechanistic understanding of much organic chemistry. We use both Lewis acids and bases and Brønsted acids and bases as the foundations for mechanistic reasoning. Although students have memorized the appropriate definitions in general chemistry, few have developed real insight about the implications of these concepts for a broader range of chemistry. We have dedicated Chapter 3 to these fundamental acid–base concepts. The terms "nucleophile," "electrophile," and "leaving group" then spring easily from Lewis and Brønsted acid–base concepts, and the curved-arrow notation makes sense. We have provided a substantial number of drill problems to test how well students have mastered these principles. We have reinforced these ideas repeatedly with each new reaction type. We also cover free-radical reactions, but not until the electron-pair concepts are fully established.

Tiered Topic Development Provides Reinforcement of Important Ideas

We have introduced complex subjects in "tiers." In other words, students will see many concepts introduced initially in a fairly simple way, then reviewed with another layer of complexity added, and later reviewed again at a greater level of sophistication.

Acid–base chemistry is an example of tiered development. After the initial chapter on acid–base chemistry and the curved-arrow notation, these concepts are revisited in detail as they are used in the early examples of reactions and mechanisms, and again with the introduction of each new reaction type.

The presentation of stereochemistry is another example of the tiered approach. The concept of stereoisomerism is introduced in Chapter 4 (Alkenes). A full chapter on stereochemistry comes two chapters later. Cyclic compounds and the stereochemistry of reactions follow in the next chapter. Then the ideas of group equivalence and nonequivalence are introduced even later, both in the context of enzyme catalysis (Chapter 10) and NMR spectroscopy (Chapter 13).

The approach to organic synthesis is yet another example. We start with simple reactions and then show students how to think about them retrosynthetically. Then, later, we introduce the idea of multistep synthesis using relatively simple two- and three-step sequences. Later still, we have another discussion in which stereochemistry comes into play. Even later, the use of protecting groups is introduced.

Another example is the presentation of resonance. We begin in Chapter 1 with resonance as a simple explanation for hybrid structures. Then, in Chapter 3, students learn to derive resonance structures with the curved-arrow notation, and they learn that resonance has implications for stability. In Chapter 15, the whole issue of resonance is reviewed and the justification of resonance with molecular orbital theory is presented.

This tiered presentation of key topics requires some repetition. Although the repetition of key points might be considered inefficient, we believe that it is crucial to the learning process. It is tempting for the instructor to cover every aspect of a topic when introducing it for the first time. However, in our experience, some students are easily overwhelmed by single, comprehensive presentations. Education research has proven that when material is revisited frequently, with new details being added each time, learning is enhanced and concepts are internalized. When a topic is considered after its first introduction, we have provided detailed cross-referencing to the original material. Students are never cast adrift with terminology that has not been completely defined and reinforced.

Everyday Analogies Help Students to Construct Their Own Knowledge

We believe in the *constructivist* theory of learning, which holds that students construct learning in their own minds by relating each new idea to something they already know. This is why the relational approach to learning organic chemistry is so important. For the same reason, we have provided common analogies from everyday experience for many of the discussions of chemical principles so that students can relate a new idea to something they already know. Many of these are found in illustrated sidebars throughout the text; examples can be found in the sidebars on pp. 163, 165, and 167.

An Increased Focus on Biological Applications Motivates Students Interested in the Allied Health Sciences

Many organic chemistry classes are populated largely by premedical students, prepharmacy students, and other students interested in the life sciences. Research has shown that teaching to the interests of students helps them to learn. One of us (ML) was involved from the beginning in the Howard Hughes Medical Institute (HHMI)–NEXUS program (2010–2014), which sprang from the "Scientific Foundations of Future Physicians" document of the American Association of Medical Colleges and HHMI. This document provided key outcomes that the medical community believed to be desirable in the undergraduate education of premedical students. At the same time that this document appeared, substantial published argumentation also appeared from the medical and biological-science communities that the foundational courses (mathematics, physics, and chemistry, and *especially* organic chemistry) should be substantially revised to incorporate greater medical and biological relevance with a reduced emphasis on synthesis.

We wanted to provide a textbook resource for those who might want to make changes along this line. Therefore, many of the changes in this edition have been to introduce a number of chemical topics in the context of biology, while still retaining the chemical rigor that has become such a staple of organic chemistry courses. To assist those who want to make this transformation, we offer a detailed course outline correlated with readings in the text to show how the text might be used in a more biology-focused organic chemistry course.

Among the unique topics relevant to biology in this edition is a full chapter on noncovalent interactions (Chapter 8) with follow-up topics in Chapter 15. Many sections on coenzymes fall within the chapters most relevant to their chemistry. For example, a section on NAD^+-promoted oxidation occurs after the section on alcohol oxidation, and another section on NADH reductions appears after the section on hydride reductions in aldehyde and ketone chemistry. Another offering to support the biologically oriented course is the all-new Chapter 25 that presents the chemistry of phosphate esters, phosphate anhydrides, and thioesters in the context of their biological importance. Throughout the text, such topics have maintained a chemical focus. Much recent research on the catalytic mechanisms of enzymes and the involvement of coenzymes is tapped for crucial examples to illustrate chemical principles. We have retained and enhanced the section in Chapter 11 on intramolecular reactions to show that intramolecularity is the basis for understanding the spectacular rate accelerations associated with enzyme catalysis. We have retained and enhanced the chapter (now Chapter 27) on amino acids and proteins that was completely rewritten for the last edition in the light of modern developments. We have retained and enhanced our chapters on carbohydrates and nucleic acids. None of these chapters have been exiled to online status.

The Textbook Is a Flexible Resource

While accommodating the need for a biologically focused course, we tried to develop a *flexible* resource. Those who want to teach a more traditional, synthetically oriented course will find everything they need in this text to continue doing so. They can incorporate as much or as little of the biology as desired.

This versatility has resulted in a text that is longer than the previous editions. One cannot simply start at the beginning of the text, go to the end, and stop. Therefore, choices in coverage will have to be made, but we leave these choices up to the instructors.

We hope that this longer text will also provide a valuable reference for students who, after taking organic chemistry, might want to broaden their perspective on the subject with further reading and study.

Solving Problems Is an Essential Component of the Learning Process

All instructors—and successful students of organic chemistry—know that solving problems is a key to learning organic chemistry. We have provided 1782 problems, many of them multipart, ranging from drill problems to problems that will challenge the most astute students. Many are based directly on material in the literature. The 885 problems within the body of the text are typically drill problems that test whether students understand the current material. The 897 problems at the end of the chapters cover material from the entire chapter and in many cases, integrate material from earlier chapters.

In addition, we have interspersed 128 Study Problems throughout the text. Each of these problems has a worked-out solution that carefully shows students the logic involved in the problem-solving process.

We will also provide for instructors a detailed breakdown of problem numbers, correlated with the fifth edition problem numbers for previous adopters, and a list of problems in this edition organized by topic and difficulty.

A Full-Color Presentation Improves Pedagogy

As we designed the art program to make use of color, we have kept one point uppermost in our minds: color should be used solely for functional, pedagogical purposes. Students have enough to worry about without having to figure out what's important in a textbook. The use of color, the presentation of the art, and the text design itself flow from the ideas in *The Psychology of Everyday Things*, a book by Don Norman. The core idea is that these elements in the text should provide subliminal cues to students that facilitate the learning process.

Illustrated Vignettes Add Enrichment

Short illustrated vignettes, or sidebars, serve a variety of purposes. Some provide analogies, some provide historical context, some provide biological relevance, some discuss the Nobel Prize–winning contributions of individual scientists, and some provide insight on current events such as climate change and biofuels. We hope that a few will evoke a laugh or two.

Supplements Provide Additional Help for Both Students and Instructors

1. The *Study Guide and Solutions Manual* presents reaction summaries, solutions to every problem in the text, Study Guide Links, and Further Explorations. The Study Guide Links, which are called out with margin icons in the text, are additional discussions of certain topics with which, in our experience, many students require additional assistance. Examples are "How to Study Organic Reactions" and "Solving Structure Problems." The Further Explorations, also called out with margin icons in the text, are short discussions that move beyond the text material. An example is "Fourier-Transform NMR." Reaction Reviews organize and summarize every reaction's details and mechanisms. The Solutions to Problems section addresses *every* problem given in the text and at the end of every chapter. The solutions are not simply answers; they contain detailed explanations, and they are fully cross-referenced to the main text so that a student can review and discover solutions on their own.

   Professor Jim Parise, teaching professor in organic chemistry at the University of Notre Dame, joins our authoring team in this edition. Jim has assumed primary responsibility for the *Study Guide and Solutions Manual*, has revised Chapters 23 and 24 of the text, and has critically reviewed and assisted in the revision of the remaining chapters.

2. **Molecular models** are available as a bundle with the text. Please contact our editor, Ben Roberts, at bwr@roberts-publishers.com for details.

3. We will maintain an up-to-date list of **errata** for both the text and the *Study Guide and Solutions Manual* supplement on the Internet. These lists of errors will be generally available to instructors and students alike.

4. The **book is offered electronically** inside of Sapling Learning's online homework platform. The platform offers automatic grading, an easy-to-use interface, and instructive feedback. Instructors can select from a variety of existing problem sets or they can modify the questions or author them from scratch. Not only does the software allow students to easily draw and interact with structures, it allows them to draw entire reaction mechanisms, including showing the movement of electrons with curved electron arrows.

   The book will also be offered in other ebook formats. Please visit the book's site at www.roberts-publishers.com for the relevant and latest details.

5. We also offer a host of other **instructor supplements**, including PowerPoint slides, and the art and photos in JPEG format. Please contact our editor, Ben Roberts, at bwr@roberts-publishers.com for details.

A BOOK WITH A SCHOLARLY HISTORY

The first edition of *Organic Chemistry*, published in 1984, required seven and a half years of development, because each topic was researched back to the original or review literature. Subsequent editions, including this one, have continued this scholarly development process. Almost every reaction example is taken from the literature. Each edition has benefited from a thorough peer review.

ACKNOWLEDGMENTS

The electronic resources of the Purdue Library have streamlined the research process for this text in a way that was unimaginable 25 years ago, and we thank Emily Mobley, Purdue's Dean of Libraries Emerita, for bringing the electronic library to fruition and Jim Mullins, the present dean, for fostering its continued improvement. We are grateful to Jeremy Garritano, former Chemistry Librarian (now at the University of Maryland), and Vicki Killion, Health Sciences Librarian, for their crucial assistance throughout this project. Thanks to ML's faculty and staff colleagues at Purdue and beyond—John Grutzner, George Bodner, Don Bergstrom, Mark Green, Chris Rochet, Carol Post, Rodolfo Pinal, Markus Lill, David Nichols, Mark Cushman, Casey Krusemark, and Karl Wood, for advice, assistance, and suggestions. ML's faculty colleagues in the HHMI–NEXUS program, professors Chris Hrycyna, Jean Chmielewski, and Marcy Towns, have provided special inspiration for some of the changes in this text. Particular thanks go to a very special colleague, Professor Animesh Aditya, who is co-teaching organic chemistry with ML, and who has provided detailed suggestions and reviews for many of the chapters in both the text and the solutions manual. The reviewers named in the list that follows this preface provided invaluable assistance in polishing this text. We would also like to thank the many students and faculty from all over the country who made suggestions, offered comments, and reported errors. We welcome correspondence with the students using this edition. We can be reached by email at loudonm@purdue.edu and jparise@nd.edu.

Ben Roberts at Roberts and Company Publishers has not only been an exceptionally energetic publisher, but a good friend. Our relationship with the composition professionals at TechArts in Boulder, Colorado, led by Kathi Townes, has been particularly gratifying. We very much appreciate the hard work and assistance of the project manager, Julianna Scott Fein; the advice and attention to detail of the copyeditor, John Murdzek; the skilled work of our exceptionally able proofreader, Kate St.Clair; and the resourcefulness of our photo and permissions editor, Sharon Donahue. We would also like to thank those acknowledged separately in the credits section for their willingness to allow me to reproduce their materials.

ML could not have completed this project without the love and support of Judy and his family, for which he is grateful beyond words.

JP would like to thank his wife, Kathryn, and his family for their support and encouragement. He is indebted to his own educators and his current and former colleagues at Oswego State, Duke, and Notre Dame. He is especially grateful to work with and learn from his coauthor ML and for his trust and guidance.

Our wish is that the students who use this text will see the amazing diversity and beauty of science through their study of organic chemistry, and that they will benefit from using this book as much as we have enjoyed writing it.

MARC LOUDON
April 2015
West Lafayette, Indiana

JIM PARISE
April 2015
Notre Dame, Indiana

Reviewers and Consultants

The author and publisher wish to acknowledge with gratitude the extensive support received from the organic chemistry community in the development of this edition. Our reviewers and consultants are listed below; some people served in multiple roles.

Sixth Edition Reviewers

Animesh V. Aditya, Purdue University
Igor Alabugin, Florida State University
John Bartmess, University of Tennessee
Peter Beak, University of Illinois
Jason Belitsky, Oberlin College
Thomas Berke, Brookdale, The County College of Monmouth
Daniel Bernier, Riverside Community College
Michael Best, University of Tennessee
Caitlin Binder, California State University, Monterey Bay
Dan Blanchard, Kutztown University of Pennsylvania
Lisa Bonner, Eckerd College
Paul Bonvallet, College of Wooster
Ned B. Bowden, The University of Iowa
Stephen G. Boyes, Colorado School of Mines
David Brown, Davidson College
Rebecca Broyer, University of Southern California
Paul Carlier, Virginia Tech
David Cartrette, South Dakota State University
Allen Clauss, University of Wisconsin
Geoffrey W. Coates, Cornell University
Bryan Cowen, University of Denver
Michael Danahy, Bowdoin College
William Daub, Harvey Mudd College
William Dichtel , Cornell University
Sally Dixon, University of Southampton
Andrew Duncan, Willamette University
Jason Dunham, Ball State University
Mark Elliott, Cardiff University
Brian Esselman, University of Wisconsin-Madison
Ed Fenlon, Franklin & Marshall College
Marcia France, Washington and Lee University
Lee Friedman, University of Maryland
Bruce Ganem, Cornell University
Sarah Goh, Williams College
Bobbie Grey, Riverside City College
Nicholas J. Hill, University of Wisconsin-Madison

John Hoberg, University of Wyoming
Joseph Houck, University of Maryland
Lyle Isaacs, University of Maryland
Madeleine Joullie, University of Pennsylvania
Sarah Kirk, Willamette University
Riz Klausmeyer, Baylor University
Brian Long, University of Tennessee
Leonard MacGillivray, University of Iowa
Charles Marth, Western Carolina University
Dan Mattern, University of Mississippi
James McKee, University of the Sciences in Philadelphia
John Medley, Centre College
Kevin P. C. Minbiole, Villanova University
Timothy Minger, Mesa Community College
Michael P. Montague-Smith, University of Maryland
Michael Nee, Oberlin College
Donna Nelson, University of Oklahoma
James Nowick, University of California, Irvine
Kimberly Pacheco, University of Northern Colorado
Laura Parmentier, Beloit College
Joshua Pierce, North Carolina State University
David P. Richardson, Williams College
Robert E. Sammelson, Ball State University
Paul Sampson, Kent State University
Nicole L. Snyder, Davidson College
Gary Spessard, University of Oregon
Brian M. Stoltz, California Institute of Technology
Scott Stoudt, Coe College
Jennifer Swift, Georgetown University
Eric Tillman, Bucknell University
Mark M. Turnbull, Clark University
David A. Vosburg, Harvey Mudd College
Ross Weatherman, Rose-Hulman Institute of Technology
Carolyn Kraebel Weinreb, Saint Anselm College
Travis Williams, University of Southern California
Jimmy Wu, Dartmouth College
Hubert Yin, University of Colorado

Fifth Edition Editorial Consultants

The following editorial consultants read chapters, evaluated the art for style and pedagogy, and advised on many important, large-scale editorial decisions.

Carolyn Anderson, Calvin College
John Bartmess, University of Tennessee

Marcia B. France, Washington and Lee University
Robert Hammer, Louisiana State University
David Hansen, Amherst College
Ahamindra Jain, Harvard University
James Nowick, University of California, Irvine
Paul R. Rablen, Swarthmore College

Fifth Edition General Reviewers

The following general reviewers helped us generate a detailed plan for changes from the fourth edition to the fifth edition.

Angela Allen, University of Michigan, Dearborn
Don Bergstrom, Purdue University
David Brown, Davidson College
Scott Bur, Gustavus Adolphus College
Joyce Blair Easter, Virginia Wesleyan University
Tom Evans, Denison College
Natia Frank, University of Victoria
Phillip Fuchs, Purdue University
Ronald Magid, University of Tennessee
Tim Minger, Mesa Community College
William Ojala, University of St. Thomas
Andy Phillips, University of Colorado, Boulder
Jetze J. Tepe, Michigan State University
Scott Ulrich, Ithaca College

Fifth Edition Chapter Reviewers

The following reviewers read chapters in detail.

Carolyn E. Anderson, Calvin College
John Bartmess, University of Tennessee

Marcia B. France, Washington and Lee University
David Hansen, Amherst College
Ahamindra Jain, Harvard University
Joseph Konopelski, University of California, Santa Cruz
Paul LePlae, Wabash College
Dewey G. McCafferty, Duke University (and his students)
Nasri Nesnas, Florida Institute of Technology
Paul R. Rablen, Swarthmore College
Christian M. Rojas, Barnard College
Barry Snider, Brandeis University
Scott A. Snyder, Columbia University
Dasan M. Thamattoor, Colby College
Lawrence T. Scott, Boston College
Scott Ulrich, Ithaca College
Richard G. Weiss, Georgetown University

Fifth Edition Focus Group Participants

The following consultants took part in an important focus group at which we discussed the price of textbooks, the trends in student learning, and the value of supplements.

David Hansen, Amherst College
Robert Hanson, Northeastern University
Ahamindra Jain, Harvard University
Cynthia McGowan, Merrimack College
Eriks Rozners, Northeastern University
Bela Torok, University of Massachusetts, Boston

About the Authors

Marc Loudon received his BS (magna cum laude) in chemistry from Louisiana State University in Baton Rouge and his PhD in organic chemistry in 1968 from the University of California, Berkeley, where he worked with Professor Don Noyce. After a two-year postdoctoral fellowship with Professor Dan Koshland in biochemistry at Berkeley, Dr. Loudon joined the Department of Chemistry at Cornell University, where he taught organic chemistry to both preprofessional students and science majors. In 1977, he joined the Department of Medicinal Chemistry and Pharmacognosy (now the Department of Medicinal Chemistry and Molecular Pharmacology) in the College of Pharmacy at Purdue University, where he has taught organic chemistry to prepharmacy students. Dr. Loudon served as the associate dean for Research and Graduate Programs for the College of Pharmacy, Nursing, and Health Sciences from 1988 to 2007. He has received numerous teaching awards: the Clark Award of the College of Arts and Sciences at Cornell (1976); the Heine Award of the School of Pharmacy at Purdue (1980 and 1985); Purdue's Class of 1922 Helping Students Learn Award (1988); the Charles B. Murphy Award that recognizes the best teachers at Purdue (1999); and the Indiana "Professor of the Year" Award of the Carnegie Foundation (2000). Dr. Loudon was named the Gustav E. Cwalina Distinguished Professor in 1996, one of the first three faculty members to be recognized by Purdue as distinguished professors for teaching and teaching scholarship. He was inducted into Purdue's Teaching Academy in 1997, and he was listed in Purdue's permanent "Book of Great Teachers" in 1999.

In collaboration with Professor George Bodner, and more recently, Prof. Animesh Aditya, Dr. Loudon has developed and implemented collaborative-learning methods for large organic chemistry classes from 1993 to the present. More recently, he was part of the HHMI-sponsored NEXUS project from 2010–2014, which resulted in the complete redesign of the first organic chemistry course to make it more relevant to the interests of students in pre–health profession programs. Dr. Loudon was named Cwalina Distinguished Professor Emeritus following his retirement from Purdue in 2015. Dr. Loudon is the author of numerous research articles, co-author of an in-house laboratory manual, and he is the co-author of the *Study Guide and Solutions Manual* for this text. The first edition of this text was published in 1984.

Dr. Loudon and his wife, Judy, celebrated their 50th anniversary in 2014, and have two grown sons and four grandchildren ranging in age from 6 to 20. Dr. Loudon is an accomplished pianist and organist who performs professionally. Dr. Loudon and his wife both enjoy tennis and travel.

Jim Parise received his BS in chemistry from the State University of New York at Oswego in 2000 and his PhD in organic chemistry in 2007 from Duke University, where he worked with Professor Eric Toone. After a postdoctoral fellowship with Professor David Lawrence at the University of North Carolina at Chapel Hill, Dr. Parise joined the Department of Chemistry at Duke University, where he taught organic chemistry to both preprofessional students and science majors, and coordinated the organic chemistry laboratory courses. In 2011, he joined the Department of Chemistry and Biochemistry at the University of Notre Dame. He teaches primarily to preprofessional students and also oversees the pedagogy of the accompanying laboratory program. Recently he was awarded the Thomas P. Madden award for outstanding teaching of first-year students at Notre Dame (2015). His research focuses on pedagogical techniques and the integration of classroom technology. He has also developed a teaching mentorship program for new instructors. Dr. Parise has authored and co-authored laboratory manuals and a peer-reviewed book chapter on writing in the laboratory, and he is co-author of the *Study Guide and Solutions Manual* for this text.

Organic Chemistry

Chemical Bonding and Chemical Structure

1.1   INTRODUCTION

A. What Is Organic Chemistry?

Organic chemistry is the branch of science that deals generally with compounds of carbon. There are roughly 15 *million* known organic compounds. The number of *possible* organic compounds is essentially infinite. But all of these have one thing in common: they contain carbon. The large number of organic compounds results from carbon's singular ability to combine with other carbons to form long chains (as you will learn).

B. How Is Organic Chemistry Useful?

This is the structure of the anticancer drug imatinib (marketed as Gleevec®).

imatinib (Gleevec®)

Chances are, you might not know how to interpret this structure, but you will learn. Imatinib illustrates the utility of organic chemistry. Prior to 2001, a medical diagnosis of chronic myelogenous leukemia (CML), a relatively uncommon cancer of the blood and bone marrow, was a death sentence. However, an oncologist, Brian Drucker, and a biochemist, Nicholas Lydon, using results on the genetic basis of CML, were able to screen a number of organic compounds for their ability to inhibit a key enzyme, *bcr–abl*. This means that they found a way to block the progression of CML. Working with organic chemists, they produced analogs of their successful compounds (compounds of similar structure), ultimately landing on imatinib as their drug of choice. Clinical trials, conducted with physician Charles Sawyers, led to approval of imatinib in 2001 by the Food and Drug Administration (FDA). Imatinib cured CML in most cases, and it has proven useful in the treatment of other cancers as well. The ability to rapidly prepare and characterize a large number of organic compounds was crucial in the development of this drug, and this required a knowledge of organic chemistry. In the larger picture, the alliance of organic chemistry, molecular biology, and medicine is clearly the way of the future in the development of effective new drugs, because *most drugs are organic compounds.*

By the time you have worked your way through this text, you will certainly understand the structure and chemical properties of imatinib and many other important molecules. Should you want to participate in the excitement of drug discovery, you will be prepared for advanced work that can set you on that road. If you are headed for a career as a practicing health-care professional, you will be prepared to understand the chemical basis of biochemistry, which is fundamental to all life sciences.

Organic chemistry is not only useful in medicine. Many useful materials come from organic chemistry. Things as diverse as textiles, body armor, artificial sweeteners, sports equipment, and computers are materials, or are based on materials, that come from organic chemistry.

Apart from its practical utility, organic chemistry is an intellectual discipline that has both theoretical and experimental aspects. You can use the study of organic chemistry to develop and apply basic skills in problem solving and, at the same time, to learn a subject of immense practical value. Whether your goal is to be a professional chemist, to remain in the mainstream of a health profession, or to be a well-informed citizen in a technological age, you will find value in the study of organic chemistry.

In this text we have several objectives. We'll present the "nuts and bolts"—the nomenclature, classification, structure, and properties of organic compounds. We'll also cover the principal reactions and the syntheses of organic molecules. But, more than this, we'll develop underlying principles that allow us to understand, and sometimes to predict, reactions *rather than simply memorizing them.* We'll bring some order to the rather daunting array of 15 million organic molecules, their reactions, and their properties. Along the way, we'll continue to highlight some of the important applications of organic chemistry in medicine, industry, and other areas.

C. The Emergence of Organic Chemistry

Although the applications of organic chemistry, as we have seen, are not restricted to the life sciences, the name *organic* certainly implies a connection to living things. In fact, the emergence of organic chemistry as a science was closely associated with the evolution of the life sciences.

As early as the sixteenth century, scholars seem to have had some realization that the phenomenon of life has chemical attributes. Theophrastus Bombastus von Hohenheim, a Swiss physician and alchemist (ca. 1493–1541) better known as Paracelsus, sought to deal with medicine in terms of its "elements" mercury, sulfur, and salt. An ailing person was thought to be deficient in one of these elements and therefore in need of supplementation with the missing substance. Paracelsus was said to have effected some dramatic "cures" based on this idea.

By the eighteenth century, chemists were beginning to recognize the chemical aspects of life processes in a modern sense. Antoine Laurent Lavoisier (1743–1794) recognized the similarity of respiration to combustion in the uptake of oxygen and expiration of carbon dioxide.

At about the same time, it was found that certain compounds are associated with living systems and that these compounds generally contain carbon. They were thought to have arisen from, or to be a consequence of, a "vital force" responsible for the life process. The term *organic* was applied to substances isolated from living things by a Swedish chemist Jöns

Jacob Berzelius (1779–1848). Somehow, the fact that these chemical substances were organic in nature was thought to put them beyond the scope of the experimentalist. The logic of the time seems to have been that life is not understandable; organic compounds spring from life; therefore, organic compounds are not understandable.

The barrier between organic (living) and inorganic (nonliving) chemistry began to crumble in 1828 because of a serendipitous (accidental) discovery by Friedrich Wöhler (1800–1882), a German analyst originally trained in medicine. When Wöhler heated ammonium cyanate, an inorganic compound, he isolated urea, a known urinary excretion product of mammals.

$$\text{(1.1)}$$

Wöhler recognized that he had synthesized this biological material "without the use of kidneys, nor an animal, be it man or dog." Not long thereafter followed the synthesis of other organic compounds, acetic acid by Hermann Kolbe in 1845, and acetylene and methane by Marcellin Berthelot in the period 1856–1863. Although "vitalism" was not so much a widely accepted formal theory as an intuitive idea that something might be special and beyond human grasp about the chemistry of living things, Wöhler did not identify his urea synthesis with the demise of the vitalistic idea; rather, his work signaled the start of a period in which the synthesis of so-called organic compounds was no longer regarded as something outside the province of laboratory investigation. Organic chemists now investigate not only molecules of biological importance, but also intriguing molecules of bizarre structure and purely theoretical interest. Thus, organic chemistry deals with compounds of carbon regardless of their origin. Wöhler seems to have anticipated these developments when he wrote to his mentor Berzelius, "Organic chemistry appears to be like a primeval tropical forest, full of the most remarkable things."

1.2   CLASSICAL THEORIES OF CHEMICAL BONDING

To understand organic chemistry, it is necessary to have some understanding of the **chemical bond**—the forces that hold atoms together within molecules. First, we'll review some of the older, or "classical," ideas of chemical bonding—ideas that, despite their age, remain useful today. Then, in the last part of this chapter, we'll consider more modern ways of describing the chemical bond.

A. Electrons in Atoms

Chemistry happens because of the behavior of electrons in atoms and molecules. The basis of this behavior is the arrangement of electrons within atoms, an arrangement suggested by the periodic table. Consequently, let's first review the organization of the periodic table (see page facing inside back cover). The shaded elements are of greatest importance in organic chemistry; knowing their atomic numbers and relative positions will be valuable later on. For the moment, however, consider the following details of the periodic table because they were important in the development of the concepts of bonding.

A neutral atom of each element contains a number of both protons and electrons equal to its atomic number. The periodic aspect of the table—its organization into groups of elements with similar chemical properties—led to the idea that electrons reside in layers, or shells, about the nucleus. The outermost shell of electrons in an atom is called its **valence shell**, and the electrons in this shell are called **valence electrons**. *The number of valence electrons for any neutral atom in an A group of the periodic table* (except helium) *equals its group number.* Thus, lithium, sodium, and potassium (Group 1A) have one valence electron, whereas carbon (Group 4A) has four, the halogens (Group 7A) have seven, and the noble gases (except helium) have eight. Helium has two valence electrons.

Walter Kossel (1888–1956) noted in 1916 that when atoms form ions they tend to gain or lose valence electrons so as to have the same number of electrons as the noble gas of closest atomic number. Thus, potassium, with one valence electron (and 19 total electrons), tends to lose an electron to become K^+, the potassium ion, which has the same number of electrons (18) as the nearest noble gas (argon). Chlorine, with seven valence electrons (and 17 total electrons) tends to accept an electron to become the 18-electron chloride ion, Cl^-, which also has the same number of electrons as argon. Because the noble gases have an octet of electrons (that is, eight electrons) in their valence shells, the tendency of atoms to gain or lose valence electrons to form ions with the noble-gas configuration has been called the **octet rule**.

B. The Ionic Bond

A chemical compound in which the component atoms exist as ions is called an **ionic compound**. Potassium chloride, KCl, is a common ionic compound. The electronic configurations of the potassium and chloride ions obey the octet rule.

The structure of crystalline KCl is shown in Fig. 1.1. In the KCl structure, which is typical of many ionic compounds, each positive ion is surrounded by negative ions, and each negative ion is surrounded by positive ions. The crystal structure is stabilized by an interaction between ions of opposite charge. Such a stabilizing interaction between opposite charges is called an **electrostatic attraction**. An electrostatic attraction that holds ions together is called an **ionic bond**. Thus, the crystal structure of KCl is maintained by ionic bonds between potassium ions and chloride ions. The ionic bond is the same in all directions; that is, a positive ion has the same attraction for each of its neighboring negative ions, and a negative ion has the same attraction for each of its neighboring positive ions.

When an ionic compound such as KCl dissolves in water, it dissociates into free ions (each surrounded by water). (We'll consider this process further in Sec. 8.6F.) Each potassium ion moves around in solution more or less independently of each chloride ion. The conduction of electricity by KCl solutions shows that the ions are present. Thus, the ionic bond is broken when KCl dissolves in water.

To summarize, the ionic bond

1. is an electrostatic attraction between ions;
2. is the same in all directions—that is, it has no preferred orientation in space; and
3. is broken when an ionic compound dissolves in water.

PROBLEMS*

1.1   How many valence electrons are found in each of the following species?
(a) Na   (b) Ca   (c) O^{2-} (d) Br^+

1.2   When two different species have the same number of electrons, they are said to be *isoelectronic*. Name the species that satisfies each of the following criteria:
(a) the singly charged negative ion isoelectronic with neon
(b) the singly charged positive ion isoelectronic with neon
(c) the dipositive ion isoelectronic with argon
(d) the neon species that is isoelectronic with neutral fluorine

---

C. The Covalent Bond

Many compounds contain bonds that are very different from the ionic bond in KCl. Neither these compounds nor their solutions conduct electricity. This observation indicates that these compounds are *not* ionic. How are the bonding forces that hold atoms together in such compounds different from those in KCl? In 1916, G. N. Lewis (1875–1946), an American physical

---

* The solutions to all problems can be found in the *Study Guide and Solutions Manual* supplement.

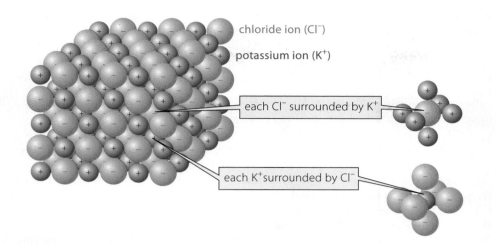

chloride ion (Cl⁻)

potassium ion (K⁺)

each Cl⁻ surrounded by K⁺

each K⁺ surrounded by Cl⁻

FIGURE 1.1 Crystal structure of KCl. The potassium and chlorine are present in this substance as K^+ and Cl^- ions, respectively. The ionic bond between the potassium ions and chloride ions is an electrostatic attraction. Each positive ion is surrounded by negative ions, and each negative ion is surrounded by positive ions. Thus, the attraction between ions in the ionic bond is the same in all directions.

chemist, proposed an electronic model for bonding in nonionic compounds. According to this model, the chemical bond in a nonionic compound is a **covalent bond**, which consists of an electron pair that is *shared* between bonded atoms. Let's examine some of the ideas associated with the covalent bond.

Lewis Structures   One of the simplest examples of a covalent bond is the bond between the two hydrogen atoms in the hydrogen molecule.

$$H \colon\! H \qquad H\!-\!H$$

covalent bond

The symbols " $\colon$ " and "$-$" are both used to denote an electron pair. *A shared electron pair is the essence of the covalent bond.* Molecular structures that use this notation for the electron-pair bond are called **Lewis structures**. In the hydrogen molecule, an electron-pair bond holds the two hydrogen atoms together. Conceptually, the bond can be envisioned to come from the pairing of the valence electrons of two hydrogen atoms:

$$H\cdot \; + \; H\cdot \; \longrightarrow \; H\colon\!H \tag{1.2}$$

Both electrons in the covalent bond are shared equally between the hydrogen atoms. Even though electrons are mutually repulsive, bonding occurs because the electron of each hydrogen atom is attracted to both hydrogen nuclei (protons) simultaneously.

An example of a covalent bond between two different atoms is provided by methane (CH_4), the simplest stable organic molecule. We can form methane conceptually by pairing each of the four carbon valence electrons with a hydrogen valence electron to make four C—H electron-pair bonds.

$$\cdot\ddot{C}\cdot \; + \; 4\,H\cdot \; \longrightarrow \; H\!:\!\overset{\overset{\displaystyle H}{\cdot\cdot}}{\underset{\underset{\displaystyle H}{\cdot\cdot}}{C}}\!:\!H \quad \text{or} \quad H\!-\!\overset{\displaystyle H}{\underset{\displaystyle H}{\overset{|}{\underset{|}{C}}}}\!-\!H \quad \text{or} \quad CH_4 \tag{1.3}$$

In the previous examples, all valence electrons of the bonded atoms are shared. In some covalent compounds, such as water (H_2O), however, some valence electrons remain unshared. In the water molecule, oxygen has six valence electrons. Two of these combine with hydrogens to make two O—H covalent bonds; four of the oxygen valence electrons are left over. These are represented in the Lewis structure of water as electron pairs on the oxygen. In general, unshared valence electrons in Lewis structures are depicted as paired dots and referred to as **unshared pairs** or **lone pairs**.

$$\text{unshared pairs} \quad \text{H} — \ddot{\text{O}} — \text{H}$$

Although we often write water as H—O—H, or even H_2O, it is a good habit to indicate all unshared pairs with paired dots until you remember instinctively that they are there.

The foregoing examples illustrate an important point: *The sum of all shared and unshared valence electrons around each atom in many stable covalent compounds is eight (two for the hydrogen atom).* This is the **octet rule** for covalent bonding. *The octet rule will prove to be extremely important for understanding chemical reactivity.* It is reminiscent of the octet rule for ion formation (Sec. 1.2A), except that in ionic compounds, valence electrons belong completely to a particular ion. In covalent compounds, *shared electrons are counted twice*, once for each of the sharing partner atoms.

Let's see how the covalent compounds we've just considered follow the octet rule. In the structure of methane (Eq. 1.3), four shared pairs surround the carbon atom—that is, eight shared electrons, an octet. Each hydrogen shares two electrons, the "octet rule" number for hydrogen. Similarly, the oxygen of the water molecule has four shared electrons and two unshared pairs for a total of eight, and again the hydrogens have two shared electrons.

Two atoms in covalent compounds may be connected by more than one covalent bond. The following compounds are common examples:

ethylene

formaldehyde

$$\text{H} — \text{C} :: \text{C} — \text{H} \quad \text{or} \quad \text{H} — \text{C} \equiv \text{C} — \text{H}$$

acetylene

Ethylene and formaldehyde each contain a **double bond**—a bond consisting of two electron pairs. Acetylene contains a **triple bond**—a bond involving three electron pairs.

Covalent bonds are especially important in organic chemistry because *all organic molecules contain covalent bonds.*

Formal Charge   The Lewis structures considered in the previous discussion are those of neutral molecules. However, many familiar ionic species, such as $[SO_4]^{2-}$, $[NH_4]^+$, and $[BF_4]^-$, also contain covalent bonds. Consider the tetrafluoroborate anion, which contains covalent B—F bonds:

tetrafluoroborate ion

Because the ion bears a negative charge, one or more of the atoms within the ion must be charged—but which one(s)? The rigorous answer is that the charge is shared by all of the atoms. However, chemists have adopted a useful and important procedure for electronic book-keeping that assigns a charge to specific atoms. The charge on each atom thus assigned is called its **formal charge**. The sum of the formal charges on the individual atoms must equal the total charge on the ion.

Computation of formal charge on an atom involves dividing the total number of valence electrons between the atom and its bonding partners. Each atom receives *all* of its unshared electrons and *half* of its bonding electrons. To assign a formal charge to an atom, then, use the following procedure:

1. Write down the *group number* of the atom from its column heading in the periodic table. This is equal to the number of valence electrons in the *neutral atom.*

2. Determine the *valence electron count* for the atom by adding the number of unshared valence electrons on the atom to the number of covalent bonds to the atom. Counting the covalent bonds in effect adds *half* the bonding electrons—one electron for each bond.

3. Subtract the valence electron count from the group number. The result is the formal charge.

This procedure is illustrated in Study Problem 1.1.

STUDY PROBLEM   1.1

Assign a formal charge to each of the atoms in the tetrafluoroborate ion, $[BF_4]^-$, which has the structure shown above.

SOLUTION   Let's first apply the procedure outlined above to fluorine:

> *Group number of fluorine:*   7
> *Valence-electron count:*   7
> (Unshared pairs contribute 6 electrons; the covalent bond contributes 1 electron.)
> *Formal charge on fluorine:* Group number − Valence-electron count = 7 − 7 = 0

Because all fluorine atoms in $[BF_4]^-$ are equivalent, they all must have the same formal charge—zero. It follows that the boron must bear the formal negative charge. Let's compute it to be sure.

> *Group number of boron:*   3
> *Valence-electron count:*   4
> (Four covalent bonds contribute 1 electron each.)
> *Formal charge on boron:* Group number − Valence-electron count = 3 − 4 = −1

STUDY GUIDE LINK 1.1*
Formal Charge

Because the formal charge of boron is −1, the structure of $[BF_4]^-$ is written with the minus charge assigned to boron:

$$:\!\ddot{F}\!:$$
$$|$$
$$:\!\ddot{F}\!-\!\overset{-}{B}\!-\!\ddot{F}\!:$$
$$|$$
$$:\!\ddot{F}\!:$$

When indicating charge on a compound, we can show the formal charges on each atom, or we can show the formal charge on the ion as a whole, *but we should not show both.*

formal charge   overall charge   don't show both

Rules for Writing Lewis Structures The previous two sections can be summarized in the following rules for writing Lewis structures.

1. Hydrogen can share no more than two electrons.
2. The sum of all bonding electrons and unshared pairs for atoms in the second period of the periodic table—the row beginning with lithium—is never greater than eight (octet rule). These atoms may, however, have fewer than eight electrons.
3. In some cases, atoms beyond the second period of the periodic table may have more than eight electrons. However, rule 2 should also be followed for these cases until exceptions are discussed later in the text.
4. Nonvalence electrons are not shown in Lewis structures.
5. The formal charge on each atom is computed by the procedure illustrated in Study Problem 1.1 and, if not equal to zero, is indicated with a plus or minus sign on the appropriate atom(s).

Here's something very important to notice: *There are two types of electron counting.* When we want to know whether an atom has a complete octet, we count all unshared valence electrons and *all bonding electrons* (rule 2 in the previous list). When we want to determine formal charge, we count all unshared valence electrons and *half of the bonding electrons.*

STUDY PROBLEM 1.2

Draw a Lewis structure for the covalent compound methanol, CH_4O. Assume that the octet rule is obeyed, and that none of the atoms have formal charges.

SOLUTION For carbon to be *both* neutral *and* consistent with the octet rule, it must have four covalent bonds:

$$-\overset{\displaystyle |}{\underset{\displaystyle |}{C}}-$$

There is also only one way each for oxygen and hydrogen to have a formal charge of zero and simultaneously not violate the octet rule:

$$-\ddot{\overset{..}{O}}- \qquad H-$$

If we connect the carbon and the oxygen, and fill in the remaining bonds with hydrogens, we obtain a structure that meets all the criteria in the problem:

$$H-\overset{\displaystyle \overset{\textstyle H}{|}}{\underset{\displaystyle \underset{\textstyle H}{|}}{C}}-\ddot{\overset{..}{O}}-H$$

correct structure of methanol

PROBLEMS

1.3 Draw a Lewis structure for each of the following species. Show all unshared pairs and the formal charges, if any. Assume that bonding follows the octet rule in all cases.

(a) $HCCl_3$ (b) NH_3 (c) $[NH_4]^+$ (d) $[H_3O]^+$
    ammonia   ammonium ion

1.4 Write two reasonable Lewis structures corresponding to the formula C_2H_6O. Assume that all bonding adheres to the octet rule, and that no atom bears a formal charge.

1.5   Compute the formal charges on each atom of the following structures. In each case, what is the charge on the entire structure?

(a)

$$\text{H}-\overset{\displaystyle \underset{|}{\overset{|}{\text{H}}}\text{H}}{\underset{\displaystyle \overset{|}{\text{H}}\ \overset{|}{\text{H}}}{\text{B}-\text{N}}}-\text{H}$$

(b)

$$:\!\ddot{\text{O}}\!-\!\text{P}\!-\!\ddot{\text{O}}\!-\!\text{H}$$

D. The Polar Covalent Bond

In many covalent bonds the electrons are not shared equally between two bonded atoms. Consider, for example, the covalent compound hydrogen chloride, HCl. (Although HCl dissolves in water to form H_3O^+ and Cl^- ions, in the gaseous state pure HCl is a covalent compound.) The electrons in the H—Cl covalent bond are *unevenly* distributed between the two atoms; they are polarized, or "pulled," toward the chlorine and away from the hydrogen. A bond in which electrons are shared unevenly is called a **polar bond**. The H—Cl bond is an example of a polar bond.

How can we determine whether a bond is polar? Think of the two atoms at each end of the bond as if they were engaging in a tug-of-war for the bonding electrons. *The tendency of an atom to attract electrons to itself in a covalent bond is indicated by its* **electronegativity**. The electronegativities of a few elements that are important in organic chemistry are shown in Table 1.1. Notice the trends in this table. Electronegativity increases to the top and to the right of the table. The more an atom attracts electrons, the more **electronegative** it is. Fluorine is the most electronegative element. Electronegativity decreases to the bottom and to the left of the periodic table. The less an atom attracts electrons, the more **electropositive** it is. Of the common stable elements, cesium is the most electropositive.

TABLE 1.1   Average Pauling Electronegativities of Some Main-Group Elements

Increasing electronegativity →

Increasing electropositivity ↓ (left margin)

Increasing electronegativity ↑ (right margin)

			H 2.20			
Li 0.98	Be 1.57	B 2.04	C 2.55	N 3.04	O 3.44	F 3.98
Na 0.93	Mg 1.31	Al 1.61	Si 1.90	P 2.19	S 2.58	Cl 3.16
K 0.82	Ca 1.00				Se 2.55	Br 2.96
Rb 0.82						I 2.66
Cs 0.79						

← Increasing electropositivity

If two bonded atoms have equal electronegativities, then the bonding electrons are shared equally. But if two bonded atoms have considerably different electronegativities, then the electrons are unequally shared, and the bond is polar. (We might think of a polar covalent bond as a covalent bond that is trying to become ionic!) Thus, *a polar bond is a bond between atoms with significantly different electronegativities.*

Sometimes we indicate the polarity of a bond in the following way:

$$\overset{\delta+}{H}\text{—}\overset{\delta-}{Cl}$$

In this notation, the delta (δ) is read as "partially" or "somewhat," so that the hydrogen atom of HCl is "partially positive," and the chlorine atom is "partially negative."

Another more graphical way that we'll use to show polarities is the *electrostatic potential map.* An **electrostatic potential map (EPM)** of a molecule starts with a map of the total electron density. This is a picture of the spatial distribution of the electrons in the molecule that comes from molecular orbital theory, which we'll learn about in Sec. 1.8. Think of this as a picture of "where the electrons are." The EPM is a map of total electron density that has been color-coded for regions of local positive and negative charge. Areas of greater negative charge are colored red, and areas of greater positive charge are colored blue. Areas of neutrality are colored green.

local positive charge   local negative charge

This is called a "potential map" because it represents the interaction of a test positive charge with the molecule at various points in the molecule. When the test positive charge encounters negative charge in the molecule, an attractive potential energy occurs; this is color-coded red. When the test positive charge encounters positive charge, a repulsive potential energy results, and this is color-coded blue.

The EPM of H—Cl shows the red region over the Cl and the blue region over the H, as we expect from the greater electronegativity of Cl versus H. In contrast, the EPM of the hydrogen molecule (dihydrogen) shows the same color on both hydrogens because the two atoms share the electrons equally. The green color indicates that neither hydrogen atom bears a net charge.

electrons pulled toward Cl

dihydrogen (H—H)

hydrogen chloride (H—Cl)

FURTHER EXPLORATION 1.1*
Dipole Moments

The uneven electron distribution in a compound containing covalent bonds is measured by a quantity called the **dipole moment**, which is abbreviated with the Greek letter $\boldsymbol{\mu}$ (mu). The dipole moment is commonly given in derived units called *debyes*, abbreviated D, and named for the physical chemist Peter Debye (1884–1966), who received the 1936 Nobel Prize

---

* *Further Explorations* are brief sections in the *Study Guide and Solutions Manual* supplement that cover the subject in greater depth.

in Chemistry. For example, the HCl molecule has a dipole moment of 1.08 D, whereas dihydrogen (H_2), which has a uniform electron distribution, has a dipole moment of zero.

The dipole moment is defined by the following equation:

$$\boldsymbol{\mu} = q\mathbf{r} \tag{1.4}$$

In this equation, q is the magnitude of the separated charge and $\mathbf{r}$ is a vector from the site of the positive charge to the site of the negative charge. For a simple molecule like HCl, the magnitude of the vector $\mathbf{r}$ is merely the length of the HCl bond, and it is oriented from the H (the positive end of the dipole) to the Cl (the negative end). The dipole moment is a *vector quantity*, and $\boldsymbol{\mu}$ and $\mathbf{r}$ *have the same direction*—from the positive to the negative end of the dipole. As a result, the dipole moment vector for the HCl molecule is oriented along the H—Cl bond from the H to the Cl:

dipole moment vector
for HCl

$$\delta+ \quad \delta- $$
$$H—Cl$$

Notice that the magnitude of the dipole moment is affected not only by the *amount* of charge that is separated (q) but also by *how far* the charges are separated ($\mathbf{r}$). Consequently, a molecule in which a relatively small amount of charge is separated by a large distance can have a dipole moment as great as one in which a large amount of charge is separated by a small distance.

> If you study electricity in physics, you may find that physicists use the convention that the dipole vector is oriented from the negative to the positive charge. In the physics convention, then, the dipole moment vector of H—Cl points from the Cl to the H. Chemists, on the other hand, view uneven distribution of *electrons* as the source of the molecular dipole moment, so they orient the dipole moment vector in the direction of excess electrons—from positive to negative. Be sure to understand that the two conventions do not differ in the location of the partial charges, but only in which charge is considered the head and which is considered the tail of the dipole moment vector. Either convention can be used in vector calculations as long as it is used consistently.

Molecules that have permanent dipole moments are called **polar molecules**. HCl is a polar molecule, whereas H_2 is a nonpolar molecule. Some molecules contain several polar bonds. Each polar bond has associated with it a dipole moment contribution, called a **bond dipole**. The net dipole moment of such a polar molecule is the vector sum of its bond dipoles. (Because HCl has only one bond, its dipole moment is equal to the H—Cl bond dipole.) Dipole moments of typical polar organic molecules are in the 1–3 D range.

The vectorial aspect of bond dipoles can be illustrated with the carbon dioxide molecule, CO_2:

EPM of carbon dioxide

As you may have learned in general chemistry, the CO_2 molecule is *linear*; therefore, the C—O bond dipoles are oriented in opposite directions. Because they have equal magnitudes, they *exactly cancel*. (Two vectors of equal magnitude oriented in opposite directions always cancel.) Consequently, CO_2 is a nonpolar molecule, *even though it has polar bonds*. In con-

**STUDY GUIDE
LINK 1.2**
Vector Addition
Review

trast, if a molecule contains several bond dipoles that do *not* cancel, the various bond dipoles add vectorially to give the overall resultant dipole moment. (See Study Guide Link 1.2.) For example, in the water molecule, which has a bond angle of 104.5°, the O—H bond dipoles add vectorially to give a resultant dipole moment of 1.84 D, which bisects the bond angle. The EPM of water shows the charge distribution suggested by the dipole vectors—a concentration of negative charge on oxygen and positive charge on hydrogen.

Polarity is an important concept because the polarity of a molecule can significantly influence its chemical and physical properties. For example, a molecule's polarity may give some indication of how it reacts chemically. Returning to HCl, we know that it dissociates in water to its ions in a manner suggested by its bond polarity.

$$\text{H}_2\text{O} + \overset{\delta+}{\text{H}}\!-\!\overset{\delta-}{\text{Cl}} \longrightarrow \text{H}_3\text{O}^+ + \text{Cl}^- \qquad (1.5)$$

We'll find many similar examples in organic chemistry in which bond polarity provides a clue to chemical reactivity.

Bond polarity is also useful because it gives us some insight that we can apply to the concept of formal charge. It's important to keep in mind that formal charge is only a book-keeping device for keeping track of charge. In some cases, formal charge corresponds to the actual charge. For example, the actual negative charge on the hydroxide ion, $^-$OH, is on the oxygen, because oxygen is much more electronegative than hydrogen. In this case, the locations of the formal charge and actual charge are the same. But in other cases the formal charge does not correspond to the actual charge. For example, in the $^-$BF$_4$ anion, fluorine is *much* more electronegative than boron (Table 1.1). So, most of the charge should actually be situated on the fluorines. In fact, the *actual* charges on the atoms of the tetrafluoroborate ion are in accord with this intuition:

$$\underset{-0.55}{\text{F}}\!-\!\!\!\underset{\substack{| \\ \\ \underset{-0.55}{\text{F}}}}{\overset{\substack{\underset{-0.55}{\text{F}} \\ |}}{\text{B}}}\!\!\!\overset{+1.20}{}\!\!\!-\!\underset{-0.55}{\text{F}}$$

actual charges in the tetrafluoroborate anion

Because the Lewis structure doesn't provide a simple way of showing this distribution, we assign the charge to the boron by the formal-charge rules. An analogy might help. Let's say a big corporation arbitrarily chalks up all of its receipts to its sales department. This is a bookkeeping device. Everyone in the company knows that the receipts are in reality due to a company-wide effort. As long as no one forgets the reality, the administrative convenience of showing receipts in one place makes keeping track of the money a little simpler. Thus, show-ing formal charge on single atoms makes our handling of Lewis structures much simpler, but applying what we know about bond polarities helps us to see where the charge really resides.

PROBLEMS

1.6   Analyze the polarity of each bond in the following organic compound. Which bond, other than the C—C bond, is the least polar one in the molecule? Which carbon has the most partial positive character?

$$
\begin{array}{c}
\text{H} \quad\; \text{O} \\
| \quad\;\; \| \\
\text{H}-\text{C}-\text{C}-\text{Cl} \\
| \\
\text{Cl}
\end{array}
$$

1.7   For which of the following ions does the formal charge give a fairly accurate picture of where the charge really is? Explain in each case.

(a) $\overset{+}{N}H_4$ (b) $:\overset{..}{\bar{N}}H_2$ (c) $\overset{+}{C}H_3$

1.3   STRUCTURES OF COVALENT COMPOUNDS

We know the **structure** of a molecule containing covalent bonds when we know its *atomic connectivity* and its *molecular geometry*. **Atomic connectivity** is the specification of how atoms in a molecule are connected. For example, we specify the atomic connectivity within the water molecule when we say that two hydrogens are bonded to an oxygen. **Molecular geometry** is the specification of how far apart the atoms are and how they are situated in space.

Chemists learned about atomic connectivity before they learned about molecular geometry. The concept of covalent compounds as three-dimensional objects emerged in the latter part of the nineteenth century on the basis of indirect chemical and physical evidence. Until the early part of the twentieth century, however, no one knew whether these concepts had any physical reality, because scientists had no techniques for viewing molecules at the atomic level. By the second decade of the twentieth century, investigators could ask two questions: (1) Do organic molecules have specific geometries and, if so, what are they? (2) How can molecular geometry be predicted?

A. Methods for Determining Molecular Geometry

Among the greatest developments of chemical physics in the early twentieth century were the discoveries of ways to deduce the structures of molecules. Such techniques include NMR spectroscopy, IR spectroscopy, UV–visible spectroscopy, and mass spectrometry, which we'll consider in Chapters 12–15. As important as these techniques are, they are used primarily to provide information about atomic connectivity. Other physical methods, however, permit the determination of molecular structures that are complete in every detail. Most complete structures today come from three sources: X-ray crystallography, electron diffraction, and microwave spectroscopy.

The arrangement of atoms in the crystalline solid state can be determined by *X-ray crystallography*. This technique, invented in 1915 and subsequently revolutionized by the availability of high-speed computers, uses the fact that X-rays are diffracted from the atoms of a crystal in precise patterns that can be mathematically deciphered to give a molecular structure. In 1930, *electron diffraction* was developed. With this technique, the diffraction of electrons by molecules of gaseous substances can be interpreted in terms of the arrangements of atoms in molecules. Following the development of radar in World War II came *microwave spectroscopy*, in which the absorption of microwave radiation by molecules in the gas phase provides detailed structural information.

Most of the spatial details of molecular structure in this book are derived from gas-phase methods: electron diffraction and microwave spectroscopy. For molecules that are not readily studied in the gas phase, X-ray crystallography is the most important source of structural information. No methods of comparable precision exist for molecules in solution, a fact that is unfortunate because most chemical reactions take place in solution. The consistency of gas-phase and crystal structures suggests, however, that molecular structures in solution probably differ little from those of molecules in the solid or gaseous state.

B. Prediction of Molecular Geometry

The way a molecule reacts is determined by the characteristics of its chemical bonds. The characteristics of the chemical bonds, in turn, are closely connected to molecular geometry. Molecular geometry is important, then, because it is a starting point for understanding chemical reactivity.

Given the connectivity of a covalent molecule, what else do we need to describe its geometry? Let's start with a simple diatomic molecule, such as HCl. The structure of such a molecule is completely defined by the **bond length**, the distance between the centers of the bonded nuclei. Bond length is usually given in angstroms; $1\ \text{Å} = 10^{-10}\ \text{m} = 10^{-8}\ \text{cm} = 100\ \text{pm}$ (picometers). Thus, the structure of HCl is completely specified by the H—Cl bond length, 1.274 Å.

When a molecule has more than two atoms, understanding its structure requires knowledge of not only each bond length, but also each **bond angle**, the angle between each pair of bonds to the same atom. The structure of water (H_2O) is completely determined, for example, when we know the O—H bond lengths and the H—O—H bond angle.

We can generalize much of the information that has been gathered about molecular structure into a few principles that allow us to analyze trends in bond length and to predict approximate bond angles.

Bond Length   The following three generalizations can be made about bond length, *in decreasing order of importance.*

1. *Bond lengths increase significantly toward higher periods (lower rows) of the periodic table.* This trend is illustrated in Fig. 1.2. For example, the H—S bond in hydrogen sulfide is longer than the other bonds to hydrogen in Fig. 1.2; sulfur is in the third period of the periodic table, whereas carbon, nitrogen, and oxygen are

FIGURE 1.2 Effect of atomic size on bond length. (Within each structure, all bonds to hydrogen are equivalent. Dashed bonds are behind the plane of page, and wedged bonds are in front; see p. 16.) Compare the bond lengths in hydrogen sulfide with those of the other molecules to see that bond lengths *increase* toward higher periods of the periodic table. Compare the bond lengths in methane, ammonia, and water to see that bond lengths *decrease* toward higher atomic number within a period (row) of the periodic table.

FIGURE 1.3 Effect of bond order on bond length. As the carbon–carbon bond order increases, the bond length decreases.

in the second period. Similarly, a C—H bond is shorter than a C—F bond, which is shorter than a C—Cl bond. These effects all reflect atomic size. Because bond length is the distance between the centers of bonded atoms, larger atoms form longer bonds.

2. *Bond lengths decrease with increasing bond order.* **Bond order** describes the number of covalent bonds shared by two atoms. For example, a C—C bond has a bond order of 1, a C=C bond has a bond order of 2, and a C≡C bond has a bond order of 3. The decrease of bond length with increasing bond order is illustrated in Fig. 1.3. Notice that the bond lengths for carbon–carbon bonds are in the order C—C > C=C > C≡C.

3. *Bonds of a given order decrease in length toward higher atomic number (that is, to the right) along a given row (period) of the periodic table.* Compare, for example, the H—C, H—N, and H—O bond lengths in Fig. 1.2. Likewise, the C—F bond in H_3C—F, at 1.39 Å, is shorter than the C—C bond in H_3C—CH_3, at 1.54 Å. Because atoms on the right of the periodic table in a given row are smaller, this trend, like that in item 1, also results from differences in atomic size. However, this effect is *much less* significant than the differences in bond length observed when atoms of different periods are compared.

Bond Angle   The bond angles within a molecule determine its *shape*. For example, in the case of a triatomic molecule such as H_2O or BeH_2, the bond angles determine whether it is bent or linear. To predict approximate bond angles, we rely on **valence-shell electron-pair repulsion theory**, or **VSEPR theory**, which you may have encountered in general chemistry. According to VSEPR theory, both the bonding electron pairs and the unshared valence electron pairs have a spatial requirement. The fundamental idea of VSEPR theory is that *bonds and unshared electron pairs are arranged about a central atom so that the bonds and unshared electron pairs are as far apart as possible.* This arrangement minimizes repulsions between electrons in the bonds.

Let's first apply VSEPR theory to three situations involving only bonding electrons: a central atom bound to four, three, and two groups, respectively.

When four groups are bonded to a central atom, the bonds are farthest apart when the central atom has tetrahedral geometry. This means that the four bound groups lie at the vertices of a **tetrahedron**. A tetrahedron is a three-dimensional object with four triangular faces (Fig. 1.4a, p. 16). Methane, CH_4, has tetrahedral geometry. The central atom is the carbon and the four groups are the hydrogens. The C—H bonds of methane are as far apart as possible when the hydrogens lie at the vertices of a tetrahedron. Because the four C—H bonds of methane are identical, the hydrogens lie at the vertices of a *regular tetrahedron*, a tetrahedron in which all edges are equal (Fig. 1.4b). The tetrahedral shape of methane requires a bond angle of 109.5° (Fig. 1.4c).

In applying VSEPR theory for the purpose of predicting bond angles, we regard all groups as identical. For example, the groups that surround carbon in CH_3Cl (chloromethane) are treated as if they were identical, even though in reality the C—Cl bond is considerably longer than the C—H bonds. Although the bond angles show minor deviations from the exact tetrahedral bond angle of 109.5°, chloromethane in fact has the general tetrahedral shape.

FIGURE 1.4 Tetrahedral geometry of methane. (a) A regular tetrahedron. (b) In methane, the carbon is in the middle of a tetrahedron and the four hydrogens lie at the vertices. (c) A ball-and-stick model of methane. The tetrahedral geometry requires a bond angle of 109.5°.

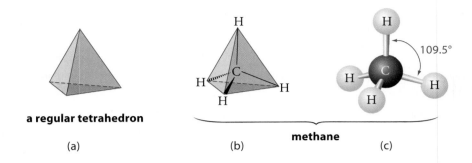

a regular tetrahedron

(a)

methane

(b)

(c)

Because you'll see tetrahedral geometry repeatedly, it is worth the effort to become familiar with it. Tetrahedral carbons are often represented by **line-and-wedge structures**, as illustrated by the following structure of dichloromethane, CH_2Cl_2.

**STUDY GUIDE
LINK 1.3**
Structure-Drawing
Conventions

The carbon, the two chlorines, and the C—Cl bonds are in the plane of the page. The C—Cl bonds are represented by lines. One of the hydrogens is behind the page. The bond to this hydrogen recedes behind the page from the carbon and is represented by a dashed wedge. The remaining hydrogen is in front of the page. The bond to this hydrogen emerges from the page and is represented by a solid wedge. (For different conventions for drawing wedges, see Study Guide Link 1.3.) Several possible line-and-wedge structures are possible for any given molecule. Which structure we draw depends on how we view the molecule. For example, we could have drawn the hydrogens in the plane of the page and the chlorines in the out-of-plane positions, or we could have drawn one hydrogen and one chlorine in the plane and the other hydrogen and chlorine out-of-plane.

A good way to become familiar with the tetrahedral shape (or any other aspect of molecular geometry) is to use **molecular models**, which are commercially available scale models from which you can construct simple organic molecules. Perhaps your instructor has required that you purchase a set of models or can recommend a set to you. *Almost all beginning students require models, at least initially, to visualize the three-dimensional aspects of organic chemistry.* Some of the types of models available are shown in Fig. 1.5. In this text, we use ball-and-stick models (Fig. 1.5a) to visualize the directionality of chemical bonds, and we use space-filling models (Fig. 1.5c) to see the consequences of atomic and molecular vol-

(a)

(b)

(c)

FIGURE 1.5 Molecular models of methane. (a) Ball-and-stick models show the atoms as balls and the bonds as connecting sticks. Most inexpensive sets of student models are of this type. (b) A wire-frame model shows a nucleus (in this case, carbon) and its attached bonds. (c) Space-filling models depict atoms as spheres with radii proportional to their covalent or atomic radii. Space-filling models are particularly effective for showing the volume occupied by atoms or molecules.

umes. You should obtain an inexpensive set of ball-and-stick molecular models and use them frequently. *Begin using them by building a model of the dichloromethane molecule discussed above and relating it to the line-and-wedge structure.*

PROBLEM

1.8   Using models if necessary, draw at least two other line-and-wedge structures of dichloromethane (p. 16).

Molecular Modeling by Computer

Scientists also use computers to depict molecular models. Computerized molecular modeling is particularly useful for very large molecules because building real molecular models in these cases can be prohibitively expensive in both time and money. The decreasing cost of computing power has made computerized molecular modeling increasingly more practical. Most of the models shown in this text were drawn to scale from the output of a molecular-modeling program on a desktop computer.

When *three* groups surround an atom, the bonds are as far apart as possible when all bonds lie in the same plane with bond angles of 120°. This is, for example, the geometry of boron trifluoride:

boron trifluoride

In such a situation the surrounded atom (in this case boron) is said to have **trigonal planar** geometry.

When an atom is surrounded by *two* groups, maximum separation of the bonds demands a bond angle of 180°. This is the situation with each carbon in acetylene, H—C≡C—H. Each carbon is surrounded by two groups: a hydrogen and another carbon. Notice that the triple bond (as well as a double bond in other compounds) is considered as one bond for purposes of VSEPR theory, because all three bonds connect the same two atoms. Atoms with 180° bond angles are said to have **linear** geometry. Thus, acetylene is a *linear* molecule.

acetylene

Now let's consider how unshared valence electron pairs are treated by VSEPR theory. *An unshared valence electron pair is treated as if it were a bond without a nucleus at one end.* For example, in VSEPR theory, the nitrogen in ammonia, :NH$_3$, is surrounded by four "bonds": three N—H bonds and the unshared valence electron pair. These "bonds" are directed to the vertices of a tetrahedron so that the hydrogens occupy three of the four tetrahedral vertices. Therefore, :NH$_3$ is essentially tetrahedral if we include the unshared electron pair. However, this geometry is called **trigonal pyramidal** because the three N—H bonds lie along the edges of a pyramid.

VSEPR theory also postulates that *unshared valence electron pairs occupy more space than an ordinary bond*. It's as if the electron pair "spreads out" because it isn't constrained by a second nucleus. As a result, the bond angle between the unshared pair and the other bonds are somewhat larger than tetrahedral, and the N—H bond angles are correspondingly smaller. In fact, the H—N—H bond angle in ammonia is 107.3°.

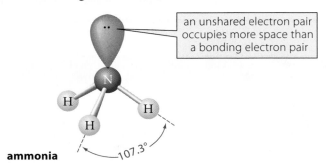

ammonia

an unshared electron pair occupies more space than a bonding electron pair

107.3°

STUDY PROBLEM 1.3

Estimate each bond angle in the following molecule, and order the bonds according to length, beginning with the shortest.

$$\underset{1}{H} \underset{(a)}{—} \underset{2}{C} \underset{(b)}{\equiv} \underset{3}{C} \underset{(c)}{—} \underset{4}{\overset{\overset{5}{\overset{\displaystyle O}{\|}{}_{(e)}}}{C}} \underset{(d)}{—} \underset{6}{Cl}$$

SOLUTION Because carbon-2 is bound to two groups (H and C), its geometry is linear. Similarly, carbon-3 also has linear geometry. The remaining carbon (carbon-4) is bound to three groups (C, O, and Cl); therefore, it has approximately trigonal planar geometry. To arrange the bonds in order of length, recall the *order of importance* of the bond-length rules. The major influence on length is the row in the periodic table from which the bonded atoms are taken. Hence, the H—C bond is shorter than *all* carbon–carbon or carbon–oxygen bonds, which are shorter than the C—Cl bond. The next major effect is the bond order. Hence, the C≡C bond is shorter than the C═O bond, which is shorter than the C—C bond. Putting these conclusions together, the required order of bond lengths is

$$(a) < (b) < (e) < (c) < (d)$$

PROBLEMS

1.9 Predict the approximate geometry in each of the following molecules.

(a) $[BF_4]^-$ (b) water   (c) $H_2C═\ddot{O}$ (d) $H_3C—C\equiv N:$

 formaldehyde   **acetonitrile**

1.10 Estimate each of the bond angles and order the bond lengths (smallest first) in the following molecule.

$$\begin{array}{c} :O: \\ \| (e) \\ H \quad C \overset{(f)}{—} C —H \\ (a) \qquad (d) \quad (g) \\ C═C \qquad H \\ (a) \quad (c) \\ H \qquad :Br: \end{array}$$

dihedral angle = 0°   dihedral angle = 90°   dihedral angle = 180°

FIGURE 1.6 The concept of dihedral angle illustrated for the hydrogen peroxide molecule, H—O—O—H. Knowledge of the bond angles does not define the dihedral angle. Three possibilities for the dihedral angle (0°, 90°, and 180°) are shown.

Dihedral Angle   To completely describe the shapes of molecules that are more complex than the ones we've just discussed, we need to specify not only the bond lengths and bond angles, but also the *spatial relationship of the bonds on adjacent atoms.*

To illustrate this problem, consider the molecule hydrogen peroxide, H—Ö—Ö—H. Both O—O—H bond angles are 96.5°. However, knowledge of these bond angles is not sufficient to describe completely the shape of the hydrogen peroxide molecule. To understand why, imagine two intersecting planes, each containing one of the oxygens and its two bonds (Fig. 1.6). To completely describe the structure of hydrogen peroxide, we need to know the angle between these two planes. This angle is called the **dihedral angle** or **torsion angle**. Three possibilities for the dihedral angle are shown in Fig. 1.6. You can also visualize these dihedral angles using a model of hydrogen peroxide by holding one O—H bond fixed and rotating the remaining oxygen and its bonded hydrogen about the O—O bond. (The actual dihedral angle in hydrogen peroxide is addressed in Problem 1.42, p. 42.) Molecules containing many bonds typically contain many dihedral angles to be specified. We'll begin to learn some of the principles that allow us to predict dihedral angles in Chapter 2.

Let's summarize: The geometry of a molecule is completely determined by its bond lengths, its bond angles, and its dihedral angles. The geometries of diatomic molecules are completely determined by their bond lengths. The geometries of molecules in which a central atom is surrounded by two or more other atoms are determined by both bond lengths and bond angles. Bond lengths, bond angles, and dihedral angles are required to specify the geometry of more complex molecules.

PROBLEM

1.11   Which of the following ions require(s) dihedral angles to specify its structure completely? Explain.

$$H_3C—\overset{+}{N}H_3 \qquad\qquad \overset{..}{\underset{..}{P}F_6} \qquad\qquad H\overset{..}{\underset{..}{O}}—\overset{..}{\underset{..}{O}}:^-$$

A   *B*   *C*

1.4   RESONANCE STRUCTURES

Some compounds are not accurately described by a single Lewis structure. Consider, for example, the structure of nitromethane, $H_3C—NO_2$.

$$H_3C—\overset{+}{N}\Big\langle\overset{\overset{..}{O}:^-}{\underset{:O:}{}}$$

nitromethane

This Lewis structure shows an N—O single bond and an N=O double bond. From the preceding section, we expect double bonds to be shorter than single bonds. However, it is found experimentally that the two nitrogen–oxygen bonds of nitromethane have the same length, and this length is intermediate between the lengths of single and double nitrogen–oxygen bonds found in other molecules. We can convey this idea by writing the structure of nitromethane as follows:

$$\left[H_3C-\overset{+}{N}\begin{smallmatrix}\ddot{O}:^-\\\\:\ddot{O}:\end{smallmatrix} \longleftrightarrow H_3C-\overset{+}{N}\begin{smallmatrix}:O:\\\\\ddot{O}:^-\end{smallmatrix} \right] \tag{1.6}$$

The double-headed arrow (⟷) means that nitromethane is a single compound that is the "average" of both structures; nitromethane is said to be a **resonance hybrid** of these two structures. Note carefully that the double-headed arrow ⟷ is different from the arrows used in chemical equilibria, ⇌. The two structures for nitromethane are *not* rapidly interconverting and they are *not* in equilibrium. Rather, they are alternative representations of *one* molecule. In this text, resonance structures will be enclosed in brackets to emphasize this point. Resonance structures are necessary because of the inadequacy of a single Lewis structure to represent nitromethane accurately.

The two resonance structures in Eq. 1.6 are *fictitious*, but nitromethane is a real molecule. Because we have no way to describe nitromethane accurately with a single Lewis structure, we must describe it as the hybrid of two fictitious structures. An analogy to this situation is a description of Fred Flatfoot, a *real* detective. Lacking words to describe Fred, we picture him as a resonance hybrid of two *fictional* characters:

Fred Flatfoot = [Sherlock Holmes ⟷ James Bond]

This suggests that Fred is a dashing, violin-playing, pipe-smoking, highly intelligent British agent with an assistant named Watson, and that Fred likes his martinis shaken, not stirred.

When two resonance structures are identical, as they are for nitromethane, they are equally important in describing the molecule. We can think of nitromethane as a 1:1 average of the structures in Eq. 1.6. For example, each oxygen bears half a negative charge, and each nitrogen–oxygen bond is neither a single bond nor a double bond, but a bond halfway in between. This hybrid character can be conveyed in a single structure in which dashed lines are used to represent partial bonds. Nitromethane, for example, can be represented in this notation in either of the following ways:

$$H_3C-\overset{+}{N}\begin{smallmatrix}O^{\delta-}\\\\O^{\delta-}\end{smallmatrix} \quad \text{or} \quad H_3C-\overset{+}{N}\begin{smallmatrix}O\\\\O\end{smallmatrix}^{-}$$

In the hybrid structure on the left, the locations of the shared negative charge are shown explicitly with partial charges. In the hybrid structure on the right, the locations of the shared negative charge are not shown. Although the use of hybrid structures is sometimes convenient, it is difficult to apply electron-counting rules to them. To avoid confusion, we'll use conventional resonance structures in these situations.

If two resonance structures are not identical, then the molecule they represent is a *weighted average* of the two. That is, one of the structures is more important than the other in describing the molecule. Such is the case, for example, with the methoxymethyl cation:

$$\left[H_2\overset{+}{C}-\ddot{O}-CH_3 \longleftrightarrow H_2C=\overset{\ddot{O}}{\underset{+}{O}}-CH_3 \right] \tag{1.7}$$

methoxymethyl cation

It turns out that the structure on the right is a better description of this cation because all atoms have complete octets. Hence, the C—O bond has significant double-bond character, and most of the formal positive charge resides on the oxygen.

Formal charge has the same limitations as a bookkeeping device in resonance structures that it does in other structures. Although most of the *formal* charge in the methoxymethyl cation resides on oxygen, the CH_2 carbon bears more of the *actual* positive charge, because oxygen is more electronegative than carbon.

A very important aspect of resonance structures is that they have implications for the stability of the molecule they represent. *A molecule represented by resonance structures is more stable than its fictional resonance contributors.* For example, the actual molecule nitromethane is more stable than either one of the fictional molecules described by the contributing resonance structures in Eq. 1.6. Nitromethane is thus said to be a *resonance-stabilized molecule*, as is the methoxymethyl cation.

How do we know when to use resonance structures, how to draw them, or how to assess their relative importance? In Chapter 3, we'll learn a technique for deriving resonance structures, and in Chapter 15, we'll return to a more detailed study of the other aspects of resonance. In the meantime, we'll draw resonance structures for you and tell you when they're important. Just try to remember the following points:

1. Resonance structures are used for compounds that are not adequately described by a single Lewis structure.

2. Resonance structures are *not* in equilibrium; that is, the compound they describe is *not* one resonance structure part of the time and the other resonance structure part of the time, but rather *a single* structure.

3. The structure of a molecule is the *weighted average* of its resonance structures. When resonance structures are identical, they are equally important descriptions of the molecule.

4. Resonance hybrids are more stable than any of the fictional structures used to describe them. Molecules described by resonance structures are said to be *resonance-stabilized.*

PROBLEMS

1.12 The compound *benzene* has only one type of carbon–carbon bond, and this bond has a length intermediate between that of a single bond and a double bond. Draw a resonance structure of benzene that, taken with the following structure, accounts for the carbon–carbon bond length.

benzene

1.13 (a) Draw a resonance structure for the allyl anion that shows, along with the following structure, that the two CH_2 carbons are equivalent and indistinguishable.

$$\left[H_2C = CH - \ddot{C}H_2 \longleftrightarrow \qquad \right]$$

allyl anion

(b) According to the resonance structures, how much negative charge is on each of the CH_2 carbons?

(c) Draw a single hybrid structure for the allyl anion that shows shared bonds as dashed lines and charges as partial charges.

1.5 THE WAVE NATURE OF THE ELECTRON

You've learned that the covalent chemical bond can be viewed as the sharing of one or more electron pairs between two atoms. Although this simple model of the chemical bond is very useful, in some situations it is inadequate. A deeper insight into the nature of the chemical bond can be obtained from an area of science called *quantum mechanics*. Quantum mechanics deals in detail with, among other things, the behavior of electrons in atoms and molecules. Although the theory involves some sophisticated mathematics, we need not explore the mathematical detail to appreciate some general conclusions of the theory. The starting point for quantum mechanics is the idea that *small particles such as electrons also have the character of waves.* How did this idea evolve?

As the twentieth century opened, it became clear that certain things about the behavior of electrons could not be explained by conventional theories. There seemed to be no doubt that the electron was a particle; after all, both its charge and mass had been measured. However, electrons could also be diffracted like light, and diffraction phenomena were associated with waves, not particles. The traditional views of the physical world treated particles and waves as unrelated phenomena. In the mid-1920s, this mode of thinking was changed by the advent of quantum mechanics. This theory holds that, in the submicroscopic world of the electron and other small particles, there is no real distinction between particles and waves. The behavior of small particles such as the electron can be described by the physics of waves. In other words, matter can be regarded as a *wave-particle duality.*

How does this wave-particle duality require us to alter our thinking about the electron? In our everyday lives, we're accustomed to a deterministic world. That is, the position of any familiar object can be measured precisely, and its velocity can be determined, for all practical purposes, to any desired degree of accuracy. For example, we can point to a baseball resting on a table and state with confidence, "That ball is at rest (its velocity is zero), and it is located exactly 1 foot from the edge of the table." Nothing in our experience indicates that we couldn't make similar measurements for an electron. The problem is that humans, chemistry books, and baseballs are of a certain scale. Electrons and other tiny objects are of a much smaller scale. A central principle of quantum mechanics, the **Heisenberg uncertainty principle**, tells us that *the accuracy with which we can determine the position and velocity of a particle is inherently limited.* For "large" objects, such as basketballs, organic chemistry textbooks, and even molecules, the uncertainty in position is small relative to the size of the object and is inconsequential. But for very small objects such as electrons, the uncertainty is significant. As a result, the position of an electron becomes "fuzzy." According to the Heisenberg uncertainty principle, we are limited to stating the *probability* that an electron is occupying a certain region of space.

In summary:

1. Electrons have wavelike properties.
2. The exact position of an electron cannot be specified; only the probability that it occupies a certain region of space can be specified.

1.6 ELECTRONIC STRUCTURE OF THE HYDROGEN ATOM

To understand the implications of quantum theory for covalent bonding, we must first understand what the theory says about the electronic structure of atoms. This section presents the applications of quantum theory to the simplest atom, hydrogen. We deal with the hydrogen atom because a very detailed description of its electronic structure has been developed, and because this description has direct applicability to more complex atoms.

A. Orbitals, Quantum Numbers, and Energy

In an earlier model of the hydrogen atom, the electron was thought to circle the nucleus in a well-defined orbit, much as Earth circles the Sun. Quantum theory replaced the orbit with the *orbital,* which, despite the similar name, is something quite different. An **atomic orbital** is a description of the wave properties of an electron in an atom. We can think of an atomic orbital of hydrogen as an *allowed state*—that is, an allowed wave motion—of an electron in the hydrogen atom.

An atomic orbital in physics is described by a mathematical function called a **wavefunction**. As an analogy, you might describe a sine wave by the function $\psi = \sin x$. This is a simple wavefunction that covers one spatial dimension. Wavefunctions for an electron in an atom are conceptually similar, except that the wavefunctions cover three spatial dimensions, and the mathematical functions are different.

Many possible orbitals, or states, are available to the electron in the hydrogen atom. This means that the wave properties of the electron can be described by any one of several wavefunctions. In the mathematics of "electron waves," each orbital is described by three **quantum numbers**. Once again, to use a simple analogy, consider the simple wave equation $\psi = \sin nx$. We get a different wave for each different value of n. If n were restricted to integers, we could think of n as a quantum number for this type of wave. (See Problem 1.14, p. 28.) Wavefunctions for the electron involve three quantum numbers. Although quantum numbers do have mathematical significance in the wave equations of the electron, for us they serve as labels, or designators, for the various orbitals, or wave motions, available to the electron. These quantum numbers can have only certain values, and the values of some quantum numbers depend on the values of others.

The **principal quantum number**, abbreviated n, can have any integral value greater than zero—that is, $n = 1, 2, 3, \ldots$.

The **angular momentum quantum number**, abbreviated l, depends on the value of n. The l quantum number can have any integral value from zero through $n - 1$, that is, $l = 0, 1, 2, \ldots, n - 1$. So that they are not confused with the principal quantum number, the values of l are encoded as letters. To $l = 0$ is assigned the letter s; to $l = 1$, the letter p; to $l = 2$, the letter d; and to $l = 3$, the letter f. The values of l are summarized in Table 1.2. It follows that there can be only one orbital, or wavefunction, with $n = 1$: this is the orbital with $l = 0$—a $1s$ orbital. However, two values of l—that is, 0 and 1—are allowed for $n = 2$. Consequently, an electron in the hydrogen atom can exist in either a $2s$ or a $2p$ orbital.

The **magnetic quantum number**, abbreviated m_l, is the third orbital quantum number. Its values depend on the value of l. The m_l quantum number can be zero as well as both positive and negative integers up to $\pm l$—that is, $0, \pm 1, \pm 2, \ldots, \pm l$. Thus, for $l = 0$ (an s orbital), m_l can only be 0. For $l = 1$ (a p orbital), m_l can have the values $-1, 0,$ and $+1$. In other words, there is one s orbital with a given principal quantum number, but (for $n > 1$) there are three p orbitals with a given principal quantum number, one corresponding to each value of m_l. Because of the multiple possibilities for l and m_l, the number of orbitals becomes increasingly large as n increases. This point is illustrated in Table 1.2 up to $n = 3$.

Just as an electron in the hydrogen atom can exist only in certain states, or orbitals, it can also have only certain allowed energies. Each orbital is associated with a characteristic

TABLE 1.2 Relationship Among the Three Orbital Quantum Numbers

n	l	m_l	n	l	m_l	n	l	m_l
1	0 (1s)	0	2	0 (2s)	0	3	0 (3s)	0
				1 (2p)	−1		1 (3p)	−1
					0			0
					+1			+1
							2 (3d)	−2
								−1
								0
								+1
								+2

electron energy. *The energy of an electron in a hydrogen atom is determined by the principal quantum number n of its orbital.* This is one of the central ideas of quantum theory. The energy of the electron is said to be *quantized*, or limited to certain values. This feature of the atomic electron is a direct consequence of its wave properties. An electron in the hydrogen atom resides in an orbital with $n = 1$ (a $1s$ orbital) and remains in that state unless the atom is subjected to the exact amount of energy (say, from light) required to increase the energy of the electron to a state with a higher n (say, $n = 2$):

If that happens, the electron absorbs energy and instantaneously assumes the new, more energetic, wave motion characteristic of the orbital with $n = 2$. (Such energy-absorption experiments gave the first clues to the quantized nature of the atom.) An analogy to this may be familiar. If you have ever blown across the opening of a soda-pop bottle (or a flute, which is a more sophisticated example of the same thing), you know that only a certain pitch can be produced by a bottle of a given size. If you blow harder, the pitch does not rise, but only becomes louder. However, if you blow hard enough, the sound suddenly jumps to a note of higher pitch. The pitch is quantized; only certain sound frequencies (pitches) are allowed. Such phenomena are observed because sound is a wave motion of the air in the bottle, and only certain pitches can exist in a cavity of given dimensions without canceling themselves out. The progressively higher pitches you hear as you blow harder (called *overtones* of the lowest pitch) are analogous to the progressively higher energy states (orbitals) of the electron in the atom. Just as each overtone in the bottle is described by a wavefunction with higher "quantum number," each orbital of higher energy is described by a wavefunction of higher principal quantum number n.

B. Spatial Characteristics of Orbitals

One of the most important aspects of atomic structure for organic chemistry is that *each orbital is characterized by a three-dimensional region of space in which the electron is most likely to exist.* That is, orbitals have *spatial* characteristics. The *size* of an orbital is governed mainly by its principal quantum number n: the larger n is, the greater the region of space occupied by the corresponding orbital. The *shape* of an orbital is governed by its angular momentum quantum number l. The *directionality* of an orbital is governed by its magnetic quantum number m_l. These points are best illustrated by example.

When an electron occupies a $1s$ orbital, it is most likely to be found in a sphere surrounding the atomic nucleus (Fig. 1.7). *We cannot say exactly where in that sphere the electron is* by the uncertainty principle; locating the electron is a matter of probability. The mathematics of quantum theory indicates that the probability is about 90% that an electron in a $1s$ orbital will be found within a sphere of radius 1.4 Å about the nucleus. This "90% probability level" is taken as the approximate size of an orbital. Thus, we can depict an electron in a $1s$ orbital as a *smear of electron density,* most of which is within 1.4 Å of the nucleus.

Because orbitals are actually mathematical functions of three spatial dimensions, it would take a fourth dimension to plot the value of the orbital (or the electron proba-

FURTHER EXPLORATION 1.2
Electron Density Distribution in Orbitals

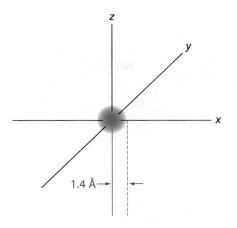

FIGURE 1.7 A 1s orbital. Most (90%) of the electron density lies in a sphere within 1.4 Å of the nucleus.

bility) at each point in space. (See Further Exploration 1.2 for an additional discussion of electron probability.) Because we are limited to three spatial dimensions, Fig. 1.7 and the other orbital pictures presented subsequently show each orbital as a geometric figure that encloses some fraction (in our case, 90%) of the electron probability. The detailed quantitative distribution of electron probability within each figure is not shown.

When an electron occupies a 2s orbital, it also lies in a sphere, but the sphere is considerably larger—about three times the radius of the 1s orbital (Fig. 1.8). A 3s orbital is even larger still. The size of the orbital reflects the fact that the electron has greater energy; a more energetic electron can escape the attraction of the positive nucleus to a greater extent.

The 2s orbital also illustrates a *node*, another very important spatial aspect of orbitals. You may be familiar with a simple wave motion, such as the wave in a vibrating string, or waves in a pool of water. If so, you know that waves have *peaks* and *troughs*, regions where the waves are at their maximum and minimum heights, respectively. As you know from trigonometry, a simple sine wave $\psi = \sin x$ has a positive sign at its peak and a negative sign at its trough (Fig. 1.9, p. 26). Because the wave is continuous, it has to have a zero value somewhere in between the peak and the trough. A **node** is a point or, in a three-dimensional wave, a *surface*, at which the wave is zero.

As you can see from Fig. 1.8 and the subsequent figures, the peaks and troughs are color-coded: peaks are blue and troughs are green. When the nodal properties of orbitals aren't important for the discussion, the orbitals will be colored gray.

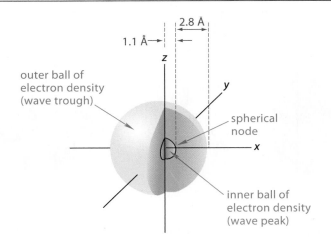

outer ball of electron density (wave trough)

spherical node

inner ball of electron density (wave peak)

FIGURE 1.8 A 2s orbital in a cutaway view, showing the positive (peak-containing) region of the electron wave in blue and the negative (trough-containing) region in green. This orbital can be described as two concentric spheres of electron density. A 2s orbital is considerably larger than a 1s orbital; most (90%) of the electron density of a 2s orbital lies within 3.9 Å of the nucleus.

FIGURE 1.9 An ordinary sine wave (a plot of $\psi = \sin x$) showing peaks, troughs, and nodes. A peak occurs in the region in which ψ is positive, and a trough occurs in a region in which ψ is negative. The nodes are points at which $\psi = 0$.

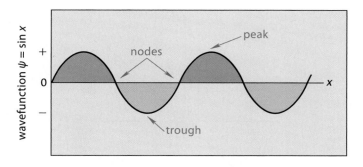

Pay very careful attention to one point of potential confusion. The *sign of the wavefunction* for an electron is *not* the same as the charge on the electron. Electrons always bear a negative charge. The sign of the wavefunction refers to the sign of the mathematical expression that describes the wave. By convention, a wave peak (color-coded blue) has a positive (+) sign and a wave trough (color-coded green) has a negative (−) sign.

As shown in Fig. 1.8, the 2*s* orbital has one node. This node separates a wave peak near the nucleus from a wave trough further out. Because the 2*s* orbital is a three-dimensional wave, its node is a *surface*. The nodal surface in the 2*s* orbital is an infinitely thin sphere. Thus, the 2*s* orbital has the characteristics of two concentric balls of electron density.

In the 2*s* orbital, the wave peak corresponds to a positive value in the 2*s* wavefunction, and the wave trough corresponds to a negative value. The node—the spherical shell of zero electron density—lies between the peak and the trough. Some students ask, "If the electron cannot exist at the node, how does it cross the node?" The answer is that the electron *is* a wave, and the node is part of its wave motion, just as the node is part of the wave in a vibrating string. The electron is not analogous to the string; it is analogous to the *wave* in the string.

We turn next to the 2*p* orbitals (Fig. 1.10), which are especially important in organic chemistry. The 2*p* orbital illustrates how the *l* quantum number governs the *shape* of an orbital. All *s* orbitals are spheres. In contrast, all *p* orbitals have dumbbell shapes and are directed in space (that is, they lie along a particular axis). One lobe of the 2*p* orbital corresponds to a wave peak, and the other to a wave trough; the electron density, or probability of finding the electron, is identical in corresponding parts of each lobe. Note that the two lobes are *parts of the same orbital*. The node in the 2*p* orbital, which passes through the nucleus and separates the two lobes, is a plane. The size of the 2*p* orbital, like that of other orbitals, is governed by its principal quantum number; it extends about the same distance from the nucleus as a 2*s* orbital.

Fig. 1.10b illustrates a drawing convention for 2*p* orbitals. Quite often the lobes of these orbitals are drawn in a less rounded, "teardrop" shape. (This shape is derived from the *square*

FIGURE 1.10 (a) A 2*p* orbital. Notice the planar node that separates the orbital into two lobes. Most (90%) of the electron density lies within 4.2 Å of the nucleus. (b) A widely used drawing style for the representation of 2*p* orbitals. (c) The three 2*p* orbitals shown together. Each orbital has a different value of the quantum number m_l.

(a)   (b)   (c)

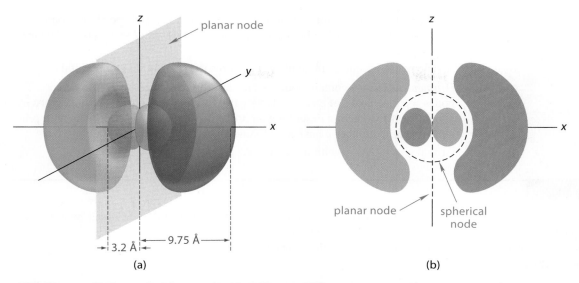

FIGURE 1.11 (a) Perspective representation of a 3*p* orbital; only the planar node is shown. There are three such orbitals, and they are mutually perpendicular. Notice that a 3*p* orbital is much larger than a 2*p* orbital in Fig 1.10. Most (90%) of the electron density lies within 9.75 Å of the nucleus; about 60% of the electron density lies in the large outer lobes. (b) Schematic planar representation of a 3*p* orbital showing both the planar and the spherical nodes.

of the wavefunction, which is proportional to the actual electron density.) This shape is useful because it emphasizes the directionality of the 2*p* orbital. This convention is so commonly adopted that we'll often use it in this text.

Recall (Table 1.2, p. 23) that there are three 2*p* orbitals, one for each allowed value of the quantum number m_l. The three 2*p* orbitals illustrate how the m_l quantum number governs the *directionality* of orbitals. The axes along which each of the 2*p* orbitals "points" are mutually perpendicular. For this reason, the three 2*p* orbitals are sometimes differentiated with the labels $2p_x$, $2p_y$, and $2p_z$. The three 2*p* orbitals are shown superimposed in Fig. 1.10c.

Let's examine one more atomic orbital, the 3*p* orbital (Fig. 1.11). First, notice the greater size of this orbital, which is a consequence of its greater principal quantum number. The 90% probability level for this orbital is almost 10 Å from the nucleus. Next, notice the *shape* of the 3*p* orbital. It is generally lobe-shaped, and it consists of four regions separated by nodes. The two inner regions resemble the lobes of a 2*p* orbital. The outer regions, however, are large and diffuse, and resemble mushroom caps. Finally, notice the number and the character of the nodes. A 3*p* orbital contains two nodes. One node is a plane through the nucleus, much like the node of a 2*p* orbital. The other is a spherical node, shown in Fig. 1.11b, that separates the inner part of each lobe from the larger outer part. *An orbital with principal quantum number n has n − 1 nodes.* Because the 3*p* orbital has $n = 3$, it has $(3 − 1) = 2$ nodes. The greater number of nodes in orbitals with higher *n* is a reflection of their higher energies. Again, the analogy to sound waves is striking: overtones of higher pitch have larger numbers of nodes.

C. Summary: Atomic Orbitals of Hydrogen

Here are the important points about orbitals in the hydrogen atom:

1. An orbital is an allowed state for the electron. It is a description of the wave motion of the electron. The mathematical description of an orbital is called a *wavefunction.*

2. Electron density within an orbital is a matter of probability, by the Heisenberg uncertainty principle. We can think of an orbital as a "smear" of electron density.

3. Orbitals are described by three quantum numbers:

   a. The principal quantum number *n* governs the energy of an orbital; orbitals of higher *n* have higher energy.

b. The angular momentum quantum number *l* governs the shape of an orbital. Orbitals with $l = 0$ (*s* orbitals) are spheres; orbitals with $l = 1$ (*p* orbitals) have lobes oriented along an axis.

c. The magnetic quantum number m_l governs the orientation of an orbital.

4. Orbitals with $n > 1$ contain nodes, which are surfaces of zero electron density. The nodes separate peaks of electron density from troughs, or, equivalently, regions in which the wavefunction describing an orbital has opposite sign. Orbitals with principal quantum number *n* have $n - 1$ nodes.

5. Orbital size increases with increasing *n*.

PROBLEMS

1.14   Sketch a plot of the wavefunction $\psi = \sin nx$ for the domain $0 \le x \le \pi$ for $n = 1, 2,$ and 3. What is the relationship between the "quantum number" *n* and the number of nodes in the wavefunction?

1.15   Use the trends in orbital shapes you've just learned to describe the general features of
(a) a 3*s* orbital   (b) a 4*s* orbital

1.7 ELECTRONIC STRUCTURES OF MORE COMPLEX ATOMS

The orbitals available to electrons in atoms with atomic number greater than 1 are, to a useful approximation, essentially like those of the hydrogen atom. This similarity includes the shapes and nodal properties of the orbitals. There is, however, one important difference: In atoms other than hydrogen, electrons with the same principal quantum number *n* but with different values of *l* have different energies. For example, carbon and oxygen, like hydrogen, have 2*s* and 2*p* orbitals, but, unlike hydrogen, electrons in these orbitals differ in energy. The ordering of energy levels for atoms with more than one electron is illustrated schematically in Fig. 1.12. As this figure shows, the gaps between energy levels become progressively smaller as the principal quantum number increases. Furthermore, the energy gap between orbitals that differ in principal quantum number is greater than the gap between two orbitals within the same principal quantum level. Thus, the difference in energy between 2*s* and 3*s* orbitals is greater than the difference in energy between 3*s* and 3*p* orbitals.

FIGURE 1.12 The relative energies of different orbitals illustrated for the first three principal quantum numbers of the carbon atom. The 1*s* orbital energy on this scale is almost five page lengths below the 2*s* energy! The energy separations differ for different atoms. The 4*s* orbital, not shown, has about the same energy as the 3*d* orbital, and the 4*s* energy drops below the 3*d* energy for atoms of higher atomic number.

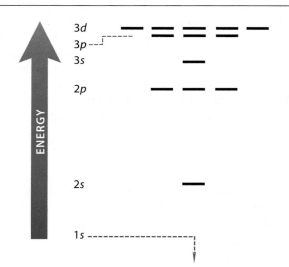

Atoms beyond hydrogen have more than one electron. Let's now consider the **electronic configurations** of these atoms—that is, the way their electrons are distributed among their atomic orbitals. To describe electronic configurations we need to introduce the concept of **electron spin**, which is a magnetic property of the electron. An electron can have only two values of spin, sometimes described as "up" and "down." Spin is characterized by a fourth quantum number m_s, which, in quantum theory, can have the values $+\frac{1}{2}$ ("up") and $-\frac{1}{2}$ ("down"). Four quantum numbers, then, are associated with any electron in an atom: the three orbital quantum numbers n, l, and m_l, and the spin quantum number m_s.

The **aufbau principle** (German, meaning "buildup principle") tells us how to determine electronic configurations. According to this principle, electrons are placed one by one into orbitals of the lowest possible energy in a manner consistent with the Pauli exclusion principle and Hund's rules. The **Pauli exclusion principle** states that no two electrons may have all four quantum numbers the same. As a consequence of this principle, a maximum of two electrons may be placed in any one orbital, and these electrons must have different spins. To illustrate, consider the electronic configuration of the helium atom, which contains two electrons. Both electrons can be placed into the $1s$ orbital as long as they have differing spin. Consequently, we can write the electronic configuration of helium as follows:

$$\text{helium, He: } (1s)^2$$

This notation means that helium has two electrons in a $1s$ orbital; and, because they occupy the same orbital, they must have opposite spins.

To illustrate Hund's rules, consider the electronic configuration of carbon, a central element in organic chemistry. A carbon atom has six electrons. The first two electrons (with opposite spins) go into the $1s$ orbital; the next two (also with opposite spins) go into the $2s$ orbital. Hund's rules tell us how to distribute the remaining two electrons among the *three equivalent 2p orbitals*. **Hund's rules** state, first, that to distribute electrons among identical orbitals of equal energy, single electrons are placed into separate orbitals before the orbitals are filled; and second, that the spins of these unpaired electrons are the same. Representing electrons as arrows, and letting their relative directions correspond to their relative spins, we can show the electronic configuration of carbon as follows:

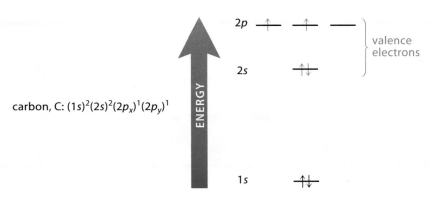

carbon, C: $(1s)^2(2s)^2(2p_x)^1(2p_y)^1$

In accordance with Hund's rules, the electrons in the carbon $2p$ orbitals are unpaired with identical spin. Placing two electrons in different $2p$ orbitals ensures that repulsions between electrons are minimized, because electrons in different $2p$ orbitals occupy different regions of space. (Recall from Fig. 1.10c that the three $2p$ orbitals are mutually perpendicular.) As shown above, we can also write the electronic configuration of carbon more concisely as $(1s)^2(2s)^2(2p_x)^1(2p_y)^1$, which shows the two $2p$ electrons in different orbitals. (The choice of x and y as subscripts is arbitrary; $2p_x$ and $2p_z$, or other combinations, are equally valid; the important point about this notation is that the two half-populated $2p$ orbitals are *different*.)

Let's now re-define the term *valence electrons*, first defined in Sec. 1.2A, in light of what we've learned about quantum theory. The **valence electrons** of an atom are the electrons that occupy the orbitals with the highest principal quantum number. (This definition applies *only* to elements in the "A" groups—that is, the nontransition groups—of the periodic table.)

For example, the $2s$ and $2p$ electrons of carbon are its valence electrons. A neutral carbon atom, therefore, has four valence electrons. The **valence orbitals** of an atom are the orbitals that contain the valence electrons. Thus, the $2s$ and $2p$ orbitals are the valence orbitals of carbon. It is important to be able to identify the valence electrons of common atoms because *chemical interactions between atoms involve their valence electrons and valence orbitals.*

STUDY PROBLEM   1.4

Describe the electronic configuration of the sulfur atom. Identify the valence electrons and valence orbitals.

SOLUTION   Because sulfur has an atomic number of 16, a neutral sulfur atom has 16 electrons. Following the aufbau principle, the first two electrons occupy the $1s$ orbital with opposite spins. The next two, again with opposite spins, occupy the $2s$ orbital. The next six occupy the three $2p$ orbitals, with each $2p$ orbital containing two electrons of opposite spin. The next two electrons go into the $3s$ orbital with paired spins. The remaining four electrons are distributed among the three $3p$ orbitals. Taking Hund's rules into account, the first three of these electrons are placed, unpaired and with identical spin, into the three equivalent $3p$ orbitals: $3p_x$, $3p_y$, and $3p_z$. The one remaining electron is then placed, with opposite spin, into the $3p_x$ orbital. To summarize:

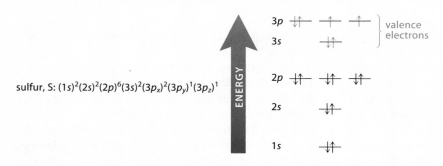

sulfur, S: $(1s)^2(2s)^2(2p)^6(3s)^2(3p_x)^2(3p_y)^1(3p_z)^1$

As shown in the diagram, the $3s$ and $3p$ electrons are the valence electrons of sulfur; the $3s$ and $3p$ orbitals are the valence orbitals.

PROBLEM

1.16   Give the electronic configurations of each of the following atoms and ions. Identify the valence electrons and valence orbitals in each.

   (a) oxygen atom   (b) chloride ion, Cl^- (c) potassium ion, K^+ (d) sodium atom

---

1.8   ANOTHER LOOK AT THE COVALENT BOND: MOLECULAR ORBITALS

A. Molecular Orbital Theory

One way to think about chemical bonding is to assume that a bond consists of two electrons localized between two specific atoms. This is the simplest view of a Lewis electron-pair bond. As useful as this picture is, it is sometimes too restrictive. When atoms combine into a molecule, the electrons contributed to the chemical bonds by each atom are no longer localized on individual atoms but "belong" to the entire molecule. Consequently, atomic orbitals are

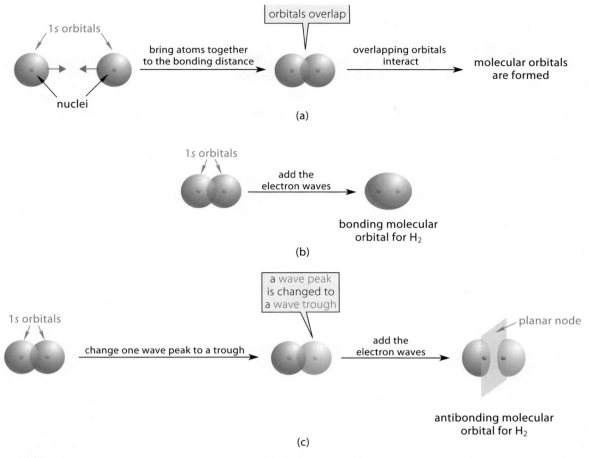

FIGURE 1.13 Formation of H_2 molecular orbitals. (a) Two hydrogen atoms are brought to the H—H bonding distance so that the $1s$ orbitals overlap. Interaction of these atomic orbitals forms the molecular orbitals. (b) To form the bonding MO of H_2, *add* the $1s$ orbital wavefunctions of the interacting hydrogen atoms. (c) To form the antibonding MO of H_2, *subtract* the $1s$ orbital wavefunctions by changing one peak to a trough and then adding. This process results in a node in the antibonding MO.

no longer appropriate descriptions for the state of electrons in molecules. Instead, **molecular orbitals**, nicknamed **MOs**, which are orbitals for the *entire molecule*, are used.

Determining the electronic configuration of a molecule is a lot like determining the electronic configuration of an atom, except that molecular orbitals are used instead of atomic orbitals. The following four steps summarize conceptually how we start with two isolated hydrogen atoms and end up with the electronic configuration of the dihydrogen molecule, H_2.

Step 1. Start with the isolated atoms of the molecule and bring them together to the positions that they have in the molecule. Their valence atomic orbitals will overlap.

For H_2, this means bringing the two hydrogen atoms together until the nuclei are separated by the length of the H—H bond (Fig. 1.13a). At this distance, the $1s$ orbitals of the atoms overlap.

Step 2. Allow the overlapping valence atomic orbitals to interact to form molecular orbitals (MOs).

This step implies that MOs of H_2 are derived by combining the $1s$ atomic orbitals of the two hydrogen atoms in a certain way. Conceptually, this is reasonable: molecules result from a combination of atoms, so molecular orbitals result from a combination of atomic orbitals. We'll learn below the process for combining atomic orbitals to form molecular orbitals.

Step 3. **Arrange the MOs in order of increasing energy.**
Steps 1 and 2 will yield two MOs for H_2 that differ in energy. We'll also learn below how to determine relative energies of these MOs.

Step 4. **Determine the electronic configuration of the molecule by redistributing the electrons from the constituent atoms into the MOs in order of increasing MO energy; the Pauli principle and Hund's rules are used.**
We redistribute the two electrons (one from each starting hydrogen atom) into the MOs of H_2 to give the electronic configuration of the molecule.

How to carry out steps 2 and 3 is the key to understanding the formation of molecular orbitals. Quantum theory gives us a few simple rules that allow us to derive the essential features of molecular orbitals without any calculations. We'll state these rules as they apply to H_2 and other cases involving the overlap of two atomic orbitals. (These rules will require only slight modification for more complex cases.)

Rules for forming molecular orbitals:

1. *The combination of two atomic orbitals gives two molecular orbitals.*
   For H_2, this rule means that the overlap of two 1s orbitals from the constituent hydrogen atoms gives two molecular orbitals. Later, we'll have situations in which we combine more than two atomic orbitals. When we combine *j* atomic orbitals, we always obtain *j* molecular orbitals.

2. *One molecular orbital is derived by the addition of the two atomic orbitals in the region of overlap.*
   To apply this to H_2, remember that the 1s orbital is a wave peak. When we add two wave peaks, they reinforce. When we add two 1s orbitals in the overlap region, they reinforce to form a continuous orbital that includes the region between the two nuclei (Fig. 1.13b). This molecular orbital is called a **bonding molecular orbital**, or **bonding MO**. The reason for the name is that, when electrons occupy this MO, they are attracted to both nuclei simultaneously. In other words, the electrons occupy not only the region around the nuclei but also *the region between the nuclei*, thus providing "electron cement" that holds the nuclei together, just as mortar between two bricks holds the bricks together.

3. *The other molecular orbital is derived by subtraction of the two atomic orbitals in the region of overlap.*
   To subtract the two 1s orbitals, we change either one of the 1s orbitals from a peak to a trough. (This is equivalent to changing the mathematical sign of the 1s wavefunction.) Then we add the two resulting orbitals. This process is illustrated in Fig. 1.13c. Adding a wave peak to a wave trough results in *cancellation* of the two waves in the region of overlap and formation of a *node*—a region in which the wave is zero. In this case, the node is a plane. The resulting orbital is called an **antibonding molecular orbital** or **antibonding MO**. Electrons that occupy this MO decrease bonding because the region between the nuclei contains no electron density.

4. *The two molecular orbitals have different energies. Orbital energy increases with the number of nodes. The bonding MO has a lower energy than the isolated 1s orbitals and the antibonding MO has a higher energy than the isolated 1s orbitals.*
   The orbital energies are summarized in an **orbital interaction diagram**, shown in Fig. 1.14. This diagram is a plot of orbital energy versus the nuclear positions of the two interacting atoms. The isolated atomic orbitals and their energies are shown on the left and right sides of the diagram, and the molecular orbitals and their energies are shown in the center, where the separation of the atoms corresponds to the bond length. *The number of nodes tells us the relative energies of*

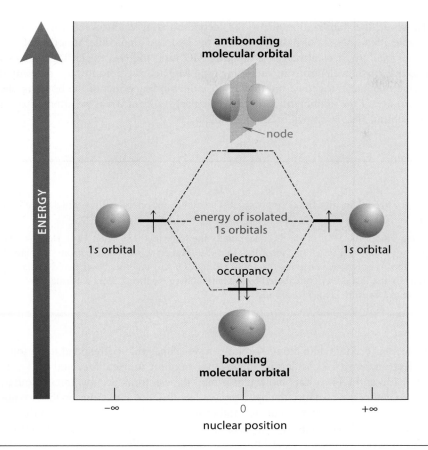

FIGURE 1.14 An orbital interaction diagram for the formation of the H_2 molecular orbitals from interacting $1s$ orbitals of two hydrogen atoms. The dashed lines show schematically how the two $1s$ orbitals interact as the internuclear distance changes from very large ($\pm \infty$) to the H—H bond length. The bonding MO has lower energy than the $1s$ orbitals and the antibonding MO has higher energy. Both electrons occupy the bonding MO.

the MOs: the more nodes an MO has, the higher is its energy. The bonding MO has no nodes and therefore has the lower energy. The antibonding MO has one node and has the higher energy. Notice that the energies of the two MOs "spread" about the energy of the isolated $1s$ orbitals—the energy of the bonding MO is lowered by a certain amount and the energy of the antibonding MO is raised by the same amount.

Now that we've described how to form the MOs and rank their energies, we're ready to populate these MOs with electrons. We apply the aufbau principle. We have two electrons—one from each hydrogen atom—to redistribute. Both can be placed in the bonding MO with opposite spins. Electron occupancy of the bonding MO is also shown in Fig. 1.14.

When we talk about the energy of an orbital, what we are really talking about is the energy of an electron that occupies the orbital. It follows, then, that the electrons in the bonding MO have lower energy than two electrons in their parent $1s$ orbitals. In other words, *chemical bonding is an energetically favorable process.* Each electron in the bonding MO of H_2 contributes about half to the stability of the H—H bond. It takes about 435 kJ (104 kcal) to dissociate a mole of H_2 into hydrogen atoms, or about 218 kJ (52 kcal) per bonding electron. This a lot of energy on a chemical scale—more than enough to raise the temperature of a kilogram of water from freezing to boiling.

According to the picture just developed, the chemical bond in a hydrogen molecule results from the occupancy of a bonding molecular orbital by two electrons. You may wonder why we concern ourselves with the antibonding molecular orbital if it is not occupied. The reason is that it *can* be occupied! If a third electron were introduced into the hydrogen molecule, then the antibonding molecular orbital would be occupied. The resulting three-electron species is the hydrogen molecule anion, H_2^- (see Prob. 1.17b, p. 34). H_2^- exists because each electron in the bonding molecular orbital of the hydrogen molecule contributes equally to the stability of the molecule. The third electron in H_2^-, the one in the antibonding molecular orbital, has a high energy that offsets the stabilization afforded by *one* of the bonding electrons. However, the stabilization due to the second bonding electron remains. Thus, H_2^- is a stable species, but only about half as stable as the hydrogen molecule. In terms of our brick-and-mortar analogy, if electrons in a bonding MO bind the two nuclei together as mortar binds two bricks, then electrons in an antibonding MO act as "anti-mortar": not only do they *not* bind the two nuclei together, but they oppose the binding effect of the bonding electrons. The importance of the antibonding MO is particularly evident when we attempt to construct diatomic helium, He_2, as shown in Study Problem 1.5.

STUDY PROBLEM 1.5

Use molecular orbital theory to explain why He_2 does not exist. The molecular orbitals of He_2 are formed in the same way as those of H_2.

SOLUTION   The orbital interaction diagram for the MOs of He_2 is conceptually the same as for H_2 (Fig. 1.14). However, He_2 contains four electrons—two from each He atom. According to the aufbau principle, two electrons are placed into the bonding MO, but the other two must occupy the antibonding MO. Any stability contributed by the bonding electrons is offset by the instability contributed by the antibonding electrons. Hence, formation of He_2 has no energetic advantage. As a result, He is monatomic.

Molecular orbitals, like atomic orbitals, have shapes that correspond to regions of significant electron density. Consider the shape of the bonding molecular orbital of H_2, shown in both Figs 1.13b and 1.14. In this molecular orbital the electrons occupy an ellipsoidal region of space. No matter how we turn the hydrogen molecule about a line joining the two nuclei, its electron density looks the same. This is another way of saying that the bond in the hydrogen molecule has **cylindrical symmetry**. Other cylindrically symmetrical objects are shown in Fig. 1.15. Bonds in which the electron density is cylindrically symmetrical about the internuclear axis are called **sigma bonds** (abbreviated σ bonds). The bond in the hydrogen molecule is thus a σ bond. The lowercase Greek letter sigma was chosen to describe the bonding molecular orbital of hydrogen because it is the Greek letter equivalent of *s*, the letter used to describe the atomic orbital of lowest energy.

PROBLEMS

1.17   Draw an orbital interaction diagram corresponding to Fig. 1.14 for each of the following species. Indicate which are likely to exist as diatomic species, and which would dissociate into monatomic fragments. Explain.

(a) the He_2^+ ion   (b) the H_2^- ion   (c) the H_2^{2-} ion   (d) the H_2^+ ion

1.18   The bond dissociation energy of H_2 is 435 kJ mol^{-1} (104 kcal mol^{-1}); that is, it takes this amount of energy to dissociate H_2 into its atoms. Estimate the bond dissociation energy of H_2^+ and explain your answer.

FIGURE 1.15 Some cylindrically symmetrical objects. Objects are cylindrically symmetrical when they appear the same no matter how they are rotated about their cylindrical axis (black line).

a barrel

a top

a goblet

bonding molecular orbital of the hydrogen molecule

B. Molecular Orbital Theory and the Lewis Structure of H₂

Let's now relate the quantum-mechanical description of H_2 to the concept of the Lewis electron-pair bond. In the Lewis structure of H_2, the bond is represented by an electron pair shared between the two nuclei. In the quantum-mechanical description, the bond is the result of the presence of two electrons in a bonding molecular orbital and the resulting electron density between the two nuclei. Both electrons are attracted to each nucleus, and these electrons thus serve as the "cement" that holds the nuclei together. Thus, for H_2, *the Lewis electron-pair bond is equivalent to the quantum-mechanical idea of a bonding molecular orbital occupied by a pair of electrons.* The Lewis picture places the electrons squarely between the nuclei. Quantum theory says that, although the electrons have a high probability of being between the bound nuclei, they can also occupy other regions of space.

The formula for bond order in MO theory reinforces the connection between an occupied bonding MO and the electron-pair bond. Recall that the *bond order* (Sec. 1.3B, p. 15) describes whether a bond between two nuclei is single (bond order = 1), double (bond order = 2), or triple (bond order = 3). The formula for bond order in MO theory is

$$\text{bond order} = \frac{\text{electrons in bonding MOs} - \text{electrons in antibonding MOs}}{2} \qquad (1.8)$$

The denominator 2 represents the fact that a full covalent bond requires two electrons. Applying this to the H_2 molecule, for example, there are two electrons in the bonding MO and no electrons in the antibonding MO; therefore, the bond order is $(2 - 0)/2 = 1$. Therefore, the covalent bond in dihydrogen is a single electron-pair bond.

Molecular orbital theory shows, however, that a chemical bond need not be an electron *pair*. For example, H_2^+ (the hydrogen molecule cation, which we might represent in the Lewis sense as H ‡ H) is a stable species in the gas phase (see Prob. 1.17d). It is not so stable as the hydrogen molecule itself because the ion has only one electron in the bonding molecular orbital, rather than the two found in a neutral hydrogen molecule. The hydrogen molecule anion, H_2^-, discussed in the previous section, might be considered to have a three-electron bond consisting of two bonding electrons and one antibonding electron. The electron in the antibonding orbital is also shared by the two nuclei, but shared in a way that reduces the energetic advantage of bonding. (H_2^- is not so stable as H_2; Sec. 1.8A.) This example demonstrates that the sharing of electrons between nuclei in some cases does *not* contribute to bonding. Nevertheless, the most stable arrangement of electrons in the dihydrogen molecular orbitals occurs when the bonding MO contains two electrons and the antibonding MO is empty—in other words, when there is an electron-pair bond.

PROBLEM

1.19 Referring to your solution to Problem 1.17d, calculate the bond order of the covalent bond in the H_2^+ ion. How does this result bear on the answer to Problem 1.18?

1.9   HYBRID ORBITALS

A. Bonding in Methane

We ultimately want to describe the chemical bonding in organic compounds, and our first step in this direction is to understand the bonding in methane, CH_4. Before quantum theory was applied to the bonding problem, it was known experimentally that the hydrogens in methane, and thus the bonds to these hydrogens, were oriented tetrahedrally about the central carbon. The valence orbitals in a carbon atom, however, are *not* directed tetrahedrally. The $2s$ orbital, as you've learned, is spherically symmetrical (see Fig. 1.8, p. 25), and the $2p$ orbitals are perpendicular (see Fig. 1.10, p. 26). If the valence orbitals of carbon aren't directed tetrahedrally, why is methane a tetrahedral molecule?

The modern solution to this problem is to apply molecular orbital theory. You can't do this with just the simple rules that we applied to H_2, but it can be done. The result is that the combination of one carbon $2s$ and three carbon $2p$ orbitals with four tetrahedrally placed hydrogen $1s$ orbitals gives four bonding MOs and four antibonding MOs. (The combination of eight atomic orbitals give eight molecular orbitals; rule 1, p. 32, with $j = 8$.) Eight electrons (four from carbon and one from each of the four hydrogens) are just sufficient to fill the four bonding MOs with electron pairs. This molecular orbital description of methane accounts accurately for its electronic properties.

The conceptual difficulty with the molecular orbital description of methane is that we can't associate a given pair of electrons in the molecule with any one bond. Instead, the electrons from all of the atoms are redistributed throughout the entire molecule. We can't even tell where atoms begin or end! If we add up all of the contributions of the electrons in the four bonding MOs of methane, we obtain a picture of the **total electron density**—that is, the probability of finding electrons in the methane molecule. (The electrostatic potential maps, or EPMs, introduced in Sec. 1.2D, are superimposed on such pictures of total electron density.)

methane
total electron density

methane
total electron density
with imbedded model

dihydrogen
total electron density
with imbedded model

In this picture of total electron density, methane looks like an "electron pudding" containing the nuclei. Although this electron density has a tetrahedral shape, there are no discrete C—H bonds. In contrast, because H_2 has only one bond, we can associate the total electron density in H_2 with the H—H bond, as shown in the previous section.

Historically, chemists have liked to think that molecules are made up of atoms connected by individual bonds. We like to build models, hold them in our hands, and manipulate chemical bonds by plucking off certain atoms and replacing them by others. Although the molecular orbital description of methane certainly describes *bonding*, it suggests that the discrete chemical bond between individual atoms is something rationally conceivable but not rigorously definable (except perhaps for simple molecules like H_2). Nevertheless, the concept of the chemical bond is so useful in organic chemistry that we can't ignore it! The problem, then, is this: is there an electronic theory of bonding in methane (other than Lewis structures) that allows us to retain the notion of discrete C—H bonds?

> To use a bonding theory that we know isn't quite right may seem inappropriate, but science works this way. A theory is a framework for unifying a body of knowledge in such a way that we can use it to make useful predictions. An example you're probably familiar with is the ideal gas law, $PV = nRT$. Most real gases don't follow this law exactly, but it can be used to make some useful predictions. For example, if you're wondering what will happen to the pressure in your automobile tires when the temperature drops in winter, this law gives a perfectly useful answer: the pressure drops. If you're interested in calculating exactly how much the pressure will drop per degree, you might need a more exact theory. We'll find it necessary to use molecular orbital theory to explain certain phenomena, but in many cases we can get by with a simpler but less accurate theory that is attractive because it allows us to think in terms of discrete chemical bonds.

An electronic description of bonding in methane that retains the C—H bonds was developed in 1928 by Linus Pauling (1901–1994), a chemist at the California Institute of Technology, who received the 1954 Nobel Prize in Chemistry for his work on chemical bonding. Pauling's theory started with the premise that the valence orbitals of the carbon *in methane* are different from the orbitals in *atomic carbon*. However, the orbitals of carbon in methane can be derived simply from those of atomic carbon. For carbon in methane, we imagine that the 2s orbital and the three 2p orbitals are mixed to give four new *equivalent* orbitals, each with a character intermediate between pure *s* and pure *p*. It's as if we mixed one dog and three cats and ended up with four identical animals, each of which is three-fourths cat and one-fourth dog. This mixing process applied to orbitals is called **hybridization**, and the new orbitals are called **hybrid orbitals**. More specifically, hybrid orbitals result from the mixing of atomic orbitals with different *l* quantum numbers. Because each of the new hybrid carbon orbitals is one part *s* and three parts *p*, it is called an sp^3 orbital (pronounced "s-p-three," not "s-p-cubed"). The six carbon electrons in this orbital picture are distributed between one 1s orbital and four equivalent sp^3 hybrid orbitals in quantum level 2. This mental transformation can be summarized as follows:

This orbital mixing can be done mathematically, and it yields the perspective drawing of an sp^3 hybrid orbital shown in Fig. 1.16a on p. 38. A simpler representation used in most texts is shown in Fig. 1.16b. As you can see from these pictures, an sp^3 orbital consists of two lobes separated by a node, much like a 2p orbital. However, one of the lobes is very small, and the other is very large. In other words, *the electron density in an sp^3 hybrid orbital is highly directed in space.* This directional character is ideal for bond formation along the axis of the large lobe.

The number of hybrid orbitals (four in this case) is the same as the number of orbitals that are mixed to obtain them. (One *s* orbital + three *p* orbitals = four sp^3 orbitals.) The large lobes of the four carbon sp^3 orbitals are directed to the corners of a regular tetrahedron, as shown in Figure 1.16c. In hybridization theory, each of the four electron-pair bonds in methane results from the overlap of a hydrogen 1s orbital containing one electron with a carbon sp^3 orbital, also containing a single electron. The resulting bond is a σ bond.

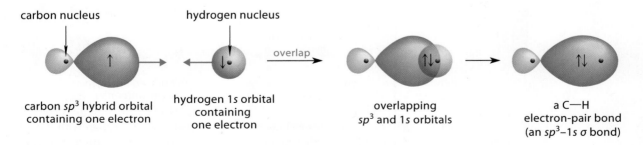

carbon sp^3 hybrid orbital
containing one electron

hydrogen 1s orbital
containing
one electron

overlapping
sp^3 and 1s orbitals

a C—H
electron-pair bond
(an sp^3–1s σ bond)

This overlap looks a lot like the overlap of two atomic orbitals that we carried out in constructing the molecular orbitals of H_2. However, *the hybrid orbital treatment is not a molecular orbital treatment because it deals with each bond in isolation.* The hybrid orbital bonding picture for methane is shown in Fig. 1.16d.

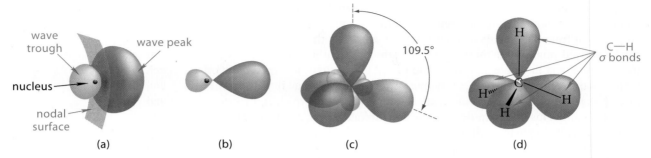

FIGURE 1.16 (a) Perspective representation of a carbon sp^3 hybrid orbital. (b) A more common representation of an sp^3 orbital used in drawings. (c) The four sp^3 orbitals of carbon shown together. (d) An orbital picture of tetrahedral methane showing the four equivalent σ bonds formed from the overlap of carbon sp^3 and hydrogen $1s$ orbitals. The rear lobes of the orbitals shown in (c) are omitted in (d) for clarity.

The hybridization of carbon itself actually costs energy. (If this weren't so, carbon atoms would exist in a hybridized configuration.) Remember, though, that this is a model for carbon *in methane*. Hybridization allows carbon to form four bonds to hydrogen that are much stronger than the bonds that would be formed without hybridization, and the strength of these bonds more than offsets the energy required for hybridization. Why does hybridization make these bonds stronger? First, the bonds are as far apart as possible, and repulsion between electron pairs in the bonds is therefore minimized. The pure s and p orbitals available on nonhybridized carbon, in contrast, are not directed tetrahedrally. Second, in each hybridized orbital, the bulk of the electron density is directed toward the bound hydrogen. This directional character provides more electron "cement" between the carbon and hydrogen nuclei, and this results in stronger (that is, more stable) bonds.

B. Bonding in Ammonia

The hybrid orbital picture is readily extended to compounds containing unshared electron pairs, such as ammonia, $:NH_3$. The valence orbitals of nitrogen in ammonia are, like the carbon in methane, hybridized to yield four sp^3 hybrid orbitals; however, unlike the corresponding carbon orbitals, one of these hybrid orbitals is fully occupied with a pair of electrons.

Each of the sp^3 orbitals on nitrogen containing one electron can overlap with the $1s$ orbital of a hydrogen atom, also containing one electron, to give one of the three N—H σ bonds of ammonia. The electrons in the filled sp^3 orbital on nitrogen become the unshared electron pair in ammonia. The unshared pair and the three N—H bonds, because they are

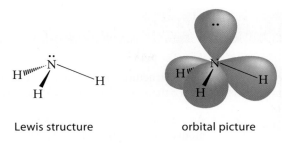

FIGURE 1.17 The hybrid orbital description of ammonia, :NH₃. As in Fig. 1.16, the small rear lobes of the hybrid orbitals are omitted for clarity.

Lewis structure   orbital picture

made up of sp^3 hybrid orbitals, are directed to the corners of a regular tetrahedron (Fig. 1.17). The advantage of orbital hybridization in ammonia is the same as in carbon: hybridization accommodates the maximum separation of the unshared pair and the three hydrogens and, at the same time, provides strong, directed N—H bonds.

You may recall from Sec. 1.3B that the H—N—H bond angle in ammonia is 107.3°, a little smaller than tetrahedral (109.5°). Our hybrid-orbital picture can accommodate this structural refinement as well. Unshared electron pairs prefer s orbitals, because s orbitals have lower energy than p orbitals. Or, to look at it another way, there's no energetic advantage to putting an unshared pair in a spatially directed orbital if it's not going to be involved in a chemical bond. But if the unshared pair were left in an unhybridized $2s$ orbital, each bond to hydrogen would have to be derived from a pure nitrogen $2p$ orbital. In such a bond, half of the electron density ("electron cement") would be directed away from the hydrogen, and the bond would be weak. In such a case, the H—N—H bond angle would be 90°, the same as the angle between the $2p$ orbitals used to form the bonds. The actual geometry of ammonia is a compromise between the preference of unshared pairs for orbitals of high s character and the preference of bonds for hybrid character. The orbital containing the unshared pair has a little more s character than the bonding orbitals. Because s orbitals cover an entire sphere (see Fig. 1.8, p. 25), orbitals with more s character occupy more space. Hence, unshared pairs have a greater spatial requirement than bonds. Hence, the angle between the unshared pair and each of the N—H bonds is somewhat *greater* than tetrahedral, and the bond angles between the N—H bonds, as a consequence, are somewhat *less* than tetrahedral. This is the same conclusion we obtained from the application of VSEPR theory to ammonia (p. 17).

A connection exists between the hybridization of an atom and the arrangement in space of the bonds around that atom. Atoms surrounded by four groups (including unshared pairs) in a tetrahedral arrangement are sp^3-hybridized. The converse is also true: sp^3-hybridized atoms always have tetrahedral bonding geometry. A trigonal planar bonding arrangement is associated with a different hybridization, and a linear bonding arrangement with yet a third type of hybridization. (These types of hybridization are discussed in Chapters 4 and 14.) In other words, *hybridization and molecular geometry are closely correlated.*

The hybridization picture of covalent bonding also drives home one of the most important differences between the ionic and covalent bond: *the covalent bond has a definite direction in space*, whereas the ionic bond is the same in all directions. *The directionality of covalent bonding is responsible for molecular shape*; and, as we shall see, molecular shape has some very important chemical consequences.

PROBLEM

1.20   (a) Construct a hybrid orbital picture for the water molecule using oxygen sp^3 hybrid orbitals.

(b) Predict any departures from tetrahedral geometry that you might expect from the presence of two unshared electron pairs. Explain your answer.

KEY IDEAS IN CHAPTER 1

- Chemical compounds can contain two types of bonds: ionic and covalent. In ionic compounds, ions are held together by electrostatic attraction (the attraction of opposite charges). In covalent compounds, atoms are held together by the sharing of electrons.

- Both the formation of ions and bonding in covalent compounds tend to follow the octet rule: each atom is surrounded by eight valence electrons (two electrons for hydrogen).

- The formal-charge convention assigns charges within a given species to its constituent atoms. The calculation of formal charge is given in Study Problem 1.1, p. 7. Formal charge is a bookkeeping device. In some cases the actual charge on an atom and the formal charge do not correspond.

- In polar covalent bonds, electrons are shared unequally between bonded atoms with different electronegativities. This unequal sharing results in a bond dipole moment. The dipole moment of a molecule is the vector sum of its individual bond dipole moments. The local charge distribution in a molecule can be described graphically with an electrostatic potential map (EPM).

- The structure of a molecule is determined by its connectivity and its geometry. The molecular geometry of a molecule is determined by its bond lengths, bond angles, and dihedral angles. Bond lengths are governed, in decreasing order of importance, by the period of the periodic table from which the bonded atoms are derived; by the bond order (whether the bond is single, double, or triple); and by the column (group) of the periodic table from which the atoms in the bond are derived. Approximate bond angles can be predicted by assuming that the groups surrounding a given atom are as far apart as possible.

- Molecules that are not adequately described by a single Lewis structure are represented as resonance hybrids, which are weighted averages of two or more fictitious Lewis structures. Resonance hybrids are more stable than any of their contributing resonance structures.

- As a consequence of their wave properties, electrons in atoms and molecules can exist only in certain allowed energy states, called orbitals. Orbitals are descriptions of the wave properties of electrons in atoms and molecules, including their spatial distribution. Orbitals are described mathematically by wavefunctions.

- Electrons in orbitals are characterized by quantum numbers, which, for atoms, are designated n, l, and m_l. Electron spin is described by a fourth quantum number m_s. The higher the principal quantum number n of an electron, the higher its energy. In atoms other than hydrogen, the energy is also a function of the l quantum number.

- Some orbitals contain nodes, which separate the wave peaks of the orbitals from the wave troughs. An atomic orbital of quantum number n has $n - 1$ nodes.

- The distribution of electron density in a given type of orbital has a characteristic arrangement in space governed by the l quantum number: all s orbitals are spheres, all p orbitals contain two equal-sized lobes, and so on. The orientation of an orbital is governed by its m_l quantum number.

- Atomic orbitals and molecular orbitals are both populated with electrons according to the aufbau principle.

- Covalent bonds are formed when the orbitals of different atoms overlap. In molecular orbital theory, covalent bonding arises from the filling of bonding molecular orbitals by electrons.

- The directional properties of bonds can be understood by the use of hybrid orbitals. The hybridization of an atom and the geometry of the atoms attached to it are closely related. All sp^3-hybridized atoms have tetrahedral geometry.

ADDITIONAL PROBLEMS

The solutions to all problems can be found in the *Study Guide and Solutions Manual* supplement.

1.21 In each of the following sets, specify the one compound that is likely to have completely ionic bonds in its solid state.

(a) CCl_4 HCl   NaAt K_2

(b) CS_2 CsF   HF XeF_2 BF_3

1.22 Which of the non-hydrogen atoms in each of the following species has a complete octet? What is the formal charge on each? Assume all unshared valence electrons are shown.

(a) CH_3 (b) $:NH_3$ (c) $:CH_3$

(d) BH_3 (e) $:\ddot{I}:$ (f) BH_4

1.23 Draw one Lewis structure for each of the following compounds; show all unshared electron pairs. None of the atoms in the compounds bears a formal charge, and all atoms have octets (hydrogens have duets).

(a) C_2H_3Cl

(b) ketene, C_2H_2O, which has a carbon–carbon double bond

(c) acetonitrile, C_2H_3N, which has a carbon–nitrogen triple bond

1.24 Draw two Lewis structures for a compound with the formula C_4H_{10}. No atom bears a charge, and all carbons have complete octets.

1.25 Give the formal charge on each atom and the net charge on each species in the following structures. All unshared valence electrons are shown.

(a)

:$\ddot{O}$:
|
:$\ddot{O}$—Cl—$\ddot{O}$:
|
:$\ddot{O}$:

perchlorate

(b)

:$\ddot{O}$:
|
N
/ | \
H_3C | CH_3
CH_3

trimethylamine oxide

(c)

$\ddot{O}$=$\overset{..}{O}$
/   \
$\ddot{O}$ $\ddot{O}$:

ozone

(d)

$\ddot{C}$
/   \
H   H

methylene

(e)

H
|
H_3C—$\overset{.}{C}$
|
H

ethyl radical

(f) :$\ddot{Cl}$—$\ddot{O}$:

hypochlorite

1.26 Give the electronic configuration of (a) the chlorine atom; (b) the silicon atom (Si); (c) the argon atom; (d) the magnesium atom. Indicate the valence electrons and the valence orbitals of Si.

1.27 Which of the following orbitals is (are) *not* permitted by the quantum theory of the hydrogen atom? Explain.

$2s$ $6s$ $5d$ $2d$ $3p$

1.28 Predict the approximate bond angles in each of the following molecules, and explain your reasoning.

(a) $:CH_2$ (b) BeH_2 (c) $^+CH_3$

(d) $:\ddot{Cl}_4Si$ (e) $\ddot{O}$=$\overset{+}{O}$—$\ddot{O}:^-$

ozone

(f) H_2C=C=CH_2 (Give H—C—C and C—C—C angles.)

allene

(g)

H_3C—$\overset{+}{N}$
/ \
$\ddot{O}:^-$:O:

1.29 Estimate each of the bond angles and order the bond lengths (smallest first) in the following molecule. State any points of ambiguity and explain.

1.30 (a) Construct a hybrid-orbital picture for the hydronium ion (H_3O^+) using oxygen sp^3-hybridized orbitals.

(b) How would you expect the H—O—H bond angles in hydronium ion to compare with those in water (larger or smaller)? Explain.

1.31 The *allyl cation* can be represented by the following resonance structures.

$$\left[\overset{H}{\underset{}{H_2\overset{+}{C}-C=CH_2}} \longleftrightarrow \overset{H}{\underset{}{H_2C=C-\overset{+}{C}H_2}} \right]$$

allyl cation

(a) What is the bond order of each carbon–carbon bond in the allyl cation?

(b) How much positive charge resides on each carbon of the allyl cation?

(c) Although the preceding structures are reasonable descriptions of the allyl cation, the following cation *cannot* be described by analogous resonance structures. Explain why the structure on the right is not a reasonable resonance structure.

$$\left[\overset{H}{\underset{}{H_2C=C-\overset{+}{N}H_3}} \quad\cancel{\longleftrightarrow}\quad \overset{H}{\underset{}{H_2\overset{+}{C}-C=NH_3}} \right]$$

1.32 Consider the resonance structures for the *carbonate ion.*

$$\left[\ \overset{:O:}{\underset{:\ddot{O}:}{\underset{|}{C}}}\overset{||}{\underset{\ddot{O}:}{}} \ \longleftrightarrow \ \overset{:\ddot{O}\bar{:}}{\underset{\ddot{O}:}{\underset{|}{C}}}\overset{}{\underset{\ddot{O}:}{}} \ \longleftrightarrow \ \overset{:\ddot{O}\bar{:}}{\underset{:\ddot{O}}{\underset{|}{C}}}\overset{}{\underset{\ddot{O}}{}} \ \right]$$

(a) How much negative charge is on each oxygen of the carbonate ion?

(b) What is the bond order of each carbon–oxygen bond in the carbonate ion?

1.33 (a) Two types of nodes occur in atomic orbitals: spherical surfaces and planes. Examine the nodes in $2s$, $2p$, and $3p$ orbitals, and show that they agree with the following statements:

    1. An orbital of principal quantum number n has $n - 1$ nodes.

    2. The value of m_l gives the number of planar nodes.

(b) How many spherical nodes does a $5s$ orbital have? A $3d$ orbital? How many nodes of all types does a $3d$ orbital have?

1.34 The shape of one of the five energetically equivalent $3d$ orbitals follows. From your answer to Problem 1.33, sketch the nodes of this $3d$ orbital, and associate a wave peak or a wave trough with each lobe of the orbital. (*Hint:* It doesn't matter where you put your first peak; you should be concerned only with the relative positions of peaks and troughs.)

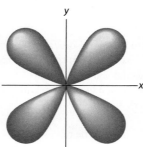

1.35 Sketch a $4p$ orbital. Show the nodes and the regions of wave peaks and wave troughs. (*Hint:* Use Fig. 1.11 and the descriptions of nodes in Problem 1.33a.)

1.36 Account for the fact that $H_3C—Cl$ (dipole moment 1.94 D) and $H_3C—F$ (dipole moment 1.82 D) have almost identical dipole moments, even though fluorine is considerably more electronegative than chlorine.

1.37 (a) Draw an appropriate bond dipole for the carbon–magnesium bond in dimethylmagnesium.

$$H_3C—Mg—CH_3$$

dimethylmagnesium

(b) What is the geometry of dimethylmagnesium? Explain.

(c) What conclusion can you draw about the dipole moment of dimethylmagnesium?

1.38 The principles for predicting bond angles do not permit a distinction between the following two conceivable forms of ethylene.

$$\overset{H}{\underset{H}{}}C=C\overset{H}{\underset{H}{}} \qquad \overset{H}{\underset{H}{}}C=C\overset{H}{\underset{H}{}}$$

      planar   staggered

The dipole moment of ethylene is zero. Does this experimental fact provide a clue to the preferred dihedral angles in ethylene? Why or why not?

1.39 (a) Give the $H—C=O$ bond angle in methyl formate.

$$H—\overset{:O:}{\underset{}{\overset{||}{C}}}—\ddot{O}—CH_3$$

methyl formate

(b) One dihedral angle in methyl formate is the angle between the plane containing the $O=C—O$ bond and the plane containing the $C—O—C$ bonds. (Notice that one bond is common to both.) Sketch the two structures of methyl formate: one in which this dihedral angle is $0°$ and the other in which it is $180°$.

1.40 A well-known chemist, Havno Szents, has heard you apply the rules for predicting molecular geometry to water; you have proposed (Problem 1.9b, p. 18) a bent geometry for this compound. Dr. Szents is unconvinced by your arguments and continues to propose that water is a linear molecule. He demands that you debate the issue with him before a distinguished academy. You must therefore come up with *experimental* data that will prove to an objective body of scientists that water indeed has a bent geometry. Explain why the dipole moment of water, 1.84 D, could be used to support your case.

1.41 Use your knowledge of vectors to explain why, even though the $C—Cl$ bond dipole is large, the dipole moment of carbon tetrachloride, CCl_4, is zero. (*Hint:* Take the resultant of any two $C—Cl$ bond dipoles; then take the resultant of the other two. Now add the two resultants to get the dipole moment of the molecule. Use models!)

1.42 Three possible dihedral angles for H_2O_2 ($0°$, $90°$, and $180°$) are shown in Fig. 1.6 on p. 19.

(a) Assume that the H_2O_2 molecule exists predominantly in one of these arrangements. Which of the dihedral angles can be ruled out by the fact that H_2O_2 has a large dipole moment (2.13 D)? Explain.

(b) The bond dipole moment of the $O—H$ bond is tabulated as 1.52 D. Use this fact and the overall dipole moment of H_2O_2 in part (a) to decide on the preferred dihedral angles in H_2O_2. Take the $H—O—O$ bond angle to be the known value (96.5°). (*Hint:* Apply the law of cosines.)

1.43 Given the dipole moment of water (1.84 D) and the $H—O—H$ bond angle (104.45°), justify the statement in

Problem 1.42b that the bond dipole moment of the O—H bond is 1.52 D.

1.44 Bring two 2s orbitals together to a bonding distance. (The wave troughs will overlap at this distance.) Form bonding and antibonding molecular orbitals, and show the electron occupancy diagram for the Li_2 molecule.

1.45 Consider two 2p orbitals, one on each of two atoms, oriented head-to-head as in Figure P1.45. Imagine bringing the nuclei closer together until the two wave peaks (the blue lobes) of the orbitals just overlap, as shown in the figure. A new system of molecular orbitals is formed by this overlap.

(a) Sketch the shape of the resulting bonding and antibonding molecular orbitals.

(b) Identify the nodes in each molecular orbital.

(c) Construct an orbital interaction diagram for molecular orbital formation.

(d) If two electrons occupy the bonding molecular orbital, is the resulting bond a σ bond? Explain.

1.46 Consider two 2p orbitals, one on each of two different atoms, oriented side-to-side, as in Figure P1.46. Imagine bringing these nuclei together so that overlap occurs as shown in the figure. This overlap results in a system of molecular orbitals.

(a) Sketch the shape of the resulting bonding and antibonding molecular orbitals.

(b) Identify the node(s) in each.

(c) Construct an orbital interaction diagram for molecular orbital formation.

(d) When two electrons occupy the bonding molecular orbital, is the resulting bond a σ bond? Explain.

1.47 In this problem we'll construct the molecular orbitals of the dioxygen molecule (O_2). Imagine bringing two oxygen atoms to within a bonding distance along the x axis (the horizontal axis in the plane of the page), and imagine the molecular orbitals that would form from overlap of the valence atomic orbitals.

(a) Sketch the bonding and antibonding combinations ($2s\sigma$ and $2s\sigma^*$, respectively) that are formed when the 2s atomic orbitals interact. Show the nodes. (*Hint:* See Problem 1.44.)

(b) Sketch the bonding and antibonding combinations ($2p\sigma$ and $2p\sigma^*$) that result when the $2p_x$ atomic orbitals overlap. (Note again that the x axis is the horizontal axis in the plane of the page.) Show the nodes. (*Hint:* See Problem 1.45.)

(c) Sketch the bonding and antibonding combinations ($2p\pi_y$ and $2p\pi_y^*$) that result when the $2p_y$ atomic orbitals overlap. (Assume that the y axis is the vertical axis in the plane of the page.) Show the nodes. (*Hint:* See Problem 1.46.)

(d) Show that the bonding and antibonding combinations ($2p\pi_z$ and $2p\pi_z^*$) that result when the $2p_z$ atomic orbitals overlap are identical to the $2p\pi_y$ and $2p\pi_y^*$ MOs in (c), but oriented at 90°. (Assume that the z axis is the vertical axis perpendicular to the plane of the page.)

(e) The energy order of these MOs and the component atomic orbitals is

$$2s\sigma < 2s \text{ (atomic orbitals)} < 2s\sigma^* < 2p\sigma < 2p\pi_y =$$

$$2p\pi_z < 2p \text{ (atomic orbitals)} < 2p\pi_y^* = 2p\pi_z^* < 2p\sigma^*$$

Construct an orbital interaction energy diagram showing the energy levels of the atomic orbitals along with the energies of these MOs. Add the available valence electrons from two oxygen atoms.

Figure P1.45

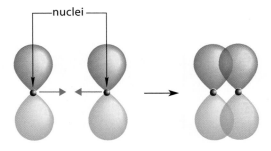

Figure P1.46

(f) Molecules that have net (nonzero) electron spin are magnetic. Explain why liquid O_2 can be trapped between the poles of a magnet.

(g) Which one of the following Lewis structures best describes the covalent bond(s) in O_2? Explain. (*Hint:* Apply Eq. 1.8.)

$$:\ddot{O}{=}\ddot{O}: \qquad \cdot\ddot{O}{-}\ddot{O}\cdot \qquad \cdot\ddot{O}{\equiv}\ddot{O}\cdot \qquad :\ddot{O}{=}\ddot{O}:$$
$$A \qquad\qquad B \qquad\qquad C \qquad\qquad D$$

$$\left[\cdot\ddot{O}{-}\ddot{O}\cdot \longleftrightarrow \cdot\ddot{O}{\equiv}\ddot{O}\cdot \right]$$
$$E$$

1.48 When a hydrogen molecule absorbs light, an electron jumps from the bonding molecular orbital to the antibonding molecular orbital. Explain why this light absorption can lead to the dissociation of the hydrogen molecule into two hydrogen atoms. (This process, called photodissociation, can sometimes be used to initiate chemical reactions.)

1.49 Suppose you take a trip to a distant universe and find that the periodic table there is derived from an arrangement of quantum numbers different from the one on Earth. The rules in that universe are

　　1. principal quantum number $n = 1, 2, \ldots$ (as on Earth);

　　2. angular momentum quantum number $l = 0, 1, 2, \ldots,$ $n - 1$ (as on Earth);

　　3. magnetic quantum number $m_l = 0, 1, 2, \ldots, l$ (that is, only *positive* integers up to and including l are allowed); and

　　4. spin quantum number $m_s = -1, 0, +1$ (that is, *three* allowed values of spin).

(a) Assuming that the Pauli exclusion principle remains valid, what is the maximum number of electrons that can populate a given orbital?

(b) Write the electronic configuration of the element with atomic number 8 in the periodic table.

(c) What is the atomic number of the second noble gas?

(d) What rule replaces the octet rule?

Alkanes

Organic compounds all contain carbon, but they can also contain a wide variety of other elements. Before we can appreciate such chemical diversity, however, we have to begin at the beginning, with the simplest organic compounds, the *hydrocarbons*. **Hydrocarbons** are compounds that contain only the elements carbon and hydrogen.

2.1 HYDROCARBONS

Methane, CH_4, is the simplest hydrocarbon. As you learned in Sec. 1.3B, all of the hydrogen atoms of methane are equivalent, occupying the corners of a regular tetrahedron. Imagine now, that instead of being bound only to hydrogens, a carbon atom could be bound to a second carbon with enough hydrogens to fulfill the octet rule. The resulting compound is *ethane*.

Lewis structures of ethane:

ball-and-stick model of ethane

space-filling model of ethane

45

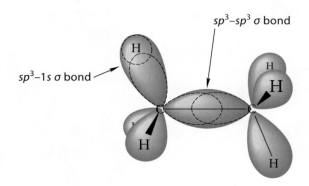

FIGURE 2.1 Hybrid-orbital description of the bonds in ethane. (The small rear lobes of the carbon sp^3 orbitals are omitted for clarity.) The component sp^3 and 1s orbitals are shown with dashed lines for the two labeled bonds.

In ethane, the bond between the two carbon atoms is longer than a C—H bond, but, like the C—H bonds, it is a covalent bond in the Lewis sense. In terms of hybrid orbitals, the carbon–carbon bond in ethane consists of two electrons in a bond formed by the overlap of two sp^3 hybrid orbitals, one from each carbon. Thus, the carbon–carbon bond in ethane is an sp^3–sp^3 σ bond (Fig. 2.1). The C—H bonds in ethane are like those of methane. They consist of covalent bonds, each of which is formed by the overlap of a carbon sp^3 orbital with a hydrogen 1s orbital; that is, they are sp^3–1s σ bonds. Both the H—C—C and H—C—H bond angles in ethane are approximately tetrahedral because each carbon bears four groups.

We can go on to envision other hydrocarbons in which any number of carbons are bonded in this way to form chains of carbons bearing their associated hydrogen atoms. Indeed, the ability of a carbon to form stable bonds to other carbons is what gives rise to the tremendous number of known organic compounds. The idea of carbon chains, a revolutionary one in the early days of chemistry, was developed independently by the German chemist August Kekulé (1829–1896) and the Scottish chemist Archibald Scott Couper (1831–1892) in about 1858. Kekulé's account of his inspiration for this idea is amusing.

> During my stay in London I resided for a considerable time in Clapham Road in the neighborhood of Clapham Common. . . . One fine summer evening I was returning by the last bus, "outside" as usual, through the deserted streets of the city that are at other times so full of life. I fell into a reverie, and lo, the atoms were gamboling before my eyes. Whenever, hitherto, these diminutive beings had appeared to me they had always been in motion. Now, however, I saw how, frequently, two smaller atoms united to form a pair. . . . I saw how the larger ones formed a chain, dragging the smaller ones after them but only at the ends of the chain. . . . The cry of the conductor, "Clapham Road," awakened me from my dreaming, but I spent a part of the night putting on paper at least sketches of these dream forms. This was the origin of the "Structure Theory."

Hydrocarbons are divided into two broad classes: *aliphatic hydrocarbons* and *aromatic hydrocarbons*. (The term *aliphatic* comes from the Greek *aleiphatos*, which means "fat." Fats contain long carbon chains that, as you will learn, are aliphatic groups.) The **aliphatic hydrocarbons** consist of three hydrocarbon families: *alkanes*, *alkenes*, and *alkynes*. We'll begin our study of aliphatic hydrocarbons with the **alkanes**, which are sometimes known as **paraffins**. Alkanes are hydrocarbons that contain only single bonds. Methane and ethane are the simplest alkanes. Later we'll consider the **alkenes**, or **olefins**, hydrocarbons that contain carbon–carbon double bonds; and the **alkynes**, or **acetylenes**, hydrocarbons that contain carbon–carbon triple bonds. The last hydrocarbons we'll study are the **aromatic hydrocarbons**, which include benzene and its substituted derivatives.

an alkane   an alkene   an alkyne   benzene

aliphatic hydrocarbons

an aromatic hydrocarbon

| 2.2 | ## UNBRANCHED ALKANES |

Carbon chains take many forms in the alkanes; they may be branched or unbranched, and they can even exist as rings (cyclic alkanes). Alkanes with unbranched carbon chains are sometimes called **normal alkanes**, or **n-alkanes**. A few of the unbranched alkanes are shown in Table 2.1, along with some of their physical properties. You should learn the names of the first 12 unbranched alkanes because they are the basis for naming many other organic compounds. The names *methane*, *ethane*, *propane*, and *butane* have their origins in the early history of organic chemistry, but the names of the higher alkanes are derived from the corresponding Greek numerical names: pentane (*pent* = five); hexane (*hex* = six); and so on.

Organic molecules are represented in different ways, which we'll illustrate using the alkane hexane. The **molecular formula** of a compound (for example, C_6H_{14} for hexane) gives its atomic composition. All noncyclic alkanes (alkanes without rings) have the general

TABLE 2.1   The Unbranched Alkanes

Compound name	Molecular formula	Condensed structural formula	Melting point (°C)	Boiling point (°C)	Density* (g mL^{-1})
methane	CH_4	CH_4	−182.5	−161.7	—
ethane	C_2H_6	CH_3CH_3	−183.3	−88.6	—
propane	C_3H_8	$CH_3CH_2CH_3$	−187.7	−42.1	—
butane	C_4H_{10}	$CH_3(CH_2)_2CH_3$	−138.3	−0.5	—
pentane	C_5H_{12}	$CH_3(CH_2)_3CH_3$	−129.8	36.1	0.6262
hexane	C_6H_{14}	$CH_3(CH_2)_4CH_3$	−95.3	68.7	0.6603
heptane	C_7H_{16}	$CH_3(CH_2)_5CH_3$	−90.6	98.4	0.6837
octane	C_8H_{18}	$CH_3(CH_2)_6CH_3$	−56.8	125.7	0.7026
nonane	C_9H_{20}	$CH_3(CH_2)_7CH_3$	−53.5	150.8	0.7177
decane	$C_{10}H_{22}$	$CH_3(CH_2)_8CH_3$	−29.7	174.0	0.7299
undecane	$C_{11}H_{24}$	$CH_3(CH_2)_9CH_3$	−25.6	195.8	0.7402
dodecane	$C_{12}H_{26}$	$CH_3(CH_2)_{10}CH_3$	−9.6	216.3	0.7487
eicosane	$C_{20}H_{42}$	$CH_3(CH_2)_{18}CH_3$	+36.8	343.0	(solid at 20°C)

* The densities tabulated in this text are of the liquids at 20 °C unless otherwise noted.

formula C_nH_{2n+2}, in which n is the number of carbon atoms. The **structural formula** of a molecule is its Lewis structure, which shows the **connectivity** of its atoms—that is, the order in which its atoms are connected. For example, a structural formula for hexane is the following:

$$
\begin{array}{ccccccccccc}
 & \text{H} & & \text{H} & & \text{H} & & \text{H} & & \text{H} & & \text{H} \\
 & | & & | & & | & & | & & | & & | \\
\text{H}- & \text{C} & - & \text{C} & - & \text{C} & - & \text{C} & - & \text{C} & - & \text{C} & -\text{H} \\
 & | & & | & & | & & | & & | & & | \\
 & \text{H} & & \text{H} & & \text{H} & & \text{H} & & \text{H} & & \text{H} \\
\end{array}
$$

hexane

(Notice that this type of formula does *not* portray the molecular geometry.) Writing each hydrogen atom in this way is very time-consuming, and a simpler representation of this molecule, called a **condensed structural formula**, conveys the same information.

$$H_3C-CH_2-CH_2-CH_2-CH_2-CH_3$$

hexane

In such a structure, the hydrogen atoms are understood to be connected to carbon atoms with single bonds, and the bonds shown explicitly are *bonds between carbon atoms*. Sometimes even these bonds are omitted, so that hexane can also be written $CH_3CH_2CH_2CH_2CH_2CH_3$. The structural formula may be further abbreviated as shown in the third column of Table 2.1. In this type of formula, for example, $(CH_2)_4$ means $-CH_2CH_2CH_2CH_2-$, and hexane can thus be written $CH_3(CH_2)_4CH_3$.

$$CH_3CH_2CH_2CH_2CH_2CH_3 \qquad CH_3(CH_2)_4CH_3$$

two other representations of hexane

The family of unbranched alkanes forms a series in which successive members differ from one another by one $-CH_2-$ group (**methylene group**) in the carbon chain. A series of compounds that differ by the addition of methylene groups is called a **homologous series**. Thus, the unbranched alkanes constitute one homologous series. Generally, physical properties within a homologous series vary in a regular way. An examination of Table 2.1, for example, reveals that the boiling points and densities of the unbranched alkanes vary regularly with increasing number of carbon atoms. This variation can be useful for quickly estimating the properties of a member of the series whose properties are not known.

The French chemist Charles Gerhardt (1816–1856) made an important chemical observation in 1845 about members of homologous series. His observation still has significant implications for learning organic chemistry. He wrote, "These (related) substances undergo reactions according to the same equations, and it is only necessary to know the reactions of one in order to predict the reactions of the others." What Gerhardt was saying, for example, is that we can study the chemical reactions of propane with the confidence that ethane, butane, or dodecane will undergo analogous reactions.

PROBLEMS

2.1   (a) How many hydrogen atoms are in the unbranched alkane with 18 carbon atoms?

(b) Is there an unbranched alkane containing 23 hydrogen atoms? If so, give its structural formula; if not, explain why not.

2.2   Give the structural formula and estimate the boiling point of tridecane, $C_{13}H_{28}$.

CONFORMATIONS OF ALKANES

In Section 1.3B, we learned that understanding the structures of many molecules requires that we specify not only their bond lengths and bond angles but also their dihedral angles. There we defined the dihedral angle as the angle between two intersecting planes. In this section, we'll learn a method to view dihedral angles easily. Then, we'll use the simple alkanes ethane and butane to develop some widely applicable simple principles that will allow us to predict the dihedral angles in more complex molecules.

A. The Conformation of Ethane

To specify the dihedral angles in ethane, we must define the relationship between the C—H bonds on one carbon and those on the other. A convenient way to do this is to view the molecule in a Newman projection, devised by Melvin S. Newman (1908–1993), who was Professor of Chemistry at The Ohio State University. A **Newman projection** is a type of planar projection along *one* bond, which we'll call the *projected bond*. For example, suppose we wish to view the ethane molecule in a Newman projection along the carbon–carbon bond, as shown in Fig. 2.2. In this projection, the carbon–carbon bond is the projected bond. To draw a Newman projection, start with a circle. The remaining bonds to the *nearer* atom in the projected bond are drawn to the center of the circle. The remaining bonds to the *farther* atom in the projected bond are drawn to the periphery of the circle. In the Newman projection of ethane (Fig. 2.2c), the three C—H bonds drawn to the center of the circle are bonds to the front carbon. The

ball-and-stick models:

line-and-wedge formulas:

(a) viewing a model of ethane from one end   (b) end-on view   (c) Newman projection (θ = dihedral angle)

FIGURE 2.2 How to derive a Newman projection for ethane using ball-and-stick models (*top*) and line-and-wedge formulas (*bottom*). (The hydrogens and C—H bonds farthest from the observer are shown in blue.) First view the ethane molecule from the end of the bond you wish to project, as in part (a). The resulting end-on view is shown in part (b). This is represented as a Newman projection (c) in the plane of the page. In the Newman projection, the bonds drawn to the center of the circle are attached to the carbon closer to the observer; the bonds drawn to the periphery of the circle (*blue*) are attached to the carbon farther from the observer. The projected bond (the carbon–carbon bond) is hidden.

C—H bonds to the periphery of the circle are the bonds to the rear carbon. *The projected bond itself, which is the fourth bond to each carbon, is hidden.*

The Newman projection of ethane makes it very easy to see the *dihedral angles θ* between its C—H bonds. When we have specified all of the dihedral angles in a molecule, we have specified its *conformation*. Thus, the **conformation** of a molecule is the spatial arrangement of its atoms when all of its dihedral angles are specified. We can also refer to conformations of parts of molecules, such as conformations about individual bonds.

Two limiting possibilities for the conformation of ethane can be seen from its Newman projections; these are termed the *staggered conformation* and the *eclipsed conformation*.

staggered conformation
of ethane

Newman projection   ball-and-stick model

eclipsed conformation
of ethane

In the **staggered conformation**, a C—H bond of one carbon bisects the angle between two C—H bonds of the other. The smallest dihedral angle in the staggered conformation is $\theta = 60°$. (The other dihedral angles are $\theta = 180°$ and $\theta = 300°$.) In the **eclipsed conformation**, the C—H bonds on the respective carbons are superimposed in the Newman projection. The smallest dihedral angle is $\theta = 0°$. (The other dihedral angles are $\theta = 120°$ and $\theta = 240°$.) Conformations intermediate between the staggered and eclipsed conformations are possible, but these two conformations will prove to be of central importance.

Which is the preferred conformation of ethane? The energies of the ethane conformations can be described by a plot of relative energy versus dihedral angle, which is shown in Fig. 2.3. In this figure, the dihedral angle is the angle between the bonds to the colored hydrogens on the different carbons. To see the relationships in Fig. 2.3, build a model of ethane, hold either carbon fixed, and turn the other carbon about the C—C bond. It doesn't matter which carbon you choose to rotate; in Fig. 2.3, we've arbitrarily chosen to rotate the front carbon. As the angle of rotation changes, the model passes alternately through three identical staggered and three identical eclipsed conformations. As shown by Fig. 2.3, identical conformations have identical energies. The graph also shows that the eclipsed conformation is characterized by an energy *maximum*, and the staggered conformation is characterized by an energy *minimum*. *The staggered conformation is therefore the more stable conformation of ethane.* The graph shows that the staggered conformation is more stable than the eclipsed conformation by about 12 kJ mol^{-1} (about 2.9 kcal mol^{-1}). This means that it would take about 12 kJ of energy to convert one mole of staggered ethane into one mole of eclipsed ethane.

The reasons for the relative stability of the staggered conformation have been debated for years. One theory is that the bonding molecular orbitals of ethane in the staggered conformation are particularly stable. A second theory holds that there is repulsion between the electrons in the C—H bonds on the two carbons. Because the bonds are closer when they have a dihedral angle of 0°, this repulsion is greater in the eclipsed form. This repulsion is termed **torsional strain**. Notice that the repulsion is not between the hydrogens but between the electrons in the bonds themselves. Recent assessments suggest that both factors are important.

One staggered conformation of ethane can convert into another by rotation of either carbon relative to the other about the carbon–carbon bond. Such a rotation about a bond is called an **internal rotation** (to differentiate it from a rotation of the entire molecule). When an internal rotation occurs, an ethane molecule must briefly pass through the eclipsed conformation. To do so, it must acquire the additional energy of the eclipsed conformation and then lose it again. What is the source of this energy?

At temperatures above absolute zero, molecules are in constant motion and therefore have kinetic energy. Heat is a manifestation of this kinetic energy. In a sample of ethane, the

FIGURE 2.3 Variation of energy with dihedral angle about the carbon–carbon bond of ethane. In this diagram, the rear carbon is held fixed and the front carbon is rotated, as shown by the green arrows. The dihedral angle plotted is the angle between the bonds to the red and blue hydrogens. Note that the staggered conformations are at the energy minima, and the eclipsed conformations are at the energy maxima.

molecules move about in a random manner, much as people might mill about in a large crowd. These moving molecules frequently collide, and molecules can gain or lose energy in such collisions. (An analogy is the collision of a bat with a ball; some of the kinetic energy of the bat is lost to the ball.) When an ethane molecule gains sufficient energy from a collision, it can undergo internal rotation, passing through the more energetic eclipsed conformation into another staggered conformation. Whether a given ethane molecule acquires sufficient energy to undergo an internal rotation is strictly a matter of *probability* (random chance). However, an internal rotation is *more probable* at higher temperature because warmer molecules have greater kinetic energy.

The probability that ethane undergoes internal rotation is reflected as its *rate* of rotation: how many times per second the molecule converts from one staggered conformation into another. This rate is determined by how much energy must be acquired for the rotation to occur: the more energy required, the smaller the rate. In the case of ethane, 12 kJ mol^{-1} (2.9 kcal mol^{-1}) is required. This amount of energy is small enough that the internal rotation of ethane is very rapid even at very low temperatures. At 25 °C, a typical ethane molecule undergoes a rotation from one staggered conformation to another at a rate of about 10^{11} times per second! This means that the interconversion between staggered conformations takes place about once every 10^{-11} second. Despite this short lifetime for any one staggered conformation, an ethane molecule spends most of its time in its staggered conformations, passing only transiently through its eclipsed conformations. Thus, an internal rotation is best characterized not as a continuous spinning but as a constant succession of jumps from one staggered conformation to another.

B. Conformations of Butane

Butane contains two distinguishable types of carbon–carbon bonds: the two terminal C—C bonds (*blue*), and the central C—C bond (*red*).

$$H_3C—CH_2—CH_2—CH_3$$

butane
two types of C—C bonds

ball-and-stick models:

line-and-wedge formulas:

(a) viewing a model of butane from one end
of the central carbon–carbon bond

(b) end-on view

(c) Newman projection

FIGURE 2.4 How to derive the Newman projection of the central carbon–carbon bond in butane using ball-and-stick models (*top*) and line-and-wedge formulas (*bottom*). The bonds and groups on the rear carbon of the projected bond are shown in blue. (Only one of the butane conformations is shown.)

We'll consider internal rotation about the central C—C bond. This rotation is a bit more complex than the ethane case, but examination of this rotation leads to important new insights about molecular conformation. As with ethane, we use Newman projections, as shown in Fig. 2.4. Remember again that the projected bond—the central C—C bond in this case—is hidden in the Newman projection.

The graph of energy as a function of dihedral angle in butane is given in Fig. 2.5. Note once again that the various rotational possibilities are generated with a model by holding either carbon fixed (the carbon away from the observer in Fig. 2.5) and rotating the other one.

Figure 2.5 shows that the staggered conformations of butane, like those of ethane, are at energy minima and are thus the stable conformations of butane. However, not all of the staggered conformations (nor all of the eclipsed conformations) of butane are alike. The different staggered conformations have been given special names. The conformations with a dihedral angle of 60° and 300° in Fig. 2.5 (or ±60°) between the two C—CH$_3$ bonds are called **gauche** (pronounced "gōsh") conformations (from the French *gauchir,* meaning "to turn aside"); the form in which the dihedral angle is 180° is called the **anti** conformation.

gauche conformation
$\theta = 60°$

anti conformation
$\theta = 180°$

gauche conformation
$\theta = 300° \, (= -60°)$

FIGURE 2.5 Variation of energy with dihedral angle about the central carbon–carbon bond of butane. In this diagram, the rear carbon is held fixed and the front carbon is rotated, as shown by the green arrows. The dihedral angle plotted is the one between the bonds to the two CH_3 groups.

The relationship between bonds also can be described with the terms *gauche* and *anti*. Two bonds that have a dihedral relationship of ±60° are said to be **gauche bonds**. Two bonds that have a dihedral relationship of 180° are said to be **anti bonds**. Notice that these terms refer to bonds on *adjacent* carbons.

Figure 2.5 shows that the gauche and anti conformations of butane have different energies. The anti conformation is more stable by 3.7 kJ mol^{-1} (0.9 kcal mol^{-1}). The gauche conformation is less stable because the CH_3 groups are very close together—so close that the hydrogens on the two groups occupy each other's space. You can see this with the aid of the space-filling model in Fig. 2.6a on p. 54.

This problem can be discussed more precisely in terms of atomic size. One measure of an atom's effective size is its **van der Waals radius**. Energy is required to force two *nonbonded* atoms together more closely than the sum of their van der Waals radii. Because the van der Waals radius of a hydrogen atom is about 1.2 Å, forcing the centers of two nonbonded hydrogens to be closer than twice this distance requires energy. Furthermore, the more the two hydrogens are pushed together, the more energy is required. The extra energy required to force two nonbonded atoms within the sum of their van der Waals radii is called a **van der Waals repulsion**. Thus, to attain the gauche conformation, butane must acquire more energy. In other words, *gauche-butane is destabilized by van der Waals repulsions between nonbonded hydrogens on the two CH_3 groups.* Such van der Waals repulsions are absent in *anti*-butane (see Fig. 2.6b). Thus, *anti*-butane is more stable than *gauche*-butane.

As with ethane, the eclipsed conformations of butane are destabilized by torsional strain. But, in the conformation in which the two C—CH_3 bonds are eclipsed, the major source of instability is van der Waals repulsions between the methyl hydrogens (Fig. 2.6c), which are forced to be even closer than they are in the gauche conformation. Notice that this is the most unstable of the eclipsed conformations ($\theta = 0°$ in Fig. 2.5).

FURTHER EXPLORATION 2.1
Atomic Radii and van der Waals Repulsion

H–H distance is less than the sum of van der Waals radii

no van der Waals repulsions

H–H distances are less than the sum of van der Waals radii

(a) *gauche*-butane

(b) *anti*-butane

(c) butane with C—CH₃ bonds eclipsed

FIGURE 2.6 Space-filling models of different butane conformations with the CH_3 (methyl) hydrogens shown in color. (a) *Gauche*-butane. A hydrogen atom from one CH_3 group is so close to a hydrogen atom of the other CH_3 group that these hydrogens, shown in pink, violate each other's van der Waals radii. The resulting van der Waals repulsions cause *gauche*-butane to have a higher energy than *anti*-butane, in which this interaction is absent. (b) *Anti*-butane. This conformation is most stable because it contains no van der Waals repulsions. (c) Butane with the C—CH₃ bonds eclipsed. In this conformation, van der Waals repulsions between the hydrogens of the two CH_3 groups (*pink*) are even greater than they are in *gauche*-butane.

It is important to understand the relative energies of the butane conformations because, when different stable conformations are in equilibrium, *the most stable conformation—the conformation of lowest energy—is present in greatest amount.* Thus, the anti conformation of butane is the predominant conformation of butane. At room temperature, there are about twice as many molecules of butane in the anti conformation as there are in the gauche conformation.

The gauche and anti conformations of butane interconvert rapidly at room temperature—almost as rapidly as the staggered forms of ethane. Because the eclipsed conformations lie at energy maxima and are unstable, they do not exist to any measurable extent.

The investigation of molecular conformations and their relative energies is called **conformational analysis**. In this section, we've learned some important principles of conformational analysis that we'll be able to apply to more complex molecules. Here is a summary of these principles:

1. Staggered conformations about single bonds are favored over eclipsed conformations.

2. Van der Waals repulsions (repulsions between nonbonded atoms) occur when atoms are "squeezed" closer together than the sum of their van der Waals radii.

3. Conformations containing van der Waals repulsions are less stable than conformations in which such repulsions are absent.

4. Rotation about C—C single bonds in most cases is so rapid that it is hard to imagine separating conformations except at very low temperature.

STUDY PROBLEM 2.1

Draw a Newman projection for the anti conformation about the C3–C4 bond of 2-methylhexane, viewing the bond so that C3 is nearest the observer.

$$H_3C—CH—CH_2—CH_2—CH_2—CH_3$$
$$|$$
$$CH_3$$

C3   C4

2-methylhexane

SOLUTION First draw a "blank" Newman projection to represent the projected bond. Remember that the projected bond itself (the C3–C4 bond) is invisible in the projection. Either template below can be used.

We arbitrarily pick the template on the left. In the view dictated by the problem, the front carbon is C3. Identify the three groups attached to C3 with bonds other than the projected bond. These groups are H, H, and the H_3C-CH- group. Put these on the *front* carbon of the Newman projection. *It doesn't matter which bonds go to*

$$CH_3$$

which groups as long as all groups are on the front carbon. It's important to understand that, because we are not examining the bonds within the large group, we can condense this group to $(CH_3)_2CH-$ or even C_3H_7-.

We then identify the groups attached to the back carbon (C4) by bonds *other than* the projected bond. These groups are H, H, and $-CH_2CH_3$. Now it *does* matter where we put these groups, because we are asked for the anti conformation. The $-CH_2CH_3$ group must be placed anti to the $(CH_3)_2CH-$ group. Remember that "anti" means a dihedral angle of 180°.

2-methylhexane
anti conformation about the C3–C4 bond

Remember that Newman projections are used to examine conformations about *a particular bond*. If we want to examine the conformations about several different bonds, we must draw a different set of Newman projections for each bond.

PROBLEMS

2.3 (a) Draw a Newman projection for each staggered and eclipsed conformation about the C2–C3 bond of isopentane, a compound containing a branched carbon chain.

$$H_3C-\overset{2}{C}H-\overset{3}{C}H_2-CH_3$$
$$\quad\quad\quad |$$
$$\quad\quad CH_3$$

isopentane

Show all staggered and eclipsed conformations.

(b) Sketch a curve of potential energy versus dihedral angle for isopentane, similar to that of butane in Fig. 2.5. Label each energy maximum and minimum with one of the conformations you drew in part (a).

(c) Which of the conformations you drew in part (a) are likely to be present in greatest amount in a sample of isopentane? Explain.

2.4 Repeat the analysis in Problem 2.3 for either one of the terminal C—C bonds of butane.

C. Methods of Drawing Conformations

In the previous section you have seen that Newman projections can be used to represent specific conformations. In this section you will learn some other ways of drawing specific conformations of molecules. At the end of this section, we will know three ways of drawing specific conformations:

1. Newman projections (already learned)
2. Line-and-wedge structures
3. Sawhorse projections

These last two methods are essentially ways of showing perspective in our structures, and we want to be able to do this systematically without attending art school! We'll illustrate each of these using the anti conformation of butane with which you are already familiar. You should approach the material that follows with *models in hand*.

Line-and-wedge structures (sometimes called **dash–wedge structures**) are extensions of the technique introduced on p. 16 for drawing the bonds to tetrahedral carbon. As before, we'll use lines to represent bonds in the plane of the page, solid wedges to represent bonds emerging in front of the page, and dashed wedges to represent bonds receding behind the page. Now, however, we'll learn how to do this for two or more carbons simultaneously. To draw a line-and-wedge structure, we draw the bond of interest and two of the adjacent bonds (in this case, the carbon–CH_3 bonds) in the plane of the page as if we're looking at the structure "side-on." If you look at a model this way, you'll see that the hydrogens behind the page are obscured by the hydrogens in front (view *A* in Eq. 2.1).

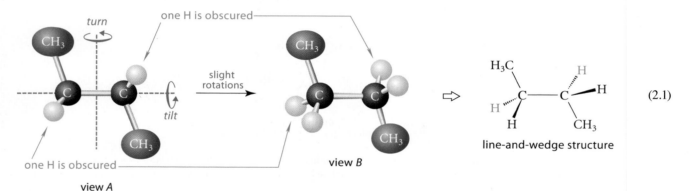

(2.1)

We then tilt and turn the model slightly as shown by the green arrows to give view *B*, and then draw this model using dashed wedges for bonds receding behind the page, solid wedges for bonds emerging from the page, and ordinary lines for bonds in the page. The "tilt and turn" about either axis can be in whichever direction brings all of the groups clearly into view, but for consistency, we'll turn the model as shown in Eq. 2.1 in most cases. The hydrogens that are obscured in view *A* are colored pink in view *B*. Notice the relationships between the solid-wedged and dashed-wedge bonds. In the view shown in Eq. 2.1, for example, the dashed-wedge (receding) bonds are above the solid-wedged (emerging) bonds on the page.

Viewing the line-and-wedge structure (or the corresponding model) end-on and projecting this view (flattening it) onto the page gives the familiar Newman projection.

$$\text{line-and-wedge structure} \qquad \Rightarrow \qquad \text{Newman projection} \qquad (2.2)$$

This procedure was also shown in more detail in Fig. 2.4 (p. 52).

A **sawhorse projection** is a view of the projected bond as if we were looking at the model from above and to the right of the model. We'll illustrate sawhorse projections with the same *anti*-butane we've been using. To draw a sawhorse projection, follow these steps:

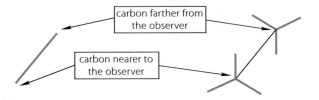

(2.3)

Step 1. Draw a long projected bond about 30° from the vertical.

Step 2. Add the bonds at either end.

Step 3. Add the atoms or groups. This is the sawhorse projection.

Some people like to draw sawhorse projections with a wedged projected bond to emphasize the perspective, but this perspective is understood whether the bond is wedged or not.

(2.4)

sawhorse projection with a wedged projected bond

To draw a line-and-wedge projection of an eclipsed conformation, we have to modify somewhat the procedure shown in Eq. 2.1. View the model side-on (view *A*) and simply tilt the structure slightly (view *B*). Draw this view on the page by exaggerating the separations and lengths of the solid wedges and dashed wedges slightly.

(2.5)

the hydrogens are obscured

view *A*

view *B*

line-and-wedge structure

The procedures for drawing sawhorse and Newman projections of eclipsed conformations are the same as for staggered conformations.

Typically, we focus on one particular bond (the projected bond) when drawing conformations, but it is possible to draw the conformations about several bonds simultaneously, provided they can all be drawn in the plane of the paper. This is the case in the following all-anti conformation of hexane:

hexane
anti conformations about
bonds C2–C3, C3–C4, and C4–C5

The most common errors committed by beginning students in drawing line-and-wedge structures are illustrated by the following examples of an *incorrectly drawn* isobutane structure:

In two of the incorrect structures (*A* and *B*), the wedge and dashed wedge are drawn on opposite sides of the in-plane carbon–carbon bonds. They must be drawn on the same side of the carbon–carbon bonds in the backbone. In structure *B*, the angle between the in-plane bonds is not "bent." In a projection, the angle between the in-plane bonds must approximate the tetrahedral angle. In structure *C*, the wedge and dashed-wedge bonds are drawn on the wrong side of the carbon backbone; they must be drawn on the convex side. The correct structures show a few of the many line-and-wedge structures of isobutane. The best way to evaluate a line-and-wedge structure is to relate it to a molecular model until you become more secure with the drawing conventions. For example, you should relate the correct structures for isobutane shown above to models.

It is possible to draw *any* structure if we are artistic enough. When we try to draw perspective views of highly branched structures, however, the task becomes difficult. Fortunately, the three techniques introduced here will meet most of our needs. We'll also learn about the conformations of some cyclic compounds in Chapter 7 and how to draw them.

PROBLEMS

2.5   (a) Use the procedures described in this section to draw the line-and-wedge and sawhorse conformations of both *gauche*-butane conformations projected about the central carbon–carbon bond. Don't hesitate to use models.

(b) Which of the following line-and-wedge structures of dibromomethane (H_2CBr_2) are correct? For the incorrect structures, tell why they are incorrect.

2.6 (a) Using models when necessary, draw a Newman projection and a sawhorse projection corresponding to the following line-and-wedge structure. Project the C2–C3 bond, with carbon-2 nearer the observer, and simplify your structures by abbreviating large groups.

(b) In your Newman projection, rotate carbon-2 clockwise by 120° about the C2–C3 bond, and draw both sawhorse and line-and-wedge structures for the resulting conformation.

(c) In your Newman projection of part (a), rotate carbon-2 180° about the C2–C3 bond, and draw both sawhorse and line-and-wedge structures for the resulting conformation.

2.4 CONSTITUTIONAL ISOMERS AND NOMENCLATURE

A. Isomers

When a carbon atom in an alkane is bound to more than two other carbon atoms, a branch in the carbon chain occurs at that position. The smallest branched alkane has four carbon atoms. As a result, there are two four-carbon alkanes; one is *butane*, and the other is *isobutane*.

These are different compounds with different properties. For example, the boiling point of butane is $-0.5\,°C$, whereas that of isobutane is $-11.7\,°C$. Yet both have the same molecular formula, C_4H_{10}. Different compounds that have the same molecular formula are said to be **isomers** or **isomeric compounds**.

There are different types of isomers. Isomers that differ in the connectivity of their atoms, such as butane and isobutane, are called **constitutional isomers** or **structural isomers**. Recall (Sec. 1.3) that *connectivity* is the order in which the atoms of the molecule are bonded. The atomic connectivities of butane and isobutane differ because in isobutane a carbon is attached to three other carbons, whereas in butane no carbon is attached to more than two other carbons.

STUDY PROBLEM 2.2

Which of the following four structures represent constitutional isomers of the same molecule, and which one is neither isomeric nor identical to the others? Explain your answers.

SOLUTION Compounds must have the same molecular formula to be either identical or isomeric. Structure *A* has a different molecular formula (C_6H_{14}) from the other structures (C_7H_{16}), and hence structure *A* represents a molecule that is neither identical nor isomeric to the others. To solve the rest of the problem, we must understand that *Lewis structures show connectivity only.* They do not represent the actual shapes of molecules unless we start adding spatial elements such as wedges and dashed wedges. This means that *we can draw a given structure many different ways.* Have you ever heard the old spiritual, "The foot-bone's connected to the ankle-bone . . . "? That's a song about connectivity of the typical human body. If the description fits you, its validity doesn't change whether you are sitting down, standing up, standing on your head, or doing yoga. Similarly, the connectivity of a molecule doesn't change whether it is drawn forward, backward, or upside-down. With that in mind, let's trace the connectivity of each structure above. Consider structures *B* and *C*. Each has two CH_3 groups connected to a CH, and that CH is connected to another CH, which in turn is connected to both a CH_3 group and a CH_2CH_3 group. In *B*, this connectivity pattern starts on the left; in *C*, it starts on the bottom. But it's the same in both. Because both structures have identical connectivities, they represent the same molecule.

Structures *D* and *B* (or *D* and *C*) have the same molecular formula C_7H_{16}; but, as you should verify, their connectivities are different, so they are constitutional isomers.

Butane and isobutane are the only constitutional isomers with the formula C_4H_{10}. However, more constitutional isomers are possible for alkanes with more carbon atoms. There are nine constitutional isomers of the heptanes (C_7H_{16}), 75 constitutional isomers of the decanes ($C_{10}H_{22}$), and 366,319 constitutional isomers of the eicosanes ($C_{20}H_{42}$)! These few examples demonstrate that millions of organic compounds are known and millions more are conceivable. It follows that organizing the body of chemical knowledge requires a system of nomenclature that can provide an unambiguous name for each compound.

B. Organic Nomenclature

An organized effort to standardize organic nomenclature dates from proposals made at a conference in Geneva in 1892. From those proposals the International Union of Pure and Applied Chemistry (IUPAC), a professional association of chemists, developed and sanctioned several accepted systems of nomenclature. The most widely applied system in use today is called **substitutive nomenclature**.

The IUPAC rules for the nomenclature of alkanes form the basis for the substitutive nomenclature of most other compound classes. Hence, it is important to learn these rules and be able to apply them.

C. Substitutive Nomenclature of Alkanes

Alkanes are named by applying the following 10 rules *in order*. This means that if one rule doesn't unambiguously determine the name of a compound of interest, we proceed down the list *in order* until we find a rule that does.

1. *The unbranched alkanes are named according to the number of carbons, as shown in Table 2.1.*

2. *For alkanes containing branched carbon chains, determine the principal chain.*

The **principal chain** is the longest continuous carbon chain in the molecule. To illustrate:

$$H_3C—CH_2—CH_2—CH—CH_2—CH_3 \quad \text{principal chain}$$
$$|$$
$$CH_3$$

When identifying the principal chain, take into account that *the condensed structure of a given molecule may be drawn in several different ways* (Study Problem 2.2). Thus, the following structures represent the *same molecule*, with the principal chain shown in red:

$$H_3C—CH_2—CH_2—CH—CH_2—CH_3 \qquad H_3C—CH—CH_2—CH_3$$
$$\qquad\qquad\qquad\quad | \qquad\qquad\qquad\qquad\qquad\quad |$$
$$\qquad\qquad\qquad\quad CH_3 \qquad\qquad\qquad\qquad\quad CH_2—CH_2—CH_3$$

(Be sure to verify that these structures have identical connectivities and thus represent the same molecule.)

3. *If two or more chains within a structure have the same length, choose as the principal chain the one with the greater number of branches.*

The following structure is an example of such a situation:

$$CH_2—CH_3 \qquad\qquad\qquad\qquad\qquad CH_2—CH_3$$
$$| \qquad\qquad\qquad\qquad\qquad\qquad\qquad |$$
$$H_3C—CH—CH—CH_2—CH_2—CH_3 \qquad H_3C—CH—CH—CH_2—CH_2—CH_3$$
$$| \qquad\qquad\qquad\qquad\qquad\qquad\qquad\quad |$$
$$CH_3 \qquad\qquad\qquad\qquad\qquad\qquad\qquad\quad CH_3$$

six-carbon chain with one branch   six-carbon chain with two branches

(this is the proper choice for principal chain)

The correct choice of principal chain is the one on the right, because it has two branches; the choice on the left has only one. (It makes no difference that the branch on the left is larger or that it has additional branching within itself.)

4. *Number the carbons of the principal chain consecutively from one end to the other in the direction that gives the lower number to the first branching point.*

In the following structure, the carbons of the principal chain are numbered consecutively from one end to give the lower number to the carbon at the $—CH_3$ branch.

$$\overset{6}{H_3C}—\overset{5}{CH_2}—\overset{4}{CH_2}—\overset{3}{CH}—\overset{2}{CH_2}—\overset{1}{CH_3} \quad \longleftarrow \text{ proper numbering}$$
$$\underset{1}{}\quad\underset{2}{}\quad\underset{3}{}\quad\underset{4}{|}\quad\underset{5}{}\quad\underset{6}{} \quad \longleftarrow \text{ improper numbering}$$
$$\qquad\qquad\qquad\qquad CH_3$$

5. *Name each branch and identify the carbon number of the principal chain at which it occurs.*

In the previous example, the branching group is a $—CH_3$ group. This group, called a *methyl group*, is located at carbon-3 of the principal chain.

Branching groups are in general called **substituents**, and substituents derived from alkanes are called **alkyl groups**. An alkyl group may contain any number of carbons. The name of an unbranched alkyl group is derived from the name of the unbranched alkane with the same number of carbons by dropping the final *ane* and adding *yl*.

$—CH_3$	methyl (= meth~~ane~~ + yl)
$—CH_2CH_3$ or $—C_2H_5$	ethyl (= eth~~ane~~ + yl)
$—CH_2CH_2CH_3$	propyl

Alkyl substituents themselves may be branched. The most common branched alkyl groups have special names, given in Table 2.2 on p. 62. These should be learned because they will be encountered frequently. Notice that the "iso" prefix is used for substituents containing two methyl groups at the end of a carbon chain. Also notice carefully the difference between an isobutyl group and a *sec*-butyl group; these two groups are frequently confused. (The use of abbreviations for substituents will be discussed in Sec. 2.5.)

**STUDY GUIDE
LINK 2.1**
Nomenclature of
Simple Branched
Compounds

TABLE 2.2   **Nomenclature and Abbreviations of Some Short Branched-Chain Alkyl Groups**

Group structure	Skeletal structure*	Condensed structure	Abbreviation*	Written name	Pronounced name
H_3C $CH-$ H_3C	⟩-}	$(CH_3)_2CH-$	*i*-Pr or *i*Pr	isopropyl	isopropyl
H_3C $CHCH_2-$ H_3C	⋎-}	$(CH_3)_2CHCH_2-$	*i*-Bu or *i*Bu	isobutyl	isobutyl
CH_3CH_2CH- $\ CH_3$	⋏-}	—	*s*-Bu or *s*Bu	*sec*-butyl	secondary butyl or "sec-butyl"
CH_3 H_3C-C- CH_3	+-}	$(CH_3)_3C-$	*t*-Bu or *t*Bu	*tert*-butyl (or *t*-butyl)	tertiary butyl or "tert-butyl"
CH_3 $H_3C-C-CH_2-$ CH_3	⋈-}	$(CH_3)_3CCH_2-$	—	neopentyl	neopentyl

* The use of skeletal structures and abbreviations is introduced in Sec. 2.5. In the skeletal structures, the bracket (}) indicates the main chain to which the substituent is attached.

6. *Construct the name by writing the carbon number of the principal chain at which the substituent occurs, a hyphen, the name of the branch, and the name of the alkane corresponding to the principal chain.*

$$H_3C-CH_2-CH_2-CH-CH_2-CH_3$$
$$\underset{CH_3}{|}$$

name: **3-methyl**hexane

↑   ↑ name of the principal chain

number and name of
the alkyl substituent

Notice that the name of the branch and the name of the principal chain are written together as one word. Notice also that the name itself has *no* relationship to the name of the isomeric unbranched alkane; that is, the preceding compound is *a constitutional isomer of heptane* because it has seven carbon atoms, but it is named as a *derivative of hexane*, because its principal chain contains six carbon atoms.

STUDY PROBLEM   2.3

Name the following compound, and give the name of the unbranched alkane of which it is a constitutional isomer.

$$H_3C-CH_2-CH_2-CH-CH_2-CH_2-CH_3$$
$$\underset{CH_3}{\overset{|}{\underset{|}{CH-CH_3}}}$$

SOLUTION   Because the principal chain has seven carbons, the compound is named as a substituted heptane. The branch is at carbon-4, and the substituent group at this branch is

$$CH-CH_3$$
$$|$$
$$CH_3$$

Table 2.2 shows that this group is an *isopropyl group*. Thus, the name of the compound is 4-isopropylheptane:

$$H_3C-CH_2-CH_2-CH-CH_2-CH_2-CH_3$$
$$|$$
$$CH-CH_3$$
$$|$$
$$CH_3$$

4-isopropylheptane

alkyl group name from Table 2.2

Because this compound has the molecular formula $C_{10}H_{22}$, it is a constitutional isomer of the unbranched alkane *decane*.

7. *If the principal chain contains multiple substituent groups, each substituent receives its own number. The prefixes di, tri, tetra, and so on, are used to indicate the number of identical substituents.*

$$CH_3$$
$$|$$
$$H_3C-C-CH_2CH_2CH_3$$
$$|$$
$$CH_3$$

2,2-dimethylpentane

two methyl substituents

shows that both methyl branches are
at carbon-2 of the principal chain

STUDY PROBLEM   2.4

Which two of the following structures represent the same compound? Name the compound.

$$H_3C-CH_2$$
$$|$$
$$CH-CH_3$$
$$|$$
$$H_3C-CH_2-CH$$
$$|$$
$$CH_3$$

$$H_3C-CH-CH_2-CH-CH_2-CH_3$$
$$|\qquad\qquad|$$
$$CH_3\qquad\ CH_3$$

$$H_3C-CH-CH-CH_2-CH_3$$
$$|\quad\ |$$
$$CH_2\ CH_3$$
$$|$$
$$CH_3$$

A   *B*   *C*

SOLUTION   The connectivities of both *A* and *C* are the same: [CH$_3$, CH$_2$, (CH connected to CH$_3$), (CH connected to CH$_3$), CH$_2$, CH$_3$]. The compound represented by these structures has six carbons in its principal chain and is therefore named as a hexane. There are methyl branches at carbons 3 and 4. Hence the name is 3, 4-dimethylhexane. (You should name compound *B* after you study the next rule.)

8. *If substituent groups occur at more than one carbon of the principal chain, alternative numbering schemes are compared number by number, and the one is chosen that gives the smaller number at the first point of difference.*

This is one of the trickiest nomenclature rules, but it is easy to handle if we are systematic. To apply this rule, write the two possible numbering schemes derived by numbering from either end of the chain. In the following example, the two schemes are 3,3,5- and 3,5,5-.

possible names:
3,3,5-trimethylheptane (correct)
3,5,5-trimethylheptane (incorrect)

A decision between the two numbering schemes is made by a *pairwise comparison* of the number sets (3,3,5) and (3,5,5).

How to do a pairwise comparison:

compare these first
compare these second

(3,3,5)   (3,5,5)

compare these last

Because the *first point of difference* in these sets occurs at the second pair—3 versus 5—the decision is made at this point, and the first scheme is chosen, because 3 is smaller than 5. If there are differences in the remaining numbers, they are ignored. The sum of the numbers is also irrelevant. Finally, it makes no difference whether the names of the substituents are the same or different; only the numerical locations are used.

The next rule deals with the order in which substituents are listed, or "cited," in the name. Don't confuse the citation order of a substituent with its numerical prefix; they aren't necessarily the same.

9. *Substituent groups are cited in alphabetical order in the name regardless of their location in the principal chain. The numerical prefixes di, tri, and so on, as well as the hyphenated prefixes tert- and sec-, are ignored in alphabetizing, but the prefixes iso, neo, and cyclo (Sec. 2.5) are considered in alphabetizing substituent groups.*

The following compounds illustrate the application of this rule:

5-ethyl-2-methylheptane
(*ethyl* is cited before *methyl* even
though it has a higher number)

3-ethyl-5,5-dimethyloctane
(note that *dimethyl* begins
with the letter *m* for
purposes of citation)

10. *If the numbering of different groups is not resolved by the other rules, the first-cited group receives the lowest number.*

In the following compound, rules 1–9 do not dictate a choice between the names 3-ethyl-5-methylheptane and 5-ethyl-3-methylheptane. Because the ethyl group is cited first in the name (rule 9), it receives the lower number, by rule 10.

$$H_3C-CH_2-CH-CH_2-CH-CH_2-CH_3$$

$$\underset{CH_3}{|} \qquad \underset{C_2H_5}{|}$$

3-ethyl-5-methylheptane

Some situations of greater complexity are not covered by these 10 rules; however, these rules will suffice for most cases.

PROBLEM

2.7   Name the following compounds.

(a)
$$\underset{|}{CH_3} \qquad \underset{|}{CH_3}$$
$$CH_3CHCHCH_2CHCH_3$$
$$\underset{|}{CH_3}$$

(b) compound *B* in Study Problem 2.4 on p. 63

(c)
$$\underset{|}{CH_2CH_2CH_3}$$
$$CH_3CH_2CHCHCH_2CH_2CH_3$$
$$\underset{|}{CH_2CH_2CH_3}$$

(d)
$$\underset{|}{CH_3} \qquad \underset{|}{CH_3}$$
$$CH_3CH_2CCH_2CH_2CHCH_3$$
$$\underset{|}{CH_3}$$

D. Highly Condensed Structures

When space is at a premium, parentheses are sometimes used to form highly condensed structures that can be written on one line, as in the following example.

$$(CH_3)_4C \quad \text{or} \quad C(CH_3)_4 \quad \text{means} \quad H_3C-\underset{\underset{CH_3}{|}}{\overset{\overset{CH_3}{|}}{C}}-CH_3$$

When such structures are complex, it is sometimes not immediately obvious, particularly to the beginner, which atom inside the parentheses is connected to the atom outside the parentheses, but a little analysis will generally solve the problem. Usually the structure is drawn so that one of the parentheses intervenes between the atoms that are connected (except for attached hydrogens). However, if in doubt, look for the atom within the parentheses that is missing its usual number of bonds. When the group inside the parentheses is CH_3, as in the previous example, the carbon has only three bonds (to the H's). Hence, it must be bound to the atom outside the parentheses. Consider as another example the CH_2OH groups in the following structure.

$$(CH_3)_2CH-CH(CH_2OH)_2 \quad \text{means} \quad \overset{H_3C}{\underset{H_3C}{\diagdown}}CH-CH\overset{CH_2OH}{\underset{CH_2OH}{\diagup}}$$

Because the oxygen is bound to a carbon and a hydrogen, it has its full complement of two bonds (the two unshared pairs are understood). The carbon of each CH_2OH group, however, is bound to only three groups inside the parentheses (two H's and the O); hence, it is the atom that is connected to the carbon outside the parentheses.

If the meaning of a condensed structure is not immediately clear, *write it out in a less condensed form*. If you will take the time to do this in a few cases, it should not be long before the interpretation of condensed structures becomes more routine.

Research in student learning strategies has shown that student success in organic chemistry is *highly correlated* with whether a student takes the time to *write out* intermediate steps in a problem. Such steps in many cases involve writing structures and partial structures. Students may be tempted to skip such steps because they see their professors working things out quickly in their heads and perhaps feel that they are expected to do the same. Professors can do this because they have years of experience. Most of them probably gained their expertise through step-by-step problem solving. In some cases, the temptation to skip steps may be a consequence of time pressure. If you are tempted in this direction, remember that a step-by-step approach applied to relatively few problems is a better expenditure of time than rushing through many problems. Study Problem 2.5 illustrates a step-by-step approach to a nomenclature problem.

STUDY PROBLEM   2.5

Write the Lewis structure of 4-*sec*-butyl-5-ethyl-3-methyloctane. Then write the structure in a condensed form.

SOLUTION   To this point, we've been giving names to structures. This problem now requires that we work "in reverse" and construct a structure from a name. Don't try to write out the structure immediately; rather, take a systematic, stepwise approach involving intermediate structures. First, write the principal chain. Because the name ends in *octane*, the principal chain contains eight carbons. Draw the principal chain without its hydrogen atoms:

$$C—C—C—C—C—C—C—C$$

Next, number the chain from either end and attach the branches indicated in the name at the appropriate positions: a *sec*-butyl group at carbon-4, an ethyl group at carbon-5, and a methyl group at carbon-3. (Use Table 2.2 to learn or relearn the structure of a *sec*-butyl group, if necessary.)

$$H_3C—CH_2—CH—CH_3 \longleftarrow sec\text{-butyl group}$$

$$\underset{1}{C}—\underset{2}{C}—\underset{3}{C}—\underset{4}{C}—\underset{5}{C}—\underset{6}{C}—\underset{7}{C}—\underset{8}{C}$$
$$\qquad\qquad CH_3 \quad CH_2—CH_3$$

Finally, fill in the proper number of hydrogens at each carbon of the principal chain so that each carbon has a total of four bonds:

$$H_3C—CH_2—CH—CH_3$$
$$H_3C—CH_2—CH—CH—CH—CH_2—CH_2—CH_3$$
$$\qquad\qquad CH_3 \qquad CH_2—CH_3$$

4-*sec*-butyl-5-ethyl-3-methyloctane

To write the structure in condensed form, put like groups attached to the same carbon within parentheses. Notice that the structure contains within it two *sec*-butyl groups (red in the following structure), even though only one is mentioned in the name; the other consists of a methyl branch and part of the principal chain.

$$H_3C—CH_2—CH—CH_3$$
$$H_3C—CH_2—CH—CH—CH—CH_2—CH_2—CH_3 \Rightarrow (CH_3CH_2CH)_2CHCHCH_2CH_2CH_3$$
$$\qquad CH_3 \qquad CH_2—CH_3 \qquad\qquad\qquad\qquad CH_3 \quad CH_2CH_3$$

Nomenclature and Chemical Indexing

The world's chemical knowledge is housed in the **chemical literature**, which is the collection of books, journals, patents, technical reports, and reviews that constitute the published record of chemical research. To find out what, if anything, is known about an organic compound of interest, we have to search the entire chemical literature. To carry out such a search, organic chemists rely on two major indexes. One is *Chemical Abstracts*, published by the Chemical Abstracts Service of the American Chemical Society, which has been the major index of the entire chemical literature since 1907. The second index is *Beilstein's Handbook of Organic Chemistry*, known to all chemists simply as *Beilstein*, which has published detailed information on organic compounds since 1881. Initially, a search of these indexes was a laborious manual process that could require hours or days in the library. Today, however, both *Chemical Abstracts* and *Beilstein* have efficient search engines (called *SciFinder*® and *Reaxys*®, respectively), that enable chemists to search for chemical information from a personal computer. Nomenclature plays a key role in locating chemical compounds, particularly in *Chemical Abstracts*, but it is also possible to search for a compound of interest by submitting its structure. A search of *Chemical Abstracts* yields a short summary, called an *abstract*, of every article that references the compound of interest, along with a detailed reference to each article. *Beilstein* yields not only the appropriate references but also detailed summaries of compound properties. One can also search for chemical reactions in both indexes.

PROBLEMS

2.8   Draw structures for all isomers of (a) hexane and (b) heptane. Give their systematic names.

2.9   Name the following compounds. Be sure to designate the principal chain properly before constructing the name.

(a) CH$_3$CH$_2$—CH——CH—CH$_2$CH$_2$CH$_3$
         | |
         CH$_2$ CH$_3$
         |
         CH$_2$—CH$_2$—CH$_3$

(b)   CH$_3$ CH$_2$CH$_3$
                        | |
     CH$_3$CH$_2$CH$_2$CH—C——CH—CH$_2$CH$_2$CH$_3$
                        | |
                        CH$_3$ CH$_3$

2.10   Draw a structure for (CH$_3$CH$_2$CH$_2$)$_2$CHCH(CH$_2$CH$_3$)$_2$ in which all carbon–carbon bonds are shown explicitly; then name the compound.

2.11   Draw the structure of 4-isopropyl-2,4,5-trimethylheptane.

E. Classification of Carbon Substitution

When we begin our study of chemical reactions, it will be important to recognize different types of carbon substitution in branched compounds. A carbon is said to be **primary**, **secondary**, **tertiary**, or **quaternary** when it is bonded to one, two, three, or four other carbons, respectively.

primary carbons

H$_3$C—CH$_2$—CH$_2$—CH——C—CH$_3$

CH$_3$ CH$_3$ quaternary carbon

CH$_3$

secondary carbons

tertiary carbon

primary carbons

Likewise, the hydrogens bonded to each type of carbon are called primary, secondary, or tertiary hydrogens, respectively.

PROBLEMS

2.12   In the structure of 4-isopropyl-2,4,5-trimethylheptane (Problem 2.11)

(a) Identify the primary, secondary, tertiary, and quaternary carbons.

(b) Identify the primary, secondary, and tertiary hydrogens.

(c) Circle one example of each of the following groups: a methyl group; an ethyl group; an isopropyl group; a *sec*-butyl group; an isobutyl group.

2.13   Identify the ethyl groups and the methyl groups in the structure of 4-*sec*-butyl-5-ethyl-3-methyloctane, the compound discussed in Study Problem 2.5 (p. 66). Note that these groups are not necessarily confined to those specifically mentioned in the name, and a particular carbon might appear in more than one group.

2.5   CYCLOALKANES, SKELETAL STRUCTURES, AND SUBSTITUENT GROUP ABBREVIATIONS

Some alkanes contain carbon chains in closed loops, or rings; these are called **cycloalkanes**. Cycloalkanes are named by adding the prefix *cyclo* to the name of the alkane. Thus, the six-membered cycloalkane is called *cyclohexane*.

cyclohexane

The names and some physical properties of the simple cycloalkanes are given in Table 2.3. The general formula for an alkane containing a single ring has two fewer hydrogens than that of the open-chain alkane with the same number of carbon atoms. For example, cyclohexane has the formula C_6H_{12}, whereas hexane has the formula C_6H_{14}. The general formula for the cycloalkanes with one ring is C_nH_{2n}.

TABLE 2.3   Physical Properties of Some Cycloalkanes

Compound	Boiling point (°C)	Melting point (°C)	Density (g mL^{-1})
cyclopropane	−32.7	−127.6	—
cyclobutane	12.5	−50.0	—
cyclopentane	49.3	−93.9	0.7457
cyclohexane	80.7	6.6	0.7786
cycloheptane	118.5	−12.0	0.8098
cyclooctane	150.0	14.3	0.8340

Because of the tetrahedral configuration of carbon in the cycloalkanes, the carbon skeletons of the cycloalkanes (except for cyclopropane) are not planar. We'll study the conformations of cycloalkanes in Chapter 7. For now, remember only that planar condensed structures for the cycloalkanes convey no information about their conformations.

Skeletal Structures   An important structure-drawing convention is the use of **skeletal structures**. For hydrocarbons, skeletal structures show only the carbon–carbon bonds. In this notation, a cycloalkane is drawn as a closed geometric figure. In a skeletal structure, it is understood that a carbon is located at each vertex of the figure, and that enough hydrogens are present on each carbon to fulfill its tetravalence. Thus, the skeletal structure of cyclohexane is drawn as follows:

a carbon and two hydrogens
are at each vertex

Skeletal structures may also be drawn for noncyclic alkanes. For example, hexane can be indicated this way:

for H_3C —CH_2—CH_2—CH_2—CH_2—CH_3

When drawing a skeletal structure for a noncyclic compound, don't forget that carbons are not only at each vertex, but also at the *ends of the structure*. Thus, the six carbons of hexane in the preceding structure are indicated by the four vertices and two ends of the skeletal structure. Here are two other examples of skeletal structures:

2,6-dimethyldecane

isopropylcyclopentane

3,3,4-triethylhexane

Use of Substituent Group Abbreviations   Another technique used to streamline structure drawing is the use of abbreviations for simple alkyl substituents. The first two letters of the alkyl group name are typically used. Therefore, Me = methyl = —CH_3; Et = ethyl = —CH_2CH_3; Pr = propyl = —$CH_2CH_2CH_3$; and Bu = butyl = —$CH_2CH_2CH_2CH_3$. Small internally branched groups can also be abbreviated; their abbreviations are given in Table 2.2 on p. 62: for example, *t*-Bu or *t*Bu = *tert*-butyl = —$C(CH_3)_3$. These abbreviations are particularly convenient when used in conjunction with skeletal structures, as in the following examples.

CH_3 CH_3

skeletal structure
with substituent
formulas

=

skeletal structure

=

Me   Me

skeletal structure
with substituent
abbreviations

2,4-dimethylhexane

—$CHMe_2$ =   —   =   —*i*Pr

isopropylcyclopentane

The isopropylcyclopentane example illustrates that there may be several different ways that group abbreviations can be used for the same structure.

Nomenclature of Cycloalkanes   The nomenclature of cycloalkanes follows essentially the same rules used for open-chain alkanes.

methylcyclobutane   **1,3-dimethyl**cyclobutane   **1-ethyl-2-methyl**cyclohexane
(Note the alphabetical citation, rule 9.)

The numerical prefix 1- is not necessary for monosubstituted cycloalkanes. For example, the first compound is methylcyclobutane, not 1-methylcyclobutane. Two or more substituents, however, must be numbered to indicate their relative positions. The lowest number is assigned in accordance with the usual rules.

Most of the cyclic compounds in this text, like those in the preceding examples, involve rings with small alkyl branches. In such cases, the ring is treated as the principal chain. However, when a noncyclic carbon chain contains more carbons than an attached ring, the ring is treated as the substituent.

$$CH_3CH_2CH_2CH_2CH_2 \!-\!\!\triangleleft$$

1-cyclopropylpentane
(not pentylcyclopropane)

STUDY PROBLEM   2.6

Name the following compound.

SOLUTION   This problem, in addition to illustrating the nomenclature of cyclic alkanes, is a good illustration of rule 8 for nomenclature, the "first point of difference" rule (p. 64). The compound is a cyclopentane with two methyl substituents and one ethyl substituent. If we number the ring carbons consecutively, the following numbering schemes (and corresponding names) are possible, depending on which carbon is designated as carbon-1:

1,2,4-	4-ethyl-1,2-dimethylcyclopentane
1,3,4-	1-ethyl-3,4-dimethylcyclopentane
1,3,5-	3-ethyl-1,5-dimethylcyclopentane

The correct name is decided by nomenclature rule 8 using the numbering schemes (*not* the names themselves). Because all numbering schemes begin with 1, the second number must be used to decide on the correct numbering. The scheme 1,2,4- has the lowest number at this point. Consequently, the correct name is 4-ethyl-1,2-dimethylcyclopentane.

Draw a skeletal structure of *tert*-butylcyclohexane.

SOLUTION   The real question in this problem is how to represent a *tert*-butyl group with a skeletal structure. The branched carbon in this group has four other bonds, three of which go to CH_3 groups. Hence:

tert-**butylcyclohexane**
(skeletal structure)

PROBLEMS

2.14   Represent each of the following compounds with a skeletal structure.

(a)

$$CH_3CH_2CH_2CH-\overset{\overset{\displaystyle CH_3}{|}}{CH}-C(CH_3)_3$$
$$\underset{\underset{\displaystyle CH_3}{|}}{}$$

(b) ethylcyclopentane

2.15   Redraw the structures in Problem 2.14 using abbreviations for substituent groups. For structure (a), draw one structure that shows as many abbreviated methyl groups as possible. (*Hint:* There are six methyl groups.) Then draw another that shows a five-carbon skeleton, two abbreviated methyl groups, and one abbreviated *tert*-butyl group.

2.16   Name the following compounds.

(a)   (b)

2.17   How many hydrogens are in an alkane of *n* carbons containing (a) two rings? (b) three rings? (c) *m* rings?

2.18   How many rings does an alkane have if its formula is (a) C_8H_{10}? (b) C_7H_{12}? Explain how you know.

2.6   PHYSICAL PROPERTIES OF ALKANES

Each time we come to a new family of organic compounds, we'll consider the trends in their boiling points, melting points, densities, and solubilities. These physical properties of an organic compound are important because they determine the conditions under which the compound is handled and used. For example, the form in which a drug is manufactured and dispensed is affected by its physical properties. In commercial agriculture, ammonia (a gas at ordinary temperatures) and urea (a crystalline solid) are both very important sources of nitrogen, but their physical properties dictate that they are handled and dispensed in very different ways.

Your goal should *not* be to memorize physical properties of individual compounds, but rather to learn to predict trends in how physical properties vary with structure.

A. Boiling Points

The **boiling point** is the temperature at which the vapor pressure of a substance equals atmospheric pressure (which, at sea level, is 760 mm Hg). Table 2.1 (p. 47) shows that, at room temperature (about 25 °C), methane, ethane, propane, and butane are gases. The unbranched alkanes with 4–17 carbons are liquids.

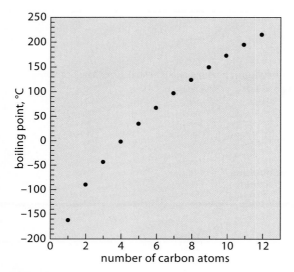

FIGURE 2.7 Boiling points of some unbranched alkanes plotted against the number of carbon atoms. Notice the steady increase with the size of the alkane, which is in the range of 20–30 °C per carbon atom.

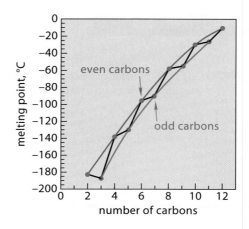

FIGURE 2.8 A plot of melting points of the unbranched alkanes against the number of carbon atoms. Notice the general increase of melting point with molecular size. Also notice that the alkanes with an even number of carbons (*red*) lie on a different curve from the alkanes with an odd number of carbons (*blue*). This trend is observed in a number of different types of organic compounds.

A regular increase in the boiling points of the unbranched alkanes occurs with increasing number of carbons (Fig. 2.7). *The regular increase in boiling point of 20–30 °C per carbon atom within a series is a general trend observed for many types of organic compounds.* The basis of this trend is the *noncovalent attractions between molecules* in the liquid state. The greater these intermolecular attractions are, the more energy (heat, higher temperature) it takes to overcome them so that the molecules escape into the gas phase, in which such attractions do not exist. *The greater the intermolecular attractions within a liquid are, the greater the boiling point is.* Now, it is important to understand that there are no covalent bonds between molecules, and furthermore, that intermolecular attractions have *nothing* to do with the strengths of the covalent bonds within the molecules themselves.

The physical basis of these attractive forces, which are important in a number of contexts, and especially in biology, will be discussed in Sec. 8.5.

B. Melting Points

The **melting point** of a substance is the temperature above which it is transformed spontaneously and completely from the solid to the liquid state. The melting point is an especially important physical property in organic chemistry because it is used both to identify organic compounds and to assess their purity. Melting points are usually depressed, or lowered, by impurities. Moreover, the melting range (the range of temperature over which a substance melts), is usually quite narrow for a pure substance and is substantially broadened by impurities. The melting point largely reflects the stabilizing intermolecular interactions between molecules in the crystal as well as the molecular symmetry, which determines the number of indistinguishable ways in which the molecule fits into the crystal. The higher the melting point, the more stable the crystal structure is relative to the liquid state. Although most alkanes are liquids or gases at room temperature and have relatively low melting points, their melting points illustrate trends that are observed in the melting points of other types of organic compounds.

One such trend is that melting points tend to increase with the number of carbons (Fig. 2.8). Another trend is that the melting points of unbranched alkanes with an even number of carbon atoms lie on a separate, higher curve from those of the alkanes with an odd number of carbons. This reflects the more effective packing of the even-carbon alkanes in the crystalline solid state. In other words, the odd-carbon alkane molecules do not "fit together" as well in the crystal as the even-carbon alkanes. Similar alternation of melting points is observed in other series of compounds, such as the cycloalkanes in Table 2.3. The effect of crystal forces on melting points is considered further in Sec. 8.5D.

Branched-chain hydrocarbons tend to have lower melting points than linear ones because the branching interferes with regular packing in the crystal. When a branched molecule has a substantial symmetry, however, its melting point is typically relatively high because of the ease with which symmetrical molecules fit together within the crystal. For example, the melting point of the very symmetrical molecule neopentane, −16.8 °C, is considerably higher than that of the less symmetrical pentane, −129.8 °C. Compare also the melting points of the compact and symmetrical molecule cyclohexane, +6.6 °C, and the extended and less symmetrical hexane, −95.3 °C.

neopentane
mp −16.8 °C

pentane
mp −129.8 °C

cyclohexane
mp +6.6 °C

hexane
mp −95.3 °C

In summary, melting points show the following general trends:

1. Melting points tend to increase with increasing molecular mass within a series.
2. In many cases, highly symmetrical molecules have unusually high melting points.
3. A sawtooth pattern of melting point behavior (see Fig. 2.8) is observed within many homologous series.

Fats and Oils

Fats and oils used in cooking illustrate the effect of structure on melting point. Most fats, such as lard, butter, and vegetable shortening, are solids at room temperature. They contain unbranched alkyl chains that pack well into a crystal lattice. Cooking oils are essentially fats (known in commerce as "unsaturated fats") that contain one or more double bonds, each of which puts a "kink" in the hydrocarbon chain. This shape makes formation of the regular crystal lattice more difficult. Oils with this structural characteristic, such as vegetable oil, canola oil, and olive oil, have much lower melting points and are therefore liquids at room temperature.

a saturated fat;
a solid at room temperature
(this structure is typical of shortening)

an unsaturated fat;
a liquid at room temperature
(this structure is typical of olive oil)

PROBLEM

2.19 In each part, match the melting point with the compound.
 (a) −109 °C, −56 °C, +100.7 °C: octane, 2-methylheptane, and 2,2,3,3-tetramethylbutane
 (b) −93 °C, +5.5 °C: benzene and toluene

benzene   toluene

C. Other Physical Properties

Among the other significant physical properties of organic compounds are dipole moments, solubilities, and densities. A molecule's dipole moment (Sec. 1.2D) determines its *polarity*, which, in turn, affects its physical properties. Because carbon and hydrogen differ little in their electronegativities, alkanes have negligible dipole moments and are therefore *nonpolar molecules*. We can see this graphically by comparing the EPMs of ethane, with a dipole moment of zero, and fluoromethane, a polar molecule with a dipole moment of 1.82 D.

EPM of ethane   EPM of fluoromethane (H_3C—F)

Solubilities are important in determining which solvents can be used to form solutions; most reactions are carried out in solution. Water solubility is particularly important for several reasons. For one thing, water is the solvent in biological systems. For this reason, water solubility is a crucial factor in the activity of drugs and other biologically important compounds. There has also been an increasing interest in the use of water as a solvent for large-scale chemical processes as part of an effort to control environmental pollution by organic solvents. The water solubility of the compounds to be used in a water-based chemical process is crucial. (We'll deal in greater depth with the important question of solubility and solvents in Chapter 8.) The alkanes are, for all practical purposes, insoluble in water—thus the saying, "Oil and water don't mix." (Alkanes are a major constituent of crude oil.)

The density of a compound is another property, like boiling point or melting point, that determines how the compound is handled. For example, whether a water-insoluble compound is more or less dense than water determines whether it will appear as a lower or upper layer when added to water. Alkanes have considerably lower densities than water. For this reason, a mixture of an alkane and water will separate into two distinct layers with the less dense alkane layer on top. An oil slick is an example of this behavior (Fig. 2.9).

PROBLEM

2.20 Gasoline consists mostly of alkanes. Explain why water is not usually very effective in extinguishing a gasoline fire.

FIGURE 2.9 The lower density of hydrocarbons and their insolubility in water allows an oil spill in flood waters to be contained by plastic tubes at a Texas refinery in the aftermath of Hurricane Rita in 2005.

2.7   COMBUSTION

A. The Combustion of Alkanes

Alkanes are among the least reactive types of organic compounds. They do not react with common acids or bases, nor do they react with common oxidizing or reducing agents.

Alkanes do, however, share one type of reactivity with many other types of organic compounds: they are flammable. This means that they react rapidly with oxygen to give carbon dioxide and water, provided that the reaction is initiated by a suitable heat source, such as a flame or the spark from a spark plug. This reaction is called **combustion**. An example is the combustion of methane, the major alkane in natural gas:

$$CH_4 + 2O_2 \longrightarrow CO_2 + 2H_2O \tag{2.6}$$

This reaction illustrates *complete combustion*: combustion in which carbon dioxide and water are the only products. Under conditions of oxygen deficiency, incomplete combustion may also occur with the formation of such byproducts as carbon monoxide, CO. Carbon monoxide is a deadly poison because it bonds to, and displaces oxygen from, hemoglobin, the protein in red blood cells that transports oxygen to tissues. It is also colorless and odorless, and is therefore difficult to detect without special equipment.

The fact that we can carry a container of gasoline in the open air without its going up in flames shows that simple mixing of alkanes and oxygen does not initiate combustion. However, once a spark is applied the combustion reaction proceeds vigorously.

Among organic compounds, alkanes are one of the best chemical sources of energy because they liberate large amounts of energy on combustion. This accounts for their importance as fuels for both transportation and heating. For example, one mole (114 g, or about 1/3 of a cup) of liquid 2,2,4-trimethylpentane, a major component of automotive gasoline, liberates 5461 kJ (1305 kcal) when it undergoes complete combustion.

$$\underset{\textbf{2,2,4-trimethylpentane}}{\text{Me}\overset{\text{Me Me}}{\wedge}\text{Me}} + \frac{25}{2}O_2 \longrightarrow 8\,CO_2 + 9\,H_2O \tag{2.7}$$

carbon dioxide

FIGURE 2.10 Atmospheric CO_2 levels for the past 1000 years. The data prior to 1958 were obtained from air bubbles trapped in dated ice core samples. More recent data were obtained from air sampling towers on Mauna Loa, Hawaii, by the Scripps Institute of Oceanography (1958–1974) and the National Oceanic and Atmospheric Administration (NOAA, 1974–present). The inset shows data obtained since 1975 in more detail. These data show the seasonal fluctuations normally observed in CO_2 levels. Notice the continuous rise in CO_2 levels since the nineteenth century.

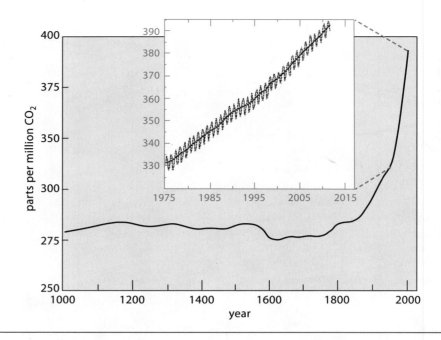

This is a *large* amount of energy; about 1/3 of a cup of this alkane liberates enough energy on combustion to propel a 3,000-pound car at conventional mileage for about half a mile, or (assuming we could capture and use all of it) to convert almost 3 *gallons* of water at 0 °C to steam at 100 °C!

Two problems with the combustion of alkanes are (1) the efficiency with which energy can be recovered from the reaction as work, and (2) the products of the reaction, specifically carbon dioxide. The efficiency of the typical automotive engine is roughly 20–25%. This is not likely to increase significantly for gasoline engines. Other, more efficient, ways of powering motor vehicles involve the generation and use of electricity in various ways, in some cases from renewable fuels, and these are under active investigation. Burning hydrocarbons for heating is very efficient, but the problem of CO_2 generation remains.

As Eq. 2.7 illustrates, every carbon atom of a hydrocarbon combines with two atoms of oxygen to generate a molar equivalent of carbon dioxide, and every pair of hydrogens combines with one oxygen atom to generate a molar equivalent of water. The atmosphere can hold a relatively small amount of water, and when that is exceeded, water returns to Earth as rain or snow. However, natural processes of removing carbon dioxide from the atmosphere are limited. After eons in which the CO_2 content of the atmosphere remained relatively constant at about 290 parts per million (ppm), the amount of CO_2 in Earth's atmosphere began to rise dramatically with the advent of the industrial age. The CO_2 level now approaches 400 ppm, an increase of more than one-third (Fig. 2.10). Most of this increase has taken place in the last 35 years. Because so much of it is produced, carbon dioxide is the most significant of several compounds known to be *greenhouse gases*, atmospheric compounds that act as a heat-reflective blanket over Earth. Most scientists are convinced that the temperature of Earth is being increased by the effect of greenhouse gases; this phenomenon is known as *global warming*. These scientists believe that global warming is beginning to have significant adverse environmental consequences, such as an increase in the intensity of hurricanes, the rapid receding of glaciers, and the extinction of animal and plant species at an increased rate. Global warming predicts that the ocean levels will rise as polar ice melts, and the resulting coastal flooding will likely displace hundreds of millions of people. As a result of these concerns, along with concerns about the political instability of the oil-producing regions of the world, the development of alternative fuels has become increasingly urgent. Ideally, the goal is to produce cheap and abundant fuels that will not, on combustion, increase the net CO_2 content of the atmosphere.

Combustion finds a minor but important use as an analytical tool for the determination of molecular formulas. In this type of analysis, the mass of CO_2 produced in the combustion of a known mass of an organic compound is used to calculate the amount of carbon in the sample. Similarly, the mass of H_2O produced is used to calculate the amount of hydrogen in the sample. (Procedures have been developed for the combustion analysis of other elements.) Combustion analysis is illustrated in Problems 2.44 and 2.45 on p. 85.

PROBLEMS

2.21 Give a general balanced reaction for
(a) the complete combustion of an alkane (formula C_nH_{2n+2}).
(b) the complete combustion of a cycloalkane containing one ring (formula C_nH_{2n}).

2.22 Calculate the number of pounds of CO_2 released into the atmosphere when 15 gallons of gasoline is burned in an automobile engine. Assume complete combustion. Also assume that gasoline is a mixture of octane isomers and that the density of gasoline is 0.692 g mL^{-1}. (This assumption ignores about 10 volume percent of oxygenated additives.) Useful conversion factors: 1 gallon = 3.785 L; 1 kg = 2.204 lb.

B. Combustion and the Chemistry of Life Processes

As the French chemist Lavoisier observed in the eighteenth century, humans and other aerobic living organisms breathe O_2 and expire CO_2, and in that sense they are carrying out combustion. The biochemical fuel is glucose, a sugar, which is obtained from foods:

$$+ \ 6\,O_2 \longrightarrow 6\,CO_2 \ + \ 6\,H_2O \tag{2.8}$$

$C_6H_{12}O_6$
D-glucopyranose
(a form of glucose)

The amount of energy available from the combustion of a mole of solid glucose, if it were released solely as heat, is 2750 kJ mol^{-1} (657 kcal mol^{-1}). [Compare this to the energy available from the combustion of the six-carbon alkane, hexane: 4163 kJ mol^{-1} (995 kcal mol^{-1})]. The biological "combustion" of glucose does not involve lighting a match and burning it. Rather, the living organism uses a series of chemical reactions that take glucose apart, one or two bonds at a time, and stores the energy liberated at each stage by forming molecules that can be tapped as energy sources when needed, such as adenosine triphosphate (ATP). We'll learn about some of these processes in this text, and, if you study biochemistry, you'll get a more thorough overview. The human body recovers the energy from glucose "combustion" with efficiencies that vary from 40–60%, depending on conditions. Given that human metabolism is 2–3 times as efficient as an automotive engine, more energy is recovered from the "combustion" of a mole of glucose than an internal-combustion engine recovers from a mole of 2,2,4-trimethylpentane!

2.8   FUNCTIONAL GROUPS, COMPOUND CLASSES, AND THE "R" NOTATION

A. Functional Groups and Compound Classes

Alkanes are the conceptual "rootstock" of organic chemistry. Replacing C—H bonds of alkanes gives the many functional groups of organic chemistry. A **functional group** is a characteristically bonded group of atoms that has about the same chemical reactivity whenever it occurs in a variety of compounds. Compounds that contain the same functional group comprise a **compound class**. Consider the following examples:

isobutylene

functional group: $\overset{\diagdown}{\underset{\diagup}{C}} = \overset{\diagup}{\underset{\diagdown}{C}}$

compound class: **alkene**

ethyl alcohol

functional group: $-\overset{|}{\underset{|}{C}}-OH$

compound class: **alcohol**

acetic acid

functional group: $-CO_2H$

compound class: **carboxylic acid**

For example, the functional group that is characteristic of the alkene compound class is the carbon–carbon double bond. Most alkenes undergo the same types of reactions, and these reactions occur at or near the double bond. Similarly, all compounds in the alcohol compound class contain an —OH group bound to the carbon atom of an alkyl group. The characteristic reactions of alcohols occur at the —OH group or the directly attached carbon, and this functional group undergoes the same general chemical transformations regardless of the structure of the remainder of the molecule. Needless to say, some compounds can contain more than one functional group. Such compounds belong to more than one compound class.

acrylic acid
contains both C=C and CO_2H functional groups
and is thus both an alkene and a carboxylic acid

The organization of this text is centered for the most part on the common functional groups and corresponding compound classes. Although you will study in detail each major functional group in subsequent chapters, you should learn to recognize the common functional groups and compound classes now. These are shown on the inside front cover.

B. "R" Notation

Sometimes we'll want to use a general structure to represent an entire class of compounds. In such a case, we can use the **R notation**, in which an R is used to represent all *alkyl groups* (Sec. 2.4C). For example, R—Cl can be used to represent an alkyl chloride.

R—Cl   could represent H_3C—Cl

R— = H_3C—   R— = $(CH_3)_2CH$—   R— = cyclohexyl—

Just as alkyl groups such as methyl, ethyl, and isopropyl are substituent groups derived from alkanes, **aryl groups** are substituent groups derived from benzene and its derivatives. The simplest aryl group is the **phenyl group**, abbreviated Ph—, which is derived from the hydrocarbon benzene. Notice that each ring carbon of an aryl group not joined to another group bears a hydrogen atom that is not shown. (This is the usual convention for skeletal structures; see Sec. 2.5.)

benzene

skeletal structure of benzene

—CH=CH₂   can be written   Ph—CH=CH₂ or Ph

phenyl group

Other aryl groups are designated by Ar—. Thus, Ar—OH could refer to any one of the following compounds, or to many others.

Ar—OH   could represent   H₃C—⟨⟩—OH or (Cl)—OH

where Ar— = H₃C—⟨⟩—   Ar— = (Cl)—

Although you will not study benzene and its derivatives until Chapter 15, before then you will see many examples in which phenyl and aryl groups are used as substituent groups.

PROBLEMS

2.23 Draw a structural formula for each of the following compounds. (Several formulas may be possible in each case.)
(a) a carboxylic acid with the molecular formula $C_2H_4O_2$
(b) an alcohol with the molecular formula $C_5H_{10}O$

2.24 A certain compound was found to have the molecular formula $C_5H_{12}O_2$. To which of the following compound classes could the compound belong? Give one example for each positive answer, and explain any negative responses.

an amide   an ether   a carboxylic acid   a phenol   an alcohol   an ester

2.9   OCCURRENCE AND USE OF ALKANES

Most alkanes come from **petroleum**, or crude oil. (The word *petroleum* comes from the ancient Greek word for "rock" (*petra*) and the Latin word for "oil" (*oleum*); thus, "oil from rocks.") Petroleum is a dark, viscous mixture composed mostly of alkanes and aromatic hydrocarbons (benzene and its derivatives) that are separated by a technique called **fractional distillation**. In fractional distillation, a mixture of compounds is slowly boiled; the vapor is then collected, cooled, and recondensed to a liquid. Because the compounds with the lowest

FIGURE 2.11 Fractionating towers such as these are used in the chemical industry to separate mixtures of compounds on the basis of their boiling points.

boiling points vaporize most readily, the condensate from a fractional distillation is enriched in the more volatile components of the mixture. As distillation continues, components with progressively higher boiling points appear in the condensate. A student who takes an organic chemistry laboratory course will almost certainly become acquainted with this technique on a laboratory scale. Industrial fractional distillations are carried out on a large scale in fractionating towers that are several stories tall (Fig. 2.11). The typical fractions obtained from distillation of petroleum are shown in Fig. 2.12.

Another important alkane source is natural gas, which is mostly methane. Natural gas comes from gas wells of various types. Recently, horizontal hydraulic fracturing ("fracking") has been used to release natural gas trapped in rock formations. This technique requires the use of large amounts of water and chemicals. Although it has contributed significantly to domestic natural gas supplies, fracking is controversial because of its environmental impact.

Significant biological sources of methane also exist that could someday be exploited commercially. For example, methane is produced by the action of certain anaerobic bacteria (bacteria that function without oxygen) on decaying organic matter (Fig. 2.13). This type of process, for example, produces "marsh gas," as methane was known before it was characterized by chemists. This same biological process can be used for the production of methane from animal and human waste. Methane produced this way is becoming practical as a local source of power (Fig. 2.13). The methane is burned to produce heat that is converted into electricity. Although this process generates carbon dioxide, the source of the carbon is the food

Fraction name	Fraction temperature	Number of carbons	Typical use
petroleum gas	up to 20 °C	C_1–C_4	bottled gas
gasoline	20–70 °C	C_5–C_{10}	automobile fuel
naphtha	70–120 °C	C_8–C_{12}	chemical feedstock
kerosene	120–240 °C	C_{10}–C_{16}	jet fuel, paraffin
light fuel oil	240–320 °C	C_{15}–C_{20}	diesel fuel
heavy oil	320–500 °C	C_{21}–C_{28}	lubricants, heating oil
asphalt, tar	above 500 °C	>C_{28}	road surfacing

DECREASING TEMPERATURE

heat

FIGURE 2.12 Schematic view of an industrial fractionating column. The temperature of the column decreases from bottom to top. Crude oil is introduced at the bottom and heated. As the vapors rise, they cool and condense to liquids. Fractions of progressively lower boiling points are collected from bottom to top of the column. The figure shows typical fractions, their boiling points, the number of carbons in the compounds collected, and the typical uses of each fraction after further processing.

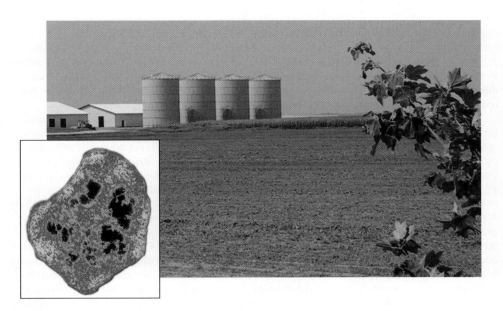

FIGURE 2.13 Manure fermenters at Fair Oaks Farms, a large commercial dairy farm in northwestern Indiana, are used to produce methane from cow manure. Electricity produced from burning the methane is fed into the local power grid. The power produced is frequently sufficient to support a large fraction of the farm's power requirements. Such fermentations are carried out by methanogens (methane-producing bacteria). The inset shows *Methanococcus voltae*, one of many known methanogen strains. This bacterium was named for Alessandro Volta, who, in 1776, collected "combustible air" (methane) that was liberated while he was exploring a marshy section of Lake Maggiore in northern Italy.

eaten by the humans and animals, and the carbon in that food comes from atmospheric carbon dioxide by photosynthesis. In other words, the CO_2 produced in this process is "recycled CO_2" and does not contribute to a net increase in atmospheric CO_2.

Alkanes of low molecular mass are in great demand for a variety of purposes—especially as motor fuels—and alkanes available directly from wells do not satisfy the demand. The petroleum industry has developed methods (called *catalytic cracking*) for converting alkanes of high molecular mass into alkanes and alkenes of lower molecular mass. The petroleum industry has also developed processes (called *reforming*) for converting unbranched alkanes into branched-chain ones, which have superior ignition properties as motor fuels.

Typically, motor fuels, fuel oils, and aviation fuels account for most of the world's hydrocarbon consumption. An Arabian oil minister once remarked, "Oil is too precious to burn." He was undoubtedly referring to the important uses for petroleum other than as fuels. Petroleum is the principal source of *carbon*, from which organic starting materials are made for such diverse products as plastics and pharmaceuticals. Petroleum is thus the basis for organic chemical *feedstocks*—the basic organic compounds from which more complex chemical substances are fabricated. However, the volatility of oil prices and the possibility that supplies will be increasingly constrained in the future have increased the interest and research in development of feedstocks from other sources such as plant-based compounds.

 Alkanes as Motor Fuels; Fuel Additives

 Alkanes vary significantly in their quality as motor fuels. Branched-chain alkanes are better motor fuels than unbranched ones. The quality of a motor fuel relates to its rate of ignition in an internal combustion engine. Premature ignition results in "engine knock," a condition that indicates poor engine performance. Severe engine knock can result in significant engine damage. The octane number is a measure of the quality of a motor fuel: the higher the octane number, the better the fuel. The octane number is the number you see associated with each grade of gasoline on the gasoline pump. Octane numbers of 100 and 0 are assigned to 2,2,4-trimethylpentane and heptane, respectively. Mixtures of the two compounds are used to define octane numbers between 0 and 100. For example, a fuel that performs as well as a 1:1 mixture of 2,2,4-trimethylpentane and heptane has an octane number of 50. The motor fuels used in modern automobiles have octane numbers in the 87–95 range.

Various additives can be used to improve the octane number of motor fuels. In the past, tetraethyllead, $(CH_3CH_2)_4Pb$, was used extensively for this purpose, but concerns over atmospheric lead pollution and the advent of catalytic converters (which are adversely affected by lead) resulted in a phase-out of tetraethyllead over the period 1976–1986 in the United States and in the European Union by 2000. This was followed by the use of methyl *tert*-butyl ether [MTBE, $(CH_3)_3C-O-CH_3$] as the major antiknock gasoline additive. After a meteoric rise in MTBE production, this compound became an object of environmental concern when its leakage from storage vessels into groundwater was discovered in several communities in the mid-1990s. Because MTBE has shown some carcinogenic (cancer-causing) activity in laboratory animals, many cities and states have enacted a phase-out of MTBE usage as a gasoline additive. Ethanol (ethyl alcohol, CH_3CH_2OH) can be used as a substitute for MTBE, and ethanol is produced by the fermentation of sugars in corn. Political action by corn-farming interests in the United States has been successful not only in substituting ethanol for MTBE as an antiknock additive, but also in the partial replacement of the hydrocarbons in gasoline by ethanol as a fuel in its own right. Ethanol production by the fermentation of sugars in corn has been subsidized. The demand for fuel ethanol was so great that the price of corn escalated sharply, and the demand for corn for ethanol production had a noticeable impact on the price of foods that depend on corn as an animal food (for example, milk, chicken, and beef). In rapidly developing Asian markets, however, which are not influenced by ethanol subsidies, MTBE continues to be used as an antiknock additive. A number of other oxygen-containing compounds can also be used for this purpose, but none of them can compete economically with MTBE and ethanol. MTBE, ethanol, and other oxygen-containing additives are collectively referred to in the fuel industry as *oxygenates*.

KEY IDEAS IN CHAPTER 2

- Alkanes are hydrocarbons that contain only carbon–carbon single bonds; alkanes may contain branched chains, unbranched chains, or rings.

- Alkanes have sp^3-hybridized carbon atoms with tetrahedral geometry. They exist in various staggered conformations that rapidly interconvert at room temperature. The conformation that minimizes van der Waals repulsions has the lowest energy and is the predominant one. In butane, the major conformation is the anti conformation; the gauche conformations exist to a lesser extent.

- Newman projections, sawhorse projections, and line-and-wedge structures can also be used to draw molecular conformations about specific bonds. A single line-and-wedge structure can be used to show the conformations about several sequential carbon–carbon bonds provided that these bonds can all be drawn in the plane of the page.

- Isomers are different compounds with the same molecular formula. Compounds that have the same molecular formula but differ in their atomic connectivities are called *constitutional isomers*.

- Alkanes are named systematically according to the substitutive nomenclature rules of the IUPAC. The name of a compound is based on its principal chain, which, for an alkane, is the longest continuous carbon chain in the molecule.

- The boiling points of alkanes and many other types of organic compounds increase within a homologous series by 20–30 °C per carbon atom.

- The melting points of alkanes and many other types of organic compounds increase with the number of carbons within a homologous series. Within this trend, however, alkanes with odd and even numbers of carbons lie on separate curves in which the melting points of even-numbered compounds have the higher values.

- Combustion is the most important reaction of alkanes. It finds practical application in the generation of much of the world's energy.

- Organic compounds are classified by their functional groups. Different compounds containing the same functional groups undergo the same types of reactions.

- The "R" notation is used as a general abbreviation for alkyl groups; Ph is the abbreviation for a phenyl group,

and Ar is the abbreviation for an aryl (substituted phenyl) group.

- Alkanes are derived from petroleum and are used mostly as fuels; however, they are also important as raw materials for the industrial preparation of other organic compounds.

ADDITIONAL PROBLEMS

2.25 Given the boiling point of the first compound in each set, estimate the boiling point of the second.

(a) $CH_3CH_2CH_2CH_2CH_2CH_2Br$ (bp 155 °C)

$CH_3CH_2CH_2CH_2CH_2CH_2CH_2Br$

(b)

$$CH_3\overset{\overset{\displaystyle O}{\|}}{C}CH_2CH_2CH_3 \qquad \text{(bp 128 °C)}$$

$$CH_3\overset{\overset{\displaystyle O}{\|}}{C}CH_2CH_2CH_2CH_2CH_3$$

(c)

$$CH_3\overset{\overset{\displaystyle O}{\|}}{C}CH_2CH_2CH_2CH_3 \quad \text{(bp 152 °C)}$$

$$CH_3CH_2\overset{\overset{\displaystyle O}{\|}}{C}CH_2CH_2CH_3$$

2.26 Draw the structures and give the names of all isomers of octane with

(a) five carbons   (b) six carbons

in their principal chains.

2.27 Label each carbon in the following molecules as primary, secondary, tertiary, or quaternary.

(a)   Me   (b)

2.28 Draw the structure of an alkane or cycloalkane that meets each of the following criteria.

(a) a compound that has more than three carbons and only primary hydrogens.

(b) a compound that has five carbons and only secondary hydrogens.

(c) a compound that has only tertiary hydrogens.

(d) a compound that has a molecular mass of 84.2.

2.29 Name each of the following compounds using IUPAC substitutive nomenclature.

(a)

(b)

$CH_2CH_2CH_3$
|
$CH_3CHCHCH_2CH_3$
|
$CH_2CH_2CH_3$

(c)

(d)

(e)

Me   Me

Et   Et

2.30 Draw structures that correspond to the following names.

(a) 4-isobutyl-2,5-dimethylheptane

(b) 2,3,5-trimethyl-4-propylheptane (skeletal structure)

(c) 5-sec-butyl-6-tert-butyl-2,2-dimethylnonane

2.31 The following labels were found on bottles of liquid hydrocarbons in the laboratory of Dr. Ima Turkey following his disappearance under mysterious circumstances. Although each name defines a structure unambiguously, some are not correct IUPAC substitutive names. Give the correct name for any compounds that are not named correctly.

(a) 2-ethyl-2,4,6-trimethylheptane

(b) 5-neopentyldecane

(c) 1-cyclopropyl-3,4-dimethylcyclohexane

(d) 3-butyl-2,2-dimethylhexane

2.32 Although compounds are indexed by their IUPAC substitutive names, sometimes chemists give whimsical names to compounds that they discover. Assist these two chemists by providing substitutive names for their compounds.

(a) Chemist Val Losipede isolated an alkane with the following skeletal structure from asphalt scrapings following a bicycle race and named it "Tourdefrançane."

(b) Chemist Slim Pickins isolated a compound with the following structure from the floor of a henhouse and dubbed it "pullane" (*pullus*, Latin for chick).

2.33 Within each set, which two structures represent the same compound?

(a)

A

B

C

(b)

CHMe₂

A   *B*   *C*

2.34 (a) Draw a skeletal structure of the compound in part (a) of Problem 2.33 that is different from the other two compounds, and name the compound.

(b) Draw a Newman projection for the most stable conformation of the compound in part (b) of Problem 2.33 that is different from the other two compounds. Draw your Newman projection about the bond between carbons 3 and 4 in the IUPAC standard numbering system, with the projection viewed from the direction of carbon-3. Describe any ambiguity you encounter in drawing this structure. Name the compound.

2.35 Sketch a diagram of potential energy versus angle of rotation about the carbon–carbon bond of chloroethane, H_3C—CH_2—Cl. The magnitude of the energy barrier to internal rotation is 15.5 kJ mol⁻¹ (3.7 kcal mol⁻¹). Label this barrier on your diagram.

2.36 Explain how you would expect the diagram of potential energy versus dihedral angle about the C2–C3 (central) carbon–carbon bond of 2,2,3,3-tetramethylbutane to differ from that for ethane (Fig. 2.3), if at all.

2.37 The anti conformation of 1,2-dichloroethane, Cl—CH_2—CH_2—Cl, is 4.81 kJ mol⁻¹ (1.15 kcal mol⁻¹) more stable than the gauche conformation. The two energy barriers (measured relative to the energy of the gauche conformation) for carbon–carbon bond rotation are 21.5 kJ mol⁻¹ (5.15 kcal mol⁻¹) and 38.9 kJ mol⁻¹ (9.3 kcal mol⁻¹).

(a) Sketch a graph of potential energy versus dihedral angle about the carbon–carbon bond. Show the energy differences on your graph and label each minimum and maximum with the appropriate conformation of 1,2-dichloroethane.

(b) Which conformation of this compound is present in greatest amount? Explain.

2.38 (a) Draw Newman projections of the most stable conformations about each of the carbon–carbon bonds in the principal chain of 2,2-dimethylpentane. *Use models!*

(b) Combine these to predict the most stable conformation of 2,2-dimethylpentane.

(c) Draw a line-and-wedge structure of the conformation you derived in (b) with the carbon–carbon bonds of the principal chain shown as lines, the bonds to all hydrogens of the principal chain shown as wedges or dashed wedges, and methyl substituents shown either as CH_3 or Me.

2.39 When the structure of compound *A* was determined in 1972, it was found to have an unusually long C—C bond and unusually large C—C—C bond angles, compared with the similar parameters for compound *B* (isobutane).

$(CH_3)_3C$ 1.611 Å H_3C 1.535 Å

∡ C—C—C = 116°   ∡ C—C—C = 110.8°

A   *B*

Explain why the indicated bond length and bond angle are larger for compound *A*.

2.40 Which of the following compounds should have the larger energy barrier to internal rotation about the indicated bond? Explain your reasoning carefully.

$(CH_3)_3C$—$C(CH_3)_3$ $(CH_3)_3Si$—$Si(CH_3)_3$

A   *B*

2.41 From what you learned in Sec. 1.3B about the relative lengths of C—C and C—O bonds, predict which of the following compounds should have the larger energy *difference* between gauche and anti conformations about the indicated bond. Explain.

CH_3O—CH_2CH_3 CH_3CH_2—CH_2CH_3

A   *B*

2.42 (a) What value is expected for the dipole moment of the anti conformation of 1,2-dibromoethane, $Br-CH_2-CH_2-Br$? Explain.

(b) The dipole moment μ of any compound that undergoes internal rotation can be expressed as a weighted average of the dipole moments of each of its conformations by the following equation:

$$\mu = \mu_1 N_1 + \mu_2 N_2 + \mu_3 N_3$$

in which μ_i is the dipole moment of conformation i, and N_i is the mole fraction of conformation i. (The mole fraction of any conformation i is the number of moles of i divided by the total moles of all conformations.) There are about 82 mole percent of anti conformation and about 9 mole percent of each gauche conformation present at equilibrium in 1,2-dibromoethane, and the observed dipole moment μ of 1,2 dibromoethane is 1.0 D. Using the preceding equation and the answer to part (a), calculate the dipole moment of a gauche conformation of 1,2-dibromoethane.

2.43 Carv and Di Oxhide drive their family car about 12,000 miles per year with an average mileage of about 25 miles per gallon of gasoline. What is the "carbon footprint" (pounds of CO_2 released into the atmosphere) of the Oxhide family car over one year? Ignoring the oxygenates present, take the density of gasoline as 0.692 g mL^{-1}. (Useful conversion factors: 1 gallon = 3.785 L; 1 kg = 2.204 lb.)

2.44 This problem illustrates how combustion can be used to determine the molecular formula of an unknown compound. A compound X (8.00 mg) undergoes combustion in a stream of oxygen to give 24.60 mg of CO_2 and 11.51 mg of H_2O.

(a) Calculate the mass of carbon and hydrogen in X.

(b) How many moles of H are present in X per mole of C? Express this as a formula C_1H_x.

(c) Multiply this formula by successive integers until the amount of H is also an integer. This is the molecular formula of X.

2.45 A hydrocarbon Y is found by combustion analysis to contain 87.17% carbon and 12.83% hydrogen by mass.

(a) Use the procedure in Problem 2.44(b) and (c) to determine the molecular formula of Y. (*Hint:* Remember that all alkanes and cycloalkanes must contain *even* numbers of hydrogens.)

(b) Draw the structure of an alkane (which may contain one or more rings) consistent with the analysis given in part (a) that has two tertiary carbons and all other carbons secondary. (More than one correct answer is possible.)

(c) Draw the structure of an alkane (which may contain one or more rings) consistent with the analysis given in part (a) that has no primary hydrogens, no tertiary carbon atoms, and one quaternary carbon atom. (More than one correct answer is possible.)

2.46 Imagine a reaction that can replace one hydrogen atom of an alkane at random with a chlorine atom.

(a) If pentane were subjected to such a reaction, how many different compounds with the formula $C_5H_{11}Cl$ would be obtained? Give their Lewis structures. Then build models (or draw line-and-wedge formulas) for each compound. Does this alter your answer in any way? Explain.

(b) Provide the same analysis as in part (a) for the same reaction carried out on 2,2-dimethylbutane.

2.47 To which compound class does each of the following compounds belong?

(a)

(b)

(c)

(d)

2.48 The α-amino acids are the building blocks of proteins. Most have the following general structure:

general structure of the α-amino acids

These amino acids differ only in their side chains —R. What functional groups are present in the side chains of each of the following amino acids? (*Hint:* See inside front cover.)

(a)

asparagine

(b)

tyrosine

(c)

threonine

2.49 Organic compounds can contain many different functional groups. Identify the functional groups (aside from the alkane carbons) present in acebutolol (Fig. P2.49), a drug that blocks a certain part of the nervous system. Name the compound class to which each group belongs.

2.50 (a) Two amides are constitutional isomers and have the formula C_4H_9NO, and each contains an isopropyl group as part of its structure. Give structures for these two isomeric amides.

(b) Draw the structure of two other amides with the formula C_4H_9NO that do *not* contain isopropyl groups.

(c) Draw the structure of a compound *X* that is a constitutional isomer of the amides in parts (a) and (b), but is *not* an amide, and contains both an amine and an alcohol functional group.

(d) Could a compound with the formula C_4H_9NO contain a nitrile functional group? Explain.

Figure P2.49   **acebutolol**

Acids and Bases
The Curved-Arrow Notation

This chapter concentrates on acid–base reactions, a topic that you have studied in earlier chemistry courses. Acid–base reactions are worth special attention in an organic chemistry course, because, first, many organic reactions are themselves acid–base reactions or are close analogs of common inorganic acid–base reactions with which you are familiar. This means that if you understand the principles behind simple acid–base reactions, you also understand the principles behind the analogous organic reactions. Second, acid–base reactions provide simple examples that can be used to illustrate some ideas that will prove useful in more complicated reactions. In particular, you'll learn in this chapter about the *curved-arrow notation*, a powerful device to help you follow, understand, and even predict organic reactions. Finally, acid–base reactions provide useful examples for discussion of some principles of chemical equilibrium.

3.1   LEWIS ACID–BASE ASSOCIATION REACTIONS

A. Electron-Deficient Compounds

In Sec. 1.2C, you learned that covalent bonding in many cases conforms to the *octet rule*, which says that the sum of the bonding and unshared valence electrons surrounding a given atom equals eight (two for hydrogen). The octet rule (or "duet" rule in the case of hydrogen) holds without exception for covalently bonded atoms from the first and second periods of the periodic table. Although the electronic octet can be exceeded when atoms from period 3 and higher are involved in covalent bonds, the rule is often obeyed for main-group elements in these periods as well.

The octet rule stipulates the *maximum* number of electrons, but it is possible for an atom to have *fewer* than an octet of electrons. In particular, some compounds contain atoms that are *short of an octet by one or more electron pairs*. Such species are called **electron-deficient compounds**. One example of an electron-deficient compound is boron trifluoride:

$$\overset{\displaystyle :\!\ddot{F}\!:}{\underset{\displaystyle :\!\ddot{F}\!:}{:\!\ddot{F}\!-\!B\!-\!\ddot{F}\!:}}$$

boron trifluoride

Boron trifluoride is electron-deficient because the boron, with six electrons in its valence shell, is two electrons, or one electron pair, short of an octet.

B. Reactions of Electron-Deficient Compounds with Lewis Bases

Electron-deficient compounds have a tendency to undergo chemical reactions that complete their valence-shell octets. In such reactions, an electron-deficient compound reacts with a species that has one or more unshared valence electron pairs. An example of such a reaction is the association of boron trifluoride and fluoride ion:

boron is electron-deficient

$$:\!\ddot{F}\!-\!B\!-\!\ddot{F}\!: \;+\; :\!\ddot{F}\!:^{-} \;\rightleftharpoons\; :\!\ddot{F}\!-\!B\!-\!\ddot{F}\!: \tag{3.1a}$$

boron trifluoride **fluoride ion** donated electron pair

tetrafluoroborate ion

In such reactions, the electron-deficient compound acts as a *Lewis acid*. A **Lewis acid** is a species that accepts an electron pair to form a new bond in a chemical reaction. Boron trifluoride is the Lewis acid in Eq. 3.1a because it accepts an electron pair from the fluoride ion to form a new B—F bond in the product, tetrafluoroborate anion. The species that donates the electron pair to a Lewis acid to form a new bond is called a **Lewis base**. Fluoride ion is the Lewis base in Eq. 3.1a. When an electron-deficient Lewis acid and a Lewis base combine to give a single product, as in this example, the reaction is called a **Lewis acid–base association reaction**.

$$:\!\ddot{F}\!-\!B\!-\!\ddot{F}\!: \;+\; :\!\ddot{F}\!:^{-} \;\underset{\text{a Lewis acid–base dissociation}}{\overset{\text{a Lewis acid–base association}}{\rightleftharpoons}}\; :\!\ddot{F}\!-\!B\!-\!\ddot{F}\!: \tag{3.1b}$$

a Lewis acid a Lewis base

(electron acceptor) (electron donor)

boron has a complete octet

As a result of this association reaction, each atom in the product tetrafluoroborate ion has a complete octet. In fact, completion of the octet provides the major driving force for this reaction.

A peculiarity in the octet-counting procedure is evident in Eqs. 3.1a and 3.1b. The fluoride ion has an octet. After it shares an electron pair with BF_3, the fluorine still has an octet in the product $^{-}BF_4$. You might ask, "How can fluorine have an octet both before and after it shares electrons?" The answer is that we count unshared pairs of electrons in the fluoride ion, but in $^{-}BF_4$, we assign to the fluorine the electrons in its unshared pairs as well as *both* electrons in the newly formed chemical bond. An apt analogy to this situation is a poor person *P* marrying a wealthy person *W*. Before the marriage, *P* is poor and *W* is wealthy; after the marriage, *W* is still wealthy, and *P*, like the boron in $^{-}BF_4$, has become wealthy by marriage! The justification for this practice of counting electrons twice is that it provides an extremely useful framework for predicting chemical reactivity. Note again that the procedure used in counting electrons for the octet differs from the one used in calculating formal charge (see Sec. 1.2C).

The reverse of a Lewis acid–base reaction is a **Lewis acid–base dissociation**. Hence, the dissociation of fluoride ion from $^-BF_4$ to give BF_3 and F^-—that is, the reverse reaction in Eqs. 3.1a and 3.1b—is an example of a Lewis acid–base dissociation.

STUDY PROBLEM 3.1

Which of the following compounds can react with the Lewis base Cl^- in a Lewis acid–base association reaction?

methane   aluminum chloride

SOLUTION For a compound to react as a Lewis acid in an association reaction, it must be able to accept an electron pair from the Lewis base Cl^-. In aluminum chloride, the aluminum is short of an octet by one pair. Hence, aluminum chloride is an electron-deficient compound and can readily accept an electron pair from chloride ion in an association reaction, as follows:

Lewis acid   **Lewis base**
(electron-deficient
compound)

In contrast, every atom in methane has the nearest noble-gas number of electrons (carbon has eight, hydrogen has two). Hence, methane is not electron-deficient and cannot undergo a Lewis acid–base association reaction.

C. The Curved-Arrow Notation for Lewis Acid–Base Association and Dissociation Reactions

Organic chemists have developed a symbolic device for keeping track of electron pairs in chemical reactions; this device is called the **curved-arrow notation**. As this notation is applied to the reactions of Lewis bases with electron-deficient Lewis acids, the formation of a chemical bond is described by a "flow" of electrons *from the electron donor* (Lewis base) *to the electron acceptor* (Lewis acid). This "electron flow" is indicated by a curved arrow drawn *from the electron source to the electron acceptor*. This notation is applied to the reaction of Eq. 3.1a in the following way:

(3.2)

The red curved arrow indicates that an unshared electron pair on the fluoride ion becomes the shared electron pair in the newly formed bond of $^-BF_4$.

The correct application of the curved-arrow notation also involves computing and properly assigning the formal charge to the products. For each reaction involving the curved-arrow notation, *the algebraic sum of the charges on the reactants must equal the algebraic sum of the charges on the products.* In other words, *total charge is conserved.* Thus, in Eq. 3.2, the reactants have a net charge of –1; hence, the products must have the same net charge. By calculating the formal charge on boron and fluorine, we determine that the charge must reside on boron.

To illustrate the application of the curved-arrow notation to a Lewis acid–base dissociation reaction, let's consider the dissociation of the ion $^-BF_4$ to give BF_3 and F^-; this reaction is the reverse of Eq. 3.2. The curved-arrow notation for this reaction is as follows:

$$
\begin{array}{ccc}
:\!\ddot{F}\!: & & :\!\ddot{F}\!: \\
| & & | \\
:\!\ddot{F}\!-\!\overset{-}{B}\!-\!\ddot{F}\!: & \longrightarrow & :\!\ddot{F}\!-\!B \;+\; :\!\ddot{F}\!:^- \\
| & & | \\
:\!\ddot{F}\!: & & :\!\ddot{F}\!:
\end{array}
\qquad (3.3)
$$

Because the B—F bond breaks in this reaction, this bond is the source of the electron pair that is transferred to a fluorine to give fluoride ion.

PROBLEM

3.1   Use the curved-arrow notation to derive a structure for the product of each of the following Lewis acid–base association reactions; be sure to assign formal charges. Label the Lewis acid and the Lewis base, and identify the atom that donates electrons in each case.

(a)
$$
\begin{array}{c}
CH_3 \\
| \\
H_3C-\overset{+}{C}\;+\;H_2\ddot{O}: \longrightarrow \\
| \\
CH_3
\end{array}
$$

(b)
$$
\begin{array}{c}
:\!\ddot{F}\!: \\
| \\
\ddot{N}H_3\;+\;B-\ddot{F}: \longrightarrow \\
| \\
:\!\ddot{F}\!:
\end{array}
$$

3.2   **ELECTRON-PAIR DISPLACEMENT REACTIONS**

A. Donation of Electrons to Atoms That Are Not Electron-Deficient

In some reactions, an electron pair is donated to an atom that is *not* electron-deficient. When this happens, another electron pair must simultaneously depart from the receiving atom so that the octet rule is not violated. The following reaction is an example of such a process.

$$ (3.4) $$

In this reaction, the carbon of bromomethane receives an electron pair from the nitrogen of ammonia. As a result, this nitrogen becomes bonded to the carbon to give the methylammonium ion, and the electron pair in the C—Br bond of bromomethane becomes an additional unshared pair in the bromide ion. If this electron pair had not departed, carbon would have ended up with more electrons than is allowed by the octet rule.

This type of reaction, in which one electron pair is displaced from an atom (in this case, from a carbon) by the donation of another electron pair from another atom, is called an **electron-pair displacement reaction**. In many such reactions, an atom is transferred between two other atoms. In this example, a carbon is transferred from the bromine to the nitrogen of ammonia.

The curved-arrow notation is particularly useful for following electron-pair displacement reactions. This usage is illustrated in Eq. 3.4. In this case, *two* arrows are required, one for the donated electron pair and one for the displaced electron pair. Notice again that each curved arrow originates at the *source* of electrons—in this case, an unshared electron pair—and terminates at the *destination* of the electron pair.

Notice also in Eq. 3.4 the *conservation of total charge* on each side of the equation, as discussed in Sec. 3.1C. The algebraic sum of the charges on the left side is zero; hence, the sum of all charges on the right side must also be zero.

STUDY GUIDE LINK 3.1
The Curved-Arrow Notation

The donated electron pairs can originate from bonds as well as unshared pairs. This is illustrated by the reaction of $^-AlH_4$ with chloromethane to give methane, AlH_3, and chloride ion:

donated
electron pair

displaced
electron pair

$$H-\overset{\underset{|}{H}}{\underset{|}{Al}}-H \qquad H_3C-\ddot{\underset{\cdot}{Cl}}: \longrightarrow H-\overset{\underset{|}{H}}{\underset{|}{Al}} \;+\; H-CH_3 \;+\; :\ddot{\underset{\cdot}{Cl}}:^- \qquad (3.5)$$

chloromethane   **methane**

In this notation, the bond corresponding to the donated electrons is "hinged" at the transferred atom (the H of the Al—H bond); it swings away from the aluminum and toward the atom that receives the electrons (the C of chloromethane).

STUDY PROBLEM 3.2

Give the curved-arrow notation for the following reaction.

$$(CH_3)_3\overset{+}{S}: \qquad ^-\!:\ddot{O}H \longrightarrow (CH_3)_2\ddot{S}: \;+\; H_3C-\ddot{O}H$$

trimethylsulfonium   **dimethyl sulfide**   **methanol**
hydroxide

SOLUTION   In this reaction, the unshared pair of the oxygen forms a bond to one of the methyl carbons of the dimethylsulfonium ion. The carbon–sulfur bond is broken, because in the product, the sulfur is bound to only two carbons. Therefore, a carbon atom (along with its three hydrogens) is transferred from the sulfur to the oxygen. Because this is an electron-pair displacement reaction, two curved arrows are required. Remember that a curved arrow is drawn from the *source* of an electron pair to its *destination*. The *source* of the donated electron pair is the ^-OH ion. The *destination* of the donated electron pair is the carbon atom. Hence, one curved arrow goes from an electron pair of the ^-OH (any one of the three pairs) to the carbon atom. Because carbon can have only eight electrons, it must lose a pair of electrons to the sulfur, which gains an electron pair in the reaction. Hence, the source of this electron pair is the C—S bond; its destination is the sulfur. The curved-arrow notation for this reaction is as follows:

$$(CH_3)_2\underset{+}{\ddot{S}}-CH_3 \quad ^-\!:\ddot{O}H \longrightarrow (CH_3)_2\ddot{S}: \;+\; H_3C-\ddot{O}H$$

(Be sure to read Study Guide Link 3.1 about the different ways that curved arrows can be drawn.) In this reaction, a methyl group is transferred from sulfur to oxygen.

Study Problem 3.2 shows how to write the curved-arrow notation for a completed reaction. Study Problem 3.3 shows how to complete a reaction for which the curved-arrow notation is given.

STUDY PROBLEM 3.3

Given the following two reactants and the curved-arrow notation for their reaction, draw the structure of the product.

$$H_3\ddot{N} \quad \overset{H}{\underset{CH_3}{\diagdown C}}{=}\ddot{O}: \quad \longrightarrow \quad ?$$

SOLUTION The bonds or unshared electron pairs at the tails of the arrows are the ones that will not be in the same place in the product. The heads of the arrows point to the places at which new bonds or unshared pairs exist in the product. Use the following steps to draw the product.

Step 1. Redraw all atoms just as they were in the reactants:

$$\overset{H}{\underset{CH_3}{H_3N \quad C \quad O}}$$

Step 2. Put in the bonds and electron pairs that do not change:

$$H_3N \quad \overset{H}{\underset{CH_3}{\diagdown C}}{-}\ddot{\underset{..}{O}}:$$

Step 3. Draw the new bonds and electron pairs indicated by the curved-arrow notation:

new electron pair

$$H_3N{-}\overset{H}{\underset{CH_3}{\diagdown C}}{-}\ddot{\underset{..}{O}}:$$

new bond

Step 4. **Complete the formal charges to give the product.** The *algebraic sum* of the formal charges in the reactants and products must be the same—zero in this case.

$$H_3\overset{+}{N}{-}\overset{H}{\underset{CH_3}{\diagdown C}}{-}\ddot{\underset{..}{\overset{..}{O}}}^{-}:$$

STUDY GUIDE
LINK 3.2
Rules for Use of the
Curved-Arrow
Notation

As you learn to use the curved-arrow notation, you will find the additional assistance in Study Guide Links 3.1 and 3.2 to be very useful. Be sure to read and study these carefully.

PROBLEMS

3.2 For each of the following cases, give the product(s) of the transformation indicated by the curved-arrow notation.

(a) $H\ddot{O}:^{-} \quad \overset{\frown}{\underset{CH_3}{C}H_2}{-}\ddot{C}l:$

(b) $(CH_3)_2C{=}C(CH_3)_2 \quad H{-}\ddot{B}r:$

(c) $H_2C \quad \overset{\ddot{O}:}{\underset{H_2C}{\parallel}} \quad \overset{+}{O}:^{-}$

3.3   Provide a curved-arrow notation for the following reaction in the left-to-right direction. (*Hint:* Use three curved arrows.)

$$CH_3\ddot{\underset{..}{O}}: \quad H-CH_2-\underset{\underset{\displaystyle CH_3}{|}}{CH}-\ddot{\underset{..}{Br}}: \quad \longrightarrow \quad CH_3\ddot{\underset{..}{O}}-H \quad H_2C=\underset{\underset{\displaystyle CH_3}{|}}{CH} \quad :\ddot{\underset{..}{Br}}:^-$$

B. Nucleophiles, Electrophiles, and Leaving Groups

In this section we'll develop a terminology that is widely used for classifying the components of an electron-pair displacement reaction. Let's return to the reaction that we used to introduce the curved-arrow notation:

$$(3.6)$$

Let's first think about the left side of this reaction from a Lewis acid–base perspective. The ammonia is a Lewis base; it is donating a pair of electrons. The carbon is accepting this electron pair and seems to be a Lewis acid. However, it is also donating a bonding electron pair to the bromine, and might also be considered simultaneously to be a Lewis base. The bromine is accepting this bonding electron pair, and might be considered to be a Lewis acid. This example shows that Lewis acid–base terminology is not very useful for describing uniquely the roles of each "actor" in this reaction.

The terms used for the components of an electron-pair displacement reaction are *nucleophile*, *electrophile*, and *leaving group*. A **nucleophile** (from the Greek word *philos*, meaning "nucleus-loving") is a species that donates an electron pair to form a new bond. In Eq. 3.6, ammonia is the nucleophile. The atom that actually donates the electron pair is called the **nucleophilic atom** or **nucleophilic center**. Nitrogen is the nucleophilic center in ammonia. An **electrophile** ("electron-loving") is a species that accepts an electron pair from the nucleophile. Bromomethane is the electrophile. The atom of the electrophile that actually accepts the electron pair is called the **electrophilic atom** or **electrophilic center**. The carbon of bromomethane is the electrophilic center. (In this case, the carbon is also giving up electrons, and this behavior may seem inconsistent for an atom that "loves electrons," but put this point aside for now.) The group that accepts electrons from the breaking bond is called, descriptively enough, a **leaving group**. The bromine is the leaving group; it becomes the bromide ion after accepting an electron pair from the breaking bond.

$$(3.7)$$

In the reverse reaction, the roles of the nucleophile and the leaving group are reversed, and the electrophilic center is the same.

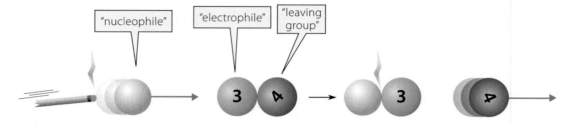

(3.8)

An analogy is what happens in a game of pool when two balls are touching, and one of them is hit by the cue ball. The cue ball is analogous to the nucleophile, the ball that is hit by the cue ball (the 3-ball below) is analogous to the electrophile, and the ball that is propelled away (the 4-ball) is analogous to the leaving group.

The same terminology can be applied to Lewis acid–base associations and dissociations. Using Eq. 3.2 (p. 89) as an example,

(3.9)

In the association (forward) direction of a Lewis acid–base association, there is a nucleophile and an electrophile, but no leaving group. In the dissociation (reverse) direction, there is a leaving group but no nucleophile.

We've already noted that in some electron-pair displacement reactions, the nucleophilic electron pair can originate from a bond rather than an unshared electron pair.

(3.10)

In this case, the nucleophilic center is the hydrogen with its pair of *bonding* electrons.

PROBLEMS

3.4   Consider the reaction analyzed in Study Problem 3.2 (p. 91), reproduced below. Identify the nucleophilic center, the electrophilic center, and the leaving group in the forward direction. (Don't hesitate to draw out the bonds between the sulfurs and each methyl group, if necessary.)

$$(CH_3)_3\overset{+}{\ddot{S}}{:} \quad \overset{..}{:}\overset{..}{O}H \longrightarrow (CH_3)_2\ddot{S}{:} \ + \ H_3C{-}\overset{..}{O}H$$

trimethylsulfonium   **dimethyl sulfide**   **methanol**
hydroxide

3.5   (a) Using the curved-arrow notation to guide you, complete the following Lewis acid–base association reaction.

$$H_3C{-}\overset{\overset{\displaystyle CH_3}{|}}{\underset{\underset{\displaystyle CH_3}{|}}{C^+}} \ + \ {:}\overset{..}{O}\underset{\underset{\displaystyle H}{|}}{CH_3} \ \rightleftharpoons$$

   (b) After you have completed the reaction, give the curved-arrow notation for the reverse direction.

   (c) Identify the nucleophilic center, the electrophilic center, and the leaving group in both forward and reverse directions of the reaction in part (a).

3.3	USING THE CURVED-ARROW NOTATION TO DERIVE RESONANCE STRUCTURES

In Sec. 1.4, you learned that resonance structures are used when the structure of a compound is not adequately represented by a single Lewis structure. *Resonance structures always differ only by the movement of electrons; nuclei do not move.* In the majority of the resonance structures you'll encounter, the electrons are moved in pairs. Because the curved-arrow notation is used to trace the flow of electron pairs, it follows that this notation can also be used to *derive* resonance structures—in other words, to show how one resonance structure can be obtained from another. Study Problem 3.4 illustrates this point with two resonance-stabilized molecules that were discussed in Sec. 1.4.

STUDY PROBLEM 3.4

In each of the following sets, show how the second resonance structure can be derived from the first by the curved-arrow notation.

(a) $$\left[CH_3\overset{..}{\underset{..}{O}}{-}\overset{+}{C}H_2 \quad \longleftrightarrow \quad CH_3\overset{+}{\underset{..}{O}}{=}CH_2 \right]$$

methoxymethyl cation

(b) $$\left[H_3C{-}\overset{\overset{\displaystyle :\overset{..}{\underset{..}{O}}{:}^-}{\diagup}}{\underset{\underset{\displaystyle \overset{..}{\underset{..}{O}}{:}}{\diagdown}}{\overset{+}{N}}} \quad \longleftrightarrow \quad H_3C{-}\overset{\overset{\displaystyle \overset{..}{O}{:}}{\diagup\!\!\!\!|}}{\underset{\underset{\displaystyle :\overset{..}{\underset{..}{O}}{:}^-}{\diagdown}}{\overset{+}{N}}} \right]$$

nitromethane

SOLUTION   (a) In the structure on the left, the positively charged carbon is electron-deficient. The structure on the right is derived by the donation of an unshared pair from the oxygen to this carbon.

$$\left[CH_3\overset{\frown}{\overset{..}{\underset{..}{O}}}{-}\overset{+}{C}H_2 \quad \longrightarrow \quad CH_3\overset{+}{\underset{..}{O}}{=}CH_2 \right]$$

This transformation resembles a Lewis acid–base association reaction, and the same curved-arrow notation is used: a single curved arrow showing the donation of the unshared pair of electrons to the electron-deficient carbon.

(b) To derive the structure on the right from the one on the left, an unshared electron pair on the upper oxygen must be used to form a bond to the nitrogen, and a bond to the lower oxygen must be used to form an unshared electron pair on the lower oxygen, as follows:

$$\left[H_3C-\overset{+}{N}\begin{matrix} :\ddot{O}:^- \\ \\ \ddot{O}: \end{matrix} \longleftrightarrow H_3C-\overset{+}{N}\begin{matrix} \ddot{O}: \\ \\ :\ddot{O}:^- \end{matrix} \right]$$

Two arrows are required because the formation of the new bond requires the *displacement* of another. Thus, we use the curved-arrow notation for electron-pair displacements.

In both of the preceding examples, the curved-arrow notation is applied in the left-to-right direction. This notation can be applied to either structure to derive the other. Thus, for part (a) in the right-to-left direction, the curved-arrow notation is as follows:

$$\left[CH_3\ddot{O}-\overset{+}{C}H_2 \longleftrightarrow CH_3\overset{+}{\ddot{O}}=CH_2 \right]$$

You should draw the curved arrow for part (b) in the right-to-left direction.

An important point is worth repeating: Even though the use of curved arrows for deriving resonance structures is identical to that for describing a reaction, the interconversion of resonance structures is *not* a reaction. The atoms involved in resonance structures do *not* move. The two structures are, taken together, a representation of a *single molecule*.

PROBLEM

3.6 (a) Using the curved-arrow notation, derive a resonance structure for the allyl cation (shown here) which shows that each carbon–carbon bond has a bond order of 1.5 and that the positive charge is shared equally by both terminal carbon atoms. (A bond with a bond order of 1.5 has the character of a single bond plus one-half of a double bond.)

$$\left[H_2\overset{+}{C}-CH=CH_2 \longleftrightarrow \quad ? \quad \right]$$

allyl cation

(b) Using the curved-arrow notation, derive a resonance structure for the allyl anion (shown here) which shows that the two carbon–carbon bonds have an identical bond order of 1.5 and that the unshared electron pair (and negative charge) is shared equally by the two terminal carbons.

$$\left[H_2\overset{..}{\overset{-}{C}}-CH=CH_2 \longleftrightarrow \quad ? \quad \right]$$

allyl anion

(c) Using the curved-arrow notation, derive a resonance structure for benzene (shown here) which shows that all carbon–carbon bonds are identical and have a bond order of 1.5.

$$\left[\bigcirc\!\!\!\hexagon \longleftrightarrow \quad ? \quad \right]$$

benzene

3.4   BRØNSTED–LOWRY ACIDS AND BASES

A. Definition of Brønsted Acids and Bases

Although less general than the Lewis concept, the *Brønsted–Lowry acid–base concept* pro-vides another way of thinking about acids and bases that is extremely important and useful in organic chemistry. The Brønsted–Lowry definition of acids and bases was published in 1923, the same year that Lewis formulated his ideas of acidity and basicity. A species that donates a proton in a chemical reaction is called a **Brønsted acid**; a species that accepts a proton in a chemical reaction is a **Brønsted base**.

The reaction of ammonium ion with hydroxide ion is an example of a Brønsted acid–base reaction.

(3.11a)

On the left side of this equation, the ammonium ion is acting as a Brønsted acid and the hydroxide ion is acting as a Brønsted base; looking at the equation from right to left, water is acting as a Brønsted acid, and ammonia as a Brønsted base.

The "classical" definition of a Brønsted acid–base reaction given above focuses on the movement of a proton. But in organic chemistry, we are always going to focus on the *movement of electrons*. As Eq. 11a illustrates, *any Brønsted acid–base reaction can be described with the curved-arrow notation for electron-pair displacement reactions*. A **Brønsted acid– base reaction** is nothing more than a *special case* of an electron-pair displacement reaction in which the electrophilic center is a proton. It's the action of the electrons that causes the net transfer of a proton from the Brønsted acid to the Brønsted base. When the electrophilic center is a proton, the electron donor is called a **Brønsted base** rather than a nucleophile. A **Brønsted acid** is the species that provides a proton to the base.

(3.11b)

Many organic electron-pair displacement reactions have Brønsted acid–base analogs. In the following examples, the only formal difference is the electrophilic center—a proton in the Brønsted acid–base reaction, and something other than a proton (a carbon in this case) in other electron-pair displacements.

A Brønsted acid–base reaction:

(3.12a)

An analogous electron-pair displacement reaction:

the electrophilic center is *something other than a* proton

leaving group $\longrightarrow$ $:\!\ddot{\text{B}}\text{r}\!-\!\text{CH}_3$ $\;\;:\!\overline{\ddot{\text{O}}\text{H}}$ $\longrightarrow$ $:\!\ddot{\text{B}}\text{r}\!:^{-}$ $+$ $\text{H}_3\text{C}\!-\!\ddot{\text{O}}\text{H}$ $\qquad$ (3.12b)

the electron donor is a *nucleophile*

Although the two types of reactions are formally similar, an important practical difference is that most Brønsted acid–base reactions are *much faster* than their organic analogs. For example, the reaction in Eq. 3.12a occurs instantaneously—about 10^9 times per second, depending on the conditions—whereas the one in Eq. 3.12b can take minutes or even hours, depending on the conditions. Despite this difference, *the fundamental similarity of the two reactions* will prove to be important and useful.

One important caution about notation: Because the traditional definition of Brønsted acids involves "proton transfer," you may sometimes be tempted to write a curved-arrow notation *incorrectly* in the following way:

Incorrect curved-arrow notation!

$\text{H}\overline{\ddot{\text{O}}}\!:^{-}$ $(\text{H})\!-\!\ddot{\text{C}}\text{l}\!:$ $\longrightarrow$ $\text{H}\ddot{\text{O}}\!-\!\text{H}$ $+$ $:\!\ddot{\text{C}}\text{l}\!:$ $\qquad$ (3.13a)

This is incorrect because it shows the movement of the proton rather than the flow of electron pairs. Someone accustomed to using the curved-arrow notation correctly would take this to imply the transfer of H^- to ^-OH, an impossible reaction! The *correct* use of the curved-arrow notation shows the flow of electron pairs:

$\text{H}\overline{\ddot{\text{O}}}\!:^{-}$ $\text{H}\!-\!\ddot{\text{C}}\text{l}\!:$ $\longrightarrow$ $\text{H}\ddot{\text{O}}\!-\!\text{H}$ $+$ $:\!\ddot{\text{C}}\text{l}\!:$ $\quad$ *Correct!* $\qquad$ (3.13b)

PROBLEMS

3.7   For each of the following electron-pair displacement reactions, give the curved-arrow notation; identify the nucleophile, the nucleophilic center, the electrophile, the electrophilic center, and the leaving group. Then write the analogous Brønsted acid–base reaction. (That is, imagine the same leaving group attached to a proton electrophilic center.) Identify the Brønsted acid and the Brønsted base in each of the resulting reactions.

(a)
$$\underset{\underset{\text{CH}_3}{|}}{\overset{\overset{\text{CH}_3}{|}}{:\!\text{O}^{+}\!\!-\!\text{CH}_3}} \;+\; {}^{-}\!:\!\ddot{\text{S}}\!-\!\text{CH}_3 \longrightarrow \underset{\underset{\text{CH}_3}{|}}{\overset{\overset{\text{CH}_3}{|}}{:\!\text{O}\!:}} \;+\; \text{H}_3\text{C}\!-\!\ddot{\text{S}}\!-\!\text{CH}_3$$

(b) $\text{H}_3\text{C}\!-\!\text{CH}_2\!-\!\ddot{\text{B}}\text{r}\!:$ $+$ $^{-}\!:\!\text{C}\!\equiv\!\text{N}\!:$ $\longrightarrow$ $\text{H}_3\text{C}\!-\!\text{CH}_2\!-\!\text{C}\!\equiv\!\text{N}\!:$ $+$ $:\!\ddot{\text{B}}\text{r}\!:^{-}$

3.8   This problem refers to the reactions shown in Eqs. 3.12a and 3.12b. When equal numbers of moles of ^-OH, H—Br, and H_3C—Br, are placed in solution together, what products are formed? (*Hint:* Which of the two possible reactions is faster?)

B. Conjugate Acids and Bases

When a Brønsted acid loses a proton, its **conjugate base** is formed; when a Brønsted base gains a proton, its **conjugate acid** is formed. When a Brønsted acid loses a proton, it becomes a Brønsted base; this acid and the resulting base constitute a **conjugate acid–base pair**. In any Brønsted acid–base reaction there are two conjugate acid–base pairs. Hence, in Eq. 3.11b, $^+NH_4$ and NH_3 are one conjugate acid–base pair, and H_2O and ^-OH are the other.

$$
\boxed{\text{conjugate acid–base pair}}
$$

$$
\overset{H}{\underset{H}{H-\overset{+}{N}-H}} + :\ddot{O}H \rightleftharpoons \overset{H}{\underset{H}{H-N:}} + H-\ddot{O}H \qquad (3.14)
$$

$$
\boxed{\text{conjugate base–acid pair}}
$$

Notice that the conjugate acid–base relationship is *across the equilibrium arrows*. For example, $^+NH_4$ and NH_3 are a conjugate acid–base pair, but $^+NH_4$ and ^-OH are *not* a conjugate acid–base pair.

The identification of a compound as an acid or a base depends on how it behaves in a specific chemical reaction. Water, for example, can act as either an acid or a base. Compounds that can act as either acids or bases are called **amphoteric compounds**. Water is the archetypal example of an amphoteric compound. In Eq. 3.14, for example, water is the conjugate *acid* in the acid–base pair $H_2O/^-OH$; in the following reaction, water is the conjugate base in the acid–base pair H_3O^+/H_2O:

STUDY GUIDE LINK 3.3
Identification of Acids and Bases

$$
\overset{H}{\underset{H}{H-\overset{+}{N}-H}} + :\overset{..}{O}-H \rightleftharpoons \overset{H}{\underset{H}{H-N:}} + H-\overset{+}{\underset{H}{O}}-H \qquad (3.15)
$$

$$
\text{acid} \qquad \text{base} \qquad\qquad \text{base} \qquad \text{acid}
$$

PROBLEMS

3.9 In the following reactions, label the conjugate acid–base pairs and specify within each pair which is the acid and which is the base. Then draw the curved-arrow notation for these reactions in the left-to-right direction.

(a) $\ddot{N}H_3 + {}^-:\ddot{O}H \rightleftharpoons {}^-:\ddot{N}H_2 + H_2\ddot{O}:$

(b) $\ddot{N}H_3 + \ddot{N}H_3 \rightleftharpoons {}^-:\ddot{N}H_2 + {}^+NH_4$

3.10 Write a Brønsted acid–base reaction in which $H_2\ddot{O}/^-:\ddot{O}H$ and $CH_3\ddot{O}H/CH_3\ddot{O}:^-$ act as conjugate acid–base pairs.

Sections 3.4A and 3.4B are about analyzing the roles of the various species in reactions. You will find that most of the reactions you will study can be analyzed in terms of these roles, and hence an understanding of these sections will help you to understand and even predict reactions. The first step in this understanding is to apply the definitions you've learned in the analysis of reactions. Study Problem 3.5 illustrates such an analysis.

STUDY PROBLEM **3.5**

Following is a series of acid–base reactions that represent the individual steps in a known organic transformation, the replacement of —Br by —OH at a carbon bearing three alkyl groups. Considering only the forward direction, classify each reaction as a Brønsted acid–base reaction or a Lewis acid–base association/dissociation. Classify each labeled species (or a group within each species) with one of the following terms: *Brønsted base, Brønsted acid, nucleophile, nucleophilic center, electrophile, electrophilic center,* and/or *leaving group.* For Brønsted acid–base reactions, show the conjugate acid–base pairs.

(3.16a)

(3.16b)

(3.16c)

SOLUTION   Classify each reaction first, and then analyze the role of each species. Reaction 3.16a is a Lewis acid–base dissociation. (Notice that a single curved arrow describes the dissociation.) In compound *A*, Br is the leaving group.

Reaction 3.16b is a Lewis acid–base association reaction. Cation *B* is an electrophile, and the electron-deficient carbon is the electrophilic center. Water molecule *D* is a nucleophile, and its oxygen is the nucleophilic center.

Reaction 3.16c is a Brønsted acid–base reaction. Ion *E* and compound *G* constitute a conjugate Brønsted acid–base pair, and compound *F* and compound *H* are a conjugate Brønsted base–acid pair. The water molecule *F* is a Brønsted base. The proton of *E* that receives an electron pair from water is an electrophilic center. The part of *E* that becomes *G* is a leaving group.

PROBLEMS

3.11   Work Study Problem 3.5 for the *reverse* of each reaction 3.16a–c.

3.12   In each of the following processes, complete the reaction using the curved arrow given; classify the process as a Brønsted acid–base reaction or a Lewis acid–base association/dissociation; and label each species (or part of each species) with one of the following terms: *Brønsted base, Brønsted acid, nucleophile, nucleophilic center, electrophile, electrophilic center,* or *leaving group.* In each part, once you complete the forward reaction, draw the curved arrow(s) for the reverse reaction and do the same exercise for it as well.

(a) $H_2C\!=\!CH_2$ $H\!-\!\overset{..}{\underset{..}{Br}}:$ $\longrightarrow$

(b) $H_2\overset{..}{O}:$ $\overset{OH}{\underset{OH}{\overset{|}{\underset{|}{B}}}}\!-\!OH$ $\longrightarrow$

C. Strengths of Brønsted Acids

Another aspect of acid–base chemistry that can be widely applied to understanding organic reactions is the strengths of Brønsted acids and bases. The strength of a Brønsted acid is determined by how well it transfers a proton to a standard Brønsted base. The standard base traditionally used for comparison is water. The transfer of a proton from an acid, HA, to water is indicated by the following equilibrium:

$$HA + H_2O \rightleftharpoons A\colon^- + H_3O^+ \tag{3.17}$$

The equilibrium constant for this reaction is given by

$$K_{eq} = \frac{[A\colon^-][H_3O^+]}{[HA][H_2O]} \tag{3.18}$$

(The quantities in brackets are molar concentrations at equilibrium.) Because water is the solvent, and its concentration remains effectively constant, regardless of the concentrations of the other species in the equilibrium, we multiply Eq. 3.18 through by $[H_2O]$ and thus define another constant K_a, called the **dissociation constant**:

$$K_a = K_{eq}[H_2O] = \frac{[A\colon^-][H_3O^+]}{[HA]} \tag{3.19}$$

Each acid has its own unique dissociation constant. The larger the dissociation constant of an acid, the more H_3O^+ ions are formed when the acid is dissolved in water at a given concentration. Thus, *the strength of a Brønsted acid is measured by the magnitude of its dissociation constant.*

Because the dissociation constants of different Brønsted acids cover a range of many powers of 10, it is useful to express acid strength in a logarithmic manner. Using p as an abbreviation for negative logarithm, we can write the following definitions:

$$pK_a = -\log K_a \tag{3.20a}$$

$$pH = -\log [H_3O^+] \tag{3.20b}$$

The pK_a values of several Brønsted acids are given in Table 3.1 on p. 102 in order of decreasing pK_a. Because stronger acids have larger K_a values, it follows from Eq. 3.20a that *stronger acids have smaller pK_a values.* Thus, HCN ($pK_a = 9.4$) is a stronger acid than water ($pK_a = 15.7$). In other words, the strengths of acids in the first column of Table 3.1 increase from the top to the bottom of the table.

PROBLEMS

3.13 What is the pK_a of an acid that has each of the following dissociation constants?
(a) 10^{-3} (b) 5.8×10^{-6} (c) 50

3.14 What is the dissociation constant of an acid that has each of the following pK_a values?
(a) 4   (b) 7.8   (c) –2

3.15 (a) Which acid is the strongest in Problem 3.13?
(b) Which acid is the strongest in Problem 3.14?

Three points about the pK_a values in Table 3.1 are worth special emphasis. The first has to do with the pK_a values for very strong and very weak acids. The direct pK_a determination of an acid in aqueous solution is limited to acids that are less acidic than H_3O^+ and more acidic than H_2O. The reason is that H_3O^+ is the strongest acid that can exist in water. If we dissolve a stronger acid in water, it immediately ionizes to H_3O^+. Similarly, ^-OH is the strongest base that can exist in water, and stronger bases react instantly with water to form ^-OH. However, pK_a values for very strong and very weak acids can be measured in other solvents, and through various methods these pK_a values can in many cases be used to estimate aqueous pK_a values. This is the basis for the estimates of the acidities of strong acids such as HCl and very weak acids such as NH_3 in Table 3.1. These approximate pK_a values will suffice for many of our applications.

TABLE 3.1 Relative Strengths of Some Acids and Bases

Conjugate acid	pK_a	Conjugate base
$\ddot{N}H_3$ (ammonia)	~35[†]	$^-{:}\ddot{N}H_2$ (amide)
$R\ddot{O}H$ (alcohol)	15–19*	$R\ddot{O}{:}^-$ (alkoxide)
$H\ddot{O}H$ (water)	15.7	$H\ddot{O}{:}^-$ (hydroxide)
HPO_4^{2-} (hydrogen phosphate)	12.3	PO_4^{3-} (phosphate)
$R\ddot{S}H$ (thiol)	10–12*	$R\ddot{S}{:}^-$ (thiolate)
$R_3\overset{+}{N}H$ (trialkylammonium ion)	9–11*	$R_3N{:}$ (trialkylamine)
$\overset{+}{N}H_4$ (ammonium ion)	9.25	$H_3N{:}$ (ammonia)
HCN (hydrocyanic acid)	9.40	$^-{:}CN$ (cyanide)
$H_2PO_4^-$ (dihydrogen phosphate)	7.21	HPO_4^{2-} (hydrogen phosphate)
$H\ddot{S}H$ (hydrosulfuric acid)	7.0	$H\ddot{S}{:}^-$ (hydrosulfide)
$R-\overset{\overset{\displaystyle O}{\|}}{C}-\ddot{O}H$ (carboxylic acid)	4–5*	$R-\overset{\overset{\displaystyle O}{\|}}{C}-\ddot{O}{:}^-$ (carboxylate)
$H\ddot{F}{:}$ (hydrofluoric acid)	3.2	$:\ddot{F}{:}^-$ (fluoride)
H_3PO_4 (phosphoric acid)	2.2	$H_2PO_4^-$ (dihydrogen phosphate)
HNO_3 (nitric acid)	–1.3	NO_3^- (nitrate)
$H_3\ddot{O}^+$ (hydronium ion)	–1.7	$H_2\ddot{O}$ (water)
$H_3C-\langle\text{ring}\rangle-SO_3H$ (p-toluene-sulfonic acid)	–2.8[†]	$H_3C-\langle\text{ring}\rangle-SO_3^-$ (p-toluene-sulfonate, or "tosylate")
H_2SO_4 (sulfuric acid)	–3[†]	HSO_4^- (hydrogen sulfate, bisulfate)
$H\ddot{C}l{:}$ (hydrochloric acid)	–6 to –7[†]	$:\ddot{C}l{:}^-$ (chloride)
$H\ddot{B}r{:}$ (hydrobromic acid)	–8 to –9.5[†]	$:\ddot{B}r{:}^-$ (bromide)
$H\ddot{I}{:}$ (hydroiodic acid)	–9.5 to –10[†]	$:\ddot{I}{:}^-$ (Iodide)
$HClO_4$ (perchloric acid)	–10[†]	ClO_4^- (perchlorate)

GREATER ACIDITY

GREATER BASICITY

* Precise value varies with the structure of R.
† Estimates; exact measurement is not possible.

The second point is that much important organic chemistry is carried out in nonaqueous solvents. In nonaqueous solvents, pK_a values typically differ substantially from pK_a values of the same acids determined in water. However, in some of these solvents, the *relative* pK_a values are roughly the same as they are in water. In other nonaqueous solvents, though, even the relative order of pK_a values is different. (We'll learn about solvent effects in Chapter 8.) Despite these differences, aqueous pK_a values such as those in Table 3.1 are the most readily available and comprehensive data on which to base a discussion of acidity and basicity.

The last point has to do with the K_a of water, which is $10^{-15.7}$, from which we obtain $pK_a = 15.7$. Don't confuse K_a with the *ion-product constant* of water, which is defined by the expression $K_w = [H_3O^+][^-OH] = 10^{-14}\ M^2$, or $-\log K_w = 14$. The ionization constant of water is defined by the expression

$$K_a = \frac{[H_3O^+][^-OH]}{[H_2O]} = \frac{K_w}{[H_2O]} = \frac{10^{-14}\ M^2}{55.6\ M} = 10^{-15.7}\ M$$

This expression has the concentration of water itself in the denominator, and thus differs from the ion-product constant of water by a factor of 1/55.6. The logarithm of this factor, –1.7, accounts for the difference between pK_a and pK_w:

$$pK_a \text{ of } H_2O = -\log K_w - \log (1/55.6) = 15.7$$

D. Strengths of Brønsted Bases

The strength of a Brønsted base is directly related to the pK_a of its conjugate acid. Thus, the base strength of fluoride ion is indicated by the pK_a of its conjugate acid, HF; the base strength of ammonia is indicated by the pK_a of its conjugate acid, the ammonium ion, $^+NH_4$. That is, when we say that a base is weak, we are also saying that its conjugate acid is strong; or, if a base is strong, its conjugate acid is weak. Thus, it is easy to tell which of two bases is stronger by looking at the pK_a values of their conjugate acids: *the stronger base has the conjugate acid with the greater (or less negative) pK_a.* For example, ^-CN, the conjugate base of HCN, is a weaker base than ^-OH, the conjugate base of water, because the pK_a of HCN is less than that of water. Therefore, the strengths of the bases in the third column of Table 3.1 increase from the bottom to the top of the table.

When we use pK_a as our measure of basicity, we are implicitly referring to the ability of the base to remove a proton from the acid H_3O^+, as shown by the reverse of Eq. 3.17. Even though the K_a of an acid AH (or its logarithmic form pK_a) is adequate to describe the basicity of its conjugate base A^-, another measure of basicity, called the *basicity constant, K_b*, is sometimes used. The basicity constant is based on the ability of a base to remove the proton from the weak acid H_2O. Again using A^- as the base, the relevant equilibrium is

$$A^- + H_2O \rightleftharpoons AH + {}^-OH \tag{3.21}$$

The equilibrium constant for this reaction is

$$K_{eq} = \frac{[^-OH][AH]}{[H_2O][A^-]} \tag{3.22a}$$

The **basicity constant K_b** is defined by multiplying this equation by $[H_2O]$:

$$[H_2O]K_{eq} = \frac{[^-OH][AH]}{[A^-]} = K_b \tag{3.22b}$$

The negative logarithm of the basicity constant is **pK_b**.

There is a simple relationship between pK_a and pK_b for any conjugate acid–base pair:

$$pK_b = pK_w - pK_a = 14 - pK_a \tag{3.23}$$

where K_w is the autoprotolysis constant of water = $10^{-14} M^2$, and $pK_w = 14$. (See Study Guide Link 3.4 for a derivation of this relationship.) In this text, we'll follow the widespread practice of using only pK_a values to describe acid and base strength, but if you encounter pK_b values, you can easily convert them into pK_a values with the simple relationship in Eq. 3.23, or vice versa. For example, the pK_a of the ammonium ion, $^+NH_4$, is 9.25 (Table 3.1). Although this number also describes the basicity of the conjugate base ammonia, we could also use the pK_b of ammonia, which is $14 - 9.25 = 4.75$, for the same purpose.

PROBLEMS

3.16   The basicities of conjugate bases A^- increase with increasing pK_a of the conjugate acids AH. How do the basicities of conjugate bases A^- change with increasing pK_b?

3.17   (a) The pK_a of acetic acid, CH_3CO_2H, is 4.67. What is the pK_b of its conjugate base acetate, $CH_3CO_2^-$?

(b) Explain why an aqueous solution of acetic acid has an acidic pH, whereas an aqueous solution of sodium acetate has a basic pH.

E. Equilibria in Acid–Base Reactions

When a Brønsted acid and base react, we can tell immediately whether the equilibrium lies to the right or left by comparing the pK_a values of the two acids involved. *The equilibrium in the reaction of an acid and a base always favors the side with the weaker acid and weaker base.* For example, in the following acid–base reaction, the equilibrium lies well to the right, because H_2O is the weaker acid and ^-CN is the weaker base.

$$
\underset{\substack{pK_a = 9.4 \\ \text{(stronger acid)}}}{HCN} + \underset{\substack{\text{(stronger} \\ \text{base)}}}{OH^-} \;\rightleftharpoons\; \underset{\substack{\text{(weaker} \\ \text{base)}}}{{}^-CN} + \underset{\substack{pK_a = 15.7 \\ \text{(weaker acid)}}}{H_2O} \tag{3.24}
$$

We'll frequently find it useful to estimate the equilibrium constants of acid–base reactions. The equilibrium constant for an acid–base reaction can be calculated in a straightforward way from the pK_a values of the two acids involved. To do this calculation, subtract the pK_a of the acid on the left side of the equation from the pK_a of the acid on the right and take the antilog of the resulting number. That is, for an acid–base reaction

$$
AH + B^- \;\rightleftharpoons\; A^- + BH \tag{3.25}
$$

in which the pK_a of AH is pK_{AH} and the pK_a of BH is pK_{BH}, the equilibrium constant can be calculated by

$$
\log K_{eq} = pK_{BH} - pK_{AH} \tag{3.26a}
$$

or

$$
K_{eq} = 10^{(pK_{BH} - pK_{AH})} \tag{3.26b}
$$

This procedure is illustrated for the reaction in Eq. 3.24 in Study Problem 3.6, and is justified in Problem 3.56 at the end of the chapter.

STUDY PROBLEM   3.6

Calculate the equilibrium constant for the reaction of HCN with hydroxide ion (see Eq. 3.24).

SOLUTION   First identify the acids on each side of the equation. The acid on the left is HCN because it loses a proton to give cyanide (^-CN), and the acid on the right is H_2O because it loses a proton to give hydroxide (^-OH). Before doing any calculation, ask whether the equilibrium should lie to the left or right. Remember that the *stronger acid and stronger base are always on one side of the equation*, and the weaker acid and weaker base are on the other side. *The equilibrium always favors the weaker acid and weaker base.* This means that the right side of Eq. 3.24 is favored and, therefore, that the equilibrium constant in the left-to-right direction is >1. This provides a quick check on whether your calculation is reasonable. Next, apply Eq. 3.26a. Subtracting the pK_a of the acid on the left of Eq. 3.24 (HCN) from the one on the right (H_2O) gives the logarithm of the desired equilibrium constant K_{eq}. (The relevant pK_a values come from Table 3.1.)

$$
\log K_{eq} = 15.7 - 9.4 = 6.3
$$

The equilibrium constant for this reaction is the antilog of this number:

$$
K_{eq} = 10^{6.3} = 2 \times 10^6
$$

This large number means that the equilibrium of Eq. 3.24 lies *far to the right*. That is, if we dissolve HCN in an equimolar solution of NaOH, a reaction occurs to give a solution in which there is *much* more ^-CN than either ^-OH or HCN. Exactly *how much* of each species is present could be determined by a detailed calculation using the equilibrium-constant expression, but in a case like this, such a calculation is unnecessary. The equilibrium constant is so large that, even with water in large excess as the solvent, the reaction lies far to the right. This also means that if we dissolve NaCN in water, only a minuscule amount of ^-CN reacts with the H_2O to give ^-OH and HCN. Typically, when K_{eq} is $\geq 10^2$, the reaction is said to lie "completely to the right"; when K_{eq} is $\leq 10^{-2}$, the reaction is said to lie "completely to the left."

PROBLEM

3.18   Using the pK_a values in Table 3.1, calculate the equilibrium constant for each of the following reactions.

   (a) NH_3 acting as a base toward the acid HCN

   (b) F^- acting as a base toward the acid HCN

Sometimes students confuse acid strength and base strength when they encounter an *amphoteric compound* (see p. 99). Water presents this sort of problem. According to the definitions just developed, the *base strength* of water is indicated by the pK_a of its *conjugate acid*, H_3O^+, whereas the *acid strength of water* (or the base strength of its conjugate base hydroxide) is indicated by the pK_a of H_2O itself. These two quantities refer to very different reactions of water:

Water acting as a base:

$$H_2O + AH \rightleftharpoons H_3O^+ + A\!:^- \qquad\qquad (3.27a)$$

$$pK_a = -1.7$$

Water acting as an acid:

$$B\!:^- + H_2O \rightleftharpoons BH + {}^-OH \qquad\qquad (3.27b)$$

$$pK_a = 15.7$$

PROBLEM

3.19   Write an equation for each of the following equilibria, and use Table 3.1 to identify the pK_a value associated with the acidic species in each equilibrium.

   (a) ammonia acting as a base toward the acid water

   (b) ammonia acting as an acid toward the base water

   Which of these reactions has the larger K_{eq} and therefore is more important in an aqueous solution of ammonia?

F. Dissociation States of Conjugate Acid–Base Pairs

When an acid or base is present in aqueous solution, its **dissociation state**—whether it is in its conjugate-acid form, its conjugate-base form, or a mixture of both—depends on the pH of the solution. It is important to know the dissociation state of an acid–base pair for several reasons. The most general reason is that the chemical reactivity of an acid–base pair depends on its dissociation state. For example, many situations in biology and medicine depend on dissociation state. For example, many biomolecules (such as enzymes) contain acidic and basic groups, and understanding the chemistry of these biomolecules requires that we know the dissociation states of these groups. The biological activities of many drugs, including their uptake into cells, depend on their dissociation states. This section will show how to determine the dissociation state of acid–base pairs.

As a specific example, consider the dissociation equilibrium of a carboxylic acid:

$$\underset{\text{AH}}{R\!-\!\overset{\overset{\displaystyle O}{\|}}{C}\!-\!OH} + H_2O \rightleftharpoons \underset{\text{A}}{R\!-\!\overset{\overset{\displaystyle O}{\|}}{C}\!-\!O^-} + H_3O^+ \qquad\qquad (3.28)$$

In this equilibrium, we'll refer to the conjugate acid as AH and its conjugate base as A, leaving the charge off of A for convenience.

Application of Le Châtelier's principle shows that in the presence of a very large H_3O^+ concentration—low pH—the equilibrium will favor the conjugate-acid form of the acid—that is, AH. Similarly, if the H_3O^+ concentration is very low—high pH—the equilibrium will favor the conjugate-base form of the acid—that is, A. Exactly what pH values are required to favor one form or the other and by how much? The answer depends on the equilibrium constant for the reaction, which in this case is the *dissociation constant*, K_a (see Sec. 3.4C and Eq. 3.19, p. 101):

$$K_a = \frac{[H_3O^+][A]}{[AH]} \tag{3.29a}$$

Because $[H_3O^+]$ and K_a are usually cited as the logarithmic quantities pH and pK_a, we'll find it useful to put this equation in logarithmic form. Taking the logarithms of both sides of this equation, we have

$$\log K_a = \log[H_3O^+] + \log \frac{[A]}{[AH]} \tag{3.29b}$$

Adopting the customary definitions, $pH = -\log[H_3O^+]$ and $pK_a = -\log K_a$,

$$-pK_a = -pH + \log \frac{[A]}{[AH]} \tag{3.29c}$$

or, rearranging,

$$pH = pK_a + \log \frac{[A]}{[AH]} \tag{3.29d}$$

Equation 3.29d is known as the **Henderson–Hasselbalch equation**, and it is nothing more than a logarithmic form of the expression for the dissociation constant (Eq. 3.29a).

Here is a very important point: the dissociation constant K_a is a property of the acid. We can't change this. However, the pH is a property of a solution that can be changed experimentally. Once we fix the pH experimentally, then, Eq. 3.29d says that the ratio [A]/[AH] is fixed. Or, conversely, if we fix the ratio [A]/[AH] experimentally, we have then fixed the pH. Therefore, the dissociation state of an acid–base pair depends on the pH of the solution (which we can change) and the pK_a of the acid (which we *cannot* change).

If we rearrange the Henderson–Hasselbalch equation to Eq. 3.29e,

$$pH - pK_a = \log \frac{[A]}{[AH]} \tag{3.29e}$$

we see that the dissociation state of the acid—the ratio [A]/[AH]—depends on the *difference* between of the pH of the solution and the pK_a of the acid.

Let's say that an acid AH is dissolved in aqueous solution, and we want to fix the ratio [A]/[AH] at a certain value. There are two ways to fix this ratio. One way is to adjust the pH by adding ^-OH (for example, by adding NaOH or KOH); or, equivalently, if we dissolve the conjugate base A in solution, we can adjust the pH by adding H_3O^+ (for example, by adding HCl). The second way to fix the ratio [A]/[AH] corresponds to a situation that is particularly relevant to biology: we dissolve either the acid AH or its conjugate base A (or both) in a *large excess* of a buffer solution that has a fixed pH value. The "large excess" of buffer is necessary because we want any acid–base reaction of AH or A with the buffer, which would change the pH, to be negligible. In human cells, which are typically buffered at pH = 7.4 ("physiological pH") by the carbonate/bicarbonate/CO_2 buffer system, most acid–base pairs are present at a much smaller concentration than the buffer. Therefore, the dissociation state of the acid–base pair present in dilute solution in the cell depends on the relationship of its pK_a to the pH of the buffer solution.

Given an acid with a certain pK_a, how does its dissociation state depend on pH? The answer is given by Eq. 3.29a or 3.29d, but we can manipulate these equations to give us a particularly convenient way of looking at this problem. The total concentration of all forms of the acid is given by [AH] + [A]. The *fraction dissociated* f_A, then, is the ratio of the dissociated form A to the total:

$$f_A = \frac{[A]}{[A] + [HA]} \tag{3.30a}$$

Now solve Eq. 3.29a for [A] to obtain $[A] = K_a[AH]/[H_3O^+]$, substitute this result into Eq. 3.30a, and cancel the common factor [HA]:

$$f_A = \frac{K_a[AH]/[H_3O^+]}{(K_a[AH]/[H_3O^+]) + [AH]} = \frac{K_a}{K_a + [H_3O^+]} \tag{3.30b}$$

Using the definitions of pH and pK_a and the property of logarithms that $x = 10^{\log x}$, Eq. 3.30b can be rewritten as

$$f_A = \frac{10^{-pK_a}}{10^{-pK_a} + 10^{-pH}} \tag{3.30c}$$

In an exactly analogous fashion, we can show that the *fraction undissociated acid* f_{AH} is given by Eq. 3.30d:

$$f_{AH} = \frac{[AH]}{[A] + [AH]} = \frac{[AH]}{(K_a[AH]/[H_3O^+]) + [AH]} = \frac{[H_3O^+]}{K_a + [H_3O^+]} = \frac{10^{-pH}}{10^{-pK_a} + 10^{-pH}} \tag{3.30d}$$

Convince yourself that the function in Eq. 3.30b goes from $f_A = 0$ at $[H_3O^+] \gg K_a$ (that is, very low pH) to $f_A = 1$ at $[H_3O^+] \ll K_a$ (very high pH). Conversely, f_{AH} (Eq. 3.30d) goes from 1 to 0 at the same extremes. Prove to yourself also that the sum $f_{AH} + f_A$ equals 1, which must be true by definition.

As we have seen—and as Eq. 3.29e shows—the dissociation state of an acid–base pair is a function of the *difference* between the pH and the pK_a of the conjugate acid. Letting $pH - pK_a = \Delta$, it can be shown (Problem 3.20, p. 109) that f_A and f_{AH} can be expressed solely as a function of Δ:

$$f_A = \frac{1}{1 + 10^{-\Delta}} \tag{3.31a}$$

$$f_{AH} = \frac{1}{1 + 10^{\Delta}} \tag{3.31b}$$

If we plot these two functions, we get the S-shaped curves shown in Fig. 3.1 on p. 108. There are several things to notice about these graphs.

1. At $pH = pK_a$, the fraction dissociated f_A = the fraction undissociated f_{AH} = 0.5. In other words, operationally, the pK_a is equal to the pH at which the acid–base pair is half-dissociated. Be sure you understand what this means. To say that an acid is "half-dissociated" does not mean that the proton is half removed from a given molecule. It means that in a population of A and AH molecules, their concentrations are equal; half of the molecules are in the A form, and half are in the AH form. Also, equilibrium is dynamic; that is, protons are rapidly jumping on and off of these molecules, but in such a way that the [A]/[AH] ratio is maintained at 1.0.

2. At pH values well below the pK_a—that is, at $pH \ll pK_a$—the acid is largely undissociated. When the pH is one unit lower than the pK_a, the acid is about 10% dissociated. When the pH is two units lower than the pK_a, the acid is about 1% dissociated.

3. At pH values well above the pK_a—that is, at $pH \gg pK_a$—the acid is largely dissociated. When the pH is one unit higher than the pK_a, the acid is about 90% dissociated. When the pH is two units higher than the pK_a, the acid is about 99% dissociated.

FIGURE 3.1. The fraction of an acid AH that is dissociated (red curve) or undissociated (black curve) as a function of the difference between the pH of the solution and the pK_a of the acid. On the left half of the curve, the pH is lower than the pK_a; on the right half, the pH is higher than the pK_a. Notice that the acid is half-dissociated when the pH and the pK_a are equal.

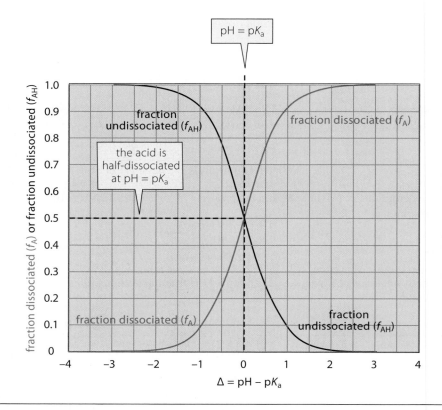

Regarding points 2 and 3, we can see that an acid is never 0% dissociated at any pH, but as the pH is lowered, 0% dissociation is approached asymptotically. Likewise, an acid is never 100% dissociated at any pH, but, as the pH is raised, 100% dissociation is approached asymptotically. As a *practical* matter, we usually say that an acid is "completely undissociated" when the pH is 2 or more units below its pK_a, and "completely dissociated" when the pH is 2 or more units above its pK_a.

If you are handy with spreadsheets, you can reproduce Fig. 3.1 for yourself. Fill column 1 with closely spaced pH values from 1 to 13 (say, 0.1 unit apart). This can be filled in automatically after you enter the first few values. Fill column 2 with the pK_a, which will be the same in each cell. (Pick any value, say 4.5.) Program column 3 to calculate $[H_3O^+] = 10^{-pH}$, column 4 to calculate $K_a = 10^{-pK_a}$, and columns 5 and 6 to calculate f_A and f_{AH} from their formulas in Eqs. 3.30c and 3.30d. Then plot columns 5 and 6 against pH (column 1) on the same set of axes. You can then change the pK_a in column 4 and see how all the numbers and the graphs change as a result.

STUDY PROBLEM 3.7

A histidine residue (B), one of the functional groups in the structure of a certain enzyme, has a conjugate-acid $pK_a = 7.8$. What is the fraction of each form (BH and B) present at physiological pH (pH = 7.4)?

Before we do a calculation, let's think about what we should expect to find. Because physiological pH (pH = 7.4) is *below* the pK_a—that is, pH – pK_a < 0—we know that the histidine residue must be less than half-dissociated—that is, [BH] > [B]. We are, then, on the left side of the curve in Fig. 3.1 at pH – pK_a = Δ = –0.4; the curve indicates that the fraction dissociation should be between 0.2 and 0.3. To calculate exactly what fraction is dissociated, use Eq. 3.31a, with the dissociated form = B:

$$f_B = \frac{1}{1 + 10^{-(-0.4)}} = \frac{1}{1 + 10^{0.4}} = \frac{1}{1 + 2.51} = \frac{1}{3.51} = 0.28$$

This shows that 28% of the enzyme molecules have histidine in the B form, and 1 – 0.28 = 0.72, or 72%, of the enzyme molecules have histidine in the BH form, and this calculation is consistent with our preliminary analysis.

PROBLEMS

3.20 Let pH – pK_a = Δ. Starting with Eqs. 3.30c and 3.30d, derive Eqs. 3.31a and 3.31b.

3.21 Ibuprofen is a drug sold as a nonprescription anti-inflammatory medication.

(a) What are the concentrations of ibuprofen and its conjugate base if 10^{-4} mole of ibuprofen is dissolved in an aqueous solution containing a large excess of a buffer at pH = 5.0?

(b) Ibuprofen is taken orally. What fraction of ibuprofen is dissociated in stomach acid? (Take the pH of stomach acid to be 2.0.)

(c) What is the dissociation state of ibuprofen in the bloodstream (pH = 7.4)?

3.22 Acetic acid, CH_3CO_2H, is a carboxylic acid with pK_a = 4.76. What is the fraction of acetic acid dissociated (f_A) if 0.1 mole of acetic acid is dissolved in pure water?

3.23 Nicotine, a habit-forming compound found in tobacco, can be protonated twice:

(a) Using intuition gained from this section, and using the pK_a values shown above, *sketch* on the same set of axes, *without performing any calculations*, plots of f_B, f_{BH}, and f_{BH_2} over the pH range 1 to 10.

(b) At what pH do you think f_{BH} is a maximum? Explain.

(c) What form of nicotine is present in greatest amount if a small amount of nicotine is dissolved in blood?

3.5   FREE ENERGY AND CHEMICAL EQUILIBRIUM

As you learned in the previous section, the equilibrium constant for a reaction tells us which species in a chemical equilibrium are present in highest concentrations. In this section, we're going to examine the connection between the equilibrium constant for a process and the *relative stabilities* of the reactants and products.

Let's start with a specific example—the dissociation equilibrium of hydrofluoric acid, a relatively weak acid:

$$H\!-\!F + H_2O \;\rightleftharpoons\; F^- + H_3O^+ \tag{3.32}$$

From Table 3.1, the pK_a of HF is 3.2. Hence, the dissociation constant K_a of HF is $10^{-3.2}$, or 6.3×10^{-4}. The small magnitude of this equilibrium constant means that HF is dissociated to only a small extent in aqueous solution. For example, in an aqueous solution containing 0.1 M HF, a detailed calculation using the equilibrium-constant expression shows that only about 7% of the acid is dissociated to fluoride ions and hydrated protons.

The dissociation constant is related to the *standard free-energy difference* between products and reactants in the following way. If K_a is the dissociation constant as defined in Eq. 3.19, then the **standard free energy of dissociation** is defined by:

$$\Delta G_a^\circ = -RT \ln K_a = -2.3RT \log K_a \tag{3.33}$$

where ln indicates natural (base-e) logarithms, log indicates common (base-10) logarithms, R is the molar gas constant (8.314×10^{-3} kJ K^{-1} mol^{-1} or 1.987×10^{-3} kcal K^{-1} mol^{-1}), and T is the absolute temperature in kelvins (K). Because $-\log K_a$ is by definition the pK_a (Eq. 3.20a), then Eq. 3.33 can be rewritten as

$$\Delta G_a^\circ = 2.3RT\,(pK_a) \tag{3.34}$$

In terms of the HF ionization, the standard free energy of dissociation ΔG_a° in Eq. 3.34 is equal to the *difference* between the standard free energies of the ionization products (H_3O^+ and F^-) and the un-ionized acid (HF). The standard free energy of the solvent (and reference base) water, because it is the same for all acids, is arbitrarily set to zero (that is, ignored).

Introducing the pK_a of HF ($= 3.2$) into Eq. 3.34, we find, at 25 °C (298 K), that

$$\Delta G_a^\circ = 18.2 \text{ kJ mol}^{-1} \text{ (4.36 kcal mol}^{-1}\text{)}$$

The meaning of this standard free-energy change is that the products of the dissociation equilibrium, H_3O^+ and F^-, have 18.2 kJ mol^{-1} (4.36 kcal mol^{-1}) more free energy than the undissociated acid HF; that is, the products are *less stable* than the reactants by 18.2 kJ mol^{-1} (4.36 kcal mol^{-1}) under standard conditions, usually taken to be 1 atm pressure (for gases) or 1 mole per liter for liquid solutions. Physically, this means that if we could somehow couple a free-energy source, such as a battery, to the HF ionization reaction, this battery would have to provide 18.2 kJ (4.36 kcal) of energy to convert one mole per liter of HF completely into one mole per liter of hydrated protons and one mole per liter of fluoride ions. Or, we can turn the idea around: if we could somehow generate a solution containing one mole per liter of hydrated protons and one mole per liter of fluoride ions, this solution would release 18.2 kJ mol^{-1} (4.36 kcal mol^{-1}) of free energy if the two reacted completely to give water and one mole per liter of HF.

Let's now generalize this result for a reaction in which the starting material is S and the product is P. The equilibrium constant K_{eq} for the interconversion of S and P is related to the standard free-energy difference ($G_P^\circ - G_S^\circ$) between P and S as follows:

$$\Delta G^\circ = G_P^\circ - G_S^\circ = -2.3RT \log K_{eq} \tag{3.35}$$

Rearranging,

$$\log K_{eq} = \frac{-\Delta G^\circ}{2.3RT} \tag{3.36a}$$

or

$$K_{eq} = 10^{-\Delta G^\circ / 2.3RT} \tag{3.36b}$$

Notice the *exponential* dependence of K_{eq} on ΔG°. This means that small changes in ΔG° result in large changes in K_{eq}. Table 3.2 shows this relationship numerically. This table

TABLE 3.2 **The Relationship Between Standard Free-Energy Changes, Equilibrium Constants, and Relative Equilibrium Concentrations at 25 °C (298 K)**

$\Delta G° = -2.3RT \log K_{eq}$ or $K_{eq} = 10^{-\Delta G°/2.3RT}$			
$\Delta G°$ (kJ mol^{-1})	$\Delta G°$ (kcal mol^{-1})	K_{eq}	[Products]:[reactants]
+34.2	+8.4	0.000001	1:1000000
+28.5	+7.0	0.00001	1:100000
+22.8	+5.6	0.0001	1:10000
+17.1	+4.2	0.001	1:1000
+11.4	+2.8	0.01	1:100
+5.71	+1.4	0.1	1:10
0	0.0	1	1:1
−5.71	−1.4	10	10:1
−11.4	−2.8	100	100:1
−17.1	−4.2	1000	1000:1
−22.8	−5.6	10000	10000:1
−28.5	−7.0	100000	100000:1
−34.2	−8.4	1000000	1000000:1

shows that a change of 5.7 kJ mol^{-1} or 1.4 kcal mol^{-1} changes the equilibrium constant K_{eq} by one order of magnitude. This free-energy change has the same effect on the product:reactant ratios at equilibrium.

In making conversions between equilibrium constant and standard free energy, the quantity 2.3RT will appear frequently. At 25 °C (298 K), 2.303RT equals 5.71 kJ mol^{-1} or 1.36 kcal mol^{-1}. These are handy numbers to remember.

Suppose that $\Delta G°$ is negative. This means that S has a greater standard free energy than P, or the product P is *more stable* than the starting material S. When S and P come to equilibrium, P will be present in greater amount. This follows from Eq. 3.36b: when $\Delta G°$ is negative, the exponent is positive, and $K_{eq} > 1$. Suppose, on the other hand, that $\Delta G°$ is positive. This means that S has a smaller standard free energy than P—that is, the product P is *less stable* than the starting material S. When S and P come to equilibrium, S will be present in greater amount. Again, this follows from Eq. 3.36b: when $\Delta G°$ is positive, the exponent is negative, and $K_{eq} < 1$. This is the situation in the H—F ionization discussed earlier. The ionization products of HF (H_3O^+ and F^-) are less stable than HF. Hence, the equilibrium constant for their formation, K_a, is very small ($10^{-3.2}$).

Let's summarize the important points of this section.

1. Chemical equilibrium favors the species of lower standard free energy.

2. The more two compounds differ in standard free energy, the greater the difference in their concentrations at equilibrium. Each 5.71 kJ mol^{-1} (1.36 kcal mol^{-1}) increment in $\Delta G°$ affects the equilibrium concentration ratio by a factor of 10.

It follows from these two points that if we can analyze relative stabilities of molecules, we can then predict equilibrium constants by applying Eq. 3.36b. Notice carefully the implication of this statement: *Knowledge of molecular stabilities can lead to an understanding of chemical phenomena*—in this case, chemical equilibrium. Molecular stabilities will form the basis for our understanding of other chemical properties as well—in particular, chemical reactivity. For this reason, we'll devote a lot of attention throughout this text to the relative stabilities of molecules.

3.24 (a) A reaction has a standard free-energy change of –14.6 kJ mol^{-1} (–3.5 kcal mol^{-1}). Calculate the equilibrium constant for the reaction at 25 °C.

  (b) Calculate the standard free-energy difference between starting materials and products for a reaction that has an equilibrium constant of 305.

3.25 (a) A reaction A + B $\rightleftharpoons$ C has a standard free-energy change of –2.93 kJ mol^{-1} (–0.7 kcal mol^{-1}) at 25 °C. What are the concentrations of A, B, and C *at equilibrium* if, at the beginning of the reaction, their concentrations are 0.1 *M*, 0.2 *M*, and 0 *M*, respectively?

  (b) Without making a calculation, tell in a qualitative sense how you would expect your answer for part (a) to change if the reaction has instead a standard free-energy change of +2.93 kJ mol^{-1} (+0.7 kcal mol^{-1}).

3.26 Complete each of the following statements with a number. Assume that the temperature is 25 °C (298 K).

  (a) Two reactions have equilibrium constants that differ by a factor of 10. Their standard free energies differ by _____ kJ mol^{-1}.

  (b) For every 1 kJ mol^{-1} in standard free energy that two reactions differ, their equilibrium constants differ by a factor of _____.

3.6   THE RELATIONSHIP OF STRUCTURE TO ACIDITY

The goal of this section is to help you learn to use the *structures* of compounds to predict trends in their chemical properties. The chemical property we are going to deal with here is Brønsted acidity, but what you learn can be brought to bear on other chemical properties. This section will answer the following question: How can we predict the relative strengths of Brønsted acids within a series? Your ability to deal with questions like this will require that you use all that you have learned in the previous sections.

A. The Element Effect

One of the most important things that determines the acidity of a Brønsted acid is *the identity of the atom to which the acidic hydrogen is attached*. For example, consider the acidities of the following two compounds:

$$CH_3CH_2-O-H \qquad CH_3CH_2-S-H$$

ethanol   **ethanethiol**
(an alcohol)   (a thiol, or mercaptan)
pK_a = 15.9   pK_a = 10.5

These two compounds are structurally similar; the sole difference between them is the element (color) to which the acidic proton is attached. The elements come from the same group in the periodic table, yet the acidity of the thiol is almost a *million times* that of the alcohol (which is about as acidic as water). Another important example of the same trend is the relative acidities of the hydrogen halides. HI is the strongest of these acids; HF is the weakest. (The relevant pK_a data are found in Table 3.1, p. 102.) These data illustrate an important trend: *Brønsted acidity increases as the atom to which the acidic hydrogen is attached has a greater atomic number within a column (group) of the periodic table.*

Now let's see how acidities vary across the periodic table within the same row, or period:

H—CH$_3$ H—NH$_2$ H—OH   H—F

pK_a: ≈55   ≈35   15.7   3.2

(3.37)

(The pK_a values of methane and ammonia are so high that they are not known with certainty.) These data demonstrate another important trend: *Brønsted acidity increases as the atom to which the acidic hydrogen is attached is farther to the right within a row (period) of the periodic table.*

The effect of changing the *directly attached* atom A on the acidity of H—A is called the **element effect** on acidity. Understanding the element effect starts by dividing the ionization process of a typical acid H—A into three imaginary steps, as shown in Eqs. 3.38a–c. This operation is allowed by the first law of thermodynamics, which states simply that the energy difference between two compounds doesn't depend on the pathway used to interconvert them, just as the height of a building doesn't depend on how one gets to the top to measure it.

Bond breaking $\quad\quad\quad\quad\quad$ H$\overset{\varsigma}{\underset{\varsigma}{-}}$A $\longrightarrow$ H$\cdot$ + A$\cdot$ $\quad\quad\quad\quad\quad\quad\quad$ (3.38a)

Loss of an electron from H· $\quad\quad\quad$ H$\cdot$ $\longrightarrow$ H$^+$ + e^- $\quad\quad\quad\quad\quad\quad$ (3.38b)

Electron transfer to A· $\quad\quad$ e^- + A$\cdot$ $\longrightarrow$ A$\overset{..}{\underset{..}{\cdot}}^{-}$ $\quad\quad\quad\quad\quad\quad\quad\quad$ (3.38c)

Sum: $\quad\quad\quad\quad\quad\quad\quad\quad$ H—A $\longrightarrow$ H$^+$ + A$\overset{..}{\underset{..}{\cdot}}^{-}$ $\quad\quad\quad\quad\quad\quad$ (3.38d)

The first step (Eq. 3.38a) is the breaking of the H—A bond "in half," with one bonding electron going to each atom. The energy required for this step is called the **bond dissociation energy**. *The bond dissociation energy is the direct measure of bond strength.* When we compare different bonds, the greater the bond dissociation energy, the stronger the bond. Because acid dissociation involves breaking the bond to hydrogen, *smaller* bond dissociation energies promote increased acidity.

The second step of acid dissociation (Eq. 3.38b) is loss of an electron from the hydrogen atom. The energy required for this step is the **ionization potential** of hydrogen. Because this is the same for all Brønsted acids, it does not enter into a comparison of different acids.

The third step of acid dissociation (Eq. 3.38c) is the transfer of an electron to A· to form the anion. The energy required for this step is the **electron affinity** of ·A. Larger electron affinities promote greater acidity. Electron affinities roughly correspond to electronegativities (Table 1.1, p. 9); this correspondence is reasonable because both are measures of electron attraction.

The acidities of a number of common acids are shown in Fig. 3.2, with the positions of the acids H—A corresponding to the positions of the elements A in the periodic table. The trends in acidity are shown with the purple arrows. These trends are a sum, or blend, of the

Decreasing bond strength

Increasing electron affinity

Increasing acidity

	H—CH$_3$	H—NH$_2$	H—OH	H—F
pK_a	~55	~35	15.7	3.2
			H—SH	H—Cl
			7.0	~ –3
				H—Br
				~ –5
				H—I
				~ –9

Increasing acidity

Decreasing bond strength

Increasing electron affinity

FIGURE 3.2 Factors affecting Brønsted acidity. The acidities of some Brønsted acids H—A are organized by the position of the elements A in the periodic table. The purple arrows show the trends in acidity along rows and columns of the periodic table. The solid arrows indicate the more important factor in each trend, and the dashed arrows show the less important factor.

trends in bond strengths and electron affinities, as shown by Eqs. 3.38a and 3.38c. Within a group, or column, of the periodic table, *bond strengths change significantly*, but electron affinities change relatively little, as indicated by the dashed arrow. Hence, the *major factor* governing the acidity increase from top to bottom within a column of the periodic table is the weaker bonds. Students are sometimes surprised that H—I is a much stronger acid than H—F. Even though fluorine is a much more potent "electron attractor" than iodine, the dominant effect governing acidity is bond strength: the H—I bond is *much* weaker than the H—F bond.

Across a row of the periodic table, bond strengths change relatively little (*dashed arrow*), but *electron affinities change significantly*. The increase in acidity across a row of the periodic table, then, is mainly controlled by the electron affinity of the elements to which the acidic hydrogen is bonded.

To summarize what we've learned about the element effect:

1. The acidities of Brønsted acids H—A increase toward higher atomic number of atom A within a group of the periodic table. The main source of this increase is the decreasing strength of the H—A bond.

2. The acidities of Brønsted acids H—A increase toward higher atomic number of the atom A within a row of the periodic table. The main source of this increase is the increasing electron-attracting ability of the atom A.

B. The Charge Effect

Another important influence on acidity is the effect of charge on the atom bonded to the acidic hydrogen. For example, the pK_a of H_3O^+ is –1.7 and the pK_a of H_2O is 15.7. The major factor responsible for this difference is that a positively charged oxygen attracts electrons much more than a neutral oxygen. Because the bond to the acidic hydrogen in both cases is an O—H bond, bond strength is not the most important factor. Such charge effects are quite general.

PROBLEMS

3.27 In each pair, which compound has the greater Brønsted acidity (smaller pK_a)?

(a) $:NH_3$ or $\overset{+}{N}H_4$ (b) $CH_3\ddot{S}H$ or $CH_3\overset{+}{S}H_2$

3.28 (a) Rank the following four acids in order of increasing Brønsted acidity.

$$CH_3\overset{+}{S}H_2 \qquad H_2\overset{+}{\ddot{F}}: \qquad CH_3\ddot{O}H \qquad CH_3\overset{+}{O}CH_3$$

$$\quad A \qquad\qquad B \qquad\qquad C \qquad\qquad\;\; | \atop H$$

$$\qquad\qquad\qquad\qquad\qquad\qquad\qquad\qquad\qquad D$$

(b) Rank the following compounds in order of increasing basicity. (*Hint:* Think about the acidity of the conjugate acids and the relationship between the acidities of acids and the basicities of their conjugate bases.)

$$CH_3\ddot{O}\overset{-}{:} \qquad CH_3\ddot{O}H \qquad \overset{-}{:}\ddot{N}H_2$$

$$\quad A \qquad\qquad B \qquad\qquad C$$

C. The Polar Effect

The previous two sections discussed effects on acidity that are associated with changes in the atom to which the acidic hydrogen is directly attached. In this section, we consider the effect on acidity that results from substitution at a more remote location in an acidic molecule. Our case study will involve *carboxylic acids*. Although carboxylic acids are classified as weak acids, they are more acidic than most other types of organic compounds. The typical carboxylic acid in aqueous solution undergoes a small degree of ionization to give its conjugate base, a *carboxylate ion*.

general structure of a carboxylic acid   general structure of a carboxylate ion   (3.39)

As shown in Eq. 3.39, carboxylate ions are resonance-stabilized. For convenience, we'll sometimes use the following hybrid structure for carboxylate ions, which shows the sharing of double-bond character and negative charge with dashed lines and partial charges:

hybrid structure of a carboxylate ion

Consider the trend in acidity indicated by the following data for acetic acid and some of its substituted derivatives:

acetic acid   **fluoroacetic acid**   **difluoroacetic acid**   **trifluoroacetic acid**   (3.40)
$pK_a = 4.76$ $pK_a = 2.66$ $pK_a = 1.24$ $pK_a = 0.23$

Within this series, the only structural difference from compound to compound is that hydrogens have been substituted by fluorines several atoms away from the acidic hydrogen. The more fluorines there are, the stronger the acid. A similar effect is observed when other electronegative atoms or groups are substituted into a carboxylic acid molecule. The following data illustrate the same type of effect:

butanoic acid   **4-chlorobutanoic acid**
$pK_a = 4.82$ $pK_a = 4.52$

3-chlorobutanoic acid   **2-chlorobutanoic acid**   (3.41)
$pK_a = 4.06$ $pK_a = 2.84$

These data show that the closer the electronegative group is to the acidic hydrogen, the greater its effect on acidity.

To understand these effects, we start with the standard free energy of the ionization process. Recall (Eq. 3.34, p. 110) that the standard free energy of ionization ΔG_a° is related to the dissociation constant K_a of an acid by the equation

$$\Delta G_a^\circ = 2.3RT\,(pK_a) \tag{3.42a}$$

FIGURE 3.3 (a) The pK_a of an acid is proportional to the standard free-energy difference between an acid and its conjugate base. (b) Lowering the standard free energy of a conjugate base reduces the pK_a of the acid and increases its acidity. The two un-ionized carboxylic acids have been arbitrarily placed at the same standard free energy to focus attention on the relative free energies of the conjugate bases.

Rearranging this equation, we have

$$pK_a = \frac{\Delta G_a^\circ}{2.3RT} \tag{3.42b}$$

These equations show that $\Delta G°$ and pK_a are *directly proportional*. Remember that the standard free energy of ionization for a carboxylic acid is equal to the *difference* between the products of ionization (the conjugate base and H_3O^+) and the carboxylic acid itself. This idea is diagrammed in Fig. 3.3a. Look at this part of the figure and think about what would happen to the pK_a of a carboxylic acid if we did something to increase the relative stability (that is, lower the relative standard free energy) of its conjugate base. This is shown in Fig. 3.3b. *Lowering the standard free energy of a conjugate base decreases $\Delta G°$ and, by Eq. 3.42b, also lowers the pKa of the acid.* In other words, *lowering the standard free energy of a conjugate base makes the conjugate acid more acidic.*

Electronegative substituent groups such as halogens increase the acidities of carboxylic acids by stabilizing their conjugate-base carboxylate ions. This stabilization originates in the polarity of the carbon–halogen bond—that is, its bond dipole. To visualize this idea, consider the *electrostatic interaction* (interaction between charges) of the negatively charged carboxylate oxygens in fluoroacetic acid with the nearby carbon–halogen bond dipole:

repulsive interaction ΔE_{like}
between like charges
destabilizes the ion
(greater distance, less important)

C—F bond dipole

attractive interaction ΔE_{unlike}
between opposite charges
stabilizes the ion
(smaller distance, more important)

$$(3.43)$$

This interaction is governed by the following equation, called the **electrostatic law**:

$$\Delta E = k \frac{q_1 q_2}{\epsilon r} \tag{3.44}$$

In this equation, q_1 and q_2 are charges, k and ϵ are constants, and r is the distance between the charges. The use of ΔE conveys the fact that this is an increment of energy added to, or subtracted from, the total energy of the molecule by the interaction between charges. The magnitudes of q_1 and q_2 in this equation include the signs of the charges. When the charges have opposite signs, $\Delta E < 0$; that is, the interaction between charges *lowers* the energy of the molecule and therefore *stabilizes* the molecule. When the charges have the same sign, then $\Delta E > 0$; that is, the interaction between the charges *raises* the energy of the molecule and therefore *destabilizes* the molecule.

Let's now inventory the various interactions between the charges within this conjugate base. The interaction of the negatively charged oxygens with the partially positive carbon in the C—F bond is *stabilizing*. Let's call this interaction ΔE_{unlike}. But the interaction of these oxygens with the partially negative fluorine is *destabilizing*. Let's call this destabilizing interaction ΔE_{like}. However—and this is the important point—the oxygens are closer to the carbon than they are to the fluorine. Therefore, r in Eq. 3.44 is smaller for the stabilizing contribution than it is for the destabilizing one. Because r is in the denominator of Eq. 3.44, then

$$\Delta E_{\text{unlike}} + \Delta E_{\text{like}} < 0 \tag{3.45}$$

That is, the overall effect of the interactions is stabilizing. Therefore, the *net* interaction of the carboxylate oxygen and the nearby C—F bond dipole is an attractive, stabilizing one. As you can see from Fig. 3.3b, this stabilization lowers the pK_a of an acid, or strengthens the acid. Because the carboxylic acid itself is uncharged, the effect of a fluorine substituent on its stability is much less important and can be ignored.

An effect on chemical properties caused by the interactions between charges, dipoles, or both is called a **polar effect**. (It is also known as an **inductive effect**.) Thus, in the present examples, halogens (or other electronegative substituents) have an *acid-strengthening polar effect* on the acidity of carboxylic acids. As the series in Eq. 3.40 shows, the more halogens there are, the greater the effect on acidity. In fact, trifluoroacetic acid is a strong acid.

Another way to describe the polar effect of halogens and other electronegative groups is to say that they exert an **electron-withdrawing polar effect** because they pull electrons toward themselves and away from the carbon to which they are attached. As we might imagine, other groups exert an opposite polar effect, called an **electron-donating polar effect** (see Problem 3.51, p. 123), and such groups raise the pK_a, or reduce the acidity, of nearby carboxylic acid groups.

The inverse relationship between the interaction energy E and distance r in Eq. 3.44 implies that the magnitude of the interaction between charges diminishes as the distance between the interacting groups increases. Hence, polar effects should be smaller for compounds in which the two interacting groups are separated by greater distances (more bonds). Indeed, within the series 3.41, you can see that the influence of a chlorine on the pK_a decreases significantly as the chlorine is more remote from the carboxylate oxygen.

Although this section has dealt specifically with polar effects on acidity, the central idea is that *stabilization of a conjugate acid or base* will alter its energy and will affect the pK_a. Stabilization of the conjugate base, as in the case discussed here, will lower its energy and therefore lower the pK_a of the conjugate acid. However, it is possible to envision a situation in which the conjugate acid is stabilized. Stabilization of a conjugate acid would *increase* ΔG_a° and lower the acidity. (Problem 3.55 on p. 124 is an example of such a case.) This type of analysis is not limited to polar effects, nor is it limited to acid–base equilibria. *Any* structural change that alters the free-energy difference between a reactant and a product will affect the equilibrium constant for the reaction. Pay careful attention to the process involved, because we will use it repeatedly:

1. Analyze the effect of a structural change on the energies of the components of an equilibrium.

FURTHER EXPLORATION 3.1 Inductive Effects

2. Then show the change on an energy diagram, as in Fig. 3.3.

3. From the diagram, the effect on the equilibrium constant follows from the relationship $\log K_{eq} = -\Delta G°/2.3RT$.

In this section, we've focused on the chemical property of Brønsted acidity. Here's what we've learned about the effects of structure on acidity:

1. The *element effect* is the effect on acidity of changing the atom to which the acidic hydrogen is bonded. The element effect has its origins in the bond energies to acidic hydrogens and the electron affinities of the elements attached to the acidic hydrogens. Trends in acidity based on the element effect can be predicted from the relationship of the elements in the periodic table.

2. The *charge effect* is the increase in acidity that results from increasing the positive charge on the atom bonded to the acidic hydrogen. The element effect and charge effect are large effects.

3. The *polar effect*, which is typically smaller in magnitude than the element and charge effects, is the effect on acidity that polar groups in an acid have through their interactions with the charged species in the acid–base equilibrium.

Two other factors that significantly affect acidity are the *hybridization* of the atom bearing the acidic proton, and *resonance stabilization* of either the conjugate acid or the conjugate base. Understanding these effects requires some additional background material, and we'll return to these effects in Chapters 14 and 18.

STUDY PROBLEM 3.8

Rank the following compounds in order of increasing basicity.

$$H_3C-\overset{\overset{\displaystyle O}{\|}}{C}-O^- \qquad H_3C-\overset{\overset{\displaystyle O}{\|}}{C}-\overset{-}{N}H \qquad H_3\overset{+}{N}-CH_2-\overset{\overset{\displaystyle O}{\|}}{C}-O^-$$

acetate ion   **acetamide anion**   **glycine** (an amino acid)

A   *B*   *C*

SOLUTION   First recognize that a problem in relative basicity is equivalent to a problem in relative acidity. If you can rank the acidities of the conjugate acids, you've solved the problem. The relevant conjugate acids are

$$H_3C-\overset{\overset{\displaystyle O}{\|}}{C}-OH \qquad H_3C-\overset{\overset{\displaystyle O}{\|}}{C}-NH_2 \qquad H_3\overset{+}{N}-CH_2-\overset{\overset{\displaystyle O}{\|}}{C}-OH$$

acetic acid   **acetamide**   **glycine conjugate acid**

AH   *BH*   *CH*

Both *AH* and *CH* are carboxylic acids; in both cases, the acidic hydrogen is bound to an oxygen. The acidic hydrogen in compound *BH* is bound to a nitrogen. The difference in acidities of *BH* and the other two compounds is therefore due primarily to the *element effect* along the first row of the periodic table, and this is the most important effect. This effect predicts that the O—H group should be more acidic than a comparably substituted N—H group, because oxygen is more electron-attracting than nitrogen. Thus, the acidities of both *AH* and *CH* are greater than the acidity of *BH*. The difference in the acidities of *AH* and *CH* is due to the *polar effect* of the $H_3\overset{+}{N}-$ group in compound *CH* on the acidity of the nearby carboxylic acid group. The full positive charge on the nitrogen has a favorable interaction with the negatively charged carboxylate oxygen. As shown in Fig. 3.3, this interaction stabilizes the conjugate base and thus enhances the acidity of *CH*. Hence, the final order of acidity is *CH* > *AH* > *BH*. Because stronger acids have weaker conjugate bases, the basicity order of the conjugate bases is *C* < *A* < *B*. Our prediction is correct: The actual pK_a values are *CH*, 2.17; *AH*, 4.76; and *BH*, ~ 16.

PROBLEMS

3.29 In each of the following sets, arrange the compounds in order of decreasing pK_a, and explain your reasoning.

(a) $ClCH_2CH_2SH$ $ClCH_2CH_2OH$ CH_3CH_2OH

(b)

$$CH_3O—CH_2—\overset{\overset{\displaystyle O}{\|}}{C}—OH \qquad H_3C—\overset{\overset{\displaystyle O}{\|}}{C}—OH \qquad \underset{\underset{\displaystyle OCH_3\ OCH_3}{|\quad\ |}}{CH_2—CH}—\overset{\overset{\displaystyle O}{\|}}{C}—OH$$

(c)

$$H_3C—\overset{\overset{\displaystyle +}{\overset{\displaystyle :OH}{\|}}}{C}—OH \qquad Cl—CH_2—\overset{\overset{\displaystyle +}{\overset{\displaystyle :OH}{\|}}}{C}—OH \qquad H_3C—\overset{\overset{\displaystyle :O:}{\|}}{C}—\overset{..}{\underset{..}{O}}H$$

A   *B*   *C*

3.30 Calculate the standard free energy for dissociation of
(a) fluoroacetic acid ($pK_a = 2.66$)
(b) acetic acid ($pK_a = 4.76$)

3.31 Rationalize your answer to the previous problem by explaining why more energy is required to ionize acetic acid than fluoro-acetic acid. (See series 3.40, p. 115, for the structures.)

KEY IDEAS IN CHAPTER 3

- There are two fundamental types of reactions that involve electron pairs:

  1. *Lewis acid–base association reactions*, in which a nucleophile donates a pair of electrons to an electron-deficient electrophile. The reverse of a Lewis acid–base association is a Lewis acid–base dissociation, in which a leaving group departs from the electrophile with a pair of electrons.

  2. *Electron-pair displacement reactions*, in which a nucleophile donates a pair of electrons to an electrophile that is not electron-deficient. In this case, a bond between the electrophilic center and a leaving group becomes an unshared pair on the leaving group, and the electrophilic atom is transferred from an atom of the leaving group to the nucleophilic center. The reverse of an electron-pair displacement reaction is another electron-pair displacement in which the roles of the nucleophile and leaving group in the forward reaction are reversed.

  All electron-pair reactions can be classified as one of these two reaction types or as combinations of them.

- The curved-arrow notation is an important symbolism for depicting the flow of electron pairs in chemical reactions. In this notation, the tail of a curved arrow indicates a source of electrons, and the head indicates the destination.

- A Lewis acid–base association reaction or its reverse requires a single curved arrow; an electron-pair displacement reaction requires two curved arrows.

- The curved-arrow notation can be used to derive resonance structures that are related by the movement of one or more electron pairs.

- A Brønsted acid–base reaction is a special case of an electron-pair displacement reaction in which the electrophilic center is a proton; a proton is transferred in the reaction, just as an electrophilic center is transferred in a general electron-pair displacement reaction. The electron donor is called a Brønsted base, and the species that donates the electrophilic proton is called the Brønsted acid.

- When a Brønsted acid loses a proton, its conjugate base is formed, and when a Brønsted base gains a proton, its conjugate acid is formed.

- The strength of a Brønsted acid is indicated by the magnitude of its dissociation constant K_a. Because dissociation constants for various acids can differ by many orders of magnitude, a logarithmic pK_a scale is used, in which $pK_a = -\log K_a$. The strength of a Brønsted acid is indicated by the magnitude of its dissociation constant K_a. Because dissociation constants for various acids can differ by many orders of magnitude, a logarithmic pK_a scale is used, in which $pK_a = -\log K_a$. The strength of a Brønsted base is inferred from the K_a (or pK_a) of its conjugate acid.

• The equilibrium constant K_{eq} for a reaction is related to the standard free-energy difference $\Delta G°$ between products and starting materials by the relationship $\Delta G° = -2.3RT \log K_{eq}$. Reactions with positive $\Delta G°$ values have $K_{eq} < 1$ and favor starting materials at equilibrium. Reactions with negative $\Delta G°$ values have $K_{eq} > 1$ and favor products at equilibrium.

• The logarithm of the equilibrium constant for any Brønsted acid–base reaction can be calculated by subtracting the pK_a of the acid on the left from the pK_a of the acid on the right. A Brønsted acid–base equilibrium always favors the weaker acid and weaker base.

• The dissociation state of an acid with dissociation constant K_a is determined by the difference $pH - pK_a$. When $pH << pK_a$, an acid is undissociated. When $pH >> pK_a$, an acid is fully dissociated. When $pH = pK_a$, half of the acid molecules are dissociated; the concentrations of an acid and its conjugate base are equal. The dissociation state is quantitatively described by the fraction dissociated, f_A, or by the fraction undissociated, f_{AH}, given in Eqs. 3.30c–d and 3.31a–b (p. 107).

• Acidity, basicity, and other chemical properties vary with structure. Three structural effects on Brønsted acidity are the *element effect*, the *charge effect*, and the *polar effect*.

• The pK_a of an acid is proportional to its standard free energy of ionization ($\Delta G_a°$). The process for analyzing the effect of structure on acidity is to assess the effect of structure on the energy of the charged species in the equilibrium and then to consider how the resulting energy affects the pK_a.

ADDITIONAL PROBLEMS

3.32 Which of the following are electron-deficient compounds? Explain.

(a)

(b)

(c)

(d)

(e)

3.33 Give the curved-arrow notation for, and predict the immediate product of, each of the following reactions. Each involves an electron-deficient Lewis acid and a Lewis base.

(a) $H_3C-\ddot{\underset{..}{O}}-CH_3 + BF_3 \longrightarrow$

(b)

(c) $H\ddot{\underset{..}{O}}-CH_2-CH_2-CH_2-\overset{+}{C}H-CH_3 \longrightarrow$

(*Hint:* This reaction forms a ring.)

(d) $(CH_3)_3B + :\bar{C}\equiv\overset{+}{O}: \longrightarrow$

(e)

3.34 For each of the Brønsted acid–base reactions shown in Fig. P3.34, label the conjugate acid–base pairs. Then give the curved-arrow notation for each reaction in the left-to-right direction.

3.35 The conversion of alcohols into alkenes, a process called *dehydration*, takes place as a succession of three simple acid–base reactions, shown in Fig. P3.35.

(a) Classify each reaction step in the forward direction with one of the following terms:

  (1) a Lewis acid–base association reaction

  (2) a Lewis acid–base dissociation reaction

  (3) an electron-pair displacement reaction

  (4) a Brønsted acid–base reaction

(b) If the step is a Brønsted acid–base reaction, indicate the conjugate acid–base pairs.

(c) Classify each species (or atoms within each species) with one of the following terms: *nucleophile, nucleophilic center, electrophile, electrophilic center, leaving group, Brønsted acid, Brønsted base*.

(d) Draw the curved-arrow notation for each step in the left-to-right direction.

3.36 Work Problem 3.35 for the *reverse* reactions in Fig. P3.35.

3.37 (a) Although we normally think of acetic acid as an acid, it is amphoteric and can also act as a weak base. The conjugate acid of acetic acid is shown below. Using the curved-arrow notation, derive a resonance structure for this ion which, taken with the structure below, shows that the two —OH groups are equivalent, the two

(a)

(b)

(c)

(d)

Figure P3.34

Step 1.

Step 2.

Step 3.

E (redrawn)

Figure P3.35

C—O bonds are equivalent, and the positive charge is shared equally by the two oxygens. Draw a single hybrid structure for this ion using dashed lines and partial charges that conveys the same idea.

conjugate acid of acetic acid

(b) The resonance structures of carbon monoxide are shown below. Show how each structure can be converted into the other using the curved-arrow notation.

$$\left[:C{=}\ddot{O}: \quad \longleftrightarrow \quad :\bar{C}{\equiv}\overset{+}{O}: \right]$$

3.38 A scientist at an Air Force Research Laboratory in California has studied "highly energetic materials" (explosive materials) and, more significantly, has lived to tell about it. In 2000, he and his equally adventurous collaborators determined the X-ray crystal structure of the N_5^+ cation; most salts of this cation are highly explosive. This species is the first ever isolated in modern times that contains more than three contiguously bonded nitrogens. The crystal structure revealed a "V-shape" for the cation, as follows:

Notice that the lines do not indicate the bonding pattern, but only the shape.

(a) Draw an acceptable Lewis structure, *including unshared electron pairs*, that accounts for the shape of the molecule and its *overall* plus charge.

Explain why your molecule meets these criteria. Be sure to show in your structure the formal charge of every atom with nonzero formal charge.

(b) Using the curved-arrow notation, derive two additional resonance structures for this cation that meet the same criteria.

3.39 Use the curved-arrow notation to derive resonance structures that convey the following ideas. In each case, draw a single hybrid structure using dashed lines and partial charges that conveys the same meaning as the resonance sructures.

(a) The outer oxygens of ozone, $:\ddot{O}=\overset{+}{\ddot{O}}-\ddot{\ddot{O}}:$, have an equal amount of negative charge.

(b) All C—O bonds in the carbonate ion are of equal length.

carbonate ion

(c) The conjugate acid of formaldehyde, $H_2C=\overset{+}{\ddot{O}}-H$, has substantial positive charge on carbon.

3.40 Draw the products of each of the following reactions indicated by the curved-arrow notation.

(a)

(b)

(c)

3.41 Use the curved-arrow notation to indicate the flow of electrons in each of the transformations given in Fig. P3.41.

3.42 Predict the products of each of the following reactions, and explain your reasoning. Use the curved-arrow notation to help you, and show the notation.

(a) $CH_3\ddot{S}H + \ddot{:}\ddot{O}CH_3 \longrightarrow$

(*Hint:* Remember that Brønsted acid–base reactions are in most cases *much* faster than nucleophilic reactions.)

(b)

(c)

(d) $\ddot{:}\ddot{O}H + H_3\overset{+}{N}-CH_2CH_3 \longrightarrow$

3.43 In each of the following processes, give the products and classify each of the atoms indicated by a colored label with one or more of the following terms: *Brønsted base, Brønsted acid, nucleophilic center, electrophilic center,* and/or *leaving group.* You may have to be a bit creative in assigning labels to some groups. If you have a problem, try to state what the difficulty is.

(a)

(b)

Figure P3.41

3.44 The examples of *incorrect* curved-arrow notation in Fig. P3.44 were found in the notebooks of Barney Bottlebrusher, a student who was known to have difficulty with organic chemistry. Explain what is wrong in each case.

3.45 Naphthalene can be described by two resonance structures in addition to the following structure. Derive these structures with the curved-arrow notation.

$$\left[\text{(naphthalene structure)} \longleftrightarrow \text{two additional structures} \right]$$

naphthalene

3.46 (a) The standard free-energy difference between 2,2-dimethylpropane and pentane is 6.86 kJ mol^{-1} (1.64 kcal mol^{-1}); 2,2-dimethylpropane is the more stable compound. If the two were present in an equilibrium mixture, what would be the percentage of each in the mixture at 25 °C?

(b) The energy difference between *anti*-butane and either one of the *gauche*-butane conformations is 2.8 kJ mol^{-1} (0.67 kcal mol^{-1}) (Fig. 2.5, p. 53). Treating this difference as a standard free energy, calculate the amounts of *gauche*- and *anti*-butane present in equilibrium in one mole of butane at 25 °C. (Remember that there are two gauche conformations.)

3.47 Arrange the compounds in each of the following sets in order of decreasing pK_a, highest first. Explain your reasoning.

(a) CH_3CH_2OH Cl_2CHCH_2OH $ClCH_2CH_2OH$

(b) $ClCH_2CH_2SH$ CH_3CH_2OH CH_3CH_2SH

(c) $H-\ddot{A}s(CH_3)_2$ $H-\overset{+}{A}s(CH_3)_2$ with H below

$H-\ddot{N}(CH_3)_2$ $H-\ddot{P}(CH_3)_2$

(d) $CH_3CH_2\ddot{O}H$ $(CH_3)_2\ddot{N}-CH_2CH_2\ddot{O}H$ $(CH_3)_3\overset{+}{N}-\ddot{O}H$

3.48 Using Table 3.1, as well as the data given below, estimate the equilibrium constants for each of the following reactions at 25 °C.

(a) $(CH_3)_3N + H-CN \rightleftharpoons (CH_3)_3\overset{+}{N}H + {}^-CN$
pK_a = 9.76

(b) $CH_3CH_2S-H + {}^-OH \rightleftharpoons CH_3CH_2S^- + H_2O$
pK_a = 10.5

3.49 (a) What is the standard free-energy change at 25 °C for reaction (a) in Problem 3.48?

(b) What is the standard free-energy change at 25 °C for reaction (b) in Problem 3.48?

(c) In reaction (a) of Problem 3.48, how much of each species will be present at equilibrium if the initial concentrations of $(CH_3)_3N$ and HCN are both zero, and $(CH_3)_3\overset{+}{N}H$ and ^-CN are present initially at concentrations of 0.1 *M*?

3.50 Phenylacetic acid has a pK_a of 4.31; acetic acid has a pK_a of 4.76.

$Ph-CH_2-\overset{\overset{O}{\|}}{C}-O-H$ $H_3C-\overset{\overset{O}{\|}}{C}-O-H$
phenylacetic acid   **acetic acid**

(a) Which acid has the more favorable (smaller) standard free energy of dissociation?

(b) What free energy would be expended to dissociate a 1 *M* solution of phenylacetic acid completely into 1 *M* of its conjugate base and 1 *M* H_3O^+?

(c) What is the fraction of phenylacetic acid that is ionized if it is dissolved in water and the pH is adjusted to 4.5 with NaOH?

(d) If equal molar amounts of phenylacetic acid and acetic acid are dissolved in water and the pH is adjusted to 4.5 with NaOH, which acid is more dissociated? Explain.

(e) According to the pK_a data, which type of polar effect is characteristic of the phenyl group: an electron-withdrawing polar effect or an electron-donating polar effect? Explain your reasoning.

3.51 Malonic acid has two carboxylic acid groups and consequently undergoes two ionization reactions. The pK_a for the first ionization of malonic acid is 2.86; the pK_a for the second is 5.70. The pK_a of acetic acid is 4.76.

$HO-\overset{\overset{O}{\|}}{C}-CH_2-\overset{\overset{O}{\|}}{C}-OH$ $H_3C-\overset{\overset{O}{\|}}{C}-OH$
malonic acid   **acetic acid**

(a) Write out the equations for the first and second ionizations of malonic acid, and label each with the appropriate pK_a value.

(b) Describe the ionization state of malonic acid if a solution of the acid is adjusted to pH 4.3 with NaOH.

(c) Use your answer to part (b) to determine the number of moles of base per mole of malonic acid required to adjust a solution of malonic acid to pH = 4.3.

(a) $H_3C-\overset{\overset{:O:}{\|}}{C}-\ddot{O}-H \,\,{}^-\!:\ddot{O}H \longrightarrow H_3C-\overset{\overset{:O:}{\|}}{C}-\ddot{O}:^- + H-\ddot{O}H$

(b) $H_3C-\ddot{O}:^- \,\, H_3C-\ddot{B}r: \longrightarrow H_3C-\ddot{O}-CH_3 + :\ddot{B}r:^-$

Figure P3.44

(d) Why is the first pK_a of malonic acid much *lower* than the pK_a of acetic acid, but the second pK_a of malonic acid is much *higher* than the pK_a of acetic acid?

(e) Malonic acid is one member of a homologous series of unbranched *dicarboxylic acids*, so-called because they have two carboxylic acid groups. Compounds in this series have the following general structure.

$$\text{HO} - \overset{\overset{O}{\|}}{C} - (CH_2)_n - \overset{\overset{O}{\|}}{C} - \text{OH}$$

How would you expect the *difference* between the first and second pK_a values to change as n increases? Explain. (*Hint:* Look at the denominator of the electrostatic law, Eq. 3.44, p. 117.)

3.52 Ascorbic acid (Vitamin C) has the following structure.

ascorbic acid
$pK_a = 4.2, 11.6$

Protons *a* and *b* are acidic; proton *a* is the more acidic of the two.

(a) Draw the structure of ascorbic acid as it exists at physiological pH (pH = 7.4). Explain.

(b) Using the curved-arrow notation, derive a resonance structure for the structure you drew in part (a); the two structures together will show that negative charge is shared between two oxygens.

3.53 In Problem 3.23 on p. 109, we considered the basicity of nicotine. Consider the pK_a of the more acidic proton in the doubly protonated (BH$_2$) form of nicotine (reproduced below). Compare this to the pK_a of the conjugate acid of 3-methylpyridine ($pK_a = 5.68$).

nicotine
(BH$_2$ form)

3-methylpyridine
(conjugate acid)

How do you account for the fact that the pK_a of the more acidic proton in the BH$_2$ form of nicotine is much lower than the pK_a of 3-methylpyridine conjugate acid? Draw an appropriate free-energy diagram as part of your answer.

3.54 Which of the following two reactions would have an equilibrium constant more favorable to the right? Explain your answer using a free-energy diagram.

(1)

(2)

3.55 From Fig. 3.3, p. 116, how would a structural effect that destabilizes the *acid* component of a conjugate acid–base pair affect its acidity? Use your analysis to predict which of the following two compounds is more basic.

$$Cl - CH_2CH_2 - \overset{..}{N}H_2 \qquad CH_3CH_2 - \overset{..}{N}H_2$$
$$A \qquad\qquad\qquad B$$

3.56 Derive Eq. 3.26b (p. 104); that is, justify the procedure used in calculating the equilibrium constant for the reaction of an acid and a base. (*Hint:* First show that $K_{eq} = K_{AH}/K_{BH}$.)

3.57 For the general acid–base reaction

$$AH + B\overset{..}{:}^- \rightleftharpoons BH + A\overset{..}{:}^-$$

derive the relationship between $\Delta G°$ for the reaction and the $\Delta G_a°$ values of the individual acids AH and BH.

3.58 (a) The acid HI is considerably stronger than HCl (see Table 3.1, p. 102). Why, then, does a 10^{-3} *M* aqueous solution of either acid in water give the same pH reading of 3?

(b) The amide ion, $^-\overset{..}{N}H_2$, whose conjugate acid ammonia has a $pK_a = 35$, is a much stronger base than hydroxide. Yet a 10^{-3} *M* solution of either base in water has pH = 11. Explain why the solution of the stronger base doesn't have a higher pH.

3.59 The borohydride anion reacts with water in the following way:

$$^-BH_4 + H_2\overset{..}{\underset{..}{O}} \longrightarrow H\overset{..}{\underset{..}{O}} - \bar{B}H_3 + H_2$$

borohydride anion

This overall transformation occurs in a sequence of two reactions. In the first reaction, a hydrogen of $^-BH_4$ with its two bonding electrons acts as a base toward water. Write out both steps and give the curved arrow notation for each.

3.60 Astatine (At) is the radioactive halogen that lies below iodine in Group 7A of the periodic table. How would you expect the following properties to compare (greater or less)?

(a) bond dissociation energy of H—At versus that of H—I

(b) electron affinity of At versus that of I

(c) dissociation constant of H—At versus that of H—I

Introduction to Alkenes
Structure and Reactivity

Alkenes are hydrocarbons that contain one or more carbon–carbon double bonds. Alkenes are sometimes called **olefins**, particularly in the chemical industry. *Ethylene* is the simplest alkene.

$$H_2C=CH_2$$

or $H_2C=CH_2$

ethylene
(substitutive name: **ethene**)

Because compounds containing double or triple bonds have fewer hydrogens than the corresponding alkanes, they are classified as **unsaturated hydrocarbons**, in contrast to alkanes, which are classified as **saturated hydrocarbons**.

This chapter covers the structure, bonding, nomenclature, and physical properties of alkenes. Then, using a few alkene reactions, some of the physical principles are discussed that are important in understanding the reactivities of organic compounds in general.

4.1   STRUCTURE AND BONDING IN ALKENES

The double-bond geometry of ethylene is typical of that found in other alkenes. Ethylene follows the rules for predicting molecular geometry (Sec. 1.3B), which require each carbon of ethylene to have trigonal planar geometry; that is, all the atoms surrounding each carbon lie in the same plane with bond angles approximating 120°. The experimentally determined structure of ethylene agrees with these expectations and shows further that *ethylene is a planar molecule*. For alkenes in general, the carbons of a double bond and the atoms directly attached to them all lie in the same plane.

FIGURE 4.1 Models of ethylene. (a) A ball-and-stick model. (b) A space-filling model. Ethylene is a planar molecule.

(a)   (b)

FIGURE 4.2 Structures of ethylene, ethane, propene, and propane. Compare the trigonal planar geometry of ethylene (bond angles near 120°) with the tetrahedral geometry of ethane (bond angles near 109.5°). All carbon–carbon double bonds are shorter than carbon–carbon single bonds. The carbon–carbon single bond in propene, moreover, is somewhat shorter than the carbon–carbon bonds of propane.

Models of ethylene are shown in Fig. 4.1, and a comparison of the geometries of ethylene and propene with those of ethane and propane is given in Fig. 4.2. Notice that the carbon–carbon double bonds of ethylene and propene (1.33 Å) are shorter than the carbon–carbon single bonds of ethane and propane (1.54 Å). This illustrates the relationship of bond length and bond order (Sec. 1.3B): double bonds are shorter than single bonds between the same atoms.

Another feature of alkene structure is apparent from a comparison of the structures of propene and propane in Fig. 4.2. Notice that the carbon–carbon *single* bond of propene (1.50 Å) is shorter than the carbon–carbon *single* bonds of propane (1.54 Å). Likewise, the bonds to the hydrogens attached to the double bonds in ethylene and propene are shorter than the C—H bonds of propane. The shortening of all these bonds is a consequence of the particular way that carbon atoms are hybridized in alkenes.

A. Carbon Hybridization in Alkenes

The carbons of an alkene double bond are hybridized differently from those of an alkane. In this hybridization (Fig. 4.3), the carbon $2s$ orbital is mixed, or hybridized, with only two of the three available $2p$ orbitals. In Fig. 4.3, we have arbitrarily chosen to hybridize the $2p_x$ and $2p_y$ orbitals. Thus, the $2p_z$ orbital is unaffected by the hybridization. Because three orbitals are mixed, the result is three hybrid orbitals and a "leftover" $2p_z$ orbital. Each hybrid orbital has one part s character and two parts p character. These hybrid orbitals are called sp^2 (pronounced "s-p-two") orbitals, and the carbon is said to be sp^2-*hybridized*. Thus, an sp^2 orbital has 33% s character (in contrast to an sp^3 orbital, which has 25% s character). A perspective drawing of an sp^2 orbital is shown in Fig. 4.4a, and a commonly used stylized representation of an sp^2 orbital is shown in Fig. 4.4b. If you compare Fig. 4.4a with Fig. 1.16a (p. 38), you can see that the shape of an individual sp^2 orbital is much like that of an sp^3 orbital. The difference between these two types of hybrid orbitals is that the electron density within an sp^2

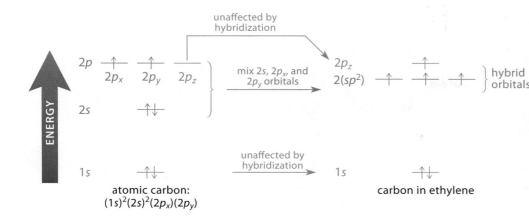

FIGURE 4.3 Orbitals of an sp^2-hybridized carbon are derived conceptually by mixing one 2s orbital and two 2p orbitals, in this case the $2p_x$ and $2p_y$ orbitals, shown in red. Three sp^2 hybrid orbitals are formed (*red*) and one $2p_z$ orbital remains unhybridized (*blue*).

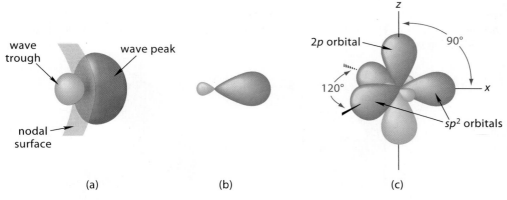

FIGURE 4.4 (a) The general shape of an sp^2 hybrid orbital is very similar to that of an sp^3 hybrid orbital, with a large and small lobe of electron density separated by a node. (Compare with Fig. 1.16a, p. 38.) (b) A common stylized representation of an sp^2 orbital. (c) Spatial distribution of orbitals on an sp^2-hybridized carbon atom. The axes of the three sp^2 orbitals lie in a common plane (the xy plane in this case) at angles of 120°, and the axis of the $2p_z$ orbital is perpendicular to this plane.

orbital is concentrated slightly closer to the nucleus. The reason for this difference is the larger amount of s character in an sp^2 orbital. Electron density in a carbon 2s orbital is concentrated a little closer to the nucleus than electron density in a carbon 2p orbital. The more s character a hybrid orbital has, then, the more "s-like" its electrons are, and the closer its electrons are to the nucleus.

Because the $2p_x$ and $2p_y$ orbitals are used for hybridization, and because the 2s orbital is spherical (that is, without direction), the axes of the three sp^2 orbitals lie in the xy plane (see Fig. 4.4c); they are oriented at the maximum angular separation of 120°. Because the "left-over" (unhybridized) 2p orbital is a $2p_z$ orbital, its axis is the z axis, which is perpendicular to the plane containing the axes of the sp^2 orbitals.

Conceptually, ethylene can be formed in the hybrid orbital model by the bonding of two sp^2-hybridized carbon atoms and four hydrogen atoms (Fig. 4.5 on p. 128). An sp^2 orbital on one carbon containing one electron overlaps with an sp^2 orbital on another to form a two-electron sp^2–sp^2 C—C σ bond. Each of the two remaining sp^2 orbitals, each containing one electron, overlaps with a hydrogen 1s orbital, also containing one electron, to form a two-electron sp^2–1s C—H σ bond. These orbitals account for the four carbon–hydrogen bonds and *one* of the two carbon–carbon bonds of ethylene, which together comprise the *sigma-bond frame-work* of ethylene. (We have not yet accounted for the 2p orbital on each carbon.) Notice carefully that the trigonal planar geometry of each carbon of ethylene is a direct consequence of the way its sp^2 orbitals are directed in space. Once again, we see that *hybridization and*

FIGURE 4.5 A hybrid orbital picture for the σ bonds of ethylene. A $2p_z$ orbital on each carbon (dashed lines) is left over after construction of σ bonds from hybrid orbitals.

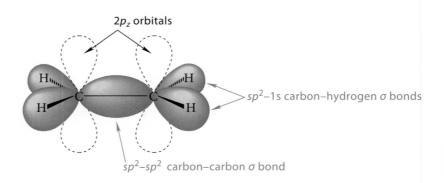

molecular geometry are related. (Sec. 1.9). Whenever a main-group atom has trigonal planar geometry, its hybridization is sp^2. Whenever such an atom has tetrahedral geometry, its hybridization is sp^3.

sp^3 orbitals are associated with tetrahedral geometry

sp^2 orbitals are associated with trigonal planar geometry

B. The π (Pi) Bond

The two $2p_z$ orbitals not used in σ-bond formation (dashed lines in Fig. 4.5) overlap side-to-side to form the second bond of the double bond. In the hybrid orbital picture, each $2p_z$ orbital contributes one electron to make an electron-pair bond. A bond formed by the side-to-side overlap of p orbitals is called a **π (pi) bond**. (The symbol π, or pi, is used because π is the Greek equivalent of the letter p and because the π bond originates from the overlap of p orbitals.)

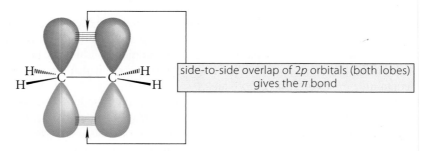

side-to-side overlap of $2p$ orbitals (both lobes) gives the π bond

To visualize electron distribution in a π bond, we'll use molecular orbital (MO) theory (Sec. 1.8). MO theory provides a richer description of the π bond, and it also forms the basis for understanding of ultraviolet spectroscopy (an important tool for molecular analysis; see Sec. 15.2) as well as a class of reactions called *pericyclic reactions* (Chapter 27). Notice that we are treating the σ-bond framework with hybrid orbital theory and the π bond with MO theory. This is justified in MO theory because the π MOs are, to a good approximation, inde-

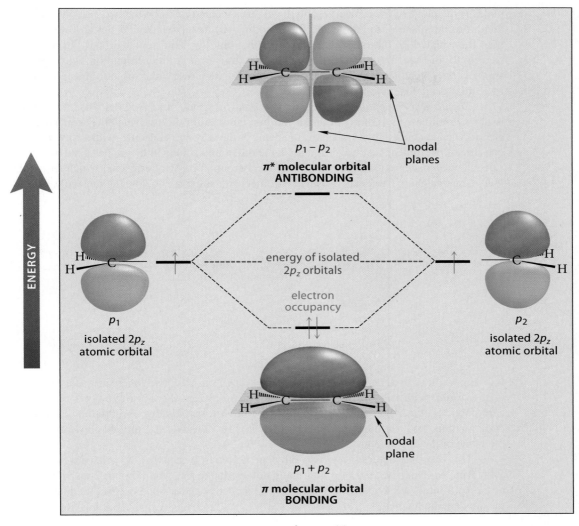

$p_1 - p_2$

**π^* molecular orbital
ANTIBONDING**

nodal
planes

ENERGY

energy of isolated
$2p_z$ orbitals

electron
occupancy

p_1

isolated $2p_z$
atomic orbital

p_2

isolated $2p_z$
atomic orbital

nodal
plane

$p_1 + p_2$

**π molecular orbital
BONDING**

nuclear position

FIGURE 4.6 An orbital interaction diagram showing the overlap of $2p$ orbitals to form bonding and antibonding π molecular orbitals of ethylene. The π bond is formed when two electrons occupy the bonding π molecular orbital. Wave peaks and wave troughs are shown with different colors. The nodal planes are perpendicular to the page.

pendent of the other MOs of an alkene molecule. This is another relatively rare situation (as in dihydrogen, H_2; Sec. 1.8B) in which molecular orbitals are associated with a particular bond that we can draw in a Lewis structure.

The interaction of two $2p_z$ orbitals of ethylene by a side-to-side overlap is shown in an *orbital interaction diagram* (Fig. 4.6). Because two atomic orbitals are used, two molecular orbitals are formed. These are formed by additive and subtractive combinations of the $2p_z$ orbitals. Remember that subtracting orbitals is the same as reversing the peaks and troughs of one orbital and then adding.

The bonding molecular orbital that results from additive overlap of the two carbon $2p$ orbitals is called a **π molecular orbital**. This molecular orbital, like the p orbitals from which it is formed, has a nodal plane (shown in Fig. 4.6); this plane coincides with the plane of the ethylene molecule. The antibonding molecular orbital, which results from subtractive overlap of the two carbon $2p$ orbitals, is called a **π^* molecular orbital**. It has two nodes. One of these nodes is the plane of the molecule, and the other is a plane between the two carbons, perpendicular to the plane of the molecule. The bonding (π) molecular orbital lies at lower energy than the isolated $2p$ orbitals, whereas the antibonding (π^*) molecular orbital lies at higher energy. By the aufbau principle, the two $2p$ electrons (one from each carbon, with opposite spin) occupy the molecular orbital of lower energy—the π molecular orbital. The antibonding molecular orbital is unoccupied.

The filled π molecular orbital is the π bond. Unlike a σ bond (Fig. 1.15, p. 34), a π bond is not cylindrically symmetrical about the line connecting the two nuclei. The π bond has electron density both above and below the plane of the ethylene molecule, with a wave peak on one side of the molecule, a wave trough on the other, and a node in the plane of the molecule. This electron distribution is particularly evident from an EPM of ethylene, which shows the local negative charge associated with electron density above and below the molecule.

π-electron density above and below the plane of the molecule

EPM of ethylene

It is important to understand that the π bond is *one bond* with two lobes, just as a $2p$ orbital is *one orbital* with two lobes. In this bonding picture, then, there are two types of carbon–carbon bonds: a σ bond, with most of its electron density relatively concentrated between the carbon atoms, and a π bond, with most of its electron density concentrated above and below the plane of the ethylene molecule.

This bonding picture shows why ethylene is planar. If the two CH_2 groups were twisted away from coplanarity, the $2p$ orbitals could not overlap to form the π bond. Thus, the overlap of the $2p$ orbitals and consequently the very existence of the π bond *require* the planarity of the ethylene molecule.

An important aspect of the π electrons is their relative energy. As Fig. 4.3 (p. 127) suggests, the $2p_z$ electrons (which become the π electrons of ethylene) have higher energy than the electrons in the hybrid orbitals. Thus, *π electrons generally have higher energy than σ electrons*, just as p electrons have higher energy than s electrons. A consequence of this higher energy is that π electrons are more easily removed than σ electrons. In fact, we'll find that electrophiles react preferentially with the π electrons in an alkene because those electrons are most easily donated. *Most of the important reactions of alkenes involve the electrons of the π bond, and many of these reactions involve the reaction of electrophiles with the π electrons.*

The π bond is also a weaker bond than typical carbon–carbon σ bonds because π overlap, which is "side-to-side," is inherently less effective than σ overlap, which is "head-to-head." It takes about 243 kJ mol^{-1} (58 kcal mol^{-1}) of energy to break a carbon–carbon π bond, whereas it takes a much greater energy—about 377 kJ mol^{-1} (90 kcal mol^{-1})—to break the carbon–carbon σ bond of ethane.

Return to the structure of propene in Fig. 4.2 (p. 126), and notice that the carbon–carbon bond to the —CH_3 group is shorter by about 0.04 Å than the carbon–carbon bonds of ethane or propane. This small but real difference is general: single bonds to an sp^2-hybridized carbon are somewhat shorter than single bonds to an sp^3-hybridized carbon. The carbon–carbon single bond of propene, for example, is derived from the overlap of a carbon sp^3 orbital of the —CH_3 group with a carbon sp^2 orbital of the alkene carbon. A carbon–carbon bond of propane is derived from the overlap of two carbon sp^3 orbitals. Because the electron density of an sp^2 orbital is somewhat closer to the nucleus than the electron density of an sp^3 orbital, a bond involving an sp^2 orbital, such as the one in propene, is shorter than one involving only

sp^3 orbitals, such as the one in propane. In other words, within bonds of a given bond order, *bonds with more s character are shorter.*

sp³–sp² single bond
(shorter)

sp³–sp³ single bond
(longer)

For exactly the same reason, sp^2–$1s$ C—H bonds are slightly shorter than sp^3–$1s$ C—H bonds, as we observed in our analysis of Fig. 4.2.

PROBLEM

4.1   Arrange the labeled bonds in the following molecule in order of increasing length, shortest first. Explain your reasoning.

C. Double-Bond Stereoisomers

The bonding in alkenes has other interesting consequences, which are illustrated by the four-carbon alkenes, the butenes. The butenes exist in isomeric forms. First, in the butenes with unbranched carbon chains, the double bond may be located either at the end or in the middle of the carbon chain.

$$H_2C{=}CH{-}CH_2{-}CH_3 \qquad H_3C{-}CH{=}CH{-}CH_3$$
1-butene   **2-butene**

Isomeric alkenes, such as these, that differ in the position of their double bonds are further examples of *constitutional isomers* (Sec. 2.4A).

The structure of 2-butene illustrates another important type of isomerism. *There are two separable, distinct 2-butenes*, each with its own characteristic properties. For example, one has a boiling point of 3.7 °C; the other has a boiling point of 0.88 °C. In the compound with the higher boiling point, called *cis*-2 butene or (*Z*)-2-butene, the methyl groups are on the same side of the double bond. In the other 2-butene, called *trans*-2-butene, or (*E*)-2-butene, the methyl groups are on opposite sides of the double bond.

cis-2-butene
(*Z*)-2-butene

trans-2-butene
(*E*)-2-butene

These isomers have identical atomic *connectivities* (CH_3 connected to CH, CH doubly bonded to CH, CH connected to CH_3). Despite their identical connectivities, *the two compounds differ in the way their constituent atoms are arranged in space.* Compounds with identical connectivities that differ in the spatial arrangement of their atoms are called **stereoisomers**. Hence, *cis*- and *trans*-2-butene are stereoisomers. [The (*E*) and (*Z*) notation has been adopted by the IUPAC as a general way of naming cis and trans isomers. This notation is discussed in Sec. 4.2B.]

The interconversion of *cis*- and *trans*-2-butene requires a 180° internal rotation about the double bond—that is, a rotation of one carbon while holding the other carbon stationary.

$$
\begin{array}{c}
\text{180°} \\
\text{internal rotation}
\end{array}
\tag{4.1}
$$

cis-**2-butene**   *trans*-**2-butene**

interconversion does not occur at ordinary temperatures

Because *cis*- and *trans*-2-butene do not interconvert, even at relatively high temperatures, it follows that this internal rotation must be very slow. For such an internal rotation to occur, the 2*p* orbitals on each carbon must be twisted away from coplanarity; that is, *the π bond must be broken* (Fig. 4.7). Because bonding is energetically favorable, lack of it is energetically costly. It takes more energy to break the π bond than is available under normal conditions; thus, the π bond in alkenes remains intact, and internal rotation about the double bond does not occur. In contrast, internal rotation about the carbon–carbon *single* bonds of ethane or butane can occur rapidly (Sec. 2.3) because no chemical bond is broken in the process.

Cis- and *trans*-2-butene are examples of *double-bond stereoisomers*. **Double-bond stereoisomers** (also called **cis–trans stereoisomers** or **E,Z stereoisomers**) are defined as compounds related by an internal rotation of 180° about the double bond. (We can always imagine such a rotation even though it does not occur at ordinary temperatures.) Another equivalent definition is that double-bond stereoisomers are different compounds related by interchange of the two groups at either carbon of a double bond.

interchange colored groups

$$ \tag{4.2} $$

different compounds, so they are *double-bond stereoisomers*

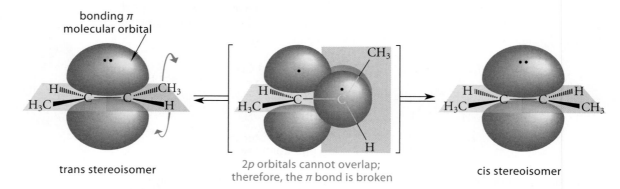

FIGURE 4.7 Internal rotation about the carbon–carbon double bond in an alkene requires breaking the π bond. This does not occur at ordinary temperatures because too much energy is required.

When an alkene can exist as double-bond stereoisomers, both carbons of the double bond are *stereocenters*. An atom is a **stereocenter** when the interchange of two bonded groups gives stereoisomers. (Other terms you might encounter that mean the same thing are **stereogenic atom** and **stereogenic center**.)

Because the exchange of the two groups at either carbon of the double bond gives stereoisomers, each of these carbons is a stereocenter.

You'll learn in Chapter 6 that double-bond stereoisomers are not the only type of stereoisomer. In every set of stereoisomers we'll be able to identify one or more stereocenters.

STUDY PROBLEM   4.1

Tell whether each of the following molecules has a double-bond stereoisomer. If so, identify its stereocenters.

SOLUTION   Apply the definition of double-bond stereoisomers as illustrated in Eq. 4.2. That is, interchange the positions of the two groups at *either* carbon of the double bond. This process will give one of two results: either the resulting molecule will be congruent to the original—that is, superimposable on the original atom-for-atom—or it will be different. If it's different, it can *only* be a stereoisomer, because its connectivity is the same.

In molecule *A*, interchanging the two groups at either carbon of the double bond gives different molecules. In the original, the methyl groups are trans; after the interchange, the methyl groups are cis. Hence, *A* has a double-bond stereoisomer:

(4.3)

(You should verify that exchanging the two groups at the other carbon of the double bond gives the same result.) The two carbons of the double bond are both stereocenters.

In the case of structure *B*, interchanging the two groups at either carbon of the double bond gives back an identical molecule.

(4.4)

**STUDY GUIDE
LINK 4.1**
Different Ways to
Draw the Same
Structure

You may have found that the structure you obtained from interchanging the two groups doesn't *look* identical to the one on the left, but it is. You can demonstrate their identity by flipping either structure 180° about a horizontal axis (*green dashed line*)—in other words, by turning it over, as shown in Eq. 4.4. But if you have difficulty seeing this, *you must build molecular models of both structures and convince yourself that the two can be superimposed atom-for-atom.* There is *no substitute* for model building when it comes to the spatial aspects of organic chemistry! After a little work with models on issues like this, you will develop the ability to see these relationships without models. Study Guide Link 4.1 offers more insights about how to achieve facility in relating alkene structures.

Because interchanging two groups in Eq. 4.4 does *not* give stereoisomers, this alkene contains no stereocenters.

PROBLEM

4.2   Which of the following alkenes can exist as double-bond stereoisomers? Identify the stereocenters in each.

(a) H_2C═$CHCH_2CH_2CH_3$ (b) CH_3CH_2CH═$CHCH_2CH_3$

      1-pentene   **3-hexene**

(c) H_2C═CH—CH═CH—CH_3 (d) CH_3CH_2CH═CCH_3 (e)

     1,3-pentadiene |

                                                     CH_3 **cyclobutene**

                             2-methyl-2-pentene

[*Hint for part (e):* Try to build a model of both stereoisomers, but don't break your models!]

4.2   NOMENCLATURE OF ALKENES

A. IUPAC Substitutive Nomenclature

The IUPAC substitutive nomenclature of alkenes is derived by modifying alkane nomenclature in a simple way. An unbranched alkene is named by replacing the *ane* suffix in the name of the corresponding alkane with the ending *ene* and specifying the location of the double bond with a number. The carbons are numbered from one end of the chain to the other so that the double bond receives the lowest number. The carbons of the double bond are numbered consecutively.

$$\overset{1}{H_2C}═\overset{2}{CH}—\overset{3}{CH_2}\overset{4}{CH_2}\overset{5}{CH_2}\overset{6}{CH_3}$$

       hex̶a̶n̶e̶ + ene = hexene   **1-hexene**

                                          ↑
                                   position of double bond

The IUPAC recognizes an exception to this rule for the name of the simplest alkene, H_2C═CH_2, which is usually called ethylene rather than ethene. [*Chemical Abstracts* (Sec. 2.4D, p. 65), however, uses the substitutive name ethene.]

The names of alkenes with branched chains are, like those of alkanes, derived from their *principal chains.* In an alkene, the principal chain is defined as *the carbon chain containing the greatest number of double bonds,* even if this is not the longest chain. If more than one candidate for the principal chain have equal numbers of double bonds, the principal chain is the longest of these. The principal chain is numbered from the end that results in the lowest numbers for the carbons of the double bonds.

When the alkene contains an alkyl substituent, the position of the double bond, not the position of the branch, determines the numbering of the chain. This is the main difference in the nomenclature of alkenes and alkanes. However, the position of the double bond is cited in the name *after* the name of the alkyl group. Study Problem 4.2 shows how these principles are implemented.

Name the following compound using IUPAC substitutive nomenclature.

$$H_2C{=}C{-}CH_2CH_2CH_3$$
$$CH_2CH_2CH_2CH_2CH_3$$

SOLUTION   The principal chain is the longest continuous carbon chain containing *both carbons* of the double bond, as shown in color in the following structure. Note in this case that the principal chain is *not* the longest carbon chain in the molecule. The principal chain is numbered from the end that gives the double bond the lowest number—in this case, 1. The substituent group is a propyl group. Hence, the name of the compound is 2-propyl-1-heptene:

position of the substituent group

position of the double bond

$$\overset{1}{H_2}C{=}\overset{2}{C}{-}CH_2CH_2CH_3$$
$$\underset{3\ \ 4\ \ 5\ \ 6\ \ 7}{CH_2CH_2CH_2CH_2CH_3}$$

2-propyl-1-heptene

principal chain (longest chain containing the double bond; double bond receives the lowest number)

If a compound contains more than one double bond, the *ane* ending of the corresponding alkane is replaced by *adiene* (if there are two double bonds), *atriene* (if there are three double bonds), and so on.

$$H_2C{=}CHCH_2CH_2CH{=}CH_2$$

1,5-hexadiene

Draw a conventional structure corresponding to the following skeletal structure, and then name it.

SOLUTION   Don't forget in a skeletal structure that there is a carbon at *each end* as well as at each vertex.

$$CH_2{-}CH{=}CH_2$$

corresponds to $CH_3CH_2CH_2CH_2{-}\overset{|}{C}{=}\overset{|}{C}{-}CH_3$

$$CH_3$$

The principal chain (color in the following structure) is the chain containing the greatest number of double bonds. One possible numbering scheme (*red*) gives the first-encountered carbons of the two double bonds the numbers 1 and 4, respectively; the other possible numbering scheme (*blue*) gives the first-encountered carbons of the double bonds the numbers 2 and 5, respectively. We compare the two possible numbering schemes pairwise—that is,

(1,4) versus (2,5). The lowest number at first point of difference (1 versus 2) determines the correct numbering. The compound is a 1,4-hexadiene, with a butyl branch at carbon-4, and a methyl branch at carbon-5:

If the name remains ambiguous after determining the correct numbers for the double bonds, then the principal chain is numbered so that the lowest numbers are given to the branches at the first point of difference.

STUDY PROBLEM   4.4

Name the following compound:

SOLUTION   Remember that Me = methyl. Two ways of numbering this compound give the double bond the numbers 1 and 2.

possible names: **1,6-dimethylcyclohexene**   **2,3-dimethylcyclohexene**
                        (correct)   (incorrect)

In this situation, choose the numbering scheme that gives the lowest number for the methyl substituents *at the first point of difference.* In comparing the substituent numbering schemes (1,6) with (2,3), the first point of difference occurs at the first number (1 versus 2). The (1,6) numbering scheme is correct because 1 is lower than 2. Notice that the number 1 for the double bond is not given explicitly in the name, because this is the only possible number. That is, when a double bond in a ring receives numerical priority, *its carbons must be numbered consecutively* with the numbers 1 and 2. That's why the following numbering scheme is incorrect. One carbon of the double bond has the number 1, but the other is not numbered consecutively.

1,2-dimethylcyclohexene
(incorrect because carbons of double bond
are not numbered consecutively)

Substituent groups may also contain double bonds. Some widely occurring groups of this type have special names that should be learned. (Both conventional and skeletal structures are shown. In these structures, the bracket indicates the point of attachment of the substituent group to the principal chain.)

$H_2C=CH-\}$ $H_2C=CH-CH_2-\}$

vinyl   **allyl**

$$CH_3$$
$$H_2C=C-\}$$

isopropenyl

Here are some examples of structures containing these substituent groups. In each case, the ring is the principal chain because it has the greater number of carbons.

—$CH=CH_2$ or

3-vinylcyclohexene

—$CH_2-CH=CH_2$ or

1-allylcyclopentene

Other substituent groups are numbered *from the point of attachment to the principal chain.*

substituent chain numbered from
its point of attachment

$-CH_2-CH_2-CH=CH_2$

1-(3-butenyl)cyclohexene

position of double bond within
the substituent

position of the substituent group
on the principal chain

The names of these groups, like the names of ordinary alkyl groups, are constructed from the name of the parent hydrocarbon by dropping the final *e* from the name of the corresponding alkene and replacing it with *yl*. Thus, the substituent in the second example above is butene + yl = butenyl. Notice the use of parentheses to set off the names of substituents with internal numbering.

Finally, some alkenes have nonsystematic traditional names that are recognized by the IUPAC. These can be learned as they are encountered. Two examples are styrene and isoprene:

$Ph-CH=CH_2$ or

styrene

$H_2C=C-CH=CH_2$ or
$$CH_3$$

isoprene

(Recall from Sec. 2.8B, p. 78, that Ph— refers to the phenyl group, a singly substituted benzene ring.)

1993 IUPAC Nomenclature Recommendations

The nomenclature in this text is based on the widely used 1979 IUPAC rules. In 1993, the IUPAC recommended an alteration in nomenclature that places the number of the double bond just before the *ene* suffix of the name. Thus, in this more recent system, 1-hexene is named hex-1-ene, and 2,4-hexadiene is named hexa-2,4-diene. This new system is being used by some chemists and not by others. While the system is logical, its general adoption would require chemical indexing systems either to recognize both old and new names or to cross-reference between them. Because *Chemical Abstracts* has not officially adopted the new system, we won't use it in this text. However, conversion between old and new names is a simple matter of moving the numerical designation.

PROBLEMS

4.3   Give the structure for each of the following:

(a) 2-methylpropene   (b) 4-methyl-1,3-hexadiene   (c) 1-isopropenylcyclopentene

(d) 5-(3-pentenyl)-1,3,6,8-decatetraene

4.4   Name the following compounds. Ignore double-bond stereochemistry.

(a)   Me

(b) $CH_3CH_2CH{=}CHCH_2CH_2CH_3$ (c)

Me

B. Nomenclature of Double-Bond Stereoisomers: The *E,Z* System

The cis and trans designations for double-bond stereoisomers are unambiguous when each carbon of a double bond has a single hydrogen, as in *cis*- and *trans*-2-butene. However, in some important situations, the use of the terms cis and trans is ambiguous. For example, is the following compound, a stereoisomer of 3-methyl-2-pentene, the cis or the trans stereoisomer?

$$H_3C \quad\quad CH_2CH_3$$
$$C{=}C$$
$$H \quad\quad CH_3$$

One person might decide that this compound is trans, because the two identical groups are on opposite sides of the double bond. Another might decide that it is cis, because the larger groups are on the same side of the double bond. Exactly this sort of ambiguity—and the use of both conventions simultaneously in the chemical literature—brought about the adoption of an unambiguous system for the nomenclature of stereoisomers. This system, first published in 1951, is part of a general system for the nomenclature of stereoisomers called the **Cahn–Ingold–Prelog system** after its inventors, Robert S. Cahn (1899–1981), then editor of the *Journal of the Chemical Society*, the most prestigious British chemistry journal; Sir Christopher K. Ingold (1893–1970), a professor at University College, London, whose work played a very important part in the development of modern organic chemistry; and Vladimir Prelog (1906–1998), a professor at the Swiss Federal Institute of Technology, who received the 1975 Nobel Prize in Chemistry for his work in organic stereochemistry. When we apply the Cahn–Ingold–Prelog system to alkene double-bond stereochemistry, we'll refer to it simply as the ***E,Z* system** for reasons that will be immediately apparent.

The *E,Z* system involves assignment of *relative priorities* to the two groups on each carbon of the double bond according to a set of *sequence rules* given in the steps to be described below. We then compare the relative locations of these groups on each alkene carbon. If the groups of higher priority are on the same side of the double bond, the compound is said to have the *Z* configuration (*Z* from the German word *zusammen*, meaning "together"). If the groups of higher priority are on opposite sides of the double bond, the compound is said to have the *E* configuration (*E* from the German *entgegen*, meaning "across"). A convenient way to remember *E* and *Z* is that *ears* begins with *E*, and your *Ears* are on opposite sides of your head.

For a compound with more than one double bond, the configuration of each double bond is specified separately.

The **sequence rules** used to assign relative priorities are the core of the Cahn–Ingold–Prelog system. To apply these rules, we must first recognize that the atoms in each group can be organized into levels. Level 1 consists of the atoms directly attached to the double bond. Level 2 consists of the atoms attached to the level-1 atoms, looking away from the double bond. Level 3 consists of the atoms attached to the level-2 atoms, and so on.

The sequence rules start with a comparison of the atoms at level 1. If no decision is possible, we proceed to the atoms at level 2, then level 3, and so on. As this diagram shows, the greater the level, the more possibilities there are for comparison at each level. The sequence rules specify both how to make the comparisons at each level and how to proceed to the next level if a decision is not possible.

To assign relative priorities, follow these steps and the accompanying rules *in order* until a difference is found. A decision must be made at the first point of difference. (Each step is illustrated with an example. Then, Study Problems 4.5–4.7 show how these steps are applied in three different cases of increasing complexity.)

Step 1. Examine the atoms directly attached to a given carbon of the double bond (level 1), and then follow the first rule that applies.

> **Rule 1a** *Assign higher priority to the group containing the atom of higher atomic number.*

> **Rule 1b** *Assign higher priority to the group containing the isotope of higher atomic mass.*

higher priority by rule 1a (C has higher atomic number than H)

higher priority by rule 1b (D has higher atomic mass than H)

(D = ^{2}H = deuterium)

Step 2. If the level-1 atoms directly attached to the double bond are the same, then, working outward from the double bond, consider within each group the set of attached atoms—that is, the level-2 set. You'll have two level-2 sets—one for each group on the double bond. Apply rule 2 to each level-2 set.

Rule 2   *Arrange the atoms within each set in descending priority order, and make a pairwise comparison of the atoms in the two sets. The higher priority is assigned to the atom of higher atomic number (or atomic mass in the case of isotopes) at the first point of difference.*

Step 3. If the level-2 sets in the two groups are identical, then, within each set, choose the atom of highest priority. Identify the level-3 set of atoms attached to it. Then compare the level-3 sets in each group by applying rule 2.

Step 3a. If no decision emerges at step 3, choose the atoms of next highest priority in the level-2 sets and repeat the process in step 3. Choose atoms of progressively lower priority in the level-2 sets until a difference is found.

Notice that we consider *only the atom directly attached* to the oxygens (C in this example); we do not consider any atoms farther out.

Step 3b. If two or more atoms in any set are the same, decide on their relative priorities for step 3 by continuing to explore outward from each, and choose as the atom of higher priority the one that gives the path of higher priority. (Step 3b is illustrated in Study Problem 4.7.)

Step 3c. If no decision is possible, move away from the double bond within each group to atoms at the next level and repeat step 3. Continue this exploration, level-by-level, until the first difference is found.

STUDY PROBLEM 4.5

What is the configuration of the following stereoisomer of 3-methyl-2-pentene? (The numbers are for reference in the solution.)

SOLUTION Assign the relative priorities of the two groups attached to each carbon. The two atoms directly attached to carbon-2 are C and H. Because C has a higher atomic number (6) than H (1), the CH_3 group is assigned the higher priority, as shown in the example for step 1. Now consider the groups attached to carbon-3. The example accompanying step 2 and rule 2 shows that ethyl has higher priority than methyl. The priority pattern is therefore as follows:

Because groups of like priority are on opposite sides of the double bond, this alkene is the *E* isomer; its complete name is (*E*)-3-methyl-2-pentene.

STUDY PROBLEM 4.6

What is the configuration (*E* or *Z*) of the following alkene? (The numbers and letters are for reference in the solution.)

6,6-dibromo-3-isobutyl-2-heptene

SOLUTION At carbon-2, the methyl group has higher priority, by rule 1a. At carbon-3, rules 1a and 1b allow no decision, because the level-1 atoms (*a1* and *b1*) are identical—both are carbons. Proceeding to step 2, the set of atoms (level-2 set) attached to either carbon *a1* or *b1* is (C,H,H); again, no decision is possible. According to step 3, we must now consider the atom of highest priority (carbon) in each level-2 set; these atoms are labeled *a2* and *b2*, respectively. The level-3 set of atoms attached to *a2* is (C,H,H); the level-3 set attached to *b2* is (C,C,H). Notice that carbons *a1* and *b1* considered in the previous step are *not considered* as members of these sets, because we always work outward, away from the double bond, by step 2. The difference in the second atoms of each set—C versus H—dictates a decision. Because the level-3 set of atoms at carbon *b2* has higher priority, the group containing carbon *b2* (the isobutyl group) also has the higher priority. The process used can be summarized as follows:

Notice that, although Br has a higher priority than H, *the decision point occurs before we reach the Br* (which is in a level-4 set). Therefore, the Br-versus-H comparison is irrelevant in this case.

Because the groups of like priority are on the same side of the double bond, this alkene has the *Z* configuration; the name is (*Z*)-6,6-dibromo-2-isobutyl-2-heptene.

Sometimes the groups to which we must assign priorities themselves contain double bonds. Double bonds are treated by a special convention, in which the double bond is rewritten as a single bond and the atoms at each end of the double bond are duplicated:

$$-CH{=}CH_2 \text{ is treated as } \overset{\displaystyle -CH-CH_2}{\underset{\displaystyle \enspace C \quad\; C}{\vert \qquad\vert}} \text{ and } -CH{=}O \text{ is treated as } \overset{\displaystyle -CH-O}{\underset{\displaystyle \enspace O \quad\; C}{\vert \qquad\vert}}$$

Notice that the duplicated atoms bear only one bond; that is, they have no other groups attached to them. The treatment of triple bonds requires triplicating the atoms involved:

$$-C{\equiv}CH \text{ is treated as } \overset{\displaystyle C \quad\; C}{\underset{\displaystyle C \quad\; C}{\overset{\vert\ \ \,\vert}{-\underset{\vert\ \ \,\vert}{C}-CH}}} \text{ and } -C{\equiv}N \text{ is treated as } \overset{\displaystyle N \quad\; C}{\underset{\displaystyle N \quad\; C}{\overset{\vert\ \ \,\vert}{-\underset{\vert\ \ \,\vert}{C}-N}}}$$

The handling of this convention is illustrated in Study Problem 4.7.

STUDY PROBLEM 4.7

Give the IUPAC name of the following compounds, including the (*E*,*Z*) designation for the double-bond stereochemistry. (The carbon numbers are for reference in the solution.)

(a) (b)

SOLUTION   (a) First, give the name without stereochemistry. By the nomenclature principles in Sec. 4.2A, the name is 3-isopropyl-1,3-pentadiene. Now assign stereochemistry. Carbons 1 and 2 are not stereocenters, but carbons 3 and 4 are. At carbon-4, the methyl group receives higher priority than H. The real challenge here is the relative priorities of the groups at carbon-3. We analyze the two groups as follows:

(The symbol ⇨ means "implies.") Notice that the carbons of the double bond are duplicated. The level-3 set (0,0,0) in the vinyl group reminds us that nothing is attached to the duplicated carbon b'. We compare carbon a of the isopropyl group with carbon a of the vinyl group. These are the same. Furthermore, the attached (level-2) sets are the same: (C,H,H). We then consider the level-2 atom of highest priority in each group and examine the level-3 sets attached to this atom. In the isopropyl group, the level-2 atom of highest priority is either of the methyl carbons. In the vinyl group, we have to choose between carbon b and its duplicated image, carbon b'. Because both atoms are the same, we use step 3b. According to step 3b, this choice is made by going out one level beyond carbons b and b'—in other words, within the vinyl group, we must compare the level-3 sets (C,H,H) for C^b with (0,0,0) for $C^{b'}$. Because any of the atoms attached to carbon b has a higher priority than "nothing" on carbon b', carbon b represents the path of higher priority for vinyl. Now we are ready to compare the vinyl and isopropyl groups again. For isopropyl, the level-3 set at carbon b is (H,H,H), and for vinyl, the level-3 set at carbon b is (C,H,H). By rule 2, vinyl receives the higher priority. The name is therefore (*E*)-3-isopropyl-1,3-pentadiene.

You should be able to work part (b) using the same tactics. Try it. Be sure to convert the skeletal structure into a conventional structure if necessary. The name is (2*E*,4*Z*)-3-isopropyl-2,4-hexadiene. (Notice that the position of the isopropyl group determines the numbering of the double bonds, which is ambiguous otherwise.) Notice also that when two or more double bonds require a stereochemical designation, the number of the double bond is included with the *E* or *Z*.

PROBLEMS

4.5   Name each of the following compounds, including the proper designation of the double-bond stereochemistry:

4.6   Give the structure of:

(a) (*E*)-4-allyl-1,5-octadiene   (b) (2*E*,7*Z*)-5-[(*E*)-1-propenyl]-2,7-nonadiene

Be sure to read Study Guide Link 4.2 if you have difficulty with this problem.

STUDY GUIDE
LINK 4.2
Drawing Structures
from Names

4.7   In each case, which group receives the higher priority?

4.3   UNSATURATION NUMBER

An alkene with one double bond has two fewer hydrogens than the alkane with the same carbon skeleton. Likewise, a compound containing a ring also has two fewer hydrogens in its molecular formula than the corresponding noncyclic compound. (Compare cyclohexane or 1-hexene, C_6H_{12}, with hexane, C_6H_{14}.)

$$H_3C—CH_2—CH_2CH_2CH_2CH_3 \qquad H_2C{=}CH—CH_2CH_2CH_2CH_3$$

hexane
(C_6H_{14})

1-hexene
(C_6H_{12})

cyclohexane
(C_6H_{12})

each double bond or ring
"costs" 2 hydrogens

As both examples illustrate, *the molecular formula of an organic compound contains "built-in" information about the number of rings and double (or triple) bonds.*

The presence of rings or double bonds within a molecule is indicated by a quantity called the **unsaturation number**, or **degree of unsaturation**, *U. The unsaturation number of a molecule is equal to the total number of its rings and multiple bonds.* The unsaturation number of a hydrocarbon is readily calculated from its molecular formula as follows. The maximum number of hydrogens possible in a hydrocarbon with *C* carbon atoms is $2C + 2$. Because *each ring or double bond reduces the number of hydrogens from this maximum by 2*, the unsaturation number is equal to half the difference between the maximum number of hydrogens and the actual number *H*:

$$U = \frac{2C + 2 - H}{2} = \text{number of rings + multiple bonds} \qquad (4.5)$$

For example, cyclohexene, C_6H_{10}, has $U = [2(6) + 2 - 10]/2 = 2$. Cyclohexene has two degrees of unsaturation: one ring and one double bond. A triple bond contributes two degrees of unsaturation. For example, 1-hexyne, $HC{\equiv}C—CH_2CH_2CH_2CH_3$, has the same formula as cyclohexene: C_6H_{10}.

How does the presence of other elements affect the unsaturation number calculation? You can readily convince yourself from common examples (for instance, ethanol, C_2H_5OH) that Eq. 4.5 remains valid when oxygen is present in an organic compound. Halogens are counted as if they were hydrogens, because halogens are monovalent, and each halogen reduces the number of hydrogens by 1. Therefore, if *X* represents the number of halogens,

$$U = \frac{2C + 2 - (H + X)}{2} \qquad (4.6)$$

Another common element found in organic compounds is nitrogen. When nitrogen is present, the number of hydrogens in a saturated compound increases by one for each nitrogen. (For example, the saturated compound methylamine, $H_3C—NH_2$, has $2C + 3$ hydrogens.) Therefore, if *N* is the number of nitrogens, the formula for the unsaturation number becomes

$$U = \frac{2C + 2 + N - (H + X)}{2} \qquad (4.7)$$

Remember that the unsaturation number is a valuable source of structural information about an unknown compound. This idea is illustrated in Problems 4.9 and 4.10.

PROBLEMS

4.8 Calculate the unsaturation number for each of the formulas in parts (a) and (b) and each of the compounds in parts (c) and (d). [Try to work parts (c) and (d) using only the compound names.]

(a) $C_3H_4Cl_4$ (b) $C_5H_8N_2$

(c) methylcyclohexane   (d) 2,4,6-octatriene

4.9 A compound has the molecular formula $C_{20}H_{34}O_2$. Certain structural evidence suggests that the compound contains two methyl groups and no carbon–carbon double bonds. Give one structure consistent with these findings in which all rings are six-membered. (Many structures are possible.)

4.10 Which of the following *cannot* be correct formula(s) for an organic compound? Explain.

(a) $\underset{A}{C_{10}H_{20}N_3}$ $\underset{B}{C_{10}H_{20}N_2O_2}$ $\underset{C}{C_{10}H_{27}N_3O_2}$ $\underset{D}{C_{10}H_{20}N_3}$

(b) Draw constitutional isomers of compounds with the formula C_4H_8O that contain (1) an alcohol functional group; (2) a ketone functional group.

4.4   PHYSICAL PROPERTIES OF ALKENES

Except for their melting points and dipole moments, many alkenes differ little in their physical properties from the corresponding alkanes.

$$H_2C{=}CH(CH_2)_3CH_3 \qquad CH_3(CH_2)_4CH_3$$

	1-hexene	**hexane**
boiling point	63.4 °C	68.7 °C
melting point	−139.8 °C	−95.3 °C
density	0.673 g mL^{-1}	0.660 g mL^{-1}
water solubility	negligible	negligible
dipole moment	0.46 D	0.085 D

Like alkanes, alkenes are flammable, nonpolar compounds that are less dense than, and insoluble in, water. The alkenes of lower molecular weight are gases at room temperature.

The dipole moments of some alkenes, though small, are greater than those of the corresponding alkanes.

$$\underset{\mu = 0.25\ D}{\underset{\overset{|}{H}\qquad\overset{|}{H}}{\overset{H_3C\qquad CH_3}{C{=}C}}} \qquad\qquad \underset{\mu = 0\ D}{H_3C{-}CH_2{-}CH_2{-}CH_3}$$

How can we account for the dipole moments of alkenes? It turns out that *sp^2-hybridized carbons are somewhat more electronegative than sp^3-hybridized carbons.* As a result, any sp^2–sp^3 carbon–carbon bond has a small bond dipole (Sec. 1.2D) in which the sp^3 carbon is the positive end of the dipole and the sp^2 carbon is the negative end.

The reason for the greater electronegativity of an sp^2-hybridized carbon is that it has an electron in a $2p$ orbital. Electron density in a $2p$ orbital is not evenly distributed around the nucleus in all directions. That is, if a $2p$ orbital is oriented along the z axis, a $2p$ electron does not screen the nucleus from other electrons situated in the xy plane (the plane of the page in the diagram at the bottom of p. 145). Therefore, the relatively unscreened positive charge of the atomic nucleus in an sp^2 carbon pulls the electrons in the sp^3–sp^2 carbon–carbon bond toward itself and creates a small partial positive charge on the attached methyl groups, and thus creates the bond dipole. (A similar effect occurs in C—H bonds; look at the EPM of ethylene on p. 130. Notice the partial positive charge on the ethylene hydrogens.)

The dipole moment of *cis*-2-butene is the vector sum of all of the H$_3$C—C and H—C bond dipoles. Although both types of bond dipole are oriented toward the alkene carbon, there is good evidence (Problem 4.63, p. 179) that the bond dipole of the H$_3$C—C bond is greater. This conclusion would be expected from the greater length of the C—C bond. (Remember that the dipole moment for a given charge separation increases with the length of the dipole; Eq. 1.4, p. 11.) Thus, *cis*-2-butene has a net dipole moment.

In summary: Bonds from alkyl groups to trigonal planar sp^2-hybridized carbons are polarized so that electrons are drawn away from alkyl groups toward the trigonal carbon. This means that a carbon–carbon double bond, when viewed as a substituent group, exerts an *electron-withdrawing polar effect* (Sec. 3.6C). The polar effect of a double bond is only about 10–15% that of a chlorine, but it's significant enough to be measurable. (See Problem 4.12.)

PROBLEMS

4.11   Which compound in each set should have the larger dipole moment? Explain.

(a) *cis*-2-butene or *trans*-2-butene   (b) propene or 2-methylpropene

4.12   Which of the following two carboxylic acids is more acidic? Explain.

4.5   RELATIVE STABILITIES OF ALKENE ISOMERS

When we ask which of two compounds is more stable, we are asking which compound has lower energy. However, energy can take different forms, and the energy we use to measure relative "stability" depends on the purpose we have in mind. We've learned that $\Delta G°$ for a reaction is the energy quantity related to the equilibrium constant, as we know from the equation $\Delta G° = -2.3RT \log K_{eq}$ (Sec. 3.5). Measuring the equilibrium constant is a good way to determine $\Delta G°$. However, if we are interested in the *total energy change* for a reaction, we use the **standard enthalpy change** for the reaction, $\Delta H°$. The $\Delta H°$ for a reaction approximates very closely the *total* energy difference between reactants and products, and it reflects the *relative stabilities of bonding arrangements in reactants and products*. The $\Delta G°$ and $\Delta H°$ for a reaction are related by the equation $\Delta G° = \Delta H° - T\Delta S°$, where $\Delta S°$ is the entropy change for the reaction and T is the absolute temperature. (For a structural interpretation of $\Delta S°$, see Further Exploration 4.1 in the Study Guide.) In other words, the $\Delta G°$ for a reaction differs from the total energy difference between reactants and products by an amount $-T\Delta S°$. In Sec. 4.5A we'll learn the conventions for presenting enthalpy data, and in Sec. 4.5B we'll use enthalpy data to investigate the relative stabilities of alkenes.

FURTHER EXPLORATION 4.1
Relationship between Free Energy and Enthalpy

A. Heats of Formation

The relative enthalpies of many organic compounds are available in standard tables as *heats of formation*. The standard **heat of formation** of a compound, abbreviated $\Delta H_f°$, is the heat change that occurs when the compound is formed from its elements in their natural state at 1 atm pressure and 25 °C. Thus, the heat of formation of *trans*-2-butene is the $\Delta H°$ of the following reaction:

$$4 \text{ H}_2 \text{ (gas)} + 4 \text{ C (solid)} \longrightarrow \textit{trans-}2\text{-butene (liquid, C}_4\text{H}_8) \qquad (4.8)$$

The sign conventions used in dealing with heats of reaction are the same as with free energies: the heat of any reaction is the *difference* between the enthalpies of the products and the reactants.

$$\Delta H°\text{(reaction)} = H°\text{(products)} - H°\text{(reactants)} \qquad (4.9)$$

A reaction in which heat is liberated is said to be an **exothermic reaction**, and one in which heat is absorbed is said to be an **endothermic reaction**. The $\Delta H°$ of an exothermic reaction, by Eq. 4.9, has a negative sign; the $\Delta H°$ of an endothermic reaction has a positive sign. The heat of formation of *trans*-2-butene (Eq. 4.8) is -11.6 kJ mol^{-1} (-2.72 kcal mol^{-1}); this means that heat is liberated in the formation of *trans*-2-butene from carbon and hydrogen, and that the alkene has lower energy than the 4 moles each of C and H$_2$ from which it is formed.

Heats of formation are used to determine the relative enthalpies of molecules—that is, which of two molecules has lower energy. How this is done is illustrated in Study Problem 4.8.

STUDY PROBLEM 4.8

Calculate the standard enthalpy difference between the cis and trans isomers of 2-butene. Specify which stereoisomer is more stable. The heats of formation are, for the cis isomer, -7.40 kJ mol^{-1}, and for the trans isomer, -11.6 kJ mol^{-1} (-1.77 and -2.72 kcal mol^{-1}, respectively).

SOLUTION The enthalpy difference requested in the problem corresponds to the $\Delta H°$ of the following hypothetical reaction:

$$
\begin{array}{lll}
\textit{cis-}2\text{-butene} & \longrightarrow & \textit{trans-}2\text{-butene} \\
\end{array}
$$

$\Delta H_f°$	-7.40	-11.6	kJ mol^{-1}	(4.10)
	-1.77	-2.72	kcal mol^{-1}	

To obtain the standard enthalpy difference, apply Eq. 4.9 *using the corresponding heats of formation in place of the $H°$ values.* Thus, $\Delta H_f°$ for the reactant, *cis*-2-butene, is subtracted from that of the product, *trans*-2-butene. The $\Delta H°$ for this reaction, then, is $-11.6 - (-7.40) = -4.2$ kJ mol^{-1} (-1.0 kcal mol^{-1}). This means that *trans*-2-butene is more stable than *cis*-2-butene by 4.2 kJ mol^{-1} (1.0 kcal mol^{-1}).

The procedure used in Study Problem 4.8 is based on the fact that *chemical reactions and their associated energies can be added algebraically.* This principle is known as **Hess's law of constant heat summation**. Hess's law is a direct consequence of the first law of thermodynamics, which requires that the energy difference between two compounds doesn't depend on the path (or reactions) used to make the measurement. Thus, what we have done in

FIGURE 4.8 Use of heats of formation to derive the relative enthalpies of two isomeric compounds. The enthalpies of both compounds are measured relative to a common reference, the elements from which they are formed. The difference between the enthalpies of formation is equal to the enthalpy difference between the two isomers.

STANDARD ENTHALPY, $H°$

4 H$_2$ + 4C

The enthalpy of the elements in their standard states is the reference point for enthalpies of formation

7.40 kJ mol^{-1} (1.77 kcal mol^{-1})

11.6 kJ mol^{-1} (2.72 kcal mol^{-1})

cis-2-butene

4.2 kJ mol^{-1} (1.0 kcal mol^{-1})

enthalpy difference between *cis*- and *trans*-2-butene

trans-2-butene

Study Problem 4.8 is to add the two formation reactions and their associated enthalpies, one in the forward direction and the other in the reverse direction:

Equations:			$\Delta H°$ (kJ mol^{-1}):	$\Delta H°$ (kcal mol^{-1}):	
4C + 4H$_2$	⟶	*trans*-2-butene	−11.6	−2.72	(4.11a)
cis-2-butene	⟶	4C + 4H$_2$	+7.4	+1.77	(4.11b)
Sum: *cis*-2-butene	⟶	*trans*-2-butene	−4.2	−1.0	(4.11c)

Because *cis*- and *trans*-2-butene are isomers, the elements from which they are formed are the same and *cancel in the comparison*. This is shown by the diagram in Fig. 4.8. Were we to compare the enthalpies of compounds that are not isomers, the two formation equations would have different quantities of carbon and hydrogen, and the sum would contain leftover C and H$_2$. This sum would not correspond to the direct comparison desired.

Using heats of formation to calculate the standard enthalpy difference between two compounds (Study Problem 4.8) is analogous to measuring the relative heights of two objects by comparing their distances from a common reference, such as the ceiling. If a table top is 5 ft below the ceiling, and an electrical outlet is 7 ft below the ceiling, then the table top is 2 ft above the outlet. The height of the ceiling can be taken arbitrarily as zero; its absolute height is irrelevant. When heats of formation are compared, the enthalpy reference point is the enthalpy of the elements in their "standard states," their normal states at 25 °C and 1 atm pressure; the enthalpies of formation of the elements in their standard states are arbitrarily taken to be zero.

Heats of formation are not measured directly, because the formation reaction is not a practical reaction. Rather, heats of formation are determined by combining the enthalpies of other, more practical reactions, such as combustion (Sec. 2.7) or catalytic hydrogenation (Sec. 4.9A), using Hess's law calculations. Heats of formation are the conventional way in which these various sources of enthalpy data are brought together and tabulated. (See Further Exploration 4.2.)

FURTHER EXPLORATION 4.2 Source of Heats of Formation

PROBLEMS

4.13 (a) Calculate the enthalpy change for the hypothetical reaction 1-butene ⟶ 2-methylpropene. The heats of formation are 1-butene, −0.30 kJ mol^{-1} (−0.07 kcal mol^{-1}); 2 methylpropene, −17.3 kJ mol^{-1} (−4.13 kcal mol^{-1}).

(b) Which butene isomer in part (a) is more stable?

4.14 (a) If the standard enthalpy change for the reaction 2-ethyl-1-butene ⟶ 1-hexene is +15.3 kJ mol^{-1} (+3.66 kcal mol^{-1}), and if $\Delta H_f°$ for 1-hexene is −40.5 kJ mol^{-1} (−9.68 kcal mol^{-1}), what is $\Delta H_f°$ for 2-ethyl-1-butene?

(b) Which isomer in part (a) is more stable?

4.15   The ΔH_f° of CO_2 is -393.51 kJ mol^{-1} (-94.05 kcal mol^{-1}), and the ΔH_f° of H_2O is -285.83 kJ mol^{-1} (-68.32 kcal mol^{-1}). Calculate the ΔH_f° of 1-heptene from its heat of combustion, -4693.1 kJ mol^{-1} (-1121.7 kcal mol^{-1}). (See Further Exploration 4.2.)

B. Relative Stabilities of Alkene Isomers

The heats of formation of alkenes can be used to determine how various structural features of alkenes affect their stabilities. We'll answer two questions using heats of formation. First, which is more stable: a cis alkene or its trans isomer? Second, how does the number of alkyl substituents at the double bond affect the stability of an alkene?

Study Problem 4.8 showed that *trans*-2-butene has a lower enthalpy of formation than *cis*-2-butene by 4.2 kJ mol^{-1} (1.0 kcal mol^{-1}) (see Eq. 4.11c). In fact, almost all trans alkenes are more stable than their cis isomers. The reason is that, in a *cis*-alkene, the larger groups are forced to occupy the same plane on the same side of the double bond. For example, a space-filling model of *cis*-2-butene (Fig. 4.9) shows that one hydrogen in each of the cis methyl groups is within a van der Waals radius of the other. Hence, van der Waals repulsions occur between the methyl groups much like those in *gauche*-butane (Fig. 2.6, p. 54). In contrast, no such repulsions occur in the trans isomer, in which the methyl groups are far apart. Not only do the heats of formation suggest the presence of van der Waals repulsions in cis alkenes, but they give us quantitative information on the magnitude of such repulsions.

Another structural aspect of alkenes that has a considerable effect on stability is the number of alkyl groups *directly attached* to the carbons of the double bond. For example, let's compare the heats of formation of the following two isomers. The first has a single alkyl group directly attached to the double bond. The second has two alkyl groups attached to the double bond.

$$\begin{array}{c}
\text{H} \qquad\qquad \text{CH(CH}_3)_2 \\
\diagdown \qquad \diagup \\
\text{C}=\text{C} \\
\diagup \qquad \diagdown \\
\text{H} \qquad\qquad \text{H}
\end{array}$$

one alkyl substituent at the double bond

$\Delta H_f^\circ = -27.4$ kJ mol^{-1}
$\qquad -6.55$ kcal mol^{-1}

(4.12a)

$$\begin{array}{c}
\text{H} \qquad\qquad \text{CH}_3 \\
\diagdown \qquad \diagup \\
\text{C}=\text{C} \\
\diagup \qquad \diagdown \\
\text{H} \qquad\qquad \text{CH}_2\text{CH}_3
\end{array}$$

two alkyl substituents at the double bond

$\Delta H_f^\circ = -35.1$ kJ mol^{-1}
$\qquad -8.39$ kcal mol^{-1}

(4.12b)

Because the isomer with two alkyl substituents at the carbon of the double bond has the more negative heat of formation, it is more stable. The data in Table 4.1 for other isomeric pairs

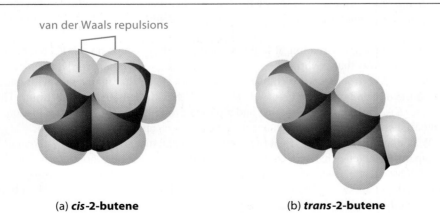

van der Waals repulsions

(a) *cis*-2-butene   (b) *trans*-2-butene

FIGURE 4.9 Space-filling models of (a) *cis*-2-butene and (b) *trans*-2-butene. *Cis*-2-butene has van der Waals repulsions between hydrogen atoms of the two methyl groups (red). In *trans*-2-butene, these van der Waals repulsions are not present.

TABLE 4.1 Effect of Branching on Alkene Stability

Alkene structure*	Number of alkyl groups on the double bond	ΔH_f°	Enthalpy difference
$H_2C{=}CH{-}CH_2CH_2CH_2CH_3$	1	-40.5 kJ mol^{-1} -9.68 kcal mol^{-1}	-10.6 kJ mol^{-1} -2.5 kcal mol^{-1}
H$_3$C, H / C=C / H, CH$_2$CH$_2$CH$_3$	2	-51.1 kJ mol^{-1} -12.2 kcal mol^{-1}	
H$_2$C=C / CH$_2$CH$_3$ \ CH$_2$CH$_3$	2	-55.8 kJ mol^{-1} -13.3 kcal mol^{-1}	-5.70 kJ mol^{-1} -1.4 kcal mol^{-1}
H$_3$C, CH$_3$ / C=C / H, CH$_2$CH$_3$	3	-61.5 kJ mol^{-1} -14.7 kcal mol^{-1}	
H$_3$C, H / C=C / H, CH(CH$_3$)$_2$	2	-60.1 kJ mol^{-1} -14.4 kcal mol^{-1}	-2.6 kJ mol^{-1} -0.6 kcal mol^{-1}
CH$_3$CH$_2$, CH$_3$ / C=C / H, CH$_3$	3	-62.7 kJ mol^{-1} -15.0 kcal mol^{-1}	
H$_3$C, CH$_3$ / C=C / H, CH(CH$_3$)$_2$	3	-88.4 kJ mol^{-1} -21.1 kcal mol^{-1}	-2.1 kJ mol^{-1} -0.5 kcal mol^{-1}
H$_3$C, CH$_3$ / C=C / CH$_3$CH$_2$, CH$_3$	4	-90.5 kJ mol^{-1} -21.6 kcal mol^{-1}	

* In each comparison, the two compounds are equally branched; they differ only in whether the branch is at the double bond.

of alkenes show that this trend continues for increasing numbers of alkyl groups directly attached to the double bond. These data show that *an alkene is stabilized by alkyl substituents on the double bond*. When we compare the stability of alkene isomers, we find that *the alkene with the greatest number of alkyl substituents on the double bond is usually the most stable one*.

To a useful approximation, it is the *number* of alkyl groups on the double bond more than their *identities* that governs the stability of an alkene. In other words, a molecule with two smaller alkyl groups on the double bond is more stable than its isomer with one larger group on the double bond. The first two entries in Table 4.1 demonstrate this point. The second entry, (*E*)-2-hexene, with a methyl and a propyl group on the double bond, is more stable than the first entry, 1-hexene, which has a single butyl group on the double bond.

Why does alkyl substitution at the double bond enhance the stability of alkenes? When we compare an alkene that has an alkyl substituent at the double bond with one that has an alkyl substituent elsewhere, we are really comparing the tradeoff of an sp^2–sp^3 carbon–carbon bond and an sp^3–$1s$ carbon–hydrogen bond with an sp^3–sp^3 carbon–carbon bond and an sp^2–$1s$ carbon–hydrogen bond.

the more stable alkene

The major effect in this tradeoff is that an sp^2–sp^3 carbon–carbon bond is stronger than an sp^3–sp^3 carbon–carbon bond. (The bonds to hydrogen have similar but smaller effects.) Increasing bond strength lowers the heat of formation.

We can go on to ask why an sp^2–sp^3 carbon–carbon bond is stronger than an sp^3–sp^3 carbon–carbon bond. Bond strength is directly related to the energy of the electrons in the bond. The lower the energy of the bonding electrons, the stronger is the bond. Because *s* electrons have lower energy than *p* electrons, a bond with more *s* character involves electrons of lower energy than one with less *s* character. A bond with more *s* character, such as an sp^2–sp^3 bond, is therefore stronger than one with less *s* character, such as an sp^3–sp^3 bond. *Bond strength increases with the fraction of s character in the component hybrid orbitals.*

Heats of formation have given us considerable information about how alkene stabilities vary with structure. To summarize:

Increasing stability:

$$R-CH=CH_2 < R-CH=CH-R \approx \underset{R}{\overset{R}{C}}=CH_2 < \underset{R}{\overset{R}{C}}=CHR < \underset{R}{\overset{R}{C}}=\underset{R}{\overset{R}{C}} \quad (4.13a)$$

and

$$\underset{H}{\overset{R}{C}}=\underset{H}{\overset{R}{C}} < \underset{H}{\overset{R}{C}}=\underset{R}{\overset{H}{C}} \quad (4.13b)$$

PROBLEMS

4.16   Within each series arrange the compounds in order of increasing stability:

(a)

(b)

4.17   Alkenes can undergo the addition of hydrogen in the presence of certain catalysts. (You will study this reaction in Sec. 4.9A.)

The $\Delta H°$ of this reaction, called the *enthalpy of hydrogenation*, can be measured very accurately and can serve as a source of heats of formation. Consider the following enthalpies of hydrogenation: (*E*)-3-hexene, -117.9 kJ mol^{-1} (28.2 kcal mol^{-1}); (*Z*)-3-hexene, -121.6 kJ mol^{-1} (29.1 kcal mol^{-1}). Calculate the heats of formation of these two alkenes, given that the $\Delta H_f°$ of hexane is -167.2 kJ mol^{-1} (40.0 kcal mol^{-1}).

4.6   ADDITION REACTIONS OF ALKENES

In the remainder of this chapter we consider three reactions of alkenes: the reaction with hydrogen halides; the reaction with hydrogen, called *catalytic hydrogenation*; and the reaction with water, called *hydration*. These reactions will be used to establish some important principles of chemical reactivity that are very useful in organic chemistry. We'll study other alkene reactions in Chapter 5.

The most characteristic type of alkene reaction is **addition** at the carbon–carbon double bond. The addition reaction can be represented generally as follows:

$$\text{C==C} + \text{X--Y} \longrightarrow -\overset{|}{\underset{X}{C}}-\overset{|}{\underset{Y}{C}}- \tag{4.14}$$

In an addition reaction, the carbon–carbon π bond of the alkene and the X—Y bond of the reagent are broken, and new C—X and C—Y bonds are formed.

PROBLEM

4.18   Give the structure of the addition product formed when ethylene reacts with each of the following reagents:

(a) H—I   (b) Br_2

(c) BH_3 (*Hint:* Each of the B—H bonds undergoes an addition to one molecule of ethylene. That is, three moles of ethylene react with one mole of BH_3.)

4.7   ADDITION OF HYDROGEN HALIDES TO ALKENES

The hydrogen halides H—F, H—Cl, H—Br, and H—I undergo addition to carbon–carbon double bonds to give products called *alkyl halides*, compounds in which a halogen is bound to a saturated carbon atom:

$$\underset{\textbf{2,3-dimethyl-2-butene}}{\overset{H_3C}{\underset{H_3C}{>}}C=C\overset{CH_3}{\underset{CH_3}{<}}} + \text{H--Br} \longrightarrow \underset{\substack{\textbf{2,3-dibromo-2,3-dimethylbutane}\\ \text{(an alkyl halide)}}}{H_3C-\overset{\overset{CH_3}{|}}{\underset{\underset{H}{|}}{C}}-\overset{\overset{CH_3}{|}}{\underset{\underset{Br}{|}}{C}}-CH_3} \tag{4.15}$$

Although the addition of HF has been used for making alkyl fluorides, HF is extremely hazardous and is avoided whenever possible. Additions of HBr and HI are generally preferred to addition of HCl because additions of HBr and HI are faster.

A. Regioselectivity of Hydrogen Halide Addition

When the alkene has an unsymmetrically located double bond, two constitutionally isomeric products are possible.

$$\underset{\textbf{1-hexene}}{H_2C=CH(CH_2)_3CH_3} + \text{HI} \longrightarrow \underset{\substack{\textbf{2-iodohexane}\\ \text{(observed)}}}{H_3C-\underset{\underset{I}{|}}{CH}(CH_2)_3CH_3} \text{ or } \underset{\substack{\textbf{1-iodohexane}\\ \text{(not observed)}}}{CH_2-\underset{\underset{I}{|}}{CH_2}(CH_2)_3CH_3} \tag{4.16}$$

As shown in Eq. 4.16, *only one* of the two possible products is formed from a 1-alkene in significant amount. Generally, *the main product is that isomer in which the halogen is bonded to the carbon of the double bond with the greater number of alkyl substituents, and the hydrogen is bonded to the carbon with the smaller number of alkyl substituents.*

$$H_2C{=}CH(CH_2)_3CH_3$$

H goes here   I goes here

(Another way to think about this result is to apply the old saying, "Them that has, gets." That is, the carbon with more hydrogens gains yet another hydrogen in the reaction.) When the products of a reaction could consist of more than one constitutional isomer, and when one of the possible isomers is formed in excess over the other, the reaction is said to be a **regioselective reaction**. Hydrogen halide addition to alkenes is a *highly regioselective reaction* because addition of the hydrogen halide across the double bond gives only one of the two possible constitutionally isomeric addition products.

When the two carbons of the alkene double bond have equal numbers of alkyl substituents, little or no regioselectivity is observed in hydrogen halide addition, even if the alkyl groups are of different size.

smaller alkyl group   larger alkyl group

$$HBr + CH_3CH{=}CHCH_2CH_3 \longrightarrow \underset{\substack{| \\ Br \quad H}}{CH_3CH{-}CHCH_2CH_3} + \underset{\substack{| \\ H \quad Br}}{CH_3CH{-}CHCH_2CH_3} \qquad (4.17)$$

2-pentene
(*E* or *Z*)

2-bromopentane   **3-bromopentane**
(nearly equal amounts)

Markovnikov's Rule

University of Moscow today

In his doctoral dissertation of 1869, the Russian chemist Vladimir Markovnikov (1838–1904; also spelled Markownikoff) proposed a "rule" for regioselective addition of hydrogen halides to alkenes. This rule, which has since become known as **Markovnikov's rule**, was originally stated as follows: "The halogen of a hydrogen halide attaches itself to the carbon of the alkene bearing the lesser number of hydrogens and greater number of carbons."

Markovnikov's higher education was in political science, economics, and law. During required organic chemistry courses in the Finance curriculum at the University of Kazan, he became infatuated with organic chemistry and eventually completed his now-famous doctoral dissertation. He was appointed to the chair of chemistry at the University of Moscow in 1873, where he was known not only for his chemistry, but also for his openness to students. He was forced to resign this position in 1893 because he would not sign an apology demanded of the faculty by a political official who had been insulted by a student. He was allowed, however, to continue working in the university for the duration of his life.

PROBLEM

4.19 Using the known regioselectivity of hydrogen halide addition to alkenes, predict the addition product that results from the reaction of:

(a) H—Cl with 2-methylpropene   (b) H—Br with 1-methylcyclohexene

B. Carbocation Intermediates in Hydrogen Halide Addition

For many years the regioselectivity of hydrogen halide addition had only an *empirical* (experimental) basis. By exploring the underlying reasons for this regioselectivity, we'll set the stage to develop a broader understanding of not only this reaction but many others as well.

A modern understanding of the regioselectivity of hydrogen halide addition begins with the fact that the overall reaction actually occurs *in two successive steps*. Let's consider each of these in turn.

In the first step, the electron pair in the π bond of an alkene is donated to the proton of the hydrogen halide. The electrons of the π bond react rather than the electrons of σ bonds because π electrons have the highest energy (Sec. 4.1A). As a result, the carbon–carbon double bond is *protonated* on a carbon atom. The other carbon becomes positively charged and electron-deficient:

$$\text{(4.18a)}$$

The species with a positively charged, electron-deficient carbon is called a **carbocation**, pronounced CAR-bo-CAT-ion. (The term **carbonium ion** was used in earlier literature.) The formation of the carbocation from the alkene is *a Brønsted acid–base reaction* (Sec. 3.4A) in which the π bond acts as a *Brønsted base* toward the *Brønsted acid* H—Br. The π bond is a *very weak* base. Nevertheless, it can be protonated to a small extent by a strong acid such as HBr.

The resulting carbocation is a powerful electron-deficient Lewis acid and is thus a potent electrophile. In the second step of hydrogen halide addition, the halide ion, which is a Lewis base, or nucleophile, reacts with the carbocation at its electron-deficient carbon atom:

$$\text{(4.18b)}$$

This is a *Lewis acid–base association reaction* (Sec. 3.1B).

The carbocations involved in hydrogen halide addition to alkenes are examples of **reactive intermediates** or **unstable intermediates**: species that react so rapidly that they never accumulate in more than very low concentration. Most carbocations are too reactive to be isolated except under special circumstances. Thus, carbocations cannot be isolated from the reactions of hydrogen halides and alkenes because they react very quickly with halide ions.

The complete description of a reaction pathway, including any reactive intermediates such as carbocations, is called the **mechanism** of the reaction. To summarize the two steps in the mechanism of hydrogen halide addition to alkenes:

1. A carbon of the π bond is protonated (a Brønsted acid–base reaction).
2. A halide ion reacts with the resulting carbocation (a Lewis acid–base association reaction).

Now that we understand the mechanism of hydrogen halide addition to alkenes, let's see how the mechanism addresses the question of regioselectivity. When the double bond of an alkene is not located symmetrically within the molecule, protonation of the double bond can occur in two distinguishable ways to give two different carbocations. For example, protonation of 2-methylpropene can give either the *tert*-butyl cation (Eq. 4.19a) or the isobutyl cation (Eq. 4.19b):

$$
\text{(4.19)}
$$

tert-butyl cation

isobutyl cation
(not formed)

(a)   (b)

These two reactions are in *competition*—that is, one can only happen at the expense of the other because the two reactions compete for the same starting material. Only the *tert*-butyl cation is formed in this reaction. The *tert*-butyl cation is formed exclusively because reaction 4.19a is *much faster* than reaction 4.19b. Because the *tert*-butyl cation is the only carbocation formed, it is the only carbocation available to react with the bromide ion. Hence, the only product of HBr addition to 2-methylpropene is *tert*-butyl bromide.

$$
\text{(4.20)}
$$

tert-butyl bromide

Notice that the bromide ion has become attached to the carbon of 2-methylpropene bearing the greater number of alkyl groups. In other words, *the regioselectivity of hydrogen halide addition is due to the formation of only one of two possible carbocations.*

To understand why the *tert*-butyl cation is formed more rapidly than the isobutyl cation in HBr addition, we need to understand the factors that influence reaction rate. The relative stability of carbocations plays an important role in understanding the rate of HBr addition. A discussion of carbocation stability, then, is an essential prelude to a more general discussion of reaction rates.

C. Structure and Stability of Carbocations

Carbocations are classified by the degree of alkyl substitution at their electron-deficient carbon atoms.

$$
\text{(4.21)}
$$

a *primary*
carbocation

a *secondary*
carbocation

a *tertiary*
carbocation

TABLE 4.2 Heats of Formation of the Isomeric Butyl Cations (Gas Phase, 25 °C)

Cation structure	Name	Heat of formation		Relative energy*	
		kJ mol⁻¹	kcal mol⁻¹	kJ mol⁻¹	kcal mol⁻¹
$CH_3CH_2CH_2\overset{+}{C}H_2$	butyl cation	845	202	155	37
$(CH_3)_2CH\overset{+}{C}H_2$	isobutyl cation	828	198	138	33
$CH_3\overset{+}{C}HCH_2CH_3$	*sec*-butyl cation	757	181	67	16
$(CH_3)_3\overset{+}{C}$	*tert*-butyl cation	690	165	(0)	(0)

* Energy difference between each carbocation and the more stable *tert*-butyl cation

That is, primary carbocations have one alkyl group bound to the electron-deficient carbon, secondary carbocations have two, and tertiary carbocations have three. For example, the isobutyl cation in Eq. 4.19b is a primary carbocation, and the *tert*-butyl cation in Eq. 4.19a is a tertiary carbocation.

The gas-phase heats of formation of the isomeric butyl carbocations are given in Table 4.2. The data in this table show that *alkyl substituents at the electron-deficient carbon strongly stabilize carbocations.* (A comparison of the first two entries shows that substituents at other carbons have a much smaller effect on stability.) The relative stability of isomeric carbocations is therefore as follows.

Stability of carbocations: tertiary > secondary > primary   (4.22)

(Remember that "greater stability" means "lower energy.")

To understand the reasons for this stability order, consider first the geometry and electronic structure of carbocations, shown in Fig. 4.10 for the *tert*-butyl cation. The electron-deficient carbon of the carbocation has *trigonal planar* geometry (Sec. 1.3B) and is therefore sp^2-hybridized, like the carbons involved in double bonds (Sec. 4.1A); however, in a carbocation, the $2p$ orbital on the electron-deficient carbon contains no electrons.

The explanation for the stabilization of carbocations by alkyl substituents is in part the same as the explanation for the stabilization of alkenes by alkyl substitution (Sec. 4.5B)—the greater number of sp^2–sp^3 carbon–carbon bonds in a carbocation with a greater number of alkyl substituents. However, if you compare the data in Tables 4.1 (p. 150) and 4.2, you'll notice that each alkyl branch stabilizes an alkene by about 7 kJ mol⁻¹, but each branch stabilizes a carbocation by nearly 70 kJ mol⁻¹. In other words, the stabilization of carbocations by alkyl substituents is considerably greater than the stabilization of alkenes.

An additional factor that accounts for the stabilization of carbocations by alkyl branching is a phenomenon called **hyperconjugation**, which is the overlap of bonding electrons from the adjacent σ bonds with the unoccupied $2p$ orbital of the carbocation.

FIGURE 4.10 Hybridization and geometry of the *tert*-butyl cation. Notice the trigonal planar geometry and the unoccupied $2p$ orbital perpendicular to the plane of the three carbons. The C—C bond length, determined in 1995 by X-ray crystallography, is less than the sp^2–sp^3 C—C bond length in propene (1.50 Å) because of hyperconjugation, which is discussed in the text.

In this diagram, the σ bond that provides the bonding electrons is a neighboring C—H bond. The energetic advantage of hyperconjugation is that it involves *additional bonding*. That is, the electrons in the C—H bonds participate in bonding not only with the C and H, but also with the electron-deficient carbon. Additional bonding is a stabilizing effect. We can show this additional bonding with resonance structures as follows:

$$\tag{4.23}$$

The shared electrons are shown in color. (Remember the meaning of resonance: the carbocation is a *single* species that has some characteristics of both resonance structures. Therefore, the proton in the right-hand structure hasn't moved; it's still part of the molecule.) The double-bond character suggested by the resonance structure on the right is reflected in the lengths of the carbon–carbon bonds in the *tert*-butyl cation. These bonds are considerably shorter (1.442 Å) than the carbon–carbon single bond in propene (1.501 Å).

We can draw analogous resonance structures for a C—H bond in each methyl group. In other words, each alkyl branch provides additional hyperconjugation and thus more stabilization. Consequently, alkyl substitution at the electron-deficient carbon stabilizes carbocations.

Let's now bring together what you've learned about carbocation stability and the mechanism of hydrogen halide addition to alkenes. The addition occurs in two steps. In the first step, protonation of the alkene double bond occurs *at the carbon with the fewer alkyl substituents* so that the more stable carbocation is formed—that is, *the one with the greater number of alkyl substituents at the electron-deficient carbon*. The reaction is completed when the halide ion, in a second step, reacts with the electron-deficient carbon.

An understanding of many organic reactions hinges on an understanding of the reactive intermediates involved. Carbocations are important reactive intermediates that occur not only in the mechanism of hydrogen halide addition, but in the mechanisms of many other reactions as well. Hence, your knowledge of carbocations will be put to use often.

FURTHER EXPLORATION 4.3
Molecular Orbital Description of Hyperconjugation

George Olah, a Holiday Party, and a Nobel Prize

Because carbocations are in most cases reactive intermediates that are too unstable to isolate, they remained hypothetical for many years after their existence was first postulated. However, their importance as reactive intermediates led to repeated, unsuccessful efforts to prepare them. In 1966–67, a team of researchers led by Professor George A. Olah (b. 1927), then at Case Western Reserve University, figured out how to prepare a number of pure carbocation salts in solution and studied their properties. For example, they formed a solution of essentially pure *tert*-butyl cation by protonation of 2-methylpropene at −80 °C. As the acid, they used HF in the presence of the powerful Lewis acid SbF_5, which actually forms the very strong acid $H^+ \, {}^-SbF_6$.

$$(CH_3)_2C\!\!=\!\!CH_2 \; + \; H^+ \, {}^-SbF_6 \quad \xrightarrow[\substack{ClSO_2F \\ (solvent)}]{-80°\,C} \quad (CH_3)_3C^+ \, {}^-SbF_6 \tag{4.24}$$

2-methylpropene   ***tert*-butyl cation**
hexafluoroantimonate salt

The fluoride ion is so tightly held within the $^-SbF_6$ complex ion that it can't act as a nucleophile towards the *tert*-butyl cation.

The discovery of this method by Olah's group was somewhat serendipitous. In 1966, his group had only recently discovered $HSbF_6$, which they called "magic acid." Following a holiday party, they put a piece of a holiday candle into magic acid. When they saw that it dissolved, they examined the solution in a nuclear magnetic resonance (NMR) instrument. (You'll learn about NMR as a powerful method for determining structure in Chapter 13.) They saw unmistakable evidence for carbocations. This was the beginning of a very productive series of investigations in which a number of carbocations were generated and examined structurally. Subsequently, Olah joined the faculty of the University of Southern California. He received the 1994 Nobel Prize in Chemistry for his work in carbocation chemistry.

PROBLEMS

4.20   Classify the isomeric carbocations in each of the following parts as primary, secondary, or tertiary, and tell which is the most stable carbocation in each part and why.

(a)

$$\underset{A}{\overset{\displaystyle CH_3}{\underset{\displaystyle CH_3}{H_3C\!-\!\overset{|}{\underset{|}{C}}\!-\!\overset{+}{C}H_2}}} \qquad \underset{B}{\overset{\displaystyle +}{\underset{\displaystyle CH_3}{H_3C\!-\!\overset{+}{\underset{|}{C}}\!-\!CH_2CH_3}}} \qquad \underset{C}{\overset{\displaystyle H}{\underset{\displaystyle CH_3}{H_3C\!-\!\overset{|}{\underset{|}{C}}\!-\!\overset{+}{C}HCH_3}}}$$

(b)

A $\qquad$ B $\qquad$ C

4.21   By writing the curved-arrow mechanism of the reaction, predict the product of the reaction of HBr with 2-methyl-1-pentene.

STUDY PROBLEM   4.9

Give the structure of an alkene that would give 2-bromopentane as the major (or sole) product of HBr addition. (The numbers are for reference in the solution.)

$$an \; alkene \; + \; HBr \; \longrightarrow \; CH_3CH_2CH_2\overset{2}{\underset{\underset{\textstyle Br}{|}}{C}}\overset{1}{H}CH_3$$

2-bromopentane

SOLUTION   The bromine of the product comes from the H—Br. However, there are many hydrogens in the product! Which ones were there to start with, and which one came from the H—Br? First, recognize that the carbon bearing the bromine must have originally been one carbon of the double bond. It then follows that the other carbon of the double bond must be an *adjacent* carbon (because two carbons involved in the same double bond must be adjacent). Use this fact to construct *all possible alkenes* that *might* be starting materials. Do this by "thinking backward": Remove the bromine and a hydrogen from *each adjacent carbon in turn*.

Remove Br from carbon-2 and H from carbon-3 $\;\Rightarrow\; CH_3CH_2CH\!\!=\!\!CHCH_3$

2-pentene (cis or trans)

Remove Br from carbon-2 and H from carbon-1 $\;\Rightarrow\; CH_3CH_2CH_2CH\!\!=\!\!CH_2$

1-pentene

(The symbol $\Rightarrow$ means "implies as starting material.") Which of these is correct? Or are they both correct? You haven't finished the problem until you've mentally carried out the addition of HBr to *each compound*. Doing this and applying the known regioselectivity of HBr addition leads to the conclusion that the desired alkyl halide could be prepared as the major product from 1-pentene. However, both carbons of the double bond of 2-pentene bear the same number of alkyl groups. Eq. 4.17 (p. 153) indicates that from this starting material we should expect not only the desired product, but also a second product:

$$CH_3CH_2CH{=}CHCH_3 \;+\; HBr \;\longrightarrow\; CH_3CH_2CH_2CHCH_3 \;+\; CH_3CH_2CHCH_2CH_3$$

2-pentene   Br   Br

                                         2-bromopentane   **3-bromopentane**

Furthermore, the two products should be formed in nearly equal amounts. This means the yield of the desired compound would be relatively low and it would be difficult to separate from its isomer, which has almost the same boiling point. Consequently, 1-pentene is the only alkene that will give the desired alkyl halide as the *major* product (that is, the one formed almost exclusively).

In solving this type of problem, it isn't enough to identify potential starting materials. You must also determine whether they really will work, given the known characteristics—in this case, the regioselectivity—of the reaction.

PROBLEM

4.22   In each case, give *two different* alkene starting materials that would react with H—Br to give the compound shown as the major (or only) addition product.

(a)   Br   (b)   Me

                                                Br

D. Carbocation Rearrangement in Hydrogen Halide Addition

In some cases, the addition of a hydrogen halide to an alkene gives an unusual product, as in the following example.

$$\underset{\substack{|\\ CH_3}}{\overset{\substack{CH_3\\|}}{H_3C{-}C}}{-}CH{=}CH_2 + HCl \;\longrightarrow\; \underset{\substack{|\\ CH_3}}{\overset{\substack{CH_3\\|}}{H_3C{-}C}}{-}\underset{\substack{|\\ Cl}}{CH}{-}CH_3 \;+\; \underset{\substack{|\\ Cl}}{\overset{\substack{CH_3\\|}}{H_3C{-}C}}{-}\underset{\substack{|\\ CH_3}}{CH}{-}CH_3 \qquad (4.25)$$

(17% of product)   (83% of product)

The minor product is the result of ordinary regioselective addition of HCl across the double bond. The origin of the major product, however, is not obvious. Examination of the carbon skeleton of the major product shows that a *rearrangement* has occurred. In a **rearrangement**, a group from the starting material has moved to a different position in the product. In this case, a methyl group of the alkene (*color*) has changed positions. As a result, the carbons of the alkyl halide product are connected differently from the carbons of the alkene starting material. Although the rearrangement leading to the second product may seem strange at first sight, it is readily understood by considering the fate of the carbocation intermediate in the reaction.

The reaction begins like a normal addition of HCl—that is, by protonation of the double bond to yield the carbocation with the greater number of alkyl substituents at the electron-deficient carbon.

$$\underset{\substack{|\\ CH_3}}{\overset{\substack{CH_3\\|}}{H_3C{-}C}}{-}CH{=}CH_2 \; H{-}\ddot{C}l{:} \;\longrightarrow\; \underset{\substack{|\\ CH_3}}{\overset{\substack{CH_3\\|}}{H_3C{-}C}}{-}\overset{+}{C}H{-}CH_3 \;+\; {:}\ddot{C}l{:}^- \qquad (4.26)$$

a secondary carbocation

Reaction of this carbocation with Cl⁻ occurs, as expected, to yield the minor product of Eq. 4.25. However, the carbocation can also undergo a second type of reaction: it can *rearrange*.

$$H_3C-\underset{\underset{CH_3}{|}}{\overset{\overset{CH_3}{|}}{C}}-\overset{+}{C}H-CH_3 \longrightarrow H_3C-\underset{\underset{CH_3}{|}}{\overset{\overset{CH_3}{|}}{\overset{+}{C}}}-CH-CH_3 \qquad (4.27a)$$

a secondary carbocation   a tertiary carbocation

FURTHER EXPLORATION 4.4
A Stepwise View of Rearrangement

In this reaction, the methyl group moves *with its pair of bonding electrons* from the carbon *adjacent* to the electron-deficient carbon. The carbon from which this group departs, as a result, becomes electron-deficient and positively charged. That is, the rearrangement converts one carbocation into another. This is essentially a *Lewis acid–base reaction* in which the electron-deficient carbon is the Lewis acid and the migrating group *with its bonding electron pair* is the Lewis base. The reaction forms a new Lewis acid—the electron-deficient carbon of the rearranged carbocation.

The major product of Eq. 4.25 is formed by the Lewis acid–base association reaction of Cl⁻ with the new carbocation.

$$H_3C-\underset{\overset{+}{|}}{\overset{\overset{CH_3}{|}}{C}}-CH(CH_3)_2 + :\overset{..}{\underset{..}{Cl}}:^- \longrightarrow H_3C-\underset{\underset{:\overset{..}{\underset{..}{Cl}}:}{|}}{\overset{\overset{CH_3}{|}}{C}}-CH(CH_3)_2 \qquad (4.27b)$$

Why does rearrangement of the carbocation occur? In the case of reaction 4.27a, a more stable tertiary carbocation is formed from a less stable secondary one. Therefore, *rearrangement is favored by the increased stability of the rearranged ion.*

$$H_3C-\underset{\underset{CH_3}{|}}{\overset{\overset{CH_3}{|}}{C}}-CH=CH_2 + HCl \longrightarrow H_3C-\underset{\underset{CH_3}{|}}{\overset{\overset{CH_3}{|}}{C}}-\overset{+}{C}H-CH_3 \quad Cl^- \qquad \text{(a secondary carbocation)}$$

(4.28)

rearrangement ⟵ competing pathways ⟶ reaction with Cl⁻

(a tertiary carbocation)
$$H_3C-\underset{\underset{Cl^-}{\overset{+}{|}}}{\overset{\overset{CH_3}{|}}{C}}——\underset{\underset{CH_3}{|}}{\overset{\overset{H}{|}}{C}}-CH_3$$

$$H_3C-\underset{\underset{CH_3}{|}}{\overset{\overset{CH_3}{|}}{C}}——\underset{\underset{Cl}{|}}{\overset{}{C}}H-CH_3$$

minor product

↓ reaction with Cl⁻

$$H_3C-\underset{\underset{Cl}{|}}{\overset{\overset{CH_3}{|}}{C}}——\underset{\underset{CH_3}{|}}{\overset{\overset{H}{|}}{C}}-CH_3$$

major product

You've now learned two pathways by which carbocations can react. They can (1) react with a nucleophile and (2) rearrange to more stable carbocations. The outcome of Eq. 4.25 represents a competition between these two pathways. In any particular case, one cannot predict exactly how much of each different product will be obtained. Nevertheless, the reactions of carbocation intermediates show why both products are reasonable.

Carbocation rearrangements are not limited to the migrations of alkyl groups. In the following reaction, the major product is also derived from the rearrangement of a carbocation intermediate. This rearrangement involves a **hydride shift**, the migration of a hydrogen *with its two bonding electrons.*

$$\underset{\substack{|\\H}}{\overset{\substack{CH_3\\|}}{H_3C-C-CH=CH_2}} + HBr \longrightarrow \underset{\substack{|\\Br}}{\overset{\substack{CH_3\\|}}{H_3C-CH-CH-CH_3}} + \underset{\substack{|\\Br}}{\overset{\substack{CH_3\\|}}{H_3C-C-CH_2CH_3}} \qquad (4.29)$$

(about 45% of product)   (about 55% of product)

The hydride migrates instead of an alkyl group because the rearranged carbocation is tertiary and thus is more stable than the starting carbocation. Migration of an alkyl group would have given another secondary carbocation.

Keep in mind the following points about the rearrangement of carbocation intermediates, all of which are illustrated by the examples in this section.

1. A rearrangement almost always occurs when a more stable carbocation can result.
2. A rearrangement that would give a less stable carbocation generally doesn't occur.
3. The group that migrates in a carbocation rearrangement comes from a carbon *directly attached* to the electron-deficient, positively charged carbon of the carbocation.
4. The group that migrates in a rearrangement is typically an alkyl group, aryl group (p. 79), or a hydrogen.
5. When there is a choice between the migration of an alkyl group (or aryl group) or a hydrogen from a particular carbon, hydride migration typically occurs because it gives the more stable carbocation.

The First Description of Carbocation Rearrangements

The first clear formulation of the involvement of carbocations in molecular rearrangements was proposed by Frank C. Whitmore (1887–1947) of Pennsylvania State University. (Such rearrangements were once called "Whitmore shifts.") Whitmore said that carbocation rearrangements result when "an atom in an electron-hungry condition seeks its missing electron pair from the next atom in the molecule." Whitmore's description emphasizes the Lewis acid–base character of the reaction.

Carbocation rearrangements are not just a laboratory curiosity; they occur extensively in living organisms, particularly in the biological pathways leading to certain cyclic compounds such as steroids (Sec. 17.6C).

PROBLEMS

4.23 Which of the following carbocations is likely to rearrange? If rearrangement occurs, give the structures of the rearranged carbocations.

(a) [cyclohexane ring with H and CH₃ groups and + charge]

(b) $\underset{+}{\overset{\substack{CH_3 \;\; CH_3\\| \qquad |}}{CH_3CH-C-CH_3}}$

(c) [branched structure with + charge]

continued

continued ──

4.24   Draw the curved-arrow mechanism for the reaction in Eq. 4.29 that accounts for the formation of both products.

4.25   Only one of the following three alkyl halides can be prepared as the *major* product of the addition of HBr to an alkene. Which compound can be prepared in this way? Explain why the other two *cannot* be prepared in this way.

$$CH_3CH_2CH_2CH_2CH_2Br \qquad\qquad \underset{\displaystyle A}{} \qquad\qquad CH_3\underset{|}{\overset{Br}{C}}HCH_2CH_2CH_3 \qquad\qquad \underset{\displaystyle B}{} \qquad\qquad H_3C\!-\!\underset{|}{\overset{Br}{C}}H\!-\!\underset{\underset{\displaystyle CH_3}{|}}{\overset{\overset{\displaystyle CH_3}{|}}{C}}\!-\!C_2H_5$$

$$\underset{\displaystyle C}{}$$

───

4.8   REACTION RATES

Whenever a reaction can give more than one possible product, two or more reactions are in competition. (You've already seen examples of competing reactions in hydrogen halide addition to alkenes.) One reaction predominates when it occurs *more rapidly* than other competing reactions. Understanding why some reactions occur in preference to others, then, is often a matter of understanding the *rates* of chemical reactions. The theoretical framework for discussing reaction rates is the subject of this section. Although we'll use hydrogen halide addition to alkenes as our example to develop the theory, the general concepts will be used throughout this text.

A. The Transition State

The **rate** of a chemical reaction can be defined for our purposes as the number of reactant molecules converted into product in a given time. The theory of reaction rates used by many organic chemists postulates that as the reactants change into products, they pass through an unstable state of maximum free energy, called the **transition state**. The transition state has a higher energy than either the reactants or products and therefore represents an **energy barrier** to their interconversion. This energy barrier is shown graphically in a **reaction free-energy diagram** (Fig. 4.11). This is a diagram of the standard free energy of a reacting system as old bonds break and new ones form along the reaction pathway. In this diagram the progress

(a) larger energy barrier, slower reaction

(b) smaller energy barrier, faster reaction

FIGURE 4.11 Reaction free-energy diagrams for two hypothetical reactions. The standard free energy of activation ($\Delta G^{\circ\ddagger}$), shown for the forward reaction, is the energy barrier that must be overcome for the reaction to occur. The reaction in part (a) is intrinsically slower because it has a larger $\Delta G^{\circ\ddagger}$ than the one in part (b).

of reactants to products is called the **reaction coordinate**. That is, the reactants define one end of the reaction coordinate, the products define the other, and the transition state is at the energy maximum somewhere in between. The energy barrier $\Delta G^{\circ\ddagger}$, called the **standard free energy of activation**, is equal to the difference between the standard free energies of the transition state and the reactants. (The double dagger, $\ddagger$, is the symbol used for transition states.) The size of the energy barrier $\Delta G^{\circ\ddagger}$ determines the rate of a reaction: *the higher the barrier, the smaller the rate.* Thus, the reaction diagrammed in Fig. 4.11a is slower than the one in Fig. 4.11b because it has a larger energy barrier. In the same sense that relative free energies of reactants and products determine the equilibrium constant, the relative free energies of the transition state and the reactants determine the reaction rate.

Notice from Fig. 4.11 that *a reaction and its reverse have the same transition state.* An analogy is that if a certain mountain pass is the shortest way to get from one town to another, then the same mountain pass is the shortest way to make the reverse journey.

If the transition state is of central importance in determining the reaction rate, it would be nice to have a way of estimating its energy. Because it lies at an energy *maximum*, a transition state can't be isolated. However, the energy of a transition state, like the energy of any ordinary molecule, depends on its structure. So, what does the transition state look like? The power of transition-state theory is that we can visualize the transition state as a *structure*. To illustrate, consider the following Brønsted acid–base reaction.

$$(CH_3)_2C \!\! = \!\! CH_2 \ + \ H \!\! - \!\! \ddot{Br}\!: \ \rightleftharpoons \ (CH_3)_2\overset{+}{C} \!\! - \!\! CH_2 \quad :\!\ddot{Br}\!:^{\!\!-} \qquad (4.30)$$
$$\qquad\qquad\qquad\qquad\qquad\qquad\qquad\qquad |$$
$$\qquad\qquad\qquad\qquad\qquad\qquad\qquad\quad H$$

You should recognize this as the first step in the addition of HBr to an alkene; see Eq. 4.18a, p. 154. In the transition state of this reaction, the H—Br bond and the carbon–carbon π bond are partially broken, the new C—H bond is partially formed, and the new charges are only partially established. We can represent this situation using dashed lines for partial bonds and using $\delta+$ and $\delta-$ for partial charges, as follows:

$$\left[(CH_3)_2\overset{\delta+}{C} \text{----} CH_2 \atop { \atop {H \atop { \atop :\!\overset{\delta-}{\ddot{Br}}\!:}}} \right]^{\ddagger} \qquad (4.31)$$

This shows the bonds breaking and forming. If we view this as an event frozen in time, we're looking at the structure of the transition state.

An Analogy for the Transition State

Suppose you do a somersault off of a high diving board at the local pool. You have someone take your picture at the height of your jump with a very fast shutter. You can think of that picture as the transition state for your dive. It's the "in-between" state that defines the point of highest potential energy between your starting point on the diving board and your finish, when you've come to rest in the water. You're never at the transition state for more than an instant, but we don't have any problem conceptualizing how you would appear at that instant. Now imagine repeating your dive an Avogadro's number of times (you never get tired) and having a picture taken of each dive at the highest point. Now you average all those pictures. Because each dive is a little different, the "averaged" picture of the transition state is a bit blurry. This averaged picture of your jump is more like what we are dealing with when we talk about the transition state for a mole of molecules, but chances are that the first picture you took is not far from the average. So, we describe the transition state with a single structure, just as we describe your large number of dives with a single picture.

4.26 Draw curved-arrow mechanisms and transition-state structures for each of the following two reactions. Each reaction occurs as a single step.

(a) $CH_3CH_2 \overset{..}{\underset{..}{Br}}: + \overset{..}{:}\overset{..}{O}CH_3 \longrightarrow CH_3CH_2 \overset{..}{\underset{..}{O}}CH_3 + :\overset{..}{\underset{..}{Br}}:^{-}$ (b) $(CH_3)_3C \overset{..}{\underset{..}{Br}}: \longrightarrow (CH_3)_3\overset{+}{C} + :\overset{..}{\underset{..}{Br}}:^{-}$

4.27 (a) Draw the transition state for the reverse reaction of Eq. 4.30. Compare it with the transition state shown in Eq. 4.31.

(b) What general statement can you make about the transition-state structures for a reaction and its reverse?

B. The Energy Barrier

We've learned that the standard free energy of the transition state (relative to the standard free energy of reactants and products) defines the free-energy barrier $\Delta G^{\circ\ddagger}$ for the reaction. Let's learn a little more about the relationship between the size of this energy barrier and rate.

The relationship between rate and standard free energy of activation is an exponential one.

$$\text{rate} \propto e^{-\Delta G^{\circ\ddagger}/RT} = 10^{-\Delta G^{\circ\ddagger}/2.3RT} \tag{4.32}$$

where R = the gas constant (8.31×10^{-3} kJ K^{-1} mol^{-1}, or 1.99 kcal K^{-1} mol^{-1}) and T is the absolute temperature (K). (The sign $\propto$ means "is proportional to.") The negative sign in the exponent means that large values of $\Delta G^{\circ\ddagger}$—that is, large energy barriers—result in a smaller rate, as shown in Fig. 4.11, p. 162. It follows that, if two reactions A and B have standard free energies of activation $\Delta G_A^{\circ\ddagger}$ and $\Delta G_B^{\circ\ddagger}$, respectively, then under standard conditions (all reactants at 1 M concentration), the relative rates of the two reactions are

$$\frac{\text{rate}_A}{\text{rate}_B} = \frac{10^{-\Delta G_A^{\circ\ddagger}/2.3RT}}{10^{-\Delta G_B^{\circ\ddagger}/2.3RT}} = 10^{(\Delta G_B^{\circ\ddagger} - \Delta G_A^{\circ\ddagger})/2.3RT} \tag{4.33a}$$

or

$$\log\left(\frac{\text{rate}_A}{\text{rate}_B}\right) = \frac{\Delta G_B^{\circ\ddagger} - \Delta G_A^{\circ\ddagger}}{2.3RT} \tag{4.33b}$$

These equations show that the rates of two reactions differ by a factor of 10 (that is, one log unit) for every increment of $2.3RT$ (5.7 kJ mol^{-1}, or 1.4 kcal mol^{-1}, at 298 K) difference in their standard free energies of activation. A factor of 10 in rate is roughly the difference in rate between a reaction that takes an hour and one that takes 10 hours. This means that reaction rates are *very sensitive* to their standard free energies of activation.

You may have learned about energy barriers if you studied reaction rates in general chemistry, and there you may have referred to the energy barrier by the name *activation energy* and the abbreviation E_a or E_{act}. The activation energy is very close to the $\Delta H^{\circ\ddagger}$ of the reaction (the *standard enthalpy difference*) rather than the standard free-energy difference between the transition state and starting materials. It is possible mathematically to relate transition-state theory to the theory of activation energy. (See Further Exploration 4.5.) However, for our discussion, the difference between the two theories is not conceptually important.

FURTHER EXPLORATION 4.5
Activation Energy

Where do molecules get enough energy to overcome the energy barrier? In general, molecules obtain this energy from thermal motions. The energy of a collection of molecules is characterized by a distribution (Fig. 4.12a), which is termed a **Maxwell–Boltzmann distribution**. An analogy is the results of an exam by a distribution of grades (Fig. 4.12b). The rate of a reaction is directly related to the fraction of molecules that has enough energy to cross the energy barrier. This fraction is shown in Fig. 4.12a as a hatched area. The smaller the barrier, the larger the hatched area will be and the greater the reaction rate will be. Analogously, the fraction of students who receive an "A" on an exam depends on the "cutoff" imposed by the instructor; the lower the "cutoff" grade is, the more students receive an "A."

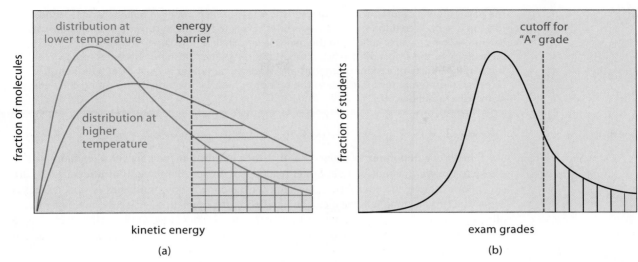

FIGURE 4.12 (a) A Maxwell–Boltzmann kinetic energy distribution at two different tempera-tures. (The right side of the distributions extend indefinitely and are cut off in the figure.) This is a plot of the number of molecules as a function of kinetic energy. The purple dashed line is the energy barrier. The fraction of molecules with enough energy to cross the barrier is given by the hatched areas. At higher temperature, the Maxwell–Boltzmann distribution is skewed to higher energy, and the fraction of molecules with enough energy to cross the barrier is greater (*red hatched area*). (b) The results of an exam can also be characterized by a distribution, which is a plot of the number of students as a function of exam grade. The "cutoff" (*purple dashed line*) defines the part of the distribution that receives an "A" grade. The fraction of students receiving an "A" is equal to the hatched area under the curve.

For a given reaction under a given set of conditions, we cannot control the size of the energy barrier; it is an intrinsic property of the reaction. Some reactions are intrinsically slow, and others are intrinsically fast. However, we can sometimes control the fraction of molecules with enough energy to cross the barrier. We can increase this fraction by *raising the temperature*. As shown in Fig. 4.12a, the Maxwell–Boltzmann distribution is skewed to higher energies at higher temperature, and, as a result, a greater fraction of molecules have the energy required to cross the barrier. In other words, reactions are faster at higher tempera-tures. Different reactions respond differently to temperature, although a *very rough* rule of thumb is that a reaction rate doubles for each 10 °C (or 10 K) increase in temperature.

Let's summarize. Two factors that govern the intrinsic reaction rate are

1. the size of the energy barrier, or standard free energy of activation $\Delta G^{\circ\ddagger}$: reactions with smaller $\Delta G^{\circ\ddagger}$ are faster (Fig. 4.11, p. 162);

2. the temperature: reactions are faster at higher temperatures.

An Analogy for Energy Barriers

An analogy that can help in visualizing these concepts is shown at left. Water in the cup would flow into the pan below if it could somehow gain enough kinetic energy to surmount the wall of the cup. The wall of the cup is a potential-energy barrier to the downhill flow of water. Likewise, molecules have to achieve a transient state of high energy—the transition state—to break stable chemical bonds and undergo reaction. An analogy to thermal motion is what happens if we shake the cup. If the cup is shallow (low energy barrier), the likelihood is good that the shaking will cause water to slosh over the sides of the cup and drop into the pan. This will occur at some characteristic rate—some number of milliliters per second. If the cup is very deep (high barrier), water is less likely to flow from cup to pan. Consequently, the rate at which water collects in the pan is lower. Shaking the cup more vigorously provides an analogy to the effect of increasing temperature. As the "sloshing" becomes more violent, the water acquires more kinetic energy, and water accumulates in the pan at a higher rate. Likewise, high temperature increases the rate of a chemical reaction by increasing the energy of the reacting molecules.

It is very important to understand that the equilibrium constant for a reaction tells us *absolutely nothing* about its rate. Some reactions with very large equilibrium constants are slow. For example, the equilibrium constant for the combustion of alkanes is very large; yet a container of gasoline (alkanes) can be handled in the open air because the reaction of gasoline with oxygen, in the absence of heat, is immeasurably slow. On the other hand, some unfavorable reactions come to equilibrium almost instantaneously. For example, the reaction of ammonia with water to give ammonium hydroxide has a very unfavorable equilibrium constant; but the small extent of reaction that does occur takes place very rapidly.

PROBLEMS

4.28   (a) The standard free energy of activation of one reaction A is 90 kJ mol^{-1} (21.5 kcal mol^{-1}). The standard free energy of activation of another reaction B is 75 kJ mol^{-1} (17.9 kcal mol^{-1}). Which reaction is faster and by what factor? Assume a temperature of 298 K.

(b) Estimate how much you would have to increase the temperature of the slower reaction so that it would have a rate equal to that of the faster reaction.

4.29   The standard free energy of activation of a reaction A is 90 kJ mol^{-1} (21.5 kcal mol^{-1}) at 298 K. Reaction B is one million times faster than reaction A at the same temperature. The products of each reaction are 10 kJ mol^{-1} (2.4 kcal mol^{-1}) more stable than the reactants.

(a) What is the standard free energy of activation of reaction B?

(b) Draw reaction free-energy diagrams for the two reactions showing the two values of $\Delta G^{\circ\ddagger}$ to scale.

(c) What is the standard free energy of activation of the reverse reaction in each case?

C. Multistep Reactions and the Rate-Limiting Step

Many chemical reactions take place with the formation of reactive intermediates. Such reactions are called **multistep reactions**. We use this terminology because, when intermediates exist in a chemical reaction, then what we commonly express as one reaction is really a sequence of two or more reactions. For example, you've already learned that the addition of hydrogen halides to alkenes involves a carbocation intermediate. This means, for example, that the following addition of HBr to 2-methylpropene

$$(CH_3)_2C{=}CH_2 + HBr \longrightarrow (CH_3)_3C{-}Br \tag{4.34}$$

is a multistep reaction involving the following two steps:

$$(CH_3)_2C{=}CH_2 + HBr \rightleftarrows (CH_3)_3C^+ + Br^- \tag{4.35a}$$

$$(CH_3)_3C^+ + Br^- \longrightarrow (CH_3)_3C{-}Br \tag{4.35b}$$

Each step of a multistep reaction has its own characteristic rate and therefore its own transition state. The energy changes in such a reaction can also be depicted in a reaction free-energy diagram. Such a diagram for the addition of HBr to 2-methylpropene is shown in Fig. 4.13. Each free-energy maximum between reactants and products represents a transition state, and the minimum represents the carbocation intermediate.

FIGURE 4.13 Reaction free-energy diagram for a multistep reaction. The rate-limiting step of a multistep reaction is the step with the transition state of highest standard free energy. In the addition of HBr to 2-methylpropene, the rate-limiting step is protonation of the double bond to give the carbocation intermediate.

Generally, the rate of a multistep reaction depends in detail on the rates of its various steps. However, it often happens that one step of a multistep reaction is considerably slower than any of the others. This slowest step in a multistep chemical reaction is called the **rate-limiting step**, or **rate-determining step**, of the reaction. In such a case, *the rate of the overall reaction is equal to the rate of the rate-limiting step*. In terms of the reaction free-energy diagram in Fig. 4.13, *the rate-limiting step is the step with the transition state of highest free energy*. This diagram indicates that in the addition of HBr to 2-methylpropene, the rate-limiting step is the first step of the reaction—the protonation of the alkene to give the carbocation. The overall rate of addition of HBr to 2-methylpropene is equal simply to the rate of this first step.

The rate-limiting step of a reaction has a special importance. Anything that increases the rate of this step increases the overall reaction rate. Conversely, if a change in the reaction conditions (for example, a change in temperature) affects the rate of the reaction, it is the effect on the rate-limiting step that is being observed. Because the rate-limiting step of a reaction has special importance, its identification receives particular emphasis when we attempt to understand the mechanism of a reaction.

An Analogy for the Rate-Limiting Step

A rate-limiting step can be illustrated by a toll station on a freeway at rush hour. We can think of the passage of cars through a toll booth as a multistep process: (1) entry of the cars into the toll area; (2) taking of the toll by the collector; and (3) exit of the cars from the toll area. Typically, paying the toll to the collector is the rate-limiting step in the passage of cars through the toll plaza. In other words, the rate of passage of cars through the toll plaza is the rate at which they pay their tolls. Cars can arrive more or less frequently, but as long as there is a line of cars, the rate of passage through the toll plaza is the same.

Installing automated change collectors generally increases the rate at which cars pass through a toll plaza. This strategy works because it increases the rate of the rate-limiting step. Increasing the speed limit at which cars can approach the toll plaza, on the other hand, would not increase the rate of passage through the toll plaza, because this change has no effect on the rate-limiting step.

Installing "E-ZPass," an electronic toll permit reader, increases the rate of toll payment even more. In fact, it's possible that with E-ZPass, toll payment is no longer the rate-limiting step. In this case the rate of passage through the toll plaza is limited by the first step, which is the rate at which cars enter the plaza. In this scenario, raising the speed limit of the approach would increase the rate, but increasing the speed of the E-Zpass monitor would have no effect.

4.30 Draw a reaction free-energy diagram for a reaction $A \rightleftharpoons B \rightleftharpoons C$ that meets the following criteria: The standard free energies are in the order $C < A < B$, and the rate-limiting step of the reaction is $B \rightleftharpoons C$.

4.31 Repeat Problem 4.30 for a case in which the standard free energies are in the order $A < C < B$, and the rate-limiting step of the reaction is $A \rightleftharpoons B$.

D. Hammond's Postulate

We've already learned that *transition states can be visualized as structures*. Recall (Eq. 4.31, p. 163) that, in the addition of HBr to an alkene, the transition state of the first step is visualized as a structure along the reaction pathway somewhere between the structures of the starting materials, the alkene and HBr, and the products of this step, the carbocation and a bromide ion:

$$(CH_3)_2C{=}CH_2 + H{-}\ddot{B}r\!: \; \rightleftharpoons \; \left[\begin{array}{c} (CH_3)_2\overset{\delta+}{C}\text{----}CH_2 \\ H \\ \overset{\delta-}{:}\!\ddot{B}r\!: \end{array}\right]^{\ddagger} \; \longrightarrow \; (CH_3)_2\overset{+}{C}{-}CH_2 \quad :\!\ddot{B}r\!:^{-} \qquad (4.36)$$

$$\underset{\text{transition state}}{} \qquad \qquad \underset{H}{}$$

transition state

What makes this transition state so unstable? First, the bonds undergoing transition are neither fully broken nor fully formed. The unstable bonding situation is why the transition state lies at an energy maximum. But additionally, a significant contribution to the high energy of the transition state comes from the same factors that account for the high energy of the carbocation. One factor is that the carbocation has one bond fewer than HBr and the alkene. Because bonding releases energy, this fact alone means that the carbocation has a considerably higher energy than starting materials (or products). The other factor is the separation of positive and negative charge. The electrostatic law (Eq. 3.44, p. 117) tells us that separation of opposite charges requires energy.

So, we've concluded that the energies of the transition state and the carbocation intermediate are very similar, and that the structural elements that account for the high energy of the carbocation also account for most of the transition-state energy. In view of these similarities, the following approximation seems justified: *The structure and energy of the transition state in Eq. 4.36 can be approximated by the structure and energy of the carbocation intermediate.* This approximation can be generalized in an important statement called *Hammond's postulate*:

> **Hammond's Postulate:** *For a reaction in which an intermediate of relatively high energy is either formed from reactants of much lower energy or converted into products of much lower energy, the structure and energy of the transition state can be approximated by the structure and energy of the intermediate itself.*

This postulate is named for George S. Hammond (1921–2005), who first stated it and applied it to organic reactions in 1955 while he was a professor of chemistry at Iowa State University. (This is not Hammond's exact statement of his postulate, but it will prove to be the most useful version of it for us.)

The utility of Hammond's postulate in dealing with reaction rates can be demonstrated by showing how we could have used it along with a knowledge of carbocation stability to predict the regioselectivity of HBr addition to 2-methylpropene. Recall (Sec. 4.8C) that the rate-limiting step in this reaction is the first step: protonation of the alkene by HBr to give a carbocation. As shown in Eqs. 4.19a and 4.19b, p. 155, this protonation could occur in two different and competing ways. Protonation of the double bond at one carbon gives the *tert*-butyl cation as the unstable intermediate; protonation of the double bond at the other carbon gives the isobutyl cation. We apply Hammond's postulate by assuming that *the structures and energies of the transition states are approximated by the structures and energies of the unstable intermediates—the carbocations—themselves.*

$$
\begin{array}{c}
H_3C \\
\qquad C{=}CH_2 + H{-}Br \\
H_3C
\end{array}
$$

similar structures and energies ← [‡] transition states [‡] → similar structures and energies

$$
\begin{array}{ccc}
H_3C & & CH_3 \\
\overset{+}{C}{-}CH_3 \quad Br^- & \qquad\qquad & H_3C{-}\overset{+}{\underset{H}{C}}{-}\overset{+}{C}H_2 \quad Br^- \\
H_3C & &
\end{array}
$$

(4.37)

tert-butyl cation
(tertiary, more stable)

isobutyl cation
(primary, much less stable)

Because the tertiary carbocation is more stable, *the transition state leading to the tertiary carbocation should also be the one of lower energy.* As a result, protonation of 2-methylpropene to give the tertiary carbocation has the transition state with the smaller free energy and is thus the faster of the two competing reactions (Fig. 4.14). Addition of HBr to alkenes is regioselective because protonation of a double bond to give a tertiary carbocation has a transition state of lower energy than the transition state for protonation to give a primary carbocation. *The stabilities of the carbocations themselves* do not determine which reaction is faster; *the relative free energies of the transition states for carbocation formation* determine the relative rates of the two processes. Only the validity of Hammond's postulate allows us to make the connection between carbocation energy and transition-state energy.

We need Hammond's postulate because the structures of transition states are uncertain, whereas the structures of reactants, products, and reactive intermediates are known. There-

FIGURE 4.14 A reaction free-energy diagram for the two possible modes of HBr addition to 2-methylpropene. Hammond's postulate states that the energy of each transition state is approximated by the energy of the corresponding carbocation. The formation of *tert*-butyl bromide (*right panel*) is faster than the formation of isobutyl bromide (*left panel*) because it involves the more stable carbocation intermediate and therefore the transition state of lower energy.

fore, knowing that a transition state resembles a particular species (for example, a carbocation) helps us to make a good guess about the transition-state structure. In this text, we'll frequently analyze or predict reaction rates by considering the structures and stabilities of reactive intermediates such as carbocations. When we do this, we are assuming that the transition states and the corresponding reactive intermediates have similar structures and energies; in other words, we are invoking Hammond's postulate.

PROBLEM

4.32   Apply Hammond's postulate to decide which reaction is faster: addition of HBr to 2-methylpropene or addition of HBr to *trans*-2-butene. Assume that the energy difference between the starting alkenes can be ignored. Why is this assumption necessary?

4.9   CATALYSIS

Some reactions take place much more rapidly in the presence of certain substances that are themselves left unchanged by the reaction. A substance that increases the rate of a reaction without being consumed is called a **catalyst**. A practical example of a catalyst occurs in the catalytic converter on the modern automobile. The catalyst in the converter brings about the rapid oxidation (combustion) of unburned hydrocarbons, the conversion of nitrogen oxides into nitrogen and oxygen, and the conversion of carbon monoxide into carbon dioxide. The catalyst increases the rates of these reactions by many orders of magnitude; they essentially do not occur in the absence of the catalyst. Despite its involvement in the reactions, the catalyst is left unchanged.

Here are some important points about catalysts.

1. A catalyst increases the reaction rate. This means that it lowers the standard free energy of activation for a reaction (Fig. 4.15).

2. A catalyst is not consumed. It may be consumed in one step of a catalyzed reaction, but if so, it is regenerated in a subsequent step.

An implication of points 1 and 2 is that a catalyst that strongly accelerates a reaction can be used in very small amounts. Many expensive catalysts are practical for this reason.

FIGURE 4.15 A reaction free-energy diagram comparing a hypothetical catalyzed reaction (*red curve*) to the uncatalyzed reaction (*blue curve*). Although both reactions are shown here as simple, one-step processes, a catalyzed reaction and the corresponding uncatalyzed reaction typically have different mechanisms and different numbers of reaction steps.

3. A catalyst does not affect the energies of reactants and products. In other words, a catalyst does not affect the $\Delta G°$ of a reaction and consequently also does not affect the equilibrium constant (Fig. 4.15).

4. A catalyst accelerates both the forward and reverse of a reaction by the same factor.

The last point follows from the fact that, at equilibrium, the rates of a reaction and its reverse are equal. If a catalyst does not affect the equilibrium constant (point 3) but increases the reaction rate in one direction, equality of rates at equilibrium requires that the rate of the reverse reaction must be increased by the same factor.

When a catalyst and the reactants exist in separate phases, the catalyst is called a **heterogeneous catalyst**. The catalyst in the catalytic converter of an automobile is a heterogeneous catalyst because it is a solid and the reactants are gases. In other cases, a reaction in solution may be catalyzed by a soluble catalyst. A catalyst that is soluble in a reaction solution is called a **homogeneous catalyst**.

A large number of organic reactions are catalyzed. In this section, we'll introduce the idea of catalysis by considering three examples of catalyzed alkene reactions. The first example, *catalytic hydrogenation*, is a very important example of heterogeneous catalysis. The second example, *hydration*, is an example of homogeneous catalysis. The last example involves catalysis of a biological reaction by an enzyme.

Catalyst Poisons, the Catalytic Converter, and Leaded Gasoline

Although in theory catalysts should function indefinitely, in practice many catalysts, particularly heterogeneous catalysts, slowly become less effective. It is as if they "wear out." One reason for this behavior is that they slowly absorb impurities, called catalyst poisons, from the surroundings; these impurities impede the functioning of the catalyst. An example of this phenomenon also occurs with the catalytic converter. Lead is a potent poison of the catalyst in a catalytic converter. This fact, as well as the desire to eliminate atmospheric lead pollution, are the major reasons why leaded gasoline is no longer used in automotive engines in the United States.

A. Catalytic Hydrogenation of Alkenes

When a solution of an alkene is stirred under an atmosphere of hydrogen, nothing happens. But if the same solution is stirred under hydrogen in the presence of a metal catalyst, the hydrogen is rapidly absorbed by the solution. The hydrogen is consumed because it undergoes an *addition* to the alkene double bond.

$$\text{cyclohexene} \;+\; H_2 \;\xrightarrow{\text{Pt/C catalyst}}\; \text{cyclohexane} \tag{4.38}$$

cyclohexene   **cyclohexane**

$$CH_3(CH_2)_5CH{=}CH_2 + H_2 \xrightarrow{\text{Pt/C catalyst}} CH_3(CH_2)_5CH_2{-}CH_3 \tag{4.39}$$

1-octene   **octane**

These reactions are examples of **catalytic hydrogenation**, an addition of hydrogen to an alkene in the presence of a catalyst. Catalytic hydrogenation is one of the best ways to convert alkenes into alkanes. Catalytic hydrogenation is an important reaction in both industry and the laboratory. The inconvenience of using a special apparatus for the handling of a flammable gas (dihydrogen) is more than offset by the great utility of the reaction.

In the preceding reactions, the catalyst is written over the reaction arrows. Pt/C is read as "Platinum supported on carbon" or simply "Platinum on carbon." This catalyst is a finely divided platinum metal that has been precipitated, or "supported," on activated charcoal. A number of noble metals, such as platinum, palladium, and nickel, are useful as hydrogenation catalysts, and they are often used in conjunction with solid support materials such as alumina (Al_2O_3), barium sulfate ($BaSO_4$), or, as in the previous examples, activated carbon. Hydrogenation can be carried out at room temperature and pressure or, for especially difficult cases, at higher temperature and pressure in a "bomb" (a closed vessel designed to withstand high pressures).

Because hydrogenation catalysts are insoluble in the reaction solution, they are examples of *heterogeneous catalysts*. (Soluble hydrogenation catalysts are also known and, although important, are not so widely used; Sec. 18.6D.) Even though they involve relatively expensive noble metals, heterogeneous hydrogenation catalysts are very practical because they can be reused. Furthermore, because they are exceedingly effective, they can be used in very small amounts. For example, typical catalytic hydrogenation reactions can be run with reactant-to-catalyst molar ratios of 100 or more.

How do hydrogenation catalysts work? Research has shown that both the hydrogen and the alkene must be adsorbed on the surface of the catalyst for a reaction to occur. The catalyst is believed to form reactive metal–carbon and metal–hydrogen bonds that ultimately are broken to form the products and to regenerate the catalyst sites. Beyond this, the chemical details of catalytic hydrogenation are not thoroughly understood. This is not a reaction for which a simple curved-arrow mechanism can be written.

The benzene ring is inert to conditions under which normal double bonds react readily:

$$\text{styrene} \xrightarrow{\text{Pt/C}} \text{ethylbenzene} \quad (4.40)$$

(Benzene rings can be hydrogenated, however, with certain catalysts under conditions of high temperature and pressure.) You will learn that many other alkene reactions do not affect the "double bonds" of a benzene ring. The relative inertness of benzene rings toward the conditions of alkene reactions was one of the great puzzles of organic chemistry that was ultimately explained by the theory of aromaticity, which is introduced in Chapter 15.

PROBLEMS

4.33   Give the product formed when each of the following alkenes reacts with a large excess of hydrogen in the presence of Pd/C.

(a) 1-pentene   (b) (*E*)-1,3-hexadiene

4.34   (a) Give the structures of five alkenes, each with the formula C_6H_{12}, that would give hexane as the product of catalytic hydrogenation.

(b) How many alkenes containing one double bond can react with H_2 over a Pt/C catalyst to give methylcyclopentane? Give their structures. (*Hint:* See Study Problem 4.9, p. 158.)

B. Hydration of Alkenes

The alkene double bond undergoes reversible addition of water in the presence of moderately concentrated strong acids such as H_2SO_4, $HClO_4$, and HNO_3.

$$\underset{\substack{\text{2-methylpropene}}}{\underset{\substack{H_3C \\ H_3C}}{\overset{H_3C}{\diagdown}}C{=}CH_2} + \underset{\substack{\text{(in excess;} \\ \text{used as solvent)}}}{H{-}OH} \quad \xrightleftharpoons{1\ M\ HNO_3} \quad \underset{\substack{\text{2-methyl-2-propanol} \\ (\textbf{\textit{tert}-butyl alcohol})}}{H_3C{-}\underset{\underset{OH}{|}}{\overset{\overset{CH_3}{|}}{C}}{-}CH_3} \qquad (4.41)$$

The addition of the elements of water is in general called **hydration**. Hence, the addition of water to the alkene double bond is called **alkene hydration**.

Hydration does not occur at a measurable rate in the absence of an acid, and the acid is not consumed in the reaction. Hence, alkene hydration is an *acid-catalyzed reaction*. Because the catalyzing acid is soluble in the reaction solution, it is a *homogeneous catalyst*.

Notice that this reaction, like the addition of HBr, is *regioselective*. As in the addition of HBr, the hydrogen adds to the carbon of the double bond with the smaller number of alkyl substituents. The more electronegative partner of the H—OH bond, the OH group, like the Br in HBr addition, adds to the carbon of the double bond with the greater number of alkyl substituents.

In this reaction, the manner in which the catalyst functions can be understood by considering the mechanism of the reaction, which is very similar to that of HBr addition. In the first step of the reaction, which is the rate-limiting step, the double bond is protonated so as to give the more stable carbocation. Because water is present, the actual acid is the hydrated proton (H_3O^+).

$$\text{(4.42a)}$$

This is a Brønsted acid–base reaction. Because this is the rate-limiting step, the rate of the hydration reaction increases when the rate of this step increases. The strong acid H_3O^+ is more effective than the considerably weaker acid water in protonating a weak base (the alkene). If a strong acid is not present, the reaction does not occur because water alone is too weak an acid to protonate the alkene.

In the next step of the hydration reaction, the nucleophile water combines with the carbocation in a Lewis acid–base association reaction:

$$(CH_3)_3C^+ \ :\!\ddot{O}H_2 \quad \xrightleftharpoons \quad (CH_3)_3C{-}\overset{+}{\ddot{O}}H_2 \qquad (4.42b)$$

Finally, a proton is lost to solvent in another Brønsted acid–base reaction to give the alcohol product and regenerate the catalyzing acid H_3O^+:

$$(CH_3)_3C{-}\overset{+}{\ddot{O}}H \quad \xrightleftharpoons \quad (CH_3)_3C{-}\ddot{O}H + H_3\overset{+}{\ddot{O}} \qquad (4.42c)$$

$pK_a = -1.7$

$pK_a \approx -2$

Notice three things about this mechanism.

1. It consists entirely of Lewis acid–base association–dissociation and Brønsted acid–base reactions.

2. Although the proton consumed in Eq. 4.42a is not the same as the one produced in Eq. 4.42c, there is no *net* consumption of protons.

3. The nucleophile in Eq. 4.42b and the Brønsted base in Eq. 4.42c is water. Some students are tempted to use hydroxide ion in a situation like this because it is a stronger base. However, there is *no* hydroxide in a 1 M nitric acid or sulfuric acid solution. Nor is hydroxide needed: the carbocation in Eq. 4.42b is *very* reactive— reactive enough to react rapidly with water; and, as we can see from the pK_a values in Eq. 4.42c, the acid on the left is strong enough to donate a proton to the weak base water. We can generalize this result as follows:

    Whenever H_3O^+ *acts as an acid, its conjugate base* H_2O *acts as the base.* (Read again about *amphoteric compounds* on p. 99 if this isn't clear.) More generally, *acids and their conjugate bases always act in tandem in acid–base catalysis.* If H_2O is the acid, then the base is ^-OH, and vice versa.

Because the hydration reaction involves carbocation intermediates, some alkenes give rearranged hydration products.

$$H_3C-\underset{\underset{CH_3}{|}}{\overset{\overset{H}{|}}{C}}-CH=CH_2 + H_2O \xrightarrow{H_3O^+} H_3C-\underset{\underset{CH_3}{|}}{\overset{\overset{OH}{|}}{C}}-CH_2-CH_3 \qquad (4.43)$$

PROBLEMS

4.35   Give the mechanism for the reaction in Eq. 4.43. Show each step of the mechanism separately with careful use of the curved-arrow notation. Explain why the rearrangement takes place.

4.36   The alkene 3,3-dimethyl-1-butene undergoes acid-catalyzed hydration with rearrangement. Use the mechanism of hydration and rearrangement to predict the structure of the hydration product of this alkene.

4.37   (a) Unlike the alcohol product Eq. 4.41, the product in Eq. 4.43 does *not* come to equilibrium with the starting alkene. However, it *does* come to equilibrium with two other alkenes. What are their structures?

  (b) Why isn't the alkene starting material in Eq. 4.43 part of the equilibrium mixture?

The equilibrium constants for many alkene hydrations are close enough to unity that the hydration reaction can be run in reverse. The reverse of alkene hydration is called *alcohol dehydration*. The direction in which the reaction is run depends on the application of **Le Châtelier's principle**, which states that if an equilibrium is disturbed, it will react so as to offset the disturbance. For example, if the alkene is a gas (as in Eq. 4.41), the reaction vessel can be pressurized with the alkene. The equilibrium reacts to the excess of alkene by forming more alcohol. Neutralization of the acid catalyst stops the reaction and permits isolation of the alcohol. This strategy is used particularly in industrial applications. One such application of alkene hydration is the commercial preparation of ethyl alcohol (ethanol) from ethylene:

$$H_2C=CH_2 + H_2O \xrightarrow[\text{300 °C}]{\substack{H_3PO_4 \\ \text{(adsorbed on a} \\ \text{solid support)}}} H_3C-CH_2-OH \qquad (4.44)$$

ethylene   **ethanol**

A high temperature is required because the hydration of ethylene is very slow at ordinary temperatures (see Problem 4.38). Recall (Sec. 4.8B) that increasing the temperature accelerates a reaction. This reaction was at one time a major source of industrial ethanol. Although it is still used, its importance has decreased as the fermentation of sugars from biomass (for example, corn) has become more prevalent (Sec. 10.13).

To run the hydration reaction in the reverse (dehydration) direction, the alkene is removed as it is formed, typically by distillation. (Alkenes have significantly lower boiling points than alcohols, as we'll further discuss in Sec. 8.5C.) The equilibrium responds by forming more alkene. Alcohol dehydration is more widely used than alkene hydration in the laboratory. We'll consider this reaction in Sec. 10.2.

Alkene hydration and alcohol dehydration illustrate two important points. First is one of the key points about catalysis: a catalyst accelerates the forward and reverse reactions of an equilibrium by the same factor. For example, because alkene hydration is acid-catalyzed, alcohol dehydration is acid-catalyzed as well. The second point is that alkene hydration and alcohol dehydration occur by the forward and reverse of the same mechanism. Generally, *if a reaction occurs by a certain mechanism, the reverse reaction under the same conditions occurs by the exact reverse of that mechanism.* This statement is called the **principle of microscopic reversibility**. Microscopic reversibility requires, for example, that if you know the mechanism of alkene hydration, then you know the mechanism of alcohol dehydration as well. A consequence of microscopic reversibility is that the rate-limiting transition states of a reaction and its reverse are the same. For example, if the rate-limiting step of alkene hydration is protonation of the double bond to form the carbocation intermediate (Eq. 4.42a), then the rate-limiting step of alcohol dehydration is the reverse of the same equation—deprotonation of the carbocation to give the alkene.

PROBLEMS

4.38   Explain why the hydration of ethylene (Eq 4.44) is a very slow reaction. (*Hint:* Think about the structure of the reactive intermediate and apply Hammond's postulate.)

4.39   Isopropyl alcohol is produced commercially by the hydration of propene. Show the mechanistic steps of this process. If you do not know the structure of isopropyl alcohol, try to deduce it by analogy from the structure of propene and the mechanism of alkene hydration.

C. Enzyme Catalysis

Catalysis is not limited to the laboratory or chemical industry. The biological processes of nature involve thousands of chemical reactions, most of which have their own unique naturally occurring catalysts. These biological catalysts are called **enzymes**. (The structures of enzymes are discussed in Sec. 27.10.) Under physiological conditions, most important biological reactions would be too slow to be useful in the absence of their enzyme catalysts. Enzyme catalysts are important not only in nature; they are used both in industry and in the laboratory.

Many of the best characterized enzymes are soluble in aqueous solution and hence are homogeneous catalysts. However, other enzymes are immobilized within biological substructures such as membranes and can be viewed as heterogeneous catalysts.

An example of an important enzyme-catalyzed addition to an alkene is the hydration of fumarate ion to malate ion.

fumarate

This reaction is catalyzed by the enzyme *fumarase*. It is one reaction in the Krebs cycle, or citric acid cycle, a series of reactions that plays a central role in the generation of energy in biological systems. Fumarase catalyzes only this reaction. The effectiveness of fumarase catalysis can be appreciated by the following comparison: At physiological pH and temperature (pH = 7, 37 °C), the enzyme-catalyzed reaction is about 10^9 (one *billion*) times faster than the same reaction in the absence of enzyme. To put the catalytic effectiveness of fumarase in greater perspective, the enzyme-catalyzed reaction occurs in a fraction of a second. The uncatalyzed reaction requires *hundreds of thousands of years*—in other words, it doesn't occur!

KEY IDEAS IN CHAPTER 4

- Alkenes are compounds containing carbon–carbon double bonds. Alkene carbon atoms, as well as other trigonal planar atoms, are sp^2-hybridized.

- The carbon–carbon double bond consists of a σ bond and a π bond. The π electrons are more reactive than the σ electrons and can be donated to Brønsted or Lewis acids.

- In the IUPAC substitutive nomenclature of alkenes, the principal chain, which is the carbon chain containing the greatest number of double bonds, is numbered so that the double bonds receive the lowest numbers.

- Because rotation about the alkene double bond does not occur under normal conditions, some alkenes can exist as double-bond stereoisomers. These are named using the *E,Z* priority system.

- The unsaturation number of a compound, which is equal to the number of rings plus double bonds in the compound, can be calculated from the molecular formula by Eq. 4.7 on p. 144.

- Heats of formation (enthalpies of formation), abbreviated ΔH_f°, can be used to determine the relative stabilities of various bonding arrangements. Heats of formation reveal that alkenes with more alkyl groups at their double bonds are more stable than isomers with fewer alkyl groups and that, in most cases, trans alkenes are more stable than their cis isomers.

- Reactants are converted into products through unstable species called transition states. A transition state of a reaction step can be approximated as a structure that is intermediate between reactants and products, and it can be drawn by using dashed bonds and partial charges.

- A reaction rate is determined by the standard free energy of activation $\Delta G^{\circ\ddagger}$, which is the standard free energy difference between the transition state and the reactants. Reactions with smaller standard free energies of activation are faster.

- The rates of reactions increase with increasing temperature.

- The rates of multistep reactions are determined by the rate of the slowest step, called the rate-limiting step. This step is the one with the transition state of highest standard free energy.

- Dipolar molecules such as H—Br and H—OH add to alkenes in a regioselective manner so that the hydrogen adds to the carbon of the double bond with the most hydrogens, and the electronegative group to the carbon of the double bond with the greater number of alkyl groups. Addition of water requires acid catalysis because water itself is too weak an acid to protonate the π bond.

- According to Hammond's postulate, the structures and energies of transition states for reactions involving unstable intermediates (such as carbocations) resemble the structures and energies of the unstable intermediates themselves.

- The regioselectivity observed in the addition reactions of hydrogen halides or water to alkenes is a consequence of two facts: (1) the rate-limiting transition states of the two competing reactions resemble carbocations; and (2) the relative stability of carbocations is in the order tertiary > secondary > primary. Application of Hammond's postulate leads to the conclusion that the reaction involving the more stable carbocation is faster.

- Reactions involving carbocation intermediates, such as hydrogen halide addition and hydration, show rearrangements in some cases. Unstable carbocations rearrange to more stable ones by a shift of an alkyl group, aryl group, or hydrogen to the electron-deficient carbon from an adjacent carbon. The group that moves brings along its bonding electron pair. As a result, the adjacent carbon becomes electron-deficient.

- A catalyst increases the rate of a reaction without being consumed in the reaction. A catalyst does not affect the equilibrium constant for a chemical equilibrium. Catalysts accelerate the forward and reverse reactions of an equilibrium equally.

- Catalysts are of two types: heterogeneous and homogeneous. Catalytic hydrogenation of alkenes is an important example of heterogeneous catalysis; acid-catalyzed hydration of alkenes involves homogeneous catalysis.

- Enzymes are biological catalysts that accelerate biological reactions by many orders of magnitude.

 REACTION REVIEW *For a summary of reactions discussed in this chapter, see the* Reaction Review *section of Chapter 4 in the* Study Guide and Solutions Manual.

ADDITIONAL PROBLEMS

4.40 Give the structures and IUPAC substitutive names of the isomeric alkenes with molecular formula C_6H_{12} containing five carbons in their principal chains.

4.41 Give the structures and the IUPAC substitutive names of the isomeric alkenes with the molecular formula C_6H_{12} containing four carbons in their principal chain.

4.42 Which alkenes in Problem 4.41 give predominantly a single constitutional isomer when treated with HBr, and which give a mixture of isomers? Explain.

4.43 Arrange the alkenes in Problem 4.40 in order of increasing heats of formation. (Some may be classified as "about the same.")

4.44 Give a structure for each of the following compounds.
(a) cyclobutene (b) 3-methyl-1-octene
(c) 5,5-dimethyl-1,3-cycloheptadiene
(d) 1-vinylcyclohexene

4.45 Give an IUPAC substitutive name for each of the following compounds. Include the *E,Z* designations where appropriate.

(a)

(b) $H_2C=CH(CH_2)_3\overset{\displaystyle CH_3}{\underset{\displaystyle CH_3}{CH}}$

(c)

(d) CH$_3$CH$_2$CH$_2$CH$_2$

(e)

(f)

4.46 A confused chemist Al Keane used the following names in a paper about alkenes. Although each name specifies a structure, in some cases the name is incorrect. Correct the names that are wrong.
(a) 3-butene (b) *trans*-1-*tert*-butylpropene
(c) (*Z*)-2-hexene (d) 6-methylcycloheptene

4.47 Specify the configuration (*E* or *Z*) of each of the following alkenes. Note that D is deuterium, or ^2H, the isotope of hydrogen with atomic mass = 2.

(a)

(b)

(c)

(d)

4.48 Classify the compounds within each of the following pairs as either identical molecules (I), constitutional isomers (C), stereoisomers (S), or none of the above (N).
(a) cyclohexane and 1-hexene
(b) cyclopentane and cyclopentene

(c) CH$_2$CH$_3$ and CH$_2$CH$_3$

(d) H$_2$C=CHCH$_2$CH$_3$ and

(e)

4.49 Use the principles of Sec. 1.3B to predict the geometry of BF_3. What hybridization of boron is suggested by this geometry? Draw an orbital diagram for hybridized boron similar to that for the carbons in ethylene shown in Fig. 4.3 (p. 127), and provide a hybrid orbital description of the bonding in BF_3.

4.50 Classify each of the labeled bonds in the following structure in terms of the bond type (σ or π) and the component orbitals that overlap to form the bond. (For example, the carbon–carbon bond in ethane is an sp^3–sp^3 σ bond.)

$$
\underset{H}{\overset{H}{\underset{|}{\overset{|}{H-C}}}}\ \underset{(a)}{\rule{1.2cm}{0.4pt}}\ \underset{(b)}{\overset{}{CH}}=\underset{(c)}{CH}\ \underset{(d)}{\rule{1cm}{0.4pt}}\ CH=\overset{\underset{(e)}{\diagup H}}{\underset{\diagdown H}{C}}
$$

4.51 (a) The following compound can be prepared by the addition of HBr to either of two alkenes; give their structures.

(b) Starting with the same two alkenes, would the products be different if DBr were used? Explain. (See note about deuterium in Problem 4.47.)

4.52 Give the structures of all the alkenes containing one double bond that would give propylcyclohexane as the product of catalytic hydrogenation.

4.53 An alkene X with molecular formula C_7H_{12} adds HBr to give a *single* alkyl halide Y with molecular formula $C_7H_{13}Br$ and undergoes catalytic hydrogenation to give 1,1-dimethylcyclopentane. Draw the structures of X and Y. (See Study Guide Link 4.3.)

STUDY GUIDE
LINK 4.3
Solving Structure
Problems

4.54 Give the structures of the two stereoisomeric alkenes with the molecular formula C_6H_{12} that react with HI to give the same *single* product and undergo catalytic hydrogenation to give hexane.

4.55 You have been called in as a consultant for the firm Alcohols Unlimited, which wants to build a plant to produce 3-methyl-1-butanol, $(CH_3)_2CHCH_2CH_2OH$. The research director, Al Keyhall, has proposed that acid-catalyzed hydration of 3-methyl-1-butene be used to prepare this compound. The company president, O. H. Gruppa, has asked you to evaluate this suggestion. Millions of dollars are on the line. What is your answer? Can 3-methyl-1-butanol be prepared in this way? Explain your answer.

4.56 A certain compound A is converted into a compound B in a reaction without intermediates. The reaction has an equilibrium constant $K_{eq} = [B]/[A] = 150$ and, with the free energy of A as a reference point, a standard free energy of activation of 96 kJ mol^{-1} (23 kcal mol^{-1}).

(a) Draw a reaction free-energy diagram for this process, showing the relative free energies of A, B, and the transition state for the reaction.

(b) What is the standard free energy of activation for the reverse reaction $B \rightarrow A$? How do you know?

4.57 A reaction $A \rightleftharpoons B \rightleftharpoons C \rightleftharpoons D$ has the reaction free-energy diagram shown in Fig. P4.57.

(a) Which compound is present in greatest amount when the reaction comes to equilibrium? In least amount?

(b) What is the rate-limiting step of the reaction?

(c) Using a vertical arrow, label the standard free energy of activation for the overall $A \rightarrow D$ reaction.

(d) Which reaction of compound C is faster: $C \rightarrow B$ or $C \rightarrow D$? How do you know?

4.58 Invoking Hammond's postulate, draw the structure of the reactive intermediate that should most closely resemble the transition state of the rate-limiting step for the hydration of 1 methylcyclohexene. (The first step in the mechanism, protonation of the double bond, is rate-limiting.)

4.59 (a) Give the product X expected when methylenecyclobutane undergoes acid-catalyzed hydration.

$$\text{[structure]}=CH_2 + H_2O \xrightarrow{\ 1\,M\,HNO_3\ } X$$

methylenecyclobutane

(b) The rate-limiting step is protonation of the double bond; use H_3O^+ as the acid catalyst. Draw the structure of the reactive intermediate formed in the rate-limiting step.

(c) Draw the transition state for the rate-limiting step.

(d) What is the rate-limiting step for dehydration of X (the reverse of the reaction shown above)?

4.60 The heat of formation of (E)-1,3-pentadiene is 75.8 kJ mol^{-1} (18.1 kcal mol^{-1}), and that of 1,4-pentadiene is 106.3 kJ mol^{-1} (25.4 kcal mol^{-1}).

reaction coordinate

Figure P4.57

(a) Which alkene has the more stable arrangement of bonds?

(b) Calculate the heat liberated when one mole of 1,3-pentadiene is burned. The heat of combustion of carbon is -393.5 kJ mol^{-1} (-94.05 kcal mol^{-1}), and that of H_2 is -285.8 kJ mol^{-1} (-68.32 kcal mol^{-1}).

4.61 The $\Delta H°$ of hydrogenation is the heat liberated when a compound undergoes catalytic hydrogenation. Consider the $\Delta H°$ values for hydrogenation of the following three alkenes: 3-methyl-1-butene, -126.8 kJ mol^{-1} (-30.3 kcal mol^{-1}); 2-methyl-1-butene, -119.2 kJ mol^{-1} (-28.5 kcal mol^{-1}); and 2-methyl-2-butene, -112.6 kJ mol^{-1} (-26.9 kcal mol^{-1}).

(a) Draw an energy diagram in which the three alkanes are placed on the same energy scale along with 2-methylbutane.

(b) Use these data to rank the three alkenes in order of stability, most stable first. Explain how you reached your conclusion.

(c) By how much do the *heats of formation* of the three alkenes differ? Explain.

(d) Explain why this stability order is expected.

4.62 Make a model of cycloheptene with the trans (or E) configuration at the double bond. Now make a model of *cis*-cycloheptene. By examining your models, determine which compound should have the greater heat of formation. Explain.

4.63 Consider the following compounds and their dipole moments:

$\mu = 2.4$ D $\mu = 1.9$ D

Assume that the C—Cl bond dipole is oriented as follows in each of these compounds.

(a) According to the preceding dipole moments, which is *more* electron-donating toward a double bond, methyl or hydrogen? Explain.

(b) Which of the following compounds should have the greater dipole moment? Explain.

4.64 Supply the curved-arrow notation for the acid-catalyzed isomerization shown in Fig. P4.64.

4.65 The curved-arrow notation can be used to understand seemingly new reactions as simple extensions of what you already know. This is the first step in developing an ability to use the notation to predict new reactions. Provide a curved-arrow mechanism for the following reaction.

To do this, follow these steps:

1. Examine the reactants and products and label corresponding atoms. If you're not sure, make a guess.

2. Describe what has happened to the functional groups in the starting material. In this case, focus on the double bond. Is this transformation similar in any way to a reaction you have seen before?

3. Make the connections you deduced in (1) with a curved-arrow mechanism, trying to use steps that are similar to mechanistic steps you've seen in other reactions. Use separate structures for each step of the mechanism; that is, don't try to write several mechanistic steps using the same structure.

4. Use a Lewis acid–base association, Lewis acid–base dissociation, or Brønsted acid–base reaction for each step.

4.66 The industrial synthesis of methyl *tert*-butyl ether involves treatment of 2-methylpropene with methanol (CH_3OH) in the presence of an acid catalyst, as shown in the following equation.

methyl *tert*-butyl ether

This ether is used commercially as an antiknock gasoline additive. Using the curved-arrow notation, propose a mechanism for this reaction.

Figure P4.64

4.67 Using the curved-arrow notation, suggest a mechanism for the reaction shown in Fig. P4.67.

[*Hints:* (1) Follow the problem-solving suggestions in Problem 4.65. (2) Use Hammond's postulate to decide which double bond should protonate first.]

4.68 The standard free energy of formation, ΔG_f°, is the free-energy change for the formation of a substance at 25 °C and 1 atm pressure from its elements in their natural states under the same conditions.

(a) Calculate the equilibrium constant for the interconversion of the following alkenes, given the standard free energy of formation of each. Indicate which compound is favored at equilibrium.

ΔG_f° 79.0 kJ mol^{-1}	75.9 kJ mol^{-1}
18.9 kcal mol^{-1}	18.1 kcal mol^{-1}

(b) What does the equilibrium constant tell us about the rate at which this interconversion takes place?

4.69 The difference in the standard free energies of formation for 1-butene and 2-methylpropene is 13.4 kJ mol^{-1} (3.2 kcal mol^{-1}). (See the previous problem for a definition of ΔG_f°.)

(a) Which compound is more stable? Why?

(b) The standard free energy of activation for the hydration of 2-methylpropene is 22.8 kJ mol^{-1} (5.5 kcal mol^{-1}) less than that for the hydration of 1-butene. Which hydration reaction is faster?

(c) Draw reaction free-energy diagrams on the same scale for the hydration reactions of these two alkenes, showing the relative free energies of both starting materials and rate-determining transition states.

(d) What is the difference in the standard free energies of the *transition states* for the two hydration reactions? Which transition state has lower energy? Using the mechanism of the reaction, suggest why it is more stable.

4.70 The standard free energy of activation ($\Delta G^{\circ\ddagger}$) for hydration of 2-methylpropene to 2-methyl-2-propanol (Eq. 4.41, p. 173) is 91.3 kJ mol^{-1} (21.8 kcal mol^{-1}). The standard free energy ΔG° for hydration of 2-methylpropene is -5.56 kJ mol^{-1} (-1.33 kcal mol^{-1}). The rate of hydration of methylenecyclobutane to give an alcohol (compound *X* in Problem 4.59) is 0.6 times the rate of hydration of 2-methylpropene. The equilibrium constant for the hydration of methylenecyclobutane is about 250 times greater (in favor of hydration) than the equilibrium constant for the hydration of 2-methylpropene. Which alcohol, *X* or 2-methyl-2-propanol, undergoes dehydration faster, and how much faster? Explain.

Figure P4.67

Addition Reactions of Alkenes

The most common reactions of alkenes are *addition reactions*. In Chapter 4, we studied hydrogen halide addition, catalytic hydrogenation, and hydration, and we learned how the curved-arrow notation and the properties of reactive intermediates can be used to understand the regioselectivity of these additions. This chapter surveys some other addition reactions of alkenes using the same approach. We'll also learn about a new type of reactive intermediate, the free radical, and we'll learn a different curved-arrow notation that we'll use for reactions involving free radicals.

5.1 AN OVERVIEW OF ELECTROPHILIC ADDITION REACTIONS

In subsequent sections, we're going to study four more alkene addition reactions in detail. First, though, we're going to look at these reactions in a more general way to see how they resemble two reactions we've already studied, the addition of HBr (Sec. 4.7) and the addition of H_2O (Sec. 4.9B). Seeing these connections will help you to learn these reactions more easily.

Here are the four reactions, each illustrated with a prototypical alkene, 2-methylpropene. In each case, first convince yourself that each of these reactions is an addition. Then determine what has been added to the double bond. Notice where each group in the product comes from.

Bromine addition:

$$\underset{\substack{\textbf{2-methylpropene} \\ \textbf{(isobutylene)}}}{\underset{H_3C}{\overset{H_3C}{\diagdown}}C=CH_2} + Br-Br \xrightarrow[\substack{CH_2Cl_2 \\ \text{(solvent)}}]{} \underset{\substack{\textbf{1,2-dibromo-} \\ \textbf{2-methylpropane}}}{H_3C-\underset{\underset{Br}{|}}{\overset{\overset{CH_3}{|}}{C}}-\underset{\underset{Br}{|}}{CH_2}} \tag{5.1}$$

Oxymercuration:

$$H_3C \underset{H_3C}{>}C{=}CH_2 + AcO{-}Hg{-}OAc + H{-}OH \longrightarrow H_3C{-}\underset{\underset{HO}{|}}{\overset{\overset{CH_3}{|}}{C}}{-}\underset{\underset{Hg{-}OAc}{|}}{CH_2} + H{-}OAc \qquad (5.2)$$

mercuric acetate   (solvent; in large excess)   **acetic acid**

In this equation, —OAc and AcO— are abbreviations for the *acetoxy group*:

$$AcO{-} = {-}OAc = {-}O{-}\overset{\overset{O}{\|}}{C}{-}CH_3$$

acetoxy group

Hydroboration:

$$H_3C \underset{H_3C}{>}C{=}CH_2 + H{-}BH_2 \xrightarrow{\text{an ether solvent}} H_3C{-}\underset{\underset{H}{|}}{\overset{\overset{CH_3}{|}}{C}}{-}\underset{\underset{BH_2}{|}}{CH_2} \qquad (5.3)$$

borane

isobutylborane

In this reaction, don't confuse the element *boron* (B) with *bromine* (Br). Some students make this erroneous subliminal association. Boron is a group 3A element and bromine is a group 7A element—a halogen.

Ozonolysis:

$$H_3C \underset{H_3C}{>}C{=}CH_2 + \ :\overset{}{O}{=}\overset{\overset{+}{O}}{\underset{}{}}\overset{}{\underset{}{}} :\ddot{O}: {}^{-} \longrightarrow H_3C{-}\overset{\overset{CH_3}{|}}{C}{-}CH_2 \qquad (5.4)$$

ozone

Ozonolysis is an example of a **cycloaddition**, which is an addition reaction that forms a ring.

When we consider the reactions in which the two groups that add to the double bond are different, and if we focus on the relative electronegativities of the two groups, the outcome of each reaction is similar to the outcomes of HBr addition and hydration. In the additions of both HBr and H_2O, the hydrogen, which is the less electronegative group in both H—Br and H—OH, adds to the CH_2 carbon of the double bond, and the more electronegative group (—Br or —OH) adds to the carbon with the methyl substituents.

less electronegative group ↘   more electronegative group ↙

$$H_3C \underset{H_3C}{>}C{=}CH_2 + H{-}X \longrightarrow H_3C{-}\underset{\underset{X}{|}}{\overset{\overset{CH_3}{|}}{C}}{-}\underset{\underset{H}{|}}{CH_2} \qquad (5.5)$$

X = Br or OH

In oxymercuration (Eq. 5.2), the Hg is a metal, which is much less electronegative than the oxygen of the —OH group. In this reaction, too, the less electronegative group adds to the carbon of the CH_2, and the more electronegative group adds to the carbon with the methyl substituents. In hydroboration (Eq. 5.3), the same pattern applies. Table 1.1 (p. 9) shows that the electronegativity of hydrogen is *greater* than that of boron, so, in this case, *hydrogen is the more electronegative group.* The boron, the less electronegative group, adds to the CH_2 carbon, and hydrogen, the more electronegative group, adds to the carbon with the methyl substituents.

To generalize these observations: In these addition reactions, when the two groups that add are different, *the carbon of the double bond with fewer alkyl substituents becomes bonded to the less electronegative group, and the carbon of the double bond with more alkyl substituents becomes bonded to the more electronegative group.* We can think of this statement as a "modified Markovnikov's rule" (p. 153).

As you'll learn in the following sections, these reactions occur by different mechanisms. But a common thread runs through all of the mechanisms. *The first step of each reaction is the donation of an electron pair from the alkene π bond, acting as a nucleophile, to an electrophilic center.* The electrophilic atom becomes bonded to the alkene carbon with fewer alkyl substituents. A nucleophilic atom that was either part of the original reactant or present in solution completes the addition by donating electrons to the alkene carbon with the greater number of alkyl substituents. Abbreviating the electrophile or electrophilic center as E (*blue*) and the nucleophile or nucleophilic center as X (*red*), this idea can be summarized as shown in Eq. 5.6. (The dashed curved arrows show the sources of electrons but aren't meant to indicate the actual reaction mechanism.)

Review the mechanisms of HBr addition (Sec. 4.7B) and hydration (Sec. 4.9B) and notice how they fit this pattern.

You may have observed that the electrophilic center is the less electronegative atom of the two atoms that add to the double bond. We might think that the more electronegative atom should be more "greedy" for electrons and for this reason should be the better electrophile. In other words, we might ask why the alkene π electrons are donated to the *less* electronegative atom. We can see why by considering the first step of the HBr addition mechanism. The hydrogen is the electrophilic atom and the Br serves as a leaving group.

This is an *electron-pair displacement reaction.* As this example illustrates, the electrophilic atom in this type of reaction both accepts *and releases* electrons, but the leaving group only accepts electrons. Therefore, electronegativity is more important for the leaving group. Although the mechanisms of the other addition reactions differ somewhat, we'll find that the same issues are present.

Because of the similarity of these addition reactions and others like them, they are grouped as a class and referred to as *electrophilic additions.* An addition reaction is an

electrophilic addition when it begins with the donation of an electron pair from a π bond to an electrophilic atom.

PROBLEM

5.1   (a) Iodine azide, I—N$_3$, adds to isobutylene in the following manner:

$$(CH_3)_2C{=}CH_2 \; + \; :\ddot{I}{-}\overset{..}{N}{=}\overset{+}{N}{=}\overset{..}{N}:^{-} \longrightarrow (CH_3)_2C{-}CH_2$$

iodine azide

$$^{-}:\overset{..}{N}{=}\overset{+}{N}{=}N: \;\; :\ddot{I}:$$

Which group is the electrophilic group to which the π bond donates electrons? How do you know? Does this result fit the electronegativity pattern for electrophilic additions? Explain.

(b) Predict the product of the following electrophilic addition reaction, and explain your reasoning.

$$(CH_3)_2C{=}CH_2 \; + \; :\ddot{I}{-}\ddot{Br}: \; \longrightarrow$$

iodine bromide

5.2   REACTIONS OF ALKENES WITH HALOGENS

A. Addition of Chlorine and Bromine

Halogens undergo addition to alkenes.

$$CH_3CH{=}CHCH_3 \; + \; Br_2 \xrightarrow[\text{(solvent)}]{CH_2Cl_2} \begin{array}{c} CH_3CH{-}CHCH_3 \\ | \quad\;\; | \\ Br \;\;\; Br \end{array} \qquad (5.8)$$

cis- **or** *trans-* **2-butene**

2,3-dibromobutane

$$\text{cyclohexene} \; + \; Cl_2 \xrightarrow[\text{(solvent)}]{CCl_4} \qquad (5.9)$$

cyclohexene

1,2-dichlorocyclohexane
(70% yield)

The products of these reactions are **vicinal dihalides.** *Vicinal* (Latin *vicinus*, for "neighborhood") means "on adjacent sites." Thus, vicinal dihalides are compounds with halogens on adjacent carbons.

Bromine and chlorine are the two halogens used most frequently in halogen addition. Fluorine is so reactive that it not only adds to the double bond but also rapidly replaces all the hydrogens with fluorines, often with considerable violence. Iodine adds to alkenes at low temperature, but most diiodides are unstable and decompose to the corresponding alkenes and I$_2$ at room temperature. Because bromine is a liquid that is more easily handled than chlorine gas, many halogen additions are carried out with bromine. Inert solvents such as methylene chloride (CH$_2$Cl$_2$) or carbon tetrachloride (CCl$_4$) are typically used for halogen additions because these solvents dissolve both halogens and alkenes. The addition of bromine to most alkenes is so fast that when bromine is added dropwise to a solution of the alkene the red bromine color disappears almost immediately. In fact, *this discharge of color is a useful qualitative test for alkenes.*

Bromine addition can occur by a variety of mechanisms, depending on the solvent, the alkene, and the reaction conditions. One of the most common mechanisms involves a reactive intermediate called a *bromonium ion.*

$$CH_3CH = CHCH_3 + Br_2 \rightleftharpoons \overset{\overset{+}{:\ddot{Br}:}}{CH_3CH - CHCH_3} \quad :\ddot{Br}:^- \qquad (5.10)$$

a bromonium ion

A **bromonium ion** is a species that contains a bromine bonded to two carbon atoms; the bromine has an octet of electrons and a positive charge. Formation of the bromonium ion occurs in a single mechanistic step involving three curved arrows. (Follow the arrows below in order.)

(2) The other bromine acts as a leaving group to form bromide ion.

(3) The electrophilic bromine also acts as a nucleophile toward the other carbon of the double bond to form the other C—Br bond.

(1) One carbon of the double bond acts as a nucleophile toward Br using the π electrons to form one C—Br bond.

$$CH_3CH = CHCH_3 \rightleftharpoons \overset{\overset{+}{:\ddot{Br}:}}{CH_3CH - CHCH_3} \quad :\ddot{Br}:^- \qquad (5.11)$$

Analogous cyclic ions form in chlorine and iodine addition.

Bromine addition is completed when the bromide ion donates an electron pair to either one of the ring carbons of the bromonium ion.

leaving group

electrophilic center

$$\overset{\overset{+}{:\ddot{Br}:}}{CH_3CH - CHCH_3} \longrightarrow \overset{:\ddot{Br}:}{CH_3CH - CHCH_3} \qquad (5.12)$$

$$:\ddot{Br}:^- \qquad :\ddot{Br}:$$

nucleophile

This is another *electron-pair displacement reaction* (Sec. 3.2), in which the nucleophile is bromide ion, the electrophilic center is the carbon that accepts an electron pair from the nucleophile, and the leaving group is the bromine of the bromonium ion. (The leaving group doesn't actually leave the molecule, because it is tethered by another bond.) This reaction occurs because the positively charged bromine is very electronegative and readily accepts an electron pair. It also occurs because, as we'll learn in Chapter 7, three-membered rings are strained, as you can see if you attempt to build a model; opening a three-membered ring releases considerable energy.

We might reasonably have written a mechanism for bromine addition that involves a carbocation intermediate, as we did for HBr addition:

$$
\begin{array}{c}
:\ddot{Br}: \\
/ \\
H \\
R \\
\quad C = CH_2 \longrightarrow \\
R'
\end{array}
\quad
\begin{array}{c}
:\ddot{Br}:^- \\
R \quad H \\
\overset{+}{C} - CH_2 \\
R'
\end{array}
\quad \Bigg| \quad
\begin{array}{c}
:\ddot{Br}: \\
/ \\
:\ddot{Br}: \\
R \\
\quad C = CH_2 \overset{\times}{\longrightarrow} \\
R'
\end{array}
\quad
\begin{array}{c}
:\ddot{Br}:^- \\
R \quad :\ddot{Br}: \\
\overset{+}{C} - CH_2 \\
R'
\end{array}
\qquad (5.13)
$$

H—Br addition   Br—Br addition

How do we know that bromonium ions rather than carbocations are reactive intermediates in bromine addition? First, rearrangements are not typically observed in bromine addition. We have seen (Sec. 4.7D) that rearrangements are observed in HBr addition. Second, bromonium ions have been isolated under special circumstances, although, like carbocations, they are not stable enough to isolate under the usual reaction conditions. Finally, there is compelling stereochemical evidence for bromonium ions that we'll consider in Sec. 7.8C after we have developed some additional background in stereochemistry.

We can then ask *why* a bromonium ion should be formed instead of a carbocation. In other words, why is a bromonium ion more stable than the corresponding carbocation? The major reason is that a bromonium ion has more covalent bonds than a carbocation, and every atom has an octet.

bromonium ion   carbocation

B. Halohydrins

In the addition of bromine, the only nucleophile available to react with the bromonium ion is the bromide ion (Eq. 5.12). When other nucleophiles are present, they, too, can react with the bromonium ion to form products other than dibromides. A common situation of this type occurs when the solvent itself can act as a nucleophile. For example, when an alkene is treated with bromine in a solvent containing a large excess of water, a water molecule rather than bromide ion reacts with the bromonium ion, because water is present in much higher concentration than bromide ion:

$$(5.14a)$$

The conjugate acid of an alcohol is very acidic—its acidity is comparable to that of H_3O^+. Therefore, the solvent H_2O can remove this acidic proton to give the product:

$$(5.14b)$$

The product is an example of a **bromohydrin**: a compound containing both an —OH and a —Br group. Bromohydrins are members of the general class of compounds called **halohydrins**, which are compounds containing both a halogen and an —OH group. In the most common type of halohydrin, the two groups occupy adjacent, or *vicinal*, positions.

general structure
of a vicinal halohydrin
(R = alkyl, aryl, or H)

X = Cl, a chlorohydrin
Br, a bromohydrin
I, an iodohydrin

Halohydrin formation involves the net addition to the double bond of the elements of a hypo-halous acid, such as hypobromous acid, HO—Br, or hypochlorous acid, HO—Cl. Although the products of I_2 addition are unstable (see Sec. 5.2A), iodohydrins can be prepared.

When the double bond of the alkene is positioned unsymmetrically, the reaction of water with the bromonium ion can give two possible products, each resulting from breakage of a different carbon–bromine bond. The reaction is highly regioselective, however, when one carbon of the alkene contains *two* alkyl substituents.

$$
\begin{array}{c}
H_3C \\
\diagdown \\
C{=}CH_2 + Br_2 + H_2O \\
\diagup \text{(solvent)} \\
H_3C
\end{array}
\longrightarrow
\begin{array}{c}
CH_3 \\
| \\
H_3C{-}C{-}CH_2 \\
| \quad\ | \\
OH \ \ Br
\end{array}
+ H_3O^+ \ Br^- \qquad (5.15)
$$

1-bromo-2-methyl-2-propanol
(77% yield)

The reason for this regioselectivity can be seen from the structure of the bromonium ion (Fig. 5.1). In this structure, about 90% of the positive charge resides on the tertiary carbon, and the bond between this carbon and the bromine is so long and weak that this species is essentially a carbocation containing a weak carbon–bromine interaction.

$$
\begin{array}{c}
\overset{\delta+}{:\!\ddot{B}r:} \\
\diagup\!\diagdown \cdot \\
H_3C{\cdots}\overset{\delta+}{C}{-}CH_2 \\
\diagup \\
H_3C
\end{array}
$$

Water reacts with the bromonium ion at the tertiary carbon, and the weaker bond to the leaving group is broken, to give the observed regioselectivity.

$$
\begin{array}{c}
\overset{\delta+}{:\!\ddot{B}r:} \\
\diagup\!\diagdown \\
\overset{\delta+}{C}{-}CH_2 \\
H_3C{-} \ \ \ | \\
\diagup \quad :\ddot{O}H_2 \\
H_3C
\end{array}
\longrightarrow
\begin{array}{c}
H_3C \ \ :\ddot{B}r: \\
| \qquad | \\
H_3C{-}C{-}CH_2 \\
| \\
:\overset{+}{O}{-}H \\
| \\
H
\end{array}
\ :\ddot{O}H_2 \underset{\longleftarrow}{\overset{\longrightarrow}{}}
\begin{array}{c}
H_3C \ \ :\ddot{B}r: \\
| \qquad | \\
H_3C{-}C{-}CH_2 + H_3\overset{..}{O}{}^+ \\
| \\
:\ddot{O}H
\end{array}
\qquad (5.16)
$$

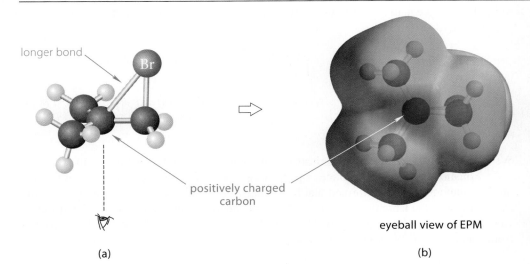

FIGURE 5.1 (a) A ball-and-stick model showing the structure of the bromonium ion involved in bromine addition to 2-methylpro-pene. Notice the very long bond between the bromine and the tertiary carbon. (b) An EPM of the bromonium ion viewed from the direction of the eyeball in part (a). Notice the concentration of positive charge on the tertiary carbon, which is indicated by the blue color. About 90% of the positive charge resides on this carbon.

longer bond

Br

positively charged carbon

eyeball view of EPM

(a)   (b)

STUDY PROBLEM 5.1

Which of the following chlorohydrins could be formed by addition of Cl_2 in water to an alkene? Explain.

$$\underset{A}{\overset{\overset{\displaystyle Cl}{|}\ \ \overset{\displaystyle OH}{|}}{H_3C - C - CH - CH_3}} \qquad \underset{B}{\text{cyclohexane with } CH_2Cl \text{ and } OH}$$

SOLUTION   The mechanistic reasoning used in this section shows that the nucleophile (water) reacts with the carbon of the double bond that has more alkyl substituents. In compound *A*, the carbon bearing the —OH group has *fewer* alkyl substituents than the one bearing the —Cl. Hence, this compound could *not* be formed in the reaction of Cl_2 and water with an alkene. Compound *B* could be formed by such a reaction, however, because the —OH group is at a carbon with more alkyl substituents than the —Cl. Don't forget that the carbons of the ring are alkyl groups even though they are part of the ring structure.

PROBLEMS

5.2   Give the products, and the mechanisms for their formation, when 2-methyl-1-hexene reacts with each of the following reagents.
(a) Br_2 (b) Br_2 in H_2O
(c) iodine azide (I—N_3) (*Hint:* See Problem 5.1a.)

5.3   Give the structure of the alkene that could be used as a starting material to form chlorohydrin *B* in Study Problem 5.1.

5.3   WRITING ORGANIC REACTIONS

As we continue with our study of organic reactions, we'll use a few widely adopted conventions for writing reactions. To avoid confusion, it's important to be aware of these conventions.

The most thorough way to write a reaction is to use a complete, balanced equation. Equations 5.8 and 5.9 on p. 184 are examples of balanced equations. Other information, such as the reaction conditions, is sometimes included in equations. For example, in Eq. 5.8 the solvent is written under the arrow, even though the solvent is not an actual reactant. Typically, solvents do not have to be learned. Also, catalysts are written over the arrow. For example, in the following equation, the H_3O^+ written over the arrow indicates that an acid catalyst is required (Sec. 4.9B).

$$\underset{H_3C}{\overset{H_3C}{\diagup}}C{=}CH_2 + H_2O \xrightarrow{\overset{\text{the catalyst is written over the arrow}}{\downarrow}\ H_3O^+} \underset{\underset{OH}{|}}{H_3C - \overset{\overset{\displaystyle CH_3}{|}}{C} - CH_3} \qquad (5.17)$$

We can also tell reactants from catalysts because a catalyst is not consumed in the reaction.

Equation 5.9 on p. 184 includes a **percentage yield**, which is the percentage of the theoretical amount of product formed that has actually been isolated from the reaction mixture by a chemist in the laboratory. Although different chemists might obtain somewhat different yields in the same reaction, the percentage yield gives a rough idea of how free the reaction is from contaminating by-products and how easily the product can be isolated from the reaction mixture. Thus, a reaction $2A + B \longrightarrow 3C + D$ should give three moles of C for every one of B and two of A used (assuming that one of these reactants is not present in excess). A 90% yield of C means that 2.7 moles of C per mole of B were *actually isolated* under these conditions. The 10% loss may have been due to separation difficulties, small amounts of by-products, or other reasons. Most of the reactions given in this book are actual laboratory examples; the percentage yield figures included in many of these reactions are not meant to be learned, but are given simply to indicate how successful a reaction actually is in practice.

Here's a convention for writing reactions that you particularly need to understand. In many cases, organic chemists abbreviate reactions by showing only the *organic starting materials* and the *major organic product(s)*. The other reactants and conditions are written over the arrow. Thus, Eq. 5.15 might have been written

$$
\begin{array}{c}
\text{H}_3\text{C} \\
\diagdown \\
\text{C}=\text{CH}_2 \\
\diagup \\
\text{H}_3\text{C}
\end{array}
\xrightarrow{\text{Br}_2/\text{H}_2\text{O (solvent)}}
\begin{array}{c}
\text{CH}_3 \\
| \\
\text{H}_3\text{C}-\text{C}-\text{CH}_2 \\
| \quad\;\; | \\
\text{OH} \;\; \text{Br}
\end{array}
\tag{5.18}
$$

1-bromo-2-methyl-2-propanol
(77% yield)

This "shorthand" way of writing organic reactions is frequently used because it saves space and time. When equations are written this way, by-products are not given and, in many cases, the equation is not balanced. This shorthand can present ambiguities for the beginner (and sometimes for the experienced chemist as well!). Are the items written over the arrow reactants, catalysts, a solvent, or something else? In Eq. 5.18 we know that Br_2 is a reactant because it is consumed in the reaction. The H_2O solvent is also a reactant (because the product contains an —OH group). The by-products H_3O^+ and Br^- are not shown.

To avoid such ambiguities, we'll present most reactions in this text initially in balanced form (when the balanced form is known). We'll also label catalysts and solvents at their first occurrence. This should help to clarify the roles of the various reaction components when abbreviated forms of the same reactions are used subsequently.

5.4   CONVERSION OF ALKENES INTO ALCOHOLS

Although the hydration of alkenes (Sec. 4.9B) is used industrially for the preparation of particular alcohols, it is rarely used for the laboratory preparation of alcohols. This section presents two reaction sequences that are especially useful in the laboratory for the conversion of alkenes into alcohols. These two processes, called *oxymercuration–reduction* and *hydroboration–oxidation*, bring about the *overall* addition of H and OH to a double bond. The two sequences, however, are complementary because they occur with opposite regiospecificities. (Notice the position of the —OH group in the product.)

$$
\begin{array}{c}
\text{R} \\
\diagdown \\
\text{C}=\text{CH}_2 \\
\diagup \\
\text{R}'
\end{array}
\left\{
\begin{array}{l}
\xrightarrow{}
\begin{array}{c}
\text{OH} \;\; \text{H} \\
| \quad\;\; | \\
\text{R}-\text{C}-\text{CH}_2 \\
| \\
\text{R}
\end{array} \\
\quad\text{oxymercuration–reduction} \\[2ex]
\xrightarrow{}
\begin{array}{c}
\text{H} \;\;\; \text{OH} \\
| \quad\;\;\; | \\
\text{R}-\text{C}-\text{CH}_2 \\
| \\
\text{R}
\end{array} \\
\quad\text{hydroboration–oxidation}
\end{array}
\right.
\tag{5.19}
$$

Each of these processes also occurs in two experimentally separate operations, which we'll consider in turn.

A. Oxymercuration–Reduction of Alkenes

Oxymercuration of Alkenes   In **oxymercuration**, alkenes react with mercuric acetate, $Hg(OAc)_2$, in aqueous solution to give addition products in which an —HgOAc (acetoxymercuri) group and an —OH (hydroxy) group derived from water have added to the double bond.

$$CH_3CH_2CH_2CH_2{-}CH{=}CH_2 + AcO{-}Hg{-}OAc + H{-}OH \xrightarrow[\text{(solvent)}]{\text{THF, H}_2O}$$

1-hexene   **mercuric acetate**

$$CH_3CH_2CH_2CH_2{-}\underset{\underset{\text{OH}}{|}}{CH}{-}\underset{\underset{\text{HgOAc}}{|}}{CH_2} + HOAc \quad (5.20)$$

acetic acid

(95% yield)

The —OAc (or AcO—) abbreviations were discussed when oxymercuration was introduced (Eq. 5.2, p. 182). Notice that the —HgOAc group goes to the carbon of the double bond with fewer alkyl substituents, and the —OH group to the carbon of the double bond with more alkyl substituents.

The solvent (written under the arrow in Eq. 5.20) is a mixture of water and THF (tetrahydrofuran), a widely used ether.

tetrahydrofuran, or THF

THF is an important solvent because it dissolves both water and many water-insoluble organic compounds. Its role in oxymercuration is to dissolve both the alkene and the aqueous mercuric acetate solution. (Recall that alkenes are not soluble in water alone; Sec. 4.4.) Water is required as both a reactant, as shown in Eq. 5.20, and as a solvent for the mercuric acetate.

The oxymercuration reaction bears a close resemblance to halohydrin formation, which was discussed in the previous section. The first step of the reaction mechanism involves the formation of a cyclic ion called a *mercurinium ion*:

**STUDY GUIDE
LINK 5.1**
Transition Elements
and the Electron-
Counting Rules

$$R{-}CH{=}CH_2 + :Hg(\ddot{O}Ac)_2 \rightleftharpoons R{-}\underset{}{CH}{-}\underset{}{CH_2}\ \ \overset{+}{Hg}{-}OAc \qquad {}^{-}:\ddot{O}Ac \qquad (5.21a)$$

a mercurinium ion

(Contrast this equation with Eq. 5.10 for the formation of a bromonium ion.) Like bromine addition, oxymercuration does not involve carbocations because carbocation rearrangements are not observed. (Stereochemical evidence that we'll discuss in Chapter 7 also supports the absence of carbocation intermediates.) Consequently, the mechanism can be viewed as a one-step process:

$$R{-}CH{=}CH_2 \rightleftharpoons R{-}CH{-}CH_2 \qquad {}^{-}:\ddot{O}Ac \qquad (5.21b)$$

This equation is analogous to Eq. 5.11 (p. 185) for bromonium ion formation.

Just as the bromonium ion in Eq. 5.14a reacts with the solvent water, which is present in large excess, the mercurinium ion also reacts with the solvent water:

$$^-OAc \quad R-CH-CH_2 \longrightarrow R-CH-CH_2 \quad ^-OAc \quad (5.21c)$$

(large excess)

Of the two carbons in the ring, the reaction of water occurs at the carbon with the greater number of alkyl substituents, just as in the reaction of water with a bromonium ion (Eq. 5.16). A difference between oxymercuration and halohydrin formation, however, is the degree of regioselectivity. In oxymercuration, the reaction of water occurs almost exclusively at the carbon with more alkyl substituents, even if that carbon has only *one* alkyl substituent (as in Eq. 5.21c). (Recall that in halohydrin formation, the reaction is highly regioselective only if one of the alkene carbons has *two* alkyl branches.)

The addition is completed by the transfer of a proton to the acetate ion formed in Eq. 5.21b.

$$R-CH-CH_2 \longrightarrow R-CH-CH_2 \quad + \quad H-\ddot{O}Ac \quad (5.21d)$$

acetate ion

$pK_a = 4.76$

$pK_a \approx -2$

The pK_a values show that the equilibrium of this last Brønsted acid–base step lies far to the right.

Conversion of Oxymercuration Adducts into Alcohols   Oxymercuration is useful because its products are easily converted into alcohols by treatment with the reducing agent sodium borohydride (NaBH$_4$) in the presence of aqueous NaOH.

$$4\,CH_3CH_2CH_2CH_2-CH-CH_2-Hg\,OAc + 4\,OH^- + NaBH_4 \longrightarrow \qquad (5.22)$$

sodium borohydride

$$4\,CH_3CH_2CH_2CH_2-CH-CH_3 + Na^+\,B(OH)_4^- + 4\,Hg^0\!\downarrow + 4\,AcO^-$$

We won't consider the mechanism of this reaction. The key thing to notice is its outcome: the carbon–mercury bond is replaced by a carbon–hydrogen bond (color in Eq. 5.22). The oxymercuration adducts are usually not isolated, but are treated directly with a basic solution of NaBH$_4$ in the same reaction vessel.

The oxymercuration and NaBH$_4$ reactions, when used sequentially, are referred to collectively as the **oxymercuration–reduction** of an alkene. (The general classification of reactions as oxidations or reductions is discussed in Sec. 10.6.) The overall result of oxymercuration–reduction is the *net* addition of the elements of water (H and OH) to an alkene double bond in a highly *regioselective* manner: the —OH group is added to the *more branched carbon of the double bond*. Here's the overall sequence applied to 1-hexene written in shorthand style. The numbers above the arrow mean that two steps are carried out *in sequence*; that is, *first*, the alkene is allowed to react with $Hg(OAc)_2$ and H_2O, and *then*, in a *separate step*, aqueous $NaBH_4$ and NaOH are added.

$$\underset{\textbf{1-hexene}}{\diagup\!\!\!\diagdown\!\!\!\diagup\!\!\!\diagdown\!\!\!\diagup\!\!\!\diagdown\!\!\!\diagup} \quad \xrightarrow[\text{2) NaBH}_4\text{/NaOH}]{\text{1) Hg(OAc)}_2\text{/H}_2\text{O}} \quad \underset{\substack{\textbf{2-hexanol}\\\text{(96\% yield)}}}{\diagup\!\!\!\diagdown\!\!\!\diagup\!\!\!\diagdown\!\!\!\overset{\text{H}}{\underset{\text{OH}}{\diagup}}} \tag{5.23}$$

Writing consecutive reactions in this manner can save lots of time and space. However, if you use this shorthand, *be sure to number the reactions*. If the numbers were left off, a reader might think that all of the reagents were added at once. Adding the reagents for both steps at the same time would *not* give the desired product!

Oxymercuration–reduction gives the same overall transformation as the hydration reaction (Sec. 4.9B). However, oxymercuration–reduction is much more convenient to run on a laboratory scale than alkene hydration, and it is free of rearrangements and other side reactions that are encountered in hydration, because carbocation intermediates are not involved in oxymercuration. For example, the alkene in the following equation gives products derived from carbocation rearrangement when it undergoes hydration (Problem 4.36, p. 174). However, no rearrangements are observed in oxymercuration–reduction:

$$\underset{\textbf{3,3-dimethyl-1-butene}}{(CH_3)_3C-CH{=}CH_2} \quad \xrightarrow[\text{2) NaBH}_4\text{/NaOH}]{\text{1) Hg(OAc)}_2\text{/H}_2\text{O}} \quad \underset{\substack{\textbf{3,3-dimethyl-2-butanol}\\\text{(94\% yield; notice:}\\\text{no rearrangement)}}}{(CH_3)_3C-\underset{\overset{|}{OH}}{CH}-CH_3} \tag{5.24}$$

The absence of rearrangements is one reason that mercurinium ions, rather than carbocations, are thought to be the reactive intermediates in oxymercuration.

PROBLEMS

5.4   Give the products expected when each of the following alkenes is subjected to oxymercuration–reduction.

(a) cyclohexene   (b) 2-methyl-2-pentene   (c) *trans*-4-methyl-2-pentene   (d) *cis*-3-hexene

5.5   Contrast the products expected when 3-methyl-1-butene undergoes (a) acid-catalyzed hydration or (b) oxymercuration–reduction. Explain any differences.

5.6   What alkenes would give each of the following alcohols as the major (or only) product as a result of oxymercuration–reduction?

(a)

$$\diagup\!\!\!\diagdown\!\!\!\overset{\diagup\!\!\!\diagdown}{\underset{OH}{\diagdown}}\!\!\!\diagup$$

(two different alkenes)

(b)

$$\triangleright\!\!-\!\!\underset{\overset{|}{OH}}{CH}\!\!-\!\!CH_3$$

(one alkene)

Laboratory Use of Mercury and Other Toxic Reagents

Mercury is a very toxic element because it can be converted in the environment into methylmercury, CH_3Hg, which can accumulate in the fatty tissues of animals such as fish. Ingestion of methylmercury leads

to neurotoxicity. The oxymercuration–reduction reaction sequence illustrates the point that chemists use a significant number of highly toxic reagents. Three issues surround the toxicity of chemical reagents. The first is the *knowledge* that they are toxic. Chemists are provided with a *Material Safety Data Sheet* (MSDS) that describes the known hazards of each reagent that they purchase, and most MSDSs are readily available on the web. The second issue is the *safe handling* of toxic or dangerous reagents in the laboratory. Part of a scientist's laboratory training is to become familiar with common laboratory hazards and to learn how they can be avoided or confronted safely. For example, if you are taking a laboratory course, you undoubtedly are required to wear safety glasses. The third issue is *protection of the environment*. Significant advances have been made in environmental protection, and there is a now a significant emphasis on developing "green" chemistry—that is, environmentally friendly chemical processes. Does this mean that a chemist must avoid the use of dangerous or environmentally unfriendly reagents? Not necessarily. The issue is to use them safely and to dispose of them properly. To take oxymercuration–reduction as an example, the ultimate mercury-containing product of the reaction (Eq. 5.22) is metallic mercury. This can be collected and recycled. Although most chemists would avoid the use of mercury where possible—for example, replacing mercury-containing thermometers in laboratories—sometimes there simply is no alternative. Oxymercuration–reduction is such an effective reaction that it is attractive despite the inconvenience of proper mercury recycling.

B. Hydroboration–Oxidation of Alkenes

We've just seen that oxymercuration–reduction brings about the addition of H and OH to a double bond so that the —OH group adds to the carbon with more alkyl substituents. Suppose, though, that we want to add H and OH to a double bond so that the —OH group adds to the carbon with *fewer* alkyl substituents. The process that we'll now discuss, *hydroboration–oxidation*, is the method that can bring about this transformation. As with oxymercuration–reduction, hydroboration–oxidation involves two separate experimental steps, which we'll discuss in turn.

Conversion of Alkenes into Organoboranes   Borane (BH_3) adds regioselectively to alkenes so that the boron becomes bonded to the carbon of the double bond with *fewer* alkyl substituents, and the hydrogen becomes bonded to the carbon with *more* alkyl substituents:

$$(CH_3)_2C\!\!=\!\!CH_2 \ + \ H\!-\!BH_2 \longrightarrow (CH_3)_2\underset{\overset{|}{H}}{C}\!-\!\underset{\overset{|}{BH_2}}{CH_2} \qquad (5.25a)$$

2-methylpropene   **borane**
(isobutylene)

isobutylborane

Because borane has *three* B—H bonds, one borane molecule can add to three alkene molecules. The first of these additions to 2-methylpropene is shown in Eq. 5.25a. The second and third additions are as follows:

Second addition:

$$(CH_3)_2C\!\!=\!\!CH_2 \ + \ (CH_3)_2\underset{\overset{|}{H}}{C}\!-\!\underset{\overset{|}{B-H}}{CH_2} \longrightarrow \ (5.25b)$$

2-methylpropene

(from Eq. 5.25a)   **diisobutylborane**

Third addition:

$$(CH_3)_2C\!\!=\!\!CH_2 \ + \ \longrightarrow \ \text{or} \ \left(\!(CH_3)_2\underset{\overset{|}{H}}{C}\!-\!CH_2\!\right)_{\!3}\!B \qquad (5.25c)$$

2-methylpropene

(from Eq. 5.25b)

triisobutylborane
(a trialkylborane)

The addition of BH_3 is called **hydroboration**. The hydroboration product of an alkene is a *trialkylborane*, such as the triisobutylborane shown in Eq. 5.25c.

Borane and Diborane

Borane actually exists as a toxic, colorless gas called *diborane*, which has the formula B_2H_6. Because borane is an electron-deficient Lewis acid, the boron has a strong tendency to acquire an additional electron pair. This tendency is satisfied by the formation of diborane, in which two hydrogens are shared between the two borons in unusual "half bonds." This bonding can be shown with resonance structures:

$$\left[\begin{array}{c} \text{H}^{\cdots}\text{B}^{\overset{\text{H}}{\diagdown}}_{\diagup}\text{B}^{\cdots}\text{H} \\ \text{H}^{\diagup}\underset{\text{H}}{\text{B}} \quad \text{B}^{\diagdown}\text{H} \end{array} \longleftrightarrow \begin{array}{c} \text{H}^{\cdots}\text{B}^{\overset{\text{H}}{\diagdown}}_{\diagup}\text{B}^{\cdots}\text{H} \\ \text{H}^{\diagup} \quad \underset{\text{H}}{} \quad \text{B}^{\diagdown}\text{H} \end{array} \right] \quad \text{or} \quad \text{H}^{\cdots}\text{B}^{\overset{\text{H}}{\diagdown}}_{\diagup}\text{B}^{\cdots}\text{H} \tag{5.26}$$

When dissolved in an ether solvent, diborane dissociates to form a borane–ether complex. Because ethers are Lewis bases, they can satisfy the electron deficiency at boron:

$$B_2H_6 + 2R\overset{\cdot\cdot}{\underset{\cdot\cdot}{O}}\text{—R} \rightleftharpoons 2H_3\bar{B}\overset{+}{\text{—}}\overset{R}{\underset{R}{O}}: \tag{5.27}$$

an ether   borane–ether complex

The following ethers are commonly used as solvents in the hydroboration reaction:

$CH_3CH_2\overset{\cdot\cdot}{O}\text{—}CH_2CH_3$ (tetrahydrofuran ring structure)

diethyl ether   **tetrahydrofuran (THF)**

$CH_3\overset{\cdot\cdot}{O}\text{—}CH_2CH_2\text{—}\overset{\cdot\cdot}{O}\text{—}CH_2CH_2\text{—}\overset{\cdot\cdot}{O}CH_3$

diethylene glycol dimethyl ether (diglyme)

Borane–ether complexes are the actual reagents involved in hydroboration reactions. For simplicity, the simple formula BH_3 often is used for borane.

Hydroboration is believed to occur in a single mechanistic step because carbocation rearrangements are not observed and because of stereochemical evidence we'll consider in Chapter 7.

$$\underset{\text{Me}_2\text{C}=\!\!=\!\!\text{CH}_2}{\text{H}\text{—}\text{BH}_2} \longrightarrow \underset{\text{Me}_2\text{C}\text{—}\text{CH}_2}{\overset{\text{H}\qquad\text{BH}_2}{|\qquad|}} \tag{5.28}$$

A reaction, like this one, that occurs in a single step without intermediates is said to occur by a **concerted mechanism** because everything happens "in concert," or simultaneously. Despite the evidence against carbocation intermediates, the concerted mechanism is consistent with the regioselectivity of the reaction only if some degree of electron deficiency is built up on the tertiary carbon in the transition state of the reaction:

$$\left[\begin{array}{c} \overset{\delta-}{\text{H}}\text{---}\text{BH}_2 \\ | \\ \overset{\delta+}{\text{Me}_2\text{C}}=\!\!=\!\!\text{CH}_2 \end{array} \right]^{\ddagger}$$

this carbon is partially electron-deficient

transition state for hydroboration

Just as alkyl substitution at the electron-deficient carbon stabilizes a carbocation, alkyl substitution at a *partially* electron-deficient carbon stabilizes a transition state. Thus, hydroboration occurs with the regioselectivity that places partial positive charge on the carbon with more alkyl substituents.

Conversion of Organoboranes into Alcohols   The utility of hydroboration lies in the many reactions of organoboranes themselves. One of the most important reactions of organoboranes is their conversion into alcohols with hydrogen peroxide (H_2O_2) and aqueous NaOH.

$$\left((CH_3)_2C{-}CH_2\right)_3 B + 3H_2O_2 + {}^-OH \longrightarrow 3\,(CH_3)_2C{-}CH_2{-}OH + {}^-B(OH)_4$$

triisobutylborane   **hydrogen peroxide**

2-methyl-1-propanol (isobutyl alcohol)
(95% yield)   (5.29)

(The mechanistic details are given in Further Exploration 5.1.) The important thing to notice about this transformation is *the replacement of the boron by an* —OH *in each alkyl group.* The oxygen of the —OH group comes from the H_2O_2.

Typically, the organoborane product of hydroboration is not isolated, but is treated directly with alkaline hydrogen peroxide to give an alcohol. The addition of borane and subsequent reaction with H_2O_2, taken together, are referred to as **hydroboration–oxidation**.

If we trace the fate of an alkene through the entire hydroboration–oxidation sequence, we find that the *net* result is addition of the elements of water (H, OH) to the double bond in a regioselective manner so that the —OH ends up at the *carbon of the double bond with the smaller number of alkyl substituents* (*red* in Eq. 5.30). Here is the hydroboration–oxidation of 2-methyl-1-butene written in our reaction shorthand. Notice again the numbered steps. Step 1 is the reaction of the alkene with borane. *After this step is complete*, a solution of hydrogen peroxide in aqueous NaOH is added in step 2.

$$(5.30)$$

(90–95% yield)

Hydroboration–oxidation is an effective way to synthesize certain alcohols from alkenes. It is particularly useful to prepare alcohols of the general structure $R_2CH{-}CH_2{-}OH$ or $R{-}CH_2{-}CH_2{-}OH$, as in Eqs. 5.29 and 5.30. Because carbocations are not involved in either the hydroboration or the oxidation reaction, the alcohol products are not contaminated by constitutional isomers arising from rearrangements.

The following example shows that the benzene ring does not react with BH_3, even though the ring apparently contains double bonds:

2-phenyl-1-propanol
(95% yield)   (5.31)

We have also seen a similar resistance of benzene rings to other addition reactions, such as catalytic hydrogenation (Sec. 4.9A). The reasons for this resistance of benzene rings to addition reactions is discussed in Chapter 15.

H. C. Brown and Hydroboration

Hydroboration was discovered accidentally in 1955 at Purdue University by Professor Herbert C. Brown (1912–2004) and his colleagues. Brown quickly realized its significance and in subsequent years carried out research demonstrating the versatility of organoboranes as intermediates in organic synthesis. Brown called the chemistry of organoboranes "a vast unexplored continent." In 1979, his research was recognized with the Nobel Prize in Chemistry, which he shared with another organic chemist, Georg Wittig (Sec. 19.13).

PROBLEMS

5.7   Give the product(s) expected from the hydroboration–oxidation of each of the following alkenes.

(a) cyclohexene   (b) 2-methyl-2-pentene   (c) *trans*-4-methyl-2-pentene   (d) *cis*-3-hexene

5.8   Contrast the answers for Problem 5.7 with the answers for the corresponding parts of Problem 5.4, p. 192. For which alkenes are the alcohol products the same? For which are they different? Explain why the same alcohols are obtained in some cases and different ones are obtained in others.

5.9   For each of the following cases, provide the structure of an alkene that would give the alcohol as the major (or only) product of hydroboration–oxidation.

(a) ⬡—CH₂OH   (b) [structure with OH]

C. Comparison of Methods for the Synthesis of Alcohols from Alkenes

Let's now compare the different ways of preparing alcohols from alkenes. The *hydration of alkenes* is a useful industrial method for the preparation of a few alcohols, but it is not a good laboratory method (Sec. 4.9B). Indeed, many industrial methods for the preparation of organic compounds are not *general*. That is, an industrial method typically works well in the specific case for which it was designed, but it cannot necessarily be applied to other related cases. The reason is that the chemical industry has gone to great effort to work out conditions that are optimal for the preparation of *particular compounds* of commercial significance (such as a few simple alcohols) using reagents that are readily available and inexpensive. These processes in many cases require high temperatures, high pressures, or elaborate reactors, conditions that are difficult to reproduce with ordinary laboratory apparatus. Moreover, there is no need to duplicate these processes in the laboratory because the relatively few compounds that are produced on large industrial scales *are* inexpensive and readily available. For laboratory work, it is impractical for chemists to design a specific procedure for each new compound. Thus, the development of general methods that work with a wide variety of compounds is important. Because laboratory synthesis is generally carried out on a relatively small scale, the expense of reagents is less of a concern.

Hydroboration–oxidation and oxymercuration–reduction are both *general laboratory methods* for the preparation of alcohols from alkenes. That is, they can be applied successfully to a wide variety of alkene starting materials. A choice between the two methods for a particular alcohol usually hinges on the difference in their regioselectivities. As shown in the following equation, hydroboration–oxidation gives an alcohol in which the —OH group has been added to the carbon of the double bond *with the smaller number of alkyl substituents*. Oxymercuration–reduction gives an alcohol in which the —OH group has been added to the carbon of the double bond *with the greater number of alkyl substituents*.

$$R—CH=CH_2 \xrightarrow[\text{of H and OH}]{\text{net addition}} \begin{cases} R—\underset{\underset{OH}{|}}{CH}—\underset{\underset{H}{|}}{CH_2} \quad \text{oxymercuration–reduction} \\ \\ R—\underset{\underset{H}{|}}{CH}—\underset{\underset{OH}{|}}{CH_2} \quad \text{hydroboration–oxidation} \end{cases}$$

(5.32)

For alkenes that yield the same alcohol by either method, such as alcohols with symmetrically located double bonds, the choice between the two is in principle arbitrary.

STUDY PROBLEM 5.2

Which of the following alcohols could be prepared free of constitutional isomers by (a) hydroboration–oxidation, (b) oxymercuration–reduction, (c) either method, or (d) neither method? Explain your answers and give the structure of the alkene starting material for the cases in which a satisfactory synthesis is possible.

$$\underset{A}{HO—CH_2CH_2CH_2CH_2CH_3} \qquad \underset{B}{CH_3\overset{\overset{OH}{|}}{C}HCH_2CH_2CH_3} \qquad \underset{C}{CH_3CH_2\overset{\overset{OH}{|}}{C}HCH_2CH_3} \qquad \underset{D}{CH_3CH_2CH_2\overset{\overset{OH}{|}}{C}HCH_2CH_3}$$

SOLUTION The steps to solving this problem are: (1) Draw the possible alkene starting materials. We should consider initially any alkene in which one carbon of the double bond is the one that bears the hydroxy group in the product. (2) Decide whether each reaction can be used on that starting material to give the desired alcohol and *only* that alcohol.

To prepare *A*, the only possible alkene starting material is 1-pentene.

$$\underset{\textbf{1-pentene}}{H_2C=CHCH_2CH_2CH_3} \xrightarrow{?} \underset{A}{HO—CH_2CH_2CH_2CH_2CH_3}$$

Hydroboration–oxidation has the correct regioselectivity to bring about this conversion, but oxymercuration–reduction does not.

To prepare *B*, the possible alkene starting materials are 1-pentene and *cis*- or *trans*-2-pentene.

$$\underset{\textbf{1-pentene}}{H_2C=CHCH_2CH_2CH_3} \text{ or } \underset{\textbf{2-pentene}}{CH_3CH=CHCH_2CH_3} \xrightarrow{?} \underset{B}{CH_3\overset{\overset{OH}{|}}{C}HCH_2CH_2CH_3}$$

Oxymercuration–reduction has the correct regiospecificity for the conversion of 1-pentene to *B*. However, the 2-pentenes would yield a mixture of *B* and *C* with either method. Hence, neither of the 2-pentenes is a satisfactory starting material. The issue is not whether the 2-pentenes would react; *the issue is whether we would obtain the constitutional isomer we want or a mixture of constitutional isomers*. Remember also that the *number* of alkyl groups on each alkene carbon, not their size, determines regioselectivity.

The (*E*)- and (*Z*)-2-pentenes are the only potential starting materials for *C*.

$$\underset{C}{CH_3CH_2\overset{\overset{OH}{|}}{C}HCH_2CH_3} \Rightarrow \underset{\text{the same compound: \textbf{2-pentene}}}{CH_3CH=CHCH_2CH_3 \text{ or } CH_3CH_2CH=CHCH_3}$$

As we just observed, neither method would yield a single compound with a 2-pentene as the starting alkene. Hence *C* cannot be prepared as a pure compound by *either* method. (There are other ways to make this alcohol.)

Finally, the possible alkene starting materials for alcohol *D* are *cis*- or *trans*-2-hexene and *cis*- or *trans*-3-hexene.

$$\underset{\textbf{2-hexene}}{CH_3CH_2CH_2CH=CHCH_3} \text{ or } \underset{\textbf{3-hexene}}{CH_3CH_2CH=CHCH_2CH_3} \xrightarrow{?} \underset{D}{CH_3CH_2CH_2\overset{\overset{OH}{|}}{C}HCH_2CH_3}$$

Because the double bond in 3-hexene is symmetrically situated, either method in principle would give alcohol *D* as the only product. Hence, either stereoisomer of 3-hexene is a satisfactory starting material. However, the reaction of a 2-hexene would, like the reaction of a 2-pentene, give a mixture of constitutional isomers by either method.

**STUDY GUIDE
LINK 5.2**
How to Study
Organic Reactions

At this point, we have studied a number of new reactions, and so, this is a good time to ask, "What is the best technique for studying and learning reactions?" Staring at the page and highlighting the reactions in the text are *not* good methods. The best methods should be *active*—that is, they should require you to think about the reactions and mentally process them, and they should require you to *write* them. They should test not only your ability to complete reactions for which you are given the starting materials, but also to provide starting materials and appropriate conditions given a desired product, as in Study Problem 5.2. You should also focus on the reaction mechanisms, using them particularly to see the similarities and differences in related reactions. (Mechanisms should not be memorized by rote.) In Study Guide Link 5.2 of the *Study Guide and Solutions Manual* we have outlined one method that seems to work fairly well. Be sure to read about this method and adopt it or some equivalent method. Then work as many problems as you can. If you apply this method as you go and not allow reactions to accumulate without learning them, you will find that you can master the large amount of material without "cramming" them right before an examination.

PROBLEMS

5.10 From what alkene and by which methods could you prepare each of the following alcohols essentially free of constitutional isomers?

(a) [cyclopentane with —OH]   (b) [branched chain with —OH]   (c) Et_3C—OH

(*Hint:* Draw out the structure!)

5.11 Which of the following alkenes would yield the same alcohol from either oxymercuration–reduction or hydroboration–oxidation, and which would give different alcohols? Explain.

(a) *cis*-2-butene   (b) 1-methylcyclohexene

5.5   OZONOLYSIS OF ALKENES

The reaction of ozone, O_3, with alkenes, like hydroboration–oxidation and oxymercuration–reduction, involves two chemically and experimentally distinct steps. In the first step, the alkene reacts with ozone to form an addition product called an *ozonide*. In the second step, the ozonide is treated with oxidizing or reducing agents to form various products. We'll consider each of these steps in turn.

Formation of Ozonides   The addition of ozone to alkenes occurs at low temperature and breaks the carbon–carbon π bond to give an unstable addition product. The spontaneous conversion of this addition product into an **ozonide** breaks the second carbon–carbon bond.

$$H_3C—CH=CH—CH_3 + \overset{..}{O}=\overset{+}{\overset{..}{O}}{}^{..}\overset{-}{\overset{..}{O}}{:}^{-} \xrightarrow[-78\,°C]{CH_2Cl_2}$$

2-butene
(cis or trans)   **ozone**

$$\underset{\text{the initial cycloaddition product}}{:\overset{..}{O}\diagup\overset{\overset{..}{O}}{}\diagdown\overset{..}{O}: \atop H_3C—HC—CH—CH_3} \longrightarrow \underset{\text{an ozonide}}{H_3C—HC\diagdown\overset{:\overset{..}{O}—\overset{..}{O}:}{\underset{\overset{..}{O}}{..}}\diagup CH—CH_3} \quad (5.33)$$

the double bond
is broken

The reaction of an alkene with ozone to yield products of double-bond cleavage is called **ozonolysis**. (The suffix *-lysis* is used for describing bond-breaking processes. Examples are *hydrolysis*, "bond breaking by water"; *thermolysis*, "bond breaking by heat"; and *ozonolysis*, "bond breaking by ozone.")

Ozone and Its Preparation

Ozone is a colorless gas that is formed in the stratosphere, the part of the atmosphere that lies about 6–30 miles above Earth's surface, by the reaction of oxygen with short-wavelength ultraviolet radiation. It is very important in shielding Earth from longer-wavelength "UV-B" radiation, which it absorbs. Although depletion of stratospheric ozone is a significant environmental concern, an increase in ozone near Earth's surface is also an environmental issue. This ozone, formed in complex reactions from nitrogen oxides and unburned hydrocarbons, is a significant contributor to smog. For example, the reaction of ozone with the double bonds in tire rubber (Sec. 15.5) accounts for the fact that tires have a shorter life in urban areas with significant ozone pollution.

Ozone can be formed from the reaction of oxygen in electrical discharges. An atmospheric example is the formation of ozone in a thunderstorm by lightning (photo). The laboratory preparation of ozone involves a similar reaction:

$$3 O_2 \xrightarrow{\text{electrical discharge}} 2 O_3$$

Ozone is produced in the laboratory by passing oxygen through an electrical discharge in a commercial apparatus called an *ozonator*. Because ozone is unstable, it cannot be stored in gas cylinders and must be produced as it is needed.

The first step in ozonolysis is another addition reaction of the alkene π bond. The central oxygen of ozone is a positively charged electronegative atom and therefore strongly attracts electrons. The curved-arrow notation shows that this oxygen can accept an electron pair when the other oxygen of the O=O bond accepts π electrons from the alkene.

$$\text{H}_3\text{C}-\text{HC}=\text{CH}-\text{CH}_3 \longrightarrow \text{H}_3\text{C}-\text{HC}-\text{CH}-\text{CH}_3 \qquad (5.34)$$

initial cycloaddition product

This reaction results in the formation of a ring because the three oxygens of the ozone molecule remain intact. Additions that give rings are called *cycloadditions*. Furthermore, the cycloaddition of ozone occurs in a single step. Hence, this is another example, like hydroboration, of a *concerted mechanism*.

The initial cycloaddition product is unstable and spontaneously forms the ozonide. In this reaction, the remaining carbon–carbon bond of the alkene is broken.

$$\text{H}_3\text{C}-\text{HC}-\text{CH}-\text{CH}_3 \longrightarrow \text{H}_3\text{C}-\text{HC} \qquad \text{CH}-\text{CH}_3 \qquad (5.35)$$

initial cycloaddition product

ozonide

This process occurs, first, by a cyclic electron flow to form an aldehyde and an aldehyde oxide. An O—O bond (a very weak bond) is broken in the process.

initial cycloaddition product   an aldehyde   an aldehyde
                                                        oxide

(5.36a)

The aldehyde flips over, and a second cycloaddition, much like ozonolysis itself—but this time to the C=O bond—completes the formation of the ozonide:

turn the
aldehyde over

(5.36b)

an aldehyde   an aldehyde   an ozonide
               oxide

products of Eq. 5.36a

Reactions of Ozonides   A few daring chemists have made careers out of isolating and studying the highly explosive ozonides. In most cases, however, the ozonides are treated further without isolation to give other compounds. Ozonides can be converted into aldehydes, ketones, or carboxylic acids, depending on the structure of the alkene starting material and the reaction conditions. When the ozonide is treated with dimethyl sulfide, $(CH_3)_2S$, the ozonide is split:

dimethyl sulfide   **acetaldehyde**
                                  (an aldehyde)

(5.37)

**FURTHER
EXPLORATION 5.2**
Mechanism of
Ozonide Conversion
into Carbonyl
Compounds

The net transformation resulting from the ozonolysis of an alkene followed by dimethyl sulfide treatment is the replacement of a $\diagup C = C \diagdown$ group by two $\diagdown C = O$ groups:

the double bond is
completely broken

$$ H_3C—CH=CH—CH_3 \xrightarrow[\text{2) } (CH_3)_2S]{\text{1) } O_3} H_3C—CH=O + O=CH—CH_3 $$

(5.38)

=O   O=
here   here

If the two ends of the double bond are identical, as in Eq. 5.38, then two equivalents of the same product are formed. If the two ends of the alkene are different, then a mixture of two different products is obtained:

$$ CH_3(CH_2)_5CH=CH_2 \xrightarrow[CH_2Cl_2]{O_3} \xrightarrow{(CH_3)_2S} CH_3(CH_2)_5CH=O + O=CH_2 $$

heptanal   **formaldehyde**
(75% yield)

(5.39)

If a carbon of the double bond in the starting alkene bears a hydrogen, then an aldehyde is formed, as in Eqs. 5.38 and 5.39. In contrast, if a carbon of the double bond bears no hydrogens, then a ketone is formed instead:

$$\begin{array}{c} H_3C \\ \diagdown \\ H_3C \end{array} C = C \begin{array}{c} H \\ \diagup \\ CH_3 \end{array} \xrightarrow[\text{2) (CH}_3)_2S]{\text{1) O}_3} \begin{array}{c} H_3C \\ \diagdown \\ H_3C \end{array} C = O \ + \ O = C \begin{array}{c} H \\ \diagup \\ CH_3 \end{array} \qquad (5.40)$$

a ketone   an aldehyde

If the ozonide is simply treated with water, hydrogen peroxide (H_2O_2) is formed as a by product. Under these conditions (or if hydrogen peroxide is added specifically), aldehydes are converted into carboxylic acids, but ketones are unaffected. Hence, the alkene in Eq. 5.40 would react as follows:

an H here . . .   . . . results in an
                                    OH here

$$\begin{array}{c} H_3C \\ \diagdown \\ H_3C \end{array} C = C \begin{array}{c} H \\ \diagup \\ CH_3 \end{array} \xrightarrow[\text{2) H}_2O \ (+ \ H_2O_2)]{\text{1) O}_3} \begin{array}{c} H_3C \\ \diagdown \\ H_3C \end{array} C = O \ + \ O = C \begin{array}{c} OH \\ \diagup \\ CH_3 \end{array} \qquad (5.41)$$

a ketone   a carboxylic acid

The different results obtained in ozonolysis are summarized in Table 5.1 on p. 202.

If the structures of its ozonolysis products are known, then the structure of an unknown alkene can be deduced. This idea is illustrated in Study Problem 5.3.

STUDY PROBLEM 5.3

Alkene X of unknown structure gives the following products after treatment with ozone followed by aqueous H_2O_2:

cyclopentanone   **propionic acid**

What is the structure of X?

SOLUTION   The structure of the alkene can be deduced by *mentally reversing* the ozonolysis reaction. To do this, rewrite the $C=O$ double bonds as "dangling" double bonds by dropping the oxygen:

Next, replace the HO— group of any carboxylic acid fragments with H—, because a carboxylic acid is formed only when there is a hydrogen on the carbon of the double bond (see Table 5.1).

Finally, connect the dangling ends of the double bonds in the two partial structures to generate the structure of the alkene:

alkene structure

PROBLEMS

5.12   Give the products (if any) expected from the treatment of each of the following compounds with ozone followed by dimethyl sulfide.

(a) 3-methyl-2 pentene   (b)   (c) cyclooctene   (d) 2-methylpentane

5.13   Give the products (if any) expected when the compounds in Problem 5.12 are treated with ozone followed by aqueous hydrogen peroxide.

5.14   In each case, give the structure of an eight-carbon alkene that would yield each of the following compounds (and no others) after treatment with ozone followed by dimethyl sulfide.

(a) $CH_3CH_2CH_2CH{=}O$ (b)   (c)

5.15   What aspect of alkene structure cannot be determined by ozonolysis?

5.16   Ozonolysis of 2-pentene gives a mixture of the following *three* ozonides. Using the mechanism in Eqs. 5.36a and 5.36b, explain the origin of all three ozonides.

A	*B*	*C*

TABLE 5.1 Summary of Ozonolysis Results Under Different Conditions

	Conditions of ozonolysis	
Alkene carbon	**O_3, then $(CH_3)_2S$**	**O_3, then H_2O_2/H_2O**
R\nC=\nR	R\nC=O\nR\nketone	R\nC=O\nR\nketone
R\nC=\nH	R\nC=O\nH\naldehyde	R\nC=O\nHO\ncarboxylic acid
H\nC=\nH	H\nC=O\nH\nformaldehyde	H\nC=O\nHO\nformic acid

5.6 FREE-RADICAL ADDITION OF HYDROGEN BROMIDE TO ALKENES

A. The Peroxide Effect

Recall that addition of HBr to alkenes is a regioselective reaction in which the bromine is directed to the carbon of a double bond with the greater number of alkyl groups (Sec. 4.7A). For example, 1-pentene reacts with HBr to give almost exclusively 2-bromopentane:

$$CH_3CH_2CH_2CH{=}CH_2 + H{-}Br \longrightarrow CH_3CH_2CH_2\overset{|}{\underset{\underset{Br}{|}}{C}}HCH_3 \qquad (5.42)$$

1-pentene

2-bromopentane
(79% yield)

For many years, results such as this were at times difficult to reproduce. Some investigators found that the addition of HBr was a highly regioselective reaction, as shown in Eq. 5.42. Others, however, obtained mixtures of constitutional isomers in which the second isomer had bromine bound at the carbon of the double bond with *fewer* alkyl groups. In the late 1920s, Morris Kharasch (1895–1957) of the University of Chicago began investigations that led to a solution of this puzzle. He found that when traces of *peroxides* (compounds of the general structure R—O—O—R) are present in the reaction mixture, *the regioselectivity of HBr addition is reversed*! In other words, 1-pentene was found to react in the presence of peroxides so that the bromine adds to the *less branched* carbon of the double bond:

$$\underset{\textbf{benzoyl peroxide}}{\underset{\text{(small amount used)}}{Ph{-}\overset{\overset{O}{\|}}{C}{-}O{-}O{-}\overset{\overset{O}{\|}}{C}{-}Ph}}$$

$$CH_3CH_2CH_2CH{=}CH_2 + H{-}Br \xrightarrow{\hspace{3cm}} CH_3CH_2CH_2CH_2CH_2{-}Br \qquad (5.43)$$

1-pentene

1-bromopentane
(96% yield)

(Contrast this result with that in Eq. 5.42.) This reversal of regioselectivity in HBr addition is called the **peroxide effect**.

> Because the regioselectivity of ordinary HBr addition is described by *Markovnikov's rule* (p. 153), the peroxide-promoted addition can be said to have *non-Markovnikov* or *anti-Markovnikov* regioselectivity. This means simply that the bromine is directed to the carbon of the alkene double bond with fewer alkyl substituents.

It was also found that light further promotes the peroxide effect. When Kharasch and his colleagues scrupulously excluded peroxides and light from the reaction, they found that HBr addition has the "normal" regioselectivity shown in Eq. 5.42.

The peroxide effect is observed with all alkenes in which alkyl substitution at the two carbons of the double bond is different. In other words, *in the presence of peroxides, the addition of HBr to alkenes occurs such that the hydrogen is bound to the carbon of the double bond bearing the greater number of alkyl substituents.* Furthermore, the peroxide-promoted reaction is faster than HBr addition in the the absence of peroxides. Very small amounts of peroxides are required to bring about this effect.

$$(5.44)$$

The regioselectivity of HI or HCl addition to alkenes is *not* affected by the presence of peroxides. For these hydrogen halides, the normal regioselectivity of addition predominates, whether peroxides are present or not. (The reason for this difference is discussed in Sec. 5.6E.)

The addition of HBr to alkenes in the presence of peroxides occurs by a mechanism that is completely different from that for normal addition. This mechanism involves reactive intermediates known as *free radicals*. To appreciate the reasons for the peroxide effect, then, let's digress and learn some basic facts about free radicals.

B. Free Radicals and the "Fishhook" Notation

In all reactions considered previously, we used a curved-arrow notation that indicates the movement of electrons in pairs. The dissociation of HBr, for example, is written

$$\text{H} \overset{\frown}{-} \overset{..}{\text{Br}}: \longrightarrow \text{H}^+ + :\overset{..}{\text{Br}}:^- \qquad \textit{heterolytic (two-electron) bond breaking} \qquad (5.45)$$

In this reaction, bromine takes *both* electrons of the H—Br covalent bond to give a bromide ion, and the hydrogen becomes an electron-deficient species, the proton. This type of bond breaking is an example of a *heterolysis*, or *heterolytic cleavage* (*hetero* = different; *lysis* = bond breaking). In a **heterolytic process**, electrons involved in the process "move" in pairs. Thus, a **heterolysis** is a bond-breaking process that occurs with electrons "moving" in pairs.

However, bond rupture can occur in another way. An electron-pair bond may also break so that each bonding partner takes one electron of the chemical bond.

$$\text{H} - \overset{..}{\text{Br}}: \longrightarrow \text{H} \cdot + \cdot \overset{..}{\text{Br}}: \qquad \textit{homolytic (one-electron) bond breaking} \qquad (5.46)$$

In this process, a hydrogen *atom* and a bromine *atom* are produced. As you should verify, these atoms are uncharged. This type of bond breaking is an example of a *homolysis*, or *homolytic cleavage* (*homo* = the same; *lysis* = bond breaking). In a **homolytic process**, electrons involved in the process "move" in an unpaired way. Thus, a **homolysis** is a bond-breaking process that occurs with electrons moving in an unpaired fashion.

A different curved-arrow notation is used for homolytic processes. In this notation, called the **fishhook notation**, *electrons move individually rather than in pairs*. This type of electron flow is represented with singly barbed arrows, or fishhooks; one fishhook is used for each electron:

homolytic bond breaking uses the *fishhook notation*

$$\text{H} \cdot \cdot \overset{..}{\text{Br}}: \quad \text{or} \quad \text{H} \overset{\frown\frown}{-} \overset{..}{\text{Br}}: \longrightarrow \text{H} \cdot + \cdot \overset{..}{\text{Br}}: \qquad (5.47)$$

Homolytic bond cleavage is not restricted to diatomic molecules. For example, peroxides, because of their very weak O—O bonds, are prone to undergo homolytic cleavage:

$$(\text{CH}_3)_3\text{C} - \overset{..}{\underset{..}{\text{O}}} \overset{\frown\frown}{-} \overset{..}{\underset{..}{\text{O}}} - \text{C}(\text{CH}_3)_3 \longrightarrow 2(\text{CH}_3)_3\text{C} - \overset{..}{\underset{..}{\text{O}}} \cdot \qquad (5.48)$$

di-*tert*-butyl peroxide   ***tert*-butoxy radical**

The fragments on the right side of this equation possess unpaired electrons. Any species with at least one unpaired electron is called a **free radical**. The hydrogen atom and the bromine atom on the right side of Eq. 5.47, as well as the *tert*-butoxy radical on the right side of Eq. 5.48, are all examples of free radicals.

Radicals Bound and Free

The "R" used in the R-group notation (Sec. 2.9B) comes from the word *radical*. In the mid-1900s, R groups were called "radicals." Thus, the CH_3 group in CH_3CH_2OH might have been referred to as the "methyl radical." Such R groups, when not bonded to anything, were then said to be "free radicals." Thus, $\cdot CH_3$ was said to be the "methyl free radical." Nowadays we simply use the word *group* to refer to a group of atoms (for example, methyl group) in a compound; we reserve the word *radical* for a species with an unpaired electron.

PROBLEMS

5.17   Draw the products of each of the following transformations shown by the fishhook notation.

(a) $(CH_3)_3C\overset{\frown\frown}{-}C(CH_3)_3 \longrightarrow$ (b) $(CH_3)_2C\!\!=\!\!CH_2$ $\cdot\ddot{B}\ddot{r}\!:$ $\longrightarrow$

(c) $\left[H_2\overset{\frown}{C}\!\!=\!\!CH\overset{\frown}{-}CH\!\!=\!\!CH\overset{\frown}{-}CH_2\cdot \quad \longleftrightarrow \quad \right]$

(d) $\overset{\frown}{R}\overset{\frown}{-}CH_2\overset{\frown}{-}CH_2\cdot \longrightarrow$

5.18   Indicate whether each of the following reactions is homolytic or heterolytic, and tell how you know. Write the appropriate fishhook or curved-arrow notation for each.

(a) $:N\!\!\equiv\!\!\overset{-}{C}\!: + \underset{\underset{\displaystyle :\ddot{B}r:}{|}}{CH_2CH_3} \longrightarrow :N\!\!\equiv\!\!C\!-\!CH_2CH_3 + :\ddot{B}\ddot{r}:^-$

(b) $CH_3CH_2OH + CH_3\ddot{O}\cdot \longrightarrow CH_3\dot{C}HOH + CH_3\ddot{O}H$

C. Free-Radical Chain Reactions

Although a few stable free radicals are known, most free radicals are very reactive. When they are generated in chemical reactions, they generally behave as *reactive intermediates*—that is, they react before they can accumulate in significant amounts. This section shows how free radicals are involved as reactive intermediates in the peroxide-promoted addition of HBr to alkenes. This discussion will provide the basis for a general understanding of free-radical reactions, as well as the peroxide effect in HBr addition, which is the subject of the next section.

Most free-radical reactions can be classified as free-radical chain reactions. A **free-radical chain reaction** involves free-radical intermediates and consists of the following three fundamental reaction steps:

1. *initiation* steps,
2. *propagation* steps, and
3. *termination* steps.

Let's examine each of these steps using the peroxide-promoted addition of HBr to alkenes as an example of a typical free-radical reaction. (The reason for the "chain reaction" terminology will become apparent.)

Initiation   In the **initiation** steps, the free radicals that take part in subsequent steps of the reaction are formed from a **free-radical initiator**, which is a molecule that undergoes homolysis with particular ease. *The initiator is the source of free radicals.* Peroxides, such as di-*tert*-butyl peroxide, are frequently used as free-radical initiators. The first initiation step in the free-radical addition of HBr to an alkene is the homolysis of the peroxide.

$$(CH_3)_3C - \overset{..}{\underset{..}{O}} - \overset{..}{\underset{..}{O}} - C(CH_3)_3 \longrightarrow 2\,(CH_3)_3C - \overset{..}{\underset{..}{O}} \cdot \qquad (5.49)$$

di-*tert*-butyl peroxide   ***tert*-butoxy radical**

Although most peroxides can serve as free-radical initiators, a notable exception is hydrogen peroxide (H_2O_2), which is *not* commonly used as a source of initiating free radicals. The reason is that the O—O bond in hydrogen peroxide is considerably stronger than the O—O bonds in most other peroxides and is therefore harder to break homolytically. The cleavage of organoboranes by hydrogen peroxide (the oxidation part of hydroboration–oxidation; Eq. 5.29, p. 195) is *not* a free-radical reaction. (See Further Exploration 5.1.)

Peroxides are not the only type of free-radical initiators. Another widely used initiator is azoisobutyronitrile, known by the acronym AIBN. This substance readily forms free radicals because the very stable molecule dinitrogen is liberated as a result of homolytic cleavage:

$$NC - \overset{\overset{\textstyle CH_3}{|}}{\underset{\underset{\textstyle CH_3}{|}}{C}} - \overset{..}{N} = \overset{..}{N} - \overset{\overset{\textstyle CH_3}{|}}{\underset{\underset{\textstyle CH_3}{|}}{C}} - CN \longrightarrow 2\,NC - \overset{\overset{\textstyle CH_3}{\diagup}}{\underset{\underset{\textstyle CH_3}{\diagdown}}{C}}\cdot \; + \; :N \equiv N: \qquad (5.50)$$

azoisobutyronitrile (AIBN)   **molecular nitrogen (dinitrogen)**

Sometimes heat or light initiates a free-radical reaction. This usually happens because the additional energy causes homolysis of the free-radical initiator—or, in some cases, the reactants themselves—into free radicals.

The effects of initiators provide some of the best clues that a reaction occurs by a free-radical mechanism. If a reaction occurs in the presence of a known free-radical initiator but does not occur in its absence, we can be fairly certain that the reaction involves free-radical intermediates. (Recall from Sec. 5.6A that Morris Kharasch proved that the change in regiospecificity of HBr addition requires peroxides. This is the type of evidence that we now take to be strongly indicative of free-radical mechanisms.)

A second initiation step occurs in the free-radical addition of HBr to alkenes: the removal of a hydrogen atom from HBr by the *tert*-butoxy free radical that was formed in the first initiation step (Eq. 5.49).

$$(CH_3)_3C - \overset{..}{\underset{..}{O}} \cdot \quad H - \overset{..}{\underset{..}{Br}}: \longrightarrow (CH_3)_3C - \overset{..}{\underset{..}{O}} - H + \cdot \overset{..}{\underset{..}{Br}}: \qquad (5.51)$$

***tert*-butoxy radical**
(from Eq. 5.49)

This is an example of another common type of free-radical process, called *atom abstraction*. In an atom abstraction reaction, a free radical removes an atom from another molecule, and a new free radical is formed (Br· in Eq. 5.51). The bromine atom is involved in the next phase of the reaction: the *propagation* steps.

Propagation   In the **propagation steps** of a free-radical reaction, radicals react with non-radical starting materials to give other radicals; starting materials are consumed and products are formed. *Propagation steps occur repeatedly.* When the propagation steps are considered together, there is no *net* formation or destruction of any of the radical species involved. This means that if a radical is formed, it must be consumed in a subsequent propagation step and another radical must be formed to take its place.

The first propagation step of free-radical addition of HBr to an alkene is the reaction of the bromine atom generated in Eq. 5.51 with the π bond.

$$R—CH{=}CH—R \quad \longrightarrow \quad R—\overset{\cdot}{C}H—CH—R \qquad (5.52a)$$
$$\cdot\ddot{B}r\text{:} \qquad\qquad\qquad :\ddot{B}r\text{:}$$

Reaction of a free radical with a carbon–carbon π bond is another common process encountered in free-radical chemistry. The π bond reacts, rather than a σ bond, because carbon–carbon π bonds are weaker than carbon–carbon σ bonds.

The second propagation step is another *atom abstraction reaction*: removal of a hydrogen atom from HBr by the free-radical product of Eq. 5.52a to give the addition product and a new bromine atom.

$$R—\overset{\cdot}{C}H—CH—R \quad \longrightarrow \quad R—CH—CH—R \;+\; :\ddot{\overset{\cdot\cdot}{B}r}\cdot \qquad (5.52b)$$
$$Br \qquad\qquad\qquad H \quad\; Br$$
$$:\ddot{B}r{—}H$$

The bromine atom, in turn, can react with another molecule of alkene (Eq. 5.52a), and this can be followed by the generation of another molecule of product along with another bromine atom (Eq. 5.52b). We can now see the basis of the term *chain reaction*. These two propagation steps continue in a chainlike fashion until the reactants are consumed. That is, the product free radical of one propagation step becomes the starting free radical for the next propagation step. For each "link in the chain"—each cycle of the two propagation steps—one molecule of the product is formed and one molecule of alkene starting material is consumed. For each free radical consumed in the propagation steps, one is produced. Because no net destruction of free radicals occurs, the initial concentration of free radicals provided by the initiator, and thus the concentration of the initiator itself, can be small. Typically, the initiator concentration is only 1–2% of the alkene concentration.

The free radicals involved in the propagation steps of a chain reaction are said to *propagate the chain*. The free radicals in Eq. 5.52a–b are the chain-propagating radicals in the peroxide-promoted free-radical addition of HBr.

> Some students think that Eq. 5.51 is part of the propagation sequence because it forms a bromine atom, which is one of the chain-propagating radicals. Rather, it is part of the initiation sequence because the *tert*-butoxy radical does not reappear in the subsequent steps of the reaction. Equation 5.51 must occur only once for the subsequent propagation steps to occur repeatedly.

An Analogy for Chain Reactions

An analogy for a chain reaction can be found in the world of business. A businessperson uses a little seed money, or capital, to purchase a small business. In time, this business produces profit that is used to buy another business. This second business, in turn, produces profit that can be used to buy yet another business, and so on. All this time, the businessperson is accumulating busi-ness property (instead of alkyl bromides), although the total amount of cash on hand is, by analogy to the chain-propagating free radicals in the reaction sequence above, small compared with the total amount of the investments.

The toppling of the dominoes shown in the photograph is another example of a chain reaction from the physical world. To initiate the process, the finger needs to provide only enough energy to topple the first domino. What "propagates the chain" once it is initiated?

Termination   In the **termination** steps of a free-radical chain reaction, two radicals react to give nonradical products. Typically, termination involves a *radical recombination reaction*, in which two radicals come together to form a covalent bond. In other words, radical recombination is the reverse of a homolysis.

The following reactions are two examples of termination reactions that can take place in the free-radical addition of HBr to alkenes. In these reactions, the chain-propagating radicals of Eqs. 5.52a and 5.52b recombine to form by-products. These by-products are present in very small amounts because they are formed *only* from free radicals, which are also present in very small amounts.

$$\text{Br·} \frown \text{·Br} \longrightarrow \text{Br}_2 \qquad (5.53)$$

$$
\begin{array}{c}
\overset{\displaystyle \text{Br}}{\underset{|}{}} \\
\text{R}-\text{CH}-\text{CH}-\text{R} \\
\text{·} \\
\text{R}-\text{CH}-\text{CH}-\text{R} \\
\underset{|}{} \\
\text{Br}
\end{array}
\longrightarrow
\begin{array}{c}
\overset{\displaystyle \text{Br}}{\underset{|}{}} \\
\text{R}-\text{CH}-\text{CH}-\text{R} \\
| \\
\text{R}-\text{CH}-\text{CH}-\text{R} \\
\underset{|}{} \\
\text{Br}
\end{array}
\qquad (5.54)
$$

Because each recombination reaction takes two free radicals "out of circulation," it terminates two propagation reactions and breaks two free-radical chains.

The recombination reactions of free radicals are in general highly exothermic; that is, they have very favorable, or negative, $\Delta H°$ values. They typically occur on every encounter of two free radicals; in other words, there is no $\Delta G°^{\ddagger}$ for radical recombination. In view of this fact, we might ask why free radicals do not simply recombine before they propagate any chains. The answer is simply a matter of the relative concentrations of the various species involved. *Free-radical intermediates are present in very low concentration*, but the other reactants are present in much higher concentration. Consequently, it is much more probable for a bromine atom to collide with an alkene molecule in the propagation reaction of Eq. 5.52a than with another bromine atom in a recombination reaction:

$$
\text{Br·}
\begin{cases}
\xrightarrow[\textit{common occurrence}]{\text{[RCH=CHR] is large;}} & \begin{array}{c} \text{Br}-\text{CH}-\overset{\displaystyle\cdot}{\text{CH}}-\text{R} \\ | \\ \text{R} \end{array} \\[2em]
\xrightarrow[\textit{rare occurrence}]{\text{[Br·] is small;}} & \text{Br}-\text{Br}
\end{cases}
\qquad (5.55)
$$

In a typical free-radical chain reaction, a termination reaction occurs once for every 10,000 propagation reactions. As the reactants are depleted, however, the probability becomes significantly greater that one free radical will survive long enough to wander into another free radical with which it can recombine. Small amounts of by-products resulting from termination reactions are typically observed in free-radical chain reactions.

In most cases, only exothermic propagation steps—steps with favorable, or negative, $\Delta H°$ values—occur rapidly enough to compete with the recombination reactions that terminate free-radical chain processes. Both of the propagation steps in the free-radical HBr addition to alkenes are exothermic, and they both occur readily. However, the first propagation step of free-radical HI addition, and the second propagation step of free-radical HCl addition, are quite *endothermic*. (This point is explored further in Sec. 5.6E.) For this reason, these processes occur to such a small extent that they cannot compete with the recombination processes that terminate these chain reactions. Consequently, the free-radical addition of neither HCl nor HI to alkenes is observed.

In summary, we've learned that many free-radical reactions occur in three phases:

1.   In the initiation steps, radicals are produced from nonradicals.

2. In propagation steps, radicals react with nonradical starting materials to give other radicals and nonradical products. Propagation steps occur repeatedly.

3. In termination steps, radicals react with each other to give small amounts of non-radical by-products.

In most free-radical reactions, products accumulate as the result of propagation steps because they occur much more frequently than termination steps.

   This section has discussed the characteristics of free-radical chain reactions using the free-radical addition of HBr to alkenes as an example. A number of other useful laboratory reactions involve free-radical chain mechanisms. Many very important industrial processes are also free-radical chain reactions (Sec. 5.7). Free-radical chain reactions of great environmental importance occur in the upper atmosphere when ozone is destroyed by chlorofluoro-carbons (Freons), the compounds that until recently have been exclusively used as coolants in air conditioners and refrigerators. (These reactions are discussed in Sec. 9.10C.) A number of free-radical processes have also been characterized in biological systems.

STUDY PROBLEM 5.4

Alkenes undergo the addition of thiols at high temperature in the presence of peroxides or other free-radical initiators. The following reaction is an example.

cyclopentene   **ethanethiol**
                     (a thiol)   a thioether or sulfide

Propose a mechanism for this reaction.

SOLUTION The fact that the reaction requires peroxides tells us that a free-radical mechanism is operating. The initiation step is abstraction of a hydrogen atom from the thiol by the *tert*-butoxy radical derived from homolysis of the free-radical initiator. (See Eq. 5.49.)

$$CH_3CH_2S-H \quad \cdot OC(CH_3)_3 \longrightarrow CH_3CH_2S\cdot + H-OC(CH_3)_3 \tag{5.56a}$$

***tert*-butoxy radical**

Notice two things about this reaction. This is *not* an acid–base reaction. Terms such as acid, base, nucleophile, electrophile, and leaving group are associated with *heterolytic* (electron-pair) reactions. In this reaction, a hydrogen *atom*, not a proton, is transferred. Second, when we write free-radical mechanisms, we can omit unshared pairs and show only the unpaired electrons. This is because free radicals in most common cases are uncharged; therefore, the calculation of formal charge and the use of the octet rule are generally not issues.

   In the first propagation step, the sulfur radical adds to the double bond of the alkene to generate a new carbon radical.

$$\tag{5.56b}$$

In the final propagation step, the carbon radical reacts with a thiol molecule to give the product plus a new sulfur radical, which propagates the chain by reacting with another alkene.

+ $\cdot SCH_2CH_3$
(reacts with
another alkene
molecule)

$$\tag{5.56c}$$

Some students are tempted to write a mechanism such as the following:

$$\tag{5.57}$$

This cannot be correct for several reasons. First, this reaction destroys free radicals. If this were the mechanism, then we would need a separate initiation step for every product molecule formed, and the initiator would then have to be present at the same concentration as the reactants. Second, there is no obvious source for the hydrogen atom. Finally, and most important, free radicals are typically present in *miniscule* concentrations. A collision of three molecules, two of which (the radicals) are present in very low concentration, is so improbable that it doesn't occur. Thus, *most free-radical reactions occur as chain reactions.* In a chain reaction, only one free radical reacts in a given step; it reacts with a molecule present in substantial concentration; and a new radical is produced. This last point means that initiators must supply only a small initial concentration of radicals to "get things started," because a small population of free radicals is maintained until the reactants run out (at which point termination reactions can compete with propagation steps).

So, here's the message: *Write chain mechanisms involving only one radical per step* when you write free-radical reactions.

PROBLEMS

5.19 In the presence of light, the addition of Br$_2$ to alkenes can occur by a free-radical mechanism rather than a bromonium-ion mechanism. Write a free-radical chain mechanism that shows the propagation steps for the following addition:

$$Br_2 + H_2C{=}CH_2 \xrightarrow{\text{light}} BrCH_2{-}CH_2Br$$

Assume that the initiation step for the reaction is the light-promoted homolysis of Br$_2$:

$$Br\frown\frown Br \xrightarrow{\text{light}} 2\,Br\cdot$$

5.20 (a) Suggest a mechanism for the free-radical addition of HBr to cyclohexene initiated by AIBN. Show the initiation and propagation steps.

   (b) In the free-radical addition of HBr to cyclohexene, suggest structures for three radical recombination products that might be formed in small amounts in the termination phase of the reaction.

D. Explanation of the Peroxide Effect

The free-radical mechanism is the basis for understanding the peroxide effect on HBr addition to alkenes—that is, why the presence of peroxides reverses the normal regioselectivity of HBr addition (Eq. 5.44, p. 203). The following example will serve as the basis for our discussion.

$$\begin{array}{c} H_3C \\ \diagdown \\ C{=}CH_2 \\ \diagup \\ H_3C \end{array} \xrightarrow{\text{HBr, peroxides}} \begin{array}{c} H_3C \\ \diagdown \\ CH{-}CH_2{-}Br \\ \diagup \\ H_3C \end{array} \tag{5.58}$$

Recall that the reaction is initiated by the formation of a bromine atom from HBr (Eq. 5.51). When the bromine atom adds to the π bond of an alkene, two reactions are in competition: the bromine atom can react at either of the two carbons of the double bond to give different free-radical intermediates.

$$\text{(5.59)}$$

a tertiary
free radical   a primary
free radical

What is the difference between these two free radicals? Free radicals, like carbocations, can be classified as primary, secondary, or tertiary.

$$R-\dot{C}H_2 \qquad R-\dot{C}H-R \qquad \underset{\underset{R}{|}}{\overset{R}{\underset{}{}}}R-\overset{R}{\underset{}{\dot{C}}}-R \qquad (5.60)$$

primary   secondary   tertiary

Equation 5.59 thus involves a competition between the formation of a primary and the formation of a tertiary free radical. *The formation of the tertiary free radical is faster.*

The tertiary free radical is formed more rapidly for two reasons. The first reason is that when the rather large bromine atom reacts at the more-branched carbon of the double bond, it experiences van der Waals repulsions with the hydrogens in the branches (Fig. 5.2a). These repulsions increase the energy of this transition state. When the bromine atom reacts at the less-branched carbon of the double bond, these van der Waals repulsions are absent (Fig. 5.2b). Because the reaction with the transition state of lower energy is the faster reaction, reaction of the bromine atom at *the alkene carbon with fewer alkyl substituents to give the more alkyl-substituted free radical* is faster. Subsequent reaction of this free radical with HBr leads to the observed products. To summarize:

(5.61)

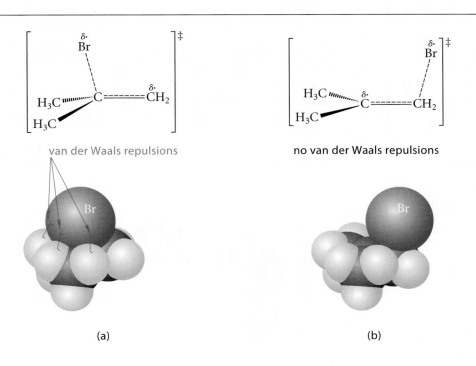

van der Waals repulsions

no van der Waals repulsions

(a)   (b)

FIGURE 5.2 Space-filling models of the alternative transition states for the addition of a bromine atom to 2-methylpropene. In part (a), the bromine is adding to the carbon of the double bond that has the two methyl substituents. This transition state contains van der Waals repulsions between the bromine and four of the six methyl hydrogens, which are shown in pink. (Three of these are shown; one is hidden from view.) In part (b), the bromine is adding to the CH_2 carbon of the double bond, and the van der Waals repulsions shown in part (a) are absent. The transition state in part (b) has lower energy and therefore leads to the observed product.

TABLE 5.2 Heats of Formation of Some Free Radicals (25 °C)

Radical	Structure	ΔH_f° (kJ mol^{-1})	ΔH_f° (kcal mol^{-1})
methyl	·CH$_3$	146.6	35.0
ethyl	·CH$_2$CH$_3$	121.3	29.0
propyl	·CH$_2$CH$_2$CH$_3$	100.4	24.0
isopropyl	CH$_3$ĊHCH$_3$	90.0	21.5
butyl	·CH$_2$CH$_2$CH$_2$CH$_3$	79.7	19.0
isobutyl	·CH$_2$CH(CH$_3$)$_2$	70	17
sec-butyl	CH$_3$ĊHCH$_2$CH$_3$	67.4	16.1
tert-butyl	·C(CH$_3$)$_3$	51.5	12.3

When a chemical phenomenon (such as a reaction) is affected by van der Waals repulsions, it is said to be influenced by a **steric effect** (from the Greek *stereos*, meaning "solid"). Thus, the regioselectivity of free-radical HBr addition to alkenes is due in part to a steric effect. Other examples of steric effects that we've already studied are the preference of butane for the anti rather than the gauche conformation (Sec. 2.3B), and the greater stability of *trans*-2-butene relative to *cis*-2-butene (Sec. 4.5B).

The second reason that the tertiary radical is formed in Eq. 5.61 has to do with its relative stability. The heats of formation of several free radicals are given in Table 5.2. Comparing the heats of formation for propyl and isopropyl radicals, or for butyl and *sec*-butyl radicals, shows that the secondary radical is more stable than the primary one by about 12 kJ mol^{-1} (about 3 kcal mol^{-1}). Similarly, the *tert*-butyl radical is more stable than the *sec*-butyl radical by about 16 kJ mol^{-1} (4 kcal mol^{-1}). Therefore:

Stability of free radicals:

$$\text{tertiary} > \text{secondary} > \text{primary} \tag{5.62}$$

Notice that free radicals have the same stability order as carbocations. However, the energy differences between isomeric free radicals are only about one-fifth the magnitude of the differences between the corresponding carbocations. (Compare Tables 5.2 and 4.2 on p. 156.)

This free-radical stability order can be understood from the geometry and hybridization of a typical carbon radical. The methyl radical ·CH$_3$ is trigonal planar (Fig. 5.3), but other carbon radicals are slightly pyramidal. However, they are close enough to planarity that we can, to a useful approximation, consider them also to be sp^2-hybridized with the unpaired electron in a $2p$ orbital. The stability order in Eq. 5.62 implies, then, that *free radicals are stabilized by alkyl-group substitution at sp^2-hybridized carbons.* The magnitude of the alkyl-group stabilization of free radicals is very similar to that observed for alkyl substitution at the sp^2-hybridized carbons of alkenes (Sec. 4.5B).

By Hammond's postulate (Sec. 4.8D), a more stable free radical should be formed more rapidly than a less stable one. Thus, when a bromine atom reacts with the π bond of an

FIGURE 5.3 Geometry of the methyl radical. The carbon is sp^2-hybridized with trigonal planar geometry, and the unshared electron occupies a $2p$ orbital. Other carbon radicals are slightly pyramidal.

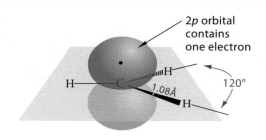

alkene, it adds to the carbon of the alkene with *fewer alkyl substituents* because this places the unpaired electron on the carbon with *more alkyl substituents*. In other words, the *more stable free radical* is formed. The product of HBr addition is formed by the subsequent reaction of this free radical with HBr (Eq. 5.52b, p. 207). Notice that whether we consider steric effects in the transition state or the relative stabilities of free radicals, the same outcome of free-radical HBr addition is predicted.

Understanding the regioselectivity of free-radical HBr addition to alkenes provides an understanding of the peroxide effect—why the regioselectivity of HBr addition to alkenes differs in the presence and absence of peroxides. Both reactions begin by attachment of an atom to the carbon of the double bond with *fewer alkyl substituents*. In the absence of peroxides, the *proton adds first* to give a carbocation at the carbon with the *greater number of alkyl substituents*. The nucleophilic reaction of the bromide ion at this carbon completes the addition. In the presence of peroxides, the free-radical mechanism occurs; a *bromine atom adds first*, thus placing the unpaired electron on the carbon with the *greater number of alkyl substituents*. A hydrogen atom is subsequently transferred to this carbon.

(5.63)

Electrophilic addition: a proton adds first to give the more stable carbocation.

Radical addition: a bromine atom adds first to give the more stable carbon radical.

The role of peroxides is to initiate the radical reaction. Any good free-radical initiator will bring about the same effect.

PROBLEMS

5.21   Give the structure the organic product(s) formed when HBr reacts with each of the following alkenes in the presence of peroxides, and explain your reasoning. If more than one product is formed, predict which one should predominate and why.
(a) 1-pentene   (b) (*E*)-4,4-dimethyl-2-pentene

5.22   Give the structures of the free radical intermediates in the peroxide-initiated reaction of HBr with each of the following alkenes.

E. Bond Dissociation Energies

How easily does a chemical bond break homolytically to form free radicals? The question can be answered by examining its *bond dissociation energy*. The **bond dissociation energy** of a bond between two atoms X—Y is defined as the standard enthalpy $\Delta H°$ of the reaction

$$X\text{—}Y \longrightarrow X\cdot + Y\cdot \tag{5.64}$$

Notice that a bond dissociation energy always corresponds to the enthalpy required to break a bond *homolytically*. Thus, the bond energy of HLBr refers to the process

$$H\text{—}Br \longrightarrow H\cdot + \cdot Br \tag{5.65}$$

and *not* to the heterolytic process

$$H\text{—}\ddot{Br}\colon \longrightarrow H^+ + \colon\ddot{Br}\colon \tag{5.66}$$

Some bond dissociation energies are collected in Table 5.3 on p. 216. *A bond dissociation energy measures the intrinsic strength of a chemical bond.* For example, breaking the H—H bond requires 435 kJ mol⁻¹ (104 kcal mol⁻¹) of energy. It then follows that forming the hydrogen molecule from two hydrogen atoms liberates 435 kJ mol⁻¹ (104 kcal mol⁻¹) of energy. Table 5.3 shows that different bonds exhibit significant differences in bond strength; even bonds of the same general type, such as the various C—H bonds, can differ in bond strength by many kilojoules per mole.

The bond dissociation energies in Table 5.3 (or others available from compilations in the literature) can be used in a number of ways. A common use of these energies is to estimate the $\Delta H°$ of a nonradical reaction by using Hess's law to treat it as the sum of fictitious radical reactions. The sum of the $\Delta H°$ values of the radical reactions, obtained from bond dissociation energies, provides the $\Delta H°$ of the overall reaction. (See Further Exploration 5.3.) For example, if you return to the discussion of the element effect on acidity (Eq. 3.38a–d, p. 113), you will see that bond dissociation energies were used to show the effect of bond strength on acidity. (Problem 5.23 on p. 217 also illustrates this idea.)

A consideration of bond dissociation energies shows why di-*tert*-butyl peroxide is an excellent free-radical initiator. The lower a bond dissociation energy, the lower the temperature required to rupture the bond in question and form free radicals at a reasonable rate. The homolysis of the O—O bond in di-*tert*-butyl peroxide requires only 159 kJ mol⁻¹ (38 kcal mol⁻¹) of energy; this is one of the smallest bond dissociation energies in Table 5.3 on p. 216. With such a small bond dissociation energy, this peroxide readily forms small amounts of free radicals when it is heated gently or when it is subjected to ultraviolet light.

An important use of bond dissociation energies is to calculate or estimate the $\Delta H°$ of free-radical reactions. As an illustration, consider the second initiation step for the free-radical addition of HBr, in which a *tert*-butoxy radical reacts with H—Br (Eq. 5.51, p. 206).

$$(CH_3)_3C—O\cdot \ + \ H—Br \ \longrightarrow \ (CH_3)_3C—O—H \ + \ \cdot Br \qquad (5.67)$$

tert-butoxy radical

In this reaction, a hydrogen is abstracted from HBr by the *tert*-butoxy radical. This is not the only reaction that might occur. Instead, the *tert*-butoxy radical might abstract a bromine atom from HBr:

$$(CH_3)_3C—O\cdot \ + \ H—Br \ \longrightarrow \ (CH_3)_3C—O—Br \ + \ \cdot H \qquad (5.68)$$

tert-butoxy radical

Why is hydrogen and not bromine abstracted? The reason lies in the relative enthalpies of the two reactions. These enthalpies are not known by direct measurement, but can be calculated using bond dissociation energies. To calculate the $\Delta H°$ for a reaction, *subtract the bond dissociation energies (BDE) of the bonds formed from the bond dissociation energies of the bonds broken.*

$$\Delta H° = BDE \ (\text{bonds broken}) - BDE \ (\text{bonds formed}) \qquad (5.69)$$

This works because BDEs are the enthalpies for bond dissociation. This procedure is illustrated in Study Problem 5.5.

FURTHER EXPLORATION 5.3
Bond Dissociation Energies and Heats of Reaction

STUDY PROBLEM 5.5

Estimate the standard enthalpies of the reactions shown in Eqs. 5.67 and 5.68.

SOLUTION To obtain the required estimates, apply Eq. 5.69. In both equations, the bond broken is the H—Br bond. From Table 5.3, the bond dissociation energy of this bond is 368 kJ mol⁻¹ (88 kcal mol⁻¹). The bond formed in Eq. 5.67 is the O—H bond in $(CH_3)_3CO$—H (*tert*-butyl alcohol). This exact compound is not found in Table 5.3; what do we do? Look for the same type of bond in as similar a compound as possible. For example, the table includes an entry for the alcohol CH_3O—H (methyl alcohol). (The O—H bond dissociation energies for methyl alcohol and *tert*-butyl alcohol differ very little.) Thus, we use 438 kJ mol⁻¹ (105 kcal mol⁻¹) for the BDE of the O—H bond.

Subtracting the enthalpy of the bond formed (the O—H bond) from that of the bond broken (the H—Br bond), we obtain $368 - 438 = -70$ kJ mol^{-1} or $88 - 105 = -17$ kcal mol^{-1}. This is the enthalpy of the reaction in Eq. 5.67. In other words, breaking the H—Br bond costs 368 kJ mol^{-1} but the formation of the O—H bond gives back 438 kJ mol^{-1}, for a net $\Delta H°$ advantage of 70 kJ mol^{-1}.

Following the same procedure for Eq. 5.68, use the bond dissociation energy of the O—Br bond in $(CH_3)_3CO$—Br, which is given in Table 5.3 as 205 kJ mol^{-1} (49 kcal mol^{-1}). The calculated enthalpy for Eq. 5.68 is then $368 - 205 = 163$ kJ mol^{-1} or $88 - 49 = 39$ kcal mol^{-1}. These $\Delta H°$ estimates are the required solution to the problem.

The calculation in Study Problem 5.5 shows that reaction 5.67 is highly exothermic (favorable) and reaction 5.68 is highly endothermic (unfavorable). Thus, it is not surprising that the abstraction of H (Eq. 5.67) is the only one that occurs. Abstraction of Br (Eq. 5.68) is so unfavorable energetically that it does not occur to any appreciable extent.

Strictly speaking, we need $\Delta G°$ values to determine whether a reaction will occur spontaneously, because $\Delta G°$ determines the equilibrium constant. However, for similar reactions, such as the two considered here, the differences between $\Delta H°$ and $\Delta G°$ tend to cancel in the comparison.

Recall that a peroxide effect is not observed when HCl and HI are added to alkenes. Bond dissociation energies can be used to help us understand this observation. Consider, for example, HI addition by a hypothetical free-radical mechanism, and compare the enthalpies for the addition of HBr and HI in the first propagation step. This propagation step involves the breaking of a π bond in each case (Eq. 5.70a, below) and the formation of a CH_2—X bond (Eq. 5.70b). Breaking the π bond requires about 243 kJ mol^{-1} (58 kcal mol^{-1}), from Table 5.3. The energy released on formation of a CH_2—X bond is approximated by the negative bond dissociation energy of the corresponding carbon–halogen bond in CH_3CH_2—X, also from Table 5.3. Because we are making the same approximation in comparing the two halogens, any error introduced tends to cancel in the comparison. The first propagation step (Eq. 5.70c) is the sum of these two processes:

	$\Delta H°$, kJ mol^{-1} (kcal mol^{-1})		
	X = Br	X = I	
$H_2C{=}CH_2 \longrightarrow H_2\dot{C}{-}\dot{C}H_2$	+243 (+58)	+243 (+58)	(5.70a)
$H_2\dot{C}{-}\dot{C}H_2 + X\cdot \longrightarrow H_2\dot{C}{-}CH_2{-}X$	−303 (−72)	−238 (−57)	(5.70b)
Sum: $X\cdot + H_2C{=}CH_2 \longrightarrow H_2\dot{C}{-}CH_2{-}X$	−60 (−14)	+5 (+1)	(5.70c)
	energetically favorable	energetically unfavorable	

This calculation shows that the first propagation step is exothermic (that is, energetically favorable) for HBr, but endothermic (that is, energetically unfavorable) for HI. Remember that the propagation steps of any free-radical chain reaction are in competition with recombination steps that terminate free-radical reactions. These recombination steps are so exothermic that they occur on every encounter of two radicals. The energy required for an endothermic propagation step, in contrast, represents an *energy barrier* that reduces the rate of this step. In effect, *only exothermic propagation steps compete successfully with recombination steps.* Hence, HI does not add to alkenes by a free-radical mechanism because the first propagation step is endothermic, and the radical chain is terminated. In Problem 5.24, you can explore from a bond-energy perspective why the addition of HCl also does not occur by a free-radical mechanism.

The use of bond dissociation energies for the calculation of $\Delta H°$ of reactions is not limited to free-radical reactions. It is necessary only that the reaction for which the calculation is

TABLE 5.3 Bond Dissociation Energies (25 °C)

Bond	$\Delta H°$ (kJ mol^{-1})	$\Delta H°$ (kcal mol^{-1})
C—H bonds		
H_3C—H	439	105
CH_3CH_2—H	423	101
$(CH_3)_2CH$—H	412	99
$(CH_3)_3C$—H	404	96
$PhCH_2$—H	378	90
$H_2C{=}CHCH_2$—H	372	89
$RCH{=}CH$—H	463	111
Ph—H	472	113
$RC{\equiv}C$—H	558	133
H—CN	528	126
C—Halogen bonds		
H_3C—F	481	115
$(CH_3)_2CH$—F	463	111
H_3C—Cl	350	84
CH_3CH_2—Cl	355	85
$(CH_3)_2CH$—Cl	356	85
$(CH_3)_3C$—Cl	355	85
H_3C—Br	302	72
CH_3CH_2—Br	303	72
$(CH_3)_2CH$—Br	309	74
$(CH_3)_3C$—Br	304	73
H_3C—I	241	58
CH_3CH_2—I	238	57
$(CH_3)_2CH$—I	238	57
$(CH_3)_3C$—I	233	56
Ph—F	531	127
Ph—Cl	406	97
Ph—Br	352	84
Ph—I	280	67
C—C bonds		
H_3C—CH_3	377	90
Ph—CH_3	433	104
$PhCH_2$—CH_3	318	78
H_3C—CN	510	122
$H_2C{=}CH_2$ (both)	728	174
$H_2C{=}CH_2$ (π bond)	243	58
$HC{\equiv}CH$	~966	~231

Bond	$\Delta H°$ (kJ mol^{-1})	$\Delta H°$ (kcal mol^{-1})
C—O bonds		
H_3C—OH	385	92
H_3C—OCH_3	347	83
Ph—OH	470	112
$H_2C{=}O$ (both)	749	179
$H_2C{=}O$ (π bond)	305	73
C—N bonds		
H_3C—NH_2	356	85
Ph—NH_2	435	104
$H_2C{=}NH$	~736	~176
$HC{\equiv}N$	~987	~236
H—X bonds		
H—OH	498	119
H—OCH_3	438	105
H—O_2CCH_3	473	113
H—OPh	377	90
H—F	569	136
H—Cl	431	103
H—Br	368	88
H—I	297	71
H—NH_2	450	108
H—SH	381	91
H—SCH_3	366	87
X—X bonds		
H—H	435	104
F—F	154	37
Cl—Cl	239	57
Br—Br	190	45
I—I	149	36
Other		
HO—OH	213	51
$(CH_3)_3CO$—$OC(CH_3)_3$	159	38
HO—Br	234	56
$(CH_3)_3CO$—Br	205	49

made does not create or destroy ions, and that the bond dissociation energies of the appropriate bonds are known or can be closely estimated. (See Problem 5.23.)

Another point worth noting is that bond dissociation energies apply to the gas phase. Bond dissociation energies can be used, however, to compare the enthalpies of two reactions in solution, provided that the effect of the solvent is either negligible or is the same for both of the reactions being compared (and thus cancels in the comparison). This assumption is valid for many free-radical reactions, including the ones in Study Problem 5.5.

PROBLEMS

5.23   Estimate the $\Delta H°$ values for each of the following gas-phase reactions using bond dissociation energies.

(a) $CH_4 + Cl_2 \longrightarrow H_3C—Cl + HCl$

(b) $H_2C{=}CH_2 + Cl_2 \longrightarrow Cl—CH_2CH_2—Cl$

5.24   (a) Consider the second propagation step for peroxide-promoted HBr addition to an alkene (Eq. 5.52b on p. 207). Calculate the $\Delta H°$ for this reaction.

(b) Calculate the $\Delta H°$ for the same step using HCl instead of HBr.

(c) Use your calculation to explain why no peroxide effect is observed for the addition of HCl to an alkene.

5.25   Consider the second propagation step of peroxide-promoted HBr addition to alkenes (Eq. 5.52b). Use bond energies to explain why hydrogen, and not bromine, is abstracted from HBr by the free-radical reactant.

5.7   POLYMERS. FREE-RADICAL POLYMERIZATION OF ALKENES

In the presence of free-radical initiators such as peroxides or AIBN, many alkenes react to form **polymers**, which are very large molecules composed of repeating units. Polymers are derived from small molecules in the same sense that a freight train is composed of boxcars. In a **polymerization reaction**, small molecules known as **monomers** react to form a polymer. For example, ethylene can be used as a monomer and polymerized with free-radical initiators or with special catalysts to yield an industrially important polymer called **polyethylene**.

$$n\,H_2C{=}CH_2 \xrightarrow[\text{or catalyst}]{\text{initiator}} -\!\!\left(CH_2—CH_2\right)\!\!\overline{}_n \qquad (5.71)$$

polyethylene
(the polymer of ethylene)

The formula for polyethylene in this equation illustrates a very important convention for representing the structures of polymers. In this formula, the subscript n means that a typical polyethylene molecule contains a very large number of repeating units, $—CH_2—CH_2—$. Typically n might be in the range of 3000 to 40,000, and a given sample of polyethylene contains molecules with a distribution of n values. Polyethylene is an example of an **addition polymer**—that is, a polymer in which no atoms of the monomer unit are lost as a result of the polymerization reaction. (Other types of polymers are discussed later in the text.)

When the polymerization of ethylene shown in Eq. 5.71 occurs by a free-radical mechanism, it is an example of **free-radical polymerization**. The reaction is initiated when a radical R·, derived from peroxides or other initiators, adds to the double bond of ethylene to form a new radical.

$$R\!\cdot\!\overset{\frown}{H_2C}\overset{\frown}{=}CH_2 \longrightarrow R—CH_2—\dot{C}H_2 \qquad (5.72a)$$

initiating
radical

The first propagation step of the reaction involves addition of the new radical to another molecule of ethylene.

$$R—CH_2—\dot{C}H_2\,\overset{\frown}{H_2C}\overset{\frown}{=}\dot{C}H_2 \longrightarrow R—CH_2—CH_2—CH_2—\dot{C}H_2 \qquad (5.72b)$$

Eqs. 5.72a and 5.72b are further examples of a typical free-radical reaction: reaction with a π bond. (Compare with Eq. 5.52a on p. 207.) This process continues indefinitely until the ethylene supply is reduced to the point that termination reactions occur.

$$R-(CH_2-CH_2)_n CH_2-\dot{C}H_2 \; + \; H_2C{=}CH_2 \longrightarrow R-(CH_2-CH_2)_{n+1} CH_2-\dot{C}H_2 \quad (5.72c)$$

An example of a termination reaction is the radical recombination of two radical chains to form a large nonradical product:

$$R-(CH_2-CH_2)_{n+1} CH_2-\dot{C}H_2 \quad \dot{C}H_2-CH_2-(CH_2-CH_2)_{n+1} R \longrightarrow$$

$$R-(CH_2-CH_2)_{n+1} CH_2-CH_2-CH_2-CH_2-(CH_2-CH_2)_{n+1} R \quad (5.72d)$$

Typically, polymers with molecular masses of 10^5 to 10^7 daltons are formed in free-radical polymerizations. The polymer chain is so long that the groups at its ends represent an insignificant part of the total structure. Hence, when we write the polymer structure as $-(CH_2-CH_2)_n$, these terminal groups are ignored, just as, by analogy, we ignore the engine and the caboose when we say that a train consists of boxcars.

The polymerization of alkenes is very important commercially. About 160 billion pounds of polyethylene valued at $112 billion is manufactured worldwide, of which about 25% is made by free-radical polymerization. The free-radical process yields a very transparent polymer, called *low-density polyethylene*, which is used in films and packaging. (Freezer bags and sandwich bags are usually made of low-density polyethylene.) The relatively low density of this polymer is due to the occurrence of significant branching in the polymer chains. Because branched chains do not pack so tightly as unbranched chains, the solid polymer contains lots of empty space. (If you've ever tried to stack a pile of brush containing highly branched tree limbs, you understand this point.) The mechanism of polymerization shown in Eqs. 5.72a–c does not explain the branching, but this is explored in Problem 5.45 (p. 226) at the end of the chapter.

Another method of polyethylene manufacture, the *Ziegler–Natta process*, employs a titanium catalyst and does not involve free-radical intermediates (Sec. 18.6D). This process yields largely unbranched polymer chains and results in *high-density polyethylene* that is used in molded plastic containers, such as milk jugs.

Many other commercially important polymers are produced from other alkene monomers by free-radical polymerization. Some of these are listed in Table 5.4. Alkene polymers surround us in many everyday articles. Cell phones, computers, automobiles, sports equipment, stereo systems, food packaging, and many other items have important components fabricated from alkene polymers.

Discovery of Teflon

In April 1938, Roy J. Plunkett (1910–1994), who had obtained his Ph.D. only two years earlier from The Ohio State University, was working in the laboratories of the DuPont company. He decided to use some tetrafluoroethylene (a gas) in the preparation of a refrigerant. When he opened the valve on the cylinder of tetrafluoroethylene, no gas escaped. Because the weight of the empty cylinder was known, Plunkett was able to determine that the cylinder had the weight expected for a full cylinder of the gas. It was at this point that Plunkett's scientific curiosity paid a handsome dividend. Rather than discard the cylinder, he checked to be sure the valve was not faulty, and then cut the cylinder open. Inside he found a polymeric material that felt slippery to the touch, could not be melted with extreme heat, and was chemically inert to almost everything. Thus, Plunkett accidentally discovered the polymer we know today as Teflon. At that time, no one imagined the commercial value of Teflon. Only with the advent of the atomic bomb project during World War II did it find a use: to form gaskets that were inert to the highly corrosive gas UF_6 used to purify the isotopes of uranium. In the 1960s, Teflon was introduced to consumers as a nonstick coating on cookware.

TABLE 5.4 Some Addition Polymers Produced by Free-Radical Polymerization

Polymer name (Trade name)	Structure of monomer	Properties of the polymer	Uses
Polyethylene	$H_2C={=}CH_2$	Flexible, semiopaque, generally inert	Containers, film
Polystyrene	$Ph{-}CH={=}CH_2$	Clear, rigid; can be foamed with air	Containers, toys, packing material and insulation
Poly(vinyl chloride) (PVC)	$H_2C={=}CH{-}Cl$	Rigid, but can be plasticized with certain additives	Plumbing, leatherette, hoses. Monomer has been implicated as a carcinogen.
Polychlorotri-fluoroethylene (Kel-F)	$F_2C={=}CF{-}Cl$	Inert	Chemically inert apparatus, fittings, and gaskets
Polytetrafluoro-ethylene (Teflon)	$F_2C={=}CF_2$	Very high melting point; chemically inert	Gaskets; chemically resistant apparatus and parts
Poly(methyl methacrylate) (Plexiglas, Lucite)	$H_2C={=}C{-}CO_2CH_3$ $\quad\quad\;\; \vert$ $\quad\quad\; CH_3$	Clear and semiflexible	Lenses and windows; fiber optics
Polyacrylonitrile (Orlon, Acrilan)	$H_2C={=}CH{-}CN$	Crystalline, strong, high luster	Fibers

---

PROBLEM

5.26   Using the monomer structure in Table 5.4, draw the structure of (a) poly(methylmethacrylate) and (b) poly(vinyl chloride) (PVC), the polymer used for the pipes in household plumbing.

5.8   ALKENES IN THE CHEMICAL INDUSTRY

More ethylene is produced industrially than any other organic compound. In the United States, it has ranked fourth in industrial production of all chemicals (behind sulfuric acid, nitrogen, and oxygen) for a number of years. About 260 billion pounds of ethylene is produced annually worldwide. Propene (known industrially by its older name *propylene*) is not far behind, with an annual world output of more than 100 billion pounds. Other important alkenes are styrene (PhCH$={=}$CH$_2$, Table 5.4), 1,3-butadiene, and 2-methylpropene (usually called isobutylene in the chemical industry).

styrene   **1,3-butadiene**   **2-methylpropene (isobutylene)**

Ethylene and propene are considered to be petroleum products, but they are not obtained directly from crude oil. Rather, they are produced industrially from alkanes in a process called **thermal cracking**. Cracking breaks larger alkanes into a mixture of dihydrogen, methane, and other small hydrocarbons, many of which are alkenes. In this process, a mixture of alkanes from the fractional distillation of petroleum (Sec. 2.9) is mixed with steam and heated in a furnace at 750–900 °C for a fraction of a second and is then quenched (rapidly cooled). The products of cracking are then separated. Specialized catalysts have also been developed that allow cracking to take place at lower temperatures ("cat cracking").

In the United States, the hydrocarbon most often used to produce ethylene is ethane, which is a component of natural gas. In the cracking of ethane, ethylene and dihydrogen (molecular hydrogen) are formed at very high temperature.

$$H_3C\!-\!CH_3 \xrightarrow{\;900\,°C\;} H_2C\!=\!CH_2 + H_2 \qquad (5.73)$$
$$\textbf{ethane} \qquad\qquad\qquad \textbf{ethylene}$$

Although petroleum is the major source of ethylene, there has been increasing interest in the production of "green" ethylene—ethylene from biological sources other than petroleum. The dehydration of ethanol produced by the fermentation of sugars in corn, sugar cane, and other sources of fermentable sugars is a promising source of "green" ethylene. Each ton of ethylene produced in this way captures 2.5 tons of atmospheric CO_2.

$$\text{plants} + CO_2 \xrightarrow{\text{photosynthesis}} \text{sugars} \xrightarrow{\text{fermentation}} CH_3CH_2OH \xrightarrow[\text{heat}]{\substack{\text{acid}\\\text{catalyst}}} H_2C\!=\!CH_2 + H_2O \quad (5.74)$$
$$\textbf{ethanol} \qquad\qquad \textbf{ethylene}$$

Currently, this process cannot compete economically with ethylene produced from natural gas, but it may become more important in the future.

Polymerization is a major end use for ethylene, propene, and styrene, which give the polymers polyethylene, polypropylene, and polystyrene, respectively (Sec. 5.7, Table 5.4). The diene 1,3-butadiene is co-polymerized with styrene to produce a synthetic rubber, styrene–butadiene rubber (SBR), which is important in the manufacture of tires. (This process is discussed in Sec. 15.5.) Ethylene is a starting material for the manufacture of ethylene glycol, HO—CH_2—CH_2—OH, which is the main ingredient of automotive antifreeze and is also a starting material in the production of polyesters. Ethylene is also used, along with benzene, to produce styrene, which, as we noted earlier, is another important alkene in the polymer industry. Propene is a key compound in the production of phenol, which is used in adhesives, and acetone, a commercially important solvent. In addition, propene is polymerized to give polypropylene, an important alkene polymer. 2-Methylpropene (isobutylene) is used to prepare octane isomers that are important components of high-octane gasoline, and it is reacted with methanol (CH_3OH) to give a gasoline additive, methyl *tert*-butyl ether (MTBE; see Problem 4.66 on p. 179). Some of these chemical interrelationships are shown in Fig. 5.4. This figure also shows how fundamental petroleum is to the chemical economy of much of the industrialized world.

 Ethylene as a Fruit Ripener

An intriguing role of ethylene in nature has been put to commercial use. Ethylene produced by plants causes fruit to ripen. That is, ethylene is a *ripening hormone*. Plants produce ethylene by the degradation of a relatively rare amino acid:

$$\tfrac{1}{2}O_2 + \underset{\substack{\textbf{1-amino-1-cyclopropane-}\\\textbf{carboxylic acid}}}{\begin{array}{c} H_2C \\[-2pt] \big| \\[-2pt] H_2C \end{array}\!\!>\!\!C\!\!\begin{array}{c} \overset{O}{\overset{\|}{C}}\!\!-\!O^- \\[4pt] NH_3 \\[-2pt] + \end{array}} \longrightarrow H_2C\!=\!CH_2 + CO_2 + HC\!\equiv\!N + H_2O$$
$$\textbf{ethylene}$$

It is ethylene, for example, that brings green tomatoes to that peak of juicy redness in the home garden. Commercial growers have made use of this knowledge by picking and transporting fruit before it is ripe, and then ripening it "on location" with ethylene!

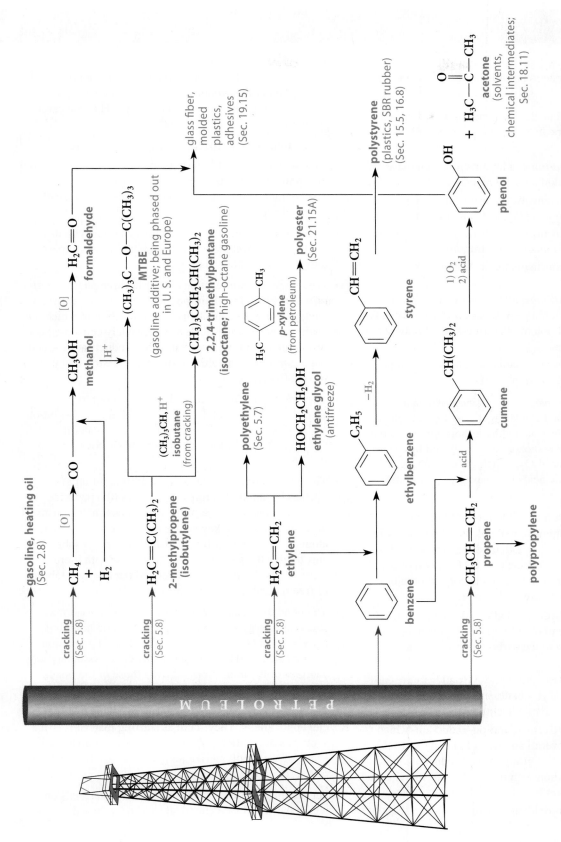

FIGURE 5.4 The organic chemical industry is a major economic force that depends strongly on the availability of petroleum. (In the reactions, [O] indicates oxidation.)

KEY IDEAS IN CHAPTER 5

- The characteristic reaction of alkenes is addition to the double bond.

- Additions to the alkene double bond discussed in this and the previous chapter can be classified in the following ways:

    1. *Electrophilic additions:* additions in which the π electrons of the double bond react with electrophiles
        a. by mechanisms involving carbocation intermediates (hydrogen halide addition, hydration);
        b. by mechanisms involving cyclic ion intermediates (halogen addition, oxymercuration);
        c. by concerted mechanisms (hydroboration, ozonolysis).
    2. *Free-radical additions:* additions in which the π electrons of the double bond react with free radicals (free-radical addition of HBr, free-radical polymerization of alkenes).
    3. Additions by other mechanisms not considered in detail (catalytic hydrogenation).

- Some useful transformations of alkenes involve additions followed by other transformations. These include oxymercuration–reduction and hydroboration–oxidation, which give alcohols; and ozonolysis followed by treatment with $(CH_3)_2S$ or H_2O_2, which gives aldehydes, ketones, or carboxylic acids by cleavage of the double bond.

- The constitutional isomer of the product formed in unsymmetrical additions to a double bond (additions in which the two groups that add are different) depends on the number of alkyl substituents at each carbon of the double bond. Reactions in which one constitutional isomer is strongly favored are said to be *regioselective*.

    1. In electrophilic hydrogen halide addition, the halogen adds to the carbon with more alkyl substituents, and the hydrogen to the carbon with fewer alkyl substituents.
    2. In free-radical HBr addition, the Br adds to the carbon with fewer alkyl substituents and the H to the carbon with more alkyl substituents.
    3. In hydration and oxymercuration–reduction, which bring about the overall addition of H and OH to the double bond, the OH group of the product alcohol is located at the carbon with more alkyl substituents and the H at the carbon with fewer alkyl substituents.
    4. In hydroboration–oxidation, which also brings about the overall addition of H and OH to the double bond, the OH of the product alcohol is located at the carbon with fewer alkyl substituents and the H at the carbon with more alkyl substituents.

- Three fundamental reactions of free radicals are
    1. reaction with a π bond;
    2. atom abstraction;
    3. recombination with another radical (the reverse of bond rupture).

- Reactions that occur by free-radical chain mechanisms are typically promoted by free-radical initiators (peroxides, AIBN), heat, or light.

- Free-radical chain reactions occur in three phases: *initiation*, in which the chain-propagating radicals are generated; *propagation*, the actual chain reaction in which the chain-propagating radicals are alternately consumed and regenerated, and the major reaction products are formed; and *termination*, in which radicals are destroyed by recombination reactions.

- The stability of free radicals is in the order tertiary > secondary > primary, but the effect of alkyl substitution on free-radical stability is considerably less significant than the effect of alkyl substitution on carbocation stability.

- The reversal of regioselectivity of HBr addition to alkenes by small amounts of peroxides (the peroxide effect) is a consequence of the free-radical mechanism of the reaction. The key step is the reaction of a bromine atom at the alkene carbon bearing fewer alkyl substituents to give the free-radical intermediate with more alkyl substituents. Both a steric effect and the relative stability of free radicals determine this outcome.

- The bond dissociation energy of a covalent bond measures the energy required to break the bond homolytically to form two free radicals. The $\Delta H°$ of a reaction can be calculated by subtracting the bond dissociation energies of the bonds formed from those of the bonds broken.

- A number of alkenes are important for many commercial applications. They serve as starting materials in the synthesis of addition polymers such as polyethylene. Many of these polymerization reactions are free-radical processes.

- Petroleum provides many important raw materials on which the worldwide chemical industry is based.

 REACTION REVIEW *For a summary of reactions discussed in this chapter, see the* Reaction Review *section of Chapter 5 in the* Study Guide and Solutions Manual.

5.27 Give the principal organic products expected when 1-butene reacts with each of the following reagents.

(a) Br_2 in CH_2Cl_2 solvent

(b) O_3, $-78°$ C

(c) product of (b) with $(CH_3)_2$ S

(d) product of (b) with H_2O_2

(e) O_2, flame

(f) HBr

(g) I_2, H_2O

(h) H_2, Pt/C

(i) HBr, peroxides

(j) BH_3 in tetrahydrofuran (THF)

(k) product of (j) with NaOH, H_2O_2

(l) $Hg(OAc)_2$, H_2O

(m) product of (l) with $NaBH_4$/NaOH

(n) HI

(o) HI, AIBN

5.28 Repeat Problem 5.27 for 1-ethylcyclopentene.

5.29 Draw the structure of

(a) a six-carbon alkene that would give the same product from reaction with HBr whether peroxides are present or not.

(b) four compounds of formula $C_{10}H_{16}$ that would undergo catalytic hydrogenation to give decalin:

decalin

(c) two alkenes that would yield 1-methylcyclohexanol when treated with $Hg(OAc)_2$ in water, then $NaBH_4$/NaOH:

1-methylcyclohexanol

(d) an alkene with one double bond that would give the following compound as the *only* product after ozonolysis followed by H_2O_2:

$$HO-\overset{O}{\overset{\|}{C}}-CH_2CH_2-\overset{O}{\overset{\|}{C}}-OH$$

(e) two stereoisomeric alkenes that would give 3-hexanol as the major product of hydroboration followed by treatment with alkaline H_2O_2:

$$\overset{OH}{\underset{|}{CH_3CH_2CHCH_2CH_2CH_3}}$$

3-hexanol

(f) an alkene of five carbons that would give the same product as a result of *either* oxymercuration–reduction *or* hydroboration–oxidation.

5.30 Draw the structure of

(a) a five-carbon alkene that would give the same product of HBr addition whether peroxides are present or not.

(b) a compound with the formula C_6H_{12} that would not undergo ozonolysis.

(c) four compounds with the formula C_7H_{12} that would undergo catalytic hydrogenation to give methylcyclohexane.

(d) two alkenes that would give the following alcohol when treated with $Hg(OAc)_2$ and H_2O in THF followed by alkaline $NaBH_4$:

$$\underset{\overset{|}{OH}}{(CH_3)_2CCH_2CH_2CH_2CH_3}$$

(e) an alkene that would give 2-pentanone as the only product of ozonolysis followed by treatment with aqueous H_2O_2:

$$CH_3\overset{O}{\overset{\|}{C}}CH_2CH_2CH_3$$

2-pentanone

(f) the alkene that would give the following alcohol after hydroboration–oxidation:

(g) an alkene with the formula C_6H_{12} that would give the same alcohol from either oxymercuration–reduction or hydroboration–oxidation.

5.31 Give the missing reactant or product in each of the following equations.

(a) ▱ + O_3 $\longrightarrow$ $\xrightarrow{H_2O, H_2O_2}$?

(b)

+ O_3 $\longrightarrow$ $\xrightarrow{(CH_3)_2S}$?

(c)

? + HBr $\xrightarrow{\text{peroxides}}$ —CH_2Br

(d)

? + HBr $\xrightarrow{\text{peroxides}}$

(the only product)

(e)

$\xrightarrow[\text{2) NaBH}_4/\text{NaOH}]{\text{1) BH}_3/\text{THF}}$?

5.32 Outline a laboratory preparation of each of the following compounds. Each should be prepared from an alkene *with the same number of carbon atoms* and any other reagents. The reactions and starting materials used should be chosen so that each compound is virtually uncontaminated by constitutional isomers.

(a)

$$CH_3CH_2-\underset{\underset{CH_2CH_3}{|}}{\overset{\overset{OH}{|}}{C}}-CH_2CH_3$$

(b) $HO-CH_2CH_2CH_2CH_2CH_2CH_3$

(c) $Br-CH_2CH_2CH_2CH_2CH_3$

(d)

$$CH_3CH_2\underset{\underset{CH_2Br}{|}}{\overset{\overset{OH}{|}}{C}}CH_2CH_3$$

(e)

$$CH_3CH-\underset{\underset{CH_3}{|}}{\overset{\overset{Br\quad Br}{|\quad\ \ |}}{C}}-CH_2CH_3$$

(f)

$$H-\overset{\overset{O}{\|}}{C}(CH_2)_4\overset{\overset{O}{\|}}{C}-CH_3$$

(g)

$$HO-\overset{\overset{O}{\|}}{C}(CH_2)_3\overset{\overset{O}{\|}}{C}-OH$$

(h)

$$CH_3\underset{\underset{Br}{|}}{\overset{}{C}}HCH_2CH_2CH_3$$

(i) $CH_3CH_2CHCH_2CH_3$
　　　　　$\underset{\underset{CH_2CH_3}{|}}{}$

5.33 Deuterium (D, or ^2H) is an isotope of hydrogen with atomic mass = 2. Deuterium can be introduced into organic compounds by using reagents in which hydrogen has been replaced by deuterium. Outline preparations of both isotopically labeled compounds from the same alkene using appropriate deuterium-containing reagents.

(a) 　　(b)

5.34 Using the mechanism of halogen addition to alkenes to guide you, predict the product(s) obtained when 2-methyl-1-butene is subjected to each of the following conditions. Explain your answers.

(a) Br_2 in CH_2Cl_2 (an inert solvent)

(b) Br_2 in H_2O

(c) Br_2 in CH_3OH solvent

(d) Br_2 in CH_3OH solvent containing concentrated $Li^+ Br^-$

5.35 Using the mechanism of the oxymercuration reaction to guide you, predict the product(s) obtained when 1-hexene is treated with mercuric acetate in each of the following solvents and the resulting products are treated with $NaBH_4/NaOH$. Explain your answers and tell what functional groups are present in each of the products.

(a) H_2O/THF　　(b) $(CH_3)_2CH-OH$
　　　　　　　　　　　　　isopropyl alcohol

(c)

$$H-\underset{\underset{\cdot\cdot}{\cdot\cdot}}{\overset{}{O}}-\overset{\overset{:O:}{\|}}{C}-CH_3$$

acetic acid

5.36 In the addition of HBr to 3,3-dimethyl-1-butene, the results observed are shown in Fig. P5.36.

(a) Explain why the different conditions give different product distributions.

$$H_3C-\underset{\underset{CH_3}{|}}{\overset{\overset{CH_3}{|}}{C}}-CH=CH_2 + HBr \longrightarrow H_3C-\underset{\underset{Br}{|}}{\overset{\overset{CH_3}{|}}{C}}-CH(CH_3)_2 + H_3C-\underset{\underset{CH_3}{|}}{\overset{\overset{CH_3}{|}}{C}}-\underset{\underset{Br}{|}}{\overset{}{C}}H-CH_3 + H_3C-\underset{\underset{CH_3}{|}}{\overset{\overset{CH_3}{|}}{C}}-CH_2CH_2Br$$

| | no peroxides: | 71% | 29% | none |
| | with peroxides: | trace | trace | 100% |

Figure P5.36

(b) Write a detailed mechanism for each reaction that explains the origin of all products.

(c) Which conditions give the faster reaction? Explain.

5.37 Give the structures of both the reactive intermediate and the product in each of the following reactions:

(a)

+ Br$_2$ ⟶

(b)

+ HBr ⟶

(c)

+ Hg(OAc)$_2$ + H$_2$O ⟶

(d)

+ HBr $\xrightarrow{\text{peroxides}}$

5.38 Trifluoroiodomethane undergoes an addition to alkenes in the presence of light by a free-radical chain mechanism.

$$CH_3CH_2CH_2CH{=}CH_2 + CF_3I \xrightarrow{\text{light}}$$

$$CH_3CH_2CH_2CH{-}CH_2{-}CF_3$$
$$\underset{I}{|}$$

The initiation step of this reaction is the light-induced homolysis of the C—I bond:

$$F_3C{-}I \xrightarrow{\text{light}} F_3C\cdot + \cdot I$$

Using the fishhook notation, write the propagation steps of a free-radical chain mechanism for this reaction.

5.39 In *thermal cracking* (Sec. 5.8), bonds generally break homolytically.

(a) In the thermal cracking of 2,2,3,3-tetramethylbutane, which bond would be most likely to break? Explain.

(b) Which compound, 2,2,3,3-tetramethylbutane or ethane, undergoes thermal cracking more rapidly at a given temperature? Explain.

(c) Calculate the $\Delta H°$ for the initial carbon–carbon bond breaking for the thermal cracking of both 2,2,3,3-tetramethylbutane and ethane. Use the $\Delta H_f°$ values in Table 5.2 as well as the $\Delta H_f°$ of 2,2,3,3-tetramethylbutane (-225.9 kJ mol^{-1}, -53.99 kcal mol^{-1}) and ethane (-84.7 kJ mol^{-1}, -20.24 kcal mol^{-1}). Use these calculations to justify your answer to part (b).

5.40 (a) What product would be obtained from the ozonolysis of natural rubber, followed by reaction with H$_2$O$_2$? (*Hint:* Write out two units of the polymer structure.)

(b) *Gutta-percha* is a natural polymer that gives the same ozonolysis product as natural rubber. Suggest a structure for gutta-percha.

5.41 Free-radical addition of thiols (molecules with the general structure RSH) to alkenes is a well-known reaction, and it is initiated by peroxides.

(a) Use information found in tables in this chapter plus the following information to calculate the C—S bond dissociation energy for ethanethiol (CH$_3$CH$_2$—SH): $\Delta H_f°$ for ethanethiol, 46.15 kJ mol^{-1}; $\Delta H_f°$ for ·SH, 143.1 kJ mol^{-1}. (To convert these energies into kcal mol^{-1}, divide by 4.184 kJ kcal^{-1}.)

(b) Using bond dissociation energies, show that the reaction of a radical such as (CH$_3$)$_3$CO· (from homolysis of a peroxide initiator) with ethanethiol should be a good source of CH$_3$CH$_2$S· radicals.

(c) Using bond dissociation energies, show that each propagation step of thiol addition to an alkene such as ethylene is exothermic and therefore favorable. (See Study Problem 5.4. on p. 209.)

5.42 Although the addition of H—CN to an alkene could be envisioned to occur by a free-radical chain mechanism, such a reaction is not observed. Justify each of the following reasons with appropriate calculations using bond dissociation energies.

(a) The reaction of H—CN with initiating (CH$_3$)$_3$C—O· radicals is *not* a good source of ·CN radicals.

(b) The second propagation step of the free-radical addition is energetically unfavorable:

$$R\dot{C}HCH_2CN + H{-}CN \longrightarrow RCH_2CH_2CN + \cdot CN$$

5.43 The halogenation of methane in the gas phase is an industrial method for the preparation of certain alkyl halides and takes place by the following equation (X = halogen):

$$X_2 + CH_4 \longrightarrow H_3C{-}X + H{-}X$$

(a) This reaction takes place readily when X = Br or X = Cl, but not when X = I. Show that these observations are expected from the $\Delta H°$ values of the reactions. Calculate the $\Delta H°$ values from appropriate bond dissociation energies.

(b) Explain why samples of methyl iodide (H$_3$C—I) that are contaminated with traces of HI darken with the color of iodine on standing a long time.

5.44 Polypropylene carbonate is considered to be a "green" polymer because it is made from CO$_2$ and is biodegradable.

polypropylene carbonate

Draw out the structure of polypropylene carbonate for $n = 3$. (Leave the "dangling bonds" at the ends.)

5.45 The mechanism for the free-radical polymerization of ethylene shown in Eqs. 5.72a–c (pp. 217–218) is somewhat simplified because it does not account for the observation that low-density polyethylene (LDPE) contains a significant number of branched chains. (The branching accounts for the low density of LDPE.) It is believed that the first step that leads to branching is an internal hydrogen abstraction reaction that occurs within the growing polyethylene chain, shown in Fig. P5.45.

(a) Use bond dissociation energies to show that this process is energetically favorable.

(b) Show how this reaction can lead to a branched polyethylene chain.

5.46 (a) Draw the structure of *polystyrene*, the polymer obtained from the free-radical polymerization of styrene.

(b) How would the structure of the polymer product differ from the one in part (a) if a few percent of 1,4-divinylbenzene were included in the reaction mixture?

$$H_2C{=}CH{-}\underset{}{\bigcirc}{-}CH{=}CH_2$$

1,4-divinylbenzene

5.47 In a laboratory a bottle was found containing a clear liquid *A*. The bottle was labeled, "$C_{10}H_{16}$, Isolated from a lemon." Because of your skills in organic chemistry, you have been hired to identify this substance. Compound *A* decolorizes Br_2 in CH_2Cl_2. When *A* is hydrogenated over a catalyst, two equivalents of H_2 are consumed and the product is found to be 1-isopropyl-4-methylcyclohexane. Ozonolysis of *A* followed by treatment of the reaction mixture with H_2O_2 gives the following compound as a major product:

$$\underset{\underset{CH_3}{\overset{\overset{O}{\|}}{\underset{\|}{\underset{C{=}O}{|}}}}{CH_3CCH_2CH_2CHCH_2}{-}\overset{\overset{O}{\|}}{C}{-}OH$$

Suggest a structure for *A* and explain all observations. (See Study Guide Link 5.3 if you need help.)

STUDY GUIDE LINK 5.3
Solving Structure Problems

5.48 A compound *A* with the molecular formula C_8H_{16} decolorized Br_2 in CH_2Cl_2. Catalytic hydrogenation of *A* gave octane. Treatment of compound *A* with O_3 followed by aqueous H_2O_2 yielded butanoic acid (*B*) as the sole product.

$$CH_3CH_2CH_2\overset{\overset{O}{\|}}{C}{-}OH$$

butanoic acid (*B*)

What is the structure of compound *A*? What aspect of the structure of *A* is not determined by the data?

5.49 Using the curved-arrow or fishhook notation, as appropriate, suggest mechanisms for each of the reactions given in Fig. P5.49.

5.50 Consider the reaction of a methyl radical (·CH_3) with the π bond of an alkene:

$$\underset{R}{\overset{R}{C}}{=}\underset{R}{\overset{R}{C}} + \cdot CH_3 \longrightarrow \underset{R}{\overset{R}{\cdot C}}{-}\underset{R}{\overset{R}{C}}{-}CH_3$$

The relative rates of the reaction shown in Fig. P5.50 were determined for various alkenes.

(a) Draw the free-radical product of the reaction in each case and explain.

(b) Explain the order of the relative rates.

5.51 Equations 5.25a–c on p. 193 show the formation of trialkylboranes from alkenes and BH_3. In the reaction of 2,3-dimethyl-2-butene with BH_3, only *two* equivalents of the alkene react, even with a large excess of alkene, to give a dialkylborane called *disiamylborane*.

$$2\,(CH_3)_2C{=}CHCH_3 + BH_3 \longrightarrow \text{"disiamylborane"}$$
$$\text{(a dialkylborane)}$$

2-methyl-2-butene

Give the structure of disiamylborane, and suggest a reason that only two equivalents of alkene react.

5.52 When *trans*-2-hexene is subjected to ozonolysis in the presence of an excess of acetaldehyde containing the isotope ^{18}O, an ozonide is isolated that contains the isotope at one of the oxygens. Use the mechanism of ozonolysis to postulate a structure for the ozonide, including the position of the isotope. (See Fig. P5.52 on p. 228.)

$$-(CH_2CH_2)_{\overline{n}}{-}CH_2\underset{\overset{}{\underset{\overset{H}{|}}{CH}}}{\overset{H_2\dot C}{\underset{CH_2}{\underset{}{}}}}\underset{CH_2}{\overset{CH_2}{}} \longrightarrow -(CH_2CH_2)_{\overline{n}}{-}CH_2\underset{\overset{\dot CH}{}}{\underset{CH_2}{}}\underset{CH_2}{\overset{H_3C}{\underset{CH_2}{}}}$$

Figure P5.45

(a)

(b)

(c) Bridged-ion intermediates are involved in the following reaction, and sulfur is the electrophile.

sulfur dichloride

1,5-cyclooctadiene

(d) For the following reaction, give the curved-arrow notation for only the reaction of the alkene with Hg(OAc)$_2$ and H$_2$O. Then show that the compounds that you obtain from this mechanistic reasoning can be converted into the observed products by the NaBH$_4$ reduction.

(54%)   (46%)

(63% total yield)

(e)

(96% yield)

(f) Use the mechanism to predict the product of the following addition. (*Hint:* See Study Problem 5.4 on p. 209.)

(g) This is a free-radical chain reaction initiated by homolysis of the O—Cl bond.

Figure P5.49

alkene:	H$_2$C=CH$_2$	(CH$_3$)$_2$CH=CH$_2$	(CH$_3$)$_2$CH=CHCH$_3$
relative rate:	1.0	1.4	0.077

Figure P5.50

5.53 Isobutylene (2-methylpropene) can be polymerized by treating it with liquid HF as shown in Fig. P5.53. A small amount of *tert*-butyl fluoride is formed in the reaction. Suggest a curved-arrow mechanism for this process, which is an example of *cationic polymerization*. (*Hint:* Carbocations are electrophiles that can react with alkene double bonds.)

5.54 In the sequence shown in Fig. P5.54, the second reaction is unfamiliar. Nevertheless, identify compounds *A* and *B* from the information provided.

The formula of compound *B* is C_6H_{12}. Compound *B* decolorizes Br_2 in CCl_4, and takes up one equivalent of H_2 over a Pt/C catalyst. Once you have identified *B*, try to give a curved-arrow notation for its formation from *A* in one step.

$$\text{trans-2-hexene} \quad + \quad O_3 \quad + \quad H_3C-CH=O^* \quad \longrightarrow \quad \text{an ozonide}$$

trans-2-hexene   **acetaldehyde-^{18}O**
                         (*O = ^{18}O)

Figure P5.52

$$H_3C-\underset{\underset{CH_3}{|}}{C}=CH_2 \quad \xrightarrow{HF} \quad \left(\underset{\underset{CH_3}{|}}{\overset{\overset{CH_3}{|}}{C}}-CH_2\right)_n \quad + \quad H_3C-\underset{\underset{CH_3}{|}}{\overset{\overset{CH_3}{|}}{C}}-F$$

2-methylpropene   **tert-butyl fluoride**
                                                 (small amount formed)

Figure P5.53

$$(CH_3)_3C-CH=CH_2 + HBr \quad \longrightarrow \quad A \quad + \quad \text{other compound(s)}$$
$$\text{(mostly)}$$

$$A + H_3C-\ddot{\underset{..}{O}}:^- \; Na^+ \quad \longrightarrow \quad H_3C-\ddot{\underset{..}{O}}H + Na^+ \; Br^- + B$$
$$\text{(a strong base)}$$

$$B + O_3 \quad \longrightarrow \quad \xrightarrow{(CH_3)_2S} \quad (CH_3)_2C=O$$
$$\text{acetone}$$
$$\text{(only compound formed)}$$

Figure P5.54

Principles of Stereochemistry

This chapter and the one that follows deal with stereoisomers and their properties. **Stereoisomers** are compounds that have the same atomic connectivity but a different arrangement of atoms in space. Recall that *E* and *Z* isomers of an alkene (Sec. 4.1C) are stereoisomers. In this chapter we'll learn about other types of stereoisomers.

The study of stereoisomers and the chemical effects of stereoisomerism is called **stereochemistry**. A few ideas of stereochemistry were introduced in Sec. 4.1C. This chapter delves more generally into stereochemistry by concentrating on the basic definitions and principles. We'll see how stereochemistry played a key role in the determination of the geometry of tetravalent carbon. Chapter 7 continues the discussion of stereochemistry by considering both the stereochemical aspects of cyclic compounds and the application of stereochemical principles to chemical reactions.

The use of molecular models during the study of this chapter is essential. Models will help you develop the ability to visualize three-dimensional structures and will make the two-dimensional pictures on the page "come to life." If you use models now, your reliance on them will gradually decrease.

In this chapter you will also be using perspective structures of molecules, particularly line-and-wedge structures. The techniques for drawing and interpreting these types of structures were introduced in Sec. 2.3C. You should review this material as a preparation for this chapter.

6.1 ENANTIOMERS, CHIRALITY, AND SYMMETRY

A. Enantiomers and Chirality

Any molecule—indeed, any object—has a mirror image. Some molecules are **congruent** to their mirror images. This means that all atoms and bonds in a molecule can be simultaneously superposed onto identical atoms and bonds in its mirror image. An example of such a mol-

FIGURE 6.1 Testing mirror-image ethanol molecules for congruence. One mirror image is shown with yellow bonds to distinguish it from the other. Aligning the central carbons, the CH₃ groups, and the OH groups on the different molecules causes the hydrogens to align as well. Notice that this alignment requires rotating one of the molecules in space.

mirror plane

180°

all groups align;
therefore, the molecules
are congruent

ecule is ethanol, or ethyl alcohol, H₃C—CH₂—OH (Fig. 6.1). Construct a model of ethanol and another model of its mirror image, and use the following procedure to show that these two models are congruent. For simplicity, use a single colored ball to represent the methyl group and a single ball of another color to represent the hydroxy (—OH) group. Place the two central carbons side by side and align the methyl and hydroxy groups, as shown in Fig. 6.1. The hydrogens should then align as well. *The congruence of an ethanol molecule and its mirror image shows that they are identical.*

Some molecules, such as 2-butanol, are *not* congruent to their mirror images (Fig. 6.2).

$$H_3C—\underset{*}{\overset{\overset{\displaystyle OH}{|}}{C}H}—C_2H_5$$

2-butanol

Build a model of 2-butanol and a second model of its mirror image. If you align the carbon with the asterisk and any two of its attached groups, the other two groups do not align. Hence, *a 2-butanol molecule and its mirror image are noncongruent and are therefore different molecules.* Because these two molecules have identical connectivities, then by definition they are *stereoisomers.* Molecules that are noncongruent mirror images are called **enantiomers**. Thus, the two 2-butanol stereoisomers are enantiomers.

Enantiomers must not only be mirror images; they must also be *noncongruent* mirror images. Thus, ethanol (Fig. 6.1) has no enantiomer because an ethanol molecule and its mirror image are congruent.

Molecules (or other objects) that can exist as enantiomers are said to be **chiral** (pronounced kī´ rŭl); they possess the property of **chirality**, or handedness. (*Chiral* comes from the Greek word for hand.) Enantiomeric molecules have the same relationship as the right and left hands—the relationship of an object and its noncongruent mirror image. Thus, 2-butanol is a chiral molecule. Molecules (or other objects) that are not chiral are said to be **achiral**—without chirality. Ethanol is an achiral molecule. Both chiral and achiral objects are matters of everyday acquaintance. A foot or a hand is chiral; the helical thread of a screw gives it chirality. Achiral objects include a ball and a soda straw.

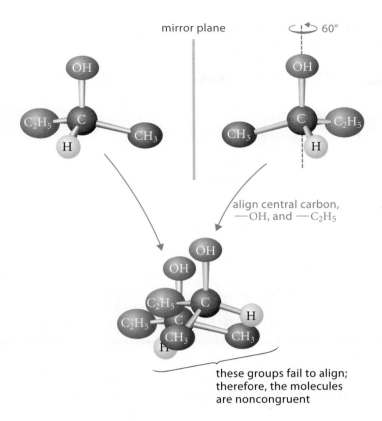

mirror plane

align central carbon,
—OH, and —C₂H₅

these groups fail to align;
therefore, the molecules
are noncongruent

FIGURE 6.2 Testing mirror-image 2-butanol molecules for congruence. As in Fig. 6.1, the bonds of one mirror image are yellow. When the central carbon and any two of the groups attached to it (OH and C_2H_5 in this figure) are aligned, the remaining groups do *not* align.

Importance of Chirality

Chiral molecules occur widely throughout all of nature. For example, glucose, an important sugar and energy source, is chiral; the enantiomer of naturally occurring glucose cannot be utilized as a food source. All sugars, proteins, and nucleic acids are chiral and occur naturally in only one enantiomeric form. Chirality is important in medicine as well. Over half of the organic compounds used as drugs are chiral, and in most cases only one enantiomer has the desired physiological activity. In rare cases, the inactive enantiomer is toxic (see the story of the drug thalidomide in Sec. 6.4B). The safety and effectiveness of synthetically prepared chiral drug molecules have been issues of concern for both pharmaceutical manufacturers and the U.S. Food and Drug Administration (FDA) for several decades.

B. Asymmetric Carbon and Stereocenters

Many chiral molecules contain one or more asymmetric carbon atoms. An **asymmetric carbon atom** is a carbon to which *four different groups* are bound. Thus, 2-butanol (see Fig. 6.2), a chiral molecule, contains an asymmetric carbon atom; this is the carbon that bears the four different groups —CH_3, —C_2H_5, —H, and —OH. In contrast, none of the carbons of ethanol, an achiral molecule, is asymmetric. *A molecule that contains only one asymmetric carbon is chiral.* No generalization can be made, however, for molecules with more than one asymmetric carbon. Although many molecules with two or more asymmetric carbons are indeed chiral, not all of them are (Sec. 6.7). Moreover, an asymmetric carbon atom (or other asymmetric atom) is not a *necessary* condition for chirality; some chiral molecules have no asymmetric carbons at all (Sec. 6.9A). Despite these caveats, it is important to recognize asymmetric carbon atoms because so many chiral organic compounds contain them.

You may encounter other terms such as *chiral carbon* and *chiral center* that mean the same thing as asymmetric carbon.

STUDY PROBLEM   6.1

Identify the asymmetric carbon(s) in 4-methyloctane:

$$CH_3$$
$$|$$
$$CH_3CH_2CH_2CHCH_2CH_2CH_2CH_3$$

4-methyloctane

SOLUTION   The asymmetric carbon is marked with an asterisk:

$$CH_3$$
$$|$$
$$CH_3CH_2CH_2CHCH_2CH_2CH_2CH_3$$
$$*$$

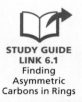

**STUDY GUIDE
LINK 6.1**
Finding
Asymmetric
Carbons in Rings

This is an asymmetric carbon because it bears four different groups: H, CH_3, $CH_3CH_2CH_2$, and $CH_2CH_2CH_2CH_3$. Notice that the propyl and butyl groups are *not* different at the point of attachment—both have CH_2 groups at that point, as well as at the next carbon removed. The difference is found at the ends of the groups. The point is that two groups are different even when the difference is remote from the carbon in question.

Although carbon is the most common asymmetric atom in organic molecules, other atoms can be asymmetric as well. For example, the following chiral compound contains an asymmetric phosphorus.

$$S$$
$$\|$$
$$CH_3CH_2\!\!\!\cdots\!\!\!P\!\!\!\diagup^{\!\!\!\!\!\!\!\!\!\text{— asymmetric phosphorus}}_{\!\!\!\diagdown OCH_3}$$
$$OH$$

Asymmetric atoms are sometimes referred to generally as **asymmetric centers**.

An asymmetric carbon (or other asymmetric atom) is another type of *stereocenter*, or *stereogenic atom*. Recall (Sec. 4.1C) that a **stereocenter** is an atom at which the interchange of two groups gives a stereoisomer. In Fig. 6.2, for example, interchanging the methyl and ethyl groups in one enantiomer of 2-butanol gives the other enantiomer. If this point is unclear from Fig. 6.2, *you need to build two models to demonstrate this to yourself.* First construct a model of either enantiomer, and then construct a model of its mirror image. Then show that the interchange of *any* two groups on one model gives the other model.

Not all carbon stereocenters are asymmetric carbons. Recall (Sec. 4.1C) that the carbons involved in the double bonds of *E* and *Z* isomers are also stereocenters. These carbons are not asymmetric carbons, though, because they are not connected to four different groups. In other words, the term *stereocenter* is not associated solely with chiral molecules. *All asymmetric atoms are stereocenters, but not all stereocenters are asymmetric atoms.*

**STUDY GUIDE
LINK 6.2**
Stereocenters and
Asymmetric Atoms

C. Chirality and Symmetry

What causes chirality? *Chiral molecules lack certain types of symmetry.* The symmetry of any object (including a molecule) can be described by certain **symmetry elements**, which are lines, points, or planes that relate equivalent parts of an object. A very important symmetry element is a **plane of symmetry**, sometimes called an **internal mirror plane**. This is a plane that divides an object into halves that are exact mirror images. For example, the mug in Fig. 6.3a has a plane of symmetry. Similarly, the ethanol molecule shown in Fig. 6.3b also has a plane of symmetry. *A molecule or other object that has a plane of symmetry is achiral.* Thus, the ethanol molecule and the mug in Fig. 6.3 are achiral. Chiral molecules and other chiral objects do *not* have planes of symmetry. The chiral molecule 2-butanol, analyzed in Fig. 6.2, has no plane of symmetry. A human hand, also a chiral object, has no plane of symmetry.

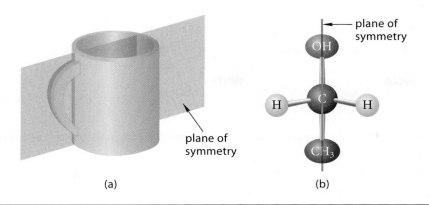

FIGURE 6.3 Examples of objects with a plane of symmetry. (a) A coffee mug. (b) An ethanol molecule. In the ethanol molecule, the plane bisects the H—C—H angle and cuts through the central carbon, the OH, and the CH$_3$.

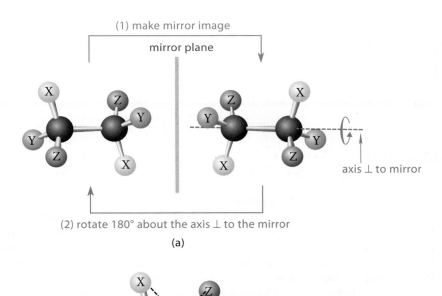

FIGURE 6.4 Center of symmetry. (a) If a molecule has a center of symmetry, forming the mirror image (1, *top arrow*) and then rotating the entire molecule 180° about an axis perpendicular to the mirror (2, *bottom arrow*) reproduces the molecule. (b) Any line through the center of symmetry (*black dot*) contacts equivalent points on the molecule at equal distances from the center.

Another important symmetry element is the **center of symmetry**, sometimes also called a *point of symmetry*. A molecule has a center of symmetry if you can reproduce it by first forming its mirror image and then rotating this mirror image by 180° about an axis perpendicular to the mirror (Fig. 6.4a). More descriptively, a center of symmetry is a point through which *any* line contacts exactly equivalent parts of the object at the same distance in both directions (Fig. 6.4b). Some objects, such as a box, can have both a center and planes of symmetry.

The plane of symmetry and the center of symmetry are the most common symmetry elements present in achiral molecules (and other achiral objects). However, some achiral molecules contain other, relatively rare, symmetry elements. How, then, can we tell whether a molecule is chiral? If a molecule has a single asymmetric carbon, it *must* be chiral. If a molecule has a plane of symmetry, a center of symmetry, or both, it is *not* chiral. If you are uncertain whether a molecule is chiral, the *most general* way to resolve the question is to build two models or draw two perspective structures, one of the molecule and the other of its mirror image, and then test the two for congruence. If the two mirror images are congruent, the molecule is achiral; if not, the molecule is chiral.

6.1 State whether each of the following molecules is achiral or chiral.

(a)
$$\underset{\substack{|\\ \text{F}}}{\overset{\substack{\text{Cl}\\|}}{\text{H}-\text{C}-\text{Br}}}$$

(b)
$$\underset{\substack{|\\ \text{Cl}}}{\overset{\substack{\text{Cl}\\|}}{\text{H}-\text{C}-\text{Br}}}$$

(c)
$$\underset{\substack{\\ \text{CH}_3}}{\overset{\substack{\text{H}\\}}{\text{Cl}-\text{C}}}\overset{\text{Cl}}{\diagup}$$

(d)
$$\text{H}_3\text{C}-\overset{\overset{\text{H}}{|}}{\underset{\underset{\text{CH}_2\text{CH}_2\text{CH}_3}{|}}{\text{N}^+}}-\text{CH}_2\text{CH}_3 \quad \text{Cl}^-$$

6.2 Ignoring specific markings, indicate whether the following objects are chiral or achiral. (State any assumptions that you make.)

(a) a shoe   (b) a book   (c) a man or woman

(d) a pair of shoes (consider the pair as one object)   (e) a pair of scissors

6.3 Show the planes and centers of symmetry (if any) in each of the following achiral objects.

(a) the methane molecule   (b) a cone

(c) the ethylene molecule   (d) the *trans*-2-butene molecule

(e) the *cis*-2-butene molecule   (f) the anti conformation of butane

6.4 Identify the asymmetric carbon(s) (if any) in each of the following molecules.

(a) $\text{CH}_3\text{CHCHCH}_3$
$$\underset{\text{Cl} \quad \text{Cl}}{\overset{|\quad|}{}}$$
(b)
(c)

6.2   NOMENCLATURE OF ENANTIOMERS: THE *R,S* SYSTEM

The existence of enantiomers poses a special problem in nomenclature. How do we indicate in the name of 2-butanol, for example, which enantiomer we have? It turns out that the same Cahn–Ingold–Prelog priority rules used to assign *E* and *Z* configurations to alkene stereoisomers (Sec. 4.2B) can be applied to enantiomers. (The Cahn–Ingold–Prelog rules were, in fact, first developed for asymmetric carbons and then later applied to double-bond stereoisomerism.) A *stereochemical configuration*, or arrangement of atoms, at each asymmetric carbon in a molecule can be assigned using the following steps, which are illustrated in Fig. 6.5.

1. Identify an asymmetric carbon and the four different groups bound to it.

2. Assign priorities to the four different groups according to the rules given in Sec. 4.2B. The convention used in this text is that the highest priority = 1 and the lowest priority = 4.

3. View the molecule along the bond *from the asymmetric carbon to the group of lowest priority*—that is, with the asymmetric carbon nearer and the lowest-priority group farther away. This is essentially a Newman projection about this bond.

4. Consider the clockwise or counterclockwise order of the remaining group priorities. If the priorities of these groups decrease in the *clockwise* direction, the asymmetric carbon is said to have the *R* configuration (*R* = Latin *rectus*, for "correct," "proper"). If the priorities of these groups decrease in the *counterclockwise* direction, the asymmetric carbon is said to have the *S* configuration (*S* = Latin *sinister*, for "left").

STUDY PROBLEM   6.2

Determine the stereochemical configuration of the following enantiomer of 3-chloro-1-pentene:

$$\text{H}_2\text{C}=\text{CH}\overset{\overset{\text{Cl}}{|}}{\underset{\underset{\text{CH}_2\text{CH}_3}{}}{\text{C}}}\cdots\text{H}$$

SOLUTION   First assign relative priorities to the four groups attached to the asymmetric carbon. These are (1) —Cl, (2) $H_2C\!\!=\!\!CH\!\!-\!\!$, (3) —CH_2CH_3, and (4) —H. Then, *using a model if necessary*, sight along the bond from the asymmetric carbon to the lowest-priority group (in this case, the H). The resulting view is essentially a Newman projection along the C—H bond:

STUDY GUIDE
LINK 6.3
Using Perspective
Structures

Because the priorities of the first three groups decrease in a counterclockwise direction, this is the *S* enantiomer of 3-chloro-1-pentene.

Study Guide Link 6.3 provides some useful shortcuts for assigning configurations.

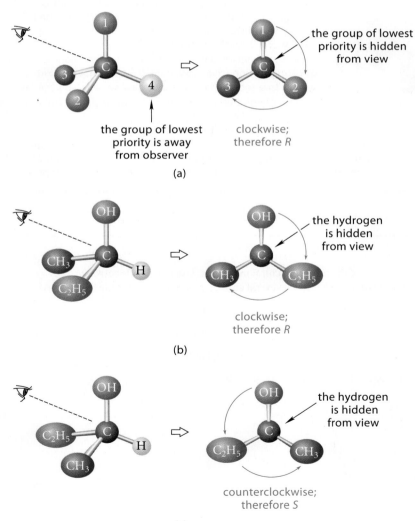

(a)

(b)

(c)

FIGURE 6.5 Use of the Cahn–Ingold–Prelog system to designate the stereochemistry of (a) a general asymmetric carbon atom, (b) (*R*)-2-butanol, (c) (*S*)-2-butanol. In each part, the direction of observation is shown by the eye, and what the observer sees is shown on the right. Priority 1 is highest and priority 4 is lowest.

A stereoisomer is named by indicating the configuration of each asymmetric carbon before the systematic name of the compound, as in the following examples:

(R)-3-methyl-1-pentene   **(3S,4S)-3,4-dimethylhexane**

(Be sure to verify these and other *R,S* assignments you find in this chapter.) As illustrated by the second example, numbers are used with the *R,S* designations when a molecule contains more than one asymmetric carbon.

The *R,S* system is not the only system used for describing stereochemical configuration. The D,L system, which predates the *R,S* system, is still used in amino acid and carbohydrate chemistry (Chapters 24 and 27). With this exception, the *R,S* system has gained virtually complete acceptance.

Is *R* Right, or Is It Proper?

Choice of the letter *R* presented a problem for Cahn, Ingold, and Prelog, the scientists who devised the *R,S* system. The letter *S* stands for *sinister*, one of the Latin words for *left*. However, the Latin word for *right* (in the directional sense) is *dexter*, and unfortunately the letter *D* was already being used in another system of configuration (the D,L system). It was difficulties with the D,L system that led to the need for a new system, and the last thing anyone needed was a system that confused the two! Fortunately, Latin provided another word for *right*: the participle *rectus*. But this "right" does not indicate direction: it means *proper*, or *correct*. (The English word *rectify* comes from the same root.) Although the Latin wasn't quite proper, it solved the problem! In passing, it might be noted that *R* and *S* are the first initials of Robert S. Cahn, one of the inventors of the *R,S* system (Sec. 4.2B). A coincidence? Perhaps.

PROBLEMS

6.5   Draw line-and-wedge representations for each of the following chiral molecules. Use models if necessary. (D = deuterium = ^{2}H, a heavy isotope of hydrogen.) Note that several correct structures are possible in each case.

(a) (S)-H₃C—CH—OH   (b) (2Z,4R)-4-methyl-2-hexene   (c)
                          |
                          D

6.6   Indicate whether the asymmetric atom in each of the following compounds has the *R* or *S* configuration.

(a)   (b)   (c)

alanine

malic acid

6.3 PHYSICAL PROPERTIES OF ENANTIOMERS. OPTICAL ACTIVITY

Recall from Sec. 2.6 that organic compounds can be characterized by their physical properties. Two properties often used for this purpose are the melting point and the boiling point. *The melting points and boiling points of a pair of enantiomers are identical.* Thus, the boiling points of (R)- and (S)-2-butanol are both 99.5 °C. Likewise, the melting points of (R)- and (S)-lactic acid are both 53 °C.

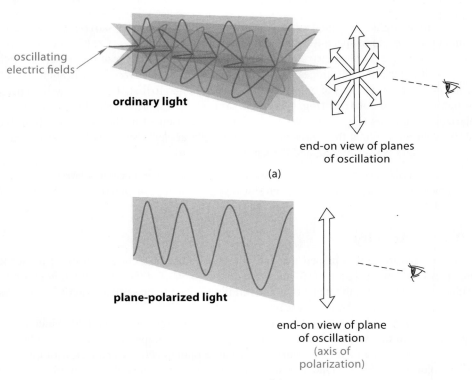

(R)-2-butanol   **(S)-2-butanol**   **(R)-lactic acid**   **(S)-lactic acid**
both have bp = 99.5 °C   both have mp = 53 °C

A pair of enantiomers also have identical densities, indices of refraction, heats of formation, standard free energies, and many other properties.

If enantiomers have so many identical properties, how can we tell one enantiomer from the other? *A compound and its enantiomer can be distinguished by their effects on polarized light.* Understanding these phenomena requires an introduction to the properties of polarized light.

A. Polarized Light

Light is a wave motion that consists of oscillating electric and magnetic fields. The electric field of ordinary light oscillates in all planes, but it is possible to obtain light with an electric field that oscillates in only one plane. This kind of light is called **plane-polarized light**, or simply, **polarized light** (Fig. 6.6).

oscillating
electric fields

ordinary light

end-on view of planes
of oscillation

(a)

FIGURE 6.6 (a) Ordinary light has electric fields oscillating in all possible planes. (Only four planes of oscillation are shown.) (b) In plane-polarized light, the oscillating electric field is confined to a single plane, which defines the axis of polarization.

plane-polarized light

end-on view of plane
of oscillation
(axis of
polarization)

(b)

FIGURE 6.7 (a) If the polarization axes of two polarizers are at right angles, no light passes through the second polarizer. (b) The same phenomenon can be observed using two pairs of polarized sunglasses.

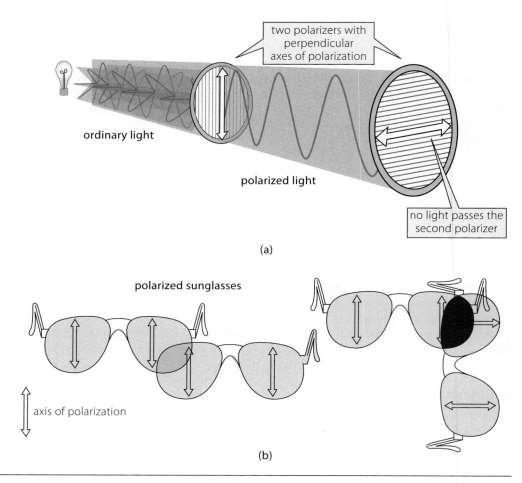

(a)

(b)

Polarized light is obtained by passing ordinary light through a polarizer, such as a Nicol prism (a prism made of specially cut and joined calcium carbonate crystals). The orientation of the polarizer's axis of polarization determines the plane of the resulting polarized light. Analysis of polarized light hinges on the fact that if plane-polarized light is subjected to a second polarizer whose axis of polarization is perpendicular to that of the first, no light passes through the second polarizer (Fig. 6.7a). This same effect can be observed with two pairs of polarized sunglasses (Fig. 6.7b). When the lenses are oriented in the same direction, light will pass through. When the lenses are turned at right angles, their axes of polarization are crossed, no light is transmitted, and the lenses appear dark.

You can see the same effect with a tablet computer, which produces a polarized image. If the image disappears when you view your tablet with polarized sunglasses, turn your tablet 90°!

B. Optical Activity

If plane-polarized light is passed through one enantiomer of a chiral substance (either the pure enantiomer or a solution of it), *the plane of polarization of the emergent light is rotated.* A substance that rotates the plane of polarized light is said to be **optically active**. *Individual enantiomers of chiral substances are optically active.*

Optical activity is measured in a device called a **polarimeter** (Fig. 6.8), which is basically the system of two polarizers shown in Fig. 6.7. The sample to be studied is placed in the light beam between the two polarizers. Because optical activity changes with the wavelength (color) of the light, monochromatic light—light of a single color—is used to measure optical activity. The yellow light from a sodium arc (the sodium D-line with a wavelength of

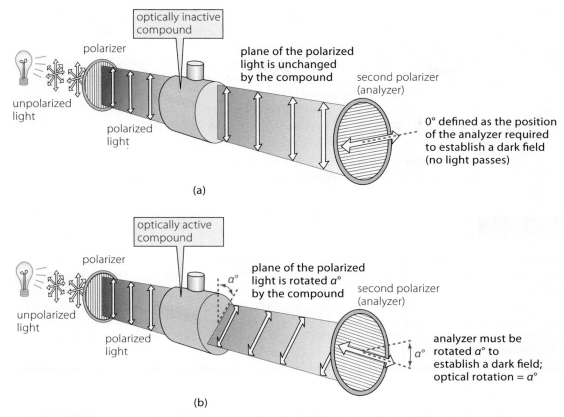

FIGURE 6.8 Determination of optical rotation in a simple polarimeter. (a) First, the reference condition of zero rotation is established as a dark field. (b) Next, the polarized light is passed through an optically active sample with observed rotation $\alpha°$. The analyzer is rotated to establish the dark-field condition again. The optical rotation $\alpha°$ can be read from the calibrated scale on the analyzer.

589.3 nm) is often used in this type of experiment. An optically inactive sample (such as air or solvent) is placed in the light beam. Light polarized by the first polarizer passes through the sample, and the analyzer is turned to establish a dark field. This setting of the analyzer defines the zero of optical rotation. Next, the sample whose optical activity is to be measured is placed in the light beam. The number of degrees α that the analyzer must be turned to reestablish the dark field is the **optical rotation** of the sample. If the sample rotates the plane of polarized light in the clockwise direction, the optical rotation is given a plus sign. Such a sample is said to be **dextrorotatory** (Latin *dexter*, meaning "right"). If the sample rotates the plane of polarized light in the counterclockwise direction, the optical rotation is given a minus sign, and the sample is said to be **levorotatory** (Latin *laevus*, meaning "left").

The optical rotation of a sample is the quantitative measure of its optical activity. The observed optical rotation α, in degrees, is proportional to the number of optically active molecules present in the path through which the light beam passes. Thus, α is proportional to both the concentration c of the optically active compound in the sample as well as the length l of the sample container:

$$\alpha = [\alpha]cl \tag{6.1}$$

The constant of proportionality, $[\alpha]$, is called the **specific rotation**. By convention, the concentration of the sample is expressed in grams per milliliter (g mL^{-1}), and the path length is in decimeters (dm). (For a pure liquid, c is taken as the density.) Thus, the specific rotation is equal to the observed rotation at a concentration of 1 g mL^{-1} and a path length of 1 dm.

Typically, the specific rotation is determined as the slope of a plot of observed rotation α against the concentration c. (See Problem 6.7.) Because the specific rotation $[\alpha]$ is independent of c and l, it is used as the standard measure of optical activity. Since the dimension of the observed rotation is degrees, the dimensions of $[\alpha]$ are degrees mL g^{-1} dm^{-1}. (Often, specific rotations are reported simply in degrees, with the other units understood.) Because the specific rotation of any compound varies with wavelength, solvent, and temperature, $[\alpha]$ is conventionally reported with a subscript that indicates the wavelength of light used and a superscript that indicates the temperature. Thus, a specific rotation reported as $[\alpha]_D^{20}$ has been determined at 20 °C using the sodium D-line.

> One decimeter = 10 centimeters. The reason for using the decimeter as a unit of length is that the length of a typical sample container used in polarimeters is 1 dm.

STUDY PROBLEM   6.3

A sample of (*S*)-2-butanol has an observed rotation of +2.18° at 20 °C. The measurement was made with a 2.0 *M* solution of (*S*)-2-butanol in methanol solvent in a sample container that is 10 cm long. What is the specific rotation $[\alpha]_D^{20}$ of (*S*)-2-butanol in this solvent?

SOLUTION   To calculate the specific rotation, the sample concentration in g mL^{-1} must be determined. Because the molecular mass of 2-butanol is 74.12, the 2.0 *M* solution contains 148.1 g L^{-1}, or 0.148 g mL^{-1}, of 2-butanol. This is the value of c used in Eq. 6.1. The value of l is 1 dm. Substituting in Eq. 6.1, $[\alpha]_D^{20}$ = (+2.18 degrees)/(0.148 g mL^{-1})(1 dm) = +14.7 degrees mL g^{-1} dm^{-1} in methanol solvent.

C. Optical Activities of Enantiomers

Enantiomers are distinguished by their optical activities because *enantiomers rotate the plane of polarized light by equal amounts in opposite directions*. Thus, if the specific rotation $[\alpha]_D^{20}$ of (*S*)-2-butanol is +14.7 degrees mL g^{-1} dm^{-1} (Study Problem 6.3), then the specific rotation of (*R*)-2-butanol is −14.7 degrees mL g^{-1} dm^{-1}. Similarly, if a particular solution of (*S*)-2-butanol has an observed rotation of +3.5°, then a solution of (*R*)-2-butanol under the same conditions will have an observed rotation of −3.5°. Another way to indicate the optical rotation of a compound is with a lower-case prefix *d* or *l*, the first letters of the words *dextrorotatory* and *levorotatory*. These are sometimes used *instead of* the plus (+) and minus (−) signs. Thus, (+)-2-butanol can also be called *d*-2-butanol; (−)-2-butanol can also be called *l*-2-butanol. We won't use this notation extensively in this text because it has the potential to be confused with the prefixes D and L, which are used in an older system of absolute stereochemical configuration that is still used with amino acids and sugars. (We'll discuss this system in Chapters 24 and 26.)

Here is a *very important point* about optical rotation: *There is no general correspondence between the sign of the optical rotation and the R or S configuration of a compound.* Thus, some compounds with the *S* configuration have positive rotations, and others have negative rotations. For example, the *S* enantiomer of 2-butanol is dextrorotatory, whereas the *S* enantiomer of 1,2-butanediol is levorotatory.

(*S*)-(+)-2-butanol
d-2-butanol
$[\alpha]_D^{20}$ = +14.7 degrees mL g^{-1} dm^{-1}

(*S*)-(−)-1,2-butanediol
l-1,2-butanediol
$[\alpha]_D^{20}$ = −15.4 degrees mL g^{-1} dm^{-1}

The only way to determine optical rotation is to measure it experimentally. Thus, the name (*S*)-(+)-2-butanol implies that someone has measured and reported the optical rotation of the

S enantiomer. We can then deduce that the *R* enantiomer should be (*R*)-(−)-2-butanol, because enantiomers have equal rotations of opposite signs.

Conversely, you can measure the optical rotation of a chiral substance without knowing its configuration. Thus, let's imagine you have isolated a compound from a natural source (let's call it "newnol") and have found that it has a positive optical rotation. Your compound is therefore (+)-newnol, or *d*-newnol. But its sign of rotation does not allow you to deduce its stereochemical configuration. (The determination of absolute stereochemical configuration is discussed in Sec. 6.5.)

It would certainly be useful to be able to deduce the sign (and the magnitude) of the optical rotation from a structure. There are ways to do this, but these methods require complex quantum-chemical calculations.

PROBLEMS

6.7 Suppose a sample of an optically active substance has an observed rotation of +10°. The scale on the analyzer of a polarimeter is circular, so +10° is the same as −350° or +370°. How would you determine whether the observed rotation is +10° or some other value?

6.8 (a) The specific rotation of sucrose (table sugar) in water is +65 degrees mL g^{-1} dm^{-1}. What is the observed optical rotation in a 1 dm path of a sucrose solution prepared from 5 g of sucrose and enough water to form 100 mL of solution?

(b) Chemist Ree N. Ventdawil has said to you that he plans to synthesize the enantiomer of sucrose so that he can measure its specific rotation. You politely inform him that you already know the result. Explain how you can make this claim.

6.4   MIXTURES OF ENANTIOMERS

A. Enantiomeric Excess

When one enantiomer of a chiral compound is uncontaminated by the other enantiomer, it is said to be **enantiomerically pure**. However, mixtures of enantiomers occur commonly. The enantiomeric composition of a mixture of enantiomers is expressed as the **enantiomeric excess**, abbreviated **EE**, which is defined as the difference between the percentages of the two enantiomers in the mixture:

$$\text{EE} = \% \text{ of the major enantiomer} - \% \text{ of the minor enantiomer} \qquad (6.2)$$

For example, if a mixture contains 80% of the (+)-enantiomer and 20% of the (−)-enantiomer, the EE is 80% − 20% = 60%. Notice that we are treating a mixture of enantiomers a little differently than we treat other mixtures. For example, if we have a mixture that contains 80% of a compound *A* and 20% of some contaminant *B*, we sometimes say that compound *A* is "80% pure." The use of enantiomeric excess comes from the effect of the contaminating enantiomer on the optical activity of the mixture. For example, if a mixture consists of 80% of the (+)-enantiomer and 20% of the (−)-enantiomer, then the optical activity of the mixture is 60% of the optical activity of the pure (+)-enantiomer. This is because the 20% of the (−)-enantiomer cancels the rotation of 20% of the (+)-enantiomer, leaving a net optical rotation of 60%. (Remember: the (+)- and (−)-enantiomers have equal rotations of opposite sign.)

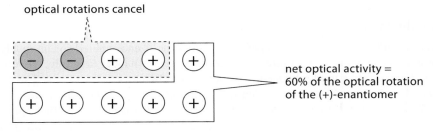

10 molecules: 20% (−)-enantiomer, 80% (+)-enantiomer   (6.3)

From this example you can see that the optical activity of the mixture has a percentage optical activity of the pure (+)-enantiomer that equals the EE. It follows that if we know the optical activity of the *pure* major enantiomer, and if the two enantiomers are the only optically active substances present in a sample, then the enantiomeric excess in the sample can be determined from the specific rotation of the mixture:

$$EE = 100\% \times \frac{[\alpha]_{\text{mixture}}}{[\alpha]_{\text{pure}}} \tag{6.4}$$

where $[\alpha]_{\text{mixture}}$ and $[\alpha]_{\text{pure}}$ are the specific rotations of the mixture and the pure major enantiomer determined under the same conditions. Conversely, if we know the optical activity of the mixture and the EE, then we can calculate the optical activity of the pure major enantiomer. When the EE is calculated from optical activities as in Eq. 6.4, it is sometimes also called **optical purity**.

Finally, the actual percentages of each enantiomer can be calculated from the enantiomeric excess. Remember that the percentages of the two enantiomers add to 100%.

$$\% \text{ minor enantiomer} = 100\% - \% \text{ major enantiomer}$$

Substituting this equation into Eq. 6.2, we have

$$EE = \% \text{ major enantiomer} - (100\% - \% \text{ major enantiomer})$$

$$= 2(\% \text{ major enantiomer}) - 100\% \tag{6.5a}$$

Solving for % major enantiomer, we find

$$\% \text{ major enantiomer} = (EE + 100\%)/2 \tag{6.5b}$$

For our initial example in which EE = 60%, the % major enantiomer is 160%/2 = 80%. The remaining 20% is the minor enantiomer.

STUDY PROBLEM   6.4

Your lab partner tells you that a sample of (+)-2-butanol has an apparent specific rotation of +13.8 degrees mL g^{-1} dm^{-1} and is known to be a mixture of the two enantiomers with EE = 94%. How much of each enantiomer is in the mixture? What is the specific rotation of enantiomerically pure (+)-2-butanol?

SOLUTION   The % of the major enantiomer (+)-2-butanol is calculated from Eq. 6.5b:

$$\% \text{ (+)-enantiomer} = (94\% + 100\%)/2 = 97\%$$

Then there is (100 – 97)% = 3% of the minor enantiomer (–)-2-butanol in the mixture. Checking our work, this means that the EE has to be 97% – 3% = 94%, as given.

Rearranging Eq. 6.4, we calculate the specific rotation of pure (+)-2-butanol as

$$[\alpha]_{\text{pure}} = \frac{100\% \times [\alpha]_{\text{mixture}}}{EE} = \frac{100\% \times (+13.8 \text{ degrees mL g}^{-1} \text{ dm}^{-1})}{94\%} = +14.2 \text{ degrees mL g}^{-1} \text{ dm}^{-1}$$

B. Racemates

A mixture containing equal amounts of two enantiomers is encountered so commonly that it is given a special name: a **racemate** or **racemic mixture**. (In older literature, the term *racemic modification* was used.) A racemate is referred to by name in two ways. The racemate of 2-butanol, for example, can be called racemic 2-butanol, (±)-2-butanol, or *d,l*-2-butanol.

Racemates typically have physical properties that are different from those of the pure enantiomers. For example, the melting point of either enantiomer of lactic acid (p. 237) is 53 °C, but the melting point of racemic lactic acid is 18 °C. The reason for the different melting points is that the crystal structures differ. Recall from Sec. 2.6B that the melting point largely reflects interactions between molecules in the crystalline solid. (Imagine packing a dozen left shoes—a "pure enantiomer"—in a box, and then imagine packing six right shoes and six left shoes—a "racemate"—in the same box. The interactions among the shoes—the way they touch each other—are different in the two cases.) The optical rotations of enantiomers and racemates are another example of differing physical properties. The optical rotation of any racemate is zero, because a racemate contains equal amounts of two enantiomers whose optical rotations of equal magnitude and opposite sign exactly cancel each other. In a racemate, the EE is also 0.

The process of forming a racemate from a pure enantiomer is called **racemization**. The simplest method of racemization is to mix equal amounts of enantiomers. As you will learn, racemization can also occur as a result of chemical reactions.

Because a pair of enantiomers have the same boiling points, melting points, and solubilities—exactly the properties that are usually exploited in designing separations—the separation of enantiomers poses a special problem. The separation of a pair of enantiomers, called an **enantiomeric resolution**, requires special methods that are discussed in Sec. 6.8.

FURTHER EXPLORATION 6.1
Terminology of Racemates

Racemates in the Pharmaceutical Industry

Over half of the pharmaceuticals sold commercially are chiral compounds. Drugs that come from natural sources (or drugs that are prepared from materials obtained from natural sources) have always been produced as pure enantiomers, because, in most cases, chiral compounds from nature occur as only one of the two possible enantiomers. (We'll explore this point in Sec. 7.8A.) Until relatively recently, however, most chiral drugs produced synthetically from *achiral* starting materials were produced and sold as racemates. The reason is that the separation of racemates into their optically pure enantiomeric components requires special procedures that add cost to the final product. (See Sec. 6.8.) The justification for selling the racemic form of a drug hinges on its lower cost and on the demonstration that the unwanted enantiomer is physiologically inactive, or at least that its side effects, if any, are tolerable. However, this is not always so.

The landmark case that dramatically demonstrated the potential pitfalls in marketing a racemic drug involved *thalidomide,* a compound first marketed as a sedative in Europe in 1958.

thalidomide

The (*R*)-(+)-enantiomer of thalidomide was found to have a higher sedative activity than the (*S*)-(−)-enantiomer, but, as was typical of the time, the drug was marketed as the

racemate for economic reasons. This drug was taken by a number of pregnant women to relieve the symptoms of morning sickness. It turned out, tragically, that thalidomide is teratogenic—that is, it causes horrible birth defects, such as deformed limbs, when taken by women in early pregnancy. An estimated 12,000 children were born with thalidomide-induced birth defects, mostly in Europe and South America. The drug was never approved for use in the United States, although some was given to doctors and dispensed for "investigational use."

Although it is believed that only the (*S*)-(−)-enantiomer of thalidomide is teratogenic, it has been shown that either enantiomer is racemized in the bloodstream. Hence, it is likely that the teratogenic effects would have been observed with even the optically pure *R* enantiomer. Nevertheless, thalidomide illustrates the point that enantiomers can, in some cases, have greatly different biological activities.

A remarkable and happier postscript to the thalidomide story is evolving. One of the reasons that thalidomide is teratogenic is that it suppresses *angiogenesis* (the growth of blood vessels), which is essential for actively dividing cells. This effect is disastrous for a developing fetus, but is likely to be beneficial for cancer patients, because the suppression of angiogenesis has been found in early trials to be effective in treating certain cancers. Thalidomide has been approved as part of a treatment for multiple myeloma and is also being used for a certain type of leprosy. Despite these potential benefits, it cannot be given to women who are pregnant or are likely to become pregnant.

The pharmaceutical industry, spurred in part by the U.S. Food and Drug Administration (FDA), has with increasing regularity developed synthetic chiral drugs as single enantiomers rather than racemates, thus ensuring that consumers will not have to contend with unanticipated side effects of therapeutically inactive stereoisomers.

PROBLEMS

6.9   (a) Identify the asymmetric carbon of thalidomide.

(b) Draw a structure of the teratogenic *S* enantiomer of thalidomide using lines, a wedge, and a dashed wedge to indicate the stereochemistry of this carbon.

6.10   A 0.1 *M* solution of an enantiomerically pure chiral compound *D* has an observed rotation of +0.20° in a 1 dm sample container. The molecular mass of the compound is 150.

(a) What is the specific rotation of *D*?

(b) What is the observed rotation if this solution is mixed with an equal volume of a solution that is 0.1 *M* in *L*, the enantiomer of *D*?

(c) What is the observed rotation if the solution of *D* is diluted with an equal volume of solvent?

(d) What is the specific rotation of *D* after the dilution described in part (c)?

(e) What is the specific rotation of *L*, the enantiomer of *D*, after the dilution described in part (c)?

(f) What is the observed rotation of 100 mL of a solution that contains 0.01 mole of *D* and 0.005 mole of *L*? (Assume a 1 dm path length.)

(g) What is the enantiomeric excess (EE) of *D* in the solution described in part (f)?

6.11   A chemist has developed a new synthesis of ibuprofen and has reported that she has prepared the (*S*)-(+)-enantiomer in 90% EE, and that this material has a measured specific rotation of +51.7 degrees mL g^{-1} dm^{-1}.

ibuprofen

(a) Draw a line-and-wedge formula of (*S*)-(+)-ibuprofen.

(b) What is the specific rotation of pure (*S*)-(+)-ibuprofen? Of pure (*R*)-(–)-ibuprofen?

(c) How much of each enantiomer is present in her sample?

6.12   What observed rotation is expected when a 1.5 *M* solution of (*R*)-2-butanol is mixed with an equal volume of a 0.75 *M* solution of racemic 2-butanol, and the resulting solution is analyzed in a sample container that is 1 dm long? The specific rotation of (*R*)-2-butanol is –14.7 degrees mL g^{-1} dm^{-1}.

6.5   STEREOCHEMICAL CORRELATION

Knowing how to assign the *R* or *S* designation to compounds with asymmetric carbons (Sec. 6.2) is one thing, but you can't apply this system to a molecule until you know the actual three-dimensional arrangement of its atoms—that is, its **absolute configuration** or **absolute stereochemistry**. If you were the first person ever to synthesize one enantiomer of a chiral compound, how would you determine its absolute configuration? (Recall that the sign of optical rotation *cannot* be used to assign an *R* or *S* configuration; Sec. 6.3C.) One way is to use a variation of X-ray crystallography called *anomalous dispersion*. Although X-ray crystallography is more widely used than it once was, it still requires specialized expensive instrumentation that is not readily available in the average laboratory. The absolute configurations of most organic compounds are determined instead by using chemical reactions to correlate them with other compounds of known absolute configurations. This process is called **stereochemical correlation**.

To illustrate a stereochemical correlation, suppose you have obtained an optically active sample of the following alkene.

$$Ph—CH—CH{=}CH_2$$
$$|$$
$$CH_3$$

$$[a]_D^{25} = -6.7 \text{ degrees mL g}^{-1} \text{ dm}^{-1}$$
R or *S* configuration unknown

You've measured its optical activity experimentally and have determined that it is levorotatory—that is, it is the (–)-enantiomer. However, you don't know its absolute configuration—whether it is R or S. Remember, the optical rotation does *not* provide this information. You go to the chemical literature and you find that no one has ever determined the absolute configuration of either the (–)- or the (+)-alkene. How would you determine its absolute configuration? This compound is a liquid, so crystallization followed by X-ray crystallography is not realistic.

Being an astute organic chemistry student, you know (Sec. 5.5) that you can convert this alkene into a carboxylic acid by ozonolysis. This reaction breaks the double bond, but *it does not break any of the bonds to the asymmetric carbon.* You carry out this reaction on your alkene and obtain a sample of the carboxylic acid, which has the common name hydratropic acid. You find by direct measurement that your sample of hydratropic acid is optically active and dextrorotatory.

$$\text{Ph}-\underset{\underset{\text{CH}_3}{|}}{\text{CH}}-\text{CH}=\text{CH}_2 \xrightarrow[\text{2) H}_2\text{O}_2/\text{H}_2\text{O}]{\text{1) O}_3} \text{Ph}-\underset{\underset{\text{CH}_3}{|}}{\text{CH}}-\overset{\overset{\text{O}}{\|}}{\text{C}}-\text{OH} \qquad (6.6)$$

(–) optical rotation   **hydratropic acid**
(+) optical rotation

Here is where things stand: You have shown that the (–)-alkene is converted by ozonolysis into (+)-hydratropic acid. Prior to running this reaction, you searched the chemical literature and found that someone in the past had determined the absolute configuration of (+)-hydratropic acid, perhaps by X-ray crystallography. (This search can be carried out rapidly by computer.) In this previous work, the (+)-enantiomer of hydrotropic acid was shown to have the S configuration. Therefore, you have converted the (–)-alkene of unknown configuration into (+)-hydrotropic acid, *known from earlier work* to have the S configuration:

$$\text{Ph}-\underset{\underset{\text{CH}_3}{|}}{\text{CH}}-\text{CH}=\text{CH}_2 \xrightarrow[\text{2) H}_2\text{O}_2/\text{H}_2\text{O}]{\text{1) O}_3} \quad \underset{\underset{\text{H}_3\text{C}}{}}{\text{H}}\overset{\overset{\text{Ph}}{|}}{\underset{}{\text{C}}}\diagdown\overset{\overset{\text{O}}{\|}}{\text{C}}-\text{OH} \qquad (6.7)$$

(–) optical rotation   **(S)-(+)-hydratropic acid**
unknown configuration   The S enantiomer is known
to have (+) optical rotation
from earlier work.

You are now in a position to deduce the absolute configuration of your alkene, because *the corresponding groups in the two compounds must be in the same relative positions.* Remember, no bonds to the asymmetric carbon were broken.

This bond is not broken by ozonolysis

$$\underset{\underset{\text{H}_3\text{C}}{}}{\text{H}}\overset{\overset{\text{Ph}}{|}}{\underset{}{\text{C}}}\diagdown\text{CH}=\text{CH}_2 \xrightarrow[\text{2) H}_2\text{O}_2/\text{H}_2\text{O}]{\text{1) O}_3} \quad \underset{\underset{\text{H}_3\text{C}}{}}{\text{H}}\overset{\overset{\text{Ph}}{|}}{\underset{}{\text{C}}}\diagdown\overset{\overset{\text{O}}{\|}}{\text{C}}-\text{OH} \qquad (6.8)$$

The relative stereochemical positions of these two carbons must be the same.

You now know the absolute configuration of the alkene—that is, you can build a three-dimensional model of it, as shown in Eq. 6.8. Assigning the configuration by the usual rules

(Sec. 6.2) shows that the alkene has the *R* configuration. You have now linked the optical rotation of the alkene to its configuration.

$$\underset{\substack{\textbf{(R)-(–)-alkene}\\ \text{(configuration deduced}\\ \text{from this correlation)}}}{\overset{\text{Ph}}{\underset{\text{H}_3\text{C}}{\text{H}\overset{}{\text{\tiny'''}}\text{C}}}\text{—CH}=\text{CH}_2} \xrightarrow[\text{2) H}_2\text{O}_2/\text{H}_2\text{O}]{\text{1) O}_3} \underset{\substack{\textbf{(S)-(+)-hydratropic acid}\\ \text{(configuration was}\\ \text{previously known)}}}{\overset{\text{Ph}}{\underset{\text{H}_3\text{C}}{\text{H}\overset{}{\text{\tiny'''}}\text{C}}}\overset{\text{O}}{\overset{\|}{\text{—C}}}\text{—OH}} \qquad (6.9)$$

Once this result is published in the chemical literature, others can use this assignment to carry out other correlations (Problem 6.13). It also follows that the dextrorotatory alkene—the enantiomer of your starting alkene—must have the *S* configuration because enantiomers must have optical rotations of opposite signs. Thus, a correlation carried out on either enantiomer establishes the configurations of both.

Although the reactant and the product in this example have different *R,S* configurations, this relationship is not true in general. It is possible for the *R,S* configurations of reactant and product to be same. The result depends on the relative priorities of the groups at the asymmetric carbon and the way these are situated in the three-dimensional model. If a reaction results in a change in the relative priorities of groups at the asymmetric carbon, as in this case, the correlated structures will have different *R,S* designations. If relative priorities don't change, the correlated structures will have the same *R,S* designations. (See Problem 6.13.)

The optical rotation of the two correlated compounds may have different signs, as in this example, or they may have different signs. The relative signs of rotation must be determined by experiment.

To summarize: We can determine the absolute configuration of one compound by converting it into another compound whose absolute configuration is known. We could also take the same approach "in reverse" and deduce the configuration of a product from a starting material of known configuration. This approach of relating reactants and products is unambiguous when the bonds to the asymmetric atom(s) are unaffected by the reaction. (A reaction that breaks these bonds can also be used if the stereochemical outcome of such a reaction has previously been carefully established.)

Stereochemical correlation has been used to establish stereochemistry throughout the history of organic chemistry. One of the most spectacular examples was the determination of the stereochemistry of glucose and other sugars, which we'll study in Sec. 24.10.

PROBLEMS

6.13   Use the known stereochemistry of the starting alkene (from this section) to assign the stereochemical configuration of the product, which was found by experiment to be levorotatory.

$$(R)\text{-}(-)\text{-Ph}\underset{\text{CH}_3}{\text{—CH—}}\text{CH}=\text{CH}_2 \xrightarrow[\text{2) H}_2\text{O}_2/\text{OH}^-]{\text{1) BH}_3} (-)\text{-Ph}\underset{\text{CH}_3}{\text{—CH—}}\text{CH}_2\text{CH}_2\text{—OH}$$

unknown configuration; the
optical rotation is known from
experimental measurement

6.14   Explain how you would use the alkene starting material in Problem 6.13 to determine the absolute configuration of the dextrorotatory enantiomer of the following hydrocarbon:

$$\text{Ph}\underset{\text{CH}_3}{\text{—CH—}}\text{CH}_2\text{—CH}_3$$

Indicate what reaction you would use. Outline the possible results of your experiment and how you would interpret them.

6.6   DIASTEREOMERS

Up to this point, our discussion has focused on molecules with only one asymmetric carbon. What happens when a molecule has two or more asymmetric carbons? This situation is illustrated by 2,3-pentanediol, in which both carbons 2 and 3 are asymmetric.

$$\underset{1}{H_3C} - \underset{2}{\overset{\overset{\displaystyle OH}{|}}{CH}} - \underset{3}{\overset{\overset{\displaystyle OH}{|}}{CH}} - \underset{4}{CH_2}\underset{5}{CH_3}$$

2,3-pentanediol

Each asymmetric carbon might have the *R* or *S* configuration. With two possible configurations at each carbon, four stereoisomers are possible:

(2*S*,3*S*)	(2*R*,3*R*)
(2*S*,3*R*)	(2*R*,3*S*)

These possibilities are shown as ball-and-stick models in Fig. 6.9. What are the relationships among these stereoisomers?

The 2*S*,3*S* and 2*R*,3*R* isomers are a pair of enantiomers because they are noncongruent mirror images; the 2*S*,3*R* and 2*R*,3*S* isomers are also an enantiomeric pair. (Demonstrate this point to yourself with models.) These structures illustrate the following generalization: *For a pair of chiral molecules with more than one asymmetric carbon to be enantiomers, they must have opposite configurations at every asymmetric carbon.*

Because neither the 2*S*,3*S* and 2*S*,3*R* pair nor the 2*R*,3*R* and 2*R*,3*S* pair are enantiomers, they must have a different stereochemical relationship. Stereoisomers that are not enantiomers are called **diastereoisomers**, or more simply, **diastereomers**. Diastereomers are not mirror images. All of the relationships among the stereoisomeric 2,3-pentanediols are shown in Fig. 6.10 (p. 248).

Diastereomers differ in all of their physical properties. Thus, diastereomers have different melting points, boiling points, heats of formation, and standard free energies. Because diastereomers differ in all of their physical properties, they can in principle be separated by conventional means, such as fractional distillation or crystallization. If diastereomers happen

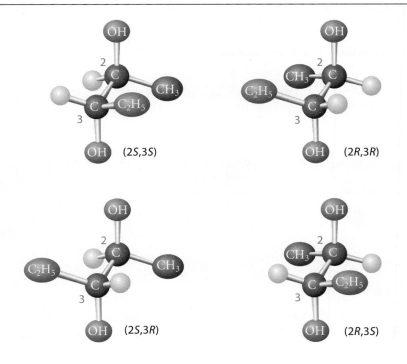

FIGURE 6.9 Stereoisomers of 2,3-pentanediol. In each model, the small unlabeled atoms are hydrogens. This illustration uses a particular conformation of each stereoisomer, but the analysis in the text is equally valid for any other conformation. (Try it!)

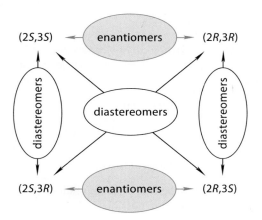

FIGURE 6.10 Relationships among the stereoisomers of 2,3-pentanediol. The pair of stereoisomers at opposite ends of any double-headed arrow have the relationship indicated within the arrow. Notice that enantiomers have different configurations at both asymmetric carbons.

to be chiral, they can be expected to be optically active, but their specific rotations will have no relationship. These points are illustrated in Table 6.1, which gives some physical properties for four stereoisomers and their racemates.

You have now seen an example of every common type of isomerism. To summarize:

1. *Isomers* have the same molecular formula.
2. *Constitutional isomers* have different atomic connectivities.
3. *Stereoisomers* have identical atomic connectivities. There are *only two* types of stereoisomers:
   a. *Enantiomers* are noncongruent mirror images.
   b. *Diastereomers* are stereoisomers that are not enantiomers.

The structural relationships among molecules are analyzed by working with *one pair of molecules at a time.* The flow chart in Fig. 6.11 provides a systematic way to determine the isomeric relationship, if one exists, between two nonidentical molecules. Study Problem 6.5 illustrates the use of Fig. 6.11.

TABLE 6.1 Properties of Four Chiral Stereoisomers

$$H_3C-\overset{\overset{\displaystyle O}{\|}}{C}-NH-\overset{2}{C}H-\overset{\overset{\displaystyle O}{\|}}{C}-OH$$
$$\underset{\underset{\displaystyle C_2H_5}{|}}{\overset{3}{C}H-CH_3}$$

Configuration	Specific rotation $[\alpha]_D^{25}$ (ethanol), degrees mL g^{-1} dm^{-1}	Melting point, °C	Relationship	
(2S,3S)	+15	150–151	enantiomers	diastereomers
(2R,3R)	−15	150–151		
(2S,3R)	+21.5	155–156	enantiomers	
(2R,3S)	−21.5	155–156		
racemate of (2S,3S) and (2R,3R)	0	117–123		
racemate of (2S,3R) and (2R,3S)	0	165–166		

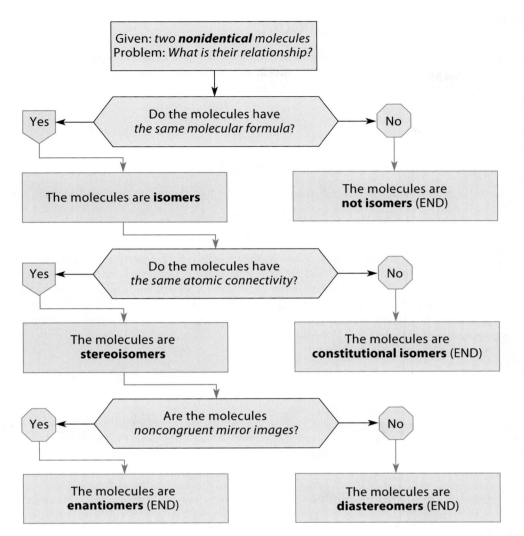

FIGURE 6.11 A systematic way to analyze the relationship between two nonidentical molecules. Given a pair of molecules, work from the top of the chart to the bottom, asking each question in order and following the appropriate branch. When you get to a red box labeled "END," the isomeric relationship is determined.

STUDY PROBLEM   6.5

Determine the isomeric relationship between the following two molecules:

SOLUTION   Work from the top of Fig. 6.11 and answer each question in turn. These two molecules have the same molecular formula; hence, they are isomers. They have the same atomic connectivity; hence, they are stereoisomers. (In fact, they are the *E* and *Z* isomers of 3-hexene.) Because the molecules are not mirror images, they must be diastereomers. Thus, (*E*)- and (*Z*)-3-hexene are diastereomers.

We can see from Study Problem 6.5 that double-bond isomerism is actually one type of diastereomeric relationship. The fact that neither (*E*)- nor (*Z*)-3-hexene is chiral shows that some diastereomers are *not* chiral. On the other hand, other diastereomers are chiral, as in the case of the diastereomeric 2,3-pentanediols (shown in Fig. 6.9).

<table>
<tr><td>**6.7**</td><td>**MESO COMPOUNDS**</td></tr>
</table>

Up to this point, each example of a molecule containing one or more asymmetric carbon atoms has been chiral. However, certain compounds containing two or more asymmetric carbons are achiral. 2,3-Butanediol provides an example:

$$\underset{\textbf{2,3-butanediol}}{H_3C\underset{1}{\text{—}}\overset{\overset{\displaystyle OH}{|}}{\underset{2}{CH}}\text{—}\overset{\overset{\displaystyle OH}{|}}{\underset{3}{CH}}\underset{4}{\text{—}}CH_3}$$

As with 2,3-pentanediol in Sec. 6.6, there appear to be four stereochemical possibilities:

(2S,3S)	(2R,3R)
(2S,3R)	(2R,3S)

Ball-and-stick models of these molecules are shown in Fig. 6.12. Consider the relationships among these structures. The 2S,3S and 2R,3R structures are noncongruent mirror images, and are thus enantiomers. However, although the 2S,3R and 2R,3S structures as drawn are mirror images, each has a center of symmetry and is therefore *achiral*. In fact, *these two structures are identical*. We can demonstrate their identity by rotating the 2R,3S structure 180° about an axis perpendicular to the C2—C3 bond:

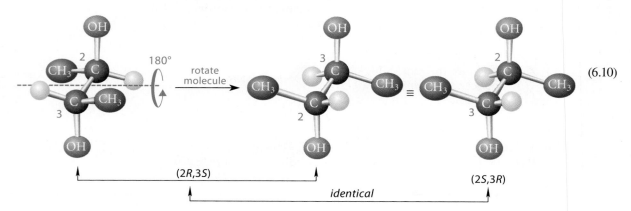

(6.10)

Hence, there are only *three* stereoisomers of 2,3-butanediol, not four as it seemed at first, because two of the possibilities are identical. Because the 2R,3S-stereoisomer of 2,3-butanediol is congruent to its mirror image, *it is achiral*. Because it is achiral, *it is optically inactive*. This stereoisomer is therefore an achiral diastereomer of both (2R,3R)- and (2S,3S)-2,3-butanediol, which, as we have seen, are chiral.

   The achiral stereoisomer of 2,3-butanediol is an example of a *meso compound*, and it is called *meso*-2,3-butanediol. A **meso compound** is an achiral (and therefore optically inactive) compound that has chiral diastereomers. In virtually all of the examples we'll encounter, *a meso compound is an achiral compound that has at least two asymmetric centers.* Although there are a few unusual exceptions, we can use this statement as the operational definition of a meso compound. For example, *cis*- and *trans*-2-butene are stereoisomers, and they are achiral; but they are not meso compounds because neither has any asymmetric carbons.

$$\underset{\textbf{\textit{cis}-2-butene}}{\overset{\displaystyle H_3C \qquad CH_3}{\underset{\displaystyle H \qquad\quad H}{C=C}}} \qquad \underset{\textbf{\textit{trans}-2-butene}}{\overset{\displaystyle H \qquad\quad CH_3}{\underset{\displaystyle H_3C \qquad H}{C=C}}}$$

Both compounds are achiral . . . BUT neither has chiral stereoisomers, and neither has asymmetric carbons. Therefore, these are *not* meso compounds.

   A summary of the relationships between the 2,3-butanediol stereoisomers is given in Fig. 6.13.

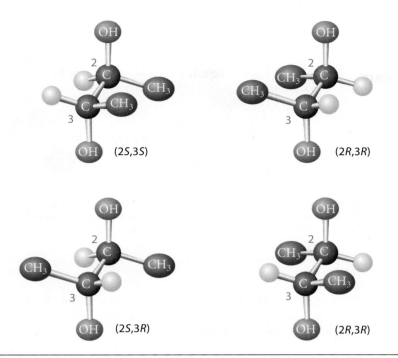

FIGURE 6.12 Stereoisomeric possibilities for 2,3-butanediol. As with Fig. 6.9, this illustration uses a particular conformation of each stereoisomer and the small unlabeled atoms on each model are hydrogens.

(2S,3S)   (2R,3R)

(2S,3R)   (2R,3R)

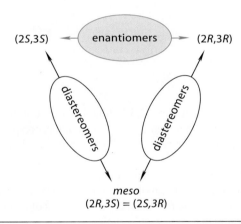

(2S,3S) ⟵ enantiomers ⟶ (2R,3R)

diastereomers   diastereomers

meso
(2R,3S) = (2S,3R)

FIGURE 6.13 Relationships among the stereoisomers of 2,3-butanediol. Any pair of compounds at opposite ends of the double-headed arrow have the relationship indicated within the arrow. The meso stereoisomer is achiral and thus has no enantiomer.

Notice carefully the difference between a meso compound and a racemate. Although both are optically inactive, a meso compound is a *single achiral compound*, but a racemate is a *mixture of chiral compounds*—specifically, an equimolar mixture of enantiomers.

The existence of meso compounds shows that *some achiral compounds have asymmetric carbons*. Thus, the presence of asymmetric carbons in a molecule is an *insufficient* condition for it to be chiral, unless it has only *one* asymmetric carbon. If a molecule contains n asymmetric carbons, then it has 2^n stereoisomers unless there are meso compounds. If there are meso compounds, then there are fewer than 2^n stereoisomers.

Suppose, now, that we have a structure that contains two or more asymmetric carbons. How can we tell whether it can exist as a meso stereoisomer? A meso compound is possible only when a molecule with two or more asymmetric atoms can be divided into halves that have the same connectivity. (The word *meso* means "in the middle.")

$$CH_3$$
$$|$$
the two halves of the molecule $\quad CH-OH$
have the same connectivity $\quad ----+----$
$$CH-OH$$
$$|$$
$$CH_3$$

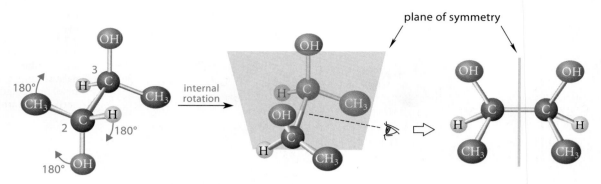

FIGURE 6.14   An eclipsed conformation of *meso*-2,3-butanediol has a plane of symmetry and is therefore achiral. (The staggered conformation on the left has a center of symmetry and is also achiral.) Finding *any* achiral conformation in a molecule with asymmetric carbons is sufficient to show that it is meso.

Once you recognize the possibility of a meso compound, how do you know which stereoisomers are meso and which are chiral? First, in a meso compound, the corresponding asymmetric atoms in each half of the molecule must have *opposite stereochemical configurations*:

$$CH_3$$
$$|$$
meso compounds have ·····> $CH—OH$
opposite configurations ·····|·····
·····> $CH—OH$
$$|$$
$$CH_3$$

Thus, one asymmetric carbon in *meso*-2,3-butanediol (see Fig. 6.12) is *R* and the other is *S*. An analogy is your two hands, held palm-to-palm so that corresponding fingers are touching. Taken as a single object, the pair is a "meso" object; its two halves (each hand) are mirror images.

Another way to identify a meso compound makes use of the following "shortcut": if you can find *any* conformation of a molecule with asymmetric carbons—even an eclipsed conformation—that is achiral, the molecule is meso. Planes of symmetry are particularly easy to spot in eclipsed conformations, and a molecule with a plane of symmetry is achiral (Sec. 6.1C). Therefore, finding an eclipsed conformation with a plane of symmetry is sufficient to show that a compound is meso, even though the compound does not exist in the eclipsed conformation. (We'll see why this works in Sec. 6.9.) For example, Fig. 6.14 shows an eclipsed conformation of *meso*-2,3-butanediol that has a plane of symmetry.

PROBLEMS

6.15   Tell whether each of the following molecules has a meso stereoisomer.

(a) [structure: CH with two Cl groups on a branched carbon chain]   (b) [structure: $CH_3CHCH_2CH_2CH_2Cl$ with Cl above]   (c) *trans*-2-hexene

6.16   Explain why the following compound has two meso stereoisomers.

$$H_3C—CH—CH—CH—CH_3$$

with OH   OH   OH on the three central carbons

(*Hint:* The plane that divides the molecule into structurally identical halves can go through one or more atoms.)

6.8 SEPARATION OF ENANTIOMERS (ENANTIOMERIC RESOLUTION)

As noted in Sec. 6.4, the separation of two enantiomers (an **enantiomeric resolution**) poses a special problem. Because a pair of enantiomers have identical melting points, boiling points, and solubilities, we cannot exploit these properties for the separation of enantiomers as we might for other compounds. How, then, are enantiomers separated?

The resolution (separation) of enantiomers takes advantage of the fact that *diastereomers, unlike enantiomers, have different physical properties.* The strategy used is to convert a mixture of enantiomers *temporarily* into a mixture of diastereomers by allowing the mixture to combine with an enantiomerically pure chiral compound called a **resolving agent**. The resulting diastereomers are separated, and the resolving agent is then removed to give the pure enantiomers.

Analogy for a Resolving Agent

Suppose you are blindfolded and asked to sort 100 gloves into separate piles of right- and left-handed gloves. (Never mind how you got into this predicament!) The gloves are identical except that 50 are right-handed and 50 are left-handed. The mixture of gloves is a "racemate." How would you separate them? You can't do it by weight, by smell, or by any other simple physical property, because right- and left-handed gloves have the same properties. The way you do it is by trying each glove on your right (or left) hand. Your hand thus acts as an "enantiomerically pure" chiral resolving agent. A right-handed glove on your right hand generates a certain feeling (which we describe by saying "it fits"), and a left-handed glove on the right hand generates a totally different feeling. In fact, we could say that the right hand wearing the right-handed glove is the diastereomer of the right hand wearing the left-handed glove. The different sensations generated by the two situations are analogous to the different physical properties of diastereomers. After classifying each glove as "right" or "left" on the basis of this sensation, you then break the hand–glove interaction (you remove the glove) and put the glove in the appropriate pile. You have thus converted the diastereomer (the hand–glove combination) into the pure enantiomer (the glove) and the chiral resolving agent (the hand).

The principle used to separate enantiomers can be stated more formally as follows:

The principle of enantiomeric differentiation: *The separation or differentiation of enantiomers requires that they interact with an enantiomerically pure chiral agent.*

This is a very important general principle that we'll use repeatedly. In the hand–glove analogy, the "chiral agent" is the hand that "interacts with" (that is, tries on) the two gloves. When the "chiral agent" is a compound, the interaction results in a pair of diastereomers, which have different properties that can be used to separate or distinguish the two enantiomers.

Many techniques have been developed for enantiomeric resolution. In this section you'll learn about three examples: *chiral chromatography, diastereomeric salt formation,* and *selective crystallization*. Try to notice how the principle of enantiomeric differentiation is applied in each example.

A. Chiral Chromatography

Chromatography is a very important technique for separating the components of mixtures. In a widely used version of this technique, called **column chromatography**, shown in Fig. 6.15a on p. 254, a mixture of compounds is introduced onto a cylindrical column containing a finely powdered solid, called the **stationary phase**. The components of the mixture adsorb (bind) reversibly to the stationary phase. This adsorption results from *noncovalent* attractions between the molecules of the stationary phase and the molecules in solution to be separated.

(a)

(b)

FIGURE 6.15 (a) In column chromatography, a mixture of two compounds is adsorbed reversibly onto a finely powdered solid (stationary phase). As solvent is passed through the column, the less strongly adsorbed component of the mixture moves through the column more rapidly and emerges first. The more strongly adsorbed component of the mixture is retained longer on the column and emerges later. (b) A plot of concentration of the mixture components vs. either time of elution or solvent volume. This is called a *chromatogram*. The area under each peak is proportional to the total amount of each component.

(Noncovalent interactions are discussed in Secs. 8.4–8.6.) The column is then eluted (washed) continuously with a solvent. If the components of the mixture adhere to the stationary phase with sufficiently different affinities, the component with the smallest affinity for the solid emerges first from the column, and the components with progressively greater affinities for the solid emerge later. A graph of the concentrations of the compounds in the mixture against time (or volume of the solvent) is called a **chromatogram** (Fig. 6.15b).

The chromatographic separation of enantiomers is called **chiral chromatography**. In a widely used type of chiral chromatography, the stationary phase typically consists of microscopic porous glass beads to which an *enantiomerically pure chiral compound* has been *covalently* attached. This material serves as the *resolving agent*. This combination of glass beads and covalently attached resolving agent is called generally a **chiral stationary phase**, abbreviated **CSP**. For example, one of many CSPs that are available commercially is shown in Eq. 6.11. (Don't be at all concerned with the detailed structure; just notice that the pendant resolving agent is a *chiral* compound and that it is *enantiomerically pure*. As an analogy, think of it as a "left hand" hanging from the glass bead.)

a chiral stationary phase
(each bead contains many
copies of the chiral molecule)

a chiral stationary phase
(molecular view)

abbreviation
(used in Eq. 6.12)

(6.11)

As Eq. 6.11 shows, each solid bead of the CSP contains many copies of the resolving agent. Now, if a mixture of enantiomers is passed through the CSP column, each of the two enantiomers forms a noncovalent complex with the immobilized resolving agent:

chiral
stationary phase

mixture of
enantiomers
to be
separated

diastereomeric complexes
have different standard free energies;
therefore, they are formed in different amounts

(6.12)

(We've used the abbreviation for the CSP in Eq. 6.11 as a specific example. Don't worry about the details of complex formation.) Because the resolving agent is enantiomerically pure, the two complexes differ in configuration at *only one* of their asymmetric carbons. In other words, the two complexes are *diastereomers*. In general, diastereomers have different free energies and different stabilities. For this reason, the *equilibrium constants for their formation differ*. This means that one of two enantiomers binds more tightly to the resolving agent than the other and, as a result, the concentrations of the two complexes are different. As solvent is passed through the column, the more strongly adsorbed enantiomer is retained longer on the column. Therefore, the solution of the less strongly binding enantiomer emerges first, and solution of the more strongly binding enantiomer next. Figure 6.16 on p. 256 shows the chromatogram for the enantiomeric resolution of Nirvanol, a synthetic drug, by the CSP in Eq. 6.11.

You may be wondering how we know what CSP to use for a particular separation. The answer is that we really don't know unless someone has done it before (as in this case). The selection of the CSP and the conditions to be used for a particular separation are sometimes matters of trial and error, but experience has led to some principles by which a resolving agent can usually be chosen rationally. The crucial point for us to notice is that the enantiomers are separated by their interaction with the CSP by *the temporary formation of diastereomers*. In terms of the analogy at the beginning of this section, the two enantiomers to be separated are the gloves, and the CSP is the hand.

Typically, chiral chromatography is used on a relatively small scale because the chiral stationary phases are fairly expensive. It is a superb method for the analysis of mixtures of enantiomers, and it is frequently used to determine enantiomeric excess (EE; Sec. 6.4A).

Practical chiral chromatography was developed by William H. Pirkle in the mid-1970s while he was professor of chemistry at the University of Illinois, and the CSPs he developed became known as *Pirkle columns*. (The CSP shown in Eq. 6.11 is an example of a Pirkle column.) A large variety of CSPs (including Pirkle columns) are now available commercially.

FIGURE 6.16 A chromatogram showing the enantiomeric resolution of Nirvanol, a synthetic anticonvulsant, on the chiral stationary phase in Eq. 6.11. (The elution solvent is an 80:20 hexane:isopropyl alcohol mixture.) Which enantiomer has the greater affinity for the chiral stationary phase?

(R)-(−)-Nirvanol

a mixture of *R*- and *S*-Nirvanol is injected here

(S)-(+)-Nirvanol

concentration

time, minutes

PROBLEM

6.17   The enantiomeric resolution in Fig. 6.16 used the chiral stationary phase (CSP) in Eq. 6.11. How would the enantiomeric resolution in Fig. 6.16 be affected if

(a) the *enantiomer* of the CSP in Eq. 6.11 were used?

(b) the *racemate* of the CSP in Eq. 6.11 were used?

B. Diastereomeric Salt Formation

Diastereomeric salt formation is a method used for the enantiomeric resolution of acidic or basic compounds. Particularly well suited for large-scale separations, this method is illustrated by the enantiomeric resolution of the racemate of α-phenethylamine:

$$:NH_2$$
$$|$$
$$Ph{-}CH{-}CH_3$$

α-phenethylamine

Amines are derivatives of ammonia in which one or more hydrogen atoms have been replaced by organic groups. Diastereomeric salt formation involving amines takes advantage of the fact that amines, like ammonia, are bases; so, they react rapidly and quantitatively with carboxylic acids to form salts:

$$R{-}\overset{\cdots}{N}H_2 \quad H{-}\overset{\cdots}{\underset{\cdots}{O}}{-}\overset{\overset{\displaystyle :O:}{\|}}{C}{-}R \quad \rightleftharpoons \quad R{-}\overset{+}{N}H_3 \quad {}^-{:}\overset{\cdots}{\underset{\cdots}{O}}{-}\overset{\overset{\displaystyle :O:}{\|}}{C}{-}R \qquad (6.13)$$

an amine	a carboxylic acid		
(a Brønsted base)	(a Brønsted acid)	$pK_a \approx 9{-}10$	
	$pK_a \approx 4{-}5$	a salt	

The resolving agent is an *enantiomerically pure* carboxylic acid. In many cases, enantiomerically pure compounds used for this purpose can be obtained from natural sources. One such compound is (2*R*,3*R*)-(+)-tartaric acid:

(2R,3R)-(+)-tartaric acid

The reaction of (+)-tartaric acid with the racemic amine as shown in Eq. 6.13 gives a mixture of two *diastereomeric* salts:

These salts are diastereomers because they differ in configuration at *only one* of their three asymmetric carbons. (Enantiomers must differ at *every* asymmetric carbon.) Because these salts are diastereomers, they have different physical properties. In this case, they have significantly different solubilities in methanol, a commonly used alcohol solvent. (This was found by trying different solvents.) The (S,R,R)-diastereomer happens to be less soluble, and it crystallizes selectively from methanol, leaving the (R,R,R)-diastereomer in solution, from which it may be recovered. Once either pure diastereomer is in hand, the salt can be decomposed with base to liberate the water-insoluble, optically active amine, leaving the tartaric acid in solution as its conjugate-base dianion.

Salt formation is such a simple and convenient reaction that it is often used for the enantiomeric resolution of amines and carboxylic acids.

PROBLEM

6.18 Which of the following amines could in principle be used as a resolving agent for a racemic carboxylic acid?

C. Selective Crystallization

Another method of enantiomeric resolution, and one used frequently in the pharmaceutical industry for the enantiomeric resolution of large quantities of chiral compounds that form crystalline solids, is *selective crystallization*. As you may know from your own laboratory work, crystallization is often a slow process, and it sometimes can be accelerated by adding a seed crystal of the compound to be crystallized. In **selective crystallization**, a solution of a mixture of enantiomers is cooled to supersaturation and a seed crystal of the desired enantiomer is added. In this case, the seed crystal serves as the resolving agent and promotes crystallization of the desired enantiomer.

How does the principle of enantiomeric differentiation operate in selective crystallization? The seed crystal contains only molecules of the pure enantiomer of the desired compound. The seed crystal can grow in two ways: it can incorporate more molecules of the *same* enantiomer or some molecules of the *opposite* enantiomer. These two possibilities generate two *diastereomeric* crystals. Because the crystals are diastereomeric, they have different properties—specifically, different solubilities. It is common that the "pure" crystal—the crystal containing molecules of only one enantiomer—has the higher melting point, and thus the lower solubility. (Given two compounds of closely related structure, the compound with the higher melting point tends to be less soluble.) Hence, the pure enantiomer crystallizes selectively.

Selective crystallization requires a small amount of the desired pure enantiomer to start with, but this can be obtained, for example, by chiral chromatography. Selective crystallization, then, provides a mechanism for "amplification" of an enantiomeric resolution to a larger scale.

Chiral chromatography, diastereomeric salt formation, and selective crystallization are only three of many methods used for enantiomeric resolutions. In any method, however, the principle of enantiomeric differentiation must operate: An enantiomerically pure resolving agent interacts with the two enantiomers in a mixture to form *temporarily* a mixture of diastereomers. It is the difference in the properties of these diastereomers that is ultimately exploited in the separation.

Chiral Recognition by Scent Receptors

In many cases, enantiomers have different odors. The enantiomers of carvone are a familiar example.

R-(−)-carvone
(spearmint)

S-(+)-carvone
(caraway)

(*R*)-(−)-Carvone gives spearmint its familiar odor. (Naturally obtained (*R*)-(−)-carvone is used as a natural flavoring, and spearmint oil production is a $100 million industry in the

United States.) Its enantiomer, (*S*)-(+)-carvone, is present in caraway seeds (actually, the fruit of the caraway plant), which give rye bread its characteristic odor.

The different odors of enantiomers provide a biological illustration of the principle of enantiomeric differentiation. Scent receptors are proteins, and they are enantiomerically pure, chiral molecules. (Humans have 347 scent receptor proteins.) Each scent receptor can therefore serve as a "chiral agent." Whether a carvone molecule interacts with one or (as is likely) several different scent receptors, the interactions of the two carvone enantiomers with any given scent receptor are diastereomeric. These diastereomeric interactions result in different neurological signals that the brain recognizes as different odors.

6.9   RAPIDLY INTERCONVERTING STEREOISOMERS

A. Stereoisomers Interconverted by Internal Rotations

Because butane, $CH_3CH_2CH_2CH_3$, has no stereocenters, you might conclude that it cannot exist in stereoisomeric forms. However, an examination of the individual conformations of butane (Sec. 2.3B) leads to a different conclusion. As shown in Fig. 6.17, the two gauche conformations of butane are noncongruent mirror images, or enantiomers; consequently,

mirror

(a) mirror images

(b) the two mirror images
are not congruent

FIGURE 6.17 (a) The two gauche conformations of butane are mirror images. The mirror images are shown with different colors for the bonds. (b) Because these mirror images are not congruent, they are enantiomers.

gauche-butane is chiral! The chirality of *gauche*-butane shows that *some chiral molecules do not contain asymmetric centers.* This example shows that the existence of an asymmetric atom is not a requirement for chirality.

The two gauche conformations of butane are **conformational enantiomers**: enantiomers that are interconverted by a conformational change. The "conformational change" in this case is an internal rotation. The anti conformation of butane, in contrast, is achiral (verify this!) and is a diastereomer of either one of the gauche conformations. *Anti*-butane and either one of the *gauche*-butanes are therefore **conformational diastereomers**: diastereomers that are interconverted by a conformational change.

Despite the chirality of any one gauche conformation of butane, the compound butane is not optically active because the two gauche conformations are present in equal amounts. The optical activity of one gauche enantiomer thus cancels the optical activity of the other. (The anti conformation, because it is achiral, would not be optically active even if it were present alone.) However, imagine an amusing experiment in which the two gauche conformations of butane are separated (by an as yet undisclosed method!) at such a low temperature that the interconversion of the gauche and anti conformations of butane is very slow. Each *gauche*-butane isomer, like any chiral molecule, would then be optically active! The two gauche isomers would have equal specific rotations of opposite signs, but many of their other properties would be the same. Because *anti*-butane is achiral, it would have zero optical rotation, and all of its properties would differ from those of its *gauche*-butane diastereomers. The isolation of individual conformations is impossible at room temperature, because the butane isomers come to equilibrium within 10^{-9} second by rotation about the central carbon–carbon bond. (This is another example of *racemization*; Sec. 6.4B.)

This discussion of butane conformations forces us to focus more closely on what we mean when we say that a molecule is chiral or achiral. Strictly speaking, the terms *chiral* and *achiral* can only be applied to a single rigid object. Thus, each gauche conformation of butane is chiral, and the anti conformation is achiral. Butane is a *mixture* of conformations, however, and is therefore a mixture of "objects." Chemists have broadened the use of the terms *chiral* and *achiral* to molecules that consist of many conformations by introducing the dimension of *time* into the definitions in the following way: *A molecule is said to be achiral when it consists of rapidly equilibrating enantiomeric conformations that cannot be separated on any reasonable time scale.* In butane, therefore, the rapidly equilibrating conformations are the two gauche conformations. We cannot isolate each conformation on any reasonable time scale because the equilibration is too fast. When we think of butane in this way, then, we are in effect considering it as one object with a *time-averaged conformation* that is achiral.

We don't necessarily have to examine every conformation of a molecule to determine whether it is achiral. If we know that the conformational equilibrium is rapid (as it is for most simple molecules), then the molecule is achiral if we can find *one* achiral conformation—even an unstable conformation such as an eclipsed conformation. This works because, once the molecule is in (or passes through) an achiral conformation, formation of either enantiomeric conformation is equally likely.

enantiomeric gauche conformations

achiral eclipsed conformation

enantiomeric gauche conformations

achiral anti conformation

(6.15)

Thus, recognizing that either the anti or eclipsed conformation of butane in Eq. 6.15 is achiral is sufficient for us to know that any chiral conformation of butane must be in rapid equilibrium with its enantiomer, and that butane is achiral as a result. We encountered the same idea with meso compounds (Fig. 6.14, p. 252). Meso compounds are like butane in the sense that they contain at least one achiral conformation and rapidly interconverting enantiomeric conformations (Problem 6.20). (They differ from butane by the presence of asymmetric carbons.) It follows that a molecule is chiral only if it has *no* achiral conformations or, equivalently, only if *all* of its conformations (even its unstable eclipsed conformations) are chiral.

   Once chemists realized that an achiral molecule could possess enantiomeric conformations, they started looking for—and found—molecules consisting of conformational enantiomers that interconvert so slowly that the individual enantiomeric conformations can be isolated. (See Further Exploration 6.2.)

FURTHER EXPLORATION 6.2
Isolation of Conformational Enantiomers

PROBLEMS

6.19   Taking the anti conformation of butane as an isolated structure, determine whether it has any stereocenters. If so, identify them.

6.20   (a) What are the stereochemical relationships among the three conformations of *meso*-2,3-butanediol (the compound discussed in Sec. 6.7)?

   (b) Explain why *meso*-2,3-butanediol is achiral even though some of its conformations are chiral.

6.21   Which of the following compounds could in principle be resolved into enantiomers at very low temperatures? Explain.

   (a) propane   (b) 2,3-dimethylbutane   (c) 2,2,3,3-tetramethylbutane

B. Asymmetric Nitrogen: Amine Inversion

Some amines, such as ethylmethylamine, undergo a rapid interconversion of stereoisomers.

$$H_3C-\overset{\displaystyle H}{\underset{\displaystyle C_2H_5}{N:}}$$

ethylmethylamine

transition state

sp^2-hybridized nitrogen

(a)

180°

enantiomers

(b)

FIGURE 6.18 Inversion of amines. (a) As the inversion takes place, the large lobe of the electron pair appears to push through the nitrogen to the other side. As this occurs, the three other
groups move first into a plane containing the nitrogen, then to the other side (*green arrows*).
(b) The mirror-image relationship of the inverted amines is shown by turning either molecule
180° in the plane of the page. Because the two mirror images are noncongruent, they are
enantiomers.

Ethylmethylamine has four different groups around the nitrogen: a hydrogen, an ethyl group,
a methyl group, and an electron pair. Because the geometry of this molecule is tetrahedral,
ethylmethylamine appears to be a chiral molecule—it should exist as two enantiomers. The
asymmetric atom is a nitrogen.

mirror plane

enantiomers of
ethylmethylamine

In fact, the two enantiomers of amines such as ethylmethylamine cannot be separated,
because they rapidly interconvert by a process called **amine inversion**, shown in Fig. 6.18. In
this process, the larger lobe of the electron pair seems to push through the nucleus to emerge
on the other side. (Imagine pulling an inflated balloon through a small hole.) The molecule
is not simply turning over; it is actually turning itself inside out! This is something like what
happens when an umbrella turns inside-out in the wind. This process occurs through a transition state in which the amine nitrogen becomes sp^2-hybridized. Figure 6.18b shows that
amine inversion interconverts the enantiomeric forms of the amine. Because this process is
rapid at room temperature, it is impossible to separate the enantiomers. Therefore, ethylmethylamine is a mixture of rapidly interconverting enantiomers. Amine inversion is yet another
example of *racemization* (Sec. 6.4B).

PROBLEM

6.22   Assume that the following compound has the *S* configuration at its asymmetric carbon.

$$CH_3CH_2-\overset{\underset{\textstyle |}{CH}}{\underset{\textstyle CH_3}{}}-\overset{\overset{\textstyle CH_3}{\diagup}}{\underset{\textstyle C_2H_5}{N:}}$$

(a) What is the isomeric relationship between the two forms of this compound that are interconverted by amine inversion?

(b) Could this compound be resolved into enantiomers?

Inversion at Other Atoms Inversion processes can occur at other atoms. When the central atom comes from the second period of the periodic table, inversion is very rapid, as it is with amines:

a carbon anion (carbanion)   an amine

an oxonium ion

All of these inversions are very fast at room temperature.

(6.16)

Therefore, if one of these atoms is the only asymmetric center in a compound, the compound cannot be resolved into enantiomers and cannot maintain optical activity.

However, when the central atom comes from the third and greater periods of the periodic table, inversion is very slow—so slow that it does not occur at room temperature, for practical purposes. (Inversion is faster and can be observed at higher temperatures.) This means that when the phosphorus atom of a phosphine, or the sulfur atom of a trialkylsulfonium ion, is an asymmetric center, such a compound can be resolved into enantiomers.

a phosphine   a sulfonium ion

Neither of these inversions occur at room temperature.
(These types of compounds can be resolved into enantiomers.)

(6.17)

The reason for the difference lies in the hybridization of the central atom. As we know (p. 18 and Sec. 1.9), the unshared electron pair on the nitrogen of ammonia (and amines) occupies an approximately sp^3-hybridized orbital. This orbital has 75% $2p$ character. In the transition state for inversion (Fig. 6.18a), the central atom is sp^2-hybridized, and the unshared pair occupies a $2p$ orbital. A relatively small amount of energy is required to add another 25% p character to the unshared pair; so, the inversion energy barrier is small, and inversion is fast.

If the central atom is from the third or greater period, the unshared electron pair occupies an orbital with a high degree of s character. It takes significant energy to convert an electron pair in a $3s$ orbital to one in a $3p$ orbital. Therefore, the inversion barrier for these atoms is larger, and inversion is slow.

Why do the unshared pairs in third-period (and higher-period) atoms have more *s* character than they do in second-period atoms? The unshared electron pair in a third-period atom is held less tightly than it is in a second-period atom, and it takes up a lot of space. According to VSEPR theory, repulsion of this unshared pair with the electrons in neighboring bonds causes the neighboring bonds to compress more than they do in second-period atoms. In fact, the R—S—R bond angles of sulfonium ions and the R—P—R bond angles of phosphines are around 100°, whereas the R—N—R bond angles in amines are typically 110°. Remember that *bond angle and hybridization are intimately related.* Bond angles closer to 90° require that the bonds have a great deal of *p* character, because *p* orbitals are oriented at 90° angles. If the bonds to sulfur or phosphorus contain most of the 3*p* character, the unshared electron pair on sulfur or phosphorus has little 3*p* character and a lot of 3*s* character.

PROBLEM

6.23 Arsenic (As) is below nitrogen and phosphorus in Group 5A of the periodic table. In an *arsine* (R₃As:) the R—As—R bond angles are about 92°. How would you expect the inversion rate of arsines to compare with that of amines and phosphines? Explain.

6.10   THE POSTULATION OF TETRAHEDRAL CARBON

Chemists recognized the tetrahedral configuration of tetracoordinate carbon almost one-half century before physical methods confirmed the idea with direct evidence. This section shows how the phenomena of optical activity and chirality played key roles in this development, which was one of the most important chapters in the history of organic chemistry.

The first chemical substance in which optical activity was observed was quartz. It was discovered that when a quartz crystal is cut in a certain way and exposed to polarized light along a particular axis, the plane of polarization of the light is rotated. In 1815, the French chemist Jean-Baptiste Biot (1774–1862) showed that quartz exists as both levorotatory and dextrorotatory crystals. The Abbé René Just Haüy (1743–1822), a French crystallographer, had earlier shown that there are two kinds of quartz crystals, which are related as object and noncongruent mirror image. Sir John F. W. Herschel (1792–1871), a British astronomer, found a correlation between these crystal forms and their optical activities: one of these forms of quartz is dextrorotatory and the other levorotatory. These were the key discoveries that clearly associated the chirality of a substance with the phenomenon of optical activity.

During the period 1815–1838, Biot examined several organic substances, both pure and in solution, for optical activity. He found that some (for example, oil of turpentine) show optical activity, and others do not. He recognized that because optical activity can be displayed by compounds in solution, *it must be a property of the molecules themselves.* (The dependence of optical activity on concentration, Eq. 6.1 (p. 239), is sometimes called *Biot's law*.) What Biot did *not* have a chance to observe is that some organic molecules exist in both dextrorotatory and levorotatory forms. The reason Biot never made this observation is undoubtedly that many optically active compounds are obtained from natural sources as single enantiomers.

The first observation of enantiomeric forms of the same organic compound involved tartaric acid:

$$\underset{\text{tartaric acid}}{\text{HO}-\overset{\overset{\text{O}}{\|}}{\text{C}}-\overset{\overset{\text{OH}}{|}}{\text{CH}}-\overset{\overset{\text{OH}}{|}}{\text{CH}}-\overset{\overset{\text{O}}{\|}}{\text{C}}-\text{OH}}$$

This substance had been known by the ancient Romans as its monopotassium salt, *tartar*, which deposits from fermenting grape juice. Tartaric acid derived from tartar was one of the compounds examined by Biot for optical activity; he found that it has a positive rotation. An isomer of tartaric acid discovered in crude tartar, called *racemic acid* (*racemus*, Latin, "a bunch of grapes"), was also studied by Biot and found to be optically inactive. The exact structural relationship of (+)-tartaric acid and its isomer racemic acid remained obscure.

All of these observations were known to Louis Pasteur (1822–1895), a French chemist and biologist. One day in 1848 the young Pasteur was viewing crystals of the sodium ammonium double salts of (+)-tartaric acid and racemic acid under the microscope. Pasteur noted that the crystals of the salt derived from (+)-tartaric acid were hemihedral (chiral). He noted, too, that the racemic acid salt was not a single type of crystal, but was actually a mixture of hemihedral crystals: some crystals were "right-handed," like those in the corresponding salt of (+)-tartaric acid, and some were "left-handed" (Fig. 6.19a; thus the name "racemic mixture"). Pasteur meticulously separated the two types of crystals with a pair of tweezers, and found that the right-handed crystals were identical in every way to the crystals of the salt of (+)-tartaric acid. When equally concentrated solutions of the two types of crystals were prepared, he found that the optical rotations of the left- and right-handed crystals were equal in magnitude, but opposite in sign. Pasteur had thus performed the first enantiomeric resolution by human hands! Racemic acid, then, was the first organic compound shown to exist as enantiomers—object and noncongruent mirror image. One of these mirror-image molecules was identical to (+)-tartaric acid, but the other was previously unknown. Pasteur's own words tell us what then took place.

> The announcement of the above facts naturally placed me in communication with Biot, who had doubts concerning their accuracy. Being charged with giving an account of them to the Academy, he made me come to him and repeat before his very eyes the decisive experiment. He handed over to me some racemic acid that he himself had studied with particular care, and that he found to be perfectly indifferent to polarized light. I prepared the double salt in his presence with soda and ammonia that he also desired to provide. The liquid was set aside for slow evaporation in one of his rooms. When it had furnished about thirty to forty grams of crystals, he asked me to call at the Collège de France in order to collect them and isolate, before his very eyes, by recognition of their crystallographic character, the right and left crystals, requesting me to state once more whether I really affirmed that the crystals that I should place at his right would really deviate [the plane of polarized light] to the right and the others to the left. This done, he told me that he would undertake the rest. He prepared the solutions with carefully measured quantities, and when ready to examine them in the polarizing apparatus, he once more invited me to come into his room. He first placed in the apparatus the more interesting solution, that which should deviate to the left [previously unknown]. Without even making a measurement, he saw by the tints of the images . . . in the analyzer that there was a strong deviation to the left. Then, very visibly affected, the illustrious old man took me by the arm and said, "My dear child, I have loved science so much all my life that this makes my heart throb!"

Pasteur's discovery of the two types of crystals of racemic acid was serendipitous (accidental). It is now known that the sodium ammonium salt of racemic acid forms separate right- and left-handed crystals only at temperatures below 26 °C. Had Pasteur's laboratories been

FIGURE 6.19 Diagrams of the crystals of the tartaric acid isomers that figured prominently in the history of stereochemistry. (a) The chiral crystals of sodium ammonium tartrate separated by Pasteur. (b) The achiral crystal of sodium ammonium racemate that crystallizes at a higher temperature.

dextrorotatory   levorotatory   optically inactive

(a)   (b)

warmer, he would not have made the discovery. Above this temperature, this salt forms only one type of crystal: a holohedral (achiral) crystal of the racemate! (Fig. 6.19b) From his discovery, and from the work of Biot, which showed that optical activity is a *molecular* property, Pasteur recognized that some molecules could, like the quartz crystals, have an enantiomeric relationship, but he was never able to deduce a structural basis for this relationship.

PROBLEMS

6.24   As described in the previous account, Pasteur discovered two stereoisomers of tartaric acid. Draw their structures [you cannot tell which is (+) and which is (−)]. Which stereoisomer of tartaric acid was yet to be discovered? (It was discovered in 1906.) What can you say about its optical activity?

6.25   Think of Pasteur's enantiomeric resolution of racemic acid in terms of the "resolving agent" idea discussed in Sec. 6.8. Did Pasteur's resolution involve a resolving agent? If so, what was it?

In 1874, Jacobus Hendricus van't Hoff (1852–1911), a professor at the Veterinary College at Utrecht, The Netherlands, and Achille Le Bel (1847–1930), a French chemist, independently arrived at the idea that if a molecule contains a carbon atom bearing four different groups, these groups can be arranged in different ways to give enantiomers. Van't Hoff suggested a tetrahedral arrangement of groups about the central carbon, but Le Bel was less specific. Van't Hoff's conclusions, published in a treatise of eleven pages entitled *La chemie dans l'espace*, were not immediately accepted. A caustic reply came from the famous German chemist Hermann Kolbe:

> A Dr. van't Hoff of the Veterinary College, Utrecht, appears to have no taste for exact chemical research. Instead, he finds it a less arduous task to mount his Pegasus (evidently borrowed from the stables of the College) and soar to his chemical Parnassus, there to reveal in his *La chemie dans l'espace* how he finds atoms situated in universal space. This paper is fanciful nonsense! What times are these, that an unknown chemist should be given such attention!

Kolbe's reply notwithstanding, van't Hoff's ideas prevailed to become a cornerstone of organic chemistry.

How can the existence of enantiomers be used to deduce a tetrahedral arrangement of groups around carbon? Let's examine some other possible carbon geometries to see the sort of reasoning that was used by van't Hoff and Le Bel. Consider a general molecule in which the carbon and its four groups lie in a single plane:

$$\text{Br}-\overset{\overset{\displaystyle \text{Cl}}{|}}{\underset{\underset{\displaystyle \text{H}}{|}}{\text{C}}}-\text{F}$$

all atoms are in the same plane

Because the mirror image of such a *planar* molecule is congruent (show this!), enantiomeric forms are impossible. The existence of enantiomers thus *rules out* this planar geometry.

However, other conceivable nonplanar nontetrahedral structures could exist as enantiomers. One structure has a pyramidal geometry:

(Convince yourself that such a structure can have an enantiomer.) This geometry could not, however, account for other facts. Consider, for example, the compound dichloromethane

(CH_2Cl_2). In the pyramidal geometry, two *diastereomers* would be known. In one, the chlorines are on opposite corners of the pyramid; in the other, the chlorines are adjacent. (Why are these diastereomers?)

pyramidal dichloromethane molecules

These molecules should be separable because diastereomers have different properties. In the entire history of chemistry, only one isomer of CH_2Cl_2, CH_2Br_2, or any similar molecule, has ever been found. Now, this is *negative* evidence. To take this evidence as conclusive would be like saying to the Wright brothers in 1902, "No one has ever seen an airplane fly; therefore, airplanes can't fly." Yet this evidence is certainly suggestive, and other experiments (Problems 6.53 and 6.54, p. 271) were subsequently carried out that could only be interpreted in terms of tetrahedral carbon. Indeed, modern methods of structure determination have shown repeatedly that van't Hoff's original proposal—tetrahedral geometry—is correct.

KEY IDEAS IN CHAPTER 6

- Stereoisomers are molecules with the same atomic connectivity but a different arrangement of their atoms in space.

- Two types of stereoisomers are:
    1. enantiomers—molecules that are related as object and noncongruent mirror image;
    2. diastereomers—stereoisomers that are not enantiomers.

- A molecule that has an enantiomer is said to be chiral. Chiral molecules lack certain symmetry elements, such as a plane of symmetry or center of symmetry.

- The absolute configurations of some compounds can be determined experimentally by correlating them chemically with other chiral compounds of previously known absolute configuration.

- The *R,S* system is used for designating absolute configuration. The system is based on the clockwise or counterclockwise arrangement of group priorities when a molecule is viewed along a bond from the asymmetric atom to the group of lowest priority. The priorities are assigned as in the *E,Z* system (Sec. 4.2B).

- Two enantiomers have the same physical properties except for their optical activities. The optical rotations of a pair of enantiomers have equal magnitudes but opposite signs.

- A mixture of enantiomers is characterized by the enantiomeric excess (EE), which is the difference in the percentages of the two enantiomers.

- An equimolar mixture of enantiomers is called a racemate, or racemic mixture.

- Diastereomers in general differ in their physical properties.

- Enantiomers can be separated or differentiated only by allowing them to interact with an enantiomerically pure chiral agent (the principle of enantiomeric differentiation).

- The separation of enantiomers is called an enantiomeric resolution. All enantiomeric resolutions involve the temporary formation of diastereomers, which have different properties. Three methods of enantiomeric resolution are chiral chromatography, diastereomeric salt formation, and selective crystallization.

- An asymmetric carbon is a carbon bonded to four different groups. All asymmetric carbons are stereocenters, but not all stereocenters are asymmetric carbons.

- A meso compound is an achiral compound that has chiral stereoisomers. Most cases of meso compounds are achiral compounds that have two or more asymmetric atoms.

- Some chiral molecules contain no asymmetric atoms.

- Some achiral molecules have enantiomeric conformations that interconvert very rapidly.

- Chiral compounds in which a central atom is bonded to three groups and an unshared electron pair (such as amines) can undergo inversion, a process that interconverts enantiomers. When the central atom is from the second period, inversion is so fast that the enantiomers cannot be isolated. When the central atom is from the

third and greater periods, stereochemical inversion is so slow that it does not occur at room temperature. The enantiomers of such compounds (such as sulfonium ions and phosphines) can be isolated.

- Optical activity and chirality are the experimental foundations for the postulate of tetrahedral bonding geometry at tetracoordinate carbon.

ADDITIONAL PROBLEMS

6.26 Point out the carbon stereocenters and the asymmetric carbons (if any) in each of the following structures.

(a) 4-methyl-1-pentene

(b) (*E*)-4-methyl-2-hexene

(c) 3-methylcyclohexene

(d) 2,4-dimethyl-2-pentene

6.27 How many stereoisomers are there of 3,4-dimethyl-2-hexene?

(a) Show all of the carbon stereocenters in the structure of this compound.

(b) Show all of the asymmetric carbons in the structure of this compound.

6.28 Identify all of the asymmetric carbon atoms (if any) in each of the following structures.

(a)
CH_3CH_2CH — cyclohexane with CH_3, CH_3

(b)
CH_3CH_2CH — cyclohexane — CH_3, with CH_3

(c) H_3C—CH—CH_2OH, CH_2OCH_3

(d) H_3C—CH—CH_2OH, NH_2

(e)
CH_3, OH

(f)
CH_3

(g)
HO

6.29 For each of the following compounds, draw a line-and-wedge structure in which all carbon–carbon bonds are in the plane of the page.

(a) (3*R*,4*R*)-3,4-hexanediol   (b) *meso*-3,4-hexanediol

HO OH

$CH_3CH_2CHCHCH_2CH_3$

3,4-hexanediol

(c) (2*S*,3*R*,4*S*)-2,3,4-hexanetriol

OH   OH

$CH_3CHCHCHCH_2CH_3$

OH

2,3,4-hexanetriol

6.30 Tell whether the configuration of each asymmetric atom in the following compounds is *R* or *S*.

(a)
O, CH_3, H_3C, OCH_3, H_3C

(b) CH_3O H, CH_3, H_3C, H OH

(c)
O^-
CH_3CH_2O ···· P^+ , OCH_3
$N(CH_3)_2$

(d) *meso*-3,4-dimethylhexane

6.31 *Ephedrine* has been known in medicine since the Chinese isolated it from natural sources in about 2800 BC. Its structure has been known since 1885. Ephedrine can be used as a bronchodilator (a compound that enlarges the air passages in the lungs), but it tends to increase blood pressure because it constricts blood vessels. *Pseudoephedrine* has the same effects, except that it is much less active in elevating blood pressure. Ephedrine is the (1*R*,2*S*)-stereoisomer of the structure below (Ph = phenyl). Pseudoephedrine is the (1*S*,2*S*)-stereoisomer of the same structure.

$$\text{Ph}-\overset{\overset{\displaystyle OH}{|}}{\underset{1}{CH}}-\overset{\overset{\displaystyle NHCH_3}{|}}{\underset{2}{CH}}-CH_3$$

(a) Draw a line-and-wedge structure for each of these two stereoisomers in which the Ph, CH_3, and the two asymmetric carbons lie in the plane of the page. In this part and part (b), do not draw out the bonds within the $-CH_3$, $-OH$, and $-NHCH_3$ groups explicitly. (More than one correct structure is possible.)

(b) Draw a sawhorse projection and a Newman projection about the C1—C2 bond for each of the structures you drew in part (a). Let the carbon nearest the observer be the one bearing the —OH group.

(c) What is the relationship between these two compounds? Choose from *enantiomers, diastereomers, identical molecules,* and *constitutional isomers.* Explain how you know.

(d) Should the melting points of these two compounds be the same or different in principle?

(e) What, if anything, can you say about the optical activities of these two compounds?

6.32 Draw the structure of the chiral alkane of lowest molecular mass not containing a ring. (No isotopes are allowed.)

6.33 Draw the structure of the chiral cyclic alkane of lowest molecular mass. (No isotopes are allowed.)

6.34 Indicate whether each of the following statements is true or false. If false, explain why.

(a) In some cases, constitutional isomers are chiral.

(b) In every case, a pair of enantiomers have a mirror-image relationship.

(c) Mirror-image molecules are in all cases enantiomers.

(d) If a compound has an enantiomer, it must be chiral.

(e) Every chiral compound has a diastereomer.

(f) If a compound has a diastereomer, it must be chiral.

(g) Every molecule containing one or more asymmetric carbons is chiral.

(h) Any molecule containing a stereocenter must be chiral.

(i) Any molecule with a stereocenter must have a stereoisomer.

(j) Some diastereomers have a mirror-image relationship.

(k) Some chiral compounds are optically inactive.

(l) Any chiral compound with a single asymmetric carbon must have a positive optical rotation if the compound has the *R* configuration.

(m) A structure is chiral if it has no plane of symmetry.

(n) All chiral molecules have no plane of symmetry.

(o) All asymmetric carbons are stereocenters.

6.35 Imagine substituting, in turn, each hydrogen atom of 3-methylpentane with a chlorine atom to give a series of isomers with molecular formula $C_6H_{13}Cl$. Give the structure of each of these isomers. Which of these are chiral? Classify the relationship of each stereoisomer with every other.

6.36 Draw the structures of all compounds with the formula $C_6H_{12}Cl_2$ that can exist as meso compounds. Indicate how many meso compounds are possible for each structure.

6.37 Construct sawhorse and Newman projections (Sec. 2.3A) of the three staggered conformations of 2-methylbutane (isopentane) that result from rotation about the C2—C3 bond.

(a) Identify the conformations that are chiral.

(b) Explain why 2-methylbutane is not a chiral compound, even though it has chiral conformations.

(c) Suppose each of the three conformations in part (a) could be isolated and their heats of formation determined. Rank these isomers in order of increasing heat of formation (that is, smallest first). Explain your choice. Indicate whether the ΔH_f° values for any of the isomers are equal and why.

6.38 (a) Draw sawhorse projections of ephedrine (Problem 6.31) about the C1—C2 bond for all three staggered and all three eclipsed conformations.

(b) Examine each conformation for chirality. How do the chiralities of these conformations relate to the overall chirality of ephedrine?

6.39 Explain why compound *A* in Fig. P6.39 can be resolved into enantiomers but compound *B* cannot.

6.40 (a) Which of the compounds shown in Fig. P6.40 can in principle be resolved into enantiomers? Explain why or why not.

$$\text{Ph}-\overset{\overset{\displaystyle CH_3}{|}}{\underset{\underset{\displaystyle CH_2Ph}{|}}{N^+}}-CH_2-CH=CH_2 \quad Cl^-$$

$$\text{Ph}-\overset{\overset{\displaystyle CH_3}{|}}{\underset{..}{N}}-CH_2-CH=CH_2$$

A   *B*

Figure P6.39

(b) Omeprazole can be separated into two enantiomers that do not interconvert at room temperature.

omeprazole

What atom is the asymmetric center in omeprazole?

(c) Esomeprazole, the *S* enantiomer of omeprazole, is a drug used to control acid reflux. Redraw the structure of omeprazole in part (b) to show it as the *S* enantiomer. (*Hint:* An unshared electron pair has lower priority than H.)

6.41 The specific rotation of the *R* enantiomer of the following alkene is $[\alpha]_D^{25} = +79$ degree mL g^{-1} dm^{-1}, and its molecular mass is 146.2.

(a) What is the observed rotation of a 0.5 *M* solution of this compound in a 5-cm sample path?

(b) What is the observed rotation of a solution formed by mixing equal volumes of the solution from part (a) and a 0.25 *M* solution of the enantiomer of the same alkene?

(c) What is the enantiomeric excess of the major enantiomer in the solution formed in part (b)?

6.42 Enantiomerically pure (*R*)-(+)-2-methyl-1,2-butanediol has a specific rotation $[\alpha]_D^{20} = +9.3$ degrees mL g^{-1} dm^{-1} in chloroform solution.

(*R*)-(+)-2-methyl-1,2-butanediol

(a) What is the EE of a mixture of (+)- and (−)-2-methyl-1,2-butanediol that has an apparent specific rotation of −6.3 degrees mL g^{-1} dm^{-1} under the same conditions?

(b) What percentage of each enantiomer is present in the mixture?

6.43 (*S*)-(+)-Aspartic acid is one of the naturally occurring α-amino acids.

(*S*)-(+)-aspartic acid
(as it exists in HCl solution)

Over time in the environment or in aqueous solution, aspartic acid can undergo racemization very slowly (by a process that we won't consider here). The racemization of aspartic acid occurs at a known rate that can be used to date tissue samples in forensic investigations. The rate of racemization in skull samples follows the equation

$$\ln\left(\frac{1 + r}{1 - r}\right) = kt$$

where $k = 6.24 \times 10^{-4}$ yr^{-1} and t = age of the sample in years. In this equation, r is the ratio of (*R*)-(−)- and (*S*)-(+)-aspartic acid.

The local coroner's office, knowing your expertise in organic chemistry, comes to you with an old skull sample that was excavated from a building site. You isolate partially racemized aspartic acid from the sample and find (by chiral chromatography) that the sample contains 6% of the (*R*)-(−)-enantiomer and 94% of the (*S*)-(+)-enantiomer.

(a) How old is the skull sample?

(b) What is the enantiomeric excess of the (*S*)-(+)-enantiomer in the sample?

(c) If (*S*)-(+)-aspartic acid has a standard rotation $[\alpha]_D^{20} = +24.5$ degrees mL g^{-1} dm^{-1} in 6 *M* aqueous HCl solution, what is the observed rotation of 100 mg of the forensic (partially racemized) sample in 10 mL of 6 *M* aqueous HCl?

A

B

C

D

E

F

Figure P6.40

6.44 The two most common forms of glucose, α-D-glucopyranose and β-D-glucopyranose, can be brought into equilibrium by dissolving them in water with a trace of an acid or base catalyst.

$$\alpha\text{-D-glucopyranose} \underset{}{\overset{H_3O^+ \text{ or } ^-OH}{\rightleftharpoons}} \beta\text{-D-glucopyranose}$$

The specific rotation of the equilibrium mixture is +52.7 deg mL g^{-1} dm^{-1}. The specific rotation of pure α-D-glucopyranose is +112 degrees mL g^{-1} dm^{-1}, and that of pure β-D-glucopyranose is +18.7 degrees mL g^{-1} dm^{-1}. What is the percentage of each form in the equilibrium mixture?

6.45 (a) 2,3,4-Trichloropentane has *two* meso stereoisomers.

2,3,4-trichloropentane

Starting with the template below for each, complete line-and-wedge structures for the two meso stereoisomers.

(b) Show the symmetry element in each meso stereoisomer that makes the compound achiral.

(c) What is the relationship (enantiomers or diastereomers) between the two meso compounds?

(d) Is carbon-3 a stereocenter in these meso compounds? How do you know?

(e) What addition to the R,S system would you have to make to assign a configuration to carbon-3? Invent a rule and then assign an R or S designation to each carbon in your two meso stereoisomers.

(f) How many stereoisomers does 2,3,4-trichloropentane have?

6.46 (a) Give the stereochemical relationship (enantiomers, diastereomers, or the same molecule) between each pair of compounds in the set shown in Figure P6.46. Assume that internal rotation is rapid.

(b) Which, if any, compounds are meso? Explain.

(c) Which compounds should be optically active? Explain.

6.47 (a) Explain why an optically inactive product is obtained when (−)-3-methyl-1-pentene undergoes catalytic hydrogenation.

(b) What is the absolute configuration of (+)-3-methylhexane if catalytic hydrogenation of (S)-(+)-3-methyl-1-hexene gives (−)-3-methylhexane?

6.48 From the outcome of the following transformation, indicate whether the levorotatory enantiomer of the product has the R or S configuration. Draw a structure of the product that shows its absolute configuration. (*Hint:* The phenyl group has a higher priority than the vinyl group in the R,S system.) (See Fig. P6.48.)

6.49 Which of the salts shown in Fig. P6.49 should have identical solubilities in methanol? Explain.

6.50 Draw the structures of the possible stereoisomers for the compound below, assuming in turn (a) tetrahedral, (b) square planar, and (c) pyramidal geometries at the

Figure P6.46

Figure P6.48

carbon atom. For each of these geometries, what is the relationship of each stereoisomer with every other?

$$Cl-\underset{\underset{I}{|}}{\overset{\overset{F}{|}}{C}}-Br$$

6.51 Two stereoisomers of the compound $(H_3N)_2Pt(Cl)_2$ with different physical properties are known. Show that this fact makes it possible to choose between the tetrahedral and square planar arrangements of these four groups around platinum.

6.52 In a structure containing a pentacoordinate phosphorus atom, the bonds to three of the groups bound to phosphorus (called *equatorial* groups) lie in a plane containing the phosphorus atom (shaded in the following structure), and the bonds to the other two groups (called *axial* groups) are perpendicular to this plane:

axial groups — equatorial groups

Cl, Br, Cl, OCH₃, OCH₃ (P)

Is this compound chiral? Explain.

6.53 In 1914, the chemist Emil Fischer carried out the following conversion in which optically active starting material was transformed into a product with an identical melting point and an optical rotation of equal magnitude and opposite sign. No bonds to the asymmetric carbon were broken in the process.

$$H-\underset{\underset{CO_2H}{|}}{\overset{\overset{CH_2CH_2CH_3}{|}}{C}}-CONH_2 \xrightarrow{\text{several reactions}} H-\underset{\underset{CONH_2}{|}}{\overset{\overset{CH_2CH_2CH_3}{|}}{C}}-CO_2H$$

Show that this result is consistent with *either* tetrahedral or pyramidal geometry at the asymmetric carbon.

6.54 Fischer also carried out the following pair of conversions. Again, no bonds to the asymmetric carbon were broken. Explain why this *pair* of conversions (but not either one alone) and the associated optical activities rule out pyramidal geometry at the asymmetric carbon, but are consistent with tetrahedral geometry.

$$H_3C-\underset{\underset{C_2H_5}{|}}{\overset{\overset{CO_2H}{|}}{C}}-CONH_2 \xrightarrow{\text{several steps}} H_3C-\underset{\underset{C_2H_5}{|}}{\overset{\overset{CO_2H}{|}}{C}}-CO_2H$$

optically active   optically inactive

$$H_3C-\underset{\underset{C_2H_5}{|}}{\overset{\overset{CO_2H}{|}}{C}}-CONH_2 \xrightarrow{\text{several steps}} H_3C-\underset{\underset{C_2H_5}{|}}{\overset{\overset{CH_3}{|}}{C}}-CONH_2$$

optically active   optically inactive

A

B

C

D

Figure P6.49

CHAPTER 7

Cyclic Compounds
Stereochemistry of Reactions

Compounds with cyclic structures present some unique aspects of stereochemistry and conformation. This chapter deals with the stereochemical and conformational issues in cyclic compounds and their derivatives followed by a discussion of how stereochemistry enters into chemical reactions. We've already learned about *regioselective reactions*, which yield one *constitutional isomer* in preference to another (for example, HBr addition to alkenes). Many reactions also yield certain *stereoisomers* to the exclusion of others. Several such reactions will be examined so that we can understand some of the principles that govern the formation of stereoisomers. We'll also see how the stereochemistry of a reaction can be used to understand its mechanism.

7.1	RELATIVE STABILITIES OF THE MONOCYCLIC ALKANES

A compound that contains a single ring is called a **monocyclic compound**. Cyclohexane, cyclopentane, and methylcyclohexane are all examples of monocyclic alkanes.

The relative stabilities of the monocyclic alkanes give us some important clues about their conformations. These relative stabilities can be determined from their heats of formation, given in Table 7.1 and shown graphically in the figure within the table. Although the monocyclic alkanes are not isomers, they have the same **empirical formula**, CH_2. This is the formula that gives the smallest whole-number proportions of the elements. When compounds have the same empirical formula, their heats of formation, and thus their stabilities, can be compared on a *per carbon basis* by dividing the heat of formation of each compound by its number of carbons. The data in Table 7.1 show that, of the cycloalkanes with 14 or fewer carbons, cyclohexane has the lowest (that is, the most negative) heat of formation per CH_2. Thus, *cyclohexane is the most stable of these cycloalkanes.*

**TABLE 7.1 Heats of Formation per —CH₂—
for Some Cycloalkanes**

(*n* = number of carbon atoms)

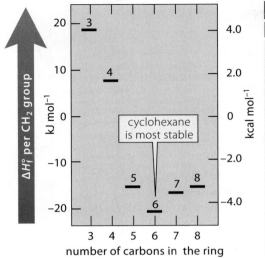

n	Compound	$\Delta H_f°/n$ kJ mol⁻¹	$\Delta H_f°/n$ kcal mol⁻¹
3	cyclopropane	+17.8	+4.25
4	cyclobutane	+7.1	+1.7
5	cyclopentane	−15.4	−3.7
6	cyclohexane	−20.7	−4.95
7	cycloheptane	−16.9	−4.0
8	cyclooctane	−15.55	−3.7
9	cyclononane	−14.8	−3.5
10	cyclodecane	−15.4	−3.7
11	cycloundecane	−16.3	−3.9
12	cyclododecane	−19.2	−4.6
13	cyclotridecane	−18.95	−4.5
14	cyclotetradecane	−17.1	−4.1

Further insight into the stability of cyclohexane comes from a comparison of its stability with that of a typical *noncyclic* alkane. The heats of formation of pentane, hexane, and heptane are −146.5, −167.1, and −187.5 kJ mol⁻¹ (−35.0, −39.9, and −44.8 kcal mol⁻¹), respectively. These data show that heats of formation, like other physical properties, change regularly within a homologous series; each CH₂ group contributes −20.7 kJ mol⁻¹ (−4.95 kcal mol⁻¹) to the heat of formation. The data for cyclohexane in Table 7.1 show that a CH₂ group in cyclohexane makes exactly the same contribution to its heat of formation (−20.7 kJ mol⁻¹ or −4.95 kcal mol⁻¹). This means that *cyclohexane has the same stability as a typical unbranched alkane.*

Cyclohexane is the most widely occurring ring in compounds of natural origin. Its prevalence, undoubtedly a consequence of its stability, makes it the most important of the cycloalkanes. Two questions emerge as we consider these data: (1) Why is cyclohexane so stable? (2) Why are the smaller rings so unstable? We'll consider the first question in Sec. 7.2 and the second question in Sec. 7.5.

7.2   CONFORMATIONS OF CYCLOHEXANE

A. The Chair Conformation

Why is cyclohexane so stable? The stability data in Table 7.1 require that the bond angles in cyclohexane must be essentially the same as the bond angles in an alkane—very close to the ideal 109.5° tetrahedral angle. If the bond angles were significantly distorted from tetrahedral, we would expect to see a greater heat of formation. The carbons of cyclohexane, then, are *sp³*-hybridized. Furthermore, cyclohexane must have a staggered conformation about each carbon–carbon bond because, otherwise, eclipsing interactions (*torsional strain*; Sec. 2.3A) would also increase the heat of formation. These two geometrical constraints can only be met if the carbon skeleton of cyclohexane assumes a nonplanar, "puckered" conformation. This conformation, shown in Fig. 7.1 (p. 274), is called the **chair conformation** because of its resemblance to a lawn chair. If you have not already done so, you should construct a model of chair cyclohexane *now* and use it to follow the subsequent discussion. If your model kit allows, leave off the hydrogens; we'll deal first only with the carbon skeleton. If the carbons of your model set include the hydrogens "built in," then simply notice the positions of the carbons in the subsequent discussion; we'll deal with the hydrogens later.

(a) ball-and-stick model

(b) space-filling model

(c) skeletal structure with hydrogens shown

(d) lawn-chair cyclohexane

FIGURE 7.1 The chair conformation of cyclohexane. (a) A ball-and-stick model. (b) A space-filling model. (c) A skeletal structure. (d) Origin of the name "chair." The different types of hydrogens are color-coded. The axial hydrogens are blue in parts (a)–(c) of the figure. The equatorial hydrogens are white in parts (a) and (b) and black in part (c).

Notice the following four points about the cyclohexane molecule and how to draw it.

1. To draw the cyclohexane ring, we use a "tilt-and-turn" technique similar to the one used for drawing line-and-wedge structures (Eq. 2.1, p. 56).

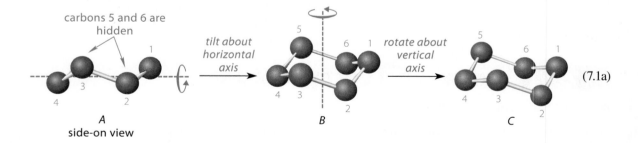

(7.1a)

First, we view the model side-on (view *A* in Eq. 7.1a). In this view, carbons 5 and 6 are obscured behind carbons 2 and 3. Then, we tilt the model about a horizontal axis to give view *B*. Finally, we turn the model slightly about a vertical axis to give the view used to draw the skeletal structure, as shown in Eq. 7.1b.

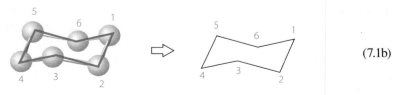

(7.1b)

view *C* from Eq. 7.1a

If we imagine carbons 1 and 4 to be in the plane of the page, then carbons 2 and 3 are in front of the page, and carbons 5 and 6 are behind the page.

(7.1c)

Remembering that the lower part of the ring is in front of the page is essential to avoiding an optical illusion.

2. Bonds on opposite sides of the ring are parallel:

3. Two perspectives are commonly used for cyclohexane rings. In one, the leftmost carbon is below the rightmost carbon; and in the other, the leftmost carbon is above the rightmost carbon:

These two perspectives are mirror images. As shown in Eqs. 7.1a–c, the perspective on the left is based on a view of the model from above and to the left of the model. The perspective on the right is based on a view of the model from above and to the right. The "tilt-and-turn" procedure for producing this structure is the same except that the rotation is in the opposite direction.

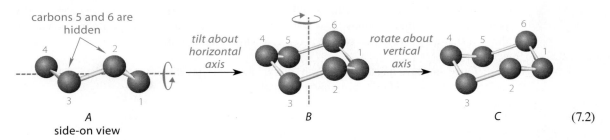

(7.2)

4. A rotation of either perspective by an odd multiple of 60° (that is, 60°, 180°, and so on), followed by the slight shift in viewing direction, gives the other perspective. *Be sure to verify this with models!*

(7.3)

It is important for you to be able to draw a cyclohexane chair conformation. Once you've examined the preceding points, practice drawing some cyclohexane rings in the two perspectives. Use the following three steps:

Step 1. Begin by drawing two parallel bonds slanted to the left for one perspective, and slanted to the right for the other. Notice that one slanted line is somewhat lower than the other in each case.

Step 2.   Connect the tops of the slanted bonds with two more bonds in a "V" arrangement.

Step 3.   Connect the bottoms of the slanted bonds with the remaining two bonds in an inverted "V" arrangement.

To summarize:

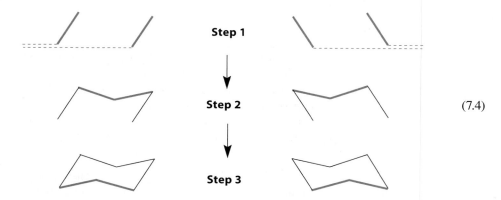

(7.4)

Now let's consider the hydrogens of cyclohexane, which are of two types. If you place your model of cyclohexane on a tabletop (you did build it, didn't you?), you'll find that six C—H bonds are perpendicular to the plane of the table. (Your model should be resting on three such hydrogens.) These hydrogens, shown in blue in Fig. 7.1a–b, are called **axial** hydrogens. The remaining C—H bonds point outward along the periphery of the ring. These hydrogens, shown in white in Fig. 7.1a–b and in black in Fig. 7.1c, are called **equatorial** hydrogens. As we might expect, other groups can be substituted for the hydrogens, and these groups also can exist in either axial or equatorial arrangements.

In the chair conformation, all bonds are staggered. You should be able to see this from your model by looking down any C—C bond, as shown in Fig. 7.2. As you learned when you studied the conformations of ethane and butane (Sec. 2.3), staggered bonds are energetically preferred over eclipsed bonds. The stability of cyclohexane (Sec. 7.1) is a consequence of the fact that all of its bonds can be staggered without compromising the tetrahedral carbon geometry.

Once you have mastered drawing the cyclohexane ring, it's time to add the C—H bonds to the ring. The axial bonds are drawn vertically.

(7.5a)

(a)   (b)   (c)

FIGURE 7.2 A demonstration that the bonds in chair cyclohexane are staggered. View a chair cyclohexane model as shown in (a); tilt the model slightly about the horizontal axis shown so that the model is viewed down the carbon–carbon bonds on opposite sides of the ring, as shown in (b). These bonds become the projected bonds in the Newman projection (c). The axial hydrogens are blue; the equatorial hydrogens are white in (a) and (b), and black in (c).

(a)

(b)

FIGURE 7.3 (a) Up and down equatorial and axial hydrogens. The up-axial hydrogens are on up carbons and the down-axial hydrogens are on down carbons. The opposite is true for equatorial hydrogens. (b) The up- and down-axial hydrogens are equivalent, and the up- and down-equatorial hydrogens are equivalent. This equivalence can be demonstrated by turning the ring "upside down" (*green arrow*). In doing so, the up carbons trade places with the down carbons, the up-axial hydrogens trade places with the down-axial hydrogens, and the up-equatorial hydrogens trade places with the down-equatorial hydrogens. (This part shows explicitly the fate of two hydrogens and the violet color shows the fate of one carbon.)

Drawing the equatorial bonds can be a little tricky. Notice that pairs of equatorial bonds are parallel to pairs of nonadjacent ring bonds (red):

 (7.5b)

(Notice also how all the equatorial bonds in Fig. 7.1 adhere to this convention.)

You should notice a few other things about the cyclohexane ring and its bonds. First, if we make a model of the cyclohexane carbon skeleton without hydrogens and place it on a tabletop, we find that every other carbon is resting on the tabletop. We'll refer to these carbons as *down carbons*. Notice that each of these carbons is at the vertex of a "V" formed by its two carbon–carbon bonds. The other three carbons lie in a plane above the tabletop. We'll refer to these carbons as *up carbons*. Each of these carbons is at the vertex of an inverted "V."

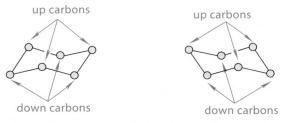

Now add the hydrogens to your model, and notice that the three axial hydrogens on up carbons point up, and the three axial hydrogens on down carbons point down. In contrast, the three equatorial hydrogens on up carbons point down, and the three equatorial hydrogens on down carbons point up (Fig. 7.3a). *The up and down hydrogens of a given type are com-*

pletely equivalent. That is, the up equatorial hydrogens are equivalent to the down equatorial hydrogens, and the up axial hydrogens are equivalent to the down axial hydrogens. You can see this equivalence by turning the ring over, as shown in Fig. 7.3b. This causes the up carbons to exchange places with the down carbons, the up-axial hydrogens to exchange places with the down-axial hydrogens, and the up-equatorial hydrogens to exchange places with the down-equatorial hydrogens. Everything looks the same as it did before turning the molecule over. (Be sure to convince yourself with models that these statements are true.)

A second useful observation is that if an axial hydrogen is up on one carbon, the two neighboring axial hydrogens are down, and vice versa. The same is true of the equatorial hydrogens.

B. Interconversion of Chair Conformations

Cycloalkanes, like noncyclic alkanes, undergo internal rotations (Sec. 2.3). However, because the carbon atoms are constrained within a ring, several internal rotations must occur at the same time. When a cyclohexane molecule undergoes internal rotations, *a change in the ring conformation occurs.* In this change, one chair conformation is converted into another, completely equivalent, chair conformation. Figure 7.4 shows this conformational interconversion and gives a four-step procedure for demonstrating it with a model. First, hold carbon-1—the rightmost carbon—so that it cannot move, and raise carbon-4 up as far as it will go. The result is a different conformation, called a **boat conformation**. (As we'll learn in Sec. 7.2C, the boat conformation is not a true intermediate in the interconversion, but it is handy as an intermediate step in the use of models.) Formation of the boat conformation involves *simultaneous internal rotations* about all carbon–carbon bonds except those to carbon-1. Now hold carbon-4 of the boat—the leftmost carbon—so it cannot move, and lower carbon-1 as far as it will go; the model returns to a chair conformation. In this case, simultaneous internal rotations have occurred about all carbon–carbon bonds except those to carbon-4. Thus, upward movement of the leftmost carbon and downward movement of the rightmost carbon changes one chair conformation into another, completely equivalent, chair conformation. But notice what has happened to the hydrogens: In this process, *the equatorial hydrogens have become axial, and the axial hydrogens have become equatorial.* In addition, up carbons have become down carbons, and vice versa.

(7.6)

(Confirm these points with your model by using groups of different colors for the axial and equatorial hydrogens.)

The interconversion of two chair forms of cyclohexane is called the **chair interconversion**. It is sometimes also called the *chair flip,* although this terminology is somewhat misleading because it implies that the ring is simply turned over. As Eq. 7.6 shows by labeling the hydrogens, this is a series of coordinated *internal* rotations rather than an overall rotation of the entire molecule. The energy barrier for the chair interconversion is about 45 kJ mol^{-1} (11 kcal mol^{-1}). This barrier is low enough that the chair interconversion is very rapid; it occurs about 10^5 times per second at room temperature.

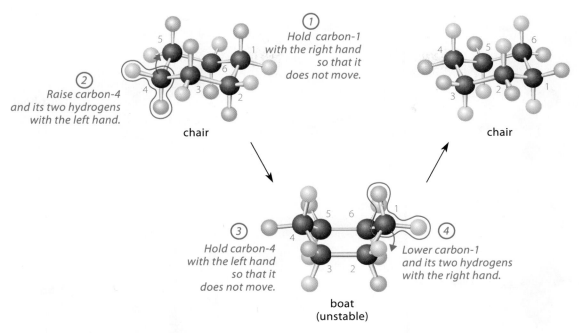

FIGURE 7.4 A four-step procedure for showing the interconversion of the two chair conformations of cyclohexane with a model. Follow the instructions in green, in order. Notice that the chair
interconversion interchanges the positions of the hydrogens: axial hydrogens in one chair conformation become equatorial hydrogens in the other.

Let's review: Although the axial hydrogens are stereochemically different from the
equatorial hydrogens in any one chair conformation, the chair interconversion causes these
hydrogens to change positions rapidly. Hence, *averaged over time*, the axial and equatorial
hydrogens of cyclohexane are equivalent and indistinguishable.

C. Boat and Twist-Boat Conformations

Figure 7.4 shows a boat conformation of cyclohexane. Let's examine this conformation in
more detail. The boat conformation is not a stable conformation of cyclohexane; it contains
two sources of instability, both of which are shown in Fig. 7.5 on p. 280. One is that certain
hydrogens (shaded in blue) are eclipsed. The second is that the two hydrogens on the "bow"
and "stern" of the boat, called *flagpole hydrogens*, experience modest van der Waals repulsion. (The flagpole hydrogens are shaded in pink in Fig. 7.5.) For these reasons, the boat
undergoes very slight internal rotations that reduce both the eclipsing interactions and the
flagpole van der Waals repulsions. The result is another stable conformation of cyclohexane
called a **twist-boat conformation**. To see the conversion of a boat into a twist-boat, view a
model of the boat conformation from above the flagpole hydrogens, as shown in Fig. 7.5b.
Grasping the model by its flagpole hydrogens, nudge one flagpole hydrogen up and the other
down to obtain a twist-boat conformation. As shown in Fig. 7.5, this motion can occur in
either of two ways, so that two twist-boats are related to any one boat conformation.

The relative enthalpies of the conformations of cyclohexane are shown in Fig. 7.6 on
p. 281. You can see from this figure that the twist-boat conformation is an intermediate in
the chair interconversion. Although the twist-boat conformation is at an energy minimum, it
is less stable than the chair conformation by about 23 kJ mol^{-1} (5.5 kcal mol^{-1}) in standard
enthalpy. The standard free-energy difference (15.9 kJ mol^{-1}, 3.8 kcal mol^{-1}) is also considerable. As Study Problem 7.1 illustrates, a sample of cyclohexane has very little twist-boat
conformation present at equilibrium. The boat conformation itself is the transition state for
the interconversion of two twist-boat conformations.

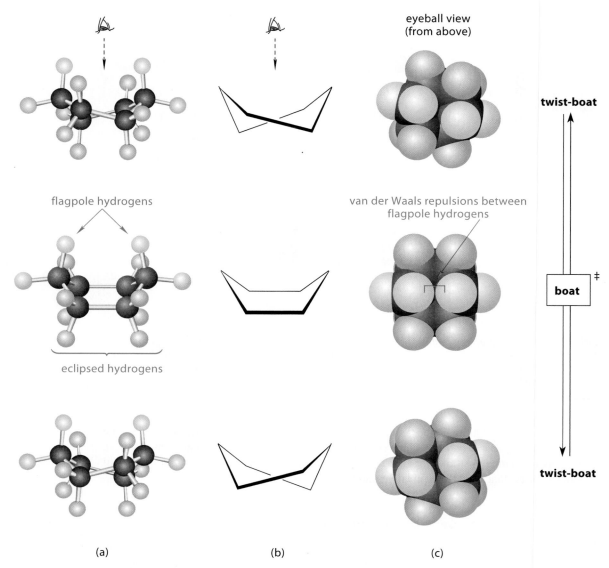

FIGURE 7.5 Boat cyclohexane (*center*) and its two related twist-boat conformations (*top* and *bottom*). The flagpole hydrogens are pink, and the hydrogens that are eclipsed in the boat conformation are blue. (a) Ball-and-stick models. Notice in the boat conformation the eclipsed relationship among the pairs of blue hydrogens. This eclipsing is reduced in the twist-boat conformation. (b) Conventional skeletal structures. (c) Space-filling models viewed from above the flagpole hydrogens (from the direction of the eyeball). Notice the van der Waals repulsion between the flagpole hydrogens in the boat conformation. This unfavorable interaction is reduced in the twist-boat conformations because the flagpole hydrogens (*pink*) are farther apart.

STUDY PROBLEM 7.1

Given that the twist-boat form is 15.9 kJ mol⁻¹ (3.8 kcal mol⁻¹) higher in standard free energy than the chair form of cyclohexane, calculate the percentages of each form present in a sample of cyclohexane.

SOLUTION What we are interested in is the equilibrium ratio of the two forms of cyclohexane—that is, the equilibrium constant for the equilibrium

$$\text{chair (C)} \quad \rightleftharpoons \quad \text{twist boat (T)}$$

This equilibrium constant can be expressed as follows:

$$K_{\text{eq}} = \frac{[\text{T}]}{[\text{C}]}$$

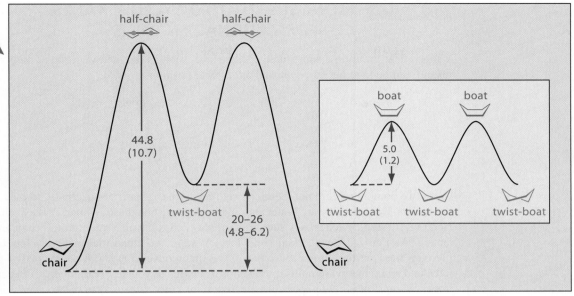

FIGURE 7.6 Relative enthalpies of cyclohexane conformations in kJ mol^{-1}. (The enthalpies in kcal mol^{-1} are in parentheses.) The inset shows the interconversion of twist-boat and boat conformations, which is much faster that the conversion of the twist-boat into either chair conformation.

The equilibrium constant is related to standard free energy by Eq. 3.35 (p. 110):

$$\Delta G° = -2.3RT \log K_{eq}$$

or its rearranged form, Eq. 3.36b,

$$K_{eq} = 10^{-\Delta G°/2.3RT}$$

Applying this equation with energies in kilojoules per mole and $R = 8.31 \times 10^{-3}$ kJ mol^{-1} K^{-1} and $T = 298$ K,

$$K_{eq} = \frac{[T]}{[C]} = 10^{-\Delta G°/2.3RT} = 10^{-15.9/5.71} = 10^{-2.79} = 1.62 \times 10^{-3}$$

Therefore, [T] = (1.62 × 10^{-3})[C]. Thus, in one mole of cyclohexane, we have

$$1 = [C] + [T] = [C] + (1.62 \times 10^{-3})[C] = 1.00162[C]$$

Solving for [C],

$$[C] = 0.998$$

and, by difference,

$$[T] = 1.000 - [C] = 0.002$$

Hence, cyclohexane contains 99.8% chair form and 0.2% twist-boat form at 25 °C.

PROBLEM

7.1 Make a model of chair cyclohexane corresponding to the leftmost model in Fig. 7.4. Raise carbon-4 so that carbons 2–5 lie in a common plane. This is the *half-chair* conformation of cyclohexane, and it is the transition state for the interconversion of the chair and twist-boat conformations. (Notice the position of this conformation on the energy diagram of Fig. 7.6.) Give two reasons why the half-chair conformation is less stable than the chair or twist-boat conformation.

7.3   MONOSUBSTITUTED CYCLOHEXANES. CONFORMATIONAL ANALYSIS

A substituent group in a substituted cyclohexane, such as the methyl group in methyl cyclohexane, can be in either an equatorial or an axial position.

These two compounds are not identical, yet they have the same connectivity, so they are stereoisomers. Because they are not enantiomers, they must be diastereomers. Like cyclohexane itself, substituted cyclohexanes such as methylcyclohexane also undergo the chair interconversion. As Fig. 7.7 shows, axial methylcyclohexane and equatorial methylcyclohexane are interconverted by this process. Notice in this interconversion that a down methyl remains down and an up methyl remains up. (Demonstrate this to yourself with models!) Because this process is rapid at room temperature, methylcyclohexane is a mixture of two *conformational diastereomers* (Sec. 6.9A). Because diastereomers have different energies, one form is more stable than the other.

Equatorial methylcyclohexane is more stable than axial methylcyclohexane. In fact, it is usually the case that *the equatorial conformation of a substituted cyclohexane is more stable than the axial conformation.* Why should this be so?

Examination of a space-filling model of axial methylcyclohexane (Fig. 7.8) shows that van der Waals repulsions occur between one of the methyl hydrogens and the two axial hydrogens on the same face of the ring. Such unfavorable interactions between axial groups are called **1,3-diaxial interactions**. These van der Waals repulsions destabilize the axial conformation relative to the equatorial conformation, in which such van der Waals repulsions are absent.

As Fig. 7.9 shows, the energy (enthalpy) difference between axial and equatorial conformations of methylcyclohexanes is 7.4 kJ mol^{-1} (1.8 kcal mol^{-1}). Because there are *two* 1,3-diaxial interactions in methylcyclohexane, each interaction is responsible for one-half of the enthalpy difference, or 3.7 kJ mol^{-1} (0.9 kcal mol^{-1}). We'll find that we can use this value in predicting the relative energies of other methyl-substituted cyclohexanes. In other words, *each methyl–hydrogen 1,3-diaxial interaction in a cyclohexane derivative raises the enthalpy by 3.7 kJ mol^{-1} (0.9 kcal mol^{-1}).*

As shown in Fig. 7.10, the 1,3-diaxial interaction of a methyl group and a hydrogen in *axial*-methylcyclohexane looks a lot like the van der Waals interaction between methyl hydrogens in *gauche*-butane. The energy cost of this interaction in *gauche*-butane is 3.7 kJ mol^{-1} (Fig. 2.5, p. 53). Because there are *two* such 1,3-diaxial interactions in *axial*-methylcyclohexane, the *gauche*-butane analogy would predict an energy cost of 2 × 3.7 = 7.4 kJ mol^{-1}.

FIGURE 7.7 The chair interconversion results in an equilibrium between equatorial (*left*) and axial (*right*) conformations of methylcyclohexane. The conversion is shown with two different ring perspectives. Notice in this interconversion that a down methyl remains down and an up methyl remains up.

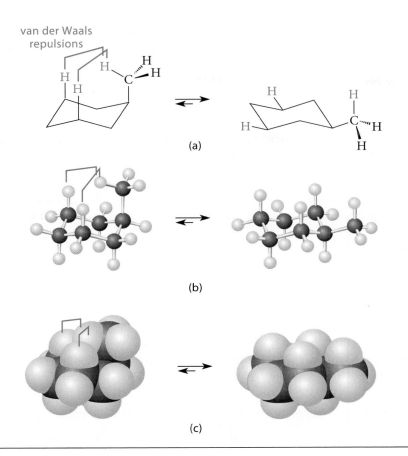

(a)

(b)

(c)

FIGURE 7.8 The equilibrium between axial and equatorial conformations of methylcyclohexane is shown with (a) Lewis structures, (b) ball-and-stick models, and (c) space-filling models. The hydrogens involved in 1,3-diaxial interactions in the axial conformation are shown in color, and the interactions themselves are indicated with red brackets.

$\Delta H°$

7.4 kJ mol⁻¹
(1.8 kcal mol⁻¹)

FIGURE 7.9 Relative enthalpies of axial and equatorial methylcyclohexane.

methylcyclohexane
(axial conformation)

gauche-butane

FIGURE 7.10 The relationship between the axial conformation of methylcyclohexane and *gauche*-butane. One *gauche*-butane part of methylcyclohexane is highlighted, and the corresponding van der Waals repulsion is shown with a red bracket. The second *gauche*-butane interaction in methylcyclohexane is shown with the blue bracket.

The actual value, 7.4 kJ mol^{-1}, is in excellent agreement with this prediction. For this reason, 1,3-diaxial methyl–hydrogen interactions in cyclohexane derivatives are sometimes called ***gauche*-butane interactions**.

   The energy cost of placing a methyl group in the axial position of a cyclohexane ring is reflected in the relative amounts of axial and equatorial methylcyclohexanes present at equilibrium. As you will see when you work Problem 7.2, methylcyclohexane contains very little of the axial conformation at equilibrium.

   The investigation of molecular conformations and their relative energies is called **conformational analysis**. We have just carried out a conformational analysis of methylcyclohexane. The conformational analyses of many different substituted cyclohexanes have been performed. As might be expected, the 1,3-diaxial interactions of large substituent groups are greater than the interactions in methylcyclohexane. For example, the equatorial conformation of *tert*-butylcyclohexane is favored over the axial conformation by about 20 kJ mol^{-1} (about 5 kcal mol^{-1}).

$$\Delta G° = 20 \text{ kJ mol}^{-1} \quad (7.7)$$
$$(5 \text{ kcal mol}^{-1})$$

***tert*-butylcyclohexane**

This means that a sample of *tert*-butylcyclohexane contains a truly minuscule amount of the axial conformation. (See Problem 7.3.)

Separation of Chair Conformations

The two chair conformations of a monosubstituted cyclohexane are diastereomers. If these conformations could be separated, they would have different physical properties. In the late 1960s, C. Hackett Bushweller, then a graduate student in the laboratory of Prof. Frederick Jensen at the University of California, Berkeley, cooled a solution of chlorocyclohexane in an inert solvent to −150 °C. Crystals suddenly appeared in the solution. He filtered the crystals at low temperature; subsequent investigations showed that he

had selectively crystallized the equatorial form of chlorocyclohexane!

selectively crystallizes
at low temperature

When the equatorial form was "heated" to −120 °C, the rate of the chair interconversion increased, and a mixture of conformations again resulted. Similar experiments have been carried out with other monosubstituted cyclohexanes.

PROBLEMS

7.2   The $\Delta G°$ difference between the axial and equatorial conformations of methylcyclohexane (7.4 kJ mol^{-1}, 1.74 kcal mol^{-1}; see Fig. 7.9) is about the same as the $\Delta H°$ difference. Calculate the percentages of axial and equatorial conformations present in one mole of methylcyclohexane at 25 °C. (*Hint:* See Study Problem 7.1.)

7.3   Using the information in the previous problem and in Eq. 7.7, contrast the relative amounts of axial conformations in samples of methylcyclohexane and *tert*-butylcyclohexane.

7.4   (a) The axial conformation of fluorocyclohexane is 1.0 kJ mol^{-1} (0.25 kcal mol^{-1}) less stable than the equatorial conformation. What is the energy cost of a 1,3-diaxial interaction between hydrogen and fluorine?

fluorocyclohexane

(b) Estimate the energy difference between the gauche and anti conformations of 1-fluoropropane.

$$H_3C—CH_2—CH_2—F$$

1-fluoropropane

7.4 DISUBSTITUTED CYCLOHEXANES

A. Cis–Trans Isomerism in Disubstituted Cyclohexanes

We'll use 1-chloro-2-methylcyclohexane to study the stereochemical and conformational aspects of disubstituted cyclohexanes. We'll begin by using structures in which the cyclohexane ring is drawn as a planar hexagon. We'll refer to this type of structure as a **planar-ring structure**.

1-chloro-2-methylcyclohexane

As we know, the cyclohexane ring isn't planar; so, a planar-ring structure is a projection of the ring into the plane of the page. In effect, it is an average of the two chair conformations. To indicate stereochemistry, we'll use solid wedges to indicate "up" bonds—bonds to substituents in front of the page—and dashed wedges to indicate "down" bonds—bonds to substituents behind the page. This compound has two asymmetric carbons and therefore four stereoisomers—two diastereomeric sets of enantiomers. In one set of enantiomers, the two substituents are both up or both down.

(7.8)

$(1R,2S)$ $(1S,2R)$

***cis*-1-chloro-2-methylcyclohexane**

We arbitrarily used the structures with both substituents up, but equivalent structures with both substituents down are equally valid. We can derive one from the other by rotating the structure 180° about the axis shown:

(7.9)

When both substituents have the same *relative* orientation—both up or both down—the substitution pattern is called **cis**. Therefore, the two compounds in Eq. 7.8 are enantiomers of *cis*-1-chloro-2-methylcyclohexane.

In the other two stereoisomers of 1-chloro-2-methylcyclohexane, one substituent is up and the other down. When both substituents have different *relative* orientations—one up and the other down—the substitution pattern is called **trans**.

(7.10)

$(1R,2R)$ $(1S,2S)$

***trans*-1-chloro-2-methylcyclohexane**

Notice that the terms *cis* and *trans*, when used with cyclic compounds, refer to the *relative* orientations of substituents (that is, their up/down relationship) and not to their absolute (that is, *R,S*) configurations.

To illustrate the relationship of the planar-ring structures to their chair conformations, we'll derive the chair conformations of (1*S*,2*S*)-1-chloro-2-methylcyclohexane. Start with two unsubstituted chairs, and then associate each of the substituted carbons in the planar structure with a carbon atom of the chairs. It is completely arbitrary where we start the numbering of the chair as long as we proceed in the same direction around the planar and chair forms. (Remember that the planar structure is a projection of the chair conformation, typically viewed from above.) It is often easiest, although not essential, to let one substituted carbon of the chair be the rightmost or leftmost carbon.

$$(7.11)$$

The chlorine is a down substituent, so we'll put it in the down position at carbon-1 in each of the two chairs. In one chair form, the down position is equatorial, and in the other it is axial. We then add the methyl substituent to the up position on each of the two chairs.

$$(7.12)$$

It may look as if carbons 1 and 2 have shifted to the left in the second chair form. That's because the viewing perspective is different for the two chair forms (p. 275). This shift in perspective is the reason that it's helpful to use one of the substituted carbons as the rightmost or leftmost carbon. For example, carbon-1 is the rightmost carbon in either perspective.

Notice two things: (1) A planar-ring structure contains information about configuration (*R*, *S*, cis, or trans) but not about conformation. Each planar-ring structure has two chair conformations. Thus, you can answer questions about chirality with planar-ring structures, but you can't answer questions about conformation—for example, whether a substituent is axial or equatorial. (2) The up and down positions in the planar-ring structures carry over to the chair conformations, because the up/down positions are not affected by the chair interconversion; that is, in the chair interconversion, up-equatorial becomes up-axial, and down-equatorial becomes down-axial, as shown in Eq. 7.12.

The two chair conformations in this case are conformational diastereomers. They have different energies, and the diequatorial form is favored at equilibrium (why?), as shown by the unequal equilibrium arrows in Eq. 7.12. You should verify that *cis*-1-chloro-2-methylcyclohexane is also a mixture of conformational diastereomers.

STUDY PROBLEM   7.2

(a) Show that *trans*-1,3-dimethylcyclohexane is chiral. (b) For the 1*R*,3*R* stereoisomer, draw the two chair conformations. (c) What is the relationship between the two chair conformations of this compound: identical, conformational enantiomers, or conformational diastereomers?

SOLUTION (a) The question about chirality is most easily answered with a planar-ring structure. Draw one of the stereoisomers; then draw its mirror image; then test for congruence. Remember: *Trans* means that the substituents have an up–down relationship; that is, one substituent is on a wedged bond, and the other is on a dashed-wedge bond.

trans-1,3-dimethylcyclohexane
(mirror images)

demonstration of noncongruence;
therefore, molecules are chiral and
are enantiomers.

Because its mirror images are noncongruent, *trans*-1,3-dimethylcyclohexane is chiral.

(b) Follow the procedure in the text for constructing the two chair conformations. (The carbons are numbered for reference.) Notice that, for ease of reference, one of the substituted carbons is chosen as the rightmost carbon.

**(1R,3R)-1,3-dimethyl-
cyclohexane**

(c) The two chair conformations are identical. They certainly don't look identical, so let's verify that they are; you should follow along with models. The goal is to rotate one of the chairs (without changing its conformation!) in various ways so as to superpose exactly one of its methyl-substituted carbons onto a methyl-substituted carbon of the other chair. The relationship of the two structures will then be easier to see. First, rotate the second chair structure 180° about the axis shown; then rotate the resulting structure 120° as shown about an axis through the ring.

identical structures

180°

120°

**STUDY GUIDE
LINK 7.1**
Relating
Cyclohexane
Conformations

Indeed, the two chair conformations are identical. Carbon-1 in one chair is equivalent to carbon-3 in the other, and vice versa.

When a ring contains more than two substituents, cis–trans nomenclature is usually cumbersome. For such cases, other systems have been developed to designate relative configuration. (See Further Exploration 7.1.)

**FURTHER
EXPLORATION 7.1**
Other Ways of
Designating Relative
Configuration

PROBLEMS

7.5   For each of the following compounds, draw the two chair conformations that are in equilibrium.

(a) *cis*-1,3-dimethylcyclohexane   (b) *trans*-1-ethyl-4-isopropylcyclohexane

7.6   For each of the compounds in Problem 7.5, draw a boat conformation.

7.7   Draw planar-ring structures for the following cyclic compounds.

(a) *cis*-1-bromo-3-methylcyclohexane (both enantiomers)

(b) (1*R*,2*S*,3*R*)-2-chloro-1-ethyl-3-methylcyclohexane

7.8   Draw the two chair conformations of each compound in Problem 7.7.

7.9   Draw the planar-ring structure that results from each rotation of the following structure.

(a) 180° about axis (a)

(b) 120° counterclockwise about an axis (b) that runs through the center of the ring and perpendicular to the page.

7.10   Redraw the following chair conformations after 180° rotations about each of the axes shown. Use models if necessary.

B. Cyclic Meso Compounds

The planar-ring structures of both *cis*-1,2-dimethylcyclohexane and *cis*-1,3-dimethylcyclo-hexane have a plane of symmetry, as indicated by the gray line:

CH₃ — plane of symmetry

CH₃

CH₃ — plane of symmetry

CH₃

(7.13)

cis-1,2-dimethylcyclohexane   **cis-1,3-dimethylcyclohexane**

The internal plane of symmetry means that these compounds are achiral. Because these compounds have asymmetric carbons and have chiral stereoisomers (the trans isomers), they are examples of meso compounds.

An examination of their chair conformations, however, shows that the chair conformations of *cis*-1,2-dimethylcyclohexane are *chiral*, but *each conformation is the enantiomer of the other*:

CH₃   chair interconversion   H₃C

H₃C   CH₃

A   *B*

(7.14a)

enantiomers

From the way they are drawn, structures *A* and *B* may not look like enantiomers, but they are. This relationship can be verified by rotating structure *B* 120° about the vertical axis shown and comparing it with structure *A*. The two are noncongruent mirror images and therefore enantiomers.

is the same as

(7.14b)

B

enantiomer of structure *A*
in Eq. 7.14a

In other words, *cis*-1,2-dimethylcyclohexane is a mixture of *conformational enantiomers* (Sec. 6.9A). If we could isolate the individual conformations (which might be possible at *very low temperature*), each would be chiral and optically active. However, because these conformations interconvert very rapidly at ordinary temperatures, *cis*-1,2-dimethylcyclohexane cannot be isolated in optically active form. Over any realistic time interval, we can think of it as the time-averaged, planar-ring structure, which is meso. This same conclusion is true for any cis-1,2-disubstituted cyclohexane in which the substituents are identical. We discussed this situation in Sec. 6.9A, where we noted that compounds with enantiomeric conformations in rapid equilibrium are considered to be achiral.

Unlike the 1,2-dimethyl isomer, *cis*-1,3-dimethylcyclohexane consists of two chair conformations that are each true meso compounds, because *each one* has an internal plane of symmetry.

(7.15a)

planes of symmetry

These two conformations are *conformational diastereomers*.

chair interconversion

(7.15b)

diastereomers

Even at very low temperature, this compound cannot be isolated in an optically active form, because each conformation is achiral.

PROBLEMS

7.11   Use their planar-ring structures to determine which of the following compounds are chiral. For any that are achiral, indicate whether their individual chair conformations are chiral. Tell how you know. For any achiral compounds, give the relationship of their two chair conformations (identical, conformational enantiomers, or conformational diastereomers).

A   *B*   *C*

continued

continued _____

7.12   Determine whether each of the following compounds can be isolated in optically active form. Explain how you know.

   (a) *trans*-1,2-dimethylcyclohexane

   (b) 1,1-dimethylcyclohexane

   (c) *cis*-1-ethyl-4-methylcyclohexane

   (d) *cis*-1-ethyl-3-methylcyclohexane

7.13   (a) Does *trans*-1,4-dimethylcyclohexane contain any asymmetric carbons? If so, identify them.

   (b) Does *trans*-1,4-dimethylcyclohexane contain any stereocenters? If so, identify them.

   (c) Is *trans*-1,4-dimethylcyclohexane chiral?

   (d) What is the relationship of the two chair conformations of *trans*-1,4-dimethylcyclohexane?

C. Conformational Analysis

Disubstituted cyclohexanes, like monosubstituted cyclohexanes, can be subjected to conformational analysis. The relative stability of the two chair conformations is determined by comparing the 1,3-diaxial interactions (or *gauche*-butane interactions) in each conformation. Such an analysis is illustrated in Study Problem 7.3.

STUDY PROBLEM   7.3

Determine the relative energies of the two chair conformations of *trans*-1,2-dimethylcyclohexane. Which conformation is more stable?

SOLUTION   The first step in solving any problem is to draw the structures of the species involved. The two chair conformations of *trans*-1,2-dimethylcyclohexane are as follows:

Conformation *A* has the greater number of axial groups and should therefore be the less stable conformation—but by how much? Conformation *A* has four 1,3-diaxial methyl–hydrogen interactions (show these!), which contribute $4 \times 3.7 = 14.8$ kJ mol^{-1} ($4 \times 0.9 = 3.6$ kcal mol^{-1}) to its energy. What about *B*? You might be tempted to say that *B* has no unfavorable interactions because it has no axial groups, but in fact *B* does have one *gauche*-butane interaction—the one between the two methyl groups themselves, which have a dihedral angle between them of 60°, just as in *gauche*-butane. This interaction can be seen in a Newman projection of the bond between the carbons bearing the methyl groups:

Newman projection

This *gauche*-butane interaction contributes 3.7 kJ mol^{-1} (0.9 kcal mol^{-1}) to the energy of conformation *B* (Fig. 2.5, p. 53). The relative energy of the two conformations is the difference between their methyl–hydrogen interactions: $14.8 - 3.7 = 11.1$ kJ mol^{-1} (or $3.6 - 0.9 = 2.7$ kcal mol^{-1}). The diaxial conformation *A* is less stable by this amount of energy.

When two groups on a substituted cyclohexane conflict in their preference for the equatorial position, the preferred conformation can usually be predicted from the relative conformational preferences of the two groups. Consider, for example, the chair interconversion of cis-1-tert-butyl-4-methylcyclohexane.

$$\qquad\qquad(7.16)$$

The tert-butyl group is so large that its van der Waals repulsions control the conformational equilibrium (see Eq. 7.7, p. 284). Hence, the chair conformation in which the tert-butyl group assumes the equatorial position is overwhelmingly favored. The methyl group is thus forced into the axial position.

> There is so little of the conformation with an axial tert-butyl group that chemists say sometimes that the conformational equilibrium is "locked." This statement is somewhat misleading because it implies that the two conformations are not at equilibrium. The equilibrium indeed occurs rapidly, but simply contains very little of the conformation in which the tert-butyl group is axial.

PROBLEMS

7.14   (a) Calculate the energy difference between the two chair conformations of trans-1,4-dimethylcyclohexane.
   (b) Calculate the energy difference between cis-1,4-dimethylcyclohexane and the more stable conformation of trans-1,4-dimethylcyclohexane.

7.15   Draw the more stable chair conformation for each of the following compounds:

7.5   CYCLOPENTANE, CYCLOBUTANE, AND CYCLOPROPANE

Recall from Sec. 7.1 that cyclopentane, cyclobutane, and cyclopropane are all less stable than cyclohexane. In this section we consider the structures of these smaller cycloalkanes and the reasons for their greater energies.

A. Cyclopentane

Cyclopentane, like cyclohexane, exists in a puckered conformation, called the **envelope conformation** (Fig. 7.11 on p. 292). This conformation undergoes very rapid conformational changes in which each carbon alternates as the "point" of the envelope.

The heats of formation in Table 7.1 (p. 273) show that cyclopentane has somewhat higher energy than cyclohexane. The higher energy of cyclopentane is due mostly to eclipsing between hydrogen atoms, which is also shown in Fig. 7.11.

Substituted cyclopentanes also exist in envelope conformations, but the substituents adopt positions that minimize van der Waals repulsions with neighboring groups. For

FIGURE 7.11 A ball-and-stick model of the envelope conformation of cyclopentane. The hydrogens shown in blue are either eclipsed or nearly eclipsed with the blue hydrogens on adjacent carbons.

eclipsed hydrogens

nearly eclipsed hydrogens

example, in methylcyclopentane, the methyl group assumes an equatorial position at the point of the envelope.

$$\text{(7.16)}$$

equatorial

axial

When a cyclopentane ring has two or more substituent groups, cis and trans relationships between the groups are possible, just as in cyclohexane.

cis-**1,2-dimethylcyclopentane**

trans-**1-ethyl-3-isopropylcyclopentane**

B. Cyclobutane and Cyclopropane

The data in Table 7.1 (p. 273) show that cyclobutane and cyclopropane are the least stable of the monocyclic alkanes. In each compound, the angles between carbon–carbon bonds are constrained by the size of the ring to be much smaller than the optimum tetrahedral angle of 109.5°. When the bond angles in a molecule deviate from their ideal values, the energy of the molecule is raised in the same sense that squeezing the handles of a hand exerciser increases the potential energy of the resisting spring. This excess energy, which is reflected in a greater heat of formation, is called **angle strain**. Hence, angle strain contributes significantly to the high energies of both cyclobutane and cyclopropane.

Puckering of the cyclobutane ring avoids complete eclipsing between hydrogens. Cyclobutane consists of two puckered conformations in rapid equilibrium (Fig. 7.12).

Because three carbons define a plane, the carbon skeleton of cyclopropane is planar; thus, neither its angle strain nor the eclipsing interactions between its hydrogens can be relieved by puckering. As the data in Table 7.1 show, cyclopropane is the least stable of the cyclic alkanes. The carbon–carbon bonds of cyclopropane are bent in a "banana" shape around the periphery of the ring. Such "bent bonds" allow for angles between the carbon orbitals that are on the order of 105°, closer to the ideal tetrahedral value of 109.5° (Fig. 7.13). Although bent bonds reduce angle strain, they do so at a cost of less effective overlap between the carbon orbitals.

FURTHER EXPLORATION 7.2
Alkenelike Behavior of Cyclopropanes

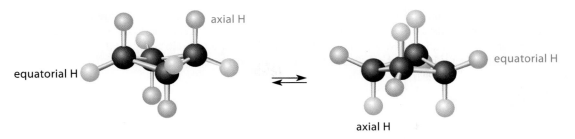

FIGURE 7.12 Cyclobutane consists of two identical puckered conformations in rapid equilibrium. As in cyclohexane, the equilibrium causes axial hydrogens (*blue on the left*) to become equatorial (*blue on the right*), and vice versa.

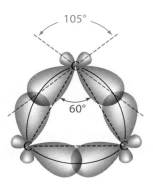

FIGURE 7.13 The orbitals that overlap to form each C—C bond in cyclopropane do not lie along the straight line between the carbon atoms. These carbon–carbon bonds (*black lines*) are sometimes called "bent" or "banana" bonds. The C—C—C angle is 60° (*purple dashed lines*), but the angles between the orbitals on each carbon are closer to 105° (*blue dashed lines*).

PROBLEMS

7.16 (a) The dipole moment of *trans*-1,3-dibromocyclobutane is 1.1 D. Explain why a nonzero dipole moment supports a puckered structure rather than a planar structure for this compound.

(b) Draw a structure for the more stable conformation of *trans*-1,2-dimethylcyclobutane.

7.17 Tell whether each of the following compounds is chiral.

(a) *cis*-1,2-dimethylcyclopropane

(b) *trans*-1,2-dimethylcyclopropane

7.6   BICYCLIC AND POLYCYCLIC COMPOUNDS

A. Classification and Nomenclature

Some cyclic compounds contain more than one ring. If two rings share two or more common atoms, the compound is called a **bicyclic compound**. If two rings have a single common atom, the compound is called a **spirocyclic compound**.

bicyclo[4.3.0]nonane   **bicyclo[2.2.1]heptane**   **spiro[4.4]nonane**
└──────── (bicyclic compounds) ────────┘   (a spirocyclic compound)

The atoms at which two rings are joined in a bicyclic compound are called **bridge-head carbons**. Bicyclic compounds are further classified according to the relationship of the bridgehead carbons. When the bridgehead carbons of a bicyclic compound are adjacent, the compound is classified as a **fused bicyclic compound**:

bridgehead carbons (*) are at adjacent positions

a fused bicyclic compound

When the bridgehead carbons are not adjacent, the compound is classified as a **bridged bicyclic compound**:

bridgehead carbons (*) are not adjacent

three views of a bridged bicyclic compound

Many students are puzzled by these unusual structures. It is important to build models of them and relate these models to the structures that are drawn on paper.

The nomenclature of bicyclic hydrocarbons is best illustrated by example:

one-carbon bridge

two-carbon bridge

three-carbon bridge

bicyclo[3.2.1]octane
(* = bridgehead carbons)

This compound is named as a bicyclooctane because it is a bicyclic compound containing a total of eight carbon atoms. The numbers in brackets represent the number of carbon atoms in the respective bridges, in order of decreasing size.

STUDY PROBLEM 7.4

Give the IUPAC name of the following compound. (Its common name is *decalin*.)

SOLUTION The compound has two fused rings that contain a total of 10 carbons, and is therefore named as a bicyclodecane. Three bridges connect the bridgehead carbons: two contain four carbons, and *one contains zero carbons*. (The bond connecting the bridgehead carbons in a fused-ring system is considered to be a bridge with zero carbons.)

zero-carbon bridge

four-carbon bridge

four-carbon bridge

bicyclo[4.4.0]decane
(* = bridgehead carbons)

The compound is named bicyclo[4.4.0]decane.

PROBLEMS

7.18   Name the following compounds and tell whether each is a bridged or a fused bicyclic compound.

(a)   (b)

7.19   Without drawing their structures, tell which of the following compounds is a fused bicyclic compound and which is a bridged bicyclic compound, and how you know.

bicyclo[2.1.1]hexane (*A*)   bicyclo[3.1.0]hexane (*B*)

Some organic compounds contain many rings joined at common atoms; these compounds are called **polycyclic compounds**. Among the more intriguing polycyclic compounds are those that have the shapes of regular geometric solids. Three of the more spectacular examples are cubane, dodecahedrane, and tetrahedrane.

cubane   **dodecahedrane**   **tetrahedrane**

Cubane, which contains eight —CH— groups at the corners of a cube, was first synthesized in 1964 by Professor Philip Eaton and his associate Thomas W. Cole at the University of Chicago. Dodecahedrane, in which 20 —CH— groups occupy the corners of a dodecahedron, was synthesized in 1982 by a team of organic chemists led by Professor Leo Paquette of the Ohio State University. Tetrahedrane itself has not yet been synthesized, although derivatives containing *tert*-butyl and Me_3Si— substituent groups at each corner were prepared in 1978, 2002, and 2011. Chemists tackle the syntheses of these unique molecules not only because they represent interesting problems in chemical bonding, but also because of the sheer challenges of the endeavors.

B. Cis and Trans Ring Fusion

Two rings in a fused bicyclic compound can be joined in more than one way. Consider, for example, bicyclo[4.4.0]decane, which has the common name *decalin*.

decalin
(bicyclo[4.4.0]decane)

There are two stereoisomers of decalin. In *cis*-decalin, two —CH_2— groups of ring *B* (circles) are cis substituents on ring *A*; likewise, two —CH_2— groups of ring *A* (squares) are cis substituents on ring *B*.

(7.17a)

ring *A*   H

ring *B*   H

chair conformation   planar-ring structure

***cis*-decalin**

The cis ring fusion can be shown in a planar-ring structure by showing the cis arrangements of the bridgehead hydrogens.

In *trans*-decalin, the —CH₂— groups adjacent to the ring fusion are in a trans-diequatorial arrangement. The bridgehead hydrogens are trans-diaxial.

(7.17b)

chair conformation   planar-ring structure

***trans*-decalin**

Both *cis*- and *trans*-decalin have two equivalent planar-ring structures:

Each cyclohexane ring in *cis*-decalin can undergo the chair interconversion. You should verify with models that when one ring changes its conformation, the other must change also. However, in *trans*-decalin, the six-membered rings can assume twist-boat conformations, but they cannot change into their alternative chair conformations. You should try the chair inter-conversion with a model of *trans*-decalin to see for yourself the validity of this point. Focus on ring *B* of the *trans*-decalin structure in Eq. 7.17b. Notice that the two circles represent carbons that are in effect *equatorial* substituents on ring *A*. If ring *A* were to convert into the other chair conformation, these two carbons in ring *B* would have to assume *axial* positions, because, in the chair interconversion, equatorial groups become axial groups. When these two carbons are in axial positions, they are much farther apart than they are in equatorial positions; the distance between them is simply too great to be spanned easily by the remaining two carbons of ring *B*.

(7.18)

As a result, the chair interconversion introduces so much ring strain into ring *B* that the inter-conversion cannot occur. Exactly the same problem occurs with ring *A* when ring *B* undergoes the chair interconversion.

PROBLEM

7.20   How many 1,3-diaxial interactions occur in *cis*-decalin? In *trans*-decalin? Which compound has the lower energy and by how much? (*Hint:* Use your models, and don't count the same 1,3-diaxial interaction twice.)

Trans-decalin is more stable than *cis*-decalin because it has fewer 1,3-diaxial interactions (Problem 7.20). Trans ring fusion, however, is not the more stable way of joining rings in all fused bicyclic molecules. In fact, if both of the rings are small, trans ring fusion is virtually impossible. For example, only the cis isomers of the following two compounds are known:

bicyclo[1.1.0]butane   **bicyclo[3.1.0]hexane**

Attempting to join two small rings with a trans ring junction introduces too much ring strain. The best way to see this is with models, using the following exercise as your guide.

PROBLEM

7.21   (a) Compare the difficulty of making models of the cis and trans isomers of bicyclo[3.1.0]hexane. (Don't break your models!) Which is easier to make? Why?

(b) Compare the difficulty of making models of *trans*-bicyclo[3.1.0]hexane and *trans*-bicyclo[5.3.0]decane. Which is easier to make? Explain.

In summary:

1. Two rings can in principle be fused in a cis or trans arrangement.

2. When the rings are small, only cis fusion is observed because trans fusion introduces too much ring strain.

3. In larger rings, both cis- and trans-fused isomers are well known, but the trans-fused ones are more stable because 1,3-diaxial interactions are minimized (as in the decalins).

Effects (2) and (3) are about equally balanced in the *hydrindanes* (bicyclo[4.3.0]nonanes); heats of combustion show that the trans isomer is only 4.46 kJ mol^{-1} (1.06 kcal mol^{-1}) more stable than the cis isomer.

hydrindane
(bicyclo[4.3.0]nonane)

C. Trans-Cycloalkenes and Bredt's Rule

Cyclohexene and other cycloalkenes with small rings have cis (or *Z*) stereochemistry at the double bond. Is there a *trans*-cyclohexene? The answer is that the trans-cycloalkenes with six or fewer carbons have never been observed. The reason becomes obvious if you try to build a model of *trans*-cyclohexene. In this molecule the carbons attached to the double bond are so far apart that it is difficult to connect them with only two other carbon atoms. To do so either introduces a great amount of strain, or requires twisting the molecule about the double bond, thus weakening the overlap of the 2*p* orbitals involved in the π bond. *Trans*-cyclooctene is the smallest trans-cycloalkene that can be isolated under ordinary conditions; however, it is 47.7 kJ mol⁻¹ (11.4 kcal mol⁻¹) less stable than its cis isomer.

Closely related to the instability of trans-cycloalkenes is the instability of any small bridged bicyclic compound that has a double bond at a bridgehead atom. The following compound, for example, is very unstable and has never been isolated:

bridgehead

bicyclo[2.2.1]hept-1(2)-ene
(unknown)

The instability of compounds with bridgehead double bonds has been generalized as **Bredt's rule**: *In a bicyclic compound, a bridgehead atom contained solely within small rings cannot be part of a double bond.* (A "small ring," for purposes of Bredt's rule, contains seven or fewer atoms within the ring.)

Julius Bredt (1855–1937) was a German chemist who, for the last 25 years of his career, was a professor and the director of the Organic Chemistry Laboratory at the Technische Hochschule of Aachen in Germany. In 1893, he proposed the correct (and then very unusual) bridged bicyclic structure of camphor, for which more than 30 incorrect structures had previously been proposed. His studies of bridged bicyclic compounds led him to formulate in 1913 the rule that bears his name.

camphor

The basis of Bredt's rule is that double bonds at bridgehead carbons within small rings are twisted; that is, the atoms directly attached to such double bonds *cannot* lie in the same plane. To see this, try to construct a model of bicyclo[2.2.1]hept-1(2)-ene, the bicyclic alkene shown above. You will see that the bicyclic ring system cannot be completed without twisting the double bond. Twisting the double bond prevents overlap of the 2*p* orbitals required to form the π bond. This is similar to the double-bond twisting that would occur in *trans*-cyclohexene. Like the corresponding trans-cycloalkenes, bicyclic compounds containing bridgehead double bonds solely within small rings are too unstable to isolate. Bicyclic compounds that have bridgehead double bonds within larger rings are more stable and can be isolated.

bicyclo[2.2.1]hept-1(2)-ene
The trans double bond
is in a 6-membered ring;
the compound is
too unstable to isolate.

bicyclo[4.4.1]undec-1(2)-ene
The trans double bond
is in a 10-membered ring;
the compound is
stable enough to isolate.

PROBLEM

7.22   Use models if necessary to help you decide which compound within each pair should have the greater heat of formation. Explain.

(a)   (b)

    A   B   A   B

D. Steroids

Of the many naturally occurring compounds with fused rings, the *steroids* are particularly important. A **steroid** is a compound with a structure derived from the following tetracyclic ring system:

Steroids have a special numbering system, which is shown in the preceding structure. The various steroids differ in the functional groups that are present on this carbon skeleton.

Two structural features are particularly common in naturally occurring steroids (Fig. 7.14 on p. 300). The first is that in many cases all ring fusions are trans. Because trans-fused cyclohexane rings cannot undergo the chair interconversion (see Eq. 7.18 and subsequent discussion), all-trans ring fusion causes a steroid to be conformationally rigid and relatively flat. This can be seen particularly with the models in Figs. 7.14c–d. Second, many steroids have methyl groups, called *angular methyls*, at carbons 10 and 13, so-called because each is located at the vertex of the angle at a bridgehead carbon. The hydrogens of these methyl groups are shown in color in Figs. 7.14c–d.

Many important hormones and other natural products are steroids. Cholesterol occurs widely and was the first steroid to be discovered (1775). The corticosteroids and the sex hormones represent two biologically important classes of steroid hormones.

cholesterol
(important component of membranes: principal component of gallstones; major constituent of atherosclerotic plaques)

cortisone
(anti-inflammatory hormone)

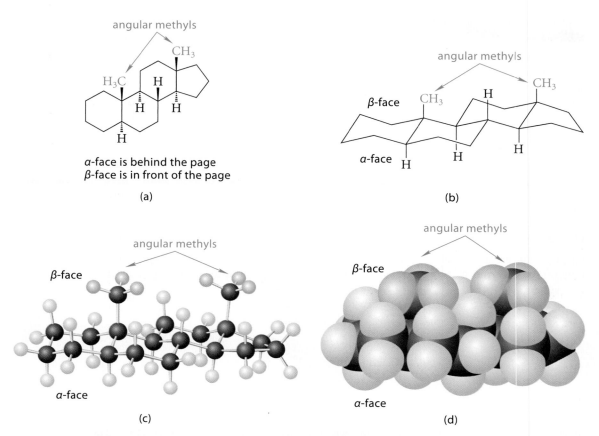

FIGURE 7.14 Four different representations of the steroid ring system. (a) A planar structure. (b) A perspective structure. (c) A ball-and-stick model. (d) A space-filling model. Notice the all-trans ring junctions and the extended, relatively flat shape. The hydrogens of the angular methyl groups are shown in blue in parts (c) and (d).

human female sex hormones

The compounds used in birth-control medications are all steroid analogs of progesterone and estradiol in which these natural steroidal female sex hormones have been synthetically modified to make them suitable for oral administration.

Sources of Steroids

Prior to 1940, steroids were obtained only from such inconvenient sources as sows' ovaries or the urine of pregnant mares, and they were scarce and expensive. In the 1940s, however, a Pennsylvania State University chemist, Russell E. Marker (1902–1995), developed a process that could bring about the conversion of a naturally occurring compound called *diosgenin* into progesterone.

$$\text{diosgenin} \xrightarrow{\text{several steps}} \textbf{progesterone}$$

diosgenin

(Various forms of this conversion, called the *Marker degradation*, are still in use.) The natural source of diosgenin is the root of a vine, *cabeza de negro*, genus *Dioscorea*, which is indigenous to Mexico (photo).

About two-thirds of modern synthetic steroid production starts with various types of *Dioscorea*, which is now grown not only in Mexico but also in Central America, India, and China. More recently, practical industrial processes have been developed that start with steroid derivatives from other sources. For example, in the United States, a process was developed to recover steroid derivatives from the by-products of soybean-oil production, and these are used to produce synthetic glucocorticoids and other steroid hormones. Some estrogens and cardiac steroids are still isolated directly from natural sources.

7.7 REACTIONS INVOLVING STEREOISOMERS

The remainder of this chapter focuses on the importance of stereochemistry in organic reactions. To begin, this section develops some general principles that apply to reactions that involve stereoisomers.

A. Reactions Involving Enantiomers

We're going to consider two important situations in this section:

1. Reactions that involve a chiral compound as a starting material
2. Reactions that form enantiomers as products from achiral starting materials

Both situations can be illustrated by a biologically important reaction, the equilibrium between malic acid and fumaric acid, a reaction from the Krebs cycle.

$$\underset{\textbf{malic acid}}{HO-\overset{O}{\overset{\|}{C}}-\overset{OH}{\underset{|}{CH}}-CH_2-\overset{O}{\overset{\|}{C}}-OH} \;\underset{\xleftarrow{\hspace{1.5em}}}{\overset{catalyst}{\xrightarrow{\hspace{1.5em}}}}\; \underset{\textbf{fumaric acid}}{\quad} + \; H_2O \qquad (7.19)$$

fumaric acid

This is an example of alcohol dehydration (in the forward reaction) and alkene hydration (in the reverse direction). We considered this type of reaction in Secs. 4.9B and 4.9C. Because the reaction has an equilibrium constant near 1, it can be studied by starting with either fumaric acid or malic acid and allowing the two compounds to come to equilibrium. In the forward direction, a chiral compound (malic acid) reacts to give an achiral compound. In the reverse direction, an achiral compound reacts to give a chiral compound. This reaction requires an acid catalyst such as H_2SO_4 and a fairly high temperature, or it can be catalyzed by an enzyme, *fumarase*, at physiological pH (7.4) and temperature (37 °C).

Do enantiomers react at the same rate or at different rates? In other words, if we carry out the reaction in Eq. 7.19 with, first, (R)-malic acid, and then again with (S)-malic acid, do the two compounds have the same reactivity or different reactivities? A general principle applies to situations like this. *Enantiomers react at identical rates with an achiral reagent.* Thus, the enantiomers of malic acid in Eq. 7.19 react with water and an acid catalyst such as H_2SO_4—both achiral reagents—at exactly the same rates to give their respective products in exactly the same yield.

An analogy from common experience can help you understand why the rates are identical. Consider your feet, an enantiomeric pair of objects. Imagine placing first your right foot, then your left, in a perfectly rectangular box—an achiral object. Each foot will fit this box in exactly the same way. If the box pinches the big toe on your right foot, it will also pinch the big toe on your left foot in the same way. It should take you the same amount of time to put one foot in the box as it does the other (if we assume equal facility with both feet). Just as your feet interact in the same way with the achiral box, so enantiomeric molecules react in exactly the same way with achiral reagents. Because water and sulfuric acid are achiral reagents, the enantiomers of malic acid in Eq. 7.20 react in exactly the same way.

The *principle of enantiomeric differentiation* (Sec. 6.8A) states that enantiomers behave differently only in the presence of a chiral agent. Neither water nor sulfuric acid is chiral; therefore, the two enantiomers of malic acid behave in exactly the same way. In terms of energies, *enantiomers have identical free energies.* That is, free energies, like boiling points and melting points, are among the properties that do not differ between enantiomers (Sec. 6.3). Both of the starting materials in Eq. 7.20 and their respective transition states are enantiomeric. The enantiomeric transition states have identical free energies, as do the enantiomeric starting materials. Because relative reactivity is determined by the difference in free energies of the transition state and starting material, and because this difference is identical for both enantiomers, enantiomers react at identical rates.

In contrast, when the reaction is catalyzed by the enzyme fumarase, the two enantiomers behave quite differently. In fact, (S)-malic acid reacts rapidly—the enzyme accelerates the reaction by a factor of about 10^9! However, the enzyme does *not* catalyze the reaction of (R)-malic acid.

(The acids are shown as their ionized forms malate and fumarate because, like typical carboxylic acids, they are ionized at pH 7.4.) Why should the enzyme catalyze the reaction of one enantiomer and not the other? Fumarase and all other enzymes are *enantiomerically pure chiral molecules.* Therefore, by the principle of enantiomeric differentiation, the two enantiomers of malate react differently with the enzyme. In terms of free energies, because the transition state for the reaction of water and malate includes the chiral enzyme, then the R and the S transition states, which are enantiomers in the *absence* of the enzyme, are diastereomers in the *presence* of the enzyme, and their free energies are different. The faster reaction—the reaction of (S)-malate—has the transition state of lower free energy.

The principle of enantiomeric differentiation is operating here exactly as it does in enantiomeric resolution (Sec. 6.8). In fact, we can think of the enzyme fumarase in principle as a chiral resolving agent. As an analogy for this situation, imagine your left and right feet interacting in turn with your left shoe. Your feet are enantiomers, and the shoe is a "chiral reagent." Your left foot "reacts" more quickly with the left shoe than your right foot does: it's easier to put on. However, your right foot "reacts" more slowly: it's more difficult to put on. Moreover, if you finally get the left shoe on your right foot, it doesn't fit as well; the interaction energy of the right foot and left shoe is higher than the interaction energy of the left foot and the left shoe. (We describe the interaction energy with the words "it doesn't fit" or "it fits.") The enzyme is analogous to the left shoe: it reacts with the two enantiomers of malic acid with different rates.

Now let's turn to the second situation: the reaction of an achiral compound to give enantiomeric products. The principle of microscopic reversibility (Sec. 4.9B, p. 175) requires that *a reaction and its reverse must have identical transition states.* Therefore, if the two enantiomers of malic acid undergo dehydration at the same rate, then the reverse reaction, hydration of fumaric acid, must give the two enantiomers of malic acid at the same rate. In general, *when chiral products are formed from achiral starting materials, both enantiomers of a pair are always formed at identical rates.* That is, the product is always the *racemate.* (This is the reason that racemates occur widely in chemistry.) Another way of expressing the same principle is that *optical activity never arises spontaneously in the reactions of achiral compounds.*

In the presence of an enzyme catalyst, as we have seen, only (S)-malate reacts to give fumarate. Therefore, in the reverse reaction, fumarate reacts in the presence of the enzyme to give only (S)-malate. The basis of this selectivity, as in Eq. 7.22, is the free energies of diastereomeric transition states. In general, *enantiomers are formed at different rates from achiral starting materials in the presence of a chiral catalyst.*

Notice that we haven't discussed the detailed molecular interactions responsible for the selectivity of fumarase. Nor, in general, can we predict which enantiomer is more reactive, or which is formed selectively, without more information. The point here is simply to show the general principles that must be operating in cases like this.

To summarize:

1. Enantiomers react at identical rates with achiral reagents or catalysts and at different rates with chiral reagents or catalysts. *How different* depends on the specific case and generally can't be predicted without more information. With enzymes as catalysts, the difference is in most cases so large that only one enantiomer reacts.

**STUDY GUIDE
LINK 7.2**
Reactions of
Chiral Molecules

**FURTHER
EXPLORATION 7.3**
Optical Activity

2. When enantiomers are formed from achiral starting materials, the product is race-mic *unless* the reaction is carried out under the influence of a chiral environment such as a chiral catalyst. In that case, the predominance of one enantiomer can be expected. Which enantiomer is preferred, and the magnitude of the preference, in general can't be predicted without more information. With enzymes as cata-lysts, however, the difference is in most cases so large that only one enantiomer is formed.

PROBLEMS

7.23   Apply the principles of this section to solve each of the following problems.

(a) Assuming equal strength in both hands, would your right and left hands differ in their ability to drive a nail? To tighten a screw with a screwdriver?

(b) Imagine that a certain Mr. D. has been visited by a certain Mr. L. from elsewhere in the universe. Mr. D. and Mr. L. are alike in every way, except that they are noncongruent mirror images! You have to introduce each of them at an international press conference, but neither will agree to give his name. How would you tell them apart? (There may be several ways.)

7.24   Tell whether the two enantiomers of alkene *A* react at the same or different rates with each of the following reagents, and explain your reasoning. Give the products of the reaction in each case.

$$Ph \overset{\displaystyle H \quad CH_3}{\underset{\displaystyle A}{\diagdown \, \diagup}}$$

(a) HBr, peroxides   (b)

$$H_3C \cdots \underset{\underset{\displaystyle H}{|}}{\underset{B}{\diagup \diagdown}} CH_3$$

(2R,5R)-2,5-dimethylborolane
(a chiral borane)

Enantiomeric Resolution in Nature

When a chiral compound occurs in nature, typically only one of its two enantiomers is found in a given natural source. That is, *nature is a source of optically active compounds.* For example, the sugar sucrose, produced by the sugar cane and sugar beet plants, occurs only as the dextrorotatory enantiomer; the naturally occurring amino acid leucine is the levorotatory enantiomer.

Many scientists hypothesize that eons ago the first chiral compounds were formed from simple, achiral starting materials, such as methane, water, and HCN. This hypothesis presents a problem. As shown in Sec. 7.7A, reactions that give chiral products from achiral starting materials always give the racemate; net optical activity cannot be generated in the reactions of achiral molecules. If the biological starting materials are all achiral, why is the world full of optically active compounds? Instead, it should be full of racemates! (This would mean that somewhere in the world your noncongruent mirror image is studying organic chemistry!) The only way out of this dilemma is to postulate that at some point in geologic time, one or more *enantiomeric resolutions* must have occurred. How could this have happened?

This question has generated much speculation. However, many scientists believe that the first enantiomeric resolution occurred purely by chance. Although we've said that a spontaneous enantiomeric resolution never occurs, a more accurate statement is that such an event is *highly improbable.* For example, you learned in Sec. 6.8C that spontaneous crystallization of one enantiomer can occur if a

(+)-sucrose

(–)-leucine

supersaturated solution of a racemate is seeded by a crystal of one enantiomer. Perhaps the spontaneous crystallization of a pure enantiomer took place on prebiotic Earth, seeded by a speck of dust with just the right shape. The question is an intriguing one, and no one really knows the answer.

Given that one or more enantiomeric resolutions occurred by chance at some time during the course of natural history, it is not difficult to understand how nature continues to manufacture enantiomerically pure compounds. You've just learned that enzymes catalyze the formation of optically active compounds from achiral starting materials, and, when the starting material of an enzyme-catalyzed reaction is chiral, an enzyme will catalyze the reaction of only one enantiomer. Such catalytic discrimination between stereoisomers guarantees a high degree of enantiomeric purity in naturally occurring compounds.

B. Reactions Involving Diastereomers

In this section, we'll consider two related situations:

1. The relative reactivities of diastereomeric starting materials in chemical reactions

2. Reactions that form diastereomeric products

Diastereomeric compounds in general have different reactivities toward *any* reagent, whether the reagent is chiral or achiral. The reason is that both the starting materials and the transition states are diastereomeric, and diastereomers have different free energies. Consequently, their standard free energies of activation, and hence their reaction rates, must in principle differ. Thus, the two diastereomeric alkenes *cis-* and *trans-*2-butene react at different rates with *any* reagent. We may not be able to predict which alkene is more reactive or by how much without further information. The cis stereoisomer may be more reactive with one reagent and the trans may be more reactive with another; but we can be sure that the two alkenes will not be equally reactive.

have different reactivities with any reagent.

(7.23)

When reactions form diastereomeric products, the products are formed at different rates and therefore in different amounts. For example, when 1-methylcyclohexene undergoes HBr addition in the presence of peroxides (free-radical addition), the diastereomeric 2-methyl-1-bromocyclohexanes—the cis and trans stereoisomers—are formed in different amounts.

(±)-*trans*-1-bromo-2-methylcyclohexane   **(±)-*cis*-1-bromo-2-methylcyclohexane**
Both products are racemates!

Without knowing more about the reaction, we might not be able to predict which diastereomer is the predominant product, or by how much, but we can expect one to be formed in greater amount than the other. (In this case the cis diastereomer is the predominant one; see Problem 7.69, p. 322.)

Diastereomers are formed in different amounts because they are formed through diastereomeric transition states. In general, one transition state has a lower standard free energy than its diastereomer. The diastereomeric reaction pathways thus have different standard free energies of activation and therefore different rates, and their respective products are formed in different amounts.

Remember that when the starting materials are achiral, each diastereomer of the product will be formed as a pair of enantiomers (the racemate), as we learned in Sec. 7.7A. This is the situation, for example, in Eq. 7.24. *Here's a drawing convention you should be aware of:* For convenience we sometimes draw only one enantiomer of each product, as in this equation, but in situations like this it is understood that each of these diastereomers *must* be racemic.

STUDY PROBLEM 7.5

What stereoisomeric products could be formed in the addition of bromine to cyclohexene? Which should be formed in the same amounts? Which should be formed in different amounts?

SOLUTION   Before dealing with any issue involving the stereochemistry of any reaction, first be sure you understand the reaction itself. Bromine addition to cyclohexene gives 1,2dibromocyclohexane:

$$\text{cyclohexene} + Br_2 \longrightarrow \text{1,2-dibromocyclohexane (Br, Br)}$$

1,2-dibromocyclohexane

Next, enumerate the possible stereoisomers of the product that might be formed. The product, 1,2-dibromocyclohexane, can exist as a pair of diastereomers:

cis-**1,2-dibromocyclohexane**   *trans*-**1,2-dibromocyclohexane**

**STUDY GUIDE
LINK 7.3**
Analysis of
Reaction
Stereochemistry

The trans diastereomer can exist as a pair of enantiomers, and the cis diastereomer is meso (Sec. 7.4B). Hence, three potentially separable stereoisomers could be formed: the cis isomer and the two enantiomers of the trans isomer. Because the cis and trans isomers are diastereomers, *they are formed in different amounts.* (You can't predict at this point which one predominates, but we'll return to that issue in Sec. 7.8C.) The two enantiomers of the trans diastereomer *must be formed in identical amounts.* Thus, whatever the amount of the trans isomer we obtain from the reaction, it is obtained as the racemate—a 50 : 50 mixture of the two enantiomers.

PROBLEMS

7.25   What stereoisomeric products are possible when *cis*-2-butene undergoes bromine addition? Which are formed in different amounts? Which are formed in the same amounts?

7.26   What stereoisomeric products are possible when *trans*-2-butene undergoes hydroboration–oxidation? Which are formed in different amounts? Which are formed in the same amounts?

7.27   Write all the possible products that might form when racemic 3-methylcyclohexene reacts with Br_2. What is the relationship of each pair? Which compounds should in principle be formed in the same amounts, and which in different amounts? Explain.

7.8   STEREOCHEMISTRY OF CHEMICAL REACTIONS

We've learned that stereochemistry adds another dimension to the study and practice of organic chemistry. No chemical structure is complete without stereochemical detail, and no chemical reaction can be planned without considering problems of stereochemistry that might arise. This section examines the possible stereochemical outcomes of two general types of reaction: addition reactions and substitution reactions. Then, some addition reactions covered in Chapter 5 will be revisited with particular attention to their stereochemistry.

A. Stereochemistry of Addition Reactions

Recall that an *addition reaction* is a reaction in which a general species X—Y adds to each end of a bond. The cases we've studied so far involve addition to double bonds:

$$R \underset{R}{\overset{R}{\diagdown}} C = C \underset{R}{\overset{R}{\diagdown}} + \quad X-Y \quad \longrightarrow \quad R-\underset{\underset{X}{|}}{\overset{\overset{R}{|}}{C}}-\underset{\underset{Y}{|}}{\overset{\overset{R}{|}}{C}}-R \qquad (7.25)$$

An addition reaction can occur in either of two stereochemically different ways, called *syn-addition* and *anti-addition*. These will be illustrated with cyclohexene and a general reagent X—Y.

The stereochemistry of addition to a double bond is discussed with reference to the plane that contains the double bond and its four attached groups. The sides of this plane are called **faces**. The side of the plane nearest the observer is typically called the *top face*, and the other side is the *bottom face*.

$$(7.26)$$

In a **syn-addition**, two groups add to a double bond from the same face.

Syn-addition:

$$(7.27a)$$

Notice that the two directions of syn-addition are *enantiomeric*: that is, when X and Y are different, the products from addition at the top face are enantiomers of the products formed from addition at the bottom face.

In an **anti-addition**, two groups add to a double bond from opposite faces.

Anti-addition:

$$(7.27b)$$

The two directions of anti-addition are also enantiomeric.

It is also conceivable that an addition might occur as a mixture of syn and anti modes. In such a reaction, the products would be a mixture of all of the products in both Eqs. 7.27a–b. Examples of both syn- and anti-additions, as well as mixed additions, will be examined later in this section.

As Eqs. 7.27a–b suggest, the syn and anti modes of addition can be distinguished *by analyzing the stereochemistry of the products.* In Eq. 7.27a, for example, the cis relationship of the groups X and Y in the product would tell us that a syn-addition has occurred. *The stereochemistry of an addition can be determined only when the stereochemically different modes of addition give rise to stereochemically different products.* In contrast, when two groups X and Y add to ethylene ($H_2C{=}CH_2$), the same product ($X{-}CH_2{-}CH_2{-}Y$) results whether the reaction is a syn- or an anti-addition. Because this product does not exist as stereoisomers, we can't tell whether the addition is syn or anti. A more general way of stating the same point is to say that syn- and anti-additions give different products only when *both* carbons of the double bond become carbon stereocenters in the product. If you stop and think about it, this should make sense, because the question of syn- and anti-addition is a question of the relative stereochemistry at *both* carbons, and the relative stereochemistry is meaningless if both carbons aren't stereocenters.

B. Stereochemistry of Substitution Reactions

In a **substitution reaction**, one group is replaced by another. In the following substitution reaction, for example, the Br is replaced by OH:

$$H_3C{-}\ddot{B}r{:} + {}^-{:}\ddot{O}H \longrightarrow H_3C{-}\ddot{O}H + {:}\ddot{B}r{:}^- \qquad (7.28)$$

The oxidation step of hydroboration–oxidation is also a substitution reaction in which the boron is replaced by an OH group.

$$^-OH + 3\,HO{-}OH + (CH_3CH_2)_3B \longrightarrow 3\,CH_3CH_2{-}OH + {}^-B(OH)_4 \qquad (7.29)$$

A substitution reaction can occur in two stereochemically different ways, called *retention of configuration* and *inversion of configuration.* When a group Y replaces a group X with **retention of configuration**, then X and Y have the same relative stereochemical positions.

Substitution with retention of configuration:

$$\qquad (7.30a)$$

Substitution with retention also implies that if X and Y have the same priorities relative to R^1, R^2, and R^3 in the *R,S* system, then the carbon that undergoes substitution will have the same configuration in the reactant and the product. Thus, if this carbon has (for example) the *S* configuration in the starting material, it has the same, or *S*, configuration in the product.

When substitution occurs with **inversion of configuration**, then X and Y have different relative stereochemical positions. Specifically, the incoming group $Y{:}^-$ must form a bond to the asymmetric carbon atom from the side *opposite* the departing group $X{:}^-$. To make room for Y and to maintain the tetrahedral configuration of carbon, the three R groups must move as shown by the green arrows:

$$\qquad (7.30b)$$

This motion very much resembles what happens in amine inversion (Fig. 6.18, p. 261).

Substitution with inversion also implies that if X and Y have the same priorities relative to R^1, R^2, and R^3 in the R,S system, then the carbon that undergoes substitution must have opposite configurations in the reactant and the product. Thus, if this carbon has (for example) the R configuration in the starting material, it has the opposite, or S, configuration in the product.

$$(7.30c)$$

As with addition, it is also possible that a reaction might occur so that both retention and inversion can occur at comparable rates in a substitution reaction. In such a case, stereoisomeric products corresponding to both pathways will be formed. Examples of substitution reactions with inversion, retention, and mixed stereochemistry are all well known.

As Eqs. 7.30a–b suggest, analysis of the stereochemistry of substitution requires that the carbon that undergoes substitution must be a stereocenter in both the reactants and the products. For example, in the following situation, the stereochemistry of substitution cannot be determined.

$$(7.31)$$

Because the carbon that undergoes substitution is not a stereocenter, the same product is obtained from both the retention and inversion modes of substitution.

A reaction in which particular stereoisomers of the product are formed in significant excess over others is said to be a **stereoselective reaction**. Thus, an addition that occurs only with anti stereochemistry, as shown in Eq. 7.27b, is a stereoselective reaction because only one diastereomer is formed to the exclusion of the other. A substitution that occurs only with inversion, as shown in Eq. 7.30b, is also a stereoselective reaction because one enantiomer of the product is formed to the exclusion of the other.

This section has established the stereochemical possibilities that might be expected in two types of reactions: additions and substitutions. The remaining sections apply these ideas in discussing the stereochemical aspects of several reactions that were first introduced in Chapter 5.

C. Stereochemistry of Bromine Addition

The addition of bromine to alkenes (Sec. 5.2A) is in many cases a highly stereoselective reaction. In this section we're going to study in detail the addition of Br_2 to *cis*- and *trans*-2-butene with two objectives in mind:

1. To see how the ideas of Sec. 7.8A are applied to acyclic compounds
2. To see how stereochemistry can be used to understand a reaction mechanism

When *cis*-2-butene reacts with Br_2, the product is 2,3-dibromobutane.

$$\text{cis-2-butene} + Br_2 \longrightarrow \textbf{2,3-dibromobutane} \tag{7.32}$$

You should now realize that three stereoisomers of this product are possible: a pair of enantiomers and the meso compound (Problem 7.25). The meso compound and the enantiomeric pair should be formed in different amounts (Sec. 7.7B) because they are diastereomers. If the enantiomers are formed, they should be formed as the racemate because the starting materials are achiral (Sec. 7.7A).

When bromine addition to *cis*-2-butene is carried out in the laboratory, the only product is the racemate. Bromine addition to *trans*-2-butene, in contrast, gives exclusively the meso compound. To summarize these results:

Experimental facts:

$$H_3C—CH{=}CH—CH_3 \xrightarrow[CH_2Cl_2]{Br_2} H_3C—CH—CH—CH_3 \tag{7.33}$$

This information indicates that addition reactions of bromine to both *cis*- and *trans*-2-butene are highly stereoselective. Are these additions syn or anti? Because the alkene is not cyclic (as it is in Eq. 7.27), the answer is not obvious. Study Problem 7.6 illustrates how to analyze the result systematically to get the answer.

STUDY PROBLEM 7.6

According to the experimental results in Eq. 7.33, is the addition of bromine to *cis*-2-butene a syn- or an anti-addition?

SOLUTION   To answer this question, you should imagine *both* syn- and anti-additions to *cis*-2-butene and see what results would be obtained for each. Comparison of these results with the experimental facts then shows us which alternative is correct.

If bromine addition were syn, the Br_2 could add to either face of the double bond. (In the following structures, we are viewing the alkene edge-on as in Eq. 7.26, p. 307.)

meso-**2,3-dibromobutane**

$$(7.34)$$

This analysis shows that syn-addition from either direction gives the meso diastereomer. Because the experimental facts (Eq. 7.33) show that *cis*-2-butene does *not* give the meso isomer, the two bromine atoms *cannot* be adding from the same face of the molecule. Therefore syn-addition does not occur.

Because bromine addition is not a syn-addition, presumably it is an anti-addition. Let's verify this. Consider the anti-addition of the two bromines to *cis*-2-butene. This addition, too, can occur in two equally probable ways.

$$2S,3S \qquad 2R,3R$$

$$\textbf{(±)-2,3-dibromobutane} \qquad (7.35)$$

This analysis shows that each mode of addition gives the enantiomer of the other; that is, the two modes of anti-addition operating at the same time should give the racemate. Because the experimental facts of Eq. 7.33 show that bromine addition to *cis*-2-butene indeed gives the racemate, this reaction is an anti-addition.

It is very important that you analyze the addition of bromine to *trans*-2-butene in a similar manner to show that this addition, too, is an anti-addition.

As suggested at the end of Study Problem 7.6, you should have demonstrated to yourself that the addition of bromine to *trans*-2-butene is also a stereoselective anti-addition. In fact, the bromine addition to most simple alkenes occurs exclusively with anti stereochemistry. Bromine addition is therefore a *stereoselective anti-addition reaction.*

The study of the stereochemistry of bromine addition to the 2-butenes raises an important philosophical point. To claim that bromine addition to the 2-butenes is an anti-addition requires that the reaction be investigated on *both* the cis and trans stereoisomers of 2-butene. It is conceivable that, in the absence of experimental evidence, anti-addition might have been observed with one stereoisomer of the 2-butenes and syn-addition with the other. Had this been the result, the bromine-addition reactions would still be highly stereoselective, but we could not have made the *more general* claim that bromine addition to the 2-butenes is an anti-addition.

Reactions such as bromine addition, in which different stereoisomers of a starting material give different stereoisomers of a product, are called **stereospecific reactions**. As the discussion in the previous paragraph demonstrates, all stereospecific reactions are stereoselective, but not all stereoselective reactions are stereospecific. To put it another way, all stereospecific reactions are a *subset* of all stereoselective reactions.

Why is bromine addition a stereospecific anti-addition? The stereospecificity of bromine addition is one of the main reasons that the bromonium-ion mechanism, shown in Eqs. 5.11–5.12 on p. 185, was postulated. Let's see how this mechanism can account for the observed results. First, the bromonium ion can form at either face of the alkene. (Reaction at one face is shown in the following equation; you should show the reaction at the other face and take your structures through the subsequent discussion.)

STUDY GUIDE
LINK 7.4
Stereoselective and
Stereospecific Reactions

$$(7.36)$$

Bromonium-ion formation as represented here is a syn-addition because both of the C—Br bonds are formed at the same face of the alkene.

If formation of the bromonium ion is a syn-addition, then the anti-addition observed in the overall reaction with bromine must be established by the stereochemistry of the reaction between the bromonium ion and the bromide ion. Suppose that the bromonium ion reacts with the bromide ion by **opposite-side substitution**. This means that the bromide ion, acting as a nucleophile, donates an electron pair to a carbon at the face opposite to the bond that breaks, which in this case is the carbon–bromine bond. *An opposite-side substitution reaction must occur with inversion of configuration* (Sec. 7.8B), because, as the substitution takes place, the methyl and the hydrogen must swing upwards (*green arrows*) to maintain the tetrahedral configuration of carbon. (Compare with Eq. 7.30b.) Reaction of the bromide ion at one carbon yields one enantiomer; reaction at the other carbon yields the other enantiomer.

(±)-2,3-dibromobutane   (7.37)

Thus, formation of a bromonium ion followed by *opposite-side substitution* of bromide is a mechanism that accounts for the observed anti-addition of Br_2 to alkenes. In general, when a nucleophile reacts at a saturated carbon atom in any substitution reaction, opposite-side substitution is observed. (Opposite-side substitution is explored further in Chapter 9.)

Might other mechanisms be consistent with the anti stereochemistry of bromine addition? Let's see what sort of prediction a carbocation mechanism makes about the stereochemistry of the reaction.

Imagine the addition of Br_2 to *cis*-2-butene to give a carbocation intermediate. (Bromine addition at only the upper face is shown below; the equally probable addition to the lower face gives the enantiomeric carbocation.) If the carbocation lasts long enough to undergo at least one internal rotation, then both diastereomers of the products would be formed even if a bromonium ion formed subsequently:

(7.38)

The reaction, then, would not be stereoselective. Because this result is not observed (Eq. 7.33), a carbocation mechanism is not in accord with the data. This mechanism also is *not* in accord with the absence of rearrangements in bromine addition. The bromonium-ion mechanism, however, accounts for the results in a direct and simple way. The credibility of this mechanism has been enhanced by the direct observation of bromonium ions under special conditions. In 1985, the structure of a bromonium ion was determined by X-ray crystallography.

Does the observation of anti stereochemistry *prove* the bromonium-ion mechanism? The answer is no. *No mechanism is ever proved.* Chemists deduce a mechanism by gathering as much information as possible about a reaction, such as its stereochemistry, the presence and absence of rearrangements, and so on, and ruling out all mechanisms that do not fit the experimental facts. If someone can think of another mechanism that explains the facts, then that mechanism is just as good until someone finds a way to decide between the two by a new experiment.

PROBLEMS

7.28   Assuming the operation of the bromonium-ion mechanism, give the structure of the product(s) (including all stereoisomers) expected from bromine addition to cyclohexene. (See Study Problem 7.5 on p. 306.)

7.29   In view of the bromonium-ion mechanism for bromine addition, which of the products in your answer to Problem 7.27 (p. 306) are likely to be the major ones?

D. Stereochemistry of Hydroboration–Oxidation

Because hydroboration–oxidation involves two distinct reactions, its stereochemical outcome is a consequence of the stereochemistry of both reactions.

Hydroboration is a stereospecific syn-addition.

(racemate)

(7.39)

Notice again the structure-drawing convention used here: Even though just one enantiomer of the product is shown, the product is racemic because the starting materials are achiral (Sec. 7.7A).

The syn-addition of borane, along with the absence of rearrangements, provides the major evidence for a concerted mechanism of the reaction.

a concerted
syn-addition

(7.40a)

Occurrence of an anti-addition by the same concerted mechanism would be virtually impossible, because it would require an abnormally long B—H bond to bridge opposite faces of the alkene π bond.

a concerted anti-addition
would require an
unrealistic B—H
bond length

(7.40b)

The oxidation of organoboranes is a stereospecific *substitution reaction* that occurs with *retention of stereochemical configuration.*

$$\text{(7.41)}$$

trans-2-methylcyclohexanol

FURTHER EXPLORATION 7.4
Stereochemistry of Organoborane Oxidation

We won't consider the mechanism of this substitution in detail here, but we can certainly conclude that it does *not* involve opposite-side nucleophilic substitution. (Why?) (The mechanism is shown in Further Exploration 7.4.)

The results from Eqs. 7.39 and 7.41 taken together show that *hydroboration–oxidation of an alkene brings about the net syn-addition of the elements of* H—OH *to the double bond.*

$$\text{(7.42)}$$

1-methylcyclohexene

(±)-trans-2-methylcyclohexanol

As far as is known, *all* hydroboration–oxidation reactions of alkenes are stereospecific *syn*-additions.

Notice carefully that the —H and —OH are added in a syn manner. The trans designation in the name of the product of Eq. 7.42 has nothing to do with the groups that have added—it refers to the relationship of the methyl group, which was part of the alkene starting material, and the —OH group. Notice again the drawing convention: only one enantiomer of each chiral molecule is drawn, but it is *understood* that the racemate of each is formed.

PROBLEMS

7.30 What products, including their stereochemistry, should be obtained when each of the following alkenes is subjected to hydroboration–oxidation? (D = deuterium = ^{2}H.)

(a) H₃C, CH₃ (b) H₃C, D
C=C   C=C
D   D   D   CH₃

7.31 Give the structure of the products formed, including their stereochemistry, when (Z)-3-methyl-2-pentene undergoes hydroboration with BH₃ in THF followed by oxidation with H₂O₂/NaOH. Tell whether any stereoisomers are formed in the same or different amounts.

E. Stereochemistry of Other Addition Reactions

Catalytic Hydrogenation Catalytic hydrogenation of most alkenes (Sec. 4.9A) is a stereospecific syn-addition. The following example is illustrative; the products are shown in eclipsed conformations for ease in seeing the stereochemical relationships.

$$\text{Ph}\overset{\text{\tiny ''}}{\underset{\text{H}_3\text{C}}{}}\text{C}=\text{C}\overset{\text{\tiny ''}\text{CH}_3}{\underset{\text{Ph}}{}} + \text{H}_2 \xrightarrow[\substack{\text{acetic acid}\\\text{(solvent)}}]{\text{Pd/C}} \text{Ph}\text{—}\text{C}\text{—}\text{C}\text{—}\text{CH}_3 \qquad \text{(7.43a)}$$

E stereoisomer   racemate

$$\text{Ph}\overset{\text{\tiny ''}}{\underset{\text{H}_3\text{C}}{}}\text{C}=\text{C}\overset{\text{Ph}}{\underset{\text{CH}_3}{}} + \text{H}_2 \xrightarrow[\substack{\text{acetic acid}\\\text{(solvent)}}]{\text{Pd/C}} \text{H}_3\text{C}\text{—}\text{C}\text{—}\text{C}\text{—}\text{CH}_3 \qquad \text{(7.43b)}$$

Z stereoisomer   meso stereoisomer

Results like these show that the two hydrogen atoms are delivered from the catalyst to the same face of the double bond. The stereospecificity of catalytic hydrogenation is one reason that the reaction is so important in organic chemistry.

Oxymercuration–Reduction   Oxymercuration of alkenes (Sec. 5.4A) is typically a stereospecific anti-addition.

$$\text{H}\overset{\text{\tiny ''}}{\underset{\text{H}_3\text{C}}{}}\text{C}=\text{C}\overset{\text{\tiny ''}\text{H}}{\underset{\text{CH}_3}{}} \xrightarrow[\text{THF}]{\text{Hg(OAc)}_2, \text{H}_2\text{O}} \overset{\text{HO}\quad\text{H}}{\text{H}\text{—}\text{C}\text{—}\text{C}\text{—}\text{CH}_3} \qquad \text{(7.44)}$$

(racemic)

(What result would you expect for the same reaction of *trans*-2-butene? See Problem 7.31.) Because this reaction occurs by a cyclic-ion mechanism (Eqs. 5.21b, p. 190) much like bromine addition, it should not be surprising that the stereochemical course of the reaction is the same. In the reaction of the mercury-containing product with NaBH$_4$, however, the stereochemical results vary from case to case. In this example, a deuterium-substituted analog, NaBD$_4$, was used to investigate the stereochemistry, and it was found that mercury is replaced by hydrogen with *loss of stereochemical configuration.*

$$\overset{\text{HO}\quad\text{H}}{\text{H}\text{—}\text{C}\text{—}\text{C}\text{—}\text{CH}_3} \xrightarrow{\text{NaBD}_4, {}^-\text{OH}} \overset{\text{HO}\quad\text{H}}{\text{H}\text{—}\text{C}\text{—}\text{C}\text{—}\text{CH}_3} + \overset{\text{HO}\quad\text{D}}{\text{H}\text{—}\text{C}\text{—}\text{C}\text{—}\text{H}} \qquad \text{(7.45)}$$

(equal amounts of each)

Hence, oxymercuration–reduction is in general *not* a stereoselective reaction. Despite its lack of stereoselectivity, the reaction is highly *regioselective* and is very useful in situations in which stereoselectivity is not an issue.

STUDY GUIDE LINK 7.5
When Stereoselectivity Matters

PROBLEMS

7.32   (a) Give the product(s) and their stereochemistry when *trans*-2-butene reacts with Hg(OAc)$_2$ and H$_2$O.

   (b) What compounds result when the products of part (a) are treated with NaBD$_4$ in aqueous NaOH? Contrast these products (including their stereochemistry) with the products of Eq. 7.45.

7.33   For which of the following alkenes would oxymercuration–reduction give (a) a single compound; (b) two diastereomers; (c) more than one constitutional isomer? Explain.

        A   B   C   D

KEY IDEAS IN CHAPTER 7

- Except for cyclopropane, the cycloalkanes have puckered carbon skeletons.

- Of the cycloalkanes containing relatively small rings, cyclohexane is the most stable because it has no angle strain and it can adopt a conformation in which all bonds are staggered.

- The most stable conformation of cyclohexane is the chair conformation. In this conformation, hydrogens or substituent groups assume axial or equatorial positions. Cyclohexane and substituted cyclohexanes rapidly undergo the chair interconversion, in which equatorial groups become axial, and vice versa. The twist-boat conformation is a less stable conformation of cyclohexane derivatives. Twist-boat conformations are interconverted through boat transition states.

- A cyclohexane conformation with an axial substituent is typically less stable than a conformation with the same substituent in an equatorial position because of unfavorable van der Waals interactions (1,3-diaxial interactions) between the axial substituent and the two axial hydrogens on the same face of the ring. The 1,3-diaxial interaction of an axial methyl group and an axial ring hydrogen is very similar to the interaction of the two methyl groups in *gauche*-butane.

- Cyclopentane exists in an envelope conformation. Cyclopentane has a greater heat of formation per CH_2 than cyclohexane because of eclipsing between hydrogen atoms.

- Cyclobutane and cyclopropane contain significant angle strain because their bonds are forced to deviate significantly from the ideal tetrahedral angle. Cyclopropane has bent carbon–carbon bonds. Cyclobutane and cyclopropane are the least stable cycloalkanes.

- Cycloalkanes can be represented by planar polygons in which the stereochemistry of substituents is indicated by dashed or solid wedges. Planar line-and-wedge structures show the configurations of carbon stereocenters but contain no conformational information.

- Bicyclic compounds contain two rings joined at two common atoms, called bridgehead atoms. If the bridgehead atoms are adjacent, the compound is a fused bicyclic compound; if the bridgehead atoms are not adjacent, the compound is a bridged bicyclic compound. Either cis or trans ring fusion is possible. Trans fusion, which avoids 1,3-diaxial interactions, is the most stable way to connect larger rings; cis fusion, which minimizes angle strain, is the most stable way to connect smaller rings. Polycyclic compounds contain many fused or bridged rings (or both).

- Cycloalkenes with trans double bonds within rings containing fewer than eight members are too unstable to exist under normal circumstances.

- Bicyclic compounds consisting of small rings containing bridgehead double bonds are also unstable (Bredt's rule) because such compounds incorporate a highly twisted double bond.

- The following fundamental principles govern reactions involving stereoisomers:

  1. A pair of enantiomers have identical reactivities unless the reaction conditions cause them to be involved in diastereomeric interactions (for example, a chiral catalyst, a chiral solvent, and so on).

  2. Chiral products are always formed as racemates in a chemical reaction involving achiral starting materials, unless the reaction conditions create diastereomeric interactions (for example, a chiral catalyst, a chiral solvent, and so forth).

  3. Diastereomers in general have different reactivities.

  4. Diastereomeric products of chemical reactions are formed at different rates and in unequal amounts.

- Addition reactions can occur with syn or anti stereochemistry. Substitution reactions can occur with retention or inversion of configuration. The stereochemistry of a reaction is determined by comparing the stereochemistry of the reactants and the products. Each carbon at which a chemical change occurs must be a stereocenter in the product in order for the stereochemistry of the reaction to be determined.

- In a stereoselective reaction, some stereoisomers of the product are formed in large excess over others. A stereospecific reaction is a highly stereoselective reaction in which each stereoisomer of the reactant gives a different stereoisomer of the product. All stereospecific reactions are stereoselective, but not all stereoselective reactions are stereospecific.

- Bromine addition to simple alkenes is a stereospecific anti-addition in which the syn-addition of one bromine to give a bromonium ion is followed by the nucleophilic reaction of bromide ion at a carbon of the bromonium ion with inversion of configuration.

• The hydroboration of alkenes is a stereospecific syn-addition, and the subsequent oxidation of organoboranes is a substitution that occurs stereospecifically with retention of configuration. Thus, hydroboration–oxidation of alkenes is an overall stereospecific syn-addition of the elements of H—OH to alkenes.

• Catalytic hydrogenation is a stereospecific syn-addition. Oxymercuration–reduction is not always stereoselective (and therefore not stereospecific), because the replacement of mercury with hydrogen can occur with mixed stereochemistry.

ADDITIONAL PROBLEMS

7.34 Draw the structures of the following compounds.
    (a) a bicyclic alkane with six carbon atoms
    (b) (*S*)-4-cyclobutylcyclohexene
Name the compound whose structure you drew in part (a).

7.35 Which of the following would distinguish (in principle) between methylcyclohexane and (*E*)-4-methyl-2-hexene? Explain your reasoning.
    (a) molecular mass determination
    (b) uptake of H_2 in the presence of a catalyst
    (c) reaction with Br_2
    (d) determination of the molecular formula
    (e) determination of the heat of formation
    (f) enantiomeric resolution

7.36 State whether you would expect each of the following properties to be identical or different for the two enantiomers of 2-pentanol. Explain.

2-pentanol

    (a) boiling point   (b) optical rotation
    (c) solubility in hexane   (d) density
    (e) solubility in (*S*)-3-methylhexane
    (f) dipole moment
    (g) taste (*Hint:* Your taste buds are chiral.)

7.37 Draw the structure of each of the following molecules after it undergoes the chair interconversion.

(*Hint:* A chair interconversion of one ring requires a simultaneous chair interconversion of the other.)

7.38 Draw a structure for each of the following compounds in its more stable chair conformation. Explain your choice.

(a)
(b)

7.39 (a) Chlorocyclohexane contains 2.07 times more of the equatorial form than the axial form at equilibrium at 25 °C. What is the standard free-energy difference between the two forms? Which is more stable?
    (b) The standard free-energy difference between the two chair conformations of isopropylcyclohexane is 9.2 kJ mol^{-1} (2.2 kcal mol^{-1}). What is the ratio of concentrations of the two conformations at 25 °C?

7.40 Suggest a reason that the energy difference between the two chair conformations of ethylcyclohexane is about the same as that for methylcyclohexane, even though the ethyl group is larger than the methyl group.

7.41 Which of the following alcohols can be synthesized relatively free of constitutional isomers and diastereomers by (a) hydroboration–oxidation; (b) oxymercuration–reduction? Explain.

7.42 For each of the following reactions, provide the following information.

(a) Give the structures of all products (including stereoisomers).

(b) If more than one product is formed, give the stereochemical relationship (if any) of each pair of products.

(c) If more than one product is formed, indicate which products are formed in identical amounts and which in different amounts.

(d) If more than one product is formed, indicate which products are expected to have different physical properties (melting point or boiling point).

(1) (R)-CH₃CH₂CH—C=CH₂ $\xrightarrow[\text{THF}]{\text{BH}_3}$ $\xrightarrow[\text{NaOH}]{\text{H}_2\text{O}_2}$
         | |
        CH₃   CH₃

(2) CH₃CH₂CH₂C=CH₂ + HBr ⟶
              |
             CH₂CH₃

(3) CH₃CH₂CH=CH₂ + Br₂ $\xrightarrow{\text{CH}_2\text{Cl}_2}$

(4) (±)-CH₃CHCH=CH₂ + Br₂ $\xrightarrow{\text{CH}_2\text{Cl}_2}$
         |
        Ph

(5)

   + H₂ $\xrightarrow{\text{Pd/C}}$

(6)

   + D₂ $\xrightarrow{\text{Pd/C}}$

7.43 Draw the structures of the following compounds. (Some parts may have more than one correct answer.)

(a) an achiral tetramethylcyclohexane for which the chair interconversion results in identical molecules

(b) an achiral trimethylcyclohexane with two chair forms that are conformational diastereomers.

(c) a chiral trimethylcyclohexane with two chair forms that are conformational diastereomers

(d) a tetramethylcyclohexane with chair forms that are conformational enantiomers

7.44 Draw a conformational representation of the following steroid. Show the α- and β-faces of the steroid, and label the angular methyl groups.

7.45 Draw the two chair conformations of the sugar α-(+)-glucopyranose, which is one form of the sugar

glucose. Which of these two conformations is the major one at equilibrium? Explain.

α-(+)-glucopyranose

7.46 From your knowledge of the mechanism of bromine addition to alkenes, give the structure and stereochemistry of the product(s) expected in each of the following reactions.

(a) addition of Br₂ to (3R,5R)-3,5-dimethylcyclopentene

(b) reaction of cyclopentene with Br₂ in the presence of H₂O (*Hint:* See Sec. 5.2B.)

7.47 Anti-addition of bromine to the following bicyclic alkene gives two separable dibromides. Suggest structures for each. (Remember that *trans*-decalin derivatives cannot undergo the chair interconversion.)

7.48 When 1,4-cyclohexadiene reacts with two equivalents of Br₂, two separable compounds with different melting points are formed. Account for this observation.

7.49 An optically active compound X with molecular formula C₈H₁₄ undergoes catalytic hydrogenation to give an optically inactive product. Which of the following structures for X is (are) consistent with all of the data?

A   B   C

D   E

7.50 Draw a chair conformation for (S)-3-methylpiperidine showing the sp^3 orbital that contains the nitrogen unshared electron pair. How many chair conformations of this compound are in rapid equilibrium? (*Hint:* See Sec. 6.9B.)

(S)-3-methylpiperidine

7.51 Which of the following compounds can be resolved into enantiomers at room temperature? Explain.

(a)

(b)

(c)

(d)

7.52 Explain why 1-methylaziridine undergoes amine inversion much more slowly than 1-methylpyrrolidine. (*Hint:* What are the hybridization and bond angles at nitrogen in the transition state for inversion?)

1-methylaziridine   **1-methylpyrrolidine**

7.53 Alkaline potassium permanganate (KMnO₄) can be used to bring about the addition of two —OH groups to an alkene double bond. This reaction has been shown in several cases to be a stereospecific syn-addition. Given the stereochemistry of the product shown in Fig. P7.53, what stereoisomer of alkene A was used in the reaction? Explain.

7.54 (a) When fumarate reacts with D₂O in the presence of the enzyme *fumarase* (Secs. 4.9C and 7.7A), only one stereoisomer of deuterated malate is formed, as shown in Fig. P7.54. Is this a syn- or an anti-addition? Explain.

(b) Why is the use of D₂O instead of H₂O necessary to establish the stereochemistry of this addition?

7.55 Give the structure and stereochemistry of all products formed in each of the following reactions. Tell whether

stereoisomeric products are formed in the same or different amounts.

(a) *trans*-2-pentene + Br₂ $\longrightarrow$

(b) *trans*-3-hexene + Br₂ + H₂O $\longrightarrow$
                    excess
                    (solvent)

(c) *cis*-3-hexene + D₂ $\xrightarrow{\text{Pt/C}}$

(d) *cis*-3-hexene + BD₃ $\xrightarrow[\substack{\text{THF} \\ \text{(solvent)}}]{}$ $\xrightarrow{\text{H}_2\text{O}_2/^-\text{OH}}$

7.56 By answering the following questions, indicate the relationship between the two structures in each of the pairs in Fig. P7.56 on p. 320. Are they chair conformations of the same molecule? If so, are they conformational diastereomers, conformational enantiomers, or identical? If they are not conformations of the same molecule, what is their stereochemical relationship? (*Hint:* Use planar structures to help you.)

7.57 When 1-methylcyclohexene undergoes hydration in D₂O, the product is a mixture of diastereomers; the hydration is thus *not* a stereoselective reaction. (See Fig. P7.57 on p. 320.)

(a) Show why the accepted mechanism for this reaction is consistent with these stereochemical results.

(b) Why must D₂O (rather than H₂O) be used to investigate the stereoselectivity of this addition?

(c) What isotopic substitution could be made in the starting material, 1-methylcyclohexene, that would allow investigation of the stereoselectivity of this addition with H₂O?

7.58 Consider the following compound.

1,2,3,4,5,6-hexachlorocyclohexane

$$CH_3(CH_2)_7CH{=}CH(CH_2)_7CH_3 \xrightarrow{\text{KMnO}_4/\text{OH}^-} \overset{\quad OH \quad\ OH}{CH_3(CH_2)_7\overset{|}{C}H{-}\overset{|}{C}H(CH_2)_7CH_3}$$

                A   meso stereoisomer

Figure P7.53

$$D_2O + \underset{\substack{\ \\ ^-O_2C \qquad\ H}}{\overset{\substack{H \qquad\ CO_2^- \\ \ }}{C{=}C}} \underset{37\,°C}{\overset{\text{fumarase}}{\rightleftharpoons}} \ ^-O_2C{-}\overset{OD}{\underset{2}{\overset{|}{C}H}}{-}\overset{D}{\underset{3}{\overset{|}{C}H}}{-}CO_2^-$$

                    fumarate   **(2S,3R)-malate-3-d**

Figure P7.54

(a) Of the nine stereoisomers of this compound, only two can be isolated in optically active form under ordinary conditions. Give the structures of these enantiomers.

(b) Give the structures of the two stereoisomers for which the chair interconversion results in identical molecules.

7.59 Give the structure of every stereoisomer of 1,2,3-trimethyl-cyclohexane. Label the enantiomeric pairs and show the plane of symmetry in each achiral stereoisomer.

7.60 Which of the following statements about *cis*- and *trans*-decalin (Sec. 7.6B) are true? Explain your answers.

(a) They are different conformations of the same molecule.

(b) They are constitutional isomers.

(c) They are diastereomers.

(d) At least one chemical bond would have to be broken to convert one into the other.

(e) They are enantiomers.

(f) They interconvert rapidly.

7.61 The following compound is the sex attractant of the female fruit fly. (This is an example of a *pheromone*; Sec. 14.9.) Females secrete this compound to attract males when they are ready for mating.

olean

Which of the following statements are true about this compound? Justify your choice(s).

(a) This compound is achiral and contains a stereocenter.

(b) This compound is achiral and contains no stereocenter.

(c) This compound is chiral and contains an asymmetric atom.

(d) This compound is chiral and contains a stereocenter.

(e) This compound is chiral and contains no stereocenter.

7.62 Which of the statements (a)–(e) in Problem 7.61 are true about the following compound? Justify your choice(s).

Figure P7.56

both compounds are racemates

Figure P7.57

7.63 Identify the stereocenters (if any) in each of the following structures, and tell whether each structure is chiral.

(a)

(b)

(c)

(d)

7.64 *One* of the stereoisomers given in Fig. P7.64 exists with one of its cyclohexane rings in a twist-boat conformation. Which is it? Explain.

7.65 It has been argued that the energy difference between *cis*- and *trans*-1,3-di-*tert*-butylcyclohexane is a good approximation for the energy difference between the chair and twist-boat forms of cyclohexane. Using models to assist you, explain why this view is reasonable.

7.66 Rank the compounds given in Fig. P7.66 according to their heats of formation, lowest first, and estimate the $\Delta H°$ difference between each pair.

7.67 Rank the compounds within each of the sets shown in Fig. P7.67 on p. 322 according to their heats of formation, lowest first. Explain.

7.68 (a) What stereoisomeric products could be formed when borolane is used to hydroborate *cis*-2-butene?

borolane

(b) Are the products in (a) formed in the same or different amounts? Explain how you know.

Figure P7.64

Figure P7.66

(c) What stereoisomeric products could be formed when enantiomerically pure (2*R*,5*R*)-2,5-dimethylborolane is used to hydroborate *cis*-2-butene?

$$H_3C\text{---}\underset{\underset{H}{|}}{B}\text{---}CH_3$$

(2*R*,5*R*)-2,5-dimethylborolane

(d) The products in (c) are formed in different amounts. Why?

(e) In fact, only one of the products in (c) is formed in significant amount. Which one? (*Hint:* Build models of the borane and the alkene. Let the borane model approach the alkene model from one face of the π bond, then the other. Decide which reaction is preferred by analyzing van der Waals repulsions in the transition state in each case.)

(f) When the product borane determined in part (e) is treated with alkaline H_2O_2, mostly a single enantiomer of the product alcohol is formed. What is the absolute configuration of this alcohol (*R* or *S*)?

7.69 The peroxide-initiated addition of HBr to 1-methylcyclohexene could give two diastereomeric products, as shown in Eq. 7.24, p. 305. Investigation of the stereochemistry of this reaction shows that it is an anti-addition.

(a) Which of the two possible diastereomers is the predominant one?

(b) Remember that this is a free-radical chain reaction, not an ionic reaction. Within the context of a free-radical chain mechanism, suggest a reason that the reaction occurs with anti stereochemistry.

7.70 (a) What two diastereomeric products could be formed in the hydroboration–oxidation of the following alkene?

(b) Considering the effect of the methyl group on the approach of the borane–THF reagent to the double bond, suggest which of the two products you obtained in part (a) should be the major product.

7.71 Propose a structure for the product *X* of the following reaction and give a mechanism for its formation. Pay particular attention to the stereochemistry of each step. (*Hint:* Draw the conformation of the starting material.)

7.72 (a) The $\Delta G°$ for the equilibrium between *A* and *B* shown in Fig. P7.72a is 8.4 kJ mol^{-1} (2.0 kcal mol^{-1}). (Conformation *A* has lower energy.) Use this information to estimate the energy cost of a 1,3-diaxial interaction between two methyl groups:

methyl–methyl.1,3-diaxial interaction

(b) Using the result in part (a), estimate the $\Delta G°$ for the equilibrium between *C* and *D* given in Fig. P7.72b.

7.73 The $\Delta G°$ for the equilibrium in Fig. P7.73a is 4.73 kJ mol^{-1} (1.13 kcal mol^{-1}). (The equilibrium favors conformation *A*.)

(a) Which behaves as if it is larger, methyl or phenyl (Ph)? Why is this reasonable?

(b) Use the $\Delta G°$ given above, along with any other appropriate data, to estimate the $\Delta G°$ for the two equilibria in Fig. P7.73b.

7.74 (a) The following two tricyclic compounds are examples of *propellanes* (propeller-shaped molecules). What is

Figure P7.67

the relationship between these two molecules (identical, enantiomers, diastereomers)? Tell how you know.

    A   *B*

(b) The chemist who prepared these compounds wrote that they are *E,Z* isomers. Do you agree or disagree? Explain.

7.75 (a) In how many stereochemically different ways can the two rings in a *bridged* bicyclic compound be joined?

(b) For which one of the following bridged bicyclic compounds are all such stereoisomers likely to be stable enough to isolate? Explain.

  (1) bicyclo[2.2.2]octane

  (2) bicyclo[25.25.25]heptaheptacontane

  (A heptaheptacontane has 77 carbons.)

Figure P7.72

Figure P7.73

Noncovalent Intermolecular Interactions

This chapter contains material on two different topics. The first is *organic nomenclature*. In Sections 2.4 and 4.2 we introduced the principles of nomenclature. In this chapter, we're going to learn the more general rules of nomenclature that will apply throughout the remainder of this text. We're also going to introduce several different types of organic compounds—*alcohols*, *thiols*, *alkyl halides*, *ethers*, and *sulfides*—which are considered together because their chemical reactions are related, as we'll see in Chapters 9–11, and because many of these compound types are used as examples in the second part of this chapter.

The second topic, and the major part this chapter, is *noncovalent intermolecular interactions*: the ways that molecules can interact without forming chemical bonds. Such interactions are fundamental to an understanding of many chemical and biological phenomena.

8.1 DEFINITIONS AND CLASSIFICATION OF ALKYL HALIDES, ALCOHOLS, THIOLS, ETHERS, AND SULFIDES

An **alkyl halide** is a compound in which a halogen (—F, —Br, —Cl, or —I) is bound to the carbon of an alkyl group.

CH₃CH₂—Br

an alkyl bromide

an alkyl fluoride

an alkyl chloride

alkyl halides

An alcohol is a compound in which a **hydroxy group** (—OH) is bound to the carbon of an alkyl group, and a **thiol** is a compound in which a **mercapto group**, or **sulfhydryl group** (—SH) is bound to the carbon of an alkyl group. Thiols are sometimes called **mercaptans**.

CH_3CH_2—OH ⬠—OH

SH
(structure)

$(CH_3)_3C$—SH

|—————— alcohols ——————| |———— thiols ————|
(sometimes called *mercaptans*)

Generally, the term *thio* is used to mean "sulfur in place of oxygen." For example, a thiol is an alcohol in which a sulfur has been substituted for oxygen. The term (from the Greek word *theio*, meaning "brimstone") originates from the observation that many volatile sulfur-containing organic compounds have very unpleasant odors, in some cases like the odor of burning rubber. (The odor-causing compound of an angry skunk is a thiol.) The name *mercaptan* is derived from the fact that thiols form very stable derivatives with mercury (and other heavy metals)—that is, a mercaptan "captures mercury."

Don't confuse an *alcohol* with a *phenol* or an *enol*. In a **phenol**, the —OH group is bound to the carbon of an *aryl group* (Sec. 2.8B).

the OH is bound to a carbon of a benzene ring (an aryl group)

the OH is bound to an sp^3-hybridized carbon

H_3C— (benzene ring) —OH

(benzene ring)—CH_2—OH

a phenol

an alcohol

In an **enol**, the —OH group is bound to a carbon that is part of a double bond. In an alcohol, the —OH group is bound to an sp^3-hybridized carbon.

the OH is bound to the carbon of a double bond

the OH is bound to an sp^3-hybridized carbon

OH
|
H_2C=C—CH_3

H
|
H_2C=C—CH_2—OH

an enol

an alcohol

Specifically, in an alcohol, the carbon to which the OH is bound is *not* part of a double or triple bond. Phenols, enols, and alcohols have very different properties; we'll learn about phenols and enols in later chapters.

Ethers are compounds in which an oxygen is bound to two carbon groups, which can be alkyl or aryl. **Sulfides**, which are also called **thioethers**, are the sulfur analogs of ethers.

CH_3CH_2—O—$CH(CH_3)_2$ (benzene ring)—O—CH_3 H_3C—S—CH_2CH_3

a sulfide
(a thioether)

|————————— ethers —————————|

The carbon bonded to the halogen in an alkyl halide, or to the oxygen in an alcohol or ether, is called the **alpha-carbon**, usually written with the Greek letter as **α-carbon**.

$$\underset{\underset{CH_3}{|}}{\overset{\overset{CH_3}{|}}{H_3C-C-Br}} \qquad \underset{\underset{H}{|}}{\overset{\overset{OH}{|}}{H_3C-C-CH_2CH_3}}$$

α-carbon   α-carbon

Alkyl halides and alcohols are classified by the *number of alkyl groups attached to the α-carbon*. A methyl halide or methyl alcohol has no alkyl groups; a **primary** halide or alcohol has one alkyl group; a **secondary** halide or alcohol has two alkyl groups; and a **tertiary** halide or alcohol has three alkyl groups. In the following examples, the alkyl substituents are shown in blue and the α-carbon in red.

$$H_3C-Br \qquad \overset{\overset{I}{|}}{H_3C-CH_2} \qquad \underset{\underset{CH_2CH_3}{|}}{\overset{\overset{Cl}{|}}{H_3C-CH}} \qquad \underset{\underset{CH(CH_3)_2}{|}}{\overset{\overset{CH_2CH_3}{|}}{H_3C-C-Br}}$$

methyl bromide   a primary alkyl iodide   a secondary alkyl chloride   a tertiary alkyl bromide

$$H_3C-OH \qquad \overset{\overset{OH}{|}}{H_3C-CH_2} \qquad \underset{\underset{CH_2CH_3}{|}}{\overset{\overset{OH}{|}}{H_3C-CH}} \qquad \underset{\underset{CH(CH_3)_2}{|}}{\overset{\overset{CH_2CH_3}{|}}{H_3C-C-OH}}$$

methyl alcohol   a primary alcohol   a secondary alcohol   a tertiary alcohol

---

| **8.2** | ## NOMENCLATURE OF ALKYL HALIDES, ALCOHOLS, THIOLS, ETHERS, AND SULFIDES |

Several systems are recognized by the IUPAC for the nomenclature of organic compounds. **Substitutive nomenclature**, the most broadly applicable system, was introduced in the nomenclature of both alkanes (Sec. 2.4C) and alkenes (Sec. 4.2A), and will be applied to the compound classes in this chapter as well. Another widely used system that will be introduced in this chapter is called **radicofunctional nomenclature** by the IUPAC; for simplicity, this system will be called **common nomenclature**. Common nomenclature is generally used only for the simplest and most common compounds. Although the adoption of a single nomenclature system might seem desirable, historical usage and other factors have dictated the use of both common and substitutive names.

A. Nomenclature of Alkyl Halides

Common Nomenclature   The common name of an alkyl halide is constructed from the name of the alkyl group (see Table 2.2, p. 62) followed by the name of the halide as a separate word.

STUDY GUIDE LINK 8.1 Common Nomenclature and the *n*-Prefix

CH_3CH_2-Cl

ethyl chloride

CH_2Cl_2

methylene chloride
($-CH_2-$ group = methylene group)

$CH_3CH_2CH_2CH_2-Br$

butyl bromide

$(CH_3)_2CH-I$

isopropyl iodide

The common names of the following compounds should be learned.

$H_2C=CH-CH_2-Cl$ $Ph-CH_2-Br$ $H_2C=CH-Cl$ CCl_4

allyl chloride   **benzyl bromide**   **vinyl chloride**   **carbon tetrachloride**

(Compounds with halogens attached to alkene carbons, such as vinyl chloride, are not alkyl halides, but it is convenient to discuss their nomenclature here.)

The **allyl group**, as the structure of allyl chloride implies, is the $H_2C=CH-CH_2-$ group. This should not be confused with the **vinyl group**, $H_2C=CH-$, which lacks the additional $-CH_2-$. Similarly, the **benzyl group**, $Ph-CH_2-$, should not be confused with the **phenyl group**.

phenyl group   **benzyl group**

The **haloforms** are the methyl trihalides. Chloroform is a commonly used organic solvent.

$HCCl_3$ $HCBr_3$ HCI_3

chloroform   **bromoform**   **iodoform**

Substitutive Nomenclature   The IUPAC substitutive name of an alkyl halide is constructed by applying the rules of alkane and alkene nomenclature (Secs. 2.4C and 4.2A). Halogens are always treated as substituents; the halogen substituents are named fluoro, chloro, bromo, or iodo. Double bonds have precedence in numbering just as they do in alkenes.

CH_3CH_2-Cl 〈 〉$-Br$ $H_3C-\overset{\overset{\displaystyle F}{|}}{C}H-CH_2CH_2CH_3$

chloroethane   **2-fluoropentane**

bromocyclohexane

$H_3C-\overset{\overset{\displaystyle }{|}}{C}H-\overset{\overset{\displaystyle }{|}}{C}H-CH_2CH_2CH_3$
                Cl   CH_3

2-chloro-3-methylhexane

(E)-5-chloro-2-pentene   **3-ethyl-4-iodohexane**

PROBLEMS

8.1   Give the common name for each of the following compounds, and tell whether each is a primary, secondary, or tertiary alkyl halide.

(a) $(CH_3)_2CHCH_2-F$ (b) $CH_3CH_2CH_2CH_2CH_2CH_2-I$

(c) $-Br$ (d) $H_3C-\overset{\overset{\displaystyle CH_3}{|}}{\underset{\underset{\displaystyle CH_3}{|}}{C}}-CH_2Cl$

continued

continued ───

8.2   Give the structure of each of the following compounds.

(a) 2,2-dichloro-5-methylhexane

(b) chlorocyclopropane

(c) 6-bromo-1-chloro-3-methylcyclohexene

(d) methylene iodide

8.3   Give the substitutive name for each of the following compounds.

(a)
$$CH_3$$ ... Cl, Br (cyclopropene structure)

(b) H_3C, Cl, $C=C$, H, CH_2CH_3

(c) $H_3C-CH-CH-CCl_3$ with Br and F substituents

(d) chloroform   (e) neopentyl bromide (see Table 2.2, p. 62)

(f)   Br ... Br (cyclobutane structure)

(g) Cl CH_3 ... $CH(CH_3)_2$ (cyclohexane structure)

───

B. Nomenclature of Alcohols and Thiols. The Principal Group

Common Nomenclature   The common name of an alcohol is derived by specifying the alkyl group to which the —OH group is attached, followed by the separate word *alcohol*.

$$H_3C-OH$$
methyl alcohol

$$(CH_3)_2CH-OH$$
isopropyl alcohol

⬡—OH
cyclohexyl alcohol

$$CH_3CH_2CH_2-OH$$
propyl alcohol

$$H_2C=CH-CH_2-OH$$
allyl alcohol

$$Ph-CH_2-OH$$
benzyl alcohol

Compounds that contain two or more hydroxy groups on different carbons are called **glycols**. The simplest glycol is ethylene glycol, the main component of automotive antifreeze. A few other glycols also have widely used traditional names.

$$HO-CH_2CH_2-OH$$
ethylene glycol

$$CH_2-CH-CH_3$$
OH   OH
propylene glycol

$$CH_2-CH-CH_2$$
OH   OH   OH
glycerol (glycerin)

Thiols are named in the common system as *mercaptans*.

$$CH_3CH_2-SH$$

ethyl mercaptan

Substitutive Nomenclature   The substitutive nomenclature of alcohols and thiols involves a concept called the *principal group. This is a very important nomenclature concept* that will be used repeatedly. The **principal group** is the chemical group on which the name is based, *and it is always cited as a suffix in the name.* For example, in a simple alcohol, the —OH group is the principal group, and its suffix is *ol*. The name of an alcohol is constructed by dropping the final *e* from the name of the parent alkane and adding this suffix.

$$CH_3CH_2—OH$$

ethane + *ol* = **ethanol**

The final *e* is generally dropped when the suffix begins with a vowel; otherwise, it is retained.

For simple thiols, the —SH group is the principal group, and its suffix is *thiol*. The name is constructed by adding this suffix to the name of the parent alkane. Note that because the suffix begins with a consonant, the final *e* of the alkane name is retained.

$$CH_3CH_2—SH$$

ethane + *thiol* = **ethanethiol**

Only certain groups are cited as principal groups. The —OH and —SH groups are the only ones in the compound classes considered so far, but others will be added in later chapters. If a compound does not contain a principal group, it is named as a substituted hydrocarbon in the manner illustrated for the alkyl halides in Sec. 8.2A.

The *principal group* and the *principal chain* are the key concepts defined and used in the construction of a substitutive name according to the *general rules for substitutive nomenclature of organic compounds*, which follow. The simplest way to learn these rules is to read through the rules briefly and then concentrate on the study problems and examples that follow, letting them guide you through the application of the rules in specific cases.

1. *Identify the principal group.*

   When a structure has several candidates for the principal group, the group chosen is the one given the highest priority by the IUPAC. The IUPAC specifies that the —OH group receives precedence over the —SH group:

Priority as principal group: —OH > —SH   (8.1)

   A complete list of principal groups and their relative priorities are summarized in Appendix I. (If there is no principal group, follow rule 4b below.)

2. *Identify the principal carbon chain.*

   The **principal chain** is the carbon chain on which the name is based (Sec. 2.4C). The principal chain is identified by applying the following criteria *in order* until a decision can be made:
   a. the chain with the greatest number of principal groups;
   b. the chain with the greatest number of double and triple bonds;
   c. the chain of greatest length;
   d. the chain with the greatest number of other substituents.

   These criteria cover most of the cases you'll encounter.

3. *Number the carbons of the principal chain consecutively from one end.*

   In numbering the principal chain, apply the following criteria *in order* until there is no ambiguity:
   a. the lowest numbers for the principal groups;
   b. the lowest numbers for multiple bonds, with double bonds having priority over triple bonds in case of ambiguity;
   c. the lowest numbers for other substituents;
   d. the lowest number for the substituent cited first in the name.

4. *Begin construction of the name with the name of the hydrocarbon corresponding to the principal chain.*
   a. Cite the principal group by its suffix and number; its number is the last one cited in the name. (See the examples in Study Problem 8.1.)
   b. If there is no principal group, name the compound as a substituted hydrocarbon. (See Secs. 2.4C and 4.2A.)
   c. Cite the names and numbers of the other substituents in alphabetical order at the beginning of the name.

Provide an IUPAC substitutive name for each of the following compounds.

(a) CH$_3$CH$_2$CHCH$_3$ (b) CH$_3$CHCH═CHCHCH$_3$
         | | |
        OH   OH   CH$_2$CH$_2$SH

SOLUTION   (a) From rule 1, the principal group is the —OH group. Because there is only one possibility for the principal chain, rule 2 does not enter the picture. By applying rule 3a, we decide that the principal group is located at carbon-2. From rule 4a, the name is based on the four-carbon hydrocarbon, butane. After dropping the final *e* and adding the suffix *ol*, the name is obtained: 2-butanol.

$$\overset{4}{C}H_3\overset{3}{C}H_2\overset{2}{C}H\overset{1}{C}H_3$$
|
OH

2-butanol

(b) From rule 1, the principal group is again the —OH group, because —OH has precedence over —SH. From rules 2a–2c, the principal chain is the longest one containing both the —OH group and the double bond, and therefore it has seven carbons. Numbering the principal chain in accord with rule 3a gives the —OH group the lowest number at carbon-2 and a double bond at carbon-3:

$$\overset{1}{C}H_3\overset{2}{C}H\overset{3}{C}H═\overset{4}{C}H\overset{5}{C}HCH_3$$

principal group ⟶ OH   CH$_2$CH$_2$SH
                           6   7 ◄——— principal chain numbering

By applying rule 4a, we decide that the parent hydrocarbon is 3-heptene, from which we drop the final *e* and add the suffix *ol*, to give 3-hepten-2-ol as the final part of the name. (Notice that because we have to cite the number of the double bond, the number for the —OH principal group is located before the final suffix *ol*.) Rule 4c requires that the methyl group at carbon-5 and the —SH group at carbon-7 be cited as ordinary substituents. (The substituent name of the —SH group is the *mercapto* group.) The name is

substituent numbers; note the alphabetical
citation of substituents

7-mercapto-5-methyl-3-hepten-2-ol

                        number of the principal group
number of the double bond

To name an alcohol containing more than one —OH group, the suffixes *diol*, *triol*, and so on are added to the name of the appropriate alkane *without* dropping the final *e*.

2,3-pentanediol

Name the following compound.

SH
|
/\—OH
|
OH

SOLUTION   From rule 1, the principal groups are the —OH groups. By rule 3a, these groups are given numerical precedence; thus, they receive the numbers 1 and 3. Because two numbering schemes give these groups the numbers 1 and 3, we choose the scheme that gives the double bond the lower number, by rule 3b. From rule 4a, the parent hydrocarbon is cyclohexene, and because the suffix is *diol*, the final *e* is retained to give the partial name 4-cyclohexene-1,3-diol. Finally, the —SH group has been eliminated from consideration as the principal group, so it is treated as an ordinary substituent group by rule 4c. The completed name is thus

6-mercapto-4-cyclohexene-1,3-diol

In a sidebar on p. 138 we introduced the 1993 IUPAC nomenclature recommendations. Although we are continuing to use the 1979 recommendations for the reasons given there, conversion of most names to the 1993 recommendations is not difficult. The handling of double bonds, triple bonds, and principal groups is the major change introduced by the 1993 recommendations. In the 1993 system, the number of the double bond, triple bond, or principal group immediately precedes the citation of the group in the name. Thus, in the 1993 convention, the name of the compound in Study Problem 8.1(a) would be butan-2-ol rather than 2-butanol. The name of the compound in Study Problem 8.1(b) would be 7-mercapto-5-methylhept-3-en-2-ol. The name 2,3-butanediol would be changed to butane-2,3-diol, and the name of the compound in Study Problem 8.2 would become 6-mercaptocyclohex-4-ene-1,3-diol. As in the 1979 nomenclature, the final *e* of the hydrocarbon name is dropped when the suffix begins in a vowel.

Common and substitutive nomenclature should not be mixed. This rule is frequently disregarded in naming the following compounds:

common: *tert*-butyl alcohol
substitutive: 2-methyl-2-propanol
incorrect: *t*-butanol or *tert*-butanol

common: isopropyl alcohol
substitutive: 2-propanol
incorrect: isopropanol

PROBLEMS

8.4   Draw the structure of each of the following compounds.

(a) *sec*-butyl alcohol   (b) 3-ethylcyclopentanethiol   (c) 3-methyl-2-pentanol

(d) (*E*)-6-chloro-4-hepten-2-ol   (e) 2-cyclohexenol

8.5   Give the substitutive name for each of the following compounds.

(a) $CH_3CH_2CH_2CH_2OH$ (b) $CH_3CHCH_2CH_2OH$
                                             |
                                            Br

continued

continued

(f)   OH   (g) $CH_3CH_2CH_2CHCH_2SH$ (h)

$\quad\quad\quad\quad\quad\quad\quad\quad\quad\quad\quad\quad\quad\quad$ OH

$H_3C-\overset{\overset{\displaystyle CH_3}{|}}{\underset{\underset{\displaystyle SH}{|}}{C}}-CH_3$

C. Nomenclature of Ethers and Sulfides

Common Nomenclature   The common name of an ether is constructed by citing as separate words the two groups attached to the ether oxygen in alphabetical order, followed by the word *ether*.

$$CH_3CH_2-O-CH_2CH_3 \quad\quad H_3C-O-C_2H_5$$

diethyl ether$\quad\quad\quad\quad\quad$**ethyl methyl ether**
(also called **ethyl ether** or
simply **ether**)

A sulfide is named in a similar manner, using the word *sulfide*. (In older literature, the word *thioether* was also used.)

$$CH_3CH_2-S-CH_3 \quad\quad (CH_3)_2CH-S-CH(CH_3)_2$$

ethyl methyl sulfide$\quad\quad\quad$**diisopropyl sulfide**
(also **ethyl methyl thioether**)

Substitutive Nomenclature   In substitutive nomenclature, ethers and sulfides are never cited as principal groups. *Alkoxy groups* (RO—) and *alkylthio groups* (RS—) are always cited as substituents.

ethoxy substituent ⟶ $CH_3CH_2O \quad\quad\quad CH_3$

$\quad\quad\quad\quad\quad\quad\quad\quad\quad\quad\quad\quad\quad\quad\quad\quad\quad |\quad\quad\quad\quad\quad\quad |$

principal chain ⟶ $CH_3CHCH_2CH_2CHCH_3$

2-ethoxy-5-methylhexane

In this example, the principal chain is a six-carbon chain. Hence, the compound is named as a hexane, and the C_2H_5O— group and the methyl group are treated as substituents. The C_2H_5O— group is named by dropping the final *yl* from the name of the alkyl group and adding the suffix *oxy*. Thus, the C_2H_5O— group is the (eth*yl* + *oxy*) = ethoxy group. The numbering follows from nomenclature rule 3d on p. 329.

The nomenclature of sulfides is similar. An RS— group is named by adding the suffix *thio* to the name of the R group; the final *yl* is not dropped.

methylthio substituent ⟶ SCH_3

$\quad\quad\quad\quad\quad\quad\quad\quad\quad\quad\quad\quad\quad\quad\quad\quad\quad |$

principal chain ⟶ $CH_3CHCH_2CH_2CH_2CH_3$
$\quad\quad\quad\quad\quad\quad\quad\quad\quad\quad\quad 1 \quad 2 \quad 3 \quad 4 \quad 5 \quad 6$

2-(methylthio)hexane

The parentheses in the name are used to indicate that "thio" is associated with "methyl" rather than with "hexane."

STUDY PROBLEM 8.3

Name the following compound.

$$CH_3CH_2CH_2CH_2—O—CH_2CH_2CH_2—OH$$

SOLUTION The —OH group is cited as the principal group, and the principal chain is the chain containing this group. Consequently, the $CH_3CH_2CH_2CH_2O—$ group is cited as a butoxy substituent (butyl + oxy) at carbon-3 of the principal chain:

$$CH_3CH_2CH_2CH_2—O—\underset{\substack{\smile \\ \text{principal chain} \\ \text{(contains the principal group —OH)}}}{\overset{3\quad 2\quad 1}{CH_2CH_2CH_2}}—OH$$

3-butoxy-1-propanol

Heterocyclic Nomenclature A number of important ethers and sulfides contain an oxygen or sulfur atom within a ring. Cyclic compounds with rings that contain at least one atom other than carbon are called **heterocyclic compounds**. The names of some common heterocyclic ethers and sulfides should be learned.

furan **tetrahydrofuran** **thiophene** **1,4-dioxane** **oxirane**
 (often called **THF**) (often called simply **dioxane**) **(ethylene oxide)**

(The IUPAC name for tetrahydrofuran is *oxolane*, but this name is not commonly used.)

Oxirane is the parent compound of a special class of heterocyclic ethers, called **epoxides**, which are three-membered rings that contain an oxygen atom. A few epoxides are named traditionally as oxides of the corresponding alkenes:

$$\overset{O}{\overset{\diagup\diagdown}{H_2C—CH_2}} \qquad H_2C{=}CH_2 \qquad Ph—\overset{O}{\overset{\diagup\diagdown}{CH—CH_2}} \qquad Ph—CH{=}CH_2$$

ethylene oxide   **ethylene**   **styrene oxide**   **styrene**

The reason for this nomenclature is that epoxides frequently are prepared from alkenes (Sec. 11.3A). However, most epoxides are named substitutively as derivatives of oxirane. The atoms of the epoxide ring are numbered consecutively, with the oxygen receiving the number 1 *regardless of the substituents present.*

$$\overset{\overset{1}{O}}{\underset{3\qquad 2}{\overset{\diagup\diagdown}{H_2C—C}}}\overset{CH_3}{\underset{CH_3}{\diagup\diagdown}}$$

2,2-dimethyloxirane

PROBLEMS

8.6   Draw the structure of each of the following compounds.
  (a) ethyl propyl ether   (b) dicyclohexyl ether
  (c) *tert*-butyl isopropyl sulfide   (d) allyl benzyl ether
  (e) phenyl vinyl ether   (f) (2*R*,3*R*)-2,3-dimethyloxirane
  (g) 5-(ethylthio)-2-methylheptane

continued

continued

8.7 Give a substitutive name for each of the following compounds.

(a) $(CH_3)_3C-O-CH_3$ (b) $CH_3CH_2-O-CH_2CH_2-OH$

(c)

(d) CH_3OCH_2 H

       C=C

    H CH_2CH_2-OH

8.8 (a) A chemist used the name 3-butyl-1,4-dioxane in a paper. Although the name unambiguously describes a structure, what should the name have been? Explain.

(b) Give the structure of 2-butoxyethanol, which is an ingredient in whiteboard cleaner and kitchen cleaning sprays.

8.3   STRUCTURES OF ALKYL HALIDES, ALCOHOLS, THIOLS, ETHERS, AND SULFIDES

In all of the compounds covered in this chapter, the bond angles at carbon are very nearly tetrahedral, and the α-carbons are sp^3-hybridized. For example, in the simple methyl derivatives (the methyl halides, methanol, methanethiol, dimethyl ether, and dimethyl sulfide) the H—C—H bond angle in the methyl group does not deviate more than a degree or so from 109.5°. In an alcohol, thiol, ether, or sulfide, the bond angle at oxygen or sulfur further defines the shape of the molecule. You learned in Sec. 1.3B that the shapes of such molecules can be predicted using VSEPR theory by thinking of an unshared electron pair as a bond without an atom at the end. This means that the oxygen or sulfur has four "groups": two electron pairs and two alkyl groups or hydrogens. These molecules are therefore bent at oxygen and sulfur, as you can see from the structures in Fig. 8.1. The angle at sulfur is generally found to be closer to 90° than the angle at oxygen. One reason for this trend is that the unshared electron pairs on sulfur occupy orbitals derived from quantum level 3 that take up more space than those on oxygen, which are derived from quantum level 2. The repulsion between these unshared pairs and the electrons in the chemical bonds forces the bonds closer together than they are on oxygen.

The lengths of bonds between carbon and other atoms follow the trends discussed in Sec. 1.3B. Within a column of the periodic table, bonds to atoms of higher atomic number are longer. Thus, the C—S bond of methanethiol is longer than the C—O bond of methanol (see Fig. 8.1 and Table 8.1). Within a row, bond lengths decrease toward higher atomic number (that is, to the right). Thus, the C—O bond in methanol is longer than the C—F bond in methyl fluoride (see Table 8.1); similarly, the C—S bond in methanethiol is longer than the C—Cl bond in methyl chloride.

PROBLEMS

8.9 Using the data in Table 8.1, estimate the carbon–selenium bond length in $H_3C-Se-CH_3$.

8.10 From the data in Fig. 8.1, tell which bonds have the greater amount of p character (Secs. 1.9B and 6.9B): C—O bonds or C—S bonds. Explain.

8.4   NONCOVALENT INTERMOLECULAR INTERACTIONS: INTRODUCTION

When we think of the ways that molecules might interact, the first thing that comes to mind is *chemical reactions*: processes in which covalent bonds are broken and formed. However,

FIGURE 8.1 Bond lengths and bond angles in a simple alcohol, thiol, ether, and sulfide. Bond angles at sulfur are smaller than those at oxygen, and bonds to sulfur are longer than the corresponding bonds to oxygen.

Increasing electronegativity

Increasing atomic radius

TABLE 8.1 Bond Lengths (in Angstroms) in Some Methyl Derivatives

H_3C—CH_3 1.536	H_3C—NH_2 1.474	H_3C—OH 1.426	H_3C—F 1.391
		H_3C—SH 1.82	H_3C—Cl 1.781
			H_3C—Br 1.939
			H_3C—I 2.129

molecules can interact in other ways that do *not* involve covalent bonds; that is, molecules can interact *noncovalently*. We've already seen an example of noncovalent interaction when we studied *steric effects* in chemical reactions such as the free-radical addition of HBr to alkenes (Fig. 5.2, p. 211). The interaction of a bromine atom with alkyl branches on a double bond is an *energetically unfavorable* noncovalent interaction—that is, an **intermolecular repulsion**. Intermolecular repulsions raise the energy of the interacting species. You're probably also familiar with an example of an *energetically favorable* noncovalent interaction—that is, an **intermolecular attraction**: the attractions between ions in the crystal structure of a salt, which we often call an *ionic bond*. Intermolecular attractions lower the energy of the interacting species. Other types of noncovalent attractions can occur, and these are considered in the next few sections.

Many biological structures, as we'll see, owe their very existence to noncovalent intermolecular attractions: among these are biomolecules such as proteins and DNA, and biological structures such as membranes. The interactions of protein receptors with small molecules such as drugs are noncovalent. To catalyze the reaction of its substrates, an enzyme must first associate noncovalently with them. To understand these phenomena, we need to understand noncovalent interactions. We're going to "start small" first, with interactions between identical small molecules. Then, we'll consider solutions, which involve interactions between different molecules. Finally, we'll examine a few examples from both chemistry and biology in which noncovalent interactions play a key role.

8.5   HOMOGENEOUS NONCOVALENT INTERMOLECULAR ATTRACTIONS: BOILING POINTS AND MELTING POINTS

Any condensed state of matter (a solid or a liquid) owes its existence to noncovalent intermolecular attractions. If there were no attractions between molecules in a solid or a liquid, the substance would be a gas. (Intermolecular interactions within an ideal gas don't occur, and we'll use the ideal-gas model as our description of gases.) The attractions between molecules in a solid or a liquid are noncovalent because no chemical reaction occurs when we convert a liquid to a gas or a solid to a liquid. That is, *chemical bonds are not broken*. Because no covalent bonds are broken, these noncovalent attractions are much weaker than the attractions that hold atoms together in covalent bonds. (We'll return to this point in Sec. 8.8.)

We can learn a lot about noncovalent attractions by studying the conversion of a liquid, in which noncovalent intermolecular attractions exist, to a gas, in which noncovalent attractions have largely disappeared. Specifically, we can use the *boiling point* as a crude measure of noncovalent attractions. The **boiling point** is the temperature required to raise the vapor pressure of a liquid to atmospheric pressure (760 mm Hg at sea level). It is possible to justify thermodynamically the use of the boiling point to measure noncovalent attractions (see Further Exploration 8.1), but let's justify it intuitively instead. The boiling point is a measure of the energy required to bring a liquid to the state in which all of the molecules want to escape from the liquid into the gas. As the boiling point increases, then, *more energy is required to break the intermolecular attractions in the liquid state*. It is important to understand that there are no covalent bonds between molecules, and furthermore, that intermolecular attractions have *nothing* to do with the strengths of the covalent bonds in, or the stabilities of, the molecules themselves.

**FURTHER
EXPLORATION 8.1**
Trouton's Rule

A. Attractions between Induced Dipoles: van der Waals (Dispersion) Forces

One of the most significant observations about the boiling points of organic compounds is that *boiling points increase regularly with molecular size within a homologous series*. For example, Fig. 2.7 (p. 72) shows that there is a fairly regular increase in the boiling points of unbranched alkanes with the number of carbon atoms. Figure 8.2a shows that this regular increase occurs not only for alkanes, but also for many other compound classes, and these are only a few of many examples. We'll consider later why the trend lines of some compounds classes are displaced to higher or lower values, and why the line for alcohols (as well as a few other classes not shown) looks a little different. But generally, *the boiling points of compounds in a given class increase about 20–30 °C per carbon atom*.

Figure 8.2b shows the same data plotted against molecular mass. This plot shows that a few different compound classes have the *same* boiling points at a given molecular mass.

Why should boiling points increase with increasing molecular size? In Chapter 1 (p. 36), we learned that electrons in bonds are not confined between the nuclei but rather reside in bonding molecular orbitals that surround the nuclei. We can think of the total electron distribution as an "electron cloud." The electron clouds in some molecules (such as alkanes) are rather "squishy" and can undergo distortions. Such distortions occur rapidly and at random, and when they occur, they result in the temporary formation of regions of local positive and negative charge; that is, these distortions cause a *temporary* dipole moment within the molecule (Fig. 8.3 on p. 338). When a second molecule is located nearby, its electron cloud distorts to form a complementary dipole, called an **induced dipole**. The positive charge in one molecule is attracted to the negative charge in the other. The attraction between temporary dipoles—called a **van der Waals attraction**, or a **dispersion interaction**—is the cohesive interaction that must be overcome to vaporize a liquid. Alkanes do *not* have significant *permanent* dipole moments. The dipoles discussed here are *temporary*, and the presence of a temporary dipole in one molecule induces a temporary dipole in another. These attractions come and go. But, over time, "nearness makes the molecules grow fonder."

The *time scale* of van der Waals attractions is extremely small. Collisions between molecules occur in fractions of a nanosecond (10^{-9} s), and the electronic redistributions associated with temporary dipole formation occur in roughly a *femtosecond* (10^{-15} s). Thus, these temporary dipoles can form and dissipate many times during a molecular collision. In other words, the electron clouds in these molecules form "flickering dipoles." However, if one molecule changes its electron distribution, the other instantly follows suit so as to maintain a net attraction.

boiling point, °C

number of carbons in the alkyl chain
(a)

FIGURE 8.2 The boiling points of several compound classes with unbranched carbon chains plotted against (a) number of carbons and (b) molecular mass. In (a), carbon number = 0 indicates the parent compound with no carbons. For example, for alkyl halides (R—X) the parent compound is H—X; for alcohols (R—OH) the parent compound is water (H—OH); and so on. For nitriles, the carbon of the C≡N group is not included in the carbon number because it is part of the functional group.

boiling point, °C

molecular mass
(b)

Now we are ready to understand why larger molecules have higher boiling points. Van der Waals attractions increase with the *surface areas of the interacting electron clouds*. That is, the larger the interacting surfaces, the greater the magnitude of the induced dipoles. A larger molecule has a greater surface area of electron clouds and therefore greater van der Waals interactions with other molecules. It follows, then, that large molecules have higher boiling points.

To see that it is the *surface area* and not the *volume* of the molecule that controls boiling point, consider how the *shape* of a molecule affects its boiling point. For example, a comparison of the boiling points of the highly branched alkane neopentane (9.4 °C) and its unbranched isomer pentane (36.1 °C) is particularly striking. Neopentane has four methyl groups disposed in a tetrahedral arrangement about a central carbon. As the following space-filling models

FIGURE 8.3 A cartoon showing the origin of van der Waals attractions in two pentane molecules. The frames are labeled t_1, t_2, t_3, and t_4 to show successive points in time. The colors represent electrostatic potential maps (EPMs). The green color indicates the absence of a significant permanent dipole. Although the attraction at t_3 is temporary, it recurs frequently. This attraction, averaged over time, is the source of van der Waals attractions.

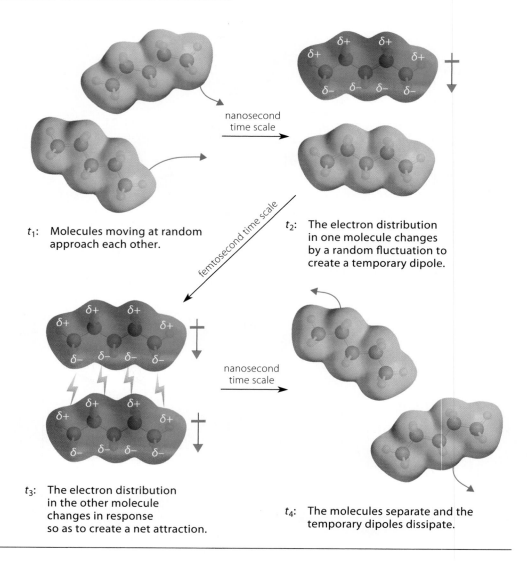

t_1: Molecules moving at random approach each other.

t_2: The electron distribution in one molecule changes by a random fluctuation to create a temporary dipole.

t_3: The electron distribution in the other molecule changes in response so as to create a net attraction.

t_4: The molecules separate and the temporary dipoles dissipate.

show, the molecule almost resembles a compact ball and could fit readily within a sphere. On the other hand, pentane is rather extended, is ellipsoidal in shape, and would *not* fit within the same sphere.

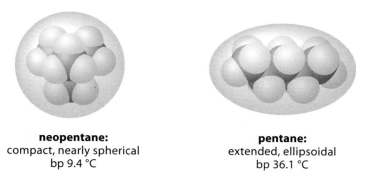

neopentane:
compact, nearly spherical
bp 9.4 °C

pentane:
extended, ellipsoidal
bp 36.1 °C

Typically, extensive branching in a compound lowers the boiling point relative to the unbranched isomer in the same class.

The more a molecule approaches spherical proportions, the less surface area it presents to other molecules, because a sphere is the three-dimensional object with the minimum surface-to-volume ratio. Because neopentane has less surface area at which van der Waals attractions with other neopentane molecules can occur, it has weaker attractions than pentane, and thus, a lower boiling point.

In summary, by analysis of the boiling points of alkanes, we have learned two general trends in the variation of boiling point with structure, and we have also learned the *mechanism* by which these noncovalent attractions occur:

1. Boiling points increase with increasing molecular weight within a homologous series—typically 20–30 °C per carbon atom. This increase is due to the greater van der Waals attractions between larger molecules.

2. Boiling points tend to be lower for highly branched molecules, because they have less molecular surface available for van der Waals attractions.

Polarizability   You've just learned that the deformability, or "squishiness," of electron clouds is what makes van der Waals attractions possible. The **polarizability** of a molecule is a direct measure of how easy it is energetically for an external charge (or dipole) to alter the electron distribution in a molecule or atom. In other words, more polarizable molecules have "squishier" electron clouds. An analogy to polarizability is the comparison of a marshmallow or a balloon to a golf ball or handball. Imagine that these objects correspond to electron clouds. It takes very little energy to deform a marshmallow or a balloon; we can do it with our hands. These objects are "squishy." Molecules or groups that have "squishy," easily deformable electron clouds are polarizable and easily develop temporary dipoles when charges or other dipoles are nearby. However, it takes lots of energy to deform a golf ball or a handball— so much that we can't do it with our hands. They're hard and not at all "squishy." By analogy, molecules and groups with electron clouds that are not easily deformed are less polarizable and do not form temporary dipoles when other charges or dipoles are nearby.

Although polarizability can be both measured and calculated, we'll keep our discussion at a more descriptive level. Molecules (or groups within molecules) that contain very electronegative atoms are typically not very polarizable, because their electrons are held tightly and pulled closer to the nuclei. Molecules or groups that contain atoms of lower electronegativity are typically more polarizable. For example, in the periodic table, the polarizability of the iodine atom is about 10 times that of the fluorine atom. Because fluorine is very electronegative, the electrons in the fluorine atom are very difficult to pull away from the nucleus. The valence electrons in iodine, however, lie in level-5 orbitals, and these orbitals are easily deformed by external charges because they are held less strongly.

From what we've learned, there ought to be a correspondence between molecular polarizability and boiling point. For a given shape and molecular mass, a liquid consisting of more polarizable molecules should have a higher boiling point than one consisting of less polarizable molecules because the van der Waals attractions are stronger between more polarizable molecules. We can make such a comparison using the boiling points of alkanes and perfluoroalkanes (alkanes in which all hydrogens have been replaced by fluorines). Because of the electronegativity of fluorine, perfluoroalkanes have significantly lower polarizabilities than alkenes of the same molecular mass. For example, compare the boiling points of hexafluoroethane with those of the hydrocarbon 2,2,3,3-tetramethylbutane:

perfluoroethane
molecular mass = 138.0
boiling point = −78 °C

2,2,3,3-tetramethylbutane
molecular mass = 114.2
boiling point = +107 °C

The boiling point of the fluorocarbon is *185° lower* even though it has about the same molecular mass as the hydrocarbon. This difference reflects the smaller polarizability of the fluorocarbon, which is about 20% that of the hydrocarbon. In other words, the van der Waals attractions in the liquid fluorocarbon are *much* weaker than those in the liquid hydrocarbon;

so, the van der Waals attractions between molecules in the liquid fluorocarbon are broken and its conversion into a gas occurs at a lower temperature. Perfluorohexane, despite its considerably greater molecular mass and size, has such a low polarizability that it actually has a boiling point that is 12° lower than the hydrocarbon hexane:

<table>
<tr><td style="text-align:center">$CF_3CF_2CF_2CF_2CF_2CF_3$</td><td style="text-align:center">$CH_3CH_2CH_2CH_2CH_2CH_3$</td></tr>
<tr><td style="text-align:center">**perfluorohexane**
molecular mass = 338.1
boiling point = 57 °C</td><td style="text-align:center">**hexane**
molecular mass = 86.2
boiling point = 69 °C</td></tr>
</table>

The polymer Teflon (polytetrafluoroethylene; Table 5.4, p. 219, and sidebar, p. 218) is perhaps the "ultimate fluorocarbon." It is valued precisely because it has weak noncovalent attractions with practically everything.

$$-\!\left[CF_2CF_2\right]_x\!-$$

polytetrafluoroethylene
(Teflon)

The weakness of its noncovalent attractions makes Teflon slippery because it does not adhere to other molecules, including surfaces. This property is the basis for its use in nonstick cookware.

It is intriguing to imagine a substance that has such a low polarizability that it would never liquefy. We don't have to go beyond element 2 of the periodic table—helium—to come close. Helium has two protons that hold its two electrons very tightly in a $1s$ orbital. The polarizability of helium is the lowest of any element. In fact, helium is a gas down to 4.2 K (−269 °C), about four degrees above absolute zero. The van der Waals attractions between helium atoms in the liquid state are so weak that "heating" to 4.2 K is all that is necessary to overcome these attractions and convert the element into a gas.

 ### Dancing on the Ceiling: The Gecko and van der Waals Attractions

The gecko (lizard), of which there are hundreds of species worldwide, is familiar to people who live in warmer climates. Geckos have the amazing ability to walk on walls or even ceilings. This capability is due to the presence of *setae* in their footpads, small hairlike structures roughly 5 micrometers (5×10^{-6} meters) in diameter. Each seta terminates with between 100 and 1000 spatulae, which are each 0.2 micro-

meters in diameter; the spatulae contact the surface on which the gecko moves. The protein in each spatula was shown in 2002 by a team from the University of California, Berkeley, and Stanford to adhere to surfaces by van der Waals forces. It has been estimated that all of the spatulae on a gecko could support a weight of 60 pounds or more. There is interest in developing synthetic spatulae that could be used to allow robots (or people) to walk on walls or ceilings.

The gecko doesn't do so well on surfaces that offer only weak van der Waals attractions. When a gecko is placed on a vertical Teflon surface, it slides right off!

B. Attractions between Permanent Dipoles

Fundamentally, van der Waals attractions are due to attractions between fluctuating dipoles. It should come as no surprise, then, that molecules with permanent dipoles also can show enhanced intermolecular attractions. Molecules with permanent dipoles can have higher boiling points than the alkanes of the same size and shape. For example, consider the boiling points of the following ethers and the alkanes of roughly the same shape and molecular mass:

	dimethyl ether	**propane**
dipole moment	1.31 D	0.08 D
boiling point	−23.7 °C	−42.1 °C

	tetrahydrofuran (THF)	**cyclopentane**
dipole moment	1.7 D	0 D
boiling point	66 °C	49.3 °C

A comparison of the EPMs of dimethyl ether and propane shows clearly the distribution of charge that leads to the permanent dipole moment: the electronegative oxygen has a partial negative charge (red), and the methyl hydrogens have partial positive charges (blue).

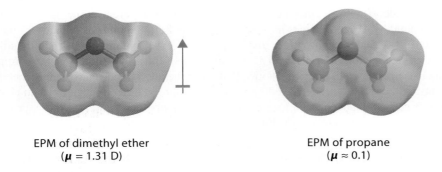

EPM of dimethyl ether
(μ = 1.31 D)

EPM of propane
($\mu \approx 0.1$)

The higher boiling point of the ether results from *greater attractions between molecules* in the liquid state. Molecules with permanent dipoles are attracted to each other because they can align part of the time in such a way that the negative end of one dipole is attracted to the positive end of the other. For example, two dimethyl ether molecules might align in the following way:

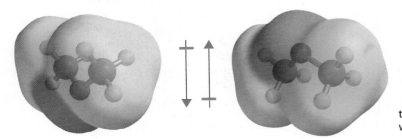

two dimethyl ether molecules
with their dipoles aligned for attraction

Although only two molecules are shown here, attractions like this can occur among many molecules at the same time. Molecules in the liquid state are in constant motion, so their relative positions are changing constantly; however, on the average, this attraction exists and raises the boiling point of a polar compound. The electronegativity of the oxygen in an ether reduces its polarizability; but this reduction in polarizability is more than offset by the attraction between the permanent dipoles.

If you examine Fig. 8.2b on p. 337, you will see that the boiling points of ethers (green line) and alkanes (black line), except at the very small molecular masses, are not very different. This is because van der Waals attractions between the alkyl groups, and the resulting induced dipoles, become the dominant source of intermolecular attractions even in molecules of modest size. However, very polar molecules show significantly higher boiling points than alkanes, even at large molecular masses. For example, Fig. 8.2b shows that nitriles (purple line) have particularly high boiling points. They also have unusually large dipole moments (typically 3.7–3.8 D) that result from the parallel alignment of two bond dipoles of the two bonds to the carbon of the triple bond. (Carbons of triple bonds, like carbons of double bonds, are relatively electronegative, but more so; see Sec. 4.4, p. 145.)

The tradeoff between molecular size and polarity is apparent in the boiling points of the alkyl halides. Alkyl chlorides have about the same boiling points as alkanes of the same molecular mass, and alkyl bromides and iodides have lower boiling points than the alkanes of about the same molecular mass. (The boiling point of the alkyl chlorides, alkyl bromides, and alkanes are shown in Fig. 8.2.) Notice that all of the alkyl halides have significant dipole moments. Although bromine and iodine are considerably less electronegative than chlorine, the longer carbon–halogen bond lengths increase the dipole moment, which is a product of charge separation and bond length (Eq. 1.4, p. 11).

	$CH_3CH_2CH_2CH_2Cl$	$CH_3CH_2CH_2CH_2CH_2CH_3$
molecular mass	92.6	86.3
dipole moment	2.0 D	0 D
boiling point	78.4 °C	68.7 °C
density	0.886 g mL^{-1}	0.660 g mL^{-1}

	CH_3CH_2Br	$CH_3CH_2CH_2CH_2CH_2CH_2CH_3$
molecular mass	109	100.2
dipole moment	2.0 D	0 D
boiling point	38.4 °C	98.4 °C
density	1.46 g mL^{-1}	0.684 g mL^{-1}

	CH_3I	$CH_3CH_2CH_2CH_2CH_2CH_2CH_2CH_2CH_2CH_3$
molecular mass	142	142
dipole moment	1.7 D	0 D
boiling point	42.5 °C	174 °C
density	2.28 g mL^{-1}	0.73 g mL^{-1}

The key to understanding these trends is to realize that although the molecules compared in each row have similar molecular masses, they have *very different molecular sizes and shapes.* Their relatively high densities show that alkyl halide molecules have large masses within relatively small volumes. Thus, *for a given molecular mass,* alkyl halide molecules have smaller volumes than alkane molecules. Remember that the attractive forces between molecules—van der Waals forces, or dispersion forces—are greater for larger molecules. Larger intermolecular attractions translate into higher boiling points. The greater molecular volumes of alkanes, then, should cause them to have *higher* boiling points than alkyl halides. The polarity of alkyl halides, in contrast, has the opposite effect on boiling points: if polarity were the only effect, alkanes would have *lower* boiling points than alkyl halides. Thus, the effects of molecular volumes and polarity oppose each other. They nearly cancel in the case of alkyl chlorides, which have about the same boiling points as alkanes of about the same molecular mass. However, alkane molecules are so much larger than alkyl bromide and alkyl iodide molecules of the same molecular mass that the size (surface area) effect trumps polarity, and alkanes have higher boiling points.

PROBLEMS

8.11   The boiling points of the 1,2-dichloroethylene stereoisomers are 47.4 °C and 60.3 °C. Which stereoisomer has each boiling point? Explain. (*Hint:* Consider their relative dipole moments.)

8.12   Octane and 2,2,3,3-tetramethylbutane have boiling points that differ by about 20 °C (106 °C and 126 °C). Which has the higher boiling point and why?

8.13   (a) The dipole moment of acetaldehyde is 2.7 D and that of propene is 0.5 D. Even though they have about the same molecular mass, they differ in boiling point by about 68 °C (–47 °C and +21 °C). Which has the higher boiling point? Why?

$$\underset{\textbf{acetaldehyde}}{H_3C-\overset{\overset{\textstyle O}{\|}}{C}-H} \qquad \underset{\textbf{propene}}{H_3C-\overset{\overset{\textstyle CH_2}{\|}}{C}-H}$$

(b) Given that alkenes and alkanes with the same branching pattern and number of carbons have about the same boiling points, show where you would expect the curve for aldehyde boiling points to fall in Fig. 8.2b. Explain.

C. Hydrogen Bonding

The boiling points of alcohols, especially alcohols of lower molecular mass, are unusually high when compared with those of other organic compounds. For example, ethanol has a much higher boiling point than other organic compounds of about the same shape and molecular mass.

	CH_3CH_2-OH	$CH_3CH_2CH_3$	$H_3C-O-CH_3$	CH_3CH_2-F
	ethanol	**propane**	**dimethyl ether**	**ethyl fluoride**
boiling point	78 °C	–42 °C	–24 °C	–38 °C
dipole moment	1.7 D	0 D	1.3 D	1.8 D

The contrast between ethanol and each of the last two compounds is particularly striking because all have similar dipole moments, and yet the boiling point of ethanol is much higher. The fact that something is unusual about the boiling points of alcohols is also apparent from a comparison of the boiling points of ethanol, methanol, and the simplest "alcohol," water.

	CH_3CH_2-OH	H_3C-OH	$H-OH$
	ethanol	**methanol**	**water**
boiling point	78 °C	65 °C	100 °C

Recall from Fig. 8.2 that each additional —CH_2— group results in a 20–30 °C increase in the boiling points of successive members of a homologous series. Yet the difference in the boiling points of methanol and ethanol is only 13 °C; and water, although the "alcohol" of lowest molecular mass, has the highest boiling point of the three compounds. These observations are reflected in the unusual shape of the alcohol curves in Fig. 8.2 at low molecular mass. These unusual phenomena are due to a very important intermolecular attraction called *hydrogen bonding.*

   Hydrogen bonding is an intermolecular attraction that results from the association of a hydrogen on one atom with an unshared electron pair on another. Hydrogen bonding can occur within the same molecule, or it can occur between molecules. For example, in the case of the simple alcohols, hydrogen bonding is a weak association of the O—H proton of one molecule with the oxygen of another.

Formation of a hydrogen bond requires two partners: the *hydrogen-bond donor* and the *hydrogen-bond acceptor*. The **hydrogen-bond donor** is the atom to which the hydrogen is fully bonded, and the **hydrogen-bond acceptor** is the atom bearing the unshared pair to which the hydrogen is partially bonded. In this case the oxygen atom serves both roles.

In a classical Lewis sense, a proton can only share two electrons. Thus, a hydrogen bond is difficult to describe with conventional Lewis structures. Consequently, hydrogen bonds are often depicted as dashed lines. The hydrogen bond results from the combination of two factors: first, a weak covalent interaction between a hydrogen on the donor atom and unshared electron pairs on the acceptor atom; and second, an electrostatic attraction between oppositely charged ends of two dipoles. Opinions differ as to the relative importance of these two factors.

The hydrogen bond between two molecules resembles the same two molecules poised to undergo a Brønsted acid–base reaction:

$$(8.2)$$

The hydrogen-bond donor is analogous to the Brønsted acid in Eq. 8.2, and the acceptor is analogous to the Brønsted base. In an acid–base reaction, the proton is fully transferred from the acid to the base; in a hydrogen bond, the proton remains covalently bound to the donor, but it interacts weakly with the acceptor.

The best hydrogen-bond donor atoms in neutral molecules are oxygens, nitrogens, and halogens. In addition, as might be expected from the similarity between hydrogen-bond interactions and Brønsted acid–base reactions, all strong Brønsted acids are also good hydrogen-bond donors. The best hydrogen-bond acceptors in neutral molecules are the electronegative first-row atoms oxygen and nitrogen. Most anions with unshared pairs and all strong Brønsted bases are also good hydrogen-bond acceptors.

Sometimes an atom can act as both a donor and an acceptor of hydrogen bonds. For example, because the oxygen atoms in water or alcohols can act as both donors and acceptors, some of the molecules in liquid water and alcohols exist in hydrogen-bonded networks. The hydrogen bonds in these networks are not static, but rather are rapidly breaking and re-forming.

In contrast, the oxygen atom of an ether is a hydrogen-bond acceptor, but it is not a donor because it has no hydrogen to donate. Finally, some atoms are donors but not acceptors. The ammonium ion, $^+NH_4$, is a good hydrogen-bond donor; but, because the nitrogen has no unshared electron pair, it is not a hydrogen-bond acceptor.

Hydrogen bonding accounts for the unusually high boiling points of alcohols. In the liquid state, hydrogen bonding is an attraction that holds molecules together. In the gas phase, hydrogen bonding is much less important (because molecules are farther apart than in a liquid or solid) and, at low pressures, it does not exist in most compounds. To vaporize a hydrogen-bonded liquid, then, the hydrogen bonds between molecules must be broken, and breaking hydrogen bonds requires energy. This energy is manifested as an unusually high boiling point for hydrogen-bonded compounds such as alcohols.

Hydrogen bonding is also important in other ways. You'll see in Sec. 8.6C how it affects the water solubility of organic compounds. It is also a very important phenomenon in biology. Hydrogen bonds have critical roles in maintaining the structures of proteins and nucleic acids. (This is covered in Secs. 27.9 and 26.5B.) Without hydrogen bonds, life as we know it would not exist.

In summary, the tendency of molecules to associate noncovalently in the liquid state increases their boiling points. The most important forces involved in these intermolecular associations are

1. *hydrogen bonding*: hydrogen-bonded molecules have greater boiling points than molecules of similar polarity that are not hydrogen-bonded;

2. *attractions between permanent dipoles*: molecules with permanent dipole moments have higher boiling points than molecules of the same size and shape with zero or small dipole moments;

3. *attractive van der Waals forces*, which are influenced by
   a. molecular surface area: molecules with greater surface area have greater boiling points;
   b. molecular shape: more extended, less spherical (less branched) molecules have greater boiling points; and
   c. polarizability: more polarizable molecules have stronger intermolecular attractions than less polarizable ones.

What is the relative importance of these effects? These effects are listed above in the approximate order of importance. For example, hydrogen bonding in small molecules usually trumps molecular polarity. In Fig. 8.2b, you can see that the curve for alcohols (the only hydrogen-bonding group in the figure) lies above all of the other curves except one. Thus, 2-propanol (isopropyl alcohol) has a higher boiling point than acetone, which has the greater dipole moment.

	acetone	2-propanol
boiling point	56.5 °C	82.3 °C
dipole moment	2.7 D	1.7 D

2-Propanol can donate and accept hydrogen bonds, but acetone lacks a donor group. (For hydrogen bonding to occur between identical molecules, they must contain *both* a donor and an acceptor.) Notice in making this comparison that we are holding size and shape virtually constant and comparing only polarity and hydrogen-bonding capability.

However, it is impossible to make a general statement about the relative importance of these effects that will hold true in every case. For example, nitriles, which contain no hydrogen-bond donor, are exceptionally polar molecules, with dipole moments in the 3.7–3.8 D range. You can see in Fig. 8.2b that they have higher boiling points than even alcohols. This is an unusual, but by no means isolated, case in which polarity trumps hydrogen bonding. We've already discussed the relative boiling points of alkyl halides and alkanes (p. 342). Because of the density of alkyl halides, molecular surface area is more important than polarity.

Your goal should be to understand the factors that affect boiling point and the trends in boiling point rather than to "split hairs." An example of the sort of reasoning to be used is shown in the Study Problem 8.4.

STUDY PROBLEM 8.4

Arrange the following compounds in order of increasing boiling point:
1-hexanol, 1-butanol, *tert*-butyl alcohol, pentane.

SOLUTION First, draw the structures!

$$CH_3CH_2CH_2CH_2CH_2CH_2—OH$$
1-hexanol

$$CH_3CH_2CH_2CH_2—OH$$
1-butanol

$$CH_3CH_2CH_2CH_2CH_3$$
pentane

***tert*-butyl alcohol**

1-Butanol and pentane have almost the same molecular mass and about the same size and shape. However, because 1-butanol is a polar molecule that can both donate and accept hydrogen bonds, it has a considerably higher boiling point than pentane. Because 1-hexanol, also a primary alcohol, is a larger molecule than 1-butanol, its boiling point is the highest of the three. So far, the order of increasing boiling points is: pentane < 1-butanol < 1-hexanol. *Tert*-butyl alcohol has about the same molecular mass as pentane, but the alcohol has a higher boiling point because of its polarity and hydrogen bonding. However, a *tert*-butyl alcohol molecule is more branched and more nearly spherical than the isomeric 1-butanol molecule; thus, the boiling point of *tert*-butyl alcohol should be lower than that of 1-butanol. Therefore, the correct order of boiling points is: pentane < *tert*-butyl alcohol < 1-butanol < 1-hexanol. (The respective boiling points in °C are 36, 82, 118, and 157.)

PROBLEMS

8.14   Within each set, arrange the compounds in order of increasing boiling point.

   (a) 4-ethylheptane, 2-bromopropane, 4-ethyloctane

   (b) 1-butanol, 1-pentene, chloromethane

8.15   Label each of the following molecules as a hydrogen-bond acceptor, donor, or both. Indicate the hydrogen that is donated or the atom that serves as the hydrogen-bond acceptor.

   (a) H—B̈r: (b) H—F̈:   (c)

$$\underset{\text{H}_3\text{C}}{\overset{\displaystyle :\text{O}:}{\underset{\displaystyle}{\|}}}\text{C}-\text{CH}_3$$

   (d)

$$\text{H}_3\text{C}-\overset{\displaystyle :\text{O}:}{\underset{\displaystyle \|}{\text{C}}}-\overset{..}{\text{N}}\text{H}-\text{CH}_3$$

   (e)

(benzene ring)—ÖH

   (f) $\text{H}_3\text{C}-\text{CH}_2-\overset{+}{\text{N}}\text{H}_3$

D. Melting Points

Melting points were discussed in Sec. 2.6B. Recall that the **melting point** is the temperature above which a solid is spontaneously transformed into a liquid. At the melting point, a solid and its liquid are in equilibrium. Melting points, like boiling points, are a reflection of non-covalent intermolecular attractions—van der Waals forces, dipole–dipole interactions, and hydrogen bonding. However, boiling points provide information about a single state—the liquid state—because (to a useful approximation) there are no intermolecular interactions in the gaseous state. Melting points, however, reflect the effects of noncovalent interactions *in both the liquid state and the crystalline solid state*. For that reason, it isn't possible to interpret melting points in terms of interactions within a single state of matter.

Nevertheless, there are two observations about melting points that are worth knowing about, and they have a common origin. We touched on both in Sec. 2.6B. The first is that, within a homologous series, we often find a "sawtooth" pattern when melting points are plotted against number of carbons (Fig. 2.8, p. 72).

The second and more useful observation is that symmetrical compounds tend to have considerably higher melting points than less symmetrical isomers. This observation has been known since the 1880s, and in some instances can be quite dramatic. For example, the melting points of the following two constitutional isomers differ by more than 30 °C.

1,3,5-trimethoxybenzene
melting point = 51–53 °C

1,2,4-trimethoxybenzene
melting point = 19–20 °C

(We've drawn the double bonds as delocalized resonance hybrids to stress the symmetry; the benzene ring is planar.) When we say that a compound is *symmetrical*, we mean that there are several ways that we can *rotate* or *reflect* the structure to produce exactly the same orientation. That is, if you turned your back and someone carried out these rotations or reflections, and then you turned around to look at the resulting structure, you wouldn't be able to tell that anything had been done. For example, we can rotate the structure of 1,3,5-trimethoxybenzene any multiple of 120° about an axis perpendicular to the page (green dot) and reproduce an indistinguishable orientation.

(8.3a)

A similar rotation of 1,2,4-trimethoxybenzene gives a distinguishable orientation; that is, we can tell that the structure has been rotated.

(8.3b)

Hence, 1,3,5-trimethoxybenzene has three-fold rotational symmetry, and its isomer does not. We can also reflect the structure of 1,3,5-trimethoxybenzene through three different planes perpendicular to the page and reproduce the same structure. (Can you find these?) We can't do that for 1,2,4-trimethoxybenzene. Therefore, 1,3,5-trimethoxybenzene also has reflection symmetry that 1,2,4-trimethoxybenzene doesn't have. The only symmetry that the two structures have in common is reflection in the plane of the page; either structure can be reflected in the page to give an identical orientation. We can conclude that 1,3,5-trimethoxybenzene is more symmetrical and thus has the higher melting point.

Because the boiling points of the two compounds are virtually identical (255 °C), the effect of symmetry on melting point has something to do with the crystalline state. (Identical boiling points show that the strengths of the intermolecular attractions within the liquid states of the two compounds are about the same.) Symmetry affects melting point because *it increases the probability of forming the crystal*. In a crystal, molecules have a regular, repeating arrangement. There are many indistinguishable ways that you can turn or reflect each symmetrical molecule to get exactly the same molecular arrangement in the crystal. By analogy, if you stack cubical blocks in a cubical box, there are many different ways we could turn or reflect each block to get exactly the same pattern within the box. Thus, there is an inherent statistical preference—a greater probability—for crystallization of symmetrical compounds. The thermodynamic measure of probability is the *entropy*, $\Delta S°$. (We'll discuss the relationship of entropy and probability in more depth in Sec. 8.6A.) In other words, symmetry *increases* the $\Delta S°$ of the crystal.

In addition, symmetrical molecules can pack into the crystal more closely than unsymmetrical ones. Returning to the previous analogy, if we stack cubical blocks in a cubical box (as we just described), and then stack irregularly shaped chunks of wood, such as twigs (each with the same volume as the cubical block) in an identical box, we find that the cubical blocks are much closer together. When molecules are closer, their noncovalent attractions are stronger; that is, the energy of the noncovalent attractions is reduced (lower energy = greater stability). These attractions are manifested in a lower enthalpy ($\Delta H°$) of the crystal. Therefore, symmetry lowers the $\Delta H°$ of the crystal.

Remember that $\Delta G° = \Delta H° - T\Delta S°$. Because $\Delta S°$ is *increased* by symmetry and $\Delta H°$ is *decreased*, then the $\Delta G°$ of the crystal is *decreased* by symmetry—in other words, the crystalline state is stabilized by symmetry. If the crystal is stabilized by symmetry, then more energy has to be expended to convert the crystal into a liquid. As a result, the crystal has a higher melting point.

The relative melting points of pentane (–129.8 °C) and neopentane (–16.8 °C) [structures in Sec. 8.4, p. 334], reflect exactly the same phenomenon.

The efficiency of crystal packing also is believed to be the cause of the "sawtooth" melting-point behavior shown in Fig. 2.8, p. 72. Unbranched alkanes with odd numbers of carbons lie on a lower curve of melting point versus carbon number than those with an even number of carbons. In other words, the crystal packing of "even-carbon" alkanes is more efficient, van der Waals attractions in the solid state are somewhat greater, and melting points are higher.

PROBLEMS

8.16   Match the structures with the following melting points. Explain. 168–172 °C; –74.5 °C; 143–147 °C; –157 °C. (*Hint:* The alkenes are liquids at room temperature, and the carboxylic acids are solids.)

8.17   If the "sawtooth" pattern of melting point behavior is explained by the efficiency of crystal packing, what would we expect to find if we were to plot the density of the *crystal* (solid) form of unbranched alkanes against carbon number?

8.6   HETEROGENEOUS INTERMOLECULAR INTERACTIONS: SOLUTIONS AND SOLUBILITY

Now we are going to examine the noncovalent interactions between different compounds. To do this, we'll examine the process of forming a solution of two liquids, in which two or more compounds coexist in the same liquid phase. First, in Sec. 8.6A, we'll discuss precisely what we mean by a solution, we'll present some terminology, and show how we can think about the solution process in terms of energies. In Sec. 8.6B, we'll learn about common solvents, and how we classify them. In Sec. 8.6C, we'll learn some of the practical aspects and "rules of thumb" governing the solubility of covalent compounds. We'll try to develop an intuition about solubilities—whether two substances will form a solution or remain in separate phases. In Sec. 8.6D, we're going to consider specifically why hydrocarbons are insoluble in water. The principles we learn there underlie a number of important biological phenomena. All of these sections will involve the solubility behavior of liquids in other liquids. In Sec. 8.6E, we'll consider briefly how we have to modify our thinking when we consider the solubility behavior of solids. Finally, in Sec. 8.6F, we'll consider the solubility of ionic compounds. In Sec. 8.7 that follows, we'll then examine a number of applications in both chemistry and biology that depend on the principles that we will have learned in this section.

A. Solutions. Definitions and Energetics

In this section we'll define what we mean by the term *solution*, and we'll explore the energetics of solution formation. We'll keep things simple by using only two components, both liquids. Imagine that a small amount of a liquid *A*, the **solute**, is added to a large amount of liquid *S*, the **solvent**. We assume that the compounds do not react with each other. If the two compounds persist as separate phases, even when in contact, we say that *A* is **insoluble** in *S*. If, however, *A* and *S* form a single, clear, liquid phase, we say that *A* has formed a **solution** in *S*, and that *A* is **soluble** in *S*. We have then *dissolved A* in *S*. A diagram of the process of forming a solution at the molecular level is shown in Fig. 8.4 on p. 350.

Figure 8.4 shows that the process of forming a solution involves the interplay of *non-covalent intermolecular interactions*. In the pure compounds, on the left, *A* molecules interact

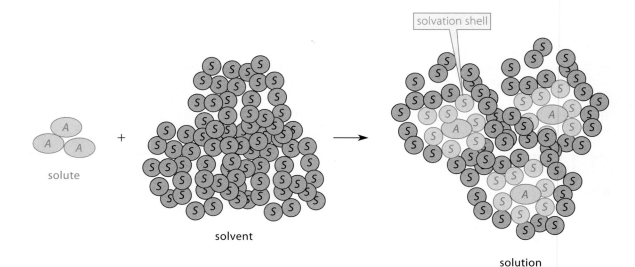

FIGURE 8.4 A schematic view of solution formation at the molecular level. Pure solute molecules *A* interact with other solute molecules, and pure solvent molecules *S* interact with other solvent molecules. When the two are mixed, solute disperses into the solvent. In dilute solution, each solute molecule is surrounded only by solvent molecules. The solvent molecules that directly interact with the solute molecules, shown in blue, constitute the solvation shell. Solvent structure is very dynamic; solvent and solvation-shell molecules exchange places rapidly.

with each other, and *S* molecules interact with each other. When the solution on the right is formed, some of the interactions between *S* molecules and *all* of the interactions between *A* molecules are replaced by interactions between *A* and *S*. The molecules of solvent *S* that are in direct contact with the solute molecules *A* (shown in blue in Fig. 8.4) are called collectively the **solvent shell**, or **solvent cage**. For simplicity, we've shown the solvent shell as a single layer of solvent molecules, although the interactions between solute and solvent can in some cases extend beyond a single layer. Although the diagram in Fig. 8.4 is necessarily static, solvent structure is very dynamic, with molecules in the solvent and molecules in the solvent shell moving around and exchanging places rapidly.

When we consider the energetics of the solution process, we can think of it much as we would a chemical reaction. The free energy change is equal to the free energy of the "products" (the solution) minus the free energy of the "reactants" (the pure liquids). When a solute *A* is dissolved in a liter of solvent *S*, the free-energy change, ΔG_s, is called the **free energy of solution**.

$$\Delta G_s = G(\text{solution}) - [G(\text{pure solute}) + G(\text{pure solvent})] \tag{8.4}$$

When $\Delta G_s < 0$, the solution process is favorable; when $\Delta G_s > 0$, the solution process is unfavorable. The magnitude of ΔG_s tells us how favorable or unfavorable the process is.

The first aspect of solution formation is that, regardless of the intermolecular interactions involved, *there is a statistical driving force for formation of the solution*, called the *entropy of mixing,* ΔS_{mixing}. The **entropy of mixing** is a quantitative description of the probability of solution formation that is completely independent of any intermolecular interactions that may be involved.

To understand the entropy of mixing, we need some intuition about entropy. **Entropy** is a measure of *probability.* If a system goes from a less probable to a more probable state, then the entropy of the system has increased. Let's consider a few examples of this idea.

Suppose you have four coins, all heads. Imagine a process in which you flip each coin once. Suppose you end up with two heads (H) and two tails (T). There is only one way that four coins can be all heads, but there are *six* ways that four coins could have two heads and two tails:

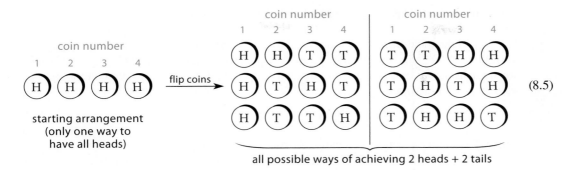

all possible ways of achieving 2 heads + 2 tails

This makes the two heads–two tails combination more probable by a factor of 6. The coin-flipping process increases the entropy of the system of coins because it takes the coins from a less probable state to a more probable state. If you do the same experiment with *six* coins, the relative probability of the three heads–three tails combination is 20. (Verify this!) If you no longer specify that the flipped coins must have half heads and half tails, and instead allow *all* possible results, the number of possibilities is even larger. There are 16 rather than 6 possible outcomes for the four coins, and 64 possible outcomes for the six coins. Imagine if you had Avogadro's number of coins (that is, a "mole of coins"), all heads or all tails, and flipped them all. The number of possible outcomes is huge.

In another analogy, let's say that your professor's office is said to be "neat" when every one of the thousand or so items (books, papers, coffee mugs, tennis racquets) is in a specific, designated place. Over time, items are moved around, they are not put back in their proper place, and the office becomes "messy." We'll define a "messy" state to be any state not equal to the "neat" state. The "messy" states are far more probable because there are so many more of them possible than there are "neat" states. Therefore, the entropy of the office spontaneously increases with use because the "messy" states are inherently more probable (there are so many more ways that the office can be messy). The only way to overcome the effects of entropy is to do work to move objects around to reconstitute the neat state. (Entropy is sometimes linked with "disorder," but, while this is often true, it's not generally true.)

Now let's say that you have a box of 10 blue balls and another box of 1000 red balls. Imagine putting the blue balls into the box containing the red balls, covering the box, and shaking it. What would you find? Would the blue balls all stay together? Certainly not. They would disperse among the red balls, and the odds are that each blue ball would be surrounded by nothing but red balls. There are a huge number of possible arrangements of balls that can give this result, but only one arrangement (or only a few arrangements, if you let the blue stack move around) in which the blue balls could, by chance, stay together. The blue balls disperse into the red balls because there are so many more ways that the balls could be arranged in the dispersed state—the dispersed state is *much* more probable. The entropy change expresses this increased probability.

The foregoing analogy is *directly related* to the formation of a solution. If we squirt some water-soluble blue ink into a beaker of water, we see the ink spontaneously disperse into the water because, in the dispersed state, there are many more ways that the ink molecules can be located relative to the water molecules than the state in which the ink molecules stay together. This is a result of entropy. The dispersed state is simply more probable. If we try to envision the reverse process starting with a solution of water-soluble blue ink, wouldn't it be amazing to see all the ink aggregate spontaneously in one corner of the beaker? It doesn't happen, because this would violate the natural tendency to higher probability (higher entropy).

The ΔS_{mixing} can be calculated for mixing any number of moles of solute with any number of moles of solvent. This formula was derived directly from probability considerations like the ones we have been discussing.

$$\Delta S_{\text{mixing}} = -2.3R(n_1 \log x_1 + n_2 \log x_2) \qquad (8.6)$$

In this equation n_1 and n_2 are the numbers of moles of the two solution components (solvent and solute) used to form the solution, x_1 and x_2 are the *mole fractions* of the two components in the solution, and R is the gas constant, 8.31 J K^{-1} mol^{-1}. For example, mixing one mole of any solute into one liter (55.6 moles) of water, the mole fraction of the solute is $1/(1 + 55.6) = 0.0177$, and the mole fraction of water is $55.6/(1 + 55.6) = 0.982$. From Eq. 8.6, $\Delta S_{mixing} = +41.9$ J mol^{-1} K^{-1}. To convert ΔS_{mixing} to ΔG_{mixing}, we recognize that $\Delta H_{mixing} = 0$ because we are considering only probabilities and not energy changes associated with intermolecular interactions. Therefore,

$$\Delta G_{mixing} = -T\Delta S_{mixing} \qquad (8.7)$$

where T is the temperature in kelvins. At room temperature (25 °C or 298 K), $\Delta G_{mixing} = (-298)(41.9) = -12{,}500$ J $= -12.5$ kJ (-2.99 kcal). Because ΔG_{mixing} is negative, there is an intrinsic tendency to form this solution, and the magnitude of ΔG_{mixing} indicates exactly what that tendency is in free-energy terms.

If the free energy of mixing were the only free-energy change involved in solution formation, every liquid would dissolve in every other liquid! We know from experience that this is not the case. The reason is that noncovalent interactions also make a contribution to the free energy. Figure 8.4 (p. 350) shows that forming a solution involves replacing some S–S interactions and all of the A–A interactions with S–A interactions. Let the free-energy change associated with these interaction changes be ΔG_{inter}. The overall free energy of solution, ΔG_s, results from the balance of ΔG_{inter} against the always-favorable free-energy of mixing:

$$\Delta G_s = \Delta G_{inter} + \Delta G_{mixing} \qquad (8.8)$$

This equation is combined with Eq. 8.7 to give

$$\Delta G_s = \Delta G_{inter} - T\Delta S_{mixing} \qquad (8.9)$$

which stresses the strictly entropic origin of the mixing contribution.

Let's use our previous blue ball–red ball analogy to illustrate this balance. Suppose our 10 blue balls contained embedded bar magnets so that they stick together. If we mix them with our 1000 red balls and shake, they would not disperse into the red balls if the magnets were strong enough. In this case, the attractions between the blue balls overcome the tendency toward spontaneous mixing. That is, *separating the blue balls would require energy.* Similarly, if the solute molecules, or the solvent molecules, have significant intermolecular attractions for each other that *are not replaced by compensating S–A interactions in solution,* then $\Delta G_{inter} > 0$. If ΔG_{inter} is large enough, it overcomes the entropy of mixing, and $\Delta G_s > 0$. As ΔG_s increases (becomes more positive), the amount of A that dissolves in S decreases.

Let's suppose that the solute and the solvent are so similar that the intermolecular interactions between solute and solvent in the solution are *exactly the same* as the interactions between solute molecules in the pure solute, and solvent molecules in the pure solvent. For example, imagine that we dissolve one mole of hexane labeled at one carbon with ^{13}C (call this material hexane*) in one liter of ordinary hexane. In this case, hexane*–hexane* interactions and some hexane–hexane interactions are replaced with hexane*–hexane interactions. The presence of a single ^{13}C in a hexane molecule has a negligible contribution to its van der Waals interactions. Therefore, in this case, $\Delta G_{inter} = 0$, and the overall ΔG_s equals the ΔG_{mixing} as calculated from Eq. 8.6. As intuition should dictate, the isotopically labeled hexane sample dissolves completely in the ordinary hexane solvent. If, however, we mix two substances that are quite different, then whether ΔG_s is < 0—whether a solution forms—depends on the balance between ΔG_{inter} and ΔG_{mixing}. Two substances can form a solution even if ΔG_{inter} is unfavorable (positive), provided that it is not *too* positive.

PROBLEMS

8.18 In each case, which distribution has the higher entropy? Explain.

(a) four coins in which two are heads and two are tails, or four coins in which one is heads and three are tails.

(b) six coins in which two are heads and four are tails, or six coins in which two are tails and four are heads.

(c) 1 mole of a solute in 1 L of water, or 5 moles of the same solute in 1 L of water; assume that $\Delta G_{inter} = 0$.

8.19   Use Eq. 8.6 to calculate the free energy of mixing of 1 mole of isotopically labeled hexane in 1 liter of ordinary hexane. The molecular mass of hexane is 114 g mol^{-1}, and the density of hexane is 0.660 g mL^{-1}.

B. Classification of Solvents

Our goal is to understand noncovalent intermolecular interactions in terms of the structures of the interacting molecules. For this purpose we'll find it useful to classify the most common solvents in terms of three broad categories. These categories are not mutually exclusive; that is, a solvent can be in more than one category.

1. A solvent can be *protic* or *aprotic.*
2. A solvent can be *polar* or *apolar.*
3. A solvent can be a *donor* or a *nondonor.*

A **protic solvent** consists of molecules that can act as *hydrogen-bond donors.* Water, alcohols, and carboxylic acids are protic solvents. Solvents that cannot act as hydrogen-bond donors are called **aprotic solvents**. Ether, dichloromethane, and hexane are aprotic solvents.

Unfortunately, when it comes to describing solvents, the word *polar* has a double usage in organic chemistry. One usage refers to whether the individual molecules of the solvent have a significant dipole moment (Sec. 1.2D). Recall (Sec. 8.5B) that attractions between dipoles of organic molecules can significantly affect their boiling points. As we'll learn, the attractions between the dipoles of solvent and solute molecules also have significant effects on solubility. We'll call solvents consisting of molecules with significant dipole moments ($\mu > 1$ D) **dipolar solvents**.

The other usage of the word *polar* has to do with the solvent *dielectric constant*, for which we use the symbol ϵ. In this usage, a **polar solvent** is a solvent that has a high dielectric constant ($\epsilon \geq 15$); an **apolar solvent** is one with a low dielectric constant. The **dielectric constant** is defined by the *electrostatic law*, which gives the interaction energy E between two ions with respective charges q_1 and q_2 separated by a distance r:

$$E = k\frac{q_1 q_2}{\epsilon r} \tag{8.10}$$

In this equation, k is a proportionality constant and ϵ is the dielectric constant of the solvent in which the two ions are imbedded. This equation shows that when the dielectric constant ϵ is large, the magnitude of E, the energy of interaction between the ions, is small. This equation means that both attractions between ions of opposite charge and repulsions between ions of like charge are weak in a polar solvent. *Thus, a polar solvent effectively separates, or shields, ions from one another.* Therefore, the tendency of oppositely charged ions to associate is less in a polar solvent than it is in an apolar solvent. Water ($\epsilon = 78$), methanol ($\epsilon = 33$), and formic acid ($\epsilon = 59$) are polar solvents. Hexane ($\epsilon = 2$), ether ($\epsilon = 4$), and acetic acid ($\epsilon = 6$) are apolar solvents. As we'll see in Sec. 8.6F, the dielectric constant contributes significantly to the ability of a solvent to dissolve ionic compounds.

The dielectric constant is a property of many molecules of a solvent acting together, whereas the dipole moment is a property of individual molecules. Fortunately, all polar solvents consist of dipolar molecules; so, when we say "polar solvent" we can assume that the solvent molecules have a significant dipole moment. However, the converse is not true. A number of *apolar* solvents (in the dielectric-constant sense) also contain dipolar molecules. The contrast between acetic acid and formic acid is a particularly striking example:

acetic acid	**formic acid**
$\mu = 1.5$–1.7 D	$\mu = 1.6$–1.8 D
$\epsilon = 6.1$	$\epsilon = 59$

These two compounds contain identical functional groups and have very similar structures and dipole moments. Both are *polar molecules* and, as solvents, they can be termed *dipolar solvents*. Yet they differ substantially in their dielectric constants and their solvent properties. Formic acid, a *polar solvent*, is much more effective in dissolving ionic compounds than acetic acid, an *apolar solvent*.

Donor solvents consist of molecules containing oxygens or nitrogens that can donate unshared electron pairs—that is, molecules that can act as Lewis bases. Ether, THF, and methanol are donor solvents. **Nondonor solvents** cannot act as Lewis bases; pentane and benzene are nondonor solvents. Although the chlorines in halogenated solvents such as dichloromethane and chloroform have unshared pairs, these solvents are very poor Lewis bases and are considered to be nondonor solvents.

Table 8.2 lists some common solvents used in organic chemistry along with their abbreviations, properties, and classifications. This table shows that a solvent can have a combination of properties, as noted at the beginning of this section. For example, some polar solvents are protic (such as water and methanol), but others are aprotic (such as acetone).

PROBLEM

8.20 Use their structures and dielectric constants to classify each of the following substances according to their solvent properties (as in Table 8.2).

(a) 2-methoxyethanol ($\epsilon = 17$)   (b) 2,2,4-trimethylpentane ($\epsilon = 2$)

(c)

$$\overset{\displaystyle O}{\underset{\displaystyle \|}{}}$$

$H_3C\!-\!\overset{O}{\overset{\|}{C}}\!-\!CH_2CH_3 \; (\epsilon = 19)$

C. Solubility of Covalent Compounds

In this section, we'll focus on applying the principles we learned in Secs. 8.6A and 8.6B to develop some intuition about the practical aspects of solubility. We start with the solubility of liquids, and then we'll extend what we learn to the solubility of solids.

In determining a solvent for a liquid covalent compound, a useful rule of thumb is that **like dissolves like**. That is, a good solvent usually has some of the molecular characteristics of the compound to be dissolved. For example, an apolar aprotic solvent is likely to be a good solvent for another apolar aprotic liquid. In contrast, a protic solvent in which significant hydrogen bonding occurs between molecules is likely to dissolve another liquid in which hydrogen bonding between molecules also occurs.

Let's try to understand the basis of the like-dissolves-like rule. To illustrate, imagine dissolving pentane in hexane. In terms of the energetics discussed in Sec. 8.6A, the pentane–pentane attractions and some of the hexane–hexane attractions in the pure liquids are replaced by hexane–pentane attractions in the solution. In both of the pure liquids, the major type of intermolecular attraction is van der Waals attractions, which we discussed in Sec. 8.5A. In the solution, the major type of attraction between hexane and pentane molecules should also be van der Waals attractions, *because both molecules are of the same type*. We expect (and find) that ΔG_{inter} is close to zero. From Eq. 8.8, the ΔG_s is then determined by the free energy of mixing ΔG_{mixing}, which is always favorable (negative). In fact, pentane and hexane are **miscible** (literally, "mixable")—they form a solution when mixed in any proportions.

Here, then, is the *physical reason for the like-dissolves-like rule*: We can confidently predict that when the attractions between molecules in the pure liquids are similar to the attractions between the solvent and solute in the solution, then ΔG_{inter} will be small, ΔG_s will be dominated by the free energy of mixing, and a solution will be formed.

Let's examine a few more examples of "like dissolves like." The smaller alcohols (methanol, ethanol, propanol) are all miscible in water. The *major* noncovalent interaction in the

TABLE 8.2 Properties of Some Common Organic Solvents
(Listed in order of increasing dielectric constant)

Solvent	Structure	Common abbrevation	Boiling point, °C	Dielectric constant ϵ^*	Dipole moment	Polar	Protic	Donor
hexane	$CH_3(CH_2)_4CH_3$	—	68.7	1.9	0.08			
1,4-dioxane[†]	(ring structure)	—	101.3	2.2	0.4			x
benzene[†]	(ring structure)	—	80.1	2.3	0.1			
diethyl ether	$(C_2H_5)_2O$	Et_2O	34.6	4.3	1.2			x
chloroform	$CHCl_3$	—	61.2	4.8	1.2			
ethyl acetate	$CH_3COC_2H_5$	EtOAc	77.1	6.0	1.6			x
acetic acid	CH_3COH	HOAc	117.9	6.1	1.6		x	x
tetrahydrofuran	(ring structure)	THF	66	7.6	1.7			x
dichloromethane	CH_2Cl_2	DCM	39.8	8.9	1.1			
acetone	CH_3CCH_3	Me_2CO	56.3	21	2.7	x		x
ethanol	C_2H_5OH	EtOH	78.3	25	1.7	x	x	x
N-methylpyrrolidone	(ring structure)	NMP	202	32	4.0	x		x
methanol	CH_3OH	MeOH	64.7	33	2.9	x	x	x
nitromethane	CH_3NO_2	$MeNO_2$	101.2	36	3.4	x		x
N,N-dimethylformamide	$HCN(CH_3)_2$	DMF	153.0	37	3.9	x		x
acetonitrile	$CH_3C{\equiv}N$	MeCN	81.6	38	3.4	x		x
sulfolane	(ring structure)	—	287 (dec)	43	4.7	x		
dimethyl sulfoxide	CH_3SCH_3	DMSO	189	47	4.0	x		x
formic acid	$HCOH$	—	100.6	59	1.7	x	x	x
water	H_2O	—	100.0	78	1.9	x	x	x
formamide	$HCNH_2$	—	211 (dec)	111	3.9	x	x	x

* Most values are at or near 25 °C   † Known carcinogen

pure substances, as we learned from boiling points, is hydrogen bonding. When (for example) methanol and water are mixed, water and methanol molecules can also form hydrogen bonds.

an alcohol can accept hydrogen bonds from water

an alcohol can donate a hydrogen bond to water

Because the interactions between solute and solvent are similar to the interactions between molecules in the pure liquids, we expect the free-energy cost of dissolving one in the other to be modest. As a result, the positive entropy (and resulting negative free energy) of mixing ensures the formation of a solution.

Hydrocarbons such as pentane and chlorinated solvents such as dichloromethane (CH_2Cl_2) are miscible. In pure pentane, the major source of intermolecular attractions is van der Waals forces. In dichloromethane, the major source of intermolecular attractions is both dipole–dipole attractions and van der Waals forces. When the two solvents are mixed, however, the dipoles of dichloromethane can *induce* temporary dipoles in nearby pentane molecules, and this interaction results in attractions as well. (Such interactions are sometimes called **dipole–induced dipole** attractions.)

pentane has no permanent dipole moment

CH_2Cl_2 has a permanent dipole moment

separated, noninteracting molecules

the permanent dipole in CH_2Cl_2 induces a temporary dipole moment in pentane

interacting molecules

Such interactions are merely variations of the van der Waals forces discussed in Sec. 8.5A in which one of the molecules has a permanent dipole. In other words, the interactions between solute and solvent molecules in solution are fundamentally the same as the interactions in the pure liquids.

One of the most important practical considerations is solubility in water. Whether we are dealing with environmental issues (such as the solubility of chemicals in ground water), the design of drugs (how to make a drug more water-soluble), or laboratory "green" chemistry (chemistry carried out in environmentally benign solvents), water solubility is important. So, let's look at some trends in water solubility. Consider, for example, the water solubilities of the following compounds of comparable size and molecular mass:

$$CH_3CH_2CH_2CH_3 \qquad CH_3CH_2Cl \qquad CH_3CH_2-O-CH_3 \qquad CH_3CH_2CH_2-OH$$

water solubility: virtually insoluble   soluble   miscible

Of these compounds, the alcohol, 1-propanol, is most soluble; in fact, it is miscible with water. Of the compounds shown, the alcohol is also most like water because it is protic. The ability both to donate a hydrogen bond to water and to accept a hydrogen bond from water is an important factor in water solubility.

an alcohol can accept
hydrogen bonds from water

$$\ddot{O}$$
$$H \qquad H$$

$$CH_3CH_2CH_2 - \ddot{O}$$
$$H$$
$$:\ddot{O}$$
$$H$$
$$H$$

an alcohol can donate a
hydrogen bond to water

The ether contains an atom (oxygen) that can accept hydrogen bonds from water, although it cannot donate a hydrogen bond; hence, it has some waterlike characteristics, but is less like water than the alcohol.

$$\ddot{O}$$
$$H \qquad H$$

$$CH_3CH_2 \qquad \ddot{O} \qquad CH_3$$

an ether can accept hydrogen bonds from water

Notice that the hydrogen-bond acceptor ability of ethers is relevant to their aqueous solubilities, *but not to their boiling points*. The boiling point is determined by the interactions of identical ether molecules. Ether molecules cannot form hydrogen bonds with each other because they contain no hydrogen-bond donor group.

Finally, the alkane (butane) and the alkyl halide (ethyl chloride) can neither donate nor accept hydrogen bonds and are therefore least like water; they are also the least water-soluble compounds on the list. The aqueous insolubility of hydrocarbons is important to many phenomena in biology, and we'll return to this topic in the next section.

The balancing of "waterlike" and "hydrocarbonlike" character is evident in the following series:

$$CH_3OH \qquad CH_3CH_2OH \qquad CH_3CH_2CH_2OH \qquad CH_3CH_2CH_2CH_2OH$$

water solubility: miscible   0.96 mole/liter

$$CH_3CH_2CH_2CH_2CH_2OH \qquad\qquad CH_3CH_2CH_2CH_2CH_2CH_2OH$$

water solubility: 0.23 mole/liter   0.0032 mole/liter

Alcohols with long hydrocarbon chains—that is, large alkyl groups—are more like alkanes than are alcohols containing small alkyl groups. Because alkanes cannot form hydrogen bonds, they are insoluble in water, but they are soluble in other apolar aprotic solvents, including other alkanes. Hence, alcohols (as well as any other organic compounds) with long hydrocarbon chains are relatively insoluble in water and are more soluble in apolar aprotic solvents than alcohols with small alkyl chains.

One of the most important generalizations about water solubility is *the importance of hydrogen bonding*. Generally, compounds that are both donors and acceptors have better water solubility than compounds that are simply acceptors; but acceptors have greater solubility than compounds that are neither donors nor acceptors. Hydrocarbon groups reduce aqueous solubility. A useful rule of thumb is that compounds containing one —OH group for every five carbons usually have significant water solubility.

In qualitative considerations of solubility, we can't be expected to make distinctions that are too fine. For example, diethyl ether and tetrahydrofuran (THF), two widely used reaction solvents, have the same number of carbons and a single oxygen.

$$CH_3CH_2 - O - CH_2CH_3$$

tetrahydrofuran (THF)
miscible with water

diethyl ether
water solubility = 6 mass % ($\approx$0.8 mole/liter)
(forms a separate phase with water)

Yet their water solubilities are significantly different. If we pour THF into water, we get a solution. If we pour diethyl ether into water, we get two layers. To be sure, the water contains a significant amount of dissolved diethyl ether, and diethyl ether contains dissolved water, but qualitatively speaking, diethyl ether and water are not soluble. Both ethers are apolar aprotic substances, and both dissolve a wide range of other compounds. But, because of its water solubility, THF is often used in reactions in which the presence of water is required, such as the oxymercuration of alkenes (Sec. 5.4A); it dissolves both water and alkenes.

What you should begin to see from this discussion are the *trends* to be expected in the solubility behavior of various compounds. You cannot be expected to remember absolute solubilities, but you should be able to make an intelligent guess about the relative solubilities of a given compound in different solvents or the relative solubilities of a series of compounds in a given solvent. This ability, for example, is required to solve the following problem.

STUDY PROBLEM 8.5

In which of the following solvents should 1-octene be least soluble: diethyl ether, dichloromethane, methanol, or 1-octanol? Explain.

SOLUTION   1-Octene is an alkene, and the intermolecular interactions in pure 1-octene should be very similar to those in octane because both are hydrocarbons. The interaction mechanism should be van der Waals attractions. (In fact, the boiling points of 1-alkenes and alkanes are nearly identical for the same carbon number.) We've learned that compounds with dipole moments can interact attractively with hydrocarbons by the dipole–induced dipole mechanism. Therefore, 1-octene should have significant solubility, if not miscibility, in diethyl ether and dichloromethane. However, for octane to dissolve in methanol, the hydrogen bonds between methanol molecules must be disrupted, and octane has no groups that can replace this interaction. The same is true in 1-octanol, except that the long alkyl chain of 1-octanol should have favorable van der Waals attractions with the alkene. Therefore, 1-octene should be least soluble in methanol.

Solubility is very important in the metabolism of xenobiotics. A **xenobiotic** (from the Greek *xenos*, meaning "alien") is any substance that is not a normal constituent of a living organism. Among the most important xenobiotics are environmental pollutants and drugs. When an organism (such as the human body) encounters a xenobiotic, it typically routes the substance into a pathway by which it can be eliminated. If the xenobiotic has very low water solubility, one of the most widely used strategies is to couple (that is, chemically attach) the xenobiotic to another group that increases its water solubility. The soluble coupling product can then be routed (for example) to the kidneys, where it can be excreted as an aqueous solution in the urine. This type of process is termed *phase II metabolism*.

One of the most common coupling strategies used in nature for this purpose is *glucuronidation*. For example, the broad-spectrum antibiotic chloramphenicol has relatively low water solubility. It is excreted largely as a glucuronide derivative. (Glucuronides are derived from glucuronic acid, a derivative of the sugar glucose.)

chloramphenicol → *phase II metabolism* → a glucuronide derivative of chloramphenicol (excreted in the urine)   (8.11)

Because of its many hydroxy groups, which have the capacity to form hydrogen bonds to water, the glucuronide has greater water solubility than the uncoupled drug. In addition, as we'll learn in Sec. 8.6E, the ionized carboxylic acid group is strongly solvated by water and also enhances aqueous solubility.

Although glucuronides are not the only derivatives used for phase II metabolism, they are among the most common. This example shows that the like-dissolves-like principle is useful in nature just as it is in the laboratory.

PROBLEMS

8.21   Into a separatory funnel is poured 200 mL of dichloromethane (density = 1.33 g mL^{-1}) and 55 mL of water. This mixture forms two layers. One milliliter of 2-octanol is added, and the mixture is shaken. After a time, two layers are again formed. Where is the 1-octanol—in the upper or lower layer?

8.22   A widely used undergraduate experiment is the recrystallization of acetanilide from water. Acetanilide (see following structure) is moderately soluble in hot water, but much less soluble in cold water. Identify one structural feature of the acetanilide molecule that would be expected to contribute positively to its solubility in water and one that would be expected to contribute negatively.

acetanilide

D. Solubility of Hydrocarbons in Water: Hydrophobic Bonding

Most of us know from experience that "oil and water don't mix." That is, if we pour any hydrocarbon into water, it forms a separate layer. This is one of the reasons that large-scale oil spills are such environmental disasters; water can't dissolve the oil and wash it away. At one level, we could say simply that the insolubility of hydrocarbons in water is a reverse manifestation of the like-dissolves-like rule: hydrocarbons (apolar, aprotic, nondonor) are not at all like water (polar, protic, donor); therefore, hydrocarbons don't dissolve in water. However, a deeper look at the insolubility of hydrocarbons in water will provide some insights that will help us to understand a number of biological phenomena, some of which we'll explore in the next section.

To say that hydrocarbons are insoluble in water is not quite accurate. Actually, hydro-carbons *do* have a *very small* solubility in water that can be measured. For example, imagine that we pour some pentane into one liter of water, stir, and let the system come to equilibrium.

(8.12)

pure pentane / 1 L of pure water   *transfer some pentane*   pure pentane / solution of pentane in water

The concentration of pentane in the aqueous layer is small, but definitely not zero: the concentration of pentane is about 5×10^{-4} M. The ΔG_s° for dissolving pentane in water (at 298 K) is +29.1 kJ mol^{-1}. This is a *standard* free energy: the free energy that would have to be expended to produce a standard solution in which one mole of pentane is dissolved in a liter of water. This large, positive ΔG_s° shows that forming such a solution is *very* unfavorable energetically. The standard enthalpy of solution ΔH_s° is –2.2 kJ mol^{-1}; $\Delta H_s^\circ = \Delta H_{inter}^\circ$ because, by definition, $\Delta H_{mixing}^\circ = 0$. The overall standard entropy of solution ΔS_s° is –105 J mol^{-1} K^{-1}. From Eq. 8.9,

$$\Delta G_s^\circ = \Delta G_{inter}^\circ - T\Delta S_{mixing}^\circ$$

We calculated $T\Delta S_{mixing}^\circ$ in our discussion of Eq. 8.7 as 12.5 kJ mol^{-1} K^{-1}. Therefore, the free-energy change due solely to intermolecular interactions, ΔG_{inter}°, is 29.1 + 12.5 = +42.6 kJ mol^{-1}. The interaction enthalpy ΔH_{inter}° contributes only –2.2 kJ mol^{-1} to this total.

Therefore, the unfavorable standard free energy for dissolving pentane in water is due to the large, negative entropy associated with intermolecular interactions.

It might seem surprising that moving pentane into water causes an enthalpy reduction. Although pentane cannot form hydrogen bonds to water, pentane and water can interact by dipole–induced dipole (van der Waals) attractions. Pentane is not an isolated case: *the small solubilities of other hydrocarbons in water are determined mostly by unfavorable entropies of solution.*

To understand the entropy effect for dissolving pentane in water, consider how the solvent water changes when pentane molecules are introduced into aqueous solution (Fig. 8.5). When molecules of a hydrocarbon dissolve, the solvent shells of the dissolved molecules (red oxygens in Fig. 8.5) must come from the solvent water (blue oxygens in Fig. 8.5). It turns out that *water molecules in the solvation shell of a pentane molecule are different from those in ordinary water.* Specifically, the water molecules at the water–pentane interface have *reduced motional freedom* relative to solvent water. Scientists once described this water as being ice-like, but more recent investigations suggest a more complex picture. A simple static picture for this "interface water" may not be possible. However, perhaps we can think of the situation in terms of the following analogy. Imagine hundreds of small children on a playground, running around at random, and a teacher comes up to some of them and says, "Form a circle around me." The children open up a "cavity" (a circle) for the teacher, and, as a result, the children on the edge of the cavity have less freedom to move around—for one thing, they have to stay out of the cavity. By analogy, the water surrounding a pentane molecule—the solvation

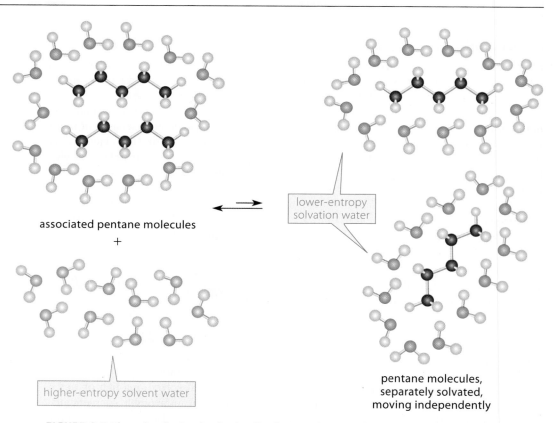

associated pentane molecules

+

higher-entropy solvent water

lower-entropy solvation water

pentane molecules, separately solvated, moving independently

FIGURE 8.5 The role of solvation in the dissolution of hydrocarbons in water. When hydrocarbons dissolve, some solvent water must be transformed into the water of the solvent shells. Because solvation water (red oxygens) has lower entropy than solvent water (blue oxygens), the process of dissolution is accompanied by a reduction in entropy. The *reverse* of this process is essentially a model for the hydrophobic bond. That is, in the reverse direction, the association of hydrocarbon molecules releases solvation water into solvent and therefore involves an entropy increase. This diagram is necessarily two-dimensional; because solvation shells surround molecules in three dimensions, more water molecules are involved than are shown here.

shell—has reduced motional freedom. *Reduced motional freedom lowers entropy.* Therefore, *the water in the hydrocarbon solvent shell has lower entropy than ordinary water.* Because of the small size of a water molecule, many water molecules are involved in cavity formation (solvation) for each hydrocarbon molecule that dissolves. Thus, the entropy *decrease* for the solvent is much greater than the positive entropy of mixing of pentane and water.

Now let's look at the solubility of hydrocarbons in reverse. Envision a situation in which hydrocarbon molecules are dispersed in aqueous solution (the reverse of Fig. 8.5). As we've just learned, there is a driving force for them to associate with each other rather than with water. The association of hydrocarbon groups in aqueous solution is called **hydrophobic bonding**. This terminology, although widely used in biology, is somewhat of a misnomer. The implication in this name ("water-fearing") is that hydrocarbons do not interact favorably with water. However, as we've seen, the enthalpy (energy) of solution of hydrocarbons in water is relatively favorable; so, hydrocarbons are really not "hydrophobic" in an energetic sense. The major reason for the hydrophobic behavior of hydrocarbons is the entropy change of the solvent water. When hydrocarbon groups associate with each other, low-entropy solvation water is released to become ordinary, higher-entropy water. In terms of our previous analogy, when many teachers join hands on the playground, many of the "immobilized" children that formed individual cavities around each teacher are released onto the playground and can move normally. The resulting $\Delta S°$ is highly positive, and this is why $\Delta G°$ is negative for hydrocarbon association.

More sophisticated models of hydrophobic bonding have shown that there is a favorable enthalpic contribution to the interaction between water and very large "hydrophobic" surfaces in addition to the favorable entropic component. This provides a further thermodynamic drive for the formation of large "hydrophobic" clusters.

Many biological processes are driven all or in part by hydrophobic bonding—the association of hydrocarbon groups. Among these processes are membrane formation (Sec. 8.7A), the "folding" of proteins into well-defined conformations (Sec. 27.9), and the binding of many enzymes to their substrates (Sec. 27.10).

PROBLEM

8.23   How would you expect the standard entropy of solution to change along a series of primary alcohols (for example, from methanol to 1-octanol) as the chain length increases? Explain.

E. Solubility of Solid Covalent Compounds

When we dissolve a solid in a liquid solvent, the process is not as simple as dissolving a liquid. Let's consider the standard free-energy change in dissolving a solid. Remember that we can choose any paths we wish for the calculation of free energy changes as long as the starting and ending points are the same. So, let's divide the process of dissolving a solid into two hypothetical steps. From a conceptual standpoint, two things have to happen. First, the solid has to become a liquid. Then, the resulting liquid has to dissolve in the solvent. (We assume a temperature of 25 °C, or 298 K.)

$$\boxed{\begin{array}{c}\text{crystalline solid solute}\\\text{at 298 K}\end{array}} \xrightarrow{\Delta G_1°} \boxed{\begin{array}{c}\text{pure liquid solute}\\\text{at 298 K}\end{array}} \xrightarrow{\Delta G_2°} \boxed{\begin{array}{c}\text{solution}\\\text{at 298 K}\end{array}} \qquad (8.13)$$

$$\boxed{\begin{array}{c}\text{pure liquid solvent}\\\text{at 298 K}\end{array}}$$

The standard free energy of solution for the solid, $\Delta G_s°(\text{solid})$, is the sum of the standard free energies of the individual steps, $\Delta G_1° + \Delta G_2°$.

The free-energy change step 1, ΔG_1°, is the free energy of fusion of the solid at 298 K, $\Delta G_{fusion,298}^\circ$. (This is the free energy required to melt a mole of the compound at 298 K.) Because we are forming a solution at a temperature below the melting point (otherwise the compound would not be a solid), *this transition is unfavorable*—that is, $\Delta G_{fusion,298}^\circ > 0$. That is, the solid is converted into a "supercooled" liquid. The further the temperature of the solution is from the melting point of the solid, the greater is $\Delta G_{fusion,298}^\circ$. Once the solid is transformed into its liquid state, we then apply the solubility principles of liquids. Therefore, $\Delta G_2^\circ = \Delta G_s^\circ(\text{liquid})$. Therefore, for the overall process in Eq. 8.13, we have

$$\Delta G_s^\circ(\text{solid}) = \Delta G_{fusion,298}^\circ + \Delta G_s^\circ(\text{liquid}) \tag{8.14}$$

The point of this equation is that *melting a solid is part of the solution process* and the energy required for melting adds an additional free-energy increment to the process of dissolving a solid. When we think about the solubility of a solid, we have to take into account not only the like-dissolves-like considerations of dissolving a liquid, but also how easy or difficult it is to melt the solid. From this discussion we conclude that, other things being equal, solids with high melting points should have greater $\Delta G_{fusion,298}^\circ$ and therefore should be less soluble in a given solvent than isomers of similar structure with a much lower melting point.

For our first example of solid solubility, recall from Sec. 8.5D that symmetrical compounds generally have higher melting points than their less symmetrical isomers. Therefore, symmetrical compounds should then have lower solubilities *in any solvent* than their less symmetrical isomers. For example, 2-methylbenzoic acid is about three times as soluble in water as its more symmetrical isomer 4-methylbenzoic acid. Notice the difference in melting points.

2-methylbenzoic acid
melting point = 105 °C
more soluble in water

4-methylbenzoic acid
melting point = 182 °C
less soluble in water

As another example, compare the drug nifedipine (used to control angina and hypertension) and the hydrocarbon 9,10-dihydroanthracene.

nifedipine
melting point = 173 °C
water solubility = 1.3×10^{-5} M

9,10-dihydroanthracene
melting point = 109 °C
water solubility = 1.4×10^{-5} M

Notice that nifedipine has a number of hydrogen-bond acceptor sites as well as an N—H hydrogen-bond donor site. The hydrocarbon is devoid of any such sites. On the basis of "like dissolves like," we would expect nifedipine to be considerably more soluble in water. In agreement with this expectation, *liquid* nifedipine is almost 30 times as soluble in water as *liquid* 9,10-dihydroanthracene. However, the water solubility of the two *solids* is about the same. The unexpectedly low solubility of *solid* nifedipine is due to its higher melting point.

The "flip side" of solubility is ease of crystallization. Practicing organic chemists know that it can sometimes be challenging to crystallize a solid with a low melting point. Solids with high melting points are generally easier to crystallize.

As you probably know from personal experience, many drugs are administered in the solid state—for example, as capsules or tablets. For such a drug to function, it has to be in solution. For this reason, the dissolution behavior of a drug is crucial in its formulation, and characterization of the solubility properties of a potential new drug is a major goal of pharmaceutical research. The melting points of drug candidates are a particular concern when solubility is considered. A drug must have a high enough melting point to be crystallized, but a low enough melting point that it will dissolve. The best compromise depends on the drug molecule, but many potentially useful drug candidates have been rejected because of low solubility resulting from a high melting point. Even different crystal forms of the same drug can have different melting points and therefore different solubilities. A few cases are known in which one crystal form of a drug is biologically active and another is so insoluble that it is inactive even though the molecules are the same! Specific crystal forms of a drug can be patented, and patent litigations have occurred over the crystal forms of commercially important drugs.

F. Solubility of Ionic Compounds

Because of the importance of both ionic reagents and ionic reactive intermediates in organic chemistry, the solubility of ionic compounds is worth special attention. Ionic compounds in solution can exist in several forms, two of which, *ion pairs* and *dissociated ions*, are shown in Fig. 8.6. In an **ion pair,** each ion is closely associated with an ion of opposite charge. In contrast, **dissociated ions** move more or less independently in solution and are surrounded by several solvent molecules, called collectively the **solvation shell** or **solvent cage** of the ion, by analogy with the similar terms for nonionic compounds (Fig. 8.4). **Solvation** refers to the favorable interaction of a dissolved molecule with solvent. When solvent molecules interact favorably with an ion, they are said to **solvate** the ion.

Most ionic inorganic compounds are solids. The ones with which we are most familiar have high melting points. For example, the melting point of solid sodium chloride ($Na^+ Cl^-$) is 801 °C. This high melting point shows that, unlike van der Waals forces or hydrogen bonds, the forces between ions in a crystal of sodium chloride are very strong. The attractions between ions

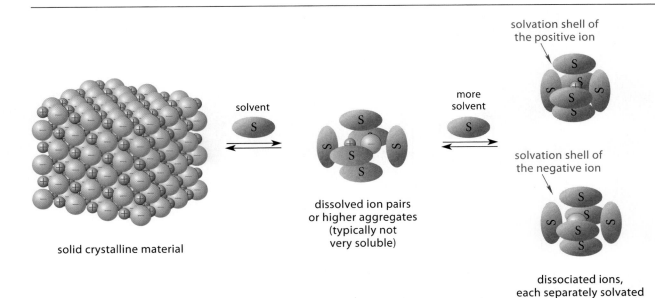

solvation shell of
the positive ion

solvent

more
solvent

solid crystalline material

dissolved ion pairs
or higher aggregates
(typically not
very soluble)

solvation shell of
the negative ion

dissociated ions,
each separately solvated

FIGURE 8.6 Ions in solution can exist as ion pairs and dissociated ions. The blue spheres are positive ions and the red spheres are negative ions. The solubility of an ionic compound depends on the ability of the solvent to break the electrostatic attractions between ions and form separate solvation shells around the dissociated ions. Solvent molecules are represents by the gray ellipses. Solvation is very dynamic; that is, solvent molecules in the solvation shells are not fixed, but are rapidly exchanging with molecules in the bulk solvent.

are called generally **electrostatic attractions**. The energies of these attractions are governed essentially by the *electrostatic law* (Eq. 8.10), although the details of the application of this equation to the entire system of ions in a crystal is complex. But this calculation has been done, and it shows that the electrostatic attraction of ions in a sodium chloride crystal averages about -770 kJ mol^{-1} per pair of ions. (The negative sign indicates an attraction.) This is a huge attraction, roughly twice the value of a covalent bond energy. For a solvent to dissolve an ionic compound, then, the solvent must provide significant energetically favorable solvation to replace the large attraction between ions in the crystal. The fact that ionic solids such as sodium chloride can be dissolved shows that the noncovalent forces involved in solvating ions must be considerable. Let's examine the factors involved in the dissolution of ions.

Ion separation and *ion solvation* are mechanisms by which ions are stabilized in solution. If you think of the ion dissolution sequence in Fig. 8.6 as an ordinary chemical equilibrium, you can see that anything that favors the right side of this equilibrium tends to make ions soluble. The separation and solvation of ions reduce the tendency of the ions to associate into aggregates and, ultimately, to precipitate as solids from solution. Hence, ionic compounds are relatively soluble in solvents in which ions are *well separated* and *well solvated*. What solvent properties contribute to the separation and solvation of ions?

The ability of a solvent to *separate* ions is measured by its dielectric constant ϵ in Eq. 8.10 on p. 353. Look carefully at this equation again. The energy of attraction of two ions of opposite charge is reduced in a solvent with a high dielectric constant. Hence, ions of opposite charge *have a reduced tendency to associate* in solvents with high dielectric constants, and thus a *greater solubility* in those solvents.

Solvent molecules solvate dissolved ions in various ways, which are illustrated in Fig. 8.7 for the interaction of the solvent water with dissolved sodium chloride. A donor solvent can act as a Lewis base to donate its unshared electron pairs to a cation, which acts as a Lewis

FIGURE 8.7 A "snapshot" of the interaction of solvent water molecules in the solvent shells of dissolved sodium and chloride ions. Although two donor interactions are shown for the cation and two hydrogen-bonding interactions for the anion, a greater number of such interactions can occur. Solvation shells can also contain more than six water molecules. Solvation is a dynamic process in which water molecules from the bulk solvent rapidly exchange with water molecules in the solvent shell.

acid. This covalent interaction is called a **donor interaction**. In addition, the dipole moments of solvent molecules can interact electrostatically with the charge of the ion. This means that the water molecule is oriented so that the negative end of its dipole moment vector is pointing toward the positive ion, thus creating a favorable electrostatic interaction by Eq. 8.10. This is called a **charge–dipole interaction**. Because the orientation of the water molecule is almost the same in both charge–dipole and donor interactions, the two interactions are sometimes considered to be different aspects of the same interaction. For solvation of the anion, a favorable charge–dipole interaction can occur in which the solvent molecules are turned so that the positive ends of their dipole moment vectors are pointing toward the negative ion. Finally, if the solvent is protic and the anion can accept hydrogen bonds, the solvent can solvate a negative ion by a **hydrogen-bonding interaction**.

Solvation is *dynamic*. That is, although the solvent shells in Figs. 8.6 and 8.7 are shown as static structures, this figure represents a "snapshot" of a rapidly changing situation. The water molecules are constantly exchanging places with water molecules from bulk solvent, and the solvation mechanism within a solvent shell can change rapidly as well.

To summarize: The *high dielectric constant* of a polar solvent reduces the attraction between ions of opposite charge; as a result, these ions are easier to separate and bring into solution. Dissolved ions are stabilized (that is, kept in solution) by three general types of interaction:

1. *Charge–dipole interactions*, by which the dipole moment vectors of the molecules in a polar solvent are oriented so as to create an attractive (stabilizing) interaction with the charge of the ion.

2. *Hydrogen-bonding interactions*, by which dissolved ions can be stabilized by hydrogen bonding with solvent molecules.

3. *Donor interactions*, by which solvent molecules with unshared electron pairs can act as Lewis bases toward dissolved cations.

These points show why water is the ideal solvent for ionic compounds, something you probably know from experience. First, because it is polar—it has a very large dielectric constant—it is effective in separating ions of opposite charge. Second, because it is protic—a good hydrogen-bond donor—it readily solvates anions. Third, because it is a Lewis base—an electron-pair donor—it can solvate cations by a donor interaction. Finally, its significant dipole moment enables water to provide stabilizing ion–dipole attractions to both cations and anions. In contrast, hydrocarbons such as hexane do not dissolve ordinary ionic compounds because such solvents are apolar, aprotic, and nondonor solvents. Some ionic compounds, however, have appreciable solubilities in *polar aprotic* solvents such as acetone or DMSO (see Table 8.2, p. 355). Although these solvents lack the protic character that solvates anions, their donor capacity solvates cations, their substantial dipole moments provide favorable charge–dipole interactions, and their high dielectric constants separate ions of opposite charge. However, it is not surprising that, because polar aprotic solvents lack the protic character that stabilizes anions, most salts are less soluble in these solvents than in water, and salts dissolved in polar aprotic solvents exist to a greater extent as ion pairs (see Fig. 8.6).

PROBLEM

8.24   Dimethyl sulfoxide (DMSO, Table 8.2, p. 355) has a very large dipole moment (4.0 D). Using structures, show the stabilizing interactions to be expected between DMSO solvent molecules and (a) a dissolved sodium ion; (b) dissolved water; (c) a dissolved chloride ion.

The effective solvation of ions by water can lead to significant solubility changes as a result of acid–base reactions. For example, benzoic acid has a very small solubility in cold water. When a base that is strong enough to ionize the carboxylic acid, such as sodium hydroxide, is added to a stirred suspension of benzoic acid in water, the benzoic acid appears to dissolve. What is happening is that the base converts the carboxylic acid into its sodium salt, an ionic compound.

benzoic acid
slightly soluble
in cold water

sodium benzoate
very soluble
in cold water

The ions of the salt are so strongly solvated in aqueous solution that the salt has a much greater aqueous solubility than the un-ionized carboxylic acid. As the acid is converted into the salt, the salt dissolves. Conversely, if we start with an aqueous solution of the salt and add concentrated HCl, the benzoate anion is protonated, and the carboxylic acid precipitates.

A biological example in which an ionic group contributes to aqueous solubility is the ionic group of glucuronic acid (Eq. 8.11, p. 358).

PROBLEM

8.25   (a) Triethylamine, Et$_3$N:, is a liquid that is not soluble in water. When HCl is added to a stirred mixture of triethylamine and water, a solution is formed. Explain.

(b) What could you add to the solution formed in (a) that would bring the triethylamine back out of solution?

8.7   APPLICATIONS OF SOLUBILITY AND SOLVATION PRINCIPLES

A. Cell Membranes and Drug Solubility

We learned in Sec. 8.6E that solubility is a crucial issue in drug action. If a drug is to be administered in an aqueous solution, it must have adequate aqueous solubility. However, water solubility is not the whole story. For drugs to act, they must get to their sites of action. For many drugs, this means that they must enter cells. The only way for a drug to get into a cell is for it to pass through the *cell membrane*, the "envelope" that surrounds the cell. Drugs and other substances pass through cell membranes by a variety of mechanisms; in some cases, transport requires carrier molecules imbedded in the membrane; and, in some cases, transport requires the expenditure of metabolic energy. However, in many cases, drugs simply pass unassisted through the cell membrane. It turns out that the ability of a molecule to penetrate a cell membrane is very much a solubility issue. To understand this issue, let's examine the structure of a cell membrane.

Cell membranes consist primarily of molecules called *phospholipids*. To understand what phospholipids are, we first need to understand what a *lipid* is. A **lipid** is a compound that shows significant solubility in apolar solvents. Because lipids are defined by a behavior rather than a precise structure, a number of different biomolecule types fit into this category. For example, steroids (Sec. 7.6D) are lipids. Even though lipids might contain polar functional groups, their solubility behavior is dominated by their significant hydrocarbon character.

Phospholipids are lipids that contain phosphate groups. However, *membrane phospholipids* have specialized structures. To understand these structures, let's build a membrane phospholipid, phosphatidylethanolamine, from its component parts. To start with, all membrane phospholipids are built on a glycerol "scaffold."

(8.15)

glycerol (1,2,3-propanetriol)

Two of the hydroxy groups of glycerol form ester linkages to *fatty acids*, which are themselves lipids consisting of carboxylic acids with long, *unbranched* hydrocarbon chains. These chains can contain 15–17 carbon atoms, and they can also contain one or more cis double bonds. For this example, we'll use a carbon chain of 17 carbons in which $-C_{17}H_{35}$ is an abbreviation for $-CH_2\,(CH_2)_{15}CH_3$.

a 1,2-diacylglycerol   (8.16)

Note that the diacylglycerol is chiral even though the starting reactants are not. The production of a single stereoisomer is assured by the catalysis of this assembly process by chiral enzymes (Sec. 7.7A). The remaining hydroxy group of the glycerol backbone is connected to a phosphoric acid molecule as a phosphate ester.

a 1,2-diacylglycerol   a 2,3-diacylglycerol-1-phosphate   (8.17)

Finally, the phosphate is connected to a protonated ethanolamine molecule in another phosphate ester linkage.

a 2,3-diacylglycerol-1-phosphate

phosphatidylethanolamine
(a membrane phospholipid)   (8.18)

A number of different compounds are utilized in this final step to give the most common membrane phospholipids. Besides ethanolamine, choline and serine are the amino alcohols used to form the most common membrane phospholipids.

$$(CH_3)_3\overset{+}{N}CH_2CH_2OH$$

choline

serine

Phosphatidylethanolamines are called *cephalins*, and phosphatidylcholines are called *lecithins*.

PROBLEM

8.26   Give the structure of (a) a phosphatidylserine; (b) a lecithin.

Two structural features of membrane phospholipids are particularly important in understanding their properties. The first is the **polar head group**, which consists of the ethanolamine, choline, or serine and the esterified phosphate. The second is the **nonpolar tails**, which are the long, unbranched hydrocarbon portions of the molecules. A space-filling model and a chemical structure of phosphatidylethanolamine are compared in Fig. 8.8a–b. The polar head group, being ionic, is well solvated by water and counterions. Groups such as the polar head group, which have stabilizing interactions with water, are sometimes called **hydrophilic groups**. The nonpolar tails are examples of **hydrophobic groups**. (More informally, hydrophobic groups might be called "greasy groups." The aqueous solvation of hydrophobic groups was discussed in Sec. 8.6D.) Thus, a membrane phospholipid has a hydrophilic part and a hydrophobic part. Molecules such as phospholipids that contain discrete hydrophilic and hydrophobic regions are said to be **amphipathic**. As a reflection of this amphipathic character, membrane phospholipids are often represented in diagrams as a circle (for the polar head group) with two "squiggly tails," as shown in Fig. 8.8c.

When membrane phospholipids are added to water, something very remarkable happens: they undergo a process called *self-assembly* in which they *spontaneously* form a **phospholipid bilayer**. A phospholipid bilayer consists of many molecules in a double layer in which the nonpolar tails interact with each other on the interior of the layer, and the polar head groups interact with water on the outside of the layer (Fig. 8.9). The spontaneous formation of this very ordered structure would appear to be, at first sight, a violation of the tendency toward increasing entropy. A random distribution of phospholipid molecules has become a seemingly improbable, ordered, bilayer structure. If we focus solely on the phospholipid molecules, their self-assembly into bilayers does indeed represent a highly negative entropy change. However, there is more to this self-assembly process than meets the eye. This is an example of *hydrophobic bonding*. The nonpolar tail of each phospholipid molecule in solution is surrounded by many waters of solvation. Remember that the water molecules in the solvation shell of a hydrocarbon have *particularly low entropy* (Sec. 8.6D, Fig. 8.5). When two (or more) hydrocarbon tails come together into the bilayer, all of the solvation water between the tails is released into the solvent. This provides a highly positive and dominating entropic contribution to the self-assembly process. This process is conceptually much like the aggregation of pentane molecules into a separate liquid phase. In other words, the self-assembly of phospholipids into bilayers is entropy-driven by the release of water.

In the phospholipid bilayer, the solubility characteristics of both parts of the molecule can be satisfied: hydrocarbon groups are next to hydrocarbon groups, and the polar head groups are near water. These double layers continue to grow as more phospholipid is added to form *phospholipid vesicles*—closed, more or less spherical structures in which a phospholipid bilayer encloses an inner aqueous region (Fig. 8.10 on p. 370). The polar head groups interact with water on both the inside and outside of the vesicle. Chemically, the living cell

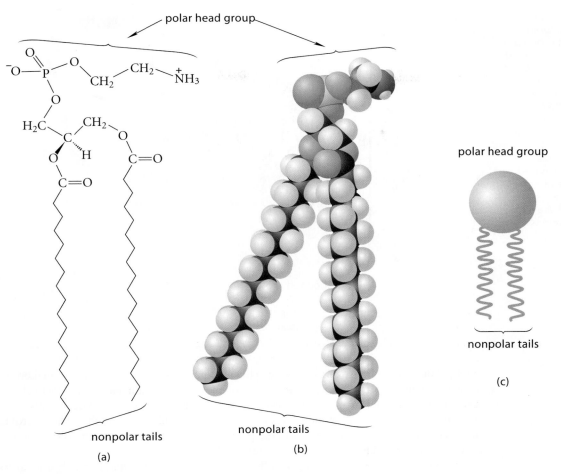

FIGURE 8.8 (a) A Lewis structure of phosphatidylethanolamine, a membrane phospholipid. (b) A space-filling model of phosphatidylethanolamine. (c) A schematic representation of a membrane phospholipid. The polar head group is represented as a circle and the nonpolar tail by "squiggles." This representation is often used in diagrams, such as Fig. 8.10.

FIGURE 8.9 Part of a phospholipid bilayer. Notice that the polar head groups interact with the solvent water, and the nonpolar tails interact with other nonpolar tails.

FIGURE 8.10 Schematic view of a cell membrane. The enlargement shows a section of the phospholipid bilayer and some imbedded proteins. The phospholipid molecules are represented as shown in Fig. 8.8c, with the polar head groups represented as circles and the hydrocarbon tails as "squiggles." The polar head groups are exposed to water, and the tails form a hydrocarbonlike region on the interior of the membrane, isolated from water.

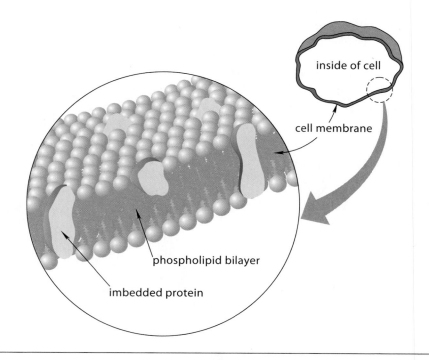

inside of cell

cell membrane

phospholipid bilayer

imbedded protein

can be thought of simplistically as a large phospholipid vesicle in which the cell membrane is a phospholipid bilayer. Actual cells are more complex: they contain nuclei and other substructures (many with their own membranes), enzymes, many different biomolecules, and so on. And cell membranes contain molecules in addition to phospholipids, such as cholesterol and imbedded proteins. Nevertheless, it is the phospholipid bilayer that is primarily responsible for the unique character of the cell membrane.

The phospholipid bilayer is generally impermeable to ions. Charged molecules, as well as inorganic ions, cannot penetrate the hydrocarbonlike interior of the lipid bilayer any more than ions can dissolve in gasoline. The insolubility of ionic compounds in hydrocarbons— specifically, the phospholipid bilayer—is crucial to the cell's ability to retain proper ion balance. The transport of ions through cell membranes requires special carriers or pores, which are proteins imbedded in the membrane. The operation of these ion-carrying systems is tightly regulated by the biochemistry of the cell. (See Sec 8.7B.)

Unlike ions, a number of *uncharged* molecules diffuse readily through the cell membrane. One of the simplest molecules of this type is molecular oxygen (O_2). Many drug molecules are also in this category. In fact, the ability of drugs to pass through the cell membranes correlates with their solubilities in hydrocarbons in the following way. Drugs that are completely insoluble in hydrocarbons do not pass through membranes. Drugs that are highly soluble in hydrocarbons don't either: they move into the phospholipid bilayer and stay there. The drugs that pass through membranes are typically those that have a moderate solubility in hydrocarbons. They are soluble enough in the membrane interior so that they can enter the membrane, but they are soluble enough in water to leave again.

In fact, simple solubility measurements have value in predicting the effectiveness of drug candidates. The potency of many drugs can be correlated, in part, with their relative solubilities in 1-octanol [$CH_3(CH_2)_6CH_2OH$], and water. This relative solubility of a drug candidate is determined by shaking it with a mixture of 1-octanol and an aqueous buffer at pH = 7.4 (physiological pH), and then measuring the concentration of the drug in each phase. If the drug contains a group that can ionize (such as a carboxylic acid) or protonate (such as an amine), the concentration of the neutral drug molecule is calculated from its pK_a value. The ratio of concentrations of the neutral molecule in the 1-octanol and aqueous phases is called the **octanol–water partition coefficient**. Hydrophobic molecules have larger partition coefficients than hydrophilic ones, and, for a given drug class, there is an optimum value for the partition coefficient. Presumably the 1-octanol phase mimics the hydrophobic environments with which a drug must interact to exert its physiological effect. Such environments might include the phospholipid bilayer in the membrane of the target cells, the phospholipid bilayer in the intestinal epithelium (the layer of cells through which an orally administered drug must pass in order to be absorbed), and the active site of a protein target to which the drug ultimately binds, thereby exerting its effect. (See Problem 8.54, p. 381, for an example.) The correlation of 1-octanol–water partition coefficients with drug activity was developed by Professor Corwin Hansch (1918–2011) of Pomona College in California.

Nicotine, the Nicotine Patch, and Cigarette Addiction

The nicotine patch is a practical example of the importance of drug transport across cell membranes. Nicotine is the addictive substance in tobacco, and thus in cigarettes. Nicotine is a base and can exist both as an uncharged free base and as the positively charged conjugate acid. The neutral form has a greater solubility in hydrocarbons than the salt form; the salt form, being ionic, is more soluble in water than the basic form.

$$\text{nicotine} + H_3O^+ \rightleftharpoons \text{nicotine} + H_2O \tag{8.19}$$

nicotine
conjugate base
(neutral)

nicotine
conjugate acid
(cationic)

The nicotine patch is used to wean smokers from cigarettes gradually by providing the addictive material in successively lower doses without requiring smoking. Nicotine within the patch is in the conjugate-base (neutral) form, which readily passes through the skin and various other membrane barriers on its way to the brain, where it exerts its neurological effects. The conjugate acid of nicotine would not be as effective in the patch, because, as an ion, it would not pass through the membrane barriers of the skin.

Cigarette manufacturers have long known that including compounds that release ammonia at high (smoking) temperatures in their cigarettes increases the addictive potential of their products. Ammonia is a base and serves to maintain nicotine in its free-base form, which is readily absorbed through the membranes of the nose, mouth, and lungs.

B. Cation-Binding Molecules

If we truly understand the mechanisms of ionic solvation, we should be able to create synthetic molecules that mimic the solvation shells of ions. Such molecules should be **ionophores**—molecules that form strong complexes with ions. (The word *ionophore* means "ion-bearing.") This goal has been largely realized in the design of *crown ethers* and *cryptands*, which we'll examine in this section. Then, as we consider the structures of ion carriers found in nature, such as *ionophore antibiotics* and *ion channels*, we'll find that the very same mechanisms of ionic solvation are important.

Crown Ethers and Cryptands   Some metal cations form stable complexes with a class of synthetic ionophores known as *crown ethers*, which were first prepared in 1967. **Crown ethers** are heterocyclic ethers containing a number of regularly spaced oxygen atoms. Some examples of crown ethers are the following:

[18]-crown-6   **[12]-crown-4**   **dibenzo[18]-crown-6**

FIGURE 8.11 Structure of the [18]-crown-6 complex of the potassium ion. (a) A Lewis structure; (b) a ball-and-stick model; and (c) a space-filling model. The oxygen atoms are shown in red. Because the outside of the complex is essentially a hydrocarbon, crown ethers and their complexes are soluble in hydrocarbon solvents. The dashed lines indicate donor and/or charge–dipole interactions of the oxygens with the ion.

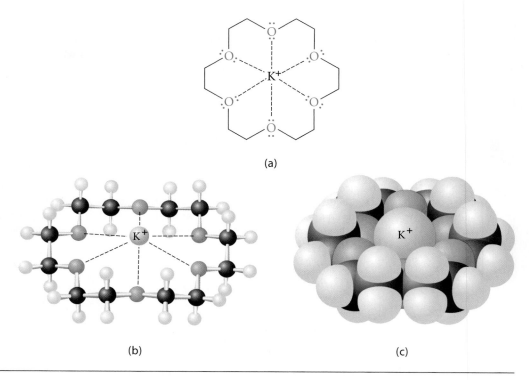

(a)

(b) (c)

(The number in brackets indicates the total number of atoms in the ring, and the number following the hyphen indicates the number of oxygens.) The term *crown* was suggested by the three-dimensional shape of these molecules, shown in Fig. 8.11 for the complex of [18]-crown-6 with the potassium ion (K^+). The oxygens of the "host" crown ether wrap around the "guest" metal cation, complexing it within the cavity of the ether using the donor and charge–dipole interactions discussed in the previous section. In fact, one can think of a crown ether molecule as a "synthetic solvation shell" for a cation. Because the metal ion must fit within the cavity, the crown ethers have some selectivity for metal ions according to size. For example, [18]-crown-6 forms the strongest complexes with potassium ion and somewhat weaker complexes with sodium, cesium, and rubidium ions. It does not form complexes with lithium ions. On the other hand, [12]-crown-4, with its smaller cavity, specifically forms complexes with the lithium ion.

Closely related to the crown ethers are the **cryptands**, which are nitrogen-containing analogs of the crown ethers. The presence of nitrogen allows for a bicyclic structure that provides an additional pair of oxygens to assist in binding the metal ion. The structure of a typical cryptand and its complex with a potassium ion is shown in Fig. 8.12. (Complexes of metal ions and cryptands are called *cryptates*.)

Because their structures contain hydrocarbon groups, crown ethers and cryptands have significant solubilities in hydrocarbon solvents such as hexane or benzene. The remarkable thing about the crown ethers and cryptands is that they can cause inorganic salts to dissolve in solvents in which these salts ordinarily have little or no solubility. For example, when potassium permanganate is added by itself to the hydrocarbon benzene, no $KMnO_4$ dissolves. Upon addition of a little dibenzo[18]-crown-6, which forms a complex with the potassium ion, the benzene takes on the purple color of a $KMnO_4$ solution, and this solution (nicknamed "purple benzene") acquires the oxidizing power typical of $KMnO_4$. What happens is that the crown ether complexes the potassium cation and dissolves it in benzene; electrical neutrality demands that the permanganate ion accompany the complexed potassium ion into solution. The stabilization of the potassium ion by the crown ether compensates for the fact that the permanganate anion is essentially unsolvated, or "naked." Other potassium salts can be dissolved in hydrocarbon solvents in a similar manner. For example, KCl and KBr can be dissolved in hydrocarbons in the presence of crown ethers to give solutions of "naked chloride" and "naked bromide," respectively.

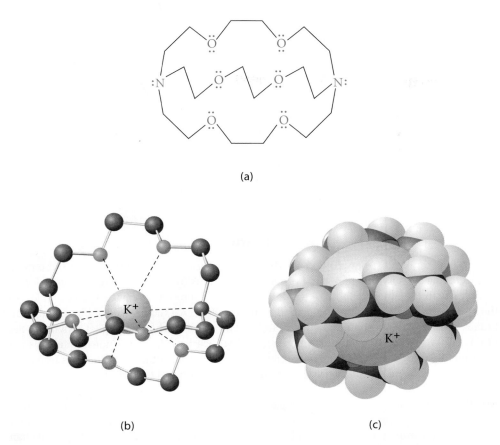

(a)

(b)   (c)

FIGURE 8.12 (a) A skeletal structure of [2.2.2]-cryptand. (The numbers refer to the number of oxygens in each of the three chains.) (b) A ball-and-stick model of a cryptate, a [2.2.2]-cryptand containing a bound potassium ion. (The hydrogen atoms are not shown.) (c) A space-filling model of the cryptate in (b) with hydrogens shown. The dashed lines indicate donor and/or charge–dipole interactions of the oxygens (*red*) or nitrogens (*blue*) with the ion.

Host–Guest Chemistry

As discussed on p. 372, crown ethers and cryptands can discriminate among various cations on the basis of *ionic size*. As a result, crown ethers and cryptands bind ions with a degree of *selectivity*. In recent years, chemists have designed other classes of molecules that can "recognize" and bind more complicated compounds on the basis of their precise structures. This type of work has been spurred, at least in part, by a desire to understand and duplicate synthetically the highly specific binding characteristic of such biological molecules as enzymes and receptors. This general field, called *host–guest chemistry*, or *molecular recognition*, was recognized with the 1987 Nobel Prize in Chemistry, which was awarded to three of its pioneers: Charles J. Pedersen (1904–1989), then a chemist with DuPont, who invented the crown ethers; Jean-Marie Lehn (b. 1939), professor of chemistry at Université Louis Pasteur in Strasbourg, France, and the Collège de France in Paris, who devised the cryptands; and Donald J. Cram (1919–2001), who was a professor of chemistry at the University of California, Los Angeles.

Ionophore Antibiotics The *ionophore antibiotics* are biologically important examples of ionophores. An *antibiotic* is a compound that interferes with the growth or survival of one or more microorganisms. The ionophore antibiotics form strong complexes with metal ions in much the same way as crown ethers and cryptands. An example of such a compound is nonactin, one of a group of antibiotics produced by a microorganism, *Streptomyces griseus.*

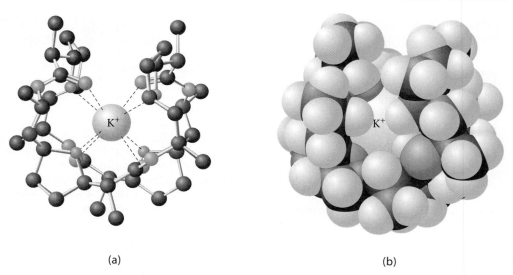

nonactin

Nonactin has a strong affinity for the potassium ion. As shown in Fig. 8.13, the molecule contains a cavity in which eight of the oxygen atoms (red in the above structure) form a complex with the ion. In contrast, the atoms on the outside of the nonactin molecule are for the most part hydrocarbon groups. Recall from Sec. 8.7A that the interior of biological membranes consists of a phospholipid bilayer, and that this hydrocarbonlike region provides a natural barrier to the passage of ions. However, the hydrocarbon surface of nonactin allows it to enter readily into, and pass through, membranes. Because nonactin binds and thus transports ions, the ion balance crucial to proper cell function is upset, and the cell dies.

Ion Channels   Ion channels, or "ion gates," provide passageways for ions into and out of cells. (Recall that ions are not soluble in membrane phospholipids.) The flow of ions is essential for the transmission of nerve impulses and for other biological processes. A typical channel is a large protein molecule imbedded in a cell membrane. Through various mechanisms, ion channels can be opened or closed to regulate the concentration of ions in the interior of the cell. Ions do not diffuse passively through an open channel; rather, an open channel contains regions that bind a specific ion. Such an ion is bound specifically within the channel at one side of the membrane and is expelled from the channel on the other side.

(a)   (b)

FIGURE 8.13 Models of the complex of the antibiotic nonactin with potassium ion. (a) A ball-and-stick model in which the hydrogens are not shown. The dashed lines indicate interaction between oxygen atoms (*red*) and the potassium ion. (b) A space-filling model in which the hydrogens are shown. The nonactin molecule wraps around the potassium ion like a hand holding a ball. Because the outside of the complex is essentially hydrocarbon in character, the complex, like the crown ether–metal ion complexes (Fig. 8.11), is soluble in nonpolar aprotic solvents.

water outside of the cell

protein   protein

K$^+$

K$^+$

K$^+$

phospholipid
bilayer

phospholipid
bilayer

water inside of the cell

(a)

K$^+$

(b)

FIGURE 8.14 An open potassium ion channel. (a) A schematic cross-sectional diagram showing the location of the channel in the cell membrane. The protein is color-coded green. The protein spans the membrane, and much of the protein exterior contacts the lipid bilayer. The red region is the selectivity filter, which is the region of the protein that binds the potassium ion selectively. (b) An atomic-level view from above (eyeball) of the selectivity filter containing one potassium ion. The oxygens (*red*) are derived from C=O bonds of the protein (double bonds are not shown). Carbons (*black*) and nitrogens (*blue*) of the protein are shown without their attached hydrogens. The dashed lines indicate the ion–dipole and donor interactions of the oxygens with the ion.

Remarkably, the structures of the ion-binding regions of these channels have much in common with the structures of ionophores such as nonactin. The first X-ray crystal structure of a potassium-ion channel was determined in 1998 by a team of scientists at Rockefeller University led by Prof. Roderick MacKinnon (b. 1956), who shared the 2003 Nobel Prize in Chemistry for this work. A schematic diagram of this protein and how it is situated in a membrane is shown in Fig. 8.14a. The exterior of the channel interacts with the hydrocarbon part of the membrane phospholipid bilayer. As we might expect from our discussion of noncovalent interactions, the groups on this part of the channel are largely hydrocarbon in character; the channel remains anchored in the membrane by hydrophobic bonding and van der Waals attractions. The entrance to the channel on the outside of the cell contains binding sites for two potassium ions; these sites are oxygen-rich, much like the interior of nonactin. The oxygens come from C=O groups within the protein structure. They are situated so that they perfectly accommodate a potassium ion and are too far apart to interact effectively with a sodium ion. This part of the channel is called the "selectivity filter." A magnified view of the selectivity filter from above at the atomic level is shown in Fig. 8.14b. The oxygens in the selectivity filter provide "solvation" for the potassium ion that replaces the solvation water of the ion in solution. Release of this solvation water undoubtedly provides an entropic drive for binding of the ion to the channel. As one ion moves down the selectivity filter, a second potassium ion can enter the channel. The repulsion between the two ions then balances the ion-binding forces, and one of the ions can then leave the channel; this is postulated to be the mechanism of ionic conduction.

PROBLEM

8.27 The crown ether [18]-crown-6 (structure on p. 372) has a strong affinity for the methylammonium ion, $CH_3\overset{+}{N}H_3$. Propose a structure for the complex between [18]-crown-6 and this ion. (Although the crown ether is bowl-shaped, you can draw a planar structure for purposes of this problem.) Show the important interactions between the crown ether and the ion.

8.8	STRENGTHS OF NONCOVALENT INTERMOLECULAR ATTRACTIONS

We've now seen an array of noncovalent attractions and some of their consequences in both chemistry and biology. Although the attractions between ions can be very strong, the other attractive forces we have discussed—van der Waals attractions, dipole–dipole attractions, and hydrogen bonds—are all relatively weak. It is worth stating again that the strengths of these noncovalent attractions have *nothing* to do with the stabilities of the molecules involved in those interactions. The stabilities of individual molecules (as measured by heats of formation) result from the stability of their covalent bonds. Because melting or boiling—processes that break noncovalent intermolecular attractions—do not involve breaking and forming covalent bonds, noncovalent attractions must be much weaker than the covalent bonds. For example, breaking a typical alkane carbon–carbon bond requires about 380 kJ mol^{-1}. Converting hexane from a liquid into a gas requires about 29 kJ mol^{-1}, or about 5 kJ mol^{-1} per carbon, and this increases with alkane size by only about 3 kJ mol^{-1} per carbon. In other words, the van der Waals attractions between hydrocarbon groups in a liquid on a per-carbon basis are less than 1% of the carbon–carbon bond energy. Although dipole–dipole attractions and hydrogen bonds are somewhat stronger, they, too, are much weaker than covalent bonds. For example, the strengths of hydrogen bonds are typically 3–10% of the strengths of covalent bonds.

Despite the magnitudes of these "weak" forces, they have chemically significant consequences, in many cases because they result from *many molecules acting together.* As you've seen, the structures of membranes, and as you'll learn, the structures of proteins and the DNA, are a consequence of many weak noncovalent attractions acting in concert. It's a bit like an organization in which all the members work together towards the same objectives: the efforts of a single individual may seem relatively insignificant, but all of the individuals working together can have powerful consequences.

KEY IDEAS IN CHAPTER 8

- Alcohols, ethers, alkyl halides, thiols, and sulfides are classified as methyl, primary, secondary, or tertiary according to the number of substituent groups at the alpha (α) carbon.

- Organic compounds are named by both common and substitutive nomenclature. In substitutive nomenclature, the name is based on the principal group and the principal chain. The principal group, specified by priority, is cited as a suffix in the name. Other groups are cited as substituents. Hydroxy (—OH) and mercapto (—SH) groups can be cited as principal groups with the suffixes *ol* and *thiol*, respectively; hydroxy has a higher priority for citation as a principal group. Halogens, alkoxy (—OR) groups, and alkylthio (—SR) groups are always cited as substituents.

- Boiling points depend on the noncovalent attractions between molecules. The types of noncovalent forces that affect boiling point are (1) van der Waals attractions, which include induced dipole–induced dipole attractions and dipole–induced dipole attractions; (2) dipole–dipole attractions; and (3) hydrogen bonding. All of these noncovalent attractions are much weaker than covalent bonds.

- The general increase of boiling points (roughly 20–30 °C per carbon) within many homologous series is due to the increasing areas of the molecular surfaces at which molecules can interact by van der Waals attractions.

- Van der Waals attractions are strongest when interacting molecules are polarizable—that is, when their electron clouds are easily deformed by interactions with external charges. Less polarizable molecules, such as perfluorocarbons, have lower boiling points than more polarizable molecules, such as hydrocarbons, of the same molecular mass and shape.

- Large amounts of branching in hydrocarbon groups reduces their surface area relative to unbranched groups of the same molecular mass; therefore, the boiling points of highly branched molecules tends to be lower than that of their unbranched isomers containing the same functional groups.

- Solubilities are a consequence of the noncovalent intermolecular interactions of unlike molecules (solute and solvent) weighed against the noncovalent intermolecular interactions of pure solute molecules and pure solvent molecules. The solubility of a solute in a solvent is a consequence of the free energy of solution, ΔG_s, which is the free energy of the solution itself minus the free energies of pure solvent and pure solute.

- The free energy of solution ΔG_s is composed of two parts: the free energy of interaction, ΔG_{inter}, and the free energy of mixing, ΔG_{mixing} (Eq. 8.8). The free energy of mixing originates solely from the entropy of mixing, ΔS_{mixing}, by the equation $\Delta G_{mixing} = -T\Delta S_{mixing}$. The ΔG_{mixing} is always favorable, just as the mixing of blue balls and red balls in a box is favorable; that is, mixing occurs spontaneously.

- Solvents are classified as protic or aprotic, depending on their ability to donate hydrogen bonds; polar or apolar, depending on the magnitude of its dielectric constant; and donor or nondonor, depending on its ability to act as a Lewis base. These categories can and do overlap.

- When the interactions between molecules in the pure substances are similar to those between the solvent and solute in the solution, then the ΔG_s is dominated by the entropy of mixing, and a solution forms. This situation is frequently observed when solute and solvent have the same classification (for example, both protic, or both apolar aprotic). This is the basis of the like-dissolves-like rule of solubility.

- The tendency of hydrocarbon groups in aqueous solution to associate is called *hydrophobic bonding*. Hydrophobic bonding is driven by the entropy increase of water when the water molecules that solvate hydrocarbon chains (lower-entropy water) are released into solvent water (higher-entropy water). The association of phospholipid molecules to form lipid bilayers and cell membranes is an example of hydrophobic bonding.

- Within many homologous series, a graph of melting points against the number of carbons in the alkyl chain shows a sawtooth pattern. Highly symmetrical compounds typically have significantly higher melting points than their less symmetrical counterparts. Both effects are caused by (a) the greater entropy associated with a crystal containing symmetrical molecules; and (b) by the more efficient crystal packing and, as a consequence, greater van der Waals attractions, in the crystals of the higher-melting compounds.

- The solubility of a solid is affected by its melting point. Solids with high melting points are typically less soluble than isomers containing the same functional groups with lower melting points. Structural symmetry, because it raises melting point, tends to reduce solubility.

- The high dielectric constant of a polar solvent contributes to the solubility of an ionic compound by reducing the attractions between oppositely charged ions, as shown by the electrostatic law (Eq. 8.10, p. 353). Ionic solubility is also enhanced by donor interactions, charge–dipole attractions, and hydrogen bonding of an ion with the solvent molecules in its solvation shell.

- Because of the effective solvation of ions, molecules with limited water solubility typically become more soluble when they can be ionized.

- Crown ethers, cryptands, and other ionophores form complexes with cations by creating artificial solvation shells for them in which the ions are stabilized by ion–dipole and donor interactions. Ionophore antibiotics and ion channels bind cations by the same principle.

ADDITIONAL PROBLEMS

8.28 (a) Give the structures of all alcohols with the molecular formula $C_5H_{11}OH$.

(b) Which of the compounds in part (a) are chiral?

(c) Name each compound using IUPAC substitutive nomenclature.

(d) Classify each as a primary, secondary, or tertiary alcohol.

(e) Show the α-carbon on each structure.

8.29 (a) Give the structures of all ethers with the molecular formula $C_5H_{12}O$.

(b) Which of the compounds in part (a) are chiral?

(c) Name each compound using both common and IUPAC substitutive nomenclature.

(d) Show the two α-carbons in each structure.

8.30 Give a structure for each of the following compounds. (In some cases, more than one answer is possible.)

(a) a chiral ether $C_5H_{10}O$ that has no double bonds

(b) a chiral alcohol C_4H_6O

(c) a vicinal glycol $C_6H_{10}O_2$ that cannot be optically active at room temperature

(d) a diol $C_4H_{10}O_2$ that exists in only three stereoisomeric forms

(e) a diol $C_4H_{10}O_2$ that exists in only two stereoisomeric forms

(f) the *six* epoxides (counting stereoisomers) with the molecular formula C_4H_8O

8.31 Give the IUPAC substitutive name for each of the following compounds, which have been used as general anesthetics.

(a)

$$H-\overset{\overset{\displaystyle Br}{|}}{\underset{\underset{\displaystyle Cl}{|}}{C}}-CF_3$$

halothane

(b) $Cl_2CH-CF_2-OCH_3$

methoxyflurane

8.32 Thiols of low molecular mass are known for their extremely foul odors. In fact, the following two thiols are the active components in the scent of the skunk. Give the IUPAC substitutive names for these compounds.

(a) $CH_3CH{=}CHCH_2SH$

(b) $(CH_3)_2CHCH_2CH_2SH$

8.33 Give an IUPAC name for each of the following compounds, which may have been isolated from the shoes of a tennis player. Ignore stereochemistry in (a).

8.34 Without consulting tables, arrange the compounds within each of the following sets in order of increasing boiling point, and give your reasoning.

(a) 1-hexanol, 2-pentanol, *tert*-butyl alcohol

(b) 1-hexanol, 1-hexene, 1-chloropentane

(c) diethyl ether, propane, 1,2-propanediol

(d) cyclooctane, chlorocyclobutane, cyclobutane

8.35 In each of the following parts, explain why the first compound has a higher boiling point than the second, despite a lower molecular mass.

(a)

$$H_3C-\overset{\overset{\displaystyle O}{\|}}{C}-OH \qquad H_3C-\overset{\overset{\displaystyle O}{\|}}{C}-OCH_2CH_3$$

(bp 118 °C)   (bp 77 °C)

(b)

$$H_3C-\overset{\overset{\displaystyle O}{\|}}{C}-NH_2 \qquad H_3C-\overset{\overset{\displaystyle O}{\|}}{C}-N(CH_3)_2$$

(bp 221 °C)   (bp 166 °C)

8.36 The molecules nitromethane and 2-propanol have roughly the same shape and molecular mass.

	2-propanol (isopropyl alcohol)	nitromethane
molecular mass	60.1	61.0
boiling point	82.4 °C	101.2 °C

Liquid 2-propanol contains hydrogen bonds, but liquid nitromethane does not. Yet nitromethane has the higher boiling point. Why does nitromethane have such a high boiling point? What physical properties of the two molecules could you look up to support your answer?

8.37 For the following problems, see Table 8.2 (p. 355) for structures and dipole moments. Explain your reasoning in each case.

(a) With which *one* of the following solvents is DMSO *not* miscible: water, acetone, hexane, or acetonitrile?

(b) With which *one* of the following solvents is hexane *not* miscible: methanol, 1-propanol, diethyl ether, or acetone?

8.38 (a) One of the following compounds is an unusual example of a salt that is soluble in hydrocarbon solvents. Which one is it? Explain your choice.

$$CH_3(CH_2)_{15}-\overset{\overset{\displaystyle CH_2CH_2CH_2CH_3}{|}}{\underset{\underset{\displaystyle CH_2CH_2CH_2CH_3}{|}}{\overset{+}{N}}}-CH_2CH_2CH_2CH_3 \quad Br^- \qquad \overset{+}{N}H_4 \quad Cl^-$$

A   *B*

(b) Which of the following would be present in greater amount in a hexane solution of the compound in part (a): separately solvated ions, or ion pairs and higher aggregates? Explain your reasoning.

8.39 The pK_a of water is 15.7. Titration of an aqueous solution containing Cu^{2+} ion suggests the presence of a species that acts as a Brønsted acid with $pK_a = 8.3$. Suggest a structure for this species. (*Hint*: Cu^{2+} is a Lewis acid.)

8.40 Normally, dibutyl ether is much more soluble in benzene than it is in water. Explain why this ether can be extracted from benzene into water if the aqueous solution contains moderately concentrated nitric acid. (*Hint:* The oxygen of ethers is weakly basic.)

8.41 Show the types of solvation interactions that might be expected in each solution. Consider the solvation of both the cation and the anion.

(a) a solution of the salt ammonium chloride in ethanol

(b) a solution of the salt sodium chloride in *N*-methylpyrrolidone

8.42 The dissociation constant K_d of the complex between a crown ether and a metal ion M^+ is given by

$$K_d = \frac{[\text{crown ether}][M^+]}{[\text{crown ether--}M^+\text{ complex}]}$$

Explain why the complex of the crown ether [18]-crown-6 with potassium ion has a much larger dissociation constant in water than it does in ether.

8.43 Nonactin (structure on p. 374) forms a strong complex with the ammonium ion, $^+NH_4$. What types of interactions are expected between the nonactin molecule and the bound ion? Contrast these interactions with those between nonactin and the bound potassium ion (Fig. 8.13, p. 374).

8.44 Suggest a structure for a constitutional isomer of the following compound that should have *greater* water solubility, and explain your reasoning. The structure should *not* be an enol (p. 325), because enols are not stable. (Several correct answers are possible.)

$$CH_3CH_2CH_2 - \overset{\overset{\displaystyle O}{\|}}{C} - \underset{\underset{\displaystyle CH_3}{|}}{N} - OCH_3$$

8.45 The effectiveness of barbiturates as sedatives has been found to be directly related to their solubility in, and thus their ability to penetrate, the lipid bilayers of membranes. Which of the following two barbiturate derivatives should be the more potent sedative? Explain.

barbital   **hexethal**

8.46 When salad oil (typical structure shown in Fig. P8.46) is mixed with water and shaken, two layers quickly separate, the oily layer on top and the water layer on the bottom. When an egg yolk (which is rich in lecithin, a phospholipid) is added and the mixture shaken, an emulsion is formed. This means that the oil is suspended (not dissolved) in water as tiny particles that no longer form a separate layer. Explain the action of the lecithin. (Addition of romaine lettuce, garlic, croutons, and Parmesan cheese yields a Caesar salad!)

8.47 *Propofol* is used as a general anesthetic in the first stages of surgery.

2,6-diisopropylphenol
(propofol)

Propofol is insoluble in water, but it has to be administered by intravenous injection. For this reason, propofol is formulated as a mixture with soybean oil and lecithin. (This milky formulation is sometimes jokingly called "milk of amnesia" by anesthesiologists.) What is the purpose of these additives? (The structure of an oil is shown in Fig. P8.46.)

8.48 Vitamins can be classified as "fat-soluble" or "water-soluble." Fat-soluble vitamins can be stored in fatty tissues, whereas water-soluble vitamins can be excreted in the urine.

(a) The structures of some vitamins are given in Fig. P8.48 on p. 380. Using their structures to guide you, classify each as fat soluble or water soluble and explain. (The structure of a fat is similar to that of an oil; see Fig. P8.46.)

(b) For which type of vitamin would an overdose be more dangerous? Why?

8.49 Ethyl alcohol in the solvent CCl_4 forms a hydrogen-bonded complex with an equilibrium constant $K_{eq} = 11$.

$$2\,C_2H_5OH \;\rightleftharpoons\; C_2H_5\ddot{O}: \text{----}H - \overset{\displaystyle \ddot{O}}{\underset{\displaystyle C_2H_5}{|}}$$
$$\qquad\qquad\quad \underset{\displaystyle H}{|}$$

(a) What happens to the concentration of the complex as the concentration of ethanol is increased? Explain.

(b) What is the standard free-energy change for this reaction at 25 °C?

(c) If one mole of ethanol is dissolved in one liter of CCl_4, what are the concentrations of free ethanol and of complex?

(d) The equilibrium constant for the analogous reaction of ethanethiol is 0.004. Which forms stronger hydrogen bonds, thiols or alcohols?

(e) Which would be more soluble in water: $CH_3OCH_2CH_2SH$ or its isomer $CH_3SCH_2CH_2OH$? Explain.

typical structure of an oil or fat
R = an unbranched chain with 15–17 carbons

an oil: R contains one or more cis double bonds
a fat: R is saturated (no double bonds)

Figure P8.46

8.50 In each of the following pairs, one compound has a melting point that is much higher than the other, but the two have very similar boiling points. Choose the compound with the greater melting point, and explain your reasoning.

(a) (reported melting points 62–65 °C, 17 °C)

A

B

(b) (reported melting points 61–62 °C, 21 °C)

A

B

8.51 The boiling points of *tert*-butyl alcohol (2-methyl-2-propanol) and 1-butanol differ by 36 °C (82° and 118°). Before proceeding, draw their structures.

(a) Which has the higher boiling point? Give a reason for this difference.

(b) How would the differences in their structures affect the ΔS_s (entropy of solution) contribution to their water solubilities? (*Hint:* Think about the surface areas of the two alkyl groups and decide which would require the greater amount of low-entropy water of solvation in aqueous solution.)

(c) One compound is miscible with water and the other has a limited solubility of 8 mass percent. Which compound is miscible? Explain. (*Hint:* What does the boiling point tell you about the intermolecular attractions within the pure liquids?)

8.52 Offer explanations for each of the following.

(a) Ethanol and 1-propanol disrupt the lipid bilayer of cell membranes.

(b) Sugars such as glucose do not diffuse freely through the cell membrane.

α-D-glucopyranose
(a form of glucose)

(Sugars require specific protein transporters to enter cells.)

(c) The $\Delta S°$ for transfer of the salt Bu_4N^+ Cl^- from water to DMSO is strongly positive (+130 J K^{-1} mol^{-1}, +31 cal K^{-1} mol^{-1}). (*Hint:* Consider solvation of the cation.)

(d) Diethyl ether (Et—O—Et) is significantly more soluble in water (about 3.4 times) than 1-pentanol ($CH_3CH_2CH_2CH_2CH_2$—OH), the primary alcohol with about the same molecular mass and shape. (*Hint:* Consider the relative boiling points and use the hint in Problem 8.51c.)

8.53 If hydrophobic bonding is driven by a highly positive $\Delta S°$, how would you expect the equilibrium constant for an association reaction driven by hydrophobic bonding to change with increasing temperature? That is, would it become greater, become smaller, or stay the same? Explain.

vitamin C
(ascorbic acid)

vitamin A

vitamin B₆
(pyridoxine)

vitamin D₃
(R = a hydrocarbon group)

vitamin B₂
(riboflavin)

Figure P8.48

8.54 (*S*)-Propranolol and (*S*)-atenolol (Fig. P8.54) are two well-known drugs used as β-blockers. They suppress the "fight-or-flight" response and are sometimes used to treat panic in examination or performance situations. Drugs such as these are often characterized by their *octanol–water partition coefficients* P_{ow} (p. 370). A small amount of drug is dissolved in a known volume of 1-octanol and then the solution is shaken with a known amount of water. The two phases are allowed to separate, and concentration of drug in each phase is determined. The octanol–water partition coefficient is calculated from the equilibrium concentrations in each phase:

$$P_{ow} = \frac{\text{drug concentration in 1-octanol phase}}{\text{drug concentration in water phase}}$$

(a) The partition coefficients of these two drugs were determined using water that was made basic to suppress protonation of the amine group. The partition coefficient of one drug is more than 2000 times (3.3 log units) greater than that of the other. Which drug has the greater partition coefficient? Explain.

(b) The partition coefficient of atenolol, expressed in logarithmic units, is log $P_{ow} = 0.22$. If 1.0 g of atenolol is partitioned between equal volumes of the two phases, how much atenolol is in each phase?

(c) What does your answer to part (a) tell you, if anything, about the relative water solubilities of crystalline atenolol and propranolol? The reported melting points of the two drugs are 152–153 °C (atenolol) and 72 °C (propranolol). (*Hint:* See Eq. 8.14 on p. 362.)

(d) The amino nitrogen is basic in both drugs. Draw the structure of the conjugate acids of each.

(e) The pK_a values of the conjugate acids of both compounds are about the same (9.5). The partition coefficient of each compound was re-determined using a water phase containing a buffer at pH 7.4. How would the partition coefficient of these drugs change (if at all) under these conditions? Would it significantly increase, significantly decrease, or stay about the same? Explain.

(f) These drugs are marketed as their hydrochloride salts (their conjugate acids with a chloride counter-ion). The drugs are administered in the solid phase orally as capsules. What is one reason that the drugs are formulated in this form instead of in their basic form?

8.55 Offer an explanation for each of the following observations.

(a) Compound *A* exists mostly in a chair conformation with an equatorial —OH group, but compound *B* prefers a chair conformation with an axial —OH group.

A   *B*

(b) The racemate of 2,2,5,5-tetramethyl-3,4-hexanediol exists with a strong intramolecular hydrogen bond, but the meso stereoisomer has no intramolecular hydrogen bond.

8.56 (a) Show that the dipole moment of 1,4-dioxane (p. 333) should be zero if the molecule exists solely in a chair conformation.

(b) Account for the fact that the dipole moment of 1,4-dioxane, although small, is definitely not zero. (It is 0.38 D.)

8.57 (a) Use the relative bond lengths of the C—C and C—O bonds to predict which of the following two equilibria lies farther to the right. (That is, predict which of the two compounds contains more of the conformation with the axial methyl group.)

(1)

(2)

(b) Which one of the following compounds contains the greater amount of gauche conformation for internal rotation about the bond shown? Explain.

CH_3CH_2———OCH_3 CH_3CH_2———CH_2CH_3

A   *B*

(S)-propranolol   **(S)-atenolol**

Figure P8.54

CHAPTER 9

The Chemistry of Alkyl Halides

The major part of this chapter is devoted to two very important types of alkyl halide reactions: *nucleophilic substitution reactions* and *β-elimination reactions*. These are among the most common and important reactions in organic chemistry, and we cover them together because they frequently occur as competing reactions. Then we'll introduce *organometallic compounds*—compounds that involve carbon–metal bonds. Alkyl halides in many cases serve as starting materials for the preparation of these compounds. This chapter also introduces another class of reactive intermediate: *carbenes*, and how they are used to prepare cyclopropanes. Finally, we'll consider some of the industrial and environmental aspects of halogen-containing organic compounds, including how alkyl chlorides and bromides can be prepared by free-radical halogenation reactions of alkanes.

Compared with the total number of organic compounds, relatively few halogen-containing compounds occur naturally—around 5000. If you are a life science or premedical student, you might wonder why you should bother learning about the chemistry of compounds that you are not likely to encounter in biology. The reason is that much of our quantitative data and current understanding of reactivity comes from studies on alkyl halides. Moreover, the chemistry of alkyl halides demonstrates in a straightforward way the types of reactivity and mechanism that we'll encounter in more complex molecules, including those that occur in biological systems. Think of a musical theme and variations: alkyl halide chemistry provides the theme, and the chemistry of alcohols, ethers, and amines will provide the variations. In other words, *alkyl halides provide simple models from which we understand the chemistry of other compound classes*. These same considerations are equally valid for the chemistry major; furthermore, alkyl halides are important starting materials used in a wide variety of laboratory reactions.

9.1 OVERVIEW OF NUCLEOPHILIC SUBSTITUTION AND β-ELIMINATION REACTIONS

A. Nucleophilic Substitution Reactions

When a *methyl halide* or a *primary alkyl halide* reacts with a nucleophile, such as sodium ethoxide, a reaction occurs in which the nucleophilic atom of the base, in this case oxygen, displaces the halogen, which is expelled as halide ion.

$$Na^+ \ CH_3CH_2\ddot{O}:^- + :\ddot{Br}-CH_2CH_3 \longrightarrow CH_3CH_2\ddot{O}-CH_2CH_3 + Na^+ :\ddot{Br}:^- \quad (9.1)$$

sodium ethoxide **ethyl bromide** **diethyl ether** **sodium bromide**

This is an example of a very general type of reaction, called a **nucleophilic substitution reaction**, or **nucleophilic displacement reaction**. In the simplest type of nucleophilic substitution reaction, a *nucleophile* donates an electron pair to an *electrophile* to displace a *leaving group*.

$$Na^+ \ CH_3CH_2\ddot{O}:^- \quad \overset{\overset{\text{leaving group}}{(:\ddot{Br}:}}{CH_2CH_3} \longrightarrow CH_3CH_2\ddot{O}-CH_2CH_3 + Na^+ :\ddot{Br}:^- \quad (9.2)$$

nucleophile   electrophile

(You should review these terms in Sec. 3.2B if necessary.) In this chapter, the electrophile will typically be a carbon and the leaving group will typically be a halide ion. As you've already seen, however, there are many examples of nucleophilic substitution reactions that involve other electrophiles and leaving groups.

The base used in Eqs. 9.1–9.2, ethoxide ($CH_3CH_2O^-$, often abbreviated as EtO^-), is the conjugate base of ethanol (CH_3CH_2OH). Ethanol is a weak acid ($pK_a = 15.9$); it has about the same acidity as water. Therefore, ethoxide is a strong base—about as strong as hydroxide ion. In general, the conjugate acids of alcohols are called **alkoxides**. You'll see many reactions in this chapter in which alkoxide bases are used. (The acidity of alcohols and the preparation of alkoxides are covered more generally in Sec. 10.1.)

In Eqs. 9.1–9.2, the sodium ion plays the role of a **spectator ion**—an ion that has no overt role in the overall reaction. If we think of a chemical equation as an algebraic entity, a spectator ion can be "subtracted" from both sides of the equation without changing the bond-making and bond-breaking chemistry. For simplicity, we often write nucleophilic substitution reactions (and elimination reactions that you'll learn about in the Sec. 9.1B) as **net ionic equations**—equations in which only the reacting ionic components are shown and the spectator ion is omitted. For example, the net ionic equation corresponding to Eq. 9.1 is as follows:

$$CH_3CH_2\ddot{O}:^- + :\ddot{Br}-CH_2CH_3 \longrightarrow CH_3CH_2\ddot{O}-CH_2CH_3 + :\ddot{Br}:^- \quad (9.3)$$

ethoxide ion **ethyl bromide** **diethyl ether** **bromide ion**

When you see a net ionic equation, you can assume always that there is a spectator ion present—in many cases Na^+ or K^+.

Although many nucleophiles are anions, others are uncharged. The following equation contains an example of an uncharged nucleophile. In addition, it illustrates an *intramolecular substitution reaction*—a reaction in which the nucleophilic center, the electrophilic center, and the leaving group are part of the same molecule. In this case, the nucleophilic substitution reaction causes a ring to form.

$$\text{(9.4)}$$

Nucleophilic substitution reactions can involve many different nucleophiles, a few of which are listed in Table 9.1. Notice from this table that nucleophilic substitution reactions can be used to transform alkyl halides into a wide variety of other functional groups. We'll discuss nucleophilic substitution reactions in more detail in Secs. 9.4 and 9.6.

PROBLEMS

9.1   What is the expected nucleophilic substitution product when
(a) methyl iodide reacts with $Na^+\ CH_3CH_2CH_2CH_2S^-$?
(b) ethyl iodide reacts with ammonia?

9.2   Write the equation for the reaction in Problem 9.1a as a net ionic equation.

B. β-Elimination Reactions

When a *tertiary alkyl halide* reacts with a Brønsted base such as sodium ethoxide, a very different type of reaction is observed.

$$\text{(9.5)}$$

sodium ethoxide   **tert-butyl bromide**   **ethanol**   **2-methylpropene (isobutylene)**   **sodium bromide**

This is an example of an **elimination reaction**: a reaction in which two or more groups (in this case H and Br) are lost from within the same molecule. We'll discuss the mechanism of this reaction in Sec. 9.5A.

In an alkyl halide, the carbon bearing the halogen is often referred to as the **α-carbon** (Sec. 8.1), and the adjacent carbons are referred to as the **β-carbons**. Notice in Eq. 9.5 that the halide is lost from the α-carbon and a proton from a β-carbon.

An elimination that involves loss of two groups from adjacent carbons to form a double bond is called a **β-elimination**. This is the most common type of elimination reaction in organic chemistry. Notice that a β-elimination reaction is conceptually the reverse of an addition to an alkene.

Strong bases promote the β-elimination reactions of alkyl halides. Among the most frequently used bases are alkoxides, which we introduced in Sec. 9.1A. The two bases you'll

TABLE 9.1 Some Nucleophilic Substitution Reactions
(X = halogen or other leaving group; R, R′ = alkyl groups)

R—$\ddot{\text{X}}$: + Nucleophile (name)	⟶	:$\ddot{\text{X}}$:⁻ + Product (name)
R—$\ddot{\text{X}}$: + :$\ddot{\text{Y}}$:⁻ (another halide)	⟶	:$\ddot{\text{X}}$:⁻ + R—$\ddot{\text{Y}}$: (another alkyl halide)
+ ⁻:C≡N: (cyanide)	⟶	+ R—C≡N: (nitrile)
+ ⁻:$\ddot{\text{O}}$H (hydroxide)	⟶	+ R—$\ddot{\text{O}}$H (alcohol)
+ ⁻:$\ddot{\text{O}}$R′ (alkoxide)	⟶	+ R—$\ddot{\text{O}}$—R′ (ether)
+ ⁻N₃ (azide = :$\ddot{\text{N}}$=$\overset{+}{\text{N}}$=$\ddot{\text{N}}$:)	⟶	+ R—N₃ (alkyl azide)
+ ⁻:$\ddot{\text{S}}$R′ (alkanethiolate)	⟶	+ R—$\ddot{\text{S}}$—R′ (thioether or sulfide)
+ :NR′₃ (amine)	⟶	R—$\overset{+}{\text{N}}$R′₃ :$\ddot{\text{X}}$:⁻ (alkylammonium salt)
+ :$\ddot{\text{O}}$H₂ (water)	⟶	R—$\overset{+}{\underset{\text{H}}{\ddot{\text{O}}}}$—H :$\ddot{\text{X}}$:⁻ ⇌ R—$\ddot{\text{O}}$—H + H$\ddot{\text{X}}$: (alcohol)
+ :$\underset{\text{H}}{\ddot{\text{O}}}$—R′ (alcohol)	⟶	R—$\overset{+}{\underset{\text{H}}{\ddot{\text{O}}}}$—R′ :$\ddot{\text{X}}$:⁻ ⇌ R—$\ddot{\text{O}}$—R′ + H$\ddot{\text{X}}$: (ether)

see most frequently in this chapter are ethoxide, which is typically used as sodium ethoxide (Na⁺ CH₃CH₂O⁻, abbreviated Na⁺ ⁻OEt or simply NaOEt) and *tert*-butoxide, which is typically used as potassium *tert*-butoxide [K⁺ (CH₃)₃C—O⁻, abbreviated K⁺ ⁻OtBu or simply KOtBu]. Often the conjugate-acid alcohols of these bases are used as solvents. Just as ⁻OH is used as a solution in its conjugate acid water, sodium ethoxide is frequently used as a solution in ethanol and potassium *tert*-butoxide in *tert*-butyl alcohol. (The preparation of these bases is covered in Sec. 10.1B.)

If the reacting alkyl halide has more than one type of β-hydrogen atom, then more than one β-elimination reaction are possible. When these different reactions occur at comparable rates, more than one alkene product are formed, as in the following example.

Note the use of a net ionic equation in this example.

PROBLEM

9.3   What product(s) are expected in the ethoxide-promoted β-elimination reaction of each of the following compounds?
(a) 2-bromo-2,3-dimethylbutane   (b) 1-chloro-1-methylcyclohexane

C. Competition between Nucleophilic Substitution and β-Elimination Reactions

In the presence of a strong base such as ethoxide, the nucleophilic substitution reaction is a typical one for primary alkyl halides, and a β-elimination reaction is observed for tertiary alkyl halides. What about secondary alkyl halides? A typical secondary alkyl halide under the same conditions undergoes both reactions.

$$\underset{\substack{\text{Br}\\ \textbf{isopropyl bromide}}}{\text{H}_3\text{C}-\overset{}{\underset{}{\text{CH}}}-\text{CH}_3} + \underset{\textbf{ethoxide}}{\text{EtO}^-} \xrightarrow[\text{EtOH}]{} \underset{\substack{\text{OEt}\\ \textbf{ethyl isopropyl ether}\\ \textit{substitution product}\\ \textit{(about 50\%)}}}{\text{H}_3\text{C}-\overset{}{\underset{}{\text{CH}}}-\text{CH}_3} + \underset{\substack{\textbf{propene}\\ \textit{elimination product}\\ \textit{(about 50\%)}}}{\text{H}_2\text{C}=\text{CH}-\text{CH}_3} \quad (9.7)$$

Because both substitution and elimination reactions are observed, the two reactions occur at comparable *rates*—in other words, the reactions are *in competition*. In fact, nucleophilic substitution and base-promoted β-elimination reactions are in competition for *all* alkyl halides with β-hydrogens, even primary and tertiary halides. It happens that in the presence of a strong Brønsted base, nucleophilic substitution is a faster reaction (it "wins the competition") for many primary alkyl halides, and in most cases β-elimination is a faster reaction for tertiary halides; that is why substitution predominates in the former case and elimination in the latter. However, under some conditions, the results of the competition can be changed. For example, we can sometimes find conditions under which some primary alkyl halides give mostly elimination products.

In the following sections, we'll focus first on nucleophilic substitution reactions, then on β-elimination reactions. We'll discuss the factors that govern the reactivities of alkyl halides in each of these reaction types. Although each type of reaction is considered in isolation, keep in mind that substitutions and eliminations are always in competition.

PROBLEM

9.4   What substitution and elimination products (if any) might be obtained when each of the following alkyl halides is treated with sodium methoxide in methanol?

(a) *trans*-1-bromo-3-methylcyclohexane   (b) methyl iodide

(c) (bromomethyl)cyclopentane

9.2   EQUILIBRIA IN NUCLEOPHILIC SUBSTITUTION REACTIONS

Table 9.1 (p. 385) shows some of the many possible nucleophilic substitution reactions. How do we know whether the equilibrium for a given substitution is favorable? This problem is illustrated by the reaction of a cyanide ion with methyl iodide, which has an equilibrium constant that favors the product acetonitrile by many powers of 10.

$$^-\text{:}\text{C}\equiv\text{N}\text{:} + \text{H}_3\text{C}-\overset{..}{\underset{..}{\text{I}}}\text{:} \longrightarrow \underset{\textbf{acetonitrile}}{\text{H}_3\text{C}-\text{C}\equiv\text{N}\text{:}} + \text{:}\overset{..}{\underset{..}{\text{I}}}\text{:}^- \qquad (9.8)$$

Other substitution reactions, however, are reversible or even unfavorable.

$$\ddot{:}\!\!\ddot{I}\!:^- + H_3C\!-\!\ddot{B}\!\ddot{r}\!: \;\rightleftarrows\; H_3C\!-\!\ddot{I}\!: + :\!\ddot{B}\!\ddot{r}\!:^- \tag{9.9a}$$

$$:\!\ddot{I}\!:^- + H_3C\!-\!\ddot{O}H \;\longleftarrow\; H_3C\!-\!\ddot{I}\!: + {}^-\!:\!\ddot{O}H \quad \text{(does not proceed to the right)} \tag{9.9b}$$

Results such as these can be predicted by recognizing that each nucleophilic substitution reaction is conceptually similar to a Brønsted acid–base reaction. That is, if the alkyl group of the alkyl halide is replaced with a hydrogen, the substitution reaction becomes an acid–base reaction.

$$^-OH + H_3C\!-\!I \;\longrightarrow\; H_3C\!-\!OH + I^- \quad \text{(substitution reaction)} \tag{9.10a}$$

$$^-OH + H\!-\!I \;\longrightarrow\; H\!-\!OH + I^- \quad \text{(acid–base reaction)} \tag{9.10b}$$

In the nucleophilic substitution reaction, ^-OH, which acts as a nucleophile, displaces the I^- leaving group from the *carbon* electrophile. In the Brønsted acid–base reaction, ^-OH, which acts as a Brønsted base, displaces the I^- leaving group from the *proton* electrophile. (This same analysis was introduced in Sec. 3.2B.) What makes this comparison especially useful is that *it can be used to predict whether the equilibrium in the nucleophilic substitution is favorable.* To do this, we determine whether the equilibrium for the Brønsted acid–base reaction is favorable using the method described in Sec. 3.4E. Thus, if the Brønsted acid–base reaction

$$H\!-\!X + Y:^- \;\rightleftarrows\; Y\!-\!H + X:^- \tag{9.11a}$$

strongly favors the right side of the equation, then the analogous nucleophilic substitution reaction

$$R\!-\!X + Y:^- \;\rightleftarrows\; Y\!-\!R + X:^- \tag{9.11b}$$

likewise favors the right side of the equation. This means that *the equilibrium in any nucleophilic substitution reaction, as in an acid–base reaction, favors release of the weaker base.* This principle, for example, shows why I^- will *not* displace ^-OH from CH_3OH in the reverse of Eq. 9.10a: I^- is a much weaker base than ^-OH (Table 3.1). In fact, the reverse reaction occurs: ^-OH readily displaces I^- from CH_3I. This example illustrates how the acid–base principles discussed in Chapter 3 can prove very useful in understanding nucleophilic substitution reactions.

The correspondence between equilibrium constants for nucleophilic substitution reactions and Brønsted acid–base reactions is not quantitatively exact because the electrophilic centers aren't the same (carbon versus hydrogen). In addition, the pK_a values used to predict the equilibrium constants for acid–base reactions are determined in water, whereas solvents other than water are often used for nucleophilic substitution reactions. Nevertheless, when the basicity difference between the nucleophile and the leaving group is large (as it will be in most alkyl halide reactions), we can be confident that our predictions about equilibria in nucleophilic substitution reactions will be qualitatively correct.

Some equilibria that are not too unfavorable can be driven to completion by applying *Le Châtelier's principle* (Sec. 4.9B, p. 174). For example, alkyl chlorides normally do not react to completion with iodide ion because iodide is a weaker base than chloride; the equilibrium favors the formation of the weaker base, iodide. However, in the solvent acetone, it happens that potassium iodide is relatively soluble and potassium chloride is relatively insoluble. Thus, when an alkyl chloride reacts with KI in acetone, KCl precipitates, and the equilibrium compensates for the loss of KCl by forming more of it, along with more alkyl iodide.

$$R\!-\!Cl + KI \;\xrightarrow[\text{acetone}]{}\; R\!-\!I + KCl \text{ (precipitates)} \tag{9.12}$$

PROBLEM

9.5   Tell whether each of the following reactions favors reactants or products at equilibrium. (Assume that all reactants and products are soluble.)

(a) $CH_3Cl + F^- \longrightarrow CH_3F + Cl^-$

(b) $CH_3Cl + N_3^- \longrightarrow CH_3N_3 + Cl^-$ (*Hint:* The pK_a of HN_3 is 4.72.)

(c) $CH_3Cl + {}^-OCH_3 \longrightarrow CH_3OCH_3 + Cl^-$

9.3   REACTION RATES

The previous section showed how to determine whether the equilibrium for a nucleophilic substitution reaction is favorable. *Knowledge of the equilibrium constant for a reaction provides no information about the rate at which the reaction takes place.* Although some substitution reactions with favorable equilibria proceed rapidly, others proceed slowly. For example, the reaction of methyl iodide with cyanide is a relatively fast reaction, whereas the reaction of cyanide with neopentyl iodide is so slow that it is virtually useless:

$$^-:C{\equiv}N: + \ H_3C{-}\ddot{\underset{..}{I}}: \longrightarrow H_3C{-}C{\equiv}N: + :\ddot{\underset{..}{I}}:^- \qquad (9.13a)$$

methyl iodide
(reacts rapidly)

$$^-:C{\equiv}N: + \ H_3C{-}\underset{\underset{CH_3}{|}}{\overset{\overset{CH_3}{|}}{C}}{-}CH_2{-}\ddot{\underset{..}{I}}: \longrightarrow H_3C{-}\underset{\underset{CH_3}{|}}{\overset{\overset{CH_3}{|}}{C}}{-}CH_2{-}C{\equiv}N: + :\ddot{\underset{..}{I}}:^- \qquad (9.13b)$$

neopentyl iodide
(reacts very slowly)

Why do reactions that are so conceptually similar differ so drastically in their rates? In other words, what determines the *reactivity* of a given alkyl halide in a nucleophilic substitution reaction? Because this question deals with reaction rates and the concept of the transition state, you should review the introduction to reaction rates and transition-state theory in Sec. 4.8.

A. Definition of Reaction Rate

The term *rate* implies that something is changing with time. For example, in the rate of travel, or the *velocity*, of a car, the car's position is the "something" that is changing.

$$\text{velocity} = v = \frac{\text{change in position}}{\text{corresponding change in time}} \qquad (9.14)$$

The quantities that change with time in a chemical reaction are the *concentrations of the reactants and products.*

$$\text{reaction rate} = \frac{\text{change in product concentration}}{\text{corresponding change in time}} \qquad (9.15a)$$

$$= -\frac{\text{change in reactant concentration}}{\text{corresponding change in time}} \qquad (9.15b)$$

(The signs in Eqs. 9.15a and 9.15b differ because the concentrations of the reactants decrease with time, and the concentrations of the products increase.)

In physics, a rate has the dimensions of length per unit time, such as meters per second. A reaction rate, by analogy, has the dimensions of concentration per unit time. When the concentration unit is the mole per liter (M), and if time is measured in seconds, then the unit of reaction rate is

$$\frac{\text{concentration}}{\text{time}} = \frac{\text{mols L}^{-1}}{\text{s}} = M \ \text{s}^{-1} \tag{9.16}$$

B. The Rate Law

For molecules to react with one another, they must "get together," or collide. Because molecules at higher concentrations are more likely to collide than molecules at lower concentrations, the rate of a reaction is a function of the concentrations of the reactants. The mathematical statement of how a reaction rate depends on concentration is called the **rate law**. A rate law is determined experimentally by varying the concentration of each reactant (including any catalysts) independently and measuring the resulting effect on the rate. Each reaction has its own characteristic rate law. For example, suppose that for the reaction $A + B \longrightarrow C$ the reaction rate doubles if *either* [A] or [B] is doubled and increases by a factor of four if *both* [A] and [B] are doubled. The rate law for this reaction is then

$$\text{rate} = k[A][B] \tag{9.17}$$

If in another reaction $D + E \longrightarrow F$, the rate doubles only if the concentration of D is doubled, and changing the concentration of E has no effect, the rate law is then

$$\text{rate} = k[D] \tag{9.18}$$

The concentrations in the rate law are the concentrations of reactants at any time during the reaction, and the rate is the velocity of the reaction at that same time. The constant of proportionality, k, is called the **rate constant**. In general, the rate constant is different for every reaction, and it is a property of each reaction under particular conditions of temperature, pressure, solvent, and so on. As Eqs. 9.17 and 9.18 show, the rate constant is numerically equal to the rate of the reaction when all reactants are present under the standard conditions of 1 M concentration. *The rates of two reactions are compared by comparing their rate constants.*

An important aspect of a reaction is its *kinetic order*. The **overall kinetic order** for a reaction is the sum of the powers of all the concentrations in the rate law. For a reaction described by the rate law in Eq. 9.17, the overall kinetic order is two; the reaction described by this rate law is said to be a **second-order reaction**. The overall kinetic order of a reaction having the rate law in Eq. 9.18 is one; such a reaction is thus a **first-order reaction**. The **kinetic order in each reactant** is the power to which its concentration is raised in the rate law. Thus, the reaction described by the rate law in Eq. 9.17 is said to be *first order in each reactant*. A reaction with the rate law in Eq. 9.18 is first order in D and zero order in E.

The *units* of the rate constant depend on the kinetic order of the reaction. With concentrations in moles per liter, and time in seconds, the rate of any reaction has the units of M s^{-1} (see Eq. 9.16). For a second-order reaction, then, dimensional consistency requires that the rate constant have the units of M^{-1} s^{-1}.

$$\text{rate} = k[A][B]$$
$$M\,\text{s}^{-1} = M^{-1}\,\text{s}^{-1} \quad M \ \ M \tag{9.19}$$

Similarly, the rate constant for a first-order reaction has units of s^{-1}.

FURTHER EXPLORATION 9.1 Reaction Rates

C. Relationship of the Rate Constant to the Standard Free Energy of Activation

According to transition-state theory, which was discussed in Sec. 4.8, the standard free energy of activation, or energy barrier, determines the rate of a reaction under standard conditions.

**FURTHER
EXPLORATION 9.2**
Absolute Rate Theory

In Sec. 9.3B we showed that the rate constant is numerically equal to the reaction rate under standard conditions—that is, when the concentrations of all the reactants are 1 M. It follows, then, that the rate constant is related to the standard free energy of activation, $\Delta G^{\circ\ddagger}$. If $\Delta G^{\circ\ddagger}$ is large for a reaction, the reaction is relatively slow, and the rate constant is small. If $\Delta G^{\circ\ddagger}$ is small, the reaction is relatively fast, and the rate constant is large. This relationship is shown in Fig. 9.1.

Table 9.2 illustrates the quantitative relationship between the rate constant and the standard free energy of activation. (For the derivation these numbers, see Further Exploration 9.2.) Table 9.2 also translates these numbers into practical terms by giving the time required for the completion of a reaction with a given rate constant. This time is approximately $7/k$. (This is also justified in Further Exploration 9.2.)

We'll most often be interested in the *relative rates* of two reactions. That is, we'll be comparing the rate of a reaction to that of a standard reaction. A **relative rate** is defined as the ratio of two rates. The relationship between the relative rate of two reactions A and B under

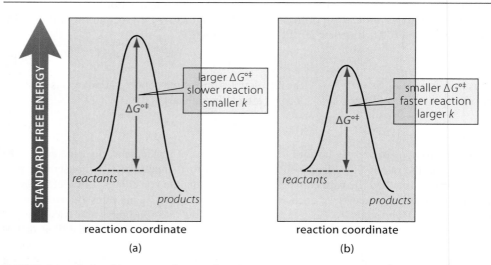

FIGURE 9.1 Relationship among the standard free energy of activation ($\Delta G^{\circ\ddagger}$), reaction rate, and rate constant (k). (a) A reaction with a larger $\Delta G^{\circ\ddagger}$ has a smaller rate and a smaller rate constant. (b) A reaction with a smaller $\Delta G^{\circ\ddagger}$ has a larger rate and a larger rate constant.

TABLE 9.2 Relationship between Rate Constants, Standard Free Energies of Activation, and Reaction Times for First-Order Reactions

Rate constant (s⁻¹) ($T = 298$ K)	Time to completion*	$\Delta G^{\circ\ddagger}$ kJ mol⁻¹	$\Delta G^{\circ\ddagger}$ kcal mol⁻¹
10^{-8}	22 years	119	28.4
10^{-6}	83 days	107	25.7
10^{-4}	20 hours	96.0	23.0
10^{-2}	12 minutes	84.6	20.2
1	7 seconds	73.2	17.5
10^{2}	70 milliseconds	61.7	14.8
10^{4}	700 microseconds	50.4	12.0
10^{6}	7 microseconds	38.9	9.3
6.2×10^{12}	0.01 nanosecond	0	0

* Time required for 99% completion of reaction ≈ $7/k$

standard conditions (that is, 1 M in all reactants) and their standard free energies of activation, first presented as Eq. 4.33a (p. 164), is given in Eq. 9.20a:

$$\text{relative rate} = \frac{\text{rate}_A}{\text{rate}_B} = 10^{(\Delta G_B^{\circ\ddagger} - \Delta G_A^{\circ\ddagger})/2.3RT} \tag{9.20a}$$

Because the rate constant is numerically equal to the rate under standard conditions, the relative rate is the ratio of rate constants:

$$\text{relative rate} = \frac{k_A}{k_B} = 10^{(\Delta G_B^{\circ\ddagger} - \Delta G_A^{\circ\ddagger})/2.3RT} \tag{9.20b}$$

or

$$\log(\text{relative rate}) = \log\left(\frac{k_A}{k_B}\right) = \frac{\Delta G_B^{\circ\ddagger} - \Delta G_A^{\circ\ddagger}}{2.3RT} \tag{9.20c}$$

This equation says that each increment of $2.3RT$ (5.7 kJ mol^{-1} or 1.4 kcal mol^{-1} at 298 K) in the $\Delta G^{\circ\ddagger}$ difference for two reactions corresponds to a one log unit (that is, 10-fold) factor in their relative rate constants.

PROBLEMS

9.6 For each of the following reactions, (1) what is the overall kinetic order of the reaction, (2) what is the order in each reactant, and (3) what are the dimensions of the rate constant?

(a) an addition reaction of bromine to an alkene with the rate law

$$\text{rate} = k[\text{alkene}][\text{Br}_2]^2$$

(b) a substitution reaction of an alkyl halide with the rate law

$$\text{rate} = k[\text{alkyl halide}]$$

9.7 (a) What is the ratio of rate constants k_A/k_B at 25 °C for two reactions A and B if the standard free energy of activation of reaction A is 14 kJ mol^{-1} (3.4 kcal mol^{-1}) less than that of reaction B?

(b) What is the difference in the standard free energies of activation at 25 °C of two reactions A and B if reaction B is 450 times faster than reaction A? Which reaction has the greater $\Delta G^{\circ\ddagger}$?

9.8 What prediction does the rate law in Eq. 9.18 make about how the rate of the reaction changes as the reactants D and E are converted into F over time? Does the rate increase, decrease, or stay the same? Explain. Use your answer to sketch a plot of the concentrations of starting materials and products against time.

9.4   THE S_N2 REACTION

A. Rate Law and Mechanism of the S_N2 Reaction

Consider now the nucleophilic substitution reaction of ethoxide ion with methyl iodide in ethanol at 25 °C.

$$\text{CH}_3\text{CH}_2\text{O}^- + \text{H}_3\text{C}\!-\!\text{I} \xrightarrow{\text{CH}_3\text{CH}_2\text{OH}} \text{CH}_3\text{CH}_2\text{O}\!-\!\text{CH}_3 + \text{I}^- \tag{9.21}$$

ethoxide ion   **ethyl methyl ether (methoxyethane)**

The following rate law was experimentally determined for this reaction:

$$\text{rate} = k[\text{CH}_3\text{I}][\text{C}_2\text{H}_5\text{O}^-] \tag{9.22}$$

with $k = 6.0 \times 10^{-4}\ M^{-1}\ \text{s}^{-1}$. That is, this is a *second-order reaction that is first order in each reactant.*

The rate law of a reaction is important because it provides fundamental information about the reaction mechanism. Specifically, the concentration terms of the rate law indicate *which species are present in the transition state of the rate-limiting step*. Hence, the rate-limiting transition state of reaction 9.21 consists of the elements of one methyl iodide molecule and one ethoxide ion. The rate law excludes some mechanisms from consideration. For example, any mechanism in which the rate-limiting step involves two molecules of ethoxide is *ruled out* by the rate law, because the rate law for such a mechanism would have to be second order in ethoxide.

The simplest possible mechanism consistent with the rate law is one in which the ethoxide ion *directly displaces* the iodide ion from the methyl carbon:

$$CH_3CH_2\overset{..}{\underset{..}{O}}:^- \quad H_3C-\overset{..}{\underset{..}{I}}: \longrightarrow \left[CH_3CH_2\overset{..}{\underset{..}{O}}\overset{\delta-}{---}\overset{\overset{\displaystyle H}{|}}{\underset{\underset{\displaystyle H\;\;H}{}}{C}}\overset{\delta-}{---}:\overset{..}{\underset{..}{I}}: \right]^{\ddagger} \longrightarrow CH_3CH_2\overset{..}{\underset{..}{O}}-CH_3 \;+\; :\overset{..}{\underset{..}{I}}:^- \quad (9.23)$$

transition state

Mechanisms like this account for many nucleophilic substitution reactions. A mechanism in which electron-pair donation by a nucleophile to an atom (usually carbon) displaces a leaving group from the same atom in a concerted manner (that is, in one step, without reactive intermediates) is called an **S_N2 mechanism**. Reactions that occur by S_N2 mechanisms are called **S_N2 reactions**. The meaning of the "nickname" S_N2 is as follows:

$$\text{substitution} \nearrow \overset{\displaystyle S_N2}{\underset{\displaystyle \text{nucleophilic}}{\uparrow}} \nwarrow \text{bimolecular}$$

**STUDY GUIDE
LINK 9.1**
Deducing
Mechanisms from
Rate Laws

(The word *bimolecular* means that the rate-limiting step of the reaction involves two species—in this case, one methyl iodide molecule and one ethoxide ion.) Notice that an S_N2 reaction, because it is concerted, involves no reactive intermediates.

The rate law does not reveal all of the details of a reaction mechanism. *Although the rate law indicates what atoms are present in the rate-limiting step, it provides no information about how they are arranged.* Thus, the following two mechanisms for the S_N2 reaction of ethoxide ion with methyl iodide are equally consistent with the rate law.

(9.24)

same-side substitution   opposite-side substitution

As far as the rate law is concerned, either mechanism is acceptable. To decide between these two possibilities, other types of experiments are needed (Sec. 9.4C).

Let's summarize the relationship between the rate law and the mechanism of a reaction.

1. The concentration terms of the rate law indicate what species are involved in the rate-limiting step.
2. Mechanisms that are inconsistent with the rate law are ruled out.
3. Of the chemically reasonable mechanisms consistent with the rate law, the simplest one is provisionally adopted.
4. The mechanism of a reaction is modified or refined if required by subsequent experiments.

Point (4) may seem disturbing because it means that a mechanism can be changed at a later time. Perhaps it seems that an "absolutely true" mechanism should exist for every reaction. However, a mechanism can never be proved; it can only be disproved. The value of a mechanism lies not in its absolute truth but rather in its validity as a conceptual framework, or theory, that generalizes the results of many experiments and predicts the outcome of others. Mechanisms allow us to place reactions into categories and thus impose a conceptual order on chemical observations. Thus, when someone observes an experimental result different from that predicted by a mechanism, the mechanism must be modified to accommodate both the previously known facts and the new facts. The evolution of mechanisms is no different from the evolution of science in general. Knowledge is dynamic: theories (mechanisms) predict the results of experiments, a test of these theories may lead to new theories, and so on.

PROBLEMS

9.9   The reaction of acetic acid with ammonia is very rapid and follows the simple rate law shown in the following equation. Propose a mechanism that is consistent with this rate law.

$$H_3C-\overset{\overset{\displaystyle O}{\|}}{C}-\overset{..}{\underset{..}{O}}-H + :NH_3 \;\rightleftharpoons\; H_3C-\overset{\overset{\displaystyle O}{\|}}{C}-\overset{..}{\underset{..}{O}}:^- + \overset{+}{N}H_4$$

acetic acid

$$\text{rate} = k\left[H_3C-\overset{\overset{\displaystyle O}{\|}}{C}-OH\right]\left[NH_3\right]$$

9.10   What rate law would be expected for the reaction of cyanide ion ($^-$:CN) with ethyl bromide by the S$_N$2 mechanism?

B. Relative Rates of S$_N$2 Reactions and Brønsted Acid–Base Reactions

In Sec. 3.2B, and again in Sec. 9.2, we learned about the close analogy between nucleophilic substitution reactions and acid–base reactions. The equilibrium constants for a nucleophilic substitution reaction and its acid–base analog are very similar, and the curved-arrow notations for an S$_N$2 reaction and its acid–base analog are identical. However, it is important to understand that their rates are very different. *Most ordinary acid–base reactions occur instantaneously*—as fast as the reacting pairs can diffuse together. The rate constants for such reactions are typically in the 10^8–10^{10} M^{-1} s^{-1} range. Although many nucleophilic substitution reactions occur at convenient rates, they are *much* slower than the analogous acid–base reactions. Thus, the reaction in Eq. 9.25a is completed in a little over an hour, but the corresponding acid–base reaction in Eq. 9.25b occurs within about a *billionth of a second*!

Nucleophilic substitution reaction:

$$CH_3CH_2\overset{..}{\underset{..}{O}}:^- + H_3C-\overset{..}{\underset{..}{I}}: \xrightarrow[CH_3CH_2OH]{} CH_3CH_2\overset{..}{\underset{..}{O}}-CH_3 + :\overset{..}{\underset{..}{I}}:^- \qquad (9.25a)$$

(complete in about an hour)

Brønsted acid–base reaction:

$$CH_3CH_2\overset{..}{\underset{..}{O}}:^- + H-\overset{..}{\underset{..}{I}}: \xrightarrow[CH_3CH_2OH]{} CH_3CH_2\overset{..}{\underset{..}{O}}-H + :\overset{..}{\underset{..}{I}}:^- \qquad (9.25b)$$

(complete in 10^{-9} second)

This means that if an alkyl halide and a Brønsted acid are in competition for a Brønsted base, the Brønsted acid reacts much more rapidly. In other words, *the Brønsted acid always wins*.

9.11 Methyl iodide (0.1 *M*) and hydriodic acid (HI, 0.1 *M*) are allowed to react in ethanol solution with 0.1 *M* sodium ethoxide. What products are observed?

9.12 Ethyl bromide (0.1*M*) and HBr (0.1 M) are allowed to react in aqueous THF with 1 *M* sodium cyanide (Na$^+$ $^-$CN). What products are observed? Are any products formed more rapidly than others? Explain.

C. Stereochemistry of the S$_N$2 Reaction

The mechanism of the S$_N$2 reaction can be described in more detail by considering its *stereochemistry*. The stereochemistry of a substitution reaction can be investigated only if the carbon at which substitution occurs is a stereocenter in both reactants and products (Sec. 7.8B). A substitution reaction can occur at a stereocenter in three stereochemically different ways:

1. with *retention of configuration* at the stereocenter;
2. with *inversion of configuration* at the stereocenter; or
3. with a combination of (1) and (2); that is, mixed retention and inversion.

If approach of the nucleophile Nuc$\overline{:}$ to an asymmetric carbon and departure of the leaving group X$\overline{:}$ occur from more or less the same direction (*same-side substitution*), then a substitution reaction would result in a product with *retention of configuration* at the asymmetric carbon.

$$\text{(9.26a)}$$

transition state

In contrast, if approach of the nucleophile and loss of the leaving group on an asymmetric carbon occur from opposite directions (*opposite-side substitution*), the other three groups on carbon must invert, or "turn inside out," to maintain the tetrahedral bond angle. This mechanism would lead to a product with *inversion of configuration* at the asymmetric carbon.

$$\text{(9.26b)}$$

transition state

We can distinguish between these two results if R^1, R^2, and R^3 are different, in which case the products of Eqs. 9.26a and 9.26b are *enantiomers*. Thus, the two types of substitution can be distinguished by subjecting one enantiomer of a chiral alkyl halide to the S$_N$2 reaction and determining which enantiomer of the product is formed. If both paths occur at equal rates, then the racemate will be formed.

What are the experimental results? The reaction of hydroxide ion with 2-bromooctane, a chiral alkyl halide, to give 2-octanol is a typical S$_N$2 reaction. The reaction follows a second-order rate law, first order in $^-$OH and first order in the alkyl halide. When (*R*)-2-bromooctane is used in the reaction, the product is (*S*)-2-octanol.

$$\text{(9.27)}$$

(R)-2-bromooctane   (S)-2-octanol

The stereochemistry of this S_N2 reaction shows that it proceeds with *inversion of configuration*. Thus, the reaction occurs by *opposite-side substitution* of hydroxide ion on the alkyl halide.

Recall that opposite-side substitution is also observed for the reaction of bromide ion and other nucleophiles with the bromonium ion intermediate in the addition of bromine to alkenes (Sec. 7.8C). As you can now appreciate, that reaction is also an S_N2 reaction. In fact, *inversion of stereochemical configuration is generally observed in all S_N2 reactions at carbon stereocenters.*

The stereochemistry of the S_N2 reaction calls to mind the *inversion of amines* (Fig. 6.18, p. 261). In the hybrid orbital description of both processes, the central atom is turned "inside out," and it is approximately sp^2-hybridized at the transition state. In the transition state for amine inversion, the $2p$ orbital on the nitrogen contains an unshared electron pair. In the transition state for an S_N2 reaction on carbon, the nucleophile and the leaving group are partially bonded to opposite lobes of the carbon $2p$ orbital (Fig. 9.2).

Why is opposite-side substitution preferred in the S_N2 reaction? The hybrid orbital description of the reaction in Fig. 9.2 provides no information on this question, but a molecular orbital analysis does, as shown in Fig. 9.3 (p. 396) for the reaction of a nucleophile (Nuc:) with methyl chloride (CH_3Cl). When a nucleophile donates electrons to an alkyl halide, the orbital containing the donated electron pair must initially interact with an *unoccupied* molecular orbital of the alkyl halide. The MO of the nucleophile that contains the donated electron pair interacts with the unoccupied alkyl halide MO of lowest energy, called the **LUMO** (for "lowest unoccupied molecular orbital"). Now, all of the bonding MOs of the alkyl halide are occupied; therefore, the alkyl halide LUMO is an antibonding MO, which is shown in Fig. 9.3. When opposite-side substitution occurs (Fig. 9.3a), *bonding overlap* of the nucleophile orbital occurs with the alkyl halide LUMO; that is, wave peaks overlap. But in same-side substitution (Fig. 9.3b), the nucleophile orbital has both bonding and antibonding overlap with the LUMO; the antibonding overlap (wave peak to wave trough) cancels the bonding overlap, and no net bonding can occur. Because only opposite-side substitution gives bonding overlap, this is *always* the observed substitution mode.

PROBLEM

9.13 What is the expected substitution product (including its stereochemical configuration) in the S_N2 reaction of potassium iodide in acetone solvent with the following compound? ($D = {}^2H$ = deuterium, an isotope of hydrogen.)

$$(R)\text{-}CH_3CH_2CH_2CH\!-\!Cl$$
$$\mid$$
$$D$$

FIGURE 9.2 Stereochemistry of the S_N2 reaction. The green arrows show how the various groups change position during the reaction. (Nuc: = a general nucleophile.) Notice that the sterochemical configuration of the asymmetric carbon is inverted by the opposite-side substitution reaction.

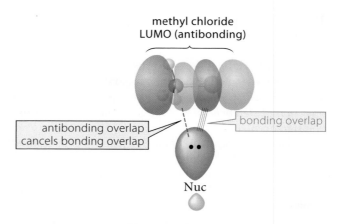

FIGURE 9.3 In the S_N2 reaction, the orbital containing the nucleophile electron pair interacts with the unoccupied molecular orbital of lowest energy (LUMO) in the alkyl halide. (a) Opposite-side substitution leads to bonding overlap. (b) Same-side substitution gives both bonding and antibonding overlaps that cancel. Therefore, opposite-side substitution is always observed.

D. Effect of Alkyl Halide Structure on the S_N2 Reaction

One of the most important aspects of the S_N2 reaction is how the reaction rate varies with the structure of the alkyl halide. (Recall Eqs. 9.13a and 9.13b, p. 388.) If an alkyl halide is very reactive, its S_N2 reactions occur rapidly under mild conditions. If an alkyl halide is relatively unreactive, then the severity of the reaction conditions (for example, the temperature) must be increased for the reaction to proceed at a reasonable rate. However, harsh conditions increase the likelihood of competing side reactions. Hence, if an alkyl halide is unreactive enough, the reaction has no practical value.

Alkyl halides differ, in some cases by many orders of magnitude, in the rates with which they undergo a given S_N2 reaction. Typical reactivity data are given in Table 9.3. To put these data in some perspective: If the reaction of a methyl halide takes about one *minute*, then the reaction of a neopentyl halide under the same conditions takes about *23 years*!

TABLE 9.3 Effect of Alkyl Substitution in the Alkyl Halide on the Rate of a Typical S_N2 Reaction

$$\text{Nuc:}^- + \text{H}_3\text{C—I} \xrightarrow[\substack{25\,°C}]{\text{acetone}} \text{Nuc—CH}_3 + \text{I}^-$$

R—	Name of R	Relative rate*
CH_3—	methyl	145
Increased alkyl substitution at the β-carbon:		
$CH_3CH_2CH_2$—	propyl	0.82
$(CH_3)_2CHCH_2$—	isobutyl	0.036
$(CH_3)_3CCH_2$—	neopentyl	0.000012
Increased alkyl substitution at the α-carbon:		
CH_3CH_2—	ethyl	1.0
$(CH_3)_2CH$—	isopropyl	0.0078
$(CH_3)_3C$—	*tert*-butyl	~0.0005[†]

* All rates are relative to that of ethyl bromide.
[†] Estimated from the rates of closely related reactions.

(a) Br$^-$ + CH$_3$—I

(b) Br$^-$ + (CH$_3$)$_3$C—CH$_2$—I

FIGURE 9.4 Transition states for S$_N$2 reactions. The upper panels show the transition states as ball-and-stick models, and the lower panels show them as space-filling models. (a) The reaction of methyl bromide with iodide ion. (b) The reaction of neopentyl bromide with iodide ion. The S$_N$2 reactions of neopentyl bromide are very slow because of the severe van der Waals repulsions of both the nucleophile and the leaving group with the pink hydrogens of the methyl substituents. These repulsions are indicated with red brackets in the models.

The data in Table 9.3 show, first, that *increased alkyl substitution at the β-carbon retards an S$_N$2 reaction*. As Fig. 9.4 shows, these data are consistent with an opposite-side substitution mechanism. When a methyl halide undergoes substitution, approach of the nucleophile and departure of the leaving group are relatively unrestricted. However, when a neopentyl halide reacts with a nucleophile, both the nucleophile and the leaving group experience severe van der Waals repulsions with hydrogens of the methyl substituents. These van der Waals repulsions raise the energy of the transition state and therefore reduce the reaction rate. This is another example of a *steric effect*. Recall from Sec. 5.6D (Fig. 5.2, p. 511) that a **steric effect** is any effect on a chemical phenomenon (such as a reaction) caused by van der Waals repulsions. Thus, S$_N$2 reactions of branched alkyl halides are retarded by a steric effect. Indeed, S$_N$2 reactions of neopentyl halides are so slow that they are not practically useful.

The data in Table 9.3 help explain why elimination reactions compete with the S$_N$2 reactions of secondary and tertiary alkyl halides (Sec. 9.1C): these halides react so slowly in S$_N$2 reactions that the rates of elimination reactions are competitive with the rates of substitution. The rates of the S$_N$2 reactions of tertiary alkyl halides are so slow that elimination is the only reaction observed. The competition between β-elimination and S$_N$2 reactions will be considered in more detail in Sec. 9.5G.

E. Nucleophilicity in the S$_N$2 Reaction

As Table 9.1 (p. 385) illustrates, the S$_N$2 reaction is especially useful because of the variety of nucleophiles that can be employed. However, nucleophiles differ significantly in their reactivities. The relative reactivity of a nucleophile—how rapidly it reacts under a defined set of conditions—is called **nucleophilicity**. What factors govern nucleophilicity in the S$_N$2 reaction and why?

Nucleophilicity is determined by three factors:

1. the *Brønsted* basicity of the nucleophile;
2. the *solvent* in which the reaction is carried out; and
3. the *polarizability* of the nucleophile.

Basicity and Solvent Effects on Nucleophilicity Because basicity and solvent effects on nucleophilicity are interdependent, we'll consider theses two effects together. Remember that the Brønsted basicity of any base is measured by the pK_a of its *conjugate acid*: the higher the conjugate-acid pK_a, the greater the basicity of the nucleophile. We might expect some correlation between *nucleophilicity* and the *Brønsted basicity* of a nucleophile because both are aspects of its Lewis basicity. That is, in either role a Lewis base donates an electron pair. (Be sure to review the definitions of these terms in Sec. 3.2A.) Let's first examine some data for the S_N2 reactions of methyl iodide with anionic nucleophiles of different basicity to see whether this expectation is met in practice. Some data for the reaction of methyl iodide with various nucleophiles in methanol solvent are given in Table 9.4 and plotted in Fig. 9.5. Notice in this table that the nucleophilic atoms are all from the second period of the periodic table. Fig. 9.5 shows a *very rough* trend toward faster reactions with the more basic nucleophiles.

Let's now consider some data for the same reaction with anionic nucleophiles from different periods (rows) of the periodic table. These data are shown in Table 9.5. If we are expecting a similar correlation of nucleophilic reactivity and basicity, we get a surprise. Notice that the sulfide nucleophile is more than three orders of magnitude *less basic* than the oxide nucleophile, and yet it is more than *four orders of magnitude more reactive*. Similarly, for the halide nucleophiles, the *least basic halide ion* (iodide) *is the best nucleophile*.

Let's generalize what we've learned so far. The following apply to *nucleophilic anions* in *polar, protic solvents* (such as water and alcohols):

1. In a series of nucleophiles in which the nucleophilic atoms are from the same period of the periodic table, there is a rough correlation of nucleophilicity with basicity.

2. In a series of nucleophiles in which the nucleophilic atoms are from the same group (column) but different periods of the periodic table, the less basic nucleophiles are more nucleophilic.

The interaction of the nucleophile with the solvent is the most significant factor that accounts for both of these generalizations. Let's start with generalization 2—the *inverse* relationship of basicity and nucleophilicity within a group of the periodic table. The solvent in all of the cases shown in Tables 9.4 and 9.5 and Fig. 9.5 is methanol, a *protic* solvent. In a protic solvent, *hydrogen bonding* occurs between the protic solvent molecules (as hydrogen bond donors) and the nucleophilic anions (as hydrogen bond acceptors). *The strongest Brønsted bases are the best hydrogen bond acceptors.* For example, fluoride ion forms much stronger hydrogen bonds than iodide ion. When the electron pairs of a nucleophile are involved in

---

TABLE 9.4 Dependence of S_N2 Reaction Rate on the Basicity of the Nucleophile in Methanol

$$\text{Nuc:}^- + \text{H}_3\text{C}-\text{I} \xrightarrow[25\,°\text{C}]{\text{CH}_3\text{OH}} \text{Nuc}-\text{CH}_3 + \text{I}^-$$

Nucleophile (name)	pK_a of conjugate acid*	k (second-order rate constant, $M^{-1}\,s^{-1}$)	log k
CH_3O^- (methoxide)	15.1	2.5×10^{-4}	−3.6
PhO^- (phenoxide)	9.95	7.9×10^{-5}	−4.1
^-CN (cyanide)	9.4	6.3×10^{-4}	−3.2
AcO^- (acetate)	4.76	2.7×10^{-6}	−5.6
N_3^- (azide)	4.72	7.8×10^{-5}	−4.1
F^- (fluoride)	3.2	5.0×10^{-8}	−7.3
SO_4^{2-} (sulfate)	2.0	4.0×10^{-7}	−6.4
NO_3^- (nitrate)	−1.2	5.0×10^{-9}	−8.3

* pK_a values in water

FIGURE 9.5 The dependence of nucleophile S_N2 reactivity on nucleophile basicity for a series of nucleophiles in methanol solvent. Reactivity is measured by log k for the reaction of the nucleophile with methyl iodide. Basicity is measured by the pK_a of the conjugate acid of the nucleophile. The blue dashed line of slope = 1 shows the trend to be expected if a change of one log unit in basicity resulted in the same change in nucleophilicity. The solid blue line shows the actual trend for a series of nucleophiles (*blue squares*) in which the reacting atom is —O⁻. The black circles show the reactivity of other nucleophilic anions in which the reacting atoms are from period 2 of the periodic table, the same period as oxygen.

TABLE 9.5 Dependence of S_N2 Reaction Rate on the Basicity of Nucleophiles from Different Periods of the Periodic Table in Methanol

$$\text{Nuc:}^- + \text{H}_3\text{C—I} \xrightarrow[25\,°C]{\text{CH}_3\text{OH}} \text{Nuc—CH}_3 + \text{I}^-$$

Nucleophile	pK_a of conjugate acid*	k (second-order rate constant, $M^{-1}\,s^{-1}$)	log k
Group 6A Nucleophiles			
PhS⁻	6.52	1.1	+0.03
PhO⁻	9.95	7.9×10^{-5}	−4.1
Group 7A Nucleophiles			
I⁻	−10	3.4×10^{-3}	−2.5
Br⁻	−8	8.0×10^{-5}	−4.1
Cl⁻	−6	3.0×10^{-6}	−5.5
F⁻	3.2	5.0×10^{-8}	−7.3

* pK_a values in water

hydrogen bonding, they are unavailable for donation to carbon in an S_N2 reaction. For the S_N2 reaction to take place, *a hydrogen bond between the solvent and the nucleophile must be broken* (Fig. 9.6, p. 400). More energy is required to break a strong hydrogen bond to fluoride ion than is required to break a relatively weak hydrogen bond to iodide ion. This extra energy is reflected in a greater free energy of activation—the energy barrier—and, as a result, the reaction of fluoride ion is slower. To use a football analogy, the nucleophilic reaction of a strongly

FIGURE 9.6 An S_N2 reaction of methyl iodide involving a nucleophile ($:\ddot{X}:^-$) in a protic solvent requires breaking a hydrogen bond between the solvent and the nucleophile. The energy required to break this hydrogen bond becomes part of the standard free energy of activation of the substitution reaction and thus retards the reaction.

transition state

hydrogen-bonded anion with an alkyl halide is about as likely as a tackler bringing down a ball carrier when both of the tackler's arms are being held by opposing linemen.

The data in Fig. 9.5 and generalization 1 can be understood with a similar argument. If nucleophilicity and basicity were exactly correlated, the graph would follow the dashed blue line of slope = 1. Focus on the blue curve, which shows the trend for nucleophiles that all have —O⁻ as the reacting atom (*blue squares*). The downward curvature shows that the nucleophiles of higher basicity do not react as rapidly with an alkyl halide as their basicity predicts, and the deviation from the line of unit slope is greatest for the most basic nucleophiles. The strongest bases form the strongest hydrogen bonds with the protic solvent methanol, and one of these hydrogen bonds has to be broken for the nucleophilic reaction to occur. As the hydrogen bond to solvent increases in strength, the rate-retarding effect on nucleophilicity also increases.

The data for nucleophiles shown with the black circles in Fig. 9.5 reflect the effects of hydrogen bonding to nucleophilic atoms that come from different groups within the same period (row) of the periodic table. For example, fluoride ion lies *below* the trend line for the oxygen nucleophiles. That is, fluoride ion is a worse nucleophile than an oxygen anion with the same basicity. The hydrogen bonds of fluoride with protic solvents are exceptionally strong, and hence its nucleophilicity is correspondingly reduced. Conversely, the hydrogen bonds of azide ion and the carbon of cyanide ion with protic solvents are weaker than those of the oxygen anions, and their nucleophilicities are somewhat greater.

If hydrogen bonding by the solvent tends to reduce the reactivity of very basic nucleophiles, it follows that S_N2 reactions might be considerably accelerated if they could be carried out in solvents in which such hydrogen bonding is not possible. Let's examine this proposition with the aid of some data shown in Table 9.6. The two solvents, methanol ($\epsilon = 33$) and *N,N*-dimethylformamide (DMF, $\epsilon = 37$), were chosen for the comparison because their dielectric constants are nearly the same; that is, their polarities are very similar.

$$H_3C—\ddot{O}—H$$

methanol
a polar protic solvent
$\epsilon = 33$

$$H—\overset{:O:}{\overset{\|}{C}}—\ddot{N}(CH_3)_2$$

***N,N*-dimethylformamide (DMF)**
a polar aprotic solvent
$\epsilon = 37$

As you can see from the data in this table, changing from a protic solvent to a polar aprotic solvent accelerates the reactions of all nucleophiles, but the increase of the reaction rate for fluoride ion is particularly noteworthy—a factor of 10^8. In fact, the acceleration of the reaction with fluoride ion is so dramatic that an S_N2 reaction with fluoride ion as the nucleophile is converted from an essentially useless reaction in a protic solvent—one that takes years—to

TABLE 9.6   Solvent Dependence of Nucleophilicity in the S$_N$2 Reaction

$$Nuc:^- + H_3C-I \xrightarrow[\substack{CH_3OH \\ \text{or DMF} \\ 25\,°C}]{} Nuc-CH_3 + I^-$$

Nucleophile	pK$_a$*	In methanol		In DMF‡	
		k, M^{-1} s^{-1}	Reaction is over in—†	k, M^{-1} s^{-1}	Reaction is over in—†
I$^-$	−10	3.4×10^{-3}	17 min	4.0×10^{-1}	8.7 s
Br$^-$	−8	8.0×10^{-5}	12 h	1.3	2.7 s
Cl$^-$	−6	3.0×10^{-6}	13 days	2.5	1.4 s
F$^-$	3.2	5.0×10^{-8}	2.2 years	>3	<1.2 s
$^-$CN	9.4	6.3×10^{-4}	1.5 h	3.2×10^2	0.011 s

* pK$_a$ values of the conjugate acid in water
† Time required for 97% completion of the reaction
‡ DMF = N,N-dimethylformamide

a very rapid reaction in the polar aprotic solvent. Other polar aprotic solvents have effects of a similar magnitude, and similar accelerations occur in the S$_N$2 reactions of other alkyl halides. The effect on rate is due mostly to the *solvent proticity*—whether the solvent is protic. Fluoride ion is *by far* the most strongly hydrogen-bonded halide anion in Table 9.5; consequently, the change of solvent has the greatest effect on the rates of its S$_N$2 reactions. As the data demonstrate, *eliminating the possibility of hydrogen bonding to nucleophiles strongly accelerates their S$_N$2 reactions.*

What we've learned, then, is that S$_N$2 reactions of nucleophilic anions with alkyl halides are much faster in polar aprotic solvents than they are in protic solvents. If this is so, why not use polar aprotic solvents for all such S$_N$2 reactions? Here we must be concerned with an element of practicality. To run an S$_N$2 reaction in solution, we must find a solvent that dissolves a salt that contains the nucleophilic anion of interest. We must also remove the solvent from the products when the reaction is over. Protic solvents, precisely because they are protic, dissolve significant quantities of salts. Methanol and ethanol, two of the most commonly used protic solvents, are cheap, are easily removed because they have relatively low boiling points, and are relatively safe to use. When the S$_N$2 reaction is rapid enough, or if a higher temperature can be used without introducing side reactions, the use of protic solvents is often the most practical solvent for an S$_N$2 reaction. Except for acetone and acetonitrile (which dissolve relatively few salts), many of the commonly used polar aprotic solvents have very high boiling points and are difficult to remove from the reaction products. Furthermore, the solubility of salts in polar aprotic solvents is much more limited because they lack the protic character that solvates anions. However, for the less reactive alkyl halides, or for the S$_N$2 reactions of fluoride ion, polar aprotic solvents are in some cases the only practical alternative.

The S$_N$2 Solvent Effect in Cancer Diagnosis

Positron emission tomography, or "PET," is a widely used technique for cancer detection. In PET, a glucose derivative containing an isotope that emits positrons is injected into the patient. A glucose derivative is used because rapidly growing tumors have a high glucose requirement and therefore take up glucose to a greater extent than normal tissue. The emission of positrons (β^+ particles, or positive electrons) is detected when they collide with nearby electrons (β^-particles). This antimatter–matter reaction results in annihilation of the two particles and the production of two gamma rays that retreat from the site of collision in opposite directions, and these are detected ultimately as light. The light emission pinpoints the site of glucose uptake—that is, the tumor.

The glucose derivative used in PET is 2-^{18}fluoro-2-deoxy-D-glucopyranose, or FDG, which contains the positron-emitting isotope ^{18}F ("fluorine-18"). The structure of FDG is so similar to the structure of glucose that FDG is also taken up by cancer cells.

2-(^{18}F)-fluoro-2-deoxy-D-glucopyranose (FDG)

D-glucopyranose (glucose)

The half-life of ^{18}F is only about 110 minutes. This means that half of it has decayed after 110 minutes, 75% has decayed after 220 minutes, and so on. This short half-life is good for the patient because the emitting isotope doesn't last very long in the body. But it places constraints on the chemistry used to prepare FDG. Thus, ^{18}F, which is generated from $H_2^{18}O$ as an aqueous solution of K^+ $^{18}F^-$, must be produced at or near the PET facility and used to prepare FDG quickly in the PET facility. An S_N2 reaction is used to prepare an FDG derivative using ^{18}F-fluoride as the nucleophile. Like other S_N2 reactions, this reaction occurs with inversion of configuration.

$$AcO— = acetate = H_3C—\overset{\overset{\displaystyle O}{\|}}{C}—O—$$

(9.28)

(The leaving group is a *triflate* group, which we'll discuss in Sec. 10.4A.) This synthesis cannot be carried out in water as a solvent because fluoride ion in protic solvents is virtually unreactive as a nucleophile. To solve this problem, water is completely removed from the aqueous fluoride solution and is replaced by acetonitrile, a polar aprotic solvent (Table 8.2, p. 355). Fluoride ion in anhydrous acetonitrile is a potent nucleophile, and to make it even more nucleophilic, a cryptand (Fig. 8.12, p. 373) is added to sequester the potassium counterion. This prevents the potassium ion from forming ion pairs with the fluoride ion. The "naked" and highly nucleophilic fluoride ion reacts rapidly with mannose triflate tetraacetate to form FDG tetraacetate, as shown in Eq. 9.28.

The acetate (—OAc) groups are used for several reasons. One reason is that they make the mannose derivative more soluble in acetonitrile than it would be if —OH groups were present. But the most important reason they are used is that if O—H groups were present they would themselves form hydrogen bonds with $^{18}F^-$, thus reducing its nucleophilicity and preventing the nucleophilic reaction from taking place. The acetate groups are rapidly removed in a subsequent ester hydrolysis reaction (Sec. 21.7A) to give FDG itself.

FDG 1,3,4,6-tetraacetate

hydrolytic removal of acetate groups

FDG

(9.29)

malignancy as visualized by PET

FIGURE 9.7 The PET image of a malignant lung tumor. The positron-emitting ^{18}F is incorporated in the structure of FDG, a glucose derivative. FDG uptake, like glucose uptake, is enhanced in malignant tumors because they are rapidly growing and require more glucose than normal tissues.

Figure 9.7 shows the PET image of a malignant lung tumor. PET is so sensitive that it has led to the detection of some cancers at an earlier and less invasive stage than previously possible. As we've seen, PET hinges on the rapid synthesis of FDG, which in turn hinges on the clever use of polar aprotic solvents and ion-complexing agents to enhance the nucleophilicity of fluoride ion.

Polarizability Effects on Nucleophilicity Examine the relative nucleophilicity of the halide ions in DMF (Table 9.6). The most basic anions are the most nucleophilic ones, as we expect in a polar aprotic solvent, but if we compare basicity and nucleophilicity *quantitatively*, we find that the variation of nucleophilicity with basicity is not very large. Although the chloride ion is about 10,000 times as basic as iodide ion, it is only 6 times as nucleophilic. Cyanide ion is about 10^{14} times (that is, 14 pK units) as basic as bromide ion, but it is only 300 times as nucleophilic. In other words, iodide ion and bromide ion are much more nucleophilic than their basicities suggest. Why should such weak bases be good nucleophiles?

The reason is that nucleophilic atoms from higher-numbered periods of the periodic table are very *polarizable*. Recall (Sec. 8.5A) that *polarizability* is a measurement of how easily an electron cloud is distorted by an external charge—or, more intuitively, how "squishy" an electron cloud is. In nucleophiles derived from higher-numbered rows of the periodic table, the valence electron clouds are polarizable because they are screened from the nucleus by nonvalence electrons and are easily pulled away from the nucleus. For example, iodide ion is 3.1 times as polarizable as chloride ion and 6.9 times as polarizable as fluoride ion; sulfur is 3.6 times as polarizable as oxygen, and phosphorus is 3.3 times as polarizable as nitrogen. The distortion of a nucleophile's electron cloud is important as it forms a partial bond to the electrophile in a transition state; more bonding can occur at longer distances if the nucleophile is polarizable, and this leads to a transition state of lower energy.

In summary, then, *weak bases can be good nucleophiles if the nucleophilic atom is highly polarizable.* For example, in the following equation, iodide ion displaces bromide ion rapidly in the polar aprotic solvent acetone because, despite its very low basicity, iodide, a highly polarizable anion, is a good nucleophile.

$$\begin{array}{c} H_3C \\ \\ H_3C \end{array}\!\!\!CH-CH_2-Br \ + \ K^+ \ I^- \ \underset{\text{acetone}}{\overset{}{\rightleftarrows}} \ \begin{array}{c} H_3C \\ \\ H_3C \end{array}\!\!\!CH-CH_2-I \ + \ K^+ \ Br^- \downarrow \quad (9.30)$$

Although the equilibrium constant for the reaction is unfavorable, it is driven to the right by the insolubility of potassium bromide in acetone.

As we've seen, *nucleophiles* and *bases* both donate electrons in their reactions. However, there are important differences between *nucleophilicity* and *basicity*. To summarize:

1. Nucleophilicity involves bond formation to atoms other than hydrogen, whereas basicity involves bond formation to hydrogen.
2. Nucleophilicities are measured with relative rates, but basicities are measured with equilibrium constants—that is, pK_a values. Polarizability has a greater effect on nucleophilicity than on basicity.

PROBLEMS

9.14   When methyl bromide is dissolved in ethanol, no reaction occurs at 25 °C. When excess sodium ethoxide is added, a good yield of ethyl methyl ether is obtained. Explain.

9.15   (a) Give the structure of the S_N2 reaction product between ethyl iodide and potassium acetate.

potassium acetate

(b) In which solvent would you expect the reaction to be faster: acetone or ethanol? Explain.

9.16   Which nucleophile, $:N(C_2H_5)_3$ or $:P(C_2H_5)_3$, reacts most rapidly with methyl iodide in ethanol solvent? Explain, and give the product formed in each case.

F. Leaving-Group Effects in the S_N2 Reaction

In many cases, when an alkyl halide is to be used as a starting material in an S_N2 reaction, a choice of leaving group is possible. That is, an alkyl halide might be readily available as an alkyl chloride, alkyl bromide, or alkyl iodide. In such a case, the halide that reacts most rapidly is usually preferred. The reactivities of alkyl halides can be predicted from the close analogy between S_N2 reactions and Brønsted acid–base reactions. Recall that the ease of dissociating an H—X bond within the series of hydrogen halides depends mostly on the H—X bond energy (Sec. 3.6A), and, for this reason, H—I is the strongest acid among the hydrogen halides. Likewise, S_N2 reactivity depends primarily on the *carbon*–halogen bond energy, which follows the same trend: Alkyl iodides are the most reactive alkyl halides, and alkyl fluorides are the least reactive.

Relative reactivities in S_N2 reactions:

$$R\text{—}I > R\text{—}Br > R\text{—}Cl \gg R\text{—}F \tag{9.31}$$

In other words, *the best leaving groups in the S_N2 reaction are those that give the weakest bases as products.* Fluoride is the strongest base of the halide ions; consequently, alkyl fluorides are the least reactive of the alkyl halides in S_N2 reactions. In fact, alkyl fluorides react so slowly that they are useless as leaving groups in most S_N2 reactions. In contrast, chloride, bromide, and iodide ions are much less basic than fluoride ion; alkyl chlorides, alkyl bromides, and alkyl iodides all have acceptable reactivities in typical S_N2 reactions, and alkyl iodides are the most reactive of these. On a laboratory scale, alkyl bromides, which are in most cases less expensive than alkyl iodides, usually represent the best compromise between expense and reactivity. For reactions carried out on a large scale, the lower cost of alkyl chlorides offsets the disadvantage of their lower reactivity.

Halides are not the only groups that can be used as leaving groups in S_N2 reactions. Section 10.4 will introduce a variety of alcohol derivatives that can also be used as starting materials for S_N2 reactions.

G. Summary of the S$_N$2 Reaction

Most primary and some secondary alkyl halides undergo nucleophilic substitution by the S$_N$2 mechanism. Let's summarize six of the characteristic features of this mechanism.

1. The reaction rate is second order overall: first order in the nucleophile and first order in the alkyl halide.

2. The mechanism involves an opposite-side substitution reaction of the nucleophile with the alkyl halide and inversion of stereochemical configuration.

3. The reaction rate is decreased by alkyl substitution at both the α- and β-carbon atoms; alkyl halides with three β-branches are unreactive.

4. Nucleophilicity depends on the basicity of the nucleophile, the polarizability of the nucleophile, and the solvent.

   a. S$_N$2 reactions are much faster in polar aprotic solvents than in protic solvents provided that the nucleophile is soluble enough for the reaction to be practical. Protic solvents are useful if the reaction is fast enough.

   b. Nucleophiles in which the nucleophilic center is from periods $\geqslant 3$ of the periodic table have enhanced nucleophilicity because they are highly polarizable.

   c. In polar aprotic solvents, nucleophilicity increases with the basicity of the nucleophile.

   d. In protic solvents, nucleophilicity increases with the basicity of the nucleophile if the nucleophilic atom is the same.

   e. In protic solvents, nucleophilicity is significantly reduced when the nucleophilic atom is a good hydrogen-bond acceptor. For this reason, nucleophiles with period-2 nucleophilic atoms are less reactive in protic solvents than nucleophiles in which the nucleophilic atoms come from higher-numbered periods within the same group.

5. The fastest S$_N$2 reactions involve leaving groups that give the weakest bases as products.

9.5   THE E2 REACTION

This section discusses base-promoted β-elimination, which is a second important reaction of alkyl halides. An example of such a reaction is the elimination of the elements of HBr from *tert*-butyl bromide:

$$\underset{\underset{CH_3}{|}}{\overset{\overset{CH_3}{|}}{H_3C-C}}-Br + Na^+\ CH_3CH_2O^- \xrightarrow[25\,°C]{CH_3CH_2OH} H_2C=\underset{CH_3}{\overset{CH_3}{C}} + CH_3CH_2O-H + Na^+\ Br^- \qquad (9.32)$$

Recall from Sec. 9.1B that this type of elimination is a dominant reaction of tertiary alkyl halides in the presence of a strong base, and it competes with the S$_N$2 reaction in the case of secondary and primary alkyl halides.

A. Rate Law and Mechanism of the E2 Reaction

Base-promoted β-elimination reactions typically follow a rate law that is second order overall and first order in each reactant:

$$\text{rate} = k[(CH_3)_3C-Br][CH_3CH_2O^-] \qquad (9.33)$$

A mechanism consistent with this rate law is the following:

(9.34)

This type of mechanism, involving concerted removal of a β-proton by a base and loss of a halide ion, is called an **E2 mechanism**. Reactions that occur by the E2 mechanism are called **E2 reactions**. The meaning of the "nickname" E2 is as follows:

Remember that *bimolecular* means that two molecules are involved in the rate-limiting step of the reaction. In this case, the two molecules are the base and the alkyl halide.

B. Why the E2 Reaction Is Concerted

The curved-arrow notation for the E2 mechanism in Eq. 9.34 is worth some attention. The simplest electron-pair displacement reactions we've encountered have involved the donation of an electron pair from either a Brønsted base or a nucleophile to an electrophile and simultaneous loss of a leaving group; the process is fully described by two curved arrows. However, the E2 reaction involves three curved arrows. In the E2 reaction, the base acts as a Brønsted base to remove the β-proton, and the halide acts as a leaving group. How do we analyze the middle curved arrow? This arrow shows that the β-carbon acts *simultaneously* as a leaving group and as a nucleophile that reacts at the α-carbon. That is, the electron pair that departs from the β-hydrogen is donated to the α-carbon to expel the bromide ion.

(9.35)

Let's separate this concerted mechanism into two *fictional* but more conventional two-curved-arrow steps. This will help us to understand why the reaction is concerted. Suppose that in the first step of the elimination the base abstracts a proton to give a *carbon anion* as the product. In this step, the β-carbon acts as a leaving group.

(9.36a)

Then, in the second step, the electron pair of the carbon anion acts as a nucleophile by reacting at the α-carbon, displacing bromide ion:

(9.36b)

Let's calculate the approximate equilibrium constant for the first step (Eq. 9.36a) using the method of Sec. 3.4E. The pK_a of the β-proton should be a little less than the pK_a of an alkane—perhaps about 50. The pK_a of ethanol is 15.9. The equilibrium constant for the first step is then $10^{(15.9-50)}$ or about 10^{-34}. The corresponding standard free-energy change is about 194 kJ mol^{-1}. This means that if the reaction were to occur by this stepwise mechanism, the standard free energy of activation for the first step of this reaction must be at least 194 kJ mol^{-1}, because this is the amount of energy required to form the carbon–anion intermediate. The rate of such a reaction is unimaginably small: the reaction would take approximately 10^{15} years at room temperature! In other words, the elimination would not occur. In fact, typical E2 reactions occur in minutes to a few hours and have standard free energies of activation typically in the 85–95 kJ mol^{-1} range.

The concerted mechanism, then, avoids the formation of a very unstable, strongly basic, carbon–anion intermediate. The concerted mechanism brings about a net transfer of electrons from the oxygen of ethoxide to bromine to form the much weaker base bromide ion. And that is why the middle curved arrow doesn't "pause" at carbon as an electron pair and "hang around" before it reacts at the α-carbon.

In later sections of this text, we'll learn about β-eliminations that *do* involve carbon–anion intermediates. As we might expect, these reactions can take place only if the carbon anion is stabilized in some way. To say that the carbon anion is more stable is to say that the β-proton is much more acidic. Hence, the stepwise β-elimination mechanism will be observed only with compounds in which the β-proton is unusually acidic.

PROBLEMS

9.17 The following hydroxide-catalyzed β-elimination takes place by a carbon–anion stepwise mechanism. Show the carbon–anion intermediate and explain its stability. Think in terms of a polar effect (Sec. 3.6C). Recalling also that resonance structures imply heightened stability (Sec. 1.4), draw a resonance structure for this anion as well.

$$\text{HO—CH}_2\text{—CH}_2\text{—}\overset{\overset{\text{O}}{\|}}{\text{C}}\text{—CH}_3 \underset{}{\overset{\text{HO}^-}{\rightleftharpoons}} \text{H}_2\text{C=CH—}\overset{\overset{\text{O}}{\|}}{\text{C}}\text{—CH}_3 + \text{H}_2\text{O}$$

9.18 We can conceive of a stepwise version of the S_N2 reaction consisting of a Lewis acid–base dissociation followed by a Lewis acid–base association. (Nuc:$^-$ = a nucleophile.)

$$\text{Nuc:}^- \quad \text{CH}_3\text{—}\ddot{\text{I}}\text{:} \longrightarrow \text{Nuc:}^- \quad ^+\text{CH}_3 \quad \text{:}\ddot{\text{I}}\text{:}^- \longrightarrow \text{Nuc—CH}_3 \quad \text{:}\ddot{\text{I}}\text{:}^-$$

(a) Why should the stepwise process be slower than the concerted process?

(b) For what type of alkyl halide is the stepwise process likely to be observed?

C. Leaving-Group Effects on the E2 Reaction

In the mechanism of the E2 reaction, the role of the leaving halide is much the same as it is in the S_N2 reaction: Its bond to carbon breaks and it takes on an additional electron pair to become a halide ion. Consequently, it should not be surprising to find that the rates of S_N2 and E2 reactions are affected in similar ways by changing the halide leaving group:

Relative rates of E2 reactions:

$$\text{R—I} > \text{R—Br} > \text{R—Cl} \tag{9.37}$$

As in the S_N2 reaction, the reactivity difference between alkyl bromides and iodides is not great. Alkyl bromides are usually used in the laboratory for E2 reactions as the best compromise of reactivity and expense, and, when possible, the less expensive alkyl chlorides are used in large-scale reactions.

D. Deuterium Kinetic Isotope Effects in the E2 Reaction

The mechanism in Eq. 9.34 implies that a proton is removed in the transition state of the E2 reaction. This aspect of the mechanism can be tested in an interesting way. When a hydrogen is transferred in the rate-limiting step of a reaction, a compound in which that hydrogen is replaced by its isotope deuterium will react more slowly in the same reaction. This effect of isotopic substitution on reaction rates is called a **primary deuterium kinetic isotope effect**. For example, suppose the rate constant for the following E2 reaction of 2-phenyl-1-bromoethane is k_H, and the rate constant for the reaction of its β-deuterium analog is k_D:

$$\text{Ph}-\text{CH}_2-\text{CH}_2-\text{Br} + \text{EtO}^- \xrightarrow[\text{EtOH}]{\text{rate constant } k_H} \text{Ph}-\text{CH}=\text{CH}_2 + \text{Br}^- + \text{EtOH} \qquad (9.38a)$$

$$\text{Ph}-\text{CD}_2-\text{CH}_2-\text{Br} + \text{EtO}^- \xrightarrow[\text{EtOH}]{\text{rate constant } k_D} \text{Ph}-\text{CD}=\text{CH}_2 + \text{Br}^- + \text{EtOD} \qquad (9.38b)$$

The primary deuterium kinetic isotope effect is the ratio of the rates for the two reactions—that is, k_H/k_D; typically such isotope effects are in the range 2.5–8. For example, k_H/k_D for the reactions in Eq. 9.38 is 7.1. The observation of a kinetic isotope effect of this magnitude shows that the bond to a β-hydrogen is broken in the rate-limiting step of this reaction.

The theoretical basis for the primary kinetic isotope effect lies in the comparative strengths of C—H and C—D bonds. In the starting material, the bond to the heavier isotope D is slightly stronger (and thus requires more energy to break; Sec. 5.6E) than the bond to the lighter isotope H. However, in the transition states for both reactions, the bond from H or D to carbon is partly broken, and the bond from H or D to the base is partly formed. To a crude approximation, the isotope undergoing transfer is not bonded to anything—it is "in flight." Because there is no bond, there is no bond-energy difference between the two isotopes in the transition state. Therefore, the compound with the C—D bond starts out at a lower energy than the compound with the C—H bond and requires more energy to achieve the transition state (Fig. 9.8). In other words, the energy barrier, or free energy of activation, for the compound with the C—D bond is greater; as a result, its rate of reaction is smaller.

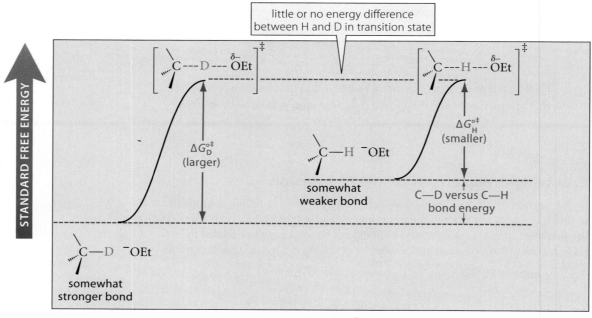

FIGURE 9.8 The source of the primary kinetic deuterium isotope effect is the stronger carbon–deuterium bond. (The difference between the bond energies of the C—H and C—D bonds is greatly exaggerated for purposes of illustration.)

A primary kinetic deuterium isotope effect is observed only when the *hydrogen that is transferred* in the rate-determining step is substituted by deuterium. Substitution of other hydrogens with deuterium usually has little or no effect on the rate of the reaction.

PROBLEMS

9.19 In each of the following series, arrange the compounds in order of increasing reactivity in the E2 reaction with $Na^+ EtO^-$.

(a)

A B C

(b)

A B

9.20 (a) The rate-limiting step in the hydration of styrene ($Ph—CH{=}CH_2$) is the initial transfer of the proton from H_3O^+ to the alkene (Sec. 4.9B). How would you expect the rate of the reaction to change if the reaction were run in D_2O/D_3O^+ instead of H_2O/H_3O^+? Would the product be the same?

(b) How would the rate of styrene hydration in H_2O/H_3O^+ differ from that of an isotopically substituted styrene $Ph—CH{=}CD_2$? Explain.

E. Stereochemistry of the E2 Reaction

When an E2 reaction occurs, the tetrahedral α- and β-carbons become trigonal when the β-proton is removed and the halide leaves. The R-groups on these two carbons move into a common plane that also contains the alkene carbons. This motion is shown in Fig. 9.9.

The stereochemistry of the E2 reaction uses this plane as a frame of reference. The E2 reaction can occur in two stereochemically distinct ways, illustrated as follows for the elimination of the elements of H—X from a general alkyl halide:

syn-elimination:

(9.39a)

anti-elimination:

(9.39b)

FIGURE 9.9 The stereochemical changes that occur during an E2 elimination. The α- and β-carbons are rehybridized from sp^3 to sp^2, and the R-groups attached to these carbons move into a common plane. In this view, this plane is perpendicular to the page and tilted slightly downward.

In a **syn-elimination**, the dihedral angle between the C—H and C—X bonds is 0°; that is, the H and X groups leave from the same side of the reference plane. In an **anti-elimination**, the dihedral angle between the C—H and C—X bonds is 180°; that is, the H and X groups leave from opposite sides of the reference plane. (The elimination shown in Fig. 9.9 is anti.) A Newman projection or a sawhorse projection of the transition state for anti-elimination is another way to see the anti relationship of the proton that is removed and the leaving group:

Newman projection
for anti-elimination

sawhorse projection
for anti-elimination

sawhorse projection
for the alkene product
of anti-elimination

Only syn- and anti-eliminations are possible because only these geometries result in the planar alkene geometry that is required for π-orbital overlap. Recall from Sec. 7.8A that the terms *syn* and *anti* were used in discussing the stereochemistry of additions to double bonds. Notice that syn-elimination is conceptually the reverse of a syn-addition, and anti-elimination is conceptually the reverse of an anti-addition.

Investigation of the stereochemistry of an elimination reaction requires the α- and β-carbons to be stereocenters in both the starting alkyl halide and the product alkene. In such cases, it is found experimentally that most E2 reactions are stereoselective anti-eliminations, as in the following example.

bonds to H and Br are anti

(Z)-α-methylstilbene
(only product observed)

(9.40a)

When the hydrogen and halogen are eliminated from a conformation in which they are anti, the phenyl groups (Ph) are on the same side of the molecule and therefore must end up in a cis relationship in the product alkene. A syn-elimination would give the other alkene stereoisomer:

bonds to H and Br are eclipsed

(E)-α-methylstilbene
(not observed)

(9.40b)

Anti-elimination is preferred for three reasons. First, syn-elimination occurs through a transition state that has an eclipsed conformation, whereas anti-elimination occurs through a transition state that has a staggered conformation.

syn-elimination:
the molecule is in
an eclipsed conformation

anti-elimination:
molecule is in
a staggered conformation

Because eclipsed conformations are unstable, the transition state for syn-elimination is less stable than the transition state for anti-elimination. As a consequence, anti-elimination is faster. The second reason that anti-elimination is preferred is that the base and leaving group are on opposite sides of the molecule, out of each other's way. In syn-elimination, they are on the same side of the molecule and can interfere sterically with each other. Finally, calculations of transition-state energies using molecular-orbital theory show that anti-elimination is more favorable; the reasoning relates to the fact that an anti-elimination involves all-opposite-side electron displacements, as in the S_N2 reaction.

this electron pair approaches
from the opposite side
of the C—X bond

this electron pair approaches
from the same side
as the C—X bond

anti

syn

PROBLEMS

9.21 Predict the products, including their stereochemistry, from the E2 reactions of the following diastereomers of stilbene dibromide with sodium ethoxide in ethanol. Assume that one equivalent of HBr is eliminated in each case.

(a) $(\pm)$-Ph—CH—CH—Ph
         | |
         Br   Br

(b) *meso*-Ph—CH—CH—Ph
              | |
              Br   Br

9.22 Draw the structure of the starting material that would undergo anti-elimination to give the *E* isomer of the alkene product in the E2 reaction of Eq. 9.40a.

F. Regioselectivity of the E2 Reaction

When an alkyl halide has more than one type of β-hydrogen, more than one alkene product can be formed (Sec. 9.1B).

(9.41)

2-bromobutane   ***cis*-2-butene**   ***trans*-2-butene**   **1-butene**

This section focuses on which of the possible products is preferred and why.

When simple alkoxide bases such as methoxide and ethoxide are used, *the predominant product of an E2 reaction is usually the most stable alkene isomer.* Recall that the most stable alkene isomers are generally those with the most alkyl substituents at the carbons of the double bond (Sec. 4.5B). These isomers, then, are the ones formed in greatest amount.

$$CH_3CH_2C(CH_3)_2 \xrightarrow[\text{EtOH}]{K^+ \ ^-OEt} CH_3CH{=}C(CH_3)_2 + CH_3CH_2C \begin{matrix} CH_2 \\ \| \\ \backslash \\ CH_3 \end{matrix} \qquad (9.42)$$

In this reaction, the alkene isomer formed in smaller amount would actually be favored statistically: six equivalent hydrogens can be lost from the alkyl halide to give this alkene, but only two can be lost to give the other alkene. In the absence of a structural effect on the product distribution, three times as much of the 1-alkene would have been formed. The fact that the other alkene is the major one shows that some other factor is operating.

The predominance of the more stable alkene isomer does *not* result from equilibration of the alkenes themselves, because *the alkene products are stable under the conditions of the reaction.* Because the product mixture, once formed, does not change, the distribution of products must reflect the relative rates at which they are formed. Hence, we look for the explanation in transition-state theory.

The transition state for the E2 reaction can be visualized as a structure that lies somewhere between alkyl halide and alkene (plus the other species present). To the extent that the transition state resembles the alkene product, it is stabilized by the same factors that stabilize alkenes—and one such factor is alkyl substitution at the double bond. A reaction that can give two alkene products is really two reactions in competition, each with its own transition state. The reaction with the transition state of lower energy—the one with more alkyl substitution at the developing double bond—is the faster reaction. Hence, more product is formed through this transition state (Fig. 9.10)

Zaitsev's Rule

An elimination reaction that forms predominantly the most stable alkene isomers is sometimes called a *Zaitsev elimination*, after Alexander M. Zaitsev (1841–1910), a Russian chemist who observed this phenomenon in 1875. Just as the Markovnikov rule describes the regioselectivity of hydrogen halide addition to alkenes, the Zaitsev rule describes the regioselectivity of elimination reactions. And, like the Markovnikov rule, the Zaitsev rule is purely descriptive; it does not attempt to explain the reasons behind the observations.

When an alkyl halide has more than one type of β-hydrogen, a mixture of alkenes is generally formed in its E2 reaction, as Eq. 9.41 illustrates. The formation of a mixture means that the yield of the desired alkene isomer is reduced. Furthermore, because the alkenes in such mixtures are isomers, they generally have similar boiling points and are therefore difficult to separate. Consequently, the greatest use of the E2 elimination for the preparation of alkenes occurs when the alkyl halide has only one type of β-hydrogen, and only one alkene product is possible.

The following study problem gives you some practice in integrating the principles involved in stereochemistry and regioselectivity of eliminations with what you learned about cyclohexane conformations in Chapter 7.

FIGURE 9.10 The alkene with more alkyl substituents on the double bond is formed more rapidly and in greater amount than the alkene with fewer substituents because the free energy of the transition state, like that of the alkene, is lowered by alkyl substitution.

STUDY PROBLEM   9.1

Give the predominant product of the E2 reaction when each of the following diastereomers is allowed to react with potassium *tert*-butoxide in DMSO. Explain your reasoning.

A   *B*

SOLUTION   Two principles are important in solving this problem. The first is that the stereochemistry of the E2 reaction is anti. In a cyclohexane derivative, the required anti relationship of the proton that is lost and the leaving group is possible only if the two are trans. In compound *A*, there are two hydrogens that are trans to the Cl: H^x and H^y. Therefore, elimination of either proton could occur to give products *X* and *Y*, respectively.

The second principle is that the predominant product is the one with the greater number of alkyl substituents on the double bond—product *X*. Therefore, this is the major product, as shown in the equation.

In compound *B*, only hydrogen H^p is trans to the leaving group. Hydrogen H^q is cis to the leaving group and therefore cannot be anti. Consequently, only compound *Y* is formed. Although isomer *X* is more stable, there is no stereochemically acceptable path by which it can be formed.

Notice that in both cyclohexane derivatives, the Cl leaving group and the trans hydrogen are anti in *only one* of the two chair conformations of compound *A*; they are gauche in the other. Therefore, the elimination must occur from the anti conformation, shown here for the formation of the major product *X* from compound *A*:

The H-anti conformation is depleted as it reacts, but it is restored rapidly by the conformational equilibrium, by Le Châtelier's principle. The same is true for the formation of product *Y* from both isomers *A* and *B* (show this). (Problems 9.77, 9.78, and 9.80 at the end of the chapter will provide more practice in solving this type of problem.)

G. Competition between the E2 and S$_N$2 Reactions: A Closer Look

Nucleophilic substitution reactions and base-promoted elimination reactions are *competing* processes (Sec 9.1C). In other words, whenever an S$_N$2 reaction is carried out, there is the possibility that an E2 reaction can also occur (if the alkyl halide has β-hydrogens), and vice versa.

(9.43)

This competition is a matter of *relative rates*: The reaction pathway that occurs more rapidly is the one that predominates.

Two variables determine which reaction—the S$_N$2 reaction or the E2 reaction—will be the major process observed in a given case: (1) the structure of the alkyl halide; and (2) the structure of the base.

The feature of an alkyl halide's structure that determines the amount of elimination versus substitution is the *number of alkyl substituents* at both the α- and β-carbons. For the S$_N$2 reaction to occur on an alkyl halide with α- or β-substituents, the nucleophile must approach through a thicket of interfering hydrogen atoms on the substituents that impede its access to the α-carbon. The resulting van der Waals repulsions create an energy barrier to the

S_N2 reaction that decreases its rate. On the other hand, when a Brønsted base initiates the E2 reaction, it reacts with a β-proton that lies near the periphery of the molecule. Reaction at the β-proton is much less affected by steric repulsions than reaction at the α-carbon atom.

reaction at the α-carbon is unhindered;
substitution occurs

reaction at the α-carbon is blocked;
reaction at the β-proton
and elimination occurs instead

(9.44)

Another reason that alkyl substitution promotes the E2 reaction is that the standard free energy of the E2 transition state, like that of an alkene, is lowered by alkyl substitution (Sec. 9.5F). Consequently, the rate of the E2 reaction is *increased by alkyl substitution*. Two effects of alkyl substitution, then, favor the E2 reaction: the rate of the S_N2 reaction is *decreased*, and the rate of the E2 reaction is *increased*.

These same effects can be seen not only in tertiary alkyl halides, but also in secondary and even primary alkyl halides. Notice in the following examples that the alkyl halides with more β-alkyl substituents show a greater proportion of elimination.

Secondary alkyl halides:

(9.45a)

(9.45b)

Primary alkyl halides:

(9.46a)

(9.46b)

(9.46c)

The structure of the base is the second variable that determines whether the E2 reaction or the S_N2 reaction is faster in a given case. First of all, a *highly branched base*, such as *tert*-butoxide, increases the proportion of elimination relative to substitution.

$$(CH_3)_2CHCH_2\!-\!Br \; + \; {}^-OCH_2CH_3 \;\; \xrightarrow{\;\;CH_3CH_2OH\;\;} \;\; (CH_3)_2C\!=\!CH_2 \; + \; (CH_3)_2CHCH_2\!-\!OCH_2CH_3 \qquad (9.47a)$$

ethoxide
(a primary,
unbranched
alkoxide base)

(62% elimination)   (38% substitution)

$$(CH_3)_2CHCH_2\!-\!Br \; + \; {}^-O\!-\!\underset{\underset{CH_3}{|}}{\overset{\overset{CH_3}{|}}{C}}\!-\!CH_3 \;\; \xrightarrow{\;\;(CH_3)_3COH\;\;} \;\; (CH_3)_2C\!=\!CH_2 \; + \; (CH_3)_2CHCH_2\!-\!O\!-\!\underset{\underset{CH_3}{|}}{\overset{\overset{CH_3}{|}}{C}}\!-\!CH_3 \qquad (9.47b)$$

tert-butoxide
(a tertiary, branched
alkoxide base)

(92% elimination)   (8% substitution)

When a highly branched base reacts at the *α-carbon* to give a substitution product, the alkyl branches of the base suffer van der Waals repulsions with the surrounding hydrogens in the alkyl halide molecule; these repulsions raise the energy of the transition state for substitution. When such a base reacts at a *β-proton* to give the elimination product, the base is further removed from the offending hydrogens in the alkyl halide, and van der Waals repulsions are less severe, as shown in Eq. 9.44. Consequently, the S_N2 reaction is retarded more than the E2 reaction by branching in the base, and elimination becomes the predominant reaction. In summary, with a highly branched base, a *steric effect* selectively retards the S_N2 reaction.

A further effect of base structure on the E2–S_N2 competition has to do with its Brønsted basicity versus its nucleophilicity. Recall from Sec. 9.4E that the reactivity of a nucleophile—its nucleophilicity—affects the rate of its S_N2 reactions, whereas its Brønsted basicity affects the rate of its E2 reactions (because the base is reacting with a proton). Recall also that species with nucleophilic atoms from higher periods of the periodic table, such as iodide ion, are excellent nucleophiles even though they are relatively weak Brønsted bases. A greater fraction of S_N2 reaction is observed in the reactions of such nucleophiles. For example, the reaction of potassium iodide with isobutyl bromide in acetone gives mostly substitution product and little elimination, because iodide is an excellent nucleophile but a weak Brønsted base:

$$\underset{H_3C}{\overset{H_3C}{>}}\!CH\!-\!CH_2\!-\!Br \; + \; K^+ \; I^- \;\; \underset{acetone}{\overset{\longrightarrow}{\longleftarrow}} \;\; \underset{H_3C}{\overset{H_3C}{>}}\!CH\!-\!CH_2\!-\!I \; + \; K^+ \; Br^- \downarrow \qquad (9.48)$$

Contrast this reaction with that in Eq. 9.46c, in which sodium ethoxide reacts with the same alkyl halide. Ethoxide, a strong Brønsted base, gives a significant percentage of alkene and a smaller percentage of substitution product.

Let's summarize the effects that govern the competition between the S_N2 and E2 reactions.

1. *Structure of the alkyl halide:*

   a. Alkyl halides with greater numbers of alkyl substituents at the *α*-carbon give greater amounts of elimination. Consequently, tertiary alkyl halides give more elimination than secondary alkyl halides, which give more than primary alkyl halides.

   b. Alkyl halides with greater numbers of alkyl substituents at the *β*-carbon give greater amounts of elimination.

   c. Alkyl halides that have no *β*-hydrogens cannot undergo *β*-elimination.

2. *Structure of the base:*

   a. In a comparison of alkoxide bases with similar strengths, tertiary alkoxide bases such as *tert*-butoxide give a greater fraction of elimination than primary alkoxide bases.

   b. Weaker bases that are good nucleophiles give a greater fraction of substitution.

The application of these ideas is illustrated in Study Problem 9.2.

STUDY PROBLEM 9.2

Which alkyl halide and what conditions should be used to prepare the following alkene in good yield by an E2 elimination?

methylenecyclohexane

SOLUTION If this alkene is to be produced in an E2 reaction from an alkyl halide, the halide must be located at one of the two carbons that eventually become carbons of the double bond. This means that there are two choices for the starting alkyl halide:

A   and   *B*

The advantage of alkyl halide *A* is that, because it is tertiary, it poses no significant competition from the S_N2 reaction. The disadvantage of this alkyl halide is that it contains more than one type of β-hydrogen, and, consequently, more than one alkene product could be formed:

β-hydrogens (*a*)

β-hydrogens (*b*)

β-elimination ⟶   *C*   +   *D*   (9.49)

β-hydrogens (*a*)

A   *C*   *D*
                                          from loss of H(*a*)   from loss of H(*b*)

STUDY GUIDE LINK 9.2
Ring Carbons as Alkyl Substituents

Product *C* is the more stable alkene because its double bond has three alkyl substituents; hence, if *A* is used as the starting material, a major amount of this undesired alkene will be formed. If alkyl halide *B* is the starting material, then the desired product *D* is the *only* possible product of β-elimination. Because this alkyl halide is primary, however, it is possible that some by-product derived from the S_N2 reaction will be formed. The way to minimize the S_N2 reaction is to use a tertiary alkoxide base such as *tert*-butoxide. In addition, the extensive β-substitution in alkyl halide *B* should also minimize the substitution reaction. Hence, a reasonable preparation of the desired alkene is the following:

$$\xrightarrow[\text{(CH}_3\text{)}_3\text{C}-\text{OH}]{\text{K}^+ \text{ (CH}_3\text{)}_3\text{C}-\text{O}^-}$$ (9.50)

PROBLEMS

9.23 What nucleophile or base and what type of solvent could be used for the conversion of isobutyl bromide into each of the following compounds?

(a) $(CH_3)_2CHCH_2\overset{+}{S}(CH_3)_2$ Br^- (b) $(CH_3)_2CHCH_2SCH_2CH_3$ (c) $(CH_3)_2C{=}CH_2$

9.24 Arrange the following four alkyl halides in descending order with respect to the E2 elimination to S_N2 substitution product ratio expected in their reactions with sodium ethoxide in ethyl alcohol. Explain your answers.

$$CH_3I \qquad (CH_3)_2CHCH_2{-}Br \qquad (CH_3)_3CCH_2CH_2CH_2{-}Br \qquad (CH_3)_2CHCH{-}Br$$

$$\qquad\qquad\qquad\qquad\qquad\qquad\qquad\qquad\qquad\qquad\qquad\qquad\qquad\qquad\qquad | $$

$$\qquad A \qquad\qquad\quad B \qquad\qquad\qquad\qquad\qquad C \qquad\qquad\qquad\qquad\qquad CH_3$$

$$\qquad\qquad\qquad\qquad\qquad\qquad\qquad\qquad\qquad\qquad\qquad\qquad\qquad\qquad\qquad\qquad D$$

9.25 Arrange the following four bases in descending order with respect to the E2 elimination to S_N2 substitution product ratio expected when they react with isobutyl bromide. Explain your answers.

$$(CH_3)_2CH{-}O^- \qquad CH_3O^- \qquad (C_2H_5)_3C{-}O^- \qquad Cl^-$$

$$\qquad A \qquad\qquad\qquad B \qquad\qquad\quad C \qquad\qquad\quad D$$

H. Summary of the E2 Reaction

The E2 reaction is a β-elimination reaction of alkyl halides that is promoted by strong bases. The following list summarizes the key points about this reaction:

1. The rates of E2 reactions are second order overall: first order in base and first order in the alkyl halide.

2. E2 reactions normally occur with anti stereochemistry.

3. The E2 reaction is faster with better leaving groups—that is, those that give the weakest bases as products.

4. The rates of E2 reactions show substantial primary deuterium isotope effects at the β-hydrogen atoms.

5. When an alkyl halide has more than one type of β-hydrogen, more than one alkene product can be formed; the most stable alkenes (the alkenes with the greatest numbers of alkyl substituents at their double bonds) are formed in greatest amount.

6. E2 reactions compete with S_N2 reactions. Elimination is favored by alkyl substitution in the alkyl halide at the α- or β-carbon atoms, by alkyl substituents at the α-carbon of the base, and by highly branched bases.

9.6   THE S_N1 AND E1 REACTIONS

Until now, the discussion has stressed the reactions of alkyl halides with species that are either strong bases or good nucleophiles. When a primary alkyl halide is dissolved in a protic solvent such as ethanol with no added base, the S_N2 reaction that occurs takes two weeks or more (depending on the temperature and the alkyl halide), because a neutral, un-ionized alcohol is a weak base and therefore a poor nucleophile. When a tertiary alkyl halide such as *tert*-butyl bromide is subjected to the same conditions, however, both substitution and elimination reactions occur readily.

$$
\begin{array}{c}
CH_3 \\
| \\
H_3C{-}C{-}Br \\
| \\
CH_3
\end{array}
+ HO{-}Et \xrightarrow[55\,°C]{}
\begin{array}{c}
CH_3 \\
| \\
H_3C{-}C{-}O{-}Et \\
| \\
CH_3
\end{array}
+
\begin{array}{c}
H_3C \\
\diagdown \\
C{=}CH_2 \\
\diagup \\
H_3C
\end{array}
+ \overset{+}{Et\,O}H_2\ Br^- \qquad (9.51)
$$

tert-butyl bromide   ethanol (solvent)   *tert*-butyl ethyl ether   2-methylpropene   (ionized form of HBr in ethanol)
                                                        (72%)   (28%)

The reaction of an alkyl halide with a solvent in which no other base or nucleophile has been added is called a **solvolysis** (literally, bond breaking by solvent). The substitution that occurs in the solvolysis of *tert*-butyl bromide cannot involve an S$_N$2 mechanism because chain branching at the α-carbon retards the S$_N$2 reaction. That is, if the solvolysis of a primary alkyl halide by an S$_N$2 mechanism is very slow, then the solvolysis of a tertiary alkyl halide by the same mechanism should be *even slower*. The elimination that occurs in this solvolysis cannot occur by an E2 mechanism because a strong base is not present. Because both substitution and elimination reactions occur readily, they must then involve mechanisms that are different from the S$_N$2 and E2 mechanisms. These new mechanisms are the subject of this section.

A. Rate Law and Mechanism of the S$_N$1 and E1 Reactions

The solvolysis of *tert*-butyl bromide follows a *first-order rate law*:

$$\text{rate} = k[(CH_3)_3CBr] \tag{9.52}$$

Any involvement of solvent in the reaction cannot be detected in the rate law because the concentration of the solvent cannot be changed. However, the nature of the solvent does play a critical role in this reaction. The solvolysis reactions of tertiary alkyl halides are fastest in *polar, protic, donor solvents*, such as alcohols, formic acid, and mixtures of water with solvents in which the alkyl halide is soluble (for example, aqueous acetone). Notice that *these solvents are the ones that are best at solvating ions* (Sec. 8.6F).

The occurrence of both substitution and elimination products shows that two *competing reactions* are involved. The first step in *both* reactions involves the ionization of the alkyl halide to a carbocation and a halide ion:

$$(CH_3)_3C \overset{\frown}{-} \overset{..}{\underset{..}{Br}} : \rightleftharpoons (CH_3)_3C^+ \quad :\overset{..}{\underset{..}{Br}} :^- \quad \text{(rate-limiting step)} \tag{9.53a}$$

$$\text{carbocation}$$
$$\text{intermediate}$$

This step, which is a *Lewis acid–base dissociation* (Sec. 3.1A), is the rate-limiting step of both the substitution and elimination reactions. In other words, when a tertiary alkyl halide is dissolved in a polar, protic solvent such as ethanol, it reacts by dissociating slowly into a carbocation and a halide ion; the carbocation then rapidly reacts to give both substitution and elimination products. Thus, substitution and elimination products arise from *competing reactions of the carbocation*.

Consider first the formation of the substitution product. This product is formed by the Lewis acid–base association of a solvent molecule with the carbocation. Even though the solvent is a poor nucleophile, the reaction occurs rapidly because the solvent is present in very high concentration and because the carbocation is a very powerful Lewis acid.

$$(CH_3)_3C^+ \quad H\overset{..}{\underset{..}{O}}Et \rightleftharpoons (CH_3)_3C \overset{+}{\underset{\underset{H}{|}}{-\overset{..}{O}Et}} \quad :\overset{..}{\underset{..}{Br}} :^- \tag{9.53b}$$
$$:\overset{..}{\underset{..}{Br}} :^-$$

The nucleophile that reacts with the carbocation is ethanol, *not* ethoxide ion; such a strong base is *not present* in a solvolysis reaction; furthermore, if significant amounts of such a base were added, elimination by the E2 mechanism would be observed exclusively. The product of Eq. 9.53b is the conjugate acid of an ether, which it is a strong acid.

The final step of the substitution reaction is a Brønsted acid–base reaction in which the protonated ether (the Brønsted acid) loses a proton to solvent (the Brønsted base) to give the ether and the conjugate acid of the solvent.

$$(CH_3)_3C \overset{+\overset{..}{O}Et}{\underset{\underset{H}{\overset{\frown}{|}}}{}} \quad Br^- \rightleftharpoons (CH_3)_3C \overset{..}{-\underset{..}{O}}Et \quad + \quad H\overset{\overset{H}{\overset{+|}{}}}{\underset{..}{O}}Et \quad Br^- \tag{9.53c}$$
$$H\overset{..}{\underset{..}{O}}Et$$
$$\text{(ionized form}$$
$$\text{of HBr in ethanol)}$$

The Brønsted base involved in this reaction is ethanol, not ethoxide ion. As we noted in discussing the previous step of the reaction, ethoxide ion is not present; nor is it necessary, because the protonated ether is a strong acid. Notice that the protonated solvent plus bromide ion (that is, $Et\overset{+}{O}H_2 \ Br^-$) is the form of ionized HBr in ethanol solvent, just as $H_3O^+Br^-$ is the ionized form of HBr in water.

A substitution mechanism that involves a carbocation intermediate is called an **S_N1 mechanism**. Substitution reactions that take place by the S_N1 mechanism are called **S_N1 reactions**. The meaning of the S_N1 "nickname" is as follows:

$$\text{substitution} \nearrow \quad \underset{\text{nucleophilic}}{\overset{S_N1}{\uparrow}} \quad \nwarrow \text{unimolecular}$$

The word *unimolecular* means that a single molecule—the alkyl halide—is involved in the rate-limiting step.

Now consider the formation of the elimination product of Eq. 9.51, which involves a different reaction of the carbocation intermediate. Loss of a β-proton (a proton from the carbon adjacent to the electron-deficient carbon) gives the alkene.

$$(9.54)$$

(ionized form of HBr in ethanol)

The base that removes a β-proton from the carbocation is typically a solvent molecule. Although ethanol is a very weak base, the reaction occurs readily because ethanol, as the solvent, is present in very high concentration and because the carbocation is a *very strong* Brønsted acid (its pK_a has been estimated to be about -8). The base is *not* ethoxide ion; no ethoxide ion is present. Notice that ionized HBr is produced in this reaction as well.

A β-elimination mechanism that involves carbocation intermediates is called an **E1 mechanism**; reactions that occur by E1 mechanisms are called **E1 reactions**. The meaning of the E1 "nickname" is as follows:

$$\text{elimination} \nearrow \quad \overset{E1}{} \quad \nwarrow \text{unimolecular}$$

B. Rate-Limiting and Product-Determining Steps

The S_N1 and E1 reactions have a *common rate-limiting step*. That is, the rate at which the alkyl halide disappears as it undergoes both competing reactions is determined by its *rate of ionization*—the rate at which it forms the carbocation. The relative amounts of substitution and elimination products are determined by the relative rates of the steps that *follow* the rate-limiting step: reaction of the solvent as a nucleophile with the carbocation to give a substitution product, and loss of a β-proton to solvent from the carbocation to give the elimination product. For example, more substitution than elimination product is formed in Eq. 9.51. This means that the rate of formation of the substitution product from the carbocation is greater than the rate of formation of the elimination product. Because the relative rates of these steps determine the ratio of products, they are said to be the **product-determining steps**. Notice that *the relative rates of the product-determining steps have nothing to do with the rate at which the alkyl halide dissociates into ions.*

$$\text{(9.55)}$$

rate-limiting step: the rate of this step is the rate at which alkyl halide disappears

(two steps) EtOH

EtOH

product-determining steps: the relative rates of these steps determine the relative amounts of different products

The reaction free-energy diagram in Fig. 9.11 on p. 422 summarizes these ideas. The first step, ionization of the alkyl halide to a carbocation, is the rate-limiting step and thus has the transition state of highest free energy. The rate of this step is the rate at which the alkyl halide reacts. The relative free-energy barriers for the product-determining steps determine the relative amounts of products formed.

An Analogy for Product-Determining Steps

Imagine that we are in the Southwest at a busy "T" intersection controlled by a traffic light. Traffic backs up during the red light, and a given car perhaps has to wait for several signal changes to get through the intersection. Passage through the intersection is the *rate-limiting step* on this highway. After leaving the intersection quickly, one can turn left (to Albuquerque) or right (to Phoenix). The total rate that cars pass through the intersection is controlled by the traffic light, not by how fast the cars make the turn after the light turns green. The relative numbers of cars that take the left (eastern) or right (western) turn determine the relative numbers of cars that end up on the highways to the two destinations. Entering a turn is analogous to the product-determining step. If more cars turn west, then the rate of cars that turn west is greater than the rate at which cars turn east. For example, if three times as many cars turn west as turn east, then the number of cars entering the western highway per unit time is three times the number entering the eastern highway. Similarly, if two products *A* and *B* are formed and if the rate of formation of *A* is three times the rate of formation of *B*, then 75% of the product is *A* and 25% is *B*. However, just as the total rate at which cars pass through the intersection is controlled by the traffic light, the total rate of a reaction is controlled by the rate of the rate-limiting step.

The competition between the S$_N$1 and E1 reactions is different from the competition between the S$_N$2 and E2 reactions. The latter two reactions share nothing in common but starting materials; they follow completely separate reaction pathways with no common intermediates.

alkyl halide + Lewis base

S$_N$2
the Lewis base acts as a nucleophile
→ substitution product

E2
the Lewis base acts as a Brønsted base
→ elimination products (alkenes)

$$\text{(9.56)}$$

FIGURE 9.11 Reaction free-energy diagram for the S_N1–E1 solvolysis reaction of $(CH_3)_3CBr$ with ethanol. The rate-limiting step, ionization of the alkyl halide (*red curve*), has the transition state of highest standard free energy. The relative rates of the product-determining steps (*blue curves*) determine the relative amounts of substitution and elimination products. In this example, the energy barrier for the substitution reaction is lower; hence, the substitution reaction is faster than the elimination reaction, and more substitution than elimination product is observed. (The final proton transfer required to form the substitution product is not shown explicitly.)

In contrast, the S_N1 and E1 reactions of an alkyl halide share not only common starting materials, but also a common rate-limiting step, and hence *a common intermediate*—the carbocation.

$$\text{alkyl halide} \xrightarrow[\text{step}]{\text{rate-limiting}} \boxed{\text{carbocation}} \begin{array}{c} \xrightarrow{S_N1} \text{substitution products} \\ \text{product-determining} \\ \text{steps} \\ \xrightarrow{E1} \text{elimination products} \\ \text{(alkenes)} \end{array} \qquad (9.57)$$

In the E1 reaction, the proton is not removed from the alkyl halide, as it is in the E2 reaction, but from the *carbocation*. Because the carbocation is a strong acid, a strong base is not required for the E1 reaction as it is for the E2 reaction.

C. Reactivity and Product Distributions in S_N1–E1 Reactions

S_N1–E1 reactions are most rapid with tertiary alkyl halides, they occur more slowly with secondary alkyl halides, and they are never observed with primary alkyl halides.

Reactivity of alkyl halides in S_N1 or E1 reactions:

$$\text{tertiary} \gg \text{secondary} \gg \text{primary} \qquad (9.58)$$

The exact relative reactivities vary with conditions, particularly the solvent; as an example, *tert*-butyl bromide is 1150 times more reactive than isopropyl bromide in ethanol, and about 10^5 times more reactive in water. The relative reactivity of primary alkyl halides in S_N1 or E1 reactions is not known with certainty because they do not react at all by this mechanism.

Notice that this reactivity order is expected from the relative stability of the corresponding carbocation intermediates. Hammond's postulate (Sec. 4.8D) suggests that the rate-limiting transition state of an S_N1 or E1 reaction should closely resemble a carbocation.

The reactivity order of the alkyl halides in S_N1–E1 reactions is iodides $\gg$ bromides $>$ chlorides $\gg$ fluorides. This is the same reactivity order observed in S_N2 and E2 reactions. This relative reactivity is expected because the leaving group in the S_N1–E1 reaction has much the same role as it does in the E2 and S$_N$2 reactions. That is, the bond to the halide is breaking in the rate-limiting step, and the halide is taking on a negative charge and an additional unshared electron pair.

S$_N$1–E1 reactions are fastest in polar, protic, donor solvents. This is the result expected in a reaction for which the rate-limiting step is a dissociation of a neutral molecule into ions of opposite charge. Ionic dissociation is favored by solvents that *separate* ions (that is, polar solvents—solvents with a high dielectric constant), and by solvents that *solvate* ions (that is, protic, donor solvents). The rate-limiting step of an S$_N$1–E1 reaction is not very different conceptually from the dissolution of an ionic compound (Sec. 8.6F); both processes hinge on the stabilization of ionic species by the solvent. The critical role of solvent shows that S$_N$1 and E1 reactions cannot truly be the unimolecular processes suggested by their nicknames. In the transition states of these reactions, solvent molecules must be actively involved in solvating the developing ions.

When an alkyl halide contains more than one type of β-hydrogen, more than one type of elimination product can be formed. As in the E2 reaction, the alkene with the greatest number of alkyl substituents at the double bond is usually formed in greatest amount; and the ratio of alkene (E1) to substitution product (S$_N$1) is greater when the alkene formed contains more than two alkyl substituents at the double bond. The following examples illustrate both of these points.

In Eq. 9.59a, relatively little alkene is formed. In Eq. 9.59b, a greater proportion of alkene is formed. In addition, in Eq. 9.59b, two alkenes are formed corresponding to loss of the two types of β-hydrogens in the alkyl halide starting material; and the alkene formed in major amount is the one with the greatest number of alkyl substituents on the double bond.

Finally, rearrangements are observed in certain solvolysis reactions.

Recall that rearrangements are a telltale sign of carbocation intermediates (Sec. 4.7D). For example, the secondary carbocation intermediate initially formed in Eq. 9.60 rearranges to a more stable tertiary carbocation; the nucleophilic reaction of solvent with this carbocation accounts for the product shown (Problem 9.27).

The different products that can be formed in S_N1–E1 reactions reflect three reactions of carbocation intermediates that you have now studied:

1. reaction with a nucleophile;
2. loss of a β-proton; and
3. rearrangement to a new carbocation followed by (1) or (2).

Although solvolysis reactions of alkyl halides and related compounds have been extensively studied because of their central role in the development of carbocation theory, as a practical matter S_N1–E1 reactions of alkyl halides are not very useful for preparative purposes because mixtures of products are invariably formed (unless the alkyl halide has no β-hydrogens). However, an understanding of the S_N1 and E1 mechanisms is important because these mechanisms occur in many reactions of alcohols, ethers, and amines that *are* very useful.

PROBLEMS

9.26   Give all the products that might be formed when 3-chloro-2,2-dimethylbutane (the alkyl halide in Eq. 9.60) undergoes solvolysis in aqueous ethanol. Of the alkenes formed, which should be the major one(s)?

9.27   Write a curved-arrow mechanism for formation of the rearrangement product shown in Eq. 9.60.

D. Stereochemistry of the S$_N$1 Reaction

Let's try to predict the stereochemistry of the S_N1 reaction of a chiral alkyl halide using the reaction mechanism. Because carbocations have trigonal planar geometry, they are achiral (assuming no asymmetric carbons elsewhere in the molecule). Hence, the products that result from the reaction of nucleophiles with a carbocation *must* be racemic (Sec. 7.7A). The way racemic products would form mechanistically is shown in Fig. 9.12.

Let's see whether the experimental facts are in accord with this prediction. When (*R*)-6-chloro-2,6-dimethyloctane, a chiral tertiary alkyl halide, undergoes solvolysis in aqueous acetone, the substitution products are only *partially* racemized, and net inversion of configuration is also observed.

(R)-6-chloro-2,6-dimethyloctane

(9.61)

As Eq. 9.61 shows, the inverted product is formed in 21% *enantiomeric excess* (EE; Sec. 6.4A). This corresponds to the percent inversion. The remaining percentage—79%, or (100% – EE)—corresponds to the percentage of the racemate because, by definition, it consists of equal amounts of the *R* and *S* products. How can we account for inverted product if a free carbocation is a reactive intermediate?

FIGURE 9.12 The stereochemical consequences of the S$_N$1 reaction of a chiral alkyl chloride. If the reaction proceeds through a free carbocation, the carbocation is achiral and the product of nucleophilic reaction with the carbocation must be racemic (barring asymmetric carbons in any of the three substituent groups R^1, R^2, or R^3). The racemic product results from equal probability of reaction of the nucleophile (water in this example) at each of the two lobes of the 2p orbital of the carbocation.

First, be sure you understand that the stereochemical inversion *cannot* result from 21% of an S$_N$2 reaction because the reaction in Eq. 9.61 proceeds thousands of times faster than the S$_N$2 reactions of primary alkyl halides in the same solvent, and α-branching retards S$_N$2 reactions.

This result actually tells us something important about both the role of the solvent and the lifetime of the carbocation (Fig. 9.13, p. 426). A mechanism that can account for this result assumes that the first reactive intermediate in the S$_N$1 reaction is an *ion pair* (Sec. 8.6F)—a carbocation intimately associated with its counterion, which, in this case, is the chloride ion. Notice that this ion pair (which includes the chloride ion) is still a chiral species. The chloride ion blocks the access of solvent to the front side of the carbocation. Solvation of the carbocation in this ion pair occurs from the opposite side only; opposite-side substitution by the solvent molecule involved in this interaction results in inversion. However, the chloride ion might also escape from the carbocation into the surrounding solvent, leaving the carbocation solvated on both sides by solvent. This symmetrically solvated carbocation is achiral and can, with equal probability, react at either face with solvent to give racemic product. The occurrence of both racemization and inversion in Eq. 9.61 shows that both types of carbocations—ion pairs and free ions—are important in determining the products of S$_N$1 reactions. According to this mechanism, 21% of the product comes from the ion pair, whereas 79% (half R, half S) comes from the symmetrically solvated ion. The exact percentages of each vary from case to case.

The occurrence of some inversion also shows that the lifetime of a tertiary carbocation is very small. It takes about 10^{-8} second for a chloride counterion to diffuse away from a carbocation and be replaced by solvent. The carbocations that undergo inversion do not last long enough for this process to take place. The competition of opposite-side substitution (which gives inversion) with racemization shows that the lifetime of the carbocation is approximately in this range. In other words, a typical tertiary carbocation exists for about 10^{-8} second before it is consumed by its reaction with solvent. This very small lifetime provides a graphic illustration just how reactive carbocations are.

FIGURE 9.13 The ion-pair mechanism for carbocation formation in the S_N1 reaction. The reaction of the solvating solvent molecule with the carbocation in the ion pair occurs from the side opposite the departing chloride and gives inverted product. (Notice that the ion pair is chiral.) The fully solvated ion is formed when the chloride counterion diffuses away and itself becomes fully solvated. (Its solvation shell is not shown explicitly.) The fully solvated carbocation is achiral and gives racemic products.

PROBLEMS

9.28   The optically active alkyl halide in Eq. 9.61 reacts at 60 °C in anhydrous methanol solvent to give a methyl ether *A* plus alkenes. The substitution reaction is reported to occur with 66% racemization and 34% inversion. Give the structure of ether *A* and state how much of each enantiomer of *A* is formed.

9.29   In light of the ion-pair hypothesis, how would you expect the stereochemical outcome of an S_N1 reaction (percent racemization and inversion) to differ from the result discussed in this section for an alkyl halide that gives a carbocation intermediate which is considerably (a) more stable or (b) less stable than the one involved in Eq. 9.61?

E. Summary of the S_N1 and E1 Reactions

Let's summarize the important characteristics of the S_N1 and E1 reactions.

1. Tertiary and secondary alkyl halides undergo solvolysis reactions by the S_N1 and E1 mechanisms; tertiary alkyl halides are much more reactive.

2. If an alkyl halide has β-hydrogens, elimination products formed by the E1 reaction accompany substitution products formed by the S_N1 mechanism.

3. Both S_N1 and E1 reactions of a given alkyl halide share the same rate-limiting step: ionization of the alkyl halide to form a carbocation.

4. The S_N1 and E1 reactions are first order in the alkyl halide.

5. S_N1 and E1 reactions differ in their product-determining steps. The product-determining step in the S_N1 reaction is reaction of a nucleophile with the carbocation intermediate, and in the E1 reaction, loss of a β-proton from the carbocation intermediate.

6. Carbocation rearrangements occur when the initially formed carbocation intermediate can rearrange to a more stable carbocation.

7. The best leaving groups are those that give the weakest bases as products.

8. The reactions are accelerated by polar, protic, donor solvents.

9. S_N1 reactions of chiral alkyl halides give largely racemized products, but some inversion of configuration is also observed.

9.7 SUMMARY OF SUBSTITUTION AND ELIMINATION REACTIONS OF ALKYL HALIDES

This chapter has shown that substitution and elimination reactions of alkyl halides can occur by a variety of mechanisms. Although each type of reaction has been considered separately, a practical question to ask is what type of reaction is likely to occur when a given alkyl halide is subjected to a particular set of conditions.

When asked to predict how a given alkyl halide will react, you must first answer three major questions.

1. *Is the alkyl halide primary, secondary, or tertiary? If primary or secondary, is there a significant amount of alkyl substitution at the β-carbon?*

2. *Is a Lewis base present? If so, is it a good nucleophile, a strong Brønsted base, or both?* Most strong Brønsted bases, such as ethoxide, are good nucleophiles; but some excellent nucleophiles, such as iodide ion, are relatively weak Brønsted bases.

3. *What is the solvent?* The practical choices are limited for the most part to polar protic solvents, polar aprotic solvents, or mixtures of both.

Once these questions have been answered, a satisfactory prediction in most cases can be obtained from Table 9.7, which is in essence a summary of this chapter. *Before using*

TABLE 9.7 Predicting Substitution and Elimination Reactions of Alkyl Halides

Entry no.	Alkyl halide structure	Good nucleophile?	Strong Brønsted base?	Type of solvent?*	Major reaction(s) expected
1	Methyl	Yes	Yes or No	PP or PA	S_N2
2	Primary, unbranched	Yes	No	PP or PA	S_N2
3		Yes	Yes , unbranched	PP or PA	S_N2
4	Primary with β-substitution	Yes	Yes, unbranched	PP or PA	$E2 + S_N2$
5	Any primary	Yes	Yes, branched	PP or PA	$E2 + S_N2$
6		No	No	PP or PA	No reaction
7	Secondary	Yes	Yes	PP or PA	E2; some SN2 with isopropyl halides; only E2 with a branched base
8		Yes	No	PA	S_N2
9		No	No	PP	S_N1–E1 (solvolysis)
10		No	No	PA	No reaction
11	Tertiary	Yes	Yes	PP or PA	E2
12		Yes	No	PP	S_N1–E1 (solvolysis)
13		Yes	No	PA	no reaction, or very slow S_N2
14		No	No	PP	S_N1–E1
15		No	No	PA	No reaction

* Solvent types are PP = polar protic; PA = polar aprotic. The S_N2, E2, S_N1, and E1 reactions are rarely if ever run in apolar aprotic solvents except with the most reactive alkyl halides. In these cases, the results to be expected are similar to those above with polar aprotic (PA) solvents.

this table, you should consider each case and why the conclusions are reasonable, returning to review the material in this chapter when necessary. Study Problem 9.3 illustrates the practical application of the table.

STUDY PROBLEM 9.3

What products are formed, and by what mechanisms, in each of the following cases?
(a) methyl iodide and sodium cyanide (NaCN) in ethanol
(b) 2-bromo-3-methylbutane in hot ethanol
(c) 2-bromo-3-methylbutane in anhydrous acetone at room temperature
(d) 2-bromo-3-methylbutane in ethanol containing an excess of sodium ethoxide
(e) 2-bromo-2-methylbutane in ethanol containing an excess of sodium iodide
(f) neopentyl bromide in ethanol containing an excess of sodium ethoxide

SOLUTION (a) Methyl iodide and sodium cyanide (NaCN) in ethanol. This case corresponds to entry 1 in Table 9.7. Because a methyl halide has no β-hydrogens, it cannot undergo a β-elimination reaction. Consequently, the only possible reaction is an S_N2 reaction. Because a good nucleophile (cyanide) is present (see Table 9.4, Fig. 9.5), the product is H_3C—CN (acetonitrile), which is formed by the S_N2 mechanism. Although protic solvents are not as effective as polar aprotic ones for the S_N2 reaction, they are useful for reactive alkyl halides such as methyl iodide. However, the reaction would be faster if it were carried out in a polar aprotic solvent (Table 9.6).

(b) 2-Bromo-3-methylbutane in hot ethanol. This is a secondary alkyl halide. (Draw its structure if you have not done so!) The conditions involve no nucleophile or base other than the solvent, which is polar and protic. This situation is covered by entry 9 in Table 9.7. Because the solvent ethanol is a poor nucleophile and a weak base, neither S_N2 nor E2 reactions can occur. Because polar protic solvents promote the S_N1 and E1 reactions, these will be the only reactions observed:

Notice the rearrangement products. (You should show how these arise from the initially formed carbocation intermediate.) Any time the S_N1 or E1 reaction is expected, the possibility of rearrangements should be considered, especially when the initially formed carbocation is secondary. Finally, "hot" ethanol is necessary because the alkyl halide is secondary and is less reactive in the S_N1–E1 reaction than a tertiary alkyl halide would be.

(c) 2-Bromo-3-methylbutane in anhydrous acetone. The alkyl halide from part (b) is subjected to conditions in which a good nucleophile is not present (no S_N2 possible), no strong base has been added (no E2 possible), and a polar aprotic solvent is used. In this type of solvent, carbocations do not form under mild conditions; hence, the S_N1 and E1 reactions cannot take place. Entry 10 in Table 9.7 predicts that no reaction will occur.

(d) 2-Bromo-3-methylbutane in ethanol containing an excess of sodium ethoxide. The alkyl halide from parts (b) and (c) is subjected to a strong base in a protic solvent. This situation is covered by entry 7 in Table 9.7. The S_N2 reaction is retarded by both α- and β-alkyl substitution, but the E2 reaction can take place. Although an S_N1–E1 reaction is promoted by the protic solvent, the rate of the E2 reaction is greater because of the high base concentration. The rate of the E2 reaction is first order in base (Eq. 9.33, p. 405), whereas the rates of the SN1 and E1 reactions are unaffected by the base concentration (Eq. 9.52, p. 419). The products are the following two alkenes:

The second of these predominates because of the greater number of alkyl substituents at the double bond.

(e) 2-Bromo-2-methylbutane in ethanol containing an excess of sodium iodide. This is a tertiary alkyl halide in a polar protic solvent containing a good nucleophile but a weak base (iodide ion). Entry 12 of Table 9.7 covers this situation. The polar protic solvent promotes carbocation formation, and hence the S_N1 and E1 reactions are observed. The S_N1 products are the following:

$$\text{(structure)}\ \underset{Br}{} \ +\ Na^+\ I^- \xrightarrow{EtOH} \text{(structure)}\ I \ +\ \text{(structure)}\ OEt \ +\ EtOH_2^+ Br^- \ +\ Na^+$$

(Na⁺ is a spectator ion) (ionized form of HBr in EtOH) (spectator ion)

Product *A* arises from the Lewis acid–base association reaction of the nucleophile I⁻ with the carbocation intermediate, and product *B* is the solvolysis product that results from the reaction of the solvent with the same carbocation; ionized HBr is formed as a byproduct. Which product (*A* or *B*) is formed in greater amount? It depends on how much iodide ion is present. The more iodide there is, the more effective it will be in competing with the solvent ethanol for the carbocation. Furthermore, because iodide is also a good leaving group, compound *A* could react further to give compound *B*, also by an S_N1 reaction. You would have to monitor the reaction carefully to maximize the yield of *A*, if that were your objective. Because the E1 reaction always accompanies the S_N1 reaction of an alkyl halide with β-hydrogens, some alkenes are also formed; you should draw their structures. No rearrangement products are predicted, because the carbocation intermediate is tertiary.

STUDY GUIDE LINK 9.3
Diagnosing Reactivity Patterns in Substitution and Elimination Reactions

(f) Neopentyl bromide in ethanol containing an excess of sodium ethoxide. This is a primary alkyl halide with *three* β-alkyl groups [$(CH_3)_3C$—CH_2—Br]. Without thinking further about the structure of this alkyl halide, you might conclude that entry 4 of Table 9.7 would cover this case. However, because there are no β-hydrogens, no elimination is possible. Neopentyl halides are essentially unreactive in S_N2 reactions (Table 9.3, p. 396), and, because primary alkyl halides do not form carbocations, neither an E1 nor an S_N1 reaction is possible. Thus, this alkyl halide is essentially inert. If the reaction mixture were heated strongly, a reaction might occur after a few days, but the correct prediction is "no reaction."

PROBLEM

9.30 Predict the products expected in each of the following situations, and show the mechanism of any reaction that takes place using the curved-arrow notation.

(a) 1-bromobutane in methanol containing a large excess of sodium methoxide

(b) 2-bromobutane in *tert*-butyl alcohol containing a large excess of potassium *tert*-butoxide

(c) 2-bromo-1,1-dimethylcyclopentane in ethanol

(d) bromocyclohexane in methanol, heat

9.8 ORGANOMETALLIC COMPOUNDS. GRIGNARD REAGENTS AND ORGANOLITHIUM REAGENTS

Compounds that contain carbon–metal bonds are called **organometallic compounds**. We've already seen two examples of such compounds: the oxymercuration adducts formed when aqueous mercuric acetate reacts with alkenes (Sec. 5.4A) and the organoboranes formed by the addition of BH_3 to alkenes (Sec. 5.4B). In this section, we're going to consider two of the most useful types of organometallic compounds: Grignard reagents and organolithium reagents. We consider them here because they are most often formed from alkyl and aryl halides. We'll discuss their preparation and their use in forming alkanes and isotopically substituted compounds. However, their most important uses are their reactions with epoxides (Sec. 11.5C) and, especially, carbonyl compounds (Sec. 19.9, 21.10, and 22.11). Your instructor may defer asking you to read this section until you cover one of those uses.

A. Grignard Reagents and Organolithium Reagents

A **Grignard reagent** is a compound of the form R—Mg—X, where X = Br, Cl, or I.

Examples of Grignard reagents:

carbon–metal bond

$$CH_3CH_2 \overset{\downarrow}{—} Mg — Br$$

ethylmagnesium bromide

—MgCl

cyclopentylmagnesium chloride

—MgBr

phenylmagnesium bromide

Development of Grignard Reagents

Grignard reagents are among the most versatile and important reagents in organic chemistry. The utility of these reagents was originally investigated by Professor François Phillipe Antoine Barbier (1848–1922) of the University of Lyon in France. However, it was Barbier's successor at Lyon, Victor Grignard (1871–1935), who developed many applications of organomagnesium halides during the early part of the twentieth century. For this work, Grignard received the Nobel Prize in Chemistry in 1912.

Organolithium reagents are compounds of the form R—Li.

Examples of organolithium reagents:

carbon–metal bond

$$CH_3CH_2CH_2CH_2 \overset{\downarrow}{—} Li$$

butyllithium

—Li

phenyllithium

Although the organolithium reagents are pictured for convenience as R—Li, many studies have shown that these reagents in solution are aggregates of several molecules [that is, $(RLi)_n$], and that the aggregation state depends on the solvent.

B. Formation of Grignard Reagents and Organolithium Reagents

Both Grignard and organolithium reagents are formed by adding the corresponding alkyl or aryl halides to rapidly stirred suspensions of the appropriate metal. Anhydrous ether solvents must be used for the formation of Grignard reagents:

$$CH_3CH_2 — Br + Mg \xrightarrow{Et_2O} CH_3CH_2 — Mg — Br \qquad (9.62)$$

bromoethane   **ethylmagnesium bromide**

$$\text{—Cl} + Mg \xrightarrow{THF} \text{—Mg—Cl} \qquad (9.63)$$

chlorocyclohexane   **cyclohexylmagnesium chloride**

FURTHER EXPLORATION 9.3
Mechanism of Formation of Grignard Reagents

The solubility of Grignard reagents in ether solvents plays a crucial role in their formation. Grignard reagents are formed on the surface of the magnesium metal (Further Exploration 9.3). As they form, these reagents are dissolved from the metal surface by the ether solvent. As a result, a fresh metal surface is continuously exposed to the alkyl halide. Grignard reagents are soluble in ether solvents because the ether solvates the metal in a *Lewis acid–base interaction.*

$$\tag{9.64}$$

The magnesium of the Grignard reagent is two electron pairs short of an octet, and the oxygen of each ether molecule can donate an electron pair to the metal. (This interaction is very similar to the donor interactions that stabilize cations in solution; Sec. 8.6F.)

Organolithium reagents are typically formed in hydrocarbon solvents such as hexane:

$$CH_3CH_2CH_2CH_2-Cl + 2\,Li \xrightarrow{\text{hexane}} CH_3CH_2CH_2CH_2-Li + LiCl \tag{9.65}$$

1-chlorobutane   **butyllithium**

Because organolithium reagents are soluble in hydrocarbons, ether solvents are not required for their formation.

Grignard and organolithium reagents react violently with oxygen and (as shown in the next section) vigorously with water. For this reason these reagents must be prepared under rigorously oxygen-free and moisture-free conditions. In the case of Grignard reagents, exclusion of oxygen is easily ensured by the low boiling points of the ether solvents that are normally used. As the Grignard reagent begins to form, heat is liberated and the ether boils. Because the reaction flask is filled with ether vapor, oxygen is excluded.

PROBLEMS

9.31 Give an equation showing the preparation of each of the following organometallic compounds.
(a) $(CH_3)_2CH-MgBr$ (b) $Ph-Li$ (c) $[(CH_3)_2CH-CH_2]_3B$

9.32 Complete each of the following equations.
(a)

(b) $(CH_3)_3C-Cl + Li \xrightarrow{\text{hexane}}$

+ Mg $\xrightarrow{\text{THF}}$

C. Protonolysis of Grignard Reagents and Organolithium Reagents

All reactions of Grignard and organolithium reagents can be understood in terms of the polarity of the carbon–metal bond. Because carbon is more electronegative than either magnesium or lithium, *the negative end of the carbon–metal bond is the carbon atom.* This is illustrated graphically by the EPM of the Grignard reagent methylmagnesium iodide, which shows the high degree of electron density on the carbon and the halogen.

$$H_3C \overset{\longleftarrow}{\rule{2cm}{0.4pt}} Mg \rule{1.5cm}{0.4pt} I$$

Lewis structure and bond dipoles
of methylmagnesium iodide

EPM of methylmagnesium iodide

Imagine now carrying this picture of bond polarity to the extreme by breaking the carbon–metal bond of a Grignard or organolithium reagent so that the metal becomes positively charged and electron-deficient, and the pair of electrons in the bond ends up on carbon. Such a carbon, bearing three bonds, an unshared electron pair, and a negative formal charge, is a carbon anion, or **carbanion**. *Grignard and organolithium reagents react as if they were carbanions:*

$$
\begin{array}{ccc}
\underset{\underset{R}{\overset{R}{\big|}}}{R-C-MgX} & \text{reacts as if it were} & \underset{\underset{R}{\overset{R}{\big|}}}{R-C:^{-}}\quad \overset{+}{MgX} \\
& & \text{a carbon anion,} \\
& & \text{or carbanion}
\end{array}
\qquad (9.66)
$$

Grignard and organolithium reagents are not *true* carbanions because they have covalent carbon–metal bonds. However, we can predict their reactivity by treating them *conceptually* as carbanions.

For example, the view of Grignard and organolithium reagents as carbanions predicts the outcome of simple Brønsted acid–base reactions. Carbanions are powerful Brønsted bases because their conjugate acids, the corresponding alkanes, are extremely weak acids, with pK_a values estimated to be in the 55–60 range. The logic, then, is

1. R—H is a very weak acid (pK_a = 55–60); therefore,
2. R:⁻ is a very strong base; therefore,
3. R—MgX and R—Li are also strong bases.

Grignard and lithium reagents are such strong bases that they react instantaneously with even weak acids such as water or alcohols. The products of such a reaction are the hydrocarbon—the conjugate acid of the organometallic "carbanion"—and the conjugate base of the proton source—hydroxide ion (if the acid is water) or an alkoxide ion (if the acid is an alcohol).

$$
CH_3CH_2-MgBr + H-OH \longrightarrow CH_3CH_2-H + HO^{-}\ {}^{+}MgBr \qquad (9.67)
$$

$$
(CH_3)_3C-Li + H_2O \longrightarrow (CH_3)_3C-H + Li^{+}\ {}^{-}OH \qquad (9.68)
$$

$$
CH_3CH_2CH_2-MgBr + CH_3OH \longrightarrow CH_3CH_2CH_2-H \quad + \quad CH_3O^{-}\ {}^{+}MgBr \qquad (9.69)
$$

$$
\begin{array}{cc}
\textbf{methanol} & \textbf{bromomagnesium methoxide} \\
\text{(an alcohol)} & \text{(the alkoxide conjugate base} \\
& \text{of methanol)}
\end{array}
$$

Each of these reactions can be represented with the curved-arrow notation as the reaction of a carbanion base with the proton of water or alcohol:

$$
CH_3\ddot{C}H_2 \quad \overset{+}{MgX} \quad H-\ddot{O}R \longrightarrow CH_3CH_2-H \quad + \quad R\ddot{O}:^{-}\ {}^{+}MgX \qquad (9.70)
$$

$$
\begin{array}{cc}
\text{conjugate acid} & \text{conjugate base} \\
\text{of } CH_3CH_2:^{-} & \text{of } RO-H
\end{array}
$$

Reactions 9.67–9.70 are examples of *protonolysis*. A **protonolysis** is any reaction with the proton of an acid that breaks chemical bonds. For example, in the protonolysis of a Grignard reagent, the carbon–metal bond of the Grignard reagent is broken. The protonolysis reaction can be an annoyance, since, because of it, Grignard and organolithium reagents must be prepared in the absence of moisture. However, the protonolysis reaction is also useful, because it provides *a method for the preparation of hydrocarbons from alkyl halides.* Notice, for example, in Eq. 9.70 that ethane (a hydrocarbon) is produced from ethylmagnesium bromide, which, in turn, comes from ethyl bromide (an alkyl halide). Although one would not normally prepare an ordinary hydrocarbon by protonolysis, a particularly useful variation of this reaction is the preparation of hydrocarbons labeled with the hydrogen isotopes deuterium

(D, or ^{2}H) or tritium (T, or ^{3}H) by reaction of a Grignard reagent with the corresponding iso-topically labeled water.

$$(CH_3)_3CCH_2 \!-\! Br \xrightarrow[\text{ether}]{Mg} (CH_3)_3CCH_2 \!-\! MgBr \xrightarrow{D_2O} (CH_3)_3CCH_2 \!-\! D \quad (9.71)$$

PROBLEMS

9.33   Give the products of the following reactions. Show the curved-arrow notation for each.

(a) $H_3C \!-\! Li + CH_3OH \longrightarrow$ (b) $(CH_3)_2CHCH_2 \!-\! MgCl + H_2O \longrightarrow$

9.34   (a) Give the structures of two isomeric alkylmagnesium bromides that would react with water to give propane.

(b) What compounds would be formed from the reactions of the reagents in (a) with D_2O?

9.9   CARBENES AND CARBENOIDS

A. α-Elimination Reactions

You've learned that β-elimination is one of the reactions that can occur when certain alkyl halides containing β-hydrogens are treated with base. When an alkyl halide contains no β-hydrogens but has an α-hydrogen, a different sort of base-promoted elimination is sometimes observed. Chloroform ($H \!-\! CCl_3$) is an alkyl halide that undergoes such a reaction.

Chloroform, although a weak acid ($pK_a \approx 25$), is a much stronger acid than an alkane because of the polar effect of the three chlorines (Sec. 3.6C). Its acidity is high enough that it can react as an acid with strong bases. Therefore, when chloroform is treated with an alkoxide base such as potassium *tert*-butoxide, a small amount of its conjugate-base anion is formed.

$$(CH_3)_3C \!-\! \ddot{O}\!:^- \quad H \!-\! CCl_3 \rightleftharpoons (CH_3)_3C \!-\! OH + {}^-\!:CCl_3 \quad (9.72a)$$

***tert*-butoxide**	**chloroform**	***tert*-butyl alcohol**	**trichloromethyl anion**
	($pK_a \approx 25$)	($pK_a \approx 19$)	

This anion can lose a chloride ion to give a neutral species called *dichloromethylene*.

$$\overset{\ddot{C}l:}{\underset{Cl}{\overset{|}{{}^-\!:C} \diagdown Cl}} \quad \longleftrightarrow \quad :C \overset{Cl}{\underset{Cl}{\diagdown}} + :\ddot{C}l:^- \quad (9.72b)$$

trichloromethyl anion	**dichloromethylene**

Dichloromethylene is an example of a **carbene**—a species with a divalent carbon atom. Dichloromethylene has only six valence electrons on carbon; in other words, its carbon is two electrons short of an octet. Carbenes are unstable and highly reactive species.

The formation of dichloromethylene shown in Eqs. 9.72a and 9.72b involves an elimination of the elements of HCl from the *same* carbon atom. An elimination of two groups from the same atom is called an **α-elimination**.

$$\overset{R^1}{\underset{R^2}{\diagup}} C \overset{H}{\underset{\ddot{X}:}{\diagdown}} \longrightarrow \overset{R^1}{\underset{R^2}{\diagup}} C: + H \!-\! \ddot{X}: \quad (\textit{α-elimination}) \quad (9.73)$$

Chloroform cannot undergo a β-elimination because it has no β-hydrogens. When an alkyl halide has β-hydrogens, β-elimination occurs in preference to α-elimination because alkenes, the products of β-elimination, are much more stable than carbenes, the products of

α-elimination. For example, CH_3CHCl_2 reacts with base to form the alkene $H_2C=CHCl$ rather than the carbene $H_3C-\ddot{C}-Cl$.

The reactivity of dichloromethylene follows from its electronic structure. The carbon atom of dichloromethylene bears three groups (two chlorines and the lone pair) and therefore has approximately trigonal planar geometry. Because trigonal planar carbon atoms are sp^2-hybridized, the Cl—C—Cl bond angle is bent rather than linear, the unshared pair of electrons occupies an sp^2 orbital, and the $2p$ orbital is vacant:

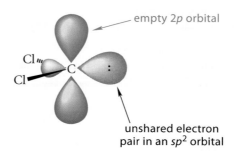

empty $2p$ orbital

unshared electron
pair in an sp^2 orbital

Because dichloromethylene lacks an electronic octet, it is an *electron-deficient compound* and can accept an electron pair; in other words, dichloromethylene is a powerful electrophile. On the other hand, an atom with an unshared electron pair can react as a nucleophile. The divalent carbon of dichloromethylene, with its unshared electron pair, fits into this category as well. Indeed, the divalent carbon of a carbene can act as a nucleophile and an electrophile at the same time!

An important reaction of carbenes that fits this analysis is cyclopropane formation. When dichloromethylene is generated in the presence of an alkene, a cyclopropane is formed.

$$HCCl_3 \ + \ (CH_3)_3C-\ddot{\underset{\cdot\cdot}{O}}:^- \ K^+ \ + \ (CH_3)_2C=CH_2 \ \longrightarrow$$

chloroform   **potassium**   **2-methylpropene**
                  ***tert*-butoxide**

$$+ \ KCl \ + \ (CH_3)_3C-OH$$

**1,1-dichloro-
2,2-dimethylcyclopropane**   (9.74)

In general, reaction of a haloform with base in the presence of an alkene yields a 1,1-dihalocyclopropane.

Mechanistically, the reaction is a concerted syn-addition. (Electron flow is shown in red, and atomic motion is shown in green.)

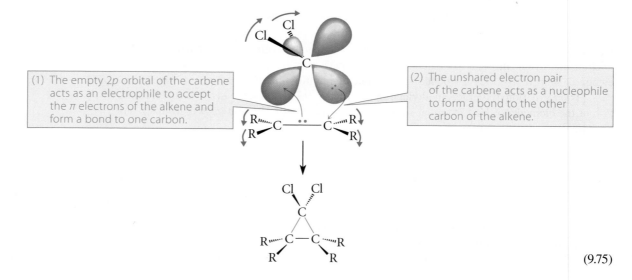

(1) The empty $2p$ orbital of the carbene acts as an electrophile to accept the π electrons of the alkene and form a bond to one carbon.

(2) The unshared electron pair of the carbene acts as a nucleophile to form a bond to the other carbon of the alkene.

(9.75)

The empty $2p$ orbital of the carbene is electron-deficient and therefore acts as an *electrophile*. The π electrons of the alkene are donated to this orbital and a bond is formed between the carbene and one carbon of the alkene. This nucleophilic reaction of one alkene carbon produces electron deficiency at the other alkene carbon. This electron deficiency is satisfied by a simultaneous reaction with the unshared electron pair of the carbene, which acts as a *nucleophile*, forming the other bond to the alkene.

The reaction is stereospecific.

(9.76a)

cis-2-butene

meso-1,1-dichloro-2,3-dimethylcyclopropane

(9.76b)

trans-2-butene

(±)-1,1-dichloro-2,3-dimethylcyclopropane

The clean syn stereochemistry of the reaction not only is synthetically useful but also provides major evidence that the reaction is a concerted process. (The relationship between stereochemistry and mechanism was discussed in Sec. 7.8C.) A concerted anti-addition is impossible because it would require the carbene carbon to add simultaneously to opposite faces of the alkene.

PROBLEMS

9.35   What alkyl halide and what alkene would yield each of the following cyclopropane derivatives in the presence of a strong base?

(a)

(b)

[*Hint for part (b):* The hydrogens on a carbon next to a benzene ring, or Ph group, are particularly acidic.]

9.36   Predict the products that result when each of the following alkenes reacts with chloroform and potassium *tert*-butoxide. Give the structures of all product stereoisomers, and, if more than one stereoisomer is formed, indicate whether they are formed in the same or different amounts.

(a) cyclopentene   (b) (*R*)-3-methylcyclohexene

B. The Simmons–Smith Reaction

Cyclopropanes without halogen atoms can be prepared by allowing alkenes to react with methylene iodide (CH_2I_2) in the presence of a copper–zinc alloy called a *zinc–copper couple*.

$$\text{cyclohexene} \quad + \quad CH_2I_2 \quad \xrightarrow{\text{Zn–Cu couple}} \quad \text{bicyclo[4.1.0]heptane (norcarane)} \quad + \quad ZnI_2 \qquad (9.77)$$

methylene iodide (diiodomethane)

bicyclo[4.1.0]heptane (norcarane)
(59% yield)

This reaction is called the **Simmons–Smith reaction** to recognize the two DuPont chemists who developed it in 1959, Howard E. Simmons and Ronald D. Smith.

Although the role of the copper is not understood, the active reagent in the Simmons–Smith reaction is believed to be an α-halo organometallic compound, a compound with a halogen and a metal on the same carbon. This species can form by a reaction analogous to the formation of a Grignard reagent (Sec. 9.8B):

$$CH_2I_2 + Zn \longrightarrow I-CH_2-ZnI \qquad (9.78)$$

Simmons–Smith reagent

From the discussion of the reactivity of carbenes with alkenes in the previous section, the cyclopropane product of the Simmons–Smith reaction is what would be expected if the parent carbene **methylene** ($:CH_2$) were a reactive intermediate. *Free* methylene is not involved in the reaction because free methylene generated in other ways gives not only cyclopropanes, but other products as well. However, the Simmons–Smith reagent can be conceptualized as methylene that is coordinated (loosely bound) to the Zn atom. This view is reasonable, first, because the carbon–zinc bond polarity is the same as the carbon–magnesium bond polarity in a Grignard reagent (Sec. 9.8C):

$$I-\overset{\frown}{C}H_2-ZnI \quad \text{reacts as if it were} \quad I-\overset{..}{\overset{-}{C}}H_2 \quad \overset{+}{Z}nI \qquad (9.79)$$

an α-halo carbanion

and second, because an α-halo carbanion loses halide ion to give a carbene (see Eq. 9.72b):

$$:\overset{..}{\underset{..}{I}}\overset{\frown}{-}\overset{..}{\overset{-}{C}}H_2 \quad \overset{+}{Z}nI \longrightarrow \quad :\overset{..}{\underset{..}{I}}:^- \quad \overset{..}{C}H_2 \quad \overset{+}{Z}nI \qquad (9.80)$$

methylene

coordinated to the Zn

Reaction of this "coordinated methylene" with the alkene double bond gives a cyclopropane. Because they show carbenelike reactivity, α-halo organometallic compounds are sometimes called *carbenoids*. A **carbenoid** is a reagent that is not a free carbene but has carbenelike reactivity.

Addition reactions of methylene from Simmons–Smith reagents to alkenes, like the reactions of dichloromethylene, are stereospecific syn-additions.

$$\text{cis-3-hexene} \quad + \quad CH_2I_2 \quad \xrightarrow{\text{Zn–Cu}} \quad \text{cis-1,2-diethylcyclopropane} \qquad (9.81a)$$

methylene iodide

cis-3-hexene

cis-1,2-diethylcyclopropane

$$\text{trans-3-hexene} + \text{CH}_2\text{I}_2 \xrightarrow{\text{Zn–Cu}} \text{trans-1,2-diethylcyclopropane} \quad (9.81b)$$

Addition of carbenes or carbenoids to alkenes to yield cyclopropanes is *a reaction that forms new carbon–carbon bonds*. Reactions that form carbon–carbon bonds are especially important in organic chemistry because they can be used to build up larger carbon skeletons from smaller ones.

PROBLEMS

9.37 Give the structure of the organic product expected when CH_2I_2 reacts with each of the following alkenes in the presence of a Zn–Cu couple:

(a) (Z)-3-methyl-2-pentene   (b)

9.38 From which alkene could each of the following cyclopropane derivatives be prepared using the Simmons–Smith reaction?

(a)   (b)

9.10   INDUSTRIAL PREPARATION AND USE OF ALKYL HALIDES

A. Free-Radical Halogenation of Alkanes

Among the methods used in industry, and occasionally in the laboratory, to produce simple alkyl halides is direct halogenation of alkanes. When an alkane such as methane is treated with Cl_2 or Br_2 in the presence of heat or light, a mixture of alkyl halides is formed by successive chlorination reactions.

$$\text{CH}_4 + \text{Cl}_2 \xrightarrow{\text{heat or light}} \text{CH}_3\text{Cl} + \text{HCl} \quad (9.82a)$$

$$\text{CH}_3\text{Cl} + \text{Cl}_2 \xrightarrow{\text{heat or light}} \text{CH}_2\text{Cl}_2 + \text{HCl} \quad (9.82b)$$

$$\text{CH}_2\text{Cl}_2 + \text{Cl}_2 \xrightarrow{\text{heat or light}} \text{CHCl}_3 + \text{HCl} \quad (9.82c)$$

$$\text{CHCl}_3 + \text{Cl}_2 \xrightarrow{\text{heat or light}} \text{CCl}_4 + \text{HCl} \quad (9.82d)$$

The relative amounts of the various products can be controlled by varying the reaction conditions, but mixtures of them are formed, and each compound must be isolated by fractional distillation (Figs. 2.11–2.12, p. 80).

The products in Eqs. 9.82a–d are formed in a series of *substitution* reactions (Sec. 7.8B). For example, CH_3Cl is formed by the substitution of a hydrogen atom in methane by a chlorine atom:

$$H-\overset{\displaystyle H}{\underset{\displaystyle H}{C}}-H + Cl_2 \xrightarrow{\text{heat or light}} H-\overset{\displaystyle H}{\underset{\displaystyle H}{C}}-Cl + HCl \qquad (9.83)$$

The conditions of this reaction (initiation by heat or light) suggest the involvement of free-radical intermediates (Sec. 5.6C). The mechanism of this reaction in fact follows the typical pattern of other free-radical chain reactions; it has initiation, propagation, and termination steps. The reaction is initiated when a small number of halogen molecules absorb energy from the heat or light and dissociate homolytically into halogen atoms:

$$:\ddot{\text{Cl}}-\ddot{\text{Cl}}: \underset{\xrightarrow{\hspace{1cm}}}{\overset{\text{light}}{\xleftarrow{\hspace{1cm}}}} :\ddot{\text{Cl}}\cdot + \cdot\ddot{\text{Cl}}: \qquad (9.84)$$

The ensuing chain reaction has the following propagation steps:

$$:\ddot{\text{Cl}}\cdot \quad H-CH_3 \longrightarrow :\ddot{\text{Cl}}-H + \cdot CH_3 \qquad (9.85a)$$
$$\textbf{methyl radical}$$

$$:\ddot{\text{Cl}}-\ddot{\text{Cl}}: \quad \cdot CH_3 \longrightarrow :\ddot{\text{Cl}}\cdot + :\ddot{\text{Cl}}-CH_3 \qquad (9.85b)$$

The chlorine radical formed in Eq. 9.85b reacts with another CH_4 as shown in Eq. 9.85a, and thus the chain reaction continues. Termination steps result from the recombination of radical species (Problem 9.40) after the methane and Cl_2 concentrations are depleted.

The halogenation of alkanes by a free-radical mechanism is an example of a **free-radical substitution** reaction: a substitution reaction that occurs by a free-radical chain mechanism. (Contrast this with the *free-radical addition* mechanism for peroxide-mediated addition to alkenes in Sec. 5.6C.)

Free-radical halogenations with chlorine and bromine proceed smoothly, halogenation with fluorine is violent, and halogenation with iodine does not occur. These observations correlate with the $\Delta H°$ values for halogenation of methane by each halogen (see Problem 5.43 on p. 225). Halogenation by fluorine is so strongly exothermic ($\Delta H° = -424$ kJ mol^{-1}, -101 kcal mol^{-1}) that the reaction is difficult to control; that is, the temperature of the reaction mixture rises more rapidly than the heat can be dissipated. Iodination is endothermic ($\Delta H° = +54$ kJ mol^{-1}, $+13$ kcal mol^{-1}); the reaction is so unfavorable energetically that it does not proceed to a useful extent. Chlorination ($\Delta H° = -106$ kJ mol^{-1}, -25 kcal mol^{-1}) and bromination ($\Delta H° = -30$ kJ mol^{-1}, -7 kcal mol^{-1}) are mildly exothermic and proceed to completion without becoming violent.

PROBLEMS

9.39   Give the free-radical chain mechanism for the formation of ethyl bromide from ethane and bromine in the presence of light.

9.40   Explain why butane is formed as a minor by-product in the free-radical bromination of ethane.

Regioselectivity of Free-Radical Halogenation When free-radical halogenation takes place on a hydrocarbon with more than one type of hydrogen, more than one product can be obtained. For example, bromination of isobutane could give both tertiary and primary alkyl bromides.

$$H_3C-\underset{\underset{CH_3}{|}}{\overset{\overset{CH_3}{|}}{C}}-H + Br_2 \xrightarrow[20\,°C]{light} H_3C-\underset{\underset{CH_3}{|}}{\overset{\overset{CH_3}{|}}{C}}-Br + H_3C-\underset{\underset{CH_3}{|}}{\overset{\overset{CH_2Br}{|}}{C}}-H + H-Br \quad (9.86)$$

isobutane	**tert-butyl bromide**	**isobutyl bromide**
(large excess)	(99.5% of product)	(0.5% of product)

(A large excess of the hydrocarbon is used to avoid additional bromination.) As this equation indicates, the product of the reaction is almost entirely the tertiary alkyl bromide. Notice that there are nine primary hydrogens and only one tertiary hydrogen in isobutane. On a purely statistical basis, then, we would expect 1/9 as much of the tertiary product as the primary product. In fact, about 200 times as much of the tertiary product is formed. Therefore, on a per-hydrogen basis, a tertiary hydrogen in isobutane is about $(200 \div 1/9) = 1800$ times as reactive as a primary hydrogen.

This difference in reactivity is a consequence of the relative stabilities of the two possible free-radical intermediates. The rate-limiting step of the bromination reaction is the first propagation step, which, by analogy to Eq. 9.85a, is the step that forms the carbon radicals:

$$H_3C-\underset{\underset{CH_3}{|}}{\overset{\overset{CH_3}{|}}{C}}-H \quad \cdot Br \longrightarrow H_3C-\underset{\underset{CH_3}{|}}{\overset{\overset{CH_3}{|}}{C}}\cdot + H-Br \quad (9.87a)$$

tert-butyl radical
$\Delta H_f^\circ = 51.5$ kJ mol^{-1}

$$Br\cdot \quad H-CH_2 \atop H_3C-\underset{\underset{CH_3}{|}}{CH} \longrightarrow H_3C-\underset{\underset{CH_3}{|}}{\overset{\overset{\cdot CH_2}{|}}{CH}} + H-Br \quad (9.87b)$$

isobutyl radical
$\Delta H_f^\circ = 70$ kJ mol^{-1}

The only difference between the two steps is the free-radical intermediate. The tert-butyl radical is much more stable than the isobutyl radical—by 18.5 kJ mol^{-1} (Table 5.2, p. 212). We invoke *Hammond's postulate* (Sec. 4.8D): the transition states for the two reactions resemble the reactive intermediates, which are the free radicals. Because the tertiary radical is more stable than the primary radical, the transition state leading to the tertiary radical has the lower energy. Therefore, the tertiary free radical is formed more rapidly.

Chlorination is much less selective than bromination. The light-promoted chlorination of isobutane actually gives a little more of the primary alkyl chloride:

$$H_3C-\underset{\underset{CH_3}{|}}{\overset{\overset{CH_3}{|}}{C}}-H + Cl_2 \xrightarrow[20\,°C]{light} H_3C-\underset{\underset{CH_3}{|}}{\overset{\overset{CH_3}{|}}{C}}-Cl + H_3C-\underset{\underset{CH_3}{|}}{\overset{\overset{CH_2Cl}{|}}{C}}-H + H-Cl \quad (9.88)$$

isobutane	**tert-butyl chloride**	**isobutyl chloride**
(large excess)	(36% of product)	(64% of product)

Taking into account the relative number of tertiary and primary hydrogens, we find (Problem 9.41) that abstraction of the tertiary hydrogen is still preferred, but only by a factor of 5.1. Because of its greater selectivity, bromination is often the preferred method of halogenation in the laboratory. (Another selective free-radical bromination method is discussed in Sec. 17.2.)

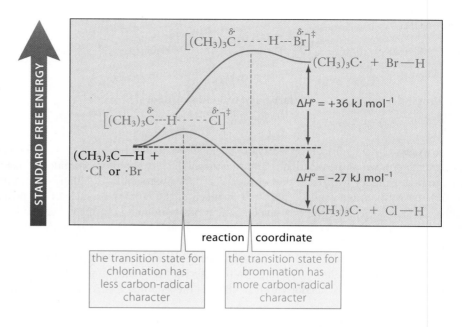

FIGURE 9.14 A free-energy versus reaction-coordinate diagram for the first propagation step of free-radical halogenation. [The relative energies of the products (*violet*) are shown as enthalpies because they are well known, but the relative standard free energies are about the same.] The transition state for the first propagation step of halogenation resembles the less stable species along the reaction coordinate. Because the transition state for bromination resembles the carbon radical and HBr, carbon-radical stability is important in determining the regioselectivity of bromination. Because the transition state for chlorination resembles isobutane and Cl·, it has very little carbon-radical character, and carbon-radical stability is much less important in determining regioselectivity.

Why is bromination so much more selective than chlorination? An exploration of this question provides a graphic illustration of Hammond's postulate. As shown in Fig. 9.14, the hydrogen abstraction step is *exothermic* for chlorination but is *endothermic* for bromination. (This difference between the two halogens is a direct consequence of the much greater bond dissociation energy of the H—Cl bond than the H—Br bond. Bromination goes to completion only because the second propagation step, and thus the overall reaction, is very exothermic.) In two such closely related reactions, it is reasonable to suppose that the relative energies of the products should have some effect on the relative energy barriers that control the rate of the reaction, and this is observed experimentally: the exothermic reaction—chlorination—is much faster than the endothermic one—bromination. (Reaction 9.88 is much faster than Reaction 9.86.) In other words, a chlorine atom is *much* more reactive with hydrocarbons than a bromine atom is. When we invoked Hammond's postulate to explain the selectivity of bromination, it worked because the transition state is very close in energy to the carbon radical along the reaction coordinate (*red* curve, Fig. 9.14). However, in chlorination, the position of the transition state along the reaction coordinate is closer to isobutane because isobutane and Cl· have a much higher energy than the carbon radical and HCl (blue curve, Fig. 9.14). In effect, isobutane and Cl· are the "unstable intermediates" in chlorination; and the transition state, by Hammond's postulate, should therefore resemble this pair more than it resembles the carbon radical. Because the transition state has very little carbon-radical character, carbon-radical stability is not very important in determining the position of reaction.

In this example, we see that a *more reactive species* (the chlorine atom) *is less selective, and a less reactive species* (the bromine atom) *is more selective.* This "inverse" relationship of reactivity and selectivity is sometimes called the **reactivity–selectivity principle**. A similar relationship is observed in many other reactions—a more reactive reagent is less selective when two or more constitutionally isomeric products are possible.

An Analogy for the Reactivity–Selectivity Principle

Suppose you are very hungry—famished—and someone offers you a chocolate from a box containing different types of chocolates. You are so hungry that you cast good manners to the wind and rapidly grab the first chocolate (or chocolates) you put your hand on. Now imagine, instead, that you have finished a large and satisfying meal, and you are allowed to select a chocolate from the same box. You are now more likely to pore slowly over the different chocolates and choose a type you like the best. You are more selective when you are not so hungry. Likewise "hungry" (reactive) reagents are in many cases less selective than "less hungry" (less reactive) reagents when they can react in different ways. In halogenation, the chlorine atom is "famished," and the bromine atom is more selective.

PROBLEMS

9.41   Given the product distribution in Eq. 9.88, calculate the relative rate of abstraction (per hydrogen) of a tertiary and a primary hydrogen by a chlorine atom.

9.42   Two monobromination products are obtained in roughly equal amounts when pentane is treated with bromine and light. What are they? Explain your reasoning.

9.43   (a) What is the major monobromination product obtained when methylcyclohexane is treated with bromine and light? Explain your reasoning.

   (b) The tertiary/primary relative reactivity per hydrogen at $-15\ °C$ in the photochlorination of triptane (2,2,3-trimethylbutane) is 4.5. What are the relative amounts formed of the three possible monochlorination products?

B. Uses of Halogen-Containing Compounds

Alkyl halides and other halogen-containing organic compounds have many practical uses. Methylene chloride and chloroform are important solvents (see Table 8.2, p. 355) that are much less flammable than ethers. (Carbon tetrachloride was also important until its toxicity was recognized.) Tetrachloroethylene, trichlorofluoroethane, and trichloroethylene are used industrially as dry-cleaning solvents. A number of halogen-containing alkenes serve as monomers for the synthesis of useful polymers, such as PVC, Teflon, and Kel-F (see Table 5.4, p. 219). A number of other brominated organic compounds are used as commercial flame retardants. The compound (2,4-dichlorophenoxy)acetic acid (sold as 2,4-D) mimics a plant growth hormone and causes broadleaf weeds to overgrow and eventually die. This is the dandelion killer used in commercial lawn fertilizers.

$$Cl \quad\quad\quad\quad\quad O$$
$$Cl{-}{-}O{-}CH_2{-}\overset{\|}{C}{-}OH$$

(2,4-dichlorophenoxy)acetic acid
(2,4-D)

A few alkyl halides have medical uses. Desflurane [$CF_3{-}CHF{-}O{-}CHF_2$] and sevoflurane [$(CF_3)_2CH{-}O{-}CH_2F$], are safe and inert general anesthetics that have replaced the highly flammable compounds ether and cyclopropane, which were once used widely. Certain fluorocarbons dissolve substantial amounts of oxygen, and some of these have been the subject of ongoing research as artificial blood in surgical applications.

C. Environmental Issues

Because alkyl halides are rarely found in nature, and because many are not biologically degraded, it is perhaps not surprising that some alkyl halides released into the environment have become issues of concern. The chlorofluorocarbons (CFCs, or freons) such as F_2CCl_2, and the hydrochlorofluorocarbons (HCFCs) such as $HCClF_2$ and $HCCl_2F$, are among the most noteworthy examples. Until relatively recently, these compounds were the only ones used as refrigerants in commercial cooling systems, and they were also widely used as propellants in aerosol products. Nontoxic and nonflammable, and with properties ideally suited to their applications, they seemed to be ideal industrial chemicals. During the 1970s, a number of studies implicated them in the destruction of stratospheric ozone. (The ozone layer provides an important shield against harmful ultraviolet solar radiation.) In October 1978, the United States government banned their use in virtually all aerosol products except those that are medically essential. In 1987, a number of countries, including the United States, initialed the "Montreal Protocol on Substances That Deplete the Ozone Layer," under which industrial nations agreed to phase out the production of CFCs, carbon tetrachloride, and certain other substances by 2010. As a direct result, the production of CFCs by 1996 had dropped to about 16% of its pretreaty value. CFCs in existing refrigeration systems are recycled. HCFCs were phased out in 2015.

The problem with CFCs stems from their chlorine content. Chlorine atoms are liberated from these compounds in upper-atmosphere *photodissociation reactions* (bond-homolysis reactions initiated by light).

$$\underset{\text{a freon}}{F_2C\!-\!Cl} \ \xrightarrow{\ \text{light}\ }\ \underset{\text{a chlorine atom}}{F_2\dot{C} \ + \ \cdot Cl} \tag{9.89a}$$

(with Cl on top of each carbon)

A chlorine atom reacts with ozone to give $ClO\cdot$ and O_2:

$$\cdot Cl + O_3 \longrightarrow ClO\cdot + O_2 \tag{9.89b}$$

The $ClO\cdot$ produced in Eq. 9.89b reacts with oxygen atoms (O) produced in the upper atmosphere by the normal photodissociation of O_2:

$$\cdot ClO + \underset{\substack{\text{(from} \\ \text{photodissociation} \\ \text{of } O_2)}}{O} \longrightarrow Cl\cdot + O_2 \tag{9.89c}$$

In this process, a chlorine atom is regenerated and is thus available to repeat the cycle. The sum of Eqs. 9.89b and 9.89c is

$$O + O_3 \longrightarrow 2O_2 \tag{9.89d}$$

In effect, then, chlorine atoms *catalyze* the destruction of ozone; it has been estimated that a single chlorine atom can promote the destruction of 10^5 molecules of ozone.

> The 1995 Nobel Prize in Chemistry was awarded for research into the chemical reactions that lead to the destruction of stratospheric ozone. The recipients of the prize were Mario Molina (b. 1943), a chemist then at the Massachusetts Institute of Technology; F. Sherwood Rowland (1927–2012), a chemist from The University of California, Irvine; and Paul Crutzen (b. 1933), a meterologist-chemist from the Max-Planck Institute for Chemistry in Mainz, Germany.

One solution to this problem is to replace CFCs with related compounds that contain no chlorine. Indeed, one of the most common replacements for CFCs is the family of hydrofluorocarbons (HFCs) such as 1,1,1,2,2-pentafluoroethane (CF_3CHF_2). Although this class of compounds is less harmful to the ozone layer, HFCs nevertheless have adverse effects as greenhouse gases and ultimately exacerbate global warming.

Some potent and effective insecticides are organohalogen compounds.

DDT

chlordane
(a mixture of isomers)

DDT was first synthesized in 1873, but it was introduced in 1939 as a pesticide by Paul Müller (1899–1965), a Swiss chemist at the Laboratorium der Farben-Fabriken J.R. Geigy A.G., Basel. So effective was this insecticide that it was viewed for about 25 years as a savior of humankind. (For example, it virtually eliminated malaria in many areas of the world, including parts of the southern United States.) Müller received the Nobel Prize Physiology or Medicine in 1948. Unfortunately, DDT, chlordane, and a number of other chlorinated broad-spectrum insecticides were subsequently found to accumulate in the fatty tissues of birds and fish, to be passed up the food chain, and to have harmful physiological effects. Hence, their use has been banned or severely curtailed. The reduction in use of DDT was followed by the recovery of a number of DDT-susceptible endangered species, such as the bald eagle. However, malaria is such a large problem in certain parts of the world, and DDT is so effective against it, that it is still used in World Health Organization-endorsed spraying strategies that are designed to minimize environmental impact.

The conflict between the use of chemistry to improve humanity's living conditions and the generation of new problems caused by the release of chemicals into the environment finds real focus in the controversies surrounding the use of many organohalogen compounds. The great promise and public optimism that chemistry offered following World War II has given way to a public skepticism—or, at least, a period of public reflection and debate—as an increasing number of problems related to synthetic chemicals have surfaced. Is commercial organic chemistry in the end to be nothing but a Pandora's box of problems? Perhaps a more realistic view is that few if any human technological endeavors are without risk, and chemistry is no exception. Each new generation of useful organic chemicals—whether they be pharmaceuticals, refrigerants, or insecticides—will likely bring with their benefits some new problems. These problems will provide a *great research opportunity* for chemists of the future who will take up the challenge of using their knowledge to improve further the benefits and to reduce or eliminate the problems.

A Nineteenth-Century Ball Ended by an Alkyl Halide Reaction

Perhaps the first recorded instance of an adverse environmental effect caused by alkyl halides occurred during the reign of Charles X of France. The French chemist Jean-Baptiste André Dumas (1800–1884) was asked to investigate something unusual that occurred during a ball given at the Tuileries. The candles used at the ball had sputtered and had given off noxious fumes, driving the guests from the ballroom. Dumas found that the beeswax used to make the candles had been bleached with chlorine gas. (Beeswax contains large numbers of double bonds. What reaction with chlorine took place?) The heat from the candle flame caused the chlorinated beeswax to decompose, liberating HCl gas—the noxious fumes.

- Two of the most important types of alkyl halide reactions are nucleophilic substitution and β-elimination.

- Nucleophilic substitution reactions occur by two mechanisms.

  1. The S_N2 reaction occurs in a single step with inversion of stereochemical configuration and is characterized by a second-order rate law. It occurs when an alkyl halide reacts with a good nucleophile. It is especially rapid in polar aprotic solvents. Alkyl substitution in the alkyl halide at the α- or β-carbons retards the S_N2 reaction and, if a good Brønsted base is present, favors competing elimination by the E2 mechanism. The S_N2 reaction is most commonly observed with methyl, primary, and unbranched secondary alkyl halides.

  2. The S_N1 reaction is characterized by a first-order rate law that contains only a term in alkyl halide concentration. This type of reaction occurs mostly in polar protic solvents. A very common example of this reaction is solvolysis, in which the solvent is the nucleophile. Because the reaction involves a carbocation intermediate, it is promoted by alkyl substitution at the α-carbon. Consequently, the S_N1 reaction is observed mostly with tertiary and secondary alkyl halides. When the alkyl halide is chiral, the products of the S_N1 reaction are mostly racemic, although some products of inverted configuration are observed in many cases.

- β-Elimination reactions also occur by two mechanisms.

  1. The E2 mechanism competes with the S_N2 mechanism, has a second-order rate law (first order in the base), and occurs with anti stereochemistry. It is favored both by use of a strong Brønsted base and by α- and β-alkyl substitution in both the alkyl halide and the base. It is the major reaction of tertiary and secondary alkyl halides in the presence of a strong Brønsted base. When an alkyl halide has more than one type of β-hydrogen, more than one alkene product is generally obtained. The alkene with the most alkyl substitution at the double bond is generally the predominant product.

  2. The E1 mechanism is an alternative product-determining step of the S_N1 mechanism in which a carbocation intermediate loses a β-proton to form an alkene. The alkene with the greatest number of alkyl substituents on the double bond predominates.

- Nucleophilicity in the S_N2 reaction is affected by

  1. base strength: Stronger bases are better nucleophiles in the absence of hydrogen-bonding effects. Correlations with basicity are generally most straightforward when comparing nucleophiles that contain the same nucleophilic atom.

  2. the solvent: Solvents that can donate hydrogen bonds reduce nucleophilicity. Overall nucleophilicity is a compromise of basicity and hydrogen bonding.

  3. polarizability: Nucleophilicity increases with polarizability of the nucleophilic atom. Highly polarizable nucleophiles (such as sulfide and iodide ions) typically contain nucleophilic atoms from higher-numbered periods of the periodic table. Some weak bases are good nucleophiles because they are polarizable, and these show a smaller percentage of E2 reaction in the E2–S_N2 competition.

- The rate law indicates the species (except for solvent) involved in the rate-limiting transition state of a reaction, but not how they are arranged.

- A primary deuterium isotope effect on the reaction rate indicates that a proton transfer takes place in the rate-limiting step of a reaction.

- Alkyl halides react with magnesium metal in ether solvents to give Grignard reagents, and with lithium in hexane to yield organolithium reagents. Both types of reagents react as strong Brønsted bases with acids, including water and alcohols, to give alkanes. This reaction can be used to introduce hydrogen isotopes from isotopically enriched water or other acids.

- Carbenes are unstable species containing divalent carbon. Carbenes can act simultaneously as electrophiles (because they have an empty valence orbital) and nucleophiles (because they have unshared electron pairs). A haloform reacts with base in an α-elimination to give dihalomethylene, which undergoes syn-additions with alkenes to give dihalocyclopropanes.

- Methylene iodide (diiodomethane) reacts with a zinc–copper couple to give a carbenoid (carbene-like) organometallic reagent (the Simmons–Smith reagent). This reagent undergoes syn-additions with alkenes to give cyclopropanes.

- Alkanes react with bromine and chlorine in the presence of heat or light in free-radical substitution reactions to give alkyl halides. Bromination is the more selective reaction; its selectivity is governed by the relative stabilities of the possible carbon-radical intermediates.

REACTION REVIEW *For a summary of reactions discussed in this chapter, see the* Reaction Review *section of Chapter 9 in the* Study Guide and Solutions Manual.

ADDITIONAL PROBLEMS

9.44 Choose the alkyl halide(s) from the following list of $C_6H_{13}Br$ isomers that meet each criterion below.

(1) 1-bromohexane

(2) 3-bromo-3-methylpentane

(3) 1-bromo-2,2-dimethylbutane

(4) 3-bromo-2-methylpentane

(5) 2-bromo-3-methylpentane

(a) the compound(s) that can exist as enantiomers

(b) the compound(s) that can exist as diastereomers

(c) the compound that gives the fastest S_N2 reaction with sodium methoxide

(d) the compound that is least reactive to sodium methoxide in methanol

(e) the compound(s) that give only one alkene in the E2 reaction with K^+ ^-OtBu

(f) the compound(s) that give an E2 but no S_N2 reaction with sodium methoxide in methanol

(g) the compound(s) that undergo an S_N1 reaction to give rearranged products

(h) the compound that gives the fastest S_N1 reaction

9.45 Give the products expected when isopentyl bromide (1-bromo-3-methylbutane) or the other substances indicated react with the following reagents.

(a) KI in aqueous acetone

(b) KOH in aqueous ethanol

(c) K^+ $(CH_3)_3C$—O^- in $(CH_3)_3C$—OH

(d) product of part (c) + HBr

(e) CsF in *N,N*-dimethylformamide (a polar aprotic solvent)

(f) product of part (c) + chloroform + potassium *tert*-butoxide

(g) product of part (c) + CH_2CI_2 in the presence of a Zn–Cu couple

(h) Li in hexane, then ethanol

(i) sodium methoxide in methanol

(j) Mg and anhydrous ether, then D_2O

9.46 Give the products expected when 2-bromo-2-methylhexane or the other substances indicated react with the following reagents.

(a) warm 1:1 ethanol–water

(b) sodium ethoxide in ethanol

(c) KI in aqueous acetone

(d) product(s) of part (b) + HBr in the presence of peroxides

(e) product(s) of part (b) + $Hg(OAc)_2$ in THF–water, followed by $NaBH_4$

(f) product(s) of part (b) + BH_3 in THF, followed by alkaline H_2O_2

9.47 Identify the gas evolved in each of the following reactions.

(a) $CH_3CH_2MgBr + H_2O \longrightarrow$

(b)

$$\underset{\overset{|}{MgBr}}{H_3C-CH-CH_3} + \underset{\overset{|}{OH}}{H_3C-CH-CH_3} \longrightarrow$$

9.48 Give the structure that meets the criteria given in each of the following cases:

(a) A compound C_6H_{14} that gives only two products of monochlorination, one of which is chiral.

(b) Four stereoisomeric compounds C_4H_8O, all optically active, that contain no double bonds and evolve a gas when treated with EtMgBr.

9.49 Rank the following compounds in order of increasing S_N2 reaction rate with KI in acetone.

$$\underset{A}{(CH_3)_3CCl} \quad \underset{B}{(CH_3)_2CHCl} \quad \underset{C}{(CH_3)_2CHCH_2Cl}$$

$$\underset{D}{CH_3CH_2CH_2CH_2Br} \quad \underset{E}{CH_3CH_2CH_2CH_2Cl}$$

9.50 In each of the following series, order the atoms, compounds, or ions in order of increasing polarizability, and explain your choices.

(a) Se, O, S

(b) chloroform, fluoroform, iodoform

(c) I^-, Br^-, Cl^-, F^-

9.51 Rank the following compounds in order of increasing S_N2 reaction rate with KI in acetone.

methyl bromide   *sec*-butyl bromide
          A   *B*

3-(bromomethyl)-3-methylpentane
                    C

1-bromopentane   1-bromo-2-methylbutane
          D   *E*

9.52 Give the structure of the nucleophile that could be used to convert iodoethane into each of the following compounds in an S_N2 reaction.

(a) 1-ethoxypropane

(b) CH_3CH_2CN

(c)

(d)
CH_3CH_2S—   ‒‒CH_3

(e) $(CH_3)_3\overset{+}{N}CH_2CH_3$ I^-

9.53 Give all of the product(s) expected, including pertinent stereochemistry, when each of the following compounds reacts with sodium ethoxide in ethanol. (D = deuterium = ^2H, an isotope of hydrogen.)

(a) (*R*)-2-bromopentane   (b)

9.54 Tell which of the following alkyl halides can give only one alkene, and which can give a mixture of alkenes, in the E2 reaction.

(a)

(b)

(c)

(d) $CH_3CH_2CHCH_2Br$ with CH_3

9.55 In the *Williamson ether synthesis*, an alkoxide reacts with an alkyl halide to give an ether.

$$R—\overset{..}{\underset{..}{O}}{:}^- + R'—\overset{..}{\underset{..}{X}}{:} \longrightarrow R—\overset{..}{\underset{..}{O}}—R' + {:}\overset{..}{\underset{..}{X}}{:}^-$$

You are in charge of a research group for a large company, Ethers Unlimited, and you have been assigned the task of synthesizing *tert*-butyl methyl ether, $(CH_3)_3C—O—CH_3$. You have decided to delegate this task to two of your staff chemists. One chemist, Ima Smart, allows $(CH_3)_3C—O^-$ K^+ to react with $H_3C—I$ and indeed obtains a good yield of the desired ether. The other chemist, Dimma Light, allows CH_3O^- Na^+ to react with $(CH_3)_3C—Br$. To his surprise, no ether was obtained, although the alkyl halide could not be recovered from the reaction. Explain why Dimma Light's reaction failed.

9.56 Propose a synthesis of ethyl neopentyl ether, $C_2H_5OCH_2C(CH_3)_3$, from an alkyl bromide and any other reagents.

9.57 Three alkyl halides, each with the formula $C_7H_{15}Br$, have different boiling points. One of the compounds is optically active. Following reaction with Mg in ether, then with water, each compound gives 2,4-dimethylpentane. After the same reaction with D_2O, a different product is obtained from each compound. Suggest a structure for each of the three alkyl halides.

9.58 When 2,3-dimethylbutane is treated with Br_2 in the presence of light, the bromine-containing compounds obtained in greatest amount are compound *A* ($C_6H_{13}Br$) and compound *B* ($C_6H_{12}Br_2$). Propose structures for these compounds and explain your reasoning.

9.59 The banned insecticide chlordane is reported to lose some of its chlorine and to be converted into other compounds when exposed to alkaline conditions. Explain.

principal component
of chlordane

9.60 What products are expected, including their stereochemistry, when (2*S*,3*R*)-2-bromo-3-methylpentane is subjected to each of the following conditions? Explain.

(a) methanol containing an excess of sodium methoxide

(b) hot methanol containing no sodium methoxide

9.61 (a) Explain why the compound given in part (a) of Fig. P9.61 reacts to give a mixture of 2-butene stereoisomers in which only the *Z* isomer contains deuterium.

(b) Explain why the compound given in part (b) of Fig. P9.61 reacts to give a mixture of 2-butene stereoisomers in which only the *E* isomer contains deuterium.

(*Hint:* Convert the conformations shown into a conformation that allows the elimination to occur with proper stereochemistry.)

9.62 Which of the following secondary alkyl halides reacts faster with ⁻CN in the S_N2 reaction? (*Hint:* Consider the hybridization and geometry of the S_N2 transition state.)

$$\triangleright\!\!-I \qquad (CH_3)_2CH\!-\!I$$

$$A \qquad\qquad B$$

9.63 Explain why 1-chlorobicyclo[2.2.1]heptane, even though it is a tertiary alkyl halide, is virtually unreactive in the S_N1 reaction. (It has been estimated that it is 10^{-13} times as reactive as *tert*-butyl chloride!) (*Hint:* Consider the preferred geometry of the reactive intermediate.)

1-chlorobicyclo[2.2.1]heptane

9.64 Explain each of the following observations.

(a) When benzyl bromide (Ph—CH_2—Br) is added to a suspension of potassium fluoride in benzene, no reaction occurs. However, when a *catalytic* amount of the crown ether [18]-crown-6 (Sec. 8.7B) is added to the solution, benzyl fluoride can be isolated in high yield.

(b) If lithium fluoride is substituted for potassium fluoride, no reaction occurs even in the presence of the crown ether.

9.65 *Tert*-Butyl chloride undergoes solvolysis in either acetic acid or formic acid.

$$\underset{\substack{\textbf{acetic acid}\\ \epsilon=6}}{H_3C\!-\!\overset{\overset{\displaystyle O}{\|}}{C}\!-\!OH} \qquad \underset{\substack{\textbf{formic acid}\\ \epsilon=59}}{H\!-\!\overset{\overset{\displaystyle O}{\|}}{C}\!-\!OH}$$

Both solvents are protic, donor solvents, but they differ substantially in their dielectric constants ϵ.

(a) What is the S_N1 solvolysis product in each solvent?

(b) In one solvent, the S_N1 reaction is 5000 times faster than it is in the other. In which solvent is the reaction more rapid, and why?

9.66 Suppose that CH_3I is added to an ethanol solution containing an excess of both Na^+ $CH_3CH_2O^-$ and K^+ $CH_3CH_2S^-$ in equimolar amounts.

(a) What is the major product that will be isolated from the reaction? Explain.

(b) How would your answer change (if at all) if the experiment were conducted in anhydrous DMSO, a polar aprotic solvent?

9.67 (a) Two isomeric S_N2 products are possible when sodium thiosulfate is allowed to react with one equivalent of methyl iodide in methanol solution. Give the structures of the two products.

$$Na^+ \quad ^-\!\!:\!\ddot{S}\!-\!\overset{\overset{\displaystyle :\ddot{O}:^-}{|}}{\underset{\underset{\displaystyle :\ddot{O}:^-}{|}}{S}}{}^{+2}\ddot{O}\!:^- \quad Na^+$$

sodium thiosulfate

(Thiosulfate is an example of an *ambident*, or "two-toothed," nucleophile.)

(b) In fact, only one of the two possible products is formed. Which one is formed, and why?

9.68 Consider the following equilibrium:

$$(CH_3)_2\ddot{S}: + \; H_3C\!-\!\ddot{B}r: \; \rightleftharpoons \; (CH_3)_2\overset{+}{\ddot{S}}\!-\!CH_3 + :\ddot{B}r:^-$$

In each case (a) and (b), choose the solvent in which the equilibrium would lie farther *to the right*. Explain. (Assume that the products are soluble in all solvents considered.)

(a) ethanol or diethyl ether

(b) *N,N*-dimethylacetamide (a polar, aprotic solvent, $\epsilon = 38$) or a mixture of water and methanol that has the same dielectric constant

9.69 Although Grignard reagents are normally insoluble in hydrocarbon solvents, they can be dissolved in such solvents if a tertiary amine (a compound with the general structure R_3N:) is added. Explain.

9.70 When methyl iodide at 0.1 *M* concentration is allowed to react with sodium ethoxide at 0.1 *M* concentration in

Figure P9.61

ethanol solution, the product ethyl methyl ether is obtained in good yield. Explain why the reaction is over much more quickly, but about the same yield of the ether is obtained, when the reaction is run with an excess (0.5 M) of sodium ethoxide.

9.71   When methyl bromide is dissolved in methanol and an equimolar amount of sodium iodide is added, the concentration of iodide ion quickly decreases, and then slowly returns to its original value. Explain.

9.72   Consider the following experiments with trityl chloride, Ph_3C—Cl, a very reactive tertiary alkyl halide:

(1) In aqueous acetone, the reaction of trityl chloride follows a rate law that is first order in the alkyl halide, and the product is trityl alcohol, Ph_3C—OH.

(2) In another reaction, when one equivalent of sodium azide (Na$^+$ N$_3^-$; see Table 9.3, p. 396) is added to a solution that is otherwise identical to that used in experiment (1), the reaction rate is the same as in (1); however, the product isolated in good yield is trityl azide, Ph_3C—N$_3$.

(3) In a reaction mixture in which both sodium azide and sodium hydroxide are present in equal concentrations, *both* trityl alcohol and trityl azide are formed, but the reaction rate is again unchanged.

Explain why the reaction rate is the same but the products are different in these three experiments.

9.73   The first demonstration of the stereochemistry of the S_N2 reaction was carried out in 1935 by Prof. E. D. Hughes and his colleagues at the University of London. They allowed (R)-2-iodooctane to react with radioactive iodide ion (*I$^-$).

$$CH_3CH(CH_2)_5CH_3 + *I^- \rightleftharpoons CH_3CH(CH_2)_5CH_3 + I^-$$
$$\quad | \qquad\qquad\qquad\qquad\qquad | $$
$$\quad I \qquad\qquad\qquad\qquad\qquad *I $$

2-iodooctane   **2-iodooctane**
                          (radioactive)

The rate of substitution (rate constant k_S) was determined by measuring the rate of incorporation of radioactivity into the alkyl halide. The rate of loss of optical activity from the alkyl halide (rate constant $k°$) was also determined under the same conditions.

(a) What ratio $k°/k_S$ is predicted for each of the following stereochemical scenarios:
(1) retention; (2) inversion; (3) equal amounts of both retention and inversion? Explain.

(b) The experimental rate constants were found to be as follows:
$k_S = (13.6 \pm 1.1) \times 10^{-4} \ M^{-1} \ s^{-1}$
$k° = (26.2 \pm 1.1) \times 10^{-4} \ M^{-1} \ s^{-1}$
Which scenario in part (a) is consistent with the data?

9.74   In a laboratory, two liquids, A and B, were found in a box labeled only "isomeric alkyl halides $C_5H_{11}Br$." You have been employed to deduce the structures of these compounds from the following data left in an accompanying

laboratory notebook. Reaction of each compound with Mg in ether, followed by water, gives the same hydrocarbon. Compound A, when dissolved in warm ethanol, reacts to give an ethyl ether C and an acidic solution in a few minutes. Compound B reacts more slowly but eventually gives the *same* ether C and an acidic solution under the same conditions. Both acidic solutions, when tested with AgNO$_3$ solution, give a light yellow precipitate of AgBr. Reaction of compound B with sodium ethoxide in ethanol gives two alkenes, one of which reacts with O$_3$, then aqueous H$_2$O$_2$, to give acetone $(CH_3)_2C$=O as one product. Give the structures of compounds A, B, and C, and explain your reasoning.

9.75   An optically active compound A has the formula $C_8H_{13}Br$. Compound A gives no reaction with Br$_2$ in CH_2C_{12}, but it reacts with K$^+$(CH$_3$)$_3$C—O$^-$ to give a single new compound B in good yield. Compound B decolorizes Br$_2$ in CH_2Cl_2 and takes up hydrogen over a catalyst. When compound B is treated with ozone followed by aqueous H$_2$O$_2$, dicarboxylic acid C is isolated in excellent yield; notice its cis stereochemistry.

C

Identify compounds A and B, and account for all observations. (If you need a refresher on how to solve this type of problem, see Study Guide Links 4.3 and 5.3.)

9.76   In the laboratories of the firm "Halides 'R' Us," a compound A has been found in a vial labeled only "achiral alkyl halide $C_{10}H_{17}Br$." The management feels that the compound might be useful as a pesticide, but they need to know its structure. You have been called in as a consultant at a handsome fee. Compound A, when treated with KOH in warm ethanol, yields two compounds (B and C), each with the molecular formula $C_{10}H_{16}$. Compound A rapidly reacts in aqueous ethanol to give an acidic solution, which, in turn, gives a precipitate of AgBr when tested with AgNO$_3$ solution. Ozonolysis of A followed by treatment with (CH$_3$)$_2$S affords (CH$_3$)$_2$C=O (acetone) as one of the products plus unidentified halogen-containing material. Catalytic hydrogenation of either B or C gives a mixture of both *trans*- and *cis*-1-isopropyl-4-methylcyclohexane. Compound A reacts with one equivalent of Br$_2$ to give a mixture of two separable compounds, D and E, both of which can be shown to be achiral compounds. Finally, ozonolysis of compound B followed by treatment with aqueous H$_2$O$_2$ gives acetone and the diketone F.

F

Propose structures for compounds A through E that best fit the data (and collect your fee).

9.77 Which one of the following stereoisomers should undergo β-elimination most rapidly with sodium ethoxide in ethanol? Explain your reasoning.

A   *B*

9.78 When menthyl chloride (see Fig. P9.78) is treated with sodium ethoxide in ethanol, 2-menthene is the only alkene product observed. When neomenthyl chloride is subjected to the same conditions, the alkene products are mostly 3-menthene (78%) along with some 2-menthene (22%). Explain why different alkene products are formed from the different alkyl halides, and why 3-menthene is the major product in the second reaction. (*Hint:* Draw the chair conformations of the starting materials, remember the stereochemistry of the E2 reaction, and don't forget about the chair interconversion of cyclohexanes.)

9.79 (a) Tell whether each of the eliminations shown in Fig. P9.79 is syn or anti.
(b) Reaction (2) shows first-order kinetics. Draw a curved-arrow mechanism for this reaction that is consistent with its kinetic order and with its stereochem-istry. (*Hint:* Be sure to draw out the structure of the acetate group.)

9.80 Explain why each alkyl halide stereoisomer gives a different alkene in the E2 reactions shown in Fig. P9.80 on p. 450. It will probably help to build models or draw out the conformations of the two starting materials.

9.81 Knowing that carbocations rearrange, chemists H. C. Brown and Glenn A. Russell at Purdue University in the early 1950s investigated whether free-radical rearrangements might also occur:

They carried out light-promoted free-radical chlorination of isobutane-2d, in which the tertiary hydrogen of isobutane was replaced by deuterium ($^2H = D$), as shown in Fig. P9.81 on p. 450. The ratio of DCl to HCl produced in the reaction was found to be exactly the same as the ratio of *A* to *B*. (Any isotopically substituted *A*, if present, would not be differentiated from *A*.)

(a) This result was cited as evidence that free radicals do *not* rearrange. Explain the logic of this conclusion. (*Hint:* Compare this result with the result that would occur with rearrangement.)

menthyl chloride

2-menthene
(100%)

neomenthyl chloride

2-menthene +
(22%)

3-menthene
(78%)

Figure P9.78

(1)

(2)

Figure P9.79

(b) Assuming the same reaction conditions, how would the relative percentages of A and B compare with the percentages of *tert*-butyl chloride and isobutyl chloride in Eq. 9.88 (greater or less)? Explain. (*Hint:* See Sec. 9.5D.)

(c) Suggest a method for the preparation of the isotopically labeled starting material, isobutane-2d. (*Hint:* See Sec. 9.8C.)

9.82 (a) The reagent tributyltin hydride, Bu_3Sn—H, brings about the rapid conversion of 1-bromo-1-methylcyclohexane into methylcyclohexane. The reaction is particularly fast in the presence of AIBN (Sec. 5.6C). Suggest a mechanism for this reaction. (*Hint:* The Sn—H bond is relatively weak.)

(b) Suggest two other reaction sequences using other reagents that would bring about the same overall transformation.

9.83 The cis and trans stereoisomers of 4-chlorocyclohexanol give different products when they react with OH^-, as shown in the reactions given in Fig. P9.83.

(a) Give a curved-arrow mechanism for the formation of each product.

(b) Explain why the bicyclic material B is observed in the reaction of the trans isomer, but not in the reaction of the cis isomer.

9.84 The reaction of butylamine, $CH_3(CH_2)_3\overset{..}{N}H_2$, with 1-bromobutane in 60% aqueous ethanol follows the rate law

$$rate = k[\text{butylamine}][\text{1-bromobutane}]$$

The product of the reaction is $(CH_3CH_2CH_2CH_2)_2\overset{+}{N}H_2$ Br^-. The following very similar reaction, however, has a first-order rate law:

rate = $k[A]$

A

Give a mechanism for each reaction that is consistent with its rate law and with the other facts about nucleophilic substitution reactions. Use the curved-arrow notation.

Figure P9.80

isobutane-2d   *A*   *B*

Figure P9.81

***trans*-4-chlorocyclohexanol**   *A*   *B* (alternate views)

***cis*-4-chlorocyclohexanol**   *A*   *C*

Figure P9.83

9.85 In 1975, a report was published in which the reaction given in Fig. P9.85 was observed. The —OBs (brosylate) group is a leaving group conceptually like halide. (Think of this group as you would —Br.) Notice that the reaction conditions favor an S_N2 reaction.

(a) This result created quite a stir among chemists because it seemed to question a fundamental principle of the S_N2 reaction. Explain.

(b) Because the result was potentially very significant, the work was reinvestigated very soon after it was published. In this reinvestigation, it was found that after about 10 hours' reaction time, the product consisted almost completely of *trans-P*. Only on standing for a much longer time under the reaction conditions did *cis-P* form (and *trans-P* disappear) to give the product mixture shown in Fig. P9.85. Furthermore, when the trans isomer of *S* was subjected to the same conditions, mostly *cis-P* was formed after 10 h, but, after 5 days, the same 75:25 cis:trans product mixture was formed as in Fig. P9.85. Finally, subjecting pure *cis-P* or pure

trans-P to the reaction conditions gave, after five days, the same 75:25 mixture. Explain these results.

(c) Why is *cis-P* favored in the product mixture?

9.86 Consider the reaction sequence given in Fig. P9.86. (Bu— = butyl group = $CH_3CH_2CH_2CH_2$—.)

(a) Use what you know about the stereochemistry of bromine addition to propose the stereochemistry of compound *B*.

(b) Is the $B \rightarrow C$ reaction a syn- or an anti-elimination? Show your analysis.

(c) How would the stereochemistry of products change if the E stereoisomer of compound *A* were carried through the same sequence of reactions? Explain.

9.87 Account for each of the results, shown in Fig. P9.87, with a mechanism. In part (a), note that the reaction is not observed in the absence of NaOH. In part (b), note that organolithium reagents are strong bases and that the hydrogens on a carbon adjacent to a benzene ring are relatively acidic.

Figure P9.85

Figure P9.86

(a) $HCCl_3 + Na^+\ I^- \xrightarrow[35\ °C]{NaOH/H_2O} HCCl_2I + Na^+\ Cl^-$

(b) $Ph—CH_2—Cl + CH_3CH_2CH_2CH_2—Li +$ $\longrightarrow$ $—Ph + LiCl + CH_3CH_2CH_2CH_3$

Figure P9.87

The Chemistry of Alcohols and Thiols

This chapter focuses on the reactions of alcohols and thiols. The nomenclature and classification of alcohols and thiols were discussed in Sec. 8.1 and Sec. 8.2B. We begin this chapter with some of the simplest but most important reactions of alcohols and thiols: their Brønsted acid–base reactions. Next we consider some reactions that alcohols and thiols have in common with alkyl halides: substitution and elimination reactions. Then we'll study *oxidation reactions* of alcohols, which have no simple parallel in alkyl halide chemistry. We'll learn how to recognize oxidations and reductions and how the oxidations of alcohols and thiols are carried out in the laboratory. A consideration of alcohol oxidation in nature leads to a discussion of the stereochemical relationships of groups within molecules. Finally, the strategy used in planning the preparation of organic compounds—*organic synthesis*—will be introduced.

10.1 ALCOHOLS AND THIOLS AS BRØNSTED ACIDS AND BASES

A. Acidity of Alcohols and Thiols

Alcohols and thiols are weak acids. In view of the similarity between the structures of water and alcohols, it may come as no surprise that their acidities are about the same.

$$CH_3CH_2-O-H \qquad H-O-H$$

$$pK_a \qquad 15.9 \qquad \qquad 15.7$$

For example, we've already encountered alkoxide bases as nucleophiles in S_N2 reactions and as bases in E2 reactions (Secs. 9.1A, 9.1B). The conjugate bases of alcohols are generally called *alkoxides*.

The common name of an alkoxide is constructed by deleting the final *yl* from the name of the alkyl group and adding the suffix *oxide*. In substitutive nomenclature, the suffix *ate* is simply added to the name of the alcohol.

$$CH_3CH_2\ddot{O}:^- \ Na^+$$

common: **sodium ethoxide**
substitutive: **sodium ethanolate**

Thiols, although weak acids, are much more acidic than alcohols.

$$CH_3CH_2 \overset{S}{\diagup} \diagdown H \qquad CH_3CH_2 \overset{O}{\diagup} \diagdown H$$

pK_a 10.5   15.9

The relative acidities of alcohols and thiols are a reflection of the *element effect* (Sec. 3.6A).

The conjugate bases of thiols are called *mercaptides* in common nomenclature and *thiolates* in substitutive nomenclature.

$$CH_3\ddot{S}:^- \ Na^+$$

common: **sodium methyl mercaptide**
substitutive: **sodium methanethiolate**

PROBLEMS

10.1 Give the structure of each of the following compounds.
   (a) sodium isopropoxide   (b) potassium *tert*-butoxide
   (c) magnesium 2,2-dimethyl-1-butanolate

10.2 Name the following compounds.
   (a) $Ca(OCH_3)_2$ (b) $CuSCH_2CH_3$

B. Formation of Alkoxides and Mercaptides

Because the pK_a of a typical alcohol is about the same as that of water, an alcohol *cannot* be converted completely into its alkoxide conjugate base in an aqueous NaOH solution.

$$CH_3CH_2\ddot{O}-H + {}^-:\!\ddot{O}H \rightleftharpoons CH_3CH_2\ddot{O}:^- + H_2\ddot{O}: \qquad (10.1)$$

$pK_a = 15.9$ $pK_a = 15.7$

Because the relative pK_a values for ethanol and water are nearly the same, both sides of the equation contribute significantly at equilibrium. In other words, *hydroxide is not a strong enough base to convert an alcohol completely into its conjugate-base alkoxide.*

Alkoxides can be formed irreversibly from alcohols with stronger bases. One convenient base used for this purpose is sodium hydride, NaH, which is a source of the *hydride ion*, $H:^-$. Hydride ion is a very strong base; the pK_a of its conjugate acid, H_2, is about 37. Hence, its reactions with alcohols go essentially to completion. In addition, when NaH reacts with an alcohol, the reaction cannot be reversed because the by-product, hydrogen gas, simply bubbles out of the solution.

$$Na^+ \ H:^- + \ H-\ddot{O}-CHCH_2CH_3 \longrightarrow Na^+ \ ^-:\!\ddot{O}-CHCH_2CH_3 + H_2\uparrow \qquad (10.2)$$
$$\qquad\qquad\qquad\qquad | \qquad\qquad\qquad\qquad\qquad\qquad\qquad |$$
$$\qquad\qquad\qquad\qquad CH_3 \qquad\qquad\qquad\qquad\qquad\qquad CH_3$$

quantitative yield

Potassium hydride and sodium hydride are supplied as dispersions in mineral oil to protect them from reaction with moisture. When these compounds are used to convert an alcohol into an alkoxide, the mineral oil is rinsed away with pentane, a solvent such as ether or THF is added, and the alcohol is introduced cautiously with stirring. Hydrogen is evolved vigorously and a solution or suspension of the pure potassium or sodium alkoxide is formed.

Solutions of alkoxides in their conjugate-acid alcohols find wide use in organic chemistry. The reaction used to prepare such solutions is analogous to a reaction of water you may have observed. Sodium reacts with water to give an aqueous sodium hydroxide solution:

$$2\,H\!-\!\ddot{\underset{\cdot\cdot}{O}}H + 2\,Na \longrightarrow 2\,Na^+ \; {}^-\!:\!\ddot{\underset{\cdot\cdot}{O}}H + H_2\uparrow \tag{10.3a}$$

The analogous reaction occurs with many alcohols. Thus, sodium metal reacts with an alcohol to afford a solution of the corresponding sodium alkoxide:

$$2\,H\!-\!\ddot{\underset{\cdot\cdot}{O}}R + 2\,Na \longrightarrow 2\,Na^+ \; {}^-\!:\!\ddot{\underset{\cdot\cdot}{O}}R + H_2\uparrow \tag{10.3b}$$

sodium alkoxide

The rate of this reaction depends strongly on the alcohol. The reactions of sodium with anhydrous (water-free) ethanol and methanol are vigorous, but not violent. However, the reactions of sodium with some alcohols, such as *tert*-butyl alcohol, are rather slow. The alkoxides of such alcohols can be formed more rapidly with the more reactive potassium metal.

Because thiols are much more acidic than water or alcohols, they, unlike alcohols, can be converted completely into their conjugate-base mercaptide anions by reaction with one equivalent of hydroxide or alkoxide. In fact, a common method of forming alkali-metal mercaptides is to dissolve them in ethanol containing one equivalent of sodium ethoxide:

$$CH_3CH_2\ddot{\underset{\cdot\cdot}{S}}H + CH_3CH_2\ddot{\underset{\cdot\cdot}{O}}\!:^- \; \rightleftharpoons \; CH_3CH_2\ddot{\underset{\cdot\cdot}{S}}\!:^- + CH_3CH_2\ddot{\underset{\cdot\cdot}{O}}H \tag{10.4}$$

| **ethanethiol** | **ethoxide ion** | **ethanethiolate** | **ethanol** |
| $pK_a = 10.5$ | | **ion** | $pK_a = 15.9$ |

Because the equilibrium constant for this reaction is $>10^5$ (Sec. 3.4E), the reaction goes essentially to completion.

Although alkali-metal mercaptides are soluble in water and alcohols, thiols form insoluble mercaptides with many heavy-metal ions, such as Hg^{2+}, Cu^{2+}, and Pb^{2+}.

$$2\,CH_3(CH_2)_9S\!-\!H + PbCl_2 \; \xrightarrow{\;EtOH\;} \; [CH_3(CH_2)_9S]_2Pb + 2\,HCl \tag{10.5}$$

decanethiol   **lead(II) decanethiolate**
(87% yield)

$$2\,PhS\!-\!H + HgCl_2 \longrightarrow (PhS)_2Hg + 2\,HCl \tag{10.6}$$

(98% yield)

The insolubility of heavy-metal mercaptides is analogous to the insolubility of heavy-metal sulfides (for example, lead(II) sulfide, PbS), which are among the most insoluble inorganic compounds known. One reason for the toxicity of lead salts is that the lead forms very strong (stable) mercaptide complexes with the thiol groups of important biomolecules (see sidebar).

Curing a Disease with Mercaptides

A relatively rare inherited disease of copper metabolism, Wilson's disease, can be treated by using the tendency of thiols to form complexes with copper ions. Accumulation of toxic levels of copper in the brain and liver causes the disease, which can be diagnosed by a brown ring around the iris of the eye (photo). Penicillamine is administered to form a complex with the Cu^{2+} ions:

penicillamine

complex of two penicillamine molecules
with Cu^{2+}

The penicillamine–copper complex, unlike ordinary cupric thiolates, is relatively soluble in water because of the ionized carboxylic acid groups (Sec. 8.6F), and its solubility allows it to be excreted by the kidneys.

C. Polar Effects on Alcohol Acidity

Substituted alcohols and thiols show the same type of polar effect on acidity as do substituted carboxylic acids (Sec. 3.6C). For example, alcohols containing electronegative substituent groups have enhanced acidity. Thus, 2,2,2-trifluoroethanol is more than three pK_a units more acidic than ethanol itself.

Relative acidity:

$$H_3C—CH_2—OH < F_3C—CH_2—OH \qquad (10.7)$$

pK_a 15.9   12.4

The polar effects of electronegative groups are more important when the groups are closer to the —OH group:

Relative acidity:

$$F_3C—CH_2—CH_2—CH_2—OH < F_3C—CH_2—CH_2—OH < F_3C—CH_2—OH \qquad (10.8)$$

pK_a 15.4   14.6   12.4

Notice that the fluorines have a negligible effect on acidity when they are separated from the —OH group by four or more carbons.

PROBLEM

10.3 In each of the following sets, arrange the compounds in order of increasing acidity (decreasing pK_a). Explain your choices.

(a) $ClCH_2CH_2OH$, Cl_2CHCH_2OH, $Cl(CH_2)_3OH$

(b) $ClCH_2CH_2SH$, $ClCH_2CH_2OH$, CH_3CH_2OH

(c) $CH_3CH_2CH_2CH_2OH$, $CH_3OCH_2CH_2OH$

D. Role of the Solvent in Alcohol Acidity

Primary, secondary, and tertiary alcohols differ significantly in acidity; some relevant pK_a values are shown in Table 10.1. The data in this table show that the acidities of alcohols are in the order methyl > primary > secondary > tertiary. For many years chemists thought that this order was due to some sort of polar effect (Sec. 3.6C) of the alkyl groups around the alcohol oxygen. However, chemists were fascinated when they learned that in the *gas phase*—in the absence of solvent—the order of acidity of alcohols is exactly reversed.

Relative gas-phase acidity:

$$(CH_3)_3COH > (CH_3)_2CHOH > CH_3CH_2OH > CH_3OH \tag{10.9}$$

Notice carefully what is being stated here. The *relative order* of acidity of different types of alcohols is reversed in the gas phase compared with the *relative order* of acidity in solution. It is *not* true that alcohols are more acidic in the gas phase than they are in solution; rather, all alcohols are *much* more acidic in solution than they are in the gas phase.

Branched alcohols are more acidic than unbranched ones in the *gas phase* because α-alkyl substituents stabilize alkoxide ions more effectively than hydrogens. (Recall that stabilization of a conjugate-base anion increases acidity; Fig. 3.3, p. 116). This stabilization occurs by a *polarization* mechanism. That is, the electron clouds of each alkyl group, which are fairly polarizable, distort so that electron density moves away from the negative charge on the alkoxide oxygen, leaving a partial positive charge on the central carbon. The anion is stabilized by its favorable electrostatic interaction with this partial positive charge.

Because a tertiary alcohol has more α-alkyl substituents than a primary alcohol, a tertiary alkoxide is stabilized by this polarization effect more than a primary alkoxide. Consequently, tertiary alcohols are more acidic in the gas phase.

The same polarization effect is present in solution, but the different acidity order in solution shows that another, more important, effect is operating as well. The acidity order in solution is due to the effectiveness with which alcohol molecules *solvate* their conjugate-base anions. Recall from Sec. 8.6F that anions are solvated, or stabilized in solution, by hydrogen bonding with the solvent. Such hydrogen bonding is nonexistent in the gas phase. It is thought that the alkyl groups of a tertiary alkoxide somehow adversely affect the solvation of the alkoxide oxygen, although a precise description of the mechanism is unclear. (It is known *not* to be a simple steric effect.) Reducing the solvation of the tertiary alkoxide increases its energy and therefore increases its basicity. Because primary alkoxides do not have so many alkyl branches, the solvation of primary alkoxides is more effective. Consequently, their solution basicities are lower. To summarize: *tertiary alkoxides are more basic in solution than*

FURTHER EXPLORATION 10.1
Solvation of Tertiary Alkoxides

TABLE 10.1 Acidities of Alcohols in Aqueous Solution

Alcohol	pK_a		Alcohol	pK_a	
CH₃OH	15.1		(CH₃)₂CHOH	17.1	
CH₃CH₂OH	15.9		(CH₃)₃COH	19.2	

primary alkoxides. An equivalent statement is that *primary alcohols are more acidic in solution than tertiary alcohols.*

The essential point of this discussion is that the solvent is not an idle bystander in the acid–base reaction; rather, it takes an active role in stabilizing the molecules involved, especially the charged species.

E. Basicity of Alcohols and Thiols

Just as water can accept a proton to form the hydronium ion, alcohols and thiols can also be protonated to form positively charged conjugate acids. Alcohols do not differ greatly from water in their basicities; thiols, however, are much less basic.

$$
\begin{array}{ccc}
\text{H} & \text{H} & \text{H} \\
| & | & | \\
\text{H—O—H} & \text{Et—O—H} & \text{Et—S—H} \\
\overset{+}{} & \overset{+}{} & \overset{+}{} \\
\textbf{hydronium ion} & \textbf{conjugate acid} & \textbf{conjugate acid} \\
& \textbf{of ethanol} & \textbf{of ethanethiol} \\
\text{p}K_{\text{a}} \quad -1.74 & -2 \text{ to } -3 & -5 \text{ to } -7
\end{array}
$$

The negative $\text{p}K_{\text{a}}$ values, which reflect the *charge effect* on $\text{p}K_{\text{a}}$ (Sec. 3.6B), mean that these protonated species are very strong acids, and that their neutral conjugate bases are rather weak. Nevertheless, the ability of alcohols and thiols to accept a proton plays a very important role in many of their reactions, particularly those that take place in acidic solutions.

Notice carefully that the $\text{p}K_{\text{a}}$ values above refer in each case to the conjugate acid of the *neutral* base. Alcohols and thiols, like water, are *amphoteric* substances—that is, they can either gain or lose a proton. Thus, two acid–base equilibria are associated with an alcohol:

Loss of proton:

$$
\text{Et}\ddot{\text{O}}\text{—H} + {}^{-}{:}\ddot{\text{O}}\text{H} \rightleftharpoons \text{Et}\ddot{\text{O}}{:}^{-} + \text{H—}\ddot{\text{O}}\text{H} \tag{10.10a}
$$

$$\text{p}K_{\text{a}} = 15.9 \qquad\qquad\qquad \text{p}K_{\text{a}} = 15.7$$

Gain of a proton:

$$
\text{Et}\ddot{\text{O}}\text{—H} + \text{H}_3\ddot{\text{O}}^{+} \rightleftharpoons \overset{\text{H}}{\underset{+}{\text{Et}\ddot{\text{O}}\text{—H}}} + \text{H}_2\ddot{\text{O}}{:} \tag{10.10b}
$$

$$\text{p}K_{\text{a}} = -1.74 \qquad \text{p}K_{\text{a}} = -2 \text{ to } -3$$

The acidity of an alcohol—the loss of a proton—is exemplified by the reaction in Eq. 10.10a. Because alcohols are weak acids, this reaction occurs usually only in the presence of strong bases. The basicity of alcohols—the gain of a proton—is exemplified by the reaction in Eq. 10.10b. Because alcohols are weak bases, this reaction usually occurs significantly only in the presence of strong acids.

Thiols, too, are amphoteric: they can also act as both acids and bases. Recall (Sec. 10.1A) that thiols are much more acidic than alcohols because of the element effect. The *conjugate acids* of thiols are also more acidic than the *conjugate acids* of alcohols for the same reason. In other words, *thiols are less basic than alcohols.*

Loss of a proton:

$$
\text{Et}\ddot{\text{S}}\text{—H} + {}^{-}{:}\ddot{\text{O}}\text{H} \rightleftharpoons \text{Et}\ddot{\text{S}}{:}^{-} + \text{H—}\ddot{\text{O}}\text{H} \tag{10.11a}
$$

$$\text{p}K_{\text{a}} = 10.5 \qquad\qquad\qquad \text{p}K_{\text{a}} = 15.7$$

Gain of a proton:

$$\text{Et}\overset{..}{\underset{..}{\text{S}}}\text{—H} \;+\; \text{H}_3\overset{..}{\text{O}}^+ \;\;\underset{\longrightarrow}{\overleftarrow{}}\;\; \text{Et}\overset{\overset{\displaystyle H}{|}}{\underset{+}{\overset{}{\text{S}}}}\text{—H} \;+\; \text{H}_2\overset{..}{\text{O}}\text{:} \tag{10.11b}$$

$$\text{p}K_a = -1.74 \qquad\qquad\qquad \text{p}K_a \approx -6$$

10.2 DEHYDRATION OF ALCOHOLS

Strong acids such as H_2SO_4 and H_3PO_4 catalyze a β-elimination reaction in which water is lost from a secondary or tertiary alcohol to give an alkene. The conversion of cyclohexanol into cyclohexene is typical:

cyclohexanol   **cyclohexene**
                    (79–84% yield)

$$\tag{10.12}$$

A reaction such as this, in which the elements of water are lost from the starting material, is called a **dehydration**. Thus, in Eq. 10.12, cyclohexanol is said to be *dehydrated* to cyclohexene.

Most acid-catalyzed dehydrations of alcohols are reversible reactions. However, these reactions can easily be driven toward the alkene products by applying Le Châtelier's principle (Sec. 4.9B). For example, in Eq. 10.12, the equilibrium is driven toward the alkene product because the water produced as a by-product forms a strong hydrogen-bonded complex with the catalyzing acid H_3PO_4, and the cyclohexene product is distilled from the reaction mixture. (Alkenes can be removed by distillation because they have considerably lower boiling points than alcohols with the same carbon skeleton; Sec. 8.5C.) The dehydration of alcohols to alkenes is easily carried out in the laboratory and is an important procedure for the preparation of some alkenes.

The role of the acid catalyst in dehydration is to convert the —OH group, a poor leaving group, into the —$\overset{+}{\text{O}}\text{H}_2$ group, a good leaving group (because H_2O is a weak base). *The use of Brønsted or Lewis acids to activate the —OH group as a leaving group will prove to be a central theme in alcohol chemistry.*

Alcohol dehydration occurs by a three-step mechanism that consists entirely of acid–base reactions and involves a carbocation intermediate. In the first mechanistic step, the —OH is activated as a leaving group by acting as a Brønsted base (Sec. 10.1E) to accept a proton from the catalyzing acid:

phosphoric acid   **dihydrogen**
(the catalyst)   **phosphate**
                        (conjugate base of
                          the catalyst)

$$\tag{10.13a}$$

Thus, the basicity of alcohols is important to the success of the dehydration reaction. Next, the carbon–oxygen bond of the alcohol breaks in a Lewis acid–base dissociation to give water and a carbocation:

a carbocation

(10.13b)

Finally, the conjugate base $^-OPO_3H_2$ of the catalyzing acid removes a β-proton from the carbocation in another Brønsted acid–base reaction:

(10.13c)

This step generates the alkene product and regenerates the catalyzing acid H_3PO_4. Alternatively, the H_2O by-product generated in Eq. 10.13b can serve as the base that removes a β-proton from the carbocation, and the H_3O^+ formed in this reaction can also serve as an acid catalyst in the dehydration.

(10.13d)

Let's focus briefly on the acid–base reaction in the last step of the mechanism (Eqs. 10.13c–d). A common error in thinking about the mechanism is to use hydroxide ion (^-OH), formed by dissociation of the product H_2O, as the base; after all, hydroxide is a stronger base than dihydrogen phosphate or water, the weak bases present in the reaction mixture. However, hydroxide cannot be the base, because the reaction is carried out in strongly acidic solution; hydroxide cannot survive in strong acid because it protonates instantaneously to form water. *Nor is hydroxide necessary as a base,* because the carbocation is a very strong acid; the pK_a of a carbocation β-proton is -8 to -10. Dihydrogen phosphate is perfectly adequate as a base to remove such an acidic proton because its conjugate acid has a pK_a of 2.2. (See Sec. 3.4E.) As shown in Eq. 10.13d, even the by-product water (conjugate-acid $pK_a = -1.74$) is basic enough remove the very acidic β-proton.

This mechanism illustrates an important principle of acid–base catalysis: *An acid and its conjugate base always act in tandem in a mechanism.* If H_3O^+ is a catalyzing acid, then its conjugate base H_2O will act as the base. In base-catalyzed reactions that involve ^-OH as a catalyst, then H_2O, the conjugate acid of ^-OH, acts as the acid—not H_3O^+, which does not exist in significant concentration in basic solution. (If necessary, re-read the discussion of amphoteric compounds on p. 99, Sec. 3.4B.)

Let's now return to the dehydration mechanism shown in Eqs. 10.13a–c. We've seen a mechanism like this twice before. First, alcohol dehydration is an E1 reaction (Sec. 9.6). Once the —OH group of the alcohol is protonated, it becomes a very good leaving group (water). Like a halide leaving group in the E1 reaction, the protonated —OH departs to give a carbocation, which then loses a β-proton to give an alkene.

Alcohol dehydration: *E1 reaction of an alkyl halide:*

$$(10.14)$$

Second, the dehydration of alcohols is the reverse of the hydration of alkenes (Sec. 4.9B). *Hydration of alkenes and dehydration of alcohols are the forward and reverse of the same reaction.*

Recall from the *principle of microscopic reversibility* (Sec. 4.9B) that the forward and reverse of the same reaction must have the same intermediates and the same rate-limiting transition states. Thus, because protonation of the alkene is the rate-limiting step in alkene hydration, the reverse of this step—loss of the proton from the carbocation intermediate (Eq. 10.13c or 10.13d)—is rate-limiting in alcohol dehydration. This principle also requires that if a catalyst accelerates a reaction in one direction, it also accelerates the reaction in the reverse direction. Thus, both the hydration of alkenes to alcohols and the dehydration of alcohols to alkenes are catalyzed by acids.

The involvement of carbocation intermediates explains several experimental facts about alcohol dehydration. First, the relative rates of alcohol dehydration are in the order tertiary > secondary >> primary. Application of Hammond's postulate (Sec. 4.8D) suggests that the rate-limiting transition state of a dehydration reaction should closely resemble the corresponding carbocation intermediate. Because tertiary carbocations are the most stable carbocations, dehydration reactions involving tertiary carbocations should be faster than those involving either secondary or primary carbocations, as observed. In fact, the dehydration of primary alcohols is generally not a useful laboratory procedure for the preparation of alkenes. (Primary alcohols react in other ways with H_2SO_4; see Problem 10.66, p. 509.)

Second, if the alcohol has more than one type of β-hydrogen, then a mixture of alkene products can be expected. As in the E1 reaction of alkyl halides, the most stable alkene—the one with the greatest number of branches at the double bond—is the alkene formed in greatest amount:

$$(10.15)$$

2-methyl-2-butanol   **2-methyl-2-butene** **2-methyl-1-butene**
                                major product   minor product

Finally, alcohols that react to give rearrangement-prone carbocation intermediates yield rearranged alkenes:

$$(10.16)$$

3,3-dimethyl-2-butanol   **2,3-dimethyl-1-butene** **2,3-dimethyl-2-butene**
                                    (29%)   (71%)

1-cyclobutylethanol   **1-methylcyclopentene**   (10.17)

The mechanism of the rearrangement in Eq. 10.17 is worked through in Study Problem 10.1.

STUDY PROBLEM   10.1

Provide a curved-arrow mechanism and a rationale for the rearrangement shown in Eq. 10.17. (Use H—A as a general abbreviation for the catalyzing acid.)

SOLUTION   A rearrangement suggests the involvement of a carbocation intermediate. Therefore, the first part of the mechanism is the generation of a carbocation. This reaction is like the one shown in Eq. 10.13a. The resulting carbocation is secondary.

carbocation *A*

The rearrangement involves a shift of one of the *ring carbons*; this shift enlarges the ring. Notice that we have drawn out the ring structure to see this shift more clearly. (Remember that a ring carbon is just another alkyl group that just happens to be "tied back.")

carbocation intermediate *A*
with all carbons shown

rearranged
carbocation intermediate
with all carbons shown

skeletal structure
of the rearranged
carbocation intermediate

This rearrangement is analogous to the expansion of a noose in a rope:

Why should a secondary carbocation rearrange to another secondary carbocation? The answer is that the four-membered ring is *strained* (Sec. 7.5B), and the expansion to a five-membered ring relieves some of its ring strain. Finally, the rearranged carbocation loses a β-proton to give the product alkene and regenerate the catalyzing acid:

(There are other β-protons; why aren't these lost instead?)

10.4   What alkene(s) are formed in the acid-catalyzed dehydration of each of the following alcohols?

(a) 3-methyl-3-heptanol   (b) $\underset{\text{PhCHCH}_2\text{Ph}}{\overset{\text{OH}}{|}}$

10.5   Write the curved-arrow mechanism for the reaction in Problem 10.4a. (Abbreviate the catalyzing acid as H—A.) In each step, identify all Brønsted acids and bases, all electrophiles and nucleophiles, and all leaving groups.

10.6   Write the structure of the carbocation intermediate involved in the acid-catalyzed dehydration of 3-ethyl-3-pentanol.

10.7   Identify the *major* alkene product(s) in part (a) of Problem 10.4.

10.8   Draw the structures of two alcohols, one secondary and one tertiary, that could give each of the following alkenes as a major acid-catalyzed dehydration product. In each case, which alcohol would dehydrate most rapidly?

(a) 1-methylcyclohexene   (b) 3-methyl-2-pentene

10.9   Write a curved-arrow mechanism for the reaction in Eq. 10.16.

10.10 A certain reaction is carried out in methanol with H_2SO_4 as a catalyst.

(a) What Brønsted acid is present in highest concentration in such a solution? (*Hint:* H_2SO_4 is completely dissociated in methanol, just as it is in water.)

(b) If a base is involved in the reaction mechanism, what is the basic species?

10.3   REACTIONS OF ALCOHOLS WITH HYDROGEN HALIDES

Alcohols react with hydrogen halides to give alkyl halides:

3-methyl-1-butanol   **1-bromo-3-methylbutane**
                                  (93% yield)   (10.18)

$$(CH_3)_3C—OH + HCl \xrightarrow[\substack{25\,°C,\ 20\ min}]{H_2O} (CH_3)_3C—Cl + H_2O \quad (10.19)$$

***tert*-butyl alcohol**   ***tert*-butyl chloride**
                                  (almost quantitative)

The equilibrium constant for the formation of alkyl halides from alcohols is not large; hence, the successful preparation of alkyl halides from alcohols, like the dehydration of alcohols to alkenes, usually depends on the application of Le Châtelier's principle (Sec. 4.9B). For example, in both Eqs. 10.18 and 10.19, the reactant alcohols are soluble in the reaction solvent, which is an aqueous acid, but the product alkyl halides are not. Separation of the alkyl halide products from the reaction mixture as water-insoluble layers drives both reactions to completion. For alcohols that are not water-soluble, a large excess of gaseous HBr, one of the reactants, can be used to drive the reaction to completion, by Le Châtelier's principle.

The mechanism of alkyl halide formation depends on the type of alcohol used as the starting material. In the reactions of tertiary alcohols, protonation of the alcohol oxygen is followed by carbocation formation. The carbocation reacts with the halide ion, which is formed by ionization of strong acid HCl, and which is present in great excess:

$$(CH_3)_3C—\overset{..}{\underset{..}{O}}H + H\overset{\frown}{—}\overset{..}{\underset{..}{Cl}}: \rightleftharpoons (CH_3)_3C—\overset{\overset{\displaystyle H}{|}}{\underset{..}{O}}{}^{+}—H + :\overset{..}{\underset{..}{Cl}}:^{-} \quad (10.20a)$$

(excess)

$$(CH_3)_3C-\overset{\overset{H}{|}}{\underset{\cdot\cdot}{O}}{}^{+}-H \quad \rightleftarrows \quad (CH_3)_3C^+ + H_2\overset{\cdot\cdot}{O}: \quad \left.\begin{array}{c} \\ \end{array}\right\} \quad (10.20b)$$

$$(CH_3)_3\overset{+}{C} \quad :\overset{\cdot\cdot}{\underset{\cdot\cdot}{Cl}}:^{-} \quad \rightleftarrows \quad (CH_3)_3C-\overset{\cdot\cdot}{\underset{\cdot\cdot}{Cl}}: \quad \left.\begin{array}{c} \\ \end{array}\right\} S_N1 \text{ reaction} \quad (10.20c)$$

Once the alcohol is protonated, *the reaction is an S_N1 reaction with H_2O as the leaving group.*

When a primary alcohol is the starting material, the reaction occurs as a concerted displacement of water xfrom the protonated alcohol by halide ion. In other words, *the reaction is an S_N2 reaction in which water is the leaving group.*

Notice that the initial step in both of these S_N1 and S_N2 mechanisms is protonation of the —OH group.

As the conditions of Eqs. 10.18 and 10.19 suggest, the reactions of tertiary alcohols with hydrogen halides are much faster than the reactions of primary alcohols. Typically, tertiary alcohols react with hydrogen halides rapidly at room temperature, whereas the reactions of primary alcohols require heating for several hours. The reactions of primary alcohols with HBr and HI are satisfactory, but their reactions with HCl are very slow. Although reactions of alcohols with HCl can be accelerated with certain catalysts, other methods for preparing primary alkyl chlorides (discussed in the following section) are better.

The reactions of secondary alcohols with hydrogen halides tend to occur by the S_N1 mechanism. This means that carbocations are involved as reactive intermediates; consequently, rearrangements can occur in many cases:

2-methyl-3-pentanol   **2-bromo-2-methylpentane**

PROBLEMS

10.11 Suggest an alcohol starting material and the conditions for the preparation of each of the following alkyl halides.

(a) $\underset{CH_3CHCH_2CH_3}{\overset{Br}{|}}$ (b)   (c) $I-CH_2CH_2CH_2CH_2CH_2-I$

10.12 Write a curved-arrow mechanism for the rearrangement shown in Eq. 10.22.

10.13 Draw the structure of the alkyl halide product expected (if any) in each of the following reactions.

(a) 1-propanol + HBr in the presence of H_2SO_4 catalyst

(b) $HOCH_2CH_2CH_2OH$ + excess HI $\xrightarrow{\text{heat}}$

(c) $\underset{Me_3C-CH-CH_3}{\overset{OH}{|}}$ + excess HBr $\xrightarrow{\text{heat}}$

(d) $(CH_3)_3CCH_2OH$ + HCl $\xrightarrow{25\,°C}$

(*Hint:* See Fig. 9.4, p. 397.)

The dehydration of alcohols to alkenes and the reactions of alcohols with hydrogen halides have some important things in common. Both take place in very acidic solution; in both reactions, the acid converts the —OH group into a good leaving group. We've already discussed this point for dehydrations in Sec. 10.2. For substitution reactions, if acid were not present, the halide ion would have to displace ⁻OH to form the alkyl halide. This reaction does not take place because ⁻OH is a much stronger base than any halide ion (Table 3.1, p. 102), and strong bases are poor leaving groups (Sec. 9.4F).

$$:\ddot{Br}:^- \; + \; H_3C—\ddot{O}H \quad \times\!\!\!\longrightarrow \quad :\ddot{Br}—CH_3 \; + \; {}^-:\ddot{O}H \qquad (10.23a)$$

a weak base   a strong base
(a poor leaving group)

$$:\ddot{Br}:^- \; + \; H_3C—\overset{H}{\underset{+}{\ddot{O}}}—H \quad\rightleftharpoons\quad :\ddot{Br}—CH_3 \; + \; H_2\ddot{O}: \qquad (10.23b)$$

a weak base
(a good leaving group)

Remember: *Substitution and elimination reactions of alcohols require the —OH group to be converted into a better leaving group.*

The formation of secondary and tertiary alkyl halides and the dehydration of secondary and tertiary alcohols have the same initial steps: protonation of the alcohol oxygen and formation of a carbocation.

$$\underset{H}{\overset{:\ddot{O}H}{\underset{|}{R_2C—CR_2}}} + HX \rightleftharpoons \boxed{\text{protonation of the OH}} \; \underset{H}{\overset{\overset{+}{:\ddot{O}H_2} \; X^-}{\underset{|}{R_2C—CR_2}}} \rightleftharpoons \boxed{\text{formation of a carbocation}} \; \underset{H}{\overset{X^-}{\underset{|}{\overset{+}{R_2C}—CR_2}}} + H_2\ddot{O}: \qquad (10.24a)$$

The two reactions differ in the fate of this carbocation, which in turn is governed by the conditions of the reaction. In the presence of a hydrogen halide, the halide ion is present in excess and reacts with the carbocation. In dehydration, no halide ion is present, and when the alkene forms by loss of a β-proton from the carbocation, the conditions of the dehydration reaction force the removal of the alkene product and the water by-product from the reaction mixture. It follows, then, that *alkyl halide formation and dehydration to alkenes are alternative branches of a common mechanism:*

$$\underset{H}{\overset{\overset{+}{R_2C}—CR_2}{\underset{|}{}}}$$

carbocation

| Lewis acid–base association | ⟵ X⁻ (halide ion)   H₂O ⟶ | Brønsted acid–base reaction |

$$\underset{X \quad H}{R_2C—CR_2} \qquad\qquad \underset{R \quad\quad R}{\overset{R \quad\quad R}{C=C}} + H_3O^+ \qquad (10.24b)$$

alkyl halide
(the S_N1 product)   alkene
(the E1 product)

Notice that the principles you've studied in Chapter 9 for the substitutions and eliminations of alkyl halides are valid for other functional groups—in this case, alcohols.

10.4   ALCOHOL-DERIVED LEAVING GROUPS

When an alkyl halide is prepared from an alcohol and a hydrogen halide, protonation converts the —OH group into a good leaving group. However, if the alcohol molecule contains a group that might be sensitive to strongly acidic conditions, or if milder or even nonacidic conditions must be used for other reasons, different ways of converting the —OH group into a good leaving group are required. Laboratory methods for accomplishing this objective are the subject of this section. In addition, we'll learn about some of the leaving groups that are found in naturally occurring, biologically important reactions.

A. Sulfonate Ester Derivatives of Alcohols

Structures of Sulfonate Esters   An important method of activating alcohols toward nucleophilic substitution and β-elimination reactions is to convert them into *sulfonate esters*. Sulfonate esters are derivatives of **sulfonic acids**, which are compounds of the form R—SO_3H. Some typical sulfonic acids are the following:

methanesulfonic acid   **benzenesulfonic acid**   ***p*-toluenesulfonic acid**

(The *p* in the name of the last compound stands for *para*, which indicates the relative positions (1,4) of the two groups on the benzene ring. This type of nomenclature is discussed in Sec. 16.1.) A **sulfonate ester** is a compound in which the acidic hydrogen of a sulfonic acid is replaced by an alkyl or aryl group. Thus, in ethyl benzenesulfonate, the acidic hydrogen of benzenesulfonic acid is replaced by an ethyl group.

benzenesulfonic acid

ethyl benzenesulfonate
(a sulfonate ester)

Sulfur has more than the octet of electrons in these Lewis structures. Such "octet expansion" is common for atoms in the third and higher periods of the periodic table. Bonding in sulfonic acids and their derivatives is discussed further in Sec. 10.10B.

Organic chemists often use abbreviated structures and names for certain sulfonate esters. Esters of methanesulfonic acid are called *mesylates* (abbreviated R—OMs), and esters of *p*-toluenesulfonic acid are called *tosylates* (abbreviated R—OTs).

CH_3CH_2—O—S—CH_3 is the same as CH_3CH_2—OMs

ethyl mesylate

ethyl methanesulfonate

is the same as

sec-butyl *p*-toluenesulfonate   **sec-butyl tosylate**

10.14 Draw both the complete structure and the abbreviated structure, and give another name for each of the following compounds.

(a) isopropyl methanesulfonate (b) methyl *p*-toluenesulfonate

(c) phenyl tosylate (d) cyclohexyl mesylate

Preparation of Sulfonate Esters Sulfonate esters are prepared from alcohols and other sulfonic acid derivatives called sulfonyl chlorides. For example, *p*-toluenesulfonyl chloride, often known as *tosyl chloride* and abbreviated TsCl, is the sulfonyl chloride used to prepare tosylate esters.

**FURTHER
EXPLORATION 10.2**
Mechanism of
Sulfonate Ester
Formation

This is a nucleophilic substitution reaction in which the oxygen of the alcohol displaces chloride ion from the tosyl chloride. The pyridine used as the solvent is a base. Besides catalyzing the reaction, it also neutralizes the HCl that would otherwise form in the reaction (color in Eq. 10.25).

10.15 Suggest a preparation of cyclohexyl mesylate from the appropriate alcohol.

Reactivity of Sulfonate Esters Sulfonate esters, such as tosylates and mesylates, are useful because *they have approximately the same reactivities as the corresponding alkyl bromides in substitution and elimination reactions.* (In other words, you can think of a tosylate or mesylate ester group roughly as a "fat" bromo group.) The reason for this similarity is that *sulfonate anions, like bromide ions, are good leaving groups.* Recall that, among the halides, the weakest bases, bromide and iodide, are the best leaving groups (Sec. 9.4F). In general, *good leaving groups are weak bases.* Sulfonate anions are weak bases; they are the conjugate bases of sulfonic acids, which are strong acids.

p-toluenesulfonic acid:
a strong acid
($pK_a \approx -3$)

p-toluenesulfonate anion
(tosylate anion):
a weak base

Thus, sulfonate esters prepared from primary and secondary alcohols, like primary and secondary alkyl halides, undergo S_N2 reactions in which a sulfonate ion serves as the leaving group.

$$\text{Nuc:}^- \quad \text{CH}_2\!-\!\ddot{\text{O}}\text{Ts} \longrightarrow \text{Nuc}\!-\!\text{CH}_2 + \text{:}\ddot{\text{O}}\text{Ts} \qquad (10.26)$$

nucleophile

tosylate leaving group

Similarly, secondary and tertiary sulfonate esters, like the corresponding alkyl halides, also undergo E2 reactions with strong bases, and they undergo S_N1–E1 solvolysis reactions in polar protic solvents.

Occasionally we'll need a sulfonate ester that is much more reactive than a tosylate or mesylate. In such a case a *trifluoromethanesulfonate ester* is used. The trifluoromethanesulfonate group is nicknamed the **triflate** group and it is abbreviated —OTf.

the triflate group, a very good leaving group

a triflate ester

abbreviation for a triflate ester

triflate anion
a very weak base
(conjugate-acid $pK_a \approx -13$)

The triflate anion is an exceptionally weak base; the pK_a of its conjugate acid is about –13. (See Problem 10.51 on p. 506.) Hence, the triflate group is an exceptionally good leaving group, and triflate esters are highly reactive. For example, consider again the S_N2 reaction used to prepare FDG, an imaging agent used in positron emission tomography (PET). (See Eq. 9.28 on p. 402.) This reaction is far too slow to be useful with a tosylate leaving group. However, the triflate leaving group has considerably greater reactivity and is an ideal leaving group for this reaction, which must be carried out quickly.

Triflate esters are prepared in the same manner as tosylate esters (Eq. 10.25), except that triflic anhydride is used instead of tosyl chloride.

$\text{CH}_3(\text{CH}_2)_4\text{OH}$ + triflic anhydride + pyridine

1-pentanol

triflic anhydride   **pyridine**

abbreviated: $\text{CH}_3(\text{CH}_2)_4\text{OTf}$

1-pentyl triflate
(90% yield)

(10.27)

The use of sulfonate esters in S_N2 reactions is illustrated in Study Problem 10.2.

STUDY PROBLEM   10.2

Outline a sequence of reactions for the conversion of 3-pentanol into 3-bromopentane.

SOLUTION   Before doing *anything* else, write the problem in terms of structures.

$$\underset{\text{CH}_3\text{CH}_2\text{CHCH}_2\text{CH}_3}{\overset{\text{OH}}{|}} \xrightarrow{\ ?\ } \underset{\text{CH}_3\text{CH}_2\text{CHCH}_2\text{CH}_3}{\overset{\text{Br}}{|}}$$

Alcohols can be converted into alkyl bromides using HBr and heat (Sec. 10.3). However, because secondary alcohols are prone to carbocation rearrangements, the HBr method is likely to give byproducts. However, if conditions can be chosen so that the reaction will occur by the S_N2 mechanism, carbocation rearrangements will not be an issue. To accomplish this objective, first convert the alcohol into a tosylate or mesylate.

$$\underset{\text{CH}_3\text{CH}_2\text{CHCH}_2\text{CH}_3}{\overset{\text{OH}}{|}} \xrightarrow[\text{pyridine}]{\text{TsCl}} \underset{\text{CH}_3\text{CH}_2\text{CHCH}_2\text{CH}_3}{\overset{\text{OTs}}{|}} \tag{10.28a}$$

Next, displace the tosylate group with bromide ion in a polar aprotic solvent such as DMSO (Table 8.2, p. 355).

$$\underset{\text{CH}_3\text{CH}_2\text{CHCH}_2\text{CH}_3}{\overset{\text{OTs}}{|}} + \text{Na}^+\ \text{Br}^- \xrightarrow[\text{DMSO}]{} \underset{\text{CH}_3\text{CH}_2\text{CHCH}_2\text{CH}_3}{\overset{\text{Br}}{|}} + \text{Na}^+\ ^-\text{OTs} \tag{10.28b}$$

Because secondary alkyl tosylates, like secondary alkyl halides, are not as reactive as primary ones in the S_N2 reaction, use of a polar aprotic solvent ensures a reasonable rate of reaction (Sec. 9.4E). This type of solvent also suppresses carbocation formation, which would be more likely to occur in a protic solvent. (The transformation in Eq. 10.28b takes place in 85% yield.)

The E2 reactions of sulfonate esters, like the analogous reactions of alkyl halides, can be used to prepare alkenes:

$$\text{(cyclohexyl—OTs with H)} + \text{K}^+\ ^-\text{O}t\text{Bu} \xrightarrow[\text{DMSO(solvent)}]{20\text{–}25\ ^\circ\text{C, 30 min}} \text{(cyclohexene)} + \text{K}^+\ ^-\text{OTs} + t\text{BuOH} \tag{10.29}$$

(83% yield)

This reaction is especially useful when the acidic conditions of alcohol dehydration lead to rearrangements or other side reactions, or for primary alcohols in which dehydration is not an option.

To summarize: An alcohol can be made to undergo substitution and elimination reactions typical of the corresponding alkyl halides by converting it into a good leaving group such as a sulfonate ester.

PROBLEMS

10.16   Design a preparation of each of the following compounds from an alcohol using sulfonate ester methodology.

(a) (branched chain structure ending in I)   (b) (cyclopentyl)—$\text{CH}_2\text{CH}_2\text{CH}_2$—$\text{SCH}_3$

10.17   Give the product that results from each of the following sequences of reactions.

(a)
$$\underset{\text{CH}_3\text{CHCH}_2\text{CH}_3}{\overset{\text{OH}}{|}} \xrightarrow[\text{pyridine}]{\text{TsCl}} \xrightarrow[\text{DMSO}]{\text{NaCN}}$$

(b)
$$\overset{\text{OH}}{\text{(structure)}} \xrightarrow[\text{pyridine}]{\text{triflic anhydride}} \xrightarrow[\substack{\text{[18]-crown-6} \\ \text{anhydrous acetonitrile}}]{\text{K}^+\ \text{F}^-}$$

B. Alkylating Agents

As you've learned, alkyl halides, alkyl tosylates, and other sulfonate esters are reactive in nucleophilic substitution reactions. In a nucleophilic substitution, an alkyl group is transferred from the leaving group to the nucleophile.

The nucleophile is said to be **alkylated** by the alkyl halide or the sulfonate ester in the same sense that a Brønsted base is *protonated* by a strong acid. For this reason, alkyl halides, sulfonate esters, and related compounds containing good leaving groups are sometimes referred to as *alkylating agents*. To say that a compound is a good **alkylating agent** means that it reacts rapidly with nucleophiles in S_N2 or S_N1 reactions to transfer an alkyl group.

C. Ester Derivatives of Strong Inorganic Acids

The esters of strong inorganic acids exemplify another type of *alkylating agent* (Sec. 10.4B). Like tosylates and mesylates, these compounds are derived conceptually by replacing the acidic hydrogen of a strong acid (in this case an inorganic acid) with an alkyl group. For example, dimethyl sulfate is an ester in which the acidic hydrogens of sulfuric acid are replaced by methyl groups.

Alkyl esters of strong inorganic acids are typically very potent alkylating agents because they contain leaving groups that are very weak bases. For example, dimethyl sulfate is a very effective methylating agent, as shown in the following example.

Dimethyl sulfate and diethyl sulfate are available commercially. These reagents, like other reactive alkylating agents, are toxic because they react with nucleophilic functional groups on proteins and nucleic acids (Sec. 26.5C).

10.18   Phosphoric acid, H_3PO_4, has the following structure.

$$
\begin{array}{c}
\text{O} \\
\parallel \\
\text{HO} - \text{P} - \text{OH} \\
\mid \\
\text{OH}
\end{array}
$$

(a) Draw the structure of trimethyl phosphate.

(b) Draw the structure of the monoethyl ester of phosphoric acid.

10.19   Predict the products in the reaction of dimethyl sulfate with each of the following nucleophiles.

(a) $CH_3\ddot{N}H_2$ (b) water   (c) sodium ethoxide   (d) sodium 1-propanethiolate

methylamine

D. Reactions of Alcohols with Thionyl Chloride and Triphenylphosphine Dibromide

In most cases, the preparation of primary alkyl chlorides from alcohols with HCl is not as satisfactory as the preparation of the analogous alkyl bromides with HBr (Sec. 10.3). A better method for the preparation of primary alkyl chlorides is the reaction of alcohols with thionyl chloride:

$$CH_3(CH_2)_6CH_2OH + SOCl_2 \xrightarrow{\text{pyridine}} CH_3(CH_2)_6CH_2Cl + SO_2\uparrow + HCl \qquad (10.32)$$

1-octanol   **thionyl**   **1-chlorooctane**   (reacts with
                    chloride   (80% yield)   pyridine)

Thionyl chloride is a dense, fuming liquid (bp 75–76°C). One advantage of using thionyl chloride for the preparation of alkyl chlorides is that the by-products of the reaction are HCl, which reacts with the base pyridine, and sulfur dioxide (SO_2), a gas. Consequently, in many cases, there are no separation problems in the purification of the product alkyl chlorides.

The preparation of an alkyl chloride from an alcohol with thionyl chloride, like the use of a sulfonate ester, involves the conversion of the alcohol —OH group into a good leaving group. When an alcohol reacts with thionyl chloride, a *chlorosulfite ester* intermediate is formed. (This reaction is analogous to Eq. 10.25.)

$$
RCH_2OH + \underset{\substack{\textbf{thionyl} \\ \textbf{chloride}}}{Cl-\overset{\overset{\textstyle O}{\parallel}}{S}-Cl} + \underset{\substack{\textbf{pyridine} \\ \text{(solvent)}}}{\text{[pyridine]}} \longrightarrow \underset{\substack{\text{a chlorosulfite} \\ \text{ester}}}{RCH_2O\overset{\overset{\textstyle O}{\parallel}}{S}Cl} + \text{[pyridinium]}\ Cl^- \qquad (10.33)
$$

The chlorosulfite ester reacts readily with nucleophiles because the chlorosulfite group, —O—SO—Cl, is a very weak base and thus a very good leaving group. The chlorosulfite ester is usually not isolated, but reacts with the chloride ion formed in Eq. 10.33 to give the alkyl chloride. The displaced $^-$O—SO—Cl ion is unstable and decomposes to SO_2 and Cl^-.

$$
R - CH_2 - \overset{..}{\underset{..}{O}} - \overset{\overset{\textstyle :O:}{\parallel}}{S} - Cl \longrightarrow R - CH_2 + \ ^-\overset{..}{\underset{..}{O}} - \overset{\overset{\textstyle :O:}{\parallel}}{S} - \overset{..}{\underset{..}{Cl}}: \longrightarrow \ \overset{..}{\underset{..}{O}} = S = \overset{..}{\underset{..}{O}} + :\overset{..}{\underset{..}{Cl}}:^- \qquad (10.34)
$$

Although the thionyl chloride method is most useful with primary alcohols, it can also be used with secondary alcohols, although rearrangements in such cases have been known to occur. Rearrangements are best avoided in the preparation of secondary alkyl halides by using S_N2 conditions: the reaction of a halide ion with a sulfonate ester in a polar aprotic solvent (as in Study Problem 10.2).

A related method for the conversion of alcohols into alkyl bromides involves the use of Ph_3PBr_2 (dibromotriphenylphosphorane; this compound is universally known to organic chemists as *triphenylphosphine dibromide*.)

$$\text{cyclopentanol} \quad \text{—OH} \quad + \quad Ph_3PBr_2 \quad \xrightarrow{\text{DMF}} \quad \text{—Br} \quad + \quad \overset{+}{Ph_3P}\text{—}\overset{-}{O} \quad + \quad H\text{—}Br \qquad (10.35)$$

cyclopentanol   **dibromotriphenyl-phosphorane (triphenylphosphine dibromide)**   **bromocyclopentane** (83% yield)   **triphenylphosphine oxide**

Triphenylphosphine dibromide is actually an ionic compound, as shown in Eq. 10.36a below. The first mechanistic step of the reaction is a Lewis acid–base association reaction in which the oxygen of the alcohol acts as a nucleophilic center and phosphorus as the electrophilic center. (As we'll learn in Sec. 10.10A, phosphorus can violate the octet rule and form five covalent bonds.) The bromide ion of the reagent acts then as a base to bring about a β-elimination of HBr and form a resonance-stabilized intermediate:

$$(10.36a)$$

The bromide ion that was displaced then acts as a nucleophile at the α-carbon to displace $^-O\text{—}\overset{+}{P}Ph_3$ (triphenylphosphine oxide) as a leaving group.

$$(10.36b)$$

triphenylphosphine oxide
conjugate-acid $pK_a \approx -2.1$

This reaction occurs very rapidly for three reasons:

1. triphenylphosphine oxide is a *very* weak base and therefore a good leaving group;
2. the reaction is typically carried out in polar aprotic solvents such as acetonitrile or DMF, which accelerate S_N2 reactions (Sec. 9.4E); and
3. the reaction of the bromide ion nucleophile is essentially intramolecular. (That is, the bromide leaving group in Eq. 10.36a reacts as a nucleophile in Eq. 10.36b before it can diffuse away.)

The reaction of alcohols with triphenylphosphine dibromide is so fast that it can even be carried out successfully with neopentyl alcohol. Recall that neopentyl derivatives are very unreactive in S_N2 reactions (Table 9.3, Fig. 9.4).

$$H_3C\text{—}\overset{\overset{\displaystyle CH_3}{|}}{\underset{\underset{\displaystyle CH_3}{|}}{C}}\text{—}CH_2OH \quad \xrightarrow[\text{DMF}]{Ph_3PBr_2} \quad H_3C\text{—}\overset{\overset{\displaystyle CH_3}{|}}{\underset{\underset{\displaystyle CH_3}{|}}{C}}\text{—}CH_2Br \qquad (10.37)$$

2,2-dimethyl-1-propanol (neopentyl alcohol)   **1-bromo-2,2-dimethylpropane (neopentyl bromide)** (91% yield)

The triphenylphosphine dibromide reaction is particularly useful for the preparation secondary bromides, as shown in Eq. 10.35. As expected for the S_N2 mechanism, this reaction occurs without rearrangement. (See Problem 10.25.) An analogous reagent, triphenylphosphine dichloride (Ph_3PCl_2), can be used for the preparation of alkyl chlorides.

PROBLEMS

10.20   Give three reactions that illustrate the preparation of 1-bromobutane from 1-butanol.

10.21   (a) According to the mechanism of the reaction shown in Eq. 10.34, what would be the absolute configuration of the alkyl chloride obtained from the reaction of thionyl chloride with (S)-CH$_3$CH$_2$CH$_2$CHD—OH? Explain.

(b) According to the mechanism shown in Eqs. 10.36a and 10.36b, what would be the absolute configuration of 2-bromopentane obtained from the reaction of Ph_3PBr_2 with the R enantiomer of 2-pentanol? Explain.

E. Biological Leaving Groups: Phosphate and Pyrophosphate

Most of what is known about nucleophilic substitution reactions in organic chemistry has been learned from studying the substitution reactions of alkyl halides, sulfonate esters, and related compounds. Halogen-containing compounds are relatively rare in nature, and alkyl halides are particularly rare. Sulfonate esters don't occur at all. However, the key concept that underlies the use of sulfonate esters—*conversion of the oxygen of an alcohol into a good leaving group*—is also operative in two of the most important leaving groups found in biological substitution reactions: phosphate and pyrophosphate, shown below in the ionization state that occurs at physiological pH.

phosphate
(conjugate-acid pK_a = 7.21)

pyrophosphate
(conjugate-acid pK_a = 6.6)

(Phosphorus can form five covalent bonds, as we'll discuss in Sec. 10.10A.) Just as sulfonate esters (Sec. 10.4A) can serve as alkylating agents in laboratory reactions by loss of sulfonate groups, **pyrophosphate esters** can serve as alkylating agents by loss of the pyrophosphate leaving group.

an alkyl pyrophosphate
(a pyrophosphate ester)

(Phosphate esters are important in other types of nucleophilic substitutions; we'll consider phosphate esters more generally in Chapter 25.)

Farnesylation is an example of a biological alkylation reaction that involves a pyrophosphate leaving group. In this reaction, a 15-carbon alkyl pyrophosphate, *farnesyl pyrophosphate*, reacts in an S_N2 process with a thiolate group of a protein called *Ras*. (*Ras* is a protein that regulates cell growth. Mutations of *Ras* have been implicated in pancreatic and other cancers.) This reaction is catalyzed by a *farnesylating enzyme*, a protein distinct from *Ras* itself. The thiolate nucleophile comes from the side chain of a cysteine, one of the amino acids of *Ras*. (See Table 27.1, p. 1377.) Weak coordination of the thiolate to a Zn^{2+} ion on the farnesylating enzyme ensures that it remains ionized.

farnesyl pyrophosphate

farnesylated *Ras*

to membrane

pyrophosphate (10.38)

Through the use of deuterium substitution for one of the α-hydrogens, an isotopically chiral farnesyl pyrophosphate was synthesized. With this chiral derivative containing an asymmetric α-carbon, substitution was shown to occur with inversion of configuration, as we would expect for an S_N2 reaction. The farnesyl group, a large hydrocarbon group, moves to the cell membrane and becomes anchored in the lipid bilayer, and thus *Ras* is tethered to the membrane as well. This event is required for *Ras* to become active. (Interfering with this process has become an attractive target for anticancer drug discovery.)

We learned in our study of the S_N2 reaction that *leaving group effectiveness is inversely correlated with basicity* (Sec. 9.4F). That is, *the weakest bases make the best leaving groups*. Bromide (Br^-) and iodide (I^-) are excellent leaving groups for the same reason that H—Br and H—I are strong acids: the bond energies of H—Br and H—I bonds are relatively low, as are the C—Br and C—I bond energies. How basic, then, is pyrophosphate? The answer is that the pK_a of the pyrophosphate di-anion $H_2P_2O_7^{2-}$ is 6.6. This group is *much* more basic than halide ions and tosylate ions. Therefore, the pyrophosphate group is a *very poor* leaving group. How, then, can it act as a leaving group in biological systems?

The answer is that pyrophosphate *itself* is not a leaving group. Rather, it is activated within enzyme active sites by binding to divalent metal ions—in many cases, Mg^{2+}—or by hydrogen bonding to acidic groups on the side chains of enzyme amino acids.

(Magnesium binding is shown in Eq. 10.38.) We can think of these interactions as examples of ionic solvation (Sec. 8.6F), except that the interactions occur within an enzyme rather than in the solvent.

When the pyrophosphate is bound to the active site of the catalyzing enzyme—and *only* when it is bound—the metal ion and/or proton sources in the active site virtually neutralize two of the negative charges on the pyrophosphate. When this happens, the pyrophosphate leaving group becomes more like pyrophosphoric acid. (The pK_{a2} of $H_3P_2O_7^-$ is 2.0.) It is also likely that noncovalent binding of the pyrophosphate group to the enzyme is used to stretch, and therefore weaken, the C—O bond. As the bond to the leaving group gets weaker, the leaving group gets better. An analogy is what happens when you put your arms in a coat that is too small. When you put the coat on, you rip it apart. (Other reasons for rate enhancements in enzyme catalysis are discussed in Sec. 11.8D, p. 549.)

A moment's reflection will convince you that enzyme-mediated enhancement of leaving-group reactivity is exactly what is needed for biological leaving groups. Alkyl phosphates and pyrophosphates are fairly unreactive as they meander around the cell. We would not want a compound with a very reactive leaving group to be loose in the cell because it could react indiscriminately with many different nucleophiles. In fact, when potent alkylating agents *are* introduced into biological systems, they wreak havoc! For example, a number of strong alkylating agents alkylate nucleic acids in DNA and are known carcinogens (Sec. 26.5C). The optimal situation is for the leaving group to become effective *only* when it is needed—and that is when it is bound to an enzyme that catalyzes a *particular* reaction. Phosphates and pyrophosphates ideally meet this criterion of *adjustable reactivity*.

PROBLEM

10.22 An important class of substitution reactions that you will study in Chapter 21 is called *nucleophilic acyl substitution*. Substitution reactions of this type occur at the carbon of a carbonyl (C=O) group. An important laboratory example is the reaction of nucleophiles (ammonia in the example below) with acyl chlorides (also called acid chlorides).

$$R-\overset{\overset{\text{O}}{\|}}{C}-Cl \ + \ \overset{..}{N}H_3 \ \longrightarrow \ R-\overset{\overset{\text{O}}{\|}}{C}-NH_2 \ + \ H-Cl$$

an acyl chloride

The corresponding derivatives found in biology are *acyl phosphates*. Draw the general structure of an acyl phosphate. Be sure to show its ionization state at physiological pH. Show how Mg^{2+} ions or hydrogen-bond-donating groups in an enzyme active site might enhance the reactivity of acyl phosphates.

10.5   CONVERSION OF ALCOHOLS INTO ALKYL HALIDES: SUMMARY

You have now studied a variety of reactions that can be used to convert alcohols into alkyl halides. These are

1. reaction with hydrogen halides (Sec. 10.3);
2. formation of sulfonate esters followed by S_N2 reaction with halide ions (Sec. 10.4A, Study Problem 10.1); and
3. reaction with thionyl chloride ($SOCl_2$) or triphenylphosphine dibromide (Ph_3PBr_2) (Sec. 10.4D).

Now that we've considered these methods individually, let's now view these methods holistically by asking which method should be used in a given situation. The method of choice depends on the structure of the alcohol and on the type of alkyl halide (chloride, bromide, iodide) to be prepared.

Primary Alcohols: Alkyl bromides are prepared from primary alcohols by the reaction of the alcohol with concentrated HBr or with Ph_3PBr_2. HBr is often chosen for convenience and because the reagent is relatively inexpensive. The reaction with Ph$_3$PBr$_2$ is quite general, but it is particularly useful when the alcohol contains another functional group that would be adversely affected by the strongly acidic conditions of the HBr reaction. (You'll learn about such functional groups in later chapters.) Primary alkyl iodides can be prepared with HI, which is usually supplied by mixing an iodide salt such as KI with a strong acid such as phosphoric acid. Thionyl chloride is the method of choice for the preparation of primary alkyl chlorides because the reactions of primary alcohols with HCl are slow. The sulfonate ester method works well with primary alcohols, but it requires two separate reactions (formation of the sulfonate ester, then reaction of the ester with halide ion). Because all of these methods have an S$_N$2 mechanism as their basis, alcohols with several β-alkyl substituents, such as neopentyl alcohol, do not react under the usual conditions.

Tertiary Alcohols: Tertiary alcohols react rapidly with HCl or HBr under mild conditions to give the corresponding alkyl halides. The sulfonate ester method shown in Study Problem 10.2 (p. 468) is not used with tertiary alcohols because tertiary sulfonates, like tertiary alkyl halides, do not undergo S$_N$2 reactions.

Secondary Alcohols: If the secondary alcohol has no β-alkyl substitution, the thionyl chloride method can be used to prepare alkyl chlorides. To avoid rearrangements completely, the alcohol can be converted into a sulfonate ester, which, in turn, can be treated with the appropriate halide ion (Cl$^-$, Br$^-$, or I$^-$) in a polar aprotic solvent. This type of solvent provides the enhanced nucleophilicity of the halide ion necessary to overcome the relatively low S$_N$2 reaction rate of a secondary sulfonate ester (Sec. 9.4E). Less reactive secondary alcohols can be converted into triflates, which are much more reactive than tosylates or mesylates toward halide ions in polar aprotic solvents. The HBr method can be expected to lead to rearrangements and is thus not very satisfactory (unless rearranged products are desired). The Ph$_3$PBr$_2$ method can be used to form alkyl bromides without rearrangement from primary and secondary alcohols that have significant β-alkyl substitution.

Let's also remind ourselves what we have learned mechanistically about the substitution and elimination reactions of alcohols. The —OH group itself cannot act as a leaving group because $^-$OH is far too basic. To break the carbon–oxygen bond, the —OH group must first be converted into a good leaving group. Two general strategies can be used for this purpose:

1. *Protonation:* Protonated alcohols are intermediates in both dehydration to alkenes and the reaction with hydrogen halides to give alkyl halides.

2. *Conversion into sulfonate esters, inorganic esters, or related leaving groups:* Sulfonate esters, to a useful approximation, react like alkyl halides. That is, the principles of alkyl halide reactivity you learned in Chapter 9 are equally applicable to sulfonate esters. Thionyl chloride and triphenylphosphine dibromide are additional examples of this approach in which the reagent both converts the alcohol —OH into a good leaving group and provides the displacing nucleophile.

PROBLEMS

10.23 Suggest conditions for carrying out each of the following conversions to yield a product that is as free of isomers as possible.

(a)

(b) $(CH_3)_2CH(CH_2)_4OH \longrightarrow (CH_3)_2CH(CH_2)_4Cl$

(c)

(d)

continued

continued _____

10.24   Give the structure of two secondary alcohols that could be converted by HBr/H_2SO_4 into the corresponding alkyl bromide without rearrangement.

10.25   Contrast the products expected when 2-methyl-3-pentanol is treated with (a) HBr/H_2SO_4 or (b) Ph_3PBr_2. Explain.

10.6   OXIDATION AND REDUCTION IN ORGANIC CHEMISTRY

The previous sections have discussed the substitution and elimination reactions of alcohols and their derivatives. These reactions have much in common with the analogous reactions of alkyl halides. Now we turn to a different type of reaction: oxidation. Oxidation is a reaction of alcohols that has no simple analogy in alkyl halide chemistry.

A. Half-Reactions and Oxidation Numbers

An **oxidation** is a transformation in which electrons are lost; a **reduction** is a transformation in which electrons are gained. Each oxidation is accompanied by a reduction, and vice versa. The gain or loss of electrons can be illustrated with a **half-reaction**, which shows either the oxidation or the reduction but not both. An example of such a half-reaction in organic chemistry is the oxidation of ethanol to acetic acid:

$$H_3C-CH_2-OH \longrightarrow H_3C-\overset{\overset{\displaystyle O}{\|}}{C}-OH \qquad (10.39)$$

ethanol   **acetic acid**

This is a half-reaction because the reagent that brings about this oxidation (and is itself reduced) is not included. That this is an oxidation can be demonstrated by balancing the half reaction using protons and free electrons, a technique that you may have learned in general chemistry. This process involves three steps:

Step 1.   Use H_2O to balance missing oxygens.

Step 2.   Use protons (that is, H^+) to balance missing hydrogens.

Step 3.   Use electrons to balance charges.

This process is illustrated in Study Problem 10.3.

STUDY PROBLEM   10.3

Write the transformation of Eq. 10.39 as a balanced half-reaction.

SOLUTION

Step 1.   Balance the extra oxygen on the right with a water on the left:

$$CH_3CH_2OH + H_2O \longrightarrow H_3C-\overset{\overset{\displaystyle O}{\|}}{C}-OH \qquad \text{(oxygens are balanced)} \qquad (10.40a)$$

Step 2.   Balance the extra hydrogens on the left with four protons on the right:

$$CH_3CH_2OH + H_2O \longrightarrow H_3C-\overset{\overset{\displaystyle O}{\|}}{C}-OH + 4H^+ \qquad \text{(hydrogens and oxygens are balanced)} \qquad (10.40b)$$

Step 3. Balance the extra positive charges on the right with electrons so that the charges on both sides of the equation are equal:

$$CH_3CH_2OH + H_2O \longrightarrow H_3C-\overset{\overset{\displaystyle O}{\|}}{C}-OH + 4H^+ + 4e^- \qquad \text{(everything is balanced)} \qquad (10.40c)$$

The result is the balanced half-reaction.

According to this half-reaction, *four electrons are lost from the ethanol molecule when acetic acid is formed*. The loss of electrons means physically that this half-reaction can be carried out at the anode of an electrochemical cell. In most cases, though, we carry out oxidations with *reagents* (rather than anodes) that accept electrons, called *oxidizing agents*, which are discussed in Sec. 10.6B. Nevertheless, on the basis of this half-reaction, it can be said that *the oxidation of ethanol to acetic acid is a four-electron oxidation*. You will see this type of terminology used frequently if you study biochemistry.

A quicker way to determine whether a transformation is an oxidation or a reduction, as well as how many electrons are involved, is to determine *oxidation numbers* for the reactant and product. This is a three-step "bookkeeping" process that focuses on *individual carbon atoms* involved in the transformation. After glancing over these steps, read carefully through Study Problem 10.4, which takes you through this process.

Step 1. Assign an *oxidation level* to each carbon that undergoes a change between reactant and product by the following method:

    a. For every bond from the carbon to a less electronegative element (including hydrogen), and for every negative charge on the carbon, assign a -1.

    b. For every bond from the carbon to another carbon atom, and for every unpaired electron on the carbon, assign a 0.

    c. For every bond from the carbon to a more electronegative element, and for every positive charge on the carbon, assign a $+1$.

    d. Add the numbers assigned under parts (a), (b), and (c) to obtain the oxidation level of the carbon under consideration.

Step 2. Determine the *oxidation number N_{ox} of both the reactant and product by adding, within each compound, the oxidation levels of all the carbons computed in step 1.* Remember: Consider only the carbons that undergo a change in the reaction.

Step 3. Compute the difference N_{ox}(product) $- N_{ox}$(reactant) to determine whether the transformation is an oxidation, reduction, or neither.

    a. If the difference is a positive number, the transformation is an *oxidation*.

    b. If the difference is a negative number, the transformation is a *reduction*.

    c. If the difference is zero, the transformation is neither an oxidation nor a reduction.

STUDY PROBLEM 10.4

Use oxidation numbers to verify that the transformation in Eq. 10.39 (also shown below) is an oxidation.

$$H_3C-CH_2-OH \longrightarrow H_3C-\overset{\overset{\displaystyle O}{\|}}{C}-OH \qquad (10.39)$$

                ethanol   **acetic acid**

SOLUTION

Step 1. For both the reactant and the product, compute the oxidation level of each carbon that undergoes a change. Because the methyl group is unchanged, you don't need to assign an oxidation level to its carbon. Only one carbon is changed. For this carbon, −1 is assigned for each bond to hydrogen; 0 is assigned to the bonds to carbon; and +1 is assigned to the bonds to oxygen. The carbon–oxygen double bond of the product is treated as *two bonds* and makes a contribution of +2. Add the resulting numbers.

Reactant:

$$\underset{\underset{H}{\overset{|}{_{-1}}}}{H_3C \underset{0}{-} C \overset{+1|\ OH}{\underset{}{-}} {}_{-1}\!\!H}$$

Product:

$$H_3C \underset{0}{-} C \overset{O\ ||+2}{\underset{+1}{-}} OH$$

Step 2. Add the oxidation levels for each carbon that changes to determine the oxidation number. Because only one carbon changes, the oxidation level of this carbon, computed in step 1, is the oxidation number of the compound. Therefore, N_{ox}(reactant), the oxidation number of the reactant ethanol, is −1; and N_{ox}(product), the oxidation number of the product acetic acid, is +3.

Sum for the reactant: *Sum for the product:*

$(+1) + 0 + (-1) + (-1) = -1$ $0 + (+1) + (+2) = +3$

Step 3. Compute the difference N_{ox}(product) − N_{ox}(reactant), which is $+3 − (−1) = +4$. Because this difference is positive, the transformation of ethanol to acetic acid is an oxidation.

Notice that the change in oxidation number, +4, determined in Step 3 of Study Problem 10.4, is the same as the number of electrons lost, determined in Study Problem 10.3 from the balanced half-reaction. *This correspondence is general.* That is, *the change in oxidation number is always equal to the number of electrons lost or gained in the half-reaction.* If the number is positive, as in this example, the transformation is an oxidation. If it is negative, the transformation is a reduction.

In some transformations that are neither oxidations nor reductions, the oxidation level of one carbon in a molecule can be increased and the oxidation level of another carbon can be decreased by the same amount. It is the sum of *all* of the changes in carbon oxidation levels that determines whether a net oxidation or reduction has occurred. This situation is illustrated in Study Problem 10.5.

STUDY PROBLEM 10.5

Verify that the acid-catalyzed hydration of 2-methylpropene is neither an oxidation nor a reduction.

SOLUTION   First, write the structures involved in the transformation:

$$\begin{array}{c} H_3C \\ \diagdown \\ C{=}CH_2 \\ \diagup \\ H_3C \end{array} \xrightarrow{\text{H}_2\text{O, acid}} \begin{array}{c} CH_3 \\ | \\ H_3C{-}C{-}CH_3 \\ | \\ OH \end{array}$$

2-methylpropene   ***tert*-butyl alcohol**

The oxidation number of the organic reactant, 2-methylpropene, is −2.

$$\begin{array}{c} \overset{0}{} \; \overset{-2}{} \\ H_3C \; \downarrow \;\; \downarrow \\ \diagdown \\ C{=}CH_2 \qquad N_{ox} = 0 + (-2) = -2 \\ \diagup \\ H_3C \end{array}$$

The oxidation number of the organic product, *tert*-butyl alcohol, is also -2:

$$H_3C-\overset{\overset{\displaystyle CH_3}{|}}{\underset{\underset{\displaystyle OH}{|}}{C}}-CH_3 \qquad N_{ox} = +1 + (-3) = -2$$

Notice that an oxidation level is computed for only the one methyl group that was formed as a result of the transformation. Because the oxidation numbers of the reactant and product are equal, the hydration reaction is neither an oxidation nor a reduction. The same conclusion must apply to the reverse reaction, the dehydration of the alcohol to the alkene.

The methods described here show that the addition of Br_2 to an alkene is an oxidation (the change in oxidation number is $+2$):

$$R-CH=CH-R + Br_2 \longrightarrow R-\overset{\overset{\displaystyle Br}{|}}{C}H-\overset{\overset{\displaystyle Br}{|}}{C}H-R \qquad (10.41)$$

$$N_{ox} = -2 \qquad\qquad\qquad N_{ox} = 0$$

Thus, whether a reaction is an oxidation or a reduction does not necessarily depend on the introduction or loss of oxygen. However, in most oxidations of organic compounds, either a hydrogen in a C—H bond or a carbon in a C—C bond is replaced by a more electronegative element, which *may* be oxygen, but *which may also be another element such as a halogen.*

PROBLEMS

10.26 Considering the organic compound, classify each of the following transformations, some of which may be unfamiliar, as an oxidation, a reduction, or neither. For those that are oxidations or reductions, tell how many electrons are gained or lost.

(a) $CH_4 \xrightarrow{Br_2, \text{light}} CH_3Br$

(b)

$$Ph-CH_3 \xrightarrow[H_2O]{Cr^{6+}} Ph-\overset{\overset{\displaystyle O}{||}}{C}-OH$$

(c) $CH_3CH_2CH_2I \xrightarrow{LiAlH_4} CH_3CH_2CH_3 + I^-$

(d)

(e)

(f)

(g)

10.27 Write the transformation in Problem 10.26b as a balanced half-reaction. Complete the following sentence: This reaction is a _____ (how many)-electron _____ (oxidation or reduction).

B. Oxidizing and Reducing Agents

Oxidations and reductions always occur in pairs. Therefore, *whenever something is oxidized, something else is reduced.* When an organic compound is oxidized, the reagent that brings about the transformation is called an **oxidizing agent**. Similarly, when an organic compound is reduced, the reagent that effects the transformation is called a **reducing agent**.

For example, chromate ion (CrO_4^{2-}) can be used to bring about the oxidation of ethanol to acetic acid in Eq. 10.39; in this reaction, chromate ion is reduced to Cr^{3+}. We calculate the change in oxidation state for chromium in this reaction by calculating the oxidation states of Cr in the reactant and product and then taking the difference. To determine the oxidation state of Cr, we apply essentially the same technique we used for determining the oxidation state of carbon. Let's start with the oxidation state of Cr in chromate ion.

Sum = oxidation number of Cr = +6

Each single bond to oxygen makes a contribution of +1 and each double bond to oxygen a contribution of +2. (The negative charges do not enter into the calculation because they are on oxygen, not chromium.) The oxidation number of Cr is therefore +6. We indicate this by saying that Cr is in the +6 oxidation state and abbreviate this oxidation state as Cr(VI).

When CrO_4^{2-} is used to oxidize ethanol, Cr^{3+} is formed. To compute the oxidation state of Cr^{3+}, we count +1 for each positive charge. Thus, Cr^{3+} has an oxidation state of +3. Hence, chromium changes oxidation state from Cr(VI) to Cr(III), and consequently, chromate ion undergoes a *three-electron reduction.* We can verify this by balancing the corresponding half-reaction:

$$8H^+ + 3e^- + CrO_4^{2-} \longrightarrow Cr^{3+} + 4H_2O \qquad (10.42)$$

A complete, balanced reaction for the oxidation of ethanol to acetic acid by chromate is obtained by reconciling the electrons in the half-reactions given by Eqs. 10.40c (p. 477) and Eq. 10.42. That is, every mole of Cr(VI) ($3e^-$ gained) can oxidize three-fourths of a mole of ethanol ($0.75 \times 4e^-$ lost). A more structured process for balancing the overall reaction is illustrated in Study Problem 10.6.

STUDY PROBLEM 10.6

Give a complete balanced equation for the oxidation of ethanol to acetic acid by chromate ion.

SOLUTION   The two half-reactions are

$$CH_3CH_2OH + H_2O \longrightarrow CH_3CO_2H + 4H^+ + 4e^- \qquad (10.43a)$$

$$8H^+ + 3e^- + CrO_4^{2-} \longrightarrow Cr^{3+} + 4H_2O \qquad (10.43b)$$

Multiply each equation by a factor that gives the same number of "free" electrons in both half-reactions. Thus, multiplying Eq. 10.43a by 3 and Eq. 10.43b by 4 gives 12 electrons in both reactions:

$$3CH_3CH_2OH + 3H_2O \longrightarrow 3CH_3CO_2H + 12H^+ + 12e^- \qquad (10.44a)$$

$$32H^+ + 12e^- + 4CrO_4^{2-} \longrightarrow 4Cr^{3+} + 16H_2O \qquad (10.44b)$$

Add these equations, canceling like terms on both sides. Thus, all electrons cancel; the three water molecules on the left are canceled by three of those on the right to leave 13 water molecules on the right; and the 12 protons

**STUDY GUIDE
LINK 10.1**
More on
Half-Reactions

on the right are canceled by 12 of those on the left, leaving 20 protons on the left. Hence, the fully balanced equation is

$$20\,H^+ \;+\; 4\,CrO_4^{2-} \;+\; 3\,CH_3CH_2OH \;\longrightarrow\; 4\,Cr^{3+} \;+\; 3\,CH_3CO_2H \;+\; 13\,H_2O \qquad (10.45)$$

This equation shows that three ethanol molecules are oxidized for every four chromate ions reduced, or, as noted earlier, three-fourths of a mole of ethanol per mole of chromate ion.

By considering the change in oxidation number for a transformation, you can tell whether an oxidizing or reducing agent is required to bring about the reaction. For example, the following unfamiliar transformation is neither an overall oxidation nor reduction (verify this statement):

$$H_3C-\underset{\underset{OH}{|}}{\overset{\overset{CH_3}{|}}{C}}-\underset{\underset{OH}{|}}{\overset{\overset{CH_3}{|}}{C}}-CH_3 \;\longrightarrow\; H_3C-\underset{\underset{CH_3}{|}}{\overset{\overset{CH_3}{|}}{C}}-\overset{\overset{O}{\|}}{C}-CH_3 \qquad (10.46)$$

Although one carbon is oxidized, another is reduced. Even though you might know nothing else about the reaction, an oxidizing or reducing agent alone would not effect this transformation. (In fact, the reaction is brought about by strong acid.)

The oxidation-number concept can be used to organize organic compounds into functional groups with the same oxidation level, as shown in Table 10.2 on p. 482. Compounds *within* a given box are generally interconverted by reagents that are neither oxidizing nor reducing agents. For example, alcohols can be converted into alkyl halides with HBr, which is neither an oxidizing nor a reducing agent. On the other hand, conversion of an alcohol into a carboxylic acid involves an increase in oxidation level, and indeed this transformation requires an oxidizing agent. Notice also in Table 10.2 that carbons with larger numbers of hydrogens have a greater number of possible oxidation states. Thus, a tertiary alcohol cannot be oxidized at the α-carbon (without breaking carbon–carbon bonds) because this carbon bears no hydrogens. Methane, on the other hand, can be oxidized to CO_2. (Any hydrocarbon can be oxidized to CO_2 if carbon–carbon bonds are broken; Sec. 2.7.)

PROBLEMS

10.28 For each of the following balanced oxidation–reduction reactions, indicate which compound(s) are oxidized and which are reduced. (*Hint:* Consider the change in the organic compounds in each reaction first.)

(a) $H_2C{=}CH_2 \;+\; H_2 \;\xrightarrow{\text{Pd/C}}\; H_3C{-}CH_3$

(b) $CH_3CH_2Br \;+\; Li^+ \;^-AlH_4 \;\longrightarrow\; CH_3CH_3 \;+\; Li^+\,Br^- \;+\; AlH_3$

(c) $H_3C{-}CH{=}CH_2 \;+\; Br_2 \;\longrightarrow\; H_3C-\underset{\underset{Br}{|}}{CH}-\underset{\underset{Br}{|}}{CH_2}$

(d) $Ph-\underset{\underset{CH_3}{|}}{\overset{\overset{CH_3}{|}}{CH}}-CH_3 \;+\; O_2 \;\longrightarrow\; Ph-\underset{\underset{O-O-H}{|}}{\overset{\overset{CH_3}{|}}{C}}-CH_3$

10.29 How many moles of permanganate are required to oxidize one mole of toluene to benzoic acid? (Use H_2O and protons to balance the equation.)

$$\underset{\textbf{toluene}}{Ph{-}CH_3} \;+\; \underset{\textbf{permanganate}}{MnO_4^-} \;\longrightarrow\; \underset{\textbf{benzoic acid}}{Ph{-}CO_2H} \;+\; \underset{\substack{\textbf{manganese}\\\textbf{dioxide}}}{MnO_2}$$

TABLE 10.2 Comparison of Oxidation States of Various Functional Groups

All molecules in the same box have the same oxidation number.
X = an electronegative group such as halogen

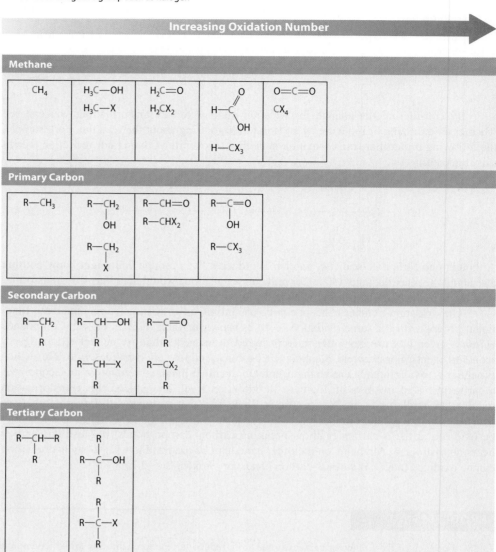

10.7   OXIDATION OF ALCOHOLS

A. Oxidation to Aldehydes and Ketones

Primary and secondary alcohols can be oxidized by reagents containing Cr(VI)—that is, chromium in the +6 oxidation state—to give certain types of *carbonyl compounds* (compounds that contain the carbonyl group, $\backslash$C$=$O). For example, secondary alcohols are oxidized to ketones:

$$\text{2-octanol} \xrightarrow[\text{aqueous } H_2SO_4]{Na_2Cr_2O_7} \text{2-octanone} \qquad (10.47)$$

2-octanol

2-octanone
(94% yield)

(10.48)

(89% yield)

Several forms of Cr(VI) can be used to convert secondary alcohols into ketones. Three of these are chromate (CrO_4^{2-}), dichromate ($Cr_2O_7^{2-}$), and chromic anhydride or chromium trioxide (CrO_3). The first two reagents are customarily used under strongly acidic conditions; the last is often used in pyridine. In all cases, the chromium is reduced to a form of Cr(III) such as Cr^{3+}.

Primary alcohols react with Cr(VI) reagents to give aldehydes. However, if water is present, the reaction cannot be stopped at the aldehyde stage because aldehydes are further oxidized to carboxylic acids:

(10.49)

2-methyl-1-butanol   **2-methylbutanal**   **2-methylbutanoic acid**

For this reason, *anhydrous* preparations of Cr(VI) are generally used for the laboratory preparation of aldehydes from primary alcohols. One commonly used reagent of this type is a complex of pyridine, HCl, and chromium trioxide called *pyridinium chlorochromate*, which goes by the abbreviation "PCC." This reagent is typically used in methylene chloride solvent.

(10.50)

1-decanol   **decanal**
(92% yield)

Water promotes the transformation of aldehydes into carboxylic acids because, in water, aldehydes are in equilibrium with hydrates formed by the addition of water across the C=O double bond. (Hydration is discussed in Sec. 19.7.)

(10.51)

an aldehyde   an aldehyde hydrate   a carboxylic acid

Aldehyde hydrates are really alcohols and therefore can be oxidized just like secondary alcohols. Because of the absence of water in anhydrous reagents such as PCC, a hydrate does not form, and the reaction stops at the aldehyde.

Tertiary alcohols are not oxidized under the usual conditions. For oxidation of alcohols at the α-carbon to occur, *the α-carbon atom must bear one or more hydrogen atoms.*

The mechanism of alcohol oxidation by Cr(VI) involves several steps that have close analogies to other reactions. Consider, for example, the oxidation of isopropyl alcohol to the ketone acetone by chromic acid (H_2CrO_4). The first steps of the reaction involve an acid-catalyzed displacement of water from chromic acid by the alcohol to form a *chromate ester*. (This ester is analogous to ester derivatives of other strong acids; Sec. 10.4C.)

$$\underset{\textbf{isopropyl alcohol}}{\underset{H_3C}{\overset{H_3C}{>}}CH-OH} \;+\; \underset{\textbf{chromic acid}}{\underset{HO}{\overset{O\;\;\;\;O}{Cr}}OH} \quad\xrightarrow[\text{(several steps)}]{H_3O^+}\quad \underset{\text{a chromate ester}}{\underset{H_3C}{\overset{H_3C}{>}}CH-O-\overset{O}{Cr}=O} \;+\; H_2O \qquad (10.52a)$$

After protonation of the chromate ester (Eq. 10.52b), it decomposes in a β-elimination reaction (Eq. 10.52c).

$$\underset{H_3C}{\overset{H_3C}{>}}CH-O-\overset{O}{\underset{OH}{Cr}}=\ddot{O}\;\;H-\overset{+}{O}H_2 \;\;\rightleftharpoons\;\; \underset{H_3C}{\overset{H_3C}{>}}CH-O-\overset{O}{\underset{OH}{Cr}}=\overset{+}{O}H \;+\; \ddot{O}H_2 \qquad (10.52b)$$

$$\underset{\substack{H-\ddot{O}: \\ | \\ H}}{H_3C-\overset{CH_3}{\underset{H}{C}}-O-\overset{O}{\underset{OH}{Cr}}=\overset{+}{O}H} \quad\xrightarrow{E2}\quad \underset{\textbf{acetone}}{\underset{H_3C}{\overset{H_3C}{>}}C=O} \;+\; \left[\underset{\substack{\\OH}}{-:\overset{O}{Cr}=\overset{+}{O}H} \;\longleftrightarrow\; \underset{\substack{\\OH}}{:\overset{O}{Cr}-\ddot{O}H} \right] \;+\; H_3\ddot{O}^+ \qquad (10.52c)$$

chromium accepts electrons

Cr(IV)

This last step is essentially an E2 reaction. Because Cr(VI) is particularly electronegative, especially when protonated, the chromium readily accepts an electron pair and is thus reduced. The reaction is so rapid that the Brønsted base in the reaction can be very weak; in fact, water is the base in Eq. 10.52c. In the resulting H_2CrO_3 by-product, chromium is in a +4 oxidation state. The ultimate by-product is Cr^{3+} because, in subsequent reactions, Cr(IV) and Cr(VI) react to give two equivalents of a Cr(V) species, which then oxidizes an additional molecule of alcohol by a similar mechanism.

$$Cr(IV) + Cr(VI) \longrightarrow 2Cr(V) \qquad (10.53a)$$

$$Cr(V) + (CH_3)_2CH-OH \longrightarrow Cr(III) + (CH_3)_2C=O + 2H^+ \qquad (10.53b)$$

The Breath Test for Ethanol

Law enforcement officers can use any of several devices to estimate a person's blood alcohol content (BAC), and two of these are based on ethanol oxidation to acetic acid. All of the devices are based on the fact that, in the lungs, a known small fraction of ethanol in the blood escapes into the air that is subsequently exhaled, and it is the alcohol content of this exhaled vapor that is analyzed. In the original measuring device, called a "breathalyzer," the exhaled air is collected and allowed to react with acidic potassium dichromate ($K_2Cr_2O_7$). The alcohol reduces the Cr(VI) to Cr^{3+}. The resulting change in color of the chromium from the yellow-orange of the Cr(VI) oxidation state to the blue-green of Cr^{3+} is detected by a simple spectrometer. The amount of Cr(VI) reduced is calibrated in terms of percent blood alcohol. (See Problem 10.54, p. 506, at the end of the chapter.) Most modern devices use electrochemical technology. The alcohol is oxidized at a platinum electrode to produce acetic acid along with four protons and four electrons. (See Study Problem 10.3 on p. 476.) The resulting electron flow in the electrochemical cell is detected and calibrated in terms of blood alcohol concentration. A third device uses infrared spectroscopy (Chapter 12) to detect ethanol.

B. Oxidation to Carboxylic Acids

As noted in the previous section (Eq. 10.49), primary alcohols can be oxidized to carboxylic acids using *aqueous* solutions of Cr(VI), such as aqueous potassium dichromate ($K_2Cr_2O_7$) in acid. Another useful reagent for oxidizing primary alcohols to carboxylic acids is potassium permanganate ($KMnO_4$) in basic solution:

$$\text{(10.54)}$$

2-ethyl-1-hexanol

a carboxylate ion
(conjugate base of the
carboxylic acid)

2-ethylhexanoic acid
(74% yield)

As shown in this equation, the immediate product of the permanganate oxidation is the conjugate base of the carboxylic acid because the reaction is run in alkaline solution. Isolation of the carboxylic acid itself requires addition of a strong acid such as H_2SO_4 in a second step.

The manganese in $KMnO_4$ is in the Mn(VII) oxidation state; in the oxidation of alcohols, it is reduced to MnO_2, a common form of Mn(IV). Because $KMnO_4$ reacts with alkene double bonds (Sec. 11.6A), Cr(VI) is required for the oxidation of alcohols that contain double or triple bonds (see Eq. 10.48).

Potassium permanganate is not used for the oxidation of secondary alcohols to ketones because many ketones react further with the alkaline permanganate reagent.

PROBLEMS

10.30 Give the product expected when each of the following alcohols reacts with pyridinium chlorochromate (PCC).

(a) $HO-CH_2CH_2CH_2-OH$ (b)

10.31 From which alcohol and by what method would each of the following compounds best be prepared by an oxidation?

(a) $(CH_3)_2CHCH_2CH_2CH_2CO_2H$ (b)

(c)

(d)

10.8   BIOLOGICAL OXIDATION OF ETHANOL

Oxidation and reduction reactions are very important in living systems. A typical biological oxidation is the conversion of ethanol into acetaldehyde, the principal reaction by which ethanol is removed from the bloodstream.

$$CH_3CH_2OH \xrightarrow[\text{(an enzyme)}]{\text{alcohol dehydrogenase}} CH_3CH{=}O \qquad \text{(10.55)}$$

ethanol   **acetaldehyde**

The reaction is carried out in the liver and is catalyzed by an enzyme called *alcohol dehydrogenase*. (Recall from Sec. 4.9C that enzymes are biological catalysts.) The actual oxidizing

FIGURE 10.1 Abbreviated and full structures of NAD+. The colored portion of the full structure is abbreviated as an R-group. In some biological oxidations, the NAD+ is phosphorylated as shown. The phosphate-ester derivative is called NADP+. The chemical mechanisms of oxidation by NAD+ and NADP+ are the same.

agent is not the enzyme, but a large molecule called *nicotinamide adenine dinucleotide*, abbreviated NAD+; the structure of NAD+ and a convenient abbreviated structure for it are shown in Fig. 10.1. When ethanol is oxidized, NAD+ is reduced to a product called NADH. The hydrogen removed from carbon-1 of the ethanol ends up in the NADH; the —OH hydrogen is lost as a proton.

$$(10.56)$$

The compound NAD+ is one of nature's most important oxidizing agents. [From the perspective of laboratory chemistry, it might be called "nature's substitute for Cr(VI)."] NAD+ is an example of a *coenzyme*. **Coenzymes** are molecules required, along with enzymes, for certain biological reactions to occur. For example, ethanol cannot be oxidized by an enzyme unless the coenzyme NAD+ is also present, because NAD+ is one of the reactants. Thus, an ethanol molecule and an NAD+ molecule are juxtaposed when they bind noncovalently to alcohol dehydrogenase, the enzyme that catalyzes ethanol oxidation. It is within the complex of these three molecules that ethanol is oxidized to acetaldehyde and NAD+ is reduced to NADH.

The coenzymes NAD+ and NADH are derived from the vitamin *niacin*, a deficiency of which is associated with the disease pellagra (black tongue). Many biochemical processes employ the NAD+ ⇌ NADH interconversion, some of which reoxidize the NADH formed in ethanol oxidation back to NAD+.

Fermentation

Fermenting pinot noir grapes

The human body uses the ethanol-to-acetaldehyde reaction of Eq. 10.56 to remove ethanol, but yeast cells use the reaction in reverse as the last step in the production of ethanol. Thus, yeast added to dough produces ethanol by the reduction of acetaldehyde, which, in turn, is produced in other reactions from sugars in the dough. Ethanol vapors, wafted away from the bread by CO_2 produced in other reactions, give rising bread its pleasant odor. Special strains of yeast ferment the sugars in grape juice or barley in the production of wine, whiskey, or beer (photo). For the fermentation reaction to take place, the reaction must occur in the absence of oxygen. Otherwise, acetic acid, CH_3CO_2H (vinegar, or "spoiled wine"), is formed instead by other reactions. In winemaking, air is excluded by trapping the CO_2 formed during fermentation as a blanket in the fermentation vessel. Because the production of alcohol by yeast occurs in the absence of air, it is called *anaerobic fermentation*. This is one of the oldest chemical reactions known to civilization.

How does NAD$^+$ work as an oxidizing agent? The resonance structure of NAD$^+$ shows that because the nitrogen in the ring can accept electrons, the molecule takes on the character of a carbocation:

$$(10.57)$$

The electron-deficient carbon of NAD$^+$ and an α-hydrogen of ethanol (color) are held in proximity by the enzyme. The carbocation removes a *hydride* (a hydrogen with two electrons) from the α-carbon of ethanol:

$$(10.58a)$$

Although this reaction may initially look strange, it is really just like a carbocation rearrangement involving the migration of a hydride (Sec. 4.7D), except that in this case the hydride moves to a different molecule. As a result, NADH and a new carbocation are formed. By loss of the proton bound to oxygen, acetaldehyde is formed.

$$H_3C-\overset{\overset{\displaystyle H}{|}}{C}=\overset{..}{\underset{..}{O}}-H \quad :\overset{..}{O}H_2 \longrightarrow H_3C-\overset{\overset{\displaystyle H}{|}}{C}=\overset{..}{\underset{..}{O}} + H_3\overset{+}{O}: \tag{10.58b}$$

The acetaldehyde and NADH dissociate from the enzyme, which is then ready for another round of catalysis. (NADH acts as a reducing agent in other reactions, by which it is converted back into NAD^+.)

Despite the fact that the NAD^+ molecule is large, the chemical changes that occur when it serves as an oxidizing agent take place in a relatively small part of the molecule. A number of other coenzymes also have complex structures but undergo simple reactions. The part of the molecule abbreviated by "R" in Fig. 10.1 (p. 486) provides the groups that cause it to bind tightly to the enzyme catalyst by noncovalent attractions but remain unchanged in the oxidation reaction.

This section has shown that the chemical changes that occur in NAD^+-promoted oxidations have analogies in common laboratory reactions. Most other biochemical reactions have common laboratory analogies as well. Even though the molecules involved may be complex, their chemical transformations are in most cases relatively simple. Thus, *an understanding of the fundamental types of organic reactions and their mechanisms is essential in the study of biochemical processes.*

PROBLEM

10.32   Write a curved-arrow mechanism for the following oxidation of 2-heptanol, which proceeds in 82% yield.

In this section we've considered the chemical transformations associated with the biological oxidation of ethanol. This reaction also has some interesting *stereochemical* aspects, which we'll consider in Sec. 10.9B. To understand that discussion, we'll need to learn about the stereochemical relationships of groups within molecules, which is the subject of Sec. 10.9A.

10.9   CHEMICAL AND STEREOCHEMICAL GROUP RELATIONSHIPS

Different molecules with the same molecular formula—isomers—can have various relationships: they can be constitutional isomers, or they can be stereoisomers, which in turn can be diastereomers or enantiomers (Chapter 6). The subject of this section is the relationships that groups *within* molecules can have. This subject is particularly important in two areas. First, it is important in understanding the stereochemical aspects of enzyme catalysis. We'll demonstrate this by returning to the biological oxidation of ethanol, the reaction discussed in

Sec. 10.8. Second, the subject of group relationships is important in spectroscopy, particularly nuclear magnetic resonance spectroscopy, which is discussed in Chapter 13.

A. Chemical Equivalence and Nonequivalence

It is sometimes important to know when two groups within a molecule are *chemically equivalent*. When two groups are **chemically equivalent**, they behave in exactly the same way toward a chemical reagent. Otherwise, they are **chemically nonequivalent**.

An understanding of chemical equivalence hinges, first, on the concept of *constitutional equivalence*. Groups within a molecule are **constitutionally equivalent** when they have the same connectivity relationship to all other atoms in the molecule. Thus, within each of the following molecules, the hydrogens shown in color are constitutionally equivalent:

For example, in compound *C*, each of the red hydrogens is connected to a carbon that is connected to a chlorine and to a CH_3CHCl— group. Each of these hydrogens has the same connectivity relationship to the other atoms in the molecule. On the other hand, the H_3C— hydrogens in ethanol (molecule *B*) are constitutionally nonequivalent to the —CH_2— hydrogens. The H_3C— hydrogens are connected to a carbon that is connected to a —CH_2OH, but the —CH_2— hydrogens are connected to a carbon that is connected to an —OH and a —CH_3. However, the two —CH_2— hydrogens of ethanol (*B*, red) are constitutionally equivalent to each other, as are the three H_3C— hydrogens.

In general, *constitutionally nonequivalent groups are chemically nonequivalent.* This means that constitutionally nonequivalent groups have different chemical behaviors. Thus, the H_3C— and —CH_2— hydrogens of ethyl alcohol, which are constitutionally nonequivalent, have different reactivities toward chemical reagents. A reagent that reacts with one type of hydrogen will in general have a different reactivity (or perhaps none at all) with the other type. For example, the oxidation of ethanol with Cr(VI) reagents results in the loss of a —CH_2— hydrogen, but the H_3C— hydrogens are unaffected. Consequently, the two types of hydrogens have very different reactivities with the oxidizing agent.

Constitutional nonequivalence is a sufficient but not a necessary condition for chemical nonequivalence. That is, some constitutionally equivalent groups are chemically nonequivalent. *Whether two constitutionally equivalent groups are chemically equivalent depends on their stereochemical relationship.* Therefore, to understand chemical equivalence, we need to understand the various *stereochemical relationships* that are possible between constitutionally equivalent groups.

The stereochemical relationship between constitutionally equivalent groups is revealed by a **substitution test**. In this test, we substitute each constitutionally equivalent group in turn with a fictitious circled group and compare the resulting molecules. Their stereochemical relationship determines the relationship of the circled groups. This process is best illustrated by example, starting with the molecules *A*, *B*, and *C* shown earlier in this section. Substitute each hydrogen of molecule *A* with a circled hydrogen:

(10.59)

Each of these "new" molecules is identical to the other. For example, the identity of *A1* and *A2* is shown in the following way:

(10.60)

When the substitution test gives identical molecules, as in this example, the constitutionally equivalent groups are said to be **homotopic**. Thus, the three hydrogens of methyl chloride (compound *A*) are homotopic. *Homotopic groups are chemically equivalent and indistinguishable under all circumstances.* Thus, the homotopic hydrogens of compound *A* (methyl chloride) all have the same reactivity toward any chemical reagent; it is impossible to distinguish among these hydrogens.

Substitution of each of the constitutionally equivalent —CH$_2$— hydrogens in molecule *B* (ethanol) gives *enantiomers*:

(10.61)

When the substitution test gives enantiomers, the constitutionally equivalent groups are said to be **enantiotopic**. Thus, the two —CH$_2$— hydrogens of compound *B* (ethanol) are enantiotopic. *Enantiotopic groups are chemically nonequivalent toward chiral reagents, but are chemically equivalent toward achiral reagents.*

Because enzymes are chiral, they can generally distinguish between enantiotopic groups within a molecule. For example, in the enzyme-catalyzed oxidation of ethanol, one of the two enantiotopic α-hydrogens is selectively removed. (This point is further explored in the next section.) Achiral reagents, however, cannot distinguish between enantiotopic groups. Thus, in the oxidation of ethanol to acetaldehyde by chromic acid, an achiral reagent, the α-hydrogens of ethanol are removed indiscriminately.

Finally, substitution of each of the constitutionally equivalent —CH$_2$— hydrogens in molecule *C* gives *diastereomers*.

When the substitution test gives diastereomers, the constitutionally equivalent groups are said to be **diastereotopic**. Thus, the two —CH$_2$— hydrogens of compound *C* are diastereotopic. *Diastereotopic groups are chemically nonequivalent under all conditions.* For example, the hydrogens labeled H^a and H^b in 2-bromobutane (Eq. 10.62; below) are also diastereotopic. In the E2 reaction of this compound, anti-elimination of H^b and Br gives *cis*-2-butene, and anti-elimination of H^a and Br gives *trans*-2-butene. (Verify these statements using models, if necessary.)

$$(10.62)$$

Different amounts of these two alkenes are formed in the elimination reaction precisely because the two diastereotopic hydrogens are removed at different rates—that is, these hydrogens are distinguished by the base that promotes the elimination.

Diastereotopic groups are easily recognized at a glance in two important situations. The first occurs when two constitutionally equivalent groups are present in a molecule that contains an asymmetric carbon:

The carbon that bears the diastereotopic hydrogens need not be adjacent to the asymmetric carbon. For example, in the following compound, the two hydrogens at each of the carbons marked with a diamond (◊) are diastereotopic, even though some of these carbons are not adjacent to asymmetric carbons.

* = asymmetric carbons

◊ = the two hydrogens at each of these carbons are diastereotopic

The second common situation occurs when two groups on one carbon of a double bond are the same and the two groups on the other carbon are different:

As you should verify, the substitution test on the red hydrogens gives *E,Z* isomers, which are diastereomers.

Just as Fig. 6.11 (p. 249) can be used to summarize the relationships between isomeric molecules, Fig. 10.2 can be used to summarize the relationships of groups within a molecule. Notice the close analogy between the relationships between *different molecules* and the relationships between *groups within a molecule*. Just as two broad classes of isomers are based on connectivity—constitutional isomers (isomers with different connectivities) and stereoisomers (isomers with the same connectivities)—the two broad classes of groups *within* a molecule are also based on connectivity: constitutionally equivalent groups and constitutionally nonequivalent groups. Just as two different *structures* can be identical, two *constitutionally equivalent groups* within the same molecule can be homotopic. Just as there are classes of stereoisomeric relationships between *molecules*—enantiomers and diastereomers—there are corresponding relationships between *constitutionally equivalent groups within molecules*—enantiotopic and diastereotopic relationships. Just as *enantiomers* have different reactivities only with chiral reagents, *enantiotopic groups* also have different reactivities only with chiral reagents. Just as *diastereomers* have different reactivities with any reagent, *diastereotopic groups* have different reactivities with any reagent.

FURTHER EXPLORATION 10.3
Symmetry Relationships among Constitutionally Equivalent Groups

FIGURE 10.2 A flowchart for classifying groups within molecules.

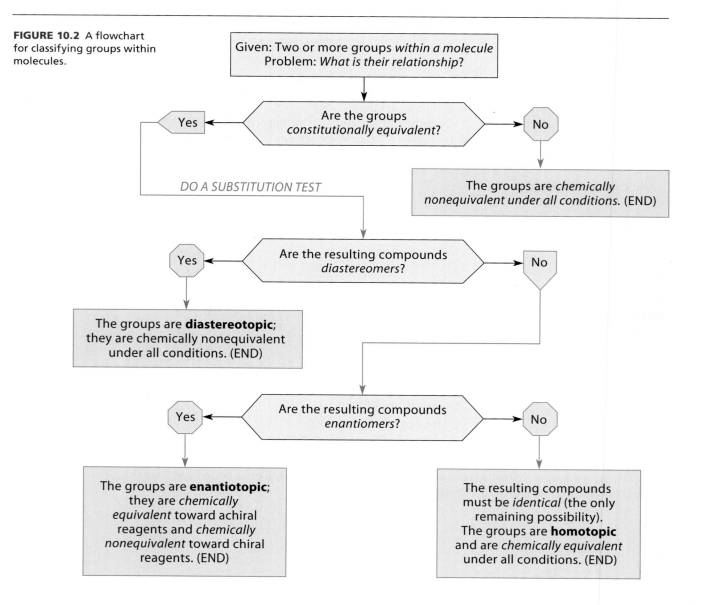

Let's summarize the answer to the question posed at the beginning of this section: When are two groups in a molecule chemically equivalent?

1. Constitutionally nonequivalent groups are chemically nonequivalent in all situations.
2. Homotopic groups are chemically equivalent in all situations.
3. Enantiotopic groups are chemically equivalent toward achiral reagents, but are chemically nonequivalent toward chiral reagents (such as enzymes).
4. Diastereotopic groups are chemically nonequivalent in all situations.

PROBLEM

10.33 For each of the following molecules, state whether the groups indicated by italic letters are constitutionally equivalent or nonequivalent. If they are constitutionally equivalent, classify them as homotopic, enantiotopic, or diastereotopic. (For cases in which more than two groups are designated, consider the relationships within each pair of groups.)

(a)

$$H_3C-\overset{\overset{\displaystyle H^a}{|}}{\underset{\underset{\displaystyle H^b}{|}}{C}}-Br$$

(b)

$$\underset{\underset{\displaystyle c}{H_3C}}{\overset{\overset{\displaystyle H}{}}{}}C=C\overset{\overset{\displaystyle a}{CH_3}}{\underset{\underset{\displaystyle b}{CH_3}}{}}$$

Classify each pair of methyl groups.

(c)

$$Ph-\overset{\overset{\displaystyle F^a}{|}}{\underset{\underset{\displaystyle F^b}{|}}{C}}-\overset{Me}{\underset{Et}{CH}}$$

(d) cyclohexane with H^a and H^b

(e)

$$H_3C-\overset{\overset{\displaystyle H^a}{}}{\underset{\underset{\displaystyle H^d}{}}{C}}-\overset{\overset{\displaystyle H^b}{}}{\underset{\underset{\displaystyle H^c}{}}{C}}-CH_3$$

(f) citric acid

$$HO_2C\overset{5}{}-\overset{\overset{\displaystyle HO}{|}}{\underset{\underset{\displaystyle H^d\ H^c}{}}{\underset{4}{C}}}\overset{}{\underset{3}{C}}\overset{\overset{\displaystyle CO_2H}{|}}{\underset{\underset{\displaystyle H^b\ H^a}{}}{\underset{2}{C}}}CO_2H$$

citric acid

Classify the relationship between each pair of labeled hydrogens as well as the relationship between carbon-2 and carbon-4.

B. Stereochemistry of the Alcohol Dehydrogenase Reaction

In this section we'll use the alcohol dehydrogenase reaction, which was introduced in Sec. 10.8, to demonstrate not only that a chiral reagent can distinguish between enantiotopic groups, but also *how this discrimination can be detected*. Suppose each of the enantiotopic α-hydrogens of ethanol is replaced in turn with the isotope deuterium (^2H, or D). Replacing one hydrogen, called the pro-(*R*)-hydrogen, with deuterium gives (*R*)-1-deuterioethanol; replacing the other, called the pro-(*S*)-hydrogen, gives (*S*)-1-deuterioethanol. Although ethanol itself is *not* chiral, the deuterium-substituted analogs *are* chiral; they are a pair of enantiomers:

H ◀——— pro-(*S*)-hydrogen

$$HO\overset{\overset{\displaystyle H}{|}}{\underset{\underset{\displaystyle H_3C}{}}{C}}-H ◀——— pro-(R)-hydrogen$$

ethanol

$$HO\overset{\overset{\displaystyle H}{|}}{\underset{\underset{\displaystyle H_3C}{}}{C}}-D$$

$$HO\overset{\overset{\displaystyle D}{|}}{\underset{\underset{\displaystyle H_3C}{}}{C}}-H$$

(R)-(+)-1-deuterioethanol   **(S)-(−)-1-deuterioethanol**

The deuterium isotope, then, provides a subtle way for the experimentalist to "label" the enantiotopic α-hydrogens of ethanol and thus to distinguish between them.

If the alcohol dehydrogenase reaction is carried out with (R)-1-deuterioethanol, *only the deuterium* is transferred from the alcohol to the NAD⁺:

$$+ \text{H}_2\text{O} \xrightleftharpoons[\text{enzyme}]{} + \text{H}_3\text{C}-\overset{\text{O}}{\overset{\|}{\text{C}}}-\text{H} + \text{H}_3\text{O}^+ \quad (10.63)$$

(R)-(+)-1-deuterioethanol   **NAD⁺**   **NADD**   **acetaldehyde**

However, if the alcohol dehydrogenase reaction is carried out on (S)-1-deuterioethanol, *only the hydrogen* is transferred.

$$+ \text{H}_2\text{O} \xrightleftharpoons[\text{enzyme}]{} + \text{H}_3\text{C}-\overset{\text{O}}{\overset{\|}{\text{C}}}-\text{D} + \text{H}_3\text{O}^+ \quad (10.64)$$

(S)-(–)-1-deuterioethanol   **NAD⁺**   **NADH**   **1-deuterioacetaldehyde**

These two experiments show that *the enzyme distinguishes between the two α-hydrogens of ethanol*. These results cannot be attributed to a primary deuterium isotope effect (Sec. 9.5D) because an isotope effect would cause the enzyme to transfer the hydrogen in preference to the deuterium in both cases. Although the isotope is used to detect the preference for transfer of one hydrogen and not the other, this experiment requires that even in the absence of the isotope the enzyme transfers the pro-(R)-hydrogen of ethanol.

The enzyme can distinguish between the two α-hydrogens of ethanol because the enzyme is chiral, and the two α-hydrogens are *enantiotopic*. This case, then, is an example of the principle that *enantiotopic groups react differently with chiral reagents* (Sec. 10.9A).

Another interesting point about the stereochemistry of the alcohol dehydrogenase reaction is shown in Eqs. 10.63 and 10.64: the deuterium (or hydrogen) that is removed from the isotopically substituted ethanol molecule is transferred specifically to one particular face, or side, of the NAD⁺ molecule. That is, the deuterium in the product NADD (color) occupies the position above the plane of the page. This result and the principle of microscopic reversibility (Sec. 4.9B, p. 175) require that if acetaldehyde and the NADD stereoisomer shown on the right of Eq. 10.63 were used as starting materials and the reaction run in reverse, only the deuterium should be transferred to the acetaldehyde, and (R)-1-deuterioethanol should be formed. Indeed, Eq. 10.63 *can* be run in reverse, and the experimental result is as predicted. No matter how many times the reaction runs back and forth, the H and the D on both the ethanol and the NADD molecules are never "scrambled"; they maintain their respective stereochemical positions. Because the R-group in NADH contains asymmetric carbons (Fig. 10.1, p. 486), the two CH₂ hydrogens in NADH are in fact *diastereotopic*; they are distinguished not only by the enzyme, but also in the absence of the enzyme (at least in principle), although without the enzyme they might not be distinguished as thoroughly.

PROBLEM

10.34   In each of the following cases, imagine that the two reactants shown are allowed to react in the presence of alcohol dehydrogenase. Tell whether the ethanol formed is chiral. If the ethanol is chiral, draw a line-and-wedge structure of the enantiomer that is formed.

(a) $H_3C-\overset{\displaystyle O}{\overset{\|}{C}}-D$ +   [structure with H, D, CONH$_2$, N, R]

(b) $H_3C-\overset{\displaystyle O}{\overset{\|}{C}}-D$ +   [structure with D, H, CONH$_2$, N, R]

10.10   OCTET EXPANSION AND OXIDATION OF THIOLS

A. Octet Expansion

Some of the chemistry of thiols is closely analogous to the chemistry of alcohols because sulfur and oxygen are in the same group of the periodic table. For oxidation reactions, however, this similarity disappears. Oxidation of an alcohol (Sec. 10.7) occurs at the *carbon* atom bearing the —OH group.

$$RCH_2OH \xrightarrow{\text{oxidation}} RCH{=}O \xrightarrow{\text{oxidation}} RCO_2H \qquad (10.65a)$$

However, oxidation of a thiol takes place at the *sulfur*. Although sulfur analogs of aldehydes, ketones, and carboxylic acids are known, they are *not* obtained by the simple oxidation of thiols:

$$RCH_2SH \xrightarrow{\text{oxidation}} \!\!\!\times\!\!\! RCH{=}S \xrightarrow{\text{oxidation}} \!\!\!\times\!\!\! R-\overset{\displaystyle O}{\overset{\|}{C}}-SH,\ \ R-\overset{\displaystyle S}{\overset{\|}{C}}-OH \quad (10.65b)$$

Oxidation occurs at sulfur because sulfur can have multiple oxidation states. Multiple oxidation states are common for elements in the third and greater periods of the periodic table. Sulfur and phosphorus are two important examples. Figure 10.3 (p. 496) shows Lewis structures of organic compounds containing these two elements in various oxidation states, along with the structures of the important inorganic compounds sulfuric acid and phosphoric acid.

The Lewis structures of some of the compounds in Fig. 10.3 have more than an octet of electrons around sulfur or phosphorus. Consider, for example, the structure of a sulfonic acid. Counting two electrons for each bond, a sulfonic acid has a total of 12 electrons around sulfur, although we can also draw a resonance structure that retains the octet. The octet structure, however, has a great deal of "separated" formal charge—that is, formal charges of opposite sign on adjacent atoms.

$$\left[\ R-\overset{\displaystyle :\!O\!:}{\underset{\displaystyle :\!O\!:}{\overset{\|}{\underset{\|}{S}}}}-\ddot{O}H \quad\longleftrightarrow\quad R-\overset{\displaystyle :\!\ddot{O}\!:^{-}}{\underset{\displaystyle :\!\ddot{O}\!:_{-}}{\overset{|}{\underset{|}{S^{2+}}}}}-\ddot{O}H\ \right] \qquad (10.66)$$

the uncharged structure   the octet structure has
has 12 electrons   formal charges
around sulfur   of opposite sign
on adjacent atoms

$$R-\overset{\cdot\cdot}{\underset{\cdot\cdot}{S}}H \qquad R-\overset{\cdot\cdot}{\underset{\cdot\cdot}{S}}-\overset{\cdot\cdot}{\underset{\cdot\cdot}{S}}-R \qquad R-\overset{\cdot\cdot}{\underset{\cdot\cdot}{S}}-\overset{\cdot\cdot}{\underset{\cdot\cdot}{O}}H$$

thiol   disulfide   sulfenic
                                   acid

$$R-\overset{\displaystyle :O:}{\overset{\|}{S}}-\overset{\cdot\cdot}{\underset{\cdot\cdot}{O}}H \qquad R-\overset{\displaystyle :O:}{\underset{:O:}{\overset{\|}{S}}}-\overset{\cdot\cdot}{\underset{\cdot\cdot}{O}}H \qquad HO-\overset{\displaystyle :O:}{\underset{:O:}{\overset{\|}{S}}}-\overset{\cdot\cdot}{\underset{\cdot\cdot}{O}}H$$

sulfinic   sulfonic   sulfuric
acid   acid   acid

Increasing sulfur oxidation level

$$R_3P: \qquad R_3P=\overset{\cdot\cdot}{\underset{\cdot\cdot}{O}}$$

phosphine   phosphine oxide

$$R-\overset{\displaystyle :O:}{\underset{:OH}{\overset{\|}{P}}}-\overset{\cdot\cdot}{O}H \qquad HO-\overset{\displaystyle :O:}{\underset{:OH}{\overset{\|}{P}}}-\overset{\cdot\cdot}{O}H$$

phosphonic acid   phosphoric acid

Increasing phosphorus oxidation level

FIGURE 10.3 Organic derivatives of sulfur and phosphorus along with the inorganic compounds sulfuric acid and phosphoric acid. These are arranged in order of increasing oxidation state of the sulfur or phosphorus. Sulfenic and sulfinic acids are not very common in organic chemistry. Sulfuric acid and phosphoric acid represent the highest oxidation states of sulfur (+6) and phosphorus (+5). One or more hydrogens of these acids can be replaced by R-groups to give organic esters; see, for example, Secs. 10.4C (sulfate esters) and 10.4E (phosphate esters).

The octet structure helps to explain the acidity of sulfonic acids. The large amount of positive charge on sulfur, for example, is one reason that sulfonic acids are strong acids ($pK_a \approx -3$). This positive charge stabilizes the conjugate-base anion (polar effect; Sec. 3.6C). In contrast, the importance of the uncharged resonance structure is indicated by the bond strength and bond length of the sulfur–oxygen bonds, which are stronger and shorter than sulfur–oxygen single bonds. Typically, we use the uncharged structure for convenience, but both structures are important.

An atom surrounded by more than an octet of electrons is said to undergo **octet expansion**. Why can octet expansion occur with elements in periods ≥ 3? Sulfur and phosphorus can accommodate more than eight valence electrons because, in addition to their occupied $3s$ and $3p$ orbitals, they have unoccupied $3d$ orbitals of relatively low energy. The overlap of an oxygen electron pair in a sulfonic acid with a sulfur $3d$ orbital is shown in Fig. 10.4; this

FIGURE 10.4 Bonding in the higher oxidation states of sulfur involves sulfur $3d$ orbitals. Wave peaks and troughs are shown in blue and green, respectively. In a sulfonic acid (RSO$_3$H; see Fig. 10.3), an electron pair on oxygen overlaps with one of several sulfur $3d$ orbitals. The overlap is indicated with gray lines. Notice that this overlap is not very efficient because the orbitals have different sizes and because half of the sulfur $3d$ orbital is directed away from the bond.

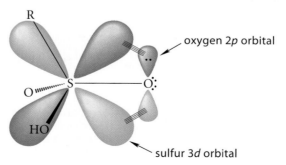

is essentially an orbital picture of the $S=O$ double bond. Notice that much of the sulfur $3d$ orbital is directed away from the oxygen $2p$ orbital; thus, this additional bonding, although significant, is not as strong as a π bond formed by the overlap of two $2p$ orbitals. For this reason, the charge-separated resonance structure in Eq. 10.66 also has some importance.

Because principal quantum level 2 has no d orbitals, the $2p$–$3d$ type of bonding is not possible in second-period atoms such as carbon, nitrogen, oxygen, and fluorine. Therefore, these atoms cannot expand their octets, and violation of the octet rule cannot occur.

PROBLEMS

10.35 How many electrons are involved in the oxidation of 1-propanethiol to each of the following compounds. (See Fig. 10.3 for detailed Lewis structures.)

(a) 1-propanesulfonic acid, $CH_3CH_2CH_2SO_3H$

(b) 1-propanesulfenic acid, $CH_3CH_2CH_2S-OH$

10.36 (a) How many electrons are involved in the oxidation of triphenylphosphine ($Ph_3P:$) to triphenylphosphine oxide ($Ph_3P=O$)? Show your reasoning.

(b) Draw a resonance structure for triphenylphosphine oxide in which phosphorus obeys the octet rule.

(c) Draw a Lewis structure for trimethylamine oxide (Me_3NO). How is the bond order of the $N-O$ bond in this compound different from the bond order of the $P-O$ bond in the phosphorus analog, trimethylphosphine oxide? Justify your answer with appropriate resonance structures.

B. Oxidation of Thiols

The vigorous oxidation of thiols or disulfides with $KMnO_4$ or nitric acid (HNO_3) gives sulfonic acids.

$$CH_3CH_2CH_2CH_2-SH \xrightarrow{\text{conc. } HNO_3} CH_3CH_2CH_2CH_2-SO_3H \qquad (10.67)$$

1-butanethiol   **1-butanesulfonic acid**
(72–96% yield)

We have seen sulfonic acids before: recall that sulfonate esters (Sec. 10.4A) are derivatives of sulfonic acids. Other sulfonic acid chemistry is considered in Chapters 16 and 20.

Many thiols spontaneously oxidize to disulfides merely on standing in air (O_2). Thiols can also be converted into disulfides by mild oxidants such as I_2 in base or Br_2 in CCl_4:

$$2\,CH_3(CH_2)_4SH + I_2 + 2\,NaOH \longrightarrow CH_3(CH_2)_4S-S(CH_2)_4CH_3 + 2\,NaI + 2\,H_2O \qquad (10.68)$$

(70% yield)

$$2\,C_2H_5SH + Br_2 \longrightarrow C_2H_5S-SC_2H_5 + 2\,HBr \qquad (10.69)$$

ethanethiol   **diethyl disulfide**
(nearly quantitative yield)

A reaction like Eq. 10.68 can be viewed as a series of S_N2 reactions in which thiolate-anion nucleophiles, formed by thiol ionizations, react as nucleophiles first toward a halogen and then toward a sulfur electrophile.

$$R-\overset{..}{\underset{..}{S}}-H + {}^-:\!\overset{..}{\underset{..}{O}}H \rightleftharpoons R-\overset{..}{\underset{..}{S}}{}^{:-} + H_2\overset{..}{\underset{..}{O}}: \qquad (10.70a)$$

thiol ionization

$$R—\overset{..}{\underset{..}{S}}{:}^{-} \quad \overset{\frown}{:}\overset{..}{\underset{..}{I}}—\overset{..}{\underset{..}{I}}{:} \quad \longrightarrow \quad R—\overset{..}{\underset{..}{S}}—\overset{..}{\underset{..}{I}}{:} \; + \; :\overset{..}{\underset{..}{I}}{:}^{-} \qquad (10.70b)$$

nucleophile / electrophile

$$R—\overset{..}{\underset{..}{S}}{:}^{-} \quad R—\overset{..}{\underset{..}{S}}\overset{\frown}{—}\overset{..}{\underset{..}{I}}{:} \quad \longrightarrow \quad R—\overset{..}{\underset{..}{S}}—\overset{..}{\underset{..}{S}}—R \; + \; :\overset{..}{\underset{..}{I}}{:}^{-} \qquad (10.70c)$$

nucleophile / electrophile

When thiols and disulfides are present together in the same solution, an equilibrium among them is rapidly established. For example, if ethanethiol and dipropyl disulfide are combined, they react to give a mixture of all possible thiols and disulfides:

$$\text{EtSH} \quad + \quad \text{PrS—SPr} \quad \rightleftharpoons \quad \text{EtS—SPr} \quad + \quad \text{PrSH} \qquad (10.71a)$$

ethanethiol   **dipropyl disulfide**   **ethyl propyl disulfide**   **propanethiol**

$$\text{EtSH} \quad + \quad \text{EtS—SPr} \quad \rightleftharpoons \quad \text{EtS—SPr} \quad + \quad \text{PrSH} \qquad (10.71b)$$

ethanethiol   **ethyl propyl disulfide**   **diethyl disulfide**   **propanethiol**

Thiols and sulfides are very important in biology. Many enzymes contain thiol groups that have catalytically essential functions, and disulfide bonds in proteins help to stabilize their three-dimensional structures (Sec. 27.8A). The vulcanization of rubber (curing in the presence of sulfur) introduces disulfide bonds that increase the strength and rigidity of the rubber (Sec. 15.5).

PROBLEM

10.37 The rates of the reactions in Eqs. 10.71a–b are increased when the thiol is ionized by a base such as sodium ethoxide. Suggest a mechanism for Eq. 10.71a that is consistent with this observation, and explain why the presence of base makes the reaction faster.

10.11   SYNTHESIS OF ALCOHOLS

The preparation of organic compounds from other organic compounds by the use of one or more reactions is called **organic synthesis**. This section reviews the methods presented in earlier chapters for the synthesis of alcohols. All of these methods begin with alkene starting materials:

     1. Hydroboration–oxidation of alkenes (Sec. 5.4B).

$$3\,RCH{=}CH_2 + BH_3 \xrightarrow[\text{THF}]{} (RCH_2CH_2)_3B \xrightarrow{H_2O_2/NaOH} 3\,RCH_2CH_2OH \qquad (10.72)$$

     2. Oxymercuration-reduction of alkenes (Sec. 5.4A).

$$R{-}CH{=}CH_2 + Hg(OAc)_2 \xrightarrow{H_2O} R{-}\overset{\overset{\displaystyle OH}{|}}{CH}{-}CH_2{-}HgOAc \xrightarrow{NaBH_4} R{-}\overset{\overset{\displaystyle OH}{|}}{CH}{-}CH_3 \qquad (10.73)$$

Acid-catalyzed hydration of alkenes (Sec. 4.9B) is used industrially to prepare certain alcohols, but this is not an important laboratory method. In principle, the S_N2 reaction of ^-OH with primary alkyl halides can also be used to prepare primary alcohols. This method is of little practical importance, however, because alkyl halides are generally prepared from alcohols themselves.

Some of the most important methods for the synthesis of alcohols involve the reduction of carbonyl compounds (aldehydes, ketones, or carboxylic acids and their derivatives), as well as the reactions of carbonyl compounds with Grignard or organolithium reagents. These methods are presented in Chapters 19, 20, and 21. A summary of methods used to prepare alcohols is found in Appendix V.

10.12 PLANNING AN ORGANIC SYNTHESIS: RETROSYNTHETIC ANALYSIS

Sometimes the synthesis of a compound from a given starting material can be completed with a single reaction. More often, however, the conversion of one compound into another requires more than one reaction. A synthesis involving a sequence of several reactions is called a **multistep synthesis**. Planning a multistep synthesis involves a type of reasoning that we're going to examine here.

The molecule to be synthesized is called the **target molecule**. To assess the best route to the target molecule from the starting material, you should take the same approach that a military officer might take in planning an assault—namely, *work backward from the target toward the starting material*. The officer does *not* send out groups of soldiers from their present position in random directions, hoping that one group will eventually capture the objective. Rather, the officer considers the most useful point from which the *final* assault on the target can be carried out—say, a hill or a farmhouse—and plans the approach to that objective, again by working backward. Similarly, in planning an organic synthesis, you should *not* try reactions at random. Rather, you should first assess what compound can be used as the immediate precursor of the target. You should then continue to work backward from this precursor step-by-step until the route from the starting material becomes clear. Sometimes more than one synthetic route will be possible. In such a case, each synthesis is evaluated in terms of yield, limitations, expense, and so on. It is not unusual to find (both in practice and on examinations) that one synthesis is as good as another.

The process of working backward from the target is called **retrosynthetic analysis**. Study Problem 10.7 illustrates this process.

STUDY PROBLEM 10.7

Outline a synthesis of hexanal from 1-hexene.

$$CH_3CH_2CH_2CH_2CH{=}CH_2 \longrightarrow CH_3CH_2CH_2CH_2CH_2CH{=}O$$

1-hexene   **hexanal**

SOLUTION To "outline a synthesis" means to suggest the reagents and conditions required for each step of the synthesis, along with the structure of each intermediate compound. Begin by working backward from hexanal, the target molecule. First, ask whether aldehydes can be directly prepared from alkenes. The answer is yes. Ozonolysis (Sec. 5.5) can be used to transform alkenes into aldehydes and ketones. However, ozonolysis breaks a carbon–carbon double bond and certainly would not work for preparing an aldehyde from an alkene *with the same number of carbon atoms*, because at least one carbon is lost when the double bond is broken. No other ways of preparing aldehydes directly from alkenes have been covered. The next step is to ask how aldehydes can be prepared from other starting materials. Only one way has been presented: the oxidation of primary alcohols (Sec. 10.7A). Indeed, the oxidation of a primary alcohol could be a final step in a satisfactory synthesis:

$$CH_3CH_2CH_2CH_2CH_2CH_2-OH \xrightarrow[CH_2Cl_2]{PCC} CH_3CH_2CH_2CH_2CH_2CH=O$$

1-hexanol   **hexanal**

Now ask whether it is possible to prepare 1-hexanol from 1-hexene; the answer is yes. Hydroboration–oxidation will convert 1-hexene into 1-hexanol. The synthesis is now complete:

$$CH_3CH_2CH_2CH_2CH=CH_2 \xrightarrow[THF]{BH_3} \xrightarrow[NaOH]{H_2O_2} CH_3CH_2CH_2CH_2CH_2CH_2-OH \xrightarrow[CH_2Cl_2]{PCC} CH_3CH_2CH_2CH_2CH_2CH=O$$

1-hexene   **1-hexanol**   **hexanal**

Notice carefully the use of retrosynthetic analysis—the process of working backward from the target molecule one step at a time. We can summarize the retrosynthetic process we've used as follows:

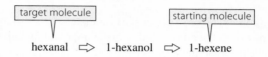

where the symbol ⇨ means "implies as starting material." Although this example, with a single intermediate compound, is relatively simple, the same process can be used to develop more complex syntheses, as we'll see in later chapters.

Working problems in organic synthesis is one of the best ways to master organic chemistry. It is akin to mastering a language: it is relatively easy to learn to read a language (be it a foreign language, English, or even a computer language), but writing it requires a more thorough understanding. Similarly, it is relatively easy to follow individual organic reactions, but to integrate them and use them out of context requires more understanding. One way to bring together organic reactions and study them systematically is to go back through the text and write a representative reaction for each of the methods used for preparing each functional group. (See Study Guide Link 5.2.) For example, which reactions can be used to prepare alkanes? To prepare carboxylic acids? Jot down some notes describing the stereochemistry of each reaction (if known) as well as its limitations—that is, the situations in which the reaction would not be expected to work. For example, dehydration of tertiary and secondary alcohols is a good laboratory method for preparing alkenes, but dehydration of primary alcohols is not. (Do you understand the reason for this limitation?) This process should be continued throughout future chapters.

The summary in Appendix V can be a starting point for this type of study. In Appendix V are presented lists of reactions, in the order that they occur in the text, which can be used to prepare compounds containing each functional group. In this summary the reactions have *not* been written out, because it is important for you to create your own summaries; such an exercise will help you to learn the reactions.

PROBLEM

10.38   Outline a synthesis of each of the following compounds from the indicated starting material. Begin each synthesis with a retrosynthetic analysis.

(a) 2-methyl-3-pentanol from 2-methyl-2-pentanol

(b) $CH_3CH_2CH_2CH_2CH_2CH_2D$ from 1-hexanol

(c) ⬡—CO_2H from ⬡=CH_2

(d) $CH_3CH_2CH_2\underset{\underset{CH_3}{|}}{C}HCH=O$ from $CH_3CH_2CH_2\underset{\underset{CH_3}{|}}{C}HCH_2Br$

10.13 PRODUCTION AND USE OF ETHANOL AND METHANOL

Ethanol A number of alcohols are important in commerce. None, however, has been affected by recent events as dramatically as ethanol. In 1980, the U.S. production of ethanol was 175 million gallons. Much of it was made industrially by the hydration of ethylene (Sec. 4.9B), which in turn comes from petroleum (Sec. 5.8).

$$H_2C=CH_2 + H-OH \xrightarrow[300\ °C]{H_3PO_4} CH_3CH_2OH \qquad (10.74)$$

ethylene   **ethanol**

In recent years, the U.S. production of ethanol has climbed rapidly, reaching 15 billion gallons in 2013 and likely to climb significantly higher. Most of this ethanol is produced by fermentation of the sugars in grain—mostly corn (Fig. 10.5). In 2013, the world production of ethanol was more than 20 billion gallons, with the United States providing about 57% and Brazil about 27% of this total. (Brazil, where plants such as sugar cane that produce fermentable sugars grow rapidly, has long operated on an ethanol-based fuel economy.) The increase in ethanol production in the United States has been driven by the demand for motor fuels that are not derived from hydrocarbons, by the phase-out of methyl *tert*-butyl ether (MTBE) and its replacement by ethanol as the main oxygen-containing component of gasoline, and by the notion that the use of ethanol and other plant-derived fuels will reduce the amount of CO_2 released into the atmosphere. (See the sidebar on p. 502.)

The relatively small amount of ethanol not used for fuel is used as a starting material for the preparation of other compounds and as a solvent for inks, fragrances, and the like. Chemically pure ethanol is subject to tight federal controls to ensure that it will not be used in beverages. In many cases the ethanol used in solvent applications is *denatured alcohol,* which is ethanol made unfit for human consumption by the addition of certain toxic additives such as methanol.

Beverage alcohol is produced by the fermentation of barley, grape juice, corn mash, or other sources of natural sugar. Beverage alcohol is not isolated; rather, alcoholic beverages are the mixtures of ethanol, water, and the natural colors and flavorings produced in the fermentation process and purified by sedimentation (as in wine) or distillation (as in brandy or whiskey). Industrial alcohol cannot be used legally to alter the alcoholic composition of beverages.

Ethanol is a drug, and, like many useful drugs, is toxic when consumed in excess. Ethanol is the most abused drug in the world.

FIGURE 10.5 The fermentation of the sugars in corn to ethanol is carried out in plants such as this one in Iowa. Many such facilities have come on line in the last 10–15 years.

Environmental Change and Biofuels

Global warming, caused by a rapid increase in atmospheric CO_2 levels, was discussed in Sec. 2.7 (see Fig. 2.10, p. 76). The increase in CO_2 levels has resulted from the combustion of fossil fuels—oil, coal, and natural gas. The environmental impact of fossil fuels and the political instability of many of the world's oil-producing regions have conspired to make the development of alternative fuels increasingly urgent. Ideally, the goal is to produce cheap and abundant fuels that will not, on combustion, increase the net CO_2 content of the atmosphere. For this reason, fuels derived from plants (biomass), termed generally *biofuels*, are attractive. Plants are rich sources of glucose in various polymerized forms, such as starch (from corn) and cellulose (from stalks and grasses). Cellulose is the single most abundant organic compound on Earth. When plants produce glucose in its various forms, they remove CO_2 from the atmosphere. The energy for the plant synthesis of glucose comes from the Sun through photosynthesis. If we add the equations for the photosynthesis of glucose, the fermentation of glucose to ethanol, and the combustion of ethanol as a fuel, the result is no net change in atmospheric CO_2:

$$6\,CO_2 + 6\,H_2O \xrightarrow{\text{light}} C_6H_{12}O_6 + 6\,O_2 \quad \text{(biosynthesis of glucose, } C_6H_{12}O_6\text{)}$$

$$C_6H_{12}O_6 \longrightarrow 2\,C_2H_5OH + 2\,CO_2 \quad \text{(anaerobic fermentation)}$$

$$2\,C_2H_5OH + 6\,O_2 \longrightarrow 4\,CO_2 + 6\,H_2O \quad \text{(combustion of ethanol)}$$

Sum: No Net Change

(10.75)

It is interesting to think of photosynthesis in the context of solar energy as a power source, which has also been of considerable interest. Photosynthesis harnessed to produce cellulose is nature's solar-energy collection and storage mechanism. However, this simple picture does not take into account the energy required to produce and transport the agricultural products, to run the fermenters, and to deliver the ethanol-containing products to the end users. Accordingly, the net energy gain from the use of various biofuels has been a matter of considerable debate.

Putting the carbon cycle of Eq. 10.75 into practice requires decisions on which fuels to use and what plants are to be their major sources. These issues are fraught with huge political and economic ramifications. For example, the diversion of grain to ethanol production has had a significant impact on the cost of food, because producers who raise poultry, pork, and beef on feed grain have seen their feed costs rise dramatically, and these costs are passed on to the consumer. Hence, finding other sources of fermentable cellulose, such as silage, waste products from wood, and grasses such as switchgrass (photo) is essential if ethanol is to become a major fuel source. Fermentation of switchgrass and related cellulose sources currently produces less than 1 million of the 15 billion gallons of ethanol produced in the United States.

The problem of energy is one of the greatest challenges ever to face the human race. Cheap and abundant energy—and ways to store it—would contribute significantly to solving many of the world's problems, such as poverty, the scarcity of food, and inadequate shelter. Increasingly expensive and scarce energy will lead to wars over limited resources and to environmental disaster. Biofuels will undoubtedly be only a part of the solution to the energy problem, but they will probably fill an important niche in the ultimate solution. Although it is tempting to be discouraged by the vast scale of this problem, it is encouraging to realize that immense opportunities undoubtedly await the scientifically trained citizens who can think in new, creative ways about the solutions.

Methanol Methanol is formed from a mixture of carbon monoxide and hydrogen, called synthesis gas, at high temperature over a special catalyst.

$$\underbrace{CO + H_2}_{\text{synthesis gas}} \xrightarrow[\substack{100\text{–}600\text{ atm} \\ 250\text{–}400\,°C}]{\text{Cr-Zn catalyst}} \underset{\textbf{methanol}}{CH_3OH}$$

(10.76)

Synthesis gas comes from the partial oxidation of methane, which is, in turn, derived from natural gas, the cracking of hydrocarbons (Sec. 5.8), or the gasification of coal. As oil supplies dwindle, coal will assume increasing importance as a fossil fuel and could become a major source for methane. Methane produced by the fermentation of biological waste can also provide a link for methanol production to biofuels.

In 2012, about 17 billion gallons of methanol were produced globally. Important uses of methanol include its oxidation to formaldehyde ($H_2C{=}O$) and its reaction with carbon monoxide over special catalysts to give acetic acid (CH_3CO_2H). A rapidly growing use of methanol is new processes for the production of alkenes, so-called MTO (methanol-to-olefin) technologies. A bright spot in methanol's future may be its use in the production of biodiesel fuel. With an octane rating of 116, methanol itself also has a largely unrealized potential for use as a motor fuel. (It has been used as a fuel in Formula One racing engines for years.)

In the 1990s, methanol became an important compound in the fight against urban automotive air pollution. Prior to 1990, efforts to control automotive air pollution were focused on the automobile itself—thus the "catalytic converter." In 1990, a new strategy for reducing automotive air pollution was mandated by the Clean Air Act amendments: to add chemicals ("additives") to gasoline itself. Chief among these additives were the so-called oxygenates, and the two most important of these were ethanol and methyl *tert*-butyl ether (MTBE). Methanol is one of the two starting materials in the industrial synthesis of MTBE.

$$\underset{\textbf{2-methylpropene}}{\begin{array}{c}H_3C\\\diagdown\\C{=}CH_2\\\diagup\\H_3C\end{array}} + \underset{\textbf{methanol}}{CH_3OH} \xrightarrow{\;H_2SO_4\;} \underset{\substack{\textbf{methyl }\textit{tert}\textbf{-butyl ether}\\\textbf{(MTBE)}}}{H_3C-\overset{\displaystyle OCH_3}{\underset{\displaystyle CH_3}{\overset{|}{\underset{|}{C}}}}-CH_3} \qquad (10.77)$$

The use of MTBE in gasoline significantly reduced urban air pollution from automobile exhausts. It rapidly became one of the top ten industrial organic chemicals, and it took methanol along for the ride. New methanol plants were built to feed the demand for MTBE. Then trouble started for MTBE. It was found in groundwater in California and Maine, and the source of the chemical was leakage from underground storage tanks. An advisory panel of the Environmental Protection Agency (EPA) recommended in August 1999 that Congress move to reduce substantially the use of MTBE in gasoline. A controversy developed about whether MTBE should be banned, but California mandated a phase-out by 2002. Many other states followed suit, and MTBE has largely been discontinued as a gasoline additive in the United States and Western Europe, leaving ethanol as the primary oxygenate used in gasoline. Ethanol producers are happy with this turn of events, but methanol producers were left with excess capacity. Many methanol plants were closed, and production in the United States decreased over the ensuing period. It is now rebounding somewhat because of the other uses for methanol.

KEY IDEAS IN CHAPTER 10

- Alcohols ($pK_a \approx 15$–16) and thiols ($pK_a \approx 10$) are weak acids. The conjugate bases of alcohols are called alkoxides or alcoholates, and the conjugate bases of thiols are called mercaptides or thiolates.

- Alkoxides are formed by the reaction of alcohols with strong bases such as sodium hydride (NaH) or by reaction with alkali metals. Solutions of thiolates can be formed by the reaction of thiols with NaOH in alcohol solvents.

- The acidity of alcohols in solution is reduced (the pK_a is increased) by branching near the —OH group and increased by electron-withdrawing substituents.

- Alcohols and thiols are weak Brønsted bases and react with strong acids to form positively charged conjugate-acid cations that have negative pK_a values. Alcohols are considerably more basic than thiols in solution.

- Several reactions of alcohols involve breaking the C—O bond. A unifying principle in all these reactions is that the —OH group, itself a very poor leaving group, is converted into a good leaving group.

1. In dehydration and in the reaction with hydrogen halides, the —OH group of an alcohol is converted by protonation into a good leaving group. The protonated —OH is eliminated as water (in dehydration) to give an alkene, or is displaced by halide (in the reaction with hydrogen halides) to give an alkyl halide.

2. In the reactions of alcohols with $SOCl_2$ and with Ph_3PBr_2, the reagents themselves convert the —OH into a good leaving group, which is displaced in a subsequent substitution reaction to form an alkyl halide.

3. In the reaction with a sulfonyl chloride, the alcohol is converted into a sulfonate ester, such as a tosylate or a mesylate. The sulfonate group serves as an excellent leaving group in substitution or elimination reactions.

• Phosphates and pyrophosphates are in many cases utilized as leaving groups in biology. They are relatively inert in solution but are activated by hydrogen bonding or metal-ion bonding within enzyme active sites.

• In an organic reaction, a compound has been oxidized when the product into which it is converted has a greater (more positive, less negative) oxidation number. A compound has been reduced when the product into which it is converted has a smaller (less positive, more negative) oxidation number. The change in oxidation number from a reactant to the corresponding product is the same as the number of electrons lost or gained in the corresponding half-reaction.

• Alcohols can be oxidized to carbonyl compounds with Cr(VI). Primary alcohols are oxidized to aldehydes (in the absence of water) or carboxylic acids (in the presence of water), secondary alcohols are oxidized to ketones, and tertiary alcohols are not oxidized. Primary alcohols are oxidized to carboxylic acids with alkaline $KMnO_4$ followed by acid.

• Thiols are oxidized at sulfur rather than at the α-carbon. Disulfides and sulfonic acids are two common oxidation products of thiols.

• Sulfur, phosphorus, and other atoms in periods higher than 3 can violate the octet rule by octet expansion, which involves additional bonding between a d orbital and a lone pair on an attached atom. Sulfonic acid and phosphoric acid derivatives are examples of compounds in which sulfur and phosphorus have expanded octets.

• Just as two isomeric molecules can be classified as constitutional isomers or stereoisomers by analyzing their connectivities, two groups within a molecule can be classified as constitutionally equivalent or constitutionally nonequivalent. In general, constitutionally nonequivalent groups are chemically distinguishable. Constitutionally equivalent groups are of three types.

1. Homotopic groups, which are chemically equivalent under all conditions.

2. Enantiotopic groups, which are chemically nonequivalent in reactions with chiral reagents such as enzymes, but are chemically equivalent in reactions with achiral reagents.

3. Diastereotopic groups, which are chemically nonequivalent under all conditions.

• An example of a naturally occurring oxidation is the conversion of ethanol into acetaldehyde by NAD^+; this reaction is catalyzed by the enzyme alcohol dehydrogenase. Because the enzyme and coenzyme are chiral, one of the enantiotopic α-hydrogens of ethanol is selectively removed in this reaction.

• A useful strategy for organic synthesis is to work backward from the target compound systematically one step at a time (retrosynthetic analysis).

 REACTION REVIEW *For a summary of reactions discussed in this chapter, see the* **Reaction Review** *section of Chapter 10 in the* Study Guide and Solutions Manual.

ADDITIONAL PROBLEMS

10.39 Give the product expected, if any, when 1-butanol (or other compound indicated) reacts with each of the following reagents.

(a) concentrated aqueous HBr, H_2SO_4 catalyst, heat

(b) cold aqueous H_2SO_4

(c) pyridinium chlorochromate (PCC) in CH_2Cl_2

(d) NaH

(e) product of part (d) + CH_3I in DMSO

(f) *p*-toluenesulfonyl chloride in pyridine

(g) PrMgBr in anhydrous ether

(h) $SOCl_2$ in pyridine

(i) Ph_3PBr_2 in DMF

(j) product of part (a) + Mg in dry ether

(k) product of part (f) + K^+ $(CH_3)_3C$—O^- in $(CH_3)_3COH$

(l) triflic anhydride and pyridine

(m) product of (l) followed by anhydrous K^+ F^- in acetonitrile

10.40 Give the product expected, if any, when 2-methyl-2-propanol (or other compound indicated) reacts with each of the following reagents.

(a) concentrated aqueous HCl

(b) CrO_3 in pyridine

(c) H_2SO_4, heat

(d) Br_2 in CH_2Cl_2 (dark)

(e) potassium metal

(f) methanesulfonyl chloride in pyridine

(g) product of part (f) + NaOH in DMSO

(h) product of part (e) + product of part (a)

10.41 Give the structure of a compound that satisfies the criterion given in each case. (There may be more than one correct answer.)

(a) a seven-carbon tertiary alcohol that yields a *single* alkene after acid-catalyzed dehydration

(b) an alcohol that, after acid-catalyzed dehydration, yields an alkene that, in turn, on ozonolysis and treatment with $(CH_3)_2S$, gives only benzaldehyde, Ph—CH=O

(c) an alcohol that gives the same product when it reacts with $KMnO_4$ as is obtained from the ozonolysis of *trans*-3,6-dimethyl-4-octene followed by treatment with H_2O_2.

10.42 The following triester is a powerful explosive, but is also a medication for angina pectoris (chest pain). From what inorganic acid and what alcohol is it derived?

$$\underset{CH_2\text{——}CH\text{——}CH_2}{\overset{ONO_2\quad ONO_2\quad ONO_2}{|\qquad\quad |\qquad\quad |}}$$

10.43 In each compound, identify (1) the diastereotopic fluorines, (2) the enantiotopic fluorines, (3) the homotopic fluorines, and (4) the constitutionally nonequivalent fluorines.

(a)
$$\underset{\underset{F}{|}\quad\underset{F}{|}}{\overset{\overset{F}{|}\quad\overset{F}{|}}{F\text{—}C\text{—}C\text{—}H}}$$

(b)
$$\underset{F}{\overset{F}{}}C=C\underset{CH_3}{\overset{F}{}}$$

(c)
$$\underset{\underset{F}{|}\quad\underset{CH_3}{|}}{\overset{\overset{F}{|}\quad\overset{F}{|}}{H\text{—}C\text{—}C\text{—}H}}$$

10.44 How many chemically nonequivalent sets of hydrogens are in each of the following structures?

(a)
$$\underset{H}{\overset{CH_3CH_2}{}}C=C\underset{CH_2CH_3}{\overset{H}{}}$$

(b)
$$CH_3O\text{—}CH_2\text{—}\underset{\underset{Cl}{|}}{\overset{\overset{Br}{|}}{CH}}$$

(c) [cyclohexane with Cl and Cl substituents]

(d)
$$CH_3CH_2O\text{—}\underset{\underset{OCH_2CH_3}{|}}{CH}\text{—}CH_3$$

10.45 When *tert*-butyl alcohol is treated with $H_2^{18}O$ (water containing the heavy oxygen isotope ^{18}O) in the presence of a small amount of acid, and the *tert*-butyl alcohol is re-isolated, it is found to contain ^{18}O. Write a curved-arrow mechanism consisting of Brønsted acid–base reactions, Lewis acid–base associations, and Lewis acid–base dissociations that explains how the isotope is incorporated into the alcohol.

10.46 Arrange the compounds or ions within each set in order of increasing acidity (decreasing pK_a) in solution. Explain your reasoning.

(a) propyl alcohol, isopropyl alcohol, *tert*-butyl alcohol, 1-propanethiol

(b) 2-chloro-1-propanethiol, 2-chloroethanol, 3-chloro-1-propanethiol

(c)
$CH_3\ddot{N}H\text{—}CH_2CH_2\text{—}\ddot{O}H$, $CH_3\ddot{N}H\text{—}CH_2CH_2CH_2\text{—}\ddot{O}H$,

$(CH_3)_3\overset{+}{N}\text{—}CH_2CH_2\text{—}\ddot{O}H$

(d)
$CH_3\ddot{O}\text{—}CH_2CH_2\text{—}\ddot{O}H$, $CH_3\ddot{O}\text{—}CH_2CH_2CH_2\text{—}\overset{+}{O}H$,

$\overset{-}{:}\ddot{O}\text{—}CH_2CH_2\text{—}\ddot{O}H$, $CH_3CH_2CH_2\text{—}\ddot{O}H$

(e)
$CH_3CH_2\ddot{O}H$, $CH_3\underset{\overset{|}{+}}{\overset{\overset{H}{|}}{S}}CH_3$, $CH_3CH_2\underset{\overset{|}{+}}{\overset{\overset{H}{|}}{O}}CH_2CH_3$

10.47 (a) When 1-propanol containing deuterium (D, or ^2H) rather than hydrogen at the oxygen, PrOD, is treated with an excess of H_2O containing a catalytic amount of NaOH, 1-propanol is formed rapidly. Write a curved-arrow mechanism for this reaction. Where does the deuterium go?

(b) How would one prepare PrOD from 1-propanol?

10.48 Indicate whether each of the following transformations is an oxidation, a reduction, or neither, and how many electrons are involved in each oxidation or reduction process.

(a)
$$\underset{CH_3CH_2CHCH=O}{\overset{\overset{OH}{|}}{}} \longrightarrow \underset{CH_3CH_2CCH_2OH}{\overset{\overset{O}{\|}}{}}$$

(b) [benzene ring with OCH3] $\longrightarrow$ [cyclohexenone] + CH_3OH

(c)

$$H_3C-\overset{\overset{\displaystyle O}{\|}}{C}-NH_2 \longrightarrow CH_3NH_2 + O=C=O$$

(d)

$$2\,Ph-\overset{\overset{\displaystyle CH_3}{|}}{\underset{\underset{\displaystyle CH_3}{|}}{C}}-H \longrightarrow Ph-\overset{\overset{\displaystyle CH_3}{|}}{\underset{\underset{\displaystyle CH_3}{|}}{C}}-\overset{\overset{\displaystyle CH_3}{|}}{\underset{\underset{\displaystyle CH_3}{|}}{C}}-Ph$$

(e) $(CH_3)_3C-Cl \longrightarrow (CH_3)_3C^+ + Cl^-$

10.49 Outline a synthesis for the conversion of enantiomerically pure (R)-CH₃CH₂CHD—OH into each of the following isotopically labeled compounds. Assume that Na¹⁸OH or H₂¹⁸O is available as needed.

(a) (S)-CH₃CH₂CHD—¹⁸OH

(b) (R)-CH₃CH₂CHD—¹⁸OH (*Hint:* Two inversions of configuration correspond to a retention of configuration.)

10.50 Outline a synthesis for each of the following compounds from the indicated starting material and any other reagents.

(a)

$$H_3C-\overset{\overset{\displaystyle CH_3}{|}}{\underset{\underset{\displaystyle D}{|}}{C}}-CH_2CH_2CH_3 \quad \text{from 2-methyl-1-pentanol}$$

(b) cyclopentyl—CH₂CO₂H from cyclopentyl—CH=CH₂

cyclopentylethylene

(c) cyclopentyl—CH₂CH=O from cyclopentylethylene

(d) cyclopentyl—CO₂H from cyclopentylethylene

(e) (±)- [CH₃, Br cyclopentane] from [CH₃ cyclopentene]

(f) ethylcyclopentane from cyclopentylethylene

10.51 Ethyl triflate is much more reactive than ethyl mesylate toward nucleophiles in Sₙ2 reactions.

$$CH_3CH_2-O-\overset{\overset{\displaystyle O}{\|}}{\underset{\underset{\displaystyle O}{\|}}{S}}-CF_3 \qquad CH_3CH_2-O-\overset{\overset{\displaystyle O}{\|}}{\underset{\underset{\displaystyle O}{\|}}{S}}-CH_3$$

ethyl triflate (more reactive)   **ethyl mesylate** (less reactive)

(a) Give the structures of all of the products formed when each compound reacts with potassium iodide in acetone (a polar aprotic solvent).

(b) Explain in detail why the triflate anion is a much weaker base than the mesylate anion.

(c) Explain why the relative reactivities of ethyl mesylate and ethyl triflate correlate inversely with the relative basicities of the two anions in (b).

10.52 How many grams of CrO₃ are required to oxidize 10 g of 2-heptanol to the corresponding ketone?

10.53 The primary alcohol 2-methoxyethanol, CH₃O—CH₂CH₂—OH, can be oxidized to the corresponding carboxylic acid with aqueous nitric acid (HNO₃). The by-product of the oxidation is nitric oxide, NO. How many moles of HNO₃ are required to oxidize 0.1 mole of the alcohol?

10.54 A police officer, Lawin Order, has detained a driver, Bobbin Weaver, after observing erratic driving behavior. Administering a breathalyzer test, Officer Order collects 52.5 mL of expired air from Weaver and finds that the air reduces 0.507 × 10⁻⁶ mole of K₂Cr₂O₇ to Cr³⁺. Assuming that 2100 mL of air contains the same amount of ethanol as 1 mL of blood, calculate the "percent blood alcohol content" (BAC), expressed as (grams of ethanol per mL of blood) × 100. If 0.08% BAC is the lower limit of legal intoxication, should Officer Order make an arrest?

10.55 Consider the following well-known reaction of glycols (vicinal diols).

$$R-\overset{\overset{\displaystyle OH}{|}}{CH}-\overset{\overset{\displaystyle OH}{|}}{CH}-R + IO_4^- \longrightarrow 2\,R\overset{\overset{\displaystyle O}{\|}}{CH} + IO_3^- + H_2O$$

a glycol   **periodate**

(a) In this reaction, what species is oxidized? What species is reduced? Explain how you know.

(b) How many electrons are involved in the oxidation half-reaction? In the reduction half-reaction? Explain how you arrived at your answer.

(c) How many moles of periodate are required to react completely with 0.1 mole of a glycol?

10.56 Chemist Stench Thiall, intending to prepare the disulfide A, has mixed one mole each of 1-butanethiol and 2-octanethiol with I₂ and base. Stench is surprised at the low yield of the desired compound and has come to you for an explanation. Explain why Stench should not have expected a good yield in this reaction.

$$CH_3CH_2CH_2CH_2-S-S-\overset{\overset{\displaystyle }{|}}{\underset{\underset{\displaystyle CH_3}{|}}{CH}}(CH_2)_5CH_3$$

A

10.57 Compound A, C₇H₁₄, decolorizes Br₂ in CH₂Cl₂ and reacts with BH₃ in THF followed by H₂O₂/OH⁻ to yield compound B. When treated with KMnO₄, then H₃O⁺, B is oxidized to a carboxylic acid C that can be resolved into enantiomers. Compound A, after ozonolysis and workup with H₂O₂, yields the same compound D as is formed by the oxidation of 3-hexanol with chromic acid. Identify compounds *A*, *B*, *C*, and *D*.

10.58 In a laboratory are found two different compounds: *A* (melting point −4.7 °C) and *B* (melting point −1 °C). Both compounds have the same molecular formula ($C_7H_{14}O$), and both can be resolved into enantiomers. Both compounds give off a gas when treated with NaH. Treatment of either *A* or *B* with tosyl chloride in pyridine yields a tosylate ester, and treatment of either tosylate with potassium *tert*-butoxide gives a mixture of the same two alkenes, *C* and *D*. However, reaction of the tosylate of *A* with potassium *tert*-butoxide to give these alkenes is noticeably slower than the corresponding reaction of the tosylate of *B*. When either *optically active A* or *optically active B* is subjected to the same treatment, *both* alkene products *C* and *D* are optically active. Treatment of either *C* or *D* with H_2 over a catalyst yields methylcyclohexane. Identify all unknown compounds and explain your reasoning.

10.59 Complete each of the following reactions by giving the principal organic product(s) formed in each case. Explain your reasoning.

(a)

$$\xrightarrow[\text{pyridine}]{\text{tosyl chloride}} \xrightarrow[\text{DMSO}]{\text{NaBr}}$$

(b)

$$-CH_2CH_2CH_2OH + HI \xrightarrow{H_3PO_4}$$

(c)

$$i\text{-Pr}-SH + CH_3O^- \xrightarrow[\text{CH}_3\text{OH}]{} \xrightarrow{\begin{array}{c}\text{dimethyl}\\\text{sulfate}\end{array}}$$

(d) $CH_3CHCH_2CH_2Ph + SOCl_2 \longrightarrow$
    $\quad\;|$
    $\quad OH$

(e)

$$+ Ph_3PCl_2 \longrightarrow$$

menthol

(Give the stereochemistry of the product.)

(f)

$$\xrightarrow[\text{heat}]{H_3PO_4}$$

(g)

$$\xrightarrow{\text{conc. HBr}} \xrightarrow[\text{(CH}_3)_3\text{COH}]{K^+ \ (CH_3)_3CO^-}$$

(h)

$$\text{3-methyl-1-butanethiol} + \text{Et}-S-S-\text{Et} \xrightarrow[\text{(catalyst)}]{^{-}\text{OH}}$$

10.60 (a) When the rate of oxidation of isopropyl alcohol to acetone is compared with the rate of oxidation of a deuterated derivative, a primary isotope effect (Sec. 9.5D) is observed (see part (a) of Fig. P10.60). Assuming that the mechanism is the same as the one shown in Eqs. 10.52a–c (p. 484), which step of this mechanism is rate-limiting?

(b) When either (*S*)- or (*R*)-1-deuterioethanol is oxidized with PCC in CH_2Cl_2, the same product mixture results and it contains significantly more of the deuterated aldehyde *A* than the undeuterated aldehyde *B* (see part (b) of Fig. P10.60). Explain why the deuterated aldehyde is the major product.

(c) Explain why, in the oxidation of (*R*)-1-deuterioethanol with alcohol dehydrogenase and NAD^+, *none* of the deuterated aldehyde is obtained.

(a)

$$H_3C-\underset{\underset{H}{|}}{\overset{\overset{CH_3}{|}}{C}}-OH \xrightarrow{H_2CrO_4} \underset{H_3C}{\overset{H_3C}{>}}C=O \qquad \text{relative rate} = 6.6$$

$$H_3C-\underset{\underset{D}{|}}{\overset{\overset{CH_3}{|}}{C}}-OH \xrightarrow{H_2CrO_4} \underset{H_3C}{\overset{H_3C}{>}}C=O \qquad \text{relative rate} = 1.0$$

(b)

$$H_3C-\underset{\underset{D}{|}}{\overset{\overset{H}{|}}{C}}-OH \xrightarrow[\text{CH}_2\text{Cl}_2]{\text{PCC}} H_3C-\underset{\underset{D}{|}}{C}=O + H_3C-\underset{\underset{H}{|}}{C}=O$$

1-deuterioethanol $\qquad\qquad\quad$ *A* $\qquad\quad$ *B*
$\qquad\qquad\qquad\qquad\qquad\qquad$ (mostly)

Figure P10.60

10.61 The *Swern oxidation*, shown in Fig. P10.61, is a very mild two-step procedure for the oxidation of primary and secondary alcohols.

(a) How many electrons are involved in this oxidation? Explain.

(b) What is the oxidizing agent?

(c) The following intermediate *A* is formed prior to the addition of the base triethylamine.

$$RCH_2-O-\overset{+}{\underset{..}{S}}(CH_3)_2 \ Cl^-$$

A

Triethylamine reacts with intermediate *A* to give the product aldehyde and the by-product dimethyl sulfide. When an isotopically substituted alcohol RCD_2OH is subjected to the Swern oxidation, a deuterium-substituted aldehyde $R-CD=O$ is formed, and the by-product dimethyl sulfide contains one atom of deuterium per molecule, $H_3C-S-CH_2D$. Write a curved-arrow mechanism for the reaction of intermediate *A* and the base triethylamine that accounts for the isotopic distribution in the products.

10.62 The enzyme *aconitase* catalyzes a reaction of the Krebs cycle (see Fig. P10.62), an important biochemical pathway. In this equation, $^*C = {}^{14}C$, a radioactive isotope of carbon, and each $-CO_2^-$ group is the conjugate base of a carboxylic acid group. Notice that the dehydration occurs toward the branch of citrate that does not contain the radioactive carbon. Imagine carrying out this dehydration in the laboratory using a strong acid such as H_3PO_4 or H_2SO_4. How would the product(s) of the reaction be expected to differ, if at all, from those of the enzyme-catalyzed reaction, and why?

10.63 When the hydration of fumarate is catalyzed by the enzyme *fumarase* in D_2O, only (2S,3R)-3-deuteriomalate is formed as the product. (Each $-CO_2^-$ group is the conjugate base of a carboxylic acid group.)

This reaction can also be run in reverse. By applying the principle of microscopic reversibility, predict the product (if any) when each of the following compounds is treated with fumarase in H_2O:

10.64 Monoamine oxidase (MAO) is an enzyme that catalyzes the oxidation of certain biologically important amines. One form of the enzyme catalyzes the following oxidation

Figure P10.61

Figure P10.62

of serotonin, an important neurotransmitter (Fig. P10.64; a number of antidepressants inhibit this reaction).

(a) How many electrons are lost in this oxidation?

(b) The pro-*R* hydrogen is removed from carbon in this reaction. Give the product formed when each of the following compounds undergoes this oxidation.

A   B

(c) One of the compounds in part (b) reacts about 10 times faster than the other. Which compound reacts faster? Why?

10.65 Buster Bluelip, a student repeating organic chemistry for the fifth time, has observed that alcohols can be converted into alkyl bromides by treatment with concentrated HBr. He has proposed that, by analogy, alcohols should be converted into nitriles (organic cyanides, R—C≡N) by treatment with concentrated HC≡N. Upon running the reaction, Bluelip finds that the alcohol does not react. Another student has suggested that the reason the reaction

failed is the absence of a strong acid catalyst. (HCN is a weak acid.) Following this suggestion, Bluelip runs the reaction in the presence of H_2SO_4 and again observes no reaction of the alcohol. Explain the difference in the reaction of alcohols with HBr and HCN—that is, why the latter fails but the former succeeds.

10.66 Primary alcohols, when treated with H_2SO_4, do not dehydrate to alkenes under the usual conditions. However, they do undergo another type of "dehydration" to form ethers if heated strongly in the presence of H_2SO_4. The reaction of ethyl alcohol to give diethyl ether is typical:

$$2\,\text{EtOH} \xrightarrow[140\,°C]{H_2SO_4} \text{Et—O—Et} + H_2O$$

ethanol   **diethyl ether**

Using the curved-arrow notation, show mechanistically how this reaction takes place.

10.67 The reactions shown in Fig. P10.67 all involve conversion of the alcohol oxygen into a good leaving group, followed by the reaction of the resulting compound with a nucleophile provided by the reagent. Write curved-arrow mechanisms for each reaction.

[*Hints:* In reaction (b), the phosphorus of triphenylphosphite is nucleophilic. Identify the intermediate *X* and show how it leads to the products. In reaction (c), let triphenyl-

---

HO— (structure of serotonin with indole ring and CH₂ side chain)

$\ddot{N}H_2$

H H

$\xrightarrow{\text{MAO} + \text{a coenzyme}}$

HO— (structure of oxidized product)

$\overset{+}{N}H_2$

H

serotonin

Figure P10.64

(a)   OH
         |
CH₃CH₂CHCH₂CH₃ + PBr₃ ⟶

         Br
         |
CH₃CH₂CHCH₂CH₃ + HO—PBr₂

(b)

(PhO)₃P: + H₃C—I ⟶ [*X*] $\xrightarrow{\text{(cyclohexanol)}}$ (cyclohexyl)—I + (PhO)₂P—CH₃ + PhOH

triphenyl-   **an ionic** $:\ddot{O}:^-$ $pK_a = 10$
phosphite   **compound**

(c)

Ph₃P: + CBr₄ + (pentyl chain)—OH ⟶ (pentyl chain)—Br + Ph₃P⁺ + HCBr₃

triphenyl-   **triphenyl-** **bromoform**
phosphine   **phosphine** $(pK_a \approx 24)$
                    oxide

Figure P10.67

phosphine act as a nucleophile and a bromine of CBr_4 as the electrophile to form an ionic intermediate, which then reacts with the alcohol. Note that bromoform ($HCBr_3$) has $pK_a \approx 24$.]

10.68 Using the curved-arrow notation, give a mechanism for each of the known conversions shown in Fig. P10.68. (If necessary, re-read Study Problem 10.1, p. 461.)

   The reaction given in part (c) of Fig. P10.68 is very important in the manufacture of high-octane gasoline. [*Hint:* If deuterated isobutane $(CH_3)_3C$—D is used, the product is $(CH_3)_2CDCH_2C(CH_3)_3$.]

10.69 (a) Describe the π bond for the double bond in Ph_3P=CH_2; that is, what orbitals are involved on carbon and phosphorus?

   (b) Draw a resonance structure for the compound in part (a) that maintains the octet rule on both carbon and phosphorus.

(c) This compound is a strong base that reacts rapidly with water. Explain.

10.70 The trichloromethyl anion, $^-$:CCl_3, which is the conjugate-base anion of chloroform ($HCCl_3$), is stabilized not only by the polar effect of the chlorines but also by resonance:

$$\left[:\ddot{\underset{\underset{:\ddot{Cl}:}{|}}{Cl}}-\overset{-}{\ddot{C}}-\ddot{Cl}: \longleftrightarrow :\ddot{\underset{\underset{:\ddot{Cl}:}{|}}{Cl}}-C=\ddot{Cl}: \longleftrightarrow \begin{array}{l}\text{(similar structures}\\ \text{for the other}\\ \text{chlorines)}\end{array} \right]$$

Show the orbital overlap between carbon and chlorine that is implied by the resonance structures. (*Hint:* The chlorine provides the *empty* orbital.)

(a)

(74% yield)

(b)

(one of several alkene products)

(c)

$$(CH_3)_2C{=}CH_2 + (CH_3)_3C{-}H \xrightarrow[\text{(catalyst)}]{\text{HF}} (CH_3)_2CHCH_2C(CH_3)_3$$

2-methylpropene   **isobutane**   **2,2,4-trimethylpentane**

(d)

$$CH_3CH{=}CH_2 + HO-\overset{\overset{\displaystyle O}{\|}}{\underset{\underset{\displaystyle O}{\|}}{S}}-CF_3 \longrightarrow (CH_3)_2CH-O-\overset{\overset{\displaystyle O}{\|}}{\underset{\underset{\displaystyle O}{\|}}{S}}-CF_3$$

Figure P10.68

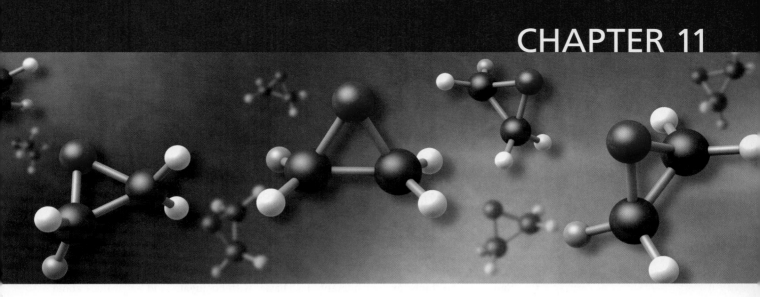

The Chemistry of Ethers, Epoxides, Glycols, and Sulfides

The compound classes discussed in this chapter were introduced in Sec. 8.1. The chemistry of ethers is closely intertwined with the chemistry of alkyl halides, alcohols, and alkenes. Ethers, however, are considerably less reactive than these other types of compounds. This chapter covers the synthesis of ethers and shows why the ether functional group is relatively unreactive.

Epoxides are heterocyclic compounds in which the ether linkage is part of a three-membered ring. Unlike ordinary ethers, epoxides are very reactive. This chapter also presents the synthesis and reactions of epoxides.

Because glycols are diols, it might seem more appropriate to consider them along with alcohols. Although glycols do undergo some reactions of alcohols, they have unique chemistry that is related to that of epoxides. For example, we'll see that epoxides are easily converted into glycols; and both epoxides and glycols can be easily prepared by the oxidation of alkenes.

Sulfides (thioethers), sulfur analogs of ethers, are also discussed briefly in this chapter. Although sulfides share some chemistry with their ether counterparts, they differ from ethers in the way they react in oxidation reactions, just as thiols differ from alcohols.

In this chapter we'll also learn the principles governing intramolecular reactions: reactions that take place between groups in the same molecule. These principles are central to an understanding of enzyme catalysis.

Finally, the strategy of organic synthesis will be revisited with a classification of reactions according to the way they are used in synthesis, and a further discussion of how to plan multistep syntheses.

11.1 BASICITY OF ETHERS AND SULFIDES

Ethers are not appreciably acidic. However, ethers are weakly basic and can accept a proton to form a conjugate-acid cation. Ethers are about as basic as alcohols, with conjugate-acid pK_a values about -2 to -3.

hydronium ion   **conjugate acid of ethanol**   **conjugate acid of diethyl ether**

pK_a -1.74 -2 to -3

Although ethers are weak bases, their basicities are important in many of their reactions that take place in acidic solution.

Sulfides, like thiols, are considerably less basic, with conjugate-acid pK_a values about -6 to -7. This reduced basicity is a consequence of the element effect: the S—H bond is weaker than the O—H bond.

conjugate acid of ethanethiol   **conjugate acid of diethyl sulfide**

$pK_a \approx -6$ to -7

Ethers are also important Lewis bases. For example, the Lewis acid–Lewis base complex of boron trifluoride and diethyl ether is stable enough that it can be distilled (bp 126 °C). This complex is a convenient way to handle BF_3.

(11.1)

boron trifluoride etherate

The solvation of Grignard reagents by ether solvents (Eq. 9.64, p. 431) is another example of the Lewis basicity of ethers.

Water and alcohols are also excellent Lewis bases, but their complexes with Lewis acids are typically unstable. The reason is that the protons on water and alcohols can react further and, as a result, the complex is destroyed.

(reacts further with EtOH in the same way)   (11.2)

PROBLEM

11.1 Arrange the ions in the following list in order of increasing acidity, and explain your reasoning.

      A   *B*   *C*   *D*

11.2   SYNTHESIS OF ETHERS AND SULFIDES

A. Williamson Ether Synthesis

Some ethers can be prepared from alcohols and alkyl halides. First, the alcohol is converted into an alkoxide (Sec. 10.1B):

$$Ph-\overset{\overset{\displaystyle \ddot{O}-H}{|}}{CH}-CH_3 \;+\; Na-H \;\xrightarrow[\text{THF}]{}\; Ph-\overset{\overset{\displaystyle :\ddot{O}^{\bar{}}\; Na^+}{|}}{CH}-CH_3 \;+\; H-H \qquad (11.3a)$$

an alkoxide
(conjugate base of
the alcohol)   **dihydrogen**

Then, the alkoxide is allowed to react as a nucleophile with a methyl halide, primary alkyl halide, or the corresponding sulfonate ester to give an ether.

$$Ph-\overset{\overset{\displaystyle :\ddot{O}^{\bar{}}\; Na^+}{|}}{CH}-CH_3 \;+\; :\ddot{I}-CH_3 \;\longrightarrow\; Ph-\overset{\overset{\displaystyle :\ddot{O}-CH_3}{|}}{CH}-CH_3 \;+\; Na^+ :\ddot{I}:^{\bar{}} \qquad (11.3b)$$

an alkoxide   an ether
(90% yield)

Some sulfides can be prepared in a similar manner from thiolates, the conjugate bases of thiols. In the following equation, a tosylate (Sec. 10.4A) is used as the alkylating agent.

$$CH_3(CH_2)_3-\ddot{\underset{\cdot\cdot}{S}}H \;\xrightarrow[\text{CH}_3\text{OH}]{^-\text{OH}}\; CH_3(CH_2)_3-\ddot{\underset{\cdot\cdot}{S}}^{\bar{}} \;\xrightarrow[\text{ethyl tosylate}]{CH_3CH_2-\ddot{O}Ts}\; CH_3(CH_2)_3-\ddot{\underset{\cdot\cdot}{S}}-CH_2CH_3 \;+\; ^{\bar{}}\ddot{O}Ts \qquad (11.4)$$

1-butanethiol   **1-butanethiolate**   **butyl ethyl sulfide, or
(1-ethylthio)butane**
(78% yield)

Both of these reactions are examples of the **Williamson ether synthesis**, which is the preparation of an ether by the alkylation of an alkoxide (and, by extension, a sulfide by the alkylation of a thiolate). This synthesis is named for Alexander William Williamson (1824–1904), who was professor of chemistry at the University of London.

The Williamson ether synthesis is an important practical example of the S_N2 reaction (Table 9.1, p. 385). In this reaction the conjugate base of an alcohol or thiol acts as a nucleophile toward the α-carbon of the alkyl halide; an ether is formed by the displacement of a halide or other leaving group.

$$R-\ddot{O}:^{\bar{}} \;+\; H_3C-\ddot{I}: \;\longrightarrow\; R-\ddot{O}-CH_3 \;+\; :\ddot{I}:^{\bar{}} \qquad (11.5)$$

Tertiary and many secondary alkyl halides cannot be used in this reaction. (Why?)

In principle, two different Williamson syntheses are possible for any ether with two different alkyl groups.

$$\left.\begin{array}{c} R^1-\ddot{O}:^{\bar{}} \;+\; R^2-X \\[4pt] \text{or} \\[4pt] R^2-\ddot{O}:^{\bar{}} \;+\; R^1-X \end{array}\right\} \;\longrightarrow\; R^1-\ddot{O}-R^2 \;+\; X^- \qquad (11.6)$$

The preferred synthesis is usually the one that involves the alkyl halide with the greater S_N2 reactivity. This point is illustrated by Study Problem 11.1.

STUDY PROBLEM   11.1

Outline a Williamson ether synthesis for *tert*-butyl methyl ether.

$$H_3C - \underset{\underset{CH_3}{|}}{\overset{\overset{CH_3}{|}}{C}} - O - CH_3$$

tert-butyl methyl ether

SOLUTION   From Eq. 11.6, two possibilities for preparing this compound are the reaction of methyl bromide with potassium *tert*-butoxide and the reaction of *tert*-butyl bromide with sodium methoxide. Only the former combination will work.

$$(CH_3)_3C - O^- \ K^+ + H_3C - Br \qquad\qquad CH_3O^- \ Na^+ + (CH_3)_3C - Br \qquad (11.7)$$

satisfactory reaction

does not occur; why?

$$(CH_3)_3C - O - CH_3$$

Do you know why sodium methoxide and *tert*-butyl bromide would not work? (See Sec. 9.5G.)

PROBLEMS

11.2   Complete the following reactions. If no reaction is likely, explain why.

(a) $(CH_3)_2CH - OH + Na \xrightarrow{\quad\quad} \overset{CH_3I}{\longrightarrow}$

(b)

$CH_3SH + NaOH \xrightarrow{\quad\quad} \overset{\diagdown\!\diagup\diagdown Cl}{\longrightarrow}$
    (1 equiv.)

(c) $CH_3O^- \ Na^+ + (CH_3)_3C - Br \xrightarrow[CH_3OH]{\quad\quad}$

(d) $EtO^- \ K^+ + (CH_3)_3CCH_2 - OTs \xrightarrow[\substack{25\ °C \\ EtOH}]{\quad\quad}$

11.3   Suggest a Williamson ether synthesis, if one is possible, for each of the following compounds. If no Williamson ether synthesis is possible, explain why.

(a)

$\bigcirc\!\!-CH_2CH_2 - O - CH_2CH_3$

(b) $(CH_3)_2CH - S - CH_3$

(c) $(CH_3)_3C - O - C(CH_3)_3$

B. Alkoxymercuration–Reduction of Alkenes

Another method for the preparation of ethers is a variation of oxymercuration–reduction, which is used to prepare alcohols from alkenes (Sec. 5.4A). If the *aqueous* solvent used in the oxymercuration step is replaced by an *alcohol* solvent, an ether instead of an alcohol is formed after the reduction step. This process is called **alkoxymercuration–reduction**:

$$\diagup\!\!\diagdown\!\!\diagup\!\!\diagdown + Hg(OAc)_2 + (CH_3)_2CHOH \longrightarrow AcOHg\!\!\diagdown\!\!\underset{\underset{}{}}{\overset{\overset{OCH(CH_3)_2}{|}}{\diagup}}\!\!\diagdown\!\!\diagup\!\!\diagdown + H - OAc \qquad (11.8a)$$

1-hexene   (solvent)   **1-acetoxymercuri-2-isopropoxyhexane**

$$\text{AcOHg}\diagdown\!\!\!\!\overset{\overset{\displaystyle OCH(CH_3)_2}{|}}{\diagup}\!\!\!\!\diagdown\!\!\!\!\diagup\!\!\!\!\diagdown + NaBH_4 \longrightarrow H\diagdown\!\!\!\!\overset{\overset{\displaystyle OCH(CH_3)_2}{|}}{\diagup}\!\!\!\!\diagdown\!\!\!\!\diagup\!\!\!\!\diagdown + Hg + \text{borates} \qquad (11.8b)$$

2-isopropoxyhexane
(91% yield)

Contrast:

$$H_2C{=}C\underset{\displaystyle H}{\overset{\displaystyle R'}{\Big\langle}}
\begin{cases}
+ \; H{-}OH & \xrightarrow[\text{THF/HOH}]{Hg(OAc)_2} \quad \xrightarrow{NaBH_4} \quad H_3C{-}\underset{\underset{\displaystyle OH}{|}}{CHR'} \quad \text{(oxymercuration–reduction)} \\[3mm]
& \qquad\qquad\qquad\qquad\qquad\qquad\quad \text{an alcohol} \\[3mm]
+ \; H{-}OR & \xrightarrow[\text{HOR}]{Hg(OAc)_2} \quad \xrightarrow{NaBH_4} \quad H_3C{-}\underset{\underset{\displaystyle OR}{|}}{CHR'} \quad \text{(alkoxymercuration–reduction)} \\[3mm]
& \qquad\qquad\qquad\qquad\qquad\qquad\quad \text{an ether}
\end{cases}$$

(11.9)

After reviewing the mechanism of oxymercuration in Eqs. 5.21a–d, pp. 190–191, you should be able to write the mechanism of the reaction in Eq. 11.8a. The two mechanisms are essentially identical, except that an alcohol instead of water is the nucleophile that reacts with the mercurinium ion intermediate.

**STUDY GUIDE
LINK 11.1**
Learning New
Reactions from
Earlier Reactions

PROBLEMS

11.4   (a) Write the mechanism of Eq. 11.8a and account for the regioselectivity of the reaction.
(b) Explain what would happen in an attempt to synthesize the ether product of Eq. 11.8b by a Williamson ether synthesis.

11.5   Complete the following reaction:

$$(CH_3)_2CH{-}CH{=}CH_2 + CH_3CH_2OH + Hg(OAc)_2 \longrightarrow \xrightarrow{NaBH_4}$$

11.6   Explain why a mixture of two isomeric ethers is formed in the following reaction.

$$CH_3CH_2CH{=}CHCH_3 + MeOH + Hg(OAc)_2 \longrightarrow \xrightarrow{NaBH_4}$$

11.7   Outline a synthesis of each of the following ethers using alkoxymercuration–reduction:
(a) dicyclohexyl ether   (b) *tert*-butyl isobutyl ether

C. Ethers from Alcohol Dehydration and Alkene Addition

In some cases, two molecules of a primary alcohol can react with loss of one molecule of water to give an ether. This dehydration reaction requires relatively harsh conditions: strong acid and heat.

$$2\,CH_3CH_2{-}OH \xrightarrow[140\,°C]{H_2SO_4} CH_3CH_2{-}O{-}CH_2CH_3 + H_2O \qquad (11.10)$$

ethanol   **diethyl ether**

This method is used industrially for the preparation of diethyl ether, and it can be used in the laboratory. However, it is generally restricted to the preparation of *symmetrical* ethers derived from *primary* alcohols. (A symmetrical ether is one in which both alkyl groups are the same.) Secondary and tertiary alcohols cannot be used because they undergo dehydration to alkenes (Sec. 10.2).

The formation of ethers from primary alcohols is an S_N2 reaction in which one alcohol displaces water from another molecule of *protonated* alcohol (see Problem 10.66, p. 509).

$$CH_3CH_2-\ddot{O}H \quad HO-CH_2CH_3 \longrightarrow$$
$$\overset{H}{\underset{|}{}}$$

$$CH_3CH_2-\overset{+}{\underset{\cdot\cdot}{O}}-CH_2CH_3 + H_2\ddot{O} \underset{\overset{H\ddot{O}CH_2CH_3}{\text{(solvent)}}}{\rightleftharpoons} CH_3CH_2-\ddot{O}-CH_2CH_3 + H_2\overset{+}{O}CH_2CH_3 \quad (11.11)$$

(protonated solvent molecule)

High temperature is required because alcohols are relatively poor nucleophiles in the S_N2 reaction.

Tertiary alcohols can be converted into *unsymmetrical* ethers by treating them with dilute strong acids in an alcohol solvent. The conditions are much milder than those required for ether formation from primary alcohols. For example, ethyl *tert*-butyl ether can be prepared when *tert*-butyl alcohol is treated with ethanol (as the solvent) in the presence of an acid catalyst:

$$H_3C-\overset{CH_3}{\underset{CH_3}{\overset{|}{C}}}-OH + EtOH \xrightarrow{\text{dilute } H_2SO_4} H_3C-\overset{CH_3}{\underset{CH_3}{\overset{|}{C}}}-OEt + H_2O \quad (11.12)$$

ethanol
(excess, solvent)

***tert*-butyl alcohol**

ethyl *tert*-butyl ether
(95% yield)

The key to using this type of reaction successfully is that only one of the alcohol starting materials (in this case, *tert*-butyl alcohol) can readily lose water after protonation to form a relatively stable carbocation. The alcohol that is used in excess (in this case, ethanol) must be one that either *cannot* form a carbocation by loss of water or should form a carbocation much less readily.

$$(CH_3)_3C-\ddot{O}H \underset{\overset{H_2SO_4}{}}{\rightleftharpoons} (CH_3)_3C-\overset{\cdot\cdot}{\underset{|}{\overset{+}{O}H}} \quad | \quad Et-\ddot{O}H \underset{\overset{H_2SO_4}{}}{\rightleftharpoons} Et-\overset{\cdot\cdot}{\underset{|}{\overset{+}{O}H}}$$
$$\quad\quad\quad\quad\quad\quad\quad\quad\quad\quad\quad H \quad\quad\quad\quad\quad\quad\quad\quad\quad\quad\quad\quad\quad\quad\quad H$$

$$\updownarrow \quad\quad\quad\quad\quad\quad\quad\quad\quad\quad\quad\quad\quad\quad\quad\quad \times$$

$$H_2\ddot{O} + (CH_3)_3C^+ \quad\quad\quad\quad\quad CH_3\overset{+}{C}H_2 + H_2\ddot{O} \quad (11.13)$$

a tertiary
carbocation

a primary
carbocation
(does not form)

When the carbocation derived from the tertiary alcohol is formed, it reacts rapidly with ethanol, which is present in large excess because it is the solvent.

$$(CH_3)_3C^+ \quad H\ddot{O}-Et \longrightarrow (CH_3)_3C-\overset{\cdot\cdot}{\underset{|}{\overset{+}{O}}}-Et \quad (11.14)$$
$$\quad\quad\quad\quad\quad\quad\quad\quad\quad\quad\quad\quad\quad\quad H$$

(loses a proton to solvent
to give the product)

There is an important relationship between this reaction and alkene formation by alcohol dehydration. Alcohols, especially tertiary alcohols, undergo dehydration to alkenes in the presence of strong acids (Sec. 10.2). Ether formation from tertiary alcohols and the dehydration of tertiary alcohols are *alternate branches of a common mechanism*. Both ether for-mation and alkene formation involve carbocation intermediates; the conditions dictate which product is obtained. The dehydration of alcohols to alkenes involves relatively high tempera-tures and *removal of the alkene and water products* as they are formed. Ether formation from tertiary alcohols involves milder conditions under which alkenes are *not* removed from the reaction mixture. In addition, *a large excess of the other alcohol* (ethanol in Eq. 11.12) *is used as the solvent*, so that the major reaction of the carbocation intermediate is with this alcohol. Any alkene that does form is not removed but is reprotonated to give back the same carboca-tion, which eventually reacts with the alcohol solvent:

STUDY GUIDE LINK 11.2
Common Intermediates from Different Starting Materials

$$(CH_3)_3C\text{—}OH$$

$$\downarrow H_2SO_4, -H_2O$$

$$\underset{HSO_4^-}{\overset{H_2SO_4}{\rightleftarrows}} \quad \underset{\text{carbocation intermediate}}{\overset{H_3C}{\underset{H_3C}{>}}C^+\text{—}CH_3} \quad \xrightarrow[\text{(solvent)}]{\text{EtOH}} \quad H_3C\text{—}\overset{CH_3}{\underset{CH_3}{\overset{|}{\underset{|}{C}}}}\text{—}O\text{—}Et \quad (11.15)$$

with left structure:

$$\underset{H_3C}{\overset{H_3C}{>}}C{=}CH_2$$

This analysis suggests that the treatment of an alkene with a large excess of alcohol in the presence of an acid catalyst should also give an ether, provided that a relatively stable carbocation intermediate is involved. Indeed, such is the case; for example, the acid-catalyzed additions of methanol or ethanol to 2-methylpropene to give, respectively, methyl *tert*-butyl ether and ethyl *tert*-butyl ether are important industrial processes for the synthesis of these gasoline additives (Eq. 10.77, p. 503).

$$\underset{\underset{\textbf{2-methylpropene}}{H_3C}}{\overset{H_3C}{>}}C{=}CH_2 \; + \; \underset{\textbf{methanol}}{CH_3OH} \quad \xrightarrow{\text{dilute } H_2SO_4} \quad H_3C\text{—}\overset{CH_3}{\underset{CH_3}{\overset{|}{\underset{|}{C}}}}\text{—}OCH_3 \quad (11.16)$$

methyl *tert*-butyl ether (MTBE)

Eqs. 11.12, 11.15, and 11.16 show that for the preparation of tertiary ethers, it makes no difference in principle whether the starting material from which the tertiary group is derived is an alkene or a tertiary alcohol.

PROBLEMS

11.8 Explain why the dehydration of primary alcohols can only be used for preparing *symmetrical* ethers. What would happen if a mixture of two different alcohols were used as the starting material in this reaction?

11.9 Complete the following reaction by giving the major organic product.

$$\text{(cyclohexane ring)}\overset{OH}{\underset{}{<}}CH_3 \; + \; \underset{\text{(solvent)}}{EtOH} \quad \xrightarrow{\text{dilute } H_2SO_4}$$

11.10 Draw the structure of two alkenes, either of which when treated with dilute H_2SO_4 and ethanol will give the same ether product as the reaction in Problem 11.9.

11.11 Outline a synthesis of each ether using either alcohol dehydration or alkene addition, as appropriate.
  (a) $ClCH_2CH_2OCH_2CH_2Cl$ (b) 2-methoxy-2-methylbutane
  (c) *tert*-butyl isopropyl ether   (d) dibutyl ether

11.3   SYNTHESIS OF EPOXIDES

A. Oxidation of Alkenes with Peroxycarboxylic Acids

One of the best laboratory preparations of epoxides involves the direct oxidation of alkenes with peroxycarboxylic acids.

$$CH_3(CH_2)_5CH\!=\!CH_2 \ + \quad \text{mCPBA} \quad \xrightarrow[\text{benzene}]{25\,°C} \quad CH_3(CH_2)_5CH\!-\!CH_2 \ + \quad \text{m-chlorobenzoic acid}$$

1-octene

meta-chloroperoxybenzoic acid (mCPBA)
a peroxycarboxylic acid

2-hexyloxirane
(81% yield)

meta-chlorobenzoic acid

(11.17)

The use of alkenes as starting materials for epoxide synthesis is one reason that certain epoxides are named traditionally as oxidation products of the corresponding alkenes (Sec. 8.1C).

The oxidizing agent in Eq. 11.17, *meta*-chloroperoxybenzoic acid (abbreviated mCPBA), is an example of a **peroxycarboxylic acid**, which is a carboxylic acid that contains an —O—O—H (hydroperoxy) group instead of an —OH (hydroxy) group.

hydroperoxy group

$$R\!-\!\overset{\displaystyle :O:}{\underset{\displaystyle \|}{C}}\!-\!\ddot{O}H \quad \text{or} \quad RCO_2H \qquad\qquad R\!-\!\overset{\displaystyle :O:}{\underset{\displaystyle \|}{C}}\!-\!\ddot{O}\!-\!\ddot{O}H \quad \text{or} \quad RCO_3H$$

a carboxylic acid

a peroxycarboxylic acid

(The terms **peroxyacid** or **peracid** are sometimes used instead of *peroxycarboxylic acid*. These are actually more general terms that refer not only to peroxycarboxylic acids, but also to *any* acid containing an —O—O—H group instead of an —OH group.) Many peroxycarboxylic acids are unstable, but they can be formed just prior to use by mixing a carboxylic acid and hydrogen peroxide. In principle, any one of several peroxycarboxylic acids can be used for the epoxidation of alkenes. The peroxyacid used in Eq. 11.17, mCPBA, has been popular because it is a crystalline solid that can be shipped commercially and stored in the laboratory. However, mCPBA, like most other peroxides, will detonate if it is not handled carefully. A less hazardous peroxycarboxylic acid that has essentially the same reactivity is the magnesium salt of monoperoxyphthalic acid, abbreviated MMPP.

$$\left[\quad \right]_2 Mg^{2+}$$

magnesium monoperoxyphthalate (MMPP)

Epoxidation is a *concerted electrophilic addition*. The mechanism is very similar to the mechanism of bromonium-ion formation in electrophilic addition (Eq. 5.11, p. 185):

(11.18a)

The protonated epoxide formed in this addition, like other ethers (Sec. 11.1), has a negative pK_a value. This very acidic proton is removed in a Brønsted acid–base reaction by the carboxylate ion to form the epoxide and a carboxylic acid. You can convince yourself using the method described in Sec. 3.4E that the equilibrium for this final step is highly favorable.

(11.18b)

We know this process is concerted because (1) carbocation rearrangements do not occur, and (2) the reaction is a *stereospecific* syn-addition. That is, the reaction takes place with complete retention of alkene stereochemistry. (Recall from Eq. 7.38, p. 312, that a stepwise process would not be stereospecific.) This means that a cis alkene gives a cis-substituted epoxide and a trans alkene gives a trans-substituted epoxide.

(11.19a)

(11.19b)

The reaction is a syn-addition because, in an anti-addition, the epoxide oxygen would have to bridge opposite faces of the two alkene carbons simultaneously. For this to happen, the double bond in the transition state for anti-addition would have to be significantly twisted and the transition state would be highly strained. As we shall see, the stereospecificity of this reaction is one reason why epoxide formation is a highly valuable synthetic reaction.

Because epoxides contain three-membered rings, they, like cyclopropanes (Sec. 7.5B), have significant angle strain. As we'll see in Sec. 11.5, this strain imparts valuable reactivity to epoxides. It is possible to form such strained compounds so easily because the O—O bond of a peroxycarboxylic acid is *very* weak. In other words, it is the high energy of the peroxycarboxylic acid that drives epoxide formation.

PROBLEMS

11.12 Give the structure of the alkene that would react with mCPBA to give each of the following epoxides.

11.13 Give the product expected when each of the following alkenes is treated with MMPP.

(a) *trans*-3-hexene   (b) ▷=CH$_2$

B. Cyclization of Halohydrins

Epoxides can also be synthesized by the treatment of halohydrins (Sec. 5.2B) with base:

$$
\underset{\substack{\textbf{1-bromo-2-methyl-2-propanol}\\ \text{(a halohydrin)}}}{\overset{\overset{\displaystyle OH}{|}}{(CH_3)_2C-CH_2-Br}} + Na^+ OH^- \xrightarrow{60\,°C} \underset{\substack{\textbf{2,2-dimethyloxirane}\\ \text{(81% yield)}}}{\overset{O}{(CH_3)_2C-CH_2}} + Na^+ Br^- + H-OH \quad (11.20)
$$

This reaction is an intramolecular variation of the Williamson ether synthesis (Sec. 11.2A); in this case, the alcohol and the alkyl halide are part of the same molecule. The alkoxide anion, formed reversibly by reaction of the alcohol with NaOH, displaces halide ion from the neighboring carbon:

$$
\overset{\overset{\displaystyle \overset{\frown}{:}\ddot{O}H}{}}{\underset{(CH_3)_2C-CH_2-\ddot{Br}:}{:\ddot{O}-H}} \rightleftharpoons \overset{H-\ddot{O}H}{\underset{(CH_3)_2C-CH_2\overset{\frown}{-}\ddot{Br}:}{:\ddot{O}:^-}} \longrightarrow \overset{:O:}{\underset{(CH_3)_2C-CH_2}{\triangle}} + :\ddot{Br}:^- \quad (11.21)
$$

Like bimolecular S$_N$2 reactions, this reaction takes place by *opposite-side substitution* of the nucleophile—in this case, the oxygen anion—at the halide-bearing carbon (Sec. 9.4C). Such an opposite-side substitution requires that the nucleophilic oxygen and the leaving halide assume an anti relationship in the transition state of the reaction. In most noncyclic halohydrins, this relationship can generally be achieved through a simple internal rotation.

(11.22a)

C—O and C—Br bonds
are gauche;
opposite-side substitution
is not possible

C—O and C—Br bonds
are anti;
opposite-side substitution
is possible

(11.22b)

Halohydrins derived from cyclic compounds must be able to assume the required anti relationship through a conformational change if epoxide formation is to succeed. The following cyclohexane derivative, for example, must undergo the chair interconversion before epoxide formation can occur.

diequatorial
conformation

diaxial conformation

(11.23)

Even though the diaxial conformation of the halohydrin is less stable than the diequatorial conformation, the two conformations are in rapid equilibrium. As the diaxial conformation reacts to give epoxide, it is replenished by the rapidly established conformational equilibrium.

PROBLEMS

11.14 From models of the transition states for their reactions, predict which of the following two diastereomers of 3-bromo-2-butanol should form an epoxide at the greater rate when treated with base, and explain your reasoning. (*Hint:* Draw the conformations required for reaction and consider their relative energies.)

stereoisomer *A*: 2*R*,3*S*
stereoisomer *B*: 2*R*,3*R*

3-bromo-2-butanol

11.15 The chlorohydrin *trans*-2-chlorocyclohexanol reacts rapidly in base to form an epoxide. The cis stereoisomer, however, is relatively unreactive and does not give an epoxide. Explain why the two stereoisomers behave so differently.

11.4   CLEAVAGE OF ETHERS

The ether linkage is relatively unreactive under a wide variety of conditions. This is one reason ethers are widely used as solvents; that is, a great many reactions can be carried out in ether solvents without affecting the ether linkage.

Ethers do not react with nucleophiles for the same reason that alcohols do not react: an S_N2 reaction would result in a very basic leaving group. Alkoxide ions, like hydroxide ion, are strong bases.

$$\text{Nuc}:^- \quad \overset{}{\underset{\underset{\text{CH}_3}{|}}{\text{CH}_2}}-\ddot{\text{O}}\text{R} \quad \xrightarrow{\times} \quad \text{Nuc}-\overset{}{\underset{\underset{\text{CH}_3}{|}}{\text{CH}_2}} \quad + \quad {}^-:\ddot{\text{O}}\text{R} \tag{11.24}$$

a nucleophile — an alkoxide ion (a strong base and a poor leaving group)

Remember that the S_N2 reactions of compounds with strongly basic leaving groups are very slow. In this case, the reaction is so slow that ethers in general are stable toward reactions with bases. For example, ethers do not react with NaOH.

Ethers do react with HI or HBr to give alcohols and alkyl halides. This reaction is called *ether cleavage*. The conditions required for ether cleavage vary with the type of ether. Ethers containing only primary alkyl groups are cleaved only under relatively harsh conditions, such as concentrated HBr or HI and heat.

$$\text{CH}_3\text{CH}_2-\text{O}-\text{CH}_2\text{CH}_3 + \text{H}-\text{I} \xrightarrow{\text{heat}} \text{CH}_3\text{CH}_2-\text{OH} + \text{I}-\text{CH}_2\text{CH}_3 \tag{11.25}$$

diethyl ether   **ethanol**   **ethyl iodide**

The alcohol formed in the cleavage of an ether (ethanol in Eq. 11.25) can go on to react with HI to give a second molecule of alkyl halide (Sec. 10.3).

The mechanism of ether cleavage involves, first, protonation of the ether oxygen:

$$\text{CH}_3\text{CH}_2-\underset{\cdot\cdot}{\overset{}{\text{O}}}-\text{CH}_2\text{CH}_3 \rightleftharpoons \text{CH}_3\text{CH}_2-\underset{}{\overset{\text{H}}{\text{O}}}{}^+-\text{CH}_2\text{CH}_3 \quad :\ddot{\text{I}}:^- \tag{11.26a}$$

Then the iodide ion, which is a good nucleophile (Sec. 9.4E), reacts with the protonated ether in an S_N2 reaction to form an alkyl halide and liberate an alcohol as a leaving group.

$$\text{CH}_3\text{CH}_2-\underset{}{\overset{\text{H}}{\text{O}}}{}^+-\text{CH}_2\text{CH}_3 \quad {}^-:\ddot{\text{I}}: \longrightarrow \text{CH}_3\text{CH}_2-\ddot{\text{O}}\text{H} + \ddot{\text{I}}-\text{CH}_2\text{CH}_3 \tag{11.26b}$$

If the ether is tertiary, the cleavage occurs under milder conditions (lower temperatures, more dilute acid). The first step of the mechanism is the same—protonation of the ether linkage:

$$\underset{\underset{\text{CH}_3}{|}}{\overset{\overset{\text{CH}_3}{|}}{\text{H}_3\text{C}-\text{C}}}-\ddot{\text{O}}-\text{Et} \quad \overset{}{\text{H}-\ddot{\text{I}}:} \rightleftharpoons \underset{\underset{\text{CH}_3}{|}}{\overset{\overset{\text{CH}_3}{|}}{\text{H}_3\text{C}-\text{C}}}-\overset{\text{H}}{\underset{}{\text{O}}}{}^+-\text{Et} \quad :\ddot{\text{I}}:^- \tag{11.27a}$$

In this case, the formation of the alkyl iodide occurs by an S_N1 mechanism. A *tertiary* carbocation, formed by loss of the *primary* alcohol leaving group, reacts with the iodide ion.

$$\underset{\underset{\text{CH}_3}{|}}{\overset{\overset{\text{CH}_3\ \ \text{H}}{|\ \ \ |}}{\text{H}_3\text{C}-\text{C}}}-\text{O}^+-\text{Et} \rightleftharpoons \underset{\underset{\text{CH}_3}{|}}{\overset{\overset{\text{CH}_3}{|}}{\text{H}_3\text{C}-\text{C}^+}} \quad :\ddot{\text{I}}:^- \longrightarrow \underset{\underset{\text{CH}_3}{|}}{\overset{\overset{\text{CH}_3}{|}}{\text{H}_3\text{C}-\text{C}}}-\ddot{\text{I}}: \tag{11.27b}$$

:Ï:⁻

a tertiary carbocation

$$+ \text{H}\ddot{\text{O}}-\text{Et}$$

Notice that the *tertiary* alkyl halide is formed along with the *primary* alcohol. Because the S_N1 reaction is faster than competing S_N2 processes, none of the primary alkyl iodide is formed.

Notice the great similarity in the reactions of ethers and alcohols with halogen acids (Sec. 10.3). In both cases, protonation converts a poor leaving group (—OH or —OR) into a good leaving group. In the reaction of an alcohol, *water* is the leaving group. In the reaction of an ether, an *alcohol* is the leaving group. Otherwise, the reactions are essentially the same.

Methyl or primary alcohol: *Methyl or primary ether:*

(11.28)

Tertiary alcohol: *Tertiary ether:*

(11.29)

Although the cleavage of alkyl ethers gives alkyl halides and alcohols as products, this reaction is rarely used to prepare these compounds because ethers themselves are most often prepared from alkyl halides or alcohols, as shown in Sec. 11.2. However, it is important to appreciate these reactions because they explain the instability of ethers under acidic conditions.

PROBLEMS

11.16   Explain each of the following facts with a mechanistic argument.

   (a) When butyl methyl ether (1-methoxybutane) is treated with HI and heat, the initially formed products are mainly methyl iodide and 1-butanol; little or no methanol and 1-iodobutane are formed.

   (b) When the reaction mixture in part (a) is heated for longer times, 1-iodobutane is also formed.

continued

continued _____

(c) When *tert*-butyl methyl ether is treated with HI, the products formed are *tert*-butyl iodide and methanol.

(d) *Tert*-butyl methyl ether cleaves much faster in HBr than its sulfur analog, *tert*-butyl methyl sulfide. (*Hint:* See Sec. 11.1.)

(e) When enantiomerically pure (*S*)-2-methoxybutane is treated with HBr, the products are enantiomerically pure (*S*)-2-butanol and methyl bromide.

11.17 What products are formed when each of the following ethers reacts with concentrated aqueous HI?

(a) diisopropyl ether   (b) 2-ethoxy-2,3-dimethylbutane

11.5   NUCLEOPHILIC SUBSTITUTION REACTIONS OF EPOXIDES

A. Ring-Opening Reactions under Basic Conditions

Epoxides readily undergo reactions in which the epoxide ring is opened by nucleophiles.

$$(CH_3)_2C \overset{O}{\overset{\triangle}{-}} CH_2 \; + \; EtOH \; \xrightarrow[\text{5 h, 80 °C}]{Na^+ EtO^-} \; (CH_3)_2C \overset{OH}{\underset{|}{-}} CH_2 - OEt \qquad (11.30)$$

2,2-dimethyloxirane   **ethanol**   **1-ethoxy-2-methyl-2-propanol**
(isobutylene oxide)   (solvent)   (83% yield)

A reaction of this type is *an S_N2 reaction in which the epoxide oxygen serves as the leaving group*. In this reaction, though, the leaving group does not depart as a separate entity, but rather remains within the same product molecule.

Notice that the base ethoxide is the nucleophile. Because protonation of the ring-opened alkoxide regenerates the ethoxide, it is a catalyst and can be used in catalytic amounts when ethanol is the solvent.

Because an epoxide is a type of ether, the ring opening of epoxides is an ether cleavage. Recall that ordinary ethers do *not* undergo cleavage in base (Eq. 11.24, p. 522). Epoxides, however, are opened readily by basic reagents. Epoxides are reactive because they, like their carbon analogs, the cyclopropanes, possess significant *angle strain* (Sec. 7.5B). Because of this strain, the bonds of an epoxide are weaker than those of an ordinary ether, and are thus more easily broken. The opening of an epoxide relieves the strain of the three-membered ring just as the snapping of a twig relieves the strain of its bending.

In an unsymmetrical epoxide, two ring-opening products could be formed corresponding to the reaction of the nucleophile at the two different carbons of the ring. As Eq. 11.31 illustrates, *nucleophiles typically react with unsymmetrical epoxides at the carbon with fewer alkyl substituents*. This regioselectivity is expected from the effect of alkyl substitution on the rates of S_N2 reactions (Sec. 9.4D). Because alkyl substitution retards the S_N2 reaction, the reaction of a nucleophile at the unsubstituted carbon is faster and leads to the observed product.

Like other S_N2 reactions, the ring opening of epoxides by bases involves *opposite-side substitution* of the nucleophile on the epoxide carbon. When this carbon is a stereocenter, inversion of configuration occurs, as illustrated by Study Problem 11.2.

STUDY PROBLEM   11.2

What is the stereochemistry of the 2,3-butanediol formed when *meso*-2,3-dimethyloxirane reacts with aqueous sodium hydroxide?

SOLUTION   First draw the structure of the epoxide. The meso stereoisomer of 2,3-dimethyloxirane has an internal plane of symmetry, and its two asymmetric carbons have opposite configurations.

meso-2,3-dimethyloxirane

Because the two different carbons of the epoxide ring are enantiotopic (Sec. 10.9A), the hydroxide ion reacts at either one at the same rate. Opposite-side substitution on each carbon should occur with inversion of configuration.

The product shown is the 2*S*,3*S* stereoisomer. Reaction at the other carbon gives the 2*R*,3*R* stereoisomer. (Verify this point!) Because the starting materials are achiral, the two enantiomers of the product must be formed in equal amounts (Sec. 7.7A). Therefore, the product of the reaction is racemic 2,3-butanediol. (This predicted result is in fact observed in the laboratory.)

Although the examples in this section have involved hydroxide and alkoxides as nucleophiles, the pattern of reactivity is the same with *any* nucleophile: The nucleophile reacts at the carbon with no alkyl substituents and opens the epoxide to form an alkoxide, which then reacts in a Brønsted acid–base reaction with a proton source to give an alcohol. Letting Nuc:⁻ be a general nucleophile, we can summarize this pattern of reactivity with Eq. 11.32:

PROBLEMS

11.18   Predict the products of the following reactions by drawing a curved-arrow mechanism for each.

(a)

continued

continued

(b)

$$H_3C\overset{\overset{\displaystyle O}{\diagup\!\!\diagdown}}{\underset{\displaystyle CH_3CH_2}{C}}-CH_2 \ + \ Na^+ \ N_3^- \ \xrightarrow[\text{EtOH/H}_2\text{O}]{}$$

sodium azide

11.19 From what epoxide and what nucleophile could each of the following compounds be prepared? (Assume each is racemic.)

(a)

(b)

$$CH_3(CH_2)_4\overset{\displaystyle OH}{\underset{|}{C}}HCH_2CN$$

B. Ring-Opening Reactions under Acidic Conditions

Ring-opening reactions of epoxides, like those of ordinary ethers, can be catalyzed by acids. However, epoxides are *much* more reactive than ethers under acidic conditions because of their angle strain. Hence, milder conditions can be used for the ring-opening reactions of epoxides than are required for the cleavage of ordinary ethers. For example, very low concentrations of acid catalysts are required in ring-opening reactions of epoxides.

$$(CH_3)_2\overset{\overset{\displaystyle O}{\diagup\!\!\diagdown}}{C}-CH_2 \ + \ CH_3OH \ \xrightarrow{\text{H}_2\text{SO}_4\ \text{(trace)}} \ (CH_3)_2C-CH_2-OH \qquad (11.33)$$

2,2-dimethyloxirane   **methanol** $|$
(isobutylene oxide)   (solvent) OCH_3

2-methoxy-2-methyl-1-propanol
(76% yield)

The regioselectivity of the ring-opening reaction is different under acidic and basic conditions. The structure of the product in Eq. 11.33 shows that the nucleophile methanol reacts at the *more substituted carbon* of the epoxide. Contrast this with the result in Eq. 11.32, in which the nucleophile reacts at the *less-substituted carbon* under basic conditions. In general, if one of the carbons of an unsymmetrical epoxide is tertiary, nucleophiles react at this carbon under acidic conditions.

Some insight into why different regioselectivities are observed under different conditions comes from mechanistic considerations. The first step in the mechanism of Eq. 11.33, like the first step of ether cleavage, is protonation of the oxygen.

the proton comes from a
protonated solvent molecule

$$\qquad (11.34a)$$

protonated epoxide

The structural properties of the protonated epoxide show that it can be expected to behave like a tertiary carbocation.

long, weak bond

nearly trigonal
planar geometry

$$\qquad (11.34b)$$

a large amount of
positive charge

First, calculations show that the tertiary carbon bears about 0.7 of a positive charge. Second, the geometry at the tertiary carbon is nearly trigonal planar. This means that the tertiary carbon and the groups around it are very nearly flattened into a common plane so that little or no steric hindrance prevents the approach of a nucleophile to this carbon. Finally, the bond between the tertiary carbon and the —OH group is unusually long and weak. This means that this bond is more easily broken than the other C—O bond. In fact, this cation resembles a carbocation solvated by the leaving group (see Fig. 9.13, p. 426). The —OH leaving group blocks the front side of the carbocation so that the nucleophilic reaction must occur at the opposite side with inversion of stereochemistry. In other words, we can think of this reaction as an S_N1 reaction with stereochemical inversion.

Thus, a solvent molecule reacts with the protonated epoxide at the tertiary carbon, and loss of a proton to solvent gives the product.

$$(CH_3)_2\overset{+}{C}—CH_2 \longrightarrow (CH_3)_2C—CH_2 \rightleftharpoons (CH_3)_2C—CH_2 \qquad (11.34c)$$

It is a solvent molecule, not the alkoxide conjugate base of the solvent, that reacts with the protonated epoxide. The alkoxide conjugate base cannot exist in acidic solution; nor is it necessary, because the protonated epoxide is very reactive and because the nucleophile is also the solvent and is thus present in great excess.

When the carbons of an unsymmetrical epoxide are secondary or primary, there is much less carbocation character at either carbon in the protonated epoxide, and acid-catalyzed ring-opening reactions tend to give mixtures of products; the exact compositions of the mixtures vary from case to case.

$$H_3C—CH—CH_2 \xrightarrow[\text{EtOH (solvent)}]{0.8\% \ H_2SO_4} H_3C—\underset{OEt}{CH}—CH_2—OH + H_3C—\underset{OH}{CH}—CH_2—OEt \qquad (11.35)$$

$$\text{secondary} \qquad \text{primary} \qquad\qquad\qquad \text{37\% of the product} \qquad \text{63\% of the product}$$

The mixture reflects the balance between opening of the weaker bond, which favors reaction at the carbon with more substituents, and van der Waals repulsions with the nucleophile, which favors reaction at the carbon with fewer substituents.

The regioselectivities of acid-catalyzed epoxide ring opening and the reactions of solvent nucleophiles with bromonium ions are very similar (see Eq. 5.15, p. 187). This is not surprising, because both types of reactions involve the opening of strained rings containing positively charged, electronegative leaving groups.

Acid-catalyzed ring-opening reactions of epoxides, like base-catalyzed ring-opening reactions, occur with *inversion of stereochemical configuration*.

$$\text{(cyclohexene oxide)} + CH_3OH \xrightarrow[\text{(solvent)}]{H_2SO_4 \text{ catalyst}} \qquad (11.36)$$

inversion of configuration

cyclohexene oxide

(±)-*trans*-2-methoxycyclohexanol
(82% yield)

When water is used as a nucleophile in acid-catalyzed epoxide ring opening, the product is a 1,2-diol, or *glycol*. Acid-catalyzed epoxide hydrolysis is generally a useful way to prepare glycols.

$$\text{cyclohexene oxide} + H_2O \xrightarrow[\text{30 min}]{HClO_4\text{ (trace)}} (\pm)\text{-}trans\text{-1,2-cyclohexanediol}$$

(11.37)

cyclohexene oxide   **(±)-*trans*-1,2-cyclohexanediol**
(a glycol; 80% yield)

Notice the trans relationship of the two hydroxy groups in the product, which results from the inversion of configuration that occurs when water reacts with the protonated epoxide. It follows that *cis*-1,2-cyclohexanediol *cannot* be prepared by epoxide opening. However, in Sec. 11.6A, you will learn how the cis stereoisomer can be prepared by another method.

Although base-catalyzed hydrolysis of epoxides also gives glycols (see Study Problem 11.2), polymerization sometimes occurs as a side reaction under basic conditions (see end-of-chapter Problem 11.69). Consequently, acid-catalyzed hydrolysis of epoxides is generally preferred for the preparation of glycols.

PROBLEM

11.20   Predict the major product(s) of each of the following transformations.

(a)

(optically active)

(b) The enantiomer of the epoxide in part (a) + $CH_3OH \xrightarrow[\text{(trace)}]{H_2SO_4}$
(solvent)

Let's summarize the facts about the regioselectivity and stereoselectivity of epoxide ring-opening reactions:

1.   Nucleophiles react with unsymmetrical epoxides under basic conditions at the less branched carbon, and inversion of configuration is observed if reaction occurs at a stereocenter.

2.   Nucleophiles react with unsymmetrical epoxides under acidic conditions at the tertiary carbon. If neither carbon is tertiary, a mixture of products is formed in most cases. Inversion of configuration is observed if reaction occurs at a stereocenter.

These facts are applied in Study Problem 11.3.

STUDY PROBLEM   **11.3**

Predict the major product in each case that would be obtained when the following epoxide reacts with water under (a) basic conditions; (b) acidic conditions. (The epoxide carbons are numbered for reference in the solution.)

SOLUTION As the preceding summary suggests, when attempting to predict the products of an epoxide ring-opening reaction, first decide whether the conditions of the reaction are basic or acidic. If basic, the nucleophile reacts at the *less-substituted* carbon of the epoxide; if acidic, the nucleophile reacts the *tertiary carbon* of the epoxide. Then determine whether the carbon at which the reaction occurs is a stereocenter. If so, make sure to predict the product that results from inversion of configuration.

(a) Under basic conditions, hydroxide ion is the nucleophile. It will react at the *less-substituted carbon* (carbon-1) of the epoxide. (If you have difficulty seeing why this is the less-substituted carbon, re-read Study Guide Link 9.2.) Because this carbon is not a stereocenter, the stereochemistry of the substitution does not matter. Consequently, the reaction is

$$(CH_3)_3C \underset{1}{\overset{2}{\bigwedge}} CH_2 + H_2O \xrightarrow{\ ^-OH\ } (CH_3)_3C \underset{}{\bigwedge} CH_2OH \qquad (11.38)$$

(b) Under acidic conditions, water is the nucleophile. It reacts with the *protonated* epoxide at the more-branched carbon (carbon-2). Notice that carbon-2 is a stereocenter (even though it is *not* an asymmetric carbon); reaction of the nucleophile at carbon-2 occurs with inversion of configuration. Consequently, the product of the reaction under acidic conditions is a diastereomer of the product obtained under basic conditions.

$$(CH_3)_3C \underset{1}{\overset{2}{\bigwedge}} CH_2 + H_2O \xrightarrow[\text{(catalyst)}]{H_2SO_4} (CH_3)_3C \underset{}{\bigwedge} \overset{CH_2OH}{\underset{OH}{}} \qquad (11.39)$$

PROBLEM

11.21 (a) Suppose 2,2-dimethyloxirane reacts with water that has been enriched with the oxygen isotope ^{18}O. Indicate how the hydrolysis product would differ under acidic and basic conditions.

(b) Show how the stereochemistry of the products will differ (if at all) when the following enantiomerically pure epoxide reacts with water under acidic and basic conditions.

$$H_3C\overset{O}{\underset{D_3C}{\overset{\diagup\!\!\diagdown}{\cdots C}}}\overset{}{\underset{D}{\overset{}{\cdots H}}}$$

C. Reactions of Epoxides with Organometallic Reagents

Grignard reagents (Sec. 9.8) react with ethylene oxide to give, after a protonation step, primary alcohols:

$$CH_3(CH_2)_4CH_2MgBr + H_2C\overset{O}{\overset{\diagup\!\!\diagdown}{-}}CH_2 \xrightarrow[\text{2) } H_3O^+]{\text{1) ether, heat}} CH_3(CH_2)_4CH_2CH_2CH_2OH \qquad (11.40)$$

hexylmagnesium bromide   **ethylene oxide**   **1-octanol**
(a Grignard reagent)   (71% yield)

This reaction is another epoxide ring-opening reaction. To understand this reaction, recall that the carbon in the C—Mg bond of the Grignard reagent has *carbon-anion* character and is therefore a very *basic* carbon (Sec. 9.8B). This carbon is the nucleophile that reacts with the epoxide. At the same time, the magnesium of the Grignard reagent, which is a Lewis acid, coordinates to the epoxide oxygen. (Recall that Grignard reagents associate strongly with ether oxygens; see Eq. 9.64, p. 431.) Just as protonation of an oxygen makes it a better leaving group, coordination of an oxygen to a Lewis acid also makes it a better leaving group.

Consequently, this coordination assists the ring opening of the epoxide in much the same way that Brønsted acids catalyze ring opening (Sec. 11.5B).

$$R\text{—}CH_2\text{—}CH_2\text{—}\ddot{\underset{..}{O}}\text{:}^{-}\text{----}Mg\text{—}R \rightleftharpoons R\text{—}CH_2\text{—}CH_2\text{—}\ddot{\underset{..}{O}}\text{:}^{-}\text{----}{}^{+}MgBr + RMgBr \quad (11.41a)$$

a bromomagnesium alkoxide

As Eq. 11.41a shows, this reaction yields an alkoxide, which is the conjugate base of an alcohol (Sec. 10.1A). After the Grignard reagent has reacted, the alkoxide is converted into the alcohol product in a separate step by the addition of water or dilute acid:

$$R\text{—}CH_2CH_2\text{—}\ddot{\underset{..}{O}}\text{:}^{-}\ {}^{+}MgBr \quad H\text{—}\overset{+}{\underset{..}{O}}H_2 \longrightarrow R\text{—}CH_2CH_2\text{—}\ddot{\underset{..}{O}}H + H_2\ddot{\underset{..}{O}} + Mg^{2+} + Br^{-} \quad (11.41b)$$

(See Eq. 11.79, p. 554, for the use of this reaction in organic synthesis.)

It would be reasonable to suppose that Grignard and organolithium reagents would react with epoxides other than ethylene oxide, and they do. However, many reactions of Grignard and lithium reagents with epoxides are unsatisfactory because they give not only the expected products of ring opening but also rearrangements and other side reactions as well. (Grignard reagents and organolithium reagents have some Lewis acid character that promotes such side reactions.) However, another type of organometallic reagent, the lithium organocuprate, undergoes useful ring-opening reactions with epoxides.

Two types of organocuprates are used most commonly in organic chemistry. The first type is formed from the reaction of two equivalents of an alkyllithium reagent with copper(I) halide in an ether solvent. The first equivalent reacts to form an alkylcopper reagent plus a lithium halide. The driving force for the reaction is the greater tendency of lithium, the more electropositive metal, to exist as an ion.

$$CH_3CH_2\overset{Li}{|} \quad Cu\text{—}Cl \longrightarrow CH_3CH_2\overset{Cu}{|} + Li^{+}Cl^{-} \quad (11.42)$$

Because the copper is a Lewis acid, the alkylcopper reagent reacts with a second equivalent of the alkyllithium to give a **lithium dialkylcuprate**.

$$CH_3CH_2\overset{Li}{|} \quad CuCH_2CH_3 \longrightarrow Li^{+}\ \bar{C}u(CH_2CH_3)_2 \quad (11.43)$$

lithium diethylcuprate
(a lithium dialkylcuprate)

(Aryllithium reagents such as phenyllithium, Ph—Li, can also be used to prepare lithium diarylcuprates.) Lithium dialkylcuprate or diarylcuprate reagents are sometimes called *Gilman reagents* after Henry Gilman (1893–1986), who was Professor of Chemistry at Iowa State University when he discovered the reagents in 1952.

If copper(I) cyanide, CuCN, is used instead of a copper(I) halide, the cyanide group, which is much more basic than halide, remains bound to the copper, and a more complex reagent is formed:

$$2\ CH_3CH_2Li + CuCN \longrightarrow (CH_3CH_2)_2Cu(CN)Li_2 \quad (11.44)$$

a higher-order organocuprate

Although Eq. 11.44 describes the stoichiometry of the reagent, it exists in a state (or states) of higher aggregation. Such reagents are called **higher-order organocuprates**.

Both types of organocuprate reagents are useful in organic chemistry, and both react with epoxides. However, the higher-order organocuprates are the preferred reagents for use with epoxides because they react with a wider variety of epoxides and give fewer side reactions. (We'll see some important uses of lithium dialkylcuprates in later chapters.)

An organocuprate reagent reacts at the carbon of the epoxide with fewer alkyl substituents to give products of ring opening. Protonolysis gives the alcohol.

$$(CH_3CH_2)_2Cu(CN)Li_2 + \quad \xrightarrow[\text{$-20\ °C$}]{\text{THF}}$$

(1S,2S)-2-ethyl-1-methyl-1-cyclopentanol
(96% yield)

$$+ \; CH_3CH_2\overset{-}{C}u(CN) \; Li^+ \qquad (11.45)$$

Notice that the alkyl group from the reagent reacts at the epoxide carbon with *inversion of stereochemical configuration.*

We can think of the reaction as an S_N2 process in which a "carbon–anion" nucleophile is delivered from the copper to the epoxide carbon electrophile with stereochemical inversion. Epoxide opening is assisted by lithium ion, which is a "built-in" Lewis acid:

$$\longrightarrow \qquad + \; CH_3CH_2\overset{-}{C}u(CN) \; Li^+ \qquad (11.46)$$

Although this mechanism doesn't show the aggregated structure of the reagent, it correctly predicts the chemical and stereochemical outcome of the reaction.

The reactions of organometallic reagents with epoxides provide other methods for the synthesis of alcohols that can be added to the list in Sec. 10.11. You should ask yourself what limits the types of alcohols that can be prepared by each method.

These reactions also provide methods for the *formation of carbon–carbon bonds.* Reactions that form carbon–carbon bonds are especially important in organic chemistry because they can be used to lengthen carbon chains. We'll explore this point further in Sec. 11.10.

PROBLEMS

11.22 (a) From what Grignard reagent can 3-methyl-1-pentanol be prepared by reaction with ethylene oxide, then aqueous acid?

(b) From what epoxide and what higher-order cuprate reagent can 3-ethyl-3-heptanol be prepared?

11.23 Complete the following reactions by giving the structures of the alcohol products. In part (b), show the stereochemistry of the product as well.

(a)

$$\text{bromocyclopentane} \xrightarrow[\text{ether}]{\text{Mg}} \xrightarrow{} \xrightarrow{\text{H}_3\text{O}^+}$$

(b)

$$2\ \text{Ph}\!-\!\text{Li} \; + \; \text{CuCN} \xrightarrow[\text{ether}]{} \xrightarrow{\text{H}_3\text{O}^+}$$

11.6   PREPARATION AND OXIDATIVE CLEAVAGE OF GLYCOLS

Glycols are compounds that contain hydroxy groups on different carbon atoms. In practice (and in this text), the term *glycol* typically refers to *vicinal* diols—diols in which the two hydroxy groups are on adjacent carbons. (*Vicinal* comes from the Latin word *vicinus* = neighborhood.)

$$
\begin{array}{c}
\text{OH} \quad \text{OH} \\
| \qquad | \\
\text{R}-\text{C}-\text{C}-\text{R} \\
| \qquad | \\
\text{R} \qquad \text{R}
\end{array}
\qquad
\textit{Example:} \quad
\begin{array}{c}
\text{OH} \\
| \\
\text{H}_3\text{C}-\text{CH}-\text{CH}_2\text{OH}
\end{array}
$$

general structure of a vicinal glycol
(R = alkyl, aryl, or H)

**1,2-propanediol
(propylene glycol)**

Although glycols are alcohols, some glycol chemistry is quite different from the chemistry of alcohols. Some of this unique chemistry is the subject of this section.

A. Preparation of Glycols

You have already learned that some glycols can be prepared by the acid-catalyzed reaction of water with epoxides (Eq. 11.37). This is one of two important methods for the preparation of glycols.

The other important method for the preparation of glycols is the oxidation of alkenes with osmium tetroxide (OsO_4).

**osmium(VIII)
tetroxide**

an osmate ester
(typically not isolated)

**2-methyl-2-phenyl-
1,2-propanediol**
a glycol
(90–95% yield)

$$(11.47)$$

The osmium in OsO_4 is in a +8 oxidation state. Metals in high oxidation states [such as Mn(VII) and Cr(VI), as you've learned] are oxidizing agents because they attract electrons. This electron-attracting ability of Os(VIII) results in a concerted (that is, one-step) cycloaddition reaction between OsO_4 and an alkene to give the intermediate osmate ester:

$$(11.48a)$$

osmate ester

(The osmate ester is another example of an organic ester derivative of an inorganic acid; Sec. 10.4C.) The curved-arrow notation shows that in this reaction osmium accepts an electron pair. As a result, its oxidation state is decreased to +6.

A glycol is formed when the cyclic osmate ester is treated with water. Two water molecules, acting as nucleophiles, displace the glycol oxygens from the osmium. A mild reducing agent such as sodium bisulfite, $NaHSO_3$, is often added to convert the osmium-containing by-products into reduced forms of osmium that are easy to remove by filtration. (The $NaHSO_3$ is converted into sodium sulfate, Na_2SO_4.)

$$R_2C - CR_2 + 2H_2O \longrightarrow R_2C - CR_2 + \text{(Os compound)} \xrightarrow{NaHSO_3} \begin{array}{c} \text{reduced forms} \\ \text{of osmium} \end{array} \quad (11.48b)$$

Two practical drawbacks to the use of the OsO_4 oxidation are that osmium and its compounds are very toxic, and they are quite expensive. However, the reaction of OsO_4 with alkenes is so useful that chemists have devised ways for it to be used with very small amounts of OsO_4. This is done by including in the reaction mixture an oxidant that "recycles" the Os(VI) by-product back into OsO_4. Among the common oxidants used for this purpose are *amine oxides*, which are compounds of the form $R_3\overset{+}{N}-O^-$. Two amine oxides used commonly are the following:

trimethylamine-*N*-oxide
(TMAO)

N-methylmorpholine-*N*-oxide
(NMMO)

In other words, once a small amount of OsO_4 is used up, the Os(VI) by-product is oxidized within the reaction mixture by the amine oxide to re-form OsO_4. Thus, a catalytic amount of OsO_4 can be used and the amine oxide acts as the ultimate oxidant.

2,3-dimethyl-2-butene
(0.025 mole)

TMAO
(0.034 mole)

2,3-dimethyl-2,3-butanediol
(85% yield)

The OsO_4 oxidation is particularly useful because of its stereochemistry. The formation of glycols from alkenes is a stereospecific syn-addition.

cyclohexene   **NMMO**   ***cis*-1,2-cyclohexanediol**
(89% yield)

The mechanism of this reaction provides a simple explanation for the syn stereochemistry. The five-membered osmate ester ring is easily formed when two oxygens of OsO_4 are added to the same face of the double bond by a concerted mechanism. This process closely resembles the concerted cycloaddition mechanism of ozonolysis. (See Eq. 5.34, p. 199.) Hydrolysis of the osmate ester gives the glycol.

an osmate ester   a 1,2-glycol

On the other hand, an anti-addition by a concerted mechanism would be very difficult, if not impossible: the two reacting oxygens of OsO_4 cannot simultaneously reach opposite faces of the π bond.

The hydrolysis of epoxides and the OsO_4 oxidation are complementary reactions because they provide glycols of different stereochemistry. This point is explored in Study Problem 11.4.

STUDY PROBLEM 11.4

Outline preparations of *cis*-1,2-cyclohexanediol and (±)-*trans*-1,2-cyclohexanediol from cyclohexene.

SOLUTION   As Eq. 11.50 shows, the direct oxidation of cyclohexene by OsO_4 yields *cis*-1,2-cyclohexanediol by a syn-addition. In contrast, conversion of cyclohexene into the epoxide with a peroxycarboxylic acid (see Problem 11.12a), followed by acid-catalyzed hydrolysis (Eq. 11.37), gives the trans-diol. Epoxide hydrolysis gives the trans-diol because it occurs with *inversion of configuration*.

$$(11.52)$$

(Remember the following convention: Although we draw a single enantiomer of the product for convenience, it should be understood to be the racemate; Sec. 7.7A.)

Glycol formation from alkenes can also be carried out with potassium permanganate ($KMnO_4$), usually under aqueous alkaline conditions. This reaction is also a stereospecific syn-addition, and its mechanism is probably similar to that of OsO_4 addition.

$$(11.53)$$

Although the use of $KMnO_4$ avoids the expense of OsO_4, a problem with the use of $KMnO_4$ is that yields are low in many cases because over-oxidation occurs; that is, the glycol product is oxidized further. Conditions have to be carefully worked out in each case to avoid this side reaction.

The manganese in MnO_4^- is in the +7 oxidation state. It is converted into Mn(IV) as a result of the reaction. Visually, when oxidation occurs, the brilliant purple color of the permanganate ion is replaced by a muddy-looking brown precipitate of manganese dioxide (MnO_2). This color change can be used as a test for functional groups that can be oxidized by $KMnO_4$.

PROBLEMS

11.24   What organic product is formed (including its stereochemistry) when each of the following alkenes is treated with NMMO in the presence of H_2O and a catalytic amount of OsO_4?

(a) 1-methylcyclopentene   (b) *trans*-2-butene

11.25   From what alkene could each of the following glycols be prepared by the OsO_4 or $KMnO_4$ method?

(a) $CH_3CH_2OCH_2CH_2CHCH_2OH$ with OH

(b) with OH and CH_2OH

(c) *meso*-4,5-octanediol   (d) (±)-4,5-octanediol

11.26 Show a curved-arrow mechanism for the first step, and the structure of the cyclic intermediate formed, when an alkene is treated with $KMnO_4$. A Lewis structure for the permanganate ion is as follows:

permanganate ion

B. Oxidative Cleavage of Glycols

The carbon–carbon bond between the —OH groups of a glycol can be cleaved with periodic acid to give two carbonyl compounds:

$$\tag{11.54}$$

Periodic acid (pronounced PURR-eye-OH-dik) is the iodine analog of perchloric acid.

$$HClO_4 \qquad\qquad HIO_4$$

perchloric acid   periodic acid

Periodic acid is commercially available as the dihydrate, $HIO_4 \cdot 2H_2O$, often abbreviated, as in Eq. 11.54, as H_5IO_6 (sometimes called *para-periodic acid*). Its sodium salt, $NaIO_4$ (sodium metaperiodate), is sometimes also used. Periodic acid is a fairly strong acid ($pK_a = -1.6$). The periodate cleavage reaction has been used as a test for glycols as well as for synthesis. The formulas HIO_4 or H_5IO_6 are used interchangeably for periodic acid.

The cleavage of glycols with periodic acid takes place through a cyclic periodate ester intermediate (Sec. 10.4C) that forms when the glycol displaces two —OH groups from H_5IO_6.

$$\tag{11.55a}$$

The cyclic ester spontaneously breaks down by a cyclic flow of electrons in which the iodine accepts an electron pair. (The clockwise direction of electron flow is arbitrary.)

$$\tag{11.55b}$$

A glycol that cannot form a cyclic ester intermediate is not cleaved by periodic acid. For example, the following compound is not cleaved because it is impossible for both oxygens to be part of the same cyclic periodate ester. (If you can't see why, build a model and try connecting the two oxygens with one other atom.)

Do not confuse osmium tetroxide, permanganate, and periodate oxidations, all of which occur through cyclic ester intermediates (Sec. 11.5A). Periodate oxidizes *glycols*, but the other two reagents oxidize *alkenes* to give glycols. In all of these reactions, oxidation occurs because an atom in a highly positive oxidation state can accept an additional pair of electrons. In the periodate oxidation, the reduction of the iodine occurs during the *breakdown* of a cyclic ester; in the permanganate and osmium tetroxide oxidations, the metals are reduced during the *formation* of a cyclic ester.

PROBLEMS

11.27   Give the product(s) expected when each of the following compounds is treated with periodic acid.

(a)

(b)

$$PhCH_2CHCH_2OH$$

(c)

11.28   What glycol undergoes oxidation to give each of the following sets of products?

(a)

(b)

11.7   OXONIUM AND SULFONIUM SALTS

A. Reactions of Oxonium and Sulfonium Salts

If the acidic hydrogen of a protonated ether is replaced with an alkyl group, the resulting compound is called an **oxonium salt**. The sulfur analog of an oxonium salt is a **sulfonium salt**:

Oxonium and sulfonium salts react with nucleophiles in S_N2 reactions:

$$HO:^- + H_3C\overset{CH_3}{\underset{CH_3}{\overset{|}{\underset{|}{+O}}}}: BF_4^- \longrightarrow HOCH_3 + (CH_3)_2O: + BF_4^- \quad (11.56)$$

(89% yield)

$$(CH_3)_3N: \quad H_3C\overset{CH_3}{\underset{CH_3}{\overset{|}{\underset{|}{+S}}}}: NO_3^- \longrightarrow (CH_3)_4\overset{+}{N} NO_3^- + (CH_3)_2S: \quad (11.57)$$

Oxonium salts are among the most reactive alkylating agents known, and they react very rapidly with most nucleophiles. Because of their reactivity, oxonium salts must be stored in the absence of moisture. For the same reason, these salts are stable only when they contain counterions that are not nucleophilic, such as tetrafluoroborate ($^-BF_4$). (Tetrafluoroborate ion is not nucleophilic because, even though boron bears a negative charge, it has no unshared electron pairs.) *Sulfonium salts* are considerably less reactive and therefore are handled more easily. Sulfonium salts are somewhat less reactive than the corresponding alkyl chlorides in S_N2 reactions.

PROBLEMS

11.29 Explain why all attempts to isolate trimethyloxonium iodide lead instead to methyl iodide and dimethyl ether.

11.30 Complete the following reactions.

(a) [pyridine structure] $+ (CH_3)_3O^+ BF_4^- \longrightarrow$

(b) $(CH_3)_2S: + (CH_3)_3\overset{+}{O} BF_4^- \longrightarrow$

B. *S*-Adenosylmethionine: Nature's Methylating Agent

A sulfonium salt, *S*-adenosylmethionine (SAM), is important in biology as a methylating agent for nucleophiles. The structure of SAM is shown in Fig. 11.1. SAM is an interesting example of a compound containing asymmetric sulfur. Recall (Sec. 6.9B) that sulfonium salts, unlike amines, undergo inversion so slowly that individual stereoisomers can be isolated. SAM has the *S* configuration at sulfur. (The electron pair has lowest priority in the *R,S* system.)

sulfur has the *S* configuration

FIGURE 11.1 The structure of *S*-adenosylmethionine (SAM). The boxed parts of the structure are abbreviated R[1] and R[2] in the text. Notice that the sulfur of SAM is an asymmetric center. The solid and dashed wedges at sulfur show the orientation of the methyl group and the orbital containing the unshared electron pair.

Although the structure of SAM seems complicated, the chemistry of SAM arises solely from its sulfonium ion functional group. (The rich functionality of the R^1 and R^2 groups is utilized to form noncovalent interactions that strengthen the binding of SAM to the active sites of enzymes that catalyze its reactions.) Like the sulfonium salts in Eq. 11.57, SAM reacts with nucleophiles at the methyl carbon, liberating a sulfide leaving group:

$$R^3-\ddot{\underset{..}{O}}:^- \quad H_3C-\overset{R^1}{\underset{R^2}{\overset{+}{S}}} \quad \xrightarrow{\text{enzyme}} \quad R^3-\ddot{\underset{..}{O}}-CH_3 \; + \; :\overset{R^1}{\underset{R^2}{\ddot{S}}} \qquad (11.58)$$

a biological nucleophile

sulfide leaving group

SAM is stable enough to survive in aqueous solution, but it is reactive enough to undergo enzyme-catalyzed S_N2 reactions. Evidence supporting an S_N2 mechanism in methylation reactions involving SAM was obtained by a very elegant experiment. The methyl carbon of SAM was made asymmetric by using the two hydrogen isotopes deuterium (D, or ^2H) and tritium (T, or ^3H). It was found that substitutions on this methyl group proceed with inversion of configuration, exactly as expected for the S_N2 mechanism.

$$R^3-\ddot{\underset{..}{O}}:^- \quad \underset{D}{\overset{H}{\underset{T}{C}}}-\overset{R^1}{\underset{R^2}{\overset{+}{\ddot{S}}}} \quad \longrightarrow \quad R^3-\ddot{\underset{..}{O}}-\underset{D}{\overset{H}{C}}\cdots T \; + \; :\overset{R^1}{\underset{R^2}{\ddot{S}}} \qquad (11.59)$$

inversion of configuration

The compound *S*-adenosylmethionine, like NAD$^+$ (Sec. 10.8), is another example of a complex biological molecule that undergoes transformations which are readily understood in terms of common analogies from organic chemistry.

The Role of SAM in the Regulation of Gene Expression

SAM has been shown to play a pivotal role in regulating the expression of genes. Within the nucleus, DNA, which encodes genetic information, is tightly packaged in *nucleosomes*. These are complexes in which the DNA is wrapped around a core of proteins called *histones*, in the same sense that thread is wrapped on a spool (see photo). In such tightly wound DNA, the DNA is "silent"—that is, it is not expressed. The histones contain large amounts of the amino acids lysine and arginine:

lysine (Lys) residue of a protein

arginine (Arg) residue of a protein

The methylation of certain lysines and arginines in histones by SAM has been shown to regulate whether a gene is actually expressed within the cell, or whether it remains "silent." This methylation probably affects the noncovalent interactions between the histones and DNA so that the DNA-translating machinery of the cell can gain access to the DNA. The precise mechanism of this regulation is yet to be unraveled. Lysine residues in histones can be methylated, once, twice, or even three times, and arginine residues can be methylated once or twice.

$$R-\overset{+}{N}H_3 + H_3C-\overset{+}{\underset{R^2}{S}}-R^1 \xrightarrow{\text{enzymes}} R-\overset{+}{N}H_2-CH_3 \quad R-\overset{+}{\underset{CH_3}{N}}H-CH_3 \quad R-\overset{CH_3}{\underset{CH_3}{\overset{+}{N}}}-CH_3 + \overset{R^1}{\underset{R^2}{S}} \tag{11.60a}$$

lysine   **SAM**   **N-methyllysine**

N,N-dimethyllysine   **N,N,N-trimethyllysine**

$$R-NH-\overset{\overset{+}{N}H_2}{\underset{NH_2}{C}} + H_3C-\overset{+}{\underset{R^2}{S}}-R^1 \xrightarrow{\text{enzymes}} R-NH-\overset{\overset{+}{N}H-CH_3}{\underset{NH_2}{C}} + R-NH-\overset{\overset{+}{N}H-CH_3}{\underset{NH-CH_3}{C}} + \overset{R^1}{\underset{R^2}{S}} \tag{11.60b}$$

arginine   **SAM**   **methylarginine**   **dimethylarginine**

The degree of methylation is controlled by the catalyzing enzymes. Because cancer is a disease characterized by aberrant gene expression, these methylations are of interest in the fight against cancer.

PROBLEM

11.31   Using the abbreviations for lysine and SAM shown in Eq. 11.60a, write a curved-arrow mechanism for the formation of *N,N*-dimethyllysine. Assume that acids ($^+$BH) and bases (:B) are available as necessary. (*Hint:* A Brønsted acid–base reaction must precede each methylation; why?)

11.8   INTRAMOLECULAR REACTIONS AND THE PROXIMITY EFFECT

In this section we're going to consider **intramolecular reactions**: reactions that occur between groups within the *same* molecule. Reactions between groups on *different* molecules are called **intermolecular reactions**. We've seen some examples of intramolecular reactions in previous sections, but we haven't really stopped to think about how intramolecular and intermolecular reactions are different—after all, it might seem that a reaction between (for example) a nucleophile and an electrophile should be the same whether the reactions are intramolecular or intermolecular. Indeed, the *outcome* of the reaction might be the same. But what is different is that many intramolecular reactions are *faster* than their intramolecular counterparts—in some cases, *thousands of times* faster. We're going to consider some examples of such reactions and try to understand why they are so fast. The reason that we want to understand this phenomenon is that the catalytic effectiveness of enzymes—that is, biological catalysis—is largely explained by the theory of intramolecularity.

A. The Kinetic Advantage of Intramolecular Reactions

Consider the remarkable difference in the rates of the following two substitution reactions, which are superficially very similar:

relative rate:

$$CH_3CH_2CH_2CH_2CH_2CH_2-Cl + H_2O \xrightarrow[\substack{\text{dioxane} \\ 100\,°C}]{20\,M\,\text{water}} CH_3CH_2CH_2CH_2CH_2CH_2-OH + HCl \qquad 1 \tag{11.61a}$$

hexyl chloride   **1-hexanol**
(1-chlorohexane)

$$Et\overset{..}{\underset{..}{S}}CH_2CH_2-Cl + H_2O \xrightarrow[\substack{\text{dioxane} \\ 100\,°C}]{20\,M\,\text{water}} Et\overset{..}{\underset{..}{S}}CH_2CH_2-OH + HCl \qquad 3200 \tag{11.61b}$$

β-chloroethyl   **2-(ethylthio)ethanol**
ethyl sulfide

At first sight, both reactions appear to be simple S_N2 reactions in which chloride is displaced as a leaving group by water. In fact, this *is* the mechanism by which hexyl chloride reacts:

$$CH_3CH_2CH_2CH_2CH_2CH_2 - \ddot{\underset{\cdot\cdot}{Cl}}: \xrightarrow[\substack{\text{dioxane} \\ 100\,°C}]{20\,M\,\text{water}} CH_3CH_2CH_2CH_2CH_2CH_2 - \overset{+}{\underset{\underset{H}{|}}{\ddot{O}H}} \quad :\ddot{\underset{\cdot\cdot}{Cl}}:^-$$

$$CH_3CH_2CH_2CH_2CH_2CH_2 - \ddot{O}H + H_3\ddot{O}^+ \qquad (11.62)$$

This reaction requires high temperature because water is such a poor nucleophile in the S_N2 reaction that the reaction is extremely slow at lower temperatures. The presence of sulfur in the alkyl halide molecule should have little effect on the rate of the S_N2 reaction, because the S_N2 mechanism is not very sensitive to the electronegativities of substituent groups. (In fact, electronegative substituents are known to *retard* S_N2 reactions slightly.) Yet the reaction in Eq. 11.61b is thousands of times faster than the reaction in Eq. 11.61a. The reaction in Eq. 11.61a takes almost two *months,* whereas the reaction in Eq. 11.61b takes about 30 *minutes.* This huge difference is due solely to the presence of sulfur in the molecule.

The rate of Eq. 11.61b is unusually large because a special mechanism facilitates the reaction, a mechanism not available to hexyl chloride. In the first and rate-limiting step of the mechanism, the nearby sulfur displaces the chloride *within the same molecule*:

$$\boxed{\begin{array}{c}\text{an intramolecular} \\ \text{reaction}\end{array}}$$

$$Et - \ddot{\underset{\cdot\cdot}{S}} - CH_2CH_2 - Cl \longrightarrow \underset{H_2C - CH_2}{\overset{\overset{\displaystyle Et}{\overset{\displaystyle |}{:\overset{+}{S}}}}{}} \quad Cl^- \qquad (11.63a)$$

$$\text{an episulfonium salt}$$

The episulfonium ion that results from this internal nucleophilic substitution reaction is structurally similar to a protonated epoxide (Sec. 11.5B) or a bromonium ion (Secs. 5.2A and 7.8C). It is very reactive because it contains a strained three-membered ring and a good leaving group. Water rapidly reacts with this intermediate as it would with a protonated epoxide or bromonium ion to give the observed substitution product.

$$\underset{H_2C - CH_2}{\overset{\overset{\displaystyle Et}{\overset{\displaystyle |}{:\overset{+}{S}}}}{}} \longrightarrow Et - \ddot{\underset{\cdot\cdot}{S}} - CH_2CH_2 - \overset{+}{\ddot{O}}H \xrightarrow{H_2O} Et - \ddot{\underset{\cdot\cdot}{S}} - CH_2CH_2 - \ddot{O}H + H_3\ddot{O}^+ \qquad (11.63b)$$

Notice that this product is identical to the one that would have been formed in an ordinary S_N2 reaction in which the sulfur played no active role. Thus, in this case, the role of the sulfur is not apparent from the identity of the product. Only *the rate of the reaction* suggests that the sulfur has a special role in the mechanism.

The covalent involvement of a neighboring group—a group within the same molecule—in a chemical reaction is called **neighboring-group participation** or **anchimeric assistance** (from the Greek word *anchi,* meaning "near"). The neighboring-group mechanism of Eq. 11.63a is *in competition* with an ordinary S_N2 mechanism (Eq. 11.62) in which water reacts with the alkyl halide directly. Because the faster of two competing reactions is always the observed reaction, a reaction that involves neighboring-group participation, in order to be observed, *must* be faster than the same overall reaction that occurs by other competing mechanisms. The rate of the reaction in Eq. 11.61a provides the basis of comparison—that is, a rough idea of what rate to expect in this case for a reaction that occurs by direct substitution of water in the absence of neighboring-group participation. A large rate acceleration, such as the one in Eq. 11.61b, is typical of the experimental evidence used to diagnose the involvement of a neighboring group in a chemical reaction. Professor Saul Winstein (1912–1969) of the University of California, Los Angeles, discovered numerous examples of neighboring-group participation and showed that all of these are associated with significant rate accelerations. (Other evidence for neighboring-group participation is discussed in Sec. 11.8C.)

You studied a similar intramolecular substitution reaction earlier in this chapter: the cyclization of halohydrins (Sec. 11.3B). The alkoxide of a bromohydrin has two competing possibilities for reaction. First, it can undergo an *intramolecular reaction* to form an epoxide:

$$(11.64a)$$

an epoxide

This is the observed reaction. However, another possible reaction is for the alkoxide to react as a nucleophile in an *intermolecular reaction* with a *second* molecule of bromohydrin:

$$(11.64b)$$

Because the epoxide—the *intramolecular* reaction product—is observed, the intramolecular reaction must be significantly faster.

Why should an intramolecular reaction be faster than an intermolecular reaction? The answer has to do with the *probability* that the reaction will occur. In Sec. 8.6A, you learned that the thermodynamic expression of the probability of a process is the *entropy change*, $\Delta S°$. Because we are dealing with a question of relative rates, the answer lies in the entropic component of $\Delta G°^{\ddagger}$, the standard free energy of activation, which is the energy barrier for a reaction that determines the reaction rate (Sec. 4.8B).

$$\Delta G°^{\ddagger} = \Delta H°^{\ddagger} - T\Delta S°^{\ddagger} \qquad (11.65)$$

In this equation, $\Delta H°^{\ddagger}$ is the **standard enthalpy of activation**, $\Delta S°^{\ddagger}$ is the **standard entropy of activation**, and T is the absolute temperature in kelvins. The *enthalpy of activation* is determined by the strengths of the bonds broken and formed, van der Waals repulsions, and other energies that result from atomic and molecular interactions. *The standard entropy of activation reflects the intrinsic probability of forming a transition state for a reaction. A positive $\Delta S°^{\ddagger}$, which corresponds to a high probability of reaction, makes $-T\Delta S°^{\ddagger}$ negative, lowers $\Delta G°^{\ddagger}$, and increases the rate of a reaction. A negative $\Delta S°^{\ddagger}$, which corresponds to a low probability of reaction, makes $-T\Delta S°^{\ddagger}$ more positive, raises $\Delta G°^{\ddagger}$, and decreases the rate of a reaction.*

Let's analyze the entropic factors that are important in comparing intermolecular and intramolecular reactions. The first is the *translational entropy* change. When two molecules react to become one molecule—an intermolecular reaction—each reactant starts with three degrees of translational freedom (that is, each molecule can move freely in the three spatial coordinates *x, y,* and *z*). Therefore two reactants start with *six* degrees of translational freedom. As they come together to form the product—or the transition state leading to product— the pair have to move together in a coordinated way. Therefore, in this process three degrees of translational freedom have been lost. Bringing two molecules together into one and having them *stay* together is inherently improbable. This improbability is reflected in a high entropy cost for an intermolecular reaction.

The next entropy factor is the *rotational entropy*. When each molecule moves randomly, the entire molecule can rotate about the three spatial axes even when it is not "translating." When two molecules come together to form one species, then, three degrees of rotational freedom are lost. The loss of rotational entropy is not as large as the loss of translational entropy because rotation does not require as much space, but it is significant.

Dancers, Airports, and Entropy

Consider two dancers alone on a crowded dance floor who decide to become a dancing couple. Becoming a couple requires a conscious decision to lower their collective entropy—to overcome the negligible probability that they might bump into each other at random. When the two dancers come together and begin moving as a couple, the six total degrees of translational freedom for the individuals become three degrees of translational freedom for the couple. In becoming a dancing couple, the two dancers have lost translational entropy.

Similarly, when each dancer is alone, he or she can spin and turn independently. When the dancers become a couple, they turn and spin as a couple (assuming they are holding each other). Becoming a couple has resulted in the loss of three degrees of rotational freedom as well. The couple has also lost rotational entropy.

Now consider the following analogy for rates: Suppose you are in a crowded airport terminal and told to find a particular person and shake hands. This would take you a very long time if you had to search at random throughout the terminal. However, this "reaction" would be very rapid if the person you are looking for were tied to you by a very short rope! Tying the two of you together has significantly increased the probability that you will make contact.

The third entropy factor is the *entropy of internal rotation*. Intramolecular reactions have a *disadvantage* in this type of entropy. Intramolecular reactions by definition form rings. When a ring is formed, internal rotations are lost. The larger the ring, the more internal rotations are lost. For example, when a five-membered ring is formed in a nucleophilic reaction, four internal rotations are lost.

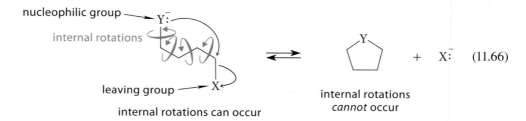

$$\text{(11.66)}$$

Internal rotations are not lost in intermolecular reactions because rings aren't formed. The loss of internal rotations lowers the entropy of a process because loss of internal rotation is a constraint on molecular freedom.

An analogy is trying to swat a fly that is tied to you with a string. The longer the string, the less probable it is that a random swat will hit the fly. Similarly, if a nucleophile is tied to an electrophile in the same molecule with a very long carbon chain, it is less probable that the molecule will achieve exactly the right conformation for reaction to occur. The probability is greater as the carbon chain becomes shorter.

When rings have more than three members, the rings can undergo conformational changes (for example, the chair interconversion). Conformational changes are coordinated partial internal rotations. The entropy of conformational changes partially offsets the loss of entropy of internal rotation, but not completely.

The $\Delta S^{\circ\ddagger}$ advantage of intramolecular reactions corresponds to the translational and rotational entropy advantage minus the internal rotation disadvantage. The translational and rotational entropy advantage is roughly 150 J K^{-1} (36 cal K^{-1}). (This is a rough estimate that varies with reaction type, but we can use it as a benchmark.) The corresponding effect on $\Delta G^{\circ\ddagger}$ is $-T\Delta S^{\circ\ddagger} = -(298)(150 \times 10^{-3})$ kJ mol^{-1} = 45 kJ mol^{-1} (10.7 kcal mol^{-1}). Now, if this entire entropic advantage is reflected in the standard free energy of activation—the energy barrier for the reaction—the effect on rate can be calculated as a ratio of rate constants from Eq. 9.20b (p. 391), with k_1 = the rate constant for the intramolecular reaction and k_2 = the rate constant for the intermolecular reaction.

$$\frac{k_1}{k_2} = 10^{(\Delta G_2^{\circ\ddagger} - \Delta G_1^{\circ\ddagger})/2.3RT} = 10^{45/5.71} \approx 10^8 \tag{11.67}$$

This means that intramolecular reactions are accelerated by factors of up to 10^8 over the corresponding intermolecular reactions. This advantage results, as we have seen, from the fact that, in an intramolecular reaction, the reactants are already "tied together" and do not have to find each other by random diffusion and rotation.

As we noted above, intramolecular reactions can have an entropic disadvantage that results from loss of internal rotations. Each internal rotation that is lost when an intramolecular reaction occurs reduces the rate of the intramolecular reaction by a factor of about 5. For example, if a five-membered ring is formed in an intramolecular reaction that results in a loss of four internal rotations (Eq. 11.66), the maximum intramolecular advantage is reduced by a factor of about $5^4 = 3125$. The net intramolecular rate advantage would then be $10^8/3125$ = about 32,000—a much smaller, but still considerable, intramolecular advantage. Notice, though, that as rings get larger, the intramolecular advantage decreases. In practice, intramolecular reactions involving the formation of seven-membered or larger rings are not very common.

As Eq. 11.65 shows, the reaction probability, or $\Delta S^{\circ\ddagger}$, is balanced against the $\Delta H^{\circ\ddagger}$ for the reaction. Thus, when a cyclic species such as the episulfonium ion in Eq. 11.63a is formed in an intramolecular reaction, its angle strain increases the $\Delta H^{\circ\ddagger}$ of the reaction. This opposes the entropic advantage of the reaction. This is why the acceleration for Eq. 11.61b is less than the theoretical maximum. Nevertheless, the entropic advantage of the intramolecular reaction is so great that *it occurs despite the strain in the three-membered ring that is formed*. In fact, the strain in the episulfonium ion in Eq. 11.63a is the reason that it reacts rapidly with water (Eq. 11.63b). If a neighboring sulfur were located four carbons away from the α-carbon of the alkyl chloride (end-of-chapter Problem 11.75), the $\Delta S^{\circ\ddagger}$ would be smaller (less positive, more negative) because of the larger number of internal rotations that are lost on forming a larger ring, but the $\Delta H^{\circ\ddagger}$ would not be increased by ring strain.

PROBLEMS

11.32   Give the structure of an *intramolecular* substitution product and an *intermolecular* substitution product that might be obtained from 4-bromo-1-butanol on treatment with one equivalent of NaOH. Which product do you think would be the major one? Why?

11.33   Two reactions, A and B, have the same $\Delta H^{\circ\ddagger}$, but the $\Delta S^{\circ\ddagger}$ of reaction A is -30 J deg^{-1} mol^{-1}, and the $\Delta S^{\circ\ddagger}$ of reaction B is -180 J deg^{-1} mol^{-1}. At 25 °C (298 K), which reaction is faster and by what factor? (*Hint:* Apply Eq. 9.20b, p. 391.)

11.34   Indicate which reaction in each of the following pairs should have the greater (less negative, more positive) standard entropy of activation. Explain.

(a) Formation of product A or formation of product B. (*Hint:* Remember that loss of internal rotations decreases freedom of motion and thus lowers entropy.)

continued

continued

B. The Proximity Effect and Effective Molarity

The 10^8 intramolecular–intermolecular rate ratio calculated in Eq. 11.67 is the *maximum* theoretical kinetic (rate) advantage of intramolecular reactions. The actual ratio varies from reaction to reaction. The actual rate acceleration of an intramolecular reaction over its inter-molecular counterpart is called the **proximity effect**. The proximity effect is expressed quantitatively as the ratio of rate constants for an intramolecular reaction (rate constant k_1) and its intermolecular counterpart (rate constant k_2).

$$\text{proximity effect} = \frac{k_1}{k_2} \qquad (11.68)$$

An intramolecular process is typically unimolecular and follows a first-order rate law. An intermolecular process is typically bimolecular and follows a second-order rate law. Because a first-order rate constant k_1 has units of s^{-1} and a second-order rate constant has units of $M^{-1}\,s^{-1}$ (Sec. 9.3B), the proximity effect k_1/k_2 has units of concentration (that is, M). For this reason, the proximity effect is sometimes called the **effective molarity**. Interpretation of the effective molarity provides us with additional insight on the advantages of intramolecular reactions. Calculation of a proximity effect and the significance of the effective molarity are explored in Study Problem 11.5.

STUDY PROBLEM   11.5

Calculate the proximity effect for the intramolecular reaction in Eq. 11.61b given that its rate is 3200 times greater than its intermolecular counterpart.

SOLUTION   The intramolecular reaction (Eq. 11.61b) is 3200 times faster than the intermolecular reaction (Eq. 11.61a). Notice that the solvolysis rate of hexyl chloride is taken as an approximation of the intermolecular solvolysis rate of the sulfide. This approximation is necessary because the intramolecular reaction is so fast that the sulfide does not undergo the intermolecular reaction. The factor of 3200 is the ratio of rates. Taking the ratio of the rate laws for the two reactions, noting that the water concentration is 20 M, and letting the alkyl halides be at the same concentration for comparison,

$$\frac{\text{rate of reaction 11.60b}}{\text{rate of reaction 11.60a}} = 3200 = \frac{k_1[\text{alkyl halide}]}{k_2[\text{alkyl halide}][\text{H}_2\text{O}]} = \frac{k_1}{k_2(20\ M)} \qquad (11.69a)$$

from which we calculate the proximity effect as

$$\text{proximity effect} = \frac{k_1}{k_2} = (3200)(20\ M) = 64{,}000\ M \qquad (11.69b)$$

The effective molarity is therefore 64,000 M. The effective molarity is the concentration of the nucleophile (water in this case) required for the rate of the intermolecular reaction to equal that of the intramolecular reaction. Because the concentration of water is 55.6 M in pure water, 64,000 M is an impossibly large concentration. What this large effective molarity means, then, is that it is impossible to make the water concentration high enough for the intermolecular reaction to take place; the intramolecular reaction is orders of magnitude faster than the corresponding intermolecular one regardless of the nucleophile concentration. The intermolecular reaction doesn't stand a chance in the competition with its intramolecular counterpart.

As we have seen, the rate accelerations of intramolecular reactions are described by the proximity effect. The ratio of 10^8 M calculated in Eq. 11.67 is the maximum value of the proximity effect. However, as Study Problem 11.5 shows, the actual proximity effect, although still considerable, can be much less.

Let's summarize what we've learned about the proximity effect:

1. Intramolecular nucleophilic substitution reactions are particularly common for cases involving the formation of three-, five-, and six-membered rings. Intramolecular nucleophilic substitution reactions that form seven-membered or larger rings are much less common.

2. The main reason that intramolecular nucleophilic substitutions are favored over their intermolecular counterparts is a more favorable (less negative or more positive) $\Delta S^{\circ\ddagger}$—the reaction probability—for intramolecular reactions.

3. The rate acceleration of intramolecular reactions over competing intermolecular reactions is measured as the ratio of rate constants for the two processes. This ratio is the *proximity effect* or the *effective molarity.*

4. The magnitude of the proximity effect varies from reaction to reaction, but the $\Delta S^{\circ\ddagger}$ contribution can theoretically be as high as 10^8 M.

The Bombing of Bari Harbor, Intramolecular Reactions, and Cancer Chemotherapy

During the later part of World War II, the port of Bari in southeastern Italy was used as a logistics hub for the Allied invasion of Italy. The British commander, Sir Arthur Coningham, at a press conference on December 2, 1943, declared that the German air force had been defeated and posed no threat to Allied operations. That very evening, the Germans staged a surprise air raid on the well-lit port that destroyed 30 ships and damaged several more. (The port was disabled for several months; the raid became known as "Little Pearl Harbor.") The raid also burst a fuel line and the harbor was set ablaze with burning fuel. In addition, vast amounts of oil from the sinking ships covered the water. One of the ships, the *John Harvey*, carried a secret load of mustard gas ($Cl—CH_2CH_2—S—CH_2CH_2—Cl$) that was intended for retaliatory use should the Germans resort to chemical warfare. Mustard was spread all over the harbor, readily dissolving in the oil that covered the water, and a cloud of gas seeped into the town, which had a population of 250,000. Mustard gas causes severe burns, and hundreds of both soldiers and townspeople, especially soldiers who were covered with oil containing dissolved mustard, suffered the effects of acute mustard poisoning. Over a thousand people eventually died from mustard poisoning.

Some of the survivors of this raid who had been exposed to mustard gas were subsequently found to have a greatly depleted lymphocyte (white blood cell) count and a scorched bone marrow. At about the same time, the U.S. Army had awarded a research contract to two Yale University pharmacologists, Louis Goodman and Alfred Gilman, who were studying the effects of mustards and similar compounds on laboratory animals. They found that animals injected with mustards did not show any of the burn symptoms, but the white cells in their blood and bone marrow virtually disappeared. Goodman and Gilman wondered whether small doses of mustard might actually bring about remissions in lymphoma, a type of cancer in which abnormal white cells proliferate rapidly. A subsequent trial on a single human lymphoma patient brought about a remission. A nitrogen-containing version of mustard gas (mechlorethamine) was developed. Lymphomas and other cancers are treated today with newer, less toxic versions of the nitrogen mustards along with other drugs. Chlorambucil and cyclophosphamide are examples. Notice the "nitrogen mustard" functionality common to all of these drugs.

mechlorethamine (mustine)   **chlorambucil**   **cyclophosphamide**

It was subsequently shown that the mustards kill cells because they react with bases on DNA. The mustards, being very hydrophobic molecules, readily traverse the cell and nuclear membranes, and, once in the nucleus, react with a nitrogen (*N*-7) on a guanine (G) base of DNA.

guanine
(a DNA base)

N-7; this is the nucleophile that reacts with the mustards

DNA

We've already learned why mustard gas molecules are so reactive: they rapidly form three-membered sulfonium ion rings that are then opened rapidly by nucleophiles.

(11.70)

N-7 of a guanidine

alkylated N-7 of a guanidine

This would be bad enough for the DNA, but the real damage hasn't yet been done. Notice that the mustard molecule contains two reactive alkyl halide groups. If there is another guanine nearby on the *opposite* strand of DNA in the double helix, it reacts at its *N*-7 with the second group on the mustard. This reaction is *very* fast, because the second guanine is perfectly positioned to react, and its reaction is truly intramolecular. As a result, the DNA strands become tethered together, or *crosslinked*.

alkylated N-7 of a guanidine

(11.71)

N-7 of another guanidine on the other strand of the double helix

crosslinked DNA strands

During cell division, the two strands of the DNA double helix must separate. Crosslinking completely prevents strand separation and therefore shuts down DNA proliferation and white-cell proliferation. Furthermore, the cells with damaged DNA are destroyed by biological mechanisms that have evolved for this purpose. This is the desired result in cancer treatment.

PROBLEM

11.35 The nucleophilic substitution reaction of sodium 2-bromopropanoate with water and/or ⁻OH can occur by both an S_N2 (intermolecular) mechanism and a mechanism that involves neighboring-group participation.

sodium 2-bromopropanoate   **sodium lactate**

(a) Give the curved-arrow notation for the S_N2 mechanism with ⁻OH as the nucleophile.

(b) Give the curved-arrow notation for an intramolecular mechanism. This mechanism should lead you to the structure of an unstable intermediate, which then reacts with ⁻OH to give the product.

(c) The first-order rate constant k_1 for the intramolecular reaction is 1.2×10^{-4} s^{-1}. The second-order rate constant for the S_N2 reaction is 6.4×10^{-4} M^{-1} s^{-1}. Calculate the proximity effect for the intramolecular reaction.

(d) At what NaOH concentration does this reaction proceed by the two mechanisms at the same rate?

(e) What is the predominant mechanism in 1 M NaOH?

(f) Consider the structure of the intermediate you derived in part (b). The reason for the small proximity effect is that this intermediate is very unstable. Explain why this intermediate is more unstable than an ordinary epoxide. (*Hint:* Think about the preferred bond angles.)

C. Stereochemical Consequences of Neighboring-Group Participation

Neighboring-group participation can sometimes be diagnosed by considering its stereochemical outcome. Notice that the mechanism of the neighboring-group reaction given in Eqs. 11.63a–b involves *two* substitution reactions. The first is the intramolecular substitution by the neighboring sulfur nucleophile to give the episulfonium ion intermediate. The second is the ring-opening reaction of water with the episulfonium ion. This double-substitution mechanism has stereochemical consequences if the reacting molecule contains appropriately situated stereocenters. This point is illustrated in Study Problem 11.6.

STUDY PROBLEM   11.6

Indicate the stereochemical outcome of the following substitution reaction (a) if neighboring-group participation does not occur, and (b) if neighboring-group participation takes place.

(2R,3S)

SOLUTION

(a) If the reaction were a simple S_N2 reaction, the nucleophile would displace the chloride leaving group with inversion of configuration (Sec. 9.4C). Thus, the (2R,3S)-stereoisomer of the starting material should give the (2S,3S)-stereoisomer of the product:

(11.72)

(b) If the reaction involves neighboring-group participation, the intramolecular substitution occurs with inversion of configuration. Notice in this case that the resulting episulfonium ion has the 2S,3S configuration.

$$(11.73a)$$

The two carbons of the episulfonium ion are homotopic. Reaction of the nucleophile water at either carbon of the episulfonium ion gives the *same product*:

$$(11.73b)$$

If we compare the product stereochemistry with that of the starting material, we find that the neighboring-group mechanism gives net *retention of stereochemistry*.

$$(11.73c)$$

Because we know the neighboring-group mechanism to be correct, this is the expected stereochemical result.

When a substitution reaction occurs with overall retention of stereochemistry at an electrophilic center, we should strongly suspect that two substitutions have occurred, because all known instances of concerted (one-step) S_N2 reactions occur with inversion. If a potential nucleophile is present in the starting material so that it can form a small ring, a cyclic intermediate is likely.

PROBLEMS

11.36   Carry out an analysis similar to Study Problem 11.6 of the stereochemical result expected when the (2R,3R)-stereoisomer of the starting material is used.

11.37   The nucleophilic substitution reaction of sodium 2-bromopropanoate with water shown in Problem 11.35 occurs with retention of configuration at very low NaOH concentrations, but occurs with inversion of configuration at 1 M NaOH. Relate this finding to your answers for Problem 11.35.

11.38   In the nucleophilic substitution reaction of the following radioactively labeled compound with water, what labeling pattern should be observed in the product (a) if the neighboring-group participation does not occur and (b) if neighboring-group participation does occur?

$$Et\ddot{S}—CH_2\overset{*}{C}H_2—Cl \qquad \overset{*}{C} = {}^{14}C$$

11.39 Explain why the following two alcohols each react with HCl to give the same alkyl chloride.

$$Et\ddot{S}—CH—CH_2—OH \xrightarrow[\text{(81\% yield)}]{HCl}$$
$$\qquad\qquad |$$
$$\qquad\qquad CH_3$$

$$Et\ddot{S}—CH_2—CH—OH \xrightarrow[\text{(72\% yield)}]{HCl}$$
$$\qquad\qquad\qquad |$$
$$\qquad\qquad\qquad CH_3$$

$$\xrightarrow{\qquad} \quad Et\ddot{S}—CH_2—\overset{\overset{\displaystyle Cl}{|}}{CH}—CH_3 + H_2O$$

D. Intramolecular Reactions and Enzyme Catalysis

Enzymes, nature's catalysts, bring about rate accelerations of many orders of magnitude in the reactions that they catalyze. Enzymes are large protein molecules; the smallest enzymes have molecular masses of about 10,000 g mol^{-1}, and most are considerably larger. The molecules on which they act are called **substrates**. When an enzyme acts on its substrate, the first step is to bind the substrate tightly and *noncovalently* into a specific part of the enzyme called the *active site*. The enzyme with its bound substrate is called the **enzyme–substrate complex**. *Binding is an obligatory step in catalysis by all enzymes.* If E is the enzyme, S is a substrate, and P is the reaction product, this scheme is as follows.

$$E + S \underset{k_2}{\overset{k_1}{\rightleftharpoons}} E{\cdot}S \overset{k_3}{\longrightarrow} E{\cdot}P \rightleftharpoons E + P \qquad (11.74)$$

binding step | catalysis step | enzyme–substrate complex

$$K_s = \text{dissociation constant of the E{\cdot}S complex} = k_2/k_1$$

A diagram of this process is shown in Fig. 11.2 on p. 550. Binding is an essential part of catalysis because the binding event positions the substrate in an optimal configuration relative to the chemical groups on the enzyme that bring about the actual reaction. The enzyme contains, as part of its structure, groups that can serve as acid catalysts, base catalysts, and nucleophiles. Metal ions, held in place by coordination with groups on the enzyme, can additionally act as Lewis acid catalysts. Here is the key point: Once the substrate is bound, *the subsequent catalytic step (k_3 in Eq. 11.74) is an intramolecular reaction* within the E·S complex. We've just learned about the kinetic advantage of intramolecular reactions. *The major part of the catalytic efficiency of enzymes resides in the fact that their reactions are intramolecular.* Moreover, we sometimes find that a mechanism occurring in the active site of an enzyme is not observed in nonenzymatic reactions because, in a nonenzymatic reaction, the simultaneous positioning of the reactant and several catalytic molecules is simply too improbable (would have a very large, negative $\Delta S^{\circ\ddagger}$). In contrast, in an enzyme-catalyzed reaction, the substrate and all of the catalytic groups are "pre-positioned" by the binding event so that a "multi-catalyst" mechanism is strongly accelerated.

Perhaps you've realized that the process of bringing the substrate (or, in some cases, several substrates) together into an enzyme active site must overcome the highly unfavorable translational and rotational entropies associated with converting two species (E and S) into one (the E·S complex). How is this accomplished? Enzymes use noncovalent forces of the type that we learned about in Chapter 8—hydrogen bonding, hydrophobic bonding, and, in some cases, electrostatic and ion–dipole attractions—to pay the entropic price of binding. Electrostatic attractions—attractions between oppositely charged ionic groups on the enzyme and substrate, when they are present, are particularly strong. The noncovalent attractions involved in enzyme–substrate binding are significant enough that, even after paying the entropic cost,

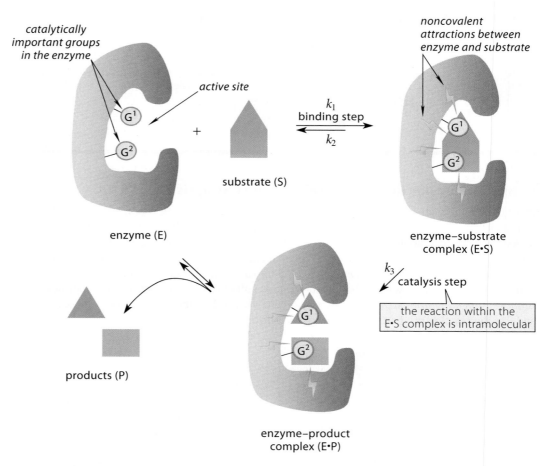

FIGURE 11.2 A diagram of enzyme catalysis that follows Eq. 11.74. The substrate S is bound to the enzyme E by noncovalent attractions to give the enzyme–substrate complex E·S. The conversion of bound substrate to bound product (E·S to E·P) is intramolecular. Release of product P from the enzyme–product complex regenerates the enzyme for further catalysis.

the binding of E and S is fairly strong. The enzyme–substrate binding is described by the dissociation constant K_s of the E·S complex (Eq. 11.74); stronger binding is characterized by smaller values of K_s. The K_s values of typical E·S complexes are in the 10^{-3} to 10^{-7} M range. Because binding is noncovalent, chemical bonds do not have to be broken and formed; so, the enzymes and substrates form E·S complexes very rapidly, in many cases with rate constants k_1 in Eq. 11.74 that are at or near the rates of diffusion (10^9–10^{10} M^{-1} s^{-1}).

Fig. 11.2 is a "conceptual framework" for understanding enzyme catalysis. In later chapters, and particularly in Sec. 27.10, we'll be examining actual cases of enzyme binding and catalysis at the molecular level.

In summary, enzymes use noncovalent forces to overcome the entropic cost of binding their substrates. The structure of each enzyme incorporates one or more catalytic groups in proximity to the substrate. For this reason, enzymes have been referred to as "entropy traps." Once the substrate is bound, the rates of catalytic reactions are strongly enhanced because they are intramolecular.

The binding event is the "Achilles heel" of the enzyme. If we can find or design small molecules that undergo strong binding to an enzyme active site without undergoing the subsequent catalytic reaction, such molecules can be used to compete with substrate molecules and, if they bind tightly enough, to shut down enzyme activity. Such molecules are called **competitive inhibitors**. If the enzyme activity is harmful, competitive inhibitors can be used as drugs. The design of competitive inhibitors is a significant part of modern drug development.

11.9 OXIDATION OF ETHERS AND SULFIDES

A. Oxidation of Ethers as Safety Hazards

Ethers are relatively inert toward many of the common oxidants used in organic chemistry if the reaction conditions are not too vigorous. For example, diethyl ether can be used as a solvent for oxidations with Cr(VI). On standing in air, however, ethers undergo slow **autoxidation**, the spontaneous oxidation by oxygen in air. Samples of ethers can accumulate dangerous quantities of explosive peroxides and hydroperoxides by autoxidation. This reaction is known to occur in two of the most common ethers used in the laboratory, tetrahydrofuran (THF) and diethyl ether.

$$CH_3CH_2—O—CH_2CH_3 + O_2 \longrightarrow CH_3CH_2O—\underset{\underset{O—O—H}{|}}{CH}—CH_3 \longrightarrow \text{other polymeric peroxides} \qquad (11.75)$$

diethyl ether

a hydroperoxide

$$CH_3CH_2—O—O—CH_2CH_3$$

diethyl peroxide

These peroxides can form by free-radical processes in samples of anhydrous diethyl ether, THF, and other ethers within less than two weeks. For this reason, some ethers are sold with small amounts of free-radical inhibitors, which can be removed by distilling the ether. Because peroxides are particularly explosive when heated, it is a good practice not to distill ethers to dryness. Peroxides in an ether can be detected by shaking a portion of the ether with 10% aqueous potassium iodide solution. If peroxides are present, they oxidize the iodide to iodine, which imparts a yellow tinge to the solution. Small amounts of peroxides can be removed by distillation of the ethers from lithium aluminum hydride ($LiAlH_4$), which both reduces the peroxides and removes contaminating water and alcohols.

A second oxidation reaction—combustion—is a particular hazard of diethyl ether. Its flammability is indicated by its very low flash point of –45 °C. The **flash point** of a material is the minimum temperature at which it is ignited by a small flame under certain standard conditions. In contrast, the flash point of THF is –14 °C. Compounding the flammability hazard of diethyl ether is the fact that its vapor is 2.6 times more dense than air. This means that vapors of diethyl ether from an open vessel will accumulate in a heavier-than-air layer along a laboratory floor or benchtop. For this reason flames can ignite diethyl ether vapors that have spread from a remote source. Good safety practice demands that open flames or sparks not be permitted anywhere in a laboratory in which diethyl ether is in active use. Even the spark from an electric switch (such as that on a hot plate) can ignite diethyl ether vapors. A steam bath is therefore one of the safest ways to heat this ether.

The sulfur analogs of peroxides are *disulfides* (R—S—S—R), which are the oxidation products of thiols (Sec. 10.10B). Disulfides are not explosive, and they in fact occur widely in nature within the structures of proteins (Sec. 27.8A).

B. Oxidation of Sulfides

Like thiols, sulfides oxidize at *sulfur* rather than carbon when they react with common oxidizing agents. Sulfides can be oxidized to **sulfoxides** and **sulfones**:

$$R—\overset{..}{\underset{..}{S}}—R \xrightarrow{oxidation} \left[R—\overset{\overset{:O:}{||}}{\underset{..}{S}}—R \longleftrightarrow R—\overset{\overset{:\overset{..}{O}:^-}{|}}{\underset{+}{S}}—R \right] \xrightarrow{oxidation} \left[R—\overset{\overset{:O:}{||}}{\underset{\underset{:O:}{||}}{S}}—R \longleftrightarrow R—\overset{\overset{:\overset{..}{O}:^-}{|}}{\underset{\underset{:\overset{..}{O}:^-}{|}}{S^{2+}}}—R \right] \qquad (11.76)$$

a sulfoxide   a sulfone

Dimethyl sulfoxide (DMSO) and sulfolane are well-known examples of a sulfoxide and a sulfone, respectively. (Both compounds are excellent dipolar aprotic solvents; see Table 8.2, p. 355.)

$$
\begin{array}{cc}
\underset{\substack{\textbf{dimethyl sulfoxide}\\ \textbf{(DMSO)}}}{H_3C-\overset{\overset{O}{\|}}{\underset{\cdot\cdot}{S}}-CH_3} &
\underset{\substack{\textbf{tetramethylene sulfone, or}\\ \textbf{sulfolane}}}{\underset{O\quad O}{\overset{S}{\diamond}}}
\end{array}
$$

Notice that nonionic Lewis structures for sulfoxides and sulfones have more than an octet of electrons on the sulfur. The bonding at sulfur in such situations was discussed in Sec. 10.10A.

Sulfoxides and sulfones can be prepared by the direct oxidation of sulfides with one and two equivalents, respectively, of hydrogen peroxide, H_2O_2:

$$
\text{(structure)} \xrightarrow[\substack{H_2O/acetone\\25\,°C,\,48\,h}]{1\ equiv.\ H_2O_2} \text{(structure)} \quad \text{(88\% yield)}
$$

2 equiv. H_2O_2
heat, 4 h

(97% yield)

(11.77)

Other common oxidizing agents, such as $KMnO_4$, HNO_3, and peroxyacids (Sec. 11.3A), also readily oxidize sulfides.

Sulfide Oxidation and Protein-Aggregation Diseases

Alzheimer's disease (AD) and Parkinson's disease (PD) are two diseases that occur mostly in the elderly. AD is characterized by increasing dementia, and PD by tremors and debilitating movement disorders. Both increase over time. As people are living longer, the incidence of both diseases is increasing. Both diseases are characterized by the aggregation of proteins into tangles or fibrils that affect neuronal function. For example, in PD, an aggregated form of a common protein α-synuclein accumulates in neuronal cells of the brain; these accumulated fibrils, when stained, appear under the microscope as "Lewy bodies" (photo). When dopamine-producing cells begin to accumulate α-synuclein fibrils, the cells begin to die. Whether the α-synuclein fibrils are causative has not been firmly established, but the process of "fibrillization" seems to be toxic in some way. An intensive research effort is underway to understand what causes formation of these fibrils.

There is some evidence that oxidation of the sulfide group of the α-amino acid methionine in certain proteins might be a primary event leading to the formation of fibrils. (Proteins are chains of α-amino acids with the general structure $\overset{+}{H_3N}-CHR-CO_2^-$ connected by amide bonds, as shown below. The various α-amino acids differ in the structures of their R-groups.) Methionine is oxidized to a sulfoxide.

(11.78)

methionine
(the amino acid)

a methionine residue
in a peptide or protein

a methionine sulfoxide residue
in a peptide or protein

This hypothesis has led to the idea that a diet rich in antioxidants might offer some protection against AD and PD, but there is no firm evidence yet to support this idea.

11.10   THE THREE FUNDAMENTAL OPERATIONS OF ORGANIC SYNTHESIS

Section 10.12 introduced organic synthesis with a systematic approach to solving synthesis problems. This section continues this approach by classifying the operations involved in a typical synthesis. Most reactions used in organic synthesis involve one or more of *three fundamental operations*:

1. functional-group transformation
2. control of stereochemistry
3. formation of carbon–carbon bonds

Functional-group transformation—the conversion of one functional group into another—is the most common type of synthetic operation. Most of the reactions you've studied so far involve functional-group transformation. For example, the hydrolysis of epoxides transforms epoxides into glycols; hydroboration–oxidation converts alkenes into alcohols.

Control of stereochemistry is accomplished with stereoselective reactions. Whenever you have to prepare a compound that can exist as several stereoisomers, you should think in terms of these reactions. Examples of stereoselective reactions include hydroboration–oxidation, which is a syn-addition, and S_N2 reactions, which occur with inversion of configuration.

Reactions that bring about the *formation of carbon–carbon bonds* are particularly important, because these reactions must be used to add carbon atoms, and thus "grow" larger carbon chains from smaller ones. Only two reactions of this type have been presented:

1. cyclopropane formation from carbenes or carbenoids and alkenes (Sec. 9.9)
2. reaction of Grignard and organocuprate reagents with epoxides (Sec. 11.5C)

Most reactions involve combinations of at least two of the three fundamental operations. For example, hydroboration–oxidation is a functional-group transformation (alkene $\longrightarrow$ alcohol) that also allows, at the same time, the control of stereochemistry. The reaction of an epoxide with a Grignard or organocuprate reagent effects both carbon–carbon bond formation and a functional-group transformation (epoxide $\longrightarrow$ alcohol).

Study Problems 11.7 and 11.8 demonstrate how to use the three fundamental operations in planning an organic synthesis.

STUDY PROBLEM   11.7

Outline a synthesis of 1-hexanol from 1-butanol and any other reagents.

SOLUTION   As usual, first write the problem in terms of structures:

$$CH_3CH_2CH_2CH_2-OH \xrightarrow{\ ?\ } CH_3CH_2CH_2CH_2CH_2CH_2-OH$$

Next, analyze the types of operations needed. Two carbons must be added, but no issues of stereochemistry are involved. You should *not* assume that because both the starting material and final product are alcohols, no functional group transformations will be necessary. As shown in the following reactions, the alcohol group is transformed into other groups during the synthesis.

Now work backward from the product. The reaction of a Grignard reagent with ethylene oxide followed by protonolysis would add the required two carbon atoms and would form the desired alcohol:

$$CH_3CH_2CH_2CH_2\!-\!MgBr + H_2C\!\overset{\displaystyle O}{\diagdown\!\!\diagup}\!CH_2 \longrightarrow \xrightarrow{\;H_3O^+\;} CH_3CH_2CH_2CH_2CH_2CH_2\!-\!OH$$

Next, decide how to prepare the Grignard reagent. We've learned only one way:

$$CH_3CH_2CH_2CH_2\!-\!Br + Mg \xrightarrow[\text{ether}]{} CH_3CH_2CH_2CH_2\!-\!MgBr$$

Because the alkyl halide required for this step has the same number of carbons as the starting alcohol, one functional-group transformation remains to complete the synthesis. A primary alcohol can be converted into the required primary alkyl halide with concentrated HBr. Summarizing the completed synthesis:

$$CH_3CH_2CH_2CH_2\!-\!OH \xrightarrow[\text{heat}]{HBr,\ H_2SO_4} CH_3CH_2CH_2CH_2\!-\!Br \xrightarrow[\text{ether}]{Mg}$$

$$CH_3CH_2CH_2CH_2\!-\!MgBr \xrightarrow{H_2C\overset{O}{\diagdown\!\!\diagup}CH_2} \xrightarrow{H_3O^+} CH_3CH_2CH_2CH_2CH_2CH_2\!-\!OH$$

Notice that the sequence of reactions used in Study Problem 11.7 is a general one for the net *two-carbon chain extension* of a primary alcohol.

$$R\!-\!OH \longrightarrow R\!-\!Br \longrightarrow RMgBr \xrightarrow{H_2C\overset{O}{\diagdown\!\!\diagup}CH_2} \xrightarrow{H_3O^+} R\!-\!CH_2CH_2\!-\!OH \qquad (11.79)$$

net carbon chain extension by two carbons

Outline a synthesis of (±)-*trans*-2-methoxycyclohexanol from cyclohexene.

SOLUTION   Notice that no new carbon–carbon bonds are joined to the cyclohexene ring, so reactions that form carbon–carbon bonds are not likely to be useful. There is, however, a stereochemical problem: the two oxygens must be introduced in a trans arrangement. Finally, notice that a net addition of $CH_3O\!-$ and $HO\!-$ to the carbon–carbon double bond is required. Such an addition cannot be completed in one step. However, in the opening of epoxides, the epoxide oxygen becomes an $-OH$ group, and this transformation occurs with inversion of stereochemistry at the broken bond so that the resulting groups end up trans. Opening an epoxide with CH_3O^- in CH_3OH, or with CH_3OH and an acid catalyst, is thus a good last step to the synthesis.

Completion of the synthesis requires only the preparation of the epoxide from cyclohexene. (How is this accomplished? See Problem 11.12a.)

PROBLEM

11.40 Outline a synthesis for each of the following compounds from the indicated starting materials and any other reagents:

(a) $(CH_3)_2CHCH_2CH_2CO_2H$ from $(CH_3)_2C{=}CH_2$ (2-methylpropene)

(b) $(CH_3)_2CHCO_2H$ from 2-methylpropene

(c) dibutyl sulfone from 1-butanethiol

(d) ($\pm$)-*trans*-1-ethoxy-2-methylcyclopentane from cyclopentene

---

11.11 SYNTHESIS OF ENANTIOMERICALLY PURE COMPOUNDS: ASYMMETRIC EPOXIDATION

As discussed in Sec. 11.10, control of stereochemistry is one of the important elements of an organic synthesis. Conventionally, the control of stereochemistry is limited to the synthesis of individual diastereomers. If a synthesis begins with achiral starting materials, as many syntheses do, chiral products are obtained as racemates (Sec. 7.7). In such a case, the only way to obtain pure enantiomers is to carry out an optical resolution at some stage of the synthesis. When such an enantiomeric resolution is necessary, half of the material—the unwanted enantiomer—is wasted. An optical resolution can be avoided if a chiral starting material can be obtained as a single enantiomer. The enantiomerically pure chiral compounds found in nature can serve as one source of such starting materials. Occasionally, an enzyme can be found to catalyze the formation of an enantiomerically pure compound from achiral starting materials. However, such sources of pure enantiomers are relatively limited. Yet, the synthesis of pure enantiomers without enantiomeric resolution has taken on special importance as regulatory agencies have mandated that chiral pharmaceuticals be produced as pure enantiomers rather than racemates.

In recent years, chemists have learned to "custom-design" chiral catalysts that can bring about the formation of enantiomerically pure chiral compounds from achiral starting materials. This section discusses one of the most useful catalysts, which brings about the epoxidation of allylic alcohols to give enantiomerically pure epoxides.

An allylic alcohol contains an —OH group on a carbon *adjacent to* (not part of) a double bond. The simplest compound of this type, 2-propen-1-ol, has the common name *allyl alcohol*.

general structure of an allylic alcohol

$H_2C{=}CH{-}CH_2{-}OH$

2-propen-1-ol
(allyl alcohol)

When an allylic alcohol is treated with titanium(IV) isopropoxide catalyst, which we'll abbreviate as $Ti(OiPr)_4$, along with *tert*-butyl hydroperoxide, an epoxide is formed at the double bond of the allylic alcohol functionality. Aqueous NaOH is added to destroy the catalyst and extract the by-products; the epoxide is isolated from the CH_2Cl_2 solvent. Although the reaction is a stereospecific syn-addition, the product epoxide is racemic.

(11.80)

(E)-2-hexen-1-ol

tert-butyl hydroperoxide

trans-3-propyloxiranemethanol
(racemic)

But when (2R,3R)-(+)-diethyl tartrate is added to the reaction mixture, the product, obtained in 97% enantiomeric purity, is the 2S,3S enantiomer, and the (+)-diethyl tartrate can be recovered unchanged!

(2R,3R)-(+)-diethyl tartrate

(E)-2-hexen-1-ol

(2S,3S)-3-propyloxiranemethanol
(80% yield; 97% enantiomerically pure)

(11.81a)

When (2S,3S)-(−)-diethyl tartrate is used instead, the other enantiomer—the 2R,3R enantiomer—of the product is obtained.

(2S,3S)-(−)-diethyl tartrate

(E)-2-hexen-1-ol

(2R,3R)-3-propyloxiranemethanol

(11.81b)

The two enantiomeric tartrate esters can be easily prepared from the enantiomers of tartaric acid, an inexpensive and readily available, naturally occurring compound, and they are commercially available. (You may recall that tartaric acid is the compound that was the object of the first enantiomeric resolution by Louis Pasteur; Sec. 6.10). These tartrate esters will be abbreviated as (+)-DET and (−)-DET, respectively.

These results are general for a large number of allylic alcohols. If we draw the allylic alcohol with the —CH$_2$OH group in the upper-right position on the double bond, the results can be generalized as follows:

"O" adds from below

"O" adds from above

(11.82)

Only double bonds with an allylic —OH group form an epoxide; others usually don't react.

(11.83)

(80% yield)

This reaction, called **asymmetric epoxidation**, was discovered by Prof. K. Barry Sharpless (b. 1941) and his student Tsutomu Katsuki in 1980 at the Massachusetts Institute of Technology. Prof. Sharpless, now at The Scripps Research Institute, shared the 2001 Nobel Prize in Chemistry for this discovery. (The reaction is often called the **Sharpless epoxidation**.) Its importance hinges on the many stereospecific ring-opening reactions that epoxides can undergo, as illustrated in Study Problem 11.8. Following asymmetric epoxidation, stereospecific ring opening of the epoxide can introduce a wide variety of enantiomerically pure organic compounds. Since its introduction, asymmetric epoxidation has been a key step the synthesis of many important, optically pure, chiral compounds.

STUDY PROBLEM 11.9

Outline a synthesis of the following compound as a single enantiomer.

SOLUTION   When we see an —OH and another functional group, in this case the $(CH_3)_2N$— (dimethylamino) group, on adjacent carbons in a trans stereochemical relationship, we should think of an epoxide opening as a way to introduce both groups, as in Study Problem 11.8. The following reaction would work.

The epoxide is one that can be made by allylic epoxidation; the pattern in Eq. 11.82 shows that (+)-DET should be the chiral additive.

An intriguing question is how the chirality of a diethyl tartrate enantiomer determines the stereochemical outcome of the reaction. First of all, titanium(IV) alkoxides can readily exchange any or all of their alkoxide groups with other alcohols.

$$
iPrO-\underset{\underset{OiPr}{|}}{\overset{\overset{OiPr}{|}}{Ti}}-OiPr \xrightarrow{ROH} RO-\underset{\underset{OiPr}{|}}{\overset{\overset{OiPr}{|}}{Ti}}-OiPr \xrightarrow{ROH} \text{further exchanges} \qquad (11.84)
$$

$$+ \; iPrOH$$

The hydroxy groups of DET, the *tert*-butyl hydroperoxide, and the allylic alcohol can thus replace the isopropyloxy groups and become bound to the same titanium. Once one of the hydroxy groups of DET reacts this way, the reaction of the second hydroxy group with the titanium becomes an *intramolecular* reaction (Sec. 11.8) and is very fast. Because DET is chiral, the addition of DET results in a chiral titanium alkoxide complex. Structural studies have shown that the complex actually involves two titaniums and two DET molecules with the structure shown in Fig. 11.3a (for (+)-DET as the additive). In this complex, the oxygen of *tert*-butyl hydroperoxide (shown in yellow) is poised below the double bond in an optimal arrangement to receive the nucleophilic donation of π electrons, as shown in Fig. 11.3c. Fig. 11.3b shows the same complex with the alkene oriented for reaction of the oxygen at the opposite face of the double bond—the reaction that is not observed. In this arrangement, the —CH$_2$— group of the allylic alcohol is forced into a more congested part of the catalyst, where it has unfavorable steric interactions (van der Waals repulsions) with one of the tartrate ester groups. This steric effect is the mechanism by which the chirality of the tartrate ester "dictates" the stereochemistry of the reaction.

The structure explains the catalyst *specificity*; that is, the catalyst is largely specific for allylic alcohols because of the relationship of the allylic —OH, the double bond, and the peroxide oxygen. Other unsaturated alcohols undoubtedly bond to the titanium, but their double bonds are not in the proper relationship to the peroxide oxygen for reaction. The reaction also doesn't work for *tertiary* allylic alcohols, because the van der Waals repulsions of the alkyl branches with the other groups bound to the catalyst prevent proper binding.

PROBLEMS

11.41 Give the product and its stereochemistry when each of the following alcohols is subjected to asymmetric epoxidation with *tert*-butyl hydroperoxide, Ti(O*i*Pr)$_4$, and the stereoisomer of diethyl tartrate (DET) indicated.

(a) Ph—CH=CH—CH$_2$OH, (−)-DET

(b)
$$\underset{H_3C}{\overset{H}{\diagdown}}C=C\underset{CH_3}{\overset{CH_2OH}{\diagup}}, \; (+)\text{-DET}$$

(c) (diene with OH) , (−)-DET

11.42 Propose a synthesis for each of the following compounds in enantiomerically pure form. Use an asymmetric epoxidation in each synthesis.

(a) (cyclopentane epoxide with H and CH$_2$OH)

(b)
$$\underset{CH_3CH_2CH_2S}{\overset{CH_3CH_2}{\diagdown}}\overset{H}{\underset{}{C}}-\overset{OH}{\underset{CH_3}{C}}\cdots CH_2OH$$

11.43 (a) Use the picture of the catalyst complex in Fig. 11.3a to explain why most *E* allylic alcohols undergo asymmetric epoxidation more rapidly than their *Z* isomers.

$$
\underset{R}{\overset{H}{\diagdown}}C=C\underset{H}{\overset{CH_2OH}{\diagup}} \qquad \underset{H}{\overset{R}{\diagdown}}C=C\underset{H}{\overset{CH_2OH}{\diagup}}
$$

an *E* allylic alcohol   a *Z* allylic alcohol

(b) Would the same phenomenon be observed with (−)-DET, the enantiomer of the DET used in Fig. 11.3? Explain.

these hydrogens are involved in
van der Waals repulsions

(a)

(b)

(c)

FIGURE 11.3 Molecular models of complexes for asymmetric epoxidation, shown with allyl alcohol as the reacting alcohol and (2*R*,3*R*)-(+)-diethyl tartrate as the chiral additive. The carbons and bonds of allyl alcohol are shown in *purple*, and the hydrogens in *blue*. The oxygen of *tert*-butyl hydroperoxide that is delivered to the alkene is shown in *yellow*. Notice that this oxygen is perfectly positioned to receive electrons from the π bond. (a) The complex that leads to the epoxide with the observed chirality. (b) The complex that leads to the epoxide with the chirality that is not observed. This complex is less stable because of van der Waals repulsions between the allylic —CH₂— hydrogens and one of the tartrate ester groups. (c) A curved-arrow mechanism corresponding to stereochemistry (a).

KEY IDEAS IN CHAPTER 11

- Ethers are weak bases with conjugate-acid pK_a values of –2 to –3. Sulfides are much weaker bases with conjugate-acid pK_a values of –6 to –7. Because ethers are Lewis bases, they donate electrons to Lewis acids such as BF_3 and Grignard reagents.

- Ethers can be synthesized by the Williamson ether synthesis (an S_N2 reaction), by alkoxymercuration–reduction, or by the dehydration of alcohols or the related acid-catalyzed addition of alcohols to alkenes.

- Epoxides can be synthesized by the syn-oxidation of alkenes with peroxycarboxylic acids or by the cyclization of halohydrins. Epoxides of allylic alcohols can be prepared in high enantiomeric purity by asymmetric epoxidation using titanium isopropoxide, *tert*-butyl hydroperoxide, and either (+)- or (−)-diethyl tartrate.

- Ordinary ethers are relatively unreactive compounds. The major reaction of ethers is cleavage, which occurs under strongly acidic conditions, but not under typical basic conditions. The acid-catalyzed cleavage of tertiary ethers occurs more readily than the cleavage of primary or methyl ethers because tertiary carbocation intermediates can be formed.

- Because of their ring strain, epoxides undergo ring-opening reactions with ease. For example, epoxides react with water to give glycols, ethylene oxide reacts with Grignard reagents to give primary alcohols, and other epoxides react with lithium organocuprate reagents to give secondary or tertiary alcohols. In acidic solution, a protonated epoxide preferentially reacts with nucleophiles at the tertiary carbon. Bases react with an epoxide at the carbon with fewer alkyl substituents.

- Ring-opening reactions of epoxides in either acid or base occur with inversion of configuration at carbon stereocenters.

- Glycols (1,2-diols) can be prepared from alkenes by treatment with OsO_4 followed by hydrolysis, by variations of this reaction in which a catalytic amount of OsO_4 is used along with other oxidants, or by treatment with aqueous alkaline $KMnO_4$. This stereospecific reaction results in a net syn-addition of two hydroxy groups to the alkene double bond.

- Glycols can be oxidatively cleaved into two carbonyl compounds (aldehydes or ketones) by treating them with periodic acid.

- Oxonium and sulfonium salts react with nucleophiles in substitution and elimination reactions; oxonium salts are more reactive than sulfonium salts. *S*-Adenosyl-methionine (SAM) is a sulfonium salt used in nature as a methylating agent.

- A number of intramolecular reactions are faster than their intermolecular counterparts because intramolecular

reactions have larger (less negative) $\Delta S^{\circ\ddagger}$ values. The rate acceleration of an intramolecular reaction is characterized by the proximity effect, or effective molarity, which is the ratio of rate constants for the intramolecular and corresponding intermolecular reactions. Proximity effects due to $\Delta S^{\circ\ddagger}$ differences can be as high as 10^8 M.

- The catalytic power of enzymes resides in the fact that they form enzyme–substrate complexes; because the reactions that take place within these complexes are intramolecular, they are strongly accelerated.

- Intramolecular nucleophilic substitution reactions are most common when the reaction forms six-membered and smaller rings.

- Alkyl halides with β-alkylthio substituents react intramolecularly by forming three-membered sulfonium ions, which are rapidly opened with nucleophiles. If the α-carbon of the alkyl halide is a stereocenter, such reactions involve two consecutive inversions of configuration, which then results in overall retention of configuration.

- Ethers, on standing in air, slowly form explosive peroxides, and diethyl ether is particularly flammable. Both oxidation and combustion reactions represent safety hazards of ethers. Ethers are largely inert toward other oxidation reactions.

- Sulfides are readily oxidized to sulfoxides and sulfones.

- The three fundamental operations of organic synthesis are (1) functional-group transformation, (2) control of stereochemistry, and (3) carbon–carbon bond formation.

 REACTION REVIEW *For a summary of reactions discussed in this chapter, see the* Reaction Review *section of Chapter 11 in the* Study Guide and Solutions Manual.

ADDITIONAL PROBLEMS

11.44   Draw the structure of each of the following. (Some parts may have more than one correct answer.)

   (a) a nine-carbon ether that *cannot* be prepared by the Williamson synthesis

   (b) a nine-carbon ether that *can* be prepared by the Williamson synthesis

   (c) a four-carbon ether that would yield 1,4-diiodobutane after heating with an excess of HI

   (d) an ether that would react with HBr to give propyl bromide as the only alkyl halide

   (e) a four-carbon alkene that would give different glycols after treatment with alkaline $KMnO_4$ or treatment

with *meta*-chloroperoxybenzoic acid followed by dilute aqueous acid

   (f) a four-carbon alkene that would give the same glycol as a result of the different reaction conditions in (e)

   (g) a diene (a compound with two double bonds) C_6H_8 that can form only one mono-epoxide and two di-epoxides (counting stereoisomers)

   (h) an alkene C_6H_{12} that would give the same glycol either from treatment with a peroxycarboxylic acid, followed by acid catalyzed hydrolysis, or from glycol formation with OsO_4

11.45 Give the major organic product of each of the following reactions. Include stereochemistry where relevant.

(a) dibutyl sulfide with 1 equivalent of H_2O_2 at 25 °C

(b) dibutyl sulfide with 2 or more equivalents of H_2O_2 and heat

(c) *cis*-3-hexene with magnesium monoperoxyphthalate (MMPP)

(d) the product of (c) with $(CH_3)_2CuCNLi_2$, then H_3O^+

(e) the product of (c) with solvent H_2O in acidic solution

(f) the product of (e) with periodic acid

(g) the product of (d) with NaH in THF followed by CH_3I

(h) (*E*)-3-methyl-3-hexene with $Hg(OAc)_2$ in ethanol solvent followed by $NaBH_4$

(i) the product of (h) with acidic methanol

(j) (*R*)-3-Methyl-1-bromopentane with Mg in ether, then with ethylene oxide, then with CH_3I

11.46 Give the products of the reaction of 2-ethyl-2-methyloxirane (or other compound indicated) with each of the following reagents.

(a) water, H_3O^+

(b) water, NaOH, heat

(c) $Na^+ CH_3O^-$ in CH_3OH

(d) CH_3OH and a catalytic amount of H_2SO_4

(e) dilithium dimethylcyanocuprate, then H_3O^+

(f) product of part (c) + HBr, 25 °C

(g) product of part (d) + HBr, 25 °C

(h) product of part (c) + NaH, then CH_3I

(i) product of part (d) + NaH, then CH_3CH_2I

(j) product of part (a) + periodic acid

(k) product of part (f) + Mg in dry ether

(l) product of part (k) + ethylene oxide, then H_3O^+

11.47 Which of the ring-opening reactions given in Fig. P11.47 should occur most readily? Explain.

11.48 Which of the ring-opening reactions given in Fig. P11.48 should occur most readily? Explain.

11.49 Explain how you could differentiate between the compounds in each of the following pairs by using simple physical or chemical tests that give readily observable results, such as obvious solubility differences, color changes, evolution of gases, or formation of precipitates.

(a) 3-ethoxypropene and 1-ethoxypropane

(b) 1-pentanol and 1-methoxybutane

(c) 1-methoxy-2-methylpropane and 2-chloro-1-methoxy-2-methylpropane

11.50 When HCl is formed as a by-product in reactions, it is usually removed from reaction mixtures by neutralization with aqueous base. At times, however, the use of base is not compatible with the products or conditions of a reaction. It has been found that propylene oxide (2-methyloxirane) can be used to remove HCl quantitatively. Explain why this procedure works.

11.51 A student has run the reactions shown in Fig. P11.51 on p. 562 and is disappointed to find that each has given none of the desired product. Explain why each reaction failed.

11.52 Tell whether each of the following compounds can be prepared by the reaction of a Grignard reagent with ethylene oxide. If so, show the reaction; if not, explain why and give a different synthesis that uses a different epoxide starting material.

(a) 2-pentanol   (b) 1-pentanol

---

(1)

$$CH_3OH + H_2C\overset{CH_2}{\underset{}{\triangle}}CH_2 \xrightarrow{CH_3O^-} CH_3O-CH_2CH_2-CH_3$$

(2)

$$CH_3OH + H_2C\overset{O}{\underset{}{\triangle}}CH_2 \xrightarrow{CH_3O^-} CH_3O-CH_2CH_2-OH$$

(3)

$$CH_3OH + H_2C\overset{S}{\underset{}{\triangle}}CH_2 \xrightarrow{CH_3O^-} CH_3O-CH_2CH_2-SH$$

Figure P11.47

---

(1)

$$CH_3OH + H_2C\overset{O}{\underset{}{\triangle}}CH_2 \xrightarrow{CH_3O^-} CH_3O-CH_2CH_2-OH$$

(2)

$$CH_3OH + \overset{}{\underset{O}{\square}} \xrightarrow{CH_3O^-} CH_3O-CH_2CH_2CH_2CH_2-OH$$

Figure P11.48

11.53 For each of the following alkenes, state whether a reaction with OsO_4 followed by aqueous $NaHSO_3$ will give a racemic mixture of products that can (in principle) be resolved into enantiomers under ordinary conditions.

(a) ethylene (b) *cis*-2-butene

(c) *trans*-2-butene (d) *cis*-2-pentene

11.54 The (+)-stereoisomer of 2-methyloxirane reacts with aqueous NaOH to give the (R)-(−)-stereoisomer of 1,2-propanediol. Use this observation to propose the absolute stereochemical configuration of (+)-2-methyloxirane.

11.55 Predict the absolute configuration of the major diol product formed by treatment of (S)-2-ethyl-2-methyloxirane with water in the presence of an acid catalyst.

11.56 Keeping in mind that many intramolecular reactions that form six-membered rings are faster than competing intermolecular reactions (Sec. 11.8), predict the product of the reaction given in Fig. P11.56.

11.57 When (3S,4S)-4-methoxy-3-methyl-1-pentene is treated with mercuric acetate in methanol solvent, then with $NaBH_4$, two isomeric compounds with the formula $C_8H_{18}O_2$ are isolated. One, compound A, is optically inactive, but the other, compound B, is optically active. Give the structures and absolute configurations of both compounds. (See Study Guide Link 7.2.)

11.58 You are a manager for a company, Weighty Matters, that specializes in the manufacture of organic compounds containing ^{18}O, a heavy isotope of oxygen. You have assigned the task of preparing ether B to a team of two staff experts, and you have stipulated that alcohol A must be used as a starting material ($^{*}O = {}^{18}O$):

A member of your staff, Homer Flaskclamper, has proposed the following two possible syntheses and has come to you for advice.

(1)

(2)

Which synthesis would you advise Flaskclamper to use and why?

11.59 Match each of the following four compounds with one of the compounds A–D on the basis of the following experimental facts. Compounds A, B, and C are optically active, but compound D is not. Compound C gives the same products as compound D on treatment with periodic acid, but compound B gives a different product. Compound A does not react with periodic acid.

(1) (2S,3S)-2,4-dimethoxy-1,3-butanediol

(2) *meso*-1,4-dimethoxy-2,3-butanediol

(3) (+)-1,4-dimethoxy-2,3-butanediol

(4) (2R,3R)-3,4-dimethoxy-1,2-butanediol

11.60 Complete the reactions given in Fig. P11.60 by giving the principal organic products. Indicate the stereochemistry of the products in parts (d), (g), (h), (i), and (k).

11.61 Outline a synthesis for each of the following compounds from the indicated starting material and any other reagents. [All chiral compounds should be prepared as racemates except in parts (1) and (m).]

(a) 2-ethoxy-3-methylbutane from 3-methyl-1-butene

(b) 2-ethoxy-2-methylbutane from 2-methyl-2-butanol

(c) 4,4-dimethyl-1-pentanol from ethylene oxide

(d)

from compounds containing ≤ 2 carbons

(Problem continues at top of p. 563.)

(a)

(b)

Figure P11.51

Figure P11.56

(Problem 11.61 continued)

(e)

from an alkene

(f)

from an alkene

(g) cyclohexyl isopropyl ether from cyclohexene

(h)

from 3-methyl-1-butene

(i)

$\text{CHCCH}_2\text{OCH}_3$ from 3-methyl-1-pentene

(j)

$\text{Me}_2\text{C}\text{—CH}\text{=}\text{O}$ from 2-methylpropene

(k)

$\text{EtO}\text{—CH}_2\text{—CH}$ from

allyl chloride

(*Hint:* Allyl chloride is very reactive in S_N2 reactions.)

(Problem continues on p. 564.)

(a) $\text{CH}_3\text{CH}_2\text{CH}_2\text{—Br} + \text{Na}^+ \text{EtO}^- \xrightarrow[\text{EtOH}]{}$

(b)

(c)

$\text{C}\text{=}\text{CH}\text{—CH}_3 + (\text{CH}_3)_2\text{CH}\text{—OH} \xrightarrow{\text{Hg(OAc)}_2} \xrightarrow{\text{NaBH}_4}$
(solvent)

(d)

(e) $\text{H}_3\text{C}\text{—CH}\text{=}\text{CH}_2 + \text{H}_2\text{O} + \text{Br}_2 \longrightarrow \xrightarrow[\text{pyridine}]{\text{CrO}_3}$

(f)

$\text{BrCH}_2\text{CH}_2\text{CH}_2\text{—HC}\text{—CH}_2 + \text{HIO}_4 \xrightarrow[\text{H}_2\text{O}]{}$

(*Hint:* Periodic acid, HIO_4, is a fairly strong acid.)

(g)

oleic acid

(h)

$+ \text{Na}^+\text{N}_3^- \xrightarrow{\text{H}_2\text{O}}$

sodium azide
(see Table 9.1, p. 385)

(i)

$+ \text{Li}_2(\text{CH}_3)_2\text{CuCN} \xrightarrow[\text{THF}]{} \xrightarrow{\text{H}_3\text{O}^+}$

(j) $\text{ClCH}_2\text{CH}_2\text{CH}_2\text{CH}_2\text{Cl} + \text{Na}_2\text{S} \xrightarrow[\substack{\text{DMF} \\ \text{(a polar aprotic} \\ \text{solvent)}}]{}$ a compound with the formula $\text{C}_4\text{H}_8\text{S}$

(*Hint:* Think of Na_2S as $\text{:}\ddot{\text{S}}\text{:}^{2-}$)

(k)

$meso\text{-CH}_3\text{CH}\text{—CHCH}_3 + \text{H}_3\text{C}$ ⬡ $\text{—SO}_2\text{Cl} \xrightarrow[\text{pyridine}]{} \xrightarrow{\text{NaOH}}$

(1 equiv.)

Figure P11.60

(Problem 11.61 continued)

(l)   H₃C   O   CH₂OH

from

enantiomerically
pure

(m) H····   O   ···CH=O   CH₂OH

from

enantiomerically
pure

11.62 Outline a synthesis for each of the following compounds in enantiomerically pure form from enantiomerically pure (2R,3R)-2,3-dimethyloxirane:

(a) (2R,3S)-3-methoxy-2-butanol

(b)   MeO   O
        | ‖
   (3S)-CH₃CH — C — CH₃
          3   2   1

(c)   EtO   OMe
        | |
   (2R,3S)-CH₃CH — CHCH₃
          2   3

(d)   EtO   OMe
        | |
   (2S,3R)-CH₃CH — CHCH₃
          2   3

11.63 Compound *A*, C_8H_{16}, undergoes catalytic hydrogenation to give octane. When treated with *meta*-chloroperoxybenzoic acid, *A* gives an epoxide *B*, which, when treated with aqueous acid, gives a compound *C*, $C_8H_{18}O_2$, which can be resolved into enantiomers. When *A* is treated with OsO_4 followed by aqueous $NaHSO_3$, an achiral compound *D* (a stereoisomer of *C*) forms. Identify all compounds, including stereochemistry where appropriate.

11.64 An alternative to the ozonolysis of alkenes is to treat an alkene with *two* molar equivalents of periodic acid and a catalytic amount of OsO_4.

(a) Explain the role of each reagent in the reaction shown in Fig. P11.64. Your explanation should account for the fact that *two* molar equivalents of periodic acid are required.

(b) Complete the following reaction.

       + 2 H₅IO₆   --OsO₄--→
                      H₂O
   CH₃

11.65 Give the structures of all epoxides that could in principle be formed when each of the following alkenes reacts with *meta*-chloroperoxybenzoic acid (mCPBA). Which epoxide should predominate in each case? Why?

(a) *cis*-4,5-dimethylcyclohexene

(b)   CH₃

          H

11.66 When $CH_3CH_2\ddot{S}CH_2CH_2\ddot{S}CH_2CH_3$ reacts with two equivalents of CH_3I, the following double sulfonium salt precipitates:

         CH₃   CH₃
         | |
   CH₃CH₂ — S — CH₂CH₂ — S — CH₂CH₃   2 I⁻
            +   +

(a) Give a curved-arrow mechanism for the formation of this salt.

(b) Upon closer examination, this compound is found to be a mixture of two isomers with melting points of 123–124 °C and 154 °C, respectively. Explain why *two* compounds of this structure are formed. What is the relationship between these isomers? (*Hint:* See Sec. 6.9B.)

11.67 When *S*-adenosylmethionine (Fig. 11.1, p. 537), isolated from natural sources, is allowed to stand in aqueous solution for several weeks at room temperature, a stereoisomeric contaminant appears in solution that can be separated by ordinary methods. Suggest a structure for this contaminant and a reason that it forms. (*Hint:* No covalent bonds are broken in this process.)

11.68 When the naturally occurring amino acid (*S*)-methionine (see Eq. 11.78, p. 552) is converted into methionine sulfoxide, two isomers with different physical properties are formed. What are their structures and what is their stereochemical relationship?

11.69 One of the side reactions that occur when epoxides react with ⁻OH is the formation of polymers. Propose a mechanism for the following polymerization reaction, using the curved-arrow notation.

                  O   CH₃
                 / \ |
   n H₃C — CH — CH₂   --⁻OH--→   (O — CH — CH₂)ₙ

H₃C   H₃C
   \   \
    C   + 2 H₅IO₆ --OsO₄--→   C=O + O=⟨ ⟩ + 2 H₃IO₄ + 2 H₂O
   /   (catalytic amount)   /
H₃C   H₂O   H₃C

Figure P11.64

11.70 (a) Give a curved-arrow mechanism for the reaction shown in Fig. P11.70. Be sure your mechanism indicates the role of the weak acid ammonium chloride.

(b) Why does the reaction of an aziridine require this weak acid? (*Hint:* The pK_a of an amine RNH_2 is about 32.)

(c) The pK_a of ammonium ion is 9.25; the conjugate-acid pK_a of an aziridine is about 7; and the pK_a of HN_3 is 4.2. A student has suggested that dilute (0.01 *M*) HCl should be an even better catalyst. Critique this suggestion.

11.71 The drug *mechlorethamine* (mustine) has been used in antitumor therapy.

$$Cl—CH_2CH_2—\overset{\overset{\displaystyle CH_3}{|}}{\underset{\displaystyle\cdot\cdot}{N}}—CH_2CH_2—Cl$$

mechlorethamine (mustine)

It is one of a family of compounds called *nitrogen mustards*, which also includes the antitumor drugs cyclophosphamide and chlorambucil.

(a) Mechlorethamine undergoes a nucleophilic substitution reaction with water that is several thousand times faster than the nucleophilic substitution reaction of 1,5-dichloropentane. Give the product of the mechlorethamine reaction and the mechanism for its formation.

(b) The antitumor effects of mechlorethamine are due to the fact that it crosslinks DNA. (See the vignette on p. 545, particularly Eqs. 11.70–11.71.) Show this crosslinking process with the curved-arrow notation, using $R_3N\colon$ to represent a base on DNA.

11.72 In each of the following pairs, *one* of the glycols is virtually inert to periodate oxidation. Which glycol is inert? Explain why. (*Hint:* Consider the structure of the intermediate in the reaction.)

(a) *A* or *B*

(b) *A* or *B*

11.73 Account for the following observations with a mechanism. (Refer to Fig. P11.73.)

(1) In 80% aqueous ethanol, compound *A* reacts to give compound *B*. Notice that *trans-B* is the only stereoisomer of this compound that is formed.

(2) Optically active *A* gives completely racemic *B*.

(3) The reaction of *A* is about 10^5 times faster than the analogous substitution reactions of both its stereoisomer *C* and chlorocyclohexane.

11.74 Provide a curved-arrow mechanism for each of the reactions in Fig. P11.74 on p. 566 that accounts for the stereochemical results. Show the structure of the unstable intermediate in each case and explain why it is unstable.

11.75 The reaction of δ-chlorobutyl phenyl sulfide in dioxane containing 20 *M* water at 100 °C gives a cyclic compound *X* that can be isolated. (See Fig. P11.75 on p. 566). This reaction is 21 times faster than the reaction of 1-chlorohexane under the same conditions.

(a) Deduce the structure of *X* and give the curved-arrow mechanism for its formation.

(b) Calculate the proximity effect for this reaction.

(c) Suggest a reason the proximity effect for this reaction is much smaller than the effect calculated for the similar reaction in Eq. 11.61b (p. 539). (This proximity effect was calculated in Study Problem 11.5.)

Figure P11.70

Figure P11.73

11.76 One of the reactions given in Fig. P11.76 is about 2000 times faster in pure water than it is in pure ethanol. Another is about 20,000 times faster in pure ethanol than it is in pure water. The rate of the third changes very little when the solvent composition is changed from ethanol to water. Which of the reactions is faster in ethanol, which is faster in water, and which has a rate that is solvent-invariant? Explain. The solvent (ethanol, water, or a mixture of the two) in the following equations is indicated by ROH. (*Hint:* Notice the difference in dielectric constants for ethanol and water in Table 8.2 on p. 355.)

11.77 Draw a curved-arrow mechanism for each of the conversions shown in Fig. P11.77 on p. 567.

Hint: The structure of the product in part (c) of Fig. P11.76 can also be redrawn as follows:

11.78 (a) As shown in Fig. P11.78(a) on p. 568, 1,5-cyclooctadiene undergoes an electrophilic addition with SCl$_2$ to give compound *A*. (Notice the conformation of *A*, also shown.) Provide a curved-arrow mechanism for this transformation that accounts for the stereochemistry. (*Hint:* Start with a simple electrophilic addition of SCl$_2$ to one double bond.)

(b) Suggest a mechanism that accounts for the reaction of *A* shown in Fig. P11.78(b).

11.79 Each of three bottles, labeled respectively *A*, *B*, and *C*, contains one of the compounds given in Fig. P11.79 on p. 568. On treatment with KOH in methanol, compound *A* gives no epoxide, compound *B* gives an epoxide *D*, and compound *C* gives an epoxide *E*. Epoxides *D* and *E* are stereoisomers. Under identical conditions, *C* gives *E* much more slowly than *B* gives *D*. Identify *A*, *B*, and *C*, and explain all observations.

11.80 Two of the compounds given in Fig. P11.80 on p. 568 form epoxides readily when treated with $^-$OH, one forms an epoxide slowly, and one does not form an epoxide at all. Identify the compound(s) in each category and explain.

11.81 (a) Account for the stereochemical results in the reaction in Fig. P11.81(a) on p. 568 with a mechanism. (*Hint:* See Study Problem 11.6, p. 547.)

(b) What stereochemical result would you expect if the 2*S*,3*S*-stereoisomer of 3-bromo-2-butanol undergoes the same reaction?

(a)   H$_3$C—N⋯ [bicyclic structure] —H + EtOH $\longrightarrow$ H$_3$C—N⋯ [bicyclic structure] —H + HCl
       Cl   (solvent)   OEt

(b)   [thiacyclohexane ring] S⋯ —Cl + MeOH $\longrightarrow$ [thiacyclohexane ring] S⋯ —OMe + HCl
                                    (solvent)

Figure P11.74

Ph—S⋯—CH$_2$CH$_2$CH$_2$CH$_2$—Cl $\xrightarrow[\substack{\text{dioxane} \\ 100\ °C}]{20\ M\ \text{water}}$ a cyclic compound *X*

δ-chlorobutyl phenyl sulfide

Figure P11.75

---

(1) $(CH_3)_3S^+$ + ^-OH $\xrightarrow{\text{ROH}}$ CH_3—OH + $(CH_3)_2S$
                                        (S$_N$2 reaction)

(2) $(CH_3)_3C$—Cl + R—OH $\longrightarrow$ $\left[\begin{array}{c}(CH_3)_3C^+ \\ Cl^-\end{array}\right]$ $\longrightarrow$ $(CH_3)_3C$—OR + HCl
              (solvent)   (S$_N$1 reaction)

(3) $(CH_3)_3C$—$\overset{+}{S}(CH_3)_2$ + ROH $\longrightarrow$ $[(CH_3)_3C^+]$ $\longrightarrow$ $(CH_3)_3C$—OR + $\overset{+}{R}OH_2$
              (solvent)   + $S(CH_3)_2$ (S$_N$1 reaction)

Figure P11.76

(a)

$$HO-\underset{\underset{CH_3}{|}}{\overset{\overset{CH_3}{|}}{C}}-CH=CH_2 \quad \xrightarrow[H_2O]{Br_2} \quad H_3C-\underset{\underset{CH_3}{|}}{C}\overset{O}{\overset{/\backslash}{}}CH-CH_2Br$$

(b)

$$\xrightarrow{\text{trace OH}^-}$$

(c)

$$\xrightarrow{\text{Na}^+ \text{ CH}_3\text{O}^-} \quad + \text{ CH}_3\text{OH} + \text{Na}^+ \text{ }^-\text{OTs}$$

(d)

$$H_3C-CH\overset{O}{\overset{/\backslash}{}}C(CH_3)_2 \quad \xrightarrow[\text{DMSO}]{t\text{BuO}^- \text{ K}^+} \quad H_2C=CH-\underset{|}{\overset{\overset{OH}{|}}{C}}(CH_3)_2 \;+\; H_3C-\underset{}{\overset{\overset{OH}{|}}{C}}H-\underset{}{\overset{\overset{CH_3}{|}}{C}}=CH_2$$

(80%)   (15%)

(e)

+ H—OSO$_2$CF$_3$ ⟶
(a strong acid)

(f)

+ H$_3$C—Br ⟶ CH$_3$S—CH$_2$CH$_2$—Br

(g)

$$\xrightarrow[\text{H}_2\text{O/ethanol}]{\text{Na}^+ \text{ }^-\text{CN}}$$

(h) PhCH$_2$—S

$$\xrightarrow[\text{CH}_3\text{OH}]{\text{NaHCO}_3}$$

(57% yield)

(i)

lead(IV) tetraacetate   **lead(II) diacetate**

(*Hint:* Lead tetraacetate cleaves diols by a mechanism that is very similar to that of periodic acid cleavage.)

Figure P11.77

(a)

1,5-cyclooctadiene

A

conformation of *A*

(b)

sodium azide
(see Table 9.1,
p. 385)

A

Figure P11.78

$(CH_3)_3C$ (1) $(CH_3)_3C$ (2) $(CH_3)_3C$ (3)

Figure P11.79

A   *B*   *C*   *D*

Figure P11.80

(a)

$$(2S,3R)\text{-}CH_3CHCHCH_3 + HBr \longrightarrow meso\text{-}CH_3CHCHCH_3 + H_2O$$

(2S,3R)-3-bromo-2-butanol   **meso-2,3-dibromobutane**

Figure P11.81

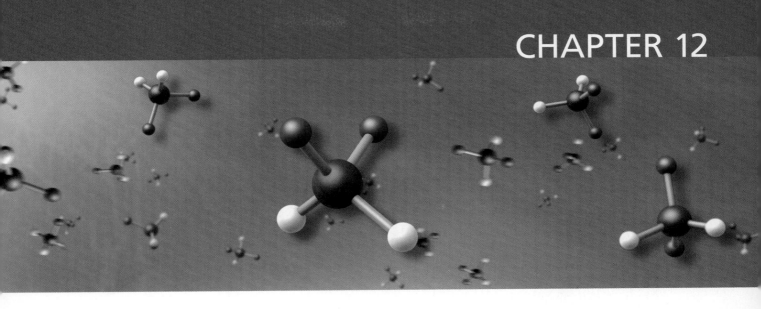

Introduction to Spectroscopy
Infrared Spectroscopy and Mass Spectrometry

Until this point in our study of organic chemistry, we have taken for granted that when a product of unknown structure is isolated from a reaction, it is possible somehow to determine its structure. At one time the structure determinations of many organic compounds required elaborate and laborious chemical-degradation studies. Although many of these proofs were ingenious, they were also very time-consuming, required relatively large amounts of compounds, and were subject to a variety of errors. In the last 50–60 years, however, physical methods have become available that allow chemists to determine molecular structures accurately, rapidly, and nondestructively, using very small quantities of material. With these methods it is not unusual for a chemist to do in 30 minutes or less a proof of structure that once took a year or more to perform! This chapter and Chapter 13 are devoted to a study of some of these methods.

12.1   INTRODUCTION TO SPECTROSCOPY

Fundamental to modern techniques of structure determination is the field of **spectroscopy**: the study of the interaction of matter and light (or other electromagnetic radiation). Spectroscopy has been immensely important to many areas of chemistry and physics. For example, much of what is known about orbitals and bonding comes from spectroscopy. But spectroscopy is also important to the laboratory organic chemist because *it can be used to determine unknown molecular structures*. Although this presentation of spectroscopy will focus largely on its applications, some fundamentals of spectroscopy theory must be considered first.

A. Electromagnetic Radiation

Visible light is one type of **electromagnetic radiation**. Most of us are familiar with the notion that light is a *wave motion*. In other words, light can be thought of as an oscillation, like a

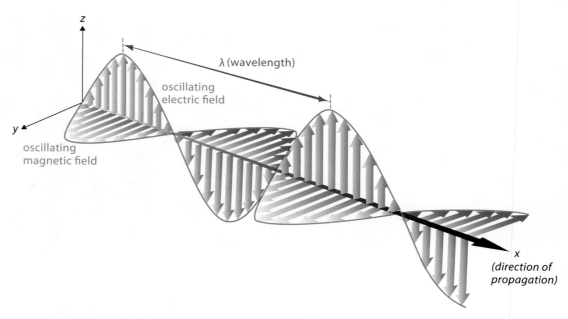

FIGURE 12.1 Electromagnetic radiation consists of electric and magnetic fields oscillating in perpendicular directions. The various forms of electromagnetic radiation differ only in their wavelengths. The wavelength (λ) is the distance between successive peaks or successive troughs.

sine or cosine wave. What is actually oscillating in a light wave? Electromagnetic radiation contains oscillating electric and magnetic field vectors that are perpendicular to each other (Fig. 12.1). The notion of a "field" may not be familiar to you. An *electric field* is capable of transferring energy to electrical charges, including charged atoms. This electrical aspect of electromagnetic radiation, as we shall see, is important in IR spectroscopy. A *magnetic field* is capable of transferring energy to magnetic dipoles, which we can think of as tiny bar magnets. The magnetic aspect of electromagnetic radiation is important in NMR spectroscopy, which we'll study in Chapter 13. Other common forms of electromagnetic radiation are X-rays; ultraviolet radiation (UV, the radiation from a tanning lamp); infrared radiation (IR, the radiation from a heat lamp); microwaves (used in radar and microwave ovens); and radiofrequency waves (rf, used to carry AM and FM radio and television signals).

Electromagnetic radiation propagates through space with a characteristic velocity, called the "speed of light," c, which is 3×10^8 m s^{-1}, or 3×10^{10} cm s^{-1}. Conceptually, this means that if we were to "ride" the wave at a particular point—say, on a particular peak—we would be moving through space at the speed of light.

A very important aspect of electromagnetic radiation is its **wavelength**, abbreviated with the Greek letter *lambda* (λ). This is the distance between successive peaks or successive troughs, also shown in Fig. 12.1. The various forms of electromagnetic radiation are fundamentally the same but differ in their wavelengths. The most common manifestation of wavelength is the phenomenon of *color*. For example, blue light has a smaller wavelength than red light. The unit of distance conventionally used to express wavelength depends on the type of radiation. For example, the wavelengths of ultraviolet and visible radiation are typically given in either Ångstroms (Å) or nanometers (nm). One nm is 10^{-9} meter, and one Å is 10^{-10} meter. Red light has $\lambda = 6800$ Å (680 nm) and blue light has $\lambda = 4800$ Å (480 nm). Ultraviolet radiation has smaller wavelengths than blue light, and microwaves have much greater wavelengths than visible light.

Closely related to the wavelength of a wave is its *frequency*. The concept of frequency is necessary because electromagnetic waves are not stationary, but propagate through space with a characteristic velocity c. The **frequency** of a wave is the number of wavelengths that pass a point per unit time when the wave is propagated through space. The conventional symbol for frequency is the Greek letter *nu* (*ν*). Notice that this is *not* the same as an italic *v*, even though the two letters look very similar. The frequency *ν* of any wave with wavelength λ is

$$\nu = \frac{c}{\lambda} \tag{12.1}$$

in which *c* = the velocity of light. Because λ has the dimensions of length, *ν* has the dimensions of s^{-1}, a unit more often called *cycles per second* (cps), or *hertz* (Hz). For example, the frequency of red light is

$$\nu = \left(\frac{3 \times 10^8 \ \cancel{m} \ s^{-1}}{6800 \ \cancel{Å}} \right) \left(10^{10} \ \frac{Å}{\cancel{m}} \right)$$

$$= 4.4 \times 10^{14} \ s^{-1} = 4.4 \times 10^{14} \ Hz \tag{12.2}$$

This means that 4.4×10^{14} wavelengths of red light pass a given point in one second.

PROBLEM

12.1   Calculate the frequency of
(a) infrared light with λ = 9×10^{-6} m   (b) blue light with λ = 4800 Å

Although light can be described as a wave, it also shows particlelike behavior. The light particle is called a **photon**. The relationship between the energy of a photon and the wavelength or frequency of light is a fundamental law of physics:

$$E = h\nu = \frac{hc}{\lambda} \tag{12.3}$$

In this equation, *h* is *Planck's constant*. Planck's constant is a universal constant that has the value

$$h = 6.625 \times 10^{-27} \ erg \ s = 6.625 \times 10^{-34} \ J \ s \tag{12.4}$$

For a mole of photons, Planck's constant has the value

$$h = 3.99 \times 10^{-13} \ kJ \ s \ mol^{-1} \quad or \quad 9.53 \times 10^{-14} \ kcal \ s \ mol^{-1} \tag{12.5}$$

Equation 12.3 shows that *the energy, frequency, and wavelength of electromagnetic radiation are interrelated*. Thus, when the frequency or wavelength of electromagnetic radiation is known, its energy is also known.

The total range of electromagnetic radiation is called the *electromagnetic spectrum*. The types of radiation within the electromagnetic spectrum are shown in Fig. 12.2 on p. 572. Note that the frequency and energy increase as the wavelength decreases, in accordance with Eqs. 12.1 and 12.3. *All electromagnetic radiation is fundamentally the same; the various forms differ in energy.*

PROBLEMS

12.2   Calculate the energy in kJ mol^{-1} of the light described in
(a) Problem 12.1(a)   (b) Problem 12.1(b)

12.3   Use Figure 12.2 to answer the following questions.
(a) How does the energy of X-rays compare with that of blue light (greater or smaller)?
(b) How does the energy of radar waves compare with that of red light (greater or smaller)?

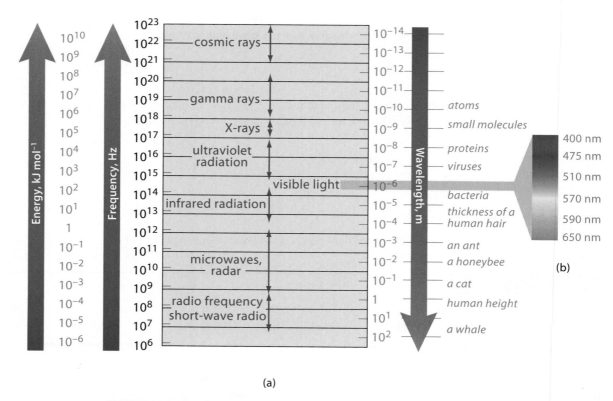

(a)

FIGURE 12.2 The electromagnetic spectrum. (a) The various forms of electromagnetic radiation as a function of energy (*red scale, red ticks*), frequency (*black scale, black lines*), and wavelength (*blue scale, blue ticks*). The wavelengths are compared with the lengths of various objects (*blue italics*). Notice the inverse relationship of wavelength with both energy and frequency; that is, longer wavelengths correspond to smaller energies and frequencies, as Eq. 12.3 indicates. (b) The range for visible light is expanded, and the numbers are the approximate wavelengths of the different colors in nanometers. (A nanometer is 10^{-9} meter.)

B. Absorption Spectroscopy

The most common type of spectroscopy used for structure determination is *absorption spectroscopy*. The basis of absorption spectroscopy is that matter can absorb energy from electromagnetic radiation, and *the amount of absorption is a function of the wavelength of the radiation used*. In an **absorption spectroscopy** experiment, the absorption of electromagnetic radiation is determined as a function of wavelength, frequency, or energy in an instrument called a **spectrophotometer** or **spectrometer**. The basic idea of an absorption spectroscopy experiment is shown schematically in Fig. 12.3. The experiment requires, first, a *source* of electromagnetic radiation. (If the experiment measures the absorption of visible light, the source could be a common light bulb.) The material to be examined, the *sample*, is placed in the radiation beam. A *detector* measures the intensity of the radiation that passes through the sample unabsorbed; when this intensity is subtracted from the intensity of the source, the amount of radiation absorbed by the sample is known. The wavelength of the radiation falling on the sample is then varied, and the radiation absorbed at each wavelength is recorded as a graph of either radiation transmitted or radiation absorbed versus wavelength or frequency. This graph is commonly called a **spectrum** of the sample.

The infrared spectrum of the hydrocarbon nonane, $CH_3(CH_2)_7CH_3$, is an example of such a graph. This spectrum is shown in Fig. 12.4. This spectrum is a plot of the radiation transmitted by a sample of nonane over a range of wavelengths in the infrared. (You'll learn how to interpret such a spectrum in more detail in the next section.)

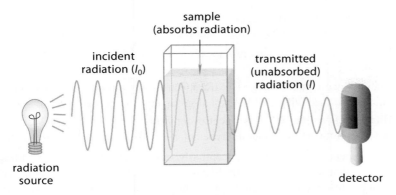

FIGURE 12.3 The conceptual absorption spectroscopy experiment. Electromagnetic radiation of a certain wavelength from a source is passed through the sample and onto a detector. If radiation is absorbed by the sample, the radiation emerging from the sample has a lower intensity than the incident radiation. A comparison of the intensities of the incident radiation and the radiation emerging from the sample tells how much radiation is absorbed by the sample. The spectrum is a plot of radiation absorbed or transmitted versus wavelength or frequency of the radiation.

FIGURE 12.4 Infrared spectrum of nonane. The light transmitted through a sample of nonane is plotted as a function of wavelength (*upper horizontal axis*) or wavenumber (*lower horizontal axis*). (Wavenumber, which is proportional to frequency, is defined by Eq. 12.6 in Sec. 12.2A.) Absorptions are indicated by the inverted peaks.

 An Everyday Analogy to a Spectroscopy Experiment

 You don't need experience with a spectrophotometer to appreciate the basic idea of a spectroscopy experiment. Imagine holding a piece of green glass up to the white light of the Sun. The Sun is the source, the glass is the sample, your eyes are the detector, and your brain provides the spectrum. White light is a mixture of all wavelengths. The glass appears green because only green light is transmitted through the glass; the other colors (wavelengths) in white light are absorbed. If you view the same green through a red glass, no light is transmitted to your eyes—the red glass looks black—because the red glass absorbs the green light.

A very important aspect of spectroscopy for the chemist is that *the spectrum of a compound is a function of its structure.* For this reason, spectroscopy can be used for structure determination. Chemists use many types of spectroscopy for this purpose. The three types of greatest use, and the general type of information each provides, are as follows:

1. Infrared (IR) spectroscopy provides information about what functional groups are present.

2. Nuclear magnetic resonance (NMR) spectroscopy provides information on the number, connectivity, and functional-group environment of carbons and hydrogens.

3. Ultraviolet–visible (UV–vis) spectroscopy (often called simply UV spectroscopy) provides information about the types of π-electron systems that are present.

These types of spectroscopy differ conceptually only in the frequency of radiation used, although the practical aspects are quite different. A fourth physical technique, mass spectrometry, which allows us to determine molecular masses, is also widely used for structure determination. Mass spectrometry is not a type of absorption spectroscopy and is thus fundamentally different from NMR, IR, and UV spectroscopy.

The remainder of this chapter is devoted to a description of IR spectroscopy and mass spectrometry. NMR spectroscopy is covered in Chapter 13 and UV spectroscopy in Chapter 15.

12.2   INFRARED SPECTROSCOPY

A. The Infrared Spectrum

An infrared spectrum, like any absorption spectrum, is a record of the light absorbed by a substance as a function of wavelength. The IR spectrum is measured in an instrument called an *infrared spectrometer*, described briefly in Sec. 12.5. In practice, the absorption of infrared radiation with wavelengths between 2.5×10^{-6} and 20×10^{-6} meter is of greatest interest to organic chemists. Let's consider the details of an IR spectrum by returning to the spectrum of nonane in Fig. 12.4.

Reexamine this spectrum. The quantity plotted on the lower horizontal axis is the **wavenumber** $\tilde{\nu}$ of the light. (The symbol $\tilde{\nu}$ is called *nu-bar*.) The wavenumber is simply the inverse of the wavelength:

$$\tilde{\nu} = \frac{1}{\lambda} \tag{12.6}$$

In this equation, the unit of wavelength is the meter, and the unit of wavenumber is therefore m^{-1}, or reciprocal meter. Physically, the wavenumber is the number of wavelengths contained in one meter. In IR spectroscopy, the conventional unit of wavelength is the *micrometer*. A **micrometer**, abbreviated μm, is 10^{-6} meter. The conventional unit for the wavenumber in IR spectroscopy is *reciprocal centimeters* or *inverse centimeters* (cm^{-1}). To apply Eq. 12.6 with these units, we must include the conversion factor 10^4 μm cm^{-1}. In conventional units, then, Eq. 12.6 becomes

$$\tilde{\nu} \ (cm^{-1}) = \frac{10^4 \ \mu m \ cm^{-1}}{\lambda \ (\mu m)} \tag{12.7a}$$

$$\lambda \ (\mu m) = \frac{10^4 \ \mu m \ cm^{-1}}{\tilde{\nu} \ (cm^{-1})} \tag{12.7b}$$

The micrometer and the reciprocal centimeter are the units used in Fig. 12.4. The wavenumber of the infrared radiation in cm^{-1} is plotted on the lower horizontal axis, and the wavelength in μm is plotted on the upper horizontal axis. (The vertical lines in the grid correspond to the wavenumber scale.) Thus, according to Eq. 12.7a, a wavelength of 10 μm corresponds to a wavenumber of 1000 cm^{-1}. Notice in Fig. 12.4 that 10 μm and 1000 cm^{-1} correspond to the same point on the horizontal axis.

Notice also in Fig. 12.4 that wavenumber, across the bottom of the spectrum, increases to the left; wavelength, across the top of the spectrum, increases to the right. Finally, notice that the wavenumber scale is divided into three distinct regions in which the linear scale is different; the changes in scale occur at 2200 cm^{-1} and 1000 cm^{-1}. The positions of the scale breaks can vary with different spectrometers.

The relationships among wavenumber, energy, and frequency can be derived by combining Eqs. 12.1 and 12.3 with Eq. 12.6, the definition of wavenumber.

$$\nu = \frac{c}{\lambda} = c\tilde{\nu} \tag{12.8a}$$

$$E = h\nu = \frac{hc}{\lambda} = hc\tilde{\nu} \tag{12.8b}$$

(These equations must be used with consistent units, such as λ in m, $\tilde{\nu}$ in m^{-1}, and c in m s^{-1}.) According to Eq. 12.8a, the frequency ν and the wavenumber $\tilde{\nu}$ are proportional. Because of this proportionality, you will often hear the wavenumber loosely referred to as a frequency.

Now let's consider the vertical axis of Fig. 12.4. The quantity plotted on the vertical axis is *percent transmittance*. The *transmittance T* is defined as the ratio of the intensity I of the light emerging the sample to the intensity I_0 of the light entering the sample (see Fig. 12.3). Transmittance by definition is a fraction that can have values between 0 and 1.

$$\text{transmittance} = T = \frac{I}{I_0} \tag{12.9a}$$

The *percent transmittance* is $100 \times T$.

$$\text{percent transmittance} = \%T = 100 \times T \tag{12.9b}$$

If the sample absorbs all of the radiation, then none is transmitted, and $I = 0$; the sample has 0% transmittance. If the sample absorbs no radiation, then all of the radiation is transmitted, and $I = I_0$; the sample has 100% transmittance. Thus, absorptions in the IR spectrum are registered as downward deflections—that is, "upside-down peaks." From Fig. 12.4, you can see that absorptions in the spectrum of nonane occur at about 2850–2980, 1470, 1380, and 720 cm^{-1}. The presentation of peaks in the transmittance mode is largely for historical reasons; that is, early instruments produced data that were most conveniently plotted this way, and the practice simply hasn't changed.

In other forms of spectroscopy, absorptions are presented as "normal peaks"—that is, upward deflections. Some IR instruments produce data plotted this way as well. The quantity plotted in this case is the *absorbance A*. Absorbance is the *negative* logarithm of transmittance.

$$\text{absorbance} = A = -\log T = -\log \frac{I}{I_0} = -\log \frac{I_0}{I} \tag{12.9c}$$

A sample that absorbs half of the incident light has $\%T = 50$, and it has an absorbance of $-\log(0.5) = 0.3$.

Sometimes an IR spectrum is not presented in graphical form, but is summarized completely or in part using descriptions of the *positions* of the important peaks. Intensities are often expressed qualitatively using the designations vs (very strong), s (strong), m (moderate), or w (weak). Some peaks are narrow (or sharp, abbreviated sh), whereas others are wide (or broad, abbreviated br). The spectrum of nonane can be summarized as follows:

$$\tilde{\nu} \text{ (cm}^{-1}\text{): } 2980\text{–}2850 \text{ (s); } 1470 \text{ (m); } 1380 \text{ (m); } 720 \text{ (w)}$$

PROBLEMS

12.4 (a) What is the wavenumber (in cm^{-1}) of infrared radiation with a wavelength of 6.0 μm?

(b) What is the wavelength of light with a wavenumber of 1720 cm^{-1}?

12.5 In the infrared spectrum of nonane in Fig. 12.4, what is the absorbance of the sharp peak at 1380 cm^{-1}?

B. The Physical Basis of IR Spectroscopy

The interpretation of an IR spectrum in terms of structure requires some understanding of why molecules absorb infrared radiation. The absorptions observed in an IR spectrum are the result of *vibrations* within a molecule. Atoms within a molecule are not stationary, but are constantly moving. Consider, for example, the C—H bonds in a typical organic compound. These bonds undergo various oscillatory stretching and bending motions, called **bond vibrations**. For example, one such vibration is the C—H stretching vibration. A useful analogy to the C—H stretching vibration is the stretching and compression of a spring (see Fig. 12.5). This vibration takes place with a certain frequency ν; that is, it occurs a certain number of times per second. Suppose that a C—H bond has a stretching frequency of 9×10^{13} times per second; this means that it undergoes a vibration every $1/(9 \times 10^{13})$ second or every 1.1×10^{-14} second.

The blue line in Fig. 12.5 shows that the stretching of the C—H bond over time describes a wave motion. It turns out that a wave of electromagnetic radiation can transfer its energy to the vibrational wave motion of the C—H bond only if *there is an exact match between the frequency of the radiation and the frequency of the vibration.* Thus, if a C—H vibration has a frequency of 9×10^{13} s^{-1}, then it will absorb energy from radiation with the same frequency. From the relationship $\lambda = c/\nu$ (Eq. 12.1), the radiation must then have a wavelength of $\lambda = (3 \times 10^8 \text{ m s}^{-1})/(9 \times 10^{13} \text{ s}^{-1}) = 3.33 \times 10^{-6}$ m = 3.33 μm. The corresponding wavenumber of this radiation (Eq. 12.7a) is 3000 cm^{-1}. When radiation of this wavelength interacts with a vibrating C—H bond, energy is absorbed and both the frequency and the intensity of the bond vibration increase. That is, after absorbing energy, the bond vibrates with a greater frequency (actually, twice the frequency) and with a larger *amplitude* (a larger stretch and tighter compression; Fig. 12.6). This absorption gives rise to the peak in the IR spectrum. Eventually, the bond returns to its normal, less intense, vibration, and when this happens, energy is released in the form of heat. This is why an infrared "heat lamp" makes your skin feel warm.

 An Analogy to Energy Absorption by a Vibrating Bond

Imagine an unopened 12-ounce can of soda suspended from your finger with a rubber band by its pull-tab. If you move your finger up and down very slowly, the soda can scarcely moves. However, if you increase the rate of oscillation of your finger gradually, at some point the soda can will start to move up and down vigorously. The motion of the soda can will decrease as you move your finger more rapidly. The motion of your finger is analogous to electromagnetic radiation, and the rubber band–soda can combination is analogous to the vibrating bond. Energy is absorbed from the motion of your finger by the rubber band–soda can oscillator only when its natural oscillation frequency matches the oscillation frequency of your finger. Likewise, a vibrating bond absorbs energy from electromagnetic radiation only when there is a frequency match between the two oscillating systems.

Although the intuitive mechanical descriptions of infrared absorption are useful, bond vibrations are subject to quantum theory. Let's contrast our "bond spring" with a mechanical spring. If you allow a mechanical spring with a weight attached to it to hang from a hook in the ceiling, you can change the energy of the weight-and-spring system continuously by increasing or decreasing the distance you pull on the weight before releasing it. With bonds, however, the allowed vibrational energies are *quantized*; that is, only certain vibrational energies are allowed; these energies are characterized by *vibrational energy levels*. Absorption of one photon of infrared radiation by a bond causes the vibrational energy to jump to the next energy level.

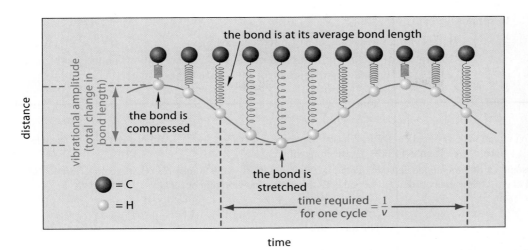

FIGURE 12.5 Chemical bonds undergo different types of vibrations. The one illustrated here is a stretching vibration. The bond, represented as a spring, is shown at various times. The bond stretches and compresses over time. The time required for one complete cycle of vibration is the reciprocal of the vibrational frequency.

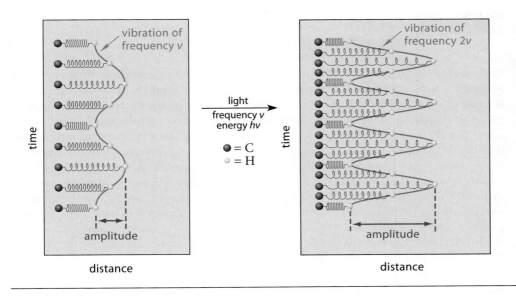

FIGURE 12.6 Light absorption by a bond vibration causes the bond (shown as a spring) to vibrate with larger amplitude and twice the frequency. The frequency of the light must exactly equal the frequency of the bond vibration for absorption to occur. (The increase in amplitude is greatly exaggerated for emphasis.)

$$\Delta E = E_1 - E_0 = h\nu_0 \qquad (12.10)$$

The energy of the photon required to bring about this transition equals the *difference* in the energy ΔE of the two vibrational levels. As the diagram in Eq. 12.10 shows, this difference equals $h\nu_0$, where ν_0 is the vibrational frequency of the bond. (See Further Exploration 12.1 in the *Study Guide and Solutions Manual* for the derivation of this result.) In other words, the frequency of the photon that brings about the vibrational transition is the same as the frequency of the bond vibration. If bond vibrations weren't quantized, a bond could absorb photons of any energy, and there would be no peaks in the spectrum. Why different bonds have different vibrational frequencies (and therefore different IR absorption frequencies) is addressed in Sec. 12.3A.

FURTHER EXPLORATION 12.1
The Vibrating Bond in Quantum Theory

In summary:

1. Bonds vibrate with characteristic frequencies.
2. Absorption of energy from infrared radiation can occur only when the wavelength of the radiation and the wavelength of the bond vibration match.

PROBLEMS

12.6 Given that the stretching vibration of a typical C—H bond has a frequency of about 9×10^{13} s^{-1}, which peak(s) in the IR spectrum of nonane (Fig. 12.4) would you assign to the C—H stretching vibrations?

12.7 The physical basis of some carbon monoxide detectors is the infrared detection of the unique C≡O stretching vibration of carbon monoxide at 2143 cm^{-1}. How many times per second does this stretching vibration occur?

$$:\bar{C}\!\equiv\!\overset{+}{O}:$$

carbon monoxide

12.3 INFRARED ABSORPTION AND CHEMICAL STRUCTURE

Each peak in the IR spectrum of a molecule corresponds to the absorption of energy by the vibration of a particular bond or group of bonds. IR spectroscopy is useful for the chemist because *in all compounds, a given type of functional group absorbs in the same general region of the IR spectrum.* The major regions of the IR spectrum are shown in Table 12.1.

The IR spectrum of a typical organic compound contains many more absorptions than can be readily interpreted. A major part of mastering IR spectroscopy is to learn which absorptions are important. Certain absorptions are *diagnostic*; that is, they indicate with reasonable certainty that a particular functional group is present. For example, an intense peak in the 1700–1750 cm^{-1} region indicates the presence of a carbonyl (C=O) group. Other peaks are *confirmatory*; that is, similar peaks can be found in other types of molecules, but their presence confirms a structural diagnosis made in other ways. For example, absorptions in the 1050–1200 cm^{-1} region of the IR spectrum due to a C—O bond could indicate the presence of an alcohol, an ether, an ester, or a carboxylic acid, among other things. However, if other evidence (perhaps obtained from other types of spectroscopy) suggests that the unknown molecule is, say, an ether, a peak in this region can serve to support this diagnosis. In the sections that follow, you'll learn about the relatively few absorptions that are important in IR spectroscopy.

Rarely if ever does an IR spectrum completely define a structure; rather, it provides information that restricts the possible structures under consideration. Once the structure for an unknown compound has been deduced, a comparison of its IR spectrum with that of an authentic sample can be used as a criterion of identity. Even subtle differences in structure generally give discernible differences in the IR spectrum, particularly in the region between 1000 cm^{-1} and 1600 cm^{-1}. This point is illustrated by the superimposed IR spectra of the two

---

TABLE 12.1 Regions of the Infrared Spectrum

Wavenumber range, cm^{-1}	Type of absorptions	Name of region
3400–2800	O—H, N—H, C—H stretching	Functional group
2250–2100	C≡N, C≡C stretching	
1850–1600	C=O, C=N, C=C stretching	
1600–1000	C—C, C—O, C—N stretching; various bending absorptions	Fingerprint
1000–600	C—H bending	C—H bending

FIGURE 12.7 Comparison of the IR spectra of the cis (*red*) and trans (*blue*) stereoisomers of ,3-dimethylcyclohexane. Although these spectra are very similar, discernible differences between them occur in the fingerprint region.

diastereomers *cis-* and *trans-*1,3-dimethylcyclohexane (Fig. 12.7). The greatest differences between the two spectra occur in this region of the spectrum. (That these spectra are different is another illustration of the principle that diastereomers have different physical properties.) Even though absorptions in this region are generally not interpreted in detail, they serve as a valuable "molecular fingerprint." That is why this region of the spectrum is called the "fingerprint region" in Table 12.1.

A. Factors That Determine IR Absorption Position

One approach to the use of IR spectroscopy is simply to memorize the wavenumbers at which characteristic functional group absorbances appear and to look for peaks at these positions in the determination of unknown structures. However, you can use IR spectroscopy much more intelligently and learn the important peak positions much more easily if you understand a little more about the physical basis of IR spectroscopy. Two aspects of IR absorption peaks are particularly important. First is the *position* of the peak—the wavenumber or wavelength at which it occurs. Second is the *intensity* of the peak—how strong it is. Let's consider each of these aspects in turn.

What factors govern the position of IR absorption? Three considerations are most important.

1. strength of the bond
2. masses of the atoms involved in the bond
3. the type of vibration being observed

Hooke's law, which comes from the treatment of the vibrating spring by classical physics, nicely accounts for the first two of these effects. Let two atoms with masses M and m, respectively, be connected by a bond, which we'll treat as a spring. The tightness of the spring (bond) is described by a *force constant, κ*: the larger the force constant, the tighter the spring, or the stronger the bond.

a spring (bond) with a force constant κ

mass M ⬤〰〰 mass m

The following equation describes the dependence of the vibrational wavenumber on the masses and the force constant:

$$\tilde{\nu} = \frac{1}{2\pi c} \sqrt{\frac{\kappa(m+M)}{mM}}$$

(12.11)

Before we use this equation, let's ask what our intuition tells us about how bond strength affects the vibrational frequency. Intuitively, a stronger bond corresponds to a tighter spring. That is, stronger bonds should have larger force constants. Objects connected by a tighter spring vibrate more rapidly—that is, with a higher frequency or wavenumber. Likewise, atoms connected by a stronger bond also vibrate at higher frequency. A simple measure of bond strength is the energy required to break the bond, which is the *bond dissociation energy* (Table 5.3 on p. 216). It follows, then, that *the higher the bond dissociation energy, the stronger the bond.* Thus, *the IR absorptions of stronger bonds—bonds with greater bond dissociation energies—occur at higher wavenumber.* Study Problem 12.1 illustrates this effect.

STUDY PROBLEM 12.1

The typical stretching frequency for a carbon–carbon double bond is 1650 cm^{-1}. Estimate the stretching frequency of a carbon–carbon triple bond.

SOLUTION   Use Eq. 12.11. We are given $\tilde{\nu}$ for a double bond, which we'll call $\tilde{\nu}_2$, and we're asked to estimate $\tilde{\nu}_3$, the stretching frequency of the triple bond. Let the force constant for a double bond be κ_2, and that of a triple bond be κ_3. Now take the ratio of $\tilde{\nu}_3$ and $\tilde{\nu}_2$ as given by Eq. 12.11:

$$\frac{\tilde{\nu}_3}{\tilde{\nu}_2} = \sqrt{\frac{\kappa_3}{\kappa_2}} \qquad (12.12a)$$

All of the mass terms cancel because they are the same in both cases—the mass of a carbon atom. We could complete the problem if we knew the force constants, but these aren't given. Let's apply some intuition. As we just learned, force constants are proportional to bond dissociation energies. We could look these up in Table 5.3, but let's simply assume that the relative strengths of triple and double bonds are in the ratio 3:2. If this were so, then Eq. 12.12a becomes

$$\frac{\tilde{\nu}_3}{\tilde{\nu}_2} = \sqrt{\frac{3}{2}} = 1.22 \qquad (12.12b)$$

With $\tilde{\nu}_2 = 1650$ cm^{-1}, we then estimate $\tilde{\nu}_3$ to be $(1.22)(1650$ cm$^{-1}) = 2013$ cm^{-1}.

How close are we? See for yourself by jumping ahead to Fig. 14.5, p. 687, which shows the C≡C stretching absorption of an alkyne.

Now let's consider the effect of mass on the stretching frequency of a bond. Eq. 12.11 also describes this effect. However, a special case of this equation is very important. Suppose the two atoms connected by a bond differ significantly in mass (for example, a carbon and a hydrogen in a C—H bond). *The vibration frequency for a bond between two atoms of different mass depends more on the mass of the lighter object than on the mass of the heavier one.* The following analogy illustrates this point.

Analogy for the Effect of Mass on Bond Vibrations

Imagine three situations: two identical light rubber balls connected by a spring; a rubber ball connected to a heavy cannonball with the same spring; and a rubber ball connected to the Empire State Building by the same spring. When the spring connecting the two rubber balls is stretched and released, both balls oscillate. That is, both masses are involved in the vibration. When the spring connecting the rubber ball and the cannonball is stretched and released, the rubber ball oscillates, and the cannonball remains almost stationary. When the spring connecting the rubber ball and the Empire State Building is stretched and released, only the rubber ball appears to oscillate; the motion of the building is imperceptible. In the last two cases, the vibrational frequencies are virtually identical, even though the larger masses differ by orders of magnitude. Hence, changing the larger mass has no effect on the vibration frequency. If we now attach a cannonball at one end of the same spring and leave the other attached to the building, the frequency is significantly

reduced, even though we haven't changed the larger mass at all. Thus, the smaller mass controls the vibration frequency.

Now let's use Eq. 12.10 to verify that the smaller mass determines the vibration frequency. Suppose in this equation that $M \gg m$. In such a case, the smaller mass can be ignored in the numerator of Eq. 12.10, and the larger mass M then cancels and vanishes from the equation, leaving only the smaller mass m in the denominator. Eq. 12.10 then becomes

$$\tilde{\nu} = \frac{1}{2\pi c}\sqrt{\frac{\kappa}{m}} \quad \text{(for } M \gg m\text{)} \tag{12.13}$$

According to this equation, the vibration frequency of a bond between a heavy and a light atom depends primarily on the mass of the light atom, as our preceeding intuitive argument suggested. This is why C—H, O—H, and N—H bonds all absorb in the same general region of the IR spectrum, and why C=O, C=N, and C=C bonds absorb in the same general region (Table 12.1, p. 578). In fact, the differences that do exist between the vibrational frequencies of these bonds are not primarily mass effects, but mostly bond-strength effects.

One of the best illustrations of the mass effect on vibrational frequency is a comparison of the frequencies of bond vibrations for an X—H bond with its deuterium-substituted analog X—D. (Deuterium, or ^{2}H, is the isotope of hydrogen with atomic mass = 2.) This effect is the subject of Problem 12.9.

PROBLEMS

12.8 The following bonds all have IR stretching absorptions in the 4000–2900 cm^{-1} region of the spectrum. Rank the following bonds in order of decreasing stretching frequencies, greatest first, and explain your reasoning. (*Hint:* Consult Table 5.3 on p. 216.)

C—H, O—H, N—H, F—H

12.9 The =C—H stretching absorption of 2-methyl-1-pentene is observed at 3090 cm^{-1}. If the hydrogen were replaced by deuterium, at what wavenumber would the =C—D stretching absorption be observed? Explain. (Assume that the force constants for the C—H and C—D bonds are identical.)

The third factor that affects the absorption frequency is the *type of vibration*. The two general types of vibrations in molecules are *stretching vibrations* and *bending vibrations*. A **stretching vibration** occurs along the line of the chemical bond. A **bending vibration** is any vibration that does not occur along the line of the chemical bond. A bending vibration can be envisioned as a ball hanging on a spring and swinging side to side. In general, *bending vibrations occur at lower frequencies (higher wavelengths) than stretching vibrations of the same groups.*

 An Analogy for Bending Vibrations

Imagine a ball attached to a stiff spring hanging from the ceiling. A gentle tap makes it swing back and forth. It takes considerably more energy to stretch the spring. Because the energy required to set the spring in motion is proportional to its frequency, the swinging (bending) motion has a lower frequency than the stretching motion.

The only possible type of vibration in a diatomic molecule (for example, H—F) is a stretching vibration. However, when a molecule contains more than two atoms, both stretching and bending vibrations are possible. The allowed vibrations of a molecule are called its

FIGURE 12.8 Normal vibrational modes of a typical CH_2 group in a structure R—CH_2—R. In each model, the white spheres are the hydrogens, the black sphere is the carbon of the CH_2, and the blue spheres represent the R groups. The center model in each normal mode represents the average position of the hydrogens. For each mode, start at the center and view left, center, right, center to see how the hydrogens move with time.

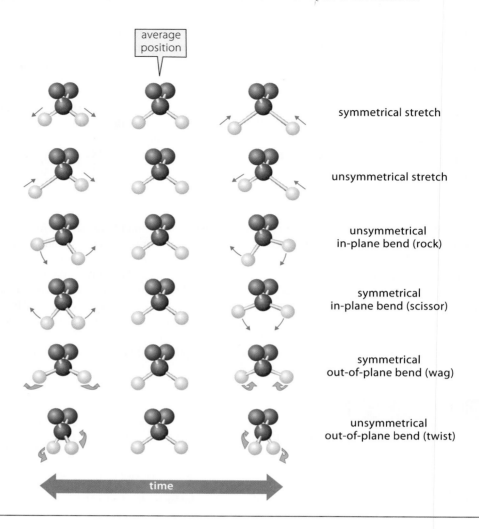

average position

symmetrical stretch

unsymmetrical stretch

unsymmetrical in-plane bend (rock)

symmetrical in-plane bend (scissor)

symmetrical out-of-plane bend (wag)

unsymmetrical out-of-plane bend (twist)

time

normal vibrational modes. The normal vibrational modes for a —CH_2— group are shown in Fig. 12.8. They serve as models for the kinds of vibrations that can be expected for other groups in organic molecules. The bending vibrations can be such that the hydrogens move *in the plane* of the —CH_2— group, or *out of the plane* of the —CH_2— group. Furthermore, stretching and bending vibrations can be *symmetrical* or *unsymmetrical* with respect to a plane between the two vibrating hydrogens. The bending motions have been given graphic names (scissoring, wagging, and so on) that describe the type of motion involved. Each of these motions occurs with a particular frequency and can have an associated peak in the IR spectrum (although some peaks are weak or absent for reasons to be considered later). The —CH_2— groups in a typical organic molecule undergo all of these motions simultaneously. That is, while the C—H bonds are stretching, they are also bending. The IR spectrum of nonane (Fig. 12.4) shows absorptions for both C—H stretching and C—H bending vibrations. The peak at 2920 cm^{-1} is due to the C—H stretching vibrations; the peaks at 1470 and 1380 cm^{-1} are due to various bending modes of both —CH_2— and CH_3— groups; and the peak at 720 cm^{-1} is due to a different bending mode, the —CH_2— rocking vibration. Notice that all of the bending vibrations absorb at lower wavenumber (and therefore lower energy) than the stretching vibrations.

B. Factors That Determine IR Absorption Intensity

The different peaks in an IR spectrum typically have very different intensities. Several factors affect absorption intensity. First, a greater number of molecules in the sample and more absorbing groups within a molecule give a more intense spectrum. Thus, a more concentrated sample gives a stronger spectrum than a less concentrated one, other things being equal. Similarly, at a given concentration, a compound such as nonane, which is rich in C—H bonds, has a stronger absorption for its C—H stretching vibrations than a compound of similar molecular mass with relatively few C—H bonds.

The dipole moment of a molecule also affects the intensity of an IR absorption. We can see why this should be so if we think about the nature of light and how it interacts with a single vibrating bond. A light wave consists of perpendicular vibrating electric and magnetic fields, as shown in Fig. 12.1, p. 570. Only the vibrating electric field is relevant to IR spectroscopy, so we can forget about the magnetic field for now. In Fig. 12.9a, the vibrating electric field of light is shown again as a vector oscillating in the plane of the page. Recall from Sec. 12.1A that *an electric field exerts a force on a charge*. This means that an electric field affects the motion of a charge. The field imposes an acceleration component on the charge in the direction of the field. In particular, if a charge is moving in the same direction as the electric field, the field increases the velocity of the charge; that is, the field makes the charge "move more."

Recall from Sec. 1.2D that a polar chemical bond has a *bond dipole*. This means that we can think of a polar bond as a system of separated positive and negative charges. If a polar bond vibrates with a particular frequency, its bond dipole vibrates with the same frequency (Fig. 12.9b). According to Eq. 1.4 (p. 11), the magnitude of the bond dipole is proportional not only to the amount of charge on each bonded atom but also to the distance between the atoms—that is, the bond length. Therefore, as the bond stretches, the bond dipole increases, and, as the bond compresses, the bond dipole decreases. To a light wave, a polar bond is basically a system of moving charges. Light interacts with a polar bond because its electric field exerts a force on the system of moving charges (the vibrating bond dipole). This interaction can occur only if the light wave and the charges in the bond are oscillating with the same frequency. When the electric field of the light wave exerts a force on the charges in the bond dipole, the bond dipole gains energy, as shown in Fig. 12.6, and consequently the light wave loses energy; this is the process of absorption.

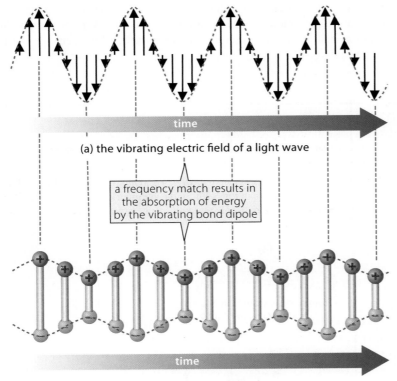

(a) the vibrating electric field of a light wave

a frequency match results in the absorption of energy by the vibrating bond dipole

(b) a vibrating bond dipole

FIGURE 12.9 (a) One component of a light wave is a vibrating electric field, here represented as a vector that oscillates in a sinusoidal manner over time. (b) A bond dipole vibrating at the same frequency as the light wave in (a). When the frequencies of the light wave and the vibrating bond dipole match (as in this figure), the bond dipole absorbs energy from the light wave. (See also Fig. 12.6 on p. 577.) The alignment of peaks and troughs of the two waves is indicated with gray dashed lines.

If a bond has no bond dipole, then there are no moving charges with which the electric field of the light wave can interact. For example, the $C=C$ bond in the alkene 2,3-dimethyl-2-butene has no bond dipole because of its symmetrical position in the molecule. Hence, its stretching vibration does not interact with the electric field of light. The $C=C$ stretching vibration occurs, but it does *not* absorb energy from light. This means that 2,3-dimethyl-2-butene does not have a $C=C$ stretching absorption in its IR spectrum in the 1600–1700 cm^{-1} region of its IR spectrum, the region in which the double bonds of many other alkenes have such an absorption.

2,3-dimethyl-2-butene:
zero dipole moment

"stretched" 2,3-dimethyl-2-butene:
zero dipole moment

no IR absorption is observed for the $C=C$ stretching vibration

(12.14)

(This compound does have other IR absorptions.) Molecular vibrations that occur but do not give rise to IR absorptions are said to be **infrared-inactive**. (IR-inactive vibrations can be observed with *Raman spectroscopy*, another type of spectroscopy.) In contrast, any vibration that gives rise to an IR absorption is said to be **infrared-active**. We often find in practice that highly symmetrical compounds have less complex IR spectra than unsymmetrical isomers because their symmetry results in the absence of a molecular dipole moment and a relatively large number of infrared-inactive vibrations.

It is possible for a molecule that has a zero dipole moment to have a molecular vibration that creates a temporary dipole moment in the distorted molecule. Such vibrations are also infrared-active. An example of this situation is found in Study Problem 12.2. This is why some vibrations, even in symmetrical molecules, are infrared-active.

Because the intensity of an IR absorption depends on the size of the dipole moment change that accompanies the corresponding vibration, IR absorptions differ widely in intensity. Chemists do not try to predict intensities; rather, they rely on collective experience to know which absorptions are weaker and which are stronger. Nevertheless, for symmetrical molecules with a zero dipole moment, we must be particularly aware of the possibility of IR-inactive vibrations that would be observed in less symmetrical molecules containing the same functional groups.

STUDY PROBLEM 12.2

Which one of the following molecular vibrations is infrared-inactive? (a) the $C=O$ symmetrical stretch of CO_2; (b) the $C=O$ unsymmetrical stretch of CO_2. (See Fig. 12.8 on p. 582.)

SOLUTION First be sure you understand what is meant by the terms *symmetrical stretch* and *unsymmetrical stretch*. These are defined by analogy to the $C—H$ stretching vibrations in Fig. 12.8. In the symmetrical stretch, the two $C=O$ bonds lengthen (or shorten) at the same time so that the molecule maintains its symmetry with respect to the symmetry plane:

symmetrical stretch: maintains
the molecular symmetry

In the unsymmetrical stretch, one $C=O$ bond shortens when the other lengthens:

unsymmetrical stretch

Which of these vibrational modes results in a change of dipole moment? Because the CO_2 molecule is linear, the two C=O bond dipoles exactly oppose each other. Stretching a bond increases its bond dipole because the size of a bond dipole is proportional not only to the magnitudes of the partial charges at each end of the bond but also to *the distance by which the charges are separated* (Eq. 1.4, p. 11). Consequently, after the symmetrical stretch, both bond dipoles are increased; but, because they are exactly equal and oppose each other, the dipole moment remains zero. Hence, the symmetrical stretching vibration is IR-inactive.

In an unsymmetrical stretch, one C=O bond is reduced in length while the other is increased. Because the "long" C=O bond has a greater bond dipole than the "short" C=O bond, the two bond dipoles no longer cancel. Thus, the unsymmetrical stretch imparts a temporary dipole moment to the CO_2 molecule. Consequently, this vibration is infrared-active—it gives rise to an IR absorption.

PROBLEM

12.10 Which of the following vibrations should be infrared-active and which should be infrared-inactive (or nearly so)?

(a) $CH_3CH_2CH_2CH_2C{\equiv}CH$ C≡C stretch

(b) $(CH_3)_2C{=}O$ C=O stretch

(c)   cyclohexane ring "breathing" (simultaneous stretch of all C—C bonds)

(d) $CH_3CH_2C{\equiv}CCH_2CH_3$ C≡C stretch

(e)   symmetrical N—O stretch

(f) $(CH_3)_3C{-}Cl$ C—Cl stretch

(g) *trans*-3-hexene   C=C stretch

12.4   FUNCTIONAL-GROUP INFRARED ABSORPTIONS

A typical IR spectrum contains many absorptions. Chemists do not try to interpret every absorption in a spectrum. Experience has shown that some absorptions are particularly useful and important in diagnosing or confirming certain functional groups. In this section, we'll focus on those. We'll show you sample spectra so that you will begin to see how these absorptions appear in actual spectra.

We'll consider here only the functional groups covered in Chapters 1–11. Subsequent chapters contain short sections that discuss the IR spectra of other functional groups. These sections, however, can be read and understood at any time with your present knowledge of infrared spectroscopy. In addition, a summary of key IR absorptions is given in Appendix II.

A. IR Spectra of Alkanes

The characteristic structural features of alkanes are the carbon–carbon and carbon–hydrogen single bonds. The stretching of the carbon–carbon single bond is infrared-inactive (or nearly so) because this vibration is associated with little or no change of the dipole moment. The stretching absorptions of alkyl C—H bonds are typically observed in the 2850–2960 cm^{-1} region. The peaks near 2920 cm^{-1} in the IR spectrum of nonane (Fig. 12.4, p. 573) are examples of such absorptions. Various bending vibrations are also observed in the fingerprint region (1380 and 1470 cm^{-1} in nonane) and in the C—H bending region (720 cm^{-1} in nonane). Absorptions in these general regions can be expected not only for alkanes but also for any compounds that contain H_3C— and —CH_2— groups. Consequently, these absorptions are not often useful, but it is important to be aware of them so that they are not mistakenly attributed to other functional groups.

B. IR Spectra of Alkyl Halides

The carbon–halogen stretching absorption of alkyl chlorides, bromides, and iodides appear in the low-wavenumber end of the spectrum, but many interfering absorptions also occur in this region. NMR spectroscopy and mass spectrometry are more useful than IR spectroscopy for determining the structures of these alkyl halides.

Alkyl fluorides, in contrast to the other alkyl halides, have useful IR absorptions. A single C—F bond typically has a very strong stretching absorption in the 1000–1100 cm^{-1} region. Multiple fluorines on the same carbon increase the frequency; for example, a CF$_3$ group typically has a stretching frequency in the 1300–1360 cm^{-1} region.

C. IR Spectra of Alkenes

Unlike the spectra of alkanes and alkyl halides, the infrared spectra of alkenes are very useful and can help determine not only whether a carbon–carbon double bond is present, but also the carbon substitution pattern at the double bond. Typical alkene absorptions are given in Table 12.2. These fall into three categories: C=C stretching absorptions, =C—H stretch-

TABLE 12.2 Important Infrared Absorptions of Alkenes

Functional group	Absorption*
C=C **stretching absorptions**	
—CH=CH$_2$ (terminal vinyl)	1640 cm^{-1} (m, sh)
C=CH$_2$ (terminal methylene)	1655 cm^{-1} (m, sh)
(cis, trans, and other disubstituted alkenes)	1660–1675 cm^{-1} (w) (absent in some compounds)
=C—H **stretching absorptions**	
=C—H =CH$_2$	3000–3100 cm^{-1} (m)
=C—H **bending absorptions**	
—CH=CH$_2$ (terminal vinyl)	910, 990 cm^{-1} (s) two absorptions
C=CH$_2$ (terminal methylene)	890 cm^{-1} (s)
(trans alkene)	960–980 cm^{-1} (s)
(cis alkene)	675–730 cm^{-1} (br) (ambiguous and variable for different compounds)
C=C (trisubstituted)	800–840 cm^{-1} (s)

* *Intensity designations:* s = strong; m = moderate; w = weak
Shape designations: sh = sharp (narrow); br = broad (wide)

FIGURE 12.10 IR spectra of (a) 1-octene and (b) *trans*-3-hexene. Be sure to correlate the key bands indicated in these spectra with the corresponding entries in Table 12.2.

ing absorptions, and =C—H bending absorptions. The stretching vibration of the carbon–carbon double bond occurs in the 1640–1675 cm^{-1} range; the frequency of this absorption tends to be greater, and its intensity smaller, with increased alkyl substitution at the double bond. The reason for the intensity variation is the dipole moment effect discussed in the previous section. Thus, the C=C stretching absorption is clearly evident in the IR spectrum of 1-octene at 1642 cm^{-1} (see Fig. 12.10a), but is virtually absent in the spectrum of the symmetrical alkene *trans*-3-hexene (Fig. 12.10b). The C=C stretching vibration is weak or absent even in unsymmetrical alkenes that have the same number of alkyl groups on each carbon of the double bond.

NMR spectroscopy is particularly useful for observing alkene hydrogens (Sec. 13.7A). Nevertheless, a =C—H stretching absorption can often be used for confirmation of an alkene functional group. In general, the stretching absorptions of C—H bonds involving sp^2-hybridized carbons occur at wavenumbers *greater than* 3000 cm^{-1}, and the stretching absorptions of C—H bonds involving sp^3-hybridized carbons occur at wavenumbers *less than* 3000 cm^{-1}. Thus, 1-octene has a =C—H stretching absorption at 3080 cm^{-1} (Fig. 12.10a), and *trans*-3-hexene has a similar absorption which is barely discernible at 3030 cm^{-1} (Fig. 12.10b). The higher frequency of =C—H stretching absorptions is a manifestation of the bond-strength effect: bonds to sp^2-hybridized carbons are stronger (Table 5.3, p. 216), and stronger bonds vibrate at higher frequencies.

The alkene =C—H bending absorptions that appear in the low-wavenumber region of the IR spectrum are in many cases very strong and can be used to determine the substitution pattern at the double bond. The first three of these absorptions in Table 12.2—the ones for terminal vinyl, terminal methylene, and trans-alkene—are the most reliable. The 910 and 990 cm^{-1} terminal vinyl absorptions are illustrated in the IR spectrum of 1-octene (Fig. 12.10a), and the trans-alkene absorption is illustrated by the 965 cm^{-1} peak in the IR spectrum of *trans*-3-hexene (Fig. 12.10b).

STUDY PROBLEM 12.3

Each of three alkenes, *A*, *B*, and *C*, has the molecular formula C_5H_{10}, and each undergoes catalytic hydrogenation to yield pentane. Alkene *A* has IR absorptions at 1642, 990, and 911 cm^{-1}; alkene *B* has an IR absorption at 964 cm^{-1} and no absorption in the 1600–1700 cm^{-1} region; and alkene *C* has absorptions at 1658 and 695 cm^{-1}. Identify the three alkenes.

SOLUTION In this problem, you can *write out all the possibilities* and then use the IR spectra to decide between them. The molecular formulas and the hydrogenation data show that the carbon chains of all of the alkenes are unbranched and that all are isomeric pentenes. Hence, the *only* possibilities for compounds *A*, *B*, and *C* are the following:

$H_2C{=}CHCH_2CH_2CH_3$

1-pentene

cis-2-pentene

trans-2-pentene

The C—H bending absorptions of *A* at 990 cm^{-1} and 911 cm^{-1} indicate that it is a 1-alkene; thus, it must be 1-pentene. The 964 cm^{-1} C—H bending absorption of *B* shows that it is *trans*-2-pentene. (Why is the C=C stretching vibration absent?) The remaining alkene *C* must be *cis*-2-pentene; the 1658 cm^{-1} C=C stretching absorption and the 695 cm^{-1} C—H bending absorption are consistent with this assignment.

Notice that you do not need the complete IR spectrum of each compound, but only the key absorptions, to solve this problem.

PROBLEMS

12.11 Five isomeric alkenes *A*–*E* each undergo catalytic hydrogenation to give 2-methylpentane. The IR spectra of these five alkenes have the following key absorptions (in cm^{-1}):

Compound *A*: 912 (s), 994 (s), 1643 (s), 3077 (m)

Compound *B*: 833 (s), 1667 (w), 3050 (weak shoulder on C—H absorption)

Compound *C*: 714 (s), 1665 (w), 3010 (m)

Compound *D*: 885 (s), 1650 (m), 3086 (m)

Compound *E*: 967 (s), no absorption 1600–1700, 3040 (m)

Propose a structure for each alkene.

12.12 One of the spectra in Fig. 12.11 is that of *trans*-2-heptene and the other is that of 2-methyl-1-hexene. Which is which? Explain.

D. IR Spectra of Alcohols and Ethers

The most characteristic absorption of alcohols is the O—H stretching absorption. The position and strength of this absorption is determined by the degree of hydrogen bonding, which depends, in turn, on the physical state of the alcohol. In the gas phase, in which O—H groups are not hydrogen-bonded to other groups, the O—H stretching absorption is a fairly sharp peak of moderate intensity that occurs near 3600 cm^{-1}. However, in most liquid and solid samples—which are more commonly used in everyday IR spectroscopy—the O—H groups are strongly hydrogen-bonded and give a broad peak of moderate to strong intensity in the 3200–3400 cm^{-1} region of the IR spectrum. Such an absorption, which is an important spectroscopic identifier for alcohols, is clearly evident in the IR spectrum of 1-hexanol (see Fig. 12.12).

The other characteristic absorption of alcohols is a strong C—O stretching peak that occurs in the 1050–1200 cm^{-1} region of the spectrum; primary alcohols absorb near the low

FIGURE 12.11 IR spectra for Problem 12.12.

FIGURE 12.12 The IR spectrum of 1-hexanol. Note particularly the broad O—H stretching absorption.

end of this range and tertiary alcohols near the high end. For example, this absorption occurs at about 1060 cm^{-1} in the spectrum of 1-hexanol. Because some other functional groups, such as ethers, esters, and carboxylic acids, also show C—O stretching absorptions in the same general region of the spectrum, the C—O stretching absorption is mainly used to support or confirm the presence of an alcohol diagnosed from the O—H absorption or from other spectroscopic evidence.

The most characteristic infrared absorption of ethers is the C—O stretching absorption, which, for the reasons just stated, is not very useful except for confirmation when an ether is already suspected from other data. For example, both dipropyl ether and an isomer 1-hexanol have strong C—O stretching absorptions near 1100 cm^{-1}.

FIGURE 12.13 The IR spectrum for Problem 12.13.

PROBLEMS

12.13 Match the IR spectrum in Fig. 12.13 to one of the following three compounds: 2-methyl-1-octene, butyl methyl ether, or 1-pentanol.

12.14 Explain why the IR spectra of some ethers have *two* C—O stretching absorptions. (*Hint:* See Fig. 12.8, p. 582.)

12.15 Explain why the frequency of the O—H stretching absorption of an alcohol in solution changes as the alcohol solution is diluted.

FURTHER EXPLORATION 12.2
FT-IR Spectroscopy

12.5   OBTAINING AN INFRARED SPECTRUM

Infrared spectra are obtained with an instrument called an **infrared spectrometer**. In its simplest concept, the IR spectrometer provides a way to carry out the absorption spectroscopy experiment shown in Fig. 12.3 (p. 573). The instrument houses a source of infrared radiation, a place to mount the sample in the infrared beam, and the optics and electronics necessary to measure the intensity of light absorbed or transmitted as a function of wavelength or wavenumber. Modern IR spectrometers, called *Fourier-transform infrared spectrometers* (FT-IR spectrometers), can provide an IR spectrum in a few seconds. (See Further Exploration 12.2.) The IR spectra in this text were obtained with an FT-IR spectrometer.

The sample containers ("sample cells") used in IR spectroscopy must be made of an infrared-transparent material. Because glass absorbs infrared radiation, it cannot be used. The conventional material used for sample cells is sodium chloride. The IR spectrum of an undiluted ("neat") liquid can be obtained by compressing the liquid to form a thin film between two optically polished salt plates. IR spectra can also be taken in solution cells, which consist of sodium chloride plates in appropriate holders equipped with syringe fittings for injecting the solution. If the sample is a solid, the finely ground solid can be dispersed ("mulled") in a mineral oil and the dispersion compressed between salt plates. Alternatively, a solid can be co-fused (melted) with KBr, another IR-transparent material, to form a clear pellet. Simple presses are available to prepare KBr pellets.

When mineral-oil dispersions or solvents are used, we have to be aware of the regions in which the oil or the solvents themselves absorb IR radiation, because these absorptions interfere with those of the sample. A number of solvents are commonly used; chloroform ($CHCl_3$), its isotopic analog ($CDCl_3$), and methylene chloride (CH_2Cl_2) are among them. As a few students learn the hard way, solvents that dissolve sodium chloride, such as water and alcohols, cannot be used.

A more recent technique for obtaining IR spectra is called *attenuated total reflectance* (ATR), illustrated in Fig. 12.14. In this technique, a thin layer of the sample (either solid or liquid) is spread across, and pressed onto, a crystal support with a high refractive index, such

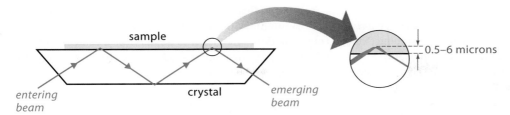

FIGURE 12.14 A diagram of the attenuated total reflectance (ATR) technique. The infrared beam is reflected from the interior surfaces of the crystal support. As the magnification shows, the IR beam at each surface actually extends for a few micrometers beyond the surface into the sample, so that the intensity of the reflected IR beam (indicated by the thickness of the red line) is reduced by sample absorption. As in other forms of absorption spectroscopy, the measurement of absorption intensity versus frequency gives the IR spectrum.

as germanium, zinc selenide, or, in research environments, diamond. The IR beam is directed through the crystal from the lower surface at an angle so that it reflects from each crystal surface. The IR beam at each surface actually extends for a few micrometers beyond each surface, so that the intensity of the reflected IR beam is reduced ("attenuated") somewhat by absorption of the sample. The spectrum, as usual, is the measurement of absorption versus wavelength. The advantage of the ATR technique is that the use of optically polished sodium chloride plates can be avoided, and substances that dissolve sodium chloride can be examined. Relatively inexpensive ATR instruments are available in some undergraduate laboratories.

12.6   INTRODUCTION TO MASS SPECTROMETRY

In contrast to other spectroscopic techniques, mass spectrometry does not involve the absorption of electromagnetic radiation, but operates on a completely different principle. As the name implies, mass spectrometry is used to determine molecular masses, and it is the most important technique used for this purpose. It also has some use in determining molecular structure.

A. Electron-Ionization (EI) Mass Spectra

The instrument used to obtain a mass spectrum is called a **mass spectrometer**. In one type of instrument, an **electron-ionization (EI) mass spectrometer**, the organic compound to be analyzed is vaporized in a vacuum chamber and bombarded with an electron beam of high energy, typically 70 eV (electron volts)—more than 6700 kJ mol^{-1}. The electron cloud of a molecule repels these electrons, so these electrons don't collide with the molecule, but veer off in what might be termed a "glancing blow." As a high-energy electron passes close by, its electric field, because of its high energy, causes the ejection of an electron from the molecule. For example, if methane is treated in this manner, it loses an electron from one of the C—H bonds.

$$\text{H:}\overset{\overset{\displaystyle H}{\displaystyle \cdot\cdot}}{\underset{\underset{\displaystyle H}{\displaystyle \cdot\cdot}}{C}}\text{:H} + e^- \longrightarrow \text{H:}\overset{\overset{\displaystyle H}{\displaystyle \cdot\cdot}}{\underset{\underset{\displaystyle H}{\displaystyle \cdot\cdot}}{\overset{+}{C}}}\text{H} + 2e^- \tag{12.15}$$

The product of this reaction is sometimes abbreviated as follows:

$$\text{H:}\overset{\overset{\displaystyle H}{\displaystyle \cdot\cdot}}{\underset{\underset{\displaystyle H}{\displaystyle \cdot\cdot}}{\overset{+}{C}}}\text{H} \quad \text{is abbreviated as} \quad CH_4^{+\cdot}$$

The symbol $^{+\cdot}$ means that the molecule is both a radical (a species with an unpaired electron) and a cation—a **radical cation**. The species $CH_4^{+\cdot}$ is called the *methane radical cation*.

Following its formation, the methane radical cation decomposes in a series of reactions called *fragmentation reactions*. In a **fragmentation reaction**, a radical cation literally comes apart. The ionic product of the fragmentation (whether it is a cation or a radical cation) is called a **fragment ion**. For example, in one fragmentation reaction, it loses a hydrogen *atom* (the radical) to generate the methyl cation, a carbocation.

$$CH_4^{\ddagger} \longrightarrow {}^+CH_3 + H\cdot \qquad (12.16)$$

mass = 16   **methyl cation**
                   mass = 15

Notice that the hydrogen atom carries the unpaired electron, and the methyl cation carries the charge; consequently, the methyl cation is the *fragment ion* in this case. The process can be represented with the free-radical (fishhook) arrow notation as follows:

$$H-\overset{\overset{\displaystyle H}{|}}{\underset{\underset{\displaystyle H}{|}}{C}}{\ddagger}H \longrightarrow H-\overset{\overset{\displaystyle H}{|}}{\underset{\underset{\displaystyle H}{|}}{C}}{}^+ + \cdot H \qquad (12.17)$$

Alternatively, the unpaired electron may remain associated with the carbon atom; in this case, the products of the fragmentation are a methyl radical and a proton.

$$CH_4^{\ddagger} \longrightarrow \cdot CH_3 + H^+ \qquad (12.18)$$

mass = 16   **methyl radical**   mass = 1

In this case the proton is the fragment ion. Further decomposition reactions give fragments of progressively smaller mass. (Show how these occur by using the fishhook notation.)

$$^+CH_3 \longrightarrow {}^{\ddagger}CH_2 + H\cdot \qquad (12.19a)$$
              mass = 14

$$^{\ddagger}CH_2 \longrightarrow {}^+CH + H\cdot \qquad (12.19b)$$
              mass = 13

$$^+CH \longrightarrow C^{\ddagger} + H\cdot \qquad (12.19c)$$
              mass = 12

The ions formed in Eqs. 12.16 and 12.19a–c are *very* unstable species. They are not the sorts of species that would be involved as reactive intermediates in a solution reaction. Recall from Sec. 9.6, for example, that methyl and primary carbocations are *never* formed as intermediates in S_N1 reactions. These ions are formed in the mass spectrometer only because of the immense energy imparted to the methane molecules by the bombarding electron beam.

Thus, methane undergoes fragmentation in the mass spectrometer to give several positively charged fragment ions of differing mass: $CH_4^{\ddagger}$, $^+CH_3$, $^{\ddagger}CH_2$, ^+CH, $C^{\ddagger}$, and H^+. In the mass spectrometer, the fragment ions are separated according to their **mass-to-charge ratio**, m/z (m = mass, z = the charge of the fragment). Because most ions formed in the electron-ionization mass spectrometer have unit charge, the m/z value can generally be taken as the mass of the ion. A **mass spectrum** is a graph of the relative amount of each ion (called the **relative abundance**) as a function of the ionic mass (or m/z). When the ions are produced by electron ionization, the mass spectrum is called an **EI mass spectrum**. The EI mass spectrum of methane is shown in Fig. 12.15. Note that only *ions* are detected by the mass spectrometer—neutral molecules and radicals do not appear as peaks in the mass spectrum. The mass spectrum of methane shows peaks at $m/z =$ 16, 15, 14, 13, 12, and 1, corresponding to the vari-

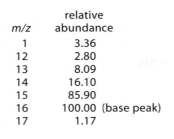

m/z	relative abundance
1	3.36
12	2.80
13	8.09
14	16.10
15	85.90
16	100.00 (base peak)
17	1.17

FIGURE 12.15 The EI mass spectrum of methane. Can you explain why there is an ion at m/z = 17? (See Sec. 12.6B for the answer.)

FIGURE 12.16 The EI mass spectrum of 1-heptene. Notice that the molecular ion is *not* the base peak.

ous ionic species that are produced from methane by electron ejection and fragmentation, as shown in Eqs. 12.16–12.19.

The mass spectrum can be determined for any molecule that can be vaporized in a high vacuum, and this includes most organic compounds. (Other techniques for vaporizing less volatile molecules have been developed and are discussed briefly in Sec. 12.6E.) Mass spectrometry is used for three purposes: (a) to determine the molecular mass of an unknown compound, (b) to determine the structure (or a partial structure) of an unknown compound by an analysis of the fragment ions in the spectrum, and (c) to confirm the structures of compounds with known or suspected structures.

The ion derived from electron ejection before any fragmentation takes place is known as the **molecular ion**, abbreviated M. *The molecular ion occurs at an m/z value equal to the molecular mass of the sample molecule.* Thus, in the mass spectrum of methane, the molecular ion occurs at m/z = 16. In the mass spectrum of 1-heptene (see Fig. 12.16), the molecular ion occurs at m/z = 98. Except for peaks due to isotopes, discussed in the next section, the molecular ion peak is the peak of highest m/z in any ordinary mass spectrum.

The **base peak** is the ion of greatest relative abundance in the mass spectrum—that is, the ion with the largest peak. The base peak is arbitrarily assigned a relative abundance of 100%, and the other peaks in the mass spectrum are scaled relative to it. In the mass spectrum of methane, the base peak is the same as the molecular ion, but in the mass spectrum of 1-heptene (Fig. 12.16), the base peak occurs at m/z = 41. In the 1-heptene spectrum and in most others, the molecular ion and the base peak are different.

B. Isotopic Peaks

Look again at the mass spectrum of methane in Fig. 12.15. This mass spectrum shows a small but real peak at m/z = 17, a mass that is one unit higher than the molecular mass. This peak is called an M + 1 peak, because it occurs one mass unit higher than the molecular ion (M). This ion occurs because chemically pure methane is really a mixture of compounds containing the various isotopes of carbon and hydrogen.

$$\text{methane} = {}^{12}CH_4, \ {}^{13}CH_4, \ {}^{12}CDH_3, \text{ and so on}$$
$$m/z = \quad 16 \qquad 17 \qquad 17$$

The isotopes of several elements and their natural abundances are given in Table 12.3.

Possible sources of the $m/z = 17$ peak for methane are ${}^{13}CH_4$ and ${}^{12}CDH_3$. *Each isotopic compound contributes a peak with a relative abundance in proportion to its amount.* In turn, the amount of each isotopic compound is directly related to the natural abundance of the isotope involved. The abundance of ${}^{13}CH_4$ methane relative to that of ${}^{12}CH_4$ methane is then given by the following equation:

$$\text{relative abundance} = \left(\frac{\text{abundance of } {}^{13}C \text{ peak}}{\text{abundance of } {}^{12}C \text{ peak}} \right) \tag{12.20a}$$

$$= (\text{number of carbons}) \times \left(\frac{\text{natural abundance of } {}^{13}C}{\text{natural abundance of } {}^{12}C} \right)$$

$$= (\text{number of carbons}) \times \left(\frac{0.0110}{0.9890} \right)$$

$$= (\text{number of carbons}) \times 0.0111 \tag{12.20b}$$

Because methane has only one carbon, the $m/z = 17$ (M + 1) peak due to ${}^{13}CH_4$ is about 1.1% of the $m/z = 16$, or M, peak. A similar calculation can be made for deuterium.

$$\text{relative abundance} = (\text{number of hydrogens}) \times \left(\frac{\text{natural abundance of } {}^{2}H}{\text{natural abundance of } {}^{1}H} \right) \tag{12.21}$$

$$= (4) \times \left(\frac{0.00015}{0.99985} \right) = 0.0006$$

Thus, the CDH_3 naturally present in methane contributes 0.06% to the isotopic peak. Because the contribution of deuterium is so small, ${}^{13}C$ is the major isotopic contributor to the M + 1 peak. (We'll ignore contributions of ${}^{2}H$ in subsequent calculations of M + 1 peak intensities.)

In a compound containing more than one carbon, the M + 1 peak is larger than 1.1% of the M peak because there is a 1.1% probability that *each carbon* in the molecule will be present as ${}^{13}C$. For example, cyclohexane has six carbons, and the abundance of its M + 1 ion relative to that of its molecular ion should be 6(1.1) = 6.6%. In the mass spectrum of cyclohexane, the molecular ion has a relative abundance of about 70%; that of the M + 1 ion is calculated to be (0.066)(70%) = 4.6%, which corresponds closely to the value observed. Not only the molecular ion peak, but also every other peak in the mass spectrum has isotopic peaks.

Several elements of importance in organic chemistry have isotopes with significant natural abundances. Table 12.3 shows that silicon has significant M + 1 and M + 2 contributions; sulfur has an M + 2 contribution; and the halogens chlorine and bromine have very important M + 2 contributions. In fact, the naturally occurring form of the element bromine consists of about equal amounts of ${}^{79}Br$ and ${}^{81}Br$. The mixture of isotopes leaves a characteristic trail in the mass spectrum that can be used to diagnose the presence of the element.

Consider, for example, the EI mass spectrum of bromomethane, shown in Fig. 12.17. The peaks at $m/z = 94$ and 96 result from the contributions of the two bromine isotopes to the molecular ion. They are in the relative abundance ratio 100 : 98 = 1.02, which is in good agreement with the ratio of the relative natural abundances of the bromine isotopes (Table 12.3). This double molecular ion, often called a "bromine doublet," is a dead giveaway

TABLE 12.3 Exact Masses and Isotopic Abundances of Several Isotopes Important in Mass Spectrometry

Element	Isotope	Exact mass	Abundance, %
hydrogen	^1H	1.007825	99.985
	^2H*	2.0140	0.015
carbon	^{12}C	12.0000	98.90
	^{13}C	13.00335	1.10
nitrogen	^{14}N	14.00307	99.63
	^{15}N	15.00011	0.37
oxygen	^{16}O	15.99491	99.759
	^{17}O	16.99913	0.037
	^{18}O	17.99916	0.204
fluorine	^{19}F	18.99840	100.
silicon	^{28}Si	27.97693	92.21
	^{29}Si	28.97649	4.67
	^{30}Si	29.97377	3.10
phosphorus	^{31}P	30.97376	100.
sulfur	^{32}S	31.97207	95.0
	^{33}S	32.97146	0.75
	^{34}S	33.96787	4.22
chlorine	^{35}Cl	34.96885	75.77
	^{37}Cl	36.96590	24.23
bromine	^{79}Br	78.91834	50.69
	^{81}Br	80.91629	49.31
iodine	^{127}I	126.90447	100.

* ^2H is commonly known as deuterium, abbreviated D.

FIGURE 12.17 The EI mass spectrum of bromomethane. The two molecular ions at $m/z = 94$ and at $m/z = 96$ have nearly equal abundance and result from the presence of the two isotopes ^{79}Br and ^{81}Br.

for a compound containing a single bromine. Notice that along with each major isotopic peak is a smaller isotopic peak one mass unit higher. These peaks are due to the isotope ^{13}C present naturally in bromomethane. For example, the $m/z = 95$ peak corresponds to bromomethane containing only ^{79}Br and one ^{13}C. The $m/z = 97$ peak arises from bromomethane that contains only ^{81}Br and one ^{13}C.

Although isotopes such as ^{13}C and ^{18}O are normally present in small amounts in organic compounds, it is possible to synthesize compounds that are selectively enriched with these and other isotopes. Isotopes are especially useful because they provide specific labels at particular atoms without significantly changing their chemical properties. One use of such compounds is to determine the fate of specific atoms in deciding between two mechanisms. Another use is to provide nonradioactive isotopes for biological metabolic studies (studies that deal with the fates of chemical compounds when they react in biological systems). When a compound has been isotopically enriched, isotopic peaks are much larger than normal. Mass spectrometry is used to measure quantitatively the amount of such isotopes present in labeled compounds.

PROBLEMS

12.16 The mass spectrum of tetramethylsilane, $(CH_3)_4Si$, has a base peak at $m/z = 73$. Calculate the relative abundances of the isotopic peaks at $m/z = 74$ and 75.

12.17 From the information in Table 12.3, predict the appearance of the molecular ion peak(s) in the mass spectrum of chloromethane. (Assume that the molecular ion is the base peak.)

C. Fragmentation

In EI mass spectrometry, the molecular ion is formed by loss of an electron. If this ion is stable, it decomposes slowly and is detected by the mass spectrometer as a peak of large relative abundance. If this ion is less stable, it decomposes, sometimes completely, into smaller pieces. Two cases of such fragmentation are most commonly observed, and in each case, two products are formed.

1. One fragmentation product can be a radical, in which case the other product is a cation with no unpaired electrons (an **even-electron ion**).

2. One fragmentation product can be a neutral molecule, in which case the other product must be, like the molecular ion, a radical cation (an **odd-electron ion**).

In either case, the cation is called a **fragment ion**. *Only the ion is detected in the mass spectrum*; the radical (case 1) or neutral molecule (case 2) is not detected.

As an example of case 1, consider the $m/z = 57$ ion in the mass spectrum of decane (Fig. 12.18). This is formed in the following way. One of several possible molecular ions is formed by ejection of an electron from a carbon–carbon bond.

$$CH_3CH_2CH_2CH_2-CH_2CH_2CH_2CH_2CH_2CH_3 \xrightarrow{-e^-} CH_3CH_2CH_2CH_2 \overset{+}{\cdot} CH_2CH_2CH_2CH_2CH_2CH_3 \quad (12.22)$$

decane   molecular ion of decane
(a radical cation, $m/z = 142$)

Next, the molecule splits at the site of electron ejection to give a carbocation with $m/z = 57$ and a radical with mass $= 85$. In this fragmentation, only the cation is detected.

$$CH_3CH_2CH_2CH_2 \overset{+}{\cdot} CH_2CH_2CH_2CH_2CH_2CH_3 \longrightarrow CH_3CH_2CH_2\overset{+}{C}H_2 + \overset{\cdot}{C}H_2CH_2CH_2CH_2CH_2CH_3 \quad (12.23)$$

molecular ion of decane   a cation   a radical
(a radical cation, $m/z = 142$) $m/z = 57$ (not detected by the
(detected by the   mass spectrometer)
mass spectrometer)

Notice that there is also a peak in Fig. 12.15 at $m/z = 85$. *This does not arise from the radical*, but rather from fragmentation of the same bond in the opposite manner to give the carbocation with $m/z = 85$ and the radical with mass $= 57$.

$$CH_3CH_2CH_2CH_2 \overset{+}{\cdot} CH_2CH_2CH_2CH_2CH_2CH_3 \longrightarrow CH_3CH_2CH_2\overset{\cdot}{C}H_2 + \overset{+}{C}H_2CH_2CH_2CH_2CH_2CH_3 \quad (12.24)$$

molecular ion of decane   a radical   a cation
(not detected by the $m/z = 85$
mass spectrometer)   (detected by the
mass spectrometer)

FIGURE 12.18 The EI mass spectrum of decane.

FIGURE 12.19 The EI mass spectrum of 1-heptanol.

Other prominent peaks in the mass spectrum of decane arise from fragmentations by the same mechanism at different bonds.

A fragmentation of type 2 is illustrated by the mass spectra of many primary alcohols. For example, in the mass spectrum of 1-heptanol (molecular mass = 116), shown in Fig. 12.19, the molecular ion is formed by electron ejection from one of the oxygen unshared pairs. (Because unshared electrons are not held in bonds, they are ejected more easily than bonding electrons.)

$$CH_3(CH_2)_4CH_2\!-\!\overset{\displaystyle:\ddot{O}H}{CH_2} \xrightarrow{\ -e^-\ } CH_3(CH_2)_4CH_2\!-\!\overset{\displaystyle+\!\cdot\!\ddot{O}H}{CH_2} \qquad (12.25)$$

molecular ion of 1-heptanol
$m/z = 116$

The molecular ion is absent in this spectrum because a type-2 fragmentation of the molecular ion occurs rapidly. An internal hydrogen-atom transfer through a six-membered transition state produces a radical cation of the same mass in which the oxygen is now poised to act as a leaving group. Loss of water (a stable neutral molecule, 18 mass units) leaves another odd-electron ion with $m/z = M - 18 = 98$.

$$ \qquad (12.26a)$$

$m/z = 116 \qquad\qquad m/z = 98$

$+\ H_2\ddot{O}\!:$ (a neutral molecule not detected by the mass spectrometer)

The $m/z = 98$ peak is a fairly minor one in the mass spectrum because it undergoes a further type-2 fragmentation to give the neutral molecule ethylene (28 mass units) and another odd-electron radical cation, this one with five carbons.

$$
\begin{array}{c}
\text{CH}_3\text{CH}_2\text{CH}_2\overset{\text{H}}{\underset{|}{\text{C}}}\cdot \\
\text{H}_2\text{C} \overset{+}{\diagdown} \text{CH}_2 \\
\underset{\text{H}_2}{\text{C}} \\
m/z = 98
\end{array}
\longrightarrow
\underset{\substack{m/z = 70 \\ \text{(base peak)}}}{\text{CH}_3\text{CH}_2\text{CH}_2\overset{\cdot}{\text{CH}}-\overset{+}{\text{CH}}_2}
+ \ \text{H}_2\text{C}=\text{CH}_2 \qquad (12.26\text{b})
$$

The radical-cation product, in which a radical and a carbocation are located on adjacent carbons, is a type of odd-electron ion that is observed frequently in fragmentation. In this structure, both the carbon bearing the unpaired electron and the carbon bearing the positive charge contain $2p$ orbitals. Therefore, the unpaired electron is delocalized between them:

$$
\left[\text{CH}_3\text{CH}_2\text{CH}_2\overset{\cdot}{\text{CH}}-\overset{+}{\text{CH}}_2 \longleftrightarrow \text{CH}_3\text{CH}_2\text{CH}_2\overset{+}{\text{CH}}-\overset{\cdot}{\text{CH}}_2 \right] \qquad (12.26\text{c})
$$

If a molecule contains only C, H, O, and halogen, its even-electron fragment ions have odd mass and its odd-electron fragment ions have even mass. You can verify this with the examples in Eqs. 12.22–12.26. Thus, from the mass of the fragment ion—odd or even—you immediately know something about its structure and its origin.

As all of the mass spectra is this section illustrate, the peaks in a mass spectrum are typically not the same height. What controls the relative abundances of ions in a mass spectrum? Typically, the most stable ions appear in greatest abundance. If an ion is relatively stable, it decomposes slowly and appears as a relatively large peak. If an ion is relatively unstable, it decomposes rapidly and appears as a relatively small peak—or perhaps not at all. The principles of carbocation stability that you already know can help you to understand why certain fragment ions in a mass spectrum are prominent and others are not. This idea is illustrated in Study Problem 12.4.

STUDY PROBLEM 12.4

The base peak in the mass spectrum of 2,2,5,5-tetramethylhexane (molecular mass = 142) is at $m/z = 57$, which corresponds to a composition C_4H_9. (a) Suggest a structure for the fragment that accounts for this peak. (b) Offer a reason that this fragment is so abundant. (c) Give a mechanism that shows the formation of this fragment.

SOLUTION The first step is to draw the structure of 2,2,5,5-tetramethylhexane:

$$(\text{CH}_3)_3\text{C}-\text{CH}_2\text{CH}_2-\text{C}(\text{CH}_3)_3$$

(a) A fragment with the composition C_4H_9 could be a *tert*-butyl cation formed by splitting the compound at the bond to either of the *tert*-butyl groups:

$$
\begin{array}{c}
\overset{\text{CH}_3}{\underset{\text{CH}_3}{\overset{|}{\underset{|}{\text{H}_3\text{C}-\text{C}-\text{CH}_2-\text{CH}_2}}}}\!\!\overset{\displaystyle\overbrace{}^{\text{C}_4\text{H}_9}}{\underset{\underset{\text{CH}_3}{|}}{\overset{\overset{\text{CH}_3}{|}}{\text{C}-\text{CH}_3}}}
\end{array}
$$

(b) The most abundant peaks in the mass spectrum result from the most stable cationic fragments. Because a *tert*-butyl cation is a relatively stable carbocation (it is tertiary), it is formed in relatively high abundance.

(c) To form this cation, one electron is ejected from the C—C bond, and the compound fragments so that the unpaired electron remains on the methylene carbon (see Eqs. 12.23–12.24):

$$\underset{\text{molecular mass} = 142}{H_3C-\overset{\overset{\displaystyle CH_3}{|}}{\underset{\underset{\displaystyle CH_3}{|}}{C}}-CH_2-CH_2\!:\!\overset{\overset{\displaystyle CH_3}{|}}{\underset{\underset{\displaystyle CH_3}{|}}{C}}-CH_3} \quad \xrightarrow{\text{electron ejection}} \quad \underset{m/z = 142;\ \text{molecular ion}}{H_3C-\overset{\overset{\displaystyle CH_3}{|}}{\underset{\underset{\displaystyle CH_3}{|}}{C}}-CH_2-CH_2\!\overset{+}{\cdot}\overset{\overset{\displaystyle CH_3}{|}}{\underset{\underset{\displaystyle CH_3}{|}}{C}}-CH_3}$$

fragmentation

$$H_3C-\overset{\overset{\displaystyle CH_3}{|}}{\underset{\underset{\displaystyle CH_3}{|}}{C}}-CH_2-\dot{C}H_2 \ + \ \underset{m/z = 57}{\overset{+}{C}-\overset{\overset{\displaystyle CH_3}{|}}{\underset{\underset{\displaystyle CH_3}{|}}{}}CH_3} \qquad (12.27)$$

Fragmentation might have occurred at the same bond so that the unpaired electron remains associated with the *tert*-butyl group and a primary carbocation with $m/z = 85$ is formed. (In other words, a *more stable* free radical and a *less stable* carbocation would be formed.) There is no peak at $m/z = 85$. That this mode of fragmentation is not observed demonstrates that *carbocation stability is more important than free-radical stability in determining fragmentation patterns.*

PROBLEMS

12.18 The peak of highest mass in the EI mass spectrum of 2,2,5,5-tetramethylhexane (the molecule discussed in Study Problem 12.4) occurs at $m/z = 71$ and has about 33% relative abundance.

(a) In a structure of the molecule, indicate the bond at which fragmentation occurs to give this ion.

(b) Give a mechanism for this fragmentation.

(c) What is the structure of the fragment ion at $m/z = 71$? (*Hint:* Apply what you know about carbocations.)

12.19 Indicate whether the following peaks in the mass spectrum of 1-heptanol are odd-electron or even-electron ions. (Don't attempt to give their structures.)

(a) $m/z = 83$ (b) $m/z = 56$ (c) $m/z = 41$

12.20 The mass spectrum of 2-chloropentane shows large and almost equally intense peaks at $m/z = 71$ and $m/z = 70$.

(a) Classify each peak as an even-electron or odd-electron ion.

(b) What stable neutral molecule can be lost to give the odd-electron ion?

(c) Write a mechanism for the origin of each fragment ion.

D. The Molecular Ion. Chemical-Ionization Mass Spectra

The molecular ion is the most important peak in the mass spectrum for two reasons. First, the m/z of the molecular ion occurs at the molecular mass, and one of the most important uses of mass spectrometry is the determination of molecular mass. Second, the mass of the molecular ion is the basis for the calculation of losses due to fragmentation.

Unfortunately, a peak due to the molecular ion is weak or absent in some mass spectra. Consider, for example, the EI mass spectrum of di-*sec*-butyl ether shown in Fig. 12.20a on p. 600. The molecular mass of this ether is 130. A peak at this mass, however, is essentially absent. The three most prominent peaks in the EI mass spectrum of di-*sec*-butyl ether occur at $m/z = 101$, $m/z = 57$, and $m/z = 45$ (base peak). The $m/z = 101$ peak correspond to a loss of

FIGURE 12.20 Mass spectra of di-*sec*-butyl ether. (a) Electron-ionization (EI) mass spectrum. (b) Chemical-ionization (CI) mass spectrum. The molecular ion at $m/z = 130$ is essentially absent in the EI spectrum, whereas the molecular ion (as the protonated ether at $m/z = 131$) is the base peak in the CI spectrum. Notice that the CI spectrum has a smaller number of fragment ions that the EI spectrum.

(a) EI mass spectrum

(b) CI mass spectrum

29 mass units (that is, an ethyl group), and occurs in the following way. First, the molecular ion is formed by loss of an electron from the oxygen unshared pair:

$$CH_3CH_2CH-\overset{..}{\underset{..}{O}}-CHCH_2CH_3 \xrightarrow{-e^-} CH_3CH_2CH-\overset{+}{\underset{..}{O}}-CHCH_2CH_3 \tag{12.28}$$

di-*sec*-**butyl ether**   molecular ion of di-*sec*-butyl ether
                                          ($m/z = 130$)

(Remember from the discussion of Eq. 12.25 that unshared electrons are more easily removed than bonding electrons.) Next, an ethyl radical is lost by a process that mass spectrometrists call *α-cleavage* and free-radical chemists call *β-scission:*

$$CH_3\overset{\frown}{CH_2}\overset{\frown}{-CH}\overset{\frown}{-\overset{+}{\underset{..}{O}}}-CHCH_2CH_3 \longrightarrow CH_3\dot{C}H_2 + CH=\overset{+}{\underset{..}{O}}-CHCH_2CH_3 \tag{12.29}$$

molecular ion of di-*sec*-butyl ether $m/z = 101$
        ($m/z = 130$)

This ion reacts further by *α*-elimination of 2-butene (56 mass units) to give the base peak at $m/z = 45$.

$$CH_3CH = \overset{+}{\underset{\ddot{}}{O}} - CHCH_3 \xrightarrow{\boxed{\beta\text{-elimination}}} CH_3CH = \overset{+}{\underset{}{O}}H + \underset{\underset{CHCH_3}{\|}}{CHCH_3} \qquad (12.30)$$

$$m/z = 101 \qquad\qquad m/z = 45$$

The $m/z = 57$ peak is formed by a process called *inductive cleavage*, which is nothing more than the radical cation version of an S_N1-like dissociation:

$$CH_3CH_2\underset{\underset{CH_3}{|}}{CH} - \overset{+}{\underset{\ddot{}}{O}} - \underset{\underset{CH_3}{|}}{CHCH_2CH_3} \xrightarrow{\boxed{\text{inductive cleavage}}} CH_3CH_2\overset{+}{\underset{\underset{CH_3}{|}}{CH}} + \cdot\overset{..}{\underset{..}{O}} - \underset{\underset{CH_3}{|}}{CHCH_2CH_3} \qquad (12.31)$$

molecular ion of di-*sec*-butyl ether
($m/z = 130$)

The decomposition mechanisms shown here—α-cleavage, β-elimination, and inductive cleavage—are very common decomposition mechanisms in the mass spectra of molecules containing atoms with unshared electron pairs. These processes lead to relatively stable cations, and this is why the molecular ion does not survive.

This example illustrates the point that we cannot be sure in many cases whether the ion of highest mass in a compound of *unknown structure* is the molecular ion or a fragment ion. The question is, then, how can we determine with certainty the molecular mass of an unknown compound?

Recall that molecular ions in EI mass spectra are formed by a highly energetic electron-bombardment process. When a molecular ion has a very high energy, it is likely to dissipate that energy by fragmentation. However, if we could form a molecular ion by a "softer" (less energetic) method, the tendency of the ion to undergo fragmentation would be decreased. An ionization method commonly used for this purpose is called *chemical ionization*, and mass spectra derived from chemical ionization are called **chemical-ionization mass spectra**, or **CI mass spectra** for short.

In **chemical ionization**, the vaporized molecule of interest (di-*sec*-butylether) is mixed with a large excess of a *reagent gas* such as methane or isobutane. (We'll illustrate this process with methane.) When this mixture is subjected to electron bombardment, the reagent gas rather than the ether is selectively ionized (to the methane radical cation) because of its much higher concentration as shown in Eq. 12.15 (p. 591). Because the reagent gas is present in high concentration, the methane radical cation almost always collides with another methane molecule. When this happens, a proton is transferred to the methane to give a species $^+CH_5$.

$$^+_{\cdot}CH_4 + CH_4 \longrightarrow \dot{C}H_3 + {}^+CH_5 \qquad (12.32a)$$

In this unusual species, *five* protons share the *four* bonding electron pairs that were in the original methane. The important thing about $^+CH_5$ is that it is *very acidic*: in fact, *it is a source of gas-phase protons*. Although $^+CH_5$ mostly collides with methane (giving no net reaction), when it eventually collides with the ether, the ether is protonated at its most basic site, which is a lone pair on the oxygen (Sec. 11.1). Thus, the ether is ionized.

$$CH_3CH_2\underset{\underset{CH_3}{|}}{CH} - \overset{..}{\underset{..}{O}} - \underset{\underset{CH_3}{|}}{CHCH_2CH_3} + {}^+CH_5 \longrightarrow CH_3CH_2\underset{\underset{CH_3}{|}}{CH} - \overset{\overset{\overset{H}{|}}{+}}{\underset{\underset{..}{}}{O}} - \underset{\underset{CH_3}{|}}{CHCH_2CH_3} + CH_4 \qquad (12.32b)$$

di-*sec*-butyl ether   (a gas-phase proton source)   conjugate acid of di-*sec*-butyl ether
$m/z = 131$

This conjugate-acid cation is an even-electron ion; it is not a radical cation. In a CI mass spectrum, the peak for this ion necessarily occurs one mass unit higher than the molecular mass of the molecule itself because of the added proton. Because this ion is formed in a relatively low-energy process, it does not fragment so readily as the molecular ion in the EI mass spectrum. The CI mass spectrum of di-*sec*-butyl ether is shown in Fig. 12.18b. This shows a prominent M + 1 ion at $m/z = 131$, which is also the base peak. The relatively small number of fragments come from the loss of various neutral molecules from this ion. For example, the largest fragment peak at $m/z = 75$ arises from loss of 2-butene in a β-elimination process analogous to the one in Eq. 12.30:

$$CH_3CH_2CH-\overset{+}{\underset{\underset{CH_3}{|}}{O}}-CHCH_2CH_3 \longrightarrow CH_3CH_2CH-\overset{+}{\underset{\underset{CH_3}{|}}{O}}-H + \overset{CHCH_3}{\underset{CHCH_3}{\|}} \quad (12.33)$$

conjugate acid of di-*sec*-butyl ether
$m/z = 131$

$m/z = 75$

Typically, the mass spectrometry of a compound with unknown structure is investigated by running both its EI and CI mass spectra. The CI mass spectrum typically gives a strong M + 1 peak that reveals the molecular mass M. The richer fragmentation pattern of the EI spectrum can then be used to deduce other aspects of the structure.

PROBLEMS

12.21 (a) The EI mass spectrum of $CH_3OCH_2CH(CH_3)_2$ (methyl isobutyl ether) contains only a trace of a molecular ion and a base peak at $m/z = 45$, which arises from α-cleavage. Show the α-cleavage process and give the structure of the ion with mass = 45.

(b) The mass spectrum of methyl isobutyl ether does *not* show a peak due to inductive cleavage, in contrast to the mass spectrum of di-*sec*-butyl ether (Eq. 12.31). Use what you know about carbocation stability to explain the absence of this peak.

(c) What major difference(s) would you expect to find when comparing the CI mass spectrum of methyl isobutyl ether with its EI spectrum?

12.22 Show the elimination reactions that account for each of the following fragments in the CI mass spectrum of di-*sec*-butyl ether (Fig. 12.18b). (*Hint:* β-Elimination reactions can also form C=O double bonds.)

(a) $m/z = 101$ (b) $m/z = 115$ (c) $m/z = 73$

E. The Mass Spectrometer

FURTHER EXPLORATION 12.3
The Mass Spectrometer

A mass spectrometer must produce gas-phase ions, sort them by mass, and detect the relative number of ions of each mass. We've already learned about two ways of producing ions: electron ionization and chemical ionization. Ion sorting in a conventional "magnetic-sector" mass spectrometer depends on the fact that the paths of moving ions are bent by a magnetic field. When subjected to a magnetic field, ions with large m/z traverse a path of larger radius than ions of smaller m/z (Fig. 12.21). Another type of ion sorting is called "time-of-flight." A time-of-flight spectrometer differentiates ions by the amount of time it takes them to move through an electric field. Ions of smaller m/z are accelerated more easily by the electric field, and thus take less time to move through the field, than ions of larger m/z. Regardless of the ion-sorting method, ions are detected as an ion current. The mass spectra in this text are actually plots of relative ion currents versus m/z, with the largest ion current (that is, the base peak) assigned a relative value of 100.

A modern mass spectrometer is an extremely sensitive instrument and can readily produce a mass spectrum from amounts of material in the range of micrograms (10^{-6} g) to picograms (10^{-12} g). For this reason, the instrument is very useful for the analysis of materials available in only trace quantities. It has played a key role in such projects as the analysis

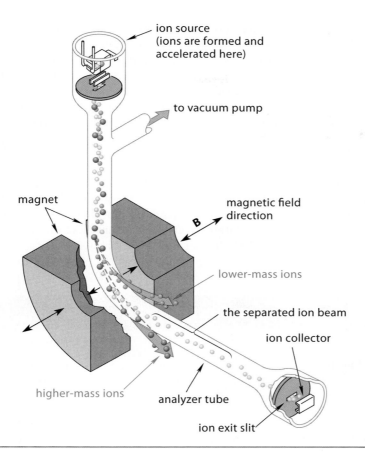

ion source
(ions are formed and
accelerated here)

to vacuum pump

magnet

magnetic field
direction

B

lower-mass ions

the separated ion beam

ion collector

higher-mass ions

analyzer tube

ion exit slit

FIGURE 12.21 Diagram of a magnetic-sector mass spectrometer. After ionization of the sample by electron bombardment, the ions are accelerated by a high voltage and are passed into a magnetic field **B** along a path perpendicular to the field. The field bends the paths of the ions; the paths of lower-mass ions (*red*) are bent more than those of higher-mass ions (*blue*). (See Further Exploration 12.3.) As the field is progressively increased, ions of increasingly higher mass attain exactly the correct path to enter the ion exit slit.

of drug levels in blood serum and the elucidation of the structures of insect pheromones (Sec. 14.9) that are available only in minuscule amounts. It is also an important tool in the modern forensics laboratory ("crime lab").

One of the operating characteristics of a mass spectrometer is its *resolution*—how well it separates ions of different mass. A relatively simple mass spectrometer readily distinguishes, over a total m/z range of several hundred, ions that differ in mass by one unit. More complex mass spectrometers, called *high-resolution mass spectrometers*, can resolve ions that are separated in mass by only a few thousandths of a mass unit. Why is such high resolution useful? Suppose an unknown compound has a molecular ion at $m/z = 124$. Two possible formulas for this ion are $C_8H_{12}O$ and C_9H_{16}. Both formulas have the same **nominal mass** (that is, the same mass to the nearest whole number). However, if the **exact mass** (the mass to four or more decimal places) is calculated for each formula (using the values of the most abundant isotopes in Table 12.2), then different results are obtained:

$$C_8H_{12}O, \text{ exact mass: } 124.0888$$

$$C_9H_{16}, \text{ exact mass: } 124.1252$$

The difference of 0.0364 mass units is easily resolved by a high-resolution mass spectrometer. Computers used with such instruments can be programmed to work backward from the exact mass and provide *an elemental analysis of the molecular ion* (and therefore the compound of interest) *as well as the elemental analysis of each fragment in the mass spectrum*! Because a modern high-resolution mass spectrometer with its associated computer and other accessories can cost several hundred thousand dollars, it is generally shared by a large number of researchers.

Before a compound can be analyzed by mass spectrometry, it must be vaporized. This presents a difficult problem for large molecules that have negligible vapor pressures. Research in mass spectrometry has focused on novel ways to produce ions in the gas phase from large nonvolatile molecules, many of which are of biological interest. In one technique, nicknamed MALDI (matrix-assisted laser desorption ionization), the material to be analyzed (analyte) is co-crystallized with a material, termed a *matrix*, that can absorb radiation from a laser. In a process that is not fully understood, bombarding the matrix–analyte mixture with light from the laser ultimately produces gas-phase ions of the analyte, which are analyzed by mass spectrometry. In another technique, nicknamed ESI (electrospray ionization), a solution of the analyte is atomized in highly charged droplets, much as we might atomize perfume in a sprayer. This process results in the formation of highly charged molecules in the gas phase, and these are analyzed by mass spectrometry. These techniques have made possible the analysis of materials with molecular masses in excess of 100,000, such as proteins, nucleic acids, and synthetic polymers. (Some examples of mass spectra obtained by these techniques are discussed in Sec. 27.8B.) For their discovery and development of these techniques, John P. Fenn, of Virginia Commonwealth University, and Koichi Tanaka, of the Shimadzu Corporation in Tokyo, shared part of the 2002 Nobel Prize in Chemistry.

KEY IDEAS IN CHAPTER 12

- Spectroscopy deals with the interaction of matter and electromagnetic radiation. Electromagnetic radiation is characterized by its energy, wavelength, and frequency, which are interrelated by Eq. 12.3.

- Infrared spectroscopy deals with the absorption of infrared radiation by molecular vibrations. An infrared spectrum is a plot of the infrared radiation transmitted through a sample as a function of the wavenumber or wavelength of the radiation.

- The frequency of an absorption in the infrared spectrum is equal to the frequency of the bond vibration involved in the absorption.

- The wavenumber or frequency of an absorption is greater for vibrations involving stronger bonds and smaller atomic masses (Eqs. 12.11 and 12.13). The smaller of two atomic masses involved in a bond vibration has the greater effect on the frequency of the vibration.

- The intensity of an absorption increases with the number of absorbing groups in the sample and the size of the dipole moment change that occurs in the molecule when the vibration occurs. Absorptions that result in no dipole moment change are infrared-inactive.

- The infrared spectrum provides information about the functional groups present in a molecule. The $=C-H$ stretching and bending absorptions and the $C=C$

stretching absorption are very useful for the identification of alkenes. The $O-H$ stretching absorption is diagnostic for alcohols.

- In electron-ionization (EI) mass spectrometry, a molecule loses an electron to form the molecular ion, a radical cation, which in most cases decomposes to fragment ions. The relative abundances of the fragment ions are recorded as a function of their mass-to-charge ratios m/z, which, for most ions, equal their masses. Both molecular masses and partial structures can be derived from the masses of these ionic fragments.

- Associated with each peak in a mass spectrum are other peaks at higher mass that arise from the presence of isotopes at their natural abundance. Such isotopic peaks are particularly useful for diagnosing the presence of elements that consist of more than one isotope with high natural abundance, such as chlorine and bromine.

- Ionic fragments are of two types: even-electron ions, which contain no unpaired electrons; and odd-electron ions, which contain an unpaired electron.

- In chemical-ionization (CI) mass spectrometry, molecules are ionized by direct protonation in the gas phase. Because this is a much gentler ionization technique than EI, a CI mass spectrum typically contains a greater proportion of molecular ion (as its conjugate acid) than the EI spectrum of the same compound.

12.23 List the factors that determine the wavenumber of an infrared absorption.

12.24 List two factors that determine the intensity of an infrared absorption.

12.25 Indicate how you would carry out each of the following chemical transformations. What are some of the changes in the infrared spectrum that could be used to indicate whether the reaction has proceeded as indicated? (Your answer can include disappearance as well as appearance of IR absorptions.)

(a) 1-methylcyclohexene $\longrightarrow$ methylcyclohexane

(b) 1-hexanol $\longrightarrow$ 1-methoxyhexane

12.26 Which of the molecules in each of the following pairs should have identical IR spectra, and which should have different IR spectra (if only slightly different)? Explain your reasoning carefully.

(a) 3-pentanol and (±)-2-pentanol

(b) (R)-2-pentanol and (S)-2-pentanol

(c)

(Assume you could determine the IR spectra of each conformation individually.)

12.27 Match each of the IR spectra in Fig. P12.27 on p. 606 to one of the following compounds. (Notice that there is no spectrum for two of the compounds.)

(a) 1,5-hexadiene   (b) 1-methylcyclopentene

(c) 1-hexen-3-ol   (d) dipropyl ether

(e) *trans*-4-octene   (f) cyclohexane

(g) 3-hexanol

12.28 A former theological student, Heavn Hardley, has turned to chemistry and, during his eighth year of graduate study, has carried out the following reaction:

Unfortunately, Hardley thinks he may have mislabeled his samples of A and B, but has wisely decided to take an IR spectrum of each sample. The spectra are reproduced in Fig. P12.28 on p. 607. Which sample goes with which spectrum? How do you know?

12.29 (a) Given the stretching frequencies for the C—H bonds shown in color, arrange the corresponding bonds in order of increasing strength. Explain your reasoning.

$$
\begin{array}{ccc}
\overset{\displaystyle H}{|} & \overset{\displaystyle H}{|} & \\
RCH{=}CH & RCH_2 & RC{\equiv}C{-}H \\
3080\ cm^{-1} & 2850\ cm^{-1} & 3300\ cm^{-1}
\end{array}
$$

(b) If the bond dissociation energy of the $\equiv$C—H bond is 558 kJ mol^{-1} (133 kcal mol^{-1}), use the stretching frequencies in part (a) to estimate the bond dissociation energy of the C—H bond in RCH$_2$—H.

12.30 Arrange the following bonds in order of increasing stretching frequencies, and explain your reasoning.

$$C{=}C \qquad C{\equiv}C \qquad C{=}O \qquad C{-}C$$

12.31 (a) The water molecule has three distinguishable molecular vibrations. Construct a diagram like Fig. 12.8 on p. 582 for the three vibrational modes of water. (*Hint:* Each vibration at its extremes must change the molecule so that it can be distinguished from the molecule before the vibration occurs.)

(b) Classify each vibration as a stretching or bending vibration.

(c) The IR spectrum of water vapor has three absorptions: 1595, 3652, and 3756 cm^{-1}. Which are stretching vibrations and which are bending vibrations? Explain.

12.32 Explain why a nitro compound has two N—O stretching vibrations. (These typically occur at about 1370 and 1550 cm^{-1}.)

$$
\left[\ R{-}\overset{+}{N}\overset{\ddot{O}:}{\underset{:\ddot{O}:^-}{}} \quad \longleftrightarrow \quad R{-}\overset{+}{N}\overset{:\ddot{O}:^-}{\underset{\ddot{O}:}{}}\ \right]
$$

12.33 (a) Explain why the S—H stretching absorption in the IR spectrum of a thiol is less intense and occurs at lower frequency (2550 cm^{-1}) than the O—H stretching absorption of an alcohol.

(b) Is the wavenumber difference between O—H and S—H absorptions caused primarily by the greater mass of sulfur or by the relative strengths of the two bonds? Explain how you know.

(c) Two unlabeled bottles, A and B, contain liquids. Laboratory notes suggest that one compound is (HSCH$_2$CH$_2$)$_2$O and the other is (HOCH$_2$CH$_2$)$_2$S.

Figure P12.27 *(continues on p. 607)*

The IR spectra of the two compounds are given in Fig. P12.33 on p. 608. Identify *A* and *B* and explain your choice.

12.34 (a) You have found in the laboratory two liquids, *C* and *D*, in unlabeled bottles. You suspect that one is deuterated chloroform ($CDCl_3$) and the other is ordinary chloroform ($CHCl_3$). Unfortunately, the mass spectrometer is not operating because the same person who failed to label the bottles has been recently using the mass spectrometer! From the IR spectra of the two compounds, shown in Fig. P12.34 on p. 608, indicate which compound is which. Explain.

(b) How would these compounds be distinguished by mass spectrometry?

12.35 Rationalize the indicated fragments in the EI mass spectrum of each of the following molecules by proposing a

Figure P12.27 (*continued from p. 606*)

Figure P12.28

Figure P12.33

Figure P12.34

structure of the fragment and a mechanism by which it is produced.

(a)

$$\underset{m/z\ =\ 72}{CH_3CH_2CH_2-\overset{\overset{\displaystyle CH_3}{|}}{\underset{\underset{\displaystyle CH_2CH_3}{|}}{C}}-\ddot{N}H_2}$$

(b) 3-methyl-3-hexanol, $m/z = 73$

(c) 1-pentanol, $m/z = 70$

(d) neopentane, $m/z = 57$

12.36 Suggest structures for the following *neutral* molecules commonly lost in mass spectral fragmentation.

(a) mass = 28 from a compound containing only C and H

(b) mass = 18 from a compound containing C, H, and O

(c) mass = 36 from a compound with an M + 2 peak about one-third the size of the molecular ion.

12.37 An alcohol *A*, when treated with NaH followed by CH_3I, gives a compound *B* with a strong M + 1 peak in its CI mass spectrum at $m/z = 117$. Compound *A*, known from other evidence to be a tertiary alcohol, has prominent fragments in its EI mass spectrum at $m/z = 87$ and $m/z = 73$ (base peak). Propose structures for compounds *A* and *B*.

12.38 A chemist, Ilov Boronin, carried out a reaction of *trans*-2-pentene with BH_3 in THF followed by treatment with $H_2O_2/{}^-OH$. Two products were separated and isolated.

Desperate to know their structures, Ilov took his compounds to the spectroscopy laboratory and found that only the mass spectrometer was operating. The mass spectra of the two products are given in Fig. P12.38. Suggest structures for the compounds, and indicate which mass spectrum goes with which compound.

12.39 Rationalize each of the following observations by postulating a structure for the fragment ion(s) and the mechanisms for their formation.

(a) The EI mass spectrum of 1-methoxybutane shows fragment ions at $m/z = 56$ and $m/z = 45$ (base peak).

(b) The EI mass spectrum of 2-methoxybutane shows a base peak at $m/z = 59$.

12.40 Explain why the mass spectrum of dibromomethane has three peaks at $m/z = 172$, 174, and 176 in the approximate relative abundances 1 : 2 : 1.

12.41 Predict the relative intensities of the three peaks in the mass spectrum of dichloromethane at $m/z = 84$, 86, and 88.

12.42 Suggest a structure for each of the ions corresponding to the following peaks in the EI mass spectrum of ethyl bromide, and give a mechanism for the formation of each ion. (The numbers in parentheses are the relative abundances.)

(a) $m/z = 110$ (98%) (b) $m/z = 108$ (100%)

(c) $m/z = 81$ (5%) (d) $m/z = 79$ (5%)

(a)

(b)

Figure P12.38

(e) $m/z = 29$ (61%)   (f) $m/z = 28$ (25%)
(g) $m/z = 27$ (53%)

12.43 A compound contains carbon, hydrogen, oxygen, and one nitrogen. Classify each of the following fragment ions derived from this compound as an odd-electron or an even-electron ion. Explain.

(a) the molecular ion

(b) a fragment ion of even mass containing one nitrogen

(c) a fragment ion of odd mass containing one nitrogen

12.44 (a) Explain why ionization of a π electron requires less energy than ionization of a σ electron.

(b) Draw the structure of the molecular ion of 1-heptene formed by ionization of a π electron.

(c) The base peak in the EI mass spectrum of 1-heptene (Fig. 12.16, p. 593) occurs at $m/z = 41$, which is believed to correspond to the allyl cation, a resonance-stabilized carbocation.

$$H_2C = CH - \overset{+}{C}H_2$$

allyl cation
$m/z = 41$

Draw a curved-arrow mechanism (or fishhook mechanism) that shows the conversion of the molecular ion you drew in part (b) into the allyl cation.

Nuclear Magnetic Resonance Spectroscopy

Infrared spectroscopy can be used to determine the functional groups present in a compound, and mass spectrometry provides the masses of a molecule and its coherent fragments. With rare exceptions, however, neither of these techniques gives enough information to define a complete structure. Another form of spectroscopy, *nuclear magnetic resonance* (NMR) spectroscopy, enables us to probe molecular structure in much greater detail. Using NMR, sometimes in conjunction with other forms of spectroscopy but often by itself, we can in many cases determine a complete molecular structure in a very short time. In the period since its commercial introduction in the 1950s, NMR spectroscopy has revolutionized organic chemistry. This chapter presents the basic principles of NMR spectroscopy and shows how it is used in structure determination.

13.1 AN OVERVIEW OF PROTON NMR SPECTROSCOPY

NMR spectroscopy is used to detect *nuclei*, but only those nuclei that have a magnetic property called *spin*, which we'll discuss further in Sec. 13.2. The proton (^1H) and a minor isotope of carbon with atomic mass = 13 (^{13}C) have spin and can therefore be detected with NMR. The common isotope of carbon ^{12}C does not have spin and cannot be detected with this technique. (^{13}C NMR is discussed in Sec. 13.9.)

Historically, the first use of NMR in organic chemistry, and still a very important use, is for the detection of protons—hydrogen nuclei—in organic compounds. This type of NMR is called **proton NMR**, or **^1H NMR**. In the first part of this chapter we'll deal with proton NMR.

The best way to begin a study of NMR is to look at a simple NMR spectrum. Consider the proton NMR spectrum of dimethoxymethane, which is shown in Figure 13.1 on p. 612.

$$CH_3O—CH_2—OCH_3$$

dimethoxymethane

FIGURE 13.1 The proton NMR spectrum of dimethoxymethane. The lower axis is the chemical shift scale in parts per million (ppm), and the upper axis is the chemical shift scale in frequency units (Hz). The peaks represent energy absorption by the protons of each chemical type. The small peak at the far right is from the protons of tetramethylsilane (TMS), a reference standard added in a small amount.

This spectrum is a plot of energy absorption on the y axis versus relative frequency of the radiation from which energy is absorbed on the x axis. The absorptions detect the *protons* in the molecule. The units on the lower horizontal axis, abbreviated δ, are called *parts per million*, or *ppm*. For now, don't be concerned about how these units are derived; you should simply view them as position markers on this axis. The numbers on the upper horizontal axis are frequency units in hertz (Hz; the Hz was defined in Sec. 12.1A). Frequency decreases from left to right, as it does in an IR spectrum.

The numbers on the ppm, or δ, scale and the numbers on the frequency scale are proportional. For most of the spectra in this book, the frequency numbers are exactly 300 times the δ numbers. That is, if the frequency numbers on the upper axis are symbolized by $\Delta\nu$, then

$$\delta = \frac{\Delta\nu}{\nu_0} \quad (\Delta\nu \text{ in Hz, } \nu_0 \text{ in MHz}) \tag{13.1}$$

where $\nu_0 = 300$ MHz. The proportionality constant ν_0, in units of megahertz, or MHz, is called the **operating frequency** of the NMR spectrometer. As the name implies, this is an operating characteristic of the NMR spectrometer. We'll learn more about this relationship in Sec. 13.3B.

Peaks in an NMR spectrum are called **resonances**, **absorptions**, or **lines**. The position of an absorption on the horizontal axis is called its **chemical shift**. We usually express the peak positions in ppm—that is, we use the lower horizontal axis. In this case, the chemical shift of an absorption is written with a δ followed by the numerical value of the peak position. When we use the δ notation, the units of ppm are implied and are not repeated. Thus, we see three peaks in Fig. 13.1: these have chemical shifts at δ 0, δ 3.35, and δ 4.56. Or, we can say that the spectrum contains peaks at 0, 3.35, and 4.56 ppm.

The rightmost absorption—the one at δ 0—is not an absorption of dimethoxymethane. Rather, this is an absorption of tetramethylsilane (TMS), a compound added to each sample to provide a reference point.

tetramethylsilane (TMS)

The absorption position of TMS defines the δ 0 position on the x axis of each spectrum. TMS is used as a standard because it has a single strong absorption, it is chemically inert, and its chemical shift is smaller than that of most common organic compounds. TMS also has a low boiling point (26.5 °C), which allows it to be removed easily if recovery of the sample is desired.

The other two peaks—the ones at δ 3.35 and δ 4.56—are the NMR absorptions of the protons of dimethoxymethane. Before reading further, can you guess why there are two absorptions, and why they have different sizes?

There are two absorptions because there are two chemically distinguishable sets of protons in dimethoxymethane, the CH_2 protons and the CH_3 protons. The resonance at δ 4.56 is the absorption of the CH_2 protons, and the resonance at δ 3.23 is the absorption of the CH_3 protons. This illustrates a very important point about NMR: *The NMR spectrum of any compound contains a separate resonance* (barring accidental overlaps) *for each chemically distinguishable* (that is, *chemically nonequivalent*) *set of nuclei.* We'll discuss this point further in Sec. 13.3D.

The chemical shift of each absorption is determined by the nature of nearby groups. Nearby oxygens (or other electronegative atoms) cause shifts to the left—to higher frequency. Thus, the carbon bearing the CH_2 protons is adjacent to *two* oxygens, whereas the carbons bearing the CH_3 protons are each adjacent to *one* oxygen. The protons nearer the two oxygens (the CH_2 protons) have the greater chemical shift. This analysis illustrates another important point about NMR: *The chemical shifts of absorptions in an NMR spectrum vary in a predictable way with the chemical environment of the corresponding protons.* We'll discuss the effect of structure on chemical shift in Sec. 13.3C.

The two peaks have different sizes because different numbers of protons contribute to each absorption. The resonance at δ 3.35 is larger because more protons (six) contribute to this resonance than to the resonance at δ 4.56 (two). In fact, you'll observe that the resonance at δ 3.35 is about three times as tall. This illustrates yet another important aspect of NMR spectra: *The size of a peak* (actually, the area under the peak) *is proportional to the number of protons contributing to the absorption.* This means that we can count the protons of each chemical type! We'll come back to this point in Sec. 13.3E.

When a compound contains hydrogens on adjacent carbons, the NMR spectrum provides additional, very powerful, information. *We can count the protons on adjacent carbons.* This aspect of NMR is not illustrated by the spectrum of dimethoxymethane, because in this molecule no two carbons are adjacent. This additional capability of NMR comes from a phenomenon called *splitting*, which we'll discuss in Sec. 13.4.

In summary, proton NMR provides four types of information:

1. the number of sets of chemically nonequivalent protons
2. the chemical environments of each set of protons (chemical shift)
3. the number of protons within each set
4. the number of protons in adjacent sets

With these four types of information we can in many cases deduce completely the structures of unknown compounds. Using these ideas, let's now consider the various aspects of NMR spectra in more detail. We begin by considering the NMR phenomenon itself.

13.2   THE PHYSICAL BASIS OF NMR SPECTROSCOPY

To understand NMR spectroscopy and to use it intelligently we must understand its physical basis. NMR spectroscopy is based on the magnetic properties of nuclei that result from a property called *nuclear spin.*

Just as electrons have two allowed spin states, designated by the quantum numbers $+\frac{1}{2}$ and $-\frac{1}{2}$, some *nuclei* also have spin. The hydrogen nucleus ^1H—the proton—has a nuclear spin that also can assume either of two values, designated by quantum numbers $+\frac{1}{2}$ and $-\frac{1}{2}$.

The physical significance of nuclear spin is that the nucleus acts like a tiny magnet. You know from experience that magnets assume a preferred orientation in the presence of a magnetic field. (An example is the orientation of a compass needle in Earth's magnetic field.) The same is true of nuclei, which can be thought of as tiny magnets. Thus, the magnetic poles of hydrogen nuclei become oriented in a magnetic field. That is, when a compound containing hydrogens is placed in a magnetic field, its hydrogen nuclei become magnetized.

Let's represent the hydrogen nuclei in a chemical sample with arrows indicating their magnetic ("north–south") polarity. In the absence of a magnetic field, the nuclear magnetic poles are oriented randomly. After a magnetic field is applied, the magnetic poles of nuclei with spin of $+\frac{1}{2}$ are oriented parallel to the magnetic field, and those of nuclei with spin of $-\frac{1}{2}$ are oriented antiparallel to the field.

The most important effect of the magnetic field for NMR is how it affects the energies of the two spin states of the proton. In the absence of a field, the two spin states have the same energy. But when a magnetic field is applied, the two spin states have *different* energies: the $+\frac{1}{2}$ spin state has lower energy than the $-\frac{1}{2}$ spin state. The energy difference between the two spin states of a proton p, $\Delta\epsilon_p$, is given by the **fundamental equation of NMR**:

$$\Delta\epsilon_p = \frac{h\gamma_H}{2\pi}\mathbf{B_p} \tag{13.2}$$

In this equation, h is Planck's constant (Sec. 12.1A), 3.99×10^{-13} kJ s mol^{-1}; $\mathbf{B_p}$ is the magnitude of the magnetic field at the proton, in gauss (rhymes with house); and γ_H is a fundamental constant of the proton, called the **gyromagnetic ratio**. The value of this constant is 26,753 radians gauss^{-1} s^{-1}. This equation shows that when the magnetic field is zero, there is no energy difference between the spin states, as you just learned; and as the magnetic field is increased, the energy difference between the two spin states grows, as shown in Fig. 13.2.

A typical field strength used in modern NMR spectrometers is 70,500 gauss. This, by the way, is a *very strong* magnetic field! If you insert this value into Eq. 13.2, you can calculate that the energy separation $\Delta\epsilon_p$ is about 0.00012 kJ mol^{-1}. This is a *very small* energy! If we treat this as if it were a $\Delta G°$ and calculate the equilibrium constant between the two spin states using Eq. 3.36b, p. 110, we find that this energy corresponds to an equilibrium constant of 0.9999516. This equilibrium constant is so close to unity that in any sample of *one million*

FIGURE 13.2 Effect of increasing magnetic field at a proton on the energy difference between its $+\frac{1}{2}$ and $-\frac{1}{2}$ spin states (Eq. 13.2). The two spin states have identical energies when the field is absent, and the energy difference between the two spin states grows with increasing field.

increasing magnetic field $\mathbf{B}_p$

protons, the difference in the populations of the two spin energy states is very small; the lower-energy spin state has roughly 20 more protons than the higher-energy spin state. Even though this difference is *minuscule*, it is physically significant and is the basis for NMR. It takes a very large magnetic field to induce even this tiny energy difference between spin states.

Here is where things stand: Molecules of a sample are situated in a magnetic field; each proton is in one of two spin states that differ in energy by an amount $\Delta\epsilon_p$; and a small excess of protons have spin $+\frac{1}{2}$. If the sample is now subjected to electromagnetic radiation with energy E_p *exactly equal* to $\Delta\epsilon_p$, this energy is absorbed by some of the protons in the $+\frac{1}{2}$ spin state. The absorbed energy causes these protons to invert or "flip" their spins and assume a more energetic state with spin $-\frac{1}{2}$.

randomly oriented nuclear spins of equal energy

add a magnetic field

ϵ_p

the lower energy state has an excess of spins (a few spins per million)

a few $+\frac{1}{2}$ spins absorb energy and become $-\frac{1}{2}$ spins

add radiation with energy $E_p = \epsilon_p$

$-\frac{1}{2}$

$+\frac{1}{2}$

(13.3)

This absorption phenomenon, called **nuclear magnetic resonance**, can be detected in a type of absorption spectrometer called a *nuclear magnetic resonance spectrometer*, or **NMR spectrometer** (Sec. 13.11). The study of this absorption is called **NMR spectroscopy**. This absorption results in the peaks we see in an NMR spectrum, such as the one in Fig. 13.1. (Note that the resonance phenomenon in NMR has *nothing* whatsoever to do with resonance structures discussed in Sec. 1.4.)

> No radioactivity is involved in an NMR experiment. The popular association of the word *nuclear* with the phenomenon of radioactivity is why magnetic resonance imaging (MRI), used extensively in medicine, was not called *nuclear* magnetic resonance imaging. MRI relies on the same NMR phenomenon used for determining molecular structures (Sec. 13.12).

To summarize: For nuclei to absorb energy, they must have a nuclear spin and must be situated in a magnetic field. Once these two conditions are met, then the nuclei can be examined by an absorption spectroscopy experiment that is conceptually the same as the simple experiment shown in Fig. 12.3. The absorption of energy corresponds physically to the "flipping" of nuclear spins from a spin state of lower energy to one of higher energy.

The frequency of the electromagnetic radiation required for "spin flipping" of a set of protons p can be calculated from the equation $E_p = h\nu_p$ and the energy derived from Eq. 13.2:

$$\text{radiofrequency required for absorption} = \nu_p = \frac{E_p}{h} = \frac{\Delta\epsilon_p}{h} = \frac{\gamma_H}{2\pi}\mathbf{B_p} \qquad (13.4)$$

Using this equation, we can verify that for a set of protons p that experience a magnetic field $\mathbf{B_p}$ of 70,500 gauss, the frequency ν_p required to "spin-flip" protons in that set is 300×10^6 Hz, or 300 megahertz (MHz). This frequency is near the FM and ham radio bands; thus, the electromagnetic radiation used in NMR spectroscopy consists essentially of radio waves. Typical values of the frequency used in NMR experiments are between 60 MHz and 950 MHz, and the magnetic fields required vary proportionately in accord with Eq. 13.4. Indeed, an ordinary radio receiver located near an NMR spectrometer and tuned to the appropriate frequency can produce audible sounds associated with an NMR experiment.

The NMR phenomenon was demonstrated in 1945–1946 simultaneously in the laboratories of physicists Felix Bloch (1905–1983) at Stanford University and Edward M. Purcell (1912–1997) at Harvard University. Bloch and Purcell jointly received the 1952 Nobel Prize in Physics for their work in NMR.

PROBLEM

13.1   (a) What frequency would be required to cause "spin-flipping" in an NMR spectrometer in which the magnetic field at the proton is 117,400 gauss?

   (b) What magnetic field at the proton would be required to cause "spin-flipping" in an NMR experiment in which the frequency imposed on the sample is 900 MHz?

13.3   THE NMR SPECTRUM: CHEMICAL SHIFT AND INTEGRAL

A. Chemical Shift

In Fig. 13.1, you saw that the two chemically distinguishable types of protons in dimethoxymethane have different *chemical shifts*. From our description of the NMR experiment in the previous section, we now understand what this means: The two types of protons situated in a magnetic field absorb electromagnetic radiation at different frequencies. Because chemical shift tells us a great deal about structure, we need to understand the basis of chemical shift.

The key point in understanding chemical shift is that *the effective, or local, magnetic field* B_p *"sensed" by a proton is different from the external magnetic field* B_0 *provided by the NMR spectrometer.* In general, B_p *is somewhat less than* B_0. The reason is that the electrons circulating in the vicinity of a proton exert their own magnetic fields that *oppose* the external field. The reduction of the local field by the circulation of nearby electrons is called **shielding**. Therefore, atoms of relatively low electronegativity, such as Si, near a proton increase the electron density at the proton and thus increase the shielding of the proton from the external field. Conversely, more electronegative atoms such as O and Cl near a proton reduce the electron density at the proton and decrease the shielding of the proton. In summary, then, the local magnetic field at a more-shielded proton is smaller than the local field at a less-shielded proton.

 An Analogy for Magnetic Shielding

If you use an umbrella when you go outside on a rainy day, the umbrella shields you from the rain. If you don't use an umbrella, you get soaked. If you use a very small umbrella, you get wet, but perhaps not soaked. Likewise, we can think of electrons as providing a "magnetic umbrella" that shields nuclei from the external field. If electron density is high in the vicinity of a proton, the "magnetic umbrella" is relatively large, and the effective field—the local field—at the proton is smaller. In this case, the chemical shift is small. If electron density is smaller at a proton, the "magnetic umbrella" is smaller, and the local field at the proton is higher. In this case, the chemical shift is larger.

Let's apply these ideas to the chemical shifts in the NMR spectrum of dimethoxymethane (Fig. 13.1).

$$\overset{a}{C}H_3O—\overset{b}{C}H_2—O\overset{a}{C}H_3$$

dimethoxymethane

Let $\mathbf{B_a}$ be the local magnetic field at protons a and $\mathbf{B_b}$ be the local field at protons b. Because protons b are adjacent to *two* oxygens and protons a are adjacent to only one oxygen, protons b are less shielded than protons a. Therefore,

$$\mathbf{B}_b > \mathbf{B}_a \tag{13.5a}$$

By multiplying both sides of this inequality by $\gamma_H/2\pi$ and applying Eq. 13.4, we obtain the relative absorption frequencies:

$$\frac{\gamma_H}{2\pi}\,\mathbf{B}_b > \frac{\gamma_H}{2\pi}\,\mathbf{B}_a \tag{13.5b}$$

$$\nu_b > \nu_a \tag{13.5c}$$

Thus, protons b undergo resonance at a greater frequency than protons a. The chemical shift of any proton in Hz is defined as the difference between its resonance frequency and that of the protons in the reference compound TMS. Therefore, when we subtract the resonance frequency of TMS from both sides of Eq. 13.5c, we obtain the relationship between the chemical shifts of protons a and b in Hz:

$$\nu_b - \nu_{TMS} > \nu_a - \nu_{TMS} \tag{13.5d}$$

$$\Delta\nu_b > \Delta\nu_a \tag{13.5e}$$

$$\text{chemical shift of } H^b > \text{chemical shift of } H^a \quad \text{(in Hz)} \tag{13.5f}$$

The frequency scale across the top of the NMR spectrum in Fig. 13.1 is the scale of these $\Delta\nu$ values. As you can see in Fig. 13.3, $\Delta\nu_b$ is 1368 Hz and $\Delta\nu_a$ is 1005 Hz. These frequency differences are due to the different chemical environments of protons a and b. These frequency differences are the **chemical shifts** in Hz of the two proton sets. Therefore, the chemical shift of protons b is greater than that of protons a.

By dividing both sides of Eq. 13.5e by the operating frequency of the spectrometer, ν_0, as shown in Eq. 13.1, we obtain the relationship between the two chemical shifts in ppm:

$$\frac{\Delta\nu_b}{\nu_0} > \frac{\Delta\nu_a}{\nu_0} \quad (\Delta\nu \text{ in Hz, } \nu_0 \text{ in MHz}) \tag{13.6a}$$

$$\delta_b > \delta_a \tag{13.6b}$$

The operating frequency of the spectrometer used in most of the spectra in this text, including Fig. 13.1, is 300×10^6 Hz, or 300 MHz. Because we express $\Delta\nu$ in Hz and ν_0 in MHz, the units of δ are *parts per million*. Thus, $\delta_a = 3.35$ ppm and $\delta_b = 4.56$ ppm. We can read chemical shifts in ppm directly from the scale on the lower horizontal axis of the spectrum. Notice that chemical shifts in Hz are a *very small fraction* of the applied frequency. It follows that the relative shieldings of protons a, protons b, and the TMS protons are very small fractions of the external applied field. (See Problem 13.2.)

In summary, the less shielded a proton is, the greater is its chemical shift. Protons b are less shielded, and thus have a greater chemical shift, than protons a. This reduction in shielding (sometimes called *deshielding*) is a consequence of decreased electron density at the b protons. These ideas are summarized in Fig. 13.3 on p. 618.

PROBLEMS

13.2 What is the reduction in shielding of (a) protons a and (b) protons b relative to the protons of TMS in magnetic field units of gauss? The applied field in a 300 MHz NMR spectrometer is 70,500 gauss.

13.3 An NMR spectrum of a compound X contains four absorptions at δ 1.3, δ 4.7, δ 4.6, and δ 5.5, respectively. Which absorption comes from the most shielded protons? From the least shielded protons? Explain.

FIGURE 13.3 The spectrum of dimethoxymethane (Fig. 13.1) showing the effect of the local field on chemical shift. Different chemical shifts are a consequence of different local fields at protons in different chemical environments.

B. Chemical Shift Scales

In the previous section, we learned that a chemical shift can be expressed as a frequency difference $\Delta \nu$ in Hz or as a δ value in ppm. Typically the δ (ppm) scale is used for citing chemical shifts for reasons that we'll explore here.

The operating characteristics of a given NMR spectrometer are determined by the strength of its magnet—that is, by its applied field strength $\mathbf{B_0}$. Closely related to the applied field strength is the *operating frequency* ν_0 of the NMR instrument. This is the resonance frequency of a proton if it were subjected to the full, unshielded magnetic field of the instrument. This is calculated from Eq. 13.4, with $\mathbf{B_p} = \mathbf{B_0}$:

$$\nu_0 = \frac{\gamma_H}{2\pi} \mathbf{B_0} \tag{13.7}$$

Notice two things about this equation. First, the operating frequency and the applied magnetic field of an instrument are proportional. Second, if you know the applied field strength, you know the operating frequency, and vice versa. For example, Eq. 13.7 shows that if a spectrometer operates at a magnetic field of 70,500 gauss, its operating frequency must be 300,000,000 Hz, or 300 MHz. A reference to a "300 MHz proton NMR spectrum" means that the spectrum was taken on an instrument with a magnetic field of 70,500 gauss.

It turns out that chemical shifts expressed as frequency differences $\Delta \nu$ are also proportional to the field strength of the spectrometer. That is, if the applied field is doubled, the chemical shifts in Hz are also doubled. It happens that a variety of NMR instruments (and a variety of applied fields) are used in NMR spectroscopy. Therefore, if chemical shifts were cited as frequency differences, we would also have to know the applied field or operating frequency for these shifts to be meaningful. It is more convenient to tabulate chemical shifts in a way that doesn't depend on the type of spectrometer used to obtain them. This is why the δ scale is used.

$$\delta \text{ (parts per million)} = \frac{\Delta \nu \text{ in Hz}}{\nu_0 \text{ in MHz}} \tag{13.1}$$

This definition ensures that the field dependence is removed from δ because *both* $\Delta\nu$ and ν_0 are proportional to the applied field, and the field dependence thus cancels in the ratio. Therefore, the chemical shift of a proton in ppm in a given molecule is the same for any field strength and thus for any spectrometer (assuming the spectra are obtained under the same conditions of solvent, concentration, temperature, etc.). Thus, the $\Delta\nu$ (Hz) chemical shift of the CH_2 protons of dimethoxymethane (Fig. 13.1) depends on the field strength of the instrument, but the δ (ppm) chemical shift of these protons is 4.56 on *any* spectrometer. Therefore, *chemical shifts are conventionally cited in ppm*. The chemical shift in frequency units, if needed, can be calculated from the operating frequency ν_0 and the definition of δ in Eq. 13.1.

It is certainly not obvious from the NMR spectrum of dimethoxymethane why one would want to use a more powerful NMR instrument to obtain the spectrum, because the NMR spectrum of this compound is two single lines at any field strength. In Sec. 13.5B you will learn, however, that the use of high-field NMR instruments gives more detailed NMR spectra for many compounds. Higher magnetic fields also result in greater instrument sensitivity (that is, the ability to detect smaller concentrations). The reason is that the sensitivity of the instrument depends on the excess of spins in the lower spin-energy level, which, in turn, depends exponentially on the difference in energy between the two spin energy levels (Fig. 13.2 and the associated discussion). The problem with high-field instruments is their very high cost. "Workhorse" NMR instruments—the type used in most everyday work—represent a compromise between cost and sensitivity. The 300 MHz spectra in this text were taken on this type of instrument.

PROBLEMS

13.4   The spectrum in Fig. 13.1 was taken on a 300 MHz NMR instrument. What is the chemical shift, in Hz, of the CH_3 protons of dimethoxymethane in a spectrum taken on different instruments with the following applied magentic fields?
(a) 21,100 gauss   (b) 141,000 gauss

13.5   What is the chemical-shift difference in ppm of two resonances separated by 45 Hz at each of the following operating frequencies?
(a) 60 MHz   (b) 300 MHz

C. The Relationship of Chemical Shift to Structure

Because the chemical shift of a proton is influenced by nearby groups, the chemical shift of a proton resonance gives information about the proton's chemical environment. As we've noted in the previous section, one of the most important factors that affects a proton's chemical shift is the *electronegativities of nearby groups*. Some data that illustrate this idea are presented in Table 13.1. Examine these data using Problem 13.6 as your guide.

TABLE 13.1 Effect of Electronegativity on Proton Chemical Shift

Entry number	Compound	Chemical shift, δ	Electronegativity (Table 1.1)
1	CH_3F	4.26	F: 3.98
2	CH_3Cl	3.05	Cl: 3.16
3	CH_3Br	2.68	Br: 2.96
4	CH_3I	2.16	I: 2.66
5	CH_2Cl_2	5.30	
6	$CHCl_3$	7.27	
7	CH_3CCl_3	2.70	
8	$(CH_3)_4C$	0.86	C: 2.55
9	$(CH_3)_4Si$	0.00*	Si: 1.90

* By definition.

PROBLEM

13.6 (a) Consider entries 1 through 4 of Table 13.1. How does the chemical shift of a proton vary with the electronegativity of the neighboring halogen?

(b) Compare entries 2, 5, and 6 of Table 13.1. How does chemical shift vary with the number of neighboring halogens?

(c) Compare entries 6 and 7. How is the chemical shift of a proton affected by its distance from an electronegative group?

(d) Explain why $(CH_3)_4Si$ absorbs at lower chemical shift than the other molecules in the table. Can you think of a molecule with protons that would have a smaller chemical shift than TMS (that is, a negative δ value)?

You should have concluded from your examination of Table 13.1 that the following factors *increase* the proton chemical shift:

1. increasing *electronegativity* of nearby groups

2. increasing *number* of nearby electronegative groups

3. decreasing *distance* between the proton and nearby electronegative groups

The effect of electropositive groups (such as Si) is opposite that of electronegative groups.

The basis of these effects, as we learned in the previous section, is the different magnetic shielding of protons by surrounding electrons in different chemical environments. The spectrum of dimethoxymethane (Fig. 13.3) also illustrates these points. The CH_2 protons, which have the larger chemical shift, are adjacent to two oxygens, and are less shielded than the CH_3 protons, which are adjacent to only one oxygen.

Although methods exist for estimating chemical shifts fairly accurately (Further Exploration 13.1), it is sufficient to learn for now the general chemical-shift ranges for protons in particular environments. The chemical shifts of protons bound to carbon in various functional environments are shown in Fig. 13.4. For example, notice in Fig. 13.4 that the α-protons of an ether or alcohol have chemical shifts in the δ 3.2–4.2 range.

The chemical shifts in Fig. 13.4 support the idea that increasing the electronegativity of nearby groups increases chemical shift. The few notable exceptions to this trend are the chemical shifts of vinylic protons (protons directly attached to carbons of double bonds, δ = 4.5–6.0), allylic protons (protons attached to carbons that are directly attached to double bonds (δ = 1.6–2.8) and the related protons associated with benzene rings (phenyl protons, δ = 6.5–8.0, and benzylic protons, δ = 2.3–2.9). The chemical-shift effects of these groups are greater than would be predicted by their electronegativities. The reason for this effect is a special deshielding effect of the π electrons, which is discussed in Sec. 13.7A, Fig. 13.15 (p. 646), and Sec. 16.3B, Fig. 16.2).

Two other general observations about chemical shift are worth remembering. The first has to do with the amount of alkyl substitution on the carbon to which a proton is bound. The chemical shifts of **methyl protons** (that is, the protons of CH_3 groups) are typically at the lower end of a given chemical shift range. The chemical shifts of **methylene protons** (that is, the protons of —CH_2— groups) are a few tenths of a ppm greater and are likely to be near the middle of the chemical-shift range. Finally, the chemical shifts of **methine protons** (that is, —C̶—H protons) are typically greater still. The following chemical shifts for the α-protons of ethers illustrate this trend.

FURTHER EXPLORATION 13.1
Quantitative Estimation of Chemical Shifts

H_3C—O—$C(CH_3)_3$ CH_3CH_2—O—CH_2CH_3 H_3C—$\overset{\overset{\displaystyle CH_3}{|}}{CH}$—O—$\overset{\overset{\displaystyle CH_3}{|}}{CH}$—$CH_3$

methyl protons
δ 3.22

methylene protons
δ 3.45

methine protons
δ 3.67

FIGURE 13.4 Approximate chemical-shift ranges for protons bound to carbon in various chemical environments.

The second observation about chemical shift applies when a proton is near more than one functional group. In such a case, chemical shifts are affected by both groups. The following examples illustrate this point. (Remember in this context that α means "next to.")

$$CH_3CH_2-O-CH_2CH_3 \qquad H_3CO-CH_2-OCH_3$$

methylene protons
α to one oxygen
δ 3.43

methylene protons
α to two oxygens
δ 4.56

$$H_3C-\overset{\overset{\text{Cl}}{|}}{CH}-CH_2-CH_3 \qquad H_3C-\overset{\overset{\text{CH}_3}{|}}{CH}-CH=CH_2 \qquad H_3C-\overset{\overset{\text{Cl}}{|}}{CH}-CH=CH_2$$

methine proton
α to a chlorine
δ 3.95

methine proton
α to a double bond
(allylic proton)
δ 2.63

methine proton
α to both a chlorine
and a double bond
δ 4.53

Entries 2, 5, and 6 in Table 13.1 also illustrate the same point.

13.7   In each the following sets, the NMR spectra of the compounds shown consist of a single resonance. Arrange the compounds in order of increasing chemical shift, with the smallest first.

(a) CH_2Cl_2 CH_2I_2 CH_3I

       A   *B*   *C*

(b)

$$Cl—CH_2CH_2—Cl \qquad \underset{\underset{CH_3 \; CH_3}{|\quad|}}{\overset{\overset{CH_3 \; CH_3}{|\quad|}}{Cl—C—C—Cl}} \qquad \underset{\underset{Cl \quad Cl}{|\quad|}}{Cl—CH—CH—Cl} \qquad Cl—CH_2—Cl$$

           A   *B*   *C*   *D*

(c) $(CH_3)_4C$ $(CH_3)_4Sn$ $(CH_3)_4Si$

       A   *B*   *C*

(*Hint:* Consider trends in electronegativity.)

D. The Number of Absorptions in an NMR Spectrum

How do we know whether the different protons in a molecule will show different absorptions? This is equivalent to asking whether different protons have different chemical shifts. *Protons have different chemical shifts when they are in different chemical environments.* In many cases, deciding whether two protons are in different chemical environments is nearly intuitive. For example, in dimethoxymethane, $CH_3O—CH_2—OCH_3$, the chemical environment of the CH_2 protons is different from that of the CH_3 protons. But this distinction is not so intuitive in every case. The discussion in this section will allow us to decide *rigorously* whether we can expect two protons to have different chemical shifts.

    Predicting chemical-shift nonequivalence is the same as predicting chemical nonequivalence. If you have read and understood Sec. 10.9A, you already know how to do this. (If you haven't studied Sec. 10.9A, you should do so before reading further.) *Chemically nonequivalent protons in principle have different chemical shifts.* (The qualifier "in principle" is used because it is possible for chemical-shift differences to be so small that they are undetectable.) *Chemically equivalent protons have identical chemical shifts.*

    Recall from Sec. 10.9A that constitutionally nonequivalent protons are *chemically nonequivalent.* You can tell whether two protons are constitutionally equivalent by tracing their connectivity relationships to the rest of the molecule. For example, the CH_3 protons and the CH_2 protons of dimethoxymethane (Fig. 13.1) have a different connectivity relationship; consequently, they are constitutionally nonequivalent. This means that they are chemically nonequivalent and thus have different chemical shifts, as we have seen (Fig. 13.1).

    Recall also from Sec. 10.9A that *diastereotopic groups* are constitutionally equivalent but are *chemically nonequivalent.* It follows, then, that *diastereotopic protons in principle have different chemical shifts.*

    In contrast, *enantiotopic protons are chemically equivalent* as long as they are in an achiral environment. Thus, in an achiral solvent such as CCl_4 or $HCCl_3$, enantiotopic protons have identical chemical shifts; but in an enantiomerically pure chiral solvent, or in any chiral environment such as an enzyme active site, enantiotopic protons in principle have different chemical shifts.

    Finally, you learned in Sec. 10.9A that *homotopic protons are chemically equivalent.* Thus, homotopic protons have identical chemical shifts under all circumstances.

    Study Problem 13.1 shows how to apply the results of Sec. 10.9A to the determination of chemical-shift equivalence and nonequivalence in some cases involving constitutionally equivalent protons.

STUDY PROBLEM   13.1

In each of the following cases, the labeled protons are constitutionally equivalent. Determine whether the labeled protons in each case are expected to have identical or different chemical shifts.

(a)

H^a Cl
  \   /
   C=C
  /   \
H^b CH_3

(b)

        H^a
         |
H_3C—C—OH
         |
        H^b

(c)

        H^a Cl
         | |
H_3C—C—CH—CH_3
         |
        H^b

(d)

           H^a
            |
CH_3O—C—OCH_3
            |
           H^b

SOLUTION   Apply the principles of Sec. 10.9A to determine whether the protons in question are chemically equivalent. If the two protons are chemically equivalent, they have the same chemical shift. If not, their chemical shifts are different in principle. Because it is given that the protons in question are constitutionally equivalent, we need to determine only whether the protons are diastereotopic, enantiotopic, or homotopic.

(a) Perform the *substitution test* (Sec. 10.9A) on protons H^a and H^b; that is, replace H^a and H^b in turn with a circled proton, and examine the relationship between the resulting structures:

H^a Cl
  \   /
   C=C
  /   \
H^b CH_3

replace H^a replace H^b

(H)   Cl   H   Cl
  \   /   \   /
   C=C   ←— diastereomers —→   C=C
  /   \   /   \
 H CH_3 (H) CH_3

Remember to think of an "H" and a "circled H" as different atoms. The two structures that result are *E,Z* isomers and are therefore diastereomers. Consequently, H^a and H^b are diastereotopic and are therefore *chemically nonequivalent*. They are expected to have different chemical shifts.

(b) Carry out the substitution test on the two α-protons of ethanol:

        OH
         |
H^a,,,C—CH_3
         |
        H^b

replace H^a replace H^b

    OH   OH   OH
     | | |
(H),,,C—CH_3 H,,,C—CH_3 —rotate 180°→ H_3C—C,,,(H)
     | | |
     H   (H)   H

|———————— enantiomers ————————|

Because the resulting structures are enantiomers, the two hydrogens are enantiotopic and have the same chemical shift (in an ordinary achiral solvent).

(c) Carry out the substitution test on protons H^a and H^b of 2-chlorobutane. Notice that the molecule prior to the substitution test contains one asymmetric carbon. Choose a particular configuration (2R or 2S) for this carbon (2S is used below) and maintain whatever configuration you choose throughout the substitution test. (A circled hydrogen is assigned a higher priority than an uncircled one.)

Because diastereomers are obtained in the substitution test, these protons are diastereotopic and hence chemically nonequivalent. These protons have different chemical shifts.

(d) This is dimethoxymethane (Fig. 13.1). A substitution test shows that these two protons are homotopic. (Do this test!) Thus, they have identical chemical shifts. That is why a single resonance is observed for these protons in the NMR spectrum. Likewise, the two CH_3 groups are homotopic, and, within each of these groups, the three protons are also homotopic. Thus, a single resonance is observed for all six CH_3 protons.

Two of the most common situations in which diastereotopic protons are encountered are diastereotopic protons on a double bond, as in part (a) of Study Problem 13.1; and diastereotopic protons of methylene groups within a molecule containing an asymmetric carbon, as in part (c) of Study Problem 13.1. These protons are chemically nonequivalent and have different chemical shifts. Unless you are particularly alert to these situations, it is easy to regard such groups mistakenly as chemically equivalent.

It is important to understand chemical-shift equivalence and nonequivalence because *the minimum number of chemically nonequivalent sets of protons in a compound of unknown structure can be determined by counting the number of different resonances in its NMR spectrum.* (The "minimum" qualifier is used because of the possibility that the resonances of chemically different groups might overlap.) Hence, the simple act of counting resonances tells you a great deal about chemical structure!

PROBLEMS

13.8 Specify whether the labeled protons in each of the following structures would be expected to have the same or different chemical shifts.

13.9   How many different absorptions are observed in the NMR spectrum of each of the following compounds? Explain.

(a)

H₃C   CH₃
   \\   /
    C = C
   /   \\
(CH₃)₃C   CH₃

(b) $(CH_3)_3C—C(CH_3)_2$
$|$
Cl

(c)

H₃C   Cl
  | |
H₃C—C—CH
  | |
  Cl   Br

E. Counting Protons with the Integral

The two resonances of dimethoxymethane in Fig. 13.1 are not the same size because the size of an NMR absorption depends on the number of protons contributing to it. The intensity of an NMR absorption is determined from the *total area under the absorption peak*, called the **integral**. This quantity can be determined by mathematical integration of the peak using more or less the same integration procedures used in calculus to determine the area under a curve. NMR instruments (or the associated computers) can display the integral on the spectrum. Such a spectrum integral is illustrated in Fig. 13.5 for the dimethoxymethane spectrum as the red curve superimposed on each peak. *The relative height of the integral (in any convenient units, such as millimeters or inches) is proportional to the number of protons contributing to the peak.* You can verify with a ruler or with the chart grid spaces that the relative heights of the integrals in Fig. 13.5 are in the ratio 1 : 3, and the relative numbers of protons are 2 and 6, respectively, also in the ratio 1 : 3. Be aware that errors of a few percent in the integrals are quite common.

Don't forget that the integral gives us the *ratios* of hydrogens of each type, not the absolute number. In some cases the absolute number of hydrogens is ambiguous. Thus, if the spectrum in Fig. 13.5 had been that of an unknown compound, the integrals in Fig. 13.5 could have corresponded to 8 hydrogens in the ratio 2H : 6H, 12 hydrogens in the ratio 3H : 9H, or to some other multiple of 4 hydrogens. In that situation, we would have needed a molecular formula to decide between these alternatives.

In this text, integrals of NMR spectra will be presented either with the integral curve, as in Fig. 13.5, or with the actual number of hydrogens indicated in red over each resonance, as in Fig. 13.7 on p. 631.

If a sample contains more than one compound, the spectrum intensity of each compound is proportional to its concentration. Thus, the intensity of the TMS peak in most spectra is very small because it is added in very low concentration. For example, in Fig. 13.5, if the

---

FIGURE 13.5 The NMR spectrum of Fig. 13.1 with superimposed integral. The integrals are the red curves. The heights of the integrals are indicated by the purple arrows. The integral heights have no absolute significance; their relative heights are proportional to the number of protons in each set.

TMS were in the same concentration as dimethoxymethane, its resonance would be twice the intensity of the δ 3.35 resonance because it represents twice the number of protons—that is, 12 equivalent hydrogens. In fact, another advantage of TMS is that, because of its large number of equivalent hydrogens, very little of it must be added to obtain a measurable reference line.

F. Using the Chemical Shift and Integral to Determine Unknown Structures

**STUDY GUIDE
LINK 13.1**
Approaches to
Problem Solving

Let's summarize the main ideas of the previous sections that will be useful in applying NMR spectroscopy to the solution of unknown structures.

First, each chemically nonequivalent set of protons in a molecule gives (in principle) a different resonance in the NMR spectrum. Thus, the number of absorptions in principle indicates the number of chemically nonequivalent sets of protons. Next, the chemical shift of a set of protons provides information about what groups are nearby. Finally, the integral of each absorption is proportional to the number of contributing protons. Thus, the integral indicates the relative numbers of protons in each nonequivalent set. If the total number of protons is known (from a molecular formula), then the absolute number of protons in each set can be calculated.

These ideas are incorporated into the following systematic approach for the determination of unknown structures by NMR spectroscopy:

1. Write down everything about the molecular structure that is known from the molecular formula, including the unsaturation number and the functional groups that might be present.

2. From the number of absorptions, determine how many chemically nonequivalent sets of hydrogens are in the unknown.

3. Use the total integral of the entire spectrum and the molecular formula to determine the number of integral units per proton.

4. Use the integral of each absorption and the result from step 3 to determine the number of protons in each set.

5. From the chemical shift of each set, determine which set must be closest to each of the functional groups that are present.

6. Write down partial structures that are consistent with each piece of evidence, and then write down *all possible structures* that are consistent with *all* of the evidence.

7. Use Fig. 13.4 to estimate the chemical shifts of the protons in each structure, and, if possible, choose the structure that best reconciles the predicted and observed chemical shifts.

This approach is illustrated in Study Problem 13.2.

STUDY PROBLEM 13.2

An unknown compound with the molecular formula $C_5H_{11}Br$ has an NMR spectrum consisting of two resonances, one at δ 1.02 (relative integral 8378 units), and the other at δ 3.15 (relative integral 1807 units). Propose a structure for this compound.

SOLUTION Follow the seven steps.

Step 1. Because the unsaturation number is zero (Eq. 4.7, p. 144), the compound has no rings or double bonds; it is a simple alkyl halide.

Step 2. Because the spectrum consists of two lines (absorptions), the compound contains (barring accidental overlaps) only two chemically nonequivalent sets of hydrogens.

Step 3. The total integral is (8378 + 1807) = 10,185 units; because the molecular formula indicates 11 hydrogens, the integral corresponds to 10,185/11 = 926 units per hydrogen.

Step 4. The larger peak accounts for $(8378/926) = 9.05$ protons; the smaller peak accounts for $(1807/926) = 1.95$ protons. Rounding these to whole numbers, the two peaks are in the ratio $9 : 2$. (Remember, integrals in many cases contain errors of a few percent.) Notice that all 11 protons have been accounted for.

Step 5. Because of its greater chemical shift, the two-proton set must be closer to the bromine than the nine-proton set.

Step 6. A partial structure consistent with the conclusion of step 5 is

$$\overset{\displaystyle |}{\underset{\displaystyle |}{-\text{C}}}-\text{CH}_2-\text{Br}$$

Comparison of this partial structure with the molecular formula shows that three carbons and nine protons are missing from this partial structure. The nine protons are chemically equivalent. Adding three methyl groups to the preceding partial structure gives the correct structure:

$$(\text{CH}_3)_3\text{C}-\text{CH}_2-\text{Br}$$

$\delta\ 1.02$ (9 hydrogens) $\delta\ 3.15$ (2 hydrogens)

neopentyl bromide

Step 7. According to the discussion of chemical shifts, methylene protons α to a bromine should have a chemical shift in the middle of the range shown in Fig. 13.4, about $\delta\ 3.4$; the observed shift is very close to this value. The shift of the methyl protons should be in the alkyl region, about $\delta\ 1.2$; the observed shift is also quite close to this value.

PROBLEMS

13.10 In each case, give a single structure that fits the data provided.
   (a) A compound $C_7H_{15}Cl$ has two NMR absorptions at $\delta\ 1.08$ and $\delta\ 1.59$, with relative integrals of $3 : 2$, respectively.
   (b) A compound $C_5H_9Cl_3$ has three NMR absorptions at $\delta\ 1.99$, $\delta\ 4.31$, and $\delta\ 6.55$ with relative integrals of $6 : 2 : 1$, respectively.

13.11 Predict the NMR spectrum, including approximate chemical shifts, of the following compound. Explain your reasoning.

$$\text{CH}_3\text{O}-\text{CH}_2-\overset{\displaystyle \text{OCH}_3}{\underset{\displaystyle \underset{\displaystyle \text{OCH}_3}{\overset{\displaystyle |}{\text{CH}_2}}}{\overset{\displaystyle |}{\underset{\displaystyle |}{\text{C}}}}}-\text{CH}_2-\text{OCH}_3$$

13.12 Strange results in the undergraduate organic laboratory have led to the admission by a teaching assistant, Thumbs Throckmorton, that he has accidentally mixed some *tert*-butyl bromide with the methyl iodide. The NMR spectrum of this mixture indeed contains two single resonances at $\delta\ 2.2$ and $\delta\ 1.8$ with relative integrals of $5 : 1$.
   (a) What is the mole percent of each compound in the mixture? (*Hint:* Be sure to assign each absorption to a compound before doing the analysis.)
   (b) Which would be more easily detected by NMR: 1 mole percent CH_3I impurity in $(CH_3)_3C$—Br, or 1 mole percent $(CH_3)_3C$—Br impurity in CH_3I? Explain.

13.4   THE NMR SPECTRUM: SPIN–SPIN SPLITTING

Although substantial information can be gleaned from the chemical shift and integral, another aspect of NMR spectra provides the most detailed information about chemical structures. Consider the compound bromoethane (ethyl bromide):

$$\overset{a}{\text{H}_3}\text{C}-\overset{b}{\text{CH}_2}-\text{Br}$$

bromoethane
(ethyl bromide)

FIGURE 13.6 The NMR spectrum of bromoethane illustrates splitting. The resonance for the CH₃ protons is split into a three-line packet, or triplet, because the *adjacent carbon* has two protons. The resonance for the CH₂ protons is split into a four-line packet, or quartet, because the *adjacent carbon* has three protons. Notice that splitting gives the number of protons on *adjacent* atoms. Splitting patterns contain $n + 1$ lines, where n is the number of protons on adjacent atoms. The separation between lines within each packet (in this case, 7.2 Hz) is J, the coupling constant.

This molecule has two chemically different sets of hydrogens, labeled *a* and *b*. We expect to find NMR absorptions for these two sets of protons in the integral ratio 3 : 2, respectively, with the absorption of protons *a* at smaller δ. The NMR spectrum of bromoethane is shown in Fig. 13.6. This spectrum contains more lines than you might have expected—seven lines in all. Moreover, the lines fall into two distinct groups: a packet of three lines, or *triplet,* at smaller δ; and a packet of four lines, or *quartet,* at higher δ. These packets are expanded horizontally in boxes on the spectrum so that their details can be seen more clearly. It turns out that all three lines of the triplet are the absorption for the CH₃ protons, and all four lines of the quartet are the absorption for the CH₂ protons. (The relative integrals of the triplet and quartet, shown in red, are 3 : 2, respectively.) The chemical shift of each packet of lines, taken at its center, is in agreement with the predictions of Fig. 13.4. The quartet and the triplet have total integrals, respectively, in the ratio 2 : 3.

When an NMR resonance for a set of equivalent nuclei appears as more than one line, the resonance is said to be *split*. **Splitting** *arises from the effect that one set of protons has on the NMR absorption of neighboring protons.* The physical reasons for splitting are considered in Sec. 13.4B. First, let's focus on the appearance of the splitting pattern and the information it provides about structure.

A. The $n + 1$ Splitting Rule

While the integral gives a proton count for each resonance, the splitting pattern gives a different proton count, namely, the number of protons *adjacent* to the protons being observed. The relationship between the number of lines in the splitting pattern for an observed proton and the number of adjacent protons is given by the **$n + 1$ rule**: *n adjacent protons cause the resonance of an observed proton to be split into $n + 1$ lines.*

Let's see how the $n + 1$ rule accounts for the splitting patterns in the spectrum of bromoethane. Consider first the resonance for the methyl (CH₃) protons. Because the carbon *adjacent* to the CH₃ group has two protons, the resonance for the CH₃ protons themselves is split into a pattern of $2 + 1 = 3$ lines—that is, a *triplet*. (The fact that there are also three methyl protons is a coincidence; the number of protons determined by the integral has *nothing* to do with their splitting.)

Now consider the resonance of the methylene (CH_2) protons. Because the carbon *adjacent* to the CH_2 group has three equivalent protons, the resonance for the CH_2 protons is split into a pattern of $3 + 1 = 4$ lines—that is, a *quartet*.

Splitting is always mutual; that is, if protons *a* split protons *b*, then protons *b* split protons *a*. Thus, in the bromoethane spectrum, the CH_3 resonance is split by the CH_2 protons, and the CH_2 resonance is split by the CH_3 protons. When two sets of protons split each other, they are said to be **coupled**. Hence, the CH_3 and the CH_2 protons of bromoethane are coupled.

3 H's on *adjacent* carbon; 3 + 1 = 4 lines (a quartet)

$$H_3C—CH_2—Br$$

2 H's on *adjacent* carbon; 2 + 1 = 3 lines (a triplet)

Why is no splitting observed in our previous examples of NMR spectra? First of all, *splitting is not observed between chemically equivalent hydrogens.* Thus, the three hydrogens of iodomethane appear as a singlet because these hydrogens are chemically equivalent. Likewise, the two hydrogens of 1,1,2,2-tetrachloroethane also appear as a singlet because these two hydrogens, *even though they are on different carbons*, are chemically equivalent.

$$H_3C—I \qquad\qquad Cl_2CH—CHCl_2$$

iodomethane **1,1,2,2-tetrachloroethane**
hydrogens are chemically   hydrogens are chemically
equivalent; no splitting   equivalent; no splitting
is observed   is observed

Second, *with saturated carbon atoms, splitting is normally not observed between protons on nonadjacent carbon atoms.* Thus, because the protons in dimethoxymethane (Fig. 13.1, p. 612) are on *nonadjacent* carbons, their splitting is negligible; the two absorptions in the NMR spectrum of this compound are *singlets* (unsplit single lines).

Splitting provides *connectivity information*: when you observe the resonance of one proton, its splitting tells you how many protons are on *adjacent* atoms. As an analogy, suppose you were describing a puppy to a person who has never seen one. It's one thing to say that a puppy has four legs in two nonequivalent sets of two; it's much more revealing when you say that one set is attached to the body at the end near the head and the other is attached to the body at the end near the tail. It's the connectivity information that allows the more complete description of the puppy.

Now let's consider some of the details of splitting. The spacing between adjacent peaks of a splitting pattern, measured in Hz, is called the **coupling constant** (symbolized by *J*). This spacing can be measured approximately with a ruler using the Hz scale on the upper horizontal axis of the spectrum, but the exact value is determined from analysis of the spectrum by computer. For bromoethane (Fig. 13.6), this spacing is 7.2 Hz. *Two coupled protons must have the same value of J.* Thus, the coupling constants for both the CH_2 protons and the CH_3 protons of bromoethane are the same because these protons split each other. Letting the CH_3 protons be *a* and the CH_2 protons be *b*, then $J_{ab} = J_{ba}$. The coupling constant, unlike the chemical shift in Hz (Eq. 13.5e, p. 617), does not vary with the operating frequency or the applied magnetic field strength. Thus, the value of *J* for bromoethane is 7.2 Hz whether the spectrum is taken at 60 MHz or 300 MHz.

The chemical shift of a split resonance in most cases occurs at or near the midpoint of the splitting pattern. Thus, in the bromoethane spectrum, the chemical shift of the CH_2 protons can be taken to be the midpoint of the quartet, and that of the CH_3 protons is at the middle line of the triplet.

How do you know whether a group of lines is a single split resonance rather than several individual resonances with different intensities? Sometimes there is ambiguity, but in most cases a splitting pattern can be discerned by the *relative intensities of its component lines.*

TABLE 13.2 Relative Intensities of Lines within Common NMR Splitting Patterns

Number of equivalent adjacent protons	Number of lines in splitting pattern (name)	Relative line intensity within splitting pattern						
0	1 (singlet)				1			
1	2 (doublet)			1		1		
2	3 (triplet)			1	2	1		
3	4 (quartet)		1	3		3	1	
4	5 (quintet)		1	4	6	4	1	
5	6 (sextet)	1	5	10		10	5	1
6	7 (septet)	1	6	15	20	15	6	1

These intensities have well-defined ratios, shown in Table 13.2. For example, the relative intensities of a triplet, such as the one in bromoethane, are in the ratio 1 : 2 : 1; the relative intensities of a quartet are 1 : 3 : 3 : 1. In reality, slight deviations from these ratios occur, but these deviations become significant only when the chemical shifts of the coupled protons are very similar. We'll see an example of this behavior in Sec. 13.7.

A set of protons can be split by protons on more than one adjacent carbon. An example of this situation occurs in 1,3-dichloropropane, the NMR spectrum of which is shown in Fig. 13.7.

$$\overset{b}{Cl-CH_2}-\overset{a}{CH_2}-\overset{b}{CH_2}-Cl$$

1,3-dichloropropane

This molecule has two chemically nonequivalent sets of protons, labeled a and b. The key to understanding this spectrum is to recognize that all four protons H^b are chemically equivalent. The absorption for H^a is therefore split into a quintet because there are four protons on adjacent carbons; the fact that two of the protons are on one carbon and two are on the other makes no difference. The absorption for H^b is a triplet that appears at larger chemical shift.

PROBLEM

13.13 Predict the NMR spectrum of each of the following compounds, including splitting and approximate chemical shifts. (Assume that the coupling constants are about the same as those for bromoethane.)

(a) $H_3C-CHCl_2$ (b) $ClCH_2-CHCl_2$ (c)

$$\begin{array}{c} CH_3 \\ | \\ H_3C-CH-I \end{array}$$

(d) $CH_3O-CH_2CH_2CH_2-OCH_3$ (e) ⬠ (with O)   (f) $Cl-CH_2-\underset{\underset{Cl}{|}}{C}(CH_3)_2$

B. Why Splitting Occurs

Splitting occurs because the magnetic field caused by the spin of a neighboring proton affects the total field experienced by an observed proton. To understand, let's analyze the absorption of a set of equivalent protons a (such as the methyl protons of bromoethane) adjacent to *two* equivalent neighboring protons b. What are the spin possibilities for the neighboring protons b? In any one molecule, both of these protons could have spin $+\frac{1}{2}$; both could have spin $-\frac{1}{2}$; or they could have differing spins.

FIGURE 13.7 The NMR spectrum of 1,3-dichloropropane. (Integrals are shown in red.) Hydrogens *a* are split by all four hydrogens *b*.

$$H^a - \underset{\underset{H^a}{|}}{\overset{\overset{H^a}{|}}{C}} - \underset{\underset{H^b}{|}}{\overset{\overset{H^b}{|}}{C}} - Br \qquad (13.8)$$

	b protons					
	1	2	1	2	1	2
spin combinations	+½	+½	+½	+½	−½	−½
			−½	+½		

(The spins of the *b* protons can differ in two ways: proton 1 can have spin +½ and proton 2 spin −½, or vice versa.)

Suppose that the *a* protons are next to two *b* protons with a spin of +½. Because the spins of these *b* protons are parallel to the applied field, they add to, or enhance, the applied field, and thus the *a* protons are subjected to a slightly greater field than they would be in the absence of the *b* protons. Hence, radiation of higher energy (and frequency) is required to bring the *a* protons into resonance. The result is a line in the splitting pattern at *higher* frequency (Fig. 13.8, p. 632). Now suppose that the *a* protons are next to two *b* protons with a spin of −½. Because the spins of these *b* protons are antiparallel to the applied field, they subtract from the applied field, and as a result the *a* protons are subjected to a slightly smaller field than they would be in the absence of the *b* protons. Hence, radiation of lower frequency is required to bring the *a* protons into resonance. The result is a line in the splitting pattern at *lower* frequency. Finally, suppose that the two *b* protons have opposite spins; in this case, the effects of the two *b* proton spins cancel, and protons *a* are subjected to the same field that they would be in the absence of protons *b*. The result is the line in the center of the splitting pattern. Thus, there are three lines in the splitting pattern.

What about the intensities of these lines? The likelihood that each proton has a spin of +½ is about the same as the likelihood that it has a spin of −½; the excess of spins in the +½

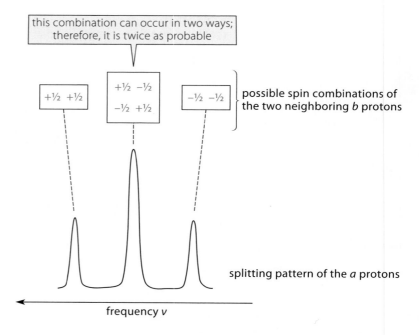

FIGURE 13.8 Analysis of the triplet splitting of a set of protons *a* by two neighboring protons *b*. When the spins of both *b* protons are aligned parallel to the applied field (+½ spin), they augment the local field; hence, a greater frequency is required to bring the *a* protons into resonance. When the spins of both *b* protons are aligned against the applied field (−½ spin), they decrease the local field; hence, a smaller frequency is required to bring the *a* protons into resonance. When neighboring *b* protons have opposite spin, the local field is unaffected; hence, the resonance frequency of the *a* protons is unchanged. The center line of a triplet has twice the intensity of the outer lines because the corresponding spin combinations of neighboring *b* protons are twice as probable.

---

state (Eq. 13.3) is so small that it can be ignored in the analysis of splitting. The intensities then follow from the relative probabilities of each spin combination. The probability that both *b* protons have spin +½ is equal to the probability that they both have spin −½, but the probability that these protons have opposite spins is twice as high because this situation can occur in *two ways* (Eq. 13.8). Hence, the center line of the splitting pattern is twice as large as the outer lines, and a 1 : 2 : 1 triplet is observed for the absorption of protons *a*.

> An analogy to the spin combinations is the combinations that can be rolled with a pair of dice. A 3 is twice as probable as a 2 because it can be rolled in two ways (2 + 1, 1 + 2), but a 2 can only be rolled in one way (1 + 1).

You should be able to analyze the splitting of the *b* protons of bromoethane in a similar manner (Problem 13.14).

How does one proton "know" about the spin of adjacent protons? One of the most important ways that proton spins interact is *through the electrons in the intervening chemical bonds.* This interaction is weaker as the protons are separated by more chemical bonds. Thus, the coupling constant between hydrogens on adjacent saturated carbons in acyclic compounds is typically 5–8 Hz, but the coupling constant between more widely separated hydrogens is normally so small that it is not observed. This is why splitting is usually not observed between protons on nonadjacent carbon atoms.

> An analogy to the effect of distance on proton coupling can be observed with an ordinary magnet and a few paper clips. If one paper clip is held to the magnet, it may be used to hold a second paper clip, and so on. The magnetic field of the magnet dies off

with distance, however, so that typically the third or fourth paper clip is not magnetized.

PROBLEM

13.14   Analyze the quartet splitting pattern for a set of equivalent *b* protons in the presence of *three* equivalent adjacent *a* protons. Include an analysis of the relative intensity of each line of the splitting pattern. (This is the splitting pattern for the CH_2 protons of bromoethane.)

C. Solving Unknown Structures with NMR Spectra Involving Splitting

We've now learned all of the basics of NMR. Let's summarize the type of information available from NMR spectra.

1. The *number of resonances* tells us the number of sets, or types, of chemically distinct hydrogens.

2. The *chemical shift* provides information about functional groups that are near an observed proton. The number of resonances (barring accidental overlaps)—that is, the number of different chemical shifts represented in the spectrum—equals the number of chemically nonequivalent sets of protons.

3. The *integral* indicates the relative number of protons contributing to a given resonance.

4. The *splitting pattern* indicates the number of protons adjacent to an observed proton.

These four elements of an NMR spectrum can be put together like pieces of a puzzle to deduce a great deal about chemical structure; it is not unusual for a complete structure to be determined from the NMR spectrum alone.

Because NMR spectra consume a large amount of space, it is common to see NMR spectra recorded in books and journals in an abbreviated form. In the form used in this text, the chemical shift of each resonance is followed by its integral, its splitting, and (if split) its coupling constant, if known. Abbreviations used to indicate splitting patterns are s (singlet), d (doublet), t (triplet), and q (quartet); complex patterns in which the nature of the splitting is not clear are designated m (multiplet). It is assumed that the relative intensities of each splitting pattern approximately match those in Table 13.2. For example, the spectrum of bromoethane (Fig. 13.6) would be summarized as follows:

$$\delta\ 1.67\ (3H,\ \text{t},\ J = 7.2\ \text{Hz});\ \delta\ 3.43\ (2H,\ \text{q},\ J = 7.2\ \text{Hz}) \tag{13.9}$$

coupling constant
type of splitting pattern (triplet; intensities in Table 13.2)
integral
chemical shift (ppm relative to TMS)

You may find that you can interpret a spectrum more easily if you can see it. If so, do not hesitate to sketch the spectrum. You can do this quickly by using vertical bars for the individual lines.

You should now have the tools needed to determine some structures using NMR spectra that contain some splitting information. The general method of problem solving given in Sec. 13.3F remains valid when splitting information is involved. Just remember to take into account splitting information when writing out partial structures (step 6). Study Problem 13.3 illustrates this approach.

STUDY PROBLEM 13.3

Give the structure of a compound $C_7H_{16}O_3$ with the following NMR spectrum: δ 1.30 (3*H*, s); δ 1.93 (2*H*, t, *J* = 7.3 Hz); δ 3.18 (6*H*, s); δ 3.33 (3*H*, s); δ 3.43 (2*H*, t, *J* = 7.3 Hz). Its IR spectrum shows no O—H stretching absorption.

SOLUTION   To solve the problem, apply the procedure in Sec. 13.3F.

Step 1. The unsaturation number is zero; hence, the unknown contains no rings and no double or triple bonds. The IR spectrum rules out an alcohol; therefore, all of the oxygens must be involved in ether linkages.

Step 2. The unknown contains five sets of chemically nonequivalent protons.

Steps 3–4. The numbers of protons in each resonance is given in the problem. Hence, we can skip steps 3 and 4.

Step 5. The protons at δ 3.18, δ 3.33, and δ 3.43 are on carbons bonded to oxygens. (See Fig. 13.4.)

Step 6. The resonance at δ 3.33 integrates for three equivalent protons and must therefore be a methyl group. Because the methyl group is attached to an oxygen (step 5), it must be a methoxy (—OCH₃) group. The resonance at δ 3.18 integrates for six equivalent protons, and these protons, too, are on carbons bound to oxygens. Two possibilities are three equivalent —CH₂O— groups with no hydrogens on adjacent carbons, or two equivalent —OCH₃ groups. Three chemically equivalent —CH₂O— groups would account for three oxygens in addition to the one previously accounted for. Because the molecule contains only three oxygens, this possibility cannot be correct. Hence, the δ 3.18 resonance corresponds to two equivalent —OCH₃ groups. The two-proton resonance at δ 3.43 must be a —CH₂O— group, and, from its triplet splitting, there must be an adjacent —CH₂— group. A partial structure consistent with the observations thus far is the following:

From the chemical shift, the three δ 1.30 protons must be part of another methyl group that is *not* attached to an oxygen and thus must be attached to a carbon. Because these methyl protons are a singlet, the attached carbon must contain no hydrogens. Completing the bonds to the remaining carbon with a methyl group and the two equivalent methoxy groups gives the final structure.

1,3,3-trimethoxybutane

**STUDY GUIDE
LINK 13.2
More NMR
Problem-Solving
Hints**

Step 7. The chemical shifts are all consistent with those given in Fig. 13.4. The shifts of the —CH₂— and —CH₃ groups at δ 1.93 and δ 1.30 illustrate another trend in chemical shift: oxygens have a discernible chemical-shift effect roughly equal to 0.2–0.3 ppm on β-protons (protons two carbons removed from the oxygen). Typically a —CH₂— has a chemical shift of about δ 1.2, but the δ 1.94 protons are shifted another 0.7–0.8 ppm by the three β-oxygens. Similarly, a methyl group typically has a chemical shift of δ 0.9, but the δ 1.30 methyl protons are shifted by the two β-oxygens of the methoxy groups. (See Further Exploration 13.1 for a discussion of these effects.)

PROBLEMS

13.15 Explain why the following two structures are ruled out by the data in Study Problem 13.3.

(a)

$$\underset{\underset{OCH_3}{|}}{CH_3OCHCH_2\overset{\overset{OCH_3}{|}}{CHCH_3}}$$

(b)

$$\underset{\underset{OCH_3}{|}}{CH_3OCH_2\overset{\overset{OCH_3}{|}}{CHCHOCH_3}}$$

13.16 Give structures for each of the following compounds.

(a) C_3H_7Br: δ 1.03 (3H, t, J = 7 Hz); δ 1.88 (2H, sextet, J = 7 Hz); δ 3.40 (2H, t, J = 7 Hz)

(b) $C_2H_3Cl_3$: δ 3.98 (2H, d, J = 7 Hz); δ 5.87 (1H, t, J = 7 Hz)

(c) $C_5H_8Br_4$: δ 3.6 (s; only resonance in the spectrum)

13.17 A compound X with the molecular formula $C_{11}H_{24}O_3$ and no O—H stretching absorption in its IR spectrum has the following proton NMR spectrum: δ 1.2 (3H, s); δ 1.4 (6H, t, J = 7 Hz); δ 3.2 (9H, s); δ 3.4 (6H, t, J = 7 Hz)

Deduce the structure of X and explain your reasoning.

An important use of spectroscopy is to confirm structures that are suspected from other information. For example, if a well-known reaction is run on a known starting material, the structure of the major product can often be predicted from a knowledge of the reaction. NMR spectroscopy can be used to confirm (or refute) the predicted structure, as shown in the following problem.

PROBLEM

13.18 When 3-bromopropene is allowed to react with HBr in the presence of peroxides, a compound A is formed that has the following NMR spectrum: δ 3.60 (4H, t, J = 6 Hz); δ 2.38 (2H, quintet, J = 6 Hz).

(a) From the reaction, what do you think A is?

(b) Use the NMR spectrum to confirm or refute your hypothesis. Identify A.

13.5   COMPLEX NMR SPECTRA

The NMR spectra of some compounds contain splitting patterns that do not appear to be the simple ones predicted by the $n + 1$ rule. Two common sources of such complex spectra are, first, the splitting of one set of protons by more than one other set, called *multiplicative splitting*, and, second, the breakdown of the $n + 1$ rule itself in certain cases. This section discusses each of these situations and shows how to deal with them.

A. Multiplicative Splitting

To understand multiplicative splitting, we first have to understand what controls the size of the coupling constants J between protons on adjacent carbons, called **vicinal coupling constants**. (Remember, the coupling constant is the separation between the individual lines of a splitting pattern.) One of the major factors that govern the size of vicinal coupling constants is the dihedral angle θ between the bonds to the coupled protons. This relationship is shown by the curve in Fig. 13.9 on p. 636, called the **Karplus curve**, after Martin Karplus (b. 1930), a chemist on the faculty of Harvard University who was awarded the Nobel Prize in Chemistry in 2013. The equation describing this curve is

$$J = J_0 \cos^2 \theta \qquad (13.10)$$

where J_0 has two different values that depend on the domain of θ. For the plot in Fig. 13.9, $J_0 = 10$ for $\theta \leqslant 90°$, and $J_0 = 14$ for $\theta \geqslant 90°$. (The value of J_0 at $\theta = 90°$ doesn't matter because $\cos^2 90° = 0$.) The exact value of J_0 depends on the nature of the substituents at the carbons bearing the coupled protons and other structural features; for example, electronegative substituents reduce this value somewhat. The important thing to notice about this curve, however, is not the exact size of J_0 but rather that (1) J is greatest when $\theta = 180°$—that is, when two

FIGURE 13.9 The Karplus curve plotted for Eq. 13.10 with $J_0 = 10$ ($\theta \leqslant 90°$) and $J_0 = 14$ ($90° \leqslant \theta \leqslant 180°$). This curve describes the coupling constant between two vicinal protons as a function of dihedral angle. The values for gauche protons ($\theta = 60°$) and anti protons ($\theta = 180°$) are marked on the graph. Notice that anti splitting is much greater than gauche splitting.

protons have an anti relationship; and (2) the value of J is much smaller when $\theta = 60°$—that is, when two protons have a gauche relationship. The following ranges for J have been observed:

These ranges show that anti and gauche relationships between C—H bonds can be clearly differentiated by the coupling constants between the protons.

What is the reason for this angular dependence of splitting? Recall (p. 632) that splitting between protons is mediated by the electrons in the bonds between the protons. The size of the coupling constant depends on the strength of the interactions between the bond orbitals containing these electrons. The sp^3 orbitals of C—H bonds interact most strongly when they are coplanar—that is, when they have an anti or eclipsed relationship. Orbitals at right angles do not interact at all, and their interaction is weak when they have a gauche relationship.

The best illustrations of the Karplus relationship involve molecules that have one predominant conformation, such as the following two diastereomers. (The asterisks indicate corresponding carbons in the chair structure and the Newman projection.)

A
H^a and H^b are anti
$J_{ab} \approx 12$ Hz

B
H^a and H^b are gauche
$J_{ab} \approx 3$ Hz

Notice that these molecules exist almost completely in the single chair conformations shown above because the large *tert*-butyl group and the methyl groups occupy equatorial positions (Eq. 7.7, p. 284). Let's focus only on the resonance of H^a in the NMR spectrum of each stereoisomer. This resonance is well separated from the other resonances in the spectrum and easy to identify because it has the greatest chemical shift. (We'll ignore the remaining parts of the spectra.) Specifically, let's focus on the splitting of the resonance of proton H^a by the two equivalent protons H^b in each case. In the first molecule A, the $C—H^a$ bond is anti to each of the equivalent $C—H^b$ bonds. The coupling constant between these two protons is predicted by the Karplus curve to be large. (In fact, it is about 12 Hz.) Therefore, the resonance for H^a is a triplet with J_{ab} = 12 Hz. In the second molecule B, the $C—H^a$ bond is gauche to each of the equivalent $C—H^b$ bonds. The Karplus curve predicts a much smaller coupling constant; in fact it is about 3 Hz. Therefore, in this compound, the resonance for H^a is a triplet with J_{ab} = 3 Hz. Had the relative stereochemistry of these two compounds not been known, it could have been confidently assigned by analyzing the coupling constants.

Now we are ready to consider the issue of multiplicative splitting. In the following compound (which is much like compound A above but without the methyl groups), the resonance of proton H^a is split by the two equivalent protons H^b and by the two equivalent protons H^c.

trans-1-tert-butyl-4-chlorocyclohexane
(* indicates corresponding carbons)

Notice that H^b and H^c are *not* equivalent. Because H^a is anti to H^b but gauche to H^c, the Karplus curve indicates that J_{ab} and J_{ac} should be quite different. When a proton is coupled to two different protons with *different coupling constants*, the splitting of each is applied successively. The successive application of different splittings to the same proton is called **multiplicative splitting**. In this case, the two splittings are observed to be J_{ab} = 11.8 Hz and J_{ac} = 4.3 Hz. To see the resulting spectrum, we construct a **splitting diagram**. In this type of diagram, shown in Fig. 13.10(a) on p. 638, we use single lines of the appropriate height to represent the individual resonances of a splitting pattern. Start with a single line at the chemical shift of H^a, as shown at the top of the figure. Now apply either of the two splittings—we'll start with J_{ab}. This results in a triplet (with 1 : 2 : 1 line intensity) in which the lines of the triplet are separated by 11.8 Hz, and the total intensity of the three lines equals the intensity of the starting resonance. Now we apply the second splitting J_{ac} = 4.3 Hz to *each* line of the splitting pattern we just created, remembering to divide the intensities appropriately. We thus obtain a *triplet of triplets*, or a nine-line pattern. The actual resonance of H^a in this compound is shown in Fig. 13.10(b). (The order in which we apply the two splittings doesn't matter—the end result is the same.)

Consider now the splitting of H^a in the cis stereoisomer of the same compound.

cis-1-tert-butyl-4-chlorocyclohexane
(* indicates corresponding carbons)

In this compound, H^a has a gauche relationship to *both* H^b and H^c. It turns out that the two splittings in this case, J_{ab} and J_{bc}, are the same (2.9 Hz). This is reasonable because the two dihedral angles ($C—H^b$ with $C—H^a$ and $C—H^c$ with $C—H^a$) are essentially the same. A splitting diagram for this situation is shown in Fig. 13.11(a) on p. 639. Successive application of two equal splittings results in the exact overlap of several lines, and the result is an apparent quintet. This is exactly the same result that we would have expected by applying the $n + 1$ rule for n = the total number of neighboring protons. Proton H^a has four neighboring protons, and the result is that the resonance for H^a is 4 + 1 = 5 lines with the relative intensities 1 : 4 : 6 : 4 : 1 as shown in Table 13.2. The experimental spectrum is shown in Fig. 13.11b.

FIGURE 13.10 (a) A splitting diagram illustrating the application of two successive triplet splittings with different coupling constants to a single resonance. Resonances are depicted as vertical lines with relative intensity proportional to height. The splitting is applied to each line of a pattern, and the intensity of that line is divided 1 : 2 : 1 in the daughter lines. Although the larger splitting has been applied first, the order makes no difference. (b) The actual spectrum corresponding to this case.

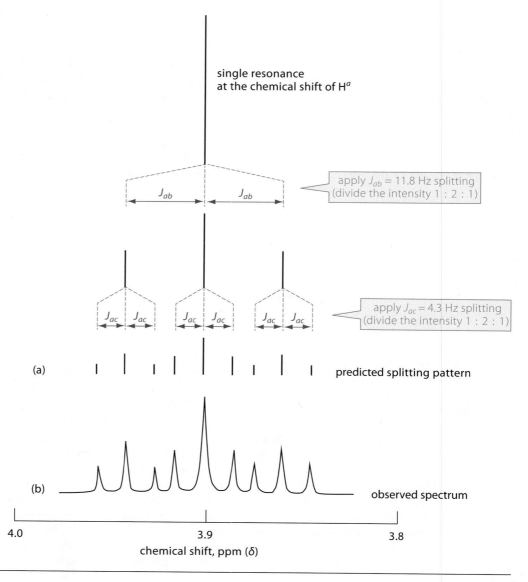

What we learn from this case is that *when two identical splittings are applied multiplicatively, the result is the same as predicted by the n + 1 rule for the total number of neighboring protons.*

The previous two examples involve compounds that exist in one major conformation. What happens when a molecule exists in more than one conformation, so that the angular relationship between protons is constantly changing? If the conformations are interconverting very rapidly—as in internal rotation around single bonds—the coupling constants are averaged, and multiplicative splitting does not enter the picture. For example, in bromoethane (NMR spectrum in Fig. 13.6), if we were to "freeze" the internal rotation and analyze the splittings in one instant of time, we would obtain the following:

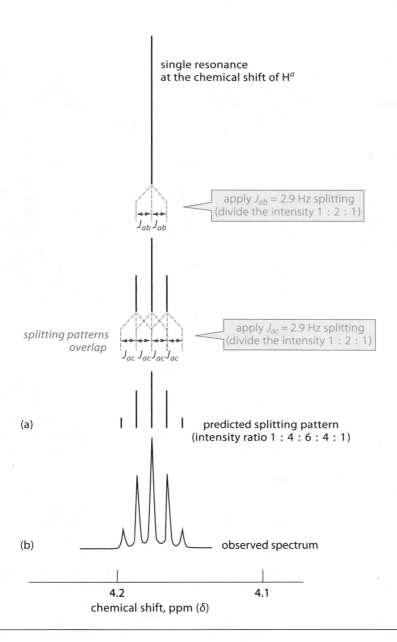

single resonance
at the chemical shift of H^a

apply J_{ab} = 2.9 Hz splitting
(divide the intensity 1 : 2 : 1)

$J_{ab} J_{ab}$

splitting patterns overlap

apply J_{ac} = 2.9 Hz splitting
(divide the intensity 1 : 2 : 1)

$J_{ac} J_{ac} J_{ac} J_{ac}$

(a)   predicted splitting pattern
(intensity ratio 1 : 4 : 6 : 4 : 1)

(b)   observed spectrum

4.2   4.1
chemical shift, ppm (δ)

FIGURE 13.11 (a) Application of two successive equal triplet splittings to a resonance. The process used is the same as in Fig. 13.10a. In this case, overlap occurs within the inner lines; the intensities of the overlapping lines are added. As a result, the splitting pattern contains five rather than nine lines in the ratio expected from application of the $n + 1$ rule to all four of the neighboring protons. (b) The actual spectrum corresponding to this case.

The methyl protons (red) experience two different environments, one of which is twice as likely as the other. Each environment is associated with a different coupling constant with the methylene protons. Over time, however, the methyl protons rapidly exchange places by internal rotation, and, as a result, the splittings of the protons on a given carbon are averaged. For example, the six different splittings of the methyl protons are averaged to

$$J(\text{methyl, average}) = \frac{(2J_{\text{gauche}}) + (J_{\text{gauche}} + J_{\text{anti}}) + (J_{\text{gauche}} + J_{\text{anti}})}{6} = \frac{4J_{\text{gauche}} + 2J_{\text{anti}}}{6} \quad (13.11a)$$

Likewise, the six splittings of the methylene protons (blue) are averaged to

$$J(\text{methylene, average}) = \frac{(2J_{\text{gauche}} + J_{\text{anti}}) + (2J_{\text{gauche}} + J_{\text{anti}})}{6} = \frac{4J_{\text{gauche}} + 2J_{\text{anti}}}{6} \quad (13.11b)$$

The average coupling constants are the same for the two protons, as they must be. Notice in Fig. 13.6 that the observed coupling constant in bromoethane is 7.2 Hz. From the equations

FIGURE 13.12 An NMR spectrum illustrating splitting by nonequivalent protons with nearly identical coupling constants.

above, this average value could result, for example, from $J_{gauche} = 3.9$ Hz and $J_{anti} = 13.9$ Hz, values well within the known ranges for gauche and anti splitting. We often find that, for protons on adjacent carbons undergoing rapid internal rotation, coupling constants are typically in the 6–8 Hz range. This is the result of conformational averaging. Fortunately, we don't have to worry about multiplicative splitting between the methylene protons and the two different types of methyl protons in bromoethane because of averaging by rapid internal rotation. To summarize: *In molecules that contain multiple rapidly interchanging conformations, the different possible coupling constants are averaged to a single value.*

What we've learned here is that splittings between protons on adjacent carbons can be multiplicative or they can be the same. A frustration for the beginning student is how to know what type of splitting to expect when working with the NMR spectrum of an unknown compound. Actually, the spectrum itself of the unknown compound helps us to decide. If the spectrum consists of singlets and/or simple splitting patterns—doublets, triplets, quartets, quintets, etc., with the intensities shown in Table 13.2—*we can use the n + 1 rule without worrying about the complications of multiplicative splitting.* To illustrate this point, Fig. 13.12 shows the NMR spectrum of a compound C_3H_6BrCl. The spectrum has three simple splitting patterns—two triplets and a quintet. *The simplicity of the splitting patterns shows that we don't need to be concerned with multiplicative splitting.* With an integral of two protons for each pattern, the only possible structure is

$$Br - \overset{a}{C}H_2 - \overset{b}{C}H_2 - \overset{c}{C}H_2 - Cl$$

1-bromo-3-chloropropane

Notice that the resonance for protons H^b is a quintet, as we would expect for a total of four neighboring protons.

Looking at the structure, we might have worried about the possibility of unequal values for J_{ab} and J_{bc} because protons H^a and H^c are nonequivalent. In fact, these coupling constants *are* unequal—but only slightly. (Very small differences in the coupling constants don't cause multiplicative splittings.) If the coupling constants had been significantly different, however, the splitting of H^b would multiplicative, and we would have seen a triplet of triplets for H^b (up to nine lines). Another way to look at this situation is that the resonance for H^b is multiplicative with nearly identical coupling constants. Recall (Fig. 13.11) that when the coupling constants are identical, we can apply the $n + 1$ rule to the total number of neighboring protons.

In summary: Let the spectrum of an unknown compound be your guide. If it consists of common splitting patterns, apply the $n + 1$ rule without considering the possibility of multiplicative splitting.

What if splitting is complex? Beginning students are not expected to interpret complex splitting patterns. More experienced students in organic chemistry will begin to recognize some of the more common multiplicative patterns. Splitting diagrams can be constructed "in reverse" to analyze simpler multiplicative splittings, and there are computer-simulation programs that enable the detailed analysis of complex patterns. However, even if a spectrum is complex, chances are that it will contain some simple nonmultiplicative splitting patterns that you can readily interpret with the $n + 1$ rule, and you can also use the chemical shift and integral information. Sometimes this information is sufficient to solve a structure without a detailed interpretation of all the splitting. In the next section, we'll illustrate this approach.

PROBLEMS

13.19 The following compound is unknown, but you are contemplating its synthesis and characterization. Predict its NMR spectrum under each of the following assumptions:

(a) $J_{ab} = J_{bc}$ (b) $J_{ab} \neq J_{bc}$

13.20 The three absorptions in the NMR spectrum of 1,1,2-trichloropropane have the following characteristics:

$$Cl_2\overset{a}{C}H - \overset{b}{C}H - \overset{c}{C}H_3$$
$$\underset{Cl}{|}$$

H^a: δ 5.82, J_{ab} = 3.6 Hz

H^b: δ 4.40, J_{ab} = 3.6 Hz, J_{bc} = 6.6 Hz

H^c: δ 1.78, J_{bc} = 6.6 Hz.

Using bars to represent lines in the spectrum and a splitting diagram to determine the appearance of the H^b absorption, sketch the appearance of the spectrum. (Graph paper is useful in constructing splitting diagrams.)

13.21 Predict the complete NMR spectrum of 1,2-dichloropropane under each of the following assumptions. Notice that protons H^b and H^c are diastereotopic and chemically nonequivalent.

(a) $J_{ab} = J_{ac}$ (b) $J_{ab} \neq J_{ac}$

1,2-dichloropropane

B. Breakdown of the *n* + 1 Rule

NMR spectra in which all resonances conform to the $n + 1$ rule are called **first-order spectra**. In all of the spectra discussed to this point, even the complex multiplicative patterns discussed in the previous section, splitting patterns have been first-order. The spectra of some compounds, however, contain splitting patterns that are more complex than predicted by the $n + 1$ rule. Although such spectra can be analyzed rigorously (in many cases) by special mathematical or instrumental techniques, a great deal of information can be obtained from them without such methods. Study Problem 13.4 illustrates a situation of this sort.

STUDY PROBLEM 13.4

Determine the structure of the compound with the formula $C_6H_{13}Cl$ that has the NMR spectrum shown in Fig. 13.13.

SOLUTION The unknown compound has an unsaturation number of zero and is therefore an alkyl chloride. The spectrum contains a very complex splitting pattern in the δ 1.2–1.5 region that cannot be readily interpreted. Three first-order features appear in the spectrum: the triplet at δ 3.52, which integrates for two protons; the triplet at δ 0.9, which integrates for three protons; and the quintet at δ 1.77, which integrates for two protons. The complex pattern in the δ 1.2–1.5 region accounts for the remaining hydrogens. The chemical shift of the δ 3.52 resonance indicates that this triplet must arise from protons on the carbon that bears the chlorine. Because it accounts for two protons, we can *immediately* write the partial structure —CH_2Cl. Its splitting shows that two protons are on the *adjacent* carbon. This information gives the partial structure —CH_2CH_2Cl. The three-proton resonance at δ 0.9 must be a methyl group, and its triplet splitting indicates the partial structure CH_3CH_2—. Forget about the quintet at δ 1.77: we have enough information to solve the structure. A compound with the formula $C_6H_{13}Cl$ and the partial structures above can *only* be 1-chlorohexane, $CH_3CH_2CH_2CH_2CH_2CH_2Cl$. The quintet gives extra information. It integrates for two protons, and must be a —CH_2— group; the quintet splitting suggests four neighboring protons—that is, —$CH_2CH_2CH_2$—; and finally, it has the second largest chemical shift, and, except for the δ 3.52 protons, must be closest to the chlorine. This gives the partial structure —$CH_2\mathbf{CH_2}CH_2Cl$, which is completely consistent with the proposed structure.

Study Problem 13.4 shows that you don't have to interpret *every* splitting pattern in a spectrum to solve a structure because *most spectra contain redundant information.*

Something very important about NMR, however, can be learned by asking why the NMR spectrum of 1-chlorohexane is so complex—why it is not first-order. It turns out that first-order NMR spectra are generally observed when *the chemical shift difference, in Hz, between coupled protons is much greater than their coupling constant.* If the difference in chemical shift of two resonances a and b is $\Delta\nu_{ab}$ (in Hz) and their coupling constant is J_{ab}, then this condition is simply expressed as follows:

Condition for first-order splitting: $\qquad \Delta\nu_{ab} \gg J_{ab}$ (13.12)

In practice, we can interpret this condition to mean that *first-order splitting can be expected when the splitting patterns of the two coupled protons do not overlap.* For example, in the spectrum of 1-chlorohexane (Fig. 13.13), the splitting pattern of the δ 0.93 resonance is

---

FIGURE 13.13 The NMR spectrum for Study Problem 13.4.

first-order—a simple triplet—because it does not overlap with the splitting pattern of the coupled proton at δ 1.3. However, in the δ 1.3–1.5 region, the splitting patterns of three different sets of protons overlap. Consequently, the splittings of these protons are not first-order, and we can't use the $n + 1$ rule to interpret them.

splitting patterns overlap;
splitting is not first-order

$\overbrace{}$

H_3C—CH_2—CH_2—CH_2—CH_2—CH_2—Cl
    a b c d e f

splitting patterns do not overlap;   splitting patterns do not overlap;
splitting of H^a is first-order   splittings of H^e and H^f
are first-order

Similarly, the splitting patterns of H^e and H^f are first-order because they are well separated from each other and from the pattern for H^d.

An important aspect of NMR provides an experimental way to simplify splitting patterns that are not first-order: *Coupling constants do not vary with the magnitude of the operating frequency or the applied magnetic field.* Recall from Sec. 13.3B that NMR experiments can be run at a variety of operating frequencies (ν_0 in Eq. 13.7, p. 618) and corresponding magnetic field strengths $\mathbf{B_0}$. Recall also that chemical shifts in Hz vary in proportion to the operating frequency used. Consequently, if a very large magnetic field and a correspondingly large operating frequency are used, the chemical shifts in Hz are much greater, but the coupling constants are unchanged. Consequently, the condition for first-order behavior in Eq. 13.12 is more likely to be met. This point is illustrated in Fig. 13.14 on p. 644 by comparing the δ 1.2–1.8 regions of the NMR spectra of 1-chloropentane taken at two different magnetic field strengths (and hence, two different operating frequencies). Notice the greater resolution and simplification of the spectrum in the H^b/H^c region at higher field (Fig. 13.14b) than at lower field (Fig. 13.14a). In the 300-MHz spectrum (Fig. 13.14a), the splitting patterns of H^b and H^c overlap extensively. Therefore, the splitting is not first-order. However, in the 600-MHz spectrum (Fig. 13.14b), the chemical-shift difference in Hz is doubled, but the coupling constants are unchanged. As a result, the splitting patterns of H^b and H^c are well separated, and the $n + 1$ rule can be applied. (The apparent complexity of the H^c pattern is the result of multiplicative splitting; Sec. 13.5A.)

Instruments that employ very high magnetic fields are very expensive to purchase and maintain. Although we have illustrated the advantages of such an instrument with a relatively simple molecule, the major use of such instruments is in unraveling the structures of complex molecules whose NMR spectra would be hopelessly complicated when taken at lower field.

PROBLEM

13.22 Identify the following two isomeric alkyl halides ($C_5H_{11}Br$) from their 300-MHz NMR spectra, which are as follows:

Compound A: δ 0.91 (6*H*, d, *J* = 6 Hz); δ 1.7–1.8 (3*H*, complex); δ 3.42 (2*H*, t, *J* = 6 Hz)

Compound B: δ 1.07 (3*H*, t, *J* = 6.5 Hz); δ 1.75 (6*H*, s); δ 1.84 (2*H*, q, *J* = 6.5 Hz)

(a) Give the structure of each compound and explain your reasoning.
(b) Predict how the spectrum of compound A might change if it were taken at 600 MHz.

13.6   USING DEUTERIUM SUBSTITUTION IN PROTON NMR

Deuterium (^2H, or D) finds special use in proton (^1H) NMR. Although deuterium has a nuclear spin, deuterium NMR and proton NMR require greatly different operating frequencies at a

(a) Spectrum taken at $\nu_0 = 300$ MHz (field strength $\mathbf{B}_0 = 70{,}500$ gauss)

(b) Spectrum taken at $\nu_0 = 600$ MHz (field strength $\mathbf{B}_0 = 141{,}000$ gauss)

FIGURE 13.14 A contrast of the H^b, H^c, and H^d regions (δ 1.2–1.8) in the NMR spectrum of 1-chloropentane. (a) The 300-MHz spectrum. (b) The 600-MHz spectrum. In (a), the splitting patterns of H^b and H^c overlap. As a result, the splitting between these protons is not first-order. In (b), the splitting patterns of H^b and H^c are well separated, and their splitting patterns are first-order. (The splitting of H^c is a case of multiplicative splitting that results from the different coupling constants J_{cb} and J_{cd}; Sec. 13.5A.)

given magnetic field strength. Consequently, deuterium NMR absorptions are not detected under the conditions used for proton NMR, so deuterium is effectively "silent" in proton NMR.

One important practical application of this fact is the use of deuterated solvents in NMR experiments. (Solvents are needed for solid and viscous liquid samples because, in the usual type of proton NMR experiment, the sample must be in a free-flowing liquid state.) To ensure that the solvent does not interfere with the NMR spectrum of the sample, it must either be devoid of protons, or its protons must not have NMR absorptions that obscure the sample absorptions. Carbon tetrachloride (CCl_4) is a useful solvent because it has no protons, and therefore has no ^1H NMR absorption. However, many organic compounds are not dissolved by carbon tetrachloride. In addition, this solvent has fallen out of use because of its toxicity. Many of the most useful organic solvents contain hydrogens, which have interfering absorptions. Fortunately, many such solvents are available with their hydrogens substituted by deuteriums; these "deuterated" solvents have no interfering NMR absorptions. The most widely used example of such a solvent is $CDCl_3$ (chloroform-d, or "deuterochloroform"), the deuterium analog of chloroform, $CHCl_3$. Most of the spectra in this text were taken in $CDCl_3$.

In these spectra, you may see a tiny resonance near δ 7.3. This is due to the very small amount of $CHCl_3$ present in commercial $CDCl_3$.

The coupling constants for proton–deuterium splitting are very small. Even when H and D are on adjacent carbons, the H–D coupling is negligible. For this reason, deuterium substitution can be used to simplify NMR spectra and assign resonances. Although deuterium substitution is normally most useful in more complex molecules, let's see how it might be used to assign the resonances of bromoethane (Fig. 13.6). If you were to synthesize CH_3CD_2Br and record its NMR spectrum, you would find that the quartet of bromoethane has disappeared from the NMR spectrum, and the remaining resonance is a singlet. This experiment would establish that the quartet is the resonance of the CH_2 group and that the triplet is that of the CH_3 group.

The use of deuterium to simplify NMR spectra is particularly important for alcohols. Simply shaking the solution of an alcohol with a little D_2O results in very rapid exchange of the O—H proton for deuterium. This strategy is called **D_2O exchange** or, more colloquially, the "**D_2O shake.**" This exchange eliminates the O—H resonance (thus identifying it) and also eliminates any splitting between the α-protons and the O—H proton. The only splitting remaining is then the splitting with any β-protons. For example, the spectrum of dry ethanol (shown in Fig. 13.19a, p. 652) is transformed by the D_2O shake in the following way:

$$H_3C-CH_2-OH \xrightarrow{\text{D}_2\text{O shake}} H_3C-CH_2-OD \qquad (13.13)$$

To summarize: Substitution of a hydrogen by deuterium eliminates its resonance from the proton NMR spectrum and removes any splitting that it causes.

PROBLEMS

13.23 The δ 1.2–1.5 region of the 300-MHz NMR spectrum of 1-chlorohexane, given in Fig. 13.13, is complex and not first-order. Assuming you could synthesize the needed compounds, explain how to use deuterium substitution to determine the chemical shifts of the protons that absorb in this region of the spectrum. Explain what you would see and how you would interpret the results.

13.24 Explain how the NMR spectra of (a) 3-methyl-2-buten-1-ol and (b) 1,2,2-trimethyl-1-propanol would change following a D_2O shake.

13.7 CHARACTERISTIC FUNCTIONAL-GROUP NMR ABSORPTIONS

This section surveys the important NMR absorptions of the major functional groups that we've already studied. The NMR spectra of other functional groups will be considered in the chapters devoted to those groups. A summary table of chemical shift information is given in Appendix III.

A. NMR Spectra of Alkenes

Two characteristic proton NMR absorptions for alkenes are the absorptions for the protons on the double bond, called **vinylic protons** (red in the following structures), and the protons on carbons *adjacent* to the double bond, called **allylic protons** (blue in the following structures). Don't confuse these two types of protons. Typical alkene chemical shifts are illustrated in the following structures and are summarized in Fig. 13.4.

vinylic proton

H δ 5.58

H δ 1.97

allylic proton

terminal vinylic protons

allylic protons

δ 4.92 δ 1.99 δ 0.88

H $CH_2 - CH_2 - CH_3$

δ 1.32

$C = C$

δ 4.88 H   H δ 5.68

internal vinylic proton

(13.14)

In these structures, allylic protons have greater chemical shifts than ordinary alkyl protons, but considerably smaller chemical shifts than vinylic protons. Additionally, the chemical shifts of internal vinylic protons are greater than those of terminal vinylic protons. Recall from Sec. 13.3C that the same trend of chemical shift with branching is evident in the relative shifts of methyl, methylene, and methine protons on saturated carbon atoms.

The chemical shifts of vinylic protons are much greater than would be predicted from the electronegativity of the alkene functional group and can be understood in the following way. Imagine that an alkene molecule in an NMR spectrometer is oriented with respect to the external applied field B_0 as shown in Figure 13.15. The applied field induces a circulation of the π electrons in closed loops above and below the plane of the alkene. This electron circulation gives rise to an induced magnetic field B_i that *opposes* the applied field B_0 at the center of the loop. This induced field can be described as contours of closed circles. Although the induced field opposes the applied field B_0 in the region of the π bond, the curvature of the induced field causes it to lie in the same direction as B_0 at the vinylic protons. The induced field, therefore, adds to, or enhances, the local field at the vinylic protons. As a result, the vinylic protons are subjected to a *greater* local field. This means that a greater frequency is required to bring them into resonance (Eq. 13.4). Consequently, their NMR absorptions occur at relatively high chemical shift.

Because molecules in solution are constantly in motion and tumbling rapidly, at any given time only a small fraction of the alkene molecules are oriented with respect to

FIGURE 13.15 In an alkene, the induced field B_i (*red*) of the circulating π electrons augments the external applied field at the vinylic protons. As a result, vinylic protons have NMR absorptions at relatively large chemical shift (high frequency).

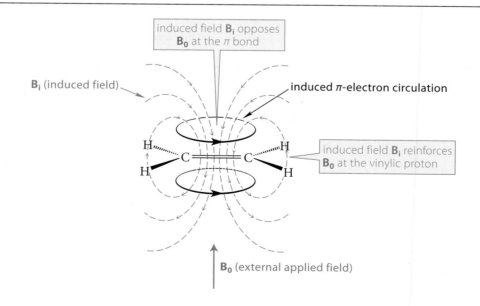

induced field B_i opposes B_0 at the π bond

B_i (induced field)

induced π-electron circulation

induced field B_i reinforces B_0 at the vinylic proton

B_0 (external applied field)

TABLE 13.3 Coupling Constants for Proton Splitting in Alkenes

Relationship of protons	Name of relationship	Coupling constant J, Hz
	cis	6–14
	trans	11–18
	geminal	0–3.5
	vicinal	4–10
 (double bond can be cis or trans)	four-bond (allylic)	0–3.0
 (double bond can be cis or trans)	five-bond	0–1.5

the external applied field as shown in Fig. 13.15. The chemical shift of a vinylic proton is an average over all orientations of the molecule. However, this particular orientation makes such a large contribution that it dominates the chemical shift.

Splitting between vinylic protons in alkenes depends strongly on the geometrical relationship of the coupled protons. Typical coupling constants are given in Table 13.3. Three of the most important splitting relationships are those between trans protons, cis protons, and geminal protons. Splitting between trans protons is largest; splitting between cis protons is intermediate; and splitting between geminal protons is very small. These coupling constants, along with the characteristic $=$C$-$H bending bands from IR spectroscopy (Sec. 12.4C), provide important ways to determine alkene stereochemistry.

Vinylic splitting is illustrated by Figure 13.16 on p. 648, which is the NMR spectrum of the following compound.

vinyl 2,2-dimethylpropanoate
(vinyl pivalate)

This spectrum illustrates all three types of splitting in alkenes and, in addition, provides a useful and relatively straightforward case study for multiplicative splitting. (Do not be con-

FIGURE 13.16 NMR spectrum of vinyl pivalate showing expansion of the vinylic region. The resonances for these protons show multiplicative splitting, which is analyzed in Fig. 13.17.

cerned that this molecule has an unfamiliar functional group; the principles are the same.) The nine equivalent *tert*-butyl protons of vinyl pivalate (H^a) give the large singlet at δ 1.22. The interesting part of this spectrum is the region containing the resonances of the alkene protons (which generally have chemical shifts greater than 4.5 ppm; Fig. 13.4). The protons H^b and H^c are farthest from the electronegative oxygen and therefore have the smallest chemical shifts; the complex resonances in the δ 4.5–5.0 region are from these protons. The four lines in the δ 7.0–7.5 region are all resonances of the one proton H^d.

Each vinylic proton is split by the other two with different coupling constants. As a result, the splitting is multiplicative, and the resonance of each proton is a doublet of doublets.

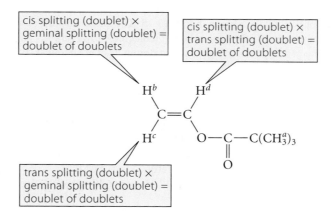

A splitting diagram for the three protons is shown in Fig. 13.17. It is easy to identify the proton that corresponds to each of the resonances by its coupling constants. For example, the resonance of proton H^d has the two largest (cis and trans) splittings; the resonance of proton H^b has the two smallest (geminal and cis) splittings; and the resonance of H^c has a small (geminal) and large (trans) splitting. This splitting pattern is almost always observed for a terminal vinyl group (H_2C=CH—).

In principle, the intensities of these lines should be equal (Table 13.2, p. 630). However, this is a case in which the chemical shifts of the coupled protons are very similar. In such a case, deviations from the ideal ratios in Table 13.2 are observed. The closer the chemical shifts of the coupled protons, the greater the deviations. (Such deviations are sometimes

FIGURE 13.17 A splitting diagram for analysis of the multiplicative splittings in Fig. 13.16. Notice that the trans, cis, and geminal coupling constants are consistent with the values in Table 13.3.

called *leaning*.) Each four-line pattern is readily distinguishable from a quartet, however, because in a quartet the spacings between the lines of the splitting pattern must be identical.

The last two entries in Table 13.3 show that small splitting in alkenes is sometimes observed between protons separated by more than three bonds. Recall that splitting over these distances is usually *not* observed in saturated compounds. These long-distance interactions between protons are transmitted by the π electrons.

In many spectra, geminal, four-bond, and five-bond splittings are not readily discernible as clearly separated lines, but instead appear as perceptibly broadened peaks. Such is the case, for example, in the NMR spectrum in Fig. 13.18 (Problem 13.25).

FIGURE 13.18 The NMR spectrum for Problem 13.25.

13.25 Propose a structure for a compound with the formula C_7H_{14} with the NMR spectrum shown in Fig. 13.18. Explain in detail how you arrived at your structure.

B. NMR Spectra of Alkanes and Cycloalkanes

Because all of the protons in a typical alkane are in very similar chemical environments, the NMR spectra of most alkanes and cycloalkanes cover a very narrow range of chemical shifts, typically δ 0.7–1.7.

An interesting exception is the chemical shifts of protons on a cyclopropane ring, which are unusual for alkanes; they absorb at unusually low chemical shifts, typically δ 0–0.5. Some even have resonances at *smaller* chemical shifts than TMS (that is, negative δ values). For example, the chemical shifts of the ring protons of *cis*-1,2dimethylcyclopropane shown in red are δ (−0.11).

The cause of this unusual chemical shift is an induced electron current in the cyclopropane ring that is oriented so as to shield the cyclopropane protons from the applied field. As a result, these protons are subjected to a smaller local field, and their chemical shifts are *decreased*.

C. NMR Spectra of Alkyl Halides and Ethers

Several NMR spectra of alkyl halides and ethers were presented in developing the principles of NMR earlier in this chapter. The chemical shifts caused by the halogens are usually in proportion to their electronegativities. For the most part, chloro groups and ether oxygens have about the same chemical-shift effect on neighboring protons (Fig. 13.4). However, epoxides, like cyclopropanes, have considerably smaller chemical shifts than their open-chain analogs.

An interesting type of splitting is observed in the NMR spectra of compounds containing fluorine. The common isotope of fluorine (^{19}F) has a nuclear spin. Proton resonances are split by neighboring fluorine in the same general way that they are split by neighboring protons; the same $n + 1$ splitting rule applies. For example, the proton in $HCCl_2F$ appears as a doublet centered at δ 7.43 with a large coupling constant J_{HF} of 54 Hz. This is *not* the NMR spectrum of the fluorine; it is the *splitting of the proton spectrum caused by the fluorine*. (It is also possible to do fluorine NMR, but this requires, for the same magnetic field, a different operating frequency; the spectra of ^{1}H and ^{19}F do not overlap.) Values of H–F coupling constants are larger than H–H coupling constants. The J_{HF} value in $(CH_3)_3C$—F is 20 Hz; a typical J_{HH} value over the same number of bonds is 6–8 Hz. Because J_{HF} values are so large, coupling between protons and fluorines can sometimes be observed over as many as four single bonds.

PROBLEMS

13.26 Suggest structures for compounds with the following proton NMR spectra.

(a) $C_4H_{10}O$: δ 1.13 (3*H*, t, *J* = 7 Hz); δ 3.38 (2*H*, q, *J* = 7 Hz)

(b) $C_3H_5F_2Cl$: δ 1.75 (3*H*, t, *J* = 17.5 Hz); δ 3.63 (2*H*, t, *J* = 13 Hz)

13.27 How would the NMR spectrum of ethyl fluoride differ from that of ethyl chloride?

D. NMR Spectra of Alcohols

Protons on the α-carbons of primary and secondary alcohols generally have chemical shifts in the same range as ethers, from δ 3.2 to δ 4.2 (see Fig. 13.4). Because tertiary alcohols have no α-protons, the observation of an O—H stretching absorption in the IR spectrum accompanied by the *absence* of the —CH—O absorption in the NMR is good evidence for a tertiary alcohol (or a phenol; see Sec. 16.3B).

$$H_3C—OH \qquad H_3C—CH_2—OH \qquad (CH_3)_2CH—OH \qquad (CH_3)_3C—OH$$

$$\delta\ 3.5 \qquad\qquad \delta\ 3.6 \qquad\qquad\qquad \delta\ 4.0 \qquad\qquad \text{no proton absorption in } \delta\ 3\text{--}4 \text{ region}$$

The chemical shift of the OH proton in an alcohol is difficult to predict because it depends on the degree to which the alcohol is involved in hydrogen bonding under the conditions used to determine the spectrum. For example, in pure ethanol, in which the alcohol molecules are extensively hydrogen-bonded, the chemical shift of the OH proton is δ 5.3. When a small amount of ethanol is dissolved in CCl_4, the ethanol molecules are more dilute and less extensively hydrogen bonded, and the OH absorption occurs at δ 2–3. In the gas phase, there is almost no hydrogen bonding, and the OH resonance of ethanol occurs at δ 0.8.

> The chemical shift of the O—H proton in the gas phase is not as large as might be expected for a proton bound to an electronegative atom such as oxygen. The surprisingly small chemical shift of unassociated OH protons is probably due to the induced field caused by circulation of the unshared electron pairs on oxygen. This field shields the OH proton from the external applied field (Sec. 13.3A). Hydrogen-bonded protons, on the other hand, have greater chemical shifts because they bear less electron density and more positive charge (Sec. 8.5C).

The splitting between the OH proton and the α-protons of alcohols is interesting. The NMR spectrum of dry ethanol is shown in Fig. 13.19a on p. 652. By the $n + 1$ splitting rule, the OH resonance of ethanol is a triplet, and the CH_3 resonance is also a triplet. However, the coupling constants for the two splittings are significantly different. For this reason, the resonance for the CH_2 protons, at δ 3.7 in the spectrum, shows multiplicative splitting (Sec. 13.5A) by both the adjacent CH_3 and OH protons and consists of $4 \times 2 = 8$ lines.

multiplicative splitting:
4 × 2 = 8 lines

$$H_3C—CH_2—OH$$

triplet,
J = 7.1 Hz

triplet,
J = 5.1 Hz

However, when a trace of water, acid, or base is added to the ethanol, the spectrum changes, as shown in Fig. 13.19b. *The presence of water, acid, or base causes collapse of the* O—H *resonance to a single line and obliterates all splitting associated with this proton.* Thus, the

FIGURE 13.19 The NMR spectra of ethanol. (a) Absolute, or very dry, ethanol. The CH_2 resonance is split by both the CH_3 and OH protons. (b) Wet acidified ethanol. The CH_2 resonance is split only by the CH_3 protons. Notice also the shift of the OH resonance under wet and dry conditions. The more extensive hydrogen bonding under wet conditions causes a larger chemical shift. Following a "D_2O shake" (Eq. 13.13, p. 645), the spectrum in (b) is also obtained except that the O—H resonance is eliminated.

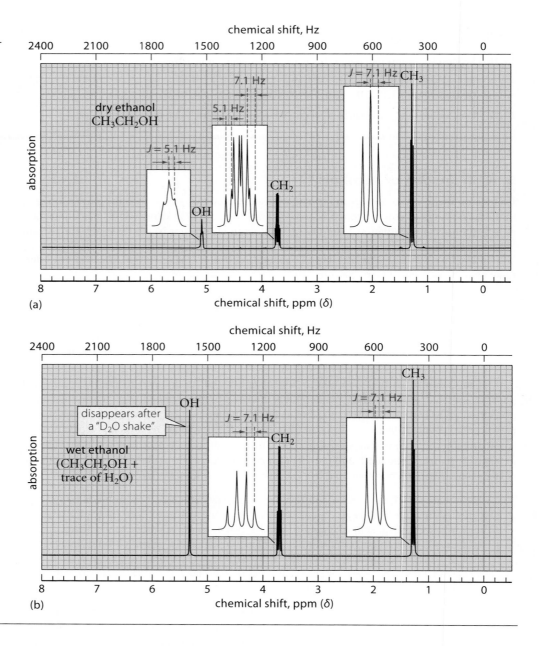

CH_2 proton resonance becomes a quartet, apparently split only by the CH_3 protons. This type of behavior is quite general for alcohols, amines, and other compounds with a proton bonded to an electronegative atom.

This effect of moisture on splitting is caused by a phenomenon called **chemical exchange**: an equilibrium involving chemical reactions that take place very rapidly as the NMR spectrum is being determined. In this case, the chemical reaction is proton exchange between the protons of the alcohol and those of water (or other alcohol molecules). For example, acid-catalyzed proton exchange occurs as two successive acid–base reactions:

$$R—\overset{..}{\underset{\underset{H}{|}}{O}}: + H—\overset{+}{\overset{..}{O}}H_2 \rightleftharpoons R—\overset{..}{\underset{\underset{H}{|}}{O}}—H + :\overset{..}{O}H_2 \qquad (13.15a)$$

$$R—\overset{+}{\overset{..}{\underset{\underset{H}{|}\ :\overset{..}{O}H_2}{O}}}—H \rightleftharpoons R—\overset{..}{\underset{..}{O}}—H + H—\overset{+}{\overset{..}{O}}H_2 \qquad (13.15b)$$

(Write the mechanism for ⁻OH-catalyzed exchange.) For reasons that are discussed in Sec. 13.8, *rapidly exchanging protons do not show spin–spin splitting with neighboring protons.* Acid and base catalyze this exchange reaction, accelerating it enough that splitting is obliterated. In the absence of acid or base, this exchange is much slower, and splitting of the OH proton and neighboring protons is observed.

As a practical matter, you have to be alert to the possibility of either fast or slow exchange when dealing with an NMR spectrum of an unknown that might be an alcohol. An intermediate situation is also common, in which the OH proton resonance is broadened but the α-protons show the splitting characteristic of fast exchange. The assignment of the OH proton can be confirmed in either of two ways. The first is by addition of a trace of acid to the NMR tube. If the α- and O—H protons are involved in splitting, the acid will obliterate this splitting and will simplify the resonances for these two protons. The second way is to use the "D$_2$O shake," discussed in Sec. 13.6. If a drop of D$_2$O is added to the NMR sample tube and the tube is shaken, the OH protons rapidly exchange with the protons of D$_2$O to form OD groups on the alcohol. As a result, the O—H resonance disappears when the spectrum is rerun. Any splitting of the α-proton caused by the O—H proton will also be obliterated because the O—H proton is no longer present.

PROBLEM

13.28 Suggest structures for each of the following compounds.
   (a) $C_4H_{10}O$; δ 1.27 (9H, s); δ 1.92 (1H, broad s; disappears after D$_2$O shake)
   (b) $C_5H_{10}O$: δ 1.78 (3H, s); δ 1.83 (3H, s); δ 2.18 (1H, broad s; disappears after D$_2$O shake); δ 4.10 (2H, d, J = 7 Hz); δ 5.40 (1H, t, J = 7 Hz)

13.8   NMR SPECTROSCOPY OF DYNAMIC SYSTEMS

The NMR spectrum of cyclohexane consists of a singlet at δ 1.4. Yet cyclohexane has two diastereotopic, and therefore chemically nonequivalent, sets of hydrogens: the *axial* hydrogens and the *equatorial* hydrogens. Why shouldn't cyclohexane have two resonances, one for each type of hydrogen? Recall that cyclohexane undergoes a rapid conformational equilibrium, the *chair interconversion* from Sec. 7.2B. The reason that the NMR spectrum of cyclohexane shows only one resonance has to do with the *rate* of the chair interconversion, which is so rapid that the NMR instrument detects only the average of the two conformations. Because the chair interconversion interchanges the positions of axial and equatorial protons (Eq. 7.6, p. 278), only the resonance of the "average" proton in cyclohexane is observed—a proton that is axial half the time and equatorial half the time. This example illustrates an important aspect of NMR spectroscopy: *the spectrum of a compound involved in a rapid equilibrium is a single spectrum that is the time-average of all species involved in the equilibrium.* In other words, *the NMR spectrometer is intrinsically limited to resolve events in time.*

Although some equations describe this phenomenon exactly, it can be understood by the use of an analogy from common experience. Imagine looking at a three-blade fan or propeller that is rotating at a speed of about 100 times per second (see Fig. 13.20 on p. 654). Our eyes do not see the individual blades, but only a blur. The appearance of the blur is a time-average of the blades and the empty space between them. If we photograph the fan using a shutter speed of about 0.1 second, the fan appears as a blur in the resulting picture for the same reason: during the time the camera shutter is open (0.1 s) the blades make 10 full revolutions (Fig. 13.20a). Now imagine that we slow the fan to about 1 rotation per second. While the shutter is open, the fan blades make only 0.1 revolution—about 36°. The fan blades are more distinct, but still somewhat blurred (Fig. 13.20b). Finally, imagine that the fan is rotating very slowly, say, one rotation every hundred seconds. While the shutter is open, the fan blades traverse only 1/1000 of a circle—about 0.36°. In the resulting picture the individual blades

(a)   100 rotations per second;
image totally blurred

(b)   1 rotation per second;
individual blades visible
but blurred

(c)   0.01 rotation per second;
individual blades visible
and in sharp focus

FIGURE 13.20 What we see when a three-blade propeller is rotated at various speeds and pho-
tographed for a duration of 0.1 second. (a) Propeller speed = 100 rotations per second (100 Hz);
(b) Propeller speed = 1 Hz; (c) Propeller speed = 0.01 Hz. Notice that the individual blades of the
propeller lose their identity, or blur, as the rotation rate increases.

are visible and in relatively sharp focus (Fig. 13.20c). The rapid conformational equilibrium of cyclohexane is to the NMR spectrometer roughly what the rapidly rotating
propeller is to the slow camera shutter.

Both types of cyclohexane protons can be observed if the rate of the chair interconversion is reduced by lowering the temperature. Imagine cooling a sample of cyclohexane in
which all protons but one have been replaced by deuterium. (The use of deuterium virtually
eliminates splitting with neighboring protons, because splitting between H and D is very
small; Sec. 13.6.)

$$\text{(13.16)}$$

cyclohexane-d_{11}

In the chair interconversion, the single remaining proton alternates between axial and equatorial positions.

The NMR spectrum of cyclohexane-d_{11} at various temperatures is shown in Fig. 13.21.
At room temperature, the spectrum consists of a single line, as in cyclohexane itself. As the
temperature is lowered progressively, the resonance becomes broader until, near -60 °C, it
divides into two broad resonances equally spaced about the original one. When the temperature is lowered still further, the spectrum becomes two sharp single lines. Thus, lowering the
temperature progressively retards the chair interconversion until, at low temperature, NMR
spectrometry can detect both chair forms independently. This is analogous to taking pictures
of the propeller in Fig. 13.20 with a constant shutter speed and slowing down the propeller
until the blur disappears and the individual blades become clearly separated.

It is possible to use the information from these spectra at different temperatures to calculate the rate of the chair interconversion. The energy barrier for the chair interconversion
shown in Fig. 7.6 (p. 281) was obtained from this type of calculation.

Just as NMR spectroscopy detects a time-average of the two chair conformations of cyclohexane at room temperature, it also detects an average of all conformations of any molecule undergoing rapid conformational equilibria. Thus, the CH_3 protons of bromoethane (CH_3CH_2Br) give a single resonance and a single coupling constant for splitting by the CH_2 protons because the molecule undergoes rapid internal rotation about the carbon–carbon bond. (The averaging of coupling constants was discussed in Sec. 13.5A.) If this rotation were so slow that the NMR spectrometer could resolve individual conformations, the NMR spectrum of bromoethane would be more complex.

The time-averaging effect of NMR is not limited simply to conformational equilibria. The spectra of molecules undergoing any rapid process, even a chemical reaction, are also averaged by NMR spectroscopy. This is the reason, for example, that the splitting associated with the OH protons of an alcohol is obliterated by chemical exchange (Sec. 13.7D). For example, consider the effects of chemical exchange on the spectrum of the CH_3 protons of methanol. In absolutely dry methanol, the resonance of these protons is split by the OH proton into a doublet.

doublet in dry methanol; singlet in wet methanol ⟶ H₃C—OH ⟵ quartet in dry methanol; singlet in wet methanol

Recall from Fig. 13.8 that this splitting occurs because the adjacent OH proton can have either of two spins. If acid or base is added to the methanol, causing the OH protons to exchange rapidly, protons of different spins jump quickly on and off the OH. Thus, the CH_3 protons *on any one molecule* are next to an OH proton with spin $+\frac{1}{2}$ half of the time and an OH proton with spin $-\frac{1}{2}$ half of the time. (The minuscule difference between the numbers of protons in the two spin states can be ignored.) In other words, the CH_3 protons "see" an adjacent OH proton with a spin that averages to zero over time. Because a proton is not split by an adjacent nucleus with zero spin, rapid exchange eliminates splitting of the CH_3 protons. Similar reasoning can be applied to the spectrum of the methanol OH proton, which, in a dry sample, is a quartet, but is a singlet in a sample containing traces of moisture.

PROBLEMS

13.29 Suppose you were able to cool a sample of 1-bromo-1,1,2-trichloroethane enough that rotation about the carbon–carbon bond becomes slow on the NMR time scale. What changes in the NMR spectrum would you anticipate? Be explicit.

13.30 Describe in detail what changes you would expect to see in the NMR resonance of the methyl group as 1-chloro-1-methylcyclohexane is cooled from room temperature to very low temperature.

13.9 NMR SPECTROSCOPY OF OTHER NUCLEI. CARBON NMR

A. NMR Spectroscopy of Other Nuclei

Although we've concentrated our attention on proton NMR, any nucleus with a nuclear spin can be studied by NMR spectroscopy. Table 13.4 lists a few other nuclei with spin $= \pm\frac{1}{2}$. For a given magnetic field strength, different nuclei absorb energy in different frequency ranges. For a given field strength $\mathbf{B_0}$, the absorption frequency can be calculated from Eq. 13.4 using the appropriate gyromagnetic ratio:

$$\nu_n = \frac{\Delta\epsilon_n}{h} = \frac{\gamma_n}{2\pi}\mathbf{B_0} \tag{13.17}$$

In this equation, $\Delta\epsilon_n$ is the energy separation between spin energy levels for nucleus n, γ_n is the gyromagnetic ratio of nucleus n, and $\mathbf{B_0}$ is the applied magnetic field. This equation shows that the absorption frequency of any nucleus at a given magnetic field strength depends on its gyromagnetic ratio. For example, the gyromagnetic ratio of the proton, γ_H, is 26,753 rad gauss^{-1} s^{-1}. If the applied magnetic field, for example, is 70,500 gauss, an operating frequency of 300 MHz is required for proton NMR. Because the gyromagnetic ratio for ^{13}C is 6728 rad gauss^{-1} s^{-1}, or about one-quarter of that for a proton, the operating frequency required for ^{13}C NMR at the same field strength is also one-quarter of that for a proton, or about 75 MHz.

Suppose we have a molecule containing two different magnetically active nuclei, such as an alkyl fluoride that contains both ^1H and ^{19}F. In the *proton* NMR spectrum of such a molecule, the NMR signals of protons are observed, but not those of the fluorines. (The proton splitting *caused* by the fluorines *is* observed, however; Sec. 13.7C.) To observe *fluorine NMR*, a different frequency range is used, in which case the fluorine resonances but not the proton resonances are observed. (In this situation, the splitting of fluorine signals caused by nearby protons would be observed.)

B. Carbon-13 NMR Spectroscopy

Because all organic compounds contain carbon, the NMR spectroscopy of carbon is very important. However, the most abundant nucleus of carbon (^{12}C) does not have a nuclear spin and therefore cannot be detected by NMR. As Table 13.4 shows, the only isotope of carbon that has a nuclear spin is ^{13}C. The NMR spectroscopy of ^{13}C is called **^{13}C NMR spectroscopy**, which is often shortened simply to **carbon NMR**.

TABLE 13.4 Properties of Some Nuclei with Spin $\pm\frac{1}{2}$

Isotope	Relative sensitivity	Natural abundance, %	Observation frequency ν_n, MHz*	Gyromagnetic ratio[‡]
^1H	(1.00)	99.98	300	26,753
^{13}C	0.0159	1.10	75	6728
^{19}F	0.834	100	282	25,179
^{31}P	0.0665	100	122	10,840

* At magnetic field $\mathbf{B_0}$ = 70,500 gauss. [‡] In radians gauss^{-1} s^{-1} defined in Eq. 13.17.

One problem with ^{13}C NMR spectroscopy is that the resonance of a ^{13}C nucleus is *intrinsically* weaker than that of a proton because of the magnetic properties of the carbon nucleus. The intrinsic intensity of the NMR signal from each nucleus is proportional to the *cube* of its gyromagnetic ratio. Thus, the relative intensity of a proton signal versus that from the same number of ^{13}C atoms is $(\gamma_H/\gamma_C)^3 = (26{,}753/6728)^3$, or 62.9. In other words, a ^{13}C NMR resonance is about $1/62.9 = 0.0159$ times as intense as a proton resonance. Another problem is the low natural abundance of the ^{13}C isotope: organic compounds contain only about 1.1% of ^{13}C at each carbon position. The cumulative effect of intrinsic intensity and low natural abundance, therefore, is that carbon resonances are only $(0.0159)(0.011) = 0.000175$ times as intense as proton resonances.

Although the weak ^{13}C NMR resonance at one time presented a serious obstacle to detection, advances in instrumentation (Sec. 13.11) and computer power have made it possible to obtain ^{13}C NMR spectra on compounds containing the *natural abundance* of ^{13}C on a routine basis. These advances (Sec. 13.11) made it possible to acquire a single NMR spectrum in a fraction of a second and to store it digitally in a computer. A useful ^{13}C NMR spectrum of a compound is obtained by taking a few thousand individual ^{13}C NMR carbon spectra of the compound and adding them together. Because electronic noise is random, it is reduced when many spectra are added together, whereas the resonances themselves are enhanced. Almost 6000 ^{13}C NMR spectra must be added together in this manner to get the same intensity that we would obtain in a single proton NMR spectrum of the same compound at the same concentration. ^{13}C NMR spectroscopy is a very important analytical technique in organic chemistry.

Although the principles of ^{13}C NMR and proton NMR are essentially the same, some aspects of ^{13}C NMR are unique. First, coupling (splitting) between carbons is *not* generally observed. The reason is the low natural abundance of ^{13}C. Recall that ^{13}C NMR measures the resonance of ^{13}C, not the common isotope ^{12}C. If the probability of finding a ^{13}C at a given carbon is 0.0110, then the probability of finding ^{13}C at any two carbons in the same molecule is $(0.0110)^2$, or 0.00012. This means that *two ^{13}C atoms almost never occur together within the same molecule.* (The two ^{13}C atoms would have to occur in the *same* molecule for coupling to be observed.) It is possible, though, to prepare compounds that are isotopically enriched in ^{13}C, in which case the usual splitting rules apply (see Problem 13.63, p. 679).

A second important aspect of ^{13}C NMR is that the range of chemical shifts is very large compared with that in proton NMR. Typical carbon chemical shifts, shown in Fig. 13.22 on p. 658, cover a range of about 200 ppm. With a few exceptions, trends in carbon chemical shifts parallel those for proton chemical shifts, but chemical shifts in ^{13}C NMR are more sensitive to small changes in chemical environment. As a result, it is often possible to observe distinct resonances for two carbons in very similar chemical environments. This point is illustrated in the ^{13}C NMR spectrum of 3-methylpentane, in which each chemically nonequivalent set of carbons gives a separate, clearly discernible resonance:

δ 18.8

$$H_3C \underset{\delta\ 11.5}{\text{——}} CH_2 \underset{\delta\ 29.2}{\text{——}} \overset{\overset{\textstyle CH_3}{|}}{\underset{\delta\ 36.2}{CH}} \text{——} CH_2 \text{——} CH_3$$

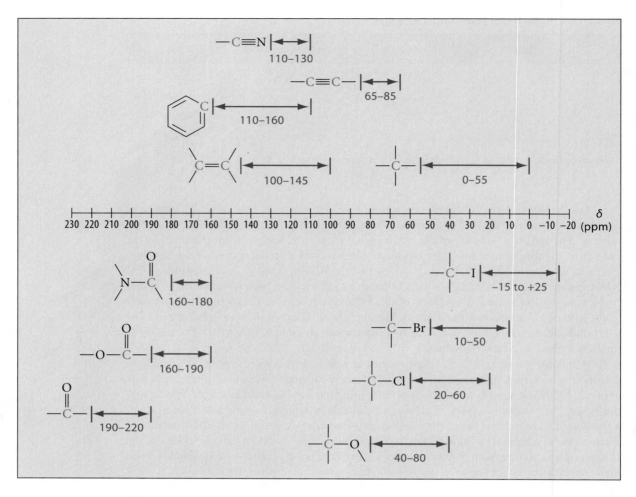

FIGURE 13.22 A carbon chemical-shift chart for common functional groups. The chemical-shift range for carbons is more than 10 times that of protons. (Compare with Figure 13.4.) For that reason, carbons in very similar chemical environments usually give distinguishable resonances.

As this example illustrates, the chemical shift for a carbon depends in many cases on the number of attached carbons in the order of δ (tertiary) $> \delta$ (secondary) $> \delta$ (methyl).

A third unique aspect of ^{13}C NMR is that the splitting of ^{13}C resonances by protons (^{13}C—^{1}H splitting) is large; typical coupling constants are 120–200 Hz for directly attached protons. Furthermore, carbon NMR signals are also split by more remote protons. Although such splitting can sometimes be useful, more typically it presents a serious complication in the interpretation of ^{13}C NMR spectra, because the ^{13}C—H splitting patterns overlap. In most ^{13}C NMR work, splitting is eliminated by a special instrumental technique called *proton spin decoupling*. Spectra in which proton coupling has been eliminated are called **proton-decoupled ^{13}C NMR spectra**. In such spectra a *single unsplit line* is observed for each chemically nonequivalent set of carbon atoms.

These points are illustrated by the proton-decoupled ^{13}C NMR spectrum of 1-chlorohexane, shown in Fig. 13.23. The carbon spectrum consists of six single lines, one for each carbon of the molecule. The assignment of the lines in Fig. 13.23 shows that carbon chemical shifts, like proton chemical shifts, decrease with distance from the electronegative chlorine.

^{13}C NMR is particularly useful in differentiating closely related compounds on the basis of their molecular symmetry. The basis of this idea is that symmetrical compounds have fewer chemically nonequivalent sets of carbons than less symmetrical isomers. This point is illustrated in Study Problem 13.5.

FIGURE 13.23 The proton-decoupled ^{13}C NMR spectrum of 1-chlorohexane. Notice that the resonance of each carbon is visible. (The peaks labeled "CDCl$_3$" are due to the ^{13}C resonance of the solvent. The reason that the CDCl$_3$ carbon signal is a triplet is considered in Problem 13.50b.)

STUDY PROBLEM 13.5

How would you use ^{13}C NMR spectroscopy to differentiate the two isomers 1-chloropentane and 3-chloropentane?

SOLUTION First, draw the structures of the two compounds.

$$CH_3CH_2CH_2CH_2CH_2Cl \qquad CH_3CH_2\overset{\overset{\displaystyle Cl}{|}}{C}HCH_2CH_3$$

1-chloropentane   **3-chloropentane**

If we assume that a separate resonance is observed for each chemically nonequivalent set of carbons, then the proton-decoupled ^{13}C NMR spectrum of 1-chloropentane should consist of five lines, but that of 3-chloropentane should consist of only three lines; the two CH$_3$ carbons are chemically equivalent, and the two CH$_2$ carbons are chemically equivalent. As this example shows, if a molecule has symmetry, it will have fewer absorptions than there are carbons.

PROBLEMS

13.31 The proton-decoupled ^{13}C NMR spectra of 3-heptanol (*A*) and 4-heptanol (*B*) are given in Fig. 13.24 on p. 660. Indicate which compound goes with each spectrum, and explain your reasoning.

13.32 Indicate two things you would look for in their ^{13}C NMR spectra to distinguish between 1,1-dichlorocyclohexane and *cis*-1,2-dichlorocyclohexane.

---

^{13}C NMR spectra are generally not integrated because the instrumental technique used for taking the spectra (Sec. 13.11) gives relative peak integrals that are governed by factors other than the number of carbons. However, even this fact can be useful. For example, the proton decoupling technique enhances the peaks of carbons that bear hydrogens; hence, resonances for carbons that bear *no* hydrogens, such as quaternary carbons, carbons of carbonyl groups, and the α-carbons of tertiary alcohols, are usually smaller than those for other carbons.

A number of techniques enhance the utility of ^{13}C NMR by providing a count of the protons directly attached to each carbon. In other words, it is possible to determine which of the carbon signals in a ^{13}C NMR spectrum come from methyl, methylene, methine, or quaternary carbons. One technique for making such a determination is known by the acronym **DEPT** (for Distortionless Enhancement with Polarization Transfer). The DEPT technique

FIGURE 13.24 Proton-decoupled ^{13}C NMR spectra for Problem 13.31.

yields separate spectra for methyl, methylene, and methine carbons, and each line in these spectra corresponds to a line in the complete ^{13}C NMR spectrum. Lines in the complete ^{13}C NMR spectrum that do not appear in the DEPT spectra arise from carbons that have no attached hydrogens. This technique is illustrated with the DEPT spectra of camphor (see Fig. 13.25).

Notice two other things in the full spectrum of camphor (Fig. 13.25d). First, the intensities of the quaternary carbons are smaller, a point made previously. Second, the chemical shifts of the carbons remote from the carbonyl group (carbons 4–9) are in the order quaternary > tertiary > secondary > methyl. This trend was also mentioned previously in this section. (The shifts of the carbons that are part of, or near, the carbonyl group are increased by the electronegativity of oxygen and the deshielding effects of π-electron circulation, effects that are also observed in proton NMR spectroscopy; see Fig. 13.15.)

STUDY PROBLEM 13.6

A compound $C_7H_{16}O_3$ has the following ^{13}C NMR-DEPT spectrum (the numbers in parentheses indicate the number of attached hydrogens):

$$\delta\ 15.2\ (3),\ \delta\ 59.5\ (2),\ \delta\ 112.9\ (1)$$

Propose a structure for this compound.

SOLUTION The compound has no rings or double bonds because its unsaturation number is zero. The simplest assumption from the ^{13}C NMR spectrum is that the compound has three chemically nonequivalent sets of carbons, because there are three lines. One set (δ 15.2) consists of methyl groups (three attached hydrogens) which, from their chemical shift, are not very close to the oxygens.

$$H_3C—\overset{|}{\underset{|}{C}}—$$

$$\uparrow$$

$$\delta\ 15.2$$

Another set consists of methylene (CH_2) groups, which are within the chemical shift range for the α-carbons of ethers (Fig. 13.22).

$$—CH_2—O—$$

$$\uparrow$$

$$\delta\ 59.5$$

(a) Methyl (CH_3) carbons

(b) Methylene (CH_2) carbons

(c) Methine (CH) carbon

(d) Complete ^{13}C spectrum. The methyl carbons are labeled p (for primary), the methylene carbons s (for secondary), the methine carbon t (for tertiary), and the quaternary carbons q.

chemical shift, ppm (δ)

FIGURE 13.25 The ^{13}C NMR spectrum of camphor (structure in the box at the left) edited by the DEPT technique. The absorptions for the methyl (CH_3) carbons are given in part (a), the methylene (CH_2) carbons in part (b), and the methine (CH) carbon in part (c). Each peak in these three spectra corresponds to a peak in the full spectrum, shown in part (d). Absorptions in the full spectrum that do not appear in parts (a), (b), or (c) are due to the carbonyl carbon or the quaternary carbons. The number over each peak is the assignment using the carbon number in the camphor structure. Notice in the full spectrum that the intensities of the resonances for the carbonyl (carbon 2) and quaternary (carbons 1 and 7) carbons are lower than the intensities of carbons with attached hydrogens. (Courtesy John Kozlowski, Purdue University.)

The last set consists of one or more methine (CH) groups, which, from the chemical shift, must be bound to more than one oxygen.

$$
\begin{array}{ccc}
& \text{O}{-} & & \text{O}{-} \\
& | & & | \\
{-}\text{O}{-}\text{CH} & & \text{or} & \text{HC}{-}\text{O}{-} \\
& | & & | \\
& \text{O}{-} & & \\
\end{array}
$$

δ 112.9

Only the triethoxymethane structure gives only three absorptions while accommodating these partial structures:

$$
\begin{array}{c}
\text{OCH}_2\text{CH}_3 \\
| \\
\text{CH}_3\text{CH}_2\text{O}{-}\text{C}{-}\text{H} \\
| \\
\text{OCH}_2\text{CH}_3
\end{array}
$$

triethoxymethane

PROBLEM

13.33 Explain why each of the following structures is *not* consistent with the ^{13}C NMR data in Study Problem 13.6.

(a)
$$
\begin{array}{c}
\text{OCH}_2\text{CH}_3 \\
| \\
\text{CH}_3\text{CH}_2\text{O}{-}\text{C}{-}\text{CH}_3 \\
| \\
\text{OCH}_3
\end{array}
$$

(b)
$$
\begin{array}{c}
\text{CH}_2\text{OCH}_3 \\
| \\
\text{CH}_3\text{OCH}_2{-}\text{C}{-}\text{H} \\
| \\
\text{CH}_2\text{OCH}_3
\end{array}
$$

13.10   SOLVING STRUCTURE PROBLEMS WITH SPECTROSCOPY

You are now ready to use what you know about IR, NMR, and mass spectrometry to solve some problems that require more than one of these techniques. Study Problems 13.7 and 13.8 illustrate the techniques involved. Although no single method works in every case, the following suggestions should prove useful.

1. From the mass spectrum determine, if possible, the molecular mass.

2. If an elemental analysis is given, calculate the molecular formula and determine the unsaturation number.

3. Look for evidence in both the IR and NMR spectra for any functional groups that are consistent with the molecular formula: OH groups, alkenes, and so on. Write down any structural fragments indicated by the spectra.

4. Use the ^{13}C NMR spectrum and, if possible, the proton NMR spectrum, to determine the number of nonequivalent sets of carbons or protons (or both). If the proton NMR spectrum is complex, this may not be possible, but you should be able to set some limits.

5. Apply the suggestions in both Sec. 13.3F and Sec. 13.4C to complete your analysis by NMR. *Be sure to write out partial structures and all possible complete structures that are consistent with your spectra.* As you write out partial struc-

tures, notice how many carbons are unaccounted for; different partial structures may have carbons in common. Decide between possible structures by asking what features of the different spectra would be expected for each, and look for those features; it is sometimes easy to overlook some feature of a spectrum that will decide between structures.

6. Finally, rationalize all spectra for consistency with the proposed structure.

STUDY PROBLEM 13.7

Propose a structure for the compound with the IR, NMR, and EI mass spectra shown in Fig. 13.26 on p. 664.

SOLUTION The EI mass spectrum of this compound shows a pair of peaks at $m/z = 90$ and 92, with the latter peak about one-third the size of the former. This pattern indicates the presence of chlorine. Furthermore, the base peak at $m/z = 55$ corresponds to a loss of Cl (35 and 37 mass units, respectively). Let's adopt the hypothesis that this is a chlorine-containing compound with molecular mass of 90 (for the ^{35}Cl isotope). In the IR spectrum, the peak at 1642 cm^{-1} suggests a C=C stretch, and, in the NMR spectrum, there is a complex signal in the vinylic proton region. Evidently this compound is a chlorine-containing alkene.

In the NMR spectrum, the vinylic protons at δ 5–6 account for three hydrogens; the quintet at δ 4.6 accounts for one hydrogen; and the doublet at δ 1.6 accounts for two hydrogens, for a total of seven protons. If the compound has seven protons (7 mass units) and one chlorine (35 mass units), then the remaining 48 mass units can be accounted for by four carbons, two of which are part of an alkene double bond. A possible molecular formula is then C_4H_7Cl.

The unsaturation number for this formula is 1. Hence, the molecule contains only one double bond. Because the NMR integral indicates three vinylic protons, the molecule must contain a —CH=CH$_2$ group. In the IR spectrum, the peaks at 930 and 990 cm^{-1} are consistent with such a group, although the former peak is at somewhat higher wavenumber than is usual for this type of alkene. The three-proton doublet at δ 1.6 suggests a methyl group adjacent to a CH group.

$$H_3C-\overset{|}{C}H-$$

The δ 4.6 absorption accounts for one proton, and its coupling constant ($J = 6.6$ Hz) matches that of the absorption at δ 1.6. The splitting and chemical shift of the δ 4.6 absorption fit the partial structure

$$H_3C-\overset{\overset{\displaystyle Cl}{|}}{C}H-CH=$$

With a molecular mass of 90 and three vinylic protons, the only possible complete structure is

$$\delta\ 1.60\ \ H_3C-\overset{\overset{\displaystyle Cl}{|}\ \delta\ 5.9\text{--}6.0}{C}H-CH=CH_2\ \ \ \delta\ 5.0\text{--}5.3$$
$$\delta\ 4.60$$

STUDY PROBLEM 13.8

A compound $C_8H_{18}O_2$ with a strong, broad infrared absorption at 3293 cm^{-1} has the following proton NMR spectrum:

$$\delta\ 1.22\ (12H,\ s);\ \delta\ 1.57\ (4H,\ s);\ \delta\ 1.96\ (2H,\ s)$$

(The resonance at δ 1.96 disappears when the sample is shaken with D_2O.) The proton-decoupled ^{13}C NMR spectrum of this compound consists of three lines, with the following chemical shifts and DEPT data (in parentheses) for attached protons:

$$\delta\ 29.4\ (3),\ \delta\ 37.8\ (2),\ \delta\ 70.5\ (0)$$

Identify the compound.

SOLUTION The IR spectrum indicates the presence of an alcohol, and the disappearance of the δ 1.96 NMR absorption after the D$_2$O shake (Secs. 13.6 and 13.7D) provides confirmation. Furthermore, because this absorption integrates for two protons, and because the formula contains two oxygens, the compound is a diol. Because the proton NMR spectrum contains no α-hydrogen absorptions in the δ 3–4 region, both alcohols must be tertiary. The proton NMR indicates only three chemically nonequivalent sets of hydrogens, and the ^{13}C NMR indicates only three chemically nonequivalent sets of carbons, one of which must be the two α-carbons of the tertiary alcohol groups. The DEPT data confirm that one set of carbons indeed has no attached protons, as expected for a tertiary alcohol,

FIGURE 13.26 Proton NMR, IR, and EI mass spectra for Study Problem 13.7. The integral of each resonance in the NMR spectrum is shown in red.

and the chemical shift is consistent with that expected for the α-carbon of an alcohol. The presence of only *three* nonequivalent sets of protons and *three* nonequivalent sets of carbons requires a structure of considerable symmetry. The *only* structure that fits these data is

$$\underset{\substack{\displaystyle | \\ \displaystyle OH}}{\overset{\substack{\displaystyle CH_3 \\ \displaystyle |}}{H_3C-C}}-CH_2CH_2-\underset{\substack{\displaystyle | \\ \displaystyle OH}}{\overset{\substack{\displaystyle CH_3 \\ \displaystyle |}}{C}}-CH_3$$

2,5-dimethyl-2,5-hexanediol

PROBLEMS

13.34 (a) Tell why each of the following structures is inconsistent with the data in Study Problem 13.7.

trans-1-chloro-2-butene   2-chloro-1-butene

A   *B*

(b) Although we did not have to analyze the vinylic proton resonances in detail to determine the structure at the end of Study Problem 13.7, it is interesting to consider these resonances further. First, justify the assignments given at the end of the Study Problem for the resonances of the vinylic protons. Then notice in the NMR spectrum (Fig. 13.26) that the =CH$_2$ proton resonances do not split each other detectably (Table 13.3, geminal protons). Next, draw out the structure of this compound to show the stereochemical relationships of the =CH$_2$ protons to the other vinylic proton. Which resonances in the δ 5.0–5.3 region go with which =CH$_2$ proton? How do you know? (*Hint*: See Fig. 13.15, p. 646.)

13.35 Tell why each of the following structures is *not* consistent with the spectroscopic data in Study Problem 13.8.

13.11 THE NMR SPECTROMETER

The basic components of an NMR spectrometer are shown in Fig. 13.27 on p. 666. An NMR instrument requires, first, a strong magnetic field to establish the tiny energy differences between nuclear spin states. Early NMR instruments employed electromagnets or permanent magnets that generated fields in the range of 7000–23,000 gauss. Modern instruments utilize large solenoids—essentially doughnut-shaped wire coils—fabricated of superconducting wire. Current flowing in the coil generates the magnetic field. In a superconducting wire, electric current, once established, persists indefinitely and flows without electrical resistance. Superconducting solenoids are required because the electric current required for large magnetic fields would generate far too much resistance (and therefore heat) in a conventional, non-superconducting solenoid. Most metals that exhibit superconductivity do so only at very low temperatures. Because liquid helium is used to maintain these low temperatures, the solenoid is housed in an elaborate cryostat (essentially, a multiwall thermos bottle). It is possible to construct superconducting solenoids that can develop magnetic fields greater than 200,000 gauss.

The second instrumental component of the NMR experiment is the radiofrequency (rf) radiation. Application of rf radiation and detection of its absorption are managed through the use of a small wire coil surrounding the sample, which is held in a glass tube that is rapidly spun about its longitudinal axis. The sample and the rf coils are housed in a precisely con-

FIGURE 13.27 A 300-MHz NMR spectrometer with a cross-section of the solenoid vessel. The height of the solenoid is about one meter. (The size of the sample and rf coil are exaggerated.) The heart of the solenoid is a coil of superconducting wire. Because superconductivity requires extremely low temperatures, the coil is immersed in liquid helium, which has a temperature of about 4 K. This is insulated from the surroundings by an evacuated container (essentially a large thermos bottle), which itself is contained within another "thermos bottle" of liquid nitrogen. When the solenoid is energized, a large current flows around the coil with zero resistance. This current generates the magnetic field of 70,500 gauss (*red dashed lines*). The large current required for a field of this magnitude is possible only with superconducting alloys; too much heat would be generated with conventional wire. The sample, contained in a small glass tube, is placed inside the sample probe within the gap in the center of the solenoid. Within the probe the sample is surrounded by the radiofrequency (rf) coil, which is used to transmit rf radiation and to measure rf emission. The sample tube is spun rapidly to average out small differences in the field throughout the sample. Many of the spectra in this text were obtained with this instrument.

structed probe that can be inserted into the center of the solenoid (that is, into the "hole" in the wire "doughnut"). Except for the magnet or solenoid, the NMR instrument is in essence a radio transmitter and receiver.

NMR experiments in early instruments involved varying the frequency of the rf radiation slowly and detection of each resonance separately. With this technique, a spectrum was obtained in a few minutes. Although this technique was used in spectrometers through the early 1970s, modern NMR spectrometers employ a technique for taking spectra called **pulse-Fourier-transform NMR (FT-NMR)**. In FT-NMR, all of the proton spins are excited instantaneously with an rf pulse containing a broad band of frequencies, and the spectrum is obtained by analyzing the emission of rf energy as the spins return to equilibrium. With FT-NMR, an entire proton NMR spectrum can be obtained in less than a second. Consequently, a large number of spectra of a given sample (anywhere from 50 to 20,000, depending on the sample concentration and the isotope) can be recorded in a relatively short time. A computer stores, analyzes, and mathematically sums the spectra. Because electronic noise is random, it sums to zero when averaged over many spectra, whereas the resonances of the sample reinforce to give a much stronger spectrum than could be obtained in a single experiment. The FT-NMR technique has made possible the routine use of ^{13}C NMR for structure determination. FT-NMR instruments became truly practical with the advent of relatively inexpensive dedicated small computers that are required for application of the FT-NMR technique.

The FT-NMR method was conceived by Richard R. Ernst (b. 1933) of the ETH (Federal Technical Institute) in Zürich, Switzerland; for this contribution, he was honored with the 1991 Nobel Prize in Chemistry.

13.12 MAGNETIC RESONANCE IMAGING

One of the most important medical uses of NMR is *magnetic resonance imaging (MRI)*. MRI is a noninvasive method of biomedical imaging that is particularly effective for visualizing soft tissue. Like the NMR techniques for determining structure that we've considered in this chapter, MRI involves the use of high magnetic fields and radiofrequency radiation, which are much less hazardous than X-rays. However, unlike structural NMR, MRI does not involve chemical shifts or splitting. Rather, MRI monitors the protons of a single compound—water—in various physiological environments by utilizing a physical phenomenon called *nuclear relaxation*. To understand nuclear relaxation conceptually, we have to expand our view of nuclear spins somewhat (Fig. 13.28, p. 668). First, the term "spin" refers to the fact that nuclei behave as if they are spinning charges. Spinning charges produce magnetic fields, which have magnitude and direction. Therefore, we represent the spinning charge as a vector. This means that we can think of the spinning charge as a tiny bar magnet, with the vector running from "south" to "north." In other words, the direction of the spin vector is the direction of the bar magnet. In our previous discussions, we have characterized nuclear spins merely as "up" ($+\frac{1}{2}$) or "down" ($-\frac{1}{2}$) relative to the magnetic field direction. However, the spin vector is not static, but rather *precesses* about the magnetic field direction. This precessional motion is like that of a spinning gyroscope or top (Fig. 13.28a). Not only is the top spinning on its own axis, but the axis of the top is also rotating about the gravitational axis. Similarly (Fig. 13.28b), the nuclear spin vector M precesses about the direction of the magnetic field, which we define as the z axis. The frequency of the precession—the number of times per second it precesses—is the frequency ν_0 defined by Eq. 13.7, p. 618. Now let's resolve the spin vector M into two components: one along the z axis, which we'll call M_z, and one in the xy plane, which we'll call M_{xy} (Fig. 13.28c). Notice that M_z is stationary, whereas M_{xy} is rotating, also at the frequency ν_0. Here is the connection with what we've learned previously: "up" and "down" refer to the direction of M_z. Fig. 13.28c shows an "up" ($+\frac{1}{2}$) spin; a down ($-\frac{1}{2}$) spin would be described by the same figure simply "flipped" 180° relative to the field direction.

Let's now imagine many water molecules in a magnetic field. As we learned in Eq. 13.3, p. 615, some of the protons of these water molecules have spin = $+\frac{1}{2}$, and others have spin = $-\frac{1}{2}$, and the two spin states are in equilibrium, with the lower-energy (spin = $+\frac{1}{2}$) protons in slight excess. Their M_z components of magnetization sum to a total value $M_{z,eq}$, which is in the "up" direction, because there is an excess of spins in the "up" (lower-energy) state. If this system of water molecules is pulsed briefly with a strong burst of radiofrequency (rf) radiation at the proton resonance frequency, lower-energy "up" proton spins are "flipped" and become higher-energy "down" spins. After the pulse, there is a net decrease in the value of M_z, because there has been an increase in the number of "down" spins. The rf radiation is turned off and the proton spins are allowed to return to equilibrium. **Nuclear relaxation** is nothing more than this return to equilibrium. In returning to equilibrium, the spins in the higher-energy ($-\frac{1}{2}$) state lose energy to their surroundings by various mechanisms that we

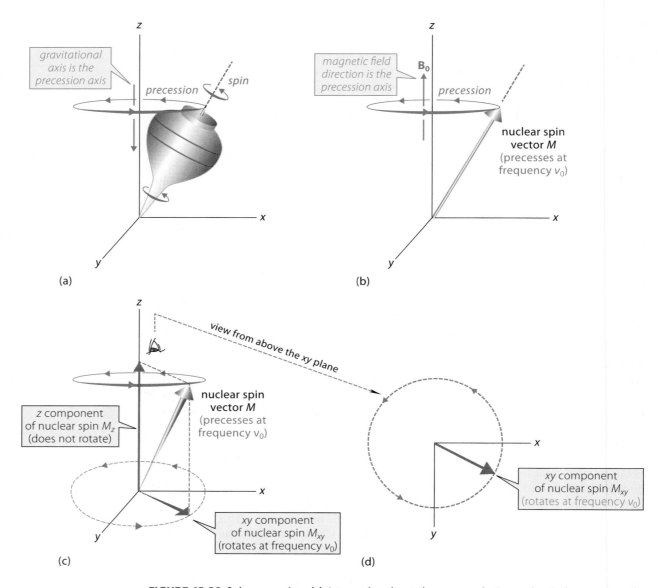

FIGURE 13.28 Spin precession. (a) A top spins about the green axis. Precession is the rotation of the spin axis itself about the z axis, which is the direction of gravity. (b) The precession of nuclear spins occurs at the resonance frequency ν_0 about the axis defined by the direction of the external magnetic field. (c) Resolution of the nuclear spin vector M into perpendicular components: a stationary component M_z along the z axis, and a rotating component M_{xy} in the xy plane. (d) A view of the rotating component M_{xy} from above the xy plane.

don't need to consider here. At this point, however, we need to consider the relaxation of the spin components M_z and M_{xy} separately.

The relaxation of M_z occurs with a characteristic rate that has a rate constant $= 1/T_1$, where the rate constant, like a first-order reaction rate constant, has the dimension of seconds^{-1}. The reciprocal of the rate constant, T_1, then, has the dimension of seconds. The quantity T_1 is called the **longitudinal relaxation time**. (It is sometimes also called the *spin-lattice relaxation time*.) The T_1 is a measure of how long it takes the proton spins of water to return (by "flipping") back to equilibrium. The pulse–relaxation process is shown in the following diagram:

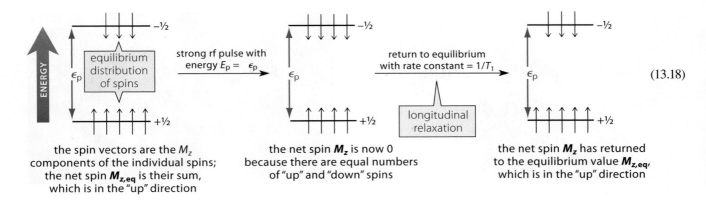

(13.18)

the spin vectors are the M_z components of the individual spins; the net spin $M_{z,eq}$ is their sum, which is in the "up" direction

the net spin M_z is now 0 because there are equal numbers of "up" and "down" spins

the net spin M_z has returned to the equilibrium value $M_{z,eq}$, which is in the "up" direction

(The excess of spins at equilibrium is indicated in this example as 2/8, or 25%, but this is *greatly* exaggerated for illustration; the excess is only a few parts per million.)

Now let's consider the component of spin in the xy plane. Before the rf pulse occurs, the xy components of the spin vectors are randomly oriented in the xy plane. That is, the precession angles of the spins are randomly distributed. Therefore, the sum of all the xy components of many water spin vectors M_{xy} is 0. When the rf pulse occurs, the precession of the nuclear spins is brought into phase coherence. (We won't discuss the physics of this process.) That is, after the rf pulse, the M_{xy} components for the many spins all coincide at the same rotational angle relative to the x and y axes, and their sum M_{xy} is no longer 0. Over time these rotations begin to fall out of phase; that is, the rotation rates change slightly so that the angles of rotation of the different spins become randomized as they were before the rf pulse, and M_{xy} ultimately decays back to 0. This process occurs with a rate constant $= 1/T_2$. The reciprocal of this rate constant, T_2, is called the **transverse relaxation time**. Transverse relaxation is typically faster than longitudinal relaxation.

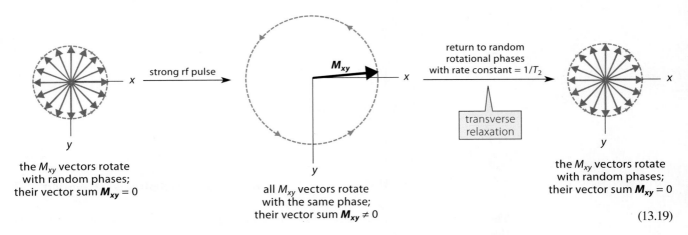

the M_{xy} vectors rotate with random phases; their vector sum $M_{xy} = 0$

all M_{xy} vectors rotate with the same phase; their vector sum $M_{xy} \neq 0$

the M_{xy} vectors rotate with random phases; their vector sum $M_{xy} = 0$

(13.19)

Determination of the MRI image involves the measurement of T_1 and T_2 values in various parts of the body. We don't need to understand exactly how this analysis is done, except that it requires a sequence of rf pulses. Each sequence is repeated many times and the results are averaged. The various rf pulses are actually audible to the patient.

The MRI image (Fig. 13.29, p. 670) is a map of relaxation times in various parts of the body. The central principle of MRI is that *the values of T_1 and T_2 depend on the physiological environment of water.* For example, T_1 for fluids is 1.5–2 s; for water-based tissues, it is 0.4–1.2 s; and for fats, it is 0.1–0.15 s. So, what we're viewing in an MRI image is *not* a "photograph" of, say, a brain, but rather a map of water relaxation rates in various parts the brain. Since these rates vary with location, the MRI image is, in effect, an image of the physiological environment. When a patient undergoes MRI imaging, typically various images are obtained in which T_1 values are weighted in some images and T_2 values in others.

(a)   (b)

FIGURE 13.29 (a) A modern MRI instrument. The patient, lying on the bed, is moved into the cavity of the magnet for determination of the image. (b) T_1-weighted MRI images both with (*left*) and without (*right*) Gd^{3+} contrast. The Gd^{3+} scan has revealed abnormal blood penetration into the brain following a stroke (*yellow arrow*).

Nuclear relaxation involves loss of energy from the nuclear spins to the surroundings. This process is rather inefficient, and the T_1 process is particularly so. However, longitudinal relaxation can be accelerated by the addition of paramagnetic substances—substances with unpaired electrons—which are used as *contrast agents* in MRI. One of the most popular substances used for this purpose is the rare-earth ion gadolinium(III), provided in the form of various coordination compounds such as the following.

Gadopentetic acid (Optimark®, Magnevist®)

This ion has seven unpaired electrons (a record in the periodic table!). As a water molecule (*blue*) from solvent enters the coordination sphere of the Gd^{3+} ion, its nuclear spins can gain or lose energy by interacting with the spins of the Gd^{3+} electrons. Because water molecules interchange rapidly between solvent and the Gd^{3+}, this process in effect enhances water T_1 relaxation—that is, T_1 values are reduced. The magnitude of the effect depends on the Gd^{3+} concentration, but effects of 3- to 8-fold are common in practice with low concentrations of Gd^{3+}. Because the imaging agent is water-soluble, this enhancement is greatest in highly aqueous physiological regions that are penetrated well by the imaging agent. For example, the image in Fig. 13.29b shows the exposure of an aneurism in which blood has leaked into the surrounding brain tissue.

For their research on MRI, Paul Lauterbur (1929–2007) of the University of Illinois (research done at State University of New York, Stony Brook) and Sir Peter Mansfield (b. 1933) of the University of Nottingham shared the 2003 Nobel Prize in Physiology or Medicine.

Other applications of NMR are worth noting. *Solid-state NMR* is being used to study the properties of important solid substances as diverse as drugs, coal, and industrial polymers.

Phosphorus NMR (^{31}P NMR) is being used to study biological processes, in some cases using intact cells or even whole organisms. *Functional MRI* (fMRI) is a variation of MRI in which mental operations can be mapped by observing differential blood flow to various areas of the brain. This is possible because deoxygenated blood has magnetic properties that impart relatively high contrast to MRI images. The reduction in contrast that is produced with a relatively higher level of oxygenated blood can be translated into maps of brain activity associated with such processes as learning or thinking about a loved one. Functional MRI also promises to open new windows on mental illness.

Once a curiosity in a physics laboratory, nuclear magnetic resonance has revolutionized not only chemistry but also human medicine, and the end is not in sight on either frontier.

KEY IDEAS IN CHAPTER 13

- The NMR spectrum records the absorption of energy by nuclei from a radiofrequency (rf) source in the presence of a magnetic field. Absorption results in "spin-flipping"—that is, promotion of nuclear spins to a higher energy level.

- Only nuclei with spin give NMR spectra; protons (^{1}H) and carbon-13 (^{13}C) are two of the nuclei that have spin $\pm\frac{1}{2}$.

- The three elements of a proton NMR spectrum are the chemical shift, which provides information on the chemical environment of the observed protons; the integral, which indicates the relative number of protons being observed; and the splitting, which gives information about the number of protons (or other spin-active nuclei) on adjacent atoms.

- Proton and carbon-13 chemical shifts can be estimated using Fig. 13.4 and Fig. 13.22, respectively.

- Chemically nonequivalent nuclei in principle have different chemical shifts. Constitutionally nonequivalent nuclei and diastereotopic nuclei are chemically nonequivalent.

- The $n + 1$ rule determines the splitting observed in many spectra. When a nucleus is split by more than one chemically nonequivalent set of nuclei, the observed splitting is multiplicative; that is, it is the result of successive applications in any order of the splittings caused by each set.

- Splitting by nuclei of spin $\pm\frac{1}{2}$ such as the proton, results in splitting patterns in which the individual lines ideally have the relative intensities shown in Table 13.2. In practice, splitting patterns show leaning: they deviate from these intensities, and the deviation is greater as the difference in the chemical shifts of the mutually split protons is smaller.

- The magnitude of the splitting between two nuclei is given by the coupling constant *J*. For splitting between vicinal protons, the magnitude of *J* varies with the dihedral angle between the bonds to the coupled protons according to the Karplus relationship (Eq. 13.10, p. 635). For example, the splitting between anti protons is much greater than that between gauche protons. This relationship can be used to determine molecular conformations in some cases. When the dihedral angle between coupled protons varies rapidly because of conformational interconversion, the observed coupling constants are weighted averages.

- Splitting more complex than that predicted by the $n + 1$ rule is observed when the resonances of two coupled protons (in Hz) differ in chemical shift by an amount that is comparable to their mutual coupling constant. Because the chemical shift in Hz increases in proportion to the operating frequency, but coupling constants do not, many compounds that give non-first-order spectra at a lower field give first-order spectra at a higher field.

- Deuterium resonances are not observed in a proton NMR spectrum. Thus, shaking the solution of an alcohol with D_2O (the "D_2O shake") removes the resonance of the OH proton because of its rapid exchange for deuterium.

- A time-averaged NMR spectrum is observed for species involved in rapid equilibria. It is possible to observe absorptions for the individual species by retarding the processes involved in the equilibria (for example, by lowering the temperature).

- The NMR of carbon nuclei in a compound can be observed as a ^{13}C NMR spectrum, despite the low natural abundance of this isotope.

- In a proton-decoupled ^{13}C NMR spectrum, the carbon–proton couplings are removed; each chemically nonequivalent set of carbons appears as a single line. The number of protons attached to each carbon can be determined using the DEPT technique.

• Magnetic resonance imaging (MRI) is a safe, noninvasive imaging technique used in medicine. MRI images rely on differences in the proton relaxation times of water molecules in different physiological environments. Two relaxation times are observed: the longitudinal relaxation time T_1, which measures the rate of return of z magnetization to equilibrium; and the transverse relaxation time T_2, which measures the rate of randomization of nuclear precessional phase angles (xy magnetization). Coordination compounds of the paramagnetic ion Gd^{3+} significantly reduce T_1 values; these compounds are widely used as contrast agents in MRI.

ADDITIONAL PROBLEMS

Note: In these problems, assume that the term NMR refers to *proton* NMR unless otherwise indicated. Assume that all ^{13}C NMR spectra are proton-decoupled.

13.36 What four primary types of information are available from an NMR spectrum? How is each used?

13.37 How would you distinguish among the compounds within each of the following sets using their NMR spectra? Explain carefully and explicitly what features of the NMR spectrum you would use.

(a) cyclohexane and *trans*-2-hexene

(b) *trans*-3-hexene and 1-hexene

(c) 1,1-dichlorohexane, 1,6-dichlorohexane, and 1,2-dichlorohexane

(d) *tert*-butyl methyl ether and isopropyl methyl ether

(e) $Cl_3C—CH_2—CH_2—CHF_2$ and $H_3C—CH_2—CCl_2—CClF_2$

13.38 Answer each of the following questions as briefly as possible.

(a) How does NMR spectroscopy differ conceptually from other forms of absorption spectroscopy?

(b) What happens physically when energy is absorbed by nuclei in an NMR experiment?

(c) How does chemical shift (in frequency units) of a proton change with the size of the field imposed by the NMR instrument?

(d) What is the relationship between coupling constant J and the size of the field imposed by the NMR instrument?

(e) What is the relationship between the coupling constants of vicinal protons and the dihedral angle of their bonds?

(f) Why does the chemical shift in ppm not change with operating frequency?

(g) What condition must be met for an NMR spectrum to be first-order?

13.39 Give the structure of each of the following compounds. (In some cases, more than one correct answer is possible.)

(a) a six-carbon hydrocarbon, not an alkene, whose proton NMR spectrum consists of one singlet

(b) a six-carbon alkene whose proton NMR spectrum consists of one singlet

(c) an eight-carbon ether whose proton NMR spectrum consists of one singlet and whose proton-decoupled ^{13}C NMR spectrum consists of two lines

(d) a nine-carbon hydrocarbon whose proton NMR spectrum consists of two singlets

(e) a seven-carbon hydrocarbon whose proton NMR spectrum consists of two singlets at δ 0.23 and δ 1.21 (relative integral 1 : 6) and whose proton-decoupled ^{13}C NMR spectrum consists of three absorptions

13.40 Give the structure that corresponds to each of the following molecular formulas and NMR spectra:

(a) C_5H_{12}; δ 0.93, s   (b) C_5H_{10}; δ 1.5, s

(c) $C_4H_{10}O_2$: δ 1.36 (3H, d, J = 5.5 Hz); δ 3.32 (6H, s); δ 4.63 (1H, q, J = 5.5 Hz)

(d) $C_7H_{16}O$; NMR spectrum in Fig. P13.40a.

(e) C_8H_{16}: NMR spectrum in Fig. P13.40b. This compound undergoes catalytic hydrogenation to give 2,2,4-trimethylpentane.

(f) $C_7H_{12}Cl_2$: δ 1.07 (9H, s); δ 2.28 (2H, d, J = 6 Hz); δ 5.77 (1H, t, J = 6 Hz).

(g) $C_2H_2Br_2F_2$: δ 4.02 (t, J = 16 Hz)

(h) $C_2H_2F_3I$: δ 3.56 (q, J = 10 Hz)

(i) $C_7H_{16}O_4$: δ 1.93 (t, J = 6 Hz); δ 3.35 (s); δ 4.49 (t, J = 6 Hz); relative integral 1 : 6 : 1

(j) $C_6H_{14}O$: δ 0.91 (6H, d, J = 7 Hz); δ 1.17 (6H, s; disappears following D_2O shake); δ 1.65 (1H, septet, J = 7 Hz) ^{13}C NMR: δ 17.6, δ 26.5, δ 38.7, δ 73.2

13.41 Suppose you wish to carry out the following reactions and you have the NMR spectrum of each starting material. In each case, explain what evidence you would look for in the NMR spectra to verify that the reactions have proceeded as shown.

(a) $(CH_3)_2C{=}CH_2 + HBr \longrightarrow (CH_3)_3CBr$

(b) $(CH_3)_2C{=}C(CH_3)_2 + HCl \longrightarrow$

$$(CH_3)_2CHC(CH_3)_2$$
$$|$$
$$Cl$$

13.42 A compound *A* reacts with H_2 over Pd/C to give methyl-cyclohexane. A colleague, Al Keen, has deduced that the compound must be either 1-methylcyclohexene or 3-methylcyclohexene. You have been called in as a consultant to help Keen decide between these two structures. What evidence would you look for in the proton NMR spectrum to decide between these two possibilities?

13.43 How many absorptions should be observed in the ^{13}C NMR spectrum of each of the following compounds? (Assume that the chair interconversion is rapid.)

(a)

(b)

(a)

(b)

Figure P13.40 NMR spectra for (a) Problem 13.40d and (b) 13.40e. In (b), the violet numbers are coupling constants.

13.44 Explain how the proton NMR spectra of the compounds within each of the following sets would differ, if at all.

(a) $(CH_3)_2CH—Cl$ and $(CH_3)_2CD—Cl$

(b)

$Cl—CD_2CH_2CH_2—Cl$ and $Cl—CH_2CH_2CH_2—Cl$

(c)

$(1R,2R)\text{-}D—\underset{\underset{Cl}{|}}{CH}—\underset{\underset{Cl}{|}}{CH}—CH_3$

$(1S,2R)\text{-}D—\underset{\underset{Cl}{|}}{CH}—\underset{\underset{Cl}{|}}{CH}—CH_3$ $ClCH_2—\underset{\underset{CH_3}{|}}{\overset{\overset{Cl}{|}}{C}}—D$

13.45 Each of four bottles, *A*, *B*, *C*, and *D*, is labeled only "C_6H_{12}" and contains a colorless liquid. You have been called in as an expert to identify these compounds from their spectra:

Compound *A*:

NMR: one line only at δ 1.66 (s); IR: no absorption in the range 1620–1700 cm^{-1}; reacts with Br_2 in CCl_4

Compound *B*:

IR: 3080, 1646, 888 cm^{-1}; NMR spectrum in Fig. P13.45a

Compound *C*:

IR: 3090, 1642, 911, 999 cm^{-1}; NMR spectrum in Fig. P13.45b

Compound *D*:

NMR: one line only at δ 1.40 (s); does not react with Br_2 in CCl_4

Figure P13.45 NMR spectra for Problem 13.45. (a) Spectrum of compound *B*. (b) Spectrum of compound *C*. The violet numbers are coupling constants.

13.46 To which of the following compounds does the NMR spectrum shown in Fig. P13.46 belong? Explain your choice carefully. Once you have made your choice, explain why the resonance at δ 3.7 is so complex.

cis-3-hexene   (Z)-1-ethoxy-1-butene

       A   B

2-ethyl-1-butene $Cl_2CH—CH(OCH_2CH_3)_2$

       C   D

$$Cl—CH—CH—Cl$$
$$CH_3CH_2O \quad\quad OCH_2CH_3$$
$$E$$

13.47 The two protons H^a and H^b in 1,2,3-trichloropropane have slightly different chemical shifts, and the splitting pattern of each is a doublet of doublets. For one proton, $J = 9.0$ Hz and 4.9 Hz; for the other, $J = 9.0$ Hz and 6.0 Hz.

$$H^a \; H^b \quad H^a \; H^b$$
$$Cl \quad\quad\quad\quad Cl$$
$$H \quad Cl$$

1,2,3-trichloropropane

(a) Explain why H^a and H^b have different chemical shifts. (*Hint:* What is the relationship between H^a and H^b?)

(b) Explain why the splitting pattern for each proton is a doublet of doublets.

13.48 The proton NMR spectrum of valine methyl ester hydrochloride is summarized at the top of the next column.

δ 1.16
(doublet, $J = 7.0$ Hz)

δ 1.15
(doublet, $J = 7.0$ Hz) ⟶ $H_3^aC \quad CH_3^b$ O

H^c C—C
δ 2.49 OCH_3 δ 3.83 (singlet)

H^d NH_3^e δ 8.85 (broad singlet)
δ 4.02 Cl^-
doublet ($J = 4.3$ Hz)

valine methyl ester hydrochloride

Note that protons H^e are not split by H^d (and vice versa) because protons H^e are rapidly exchanging. Also, the chemical shifts of protons H^a and H^b could be reversed; the important point is that they are different.

(a) Explain why protons H^a and H^b have different chemical shifts and why the splitting of each is a doublet.

(b) Describe the splitting of proton H^c by constructing a splitting diagram. (*Hint:* This is a case of multiplicative splitting; the coupling constants are shown in the structure.)

13.49 To which of the following compounds does the following ^{13}C NMR-DEPT spectrum belong (attached protons in parentheses):

δ 15.5 (3), δ 20.1 (3), δ 60.7 (2), δ 99.6 (1)

$$CH_3CH_2O—CH—OCH_2CH_3 \quad\quad CH_3OCH_2—CH—CH_2OCH_3$$
$$CH_3 \quad\quad\quad\quad\quad\quad\quad\quad\quad\quad CH_3$$
$$A \quad\quad\quad\quad\quad\quad\quad\quad\quad\quad\quad\quad\quad B$$

$$CH_3CHO—CH_2CH_2—OCHCH_3$$
$$CH_3 \quad\quad\quad\quad\quad\quad\quad\quad CH_3$$
$$C$$

Figure P13.46 The NMR spectrum for Problem 13.46.

13.50 Although this chapter has discussed only nuclei that have spin $\pm\frac{1}{2}$, several common nuclei such as ^{14}N and deuterium (^2H, or D) have a spin of 1. This means that the spin has three equally probable possibilities: +1, 0, and −1.

   (a) How many lines would you expect to observe in the proton NMR of $^+NH_4$? What is the theoretical relative intensity of each line?

   (b) How many lines would you expect to observe in the ^{13}C NMR spectrum of $CDCl_3$? (For the answer, see Fig. 13.23 on p. 659.)

   (c) Although the splitting of protons by deuteriums on *adjacent carbons* is generally negligible, the splitting of protons by deuteriums on the *same* carbon can be significant. Explain how you could tell samples of $H_2CD—I$, $D_2CH—I$, and $D_3C—I$ apart by proton NMR. What other technique could be used for this determination?

13.51 A compound A has a strong, broad IR absorption at 3200–3500 cm^{-1} and the proton NMR spectrum shown in Fig. P13.51a. Treatment of compound A with H_2SO_4 gives compound B, which has the NMR spectrum shown in Fig. P13.51b and a molecular ion at $m/z = 84$ in its EI mass spectrum. Identify compounds A and B.

13.52 Give structures for each of the compounds below.

   (a) Compound A, $C_6H_{14}O_2$; IR: 3200–3600 cm^{-1} (broad, strong), 1090 (strong); NMR spectrum in Fig. P13.52a. (*Note:* This is a very dry sample.)

   (b) Compound B, $C_5H_{10}O$; IR: 3200–3600 cm^{-1} (broad, strong), 3085 cm^{-1} (sharp), 1658 cm^{-1} (sharp), 1055 cm^{-1}, 875 cm^{-1}; NMR spectrum in Fig. P13.52b.

13.53 Identify the compound A ($C_5H_{10}O$) with the proton NMR spectrum shown in Fig. P13.53 on p. 678. Compound A

Figure P13.51 NMR spectra for Problem 13.51. (a) Spectrum of compound A. (b) Spectrum of compound B. In the spectrum of compound B, the expansion of the resonance at δ 5.1 is on a larger horizontal scale than the expansions of the other resonances. The violet numbers in spectrum (b) are coupling constants.

has IR absorptions at 3200–3600 cm⁻¹ (strong, broad), 1676 cm⁻¹ (weak), and 965 cm⁻¹, and also has ^{13}C NMR absorptions (attached protons in parentheses) at δ 17.5 (3), δ 23.3 (3), δ 68.8 (1), δ 125.5 (1), and δ 135.5 (1). Compound *A* is optically inactive, but it can be resolved into enantiomers.

13.54 Propose a structure for the compound that has the following spectra:

NMR: δ 1.28 (3*H*, t, *J* = 7 Hz); δ 3.91 (2*H*, q, *J* = 7 Hz); δ 5.0 (1*H*, d, *J* = 4 Hz); δ 6.49 (1*H*, d, *J* = 4 Hz)

IR: 3100, 1644 (strong), 1104, 1166, 694 cm⁻¹ (strong); no IR absorptions in the range 700–1100 cm⁻¹ or above 3100 cm⁻¹

Mass spectrum: *m/z* = 152, 150 (equal intensity; double molecular ion)

13.55 You work for a reputable chemical supply house. An angry customer, Fly Ofterhandle, has called, alleging that a sample of 2,5-hexanediol he purchased cannot be the correct compound. As evidence, he cites its ^{13}C NMR spectrum:

δ 23.2, δ 23.5, δ 35.1, δ 35.8, δ 67.4, δ 67.8

(Notice that the spectrum contains three sets of two closely spaced absorptions.) After verifying the ^{13}C NMR spectrum, you can confidently assure him that the sample he purchased contains *only* 2,5-hexanediol. Explain why the ^{13}C NMR spectrum is consistent with this claim.

Figure P13.52 NMR spectra for Problem 13.52.

13.56 (a) How many different sets of proton absorptions (ignoring splitting) should be observed in the proton NMR spectrum of 4-methyl-1-penten-3-ol?

(b) How many absorptions should there be in the ^{13}C NMR spectrum?

13.57 (a) How would the proton NMR spectrum of very dry 2-chloroethanol differ from that of the same compound containing a trace of aqueous acid? Explain your answer.

(b) How would the proton NMR spectrum of the compound in part (a) change following a D_2O shake?

13.58 The NMR spectrum of vitamin D_3 is given in Fig. P13.58.

Interpret the resonances marked with asterisks (*) by indicating the part of the structure to which they correspond. (Do not try to assign the individual resonances within the groups.) Explain your choices.

13.59 A compound X with the molecular formula $C_5H_{10}O_2$ has an IR spectrum with strong absorption in the

1000–1100 cm^{-1} region; very strong, broad absorption in the 3000–3600 cm^{-1} region; and no absorption in the 1600–1700 cm^{-1} region. The proton NMR spectrum of X is given in Fig. P13.59. When the sample is shaken with D_2O, the triplet at δ 3.5 disappears and the doublet at δ 3.7 becomes a singlet. Propose a structure for this compound, and explain your reasoning carefully.

13.60 A reaction has yielded a mixture of the following two stereoisomers. Explain how you would use proton NMR to determine which is which.

13.61 ^{17}O is a rare isotope that has a nuclear spin. The ^{17}O NMR of a small amount of water dissolved in CCl_4 is a triplet (intensity ratio 1 : 2 : 1). When water is dissolved in the strongly acidic HF-SbF$_5$ solvent, its ^{17}O NMR becomes a 1 : 3 : 3 : 1 quartet. Suggest a reason for these observations. (*Hint:* Think of this solvent as H$^+$ SbF$_6^-$.)

13.62 Imagine taking the NMR spectrum of a sample of "naked" protons—that is, H$^+$ in the gas phase not chemically bound to anything. In which of the following ranges of chemical shifts would you expect to find the resonance for these protons, and why?

$\delta > 8$ $8 > \delta > 0$ $\delta < 0$ (that is, negative δ)

Figure P13.53 NMR spectrum for Problem 13.53. The integral is shown above each resonance in red as the actual number of hydrogens, and the violet numbers are coupling constants.

13.63 Carbon–carbon splitting is not apparent in natural-abundance ^{13}C NMR spectra because of the rarity of the ^{13}C isotope. However, it can be observed in compounds that are enriched in ^{13}C. A chemist, Buster Magnet, has just completed a synthesis of CH_3CH_2Br that contains 50% ^{13}C at each position. (What this really means is that some of the molecules contain no ^{13}C, some contain ^{13}C at one position, and some contain ^{13}C at both positions.) Buster does not know what to expect for the spectrum of this compound and has come to you for assistance. Describe the proton-decoupled ^{13}C NMR spectrum of this compound. (*Hint:* List all of the species present and decide on their relative amounts. To determine their relative amounts, remember that the probability that two events will occur simultaneously is the product of their individual probabilities. The spectrum of a mixture shows peaks for each compound in the mixture.)

Figure P13.58 The NMR spectrum of vitamin D_3. The violet numbers are coupling constants. The expansion of the δ 6.0–6.4 region is on a different scale than the other expansion.

Figure P13.59 The proton NMR spectrum for Problem 13.59. The integrals (*red*) are the actual numbers of hydrogens, and the violet numbers are coupling constants.

13.64 (a) Because electrons have spin, they can also undergo magnetic resonance. *Electron spin resonance spectroscopy* (ESR spectroscopy) is used to study the magnetic resonance of unpaired electrons in free radicals. (ESR spectroscopy is to unpaired electrons what NMR spectroscopy is to protons.) Explain why the ESR spectrum of the unpaired electron in the methyl radical, $\cdot CH_3$, is a quartet of four lines in a $1 : 3 : 3 : 1$ ratio.

(b) The gyromagnetic ratio of the electron is 17.60×10^6 rad gauss^{-1} s^{-1}, 658 times greater than that of the proton. What operating frequency would be required to detect the magnetic resonance of an unpaired electron in a magnetic field of 3400 gauss (a common field used in ESR spectrometers)? In what region of the electromagnetic spectrum does this frequency lie? (Consult Fig. 12.2 on p. 572.)

13.65 The 60-MHz proton NMR spectrum of 2,2,3,3-tetrachlorobutane consists of a sharp singlet at 25 °C, but at −45 °C consists of two singlets of different intensities separated by about 10 Hz.

(a) Explain the changes in the spectrum as a function of temperature.

(b) Explain why the two lines observed at low temperature have different intensities.

13.66 The NMR spectrum of iodocyclohexane was taken at −80 °C. At that temperature, the chair interconversion is slow and each chair conformation can be observed separately.

The resonance of proton H^a is well resolved from the rest of the spectrum because of its unique chemical shift. The resonance of proton H^a in *one* of the conformations at −80 °C is shown in Fig. P13.66.

(a) Which conformation has the H^a resonance shown in Fig. P13.66? Explain.

(b) How would you expect the resonance of H^a in the other conformation (not shown) to differ from the one shown in Fig. P13.66?

(c) The integrals of the H^a resonance in the two conformations at −80 °C are in the ratio 3.39 : 1. What is the ΔG for the chair interconversion of the two conformations? From what you know about conformational analysis, which conformation is the major one?

13.67 What changes would you expect in the ^{13}C NMR spectrum of 1-bromopropane upon cooling the compound to very low temperature?

Figure P13.66 The proton NMR resonance of H^a in one of the two chair conformations of iodocyclohexane at −80 °C.

The Chemistry of Alkynes

An **alkyne** is a hydrocarbon containing a carbon–carbon triple bond; the simplest member of this family is *acetylene*, H—C≡C—H. The chemistry of the carbon–carbon triple bond is similar in many respects to that of the carbon–carbon double bond; indeed, alkynes and alkenes undergo many of the same addition reactions. But alkynes also have some unique chemistry, most of it associated with the bond between hydrogen and the triply bonded carbon, the ≡C—H bond. Alkynes are also useful in the preparation of other types of organic compounds; in this chapter, we'll also explore the synthetic utility of alkynes.

14.1 STRUCTURE AND BONDING IN ALKYNES

Because each carbon of acetylene is connected to two groups—a hydrogen and another carbon—the H—C≡C bond angle in acetylene is 180° (Sec. 1.3B); thus, the acetylene molecule is *linear.*

$$\text{H}-\overset{\underset{180°}{}}{\text{C}}\equiv\overset{\underset{}{}}{\text{C}}-\text{H}$$

The C≡C bond, with a bond length of 1.20 Å, is shorter than the C=C and C—C bonds, which have bond lengths of 1.33 Å and 1.54 Å, respectively.

Because of the 180° bond angles at the carbon–carbon triple bond, cis–trans isomerism cannot occur in alkynes. Thus, although 2-butene exists as cis and trans stereoisomers, 2-butyne does not. Another consequence of this linear geometry is that cycloalkynes smaller than cyclooctyne cannot be isolated under ordinary conditions (see Problem 14.1).

The hybrid-orbital model for bonding provides a useful description of bonding in alkynes. We learned in Sec. 4.1A (pp. 127–128) that carbon hybridization and geometry are

FIGURE 14.1 Orbitals of an *sp*-hybridized carbon are derived conceptually by mixing one 2s orbital and one 2p orbital, shown in red. Two *sp* hybrid orbitals are formed (*red*) and two 2p orbitals remain unhybridized (*blue*).

correlated: tetrahedral carbon is sp^3-hybridized, and trigonal planar carbon is sp^2-hybridized. The linear geometry found in alkynes is characterized by a third type of carbon hybridization, called *sp hybridization*, diagrammed in Fig. 14.1. Imagine that the 2s orbital and *one* 2p orbital (say, the $2p_x$ orbital) on carbon mix to form two new hybrid orbitals. Because these two new orbitals are each one part s and one part p, they are called *sp hybrid orbitals*. Two of the 2p orbitals ($2p_y$ and $2p_z$) are not included in the hybridization. An ***sp* hybrid orbital**, then, is an orbital derived from the mixing of one s orbital and a p orbital of the same principal quantum number.

An *sp* orbital has much the same shape as an sp^2 or sp^3 orbital (Fig. 14.2; compare with Figs. 1.16a and 4.4a). However, electrons in an *sp* hybrid orbital are, on the average, somewhat closer to the carbon nucleus than they are in sp^2 or sp^3 hybrid orbitals. In other words, *sp* orbitals are more compact than sp^2 or sp^3 hybrid orbitals. The reason is that an *sp* orbital contains a greater fraction of s character than an sp^2 or an sp^3 orbital, and 2s electrons are, on the average, closer to the nucleus than 2p electrons. An *sp*-hybridized carbon atom, shown in Fig. 14.2c, has two *sp* orbitals at a relative orientation of 180°. The two remaining unhybridized 2p orbitals lie along axes that are at right angles both to each other and to the *sp* orbitals.

The σ bonds in acetylene result from the combination of two *sp*-hybridized carbon atoms and two hydrogen atoms (Fig. 14.3). One bond between the carbon atoms is a σ bond resulting from the overlap of two *sp* hybrid orbitals, each containing one electron. This bond is an *sp–sp* σ *bond*. The remaining *sp* orbital on each carbon overlaps with a hydrogen 1s orbital to form a carbon–hydrogen σ bond. These bonds are *sp–1s* σ *bonds*. Because electron density in an *sp* hybrid orbital is closer to the nucleus than electron density in other hybrid orbitals, the C—H bond in acetylene is shorter (1.06 Å) than the C—H bonds in ethylene

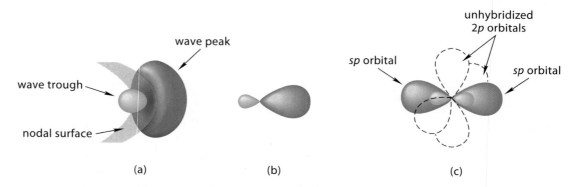

FIGURE 14.2 (a) A perspective representation of an *sp* hybrid orbital. (b) A more common representation of an *sp* hybrid orbital used in drawings. (c) The two *sp* hybrid orbitals shown together. The "leftover" (unhybridized) 2p orbitals are shown with dashed lines. Notice that the *sp* hybrid orbitals are oriented at 180°. The blue and green colors represent wave peaks and wave troughs.

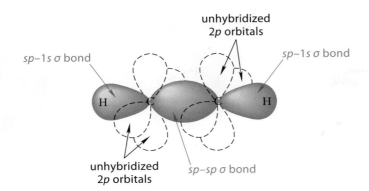

FIGURE 14.3 The σ-bond framework of acetylene (shown in blue). Overlap of carbon sp hybrid orbitals gives the carbon–carbon σ bond, and the overlap of carbon sp hybrid orbitals with hydrogen $1s$ orbitals gives the carbon–hydrogen σ bonds. Two $2p$ orbitals on each carbon, shown as dashed lines, do not participate in σ bonding. (See Fig. 14.4, p.684.)

(1.08 Å) and ethane (1.11 Å). Table 5.3 (p. 216) shows that the C—H bond in acetylene, with a bond dissociation energy of 558 kJ mol^{-1} (133 kcal mol^{-1}), is also stronger than the C—H bonds of ethylene (463 kJ mol^{-1}, 111 kcal mol^{-1}) or ethane (423 kJ mol^{-1}, 101 kcal mol^{-1}). This bond-strength effect occurs because the C—H bond in acetylene contains a greater percentage of the lower-energy $2s$ orbital than the bonds derived from sp^2 or sp^3 hybrid orbitals, which, in contrast, contain progressively more high-energy $2p$ character. Notice that the linear geometry of acetylene results from the 180° orientation of the sp orbitals on each carbon. Again: *hybridization and geometry are correlated.*

The leftover $2p$ orbitals on each carbon overlap to form π bonds. Because each carbon of acetylene has *two* $2p$ orbitals, *two* π bonds are formed. Like the $2p$ orbitals from which they are formed, they are mutually perpendicular. The two bonding π molecular orbitals that result from this overlap are shown in Fig. 14.4 on p. 684. Notice that the acetylene molecule is literally surrounded by π electrons. The total electron density from all of the π electrons taken together forms a cylinder, or barrel, about the axis of the molecule (Fig. 14.4c). This cylinder of π-electron density is particularly evident in the electrostatic potential map (EPM) of acetylene. Compare this with the EPM of ethylene, which has π-electron density above and below the plane of the molecule.

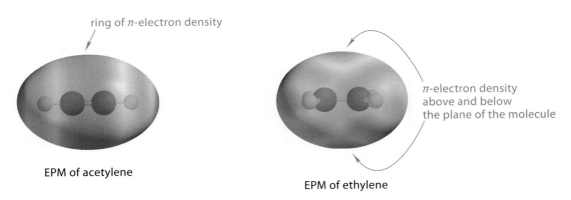

ring of π-electron density

π-electron density above and below the plane of the molecule

EPM of acetylene

EPM of ethylene

The following heats of formation show that alkynes are less stable than isomeric dienes:

H—C≡C—CH₂CH₂CH₃   H₃C—C≡C—CH₂CH₃   H₂C=CH—CH₂—CH=CH₂

	1-pentyne	**2-pentyne**	**1,4-pentadiene**
ΔH_f°	+144 kJ mol^{-1}	+129 kJ mol^{-1}	+106 kJ mol^{-1}
	(34.5 kcal mol^{-1})	(30.8 kcal mol^{-1})	(25.4 kcal mol^{-1})

In other words, the sp hybridization state is inherently less stable than the sp^2 hybridization state, other things being equal. These heats of formation also show that a triple bond, like a double bond, is more stable in the interior of a carbon chain than at the end.

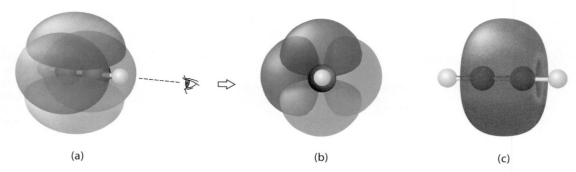

(a)   (b)   (c)

FIGURE 14.4 The two bonding π molecular orbitals in acetylene. (a) A perspective view. (b) An "end-on" view as indicated by the eyeball in (a). The blue and green colors in (a) and (b) represent wave peaks and wave troughs. (c) The total π-electron density in acetylene. Acetylene is completely surrounded by π electrons.

PROBLEM

14.1 (a) Attempt to build a model of cyclohexyne. Explain why this compound is unstable.

(b) Build a model of cyclodecyne. Compare its stability qualitatively to that of cyclohexyne; explain your answer.

14.2 NOMENCLATURE OF ALKYNES

In common nomenclature, simple alkynes are named as derivatives of the parent compound acetylene:

$$H_3C—C\equiv C—H \qquad H_3C—C\equiv C—CH_3 \qquad CH_3CH_2—C\equiv C—CH_3$$

methylacetylene   **dimethylacetylene**   **ethylmethylacetylene**

Certain compounds are named as derivatives of the **propargyl group**, $HC\equiv C—CH_2—$, in the common system. The propargyl group is the triple-bond analog of the allyl group.

$$HC\equiv C—CH_2—Cl \qquad H_2C\equiv CH—CH_2—Cl$$

propargyl chloride   **allyl chloride**

As we might expect, the substitutive nomenclature of alkynes is much like that of alkenes. The suffix *ane* in the name of the corresponding alkane is replaced by the suffix *yne*, and the triple bond is given the lowest possible number.

$$H_3C—C\equiv C—H \qquad CH_3CH_2CH_2CH_2—C\equiv C—CH_3 \qquad H_3C—CH_2—C\equiv C—H$$

propyne   **2-heptyne**   **1-butyne**

$$H_3C—\underset{\underset{CH_3}{|}}{CH}—C\equiv C—CH_3 \qquad HC\equiv C—CH_2—CH_2—C\equiv C—CH_3$$

                                   1,5-heptadiyne

4-methyl-2-pentyne

Substituent groups that contain a triple bond (called *alkynyl groups*) are named by replacing the final *e* in the name of the corresponding alkyne with the suffix *yl*. (This is exactly analogous to the nomenclature of substituent groups containing double bonds; see Sec. 4.2A.) The alkynyl group is numbered from its point of attachment to the main chain:

$HC\equiv C-$

ethynyl group
(ethyne + yl)

$HC\equiv C-CH_2-$

2-propynyl group

3-(2-**propynyl)cyclohexanol**

position of the triple bond within the substituent

position of the 2-propynyl group on the ring

As with alkenes, groups that can be cited as principal groups, such as the —OH group in the following example (as well as in the previous one), are given numerical precedence over the triple bond. (See Appendix I for a summary of nomenclature rules.)

$$HC\underset{5}{\equiv}C\underset{4}{-}CH_2\underset{3}{-}\underset{2}{CH}-\underset{1}{CH_3}$$

4-pentyn-2-ol

the OH group receives numerical priority

When a molecule contains both double and triple bonds, the bond that has the lower number at first point of difference receives numerical precedence. However, if this rule is ambiguous, a double bond receives numerical precedence over a triple bond.

$$\underset{1}{HC}\equiv\underset{2}{C}-\underset{3}{CH}=\underset{4}{CH}\underset{5}{CH_3}$$

3-penten-1-yne

$$\underset{5}{CH_3}\underset{4}{C}\equiv\underset{3}{C}-\underset{2}{CH}=\underset{1}{CH_2}$$

1-penten-3-yne

precedence is given to the bond that has lower number at first point of difference

$$\underset{1}{H_2C}=\underset{2}{CH}\underset{3}{CH_2}\underset{4}{C}\equiv\underset{5}{CH}$$

1-penten-4-yne

precedence is given to the double bond when numbering is ambiguous

PROBLEMS

14.2 Draw a Lewis structure for each of the following alkynes.
   (a) isopropylacetylene   (b) cyclononyne   (c) 4-methyl-1-pentyne
   (d) 1-ethynylcyclohexanol   (e) 2-butoxy-3-heptyne   (f) 1,3-hexadiyne

14.3 Provide the substitutive name for each of the following compounds. Also provide common names for (a) and (b).
   (a) $CH_3CH_2CH_2CH_2C\equiv CH$ (b) $CH_3CH_2CH_2CH_2C\equiv CCH_2CH_2CH_2CH_3$
   (c)

   $$H_3C-\underset{\underset{CH_3}{|}}{\overset{\overset{OH}{|}}{C}}-C\equiv C-CH_3$$

   (d) $HC\equiv CCHCH_2CH_2CH_3$

   (e)

   $$HC\equiv C-\overset{\overset{OH}{|}}{CH}-CH=CH_2$$

14.3   PHYSICAL PROPERTIES OF ALKYNES

A. Boiling Points and Solubilities

The boiling points of most alkynes are not very different from those of analogous alkenes and alkanes:

$$HC{\equiv}C(CH_2)_3CH_3 \qquad H_2C{=}CH(CH_2)_3CH_3 \qquad H_3C{-}CH_2(CH_2)_3CH_3$$

	1-hexyne	**1-hexene**	**hexane**
boiling point:	71.3 °C	63.4 °C	68.7 °C
density:	0.7155 g mL^{-1}	0.6731 g mL^{-1}	0.6603 g mL^{-1}

Like alkanes and alkenes, alkynes have much lower densities than water and are also insoluble in water.

B. IR Spectroscopy of Alkynes

Many alkynes have a $C{\equiv}C$ stretching absorption in the 2100–2200 cm^{-1} region of the infrared spectrum. This absorption is clearly evident, for example, at 2120 cm^{-1} in the IR spectrum of 1-octyne (Fig. 14.5). However, this absorption is very weak or absent in the IR spectra of many symmetrical, or nearly symmetrical, alkynes because of the dipole moment effect (Sec. 12.3B). For example, 4-octyne has no $C{\equiv}C$ stretching absorption at all.

The $C{\equiv}C$ stretching absorption (2120 cm^{-1}) lies at considerably higher frequency than the $C{=}C$ stretching frequency (1640–1675 cm^{-1}). This is a clear manifestation of the bond-strength effect on absorption frequency. (See Sec. 12.3A and Study Problem 12.1, p. 580.)

A very useful absorption of 1-alkynes is the ${\equiv}C{-}H$ stretching absorption, which occurs at about 3300 cm^{-1}. This strong, sharp absorption, very prominent in the spectrum of 1octyne (Fig. 14.5), is well separated from other $C{-}H$ stretching absorptions. Because alkynes other than 1-alkynes lack the unique ${\equiv}C{-}H$ bond, they do not show this absorption.

C. NMR Spectroscopy of Alkynes

Proton NMR Spectroscopy   Compare the typical chemical shifts observed in the proton NMR spectra of alkynes with the analogous shifts for alkenes:

Although the chemical shifts of allylic and propargylic protons are very similar (as might be expected from the fact that both double and triple bonds involve π electrons), the chemical shifts of acetylenic protons are much smaller than those of vinylic protons.

The explanation for the unusual proton chemical shifts observed in alkynes is closely related to the explanation for the chemical shifts of vinylic protons (Fig. 13.15, p. 646), although *the effect is in the opposite direction*. An alkyne molecule in solution is tumbling rapidly, but alkyne chemical shifts are dominated by the effects resulting from one particular orientation of the alkyne molecule relative to the magnetic field, as shown in Fig. 14.6. When an alkyne molecule is oriented in the applied field $\mathbf{B_0}$ as shown in this figure, an induced electron circulation is set up in the cylinder of π electrons (Fig. 14.4c) that encircles the molecule. The resulting induced field $\mathbf{B_i}$ *opposes* the applied field along the axis of this cylinder. Because the acetylenic proton lies along this axis, the local field at this proton is reduced. Consequently, by Eq. 13.4, p. 615, acetylenic protons have NMR absorptions at smaller chemical shift than they would have in the absence of this effect.

FIGURE 14.5 The IR spectrum of 1-octyne. The two key absorptions indicated are absent in the spectrum of 4-octyne.

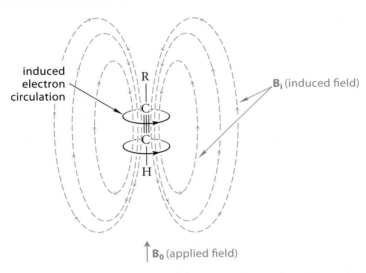

FIGURE 14.6 Explanation of the chemical shift of acetylenic protons. The induced field $\mathbf{B_i}$ of the circulating π electrons (*red*) opposes the applied field $\mathbf{B_0}$ (*blue*) from the spectrometer in the region of space occupied by acetylenic protons. As a result, the local field at an acetylenic proton is reduced. Hence, acetylenic protons have NMR absorptions at relatively small chemical shift. The same effect accounts for the chemical shifts of acetylenic and propargylic carbons in the ^{13}C NMR spectra of alkynes.

Carbon-13 NMR Spectroscopy   Chemical shifts of alkynes in ^{13}C NMR are subject to the same influences as proton chemical shifts. Although carbons involved in double bonds have chemical shifts in the δ 100–145 range, carbons involved in triple bonds absorb at considerably lower chemical shift, in the δ 65–85 range. Propargylic carbons, like acetylenic hydrogens, also have smaller chemical shifts, typically by 5–15 ppm. The chemical shifts in 2-heptyne are typical:

propargylic carbons

δ 3.3   75.2   79.2   18.7   31.7   22.3   13.8

$$H_3C—C\equiv C—CH_2—CH_2—CH_2—CH_3$$

acetylenic carbons   **2-heptyne**

Compare, for example, the chemical shift of the propargylic methyl carbon (δ 3.3) with that of the other methyl carbon (δ 13.8), which is much like that of an alkane methyl group.

FIGURE 14.7 The IR spectrum for Problem 14.4.

The explanation for these chemical-shift effects is the same one (Fig. 14.6) discussed for the proton chemical shifts in alkynes.

PROBLEMS

14.4 Identify the compound with a molecular mass of 82 that has the IR spectrum shown in Fig. 14.7 and the following NMR spectrum: δ 1.90 (1H, s); δ 1.21 (9H, s)

14.5 (a) Match each of the following ^{13}C NMR spectra to either 2-hexyne or 3-hexyne. Explain.

Spectrum A: δ 3.3, 13.6, 21.1, 22.9, 75.4, 79.1

Spectrum B: δ 12.7, 14.6, 81.0

(b) Assign each of the resonances in the two spectra to the appropriate carbon atoms.

14.6 A student consulted a well-known compilation of reference spectra for the proton NMR spectrum of propyne and was surprised to find that this spectrum consists of a single unsplit resonance at δ 1.8. Believing this to be an error, he comes to you for an explanation. Explain to him why it is reasonable that propyne could have this spectrum.

14.4 INTRODUCTION TO ADDITION REACTIONS OF THE TRIPLE BOND

In Chapters 4 and 5 we learned that the most common reactions of alkenes involve additions to the double bond. Additions to the triple bond also occur, although in most cases they are somewhat slower than the same reactions of comparably substituted alkenes. For example, HBr can be added to the triple bond.

$$CH_3(CH_2)_3C\equiv CH + HBr \xrightarrow[CH_2Cl_2]{(Et)_4N^+ Br^-} CH_3(CH_2)_3C=CH_2 \qquad (14.1)$$

1-hexyne

2-bromo-1-hexene
(89% yield)
Br

The regioselectivity of the addition is analogous to that found in the addition of HBr to alkenes (Sec. 4.7A): the bromine adds to the carbon of the triple bond that bears the alkyl substituent. As in alkene additions, the regioselectivity is reversed in the presence of peroxides because free-radical intermediates are involved (Sec. 5.6).

$$CH_3(CH_2)_3C{\equiv}CH + HBr \xrightarrow[\text{0–5 °C, 1 h}]{\text{peroxides}} CH_3(CH_2)_3CH{=}CHBr \qquad (14.2)$$

1-hexyne   **1-bromo-1-hexene;**
stereochemistry not determined
(74% yield)

Because addition to an alkyne gives a substituted alkene, a second addition can occur in many cases.

$$H_3C{-}C{\equiv}C{-}CH_3 + HBr \longrightarrow H_3C{-}\underset{}{\overset{Br}{C}}{=}CH{-}CH_3 \xrightarrow{HBr} H_3C{-}\underset{Br}{\overset{Br}{C}}{-}CH_2CH_3 \qquad (14.3a)$$

2-butyne   (excess)   (not isolated)

2,2-dibromobutane
(60% yield)

The regioselectivity of this addition reaction is determined by the relative stabilities of the two possible carbocation intermediates. One of the two possible carbocations (*A* in the following equation) is stabilized by resonance. By Hammond's postulate (Sec. 4.8D), this carbocation is formed more rapidly.

resonance-stabilized carbocation *A*   less stable carbocation *B*

observed product   not formed

    In the addition of a hydrogen halide or a halogen to an alkyne, the second addition is usually slower than the first. The reason is that the halogen that enters the molecule in the first addition exerts a rate-retarding polar effect (Sec. 3.6C) on carbocation formation in the second addition. In other words, both carbocations *A* and *B* in Eq. 14.3b are destabilized by the polar effect of bromine, and this polar effect is only partially counterbalanced by the resonance stabilization in carbocation *A*. Because the second addition is slower, it is possible to isolate the product of the first addition if one equivalent of HBr is used, as in Eq. 14.1.

PROBLEMS

14.7   Give the product that results from the addition of one equivalent of Br$_2$ to 3-hexyne. What are the possible stereoisomers that could be formed?

14.8   The addition of HCl to 3-hexyne occurs as an anti-addition. Give the structure, stereochemistry, and name of the product.

14.5   CONVERSION OF ALKYNES INTO ALDEHYDES AND KETONES

A. Hydration of Alkynes

Water can be added to the triple bond. Although the reaction can be catalyzed by a strong acid, it is faster, and yields are higher, when a combination of dilute acid and mercuric ion (Hg^{2+}) catalysts is used.

$$\text{cyclohexyl—C} \equiv \text{CH} + H_2O \xrightarrow{\ Hg^{2+},\ H_2SO_4\ (\text{dilute})\ } \text{cyclohexyl—}\overset{\displaystyle O}{\overset{\|}{C}}\text{—CH}_3 \qquad (14.4)$$

cyclohexylacetylene   **cyclohexyl methyl ketone**
                                              (91% yield)

The addition of water to a triple bond, like the corresponding addition to a double bond, is called **hydration**. The hydration of alkynes gives ketones (except in the case of acetylene itself, which gives an aldehyde; see Study Problem 14.1, p. 692). Unlike alkene hydration, *alkyne hydration is irreversible.* (We'll see why shortly.)

Although the Hg^{2+}-catalyzed process is more useful synthetically, let's focus mechanistically on the acid-catalyzed hydration so that we can contrast this process with the hydration of alkenes (Sec. 4.9B). The hydration of alkynes and alkenes begins with exactly the same first step: protonation of a carbon–carbon π bond to give a carbocation:

$$R\text{—C} \equiv \text{CH} \quad H\text{—}\overset{+}{\overset{..}{O}}H_2 \longrightarrow R\text{—}\overset{+}{C}\text{=}CH_2 + :\overset{..}{O}H_2 \qquad (14.5a)$$

a vinylic cation

This type of carbocation is called a **vinylic cation**. In a vinylic cation, the electron-deficient carbon is sp-hybridized, and it contains two $2p$ orbitals (Fig. 14.8). One $2p$ orbital is empty, and the other is used to form the π bond to the other carbon of the double bond. Notice that we form the carbocation at the carbon with the alkyl branch (—R); why?

Like other carbocations, a vinylic cation can undergo a Lewis acid–base association reaction with a nucleophile, which in this case is water. A subsequent Brønsted acid–base reaction produces the product, which is a vinylic alcohol, or *enol.*

$$R\text{—}\overset{+}{C}\text{=}CH_2 \quad :\overset{..}{O}H_2 \longrightarrow R\text{—}\underset{\overset{|}{\overset{+}{O}}\text{—}H}{\overset{\overset{\displaystyle H}{\overset{|}{}}}{C}}\text{=}CH_2 \quad \underset{\overset{\longleftarrow}{\longrightarrow}}{:\overset{..}{O}H_2} \quad R\text{—}\underset{}{\overset{:\overset{..}{O}H}{\overset{|}{C}}}\text{=}CH_2 + H_3\overset{+}{\overset{..}{O}} \qquad (14.5b)$$

enol

An **enol** (pronounced ēn´-ôl) is a special type of alcohol in which the OH group is on a carbon of a double bond. *Most enols cannot be isolated because they are unstable and are rapidly converted into the corresponding aldehydes or ketones.*

$$R\text{—}\underset{}{\overset{OH}{\overset{|}{C}}}\text{=}CH_2 \quad \underset{\overset{\longleftarrow}{\longrightarrow}}{} \quad R\text{—}\overset{\displaystyle O}{\overset{\|}{C}}\text{—CH}_3 \qquad (14.5c)$$

an enol   a ketone

Eq. 14.5c shows why alkynes give ketones as products: the enol that is formed initially is rapidly converted into a ketone.

$$H_2C{=}\overset{+}{C}{-}R$$

(a) Lewis structure

2p orbitals of the double bond

H

empty 2p orbital

sp-hybridized carbon

(b) 2p orbitals

FIGURE 14.8 Vinylic cations. (a) Lewis structure. (b) The arrangement of 2p orbitals. On the sp-hybridized carbon atom, one 2p orbital (*blue*) is empty. The other is involved in π bond formation with the 2p orbital on the adjacent carbon.

Equation 14.5c also shows why alkyne hydration, unlike alkene hydration, is not reversible. The equilibrium between an aldehyde or ketone and its enol isomers *strongly* favors the aldehyde or ketone. The equilibrium concentrations of enols are in most cases minuscule— typically, one part in 10^8 or less. (The relationship between aldehydes, ketones, and their enol isomers is explored in more detail in Chapter 22.) The important point here is that, because the equilibrium in Eq. 14.5c lies so far to the right, the hydration of alkynes is effectively irreversible.

What makes enols so unstable? Actually, nothing is particularly unfavorable about the structures of enols. They are unstable only because they can be rapidly converted into *even more stable* ketone isomers. (Ketones are more stable than enols largely because the C=O bond is considerably stronger than the C=C bond.)

Let's see how the acid-catalyzed conversion of enols into ketones takes place. We've already learned that alkynes sometimes undergo two addition reactions because the first addition product contains a double bond, which can itself undergo an addition. We should not be surprised, then, that the enol reacts further, because it is an alkene "at heart" and its π bond can be protonated. The carbocation produced by protonation of the enol is *resonance-stabilized*:

$$R{-}\underset{:\ddot{O}H}{C}{=}CH_2 \quad H{-}\overset{+}{\ddot{O}}H_2 \quad \rightleftharpoons \quad \left[R{-}\underset{+}{\overset{:\ddot{O}H}{C}}{-}CH_3 \quad \longleftrightarrow \quad R{-}\overset{+\ddot{O}H}{C}{-}CH_3 \right] + :\ddot{O}H_2 \quad (14.6a)$$

a resonance-stabilized carbocation

The resonance structure on the right shows that this carbocation is the conjugate acid of a ketone. Loss of a proton gives the ketone product.

$$R{-}\overset{\overset{+}{:}O}{C}{-}CH_3 \quad H \quad :OH_2 \quad \rightleftharpoons \quad R{-}\overset{:\ddot{O}}{C}{-}CH_3 + H{-}\overset{+}{O}H_2 \quad (14.6b)$$

In summary, then, alkynes undergo acid-catalyzed hydration to give enols, which are rapidly and irreversibly transformed into ketones under the reaction conditions.

The Hg^{2+}-catalyzed hydration of alkynes is a variation of oxymercuration (Sec. 5.4A, p. 190). As in alkene oxymercuration, carbocation rearrangements do not occur. The absence of rearrangements is a desirable aspect of Hg^{2+}-catalyzed hydration for its use in organic synthesis. A difference in the two reactions is that, in the case of alkenes, a reducing agent is required to remove the mercury (p. 191), and a full equivalent of Hg^{2+} is required. However, in the hydration of alkynes, the mercuric ion is a catalyst; it is regenerated by a reaction of the oxymercuration addition product with the acid co-catalyst.

$$R-C{\equiv}CH + Hg^{2+} + 2H_2O \longrightarrow \underset{\substack{\text{the oxymercuration}\\\text{addition product}}}{\overset{\displaystyle RHg^+}{\underset{HOH}{C=C}}} + H_3O^+ \qquad (14.7a)$$

$$\underset{Hg^+}{\overset{:\ddot{O}H}{R-\overset{|}{C}=CH}} \xrightarrow{\;H-\overset{+}{\ddot{O}}H_2\;} \underset{\substack{Hg^+\\+\ :\ddot{O}H_2}}{\overset{:\ddot{O}H}{R-\overset{|}{\underset{+}{C}}-CH_2}} \longrightarrow \underset{\substack{\text{(the catalyst is}\\\text{regenerated)}}}{\overset{:\ddot{O}H}{R-\overset{|}{C}=CH_2}} + Hg^{2+} \qquad (14.7b)$$

Because mercuric ion is a catalyst in the hydration of alkynes, it can be used in relatively small amounts.

The hydration of alkynes is a useful way to prepare ketones provided that the starting material is a 1-alkyne or a symmetrical alkyne (an alkyne with identical groups on each end of the triple bond). This point is explored in Study Problem 14.1.

STUDY PROBLEM 14.1

Which one of the following compounds could be prepared by the hydration of alkynes so that it is uncontaminated by constitutional isomers? Explain your answer.

(a)
$$\overset{O}{\overset{\|}{CH_3CH}}$$
acetaldehyde

(b)
$$\overset{O}{\overset{\|}{CH_3CH_2CCH_2CH_3}}$$
3-pentanone

SOLUTION First, what alkyne starting materials, if any, would give the desired products? The equations in the text show that the two carbons of the triple bond in the starting material correspond within the product to the carbon of the C=O group and an adjacent carbon. Thus, for part (a), the only possible alkyne starting material is acetylene itself, HC≡CH. For part (b), the only possible alkyne starting material is 2-pentyne, CH₃C≡CCH₂CH₃.

Next, it remains to be shown whether hydration of these alkynes gives *only* the products in the problem. Remember, a good synthesis gives relatively pure compounds. The hydration of acetylene indeed gives only acetaldehyde. (In fact, acetaldehyde is the only aldehyde that can be prepared by the hydration of an alkyne.) However, hydration of 2-pentyne gives a mixture consisting of comparable amounts of 2-pentanone and 3-pentanone, *because the carbons of 2-pentyne both have one alkyl substituent.* Thus, there is no reason that the reaction of water at either carbon should be strongly favored.

Hence, hydration would give a mixture of constitutional isomers that would have to be separated, and the yield of the desired product would be low. Consequently, hydration would *not* be a good way to prepare 3-pentanone. (However, 2-pentanone could be prepared by hydration of a different alkyne; see Problem 14.9a).

PROBLEMS

14.9 From which alkyne could each of the following compounds be prepared by acid-catalyzed hydration?

(a)

$$CH_3\overset{\overset{\displaystyle O}{\|}}{C}CH_2CH_2CH_3$$

(b)

$$(CH_3)_3C\overset{\overset{\displaystyle O}{\|}}{-C}-CH_3$$

(c)

14.10 The hydration of an alkyne is *not* a reasonable preparative method for each of the following compounds. Explain why.

(a) $CH_3CH_2CH{=}O$

(b)

$$(CH_3)_3C\overset{\overset{\displaystyle O}{\|}}{-C}-C(CH_3)_3$$

(c)

14.11 Which of the following compounds are enols? For those that are enols, show the ketone into which they would be converted.

A   *B*   *C*   *D*

B. Hydroboration–Oxidation of Alkynes

The hydroboration of alkynes is analogous to the same reaction of alkenes (Sec. 5.4B).

$$3\ CH_3CH_2{-}C{\equiv}C{-}CH_2CH_3\ +\ BH_3\ \xrightarrow[THF]{}\ \left(\begin{array}{c} CH_3CH_2\ \ \ \ CH_3CH_2 \\ \diagdown\ \ \ / \\ C{=}C \\ /\ \ \ \diagdown \\ H\ \ \ \ B \end{array}\right)_3 \quad (14.8a)$$

3-hexyne

As in the similar reaction of alkenes, oxidation of the organoborane with alkaline hydrogen peroxide yields the corresponding "alcohol," which in this case is an *enol*. As shown in Sec. 14.5A, enols react further to give the corresponding aldehydes or ketones.

$$\xrightarrow{H_2O_2/OH^-} \quad \text{an enol} \quad \longrightarrow\quad CH_3CH_2CH_2{-}\overset{\overset{\displaystyle O}{\|}}{C}{-}CH_2CH_3 \quad (14.8b)$$

3-hexanone

Because the organoborane product of Eq. 14.8a has a double bond, a second addition of BH_3 is in principle possible. However, the reaction conditions can be controlled so that only one addition takes place, as shown, provided that the alkyne is not a 1-alkyne.

If the alkyne is a 1-alkyne (that is, if it has a triple bond at the end of a carbon chain), a second addition of BH_3 cannot be prevented.

$$R{-}C{\equiv}CH\ +\ BH_3\ \longrightarrow\ \text{multiple addition reactions}$$

a 1-alkyne

However, the hydroboration of 1-alkynes can be stopped after a single addition provided that an organoborane containing highly branched groups is used instead of BH_3. One reagent developed for this purpose is disiamylborane, represented with the skeletal structure shown in Eq. 14.9. (How would you synthesize disiamylborane? See Sec. 5.4B.)

$$\left(H-\underset{\underset{CH_3}{|}}{\overset{\overset{CH_3}{|}}{C}} - \underset{\underset{H}{|}}{\overset{\overset{CH_3}{|}}{C}} \right)_2 B-H \quad \text{represented as} \quad \left(\vdash\!\!\!\dashv \right)_2 BH \qquad (14.9)$$

disiamylborane

The disiamylborane molecule is so large and highly branched that only one equivalent can react with a 1-alkyne; addition of a second molecule results in severe van der Waals repulsions in the product. In many cases, van der Waals repulsions, or *steric effects*, interfere with a *desired* reaction; in this case, however, van der Waals repulsions are used to advantage, to prevent an *undesired* second addition from occurring:

$$CH_3(CH_2)_5-C\equiv CH \xrightarrow[\text{THF}]{\left(\vdash\!\!\!\dashv\right)_2 BH} \underset{\underset{H}{}}{\overset{CH_3(CH_2)_5}{}}C=C\overset{H}{\underset{B-(\vdash\!\!\!\dashv)_2}{}} \xrightarrow[OH^-]{H_2O_2}$$

$$\underset{\underset{H}{}}{\overset{CH_3(CH_2)_5}{}}C=C\overset{H}{\underset{OH}{}} \longrightarrow CH_3(CH_2)_5-CH_2-CH=O \qquad (14.10)$$

(an enol)   **octanal**
                          (an aldehyde; 70% yield)

Notice from this example that the regioselectivity of alkyne hydroboration is similar to that observed in alkene hydroboration (Sec. 5.4B): boron adds to the unbranched carbon atom of the triple bond, and hydrogen adds to the branched carbon.

Because hydroboration–oxidation and acid-catalyzed hydration give different products when a 1-alkyne is used as the starting material (why?), these are *complementary* methods for the preparation of aldehydes and ketones in the same sense that hydroboration–oxidation and oxymercuration–reduction are complementary methods for the preparation of alcohols from alkenes.

STUDY GUIDE LINK 14.1
Functional Group Preparations

$$CH_3(CH_2)_5-C\equiv C-H \underset{\xrightarrow[H_2O,\,Hg^{2+},\,H_3O^+]{}}{\overset{\xrightarrow[2)\,H_2O_2/OH^-]{1)\,\left(\vdash\!\!\!\dashv\right)_2 BH}}{}} \begin{array}{l} CH_3(CH_2)_5CH_2-\overset{\overset{O}{\|}}{C}-H \\[1.5em] CH_3(CH_2)_5-\overset{\overset{O}{\|}}{C}-CH_3 \end{array} \qquad (14.11)$$

Notice that hydroboration–oxidation of a 1-alkyne gives an *aldehyde*; hydration of any 1-alkyne (other than acetylene itself) gives a *ketone* with a methyl group as one of the alkyl branches.

PROBLEM

14.12   Compare the results of hydroboration–oxidation and mercuric-ion-catalyzed hydration for (a) cyclohexylacetylene and (b) 2-butyne.

14.6 ## REDUCTION OF ALKYNES

A. Catalytic Hydrogenation of Alkynes

Alkynes, like alkenes (Sec. 4.9A), undergo catalytic hydrogenation. The first addition of hydrogen yields an alkene; a second addition of hydrogen gives an alkane.

$$R-C\equiv C-R \xrightarrow[\text{catalyst}]{H_2,} R-CH=CH-R \xrightarrow[\text{catalyst}]{H_2,} R-CH_2-CH_2-R \qquad (14.12)$$

The utility of catalytic hydrogenation is enhanced considerably by the fact that hydrogenation of an alkyne may be stopped at the alkene stage if the reaction mixture contains a **catalyst poison**: a compound that disrupts the action of a catalyst. Among the useful catalyst poisons are salts of Pb^{2+}, and certain nitrogen compounds, such as pyridine, quinoline, or other amines.

pyridine   **quinoline**

These compounds *selectively* block the hydrogenation of alkenes without preventing the hydrogenation of alkynes to alkenes. For example, a $Pd/CaCO_3$ catalyst can be washed with $Pb(OAc)_2$ to give a poisoned catalyst known as **Lindlar catalyst**. In the presence of Lindlar catalyst, an alkyne is hydrogenated to the corresponding alkene:

$$H_2 + \quad \text{4-octyne} \quad \xrightarrow[\text{ethanol}]{\substack{\text{Lindlar catalyst or} \\ \text{Pd/C, pyridine}}} \quad \textit{cis}\text{-4-octene} \qquad (14.13)$$

As Eq. 14.13 shows, the hydrogenation of alkynes, like the hydrogenation of alkenes (Sec. 7.8E), is a stereoselective syn-addition. Thus, in the presence of a poisoned catalyst, hydrogenation of appropriate alkynes gives cis alkenes. In fact, *catalytic hydrogenation of alkynes is one of the best ways to prepare cis alkenes.*

In the absence of a catalyst poison, two equivalents of H_2 are added to the triple bond.

$$2H_2 + \quad \text{4-octyne} \quad \xrightarrow[\text{no poison}]{Pd/C} \quad \text{octane} \qquad (14.14)$$

The catalytic hydrogenation of alkynes can therefore be used to prepare alkenes or alkanes by either including or omitting the catalyst poison. How catalyst poisons exert their inhibitory effect on the hydrogenation of alkenes is not well understood.

14.13   Give the principal organic product formed in each of the following reactions.

(a) $CH_3(CH_2)_5C\equiv CH$ + H_2 $\xrightarrow{\text{Lindlar catalyst}}$ (b) Same as part (a) with no poison
1-octyne

(c)

$+ H_2$ $\xrightarrow{\text{Pd/C}}$ (d) $H_3C-C\equiv C-CH_2CH_3$ + D_2 $\xrightarrow{\text{Lindlar catalyst}}$
2-pentyne

B. Reduction of Alkynes with Sodium in Liquid Ammonia

Reaction of an alkyne with a solution of an alkali metal (usually sodium) in liquid ammonia gives a trans alkene.

4-octyne   **trans-4-octene**
(97% yield)

(14.15)

The reduction of alkynes with sodium in liquid ammonia is complementary to the catalytic hydrogenation of alkynes, which is used to prepare cis alkenes (Sec. 14.6A).

(14.16)

a trans alkene   a cis alkene

The stereochemistry of the Na/NH_3 reduction follows from its mechanism. If sodium or other alkali metals are dissolved in pure liquid ammonia, a deep blue solution forms that contains electrons complexed with ammonia (*solvated electrons*).

$$Na\cdot + n\,NH_3\,(liq) \longrightarrow Na^+ + e^-(NH_3)_n \qquad (14.17)$$

solvated electron

The solvated electron can be thought of as the simplest free radical. Remember that free radicals add to triple bonds (Eq. 14.2, p. 689). The reaction of solvated electrons with the alkynes begins with the addition of an electron to the triple bond. The resulting species has both an unpaired electron and a negative charge. Such a species is called a **radical anion**:

a radical anion

$(14.18a)$

The radical anion is such a strong base that it readily removes a proton from ammonia to give a *vinylic radical*—a radical in which the unpaired electron is associated with one carbon of a double bond. Notice that this reaction is a Brønsted acid–base reaction and not a radical reaction.

(14.18b)

a vinylic radical

The destruction of the radical anion in this manner pulls the unfavorable equilibrium in Eq. 14.18a to the right. The vinylic radical, like the unshared electron pair of an amine (Sec. 6.9B), rapidly undergoes inversion, and the equilibrium between the cis and trans radicals favors the trans radical for the same reason that trans alkenes are more stable than cis alkenes: repulsions between the R groups are reduced.

(14.18c)

| trans vinylic radical (strongly favored at equilibrium) | transition state for inversion | cis vinylic radical |

Next, the vinylic radical accepts an electron to form an anion:

(14.18d)

solvated
electron

This step of the mechanism is the *product-determining step* of the reaction (Sec. 9.6B). The *rate constants* for the reactions of the cis and trans vinylic radicals with the solvated electron are probably the same. However, the actual *rate* of the reaction of each radical is determined by the product of the rate constant and the concentration of the radical. Because the trans vinylic radical is present in much higher concentration, the ultimate product of the reaction, the trans alkene, is derived from this radical.

The anion formed in Eq. 14.18d is also more basic than the amide anion and readily removes a proton from ammonia in another Brønsted acid–base reaction to complete the addition.

(14.18e)

trans alkene
$pK_a = 42$

$pK_a = 35$

Because ordinary alkenes do not react with the solvated electron (the initial equilibrium analogous to Eq. 14.18a is too unfavorable), the reaction stops at the trans alkene stage.

The Na/NH$_3$ reduction of alkynes does not work well on 1-alkynes unless certain modifications are made in the reaction conditions. (This is explored in Problem 14.40, p. 710.) However, this is not a serious limitation for the reaction, because the reduction of 1-alkynes to 1-alkenes is easily accomplished by catalytic hydrogenation (Sec. 14.6A).

PROBLEM

14.14   What product is obtained in each case when 3-hexyne is treated in each of the following ways? (*Hint:* The products of the two reactions are stereoisomers.)

(a) with sodium in liquid ammonia and the product of that reaction with D$_2$ over Pd/C

(b) with H$_2$ over Pd/C and quinoline and the product of that reaction with D$_2$ over Pd/C

14.7   ACIDITY OF 1-ALKYNES

A. Acetylenic Anions

Most hydrocarbons do not react as Brønsted acids. Nevertheless, let's imagine such a reaction in which a proton is removed from a hydrocarbon by a very strong base B:⁻.

$$\overset{\diagdown}{\underset{\diagup}{C}}\!-\!H + B\!:^- \rightleftharpoons \overset{\diagdown}{\underset{\diagup}{C}}\!:^- + B\!-\!H \qquad (14.19)$$

a carbanion

In this equation, the conjugate base of the hydrocarbon is a carbon anion, or *carbanion*. Recall from Sec. 9.8C that a carbanion is a species with an unshared electron pair and a negative charge on carbon.

The conjugate base of an alkane, called generally an *alkyl anion*, has an electron pair in an *sp*^{3} orbital. An example of such an ion is the 2-propanide anion:

2-propanide anion
(an alkyl anion)

The conjugate base of an alkene, called generally a *vinylic anion*, has an electron pair in an *sp*^{2} orbital. An example of this type of carbanion is the 1-propenide anion:

1-propenide anion
(a vinylic anion)

The anion derived from the ionization of a 1-alkyne, generally called an *acetylenic anion*, has an electron pair in an *sp* orbital. An example of this type of anion is the 1-propynide anion:

sp orbital

$$H_3C—C{\equiv}\bar{C}\,:$$

1-propynide anion
(an acetylenic anion)

The approximate acidities of the different types of aliphatic hydrocarbons have been measured or estimated:

$$R_3C—H \qquad R_2C{=}C\!\!\begin{array}{c}R\\ \\H\end{array} \qquad R—C{\equiv}C—H \qquad (14.20)$$

type of hydrocarbon	alkane	alkene	alkyne
approximate pK_a	55	42	25

These data show, first, that carbanions are extremely strong bases (that is, hydrocarbons are very weak acids); and second, that alkynes are the most acidic of the aliphatic hydrocarbons.

Alkyl anions and vinylic anions are seldom if ever formed by proton removal from the corresponding hydrocarbons; the hydrocarbons are simply not acidic enough. However, alkynes are sufficiently acidic that their conjugate-base acetylenic anions can be formed with strong bases. One base commonly used for this purpose is sodium amide, or sodamide, Na^+ $^-:\ddot{N}H_2$, dissolved in its conjugate acid, liquid ammonia. The amide ion, $^-:\ddot{N}H_2$, is the conjugate base of ammonia, which, as an *acid*, has a pK_a of about 35.

$$\underset{\text{p}K_a = 35}{B:^-} \;+\; :NH_3 \;\rightleftarrows\; \underset{\textbf{amide ion}}{^-:\ddot{N}H_2} \;+\; B—H \qquad (14.21)$$

STUDY GUIDE LINK 14.2
Ammonia, Solvated Electrons, and Amide Anion

(Don't confuse this pK_a with that of the *ammonium ion*, $^+NH_4$ (9.25), which is the *conjugate acid* of ammonia.)

Because the amide ion is a much stronger base than an acetylenic anion, the equilibrium for removal of the acetylenic proton by amide ion is very favorable:

$$R—C{\equiv}\overset{\frown}{C}—H \;\; ^-:\ddot{N}H_2\,Na^+ \;\underset{NH_3\,(liq)}{\rightleftharpoons}\; R—C{\equiv}\bar{C}:\,Na^+ \;+\; \ddot{N}H_3 \qquad (14.22)$$

In fact, the sodium salt of an alkyne can be formed from a 1-alkyne quantitatively (that is, in 100% yield) with $NaNH_2$. Because the amide ion is a much *weaker* base than either a vinylic anion or an alkyl anion, these ions *cannot* be prepared using sodium amide (Problem 14.17).

The relative acidity of alkynes plays a role in the method usually used to prepare *acetylenic Grignard reagents*, which are reagents with the general structure $R—C{\equiv}C—MgBr$. Recall from Sec. 9.8B that Grignard reagents are generally prepared by the reactions of alkyl halides with magnesium. The "alkyl halide" starting material for the preparation of an acetylenic Grignard reagent by this method would be a 1-bromoalkyne—that is, $R—C{\equiv}C—Br$. Such compounds are not generally available commercially and are difficult to prepare and store. Fortunately, acetylenic Grignard reagents are accessible by the acid–base reaction between a 1-alkyne and another Grignard reagent. Methylmagnesium bromide or ethylmagnesium bromide are often used for this purpose.

$$Bu—C{\equiv}C—H + CH_3CH_2—MgBr \;\xrightarrow{\text{THF}}\; \underset{\substack{\text{an acetylenic}\\\text{Grignard reagent}}}{Bu—C{\equiv}C—MgBr} \;+\; \underset{\substack{\textbf{ethane}\\\text{(a gas)}}}{CH_3CH_3} \quad (14.23)$$

$$\underset{\text{(excess)}}{H—C{\equiv}C—H} + CH_3CH_2—MgBr \;\xrightarrow{\text{THF}}\; \underset{\textbf{ethynylmagnesium bromide}}{H—C{\equiv}C—MgBr} \;+\; CH_3CH_3 \quad (14.24)$$

This reaction is extremely rapid and is driven to completion by the formation of ethane gas (when CH_3CH_2MgBr is used as the Grignard reagent). This reaction is an example of a **transmetallation**: a reaction in which a metal is transferred from one carbon to another. However, it is really just another Brønsted acid–base reaction:

$$BrMg^+ :\!\bar{C}H_2CH_3 \quad H-\!C\!\equiv\!C-\!R \longrightarrow H-\!CH_2CH_3 + BrMg^+ :\!\bar{C}\!\equiv\!C-\!R \qquad (14.25)$$

conjugate acid–base pair

conjugate base–acid pair

Although Grignard reagents are covalent compounds, the two Grignard reagents in this equation are represented as ionic compounds to stress the acid–base character of the equilibrium. This reaction is similar in principle to the reaction of a Grignard reagent with water or alcohols (Eq. 9.67, p. 432). Like all Brønsted acid–base equilibria, this one favors formation of the weaker base, which, in this case, is the acetylenic Grignard reagent. The release of ethane gas in the reaction with ethylmagnesium bromide makes the reaction irreversible and at one time was also a useful test for 1-alkynes. Alkynes with an internal triple bond do not react because they lack an acidic acetylenic hydrogen.

Other organometallic reagents, such as organolithium reagents, can also be prepared by analogous transmetallation reactions.

$$-\!C\!\equiv\!CH + Bu\!-\!Li \xrightarrow[\text{hexane, } -70\,°C]{} -\!C\!\equiv\!C\!-\!Li + Bu\!-\!H \qquad (14.26)$$

1-heptyne   **butyllithium**   an alkynyllithium   **butane**

What is the reason for the relative acidities of the hydrocarbons? Sec. 3.6A discussed two important factors that affect the acidity of an acid A—H: the A—H bond strength and the electronegativity of the group A. Bond dissociation energies show that acetylenic C—H bonds are the strongest of all the C—H bonds in the aliphatic hydrocarbons:

bond dissociation energies:

$$R-\!C\!\equiv\!C-\!H \quad > \quad R_2C\!=\!\overset{\overset{\displaystyle R}{|}}{C}-\!H \quad > \quad R_3C-\!H \qquad (14.27)$$

acetylenic C—H   vinylic C—H   alkyl C—H
($548\ kJ\ mol^{-1}$,   ($460\ kJ\ mol^{-1}$,   (402–$418\ kJ\ mol^{-1}$,
$131\ kcal\ mol^{-1}$) $110\ kcal\ mol^{-1}$) 96–$100\ kcal\ mol^{-1}$)

If bond strength were the major factor controlling hydrocarbon acidity, then alkynes would be the *least* acidic hydrocarbons. Because they are in fact the *most* acidic hydrocarbons, the electronegativities of the *carbons themselves* must govern acidity. Thus, the relative electronegativities of carbon atoms increase in the order $sp^3 < sp^2 < sp$, and the electronegativity differences on acidity must outweigh the effects of bond strength.

The explanation for this trend in electronegativity with hybridization is similar to the explanation for the dipole moments of alkenes (Sec. 4.4). If we let the acetylenic C—H bond lie along the x axis, the axes of the $2p$ orbitals of an acetylenic carbon lie along the y and z axes. The positive nucleus of an acetylenic carbon, then, is not screened by the $2p$ electrons along the x axis. Therefore, the sp carbon nucleus attracts electrons very strongly along this axis. Thus, *acetylenic carbons are particularly electronegative along the x axis.*

Here is a summary of what we have learned about hybridization, electronegativity, and acidity:

1. The electronegativity of carbons *increases* with increasing *s* character in the hybrid orbitals; that is,

    electronegativity order of carbons: $sp^3 < sp^2 < sp$

2. The acidity of attached hydrogens *increases* with increasing *s* character in the hybrid orbitals; that is,

    $$pK_a(C_{sp^3}\!-\!H) > pK_a(C_{sp^2}\!-\!H) > pK_a(C_{sp}\!-\!H)$$

    The basicity of the conjugate-base anions, then, must decrease in the same order.

Although the focus in this discussion has been carbon hybridization, the same generalizations will apply to any atom. For example, the difference in hybridization accounts for the greatly different basicities of the nitrogens in pyridine and piperidine:

In this case, the bases are neutral, not anionic, and the conjugate acids are cations; but the same generalizations hold nevertheless.

PROBLEMS

14.15 Each of the following compounds protonates on nitrogen. Draw the conjugate acid of each. Which compound is more basic? Explain.

$$H_3C\!-\!CH\!\!=\!\!\ddot{N}H \qquad H_3C\!-\!C\!\!\equiv\!\!N\!:$$
$$\qquad\quad A \qquad\qquad\qquad\quad B$$

14.16 (a) Ion A is more acidic than ion B in the gas phase. Is this the acidity order predicted by hybridization arguments? Explain.

$$H_3C\!-\!\overset{+}{O}H_2 \qquad H_3C\!-\!CH\!\!=\!\!\overset{+}{\underset{\cdot\cdot}{O}}H$$
$$\quad A \qquad\qquad\qquad B$$

(b) Ion B is less acidic because it is stabilized by resonance, whereas ion A is not. Show the resonance structure for ion B, and, with the aid of an energy diagram, show why stabilization of ion B should reduce its acidity.

(c) In aqueous solution, ion A is less acidic than ion B. Explain. (*Hint:* See Sec. 8.5C.)

14.17 (a) Using the pK_a values of the hydrocarbons and ammonia, estimate the equilibrium constant for (1) the reaction in Eq. 14.22 and (2) the analogous reaction of an alkane with amide ion. (*Hint:* See Study Problem 3.6, p. 104)

(b) Use your calculation to explain why sodium amide cannot be used to form alkyl anions from alkanes.

B. Acetylenic Anions as Nucleophiles

Although acetylenic anions are the weakest bases of the simple hydrocarbon anions, they are nevertheless strong bases—much stronger, for example, than hydroxide or alkoxides. They undergo many of the characteristic reactions of strong bases, such as S_N2 reactions with alkyl halides or alkyl sulfonates (Secs. 9.4, 10.4A). Thus, acetylenic anions can be used as nucleophiles in S_N2 reactions to prepare other alkynes.

$$CH_3CH_2CH_2CH_2\!-\!\ddot{B}r\!: + \; Na^+ \; :\bar{C}\!\equiv\!CH \; \xrightarrow{\;NH_3 \,(liq)\;} \; CH_3CH_2CH_2CH_2\!-\!C\!\equiv\!CH + Na^+ \; :\ddot{\underset{..}{B}}r\!:^- \quad (14.28)$$

electrophile   nucleophile

1-bromobutane   **sodium acetylide**   **1-hexyne**
leaving group   (64% yield)

$$CH_3CH_2CH_2CH_2\!-\!C\!\equiv\!\bar{C}\!: \; Na^+ + H_3C\!-\!\ddot{B}r\!: \; \longrightarrow \; CH_3CH_2CH_2CH_2\!-\!C\!\equiv\!C\!-\!CH_3 + Na^+ \; :\ddot{\underset{..}{B}}r\!:^- \quad (14.29)$$

2-heptyne

The acetylenic anions in these reactions are formed by the reactions of the appropriate 1-alkynes with $NaNH_2$ in liquid ammonia (Sec. 14.7A). The alkyl halides and sulfonates, as in most other S_N2 reactions, must be unhindered primary compounds. (Why? See Secs. 9.4D and 9.5G.)

The reaction of acetylenic anions with alkyl halides or sulfonates is important because *it is another method of carbon–carbon bond formation*. Let's review the methods covered so far:

1. cyclopropane formation by the addition of carbenes to alkenes (Sec. 9.9)
2. reaction of Grignard reagents with ethylene oxide and lithium organocuprate reagents with epoxides (Sec. 11.5C)
3. reaction of acetylenic anions with alkyl halides or sulfonates (this section)

PROBLEMS

14.18 Give the structures of the products in each of the following reactions.

(a) $H_3C\!-\!C\!\equiv\!\bar{C}\!: \; Na^+ + CH_3CH_2\!-\!I \; \longrightarrow$

(b) butyl tosylate $+ \; Ph\!-\!C\!\equiv\!\bar{C}\!: \; Na^+ \; \longrightarrow$

(c) $H_3C\!-\!C\!\equiv\!C\!-\!MgBr + $ ethylene oxide $\longrightarrow \; \xrightarrow{\;H_3O^+\;}$

(d) $Br(CH_2)_5Br \; + \; HC\!\equiv\!\bar{C}\!: \; Na^+ \text{(excess)} \; \longrightarrow$

14.19 Explain why graduate student Choke Fumely, in attempting to synthesize 4,4-dimethyl-2-pentyne using the reaction of $H_3C\!-\!C\!\equiv\!\bar{C}\!: \; Na^+$ with *tert*-butyl bromide, obtained none of the desired product. What product did he obtain?

14.20 Propose a synthesis of 4,4-dimethyl-2-pentyne (the compound in Problem 14.19) from an alkyl halide and an alkyne.

14.21 Outline two different preparations of 2-pentyne that involve an alkyne and an alkyl halide.

14.22 Propose another pair of reactants that could be used to prepare 2-heptyne (the product in Eq. 14.29).

14.8   ORGANIC SYNTHESIS USING ALKYNES

Let's tie together what we've learned about alkyne reactions and organic synthesis. The solution to Study Problem 14.2 requires all of the fundamental operations of organic synthesis: the formation of carbon–carbon bonds, the transformation of functional groups, and the establishment of stereochemistry (Sec. 11.10).

Notice that this problem stipulates the use of starting materials containing five or fewer carbons. This stipulation is made because such compounds are readily available from commercial sources and are relatively inexpensive.

STUDY PROBLEM 14.2

Outline a synthesis of the following compound from acetylene and any other compounds containing no more than five carbons:

$$CH_3(CH_2)_6 \quad CH_2CH_2CH(CH_3)_2$$
$$C=C$$
$$H \qquad H$$

***cis*-2-methyl-5-tridecene**

SOLUTION As usual, we start with the target molecule and work backward. First, notice the stereochemistry of the target molecule: it is a cis alkene. We've covered only one method of preparing cis alkenes free of their trans isomers: the catalytic hydrogenation of alkynes (Sec. 14.6A). This reaction, then, is used in the last step of the synthesis:

$$CH_3(CH_2)_6—C\equiv C—CH_2CH_2CH(CH_3)_2 \xrightarrow[\text{Lindlar catalyst}]{H_2}$$

2-methyl-5-tridecyne

$$CH_3(CH_2)_6 \quad CH_2CH_2CH(CH_3)_2$$
$$C=C \qquad\qquad (14.30a)$$
$$H \qquad H$$

(target molecule)

The next task is to prepare the alkyne used as the starting material in Eq. 14.30a. Because the desired alkyne contains 14 carbons and the problem stipulates the use of compounds with five or fewer carbons, we'll have to use several reactions that form carbon–carbon bonds. There are two primary alkyl groups on the triple bond; both could be introduced by the reaction of a primary alkyl halide with an acetylenic anion. The order in which they are introduced is arbitrary. Let's introduce the five-carbon fragment on the right-hand side of this alkyne in the last step of the alkyne synthesis. This is accomplished by forming the conjugate-base acetylenic anion of 1-nonyne and allowing it to react with the appropriate commercially available five-carbon alkyl halide, 1-bromo-3-methylbutane (Sec. 14.7B):

$$CH_3(CH_2)_6—C\equiv C—H \xrightarrow[\text{NH}_3 \text{ (liq)}]{\text{NaNH}_2} CH_3(CH_2)_6C\equiv\bar{C}{:}$$

1-nonyne

$$\xrightarrow{\overset{\textbf{1-bromo-3-methylbutane}}{\text{Br}—CH_2CH_2CH(CH_3)_2}}$$

$$CH_3(CH_2)_6—C\equiv C—CH_2CH_2CH(CH_3)_2 \quad (14.30b)$$

The starting material for this reaction, 1-nonyne, is prepared by the reaction of 1-bromoheptane with the sodium salt of acetylene itself.

$$H—C\equiv C—H \xrightarrow[\text{NH}_3 \text{ (liq)}]{\text{NaNH}_2} Na^+ {:}\bar{C}\equiv C—H \xrightarrow{CH_3(CH_2)_6Br} CH_3(CH_2)_6—C\equiv C—H \quad (14.30c)$$

(large excess
relative to NaNH₂)

The large excess of acetylene relative to sodium amide is required to ensure formation of the *monoanion*—that is, the anion derived from the removal of *only one* acetylene proton. If there were more sodium amide than acetylene, some *dianion* :C̄≡C̄: could form, and other reactions would occur. (What are they?) Because acetylene is cheap and is easily separated from the products (it is a gas), use of a large excess presents no practical problem.

Because the 1-bromoheptane used in Eq. 14.30c has more than five carbons, it must be prepared as well. The following sequence of reactions will accomplish this objective.

$$CH_3CH_2CH_2CH_2CH_2—Br \xrightarrow[\text{ether}]{Mg} \xrightarrow{H_2C—CH_2} \xrightarrow{H_3O^+}$$

1-bromopentane

$$CH_3CH_2CH_2CH_2CH_2CH_2CH_2—OH \xrightarrow[H_2SO_4]{\text{conc. HBr}} CH_3CH_2CH_2CH_2CH_2CH_2CH_2—Br \quad (14.30d)$$

1-heptanol   **1-bromoheptane**

The synthesis is now complete. To summarize:

$$HC≡CH \text{ (excess)} \xrightarrow[NH_3 \text{ (liq)}]{NaNH_2} \xrightarrow{CH_3(CH_2)_6Br \text{ (Eq. 14.30d)}} CH_3(CH_2)_6C≡CH \xrightarrow[NH_3 \text{ (liq)}]{NaNH_2}$$

$$\xrightarrow{BrCH_2CH_2CH(CH_3)_2} CH_3(CH_2)_6C≡CCH_2CH_2CH(CH_3)_2 \xrightarrow[\text{Lindlar catalyst}]{H_2}$$

$$\underset{H \quad\quad H}{\overset{CH_3(CH_2)_6 \quad\quad CH_2CH_2CH(CH_3)_2}{C=C}} \quad (14.30e)$$

As you gain experience working problems in organic synthesis, you should begin to think about how each reaction you study can be linked to others. For example, because alkenes can be prepared from alkynes, so alkynes can be used as starting materials for compounds that can be prepared from alkenes, such as epoxides and alcohols. Figure 14.9 shows the large number of compound types that are only two synthetic steps away from alkynes by way of alkenes. If we were willing to add additional steps, the diagram would have even more possibilities. As you build up more synthetically useful reactions, you should diagram such "synthetic networks" and how they interconnect compounds with different functional groups.

PROBLEM

14.23 Outline a synthesis of each of the following compounds from acetylene and any other compounds containing five or fewer carbons.

(a) $CH_3CH_2CH_2CH_2CH_2CH_2CH_2CH_2CH_2—OH$

1-nonanol

(b) $$H_3C—\overset{\overset{\displaystyle O}{\|}}{C}—(CH_2)_8CH_3$$

2-undecanone

(c)
$$\underset{H}{\overset{H}{\diagup}}C=C\diagdown$$

trans-2-heptene

FIGURE 14.9 A synthetic network linking alkene reactions to alkynes as starting materials.

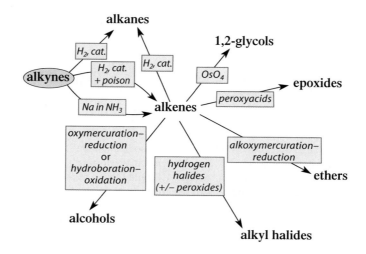

14.9   PHEROMONES

As Problems 14.24 and 14.25 on p. 706 illustrate, the chemistry of alkynes can be applied to the synthesis of a number of **pheromones**—chemical substances used in nature for communication or signaling. An example of a pheromone is a compound or group of compounds that the female of an insect species secretes to signal her readiness for mating. The sex attractant of the female Indian meal moth (*Plodia interpunctella*, a common pantry moth in the United States) is such a compound:

(9*Z*,12*E*)-9,12-tetradecadienyl acetate
(mating pheromone of the female Indian meal moth)

Pheromones are also used for defense, to mark trails, and for many other purposes. It was discovered not long ago that the traditional use of sows in France and Italy to discover buried truffles owes its success to the fact that truffles contain a steroid that happens to be identical to a sex attractant secreted in the saliva of boars during premating behavior!

About three decades ago, scientists became intrigued with the idea that pheromones might be used as a species-specific form of insect control. The thinking was that a sex attractant, for example, might be used to attract and trap the male of an insect species selectively without affecting other insect populations. Alternatively, the males of a species might become confused by a blanket of sex attractant and not be able to locate a suitable female. When used successfully, this strategy would break the reproductive cycle of the insect. The harmful environmental effects and consequent banning of such pesticides as DDT stimulated interest in such highly specific biological methods.

Although experimentation has shown that these strategies are not successful for the broad control of insect populations, they are successful in specific cases. For example, local infestations of the common pantry moth can be eradicated with commercially available traps that utilize the female sex attractant (Fig. 14.10). In large-scale agriculture, traps employing mating pheromones are useful as "early-warning" systems for insect infestations. When this approach is used, conventional pesticides need be applied only when the target insects appear

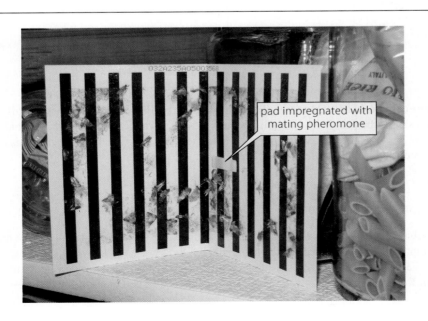

pad impregnated with mating pheromone

FIGURE 14.10 An infestation of the Indian meal moth, a common pantry moth, is controlled with a commercially available trap. The pad on the trap is impregnated with the female mating pheromone. Males attracted to the pheromone are immobilized and die on the sticky surface of the trap. The mating cycle of the moth is thus broken. Through use of these traps, fumigation of the pantry with insecticides is unnecessary.

in the traps. This strategy has brought about reductions in the use of conventional pesticides by as much as 70% in many parts of the United States. Sex pheromones, aggregation pheromones (pheromones that summon insects for coordinated attack on a plant species), and kairomones (plant-derived compounds that function as interspecies signals for host plant selection) are used commercially in this manner. Attractants for more than 250 different species of insect pests are now available commercially.

PROBLEMS

14.24 In the course of the synthesis of the sex attractant of the grape berry moth, both the cis and trans isomers of the following alkene were needed.

$$CH_3CH_2CH{=}CHCH_2CH_2CH_2CH_2CH_2CH_2CH_2CH_2{-}O{-}\bigcirc_O$$

(a) Outline a synthesis of the cis isomer of this alkene from the following alkyl halide and any other organic compounds.

$$Br{-}CH_2CH_2CH_2CH_2CH_2CH_2CH_2CH_2{-}O{-}\bigcirc_O$$

(b) Outline a synthesis of the trans isomer of the same alkene from the same alkyl halide and any other organic compounds.

14.25 The following compound is an intermediate in one synthesis of the mating pheromone of the female Indian meal moth (structure on p. 705). Show how this compound can be converted into the pheromone in a single reaction.

$$\underset{H_3C}{\overset{H}{>}}C{=}C\underset{H}{\overset{CH_2-C{\equiv}C-CH_2CH_2CH_2CH_2CH_2CH_2CH_2CH_2-O-\overset{O}{\overset{\|}{C}}-CH_3}{}}$$

14.10   OCCURRENCE AND USE OF ALKYNES

Naturally occurring alkynes are relatively rare. Alkynes do not occur as constituents of petroleum, but instead are synthesized from other compounds.

Acetylene itself comes from two common sources. Acetylene can be produced by heating coke (carbon from coal) with calcium oxide in an electric furnace to yield calcium carbide, CaC_2.

$$CaO + 3C \xrightarrow{\text{heat}} CaC_2 + CO \qquad (14.31a)$$

calcium oxide   **calcium carbide** **carbon monoxide**

Calcium carbide is an organometallic compound that can be regarded conceptually as the calcium salt of the acetylene dianion. Like any other acetylenic anion, calcium carbide reacts vigorously with water to yield the hydrocarbon; the calcium oxide by-product of this reaction can be recycled in Eq. 14.31a.

$$Ca^{2+} :\bar{C}{\equiv}\bar{C}: + H_2O \longrightarrow HC{\equiv}CH + CaO \qquad (14.31b)$$

calcium carbide   **acetylene**

The carbide process is widely used in Japan and eastern Europe, and it may become more important in the United States as the use of coal as a carbon source grows.

The second process for the manufacture of acetylene, and the predominant process used in the United States, is the thermal "cracking" (that is, decomposition) of ethylene at temperatures above 1400 °C to give acetylene and H_2. (This process is thermodynamically unfavorable at lower temperatures.)

The most important general use of acetylene is for a chemical feedstock (starting material), as illustrated by the following examples:

$$HC \equiv CH + HCl \xrightarrow{\text{HgCl}_2} H_2C = CHCl \tag{14.32}$$

acetylene   **vinyl chloride**
a monomer used in the manufacture
of poly(vinyl chloride), PVC

$$2\,HC \equiv CH \xrightarrow{\text{catalyst}} HC \equiv C - CH = CH_2 \xrightarrow{\text{HCl}} H_2C = \underset{\underset{Cl}{|}}{C} - CH = CH_2 \tag{14.33}$$

acetylene   **vinylacetylene**

chloroprene
(can be polymerized to give
neoprene rubber; Sec. 15.5)

Oxygen–acetylene welding is an important use of acetylene, although it accounts for a relatively small percentage of acetylene consumption. The acetylene used for this purpose is supplied in cylinders, but it is hazardous because, at concentrations of 2.5–80% in air, it is explosive. Furthermore, because gaseous acetylene at even moderate pressures is unstable, this substance is not sold simply as a compressed gas. Acetylene cylinders contain a porous material saturated with a solvent such as acetone. Acetylene is so soluble in acetone that most of it actually dissolves. As acetylene gas is drawn off, more of the material escapes from solution as the gas is needed—another example of Le Châtelier's principle in action!

KEY IDEAS IN CHAPTER 14

- Alkynes are compounds containing carbon–carbon triple bonds. The carbons of the triple bond are *sp*-hybridized. Electrons in *sp* orbitals are held somewhat closer to the nucleus than those in *sp*² or *sp*³ orbitals.

- The carbon–carbon triple bond in an alkyne consists of one σ bond and two mutually perpendicular π bonds. The electron density associated with the π bonds resides in a cylinder surrounding the triple bond. The induced circulation of these π electrons in a magnetic field shields acetylenic protons as well as acetylenic and propargylic carbons, and results in the relatively small chemical shifts observed in NMR spectra.

- The *sp* hybridization state is less stable than the *sp*² or *sp*³ state. For this reason, alkynes have greater heats of formation than isomeric alkenes.

- Alkynes have two general types of reactivity:
  1. addition to the triple bond
  2. reactions at the acetylenic ≡C—H bond

- Useful additions to the triple bond include Hg^{2+}-catalyzed hydration, hydroboration, catalytic hydrogenation, and reduction with sodium in liquid ammonia.

- Both the hydration and hydroboration–oxidation of alkynes yield enols, which spontaneously form the isomeric aldehydes or ketones.

- Catalytic hydrogenation of alkynes gives cis alkenes when a poisoned catalyst is used. When a poison is not used, hydrogenation to alkanes occurs. The reduction of alkynes with alkali metals in liquid ammonia, a reaction that involves radical anion intermediates, gives the corresponding trans alkenes.

- 1-Alkynes, with pK_a values near 25, are the most acidic of the aliphatic hydrocarbons. This acidity is due to the electronegativity of the *sp*-hybridized carbon of the 1-alkyne. C—H acidity increases with increasing *s* character: $pK_a(C_{sp^3}-H) > pK_a(C_{sp^2}-H) > pK_a(C_{sp}-H)$, and a similar trend is observed with other atoms.

- Carbon electronegativity increases with increasing *s* character in the order $sp^3 < sp^2 < sp$.

- Removal of the acidic proton of a 1-alkyne gives an acetylenic anion. Acetylenic anions are formed by the reactions of 1-alkynes with the strong base sodium amide ($NaNH_2$). In related transmetallation reactions, acetylenic Grignard or organolithium reagents can be formed by the reactions of 1-alkynes with alkylmagnesium halides and alkyllithium reagents, respectively.

• Acetylenic anions are good nucleophiles and react with primary alkyl halides and sulfonates in S_N2 reactions to form new carbon–carbon bonds.

 REACTION REVIEW *For a summary of reactions discussed in this chapter, see the* Reaction Review *section of Chapter 14 in the* Study Guide and Solutions Manual.

ADDITIONAL PROBLEMS

14.26 Give the principal product(s) expected when 1-hexyne or the other compounds indicated are treated with each of the following reagents:

  (a) HBr   (b) H_2, Pd/C

  (c) H_2, Pd/C, Lindlar catalyst

  (d) product of part (c) + O_3, then $(CH_3)_2S$

  (e) product of part (c) + BH_3 in THF, then $H_2O_2/{}^-OH$

  (f) product of part (c) + Br_2

  (g) $NaNH_2$ in liquid ammonia

  (h) product of part (g) + CH_3CH_2I

  (i) Hg^{2+}, H_2SO_4, H_2O

  (j) disiamyl borane, then $H_2O_2/{}^-OH$

  (k) CH_3CH_2MgBr in ether

  (l) product of part (k) with ethylene oxide, then H_3O^+

14.27 Give the principal products expected when 4-octyne or the other compounds indicated are treated with each of the following reagents:

  (a) H_2, Pd/C catalyst

  (b) H_2, Lindlar catalyst

  (c) product of (b) + O_3, then H_2O_2/H_2O

  (d) Na metal in liquid NH_3

  (e) Hg^{2+}, H_2SO_4, H_2O

  (f) BH_3, then $H_2O_2/{}^-OH$

14.28 In its latest catalog, Blarneystyne, Inc., a chemical company of dubious reputation specializing in alkynes, has offered some compounds for sale under the following names. Although each name unambiguously specifies a structure, all are incorrect. Propose a correct name for each compound.

  (a) 2-hexyn-4-ol

  (b) 6-methoxy-1,5-hexadiyne

  (c) 1-butyn-3-ene

  (d) 5-hexyne

14.29 In each case, draw a structure containing only carbon and hydrogen that satisfies the indicated criterion.

  (a) a stable alkyne of five carbons containing a ring

  (b) a chiral alkyne of six carbon atoms

  (c) an alkyne of six carbon atoms that gives the *same single* product in its reaction either with BH_3 in THF followed by $H_2O_2/{}^-OH$ or with $H_2O/Hg^{2+}/H_3O^+$

  (d) a six-carbon alkyne that can exist as diastereomers

14.30 On the basis of the hybrid orbitals involved in the bonds, arrange the bonds in each of the following sets in order of increasing length.

  (a) C—H bonds of ethylene; C—H bonds of ethane; C—H bonds of acetylene

  (b) C—C single bond of propane; C—C single bond of propyne; C—C single bond of propene

14.31 Rank the anions within each series in order of increasing basicity, lowest first. Explain.

  (a) $CH_3CH_2\ddot{O}{:}^-$, $HC{\equiv}\bar{C}{:}$, $:\ddot{F}{:}^-$

  (b) $CH_3(CH_2)_3C{\equiv}\bar{C}{:}$, $CH_3(CH_2)_4\bar{\ddot{C}}H_2$, $CH_3(CH_2)_3CH{=}\bar{\ddot{C}}H$

14.32 Using simple observations or chemical tests with readily observable results, show how you would distinguish between the compounds in each of the following pairs. (Don't use spectroscopy.)

  (a) *cis*-2-hexene and 1-hexyne

  (b) 1-hexyne and 2-hexyne

  (c) 4,4-dimethyl-2-hexyne and 3,3-dimethylhexane

  (d) propyne and 1-decyne

14.33 Outline a preparation of each of the following compounds from acetylene and any other reagents.

  (a) $CH_3CH_2CD_2CD_2CH_2CH_3$ (b) 1-hexene

  (c) 3-hexanol   (d) 1-hexyne

  (e)
$$CH_3(CH_2)_7{-}\overset{\overset{\displaystyle O}{\|}}{C}{-}OH$$

  (f) $(CH_3)_2CHCH_2CH_2CH_2CH{=}O$

  (g) *cis*-2-pentene   (h) *trans*-3-decene

  (i) *meso*-4,5-octanediol   (j) (Z)-3-hexen-1-ol

14.34 Using 1-butyne as the only source of carbon in the reactants, propose a synthesis for each of the following compounds.

(a) $CH_3CH_2C\equiv C—D$ (b) $CH_3CH_2CD_2CD_3$

(c)
$$CH_3CH_2CH_2\overset{\overset{\displaystyle O}{\|}}{C}OH$$

(d) 1-butoxybutane (dibutyl ether)

(e) the racemate of
$(3R,4S)\text{-}CH_3CH_2\underset{\underset{\displaystyle D}{|}}{C}H\underset{\underset{\displaystyle D}{|}}{C}HCH_2CH_2CH_2CH_3$

(f) octane (g)
$$CH_3CH_2\overset{\overset{\displaystyle O}{\|}}{C}CH_3$$

14.35 (a) Draw the structures of *all* enols that would spontaneously form the following ketone, including stereoisomers.

$$CH_3CH_2\overset{\overset{\displaystyle O}{\|}}{—C—}CH(CH_3)_2$$

(b) Would alkyne hydration be a good preparative method for this compound? If so, give the reaction. If not, explain why.

14.36 A box labeled "C_6H_{10} isomers" contains samples of three compounds: *A*, *B*, and *C*. Along with the compounds are the IR spectra of *A* and *B*, shown in Fig. P14.36. Fragmentary data in a laboratory notebook suggest that the compounds are 1hexyne, 2-hexyne, and 3-methyl-1,4-pentadiene. Identify the three compounds.

14.37 You have just been hired by Triple Bond, Inc., a company that specializes in the manufacture of alkynes containing five or fewer carbons. The President, Mr. Al Kyne, needs an outlet for the company's products. You have been asked to develop a synthesis of the housefly sex pheromone, *muscalure*, with the stipulation that all of the carbon in the product must come only from the company's alkynes. The muscalure will subsequently be used in a household fly trap. You will be equipped with a laboratory containing all of the company's alkynes, requisition forms for other reagents, and one gross of fly swatters in case you are successful. Outline a preparation of racemic muscalure that meets the company's needs.

$$\underset{H}{\overset{CH_3(CH_2)_7}{\diagdown}}C=C\underset{H}{\overset{(CH_2)_{12}CH_3}{\diagup}}$$

muscalure

Figure P14.36 IR spectra for Problem 14.36.

14.38 Outline a preparation of racemic *disparlure*, a pheromone of the gypsy moth, from acetylene and any other compounds containing not more than five carbon atoms.

$$CH_3(CH_2)_9 \overset{O}{\underset{H}{\underset{|}{C}}} \overset{}{\underset{H}{\underset{|}{C}}} (CH_2)_4CH(CH_3)_2$$

disparlure

14.39 In the preparation of ethynylmagnesium bromide by the transmetallation reaction of Eq. 14.24, ethylmagnesium bromide is added to a large excess of acetylene in THF solution. Two side reactions that can occur in this procedure are shown in Fig. P14.39.

(a) Suggest a mechanism for reaction (1), and explain why an excess of acetylene is important for avoiding this reaction.

(b) Suggest a mechanism for reaction (2), and explain why an excess of acetylene is important for avoiding this reaction.

(c) Tetrahydrofuran (THF) is used as a solvent because the undesired by-product, $BrMg—C\equiv C—MgBr$, is relatively soluble in this solvent. Explain why it is important for this by-product to be soluble if both side reactions are to be minimized.

14.40 (a) When the reduction of alkynes to alkenes by Na in liquid ammonia is attempted with a 1-alkyne, every three moles of 1-alkyne give only one mole of alkene and two moles of the acetylenic anion:

$$3\,RC\equiv CH \xrightarrow[NH_3\,(liq)]{Na} RCH=CH_2 + 2\,RC\equiv \overset{-}{C}:\,Na^+$$

Explain this result using the mechanism of this reduction and what you know about the acidity of 1-alkynes.

(b) When $(NH_4)_2SO_4$ is added to the reaction mixture, the 1-alkyne is converted completely into the alkene. Explain. (*Hint:* The pK_a of the ammonium ion is 9.25.)

14.41 Arrange the following three carboxylic acids in order of increasing acidity (decreasing pK_a). Explain your reasoning. (*Hint:* Consider the electronegativity of the β-carbon.)

$$H_2C=CHCH_2CO_2H \quad HC\equiv CCH_2CO_2H \quad CH_3CH_2CH_2CO_2H$$
$$A \qquad\qquad\qquad B \qquad\qquad\qquad C$$

14.42 Identify the following compounds from their IR and proton NMR spectra.

(a) $C_6H_{10}O$:

NMR: δ 3.31 (3H, s); δ 2.41 (1H, s); δ 1.43 (6H, s)
IR: 2110, 3300 cm^{-1} (sharp)

(b) C_4H_6O: liberates a gas when treated with EtMgBr

NMR: δ 2.43 (1H, t, J = 2 Hz); δ 3.41 (3H, s); δ 4.10 (2H, d, J = 2 Hz)
IR: 2125, 3300 cm^{-1}

(c) C_4H_6O:

NMR in Fig. P14.42
IR: 2100, 3300 cm^{-1} (sharp), superimposed on a broad, strong band at 3350 cm^{-1}

(d) C_5H_6O

IR: 3300, 2102, 1634 cm^{-1}
NMR: δ 3.10 (1H, d, J = 2 Hz); δ 3.79 (3H, s); δ 4.52 (1H, doublet of doublets, J = 6 Hz and 2 Hz); δ 6.38 (1H, d, J = 6 Hz)

14.43 (a) Identify the compound C_6H_{10} that shows IR absorptions at 3300 cm^{-1} and 2100 cm^{-1} and has the following ^{13}C NMR spectrum: δ 27.3, 31.0, 66.7, 92.8.

(b) Explain how you could distinguish between 1-hexyne and 4-methyl-2-pentyne by ^{13}C NMR.

14.44 Propose mechanisms for each of the following known transformations; use the curved-arrow notation where possible.

(a)

$$Ph—C\equiv C—H \xrightarrow[THF]{\substack{NaOD, \\ D_2O\,(large\,excess)}} Ph—C\equiv C—D$$

(b)

$$Ph—C\equiv CH + Br_2 \xrightarrow{H_2O} Ph—\overset{O}{\overset{||}{C}}—CH_2Br$$
$$+\ HBr$$

14.45 A compound A (C_6H_6) undergoes catalytic hydrogenation over Lindlar catalyst to give a compound B, which in turn undergoes ozonolysis followed by workup with aqueous H_2O_2 to yield succinic acid and two equivalents of formic acid. In the absence of a catalyst poison, hydrogenation of A gives hexane. Propose a structure for compound A.

$$HO\overset{O}{\overset{||}{C}}—CH_2CH_2—\overset{O}{\overset{||}{C}}OH \qquad H—\overset{O}{\overset{||}{C}}OH$$

succinic acid   **formic acid**

14.46 An optically active alkyne A ($C_{10}H_{14}$) can be catalytically hydrogenated to butylcyclohexane. Treatment of A with EtMgBr liberates no gas. Catalytic hydrogenation of A over Pd/C in the presence of quinoline poison and treatment of the product with O_3 and then H_2O_2 gives an *optically active* tricarboxylic acid $C_8H_{12}O_6$. (A tricarboxylic

---

(1) $H—C\equiv C—MgBr + CH_3CH_2—MgBr \longrightarrow CH_3CH_3 + BrMg—C\equiv C—MgBr$

(2) $2\,H—C\equiv C—MgBr \rightleftharpoons H—C\equiv C—H + BrMg—C\equiv C—MgBr$

Figure P14.39

acid is a compound with three —CO$_2$H groups.) Give the structure of A, and account for all observations.

14.47 Complete the reactions given in Fig. P14.47 using knowledge or intuition developed from this or previous chapters.

Figure P14.42 The NMR spectrum for Problem 14.42c. A trace of aqueous acid was added to the compound before the spectrum was obtained. The integral (as the number of protons) is shown in red over each absorption. The absorptions labeled "dry" were obtained on a very dry sample before addition of the acid. The absorptions labeled "wet" were obtained in the presence of aqueous acid.

(a) CH$_3$CH$_2$CH$_2$CH$_2$C≡CH $\xrightarrow{\text{EtMgBr}}$ $\xrightarrow{\text{D}_2\text{O}}$

(b) CH$_3$CH$_2$CH$_2$CH$_2$—C≡C—H + Bu—Li $\longrightarrow$ $\xrightarrow{\text{Me}_3\text{SiCl}}$

(*Hint:* Tertiary silyl halides, unlike tertiary alkyl halides, undergo nucleophilic substitution reactions that are not complicated by competing elimination reactions.)

(c)

$$CH_3CH_2-O-\overset{\overset{\displaystyle O}{\|}}{\underset{\underset{\displaystyle O}{\|}}{S}}-O-CH_2CH_3$$

CH$_3$(CH$_2$)$_6$—C≡CH $\xrightarrow{\text{NaNH}_2}$ $\xrightarrow{\textbf{diethyl sulfate}}$

(*Hint:* See Sec. 10.4C)

(d) Li—C≡CH + F⁀⁀⁀⁀Cl $\longrightarrow$

(1 equivalent)

(e)

Ph—CH=CH$_2$ + Br$_2$ $\longrightarrow$ $\xrightarrow{\text{NaNH}_2}$ $\xrightarrow[\text{(H}_3\text{O}^+)]{\text{neutralize}}$ (a hydrocarbon not containing bromine)

(*Hint:* See Sec. 9.5)

(f) Ph—C≡C—Ph + HCCl$_3$ $\xrightarrow{\text{K}^+\text{Me}_3\text{CO}^-}$

(*Hint:* See Sec. 9.9A)

Figure P14.47

CHAPTER 15

Dienes, Resonance, and Aromaticity

Dienes are compounds with two carbon–carbon double bonds. Their nomenclature was discussed along with the nomenclature of other alkenes in Sec. 4.2A. Dienes are classified according to the relationship of their double bonds. In **conjugated dienes**, two double bonds are separated by one single bond. These double bonds are called **conjugated double bonds**.

conjugated double bonds

$$H_2C{=}CH{-}CH{=}CH_2$$

1,3-butadiene
(a conjugated diene)

Cumulenes are compounds in which one carbon participates in two carbon–carbon double bonds; these double bonds are called **cumulated double bonds**. Propadiene (common name allene) is the simplest cumulene. The term *allene* is also sometimes used as a family name for compounds containing only two cumulated double bonds.

one carbon involved in two double bonds

$$H_2C{=}C{=}CH_2$$

propadiene
(allene)

Conjugated dienes and allenes have unique structures and chemical properties that are the basis for much of the discussion in this chapter.

Dienes in which the double bonds are separated by two or more single bonds have structures and chemical properties more or less like those of simple alkenes and do not require special discussion. These dienes will be called "ordinary" dienes.

$$H_2C\!\!=\!\!CHCH_2CH_2CH_2CH\!\!=\!\!CH_2$$
1,6-heptadiene
(an ordinary diene)

In this chapter, you'll see that the interaction of two functional groups within the same molecule—in this case two carbon–carbon double bonds—can result in special reactivity. In particular, you'll learn how conjugated double bonds differ in their reactivity from ordinary double bonds. This discussion will lead to a consideration of benzene, a cyclic hydrocarbon in which the effects of conjugation are particularly dramatic. The chemistry of benzene and the effects of conjugation on chemical properties will continue as central themes through Chapter 18.

15.1   STRUCTURE AND STABILITY OF DIENES

A. Stability of Conjugated Dienes. Molecular Orbitals

The heats of formation listed in Table 15.1 provide information about the relative stabilities of dienes. The effect of conjugation on the stability of dienes can be deduced from a comparison of the heats of formation for (E)-1,3-hexadiene, a conjugated diene, and (E)-1,4-hexadiene, an unconjugated isomer. Notice from the heats of formation that the conjugated diene is 19.7 kJ mol^{-1} (4.7 kcal mol^{-1}) more stable than its unconjugated isomer. Because the double bonds in these two compounds have the same number of branches and the same stereochemistry, this stabilization of nearly 20 kJ mol^{-1} (5 kcal mol^{-1}) is due to conjugation.

One possible reason for this additional stability could be the differences in the σ bonds between isomeric conjugated and unconjugated dienes. For example, comparing the isomers

TABLE 15.1 Heats of Formation of Dienes and Alkynes

Compound	Structure	ΔH_f° (25 °C, gas phase)	
		kJ mol^{-1}	kcal mol^{-1}
(E)-1,3-hexadiene	$H_2C\!\!=\!\!CH$ ... $C\!\!=\!\!C$ ... H, CH_2CH_3	54.4	13.0
(E)-1,4-hexadiene	$H_2C\!\!=\!\!CHCH_2$ ... $C\!\!=\!\!C$ ... H, CH_3	74.1	17.7
1-pentyne	$HC\!\!\equiv\!\!CCH_2CH_2CH_3$	144	34.5
2-pentyne	$CH_3C\!\!\equiv\!\!CCH_2CH_3$	129	30.8
(E)-1,3-pentadiene	$H_2C\!\!=\!\!CH$ ... $C\!\!=\!\!C$ ... H, CH_3	75.8	18.1
1,4-pentadiene	$H_2C\!\!=\!\!CHCH_2CH\!\!=\!\!CH_2$	106	25.4
1,2-pentadiene	$H_2C\!\!=\!\!C\!\!=\!\!CHCH_2CH_3$	141	33.6
2,3-pentadiene	$CH_3CH\!\!=\!\!C\!\!=\!\!CHCH_3$	133	31.8

1,4-hexadiene and 1,3-hexadiene, we find that two sp^2–sp^3 σ bonds in the unconjugated diene are traded for one sp^2–sp^2 and one sp^3–sp^3 σ bond in the conjugated diene.

$$sp^2\text{–}sp^3 \quad sp^3\text{–}sp^2 \qquad sp^2\text{–}sp^3$$
$$\downarrow \qquad \downarrow \qquad\qquad \downarrow$$

H₂C＝CH—CH₂—CH＝CH—CH₃

$$sp^2\text{–}sp^2 \qquad sp^2\text{–}sp^3 \quad sp^3\text{–}sp^3$$
$$\downarrow \qquad\qquad \downarrow \qquad \downarrow$$

H₂C＝CH—CH＝CH—CH₂—CH₃

1,4-hexadiene
(unconjugated)

1,3-hexadiene
(conjugated)

We learned in Sec. 14.2 that σ bonds with more s character are stronger than those with less s character. Two sp^2–sp^3 σ bonds in the unconjugated diene, then, are "traded" for a *stronger* bond (the sp^2–sp^2 bond) and a *weaker* bond (the sp^3–sp^3 bond). These bond-strength effects almost cancel; one estimate is that these effects account at most for 5–6 kJ mol^{-1} (1.2–1.4 kcal mol^{-1}) of the enhanced stability of the conjugated diene.

The second and major reason for the greater stability of conjugated dienes is the over-lap of $2p$ orbitals across the carbon–carbon bond connecting the two alkene units. That is, not only does π bonding occur *within* each of the alkene units, but *between* them as well. Fig. 15.1a shows the alignment of carbon $2p$ orbitals in 1,3-butadiene, the simplest conjugated diene. Notice that the $2p$ orbitals on the central carbons are in the parallel alignment neces-sary for overlap.

As we learned when we considered π bonding in ethylene (Sec. 4.1B), the overlap of $2p$ orbitals results in the formation of π molecular orbitals. The overlap of j $2p$ orbitals results in the formation of j molecular orbitals. In the case of a conjugated diene, $j = 4$. Therefore, four molecular orbitals (MOs) are formed. For a conjugated diene, half of the MOs are bonding—they have a lower energy than an isolated $2p$ atomic orbital. The other half are antibonding—they have a higher energy than an isolated $2p$ orbital. These four MOs for 1,3-butadiene, the simplest conjugated diene, are shown in Fig. 15.1b. Each MO of successively higher energy has one additional node, and the nodes are symmetrically arranged within the π system. The MO of lowest energy, π_1, has no node, and the second bonding MO, π_2, has one node between the two interior carbons. The antibonding MOs, π_3^* and π_4^*, have two and three nodes, respec-tively. (The asterisk indicates their antibonding character.)

1,3-Butadiene has four $2p$ electrons; these electrons are distributed into the four MOs. Because each MO can accommodate two electrons, two electrons are placed into π_1 and two into π_2. These two bonding MOs, then, are the ones we want to examine to understand the bonding and stability of conjugated dienes. Consider first the energies of these bonding MOs. These are shown to scale relative to the energies of the ethylene MOs (Fig. 4.6, p. 129). The energy unit conventionally used with π MOs is called **beta (β)**, which, for conjugated alkenes, has a value of roughly –50 kJ mol^{-1} (–12 kcal mol^{-1}). By convention, β is a negative number. The π_1 MO of butadiene has a relative energy of 1.62β, and π_2 has a relative energy of 0.62β. Each π electron in butadiene contributes to the molecule the energy of its MO. Therefore, the two electrons in π_1 contribute $2 \times (1.62\beta) = 3.24\beta$, and the two electrons in π_2 contribute $2 \times (0.62\beta) = 1.24\beta$. The total π electron energy for 1,3-butadiene, then, is 4.48β.

To calculate the bonding advantage of a conjugated diene, then, we compare it to the π-electron energy of two isolated ethylene molecules—that is, two π-electron systems in which there is no overlap between the double bonds. As Fig. 15.1 shows, the bonding MO of ethylene lies at 1.00β; the two bonding π electrons of ethylene contribute a π-electron energy of 2.00β, and the π electrons of two isolated ethylenes contribute 4.00β. It follows that the ener-getic advantage of conjugation—orbital overlap—in 1,3-butadiene is $4.48\beta - 4.0\beta = 0.48\beta$. This energetic advantage must result from π_1, which is the MO with lower energy than the bonding MO of ethylene.

Half of the total π-electron density in 1,3-butadiene is contributed by the two electrons in π_1 and half by the two electrons in π_2. Consider now the nodal structure of these two molecular orbitals. The π_2 MO has a node that divides the molecule into two isolated "eth-ylene halves;" thus, the electrons in this MO contribute some isolated double-bond character to the π-electron structure of 1,3-butadiene. However, the π_1 MO has no node; consequently, the electron density in this MO is spread across the entire molecule. The electrons in π_1 for this reason are said to be **delocalized**. In particular, this MO contributes to bonding between

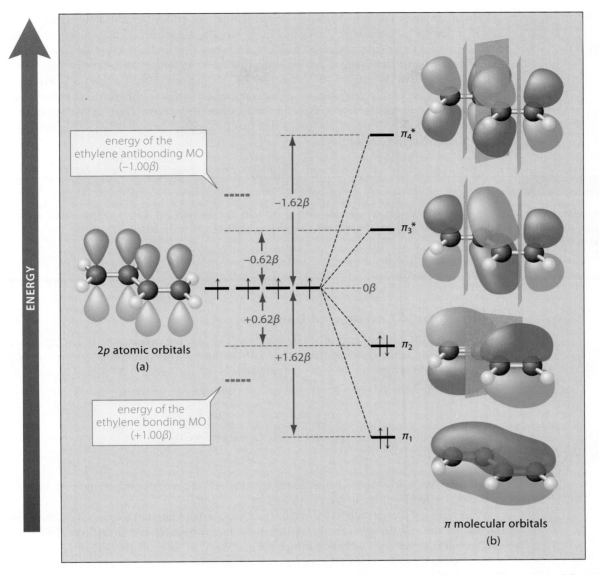

FIGURE 15.1 An orbital interaction diagram showing π molecular orbital (MO) formation in 1,3-butadiene. (a) Arrangement of 2p orbitals in 1,3-butadiene, the simplest conjugated diene. Notice that the axes of the 2p orbitals are properly aligned for overlap. (b) Interaction of the four 2p orbitals (*dashed black lines*) gives four π MOs. Nodal planes are shown in gray. Notice that nodes occur between peaks and troughs in the MOs, indicated by blue and green, respectively. (The original nodal plane of the starting 2p orbitals is not shown.) The four 2p electrons both go into π_1 and π_2, the bonding MOs. The violet arrows and numbers show the relative energies of the MOs in β units. (Remember that β is a negative number.) The relative energies of the ethylene MOs are shown in red.

the two central carbons—the carbons connected by the "single bond." The delocalization of π electrons across the central single bond is also evident from the EPM of 1,3-butadiene.

π-electron density across the C—C single bond

EPM of 1,3-butadiene

This analysis shows that electron delocalization, which is not adequately conveyed by Lewis structures, is responsible for the additional stability associated with conjugation. To put it another way, *conjugation results in additional bonding that makes a molecule more stable.*

The energetic advantage of conjugation is called the **delocalization energy**. This name colorfully describes its origin—the delocalization of electrons in π_1. Because β is negative, the delocalization energy describes the *reduction* in energy (that is, the increased stability) of a conjugated diene relative to two isolated, unconjugated ethylenes. Thus, the delocalization energy of 0.48β for 1,3-butadiene means that this conjugated diene is more stable than two unconjugated ethylene molecules by 0.48β.

PROBLEMS

15.1   The conjugated triene (E)-1,3,5-hexatriene has six π molecular orbitals with relative energies $\pm1.80\beta$, $\pm1.25\beta$, and $\pm0.44\beta$. (a) Sketch these MOs. Indicate which are bonding and which are antibonding. (b) Tell how many nodes each has. (c) Show the position of the nodes in π_1, π_2, and π_6^*.

15.2   Calculate the delocalization energy for (E)-1,3,5-hexatriene.

B. Structure of Conjugated Dienes

The length of the carbon–carbon single bond in 1,3-butadiene reflects the hybridization of the orbitals from which it is constructed. At 1.46 Å, this sp^2–sp^2 single bond is considerably shorter than both the sp^2–sp^3 carbon–carbon single bond in propene (1.50 Å) and the sp^3–sp^3 carbon–carbon bond in ethane (1.54 Å).

$$\overset{\overset{\displaystyle \vert\!\!-1.34\text{ Å}-\!\!\vert}{}}{H_2C} = \overset{}{CH} - \overset{\overset{\displaystyle \vert\!\!-1.34\text{ Å}-\!\!\vert}{}}{CH} = CH_2$$

Recall from Secs. 4.1A and 4.2 that, as the fraction of s character in the component orbitals increases, the length of the bond decreases.

Conjugated dienes such as 1,3-butadiene undergo rapid internal rotation about the central single bond of the diene unit. 1,3-Butadiene has two stable conformations. The most stable conformation is the **s-trans conformation**. (The *s*-prefix emphasizes that this refers to rotation about a *single* bond.) This conformation is sometimes called the **anti conformation**. The second conformation is the **gauche** or **skew conformation**. These conformations and their relative standard free energies are shown in Fig. 15.2a; Newman projections are shown in Fig. 15.2b. (The *s*-trans conformation is shown in Fig. 15.1 as well.) In the *s*-trans conformation, the $2p$ orbitals of all carbons are coplanar and can overlap. In the gauche conformation, the $2p$ orbitals of one double bond are twisted 38° relative to those of the other, at the cost of some orbital overlap. The partial loss of overlap accounts for the higher energy of the gauche conformation. The energy barrier between the two conformations, which is greatest at 102°, largely reflects the complete loss of overlap at this angle. The third conformation shown in Fig. 15.2a, the **s-cis conformation**, is unstable. In this conformation, the $2p$ orbitals are coplanar, but van der Waals repulsions between two of the hydrogens (shown in Fig. 15.2a) destabilize this conformation; in the gauche conformation, the offending hydrogens are further apart. Despite the instability of the s-cis conformation, it is important in some reactions of conjugated dienes (Sec. 15.3).

PROBLEM

15.3   Draw the *s*-cis and *s*-trans conformations of $(2E,4E)$-2,4-hexadiene and $(2E,4Z)$-2,4-hexadiene. Which diene contains the *greater* proportion of the gauche conformation? Why? (Use the *s*-cis conformation as an approximation of the gauche conformation.)

FIGURE 15.2 (a) The conformations of 1,3-butadiene and their relative standard free energies. Internal rotation occurs about the central carbon–carbon single bond (*green arrow*; rotation angles are shown in green along the horizontal axis). (b) Newman projections of the two stable conformations obtained by sighting along the central carbon–carbon single bond, as shown by the eyeball in (a).

C. Structure and Stability of Cumulated Dienes

The structure of allene is shown in Fig. 15.3. Because the central carbon of allene is bound to two groups, the carbon skeleton of this molecule is linear (Sec. 1.3B). A carbon atom with 180° bond angles is *sp*-hybridized (Sec. 14.1). Therefore, the central carbon of allene, like the carbons in an alkyne triple bond, is *sp*-hybridized. The two remaining carbons of the cumulated diene are *sp*²-hybridized and have trigonal planar geometry.

FIGURE 15.3 The structure of allene, the simplest cumulated diene. (a) Lewis structure showing the bond angles and bond lengths. (b) A Newman projection along the carbon–carbon double bonds as seen by the eyeball. The CH₂ groups at opposite ends of the molecule lie in perpendicular planes.

FIGURE 15.4
The π-electron structure of allene. The blue and green orbital colors represent wave peaks and wave troughs. (a) The component 2p orbitals of the double bonds. Because the central carbon is sp-hybridized, it has two mutually perpendicular 2p orbitals. (b) The π molecular orbitals that result from overlap of the 2p orbitals are mutually perpendicular and do not overlap.

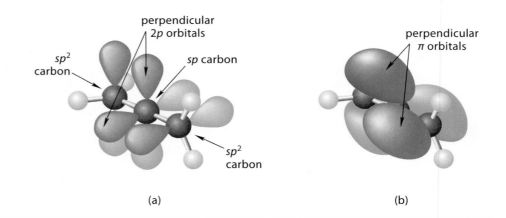

(a)   (b)

The two π bonds in allenes are mutually perpendicular, as required by the *sp* hybridization of the central carbon atom (Fig. 15.4). Consequently, *the H—C—H plane at one end of the allene molecule is perpendicular to the H—C—H plane at the other end*, as shown by the Newman projection in Fig. 15.3. Note carefully the difference in the bonding arrangements in allene and the conjugated diene 1,3-butadiene. In the conjugated diene, the π-electron systems of the two double bonds are coplanar and can overlap; all carbon atoms are sp²-hybridized. In contrast, allene contains two mutually perpendicular π systems, each spanning two carbons; the central carbon is part of both. Because these two π systems are perpendicular, they do *not* overlap. The perpendicular π orbitals of allene are reflected in the EPM of allene, which shows areas of π-electron density above and below each double bond.

regions of π-electron density lie on perpendicular axes

EPM of allene

Because of their geometries, some allenes are chiral even though they do not contain an asymmetric carbon atom. The following molecule, 2,3-pentadiene, is an example of a chiral allene.

(Using models if necessary, verify the chirality of 2,3-pentadiene by showing that these two structures are not congruent.) The two sp²-hybridized carbons are stereocenters. Thus, the enantiomers of 2,3-pentadiene differ by an internal rotation about either double bond. Because internal rotation about a double bond does not occur under normal conditions, enantiomeric allenes can be isolated.

Chiral allenes illustrate the point that *some chiral molecules do not contain asymmetric centers*. We first came across this situation when we considered the chirality of individual molecular conformations, such as *gauche*-butane (Sec. 6.9A). The chiral allenes are stable, isolable molecules that illustrate the same point.

The *sp* hybridization of allenes is reflected in their C=C stretching absorptions in the infrared spectrum. This absorption occurs near 1950 cm^{-1}, not far from the C≡C stretching absorption of alkynes.

The data in Table 15.1 (p. 713) show that allenes have greater heats of formation than other types of isomeric dienes. For example, 1,2-pentadiene is considerably less stable than 1,3-pentadiene or 1,4-pentadiene. Thus, the cumulated arrangement is the least stable arrangement of two double bonds. A comparison of the heats of formation of 2-pentyne and 2,3-pentadiene shows that allenes are somewhat less stable than isomeric alkynes as well. In fact, a common reaction of allenes is isomerization to alkynes.

Although a few naturally occurring allenes are known, allenes are relatively rare in nature.

PROBLEMS

15.4 Explain why there is a larger *difference* between the heats of formation of (*E*)-1,3-pentadiene and 1,4-pentadiene (29.3 kJ mol^{-1} or 7.1 kcal mol^{-1}) than between (*E*)-1,3-hexadiene and (*E*)-1,4-hexadiene (19.7 kJ mol^{-1} or 4.7 kcal mol^{-1}).

15.5 (a) Draw line-and-wedge structures for the two enantiomers of the following allene.

$$H_3C-CH=C=C \begin{array}{c} CH_2CH_2CH_2CH_3 \\ \\ CO_2H \end{array}$$

(b) One enantiomer of this compound has a specific rotation of $-30.7°$. What is the specific rotation of the other?

15.2 ULTRAVIOLET–VISIBLE SPECTROSCOPY AND FLUORESCENCE

The IR and NMR spectra of conjugated dienes are very similar to the spectra of ordinary alkenes. However, another type of spectroscopy can be used to analyze and identify organic compounds containing conjugated π-electron systems. In this type of spectroscopy, called **ultraviolet–visible spectroscopy**, the absorption of radiation in the ultraviolet or visible region of the spectrum is recorded as a function of wavelength. The part of the ultraviolet spectrum of greatest interest to organic chemists is the *near ultraviolet* (wavelength range 200×10^{-9} to 400×10^{-9} m). Visible light, as the name implies, is electromagnetic radiation visible to the human eye (wavelengths from 400×10^{-9} to 750×10^{-9} m). Because there is a common physical basis for the absorption of both ultraviolet and visible radiation by chemical compounds, both ultraviolet and visible spectroscopy are considered together as one type of spectroscopy, often called simply **UV–vis spectroscopy**.

Some molecules, after absorption of UV or visible radiation, can lose some of the energy gained during absorption by emission of light. We'll conclude this section with a description of one type of light emission, called *fluorescence*, which is a particularly important analytical technique in biology.

A. The UV–Vis Spectrum

Like any other absorption spectrum, the **UV–vis spectrum** of a substance is the graph of radiation absorption by the substance versus the wavelength of the radiation. The instrument

FIGURE 15.5 Ultraviolet–visible spectrum of isoprene in methanol. The λ_{max} (*red*) is the wavelength at which the absorption maximum occurs; for isoprene, the λ_{max} is 222.5 nm.

used to measure a UV–vis spectrum is called a **UV–vis spectrophotometer**. Except for the fact that it is designed to operate in a different part of the electromagnetic spectrum, it is conceptually much like any other absorption spectrometer (Fig. 12.3, p. 573).

A typical UV spectrum, that of 2-methyl-1,3-butadiene (isoprene), is shown in Fig. 15.5. Because isoprene does not absorb visible light, only the ultraviolet region of the spectrum is shown. On the horizontal axis of the UV spectrum is plotted the wavelength λ of the radiation. In UV spectroscopy, the conventional unit of wavelength is the *nanometer* (abbreviated nm). One **nanometer** equals 10^{-9} meter. (In older literature, the term *millimicron*, abbreviated mμ, was used; a millimicron is the same as a nanometer.) The relationship between the energy of the electromagnetic radiation and its frequency or wavelength should be reviewed again (Sec. 12.1A).

The vertical axis of a UV spectrum shows the **absorbance**. (Absorbance is sometimes called **optical density**, abbreviated OD.) The absorbance is a measure of the amount of radiant energy absorbed. Suppose the radiation entering a sample has intensity I_0, and the light emerging from the sample has intensity I. The absorbance A is defined as the logarithm of the ratio I_0/I:

$$A = \log (I_0/I) \tag{15.1}$$

According to Eq. 15.1, then, the more radiant energy is absorbed, the larger is the ratio I_0/I, and the greater the absorbance.

PROBLEMS

15.6 What is the energy of light (in kJ mol^{-1} or kcal mol^{-1}) with a wavelength of
(a) 450 nm?   (b) 250 nm?

15.7 (a) What percent of the incident radiation is transmitted by a sample when its absorbance is 1.0? When its absorbance is 0?
(b) What is the absorbance of a sample that transmits one-half of the incident radiation intensity?

15.8 A thin piece of red glass held up to white light appears brighter to the eye than a piece of the same glass that is twice as thick. Which piece has the greater absorbance?

In the UV–vis spectra used in this text, absorbance increases from the bottom to the top of the spectrum. Therefore, absorption maxima occur as high points or peaks in the spectrum. Notice the difference in how UV–vis and IR spectra are presented. (Absorptions in IR spectra increase from top to bottom because IR spectra are conventionally presented as plots of *trans-*

mittance, or percentage of light transmitted.) In the UV–vis spectrum shown in Fig. 15.5, the absorbance maximum occurs at a wavelength of 222.5 nm. The wavelength at the maximum of an absorption peak is called the λ_{max} (pronounced "lambda-max"). Some compounds have several absorption peaks and a corresponding number of λ_{max} values. Absorption peaks in the UV–vis spectra of compounds in solution are generally quite broad. That is, peak widths span a considerable range of wavelength, typically 50 nm or more. (The reason is discussed in Further Exploration 15.1 in the *Study Guide*.)

FURTHER EXPLORATION 15.1 More on UV Spectroscopy

The absorbance at a given wavelength depends on the number of molecules in the light path. If a sample is contained in a vessel with a thickness along the light path of l cm, and the absorbing compound is present at a concentration of c moles per liter, then the absorbance is proportional to the product lc.

$$A = \epsilon lc \qquad (15.2)$$

This equation is called the *Beer–Lambert law* or simply **Beer's law**. The constant of proportionality ϵ is called the **molar extinction coefficient** or **molar absorptivity**. The units of ϵ are L mol^{-1} cm^{-1}, or M^{-1} cm^{-1}; these units are sometimes omitted when values of ϵ are cited. Each absorption in a given spectrum has a unique extinction coefficient that depends on wavelength, solvent, and temperature. The larger is ϵ, the greater is the light absorption at a given concentration c and path length l. For example, the extinction coefficient of isoprene (Fig. 15.5) at its λ_{max} of 222.5 nm is 10,750 M^{-1} cm^{-1} in methanol solvent at 25 °C; its extinction coefficient in alkane solvents is nearly twice as large.

Extinction coefficients of 10^4–10^5 M^{-1} cm^{-1} are common for molecules with conjugated π-electron systems. This means that strong absorptions can be obtained from very dilute solutions—solutions with concentrations on the order of 10^{-4} to 10^{-6} M with a typical path length of 1 cm. Because of its intrinsic sensitivity and its relatively simple instrumentation, UV–vis spectroscopy was one of the earliest forms of spectroscopy to be used routinely in the laboratory; adequate spectra could be obtained on even the most primitive spectrometers. UV–vis spectroscopy remains a very important method for quantitative analysis.

Some UV–vis spectra are presented in abbreviated form by citing the λ_{max} values of their principal peaks, the solvent used, and the extinction coefficients. For example, the spectrum in Fig. 15.5 is summarized as follows:

$$\lambda_{max}(CH_3OH) = 222.5 \text{ nm } (\epsilon = 10{,}750)$$

or

$$\lambda_{max}(CH_3OH) = 222.5 \text{ nm } (\log \epsilon = 4.03)$$

PROBLEM

15.9 (a) From the extinction coefficient of isoprene (10,750 M^{-1} cm^{-1}) and its observed absorbance at 222.5 nm (Fig. 15.5), calculate the concentration of isoprene in mol L^{-1} (assume a 1 cm light path).

(b) From the results of part (a) and Fig. 15.5, calculate the extinction coefficient of isoprene at 235 nm.

B. Physical Basis of UV–Vis Spectroscopy

What determines whether an organic compound will absorb UV or visible radiation? Ultraviolet and/or visible radiation is absorbed by the π electrons and, in some cases, by the unshared electron pairs in organic compounds. For this reason, UV–vis spectra are sometimes called *electronic spectra*. (The electrons of σ bonds absorb at much shorter wavelengths, in the far ultraviolet.) Absorptions by compounds containing only single bonds and unshared electron pairs are generally quite weak (that is, their extinction coefficients are small). However, intense absorption of UV and visible radiation occurs when a compound contains π electrons. The simplest hydrocarbon containing π electrons, ethylene, absorbs UV radiation at $\lambda_{max} = 165$ nm ($\epsilon = 15{,}000$). Although this is a strong absorption, the λ_{max} of ethylene and other simple alkenes is below the usual working wavelength range of most conventional

UV–vis spectrophotometers; the lower end of this range is about 200 nm. However, molecules with *conjugated* double or triple bonds (for example, isoprene, Fig. 15.5) have λ_{max} values greater than 200 nm. Therefore, *UV–vis spectroscopy is especially useful for the diagnosis of conjugated double or triple bonds.*

$$H_2C=CHCH_2CH_2CH=CH_2$$

the double bonds are not conjugated; no λ_{max} above 200 nm

the double bonds are conjugated; this compound has a λ_{max} above 200 nm: λ_{max} = 227 nm

The structural feature of a molecule responsible for its UV–vis absorption is called a **chromophore**, from Greek words meaning "to bear color." Thus, the chromophore in isoprene (Fig. 15.5) is the system of conjugated double bonds. Because many important compounds do not contain conjugated double bonds or other chromophores, UV–vis spectroscopy has limited utility in structure determination compared with NMR and IR spectroscopy. However, the technique is widely used for quantitative analysis in both chemistry and biology; and, when compounds do contain conjugated multiple bonds, the UV–vis spectrum can be an important element in a structure proof.

The physical phenomenon responsible for the absorption of energy in the UV–vis spectroscopy experiment can be understood from a consideration of what happens when ethylene absorbs UV–vis radiation at 165 nm. The π-electron structure of ethylene was discussed in Sec. 4.1B, and is shown in Fig. 15.6. In the normal state of the ethylene molecule, called the *ground state*, the two π electrons occupy a *bonding* π molecular orbital. When ethylene absorbs energy from light, one π electron is elevated from this bonding molecular orbital to the *antibonding* or π^* molecular orbital. This means that the electron assumes the more energetic wave motion characteristic of the π^* orbital, which includes a node between the two carbon atoms. The resulting state of the ethylene molecule, in which there is one electron in each molecular orbital, is called an *excited state*. The energy required for this absorption must match ΔE, the difference in the energies of the π and π^* orbitals (Fig. 15.6). As a result, the 165 nm absorption of ethylene is called a $\pi \rightarrow \pi^*$ **transition** (read "pi to pi star"). The UV absorptions of conjugated alkenes are also due to $\pi \rightarrow \pi^*$ transitions.

FIGURE 15.6 Absorption of UV radiation by ethylene. The molecular orbitals of ethylene are shown on the left, and the energy difference between these orbitals is shown as ΔE. In the ground state of ethylene, two electrons of opposite spin occupy the bonding (π) molecular orbital. When ethylene is subjected to UV radiation of energy = ΔE, an electron (shown in *red*) is promoted from the bonding molecular orbital to the antibonding (π^*) molecular orbital. The product of this energy absorption is an excited state of ethylene.

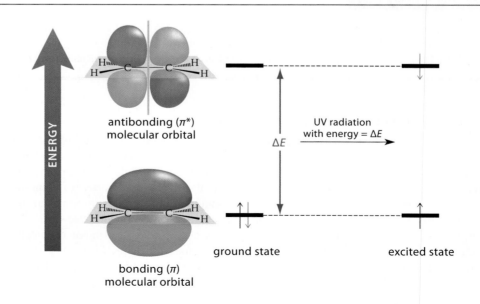

antibonding (π^*) molecular orbital

bonding (π) molecular orbital

ENERGY

ΔE

UV radiation with energy = ΔE

ground state

excited state

C. UV–Vis Spectroscopy of Conjugated Alkenes

When UV–vis spectroscopy is used to determine chemical structure, the most important aspect of a spectrum is the λ_{max} values. The structural feature of a compound that is most important in determining the λ_{max} is the number of consecutive conjugated double (or triple) bonds. *The longer the conjugated π-electron system (that is, the more consecutive conjugated multiple bonds), the higher the wavelength of the absorption.* Molecular orbital theory provides an explanation for this effect. As shown in Fig. 15.6, the energy of the radiation required for UV–vis absorption is determined by the energy separation between the occupied (bonding) MO and the unoccupied (antibonding) MO. As you learned in Sec. 12.1, this energy is inversely proportional to the wavelength λ:

$$\Delta E = h\nu = \frac{hc}{\lambda} \tag{15.3}$$

A conjugated alkene contains more than one bonding MO and more than one antibonding MO, as we found for 1,3-butadiene (Fig. 15.1, p. 715). In a conjugated diene, the UV–vis absorption at highest wavelength results in promotion of a π electron from the bonding MO of *highest* energy, called the **HOMO** (an acronym for "highest occupied molecular orbital") to the antibonding MO of *lowest* energy, called the **LUMO** (for "lowest unoccupied molecular orbital"). In other words, the HOMO–LUMO energy gap—the energy difference between these two MOs—determines the wavelength of the absorption. The relative energies of the π MOs for ethylene and the first two conjugated alkenes are shown in Fig. 15.7. This figure shows that the HOMO–LUMO gap becomes smaller as the number of conjugated double bonds increases. The energy of the radiation required for absorption, then, becomes smaller, and the wavelength greater, as the number of double bonds increases. (Other factors in addition to the HOMO–LUMO gap also contribute to the λ_{max}; see Further Exploration 15.1 in the *Study Guide*.)

Table 15.2 (p. 724) lists the λ_{max} values for a series of conjugated alkenes. Notice that λ_{max} (as well as the extinction coefficient) increases with increasing number of conjugated

FIGURE 15.7 The relationship of the absorption energy in UV–vis spectroscopy to the number of conjugated double bonds. The energies of the π molecular orbitals for ethylene, 1,3-butadiene, and (*E*)-1,3,5-hexatriene are given in β units. The energy of the UV or visible radiation required for absorption is equal to the gap (*lavender* shading) between the highest occupied MO (HOMO) and the lowest unoccupied MO (LUMO). Absorption (indicated by the *red* "squiggly arrows") results in the promotion of an electron from the HOMO to the LUMO. As the number of double bonds increase, the size of the HOMO–LUMO gap decreases and, by Eq. 15.3, the absorption wavelength increases.

TABLE 15.2 Ultraviolet Absorptions for Ethylene and Some Conjugated Alkenes

Alkene	λ_{max}, nm	ϵ
ethylene	165	15,000
∿	217	21,000
∿∿	268	34,600
∿∿∿	364	138,000

double bonds; each additional conjugated double bond increases λ_{max} by 30 to 50 nm. Molecules that contain many conjugated double bonds, such as the last one in Table 15.2, generally have several absorption peaks. These result not only from the HOMO–LUMO transition but also from electronic transitions involving other π MOs as well. The λ_{max} usually quoted for such compounds is the one at highest wavelength, which corresponds to the HOMO–LUMO transition.

If a compound has enough double bonds in conjugation, one or more of its λ_{max} values will be large enough to fall within the visible region of the electromagnetic spectrum, and the compound will be colored. An example of a conjugated alkene that absorbs visible light is β-carotene, which is found in carrots and is known to be a biological precursor of vitamin A:

β-carotene

Because of the large number of conjugated double bonds in β-carotene, its absorption, which occurs between 400 and 500 nm, is in the visible (blue-green) part of the electromagnetic spectrum. Thus, when a sample of β-carotene is exposed to white light, blue-green light is absorbed, and the eye perceives the *unabsorbed* light, which is red-orange. In fact, β-carotene is responsible for the orange color of carrots. Similarly, flamingos (Fig. 15.8) are red-orange because of the vitamin A in their diets.

The human eye can detect visible light because the eye contains organic compounds that absorb light in the visible region of the electromagnetic spectrum. In fact, light absorption

FIGURE 15.8 The bright red-orange color of flamingos is due to the vitamin A in their diets.

by a pigment, *rhodopsin*, in the rod cells of the eye (as well as a related pigment in the cone cells) is the event that triggers the physiological response that we know as *vision*. The chromophore in rhodopsin is its group of six conjugated double bonds (red in the following structure):

**rhodopsin
(visual purple)**

Absorption of a photon by this chromophore results in a $Z \rightarrow E$ isomerization about one of the double bonds, as shown above. This isomerization drastically alters the shape of the molecule, and this change of shape, in turn, causes a large change in the conformation of the surrounding protein. These events set of a cascade of molecular signals that culminate in the visual response.

Although the number of double or triple bonds in conjugation is the most important thing that determines the λ_{max} of an organic compound, other factors are involved. One is *the conformation of a diene unit about its central single bond*—that is, whether the diene is in an *s*-cis or an *s*-trans conformation (Sec. 15.1B). Recall that an acyclic diene assumes the lower energy *s*-trans conformation. However, dienes locked into *s*-cis conformations have *higher* values of λ_{max} and *lower* extinction coefficients than comparably substituted *s*-trans compounds:

Primarily *s*-trans
λ_{max} = 227 nm
(ϵ = 14,200)

constrained by the ring to
an *s*-cis conformation
λ_{max} = 256 nm (ϵ = 8000)

constrained by the the ring to
an *s*-cis conformation
λ_{max} = 239 nm (ϵ = 3400)

A third variable that affects λ_{max} in a less dramatic yet predictable way is the presence of substituent groups on the double bond. For example, each alkyl group on a conjugated double bond adds about 5 nm to the λ_{max} of a conjugated alkene. Thus, the two methyl groups of 2,3-dimethyl-1,3-butadiene add (2×5) = 10 nm to the λ_{max} of 1,3-butadiene, which is 217 nm (Table 15.2). The predicted λ_{max} is (217 + 10) = 227 nm; the observed value is 226 nm.

CH_3 ⟵——— one alkyl group   + 5 nm

       ⟵ basic diene unit   217 nm

CH_3 ⟵——— one alkyl group   + 5 nm
                                  ———————
                                  227 nm = predicted λ_{max}

2,3-dimethyl-1,3-butadiene

(observed λ_{max} = 226 nm)

Although other structural features affect the λ_{max} of a conjugated alkene, the two most important points to remember are:

1. The λ_{max} is greater for compounds containing more conjugated double bonds.
2. The λ_{max} is affected by substituents, conformation, and other structural characteristics of the conjugated π-electron system.

PROBLEM

15.10 Predict λ_{max} for the UV absorption of each of the following compounds.

Sunscreens

Sunscreens and sunblocks are used to protect the skin from harmful ultraviolet rays of the sun. The harmful UV radiation of sunlight is often discussed in terms of the parts of the solar spectrum termed "UV-A" and "UV-B." UV-B rays, which cover roughly the 280–315 nm part of the spectrum, are mostly responsible for sunburn and have long been associated with skin cancer. UV-A rays, which cover the 315–400 nm part of the spectrum, are responsible for wrinkling and aging. More recent research suggests that UV-A rays can also contribute to skin cancer. Sunscreens, which have a $1 billion annual market in the United States, are often characterized by an "SPF" (sun-protection factor). The SPF typically measures protection from UV-B rays. The best sunscreen formulations not only have high SPF numbers but also provide protection from UV-A rays.

Sunscreens work by absorbing the UV rays of the Sun. Therefore, it should not be surprising that sunscreens have UV–visible spectra, which are direct measures of their UV-absorbing capability. Notice in the following structure of a typical sunscreen the system of conjugated double bonds (*red*), which is responsible for the UV absorption.

2-ethylhexyl *p*-methoxycinnamate

The UV spectrum of this compound is shown in Fig. 15.9. This spectrum shows that this compound is an effective absorber of UV-B rays but has only modest absorption in the UV-A region. This is why sunscreen preparations should contain UV-A blockers as well. (Most UV-A blockers also contain conjugated π-electron systems with appropriate UV absorption.)

Notice also the use of the long hydrocarbon chain in the ester group (*blue*), which reduces the water solubility of the sunscreen. This same hydrocarbon group, however, promotes absorption through the skin. The absorption of sunscreens has been a cause for concern, but their beneficial effect in reducing the incidence of skin cancer seems to outweigh the other risks of their use.

D. Fluorescence

Most of us have seen things that "glow" with bright colors under a UV lamp. This glow is caused by the emission of light in the visible region from compounds that absorb ultraviolet light. These compounds are *fluorescent*. The purpose of this section is to understand the origin of fluorescence and why it is important, especially in biology.

FIGURE 15.9 The UV spectrum of a typical sunscreen, 2-ethylhexyl *p*-methoxycinnamate, in ethanol with the UV-A and UV-B regions indicated.

To start with, we consider the origin of fluorescence by looking a little more deeply into UV–visible absorption. Recall that when any conjugated molecule absorbs UV or visible radiation, an electron is promoted from the *highest occupied molecular orbital* (HOMO) to the *lowest unoccupied molecular orbital* (LUMO) (Fig. 15.7). Before absorption, the molecule is said to be in its **ground state**. After absorption, the molecule is said to be in an **excited state**.

(15.4)

The molecule in its excited state now contains energy resulting from light absorption. What happens to that energy? One of the most common things that can occur is that the molecule can lose its excess energy by light emission; this light emission is called **fluorescence**. In fluorescence, the electron that was promoted to the LUMO by absorption returns to the HOMO, and the energy that is lost is emitted as light. At first glance, it might seem that the light emitted should have the same energy (and therefore the same wavelength) as the light absorbed. However, in most cases, the light emitted occurs at *lower energy*, and therefore *longer wavelength*, than the light absorbed. For example, a molecule that absorbs blue or violet light might emit green fluorescence. The wavelength shift that occurs in fluorescence is called the **Stokes shift**. The Stokes shift is one of the most useful aspects of fluorescence. The Stokes shift is why certain fluorescent objects glow visibly under UV light.

The origin of the Stokes shift is not apparent from the diagram in Eq. 15.4. Rather, it originates in the *molecular vibrations* of the molecule. When a molecule absorbs UV or visible radiation, the promotion of the π electron occurs so rapidly that *the bond lengths of the molecule do not change during the absorption*. However, in the excited state, the optimum bond lengths may be different from those in the ground state. For example, the electrons in a conjugated double bond that are delocalized in the ground state may be less delocalized in the excited state. If the optimum length of this double bond is shorter in the excited state, the bond that had an optimum length in the ground state may find itself suddenly longer than optimum in the excited state. Therefore, absorption produces an excited state that is not only *electronically* excited but also *vibrationally* excited (Fig. 15.10a). "Vibrationally excited" means in a mechanical sense that, immediately after absorption, the bonds in the molecule find themselves stretched or compressed relative to their optimum lengths in the excited state. Even though the electronic excited state lasts only 10^{-9} to 10^{-8} seconds, the lifetime of the vibrationally excited state is typically 10^{-11} to 10^{-12} seconds—about 1000 times smaller. Therefore, immediately following absorption, the molecule undergoes vibrational relaxation to the bond lengths that are optimum for the excited state (Fig. 15.10b). In this process, some of the energy of the excited state is lost as heat. Eventually, the excited molecule returns to the ground state when the π electron returns to the HOMO from the LUMO, and the loss of energy is seen as fluorescence (Fig. 15.10c). However, here again the vibrational energy levels come into play. The fluorescence process returns the molecule to a vibrationally excited ground state, from which the remaining excess energy is lost by vibrational relaxation as heat. From this description you can see that *the energy lost in fluorescence is less than the energy gained in absorption*. Because $E = hc/\lambda$, *the wavelength of the emitted light is greater than that of the absorbed light*. This difference in wavelength between absorbed and emitted light is the Stokes shift.

Fluorescence intensity follows Beer's law (Eq. 15.2) just as absorption does. Molecules that fluoresce strongly must absorb UV–visible radiation strongly as well—that is, they must have large extinction coefficients for absorption. However, not all strong absorbers produce significant fluorescence because excited states can sometimes undergo reactions, transitions to other types of excited states, and so on—processes that we won't be concerned with here. The efficiency of fluorescence is determined by the fraction of excited states that return to ground state by fluorescence. This efficiency is called the **quantum yield** of fluorescence. Many compounds commonly used in fluorescence have quantum yields in the 0.8 to 1.0 range.

There aren't any hard-and-fast rules for determining whether a molecule will have a high quantum yield for fluorescence. To fluoresce strongly, a molecule must have a high absorbance, because fluorescence originates from absorbance. However, a high absorbance is no guarantee that a molecule will have significant fluorescence, because there are other ways besides fluorescence that an excited state can lose energy. Many strongly fluorescent molecules have chromophores consisting of several conjugated double bonds incorporated into rigid polycyclic frameworks. Non-carbon atoms such as nitrogen and oxygen are in many cases part of the conjugated system, as you will see in the examples later in this section.

The importance of fluorescence as an analytical tool lies in its sensitivity. If fluorescence follows the same concentration dependence as absorption, why is it so sensitive? Recall (Eq. 15.1, p. 720) that UV–visible absorbance is the logarithm of the *ratio* of the light intensity emerging from the sample to the intensity of the light source; $A = \log (I_0/I)$. (This ratio is usually provided physically by the optics in the spectrophotometer.) In other words, UV absorption requires comparison of the light emerging from the sample with a large background—a small difference between large numbers. In fluorescence, however, we measure the absolute intensity of light emerging from the sample, and the standard of comparison is total darkness. Typically, fluorescence detectors are set up at 90° to the light path so that the exciting light source does not interfere with detection.

An analogy to absorption sensitivity is trying to hear a few people whispering in a room full of people talking loudly. The analogy to fluorescence sensitivity is that if the same people are placed in a totally silent room, their whispering can be heard clearly.

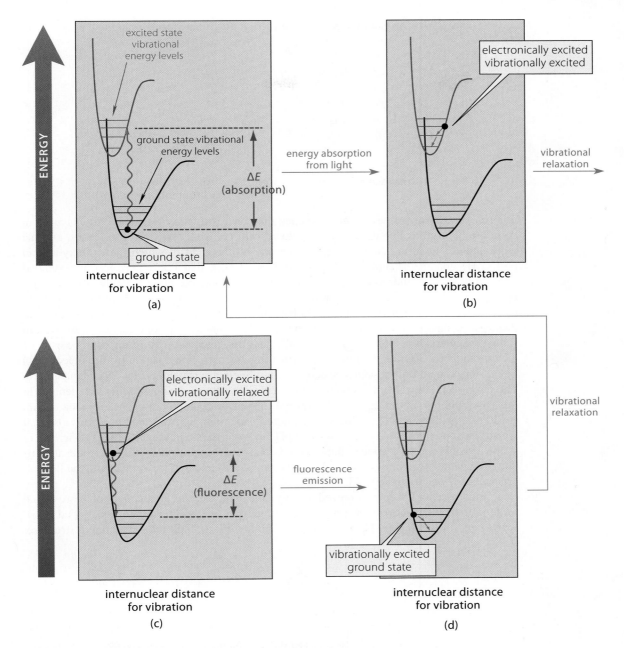

FIGURE 15.10 The absorption and fluorescence process. A molecule in its ground state (a) absorbs energy from light and is promoted to an electronically excited and vibrationally excited state (b). The electronically excited state loses vibrational energy as heat and forms an electronically excited and vibrationally relaxed state (c). This state loses energy by emitting light (fluorescence) to form a vibrationally excited ground state (d). This state then relaxes vibrationally to form the ground state (a). The energy spacing of the vibrational levels is highly exaggerated for clarity. This process is illustrated for one vibrational mode, but every vibrational mode in the molecule contributes. Notice that the energy of the light emitted, ΔE (fluorescence), is less than the energy of the light absorbed, ΔE (absorbance), and for that reason, the wavelength of the emitted light is greater than that of the light absorbed (Stokes shift).

Fluorescein is an example of a widely used, highly fluorescent compound.

fluorescein

major form at pH > 7
(chromophore in color)

Fluorescein derivatives have been developed that react with other functional groups and can therefore be used as fluorescent "tags." Fluorescein contains two ionizable groups, and the most fluorescent form, shown above, is the one on the right in which the carboxylic acid ($-CO_2H$) and the phenol ($-OH$) are ionized. Notice that fluorescein has an extended π-electron system, and, as expected, it absorbs UV–visible radiation strongly. The chromophore responsible for the absorption (and the fluorescence) at long wavelengths is relatively rigid; the attached phenyl group (black) has only a minor contribution to the absorption at these wavelengths. The unshared electron pairs of the ionized phenol oxygen are involved in resonance interaction with the double bonds, and this interaction is an important aspect of the chromophore. (Draw the resonance structures for the delocalization of these electrons.) The absorption and fluorescence spectra of fluorescein are shown in Fig. 15.11a and 15.11b. The fluorescence spectrum is obtained by exciting the molecule at a specific wavelength (in this case, 488 nm, at the λ_{max} for absorption). The peak in the fluorescence emission occurs at 510 nm. Notice the Stokes shift in the fluorescence to higher wavelength. Notice also that the absorption intensity is smaller at lower pH, and so is the fluorescence intensity. Fig. 15.11c shows a visible demonstration of fluorescein fluorescence. In this photo, the electronic absorption is activated by shining a UV light on the sample. Notice that we are viewing the fluorescence at a 90° angle to the UV light path.

Fluorescence has revolutionized biology. One of the exciting applications of fluorescence resulted from the discovery of a fluorescent protein in a jellyfish (Fig. 15.12a), called *green fluorescent protein* (GFP). The fluorescing group in GFP is "built into" the protein structure:

the fluorescent group of GFP

The protein itself wraps around the fluorescent group like a barrel and excludes it from solvent water. When this group is allowed to interact with water, its fluorescence quantum yield drops to zero because the energy of the excited state is lost to solvent vibrations rather than by fluorescence.

GFP is a fairly small and relatively stable protein. The gene for GFP can in many cases be joined to the genes of proteins of interest, and the two proteins can be co-expressed in living organisms as a single "fusion" protein. Essentially, the GFP is carried by the protein of interest as a "molecular flashlight" so that the localization of the protein of interest can be seen by viewing a cell under a fluorescence microscope (Fig. 15.12b). Even genetically modified whole animals carrying GFP have been produced (Fig. 15.12c). It is now possible to produce mutant GFPs that give fluorescence of different colors, such as red and yellow. The discovery and development of GFP as a biological tool was recognized with the 2008 Nobel Prize in Chemistry, which was awarded to Osamu Shimomura (b. 1928) of the Marine Biological Laboratory in Woods Hole, Massachusetts; Martin Chalfie (b. 1947) of Columbia University; and Roger Y. Tsien (b. 1952) of the University of California, San Diego.

FIGURE 15.11 (a) The UV–visible absorption spectrum of fluorescein at two pH values. (b) The fluorescence emission spectrum of fluorescein at the same two pH values resulting from absorption at the λ_{max} (488 nm). Notice the Stokes shift of fluorescence to greater wavelength. (c) Fluorescein sprinkled into tap water emits green fluorescence under a handheld UV light.

FIGURE 15.12 (a) The fluorescent jellyfish *Aequorea victoria*, source of the green fluorescent protein. (b) A fluorescent light micrograph of GFP-containing oligodendrocytes from a mouse. The gene for a protein localized in the oligodendrocyte was fused with the gene for GFP and the fusion protein was expressed within a genetically engineered mouse in its oligodendrocytes. These cells form the myelin sheath around neurons (nerve cells) in the central nervous system (CNS). (c) Scientists at Mayo Clinic in Rochester, Minnesota, have produced transgenic cats containing a gene that imparts resistance to feline immunodeficiency virus (FIV, the cat version of HIV). The resistance gene was fused to the GFP gene so that the two genes were incorporated together. Therefore, cats containing the resistance gene are readily detected by their fluorescence under blue light.

PROBLEMS

15.11   Refer to Fig. 15.11. At what pH does the fluorescence of fluorescein have the greater quantum yield, pH = 9 or pH = 5? How do you know?

15.12   One of the following compounds has an intense yellow fluorescence when irradiated with UV light. Which one do you think it is, and why?

15.3   THE DIELS–ALDER REACTION

A. Reaction of Conjugated Dienes with Alkenes

Conjugated dienes undergo several unique reactions. One of these was discovered in 1928, when two German chemists, Otto Diels (1876–1954) and Kurt Alder (1902–1958), showed that many conjugated dienes undergo addition reactions with certain alkenes or alkynes. The following reaction is typical:

$$ \qquad\qquad (15.5) $$

(84% yield)

This type of reaction between a conjugated diene and an alkene is called the **Diels–Alder reaction**. For their extensive work on this reaction, Diels and Alder shared the 1950 Nobel Prize in Chemistry.

The Diels–Alder reaction is an example of a **cycloaddition reaction**—an addition reaction that results in the formation of a ring. Indeed, *the Diels–Alder reaction is an important method for making rings*, as the example in Eq. 15.5 illustrates.

The Diels–Alder reaction is also an example of a **1,4-addition** or **conjugate addition**. In such a reaction, addition occurs across the outer carbons (carbons 1 and 4) of the diene. *Conjugate addition is a characteristic type of reaction of conjugated dienes.* In the Diels–Alder reaction, conjugate addition also results in the formation of a double bond between carbons 2 and 3. (The numbers indicate the relative locations of the carbons involved in the addition; it has nothing to do with the numbering of the diene used in its substitutive nomenclature.)

the diene and dienophile are joined at C1 and C4 of the diene

a new π bond is formed between C2 and C3

$$ \qquad\qquad (15.6) $$

STUDY GUIDE LINK 15.1
A Terminology Review

When discussing the reactants in the Diels–Alder reaction, we employ the following terminology, which is illustrated in Eq. 15.6. The conjugated diene reactant is referred to simply as the *diene*, and the alkene with which it reacts is called the **dienophile** (literally, "diene-loving molecule"). The diene in 1,3-butadiene is used as the diene for simplicity in Eq. 15.6, but as you'll see shortly, a wide variety of dienes can be used in this reaction.

Mechanistically, the Diels–Alder reaction occurs in a single step involving a cyclic flow of electrons. The curved arrows for this mechanism can be drawn in either a clockwise or counterclockwise direction.

or

$$ \qquad\qquad (15.7) $$

(The best evidence for a concerted rather than a stepwise mechanism for the Diels–Alder reaction comes from the stereochemistry of the reaction, which we'll consider in Secs. 15.3B and 15.3C.) A concerted reaction that involves a cyclic flow of electrons is called a **pericyclic reaction**. The Diels–Alder reaction is a pericyclic reaction, as is the hydroboration of alkenes (Sec. 5.4B). However, hydroboration is not a cycloaddition, because no ring is formed. (Pericyclic reactions as a class are discussed from the perspective of MO theory in Chapter 28.)

Some of the dienophiles that react most rapidly in the Diels–Alder reaction, as in Eq. 15.5, bear substituent groups such as esters ($—CO_2R$), nitriles ($—CN$), or certain other unsaturated, electronegative groups. However, these substituents are not strictly necessary because the reactions of many other alkenes can be promoted by heat or pressure. Some alkynes can also serve as dienophiles.

$$(15.8)$$

As you have already learned, when a simple diene is used in the Diels–Alder reaction, a new ring is formed. When the diene is cyclic, a *second* ring is formed. In other words, the Diels–Alder reaction can be used to prepare certain *bicyclic compounds* (Sec. 7.6A).

$$(15.9)$$

a cyclic diene

different representations of the bicyclic product

STUDY PROBLEM 15.1

Give the structure of the diene and dienophile that would react in a Diels–Alder reaction to give the following product:

SOLUTION   In the product of a Diels–Alder reaction, the two carbons of the double bond and the two *adjacent carbons* originate from the diene. These carbons are numbered 1 through 4 in the following structure:

The two new single bonds formed in the reaction connect carbons 1 and 4 of the diene to the carbons of the dienophile double bond, which (because they are part of the same double bond) must be adjacent in the dienophile. This analysis reveals two possibilities, A and B, for the bonds formed in the Diels–Alder reaction:

Because the product is bicyclic, the diene in either case is a *cyclic* diene. The double bonds in the diene are between carbons 1 and 2 and between carbons 3 and 4. Thus, to derive the starting materials in each case, follow these steps:

1. Disconnect the bonds between carbons 1 and 4 and their adjacent dienophile carbons.
2. Complete the diene structure by eliminating the double bond between carbons 2 and 3 and by adding the C1–C2 and C3–C4 double bonds.
3. Complete the dienophile structure by adding the double bond between its carbons.

By following these steps we find that the starting materials for possibilities *A* and *B* are as follows: (The carbon skeleton of the diene unit is first drawn exactly as it looks in the product (even though this is a distorted conformation), and then it is drawn in the more conventional way.)

In principle, either combination *A* or *B* could serve as the starting materials in a Diels–Alder reaction. Recall, however, that dienophiles with ester groups (or other electronegative groups) react faster than those without such groups. Hence, the reactants in *B* would be preferred.

It may have occurred to you that cycloaddition reactions between two alkenes or between two dienes appear to be formally possible:

These reactions do *not* occur under thermal conditions. Although we might attribute the failure of the first reaction to ring strain in the product, there is more to it than that. The basis of these observations can be found in the theory of pericyclic reactions, which is the subject of Chapter 28. Be sure you understand that the Diels–Alder reaction is the reaction of a *conjugated diene* with an *alkene*.

PROBLEMS

15.13 What products are formed in the Diels–Alder reactions of the following dienes and dienophiles?

(a)

and

(b)

and

15.14 Give the diene and dienophile that would react in a Diels–Alder reaction to give each of the following products.

(a)   (b)

15.15 (a) What product would be expected from the Diels–Alder reaction of 1,3-butadiene as the diene and ethylene as the dienophile?

(b) This product is actually not observed under ordinary conditions because 1,3-butadiene reacts with itself faster than it does with ethylene. In this reaction, one molecule of 1,3-butadiene acts as the diene component and the other as the dienophile. Give the product of this reaction.

(c) How would you alter the reaction conditions to favor the formation of the product in part (a)?

15.16 (a) Explain why two constitutional isomers are formed in the following Diels–Alder reaction:

(84% of the product)   (16% of the product)
(54% total yield)

(b) What two constitutional isomers could be formed in the following Diels–Alder reaction?

B. Effect of Diene Conformation on the Diels–Alder Reaction

Dienes that are "locked" into *s*-trans conformations are unreactive in Diels–Alder reactions:

"locked" *s*-trans dienes;
unreactive in Diels–Alder reactions

The reason is that if such dienes were to form Diels–Alder products, the *s*-trans single bond of the diene would become a trans double bond in the Diels–Alder product. This means that the

Diels–Alder product would contain a trans double bond in a six-membered ring. For example, consider the following reaction.

The product is a bicyclic compound containing a bridgehead double bond. As discussed in Sec. 7.6C, the bridgehead double bond (red in the preceding equation) has trans stereochemistry within one of the rings joined at the bridges, and therefore the product violates Bredt's rule and is too strained to exist. (For a graphic demonstration, try building a model of the product, but don't break your models.)

In contrast, dienes locked into *s*-cis conformations are considerably more reactive than the corresponding noncyclic dienes:

furan	**1,3-cyclopentadiene**	**1,3-cyclohexadiene**	**1,2-dimethylenecyclohexane**

all are "locked" *s*-cis dienes;
all are reactive in the Diels–Alder reaction

For example, 1,3-cyclopentadiene, which is locked in an *s*-cis conformation, reacts with typical dienophiles hundreds of times more rapidly than 1,3-butadiene, which exists primarily in the *s*-trans conformation.

These observations are consistent with a transition state in which the diene component of the reaction has assumed an *s*-cis conformation. This transition state is shown in Fig. 15.13 for the reaction of 1,3-butadiene and ethylene. In this transition state, the diene and the dienophile approach in parallel planes. The $2p$ orbitals on the dienophile interact with the $2p$ orbitals on the outer carbons of the diene to form the new σ bonds. Because 1,3-butadiene prefers the *s*-trans conformation (Fig. 15.1, p. 715), the energy required for it to assume the *s*-cis conformation in the transition state becomes part of the energy barrier for the reaction. In contrast, a diene that is locked by its structure into an *s*-cis conformation, such as 1,3-cyclopentadiene, does not have this additional energy barrier to climb before it can react; hence, it reacts more rapidly.

FIGURE 15.13 In the transition state for the Diels–Alder reaction, the diene (1,3-butadiene in this example) and the dieno-phile (ethylene in this example) approach in parallel planes (as shown by the green arrows) so that the $2p$ orbitals of the dienophile overlap with the $2p$ orbitals on carbons 1 and 4 of the diene to form the two new σ bonds. The developing over-lap is indicated with blue lines. Notice that the diene is in an *s*-cis conformation.

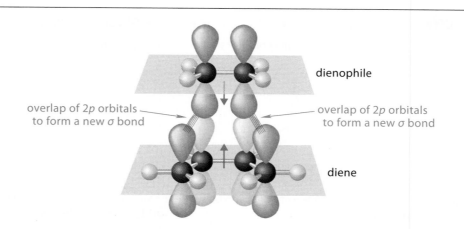

The importance of the *s*-cis diene conformation can have some fairly drastic consequences for the reactivity of some noncyclic dienes. Thus, the *E* isomer of 1,3-pentadiene reacts 12,600 times more rapidly than the *Z* isomer of the same diene with tetracyanoethylene (TCNE), a very reactive dienophile:

(15.10a)

(15.10b)

As Eq. 15.10b shows, the *s*-cis conformation of the cis diene is destabilized by a significant van der Waals repulsion between the methyl group and a diene hydrogen. The transition states for the Diels–Alder reactions of this diene, which require an *s*-cis conformation, are destabilized by the same effect. Consequently, the Diels–Alder reactions of (*Z*)-1,3-pentadiene are much slower than the corresponding reactions of (*E*)-1,3-pentadiene, in which the destabilizing repulsion in its *s*-cis conformation is between hydrogens and is much less severe.

PROBLEMS

15.17 A mixture of 0.1 mole of (2*E*,4*E*)-2,4-hexadiene and 0.1 mole of (2*E*,4*Z*)-2,4-hexadiene was allowed to react with 0.1 mole of TCNE. After the reaction, the unreacted diene was found to consist of only one of the starting 2,4-hexadiene isomers. Which isomer did not react? Explain.

15.18 Complete the following Diels–Alder reaction.

C. Stereochemistry of the Diels–Alder Reaction

If the Diels–Alder reaction takes place in a single step without reactive intermediates, and if the transition-state picture of Fig. 15.10 is correct, then the diene should undergo a syn-addition to the dienophile. Likewise, the dienophile should undergo a 1,4-syn-addition to the diene. Each component adds to the other at *one face* of the π system.

The stereochemistry of the Diels–Alder reaction is completely consistent with these predictions. If we use a dienophile that is a cis alkene, groups that are cis in the alkene starting material are also cis in the product.

$$(15.11a)$$

Use of the trans isomer of this dienophile gives the complementary result:

$$(15.11b)$$

Although one enantiomer of the product is shown in Eq. 15.11b, the product is the racemate, because both starting materials are achiral (Sec. 7.7A).

Syn-addition to the diene is revealed if the terminal carbons of the diene unit are stereocenters. To assist in the analysis of stereochemistry, let's first draw the diene in its *s*-cis conformation and then classify the groups at the terminal carbons as inner substituents (R^i) or outer substituents (R^o):

$$(15.12a)$$

A syn-addition requires that in the Diels–Alder product, the two inner substituents always have a cis relationship; the two outer substituents are also cis; and an inner substituent on one carbon is always trans to an outer substituent on the other.

$$(15.12b)$$

The following reactions of the stereoisomeric 2,4-hexadienes with the dienophile maleic anhydride demonstrate these points.

$$(15.13a)$$

(2E,4E)-2,4-hexadiene   **maleic anhydride**

$$(15.13b)$$

(2E,4Z)-2,4-hexadiene

In Eq. 15.13a, the methyl groups in the diene are both outer substituents, and they are cis in the product. In Eq. 15.13b, one methyl group in the diene is outer and the other is inner; consequently, they are trans in the product.

> Notice, incidentally, the different reaction conditions required for reactions of the two dienes in Eqs. 15.13a and 15.13b. The latter reaction requires *much* more drastic conditions. Why? (See Eq. 15.10b.)

One other stereochemical issue arises in the reactions of Eqs. 15.13a and 15.13b: the stereochemistry at the ring junction. Because maleic anhydride is a cis alkene, and because the Diels–Alder reaction is a syn-addition, we know that the stereochemistry at the ring junction must be cis. However, for a given diene and dienophile, two diastereomeric syn-addition products are possible. The reaction of Eq. 15.13a illustrates this point:

$$(15.14)$$

This issue arises when *both* the terminal carbons of the diene *and* the carbons of the dienophile are stereocenters.

Let's classify these two possibilities with a more general equation in which a cis alkene reacts with a diene:

$$(15.15)$$

endo product   exo product

Following the diagram in Eq. 15.12a, we have drawn the diene in its *s*-cis conformation and have labeled the groups at the terminal carbons as outer or inner substituents. The product in which the alkene substituents R are cis to the outer diene substituents R^o is said to have **endo** stereochemistry. The product in which the alkene substituents R are trans to the outer diene substituents R^o is said to have **exo** stereochemistry. (The terms *endo* and *exo* are from Greek roots meaning "inside" and "outside," respectively.)

In many cases, the endo products are formed more rapidly than the exo products, although exceptions occur. (A theoretical explanation for this observation exists, but we won't consider that here.) If this is so, which product would be favored in Eq. 15.14? (See Problem 15.19.)

The preference for endo products extends to cases in which cyclic dienes are used and bicyclic products are formed:

$$(15.16)$$

1,3-cyclopentadiene — endo product (76%) — exo product (24%)

Be sure you see the correspondence between Eq. 15.16 and Eq. 15.15. The $—CH_2—$ group of the diene (red) represents the inner groups R^i (tied together in one group as part of the ring); the hydrogens in blue are the outer groups R^o. You can see from this equation that, in the predominant, or endo, product, the $—CO_2CH_3$ group is cis to R^o and trans to R^i.

PROBLEMS

15.19 Give the products formed when each of the following pairs reacts in a Diels–Alder reaction; show the relative stereochemistry of the substituent groups where appropriate. In part (b), show both exo and endo products and label them.

(a)

$$(—OAc = acetoxy = —O—\overset{\overset{\textstyle O}{\|}}{C}—CH_3)$$

(b)

(c)

$HC{\equiv}C—CO_2CH_3$

15.20 Give the structures of the starting materials that would yield each of the compounds below in Diels–Alder reactions. Pay careful attention to stereochemistry, where appropriate.

(a)

(b)

(c)

(d)

15.21 (a) In the products of Eq. 15.14, the observed stereochemistry at the ring fusion is not specified. Show this stereochemistry, assuming that the Diels–Alder reaction gives the endo product.

(b) Sketch diagrams like the one in Fig. 15.10 (without the orbitals) that shows the approach of the diene and dienophile leading to both endo and exo products in part (a). Pay careful attention to the relative positions of substituent groups.

15.22 Assuming endo stereochemistry of the product, give the structure of the compound formed when 1,3-cyclohexadiene reacts with maleic anhydride. (The structure of maleic anhydride is shown in Eq. 15.13a.)

15.4 ADDITION OF HYDROGEN HALIDES TO CONJUGATED DIENES

A. 1,2- and 1,4-Additions

Conjugated dienes, like ordinary alkenes (Sec. 4.7), react with hydrogen halides; however, conjugated dienes give two types of addition product:

$$MeCH{=}CH{-}CH{=}CHMe \ + \ HBr \ \xrightarrow{-20\,°C}$$

$$
\underset{\substack{\text{1,2-addition product} \\ (65\%)}}{MeCH{-}\underset{\underset{Br}{|}}{\underset{\underset{}{|}}{CH}}{-}CH{=}CHMe} \quad + \quad \underset{\substack{\text{1,4-addition product} \\ (\text{conjugate-addition product}) \\ (35\%)}}{MeCH{-}\underset{\underset{H}{|}}{CH}{=}CH{-}\underset{\underset{Br}{|}}{CHMe}} \quad (15.17)
$$

The major product is a *1,2-addition* product. (We'll address why this is the major product in Sec. 15.4C.) **1,2-Addition** means that addition (of HBr in this case) occurs at adjacent carbons. The minor product results from *1,4-addition*, or *conjugate addition*. In a **1,4-addition**, or **conjugate addition**, addition occurs to carbons that have a 1,4-relationship. (The terms *1,2-addition* and *1,4-addition* have nothing to do with systematic nomenclature.)

The 1,2-addition reaction is analogous to the reaction of HBr with an ordinary alkene. But how can we account for the conjugate-addition product? As in HBr addition to ordinary alkenes, the first mechanistic step is protonation of a double bond. Protonation of the diene in Eq. 15.17 at either of the equivalent carbons 2 or 5 gives a resonance-stabilized carbocation:

$$MeCH{=}CH{-}CH{=}CHMe \longrightarrow$$

$$\underset{\substack{H{-}\ddot{B}r\!:}}{}$$

$$
\left[\ \underset{\underset{H}{|}}{MeCH}{-}\overset{+}{CH}{-}CH{=}CHMe \quad \longleftrightarrow \quad MeCH{-}CH{=}CH{-}\overset{+}{CH}Me \ \right] \qquad (15.18)
$$

$$:\!\ddot{B}r\!:^-$$

The resonance structures for this carbocation show that the positive charge in this ion is not localized, but is instead *shared* by two different carbons. *Two constitutional isomers are formed in Eq. 15.17 because the bromide ion can react at either of the electron-deficient carbons:*

$$
\left[\ MeCH_2{-}\overset{+}{CH}{-}CH{=}CHMe \quad \longleftrightarrow \quad MeCH_2{-}CH{=}CH{-}\overset{+}{CH}Me \ \right]
$$

$$:\!\ddot{B}r\!:^-$$

$$
\underset{\underset{:\ddot{Br}:}{|}}{MeCH_2{-}CH}{-}CH{=}CHMe \quad + \quad MeCH_2{-}CH{=}CH{-}\underset{\underset{:\ddot{Br}:}{|}}{CHMe} \qquad (15.19)
$$

Protonation at carbon-3 would give a different carbocation, which would react with bromide ion to give an alkyl halide that is different from those obtained experimentally:

$$\text{MeCH}=\text{CH}-\text{CH}=\text{CHMe} \longrightarrow \overset{+}{\text{MeCH}}-\text{CH}-\text{CH}=\text{CHMe} \longrightarrow$$

$$\text{MeCH}-\text{CH}_2-\text{CH}=\text{CHMe} \qquad (15.20)$$

(not formed)

The course of addition to conjugated alkenes is suggested by Hammond's postulate (Sec. 4.8D), which predicts that the reaction pathway that involves the more stable carbocation occurs more rapidly. Because the carbocation in Eq. 15.18 is more stable, it is formed more rapidly; therefore, the products derived from this carbocation are the ones observed. Although both possible carbocations are secondary, the carbocation in Eq. 15.18 is *resonance-stabilized*, but the carbocation in Eq. 15.20 is not:

$$\left[\text{MeCH}_2-\overset{+}{\text{CH}}-\text{CH}=\text{CHMe} \longleftrightarrow \text{MeCH}_2-\text{CH}=\text{CH}-\overset{+}{\text{CHMe}} \right]$$

more stable carbocation (Eq. 15.18); formed more rapidly

$$\overset{+}{\text{MeCH}}-\text{CH}_2-\text{CH}=\text{CHMe}$$

less stable carbocation (Eq. 15.20); not formed

In other words, resonance accounts for the greater stability of the carbocation intermediate that is formed.

In Sec. 1.4, we stated as a fact that resonance is a stabilizing effect. We've seen other examples (Eq. 14.3b, p. 689, and Eq. 14.6a, p. 691) in which the resonance stabilization of a carbocation intermediate determines the course of the reaction. Now we're going to consider why resonance should be a stabilizing effect. Understanding the connection between resonance and stability is the subject of the Sec. 15.4B.

PROBLEMS

15.23 Use the reaction mechanism, including the resonance structures of the carbocation intermediates, to predict the products of the following reactions.

(a) 1,3-butadiene + HCl $\longrightarrow$

(b) S_N1 solvolysis of 3-chloro-1-methylcyclohexene in ethanol

15.24 Suggest a mechanism for each of the following reactions that accounts for both products.

(a) $\text{CH}_3\text{CH}=\text{CHCH}_2\text{OH} + \text{HBr} \xrightarrow{\text{H}_2\text{SO}_4}$

$$\text{CH}_3\text{CH}=\text{CHCH}_2\text{Br} + \text{CH}_3\text{CHCH}=\text{CH}_2$$

(84%) $\qquad\qquad$ Br

(16%)

(b) $(\text{CH}_3)_2\text{C}=\text{CH}-\text{CH}-\text{CH}_3 \xrightarrow{\text{H}_2\text{O/acetone}}$
Cl

$$(\text{CH}_3)_2\text{C}=\text{CH}-\text{CH}-\text{CH}_3 + (\text{CH}_3)_2\text{C}-\text{CH}=\text{CH}-\text{CH}_3 + \text{HCl}$$
OH $\qquad\qquad$ OH

B. Allylic Carbocations. The Connection between Resonance and Stability

To explore the connection between resonance and stability, we'll use a carbocation similar to the one involved in HBr addition to conjugated alkenes (Eq. 15.18). This is an example of an **allylic carbocation**: a carbocation in which the positively charged, electron-deficient carbon is adjacent to a double bond.

general structure of an **allylic carbocation**

The word *allylic* is a generic term applied to any functional group at a carbon adjacent to a double bond.

Allylic carbocations are more stable than comparably branched nonallylic alkyl carbocations. Roughly speaking, an allylic carbocation is about as stable as a nonallylic alkyl carbocation with one additional alkyl branch. Thus, a secondary allylic carbocation is about as stable as a tertiary nonallylic one. To summarize:

Stability of carbocations:

The reason for the stability of allylic carbocations lies in their electronic structures. The π-electron structure of the allyl cation, $H_2C\!=\!CH\!-\!\overset{+}{C}H_2$ (the simplest allylic cation), is shown in Fig. 15.14 (p. 744). The electron-deficient carbon and the carbons of the double bond are all sp^2-hybridized; each carbon has a $2p$ orbital (Fig. 15.14a). The overlap of these three $2p$ orbitals results in three π molecular orbitals. The MO of lowest energy, π_1, is bonding with an energy of $+1.41\beta$, and it has no nodes. (Remember that β is a negative number.) The next MO, π_2, has the same energy as an isolated $2p$ orbital (0β). An MO that has the same energy as an isolated $2p$ orbital is called a **nonbonding MO** (**NBMO**). This MO has one node. Because nodes must be placed symmetrically, this node goes through the central carbon. The position of this node determines the location of the positive charge, as we'll see shortly. The MO of highest energy, π_3^*, is antibonding. It has two nodes—one between each carbon.

The allyl cation has two electrons; both reside in π_1. Each π electron contributes $+1.41\beta$ to the energy of the molecule, for a total of $+2.82\beta$. The *delocalization energy* (Sec. 15.1A) of the allyl cation is the difference between this energy and the energies of an isolated ethylene ($+2.00\beta$) and an isolated $2p$ orbital (0β). The allyl cation, then, has a delocalization energy of 0.82β. Because the source of this stabilization is the "nodeless" π_1 MO, the stabilization of the allyl cation arises from the delocalization of π electrons across the entire molecule. This delocalization, then, is a source of *additional bonding* in the allyl cation that would not be present if the alkene π bond and the carbocation could not interact.

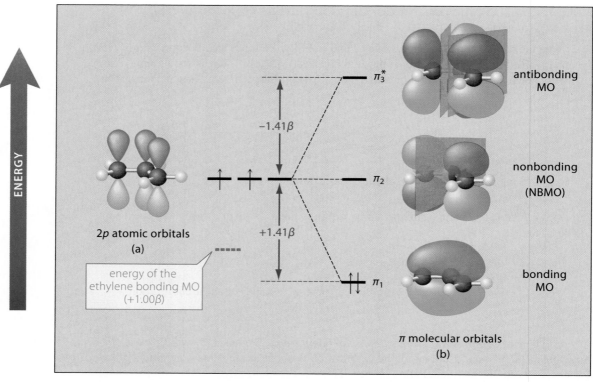

FIGURE 15.14 (a) An orbital interaction diagram that shows the arrangement of $2p$ orbitals in the allyl cation, the simplest allylic carbocation. Notice that the axes of the $2p$ orbitals are parallel and thus properly aligned for orerlap. (b) Interaction of the three $2p$ orbitals (*dashed lines*) gives three π MOs. Nodal planes are shown in gray. The two $2p$ electrons both go into π_1, the bonding MO. The violet arrows and numbers show the relative energies of the MOs in β units, and the relative energy of the ethylene bonding MO is shown in red. The absence of electrons in π_2 accounts for the positive charge. The nodal plane in π_2 cuts through the central carbon; as a result, there is no positive charge on this carbon.

Here is the connection between molecular stability and resonance: *Resonance structures provide a device to show electron delocalization with Lewis structures.* For example, the allyl cation has two equivalent resonance structures.

$$\left[\begin{array}{c} \text{C}=\text{C}-\overset{+}{\text{C}} \end{array} \longleftrightarrow \begin{array}{c} \overset{+}{\text{C}}-\text{C}=\text{C} \end{array} \right] \quad \text{or} \quad \begin{array}{c} \overset{\delta+}{\text{C}}\text{---}\text{C}\text{---}\overset{\delta+}{\text{C}} \end{array} \tag{15.22}$$

an allylic carbocation   hybrid structure

The two structures show the sharing of π-electron density and charge—in other words, electron delocalization. The hybrid structure shows the same delocalization with dashed bonds and partial charges. In summary, the logic of resonance stabilization is as follows:

1. Resonance structures symbolize electron delocalization.
2. Electron delocalization is stabilizing because it results in additional bonding associated with the formation of low-energy bonding MOs.
3. Resonance, therefore, is a stabilizing effect.

Because resonance describes electron delocalization, *delocalization energy* (Sec. 15.1B), the energetic advantage of electron delocalization, is also referred to as **resonance energy**. In Sec. 1.4 we learned that resonance-stabilized molecules are more stable than any of their fictional contributors. The resonance energy is a quantitative measure of this additional stability.

One other important aspect of resonance structures emerges from a consideration of the positive-charge distribution in the allyl cation. Notice in the resonance structures that, although the π electrons are delocalized across the entire molecule, the positive charge resides on only two of the three carbons; *there is no positive charge on the central carbon of an allylic cation.* This charge distribution is consistent with the molecular orbital description of the cation, shown in Fig. 15.14b. The charge is due to the absence of a third π electron. If a third π electron were present, it would occupy the NBMO π_2. Thus, the picture of π_2 describes the distribution of the "missing electron"—the positive charge. Because π_2 has a node on the central carbon, this carbon bears no charge. This charge distribution is shown graphically in the EPM of the allyl cation, which is calculated from MO theory. Notice that the terminal carbons have more positive charge (*blue*) than the central carbon (*red*):

* = sites of positive charge

EPM of the allyl cation

Resonance structures are useful because they give us the qualitative results of MO theory without any calculations!

PROBLEM

15.25 (a) The allyl anion has an unshared electron pair on the allylic carbon:

$$H_2C\!=\!CH\!-\!\overset{..}{\underset{}{C}}H_2$$

allyl anion

This anion has two more π electrons than the allyl cation. Use the molecular orbital diagram in Fig. 15.14b to decide which molecular orbital these "extra electrons" occupy.

(b) According to the molecular orbital description, which carbons of the allyl anion bear the negative charge?

(c) Show that the same conclusion can be reached by drawing resonance structures of the allyl anion. Use the curved-arrow notation.

C. Kinetic and Thermodynamic Control

Naively, we might expect that when a reaction can give products that differ in stability, the more stable product should be formed in greater amount. However, this is often not the case. Consider, for example, the addition of hydrogen halides to conjugated dienes. When a conjugated diene reacts with a hydrogen halide to give a mixture of 1,2- and 1,4-addition products, the 1,2-addition product predominates at low temperature:

$$H_2C\!=\!CH\!-\!CH\!=\!CH_2 \ + \ HCl \ \xrightarrow[-80\,°C]{}$$

$$\underset{\substack{\text{1,2-addition product} \\ (75\text{–}80\%)}}{H_3C\!-\!\overset{\overset{\displaystyle Cl}{\displaystyle |}}{C}H\!-\!CH\!=\!CH_2} \ + \ \underset{\substack{\text{1,4-addition product} \\ (20\text{–}25\%)}}{H_3C\!-\!CH\!=\!CH\!-\!CH_2\!-\!Cl} \qquad (15.23)$$

(See Eq.15.17, p. 741, for another example.) We learned in Sec. 4.5B that alkenes with internal double bonds are more stable than their isomers with terminal double bonds, because the internal double bonds have more alkyl branches. Hence, in Eq. 15.23, *the major product is the less stable one*. This can be demonstrated experimentally by bringing the two alkyl halide products to equilibrium with heat and Lewis acids:

$$H_3C—CH—CH=CH_2 \xrightleftharpoons[]{heat, FeCl_3} H_3C—CH=CH—CH_2—Cl \qquad (15.24)$$
$$\overset{|}{Cl}$$

major product at equilibrium

minor product at equilibrium

Because the more stable isomer always predominates in an equilibrium (Sec. 3.5), the result in Eq. 15.24 shows that the 1,4-addition product is more stable than the 1,2-addition product, as expected.

When the less stable product of a reaction is the major product, then two things must be true. First, the less stable product *must form more rapidly* than the other products. Recall from Sec. 4.8 that a reaction in which two products form from the same starting material is in reality two competing reactions. Consequently, the reaction that forms the less stable product is faster. Second, the products *must not come to equilibrium* under the reaction conditions, because otherwise the more stable compound would be present in larger amount. Thus, in the addition of HCl to conjugated dienes, the predominance of the less stable product (Eq. 15.23) shows that 1,2-addition, which gives the less stable product, is faster than 1,4-addition:

$$H_2C=CH—CH=CH_2 + HCl$$

1,2-addition
(faster reaction)

1,4-addition
(slower reaction)

$$H_3C—CH—CH=CH_2 \qquad H_3C—CH=CH—CH_2—Cl \qquad (15.25)$$
$$\overset{|}{Cl}$$

more stable product

less stable product

When the products of a reaction do not come to equilibrium under the reaction conditions, the reaction is said to be **kinetically controlled**. In a kinetically controlled reaction, the relative proportions of products are controlled solely by the relative rates at which they are formed. Thus, the addition of hydrogen halides to conjugated dienes is a kinetically controlled reaction. On the other hand, if the products of a reaction come to equilibrium under the reaction conditions, the reaction is said to be **thermodynamically controlled**.

It is possible that a given kinetically controlled reaction might give about the same distribution of products as would be obtained if the products were allowed to come to equilibrium. However, it is *impossible* for a thermodynamically controlled reaction to give a product distribution other than the equilibrium distribution. Hence, when we obtain a product distribution that is clearly different from that obtained at equilibrium (as occurs in the addition of HCl to conjugated dienes), we know immediately that the reaction must be kinetically controlled.

 An Analogy for Kinetic Control

Imagine a very disoriented steer stumbling randomly around a pasture with a shallow watering hole and a deep well with a high fence around it. Where is he likely to end up? Certainly the deep well is the state of lowest potential energy. However, because of the fence around the well, it is simply less likely that the animal will fall into the well; he is much more likely to wander into the watering hole. Now, if you imagine a large herd of similarly disoriented steers staggering around the same (very large) pasture, you should get a reasonably good image of kinetic control. Most of the animals wander into the watering hole, even though this is not the state of lowest potential energy.

Likewise with molecules: It is possible for the formation of a more stable product to have a greater standard free energy of activation (a greater energy barrier) than the formation of a less stable product. In such a case, the less stable product forms more rapidly and in greater amount.

In hydrogen halide addition to a conjugated diene, the first and rate-limiting step in the formation of both 1,2- and 1,4-addition products is the same—protonation of the double bond. Consequently, the product distribution must be determined by the relative rates of the *product-determining steps* (Sec. 9.6B): the nucleophilic reaction of the halide ion at one or the other of the electron-deficient carbons of the allylic carbocation intermediate.

$$H_2C{=}CH{-}CH{=}CH_2 + H\overset{..}{\underset{..}{Cl}}:$$

rate-limiting step

$$\left[H_3C{-}\overset{+}{C}H{-}CH{=}CH_2 \;\longleftrightarrow\; H_3C{-}CH{=}CH{-}\overset{+}{C}H_2 \right]$$

$$:\overset{..}{\underset{..}{Cl}}:^-$$

faster reaction   product-determining steps   slower reaction

$$H_3C{-}\underset{\underset{\displaystyle :\overset{..}{\underset{..}{Cl}}:}{|}}{CH}{-}CH{=}CH_2 \qquad\qquad H_3C{-}CH{=}CH{-}CH_2{-}\overset{..}{\underset{..}{Cl}}: \quad (15.26)$$

less stable product   more stable product
(major product)   (minor product)

Why is the 1,2-addition product formed more rapidly? The diene reacts with *undissociated* HCl; consequently, the carbocation and its chloride counterion, when first formed, exist as an *ion pair* (Fig. 8.6, p. 363). That is, the chloride ion and the carbocation are closely associated. The chloride ion simply finds itself closer to the positively charged carbon adjacent to the site of protonation than to the other. Addition is completed, therefore, at the nearer site of positive charge, giving the 1,2-addition product.

$$H_2C{=}CH{-}CH{=}CH_2 \longrightarrow \left[H_2C{=}CH{-}\overset{+}{C}H{-}CH_2 \;\longleftrightarrow\; H_2\overset{+}{C}{-}CH{=}CH{-}CH_2 \right]$$

$$:\overset{..}{\underset{..}{Cl}}{-}H$$

closer   farther

$$H_2C{=}CH{-}\underset{\underset{\displaystyle :\overset{..}{\underset{..}{Cl}}:}{|}}{CH}{-}CH_2 \qquad\qquad H_2C{-}CH{=}CH{-}CH_3 \quad (15.27)$$

(mostly)   (little)

(The very elegant experiment that suggested this explanation is described in Problem 15.66.)

The reason for kinetic control varies from reaction to reaction. Whatever the reason, the relative amounts of products in a kinetically controlled reaction are determined by the relative free energies of the *transition states* for each of the product-determining steps and *not* by the relative free energies of the products.

15.26 Suggest structures for the two constitutional isomers formed when 1,3-butadiene reacts with one equivalent of Br_2. (Ignore any stereochemical issues.) Which of these products would predominate if the two were brought to equilibrium?

15.5 DIENE POLYMERS

1,3-Butadiene is one of the most important raw materials of the synthetic rubber industry, which includes the production of automobile tires. The annual global production of 1,3-butadiene is about 20 billion pounds.

1,3-Butadiene can be polymerized to give polybutadiene. This polymer can result either from 1,4-addition of butadiene molecules to each other or from 1,2-addition. The 1,4-addition polymers can contain cis or trans double bonds.

1,4-addition polymer
(cis double bond)

1,4-addition polymer
(trans double bond)

1,2-addition polymer
("vinyl" polybutadiene)

These different types of polymer linkages, although shown separately in the structures above, can be present together in various proportions in a given polymer molecule. The most useful butadiene polymers contain mostly cis-1,4 units. This type of polymer is produced by the action of transition-metal organometallic catalysts of a type that we'll study in Sec. 18.5. (The cis double bond is a consequence of the way that the π bonds of the diene interact with the catalyst.) Different types of catalysts can be used to give the different types of diene polymers.

1,3-Butadiene can also be polymerized along with styrene ($H_2C{=}CH{-}Ph$), usually in about a 3 : 1 ratio, to give another type of synthetic rubber called *styrene–butadiene rubber* (SBR), most of which is used for tires and tread rubber.

3 butadiene units — 1 styrene unit

styrene–butadiene rubber (SBR)
(all-trans diene units shown)

As with polybutadiene, the butadiene units in SBR can be polymerized in either a 1,4- or 1,2-addition, and the 1,4-addition polymer can contain cis or trans double-bond stereochemistry. The example above is simplified because it shows an SBR unit with only *trans*-1,4-butadiene units, which is the predominant type of linkage in commercial SBR. For example, one typical formulation of SBR contains 54.5% *trans*-1,4-butadiene units, 9% *cis*-1,4-butadiene units, 13% "vinyl" units (resulting from 1,2-addition of butadiene), and 23.5% styrene. Moreover, the styrene units can occur at random. SBR is produced either by organometallic catalysis or by free-radical polymerization (Sec. 5.7 and Problem 15.27) in an aqueous emulsion.

SBR is an example of a **copolymer**: a polymer produced by the simultaneous polymerization of two or more monomers. The annual global demand for styrene–butadiene copolymers is about 11 billion pounds and is growing rapidly as the demand for automobiles increases around the globe.

Natural rubber is (*Z*)-polyisoprene, another diene polymer:

(Z)-polyisoprene (natural rubber)   **2-methyl-1,3-butadiene (isoprene)**

Although it is conceptually a diene polymer, natural rubber is not made in nature from isoprene. (The biosynthesis of naturally occurring isoprene derivatives is discussed in Sec. 17.6B.) Rubber hydrocarbon (polyisoprene) is obtained as a 40% aqueous emulsion from the rubber tree. Although polyisoprene can be made synthetically, the natural material is generally preferred for economic reasons. Chemists and botanists are investigating the possibility of cultivating other hydrocarbon-producing plants that could become hydrocarbon sources of the future.

Crude natural rubber, SBR, and other diene polymers do not have adequate mechanical durability for their commercial applications. For this reason, they are subjected to a process called *vulcanization*. In this process, discovered in 1840 by Charles Goodyear (1800–1860), the rubber is kneaded and heated with sulfur. The sulfur forms disulfide crosslinks between the polymer chains, which can be represented schematically as follows:

The crosslinks increase the rigidity and strength of the polymer at the cost of some flexibility.

PROBLEM

15.27 An initiation step in the free-radical co-polymerization of styrene and 1,3-butadiene is the free-radical addition of a peroxide-derived radical to a double bond of 1,3-butadiene:

(a) Use the fishhook curved-arrow notation to derive the missing resonance structure.

(b) Use this resonance structure as part of a fishhook mechanism for free-radical co-polymerization to form SBR. (Show the incorporation of one diene unit in a 1,4-addition and one styrene unit.)

(c) Suggest a reason that the diene part of the polymer has mostly trans double-bond stereochemistry.

15.6   RESONANCE

Recall from Sec. 1.4 that resonance structures are used to describe a molecule when a single Lewis structure is inadequate. Molecules that can be represented as resonance hybrids are said to be *resonance-stabilized*—that is, they are more stable than any one of their contributing structures. The reason for this additional stability is the additional bonding that results from the delocalization of electrons within low-energy bonding MOs (Sec. 15.4C). Resonance structures can be derived by the curved-arrow notation (Sec. 3.3).

The derivation and use of resonance structures is important for understanding both *molecular structure* and *molecular stability*. Because some reactive intermediates (for example, some carbocations) are resonance-stabilized, you will also find that resonance arguments can be important in analyzing reactivity. Because resonance is an essential part of the language of organic chemistry, we are now going to review the concepts of resonance and the techniques for drawing resonance structures. You'll learn to assess whether a resonance structure is significant enough to be considered seriously as an important contributor to a molecular structure. You'll also learn how resonance structures can be used to make deductions about molecular stability and, finally, chemical reactivity.

A. Drawing Resonance Structures

Resonance structures show *the delocalization of electrons*. Resonance structures are derived with the curved-arrow notation (or the fishhook notation for free radicals) *without moving any atoms*. Resonance structures are usually placed within brackets to emphasize the fact that they are being used to describe a *single species*. Here are some of the most common situations in which you can draw resonance structures:

1. *A double bond, a triple bond, or an atom with an unshared electron pair is adjacent to an electron-deficient atom.* In these cases a single curved arrow is used, as in the notation for Lewis acid–base association–dissociation reactions. (Sec. 3.1C)

    Examples:

    $$\left[H_2C\!=\!CH\!\!-\!\!\overset{+}{C}H_2 \quad\longleftrightarrow\quad H_2\overset{+}{C}\!\!-\!\!CH\!=\!CH_2 \right] \tag{15.28a}$$

    $$\left[CH_3\overset{..}{\underset{..}{O}}\!\!-\!\!\overset{+}{C}H_2 \quad\longleftrightarrow\quad CH_3\overset{..}{\underset{..}{O}}\!\!=\!\!\overset{+}{C}H_2 \right] \tag{15.28b}$$

    $$\left[CH_3\overset{..}{\underset{..}{O}}\!\!-\!\!\underset{\underset{OCH_3}{|}}{B}\!\!-\!\!OCH_3 \quad\longleftrightarrow\quad CH_3\overset{+}{\underset{..}{O}}\!\!=\!\!\underset{\underset{OCH_3}{|}}{\overset{-}{B}}\!\!-\!\!OCH_3 \right] \tag{15.28c}$$

2. *A double or triple bond is adjacent to an atom with an unshared electron pair.* In these cases two curved arrows are needed, as in the electron-pair displacement notation. (Sec. 3.2A)

    Examples:

    $$\left[H_2\overset{\frown}{C}\!=\!\overset{\frown}{C}\!\!-\!\!\underset{\underset{H}{|}}{\overset{..}{O}}\!\!:^- \quad\longleftrightarrow\quad H_2\overset{..}{\underset{..}{C}}\!\!-\!\!\underset{\underset{H}{|}}{C}\!=\!\overset{..}{O}\!: \right] \tag{15.29a}$$

    $$\left[H_3C\!\!-\!\!C\overset{\overset{\displaystyle\overset{..}{O}:}{\parallel}}{\underset{\underset{..}{\overset{..}{O}}:}{}} \quad\longleftrightarrow\quad H_3C\!\!-\!\!C\overset{\overset{\displaystyle:\overset{..}{O}^{\overline{}}}{}}{\underset{\underset{..}{O}:}{\parallel}} \right] \tag{15.29b}$$

$$\left[H_3C - \overset{+}{N} \equiv C - \overset{\cdot\cdot}{N}H_2 \longleftrightarrow H_3C - \overset{\cdot\cdot}{N} = C = \overset{+}{N}H_2 \right]$$ (15.29c)

3. *A double or triple bond is adjacent to an atom with an unpaired electron.* Any time an unpaired electron is delocalized, the fishhook notation (Sec. 5.6B) is used.

$$\left[H_2C = CH - \overset{\cdot}{C}H_2 \longleftrightarrow H_2\overset{\cdot}{C} - CH = CH_2 \right]$$ (15.30)

4. *Double bonds can be moved in a cycle around a six-membered ring.*

   The archetypal example is benzene. This type of resonance requires a special discussion, which is the subject of Sec. 15.7.

$$\text{(15.31)}$$

benzene

More extensive delocalization of electrons involving more curved arrows can always be broken down into simpler individual steps involving one or two curved arrows, as in the following example:

is equivalent to   (15.32)

Notice also in these examples that the delocalization of electrons by resonance can also result in the delocalization of charge (Eqs. 15.28 and 15.29) or the delocalization of an unpaired electron (Eq. 15.30). Remember, though, that the *delocalization of electrons* is the primary issue in resonance; the delocalization of a charge or an unpaired electron is a *consequence* of electron delocalization.

It is important to re-emphasize that *atoms do not move* between resonance structures. The following structures, although reasonable Lewis structures, are *not* resonance structures, because the movement of an atom takes place; the location of the chlorine is different in the two structures:

$$H_2C - CH = CH - CH_3 \quad \times \quad H_2C = CH - CH - CH_3 \qquad \text{(15.33a)}$$
$$\quad\quad\quad | \qquad\qquad\qquad\qquad\qquad\qquad\qquad\quad | $$
$$\quad\quad\quad Cl \qquad\qquad\qquad\qquad\qquad\qquad\qquad\quad Cl$$

In fact, these are separate compounds:

$$H_2C - CH = CH - CH_3 \quad \rightleftharpoons \quad H_2C = CH - CH - CH_3 \qquad \text{(15.33b)}$$
$$\quad\quad\quad | \qquad\qquad\qquad\qquad\qquad\qquad\qquad\quad | $$
$$\quad\quad\quad Cl \qquad\qquad\qquad\qquad\qquad\qquad\qquad\quad Cl$$

If two structures can exist as different compounds, they cannot be resonance structures. Resonance structures, in contrast, are used to describe a *single* molecule; the molecule is an average of its resonance structures. That is, resonance contributors are *fictitious structures* used to help us understand the structures of *real molecules* for which single Lewis structures are inadequate. Thus, the equilibrium double arrows $\rightleftharpoons$ and the resonance double-headed arrow $\longleftrightarrow$ have quite different meanings. *You must be careful not to use one symbol in a situation in which the other is appropriate.*

B. Evaluating the Relative Importance of Resonance Structures

In many cases, not all resonance structures are of equal importance; that is, the structure of a molecule is most closely approximated by its most important resonance structures. This section shows how to assess the *relative importance* of resonance structures.

To evaluate the relative importance of resonance structures, compare the stabilities of all of the resonance structures for a given molecule *as if each structure were a separate molecule.* That is, even though each structure is fictitious, *imagine* that each structure is real. Then use the relative stabilities of the different structures to determine their relative importance. *The most stable structures are the most important ones.* The following guidelines for evaluating resonance structures emerge from this type of analysis:

1. *Identical structures are equally important descriptions of a molecule.*

   Example:

$$H_2\overset{+}{C}-CH=CH_2 \quad\longleftrightarrow\quad H_2C=CH-\overset{+}{C}H_2 \Big] \tag{15.34}$$

   allyl cation

Because these structures of the allyl carbocation are indistinguishable, they are equally important in describing this species.

2. *Structures that have complete octets on each atom are more important than those that do not.*

   Example:

$$\Big[H_2\overset{+}{C}-\overset{..}{\overset{..}{O}}-H \quad\longleftrightarrow\quad H_2C=\overset{+}{\overset{..}{O}}-H\Big] \tag{15.35}$$

   more important because
   each atom has a complete octet

It might seem that placing a positive charge on oxygen should be avoided because oxygen is an electronegative atom. However, the octet rule is so important that a positively charged oxygen (or other electronegative atom) is perfectly acceptable if it has a complete octet.

3. *Structures that place positive charge on more electropositive (or less electronegative) atoms are more important than structures that place positive charge on more electronegative atoms. Structures that place negative charge on more electronegative atoms are more important than those that place negative charge on more electropositive (or less electronegative) atoms.*

In Eq. 15.35, guidelines 2 and 3 are in conflict. Guideline 2 is more important; but, because of guideline 3, the structure on the left has some significance: the positive charge is on the less electronegative atom. Application of guideline 3 is also illustrated in Eqs. 15.36a and 15.36b.

$$\left[(CH_3)_2\overset{+}{N}=C-\overset{..}{\overset{..}{O}}CH_3 \quad\longleftrightarrow\quad (CH_3)_2\overset{..}{N}-C=\overset{+}{\overset{..}{O}}CH_3\right] \tag{15.36a}$$
$$\qquad\qquad\quad | \qquad\qquad\qquad\qquad\qquad\quad |$$
$$\qquad\qquad\quad H \qquad\qquad\qquad\qquad\qquad\quad H$$

more important because
positive charge is on the
less electronegative atom

$$
\left[
\begin{array}{c}
\overset{\displaystyle :\ddot{O}:}{\underset{}{\overset{\parallel}{H_2\ddot{C}-C-H}}}
\end{array}
\quad\longleftrightarrow\quad
\begin{array}{c}
\overset{\displaystyle :\ddot{O}:^{-}}{\underset{}{\overset{|}{H_2C=C-H}}}
\end{array}
\right] \qquad (15.36b)
$$

more important because
negative charge is on the
more electronegative atom

As you apply guidelines 2 and 3, be aware of the difference between positively charged, electronegative atoms (such as nitrogen, oxygen, and the halogens) that have complete octets, as they do in Eq. 15.35 and 15.36a, and positively charged, *electron-deficient*, electronegative atoms, which do *not* have complete octets. The latter are to be avoided if at all possible. For example, the structure on the right for ozone is unimportant because the positively charged oxygen is electron-deficient.

ozone
$$
\left[
:\ddot{O}=\overset{+}{\ddot{O}}-\ddot{O}:^{-}
\quad\longleftrightarrow\quad
{}^{-}:\ddot{O}-\overset{+}{\ddot{O}}=\ddot{O}:
\quad\times\quad
:\ddot{O}-\overset{\cdot\cdot}{\ddot{O}}-\overset{+}{\ddot{O}}:
\right] \qquad (15.36c)
$$

identical structures;
both are important;
all atoms have octets

this structure is unimportant
because it has an
electron-deficient
electronegative atom

4. *If the orbital overlap symbolized by a resonance structure is impossible, the reso-nance structure is not important.*

   Example:

$$
\left[
\quad\overset{}{\underset{\ddot{N}}{\bigtriangleup}}{}^{+}
\quad\times\quad
\overset{}{\underset{\overset{+}{N}}{\bigtriangleup}}
\quad
\right] \qquad (15.37)
$$

$$A \qquad\qquad\qquad B$$

unimportant because it is
geometrically impossible!

Remember that resonance structures are a simple way of portraying the delocalization of elec-trons that results from orbital overlap. If a structure is constrained so that such orbital overlap is impossible, the resonance structure is invalid. This is the case with the ion in Eq. 15.37. Structure *B* cannot be a valid contributor because the orbital containing the unshared electron pair lies at a right angle to the $2p$ orbital of the carbocation and therefore cannot overlap with it. This can be seen from a view of the molecule along the bond between the C^+ and the N:

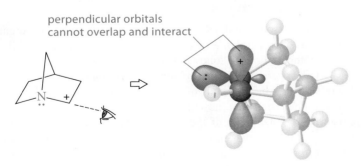

perpendicular orbitals
cannot overlap and interact

Only an energetically costly distortion of the ion would permit overlap of the two orbitals; but resonance structures cannot involve atomic motion. (Another argument against structure *B* is that it violates Bredt's rule; Sec. 7.6C.) However, in an ion containing the same functional groups in which such orbital overlap is possible, a similar resonance structure *is* important.

$$\left[Me_2\ddot{N}\!\overset{\frown}{-}\!\overset{+}{C}HCH_3 \longleftrightarrow Me_2\overset{+}{N}\!\!=\!\!CHCH_3 \right] \tag{15.38}$$
an important
resonance structure

In this example, the ion can easily adopt a conformation in which the nitrogen orbital containing the unshared electron pair is coplanar with the $2p$ orbitals of the carbocation.

2p orbitals are aligned for overlap

Guideline 4 means that it is not enough to derive a resonance structure correctly with the curved-arrow notation. We must also keep in mind the *meaning* of the structure in terms of the orbital overlap involved.

Resonance has implications for molecular geometry. Applying the VSEPR rules (Sec. 1.3B) we would predict, from the resonance structure on the left side of Eq. 15.38, that nitrogen should have tetrahedral geometry and that its unshared pair should occupy an sp^3 orbital. The same rules applied to the resonance structure on the right, however, predict that the nitrogen should have trigonal planar geometry and sp^2 hybridization, and that its unshared pair should occupy a $2p$ orbital. A $2p$ orbital on nitrogen can interact more effectively with the $2p$ orbital of the carbocation than an sp^3 orbital would because it has the same shape, size, and orientation. Because this interaction is a stabilizing effect, anything that enhances the interaction, such as the sp^2 hybridization of nitrogen, also stabilizes the molecule. Thus, the second structure in Eq. 15.38 is a more accurate description of the ion (a conclusion also reached from guideline 2 on p. 752), and, as the above structure of the ion shows, the unshared pair occupies a $2p$ orbital. Note also that the molecule doesn't "oscillate" between sp^2-hybridized nitrogen and sp^3-hybridized nitrogen. If it did, atomic motion such as changing bond lengths and angles would be involved and these structures would then not be resonance structures. Because these are resonance structures, they represent *one* molecule that has an sp^2-hybridized nitrogen.

C. Using Resonance Structures

Although, as we've seen, resonance does have a basis in quantum theory, organic chemists generally use resonance arguments in a qualitative way to compare the stabilities of molecules. To make this comparison, the following principle is applied: *All other things being equal, the molecule with the greater number of important resonance structures is more stable.* The qualifier "all other things being equal" means that the other aspects of the molecules that affect their stabilities should be about the same.

A comparison of the stabilities of the following two carbocations illustrates the use of resonance arguments. (These two carbocations were compared in the addition of HBr in Eqs. 15.18–15.20.) Both carbocations are secondary, and in fact they are isomeric. The two carbocations differ in the number of resonance structures that can be written for each. The more stable carbocation has two resonance structures. The other carbocation has only one Lewis structure and is less stable.

$$\left[Et\overset{+}{C}H—CH{=}CH—Me \longleftrightarrow EtCH{=}CH—\overset{+}{C}H—Me \right] \quad \begin{array}{l} \text{more stable} \\ \text{carbocation} \end{array}$$

$$Me\overset{+}{C}HCH_2—CH{=}CH—Me \quad \text{less stable carbocation}$$

Applying this principle broadens the meaning of the term *resonance stabilization*. Resonance-stabilized molecules are not only more stable than their individual contributing structures, they are also more stable than *other isomeric molecules* that have only one Lewis contributor (other things being equal).

The reason that the number of resonance structures is related to the stability of a molecule follows from the electronic basis of resonance itself, which was discussed in Sec. 15.4B. Molecules with many resonance structures have extensive electron delocalization and additional bonding that result from the overlap of orbitals. (In molecular orbital terms, a number of bonding molecular orbitals of low energy are formed and occupied by the π electrons.) This additional bonding is a stabilizing effect.

STUDY PROBLEM   15.2

Which of the following carbocations is more stable?

$$CH_3\ddot{\underset{..}{O}}—CH{=}CH—\overset{+}{C}H_2 \quad \text{or} \quad \begin{array}{c} :\ddot{O}CH_3 \\ | \\ H_2C{=}C—\overset{+}{C}H_2 \end{array}$$

$$\qquad\qquad A \qquad\qquad\qquad\qquad\qquad B$$

SOLUTION   The solution to this problem involves determining which carbocation has the greater number of important resonance structures. Carbocation *A* has the following important resonance structures:

$$\left[CH_3\ddot{\underset{..}{O}}—CH{=}CH—\overset{+}{C}H_2 \longleftrightarrow CH_3\ddot{\underset{..}{O}}—\overset{+}{C}H—CH{=}CH_2 \longleftrightarrow CH_3\overset{+}{\underset{..}{O}}{=}CH—CH{=}CH_2 \right]$$

Carbocation *B* has only two reasonable structures. In particular, the charge can be delocalized onto the oxygen in ion *A* but not in ion *B*:

$$\left[\begin{array}{c} :\ddot{O}CH_3 \\ | \\ H_2C{=}C—\overset{+}{C}H_2 \end{array} \longleftrightarrow \begin{array}{c} :\ddot{O}CH_3 \\ | \\ H_2\overset{+}{C}—C{=}CH_2 \end{array} \right]$$

Thus, ion *A* is more stable because it has the larger number of important resonance structures.

Resonance structures can in some cases be used to predict reactivity. Recall that Hammond's postulate provides the connection between transition-state stability and the stability of actual chemical species. When asked to predict reactivity, do the following:

1. For each reaction, identify the reactive intermediate that is structurally similar to the rate-limiting transition state.

2. Use what you know about relative stabilities to predict reactivity. Reactions that involve more stable reactive intermediates are faster. Use resonance arguments, if appropriate, in your reasoning about relative stability.

This process was first introduced in Sec. 4.8C and should be reviewed again. Study Problem 15.3 introduces resonance arguments into this process.

STUDY PROBLEM 15.3

Predict the relative reactivity of the following two compounds in an S_N1 solvolysis reaction.

$$CH_3OCH_2CH_2Cl \qquad CH_3CH_2OCH_2Cl$$

1-chloro-2-methoxyethane   (chloromethoxy)ethane

A   *B*

SOLUTION   Because the question asks about an S_N1 reaction, the rate-limiting step is ionization of the alkyl halide, and the relevant reactive intermediate is a carbocation. The carbocation formed from *A* is a primary carbocation:

$$CH_3OCH_2CH_2{\overset{\frown}{-}}\ddot{\underset{\cdot\cdot}{C}}l: \longrightarrow CH_3OCH_2\overset{+}{C}H_2 \quad :\ddot{\underset{\cdot\cdot}{C}}l:^- \tag{15.39}$$

A

This cation has no resonance structures and is destabilized by the polar effect of the oxygen. The carbocation formed from *B* is also a primary carbocation, but it has an important resonance structure:

$$CH_3CH_2\ddot{\underset{\cdot\cdot}{O}}CH_2{\overset{\frown}{-}}\ddot{\underset{\cdot\cdot}{C}}l: \longrightarrow \left[CH_3CH_2\ddot{\underset{\cdot\cdot}{O}}{\overset{\frown}{-}}\overset{+}{C}H_2 \longleftrightarrow CH_3CH_2\overset{+}{\underset{\cdot\cdot}{O}}{=}CH_2 \right] :\ddot{\underset{\cdot\cdot}{C}}l:^- \tag{15.40}$$

B

From this analysis, we deduce that the carbocation derived from *B* is more stable than the one derived from *A*. Invoking Hammond's postulate, we deduce that the transition states for the two S_N1 solvolysis reactions resemble the carbocations, and we thus conclude that the transition state for the S_N1 solvolysis of *B* has a lower energy than the transition state for solvolysis of *A*. Hence, the reaction of *B* is faster.

> Remember that the rate of a reaction is related to the *difference* in standard free energies of the transition state and the starting materials. Hence, in our analysis, we are *assuming* that the energy differences between *A* and *B* are much less important than the energy differences between the carbocations. This assumption is usually justified because the energy differences associated with the presence and absence of resonance structures are *much* greater than the energy differences between molecules of closely related structure.

How good is our prediction? Experimentally, it is found that *B* reacts 5×10^9 (5 *billion*) times faster than *A* in 36% aqueous dioxane at 100 °C. (In fact, the very slow reaction of *A* probably occurs by the S_N2 mechanism rather than the S_N1 mechanism; this means that the S_N1 reaction of *A* is even slower!) As you can see, the use of resonance arguments correctly predicts the relative order of reactivities in this solvolysis reaction.

Resonance arguments can in some cases be used to predict or understand relative acidities or basicities. These arguments will be considered in Secs. 18.7, 20.4, 22.1, and 23.5.

Resonance-stabilized compounds or intermediates in some cases lead to more than one product. The principle is that *in predicting products, we can treat each resonance structure as if it were a separate compound.* For example, in the addition of HBr to 1,3-butadiene, the carbocation intermediate has two resonance structures. The reaction of Br^- at the electron-deficient carbon of each gives the two products:

$$H_2C{=}CH{-}CH{=}CH_2 \longrightarrow \left[H_2C{=}CH{-}\overset{+}{C}H{-}CH_3 \longleftrightarrow H_2\overset{+}{C}{-}CH{=}CH{-}CH_3 \right]$$

$$\searrow H{-}\ddot{\underset{\cdot\cdot}{B}}r:$$

A   :$\ddot{\underset{\cdot\cdot}{B}}$r:$^-$ *B*

$$H_2C{=}CH{-}CH{-}CH_3 \qquad H_2C{-}CH{=}CH{-}CH_3 \tag{15.41}$$
$$\underset{:\ddot{\underset{\cdot\cdot}{B}}r:}{|} \qquad\qquad\qquad \underset{:\ddot{\underset{\cdot\cdot}{B}}r:}{|}$$

1,2-addition product   **1,4-addition product**
(from reaction of Br^- (from reaction of Br^-
with structure *A*)   with structure *B*)

The two resonance structures don't tell us how much of each product is formed, but they do predict that *some of each* could be formed. Remember that there is only *one* carbocation; the resonance structures show that the electron deficiency is shared between two carbons. This technique of using each structure to predict a product works because the Br⁻ can react at either site of electron deficiency.

PROBLEMS

15.28 In each of the following sets, show by the curved-arrow notation how each resonance structure is derived from the other one, and indicate which structure is more important and why.

(a) $\left[:C{=}\ddot{O} \longleftrightarrow :\bar{C}{\equiv}\overset{+}{O} \right]$

carbon monoxide

(b) $\left[H_3C{-}\overset{+}{C}{=}\ddot{N}H \longleftrightarrow H_3C{-}C{\equiv}\overset{+}{N}H \right]$

(c) $\left[(CH_3)_2C{=}CH{-}\overset{+}{C}H_2 \longleftrightarrow (CH_3)_2\overset{+}{C}{-}CH{=}CH_2 \right]$

(d)

15.29 Show the 2p orbitals, and indicate the orbital overlap symbolized by the resonance structures for the carbocation in Eq. 15.35 on p. 752.

15.30 (a) Break down the following resonance interaction into four separate resonance structures (counting the two below), each connected to the next by *one* curved arrow.

(b) Draw a hybrid structure corresponding to the ion in part (a) containing dashed lines and partial charges. (*Hint:* Let your answer to part (a) tell you where to place the partial charges.)

15.31 Using resonance arguments, state which ion or radical within each set is more stable. Explain.

(a)

$$H_3C{-}\overset{\overset{\displaystyle :\ddot{O}:^-}{|}}{C}H{-}CH{=}CH_2 \quad \text{or} \quad H_3C{-}\overset{\overset{\displaystyle :\ddot{O}:^-}{|}}{C}{=}CH{-}CH_3$$

(b)

$$H_2C{=}CH{-}CH{=}CH{-}\dot{C}H_2 \quad \text{or} \quad H_2\dot{C}{-}\overset{\overset{\displaystyle CH_2}{\|}}{C}{-}CH{=}CH_2$$

(c)

$$H_2C{=}\overset{\overset{\displaystyle CH_3}{|}}{C}{-}\overset{+}{C}H_2 \quad \text{or} \quad H_3C{-}CH{=}CH{-}\overset{+}{C}H_2$$

15.32 The following isomers do not differ greatly in stability. Predict which one should react more rapidly in an S_N1 solvolysis reaction in aqueous acetone. Explain.

A   *B*

continued

continued

15.33 (a) The following resonance-stabilized ion can protonate to give two different constitutional isomers. Give their structures and give the curved-arrow notation for their formation.

$$\left[H_2C\!\!=\!\!\overset{+}{N}\overset{\ddot{\ddot{O}}\!:^-}{\underset{\ddot{\ddot{O}}\!:^-}{\diagdown}} \quad\longleftrightarrow\quad H_2\ddot{C}\!-\!\overset{+}{N}\overset{\ddot{O}:}{\underset{\ddot{\ddot{O}}\!:^-}{\diagdown}} \right]$$

(b) One of the compounds in Problem 15.32 gives two different constitutional isomers as a product of its S_N1 reaction with H_2O. Which compound is it? Explain and give the structures of the two isomers.

15.7 INTRODUCTION TO AROMATIC COMPOUNDS

The term *aromatic*, as we'll come to understand it in this section, is a precisely defined *structural* term that applies to cyclic conjugated molecules that meet certain criteria. Benzene and its derivatives are the best known examples of aromatic compounds.

benzene

The origin of the term *aromatic* is historical: many fragrant compounds known from earliest times, such as the following ones, proved to be derivatives of benzene.

vanillin
(vanilla)

methyl salicylate
(oil of wintergreen)

***para*-cymene**
(cumin and thyme)

saffrole
(oil of sassafras)

Although it is known today that benzene derivatives are not distinguished by unique odors, the term *aromatic*—which has nothing to do with odor—has stuck, and it is now a class name for benzene, its derivatives, and a number of other organic compounds.

Development of the theory of aromaticity was a major theoretical advance in organic chemistry because it solved a number of intriguing problems that centered on the structure and reactivity of benzene. Before considering this theory, let's see what some of these problems were.

A. Benzene, a Puzzling "Alkene"

The structure used today for benzene was proposed in 1865 by August Kekulé (p. 46), who claimed later that it came to him in a dream. The Kekulé structure portrays benzene as a cyclic conjugated triene. Yet benzene does not undergo any of the addition reactions that are associated with either conjugated dienes or ordinary alkenes. Benzene itself, as well as benzene rings in other compounds, are inert to the usual conditions of alkene addition reactions. This property of the benzene ring is illustrated by the addition of bromine to styrene, a compound that contains both a benzene ring and one additional double bond:

$$\text{styrene}\quad \langle \!\!\bigcirc\!\!\rangle\text{—CH}\!\!=\!\!\text{CH}_2 + \text{Br}_2 \longrightarrow \langle \!\!\bigcirc\!\!\rangle\text{—CH—CH}_2 \quad\quad (15.42)$$
$$\underset{\text{Br}}{|}\quad\underset{\text{Br}}{|}$$

The noncyclic double bond in styrene rapidly adds bromine, but the benzene ring remains unaffected, even if excess bromine is used. To early chemists, *this lack of alkenelike reactivity defined the uniqueness of benzene and its derivatives.*

Does benzene's lack of reactivity have something to do with its cyclic structure? Cyclohexene, however, adds bromine readily. Perhaps, then, it is the cyclic structure and the conjugated double bonds that *together* account for the unusual behavior of benzene. However, 1,3,5,7-cyclooctatetraene (abbreviated in this text as COT) adds bromine smoothly even at low temperature.

$$\text{1,3,5,7-cyclooctatetraene}\atop\text{(COT)} \quad + \quad \text{Br}_2 \xrightarrow[\text{CHCl}_3]{-55\,^\circ\text{C}} \quad \text{(100\% yield)} \quad\quad (15.43)$$

Thus, the Kekulé structure clearly had difficulties that could not be easily explained away, but there were some ingenious attempts. In 1869, Albert Ladenburg proposed a structure for benzene, called both *Ladenburg benzene* and *prismane*, that seemed to overcome these objections.

Ladenburg benzene
(prismane)

Although Ladenburg benzene is recognized today as a highly strained molecule (it has been described as a "caged tiger"), an attractive feature of this structure to nineteenth-century chemists was its lack of double bonds.

Several facts, however, ultimately led to the adoption of the Kekulé structure. One of the most compelling arguments was that all efforts to prepare the alkene 1,3,5-cyclohexatriene using standard alkene syntheses led to benzene. The argument was, then, that benzene and 1,3,5-cyclohexatriene must be one and the same compound. The reactions used in these routes received additional credibility because they were also used to prepare COT, which, as Eq. 15.43 illustrates, has the reactivity of an ordinary alkene.

Although the Ladenburg benzene structure had been discarded for all practical purposes decades earlier, its final refutation came in 1973 with its synthesis by Professor Thomas J. Katz and his colleagues at Columbia University. These chemists found that Ladenburg benzene is an explosive liquid with properties that are quite different from those of benzene.

How, then, can the Kekulé "cyclic triene" structure for benzene be reconciled with the fact that benzene is inert to the usual reactions of alkenes? The answer to this question will occupy our attention in the next three parts of this section.

B. The Structure of Benzene

The structure of benzene is given in Fig. 15.15a. This structure shows that benzene has *one* type of carbon–carbon bond with a bond length (1.395 Å) that is the average of the lengths of sp^2–sp^2 single bonds (1.46 Å) and double bonds (1.33 Å, Fig. 15.15c). All atoms in the benzene molecule lie in one plane. The Kekulé structure for benzene shows *two* types of carbon–carbon bond: single bonds and double bonds. This inadequacy of the Kekulé structure can be remedied, however, by depicting benzene as *the hybrid of two equally contributing resonance structures*:

$$(15.44)$$

Kekulé structure   resonance hybrid

Benzene is an average of these two structures; it is *one* compound with *one* type of carbon–carbon bond with a bond order of 1.5—neither a single bond nor a double bond, but something halfway in between. A benzene ring is often represented with either of the following hybrid structures, which show the "smearing out" of double-bond character:

hybrid structures of benzene

FIGURE 15.15 Comparison of the structures of benzene, 1,3-butadiene, and COT. (a) The structure of benzene. (The hybrid structure is shown.) (b) The structure of 1,3-butadiene, a conjugated diene. (c) The structure of 1,3,5,7-cyclooctatetraene (COT). (d) A ball-and-stick model of COT. The carbon skeleton of benzene is a planar hexagon and all of the carbon–carbon bonds are equivalent with a bond length that is the average of the lengths of carbon–carbon single and double bonds in COT. In contrast, COT has distinct single and double bonds with lengths that are almost the same as those in 1,3-butadiene, and COT is tub-shaped rather than planar.

benzene

(a)

1,3-butadiene

(b)

1,3,5,7-cyclooctatetraene (COT)

(c)

ball-and-stick model of COT

(d)

As with other resonance-stabilized molecules, we'll continue to represent benzene as one of its resonance contributors because the curved-arrow notation and electronic bookkeeping devices are easier to apply to structures with fixed bonds.

It is interesting to compare the structures of benzene and 1,3,5,7-cyclooctatetraene (COT) in view of their greatly different chemical reactivities (Eqs. 15.42 and 15.43). Their structures are remarkably different (Fig. 15.15). First, although benzene has a single type of carbon–carbon bond, COT has alternating single and double bonds, which have almost the same lengths as the single and double bonds in 1,3-butadiene. Second, COT is not planar like benzene, but instead is tub-shaped.

The π bonds of benzene and COT are also different (Fig. 15.16). The Kekulé structures for benzene suggest that each carbon atom should be trigonal, and therefore sp^2-hybridized. This means each carbon atom has a $2p$ orbital (Fig. 15.16a). Because the benzene molecule is planar, and the axes of all six $2p$ orbitals of benzene are parallel, these $2p$ orbitals can overlap to form six π molecular orbitals. The bonding π molecular orbital of lowest energy is shown in Fig. 15.16b. (The other five π molecular orbitals of benzene are shown in Further Exploration 15.2 in the *Study Guide*.) This molecular orbital shows that π-electron density in benzene lies in doughnut-shaped regions both above and below the plane of the ring. *This overlap is symbolized by the resonance structures of benzene.* The π-electron overlap above and below

FURTHER EXPLORATION 15.2
The π Molecular Orbitals of Benzene

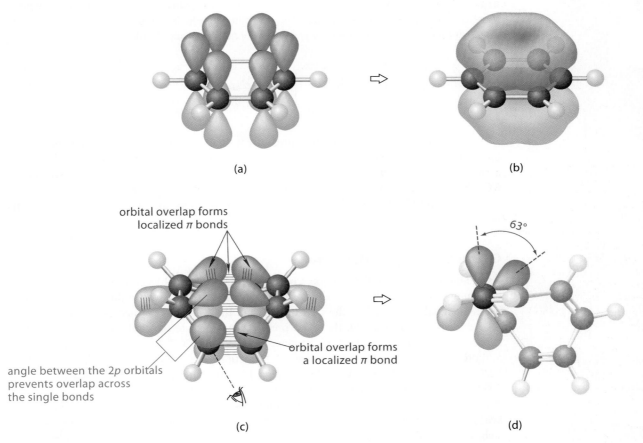

(a)   (b)

orbital overlap forms
localized π bonds

                                63°

orbital overlap forms
a localized π bond

angle between the 2p orbitals
prevents overlap across
the single bonds

(c)   (d)

FIGURE 15.16 Comparison of the p bonds in benzene and 1,3,5,7-cyclooctatetraene (COT). (a) The carbon 2p orbitals in benzene. These orbitals are properly aligned for overlap. (b) The bonding π molecular orbital of lowest energy in benzene. This molecular orbital illustrates that π-electron density lies in a doughnut-shaped region above and below the plane of the benzene ring. (Benzene has two other occupied π molecular orbitals as well as three antibonding π molecular orbitals not shown here.) (c) The carbon 2p orbitals of COT. These orbitals can overlap in pairs to form isolated π bonds, but the tub shape prevents the overlap of 2p orbitals across the single bonds. (d) The view down a single bond indicated by the eyeball in (c). The angle between 2p orbitals, indicated by the red bracket in (c), is about 63°, which is too large for effective overlap.

the plane of the benzene ring is also reflected in the EPM of benzene, which shows a concentration of negative potential in these regions.

π-electron density

EPM of benzene

In contrast, the carbon atoms of COT are not all coplanar, but they are nevertheless all trigonal. This means that there is a 2*p* orbital on each carbon atom of COT (Fig. 15.16c). The tub shape of COT forces the 2*p* orbitals on the ends of each single bond to be oriented at a 63° angle, which is too far from coplanarity for effective interaction and overlap (Fig. 15.16d). Thus, the 2*p* orbitals in COT cannot form a continuous π molecular orbital analogous to the one in benzene. Instead, COT contains four π-electron systems of two carbons each. As far as the π electrons are concerned, *COT looks like four isolated ethylene molecules.* Because there is no electronic overlap between the π orbitals of adjacent double bonds, *COT does not have resonance structures analogous to those of benzene* (Sec. 15.6B, guideline 5).

$$(15.45)$$

To summarize: resonance structures can be written for benzene, because the carbon 2*p* orbitals of benzene can overlap to provide the additional bonding and additional stability associated with *filled bonding molecular orbitals.* Resonance structures *cannot* be written for COT because there is no overlap between 2*p* orbitals on adjacent double bonds.

Why doesn't COT flatten itself to allow overlap of all its 2*p* orbitals? We'll return to this point in Sec. 15.7E.

C. The Stability of Benzene

As mentioned earlier in this section, chemists of the nineteenth century considered benzene to be unusually stable because it is inert to reagents that react with ordinary alkenes. However, chemical reactivity (or the lack of it) is not the way that we measure energy content. As we have already learned, the more precise way to relate molecular energies is by their *standard heats of formation* ΔH_f°. Because benzene and COT have the same empirical formula (CH), we can compare their heats of formation per CH group. The ΔH_f° of benzene is 82.93 kJ mol^{-1} or 82.93/6 = 13.8 kJ mol^{-1} per CH group. The ΔH_f° of COT is 298.0 kJ mol^{-1} or 298.0/8 = 37.3 kJ mol^{-1} per CH group. Thus, benzene, per CH group, is (37.3 − 13.8) = 23.5 kJ mol^{-1} more stable than COT. It follows that benzene is 23.5 × 6 = 141 kJ mol^{-1} (33.6 kcal mol^{-1}) more stable than a hypothetical six-carbon cyclic conjugated triene with the same stability as COT.

This energy difference of about 141 kJ mol^{-1} or 34 kcal mol^{-1} is called the **empirical resonance energy** of benzene. The empirical resonance energy is an experimental estimate of just how much special stability is implied by the resonance structures for benzene—thus the name "resonance energy."

The resonance energy is the energy by which benzene is *stabilized*; it is therefore an energy that benzene "doesn't have." The empirical resonance energy of benzene has been estimated in several different ways; these estimates range from 126 to 172 kJ mol^{-1}

(30 to 41 kcal mol^{-1}). (Another estimate is discussed in Sec. 16.6.) The important point, however, is not the exact value of this number, but the fact that it is *large*.

D. Aromaticity and the Hückel 4*n* + 2 Rule

We've now learned that benzene is unusually stable, and that this stability seems to be correlated with the overlap of its carbon 2*p* orbitals to form π molecular orbitals. In 1931, Erich Hückel (1896–1980), a German chemical physicist, elucidated with molecular orbital arguments the criteria for this sort of stability, which has come to be called *aromaticity*. Using Hückel's criteria, we can define aromaticity more precisely. (Remember again that aromaticity in this context has nothing to do with odor.) This definition has allowed chemists to recognize the aromaticity of many compounds in addition to benzene.

A compound is said to be **aromatic** when it meets *all* of the following criteria.

Criteria for aromaticity:

1. Aromatic compounds contain one or more rings that have a cyclic arrangement of *p* orbitals. Thus, aromaticity is a property of certain *cyclic* compounds.

2. *Every* atom of an aromatic ring has a *p* orbital.

3. Aromatic rings are *planar*.

4. The cyclic arrangement of *p* orbitals in an aromatic compound must contain 4*n* + 2 π electrons, where *n* is any positive integer (0, 1, 2, . . .). In other words, an aromatic ring must contain 2, 6, 10, . . . π electrons.

These criteria are often called collectively the **Hückel 4*n* + 2 rule** or simply the **4*n* + 2 rule**.

The basis of the 4*n* + 2 rule lies in the molecular orbital theory of *cyclic* π-electron systems. The theory holds that aromatic stability is observed only with *continuous cycles* of *p* orbitals—thus, criteria 1 and 2. The theory also requires that the *p* orbitals must overlap to form π molecular orbitals. This overlap requires that an aromatic ring must be planar; *p* orbitals cannot overlap in rings significantly distorted from planarity—thus, criterion 3. The last criterion has to do with the number of π molecular orbitals and the number of electrons they contain. Therefore, to understand criterion 4, we need to know the energies and electron occupancies of the various π molecular orbitals. Two Northwestern University physical chemists, A. A. Frost and Boris Musulin, described in 1953 a simple graphical method for deriving the π-molecular orbital energies of cyclic π-electron systems without resorting to any of the mathematics of quantum theory. This method has come to be known as the **Frost circle**. The steps used in constructing a Frost circle follow, and they are illustrated in Study Problem 15.4.

Steps for constructing a Frost circle:

1. For a cyclic conjugated hydrocarbon or ion with *j* sides (and therefore *j* overlapping 2*p* orbitals), inscribe a regular polygon of *j* sides within a circle of radius 2β with one vertex of the polygon in the vertical position. (Remember from Sec. 15.1B that β is an energy unit used with molecular orbitals.)

2. Place one MO energy level at each vertex of the polygon. Because there are *j* vertices, there will be *j* MOs.

3. Draw a horizontal line that bisects the polygon. All MOs below the line are bonding; all MOs above the line are antibonding; and any MOs on the line are nonbonding (that is, they have the same energy as the isolated 2*p* orbital).

4. The lowest energy level must lie at 2β because it is at the lowest vertex, which is at the end of the vertical radius. The energies of the other levels are calculated by determining their positions along the vertical radius by trigonometry.

5. Add the π electrons to the energy levels in accordance with the Pauli principle and Hund's rules.

Study Problem 15.4 illustrates the application of the Frost circle to benzene.

STUDY PROBLEM   15.4

Use the Frost circle to determine the energy levels and electron occupancies for the π MOs of benzene.

Step 1. For benzene, $j = 6$. Therefore, we inscribe a regular hexagon into a Frost circle of radius 2β with one vertex pointing down (Fig. 15.17a).

Step 2. Place an energy level at each vertex. This gives six energy levels.

Step 3. The horizontal dashed line in Fig. 15.17a bisects the polygon. The three MOs below the dashed line are bonding, and the three MOs above the dashed line are antibonding. There are no nonbonding MOs. (In Fig. 15.7b, the two MOs on the line are nonbonding.)

Step 4. Calculate the energies. The energies of π_1 and π_6^* clearly lie on the circle, so their energies must be 2β and -2β, respectively. (Remember that β is a negative number.) Then draw a perpendicular from the π_2 (or π_3) vertex to the vertical radius and calculate the $E(\pi_2)$, the energy of π_2, as follows:

$$E(\pi_2) = r \cos 60° = (2\beta)(0.50) = 1\beta$$

From this calculation, the energies of π_2 and π_3, which are identical, equal $+1.0\beta$. By symmetry, the energies of π_4^* and π_5^* are -1.0β.

Step 5. Add the π electrons to the energy levels. Benzene has six π electrons. Because each bonding MO can accommodate two electrons, the available electrons exactly fill the bonding MOs.

This method shows that benzene has six MOs—something we already knew. But it also shows that two bonding MOs, π_2 and π_3, have identical energies, and two antibonding MOs, π_4^* and π_5^*, also have identical energies. When orbitals have the same energy, they are said to be **degenerate**. Hence, π_2 and π_3 are degenerate MOs and π_4^* and π_5^* are degenerate MOs.

The π-electron energy of benzene is 8.0β. The π-electron energy of three isolated ethylenes (with a bonding-MO energy of 1.0β) is 6.0β. The delocalization energy, or resonance energy, of benzene is then 2.0β. If we equate this to the empirical resonance energy of benzene (Sec. 15.7C), which is 141–150 kJ mol^{-1} or 34–36 kcal mol^{-1}, we find that $\beta = 70$–75 kJ mol^{-1} (17–18 kcal mol^{-1}). The resonance energy is a consequence of the very low-lying π_1 MO; the other bonding MOs, π_2 and π_3, have the same energy (1.0β) as the bonding MO of ethylene. The π_1 MO, shown in Fig. 15.16b, has no nodes, and it most closely corresponds to the resonance-hybrid structure of benzene, which shows the π electrons spread evenly around the entire molecule.

PROBLEMS

15.34 Use a Frost circle to determine the π-electron structure of (a) the cyclopentadienyl anion, which has a planar structure and six π electrons; and (b) the cyclopropenyl cation, which has two π electrons.

cyclopentadienyl anion   **cyclopropenyl cation**

15.35 How many bonding MOs are there in a planar, cyclic, conjugated hydrocarbon that contains a ring of 10 carbon atoms? How many π electrons does it have? How many of the π electrons can be accommodated in the bonding MOs?

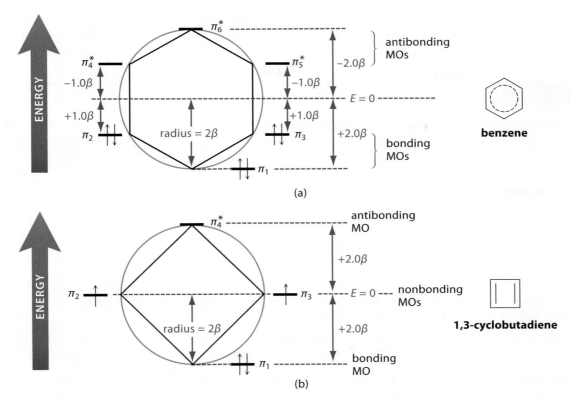

FIGURE 15.17 Application of the Frost circle to find the MOs and their energies for cyclic conjugated hydrocarbons. The Frost circle is in blue. (a) The Frost circle for benzene. (b) The Frost circle for 1,3-cyclobutadiene, an antiaromatic compound. (Antiaromatic compounds are discussed in Sec. 15.7E.) Notice that π_2 and π_3 are nonbonding and are only half filled.

Compounds (and ions) with $4n + 2$ π electrons contain exactly the number of electrons required to fill the bonding MOs. Benzene has six π electrons ($4n + 2 = 6$ for $n = 1$); as we found in Study Problem 15.4, these exactly fill the bonding MOs of benzene. A planar, cyclic conjugated hydrocarbon with 10 π electrons (see Problem 15.35) has five bonding MOs, which can accommodate all 10 electrons ($4n + 2 = 10$ for $n = 2$). A molecule that contains *more* than $4n + 2$ π electrons, even if it could meet all of the other criteria for aromaticity, must have one or more electrons in nonbonding or antibonding MOs. If a molecule contains fewer than $4n + 2$ electrons, its bonding molecular orbitals are not fully populated, and its resonance energy (delocalization energy) is reduced. But there's more to aromaticity than just fully occupied bonding MOs; after all, 1,3-butadiene and other conjugated *acyclic* hydrocarbons also have fully occupied bonding MOs (Fig. 15.1). *The bonding molecular orbitals in aromatic compounds have particularly low energy,* especially the MO at $E = 2.0\beta$. The resonance energy of benzene is 2.0β, but the resonance energy of (E)-1,3,5-hexatriene, the acyclic conjugated triene, is 1.0β (Problem 15.2, p. 716, or Fig. 15.7, p. 723). Moreover, the magnitude of β for acyclic conjugated hydrocarbons (-50 kJ mol^{-1}) is only two-thirds of that for cyclic conjugated hydrocarbons. (Simple MO theory does not account for this difference, but more advanced theories do.) This difference further increases the energetic advantage of aromaticity. To summarize the basis of the $4n + 2$ rule:

1. Cyclic conjugated molecules and ions with $4n + 2$ π electrons have exactly the right number of π electrons to fill the bonding MOs.

2. The bonding MOs in cyclic conjugated molecules and ions, especially the bonding MO of lowest energy, have a very low energy. For this reason, cyclic conjugated molecules and ions have a large resonance energy.

Recognizing aromatic compounds is a matter of applying the four criteria for aromaticity. This objective is addressed in Study Problem 15.5 and the discussion that follows it.

STUDY PROBLEM 15.5

Decide whether each of the following compounds is aromatic. Explain your reasoning.

(a) toluene

(b) $H_2C=CH—CH=CH—CH=CH_2$
1,3,5-hexatriene

(c) biphenyl

(d) **1,3,5-cycloheptatriene**

(e) **1,3-cyclobutadiene**

SOLUTION In each example, first count the π electrons by applying the following rule: *Each double bond contributes two π electrons.* Then apply *all* of the criteria for aromaticity.

(a) The ring in toluene, like the ring in benzene, is a continuous planar cycle of six π electrons. Hence, the ring in toluene is aromatic. The methyl group is a substituent group on the ring and is not part of the ring system. Because toluene contains an aromatic ring, it is considered to be an aromatic compound. This example shows that *parts of molecules* can be aromatic, or, equivalently, that aromatic rings can have nonaromatic substituents.

(b) Although 1,3,5-hexatriene contains six π electrons, it is not aromatic, because it fails criterion 1 for aromaticity: it is not cyclic. Aromatic species must be cyclic.

(c) Biphenyl has two rings, each of which is separately aromatic. Hence, biphenyl is an aromatic compound.

(d) Although 1,3,5-cycloheptatriene has six π electrons, it is not aromatic, because it fails criterion 2 for aromaticity: one carbon of the ring does not have a p orbital. In other words, the π-electron system is not continuous, but is interrupted by the sp^3-hybridized carbon of the CH_2 group.

(e) 1,3-Cyclobutadiene is not aromatic. Even though it is a continuous cyclic system of $2p$ orbitals, it fails criterion 4 for aromaticity: it does not have $4n + 2$ π electrons.

Aromatic Heterocycles Aromaticity is not confined solely to hydrocarbons. Some *heterocyclic compounds* (Sec. 8.2C) are aromatic; for example, pyridine and pyrrole are both aromatic nitrogen-containing heterocycles.

pyridine
(aromatic)

pyrrole
(aromatic)

Except for the nitrogen in the ring, the structure of pyridine closely resembles that of benzene. Each atom in the ring, including the nitrogen, is part of a double bond and therefore contributes one π electron. How does the electron pair on nitrogen figure in the π-electron count? This electron pair resides in an sp^2 orbital in the plane of the ring (see Fig. 15.18a). (It has the same relationship to the pyridine ring that any one of the C—H bonds has.) Because the nitrogen unshared pair does not overlap with the ring's π-electron system, it is not included

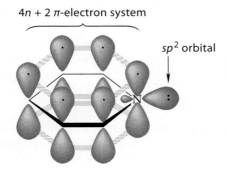

4*n* + 2 π-electron system

*sp*² orbital

(a) **pyridine**
unshared pair *is not* part of the
4*n* + 2 π-electron system

4*n* + 2 π-electron system

2*p* orbital

N—H

(b) **pyrrole**
unshared pair *is* part of the
4*n* + 2 π-electron system

FIGURE 15.18 The 2*p* orbitals in pyridine and pyrrole. The gray lines represent orbital overlap. (a) The unshared electron pair in pyridine is vinylic and is therefore in an *sp*² orbital (*blue*) and is not part of the aromatic π-electron system. (b) The unshared electron pair in pyrrole is allylic and can occupy a 2*p* orbital (*blue*) that is part of the aromatic π-electron system.

in the π-electron count. Thus, *vinylic electrons (electrons on doubly bonded atoms) are not counted as π electrons.*

In pyrrole, the electron pair on nitrogen is *allylic* (Fig. 15.18b). The nitrogen has a trigonal geometry and *sp*² hybridization that allow its electron pair to occupy a 2*p* orbital and contribute to the π-electron count. The N—H hydrogen lies in the plane of the ring. In general, *allylic electrons are counted as π electrons when they reside in orbitals that are properly situated for overlap with the other p orbitals in the molecule.* Therefore, pyrrole has six π electrons—four from the double bonds and two from the nitrogen—and it is aromatic.

Note carefully the different ways in which we handle the electron pairs on the nitrogens of pyridine and pyrrole. The nitrogen in pyridine is part of a double bond, and the electron pair *is not* part of the π-electron system. The nitrogen in pyrrole is allylic and its electron pair *is* part of the π-electron system.

Aromatic Ions   Aromaticity is not restricted to neutral molecules; a number of ions are aromatic. One of the best characterized aromatic ions is the cyclopentadienyl anion:

2,4-cyclopentadien-1-ide anion
(**cyclopentadienyl anion**; aromatic)

(The Frost circle for this ion was considered in Problem 15.34a.) The cyclopentadienyl anion resembles pyrrole; however, because the atom bearing the allylic electron pair is carbon rather than nitrogen, its charge is −1. One way to form this ion is by the reaction of sodium with the conjugate acid hydrocarbon, 1,3-cyclopentadiene; notice the analogy to the reaction of Na with H_2O.

$$2 \quad \text{(cyclopentadiene)} + 2\,\text{Na} \xrightarrow[\text{THF}]{\text{2–3 h; 0 °C}} 2\,\text{Na}^+ \quad \text{(cyclopentadienyl anion)}^- + H_2 \qquad (15.46)$$

1,3-cyclopentadiene
not aromatic

cyclopentadienyl anion
aromatic

The cyclopentadienyl anion has five equivalent resonance structures; the negative charge can be delocalized to each carbon atom:

 (15.47)

These structures show that all carbon atoms of the cyclopentadienyl anion are equivalent. For this reason, the cyclopentadienyl anion is sometimes represented with a hybrid structure:

hybrid structure of the cyclopentadienyl anion

Because of the stability of this anion, its conjugate acid, 1,3-cyclopentadiene, is an unusually strong hydrocarbon acid. (Remember: The more stable the conjugate base, the more acidic is the conjugate acid; Fig. 3.3, p. 116.) With a pK_a of 15, this compound is 10^{10} times more acidic than a 1-alkyne, and about as acidic as water!

Cations, too, may be aromatic. (See Problem 15.34b.)

$$\text{(a Lewis acid)} \quad \text{Cl} + \text{SbCl}_5 \longrightarrow \left[\cdots \right] \text{SbCl}_6^- \quad (15.48)$$

cyclopropenyl cation
(aromatic)

This example illustrates another point about counting electrons for aromaticity: *atoms with empty p orbitals are part of the π-electron system, but they contribute no electrons to the π-electron count.* Because this cation has two π electrons, it is aromatic ($4n + 2 = 2$ for $n = 0$). The stability of the cyclopropenyl cation, despite its considerable angle strain, is a particularly strong testament to the stabilizing effect of aromaticity.

Counting π electrons accurately is crucial for successfully applying the $4n + 2$ rule. Let's summarize the rules for π-electron counting.

1. Each atom that is part of a double bond contributes one π electron.
2. Vinylic unshared electron pairs do not contribute to the π-electron count.
3. Allylic unshared electron pairs contribute two electrons to the π-electron count if they occupy an orbital that is parallel to the other *p* orbitals in the molecule.
4. An atom with an empty *p* orbital can be part of a continuous aromatic π-electron system, but contributes no π electrons.

Aromatic Polycyclic Compounds The Hückel $4n + 2$ rule applies strictly to single rings. However, a number of common fused bicyclic and polycyclic compounds, such as naphthalene, quinoline, and indole are also aromatic:

two fused benzene rings

fused benzene and pyridine rings

fused benzene and pyrrole rings

naphthalene　　**quinoline**　　**indole**

Although rules have been devised to predict the aromaticity of fused-ring compounds, these rules are rather complex, and we need not be concerned with them. However, it shouldn't be difficult to see the resemblance of these compounds to monocyclic aromatic compounds—naphthalene to benzene, quinoline to benzene and pyridine, and indole to benzene and pyrrole.

Indole is particularly important in biology because of its presence in proteins within the structure of the amino acid *tryptophan*.

tryptophan
(as it occurs within proteins)

(The structures of proteins are covered in Chapter 27.) Quinoline is important because of its presence in a number of naturally occurring compounds.

Fused-ring aromatic hydrocarbons with more than two rings are well known. Some of these are shown in Fig. 15.19a (p. 770). Two of the most spectacular examples of fused-ring aromatic compounds are graphite (Fig. 15.19b) and buckminsterfullerene (Fig. 15.19c). *Graphite* is a form of elemental carbon that consists of layers of fused benzene rings. The softness of graphite and its ability to act as a lubricant can be attributed to the ease with which the layers slide past one another. Even though graphite is not an ionic compound, it is an excellent electrical conductor because of the ease with which the π electrons can be delocalized across its structure. Graphite is used in arc-welding electrodes and in lithium-ion and nickel–metal-hydride batteries. Single layers of graphite (called *graphene*) have been isolated and studied, and the 2010 Nobel Prize in Physics was awarded to Andre Geim (b. 1958) and Konstantin Novoselov (b. 1974) of the University of Manchester, U.K., for their experiments that characterized this material. (You have probably produced graphene when you have written with a lead pencil on paper. Pencil "lead" is actually graphite.)

Buckminsterfullerene was discovered as one component of soot and interstellar gas. The structure, which corresponds to the seams on a modern soccer ball, was proposed in 1985 by Sir Harold W. Kroto (b. 1939) of the University of Sussex, Brighton, U.K., along with Richard E. Smalley (1943–2005) and Robert E. Curl (b. 1933) of Rice University. All 60 of the carbons in buckminsterfullerene are equivalent, as its single ^{13}C-NMR resonance at δ 143 shows. The intriguing name of the compound (also sometimes nicknamed "buckyball") came from its resemblance to the geodesic dome designed by American architect Buckminster Fuller (1895–1983). A variety of related "buckyballs" and "buckytubes" have since been discovered. Kroto, Smalley, and Curl were recognized for this and related discoveries with the 1996 Nobel Prize in Chemistry.

Aromatic Organometallic Compounds   Some remarkable organometallic compounds have aromatic character. For example, the cyclopentadienyl anion, discussed previously in this section as one example of an aromatic anion, forms stable complexes with a number of transition-metal cations. One of the best known of these complexes is *ferrocene*, which is synthesized by the reaction of two equivalents of cyclopentadienyl anion with one equivalent of ferrous ion (Fe^{2+}).

$$2 \; \underset{\ddot{\,}\;\; Na^+}{\text{⬠}} + FeCl_2 \longrightarrow Fe^{2+} \left(\ddot{\;} \text{⬠} \right)_2 + 2\,NaCl \qquad (15.49)$$

ferrocene
(90% yield)

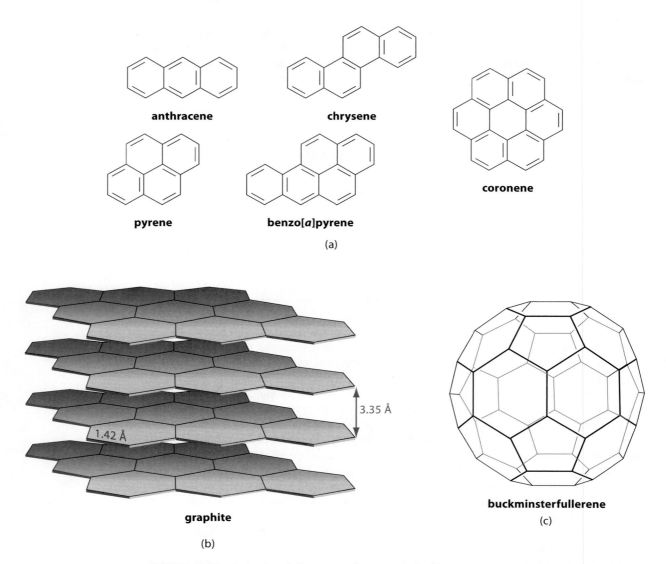

FIGURE 15.19 Some fused-ring aromatic compounds. (a) Some common fused-ring aromatic hydrocarbons containing more than two rings. (The role of benzo[a]pyrene in cancer is discussed in Sec. 16.7.) (b) Graphite. In this structure, the delocalized double bonds are not shown. Notice that the carbon–carbon bond length is very similar to that in benzene. (c) Buckminsterfullerene (C_{60}). In this structure the delocalized double bonds are not shown.

Although this synthesis resembles a metathesis (exchange) reaction in which two salts are formed from two other salts, ferrocene is not a salt but is a remarkable "molecular sandwich" in which a ferrous ion is imbedded between two cyclopentadienyl anions.

The red dashed lines mean that the electrons of the cyclopentadienyl anions are shared not only by the ring carbons but also by the ferrous ion; each carbon is bonded equally to the iron.

Let's now return to the question posed near the beginning of this section: Why is benzene inert in the usual reactions of alkenes? The *aromaticity* of benzene is responsible for its unique chemical behavior. If benzene were to undergo the addition reactions typical of alkenes, its continuous cycle of $4n + 2$ π electrons would be broken; it would lose its aromatic character and much of its stability.

This is not to say, however, that benzene is unreactive under all conditions. Indeed, benzene and many other aromatic compounds undergo a number of characteristic reactions that are presented in Chapter 16. However, the conditions required for these reactions are typically much harsher than those used with alkenes, precisely because benzene is so stable. As you will also see, the reactions of benzene give very different kinds of products from the reactions of alkenes.

PROBLEMS

15.36 Furan is an aromatic compound. Discuss the hybridization of its oxygen and the geometry of its two electron pairs.

furan
(aromatic)

15.37 Do you think it would be possible to have an aromatic free radical? Why or why not?

15.38 Which of the following species should be aromatic by the Hückel $4n + 2$ rule?

(a)

thiophene

(b)

(c)

H H

(d)

isoxazole

(e)

(f)

Et

E. Antiaromatic Compounds

Compounds that contain *planar*, continuous rings of $4n$ π electrons, in stark contrast to aromatic compounds, are especially *unstable*; such compounds are said to be **antiaromatic**. 1,3-Cyclobutadiene (which we'll call simply cyclobutadiene) is such a compound; its small ring size and the sp^2 hybridization of its carbon atoms constrain it to planarity. This compound is so unstable that it can only be isolated at very low temperature, 4 K.

1,3-cyclobutadiene

The Frost circle for 1,3-cyclobutadiene is shown in Fig. 15.17b, p. 765. Two of the π MOs, π_2 and π_3, lie at $E = 0$. The π-electron energy of cyclobutadiene is 4.0β, which is *the same* as the π-electron energy of two isolated ethylenes. In other words, *cyclobutadiene has no resonance energy.* Moreover, Hund's rules requires that the two degenerate MOs, π_2 and π_3, be half occupied. This means that cyclobutadiene has two unpaired electrons and is therefore a double free radical! Finally, cyclobutadiene has considerable angle strain.

The overlap of p orbitals in molecules with cyclic arrays of $4n$ π electrons is a *destabilizing* effect. (It could be said that antiaromatic molecules are "destabilized by resonance.")

More advanced MO calculations show that cyclobutadiene, in an effort to escape this high-energy situation, distorts by lengthening its single bonds and shortening its double bonds:

1.54–1.55 Å

1.35–1.36 Å

As a result of this distortion, the degeneracy of π_2 and π_3 is removed; so, one MO lies at slightly lower energy than the other and is doubly occupied. Cyclobutadiene, in effect, contains *localized* double bonds. This distortion, while minimizing antiaromatic overlap, introduces even more strain than the molecule would contain otherwise. The molecule can't win; it is too unstable to exist under normal circumstances.

Although cyclobutadiene is itself very unstable, it forms a very stable complex with Fe(0):

$$\left[\underset{\substack{\text{Fe(CO)}_3}}{} \longleftrightarrow \underset{\substack{\text{Fe}^{2+}\text{(CO)}_3}}{} \xleftarrow{} \begin{array}{c}\text{cyclobutadienyl}\\\text{dianion}\end{array} \right] \tag{15.50}$$

**cyclobutadieneiron
tricarbonyl**

(In this structure, the CO groups are neutral carbon monoxide ligands.) 1,3-Cyclobutadiene has four π electrons and is thus two electrons short of the number (six) required for aromatic stability. These two missing electrons are provided by the iron. As the resonance structure on the right in Eq. 15.50 suggests, this complex in effect consists of a 1,3-cyclobutadiene with two additional electrons—a cyclobutadienyl *dianion*, a six π-electron aromatic system—combined with an iron minus two electrons—that is, Fe^{2+}. In effect, the iron stabilizes the antiaromatic diene by donating two electrons, thus making it aromatic.

This section began with a comparison of the stabilities of benzene and 1,3,5,7-cyclooctatetraene (COT). You can now recognize that COT contains a continuous cycle of $4n$ π electrons.

**1,3,5,7-cyclooctatetraene
(COT)**

Is COT antiaromatic? It would be if it were planar. However, this molecule is large and flexible enough that it can escape unfavorable antiaromatic overlap by folding into a tub conformation, as shown in Figs. 15.15d and 15.16c. It is believed that *planar* cyclooctatetraene, which *is* antiaromatic, is more than 58 kJ mol^{-1} (14 kcal mol^{-1}) less stable than the tub conformation.

PROBLEMS

15.39   Using the theory of aromaticity, explain the finding that *A* and *B* are different compounds, but *C* and *D* are identical. (That *A* and *B* are different molecules was established by Prof. Barry Carpenter and his students at Cornell University in 1980.)

A   *B*   *C*   *D*

15.40   Which of the compounds or ions in Problem 15.38 (p. 771) are likely to be antiaromatic? Explain.

15.8 NONCOVALENT INTERACTIONS OF AROMATIC RINGS

In Sections 8.5–8.8 we learned about a number of noncovalent interactions:

1. Van der Waals interactions: interactions between fluctuating dipoles and interactions between dipoles and induced dipoles;
2. Electrostatic interactions: charge–charge interactions, charge–dipole interactions, and dipole–dipole interactions;
3. Hydrogen bonding: the association of an O—H or N—H donor with a Lewis-base acceptor; and
4. "Hydrophobic bonding": the association of nonpolar groups driven by entropy changes in water.

As we might imagine, benzene and its derivatives can interact noncovalently with other hydrocarbons, both aromatic and nonaromatic. Two noncovalent interactions of aromatic rings are rather unique, and they are very important in many biomolecules and in their interactions with drug molecules. The first is the noncovalent interaction of aromatic rings with each other. The second is the noncovalent interaction of aromatic rings with cations. This section describes these noncovalent interactions and provides examples of each.

The basis of both types of interaction is the charge distribution in aromatic compounds, which we'll illustrate with benzene. Although benzene, because of its symmetry, has an overall zero dipole moment, it does have local regions of positive and negative charge. As you've already learned, the π electrons in benzene and other aromatic compounds are concentrated above and below the ring plane in the bonding π molecular orbitals. Associated with this π-electron density is a localized partial negative charge in these regions, illustrated by the electrostatic potential maps (EPMs) below. The red color is associated with regions of partial negative charge.

EPMs of benzene illustrating regions of
partial positive and partial negative charge

Benzene is a neutral molecule; so, corresponding to the localized partial negative charge, there must be regions of localized partial positive charge. These regions occur around the periphery of the ring, as shown by the blue areas in the EPMs. Because the π electrons are concentrated *above and below* the ring plane, the compensating partial positive charges are concentrated *in* the ring plane. This occurs because the positive charge of the carbon nucleus is screened from outside groups *in the ring plane* by only the carbon sp^2 electrons but not the π electrons. This effect was also discussed in Sec. 4.4 (p. 145), which explains why sp^2-hybridized carbons (including those of aromatic rings) are electronegative and the attached hydrogens are relatively positive.

A. Noncovalent Interactions between Aromatic Rings

Aromatic rings can interact noncovalently with each other in two major ways. The first is that they can "stack" with their ring planes parallel.

offset stacking interaction

the offset allows attractions between oppositely charged regions to occur

The centers of the rings involved in a stacking interaction are offset to avoid the juxtaposition of like charges in the two interacting rings. We'll call this type of interaction **offset stacking**. The offset allows the positively charged region of one ring to interact attractively with the negatively charged region of the other.

In the second type of interaction between rings, the planes of the two interacting rings are perpendicular. This orientation also allows the positively charged region of one ring to interact attractively with the negatively charged region of the other. This type of interaction is called an **edge-to-face** ring interaction.

"T-type" edge-to-face interaction

attractions between oppositely charged regions

"Y-type" edge-to-face interaction

edge-to-face interactions

As the diagram shows, the edge-to-face interaction can occur in two ways; one is called a "T interaction" and the other is called a "Y interaction." Opinions differ as to the relative importance of these two orientations.

Both physical calculations and experimental evidence support the idea that the edge-to-face interaction is the more stable of the two attractions. However, there are many examples of both offset stacking and edge-to-face interactions in chemistry and biology; so, both are important.

The interactions described above are somewhat different from the "hydrophobic bonds" discussed in Sec. 8.6D, which are driven largely by a favorable entropy change associated with the solvent organization. When two aromatic rings come together in water, a negative enthalpy change $\Delta H°$ drives the association. The negative enthalpy is caused, at least in part, by a significant attraction between the rings, which is described in the foregoing discussion.

Aromatic rings can also interact with other polarizable but nonpolar groups (for example, alkyl groups) in conventional van der Waals attractions. Examples are known in which an alkyl group lies in the face of the ring, close to the π electrons, and other examples are known in which the hydrocarbon group lies along the edge of the ring. In either case, the localized charge in the aromatic ring induces a temporary dipole in the alkyl group, and, as a result, an attractive interaction develops between the two groups.

B. The Noncovalent Interaction of Aromatic Rings with Cations

Once we understand the nature of the charge distribution in aromatic rings, we can understand that an aromatic ring can interact favorably with cations. This type of noncovalent interaction is called a **pi–cation interaction** (Fig. 15.20). As we have learned, the π-electron cloud of a benzene ring or other aromatic ring is a rich source of electrons. In a pi–cation interaction, a cation is centered above or below the π-electron cloud of an aromatic ring and is attracted to the π electrons (Fig. 15.20a). This attraction was first discovered in a study of gas-phase interactions. In the gas phase, one or more benzene molecules can interact attractively with a potassium ion. Complexes of one to four benzene molecules with a potassium ion in the gas phase are known. Fig. 15.20b shows the gas-phase pi–cation interaction of four benzene rings with a potassium ion. The formation of the complex between a potassium ion and a single benzene molecule in the gas phase has a large favorable $\Delta G°$ value of -48.5 kJ mol^{-1} (-11.6 kcal mol^{-1}), which includes the loss of entropy that typically accompanies an association reaction. The interaction energy, which has the unfavorable entropy removed, is approximated by the $\Delta H°$ value of -76.6 kJ mol^{-1} (-18.3 kcal mol^{-1}).

C. Noncovalent Interactions of Aromatic Rings in Biology

Offset stacking, edge-to-face attractions, and pi–cation attractions are all very important in biology. These interactions provide mechanisms that stabilize protein structures and contribute to the formation of stable complexes between proteins (enzymes, receptors) and small molecules (substrates, inhibitors). This section provides a few biological examples of these interactions.

The aromatic rings in proteins are the groups in the side chains of the amino acid residues phenylalanine, tyrosine, and tryptophan.

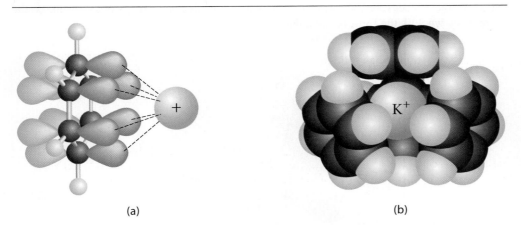

(a)   (b)

FIGURE 15.20 Pi–cation interactions. (a) An orbital description of the pi–cation interaction. (b) A space-filling model showing the interaction of four benzene molecules with a potassium cation in the gas phase. In both, the cation is centered on the face of an aromatic ring. The high π-electron density in this direction interacts attractively with the cation.

The amino acid histidine also has an aromatic imidazole ring in its side chain, but this ring (which is often found as its conjugate-acid cation) more often serves an acid–base role than a source of aromatic-ring interactions.

side chain of **histidine** (His, H)
conjugate-acid form

Double-helical DNA provides a spectacular example of offset stacking. If you are familiar with the double-helical structure of DNA, you know that the bases in each strand are "stacked" perpendicular to the axis of the double helix. (The double-helical structure of DNA is shown in Fig. 26.5, p. 1358.) These bases are aromatic compounds. Their stacking is offset by the turn in the helix. It is known that the favorable interaction between the aromatic rings is an important element—possibly the *most important* element—that stabilizes the helical structure.

The structures of many proteins also contain stabilizing interactions between the rings of aromatic amino acid residues.

An important biological example of a pi–cation interaction occurs in the binding of the cationic molecule acetylcholine to the *acetylcholine receptor*, often abbreviated AcChR. One important form of AcChR is a large protein associated with cell membranes at the neuromuscular junction and elsewhere. Acetylcholine is a cationic molecule.

acetylcholine
(AcCh)

The binding of acetylcholine to the AcChR causes a large conformational change in the protein that opens a pore in the protein and allows sodium and potassium ions to flow through the cell membrane. At the neuromuscular junction, this ion transmission potentiates muscle contraction. A major attractive interaction between acetylcholine and the receptor is a pi–cation interaction between the cationic group of acetylcholine and a tryptophan residue on the protein, shown with the EPM of the indole ring:

tryptophan residue of the protein

acetylcholine

pi–cation interaction

Binding of nicotine in its cationic, conjugate-acid form at the acetylcholine site in the AcChR of the brain is involved in the mechanism of nicotine addiction.

nicotine
(conjugate acid)

Many protein structures also contain stabilizing pi–cation interactions of the aromatic amino acid rings with the positively charged amino acid side chains of lysine and arginine.

R— =   **lysine**   **arginine**

└── positively charged amino acid residues ──┘

Aricept: Drug Therapy for Alzheimer's Disease

Alzheimer's disease (AD) is a condition in which a progressive and ultimately severe dementia is caused by protein misaggregation in the brain. Although largely a disease of the elderly, it can also occur in younger adults. Although much current research is focused on the prevention of AD, current drug therapy is focused largely to delaying the onset of cognitive impairment associated with the disease. One of the most widely used drugs for this purpose is *donepezil*, marketed as its hydrochloride salt under the trade name Aricept®.

donepezil hydrochloride (Aricept®)

The biological target of donepezil is the brain enzyme *acetylcholinesterase*. The normal function of this enzyme is to catalyze the hydrolysis of acetylcholine, an important neurotransmitter.

acetylcholine
(AcCh)   **choline**   **acetate**
(conjugate base
of acetic acid)

This hydrolysis terminates nerve transmission mediated by AcCh. Donepezil binds to acetylcholinesterase near the active site and thereby prevents the enzyme from binding AcCh. The inhibition of acetylcholine hydrolysis is believed to intensify neural activity in the brain.

As in all enzyme-catalyzed reactions, the hydrolysis reaction is preceded by binding of the substrate AcCh to the enzyme. Because donepezil binds close to the AcCh site in the enzyme, it blocks the binding of AcCh to the enzyme and thereby inhibits the hydrolysis of the neurotransmitter. The binding of donepezil to acetylcholinesterase is shown in Fig. 15.21 on p. 778. This binding involves several examples of the interactions discussed in this section. The drug molecule is shown in green, and various aromatic amino-acid side chains (identified by their three-letter abbreviations) are in gray. Identifiable offset stacking and pi–cation interactions are indicated. An edge-to-face interaction and offset interactions of groups on the enzyme with each other is also present; can you find them?

FIGURE 15.21 The binding of Aricept® (donepezil, *green*) within the active site of the enzyme acetylcholinesterase is stabilized by both offset stacking and pi–cation interactions. Aromatic side chains of the enzyme are shown in gray. Several interactions of aromatic rings within the enzyme structure are also shown; can you identify some of them?

PROBLEM

15.41   The following molecule has a barrel shape (in which the benzene rings are the "walls" of the barrel). It forms a noncovalent complex with the iodide salt of acetylcholine in chloroform solvent.

Describe the orientation of the acetylcholine molecule within the complex.

KEY IDEAS IN CHAPTER 15

- Molecules containing conjugated double bonds have additional stability, relative to unconjugated isomers, that can be attributed to the continuous overlap of their carbon $2p$ orbitals to form π molecular orbitals.

- The delocalization energy, or resonance energy, of a conjugated π-electron system containing j double bonds is the difference between its π-electron energy and the π-electron energy of j ethylenes.

- The most stable conformation of 1,3-butadiene and other conjugated dienes is the *s*-trans conformation, which is

an anti conformation about the central single bond of the diene unit.

- A cumulene is a compound with one or more *sp*-hybridized carbon atoms that are part of two double bonds. An allene is a cumulene with two cumulated double bonds. Adjacent π bonds in a cumulene are mutually perpendicular; appropriately substituted allenes are chiral.

- Heats of formation for isomers are generally in the following order: conjugated dienes < ordinary dienes < alkynes < cumulenes.

- Compounds with conjugated double or triple bonds have UV–visible absorptions at $\lambda_{max} > 200$ nm.

- Each conjugated double or triple bond in a molecule contributes 30–50 nm to its λ_{max}. When a compound contains many conjugated double or triple bonds, it absorbs visible light and appears colored.

- The intensity of the UV–vis absorption of a compound is proportional to its concentration (Beer's law). The constant of proportionality ϵ, called the molar extinction coefficient, is the intrinsic intensity of an absorption.

- Some molecules that absorb energy from UV or visible radiation can lose some of that energy by emitting light (fluorescence) at greater wavelength. Many fluorescent molecules have rigid chromophores with extended π-electron systems. Fluorescence is a very sensitive analytical method that has important applications in biology, such as the green fluorescent protein (GFP).

- The Diels–Alder reaction is a pericyclic reaction that involves the cycloaddition of a conjugated diene and a dienophile (usually an alkene). When the diene is cyclic, bicyclic products are produced.

- The diene assumes an *s*-cis conformation in the transition state of the Diels–Alder reaction; dienes that are locked into *s*-trans conformations are unreactive.

- Each component of the Diels–Alder reaction undergoes a syn-addition to the other. In many cases the endo mode of addition is kinetically favored over the exo mode.

- Conjugated dienes react with hydrogen halides to give mixtures of 1,2- and 1,4-addition (conjugate addition) products. Such a mixture of products is accounted for by the formation of a resonance-stabilized allylic carbocation intermediate, which can react with halide ion at either of two positively charged carbons.

- When the products of a reaction do not come to equilibrium under the reaction conditions, the reaction is said to be kinetically controlled. A kinetically controlled reaction can give a mixture of products that is substantially different from the mixture obtained if the products were allowed to come to equilibrium. The predominance of the 1,2-addition product in the reaction of hydrogen halides with conjugated alkenes is an example of kinetic control. If the products of a reaction come to equilibrium under the reaction conditions, the reaction is said to be thermodynamically controlled.

- Resonance structures are derived by the curved-arrow notation. A molecule is the weighted average of its resonance structures. That is, the structure of the molecule is most accurately approximated by its most important resonance structures.

- Other things being equal, the species with the greatest number of important resonance structures is most stable.

- Benzene is the prototype of a class of compounds, including some ions, that have a special stability called aromaticity. All aromatic compounds contain $4n + 2$ π electrons in a continuous, planar, cyclic array.

- The basis of the $4n + 2$ rule is that bonding MOs are completely filled in molecules with $4n + 2$ π electrons, and that the bonding MOs have very low energy.

- Compounds that contain $4n$ π electrons in a continuous, planar, cyclic array are antiaromatic and are especially unstable.

- Aromatic rings can have noncovalent attractions with other aromatic rings. Two configurations are commonly observed: a parallel-offset stacking interaction and an edge-to-face interaction. Aromatic rings can also have noncovalent attractions with positive ions. In this type of attraction, called a pi–cation interaction, a positive ion is situated at the face of the ring where it can interact favorably with the ring π electrons. These interactions are very important in some protein structures and in the binding of substrates and inhibitors to some enzymes and receptors.

 REACTION REVIEW *For a summary of reactions discussed in this chapter, see the* Reaction Review *section of Chapter 15 in the* Study Guide and Solutions Manual.

ADDITIONAL PROBLEMS

15.42 Use the curved-arrow or fishhook notation to derive the major resonance structures for each of the following species. Determine which, if any, structure is the most important one in each case.

(a)

(b)

(c)

$$H_3C - \overset{\overset{\displaystyle :O:}{\|}}{C} - \overset{..}{\underset{..}{C}}H - CH = CH_2$$

(d)   (e)   (f)

15.43 Give the principal product(s) expected, if any, when *trans*-1,3-pentadiene reacts under the following conditions. Assume one equivalent of each reagent reacts unless noted otherwise.

(a) Br_2 (dark) in CH_2Cl_2 (b) HBr

(c) H_2 (two molar equivalents), Pd/C

(d) H_2O, H_3O^+ (e) Na^+ EtO^- in EtOH

(f) maleic anhydride (see Eq. 15.13a, p. 739), heat

15.44 What six-carbon conjugated diene would give the same *single* product from either 1,2- or 1,4-addition of HBr?

15.45 Explain each of the following observations.

(a) The allene 2,3-heptadiene can be resolved into enantiomers, but the cumulene 2,3,4-heptatriene cannot.

(b) The cumulene in part (a) can exist as diastereomers, but the allene in part (a) cannot.

15.46 Using the Hückel $4n + 2$ rule, determine whether each of the following compounds is likely to be aromatic. Explain how you arrived at the π-electron count in each case.

(a)   (b)

(c)   (d)

15.47 Which of the following molecules is likely to be planar and which nonplanar? Explain.

oxepin   **tropylium ion**   **cyclooctatetraenyl dianion**

15.48 The following compound is not aromatic even though it has $4n + 2$ π electrons in a continuous cyclic array. Explain why this compound is not aromatic. (*Hint:* Draw out the hydrogens.)

15.49 Rank the isomers within each set in order of increasing heat of formation (lowest first).

(a)

(1)   (2)

$$Et - CH = C = CH - Et$$

(3)

(4)

(b)

(1)   (2)

(3)

(c)

(1)   (2)   (3)

15.50 Assume you have unlabeled samples of the compounds within each of the following sets. Explain how UV–vis spectroscopy could be used to distinguish each compound in the set from the other(s).

(a) 1,4-cyclohexadiene and 1,3-cyclohexadiene

(b)

and

(c)

and

(d)

and

(e)

and

15.51 Two of the compounds in Fig. P15.51 are used in sunscreens, and one is not. Identify the compound that does *not* act as a sunscreen; explain.

15.52 A colleague, Ima Hack, has subjected isoprene (Fig. 15.5, p. 720) to catalytic hydrogenation to give isopentane. Hack has inadvertently stopped the hydrogenation prematurely and wants to know how much unreacted isoprene remains in the sample. The mixture of isoprene and 2-methylbutane (75 mg total) is diluted to one liter with pure methanol and found to have an absorption at 222.5 nm (1 cm path length) of 0.356. Given an extinction coefficient of 10,750 at this wavelength, what mass percent of the sample is unreacted isoprene?

15.53 How would the color of β-carotene (structure on p. 724) be affected by treatment of the compound with a large excess of H_2 over a Pt/C catalyst? Explain.

15.54 Fluorescein was once used to color the Chicago River green on St. Patrick's Day until it was subsequently replaced with a vegetable dye. One problem with fluorescein was that the green color required bright sunshine for maximum effect—definitely a problem for Chicago in March! Explain this observation using the theory of fluorescence.

15.55 (Refer to Fig. 15.11a–b.) If a solution of fluorescein at pH = 9 is subjected to visible light at 488 nm, it has maximum fluorescence at 510 nm. Calculate the energy difference between the absorbed and fluorescing radiation in kJ mol^{-1}.

15.56 Draw as many resonance structures as you can for (a) the form of fluorescein present at pH = 9 (p. 730), and (b) the fluorescent group of the green fluorescent protein (p. 730). Use the curved-arrow notation to derive your structures. Be sure in both cases that an unshared electron pair on the anionic oxygen is involved in the resonance interaction.

15.57 A chemist, I. M. Shoddy, has just purchased some compounds in a going-out-of-business sale from Pybond, Inc., a cut-rate chemical supply house. The company, whose motto is "You get what you pay for," has sent Shoddy a compound A at a bargain price in a bottle labeled only "C_6H_{10}." Unfortunately, Shoddy cannot remember what he ordered, and he has come to ask your help in identifying the compound. Compound A is optically active and has an IR absorption at 2083 cm^{-1}. Partial hydrogenation of A with 0.2 equivalent of H_2 over a catalyst gives, in addition to recovered A, a mixture of *cis*-2-hexene and *cis*-3-hexene. Identify compound A, and explain your reasoning.

15.58 Account for the fact that the antibiotic *mycomycin* is optically active (see Fig. P15.58).

15.59 If a lysine residue and a phenylalanine residue are located close to each other in a protein structure, describe how would you expect them to be oriented for the most favorable interaction?

15.60 (Refer to Fig. P15.60 on p. 782.) The *N*-methylquinolinium ion forms a noncovalent complex with molecule A in water that has a standard free energy of dissociation $\Delta G_d^\circ = 28.9$ kJ mol^{-1} (6.9 kcal mol^{-1}). The neutral

Figure P15.51

mycomycin

Figure P15.58

molecule 4-methylquinoline forms a noncovalent complex with molecule A in water with $\Delta G_d^\circ = 22.2$ kJ mol^{-1} (5.3 kcal mol^{-1}).

(a) Calculate the dissociation constant for each complex.

(b) Suggest a reason that the binding of the ion to A is stronger than the binding of the neutral molecule to A.

(c) The *N*-methylquinolinium ion forms a noncovalent complex with molecule B with $\Delta G_d^\circ = 35.2$ kJ mol^{-1} (8.4 kcal mol^{-1}). Suggest a reason that the complex with molecule B has a smaller dissociation constant than the complex with molecule A.

15.61 Explain the fact that 2,3-dimethyl-1,3-butadiene and maleic anhydride (structure in Problem 15.63) readily react to give a Diels–Alder adduct, but 2,3-di-*tert*-butyl-1,3-butadiene and maleic anhydride do not.

15.62 The following natural product readily gives a Diels–Alder adduct with maleic anhydride (structure in Problem 15.63) under mild conditions. What is the most likely configuration of the two double bonds (cis or trans)?

$$CH_3(CH_2)_5CH=CH-CH=CH(CH_2)_7\overset{\overset{\displaystyle O}{\|}}{C}OH$$

15.63 Knowing that conjugated dienes react in the Diels–Alder reaction, a student, M. T. Brainpan, has come to you with an original research idea: to use conjugated alkynes as the diene component in the Diels–Alder reaction (such as the following). Would Brainpan's idea work? Explain.

$$H_3C-C\equiv C-C\equiv C-CH_3 +$$

maleic anhydride

15.64 Explain why 4-methyl-1,3-pentadiene is *much* less reactive as a diene in Diels–Alder reactions than (*E*)-1,3-pentadiene, but its reactivity is similar to that of (*Z*)-1,3-pentadiene.

15.65 (a) Which carbocation is more stable: the carbocation formed by protonation of isoprene at carbon-1 or the carbocation formed by protonation of isoprene at carbon-4? Explain.

isoprene

(b) Predict the products expected from the addition of one equivalent of HBr to isoprene; explain your reasoning.

(c) Predict the products expected from the addition of one equivalent of HBr to *trans*-1,3,5-hexatriene; explain your reasoning.

A

B

N-methylquinolinium ion

4-methylquinoline

Figure P15.60

(d) In parts (b) and (c), which are likely to be the kinetically controlled products and which are likely to be the thermodynamically controlled ones? Explain.

15.66 This problem describes the result that established the intrinsic preference for 1,2-addition in the reaction of hydrogen halides with conjugated dienes.

(a) What is the relationship between the products of 1,2- and 1,4-addition in the following reaction?

$(E)\text{-}H_2C\text{=\!=}CH\text{—}CH\text{=\!=}CH\text{—}CH_3 + HCl \longrightarrow$

(b) How does the use of DCl change this relationship, if at all?

$(E)\text{-}H_2C\text{=\!=}CH\text{—}CH\text{=\!=}CH\text{—}CH_3 + DCl \longrightarrow$

(c) The reaction with DCl gives mostly the kinetically controlled product. Give the structure of this product.

15.67 When the alcohol A undergoes acid-catalyzed dehydration, two isomeric alkenes are formed: B and C (see Fig. P15.67a). The relative percentage of each alkene formed is shown as a function of time in Fig. P15.67b. The composition of the alkene mixture at very long times is the equilibrium composition. Furthermore, if either alkene is subjected to the conditions of the reaction, the equilibrium mixture of alkenes is obtained.

(a) Is the dehydration a kinetically controlled or thermodynamically controlled reaction? Explain.

(b) Give a structural reason why compound C is favored at equilibrium.

(c) Suggest one reason why alkene B is formed more rapidly.

15.68 When 1,3-cyclopentadiene and maleic anhydride (Problem 15.63) are allowed to react at room temperature, a Diels–Alder reaction takes place in which the endo product is formed as the major product. When this product is heated above its melting point of 165 °C, it is transformed into an equilibrium mixture that contains about 57% of the exo stereoisomer and 43% of the endo stereoisomer. (The equilibrium constant for interconversion of the two stereoisomers probably does not vary greatly with temperature.)

(a) Show these transformations with equations, including the structures of all compounds.

(b) According to these observations, is the Diels–Alder reaction of maleic anhydride and 1,3-cyclopentadiene at room temperature a kinetically controlled or a thermodynamically controlled reaction?

(c) Sketch two diagrams like Fig. 15.13, p. 736, one showing the transition state that leads to the endo product, and the other showing the transition state that leads to the exo product. According to the data in this problem, which diagram portrays the transition state of the reaction at low temperature?

15.69 Consider the bromine addition shown in Figure P15.69 on p. 784. Product A is the predominant product formed at low temperature. If the products are allowed to stand under the reaction conditions or are brought to equilib-

(a)

(b)

Figure P15.67 The relative amounts of alkene products B and C in Problem 15.67 as a function of time rium at higher temperature, product *B* is the only product formed.

(a) Which is the kinetic product and which is the thermodynamic product?

(b) Give a structural reason that the thermodynamic product is more stable than the kinetic product.

(c) Propose a mechanism that explains why the kinetic product is formed more rapidly even though it is less stable. (*Hint:* The rate-limiting step of bromine addition is formation of the bromonium ion.)

(d) Propose a mechanism for the equilibration of the two compounds that does not involve the alkene starting material.

15.70 The 1,2-addition of one equivalent of HCl to the triple bond of vinylacetylene, $HC\equiv C-CH=CH_2$, gives a chlorine-containing conjugated diene called *chloroprene*. Chloroprene can be polymerized to give *neoprene*, valued for its resistance to oils, oxidative breakdown, and other deterioration. Give the structures of chloroprene and neoprene. (Assume a 1,4-addition polymerization for purposes of this problem.)

15.71 When an excess of 1,3-butadiene reacts with Cl_2 in chloroform solvent, two compounds, *A* and *B*, both with the formula $C_4H_6Cl_2$, are formed. Compound *B* reacts with more Cl_2 to form compound *C*, $C_4H_6Cl_4$, which proves to be a meso compound. Compound *A* reacts with more Cl_2 to form both *C* and a diastereomer *D*. Propose structures for *A*, *B*, *C*, and *D*, and explain your reasoning.

15.72 When 1,3-cyclopentadiene containing carbon-13 (^{13}C) *only* at carbon-5 (as indicated by the asterisk in Fig. P15.72) is treated with potassium hydride (KH), a species *X* is formed and a gas is evolved. When the resulting mixture is added to water, a mixture of ^{13}C-labeled 1,3-cyclopentadienes is formed as shown in the equation. Identify *X*, and explain both the origin and the percentages of the three labeled cyclopentadienes.

15.73 Explain why borazole (sometimes called *inorganic benzene*) is a very stable compound.

borazole

15.74 An amine R_2NH is typically more than 20 pK_a units more acidic than the hydrogens of the carbon analog, R_2CH_2 (the element effect; Sec. 3.6A). However, the acidities of 1,3-cyclopentadiene and pyrrole are an exception.

1,3-cyclopentadiene
pK_a = 15

pyrrole
pK_a = 17

Use the theory of aromaticity to explain this exception. (*Hint:* Remember that the pK_a of a compound is proportional to the free energy *difference* between an acid and its conjugate base.)

15.75 (a) Although aldehydes and ketones are weak acids, their α-hydrogens are more than 30 pK_a units more acidic than the hydrogens of alkanes.

α-hydrogens

$$H_3C-\overset{\overset{O}{\|}}{C}-R \ + \ ^-:B \ \rightleftharpoons \ H_2\overset{\overset{O}{\|}}{\overset{..}{C}}-C-R \ + \ H-B$$

an aldehyde or ketone   a base   an enolate ion
p$K_a \approx 19$

Figure P15.69

Figure P15.72

Using polar effects and resonance effects in your argument, explain the enhanced acidity of aldehydes and ketones.

(b) Which α-hydrogen of the following ketone, Hᵃ or Hᵇ, should be most acidic? Explain.

15.76 Which of the following two alkyl halides would react most rapidly in a solvolysis reaction by the S_N1 mechanism? Explain your reasoning.

$$CH_3\ddot{O}-CH=CH-CH_2-Cl$$

A (trans isomer)

$$CH_3\ddot{O}-\underset{\underset{CH_2}{\|}}{C}-CH_2-Cl$$

B

15.77 The S_N1 solvolysis of cinnamyl chloride in water gives two structurally isomeric alcohols (neglect stereoisomers).

+ H₂O $\xrightarrow[\text{water}]{\text{dioxane–}}$ two alcohols + HCl
(constitutional isomers)

***trans*-cinnamyl chloride**

(a) Show *five* resonance structures of the carbocation intermediate. In each of your structures, the positive charge should be on a different carbon.

(b) Even though the positive charge is delocalized to five different carbons, only two structurally isomeric products are formed. (Neglect stereoisomers.) What are the two products? Why are more products not formed?

(c) If you have the two alcohol products but don't know which is which, how would you use UV spectroscopy to tell them apart?

(d) A constitutional isomer *A* of cinnamyl chloride gives the same two products in a solvolysis reaction under the same conditions. Give the structure of *A* and explain.

15.78 Invoking Hammond's postulate and the properties of the carbocation intermediates, explain why the doubly allylic alkyl halide *A* undergoes much more rapid solvolysis in aqueous acetone than compound *B*. Then explain why compound *C*, which is also a doubly allylic alkyl halide, is solvolytically inert.

A   *B*   *C*

15.79 Most alkyl bromides are water-insoluble liquids. Yet, when 7-bromo-1,3,5-cycloheptatriene was first isolated, its high melting point of 203 °C and its water solubility led its discoverers to comment that it behaves more like a salt. Explain the salt-like behavior of this compound.

**7-bromo-1,3,5-cycloheptatriene
(tropylium bromide)**

15.80 Complete the reactions given in Fig. P15.80 on p. 786, giving the structures of all reasonable products and the reasoning used to obtain them.

15.81 One interesting use of Diels–Alder reactions is to trap very reactive alkenes that cannot be isolated and studied directly. One compound used as a diene for this purpose is diphenylisobenzofuran, which reacts as shown in Fig. P15.81a on p. 787. (Notice that the formation of an aromatic ring in the product helps ensure that the Diels–Alder reaction is driven to completion.) In the reaction given in Fig. P15.81b, use the structure of the Diels–Alder product to deduce the structure of the reactive species formed in the reaction. Show by the curved-arrow notation how the reactive species is formed, and explain what makes it particularly unstable.

15.82 Use the structure of the Diels–Alder adduct to deduce the structure of the product *X* in the reaction given in Fig. P15.82 on p. 787. Then give a curved-arrow mechanism for the formation of *X*.

15.83 In 1991, chemists at Rice University reported that they had trapped an unstable compound called *spiropentadiene* using its Diels–Alder reaction with excess 1,3-cyclopentadiene, giving the product in the reaction shown in Fig. P15.83 on p. 787. Use the structure of this product to deduce the structure of spiropentadiene.

15.84 Account for each of the transformations shown in Fig. P15.84 on p. 787 with a curved-arrow mechanism. (Don't try to explain any percentages.)

In part (d), identify *X*; give the mechanisms for both the formation and the subsequent reaction of *X*; and explain why the equilibrium for the reaction of *X* strongly favors the products.

15.85 When the following compound is treated with a strong Brønsted acid, a stable carbocation *A* is formed.

$\xrightarrow{\text{acid}}$ *A* (a carbocation)

(a) Propose a structure for carbocation *A*, and draw its resonance structures.

(b) The proton NMR spectrum of carbocation *A* at −10 °C consists of four singlets at δ 1.54, δ 2.36, δ 2.63, and δ 2.82 (relative integral 2 : 2 : 2 : 1). Explain why the structure of *A* is consistent with this spectrum by assigning each resonance.

(c) Explain why the NMR spectrum of *A* becomes a single broad line when the temperature is raised to 113 °C. (*Hint:* See Sec. 13.8.)

15.86 Account for the fact that the central "benzene ring" of [4]phenylene (Fig. P15.86 on p. 788) undergoes catalytic hydrogenation readily under conditions usually used for ordinary alkenes, but the other benzene rings do not.

(a)

(b)

(c)

Ph—CH=CH—CH=CH—Ph + CH₃OC—C≡C—COCH₃ →(heat)
(both double bonds are trans)

(d)

H₂C=C=CH—CH=CH₂ + benzoquinone →
(2 equivalents)

benzoquinone

(e)

maleic anhydride

(f)

H₂C=CHCCH₂CH₂CH₂CCH=CH₂ →(heat) (a compound with 10 carbon atoms)

(g)

NiCl₂ + 2 [cyclopentadienyl]:⁻ Na⁺ →

Figure P15.80

(a)

diphenylisobenzofuran

(b)

Figure P15.81

Figure P15.82

spiropentadiene +

(excess)

Figure P15.83

(a) $CH_3CH{=}CHCH_2{-}OH$ + conc. HBr $\xrightarrow[{-15\,°C}]{H_2SO_4}$ $CH_3CH{=}CHCH_2{-}Br$ + $CH_3CHCH{=}CH_2$ + H_2O

(84%) $\underset{Br}{|}$

(16%)

(b)

dextropimaric acid   **abietic acid**

(Dextropimaric acid is isolated from the exudate resin of the cluster pine.)

Figure P15.84 *(continues on p. 788)*

(c)

(d)

α-phellandrene

Figure P15.84 *(continued from p. 787)*

[4]phenylene

Figure P15.86

The Chemistry of Benzene and Its Derivatives

As we learned in Chapter 15, benzene and its derivatives are *aromatic compounds*. In this chapter, we'll learn how aromaticity affects the spectroscopic properties and the reactivity of benzene and its derivatives. In particular, we'll learn that benzene and its derivatives do not undergo most of the usual addition reactions of alkenes. Instead, they undergo a type of reaction called *electrophilic aromatic substitution*, in which a ring hydrogen is substituted by another group. Such substitution reactions can be used to prepare a variety of substituted benzenes from benzene itself. Most of this chapter is concerned with the substitution reactions of benzene and its derivatives, including some biological aspects of aromatic chemistry. Chapters 17 and 18 deal with other aspects of aromatic chemistry.

16.1 NOMENCLATURE OF BENZENE DERIVATIVES

The nomenclature of benzene derivatives follows the same rules used for other substituted hydrocarbons:

chlorobenzene   **nitrobenzene**   **ethylbenzene**

The *nitro group*, abbreviated —NO_2, which is a part of the nitrobenzene structure shown here, may be less familiar than the other substituent groups. The nitro group can be represented in more detail as a resonance hybrid of two equivalent dipolar structures:

$$\left[\begin{array}{c} \overset{\ddot{O}:}{\underset{:\ddot{O}:^-}{R - \overset{+}{N}}} \end{array} \longleftrightarrow \begin{array}{c} \overset{:\ddot{O}:^-}{\underset{:\ddot{O}:}{R - \overset{+}{N}}} \end{array} \right] = R - NO_2$$

Some monosubstituted benzene derivatives have well-established common names that should be learned.

| **toluene** | **styrene** | **phenol** | **anisole** | **aniline** |

The positions of substituent groups in disubstituted benzenes can be designated in two ways. Modern substitutive nomenclature utilizes numerical designations in the same manner as that for other compound classes. However, an older system, which is still used, employs special letter prefixes. The prefix *o* (for *ortho*) is used for substituents in a 1,2-relationship; *m* (for *meta*) for substituents in a 1,3-relationship; and *p* (for *para*) for substituents in a 1,4-relationship.

o-dichlorobenzene   **m-bromonitrobenzene**   **p-fluoroiodobenzene**
1,2-dichlorobenzene   **1-bromo-3-nitrobenzene**   **1-fluoro-4-iodobenzene**

As these examples illustrate, when none of the substituents qualifies as a principal group, the substituents are cited and numbered in alphabetical order. In contrast, if a substituent is eligible for citation as a principal group, it is assumed to be at carbon-1 of the ring.

m-nitrophenol (3-nitrophenol)
—OH group is the principal group

Some disubstituted benzene derivatives also have time-honored common names. The dimethylbenzenes are called *xylenes*, and the methylphenols are called *cresols*.

o-xylene   **m-cresol**

The phenols with two hydroxy groups also have important common names.

catechol   **resorcinol**   **hydroquinone**

When a benzene derivative contains more than two substituents on the ring, the *o, m,* and *p* designations are not appropriate; only numbers may be used to designate the positions of substituents. The usual nomenclature rules are followed (Secs. 2.4C, 4.2A, 8.2).

alphabetical citation: **bromodifluoro**
numbering: **1,2,3**
name: **1-bromo-2,3-difluorobenzene**

2-ethoxy-5-nitrophenol

Sometimes it is simpler to name a benzene ring as a substituent group. A benzene ring or substituted benzene ring cited as a substituent is referred to generally as an **aryl group**; this term is analogous to *alkyl group* in nonaromatic compounds (Sec. 2.8B). When an unsubstituted benzene ring is a substituent, it is called a **phenyl group**. This group can be abbreviated Ph—. It is also sometimes abbreviated by its group formula, C_6H_5—.

$C_6H_5—O—C_6H_5$ Ph—O—Ph

diphenyl ether (phenoxybenzene)
three different ways to write the structure

The Ph—CH_2— group is called the **benzyl group**.

$PhCH_2—Cl$

benzyl chloride or (chloromethyl)benzene

Carefully note the difference between the *phenyl* group, Ph—, and the *benzyl* group, Ph—CH_2—. Some students erroneously think both of these names refer to the phenyl group.

PROBLEMS

16.1 Name the following compounds.

16.2 Draw the structure of each of the following compounds.

(a) *p*-chlorophenol   (b) *m*-nitrotoluene   (c) 3,4-dichlorotoluene
(d) 1-bromo-2-propylbenzene   (e) methyl phenyl ether
(f) benzyl methyl ether   (g) *p*-xylene   (h) *o*-cresol
(i) 2,4,6-trichloroanisole (a compound present in tainted wine corks)

16.2 PHYSICAL PROPERTIES OF BENZENE DERIVATIVES

The boiling points of benzene derivatives are similar to those of other hydrocarbons with similar shapes and molecular masses.

	benzene	cyclohexane	toluene
bp	80.1 °C	80.7 °C	110.6 °C
mp	5.5 °C	6.6 °C	−95 °C

The melting points of benzene and cyclohexane are unusually high because of their symmetry (Sec. 8.5D). Notice that the boiling points of toluene and benzene fit the general trend (Sec. 8.5A) that addition of a carbon atom adds 20–30 °C to the boiling point.

The melting points of para-disubstituted benzene derivatives are typically much higher than those of the corresponding ortho or meta isomers—another effect of symmetry (Sec. 8.5D).

	p-nitrotoluene	*o*-nitrotoluene	*p*-dibromobenzene	*m*-dibromobenzene
mp	54.5 °C	−9.6 °C	87.3 °C	−7 °C

This trend can be useful in purifying the para isomer of a benzene derivative from mixtures containing other isomers. (This point will prove to be very important in the reactions of some benzene derivatives.) Because the isomer with the highest melting point is usually the one that is most easily crystallized, many para-substituted compounds can be separated from their ortho and meta isomers by recrystallization.

Benzene and other aromatic hydrocarbons are not as dense as water but are more dense than alkanes and alkenes of about the same molecular mass. Like other hydrocarbons, benzene and its hydrocarbon derivatives are insoluble in water. As we might expect, benzene derivatives with substituents that form hydrogen bonds to water are more soluble. For example, phenol has substantial water solubility.

PROBLEM

16.3 Predict the approximate boiling point of
(a) ethylbenzene   (b) propylbenzene   (c) *p*-xylene

16.3 SPECTROSCOPY OF BENZENE DERIVATIVES

A. IR Spectroscopy

The most useful absorptions in the infrared spectra of benzene derivatives are the carbon–carbon stretching absorptions of the ring, which occur at lower frequency than the C=C absorption of alkenes. Two such absorptions are typical: one near 1600 cm^{-1} and the other

FIGURE 16.1 (a) The IR spectrum of toluene. The carbon–carbon stretching absorption is the major absorption used for diagnosing the presence of benzene rings. (b) The IR spectrum of phenol. The strong O—H and C—O stretching absorptions are much like the same absorptions of tertiary alcohols.

near 1500 cm^{-1}. These absorptions, illustrated in the IR spectrum of toluene (Fig. 16.1a), occur at lower frequency than alkene C=C stretching absorptions because the carbon–carbon bonds in benzene rings have a bond order of 1.5—that is, they are intermediate between single bonds and double bonds. Other characteristic absorptions are also shown in Fig. 16.1a. For example, the *overtone and combination bands* in the 1660–2000 cm^{-1} region were once used to determine the substitution patterns of aromatic compounds. However, NMR spectroscopy is now a more reliable tool for this purpose.

Phenols have not only the characteristic aromatic absorptions, but also O—H and C—O stretching absorptions, which are very much like those of tertiary alcohols. The IR spectrum of phenol is shown in Fig. 16.1b.

B. NMR Spectroscopy

The proton NMR spectrum of benzene consists of a singlet at a chemical shift of δ 7.4. Typical alkenes, in contrast, have chemical shifts for internal vinylic protons of δ 5.0–5.7. Thus, the chemical shifts are greater than those of alkenes by about 1.5–2 ppm. (See also Fig. 13.4, p. 621.) *NMR absorptions at large chemical shifts are particularly characteristic of most benzene derivatives.* Notice also that a benzene ring contributes four degrees of unsaturation. Thus, when dealing with an unknown for which you have deduced an unsaturation number ≥4 from the molecular formula, your eyes should move immediately to the δ 7–8 region of the NMR spectrum. Absorptions in this region immediately alert you to the likelihood of a benzene ring.

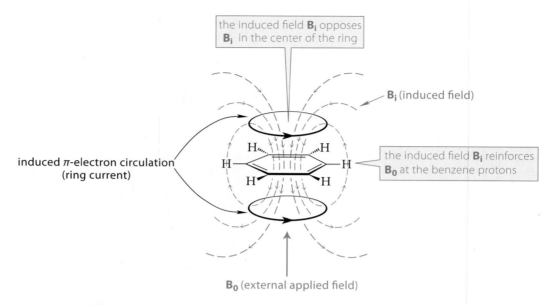

FIGURE 16.2 Origin of the large chemical shift of benzene protons. The field (*red dashed lines*, **B$_i$**) induced by the π-electron ring current opposes the applied field **B$_0$** (*blue*) in the center of the ring. However, because **B$_i$** forms closed loops, it lies in the same direction as **B$_0$** in the region occupied by the benzene protons. As a result, the induced field increases the local field at the benzene protons. Consequently, these protons require a higher frequency to meet the condition for resonance. The higher resonance frequency translates into a greater chemical shift (Eq. 13.4, p. 615, and related discussion).

What is the reason for the unusual chemical shift of benzene? Recall that the π-electron density in benzene lies in two doughnut-shaped regions above and below the plane of the ring (Fig. 15.16b, p. 761). In an NMR experiment, benzene molecules in solution are moving about randomly and thus can assume all possible orientations relative to the applied field **B$_0$**. However, a particular orientation dominates the chemical shift, as shown in Fig. 16.2. In this orientation, a circulation of π electrons around the ring, called a **ring current**, is induced. The ring current, in turn, induces a magnetic field **B$_i$** that forms closed loops through the ring. This induced field opposes the applied field along the axis of the ring, but it *adds to the applied field outside of the ring*, in the region occupied by the benzene protons. Thus, the net field at these protons is higher than it would be in the absence of the ring current. As a result, a correspondingly higher frequency is required for absorption, and the chemical shifts of aromatic protons are increased (Eq. 13.4, p. 615). This explanation is similar to that for the chemical shifts of vinylic protons in alkenes (Fig. 13.15, p. 646), except that the effect is larger for benzene protons.

The ring current and the large chemical shift are characteristic of compounds that are aromatic by the Hückel $4n + 2$ rule (Sec. 15.7D). This is reasonable because the basis of both the ring current and aromaticity is the overlap of p orbitals in a continuous cyclic array. Many chemists believe that the existence of the ring current (detected by unusually large chemical shifts) is the best *experimental* evidence of aromatic character.

PROBLEMS

16.4   Within each set, which compound should show NMR absorptions with the greater chemical shifts? Explain your choices.

16.5   (a) Verify that the following compound meets the Hückel criteria for aromaticity.

(b) The NMR spectrum of this compound consists of two sets of multiplets: one at δ 9.28 and the other at δ (−2.99); the latter resonance is at 3 ppm *lower* chemical shift than that of TMS—that is, to the right of TMS in a conventional NMR spectrum. The relative integral of the two resonances is 2 : 1, respectively. Assign the two sets of resonances, and explain why their chemical shifts are so different; in particular, explain why one of the chemical shifts is so small. (*Hint:* Look carefully at the direction of the induced field in Fig. 16.2.)

When the protons in a substituted benzene derivative are nonequivalent, they split each other, and their coupling constants depend on their positional relationships, as shown in Table 16.1. Notice that splitting can occur across more than one carbon–carbon bond. Because of this splitting, the NMR spectra of many monosubstituted benzene derivatives have complex absorptions in the aromatic region. For rings with higher degrees of substitution, the substitution pattern can in many cases be deduced directly from the splitting patterns of the aromatic protons.

One particular splitting pattern in the NMR spectra of aromatic compounds occurs often enough that it is worth remembering. This pattern is illustrated by the NMR spectrum of 1-bromo-4-ethylbenzene in Fig. 16.3 on p. 796. This spectrum consists of two apparent doublets, centered near δ 7.0 and δ 7.4. Ideally, the lines in these doublets should have equal intensities (Table 13.2, p. 630), but because the chemical shifts of the two coupled protons are similar, the intensities differ from the 1 : 1 ideal. (This phenomenon is called *leaning*.) In each doublet, the major coupling constant, J = 8.4 Hz, reflects the large ortho coupling. A superimposed, very small, para coupling causes the additional fine structure visible in the expansions of these absorptions. *Such a "two-leaning-doublet" pattern is very typical of disubstituted benzene rings in which two different ring substituents have a para relationship.*

The spectrum of 1-bromo-4-ethylbenzene also shows how substituent groups can affect the chemical shifts of ring protons. The bromo group is the more electronegative group. The protons ortho to this group (H^a in Fig. 16.3) have a greater chemical shift at δ 7.4. The protons

STUDY GUIDE LINK 16.1
NMR of Para-Substituted Benzene Derivatives

TABLE 16.1 Typical Coupling Constants of Aromatic Protons

Relationship of protons		Coupling constant
ortho		J_{ortho} = 6–10 Hz
meta		J_{meta} = 1–3 Hz
para	H— —H	J_{para} = 0–1 Hz

FIGURE 16.3 The proton NMR spectrum of 1-bromo-4-ethylbenzene. Notice three things about this spectrum. First, the "two-leaning-doublet" pattern near δ 7 is very typical of para-disubstituted benzene derivatives in which the ring substituents are different. Second, the chemical shifts of the ring protons reflect the electronegativities of nearby groups. Thus, the protons H^a, which are ortho to the electronegative bromine, have greater chemical shifts than protons H^b, which are ortho to the more electropositive ethyl group. Finally, notice that the chemical shifts of the benzylic protons H^c are slightly greater than the chemical shifts of allylic protons. (See also Fig. 13.4, p. 621.)

ortho to the more electropositive ethyl group (H^b in Fig. 16.3) have a smaller chemical shift at δ 7.0. (Resonance also affects ring-proton shifts; see Problem 16.42, p. 831.)

The chemical shifts of *benzylic protons*—protons on carbons adjacent to benzene rings—are in the δ 2–3 region. These chemical shifts are slightly greater than those of allylic protons (see Fig. 13.4, p. 621). The chemical shifts of benzylic protons in toluene and ethylbenzene are typical.

benzylic protons (δ 2.29)   benzylic protons (δ 2.53, q)   δ 1.22(t)

ring protons δ 7.1

Notice also the chemical shifts of the benzylic protons of 1-bromo-4-ethylbenzene in Fig. 16.3.

The O—H absorptions of phenols are typically observed at lower field (about δ 5–6) than those of alcohols (δ 2–3). The O—H protons of phenols, like those of alcohols, undergo exchange in D_2O.

PROBLEMS

16.6   Explain how to use NMR spectroscopy to differentiate the isomers within each of the following sets.

(a) mesitylene (1,3,5-trimethylbenzene) and *p*-ethyltoluene

(b) 1-bromo-4-ethylbenzene (Fig. 16.3) and (2-bromoethyl)benzene ($BrCH_2CH_2Ph$)

16.7 Give structures for each of the following compounds.

(a) $C_9H_{12}O$: NMR δ 1.27 (3H, d, J = 7 Hz); δ 2.26 (3H, s); δ 3.76 (1H, broad s, disappears after D_2O shake); δ 4.60 (1H, q, J = 7 Hz); δ 6.95, δ 7.10 (4H, apparent pair of doublets, J = 10 Hz)

(b) $C_8H_{10}O$: IR, 3150–3600 cm^{-1} (strong, broad); NMR, δ 1.17 (3H, t, J = 8 Hz); δ 2.58 (2H, q, J = 8 Hz); δ 6.0 (1H, broad singlet, disappears with D_2O shake); δ 6.79 (2H, d, J = 10 Hz); δ 7.13 (2H, d, J = 10 Hz)

C. ^{13}C NMR Spectroscopy

In ^{13}C NMR spectra the chemical shifts of aromatic carbons are in the carbon–carbon double bond region (δ 110–160); the exact values depend on the ring substituents that are present. The chemical shift of benzene itself is δ 128.5. The chemical shifts of the carbons in ethylbenzene are typical:

ethylbenzene

Notice the higher chemical shift for the quaternary ring carbon. This fits the pattern of larger chemical shifts for carbons that bear no hydrogens (Sec. 13.9, p. 659). Also, because the proton-decoupling technique enhances the size of peaks of carbons that bear hydrogens, the peaks for carbons that do *not* bear hydrogens are considerably smaller. Thus, the δ 144.1 resonance of ethylbenzene is the smallest peak in the spectrum.

The chemical shifts of benzylic carbons are in the δ 18–30 region—not appreciably different from the chemical shifts of ordinary alkyl carbons. The ^{13}C chemical shift for the benzylic carbon of ethylbenzene, δ 29.2, is typical.

PROBLEMS

16.8 A benzene derivative known to be a methyl ether with the formula $C_7H_6OCl_2$ has five lines in its proton-decoupled ^{13}C NMR spectrum. Propose two possible structures for this compound that fit these facts.

16.9 How would you distinguish mesitylene (1,3,5-trimethylbenzene) from isopropylbenzene (cumene) by ^{13}C NMR spectroscopy?

D. UV Spectroscopy

Simple aromatic hydrocarbons have two absorption bands in their UV spectra: a relatively strong band near 210 nm and a much weaker one near 260 nm. The spectrum of ethylbenzene in methanol solvent (Fig. 16.4, p. 798) is typical: λ_{max} = 208 nm (ϵ = 7520); 261 nm (ϵ = 200). Substituent groups on the ring alter both the λ_{max} values and the intensities of both peaks, particularly if the substituent has an unshared electron pair or $2p$ orbitals that can overlap with the π-electron system of the aromatic ring. As is also the case in alkenes, more extensive conjugation is associated with an increase in both λ_{max} and intensity. For example, 1-ethyl-4-methoxybenzene (*p*-ethylanisole) in methanol solvent has absorptions at λ_{max} = 224 nm (ϵ = 10,100) and 276 nm (ϵ = 1,930); both absorptions occur at greater wavelengths and have greater intensities than the analogous absorptions of ethylbenzene (Fig. 16.4) because the

FIGURE 16.4 Comparison of the UV spectra of ethylbenzene (*blue*) and *p*-ethylanisole (*black*). The solid lines are spectra taken at the same concentrations. The dashed line is the spectrum of ethylbenzene at 50-fold higher concentration. This comparison shows that the UV spectrum of *p*-ethylanisole is generally more intense. Notice also that the λ_{max} in the *p*-ethylanisole spectrum occurs at higher wavelength.

—OCH$_3$ group has electron pairs in orbitals that overlap with the 2*p* orbitals of the benzene ring.

_p_-ethylanisole

Notice that the VSEPR rules for determining geometry (Sec. 1.3B) predict that the oxygen of *p*-ethylanisole, like the oxygen of water, should be approximately tetrahedral (if we include the electron pairs) and should therefore be sp^3-hybridized. However, the oxygen is sp^2-hybridized. This hybridization allows one of its electron pairs to occupy a 2*p* orbital, which has the same size, shape, and orientation as the carbon 2*p* orbitals of the ring. In other words, an oxygen 2*p* orbital overlaps more effectively with the carbon 2*p* orbitals of the ring than an oxygen sp^3 orbital would. (We learned about the same effect in resonance-stabilized allylic systems; p. 754). The UV spectrum of anisole is a direct consequence of this overlap.

PROBLEMS

16.10 (a) Explain why compound *A* has a UV spectrum with considerably greater λ_{max} values and intensities than are observed for ethylbenzene.

$$\lambda_{max} = \; 256 \text{ nm } (\epsilon = 20{,}000)$$
$$283 \text{ nm } (\epsilon = 5{,}100)$$

A

(b) In view of your answer to part (a), explain why the UV spectra of compounds *B* and *C* are virtually identical.

B
bimesityl
λ_{max} = 266 nm (ϵ = 700)

C
mesitylene
λ_{max} = 266 nm (ϵ = 200)

16.11 How could you distinguish styrene (Ph—CH=CH$_2$) from ethylbenzene by UV spectroscopy?

ELECTROPHILIC AROMATIC SUBSTITUTION REACTIONS OF BENZENE

The most characteristic reaction of benzene and many of its derivatives is *electrophilic aromatic substitution*. In an **electrophilic aromatic substitution** reaction, a hydrogen of an aromatic ring is substituted by an electrophile—that is, by a Lewis acid. The general pattern of an electrophilic aromatic substitution reaction is as follows, where E is the electrophile:

$$\text{—H} + \text{E—Y} \longrightarrow \text{—E} + \text{H—Y} \qquad (16.1)$$

(Note that in this reaction and in others that follow, only one of the six benzene hydrogens is shown explicitly to emphasize that one hydrogen is lost in the reaction.)

All electrophilic aromatic substitution reactions occur by similar mechanisms. This section surveys some of the most common electrophilic aromatic substitution reactions and their mechanisms.

A. Halogenation of Benzene

When benzene reacts with bromine under harsh conditions—liquid bromine, no solvent, and the Lewis acid FeBr$_3$ as a catalyst—a reaction occurs in which *one* bromine is substituted for a ring hydrogen.

$$\text{—H} + \text{Br}_2 \xrightarrow[\text{(0.2 equiv.)}]{\text{FeBr}_3 \text{ or Fe}} \text{—Br} + \text{HBr} \qquad (16.2)$$

benzene

bromobenzene
(50% yield)

(Because iron reacts with Br$_2$ to give FeBr$_3$, iron filings can be used in place of FeBr$_3$.) An analogous chlorination reaction using Cl$_2$ and FeCl$_3$ gives chlorobenzene.

This reaction of benzene with halogens differs from the reaction of alkenes with halogens in two important ways. First is the type of product obtained. Alkenes react spontaneously with bromine and chlorine, even in dilute solution, to give *addition products*.

$$+ \text{Br}_2 \longrightarrow \qquad (16.3)$$

Halogenation of benzene, however, is a *substitution reaction*; a ring hydrogen is *replaced* by a halogen. Second, the reaction conditions for benzene halogenation are *much* more severe than the conditions for addition of halogens to an alkene.

The first step in the mechanism of benzene bromination is the formation of a complex between Br_2 and the Lewis acid $FeBr_3$ by a Lewis acid–base association.

$$:\!\ddot{B}r\!-\!\ddot{B}r\!:\!\frown FeBr_3 \quad\rightleftharpoons\quad :\!\ddot{B}r\!-\!\overset{+}{\underset{..}{B}r}\!-\!\bar{F}eBr_3 \tag{16.4}$$

Formation of this complex results in a formal positive charge on one of the bromines. A positively charged bromine is a better electron acceptor, and thus a better leaving group, than a bromine in Br_2 itself. Another (and equivalent) explanation of the leaving-group effect is that $^-FeBr_4$ is a weaker base than Br^-. (Remember from Sec. 9.4F that weaker bases are better leaving groups.) $^-FeBr_4$ is essentially the product of a Lewis acid–base association reaction of Br^- with $FeBr_3$. Therefore, in $^-FeBr_4$, an electron pair on Br^- has already been donated to Fe, and is thus less available to act as a base, than a "naked" electron pair on Br^- itself.

$$(16.5)$$

As you learned in Sec. 9.4F, —Br is a good leaving group. The fact that a much better leaving group than —Br is required for electrophilic aromatic substitution illustrates how unreactive the benzene ring is.

$$(16.6)$$

a resonance-stabilized carbocation
(an arenium ion)

This carbocation is an example of an **arenium ion**: a carbocation that is formed by the reaction of an electrophile with a double bond of an aromatic ring. Although the arenium ion is resonance-stabilized, it is *not* aromatic. Its formation disrupts the aromatic stability of the benzene ring. For that reason, harsh conditions (high temperature and a strong Lewis acid catalyst) are required for this reaction to proceed at a useful rate. These conditions are *much* harsher than those required for bromine addition to an ordinary alkene double bond (that is, bromine dissolved in an inert solvent, no catalyst, room temperature or low temperature).

The reaction is completed when a bromide ion (complexed to $FeBr_3$) acts as a base to remove the ring proton from the arenium ion, regenerate the catalyst $FeBr_3$, and give the products bromobenzene and HBr.

$$(16.7)$$

Recall that loss of a β-proton is one of the characteristic reactions of carbocations (Sec. 9.6B). Another typical reaction of carbocations—reaction of bromide ion at the electron-deficient carbon itself—doesn't occur because the resulting addition product would not be aromatic:

$$(16.8)$$

(not aromatic)

By losing a β-proton instead (Eq. 16.7), the carbocation can form bromobenzene, a stable aromatic compound.

PROBLEM

16.12 A small amount of a by-product, *p*-dibromobenzene, is also formed in the bromination of benzene shown in Eq. 16.2. Write a curved-arrow mechanism for the formation of this compound.

B. The Mechanistic Steps of Electrophilic Aromatic Substitution

Halogenation of benzene is one of many **electrophilic aromatic substitution** reactions. The bromination of benzene, for example, is an *aromatic substitution* because a hydrogen of benzene (the aromatic compound that undergoes substitution) is replaced by another group (bromine). The reaction is *electrophilic* because the substituting group reacts as an electrophile toward the benzene π electrons. In bromination, the Lewis acid is a bromine in the complex of bromine and the $FeBr_3$ catalyst (Eq. 16.6).

We've considered two other types of substitution reactions: *nucleophilic substitution* (the S_N2 and S_N1 reactions, Secs. 9.4 and 9.6) and *free-radical substitution* (halogenation of alkanes, Sec. 9.10A). In a nucleophilic substitution reaction, the substituting group acts as a nucleophile; and in free-radical substitution, free-radical intermediates are involved.

Electrophilic aromatic substitution is the most typical reaction of benzene and its derivatives. As you learn about other electrophilic substitution reactions, it will help you to understand them if you can identify in each reaction the following three mechanistic steps:

Step 1. Generation of an electrophile. The electrophile in bromination is the complex of bromine with $FeBr_3$, formed as shown in Eq. 16.4.

Step 2. **Nucleophilic reaction of the π electrons of the aromatic ring with the electrophile to form a resonance-stabilized carbocation intermediate (an arenium ion).**

> a tetrahedral, sp^3-hybridized, carbon; no longer part of the π-electron system

a resonance-stabilized, but nonaromatic, carbocation intermediate (an arenium ion)

(16.9a)

The electrophile approaches the π-electron cloud of the ring above or below the plane of the molecule. In the arenium-ion intermediate, the carbon at which the electrophile reacts becomes sp^3-hybridized and tetrahedral. This step in the bromination mechanism is Eq. 16.6.

Step 3. **Loss of a proton from the carbocation intermediate to form the substituted aromatic compound.** The proton is lost from the carbon at which substitution occurs. This carbon again becomes part of the aromatic π-electron system.

(16.9b)

This step in the bromination mechanism is Eq. 16.7.

This sequence is classified as *electrophilic* substitution because we focus on the nature of the group—an electrophile—that reacts with the aromatic ring. However, the mechanism really involves nothing fundamentally new: like both *electrophilic addition* (Sec. 5.1) and *nucleophilic substitution* (Sec. 9.1), the reaction involves the reactions of nucleophiles, electrophiles, and leaving groups.

STUDY PROBLEM 16.1

Give a curved-arrow mechanism for the following electrophilic substitution reaction.

SOLUTION Construct the mechanism in terms of the three steps given in this section.

Step 1. In this reaction, a hydrogen of the benzene ring has been replaced by an isotope D, which must come from the D_2SO_4. Because protons (in the form of Brønsted acids) are good electrophiles, the D_2SO_4 itself can serve as the electrophile.

Step 2. Reaction of the benzene π electrons with the electrophile involves protonation of the benzene ring by the isotopically substituted acid:

resonance-stabilized
carbocation intermediate

(If you're asking where that "extra" hydrogen in the carbocation came from, don't forget that each carbon of the benzene ring has a single hydrogen that is not shown explicitly in the skeletal structure. One of these is shown in the carbocation because it is involved in the next step.) You should draw the resonance structures of the carbocation intermediate.

Step 3. Removal of the proton gives the final product:

The base can remove either the deuterium or the hydrogen, but removal of the hydrogen is faster because of the primary deuterium kinetic isotope effect (Sec. 9.5D).

C. Nitration of Benzene

Benzene reacts with concentrated nitric acid, usually in the presence of a sulfuric acid catalyst, to form nitrobenzene. In this reaction, called *nitration*, the nitro group, —NO$_2$, is introduced into the benzene ring by electrophilic substitution.

$$\text{benzene} + \text{HONO}_2 \xrightarrow{\text{H}_2\text{SO}_4} \text{nitrobenzene} + \text{H}_2\text{O} \qquad (16.10)$$

benzene   **nitric acid**   **nitrobenzene** (81% yield)

This reaction fits the mechanistic pattern of the electrophilic aromatic substitution reaction outlined in the previous section:

Step 1.   Generation of the electrophile. In nitration, the electrophile is $^{+}$NO$_2$, the *nitronium ion*. This ion is formed by the acid-catalyzed removal of the elements of water from HNO$_3$.

$$(16.11a)$$

$$(16.11b)$$

nitronium ion

Step 2.   Reaction of the benzene π electrons with the electrophile to form a resonance-stabilized carbocation intermediate (an arenium ion).

$$(16.11c)$$

a resonance-stabilized carbocation

(Notice that either of the oxygens can accept the electron pair.)

Step 3. Loss of a proton from the carbocation to give a new aromatic compound.

(from Eq. 16.11a)

$$\text{(from Eq. 16.11a)}$$

(16.11d)

Nitration is the usual way that nitro groups are introduced into aromatic rings.

D. Sulfonation of Benzene

Another electrophilic substitution reaction of benzene is its conversion into benzenesulfonic acid.

benzene

sulfur trioxide

benzenesulfonic acid
(52% yield)

(16.12)

(Sulfonic acids were introduced in Sec. 10.4A as their sulfonate ester derivatives.) This reaction, called *sulfonation*, occurs by two mechanisms that operate simultaneously. Both mechanisms involve sulfur trioxide, a fuming liquid that reacts violently with water to give H_2SO_4. The source of SO_3 for sulfonation is usually a solution of SO_3 in concentrated H_2SO_4 called *fuming sulfuric acid* or *oleum*. This material is one of the most acidic Brønsted acids available commercially.

In one sulfonation mechanism, the electrophile is neutral sulfur trioxide. When sulfur trioxide reacts with the benzene ring π electrons, an oxygen accepts the electron pair displaced from sulfur.

(from solvent)

(16.13)

Sulfonic acids such as benzenesulfonic acid are rather strong acids. (Notice the last equilibrium in Eq. 16.13 and the structural resemblance of benzenesulfonic acid to another strong acid, sulfuric acid.)

Sulfonation, unlike many electrophilic aromatic substitution reactions, is reversible. The —SO_3H (sulfonic acid) group is replaced by a hydrogen when sulfonic acids are heated with steam (Problem 16.53, p. 832).

16.13 A second sulfonation mechanism involves protonated sulfur trioxide as the electrophile. Show the protonation of SO_3, and draw a curved-arrow mechanism for the reaction of protonated SO_3 with benzene to give benzenesulfonic acid.

16.14 A compound called *p*-toluenesulfonic acid is formed when toluene is sulfonated at the para position. Draw the structure of this compound, and give the curved-arrow mechanism for its formation, including resonance structures for the carbocation intermediate.

E. Friedel–Crafts Alkylation of Benzene

The reaction of an alkyl halide with benzene in the presence of a Lewis acid catalyst gives an alkylbenzene.

benzene
(large excess)   **sec-butyl chloride**

sec-butylbenzene
(71% yield)

(16.14)

This reaction is an example of a *Friedel–Crafts alkylation*. Recall that an *alkylation* is a reaction that results in the transfer of an alkyl group (Sec. 10.4B). In a **Friedel–Crafts alkylation**, an alkyl group is transferred to an aromatic ring in the presence of an acid catalyst. In the preceding example, the alkyl group comes from an alkyl halide and the catalyst is the Lewis acid aluminum trichloride, $AlCl_3$.

The electrophile in a Friedel–Crafts alkylation is formed by the complexation of the Lewis acid $AlCl_3$ with the halogen of an alkyl halide in much the same way that the electrophile in the bromination of benzene is formed by the complexation of $FeBr_3$ with Br_2 (Eq. 16.4, p. 800). If the alkyl halide is secondary or tertiary, this complex can further react to form carbocation intermediates.

carbocation   (16.15a)

Either the alkyl halide–Lewis acid complex, or the carbocation derived from it, can serve as the electrophile in a Friedel–Crafts alkylation.

alkylation by the complex

or

alkylation by the carbocation

(16.15b)

Compare the role of AlCl$_3$ in enhancing the effectiveness of the chloride leaving group with the similar role of FeBr$_3$ in the bromination of benzene (Eq. 16.5, p. 800) or FeCl$_3$ in the chlorination of benzene:

(16.15c)

We've learned three general reactions of carbocations:

1. reaction with nucleophiles (Secs. 4.7B, 9.6B)
2. rearrangement to other carbocations (Sec. 4.7D)
3. loss of a β-proton to give an alkene (Sec. 9.6B) or aromatic ring (Eq. 16.7, Sec. 16.4A)

The reaction of a carbocation with the benzene π electrons is an example of reaction 1. Loss of a β-proton to chloride ion completes the alkylation.

(16.15d)

Because some carbocations can rearrange, it is not surprising that rearrangements of alkyl groups are observed in some Friedel–Crafts alkylations:

benzene + CH$_3$CH$_2$CH$_2$CH$_2$Cl $\xrightarrow[\text{0 °C}]{\text{AlCl}_3}$ HCl + butylbenzene (27% yield) + *sec*-butylbenzene (49% yield) (16.16)

1-chlorobutane

the alkyl group has rearranged

In this example, the alkyl group in the *sec*-butylbenzene product has rearranged. Because primary carbocations are too unstable to be involved as intermediates, it is probably the complex of the alkyl halide and AlCl$_3$ that rearranges. This complex has enough carbocation character that it behaves like a carbocation.

CH$_3$CH$_2$CHCH$_2$—$\overset{+}{\ddot{\text{C}}}$l—$\bar{\text{A}}lCl_3$ $\longrightarrow$ CH$_3$CH$_2$$\overset{+}{\text{C}}HCH_2$:$\ddot{\text{C}}$l—$\bar{\text{A}}lCl_3$ (16.17)

***sec*-butyl cation**
alkylates benzene to give
sec-butylbenzene

As we might expect, rearrangement in the Friedel–Crafts alkylation is not observed if the carbocation intermediate is not prone to rearrangement.

benzene
(threefold excess)
tert-butylbenzene
(66% yield)
1,4-di-tert-butylbenzene
(small amount)
(16.18)

In this example, the alkylating cation is the *tert*-butyl cation; because it is tertiary, this carbocation does not rearrange.

Alkylbenzenes, such as butylbenzene (Eq. 16.16) that are derived from rearrangement-prone alkyl halides, are generally not prepared by the Friedel–Crafts alkylation, but rather by other methods that we'll consider in Secs. 19.12 and 18.10B.

Another complication in Friedel–Crafts alkylation is that the alkylbenzene products are more reactive than benzene itself (for reasons to be discussed in Sec. 16.5B). This means that the product can undergo further alkylation, and mixtures of products alkylated to different extents are observed along with unreacted benzene.

(equimolar amounts)
toluene
p-xylene
o-xylene

1,2,4-trimethylbenzene
(pseudocumene)
1,2,4,5-tetramethylbenzene
(durene)
+ other compounds   (16.19)

(Notice also the product of double alkylation in Eq. 16.18.) However, a monoalkylation product can be obtained in good yield if a large excess of the aromatic starting material is used. For example, in the following equation the 15-fold molar excess of benzene ensures that a molecule of alkylating agent is much more likely to encounter a molecule of benzene in the reaction mixture than a molecule of the ethylbenzene product.

benzene
(15-fold excess)
chloroethane
ethylbenzene
(83% yield)
(16.20)

(Notice also the use of excess starting material in Eqs. 16.14 and 16.18.) This strategy is practical only if the starting material is cheap, and if it can be readily separated from the product.

Alkenes and alcohols can also be used as the alkylating agents in Friedel–Crafts alkylation reactions. The carbocation electrophiles in such reactions are generated from alkenes by protonation and from alcohols by dehydration. (Recall that carbocation intermediates are formed in the protonation of alkenes and the dehydration of alcohols; Secs. 4.7B, 4.9B, 10.2.)

STUDY GUIDE
LINK 16.2
Different Sources of
the Same Reactive
Intermediate

benzene
(excess)
cyclohexene
cyclohexylbenzene
(65–68% yield)
(16.21a)

benzene (excess) + cyclohexanol → cyclohexylbenzene (90% yield) + H_2O (16.21b)

$$\text{benzene} + \text{HO} \overset{H_2SO_4}{\underset{\text{heat}}{\longrightarrow}} \text{cyclohexylbenzene} + H_2O \quad (16.21b)$$

PROBLEMS

16.15 (a) Draw a curved-arrow mechanism for the reaction in Eq. 16.21a.

(b) Draw a curved-arrow mechanism for the reaction in Eq. 16.21b.

16.16 What electrophilic substitution product is formed when 2-methylpropene is added to a large excess of benzene containing HF and the Lewis acid BF_3? By what mechanism is it formed?

16.17 Predict the product of the following reaction and draw the curved-arrow mechanism for its formation. (*Hint:* Friedel–Crafts alkylations can be used to form rings.)

$$\text{(phenyl-(CH}_2)_4\text{-Cl)} \overset{AlCl_3}{\longrightarrow} \text{(a compound } C_{10}H_{12} + HCl)$$

F. Friedel–Crafts Acylation of Benzene

When benzene reacts with an acid chloride in the presence of a Lewis acid such as aluminum trichloride ($AlCl_3$), a ketone is formed.

$$\text{benzene} - H + Cl - \overset{O}{\overset{\|}{C}} - CH_3 \xrightarrow[\text{2) H}_2\text{O}]{\text{1) AlCl}_3 \text{ (1.1 equiv.)}} \text{acetophenone} - \overset{O}{\overset{\|}{C}} - CH_3 + HCl \quad (16.22)$$

benzene | acetyl chloride (an acid chloride) → acetophenone (a ketone) (97% yield)

This reaction is an example of a *Friedel–Crafts acylation* (pronounced AY-suh-LAY-shun). In an **acylation reaction**, an *acyl* (pronounced AY-sil) *group* is transferred from one group to another. In the **Friedel–Crafts acylation**, an acyl group, typically derived from an acid chloride, is introduced into an aromatic ring in the presence of a Lewis acid.

$$R - \overset{O}{\overset{\|}{C}} - Cl \qquad R - \overset{O}{\overset{\|}{C}} -$$

an acid chloride   an acyl group

The electrophile in the Friedel–Crafts acylation reaction is a carbocation called an *acylium ion*. This ion is formed when the acid chloride reacts with the Lewis acid $AlCl_3$.

$$R - \overset{:O:}{\overset{\|}{C}} - Cl \quad AlCl_3 \longrightarrow R - C \equiv \overset{+}{O}: \quad \overset{-}{AlCl_4} \quad (16.23)$$

acylium ion

Weaker Lewis acids, such as $FeCl_3$ and $ZnCl_2$, can be used to form acylium ions in Friedel–Crafts acylations of aromatic compounds that are more reactive than benzene.

The acylation reaction is completed by the usual steps of electrophilic aromatic substitution (Sec. 16.4B):

$$\text{(16.24)}$$

As we'll learn in Sec. 19.6, ketones are weakly basic. Because of this basicity, the ketone product of the Friedel–Crafts acylation reacts with the Lewis acid in a Lewis acid–base association to form a complex that is catalytically inactive. This complex is the actual product of the acylation reaction. The formation of this complex has two consequences. First, at least one equivalent of the Lewis acid must be used to ensure its presence throughout the reaction. (Notice, for example, that 1.1 equivalent of $AlCl_3$ is used in Eq. 16.22.) This is in contrast to the Friedel–Crafts *alkylation*, in which $AlCl_3$ can be used in catalytic amounts. Second, the complex must be destroyed before the ketone product can be isolated. This is usually accomplished by pouring the reaction mixture into ice water.

complex of $AlCl_3$
with the ketone product

$$\text{(16.25)}$$

Both Friedel–Crafts alkylation and acylation reactions can occur *intramolecularly* when the product contains a five- or six-membered ring. (See also Problem 16.17.)

$$\text{(16.26)}$$

4-phenylbutanoyl chloride

α-tetralone
(74–91% yield)

In this reaction, the phenyl ring "bites back" on the acylium ion within the same molecule to form a bicyclic compound. This type of reaction can only occur at an adjacent ortho position because reaction at other positions would produce highly strained products. When five- or six-membered rings are involved, this process is much faster than reaction of the acylium ion with the phenyl ring of another molecule. (Sometimes this reaction can be used to form larger rings as well.) This is another illustration of the *proximity effect*: the kinetic advantage of intramolecular reactions (Sec. 11.8).

The multiple-substitution products observed in Friedel–Crafts *alkylation* (Sec. 16.4E) are not a problem in Friedel–Crafts *acylation* because the ketone products of acylation are much less reactive than the benzene starting material, for reasons to be discussed in Sec. 16.5B.

The Friedel–Crafts alkylation and acylation reactions are important for two reasons. First, the alkylation reaction is useful for preparing certain alkylbenzenes, and the acylation reaction is an excellent method for the synthesis of aromatic ketones. Second, they provide other ways to form carbon–carbon bonds. Here is an updated list of reactions that form carbon–carbon bonds.

1. addition of carbenes and carbenoids to alkenes (Sec. 9.9)
2. reaction of Grignard reagents with ethylene oxide and lithium organocuprate reagents with epoxides (Sec. 11.5C)
3. reaction of acetylenic anions with alkyl halides or sulfonates (Sec. 14.7B)
4. Diels–Alder reaction (Sec. 15.3)
5. Friedel–Crafts reactions (Secs. 16.4E and 16.4F)

Charles Friedel and James Mason Crafts

The Friedel–Crafts acylation and alkylation (Sec. 16.4E) reactions are named for their discoverers, the French chemist Charles Friedel (1832–1899) and the American chemist James Mason Crafts (1839–1917). The two men met in Paris while in the laboratory of

Charles-Adolphe Wurtz (1817–1884) at the École de Médecine in Paris (photo). Wurtz was one of the most famous chemists of that time. In 1877, Friedel and Crafts began their collaboration on the reactions that were to bear their names. Friedel subsequently became a very active figure in the development of chemistry in France, and Crafts served as professor and chair in the Chemistry Department at Cornell University, and then as president of the Massachusetts Institute of Technology.

PROBLEMS

16.18 Give the structure of the product expected from the reaction of each of the following compounds with benzene in the presence of one equivalent of $AlCl_3$, followed by treatment with water.

(a)

isobutyryl chloride

(b)

benzoyl chloride

16.19 Show two different Friedel–Crafts acylation reactions that can be used to prepare the following compound.

16.20 The following compound reacts with $AlCl_3$ followed by water to give a ketone A with the formula $C_{10}H_{10}O$. Give the structure of A and draw a curved-arrow mechanism for its formation.

$$H_3C-\text{⟨benzene⟩}-CH_2CH_2-\overset{\overset{\displaystyle O}{\|}}{C}-Cl$$

16.5 ELECTROPHILIC AROMATIC SUBSTITUTION REACTIONS OF SUBSTITUTED BENZENES

A. Directing Effects of Substituents

When a monosubstituted benzene undergoes an electrophilic aromatic substitution reaction, three possible disubstitution products might be obtained. For example, nitration of bromobenzene could in principle give *ortho-*, *meta-*, or *para*-bromonitrobenzene. If substitution were totally random, an ortho:meta:para product ratio of 2 : 2 : 1 would be expected. (Why?) It is found experimentally that this substitution is *not* random, but is *regioselective.*

bromobenzene → HNO₃ / acetic acid → *o*-bromonitrobenzene (36%) + *p*-bromonitrobenzene (62%) + *m*-bromonitrobenzene (2%)   (16.27)

Other electrophilic substitution reactions of bromobenzene also give mostly ortho and para isomers. If a substituted benzene undergoes further substitution mostly at the ortho and para positions, the original substituent is called an **ortho, para-directing group**. Thus, bromine is an ortho, para-directing group, because all electrophilic substitution reactions of bromobenzene occur at the ortho and para positions.

In contrast, some substituted benzenes react in electrophilic aromatic substitution to give mostly the meta disubstitution product. For example, the bromination of nitrobenzene gives only the meta isomer.

nitrobenzene → Br₂ / FeBr₃ / heat → *m*-bromonitrobenzene (only product observed)   (16.28)

Other electrophilic substitution reactions of nitrobenzene also give mostly the meta isomers. If a substituted benzene undergoes further substitution mainly at the meta position, the original substituent group is called a **meta-directing group**. Thus, the nitro group is a meta-directing group because all electrophilic substitution reactions of nitrobenzene occur at the meta position.

A substituent group is either an ortho, para-directing group or a meta-directing group in all electrophilic aromatic substitution reactions; that is, no substituent is ortho, para directing in one reaction and meta directing in another. A summary of the directing effects of common substituent groups is given in the third column of Table 16.2 on p. 812.

PROBLEM

16.21 Using the information in Table 16.2, predict the product(s) of
(a) Friedel–Crafts acylation of anisole (methoxybenzene) with acetyl chloride (structure in Eq. 16.22, p. 808) in the presence of one equivalent of AlCl₃ followed by H₂O.
(b) Friedel–Crafts alkylation of a large excess of ethylbenzene with chloromethane in the presence of AlCl₃.

These directing effects occur because *electrophilic substitution reactions at one position of a benzene derivative are much faster than the same reactions at another position.* That is, the substitution reactions at the different ring positions are *in competition.* For example, in Eq. 16.27, *o-* and *p*-bromonitrobenzenes are the major products because the rate of nitration is greater at the ortho and para positions of bromobenzene than it is at the meta position. Understanding these effects thus requires an understanding of the factors that control the *rates* of aromatic substitution at each position.

TABLE 16.2 Summary of Directing and Activating or Deactivating Effects of Some Common Functional Groups

(Groups are listed in decreasing order of activation.)

Substituent group	Name of group	Directing effect	Activating or deactivating
—NH₂, —NR₂	amino, alkylamino		
—ÖH	hydroxy		
—ÖR	alkoxy		
—ṄH—C(=O)R	acylamino	ortho, para directors	activating substituents
—R	alkyl		
—Ö—C(=O)R	acyloxy		
(phenyl ring)	phenyl		
—F̈:, —B̈r:, —C̈l:, —Ï:,	halogens		
—C(=O)OH , —C(=O)NH₂ , —C(=O)OR	carboxy, carboxamido, carboalkoxy		deactivating substituents
—C(=O)R	acyl	meta directors	
—SO₃H	sulfonic acid		
—CN	cyano		
—NO₂	nitro		

Ortho, Para-Directing Groups All of the ortho, para-directing substituents in Table 16.2 are either *alkyl groups* or *groups that have unshared electron pairs on atoms directly attached to the benzene ring.* Although other types of ortho, para-directing groups are known, the principles on which ortho, para-directing effects are based can be understood by considering electrophilic substitution reactions of benzene derivatives containing these types of substituents.

First imagine the reaction of a general electrophile E^+ with anisole (methoxybenzene). Notice that the atom directly attached to the benzene ring (the oxygen of the methoxy group) has unshared electron pairs. Reaction of E^+ at the para position of anisole gives a carbocation (an arenium ion) intermediate with the following four important resonance structures:

$$(16.29)$$

The colored structure shows that the unshared electron pair of the methoxy group can delocalize the positive charge on the carbocation. This is an especially important structure because it contains more bonds than the others, and every atom has an octet.

PROBLEM

16.22 Draw the carbocation that results from the reaction of the electrophile at the ortho position of anisole; show that this ion also has four resonance structures.

If the electrophile reacts with anisole at the meta position, the carbocation intermediate that is formed has fewer resonance structures than the ion in Eq. 16.29. *In particular, the charge cannot be delocalized onto the* —OCH_3 *group when reaction occurs at the meta position.* There is no structure that corresponds to the colored structure in Eq. 16.29.

$$(16.30)$$

For the oxygen to delocalize the charge, it must be adjacent to an electron-deficient carbon, as in Eq. 16.29. The resonance structures show that the positive charge is shared on *alternate* carbons of the ring. When meta substitution occurs, the positive charge is not shared by the carbon adjacent to the oxygen.

We now use the resonance structures in Eqs. 16.29 and 16.30 (as well as those you drew in Problem 16.22) to assess relative rates. The logic to be used follows the general outline given in Study Problem 15.3, page 756. A comparison of Eq. 16.29 and the structures you drew for Problem 16.22 with Eq. 16.30 shows that the reaction of an electrophile at either the ortho or para positions of anisole gives a carbocation with more resonance structures—that is, a more stable carbocation. The rate-limiting step in many electrophilic aromatic substitution reactions is *formation of the carbocation intermediate.* Hammond's postulate (Sec. 4.8D) indicates that the more stable carbocation should be formed more rapidly. Hence, the products derived from the more rapidly formed carbocation—the more stable carbocation—are the ones observed. Because the reaction of the electrophile at an ortho or para position of anisole gives a more stable carbocation than the reaction at a meta position, the products of ortho, para substitution are formed more rapidly, and are thus the products observed (see Fig. 16.5, p. 814). This is why the —OCH_3 group is an ortho, para-directing group.

To summarize: Substituents containing atoms with unshared electron pairs adjacent to the benzene ring are ortho, para directors in electrophilic aromatic substitution reactions because their electron pairs can be involved in the resonance stabilization of the carbocation intermediates.

Now imagine the reaction of an electrophile E^+ with an alkyl-substituted benzene such as toluene. Alkyl groups such as a methyl group have no unshared electrons, but the explanation for the directing effects of these groups is similar. Reaction of E^+ at a position that is ortho or para to an alkyl group gives an ion that has one tertiary carbocation resonance structure (colored structure in the following equation).

tertiary carbocation

$$(16.31)$$

FIGURE 16.5 Basis of the directing effect of the methoxy group in the electrophilic aromatic substitution reactions of anisole. Substitution of anisole by an electrophile E⁺ occurs more rapidly at (a) the para position than at (b) the meta posi-

tion because a more stable carbocation intermediate is involved in para substitution. The carbocation and transition state structures are shown as hybrid structures; that is, the dashed lines within the structures symbolize the delocalization of electrons.

Reaction of the electrophile meta to the alkyl group also gives an ion with three resonance structures, but all resonance forms are secondary carbocations.

$$\text{(16.32)}$$

Because reaction at the ortho or para position gives the more stable carbocation, alkyl groups are ortho, para-directing groups.

Meta-Directing Groups The meta-directing groups in Table 16.2 are all *polar groups that do not have an unshared electron pair on an atom adjacent to the benzene ring.* The directing effect of these groups can be understood by considering as an example the reactions of a general electrophile E⁺ with nitrobenzene at the meta and para positions.

$$\text{(16.33)}$$

(16.34)

positive charges are on
adjacent atoms

Both reactions give carbocations that have three resonance structures, but reaction at the para position gives an ion with one particularly unfavorable structure (*red*). In this structure, positive charges are situated on adjacent atoms. Because repulsion between two like charges, and consequently their energy of interaction, increases with decreasing separation, the red resonance structure in Eq. 16.34 is less important than the others. Thus, the carbocation in Eq. 16.33, with the greater separation of like charges, is more stable than the carbocation in Eq. 16.34. By Hammond's postulate (Sec. 4.8D), the more stable carbocation intermediate should be formed more rapidly. Consequently, the nitro group is a meta director because the ion that results from meta substitution (Eq. 16.33) is more stable than the one that results from para substitution (Eq. 16.34).

In summary, substituents that have positive charges adjacent to the aromatic ring are meta directors because meta substitution gives the carbocation intermediate in which like charges are farther apart. Notice that not all meta-directing groups have full positive charges like the nitro group, but all of them have bond dipoles that place a substantial amount of positive charge next to the benzene ring.

acyl group

sulfonic acid
group

PROBLEMS

16.23 Biphenyl (phenylbenzene) undergoes the Friedel–Crafts acylation reaction, as shown by the following example.

biphenyl

p-phenylacetophenone

(a) On the basis of this result, what is the directing effect of the phenyl group?

(b) Using resonance arguments, explain the directing effect of the phenyl group.

continued

continued

16.24 Predict the predominant products that would result from bromination of each of the following compounds. Classify each substituent group as an ortho, para director or a meta director, and explain your reasoning.

(a)

⟨benzene ring⟩—NH—C(=O)—CH₃

(b)

⟨benzene ring⟩—CF₃

(c)

⟨benzene ring⟩—N⁺(CH₃)₃

(d)

⟨benzene ring⟩—C(CH₃)₃

(e)

⟨benzene ring⟩—Ö:⁻

The Ortho, Para Ratio   An aromatic substitution reaction of a benzene derivative bearing an ortho, para-directing group would give twice as much ortho as para product if substitution were completely random, because there are two ortho positions and only one para position available for substitution. However, this situation is rarely observed in practice; instead, it is often found that the para substitution product is the major one in the reaction mixture. In some cases this result can be explained by the spatial demands of the electrophile. For example, Friedel–Crafts acylation of toluene gives essentially all para substitution product and almost no ortho product. The electrophile cannot react at the ortho position without developing van der Waals repulsions with the methyl group that is already on the ring. Consequently, reaction occurs at the para position, where such repulsions cannot occur.

Typically, para substitution predominates over ortho substitution, but not always. For example, nitration of toluene gives twice as much *o*-nitrotoluene as *p*-nitrotoluene. This result occurs because the nitration of toluene at either the ortho or para position is so fast that it occurs on *every encounter* of the reagents; that is, the energy barrier for the reaction is insignificant. Hence, the product distribution corresponds simply to the relative probability of the reactions. Because the ratio of ortho and para positions is 2 : 1, the product distribution is 2 : 1. In fact, the ready availability of *o*-nitrotoluene makes it a good starting material for certain other ortho-substituted benzene derivatives.

In summary, the reasons for the ortho, para ratio vary from case to case, and in some cases these reasons are not well understood.

Whatever the reasons for the ortho, para ratio, if an electrophilic aromatic substitution reaction yields a mixture of ortho and para isomers, a problem of isomer separation arises that must be solved if the reaction is to be useful. Usually, syntheses that give mixtures of isomers are avoided because, in many cases, isomers are difficult to separate. However, the ortho and para isomers obtained in many electrophilic aromatic substitution reactions have sufficiently different physical properties that they are readily separated (Sec. 16.2). For example, the boiling points of *o*- and *p*-nitrotoluene, 220 °C and 238 °C, respectively, are sufficiently different that these isomers can be separated by careful fractional distillation. Thus, either isomer can be obtained relatively pure from the nitration of toluene. The melting points of *o*- and *p*-chloronitrobenzene, 34 °C and 84 °C, respectively, are so different that the para isomer can be selectively crystallized. As you learned in Sec. 16.2, the para isomer of an ortho, para pair typically has the higher melting point, in many cases, *considerably* higher. Most aromatic substitution reactions are so simple and inexpensive to run that when the separation of isomeric products is not difficult, these reactions are useful for organic synthesis despite the product mixtures obtained. Thus, you may assume in working problems involving electrophilic aromatic substitution on compounds containing ortho, para-directing groups that the para isomer can be isolated in useful amounts. For the reasons pointed out in the previous paragraph, *o*-nitrotoluene is a relatively rare example of a readily obtained ortho-substituted benzene derivative.

B. Activating and Deactivating Effects of Substituents

Different benzene derivatives have greatly different reactivities in electrophilic aromatic substitution reactions. If a substituted benzene derivative reacts more rapidly than benzene itself, then the substituent group is said to be an **activating group**. The Friedel–Crafts acylation of anisole (methoxybenzene), for example, is 300,000 times faster than the same reaction of benzene under comparable conditions. Furthermore, anisole shows a similar enhanced reactivity relative to benzene in all other electrophilic substitution reactions. Thus, the methoxy group is an *activating group*.

On the other hand, if a substituted benzene derivative reacts more slowly than benzene itself, then the substituent is called a **deactivating group**. For example, the rate for the bromination of nitrobenzene is less than 10^{-5} times the rate for the bromination of benzene; furthermore, nitrobenzene reacts much more slowly than benzene in all other electrophilic aromatic substitution reactions. Thus, the nitro group is a *deactivating group*.

A given substituent group is either activating in all electrophilic aromatic substitution reactions or deactivating in all such reactions. Whether a substituent is activating or deactivating is shown in the last column of Table 16.2, p. 812. In this table the most activating substituent groups are near the top of the table. Three generalizations emerge from examining this table.

1. All meta-directing groups are deactivating groups.
2. All ortho, para-directing groups except for the halogens are activating groups.
3. The halogens are deactivating groups.

Thus, except for the halogens, there appears to be a correlation between the activating and directing effects of substituents.

In view of this correlation, it is not surprising that the explanation of activating and deactivating effects is closely related to the explanation for directing effects. A key to understanding these effects is the realization that directing effects are concerned with the relative rates of substitution at different positions of the *same* compound, whereas activating or deactivating effects are concerned with the relative rates of substitution of *different* compounds—a substituted benzene compared with benzene itself. As in the discussion of directing effects, we consider the effect of the substituent on the stability of the intermediate carbocation, and we then apply Hammond's postulate by assuming that the stability of this carbocation is related to the stability of the transition state for its formation.

Two properties of substituents must be considered to understand activating and deactivating effects. First is the *resonance effect* of the substituent. The **resonance effect** of a substituent group is the ability of the substituent to stabilize the carbocation intermediate in electrophilic substitution by delocalization of electrons from the substituent into the ring. The resonance effect is the same effect responsible for the ortho, para-directing effects of substituents with unshared electron pairs, such as —OCH_3 and halogen (colored structure in Eq. 16.29, p. 812). We can summarize this effect with the following two of the four resonance structures for the carbocation intermediate in Eq. 16.29.

the resonance effect of the methoxy group *stabilizes* the carbocation

(two of the four important resonance structures)

The second property is the *polar effect* of the substituent. The **polar effect** is the tendency of the substituent group, by virtue of its electronegativity, to pull electrons away from the ring. This is the same effect discussed in connection with substituent effects on acidity (Sec. 3.6C). When a ring substituent is electronegative, it pulls the electrons of the ring toward itself and creates an electron deficiency, or positive charge, in the ring. In the carbocation intermediate of an electrophilic substitution reaction, the positive end of the bond dipole interacts repulsively with the positive charge in the ring, thus raising the energy of the ion:

$\overset{\delta-}{:\ddot{O}CH_3}$

repulsive interaction

the polar effect of the methoxy group *destabilizes* the carbocation

Thus, the electron-donating resonance effect of a substituent group with unshared electron pairs, if it were dominant, would *stabilize* positive charge and would *activate* further substitution. If such a group is electronegative, its electron-withdrawing polar effect, if dominant, would *destabilize* positive charge and would *deactivate* further substitution. These two effects operate simultaneously and in opposite directions. *Whether a substituted derivative of benzene is activated or deactivated toward further substitution depends on the balance of the resonance and polar effects of the substituent group.*

Anisole (methoxybenzene) undergoes electrophilic substitution much more rapidly than benzene because the resonance effect of the methoxy group far outweighs its polar effect. The benzene molecule, in contrast, has no substituent to help stabilize the carbocation intermediate by resonance. Hence, the carbocation intermediate (and the transition state) derived from the electrophilic substitution of anisole is more stable relative to starting materials than the carbocation (and transition state) derived from the electrophilic substitution of benzene. Thus, in a given reaction, the ortho and para substitution of anisole are faster than the substitution of benzene. In other words, the methoxy group activates the benzene ring toward ortho and para substitution.

There is also an important subtlety here. Although the ortho and para positions of anisole are highly activated toward substitution, the meta position is deactivated. When substitution

FIGURE 16.6 Basis of the activating effect of the methoxy group on electrophilic aromatic substitution in anisole. (a) The energy barrier for substitution of benzene by an electrophile E⁺. (b) The energy barrier for substitution of anisole by E⁺ at the para position. (c) The energy barrier for substitution of anisole by E⁺ at the meta position. (Notice that the diagrams for parts (b) and (c) are the same as parts (a) and (b) of Fig. 16.5, p. 814.) The substitution of anisole at the para position is faster than the substitution of benzene; the substitution of anisole at the meta position is slower than the substitution of benzene. The methoxy group is an activating group because the observed reaction of anisole—substitution at the para position—is faster than the substitution of benzene.

overlap of carbon and oxygen 2*p* orbitals

(a)

overlap of carbon 2*p* and chlorine 3*p* orbitals

(b)

FIGURE 16.7 The carbocation (arenium ion) resulting from the reaction of an electrophile E$^+$ at the para position of a benzene ring substituted with (a) a methoxy group and (b) a chlorine. Below each ion is shown the orbital overlap between (a) oxygen and carbon and (b) chlorine and carbon. This orbital overlap stabilizes the ion. The blue and green parts of the orbitals represent wave peaks and wave troughs, respectively. The overlap of carbon and oxygen 2*p* orbitals, shown in (a), is more effective than the overlap of carbon 2*p* and chlorine 3*p* orbitals, shown in (b). Bonding overlap occurs only when peaks overlap with peaks and troughs overlap with troughs. Because orbitals with different quantum numbers have different sizes and different numbers of nodes, part of the chlorine 3*p* orbital cannot overlap with the carbon 2*p* orbital.

occurs in the meta position, the methoxy group cannot exert its resonance effect (Eq. 16.30), and only its rate-retarding polar effect is operative. Thus, whether a group activates or deactivates further substitution really depends on the *position* on the ring being considered. Thus, the methoxy group activates ortho, para substitution and deactivates meta substitution. But this is just another way of saying that the methoxy group is an ortho, para director. Because ortho, para substitution is the *observed* mode of substitution, the methoxy group is considered to be an activating group. These ideas are summarized in the reaction free-energy diagrams shown in Fig. 16.6.

The deactivating effects of halogen substituents reflect a different balance of resonance and polar effects. Consider the chloro group, for example. Because chlorine and oxygen have similar electronegativities, the polar effects of the chloro and methoxy groups are similar. However, the resonance interaction of chlorine electron pairs with the ring is much less effective than the interaction of oxygen electron pairs because the chlorine valence electrons reside in orbitals with higher quantum numbers. Because these orbitals and the carbon 2*p* orbitals of the benzene ring have *different sizes* and *different numbers of nodes*, they do not overlap so effectively (Fig. 16.7). Because this overlap is the basis of the resonance effect, the resonance effect of chlorine is weak. With a weak rate-enhancing resonance effect and a strong rate-retarding polar effect, chlorine is a deactivating group. Bromine and iodine exert weaker polar effects than chlorine, but their resonance effects are also weaker. (Why?) Hence, these groups, too, are deactivating groups. Fluorine, as a second-period element, has a stronger resonance effect than the other halogens, but, as the most electronegative element, it has a stronger polar effect as well. Fluorine is also a deactivating group.

The deactivating, rate-retarding polar effects of the halogens are similar at all ring positions but are offset somewhat by their resonance effects when substitution occurs para to the halogen. However, the resonance effect of a halogen cannot come into play at all when substitution occurs at the meta position of a halobenzene. (Why?) Hence, meta substitution in halobenzenes is deactivated even more than para substitution is. This is another way of saying that halogens are ortho, para-directing groups.

Alkyl substituents such as the methyl group have no resonance effect, but the polar effect of any alkyl group toward electron-deficient carbons is an electropositive, stabilizing effect (Sec. 4.7C). It follows that alkyl substituents on a benzene ring stabilize carbocation intermediates in electrophilic substitution, and for this reason, they are activating groups. It turns out that alkyl groups activate substitution at all ring positions, but they are ortho, para directors because they activate ortho, para substitution more than they activate meta substitution (Eqs. 16.31, p. 813; 16.32, p. 814).

Finally, consider the deactivating effects of meta-directing groups such as the nitro group. Because a nitro group has no electron-donating resonance effect, the polar effect of this electronegative group destabilizes the carbocation intermediate and retards electrophilic substitution at *all* positions of the ring. The nitro group is a meta-directing group because substitution is retarded more at the ortho and para positions than at the meta positions (Eqs. 16.33 and 16.34, p. 815). In other words, the meta-directing effect of the nitro group is not due to selective activation of the meta positions, but rather to greater *deactivation* of the ortho and para positions. For this reason, the nitro group and the other meta-directing groups might be called *meta-allowing groups*.

PROBLEMS

16.25 Draw reaction-free energy profiles analogous to that in Fig. 16.6 in which substitution on benzene by a general electrophile E^+ is compared with substitution at the para and meta positions of (a) chlorobenzene; (b) nitrobenzene.

16.26 Explain why the nitration of anisole is much faster than the nitration of thioanisole under the same conditions.

anisole   **thioanisole**

16.27 Which should be faster: bromination of benzene or bromination of *N,N*-dimethylaniline? Explain your answer carefully.

***N,N*-dimethylaniline**

C. Electrophilic Aromatic Substitution in Biology: Biosynthesis of Thyroid Hormones

Electrophilic aromatic substitution in biology is illustrated by one of the steps in the biosynthesis of thyroid hormones. The thyroid hormones T3 and T4 are produced by the thyroid gland. These hormones are involved in the regulation of metabolism. The formation of T3 and T4 accounts for the dietary requirement for iodine (as iodide).

(S)-triiodothyronine (T3)
a thyroid hormone

(S)-thyroxine (T4)
a thyroid hormone

A key step in the production of these hormones is the iodination of several tyrosine residues in a protein called *thyroglobulin*.

tyrosine residues in thyroglobulin   3′,5′-diiodotyrosine residues in thyroglobulin

$$(16.35)$$

As with other electrophilic aromatic substitution reactions, the first step is formation of the electrophile. The thyroid contains an iodide transporter protein that concentrates sodium iodide within the appropriate part of the thyroid tissue. Iodide ion is not electrophilic, and it must be oxidized to become an electrophile. A thyroid enzyme called *thyroid peroxidase* catalyzes the oxidation reaction between the iodide ion and hydrogen peroxide to give the actual electrophile, which is probably hypoiodous acid (I—OH), molecular iodine (I_2), or an equivalent species. The formation of I—OH is shown here:

$$H_3O^+ \ + \ I^- \ + \ H_2O_2 \ \xrightarrow[\text{(enzyme)}]{\text{thyroid peroxidase}} \ I-OH \ + \ 2\,H_2O \qquad (16.36)$$

iodide ion **hydrogen peroxide**   **hypoiodous acid**

The electrophile reacts rapidly with the aromatic ring of a tyrosine residue in an electrophilic aromatic substitution reaction.

tyrosine residue in thyroglobulin

$$(16.37a)$$

"carbocation" intermediate

In this mechanism, the tyrosine —OH group, which is weakly acidic, protonates the —OH leaving group in hypoiodous acid. This protonation improves the leaving group (as in the protonation of alcohol —OH groups) and results in a negative charge on the tyrosine oxygen that stabilizes the arenium-ion intermediate. As a result, this intermediate is actually not a carbocation but a neutral species, as the second resonance structure in Eq. 16.37a shows. The water molecule formed as a leaving group then facilitates the proton transfers to regenerate aromaticity in the substitution product.

$$(16.37b)$$

iodinated tyrosine residue

The second iodine atom (Eq. 16.35) is introduced into the ring by the same mechanism.

Two points are especially pertinent. First, the hydroxy substituent on the tyrosine ring is strongly ortho, para directing. Because the para position is already occupied in tyrosine, the two ortho positions are iodinated. Therefore, the directing effect of the hydroxy group accounts nicely for the position of iodination. Second, the iodination of tyrosines in thyroglobulin is very rapid. We expect a very rapid reaction from the strong activating effect of the hydroxy substituent (Table 16.2, p. 812).

After the tyrosine residues in thyroglobulin are iodinated, subsequent enzyme-catalyzed oxidative reactions of thyroglobulin (which we won't consider here) form the T3 and T4 structures within the protein. In subsequent enzyme-catalyzed reactions, these are excised from the thyroglobulin protein as the free amino acids.

PROBLEMS

16.28 Write the two half-reactions that correspond to the oxidation of iodide ion in Eq. 16.36. (Review Sec. 10.6A if you need help.)

16.29 The iodination reaction discussed in this section can be carried out on the amino acid tyrosine and related compounds in the laboratory with iodide ion in the presence of the enzyme thyroid oxidase. Which would undergo iodination more rapidly: tyrosine (*A*) or 3,3-difluorotyrosine (*B*)? Explain.

tyrosine
A

3,3-difluorotyrosine
B

D. Use of Electrophilic Aromatic Substitution in Organic Synthesis

Both activating/deactivating and directing effects of substituents can come into play in planning an organic synthesis that involves electrophilic substitution reactions. The importance of directing effects is illustrated in Study Problem 16.2.

STUDY PROBLEM 16.2

Outline a synthesis of *p*-bromonitrobenzene from benzene.

SOLUTION The key to this problem is whether the bromine or the nitro group should be the first ring substituent introduced. Introduction of the bromine first takes advantage of its directing effect in the subsequent nitration reaction:

benzene   bromobenzene   *p*-bromonitrobenzene   (16.38)

Introduction of the nitro group first followed by bromination would give instead *m*-bromonitrobenzene, because the nitro group is a meta-directing group.

benzene   nitrobenzene   *m*-bromonitrobenzene   (16.39)

Hence, to prepare the desired compound, brominate first and *then* nitrate the resulting bromobenzene, as shown in Eq. 16.38.

When an electrophilic substitution reaction is carried out on a benzene derivative with more than one substituent, the activating and directing effects are roughly the sum of the effects of the separate substituents. First, let's consider directing effects. In the Friedel–Crafts acylation of *m*-xylene, for example, both methyl groups direct the substitution to the same positions.

$$H_3C{-}\langle\rangle{-}CH_3 \ +\ Cl{-}\overset{O}{\overset{\|}{C}}{-}CH_3 \ \xrightarrow{AlCl_3}\ H_3C{-}\langle\rangle{-}\overset{O}{\overset{\|}{C}}{-}CH_3\ +\ HCl \qquad (16.40)$$

substitution at this position is hindered by two ortho methyl groups

m-xylene

(80% yield)

Methyl groups are ortho, para directors. Substitution at the position ortho to both methyl groups is difficult because van der Waals repulsions between both methyls and the electrophile would be present in the transition state. Consequently, substitution occurs at a ring position that is para to one methyl and, of necessity, ortho to the other, as shown in Eq. 16.40.

Two meta-directing groups on a ring, such as the carboxylic acid ($-CO_2H$) groups in the following example, direct further substitution to the remaining open meta position:

$$+ \ HNO_3 \ \xrightarrow{H_2SO_4} \qquad\qquad + \ H_2O \qquad (16.41)$$

1,3-benzenedicarboxylic acid

5-nitro-1,3-benzenedicarboxylic acid
(96% of product)

In each of the previous two examples, both substituents direct the incoming group to the same position. What happens when the directing effects of the two groups are in conflict? If one group is much more strongly activating than the other, the directing effect of the more powerful activating group generally predominates. For example, the $-OH$ group is such a powerful activating group that phenol can be brominated three times, even without a Lewis acid catalyst. (Notice that the $-OH$ group is near the top of Table 16.2, p. 812.)

$$+ \ 3\,Br_2 \ \xrightarrow{H_2O} \qquad\qquad + \ 3\,HBr \qquad (16.42)$$

phenol

2,4,6-tribromophenol
quantitative;
virtually instantaneous

(This reaction is discussed in more detail in Chapter 18, pp. 925–926.) After the first bromination, the $-OH$ and $-Br$ groups direct subsequent brominations to different positions. The strong activating and directing effect of the $-OH$ group at the ortho and para positions overrides the weaker directing effect of the $-Br$ group. The same principle operates in the iodination of tyrosine in thyroglobulin (Eq. 16.35, p. 821).

In other cases, mixtures of isomers are typically obtained.

4-chlorotoluene   **4-chloro-3-nitrotoluene**   **4-chloro-2-nitrotoluene**
(42%)   (58%)

(16.43)

PROBLEM

16.30   Predict the predominant product(s) from:

(a) monosulfonation of *m*-bromotoluene   (b) mononitration of *m*-bromoiodobenzene

---

**STUDY GUIDE
LINK 16.3
Reaction Conditions
and Reaction Rate**

You've just learned that the activating and directing effects of substituents must be taken into account in developing the strategy for an organic synthesis that involves a substitution reaction on an already-substituted benzene ring. The activating or deactivating effects of substituents in an aromatic compound also determine the *conditions* that must be used in an electrophilic substitution reaction. The bromination of nitrobenzene, for example (Eq. 16.28, p. 811), requires relatively harsh conditions of heat and a Lewis acid catalyst because the nitro group deactivates the ring toward electrophilic substitution. The conditions in Eq. 16.28 are more severe than the conditions required for the bromination of benzene itself, because benzene is the more reactive compound. An even more dramatic example in the opposite direction is provided by the bromination of mesitylene (1,3,5-trimethylbenzene). Mesitylene can be brominated under *very* mild conditions, because the ring is activated by three methyl groups; a Lewis acid catalyst is not even necessary.

$$H_3C \diagdown \diagup CH_3 + Br_2 \xrightarrow[CCl_4]{0-10\ °C} H_3C \diagdown \diagup CH_3 + HBr \qquad (16.44)$$

mesitylene   (80% yield)

A similar contrast is apparent in the conditions required to sulfonate benzene and toluene. Sulfonation of benzene requires fuming sulfuric acid (Eq. 16.12, p. 804). However, because toluene is more reactive than benzene, toluene can be sulfonated with concentrated sulfuric acid, a milder reagent than fuming sulfuric acid.

$$H_3C \diagdown \!\!\!\!\bigcirc\!\!\!\! + H_2SO_4 \longrightarrow H_3C \diagdown \!\!\!\!\bigcirc\!\!\!\! -SO_3H + H_2O \qquad (16.45)$$

toluene   ***p*-toluenesulfonic acid**

Another very important consequence of activating and deactivating effects is that when a deactivating group—for example, a nitro group—is being introduced by an electrophilic substitution reaction, it is easy to introduce one group at a time, because the products are *less reactive* than the reactants. Thus, toluene can be nitrated only once because the nitro group that is introduced retards a second nitration on the same ring. The following three equations show the conditions required for successive nitrations. The actual amounts of nitric and sulfuric acid are given to show that each additional nitration requires harsher conditions.

toluene
50 g

4-nitrotoluene   (16.46a)

4-nitrotoluene
50 g

2,4-dinitrotoluene
(90% yield)   (16.46b)

2,4-dinitrotoluene
50 g

2,4,6-trinitrotoluene ("TNT")
(90% yield)   (16.46c)

Fuming nitric acid (Eq. 16.46c) is an especially concentrated form of nitric acid. Ordinary nitric acid contains 68% by weight of nitric acid; fuming nitric acid is 95% by weight nitric acid. It owes its name to the layer of colored fumes usually present in the bottle of the commercial product. Fuming nitric acid is a much harsher (that is, more reactive) nitrating reagent than nitric acid itself.

In contrast, when an activating group is introduced by electrophilic substitution, the products are *more reactive* than the reactants; consequently, additional substitutions can occur easily under the conditions of the first substitution and, as a result, mixtures of products are obtained. This is the situation in Friedel–Crafts alkylation. As noted in the discussion of Eq. 16.19 (p. 807), one way to avoid multiple substitution in such cases is to use a large excess of the starting material. (Friedel–Crafts alkylation is the only electrophilic aromatic substitution reaction discussed in this chapter that introduces an activating substituent.)

Some deactivating substituents retard some reactions to the point that they are not useful. For example, Friedel–Crafts *acylation* (Sec. 16.4F) does not occur on a benzene ring substituted *solely* with one or more meta-directing groups. In fact, nitrobenzene is so unreactive in the Friedel–Crafts acylation that it can be used as the solvent in the acylation of other aromatic compounds! Similarly, the Friedel–Crafts *alkylation* (Sec. 16.4E) is generally too slow to be useful on compounds that are more deactivated than benzene itself, even halobenzenes.

PROBLEMS

16.31 In each of the following sets, rank the compounds in order of increasing harshness of the reaction conditions required to accomplish the indicated reaction.

(a) sulfonation of benzene, *m*-xylene, or *p*-dichlorobenzene

(b) Friedel–Crafts acylation of chlorobenzene, anisole, or toluene.

16.32 Outline a synthesis of *m*-nitroacetophenone from benzene; explain your reasoning.

***m*-nitroacetophenone**

16.6   HYDROGENATION OF BENZENE DERIVATIVES

Because of its aromatic stability, the benzene ring is resistant to conditions used to hydrogenate ordinary double bonds.

$$
\text{stilbene} \; (\text{CH}{=}\text{CH}) + \text{H}_2 \xrightarrow[\text{25 °C}]{\text{Pd/C}} \text{(CH}_2{-}\text{CH}_2) \tag{16.47}
$$

stilbene
(cis or trans)

(2-phenylethyl)benzene
(bibenzyl)
(95% yield)

Nevertheless, aromatic rings can be hydrogenated under more extreme conditions of temperature or pressure (or both), and practical laboratory apparatus that can accommodate these conditions is readily available. Typical conditions for carrying out the hydrogenation of benzene derivatives include Rh or Pt catalysts at 5–10 atm of hydrogen pressure and 50–100 °C, or Ni or Pd catalysts at 100–200 atm and 100–200 °C. For example, compare the conditions for the following hydrogenation with those for the hydrogenation in Eq. 16.47.

$$
\text{Et} + 3\,\text{H}_2 \xrightarrow[\substack{175\,°\text{C} \\ 180\ \text{atm}}]{\text{Ni}} \text{Et} \tag{16.48}
$$

ethylbenzene   **ethylcyclohexane**
(93% yield)

As this example illustrates, a good way to prepare a substituted cyclohexane in many cases is to prepare the corresponding benzene derivative and then hydrogenate it.

Catalytic hydrogenation of benzene derivatives gives the corresponding cyclohexanes and cannot be stopped at the cyclohexadiene or cyclohexene stage. The reason follows from the enthalpies of hydrogenation of benzene, 1,3-cyclohexadiene, and cyclohexene.

$$
\text{benzene} + \text{H}_2 \longrightarrow \text{1,3-cyclohexadiene} \qquad H° = +24.3\ \text{kJ mol}^{-1}\ (+5.8\ \text{kcal mol}^{-1}) \tag{16.49a}
$$

$$
\text{1,3-cyclohexadiene} + \text{H}_2 \longrightarrow \text{cyclohexene} \qquad H° = -111\ \text{kJ mol}^{-1}\ (-26.5\ \text{kcal mol}^{-1}) \tag{16.49b}
$$

$$
\text{cyclohexene} + \text{H}_2 \longrightarrow \text{cyclohexane} \qquad H° = -118\ \text{kJ mol}^{-1}\ (-28.2\ \text{kcal mol}^{-1}) \tag{16.49c}
$$

The hydrogenation of most ordinary alkenes is *exothermic* by 113–126 kJ mol^{-1} (27–30 kcal mol^{-1}); yet the reaction in Eq. 16.49a is *endothermic*. The unusual $\Delta H°$ of this reaction reflects the aromatic stability of benzene. In fact, the $\Delta H°$ of hydrogenation of benzene can be used to provide another estimate of the *empirical resonance energy* of benzene—the aromatic stabilization energy of benzene (Sec. 15.7C). If benzene were an "ordinary" alkene, the $\Delta H°$ of hydrogenation of its three "ordinary" double bonds should be about the same as that of three cyclohexenes, which, from Eq. 16.49c, is $3 \times (-118) = -354$ kJ mol^{-1} (−85 kcal mol^{-1}). The actual $\Delta H°$ of hydrogenation of benzene is obtained by adding the $\Delta H°$ values of Eqs. 16.49a–c to obtain −205 kJ mol^{-1} (−49 kcal mol^{-1}). The difference, which is the empirical resonance energy of benzene, is 149 kJ mol^{-1} (36 kcal mol^{-1}). This is very similar to the estimate obtained in Sec. 15.7C (141 kJ mol^{-1}, 34 kcal mol^{-1}) by comparing the heats of formation of benzene and cyclooctatetraene (COT).

Because the first hydrogenation reaction of benzene is endothermic, energy must be added for it to take place—thus, the harsh conditions required for the hydrogenation of benzene derivatives. The hydrogenations of 1,3-cyclohexadiene and cyclohexene proceed so rapidly under these vigorous conditions that once these compounds are formed in the hydrogenation of benzene, they react instantaneously.

PROBLEM

16.33 Using benzene and any other reagents, outline a synthesis of each of the following compounds.
(a) cyclohexylcyclohexane   (b) *tert*-butylcyclohexane

16.7 POLYCYCLIC AROMATIC HYDROCARBONS AND CANCER

Most people understand that certain chemicals are hazardous. Among the most worrisome chemical hazards is *carcinogenicity*—the proclivity of a substance to cause cancer. Cancer-causing materials are termed **carcinogens**. A few aromatic compounds are potent carcinogens. Both the historical aspects of this finding and the reasons underlying it are interesting.

After the great fire of London in 1666, Londoners began the practice of building homes with long and tortuous chimneys. The use of coal for heating resulted in deposits of black soot that had to be periodically removed from these chimneys, but the only people who could negotiate these narrow passages were small boys, called "sweeping boys." It was common for these boys to contract a disease that we now know is cancer of the scrotum. In 1775, Percivall Pott (1714–1788), a surgeon at London's St. Bartholomew's Hospital, identified coal dust as the source of "this noisome, painful, and fatal disease," and Pott's findings subsequently led to substantial reform in the child-labor statutes in England. In 1892, Henry T. Butlin (1845–1912), also of St. Bartholomew's, pointed out that the disease did not occur in countries in which the chimney sweeps washed thoroughly after each day's work.

The question remained: why did exposure to large amounts of soot cause cancer in the sweeping boys? The source of the problem was traced to benzo[*a*]pyrene, a compound that had been isolated in 1933 from coal tar. This compound and 7,12-dimethylbenz[*a*]anthracene, both polycyclic aromatic hydrocarbons, have been found to be two of the most potent carcinogens known.

benzo[*a*]pyrene

7,12-dimethylbenz[*a*]anthracene (7,12-DMBA)

These and related compounds are the carcinogens in soot. Materials such as these are also found in cigarette smoke, automobile exhaust, the smoke from wood and coal burning, and even in grilled meats. Several thousand tons of benzo[*a*]pyrene and related hydrocarbons per year—the exact amount is not known with certainty—are released into the environment. The association of lung cancer with cigarette smoking has been definitively linked to benzo[*a*]pyrene and related compounds in tobacco smoke.

A study of the carcinogenicity of benzo[*a*]pyrene and related aromatic hydrocarbons led to the finding that the hydrocarbons themselves are not the cancer-causing agents. Rather, they are metabolized in living systems to form certain epoxide derivatives, which are the *ultimate carcinogens* (the true carcinogens). The diol-epoxide shown in Eq. 16.50, formed when living cells attempt to metabolize benzo[*a*]pyrene, is the ultimate carcinogen derived from this hydrocarbon.

benzo[*a*]pyrene → **benzo[*a*]pyrene diol-epoxide**

living tissue (enzymes, O_2)

(16.50)

This transformation is particularly remarkable in view of the usual resistance of benzene rings to undergo addition reactions. This metabolism is an example of *phase I metabolism*—enzyme-catalyzed reactions in which molecules are derivatized to produce more water-soluble products. Recall that in phase II metabolism (Sec. 8.6C), compounds are chemically coupled to water-soluble "handles," such as glucuronides, that increase their water solubility. In phase I metabolism, the compound itself undergoes a functional-group transformation that ultimately has the same metabolic purpose—an increase in water solubility. You can see that the diol-epoxide in Eq. 16.50 contains groups that should enhance water solubility—particularly the two alcohol groups. The problem is the epoxide functional group. The diol-epoxide survives long enough that it can migrate into the nuclei of cells, where the epoxide group reacts with nucleophilic groups on DNA. (See Problem 16.34.) When this reaction occurs in certain genes, the regulation of cell division is disrupted and cancer results. In other words, the metabolic process, in an effort to clear the polycyclic aromatic hydrocarbons from the cell, actually activates these hydrocarbons to become the ultimate carcinogens.

Benzene has also been found to be carcinogenic, and it has been supplanted for many uses by toluene, which is not carcinogenic. (Not all aromatic compounds are carcinogens.) However, benzene is much less carcinogenic than the polycyclic hydrocarbons shown here, and it continues to be used with due caution in applications for which it cannot be readily replaced, particularly in the chemical industry (see Sec. 16.8).

PROBLEM

16.34 The group in DNA that reacts with the diol-epoxide has the following general structure. Using mechanistic reasoning, show how the amino group indicated by the asterisk (*) might react with the epoxide group of the diol-epoxide in Eq. 16.50. (*Hint:* Abbreviate this structure as R—N̈H$_2$ and then fill in the completed structure after you have completed the mechanism.)

16.8 THE SOURCE AND INDUSTRIAL USE OF AROMATIC HYDROCARBONS

The most common source of aromatic hydrocarbons is petroleum. Some petroleum sources are relatively rich in aromatic hydrocarbons, and aromatic hydrocarbons can be obtained by catalytic re-forming of the hydrocarbons from other sources. Another potentially important, but currently minor, source of aromatic hydrocarbons is *coal tar*, the tarry residue obtained when coal is heated in the absence of oxygen. Once a major source of aromatic hydrocarbons, coal tar may increase in importance as a source of aromatic compounds if the use of coal increases.

Benzene itself is obtained by separation from petroleum fractions and by demethylation of toluene. The worldwide annual production of benzene is about 12 billion gallons. Benzene serves as a principal source of ethylbenzene, styrene, and cumene (see Eq. 16.51), and as one of the sources of cyclohexane. Because cyclohexane is an important intermediate in the production of nylon (Sec. 21.12A), benzene has substantial importance to the nylon industry.

Toluene is also obtained by separation from *reformates*, the products of hydrocarbon interconversion over certain catalysts. As noted in the previous paragraph, some toluene is used in the production of benzene. Toluene is also used as an octane booster for gasoline and as a starting material in the polyurethane industry.

Ethylbenzene and cumene are obtained by the alkylation of benzene with ethylene and propene, respectively, in the presence of acid catalysts (Friedel–Crafts alkylation).

$$\text{benzene} + \underset{\text{propene}}{CH_3CH=CH_2} \xrightarrow[\text{or } H_2SO_4]{AlCl_3/HCl} \text{cumene}-CH(CH_3)_2 \qquad (16.51)$$

Cumene is an important intermediate in the manufacture of phenol and acetone (Sec. 18.11). The major use of ethylbenzene is for dehydrogenation to styrene ($PhCH=CH_2$), one of the most commercially important aromatic hydrocarbons. Its principal uses are in the manufacture of polystyrene (Sec. 5.7) and styrene–butadiene rubber (Sec. 15.5).

The xylenes (dimethylbenzenes) are obtained by separation from petroleum and by re-forming C_8 petroleum fractions. Of the xylenes, *p*-xylene is the most important commercially. Virtually the entire production of *p*-xylene is used for oxidation to terephthalic acid (Eq. 16.52), an important intermediate in polyester synthesis (for example, Dacron; Sec. 21.12A). (Oxidation of alkylbenzenes is discussed in Sec. 17.5C.)

$$H_3C-\text{(ring)}-CH_3 + O_2 \xrightarrow[\text{heat}]{\text{Co–Mn catalyst}} HO_2C-\text{(ring)}-CO_2H \qquad (16.52)$$

p-xylene   terephthalic acid

KEY IDEAS IN CHAPTER 16

- Benzene derivatives are distinguished spectroscopically by the NMR absorptions of their ring protons, which occur at greater chemical shift than the absorptions of vinylic protons. The unusual chemical shifts of aromatic protons are caused by the ring-current effect. The ^{13}C NMR absorptions of ring carbons are observed in about the same part of the spectrum as the absorptions of the vinylic carbons of alkenes.

- The most characteristic reaction of aromatic compounds is electrophilic aromatic substitution. In this type of reaction, an electrophile is attacked by the π electrons of a benzene ring to form a resonance-stabilized carbocation. Loss of a proton from this ion gives a new aromatic compound.

- Examples of electrophilic aromatic substitution reactions discussed in this chapter are halogenation, used to prepare halobenzenes; nitration, used to prepare nitrobenzene derivatives; sulfonation, used to prepare benzenesulfonic acid derivatives; Friedel–Crafts alkyla-

tion, used to prepare alkylbenzenes; and Friedel–Crafts acylation, used to prepare aryl ketones.

- Electrophilic iodination of tyrosine, which occurs in the biosynthesis of thyroid hormones, is an example of electrophilic substitution in biology.

- Derivatives containing substituted benzene rings can undergo further substitution either at the ortho and para positions or at the meta position, depending on the ring substituent.

- Benzene rings with alkyl substituents or substituent groups that delocalize positive charge by resonance typically undergo substitution at the ortho and para positions; these substituent groups are called ortho, para-directing groups.

- Benzene rings with electronegative substituents that cannot stabilize carbocations or delocalize positive charge by resonance typically undergo substitution at the meta position. These substituents are called meta-directing groups.

• Whether a substituted benzene undergoes substitution more rapidly or more slowly than benzene itself is determined by the balance of resonance and polar effects of the substituents. Monosubstituted benzene rings containing an ortho, para-directing group other than halogen react more rapidly in electrophilic aromatic substitution than benzene itself. Monosubstituted benzene rings containing a halogen substituent or any meta-directing group react more slowly in electrophilic aromatic substitution than benzene itself. The effects of multiple substituents are roughly additive.

• The activating and directing effects of substituent groups must be taken into account when planning a synthesis.

• Alkene double bonds can generally be hydrogenated without affecting the benzene ring. Benzene derivatives, however, can be hydrogenated under relatively harsh conditions to cyclohexane derivatives.

• Polycyclic aromatic hydrocarbons are metabolized in phase I metabolism to epoxides and diol-epoxides, which can, in some cases, alkylate DNA. Benzo[a]pyrene is a potent environmental carcinogen that is metabolized this way.

 REACTION REVIEW *For a summary of reactions discussed in this chapter, see the* Reaction Review *section of Chapter 16 in the* Study Guide and Solutions Manual.

ADDITIONAL PROBLEMS

16.35 Give the products expected (if any) when ethylbenzene reacts under the following conditions.

(a) Br_2 in CCl_4 (dark)   (b) HNO_3, H_2SO_4

(c) concd. H_2SO_4

(d)
$$Et-\overset{\overset{\textstyle O}{\|}}{C}-Cl, AlCl_3 \text{ (1.1 equiv.), then } H_2O$$

(e) CH_3Br, $AlCl_3$ (f) Br_2, $FeBr_3$

16.36 Give the products expected (if any) when nitrobenzene reacts under the following conditions.

(a) Cl_2, $FeCl_3$, heat   (b) fuming HNO_3, H_2SO_4

(c)
$$H_3C-\overset{\overset{\textstyle O}{\|}}{C}-Cl, AlCl_3 \text{ (1.1 equiv.), then } H_2O$$

16.37 Which of the following compounds *cannot* contain a benzene ring? How do you know?

$C_{10}H_{16}$ $C_8H_6Cl_2$ C_5H_4 $C_{10}H_{15}N$

  A   B   C   D

16.38 (a) Arrange the three isomeric dichlorobenzenes in order of increasing dipole moment (smallest first).

(b) Assuming that the dipole moment is the principal factor governing their relative boiling points, arrange the compounds from part (a) in order of increasing boiling point (smallest first). Explain your reasoning.

16.39 Explain how you would distinguish each of the following isomeric compounds from the others using NMR spectroscopy. Be explicit.

A

B

C

16.40 Explain how you would distinguish between ethylbenzene, *p*-xylene, and styrene solely by NMR spectroscopy.

16.41 Explain why

(a) the NMR spectrum of the sodium salt of cyclopentadiene consists of a singlet.

$: Na^+$

sodium salt of cyclopentadiene

(b) the methyl group in the following compound has an unusual chemical shift of δ (−1.67), about 4 ppm lower than the chemical shift of a typical allylic methyl group.

16.42 Show how resonance interaction of the electron pairs on the oxygen with the ring π electrons can account for the fact that the chemical shift of protons H^a in *p*-methoxytoluene is smaller than that of protons H^b in spite of the fact that the oxygen has a greater electronegativity than the methyl carbon.

16.43 Outline laboratory syntheses of each of the following compounds, starting with benzene and any other reagents. (The references to equations will assist you with nomenclature.)

(a) *p*-nitrotoluene

(b) *p*-dibromobenzene

(c) *p*-chloroacetophenone (Eq. 16.22, p. 808)

(d) *m*-nitrobenzenesulfonic acid (Eq. 16.12, p. 804)

(e) *p*-chloronitrobenzene

(f) 1,3,5-trinitrobenzene

(g) 2,6-dibromo-4-nitrotoluene

(h) 2,4-dibromo-6-nitrotoluene

(i) 4-ethyl-3-nitroacetophenone (Eq. 16.22, p. 808)

(j) cyclopentylbenzene

16.44 Arrange the following compounds in order of increasing reactivity toward HNO_3 in H_2SO_4. (The references to equations will assist you with nomenclature.)

(a) mesitylene (Eq. 16.44, p. 824), toluene, 1,2,4-trimethylbenzene

(b) chlorobenzene, benzene, nitrobenzene

(c) *m*-chloroanisole, *p*-chloroanisole, anisole

(d) acetophenone (Eq. 16.22, p. 808), *p*-methoxyacetophenone, *p*-bromoacetophenone

16.45 Indicate whether each of the following compounds should be nitrated more rapidly or more slowly than benzene, and give the structure of the principal mononitration product in each case. Explain your reasoning.

(a)

benzeneboronic acid

(b)

(c)

biphenyl

(d)

16.46 Rank the following compounds in order of increasing reactivity in electrophilic bromination. In each case, indicate whether the principal monobromination products will be the ortho and para isomers or the meta isomer, and whether the compound will be more or less reactive than benzene. Explain carefully the points that cause any uncertainty.

16.47 Nitration of phenyl acetate (compound *A*) results in para substitution of the nitro group. However, nitration of dimethyl phenyl phosphate (compound *B*) results in meta substitution of the nitro group. Suggest a reason that the two compounds nitrate in different positions. (*Hint:* Draw an octet structure for the phosphorus.)

16.48 Give two Friedel–Crafts acylation reactions that could be used to prepare 4-methoxybenzophenone. Which reaction would be faster or occur under milder conditions? Explain.

4-methoxybenzophenone

16.49 Two alcohols, *A* and *B*, have the same molecular formula $C_9H_{10}O$ and react with sulfuric acid to give the same hydrocarbon *C*. Compound *A* is optically active, and compound *B* is not. Catalytic hydrogenation of *C* gives a hydrocarbon *D*, C_9H_{10}, which gives two and only two products when nitrated once with HNO_3 in H_2SO_4. Give the structures of *A*, *B*, *C*, and *D*.

16.50 Give the structures of all the hydrocarbons $C_{10}H_{10}$ that would undergo catalytic hydrogenation to give *p*-diethylbenzene.

16.51 Suggest a reason that the λ_{max} values and intensities of the UV absorptions of styrene (PhCH=CH$_2$) and phenylacetylene (PhC≡CH) are essentially identical even though phenylacetylene contains an additional π bond.

16.52 Diphenylsulfone is a by-product that is formed in the sulfonation of benzene. Give a curved-arrow mechanism for its formation.

diphenylsulfone

16.53 Sulfonation, unlike most other electrophilic aromatic substitution reactions, is reversible. Benzenesulfonic acid (structure in Eq. 16.12, p. 804) can be converted into benzene and H$_2$SO$_4$ with hot water (steam). Write a curved-arrow mechanism for this reaction.

16.54 When the following compound is treated with H$_2$SO$_4$, the product of the resulting reaction has the formula C$_{15}$H$_{20}$ and does not decolorize Br$_2$ in CCl$_4$. Suggest a structure for this product and give a curved-arrow mechanism for its formation.

16.55 Celestolide, a perfuming agent with a musk odor, is prepared by the sequence of reactions given in Fig. P16.55.
  (a) Give the curved-arrow mechanism for the formation of compound A (reaction a).
  (b) In your mechanism, identify the three basic steps of electrophilic aromatic substitution discussed in Sec. 16.4B.
  (c) What product would be obtained in reaction b if this reaction followed the usual directing effects of alkyl substituents? Suggest a reason that celestolide is formed instead.

16.56 When styrene is treated with a sulfonic acid catalyst (RSO$_3$H) in cyclohexane solvent, an alkene X (C$_{16}$H$_{16}$) is formed that is slowly transformed into isomeric compounds Y and Z (Fig. P16.56). Provide a curved-arrow mechanism for the formation of Y and Z, which should include a structure for alkene X.

16.57 The solvolysis reaction of 2-bromooctane in ethanol is relatively slow. However, this reaction is accelerated by the addition of silver ion (as AgNO$_3$), and one of the products is AgBr. Explain how Ag$^+$ accelerates the reaction. (*Hint:* See Eq. 16.15c, p. 806.)

16.58 An optically active compound A (C$_9$H$_{11}$Br) reacts with sodium ethoxide in ethanol to give an optically inactive hydrocarbon B (NMR spectrum in Fig. P16.58). Compound B undergoes hydrogenation over a Pd/C catalyst at room temperature to give a compound C, which has the formula C$_9$H$_{12}$. Give the structures of A, B, and C.

Figure P16.55

Figure P16.56

16.59 Identify each of the following compounds.

    (a) Compound *A*: IR 1605 cm^{-1}, no O—H stretch NMR: δ 3.72 (3*H*, s); δ 6.72 (2*H*, apparent doublet, *J* = 9 Hz); δ 7.15 (2*H*, apparent doublet, *J* = 9 Hz). Mass spectrum in Fig. P16.59a, on p. 834. (*Hint:* Notice the M + 2 peak.)

    (b) Compound *B* (C$_{10}$H$_{12}$O): IR 965, 1175, 1247, 1608, 1640 cm^{-1}; no O—H stretch. NMR in Fig. 16.59b. UV: λ_{max} (ethanol) 260 (ϵ = 18,200); this is a greater wavelength and about the same intensity as the UV spectrum of styrene.

16.60 A method for determining the structures of disubstituted benzene derivatives was proposed in 1874 by Wilhelm Körner of the University of Milan. Körner had in hand three dibromobenzenes, *A*, *B*, and *C*, with melting points of 89, 6.7, and −6.5 °C, respectively. He nitrated each isomer in turn and meticulously isolated *all* of the mononitro derivatives of each. Compound *A* gave one mononitro derivative; compound *B* gave two mononitro derivatives; and compound *C* gave three. These experiments gave him enough information to assign the structures of *o*-, *m*-, and *p*-dibromobenzene.

    (a) Assuming the correctness of the Kekulé structure for benzene, assign the structures of the dibromobenzene derivatives.

    (b) Körner had no way of knowing whether the Kekulé or Ladenburg benzene structure (p. 759) was correct. Assuming the correctness of the Ladenburg benzene structure, assign the structures of the dibromobenzene derivatives.

    (c) It is a testament to Körner's experimental skill that he could isolate all of the mononitration products. Of all the mononitration products that he isolated, which one(s) were formed in smallest amount? Explain.

    (d) Jack Körner, Wilhelm's grandnephew twice removed, has decided to verify great-uncle Wilhelm's results by using ^{13}C NMR to identify the dibromobenzene isomers. What differences can he expect in the ^{13}C NMR spectra of these compounds?

16.61 Give the structures of the principal organic product(s) expected in each of the reactions given in Fig. P16.61 on p. 835, and explain your reasoning.

16.62 Would 1-methoxynaphthalene nitrate more rapidly or more slowly than naphthalene at (a) carbon-4; (b) carbon-5; (c) carbon-6? Explain your reasoning.

1-methoxynaphthalene

16.63 Furan is an aromatic heterocyclic compound that undergoes electrophilic aromatic substitution. By drawing resonance structures for the carbocation intermediates involved, deduce whether furan should undergo Friedel–Crafts acylation more rapidly at carbon-2 or carbon-3.

furan

Figure P16.58 The NMR spectrum for Problem 16.58. The integrals are shown in red over the peaks.

16.64 Given that anisole (methoxybenzene) protonates primarily on oxygen in concentrated H_2SO_4, explain why 1,3,5-trimethoxybenzene protonates primarily on a carbon of the ring. As part of your reasoning, draw the structure of each conjugate acid.

16.65 A Diels–Alder reaction of 2,5-dimethylfuran and maleic anhydride gives a compound A that undergoes acid-catalyzed dehydration to give 3,6-dimethylphthalic anhydride (see Fig. P16.65).

(a) Deduce the structure of compound A.

(b) Give a curved-arrow mechanism for the conversion of A into 3,6-dimethylphthalic anhydride.

16.66 Propose a curved-arrow mechanism for the reaction given in Fig. P16.66. (*Hint:* Modify Step 3 of the usual aromatic substitution mechanism in Sec. 16.4B).

16.67 At 36 °C the NMR resonances for the ring methyl groups of "isopropylmesitylene" (protons H^a and H^b in the following structure) are two singlets at δ 2.25 and δ 2.13 with a 2 : 1 intensity ratio, respectively. When the spectrum is taken at −60 °C, however, it shows three singlets of equal intensity for these groups at δ 2.25, δ 2.17, and δ 2.11. Explain these results.

"isopropylmesitylene"

Figure P16.59 (a) The mass spectrum of compound A in Problem 16.59a. (b) The NMR spectrum of compound B in Problem 16.59b. The integrals are shown in red over the peaks, and the coupling constants are shown in violet.

16.68 Each of the following compounds can be resolved into enantiomers. Explain why each is chiral, and why compound (b) racemizes when it is heated.

(a)

hexahelicene
$[\alpha]_D^{25} = 3700$ degrees mL g^{-1} dm^{-1}

(b)

SO$_3$H

SO$_3$H

(a) benzene (large excess) + ClCH$_2$—〈 〉—CH$_2$Cl $\xrightarrow{\text{AlCl}_3}$

(b) benzene (large excess) + CHCl$_3$ $\xrightarrow{\text{AlCl}_3}$ (a hydrocarbon C$_{19}$H$_{16}$)

(c) 〈 〉—O—CH$_2$CH$_2$CH—Cl $\xrightarrow{\text{AlCl}_3}$ (a compound with 10 carbons)
        |
        CH$_3$

(d) naphthalene + Cl—C—C—C—Cl $\xrightarrow[\text{2) H}_2\text{O}]{\text{1) AlCl}_3}$ (three products, all isomers with formula C$_{15}$H$_{12}$O$_2$)
    naphthalene   H$_3$C CH$_3$
    ***α,α*-dimethylmalonyl dichloride**

(e) HO—〈 〉—CH$_3$ + I—OH $\longrightarrow$ (*Hint*: See Sec. 16.5C.)
        (large excess)

(f) 〈 〉 + HNO$_3$ $\xrightarrow[0\,°C]{\text{H}_2\text{SO}_4}$

(g) ferrocene (p. 769) + H$_3$C—C—Cl $\xrightarrow[\text{2) H}_2\text{O}]{\text{1) AlCl}_3}$ (C$_{12}$H$_{12}$OFe)

(h) CH$_3$O—〈 〉—SO$_3$H $\xrightarrow{\text{HNO}_3}$ $\xrightarrow{\text{Br}_2,\text{Fe}}$ (a compound containing one nitro group and one bromine)

Figure P16.61

H$_3$C—〈O〉—CH$_3$ + maleic anhydride $\longrightarrow$ A $\xrightarrow{\text{H}_2\text{SO}_4}$ **3,6-dimethylphthalic anhydride** + H$_2$O

2,5-dimethylfuran

maleic anhydride

Figure P16.65

(CH$_3$)$_3$C—〈 〉—C(CH$_3$)$_3$ + HNO$_3$ $\longrightarrow$ (CH$_3$)$_3$C—〈 〉—NO$_2$ + (CH$_3$)$_2$C=CH$_2$

Figure P16.66

Allylic and Benzylic Reactivity

An **allylic group** is a group on a carbon adjacent to a double bond. A **benzylic group** is a group on a carbon adjacent to a benzene ring or substituted benzene ring.

allylic chlorine

allylic carbon

$H_2C=CH-\overset{\displaystyle \overset{Cl}{|}}{CH}-CH_3$

allylic hydrogen

benzylic hydroxy group

benzylic carbon

$\overset{\displaystyle OH}{CH}-CH_3$

benzylic hydrogen

In many situations *allylic and benzylic groups are unusually reactive.* This chapter examines what happens when some familiar reactions occur at allylic and benzylic positions and discusses the reasons for allylic and benzylic reactivity. This chapter also presents some reactions that occur *only* at the allylic and benzylic positions. Finally, Secs. 17.5B and 17.6 will show that allylic reactivity is also important in some chemistry that occurs in living organisms.

PROBLEMS

17.1   Identify the allylic carbons in each of the following structures.

(a)   (b)

17.2 Identify the benzylic carbons in each of the following structures.

(a) ![structure a] (b) ![structure b]

17.1 REACTIONS INVOLVING ALLYLIC AND BENZYLIC CARBOCATIONS

Recall that allylic carbocations are resonance-stabilized (Sec. 15.4B). The simplest example of an allylic cation is the *allyl cation* itself:

$$\left[H_2C \overset{\frown}{=} CH \overset{+}{-} \overset{+}{C}H_2 \quad \longleftrightarrow \quad H_2\overset{+}{C} - CH = CH_2 \right] \tag{17.1}$$

resonance structures of the allyl cation

These resonance structures symbolize the delocalization of electrons and electron deficiency (along with the associated positive charge) that result from the overlap of $2p$ orbitals to form π molecular orbitals, as shown in Fig. 15.14 on p. 744.

Benzylic carbocations are also resonance-stabilized. The *benzyl cation* is the simplest example of a benzylic cation:

$$\left[\text{resonance structures} \right] \tag{17.2}$$

resonance structures of the benzyl cation

The resonance structures of the benzyl cation symbolize the overlap of $2p$ orbitals of the benzylic carbon and the benzene ring to form bonding MOs. As these structures show, the electron deficiency and resulting positive charge on a benzylic carbocation are shared not only by the benzylic carbon, but also by alternate carbons of the ring. As with the allyl cation (Sec. 15.4B), the resonance structures of the benzyl cation correctly account for the distribution of positive charge that is calculated from MO theory and shown by the EPM:

EPM of the allyl cation

EPM of benzyl cation

✳ = sites of partial positive charge

The resonance energy (p. 744) of the benzyl cation calculated from MO theory (over and above that of the benzene ring itself) is about 0.72β, which is about 90% of the resonance energy of the allyl cation. In other words, the resonance stabilizations of the benzyl and allyl cations are about the same.

The structures and stabilities of allylic and benzylic carbocations have important consequences for reactions in which they are involved as reactive intermediates. First, *reactions in which benzylic or allylic carbocations are formed as intermediates are generally considerably faster than analogous reactions involving comparably substituted nonallylic or*

TABLE 17.1 Comparison of S_N1 Solvolysis Rates of Allylic and Nonallylic Alkyl Halides

$$R—Cl + C_2H_5OH + H_2O \xrightarrow[\text{50\% aqueous ethanol}]{\text{44.6 °C}} R—OC_2H_5 + R—OH + HCl$$

Alkyl chloride R—Cl	Relative rate
H₂C=CH—C(CH₃)₂—Cl (with CH₃ groups)	162
(H₃C)₂C=CH—CH₂—Cl	38
CH₃CH₂—C(CH₃)₂—Cl	(1.00)
(H₃C)₂CH—CH₂—CH₂—Cl	<0.00002

TABLE 17.2 Comparison of S_N1 Solvolysis Rates of Benzylic and Nonbenzylic Alkyl Halides

$$R—Cl + H_2O \xrightarrow[\text{90\% aqueous acetone}]{\text{25 °C}} R—OH + HCl$$

Alkyl chloride R—Cl	Common name	Relative rate
(CH₃)₃C—Cl	*tert*-butyl chloride	(1.0)
Ph—CH(CH₃)—Cl	α-phenethyl chloride	1.0
Ph—C(CH₃)₂—Cl	*tert*-cumyl chloride	620
Ph₂CH—Cl	benzhydryl chloride	200*
Ph₃C—Cl	trityl chloride	>600,000

* In 80% aqueous ethanol.

nonbenzylic carbocations. This point is illustrated by the relative rates of S_N1 solvolysis reactions, shown in Tables 17.1 and 17.2. For example, the tertiary allylic alkyl halide in the first entry of Table 17.1 reacts more than 100 times faster than the tertiary nonallylic alkyl halide in the third entry. A comparison of the first and third entries of Table 17.2 shows the effect of benzylic substitution. *Tert*-cumyl chloride, the third entry, reacts more than 600 times faster than *tert*-butyl chloride, the first entry.

The greater reactivities of allylic and benzylic halides result from the stabilities of the carbocation intermediates that are formed when they react. For example, *tert*-cumyl chloride (the third entry of Table 17.2) ionizes to a carbocation with four important resonance structures:

$$(17.3)$$

resonance-stabilized carbocation

Ionization of *tert*-butyl chloride, on the other hand, gives the *tert*-butyl cation, a carbocation with only one important contributing structure.

$$(17.4)$$

tert-butyl chloride

The benzylic cation is more stable relative to its alkyl halide starting material than is the *tert*-butyl cation, and application of Hammond's postulate predicts that the more stable carbocation should be formed more rapidly. A similar analysis explains the reactivity of allylic alkyl halides.

Because of the possibility of resonance, ortho and para substituent groups on the benzene ring that activate electrophilic aromatic substitution further accelerate S_N1 reactions at the benzylic position:

$$(17.5)$$

relative solvolysis rates   1   3400
(90% aqueous acetone, 25 °C)

The carbocation derived from the ionization of the *p*-methoxy derivative in Eq. 17.5 not only has the same types of resonance structures as the unsubstituted compound, shown in Eq. 17.3, but also an additional structure (*red*) that results from the delocalization of an oxygen lone pair (*red arrow*) and the sharing of the positive charge by the substituent oxygen.

the oxygen electrons are delocalized

$$(17.6)$$

the positive charge is shared by oxygen;
all atoms have octets

FIGURE 17.1 The delocalization of oxygen electrons in Eq. 17.6 results from the overlap of the oxygen 2*p* orbital containing a lone pair with the carbon 2*p* orbitals of the carbocation. (Compare with Fig. 16.7, p. 819.)

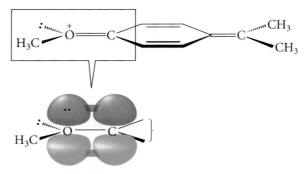

overlap of oxygen and carbon 2*p* orbitals

This delocalization is the result of the overlap of a lone pair in a 2*p* orbital of the *sp²*-hybridized oxygen with the 2*p* orbitals of the ring (Fig. 17.1). This overlap is identical to the overlap described in Fig. 16.7 on p. 819.

Other reactions that involve carbocation intermediates are accelerated when the carbocations are allylic or benzylic. Thus, the dehydration of an alcohol (Sec. 10.2) and the reaction of an alcohol with a hydrogen halide (Sec. 10.3) are also faster when the alcohol is allylic or benzylic. For example, most alcohols require forcing conditions or Lewis acid catalysts to react with HCl to give alkyl chlorides, but such conditions are unnecessary when benzylic alcohols react with HCl. The addition of hydrogen halides to conjugated dienes also reflects the stability of allylic carbocations. Recall that protonation of a conjugated diene gives the allylic carbocation rather than its nonallylic isomer because the allylic carbocation is formed more rapidly (Sec. 15.4A).

A second consequence of the involvement of allylic carbocations as reactive intermediates is that in many cases *more than one product can be formed*. More than one product is possible because the positive charge (and electron deficiency) is shared between two carbons. Nucleophiles can react at either of the electron-deficient carbon atoms and, if the two carbons are not equivalent, two different products result.

$$(CH_3)_2C{=}CH{-}CH_2{-}Cl \longrightarrow \left[(CH_3)_2C{=}CH{-}\overset{+}{C}H_2 \longleftrightarrow (CH_3)_2\overset{+}{C}{-}CH{=}CH_2\right] + Cl^-$$

$$:\overset{..}{O}H_2$$

$$(CH_3)_2C{=}CH{-}CH_2 + (CH_3)_2C{-}CH{=}CH_2$$

$$:\overset{+}{O}{-}H \qquad :\overset{+}{O}{-}H$$

$$H \qquad H$$

$$\Big\downarrow H_2\overset{..}{O} \qquad\qquad \Big\downarrow H_2\overset{..}{O}$$

$$H_3\overset{..}{O}{}^+ + (CH_3)_2C{=}CH{-}CH_2 \qquad (CH_3)_2C{-}CH{=}CH_2$$

$$:\overset{..}{O}H \qquad\qquad :\overset{..}{O}H$$

(15% of product)   (85% of product)   (17.7)

The two products are derived from *one* allylic carbocation that has two resonance forms. Recall that similar reasoning explains why a mixture of products (1,2- and 1,4-addition products) is obtained in the reactions of hydrogen halides with conjugated alkenes (Sec. 15.4A; Eq. 15.41, p. 756).

We might expect that several substitution products in the S_N1 reactions of benzylic alkyl halides might be formed for the same reason.

$$\text{(17.8)}$$

As Eq. 17.8 shows, the products derived from the reactions of water at the ring carbons are not formed. The reason is that *these products are not aromatic* and thus lack the stability associated with the aromatic ring. Aromaticity is such an important stabilizing factor that only the aromatic product (*red*) is formed.

PROBLEMS

17.3 Predict the order of relative reactivities of the compounds within each series in S_N1 solvolysis reactions, and explain your answers carefully.

17.4 Give the structure of an isomer of the allylic halide reactant in Eq. 17.7 that would react with water in an S_N1 solvolysis reaction to give the same two products. Explain your reasoning.

17.5 Why is trityl chloride much more reactive than the other alkyl halides in Table 17.2?

17.2   REACTIONS INVOLVING ALLYLIC AND BENZYLIC RADICALS

An **allylic radical** has an unpaired electron at an allylic position. Allylic radicals are resonance-stabilized and are more stable than comparably substituted nonallylic radicals. The simplest allylic radical is the *allyl radical* itself:

$$\left[H_2C{=}CH{-}\dot{C}H_2 \quad \longleftrightarrow \quad H_2\dot{C}{-}CH{=}CH_2 \right] \qquad \text{(17.9)}$$

resonance structures of the allyl radical

FIGURE 17.2 Use of bond dissociation energies to determine the stabilization of an allylic radical. The stabilization of the allylic radical results from the lower energy required (50 kJ mol⁻¹, 12 kcal mol⁻¹) to remove an allylic hydrogen (*red*) compared with a nonallylic one (*blue*).

Similarly, a **benzylic radical**, which has an unpaired electron at a benzylic position, is also resonance-stabilized. The *benzyl radical* is the prototype:

$$\left[\quad \longleftrightarrow \quad \longleftrightarrow \quad \longleftrightarrow \quad \right] \quad (17.10)$$

resonance structures of the benzyl radical

These resonance structures symbolize the delocalization (sharing) of the unpaired electron that results from the overlap of carbon $2p$ orbitals to form molecular orbitals.

The enhanced stabilities of allylic and benzylic radicals can be experimentally demonstrated with bond dissociation energies. The bond dissociation energies of allylic and nonallylic hydrogens in 2-pentene are compared in Fig. 17.2. One set of methyl hydrogens is allylic and the other is not. It takes 50 kJ mol⁻¹ (12 kcal mol⁻¹) less energy to remove the allylic hydrogen H^a than the nonallylic one H^b. As Fig. 17.2 shows, the difference in bond dissociation energies is a direct measure of the relative energies of the two radicals. This comparison shows that the allylic radical is stabilized by 50 kJ mol⁻¹ (12 kcal mol⁻¹) relative to the nonallylic radical. The same type of comparison shows that benzylic radicals are about 42 kJ mol⁻¹ (10 kcal mol⁻¹) more stable than comparably substituted nonbenzylic radicals.

Because allylic and benzylic radicals are especially stable, they are more readily formed as reactive intermediates than ordinary alkyl radicals. Consider what happens, for example, in the bromination of cumene:

$$(17.11)$$

cumene   ***tert*-cumyl bromide**
                              (nearly quantitative)

This is a free-radical chain reaction (Secs. 5.6C and 9.10A). Notice that *only the benzylic hydrogen is substituted.*

The initiation step in this reaction is the dissociation of molecular bromine into bromine atoms; this reaction is promoted by heat or light.

$$:\!\ddot{Br}\!-\!\ddot{Br}\!: \longrightarrow 2\,:\!\ddot{Br}\cdot \qquad (17.12a)$$

In the first propagation step, a bromine atom abstracts the one benzylic hydrogen in preference to either the six nonbenzylic methyl hydrogens or the five hydrogens of the aromatic ring. It is in this propagation step that the selectivity for substitution of the benzylic hydrogen occurs.

$$(17.12b)$$

The reason for this selectivity is the relative weakness of the benzylic C—H bond that is broken (Fig. 17.1) or, equivalently, the greater stability of the benzylic radical that is formed.

In the second propagation step, the benzylic radical reacts with another molecule of bromine to generate a molecule of product as well as another bromine atom, which can react again in Eq. 17.12b.

$$(17.12c)$$

Free-radical halogenation is used to halogenate alkanes industrially (Sec. 9.10A). Because free-radical halogenation of alkanes with different types of hydrogens gives mixtures of products, this reaction is ordinarily not very useful in the laboratory. (It can be used industrially because industry has developed efficient fractional distillation methods that can separate liquids of similar boiling points.) However, when a benzylic hydrogen is present, it undergoes substitution so much more rapidly than an ordinary hydrogen that a single product is obtained. Consequently, free-radical halogenation can be used for the laboratory preparation of benzylic halides.

Because the allylic radical is also relatively stable, a similar substitution occurs preferentially at the allylic positions of an alkene. But a competing reaction occurs in the case of an alkene that is not observed with benzylic substitution: addition of halogen to the alkene double bond by an ionic mechanism (Sec. 5.2A).

$$(17.13)$$

(Such a competing addition is not a problem in benzylic bromination because bromine doesn't add to the benzene ring in Eq. 17.11. Why?)

One reaction can be promoted over the other if the reaction conditions are chosen carefully. *Addition* of bromine is the predominant reaction if (1) free-radical substitution is suppressed by avoiding conditions that promote free-radical reactions (heat, light, or free-radical initiators); and if (2) the reaction is carried out in solvents of even slight polarity that promote the ionic mechanism for bromine addition. Thus, addition is observed at 25 °C if the reaction is run in the dark in dichloromethane, CH_2Cl_2. On the other hand, free-radical *substitution* occurs when the reaction is promoted by heat, light, or free-radical initiators, an apolar solvent such as CCl_4 is used, and *the bromine is added slowly so that its concentration remains very low.* To summarize:

Addition:

$$\text{(cyclohexene)} + Br_2 \xrightarrow[\substack{25\ °C \\ CH_2Cl_2}]{\text{dark}} \text{(trans-1,2-dibromocyclohexane)}$$ (17.14a)

Substitution:

$$\text{(cyclohexene)} + Br_2 \xrightarrow[CCl_4]{\text{heat or light}} \text{(3-bromocyclohexene)} + HBr$$ (17.14b)

added slowly; concentration kept low

FURTHER EXPLORATION 17.1
Addition versus Substitution with Bromine

The effect of bromine concentration results from the rate laws for the competing reactions. Addition has a higher kinetic order in $[Br_2]$ than substitution. Hence, the rate of addition is decreased more than the rate of substitution by lowering the bromine concentration. This effect is discussed in Further Exploration 17.1.

Adding bromine to a reaction so slowly that it remains at very low concentration is experimentally inconvenient, but a very useful reagent can be employed to accomplish the same objective: *N*-bromosuccinimide (abbreviated NBS). When a compound with allylic hydrogens is treated with *N*-bromosuccinimide in CCl_4 under free-radical conditions (heat or light and peroxides), allylic bromination takes place, and addition to the double bond is not observed.

$$\text{cyclohexene}{-}H + \text{NBS (}N{-}Br\text{)} \xrightarrow[CCl_4]{\substack{\text{heat or light} \\ \text{peroxides}}} \text{3-bromocyclohexene}{-}Br + \text{succinimide (}N{-}H\text{)}$$ (17.15)

cyclohexene   **N-bromosuccinimide (NBS)**   **3-bromocyclohexene** (82–87% yield)   **succinimide**

As we might expect, *N*-bromosuccinimide can also be used for benzylic bromination.

$$\text{Ph}{-}CH_2{-}CH_3 + \text{NBS (}N{-}Br\text{)} \xrightarrow[CCl_4]{\text{peroxides}} \text{Ph}{-}\underset{\underset{Br}{|}}{CH}{-}CH_3 + \text{succinimide (}N{-}H\text{)}$$ (17.16)

(80% yield)

The initiation step in allylic and benzylic bromination with NBS is the formation of a bromine atom by homolytic cleavage of the N—Br bond in NBS itself. The ensuing substitution reaction has three propagation steps, which we'll illustrate for allylic bromination. First, the bromine atom abstracts an allylic hydrogen from the alkene molecule:

$$\text{Br}\cdot \quad \text{H} \quad \underset{|}{\overset{|}{\text{C}}}-\text{CH}=\text{CH}_2 \longrightarrow \text{H}-\text{Br} + \cdot\underset{|}{\overset{|}{\text{C}}}-\text{CH}=\text{CH}_2 \qquad (17.17a)$$

allylic H   alkene   allylic radical

The HBr thus formed reacts with the NBS by an ionic mechanism to produce a Br_2 molecule.

$$\text{HBr} + \text{NBS} \longrightarrow \text{N}-\text{H} + \text{Br}_2 \qquad (17.17b)$$

NBS

The last propagation step is the reaction of this bromine molecule with the radical formed in Eq. 17.17a. A new bromine atom is produced that can begin the cycle anew.

$$\text{Br}-\text{Br} \quad \cdot\underset{|}{\overset{|}{\text{C}}}-\text{CH}=\text{CH}_2 \longrightarrow \text{Br}\cdot + \text{Br}-\underset{|}{\overset{|}{\text{C}}}-\text{CH}=\text{CH}_2 \qquad (17.17c)$$

The first and last propagation steps are identical to those for free-radical substitution with Br_2 itself (Eqs. 17.12b,c). The unique role of NBS is to maintain the very low concentration of bromine by reacting with HBr in Eq. 17.17b. The Br_2 concentration remains low because it can be generated no faster than an HBr molecule and an allylic radical are generated in Eq. 17.17a. Thus, every time a bromine molecule is formed, an allylic radical is also formed with which the bromine can react.

The low solubility of NBS in CCl_4 ($\leq 0.005\ M$) is crucial to the success of allylic bromination with NBS. When solvents that dissolve NBS are used, different reactions are observed. Hence, CCl_4 *must* be used as the solvent in allylic or benzylic bromination with NBS. During the reaction, the insoluble NBS, which is more dense than CCl_4, is consumed from the bottom of the flask and is replaced by the less dense by-product succinimide (Eq. 17.15), which forms a layer on the surface of the CCl_4. Equation 17.17b, and possibly other steps of the mechanism, occur at the surface of the insoluble NBS. (These very specific aspects of the NBS allylic bromination reaction were known many years before the reasons for them were understood.)

Mixtures of products are formed in some allylic bromination reactions because, as resonance structures indicate, the unpaired electron in the free-radical intermediate is shared by two different carbons. This point is explored in Study Problem 17.1.

STUDY PROBLEM   17.1

What products are expected in the reaction of $H_2C\!=\!CHCH_2CH_2CH_2CH_3$ (1-hexene) with NBS in CCl_4 in the presence of peroxides? Explain your answer.

SOLUTION Work through the NBS mechanism with 1-hexene. In the step corresponding to Eq. 17.17a, the following resonance-stabilized allylic free radical is formed as an intermediate:

$$\left[H_2C\!=\!CH-\dot{C}H-CH_2CH_2CH_3 \quad \longleftrightarrow \quad H_2\dot{C}-CH\!=\!CH-CH_2CH_2CH_3 \right]$$

A $\qquad\qquad\qquad\qquad\qquad\qquad$ B

Because the unpaired electron is shared by *two different carbons*, this radical can react in the final propagation step to give *two different products*. Reaction of Br_2 at the radical site shown in structure *A* gives product (1), and reaction at the radical site shown in structure *B* gives product (2):

$$H_2C{=}CH{-}\underset{\underset{Br}{|}}{CH}{-}CH_2CH_2CH_3 \qquad H_2\underset{\underset{Br}{|}}{C}{-}CH{=}CH{-}CH_2CH_2CH_3$$

$$\text{(1)} \qquad\qquad\qquad\qquad \text{(2)}$$

Notice that product (1) is chiral, and product (2) can exist as both cis and trans stereoisomers. Hence, bromination of 1-hexene gives racemic (1) as well as *cis*- and *trans*-(2), although the trans isomer should predominate because of its greater stability.

PROBLEM

17.6 What product(s) are expected when each of the following compounds reacts with one equivalent of NBS in CCl_4 in the presence of light and peroxides? Explain your answers.

(a) cyclohexene

(b) 3,3-dimethylcyclohexene

(c) *trans*-2-pentene

(d) 4-*tert*-butyltoluene

(e) 1-isopropyl-4-nitrobenzene

17.3 REACTIONS INVOLVING ALLYLIC AND BENZYLIC ANIONS

The prototype for allylic anions is the *allyl anion*, and the simplest benzylic anion is the *benzyl anion*.

$$\left[H_2C{=}CH{-}\ddot{C}H_2 \longleftrightarrow H_2\ddot{C}{-}CH{=}CH_2 \right] \qquad (17.18)$$

allyl anion

benzyl anion

$$(17.19)$$

Allylic and benzylic anions are about 59 kJ mol^{-1} (14 kcal mol^{-1}) more stable than their nonallylic and nonbenzylic counterparts. There are two reasons for the stabilities of these anions. The first is resonance stabilization, as indicated by the preceding resonance structures. The second reason is the *polar effect* (Sec. 3.6C) of the double bond (in the allyl anion) or the phenyl ring (in the benzyl anion). The polar effect of both groups stabilizes anions. (Opinions differ about the relative importance of resonance and polar effects.) The reason for the polar effect of a double bond is the electronegativity of an sp^2-hybridized carbon, discussed in Secs. 4.4 and 14.7A. The polar effect of a phenyl ring has a similar explanation.

The enhanced stability of allylic and benzylic anions is reflected in the pK_a values of propene and toluene ($B{:}^- = $ a base):

$$H_2C{=}CH{-}CH_2{-}H + B{:}^- \rightleftarrows H_2C{=}CH{-}\overset{..}{\overset{..}{C}}H_2 + B{-}H \qquad (17.20)$$

propene
$pK_a \approx 43$

$+ B{:}^- \rightleftarrows$ $+ B{-}H \qquad (17.21)$

toluene
$pK_a \approx 41$

Although these compounds are very weak acids, their acidities are much greater than the acidities of alkanes that do not contain allylic or benzylic hydrogens. Recall from Sec. 14.7A that ordinary alkanes have pK_a values of about 55.

Free benzylic or allylic carbanions are rarely involved as reactive intermediates. However, a number of reactions involve species that have *carbanion character*. Two of these are the reactions of Grignard and related organometallic reagents, and E2 eliminations. The following sections show how these reactions are affected when carbanion character occurs at benzylic or allylic positions.

A. Allylic Grignard Reagents

Recall that Grignard reagents have many of the properties expected of *carbanions* (Sec. 9.8C). Thus, allylic Grignard reagents resemble allylic carbanions.

$$H_2C{=}CH{-}CH_2{-}MgBr \quad \text{resembles} \quad H_2C{=}CH{-}\overset{..}{\overset{..}{C}}H_2 \ \overset{+}{MgBr} \quad (17.22)$$

Allylic Grignard reagents undergo a rapid equilibration in which the —MgBr group moves back and forth between the two partially negative carbons at a rate of about 1000 times per second.

(The right side of the equilibrium is favored because the double bond has more alkyl substituents; Sec. 4.5B.) The transition state for this reaction can be envisioned as an ion pair consisting of an allylic carbanion and a ^+MgBr cation.

Because the allylic carbanion is resonance-stabilized, this transition state has relatively low energy, and consequently, the equilibration occurs rapidly.

The equilibration in Eq. 17.23 is an example of an *allylic rearrangement*. An **allylic rearrangement** involves the simultaneous movement of a group G and a double bond so that one allylic isomer is converted into another.

G = any group $\qquad (17.24)$

allylic rearrangement

Notice that these two structures are *not* resonance structures; they are two *distinct isomeric* species in rapid equilibrium.

The rapid allylic rearrangement of an unsymmetrical Grignard reagent, such as the one shown in Eq. 17.23, means that the reagent is actually a mixture of two different species. This has two consequences. First, the same mixture of species is obtained from either of two allylically related alkyl halides:

$$H_3C-\overset{\overset{\displaystyle CH_3}{|}}{C}=CH-CH_2-Br \qquad\qquad H_3C-\overset{\overset{\displaystyle CH_3}{|}}{\underset{\underset{\displaystyle Br}{|}}{C}}-CH=CH_2$$

$$\downarrow Mg \qquad\qquad\qquad\qquad\qquad\qquad \downarrow Mg$$

$$H_3C-\overset{\overset{\displaystyle CH_3}{|}}{C}=CH-CH_2MgBr \quad\overset{fast}{\underset{}{\rightleftarrows}}\quad H_3C-\overset{\overset{\displaystyle CH_3}{|}}{\underset{\underset{\displaystyle MgBr}{|}}{C}}-CH=CH_2 \qquad (17.25)$$

Second, when the Grignard reagents undergo a subsequent reaction, a mixture of products is usually obtained, and the same mixture of products is obtained regardless of the alkyl halide used to form the Grignard reagent. For example, protonolysis of the mixture of equilibrating Grignard reagents in Eq. 17.25 gives the following result:

$$H_3C-\overset{\overset{\displaystyle CH_3}{|}}{C}=CH-CH_2MgBr \quad\overset{fast}{\underset{}{\rightleftarrows}}\quad H_3C-\overset{\overset{\displaystyle CH_3}{|}}{\underset{\underset{\displaystyle MgBr}{|}}{C}}-CH=CH_2$$

$$\downarrow H_2O$$

$$H_3C-\overset{\overset{\displaystyle CH_3}{|}}{C}=CH-CH_3 \;+\; H_3C-\overset{\overset{\displaystyle CH_3}{|}}{\underset{\underset{\displaystyle H}{|}}{C}}-CH=CH_2 \;+\; HO-MgBr \qquad (17.26)$$

mixture of products

PROBLEM

17.7   What product(s) are formed when a Grignard reagent prepared from each of the following alkyl halides is treated with D_2O?

(a)   [structure: cyclohexene with CH_2Br substituent]

1-(bromomethyl)cyclohexene

(b)   [structure: cyclohexene with CH_3 substituent and Br]

6-bromo-1-methylcyclohexene

B. E2 Eliminations Involving Allylic or Benzylic Hydrogens

Recall that the S_N2 (bimolecular substitution) and E2 (bimolecular elimination) reactions of alkyl halides are *competing reactions*, and that the structure of the alkyl halide is one of the major factors that determine which reaction is the dominant one (Sec. 9.5G). A structural effect in the alkyl halide that tends to promote a greater fraction of elimination is *enhanced acidity of the β-hydrogens.* It is found that a greater ratio of elimination to substitution is observed when the β-hydrogens of the alkyl halide have higher than normal acidity. Such a situation can occur when the β-hydrogens are allylic or benzylic. (Recall from the introduction to this section that allylic and benzylic hydrogens are more acidic than ordinary alkyl hydrogens.) For example, the E2 reaction of the alkyl bromide in Eq. 17.27 is more than 100 times faster than the E2 reaction of isopentyl bromide [$(CH_3)_2CHCH_2CH_2Br$], a comparably branched alkyl halide.

$$\text{(95\% elimination)} \qquad \text{(5\% substitution)} \qquad (17.27)$$

(Elimination predominates because the E2 reaction is particularly fast; the S_N2 component of the competition occurs at a normal rate.)

Why should an acidic β-hydrogen increase the rate of an E2 reaction? In the transition state of the E2 reaction, the base is removing a β-proton, and the transition state of the reaction has *carbanion character* at the β-carbon atom.

transition state for the E2 reaction

This partially formed carbanion is stabilized in the same way that a fully formed carbanion is; a more stable transition state results in a faster reaction. Another reason that benzylic E2 reactions are faster is that the alkene double bond, which is partially formed in the transition state, is conjugated with the benzene ring; recall that conjugated double bonds are more stable than unconjugated double bonds (Sec. 15.1A).

Let's summarize the structural characteristics of alkyl halides or sulfonate esters that favor E2 reactions over S_N2 reactions. Elimination reactions are favored by:

1. branching at the α-carbon (Sec. 9.5G)
2. branching at the β-carbon (Sec. 9.5G)
3. greater acidity of the β-hydrogens (this section)

PROBLEM

17.8   Predict the major product that is obtained when each of the following alkyl halides is treated with potassium *tert*-butoxide. Explain your reasoning.

17.4 ALLYLIC AND BENZYLIC S_N2 REACTIONS

S_N2 reactions of allylic and benzylic halides are relatively fast, even though they do not involve reactive intermediates. The following data for allyl chloride are typical:

$$H_2C=CH-CH_2-Cl + I^- \xrightarrow[\text{acetone}]{50\,°C} H_2C=CH-CH_2-I + Cl^- \qquad \begin{array}{c}\text{relative rate}\\73\end{array} \qquad (17.28a)$$

$$H_3C-CH_2-CH_2-Cl + I^- \xrightarrow[\text{acetone}]{50\,°C} H_3C-CH_2-CH_2-I + Cl^- \qquad 1 \qquad (17.28b)$$

An even greater acceleration is observed for benzylic halides.

$$\text{Ph}-CH_2-Cl + I^- \xrightarrow[\text{acetone}]{60\,°C} \text{Ph}-CH_2-I + Cl^- \qquad \begin{array}{c}\text{relative rate}\\\sim100{,}000\end{array} \qquad (17.29a)$$

$$\begin{array}{c}H_3C\\ \diagdown\\CH-CH_2-Cl\\ \diagup\\H_3C\end{array} + I^- \xrightarrow[\text{acetone}]{60\,°C} \begin{array}{c}H_3C\\ \diagdown\\CH-CH_2-I\\ \diagup\\H_3C\end{array} + Cl^- \qquad 1 \qquad (17.29b)$$

Allylic and benzylic S_N2 reactions are accelerated because the energies of their transition states are reduced by $2p$-orbital overlap, shown in Fig. 17.3, for an allylic S_N2 reaction. In the transition state of the S_N2 reaction, the carbon at which substitution occurs is sp^2-hybridized (Fig. 9.2, p. 395); the incoming nucleophile and the departing leaving group are partially bonded to a $2p$ orbital on this carbon. Overlap of this $2p$ orbital with the $2p$ orbitals of an adjacent double bond or phenyl ring provides additional bonding that lowers the energy of the transition state and accelerates the reaction.

PROBLEM

17.9 Explain how and why the product(s) would differ in the following reactions of *trans*-2-buten-1-ol.

(1) Reaction with concentrated aqueous HBr

(2) Conversion into the tosylate, then reaction with NaBr in acetone

FIGURE 17.3 Transition states for S_N2 reactions at (a) an allylic carbon and (b) a nonallylic carbon. Nuc and X are the nucleophile and leaving group, respectively. The allylic substitution is faster because the transition state is stabilized by overlap (*blue lines*) of the $2p$ orbital at the site of substitution with the adjacent π bond.

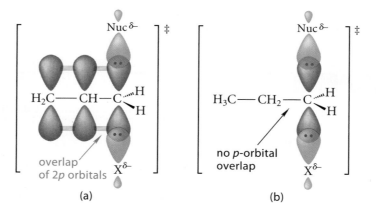

(a)   (b)

ALLYLIC AND BENZYLIC OXIDATION

A. Oxidation of Allylic and Benzylic Alcohols with Manganese Dioxide

Allylic and benzylic alcohols are oxidized selectively by a suspension of activated manganese(IV) dioxide, MnO_2. Primary allylic alcohols are oxidized to aldehydes and secondary allylic alcohols are oxidized to ketones.

$$CH_3O-\underset{}{\bigcirc}-CH_2OH + \underset{\text{(insoluble)}}{MnO_2} \xrightarrow[\text{(solvent)}]{CH_2Cl_2} CH_3O-\underset{}{\bigcirc}-CH{=}O + Mn(OH)_2 \quad (17.30)$$

(4-methoxyphenyl)methanol　　　　　　　　　　**4-methoxybenzaldehyde**
(*p*-methoxybenzyl alcohol)　　　　　　　　　　　(81% yield)

$$\underset{CH_3CH_2}{\overset{H}{\diagdown}}C{=}C\underset{H}{\overset{CH_2OH}{\diagup}} + \underset{\text{(insoluble)}}{MnO_2} \xrightarrow[\text{(solvent)}]{CH_2Cl_2} \underset{CH_3CH_2}{\overset{H}{\diagdown}}C{=}C\underset{H}{\overset{CH{=}O}{\diagup}} + Mn(OH)_2 \quad (17.31)$$

(*E*)-2-penten-1-ol　　　　　　　　　　　**(*E*)-2-pentenal**
　　　　　　　　　　　　　　　　　　　(83% yield)

"Activated MnO_2" is obtained by the oxidation–reduction reaction of potassium permanganate ($KMnO_4$) with an Mn^{2+} salt such as $MnSO_4$ under either alkaline or acidic conditions followed by thorough drying.

$$2\,H_2O + 3\,MnSO_4 + 2\,KMnO_4 \xrightarrow{H_2O} 5\,MnO_2 + 2\,H_2SO_4 + K_2SO_4 \quad (17.32)$$

**manganese
dioxide**

Allylic and benzylic oxidation of alcohols takes place on the surface of the MnO_2, which is insoluble in the solvents used for the reaction. Water competes with alcohol for the sites on the MnO_2 and thus must be removed by drying to produce an active oxidant.

The selectivity of MnO_2 oxidation for allylic and benzylic alcohols is illustrated by the following example, in which the ordinary (nonbenzylic) alcohol is unaffected.

the benzylic alcohol
is selectively oxidized

$$CH_3O-\underset{CH_3O}{\bigcirc}-\underset{OH}{\underset{|}{CH}}CH_2CH_2OH \xrightarrow[\text{acetone}]{MnO_2}{\text{(solvent)}} CH_3O-\underset{CH_3O}{\bigcirc}-\overset{O}{\overset{\|}{C}}-CH_2CH_2OH \quad (17.33)$$

3-(3,4-dimethoxyphenyl)-1,3-propanediol　　**1-(3,4-dimethoxyphenyl)-3-hydroxy-1-propanone**
　　　　　　　　　　　　　　　　　　　　　　(94% yield)

(This example illustrates the selectivity for benzylic alcohols; a corresponding selectivity is observed for allylic alcohols.)

The reaction is selective because allylic and benzylic alcohols react much more rapidly than "ordinary" alcohols. An understanding of this selectivity comes from the mechanism. In the first step of the mechanism, the O—H group of the alcohol rapidly adds to MnO_2 to give an ester (Sec. 10.4C).

$$R-CH_2-OH + Mn=O \rightleftharpoons R-CH_2-O-Mn-OH \qquad (17.34a)$$

(in solution) (on the MnO$_2$ surface)

(The solid-state structures of the Mn-containing species are simplified in these equations.) In the next step, which is rate-limiting, Mn(IV) accepts an electron to become Mn(III), and, at the same time, a hydrogen atom is transferred from the allylic or benzylic carbon to an oxygen of the oxidant. The product has an unpaired electron on the allylic or benzylic carbon and is therefore a resonance-stabilized radical.

(17.34b)

an allylic or benzylic radical is resonance-stabilized

The stability of the radical intermediate, by Hammond's postulate (Sec. 4.8D), increases the rate of this step. The allylic/benzylic selectivity occurs because the analogous radical intermediate in the oxidation of an alcohol that is not allylic or benzylic is less stable and is formed more slowly. In the rapid final step, Mn(III) is reduced to the more stable Mn(II), and a strong C=O double bond is formed to give the aldehyde product, which is washed away from the oxidant surface by the solvent.

(17.34c)

PROBLEMS

17.10 Give the structure of the product expected when each of the following alcohols is subjected to MnO$_2$ oxidation.

(a)

(b)

(c)

(d)

17.11 In each case, give the structure of a starting material that would give the product shown by MnO$_2$ oxidation.

(a)

(b)

(c) HOCH$_2$—

17.12 In each case, tell whether oxidation with pyridinium chlorochromate (PCC) and oxidation with MnO_2 would give the same product, different products, or no reaction. Explain.

(a) [structure: cyclohexene with OH group]

(b) [structure: HO-cyclohexene]

(c) [structure: benzene ring attached to C with OH, CH₃, and CH₃ groups]

$$\text{OH}$$
$$| $$
$$\text{C—CH}_3$$
$$| $$
$$\text{CH}_3$$

(d) [structure: eight-membered ring with CH₂OH and HOCH₂ substituents across a double bond]

B. Oxidation with Cytochrome P450

In Sec. 8.6C (Eq. 8.11, p. 358) we learned that many *xenobiotics*—foreign substances introduced into biological systems—are metabolized by living organisms. There we learned that one metabolic strategy is *conjugation*, in which an existing functional group of the xenobiotic is joined to another group that makes the xenobiotic more soluble in water. The modified xenobiotic can then be excreted in the urine. Glucuronidation is one example of conjugation. We learned that the conjugation of xenobiotics is generally termed **phase II metabolism**.

A different type of metabolism occurs with some xenobiotics. In this type, called **phase I metabolism**, a xenobiotic undergoes an enzyme-catalyzed chemical transformation in which a new functional group is introduced. Typically, such reactions include hydrolysis, oxidation, reduction, and other reactions. (An example of phase I metabolism—the oxidation of polycyclic aromatic hydrocarbons—was discussed in Sec. 16.7.) Although many types of phase I transformations occur in biology, oxidations play a central role. Why should this be? Oxidation usually introduces oxygen, often in the form of a hydroxy group, which both enhances water solubility and provides a site for glucuronidation.

A number of these phase I oxidations are catalyzed by an enzyme called **cytochrome P450**, which we'll abbreviate as **CyP450**. This enzyme (called P450 because of its spectroscopic properties) exists in a very large number of genetic variants, and humans differ in their distribution of the different variants. One reason that drugs affect people differently is the different activities of their cytochromes. Cases of drug toxicity are known that result from a deficiency in CyP450. For example, the metabolism of the cholesterol-reducing drug atorvastatin by CyP450 is inhibited by a compound found in grapefruit; under certain circumstances, eating grapefruit while taking atorvastatin can lead to dangerously high levels of the drug that result in destruction of muscle tissue (rhabdomyolysis).

CyP450 mediates a number of different oxidation reactions. One common reaction is the epoxidation of double bonds. For example, the epoxidation oxidation of aromatic hydrocarbons (Sec. 16.7) is promoted by CyP450. Another common reaction, and the one that we'll consider here, is the hydroxylation of C—H bonds; that is,

$$\text{R—}\underset{|}{\overset{|}{\text{C}}}\text{—H} \xrightarrow{\text{CyP450}} \text{R—}\underset{|}{\overset{|}{\text{C}}}\text{—OH} \qquad (17.35)$$

comes from O_2

In CyP450-promoted oxidations, it is common to find that allylic and benzylic hydrogens are replaced selectively. For example, in "THC," the active principle in marijuana, the primary sites of oxidation are allylic hydrogens.

(−)-*trans*-Δ⁹-tetrahydrocannabinol
("THC")

(17.36)

In this case, not all of the allylic hydrogens, and none of the benzylic hydrogens, are replaced. Presumably, this selectivity is due to the proximity of the reactive hydrogens to the active species in the CyP450 active site (discussed below). Moreover, many cases are known in which "ordinary" hydrogens are replaced, as in the metabolism of the common analgesic ibuprofen:

ibuprofen

(17.37)

major metabolites of ibuprofen

This is a remarkable process in the sense that it is not very common in the laboratory. How does this hydroxylation occur?

The active site of a cytochrome P450 contains an iron in a high oxidation state that is held in place by coordination to the nitrogens of a large heterocyclic compound called *protoporphyrin IX*, which is, in turn, tightly held within the enzyme active site by both noncovalent attractions and by covalent bonding of the sulfur atom of a cysteine residue to the iron. The iron is coordinated to an oxygen atom that originates from molecular oxygen. (The other oxygen of O_2 is converted into water in a reductive process preceding the formation of this species.) Fig. 17.4a shows the detailed structure of this species, and Fig. 17.4b shows a convenient abbreviation that we'll use in discussing the mechanism. (We'll learn more about metal oxidation states in such compounds in Sec. 18.5B.) The protoporphyrin IX–Fe–oxygen group is the actual oxidizing agent in many cases of CyP450-mediated oxidation. The enzyme itself binds the substrate close to this group, and the reaction then takes place within the enzyme–substrate complex.

The mechanism of oxidation by CyP450 has been widely debated, but the currently accepted mechanism for hydroxylation involves a radical mechanism. This is *not* a radical *chain* mechanism, because the radicals involved never exit the enzyme active site; so, they are not "free" radicals. Rather, the mechanism is more like the initiation and termination steps of a radical chain reaction, with no propagation steps in between.

FIGURE 17.4 (a) The structure of protoporphyrin IX at the active site of CyP450, along with the coordinated oxo-iron(V) group. The cysteine bound to the iron is part of the CyP450 protein. (b) The abbreviation of CyP450 used in the text discussion.

The iron in the $Fe^V{=}O$ form of the enzyme has unpaired electron spins, and this unpaired-spin character is shared by the oxygen. In effect, we can think of the oxygen as a bound *oxy radical*:

$$(17.38a)$$

The oxy radical abstracts a nearby hydrogen atom from the bound drug molecule within the CyP450–substrate complex, and the resulting carbon radical then abstracts an OH radical instantly from the iron to give the product alcohol. This is called a "rebound mechanism."

$$(17.38b)$$

a series of transformations converts CyP450 back to the oxidized form

As shown in the equation, the iron is reduced; therefore, it must then be re-oxidized to give the active form of the enzyme. This oxidation is a multistep, enzyme-catalyzed process that we won't discuss here. Therefore, CyP450 alone is not a catalyst in Eq. 17.38b, but is actually a reactant. The catalytic cycle for hydrocarbon oxidation therefore also includes the reactions by which CyP450 is re-oxidized.

The exact hydrogen that is removed depends on which carbon atom is positioned closest to the reactive oxygen within the protein–substrate complex and on which genetic variant of CyP450 is involved. Thus, it is sometimes difficult to predict in a new compound which hydrogen will be abstracted. When the substrate is bound so that an allylic or benzylic hydrogen is closest to the oxygen, it is selectively removed, as in Eq. 17.36. However, if the binding of the substrate positions a nonallylic or nonbenzylic hydrogen close to the oxygen, this type of "ordinary" hydrogen can be removed as well (Eq. 17.37). Even though an ordinary C—H bond (typical bond dissociation energy of 422 kJ mol^{-1} [101 kcal mol^{-1}]) is much stronger than an allylic C—H bond (typical bond dissociation energy of 372 kJ mol^{-1} [89 kcal mol^{-1}]), an O—H bond is somewhat stronger than either type of C—H bond, and a rapid rate of the abstraction reaction is ensured by the proximity of the oxygen and the C—H at the enzyme active site. This peculiar substrate-dependent selectivity demonstrates the importance of proximity (Sec. 11.8C) in determining relative reaction rates.

The level of a drug in the body is determined by a balance of its dose size, its absorption rate, and its metabolic rate. Strategies that reduce the rate of CyP450-promoted oxidation of drugs have the potential to reduce the size of the drug dose required for effectiveness and therefore the potential side effects of the drug. One approach to this problem uses drugs in which deuterium has been substituted for hydrogen at the sites of CyP450-catalyzed oxidation. Such deuterium-substituted drugs should have a negligible difference in their binding to proteins or other macromolecules involved in their pharmacological activity. However, the metabolism of such a deuterium-substituted drug might be retarded if a *primary kinetic deuterium isotope effect* (Sec. 9.5D) occurs when a deuterium atom is removed at the site of metabolism.

This strategy has been realized in a number of drugs. For example, the antidepressant venlafaxine (Effexor®) is metabolized by CyP450-mediated removal of the hydrogens next to O and N. (This is a common site of action for CyP450.)

The deuterium-substituted analog is in fact metabolized more slowly than the hydrogen analog, presumably because the removal of a hydrogen atom in the first step of the mechanism (Eq. 17.38b) is rate-limiting. Substitution of a hydrogen by deuterium results in a rate-retarding primary kinetic isotope effect.

PROBLEMS

17.13 What is the major CyP450-promoted hydroxylation product of each of the following compounds? Assume that benzylic hydroxylation occurs, and neglect stereochemistry.

(a) toluene   (b)

imipramine
(an antidepressant)

17.14 When a deuterium-substituted cyclohexene (shown below) is subjected to CyP450-promoted oxidation, two different allylic alcohols are formed. Explain. (*Hint:* See Eq. 17.38b and think about resonance structures for the intermediate.)

C. Benzylic Oxidation of Alkylbenzenes

Treatment of alkylbenzene derivatives with strong oxidizing agents under vigorous conditions converts the alkyl side chain into a carboxylic acid group. Oxidants commonly used for this purpose are Cr(VI) derivatives, such as $Na_2Cr_2O_7$ (sodium dichromate) or CrO_3; the Mn(VII) reagent $KMnO_4$ (potassium permanganate); or O_2 and special catalysts, a procedure that is used industrially (Eq. 16.52, p. 829).

o-chlorotoluene $\xrightarrow[\substack{100\ °C,\ 3-4\ h \\ H_2O}]{KMnO_4}$ 2-chlorobenzoic acid (17.39)

o-chlorotoluene

2-chlorobenzoic acid
(77% yield)

propylbenzene $\xrightarrow[\substack{48\ h,\ 100\ °C}]{CrO_3,\ 40\%\ H_2SO_4}$ benzoic acid (17.40)

propylbenzene

benzoic acid
(55% yield)

STUDY GUIDE LINK 17.1 Synthetic Equivalence

Notice that the benzene ring is left intact, and notice from Eq. 17.40 that the alkyl side chain, *regardless of length*, is converted into a carboxylic acid group. This reaction is useful for the preparation of some carboxylic acids from alkylbenzenes.

Oxidation of alkyl side chains requires the presence of a benzylic hydrogen. Consequently, *tert*-butylbenzene, which has no benzylic hydrogen, is resistant to benzylic oxidation. Although we won't consider the mechanisms of these benzylic oxidations, they occur in many cases because resonance-stabilized benzylic intermediates such as benzylic radicals are involved.

The conditions for this side-chain oxidation are generally vigorous: heat, high concentrations of oxidant, and/or long reaction times. It is also possible to effect less extensive oxidations of side-chain groups. Thus, 1-phenylethanol is readily oxidized to acetophenone under milder conditions—the normal oxidation of secondary alcohols to ketones (Sec. 10.7A)—but it is converted into benzoic acid under more vigorous conditions.

1-phenylethanol

benzoic acid

acetophenone

(17.41)

You do not need to be concerned with learning the exact conditions for these reactions; rather, it is important simply to be aware that it is usually possible to find appropriate conditions for each type of oxidation. (See Study Guide Link 16.3.)

17.15   Give the products of vigorous $KMnO_4$ oxidation of each of the following compounds.
(a) *p*-nitrobenzyl alcohol   (b) 1-butyl-4-*tert*-butylbenzene

17.16   (a) A compound *A* has the formula C_8H_{10}. After vigorous oxidation, it yields phthalic acid. What is the structure of *A*?

phthalic acid

(b) A compound *B* has the formula C_8H_{10}. After vigorous oxidation, it yields benzoic acid (structure in Eq. 17.40). What is the structure of *B*?

17.6   BIOSYNTHESIS OF TERPENES AND STEROIDS

In this section, we'll see how allylic reactivity is utilized by living systems in the *biosynthesis* of a class of naturally occurring organic compounds called *terpenes*. **Biosynthesis** is the synthesis of chemical compounds by living organisms. The study of biosynthesis is an active area of research that lies at the interface of chemistry and biochemistry. This area is seeing renewed interest because biologists and chemists are collaborating to use genetic engineering technology to alter biosynthetic pathways. This technology holds the promise of using microorganisms as microscopic factories to turn out specially engineered molecules such as drugs.

To begin with, the structure and classification of terpenes is considered in Sec. 17.6A. Then, in Sec. 17.6B, we consider how allylic reactivity is brought to bear on terpene biosynthesis. Then, in Sec. 17.6C, we consider how carbocation chemistry is used in the biosynthetic conversion of terpenes into *steroids*. (The structure of steroids was introduced in Sec. 7.6D.)

A. Terpenes and the Isoprene Rule

People have long been fascinated with the pleasant-smelling substances found in plants—for example, the perfume of a rose—and have been curious to learn more about these materials, which have come to be called *essential oils*. An **essential oil** is a substance that possesses a key characteristic, such as an odor or flavor, of the natural material from which it comes. (See sidebar.)

Essential Oils

The history of the essential oils is an important part of the early history of both chemistry and medicine. A Swiss alchemist, Theophrastus Bombastus von Hohenheim, better known as Paracelsus (*ca.* 1493–1541), believed that everything had a chemical essence. For example, he could demonstrate that the odor of spearmint could be liberated from the plant as a volatile oil on heating; this was the "essence of spearmint"—its

quintessence, or "reason for being." (Paracelsus tried unsuccessfully to find the essences of rocks and other refractory objects.) He believed that a person who was ill was missing part of his or her essence—and Paracelsus sought to restore the essence with chemical cures, among them mercury and sulfur, which, we now know, kill microorganisms (but with significant toxicity to the host). Paracelsus was reputed to have effected some very dramatic cures. His firm belief in the chemical essence of all things was reflected in public tirades against physicians of the day, who (he said) "...strut about with haughty gait, dressed in silk with rings upon their fingers displayed ostentatiously, or silver poniards fixed upon their loins and sleek gloves upon their hands..." while chemists "sweat whole nights and days over fiery furnaces, do not waste time with empty talk, but express delight in the laboratory." Paracelsus was driven out of Basel when he angered some local clergy during a quarrel over his fees; he died in exile.

Some of the most familiar essential oils turned out to be terpenes. The perfume industry in southern France developed around the isolation of fragrant essential oils from flowers, and this industry flourishes today. (The photo shows an old distillation apparatus on display at the Fragonard perfumery in Grasse, France.) The spearmint industry in the northern states of the western and midwestern U.S. is also based on the isolation of the "essence of spearmint," which is *R*-(–)-carvone and other terpenes. (See the sidebar, p. 258.) It was the fascination with, and curiosity about, the essential oils that ultimately led chemists to their understanding of how chemical substances are synthesized in the natural world.

Essential oils, particularly oil of turpentine, were known to the ancient Egyptians. However, not until early in the nineteenth century was an effort made to determine the chemical constitution of the essential oils. In 1818, it was found that the C : H ratio in oil of turpentine was 5 : 8. This same ratio was subsequently found for a wide variety of natural products. These related natural products became known collectively as **terpenes**, a name coined by August Kekulé (p. 46). The similarity in the atomic compositions of the many terpenes led to the idea that they might possess some unifying structural element.

In 1887, the German chemist Otto Wallach (1847–1931), who received the 1910 Nobel Prize in Chemistry, pointed out the common structural feature of the terpenes: they all consist of repeating units that have the same carbon skeleton as the five-carbon diene isoprene. This generalization subsequently became known as the **isoprene rule**.

$$\text{isoprene} \qquad \text{the carbon skeleton of isoprene} \qquad\qquad (17.42)$$

For example, citronellol (from oil of roses and other sources) incorporates two isoprene units:

citronellol

Because of this relationship to isoprene, terpenes are also called **isoprenoids**. Notice carefully that the basis of the terpene or isoprenoid classification is only *the connectivity of the carbon skeleton.* The presence or the positions of double bonds and other functional groups, or the configurations of double bonds and asymmetric carbons, have nothing to do with the terpene classification.

Some notational conventions in terpene chemistry are important. As illustrated by the isoprene structure in Eq. 17.42, the carbons at the ends of the isoprene skeleton are classified as carbon-1 and carbon-4, carbon-4 being either carbon of the dimethyl branch. (Note that these numbers are not the same numbers used in IUPAC nomenclature.) These carbons used to be called "head" and "tail," but British and American chemists confused the issue by adopting different conventions for head and tail. The Americans called carbon-1 the "head" and carbon-4 the "tail," whereas the British adopted the reverse convention. To settle the issue, Dale Poulter, a professor at the University of Utah, proposed that the carbons simply be referred to by numbers, and that is what we'll do.

In many terpenes, the isoprene units are connected in a 1′–4 arrangement (formerly called a "head-to-tail" or "tail-to-head" arrangement). This means that carbon-4 of one skeleton is connected to carbon-1 of the other. The prime (′) on one number and its absence on the other mean that the connection is between *different* isoprene units. For example, this connection is readily apparent in the terpenes geraniol (from oil of geraniums) and limonene (from oil of lemons). Cyclic terpenes such as limonene have additional connections between the isoprene units (in the case of limonene, a C1–C2′ connection) that close the ring.

geraniol

limonene

As you can see, citronellol (shown on p. 859) has a 1′–4 connection between isoprene units as well. Because this arrangement is so common, Wallach assumed the generality of 1′–4 connectivity in his original statement of the isoprene rule. However, many examples are now known in which the isoprene units have a 1′–1 connectivity. Furthermore, some compounds are derived from the conventional terpene structures by skeletal rearrangements. Although these compounds do not have the exact terpene connectivity, they are nevertheless classified as terpenes. For our purposes, though, it will be sufficient to recognize terpenes by two criteria.

1. a multiple of five carbon atoms in the main carbon skeleton
2. the carbon connectivity of the isoprene carbon skeleton within each five-carbon unit

Because terpenes are assembled from five-carbon units, their carbon skeletons contain multiples of five carbon atoms (10, 15, 20, . . . , 5n). Terpenes with 10 carbon atoms in their carbon chains are classified as **monoterpenes**, those with 15 carbons **sesquiterpenes**, those with 20 carbons **diterpenes**, and so on. Some examples of terpenes are given in Fig. 17.5 on p. 862. Many of these compounds are familiar natural flavorings or fragrances.

STUDY PROBLEM   7.2

Determine whether the following compound, isolated from the frontal gland secretion of a termite soldier, is a terpene.

SOLUTION   Because stereochemistry is not an issue, delete all stereochemical details for simplicity. First, count the number of carbons. Because the compound has a multiple of five carbon atoms, it could be a terpene. To check for terpene connectivity, look first for a methyl branch. One is at the end of the long side chain. Identify within this group a chain of four carbons with a methyl branch at the second carbon:

Starting at the next carbon, look for the same pattern. Remember that a bond must connect each isoprene skeleton.

(We arbitrarily chose to proceed clockwise around the ring; you should convince yourself that in this case a counterclockwise path also works.) Continue in this fashion until either the pattern is broken or, as in this case, all carbons are included:

This compound incorporates four isoprene skeletons and is therefore a diterpene.

PROBLEM

17.17 Show the isoprene skeletons within the following compounds of Fig. 17.5.

(a) vitamin A   (b) caryophyllene

B. Biosynthesis of Terpenes

How are terpenes synthesized in nature? What is responsible for the regular repetition of isoprene skeletons? To answer these questions, chemists have studied the biosynthesis of terpenes. In a biosynthetic study, chemists and biologists synthesize and use molecules hypothesized to be the biological starting materials that contain isotopically substituted atoms—for example, ^{14}C or ^{13}C substituted for natural-abundance carbon (mostly ^{12}C)—and determine by chemical degradations, mass spectrometry, and/or NMR whether the biosynthetic products contain the isotopic substituents and, if so, where in the products the isotopes are located. These studies can be followed in some cases by isolation and characterization of the enzymes that catalyze the transformations of interest. Eventually, plausible mechanisms are formulated for the transformations that are consistent with all of the findings. We won't be concerned here with the methodology that was used (over many years) to determine the mechanism of terpene biosynthesis; rather, we'll see the end result and how it fits the notions of allylic reactivity that we have learned.

FIGURE 17.5 Examples of
terpenes. In the monoter-
penes, the isoprene skeletons are shown in red.

limonene
(from oranges
and lemons)

(–)-α-pinene
(from turpentine)

menthol
(from peppermint)

monoterpenes

caryophyllene
(from cloves, hemp, or black pepper)
a sesquiterpene

vitamin A
a diterpene

head-to-head connection

β-carotene
(the orange pigment in carrots;
converted into vitamin A by the human liver)

natural rubber
(a polyterpene)

The repetitive isoprene skeleton in all terpenes has its origin in two simple five-carbon
compounds.

isopentenyl pyrophosphate
(IPP)

γ,γ-dimethylallyl pyrophosphate
(DMAP)

(17.43)

IPP and DMAP are rapidly interconverted in biological systems by an isomerization reaction
(Eq. 17.43) catalyzed by an enzyme, IPP:DMAP isomerase. (For the mechanism, see Problem
17.39a.) Because of this interconversion, the presence of one compound ensures the presence
of the other.

The —OPP in Eq. 17.43 is an abbreviation for the *pyrophosphate group* (*red* in the following structure), which, in nature, is usually complexed to a metal ion such as Mg^{2+} or Mn^{2+}
(*blue*).

a metal-complexed alkyl pyrophosphate

Recall that phosphate and pyrophosphate were discussed in Sec. 10.4E as two of the most important biological leaving groups. The important role of the metal ion and enzyme active-site residues in modulating the reactivity of these groups was discussed there; this should be reviewed, if necessary.

The biosynthesis of the simple monoterpene *geraniol* illustrates the general pattern of terpene biosynthesis. In the first step of geraniol biosynthesis, IPP and DMAP (Eq. 17.43) are bound to the enzyme *prenyl transferase*. The DMAP loses its pyrophosphate leaving group in an S_N1-like process.

(17.44)

The carbocation formed in Eq. 17.44 is a relatively stable allylic cation (Sec. 17.1). Carbocations, like other electrophiles, can react with the π electrons of a double bond, which act as a nucleophile. (This same type of reaction is involved in Friedel–Crafts acylations and alkylations; see Secs. 16.4E–F.) The reaction of this carbocation with the double bond of IPP gives a new carbocation. Loss of a proton from a β-carbon of this carbocation gives the monoterpene geranyl pyrophosphate. (A:$^-$ and AH are basic and acidic groups, respectively, of the enzyme catalyst.)

(17.45)

geranyl pyrophosphate

Geraniol is formed in an S_N2 reaction between water and geranyl pyrophosphate.

geranyl pyrophosphate

(17.46)

geraniol

The S_N2 nature of the reaction was suggested by the observation of stereochemical inversion at the electrophilic carbon. (See Problem 17.40, p. 873.)

All 1′–4 terpenes are formed by reactions analogous to the ones shown in Eqs. 17.45 and 17.46. A large body of evidence for the carbocation character of these reactions was developed in an elegant series of investigations by Profs. Dale Poulter (b. 1942) and Hans C. Rilling (b. 1933) and their students at the University of Utah in the period 1975–1980. As those investigations showed, and as these examples illustrate, the biosynthesis of terpenes can be understood in terms of carbocation intermediates that are like those involved in laboratory chemistry. As we have noted previously (Secs. 10.8 and 11.7B), the organic chemistry of living systems is understandable in terms of laboratory analogies.

The biosynthesis of terpenes also illustrates the economy of nature: A remarkable array of substances is generated from a common starting material. This economy is evident also in other families of natural products. For example, terpenes also serve as the starting point for the biosynthesis of *steroids* (Sec. 7.6D). The isoprene rule is thus one of the unifying principles that underlie the chemical diversity of nature.

PROBLEMS

17.18   (a) Give a biosynthetic mechanism for the formation of the cyclic terpene limonene (Fig. 17.4) beginning with an intramolecular reaction of the following carbocation. (Assume acids and bases are present as necessary.)

$$^+CH_2$$

(b) Draw a curved-arrow mechanism for the biosynthesis of the carbocation intermediate in part (a) from geranyl pyrophosphate.

17.19   Propose a biosynthetic pathway for each of the following natural products. Assume acids and bases are present as necessary.

(a)

farnesol

(*Hint:* Start with geranyl pyrophosphate, Eq. 17.45.)

(b)

α-**pinene**

(*Hint:* Start with the carbocation intermediate in Problem 17.18a.)

C. Biosynthesis of Steroids

The biosynthesis of steroids has been a very important research area for decades because atherosclerosis (cholesterol deposits in arteries) is the leading cause of death in the United States. (See sidebar in Sec. 25.5C, p. 1295.) The starting material for the biosynthesis of steroids is the triterpene *squalene*, which comes from the head-to-head (1–1′) joining of two farnesyl pyrophosphates. (This transformation requires a reductive step that we won't show here.)

farnesyl pyrophosphate

$2e^-$ *reduction*

head-to-head bond

$+ \ 2 \ \bar{O}PP$ (17.47)

squalene

The biosynthesis of farnesyl pyrophosphate is an extension of the biosynthesis of geranyl pyrophosphate shown in Sec. 17.6B (see Problem 17.19a).

In mammals, squalene is converted into a steroid called *lanosterol* in just two enzymatic steps:

two enzyme-catalyzed steps

(17.48)

squalene
(0 rings, 0 asymmetric carbons)

lanosterol
(4 rings, 7 asymmetric carbons)

Notice that this conversion introduces not only four rings, but also seven asymmetric carbons, in a completely stereoselective manner.

The first of the enzymatic steps is an epoxidation catalyzed by the enzyme *squalene oxidase*. The actual oxidizing agent is *flavin hydroperoxide*, which, in turn, is derived ultimately from riboflavin, or vitamin B$_2$. The structure of flavin hydroperoxide is given in Fig. 17.6 on p. 866. As in many biomolecules, the reactive part of the molecule is a small part of the structure, and it will suffice for us to abbreviate flavin hydroperoxide as R—O—O—H. The source of the oxygen in flavin hydroperoxide is O$_2$. We can think of flavin hydroperoxide in this reaction as "nature's peroxy acid," because its role here is exactly the same as that of a peroxy acid in the laboratory epoxidation of alkenes (Sec. 11.3A). The formation of the epoxide is a concerted electrophilic addition. (Compare with Eq. 11.18a, p. 519.) This establishes the stereochemistry at one of the seven asymmetric carbons.

flavin hydroperoxide

an acidic group of the enzyme

squalene oxidase (enzyme)

squalene

squalene epoxide (2,3-oxidosqualene) (conjugate-acid form)

squalene epoxide (2,3-oxidosqualene)

(17.49)

FIGURE 17.6 (a) The structure of the coenzyme flavin adenine dinucleotide (FAD). The boxed structure (with an O—H rather than an O—P bond) is vitamin B$_2$ (riboflavin). FAD and a reduced derivative, FADH$_2$, are important in a number of biological oxidation–reduction reactions. (b) Flavin hydroperoxide is a modified form of FAD. (The part of the structure above the bracket is the same as in FAD.) This can be thought of in the context of steroid biosynthesis as the natural equivalent of peroxy acids. The abbreviation used in the text is shown at right.

The second, and more remarkable, enzymatic step is the cyclization of squalene epoxide, which is catalyzed by the enzyme *2,3-oxidosqualene:lanosterol cyclase*. To understand what happens in this step, we have to consider the conformation of the epoxide 2,3-oxidosqualene as it exists in the enzyme active site. Noncovalent interactions are used to constrain 2,3-oxidosqualene to adopt the following conformation in the enzyme active site.

In this conformation, the 2*p* orbitals of successive double bonds are positioned so that they can react sequentially with little or no molecular realignment. The enzyme provides an acidic group that protonates the epoxide. As we learned in Sec. 11.5B, protonated tertiary epoxides are essentially shielded carbocations, and they undergo reactions with nucleophiles at the more-branched carbon atom. So it is in this case. The carbocation undergoes an electrophilic addition to the neighboring double bond. This generates electron deficiency at the other carbon of the double bond, which undergoes an electrophilic addition to the next double bond, and so on. The first intermediate is a tertiary carbocation in which all the ring bonds are formed stereospecifically.

(17.50)

This reaction is a spectacular example of a *proximity effect* (Sec. 11.8). There is good evidence that the proximity of the double bonds to each other accelerates the reaction.

To understand this addition, we can think of it as a stepwise process. The carbocation resulting from opening of the protonated epoxide undergoes an electrophilic addition to the neighboring double bond; the resulting carbocation undergoes an electrophilic addition to the next double bond; and so on.

etc.

(17.51)

Although it may be helpful to view the reaction this way, the reaction appears to be a concerted process.

What is the energetic advantage of converting one tertiary carbocation into another? In this process, several π bonds are converted into σ bonds. Each π bond has a bond dissociation energy of about 243 kJ mol^{-1} (58 kcal mol^{-1}), whereas a carbon–carbon σ bond has a bond dissociation energy of about 377 kJ mol^{-1} (90 kcal mol^{-1}) (Table 5.3, p. 216). Therefore, the conversion of *each* π bond has a $\Delta H°$ of roughly –134 kJ mol^{-1} (–32 kcal mol^{-1}). The magnitude of this energy is reduced by the boat conformation of one ring and by the entropic cost of eliminating a number of bond rotations, but the reaction is highly favorable nevertheless. In addition, the carbocation product is stabilized by pi–cation interactions (Sec. 15.8B) with aromatic amino acid side chains of the enzyme catalyst.

The carbocation formed in Eq. 17.49 then undergoes a series of rearrangements and loss of a proton to give lanosterol. Each rearranging group is anti to the next; we can think of this as a substitution of one rearranging group with its electron pair by the next. Like all nucleophilic substitutions, each occurs by an opposite-side substitution.

(17.52)

base in the enzyme active site

lanosterol

As with the additions, we could rewrite the rearrangements in individual steps:

etc. (17.53)

However, it is believed that these take place in a concerted manner.

The rearrangement alters the isoprenoid carbon skeleton, and this is why lanosterol, although derived from a triterpene, is not itself a terpene.

Lanosterol is converted into the steroid cholesterol in 19 additional biosynthetic steps. Lanosterol also serves as the biosynthetic precursor of other steroids. Therefore, the biosynthesis of steroids that we've considered in this and the previous section can be summarized:

isopentenyl pyrophosphate $\longrightarrow$ squalene $\longrightarrow$ squalene epoxide $\longrightarrow$

$$\text{lanosterol} \xrightarrow{\textit{19 steps}} \text{cholesterol} \qquad (17.54)$$

Other important steps in the biosynthesis of cholesterol lead to the formation of isopentenyl pyrophosphate. In Sec. 25.5C, we'll encounter this topic again and show how an understanding of cholesterol biosynthesis has led to the development of a number of important drugs to fight atherosclerosis.

The biosynthesis of cholesterol was elucidated by Konrad Bloch (1912–2000), a Harvard professor of chemistry. For this work, Bloch shared the 1964 Nobel Prize in Physiology or Medicine with Feodor Lynen (1911–1974), a professor at the Max Planck Institute for Cellular Chemistry in Munich.

PROBLEMS

17.20 (a) Complete the stepwise mechanism shown in Eq. 17.51 to give the product carbocation of Eq. 17.50.

  (b) Examine each of the carbocations in the stepwise process you drew in part (a). What problem do you see with the stepwise mechanism? (*Hint:* Classify the carbocation formed in each step as primary, secondary, or tertiary.)

17.21 Complete the stepwise process shown in Eq. 17.53 to give lanosterol, the product of Eq. 17.52.

KEY IDEAS IN CHAPTER 17

- Functional groups at allylic and benzylic positions are in many cases unusually reactive.

- The addition of hydrogen halides to dienes, and the solvolysis of allylic and benzylic alkyl halides are reactions that involve allylic or benzylic carbocations. The acceleration of reactions that involve these carbocations can be attributed to the resonance stabilization of the carbocations.

- A mixture of isomeric products is typically obtained from a reaction involving an unsymmetrical allylic carbocation as a reactive intermediate because charge is shared by more than one carbon in the ion, and because a nucleophile can react with each charged carbon. A reaction involving a benzylic carbocation, in contrast, typically gives only the product derived from the reaction of a nucleophile at the benzylic position because only in this product is the aromaticity of the ring not disrupted.

- Free-radical halogenation is selective for allylic and benzylic hydrogens because the allylic or benzylic free-radical intermediates that are involved are resonance-stabilized.

- *N*-Bromosuccinimide (NBS) in CCl_4 solution is used to carry out allylic and benzylic brominations. Benzylic bromination can also be carried out with bromine and light.

- Because the unpaired electron in allylic radicals is shared on different carbons, some reactions that involve allylic radicals give more than one product.

- The substitution of a hydrogen in a C—H bond by —OH is an important reaction of phase I metabolism promoted by cytochrome P450. This reaction occurs by a radical mechanism and is especially prevalent at allylic and benzylic C—H bonds, although ordinary C—H bonds sometimes undergo oxidation as well.

- Allylic and benzylic anions are stabilized by resonance and by the polar effect of double bonds.

- Allylic Grignard and organolithium reagents undergo rapid allylic rearrangements. The transition state for this rearrangement is stabilized because it has the character of a resonance-stabilized allylic carbanion.

- A β-elimination reaction involving allylic or benzylic β-hydrogens is accelerated because the anionic character in the transition state is stabilized in the same manner as an allylic or benzylic anion, and because the developing double bond in the transition state is conjugated.

- S_N2 reactions at allylic and benzylic positions are accelerated because their transition states are stabilized by the overlap of $2p$ orbitals.

- Allylic and benzylic alcohols can be oxidized selectively by a suspension of manganese dioxide. The selectivity of the reaction is due to the formation of an allylic or benzylic radical in the reaction mechanism.

- In aromatic compounds, alkyl side chains that contain benzylic hydrogens can be oxidized to carboxylic acid groups under vigorous conditions.

- Terpenes, or isoprenoids, are natural products with carbon skeletons characterized by repetition of the five-carbon isoprene unit.

- Terpenes are synthesized in nature by enzyme-catalyzed processes involving the reaction of allylic carbocations with double bonds. The biosynthetic precursor of all terpenes is isopentenyl pyrophosphate (IPP).

- Isopentenyl pyrophosphate is the biosynthetic precursor of steroids. Squalene, a triterpene containing a head-to-head linkage, is converted into an epoxide, which, after protonation, undergoes a concerted cyclization that can be viewed as a series of electrophilic additions of carbocations to successive double bonds. The carbocation product of the cyclization undergoes a series of rearrangements and loss of a proton to give lanosterol, in which the typical steroid ring system is established. Lanosterol is a biosynthetic precursor of cholesterol.

 REACTION REVIEW *For a summary of reactions discussed in this chapter, see the* Reaction Review *section of Chapter 17 in the* Study Guide and Solutions Manual.

ADDITIONAL PROBLEMS

Note: Problems 15.76–15.78 and 15.85 (Chapter 15) are also related to the material in this chapter.

17.22 Give the principal organic product(s) expected when *trans-2-butene* or another compound indicated reacts under the following conditions. Assume one equivalent of each reagent reacts in each case.
   (a) Br_2 in CH_2Cl_2, dark
   (b) *N*-bromosuccinimide in CCl_4, light
   (c) product(s) of part (b), solvolysis in aqueous acetone
   (d) product(s) of part (b) + Mg in ether
   (e) product(s) of part (d) + D_2O

17.23 Give the principal product(s) expected when 4-methylcyclohexene or other compound indicated reacts under the conditions in Problem 17.22.

17.24 Which of the following compounds, all known in nature, can be classified as terpenes? Show the isoprene skeletons in each terpene.

(a)

ipsdienol
(one component of the pheromone of the Norwegian spruce beetle)

(b)

saffrole
(oil of sassafras)

(c)

modhephene
(from rayless goldenrod)

(d)

CH₃

H

O

H₃C   CH₃

β-thujone
(from yellow cedar)

(e)

O

O

O

H₂C

periplanone B
(pheromone of the female American cockroach)

(f)

***m*-cymene**

17.25 (a) Determine whether the following compound (zoapatanol, used as a fertility-regulating agent in Mexican folk medicine) is a terpene.

O   HO

CH₂OH

H₃C   O   H

zoapatanol

(b) What product is obtained when zoapatanol is subjected to MnO₂ oxidation?

17.26 Explain why two products are formed in the first ether synthesis, but only one in the second.

(1)

OH   OEt   OEt

+ EtOH $\xrightarrow{\text{H}_2\text{SO}_4}$ +

(2)

OH   OEt

+ NaH $\xrightarrow{\text{EtI}}$

17.27 A student Al Lillich has prepared a pure sample of 3-bromo-1-butene (*A*). Several weeks later he finds that the sample is contaminated with an isomer *B* formed by allylic rearrangement.

(a) Give the structure of *B*.

(b) Draw a curved-arrow mechanism for the formation of *B* from *A*.

(c) Which should be the major isomer at equilibrium, *A* or *B*? Explain.

17.28 Outline a synthesis of each of the following compounds from the indicated starting materials and any other reagents.

(a) benzyl methyl ether from toluene

(b) 3-phenyl-1-propanol from toluene

(c) (Z)-1,4-nonadiene from 1-hexyne

(d)

—CH₂CH=O from cyclopentene

(e) O₂N

—CO₂H from cumene
(isopropylbenzene)

(f)

O₂N—   —CO₂H from cumene

(g)   CH₃O

CH₃O—   —CH=O

from 1,2-dimethoxy-4-methylbenzene

(h)   —SCH₂CH₃ from   ·····OH

17.29 Rank the following compounds in order of increasing reactivity (least reactive first) in an S_N1 solvolysis reaction in aqueous acetone. Explain your answers. (The structure of *tert*-cumyl chloride is shown in Table 17.2.)

(1) *m*-nitro-*tert*-cumyl chloride

(2) *p*-methoxy-*tert*-cumyl chloride

(3) *p*-fluoro-*tert*-cumyl chloride

(4) *p*-nitro-*tert*-cumyl chloride

17.30 Arrange the following alcohols according to increasing rates of their acid-catalyzed dehydration to alkene (smallest rate first), and explain your reasoning.

O₂N—   —CHCH₂Ph

OH

A

CH₃

CH₃O—   —CCH₂Ph

OH

B

CH₃O—   —CHCH₂Ph

OH

C

17.31 Terfenadine is an antihistaminic drug that contains two alcohol functional groups. Suppose terfenadine were to undergo acid-catalyzed alcohol dehydration (Sec. 10.2). Which alcohol would dehydrate most rapidly? Why? What would be the dehydration product?

terfenadine

17.32 Explain why compound A reacts faster than compound B when they undergo solvolysis in aqueous acetone.

17.33 A hydrocarbon A, C_9H_{12}, is treated with N-bromosuccinimide in CCl_4 in the presence of peroxides to give a compound B, $C_9H_{11}Br$. Compound B undergoes rapid solvolysis in aqueous acetone to give an alcohol C, $C_9H_{12}O$, which cannot be oxidized with CrO_3 in pyridine. Vigorous oxidation of compound A with hot chromic acid gives benzoic acid, $PhCO_2H$. Identify compounds A–C.

17.34 A hydrocarbon A, C_9H_{10}, is treated with N-bromosuccinimide to give a single monobromo compound B. When B is dissolved in aqueous acetone it reacts to give two nonisomeric compounds: C and D. Catalytic hydrogenation of D gives back A, and C can be separated into enantiomers. When optically active C is oxidized with CrO_3 and pyridine, an optically inactive ketone E is obtained. Vigorous oxidation of A with $KMnO_4$ affords phthalic acid (structure in Problem 17.16, p. 858). Propose structures for compounds A through E, and explain your reasoning.

17.35 Predict which of the following compounds should undergo the more rapid reaction with K^+ $(CH_3)_3C—O^-$, explain your reasoning, and give the product of the reaction.

17.36 When benzyl alcohol ($\lambda_{max} = 258$ nm, $\epsilon = 520$) is dissolved in H_2SO_4, a colored solution is obtained that has a different UV spectrum: $\lambda_{max} = 442$ nm, $\epsilon = 53,000$. When this solution is added to cold NaOH, the original spectrum of benzyl alcohol is restored. Suggest a structural basis for these observations.

17.37 Match one or more of the structures below with each of the following statements.

(a) An optically active compound that is oxidized by MnO_2 to an optically inactive compound.

(b) An optically active compound that is oxidized by MnO_2 to an optically active compound.

(c) An optically inactive compound that is oxidized by MnO_2 to an optically inactive compound.

(d) A compound that is not oxidized by MnO_2.

17.38 Draw a variation on the "rebound" hydroxylation mechanism that accounts for the formation of epoxides from some alkenes by CyP450 as shown in Fig. P17.38. (*Hint:* Recall that radicals can add to a π bond.)

17.39 (a) Assuming the presence of acids (AH) and bases (A^-) as needed, give a curved-arrow mechanism for the isomerization of isopentenyl pyrophosphate into dimethylallyl pyrophosphate (Eq. 17.43).

(b) In some organisms, geranyl pyrophosphate (structure in Eq. 17.45, p. 863) is converted by hydrolysis into one of two enantiomeric *tertiary* alcohols:

linalool

(3R)-(–)-linalool
(also called licareol; has a lily-of-the-valley scent)

(3S)-(+)-linalool
(also called coriandrol; has a musty herbaceous scent)

Assuming the presence of acids (AH), bases (A^-), and water as needed, give a curved-arrow mechanism for this transformation. (Ignore stereochemistry.)

(c) Each of the stereoisomeric linalools in part (b) is converted into an epoxide A by CyP450, and the epoxide spontaneously forms the following compound B. Neglecting stereochemistry, suggest a structure for A; and, assuming the presence of H_3O^+ and H_2O, provide a curved-arrow mechanism for the conversion of A into B.

B

17.40 Consider the hydrolysis of geranyl pyrophosphate to geraniol shown in Eq. 17.46 on p. 863.

(a) What isotopically substituted derivative of geranyl pyrophosphate could be used to establish the assertion that this hydrolysis proceeds by opposite-side substitution?

(b) Given that the isotopically substituted derivative you proposed in (a) can be prepared in enantiomerically pure form with a known absolute configuration, the absolute stereochemistry of the product geraniol must be determined to establish the stereochemistry of substitution. Explain how the alcohol dehydrogenase-catalyzed oxidation of the isotopically substituted geraniol product by NAD^+ (Sec. 10.9B) can be used to establish its stereochemistry.

17.41 (a) Geranyl pyrophosphate is methylated by *S*-adenosyl-methionine (SAM; Sec. 11.7B) in the reaction shown in Fig. P17.41, which is catalyzed by the enzyme geranyl pyrophosphate *C*-methyl transferase. Assuming that the enzyme provides acidic and basic groups as needed, provide a curved-arrow mechanism for this transformation. (*Hint:* A carbocation intermediate is involved.)

(b) The product of the reaction in (a) is converted into 2-methylisoborneol in an electrophilic addition reaction catalyzed by the enzyme 2-methylisoborneol synthase. Assuming the presence of acids, bases, and water as needed, suggest a curved-arrow mechanism for the reaction. (*Hint:* Before starting, relate the carbons of the product to those of the starting material and thus establish the connections that must be made.)

Then provide a plausible mechanism for those connections that involves carbocation intermediates.)

2-methylisoborneol

17.42 Starting with isopentenyl pyrophosphate, propose a mechanism for the biosynthesis of eudesmol, a sesquiterpene obtained from eucalyptus.

eudesmol

17.43 One of the compounds responsible for the odor of moist soil is geosmin, which is produced by streptomycetes in soil from farnesyl pyrophosphate, as shown in Fig. P17.43 on p. 874. Compounds *A* and *B* are intermediates in the biosynthesis.

Using Brønsted acids (^+BH), Brønsted bases (B:), and water as needed, propose a curved-arrow mechanism for the conversion of compound *A* to geosmin with compound *B* as an intermediate.

CyP450

Figure P17.38

geranyl pyrophosphate **SAM**

geranyl pyrophosphate *C*-methyl transferase (enzyme)

2-methylgeranyl pyrophosphate

Figure P17.41

17.44 Account for each of the following facts with an explanation.

(a) 1,3-Cyclopentadiene is a considerably stronger carbon acid than 1,4-pentadiene, even though the acidic hydrogens in both cases are doubly allylic.

1,3-cyclopentadiene   **1,4-pentadiene**

(b) 3-Bromo-1,4-pentadiene undergoes solvolysis readily in protic solvents, but 5-bromo-1,3-cyclopentadiene is virtually inert.

3-bromo-1,4-pentadiene   **5-bromo-1,3-cyclopentadiene**

17.45 Complete the reactions given in Fig. P17.45 by proposing structures for the major organic products.

17.46 Propose a curved-arrow mechanism for each of the reactions given in Fig. P17.46.

17.47 (a) When 2-hexyne is treated with certain very strong bases, it undergoes the reaction given in Fig. P17.47, an example of the "acetylene zipper" reaction. Give a curved-arrow mechanism for this reaction, and explain why it is irreversible.

(b) What product(s) would be expected in the same reaction of 3-methyl-4-octyne? Explain.

17.48 Propose a curved-arrow mechanism for the reaction given in Fig. P17.48 on p. 876, and give at least two structural reasons why the equilibrium lies to the right.

17.49 Propose a curved-arrow mechanism for the following reaction. Explain why the equilibrium lies to the right.

17.50 Consider the relative rates of the two solvolysis reactions in acetic acid solvent shown in Fig. P17.50 on p. 876.

(a) Suggest a reason that compound *A* undergoes solvolysis *much* faster than compound *B*. (*Hint:* Consider anchimeric assistance by the π electrons of the double bond; see Sec. 11.8A.)

(b) Account for the retention of stereochemistry observed in reaction *A* with a mechanism. (*Hint:* See Sections 11.8A and 11.8D.)

17.51 In an effort to duplicate in the laboratory the natural processes that resemble the cyclization of 2,3-oxidosqualene, chemists in 1970 and 1980 carried out the conversions shown in Fig. P17.51 on p. 876. Give a curved-arrow mechanism for each. (For purposes of this problem, let the acid and base catalysts be a general acid A—H and its conjugate base A⁻.)

17.52 Nicotine undergoes phase I metabolism by a reaction with CyP450 that gives a compound *A*. In the presence of water, compound *A* is converted spontaneously into compound *B*, with which it is in equilibrium. (See Fig. P17.51, p. 876.)

(a) Use what you know about the reactions of CyP450 to suggest a structure for compound *A*.

(b) Assuming the presence of H₃O⁺ and H₂O, provide a curved-arrow mechanism for the conversion of *A* into *B*.

17.53 When 1-buten-3-yne undergoes HCl addition, two compounds *A* and *B* are formed in a ratio of 2.2 : 1. Neither compound shows a C≡C stretching absorption in its IR spectrum. Compound *B* reacts with maleic anhydride to give compound *C*, and compound *A* undergoes allylic

(*continues on p. 877*)

farnesyl pyrophosphate   **geosmin**   **acetone**

A   *B*

Figure P17.43

(a) $H_2C{=}CH{-}CH{=}CH{-}CH_2MgBr + D_2O \longrightarrow$

(b)

$H_2C{=}CH{-}CH{=}CH{-}CH_2MgBr + H_2C{-}CH_2$ (epoxide) $\longrightarrow \xrightarrow{H_3O^+}$

(c) *trans*-$BrCH_2CH{=}CHCH(CH_3)_2 + Na^+\ CH_3CH_2S^- \xrightarrow{ethanol}$

(d) Same as part (c), but with warm ethanol only; no $Na^+\ CH_3CH_2S^-$

(e)

$+ \ N\text{-bromosuccinimide} \xrightarrow[CCl_4]{AIBN}$
(1 equiv.)

(f)

$H_3C{-}\ \ {-}C({-}CH_3)({-}H) \ + \ N\text{-bromosuccinimide} \xrightarrow[CCl_4]{peroxides} C_{10}H_{13}Br$
(1 equiv.)

(g)

$\xrightarrow[dark]{Br_2} \xrightarrow[ethanol]{KOH} C_{10}H_8$

(h)

${-}CH{=}CH{-}CH{=}CH_2 + HBr \longrightarrow$
(1 equiv.)

(i)

$\xrightarrow[heat]{KMnO_4}$

Figure P17.45

(a)

$CH_3(CH_2)_3{-}C{\equiv}C{-}CH_2{-}Br + Mg \xrightarrow[ether]{} \xrightarrow{H_2O}$

$CH_3(CH_2)_3{-}CH{=}C{=}CH_2 + CH_3(CH_2)_3{-}C{\equiv}C{-}CH_3$

(b)

$Ph{-}C{\equiv}C{-}CH_2CH_2OH \xrightarrow{p\text{-toluenesulfonic acid}}$

(c) 1-penten-4-yne $+ Na^+\ {}^-{:}C{\equiv}CH$; then allyl bromide $\longrightarrow H_2C{=}CH{-}CH{-}C{\equiv}CH$
with $CH_2{-}CH{=}CH_2$

(d)

$+ EtOH \xrightarrow[(dilute)]{H_2SO_4}$
(excess; solvent)

(e)

$H_3C{-}C({-}CH_3)({-}Cl){-}C{\equiv}C{-}H + $ $\xrightarrow{(CH_3)_3C{-}O^-\ K^+}$

Figure P17.46

$H_3C{-}C{\equiv}C{-}CH_2CH_2CH_3 + B{:}^- \longrightarrow {}^-{:}C{\equiv}C{-}CH_2CH_2CH_2CH_3 + BH$
$(B{:}^- = {}^-{:}\ddot{N}H{-}CH_2CH_2CH_2{-}\ddot{N}H_2)$ (pK_a = 35)

Figure P17.47

Figure P17.48

Figure P17.50

(a)

(b)

Figure P17.51

nicotine
(form present
at pH 7)

Figure P17.52

(*continued from p. 874*)

rearrangement to compound *B* on heating. Propose structures for compounds *A* and *B* and explain your reasoning.

maleic anhydride   *C*

17.54 Around 1900, Moses Gomberg, a pioneer in free-radical chemistry, prepared the triphenylmethyl radical, Ph$_3$C·, sometimes called the *trityl radical* (trityl = *triphenylmethyl*).

(a) Explain why the trityl radical is an unusually stable radical.

(b) The trityl radical is known to exist in equilibrium with a dimer that, for many years, was assumed to be hexaphenylethane, Ph$_3$C—CPh$_3$. Show how hexaphenylethane could be formed from the trityl radical.

(c) In 1968 the structure of this dimer was investigated using modern methods and found not to be hexaphenylethane, but rather the following compound. Using the fishhook notation, show how this compound is formed from two trityl radicals, and explain why this compound is formed instead of hexaphenylethane. (*Hint:* Can you think of any reason why hexaphenylethane might be unstable?)

dimer of the trityl radical

17.55 (a) Triphenylmethane [structure in part (b)] has a pK_a of 31.5 and, although an alkane, it is almost as acidic as a 1-alkyne. (Most alkanes have p$K_a \geq 55$.) By considering the structure of its conjugate base, suggest a reason why triphenylmethane is such a strong hydrocarbon acid.

(b) Fluoradene is structurally very similar to triphenylmethane, except that the three aromatic rings are "tied together" with single bonds. Fluoradene has a pK_a of 11. Suggest a reason why fluoradene is much more acidic than triphenylmethane.

triphenylmethane   **fluoradene**
p$K_a \approx 31.5$ p$K_a \approx 11$

17.56 The amount of anti-addition in the chlorination of alkenes varies with the structure of the alkene, as shown in the following table. (See Sec. 7.8C, p. 309.)

Structure of R—	Percent anti-addition
H$_3$C—	99
(phenyl)	88
CH$_3$O—(phenyl)—	63

Suggest a reason for the variation in the stereochemistry of addition as the alkene structure is varied. (*Hint:* What types of reactive intermediate(s) could account for the stereochemical observations?)

17.57 In the late 1970s, a graduate student at a major West Coast university began synthesizing new classes of drugs and testing them on himself. After being expelled from the university, he began making his living by illegally synthesizing and selling to heroin addicts compound *B*, a synthetic analog of meperidine (Demerol). (See Fig. P17.57, p. 878.) After shortening his synthetic procedure and self-injecting his product, he developed severe symptoms of Parkinson's disease, as did several of his young clients; one person died. Chemists found that his compound *B* contained two by-products, alcohol *C* and another compound MPTP (C$_{12}$H$_{15}$N), which, when independently prepared and injected into animals, caused the same symptoms. (Ironically, this has been one of the most significant advances in Parkinsonism research.)

Given the illicit chemistry outlined in Fig. P17.57, provide the structure of compound *A*, suggest a structure for MPTP, and show how all products are formed.

17.58 (a) For each of the two reactions shown in Fig. P17.58 on p. 878, suggest a mechanism that is consistent with all of the experimental facts given.

Experimental observations:

(1) Both reactions conform to the following rate law, although the rate constants for each reaction are different.

$$\text{rate} = k[\text{alkyl halide}][\text{Et}_2\ddot{\text{N}}\text{H}]$$

(2) The alkyl chloride starting materials do *not* interconvert under the reaction conditions.

(3) The following compound, prepared separately, is *not* converted into the observed product under the reaction conditions.

$$\text{H}_2\text{C}=\text{CH}-\overset{\text{CH}_3}{\underset{}{\text{CH}}}-\overset{+}{\text{N}}\text{HEt}_2 \quad \text{Cl}^-$$

In particular, explain the importance of facts (2) and (3) in understanding the mechanism.

(b) The mechanism of reaction 2 is called the S_N2' mechanism. Suggest a reason why this reaction occurs by the S_N2' mechanism and reaction 1 does not.

17.59 The reaction given in Fig. P17.59 occurs by a mechanism called the S_N2' mechanism, which is a bimolecular substitution that occurs by reaction of the nucleophile at an allylic carbon (see previous problem). In this reaction, the

C—Cl bond as well as the bond to the nucleophile must be perpendicular to the plane of the alkene double bond for proper orbital overlap, but this orientation can occur in two ways: The C—Cl bond can be syn to the new bond that is formed with the nucleophile, or it can be anti. Is the following S_N2' reaction syn or anti? How does this result contrast with the stereochemistry of the S_N2 reaction?

Figure P17.57

Figure P17.58

Figure P17.59

The Chemistry of Aryl Halides, Vinylic Halides, and Phenols
Transition-Metal Catalysis

An **aryl halide** is a compound in which a halogen is bound to the carbon of a benzene ring (or another aromatic ring).

1-ethyl-2-iodobenzene
(an aryl iodide)

not an aryl halide;
the halogen is not attached directly
to benzene ring

In a **vinylic halide**, a halogen is bound to a carbon of a double bond.

$$H_2C=CH-Cl$$

**vinyl chloride
(chloroethylene)**
a vinylic chloride

(E)-1-bromo-1-pentene
(a vinylic bromide)

879

Be sure to differentiate carefully between *vinylic* and *allylic* halides (p. 836). *Allylic* groups are on a carbon *adjacent* to the double bond. Likewise, be sure that the distinction between *aryl* and *benzylic* halides is clear. *Benzylic* groups are on a carbon *adjacent* to an aromatic ring.

The reactivity of aryl and vinylic halides is quite different from that of ordinary alkyl halides. In fact, one of the major points of this chapter is that aryl halides do *not* undergo nucleophilic substitution reactions by the S_N2 or S_N1 mechanisms.

In a **phenol**, a hydroxy (—OH) group is bound to an aromatic ring. As the following structures illustrate, *phenol* is also the name given to the parent compound, and a number of phenols have traditional names.

phenol   **catechol**   **resorcinol**   **hydroquinone**   **o-cresol**   **p-cresol**

Although phenols and alcohols have some reactions in common, there are also important differences in the chemical behavior of these two functional groups.

A relatively recent field of organic chemistry involves the use of transition-metal catalysts in organic reactions, particularly in reactions that involve formation of carbon–carbon bonds. The reactivity of aryl and vinylic halides in substitution reactions is dramatically increased by certain catalysts of this type, and this heightened reactivity will be the vehicle through which we can learn some of the basic principles involved in transition-metal catalysis.

The nomenclature and spectroscopy of aryl halides and phenols were discussed in Secs. 16.1 and 16.3, respectively. The nomenclature of vinylic halides follows the principles of alkene nomenclature (Sec. 4.2A), and the spectroscopy of vinylic halides, except for minor differences due to the halogen, is also similar to that of alkenes.

18.1   LACK OF REACTIVITY OF VINYLIC AND ARYL HALIDES UNDER S_N2 CONDITIONS

One of the most important differences between vinylic or aryl halides and alkyl halides is their reactivity in nucleophilic substitution reactions. The two most important mechanisms for nucleophilic substitution reactions of alkyl halides are the S_N2 (bimolecular opposite-side substitution) mechanism, and the S_N1 (unimolecular carbocation) mechanism (Secs. 9.4 and 9.6). What happens to vinylic and aryl halides under the conditions used for S_N1 or S_N2 reactions of alkyl halides?

Consider first the S_N2 reaction. One of the most dramatic contrasts between vinylic or aryl halides and alkyl halides is that simple vinylic and aryl halides are inert under S_N2 conditions. For example, when ethyl bromide is allowed to react with $Na^+ \ CH_3CH_2O^-$ in CH_3CH_2OH solvent at 55 °C, the following S_N2 reaction proceeds to completion in about an hour with excellent yield:

$$CH_3CH_2—Br + Na^+ \ CH_3CH_2—\ddot{O}{:}^- \xrightarrow[CH_3CH_2OH]{55\,°C} CH_3CH_2—\ddot{O}—CH_2CH_3 + Na^+ \ Br^- \qquad (18.1)$$

Yet when vinyl bromide or bromobenzene is subjected to the same conditions, nothing happens!

and   are inert to $Na^+ \ CH_3CH_2O^-$ in CH_3CH_2OH at 55 °C

Why don't vinylic halides undergo S$_N$2 reactions? In Sec. 9.4C (Fig. 9.2, p. 395), we learned that, in the transition state of an S$_N$2 reaction of an alkyl halide, the carbon undergoing substitution has a 2p orbital to which the nucleophile and the leaving group are partially bound, and is therefore sp^2-hybridized. In other words, this carbon rehybridizes from sp^3 in the alkyl halide to sp^2 in the transition state. The carbon undergoing substitution in a vinylic halide is sp^2-hybridized in the starting material; it contains a 2p orbital involved in the double bond. If the S$_N$2 reaction results in the development of a second 2p orbital at this carbon, then this carbon must become sp-hybridized. Therefore, an S$_N$2 reaction at a vinylic carbon involves rehybridization from sp^2 in the vinylic halide to sp in the transition state.

$$\text{(18.2)}$$

transition state

The sp hybridization state has such high energy (Sec. 14.1) that conversion of an sp^2-hybridized carbon into an sp-hybridized carbon requires about 21 kJ mol^{-1} (5 kcal mol^{-1}) more energy than is required for an sp^3 to sp^2 hybridization change. The relatively high energy of the transition state caused by sp hybridization reduces the rate of S$_N$2 reactions of vinylic halides by almost four orders of magnitude (by Eq. 9.20a, p. 391). This means that, under the conditions in which the S$_N$2 reaction of ethyl bromide takes an hour, transition-state hybridization alone would cause the same reaction of vinyl bromide to take almost 200 days.

Rehybridization, however, is not the only reason that vinylic halides are unreactive in the S$_N$2 reaction. A second reason is that the nucleophile (Nuc:$^-$ in Eq. 18.2) would have to approach the vinylic halide at the opposite side of the halogen-bearing carbon *and in the plane of the alkene*. This arrangement results in significant van der Waals repulsions (a steric effect) of both the nucleophile and the leaving group with the groups on the other vinylic carbon. This is shown for the S$_N$2 reaction of vinyl bromide in Fig. 18.1. When the groups on the other vinylic carbon are larger than hydrogen, the repulsions are even greater. These repulsions raise the energy of the transition state and decrease the reaction rate even further.

In summary, both hybridization and van der Waals repulsions (steric effects) within the transition state retard the S$_N$2 reactions of vinylic halides to such an extent that they do not occur.

S$_N$2 reactions of aryl halides have the same problems as those of vinylic halides and two others as well. First, opposite-side approach to the carbon of the carbon–halogen bond would

(a)   (b)

FIGURE 18.1 Van der Waals repulsions in the transition state of the S$_N$2 reaction of vinyl bromide with methoxide, illustrated with (a) Lewis structures and (b) space-filling models. These van der Waals repulsions raise the energy of the transition state and decrease the rate of the reaction.

place the nucleophile on a path that goes through the plane of the benzene ring—an obvious impossibility. Furthermore, because the carbon at which substitution occurs would have to undergo stereochemical inversion, the reaction would necessarily yield a benzene derivative containing a twisted and highly strained double bond.

$$\text{Nuc}:^- \quad \boxed{\text{inversion required}} \quad \overset{\cdot\cdot}{\underset{\cdot\cdot}{X}}: \quad \times \quad \overset{\text{twisted double bond}}{\underset{\text{Nuc}}{\bigtriangleup}} \quad + \quad :\overset{\cdot\cdot}{\underset{\cdot\cdot}{X}}:^- \qquad (18.3)$$

If the impossibility of this result is unclear, try to build a model of the product—but don't break your models!

PROBLEM

18.1   Within each set, rank the compounds in order of increasing rates of their S_N2 reactions. Explain your reasoning.
(a) benzyl bromide, (3-bromopropyl)benzene, *p*-bromotoluene
(b) 1-bromocyclohexene, bromocyclohexane, 1-(bromomethyl)cyclohexene

---

18.2   ELIMINATION REACTIONS OF VINYLIC HALIDES

Although S_N2 reactions of vinylic halides are unknown, base-promoted β-elimination reactions of vinylic halides do occur and can be useful in the synthesis of alkynes.

$$\text{Ph}\text{—CH}\text{=}\text{CH}\text{—Br} + \text{KOH} \xrightarrow{200\,°C} \text{Ph}\text{—C}\text{≡}\text{C}\text{—H} + \text{K}^+ \text{Br}^- + \text{H}_2\text{O} \qquad (18.4)$$

(*E* or *Z*)   **phenylacetylene**
(distills from the
reaction mixture;
67% yield)

$$\overset{\text{Br}\quad\text{Br}}{\text{Ph}\text{—CH}\text{—CH}\text{—Ph}} + 2\,\text{KOH} \xrightarrow{\text{EtOH}} \text{Ph}\text{—C}\text{≡}\text{C}\text{—Ph} + 2\,\text{K}^+ \text{Br}^- + 2\,\text{H}_2\text{O} \qquad (18.5)$$

(67% yield)

In Eq. 18.5, two successive eliminations take place. The first gives a vinylic halide and the second gives the alkyne.

Many vinylic eliminations require rather harsh conditions (heat or very strong bases), and some of the more useful examples of this reaction involve elimination of β-hydrogens with enhanced acidity. Notice, for example, that the hydrogens which are eliminated in Eqs. 18.4 and 18.5 are benzylic (Sec. 17.3B).

Can aryl halides undergo β-elimination? Try to answer this question by constructing a model of the alkyne that would be formed in such an elimination. (This possibility is explored in Problem 18.83.)

PROBLEM

18.2   Arrange the following compounds according to increasing rate of elimination with NaOEt in EtOH. What is the product in each case?

A   *B*   *C*   *D*

LACK OF REACTIVITY OF VINYLIC AND ARYL HALIDES UNDER S$_N$1 CONDITIONS

Recall that tertiary and some secondary alkyl halides undergo nucleophilic substitution and elimination reactions by the S$_N$1 and E1 mechanisms (Sec. 9.6). For example, *tert*-butyl bromide undergoes a rapid solvolysis in ethanol to give both substitution and elimination products.

$$H_3C-\underset{\underset{CH_3}{|}}{\overset{\overset{CH_3}{|}}{C}}-\ddot{B}r: \underset{\underset{55\,°C}{EtOH}}{\overset{\longrightarrow}{\longleftarrow}} \left[H_3C-\underset{\underset{CH_3}{|}}{\overset{\overset{CH_3}{|}}{C^+}} \quad :\ddot{B}r:^- \right] \overset{EtOH}{\longrightarrow} H_3C-\underset{\underset{CH_3}{|}}{\overset{\overset{CH_3}{|}}{C}}-OEt + \underset{\underset{H_3C}{}}{\overset{\overset{H_3C}{}}{}}C=CH_2 + H\ddot{B}r: \quad (18.6)$$

$$(72\%) \qquad (28\%)$$

Vinylic and aryl halides, however, are virtually inert to the conditions that promote S$_N$1 or E1 reactions of alkyl halides. Certain vinylic halides can be forced to react by the S$_N$1–E1 mechanism under extreme conditions, but such reactions are relatively uncommon.

$$H_2C=\underset{\underset{}{\overset{\overset{CH_3}{|}}{C}}}-Br + EtOH \xrightarrow{55\,°C} \text{no reaction} \qquad (18.7)$$

$$\text{⬡}-Br + EtOH \xrightarrow{55\,°C} \text{no reaction} \qquad (18.8)$$

To understand why vinylic and aryl halides are inert under S$_N$1 conditions, consider what would happen if they *were* to undergo the S$_N$1 reaction. If a vinylic halide undergoes an S$_N$1 reaction, it must ionize to form a *vinylic cation*.

$$\underset{\underset{:\ddot{B}r:}{}}{\overset{\overset{R}{}}{}}C=CH_2 \longrightarrow R-\overset{+}{C}=CH_2 \quad :\ddot{B}r:^- \qquad (18.9)$$

a vinylic cation

A **vinylic cation** is a carbocation in which the electron-deficient carbon is also part of a double bond. An orbital diagram of a vinylic cation is shown in Fig. 18.2a on p. 884. The electron-deficient carbon is connected to two groups, the R group and the other carbon of the double bond. Hence, the geometry at this carbon is *linear* (Sec. 1.3B) and the electron-deficient carbon is therefore *sp*-hybridized. Notice that the vacant 2*p* orbital is *not* conjugated with the π-electron system of the double bond; in order to be conjugated, it would have to be coplanar with the double-bond π system. Vinylic cations are considerably less stable than alkyl carbocations because their *sp* hybridization has a higher energy than the *sp*^{2} hybridization of alkyl cations (the same reason that alkynes are less stable than isomeric dienes), and because the electron-withdrawing polar effect of the double bond discourages formation of positive charge at a vinylic carbon. Hence, one reason that vinylic halides do not undergo the S$_N$1 reaction is the *instability of the vinylic cations* that would necessarily be involved as reactive intermediates.

The second reason that vinylic halides do not undergo the S$_N$1 reaction is that carbon–halogen bonds are stronger in vinylic halides than they are in alkyl halides. A vinylic carbon–halogen bond involves an *sp*^{2} carbon orbital, whereas an alkyl carbon–halogen bond involves an *sp*^{3} carbon orbital. Hence, a vinylic carbon–halogen bond has more *s* character. Recall that bonds with more *s* character are stronger (Eq. 14.27, p. 700). Consequently, it takes more energy to break the carbon–halogen bond of a vinylic halide. This additional energy is reflected in a smaller rate of ionization.

We've seen that vinylic cations are formed as intermediates in the hydration of alkynes (Eq. 14.5a and Fig. 14.8). The formation of vinylic cations in alkyne hydra-

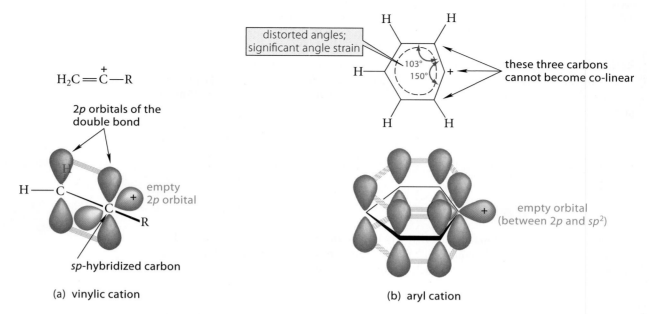

FIGURE 18.2 Lewis structures and corresponding orbital diagrams of vinylic and aryl cations. The thin gray lines indicate orbital overlap. (a) A vinylic cation. In this cation, the empty $2p$ orbital of the cation (*blue*) is oriented at right angles to the $2p$ orbitals of the aromatic π-electron system. The carbon with the empty $2p$ orbital is sp-hybridized. (b) A phenyl cation, the simplest aryl cation. The structure is a compromise between the linear geometry associated with sp hybridization and the 120° bond angles that are optimal for a planar six-membered ring. Consequently, the phenyl cation has significant angle strain, and the empty orbital (*blue*) has more s character than a $2p$ orbital. Because hybridization and geometry are connected, the hybridization of the electron-deficient carbon is somewhere between sp and sp^2.

---

tion is much less costly energetically than in S_N1 solvolysis. The reason is that the alkyne starting material is itself sp-hybridized, so the energy cost of rehybridization is not an issue. Furthermore, one of the carbons of the alkyne—the carbon that is protonated—goes from sp to sp^2 hybridization, which is energetically favorable.

both carbons are
sp-hybridized

$$R-C\equiv CH + H_3O^+ \longrightarrow R-\overset{+}{C}=CH_2 + H_2O \qquad (18.10)$$

goes from sp hybridization
to sp^2 hybridization
(energetically favorable)

formation of a vinylic cation intermediate
in the hydration of an alkyne

Furthermore, there is no strong bond to a leaving group to be broken to form this vinylic cation as there is in Eq. 18.9. The only energetically unfavorable aspects of the vinylic cation are the presence of an electron-deficient carbon, as in all carbocations, and the intrinsic electronegativity of the sp-hybridized carbons.

S_N1 reactions of aryl halides would involve aryl cations as reactive intermediates.

$$\text{(structure)} \qquad (18.11)$$

phenyl cation
(an aryl cation)

An **aryl cation** is a carbocation in which the electron-deficient carbon is part of an aromatic ring. An orbital diagram of phenyl cation, the simplest aryl cation, is shown in Fig. 18.2b. Because the electron-deficient carbon in an aryl cation is bonded to two groups, VSEPR theory suggests that it should have a linear geometry. Linear geometry is associated with sp hybridization. As shown in Fig. 18.2b, the geometry is distorted toward linearity with a resulting bond angle of 150°. To compensate, the neighboring angles are narrowed from optimum 120° to 103°. The empty orbital, which would be a $2p$ orbital in sp hybridization, is between $2p$ and sp^2. Because an aryl cation is forced to assume a nonoptimal geometry and hybridization, it has a very high energy. The electron-withdrawing polar effect of the ring double bonds also destabilizes an aryl cation, just as a double bond destabilizes a vinylic cation. Thus, S_N1 reactions of aryl halides do not occur because they would require the formation of carbocation intermediates—aryl cations—with very high energy.

> Note that an aryl cation is quite different from the arenium ion formed in electrophilic aromatic substitution (Eq. 16.9a, p. 802); the arenium-ion intermediate is stabilized by resonance. In an aryl cation, the empty orbital is not part of the ring π-electron system but is orthogonal (at right angles) to it. Hence, this carbocation is *not* resonance-stabilized.

The first direct observation of an aryl cation (the phenyl cation, Eq. 18.11) was reported in 2000 by chemists at the Ruhr-Universität in Bochum, Germany, who trapped the cation at 4 K and observed it spectroscopically. Thus, aryl cations are known species. However, they are *far* too unstable to form from aryl halides under typical S_N1 conditions.

PROBLEM

18.3 Within each series, arrange the compounds according to increasing rates of their reactions by the S_N1–E1 mechanism. Explain your reasoning.

18.4 NUCLEOPHILIC AROMATIC SUBSTITUTION REACTIONS OF ARYL HALIDES

Although aryl halides do not undergo nucleophilic substitution reactions by S_N1 and S_N2 mechanisms, aryl halides that have one or more nitro groups ortho or para to the halogen undergo nucleophilic substitution reactions under relatively mild conditions.

p-fluoronitrobenzene

p-nitroanisole
(93% yield)

1-chloro-2,4-dinitrobenzene   **2,4-dinitroaniline**
(70% yield)

$$\text{(18.13)}$$

These reactions are examples of **nucleophilic aromatic substitution**: substitution that occurs at a carbon of an aromatic ring by a nucleophilic mechanism.

Let's examine some of the characteristics of this mechanism. Like S_N2 reactions, nucleophilic aromatic substitution reactions involve nucleophiles and leaving groups, and they also obey second-order rate laws.

$$\text{rate} = k[\text{aryl halide}][\text{nucleophile}] \qquad (18.14)$$

However, nucleophilic aromatic substitution reactions do not involve a concerted opposite-side substitution for the reasons given in Sec. 18.1. Two clues about the reaction mechanism come from the reactivities of different aryl halides. First, the reaction is faster when there are more nitro groups ortho and para to the halogen leaving group:

Second, the effect of the halogen on the rate of this type of reaction is quite different from that in the S_N1 or S_N2 reaction of alkyl halides. In nucleophilic aromatic substitution reactions, aryl fluorides are most reactive.

Reactivities of aryl halides:

$$\text{Ar—F} \gg \text{Ar—Cl} \approx \text{Ar—Br} \approx \text{Ar—I} \qquad (18.16)$$

In S_N2 and S_N1 reactions of *alkyl* halides, the reactivity order is exactly the reverse: alkyl fluorides are the least reactive alkyl halides (Secs. 9.4F and 9.6C).

These data are consistent with a reaction mechanism in which the nucleophile reacts at the halide-bearing carbon below (or above) the plane of the aromatic ring to yield a resonance-stabilized anion called a *Meisenheimer complex*. In this anion, the negative charge is delocalized throughout the π-electron system of the ring. Formation of this anion is the rate-limiting step in many nucleophilic aromatic substitution reactions.

a Meisenheimer complex

The negative charge in this complex is also delocalized into the nitro group.

$$[\text{structures}] \qquad (18.17b)$$

The Meisenheimer complex breaks down to products by loss of the halide ion.

$$O_2N—\text{(ring)}—OCH_3 \longrightarrow O_2N—\text{(ring)}—OCH_3 + :\ddot{F}:^- \qquad (18.17c)$$

Let's see how this mechanism fits the experimental facts. Ortho and para nitro groups accelerate the reaction because the rate-limiting transition state resembles the Meisenheimer complex, and ortho and para nitro groups (but not meta nitro groups) stabilize this complex by resonance. Fluorine also stabilizes the negative charge by its electron-withdrawing *polar effect*, which is greater than the polar effect of the other halogens. Because the *loss of halide is not rate-limiting*, the basicity of the halide, or equivalently, the strength of the carbon–halogen bond, is not important in determining the reaction rate.

Although we have used aryl halides substituted with ortho and para nitro groups to illustrate nucleophilic aromatic substitution, it stands to reason that other substituents that can provide resonance stabilization to the Meisenheimer complex can also activate nucleophilic aromatic substitution. (See, for example, Problem 18.5b.)

Notice how the nucleophilic aromatic substitution reaction differs from the S_N2 reaction of alkyl halides. First, there is an actual intermediate in the nucleophilic aromatic substitution reaction—the Meisenheimer complex. (In some cases, this is sufficiently stable that it can be directly observed.) There is no evidence for an intermediate in any S_N2 reaction. Second, the nucleophilic aromatic substitution reaction is a frontside substitution; it requires no inversion of configuration. The S_N2 reaction of an alkyl halide, in contrast, is an opposite-side substitution with inversion of configuration. Finally, the effect of the halogen on the reaction rate (Eq. 18.16) is different in the two reactions. Aryl fluorides react most rapidly in nucleophilic aromatic substitution, whereas alkyl fluorides react most slowly in S_N2 reactions.

STUDY GUIDE LINK 18.1
Contrast of Aromatic Substitution Reactions

PROBLEMS

18.4 Complete the following reactions. (*No reaction* may be the correct response.)

(a)
$$O_2N—\text{(ring)}—Cl + Et\ddot{N}H_2 \xrightarrow{\text{heat}}$$
(with NO$_2$)

(b)
$$\text{(ring)}—F + Bu—\ddot{S}:^- \xrightarrow{CH_3OH}$$
(with NO$_2$)

(c)
$$CH_3O—\text{(ring)}—F + CH_3\ddot{O}:^- \xrightarrow[CH_3OH]{25\,°C}$$

18.5 Which of the two compounds in each of the following sets should react more rapidly in a nucleophilic aromatic substitution reaction with CH_3O^- in CH_3OH? Explain your answers.

(a) [structure with F and NO$_2$] or [structure with F and NO$_2$]

(b) [structure with F, NO$_2$, and $O=C—OCH_3$] or [structure with F and NO$_2$]

18.5 INTRODUCTION TO TRANSITION-METAL-CATALYZED REACTIONS

We've just learned that S_N1 and S_N2 reactions cannot be carried out on either aryl or vinylic halides. However, reactions that *look* very much like nucleophilic substitutions *can* be carried out using certain transition-metal catalysts. Here are some examples.

$$\text{o-bromotoluene} + H_2C=CH_2 \xrightarrow[\substack{\text{CH}_3\text{C}\equiv\text{N (solvent)} \\ \text{18 h, 125 °C}}]{\substack{\text{(catalyst)} \\ \text{Et}_3\text{N:}}} \text{o-methylstyrene} + HBr \quad (18.18)$$

o-bromotoluene

o-methylstyrene
(86% yield)

HBr neutralized by the Et₃N:

This reaction, called the *Heck reaction*, has become very important in organic synthesis. We'll revisit this reaction in Sec. 18.6A. Notice the formation of the carbon–carbon bond and the release of bromide as HBr. Superficially, it looks as if the conjugate-base anion of ethylene displaces bromide ion from the aromatic ring. However, this reaction occurs by a very different mechanism and does not happen without the palladium catalyst. (Only about 1 mole % of the catalyst is required.)

In the following reaction, we see the substitution of a *vinylic* bromide by a thiolate anion.

$$\begin{array}{c}\text{Ph} \quad \text{Br} \\ \text{C}=\text{C} \\ \text{H} \quad \text{H}\end{array} + \text{Li}^+ \ ^-\text{SEt} \xrightarrow[\text{benzene}]{\substack{\text{Pd(PPh}_3)_4 \\ \text{(catalyst)}}} \begin{array}{c}\text{Ph} \quad \text{SEt} \\ \text{C}=\text{C} \\ \text{H} \quad \text{H}\end{array} + \text{Li}^+ \ \text{Br}^- \quad (18.19)$$

(93% yield)

This reaction, too, looks superficially like a nucleophilic substitution reaction. But this reaction also proceeds by a different mechanism and does not take place without the catalyst, which is present in only 1 mole %. Notice also the *retention* of alkene stereochemistry, a very different result from that expected in an S_N2 reaction.

These are but two examples of thousands now known in which transition-metal catalysts bring about seemingly "impossible" reactions. The field of transition-metal catalysis has exploded in the last four decades, and it has become very important in both laboratory and industrial chemistry, as well as in some areas of biology. This field is part of the larger field of *organometallic chemistry*: the chemistry of carbon–metal bonds. (Grignard reagents and lithium dialkylcuprate reagents are examples of organometallic compounds that you encountered in Secs. 9.8 and 11.5C and will encounter again in subsequent chapters.) Our goal in this section is to understand some of the basic ideas of transition-metal catalysis. Then, in Sec. 18.6, we'll examine a few important transition-metal-catalyzed reactions in the light of these principles.

A. Transition Metals and Their Complexes

Recall from general chemistry that **transition metals** are the elements in the "*d* block" or "B" groups of the periodic table (groups 3–12 in the IUPAC numbering). These elements are shown in Fig. 18.3. In a given period *n*, elements are characterized by the progressive filling of *d* orbitals in quantum level *n* − 1 and the *s* orbital in quantum level *n*. Thus, in the fourth period, the transition elements are characterized by the filling of the one 4*s* and the five 3*d* orbitals. Because the 4*s* and 3*d* orbitals have very similar energies, it is usually convenient to

Group number	3B	4B	5B	6B	7B		8B		1B	2B
Valence electrons in the neutral atom	3	4	5	6	7	8	9	10	11	12
Period 4 →	Sc	Ti	V	Cr	Mn	Fe	Co	Ni	Cu	Zn
Period 5 →	Y	Zr	Nb	Mo	Tc	Ru	Rh	Pd	Ag	Cd
Period 6 →	Lu	Hf	Ta	W	Re	Os	Ir	Pt	Au	Hg

FIGURE 18.3 The transition metals. The red numbers indicate the number of valence electrons (outer shell *s* and *d* electrons) in the neutral atoms. (These are the same as the IUPAC group numbers in the periodic table; see the inside of the back cover.)

think of the electrons in both types of orbitals together as valence electrons. For example, Ni has the electronic configuration $[Ar]4s^2 3d^8$, but we classify Ni as a 10-valence-electron atom. ([Ar] represents the electronic configuration of the noble gas argon.)

Central to transition-metal chemistry are a wide variety of compounds containing transition metals surrounded by several groups, called **ligands**. Such compounds are called **coordination compounds** or **transition-metal complexes**. These can be neutral molecules, as in the first of the following examples, or *complex ions*, as in the second example.

cis-diamminedichloroplatinum(II)
(**cis-platin**, an antitumor drug)
a neutral complex

hexamminecobalt(III) ion
a complex ion

To deal systematically with transition-metal complexes, we must be aware of, and be able to apply, certain conventions:

1. how to classify ligands
2. how to specify formal charge on the metal
3. how to calculate the oxidation state of the metal
4. how to count electrons around the metal

In transition-metal chemistry, all ligands are *Lewis bases*. That is, ligands interact with transition metals by donating electron pairs. There are two types of ligands. The first we'll call an *L-type ligand*. If you imagine that if a ligand dissociates from the metal with its bonding electron pair and thus becomes a neutral molecule, the ligand is an **L-type ligand**. For example, any one of the NH_3 ligands in the two preceeding complexes is an L-type ligand because if we remove it with its bonding electron pair, we get $:NH_3$, the neutral molecule ammonia.

The second type of ligand is termed an *X-type ligand*. If you imagine that if a ligand dissociates from the metal with its bonding electron pair and thus becomes a negative ion, the ligand is an **X-type ligand**. Thus, Cl in *cis*-platin (the first example) is an X-type ligand, because removing it with its bonding pair of electrons gives the chloride ion, Cl^-.

The classification of ligands has implications for computing formal charge. *From a formal-charge perspective, the bonding electrons on L-type ligands are considered to "belong" completely to the ligand.* Let's see how this differs from the way we treat bonds in main-group chemistry. We know that a nitrogen with four bonds in main-group chemistry, for example, the ammonium ion, $^+NH_4$, has a positive formal charge. If we were to take a similar view with *cis*-platin, the nitrogens would each have a positive charge; and, because the complex is neutral, the Pt would have a charge of -2. A neutral transition-metal complex ML_6 bearing six L-type ligands would thus have a charge of -6 on the metal and a positive charge on each ligand. It is inconvenient to draw out all of these charges; moreover, a formal charge of -6 on a metal is highly unrealistic. Instead, we adopt the *convention* that the electron pair in an L-type ligand is assigned completely to the ligand. Sometimes this point is emphasized by leaving the bonding electron pair on the ligand and depicting the ligand–metal bond as an arrow from these electrons to the metal. This is called a *dative bond*.

$$
\begin{array}{c}
\text{Cl} \diagdown \quad \diagup :NH_3 \\
\text{Pt} \quad \text{dative bonds} \\
\text{Cl} \diagup \quad \diagdown :NH_3
\end{array}
$$

Because electrons on an L-type ligand belong to the ligand, removal of the ligand does not change the formal charge on either the ligand or the metal:

$$
\begin{array}{c}
\text{Cl} \diagdown \quad \diagup \text{NH}_3 \\
\text{Pt} \\
\text{Cl} \diagup \quad \diagdown \text{NH}_3
\end{array}
\qquad \longrightarrow \qquad
\begin{array}{c}
\text{Cl} \diagdown \quad \diagup \text{NH}_3 \\
\text{Pt} \\
\text{Cl} \diagup
\end{array}
\; + \; :NH_3
\qquad (18.20)
$$

In contrast, electrons in the bonds to X-type ligands are assigned in the same way that we assign electrons in main-group chemistry: *one electron is assigned to the ligand and one to the metal*. This means that if we remove an X-type ligand, it takes on an additional negative charge and the metal takes on a compensating positive charge:

$$
\begin{array}{c}
\text{Cl} \diagdown \quad \diagup \text{NH}_3 \\
\text{Pt} \\
:\ddot{\text{Cl}} \diagup \quad \diagdown \text{NH}_3
\end{array}
\qquad \longrightarrow \qquad
\begin{array}{c}
\text{Cl} \diagdown \quad \overset{+}{\diagup} \text{NH}_3 \\
\text{Pt} \\
\diagdown \text{NH}_3
\end{array}
\; + \; :\ddot{\text{Cl}}:^{-}
\qquad (18.21)
$$

Differentiating between X-type and L-type bonds is a very convenient bookkeeping device, but we should bear in mind that both types of bonds are covalent bonds, and the degree to which electrons (and charge) are transferred to the metal varies widely in both types of bonds, depending on the metal and the ligand.

Table 18.1 lists some of the common ligands used in transition-metal chemistry. These are classified as L-type or X-type ligands. It is worth noting two things about this table. First, alkenes or aromatic rings can act as ligands by donating their π electrons to a metal. Second, allyl and cyclopentadienyl (Cp) are classified as both L-type and X-type ligands. Let's consider the Cp case to understand this. The cyclopentadienyl anion was discussed in Sec. 15.7D as an example of an aromatic ion with six π electrons. Table 18.1 indicates that Cp is an example of an L_2X ligand. What this means is that *one* X-type bond accounts for the fact that Cp takes on *one* negative charge when removed with a bonding pair from the metal, and that the four remaining π electrons (that is, two double bonds) take part in two L-type bonds. In other words, we can think of a metal–Cp complex in the following way (M = metal):

$$(18.22)$$

We know that the π electrons in Cp are completely delocalized, and they remain delocalized in metal complexes. (See the structure of ferrocene, Cp_2Fe, p. 770, which shows this delocalization.) Consequently, a more accurate picture of such a complex would show the "L" and "X" character of the bonds parceled out equally over all five carbons, with each carbon participating in 20% of an X-type bond ($5 \times 0.20 = 1.0$ X-type bond) and 40% of an L-type bond ($5 \times 0.40 = 2.0$ L-type bonds). But this delocalization can be ignored for the bookkeeping purposes discussed in this section.

TABLE 18.1 Some Typical Ligands Used in Transition-Metal Chemistry

Ligand	Name	Abbreviation	Type	Electron count*
$H_3N:$	ammine		L	2
$H_2\ddot{O}:$	aquo		L	$2^\dagger$
$R_3P:$ (R = alkyl, aryl)	trialkylphosphino, triarylphosphino		L	2
$:C{=}\ddot{O}:$	carbonyl	CO	L	$2^\ddagger$
$H_2C{=}CH_2{}^\S$	ethylene		L	2
$CH_3C{\equiv}N:$	acetonitrile	MeCN	L	$2^{\S\S}$
benzene ring	benzene		L_3	6
F^-, Cl^-, Br^-, I^-	halo (e.g., chloro)	X	X	$2^\dagger$
H^-	hydrido		X	2
$H_3C{-}\overset{\overset{\displaystyle O}{\|}}{C}{-}O^-$	acetato	AcO	X	$2^\dagger$
$R:^-$ e.g., $H_3C:^-$	alkyl (e.g., methyl)		X	2
$^-:C{\equiv}N:$	cyano	CN	X	$2^\ddagger$
$H_2C{=}CH{-}\ddot{C}H_2$	allyl		LX	4^{**}
cyclopentadienyl ring	cyclopentadienyl	Cp	L_2X	6

* The sum of all electrons in the bond(s) between the ligand and the metal.

$\dagger$ Only one electron pair is involved in the ligand–metal interaction.

$\ddagger$ Only the electron pair on carbon is involved in the ligand–metal interaction.

$\S$ Ethylene is listed as a prototype for many alkenes.

$\S\S$ Donation of the nitrogen unshared election pair.

** Allyl can also bind to metals as an X-type ligand. In such a situation, the p bond is not involved in coordination and the electron count is 2 (as with alkyl).

PROBLEM

18.6   Noting the LX character of the allyl ligand in Table 18.1, sketch the allyl–metal interaction, showing both L-type and X-type bonds. Use M as a general metal.

B. Oxidation State

The oxidation state of the metal is an important concept in organometallic chemistry. Oxidation state can be conceptualized as the charge the metal would have if all covalently bonded atoms—that is, all X-type ligands—were dissociated with their bonding electron pairs. For example, if the oxygens of manganese dioxide, $O{=}Mn{=}O$, were to dissociate from the manganese with their bonding electron pairs, the oxygens would each take on a -2 charge, and the manganese would take on a $+4$ charge. Therefore, the manganese has an oxidation state of $+4$. [This process is essentially the same as the one used for assigning oxidation numbers to carbon atoms in oxidation–reduction reactions (Sec. 10.6A)]. Eq. 18.23 formalizes this idea:

$$\text{Oxidation state of M} = \text{number of X-type ligands} + Q_M \qquad (18.23)$$

In this equation, Q_M is the actual charge that M has *before* the fictitious dissociation of the X-type ligands. To illustrate a situation involving Q_M, consider the hexachloroplatinate dianion, $[Pt(Cl_6)]^{2-}$. For this species, $Q_M = -2$. If the six chlorines were to dissociate with their bonding electrons, the charge on Pt would be $+4$, because Pt starts out with a -2 charge before the fictitious dissociation. To be sure that the oxidation state and the actual charge Q_M are not confused, the oxidation state is sometimes indicated with Roman numerals. Hence, the name of the ion $[Pt(Cl_6)]^{2-}$ is hexachloroplatinate(IV).

L-type ligands do *not* contribute to the oxidation state. You should verify that the oxidation state of platinum in the neutral complex $Cl_2Pt(PPh_3)_2$ is $+2$.

PROBLEMS

18.7   Calculate the oxidation state of the metal in each of the following complexes.

(a)

$$O=Mn-O^-$$
with O double-bonded above and below Mn

permanganate

(b) $Pd(PPh_3)_4$

tetrakis(triphenylphosphine)palladium

(c) Cp_2Fe

ferrocene

18.8   What is the oxidation state of the metal in the starting material in the following reaction? How does it change, if at all, as a result of the reaction? Is this reaction an oxidation, a reduction, or neither?

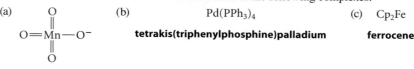

chlorotris(triphenylphosphine)rhodium

C. The d^n Notation

In understanding the reactions of main-group elements that follow the octet rule, it is important in applying acid–base concepts for us to know whether the element undergoing a transformation has unshared valence electrons. Often these unshared electrons are shown explicitly. In transition-metal chemistry, it is also important to know whether the metal has unshared valence electrons. In many cases, the metal has so many unshared valence electrons that it would be impractical or confusing to draw them all. Instead, we use a convenient algorithm to calculate the number of unshared valence electrons. The number of unshared valence electrons on the metal is the number n in a notation called d^n. For example, if the metal in a complex has eight unshared valence electrons, we say that the complex is a d^8 complex.

We calculate n by determining the number of valence electrons remaining on the metal after removing all ligands with their electron pairs. We start with the number of valence electrons in the *neutral transition element* (from Fig. 18.3). We remove an electron for each positive charge, add an electron for each negative charge, and then subtract one electron for each bond to an X-type ligand. L-type ligands have no effect on d^n.

$$n = \text{valence electrons in neutral M} - Q_M - \text{number of X-type ligands} \qquad (18.24)$$

Introducing the definition of oxidation state in Eq. 18.23, Eq. 18.24 becomes

$$n = \text{valence electrons in neutral M} - \text{oxidation state of M} \qquad (18.25)$$

STUDY PROBLEM 18.1

Calculate n in the d^n notation for ferrocene, Cp_2Fe.

SOLUTION We'll make this calculation with both Eqs. 18.24 and 18.25. From Fig. 18.3 we see that neutral Fe has eight valence electrons. The charge of the iron is zero, and Table 18.1 shows that each Cp ligand has one X-type bond; the iron thus has two bonds to X-type ligands. Hence, $n = 8 - 2 = 6$, and ferrocene is thus a d^6 complex.

From Eq. 18.25, we calculate that with two X-type ligands and zero charge, the Fe in ferrocene has a $+2$ oxidation state; hence, Eq. 18.25 gives the value of n as $8 - 2 = 6$.

PROBLEM

18.9 What is d^n for each of the following complexes?

(a) $[W(CO)_5]^{2-}$ (b) $Pd(PPh_3)_4$ (c)

D. Electron Counting. The 16- and 18-Electron Rules

In main-group chemistry, we use the octet rule as one indicator of reactivity. For example, we know that if a main-group element in a compound has fewer than an octet of electrons, it can accept an electron pair from a Lewis base in a Lewis acid–base association reaction. In other words, main-group elements have a tendency to complete their octets. Recall that counting for the octet involves adding an element's unshared valence electrons to the number of electrons in all bonds to the element. The electron count in transition-metal complexes is also important and is determined in a similar manner.

To determine the **electron count** for a transition-metal complex, we start with the n electrons in the d^n count—the unshared electrons—and add two electrons for every ligand (both L-type and X-type). Thus,

$$\text{electron count} = n + 2(\text{number of all ligands}) \tag{18.26}$$

The multiplier 2 is required because there are two electrons per bond. The rationale for this formula is that the number n is the number of unshared valence electrons on the metal; the total electron count is the unshared valence electrons plus all electrons in bonds, just as in counting for the octet rule. Using Eq. 18.25, we can rewrite this formula in terms of the oxidation state of the metal:

$$\text{electron count} = -\text{valence electrons in neutral M} - \text{oxidation state of M} + 2(\text{number of all ligands}) \tag{18.27}$$

By incorporating the definition of oxidation state (Eq. 18.28), we obtain yet another equivalent formula:

$$\text{electron count} = \text{valence electrons in neutral M} - Q_M - \text{number of X-type ligands} + 2(\text{number of all ligands}) \tag{18.28}$$

Recognizing that all ligands = X-type ligands + L-type ligands, we finally obtain a very useful formula for electron count:

$$\text{electron count} = \text{valence electrons in neutral M} - Q_M + \text{number of X-type ligands} + 2(\text{number of L-type ligands}) \tag{18.29}$$

Thus, to obtain the electron count in a complex, we start with the electron count of the neutral metal from Fig. 18.3; we subtract the charge on the metal (taking into account its algebraic sign); we add the number of X-type ligands; and we add *twice* the number of L-type ligands. Don't let the mathematical derivation of Eq. 18.29 obscure its rationale. Remember that the goal is to count all unshared and bonding electrons about the metal. Because an X-type ligand *by definition* has one electron in its M—X bond assigned to X, we have to add this electron back to obtain the total number of electrons in the bond. Because both electrons in the dative bond to an L-type ligand are assigned to the ligand, we have to multiply each L-type ligand by 2 to count both of these bonding electrons.

Let's use Eq. 18.29 to calculate some electron counts. For example, the electron count of $Ni(CO)_4$ is $10 - 0 - 0 + 2(4) = 18$. This is an 18-electron complex. (This compound, tetra-carbonylnickel(0), is a very stable complex of Ni.)

The electron count of $Cl_2Pd(PPh_3)_2$ is $10 - 0 + 2 + 2(2) = 16$. This is a 16-electron complex.

In transition-metal chemistry, the most stable complexes in many cases have electron counts of 18 electrons. This statement is called the **18-electron rule**. $Ni(CO)_4$, a very stable complex of Ni(0), is an example of the 18-electron rule. Just as the octet represents the number of valence electrons (8) in the outermost s and p orbitals of the nearest noble gas, 18 electrons is also the number of total $s + p + d$ valence electrons in the nearest noble gas.

Exceptions to the 18-electron rule occur, and an important type of exception occurs frequently with transition metals in the 8–11 valence-electron group (Fig. 18.3), which includes Ni and Pd, two metals of prime importance in the transition-metal-catalyzed reactions discussed in this section. Although a number of stable complexes of these metals have 18 electrons, others contain 16 electrons. The tendency of these metals to surround themselves with 16 electrons can be called the **16-electron rule**. The $Cl_2Pd(PPh_3)_2$ example shows the operation of the 16-electron rule.

PROBLEMS

18.10 (a) What is the electron count for the Rh complex shown in Problem 18.9c?

(b) *Sestamibi* (trade name Cardiolite®) is a complex of $^{99}Tc(I)$ (a radioactive γ-ray emitter) that is widely used for cardiac and parathyroid imaging.

$[TcL_6]^+$, where L = $\longleftarrow$ $:\overset{-}{C}\equiv\overset{+}{N}-$

"sestamibi"

OCH_3

***N*-(2-methoxyisobutyl) isocyanide (MIBI)**

Note that "MIBI" is an L-type ligand. Give the value of n for the d^n notation and the total electron count for the technetium.

18.11 How many CO ligands would be accommodated by Fe(0) if we assume that the resulting complex follows the 18-electron rule?

18.12 Using the 18-electron rule, explain why $V(CO)_6$ can be easily reduced to $[V(CO)_6]^-$.

We used hybridization arguments to understand the basis of the octet rule in main-group chemistry (Sec. 1.9). Thus, the main-group element carbon has four valence orbitals (for example, four sp^3 hybrid orbitals) that can either form two-electron bonds or house unshared electron pairs. We can justify the 18-electron rule in a similar way. Consider, for example, the complex ion $[Co(CN)_6]^{3-}$. Using Eq. 18.23, we see that the oxidation state of Co is +3, and, from Eq. 18.25, that this is a d^6 complex. This means that Co(III) in this complex has six unshared electrons. Let's imagine building this complex from a "naked" Co^{3+} ion. Start with the electronic configuration of this ion, as shown in Fig. 18.4a. Allow all the electrons to pair, as shown in Fig. 18.4b. Because this electron pairing violates Hund's rules, it requires energy. This electron pairing leaves two $3d$, one $4s$, and three $4p$ orbitals unoccupied.

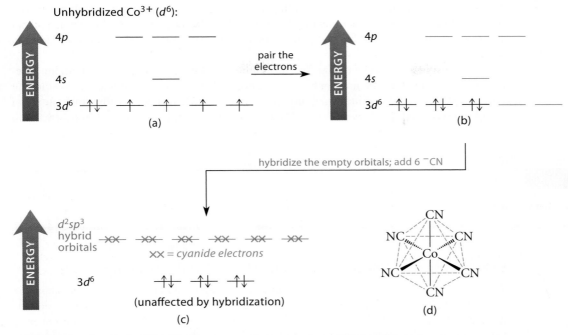

FIGURE 18.4 Development of the hybrid orbital description of $[Co(CN)_6]^{3-}$, an 18-electron complex ion. (a) The electronic configuration of Co^{3+}. (b) The Co electrons are arranged in pairs; the empty orbitals that remain (*red*) are used to form hybrid orbitals. (c) The empty orbitals are hybridized into six equivalent d^2sp^3 hybrid orbitals. Each of these orbitals can accept an electron pair (symbolized by ××) from a ^-CN ion. (d) The hybrid orbitals and, hence, the six ^-CN that bind to them are oriented to the corners of a regular octahedron. (The edges of the octahedron are indicated with blue dashed lines.)

These are hybridized, as shown in Fig. 18.4c, to give *six equivalent d^2sp^3 hybrid orbitals*. Six equivalent hybrid orbitals are directed to the corners of a regular octahedron in the same sense that sp^3 carbon orbitals in methane are directed to the corners of a regular tetrahedron. Hybridization also requires energy. Each of these empty hybrid orbitals can accept an electron pair from a cyanide ion (^-CN). Because these hybrid orbitals are directed in space, they can form stronger bonds to cyanide than unhybridized orbitals, and the strength of these bonds more than compensates for the energy cost of electron pairing and hybridization. The result is the octahedral $[Co(CN)_6]^{3-}$ complex shown in Fig. 18.4d. In other words, the 18-electron rule results from the rehybridization and maximal occupancy of all valence orbitals of the Co^{3+} ion.

The 16-electron rule is important in square planar complexes of the 10-electron elements Ni, Pd, and Pt. For example, consider the antitumor drug *cis*-platin, $Cl_2Pt(NH_3)_2$ (p. 889). This is a 16-electron d^8 complex of Pt(II) that has square planar geometry. If we start with a Pt^{2+} ion and arrange its eight electrons in pairs within four $5d$ orbitals, this leaves a single $5d$ orbital, a $6s$ orbital, and three $6p$ orbitals empty. It turns out that hybridization of four of the five empty orbitals to give four dsp^2 hybrid orbitals and one relatively high-energy $6p$ orbital is a particularly favorable hybridization:

The four hybrid orbitals are directed to the corners of a square and accept the electron pairs from the four ligands to give a square planar complex. The element platinum can also adopt 18-electron configurations, but the point is that the 16-electron configuration is reasonably stable.

As in main-group chemistry, hybridization arguments are useful for visualizing electrons in bonds, but they are inferior to molecular orbital arguments for detailed understanding of molecular energies. The branch of molecular orbital theory that deals with transition-metal complexes is called *ligand field theory*. We need not explore this theory here; but suffice it to say that this theory provides excellent support for the 16- and 18-electron rules.

PROBLEM

18.13   Use a hybridization argument to predict the geometry of (a) the $[Zn(CN)_4]^{2-}$ ion; (b) the neutral compound $Pd(PPh_3)_4$.

E. Fundamental Reactions of Transition-Metal Complexes

We have now been introduced to the preliminaries that we need to understand the mechanistic basis of some transition-metal-catalyzed reactions, and we're ready to look at these reactions in detail. It turns out that transition-metal complexes undergo a relatively small number of fundamental reaction types, and many reactions are readily understood simply as combinations of these fundamental processes. The goal of this section is to introduce a few of these.

Ligand Dissociation–Association; Ligand Substitution   One of the most common reactions of transition-metal complexes is *ligand dissociation* and its reverse, *ligand association*. In ligand dissociation, a ligand simply departs from the metal with its pair of electrons, leaving a vacant site (orbital) on the metal.

$$Pd(PPh_3)_4 \;\rightleftharpoons\; Pd(PPh_3)_3 \;+\; :PPh_3 \qquad (18.30)$$

$$\underset{\substack{Pd(0)\\ \text{an }18e^-\text{ complex}}}{} \qquad\qquad \underset{\substack{Pd(0)\\ \text{a }16e^-\text{ complex}}}{}$$

This process does not change the oxidation state of the metal, but it does change the electron count. A *ligand substitution* can occur by the dissociation of one ligand and the association of another, which is somewhat analogous to an S_N1 reaction in alkyl halides, or by a direct substitution, in which one ligand displaces another, somewhat analogous to the S_N2 reaction.

$$\underset{\substack{Pd(II)\\ \text{a }16e^-\text{ complex}}}{\underset{Ph_3P}{\overset{Ph_3P}{\diagdown}}\!\!\underset{Ph}{\overset{I}{\diagup}}\!\!Pd} \;+\; Li\!-\!CH_3 \;\longrightarrow\; \underset{\substack{Pd(II)\\ \text{a }16e^-\text{ complex}}}{\underset{Ph_3P}{\overset{Ph_3P}{\diagdown}}\!\!\underset{Ph}{\overset{CH_3}{\diagup}}\!\!Pd} \;+\; Li^+\; I^- \qquad (18.31)$$

In the most common ligand substitution reactions, ligands of the same type are exchanged: X-type ligands for X-type ligands and L-type ligands for L-type ligands.

Oxidative Addition   In **oxidative addition**, a metal M reacts with a compound X—Y to form a compound X—M—Y; the metal "inserts" into the X—Y bond. We have already studied an important reaction of this type: the formation of a Grignard reagent from Mg metal and an alkyl halide (Sec. 9.8A).

$$R\!-\!Br \;+\; \ddot{M}g \;\longrightarrow\; R\!-\!Mg\!-\!Br \qquad (18.32a)$$

$$\underset{Mg(0)}{} \qquad\qquad \underset{Mg(II)}{}$$

As the term "oxidative addition" implies, the Mg is oxidized. An electron count for Mg in this reaction is 2 for a metal and 4 in the Grignard reagent. (Remember, though, that Mg is a main-group metal and is not subject to the 16- or 18-electron rules.)

An example of oxidative addition from transition-metal chemistry is the insertion of Pd into the carbon–halogen bond of iodobenzene:

$$\text{Ph}_3\text{P} \diagdown \atop \text{Ph}_3\text{P} \diagup \text{Pd} \quad + \quad \text{Ph}\text{—}\text{I} \quad \longrightarrow \quad \text{Ph}_3\text{P} \diagdown \atop \text{Ph}_3\text{P} \diagup \text{Pd} \diagup{\text{Ph}} \atop \diagdown{\text{I}} \tag{18.32b}$$

Pd(0)
a 14e^- complex

Pd(II)
a 16e^- complex

Both of new bonds are X-type bonds. As a result of this reaction, both the electron count and the oxidation number of the metal increase by two units.

Oxidative addition is a remarkable reaction that lies at the heart of transition-metal catalysis with aryl and vinylic halides. Why is it that a metal can break a sigma bond in this way? Molecular orbital theory provides a simple way to understand this process, as shown in Fig. 18.5 on p. 898. Think of the carbon–halogen bond as a localized bond for simplicity, and imagine a molecular orbital treatment of this bond much like the molecular orbital treatment of the H—H bond in H_2 (Sec. 1.8A). The carbon–halogen bond has an associated *bonding molecular orbital*, which is occupied by the two bonding electrons, and an *antibonding molecular orbital*, which is unoccupied. The bonding molecular orbital can serve as a ligand, donating its electrons to one of the empty hybrid orbitals on the metal. At the same time, one of the filled *d* orbitals of the metal overlaps with the *antibonding* molecular orbital of the carbon–halogen bond. This additional overlap *strengthens* the metal–ligand interaction, but *weakens* the carbon–halogen bond, because addition of electrons to an antibonding molecular orbital removes the energetic advantage of bonding. (See Fig. 1.14, p. 33.) The carbon–halogen bond is weakened sufficiently that it actually breaks. Hence, electrons flow *from the aryl halide to the metal* and, at the same time, *from the metal to the aryl halide*. We can approximate the process as follows with the curved-arrow notation (L = other ligands):

$$\tag{18.33}$$

electron pair
in a *d* orbital

Oxidative addition can occur by a variety of mechanisms, but a concerted (one-step) process is fairly common.

Reductive Elimination   **Reductive elimination** is conceptually the reverse of oxidative addition, and the orbital interactions involved are the same, only in reverse. In reductive elimination, then, two ligands bond to each other and their bonds to the metal are broken; X—M—Y $\longrightarrow$ X—Y + M. An example of this process is the formation of a carbon–carbon bond between two ligands within a Ni complex:

$$\tag{18.34}$$

Ni(II)
a 16e^- complex

Ni(0)
a 14e^- complex

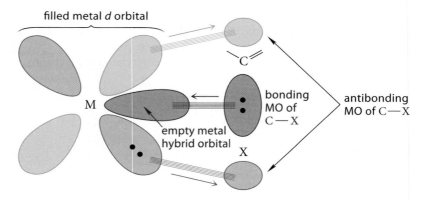

FIGURE 18.5 An orbital description of concerted oxidative addition. The bonding MO of the C—X bond donates electrons to an empty hybrid orbital on the metal. (The shape of this hybrid orbital is simplified.) These orbitals are shown in purple, the thin purple lines indicate the electronic overlap, and the purple arrow shows the direction of electron flow. At the same time, a filled d orbital on the metal donates electrons to an antibonding MO of the C—X bond. (A 3d orbital is used for simplicity.) Because the peaks and troughs of the d orbital (shown in blue and green, respectively) match the peaks and troughs of the antibonding MO of the C—X bond, this is a bonding interaction. The electronic overlap in this interaction is indicated with thin blue and green lines, and the blue and green arrows show the direction of electron flow. Because this interaction populates the antibonding MO of C—X, the C—X bond is weakened, and it breaks.

Because two X-type ligands are lost, the metal is reduced, and its electron count is decreased. Notice in this particular example that the alkene stereochemistry is retained. *Reductive elimination in general occurs with retention of stereochemistry.* Because this process is the reverse of oxidative addition, it follows that *concerted oxidative addition also occurs with retention of stereochemistry.* The electron flow in Fig. 18.5 or Eq. 18.33 is consistent with this observation.

Ligand Insertion   In this process, a ligand inserts into a metal–ligand bond; that is, L—M—R ⟶ M—L—R. Notice the difference between oxidative addition and ligand insertion. Both are insertion processes. In an oxidative addition, the metal inserts into a chemical bond within a compound not initially associated with the metal. In a ligand insertion, a ligand L inserts into the bond between the metal and a different ligand R, and the inserting ligand gains a bond.

Two types of ligand insertion are most frequently observed in transition-metal chemistry. In a *1,1-insertion*, the new bond is formed at the same atom that was bound to the metal. Insertions of CO ligands are frequently observed examples of this type. The first reaction below is an example.

$$(CO)_4Mn \longleftarrow :C{=}O \quad \xrightarrow[\text{insertion}]{\textit{1,1-ligand}} \quad (CO)_4Mn{-}\overset{\displaystyle CH_3}{\underset{}{C}}{=}O \quad \xrightarrow[\textit{ligand association}]{CO} \quad (CO)_4Mn{-}\overset{\displaystyle CH_3}{\underset{}{C}}{=}O \qquad (18.35)$$

Mn(I)   Mn(I)   Mn(I)
an 18e⁻ complex   a 16e⁻ complex   an 18e⁻ complex

In this reaction, the methyl group migrates, *with its bonding electrons*, to the carbon of the carbonyl ligand, which in turn forms an X-type bond to the metal. This migration is possible because the carbon of the carbon monoxide ligand is electron-deficient. Hence, the carbonyl carbon inserts into the Mn—CH₃ bond. Notice that this insertion leaves a vacant site (that is, an empty orbital) on the metal, as we can see from the reduction in the electron count. In the second reaction, this empty metal orbital is filled by another molecule of the ligand from solution.

Another type of ligand insertion is *1,2-insertion.* In this process, the migrating group moves to an atom adjacent to the one bound to the metal. A common example of this process is the following, in which an ethylene ligand inserts into a Pd–aryl bond.

$$(18.36)$$

Again, notice that the electron count is reduced by two; that is, the process results in an empty orbital on the metal. This orbital can then gain another electron pair by ligand association, as shown by the second step in Eq. 18.36, thus fulfilling the 16-electron rule.

We can approximate the ligand insertion process in the curved-arrow notation as follows:

$$(18.37a)$$

As Eq. 18.37a illustrates, 1,2-ligand insertion is essentially a concerted addition of the metal (Pd in this case) and the migrating ligand (Ar in this case) to the alkene π bond. Because 1,2-ligand insertion is a concerted *intramolecular* addition reaction, the two new bonds must be formed at the same face of the π bond. Hence, *ligand insertion is a syn-addition.* This becomes evident when the carbons of the alkene double bond are stereocenters, as they are in cyclohexene. In this case, syn-addition requires that the Pd and the aryl group have a cis relationship in the insertion product.

$$(18.37b)$$

β-Elimination In β-elimination, a group β to the metal migrates *with its bonding electron pair* to the metal. This process is conceptually the reverse of ligand insertion. (Run Eq. 18.37a backward mentally and you will see the β-elimination of ethylene by migration of aryl.)

It happens that many β-elimination reactions involve a hydride migration. For example, the product of Eq. 18.37a (with Ar = phenyl) can undergo β-elimination with hydride migration as follows:

$$(18.38)$$

Notice that β-elimination requires an empty orbital on the metal, because, as a result of this process, the electron count is increased by two units.

We studied another type of β-elimination, the E2 reaction, in Sec. 9.5. The elimination reaction in Eq. 18.38 looks superficially similar, but it is quite different. In the E2 reaction, a *proton* is eliminated. In the β-elimination of Eq. 18.38, a *hydride*—a hydrogen with its bonding electrons—is eliminated by migration to the metal. We can stress this point with the curved-arrow notation:

(18.39)

Because this β-elimination is *intramolecular*—within the same molecule—it must occur as a syn-elimination. This makes sense because this reaction is conceptually the reverse of 1,2-ligand insertion, which is a syn-addition (Eq. 18.37b). Contrast the stereochemistry of this β-elimination with that of the E2 elimination, which is a bimolecular reaction and occurs with anti stereochemistry (Sec. 9.5E).

STUDY PROBLEM 18.2

Consider the following mechanism for Eq. 18.19 on p. 888. Identify the process associated with each step. Counting electrons at each stage may help you.

$$Pd(PPh_3)_4 \rightleftharpoons Pd(PPh_3)_3 + PPh_3 \rightleftharpoons Pd(PPh_3)_2 + PPh_3 \qquad (18.40a)$$

(18.40b)

(18.40c)

(18.40d)

SOLUTION Step 18.40a consists of two successive *ligand dissociations* that reduce the electron count around the Pd from $18e^-$ to $14e^-$. This "makes room" for the vinylic halide, which undergoes an *oxidative addition* with retention of configuration in Step 18.40b. This step takes the electron count to $16e^-$, and oxidizes the Pd(0) to Pd(II). Step 18.40c is a *ligand substitution* of a bromo ligand with an ethylthio ligand. It might occur by a prior dissociation of the Br ligand, by association of the EtS$^-$ with the Pd to give an $18e^-$ complex followed by dissociation of Br$^-$, or by a concerted mechanism reminiscent of the S$_N$2 reaction. Finally, Step 18.40d is a *reductive elimination*, which forms the product with *retention of stereochemistry* and regenerates the catalytic Pd(0) species Pd(PPh$_3$)$_2$.

Let's use the example in Study Problem 18.2 to take stock of what the Pd is actually doing—why it makes a vinylic or aryl substitution possible. Ligation to the Pd brings two groups—the vinylic group and the nucleophile—into proximity. The oxidative addition step is the key step that makes this possible, and, as we have seen (Fig. 18.5), it is driven by the simultaneous presence of filled and empty metal orbitals that can interact with the vinylic halide so that the carbon–halogen bond is broken. The nucleophile EtS⁻ and the vinylic group are then connected by reductive elimination. The orbital interactions are essentially the same as in oxidative addition. The role of the metal finds analogy in the slider of a zipper: it brings two groups together, causes them to join and lock, and then moves on to do the same thing over again.

PROBLEMS

18.14 A student has written the following ligand substitution reaction, claiming that it changes the oxidation state of the metal by one unit. What is wrong with this reasoning?

$$Cl^- + Pd(PPh_3)_4 \longrightarrow ClPd(PPh_3)_3 + :PPh_3$$

18.15 The *Wilkinson catalyst* chlorotris(triphenylphosphine)rhodium(I), $ClRh(PPh_3)_3$, brings about the catalytic hydrogenation of an alkene in homogeneous solution:

$$\underset{\overset{|}{H}}{\overset{\overset{R}{|}}{C}} = \underset{\overset{|}{H}}{\overset{\overset{R}{|}}{C}} + H_2 \xrightarrow{\ ClRh(PPh_3)_3\ } RCH_2CH_2R \qquad (18.41)$$

(a) Using the following mechanistic steps as your guide, draw structures of the transition-metal complexes involved in each step. Give the electron count and the metal oxidation state at each step.

1. oxidative addition of H_2 to the catalyst
2. ligand substitution of one PPh_3 by the alkene
3. 1,2-insertion of the alkene into a Rh—H bond and readdition of the previously expelled PPh_3 ligand
4. reductive elimination of the alkane product to regenerate the catalyst

(b) According to the known stereochemistry of the 1,2-ligand insertion and reductive elimination steps, what would be the stereochemistry of the product if D_2 were substituted for H_2 in the reaction?

18.6 EXAMPLES OF TRANSITION-METAL-CATALYZED REACTIONS

A. The Heck Reaction

In the **Heck reaction**, an alkene is coupled to an aryl bromide or aryl iodide under the influence of a Pd(0) catalyst.

$$(18.42)$$

(86% yield)

(The aryl substituents of the phosphine ligands used in the catalyst in this case are *o*-tolyl (that is, *o*-methylphenyl) groups rather than phenyl groups, but phenyl groups are also sometimes used.) The reaction is named for Richard F. Heck (b. 1931), who discovered the reaction in the early 1970s while a professor of chemistry at the University of Delaware. (A Japanese chemist, T. Mizoroki, simultaneously discovered the reaction, but it is generally known as the Heck reaction.) The Heck reaction has proven to be one of the most useful processes for forming carbon–carbon bonds to aromatic rings and even, occasionally, to vinylic groups. Heck shared the 2010 Nobel Prize in Chemistry with Akira Suzuki (b. 1930, see Sec. 18.6B) and Ei-ichi Negishi (b. 1935) of Purdue University (see Problem 18.79) for their discovery of transition-metal-catalyzed coupling reaction.

The mechanism of the Heck reaction is outlined in the following equations. You should identify the process or processes involved in each step (L = tri-*o*-tolylphosphine ligands; the steps in Eq. 18.43b are numbered for reference).

The actual catalytically active species is believed to be PdL$_2$, which is formed by two ligand dissociations:

$$(18.43a)$$

The PdL$_2$ thus generated enters into the catalytic cycle.

$$(18.43b)$$

<hr>

PROBLEM

18.16   Characterize each step of the mechanism in Eq. 18.43b in terms of the fundamental processes discussed in the previous section. Give the electron count and the oxidation state of the metal in each complex.

<hr>

Another example of the Heck reaction illustrates two important aspects of the reaction.

$$(18.44)$$

cyclohexene   iodobenzene
(excess)

(2-cyclohexenyl)benzene
(72% yield)

First, the catalyst is not Pd(0), but rather a Pd(II) species. [Pd(OAc)$_2$ is used because it is a convenient and easily handled Pd derivative.] In some cases (typically with iodobenzenes as the aryl halides), the reaction can be run with Pd(II), but it is believed that the Pd(II) is reduced to Pd(0), perhaps by a few molecules of alkene that are converted into vinylic acetates; Pd(0) is the actual catalyst. Addition of an oxidizable ligand such as PPh$_3$ can also serve to reduce the Pd(II). Because a very small amount of Pd is used, the by-products of these reactions are also formed in very small amounts. Second, the alkene double bond in the product is *not* at the site of coupling, but rather one carbon removed. What has happened here?

This sort of product, which occurs commonly with cyclic alkenes in the Heck reaction, is a direct consequence of the stereochemistry of certain steps in the mechanism. The insertion step (step 3 in Eq. 18.43b) *must* occur in a syn manner because the reaction is intramolecular. Hence, in the initially formed insertion complex, the Pd and the phenyl group become cis substituents on a cyclohexane ring.

$$(18.45)$$

The subsequent β-elimination is also a syn process. Hence, only a hydride cis to the Pd is "eligible" for elimination. When a noncyclic alkene is used in the Heck reaction, internal rotation is possible so that the hydride on the carbon at which insertion occurs can be eliminated.

$$(18.46)$$

When the starting material is a cyclic alkene, as in Eq. 18.44, an analogous internal rotation is prevented by the ring. The only cis β-hydride available for elimination is the one (*red* in Eq. 18.45) on the *other* β-carbon. Elimination of this hydride yields an alkene in which the carbon at the insertion point—the one attached to the phenyl—is not part of the double bond, but is one carbon removed. We can summarize this in the following way, with the insertion point marked with an asterisk (*):

$$(18.47)$$

When the Heck reaction is applied to unsymmetrically substituted alkenes, such as an alkene of the form R—CH=CH$_2$, two products are in principle possible, because insertion might occur at either of the alkene carbons. It is found that when the R group is phenyl, CO$_2$R (ester), CN, or another relatively electronegative group, the aryl halide tends to react at the *unsubstituted carbon*; that is, the product is R—CH=CH—Ar, usually the *E* (or trans) stereoisomer. When R = alkyl, mixtures of products are often observed (Problem 18.17).

PROBLEMS

18.17 When iodobenzene and propene are subjected to the conditions of the Heck reaction, two constitutionally isomeric products are formed. What are they? Why are two products formed?

18.18 What *two* sets of aryl bromide and alkene starting materials would give the following compound as the product of a Heck reaction?

18.19 The product of a Heck reaction is, like the starting material, an alkene. Why doesn't a Heck reaction of the product compete with the reaction of the starting alkene?

18.20 What product is expected when cyclopentene reacts with iodobenzene in the presence of triethylamine and a Pd(0) catalyst?

B. The Suzuki Coupling

The Suzuki–Miyaura coupling reaction (usually referred to as the **Suzuki reaction** or the **Suzuki coupling**) is a Pd(0)-catalyzed process in which an aryl or vinylic boronic acid (a compound of the form RB(OH)$_2$, where R = an aryl or vinylic group) is coupled to an aryl or vinylic iodide or bromide in the presence of a base, which is in many cases aqueous sodium hydroxide or sodium carbonate. The reaction can be used to prepare three types of compounds: **biaryls**—compounds in which two aryl rings are connected by a σ bond; aryl-substituted alkenes; and conjugated alkenes. Eq. 18.48 illustrates the preparation of a biphenyl.

The Pd(0) catalyst can be Pd(PPh$_3$)$_4$, the same catalyst used in the Heck reaction, or the Pd(0) can be formed in the reaction flask from Pd(OAc)$_2$, a strategy that is also used in the Heck reaction, as in the preceding example.

Eq. 18.49 shows the preparation of an aryl-substituted alkene.

(18.49)

4-methoxybenzene-boronic acid   **(Z)-1-bromo-1-propene**

(Z)-1-methoxy-4-(1-propenyl)benzene
(*cis*-anethole; 62% yield)

As this example illustrates, the coupling occurs with retention of alkene stereochemistry. You may have noticed that this is the type of compound that can be prepared by the Heck reaction. However, the Suzuki coupling avoids issues of regiochemistry that can sometimes occur with the Heck reaction. (See Problem 18.17)

The importance of the Suzuki reaction has resulted in the commercial availability of many boronic acids and their derivatives. Two ways of preparing the required boron derivatives are, first, the reaction of Grignard or organolithium reagent with trimethyl borate:

(18.50)

In this reaction, the Grignard reagent, a strong Lewis base, donates electrons to the boron in a Lewis acid–base association. A reaction with aqueous acid results in formation of the boronic acid product. (See Problem 18.24, p. 907.) The analogous reaction can be used to form vinylic boronic acids from vinylic Grignard reagents. A second preparation of vinylic boronic acid derivatives is the hydroboration of 1-alkynes with catecholborane:

(18.51)

1-hexyne

catecholborane

This is essentially the same reaction discussed in Sec. 14.5B, in which hydroboration is carried out with disiamylborane. Recall that hydroboration occurs as a syn-addition. Both the catecholborane adducts and the disiamylborane adducts can be used in the Suzuki coupling. The following reaction illustrates both the use of vinylic catecholboranes and the formation of a conjugated alkene.

(76% yield)   (18.52)

The mechanism of the Suzuki coupling begins exactly like that of the Heck reaction (Eq. 18.43b, p. 902) with ligand dissociation to give a $14e^-$ complex, followed by oxidative addition of the aryl halide.

$$PPh_3 \rightarrow \underset{\underset{PPh_3}{\uparrow}}{\overset{\overset{PPh_3}{\downarrow}}{Pd}} \leftarrow PPh_3 \quad \underset{\longleftarrow}{\longrightarrow} \quad PPh_3 \rightarrow \underset{14e^-,\ Pd(0)}{\overset{\overset{PPh_3}{\downarrow}}{Pd}} \quad \xrightarrow[\text{oxidative addition}]{Ar-Br} \quad PPh_3 \rightarrow \underset{\underset{Ar}{|}}{\overset{\overset{PPh_3}{\downarrow}}{Pd}} - Br \qquad (18.53a)$$

$18e^-,\ Pd(0)$ $+\ 2\ PPh_3$ $16e^-,\ Pd(II)$

From this point on, the mechanism is not definitively established, but a reasonable sequence involves another ligand substitution in which the base displaces the halide ion:

$$PPh_3 \rightarrow \underset{\underset{Ar}{|}}{\overset{\overset{PPh_3}{\downarrow}}{Pd}} - Br \ +\ {}^-OH \quad \xrightarrow[\text{substitution}]{\text{ligand}} \quad PPh_3 \rightarrow \underset{\underset{Ar}{|}}{\overset{\overset{PPh_3}{\downarrow}}{Pd}} - OH \ +\ Br^- \qquad (18.53b)$$

A Lewis acid–base association brings the boron into the coordination sphere of the metal:

$$PPh_3 \rightarrow \underset{\underset{Ar}{|}}{\overset{\overset{PPh_3}{\downarrow}}{Pd}} - \ddot{O}H \quad \underset{\underset{R}{|}}{B(OH)_2} \quad \xrightarrow[\text{association}]{\text{Lewis acid–base}} \quad PPh_3 \rightarrow \underset{\underset{Ar}{|}}{\overset{\overset{PPh_3}{\downarrow}}{Pd}} - \overset{+}{\underset{\underset{R}{|}}{\ddot{O}}}\overset{H}{\underset{\bar{B}(OH)_2}{}} \qquad (18.53c)$$

This association results in a formal negative charge on boron. Remember, though, that carbon is more electronegative than boron; this means that carbon bears a significant amount of the negative charge in this complex. In other words, the Lewis acid–base association of the oxygen with boron endows the carbon in the carbon–boron bond with significant carbanion character. An intramolecular substitution of this "carbon anion," a strong base, for the weaker base HO— results in transfer of the R group to the metal. Reductive elimination then gives the coupling product and provides the catalyst for another cycle.

$$PPh_3 \rightarrow \underset{\underset{Ar}{|}}{\overset{\overset{PPh_3}{\downarrow}}{Pd}} - \overset{+}{\underset{\underset{R^{\delta-}}{|}}{\ddot{O}}}\overset{H}{\underset{B(OH)_2^{\delta-}}{}} \quad \xrightarrow[\substack{\text{ligand} \\ \text{substitution}}]{\text{intramolecular}} \quad PPh_3 \rightarrow \underset{\underset{Ar}{|}}{\overset{\overset{PPh_3}{\downarrow}}{Pd}} - R \quad \xrightarrow[\text{elimination}]{\text{reductive}} \quad PPh_3 \rightarrow \overset{\overset{PPh_3}{\downarrow}}{Pd} \ +\ Ar-R \qquad (18.53d)$$

the carbon of the carbon–boron bond has carbanion character

$+\ B(OH)_3$ regenerated catalyst   coupling product

The Suzuki coupling is named for Akira Suzuki (b. 1930), who is on the faculty of the Kirashiki University of Science and the Arts in Kirashiki, Japan. Prof. Suzuki had spent two years as a postdoctoral fellow with Herbert C. Brown (the discoverer of hydroboration), where he was immersed in the organic chemistry of boron. His intellectual synthesis of transition-metal organometallic chemistry and organoboron chemistry led to the discovery in the mid-1970s of the reaction that bears his name, in collaboration with his student Norio Miyaura, while the two were at Hokkaido University in Japan. The Suzuki coupling has become a very important reaction in both academic and industrial settings. As noted in Sec 18.6A, Suzuki shared the 2010 Nobel Prize in Chemistry.

PROBLEMS

18.21 Complete the following Pd(0)-catalyzed Suzuki reactions by giving the coupling product. For parts (b) and (c), include the stereochemistry of the products.

(a)

2-naphthaleneboronic acid

(b)

(c)

18.22 Provide two different reaction sequences that could be used to synthesize 4-methoxy-3′-methylbiphenyl. Both sequences, however, should start with both *p*-bromoanisole and *m*-bromotoluene.

4-methoxy-3′-methylbiphenyl   from   **m-bromotoluene**   and   **p-bromoanisole**

18.23 Give two different pairs of starting materials that could be used to prepare the following compound by a Suzuki coupling.

18.24 Draw a curved-arrow mechanism for the last (acid-catalyzed hydrolysis) step of Eq. 18.50.

C. Alkene Metathesis

Certain ruthenium(IV) catalysts bring about a reaction in which two alkenes are "sliced apart" at their double bonds, and the parts reassembled, to give new alkenes.

eugenol   **cis-2-buten-1,4-diol**

$$\xrightarrow[\text{CH}_2\text{Cl}_2]{\substack{\text{G2 ruthenium catalyst} \\ \text{(2 mole \%)}}}$$

(E)-4-(3-hydroxy-4-methoxyphenyl)-2-buten-1-ol
(80% yield)

$$+ \quad \text{HO}-\text{CH}_2\text{CH}=\text{CH}_2 \qquad (18.54)$$

allyl alcohol

This remarkable reaction is an example of **alkene metathesis**, also called **olefin metathesis**. A metathesis reaction (pronounced mə-tă´-thə-sĭs, from the Greek, *meta* = change, *thesis* = place) can be represented in general as follows:

$$A\!-\!B \ + \ C\!-\!D \ \longrightarrow \ A\!-\!C \ + \ B\!-\!D \qquad (18.55)$$

You are probably familiar with some inorganic examples of metathesis reactions, such as the reaction of silver nitrate with sodium chloride:

$$\overset{+}{Ag}\,\overset{-}{NO_3} \ + \ \overset{+}{Na}\,\overset{-}{Cl} \ \longrightarrow \ AgCl\!\downarrow \ + \ \overset{+}{Na}\,\overset{-}{NO_3} \qquad (18.56)$$

| **silver(I)** | **sodium** | **silver(I)** | **sodium** |
| **nitrate** | **chloride** | **chloride** | **nitrate** |

In an alkene metathesis, the groups at each end of the double bonds are interchanged. For example, the reaction in Eq. 18.54 has the following form:

$$
\underset{H}{\overset{R'}{\diagdown}}C\!\!=\!\!C\underset{H}{\overset{H}{\diagup}}
\ + \
\underset{H}{\overset{R}{\diagdown}}C\!\!=\!\!C\underset{H}{\overset{R}{\diagup}}
\ \rightleftharpoons \
\underset{H}{\overset{R'}{\diagdown}}C\!\!=\!\!C\underset{H}{\overset{R}{\diagup}}
\ + \
\underset{H}{\overset{R}{\diagdown}}C\!\!=\!\!C\underset{H}{\overset{H}{\diagup}}
\qquad (18.57)
$$

For a metathesis reaction of two unsymmetrical alkenes, 10 alkenes (not counting stereoisomers) can be formed. (See Problem 18.25.) However, the usual applications of this reaction are typically not this complex.

A number of transition-metal catalysts have been developed for alkene metathesis. These catalysts are based on tungsten, molybdenum, and especially ruthenium. The two most widely used laboratory catalysts are the ruthenium-based catalysts G1, which stands for the *Grubbs first-generation catalyst*, and G2, the *Grubbs second-generation catalyst*. The structures of these catalysts are shown in Fig. 18.6. Although we need not go into detail here, the design of these catalysts was an evolutionary process that involved a consideration of steric and electronic effects of the ligands in light of the reaction mechanism (which we'll discuss

FIGURE 18.6 The structures of two ruthenium catalysts for alkene metathesis and their abbreviated structures. Cy is the abbreviation for the cyclohexyl group, and Mes is the abbreviation for the mesityl, or 2,4,6-trimethylphenyl, group. The unusual ligand in G2 is a stabilized carbene (see Eq. 18.58), and it is abbreviated NHC (for "*N*itrogen *H*eterocycle-stabilized *C*arbene"). Formal charges in the NHC ligand are conventionally not shown. A key aspect of these catalysts is the ruthenium–carbon double bond.

Grubbs G1 catalyst
(a)

Grubbs G2 catalyst
(b)

abbreviated structure for the G2 catalyst

below), as well as some outright fortuitous discoveries! Ruthenium catalysts can be easily handled in the laboratory, and, as Eq. 18.54 illustrates, they can be used in the presence of alcohols, phenols, and other functional groups. The molybdenum and tungsten catalysts are much more air-sensitive and less tolerant of other functional groups.

The ruthenium–carbon double bond plays an important role in the operation of these catalysts. The rather unusual NHC ligand in the G2 catalyst, when "dissociated" from the metal, is actually a *carbene* (a molecule containing divalent carbon; Sec. 9.9A). Most carbenes are very unstable, but this carbene is stabilized by resonance.

resonance structures for the NHC ligand

hybrid structure for the NHC ligand   (18.58)

For electron-counting purposes, the NHC and PCy_3 ligands are L-type ligands, the chlorines are X-type ligands, and the benzylidene group is a 2X ligand because it has two bonds to the ruthenium. You should verify that both catalysts are 16-electron complexes (Eq. 18.29) and that the oxidation state of ruthenium (Eq. 18.23) is formally Ru(IV).

In many cases, the catalysts bring some or all of the possible alkenes into equilibrium. In such cases, the practical use of alkene metathesis requires the application of Le Châtelier's principle. For example, in Eq. 18.54, one of the starting materials, *cis*-2-buten-1,4-diol, is used in excess. Use of an excess of this diol is practical because it is cheap, and because both it and the by-product allyl alcohol are readily separated from the desired product.

One of the most important applications of alkene metathesis is for closing rings, and it can be used to close medium- and large-sized rings.

$$+ \ H_2C=CH_2 \qquad (18.59)$$

(72 % yield)

In this and many other ring-closing applications, the by-product is ethylene, which bubbles out of the reaction mixture, thus driving the equilibrium towards the product—Le Châtelier's principle in operation again.

The mechanism of alkene metathesis, which we'll illustrate for Eq. 18.57 using the G2 catalyst, starts with loss of the PCy_3 ligand by ligand dissociation. Because ruthenium in the resulting complex has 14 electrons, it can accept two electrons from another ligand, in this case one of the alkenes. Let the alkene RCH=CHR be present in large excess; because of its concentration, it is more likely to interact with the catalyst.

Then follows a key step in alkene metathesis, a cycloaddition to form a *metallacycle* (a cyclic compound in which the metal occupies a ring position). This reaction is essentially a ligand insertion (p. 898).

$$\underset{\substack{RCH=CHR}}{\overset{\overset{\displaystyle NHC}{\downarrow}}{Cl_2Ru=CHPh}} \xrightarrow{\boxed{cycloaddition}} \underset{\substack{RCH-CHR \\ \text{a metallacycle}}}{\overset{\overset{\displaystyle NHC}{\downarrow}}{Cl_2Ru-CHPh}} \xrightarrow{\boxed{cycloreversion}} \underset{\substack{RCH}}{\overset{\overset{\displaystyle NHC}{\downarrow}}{Cl_2Ru\leftarrow \underset{CHR}{\overset{\|}{\|}}}} \longrightarrow \underset{\substack{RCH}}{\overset{\overset{\displaystyle NHC}{\downarrow}}{Cl_2Ru}} + \underset{\substack{\text{very minor} \\ \text{by-product}}}{RCH=CHPh} \quad (18.60b)$$

As shown in this equation, the metallacycle then breaks down in the opposite sense by a *cycloreversion* (the reverse of a cycloaddition) to give a new alkene, which contains the benzylidene group. This becomes a very minor by-product, because the catalyst (and thus the benzylidene group) is typically present in 1 or 2 mole percent of the reactants. This process leaves the catalyst "primed" with the RCH= group.

The cycloaddition–cycloreversion process is now repeated. It is most probable that the process will occur with the alkene used in Eq. 18.60b, because this alkene is in excess; but the reaction with this alkene results in no change. (Be sure to demonstrate this point to yourself.) However, occasionally a molecule of the other alkene R'CH=CH$_2$ will bind to the catalyst, and this produces a metathesis product.

$$\underset{\substack{RCH \quad CHR'}}{\overset{\overset{\displaystyle NHC}{\downarrow}}{Cl_2Ru \uparrow CH_2}} \longrightarrow \underset{\substack{RCH-CHR'}}{\overset{\overset{\displaystyle NHC}{\downarrow}}{Cl_2Ru-CH_2}} \longrightarrow \underset{\substack{\text{first} \\ \text{metathesis product}}}{\overset{\overset{\displaystyle NHC}{\downarrow}}{Cl_2Ru=CH_2} + RCH=CHR'} \quad (18.60c)$$

This leaves the catalyst with a =CH$_2$ group. This catalyst molecule is most likely to react with the alkene in excess, and when that happens, the second metathesis product (which is allyl alcohol in Eq. 18.54) is formed, and the catalyst is again primed with a =CHR group.

$$\underset{\substack{RCH=CHR}}{\overset{\overset{\displaystyle NHC}{\downarrow}}{Cl_2Ru=CH_2}} \longrightarrow \underset{\substack{RCH-CHR}}{\overset{\overset{\displaystyle NHC}{\downarrow}}{Cl_2Ru-CH_2}} \longrightarrow \underset{\substack{RCH}}{\overset{\overset{\displaystyle NHC}{\downarrow}}{Cl_2Ru}} + \underset{\substack{\text{second} \\ \text{metathesis product}}}{RCH=CH_2} \quad (18.60d)$$

$$\underset{\substack{\text{catalyst enters} \\ \text{another cycle} \\ \text{(Eq. 18.60c)}}}{\downarrow}$$

The sequence in Eqs. 18.60c–d continues repeatedly until the limiting reactant is exhausted.

It is conceivable that, as the product builds up, it might undergo metathesis with itself. However, self-metathesis is largely avoided in this example because one of the starting materials is used in excess, and it intercepts the catalyst almost every time. In other words, if the product enters into the metathesis sequence and is split by the catalyst, the fragments are most likely intercepted by the alkene present in excess; and such a reaction either gives back that same alkene or the desired product.

When it is impractical to use one alkene in excess or to exploit Le Châtelier's principle in some other way, self-metathesis of the product can be a problem. However, this potential complexity of alkene metathesis is mitigated by the fact that different alkenes undergo metathesis at greatly different rates. Alkene metathesis is very sensitive to the steric environment of the alkene double bonds. For example, 2-methylpropene (isobutylene) does not undergo self-metathesis, presumably because the interaction of two (CH$_3$)$_2$C= fragments with the catalyst results in severe van der Waals repulsions (that is, steric congestion) with the bulky catalyst ligands.

$$H_3C \text{—} C(CH_3)=CH_2 + H_2C=C(CH_3)(CH_3) \xrightarrow{\ self\text{-}metathesis\ } \cancel{\ } \ (CH_3)_2C=C(CH_3)_2 + H_2C=CH_2 \quad (18.61)$$

two molecules of isobutylene   not formed

(Isobutylene can be used as a metathesis partner with less crowded alkenes, however.) In fact, the metathesis catalysts were designed to emphasize such differences in alkene reactivity.

Alkene metathesis plays a major role in the synthesis of complex organic molecules, and it is used industrially with increasing frequency in such diverse applications as alkene synthesis, polymer synthesis, and pheromone synthesis. Because ruthenium catalysts can be used in aqueous solution, they are providing options for "green chemistry"—chemistry that is environmentally friendly. The importance of alkene metathesis was recognized with the 2005 Nobel Prize in Chemistry, which was shared by three chemists: Robert H. Grubbs (b. 1942) of the California Institute of Technology, who developed the ruthenium catalysts; Richard R. Schrock (b. 1945) of the Massachusetts Institute of Technology, who developed molybdenum-based metathesis catalysts; and Yves Chauvin (1930–2015) of the Petroleum Institute of France, who first proposed the reaction mechanism.

PROBLEMS

18.25 Show that the equilibrium mixture produced by alkene metathesis of two completely different alkenes with the following general structures contains 10 different alkenes. (Assume that all alkenes have trans stereochemistry.)

$$\begin{array}{c} R^1 \\ \end{array} C=C \begin{array}{c} H \\ R^2 \end{array} \quad + \quad \begin{array}{c} R^3 \\ \end{array} C=C \begin{array}{c} H \\ R^4 \end{array}$$

18.26 Give the structure of the major product formed in each case when the reactant(s) shown undergo alkene metathesis in the presence of an appropriate ruthenium catalyst.

(a)

$$H_2C=CHCH_2\overset{\displaystyle CH_2OH}{\underset{\displaystyle |}{C}}HCH_2CH_2CH=CH_2 \longrightarrow \text{ a compound with 7 carbons}$$

(b)

a compound with 11 carbons

(c)

+

$$\begin{array}{c} HOCH_2 \\ \end{array} C=C \begin{array}{c} CH_2OH \\ \end{array}$$
       H   H

(large excess)

18.27 Suggest an alkene metathesis reaction that would yield each of the following compounds as a major product.

(a)

(–)-citronellol
(oil of roses)

(b)

(c)

18.28 Draw structures analogous to those in Eqs. 18.60a–d for the catalytic intermediates formed in the conversion of 1,7-octadiene to cyclohexene and ethylene catalyzed by the G2 catalyst.

The three coupling reactions presented in this section are additional examples of reactions that can be used to form carbon–carbon bonds. Here is a list of such reactions that we have encountered up to this point:

1. the addition of carbenes and carbenoids to alkenes (Sec. 9.9)
2. the reaction of Grignard reagents with ethylene oxide and lithium organocuprate reagents with epoxides (Sec. 11.5C)
3. the reaction of acetylenic anions with alkyl halides or sulfonate esters (Sec. 14.7B)
4. the Diels–Alder reaction (Sec. 15.3)
5. Friedel–Crafts alkylation and acylation reactions (Secs. 16.4E–F)
6. the Heck and Suzuki coupling reactions (Secs. 18.6A–B)
7. alkene metathesis (this section)

D. Other Examples of Transition-Metal-Catalyzed Reactions

One of the most important transition-metal catalysts in commerce is a catalyst formed from $TiCl_3$ and $(CH_3CH_2)_2AlCl$, called the *Ziegler–Natta catalyst*. This catalyst brings about the polymerization of ethylene and other alkenes at 25 °C and 1 atm pressure. Although free-radical polymerization of ethylene (Sec. 5.7) is very important, the Ziegler–Natta polymerization of ethylene accounts for most of the polyethylene produced; the resulting *high-density polyethylene* has different properties from the *low-density polyethylene* produced by free-radical processes. The discoverers of this catalyst, the German chemist Karl Ziegler (1898–1973) and the Italian chemist Giulio Natta (1903–1979) shared the 1963 Nobel Prize in Chemistry for their work. Although the mechanism of the polymerization has been hotly debated, the following sequence is one possibility:

$$\text{formed from } (CH_3CH_2)_2AlCl \text{ and } TiCl_3 \xrightarrow[\text{ligand association}]{H_2C=CH_2} \xrightarrow{\text{1,2-ligand insertion}}$$

$$\xrightarrow[\text{ligand association}]{H_2C=CH_2} \tag{18.62}$$

Continuation of the insertion–ligand-association sequence gives the polymer. It is believed that titanium brings about this reaction because a d^1 metal cannot undergo β-elimination. (β-Elimination requires some filled metal d orbitals for reasons that we haven't discussed.) The tendency toward β-elimination of other metals would terminate the reaction.

Hydroformylation is another commercially important process that involves a transition-metal catalyst, in this case a tetracarbonylhydridocobalt(I) catalyst. Propionaldehyde, for example, is produced by the hydroformylation of ethylene. (This is sometimes called the *oxo process*.)

$$H_2C=CH_2 + H_2 + CO \xrightarrow[100-120\,°C]{HCo(CO)_4} CH_3CH_2\overset{O}{\overset{\|}{C}}H \tag{18.63}$$

ethylene   **propionaldehyde (propanal)**

This process involves, among other things, a 1,2-insertion reaction of ethylene and a 1,1-insertion reaction of carbon monoxide (Problem 18.29).

Yet another important transition-metal-catalyzed reaction is the *homogeneous* catalytic hydrogenation of alkenes using a soluble rhodium(I) catalyst called the *Wilkinson catalyst*, $ClRh(PPh_3)_3$. This reaction was explored in Problem 18.15 (p. 901).

And let's not forget catalytic hydrogenation (Secs. 4.9A and 14.6A), a very important reaction that occurs over carbon-supported transition metals such as Ni, Pd, and Pt. The mechanism of catalytic hydrogenation is not definitively known, but it is not hard to imagine that the mechanism might involve oxidative additions and insertions much like those that take place on the Wilkinson catalyst.

Many aspects of transition-metal chemistry are beyond the scope of an introduction. How does the chemist design a catalytic system and choose a catalyst? What influences the choice of ligands and solvents? These questions are sometimes addressed with a certain degree of empiricism, but the bases for the answers to these questions are becoming better understood.

PROBLEM

18.29 Suggest a mechanism for the oxo reaction (Eq. 18.63) involving intermediates that are consistent with the 16- and 18-electron rules.

18.7 ACIDITY OF PHENOLS

A. Resonance and Polar Effects on the Acidity of Phenols

Phenols, like alcohols, can ionize.

$$\text{phenol} \rightleftharpoons \text{phenoxide ion (phenolate ion)} + H_3\ddot{O}^+ \tag{18.64}$$

The conjugate base of a phenol is named, using common nomenclature, as a *phenoxide ion* or, using substitutive nomenclature, as a *phenolate ion*. Thus, the sodium salt of phenol is called sodium phenoxide or sodium phenolate; the potassium salt of *p*-chlorophenol is called potassium *p*-chlorophenoxide or potassium 4-chlorophenolate.

Phenols are considerably more acidic than alcohols. For example, the pK_a of phenol is 9.95, but that of cyclohexanol is about 17. Thus, phenol is approximately 10^7 times more acidic than an alcohol of similar size and shape.

pK_a ≈10   ≈17

Recall from Fig. 3.2 on p. 113 that the pK_a of an acid is decreased by stabilizing its conjugate base. *The enhanced acidity of phenol is due to stabilization of its conjugate-base anion.*

What is the source of this enhanced stability? First, the phenolate anion is stabilized by resonance:

resonance structures for the phenoxide anion

Actually, phenol itself is also stabilized by resonance.

resonance structures for phenol
(notice the *separation* of charge)

$$(18.66)$$

However, the resonance structures for a phenolate ion are *more important* than the resonance structures for un-ionized phenol. Therefore, resonance in the phenolate ion has a greater effect on pK_a than resonance in un-ionized phenol.

Why is resonance in phenol less important than it is in phenolate? The resonance structures in phenol—the delocalization of electrons—involve the *separation of charge*—that is, the creation and separation of opposite charges. According to the electrostatic law of physics, pulling a negative and a positive charge apart requires energy.

$$(18.67)$$

However, the resonance structures of the phenolate ion simply delocalize the negative charge. What we learn, then, is an important principle that we can apply to acid–base reactions: *Resonance structures that involve the delocalization of charge are more important than resonance structures that involve the separation of charge.*

The second effect that stabilizes the conjugate base is the polar effect of the benzene ring, which also stabilizes negative charge (Sec. 17.3). Both resonance stabilization and polar effects are the same effects that stabilize benzylic carbanions (Sec. 17.3). Actually, a phenoxide anion *is* a benzylic anion in which the benzylic group is an oxygen instead of a carbon!

Alkoxides are stabilized neither by resonance nor by the polar effect of benzene rings or double bonds.

cyclohexanolate anion
no resonance structures;
no polar effect of double bonds

Because phenoxide ions are stabilized by both resonance and polar effects, less energy is required to form phenoxides from phenols than is required to form alkoxides from alcohols. *Because* pK_a *is directly proportional to the standard free energy of ionization* (Eq. 3.42b, p. 116), phenols have lower pK_a values, and are thus more acidic, than alcohols.

Substituent groups can also affect phenol acidity by both polar and resonance effects. For example, the relative acidities of phenol, *m*-nitrophenol, and *p*-nitrophenol reflect the operation of both effects.

	phenol	***m*-nitrophenol**	***p*-nitrophenol**
pK$_a$	9.95	8.35	7.21

m-Nitrophenol is more acidic than phenol because the nitro group is very electronegative. The polar effect of the nitro substituent stabilizes the conjugate-base anion for the same reason that it would stabilize the conjugate base of an alcohol (Sec. 10.1E). Yet *p*-nitrophenol is more acidic than *m*-nitrophenol by more than one pK$_a$ unit, even though the *p*-nitro group is farther from the phenol oxygen. This cannot be entirely the result of a polar effect, for polar effects on acidity *decrease* as the distance between the substituent and the acidic group increases. The reason for the increased acidity of *p*-nitrophenol is that the *p*-nitro group stabilizes the conjugate-base anion by resonance (red structure).

charge is delocalized
into the nitro group

(18.68)

The structure shown in red is especially important because it places charge on the electronegative oxygen atom. In *m*-nitrophenol, however, it is not possible to draw a resonance structure that delocalizes the negative charge into the nitro group.

fewer important structures than the para isomer

(18.69)

Because *p*-nitrophenoxide has more resonance structures, it is more stable relative to its corresponding phenol than is *m*-nitrophenoxide. Hence *p*-nitrophenol is the more acidic of the two nitrophenols. The acid-strengthening resonance effect of ortho and para nitro groups is so large that 2,4,6-trinitrophenol (picric acid) is actually a strong acid.

**2,4,6-trinitrophenol
(picric acid)**
pK$_a$ = 0.96

Let's summarize the factors that govern acidity as we've seen them in operation so far:

1. *Element effects*: Other things being equal, compounds are more acidic when the element to which the acidic hydrogen is bound has a higher atomic number within either a row or group of the periodic table.

   a. The effect within a row (period) of the periodic table is dominated by relative electronegativities (or electron affinities). Thus, water is much more acidic than methane, and phenol is much more acidic than toluene, because oxygen (higher atomic number) is more electronegative than carbon (lower atomic number).

   b. The effect within a column of the periodic table is dominated by relative bond energies. Thus, thiols are more acidic than alcohols because the S—H bond is weaker than the O—H bond.

2. *Charge effects*: A positive charge on the atom to which the acidic proton is attached enhances acidity. For example, H_3O^+ is more acidic than H_2O.

3. *Resonance effects*: Enhanced delocalization of electrons in the conjugate base enhances acidity.

4. *Polar effects*: Stabilization of charge in the conjugate base enhances acidity.

PROBLEM

18.30   Which of the two phenols in each set is more acidic? Explain.

(a) 2,5-dinitrophenol or 2,4-dinitrophenol

(b) phenol or *m*-chlorophenol

(c)

B. Formation and Use of Phenoxides

Alcohols are not converted completely into alkoxides by aqueous NaOH solution because the pK_a values of water and alcohols are similar (Sec. 10.1A). In contrast, the equilibrium for the reaction of phenol and NaOH lies almost completely to the right:

$$\text{phenol} \quad \rightleftharpoons \quad \text{phenoxide ion} \quad (18.70)$$

phenol
$pK_a = 9.95$

phenoxide ion

$pK_a = 15.7$

Because the difference in pK_a values of water and phenol is about 6, the equilibrium constant for this reaction is about 10^6 (Sec. 3.4E). Thus, for all practical purposes, phenols are converted completely into their conjugate-base anions by NaOH solution. Although the stronger bases used to ionize alcohols (Sec. 10.1A) can also be used for phenols, hydroxide ion or alkoxide bases such as ethoxide ion are often perfectly adequate for the purpose. Thus, when phenol is treated with a solution containing one equivalent of NaOH or NaOEt, the phenol O—H proton is titrated completely to give a solution of sodium phenoxide.

The acidities of phenols can sometimes be used to separate them from mixtures with other organic compounds. For example, suppose that we wish to separate the water-insoluble phenol, 4-chlorophenol, from the water-insoluble alcohol, 4-chlorocyclohexanol. Although the phenol itself is water-insoluble, its sodium or potassium salt, like many other alkali metal salts, has considerable solubility in water because it is an *ionic compound*. (Recall from Sec. 8.6F that water is one of the best solvents for ionic compounds.)

4-chlorophenol	**sodium 4-chlorophenolate**	**4-chlorocyclohexanol**
soluble in ether	insoluble in ether	soluble in ether
insoluble in water	soluble in water	insoluble in water
	(an ionic compound)	

Thus, when a mixture of the phenol and the alcohol in ether solution is shaken with aqueous NaOH, the phenol is selectively extracted into the aqueous solution as its sodium salt, sodium 4-chlorophenolate, while the alcohol, which is not significantly ionized by NaOH, remains in the ether. (Although alcohols of low molecular mass are soluble in water, the chlorine and hydrocarbon parts of 4-chlorocyclohexanol dominate its solubility properties.) Acidification of the aqueous solution gives the phenol, which separates from solution because, after acidi-fication, it is no longer ionized.

It is usually said that phenols are "soluble in sodium hydroxide solution." What is really meant by this statement is that if sodium hydroxide solution is added to a phenol, the phenol is converted into its conjugate-base phenoxide ion, which, because it is ionic, is the species that actually dissolves in the aqueous solution. Solubility in 5% NaOH solution is a qualitative test for phenols (and other compounds of equal or greater acidity).

Phenoxides, like alkoxides, can be used as nucleophiles. For example, aryl ethers can be prepared by the reaction of a phenoxide anion and an alkyl halide.

This is another example of the Williamson ether synthesis (Sec. 11.2A). Note that the reaction of sodium propoxide, the sodium salt of 1-propanol, with bromobenzene, would *not* be a sat-isfactory synthesis of this ether. (Why? See Sec. 18.1.)

PROBLEMS

18.31 Outline a preparation of each of the following compounds from the indicated starting material and any other reagents.

(a) *p*-nitroanisole from *p*-nitrophenol   (b) 2-phenoxyethanol from phenol

18.32 The following compound, unlike most phenols, is *soluble* in neutral aqueous solution, but *insoluble* in aqueous base. Explain this unusual behavior.

18.8   QUINONES AND SEMIQUINONES

A. Oxidation of Phenols to Quinones

Even though phenols do not have hydrogen at their α-carbon atoms, they do undergo oxida-tion. The most common oxidation products of phenols are quinones.

$$\text{(18.72)}$$

hydroquinone   **p-benzoquinone**
                            (86–92% yield)

$$\text{(18.73)}$$

2,3,6-trimethylphenol

2,3,5-trimethyl-1,4-benzoquinone
(50% yield)

$$\text{(18.74)}$$

4-methylcatechol   **4-methyl-1,2-benzoquinone**
                            (unstable red crystals)
                            (68% yield)

As Eqs. 18.72–18.74 illustrate, *p*-hydroxyphenols (hydroquinones), *o*-hydroxyphenols (catechols), and phenols with an unsubstituted position para to the hydroxy group are oxidized to quinones. A **quinone** is any compound containing either of the following structural units.

an *ortho*-quinone

a *para*-quinone

If the quinone oxygens have a 1,4 (para) relationship, the quinone is called a *para*-quinone; if the oxygens are in a 1,2 (ortho) arrangement, the quinone is called an *ortho*-quinone. The following compounds are typical quinones.

p-benzoquinone
(1,4-benzoquinone)

o-benzoquinone
(1,2-benzoquinone)

1,4-naphthoquinone

derived from

naphthalene

derived from

9,10-anthraquinone

anthracene

As illustrated by the preceding structures, the names of quinones are derived from the names of the corresponding aromatic hydrocarbons: *benzo*quinone is derived from benzene, *naph-tho*quinone from naphthalene, and so on.

Ortho-quinones, particularly *ortho*-benzoquinones, are typically considerably less stable than their *para*-quinone isomers. One reason for this difference is that in *ortho*-quinones, the ends of the C=O bond dipoles with like charges are close together and therefore have a repulsive, destabilizing interaction. In *para*-quinones these dipoles are pointed in opposite directions and are farther apart.

o-benzoquinone
like charges in the bond dipoles
are close together

p-benzoquinone
bond dipoles have opposite directions
and are farther apart

PROBLEMS

18.33 Given the structure of phenanthrene, draw structures of
(a) 9,10-phenanthroquinone   (b) 1,4-phenanthroquinone

phenanthrene

18.34 Complete the following reactions:

(a)

$$\text{OH, NO}_2 \xrightarrow[\text{H}_2\text{SO}_4]{\text{Cr(VI)}}$$

(b)

$$\text{HO} \quad \text{OH, OH} \xrightarrow[\text{H}_2\text{SO}_4]{\text{Cr(VI)}}$$

(*Hint:* The product is a hydroxy quinone; but which one?)

B. Quinones and Phenols in Biology

The amino acid *tyrosine* has a para-substituted phenol in its side chain.

tyrosine residue in a protein

Its functional role in proteins varies. In some cases the O—H acts as a hydrogen-bond donor that can be important in catalyzing chemical reactions or maintaining protein structure; it can be involved in the noncovalent interactions between rings (offset-stacking or edge-to-face interactions; Sec. 15.8A) or as the "pi" component in pi–cation interactions (Sec. 15.8B); and, in other cases, the conjugate base of the O—H group is involved as a basic group in enzyme catalysis.

The oxidation–reduction reactions and free-radical reactions of phenols are very important in biology. To understand this role, let's think of hydroquinone oxidation as a half-reaction (Sec. 10.6A).

$$+ \ 2H^+ + 2e^- \qquad (18.75)$$

(In fact, this half-reaction can be carried out reversibly in an electrochemical cell.) We can think of this oxidation mechanistically as two individual one-electron steps:

hydroquinone

a semiquinone

$$+ \ H{-}\overset{..}{\underset{+}{O}}H_2$$

$$+ \ H{-}\overset{..}{\underset{+}{O}}H_2 \qquad (18.76)$$

1,4-benzoquinone

The one-electron oxidation product of a hydroquinone is called a **semiquinone**. As shown in Eq. 18.76, semiquinones are resonance-stabilized radicals.

The view of hydroquinone oxidation (or quinone reduction) as a half-reaction is biologically significant in the process known as *oxidative phosphorylation*. In oxidative phosphorylation, O_2 is reduced to water:

$$4e^- + \ 4H^+ + \ O_2 \longrightarrow \ 2H_2O \qquad (18.77)$$

This half-reaction has a standard free energy (at pH 7) of –157.6 kJ (–37.7 kcal) per mole of O_2 consumed. This very large free energy is captured and ultimately used to drive the synthesis of ATP (Secs. 25.2B and 25.8B), the source of chemical energy in the cell. (You will study oxidative phosphorylation if you take a biochemistry course.) The free energy liberated by the reduction of O_2 is not captured in one step; rather, it is captured incrementally in a series of steps in which electrons are transferred from cellular reducing agents to oxygen. The reactions in which these electron transfers take place are called collectively the **electron-**

transport chain. A biological quinone, **ubiquinone**, sometimes called **coenzyme Q**, or simply **Q**, is involved in electron transport.

coenzyme Q
(Q, ubiquinone)

reduced coenzyme Q
(QH$_2$)
$$(18.78)$$

Q is reduced to QH$_2$ in a two-electron process, as shown in Eq. 18.78. (The reducing agents are NADH and succinate, and we won't detail this process here.) The QH$_2$ then delivers an electron to Fe^{3+}, which is part of a heterocycle–protein complex called *cytochrome c*. In this reaction, the ferric iron (Fe^{3+}) undergoes a one-electron reduction to ferrous iron (Fe^{2+}), and Q is converted into its semiquinone. Reduction of a second cytochrome-bound Fe^{3+} by the semiquinone gives QH$_2$ and another Fe^{2+}.

$$QH_2 \ + \ Fe^{3+} \ \longrightarrow \ {}^{\bullet}QH \ + \ Fe^{2+} \qquad (18.79a)$$

(in cytochrome c)   semiquinone
form of Q

$${}^{\bullet}QH \ + \ Fe^{3+} \ \longrightarrow \ Q \ + \ Fe^{2+} \qquad (18.79b)$$

Notice that *two* Fe^{3+} can be reduced by each QH$_2$, each in a one-electron step. For this reason, coenzyme Q is sometimes called a two-electron/one-electron bridge in electron transport.

    Another important quinone is **vitamin K**, which is important in a group of biochemical reactions involved in blood clotting. (Different forms of vitamin K, which are all active, differ in the detailed structure of the isoprenoid side chain).

vitamin K$_2$
(menaquinone)

Its reduction product, often called KH$_2$, undergoes oxidation to vitamin K-2,3-epoxide with oxygen as part of the blood-clotting process. The conversion of the epoxide back into KH$_2$ is essential for proper clotting activity. The widely used anticoagulant warfarin (Coumadin) inhibits this process.

vitamin K$_2$-2,3-epoxide   **reduced vitamin K$_2$ (KH$_2$)**

blocked by warfarin (Coumadin)

warfarin (Coumadin)   (18.80)

Warfarin is the active ingredient in rat poison, but is also used in human medicine as an anticoagulant.

As we noted in Sec. 5.6C, phenols are inhibitors of free-radical reactions. The basis of this effect is that a free radical (R· in Eq. 18.81) can abstract a hydrogen atom from a phenol to give a resonance-stabilized free radical. [A semiquinone (Eq. 18.76) is an example of such a radical.]

a resonance-stabilized free radical   (18.81a)

Vitamin E, a phenol, serves as a free-radical inhibitor in biological membranes and fats.

α-tocopherol
a major form of vitamin E

Because of its long hydrocarbon side chain, vitamin E is readily incorporated into the phospholipid bilayer of cell membranes, which it protects from free-radical damage by terminating radical chains as shown in Eq. 18.81a. The radical thus formed from vitamin E reacts with vitamin C, a water-soluble vitamin, at the membrane–water interface to regenerate vitamin E and form a resonance-stabilized radical derived from vitamin C.

(18.81b)

The vitamin C-derived radical is ultimately reduced back to vitamin C by other biological reducing agents. As this example illustrates, vitamin C is also a very important reducing agent in biology.

The effectiveness of several widely used food preservatives is based on reactions such as these. Examples of such preservatives are "butylated hydroxytoluene" (BHT) and "butylated hydroxyanisole" (BHA).

BHT   **BHA**

Oxidation involving free-radical processes is one way that foods discolor and spoil. A preservative such as BHT inhibits these processes by donating its OH hydrogen atom to free radicals in the food (as in Eq. 18.81). The BHT is thus transformed into a phenoxy radical, which is too stable and unreactive to propagate radical chain reactions. Although the use of BHT and BHA as food additives has generated some controversy because of their potential side effects, without such additives foods could not be stored for any appreciable length of time or transported over long distances.

Poison Ivy and Itchy Quinones

Over half of the people in the United States suffer from allergy to poison ivy, poison oak, and poison sumac. The active principle in these plants is a family of catechol derivatives known collectively as *urushiol*.

urushiol (one of several components)

(The various urushiol components differ in the number, positions, and possibly the stereochemistry of the side-chain double bonds.) The allergic reaction is not caused by the catechol itself, but rather by its oxidation product, an o-quinone:

(18.82)

The long hydrocarbon side chain of urushiol probably imbeds itself into the lipid bilayer (Sec. 8.7A) of a skin-cell membrane, thus immobilizing it near membrane-localized oxidizing enzymes. *Ortho*-quinones are examples of α,β-unsaturated ketones, which are compounds in which a ketone carbonyl group (C=O) is conjugated with a carbon–carbon double bond. As we'll explore in Sec. 22.9A, these compounds undergo rapid conjugate addition with nucleophiles, including nucleophiles available in the cell, such as the thiol and amino groups of proteins. Using the thiol group of a protein as an example, a typical addition reaction occurs as follows:

(18.83a)

A subsequent rapid reaction with Brønsted acids and bases forms a substituted catechol.

(18.83b)

Ortho-quinones are not aromatic, but the catechol addition products are; the aromatic stabilization of the product makes the reaction irreversible. These reactions result in an irreversibly modified protein, which is now sensed as "foreign" by the immune system. The resulting biological response is the all-too-familiar allergic reaction—the skin eruptions and the intense itch.

PROBLEMS

18.35 Draw the important resonance structures of the radicals formed when each of the following reacts with R·, a general free radical.
(a) vitamin E   (b) BHT

18.36 Consider the detailed structure of the semiquinone ·QH shown in Eq. 18.79a.
(a) There are two possible structures for this semiquinone; draw them both.
(b) Show the resonance structures for either of the structures you gave in part (a).

18.37 (a) Using the fishhook notation, derive the important resonance structures of the vitamin C-derived radical in Eq. 18.81b.

(b) In the laboratory, the radical derived from vitamin E can react with a second free radical R· to give the following oxidation product (among others).

vitamin E radical

Using the fishhook notation, give a mechanism for this reaction.

18.38 Electron transport takes place in the membrane of cellular organelles called *mitochondria*. What is it about the structure of ubiquinone and its reduction products that ensures their localization within membranes? Explain.

18.39 (a) Give the structure of the product formed in the reaction of urushiol with K_2CO_3 and a large excess of methyl iodide.

(b) Would this compound be likely to provoke the same allergic skin response as urushiol? Explain.

18.9 ELECTROPHILIC AROMATIC SUBSTITUTION REACTIONS OF PHENOLS

Phenols are aromatic compounds, and they undergo electrophilic aromatic substitution reactions such as those described in Sec. 16.4. In some of these reactions, the —OH group has special effects that are not common to other substituent groups.

Because the —OH group is a strongly activating substituent, phenol can be halogenated once under mild conditions that are totally ineffective for benzene itself.

$$\text{H}\!-\!\!\langle\text{benzene}\rangle\!-\!\text{OH} + \text{Br}_2 \xrightarrow[\text{CCl}_4 \text{ or CS}_2]{} \text{Br}\!-\!\!\langle\text{benzene}\rangle\!-\!\text{OH} + \text{HBr} \qquad (18.84)$$

phenol **p-bromophenol**
(82% yield)

Notice the mild conditions of this reaction. A Lewis acid such as $FeBr_3$ is not required. (A solution of Br_2 in CCl_4 is the reagent usually used for adding bromine to alkenes.) But when phenol reacts with Br_2 in H_2O (bromine water), more extensive bromination occurs and 2,4,6-tribromophenol is obtained.

$$\text{phenol} + 3\,\text{Br}_2 \xrightarrow{\text{H}_2\text{O}} \text{2,4,6-tribromophenol} + 3\,\text{HBr} \qquad (18.85)$$

2,4,6-tribromophenol
(~100% yield;
(precipitates)

This more extensive bromination occurs for two reasons. First, bromine reacts with water to give protonated hypobromous acid, a more potent electrophile than bromine itself.

$$\text{Br}_2 + \text{H}_2\ddot{\text{O}}\!: \; \rightleftarrows \; \text{Br}^- + \text{Br}\!-\!\overset{+}{\ddot{\text{O}}}\text{H}_2 \; \rightleftarrows \; \text{HBr} + \text{Br}\!-\!\ddot{\text{O}}\text{H} \qquad (18.86)$$

protonated hypobromous
hypobromous acid acid

Second, in aqueous solutions near neutrality, phenol partially ionizes to its conjugate-base phenoxide anion. Although only a small amount of this anion is present, it is very reactive and brominates instantly, thereby pulling the phenol–phenolate equilibrium to the right.

(18.87)

Phenoxide ion is much more reactive than phenol because the reactive intermediate is not a carbocation, but is instead a more stable neutral molecule (red structure).

(18.88)

p-Bromophenol is also in equilibrium with its conjugate base *p*-bromophenoxide anion, which brominates again until all ortho and para positions have been substituted. Notice in Eq. 18.87 that in the second and third substitutions the powerful ortho, para-directing and activating effects of the —$\ddot{O}$:⁻ group override the weaker deactivating and directing effects of the bromine substituents.

These same considerations apply to the iodination of tyrosine (Sec. 16.5C).

In strongly acidic solution, in which formation of the phenolate anion is suppressed, bromination can be stopped at the 2,4-dibromophenol stage.

(18.89)

phenol   **2,4-dibromophenol**
(87% yield)

Phenol is also very reactive in other electrophilic substitution reactions, such as nitration. Phenol can be nitrated once under mild conditions. (Notice that H_2SO_4 in the following reaction is not present as it is in the nitration of benzene; Eq. 16.10, p. 803.)

$$\text{phenol} \xrightarrow[\substack{15\,°C \\ CHCl_3}]{HNO_3} \text{o-nitrophenol} + \text{p-nitrophenol} + H_2O \qquad (18.90)$$

phenol

o-nitrophenol
(26% yield)

p-nitrophenol
(61% yield)

Because phenol is activated toward electrophilic substitution, it is also possible to nitrate phenol two and three times. However, direct nitration is *not* the preferred method for synthesis of di- and trinitrophenol, because the concentrated HNO_3 required for multiple nitrations is also an oxidizing agent, and phenols are easily oxidized (Sec. 18.8). Instead, 2,4-dinitrophenol is synthesized by the nucleophilic aromatic substitution reaction of 1-chloro-2,4-dinitrobenzene with ⁻OH (Sec. 18.4A).

$$\text{chlorobenzene} \xrightarrow[H_2SO_4]{HNO_3} \text{1-chloro-2,4-dinitrobenzene} \xrightarrow[\substack{1)\,^-OH \\ 2)\,H_3O^+}]{} \text{2,4-dinitrophenol} \qquad (18.91)$$

chlorobenzene

1-chloro-2,4-dinitrobenzene

2,4-dinitrophenol

The basic conditions of this reaction result in formation of the conjugate-base anion of the product; the H_3O^+ is added following the reaction to give the neutral phenol.

The great reactivity of phenol in electrophilic aromatic substitution does not extend to the Friedel–Crafts acylation reaction, because phenol reacts rapidly with the $AlCl_3$ catalyst.

$$\text{phenol} + AlCl_3 \longrightarrow \left[\cdots \longleftrightarrow \cdots \right] + HCl \qquad (18.92)$$

The adduct of phenol and $AlCl_3$ is much less reactive than phenol itself in electrophilic aromatic substitution reactions because, as shown in Eq. 18.92, the oxygen electrons are delocalized onto the electron-deficient aluminum. Because of their delocalization away from the benzene ring, these electrons are less available for resonance stabilization of the carbocation intermediate formed within the ring during Friedel–Crafts acylation (Eq. 16.24, p. 809). Thus, Friedel–Crafts acylation of phenol occurs slowly, but can be carried out successfully at elevated temperatures. Because it is not highly activated, the ring is acylated only once.

$$\text{—OH} + AlCl_3 + CH_3(CH_2)_5 - \overset{O}{\underset{\|}{C}} - Cl \xrightarrow[PhNO_2]{140\,°C} \xrightarrow{H_2O}$$

(as a complex)

(34% yield) $+ CH_3(CH_2)_5 - \overset{O}{\underset{\|}{C}}$ —OH   (18.93)

(47% yield)

**FURTHER
EXPLORATION 18.1**
The Fries Rearrangement

Friedel–Crafts alkylation of phenol is also possible.

p-tert-butylphenol
(80% yield)

$$(18.94)$$

PROBLEMS

18.40 Give the principal organic product(s) formed in each of the following reactions.

(a) o-cresol + Br_2 in CCl_4 $\longrightarrow$

(b) m-chlorophenol + HNO_3, low temperature $\longrightarrow$

(c)

18.41 Give a curved-arrow mechanism for the reaction in Eq. 18.94. Be sure to identify the electrophilic species in the reaction and to show how it is formed.

18.10   REACTIVITY OF THE ARYL–OXYGEN BOND

A. Lack of Reactivity of the Aryl–Oxygen Bond in S_N1 and S_N2 Reactions

Just as the reactions of alcohols that break the carbon–oxygen bond have close analogy to the reactions of alkyl halides that break the carbon–halogen bond, the carbon–oxygen reactivity of phenols follows the poor carbon–halogen reactivity of *aryl* halides. Recall that aryl halides do not undergo S_N1 or E1 reactions (Sec. 18.3); for the same reasons, phenols also do not react under conditions used for the S_N1 or E1 reactions of alcohols. Thus, phenols do *not* form aryl bromides with concentrated HBr; they do *not* dehydrate with concentrated H_2SO_4. (Instead, they undergo sulfonation; see Sec. 16.4D.) The reasons for these observations are exactly the same as those that explain the lack of reactivity of aryl halides (see Secs. 18.1 and 18.3).

More generally, *any* derivative of the form

in which X is a good leaving group, such as tosylate, mesylate, or even —$\overset{+}{O}H_2$, has the same lack of reactivity toward S_N1 and S_N2 conditions as aryl halides—and for the same reasons.

The lack of reactivity of the aryl–oxygen bond can be put to good use in the cleavage of aryl ethers. Recall that when ethers cleave—depending on the particular ether and the mechanism—products resulting from cleavage at either of the carbon–oxygen bonds are possible (Sec. 11.4). In the case of aryl ethers, cleavage occurs *only* at the alkyl–oxygen bond; consequently, only one set of products is formed:

this bond is broken

$$\text{(phenyl)}-O-CH_3 + HBr \xrightarrow{\text{heat}} \text{(phenyl)}-OH + CH_3Br \qquad (18.95)$$

observed products

$$\text{(phenyl)}-Br + CH_3OH$$

(not observed)

In this example, ether cleavage gives phenol and methyl bromide rather than bromobenzene and methanol because the S_N2 reaction of the bromide-ion nucleophile can only occur *at the methyl group* of the protonated ether:

$$\qquad (18.96)$$

S_N2 reaction of the Br^- cannot occur at this carbon

PROBLEM

18.42 Within each set, identify the ether that would *not* readily cleave with concentrated HBr and heat, and explain. Then give the products of ether cleavage and the mechanisms of their formation for the other ether(s) in the set.

(a)

benzyl methyl ether   ***p*-methoxytoluene**

**(2,2-dimethylpropoxy)benzene
(neopentyl phenyl ether)**

(b)

***tert*-butyl phenyl ether
(*tert*-butoxybenzene)**   **diphenyl ether
(phenoxybenzene)**

B. Substitution at the Aryl–Oxygen Bond: The Stille Reaction

We learned in Secs. 18.5 and 18.6 that certain transition-metal catalysts, particularly Pd(0), can catalyze the substitution of aryl halides at aryl carbons. Pd(0) catalysts can also catalyze substitution at the aryl–oxygen bond. The first requirement for this to occur is conversion of the phenolic —OH group into a triflate, a very reactive leaving group. (Triflates were introduced in Sec. 10.4A, p. 465.)

trifluoromethanesulfonic anhydride (triflic anhydride)

pyridine (solvent and base)

4-methoxyphenyl trifluoromethanesulfonate (*p*-methoxyphenyl triflate) (93% yield)

(18.97)

Aryl triflates react readily with organotin derivatives in the presence of Pd(0) catalysts to give coupling products.

trimethylphenylstannane

(85% yield)

(18.98)

This reaction is called the **Stille reaction**, after John K. Stille (1930–1989), who was a professor of chemistry at Colorado State University. Stille and his students developed this reaction in the early 1980s. The Stille reaction is another important application of transition-metal catalysis for the formation of carbon–carbon bonds.

A number of metals can be used to bring about similar transformations, but tin was chosen because the organotin compounds used in the Stille reaction are relatively insensitive to moisture, tolerant of other functional groups, and very easy to handle. (Some can be distilled or crystallized.) Many of them are either commercially available, or they are readily prepared from Grignard reagents and commercially available trialkyltin chlorides. The very basic "carbanion" of the Grignard reagent displaces chloride, a weak base, from the tin:

(18.99)

In Eq. 18.98, notice that the phenyl group is transferred in preference to the methyl groups in the Stille reaction. In general, vinylic groups, aryl groups, and other unsaturated groups are transferred preferentially. However, if a tetraalkylstannane is used, alkyl groups can also be transferred. This provides an excellent way to prepare alkylbenzenes. In contrast to the Friedel–Crafts alkylation reaction (Sec. 16.4E), the Stille reaction is *not* plagued by rearrangements.

$$H_3C-\overset{\displaystyle O}{\overset{\|}{C}}-\!\!\!\left\langle\bigcirc\right\rangle\!\!\!-OTf + (CH_3CH_2CH_2CH_2)_4Sn + LiCl \xrightarrow[\substack{98\ ^\circ C \\ dioxane}]{\substack{Pd(PPh_3)_4 \\ (catalyst)}}$$

tetrabutylstannane   (excess)

$$H_3C-\overset{\displaystyle O}{\overset{\|}{C}}-\!\!\!\left\langle\bigcirc\right\rangle\!\!\!-CH_2CH_2CH_2CH_3 + (CH_3CH_2CH_2CH_2)_3SnCl + LiOTf \qquad (18.100)$$

p-butylacetophenone
(82% yield)

The mechanism of the Stille reaction (Eq. 18.101) begins with oxidative addition of the aryl triflate to the 14-electron $Pd(PPh_3)_2$ (the same catalytic species as in the Heck and Suzuki reactions). The resulting complex *(1)* is very unstable, and the excess chloride ion (as LiCl) rescues the complex from decomposition by ligand substitution to form *(2)*. The aryl or alkyl R-groups on the organotin compounds have carbanion character, and they are nucleophilic enough to substitute for the chloride on the Pd. A reductive elimination completes the mechanism.

Complex *(2)* in Eq. 18.101 is the same complex that would be formed from oxidative addition of an aryl halide to PdL_2. (Compare with the product of Step *(1)* in the Heck mechanism, Eq. 18.43b, p. 902.) Hence, the Stille reaction can be carried out with aryl halides instead of aryl triflates; and the Heck reaction can be carried out with aryl triflates instead of aryl halides.

The Stille reaction adds another method to our arsenal of reactions that can be used to form carbon–carbon bonds:

1. the addition of carbenes and carbenoids to alkenes (Sec. 9.9)
2. the reaction of Grignard reagents with ethylene oxide and lithium organocuprate reagents with epoxides (Sec. 11.5C)
3. the reaction of acetylenic anions with alkyl halides or sulfonates (Sec. 14.7B)
4. the Diels–Alder reaction (Sec. 15.3)
5. Friedel–Crafts reactions (Secs. 16.4E–F)
6. the Heck and Suzuki coupling reactions (Secs. 18.6A–B)
7. alkene metathesis (Sec. 18.6C)
8. the Stille reaction (this section)

PROBLEMS

18.43   Predict the product of the Stille reaction between ethynyltrimethylstannane, $HC{\equiv}C-Sn(CH_3)_3$, and phenyl triflate, PhOTf, in the presence of $Pd(PPh_3)_4$ and excess LiCl.

18.44   What reactants would be required to form the following compound by the Stille reaction?

18.11   INDUSTRIAL PREPARATION AND USE OF PHENOL

Historically, phenol has been made in a variety of ways, but the principal method used today is an elegant example of a process that gives two industrially important compounds, phenol and acetone, $(CH_3)_2C{=}O$, from a single starting material. The starting material for the manufacture of phenol is cumene (isopropylbenzene), which comes from benzene and propene, two compounds obtained from petroleum (Sec. 16.8). The production of phenol and acetone is a two-stage process. In the first stage, cumene undergoes an *autoxidation* to form cumene hydroperoxide. (An **autoxidation** is an oxidation reaction involving molecular oxygen as the oxidizing agent).

Autoxidation:

(18.102)

Autoxidation is a free-radical chain reaction. Dioxygen is actually a *diradical*—a double free radical. (For the reason, see Problem 1.47, p. 43.) The initiation step involves abstraction of the benzylic hydrogen of cumene:

(18.103a)

In the propagation steps, the resulting benzylic radical then reacts with another molecule of dioxygen, and the resulting radical abstracts a benzylic hydrogen from another molecule of cumene to form cumene hydroperoxide.

(18.103b)

$$\text{(18.103c)}$$

cumene hydroperoxide

In the second stage of the phenol–acetone synthesis, cumene hydroperoxide is subjected to an acid-catalyzed rearrangement that yields both acetone and phenol.

Rearrangement:

phenol   **acetone** $$\text{(18.104)}$$

STUDY GUIDE LINK 18.2
The Cumene Hydroperoxide Rearrangement

Phenol is a very important commercial chemical. It is a starting material for the production of phenol–formaldehyde resins (Sec. 19.15), which are polymers that have a variety of uses, including plywood adhesives, glass fiber (fiberglass) insulation, molded phenolic plastics used in automobiles and appliances, and many others.

PROBLEM

18.45 Compound *A* is a by-product of the autoxidation of cumene, and compound *B* is a by-product of the acid-catalyzed conversion of cumene hydroperoxide to phenol and acetone.

A   *B*

Draw a curved-arrow mechanism that shows how compound *A* can react with phenol under the conditions of Eq. 18.102 to give compound *B*.

KEY IDEAS IN CHAPTER 18

- Neither aryl halides, vinylic halides, nor phenols undergo S_N2 reactions because the *sp* hybridization of the S_N2 transition state has relatively high energy and because the opposite-side substitution reaction with nucleophiles is sterically blocked.

- Neither aryl halides, vinylic halides, nor phenols undergo S_N1 reactions because of the instability of the carbocation intermediates that would be involved in such reactions.

- The aryl–oxygen bond is unreactive in both S_N1 and S_N2 reactions.

- Vinylic halides undergo β-elimination reactions in base under vigorous conditions to give alkynes.

- Aryl halides substituted with ortho or para electron-attracting groups, particularly nitro groups, undergo nucleophilic aromatic substitution. In this type of reaction, the nucleophile reacts at the ring carbon to form a resonance-stabilized anion (Meisenheimer complex), from which the halide leaving group is then expelled.

- The important concepts for electron counting in transition-metal complexes are

  1. *Oxidation state:* The number of X-type ligands plus the charge.

  2. *The d^n configuration:* The number of unshared electrons on a metal within a complex, equal to the electron count of the neutral atom minus the oxidation state.

  3. *The electron count:* The electron count of the neutral atom plus the number of X-type ligands plus twice the number of L-type ligands minus the charge.

- Some fundamental processes in transition-metal catalysis are

  1. *Ligand association and dissociation.* A succession of dissociation–association steps gives ligand substitution.

  2. *Oxidative addition.*

  3. *Reductive elimination.* (Oxidative addition and reductive elimination are conceptually the reverse of each other.)

  4. *Ligand insertion.*

  5. *β-Elimination.* (1,2-Ligand insertion and β-elimination are conceptually the reverse of each other.)

- Several transition-metal-catalyzed reactions can effect the coupling of vinylic and aryl groups. These are catalyzed by Pd(0), which is typically present as Pd(PPh$_3$)$_4$ or a related derivative. Pd(OAc)$_2$ can also be used; it is reduced in the reaction mixture to form Pd(0).

  1. The Heck reaction brings about the coupling of an alkene with an aryl bromide or iodide. The reaction is typically used to form aryl-substituted alkenes.

  2. The Suzuki reaction brings about the coupling of a vinylic or aryl bromide or iodide with an aryl or vinylic boronic acid or borane. The reaction is used to form biaryls, aryl-substituted alkenes, and conjugated alkenes.

  3. The Stille reaction is the coupling of an organic group derived from an organostannane reagent with the aryl group of a phenol in which the phenol group has been converted into a triflate group. The Stille reaction

can be used to form both alkyl- and aryl-substituted benzenes.

The key mechanistic steps in all of these reactions are an oxidative addition of the aryl halide or triflate to the metal and a reductive elimination of the coupling product.

- In alkene metathesis, two alkenes can be transformed into other alkenes by breaking the carbon–carbon double bond and scrambling the resulting fragments. The ruthenium-based Grubbs catalysts G1 and G2 can be used to bring about this transformation. A series of ruthenium metallacycles are key intermediates in alkene metathesis. Formation of cyclic alkenes is one of the most important applications of this reaction.

- Phenols are considerably more acidic than alcohols because phenoxide ions, the conjugate bases of phenols, are stabilized both by resonance and by the electron-attracting polar effect of the aromatic ring. Phenols containing substituent groups that stabilize negative charge by resonance or polar effects (or both) are even more acidic.

- Phenols can be oxidized to quinones. Two classes of quinones are *ortho*-quinones and *para*-quinones. *Para*-quinones are typically more stable.

- A hydroquinone (a 1,4-benzenediol) derivative can undergo reversible one-electron oxidation to a semiquinone, and another reversible one-electron oxidation to a quinone. Coenzyme Q is an important biological quinone that undergoes reversible conversion into its hydroquinone, QH$_2$, as part of electron transport in biology, a process used to fuel energy generation in the cell. Vitamin K is another biologically important hydroquinone that plays a crucial role in blood clotting.

- Phenols are free-radical inhibitors because abstraction of a hydrogen by a radical gives a resonance-stabilized phenoxy radical. Vitamin E serves as a biological radical scavenger in living systems.

- Some electrophilic aromatic substitution reactions of phenols show unusual effects attributable to the —OH group. Thus, phenol brominates three times in bromine water because the —OH group can ionize; and phenol is rather unreactive in Friedel–Crafts acylation reactions because the —OH group reacts with the AlCl$_3$ catalyst.

 REACTION REVIEW *For a summary of reactions discussed in this chapter, see the* Reaction Review *section of Chapter 18 in the* Study Guide and Solutions Manual.

18.46 Give the product(s) (if any) expected when *p*-iodotoluene or other compound indicated is subjected to each of the following conditions.

(a) CH_3OH, 25 °C

(b) CH_3O^- in CH_3OH, 25 °C

(c) CH_3O^-, pressure, heat

(d) Mg in THF

(e) product of part (d) + $ClSn(CH_3)_3$

(f) Li in hexane

(g) $H_2C=CH_2$, $Pd(PPh_3)_4$ catalyst, and $(Et)_3N\colon$ in CH_3CN

(h) product of part (e) with phenyl triflate, excess LiCl, and $Pd(PPh_3)_4$ catalyst in dioxane

(i) $PhB(OH)_2$, aqueous Na_2CO_3, and $Pd(PPh_3)_4$ catalyst

(j) product of (d) + $B(OCH_3)_3$, then H_3O^+/H_2O

(k) product of (j) + (*E*)-1-bromopropene, aqueous Na_2CO_3, and $Pd(PPh_3)_4$ catalyst

18.47 Give the product(s) expected (if any) when *m*-cresol or other compound indicated is subjected to each of the following conditions.

(a) concentrated H_2SO_4

(b) Br_2 in CCl_4 (dark)

(c) Br_2 (excess) in CCl_4, light

(d) dilute HCl

(e) 0.1 *M* NaOH solution

(f) HNO_3, cold

(g)
$$Et\!-\!\overset{\overset{\textstyle O}{\|}}{C}\!-\!Cl,\ AlCl_3,\ heat;\ then\ H_2O$$

(h) $Na_2Cr_2O_7$ in H_2SO_4

(i) triflic anhydride in pyridine, 0 °C

(j) product of part (i) + $(CH_3)_4Sn$, excess LiCl, and $Pd(PPh_3)_4$ catalyst in dioxane

(k) product of (i) + (*E*)-$CH_3CH=CH-B(OH)_2$, aqueous NaOH, and $Pd(PPh_3)_4$ catalyst

18.48 Give the products formed in the reaction of 1-hexene with each of the following compounds in the presence of the Grubbs G2 catalyst.

(a) an excess of allyl alcohol (2-propen-1-ol)

(b) an excess of 2-methylpropene (isobutylene)

18.49 Arrange the compounds within each set in order of increasing acidity, and explain your reasoning.

(a) cyclohexyl mercaptan, cyclohexanol, benzenethiol

(b) cyclohexanol, phenol, benzyl alcohol

(c) *p*-nitrophenol, *p*-chlorophenol,

$$H-\overset{\displaystyle \ddot{O}\colon}{\underset{\displaystyle \ddot{O}\colon^-}{\ddot{O}-\overset{+}{N}}}$$

(d) 4-nitrobenzenethiol, 4-nitrophenol, phenol

(e)

A   *B*   *C*   *D*

18.50 Although enols are unstable compounds (Sec. 14.5A), suppose that the acidity of an enol could be measured. Which would be more acidic: enol *A* or alcohol *B*? Why?

$$\underset{A}{H_3C-\overset{\displaystyle OH}{\underset{}{C}}=CH_2} \qquad \underset{B}{H_3C-\overset{\displaystyle OH}{\underset{}{CH}}-CH_3}$$

18.51 Enols, like phenols, have pK_a values of ~10–11. However, the pK_a of warfarin, a widely used anticoagulant, is 5.0.

warfarin (Coumadin)

(a) Give the structure of the conjugate base of warfarin.

(b) Use resonance arguments to justify the unusually low pK_a of warfarin.

(c) What is the predominant form of warfarin—enol or conjugate base—at physiological pH (7.4)? Explain.

18.52 Identify compounds *A*, *B*, and *C* from the following information.

(a) Compound *A*, $C_8H_{10}O$, is insoluble in water but soluble in aqueous NaOH solution, and yields 3,5-dimethylcyclohexanol when hydrogenated over a nickel catalyst at high pressure.

(b) Aromatic compound *B*, $C_8H_{10}O$, is insoluble in both water and aqueous NaOH solution. When treated successively with concentrated HBr, then Mg in THF, then water, it gives *p*-xylene.

(c) Compound *C*, $C_9H_{12}O$, is insoluble in water and in NaOH solution, but reacts with concentrated HBr and heat to give *m*-cresol and a volatile alkyl bromide.

18.53 Contrast the reactivities of cyclohexanol and phenol with each of the following reagents, and explain.

(a) aqueous NaOH solution

(b) NaH in THF

(c) triflic anhydride in pyridine, 0 °C

(d) concentrated aqueous HBr, H_2SO_4 catalyst

(e) Br_2 in CCl_4, dark

(f) $Na_2Cr_2O_7$ in H_2SO_4

(g) H_2SO_4, heat

18.54 Choose the one compound within each set (see Fig. P18.54) that meets the indicated criterion, and explain your choice.

(a) The compound that reacts with alcoholic KOH to liberate fluoride ion.

(b) The compound that *cannot* be prepared by a Williamson ether synthesis.

(c) The compound that gives an acidic solution when allowed to stand in aqueous ethanol.

(d) The ether that cleaves more rapidly in HI.

18.55 Give the products (if any) when each of the following isomers reacts with HBr and heat.

3-(hydroxymethyl)phenol   3-methoxyphenol

18.56 Explain the following observations, which were recorded in the chemical literature, concerning the reaction between *tert*-butyl bromide and potassium benzenethiolate (the potassium salt of benzenethiol): "The attempts to prepare phenyl *tert*-butyl sulfide by this route failed. If the reactants were kept at room temperature KBr was formed, but the benzenethiol was recovered unchanged."

18.57 What products (if any) are formed when 3,5-dimethylbenzenethiol is treated first with one equivalent of Na^+ EtO^- in ethanol, and then with each of the following?

(a) allyl bromide   (b) bromobenzene

18.58 Phenols, like alcohols, are Brønsted bases.

(a) Write the reaction in which the oxygen of phenol reacts as a base with the acid H_2SO_4.

(b) On the basis of resonance and polar effects, decide whether phenol or cyclohexanol should be the stronger base.

18.59 The UV spectrum of *p*-nitrophenol in aqueous solution is shown in Fig. P18.59 (spectrum A). When a few drops of concentrated NaOH are added, the solution turns yellow and the spectrum changes (spectrum B). On addition of a few drops of concentrated acid, the color disappears and spectrum A is restored. Explain these observations.

18.60 Vanillin is the active component of natural vanilla flavoring.

vanillin

When a few drops of vanilla extract (an ethanol solution of vanillin) are added to an aqueous NaOH solution, the characteristic vanilla odor is not present. Upon acidification of the solution, a strong vanilla odor develops. Explain.

18.61 A mixture of *p*-cresol (4-methylphenol), $pK_a = 10.2$, and 2,4-dinitrophenol, $pK_a = 4.11$, is dissolved in ether. The ether solution is then vigorously shaken with one of the following aqueous solutions. Which solution effects

(a) or

(b) or or

(c) or

(d) phenyl cyclohexyl ether   or   diphenyl ether

Figure P18.54

the best separation of the two phenols by dissolving one in the water layer and leaving the other in the ether solution? Explain. (*Hint:* Apply Eqs. 3.31a–b, p. 107.)

(1) a 0.1 *M* aqueous HCl solution

(2) a solution that contains a large excess of pH = 4 buffer

(3) a solution that contains a large excess of pH = 7 buffer

(4) 0.1 *M* NaOH solution

18.62 1-Haloalkynes are known compounds.

(a) Would 1-bromo-2-phenylacetylene (Br—C≡CPh) be likely to undergo an S_N2 reaction? Explain.

(b) Would the same compound be likely to undergo an S_N1 reaction? Explain.

18.63 Ferulic acid is a potent antioxidant found in tomatoes and other vegetables that protects against oxidative stress in neuronal cells, and is thus of some interest in research on Alzheimer's disease. (See the sidebar, pp. 552–553.)

4-hydroxy-3-methoxycinnamic acid (ferulic acid)

(a) Show how a free radical R· would react with ferulic acid, and include resonance structures of the product radical.

(b) Would the isomer of ferulic acid, 3-hydroxy-4-methoxycinnamic acid, be equally effective as a free-radical inhibitor? Explain.

18.64 Give the structure of the radical formed, and its resonance structures, when warfarin (structure in Problem 18.51) undergoes a one-electron oxidation.

18.65 The structure of cyanocobalamin, one of the forms of vitamin B_{12}, is given in Fig. P18.65, p. 939. Notice that cyanocobalamin is a complex of the transition metal cobalt (Co). Characterize this compound in the following ways:

(a) the oxidation state of the cobalt

(b) the d^n count (that is, the value of n)

(c) the total electron count around the metal

18.66 Provide the following information, and show how you obtained it, for the Fe(V) in the protoporphyrin IX group of cytochrome P450 (Fig. 17.4a, p. 855). In addition, verify the oxidation state of the iron.

(a) the d^n count (that is, the value of n)

(b) the total electron count around the metal

18.67 When a suspension of 2,4,6-tribromophenol is treated with an excess of bromine water, the white precipitate of 2,4,6-tribromophenol disappears and is replaced by a precipitate of a yellow compound that has the following structure. Give a curved-arrow mechanism for the formation of this compound.

18.68 It has been suggested that the solvolysis of 2-(bromomethyl)-5-nitrophenol at alkaline pH values involves the intermediate shown in brackets (see Fig. P18.68, p. 939). Give a curved-arrow mechanism for the formation of this intermediate and for its reaction with aqueous hydroxide ion to give the final product.

18.69 Outline a synthesis for each of the following compounds from the indicated starting material and any other reagents.

(a) 1-chloro-2,4-dinitrobenzene from benzene

(b) 1-chloro-3,5-dinitrobenzene from benzene

(continues on p. 938)

Figure P18.59 UV spectra for Problem 18.59. Spectrum *A* was taken in the presence of acid; spectrum *B* was taken in the presence of NaOH.

(continued from p. 937)

(c)

Ph—C≡C—Ph   from   Ph, H / C=C / H, Ph

(d)

H$_3$C—C(=O)—[ring with NO$_2$]—SCH$_2$CH$_3$ from chlorobenzene

(e) 2-chloro-4,6-dinitrophenol from chlorobenzene

(f) "butylated hydroxytoluene" (BHT; p. 923) from *p*-cresol (4-methylphenol)

(g)

(CH$_3$)$_3$C—[cyclohexadienone ring]=O   from catechol (benzene-1,2-diol)

(h) PhCH$_2$CH$_2$OH from bromobenzene

(i)

O$_2$N—[ring]—OCH$_2$CH$_2$—[ring]

from the product of part (h) and fluorobenzene

(j)

Ph / C=C(H) / H ... C—[ring with NO$_2$]

from benzene and ethylene as the only sources of carbon

(k)

[naphthalene]—C(H)=C(H)—[ring]

from [naphthalene with Br]

(l)

O$_2$N—[ring]—C(H)=C(H)—CH$_2$OH

from bromobenzene, ethylene, and any other reagents

(m)

[ring]—C≡C—[ring]—OCH$_3$

from PhC≡CH and 4-methoxyphenol. (*Hint:* See Eqs. 14.23, p. 699, and 18.99, p. 930.)

(n)

[chain]—C(H)=C(H)—[ring with OEt, OEt]

from 1-hexyne and resorcinol

(o)

Ph·····[bicyclic epoxide ring]—O   from iodobenzene

(racemate)

18.70 Complete each reaction given in Fig. P18.70, pp. 940–941, by giving the major organic product(s), and explain your reasoning. "No reaction" may be an appropriate response.

18.71 One use of alkene metathesis is to form polymers from cyclic alkenes ("ROMP," for "ring-opening metathesis polymerization"). Give the structure of the polymer formed when each of the following alkenes is polymerized with an appropriate metathesis catalyst.

(a)

[cyclooctene structure]

cyclooctene

(b)

[norbornene structure]

norbornene

18.72 (a) 1,3-Cyclopentadiene undergoes a Diels–Alder reaction with itself to give a diene known commonly as *endo*-dicyclopentadiene. Give the structure of this diene.

(b) When *endo*-dicyclopentadiene is subjected to ROMP (see Problem 18.71), a polymer is formed that is so strong that a 1-inch-thick block will stop a 9 mm bullet. Give a structure for this polymer and suggest one reason why it is so strong.

18.73 Explain why, in the reactions given in Fig. P18.73, p. 942 different stereoisomers of the starting material give different products.

18.74 The reaction given in Fig. P18.74 on p. 942 occurs readily at 95 °C (X = halogen). The relative rates of the reaction for the various halogens are 290 (X = F), 1.4 (X = Cl), and 1.0 (X = Br). When a nitro group is in the para position of each benzene ring, the reaction is substantially accelerated. Give a detailed mechanism for this reaction, and explain how it is consistent with the experimental facts.

18.75 When 1,3,5-trinitrobenzene [NMR: δ 9.1(s)] is treated with Na$^+$ CH$_3$O$^-$, an ionic compound is formed that has the following NMR spectrum: δ 3.3 (3*H*, s); δ 6.3 (1*H*, t, *J* = 1 Hz); δ 8.7 (2*H*, d, *J* = 1 Hz). Suggest a structure for this compound.

18.76 Suggest structures for *X* and *Y* in the reaction sequence given in Fig. P18.76 (p. 942), including their stereochemistry. Suggest a mechanism for the formation of *Y*. Notice that aluminum (Al) is just below boron (B) in the periodic table; to predict *X*, imagine that the Al is a B and ask how that compound would react with an alkene. (*Hint:* See Secs. 5.4B and 14.5B.)

18.77 In some Pd(0)-catalyzed reactions, a Pd(II) compound such as PdCl₂ can be used instead of Pd(0), but it is assumed that the Pd(II) is reduced to Pd(0) in the reaction. Give both the product and the curved-arrow mechanism for reduction of PdCl₂ to Pd(0) by :PPh₃. (*Hint:* If Pd is reduced, something has to be oxidized.)

18.78 Draw curved-arrow mechanisms for the reactions given in Fig. P18.78 on p. 942.

18.79 The *Negishi reaction* is a Pd(0)-catalyzed cross-coupling reaction of organozinc compounds and aryl or vinylic iodides, as in the example shown in Fig. P18.79 on p. 942.

(Ei-ichi Negishi, a Professor at Purdue University, shared the 2010 Nobel Prize with Suzuki and Heck for his research into cross-coupling reactions.)

(a) Outline a mechanism of this coupling reaction by showing the important catalytic intermediates.

(b) Show how the organozinc compound could be prepared from *o*-iodotoluene. (*Hint:* See Eq. 18.50, p. 905.)

18.80 In 1991, chemists at the University of Iowa were able to prepare trifluoromethylcopper(I), F₃C—Cu. This compound could be used to prepare trifluoromethyl-substituted aromatic compounds by reactions like the one shown in Fig. P18.80 on p. 942.

(a) Suggest a stepwise mechanism for this transformation.

In 2013, chemists at Imperial College (London) showed that they could prepare F₃C—Cu containing the isotope ¹⁸F. They could use this compound to prepare a number

Figure P18.65 The structure of cyanocobalamin, the first isolated form of Vitamin B₁₂. (a) A Lewis structure. (b) A perspective drawing that shows the position of the two ligands that are above and below the plane defined by the four ring nitrogens.

Figure P18.68

of important drugs that contain trifluoromethyl groups. Each synthesis took less than an hour.

(b) Why might this advance be medically important? (*Hint:* See sidebar, p. 401.)

(c) Why is it important that these drugs be prepared and used quickly?

18.81 Explain why the dipole moment of 4-chloronitrobenzene (2.69 D) is *less* than that of nitrobenzene (3.99 D), and the dipole moment of *p*-nitroanisole (1-methoxy-4-nitrobenzene, 4.92 D) is *greater* than that of nitrobenzene, even though the electronegativities of chlorine and oxygen are about the same.

18.82 The reaction given in Fig. P18.82 on p. 943, used to prepare the drug mephenesin (a skeletal muscle relaxant), appears to be a simple Williamson ether synthesis. During this reaction, a precipitate of NaCl forms after only about 10 minutes, but a considerably longer reaction time is required to obtain a good yield of mephenesin. Taking these facts into account, suggest a mechanism for this reaction. (*Hint:* See Sec. 11.8.)

18.83 In 1960, a group of chemists led by Prof. John D. Roberts at Caltech reported that when chlorobenzene containing a ^{14}C isotopic label at carbon-1 is treated with the very strong base potassium amide, a substitution product, aniline, is formed in which the isotopic label is equally distributed between carbon-1 and carbon-2 (Fig. P18.83, p. 943). This result was interpreted as evidence for a very interesting unstable intermediate called *benzyne*. Propose a mechanism for this substitution that accounts for the isotopic labeling result. (*Hint:* Think about a β-elimination reaction.)

(a)

O_2N—⟨benzene⟩—F + $CH_3CH_2S^-$ ⟶

(b)

O_2N—⟨benzene with Cl at top, NO_2 at right, NO_2 at bottom⟩ + ^-OH ⟶ $\xrightarrow{H_3O^+}$

(c)

CH_3CH_2—⟨benzene⟩—OH $\xrightarrow{\text{NaOH solution}}$ $\xrightarrow{\text{dimethyl sulfate}}$ $\xrightarrow[\text{high pressure}]{\text{H}_2,\ \text{catalyst}}$

(d)

⟨benzene⟩—Cl + $CH_3CH_2CH_2$—NH_2 $\xrightarrow{25\ °C}$

(e)

⟨pyridine with B(OH)$_2$⟩ + ⟨pyridine with C=O, CH$_3$⟩ $\xrightarrow[\text{NaOH}]{\text{Pd(PPh}_3)_4\ \text{catalyst}}$

(f)

⟨cyclohexene with OTf⟩ + $(HO)_2B$—⟨benzene with NO_2⟩ $\xrightarrow[\text{aqueous Na}_2\text{CO}_3]{\text{Pd(PPh}_3)_4\ \text{catalyst}}$

(g)

HO—⟨benzene with OH, OH⟩ + Br_2 $\xrightarrow{CCl_4}$ $(C_6H_5O_3Br)$

pyrogallol

Figure P18.70 (continues)

(h)

(i)

(j)

(k) *m*-chlorophenol + Na$_2$Cr$_2$O$_7$ $\xrightarrow{\text{H}_2\text{SO}_4}$

(l) 2,4-dinitrophenol $\xrightarrow[\text{pyridine}]{\text{CH}_3\text{SO}_2\text{Cl}}$ $\xrightarrow[\text{CH}_3\text{OH}]{\text{CH}_3\text{O}^-}$

(m)

(n)

(o) Give the stereochemistry of the product.

(p) Give the structure and stereochemistry of *X*, *Y*, and *Z*.

(q) Give the stereochemistry of the product and explain your reasoning. (9-BBN is a sterically hindered hydroboration reagent.)

**9-borabicyclononane
(9-BBN)**

Figure P18.70 (continued)

$$H_3C-C\equiv C-OC_2H_5$$
(only elimination product observed)

$$H_3C-C\equiv C-OC_2H_5 + H_2C=C=CH-OC_2H_5$$

Figure P18.73

Figure P18.74

**diisobutylaluminum
hydride (DIBAL)**

Figure P18.76

(a)

(b)

Figure P18.78

(78% yield)

Figure P18.79

trifluoromethylcopper(I)

Figure P18.80

18.84 In the conversion shown in Fig. P18.84, the Diels–Alder reaction is used to trap a very interesting intermediate by its reaction with anthracene. From the structure of the product, deduce the structure of the intermediate. Then write a mechanism that shows how the intermediate is formed from the starting material.

18.85 Propose a structure for the product *A* obtained in the following oxidation of 2,4,6-trimethylphenol. (Compound *A* is an example of a rather unstable type of compound called generally a *quinone methide*.)

Proton NMR of *A*: δ 1.90 (6*H*), δ 5.49 (2*H*), δ 6.76 (2*H*), all broad singlets.

18.86 For the reactions given in Fig. P18.86 (p. 944), explain why different products are obtained when different amounts of $AlBr_3$ catalyst are used. (*Hint:* See Eq. 18.92, p. 927.)

18.87 Give a mechanism for the reaction in Fig. P18.87 (p. 944) by showing the catalytic intermediates.

18.88 Two general mechanisms (or various versions of them) for alkene metathesis were originally considered. In the first (pairwise) mechanism, shown in Fig. P18.88 (p. 944), the catalyst brings about cyclobutane formation between two alkenes. The second mechanism involves metallacycle formation, as shown in Eqs. 18.60a–d on pp. 909–910.

(a) The metathesis reaction of cyclopentene and (*E*)-2-pentene gives three products (not counting stereoisomers): 2,7-nonadiene, 2,7-decadiene, and 3,8-undecadiene, in a ratio of 1 : 2 : 1. Show how this result can be used to decide between the two mechanisms. Ignore self-metathesis products of cydopentene and (*E*)-2 pentene.

(b) What result is predicted by the two mechanisms for the ratio of the three ethylene products in the reaction shown in Fig. P18.88b? Assume that once an ethylene molecule is formed, it undergoes no further reaction.

18.89 When vinylic boronic acids are treated with Br_2 and then with NaOH, vinylic bromides are formed with inversion of configuration, as shown in Fig. P18.89a on p.944. But when vinylic boronic acids are treated with I_2 in the presence of NaOH, vinylic iodides are formed with *retention* of configuration, as shown in Fig. P18.89b. Suggest mechanisms that account for these results. (*Hint:* Iodine does not add to double bonds, but it does form iodohydrins (Sec. 5.2B). Remember the stereochemistry of halogen addition.)

Figure P18.82

* = carbon-14

chlorobenzene   **potassium amide**   **aniline** (about 50% of each)

Figure P18.83

1-bromo-2-fluorobenzene   **triptycene**   + F—Mg—Br

Figure P18.84

Figure P18.86

Figure P18.87

(a) Pairwise (cyclobutane) mechanism:

cyclobutane
intermediate

(b) Metallacycle formation:

1,7-octadiene   **1,7-octadiene-1,1,8,8-d_4**   **cyclohexene**   **ethylene ethylene-1,1-d_2 ethylene-d_4**

Figure P18.88

(a)

(b)

Figure P18.89

18.90 *Bombykol* is the mating pheromone of the female silk-worm moth.

bombykol

Using the result in Problem 18.89, propose a synthesis of bombykol, using both H—C≡C(CH₂)₉OH (10-undecyne-1-ol) and 1-pentyne as starting materials.

18.91 Potassium tri(isopropoxy)borohydride sometimes finds use as a source of nucleophilic H:⁻ (hydride ion).

$$K^+ \ H—\bar{B}(OiPr)_3$$

potassium tri(isopropoxy)borohydride

Suggest a mechanism for the substitution reaction in Fig. P18.91 that accounts for the stereochemistry of the reaction. (*Hint:* See Sec. 11.8.)

18.92 Citalopram is used as an antidepressant.

Citalopram

A biologist, Heywood U. Clonum, has argued that this compound could be hazardous because it could release toxic cyanide (⁻C≡N) and fluoride ions by nucleophilic substitution reactions with biological nucleophiles such as water. Is this argument chemically reasonable? Explain.

Figure P18.91

The Chemistry of Aldehydes and Ketones
Carbonyl-Addition Reactions

This chapter begins the study of **carbonyl compounds**—compounds containing the **carbonyl group**, C=O. Aldehydes, ketones, carboxylic acids, and the carboxylic acid derivatives (esters, amides, anhydrides, and acid chlorides) are all carbonyl compounds.

This chapter focuses on the nomenclature, properties, and characteristic carbonyl-group reactions of aldehydes and ketones. Chapters 20 and 21 consider carboxylic acids and carboxylic acid derivatives, respectively. Chapter 22 deals with ionization, enolization, and condensation reactions, which are common to the chemistry of all classes of carbonyl compounds.

Many biologically important molecules contain carbonyl groups, and carbonyl-group reactivity plays an important role in many biological reactions. We'll consider a number of these reactions in Chapters 19–22 and how they relate to the laboratory reactions that you will learn.

Aldehydes and ketones have the following general structures:

$$\underset{\textbf{aldehyde}}{\overset{\overset{\displaystyle O}{\|}}{R-C-H}} \xleftarrow{\quad} \text{carbonyl group} \xrightarrow{\quad} \underset{\textbf{ketone}}{\overset{\overset{\displaystyle O}{\|}}{R-C-R'}}$$

Examples:

$$\underset{\substack{\textbf{acetaldehyde}\\ \text{(an aldehyde)}}}{\overset{\overset{\displaystyle O}{\|}}{H_3C-C-H}} \qquad \underset{\substack{\textbf{acetone}\\ \text{(a ketone)}}}{\overset{\overset{\displaystyle O}{\|}}{H_3C-C-CH_3}}$$

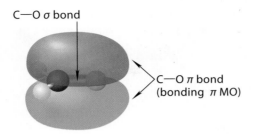

C—O σ bond

C—O π bond
(bonding π MO)

FIGURE 19.1 Bonding in formaldehyde, the simplest aldehyde, is typical of bonding in aldehydes and ketones. The carbonyl group and the two hydrogens lie in the same plane, and both a σ bond and a π bond connect the carbonyl carbon and oxygen. The π bond is a bonding π molecular orbital (MO) occupied by two electrons.

In a **ketone**, the groups bound to the carbonyl carbon (R and R′ in the preceding structures) are alkyl or aryl groups. In an **aldehyde**, at least one of the groups at the carbonyl carbon atom is a hydrogen, and the other may be alkyl, aryl, or a second hydrogen.

The carbonyl carbon of a typical aldehyde or ketone is sp^2-hybridized with bond angles approximating 120°. The carbon–oxygen double bond consists of a σ bond and a π bond, much like the double bond of an alkene (Fig. 19.1). Just as C—O single bonds are shorter than C—C single bonds, C=O bonds are shorter than C=C bonds (Sec. 1.3B). The structures of some simple aldehydes and ketones are given in Fig. 19.2.

19.1   NOMENCLATURE OF ALDEHYDES AND KETONES

A. Common Nomenclature

Common names are almost always used for the simplest aldehydes. In common nomenclature the suffix *aldehyde* is added to a prefix that indicates the chain length. A list of prefixes is given in Table 19.1 on p. 948.

$$H_2C=O \qquad H_3C—CH_2—CH_2—CH=O$$

formaldehyde   butyr + *aldehyde* = **butyraldehyde**

FIGURE 19.2 Structures of aldehydes and ketones. (a) The structures of formaldehyde, acetaldehyde, and acetone are compared with the structure of propene. The C=O bonds are shorter than the C=C bond, and the carbonyl carbon is trigonal planar with bond angles very close to 120°. (b) A ball-and-stick model of acetone. (c) A space-filling model of acetone.

TABLE 19.1 Prefixes Used in Common Nomenclature of Carbonyl Compounds

Prefix	R— in R—CH=O or R—CO$_2$H	Prefix	R— in R—CH=O or R—CO$_2$H
form	H—	isobutyr	(CH$_3$)$_2$CH—, *i*Pr
acet	H$_3$C—, Me	valer	CH$_3$CH$_2$CH$_2$CH$_2$—, Bu
propion, propi*	CH$_3$CH$_2$—, Et	isovaler	(CH$_3$)$_2$CHCH$_2$—, *i*Bu
butyr	CH$_3$CH$_2$CH$_2$—, Pr	benz, benzo†	Ph—

* Used in phenone nomenclature as discussed in the text.
† Used in carboxylic acid nomenclature (Sec. 20.1A).

Acetone is the common name for the simplest ketone, and benzaldehyde is the simplest aromatic aldehyde.

Certain aromatic ketones are named by attaching the suffix *ophenone* to the appropriate prefix from Table 19.1.

acet + *ophenone* = **acetophenone**   **benzophenone**

The common names of some ketones are constructed by citing the two groups on the carbonyl carbon followed by the word *ketone*.

cyclohexyl phenyl ketone   **dicyclohexyl ketone**

Simple substituted aldehydes and ketones can be named in the common system by designating the positions of substituents with Greek letters, beginning at the position *adjacent* to the carbonyl group.

β-bromopropionaldehyde

As suggested by this nomenclature, a carbon *adjacent* to the carbonyl group is called the **α-carbon**, and the hydrogens on the α-carbon are called **α-hydrogens**.

Many common carbonyl-containing substituent groups are named by a simple extension of the terminology in Table 19.1: the suffix *yl* is added to the appropriate prefix. The following names are examples:

formyl group   **acetyl group**   **propionyl group**   **benzoyl group**

Such groups are called in general **acyl groups**. (This is the source of the term *acylation*, used in Sec. 16.4F.) To be named as an acyl group, a substituent group must be connected to the remainder of the molecule *at its carbonyl carbon.*

$$H_3C-\overset{\overset{\displaystyle O}{\|}}{C}-\overset{}{\bigcirc}-\overset{\overset{\displaystyle O}{\|}}{C}-H$$

***p*-acetyl**benzaldehyde

Be careful not to confuse the *benzoyl* group, an acyl group, with the *benzyl* group, an alkyl group. The benzoyl group has an "o" in both the name and the structure.

$$Ph-\overset{\overset{\displaystyle O}{\|}}{C}- \qquad Ph-CH_2-$$

benzoyl group   **benzyl group**

A great many aldehydes and ketones were well known long before any system of nomenclature existed. These are known by the traditional names illustrated by the following examples:

$$Ph-CH{=}CH-CH{=}O \qquad \overset{}{\bigcirc}{-}CH{=}O \qquad H_2C{=}CH-CH{=}O$$

cinnamaldehyde   **furfural**   **acrolein**

B. Substitutive Nomenclature

The substitutive name of an aldehyde is constructed from a prefix indicating the length of the carbon chain followed by the suffix *al*. The prefix is the name of the corresponding hydrocarbon without the final *e*.

$$CH_3CH_2CH_2CH{=}O$$

butan*e* + *al* = **butanal**

In numbering the carbon chain of an aldehyde, the carbonyl carbon receives the number one.

$$\overset{4}{C}H_3\overset{3}{C}H_2\overset{2}{C}H\overset{1}{C}H{=}O$$
$$\underset{\displaystyle CH_3}{|}$$

2-methylbutanal

Note carefully the difference in chain numbering of aldehydes in common and substitutive nomenclature. In common nomenclature, numbering begins at the carbon *adjacent* to the carbonyl (the α-carbon); in substitutive nomenclature, numbering begins at the carbonyl carbon itself.

As with diols, the final *e* is not dropped when the carbon chain has more than one aldehyde group.

$$O{=}CH-CH_2CH_2CH_2CH_2-CH{=}O$$

hexanedial

When an aldehyde group is attached to a ring, the suffix *carbaldehyde* is appended to the name of the ring. (In older literature, the suffix *carboxaldehyde* was used.)

cyclohexanecarbaldehyde

In aldehydes of this type, carbon-1 is not the carbonyl carbon, but rather the ring carbon attached to the carbonyl group.

2-methylcyclohexanecarbaldehyde

The name *benzaldehyde* (Sec. 19.1A) is used in both common and substitutive nomenclature.

A ketone is named by giving the hydrocarbon name of the longest carbon chain containing the carbonyl group, dropping the final *e*, and adding the suffix *one*. The position of the carbonyl group is given the lowest possible number.

five-carbon chain: pentan*e* + *one* = **pentanone**
position of carbonyl: **2-pentanone**

cyclohexan*e* + *one* = **cyclohexanone**

3,3-dimethylcyclohexanone

As with diols and dialdehydes, the final *e* of the hydrocarbon name is not dropped in the nomenclature of diones, triones, and so on.

six-carbon chain: hexane + *dione* = **hexanedione**
positions of carbonyls: **2,4-hexanedione**

Aldehyde and ketone carbonyl groups receive higher priority than —OH or —SH groups for citation as *principal groups* (Sec. 8.2B).

Priority for citation as principal group:

$$-\text{CH (aldehyde)} > -\text{C—(ketone)} > -\text{OH} > -\text{SH} \qquad (19.1)$$

Provide a substitutive name for the following compound. (The numbers are used in the solution that follows.)

$$H_3C-\overset{\overset{O}{\|}}{\underset{1}{C}}-\underset{2}{CH_2}-\underset{3}{CH_2}-\underset{4}{\overset{\overset{OH}{|}}{CH}}-\underset{5}{CH}-\underset{}{CH_2}-CH_2-CH_3$$

$$H_3C\overset{}{\underset{6}{C}}\diagdown\underset{7}{CH_2}$$

SOLUTION   To name this compound, use the nomenclature rules in Sec. 8.2B. First, *identify the principal group.* Possible candidates are the carbonyl group at carbon-2 and the hydroxy group at carbon-4. Because ketones have a higher citation priority than hydroxy groups (Eq. 19.1), the compound is named as a ketone with the suffix *one*. Next, *identify the principal chain.* This is the longest carbon chain containing the principal group and the greatest number of double and triple bonds. This chain (numbered in the preceding structure) contains seven carbons. Notice that the longer carbon chain within the molecule is not the principal chain because the presence of double bonds takes precedence over length. Hence, the compound is named as a heptenone with hydroxy, methyl, and propyl substituents cited in alphabetical order. Finally, *number the principal chain.* Number from the end of the chain so that the principal group—the carbonyl group—receives the lowest possible number. Thus, the carbonyl carbon is carbon-2, the hydroxy carbon is carbon-4, the carbon bearing the propyl group is carbon-5, and the first alkene carbon is carbon-6 (see numbering in the preceding structure). Cite the substituent groups in alphabetical order. The name is therefore:

4-hydroxy-6-methyl-5-propyl-6-hepten-2-one

positions of
substituent groups   position of
double bond   position of
carbonyl carbon

When a ketone carbonyl group is treated as a substituent, its position is designated by the term *oxo*.

principal group: aldehyde carbonyl
name: **4-oxopentanal**

PROBLEMS

19.1   Give the structure for each of the following compounds.

(a) isobutryaldehyde   (b) valerophenone

(c) *o*-bromoacetophenone   (d) γ-chlorobutyraldehyde

(e) 3-hydroxy-2-butanone   (f) 4-(2-chlorobutyryl)benzaldehyde

(g) 3-cyclohexenone   (h) 2-oxocyclopentanecarbaldehyde

19.2   Give the substitutive name for each of the following compounds.

(a) acetone   (b) diisopropyl ketone   (c)

continued

continued

(d) EtO, H
      \ /
       C=C
      /   \
     H   CH=O

(e) CH=O
     [cyclobutane with]
     H₃C

(f) O
     [cyclohexadienone with]
     H₃C CH₃

19.2 PHYSICAL PROPERTIES OF ALDEHYDES AND KETONES

Most simple aldehydes and ketones are liquids. However, formaldehyde is a gas, and acetaldehyde has a boiling point (20.8 °C) very near room temperature, although it is usually sold as a liquid.

Aldehydes and ketones are polar molecules because of their C=O bond dipoles. The charge separation that results from this dipole is evident in the EPMs of aldehydes and ketones, illustrated here for acetone.

bond dipole
of the C=O bond   EPM of acetone

Because of their polarities, aldehydes and ketones have higher boiling points than alkenes or alkanes with similar molecular masses and shapes. But because aldehydes and ketones are not hydrogen-bond donors, their boiling points are considerably lower than those of the corresponding alcohols.

	$CH_3CH=CH_2$	$CH_3CH=O$	CH_3CH_2OH
boiling point	-47.4 °C	20.8 °C	78.3 °C
dipole moment	0.4 D	2.7 D	1.7 D

	CH_2 double bond	O (ketone)	OH
	H_3C...CH_3	H_3C...CH_3	H_3C...CH_3
boiling point	−6.9 °C	56.5 °C	82.3 °C
dipole moment	0.5 D	2.7 D	1.7 D

Aldehydes and ketones with four or fewer carbons have considerable solubilities in water because they can accept hydrogen bonds from water at the carbonyl oxygen.

HO—H ⋯ :O: ⋯ H—OH
           ‖
    $H_3C—C—CH_3$

Because they are hydrogen-bond acceptors, acetaldehyde and acetone are miscible with water (that is, soluble in all proportions). The water solubility of aldehydes and ketones along a series diminishes rapidly as the number of carbons increases.

Acetone and 2-butanone are especially valued as solvents because they dissolve not only water but also a wide variety of organic compounds. These solvents have sufficiently low boiling points that they can be easily separated from other less volatile compounds. Acetone, with a dielectric constant of 21, is a polar aprotic solvent and is often used as a solvent or co-solvent for nucleophilic substitution reactions.

19.3   SPECTROSCOPY OF ALDEHYDES AND KETONES

A. IR Spectroscopy

The principal infrared absorption of aldehydes and ketones is the C=O stretching absorption, a strong absorption that occurs in the vicinity of 1700 cm⁻¹. In fact, this is one of the most important of all infrared absorptions. Because the C=O bond is stronger than the C=C bond, the stretching frequency of the C=O bond is greater (Eq. 12.11, p. 579).

The position of the C=O stretching absorption varies predictably for different types of carbonyl compounds. It generally occurs at 1710–1715 cm⁻¹ for simple ketones and at 1720–1725 cm⁻¹ for simple aldehydes. The carbonyl absorption is clearly evident, for example, in the IR spectrum of butyraldehyde (Fig. 19.3). The stretching absorption of the carbonyl–hydrogen bond of aldehydes near 2710 cm⁻¹ is another characteristic absorption; however, NMR spectroscopy provides a more reliable way to diagnose the presence of this type of hydrogen (Sec. 19.3B).

Compounds in which the carbonyl group is conjugated with aromatic rings, double bonds, or triple bonds have lower carbonyl stretching frequencies than unconjugated carbonyl compounds.

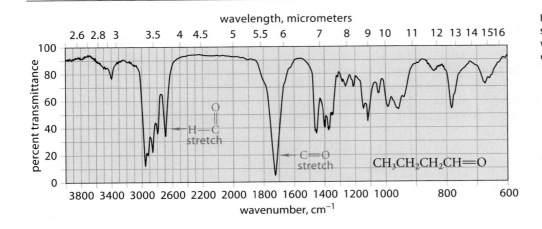

The carbon–carbon double-bond stretching frequencies are also lower in the conjugated molecules. These effects can be explained by the resonance structures for these compounds.

FIGURE 19.3 The infrared spectrum of butyraldehyde with key absorptions indicated in red.

Because the C=O and C=C bonds have some single-bond character, as indicated by the following resonance structures, they are somewhat weaker than ordinary double bonds, and therefore they absorb in the IR at lower frequency.

$$(19.3)$$

In cyclic ketones with rings containing fewer than six carbons, the carbonyl absorption frequency increases significantly as the ring size decreases. (See Further Exploration 19.1.)

FURTHER EXPLORATION 19.1
IR Absorptions of Cyclic Ketones

cyclohexanone	**cyclopentanone**	**cyclobutanone**	**cyclopropanone**	**ketene**
1715 cm^{-1} (normal)	1745 cm^{-1}	1780 cm^{-1}	1850 cm^{-1}	2150 cm^{-1}

$H_2C{=}C{=}O$ (19.4)

PROBLEM

19.3 Explain how IR spectra could be used to differentiate the isomers within each of the following pairs.

(a) cyclohexanone and hexanal   (b) 3-cyclohexenone and 2-cyclohexenone

(c) 2-butanone and 3-buten-2-ol

B. Proton NMR Spectroscopy

The characteristic NMR absorption common to both aldehydes and ketones is that of the protons on the carbons *adjacent* to the carbonyl group: the α-protons. This absorption is in the δ 2.0–2.5 region of the spectrum (see also Fig. 13.4, p. 621). This absorption occurs at somewhat greater chemical shift than the absorptions of allylic protons because the C=O group is more electronegative than the C=C group. In addition, the absorption of the aldehydic proton is quite distinctive, occurring in the δ 9–10 region of the NMR spectrum, at a greater chemical shift than most other NMR absorptions.

In general, aldehydic protons have very large chemical shifts. The explanation is the same as that for the large chemical shifts of protons on a carbon–carbon double bond (Sec. 13.7A). However, the carbonyl group has a greater effect on chemical shift than a carbon–carbon double bond because of the electronegativity of the carbonyl oxygen.

PROBLEM

19.4 Deduce the structures of the following compounds.

(a) C$_4$H$_8$O: IR 1720, 2710 cm^{-1}

NMR in Fig. 19.4

(b) C_4H_8O: IR 1717 cm^{-1}
NMR δ 0.95 (3H, t, J = 8 Hz); δ 2.03 (3H, s); δ 2.38 (2H, q, J = 8 Hz)

(c) A compound with molecular mass = 70.1, IR absorption at 1780 cm^{-1}, and the following NMR spectrum: δ 2.01 (quintuplet, J = 7 Hz); δ 3.09 (t, J = 7 Hz). The integral of the δ 3.09 resonance is twice as large as that of the δ 2.01 resonance.

(d) $C_{10}H_{12}O_2$: IR 1690 cm^{-1}, 1612 cm^{-1}
NMR δ 1.4 (3H, t, J = 8 Hz); δ 2.5 (3H, s); δ 4.1 (2H, q, J = 8 Hz); δ 6.9 (2H, d, J = 9 Hz); δ 7.9 (2H, d, J = 9 Hz)

C. Carbon NMR Spectroscopy

The most characteristic absorption of aldehydes and ketones in ^{13}C NMR spectroscopy is that of the carbonyl carbon, which occurs typically in the δ 190–220 range (see Fig. 13.22, p. 658). This large downfield shift is due to the induced electron circulation in the π bond, as in alkenes (Fig. 13.15, p. 646), and to the additional chemical-shift effect of the electronegative carbonyl oxygen. Because the carbonyl carbon of a ketone bears no hydrogens, its ^{13}C NMR absorption, like that of other quaternary carbons, is characteristically rather weak (Sec. 13.9). This effect is evident in the ^{13}C NMR spectrum of propiophenone (Fig. 19.5, p. 956). Like quaternary carbons, carbonyl carbons are absent from DEPT spectra (Sec. 13.9B; Fig. 13.25, p. 661).

The α-carbon absorptions of aldehydes and ketones show modest chemical shifts, typically in the δ 30–50 range, with, as usual, greater shifts for more branched carbons. The α-carbon shift of propiophenone, 31.7 ppm (Fig. 19.5, carbon b) is typical. Because shifts in this range are also observed for other functional groups, these absorptions are less useful than the carbonyl carbon resonances for identifying aldehydes and ketones.

PROBLEMS

19.5 Propose a structure for a compound $C_6H_{12}O$ that has IR absorption at 1705 cm^{-1}, no proton NMR absorption at a chemical shift greater than δ 3, and the following ^{13}C NMR spectrum: δ 24.4, δ 26.5, δ 44.2, and δ 212.6. The resonances at δ 44.2 and δ 212.6 have very low intensity.

19.6 The ^{13}C NMR spectrum of 2-ethylbutanal consists of the following absorptions: δ 11.5, δ 21.7, δ 55.2, and δ 204.7. Draw the structure of this aldehyde, label each chemically nonequivalent set of carbons, and assign each absorption to the appropriate carbon(s).

FIGURE 19.4 The NMR spectrum for Problem 19.4(a). The relative integrals are indicated in red over their respective resonances.

FIGURE 19.5 The ^{13}C NMR spectrum of propiophenone. Two particularly important features of the spectrum are the large chemical shift of the carbonyl carbon *g*, and the small resonances for the two carbons (*f* and *g*) that bear no protons. Recall from Sec. 13.9 that absorption intensities in ^{13}C NMR spectra generally do not accurately correspond to numbers of carbons, and quaternary carbons generally have considerably weaker absorptions than proton-bearing carbons.

D. UV Spectroscopy

The $\pi \rightarrow \pi^*$ absorptions (Sec. 15.2B) of unconjugated aldehydes and ketones occur at about 150 nm, a wavelength well below the operating range of common UV spectrometers. Simple aldehydes and ketones also have another, much weaker, absorption at higher wavelength, in the 260–290 nm region. This absorption is caused by excitation of the unshared electrons on oxygen (sometimes called the *n* electrons). This high-wavelength absorption is usually referred to as an **$n \rightarrow \pi^*$** absorption.

$$(CH_3)_2C=\ddot{O} \qquad n \rightarrow \pi^* \qquad 271 \text{ nm } (\epsilon = 16) \text{ (in ethanol)}$$

n electrons

This absorption is easily distinguished from a $\pi \rightarrow \pi^*$ absorption because it is only 10^{-2} to 10^{-3} times as strong. However, it is strong enough that aldehydes and ketones cannot be used as solvents for UV spectroscopy.

Like conjugated dienes, the π electrons of compounds in which carbonyl groups are conjugated with double or triple bonds have strong absorption in the UV spectrum. The spectrum of 1-acetylcyclohexene (Fig. 19.6) is typical. The 232-nm peak is due to light absorption by the conjugated π-electron system and is thus a $\pi \rightarrow \pi^*$ absorption. It has a very large extinction coefficient, much like that of a conjugated diene. The weak 308-nm absorption is an $n \rightarrow \pi^*$ absorption.

conjugated π-electron system

$$\lambda_{max} = 232 \text{ nm } (\epsilon = 13,200)$$
$$\lambda_{max} = 308 \text{ nm } (\epsilon = 150)$$
$$\text{(in methanol)}$$

1-acetylcyclohexene
[1-(cyclohexen-1-yl)ethanone]

The λ_{max} of a conjugated aldehyde or ketone is governed by the same variables that affect the λ_{max} values of conjugated dienes: the number of conjugated double bonds, substitution on the double bond, and so on. When an aromatic ring is conjugated with a carbonyl group, the typical aromatic absorptions are more intense and shifted to higher wavelengths than those of benzene.

FIGURE 19.6 The ultraviolet spectrum of 1-acetylcyclohexene [1-(cyclohexen-1-yl)ethanone] in methanol. The spectrum of a more concentrated solution (*red*) reveals the "forbidden" $n \rightarrow \pi^*$ absorption, which is so weak that it is not apparent in the spectrum taken on a more dilute solution (*black*).

$\lambda_{max} = 204$ nm ($\epsilon = 7900$)
254 nm ($\epsilon = 212$)

$\lambda_{max} = 240$ nm ($\epsilon = 13{,}000$)
278 nm ($\epsilon = 1100$)
319 nm ($\epsilon = 50$) ($n \rightarrow \pi^*$)

The $\pi \rightarrow \pi^*$ absorptions of conjugated carbonyl compounds, like those of conjugated alkenes, arise from the promotion of a π electron from a bonding to an antibonding (π^*) molecular orbital (Sec. 15.2B). An $n \rightarrow \pi^*$ absorption arises from promotion of one of the n (unshared) electrons on a carbonyl oxygen to a π^* molecular orbital. As stated previously in this section, $n \rightarrow \pi^*$ absorptions are weak. Spectroscopists say that these absorptions are *forbidden*. This term refers to certain physical reasons for the very low intensity of these absorptions. The 254-nm absorption of benzene, which has a very low extinction coefficient of 212, is another example of a "forbidden" absorption.

PROBLEMS

19.7 Explain how the compounds within each set can be distinguished using only UV spectroscopy.

(a) 2-cyclohexenone and 3-cyclohexenone

(b) ![structure] and ![structure]

(c) 1-phenyl-2-propanone and *p*-methylacetophenone

19.8 In neutral alcohol solution, the UV spectra of *p*-hydroxyacetophenone and *p*-methoxyacetophenone are virtually identical. When NaOH is added to the solution, the λ_{max} of *p*-hydroxyacetophenone increases by about 50 nm, but that of *p*-methoxyacetophenone is unaffected. Explain these observations.

19.9 Which one of the following compounds should have a $\pi \rightarrow \pi^*$ UV absorption at the greater λ_{max} when the compound is dissolved in NaOH solution? Explain. (*Hint:* See Problem 19.8.)

vanillin
A

isovanillin
B

E. Mass Spectrometry

Important fragmentations of aldehydes and ketones are illustrated by the electron-ionization (EI) mass spectrum of 5-methyl-2-hexanone (Fig. 19.7). The three most important peaks occur at $m/z = 71$, 58, and 43. The peaks at $m/z = 71$ and $m/z = 43$ arise from cleavage of the molecular ion at the bond between the carbonyl group and an adjacent carbon atom by two mechanisms that were discussed in Sec. 12.6C: *inductive cleavage* and *α-cleavage*. Inductive cleavage accounts for the $m/z = 71$ peak. In this cleavage, the alkyl fragment carries the charge and the carbonyl fragment carries the unpaired electron.

molecular ion from loss of unshared electron

$$(CH_3)_2CHCH_2\overset{+}{C}H_2 \;+\; \cdot\overset{:O:}{C}-CH_3 \qquad (19.5)$$
$$m/z = 71$$

α-Cleavage accounts for the $m/z = 43$ peak. In this case the same molecular ion undergoes fragmentation in such a way that the carbonyl fragment carries the charge and the alkyl fragment carries the unpaired electron:

$$(CH_3)_2CHCH_2\overset{\cdot}{C}H_2 \;+\; :\overset{+}{O}\!\!\equiv\!\!C-CH_3 \qquad (19.6)$$
$$m/z = 43$$

An analogous cleavage at the carbon–hydrogen bond accounts for the fact that many aldehydes show a strong M − 1 peak.

What accounts for the $m/z = 58$ peak? A common mechanism for the formation of odd-electron ions is hydrogen transfer followed by loss of a stable neutral molecule (p. 597). In this case, the oxygen radical in the molecular ion abstracts a hydrogen atom from a carbon *five atoms away*, and the resulting radical then undergoes α-cleavage.

$$H_3C-\overset{:\overset{+}{O}H}{C} \;+\; \overset{C(CH_3)_2}{\underset{\cdot CH_2}{CH_2}} \qquad (19.7)$$
$$m/z = 58$$

FIGURE 19.7 The EI mass spectrum of 5-methyl-2-hexanone. The odd-electron ion at $m/z = 58$ results from a McLafferty rearrangement of the molecular ion.

If we count the hydrogen that is transferred, the first step occurs through a transient six-membered ring. This process is called a **McLafferty rearrangement**, after Professor Fred McLafferty (b. 1923) of Cornell University, who investigated this type of fragmentation extensively. The McLafferty rearrangement and subsequent α-cleavage constitute a common mechanism for the production of odd-electron fragment ions in the mass spectrometry of carbonyl compounds.

As illustrated by Fig. 19.7, the EI mass spectra of many aldehydes and ketones have weak molecular ions, because relatively stable fragment ions can be formed. However, as we'll learn in Sec. 19.6, the carbonyl oxygen can be protonated. As a result, chemical-ionization (CI) mass spectra (Sec. 12.6D) of aldehydes and ketones show very strong M + 1 ions, from which accurate molecular masses can be determined.

PROBLEMS

19.10 Explain each of the following observations resulting from a comparison of the mass spectra of 2-hexanone (A) and 3,3-dimethyl-2-butanone (B).

(a) The $m/z = 57$ fragment peak is much more intense in the spectrum of B than it is in the spectrum of A.

(b) The spectrum of compound A shows a fragment at $m/z = 58$, but that of compound B does not.

19.11 Using only mass spectrometry, how would you distinguish 2-heptanone from 3-heptanone?

19.4   SYNTHESIS OF ALDEHYDES AND KETONES

Several reactions presented in previous chapters can be used for the preparation of aldehydes and ketones. The four most important of these are:

1. Oxidation of alcohols (Secs. 10.7A and 17.5A). Primary alcohols can be oxidized to aldehydes, and secondary alcohols can be oxidized to ketones.

2. Friedel–Crafts acylation (Sec. 16.4F). This reaction provides a way to synthesize aryl ketones. It also involves the formation of a carbon–carbon bond, the bond between the aryl ring and the carbonyl group.

3. Hydration of alkynes (Sec. 14.5A)

4. Hydroboration–oxidation of alkynes (Sec. 14.5B)

Two other reactions have been discussed that give aldehydes or ketones as products, but these are less important as synthetic methods:

1. Ozonolysis of alkenes (Sec. 5.5)

2. Periodate cleavage of glycols (Sec. 11.6B)

Ozonolysis and periodate cleavage are reactions that break carbon–carbon bonds. Because an important aspect of organic synthesis is the *making* of carbon–carbon bonds, use of these reactions in effect wastes some of the effort that goes into making the alkene or glycol starting materials. Nevertheless, these reactions can be used synthetically in certain cases.

Other important methods of preparing aldehydes and ketones start with carboxylic acid derivatives; these methods are discussed in Chapter 21. Appendix V gives a summary of all of the synthetic methods for aldehydes and ketones, arranged in the order in which they appear in the text.

19.5   INTRODUCTION TO ALDEHYDE AND KETONE REACTIONS

The reactions of aldehydes and ketones can be grouped into two categories: (1) reactions of the carbonyl group, which are considered in this chapter; and (2) reactions involving the α-carbon, which are presented in Chapter 22.

The great preponderance of carbonyl-group reactions of aldehydes and ketones fall into three categories:

1. *Reactions with acids.* The carbonyl oxygen is weakly basic and thus reacts with Lewis and Brønsted acids. With E^+ as a general electrophile, this reaction can be represented as follows:

$$\ddot{\underset{}{O}}: \curvearrowright E^+ \rightleftharpoons \overset{+}{\ddot{O}}{-}E \tag{19.8}$$

Carbonyl basicity is important because it plays a role in several other carbonyl-group reactions.

2. *Addition reactions.* The most important carbonyl-group reaction is addition to the $C{=}O$ double bond. With $E{-}Y$ symbolizing a general reagent, addition can be represented in the following way:

$$\underset{}{\ddot{O}}: \; + \; E{-}Y \longrightarrow \overset{\ddot{O}{-}E}{\underset{Y}{-C-}} \tag{19.9}$$

Superficially, carbonyl addition is analogous to alkene addition (Sec. 4.6).

Many reactions of aldehydes and ketones are simple additions that conform exactly to the model in Eq. 19.9. Others are multistep processes in which addition is followed by other reactions.

3. *Oxidation of aldehydes.* Aldehydes can be oxidized to carboxylic acids:

$$\overset{O}{\underset{}{-C-}}H \xrightarrow{\text{oxidation}} \overset{O}{\underset{}{-C-}}OH \tag{19.10}$$

19.6   BASICITY OF ALDEHYDES AND KETONES

Aldehydes and ketones are weakly basic and react at the carbonyl oxygen with protons or Lewis acids.

$$H_3\ddot{O}^+ + \; \underset{\underset{\textbf{acetone}}{H_3C \quad CH_3}}{\overset{\ddot{O}:}{C}} \rightleftharpoons \left[\underset{\underset{\text{all atoms have an octet}}{H_3C \quad CH_3}}{\overset{\overset{+}{\ddot{O}}{-}H}{C}} \longleftrightarrow \underset{\underset{\text{carbocation structure}}{H_3C \quad \overset{+}{} \quad CH_3}}{\overset{\ddot{O}{-}H}{C}} \right] + H_2\ddot{O}: \tag{19.11}$$

conjugate acid of acetone
$pK_a = -6$ to -7

As Eq. 19.11 shows, the conjugate acid of an aldehyde or ketone is resonance-stabilized. The resonance structure on the left is more important because all atoms have octets. However, the resonance structure on the right shows that the protonated carbonyl compound has carbocation character. In fact, in some cases, the conjugate acids of aldehydes and ketones undergo typical carbocation reactions.

The conjugate acids of aldehydes and ketones can be viewed as *α-hydroxy carbocations*. If we conceptually replace the acidic proton in a protonated aldehyde or ketone with an alkyl group, we get an *α-alkoxy carbocation*.

$$
\left[\begin{array}{ccc} \overset{+}{\ddot{\text{O}}\text{H}} & & \ddot{\text{O}}\text{H} \\ \| & & | \\ \text{H}_3\text{C}-\text{C}-\text{CH}_3 & \longleftrightarrow & \text{H}_3\text{C}-\overset{+}{\text{C}}-\text{CH}_3 \end{array} \right] \qquad \left[\begin{array}{ccc} \overset{+}{\ddot{\text{O}}\text{R}} & & \ddot{\text{O}}\text{R} \\ \| & & | \\ \text{H}_3\text{C}-\text{C}-\text{CH}_3 & \longleftrightarrow & \text{H}_3\text{C}-\overset{+}{\text{C}}-\text{CH}_3 \end{array} \right]
$$

conjugate acid of acetone   (an α-alkoxy carbocation)
(an α-hydroxy carbocation)

α-Hydroxy carbocations and α-alkoxy carbocations are considerably more stable than ordinary carbocations. For example, a comparably substituted α-alkoxy carbocation is about 100 kJ mol^{-1} (24 kcal mol^{-1}) more stable than an ordinary tertiary carbocation in the gas phase.

$$
\begin{array}{cc} \text{CH}_3\text{O} & \text{CH}_3 \\ | & | \\ \text{H}_3\text{C}-\overset{+}{\text{C}}-\text{CH} \\ | \\ \text{CH}_3 \end{array} \quad \text{is much more stable than} \quad \begin{array}{cc} \text{CH}_3\text{O} & \text{CH}_3 \\ | & | \\ \text{H}_3\text{C}-\text{CH}-\overset{+}{\text{C}} \\ | \\ \text{CH}_3 \end{array} \qquad (19.12)
$$

An α-alkoxy carbocation, like a protonated aldehyde or ketone, owes its stability to the resonance interaction of the electron-deficient carbon with the neighboring oxygen, which, as we learned in Sec. 15.4B, corresponds to the formation of bonding molecular orbitals. This resonance effect far outweighs the electron-attracting polar effect of the oxygen, which, by itself, would destabilize the carbocation.

STUDY PROBLEM 19.2

Many 1,2-diols, under the acidic conditions used for dehydration of alcohols, undergo a reaction called the *pinacol rearrangement*:

$$
\begin{array}{c} \text{OH} \quad \text{OH} \\ | \quad\quad | \\ \text{H}_3\text{C}-\text{C}-\text{C}-\text{CH}_3 \\ | \quad\quad | \\ \text{CH}_3 \quad \text{CH}_3 \end{array} \xrightarrow{\text{H}_2\text{SO}_4} \begin{array}{c} \text{CH}_3 \quad \text{O} \\ | \quad\quad \| \\ \text{H}_3\text{C}-\text{C}-\text{C}-\text{CH}_3 \\ | \\ \text{CH}_3 \end{array} + \text{H}_2\text{O} \qquad (19.13)
$$

2,3-dimethyl-2,3-butanediol   **3,3-dimethyl-2-butanone**
(pinacol)   **(pinacolone)**
(65–72% yield)

Propose a curved-arrow mechanism for this reaction, and explain why the rearrangement step is energetically favorable.

SOLUTION   First, analyze the connectivity changes that take place. A methyl group shifts to an adjacent carbon, and one of the —OH groups is lost as water. The fact that a rearrangement occurs suggests a carbocation intermediate; such a carbocation can be generated (as in the dehydration of any alcohol) by protonation of an —OH group and loss of H$_2$O:

$$
\begin{array}{c} \overset{+}{\text{H}_2\ddot{\text{O}}}-\text{H} \quad \ddot{\text{O}}\text{H} \quad \text{OH} \\ \qquad\qquad | \quad\quad | \\ \text{H}_3\text{C}-\text{C}-\text{C}-\text{CH}_3 \\ \qquad | \quad\quad | \\ \text{CH}_3 \quad \text{CH}_3 \end{array} \rightleftarrows \begin{array}{c} \overset{+}{\text{H}-\ddot{\text{O}}\text{H}} \quad \text{OH} \\ | \quad\quad | \\ \text{H}_3\text{C}-\text{C}-\text{C}-\text{CH}_3 \\ | \quad\quad | \\ \text{CH}_3 \quad \text{CH}_3 \end{array} \rightleftarrows \begin{array}{c} \text{OH} \\ | \\ \text{H}_3\text{C}-\overset{+}{\text{C}}-\text{C}-\text{CH}_3 + \text{H}_2\ddot{\text{O}} \\ | \quad\quad | \\ \text{CH}_3 \quad \text{CH}_3 \end{array} \quad (19.14\text{a})
$$

$$
+ \text{H}_2\ddot{\text{O}}
$$

The rearrangement can now take place. The product of the rearrangement is an α-hydroxy carbocation, which, as we have seen, is the same thing as the conjugate acid of a ketone.

$$\text{(19.14b)}$$

both an α-hydroxy carbocation and
a ketone conjugate acid

You've just learned that such carbocations are especially stable, a fact indicated by their resonance structures. Thus, the rearrangement step is favorable because the α-hydroxy carbocation is more stable than the tertiary carbocation. The second resonance structure in Eq. 19.14b shows that the α-hydroxy carbocation is also the conjugate acid of the ketone. Removal of a proton from the carbonyl oxygen gives the product.

$$\text{(19.14c)}$$

Aldehydes and ketones in solution are considerably less basic than alcohols (Sec. 10.1E). In other words, their conjugate acids are more acidic than those of alcohols.

$$pK_a \approx -2.5 \qquad pK_a \approx -7$$

Because protonated aldehydes and ketones are resonance-stabilized and protonated alcohols are not, we might have expected protonated carbonyl compounds to be *more stable* relative to their conjugate bases and therefore *less acidic*. The relative acidity of protonated alcohols and carbonyl compounds is an example of a solvent effect. In the *gas phase*, aldehydes and ketones *are* indeed more basic than alcohols. In solution, the solvation of a protonated alcohol by hydrogen bonding is evidently so effective that it outweighs the resonance stabilization of a protonated aldehyde or ketone.

PROBLEMS

19.12 (a) Write an S_N1 mechanism for the solvolysis of CH_3OCH_2Cl [(chloromethoxy)methane] in ethanol; draw appropriate resonance structures for the carbocation intermediate.

(b) Explain why the alkyl halide in part (a) undergoes solvolysis much more rapidly than 1-chlorobutane. (In fact, it reacts in ethanol more than 100 times more rapidly.)

19.13 Predict the product when each of the following diols undergoes the pinacol rearrangement.

(a)   (b)   (c)

19.14 Use resonance arguments to explain why

(a) *p*-methoxybenzaldehyde is more basic than *p*-nitrobenzaldehyde.

(b) 3-buten-2-one is more basic than 2-butanone.

19.7	**REVERSIBLE ADDITION REACTIONS OF ALDEHYDES AND KETONES**

One of the most typical reactions of aldehydes and ketones is *addition* to the carbon–oxygen double bond. To begin with, let's focus on two simple addition reactions: addition of hydrogen cyanide (HCN) and hydration (addition of water).

Addition of HCN:

$$H_3C-\overset{\overset{\displaystyle O}{\|}}{C}-CH_3 + H-C\equiv N \underset{pH\ 9-10}{\rightleftharpoons} H_3C-\overset{\overset{\displaystyle OH}{|}}{\underset{\underset{\displaystyle C\equiv N}{|}}{C}}-CH_3 \qquad (19.15)$$

acetone

acetone cyanohydrin
(77–78% yield)

The product of HCN addition to an aldehyde or ketone is called a **cyanohydrin**. Cyanohydrins constitute a special class of *nitriles* (compounds of the form $R-C\equiv N$). (The chemistry of nitriles is discussed in Chapter 21.) The preparation of cyanohydrins is another method of forming carbon–carbon bonds.

Addition of water (hydration):

$$H_3C-\overset{\overset{\displaystyle O}{\|}}{C}-H + H-OH \rightleftharpoons H_3C-\overset{\overset{\displaystyle OH}{|}}{\underset{\underset{\displaystyle OH}{|}}{C}}-H \qquad (19.16)$$

acetaldehyde

acetaldehyde hydrate

The product of water addition to an aldehyde or ketone is called a **hydrate**, or gem-diol. (The prefix *gem* stands for *geminal*, from the Latin word for twin, and is used in chemistry when two identical groups are present on the same carbon.)

In all carbonyl-addition reactions, the more electropositive species (for example, the hydrogen of H—CN or H—OH) adds to the carbonyl oxygen, and the more electronegative species (for example, the —CN or the —OH) adds to the carbonyl carbon.

A. Mechanisms of Carbonyl-Addition Reactions

Carbonyl-addition reactions occur by two general types of mechanisms. The first mechanism, called **nucleophilic carbonyl addition**, involves the reaction of a nucleophile at the carbonyl carbon. In cyanohydrin formation (Eq. 19.15), the nucleophile is cyanide ion, which is formed by the ionization of HCN:

$$H-CN + {}^-OH \rightleftharpoons {}^-{:}CN + H_2O \qquad (19.17a)$$

$pK_a = 9.4$

cyanide ion

Cyanide ion donates electrons to the carbonyl carbon of the aldehyde or ketone, and the carbonyl oxygen accepts the displaced electron pair and assumes a negative charge. We can think of this process in the same way that we think of nucleophilic reactions at saturated carbon atoms (Sec. 3.2A). The electrophile is the carbonyl carbon, and the "leaving group" is the π bond of the carbonyl group. Because the carbon–oxygen σ bond remains intact, the "leaving group" doesn't actually leave.

$$H_3C-C(=O)(CH_3) + {}^-:CN \quad \rightleftharpoons \quad H_3C-\underset{\underset{CN}{|}}{\overset{\overset{:\overset{..}{O}:^-}{|}}{C}}-CH_3 \qquad (19.17b)$$

the carbon–oxygen π bond serves as the "leaving group"

electrophile

nucleophile

To complete the nucleophilic addition, the negatively charged oxygen—an alkoxide ion, and a relatively strong base—is protonated by either water or HCN.

$$H_3C-\underset{\underset{CN}{|}}{\overset{\overset{:\overset{..}{O}:^-}{|}}{C}}-CH_3 \quad + \quad H-CN \quad \rightleftharpoons \quad H_3C-\underset{\underset{CN}{|}}{\overset{\overset{:\overset{..}{O}H}{|}}{C}}-CH_3 \; + \; {}^-:CN \qquad (19.17c)$$

The nucleophilic carbonyl-addition mechanism finds no analogy in additions to ordinary carbon–carbon double bonds. Yet, nucleophilic carbonyl addition occurs even though the carbon–oxygen π bond is 62 kJ mol^{-1} (15 kcal mol^{-1}) *stronger* than the carbon–carbon π bond of an alkene. The stronger bond is more reactive because the unshared electron pair (and negative charge) formed in the carbonyl-addition mechanism is transferred to a very electronegative atom, oxygen. The same reaction of an alkene would place an unshared pair and negative charge on a carbon, a much less electronegative atom.

additional unshared pair and negative charge on *oxygen*

unshared pair and negative charge on *carbon*

$$(19.18)$$

observed

not observed with ordinary alkenes

The reaction of a nucleophile with the carbonyl group, then, is driven by the ability of oxygen to accept the unshared electron pair. For this reason, a nucleophile cannot add to the carbonyl oxygen. *Nucleophiles always react with carbonyl groups at the carbonyl carbon.*

It is tempting to use resonance structures to rationalize the reactivity of a carbonyl group as follows:

$$\left[\; \overset{:O:}{\underset{}{\overset{\|}{C}}} \quad \longleftrightarrow \quad \overset{:\overset{..}{O}:^-}{\underset{+}{\overset{|}{C}}} \; \right] \qquad (19.19)$$

It is sometimes erroneously said that carbonyl compounds react with nucleophiles at the carbonyl carbon because there is positive charge at this carbon and this positive charge attracts the electron pairs of nucleophiles.

There are two fallacies in this argument. First, when two species collide, they do so randomly, not from any preferred direction. For example, a nucleophile in solution can collide randomly with any of the atoms in a carbonyl compound. In other words, a nucleophile is not selectively directed to a carbonyl carbon. Rather, reaction occurs the way it does because when the nucleophile happens to collide with the carbonyl carbon from the proper direction, *electrons (and charge) can be shifted onto the electronegative carbonyl oxygen.*

The second fallacy in this argument is its implication that resonance increases reactivity. In fact, *resonance stabilization has the opposite effect.* Carbonyl compounds are actually *less reactive* than they would be if such resonance stabilization did not exist. Many studies have shown that when molecules are prepared in which resonance stabilization of double bonds cannot occur (for example, molecules in which π-electron systems are twisted out of coplanarity), these bonds become more reactive than double bonds in similar molecules in which resonance interaction can occur. The carbon–oxygen π bond is actually about 62 kJ mol^{-1} (15 kcal mol^{-1}) *stronger* than a carbon–carbon π bond; this fact, taken alone, means that it should be *less reactive* than a carbon–carbon double bond, exactly as the resonance structures suggest.

If the carbonyl π bond is so stable, why does it react? Remember from Sec. 3.6A that when a Brønsted base reacts with a Brønsted acid, it is not only the strength of the bond to the hydrogen, but also *how well the other atom in the bond accepts electrons*, that governs how easily the reaction takes place. A nucleophilic reaction at a carbonyl carbon is no different, except that the electrophile is a carbon rather than a hydrogen, and a π bond rather than a σ bond is broken. *Carbonyl compounds react with nucleophiles at the carbonyl carbon because the electronegative oxygen readily accepts negative charge.*

It nevertheless is true that resonance structures *do* suggest ways that molecules can react. In the case of a carbonyl group, the dipolar resonance structure suggests that the carbon, with its partial positive charge, is the electrophilic site. The reason this works is that the atom which accepts electrons in the dipolar resonance structure (oxygen) is the same one that accepts electrons in the transition state for nucleophilic addition. *Such correspondences between reactivity patterns and resonance structures invariably occur throughout organic chemistry.* For this reason, organic chemists find themselves using resonance structures to predict sites of reactivity in molecules. Resonance structures are undeniably useful for this purpose; but when we use them this way, we must remember that resonance is not the *reason for reactivity*.

The geometry of nucleophilic addition is shown in Fig. 19.8. The carbonyl group and the two atoms bound to the carbonyl carbon define a reference plane. The nucleophile approaches the carbonyl carbon from above or below this plane, as shown in Fig. 19.8a. As a result of this reaction, the carbonyl carbon changes hybridization from sp^2 to sp^3, the oxygen accepts an electron pair, and the geometry at the carbonyl carbon changes from trigonal planar to

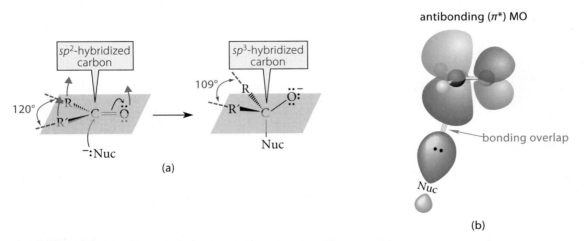

FIGURE 19.8 (a) The geometry of nucleophilic reaction with the carbonyl carbon of formaldehyde, with the nucleophile represented by Nuc:$^-$. The reference plane (*gray*) is the plane defined by the carbonyl group and the atoms attached to the carbonyl carbon. The nucleophile approaches this carbon from above or below this plane. As a result, the carbonyl carbon changes hybridization from sp^2 to sp^3, the bond angles at the carbonyl carbon compress from 120° to 109°, and the groups attached to the carbonyl carbon move as shown by the green arrows. (b) The interaction of a nucleophile with the π^* antibonding MO of formaldehyde. The lobes of this MO are concentrated above and below the plane of the molecule. The interaction of the nucleophile with this MO defines the direction of nucleophilic approach to the carbonyl carbon.

tetrahedral. In other words, the angle between the bonds to the carbonyl carbon, initially about 120°, compresses to about 109°, the tetrahedral angle. As a result of the reaction, then, the groups bound to the carbonyl carbon become closer together.

The reason for the addition geometry is similar to the reason for opposite-side substitution in the S_N2 reactions of alkyl halides (Sec. 9.4C, Fig. 9.3). The curved-arrow notation might convey the impression that the nucleophile reacts at the π bond. However, the bonding π molecular orbital of the carbonyl group (Fig. 19.1) is fully occupied with two electrons and cannot accommodate any more electrons. The electron pair of the nucleophile interacts instead with the *unoccupied* MO of lowest energy (LUMO), which, in the case of the carbonyl group, is the antibonding π^* molecular orbital. This MO is shown in Fig. 19.8b. This MO has lobes above and below the reference plane. The nucleophile, then, must begin its bonding interaction with the carbonyl carbon from the direction along which the LUMO is concentrated, as shown in Fig. 19.8a.

When the antibonding π^* MO is filled, even with electrons from another molecule, the C=O π bond is weakened. (Remember that when an antibonding molecular orbital is populated, the energetic advantage of bonding disappears.) This is the reason that the π bond breaks. The energetic "trade-off" for loss of this bonding is formation of the new bond to the nucleophile.

The second mechanism for carbonyl addition occurs under *acidic conditions* and is closely analogous to the mechanism for the addition of acids to alkenes (Secs. 4.7 and 4.9B). Acid-catalyzed hydration of aldehydes and ketones (Eq. 19.16) is an example of this mechanism. The first step in hydration is protonation of the carbonyl oxygen (Sec. 19.6).

$$(19.20a)$$

A positively charged oxygen attracts electrons even more strongly than the oxygen of an unprotonated carbonyl group. In other words, the protonated carbonyl compound is a much stronger *Lewis acid* (electron acceptor) than an unprotonated carbonyl compound. As a result, even the relatively weak base H_2O can react at the carbonyl carbon. Loss of a proton to solvent completes the reaction.

$$(19.20b)$$

Hydration of aldehydes and ketones also occurs in neutral and basic solution (Problem 19.15).

The direction of approach of the nucleophile to the protonated carbonyl group is the same as it is for approach to the unprotonated carbonyl group (Fig. 19.8)—from above or below the reference plane. This is explained by the shape of the LUMO of the protonated carbonyl group, which is very similar to the shape of the LUMO of the carbonyl group itself.

PROBLEMS

19.15 (a) Write a curved-arrow mechanism for the hydroxide-catalyzed hydration of acetaldehyde.

(b) Write a curved-arrow mechanism for the decomposition of acetone cyanohydrin (Eq. 19.15) in aqueous hydroxide. Explain why the ketone–cyanohydrin equilibrium favors the ketone at high pH.

19.16   Write a curved-arrow mechanism for

(a) the acid-catalyzed addition of methanol to benzaldehyde.

(b) the methoxide-catalyzed addition of methanol to benzaldehyde.

B. Equilibria in Carbonyl-Addition Reactions

Hydration and cyanohydrin formation are both reversible reactions. (Not all carbonyl additions are reversible.) Whether the equilibrium for a reversible addition favors the addition product or the carbonyl compound *depends strongly on the structure of the carbonyl compound*. For example, cyanohydrin formation favors the cyanohydrin addition product in the case of aldehydes and methyl ketones, but the equilibrium favors the carbonyl compound when aryl ketones are used.

The effect of aldehyde or ketone structure on the addition equilibrium for hydration is illustrated by the data in Table 19.2. Note the following trends in the table.

1. Addition is more favorable for aldehydes than for ketones.
2. Electronegative groups near the carbonyl carbon make carbonyl addition more favorable.
3. Addition is less favorable when groups are present that donate electrons by resonance to the carbonyl carbon.

The trends in this table and the reasons behind them are important for two reasons. First, the equilibria for *all* addition reactions show similar effects of structure. Second, and more important, the *rates* of carbonyl-addition reactions—that is, the *reactivities* of carbonyl compounds—follow similar trends (Sec. 19.7C).

What is the reason for the effect of structure on carbonyl addition? The relative stabilities of the carbonyl compound and the addition product govern the $\Delta G°$ for addition. This point is illustrated in Fig. 19.9 on p. 968. As shown in this figure, the primary effect on the hydration equilibrium is the *difference in the stabilities of the carbonyl compounds*. Added stability in the carbonyl compound increases the energy change $\Delta G°$, *and hence decreases the equilibrium constant*, for formation of an addition product.

The major factors that stabilize carbonyl compounds can be understood by considering the resonance structures of the carbonyl group:

$$\left[\begin{array}{ccc} \overset{\displaystyle :O:}{\underset{\displaystyle \|}{R-C-R}} & \longleftrightarrow & \overset{\displaystyle :\overset{..}{O}:^-}{\underset{\displaystyle |}{R-\underset{+}{C}-R}} \end{array} \right] \qquad (19.21)$$

TABLE 19.2 Equilibrium Constants for Hydration of Aldehydes and Ketones

$$H_2O + R-\overset{\displaystyle O}{\overset{\displaystyle \|}{C}}-R' \underset{\phantom{K_{eq}}}{\overset{K_{eq}}{\rightleftharpoons}} R-\overset{\displaystyle OH}{\underset{\displaystyle OH}{\overset{\displaystyle |}{\underset{\displaystyle |}{C}}}}-R'$$

Aldehydes	K_{eq}	Ketones	K_{eq}
$H_2C{=}O$	2.2×10^3	$(CH_3)_2C{=}O$	1.4×10^{-3}
$CH_3CH{=}O$	1.0		
$(CH_3)_2CHCH{=}O$	0.5–1.0	$Ph-\overset{\displaystyle O}{\overset{\displaystyle \|}{C}}-CH_3$	6.6×10^{-6}
$PhCH{=}O$	8.3×10^{-3}	$Ph_2C{=}O$	1.2×10^{-7}
$ClCH_2CH{=}O$	37	$(ClCH_2)_2C{=}O$	10
$Cl_3CCH{=}O$	2.8×10^4	$(CF_3)_2C{=}O$	too large to measure

FIGURE 19.9 The greater stability of a ketone relative to an aldehyde causes the ketone to have a greater standard free energy of hydration and therefore a smaller equilibrium constant for hydration. (The two hydrates have been placed at the same energy level for comparison purposes.)

The structure on the right, although not as important a contributor as the one on the left, reflects the polarity of the carbonyl group and has the characteristics of a carbocation. Therefore, anything that stabilizes carbocations also tends to stabilize carbonyl compounds. Because alkyl groups stabilize carbocations, ketones (R = alkyl) are more stable than aldehydes (R = H). This stability is reflected in the relative heats of formation of aldehydes and ketones. For example, acetone, with $\Delta H_f^\circ = -218$ kJ mol^{-1} (-52.0 kcal mol^{-1}), is 29 kJ mol^{-1} (6.9 kcal mol^{-1}) more stable than its isomer propionaldehyde, for which $\Delta H_f^\circ = -189$ kJ mol^{-1} (-45.2 kcal mol^{-1}). Because alkyl groups stabilize carbonyl compounds, the equilibria for additions to ketones are less favorable than those for additions to aldehydes (trend 1). Formaldehyde, with *two* hydrogens and no alkyl groups bound to the carbonyl, has a very large equilibrium constant for hydration.

Electronegative groups such as halogens destabilize carbocations by their polar effect and for the same reason destabilize carbonyl compounds. Thus, halogens make the equilibria for addition more favorable (trend 2). In fact, chloral hydrate (known in medicine as a hypnotic) is a stable crystalline compound.

$$
\underset{\substack{\textbf{chloral} \\ \textbf{(2,2,2-trichloroethanal)}}}{Cl_3C-\overset{\displaystyle O}{\overset{\|}{C}}H} + H_2O \longrightarrow \underset{\textbf{chloral hydrate}}{Cl_3C-\overset{\displaystyle OH}{\underset{\displaystyle OH}{C}H}} \qquad (19.22)
$$

Groups that are conjugated with the carbonyl group, such as the phenyl group of benzaldehyde, stabilize carbocations by resonance, and hence stabilize carbonyl compounds.

$$(19.23)$$

A similar resonance stabilization cannot occur in the hydrate because the carbonyl group is no longer present. Consequently, aryl aldehydes and ketones have relatively unfavorable hydration equilibria (trend 3).

A *steric effect* also operates in carbonyl addition. As the size of the groups bound to the carbonyl carbon increases, van der Waals repulsions in the corresponding addition compounds become more important. We can see why this should be so from Fig. 19.8a. The groups at the carbonyl carbon are closer together in the addition compound than in the carbonyl compound; hence, van der Waals repulsions are more pronounced in the addition compound. These van der Waals repulsions, in turn, *raise* the energy of the addition compound relative to the carbonyl compound and *increase* the $\Delta G°$ for addition.

C. Rates of Carbonyl-Addition Reactions

The trends in relative rates of addition can be predicted from the trends in equilibrium constants. That is, *compounds with the most favorable addition equilibria tend to react most rapidly in addition reactions.* Thus, *aldehydes are generally more reactive than ketones* in addition reactions; *formaldehyde is more reactive than many other simple aldehydes.*

The reason for the parallel trends in rates and equilibria is that the transition states for addition reactions resemble addition products. Thus, it is a convenient approximation to think of the addition compounds in Fig. 19.9 as transition states, and the standard free energies $\Delta G°$ as standard free energies of activation $\Delta G°^{\ddagger}$. Just as destabilization of aldehydes or ketones decreases the $\Delta G°$ for their addition reactions, the same destabilization decreases the free energies of activation $\Delta G°^{\ddagger}$ for addition and thus increases the rate of addition.

Notice that *we are analyzing carbonyl reactivity differently than we have analyzed other reactivity.* In many cases, you have learned to gauge reactivity by applying Hammond's postulate to the relative stability of *reactive intermediates* such as carbocations. The more stable the intermediate, the more reactive a compound is. However, you should remember that reactivity involves the *difference* in two standard free energies: the standard free energy of the transition state and the standard free energy of the reactant (see Fig. 4.11, p. 162). In carbonyl chemistry, it is often the relative free energies of the *reactants* that provide the major effects on reactivity. When one reactant is less stable than another relative to its transition state, the compound is more reactive because it is pushed further up the "free-energy hill" toward its transition state (Fig. 19.10). Because we'll continue to return to the point in Chapters 21 and 22, it is important for you to understand the relative reactivities of carbonyl compounds in these terms.

This section has covered two examples of addition to the carbonyl group. Subsequent sections deal with other addition reactions as well as more complex reactions that have mech-

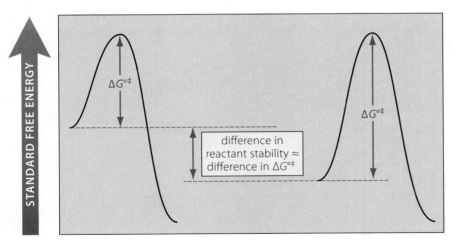

reaction coordinates

FIGURE 19.10 The effect of reactant stability on reactivity when two reactions are compared at the same transition-state free energy. In many carbonyl reactions, the relative stability of the reactants determines relative reactivity. In these cases, the less stable compound reacts more rapidly. In many carbonyl reactions, the effect of reactant stability on *rate* parallels the effect of reactant stability on *equilibrium constant* (Fig. 19.9). Therefore, we can use equilibrium data, such as that in Table 19.2, to estimate reactivity.

anisms in which the initial steps are addition reactions. These addition reactions all have mechanisms similar to the ones discussed in this section, and the trends in reactivity are the same. *Addition to the carbonyl group is a common thread that runs throughout most of aldehyde and ketone chemistry.*

PROBLEMS

19.17 Which carbonyl compound should form the greater proportion of cyanohydrin at equilibrium? Draw the structure of the cyanohydrin, and explain your reasoning.

⬡—CH=O   or $CH_3CH_2CH=O$

benzaldehyde   **propanal**

19.18 Within each set, which compound should be more reactive in carbonyl-addition reactions? Explain your choices.

(a)

$$H_3C—\overset{\overset{\displaystyle O}{\|}}{C}—CH_2CH_2Br \quad or \quad H_3C—\overset{\overset{\displaystyle O}{\|}}{C}—CH_2Br$$

(b)

$$H_3C—\overset{\overset{\displaystyle O}{\|}}{C}—\overset{\overset{\displaystyle O}{\|}}{C}—CH_3 \quad or \quad H_3C—\overset{\overset{\displaystyle O}{\|}}{C}—CH_2CH_3$$

(c)

$$O_2N—⬡—CH=O \quad or \quad CH_3\ddot{O}—⬡—CH=O$$

(d)

⬠=O   or   ▷=O

(*Hint:* Note the change in bond angles in Fig. 19.8a.)

19.8	**REDUCTION OF ALDEHYDES AND KETONES TO ALCOHOLS**

A. Reduction with Lithium Aluminum Hydride and Sodium Borohydride

Aldehydes and ketones are reduced to alcohols with either lithium aluminum hydride, $LiAlH_4$, or sodium borohydride, $NaBH_4$. These reactions result in the net *addition* of the elements of H_2 across the C=O bond.

$$4\;\square{=}O \;+\; LiAlH_4 \xrightarrow{\text{ether}} \xrightarrow{H_3O^+} 4\;\square\overset{H}{\underset{}{{-}OH}} \;+\; Li^+ \text{ and } Al^{3+} \text{ salts} \qquad (19.24)$$

cyclobutanone   **lithium aluminum hydride**   **cyclobutanol** (90% yield)

$$4\,CH_3O—⬡—CH=O \;+\; 4\,CH_3OH \;+\; NaBH_4 \xrightarrow{CH_3OH}$$

p-methoxybenzaldehyde   **sodium borohydride**

$$4\,CH_3O—⬡—\overset{H}{\underset{H}{C}}—OH \;+\; Na^+ \; {}^-B(OCH_3)_4 \qquad (19.25)$$

p-methoxybenzyl alcohol (96% yield)

As these examples illustrate, reduction of an aldehyde gives a primary alcohol, and reduction of a ketone gives a secondary alcohol.

Lithium aluminum hydride is one of the most useful reducing agents in organic chemistry. It serves generally as a source of H:$^-$, the *hydride ion*. Because hydrogen is more electronegative than aluminum (Table 1.1), the Al–H bonds of the $^-$AlH$_4$ ion carry a substantial fraction of the negative charge. In other words,

$$H—\overset{\overset{\textstyle H}{|}}{\underset{\underset{\textstyle H}{|}}{Al}}—H \quad \text{reacts as if it were} \quad H—\overset{\overset{\textstyle H}{|}}{\underset{\underset{\textstyle H}{|}}{Al}} \quad H:^- \qquad (19.26)$$

The hydride ion in LiAlH$_4$ is very basic. For this reason, LiAlH$_4$ reacts violently with water and therefore must be used in dry solvents such as anhydrous ether and THF.

$$\text{Li}^+ \quad H—\overset{\overset{\textstyle H}{|}}{\underset{\underset{\textstyle H}{|}}{Al}}—H \quad H—\overset{..}{\overset{..}{O}}H \longrightarrow H—\overset{\overset{\textstyle H}{|}}{\underset{\underset{\textstyle H}{|}}{Al}} \quad + \quad H—H \quad + \text{Li}^+ \quad ^-\!:\overset{..}{\overset{..}{O}}H \qquad (19.27)$$

$$\underset{\substack{\textbf{lithium aluminum} \\ \textbf{hydride}}}{} \qquad \underset{\substack{\text{(reacts further} \\ \text{with water)}}}{} \qquad \underset{\text{hydrogen gas}}{}$$

Like many other strong bases, the hydride ion in LiAlH$_4$ is a good nucleophile, and LiAlH$_4$ contains its own "built-in" Lewis acid, the lithium ion. The reaction of LiAlH$_4$ with aldehydes and ketones involves the nucleophilic reaction of hydride (delivered from $^-$AlH$_4$) at the carbonyl carbon. The lithium ion acts as a Lewis acid catalyst by coordinating to the carbonyl oxygen.

$$\underset{R}{\overset{:\overset{..}{O}:---\text{Li}^+}{\underset{\underset{\textstyle H—\bar{A}lH_3}{}}{\overset{\|}{C}}}}{\underset{R}{}} \longrightarrow \underset{\underset{\textstyle H}{|}}{R—\overset{:\overset{..}{O}:^-\ \text{Li}^+}{\underset{|}{C}}—R} \quad + \ \text{AlH}_3 \qquad (19.28a)$$

$$\underset{\text{a lithium alkoxide}}{}$$

The addition product, an alkoxide salt, can react with AlH$_3$, and the resulting product can also serve as a source of hydride.

$$\underset{\underset{\textstyle H}{|}}{R—\overset{\text{Li}^+ \ :\overset{..}{O}:\!\frown\!\text{AlH}_3}{\underset{|}{C}}—R} \longrightarrow \underset{\underset{\textstyle H}{|}}{R—\overset{\text{Li}^+ \ :\overset{..}{O}—\bar{A}lH_3}{\underset{|}{C}}—R} \quad \boxed{\text{these hydrides are active in further reductions}} \qquad (19.28b)$$

Similar processes occur at each stage of the reduction until all of the hydrides are consumed. Hence, as shown in the stoichiometry of Eq. 19.24, all four hydrides of LiAlH$_4$ are active in the reduction. In other words, it takes one-fourth of a mole of LiAlH$_4$ to reduce a mole of aldehyde or ketone.

After the reduction is complete, the alcohol product exists as an alkoxide addition compound with the aluminum. This is converted by protonation in a separate step into the alcohol product. The proton source can be an aqueous HCl solution or even an aqueous solution of a weak acid such as ammonium chloride.

$$\left(\underset{\underset{\textstyle R}{|}}{H—\overset{\overset{\textstyle R}{|}}{\underset{|}{C}}—O}\right)_{\!\!4}\!\!\!\overset{H—\overset{..}{\overset{+}{O}}H_2}{\bar{A}l} \ \text{Li}^+ \longrightarrow 4\,\underset{\underset{\textstyle R}{|}}{H—\overset{\overset{\textstyle R}{|}}{\underset{|}{C}}—O—H} \ + \ H_2O \ + \ \text{Li}^+, \text{Al}^{3+} \text{ salts} \qquad (19.28c)$$

$$\underset{\substack{\text{aluminum alkoxide} \\ \text{addition compound}}}{} \qquad \underset{\text{alcohol product}}{}$$

The reaction of sodium borohydride with aldehydes and ketones is conceptually similar to that of LiAlH$_4$. The sodium ion is a much weaker Lewis acid than the lithium ion. For this reason, NaBH$_4$ reductions are carried out in protic solvents such as alcohols. Hydrogen bonding between the alcohol solvent and the carbonyl group serves as a weak acid catalysis that activates the carbonyl group. Unlike LiAlH$_4$, NaBH$_4$ reacts only slowly with alcohols and can even be used in water if the solution is not acidic.

$$ \text{(19.29)} $$

As Eq. 19.25 shows, all four hydride equivalents of NaBH$_4$ are active in the reduction.

Because LiAlH$_4$ and NaBH$_4$ are hydride donors, reductions by these and related reagents are generally referred to as **hydride reductions**. The important mechanistic point about these reactions is that they are further examples of *nucleophilic addition*. Hydride ion from LiAlH$_4$ or NaBH$_4$ is the nucleophile, and the proton is delivered from acid added in a separate step (in the case of LiAlH$_4$ reductions) or solvent (in the case of NaBH$_4$ reductions).

$$ \text{(19.30)} $$

Unlike the additions discussed in Sec. 19.7, hydride reductions are *not* reversible. Reversal of carbonyl addition would require the original nucleophilic group, in this case, H:$^-$, to be expelled as a leaving group. As in S$_N$1 or S$_N$2 reactions, the best leaving groups are the weakest bases. Hydride ion is such a strong base that it is not easily expelled as a leaving group. Hence, hydride reductions of all aldehydes and ketones are irreversible—they go to completion.

Both LiAlH$_4$ and NaBH$_4$ are highly useful in the reduction of aldehydes and ketones. Lithium aluminum hydride is, however, a much more *reactive* agent than sodium borohydride. A significant number of functional groups react with LiAlH$_4$ but not with NaBH$_4$; among such groups are alkyl halides, esters, alkyl tosylates, and nitro groups. Sodium borohydride can be used as a reducing agent in the presence of these groups.

$$ \text{(19.31)} $$

3-nitrobenzaldehyde   **(3-nitrophenyl)methanol**
                                           (*m*-nitrobenzyl alcohol)
                                           (82% yield)

Sodium borohydride is also a much less hazardous reagent than lithium aluminum hydride. The greater selectivity and safety of $NaBH_4$ make it the preferred reagent in many applications, but either reagent can be used for the reduction of simple aldehydes and ketones. Both are very important in organic chemistry.

Discovery of NaBH₄ Reductions

The discovery of $NaBH_4$ reductions illustrates that interesting research findings are sometimes serendipitous (obtained by accident). In the early 1940s, the U.S. Army Signal Corps became interested in methods for generating hydrogen gas in the field. $NaBH_4$ was proposed as a relatively safe, portable source of hydrogen: addition of acidified water to $NaBH_4$ results in the evolution of hydrogen gas at a safe, moderate rate. To supply the required quantities of $NaBH_4$, a large-scale synthesis was necessary. The following reaction appeared to be suitable for this purpose.

$$4\,NaH + B(OCH_3)_3 \longrightarrow NaBH_4 + 3\,NaOCH_3 \qquad (19.32)$$

The problem with this process was that the sodium borohydride had to be separated from the sodium methoxide by-product. Several solvents were tried in the hope that a significant difference in solubilities could be found. In the course of this investigation, acetone was tried as a recrystal-

lization solvent, and it was found to react with the $NaBH_4$ to yield isopropyl alcohol. Thus was born the use of $NaBH_4$ as a reducing agent for carbonyl compounds.

These investigations, carried out by Herbert C. Brown (1912–2004) at Purdue University, were part of what was to become a major research program in the boron hydrides, shortly thereafter leading to the discovery of hydroboration (Sec. 5.4B). Brown even described his interest in the field of boron chemistry as something of an accident, because it sprung from his reading a book about boron and silicon hydrides that was given to him by his girlfriend (who later became his wife) as a graduation present. Mrs. Brown observed that the choice of this particular book was dictated by the fact that it was among the least expensive chemical titles in the bookstore; in the depression era, students had to be careful how they spent their money! For his work in organic chemistry, Brown shared the Nobel Prize in Chemistry in 1979 with Georg Wittig (Sec. 19.13).

The use of sodium borohydride has come full circle, as it is now used as a hydrogen source for experimental hydrogen-powered fuel cells.

PROBLEMS

19.19   From what aldehyde or ketone could each of the following be synthesized by reduction with either $LiAlH_4$ or $NaBH_4$?

(a) ⬡—CH_2OH

(b) $\underset{\underset{\displaystyle CH_3CHCH_2CH_3}{|}}{OH}$

(c) (decalin with OH at ring junction and OH at lower ring position)

19.20   Which of the following alcohols could *not* be synthesized by a hydride reduction of an aldehyde or ketone? Explain.

H_3C—⬡—OH   (cyclohexane with OH and CH_3)   (cyclohexane with OH and CH_3)

A   *B*   *C*

B. Hydride Reduction in Biology

In Sec. 10.8, we learned that NAD^+ is an important biological oxidizing agent. (The structure of NAD^+ is shown in Fig. 10.1, p. 486.) When NAD^+ is used for an oxidation, its reduced form, NADH, is formed. NADH is "nature's hydride reagent." It acts as a biological reducing agent in the same way that $LiAlH_4$ and $NaBH_4$ act as laboratory reducing agents, as we shall see.

Both NAD^+ and NADH exist in phosphorylated forms, $NADP^+$ and NADPH; the position of phosphorylation is shown in Fig. 10.1. These forms differ in their cellular location and the general role that they play in cellular biochemistry. You will learn these details if you study biochemistry. However, from a chemical point of view, they function in the same way.

One of the most important functions of NADH is *carbonyl reduction*. An example of this reaction is the conversion of pyruvate into (S)-lactate under anaerobic conditions in muscle tissue. The pro-(R) hydride of NADH (*red*) is transferred stereospecifically.

$$(19.33)$$

(The lactate produced in this reaction is the source of the burning in muscle that occurs after very intense exercise; the conditions of oxygen deprivation promote this reaction as a way of generating NAD^+ required in the metabolism of glucose.) This reaction is catalyzed by the enzyme *lactate dehydrogenase*. Although the enzyme is commonly called a "dehydrogenase," metabolically the reaction normally proceeds in the reductive direction as shown in Eq. 19.33. (The enzyme is "officially" named (S)-lactate:pyruvate oxidoreductase; it catalyzes the reaction in both directions, as it must by the principle of microscopic reversibility.)

In the reduction of pyruvate to lactate, the ketone carbonyl group is activated by hydrogen bonding from a protonated histidine residue of the enzyme. In this respect, the reaction resembles sodium borohydride reduction, in which the carbonyl group is activated by hydrogen-bond donation from a protic solvent (Eq. 19.29).

$$(19.34)$$

This basic theme—delivery of a hydride from NADH (or NADPH) to a carbonyl-activated substrate in an enzyme active site—is played out in all of the carbonyl reductions in which NADH or NADPH is involved.

It is worth noting that, in other reductions involving NADH, the pro-(S) hydride of NADH is delivered instead of the pro-(R) hydride. Although these hydrides are diastereotopic, their intrinsic reactivity is virtually the same. The differences among various enzymes has to do with the way the NADH or NADPH binds in the enzyme active site relative to the substrate.

PROBLEM

19.21   Yeast alcohol dehydrogenase catalyzes the reduction of acetaldehyde by NADH to ethanol in the last step of anaerobic fermentation (see sidebar, Sec. 10.8, p. 487).

(a) The structure of this enzyme shows a Zn^{2+} ion (held in place by coordination with several amino acid side chains of the enzyme) that is required for catalysis near the carbonyl of acetaldehyde. Propose a role for this metal ion in catalysis. (*Hint:* See Eq. 19.28a.)

(b) Using your answer to part (a), draw a curved-arrow mechanism for this reduction in which the role of the Zn^{2+} ion is explicitly shown.

C. Reduction by Catalytic Hydrogenation

Aldehydes and ketones can also be reduced to alcohols by catalytic hydrogenation. This reaction is analogous to the catalytic hydrogenation of alkenes (Sec. 4.9A).

(19.35)

Catalytic hydrogenation of carbonyl groups requires typically higher pressure and temperature than hydrogenation of carbon–carbon π bonds because of the greater bond strength of the carbon–oxygen π bond. For this reason, it is usually possible to use catalytic hydrogenation for the selective reduction of an alkene double bond in the presence of a carbonyl group. Palladium catalysts are particularly effective for this purpose.

(19.36)

PROBLEM

19.22   Give three different starting materials that might be used to obtain the following product by selective catalytic hydrogenation.

19.9   REACTIONS OF ALDEHYDES AND KETONES WITH GRIGNARD AND RELATED REAGENTS

Grignard reagents were introduced in Sec. 9.8. *The reaction of Grignard reagents with carbonyl groups is the most important application of the Grignard reagent in organic chemistry.* Addition of Grignard reagents to aldehydes and ketones in an ether solvent, followed by protonolysis, gives alcohols.

$$(CH_3)_2CHCH + BrMg-CH_2CH_3 \xrightarrow{\text{ether}} \xrightarrow{H_3O^+} (CH_3)_2CHCHCH_2CH_3 \qquad (19.37)$$

2-methylpropanal   **ethylmagnesium bromide**   **2-methyl-3-pentanol**
(68% yield)

$$H_3C-C-CH_3 + CH_3CH_2CH_2-MgBr \xrightarrow{\text{ether}} \xrightarrow{H_3O^+} H_3C-C-CH_3 \qquad (19.38)$$

acetone   **propylmagnesium bromide**

2-methyl-2-pentanol
(68% yield)

**STUDY GUIDE
LINK 19.1**
Lewis Acid Catalysis

The reaction of Grignard reagents with aldehydes and ketones is another example of *carbonyl addition*. In this reaction, the magnesium of the Grignard reagent, a Lewis acid, bonds to the carbonyl oxygen. This bonding, much like protonation in acid-catalyzed hydration, makes the carbonyl carbon more electrophilic (that is, makes it more reactive toward nucleophiles) by making the carbonyl oxygen a better acceptor of electrons. The carbon group of the Grignard reagent reacts as a nucleophile at the carbonyl carbon. Recall that this group is a strong base that behaves much like a *carbanion* (Secs. 9.8B and 11.5C).

$$(19.39a)$$

a bromomagnesium alkoxide

The product of this addition, a bromomagnesium alkoxide, is essentially the magnesium salt of an alcohol. Addition of dilute acid to the reaction mixture gives an alcohol.

$$(19.39b)$$

Because of the great basicity of Grignard reagents, this addition, like hydride reductions, is irreversible, and it works with just about any aldehyde or ketone.

The reactions of organolithium and sodium acetylide reagents with aldehydes and ketones are fundamentally similar to the Grignard reaction.

$$CH_3(CH_2)_3\text{—Li} + H_3C\text{—}\overset{\overset{\displaystyle O}{\|}}{C}\text{—}CH_3 \xrightarrow[\text{hexane}]{-78\,°C} \xrightarrow{H_3O^+} CH_3(CH_2)_3\text{—}\overset{\overset{\displaystyle OH}{|}}{\underset{\underset{\displaystyle CH_3}{|}}{C}}\text{—}CH_3 + Li^+ + H_2O \qquad (19.40)$$

butyllithium   **acetone**   **2-methyl-2-hexanol**
                                                                                          (80% yield)

$$\text{cyclohexanone} + HC\equiv\overset{\cdot\cdot}{C}\!:Na^+ \xrightarrow{NH_3(\text{liq.})} \xrightarrow{H_3O^+} \text{1-ethynylcyclohexanol} + Na^+ + H_2O \qquad (19.41)$$

cyclohexanone   **sodium acetylide**   **1-ethynylcyclohexanol**
                          (Sec. 14.7B)   (65–75% yield)

The reaction of Grignard and related reagents with aldehydes and ketones is important not only because it can be used to convert aldehydes or ketones into alcohols, but also because it is an excellent method of *carbon–carbon bond formation*.

STUDY GUIDE
LINK 19.2
Reactions That Form
Carbon–Carbon Bonds

$$\overset{\overset{\displaystyle O}{\|}}{C} + R\text{—MgBr} \longrightarrow \xrightarrow{H_3O^+} -\overset{\overset{\displaystyle OH}{|}}{\underset{\underset{\displaystyle R}{|}}{C}}- \qquad (19.42)$$

the carbonyl group is reduced

a new carbon–carbon bond is formed

A complete list of reactions that form carbon–carbon bonds is given in Appendix VI.

The possibilities for alcohol synthesis with the Grignard reaction are almost endless. Primary alcohols are synthesized by the addition of a Grignard reagent to formaldehyde.

$$\text{cyclohexyl—MgCl} + H_2C\!=\!O \longrightarrow \xrightarrow{H_3O^+} \text{cyclohexyl—}CH_2\text{—OH} \qquad (19.43)$$

cyclohexylmagnesium   **formaldehyde**   **cyclohexylmethanol**
chloride   (66% yield)

Because Grignard reagents are made from alkyl halides, which in many cases can be synthesized from alcohols, this reaction can be incorporated as a key element in a one-carbon chain extension of an alcohol:

$$R\text{—OH} \xrightarrow[\text{Ph}_3PBr_2]{\text{HBr or}} R\text{—Br} \xrightarrow[\text{ether}]{Mg} R\text{—MgBr} \xrightarrow[\text{ether}]{H_2C=O} \xrightarrow{H_3O^+} R\text{—}CH_2\text{—OH} \qquad (19.44)$$

net one-carbon chain extension

Addition of a Grignard reagent to an aldehyde other than formaldehyde gives a secondary alcohol (Eq. 19.37), and addition to a ketone gives a tertiary alcohol (Eq. 19.38). The Grignard synthesis of a tertiary alcohol or, in some cases, a secondary alcohol, can also be extended to an alkene synthesis by dehydration of the alcohol with strong acid during the protonolysis step (Sec. 10.2).

$$\text{(reaction)} + CH_3MgI \xrightarrow[\text{ether}]{} \xrightarrow{H_3O^+} \text{(product)} \xrightarrow{H_2SO_4} \text{(product)} + H_2O \quad (19.45)$$

(68% yield)

When you are asked to prepare an alcohol, you can determine whether it can be synthesized by the reaction of a Grignard reagent with an aldehyde or ketone if you understand that the *net effect* of the Grignard reaction, followed by protonolysis, is addition of R—H (R = an alkyl or aryl group) across the C═O double bond:

$$\underset{\text{C}}{\overset{O}{\|}} + R-MgBr \longrightarrow \xrightarrow{H-\overset{+}{O}H_2} \underset{\text{C}}{\overset{O-H}{|}}-R \quad (19.46)$$

**STUDY GUIDE
LINK 19.3**
Alcohol Syntheses

Once you grasp this relationship, you can determine the starting materials for a particular synthesis by mentally subtracting R and H from the target alcohol. This approach is illustrated in Study Problem 19.3.

STUDY PROBLEM 19.3

Propose a synthesis of 2-butanol by the reaction of a Grignard reagent with an aldehyde or ketone.

SOLUTION   The carbonyl carbon of the starting material becomes the α-carbon of the alcohol. Consequently, any alkyl group bound to this carbon in the product can be derived from a Grignard reagent. The O—H proton is derived from the water or acid used in the protonolysis step. Thus, one possible analysis of the required synthesis is as follows:

$$\underset{\text{target compound}}{H_3C-\overset{O-H}{\underset{|}{C}}H-CH_2CH_3} \xrightarrow[\text{H and } CH_2CH_3]{\text{subtract}} H_3C-\overset{O}{\overset{\|}{C}}H + BrMg-CH_2CH_3, \text{ then } H-\overset{+}{O}H_2 \quad (19.47)$$

(This type of arrow means, "Implies as starting materials." It is often used when mapping out a retrosynthetic strategy.) Another possibility for a Grignard synthesis of 2-butanol can be found by a similar analysis. What is it?

PROBLEMS

19.23   Show how ethyl bromide can be used as a starting material in the preparation of each of the following compounds. (*Hint:* How are Grignard reagents prepared?)

(a) $\underset{PhCHCH_2CH_3}{\overset{OH}{|}}$ (b) $\underset{Et_2CCH_3}{\overset{OH}{|}}$ (c) 1-butanol

(d) $CH_3CH_2CH{=}O$ (e) $\underset{Ph-\overset{Ph}{\underset{|}{C}}{=}CH-CH_3}{}$ (f) (structure: cyclopentane with OH and Et)

19.24   Outline three different Grignard syntheses for 3-methyl-3-hexanol.

19.10   ACETALS AND THEIR USE AS PROTECTING GROUPS

The preceding sections dealt with simple carbonyl-addition reactions—first, reversible additions (cyanohydrin formation and hydration); then, irreversible additions (hydride reduction and addition of Grignard reagents). This and the following sections consider some reactions that begin as additions but involve other types of mechanistic steps.

A. Preparation and Hydrolysis of Acetals

When an aldehyde or ketone reacts with a large excess of an alcohol in the presence of a trace of strong acid, an *acetal* is formed.

$$+ \ 2\,CH_3OH \quad \xrightarrow[\text{(trace)}]{H_2SO_4} \quad + \ H_2O \qquad (19.48)$$

m-nitrobenzaldehyde

m-nitrobenzaldehyde
dimethyl acetal
(76–85% yield)

$$+ \ 2\,CH_3OH \quad \xrightarrow[\text{(trace)}]{H_2SO_4} \quad + \ H_2O \qquad (19.49)$$

acetophenone

acetophenone
dimethyl acetal
(82% yield)

An **acetal** is a compound in which two ether oxygens are bound to the same carbon. In other words, acetals are the ethers of carbonyl hydrates, or gem-diols (Sec. 19.7). (Acetals derived from ketones were once called *ketals*, but this name is no longer used.)

Notice that two equivalents of alcohol are consumed in each of the preceding reactions. However, 1,2- and 1,3-diols contain two —OH groups within the same molecule. Hence, one equivalent of a 1,2- or 1,3-diol can react to form a *cyclic acetal*, in which the acetal group is part of a five- or six-membered ring, respectively.

$$\xrightarrow[\text{acid (Sec. 10.4A)}]{p\text{-toluenesulfonic}} \quad + \ H_2O \qquad (19.50)$$

cyclohexanone

ethylene
glycol

cyclohexanone
ethylene acetal
(85% yield)

The formation of acetals is readily reversible. The reaction is driven to the right either by the use of excess alcohol as the solvent or by removal of the water by-product, or both. This strategy is another application of Le Châtelier's principle. In Eq. 19.50, for example, the water can be removed by distillation as an *azeotrope* with benzene. (The benzene–water azeotrope is a mixture of benzene and water that has a lower boiling point than either benzene or water alone.)

The first step in the mechanism of acetal formation is acid-catalyzed *addition* of the alcohol to the carbonyl group to give a **hemiacetal**—a compound with an —OR and —OH group on the same carbon (*hemi* = half; *hemiacetal* = half acetal).

$$+ \ ROH \quad \underset{\text{acid}}{\rightleftarrows} \qquad (19.51a)$$

hemiacetal

Hemiacetal formation is completely analogous to acid-catalyzed hydration. (Write the stepwise mechanism of this reaction; see Problem 19.16a, p. 967.)

The hemiacetal reacts further when the —OH group is protonated and water is lost to give a relatively stable carbocation, an α-alkoxy carbocation (Sec. 19.6).

$$\text{(19.51b)}$$

STUDY GUIDE LINK 19.4
Hemiacetal Protonation

Loss of water from the hemiacetal is an S_N1 reaction analogous to the loss of water in the dehydration of an ordinary alcohol (Eq. 10.13b). The nucleophilic reaction of an alcohol molecule with the cation and deprotonation of the nucleophilic oxygen complete the mechanism.

$$\text{(19.51c)}$$

As we have just shown, the mechanism for acetal formation is really a combination of other familiar mechanisms. It involves an *acid-catalyzed carbonyl addition* followed by a *substitution* that occurs by the S_N1 mechanism.

Because the formation of acetals is reversible, acetals in the presence of acid and excess water are transformed rapidly back into the corresponding carbonyl compounds and alcohols; this process is called **acetal hydrolysis**. (A *hydrolysis* is a cleavage reaction involving water.) As expected from the principle of microscopic reversibility, the mechanism of acetal hydrolysis is the reverse of the mechanism of acetal formation. Hence, acetal hydrolysis, like hemiacetal formation, is acid-catalyzed.

The formation of *hemiacetals* is catalyzed not only by acids but by bases as well (Problem 19.16b, p. 967). However, the conversion of hemiacetals into acetals is catalyzed *only* by acids (Eqs. 19.51b and c). This is why acetal formation, which is a combination of the two reactions, is catalyzed by acids but not by bases.

$$\text{(19.51d)}$$

As expected from the principle of microscopic reversibility, the hydrolysis of hemiacetals to aldehydes and ketones is also catalyzed by bases, but the hydrolysis of acetals to hemiacetals is catalyzed *only* by acids. Hence, *acetals are stable in basic and neutral solution.*

Hemiacetals, the intermediates in acetal formation (Eq. 19.51a), in most cases cannot be isolated because they react further to yield acetals (in alcohol solution under acidic conditions) or decompose to aldehydes or ketones and an alcohol. Simple aldehydes, however, form appreciable amounts of hemiacetals in alcohol solution, just as they form appreciable amounts of hydrates in water (see Table 19.2).

$$H_3C—CH{=}O + CH_3CH_2OH \underset{\text{(solvent)}}{\rightleftarrows} \begin{array}{c} OH \\ | \\ H_3C—CH \\ | \\ OCH_2CH_3 \end{array} \qquad (19.52)$$

(97% at equilibrium)

Five- and six-membered *cyclic* hemiacetals form spontaneously from the corresponding hydroxy aldehydes, and most are stable compounds that can be isolated.

5-hydroxypentanal ⇌ a cyclic hemiacetal (94% at equilibrium)   (19.53)

4-hydroxybutanal ⇌ (89% at equilibrium)   (19.54)

You learned in Sec. 11.8 that intramolecular reactions which give six-membered or five-membered rings are faster than the corresponding intermolecular reactions. Such intramolecular reactions are also more favored thermodynamically—that is, they have larger equilibrium constants, because an intramolecular —OH group simply has a greater probability of reaction (that is, a greater $\Delta S°$) than an —OH group in a different molecule.

The five- and six-carbon sugars are important biological examples of cyclic hemiacetals.

(+)-glucose ⇌ **α-(+)-glucopyranose** + **β-(+)-glucopyranose**   (19.55)

cyclic forms of glucose

(This reaction and its stereochemistry are discussed in Sec. 24.2B.)

Storage of Aldehydes as Acetals

Some aldehydes are stored as acetals. Acetaldehyde, when treated with a trace of acid, readily forms a cyclic acetal called paraldehyde. Each molecule of paraldehyde is formed from three molecules of acetaldehyde. (Notice that an alcohol is not involved in formation of paraldehyde.) Paraldehyde, with a boiling point of 125 °C, is a particularly convenient way to store acetaldehyde, which itself boils near room temperature. Upon heating with a trace of acid, acetaldehyde can be distilled from a sample of paraldehyde. (See Problem 19.60, p. 1000.)

$$3\,CH_3CH{=}O \underset{}{\overset{\text{acid}}{\rightleftarrows}} \text{paraldehyde} \qquad (19.56)$$

acetaldehyde

paraldehyde

Formaldehyde can be stored as the acetal polymer paraformaldehyde, which precipitates from concentrated formaldehyde solutions.

$$HO \overbracket{}(CH_2\!-\!O)_n H$$

paraformaldehyde

(An alcohol is not involved in paraformaldehyde formation.) Because it is a solid, paraformaldehyde is a useful form in which to store formaldehyde, itself a gas. Formaldehyde is liberated from paraformaldehyde by heating.

PROBLEMS

19.25 Write the structure of the product formed in each of the following reactions.

(a)

$$\text{(cyclopentanone)}=O + CH_3CH_2OH \xrightarrow[\text{(solvent)}]{\text{acid}}$$

(b)

$$\underset{\substack{\displaystyle \| \\ O}}{CH_3CH_2CH_2CH} + (CH_3)_2CHOH \xrightarrow[\text{(solvent)}]{\text{acid}}$$

19.26 Propose syntheses of each of the following acetals from carbonyl compounds and alcohols.

(a)

(b)

19.27 Suggest a structure for the acetal product of each reaction.

(a)

$$\text{HO} \quad \text{H} \atop \text{O} \; + \; CH_3CH_2OH \xrightarrow[\text{(excess)}]{\text{acid}} (C_7H_{14}O_2)$$

(b)

$$\text{(cyclohexanone)}=O + HO\!-\!CH_2\!-\!\underset{\substack{\displaystyle | \\ CH_3}}{\overset{\substack{CH_3 \\ \displaystyle |}}{C}}\!-\!CH_2\!-\!OH \xrightarrow{\text{acid}}$$

(excess)

B. Protecting Groups

A common tactic of organic synthesis is the use of *protecting groups*. The method is illustrated by the following analogy. Suppose you and a friend haven't been invited to a party but are determined to attend it anyway. To avoid recognition and confrontation, you wear a disguise, which might be a wig, a false mustache, or even more drastic accoutrements. Your friend doesn't bother with such deception. The host recognizes your friend and insists that he leave the party, but, because you are not recognized, you avoid such a confrontation and can remain to enjoy the evening, removing your disguise only after the party is over.

Now, suppose two groups in a molecule, *A* and *B*, are both known to react with a certain reagent, but we want to let only group *A* react and leave group *B* unaffected. The solution to this problem is to *disguise*, or *protect*, group *B* in such a way that it cannot react. After group *A* is allowed to react, the disguise of group *B* is removed. The "chemical disguise" used with group *B* is called a **protecting group** or **protective group**. Acetals are among the most commonly used protecting groups for aldehydes and ketones. Study Problem 19.4 illustrates the use of an acetal as a protecting group.

STUDY PROBLEM 19.4

Propose a sequence of reactions for carrying out the following conversion.

(19.57)

SOLUTION   It might seem that the way to effect this conversion would be to convert the starting halide into the corresponding Grignard reagent, and then allow this reagent to react with ethylene oxide, followed by dilute aqueous acid (Sec. 11.5C). However, Grignard reagents also react with ketones (Sec. 19.9). Hence, the Grignard reagent derived from one molecule of the starting material would react with the carbonyl group of another molecule, and thus the ketone group would not survive this reaction. However, the ketone can be *protected* as an acetal, which does *not* react with Grignard reagents. (An acetal is a type of ether, and ethers are unaffected by Grignard reagents.) The following synthesis incorporates this strategy.

(19.58)

Notice in this synthesis that all steps following acetal formation involve basic or neutral conditions. Acid can be used only when removal of the acetal protecting group is desired.

Although any acetal group can in principle be used, the five-membered cyclic acetal is frequently employed as a protecting group because it forms very rapidly (proximity effect; Sec. 11.8) and it introduces relatively little steric congestion into the protected molecule.

A number of reagents that react with carbonyl groups also react with other functional groups. Acetals are commonly used to protect the carbonyl groups of aldehydes and ketones from basic, nucleophilic reagents. Once the protection is no longer needed, the acetal protecting group is easily removed, and the carbonyl group re-exposed, by treatment with dilute aqueous acid. Because acetals are unstable in acid, they do *not* protect carbonyl groups under acidic conditions.

PROBLEM

19.28   Outline a synthesis of the following compound from *p*-bromoacetophenone and any other reagents.

19.11   REACTIONS OF ALDEHYDES AND KETONES WITH AMINES

A. Reaction with Primary Amines and Other Monosubstituted Derivatives of Ammonia

A **primary amine** is an organic derivative of ammonia in which only one ammonia hydrogen is replaced by an alkyl or aryl group. An **imine** is a nitrogen analog of an aldehyde or ketone in which the C=O group is replaced by a C=NR group, where R = alkyl, aryl, or H.

$$R-\ddot{N}H_2 \qquad\qquad \underset{/}{\overset{\backslash}{C}}=\ddot{\ddot{O}} \qquad\qquad \underset{/}{\overset{\backslash}{C}}=\ddot{N}-R$$

primary amine   aldehyde or ketone   imine (Schiff base)

(Imines are sometimes called **Schiff bases** or **Schiff's bases**.) Imines are prepared by the reaction of aldehydes or ketones with primary amines.

$$\text{Ph}-CH=O + Ph-\ddot{N}H_2 \xrightarrow{\text{heat}} \text{Ph}-CH=\ddot{N}-Ph + H_2O \qquad (19.59)$$

aniline (a primary amine)   an imine (84–87% yield)   (separates from the reaction mixture)

Formation of imines is reversible and generally takes place with acid or base catalysis or with heat. Imine formation is typically driven to completion by precipitation of the imine, removal of water, or both.

The mechanism of imine formation begins as a nucleophilic addition to the carbonyl group. In this case, the nucleophile is the amine, which reacts with the aldehyde or ketone to give an unstable addition product called a *carbinolamine*. A **carbinolamine** is a compound with an amine group (—NH$_2$, —NHR, or —NR$_2$) and a hydroxy group on the same carbon.

STUDY GUIDE LINK 19.5
Mechanism of Carbinolamine Formation

$$\underset{\overset{||}{C}}{\overset{O}{\underset{/\ \ \backslash}{}}} + H_2\ddot{N}-R \;\rightleftarrows\; \underset{\underset{:NH-R}{|}}{\overset{OH}{\underset{|}{-C-}}} \qquad (19.60a)$$

carbinolamine

(You should write the detailed curved-arrow mechanism, which is analogous to the mechanism of other reversible additions.) Carbinolamines are not isolated, but undergo acid-catalyzed dehydration to form imines. This reaction is essentially an alcohol dehydration (Sec. 10.2), except that it is typically much faster than dehydration of an ordinary alcohol.

STUDY GUIDE LINK 19.6
Dehydration of Carbinolamines

$$\underset{\underset{H}{|}}{\overset{OH}{\underset{|}{-C-}}}\ddot{N}R \;\underset{}{\overset{\text{acid}}{\rightleftarrows}}\; \underset{/}{\overset{\backslash}{C}}=\ddot{N}R + H_2O \qquad (19.60b)$$

carbinolamine   imine (Schiff base)

(Write the mechanism of this reaction as well.)

Typically, the dehydration of the carbinolamine is the rate-limiting step of imine formation. This is why imine formation is catalyzed by acids. Yet the acid concentration cannot be too high because amines are basic compounds, and because protonated amines cannot act as nucleophiles.

$$R\ddot{N}H_2 + H_3\overset{+}{\ddot{O}} \rightleftharpoons R\overset{+}{\ddot{N}}H_3 + H_2\ddot{O}:$$ (19.61)

Protonation of the amine pulls the equilibrium in Eq. 19.60a to the left; consequently, if the acid concentration is high enough, carbinolamine formation cannot occur. For this reason, many imine syntheses are carried out in very dilute acid.

To summarize: Imine formation is a sequence of two reactions that have close analogies to familiar reactions—namely, *carbonyl addition* followed by *β-elimination*.

How are imines used? One important use of imines is in the preparation of amines. This process, called *reductive amination*, is discussed in Sec. 23.7B. Before the advent of spectroscopy, certain *N*-substituted imines were used to characterize or identify the aldehydes or ketones from which they were derived. (This use of imines is discussed in Further Exploration 19.2.) Some common *N*-substituted imines that were used for this purpose appear often in older chemical literature. These imines, and the *N*-substituted amines used to prepare them from carbonyl compounds, are shown in Table 19.3.

Some *N*-substituted imines are important in their own right. For example, we'll encounter hydrazones as intermediates in Sec. 19.12. Cyclohexanone oxime is an intermediate in the industrial synthesis of one type of nylon. All *N*-substituted imines, like other imines, are prepared from the parent carbonyl compound and the appropriate *N*-substituted amine. For example, cyclohexanone oxime is prepared from the reaction of cyclohexanone and hydroxylamine.

FURTHER EXPLORATION 19.2
Use of Imines for Characterization of Aldehydes and Ketones

$$\text{cyclohexanone} + \text{H}_2\text{N}-\text{OH} \xrightarrow[\text{methanol–water}]{\text{Na}^+ \text{AcO}^- \text{(base)}} \text{cyclohexanone oxime} + \text{H}_2\text{O}$$ (19.62)

hydroxylamine

cyclohexanone

cyclohexanone oxime
(92% yield)

PROBLEMS

19.29   Draw the structure of
(a) the oxime of acetone;
(b) the imine formed in the reaction between 2-methylhexanal and ethylamine (EtNH$_2$).

19.30   Write a curved-arrow mechanism for the reaction in Eq. 19.62. (This is a base-catalyzed process; the base is sodium acetate, Na$^+$ AcO$^-$.)

19.31   Write a curved-arrow mechanism for the acid-catalyzed hydrolysis of the imine derived from benzaldehyde and ethylamine (CH$_3$CH$_2$NH$_2$). Use the principle of microscopic reversibility (p. 175) to guide you.

TABLE 19.3 Some *N*-Substituted Imine Derivatives of Aldehydes and Ketones

Amine	Name	Carbonyl Derivative	Name
H$_2$$\ddot{N}$—$\ddot{O}$H	hydroxylamine	R$_2$C=$\ddot{N}$—$\ddot{O}$H	oxime
H$_2$$\ddot{N}$—$\ddot{N}H_2$	hydrazine	R$_2$C=$\ddot{N}$—$\ddot{N}$H$_2$	hydrazone
H$_2$$\ddot{N}$—$\ddot{N}$H—⟨phenyl⟩	phenylhydrazine	R$_2$C=$\ddot{N}$—$\ddot{N}$H—⟨phenyl⟩	phenylhydrazone
H$_2$$\ddot{N}$—$\ddot{N}$H—⟨2,4-dinitrophenyl⟩	2,4-dinitrophenylhydrazine (2,4-DNP)	R$_2$C=$\ddot{N}$—$\ddot{N}$H—⟨2,4-dinitrophenyl⟩	2,4-dinitrophenylhydrazone (2,4-DNP derivative)
R$_2$C=$\ddot{N}$—$\ddot{N}$H—C(=O)—$\ddot{N}$H$_2$	semicarbazide	H$_2$$\ddot{N}$—$\ddot{N}$H—C(=O)—$\ddot{N}H_2$	semicarbazone

Imines in Biology   Imines are well known in biology. Typically, imines are employed as a tether when an aldehyde group on a small molecule is connected to an enzyme or other protein. An amino group in the side chain of a lysine residue of the protein reacts with the aldehyde to give the imine linkage. An example is the connection between retinal, the visual pigment, and its protein, opsin, to give rhodopsin (see Sec. 15.2C, p. 723).

rhodopsin (an imine; shown in its conjugate-acid form)
(the light receptor in vision)   (19.63)

Because the bond between the small molecule and the protein is covalent, it is particularly strong. However, because imine formation is reversible, this linkage can be hydrolyzed back to the aldehyde and the free lysine amino group when required. Thus, after the imine of 11-(Z)-retinal is converted into the imine of 11-(E)-retinal, the retinal must be removed from the protein by imine hydrolysis. The direction in which the reaction runs depends on the conformation of the protein and, undoubtedly, the local availability of water in the vicinity of the imine linkage.

Imines are particularly important in the biological reactions of pyridoxal phosphate, a form of vitamin B_6; we'll study these in Sec. 26.4E. (See Problem 19.32.)

PROBLEM

19.32   A form of vitamin B_6 (pyridoxal phosphate) forms imine derivatives with most proteins that catalyze its reactions. Given the structure of pyridoxal phosphate below, draw the structure of its imine derivative with the lysine residue of a protein. (Use an abbreviation for the lysine residue like the one shown in Eq. 19.63.)

pyridoxal phosphate
(a form of vitamin B_6)

B. Reaction with Secondary Amines

A **secondary amine** has the general structure R_2NH, in which two ammonia hydrogens are replaced by alkyl or aryl groups. An **enamine** (pronounced ēn´-ə-mēn˝) has the following general structure:

general enamine structure

The name *enamine* is a contraction of the word *amine* (a compound of the form R_3N) and the suffix *ene*, which is used for naming alkenes. The name recognizes that an *amine* nitrogen is bonded to a carbon that is part of a double bond (that is, an alk*ene*).

Formation of an enamine occurs when a secondary amine reacts with an aldehyde or ketone, provided that the carbonyl compound has an α-hydrogen.

      (19.64)

      (19.65)

As Eq. 19.65 illustrates, the two alkyl groups of a secondary amine may be part of a ring.

Like imine formation, enamine formation is reversible and must be driven to completion by the removal of one of the reaction products (usually water; see Eq. 19.65). Enamines, like imines, revert to the corresponding carbonyl compounds and amines in aqueous acid.

The mechanism of enamine formation begins, like the mechanism of imine formation, as a nucleophilic addition to give a carbinolamine intermediate. (Write the mechanism of this reaction.)

      (16.66a)

Because no hydrogen remains on the nitrogen of this carbinolamine, imine formation cannot occur. Instead, dehydration of the carbinolamine involves loss of a hydrogen from an adjacent *carbon*.

      (16.66b)

Why don't primary amines react with aldehydes or ketones to form enamines rather than imines? The answer is the enamines bear the same relationship to imines that *enols* bear to ketones.

$$\text{an enamine} \quad\rightleftharpoons\quad \substack{\text{the isomeric imine}\\ \text{(more stable)}} \tag{19.67a}$$

$$\text{an enol} \quad\rightleftharpoons\quad \substack{\text{the isomeric ketone}\\ \text{(more stable)}} \tag{19.67b}$$

Just as most aldehydes and ketones are more stable than their corresponding enols (Sec. 14.5A), most imines are more stable than their corresponding enamines. Because secondary amines *cannot* form imines, they form enamines instead.

To summarize: Aldehydes and ketones react with primary amines (RNH_2) to give imines, and with secondary amines (R_2NH) to give enamines. In a third type of amine, a **tertiary amine** (R_3N), all hydrogens of ammonia are replaced by alkyl or aryl groups. *Tertiary amines do not react with aldehydes and ketones to form stable derivatives.* Although most tertiary amines are good nucleophiles, they have no NH hydrogens and therefore cannot even form carbinolamines. Their adducts with aldehydes and ketones are unstable and can only break down to starting materials.

$$R_3N: \;+\; \underset{\text{C}}{\overset{:O:}{\|}} \quad\rightleftharpoons\quad -\underset{\underset{R_3N^+}{|}}{\overset{\overset{:\ddot{O}:^-}{|}}{C}}- \tag{19.68}$$

PROBLEM

19.33   Give the enamine product formed when each of the following pairs reacts.

(a) acetone and   (b) $PhCH_2CH{=}O$ and $(CH_3)_2\ddot{N}H$

19.12	**REDUCTION OF CARBONYL GROUPS TO METHYLENE GROUPS**

The most common reductive transformation of aldehydes or ketones is their conversion into alcohols (Sec. 19.8). But it is also possible to reduce the carbonyl group of an aldehyde or ketone completely to a methylene ($-CH_2-$) group. One procedure for effecting this transformation involves heating the aldehyde or ketone with hydrazine (H_2N-NH_2) and strong base.

$$\underset{\substack{\textbf{propiophenone}}}{Ph-\overset{\overset{O}{\|}}{C}-CH_2CH_3} \;+\; \underset{\substack{\textbf{hydrazine}\\ \text{(85\% aqueous}\\ \text{solution)}}}{H_2N-NH_2} \quad\xrightarrow[\substack{\text{heat, 1 h}\\ \text{triethylene glycol}}]{\text{KOH}}\quad \underset{\substack{\textbf{propylbenzene}\\ \text{(82\% yield)}}}{Ph-CH_2CH_2CH_3} \;+\; H_2O \;+\; N_2 \tag{19.69}$$

3,4-dimethoxybenzaldehyde   **3,4-dimethoxytoluene**
(81% yield)

This reaction, called the **Wolff–Kishner reduction**, typically uses ethylene glycol or similar high-boiling compounds as co-solvents. (Triethylene glycol, which has the structure $HOCH_2CH_2OCH_2CH_2OCH_2CH_2OH$, and a boiling point of 278 °C, is used in Eqs. 19.69 and 19.70.) The high boiling points of these solvents allow the reaction mixtures to reach the high temperatures required for the reduction to take place at a reasonable rate.

The Wolff–Kishner reduction is an extension of imine formation (Sec. 19.11A) because a *hydrazone* (Table 19.3) is an intermediate in the reaction. A series of Brønsted acid–base reactions (see Study Guide Link 19.7) lead ultimately to expulsion of dinitrogen gas and formation of the product.

hydrazine   **a hydrazone**
$+ H_2O$

STUDY GUIDE LINK 19.7
Mechanism of the Wolff–Kishner Reaction

The Wolff–Kishner reduction takes place under strongly basic conditions. The same overall transformation can be achieved under acidic conditions by a reaction called the **Clemmensen reduction**. In this reaction, an aldehyde or ketone is reduced with zinc amalgam (a solution of zinc metal in mercury) in the presence of HCl. The reduction takes place on the surface of the Zn metal.

(93% yield)

heptanal   **heptane**
(87% yield)

The mechanism of the Clemmensen reduction is uncertain.

One of the most useful applications of the Wolff–Kishner and Clemmensen reductions is the introduction of alkyl substituents into benzene rings. This is illustrated in Study Problem 19.5.

STUDY PROBLEM 19.5

Outline a synthesis of butylbenzene from benzene and any other reagents.

SOLUTION When you are asked to prepare an alkylbenzene from benzene, Friedel–Crafts alkylation (Sec. 16.4E) should come to mind. Indeed, the Friedel–Crafts alkylation reaction is useful for introducing groups that do not rearrange, such as methyl groups, ethyl groups, and *tert*-butyl groups, into benzene rings. But when this reaction is used to prepare butylbenzene from benzene and 1-chlorobutane, a major amount of rearranged product is observed. (See Eq. 16.16, p. 806.)

$$\text{benzene} + CH_3CH_2CH_2CH_2-Cl \xrightarrow{AlCl_3} sec\text{-butylbenzene (65\%)} + \text{butylbenzene (35\%)} + HCl \qquad (19.74a)$$

Butylbenzene can be easily prepared free of isomers, however, by the Wolff–Kishner reduction of butyrophenone:

$$\text{butyrophenone} \xrightarrow[\text{heat}]{H_2NNH_2,\ ^-OH} \text{butylbenzene} \qquad (19.74b)$$

In turn, butyrophenone is readily prepared by Friedel–Crafts *acylation* (Sec. 16.4F), which is not plagued by the rearrangement problems associated with the *alkylation*.

$$\text{benzene} + \text{butyryl chloride} \xrightarrow{AlCl_3} \xrightarrow{H_3O^+} \text{butyrophenone} + HCl \qquad (19.74c)$$

(Butylbenzene can also be prepared by the Stille reaction; Sec. 18.10B.)

PROBLEMS

19.34 Draw the structures of all aldehydes or ketones that could in principle give the following product after application of either the Wolff–Kishner or Clemmensen reduction.

$$H_3C-\text{(benzene ring)}-CH_2CH(CH_3)_2$$

19.35 Outline a synthesis of 1,4-dimethoxy-2-propylbenzene from hydroquinone (*p*-hydroxyphenol) and any other reagents.

19.13   THE WITTIG ALKENE SYNTHESIS

Our tour through aldehyde and ketone chemistry started with simple additions; then addition followed by substitution (acetal formation); then additions followed by elimination (imine and enamine formation). Another addition–elimination reaction, called the **Wittig alkene synthesis**, is an important method for preparing alkenes from aldehydes and ketones.

An example of the Wittig alkene synthesis is the preparation of methylenecyclohexane from cyclohexanone.

$$\text{cyclohexanone} + :\bar{C}H_2-\overset{+}{P}Ph_3 \longrightarrow \text{methylenecyclohexane} =CH_2 + Ph_3\overset{+}{P}-\ddot{O}:^- \qquad (19.75)$$

an ylid   triphenylphosphine oxide

The Wittig synthesis is especially important because it gives alkenes in which the *position* of the double bond is unambiguous; in other words, the Wittig synthesis is completely *regioselective*. It can be used for the preparation of alkenes that would be difficult to prepare by other reactions. For example, methylenecyclohexane, which is readily prepared by the Wittig synthesis (Eq. 19.75), cannot be prepared by dehydration of 1-methylcyclohexanol; 1-methylcyclohexene is obtained instead, because alcohol dehydration gives the alkene isomer(s) in which the double bond has the greatest number of alkyl substituents (Sec. 10.2).

$$\text{(19.76)}$$

The nucleophile in the Wittig alkene synthesis is a type of *ylid* (pronounced ĭ′ ləd). An **ylid** (sometimes spelled *ylide*) is any compound with opposite charges on adjacent, covalently bound atoms, each of which has an electronic octet.

Because phosphorus, like sulfur (Sec. 10.10), can accommodate more than eight valence electrons, a phosphorus ylid has an uncharged resonance structure.

$$\left[Ph_3\overset{+}{P}\!-\!\overset{..}{C}H_2 \quad \longleftrightarrow \quad Ph_3P\!=\!CH_2 \right] \qquad \text{(19.77)}$$

Although the structures of phosphorus ylids are sometimes written with phosphorus–carbon double bonds, the charged structures, in which each atom has an octet of electrons, are very important contributors.

The mechanism of the Wittig reaction in most cases is a concerted cycloaddition.

$$\text{(19.78a)}$$

an ylid → an oxaphosphetane

We can relate this mechanism to what we have learned about carbonyl additions if we imagine it as a two-step process in which the first step is a conventional carbonyl addition to give a tetrahedral addition intermediate. The second step is the reaction of the negatively charged oxygen in the tetrahedral addition intermediate with the positively charged phosphorus to give the oxaphosphetane.

$$\text{C}=\ddot{\text{O}}: \quad \text{H}_2\bar{\text{C}}-\overset{+}{\text{P}}\text{Ph}_3 \longrightarrow \left[\begin{array}{c} \text{C}-\ddot{\text{O}}:^- \\ | \\ \text{H}_2\text{C}-\overset{+}{\text{P}}\text{Ph}_3 \end{array} \right] \longrightarrow \begin{array}{c} \text{C}-\ddot{\text{O}}: \\ | \quad\quad | \\ \text{H}_2\text{C}-\text{PPh}_3 \end{array} \tag{19.78b}$$

an ylid

tetrahedral
addition intermediate

an oxaphosphetane

Under the usual reaction conditions, the oxaphosphetane spontaneously undergoes a β-elimination to give the alkene and the by-product triphenylphosphine oxide.

$$\begin{array}{c} \text{C}-\ddot{\text{O}}: \\ || \quad\quad | \\ \text{H}_2\text{C}-\text{PPh}_3 \end{array} \longrightarrow \begin{array}{c} \text{C} \\ || \\ \text{CH}_2 \end{array} + \begin{array}{c} \ddot{\text{O}}: \\ || \\ \text{PPh}_3 \end{array} \tag{19.78c}$$

oxaphosphetane

the alkene
product

**triphenylphosphine
oxide**

The ylid starting material in the Wittig synthesis is prepared by the reaction of an alkyl halide with triphenylphosphine (Ph$_3$P) in an S$_N$2 reaction to give a *phosphonium salt*.

$$\text{Ph}_3\text{P}: \quad + \quad \text{H}_3\text{C}-\ddot{\text{B}}\text{r}: \quad \xrightarrow[\text{2 days}]{\text{benzene}} \quad \text{Ph}_3\overset{+}{\text{P}}-\text{CH}_3 \; :\ddot{\text{B}}\text{r}:^- \tag{19.79a}$$

triphenylphosphine   methyl bromide

**methyltriphenylphosphonium
bromide**
(a phosphonium salt;
99% yield)

The phosphonium salt can be converted into its conjugate base, the ylid, by reaction with a strong base such as an organolithium reagent.

$$\text{Ph}_3\overset{+}{\text{P}}-\text{CH}_2 \; \text{Br}^- \longrightarrow \text{Ph}_3\overset{+}{\text{P}}-\bar{\bar{\text{C}}}\text{H}_2 + \text{H}-\text{CH}_2\text{CH}_2\text{CH}_2\text{CH}_3 + \text{LiBr} \tag{19.79b}$$

$$\quad | \\ \quad \text{H} \\ \text{Li}-\text{CH}_2\text{CH}_2\text{CH}_2\text{CH}_3$$

an ylid   **butane**

butyllithium

To plan the preparation of an alkene by the Wittig synthesis, consider the origin of each part of the product, and then reason deductively. Thus, one carbon of the alkene double bond originates from the alkyl halide used to prepare the ylid; the other is the carbonyl carbon of the aldehyde or ketone:

$$\begin{array}{cc} \text{R}^1 & \text{R}^3 \\ \text{C}=\text{C} \\ \text{R}^2 & \text{R}^4 \end{array} \Rightarrow \begin{array}{cc} \text{R}^1 \\ \text{C}=\text{O} \\ \text{R}^2 \end{array} + \text{Ph}_3\overset{+}{\text{P}}-\bar{\bar{\text{C}}}\begin{array}{c}\text{R}^3\\ \\ \text{R}^4\end{array} \Rightarrow \text{Ph}_3\overset{+}{\text{P}}-\text{CH}\begin{array}{c}\text{R}^3\\ \\ \text{R}^4\end{array} \text{Br}^- \Rightarrow \text{Ph}_3\text{P}: + \text{Br}-\text{CH}\begin{array}{c}\text{R}^3\\ \\ \text{R}^4\end{array} \tag{19.80}$$

aldehyde or   ylid
ketone

+ base

alkyl halide

(Again, the arrows used in this retrosynthetic analysis are read "implies as starting material.") This analysis also shows that, in principle, two Wittig syntheses are possible for any given alkene; in the other possibility, the R^1 and R^2 groups could originate from the alkyl halide and the R^3 and R^4 groups from the aldehyde or ketone. However, remember that the reaction used to form the phosphonium salt is an S$_N$2 reaction; consequently, this reaction is fastest with methyl and primary alkyl halides. In other words, most Wittig syntheses are planned so that the most reactive alkyl halide can be used as one of the starting materials.

One problem with the Wittig alkene synthesis is that it gives mixtures of *E* and *Z* isomers.

$$PhCH_2Cl \xrightarrow{Ph_3P} PhCH_2\overset{+}{-}PPh_3 \; Cl^- \xrightarrow[\text{2) PhCH=O}]{\text{1) Ph—Li, ether}}$$

Ph   H   Ph   Ph
  \   /   \   /
   C=C   +   C=C   (19.81)
  /   \   /   \
 H   Ph   H   H
 (20% yield)   (62% yield)

In many cases, the *Z* isomer predominates, as in this example. However, modifications of the Wittig reaction have been discovered (the use of different bases, lower temperatures, and other modifications) that can give either nearly pure *Z* isomer or nearly pure *E* isomer, depending on the modification used. However, the details of these modifications are outside of the scope of our discussion.

STUDY PROBLEM 19.6

Outline two Wittig alkene syntheses of 2-methyl-1-hexene. Is one synthesis preferred over the other? Why?

SOLUTION   The analysis in Eq. 19.80 suggests that the "right-hand" part of the alkene can be derived from the ketone 2-hexanone:

$$H_2C=CCH_2CH_2CH_2CH_3 \Rightarrow Ph_3\overset{+}{P}-\overset{\cdot\cdot}{C}H_2 + O=CCH_2CH_2CH_2CH_3 \quad (19.82)$$

$$\underset{CH_3}{|} \qquad\qquad\qquad \underset{CH_3}{|}$$

2-methyl-1-hexene $Ph_3P: + CH_3I$ **2-hexanone**

methyl iodide

Another possibility, however, is that the "left-hand" part of the alkene is derived from formaldehyde:

$$H_2C=CCH_2CH_2CH_2CH_3 \Rightarrow Ph_3\overset{+}{P}-\overset{\cdot\cdot}{C}-CH_2CH_2CH_2CH_3 \Rightarrow Ph_3P: + Br-CHCH_2CH_2CH_2CH_3 \quad (19.83)$$

$$\underset{CH_3}{|} \qquad\qquad\qquad \underset{CH_3}{|} \qquad\qquad\qquad \underset{CH_3}{|}$$

2-methyl-1-hexene $+ H_2C=O$ **2-bromohexane**

formaldehyde

Although both syntheses seem reasonable, the latter one (Eq. 19.83) would require an S_N2 reaction of triphenylphosphine with a secondary alkyl halide, whereas the former one (Eq. 19.82) would require an S_N2 reaction of triphenylphosphine with a methyl halide. The first reaction is preferred because methyl halides are much more reactive than secondary alkyl halides (Sec. 9.4C).

Discovery of the Wittig Alkene Synthesis

The Wittig alkene synthesis is named for Georg Wittig (1897–1987), who was Professor of Chemistry at the University of Heidelberg. Wittig and his co-workers discovered the alkene synthesis in the course of other work in phosphorus chemistry; they had not set out to develop this reaction explicitly. Once the significance of the reaction was recognized, it was widely exploited. Wittig shared the 1979 Nobel Prize in Chemistry with H. C. Brown (Sec. 19.8A).

The Wittig reaction is not only important as a laboratory reaction; it has also been industrially useful. For example, it is an important reaction in the industrial synthesis of vitamin A derivatives.

PROBLEMS

19.36   Give the structure of the alkene(s) formed in each of the following reactions.

(a) CH_3CH_2I $\xrightarrow{Ph_3P}$ $\xrightarrow{\text{butyllithium}}$ $\xrightarrow{\text{acetone}}$ (b) CH_3Br $\xrightarrow{Ph_3P}$ $\xrightarrow{\text{butyllithium}}$ $\xrightarrow{\text{benzaldehyde}}$

19.37   Outline a Wittig synthesis for each of the following alkenes; give two Wittig syntheses of the compound in part (a).

(a) CH_3O—⟨benzene ring⟩—$CH{=}CH$—⟨benzene ring⟩
(mixture of cis and trans)

(b)
$$CH_3$$
$$|$$
$$H_2C{=}CCH_2CH_3$$

(c) $CH_3CH{=}$⟨cyclobutane ring⟩

19.14   OXIDATION OF ALDEHYDES TO CARBOXYLIC ACIDS

Aldehydes can be oxidized to carboxylic acids.

$$CH_3CH_2CH_2CH_2CH{-}CH{=}O \xrightarrow[H_2O]{KMnO_4/NaOH} \xrightarrow{H_3O^+} CH_3CH_2CH_2CH_2CH{-}CO_2H \quad (19.84)$$

with CH_2CH_3 substituent on each.

2-ethylhexanal   **2-ethylhexanoic acid**
                                                     (78% yield)

Other common oxidants, such as aqueous Cr(VI) reagents, also work in this reaction. These oxidizing agents are the same ones used for oxidizing alcohols (Sec. 10.7A).

Some aldehyde oxidations begin as addition reactions. For example, in the oxidation of aldehydes by Cr(VI) reagents, the *hydrate*, not the aldehyde, is actually the species oxidized. (See Eq. 10.51, p. 483.)

$$R{-}\overset{\overset{\displaystyle O}{\|}}{C}{-}H + H_2O \rightleftharpoons R{-}\overset{\overset{\displaystyle OH}{|}}{\underset{\underset{\displaystyle OH}{|}}{C}}{-}H \xrightarrow{H_2Cr_2O_7} R{-}\overset{\overset{\displaystyle O}{\|}}{C}{-}OH \quad (19.85)$$

an aldehyde   the aldehyde hydrate

That is, the "aldehyde" oxidation is really an "alcohol" oxidation, the "alcohol" being the hydrate formed by addition of water to the aldehyde carbonyl group. For this reason, some water should be present in solution so that aldehyde oxidations with Cr(VI) occur at a reasonable rate.

In the laboratory, aldehydes can be conveniently oxidized to carboxylic acids with Ag(I) reagents.

$$\text{⟨cyclohexene ring with }CH{=}O\text{⟩} + Ag_2O \xrightarrow[\text{THF–water}]{NaOH} \text{⟨cyclohexene ring with }CO_2H\text{⟩} + 2\,Ag \quad (19.86)$$

3-cyclohexenecarbaldehyde   **3-cyclohexenecarboxylic acid**
                                        (75% yield)

The expense of silver limits its use to small-scale reactions, as a rule. However, the Ag_2O oxidation is especially handy when the aldehyde to be oxidized contains double bonds or alcohol —OH groups, functional groups that react with other oxidizing reagents but do not react with Ag_2O.

Sometimes, as in Eq. 19.86, the Ag(I) is used as a slurry of brown Ag_2O, which changes to a black precipitate of silver metal as the reaction proceeds. If the silver ion is solubilized as its ammonia complex, $^+Ag(NH_3)_2$, oxidation of the aldehyde is accompanied by the deposition of a metallic silver mirror on the walls of the reaction vessel. This observation can be used as a convenient test for aldehydes, known as the **Tollens test**.

Many aldehydes are oxidized by the oxygen in air upon standing for a long time. This process, another example of *autoxidation* (Sec. 18.11), is responsible for the contamination of some aldehyde samples with appreciable amounts of carboxylic acids.

Ketones cannot be oxidized without breaking carbon–carbon bonds (see Table 10.2, p. 482). Ketones are resistant to mild oxidation with Cr(VI) reagents, and acetone can even be used as a solvent for oxidations with such reagents. Potassium permanganate, however, oxidizes ketones by breaking carbon–carbon bonds, and it is therefore not useful as an oxidizing reagent in the presence of ketones.

PROBLEMS

19.38   Give the structure of an aldehyde $C_8H_8O_2$ that would be oxidized to terephthalic acid by $KMnO_4$.

terephthalic acid

19.39   What product is formed when the following compound is treated with Ag_2O?

19.15 MANUFACTURE AND USE OF ALDEHYDES AND KETONES

The most important commercial aldehyde is formaldehyde, which is manufactured by the oxidation of methanol over a silver catalyst.

$$H_3C—OH \xrightarrow[\text{600–650 °C}]{O_2,\ \text{Ag catalyst}} H_2C{=}O \tag{19.87}$$

About 74 billion pounds of formaldehyde is produced annually worldwide.

The single most important use of formaldehyde is in the synthesis of a class of polymers known as *phenol–formaldehyde resins*. (A **resin** is a polymer with a rigid three-dimensional network of repeating units.) Although the exact structure and properties of a phenol-formaldehyde resin depend on the conditions of the reaction used to prepare it, a typical segment of such a resin can be represented schematically as follows, in which the red CH_2 groups come from formaldehyde.

Phenol–formaldehyde resins are produced by a variation of the Friedel–Crafts alkylation in which phenol and formaldehyde are heated with acidic or basic catalysts. The formaldehyde in some cases is supplied in the form of its addition product with ammonia. Various formulations of these resins are used for telephones, adhesives in exterior-grade plywood, and heat-stable bondings for brake linings. A phenol–formaldehyde resin called *Bakelite*, patented in 1909 by the Belgian immigrant Leo H. Baekeland, was the first useful synthetic polymer.

Because formaldehyde, a known carcinogen, is used extensively in the manufacture of some construction materials, some concern has developed about formaldehyde release in confined environments in which such materials have been used. This concern received particular focus when it was discovered in 2006 that the trailers provided as temporary housing to victims of Hurricane Katrina by the Federal Emergency Management Agency (FEMA) contained formaldehyde levels that were 4–7 times the federally mandated limit. Formaldehyde-based construction materials were used extensively in these trailers.

Acetone, the simplest ketone, is co-produced with phenol by the autoxidation–rearrangement of cumene (Sec. 18.11). The worldwide production of acetone is estimated at roughly 15 billion pounds. Acetone itself finds use as an important solvent, and acetone cyanohydrin, which is produced from acetone (Eq. 19.15, p. 963), is a starting material for the production of poly(methyl methacrylate), an important polymer (Table 5.4, p. 219).

KEY IDEAS IN CHAPTER 19

- The functional group in aldehydes and ketones is the carbonyl group.

- Aldehydes and ketones are polar molecules. Simple aldehydes and ketones have boiling points that are higher than those of hydrocarbons but lower than those of alcohols. The aldehydes and ketones of low molecular mass are very soluble in water.

- The carbonyl stretching absorption near 1700 cm^{-1} is the most important infrared absorption of aldehydes and ketones. The proton NMR spectra of aldehydes have distinctive low-field absorptions at δ 9–11 for the aldehydic protons, and α-protons of both aldehydes and ketones absorb near δ 2.5. The most characteristic absorptions in the ^{13}C NMR spectra of aldehydes and ketones are the carbonyl carbon resonances at δ 190–220. Aldehydes and ketones have weak $n \rightarrow \pi^*$ UV absorptions, and compounds that contain double bonds conjugated with the carbonyl group have $\pi \rightarrow \pi^*$ absorptions. α-Cleavage, inductive cleavage, and the McLafferty rearrangement are the important fragmentation modes observed in the mass spectra of aldehydes and ketones.

- Aldehydes and ketones are weak bases and are protonated on their carbonyl oxygens to give α-hydroxy carbocations. The interaction of a proton or Lewis acid with a carbonyl group activates it toward addition reactions. Proton-catalyzed additions, Grignard additions, and hydride reductions provide examples of such activation.

- The most characteristic carbonyl-group reactions of aldehydes and ketones are carbonyl-addition reactions, which,

depending on the reaction, occur under acidic or basic conditions or both. Cyanohydrin formation and hydration are examples of simple reversible carbonyl additions. Hydride reductions and Grignard reactions are examples of simple additions that are irreversible. Addition occurs by reaction of a nucleophile at the carbonyl carbon. The direction of approach of the nucleophile, from above or below the carbonyl plane, is governed by its interaction with the π^* MO of the carbonyl group.

- Hydride addition is an important process in biological reductions. NADH and its phosphorylated analog, NADPH, are the most common biological hydride reducing agents.

- Acetal formation is an example of addition to the carbonyl group followed by substitution. Acetals can be formed in acidic alcohol solvents and converted back into aldehydes or ketones in aqueous acid, but are stable to base. They make excellent protecting groups for aldehydes and ketones under basic conditions. Hemiacetals are intermediates in acetal formation.

- Imine (Schiff base) formation, enamine formation, and the Wittig alkene synthesis are examples of addition to the carbonyl group followed by elimination.

- Imines are used in biology to tether biologically important aldehydes, such as retinal and pyridoxal phosphate, to proteins; the nucleophilic amine that reacts with the aldehyde is typically the amino group of a lysine residue of the protein.

- The carbonyl group of an aldehyde or ketone can be reduced to a methylene group by either the Wolff–Kishner or the Clemmensen reduction.

- Aldehydes are readily oxidized to carboxylic acids; ketones cannot be oxidized without breaking carbon–carbon bonds.

- Aldehydes and ketones can be converted into alcohols (Grignard reactions, hydrogenation, and hydride reductions), alkenes (Wittig synthesis), and alkanes (Wolff–Kishner and Clemmensen reductions).

 REACTION REVIEW *For a summary of reactions discussed in this chapter, see the* Reaction Review *section of Chapter 19 in the* Study Guide and Solutions Manual.

ADDITIONAL PROBLEMS

19.40 Give the products expected (if any) when acetone reacts with each of the following reagents.

(a) H_3O^+

(b) $NaBH_4$ in CH_3OH, then H_2O

(c) CrO_3, pyridine

(d) NaCN, pH 10, H_2O

(e) CH_3OH (excess), H_2SO_4 (trace)

(f) , trace of acid

(g) semicarbazide, dilute acid

(h) CH_3MgI, ether, then H_3O^+

(i) product of part (b) + $Na_2Cr_2O_7$ in H_2SO_4

(j) product of part (h) + H_2SO_4

(k) H_2, PtO_2

(l) $H_2C{=}PPh_3$

(m) Zn amalgam, HCl

19.41 Give the product expected (if any) when butyraldehyde (butanal) reacts with each of the following reagents.

(a) PhMgBr, then dilute H_3O^+

(b) $LiAlH_4$ in ether, then H_3O^+

(c) alkaline $KMnO_4$, then H_3O^+

(d) aqueous $H_2Cr_2O_7$

(e) NH_2OH, pH = 5

(f) Ag_2O

(g) Zn amalgam, HCl

(h) $CH_3CH{=}PPh_3$

19.42 Sodium bisulfite adds reversibly to aldehydes and a few ketones to give *bisulfite addition products* (see Fig. P19.42).

(a) Write a curved-arrow mechanism for this addition reaction; assume water is the solvent.

(b) The reaction can be reversed by adding either H_3O^+ or ^-OH. Explain this observation using Le Châtelier's principle and your knowledge of sodium bisulfite reactions from general or inorganic chemistry.

(c) Deduce the structure of the bisulfite addition product of 2-methylpentanal.

19.43 The compound *ninhydrin* exists as a hydrate. Explain, and draw the structure of the hydrate.

ninhydrin

19.44 Each of the reactions shown in Fig. P19.44 (p. 998) gives a mixture of two separable isomers. What are the two isomers formed in each case?

19.45 Give the structures of the *four* separable isomers with the formula $C_9H_{18}O_3$ that are formed in the acid-catalyzed reaction of hexanal with glycerol (1,2,3-propanetriol).

---

sodium bisulfite

Figure P19.42

19.46 (a) What are the two constitutionally isomeric cyclic acetals that could in principle be formed in the acid-catalyzed reaction of acetone and glycerol (1,2,3-propanetriol)?

(b) Only one of the two compounds is actually formed. Given that it can be resolved into enantiomers, which isomer in part (a) is the one that is produced?

19.47 Give the structures of the two separable isomers formed in the following reaction.

19.48 Complete the reactions given in Fig. P19.48 by giving the principal organic product(s).

19.49 Using known reactions and mechanisms discussed in the text, complete the reactions given in Fig. P19.49.

19.50 (a) Complete the series of reactions in Fig. P19.50 by giving the major organic product.

(b) Show how the same product could be prepared from hydroquinone monomethyl ether (*p*-methoxyphenol).

19.51 A compound A, C_8H_8O, when treated with Zn amalgam and HCl, gives a xylene (dimethylbenzene) isomer that in turn gives only one ring monobromination product with Br_2 and Fe. Propose a structure for A.

19.52 Suggest routes by which each of the following compounds could be synthesized from the indicated starting materials and any other reagents.

(a) 1-phenyl-1-butanone (butyrophenone) from butyraldehyde

(b) 2-cyclohexyl-2-propanol from cyclohexanone

(c) cyclohexyl methyl ether (methoxycyclohexane) from cyclohexanone

(d) $PhCH_2OCH_2Ph$ (dibenzyl ether) from benzaldehyde as the only carbon source

---

(a)

(b) $PhCH{=}O + H_3C{-}CH{=}PPh_3 \longrightarrow$

Figure P19.44

(a)

phenylhydrazine
(see Table 19.3)

(b)

(c)

(d)

(e)

propiophenone

(f) $CH_3I + Ph_3P \longrightarrow \xrightarrow[\text{benzene}]{Bu{-}Li} \xrightarrow{Ph_2C{=}O}$

(g)

Figure P19.48

(e) 2,3-dimethyl-2-hexene from 3-methyl-2-hexanone

(f) 2,3-dimethyl-1-hexene from 3-methyl-2-hexanone

(g)

from (Z)-5,6-dimethyl-5-decen-1,10-diol

(h)

from the starting material for part (g)

(i) 1,6-hexanediol from cyclohexene

(j) 1-butyl-4-propylbenzene from benzene

(k)

from benzaldehyde as the only source of carbon

(l)

from

(*Hints:* 1. BH₃ in THF reduces aldehydes and ketones to alcohols; you need a protecting group.

2. Can you find an aldehyde lurking somewhere in the target molecule?)

(m)

from bromobenzene

(n)

from bromobenzene using a Heck reaction

19.53 (a) The following compound is unstable and spontaneously decomposes to acetophenone and HBr. Give a mechanism for this transformation.

(b) Use the information in part (a) to complete the following reaction:

$$H_2C{=}CH{-}Br + OsO_4 \xrightarrow{H_2O}$$

19.54 The product A of the reaction given in Fig. P19.54 (p. 1000) hydrolyzes in dilute aqueous acid to give acetophenone. Identify A, and draw a mechanism for its formation that accounts for the regioselectivity of the reaction.

19.55 Acetals can be used as protecting groups for alcohols. One such protecting group is the *tetrahydropyranyl ether* (THP ether).

a THP ether

THP ethers are introduced by treating an alcohol with dihydropyran and *p*-toluenesulfonic acid catalyst.

$$ROH + \underset{\textbf{dihydropyran}}{\text{(structure)}} \xrightarrow[\text{CH}_2\text{Cl}_2]{\substack{p\text{-toluene-}\\ \text{sulfonic acid}}} \text{(structure)}$$

THP ethers are stable to base but are rapidly removed by dilute aqueous acid.

(*Problem continues at top of p. 1000*)

(a)

$$(CH_3CH_2CH_2)_2C{-}\overset{\overset{\displaystyle O}{\|}}{C}{-}CH_2CH_3 \xrightarrow{\underset{CH_3OH}{NaBH_4}} \xrightarrow{NaH}$$

with Ph₂P=O group

(*Hint:* See Eqs. 19.78a–b, pp. 991–92.)

(b)

$$CH_3CH_2CH_2\overset{\overset{\displaystyle O}{\|}}{C}CH_3 + H_2\ddot{N}{-}Ph \longrightarrow \xrightarrow{\underset{CH_3OH}{NaBH_4}}$$

(*Hint:* The C=N bond undergoes addition much like the C=O bond.)

Figure P19.49

$$CH_3O{-}\langle\text{benzene}\rangle + H_3C{-}\overset{\overset{\displaystyle O}{\|}}{C}{-}Cl \xrightarrow{AlCl_3} \xrightarrow{H_3O^+} \xrightarrow[\substack{\text{heat}\\\text{ethylene glycol}}]{H_2NNH_2,\ NaOH}$$

Figure P19.50

(Problem 19.55 continued)

(a) Give the structure of the product formed (in addition to the alcohol ROH) when a THP ether is treated with aqueous acid.

(b) Using the THP protecting group as part of your strategy, outline a synthesis of 2-methyl-2,6-hexanediol from 4-bromo-1-butanol.

19.56 What are the starting materials for the synthesis of each of the following imines?

(a)

$$CH_3CH_2CH_2CH_2-CH{=}N-NH-\langle\bigcirc\rangle-OCH_3$$

(b)

19.57 The enzyme 3-ketobutanoyl thioester reductase catalyzes the reduction of the ketone carbonyl group of the following compound with NADPH to give the *R* stereoisomer of the product.

3-ketobutanoyl thioester

(This is an important reaction in the biosynthesis of fatty acids.)

(a) Give the structure of the product, including stereochemistry.

(b) The pro-(*R*) hydrogen of NADPH is transferred in the reaction. Show the relative positions of the NADPH and the 3-ketobutanoyl thioester molecules in the transition state of the reaction.

(c) In several different variants of this enzyme, the —OH groups of both tyrosine (Tyr) and serine (Ser) residues are positioned near the ketone carbonyl group. (For the structures of these amino acids, see Table 27.1, p. 1376–1377). What is the likely role of these groups? Add these amino acid side chains to the diagram you drew in part (b).

19.58 Compound *A*, $C_{11}H_{12}O$, which gave a negative Tollens test, was treated with $LiAlH_4$, followed by dilute acid, to give compound *B*, which could be resolved into enantiomers. When optically active *B* was treated with CrO_3 in pyridine, an optically inactive sample of *A* was obtained. Heating *A* with hydrazine in base gave hydrocarbon *C*, which, when heated with alkaline $KMnO_4$, gave carboxylic acid *D*. Identify all of the compounds and explain your reasoning.

D

19.59 Compound *A*, $C_6H_{12}O_2$, was found to be optically active, and it was slowly oxidized to an optically active carboxylic acid *B*, $C_6H_{12}O_3$, by $^+Ag(NH_3)_2$. Oxidation of *A* by anhydrous CrO_3 gave an optically inactive compound that reacted with Zn amalgam/HCl to give 3-methylpentane. With aqueous H_2CrO_4, compound *A* was oxidized to an optically inactive dicarboxylic acid *C*, $C_6H_{10}O_4$. Give structures for compounds *A*, *B*, and *C*.

19.60 (a) The insecticide DDT can be prepared by the reaction shown in Fig. P19.60a. Remembering that a protonated aldehyde or ketone is a type of carbocation, and that carbocations are electrophiles, draw a curved-arrow

---

Figure P19.54

(a)

(b)

Figure P19.60

mechanism for this electrophilic aromatic substitution reaction. (See Sec. 16.4E.)

(b) Bisphenol A is used in the manufacture of polycarbonate plastics. (Its carcinogenicity has become a cause for concern.) Bisphenol A is prepared by the acid-catalyzed reaction of acetone with two equivalents of phenol (Fig. P19.60b). Draw a curved-arrow mechanism for this reaction.

19.61 Salsolinol (Fig. P19.61) is formed in the brain when acetaldehyde reacts with dopamine. Because acetaldehyde is a biological oxidation product of ethanol (Sec. 10.8), it has been suggested that salsolinol might be used as a biological marker for alcohol consumption. Draw a curved-arrow mechanism for the formation of salsolinol from acetaldehyde and dopamine. Assume acids and bases are present as needed. (This is an example of the *Pictet–Spengler reaction.*)

19.62 From your knowledge of the reactivity of LiAlH$_4$, as well as the reactivity of epoxides with nucleophiles, predict the product (including stereochemistry, if appropriate) in each of the following reactions:

(a)

$$\text{(cyclohexene oxide)} + \text{LiAlD}_4 \longrightarrow \xrightarrow{\text{H}_3\text{O}^+}$$

(b)

$$\text{H}_3\text{C}\cdots\overset{\text{O}}{\underset{\overset{|}{\text{D}}\ \overset{|}{\text{D}}}{\triangle}}\cdots\text{CH}_3 + \text{LiAlH}_4 \longrightarrow \xrightarrow{\text{H}_3\text{O}^+}$$

19.63 Thumbs Throckmorton, a graduate student in his twelfth year of study, has designed the synthetic procedures shown in Fig. P19.63. Indicate the problems (if any) that each synthesis is likely to encounter.

19.64 Give curved-arrow mechanisms for the reactions given in Fig. P19.64 (p. 1002).

19.65 (a) You are the chief organic chemist for Bugs and Slugs, Inc., a firm that specializes in environmentally friendly pest control. You have been asked to design a synthesis of 4-methyl-3-heptanol, the aggregation pheromone of the European elm beetle (the carrier of Dutch elm disease). Propose a synthesis of this compound from starting materials containing five or fewer carbons.

(b) After successfully completing the synthesis in part (a) and delivering your compound, you are advised that it appears to be a mixture of isomers. Assuming that you have prepared the correct compound, provide an explanation.

19.66 Identify the following compounds.

(a) $C_{10}H_{10}O_2$ NMR: δ 2.82 (6*H*, s), δ 8.13 (4*H*, s)
     IR: 1681 cm^{-1}, no O—H stretch

(b) $C_5H_{10}O$ NMR: δ 9.8 (1*H*, s), δ 1.1 (9*H*, s)

(c) $C_6H_{10}O$ NMR in Fig. P19.66 (p. 1002)
     IR: 1701 cm^{-1}, 970 cm^{-1}
     UV: λ_{max} = 215 (ϵ = 17,400),
     329 (ϵ = 26)

19.67 Identify the compound with the mass spectrum and proton NMR spectrum shown in Fig. P19.67 on p. 1003. This compound has IR absorptions at 1678 cm^{-1} and 1600 cm^{-1}.

19.68 Trichloroacetaldehyde, Cl$_3$C—CH=O, forms a cyclic trimer analogous to paraldehyde (Eq. 19.56, p. 981).

(a) Account for the fact that two forms of this trimer are known (α, bp 223 °C and mp 116 °C; β, bp 250 °C and mp 152 °C).

(b) Which of your structures is likely to be the one with the higher melting point? Explain.

(c) Assuming you have in hand a sample of both forms, show how NMR spectroscopy could be used to verify your hypothesis in part (b).

---

dopamine + H$_3$C—CH(=O) ⟶ **salsolinol** + H$_2$O

Figure P19.61

(a)

$$\text{H}_3\text{C}-\overset{\text{O}}{\underset{\|}{\text{C}}}-\text{C}_6\text{H}_4-\overset{\text{O}}{\underset{\|}{\text{CH}}} \xrightarrow[\text{ether}]{\text{EtMgBr}} \xrightarrow{\text{H}_3\text{O}^+} \text{H}_3\text{C}-\overset{\text{OH}}{\underset{\underset{\text{Et}}{|}}{\overset{|}{\text{C}}}}-\text{C}_6\text{H}_4-\overset{\text{O}}{\underset{\|}{\text{CH}}}$$

(b) $\text{Ph}_3\text{P} + (\text{CH}_3)_3\text{CCH}_2\text{Br} \longrightarrow \xrightarrow{\text{CH}_3(\text{CH}_2)_3\text{Li}} \xrightarrow{\text{Ph}-\text{CH}=\text{O}} \text{PhCH}=\text{CHC}(\text{CH}_3)_3$

Figure P19.63

19.69 Starting with any organic compound you wish, outline synthetic procedures for preparing each of the following isotopically labeled materials using the indicated source of the isotope.

(a) ^{18}OH

Ph—CH—CH$_2$—Ph using H$_2{}^{18}$O

(b)   OH

Ph—CD—CH$_2$—Ph using LiAlD$_4$

19.70 Offer a rational explanation for each of the following observations.

(a) Although biacetyl (2,3-butanedione) and 1,2-cyclopentanedione have the same type of functional group, their dipole moments differ substantially.

$$H_3C-\overset{O}{\underset{||}{C}}-\overset{O}{\underset{||}{C}}-CH_3$$

biacetyl
$\mu = 1.04$ D

1,2-cyclopentanedione
$\mu = 2.21$ D

(b) When acetaldehyde is mixed with a 10-fold excess of ethanethiol, its $n \rightarrow \pi^*$ absorption at 280 nm is nearly eliminated.

(c) Compound *A* has a *much* weaker IR carbonyl absorption than compound *B*.

A   *B*

(a)

$$Ph-\overset{S}{\underset{||}{C}}-Ph \xrightarrow{H_2O} Ph-\overset{O}{\underset{||}{C}}-Ph + H_2S$$

(b)

$$(CH_3O)_3P: + H_2C-\overset{O}{\overbrace{\quad}}-CH-CH_3 \longrightarrow (CH_3O)_3P=O + H_2C=CHCH_3$$

(c)

$$CH_3CH_2-\ddot{N}H_2 + \quad \underset{H_3C\quad CH_3}{O=CH\quad CH=O} \xrightarrow{\text{acid}} \quad + 2H_2O$$

(d)

$$\xrightarrow{HCl}$$

(e)

$$3CH_3CH=O \underset{}{\overset{\text{acid}}{\rightleftharpoons}}$$

acetaldehyde   **paraldehyde**

Figure P19.64

chemical shift, Hz

Figure P19.66 The NMR spectrum for Problem 19.66(c). The relative integrals are indicated in red over their respective resonances. The horizontal scales of the insets are identical.

δ 9.51 (off scale)

chemical shift, ppm (δ)

19.71 Identify the compound $C_7H_{10}O$ that has an IR absorption at 1703 cm^{-1} and the proton NMR spectrum shown in Fig. P19.71.

19.72 Identify compound *A*, $C_6H_{12}O_3$, which has an IR absorption at 1710 cm^{-1} (no absorption in the 3200–

3400 cm^{-1} region), as well as the following ^{13}C NMR-DEPT spectrum (attached hydrogens in parentheses): δ 30.6 (3), δ 47.2 (2), δ 53.5 (3), δ 101.7 (1), δ 204.9 (0). One of the proton NMR absorptions of compound *A* is a singlet at δ 2.1.

Figure P19.67 The mass spectrum and NMR spectrum for Problem 19.67. The relative integrals in the NMR spectrum are indicated in red over their respective resonances.

Figure P19.71 The NMR spectrum for Problem 19.71. The relative integrals are indicated in color over their respective resonances.

CHAPTER 20

The Chemistry of Carboxylic Acids

The characteristic functional group in a **carboxylic acid** is the **carboxy group**.

$$H_3C-\overset{\overset{\displaystyle O}{\|}}{C}-OH \qquad\qquad H_3C-CO_2H$$

acetic acid
(a simple carboxylic acid)

carboxy group

condensed structure

Carboxylic acids and their derivatives rank with aldehydes and ketones among the most important organic compounds because they occur widely in living organisms and because they serve important roles in organic synthesis. This chapter is concerned with the structures, properties, acidities, and carbonyl-group reactions of carboxylic acids themselves, including some biologically important reactions. Chapter 21 is devoted to a study of carboxylic acid derivatives.

This chapter also surveys briefly some of the chemistry of **sulfonic acids**.

$$H_3C-\overset{\overset{\displaystyle O}{\|}}{\underset{\underset{\displaystyle O}{\|}}{S}}-OH \qquad\qquad H_3C-SO_3H$$

sulfonic acid group

methanesulfonic acid
(a simple sulfonic acid)

condensed structure

20.1   NOMENCLATURE OF CARBOXYLIC ACIDS

A. Common Nomenclature

Common nomenclature is widely used for the simpler carboxylic acids. A carboxylic acid is named by adding the suffix *ic* and the word *acid* to the prefix for the appropriate group given in Table 19.1 on p. 948.

$$H_3C-\overset{\overset{\displaystyle O}{\|}}{C}-OH$$

prefix (Table 19.1, p. 948): acet + *ic acid* = **acetic acid**

$$\langle\text{benzene ring}\rangle-\overset{\overset{\displaystyle O}{\|}}{C}-OH$$

benzo + *ic acid* = **benzoic acid**

Some of these names owe their origin to the natural source of the acid. For example, formic acid occurs in the venom of the red ant (from the Latin *formica*, meaning "ant"); acetic acid is the acidic component of vinegar (from the Latin *acetus*, meaning "vinegar"); and butyric acid is the foul-smelling component of rancid butter (from the Latin *butyrum*, meaning "butter"). The common names of carboxylic acids, given in Table 20.1 on p. 1006, are used as much or more than the substitutive names.

As with aldehydes and ketones, substitution in the common system is denoted with Greek letters rather than numbers. The position *adjacent* to the carboxy group is designated as α.

$$H_3\overset{\gamma}{C}-\overset{\beta}{C}H_2-\underset{\underset{\displaystyle Br}{|}}{\overset{\alpha}{C}H}-\overset{\overset{\displaystyle O}{\|}}{C}-OH$$

***α*-bromobutyric acid**

In common nomenclature, the position of the substituent is omitted if it is unambiguous. Thus, $ClCH_2CO_2H$ is named chloroacetic acid rather than α-chloroacetic acid.

Carboxylic acids with two carboxy groups are called **dicarboxylic acids**. The unbranched dicarboxylic acids are particularly important and are invariably known by their common names. Some important dicarboxylic acids are also listed in Table 20.1.

$$HO_2C-CH_2CH_2-CO_2H$$
succinic acid

$$HO_2C-\underset{\underset{\displaystyle CH_3}{|}}{CH}-CO_2H$$
methylmalonic acid

$$HO_2C-CH_2\underset{\underset{\displaystyle CH_3}{|}}{\overset{\overset{\displaystyle CH_3}{|}}{C}}CH_2-CO_2H$$
***β,β*-dimethylglutaric acid**

A mnemonic device used by generations of organic chemistry students for remembering the names of the dicarboxylic acids is the phrase, "*O*h, *M*y, *S*uch *G*ood *A*pple *P*ie," in which the first letter of each word corresponds to the name of successive dicarboxylic acids: oxalic, malonic, succinic, glutaric, adipic, and pimelic acids.

Phthalic acid is an important aromatic dicarboxylic acid.

$$\langle\text{benzene ring}\rangle\overset{CO_2H}{\underset{CO_2H}{}}$$

phthalic acid

TABLE 20.1 Names and Structures of Some Carboxylic Acids

Systematic name	Common name	Structure
methanoic* acid	formic acid	HCO_2H
ethanoic* acid	acetic acid	CH_3CO_2H
propanoic acid	propionic acid	$CH_3CH_2CO_2H$
butanoic acid	butyric acid	$CH_3CH_2CH_2CO_2H$
2-methylpropanoic acid	isobutyric acid	$(CH_3)_2CHCO_2H$
pentanoic acid	valeric acid	$CH_3(CH_2)_3CO_2H$
3-methylbutanoic acid	isovaleric acid	$(CH_3)_2CHCH_2CO_2H$
2,2-dimethylpropanoic acid	pivalic acid	$(CH_3)_3CCO_2H$
hexanoic acid	caproic acid	$CH_3(CH_2)_4CO_2H$
octanoic acid	caprylic acid	$CH_3(CH_2)_6CO_2H$
decanoic acid	capric acid	$CH_3(CH_2)_8CO_2H$
dodecanoic acid	lauric acid	$CH_3(CH_2)_{10}CO_2H$
tetradecanoic acid	myristic acid	$CH_3(CH_2)_{12}CO_2H$
hexadecanoic acid	palmitic acid	$CH_3(CH_2)_{14}CO_2H$
octadecanoic acid	stearic acid	$CH_3(CH_2)_{16}CO_2H$
2-propenoic* acid	acrylic acid	$H_2C{=}CHCO_2H$
2-butenoic* acid	crotonic acid	$CH_3CH{=}CHCO_2H$
benzoic acid	benzoic acid	$PhCO_2H$
Dicarboxylic acids		
ethanedioic* acid	oxalic acid	$HO_2C{-}CO_2H$
propanedioic* acid	malonic acid	$HO_2CCH_2CO_2H$
butanedioic* acid	succinic acid	$HO_2C(CH_2)_2CO_2H$
pentanedioic* acid	glutaric acid	$HO_2C(CH_2)_3CO_2H$
hexanedioic* acid	adipic acid	$HO_2C(CH_2)_4CO_2H$
heptanedioic* acid	pimelic acid	$HO_2C(CH_2)_5CO_2H$
1,2-benzenedicarboxylic* acid	phthalic acid	
(Z)-2-butenedioic* acid	maleic acid	
(E)-2-butenedioic* acid	fumaric acid	

* The common name is almost always used instead.

Many carboxylic acids were known long before any system of nomenclature existed, and their time-honored traditional names are widely used. The following are examples of these.

$$HO_2C—CH—CH—CO_2H$$
$$\underset{OH}{|} \quad \underset{OH}{|}$$

tartaric acid

$$Ph—CH\!=\!CH—CO_2H$$

cinnamic acid

salicylic acid

B. Substitutive Nomenclature

A carboxylic acid is named systematically by dropping the final *e* from the name of the hydrocarbon with the same number of carbon atoms and adding the suffix *oic* and the word *acid*.

$$CH_3CH_2—\overset{\overset{\displaystyle O}{\|}}{C}—OH$$

propan*e* + *oic acid* = **propanoic acid**

The final *e* is not dropped in the name of dicarboxylic acids.

$$HO_2C\diagup\diagdown\diagup\diagdown\diagup^{CO_2H}$$

octanedioic acid

When a carboxylic acid is derived from a cyclic hydrocarbon, the suffix *carboxylic* and the word *acid* are added to the name of the hydrocarbon. (This nomenclature is similar to that for the corresponding aldehydes; Sec. 19.1B.)

cyclohexanecarboxylic acid   **1,2,4-benzenetricarboxylic acid**

One exception to this nomenclature is benzoic acid (p. 1005), for which the IUPAC recognizes the common name.

The principal chain in substituted carboxylic acids is numbered, as in aldehydes, by assigning the number 1 to the carbonyl carbon.

$$\underset{5}{H_3C}—\underset{4}{CH_2}—\underset{3}{\overset{\overset{\displaystyle CH_3}{|}}{CH}}—\underset{2}{CH_2}—\underset{1}{CO_2H}$$

3-methylpentanoic acid

This numbering scheme should be contrasted with that used in the common system, in which numbering begins with the Greek letter α at carbon-2.

In carboxylic acids derived from cyclic hydrocarbons, numbering begins at the ring carbon bearing the carboxy group.

4-methylcyclohexanecarboxylic acid   **4-bromobenzoic acid**
or *p*-bromobenzoic acid

When carboxylic acids contain other functional groups, the carboxy groups receive priority over aldehyde and ketone carbonyl groups, hydroxy groups, and mercapto groups for citation as the principal group.

Priority for citation as principal group:

$$\underset{\substack{\| \\ O}}{-C}-OH \; > \; \underset{\substack{\| \\ O}}{-C}-H \; > \; \underset{\substack{\| \\ O}}{-C}- \; > \; -OH \; > \; -SH \tag{20.1}$$

STUDY PROBLEM   20.1

Provide a substitutive name for the following compound.

SOLUTION   First, decide on the principal group. From the order in Eq. 20.1, the carboxy group has highest priority. The aldehyde oxygen and the hydroxy group are treated as substituents. The structure has seven carbons and one double bond and hence is a heptenoic acid. The carboxy group is given the number 1; hence, the double bond is at carbon-5, and the molecule is a 5-heptenoic acid. The —OH group is named as a 4-hydroxy substituent and the aldehyde oxygen as a 7-oxo substituent. Application of the priority rules for double-bond stereochemistry (Sec. 4.2B) shows that the double bond has *E* stereochemistry. The name is therefore (*E*)-4-hydroxy-7-oxo-5-heptenoic acid. Although carbon-4 is an asymmetric carbon, its stereochemistry is omitted in the name because it is not specified in the structure.

The carboxy group is sometimes named as a substitutent:

3-(carboxymethyl)hexanedioic acid

A complete list of nomenclature priorities for all of the functional groups covered in this text is given in Appendix I.

PROBLEMS

20.1   Give the structure of each of the following compounds.
  (a) γ-hydroxybutyric acid   (b) β,β-dichloropropionic acid
  (c) (Z)-3-hexenoic acid   (d) 4-methylhexanoic acid
  (e) 1,4-cyclohexanedicarboxylic acid   (f) *p*-methoxybenzoic acid
  (g) α,α-dichloroadipic acid   (h) oxalic acid

20.2   Name each of the following compounds. Use a common name for at least one compound.

(d)

(e) $HO_2C\!-\!CH\!-\!CO_2H$
$\quad\quad\quad\quad |$
$\quad\quad\quad\quad CH_3$

(f) ▷—CO_2H

20.2 STRUCTURE AND PHYSICAL PROPERTIES OF CARBOXYLIC ACIDS

The structure of a simple carboxylic acid, acetic acid, is compared with the structures of other oxygen-containing compounds in Fig. 20.1. Carboxylic acids, like aldehydes and ketones, have trigonal geometry at their carbonyl carbons. Notice, moreover, that the two oxygens of a carboxylic acid are quite different. One, the **carbonyl oxygen**, is the oxygen involved in the C=O double bond. Figure 20.1 demonstrates that the C=O bonds of aldehydes, ketones, and carboxylic acids have the same length. The other oxygen, called the **carboxylate oxygen**, is the oxygen involved in the C—O single bond. Notice in Fig. 20.1 that the C—O bond in a carboxylic acid is considerably shorter than the C—O bond in an alcohol or ether (about 1.36 Å vs. about 1.42 Å). The reason for this difference is that the C—O bond in an acid is an sp^2–sp^3 single bond, whereas the C—O bond in an alcohol or ether is an sp^3–sp^3 single bond.

$$
\begin{array}{cc}
& O \\
& \| \\
& C \\
R & \quad OH \\
\end{array}
\qquad
\begin{array}{cc}
& H_2 \\
& C \\
R & \quad OH \\
\end{array}
\qquad (20.2)
$$

sp^2–sp^3 single bond
(more *s* character,
therefore shorter)

sp^3–sp^3 single bond
(longer)

The carboxylic acids of lower molecular mass are high-boiling liquids with acrid, piercing odors. They have considerably higher boiling points than many other organic compounds of about the same molecular mass and shape:

	acetic acid	**isopropyl alcohol**	**acetone**	**isobutylene**
boiling point	117.9° C	82.3° C	56.5° C	−6.9° C

$H_3C\!-\!\overset{\displaystyle O}{\underset{\displaystyle OH}{C}}$ $\quad$ $H_3C\!-\!\overset{\displaystyle OH}{\underset{\displaystyle CH_3}{CH}}$ $\quad$ $H_3C\!-\!\overset{\displaystyle O}{\underset{\displaystyle CH_3}{C}}$ $\quad$ $H_3C\!-\!\overset{\displaystyle CH_2}{\underset{\displaystyle CH_3}{C}}$

FIGURE 20.1 Comparison of the structures of acetic acid and other oxygen-containing compounds. The carbonyl compounds have identical C=O bond lengths, and the C—O single bond in a carboxylic acid is shorter than that in an ether or alcohol.

carbonyl oxygen

carboxylate oxygen

acetic acid — 127°, 1.21 Å, 1.52 Å, 1.36 Å, 111°

acetaldehyde — 124°, 1.21 Å, 1.51 Å, 116°

acetone — 122°, 1.21 Å, 1.52 Å, 116°

dimethyl ether — 1.42 Å, 111.5°

methanol — 1.426 Å, 109°

The high boiling points of carboxylic acids can be attributed not only to their polarity, but also to the fact that they form very strong hydrogen bonds (Sec. 8.5C). In the solid state, and under some conditions in both the gas phase and solution, carboxylic acids exist as hydrogen-bonded dimers. (A **dimer** is any structure derived from two identical smaller units.)

acetic acid dimer

The equilibrium constants for the formation of such dimers in solution are very large—on the order of 10^6 to 10^7 M^{-1}. (The equilibrium constant for hydrogen-bond dimerization of ethanol, in contrast, is 11 M^{-1}.)

Many aromatic and dicarboxylic acids are solids. For example, the melting points of benzoic acid and succinic acid are 122°C and 188°C, respectively.

The simpler carboxylic acids are very soluble in water, as expected from their hydrogen-bonding capabilities (Sec 8.6D); the unbranched carboxylic acids below pentanoic acid are miscible with water. Many dicarboxylic acids also have significant water solubilities.

PROBLEM

20.3   At a given concentration of acetic acid, in which solvent would you expect the amount of acetic acid dimer to be greater: CCl_4 or water? Explain.

20.3   SPECTROSCOPY OF CARBOXYLIC ACIDS

A. IR Spectroscopy

Two important absorptions are found in the infrared spectrum of a typical carboxylic acid. One is the C=O stretching absorption, which occurs near 1710 cm^{-1} for carboxylic acid dimers. (The IR spectra of carboxylic acids are nearly always run under conditions such that they are in the dimer form. The carbonyl absorptions of carboxylic acid monomers occur near 1760 cm^{-1} but are rarely observed.) The other important carboxylic acid absorption is the O—H stretching absorption. This absorption is much broader than the O—H stretching absorption of an alcohol or phenol and covers a very wide region of the spectrum—typically 2400–3600 cm^{-1}. (In many cases this absorption obliterates the C—H stretching absorption of the acid.) The carbonyl absorption and this broad O—H stretching absorption are illustrated in the IR spectrum of propanoic acid (Fig. 20.2a); these absorptions are hallmarks of a carboxylic acid. A conjugated carbon–carbon double bond affects the position of the carbonyl absorption much less in acids than it does in aldehydes and ketones. A substantial shift in the carbonyl absorption is observed, however, for acids in which the carboxy group is on an aromatic ring. Benzoic acid, for example, has a carbonyl absorption at 1680 cm^{-1}.

B. NMR Spectroscopy

Many aspects of the NMR spectra of carboxylic acids are illustrated by the NMR spectrum of propanoic acid, shown in Fig. 20.2b. The α-protons of carboxylic acids, like those of aldehydes and ketones, show proton NMR absorptions in the δ 2.0–2.5 chemical shift region. The O—H proton resonances of carboxylic acids occur at positions that depend on both the

(a)

(b)

FIGURE 20.2 The spectra of propanoic acid illustrate typical characteristics of carboxylic acid spectra. (a) IR spectrum of propanoic acid. Notice particularly the very broad O—H stretching absorption. (b) Proton NMR spectrum of propanoic acid. The O—H absorption occurs at very large chemical shift, and the chemical shifts of the other hydrogens are in about the same positions as shifts of the corresponding protons in aldehydes and ketones.

acidity of the acid and its concentration. Typically, the carboxylic acid OH proton resonance occurs at a very large chemical shift, in the δ 9–13 region, and in many cases it is broad. It is readily distinguished from an aldehydic proton because the acid proton, like an alcohol O—H proton, rapidly exchanges with D_2O (Sec. 13.6).

The ^{13}C NMR absorptions of carboxylic acids are similar to those of aldehydes and ketones, although the carbonyl carbon of an acid has a somewhat *smaller* chemical shift than that of an aldehyde or ketone.

This carbonyl shift is contrary to what is expected from the relative electronegativities of oxygen and carbon; electronegative atoms generally cause *greater* chemical shifts. This unusual

FURTHER EXPLORATION 20.1 Chemical Shifts of Carbonyl Carbons

chemical shift is caused by shielding effects of the unshared electron pairs on the carboxylate oxygen.

PROBLEMS

20.4   Give the structure of the compound with molecular mass = 88 and the following spectra.
Proton NMR: δ 1.2 (6H, d, J = 7 Hz); δ 2.5 (1H, septet, J = 7 Hz); δ 10 (1H, broad s)
IR: 2600–3400 cm^{-1} (broad), 1720 cm^{-1}

20.5   Give the structure of the compound $C_7H_5O_2Cl$ that has an IR absorption at 1685 cm^{-1} as well as a strong, broad O—H absorption, and the following proton NMR spectrum: δ 7.56 (2H, leaning d, J = 10 Hz); δ 8.00 (2H, leaning d, J = 10 Hz); δ 8.27 (1H, broad s, exchanges with D_2O).

20.6   Explain how you would distinguish between the isomers α,α-dimethylsuccinic acid and adipic acid by (a) ^{13}C NMR; (b) proton NMR.

20.4   ACID–BASE PROPERTIES OF CARBOXYLIC ACIDS

A. Acidity of Carboxylic and Sulfonic Acids

The acidity of carboxylic acids is one of their most important chemical properties. This acidity is due to ionization of the O—H group.

$$R—C\overset{O}{\underset{\ddot{O}—H}{}} + H_2O \;\rightleftharpoons\; R—C\overset{O}{\underset{\ddot{O}:^-}{}} + H_3O^+ \qquad (20.3)$$

carboxylic acid   carboxylate ion

The conjugate bases of carboxylic acids are called generally **carboxylate ions**. Carboxylate salts are named by replacing the *ic* in the name of the acid (in any system of nomenclature) with the suffix *ate*.

sodium acetate   **potassium benzoate**
(aceti̇c̸ + *ate* = acetate)

STUDY GUIDE LINK 20.1
Reactions of Bases with Carboxylic Acids

Carboxylic acids are among the most acidic organic compounds; acetic acid, for example, has a pK_a of 4.76. This pK_a is low enough that an aqueous solution of acetic acid gives an acid reaction with litmus or pH paper.

Carboxylic acids are more acidic than alcohols or phenols, other compounds with O—H bonds.

Increasing acidity

$CH_3CH_2—O—H$ ⟨phenol⟩—O—H $H_3C—C(=O)—O—H$

pK_a 15.9   9.95   4.76

The acidity of carboxylic acids is due to two factors. First is the *polar effect* of the carbonyl group. The carbonyl group, because of its sp^2-hybridized atoms, the partial positive charge on the carbonyl carbon, and the presence of oxygen, is a very electronegative group, much more electronegative than the phenyl ring of a phenol or the alkyl group of an alcohol. The polar effect of the carbonyl group stabilizes charge in the carboxylate ion. Remember that *stabilization of a conjugate base enhances acidity* (Fig. 3.2, p. 113).

The second factor that accounts for the acidity of carboxylic acids is the resonance stabilization of their conjugate-base carboxylate ions.

resonance structures of the acetate ion

(20.4)

STUDY GUIDE LINK 20.2
Resonance Effect on Carboxylic Acid Acidity

Although typical carboxylic acids have pK_a values in the 4–5 range, the acidities of carboxylic acids vary with structure. Recall, for example (Sec. 3.6C), that halogen substitution within the alkyl group of a carboxylic acid enhances acidity by a polar effect.

$$H_3C—CO_2H \qquad FCH_2—CO_2H \qquad F_2CH—CO_2H \qquad F_3C—CO_2H \qquad (20.5)$$

	acetic acid	**fluoroacetic acid**	**difluoroacetic acid**	**trifluoroacetic acid**
pK_a	4.76	2.66	1.24	0.23

Trifluoroacetic acid, commonly abbreviated TFA, is such a strong acid that it is often used in place of HCl and H_2SO_4 when an acid of moderate strength is required.

The pK_a values of some carboxylic acids are given in Table 20.2, and the pK_a values of the simple dicarboxylic acids in Table 20.3. The data in these tables give some idea of the range over which the acidities of carboxylic acids vary.

Sulfonic acids are much stronger than comparably substituted carboxylic acids.

***p*-toluenesulfonic acid**
(**TsOH**, or **tosic acid**)
a strong acid; $pK_a \approx -3$

TABLE 20.2 pK_a Values of Some Carboxylic Acids

Acid*	pK_a
formic	3.75
acetic	4.76
propionic	4.87
2,2-dimethylpropanoic (pivalic)	5.05
acrylic	4.26
chloroacetic	2.85
phenylacetic	4.31
benzoic	4.18
p-methylbenzoic (*p*-toluic)	4.37
p-nitrobenzoic	3.43
p-chlorobenzoic	3.98
p-methoxybenzoic (*p*-anisic)	4.47
2,4,6-trinitrobenzoic	0.65

* See Table 20.1 for structures.

TABLE 20.3 pK_a Values of Some Dicarboxylic Acids

Acid*	First pK_a	Second pK_a
carbonic	3.77‡	10.33
oxalic	1.27	4.27
malonic	2.86	5.70
succinic	4.21	5.64
glutaric	4.34	5.27
adipic	4.41	5.28
phthalic	2.95	5.41

* See Table 20.1 for structures.

‡ This value, which corrects for the amount of H_2CO_3 in aqueous CO_2, is the actual pK_a of carbonic acid. An often-cited value of 6.4 treats all dissolved CO_2 as H_2CO_3.

One reason that sulfonic acids are more acidic than carboxylic acids is the high oxidation state of sulfur. The octet structure for a sulfonate anion indicates that sulfur has considerable positive charge. This positive charge stabilizes the negative charge on the oxygens.

$$
\left[\ \ce{R-\overset{\displaystyle :\ddot{O}}{\underset{\displaystyle :\ddot{O}}{S}}-\ddot{O}:^-} \ \ \longleftrightarrow \ \ \ce{R-\overset{\displaystyle :\ddot{O}:^-}{\underset{\displaystyle :\ddot{O}:^-}{S^{2+}}}-\ddot{O}:^-} \ \ \right]
$$

expanded octet   octet at sulfur
at sulfur

└──────── a sulfonate anion ────────┘

Sulfonic acids are useful as acid catalysts in organic solvents because they are more soluble than most inorganic acids. For example, *p*-toluenesulfonic acid is moderately soluble in benzene and toluene and can be used as a strong acid catalyst in those solvents. (Sulfuric acid, in contrast, is completely insoluble in benzene and toluene.)

Many carboxylic acids of moderate molecular mass are insoluble in water. Their alkali metal salts, however, are ionic compounds, and in many cases are much more soluble in water (Sec. 8.6F; Eq. 8.15, p. 366). Therefore, many water-insoluble carboxylic acids dissolve in solutions of alkali metal hydroxides (NaOH, KOH) because the insoluble acids are converted completely into their soluble salts.

$$
\ce{R-\overset{\displaystyle O}{\overset{\|}{C}}-\ddot{O}H + Na^+\ ^-OH ->[H2O] R-\overset{\displaystyle O}{\overset{\|}{C}}-\ddot{O}:^-\ Na^+ + H2O} \tag{20.6}
$$

more soluble than the
carboxylic acid in water

Even a 5% sodium bicarbonate ($Na^+\ HCO_3^-$) solution is basic enough (pH ≈ 8.5) to dissolve a carboxylic acid. This statement follows from Eq. 3.31a, p. 107. Remember that the fraction of an acid that is ionized, f_A, is determined by the difference $\Delta = pH - pK_a$. With pH = 8.5 and $pK_a \approx 4.5$, then $\Delta = 4$, and the fraction ionized is essentially 1:

$$
f_A = \frac{1}{1+10^{-\Delta}} = \frac{1}{1+10^{-4}} = 0.9999 \approx 1 \tag{20.7}
$$

Provided that the bicarbonate is in excess (so that it is not completely consumed by the carboxylic acid), the carboxylic acid, then, is completely converted into its conjugate-base anion, which is soluble in water.

A typical carboxylic acid can be separated from mixtures with other water-insoluble, nonacidic substances by extraction with NaOH, Na_2CO_3, or $NaHCO_3$ solution. The acid dissolves in the basic aqueous solution, but nonacidic compounds do not. After separating the basic aqueous solution, it can be acidified with a strong acid to yield the carboxylic acid, which may be isolated by filtration or extraction with organic solvents. (A similar idea was used in the separation of phenols; Sec. 18.7B.) Carboxylic acids can also be separated from phenols by extraction with 5% $NaHCO_3$ if the phenol is not unusually acidic. Because the pK_a of a typical phenol is about 10, it remains largely un-ionized and thus insoluble in an aqueous solution with a pH of 8.5. (This conclusion follows from an equation for phenol ionization analogous to Eq. 20.7.)

PROBLEMS

20.7   (a) Write the equations for the first and second ionizations of succinic acid. Label each with the appropriate pK_a values from Table 20.3.

      (b) Why is the first pK_a value of succinic acid lower than the second pK_a value?

20.8 Imagine that you have just carried out a conversion of *p*-bromotoluene into *p*-bromobenzoic acid and wish to separate the product from the unreacted starting material. Design a separation of these two substances that would enable you to isolate the purified acid starting with a solution of both compounds in methylene chloride.

B. Basicity of Carboxylic Acids

Although we think of carboxylic acids primarily as acids, the carbonyl oxygens of acids, like those of aldehydes or ketones, are weakly basic.

$$R-\overset{\overset{:O:}{\|}}{C}-\ddot{O}H + H_3O^+ \ \rightleftharpoons \ \left[R-\overset{\overset{+:\ddot{O}H}{\|}}{C}-\ddot{O}H \ \longleftrightarrow \ R-C=\overset{+}{O}H \right] + H_2O \quad (20.8)$$

protonated carboxylic acid
p$K_a \approx -6$

The basicity of carboxylic acids plays a very important role in many of their reactions.

Protonation of an acid on the *carbonyl oxygen* occurs because, as Eq. 20.8 shows, a resonance-stabilized cation is formed. Protonation on the *carboxylate oxygen* is much less favorable because it does not give a resonance-stabilized cation and because the positive charge on oxygen is destabilized by the polar effect of the carbonyl group.

$$R-\overset{\overset{+\ddot{O}-H}{\|}}{C}-\ddot{O}H$$

resonance-stabilized
(Eq. 20.8)

$$R-\overset{\overset{:O:}{\|}}{C}-\overset{+}{\underset{\overset{|}{H}}{O}}-H$$

not resonance-stabilized;
does not form

20.5   **FATTY ACIDS, SOAPS, AND DETERGENTS**

Carboxylic acids with long, unbranched carbon chains are called **fatty acids** because many of them are liberated from fats and oils by a hydrolytic process called *saponification* (Sec. 21.7A). Some fatty acids contain carbon–carbon double bonds. Fatty acids with cis double bonds occur widely in nature, but those with trans double bonds are rare. The following compounds are examples of common fatty acids:

$$CH_3(CH_2)_{14}CO_2H \quad \text{or} \quad CH_3CH_2CH_2CH_2CH_2CH_2CH_2CH_2CH_2CH_2CH_2CH_2CH_2CH_2CH_2CO_2H$$

or $\diagdown\diagup\diagdown\diagup\diagdown\diagup\diagdown\diagup\diagdown\diagup\diagdown\diagup\diagdown\diagup$ CO$_2$H

palmitic acid
(from palm oil)

$\diagdown\diagup\diagdown\diagup\diagdown\diagup\diagdown\diagup\diagdown\diagup\diagdown\diagup\diagdown\diagup\diagdown$ CO$_2$H   or $CH_3(CH_2)_{16}CO_2H$

stearic acid
(Greek *stear*, meaning "tallow," or "beef fat")

$$CH_3(CH_2)_7 \diagdown \qquad \diagup (CH_2)_7CO_2H$$
$$C=C \longleftarrow \text{note the cis double bond}$$
$$\diagup \qquad \diagdown$$
$$H \qquad\quad H$$

oleic acid

The sodium and potassium salts of fatty acids, called **soaps**, are the major ingredients of commercial soap.

$$CH_3(CH_2)_{16}-\overset{\overset{\displaystyle O}{\|}}{C}-O^- \quad Na^+$$

sodium stearate
(a soap)

Closely related to soaps are **detergents**. A detergent, like a soap, has a long hydrocarbon tail and an ionic head group, but its polar head group is something other than a carboxylate group. For example, the following compound, the sodium salt of a sulfonic acid, is used in household laundry detergent formulations.

$$Na^+ \quad {}^-O-\overset{\overset{\displaystyle O}{\|}}{\underset{\underset{\displaystyle O}{\|}}{S}}-\hspace{-0.3em}\langle\bigcirc\rangle\hspace{-0.3em}-(CH_2)_{11}CH_3$$

sodium 4-dodecyl-1-benzenesulfonate
(a synthetic detergent)

Many soaps and detergents have not only cleansing properties but also germicidal properties. The basis of these properties is described below.

FURTHER EXPLORATION 20.2
More on Surfactants

Soaps and synthetic detergents are two examples of a larger class of molecules known as **surfactants**, so-called because of their effects on the surface tension of water. (See Further Exploration 20.2 for a more extensive discussion.) Surfactants are molecules with two structural parts that interact with water in different ways: *a polar head group*, which is readily solvated by water, and a *hydrocarbon tail*, which, like a long alkane, is not readily solvated by water. In Sec. 8.7A you learned about phospholipids, the major components of cell membranes, which also have polar head groups and hydrocarbon tails. Phospholipids are also surfactants. In a soap, the polar head group is the carboxylate anion, and the hydrocarbon tail is the carbon chain. The soap and the detergent shown above are examples of *anionic surfactants*—that is, surfactants with an anionic polar head group. *Cationic surfactants* are also known:

benzylcetyldimethylammonium chloride
(benzalkonium chloride)
a cationic detergent and germicide

Although small amounts of soap and detergent molecules dissolve in water, when their concentrations are raised above a certain value, called the **critical micelle concentration (CMC)**, the soap or detergent molecules spontaneously form **micelles**, which are approximately spherical aggregates of 50–150 molecules (Fig. 20.3). Think of a micelle as a large ball in which the polar head groups, along with their counterions, are exposed on the outside of the ball and the nonpolar tails are buried on the inside of the ball. The micellar structure satisfies the solvation requirements of both the polar head groups, which are close to water, and the "greasy groups"—the nonpolar tails—which associate with each other on the inside of the micelle.

The spontaneous assembly of micelles calls to mind the spontaneous formation of phospholipid bilayers in aqueous media (Sec. 8.7A). Both phenomena have a similar cause: the entropy-driven release of solvation water as the nonpolar chains come together (Sec. 8.6D). It is reasonable to ask why soaps and detergents don't form the same sort of bilayers that phospholipids do—or, to turn the question around, why phospholipids form bilayers instead

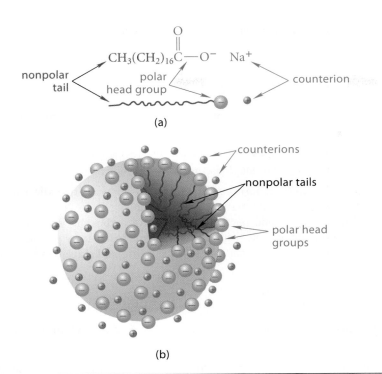

(a)

(b)

FIGURE 20.3 (a) Schematic diagram of a soap. The polar head group (the carboxylate group) is represented by a red sphere, the nonpolar tail by a wiggly line, and the counterion as a blue sphere. (b) A cutaway diagram of a micelle structure. Each micelle contains 50–150 molecules and is approximately spherical. Notice that the polar head groups within the micelle are directed outward toward the solvent water and that the nonpolar tails interact with one another and are isolated from water within the interior of the micelle.

of micelles. Recall that phospholipids contain two long hydrocarbon chains per molecule, whereas soaps and detergents have only one. In a bilayer formed from a soap, the alkyl chains would be relatively far apart, and their favorable van der Waals interactions would be relatively weak. (Van der Waals forces increase strongly with decreasing distance.) In a micelle, the hydrocarbon chains of a soap are packed more closely, and their attractive van der Waals interactions are stronger. On the other hand, in the interior of a micelle derived from a phospholipid, there isn't sufficient space to accommodate the two hydrocarbon chains per molecule. In other words, the hydrocarbon chains of phospholipid molecules would be too close together in a micelle and would experience unfavorable van der Waals interactions (steric repulsions). In the interior of the bilayer, however, there is enough room for the two hydrocarbon chains from different molecules of the phospholipid to interact favorably without steric repulsion.

The use of soaps and detergents in cleaning applications is understandable once the rationale for the formation of micelles is clear. When a fabric with greasy dirt is exposed to an aqueous solution containing micelles of a soap or detergent, the dirt associates with the "greasy" hydrocarbon chains on the interior of the micelle and is incorporated into the micellar aggregate. The dirt is thus lifted away from the surface of the fabric and carried into solution. The antiseptic action of some detergents owes its success to a similar phenomenon. When the bacterial cell is exposed to a solution containing a detergent, phospholipids of the cell membrane tend to associate with the detergent. In some cases this disrupts the membrane enough that the cell can no longer function, and it dies.

So-called *hard water*, which contains Ca^{2+} and Mg^{2+} ions, interferes with the cleaning action of soaps. Hard-water scum ("bathtub ring") is a precipitate of the calcium or magnesium salts of fatty acids, which (unlike the sodium and potassium salts) do not form micelles because they are too insoluble in water. These offending ions can be solubilized and removed by complexation with phosphates. Phosphates, however, have been found to cause excessive growth of algae in rivers and streams, and their use has been curtailed (thus, "low-phosphate detergents"). Unfortunately, no completely acceptable substitute for phosphate-containing detergents has yet been found.

Surfactants are used not only in "soap-and-water"-type cleaning operations. They are also extremely important as components of fuels and lubricating oils. For example, detergents in engine oils assist in keeping deposits suspended in the oil and thus prevent them from building up on engine surfaces.

20.6 SYNTHESIS OF CARBOXYLIC ACIDS

Two reactions covered in previous chapters are especially important for the preparation of carboxylic acids:

1. oxidation of primary alcohols and aldehydes (Secs. 10.6B and 19.14)
2. side-chain oxidation of alkylbenzenes (Sec. 17.5C)

The ozonolysis of alkenes (Sec. 5.5) can also be used to prepare carboxylic acids, although it is less important because it breaks carbon–carbon bonds.

Another important method for the preparation of carboxylic acids is the reaction of Grignard or organolithium reagents with carbon dioxide, followed by protonolysis. Typically the reaction is run by pouring an ether solution of the Grignard reagent over crushed dry ice.

$$
\underset{\textbf{\textit{sec}-butylmagnesium bromide}}{CH_3CH_2\overset{\overset{\displaystyle CH_3}{|}}{C}HMgBr} + CO_2 \longrightarrow \xrightarrow{H_3O^+} \underset{\substack{\textbf{2-methylbutanoic acid} \\ \text{(76–86\% yield)}}}{CH_3CH_2\overset{\overset{\displaystyle CH_3}{|}}{C}H-\overset{\overset{\displaystyle O}{\|}}{C}-OH} \qquad (20.9)
$$

Carbon dioxide is itself a carbonyl compound. The mechanism of this reaction is much like that for Grignard additions to other carbonyl compounds (Sec. 19.9). Addition of the Grignard reagent to carbon dioxide gives the bromomagnesium salt of a carboxylic acid. When aqueous acid is added to the reaction mixture in a separate reaction step, the neutral carboxylic acid is formed.

$$
\ddot{O}=C=\ddot{O} \longrightarrow \underset{\substack{\text{bromomagnesium salt} \\ \text{of a carboxylic acid}}}{R-\overset{\overset{\displaystyle :O:}{\|}}{C}-\ddot{O}:^- \; {}^+MgBr} \xrightarrow{H_3O^+} \underset{\substack{\text{a carboxylic} \\ \text{acid}}}{R-\overset{\overset{\displaystyle :O:}{\|}}{C}-\ddot{O}H} + Mg^{2+} + Br^- \qquad (20.10)
$$

Notice that the reaction of Grignard reagents with CO_2, unlike the other reactions listed at the beginning of this section, is another method for forming carbon–carbon bonds. (Be sure to review the others; Appendix VI.)

All of the methods used for preparing carboxylic acids are summarized in Appendix V. A number of these are discussed in Chapters 21 and 22.

PROBLEM

20.9 Outline a synthetic scheme for each of the following transformations.
(a) cyclopentanecarboxylic acid from cyclopentanol
(b) octanoic acid from 1-heptene

20.7 INTRODUCTION TO CARBOXYLIC ACID REACTIONS

The reactions of carboxylic acids can be categorized into four types.

1. reactions at the carbonyl group
2. reactions at the carboxylate oxygen
3. loss of the carboxy group as CO_2 (decarboxylation)
4. reactions involving the α-carbon

The most typical reaction at the carbonyl group is *substitution at the carbonyl carbon.* Let EY be a general reagent in which E is an electrophilic group (for example, a hydrogen) and Y is a nucleophilic group. Typically the —OH of the carboxy group is substituted by the group —Y.

$$R-\overset{\overset{\displaystyle O}{\|}}{C}-OH + E-Y \ \rightleftarrows \ R-\overset{\overset{\displaystyle O}{\|}}{C}-Y + HO-E \qquad (20.11)$$

Another reaction at the carbonyl group of carboxylic acids and their derivatives is reaction of the carbonyl oxygen with an electrophile (a Lewis acid or a Brønsted acid)—that is, the reaction of the carbonyl oxygen as a *base*:

$$R-\overset{\overset{\displaystyle :O:}{\|}}{C}-\overset{..}{O}H \ \rightleftarrows \ \left[R-\overset{\overset{\displaystyle \overset{+}{O}-E}{\|}}{C}-\overset{..}{O}H \ \longleftrightarrow \ R-\overset{\overset{\displaystyle :\overset{..}{O}-E}{|}}{C}=\overset{+}{O}H \right] + Y:^- \qquad (20.12)$$

One such reaction, protonation of the carbonyl oxygen, was discussed in Sec. 20.4B. Many substitution reactions at the carbonyl carbon are acid-catalyzed; that is, the reactions of nucleophiles at the carbonyl *carbon* are catalyzed by the reactions of acids at the carbonyl *oxygen.*

We've already studied one *reaction at the carboxylate oxygen*: the ionization of carboxylic acids (Sec. 20.4A):

$$R-\overset{\overset{\displaystyle :O:}{\|}}{C}-\overset{..}{O}-H + H_2\overset{..}{O}: \ \rightleftarrows \ \left[R-\overset{\overset{\displaystyle :O:}{\|}}{C}-\overset{..}{O}:^- \ \longleftrightarrow \ R-\overset{\overset{\displaystyle :\overset{..}{O}:^-}{|}}{C}=\overset{..}{O}: \right] + H_3\overset{+}{O}: \qquad (20.13)$$

Another general reaction involves reaction of the carboxylate oxygen as a nucleophile ($Y:^- =$ halide, sulfonate ester, or other leaving group).

$$R-\overset{\overset{\displaystyle O}{\|}}{C}-\overset{..}{O}:^- + E-Y \ \longrightarrow \ R-\overset{\overset{\displaystyle O}{\|}}{C}-\overset{..}{O}-E + Y:^- \qquad (20.14)$$

Decarboxylation is loss of the carboxy group as CO_2.

$$R-\overset{\overset{\displaystyle O}{\|}}{C}-O-H \ \longrightarrow \ R-H + O=C=O \qquad (20.15)$$

This reaction is more important for some types of carboxylic acids than for others (Sec. 20.11) and is biologically important.

This chapter concentrates on the first three types of reactions: reactions at the carbonyl group, reactions at the carboxylate oxygen, and decarboxylation. For the most part, Chapter 21 considers reactions at the carbonyl group of carboxylic acid *derivatives.* Chapter 22 takes up the fourth type of reaction—reactions involving the α-carbon—for both carboxylic acids and their derivatives.

20.8 CONVERSION OF CARBOXYLIC ACIDS INTO ESTERS

A. Acid-Catalyzed Esterification

Carboxylate esters are carboxylic acid derivatives with the following general structure:

$$R-\overset{\overset{\displaystyle O}{\|}}{C}-O-R' \qquad (R = H, \text{ alkyl, or aryl}; R' = \text{alkyl or aryl})$$

general structure
of a carboxylate ester

As you learned in Secs 10.4A and 10.4C, ester derivatives of other types of acids are well known. However, in this section, we'll use the simpler and more general term *ester* to mean a carboxylate ester.

When a carboxylic acid is treated with a large excess of an alcohol in the presence of a strong acid catalyst, an ester is formed.

$$\text{benzoic acid} + \underset{\substack{\text{methanol} \\ \text{(large excess;} \\ \text{used as solvent)}}}{\text{CH}_3\text{OH}} \xrightarrow{\text{H}_2\text{SO}_4} \underset{\substack{\text{methyl benzoate} \\ \text{(an ester; 85–95\% yield)}}}{} + \text{H}_2\text{O} \qquad (20.16)$$

This reaction is called **acid-catalyzed esterification**, or sometimes **Fischer esterification**, after the renowned German chemist Emil Fischer (1852–1919).

The equilibrium constants for esterifications with most primary alcohols, although favorable, are not large; for example, the equilibrium constant for the esterification of acetic acid with ethyl alcohol is 3.38. The reaction is driven to completion by using the reactant alcohol as the solvent. Because the alcohol is present in large excess, the equilibrium is driven toward the ester product. This is another application of Le Châtelier's principle (Sec. 4.9B).

Acid-catalyzed esterification *cannot* be applied to the synthesis of esters from phenols or tertiary alcohols. Tertiary alcohols undergo dehydration (Sec. 10.2) and other reactions under the acidic conditions of the reaction, and the equilibrium constants for the esterification of phenols are much less favorable than those for the esterification of alcohols by a factor of about 10^4. Although it is possible in principle to drive the esterification of phenols to completion, there are simpler ways for preparing esters of both phenols and tertiary alcohols that are discussed in Chapter 21.

A very important question relevant to the mechanism of acid-catalyzed esterification is whether the oxygen of the water liberated in the reaction comes from the carboxylic acid or the alcohol.

$$\text{Ph}-\overset{\overset{\displaystyle O}{\|}}{\text{C}}-\text{OH} + \text{H}_3\text{C}-\text{OH} \xrightarrow{?} \begin{cases} \text{Ph}-\overset{\overset{\displaystyle O}{\|}}{\text{C}}-\text{OCH}_3 + \text{H}_2\text{O} & (20.17a) \\ \text{or} \\ \text{Ph}-\overset{\overset{\displaystyle O}{\|}}{\text{C}}-\text{OCH}_3 + \text{H}_2\text{O} & (20.17b) \end{cases}$$

This question was answered in 1938, when it was found, using the ^{18}O isotope to label the alcohol oxygen, that the OH of the water produced comes exclusively from the carboxylic acid. Acid-catalyzed esterification is therefore a *substitution of* —OH *at the carbonyl group of the acid by the oxygen of the alcohol* (Eq. 20.17a). Thus, acid-catalyzed esterification is an example of *substitution at a carbonyl carbon.*

The mechanism of acid-catalyzed esterification is very important because it serves as a model for the mechanisms of other acid-catalyzed reactions of carboxylic acids and their derivatives. In the mechanism that follows, the formation of a methyl ester in the solvent methanol is shown. The first step of the mechanism is protonation of the carbonyl oxygen (Sec. 20.4B):

$$\underset{\substack{\text{conjugate acid} \\ \text{of the solvent}}}{} \rightleftharpoons \underset{\substack{\text{protonated} \\ \text{carboxylic acid}}}{} + \ddot{\text{H}}\ddot{\text{O}}\text{CH}_3 \qquad (20.18a)$$

The catalyzing acid is the conjugate acid of the solvent; this is the actual acid present when a strong acid such as H_2SO_4 is dissolved in methanol.

Recall from Sec. 19.7A that *protonation of a carbonyl oxygen makes the carbonyl carbon more electrophilic* because the carbonyl oxygen becomes a better electron acceptor. The carbonyl carbon of a protonated carbonyl group is electrophilic enough to react with the weakly basic methanol molecule. A nucleophilic reaction of methanol at the carbonyl carbon, followed by loss of a proton, gives a *tetrahedral addition intermediate*.

$$\text{(20.18b)}$$

A **tetrahedral addition intermediate** is simply the product of carbonyl addition. In aldehyde and ketone reactions, the product of carbonyl addition is in many cases a stable compound that can be isolated. (See, for example, Eq. 19.15 on p. 963.) In the case of carboxylic acid derivatives, it is called an *intermediate* because it reacts further, as we shall see. In esterification, formation of the tetrahedral addition intermediate is essentially the same reaction as the acid-catalyzed reaction of an alcohol with a protonated aldehyde or ketone to form a hemiacetal (Sec. 19.10A). The tetrahedral addition intermediate, after protonation, loses water to give the conjugate acid of the ester:

$$\text{(20.18c)}$$

Loss of a proton gives the ester product and regenerates the acid catalyst.

$$\text{(20.18d)}$$

The *mechanism of esterification is an extension of the mechanism of carbonyl addition.* In esterification, a nucleophile approaches the carbonyl carbon from above or below the plane of the carbonyl group and interacts with the π^* (antibonding) molecular orbital, just as in the addition reactions of aldehydes and ketones (Fig. 19.8, p. 965). Reaction of the nucleophile at the carbonyl carbon gives the addition compound. In esterification, however, the addition compound—the *tetrahedral addition intermediate*—reacts further; the —OH group from the carboxylic acid, after protonation, is expelled from the addition compound, and a carboxylic acid derivative—an ester in this case—is formed.

Esterification illustrates a very general mechanistic pattern that occurs in many substitution reactions of carboxylic acids and their derivatives. An addition intermediate is formed, and a leaving group —X is expelled from this intermediate to give a new carboxylic acid derivative.

$$
\underset{\substack{}}{\text{R}-\overset{\overset{\displaystyle O}{\|}}{\text{C}}-\text{X}} + \text{Y}-\text{H} \longrightarrow \underset{\substack{\text{tetrahedral} \\ \text{addition intermediate}}}{\text{R}-\overset{\overset{\displaystyle OH}{|}}{\underset{\underset{\displaystyle Y}{|}}{\text{C}}}-\text{X}} \longrightarrow \text{R}-\overset{\overset{\displaystyle O}{\|}}{\text{C}}-\text{Y} + \text{X}-\text{H} \qquad (20.19a)
$$

a leaving group

In other words, substitution at a carbonyl carbon is really a sequence of two processes: *addition* to the carbonyl group followed by *elimination* to regenerate the carbonyl group.

Although esterification is catalyzed only by acids, a number of other carbonyl-substitution reactions, like carbonyl-addition reactions, are base-catalyzed. Several reactions of this type are discussed in Chapters 21 and 22.

Why don't aldehydes and ketones undergo substitution at their carbonyl carbons? After a nucleophile reacts at the carbonyl carbon of an aldehyde or ketone, *neither of the groups attached to the carbonyl carbon can be a leaving group.*

$$
\text{R}-\overset{\overset{\displaystyle O}{\|}}{\text{C}}-\text{R} + \text{Y}-\text{H} \;\rightleftarrows\; \text{R}-\overset{\overset{\displaystyle OH}{|}}{\underset{\underset{\displaystyle Y}{|}}{\text{C}}}-\text{R} \qquad (20.19b)
$$

cannot be leaving groups

**FURTHER
EXPLORATION 20.3**
Orthoesters

The reason is that the H— or the R— group of an aldehyde or ketone, to be a leaving group, would be expelled as H:⁻ or R:⁻, respectively, either of which is a *very strong* base. In carbonyl-substitution reactions, as in S_N2 and S_N1 reactions, the best leaving groups are generally the *weakest* bases (Secs. 9.4F and 9.6C). In contrast, one of the groups attached to the carbonyl carbon of a carboxylic acid derivative is either a weak base, or (as in the case of esterification, Eq. 20.18c) is converted by protonation into a weak base; in either case, such a group can act as a good leaving group, and substitution results.

PROBLEMS

20.10   Give the structure of the product formed when

(a) 3-methylhexanoic acid is heated with a large excess of ethanol (as solvent) with a sulfuric acid catalyst.

(b) adipic acid (Table 20.1, p. 1006) is heated in a large excess of 1-propanol (as solvent) with a sulfuric acid catalyst.

20.11   (a) Using the principle of microscopic reversibility (Sec. 4.9B, p. 175), give a detailed mechanism for the acid-catalyzed hydrolysis of methyl benzoate (structure in Eq. 20.16, p. 1020) to benzoic acid and methanol.

(b) Given that ester formation is reversible, what reaction conditions would you use to bring about the acid-catalyzed hydrolysis of methyl benzoate to benzoic acid and methanol?

20.12   (a) You learned in Sec. 19.7 that carbonyl-addition reactions can occur under basic conditions. The hydrolysis of methyl benzoate is also promoted by ⁻OH. Write a mechanism for the hydrolysis of methyl benzoate in NaOH solution.

(b) A student has suggested the following transformation, arguing that it can be driven to completion with a large excess of methanol and sodium methoxide.

$$
\underset{}{\text{Ph}-\overset{\overset{\displaystyle O}{\|}}{\text{C}}-\text{OH}} + \underset{\text{(large excess)}}{\text{CH}_3\text{OH}} \;\underset{}{\overset{\text{CH}_3\text{O}^-}{\rightleftarrows}}\; \text{Ph}-\overset{\overset{\displaystyle O}{\|}}{\text{C}}-\text{OCH}_3 + \text{H}_2\text{O}
$$

In fact, this reaction does not occur because, under the basic conditions, the carboxylic acid undergoes a different reaction. What is that reaction?

B. Esterification by Alkylation

The esterification discussed in the previous section involves the reaction of a nucleophile at the carbonyl carbon. This section considers a different method of forming esters that illustrates another mode of carboxylic acid reactivity: nucleophilic reactivity of the *carboxylate oxygen* (Eq. 20.14).

When a carboxylic acid is treated with diazomethane in ether solution, it is rapidly converted into its methyl ester.

(E)-2-octenoic acid **methyl (E)-2-octenoate**
(91% yield)   (20.20)

Diazomethane, a toxic yellow gas (bp $= -23$ °C), is usually generated chemically as it is needed from a commercially available precursor and is co-distilled with ether into a flask containing the carboxylic acid to be esterified. Diazomethane is both explosive and allergenic and is therefore only used in small quantities under conditions that have been carefully established to maintain safety. Nevertheless, esterification with diazomethane is so mild and free of side reactions that in many cases it is the method of choice for the synthesis of methyl esters in small-scale reactions.

The acidity of the carboxylic acid is important in the mechanism of this reaction. Protonation of diazomethane by the carboxylic acid gives the methyldiazonium ion.

methyldiazonium ion

This ion contains dinitrogen, one of the best leaving groups. An S_N2 reaction of the methyldiazonium ion with the carboxylate oxygen results in the displacement of N_2 and formation of the ester.

Carboxylate ions are less basic, and therefore less nucleophilic, than alkoxides or phenoxides, but they do react with especially reactive alkylating agents. The methyldiazonium ion formed by the protonation of diazomethane is one of the most reactive alkylating agents known. Notice, though, that the carboxylic acid, not the carboxylate salt, is required for the reaction with diazomethane because protonation of diazomethane by the acid is the first step of the reaction.

The nucleophilic reactivity of carboxylates is also illustrated by the reaction of certain alkyl halides with carboxylate ions.

2-acetylbenzoic acid   **methyl 2-acetylbenzoate**
(65% yield)

This is an S_N2 reaction in which the carboxylate ion, formed by the acid–base reaction of the acid and K_2CO_3, reacts as a nucleophile with the alkyl halide. Because carboxylate ions are such weak nucleophiles, this reaction works best on alkyl halides that are especially reactive in S_N2 reactions, such as methyl iodide and benzylic or allylic halides (Sec. 17.4). This reaction is typically carried out in polar aprotic solvents that accelerate S_N2 reactions, such as acetone, as Eq. 20.22 illustrates.

Let's contrast the esterification reactions in Eqs. 20.20 and 20.22 with the acid-catalyzed esterification in Sec. 20.8A. In all of the reactions discussed in this section, the carboxylate oxygen of the acid acts as a nucleophile. This oxygen is alkylated by an alkyl halide or diazomethane. In acid-catalyzed esterification, the carbonyl carbon, after protonation of the carbonyl oxygen, acts as an electrophile (a Lewis acid). The nucleophile in acid-catalyzed esterification is the oxygen atom of the solvent alcohol molecule.

PROBLEMS

20.13 Give the structure of the ester formed when

(a) isobutyric acid reacts with diazomethane in ether.

(b) succinic acid reacts with a large excess of diazomethane in ether.

(c) isobutyric acid reacts with benzyl bromide and K_2CO_3 in acetone.

(d) benzoic acid reacts with allyl bromide and K_2CO_3 in acetone.

20.14 *Tert*-butyl esters can be prepared by the acid-catalyzed reaction of methylpropene (isobutylene) with carboxylic acids.

Suggest a mechanism for this reaction that accounts for the role of the acid catalyst. (*Hint:* See Sec. 4.9B.)

20.9 CONVERSION OF CARBOXYLIC ACIDS INTO ACID CHLORIDES AND ANHYDRIDES

Acid chlorides and acid anhydrides are very reactive carboxylic acid derivatives that play an important role in the synthesis of other carboxylic acid derivatives, such as esters and amides. This section shows how these derivatives are prepared from carboxylic acids. Their use in the synthesis of other carboxylic acid derivatives is discussed in Chapter 21 starting in Sec. 21.8.

A. Synthesis of Acid Chlorides

Acid chlorides are carboxylic acid derivatives with the following general structure:

general structure
of an acid chloride

Acid chlorides are invariably prepared from carboxylic acids. Two reagents used for this purpose are thionyl chloride, $SOCl_2$, and phosphorus pentachloride, PCl_5.

$$CH_3CH_2CH_2\overset{\displaystyle O}{\overset{\|}{C}}-OH + SOCl_2 \longrightarrow CH_3CH_2CH_2\overset{\displaystyle O}{\overset{\|}{C}}-Cl + HCl + SO_2 \qquad (20.23)$$

butyric acid **thionyl chloride** **butyryl chloride** (an acid chloride; 85% yield)

$$O_2N-\overset{\displaystyle O}{\underset{}{\overset{\|}{C}}}-OH \;+\; PCl_5 \longrightarrow O_2N-\overset{\displaystyle O}{\underset{}{\overset{\|}{C}}}-Cl + POCl_3 + HCl \qquad (20.24)$$

p-nitrobenzoic acid **phosphorus pentachloride** **p-nitrobenzoyl chloride** (90–96% yield)

Acid chloride synthesis fits the general pattern of substitution at a carbonyl group; in this case, —OH is substituted by —Cl. (Study Guide Link 20.3 provides the mechanistic details.)

$$R-\overset{\displaystyle O}{\overset{\|}{C}}-OH \xrightarrow{\boxed{\text{—Cl is substituted for —OH}}} R-\overset{\displaystyle O}{\overset{\|}{C}}-Cl \qquad (20.25)$$

STUDY GUIDE LINK 20.3
Mechanism of Acid Chloride Formation

Thionyl chloride is the same reagent used for making alkyl chlorides from alcohols (Sec. 10.4D), a reaction in which —OH is replaced by —Cl at the carbon of an alkyl group.

Recall that acid chlorides are one of the starting materials in the Friedel–Crafts acylation reaction (Sec. 16.4F), which is used for the preparation of aromatic ketones. We'll find in Chapter 21 that acid chlorides are very reactive; for this reason, they are also very useful for the synthesis of other carbonyl compounds.

STUDY GUIDE LINK 20.4
More on Synthetic Equivalents

Sulfonyl chlorides, the acid chlorides of sulfonic acids, are prepared by the treatment of sulfonic acids or their sodium salts with PCl_5.

$$H_3C-\overset{\displaystyle O}{\underset{\displaystyle O}{\overset{\|}{\underset{\|}{S}}}}-O^-\,Na^+ \;+\; PCl_5 \longrightarrow H_3C-\overset{\displaystyle O}{\underset{\displaystyle O}{\overset{\|}{\underset{\|}{S}}}}-Cl + POCl_3 + NaCl \qquad (20.26)$$

sodium methanesulfonate **phosphorus pentachloride** **methanesulfonyl chloride** (85% yield)

Aromatic sulfonyl chlorides can be prepared directly by the reaction of aromatic compounds with chlorosulfonic acid.

$$\text{benzene} + 2Cl-\overset{\displaystyle O}{\underset{\displaystyle O}{\overset{\|}{\underset{\|}{S}}}}-OH \xrightarrow{20-25\,°C} \text{benzene}-\overset{\displaystyle O}{\underset{\displaystyle O}{\overset{\|}{\underset{\|}{S}}}}-Cl + H_2SO_4 + HCl \qquad (20.27)$$

benzene **chlorosulfonic acid** **benzenesulfonyl chloride** (73–77% yield)

This reaction is a variation of aromatic sulfonation, an electrophilic aromatic substitution reaction (Sec. 16.4D). Chlorosulfonic acid, the acid chloride of sulfuric acid, acts as an electrophile in this reaction just as SO_3 does in sulfonation.

$$\text{benzene}-H + Cl-SO_3H \longrightarrow \text{benzene}-SO_3H + HCl \qquad (20.28a)$$

benzene **chlorosulfonic acid** **benzenesulfonic acid**

The sulfonic acid produced in the reaction is converted into the sulfonyl chloride by reaction with another equivalent of chlorosulfonic acid.

$$\underset{\substack{\textbf{benzenesulfonic}\\\textbf{acid}}}{\text{Ph—S(=O)(=O)—OH}} + \underset{\text{(excess)}}{\text{Cl—SO}_3\text{H}} \longrightarrow \underset{\substack{\textbf{benzenesulfonyl}\\\textbf{chloride}}}{\text{Ph—S(=O)(=O)—Cl}} + \underset{\textbf{sulfuric acid}}{\text{HO—SO}_3\text{H}} \quad (20.28b)$$

This part of the reaction is analogous to the reaction of a carboxylic acid with thionyl chloride (Eq. 20.23).

PROBLEMS

20.15   Draw the structures of the acid chlorides derived from (a) 2-methylbutanoic acid; (b) *p*-methoxybenzoic acid; (c) 1-propanesulfonic acid.

20.16   Give the structure of the acid chloride formed in each of the following transformations.

(a) sodium ethanesulfonate + PCl$_5$ $\longrightarrow$

(b) benzoic acid + SOCl$_2$ $\longrightarrow$

(c) *p*-toluenesulfonic acid + excess chlorosulfonic acid $\longrightarrow$

20.17   Outline a synthesis of the following compound from benzoic acid and any other reagents.

$$\text{CH}_3\text{O}-\!\!\!\langle\text{C}_6\text{H}_4\rangle\!\!\!-\overset{\overset{\textstyle O}{\|}}{\text{C}}-\langle\text{C}_6\text{H}_5\rangle$$

(*Hint:* See Study Guide Link 20.4.)

B. Synthesis of Anhydrides

Carboxylic acid *anhydrides* have the following general structure:

$$\underset{\substack{\text{general structure}\\\text{of an anhydride}}}{\text{R}-\overset{\overset{\textstyle O}{\|}}{\text{C}}-\text{O}-\overset{\overset{\textstyle O}{\|}}{\text{C}}-\text{R}} \quad (\text{R} = \text{H, alkyl, or aryl})$$

The name *anhydride*, which means "without water," comes from the fact that an anhydride reacts with water to give two equivalents of a carboxylic acid.

$$\underset{\text{an anhydride}}{\text{R}-\overset{\overset{\textstyle O}{\|}}{\text{C}}-\text{O}-\overset{\overset{\textstyle O}{\|}}{\text{C}}-\text{R}} + \text{H}_2\text{O} \longrightarrow \text{R}-\overset{\overset{\textstyle O}{\|}}{\text{C}}-\text{OH} + \text{HO}-\overset{\overset{\textstyle O}{\|}}{\text{C}}-\text{R} \quad (20.29)$$

The name *anhydride* also graphically describes one of the ways that anhydrides are prepared: the treatment of carboxylic acids with strong dehydrating agents.

$$2\,\underset{\substack{\textbf{trifluoroacetic}\\\textbf{acid}}}{\text{F}_3\text{C}-\overset{\overset{\textstyle O}{\|}}{\text{C}}-\text{OH}} + \underset{\substack{\textbf{phosphorus}\\\textbf{pentoxide}}}{\text{P}_2\text{O}_5} \longrightarrow \underset{\substack{\textbf{trifluoroacetic anhydride}\\(74\%\text{ yield})}}{\text{F}_3\text{C}-\overset{\overset{\textstyle O}{\|}}{\text{C}}-\text{O}-\overset{\overset{\textstyle O}{\|}}{\text{C}}-\text{CF}_3} + \text{complex phosphates} \quad (20.30)$$

Phosphorus pentoxide (actual formula P_4O_{10}) is a white powder that rapidly absorbs, and reacts violently with, water. It is also used as a potent desiccant. This compound is a complex anhydride of phosphoric acid, because it gives phosphoric acid when it reacts with an excess of water.

Most anhydrides may themselves be used to form other anhydrides. For example, P_2O_5 is an inorganic anhydride that is used to form anhydrides of carboxylic acids (Eq. 20.30). In the following example, a dicarboxylic acid reacts with acetic anhydride to form a *cyclic anhydride*—a compound in which the anhydride group is part of a ring:

β-methylglutaric acid

acetic anhydride

β-methylglutaric anhydride
(>90% yield; a cyclic anhydride)   (20.31)

Phosphorus oxychloride ($POCl_3$) and P_2O_5 (Eq. 20.30) can also be used for the formation of cyclic anhydrides. Cyclic anhydrides containing five- and six-membered anhydride rings are readily prepared from their corresponding dicarboxylic acids. Compounds containing either larger or smaller anhydride rings generally cannot be prepared in this way. Formation of cyclic anhydrides with five- and six-membered rings is so facile that in some cases it occurs on heating the dicarboxylic acid.

phthalic acid   **phthalic anhydride**

Cyclic anhydride formation is another example of a *proximity effect* (Sec. 11.8A,B). These reactions are accelerated because they are intramolecular. Furthermore, they have a particularly favorable $\Delta G°$ for the same reason. The equilibrium benefits further from a positive $\Delta S°$ component of both reactions: the release of two molecules of acetic acid in Eq. 20.31, and the expulsion of a water molecule (removed as steam from the melt) in Eq. 20.32.

The formation of anhydrides from carboxylic acids, like many other carboxylic acid reactions that have been discussed, fits the pattern of substitution at the carbonyl carbon: the —OH of one carboxylic acid molecule is substituted by the *acyloxy group* (red) of another. (The mechanistic details are provided in Study Guide Link 20.5.)

acyloxy
group

**STUDY GUIDE
LINK 20.5**
Mechanism of
Anhydride Formation

As we will find in Sec. 21.8B, anhydrides, like acid chlorides, are used in the synthesis of other carboxylic acid derivatives.

PROBLEMS

20.18   Give the structures of the products formed when (a) chloroacetic acid and (b) *p*-chlorobenzoic acid react with P_2O_5.

20.19   (a) Fumaric and maleic acids (Table 20.1) are *E,Z*-isomers. One forms a cyclic anhydride on heating and one does not. Which one forms the cyclic anhydride? Explain.

   (b) Which one of the following compounds forms a cyclic anhydride on heating: methylmalonic acid or 2,3-dimethylbutanedioic acid?

20.10   REDUCTION OF CARBOXYLIC ACIDS TO PRIMARY ALCOHOLS

When a carboxylic acid is treated with lithium aluminum hydride, $LiAlH_4$, then dilute acid, a primary alcohol is formed.

$$2\,CH_3CH_2\overset{CH_3}{\underset{}{CH}}\overset{O}{\overset{\|}{C}}\!-\!OH \;+\; LiAlH_4 \;\;\xrightarrow[\text{ether}]{}\;\; \xrightarrow{H_3O^+}\;\; 2\,CH_3CH_2\overset{CH_3}{\underset{}{CH}}CH_2\!-\!OH \qquad (20.34)$$

2-methylbutanoic acid   **2-methyl-1-butanol**
                                              (83% yield)

This is an important method for the preparation of primary alcohols.

Before the reduction itself takes place, $LiAlH_4$, a source of the very strongly basic hydride ion ($H:^-$), reacts with the acidic hydrogen of the carboxylic acid to give the lithium salt of the carboxylic acid and one equivalent of hydrogen gas (that is, dihydrogen).

$$R\!-\!\overset{\ddot{O}:}{\overset{\|}{C}}\!-\!\ddot{O}\!-\!H \quad H\!-\!\bar{A}lH_3 \;\longrightarrow\; R\!-\!\overset{:\ddot{O}:^-\;\;Li^+}{\underset{}{C}}\!=\!O \;+\; H_2 \;+\; AlH_3 \qquad (20.35a)$$

                                              dihydrogen

The lithium salt of the carboxylic acid is the species that is actually reduced.

The reduction occurs in two stages. In the first stage, the AlH_3 formed in Eq. 20.35a reduces the carboxylate ion to an aldehyde. The aldehyde is rapidly reduced further to give, after protonolysis, the primary alcohol (Sec. 19.8A).

$$R\!-\!\overset{O}{\overset{\|}{C}}\!-\!O^-\;\;Li^+ \;\;\xrightarrow[\substack{\text{(from}\\ \text{Eq. 20.35a)}}]{AlH_3}\;\; R\!-\!\overset{O}{\overset{\|}{C}}\!-\!H \;\;\xrightarrow{LiAlH_4}\;\; \xrightarrow{H_3O^+}\;\; R\!-\!CH_2OH \qquad (20.35b)$$

Because the aldehyde is more reactive than the carboxylate salt, it *cannot* be isolated. (The reactivity differences among the different carbonyl compounds, and the reasons for them, are considered in Sec. 21.7E.)

Notice that the $LiAlH_4$ reduction of a carboxylic acid incorporates two different types of carbonyl reactions. The first is a net *substitution* at the carbonyl carbon to give the aldehyde intermediate. The second is an *addition* to the aldehyde.

FURTHER EXPLORATION 20.4
Mechanism of the $LiAlH_4$ Reduction of Carboxylic Acids

$$R\!-\!\overset{O}{\overset{\|}{C}}\!-\!OH \;\;\xrightarrow{\boxed{\text{substitution}}}\;\; R\!-\!\overset{O}{\overset{\|}{C}}\!-\!H \;\;\xrightarrow{\boxed{\text{addition}}}\;\; R\!-\!\overset{OH}{\underset{H}{\overset{|}{C}}}\!-\!H \qquad (20.36)$$

Many of the reactions of carboxylic acid derivatives discussed in Chapter 21 also fit the same pattern of substitution followed by addition.

Sodium borohydride, $NaBH_4$, another important hydride reducing agent (Sec. 19.8A), does *not* reduce carboxylic acids, although it does react with the acidic hydrogens of acids in a manner analogous to Eq. 20.35a. This selectivity of $NaBH_4$ allows the reduction of aldehydes and ketones in the presence of carboxylic acids.

the carboxylic acid group is unaffected

4-formylbenzoic acid

4-(hydroxymethyl)benzoic acid
(75% yield)

(20.37)

The $LiAlH_4$ reduction of carboxylic acids can be combined with the Grignard synthesis of carboxylic acids (Sec. 20.6) to provide a one-carbon chain extension of carboxylic acids, as illustrated by Study Problem 20.2.

STUDY PROBLEM 20.2

Fatty acids containing an even number of carbon atoms are readily obtained from natural sources, but those containing an odd number of carbons are relatively rare. Outline a synthesis of the rare tridecanoic acid, $CH_3(CH_2)_{10}CH_2CO_2H$, from the readily available lauric acid (see Table 20.1).

SOLUTION The problem requires the synthesis of a carboxylic acid with the addition of one carbon atom to a carbon chain. The Grignard synthesis of carboxylic acids (Sec. 20.6) will accomplish this objective:

$$CH_3(CH_2)_{10}CH_2Br \xrightarrow[\text{2) } CO_2, \text{ then } H_3O^+]{\text{1) Mg, ether}} CH_3(CH_2)_{10}CH_2CO_2H$$

1-bromododecane   **tridecanoic acid**

The required alkyl bromide, 1-bromododecane, comes from treatment of the corresponding alcohol with concentrated HBr; and the alcohol comes, in turn, from the $LiAlH_4$ reduction of lauric acid:

$$CH_3(CH_2)_{10}CO_2H \xrightarrow[\text{2) } H_3O^+]{\text{1) } LiAlH_4 \text{ in ether}} CH_3(CH_2)_{10}CH_2OH \xrightarrow{\text{conc. HBr}} CH_3(CH_2)_{10}CH_2Br$$

lauric acid   **1-dodecanol**   **1-bromododecane**

PROBLEMS

20.20 In each case, give the structure of a compound with the indicated formula that would give the following diol in a $LiAlH_4$ reduction followed by protonolysis.

$$HOCH_2 \diagdown \bigcirc \diagdown CH_2OH$$

(a) $C_8H_6O_3$ (b) $C_8H_6O_4$

20.21 Propose reaction sequences for each of the following conversions:
(a) benzoic acid into phenylacetic acid, $PhCH_2CO_2H$
(b) benzoic acid into 3-phenylpropanoic acid

20.11 DECARBOXYLATION OF CARBOXYLIC ACIDS

The loss of carbon dioxide from a carboxylic acid is called **decarboxylation**.

$$R\text{—}\overset{\overset{\textstyle O}{\|}}{C}\text{—}O\text{—}H \longrightarrow R\text{—}H + O\text{=}C\text{=}O \qquad (20.38)$$

Although decarboxylation is not an important reaction for most ordinary carboxylic acids, certain types of carboxylic acids are readily decarboxylated. Among these are

1. β-keto acids
2. malonic acid derivatives
3. carbonic acid derivatives

In Section 20.11A we'll examine the decarboxylation reactions of each type of carboxylic acid. Then we'll use the mechanistic ideas developed there as we consider biological decarboxylation reactions in Section 20.11B.

A. Decarboxylation of β-Keto Acids, Malonic Acid Derivatives, and Carbonic Acid Derivatives

β-Keto acids—carboxylic acids with a keto group in the β-position—readily decarboxylate at room temperature in *acidic* solution.

$$\qquad (20.39a)$$

acetoacetic acid
(a β-ketoacid)

acetone

Decarboxylation of a β-keto acid involves an *enol intermediate* that is formed by an internal proton transfer from the carboxylic acid group to the carbonyl oxygen atom of the ketone. The enol is transformed spontaneously into the corresponding ketone (Sec. 14.5A).

$$\qquad (20.39b)$$

acetone enol

acetone

The *acid* form of the β-keto acid decarboxylates more readily than the conjugate-base carboxylate form because the latter has no acidic proton that can be donated to the β-carbonyl oxygen. In effect, the carboxy group promotes its own removal!

Why should this internal proton transfer promote decarboxylation? If we represent the internally hydrogen-bonded structure of the β-keto acid in Eq. 20.39 with resonance structures, we can see that the keto oxygen is partially protonated. To the extent that it is protonated, it becomes a very electronegative, positively charged atom, which attracts electrons ultimately from the carboxy group. In other words, the protonated keto carbonyl group is the destination, or "electron sink," for the electrons released by loss of the carboxy group.

a positively charged, electronegative atom is an "electron sink" for the carboxy electrons

$$\qquad (20.40)$$

The presence of an "electron sink" will also prove to be a very important aspect of biological decarboxylations (Sec. 20.11B).

Malonic acid and its derivatives readily decarboxylate upon heating in acidic solution.

$$HO_2C—CH—CO_2H \quad \xrightarrow[135\,°C]{H_3O^+} \quad HO_2C—CH_2 + CO_2 \qquad (20.41)$$

$$\underset{\text{methylmalonic acid}}{\overset{\displaystyle CH_3}{|}} \qquad\qquad \underset{\text{propionic acid}}{\overset{\displaystyle CH_3}{|}}$$

This reaction, which also does not occur in base, bears a close resemblance to the decarboxylation of β-keto acids because both malonic acids and β-keto acids have a carbonyl group β to the carboxy group.

malonic acid   a β-keto acid

Because decarboxylation of malonic acid and its derivatives requires heating, the acids themselves can be isolated at room temperature.

Carbonic acid is unstable and decarboxylates spontaneously in acidic solution to carbon dioxide and water. (Carbonic acid is formed reversibly when CO_2 is bubbled into water; carbonic acid gives carbonated beverages their acidity, and CO_2 gives them their "fizz.")

carbonic acid

$$\qquad\qquad \rightleftharpoons \quad CO_2 + H_2O \qquad (20.42a)$$

In this decarboxylation, water molecules of the solvent serve simultaneously as both acids and bases. The proton transfers are incorporated naturally within a chain of hydrogen-bonded water molecules. Loss of a water molecule as a leaving group provides the "electron sink" for the carboxy-group electrons.

water acts an acid
water acts a base
water acts a base
water acts an acid

$$(20.42b)$$

Similarly, any carbonic acid derivative with a free carboxylic acid group will also decarboxylate under acidic conditions.

methyl carbonate

$$\qquad \underset{}{\overset{H_3O^+}{\rightleftharpoons}} \quad CH_3OH + CO_2 \qquad (20.43a)$$

carbamic acid

$$\qquad \underset{}{\overset{H_3O^+}{\rightleftharpoons}} \quad CO_2 + NH_3 \quad \underset{}{\overset{H_3O^+}{\rightleftharpoons}} \quad \overset{+}{N}H_4 \qquad (20.43b)$$

Under basic conditions, carbonic acid and its derivatives exist as carboxylate salts and do not decarboxylate. For example, the sodium salts of carbonic acid, such as sodium bicarbonate ($NaHCO_3$) and sodium carbonate (Na_2CO_3), are familiar stable compounds.

Carbonic acid diesters and diamides are stable. Dimethyl carbonate (a diester of carbonic acid) and urea (the diamide of carbonic acid) are examples of such stable compounds. Likewise, the acid chloride phosgene is also stable.

dimethyl carbonate   **urea**   **phosgene**

PROBLEMS

20.22   Give the product expected when each of the following compounds is treated with acid.

(a)

(b)

(+ heat)

(c) $CH_3CH_2NHCO_2^-$ Na^+

20.23   Give the structures of all of the β-keto acids that will decarboxylate to yield 2-methylcyclohexanone.

20.24   One piece of evidence supporting the enol mechanism in Eq. 20.39 is that β-keto acids that *cannot* form enols are stable to decarboxylation. For example, the following β-keto acid can be distilled at 310 °C without decomposition. Attempt to construct a model of the enol that would be formed when this compound decarboxylates. Use your model to explain why this β-keto acid resists decarboxylation. (*Hint:* See Sec. 7.6C.)

B. Decarboxylations in Biology: Thiamin Pyrophosphate

Decarboxylation reactions occur in a number of biological pathways. Although they occur in many different situations, we'll examine a specific decarboxylation reaction to illustrate how the mechanistic ideas of Sec. 20.11A apply in biological decarboxylations.

Pyruvate is the conjugate base of pyruvic acid. Pyruvate is formed in the metabolism of glucose. In yeast, pyruvate is an important intermediate in the fermentation of sugars to ethanol, a process that is central to the manufacture of alcoholic beverages such as beer and wine. The decarboxylation of pyruvate to acetaldehyde in yeast is catalyzed by the yeast enzyme *pyruvate decarboxylase*. The acetaldehyde is subsequently reduced to ethanol by the action of alcohol dehydrogenase and NADH, a reaction discussed in Sec. 19.8B.

In the light of what we learned in the previous section, the decarboxylation of pyruvate presents a mechanistic problem, because the keto group of pyruvate is *not* a β-keto group, but rather an α-keto group. The ketone is therefore in the wrong position to serve as the "electron sink" for the electrons provided by decarboxylation. This problem is solved by the involvement of a coenzyme, **thiamin pyrophosphate** (abbreviated **TPP**) in the decarboxyl-

FIGURE 20.4 The structure of thiamin pyrophosphate (TPP) and the abbreviation for TPP used in the text. TPP is a form of vitamin B_1; vitamin B_1 itself is the structure without the pyrophosphate group—that is, —R^2 = —CH_2CH_2OH.

ation reaction. (This is sometimes called *thiamin diphosphate*, and you will also sometimes see "thiamin" spelled "thiamine.") The structure of TPP is shown in Fig. 20.4. TPP is a form of vitamin B_1.

TPP is weakly acidic ($pK_a = 18$) and can ionize to form an *ylid* in the active site of the alcohol dehydrogenase enzyme. Recall (Sec. 19.13) that an **ylid** is a species in which adjacent atoms have opposite charges and complete octets. Because both charged atoms have octets, the charges can't neutralize each other by resonance (Problem 20.25). The ylid is stabilized by the positive charge on the adjacent nitrogen and by overlap of the unshared electron pair with the $3d$ orbitals on the sulfur.

The TPP ylid reacts instantly by adding to the ketone carbonyl group of pyruvate, which is bound nearby on the enzyme. (In Sec. 21.7E you'll learn that ketone carbonyl groups are *much* more reactive than carboxylate carbonyl groups.) This addition is assisted by an acidic group of the enzyme present within the active site.

This addition provides the "electron sink" that can receive the carboxylate electrons produced in the decarboxylation process. Like the internally hydrogen-bonded β-carbonyl of a β-keto acid, the addition product contains a double bond to a positively charged, electronegative atom at the β-carbon.

Loss of CO_2 gives the addition product of TPP ylid and acetaldehyde.

$$(20.47a)$$

another acid group
in the enzyme active site

**TPP ylid–aldehyde
addition product**

This addition product breaks down to acetaldehyde and the TPP ylid.

$$(20.47b)$$

**TPP ylid–aldehyde
addition product**

TPP ylid

acetaldehyde
(reduced to ethanol;
Eq. 20.44)

The acetaldehyde is reduced to ethanol by NADH (Eq. 20.44), and the TPP ylid is ready for another cycle of catalysis by reacting with pyruvate (Eq. 20.46).

Malolactic Fermentation in Winemaking

After producing a young wine, winemakers in some cases subject the wine to a second fermentation, called a *malolactic fermentation*. This fermentation converts sharp-tasting malic acid in the wine to softer-tasting lactic acid. This is another enzyme-catalyzed decarboxylation reaction.

malic acid → **lactic acid** + CO_2

$$(20.48)$$

The progress of this reaction is typically followed by paper chromatography (illustration), in which a drop of each wine is spotted at the bottom of absorbant paper, and a solvent, moving up the paper by capillary action, separates the malic acid from lactic acid and tartaric acid (another dicarboxylic acid present in grape juice). The solvent contains bromcresol green, an acid–base indicator that allows visualization of the carboxylic acids as yellow spots against a blue-green background. The illustration shows that the wines in barrels 1 and 2 have undergone complete malolactic fermentation; the wine in barrel 3 has undergone partial malolactic fermentation; and the wine in barrel 4 has not undergone malolactic fermentation. Lactic acid has a higher pK_a than malic acid (why?), and therefore the pH of the wine is somewhat greater after the malolactic fermentation. (Besides reducing wine acidity, the fermentation also brings about other subtle changes in the wine that affect its flavor.) As Eq. 20.48 on the preceding page shows, malic acid lacks the β-keto group necessary for decarboxylation; therefore, we might wonder why the decarboxylation reaction should occur. Problem 20.26 addresses this issue.

PROBLEMS

20.25 Suppose that the following curved-arrow notation and resonance structures for TPP ylid have been proposed in an attempt to demonstrate that the charges in the ylid can be neutralized by resonance. Explain what is wrong with each resonance structure.

20.26 The enzyme malate decarboxylase, which catalyzes the reaction shown in Eq. 20.48, contains a molecule of tightly bound NAD^+ (Secs. 10.8, 19.9B, Fig. 10.1), which undergoes no *net* change as a result of the reaction. Without the NAD^+, the enzyme is completely inactive as a catalyst. Explain the role of the NAD^+ in facilitating the decarboxylation of malic acid. (*Hint:* Remember that NAD^+ is an oxidizing agent.)

KEY IDEAS IN CHAPTER 20

- The carboxy group is the characteristic functional group of carboxylic acids.

- Carboxylic acids are solids or high-boiling liquids, and carboxylic acids of relatively low molecular mass are very soluble in water.

- The carbonyl and OH absorptions are the most important infrared absorptions of carboxylic acids. In proton NMR spectra the α-hydrogens of carboxylic acids absorb in the δ 2.0–2.5 region, and the OH protons, which can be exchanged with D_2O, absorb in the δ 9–13 region. The carbonyl carbon resonances in the ^{13}C NMR spectra of carboxylic acids occur at δ 170–180, about 20 ppm higher field than the carbonyl carbon resonances of aldehydes and ketones.

- Typical carboxylic acids have pK_a values between 4 and 5, although pK_a values are influenced significantly by polar effects. Sulfonic acids are even more acidic. The acidity of carboxylic acids is due to a combination of polar and resonance effects. The conjugate base of a carboxylic acid is a carboxylate ion.

- Carboxylic acids with long unbranched carbon chains are called fatty acids. The alkali-metal salts of fatty acids are soaps. Soaps and other detergents are surfactants; they form micelles in aqueous solution.

- Because of their acidities, carboxylic acids dissolve not only in aqueous NaOH but also in aqueous solutions of weaker bases such as sodium bicarbonate.

- The reaction of Grignard reagents with CO_2 serves as both a synthesis of carboxylic acids and a method of carbon–carbon bond formation.

- The reactivity of the carbonyl carbon toward nucleophiles plays an important role in many reactions of carboxylic acids and their derivatives. In these reactions, a nucleophile approaches the carbonyl carbon from above or below the plane of the carbonyl group and reacts to form a tetrahedral addition intermediate, which then breaks down by loss of a leaving group. The result is a net substitution reaction at the carbonyl carbon. Acid-catalyzed esterification and lithium aluminum hydride reduction are two examples of nucleophilic carbonyl substitution. In lithium aluminum hydride reduction, the substitution product, an aldehyde, reacts further in an addition reaction to give, after protonolysis, an alcohol.

• The nucleophilic reactivity of the carboxylate oxygen is important in some reactions of carboxylic acids, such as ester formation by the alkylation of carboxylic acids with diazomethane or the alkylation of carboxylate salts with alkyl halides.

• Carboxylic acids are readily converted into acid chlorides with phosphorus pentachloride or thionyl chloride, and into anhydrides with P_2O_5. Cyclic anhydrides containing five- and six-membered rings are readily formed on heating the corresponding dicarboxylic acids.

• β-Keto acids, as well as derivatives of malonic acid and carbonic acid, decarboxylate in acidic solution; most malonic acid derivatives require heating. The decarboxylation reactions of β-keto acids, and probably those of malonic acids as well, involve enol intermediates.

• Decarboxylation is an important reaction in biology. Some decarboxylations involve thiamin pyrophosphate (TPP) as a coenzyme. The positively charged nitrogen of TPP serves as the ultimate destination ("electron sink") for the electrons provided by the decarboxylation reaction.

 REACTION REVIEW *For a summary of reactions discussed in this chapter, see the* Reaction Review *section of Chapter 20 in the* Study Guide and Solutions Manual.

ADDITIONAL PROBLEMS

20.27 Give the product expected when butyric acid (or other compound indicated) reacts with each of the following reagents.
  (a) ethanol (solvent), H_2SO_4 catalyst
  (b) aqueous NaOH solution
  (c) $LiAlH_4$ (excess), then H_3O^+
  (d) heat
  (e) $SOCl_2$
  (f) diazomethane in ether
  (g) product of part (c) (excess) + $CH_3CH{=}O$, HCl (catalyst)
  (h) product of part (e), $AlCl_3$, benzene, then H_2O
  (i) product of part (h), H_2NNH_2, KOH, ethylene glycol (solvent), heat

20.28 Give the product expected when benzoic acid reacts with each of the following reagents.
  (a) CH_3I, K_2CO_3 (b) concentrated HNO_3, H_2SO_4
  (c) PCl_5 (d) P_2O_5, heat

20.29 Draw the structures and give the names of all the dicarboxylic acids with the formula $C_6H_{10}O_4$. Indicate which are chiral, which would readily form cyclic anhydrides on heating, and which would decarboxylate on heating.

20.30 (a) What is the molecular mass of a carboxylic acid containing a single carboxylic acid group if 8.61 mL of 0.1 *M* NaOH solution is required to neutralize 100 mg of the acid?
  (b) How many milliliters of aqueous 0.1 *M* NaOH are required to form the disodium salt from 100 mg of succinic acid?

20.31 Give the product(s) formed and the curved-arrow notation for the reaction of 0.01 mole of each reagent below with 0.01 mole of acetic acid.
  (a) $Na^+ CH_3O^-$ (b) $Cs^+ \ ^-OH$
  (c) $CH_3CH_2{-}MgBr$ (d) $H_3C{-}Li$
  (e) $:NH_3$ (f) NaH
  (g)

$$Na^+ \ ^-O{-}\overset{\overset{\displaystyle O}{\|}}{C}{-}OH$$

20.32 Draw the structure of the major species present in solution when 0.01 mole of the following acid in aqueous solution is treated with 0.01 mole of NaOH. Explain.

$$HO{-}\overset{\overset{\displaystyle O}{\|}}{C}{-}\underset{\underset{\displaystyle Cl}{|}}{\overset{\overset{\displaystyle Cl}{|}}{C}}{-}CH_2{-}CH_2{-}\overset{\overset{\displaystyle O}{\|}}{C}{-}OH$$

20.33 Explain why the differences between the first and second pK_a values of the dicarboxylic acids become smaller as the lengths of their carbon chains increase (Table 20.3).

20.34 Rank the following compounds in order of increasing acidity. Explain your answers.

20.35 Ordinary litmus paper turns red at pH values below about 3. Show that a 0.1 M solution of acetic acid ($pK_a = 4.76$) will turn litmus red, and that a 0.1 M solution of phenol ($pK_a = 9.95$) will not.

20.36 (a) Explain why the most acidic species that can exist in significant concentration in any solvent is the conjugate acid of the solvent.

(b) Show why HBr is a stronger acid in acetic acid solvent than it is in water.

20.37 Explain why all efforts to synthesize a carboxylic acid containing the isotope oxygen-18 at *only* the carbonyl oxygen fail and yield instead a carboxylic acid in which the labeled oxygen is distributed equally between both oxygens of the carboxy group. ($^*O = {}^{18}O$)

(50% of label on each oxygen)

20.38 Outline a synthesis of each of the following compounds from isobutyric acid (2-methylpropanoic acid) and any other necessary reagents.

(a)

$$(CH_3)_2CH-\overset{\overset{\displaystyle O}{\|}}{C}-OCH_2CH_2CH_3$$

(b)

$$(CH_3)_2CH-\overset{\overset{\displaystyle O}{\|}}{C}-OCH_3$$

(c)

$$(CH_3)_2CHCH_2CH_2\overset{\overset{\displaystyle O}{\|}}{C}-OH$$

(d)

isobutyrophenone

(e) 3-methyl-2-butanone

(f) $(CH_3)_2CHCH=CH_2$

20.39 A graduate student, Al Kane, has been given by his professor a very precious sample of (−)-3-methylhexane, along with optically active samples of both enantiomers of 4-methylhexanoic acid, each of known absolute configuration. Kane has been instructed to determine the absolute configuration of (−)-3-methylhexane. Kane has come to you for assistance. Show what he should do to deduce the

configuration of the optically active hydrocarbon from the acids of known configuration. Be specific. (See Sec. 6.5.)

20.40 The sodium salt of *valproic acid* is a drug that has been used in the treatment of epilepsy. (Valproic acid is a name used in medicine.)

valproic acid

(a) Give the substitutive name of valproic acid.

(b) Give the common name of valproic acid.

(c) Outline a synthesis of valproic acid from carbon sources containing fewer than five carbons and any other necessary reagents.

20.41 Because the radioactive isotope carbon-14 is used at very low ("tracer") levels, its presence cannot be detected by spectroscopy. It is generally detected by counting its radioactive decay in a device called a scintillation counter. The location of carbon-14 in a chemical compound must be determined by carrying out chemical degradations, isolating the resulting fragments that represent different carbons in the molecule, and counting them.

A well-known biologist, Fizzi O. Logicle, has purchased a sample of phenylacetic acid ($PhCH_2CO_2H$) advertised to be labeled with the radioactive isotope carbon-14 *only* at the carbonyl carbon. Before using this compound in experiments designed to test a promising theory of biosynthesis, she has wisely decided to be sure that the radiolabel is located only at the carbonyl carbon as claimed. Knowing your expertise in organic chemistry, she has asked that you devise a way to determine what fraction of the ^{14}C is at the carbonyl carbon and what fraction is elsewhere in the molecule. Outline a reaction scheme that could be used to make this determination.

20.42 You have been employed by a biochemist, Fungus P. Gildersleeve, who has given you a very expensive sample of benzoic acid labeled equally in both oxygens with ^{18}O. He asks you to prepare methyl benzoate (structure in Eq. 20.16, p. 1020), preserving as much ^{18}O label in the ester as possible. Which method of ester synthesis would you choose to carry out this assignment? Why?

20.43 You are a chemist for Chlorganics, Inc., a company specializing in chlorinated organic compounds. A process engineer, Turner Switchback, has accidentally mixed the contents of four vats containing, respectively, *p*-chlorophenol, 4-chlorocyclohexanol, *p*-chlorobenzoic acid, and chlorocyclohexane. The president of the company, Hal Ojinn (green with anger), has ordered you to design an expeditious separation of these four compounds. Success guarantees you a promotion; accommodate him.

20.44 Penicillin-G is one of the penicillin family of drugs. In which fluid would you expect penicillin-G to be more soluble: stomach acid (pH = 2) or the bloodstream (pH 7.4)? Explain.

penicillin-G

20.45 (a) The relatively stable carbocation *crystal violet* has a deep blue-violet color in aqueous solution. When NaOH is added to the solution, the blue color fades because the carbocation reacts with sodium hydroxide in about 1–2 min to give a colorless product. Show the reaction of crystal violet with NaOH and explain why the color changes.

crystal violet

(b) When the detergent sodium dodecyl sulfate (SDS) is present in solution above its critical micelle concentration, the bleaching of crystal violet with NaOH takes several days. Account for the effect of SDS on the rate of this bleaching reaction.

$$CH_3CH_2CH_2CH_2CH_2CH_2CH_2CH_2CH_2CH_2CH_2CH_2OSO_3^- \ Na^+$$

sodium dodecyl sulfate (SDS)

20.46 In each case, draw the structure of the cyclic anhydride that forms when the dicarboxylic acid is heated.

(a)

(b) *meso*-α,β-dimethylsuccinic acid

(c)

(*Hint:* Don't forget about the chair interconversion.)

20.47 Propose a synthesis of each of the following compounds from the indicated starting material(s) and any other necessary reagents.

(a) 2-pentanol from propanoic acid

(b)

from allyl alcohol as the only carbon source

(c) 2-methylheptane from pentanoic acid

(d) *m*-nitrobenzoic acid from toluene

(e) PhCH$_2$CH$_2$CH$_2$Ph from benzoic acid

(f)

from

(*Hint:* γ-Lactones form spontaneously from γ-hydroxy acids.)

(g)

from

norbornene

(h) 5-oxohexanoic acid from 5-bromo-2-pentanone (*Hint:* Use a protecting group.)

(i)

from Ph—C≡C—H

***cis*-cinnamic acid**

20.48 (a) Decarboxylation of compound *A* gives *two* separable products; draw their structures and explain.

(b) How many products are formed when compound *B* is decarboxylated?

A   *B*

20.49 (a) Squaric acid (see following structure) has pK_a values of about 1 and 3.5. Draw the reactions corresponding to the two successive ionizations of squaric acid and label each with the appropriate pK_a value.

squaric acid

(b) Most enols have pK_a values in the 10–12 range. Using polar and resonance effects, explain why squaric acid is much more acidic.

(c) Given that squaric acid behaves like a dicarboxylic acid, draw structures for the products formed when it reacts with excess SOCl$_2$; with ethanol solvent in the presence of an acid catalyst.

20.50 Complete each of the reactions given in Fig. P20.50 by giving the principal organic product(s). Give the reasons for your answers.

20.51 (a) Organolithium reagents such as methyllithium (CH$_3$Li) react with carboxylic acids to give ketones [see part (a) of Fig. P20.51]. *Two* equivalents of the lithium reagent are required, and the ketone does not react further. Suggest a mechanism for this reaction that accounts for these facts. (*Hint:* Start by looking at Problem 20.31d.)

    (b) Give the product of the reaction given in part (b) of Fig. P20.51.

20.52 Give a curved-arrow mechanism for each of the reactions given in Fig. P20.52 on p.1040.

20.53 β-Imino carboxylic acids such as the one in Fig. P20.53 (p. 1040) decarboxylate very rapidly provided that the pH is neither too low nor too high.

    (a) Give a curved-arrow mechanism for the decarboxylation reaction shown in Fig. P20.53.

    (b) Explain why very low or very high pH values reduce the rate of this decarboxylation.

(c) The decarboxylation of acetoacetic acid is catalyzed by primary amines (amines of the form R—NH$_2$). Explain. (*Hint:* See Sec. 19.11A.)

acetoacetic acid

20.54 Pyridoxal phosphate (PLP) is a form of vitamin B$_6$.

pyridoxal phosphate (PLP)
(zwitterion form)

abbreviated structure
of PLP

PLP serves as a coenzyme in the enzyme-catalyzed decarboxylation of *meso*-2,6-diaminopimelic acid, which is the

(*continues on p. 1041*)

(a)

(b)

(c)

(d)

(e)

(f)

(g)

(h)

Figure P20.50

(a)

(b)

Figure P20.51

(a)

$$H_3C-\underset{\underset{OEt}{|}}{\overset{\overset{OEt}{|}}{C}}-OEt + H_2O \xrightarrow[\text{(catalyst)}]{\text{dil. HCl}} H_3C-\overset{\overset{O}{\|}}{C}-OEt + 2\,EtOH$$

an orthoester

(b)

$$\xrightarrow[\text{CH}_3\text{OH}]{\substack{\text{dil. HCl} \\ \text{(catalyst)}}}$$

$+ H_2O$

(c)

$$\xrightarrow{H_2SO_4} \xrightarrow{H_2O}$$

carbon monoxide

$$H_3C-\underset{\underset{CH_3}{|}}{\overset{\overset{CH_3}{|}}{C}}-CO_2H$$

(d)

$$CH_3CO_2H + \underset{\underset{Ph}{|}}{\overset{\overset{Ph}{|}}{C}}=CH_2 \xrightarrow{H_2SO_4 \text{ (cat.)}} Ph-\underset{\underset{CH_3}{|}}{\overset{\overset{Ph}{|}}{C}}-O-\overset{\overset{O}{\|}}{C}-CH_3$$

(e)

$$Ph-\overset{\overset{O}{\diagdown}}{\underset{\underset{CH_3}{|}}{C}}-CH-CO_2H \xrightarrow[\text{warm}]{\text{HCl}} Ph-\underset{\underset{CH_3}{|}}{CH}-CH=O + CO_2$$

(f)

$$+ H_2\overset{..}{\overset{-}{C}}-\overset{+}{N}\equiv N: \longrightarrow$$

(63% yield)   (15% yield) $+ N_2$

(g)

$$\xrightarrow{\text{NaOH, I}_2}$$

$+ Na^+I^- + H_2O$

Figure P20.52

$$H_2O + \underset{\underset{H_2}{\overset{|}{}}}{\overset{\overset{Ph-\overset{+}{N}H}{\|}}{H_3C-C}}-\overset{\overset{O^-}{|}}{\underset{\underset{O}{\|}}{C}} \xrightarrow{H_3O^+} H_3C-\overset{\overset{O}{\|}}{C}-CH_3 + CO_2 + Ph-NH_2$$

3-(N-phenylimino)butanoic acid
(a β-imino carboxylic acid)
zwitterion form

Figure P20.53

(*continued from p. 1039*)

last step in the biosynthesis of the amino acid *lysine*. The sequence of reactions involved in this process is shown in Fig. P20.54.

(a) In the first step of this reaction, a molecule of PLP forms an imine (Schiff base) adduct *A* with either of the amino groups of *meso*-2,6-diaminopimelate. Give the structure of adduct *A*.

(b) Carbon dioxide is lost from *A* to form a new imine intermediate *X*. Draw a curved-arrow mechanism for this decarboxylation reaction. This mechanism should lead you to the structure of *X*. Show that the formation of *A* provides the "electron sink" that is crucial to this decarboxylation reaction.

(c) Imine *X* is transformed by a series of acid–base reactions to a different imine derivative *Y*, which is identical to the imine that can be formed separately from PLP and one of the amino groups of lysine. Imine *Y* undergoes imine hydrolysis to PLP and lysine. Draw a structure for *Y* and a curved-arrow mechanism for its formation from *X*. (Assume acids and bases are present as necessary; these are provided by the enzyme.)

20.55 Isopentenyl pyrophosphate, the starting material for isoprenoid and steroid biosynthesis (Sec. 17.6B), is formed biosynthetically by the decarboxylation reaction of (*R*)-phosphomevalonate-5-pyrophosphate catalyzed by the enzyme mevalonate-5-pyrophosphate decarboxylase, shown in Fig. P20.55. When a different starting material is used in which the —CH_3 group is replaced with a fluoromethyl (FCH_2—) group, the reaction takes place 2500 times more slowly.

(a) Draw a two-step curved-arrow mechanism that includes the structure of the intermediate *X*. Your

mechanism should account for the large effect of fluorination on the rate of the reaction. (Assume acids and bases are available as needed.)

(b) What is the "electron sink" for the decarboxylation reaction?

20.56 Propose reasonable fragmentation mechanisms that explain why

(a) the EI mass spectrum of 2-methylpentanoic acid has a strong peak at $m/z = 74$.

(b) the EI mass spectrum of benzoic acid shows major peaks at $m/z = 105$ and $m/z = 77$.

20.57 (a) Give a structure for an optically inactive compound *A* ($C_6H_{10}O_4$), that can be resolved into enantiomers and has the following NMR spectra:
^{13}C NMR: $\delta13.5, \delta41.2, \delta177.9$
proton NMR: $\delta1.13$ (6*H*, d, $J = 7$ Hz); $\delta2.65$ (2*H*, quintet, $J = 7$ Hz); $\delta9.9$ (2*H*, broad s, disappears after D_2O shake)

(b) Give the structure of an isomer of compound *A* that has a different melting point than *A* and NMR spectra that are almost identical to those of *A*.

20.58 A water-insoluble hydrocarbon *A* decolorizes a solution of Br_2 in CH_2Cl_2. The base peak in the EI mass spectrum of *A* occurs at $m/z = 67$. The proton NMR of *A* is complex, but integration shows that about 30% of the protons have chemical shifts in the $\delta1.8$–2.2 region of the spectrum. Treatment of *A* successively with OsO_4, then periodic acid, and finally with Ag_2O, gives a single dicarboxylic acid *B* that can be resolved into enantiomers. Neutralization of a solution containing 100 mg of *B* requires 13.7 mL of 0.1 *M*

Figure P20.54

Figure P20.55

NaOH solution (see Problem 20.30). Compound *B*, when treated with POCl₃, forms a cyclic anhydride. Give the structures of *A* and *B*.

20.59 Provide a structure for each of the following compounds.

(a) $C_9H_{10}O_3$: IR 2300–3200, 1710, 1600 cm^{-1} NMR spectrum in Fig. P20.59a.

(b) $C_9H_{10}O_3$: IR 2400–3200, 1700, 1630 cm^{-1} NMR: δ 1.53 (3*H*, t, *J* = 8 Hz); δ 4.32 (2*H*, q, *J* = 8 Hz); δ 7.08, δ 8.13 (4*H*, pair of leaning doublets, *J* = 10 Hz); δ 10 (1*H*, broad, disappears with D₂O shake)

(c) IR 3000–3580 (broad), 1698, 981 cm^{-1} UV: λ_{max} = 212 nm (ε = 10,800)
CI mass spectrum: M + 1 peak at *m/z* = 115
NMR spectrum in Fig. P20.59b.

20.60 You are employed at Phenomenal Phenols, Inc., and have been asked by your Supervisor, O. H. Gruppa, to identify a compound *A* that was isolated from natural sources. Compound *A*, mp 129–130 °C, is soluble in NaOH solution and in hot water. The IR spectrum of *A* shows prominent absorptions at 3300–3600 cm^{-1} (broad) and 1680 cm^{-1}; the

Figure P20.59 (a) The NMR spectrum for Problem 20.59a. (b) The NMR spectrum for Problem 20.59c. In both NMR spectra, the relative values of the integrals are given in red over their respective resonances.

EI mass spectrum of A has prominent peaks at $m/z = 166$ and 107, and the CI mass spectrum has an M + 1 peak at $m/z = 167$. The NMR spectrum of A is given in Fig. P20.60. It is determined by titration that A has two acidic groups with pK_a values of 4.7 and 10.4, respectively. The UV spec-

trum of A is virtually unchanged as the pH of a solution of A is raised from 2 to 7, but the λ_{max} shifts to much higher wavelength when A is dissolved in 0.1 M NaOH solution. Propose a structure for A. Rationalize the two peaks in its mass spectrum.

Figure P20.60 The NMR spectrum for Problem 20.60. Notice that part of the spectrum is beyond δ 8 and is traced separately. The relative values of the integrals are given in red over their respective resonances.

CHAPTER 21

The Chemistry of Carboxylic Acid Derivatives

A **carboxylic acid derivative** is a compound that can be hydrolyzed under acidic or basic conditions to give the parent carboxylic acid. All carboxylic acid derivatives can be conceptually derived by replacing a portion of the carboxylic acid structure with other groups, as shown in Table 21.1.

Carboxylic acids and their derivatives have not only structural similarities, but also close relationships in their chemistry. With the exception of nitriles, all carboxylic acid derivatives contain a *carbonyl group*. Many important reactions of these compounds occur at the carbonyl group. Furthermore, the —C≡N (cyano) group of nitriles has reactivity that resembles that of a carbonyl group. Thus, the chemistry of carboxylic acid derivatives, like that of aldehydes, ketones, and carboxylic acids, involves the chemistry of the carbonyl group.

21.1 NOMENCLATURE AND CLASSIFICATION OF CARBOXYLIC ACID DERIVATIVES

A. Esters and Lactones

Carboxylic esters are named as derivatives of their parent carboxylic acids by applying a variation of the system used in naming carboxylate salts (Sec. 20.4A). The group attached to the carboxylate oxygen is named first as a simple alkyl or aryl group. This name is followed by the name of the parent carboxylate, which, as you have learned, is constructed by dropping the final *ic* from the name of the acid and adding the suffix *ate*. This procedure is used in both common and substitutive nomenclature.

TABLE 21.1 Structures of Carboxylic Acid Derivatives

| General structure, name of derivative | Condensed structure | Derivation* | | Example |
		Replace—	With—	
ester $R-\overset{\overset{\displaystyle O}{\|\|}}{C}-O-R'$	$R-CO_2R'$	—H	—R'	$H_3C-\overset{\overset{\displaystyle O}{\|\|}}{C}-O-CH_2CH_3$ **ethyl acetate**
thioester $R-\overset{\overset{\displaystyle O}{\|\|}}{C}-S-R'$	$R-CO-SR'$	—OH	—SR	$H_3C-\overset{\overset{\displaystyle O}{\|\|}}{C}-S-CH_2CH_3$ **ethyl thioacetate (S-ethyl thioacetate)**
anhydride $R-\overset{\overset{\displaystyle O}{\|\|}}{C}-O-\overset{\overset{\displaystyle O}{\|\|}}{C}-R$	$\left(R-\overset{\overset{\displaystyle O}{\|\|}}{C}\right)_2 O$	—H	$-\overset{\overset{\displaystyle O}{\|\|}}{C}-R$ (acyl group)	$H_3C-\overset{\overset{\displaystyle O}{\|\|}}{C}-O-\overset{\overset{\displaystyle O}{\|\|}}{C}-CH_3$ **acetic anhydride**
acid halide $R-\overset{\overset{\displaystyle O}{\|\|}}{C}-X$	$R-CO-X$	—OH	—X (halogen)	$H_3C-\overset{\overset{\displaystyle O}{\|\|}}{C}-Cl$ **acetyl chloride**
amide $R-\overset{\overset{\displaystyle O}{\|\|}}{C}-\overset{\overset{\displaystyle R'}{}}{\underset{\displaystyle R'}{N}}$	$R-CO-NR'_2$	—OH	$-\overset{\displaystyle R'}{\underset{\displaystyle R'}{N}}$	$H_3C-\overset{\overset{\displaystyle O}{\|\|}}{C}-NH_2$ **acetamide**
nitrile $R-C\equiv N$	$R-CN$	—CO_2H	—CN (cyano group)	$H_3C-C\equiv N$ **acetonitrile**

* Within the carboxylic acid structure $R-\overset{\overset{\displaystyle O}{\|\|}}{C}-OH$, replace the group in column 3 with the group in column 4 to obtain the derivative. (Note that this shows the relationship of structures, but not necessarily how they are interconverted chemically.)

$H_3C-\overset{\overset{\displaystyle O}{\|\|}}{C}-O-CH_2CH_3$

acetic + ate = acetate ethyl group

ethyl acetate (common)

$CH_3CH_2CH_2CH_2CH_2-\overset{\overset{\displaystyle O}{\|\|}}{C}-O-$⬡

phenyl hexanoate (substitutive)

Esters of acetic acid (acetate esters) are so common that a common abbreviation for acetate is often used.

$$acetate = -OAc = -O-\overset{\overset{\displaystyle O}{\|\|}}{C}-CH_3$$

Thus, ethyl acetate could be abbreviated CH_3CH_2OAc or even EtOAc. (This group is called *acetoxy* when it is a substituent; see Table 21.2 on p. 1050).

In ester nomenclature, substituents can occur in the acyl part of the ester or in the carboxylate part. In substitutive nomenclature, the position of a substituent in the acyl part of an ester is indicated by number as it is in carboxylic acids, in which the carbonyl carbon receives the number 1. In common nomenclature, the position of a substituent is indicated by a Greek letter, and numbering starts at the carbon *adjacent* to the carbonyl carbon.

acyl part of the ester   *carboxylate part of the ester*

common numbering
substitutive numbering

$$\underset{\gamma\,\,\,\,\,\,\,\,\,\,\beta\,\,\,\,\,\,\,\,\,\,\alpha\,\alpha\,\,\,\,\,\,\,\,\,\,\beta\,\,\,\,\,\,\,\,\,\,\gamma}{\underset{4\,\,\,\,\,\,\,\,\,\,3\,\,\,\,\,\,\,\,\,\,2\,\,\,\,\,\,\,\,\,\,1\,\,\,\,\,\,\,\,\,\,\,\,\,\,1\,\,\,\,\,\,\,\,\,\,2\,\,\,\,\,\,\,\,\,\,3}{H_3C—CH—CH_2—C—O—CH_2—CH—CH_3}}$$

common name: **β-bromopropyl β-chlorobutyrate**
substitutive name: **2-bromopropyl 3-chlorobutanoate**

As shown in this example, substitution in the carboxylate part of the ester is also indicated by number (in substitutive nomenclature) or by Greek letter (in common nomenclature). In this part of the ester, numbering starts at the carbon attached to the ester oxygen.

Esters of other acids are named by analogous extensions of acid nomenclature.

methyl 2-bromocyclohexanecarboxylate

diisopropyl succinate

$$(CH_3)_2CH—O—C—CH_2CH_2—C—O—CH(CH_3)_2$$

Cyclic esters are called **lactones**.

β-butyrolactone
(a β-lactone)

γ-butyrolactone
(a γ-lactone)

In common nomenclature, illustrated in these examples, the *name* of a lactone is derived from the acid with the same number of carbons in its principal chain; the *ring size* is denoted by a Greek letter corresponding to the point of attachment of the lactone ring oxygen to the carbon chain. Thus, in a β-lactone, the ring oxygen is attached at the β-carbon to form a four-membered ring.

The substitutive nomenclature of lactones is a specialized extension of heterocyclic nomenclature that we will not consider.

B. Acid Halides

Acid halides are named in any system of nomenclature by replacing the *ic* ending of the acid with the suffix *yl*, followed by the name of the halide.

$$CH_3CH_2—C—Cl$$

propion*ic* + yl =
propionyl chloride
(common)
propanoyl chloride
(substitutive)

$$H_3C—C—C—Br$$

α-bromo-α-methylbutyryl bromide (common)
2-bromo-2-methylbutanoyl bromide (substitutive)

$$Cl—C—CH_2—C—Cl$$

malonyl dichloride

cyclohexanecarbonyl chloride

Notice the special nomenclature required when the acid halide group is attached to a ring: the compound is named as a cycloalkanecarbonyl halide.

C. Anhydrides

To name an anhydride, the name of the parent acid is followed by the word *anhydride*.

benzoic anhydride

valeric anhydride (common)
pentanoic anhydride (substitutive)

acetic formic anhydride
(a mixed anhydride)

phthalic anhydride
(a cyclic anhydride)

Acetic formic anhydride is an example of a **mixed anhydride**, an anhydride derived from two different carboxylic acids. Mixed anhydrides are named by citing the two parent acids in alphabetical order. Phthalic anhydride is an example of a **cyclic anhydride**, an anhydride derived from two carboxylic acid groups within the same molecule, in this case, phthalic acid (Table 20.1).

D. Nitriles

Nitriles may not appear to be related to carboxylic acids at first glance. However, notice that in carboxylic acids and the carboxylic acid derivatives that we've seen, each one possesses a carbon that has three bonds to a more electronegative element (oxygen, a halogen, etc.). Carboxylic acids, for example, have a carbon double-bonded to an oxygen and single-bonded to another oxygen. Nitriles, then, fit the definition because the carbon has three bonds to nitrogen. In other words, nitriles have the same oxidation state as carboxylic acids (Sec. 10.6A).

Nitriles are named in the common system by dropping the *ic* or *oic* from the name of the acid *with the same number of carbon atoms* (counting the nitrile carbon) and adding the suffix *onitrile*. In substitutive nomenclature, the suffix *nitrile* is added to the name of the hydrocarbon with the same number of carbon atoms.

$$Ph—C≡N:$$

benzonitrile (benz*oic* + onitrile)

$$H_3C—C≡N:$$

acetonitrile (acet*ic* + onitrile)

$$H_3C—CH—CH_2—C≡N:$$
$$|$$
$$CH_3$$

isovaleronitrile (common)
3-methylbutanenitrile (substitutive)

$$:N≡C—CH_2—CH_2—C≡N:$$

succinonitrile (common)
butanedinitrile (substitutive)

The name of the three-carbon nitrile is shortened in common nomenclature:

$$CH_3CH_2—C≡N$$

propionitrile (*not* propiononitrile)

When the nitrile group is attached to a ring, a special *carbonitrile* nomenclature is used.

2-methylcyclobutanecarbonitrile

E. Amides, Lactams, and Imides

Simple amides are named in any system by replacing the *ic* or *oic* suffix of the acid name with the suffix *amide*.

benzamide (benzoic + amide)

γ-chlorovaleramide (common)
4-chloropentanamide (substitutive)

When the amide functional group is attached to a ring, the suffix *carboxamide* is used.

2-methylcyclopentanecarboxamide

Like amines (Sec. 19.11A, B; Chapter 23, p. 1183), amides are classified as *primary*, *secondary*, or *tertiary* according to the number of hydrogens on the amide nitrogen.

primary amide   secondary amide   tertiary amide

This classification, *unlike that of alkyl halides and alcohols*, refers to substitution *at nitrogen* rather than substitution at carbon. Thus, the following compound is a *secondary amide*, even though a tertiary alkyl group is bound to nitrogen.

a tertiary alkyl group

a secondary amide

Substitution on nitrogen in secondary and tertiary amides is designated with the letter *N* (italicized or underlined).

N,N-diethylacetamide
(the double *N* designation means that
both ethyl groups are on nitrogen)

4-chloro-N-methylcyclohexanecarboxamide

Cyclic amides are called **lactams**, and the common nomenclature of the simple lactams is analogous to that of lactones. Lactams, like lactones, are classified by ring size as γ-lactams (five-membered lactam ring), β-lactams (four-membered lactam ring), and so on.

γ-butyrolactam
(a γ-lactam)

penicillin-G
(a β-lactam)

Imides can be thought of as the nitrogen analogs of anhydrides. Cyclic imides, of which the following two compounds are examples, are of greater importance than open-chain imides, although the latter are also known compounds.

succinimide   **phthalimide**

F. Nomenclature of Substituent Groups

The priorities for citing principal groups in a carboxylic acid derivative are as follows:

$$\text{acid} > \text{anhydride} > \text{ester} > \text{acid halide} > \text{amide} > \text{nitrile} \qquad (21.1)$$

All of these groups have citation priority over aldehydes and ketones, as well as the other functional groups considered in previous chapters. (A complete list of group priorities is given in Appendix I.) The names used for citing these groups as substituents are given in Table 21.2 (p. 1050). The following compounds illustrate the use of these names:

p-acetamidobenzoic acid
4-(acetylamino)benzoic acid

5-chloroformyl-4-cyano-
2-methoxycarbonylbenzoic acid

G. Carbonic Acid Derivatives

Esters of carbonic acid (Sec. 20.11A) are named like any other ester, but other important carbonic acid derivatives have special names that should be learned.

dimethyl carbonate   **phosgene**   **urea**   **carbamic acid**
(unstable, but has
many stable derivatives)   **methyl carbamate**
(a stable carbamic acid
derivative)

TABLE 21.2 Names of Carboxylic Acid Derivatives When Used as Substituent Groups

Group	Name	Group	Name
$-\overset{\overset{\displaystyle O}{\|\|}}{C}-OH$	carboxy	$-\overset{\overset{\displaystyle O}{\|\|}}{C}-Cl$	chloroformyl
$-\overset{\overset{\displaystyle O}{\|\|}}{C}-OCH_3$	methoxycarbonyl	$-\overset{\overset{\displaystyle O}{\|\|}}{C}-NH_2$	carbamoyl
$-\overset{\overset{\displaystyle O}{\|\|}}{C}-OCH_2CH_3$	ethoxycarbonyl	$-NH-\overset{\overset{\displaystyle O}{\|\|}}{C}-CH_3$	acetamido or acetylamino*
$-CH_2-\overset{\overset{\displaystyle O}{\|\|}}{C}-OH$	carboxymethyl	$-C\equiv N$	cyano
$-O-\overset{\overset{\displaystyle O}{\|\|}}{C}-CH_3$	acetoxy or acetyloxy*		

* Used by Chemical Abstracts.

PROBLEMS

21.1 Give a structure for each of the following compounds. (Refer to Table 20.1 on p. 1006 for the common names of carboxylic acids.)

(a) 5-cyanopentanoic acid   (b) isopropyl valerate   (c) ethyl methyl malonate

(d) cyclohexyl acetate   (e) *N,N*-dimethylformamide   (f) γ-valerolactone

(g) glutarimide   (h) α-chloroisobutyryl chloride   (i) 3-ethoxycarbonylhexanedioic acid

21.2 Name the following compounds.

(a) $CH_3CH_2CH_2CN$ (b)   (c)

(d) $Ph-\overset{\overset{\displaystyle O}{\|\|}}{C}-N(CH_3)_2$ (e)   (f) $EtO-\overset{\overset{\displaystyle O}{\|\|}}{C}-CH_2-\overset{\overset{\displaystyle O}{\|\|}}{C}-CH_2CH_3$

(g) $CH_3CH_2-\overset{\overset{\displaystyle O}{\|\|}}{C}-O-\underset{\underset{\displaystyle CH_3}{\|}}{CH}CH_2CH=CH_2$

21.2   STRUCTURES OF CARBOXYLIC ACID DERIVATIVES

The structures of many carboxylic acid derivatives are very similar to the structures of other carbonyl compounds. For example, the C=O bond length is about 1.21 Å, and the carbonyl group and its two attached atoms are planar. The nitrile C≡N bond length, 1.16 Å, is significantly shorter than the acetylene C≡C bond length, 1.20 Å. This is another example of the shortening of bonds to smaller atoms (Sec. 1.3B).

both the carbonyl carbon
and the nitrogen have
trigonal planar geometry

FIGURE 21.1 A ball-and-stick model of *N*-methylacetamide, shown in two perspectives. The perspective on the right is obtained from the one on the left by rotation about the axis shown. The trigonal planar geometry of the carbonyl carbon and the nitrogen requires that the labeled atoms lie in the same plane.

In an amide, both the carbonyl carbon and the amide nitrogen have trigonal planar bonding (Fig. 21.1). This geometry can be understood from the following resonance structures, which show that the bond between the nitrogen and the carbonyl carbon has considerable double-bond character.

$$ (21.2) $$

amide resonance structures

Because of the trigonal planar geometry at nitrogen, secondary and tertiary amides can exist in both *E* and *Z* conformations about the carbonyl–nitrogen bond; the *Z* conformation predominates in most secondary amides because, in this form, van der Waals repulsions between the largest groups are avoided.

$$ (21.3) $$

E and *Z* conformations of *N*-methylacetamide

The interconversion of the *E* and *Z* forms of amides is too rapid at room temperature to permit their separate isolation, but it is very slow compared with rotation about ordinary carbon–carbon single bonds. A typical energy barrier for rotation about the carbonyl–nitrogen bond of an amide is 71 kJ mol^{-1} (17 kcal mol^{-1}), which results in an internal rotation rate of about 10 times per second. (In contrast, internal rotation in butane occurs about 10^{11} times per second.) The relatively low rate of internal rotation is caused by the significant double-bond character in the carbon–nitrogen bond; recall from Sec. 4.1C that rotation about carbon–carbon double bonds does not occur because rotation would require breaking the π bond. Rotation about a "partial double bond" is retarded for the same reason.

**FURTHER
EXPLORATION 21.1**
NMR Evidence for
Internal Rotation
in Amides

PROBLEMS

21.3 Shown below is one conformation of the amino acid derivative *N*-acetylproline about the amide bond.

N-acetylproline

(a) Is this the *E* or the *Z* conformation about the amide bond?

(b) Draw the other conformation about the amide bond.

21.4 Draw the structure of an amide that *must* exist in an *E* conformation about the carbonyl–nitrogen bond.

21.3 PHYSICAL PROPERTIES OF CARBOXYLIC ACID DERIVATIVES

A. Esters

Esters are polar molecules, but they lack the capability to donate hydrogen bonds that carboxylic acids have. The smaller esters are typically volatile, fragrant liquids that have lower densities than water. Most esters are insoluble in water. The low boiling point of a typical ester (*red*) is illustrated by the following comparison:

	propionic acid	2-butanone	methyl acetate	2-methyl-1-butene
boiling point	141 °C	80 °C	57 °C	31.2 °C

PROBLEMS

21.5 Pentanoic acid and methyl butyrate are constitutional isomers. Which has the higher boiling point and why?

21.6 (a) Assuming that the difference in the relative boiling points of methyl acetate and 2-butanone (see display above) is caused by the difference in their dipole moments, predict which compound has the greater dipole moment.

(b) Use a vector analysis of bond dipoles to show why your answer to part (a) is reasonable.

B. Anhydrides and Acid Chlorides

Most of the lower anhydrides and acid chlorides are dense, water-insoluble liquids with acrid, piercing odors. Their boiling points are not very different from those of other polar molecules of about the same molecular mass and shape.

	acetic anhydride	4-methyl-3-penten-2-one
boiling point	139.6 °C	129.8 °C
density	1.082	0.86

	acetyl chloride	methyl acetate	benzoyl chloride	methyl benzoate
	H_3C—$\overset{\overset{O}{\|\|}}{C}$—$Cl$	H_3C—$\overset{\overset{O}{\|\|}}{C}$—$OCH_3$	Ph—$\overset{\overset{O}{\|\|}}{C}$—$Cl$	Ph—$\overset{\overset{O}{\|\|}}{C}$—$OCH_3$
boiling point	50.9 °C	57 °C	197.2 °C	213 °C
density	1.051	0.93	1.212	1.09

The simplest anhydride, formic anhydride, and the simplest acid chloride, formyl chloride, are unstable and cannot be isolated under ordinary conditions.

C. Nitriles

Nitriles are among the most polar organic compounds. Acetonitrile, for example, has a dipole moment of 3.4 D. The polarity of nitriles is reflected in their boiling points, which are rather high despite the absence of hydrogen bonding. (See Fig. 8.2, p. 337.)

	acetonitrile	propionitrile	propyne
	H_3C—$C{\equiv}N\!:$	CH_3CH_2—$C{\equiv}N\!:$	H_3C—$C{\equiv}C$—H
boiling point	81.6 °C	97.4 °C	−23.3 °C

Although nitriles are very poor hydrogen-bond acceptors (because they are very weak bases; see Sec. 21.6), acetonitrile is miscible with water and propionitrile has a moderate solubility in water. Higher nitriles are insoluble in water. Acetonitrile serves in some cases as a useful polar aprotic solvent because of its moderate boiling point and its relatively high dielectric constant of 38 (Table 8.2, p. 355).

D. Amides

The amides of lower molecular mass are water-soluble, polar molecules with high boiling points. Primary and secondary amides, like carboxylic acids (Sec. 20.2), tend to associate into hydrogen-bonded dimers or higher aggregates in the solid state, in the pure liquid state, or in solvents that do not form hydrogen bonds. This association has a noticeable effect on the properties of amides and is of substantial biological importance in the structures of proteins (Sec. 27.9A). For example, simple amides have very high boiling points; many are solids.

	acetamide	acetic acid	acetone
boiling point	221.2 °C	117.9 °C	56.5 °C
melting point	82.3 °C	16.7 °C	−94 °C

Primary amides have two hydrogens on the amide nitrogen that can form hydrogen bonds. Along a series in which these hydrogens are replaced by methyl groups, the capacity for hydrogen bonding is reduced, and boiling points decrease in spite of the increase in molecular mass.

	acetamide	N-methylacetamide	N,N-dimethylacetamide
boiling point	221.2 °C	204–206 °C	166.1 °C
melting point	82.3 °C	28 °C	−20 °C

A number of amides have high dielectric constants (Table 8.2, p. 355). N,N-Dimethylformamide (DMF), which has a dielectric constant of 37, for example, dissolves a number of inorganic salts and is widely used as a polar aprotic solvent, despite its high boiling point.

21.4 SPECTROSCOPY OF CARBOXYLIC ACID DERIVATIVES

A. IR Spectroscopy

The most important feature in the IR spectra of most carboxylic acid derivatives is the C=O stretching absorption. For nitriles, the most important feature in the IR spectrum is the C≡N stretching absorption. These absorptions are summarized in Table 21.3, along with the absorptions of other carbonyl compounds. Some of the noteworthy trends in this table are the following:

1. Esters are readily differentiated from carboxylic acids, aldehydes, or ketones by the unique ester carbonyl absorption at 1735–1745 cm^{-1}.

2. Lactones, lactams, and cyclic anhydrides, like cyclic ketones, have carbonyl absorption frequencies that increase significantly as the ring size decreases. (See Further Exploration 19.1 in the *Study Guide* for the explanation.)

TABLE 21.3 Important Infrared Absorptions of Carbonyl Compounds and Nitriles

Compound	Carbonyl absorption, cm^{-1}	Other absorptions, cm^{-1}
ketone	1710–1715	
α,β-unsaturated ketone	1670–1680	
aryl ketone	1680–1690	
cyclopentanone	1745	
cyclobutanone	1780	
aldehyde	1720–1725	aldehydic C—H stretch at 2720
α,β-unsaturated aldehyde	1680–1690	
aryl aldehyde	1700	
carboxylic acid (dimer)	1710	OH stretch at 2400–3000 (strong, broad); C—O stretch at 1200–1300
aryl carboxylic acid	1680–1690	
ester or six-membered		
lactone (δ-lactone)	1735–1745	C—O stretch at 1000–1300
α,β-unsaturated ester	1720–1725	
5-membered lactone (γ-lactone)	1770	
4-membered lactone (β-lactone)	1840	
acid chloride	1800	a second weaker band is sometimes observed at 1700–1750
anhydride	1760, 1820 (two absorptions)	C—O stretch as in an ester
6-membered cyclic anhydride	1750, 1800	
5-membered cyclic anhydride	1785, 1865	
amide	1650–1655	N—H bend at 1640 N—H stretch at 3200–3400; double absorption for a primary amide
6-membered lactam (δ-lactam)	1670	
5-membered lactam (γ-lactam)	1700	
4-membered lactam (β-lactam)	1745	
nitrile		C≡N stretch at 2200–2250

3. Anhydrides and some acid chlorides have two carbonyl absorptions. The two carbonyl absorptions of anhydrides are due to the symmetrical and unsymmetrical stretching vibrations of the carbonyl groups (Fig. 12.8, p. 582). (The reason for the double absorption of acid chlorides is more obscure.)

4. The carbonyl absorptions of amides occur at much lower frequencies than those of other carbonyl compounds.

5. The C≡N stretching absorptions of nitriles generally occur in the triple-bond region of the spectrum. These absorptions are stronger than the C≡C absorptions of alkynes because of the large bond dipole of the carbon–nitrogen triple bond, and they occur at higher frequencies.

The IR spectra of some carboxylic acid derivatives are shown in Fig. 21.2a–c on p. 1056.

Other useful absorptions in the IR spectra of carboxylic acid derivatives are also summarized in Table 21.3. For example, primary and secondary amides show an N—H stretching absorption in the 3200–3400 cm^{-1} region of the spectrum. Many primary amides show two N—H absorptions, and secondary amides show a single strong N—H absorption. In addition, a strong N—H bending absorption occurs in the vicinity of 1640 cm^{-1}, typically appearing as a shoulder on the low-frequency side of the amide carbonyl absorption. Tertiary amides lack both of these NH vibrations. The presence of these absorptions in a primary amide and their absence in a tertiary amide are evident in the comparison of the two spectra in Fig. 21.2d–e on p. 1057.

B. NMR Spectroscopy

Proton NMR Spectroscopy The α-proton resonances of all carboxylic acid derivatives are observed in the δ 1.9–3 region of the proton NMR spectrum (see Fig. 13.4, p. 621). In esters, the chemical shifts of protons on the alkyl carbon adjacent to the carboxylate oxygen are 0.6 ppm greater than the chemical shifts of the analogous protons in alcohols and ethers. This shift is attributable to the electronegative character of the carbonyl group.

$$
\begin{array}{ccc}
\delta\,1.22(\text{t}) & \delta\,3.4(\text{q}) & \\
\text{H}_3\text{C}-\overset{\overset{\displaystyle O}{\|}}{\text{C}}-\text{O}-\text{CH}_2-\text{CH}_3 & \text{H}_3\text{C}-\text{CH}_2-\text{O}-\text{CH}_2-\text{CH}_3 & \text{H}_3\text{C}-\text{C}\equiv\text{N} \\
\delta\,1.94(\text{s}) \quad \delta\,4.02(\text{q}) & \textbf{diethyl ether} & \delta\,2.00 \\
\textbf{ethyl acetate} & & \textbf{acetonitrile}
\end{array}
$$

The *N*-alkyl protons of amides have chemical shifts in the δ 2.6–3 chemical shift region, and the NH proton resonances of primary and secondary amides are observed in the δ 7.5–8.5 region. The resonances for these protons, like those of carboxylic acid OH protons, are sometimes broad. This broadening is caused by a slow chemical exchange with the protons of other protic substances (such as traces of moisture) and by unresolved splitting with ^{14}N, which has a nuclear spin. Amide NH resonances, like the OH signals of acids and alcohols, can be eliminated by exchange with D_2O ("D_2O shake"; Sec. 13.7).

**STUDY GUIDE
LINK 21.1**
Solving Structure Problems Involving Nitrogen-Containing Compounds

$$
\begin{array}{c}
\delta\,2.74(\text{d}) \\
\text{H}_3\text{C}-\overset{\overset{\displaystyle O}{\|}}{\text{C}}-\text{NH}-\text{CH}_3 \\
\delta\,1.97(\text{s}) \qquad \delta\,8.18\ (\text{broad}),\ \text{exchanges with }D_2O \\
\textit{N}\textbf{-methylacetamide}
\end{array}
$$

FIGURE 21.2 Infrared spectra of some carboxylic acid derivatives. (a) Ethyl acetate. The distinguishing feature in the IR spectra of esters is the position of the carbonyl stretching absorption. (b) Butyronitrile. The distinguishing feature in the IR spectra of nitriles is the C≡N absorption. (c) Propionic anhydride. The distinguishing feature in the IR spectra of anhydrides is the double carbonyl stretching absorption.

An interesting aspect of amide structure is revealed by NMR spectroscopy. For example, the two *N*-methyl groups in *N*,*N*-dimethylacetamide have different chemical shifts and appear as two closely spaced singlets:

N,N-dimethylacetamide

FIGURE 21.2 (*continued*) Infrared spectra of amides. (d) 2,2-Dimethylpropanamide (pivalamide); primary amides typically have two N—H absorptions. (e) *N,N*-Dimethylpropanamide. The N—H stretching and bending absorptions seen in primary amides are absent in tertiary amides.

The different chemical shifts show that the two *N*-methyl groups are *chemically nonequivalent*. Why should this be so?

As discussed on p. 1051, there is a significant amount of double-bond character in the bond between the nitrogen and the carbonyl group, and this leads to a considerably smaller rate of internal rotation about this bond. Although the internal rotation occurs about 10 times per second, a rate that is large on the human time scale, this rate is very small in the context of an NMR experiment. That is, the time scale of the NMR measurement is so small that the internal rotation about the carbonyl–nitrogen bond appears to be frozen. (See Sec. 13.8 for a discussion of the effect of slow internal rotations on NMR spectra.) Thus, the *N*-methyls behave in the NMR experiment like substituents on a double bond. The *N*-methyls have different chemical shifts because one of them is cis to the carbonyl oxygen and the other is trans; that is, the two *N*-methyl groups are *diastereotopic*. (See Further Exploration 21.1.)

¹³C NMR Spectroscopy   In ¹³C NMR spectra, the carbonyl chemical shifts of carboxylic acid derivatives are in the range δ 165–180, very much like those of carboxylic acids.

The chemical shifts of nitrile carbons are considerably smaller, occurring in the δ 115–120 range. These shifts are much greater, however, than those of acetylenic carbons.

$$\delta\ 1.3 \qquad\qquad \delta\ 1.7 \quad \delta\ 76.9$$

$$H_3C-C\equiv N \qquad H_3C-C\equiv C-CH_2CH_2CH_3$$

$$\delta\ 117.7 \qquad\qquad \delta\ 73.7 \quad \delta\ 19.6$$

PROBLEMS

21.7   How would you differentiate between the compounds in each of the following pairs?

(a) *p*-ethylbenzoic acid and ethyl benzoate by IR spectroscopy

(b) 2,4-dimethylbenzonitrile and *N*-methylbenzamide by proton NMR spectroscopy

(c) methyl propionate and ethyl acetate by proton NMR spectroscopy

(d) *N*-methylpropanamide and *N*-ethylacetamide by proton NMR spectroscopy

(e) ethyl butyrate and ethyl isobutyrate by ^{13}C NMR spectroscopy

21.8   Identify the compound C_4H_9NO with the proton NMR spectrum given in Fig. 21.3. This compound has IR absorptions at 3300 and $1650\ cm^{-1}$.

21.5   BASICITY OF CARBOXYLIC ACID DERIVATIVES

Like carboxylic acids themselves, carboxylic acid derivatives are weakly basic and can be protonated on the carbonyl oxygen by strong acids. Similarly, nitriles are weakly basic at nitrogen. These basicities are particularly important in some of the acid-catalyzed reactions of esters, amides, and nitriles.

The basicity of an ester is about the same as the basicity of the corresponding carboxylic acid.

$$H_3C-\overset{\overset{:O:}{\|}}{C}-\overset{..}{\underset{..}{O}}CH_3\ +\ H_3O^+\ \rightleftharpoons$$

$$\left[H_3C-\overset{\overset{+\overset{..}{O}-H}{\|}}{C}-\overset{..}{\underset{..}{O}}CH_3 \longleftrightarrow H_3C-\overset{\overset{+\overset{..}{O}-H}{|}}{\underset{+}{C}}-\overset{..}{\underset{..}{O}}CH_3 \longleftrightarrow H_3C-\overset{\overset{:\overset{..}{O}-H}{|}}{C}=\overset{..}{\underset{+}{O}}CH_3 \right]\ +\ H_2O \qquad (21.4a)$$

protonated ester;
$pK_a \approx -6$

Amides are considerably more basic than other carboxylic acid derivatives. This basicity, relative to esters, is a reflection of the reduced electronegativity of nitrogen relative to oxygen. That is, the resonance structures in which positive charge is shared on nitrogen are particularly important for a protonated amide.

$$H_3C-\overset{\overset{:O:}{\|}}{C}-\overset{..}{N}H_2\ +\ H_3O^+\ \rightleftharpoons$$

$$\left[H_3C-\overset{\overset{+\overset{..}{O}-H}{\|}}{C}-\overset{..}{N}H_2 \longleftrightarrow H_3C-\overset{\overset{:\overset{..}{O}-H}{|}}{\underset{+}{C}}-\overset{..}{N}H_2 \longleftrightarrow H_3C-\overset{\overset{:\overset{..}{O}-H}{|}}{C}=\overset{}{\underset{+}{N}}H_2 \right]\ +\ H_2O \qquad (21.4b)$$

protonated amide;
$pK_a \approx -0.5$ to -1

FIGURE 21.3 The NMR spectrum for Problem 21.8. The integrals are shown in red over their respective absorptions. The broad resonance at δ 7.6 disappears after a D_2O shake, and the multiplet at δ 3.05 simplifies to a quartet.

Both esters and amides, like carboxylic acids (Sec. 20.4B), protonate on the *carbonyl oxygen*. Protonation of esters on the carboxylate oxygen, or protonation of amides on the nitrogen, would give a cation that is *not* resonance-stabilized, and additionally, one that is destabilized by the electron-attracting polar effect of the carbonyl group. The site of protonation of amides was for many years a subject of controversy, because ammonia and amines (R_3N:) are protonated on nitrogen. However, protonation of an amide on nitrogen is less favorable than carbonyl protonation by about 8 pK_a units.

Nitriles are *very* weak bases; protonated nitriles have a pK_a of about -10. To put this pK_a in perspective, a protonated nitrile is about as acidic as the strong acid HI.

STUDY GUIDE LINK 21.2
Basicity of Nitriles

$$H_3C-C\equiv N: + H_3O^+ \rightleftharpoons \left[H_3C-C\equiv\overset{+}{N}-H \longleftrightarrow H_3C-\overset{+}{C}=\overset{..}{N}-H \right] + H_2O \qquad (21.5)$$

protonated nitrile;
$pK_a \approx -10$

An Amide with a Twist

A graphic demonstration of the importance of resonance in amides and their conjugate acids was provided when, in 2006, chemists at CalTech led by Prof. Brian Stoltz synthesized the conjugate acid of the cyclic amide 2-quinuclidone.

2-quinuclidone
(conjugate acid)

They showed that this amide, unlike ordinary amides, protonates on nitrogen rather than on the carbonyl oxygen. The

reason for this difference is that the ion that would result from protonation of the carbonyl oxygen is *not* resonance-stabilized because the resonance structure of the *O*-protonated amide violates Bredt's rule (Sec. 7.6C).

twisted double bond
(violates Bredt's rule)

oxygen-protonated
2-quinuclidone

In terms of the orbitals involved, the orbitals containing the nitrogen unshared pair and the π orbital of the carbonyl group are perpendicular and cannot overlap. In other words, the amide bond is twisted because of the bicyclic ring structure, and because a rotation about the nitrogen–carbon bond that would allow orbital overlap to occur would introduce a large degree of strain (Eq. 15.37, p. 753).

perpendicular orbitals
cannot overlap and interact

As a result, nitrogen behaves more like an amine nitrogen than an amide nitrogen. (Amines are basic and are readily protonated by acids.) An attempt to form unprotonated 2-quinuclidone by treatment of its conjugate acid with base in aqueous solution resulted in its very rapid hydrolysis to form the cyclic amino acid. As you'll learn in Sec. 21.7, ordinary amides hydrolyze rather slowly because they are stabilized by the resonance interaction within the amide bond. Lacking this stabilization, 2-quinuclidone is unusually reactive.

PROBLEM

21.9 Which of the two isomers in each of the following sets should have the greater basicity at the carbonyl oxygen? Explain.

(a)

$$H_3C-CH=CH-\overset{\displaystyle O}{\overset{\|}{C}}-OCH_3 \quad \text{or} \quad H_3C-\overset{\displaystyle O}{\overset{\|}{C}}-O-CH_2-CH=CH_2$$

(b)

$$CH_3O-\left\langle\underset{}{\bigcirc}\right\rangle-\overset{\displaystyle O}{\overset{\|}{C}}-OEt \quad \text{or} \quad \left\langle\underset{CH_3O}{\bigcirc}\right\rangle-\overset{\displaystyle O}{\overset{\|}{C}}-OEt$$

(c)

$$F_3C-\overset{\displaystyle O}{\overset{\|}{C}}-OEt \quad \text{or} \quad H_3C-\overset{\displaystyle O}{\overset{\|}{C}}-OEt$$

21.6   INTRODUCTION TO THE REACTIONS OF CARBOXYLIC ACID DERIVATIVES

The reactions of carboxylic acid derivatives can be categorized as follows:

1. reactions at the carbonyl group, or cyano group of a nitrile
    a. reactions at the carbonyl oxygen or cyano nitrogen
    b. reactions at the carbonyl carbon or cyano carbon
2. reactions involving the α-carbon
3. reactions at the nitrogen of amides

The reaction of carboxylic acids and their derivatives as Brønsted bases, illustrated in the previous section, is an example of reaction type 1a. This type of reaction often serves as the first step in acid-catalyzed reactions of carboxylic acid derivatives.

As with carboxylic acids, the major carbonyl-group reaction of carboxylic acid derivatives is a reaction of type 1b. This reaction, *substitution at the carbonyl carbon*, is also called **acyl substitution**. Acyl substitution can be represented generally as follows, with E = an electrophilic group and Y = a nucleophilic group:

$$\underset{\substack{\text{a carboxylic acid} \\ \text{derivative}}}{R-\overset{\displaystyle O}{\overset{\|}{C}}-X} + E-Y \longrightarrow \underset{\substack{\text{another} \\ \text{carboxylic acid} \\ \text{derivative}}}{R-\overset{\displaystyle O}{\overset{\|}{C}}-Y} + E-X \qquad (21.6)$$

an acyl group

The term *acyl substitution* comes from the fact that substitution occurs at the carbonyl carbon of an *acyl group* (the blue group in Eq. 21.6). In other words, an acyl group is transferred in Eq. 21.8 between an —X and a —Y group. The group —X might be the —Cl of an acid chloride, the —OR of an ester, and so on; this group is substituted by another group —Y. This is precisely the same type of reaction as the esterification of carboxylic acids (—X = —OH, E—Y = H—OCH$_3$; Sec. 20.8A). Acyl substitution reactions of carboxylic acid derivatives are the major focus of this chapter.

Although nitriles are not carbonyl compounds, the C≡N bond behaves chemically much like a carbonyl group. For example, a typical reaction of nitriles is *addition*.

$$R—C≡N: + \text{E—Y} \longrightarrow R—\overset{\overset{\text{Y}}{|}}{C}=\overset{..}{N}—E \qquad (21.7)$$

(Compare this reaction with addition to the carbonyl group of an aldehyde or ketone.) Although the resulting addition products are stable in some cases, in most situations they react further.

Like aldehydes and ketones, carboxylic acid derivatives undergo certain reactions involving the α-carbon. The α-carbon reactions of all carbonyl compounds are grouped together in Chapter 22. The reactivity of amides at nitrogen is discussed in Sec. 23.11D.

21.7 HYDROLYSIS OF CARBOXYLIC ACID DERIVATIVES

All carboxylic acid derivatives have in common the fact that they undergo *hydrolysis* (a cleavage reaction with water) to yield carboxylic acids.

A. Hydrolysis of Esters and Lactones

Base-Promoted Hydrolysis (Saponification) of Esters One of the most important reactions of esters is the cleavage reaction with hydroxide ion to yield a carboxylate salt and an alcohol. Strong acid is required in a second step to form the carboxylic acid.

methyl 3-nitrobenzoate   + CH$_3$OH   **3-nitrobenzoic acid** (90–96% yield)

$$(21.8)$$

Ester hydrolysis in aqueous hydroxide is called **saponification** because it is used in the production of soaps from fats (Sec. 21.12C). Despite its association with fatty-acid esters, the term *saponification* can be used to refer to the hydrolysis in base of any carboxylic acid derivative.

The mechanism of ester saponification involves the reaction of the nucleophilic hydroxide ion at the carbonyl carbon to give a tetrahedral addition intermediate from which an alkoxide ion is expelled.

tetrahedral
addition intermediate

$$(21.9a)$$

The alkoxide ion, after expulsion as a leaving group (methoxide in Eq. 21.9a), reacts with the acid to give the carboxylate salt and the alcohol.

$$R—C(=O)—\overset{..}{\underset{..}{O}}—H + {}^-\!\!:\overset{..}{\underset{..}{O}}CH_3 \rightleftharpoons R—C(=O)—\overset{..}{\underset{..}{O}}\!:^- + H—\overset{..}{\underset{..}{O}}CH_3 \quad (21.9b)$$

$$\text{p}K_a = 4.5 \qquad\qquad\qquad \text{p}K_a = 15$$

The equilibrium in this reaction lies far to the right because the carboxylic acid is a much stronger acid than methanol. Le Châtelier's principle operates: The reaction in Eq. 21.9b removes the carboxylic acid from the equilibrium in Eq. 21.9a as its salt and thus drives the hydrolysis to completion. Hence, *saponification is effectively irreversible*. Although an excess of hydroxide ion is often used as a matter of convenience, many esters can be saponified with just one equivalent of ⁻OH. Saponification can also be carried out in an alcohol solvent, even though an alcohol is one of the products of the reaction. If saponification were reversible, an alcohol could not be used as the solvent because the equilibrium would be driven toward starting materials.

Acid-Catalyzed Ester Hydrolysis Because esterification of an acid with an alcohol is a reversible reaction (Sec. 20.8A), esters can be hydrolyzed to carboxylic acids in aqueous solutions of strong acids. In most cases, this reaction is slow and must be carried out with an excess of water, in which most esters are insoluble. Saponification, followed by acidification, is a much more convenient method for hydrolysis of most esters because it is faster, it is irreversible, and it can be carried out not only in water but also in a variety of solvents—even alcohols.

As expected from the principle of microscopic reversibility (Sec. 4.9B), the mechanism of acid-catalyzed hydrolysis is the exact reverse of the mechanism of acid-catalyzed esterification (Sec. 20.8A). The ester is first protonated by the acid catalyst:

$$R—C(=\overset{..}{\underset{..}{O}})—\overset{..}{\underset{..}{O}}CH_3 + H—\overset{+}{\underset{..}{O}}H_2 \rightleftharpoons \left[R—C(=\overset{+}{\underset{..}{O}}\!H)—\overset{..}{\underset{..}{O}}CH_3 \leftrightarrow R—\overset{+}{C}(—\overset{..}{O}H)—\overset{..}{\underset{..}{O}}CH_3 \leftrightarrow R—C(—\overset{..}{O}H)=\overset{+}{\underset{..}{O}}CH_3 \right] + H_2\overset{..}{\underset{..}{O}} \quad (21.10a)$$

As in other acid-catalyzed reactions at the carbonyl group, protonation makes the carbonyl carbon more electrophilic by making the carbonyl oxygen a better acceptor of electrons. Water, acting as a nucleophile, reacts at the carbonyl carbon and then loses a proton to give the tetrahedral addition intermediate:

$$R—\overset{+\,\overset{..}{O}—H}{\underset{\,:\overset{..}{O}H_2}{C}}—OCH_3 \rightleftharpoons R—\overset{:\overset{..}{O}H}{\underset{\overset{+}{O}—H\ (H)\ H_2\overset{..}{O}}{C}}—OCH_3 \xrightarrow{H_2O} R—\overset{:\overset{..}{O}H}{\underset{:\overset{..}{O}H}{C}}—OCH_3 + H_3\overset{..}{O}^+ \quad (21.10b)$$

tetrahedral addition
intermediate

Protonation of the leaving oxygen converts it into a better leaving group. Loss of this group gives a protonated carboxylic acid, from which a proton is removed to give the carboxylic acid itself.

The chemical mechanism shown at top of page, Eq. (21.10c):

$$(21.10c)$$

Let's summarize the important differences between acid-catalyzed ester hydrolysis and ester saponification. First, in acid-catalyzed hydrolysis, the carbonyl carbon can react with the relatively weak nucleophile water because the carbonyl oxygen is protonated. In base, the carbonyl oxygen is not protonated; hence, a much stronger base than water—namely, hydroxide ion—is required to react at the carbonyl carbon. Second, acid *catalyzes* ester hydrolysis, but base *is not a catalyst* because it is consumed by the reaction in Eq. 21.9b. Finally, acid-catalyzed ester hydrolysis is reversible, but saponification is irreversible, again because of the ionization in Eq. 21.9b.

Ester hydrolysis and saponification are both examples of *acyl substitution* (Sec. 21.6). Specifically, the mechanisms of these reactions are classified as **nucleophilic acyl substitution** mechanisms. In a nucleophilic acyl substitution reaction, the substituting group reacts as a nucleophile at the carbonyl carbon. As in the reactions of aldehydes and ketones (Fig. 19.8, p. 965), nucleophiles approach the carbonyl carbon from above or below the plane of the carbonyl group (Fig. 21.4), first interacting with the π^* (antibonding) molecular orbital of the carbonyl group. As the result of this reaction, a tetrahedral addition intermediate is formed. The leaving group is expelled from this intermediate, departing from above or below the plane of the new carbonyl group. In saponification, the nucleophile is ^-OH, and in acid-catalyzed hydrolysis, the nucleophile is water; in both cases, the —OR group of the ester is displaced. With the exception of the reactions of nitriles, most of the reactions of carboxylic acid derivatives in the remainder of this chapter are nucleophilic acyl substitution reactions. They follow the same pattern as saponification, the only substantial difference being the identity of the nucleophiles and the leaving groups.

STUDY GUIDE LINK 21.3
Mechanism of Ester Hydrolysis

FURTHER EXPLORATION 21.2
Cleavage of Tertiary Esters and Carbonless Carbon Paper

FIGURE 21.4 The geometry of nucleophilic acyl substitution. The nucleophile (Nuc:$^-$) approaches the carbonyl carbon above or below the carbonyl plane (*gray*) to form a tetrahedral intermediate. The leaving group (R′O$^-$) departs from above or below the plane of the newly formed carbonyl group (*blue*). The green arrows show the movement of the various groups.

Hydrolysis and Formation of Lactones Because lactones are cyclic esters, they undergo the reactions of esters, including saponification. Saponification converts a lactone completely into the carboxylate salt of the corresponding hydroxy acid.

$$\text{γ-butyrolactone} + {}^-OH \longrightarrow \text{γ-hydroxybutyrate} \tag{21.11}$$

γ-butyrolactone   **γ-hydroxybutyrate**

Upon acidification, the hydroxy acid forms. However, *if a hydroxy acid is allowed to stand in acidic solution, it comes to equilibrium with the corresponding lactone.* The formation of a lactone from a hydroxy acid is nothing more than an *intramolecular* esterification (an esterification within the same molecule) and, like esterification, the lactonization equilibrium is acid-catalyzed.

$$\xrightarrow[\text{catalyst}]{\text{acid}} \quad + H_2O \qquad K_{eq} \approx 145 \tag{21.12}$$

$$\xrightarrow[\text{catalyst}]{\text{acid}} \quad + H_2O \qquad K_{eq} \approx 4 \tag{21.13}$$

As the examples in Eqs. 21.12 and 21.13 illustrate, lactones containing five- and six-membered rings are favored at equilibrium over their corresponding hydroxy acids. Although lactones with ring sizes smaller than five or larger than six are well known, they are less stable than their corresponding hydroxy acids. Consequently, the lactonization equilibria for these compounds favor instead the hydroxy acids.

$$\xrightarrow[\text{catalyst}]{\text{acid}} \quad + H_2O \quad \text{(almost no lactone present at equilibrium)} \tag{21.14}$$

B. Hydrolysis of Amides

Amides can be hydrolyzed to carboxylic acids and ammonia or amines by heating them in acidic or basic solution.

$$\underset{\text{2-phenylbutanamide}}{CH_3CH_2\overset{Ph}{\underset{|}{C}}H\overset{O}{\underset{\|}{C}}-NH_2} + H_2O \xrightarrow[\text{heat, 2 h}]{55 \text{ wt \% } H_2SO_4} \underset{\substack{\text{2-phenylbutanoic acid} \\ \text{(88–90\% yield)}}}{CH_3CH_2\overset{Ph}{\underset{|}{C}}H\overset{O}{\underset{\|}{C}}-OH} + \overset{+}{NH_4} \ HSO_4^- \tag{21.15}$$

In acid, protonation of the ammonia or amine product drives the hydrolysis equilibrium to completion. The amine can be isolated, if desired, by addition of base to the reaction mixture following hydrolysis, as in the following example.

$$+ CH_3CO_2H \qquad (60-67\% \text{ yield}) \qquad (21.16)$$

The hydrolysis of amides in base is analogous to the saponification of esters. In base, the reaction is driven to completion by formation of the carboxylic acid salt.

$$(95-97\% \text{ yield}) \qquad (21.17)$$

The conditions for both acid- and base-promoted amide hydrolysis are considerably more severe than the corresponding reactions of esters. That is, amides are considerably *less* reactive than esters. The relative reactivities of carboxylic acid derivatives are discussed in Sec. 21.7E.

The mechanisms of amide hydrolysis are typical nucleophilic acyl substitution mechanisms; you are asked to explore this point in Problem 21.10.

PROBLEMS

21.10 Show in detail the curved-arrow mechanism for the hydrolysis of *N*-methylbenzamide (a) in acidic solution; (b) in aqueous NaOH. Assume that each mechanism involves a tetrahedral addition intermediate.

21.11 Give the structures of the hydrolysis products that result from each of the following reactions. Be sure to show the product stereochemistry in part (b).

(a)

(b)

C. Hydrolysis of Nitriles

Nitriles are hydrolyzed to carboxylic acids and ammonia by heating them in strongly acidic or strongly basic solution.

$$\underset{\textbf{phenylacetonitrile}}{PhCH_2 - C \equiv N} + 2\,H_2O + \underset{(57\text{ wt }\%)}{H_2SO_4} \xrightarrow[\text{3 h}]{\text{heat}} \underset{\substack{\textbf{phenylacetic acid}\\(78\%\text{ yield})}}{PhCH_2 - CO_2H} + NH_4^+\ HSO_4^- \qquad (21.18)$$

1-cyclohexenecarbonitrile

1-cyclohexenecarboxylic acid
(79% yield)

$$\text{(21.19)}$$

Nitriles hydrolyze more slowly than esters and amides. Consequently, the conditions required for the hydrolysis of nitriles are more severe.

The mechanism of nitrile hydrolysis in acidic solution involves, first, protonation of the nitrogen (Sec. 21.5):

$$\text{(21.20a)}$$

This protonation makes the nitrile carbon much more electrophilic, just as protonation of a carbonyl oxygen makes a carbonyl carbon more electrophilic. A nucleophilic reaction of water at the nitrile carbon and loss of a proton gives an intermediate called an *imidic acid*.

an imidic acid

$$\text{(21.20b)}$$

An imidic acid is the nitrogen analog of an enol (Sec. 14.5A). That is, an imidic acid is to an amide as an enol is to a ketone.

Just as enols are converted spontaneously into aldehydes or ketones, an imidic acid is converted under the reaction conditions into an amide:

$$\text{(21.20c)}$$

Because amide hydrolysis is faster than nitrile hydrolysis, the amide formed in Eq. 21.20c does not survive under the vigorous conditions of nitrile hydrolysis and is therefore hydrolyzed to a carboxylic acid and ammonium ion, as discussed in Sec. 21.7B. Thus, the ultimate product of nitrile hydrolysis in acid is a carboxylic acid.

Notice that nitriles behave mechanistically much like carbonyl compounds. Compare, for example, the mechanism of acid-promoted nitrile hydrolysis in Eqs. 21.20a and b with that for the acid-catalyzed hydration of an aldehyde or ketone (Sec. 19.7A). In both mechanisms, an electronegative atom is protonated (nitrogen of the C≡N bond, or oxygen of the C=O bond), and water then reacts as a nucleophile at the carbon of the resulting cation.

The parallel between nitrile and carbonyl chemistry is further illustrated by the hydrolysis of nitriles in base. The nitrile group, like a carbonyl group, reacts with basic nucleophiles and, as a result, the electronegative nitrogen assumes a negative charge. Proton transfer gives an imidic acid (which, like a carboxylic acid, ionizes in base).

$$R-C\equiv N: \rightleftharpoons R-C=\ddot{N}:^- \rightleftharpoons R-C=\ddot{N}H \rightleftharpoons R-C=\ddot{N}H + H_2\ddot{O} \quad (21.21a)$$

imidic acid

ionized imidic acid

As in acid-promoted hydrolysis, the imidic acid reacts further to give the corresponding amide, which, in turn, hydrolyzes under the reaction conditions to the carboxylate salt of the corresponding carboxylic acid (Sec. 21.7B).

$$\left[R-C=\ddot{N}H \longleftrightarrow R-\overset{:O:}{\underset{}{C}}-\ddot{N}H \right] \longrightarrow$$

$$R-\overset{:O:}{\underset{}{C}}-\ddot{N}H_2 + {}^-\ddot{O}H \xrightarrow[\text{hydrolysis}]{amide} R-\overset{:O:}{\underset{}{C}}-\ddot{O}:^- + :NH_3 \quad (21.21b)$$

D. Hydrolysis of Acid Chlorides and Anhydrides

Acid chlorides and anhydrides react *rapidly* with water, even in the absence of acids or bases.

$$\text{(maleic anhydride)} + H_2O \xrightarrow[\text{few minutes}]{\text{room temperature,}} \text{(maleic acid)} \quad (21.22)$$

(94% yield)

$$\underset{Ph}{\overset{Ph}{>}}C=CH-\overset{O}{\underset{}{C}}-Cl + H_2O \xrightarrow[0\,°C]{\substack{1) \text{ Na}_2\text{CO}_3/\text{H}_2\text{O} \\ 2) \text{ H}_3\text{O}^+}} \underset{Ph}{\overset{Ph}{>}}C=CH-\overset{O}{\underset{}{C}}-OH + Cl^- \quad (21.23)$$

(>95% yield)

However, the hydrolysis reactions of acid chlorides and anhydrides are almost never used for the preparation of carboxylic acids because these derivatives are themselves usually prepared from acids (Sec. 20.9). Rather, these reactions serve as reminders that if samples of acid chlorides and anhydrides are allowed to come into contact with moisture, they will rapidly become contaminated with the corresponding carboxylic acids.

E. Mechanisms and Reactivity in Nucleophilic Acyl Substitution Reactions

As we've seen, all carboxylic acid derivatives can be hydrolyzed to carboxylic acids; however, the *conditions* under which the different derivatives are hydrolyzed differ considerably.

Hydrolysis reactions of amides and nitriles require heat as well as acid or base; hydrolysis reactions of esters require acid or base, but require heating only briefly, if at all; and hydrolysis reactions of acid chlorides and anhydrides occur rapidly at room temperature even in the absence of acid and base. These trends in reactivity, which are observed not only in hydrolysis but in *all* nucleophilic acyl substitution reactions, can be summarized as follows:

Reactivities of carboxylic acid derivatives in nucleophilic acyl substitution reactions:

$$\text{nitriles} < \text{amides} < \text{esters, thioesters, acids} \ll \text{anhydrides} < \text{acid chlorides}$$

increasing reactivity

(21.24)

(The reactions of nitriles are additions, not substitutions, but they are included for comparison.)

The practical significance of this reactivity order is that selective reactions are possible. In other words, an ester can be hydrolyzed under conditions that will leave an amide in the same molecule unaffected; likewise, nucleophilic substitution reactions on an acid chloride can be carried out under conditions that will leave an ester group unaffected.

Understanding the trends in relative reactivity requires, first, an understanding of the mechanisms by which nucleophilic acyl substitution reactions take place. (The reactivity of nitriles is considered later.) As we have learned, the nucleophilic acyl substitution reaction mechanism typically consists of two steps: the addition step and the elimination step. These steps, and the corresponding transition states for them, can be represented as follows:

tetrahedral addition
intermediate

(21.25)

The rate-limiting steps can differ for different carboxylic acid derivatives; and, in the more reactive derivatives, the substitution might be concerted, with no tetrahedral intermediate at all. However, regardless of the details of the mechanism, we'll adopt the view that *the structure of the tetrahedral intermediate is an approximation of the transition-state structure,* and therefore, *the standard free energy of the tetrahedral intermediate is an approximation of the standard free energy of the transition state.* Essentially, we're invoking Hammond's postulate (Sec. 4.8D) to describe the transition state of the reaction.

The reactivity of carboxylic acid derivatives is affected by the standard free energies of both the carbonyl compound and the transition state. Lowering the standard free energy of the transition state *decreases* $\Delta G^{\circ\ddagger}$ and *increases* reactivity. An analogy is driving up a mountain to a mountain pass. If we begin our drive at 3000 feet above sea level, it takes less time (and a smaller increase in potential energy) to reach a pass at 8000 feet than it does to reach a pass at 10,000 feet. Lowering the standard free energy of the starting material (the carbonyl compound) *increases* the standard free energy of activation $\Delta G^{\circ\ddagger}$ and therefore *decreases* reactivity. By analogy, if we are driving to a particular mountain pass at 10,000 feet, it takes more time (and a larger gain in potential energy) if we start at 3000 feet than it does if we start at 5000 feet.

Let's first consider how the stability of the carbonyl compound varies among the different carboxylic acid derivatives. The major factor in the stability of the carbonyl compound is *the resonance interaction of the potential leaving group X with the carbonyl group.* The basis of this resonance interaction is the overlap of the unshared electrons of the group X with the carbonyl π molecular orbital.

$$\left[\begin{array}{c} \overset{\ddot{O}:}{\underset{\parallel}{R-C-\ddot{X}}} \end{array} \longleftrightarrow \begin{array}{c} :\overset{..}{\underset{\mid}{O}}:^{-} \\ R-C\overset{+}{=}\ddot{X} \end{array} \right]$$

Examples: $\left[\begin{array}{c} \overset{\ddot{O}:}{\underset{\parallel}{R-C-\ddot{N}H_2}} \end{array} \longleftrightarrow \begin{array}{c} :\overset{..}{O}:^{-} \\ R-C\overset{+}{=}\overset{+}{N}H_2 \end{array} \right]$ $\left[\begin{array}{c} \overset{\ddot{O}:}{\underset{\parallel}{R-C-\overset{..}{O}CH_3}} \end{array} \longleftrightarrow \begin{array}{c} :\overset{..}{O}:^{-} \\ R-C\overset{+}{=}\overset{+}{O}CH_3 \end{array} \right]$ (21.26)

resonance stabilization
of an amide

resonance stabilization
of an ester

Notice that the X atom takes on a positive charge as a result of this interaction. This resonance interaction is important in both esters and amides; however, it is less important in an ester than it is in an amide because the greater electronegativity of oxygen compared with nitrogen opposes electron donation by resonance.

Comparing the resonance interaction in an anhydride with that in an ester, we see that the resonance interaction in an anhydride creates repulsion between the positively charged oxygen and the bond dipole of the adjacent carbonyl group:

$$\left[\begin{array}{c} \overset{\ddot{O}:}{\underset{\parallel}{R-C-\overset{..}{\underset{..}{O}}-C-R}}\overset{O}{\underset{\parallel}{}} \end{array} \longleftrightarrow \begin{array}{c} :\overset{..}{O}:^{-} \qquad \overset{\delta-}{O} \\ R-C\overset{+}{=}\overset{..}{O}-\underset{\delta+}{C}-R \end{array} \right]$$ (21.27)

repulsion between like charges

Therefore, an anhydride is stabilized less by resonance than an ester is.

In an acid chloride, the resonance interaction between the chlorine and the carbonyl group is opposed by the electronegativity of the leaving group, as in an ester. However, an even more important factor is the weaker electronic overlap between a $3p$ orbital of the chlorine with the $2p$ orbital of the carbonyl carbon (Fig. 16.7, p. 819). Consequently, resonance stabilization in an acid chloride is less important than that in an anhydride.

Considering the stabilization of the carbonyl compound, we conclude that the reactivity order of the various derivatives should be the following:

increasing resonance stabilization of
carbonyl compound

amides < esters, acids << anhydrides < acid chlorides (21.28)

increasing reactivity

This is exactly the order observed.

Resonance stabilization of the carbonyl compound, however, is only half of the story. Let's now consider the *polar effect* of the leaving group X on the stability of the transition state. (Remember, we are using the tetrahedral addition intermediate as an approximation for the transition state.) A greater electronegativity of the X group results in a greater C—X bond dipole moment. A greater C—X bond dipole results in more partial positive charge on the carbon end of the dipole and therefore greater electrostatic stabilization of the transition state.

the favorable interaction between opposite charges lowers energy

This polar effect is exactly the same as the effect we discussed for the pK_a values of carboxylic acids (Sec. 3.6C). Recall that an electronegative substituent lowers the pK_a of a carboxylic acid because its bond dipole interacts favorably with the negative charge of the carboxylate anion (Eq. 3.43, p. 116).

carboxylate ion

tetrahedral intermediate

The polar effect on the stability of the tetrahedral intermediate should be even greater than the polar effect on the stability of a carboxylate anion because the X group is closer to the negatively charged oxygen in the tetrahedral intermediate than it is in a substituted carboxylate ion. (Recall that polar effects increase with decreasing distance.)

Therefore, the polar effect should increase reactivity in the following order:

increasing polar stabilization of transition state

amides < esters, acids << anhydrides < acid chlorides   (21.29)

increasing reactivity

Comparing Eqs. 21.28 and 21.29, we see that the resonance stabilization of the carbonyl compound and the polar-effect stabilization of the transition state *have reinforcing effects on reactivity*. Both effects contribute to the reactivity order that we observe (Fig. 21.5).

What about nitriles? Reactions of nitriles in base are slower than those of other acid derivatives because nitrogen is less electronegative than oxygen and accepts additional electrons less readily. Reactions of nitriles in acid are slower because of their extremely low basicities. It is the protonated form of a nitrile that reacts with nucleophiles in acid solution, but so little of this form is present (Sec. 21.5) that the rate of the reaction is very small.

PROBLEMS

21.12 Use an analysis of resonance effects and leaving-group basicities to explain why acid-catalyzed hydrolysis of esters is faster than acid-catalyzed hydrolysis of amides.

21.13 The hydrolysis of acetyl chloride is 7800 times faster than the hydrolysis of acetyl fluoride.

$$H_3C-\overset{\overset{\displaystyle O}{\|}}{C}-F \qquad H_3C-\overset{\overset{\displaystyle O}{\|}}{C}-Cl$$

acetyl fluoride   acetyl chloride

Which factor is more important in determining the relative hydrolysis rate: resonance stabilization of the carbonyl compound or polar stabilization of the transition state? Explain how you know.

21.14 Complete the following reactions.

(a)

$$N\equiv C-CH_2-\overset{\overset{\displaystyle O}{\|}}{C}-OCH_3 + {}^-OH \text{ (1 equiv.)} \xrightarrow[H_2O/CH_3OH]{}$$

(b)

$$F \text{—} \boxed{} \text{—} CO_2CH_3 \ + \ {}^-OH \ \xrightarrow[H_2O]{} \ \xrightarrow{H_3O^+}$$

(c)

$$H_2N\text{—}\overset{\overset{\displaystyle O}{\|}}{C}\text{—}NH_2 \ + \ H_2O \ \xrightarrow[\substack{H_2O \\ heat}]{H_3O^+}$$

21.8 REACTIONS OF CARBOXYLIC ACID DERIVATIVES WITH NUCLEOPHILES

Section 21.7 showed that all carboxylic acid derivatives hydrolyze to carboxylic acids. Water and hydroxide ion, the nucleophiles involved in hydrolysis, are only two of the nucleophiles that react with carboxylic acid derivatives. This section shows how the reactions of other nucleophiles with carboxylic acid derivatives can be used to prepare other carboxylic acid derivatives. As you proceed through this section, notice how all of the reactions fit the pattern of nucleophilic acyl substitution.

A. Reactions of Acid Chlorides with Nucleophiles

Among the most useful ways of preparing carboxylic acid derivatives are the reactions of acid chlorides with various nucleophiles. Because of the great reactivity of acid chlorides,

FIGURE 21.5 The effect of structure on the reactivity of carboxylic acid derivatives in nucleophilic carbonyl substitution. Within the diagram, the blue end of the arrow refers to lower energy and the red end to higher energy. The overall reactivity increases to the right, as shown on the label for the x axis, with red coding for greater reactivity.

such reactions are typically very rapid and can be carried out under mild conditions. Recall that acid chlorides are readily prepared from the corresponding carboxylic acids (Sec. 20.9A).

Reactions of Acid Chlorides with Ammonia and Amines Acid chlorides react rapidly and irreversibly with ammonia or amines to give amides. Reaction of an acid chloride with *ammonia* yields a primary amide:

$$
\underset{\substack{\textbf{decanoyl chloride}}}{CH_3(CH_2)_8-\overset{\overset{\textstyle O}{\|}}{C}-Cl} + 2\,\ddot{N}H_3 \longrightarrow \underset{\substack{\textbf{decanamide}\\ \text{a primary amide}\\ \text{(73\% yield)}}}{CH_3(CH_2)_8-\overset{\overset{\textstyle O}{\|}}{C}-\ddot{N}H_2} + \overset{+}{N}H_4\ Cl^- \qquad (21.30)
$$

$$\underset{\text{(conc. NH}_4\text{OH)}}{}$$

Reaction of an acid chloride with a *primary amine* (an amine of the form RNH$_2$) gives a secondary amide:

$$
\underset{\substack{\text{a primary amine}}}{Ph-\overset{\overset{\textstyle O}{\|}}{C}-Cl + PhCH_2CH_2\ddot{N}H_2 +} \bigodot_{\ddot{N}} \longrightarrow \underset{\substack{\text{a secondary amide}\\ \text{(89–98\% yield)}}}{Ph-\overset{\overset{\textstyle O}{\|}}{C}-\ddot{N}HCH_2CH_2Ph} + \underset{\substack{N\\H}}{\bigodot^+}\ Cl^- \qquad (21.31)
$$

Reaction of an acid chloride with a *secondary amine* (an amine of the form R$_2$NH) gives a tertiary amide:

$$
\underset{\substack{\text{a secondary}\\ \text{amine}}}{Ph-\overset{\overset{\textstyle O}{\|}}{C}-Cl + H-N} \bigcirc + NaOH \longrightarrow \underset{\substack{\text{a tertiary amide}\\ \text{(77–81\% yield)}}}{Ph-\overset{\overset{\textstyle O}{\|}}{C}-N} \bigcirc + H_2O + Na^+\ Cl^- \qquad (21.32)
$$

These reactions are all additional examples of nucleophilic acyl substitution in which an amine reacts as a nucleophile at the carbonyl carbon of the acid chloride. A chloride ion is lost from the tetrahedral intermediate.

$$
\overset{\ddot{O}:}{\underset{\ddot{N}H}{-C-\ddot{C}l:}} \rightleftharpoons \overset{:\ddot{O}:^-}{\underset{\underset{+}{N}H}{-C-\ddot{C}l:}} \rightleftharpoons \overset{:O:}{\underset{\underset{+}{N}-}{-C-N}} + :\ddot{C}l:^- \xrightarrow{\ddot{N}H} \overset{:O:}{\underset{\underset{+}{-\dot{N}H_2}}{-C-\ddot{N}}} \qquad (21.33)
$$

A proton is removed from the amide nitrogen in the last step of the mechanism. Unless another base is added to the reaction mixture, *the starting amine acts as the base in this step.* Hence, for each equivalent of amide that is formed, an equivalent of amine is protonated. When the amine is protonated, its electron pair is taken "out of action," and the amine is no longer nucleophilic.

$$
\underset{\substack{\text{an amine is a}\\ \text{good base and}\\ \text{a good nucleophile}}}{R\overset{\ddot{N}H}{\diagup}\diagdown R} \qquad\qquad \underset{\substack{\text{the conjugate acid}\\ \text{of an amine cannot act}\\ \text{as a nucleophile}}}{R\overset{\overset{\textstyle H\ \ H}{\diagdown\diagup}}{\underset{+}{\diagup N \diagdown}}R}
$$

Hence, if the only base present is the amine nucleophile (for example, as in Eq. 21.30), then at least *two* equivalents must be used: one equivalent as the nucleophile and one as the base in the final proton-transfer step.

The use of excess amine is practical when the amine is cheap and readily available. Another alternative is to use a *tertiary amine* (an amine of the form R$_3$N:) such as triethylamine or pyridine as the base. The reaction in Eq. 21.31 is an example of this strategy.

pyridine　　　　**triethylamine**

CH$_3$CH$_2$—N̈—CH$_2$CH$_3$
　　　　　|
　　　CH$_2$CH$_3$

The presence of a tertiary amine does not interfere with amide formation by another amine because a tertiary amine itself cannot form an amide. (Why?) The use of a tertiary amine is particularly practical if the amine used to form the amide is expensive and cannot be used in excess.

FURTHER EXPLORATION 21.3
Reaction of Tertiary Amines with Acid Chlorides

Yet another alternative for amide formation is to use the *Schotten–Baumann* technique. In this method, the reaction is run with an acid chloride in a separate layer (either alone or in a solvent) over an aqueous solution of NaOH (Eq. 21.32). Hydrolysis of the acid chloride by NaOH is avoided because acid chlorides are typically insoluble in water and therefore are not in direct contact with the water-soluble hydroxide ion. The amine, which is soluble in the acid chloride solution, reacts to yield an amide. The aqueous NaOH extracts and neutralizes the protonated amine that is formed.

occurs in the organic layer　　　　　occurs in the aqueous layer

$$\text{R}-\overset{\text{O}}{\overset{\|}{\text{C}}}-\text{Cl} + \text{R}'-\text{NH}_2 \longrightarrow \text{R}-\overset{\text{O}}{\overset{\|}{\text{C}}}-\text{NH}-\text{R}' + \text{R}'-\overset{+}{\text{NH}_3}\ \text{Cl}^- \xrightarrow{\ ^-\text{OH}\ } \text{R}'-\text{NH}_2 + \text{H}_2\text{O} \quad (21.34)$$

water-insoluble　　　　　　　　water-soluble

The important point about all of the methods for preparing amides is that either two equivalents of amine must be used, or an equivalent of base must be added to effect the final neutralization.

Reaction of Acid Chlorides with Alcohols and Phenols　Esters are formed rapidly when acid chlorides react with alcohols or phenols. In principle, the HCl liberated in the reaction need not be neutralized because alcohols and phenols are not basic enough to be extensively protonated by the acid. However, some esters (such as *tert*-butyl esters; see Further Exploration 21.2) and alcohols (such as tertiary alcohols; Secs. 10.2 and 10.3) are sensitive to acid. In practice, a tertiary amine like pyridine is added to the reaction mixture or is even used as the solvent to neutralize the HCl.

$$\text{3,5-dimethylphenol} + \text{Cl}-\overset{\text{O}}{\overset{\|}{\text{C}}}-\text{CH}_3 \xrightarrow[\text{ether}]{\text{pyridine}} \text{3,5-dimethylphenyl acetate} + \text{HCl} \quad (21.35a)$$

acetyl chloride

3,5-dimethylphenol　　　　　**3,5-dimethylphenyl acetate**　　+　HCl
(reacts with pyridine)

$$\text{Ph}-\overset{\text{O}}{\overset{\|}{\text{C}}}-\text{Cl} + \text{HO}-\text{C(CH}_3)_3 \xrightarrow{\text{quinoline}} \text{Ph}-\overset{\text{O}}{\overset{\|}{\text{C}}}-\text{O}-\text{C(CH}_3)_3 + \text{HCl} \quad (21.35b)$$

benzoyl chloride　　**tert-butyl alcohol**　　　**tert-butyl benzoate**　　+　HCl
(71–76% yield)　　(reacts with quinoline)

As these examples illustrate, esters of tertiary alcohols and phenols, which cannot be prepared by acid-catalyzed esterification, can be prepared by this method.

Sulfonate esters (esters of sulfonic acids) are prepared by the analogous reactions of sulfonyl chlorides (the acid chlorides of sulfonic acids) with alcohols. This reaction was introduced in Sec. 10.4A.

$$CH_3CH_2CH_2CH_2 \text{—OH}$$
1-butanol

+ (tosyl chloride structure) $\xrightarrow{\text{pyridine}}$

p-toluenesulfonyl chloride (tosyl chloride)

(butyl tosylate structure) $\quad + \quad$ HCl $\qquad$ (21.36)

butyl p-toluenesulfonate (butyl tosylate) (88–90% yield)

(reacts with pyridine)

Reaction of Acid Chlorides with Carboxylate Salts

Even though carboxylate salts are weak nucleophiles, acid chlorides are reactive enough to react with carboxylate salts to give anhydrides.

$$CH_3CH_2\text{—}\overset{\text{O}}{\overset{\|}{C}}\text{—Cl} \; + \; Na^+ \; {}^-\ddot{\text{O}}\text{—}\overset{\text{O}}{\overset{\|}{C}}\text{—CH}_3 \xrightarrow{\text{ether}} CH_3CH_2\text{—}\overset{\text{O}}{\overset{\|}{C}}\text{—}\ddot{\text{O}}\text{—}\overset{\text{O}}{\overset{\|}{C}}\text{—CH}_3 \; + \; Na^+ \; Cl^- \qquad (21.37)$$

propionyl chloride   **sodium acetate** (excess)   **acetic propionic anhydride** (60% yield)

This is a second general method for the synthesis of anhydrides. Although the anhydride synthesis discussed in Sec. 20.9B can only be used for the synthesis of symmetrical anhydrides, the reactions of acid chlorides with carboxylate salts can be used to prepare mixed anhydrides, as the example in Eq. 21.37 illustrates.

Summary: Use of Acid Chlorides in Organic Synthesis

One of the most important general methods for converting a carboxylic acid into an ester, amide, or anhydride is first to convert the carboxylic acid into its acid chloride (Sec. 20.9A) and then use one of the acid chloride reactions discussed in this section to form the desired carboxylic acid derivative. To summarize:

STUDY GUIDE LINK 21.4 Another Look at the Friedel-Crafts Reaction

$$R\text{—}\overset{\text{O}}{\overset{\|}{C}}\text{—OH} \xrightarrow{\text{SOCl}_2 \text{ or PCl}_5} R\text{—}\overset{\text{O}}{\overset{\|}{C}}\text{—Cl}$$

- $\xrightarrow{\text{amine}}$ amide
- $\xrightarrow{\text{alcohol or phenol}}$ ester
- $\xrightarrow{\text{carboxylate}}$ anhydride

$\qquad$ (21.38)

B. Reactions of Anhydrides with Nucleophiles

Anhydrides react with nucleophiles in much the same way as acid chlorides—that is, the reaction with an amine yields an amide, the reaction with an alcohol yields an ester, and so on.

$$\text{(21.39)}$$

N-(p-methoxyphenyl)acetamide
(75–79% yield)

$$\text{(21.40)}$$

2-hydroxybenzoic acid
(salicylic acid)

acetic anhydride

2-acetoxybenzoic acid
(aspirin)
(80–90% yield)

Because most anhydrides are prepared from the corresponding carboxylic acids, the use of an anhydride to prepare an ester or amide wastes one equivalent of the parent acid as a leaving group. (For example, acetic acid is a by-product in Eqs. 21.39 and 21.40.) Therefore, this reaction in practice is used only with inexpensive and readily available anhydrides, such as acetic anhydride. However, one exception is the formation of half-esters and half-amides from cyclic anhydrides:

$$\text{(21.41)}$$

succinic
anhydride

methanol

methyl hydrogen succinate
(95–96% yield)

Half-amides of dicarboxylic acids are produced in analogous reactions of amines and cyclic anhydrides. These compounds can be cyclized to imides by treatment with dehydrating agents such as anhydrides, or, in some cases, just by heating, when five- or six-membered rings are formed. This reaction is the nitrogen analog of cyclic anhydride formation (Sec. 20.9B).

$$\text{(21.42)}$$

maleic
anhydride

aniline

(97% yield)

N-phenylmaleimide
(75–80% yield)

C. Reactions of Esters with Nucleophiles

Just as esters are much less reactive than acid chlorides toward hydrolysis, they are also much less reactive toward amines and alcohols. Nevertheless, reactions of esters with these nucleophiles are sometimes useful. The reaction of esters with ammonia or amines yields amides.

$$N\equiv C-CH_2-\overset{\overset{\displaystyle O}{\|}}{C}-OEt \ + \ NH_3 \ \xrightarrow[H_2O]{} \ N\equiv C-CH_2-\overset{\overset{\displaystyle O}{\|}}{C}-NH_2 \ + \ EtOH \quad (21.43)$$

(86% yield)

The reaction of esters with *hydroxylamine* (NH_2OH, Table 19.3) gives *N*-hydroxy-amides; these compounds are known as **hydroxamic acids**.

$$R-\overset{\overset{\displaystyle O}{\|}}{C}-OEt \ + \ NH_2OH \ \longrightarrow \ R-\overset{\overset{\displaystyle O}{\|}}{C}-NHOH \ + \ EtOH \quad (21.44)$$

an ester   **hydroxylamine**   a hydroxamic acid

(Acid chlorides and anhydrides also react with hydroxylamine to form hydroxamic acids.) This chemistry is the basis of the *hydroxamate test*, used mostly for esters. The hydroxamic acid products are easily recognized because they form highly colored complexes with ferric ion.

When an ester reacts with an alcohol under acidic conditions, or with an alkoxide under basic conditions, a new ester is formed.

$$Ph-\overset{\overset{\displaystyle O}{\|}}{C}-OCH_3 \ + \ HO(CH_2)_3CH_3 \ \underset{}{\overset{K^+ \ CH_3(CH_2)_3O^-}{\rightleftharpoons}} \ Ph-\overset{\overset{\displaystyle O}{\|}}{C}-O(CH_2)_3CH_3 \ + \ CH_3OH \quad (21.45)$$

methyl benzoate   **1-butanol**   **butyl benzoate**   **methanol**
                       (excess)   (72% yield)

This reaction is an example of **transesterification**: the conversion of one ester into another by reaction with an alcohol. In transesterification, neither ester is strongly favored at equilibrium. The reaction is driven to completion by the use of an excess of the alcohol nucleophile or by removal of the alcohol by-product by distillation—Le Châtelier's principle in action once again.

PROBLEMS

21.15   Using an acid chloride synthesis as a first step, outline a conversion of hexanoic acid into each of the following compounds.
(a) ethyl hexanoate   (b) *N*-methylhexanamide

21.16   Complete the following reactions by giving the major organic products.

(a)
$$CH_3CH_2CO_2H \ \xrightarrow[\text{(excess)}]{SOCl_2} \ \xrightarrow[\text{(excess)}]{(CH_3)_2NH}$$

(b)

(c)

$$PhCH_2-\overset{\overset{\displaystyle O}{\|}}{C}-Cl \ + \ CH_3CH_2SH \ \longrightarrow$$

(d)
$$CH_3CH_2CH_2-\overset{\overset{\displaystyle O}{\|}}{C}-Cl \ + \ Na^+ \ ^-O-\overset{\overset{\displaystyle O}{\|}}{C}-CH_3 \ \longrightarrow$$

(e)
$$Cl-\overset{\overset{\displaystyle O}{\|}}{C}-Cl \ \text{(excess)} \ + \ CH_3OH \ \longrightarrow$$

(f)
$$Cl-\overset{\overset{\displaystyle O}{\|}}{C}-Cl \ + \ CH_3OH \ \text{(excess)} \ \longrightarrow$$

(g)
$$EtO-\overset{\overset{\displaystyle O}{\|}}{C}-OEt \ + \ HO-CH_2CH_2-OH \ \xrightarrow[\text{heat}]{\text{acid catalyst}} \ (C_3H_4O_3)$$

(h)

21.17 Give the structure of the product in the reaction of each of the following esters with isotopically labeled sodium hydroxide, Na$^+$ ^{18}OH$^-$ and explain your reasoning.

A

B

STUDY GUIDE
LINK 21.5
Esters and
Nucleophiles

21.18 How would you synthesize each of the following compounds from an acid chloride?

(a)

(b)

(c)

(d)

D. Reaction of Amides with Nucleophiles: Penicillin

Like other carboxylic acid derivatives, amides can react with nucleophiles. However, because amides, as we have learned, are very unreactive, weak nucleophiles such as alcohols typically do not react with amides at a useful rate in the laboratory. Nevertheless, the reactions of some amides with nucleophiles are catalyzed by enzymes. An interesting biological example of *enzyme-catalyzed* amide reactivity with an alcohol is found in the mode of action of *penicillin*, the first general-use antibiotic, discovered in 1928 by Sir Alexander Fleming.

Penicillin inhibits *transpeptidase* (TP), an important bacterial enzyme involved in the construction of cell walls, which protect the microorganism from bursting due to the high osmotic pressure within the cell. Transpeptidase catalyzes the formation of amide bonds in peptidoglycan, a component of the cell walls. This enzyme contains a particular amino acid, serine, in its active site that operates as a nucleophile in one of the reactions of peptidoglycan synthesis.

a serine residue in
the transpeptidase enzyme

In the first step of peptidoglycan synthesis, the —OH group of the serine acts as a nucleophile toward the terminal amide bond in a cell-wall precursor protein, peptidoglycan-1 (PG1), to form an ester in which the peptidoglycan-1 is esterified to the enzyme:

$(21.46a)$

transpeptidase
(enzyme)

ester of transpeptidase
and peptidoglycan-1

(R)-alanine

Acidic and basic groups (not shown) in the active site of the enzyme help to catalyze this reaction.

In the second step, the ester from the first step undergoes an aminolysis (Sec. 21.8C) with an amino group of a second precursor, peptidoglycan-2 (PG2). This reaction links the two peptidoglycan chains in an amide bond to generate the completed peptidoglycan linkage of the cell wall and regenerates the active-site serine.

$$(21.46b)$$

ester of transpeptidase and peptidoglycan-1   peptidoglycan-2   completed peptidoglycan   the regenerated transpeptidase

(This process occurs many times in the formation of a cell wall.)

The inhibition of transpeptidase by penicillin blocks cell wall biosynthesis; therefore, this inhibition is lethal to the bacterium. This inhibition occurs because the active-site serine of the transpeptidase reacts at the carbonyl of the β-lactam ring (the cyclic amide containing a four-membered ring) to form an ester.

$$(21.47)$$

penicillin-G   transpeptidase enzyme   an ester of penicillin and the transpeptidase (inhibited transpeptidase)

This reaction is unusually favorable thermodynamically, and it also occurs rapidly, because the β-lactam ring is *strained*; opening the ring releases free energy by relieving ring strain. The penicillin ester thus formed blocks the active site of the transpeptidase, and cell-wall biosynthesis cannot occur. Bacteria have developed penicillin resistance by evolving another enzyme, β-*lactamase* (β-Lac), which intercepts the penicillin and opens the β-lactam ring to form its own serine ester. However, β-lactamase also catalyzes the hydrolysis of its own serine ester to give an inactive form of penicillin and regenerate the enzyme.

$$(21.48)$$

penicillin-G β-lactamase (an enzyme in penicillin-resistant bacteria)   an ester of penicillin and the β-lactamase

an inactive penicillin derivative   the regenerated β-lactamase

The β-lactamase, then, protects the process of cell-wall biosynthesis by inactivating penicillin.

PROBLEM

21.19 (a) Draw the structure of the product that is formed when the following compound is heated with *one equivalent* of sodium methoxide in methanol. Explain your reasoning.

(b) What would the product be if the same compound were heated with a large excess of aqueous H_2SO_4?

 ### Antibiotic Resistance

Chemists have developed inhibitors of the β-lactamase that can be administered along with penicillins. (See Problem 22.98, p. 1181.) However, new bacterial strains have evolved that are resistant to this inhibitor as well. The more antibiotics are used, the greater is the pressure for the evolution of antibiotic-resistant bacterial strains. (The widespread use of antibiotics in feed grain has been a partic- ular area of concern.) For this reason, it is important to limit antibiotic use to cases in which they are truly needed. For example, the common cold is caused by a virus and cannot be cured with antibiotics; so, taking antibiotics for a cold has no effect. The continuous cycle of antibiotic discovery and the evolution of resistance can be expected to repeat itself indefinitely. This cycle means that the development of antibi- otics will always be a new opportunity in the pharmaceutical industry.

21.9   REDUCTION OF CARBOXYLIC ACID DERIVATIVES

A. Reduction of Esters to Primary Alcohols

Lithium aluminum hydride reduces all carboxylic acid derivatives. Reduction of esters with this reagent, like the reduction of carboxylic acids, gives primary alcohols.

$$2 \text{ (ethyl 2-methylbutanoate)} + \underset{\substack{\textbf{lithium}\\\textbf{aluminum}\\\textbf{hydride}}}{\text{LiAlH}_4} \xrightarrow[\text{ether}]{} \xrightarrow{\text{H}_3\text{O}^+} 2 \text{ (2-methyl-1-butanol)} + 2\,\text{EtOH} + \text{Li}^+, \text{Al}^{3+}\text{ salts}$$

ethyl 2-methylbutanoate   **2-methyl-1-butanol**   **ethanol**
(91% yield)   (21.49)

Two alcohols are formed in this reaction, one derived from the *acyl group* of the ester (2-methyl-1-butanol in Eq. 21.49), and one derived from the alkoxy group (ethanol in Eq. 21.49). In most cases, a methyl or ethyl ester is used in this reaction, and the by-product methanol or ethanol is discarded; the alcohol derived from the acyl portion of the ester is typically the product of interest.

As noted several times (Sec. 19.8A), the active nucleophile in $LiAlH_4$ reductions is the *hydride ion* ($H:^-$) delivered from $^-AlH_4$, and this reduction is no exception. Hydride replaces alkoxide at the carbonyl group of the ester to give an aldehyde. (Write the mechanism of this reaction, another example of nucleophilic acyl substitution.)

$$\text{Li}^+\ ^-\text{AlH}_4 + \underset{}{\text{R}-\overset{\text{O}}{\overset{\|}{\text{C}}}-\text{OEt}} \longrightarrow \underset{\text{an aldehyde}}{\text{R}-\overset{\text{O}}{\overset{\|}{\text{C}}}-\text{H}} + \text{Li}^+\ \text{EtO}^- + \text{AlH}_3 \qquad (21.50a)$$

The aldehyde reacts rapidly with $LiAlH_4$ to give the alcohol after protonolysis (Sec. 19.8A).

$$R-\overset{\overset{\displaystyle O}{\|}}{C}-H \xrightarrow{LiAlH_4} \xrightarrow{H_3O^+} R-\overset{\overset{\displaystyle OH}{|}}{\underset{\underset{\displaystyle H}{|}}{C}}-H \qquad (21.50b)$$

The reduction of esters to alcohols thus involves a *nucleophilic acyl substitution* reaction followed by a *carbonyl addition* reaction.

Sodium borohydride ($NaBH_4$), another useful hydride reducing agent, is much less reactive than lithium aluminum hydride. It reduces aldehydes and ketones, but it reacts very sluggishly with most esters; in fact, $NaBH_4$ can be used to reduce aldehydes and ketones selectively in the presence of esters.

Acid chlorides and anhydrides also react with $LiAlH_4$ to give primary alcohols. However, because acid chlorides and anhydrides are usually prepared from carboxylic acids, and because carboxylic acids themselves can be reduced to alcohols with $LiAlH_4$ (Sec. 20.10), the reduction of acid chlorides and anhydrides is seldom used.

B. Reduction of Amides to Amines

Amines are formed when amides are reduced with $LiAlH_4$.

$$LiAlH_4 + 2\,Ph-\overset{\overset{\displaystyle O}{\|}}{C}-NH_2 \longrightarrow \xrightarrow[\text{2) }^-OH]{\text{1) } H_3O^+} 2\,Ph-CH_2-NH_2 + Li^+, Al^{3+} \text{ salts} + 2\,H_2 \qquad (21.51)$$

lithium aluminum hydride   **benzamide**   **benzylamine** (80% yield)

In the workup conditions, H_3O^+ is followed by ^-OH. An aqueous acidic solution is often used to carry out the protonolysis step that follows the $LiAlH_4$ reduction (as shown in the following mechanism). The excess of acid that is typically used converts the amine, which is a base, into its conjugate-acid ammonium ion. Hydroxide is then required to neutralize this ammonium salt and thus give the neutral amine.

$$^-OH + RCH_2\overset{+}{N}H_3 \rightleftharpoons RCH_2\overset{..}{N}H_2 + H_2O \qquad (21.52)$$

conjugate-acid ammonium ion (typical $pK_a = 8$–11)   amine ($pK_a = 15.7$)

Although water itself rather than acid can be used in the protonolysis step, for practical reasons the acidic workup is more convenient. Thus, the extra neutralization step is required.

Amide reduction can be used not only to prepare primary amines from primary amides, but also to prepare secondary and tertiary amines from secondary and tertiary amides, respectively.

$$LiAlH_4 + \text{(cyclohexyl)}-\overset{\overset{\displaystyle O}{\|}}{C}-N(CH_3)_2 \longrightarrow \xrightarrow[\text{2) }^-OH]{\text{1) } H_3O^+} \text{(cyclohexyl)}-CH_2-N(CH_3)_2 + Li^+, Al^{3+} \text{ salts} \qquad (21.53)$$

(88% yield)

The reaction of $LiAlH_4$ with an amide differs from its reaction with an ester. In the reduction of an ester, the *carboxylate oxygen* is lost as a leaving group. If amide reduction were strictly analogous to ester reduction, the nitrogen would be lost, and a primary alcohol would be formed. Instead, it is the *carbonyl oxygen* that is lost in amide reduction.

Ester reduction:

$$R-\overset{\overset{\displaystyle O}{\|}}{C}-OR' \xrightarrow{\text{LiAlH}_4} \xrightarrow{\text{H}_3\text{O}^+} R-CH_2OH + R'OH \qquad (21.54a)$$

the carbonyl oxygen is retained

Amide reduction:

$$R-\overset{\overset{\displaystyle O}{\|}}{C}-NR_2' \xrightarrow{\text{LiAlH}_4} \xrightarrow[\text{2) }^-\text{OH}]{\text{1) H}_3\text{O}^+} R-CH_2NR_2 \qquad (21.54b)$$

the carbonyl oxygen is lost

Let's consider the reason for this difference, using as a case study the reduction of a secondary amide. (The mechanisms of reduction of primary and tertiary amides are somewhat different, but they have the same result.)

In the first step of the mechanism, the weakly acidic amide proton reacts with an equivalent of hydride, a strong base, to give hydrogen gas, AlH_3, and the lithium salt of the amide.

$$(21.55a)$$

The lithium salt of the amide, a Lewis base, reacts with the Lewis acid AlH_3 in a Lewis acid–base association.

$$(21.55b)$$

The resulting species is an active hydride reagent, conceptually much like $LiAlH_4$, and it can deliver hydride to the C=N double bond.

$$(21.55c)$$

The $^-$O—AlH_2 group is subsequently lost from the tetrahedral intermediate because it is less basic than the other possible leaving group, R$\ddot{\text{N}}$—Li. The resulting product is an *imine* (Sec. 19.11A).

$$(21.55d)$$

an imine

The C=N of the imine, like the C=O of an aldehyde, undergoes nucleophilic addition with "H:⁻" from ⁻AlH₄ or from one of the other hydride-containing species in the reaction mixture. Addition of acid to the reaction mixture converts the addition intermediate into an amine by protonolysis and then into its conjugate-acid ammonium ion.

$$\text{Li}^+ \, \ddot{\text{N}}\text{R} \quad \longrightarrow \quad \text{Al}{\Huge\langle} \; + \quad \text{Li}^+ \; :\ddot{\text{N}}\text{R} \quad \xrightarrow{\text{H}_3\text{O}^+} \quad \text{H}\ddot{\text{N}}\text{R} \quad \xrightarrow{\text{H}_3\text{O}^+} \quad \text{H}_2\overset{+}{\text{N}}\text{R} \qquad (21.55\text{e})$$

The ammonium ion is neutralized to the free amine when ⁻OH is added in a subsequent step (Eq. 21.50b).

C. Reduction of Nitriles to Primary Amines

Nitriles are reduced to primary amines by reaction with LiAlH₄, followed by the usual protonolysis step.

$$\underset{\textbf{2-(1-cyclohexenyl)ethanenitrile}}{\text{CH}_2\text{C}{\equiv}\text{N}} \;+\; \underset{\substack{\textbf{lithium}\\\textbf{aluminum}\\\textbf{hydride}}}{\text{LiAlH}_4} \quad \xrightarrow[\text{2) }^-\text{OH}]{\text{1) H}_3\text{O}^+} \quad \underset{\substack{\textbf{2-(1-cyclohexenyl)ethanamine}\\(74\% \text{ yield})}}{\text{CH}_2\text{CH}_2\text{NH}_2} \;+\; \text{Li}^+, \text{Al}^{3+} \text{ salts} \;+\; 2\,\text{H}{-}\text{H} \qquad (21.56)$$

As in amide reduction, isolation of the neutral amine requires addition of ⁻OH at the conclusion of the reaction.

The mechanism of this reaction illustrates again how the C≡N and C=O bonds react in similar ways. This reaction probably occurs as two successive *nucleophilic additions*.

$$\underset{\substack{\text{H}{-}\bar{\text{Al}}\text{H}_3}}{\text{R}{-}\text{C}{\equiv}\text{N:}} \quad \longrightarrow \quad \underset{\underset{\textbf{imine salt}}{\text{H}}}{\text{R}{-}\text{C}{=}\ddot{\text{N}}} \overset{\text{Li}}{} \;+\; \text{AlH}_3 \qquad (21.57\text{a})$$

In the second addition, the imine salt reacts in a similar manner with AlH₃ (or another equivalent of ⁻AlH₄).

$$\underset{\text{H}{-}\text{AlH}_2}{\text{R}{-}\text{CH}{=}\text{N:}}{\overset{\text{Li}}{}} \longrightarrow \left[\underset{\text{AlH}_2}{\text{R}{-}\text{CH}_2{-}\ddot{\text{N}}:}{\overset{\text{Li}}{}} \;\longleftrightarrow\; \underset{\bar{\text{Al}}\text{H}_2}{\text{R}{-}\text{CH}_2{-}\overset{+}{\text{N}}}{\overset{\text{Li}}{}} \right] \qquad (21.57\text{b})$$

In the resulting derivative, both the N—Li and the N—Al bonds are very polar, and the nitrogen has a great deal of anionic character. Both bonds are susceptible to protonolysis, as are the remaining Al—H bonds. Hence, an amine, and then an ammonium ion, is formed when aqueous acid is added to the reaction mixture.

$$\underset{\text{AlH}_2}{\text{RCH}_2{-}\ddot{\text{N}}}{\overset{\text{Li}}{}} \xrightarrow{\text{H}_3\text{O}^+} \underset{+\,\text{Li}^+,\,\text{Al}^{3+}\text{ salts}\,+\,2\,\text{H}_2}{\text{RCH}_2{-}\ddot{\text{N}}\text{H}_2} \xrightarrow{\text{H}_3\text{O}^+} \underset{\substack{(\text{neutralization}\\\text{with }^-\text{OH gives}\\\text{the amine})}}{\text{RCH}_2{-}\overset{+}{\text{N}}\text{H}_3} \qquad (21.57\text{c})$$

Nitriles are also reduced to primary amines by catalytic hydrogenation using Raney nickel, a type of nickel–aluminum alloy.

$$CH_3(CH_2)_4C \equiv N + 2H_2 \xrightarrow[\substack{2000 \text{ psi} \\ 120-130\,°C}]{\text{Raney Ni}} CH_3(CH_2)_4CH_2NH_2 \qquad (21.57d)$$

hexanenitrile   **1-hexanamine**

An intermediate in the reaction is the imine, which is not isolated but is hydrogenated to the amine product. (See also Problem 21.23, p. 1085.)

$$R—C \equiv N \xrightarrow{H_2,\ \text{catalyst}} [R—CH = NH] \xrightarrow{H_2,\ \text{catalyst}} R—CH_2—NH_2 \quad (21.58)$$
imine

The reductions discussed in this and the previous section allow the formation of the *amine* functional group from amides and nitriles, the nitrogen-containing carboxylic acid derivatives. Hence, any synthesis of a carboxylic acid can be used as part of an amine synthesis, but it is important to remember that the amine prepared by these methods must have the following form:

$$R—CH_2—N\diagup^{\diagup}_{\diagdown}$$

C═O or C≡N carbon of
the carboxylic acid derivative

As this diagram shows, the carbon of the carbonyl group or cyano group in the carboxylic acid derivative ends up as a —CH$_2$— group adjacent to the amine nitrogen.

STUDY PROBLEM   21.1

Outline a synthesis of (cyclohexylmethyl)methylamine from cyclohexanecarboxylic acid.

cyclohexanecarboxylic acid   **(cyclohexylmethyl)methylamine**

SOLUTION Any carboxylic acid derivative used to prepare the amine must contain nitrogen; the two such derivatives are amides and nitriles. However, the only type of amine that can be prepared directly by nitrile reduction is a primary amine of the form —CH$_2$NH$_2$. Because the desired product is not a primary amine, the reduction of nitriles must be rejected as an approach to this target.

The amide that could be reduced to the desired amine is *N*-methylcyclohexanecarboxamide:

***N*-methylcyclohexanecarboxamide**

This amide can be prepared, in turn, by reaction of the appropriate amine, in this case methylamine, with an acid chloride:

cyclohexanecarbonyl chloride

Finally, the acid chloride is prepared from the carboxylic acid (Sec. 20.9A).

D. Reduction of Acid Chlorides to Aldehydes

Acid chlorides can be reduced to aldehydes by either of two procedures. In the first, the acid chloride is hydrogenated over a catalyst that has been deactivated, or *poisoned*, with an amine, such as quinoline, that has been heated with sulfur. (Amines and sulfides are catalyst poisons.) This reaction is called the **Rosenmund reduction**.

3,4,5-trimethoxybenzoyl chloride　　　　　**3,4,5-trimethoxybenzaldehyde**
(54–83% yield)

$$+ H_2 \quad \xrightarrow[\substack{\text{sulfur} \\ 50 \text{ psi}}]{\text{Pd/C catalyst} \\ \textbf{quinoline}} \quad + HCl \qquad (21.59)$$

The poisoning of the catalyst prevents further reduction of the aldehyde product.

A second method of converting acid chlorides into aldehydes is the reaction of an acid chloride at low temperature with lithium tri(*tert*-butoxy)aluminum hydride, a "cousin" of LiAlH$_4$.

2,2-dimethylpropanoyl　　**lithium tri(*tert*-butoxy)aluminum**
chloride　　　　　　　**hydride**

$$(CH_3)_3C{-}C{-}Cl \quad + \quad Li^+ \quad H{-}\bar{Al}(OtBu)_3 \quad \xrightarrow[\substack{\text{diglyme} \\ -78\,°C}]{} \quad \xrightarrow{H_3O^+}$$

$$(CH_3)_3C{-}\overset{\overset{\displaystyle O}{\|}}{C}{-}H + 3\ HOtBu + LiCl + Al^{3+}\ \text{salts} \qquad (21.60)$$

2,2-dimethylpropanal

The hydride reagent used in this reduction is derived by the replacement of three hydrogens of lithium aluminum hydride by *tert*-butoxy groups. As the hydrides of LiAlH$_4$ are replaced successively with alkoxy groups, less reactive reagents are obtained. In fact, the preparation of LiAlH(O*t*Bu)$_3$ owes its success to the poor reactivity of its hydride: the reaction of LiAlH$_4$ with *tert*-butyl alcohol stops after three moles of alcohol have been consumed.

$$Li^+\ {}^-AlH_4 + 3\,tBu{-}O{-}H \longrightarrow Li^+\ H{-}\bar{Al}(O{-}tBu)_3 + 3H{-}H \qquad (21.61)$$

The one remaining hydride reduces only the most reactive functional groups. Because *acid chlorides are more reactive than aldehydes toward nucleophiles*, the reagent reacts preferentially with the acid chloride reactant rather than with the product aldehyde. In contrast, lithium aluminum hydride is so reactive that it fails to discriminate to a useful degree between the aldehyde and acid chloride groups, and it thus reduces acid chlorides to primary alcohols.

The reduction of acid chlorides adds another synthesis of aldehydes and ketones to those given in Sec. 19.4. A complete list of methods for preparing aldehydes and ketones is given in Appendix V.

PROBLEMS

21.20 Show how benzoyl chloride can be converted into each of the following compounds.

(a) benzaldehyde　　(b)

PhCH$_2${-}N

21.21 Complete the following reactions by giving the principal organic product(s).

(a)

$$PhCH_2C{\equiv}N \ + \ H_2 \ \xrightarrow[\text{heat}]{\begin{array}{c}\text{Raney Ni}\\\text{(catalyst)}\end{array}}$$

(b)

$$EtO-\overset{\overset{\displaystyle O}{\|}}{C}-CH_2-CN \ \xrightarrow[\text{(excess)}]{LiAlH_4} \ \xrightarrow[\text{2) }^-OH]{\text{1) }H_3O^+}$$

(c)

$$Ph-\underset{\underset{\displaystyle Ph-CH-CO_2Et}{|}}{\overset{\overset{\displaystyle O}{\|}}{O-C-CH_3}} \ + \ LiAlH_4 \text{ (excess)} \ \longrightarrow \ \xrightarrow{H_3O^+}$$

21.22 Give the structures of two compounds that would give the following amine after LiAlH$_4$ reduction.

$$\text{\Large \diagdown\diagup\diagdown\diagup\diagdown}NH_2$$

21.23 (a) In the catalytic hydrogenation of some nitriles to primary amines, secondary amines are obtained as by-products:

$$R-C{\equiv}N \ \xrightarrow{H_2/\text{catalyst}} \ RCH_2NH_2 \ + \ (RCH_2)_2NH$$
$$\text{a secondary amine}$$

Suggest a mechanism for the formation of this by-product. (*Hint:* What is the intermediate in the reduction? How can this intermediate react with an amine?)

(b) Explain why ammonia added to the reaction mixture prevents the formation of this by-product.

E. Relative Reactivities of Carbonyl Compounds

Recall that the reaction of lithium aluminum hydride with a carboxylic acid (Sec. 20.10) or ester (Sec. 21.9A) involves an aldehyde intermediate. But the product of such a reaction is a primary alcohol, not an aldehyde, because *the aldehyde intermediate is more reactive than the acid or ester.* The instant a small amount of aldehyde is formed, it is in competition with the remaining acid or ester for the LiAlH$_4$ reagent. Because it is more reactive, the aldehyde reacts faster than the remaining ester reacts. Hence, the aldehyde cannot be isolated under such circumstances. On the other hand, the lithium tri(*tert*-butoxy)aluminum hydride reduction of acid chlorides (Sec. 21.9D) can be stopped at the aldehyde because acid chlorides are more reactive than aldehydes. When the aldehyde is formed as a product, it is in competition with the remaining acid chloride for the hydride reagent. Because the acid chloride is more reactive, it is consumed before the aldehyde has a chance to react.

These examples show that the outcomes of many reactions of carboxylic acid derivatives are determined by the *relative reactivities of carbonyl compounds* toward nucleophilic reagents, which can be summarized as follows. (Nitriles are included as "honorary carbonyl compounds.")

Relative reactivities of carbonyl compounds:

nitriles < amides < esters, acids << ketones < aldehydes < acid chlorides

$$\xrightarrow{\hspace{3cm}\text{increasing reactivity}\hspace{3cm}} \tag{21.62}$$

The explanation of this reactivity order is the same one used in Sec. 21.7E. Relative reactivity is determined by the stability of each type of carbonyl compound relative to its transition state for addition or substitution. *The more a compound is stabilized, the less reactive it is; the more a transition state for nucleophilic addition or substitution is stabilized, the more reactive the compound is* (Fig. 21.5, p. 1071). For example, esters are stabilized by resonance (Eq. 21.26, p. 1069) in a way that aldehydes and ketones are not. Hence, esters are less reactive than aldehydes. In contrast, resonance stabilization of acid chlorides is much less important, and acid chlorides are destabilized by the electron-attracting polar effect of the chlorine. Moreover, the transition-state energies for nucleophilic substitution reactions of acid chlorides are lowered by the polar effect of chlorine. For these reasons, acid chlorides are more reactive than aldehydes, in which these effects of the chlorine are absent.

21.10 REACTIONS OF CARBOXYLIC ACID DERIVATIVES WITH ORGANOMETALLIC REAGENTS

A. Reaction of Esters with Grignard Reagents

Most carboxylic acid derivatives react with Grignard or organolithium reagents. One of the most important reactions of this type is the reaction of esters with Grignard reagents. In this reaction, a tertiary alcohol is formed after protonolysis. (Secondary alcohols are formed from esters of formic acid; see Problem 21.25a, p. 1088.)

$$(CH_3)_2CH-\overset{\overset{\displaystyle O}{\|}}{C}-OEt \ + \ 2\ CH_3MgI \ \underset{ether}{\longrightarrow} \ \overset{H_3O^+}{\longrightarrow} \ (CH_3)_2CH-\overset{\overset{\displaystyle OH}{|}}{\underset{\underset{\displaystyle CH_3}{|}}{C}}-CH_3 \ + \ EtOH \ + \ Mg^{2+}\ salts \qquad (21.63)$$

**ethyl
2-methylpropanoate**   **methylmagnesium
iodide**

2,3-dimethyl-2-butanol
(92% yield)

$$\overset{\overset{\displaystyle O}{\|}}{C}-OCH_3 \ + \ 2\ Li-(CH_2)_3CH_3 \ \underset{THF}{\longrightarrow} \ \overset{H_3O^+}{\longrightarrow} \ \overset{\overset{\displaystyle OH}{|}}{\underset{\underset{\displaystyle (CH_2)_3CH_3}{|}}{C}}-(CH_2)_3CH_3 \ + \ CH_3OH \ + \ Li^+ \qquad (21.64)$$

butyllithium

**methyl
2-methylpropenoate**

3-butyl-2-methyl-1-hepten-3-ol

Two equivalents of organometallic reagent react per mole of ester, and a second alcohol is produced in the reaction (ethanol and methanol in Eqs. 21.63 and 21.64, respectively). Recall from Sec. 21.9A that a similar situation occurs in the LiAlH$_4$ reduction of esters. This alcohol is typically not the one of interest and is discarded as a by-product.

Like the LiAlH$_4$ reduction of esters, this reaction is a nucleophilic acyl substitution followed by an addition. A ketone is formed in the substitution step. (Fill in the details of the mechanism.)

$$R-\overset{\overset{\displaystyle O}{\|}}{C}-OEt \ + \ H_3C-MgI \ \longrightarrow \ R-\overset{\overset{\displaystyle O}{\|}}{C}-CH_3 \ + \ I-Mg-OEt \qquad (21.65a)$$

The ketone intermediate is not isolated because *ketones are more reactive than esters toward nucleophilic reagents* (Eq. 21.62). The ketone therefore reacts with a second equivalent of the Grignard reagent to form a magnesium alkoxide, which, after protonolysis, gives the alcohol (Sec. 19.9).

$$R-\overset{\overset{\displaystyle O}{\|}}{C}-CH_3 \ + \ H_3C-MgI \ \xrightarrow{[Sec.\ 19.9]} \ \overset{H_3O^+}{\underset{H_2O}{\longrightarrow}} \ R-\overset{\overset{\displaystyle OH}{|}}{\underset{\underset{\displaystyle CH_3}{|}}{C}}-CH_3 \qquad (21.65b)$$

STUDY PROBLEM   21.2

From what ester and Grignard reagent could 3-methyl-3-pentanol be prepared by a single reaction, followed by protonolysis?

SOLUTION   Rephrase the problem in terms of structures:

$$R^1\!-\!\overset{\overset{\displaystyle O}{\|}}{C}\!-\!OR^2 + R^3\!-\!MgBr \longrightarrow \xrightarrow{H_3O^+} CH_3CH_2\!-\!\overset{\overset{\displaystyle OH}{|}}{\underset{\underset{\displaystyle CH_3}{|}}{C}}\!-\!CH_2CH_3$$

What choices should be made for R^1, R^2, and R^3? There are two keys to solving this problem. First, the carbonyl carbon of the ester starting material becomes the α-carbon of the target alcohol; therefore, R^1 must be attached to this carbon. Second, the two *identical* groups on the α-carbon of the target alcohol *must* correspond to group R^3 of the Grignard reagent.

identical groups; therefore, these
are R^3 of the Grignard reagent

$$CH_3CH_2\!-\!\overset{\overset{\displaystyle OH}{|}}{\underset{\underset{\displaystyle CH_3}{|}}{C}}\!-\!CH_2CH_3$$

R^1 of the ester $\longrightarrow$ CH_3

These deductions follow from the examples in the text or from the mechanism of the reaction. What about the group R^2 in the ester? It doesn't matter, because —OR^2 is the leaving group that becomes the by-product alcohol, which is discarded. Because methyl or ethyl esters are common, easily removed, and relatively inexpensive, R^2 = methyl or ethyl is a good choice. Hence, the reaction required to prepare the desired alcohol is

$$H_3C\!-\!\overset{\overset{\displaystyle O}{\|}}{C}\!-\!OEt + CH_3CH_2MgBr \longrightarrow \xrightarrow{H_3O^+} CH_3CH_2\!-\!\overset{\overset{\displaystyle OH}{|}}{\underset{\underset{\displaystyle CH_3}{|}}{C}}\!-\!CH_2CH_3 + EtOH$$

ethyl acetate   **ethylmagnesium**   **ethanol**
                          bromide   (discarded)

3-methyl-3-pentanol

As Study Problem 21.2 demonstrates, the reaction of a Grignard reagent with an ester is an important way to prepare alcohols in which at least two of the groups on the α-carbon are identical. (A complete list of methods for preparing alcohols is found in Appendix V.)

B. Reaction of Acid Chlorides with Lithium Dialkylcuprates

Because acid chlorides are more reactive than ketones, the reaction of an acid chloride with a Grignard reagent can in principle give a ketone without further reaction of the ketone itself.

$$R\!-\!\overset{\overset{\displaystyle O}{\|}}{C}\!-\!Cl + R'\!-\!MgBr \longrightarrow R\!-\!\overset{\overset{\displaystyle O}{\|}}{C}\!-\!R' + Cl\!-\!Mg\!-\!Br \qquad (21.66)$$

However, Grignard reagents are so reactive that this transformation is difficult to achieve in practice without careful control of the reaction conditions; that is, it is hard to prevent the further reaction of the product ketone with the Grignard reagent to give an alcohol. However, lithium dialkylcuprate reagents ($R_2Cu^- Li^+$) can bring about this transformation. Recall from Sec. 11.5C that these reagents are prepared from organolithium reagents and copper(I) chloride:

$$2R\!-\!Li + CuCl \longrightarrow R\!-\!\overset{-}{Cu}\!-\!R \; Li^+ + Li^+ Cl^- \qquad (21.67)$$

lithium dialkylcuprate

Lithium dialkylcuprates are less reactive than Grignard and organolithium reagents; they typically react readily with acid chlorides, aldehydes, and epoxides, very slowly with ketones, and not at all with esters. The reaction of lithium dialkylcuprates with acid chlorides gives ketones in excellent yield.

$$CH_3(CH_2)_4\overset{\overset{\displaystyle O}{\|}}{C}\!-\!Cl + (CH_3)_2Cu^-\,Li^+ \xrightarrow[\substack{-78\,°C \\ 15\ min}]{THF} \xrightarrow{H_2O} CH_3(CH_2)_4\overset{\overset{\displaystyle O}{\|}}{C}\!-\!CH_3 + H_3C\!-\!Cu + Li^+\,Cl^- \quad (21.68)$$

hexanoyl chloride   **lithium dimethylcuprate**   **2-heptanone** (81% yield)   (reacts with H₂O)

Because ketones are much less reactive than acid chlorides toward lithium dialkylcuprates, they do not react further.

The reaction of acid chlorides with lithium dialkylcuprates is one of several good methods for the preparation of ketones. Be sure to review the others in Appendix V. Both this reaction and the reaction of Grignard reagents with esters also provide additional methods for the formation of carbon–carbon bonds. You should also review the other reactions used for carbon–carbon bond formation, found in Appendix VI.

PROBLEMS

21.24 Suggest a sequence of reactions for carrying out each of the following conversions.

(a) benzoic acid to Ph₃C—OH (triphenylmethanol)

(b) butyric acid to 3-methyl-3-hexanol

(c) isobutyronitrile to 2,3-dimethyl-2-butanol (two ways)

(d) propionic acid to 3-pentanone

21.25 (a) What is the general structure of the alcohols obtained by the reaction of Grignard reagents of the form RMgBr with ethyl formate, followed by protonolysis?

(b) Outline a synthesis of 3-pentanol from ethyl formate and a Grignard reagent.

21.26 Predict the product when each of the following compounds reacts with one equivalent of lithium dimethylcuprate, followed by protonolysis. Explain.

(a)

$$N\!\equiv\!C(CH_2)_{10}\!-\!\overset{\overset{\displaystyle O}{\|}}{C}\!-\!Cl$$

(b)

$$CH_3(CH_2)_3O\!-\!\overset{\overset{\displaystyle O}{\|}}{C}(CH_2)_4\overset{\overset{\displaystyle O}{\|}}{C}\!-\!Cl$$

21.11   SYNTHESIS OF CARBOXYLIC ACID DERIVATIVES

Reactions in this and the previous chapter demonstrate that many syntheses of carboxylic acid derivatives begin with other carboxylic acid derivatives. Let's review the methods that have been covered:

Synthesis of esters:

- acid-catalyzed esterification of carboxylic acids (Sec. 20.8A)
- alkylation of carboxylic acids or carboxylate salts (Sec. 20.8B)
- reaction of acid chlorides and anhydrides with alcohols or phenols (Sec. 21.8A)
- transesterification of other esters (Sec. 21.8C)

Synthesis of acid chlorides:

- reaction of carboxylic acids with SOCl₂ or PCl₅ (Sec. 20.9A)

Synthesis of anhydrides:

- reaction of carboxylic acids with dehydrating agents (Sec. 20.9B)
- reaction of acid chlorides with carboxylate salts (Sec. 21.8A)

Synthesis of amides:

■ reaction of acid chlorides, anhydrides, or esters with amines (Sec. 21.8A,C)

Synthesis of nitriles:

The synthesis of nitriles is an important exception to the generalization that carboxylic acid derivatives are usually prepared from other carboxylic acid derivatives. Two syntheses of nitriles are:

■ cyanohydrin formation (Sec. 19.7)
■ S_N2 reaction of cyanide ion with alkyl halides or sulfonate esters

The S_N2 reaction was discussed thoroughly in Sec. 9.4, and the reaction of alkyl halides with cyanide ion was used as an example in Table 9.1 on p. 385. Let's now focus on that reaction as a useful organic synthesis. Recall that an S_N2 reaction of cyanide ion, like all S_N2 reactions, requires a primary or unbranched secondary alkyl halide or sulfonate ester, as in the following examples.

$$PhCH_2Cl \;\; Na^+ + \; ^-:C{\equiv}N \xrightarrow[EtOH/H_2O]{} \;\; PhCH_2C{\equiv}N \; + \; Na^+ \, Cl^- \quad (21.69)$$

benzyl chloride   **phenylacetonitrile** (80–90% yield)

$$Br(CH_2)_3Br \;\; + \;\; 2\,Na^+ \; ^-:C{\equiv}N \xrightarrow[EtOH/H_2O]{} \;\; N{\equiv}C(CH_2)_3C{\equiv}N \; + \; 2\,Na^+ \, Br^- \quad (21.70)$$

1,3-dibromopropane   **glutaronitrile** (77–86% yield)

Both the S_N2 reaction of cyanide and the synthesis of cyanohydrins are noteworthy because they provide additional ways to form carbon–carbon bonds. See how many reactions you can list that form carbon–carbon bonds; check yourself against the summary in Appendix VI.

Because nitriles are prepared from compounds other than carboxylic acid derivatives, the preparation of a nitrile can be particularly useful as an intermediate step in the preparation of a carboxylic acid. Study Problem 21.3 illustrates this approach.

STUDY PROBLEM 21.3

Outline a synthesis of pentanoic acid (valeric acid) from 1-butanol.

$$CH_3CH_2CH_2CH_2{-}OH \xrightarrow{?} CH_3CH_2CH_2CH_2{-}\overset{\overset{\textstyle O}{\|}}{C}{-}OH$$

1-butanol   **pentanoic acid**

SOLUTION   A new carbon–carbon bond must be formed at some point in this synthesis. Because nitriles can be hydrolyzed to carboxylic acids, the immediate precursor of the carboxylic acid can be the corresponding nitrile.

$$CH_3CH_2CH_2CH_2{-}C{\equiv}N \xrightarrow[heat]{H_2O, \, H_3O^+} CH_3CH_2CH_2CH_2{-}\overset{\overset{\textstyle O}{\|}}{C}{-}OH$$

The nitrile, in turn, can be prepared from the corresponding alkyl halide:

$$CH_3CH_2CH_2CH_2{-}Br \xrightarrow{Na^+ \; ^-CN} CH_3CH_2CH_2CH_2{-}C{\equiv}N$$

Notice the formation of the new carbon–carbon bond at this point. The alkyl halide is formed from the alcohol by any of the methods summarized in Sec. 10.5, such as by reaction with concentrated HBr:

$$CH_3CH_2CH_2CH_2{-}OH \xrightarrow{HBr, \, H_2SO_4} CH_3CH_2CH_2CH_2{-}Br$$

This alkyl halide could also be converted into the target carboxylic acid by converting it into a Grignard reagent, and then treating the Grignard reagent with CO_2 (Sec. 20.6.)

21.27   Outline two methods for the preparation of 5-methylhexanoic acid from 1-bromo-4-methylpentane. (See Sec. 20.6.)

21.28   From what nitrile can 2-hydroxypropanoic acid (lactic acid) be prepared? How can this nitrile be prepared from acetaldehyde? (*Hint:* See Sec. 19.7A.)

21.12   USE AND OCCURRENCE OF CARBOXYLIC ACIDS AND THEIR DERIVATIVES

A. Nylon and Polyesters

Two of the most important polymers produced on an industrial scale are *nylon* and *polyesters*. The chemistry of carboxylic acids and their derivatives plays an important role in the synthesis of these polymers.

Nylon is the general name given to a group of polymeric amides, or *polyamides*. The two most widely used are nylon-6,6 and nylon-6.

$$\left(\!\!-NH(CH_2)_6NHC(CH_2)_4C\!-\!\right)_n \qquad \left(\!\!-NH(CH_2)_5C\!-\!\right)_n$$

nylon-6,6   **nylon-6**

Nylon-6,6 was invented by a DuPont chemist, Wallace H. Carothers (1896–1937) in the early 1930s. About 8 billion pounds of nylon are produced annually worldwide. Nylon is used in tire cord, carpet, and apparel.

The starting material for the industrial synthesis of nylon-6,6 is adipic acid. In one process, adipic acid is converted into its dinitrile and then into 1,6-hexanediamine (hexamethylenediamine).

$$HO_2C-(CH_2)_4-CO_2H \xrightarrow{\text{several steps}} N\equiv C-(CH_2)_4-C\equiv N \xrightarrow[\text{catalyst}]{H_2/} H_2N-(CH_2)_6-NH_2 \qquad (21.71)$$

adipic acid   **hexamethylenediamine**

The hexamethylenediamine product is mixed with more adipic acid to form a salt. Heating the salt forms the polymeric amide.

$$-\overset{O}{\overset{\|}{C}}-O^-\ \ H_3\overset{+}{N}- \ \rightleftarrows \ -\overset{O}{\overset{\|}{C}}-OH + H_2N- \xrightarrow[\text{heat}]{-H_2O} -\overset{O}{\overset{\|}{C}}-NH- \qquad (21.72)$$

   salt   acid   amine   nylon
                                                          (an amide)

The reaction of an amine with a carboxylic acid to form an amide is analogous to the reaction of an amine with an ester (Sec. 21.8C). However, much more vigorous conditions are required because the amine is basic, and thus the equilibrium on the left of Eq. 21.72 strongly favors the salt. In the salt, the amine is protonated and therefore not nucleophilic, and the carboxylate ion is very unreactive toward nucleophiles. (Why?) The small amount of amine and carboxylic acid in equilibrium with the salt react when the salt is heated, pulling the equilibrium to the right.

The starting material for nylon-6 is ϵ-caprolactam. (For the structure of ϵ-caprolactam and its polymerization to nylon-6, see Problem 21.31.) Both adipic acid and ϵ-caprolactam are prepared from cyclohexanone (see Problem 21.61, p. 1101), which, in turn, is prepared by the oxidation of cyclohexane. Cyclohexane comes from petroleum. This is an example of the dependence of an important segment of the chemical economy on petroleum feedstocks.

Both nylons are examples of *condensation polymers*. A **condensation polymer** is a polymer formed in a reaction that liberates a small molecule. In the synthesis of nylon-6,6 in Eq. 21.72, for example, formation of each amide bond is accompanied by the loss of the small molecule H_2O. Contrast this type of polymer with an *addition polymer* such as polyethylene (Sec. 5.7). In the formation of polyethylene or other addition polymers, one molecule adds to the other without the loss of any molecular fragment.

Polyesters are condensation polymers derived from the reaction of diols and dicarboxylic acids. One widely used polyester, poly(ethylene terephthalate), can be produced by the esterification of ethylene glycol and terephthalic acid.

$$\text{(21.73)}$$

ethylene glycol

terephthalic acid

poly(ethylene terephthalate)
(a polyester)

Certain familiar polyester fibers and films are sold under the trade names Dacron and Mylar, respectively. Almost 100 billion pounds of polyester are produced annually worldwide. As the following synthetic scheme shows, polyester production also depends on raw materials derived from petroleum.

p-xylene

terephthalic acid   (21.74)

ethylene glycol

polyester

PROBLEMS

21.29 Which polymer should be more resistant to strong base: nylon-6,6 or the polyester in Eq. 21.73? Explain.

21.30 One interesting process for making nylon-6,6 demonstrates the potential of using biomass as an industrial starting material. The raw material for this process, outlined in the following reaction, is the aldehyde furfural, obtained from sugars found in oat hulls. Suggest conditions for carrying out each of the steps in this process indicated by italicized letters.

furfural

furan   **tetrahydrofuran**
(THF)

continued

continued ───────────────────────────────────

21.31 ε-Caprolactam is polymerized to nylon-6 when it is heated with a *catalytic amount* of water.

ε-caprolactam

$$\xrightarrow[\text{heat}]{\text{H}_2\text{O (catalyst)}} \text{nylon-6}$$

Write a mechanism for the polymerization, showing clearly the role of the water.

──

B. Proteins

Proteins are naturally occurring condensation polymers of **alpha-amino acids (α-amino acids)**.

an α-amino acid
(the form at pH 7.4)

a polymeric α-amino acid
(a protein)

As in nylon, the bonds joining the polymeric units are amide bonds. In fact, the impetus for developing nylon was to produce a "synthetic silk"; silk is a protein. Although a few proteins are polymers of a single α-amino acid, most proteins are **heteropolymers**. This means that each unit of the polymer contains a constant part and a variable part. The variable part of each unit is the side chain of the amino acid. There are 20 different naturally occurring α-amino acids. (See Table 27.1, page 1376–1377.)

repeating units
(residues)

variable part (20 different possibilities)

Proteins serve multiple roles in biology. In this and previous chapters, we have focused largely on their role as enzyme catalysts. Chapter 27 is devoted to the structure and stereochemistry of amino acids and proteins, including protein conformation. You have the background to read this chapter at any time.

C. Waxes, Fats, and Phospholipids

Waxes, fats, and phospholipids are all important naturally occurring ester derivatives of fatty acids. A **wax** is an ester of a fatty acid and a "fatty alcohol," a primary alcohol with a long unbranched carbon chain. For example, carnauba wax, obtained from the leaves of the Brazilian carnauba palm, and valued for its hard, brittle characteristics, consists of about 80% of esters derived from C_{24}, C_{26}, and C_{28} fatty acids and C_{30}, C_{32}, and C_{34} alcohols. The following compound is a typical constituent of carnauba wax.

$$CH_3(CH_2)_{24}-\overset{\overset{\text{O}}{\|}}{C}-O-(CH_2)_{31}CH_3$$

a constituent of carnauba wax

A **fat** is an ester derived from a molecule of glycerol and three molecules of fatty acid.

$$H_2C-O-H$$
$$HC-O-H$$
$$H_2C-O-H$$

glycerol

$$H_2C-O-\overset{\overset{\text{O}}{\|}}{C}-(CH_2)_{16}CH_3$$
$$HC-O-\overset{\overset{\text{O}}{\|}}{C}-(CH_2)_{16}CH_3$$
$$H_2C-O-\overset{\overset{\text{O}}{\|}}{C}-(CH_2)_{16}CH_3$$

glyceryl tristearate
(a typical fat)

derived from stearic
acid, a fatty acid

$$HO-\overset{\overset{\text{O}}{\|}}{C}-(CH_2)_{16}CH_3$$

stearic acid

The three acyl groups in a fat (shown in black in the preceding structure) may be the same, as in a glyceryl tristearate, or different, and they may contain unsaturation, which is typically in the form of one or more cis double bonds. Fats with no double bonds, called **saturated fats**, are typically solids; lard is a saturated fat. Fats containing cis double bonds, called **unsaturated fats**, are in many cases oily liquids; olive oil is an unsaturated fat. Fats, which are stored in highly concentrated form in the body, serve as the biological storehouse of energy reserves.

The treatment of fats with NaOH or KOH gives glycerol and the sodium or potassium salts of fatty acids (soaps; Sec. 20.5). This reaction is the origin of the term *saponification* (Sec. 21.7A). The treatment of lard (animal fat) with the ash residue from burning wood (potash, an impure form of potassium carbonate) had been used since antiquity to make soap until commercial soapmaking provided inexpensive bar soap and laundry detergents.

Phospholipids, which were discussed in Sec. 8.7A, are also esters of glycerol. (Compare the structure of phosphatidylethanolamine in Eq. 8.18, p. 367, with the structure of the fat glyceryl tristearate.) The only structural difference between a fat and a phospholipid is that, in a phospholipid, one of the glyceryl primary hydroxy groups is esterified to a polar phosphoric acid derivative rather than to a fatty acid.

$$H_2C-O-\text{fatty acid ester}$$
$$HC-O-\text{fatty acid ester}$$
$$H_2C-O-\text{fatty acid ester}$$

a fat

$$H_2C-O-\text{polar head group}$$
$$HC-O-\text{fatty acid ester}$$
$$H_2C-O-\text{fatty acid ester}$$

a phospholipid

This difference is responsible for the amphipathic behavior of phospholipids. Because fats lack polar head groups, they do not form vesicles when added to water; instead, a fat forms a separate, insoluble layer like any other water-insoluble compound. (If you have ever mixed cooking oil and water, you have observed this behavior.)

KEY IDEAS IN CHAPTER 21

- Carboxylic acid derivatives are polar molecules; except for amides, they have low water solubilities and low-to-moderate boiling points. Because of their capacity for hydrogen bonding, amides, in contrast, have high boiling points (many are solids) and moderate solubilities in water.

- The carbonyl absorptions (and the C≡N absorption of nitriles) are the most important absorptions in the infrared spectra of carboxylic acid derivatives (Table 21.3).

- The proton NMR spectra of carboxylic acid derivatives show typical $\delta 2$–3 absorptions for the α-protons, as well as other characteristic absorptions: $\delta 3.5$–4.5 for the O-alkyl groups of esters, $\delta 2.6$–3 for the N-alkyl groups of amides, and $\delta 7.5$–8.5 for the NH protons of amides.

- In ^{13}C NMR spectra, the chemical shifts of carbonyl carbons are in the $\delta 170$ region, and those of nitrile carbons are in the $\delta 115$–120 range.

- Carboxylic acid derivatives, like carboxylic acids themselves, are weak bases that can be protonated by strong acids on the carbonyl oxygen or the nitrile nitrogen.

- The most characteristic reaction of carboxylic acid derivatives is nucleophilic acyl substitution. Hydrolysis reactions and the reactions of acid chlorides, anhydrides, and esters with nucleophiles are examples of this type of reactivity. Nucleophilic acyl substitution typically involves addition of a nucleophile at the carbonyl carbon to form a tetrahedral addition intermediate, which then loses a leaving group to form product.

- In some reactions of carboxylic acid derivatives, nucleophilic acyl substitution is followed by addition. This is the case, for example, in the LiAlH$_4$ reductions and Grignard reactions of esters, in which the aldehyde and ketone intermediates, respectively, react further.

- The relative reactivities of carbonyl compounds in nucleophilic addition are determined by the relative free energies of the carbonyl compounds and the tetrahedral addition intermediates, which are used as models for the transition states for addition or substitution. For carboxylic acid derivatives, resonance stabilization of the carbonyl compound and polar stabilization of the transition state are important factors in determining relative reactivity.

- The relative reactivities of carbonyl compounds and nitriles with nucleophiles increase in the order nitriles < amides < esters << ketones < aldehydes < acid chlorides. The reactivity of anhydrides is considerably greater than that of esters but less than that of acid chlorides.

- Nitriles react by addition, much like aldehydes and ketones. In some reactions, the product of addition undergoes a second addition (as in reduction to primary amines); and in other cases it undergoes a substitution reaction (as in hydrolysis).

- Polyesters, such as poly(ethylene terephthalate), and polyamides, such as nylon, are important commercial examples of condensation polymers that are also carboxylic acid derivatives. Proteins are naturally occurring polyamides. In nature, waxes, fats, and phospholipids are important natural examples of esters.

 REACTION REVIEW *For a summary of reactions discussed in this chapter, see the* Reaction Review *section of Chapter 21 in the* Study Guide and Solutions Manual.

ADDITIONAL PROBLEMS

21.32 Give the principal organic product(s) expected when ethyl benzoate or the other compound indicated reacts with each of the following reagents.
  (a) H$_2$O, heat, acid catalyst
  (b) NaOH, H$_2$O
  (c) aqueous NH$_3$, heat
  (d) LiAlH$_4$, then H$_2$O
  (e) excess CH$_3$CH$_2$CH$_2$MgBr, then H$_2$O
  (f) product of part (e) + acetyl chloride, pyridine, 0 °C
  (g) product of part (e) + benzenesulfonyl chloride
  (h) Na$^+$ CH$_3$CH$_2$O$^-$

21.33 Give the principal organic product(s) expected when propionyl chloride reacts with each of the following reagents.

(a) H_2O

(b) ethanethiol, pyridine, 0 °C

(c) $(CH_3)_3COH$, pyridine

(d) $(CH_3)_2CuLi$, -78 °C, then H_2O

(e) H_2, Pd catalyst (quinoline/sulfur poison)

(f) $AlCl_3$, toluene, then H_2O

(g) $(CH_3)_2CHNH_2$ (2 equiv.)

(h) sodium benzoate

(i) *p*-cresol (4-methylphenol), pyridine

21.34 Give the structure of a compound that satisfies each of the following criteria.

(a) a compound C_3H_5N that liberates ammonia on treatment with hot aqueous KOH

(b) a compound C_3H_7ON that liberates ammonia on treatment with hot aqueous KOH

(c) a compound that gives 1-butanol and 2-methyl-2-propanol on treatment with excess CH_3MgI followed by protonolysis

(d) a compound that gives equal amounts of 1-hexanol and 2-hexanol on treatment with $LiAlH_4$ followed by protonolysis

(e) a compound $C_3H_2N_2$ that gives CO_2, $^+NH_4$, and acetic acid after boiling in concentrated aqueous HCl

21.35 Complete the diagram shown in Fig P21.35 by filling in all missing reagents or intermediates.

21.36 When (*R*)-(−)-mandelic acid (*α*-hydroxy-*α*-phenylacetic acid) is treated with CH_3OH and H_2SO_4, and the resulting compound is treated with excess $LiAlH_4$ in ether, then H_2O, a levorotatory product is obtained that reacts with

periodic acid. Give the structure, name, and absolute configuration of this product.

21.37 Contrast the results to be expected when levulinic acid (4-oxopentanoic acid) is treated in the following different ways: (a) with excess $LiAlH_4$, then H_3O^+; (b) with excess $NaBH_4$ in methanol, then H_3O^+.

21.38 In clinical studies of patients with atherosclerosis (hardening of the arteries) it was found that one of the metabolites of the hyperlipidemia drug (*Z*)-3-methyl-4-phenyl-3-butenamide (*A*, in the reaction given in Fig. P21.38) is a compound *B*, which has the formula $C_{11}H_{15}NO_2$. When compound *B* is heated in aqueous acid, lactone *C* is formed along with the ammonium ion ($^+NH_4$).

(a) Propose a structure for compound *B*, and explain your reasoning.

(b) Give a curved-arrow mechanism for the conversion of *B* into *C*.

21.39 A major component of olive oil is glyceryl trioleate.

glyceryl trioleate

Figure P21.35

Figure P21.38

(a) Give the structures of the products expected from the saponification of glyceryl trioleate with excess aqueous NaOH.

(b) Give the structure of the product formed when glyceryl trioleate is subjected to catalytic hydrogenation. How would the physical properties of this product differ from those of the starting material?

21.40 Propose a synthesis for each of the following compounds from the indicated starting materials and any other reagents.

(a) compound *A*, the active ingredient in some insect repellents, from 3-methylbenzaldehyde

(b) compound *B* from 2-bromobenzoic acid

A   *B*

21.41 Propose a synthesis for each of the following compounds from butyric acid and any other reagents.

(a) 4-methyl-4-heptanol

(b) 2-methyl-2-pentanol

(c) 4-heptanol

(d) $CH_3CH_2CH_2CH_2CH_2CH_2NH_2$

(e) $CH_3CH_2CH_2CH_2CH_2NH_2$

(f) $CH_3CH_2CH_2CH_2NH_2$

21.42 (a) Draw the structures of all of the products that would be obtained when ($\pm$)-α-phenylglutaric anhydride is treated with ($\pm$)-1-phenylethanol. Which products should be separable without enantiomeric resolution? Which products should be obtained in equal amounts? In different amounts?

(b) Answer the same question for the reaction of the *S* enantiomer of the same alcohol with ($\pm$)-α-phenylglutaric anhydride.

21.43 Treatment of acetic propionic anhydride with ethanol gives a mixture of two esters consisting of 36% of the higher boiling one, *A*, and 64% of the lower boiling one, *B*. Identify *A* and *B*.

21.44 Why are carboxylate salts much less reactive than esters in nucleophilic acyl substitution reactions?

21.45 Compare the base-promoted hydrolysis reaction of a thioester with that of an oxygen ester, and decide how each of the following factors affects the relative rate of the two reactions.

an ethyl thioester   an ethyl ester

(a) resonance stabilization of the carbonyl compound

(b) polar stabilization of the transition state

(The hydrolysis of thioesters is discussed in Sec. 25.5A.)

21.46 You are employed by Fibers Unlimited, a company specializing in the manufacture of specialty polymers. The vice president for research, Strong Fishlein, has asked you to design laboratory preparations of the following polymers. You are to use as starting materials the company's extensive stock of dicarboxylic acids containing six or fewer carbon atoms. Accommodate him.

(a)

nylon-4,6

(b)

21.47 (a) Kevlar® is a polymer made from the reaction of the following two compounds:

1,4-benzenedicarbonyl chloride   **1,4-benzenediamine**

Give the structure of Kevlar in polymer notation.

(b) Kevlar has such high strength that it can be used for bulletproof vests and other applications in which penetration resistance is important. This strength is due to inter-chain hydrogen bonds. Draw a few repeating units of two Kevlar chains and show the hydrogen bonds that connect the two chains. Indicate the hydrogen-bond donors and acceptors.

21.48 A compound *A* has prominent infrared absorptions at 1050, 1786, and 1852 cm^{-1} and shows a single absorption in the proton NMR at $\delta 3.00$. When heated gently with methanol, compound *B*, $C_5H_8O_4$, is obtained. Compound *B* has IR absorption at 2500–3000 (broad), 1730, and 1701 cm^{-1}, and its proton NMR spectrum in D_2O consists of resonances at $\delta 2.7$ (complex splitting) and $\delta 3.7$ (a singlet) in the intensity ratio 4:3. Identify *A* and *B*, and explain your reasoning.

21.49 Propose a structure for a compound *A* that has an infrared absorption at 1820 cm^{-1} and a single proton NMR absorption at $\delta 1.5$. Compound *A* reacts with water to give dimethylmalonic acid and with methanol to give the monomethyl ester of the same acid. (Compound *A* was unknown until its preparation in 1978 by chemists at the University of California, San Diego.)

21.50 You are a chemist working for the Imahot Pepper Company and have been asked to provide some information about *capsaicin*, the active ingredient of hot peppers.

capsaicin

How should capsaicin or the other compounds indicated react under each of the following conditions?

(a) Br_2/CH_2Cl_2

(b) dilute aqueous NaOH

(c) dilute aqueous HCl

(d) H_2, catalyst

(e) product of part (d) + 6 M HCl, heat

(f) product of part (b) + CH_3I

(g) product of part (d) + concentrated aqueous HBr, heat

21.51 Exactly 2.00 g of an ester *A* containing only C, H, and O was saponified with 15.00 mL of a 1.00 M NaOH solution. Following the saponification, the solution required 5.30 mL of 1.00 M HCl to titrate the unused NaOH. Ester *A*, as well as its acid and alcohol saponification products *B* and *C*, respectively, were all optically active. Compound *A* was not oxidized by $K_2Cr_2O_7$, nor did compound *A* decolorize Br_2 in CH_2Cl_2. Alcohol *C* was oxidized to acetophenone by $K_2Cr_2O_7$. When acetophenone was reduced with $NaBH_4$, a compound *D* was formed that reacted with

the acid chloride derived from *B* to give two optically active compounds: *A* (identical to the starting ester) and *E*. Propose a structure for each compound that is consistent with the data. (Note that the absolute stereochemical configurations of chiral substances cannot be determined from the data.)

21.52 Klutz McFingers, a graduate student in his ninth year of study, has suggested the following synthetic procedures and has come to you in the hope that you can explain why none of them works very well (or not at all).

(a) Noting the fact that primary alcohols + HBr give alkyl halides, Klutz has proposed by analogy the nitrile synthesis shown in part (a) of Fig. P21.52.

(b) Klutz has proposed the synthesis for the half-ester of adipic acid shown in part (b) of Fig. P21.52.

(c) Klutz has proposed the synthesis of acetic benzoic anhydride shown in part (c) of Fig. P21.52.

(d) Noting correctly that methyl benzoate is completely saponified by one molar equivalent of NaOH, Klutz has suggested that methyl salicylate should also undergo saponification with one equivalent of NaOH. [See structure in part (d) of Fig. P21.52.]

(e) Klutz, finally able to secure a position with a pharmaceutical company working on β-lactam antibiotics, has proposed the reaction given in part (e) of Fig. P21.52 for deamidation of a cephalosporin derivative.

21.53 An optically active compound *A*, $C_6H_{10}O_2$, when dissolved in NaOH solution, consumed one equivalent of base. On acidification, compound *A* was slowly regenerated. Treatment of *A* with $LiAlH_4$ in ether followed by protonolysis

(a) $Ph—CH_2—OH + HCN \xrightarrow{H_2O} Ph—CH_2—CN + H_2O$

(*Hint:* HCN is a weak acid; its pK_a is 9.4.)

(b) $HO_2C(CH_2)_4CO_2H + CH_3OH \xrightarrow{H_2SO_4} HO_2C(CH_2)_4CO_2CH_3 + H_2O$
    (1.0 mole)   (1.0 mole)

(c)

(d)

methyl salicylate

(e)

Figure P21.52

gave an optically inactive compound *B* that reacted with acetic anhydride to give an acetate diester derivative *C*. Compound *B* was oxidized by aqueous chromic acid to β-methylglutaric acid (3-methylpentanedioic acid). Identify compounds *A*, *B*, and *C*, and explain your reasoning. (The absolute stereochemical configurations of chiral substances cannot be determined from the data.)

21.54 Complete the reactions given in Fig. P21.54 by giving the principal organic products. Explain how you arrived at your answers.

21.55 Identify each of the following compounds from their spectra.

(a) Compound *A*: molecular mass 113; gives a positive hydroxamate test; IR 2237, 1733, 1200 cm^{-1}; proton NMR: δ 1.33 (3*H*, t, *J* = 7 Hz), δ 3.45 (2*H*, s), δ 4.27 (2*H*, q, *J* = 7 Hz).

(b) Compound *B*: $C_6H_{12}O_2$; IR: 1743 cm^{-1}; proton NMR spectrum shown in Fig. P21.55a.

(c) Compound *C*: molecular mass 71; IR: 3200 (strong, broad), 2250 cm^{-1}; no absorptions in the 1500–2250 cm^{-1} range; proton NMR: δ 2.62 (2*H*, t, *J* = 6 Hz), δ 3.42 (1*H*, broad s; eliminated by D$_2$O shake), δ 3.85 (2*H*, t, *J* = 6 Hz).

(d) Compound *D*: EI mass spectrum: two molecular ions of about equal intensity at *m/z* = 180 and 182; IR: 1740 cm^{-1}; proton NMR: δ 1.30 (3*H*, t *J* = 7 Hz); δ 1.80 (3*H*, d, *J* = 7 Hz); δ 4.23 (2*H*, q, *J* = 7 Hz); δ 4.37 (1*H*, q, *J* = 7 Hz).

(e) Compound *E*: UV spectrum: λ$_{max}$ = 272 nm (ε = 39,500); EI mass spectrum: *m/z* = 129 (molecular ion and base peak); IR: 2200, 970 cm^{-1}; proton NMR: δ 5.85 (1*H*, d, *J* = 17 Hz); δ 7.35 (1*H*, d, *J* = 17 Hz); δ 7.4 (5*H*, apparent s).

(f) Compound *F*: $C_{10}H_{13}NO_2$; IR: 3285, 1659, 1246 cm^{-1}; proton NMR spectrum shown in Fig. P21.55b.

(g) Compound *G*: molecular mass 101; IR: 3397, 3200, 1655, 1622 cm^{-1}; ^{13}C NMR: δ 27.5, δ 38.0 (weak), δ 180.5 (weak).

(a)

$$H_3C-\overset{\overset{\displaystyle O}{\|}}{C}-O-\underset{\underset{\displaystyle Ph}{|}}{C}=CH_2 + H_2O \xrightarrow{\text{NaOH}}$$

(b)

$$H_3C-\overset{\overset{\displaystyle O}{\|}}{C}-CH_2-O-\overset{\overset{\displaystyle O}{\|}}{CH} + CH_3OH \xrightarrow[\text{(solvent)}]{CH_3O^- \text{ (trace)}}$$

(c)

2-aminobenzoic acid (NH$_2$, CO$_2$H) $+ Cl-\overset{\overset{\displaystyle O}{\|}}{C}-Cl \longrightarrow (C_8H_5NO_3)$

(d)

$$H_2N-NH_2 + Ph-\overset{\overset{\displaystyle O}{\|}}{C}-Cl \text{ (excess)} \xrightarrow{\text{NaOH}}$$

(e)

$$Ph-\overset{\overset{\displaystyle O}{\|}}{C}-NH(CH_2)_5CN \xrightarrow[\text{heat}]{H_3O^+, H_2O}$$

(f)

2-ethyl pyrrolidinone (NH, CH$_2$CH$_3$) $+ LiAlH_4 \text{ (excess)} \longrightarrow \xrightarrow{H_2O}$

(g)

γ-lactone with CH$_2$(CH$_2$)$_3$CH$_3$ substituent $+ CH_3MgBr \text{ (excess)} \longrightarrow \xrightarrow{H_3O^+}$

(h)

$$EtO-\overset{\overset{\displaystyle O}{\|}}{C}-OEt + EtMgBr \text{ (large excess)} \longrightarrow \xrightarrow{H_3O^+}$$

(i)

$$(CH_3)_3C-\overset{\overset{\displaystyle O}{\|}}{C}-Cl + LiAlH_4 \text{ (excess)} \longrightarrow \xrightarrow{H_3O^+}$$

(j)

$$Ph-MgBr \text{ (1 equiv.)} + CH_3O-\overset{\overset{\displaystyle O}{\|}}{C}(CH_2)_3CH=O \longrightarrow \xrightarrow{H_3O^+}$$

(k) glyceryl tristearate (structure on p. 1093) $+ CH_3OH \xrightarrow{CH_3O^-}$

(l) $(CH_3)_2\underset{\underset{\displaystyle NH_2}{|}}{C}CH_2CH_2CO_2CH_3 \xrightarrow{\text{stand in } CH_3OH} (C_6H_{11}NO)$

Figure P21.54

21.56 Outline a synthesis of each of the following compounds from the indicated starting materials and any other reagents.

(a)

O
‖
CCH₂CH₂CH₃

CH₃

from *o*-bromotoluene

o-methylbutyrophenone

(b) 1-cyclohexyl-2-methyl-2-propanol from bromocyclohexane

(c)

NH   from

CO₂H

CO₂H

phthalic acid

(d) PhNHCH₂CH₂CH(CH₃)₂
   from   (CH₃)₂CHCH₂CO₂H (isovaleric acid)

(a)

(b)

Figure P21.55 (a) The NMR spectrum of compound *B* for Problem 21.55b. The integrals are shown in red over their respective resonances. The resonance at about δ 1.6 is coupled to both of the resonances at δ 1.0 and δ 4.0; the coupling constants of 7.4 Hz and 7.6 Hz are sufficiently close that the *n* + 1 rule is followed. (b) The NMR spectrum of compound *F* in Problem 21.55f. The integrals are shown in red over their respective resonances.

(e)

from

(f) HO CH$_2$NH$_2$

from cyclohexanone

(g)

from benzoic acid as the only carbon source

(h)

from

$$HO-\overset{O}{\underset{\|}{C}}-CH_2CH_2CH_2-\overset{O}{\underset{\|}{C}}-OCH_2CH_3$$

adipic acid monoethyl ester

(i)

$$CH_3NH-\overset{O}{\underset{\|}{C}}-\langle\rangle-OCH_3$$

from *p*-methoxytoluene

(j)

$$PhO-\overset{O}{\underset{\|}{C}}-OCH_3 \quad \begin{array}{l}\text{from phosgene}\\ \text{(Sec. 21.1G)}\end{array}$$

(k)

$$H_3C-\overset{O}{\underset{\|}{C}}-\langle\rangle-\langle\rangle-\overset{O}{\underset{\|}{C}}-CH_3$$

from *p*-bromotoluene

21.57 Rationalize each of the reactions in Fig. P21.57 with a mechanism, using the curved-arrow notation where possible. In part (d), identify compound *A* and show the mechanism for its formation. (Do not give the mechanism of the NaBH$_4$ step.)

21.58 The reaction of Grignard reagents with nitriles is another method of preparing ketones. The example of this synthesis is shown in Fig. P21.58. Identify compound *A*, and give a mechanism for its formation.

21.59 The sequence shown in Fig. P21.59 illustrates a method for the preparation of nitriles from aldehydes. Identify compound *A*, and give a curved-arrow mechanism for the conversion of *A* to the products.

21.60 The *Weinreb amide* method is a good way to prepare ketones in high yields from acid chlorides.

(a) In the first step of this method, an acid chloride is converted into an amide with *N,O*-dimethylhydroxyl-amine. The product is known generically as a *Weinreb amide*, after its inventor Steven Weinreb (b. 1941), a Pennsylvania State University chemist. Give the struc-

(a)

$$O{=}C{=}O + H_2O \underset{\xrightarrow{H_3O^+}}{\rightleftharpoons} HO-\overset{O}{\underset{\|}{C}}-OH$$

(b)

(92% yield)

(c)

(d)

(e)

Figure P21.57

ture of the Weinreb amide by completing the following reaction.

pyridine   **N,O-dimethyl-hydroxylamine**

(b) The Weinreb amide is allowed to react with a Grignard or organolithium reagent to form a *stable* tetrahedral addition intermediate. Using a generic structure for the organolithium reagent R′—Li, draw a structure for this intermediate. This intermediate is stable to decomposition for two reasons. What are they? (*Hint:* The intermediate contains the metal ion from the Grignard or lithium reagent.)

(c) Only after aqueous acid is added does the tetrahedral intermediate break down to the ketone product and other by-products. Write a curved-arrow mechanism for this process and show all the products.

(d) Give the products formed when cyclohexanecarbonyl chloride (*A*) is converted into a Weinreb amide (*B*), and this amide is allowed to react with propylmagnesium bromide (*C*) followed by protonolysis.

A
cyclohexanecarbonyl chloride

C
propylmagnesium bromide

(e) Why not simply allow the acid chloride to react with the Grignard or organolithium reagent? Why is the Weinreb amide used instead?

21.61 ϵ-Caprolactam is the starting material for nylon-6 preparation (p. 1091). ϵ-Caprolactam can be prepared from cyclohexanone in a reaction sequence that involves the *Beckmann rearrangement*, the second of the two reactions shown in Fig. P21.61.

Identify compound *A*, and suggest a curved-arrow mechanism for its conversion to ϵ-caprolactam.

21.62 (a) Build a model of mesitoic acid (2,4,6-trimethylbenzoic acid). What is the most likely conformation of the molecule at the bond between the carbonyl group and the ring? Explain.

mesitoic acid

(b) Explain why the acid-catalyzed hydrolysis of the methyl ester of mesitoic acid does not occur at a measurable rate.

(c) Which *one* of the following methods should be used to make the methyl ester of mesitoic acid: acid-catalyzed esterification in methanol or esterification with diazomethane? Explain your choice.

$Ph—C{\equiv}N + PhMgI \longrightarrow \xrightarrow{H_2O} A \xrightarrow{H_3O^+, H_2O} Ph—\overset{O}{\overset{\|}{C}}—Ph + {}^+NH_4$

Figure P21.58

Figure P21.59

---

cyclohexanone   **ϵ-caprolactam**

Figure P21.61

21.63 In aqueous solution at pH 3, the hydrolysis of phthalamic acid to phthalic acid (see the reaction in Fig. P21.63) is 10^5 times faster than the hydrolysis of benzamide under the same conditions. Furthermore, an isotope *double-labeling experiment* gives the results shown in Fig. P21.63 ($* = {}^{18}O$, $\# = {}^{13}C$). Phthalic anhydride (see p. 1047) was postulated as an intermediate in this reaction.

(a) Using the curved-arrow notation, show how phthalic anhydride is formed from the starting materials.

(b) Show how the intermediacy of phthalic anhydride can explain the double-labeling experiment.

(c) Explain why this mechanism results in a large rate acceleration. (*Hint:* See Sec. 11.8.)

phthalamic acid

equal amounts of each
phthalic acid (isotopically labeled)

Figure P21.63

The Chemistry of Enolate Ions, Enols, and α,β-Unsaturated Carbonyl Compounds

Chapters 19–21 examined the chemistry of carbonyl compounds, concentrating largely on reactions at the carbonyl group. This chapter completes the survey of carbonyl compounds by considering reactions involving the α-carbon. As we will learn, hydrogens at the α-carbon of carbonyl compounds are somewhat acidic. When an α-proton is removed, a conjugate-base anion is formed at the α-carbon. The conjugate-base anion of a carbonyl compound formed by removal of an α-hydrogen is called an **enolate ion** or sometimes simply an **enolate**.

α-hydrogen ⟶ H   :O:
α-carbon ⟶ C—C—R + ⁻:Base ⇌ [R—C̈—C̈—C—R ⟷ R—C=C—R] + H—Base   (22.1)
R   R   R
R

enolate ion

Enolate ions are bases, and, like most bases, they can act as nucleophiles. Hence, the α-carbon of a carbonyl compound, as the site of a conjugate-base enolate ion, is a site of *nucleophilic reactivity*. Much of this chapter deals with aspects of this reactivity.

The α-carbon and α-hydrogens of carbonyl compounds are also involved in the formation of *enols*, which were first introduced in Sec. 14.5A. An **enol** is any compound in which a hydroxy group is on a carbon of a carbon–carbon double bond; that is, an enol is a vinylic alcohol.

$$\underset{\substack{R \\ R}}{\overset{\substack{H \\ |}}{C}}\overset{\overset{O}{\|}}{\underset{}{C}}-R \;\rightleftharpoons\; \underset{\substack{R' \\ }}{\overset{\substack{R \\ }}{C}}=\underset{}{\overset{\overset{OH}{|}}{C}}-R \quad\longleftarrow\; \begin{array}{l}\text{hydroxy group} \\ \text{on a carbon of the} \\ \text{C}=\text{C bond}\end{array} \qquad (22.2)$$

$$\text{enol}$$

As this equation shows, an enol and the corresponding carbonyl compounds are *constitutional isomers*. Most carbonyl compounds with α-hydrogens are in equilibrium with small (in many cases, *very* small) amounts of their enol isomers. Despite their low concentration, enols are intermediates in a number of important reactions of carbonyl compounds. The chemistry of enols is another topic of this chapter.

This chapter also covers some unique chemistry of **α,β-unsaturated carbonyl compounds**—compounds in which a carbonyl group is conjugated with a carbon–carbon double bond.

$$H_3C-CH=CH-\overset{\overset{O}{\|}}{C}-CH_3 \qquad\qquad \qquad\qquad H_3C-CH=CH-\overset{\overset{O}{\|}}{C}-OEt$$

$$\begin{array}{c}\text{a,}\beta\text{-unsaturated ketones} \\ \text{(sometimes called } enones\text{)}\end{array} \qquad\qquad\qquad \text{an a,}\beta\text{-unsaturated ester}$$

Thus, as its title suggests, this chapter has three focal points: the formation of enolate ions and their chemistry; the formation of enols and their chemistry; and the chemistry of α,β-unsaturated carbonyl compounds. Following the pattern of previous chapters, we'll also see some examples of the chemistry of enolates, enols, and α,β-unsaturated carbonyl compounds in biology.

22.1   ACIDITY OF CARBONYL COMPOUNDS

A. Formation of Enolate Anions

The α-hydrogens of many carbonyl compounds, as well as those of nitriles, are weakly acidic. Ionization of an α-hydrogen gives the conjugate-base *enolate anion*.

$$\begin{array}{c}\alpha\text{-hydrogen} \\ pK_a = 18.2\end{array}$$

$$B{:}^- + H_2C\overset{\overset{H}{|}}{\underset{}{-}}\overset{\overset{:O:}{\|}}{C}-Ph \;\rightleftharpoons\; \left[\; H_2\overset{..}{C}\overset{:O:}{-}\overset{}{C}-Ph \;\longleftrightarrow\; H_2C=\overset{\overset{:\overset{..}{O}:^-}{|}}{C}-Ph \;\right] + B-H \qquad (22.3)$$

$$\text{a base} \qquad\qquad\qquad\qquad\qquad \text{conjugate-base enolate ion}$$

$$\begin{array}{c}\alpha\text{-hydrogen} \\ pK_a = 25\end{array}$$

$$B{:}^- + H_2C\overset{\overset{H}{|}}{\underset{}{-}}\overset{\overset{:O:}{\|}}{C}-OEt \;\rightleftharpoons\; \left[\; H_2\overset{..}{C}\overset{:O:}{-}\overset{}{C}-OEt \;\longleftrightarrow\; H_2C=\overset{\overset{:\overset{..}{O}:^-}{|}}{C}-OEt \;\right] + B-H \qquad (22.4)$$

$$\text{conjugate-base enolate ion}$$

The pK_a values of simple aldehydes or ketones are in the range 16–20, and the pK_a values of esters are about 25. The α-hydrogens of nitriles and tertiary amides also have acidities similar to those of esters.

Although carbonyl compounds are classified as weak acids, their α-hydrogens are much more acidic than other types of hydrogens bound to carbon. For example, the dissociation constants of carbonyl compounds are greater than those of alkanes by about 30 powers of 10! To understand the greater acidity of carbonyl compounds, first recall that the stabilization of a base lowers the pK_a of its conjugate acid (Fig. 3.3, p. 116). Enolate ions are resonance-stabilized, as shown in Eqs. 22.3 and 22.4. Hence, carbonyl compounds have lower pK_a values—greater acidities—than carbon acids that lack this stabilization. As discussed in

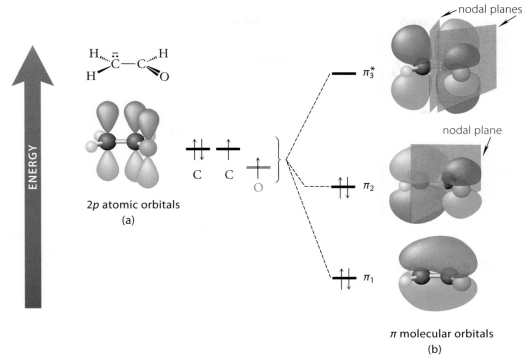

π molecular orbitals
(b)

FIGURE 22.1 An orbital interaction diagram for the conjugate-base enolate ion of acetaldehyde, which is shown in the upper left. (a) The 2*p* orbitals are shown aligned for overlap. The energy of the 2*p* orbital of oxygen (*red*) is lower than the energy of the carbon 2*p* orbitals. (b) The three *π* MOs of the ion. The presence of oxygen lowers the energy of π_2. (Compare with the orbital interaction diagram in Fig. 15.14, p. 744, for the allyl cation.) In the π_1 MO, electrons are delocalized across the entire molecule, and this MO has the lowest energy. This MO is the major source of stabilization of the ion.

Sec. 15.4B, resonance is a symbolic way of depicting orbital overlap. In an enolate ion, the anionic *α*-carbon is *sp²*-hybridized. This hybridization allows the unshared pair of electrons to occupy a 2*p* orbital, which is aligned for overlap with the 2*p* orbitals of the carbonyl group. Fig. 22.1a shows this 2*p*-orbital alignment for the conjugate-base enolate ion of acetaldehyde. This overlap results in the formation of three *π* molecular orbitals (MOs), which are shown in Fig. 22.1b. (Compare Fig. 22.1 with Fig. 15.14, p. 744, for the allyl cation.) Because the enolate ion has four *π* electrons, two of its MOs, π_1 and π_2, are fully occupied. In the occupied MO of lowest energy (π_1), the *π* electrons extend across all three constituent atoms. *This additional overlap provides additional bonding and, hence, additional stabilization.* The occupied MO of higher energy, (π_2), however, has a node that is more or less at the carbonyl carbon. The electrons in this molecular orbital are the ones involved in the chemical reactions of enolate ions. In this molecular orbital, *the α-carbon and the carbonyl oxygen are the major sites of electron density.* This is exactly the same conclusion that we reach from the resonance structures of the enolate ions in Eq. 22.1. These sites of negative charge are also revealed by the red regions in the EPM of this enolate ion:

EPM of acetaldehyde enolate

hybrid structure
of acetaldehyde enolate

If the structure of an enolate ion constrains the geometry of the component $2p$ orbitals so that they *cannot* overlap, the enolate ion is no longer stabilized. (See guideline 4 for writing resonance structures, p. 753.) For example, an enolate ion *cannot* form at the bridgehead carbon of the following bicyclic ketone because the resonance structure of the enolate ion violates Bredt's rule (Sec. 7.6C).

this structure violates
Bredt's rule (Sec. 7.6C)

In other words, the bicyclic structure forces the sp^2 orbital containing the unshared pair of the enolate to be oriented so that it cannot overlap with the π-electron system of the carbonyl group.

perpendicular orbitals
cannot overlap and interact

The second reason for the acidity of α-hydrogens is that the negative charge in an enolate ion is delocalized onto oxygen, an electronegative atom. Thus, the α-hydrogens of carbonyl compounds are much more acidic than the allylic hydrogens of alkenes, even though the conjugate-base anions of both types of compounds are resonance-stabilized.

allylic hydrogen
$pK_a \approx 42$

α-hydrogen
$pK_a = 16.7$

A third reason for the acidity of α-hydrogens is that the *polar effect* of the carbonyl group stabilizes enolate anions, just as it stabilizes carboxylate anions (Sec. 3.6C). This stabilization results from the favorable interaction of the positive end of the C=O bond dipole with the negative charge of the ion:

favorable charge–dipole interaction

A comparison of the pK_a values in Eqs. 22.3 and 22.4 shows that aldehydes and ketones are about 10 million times (seven pK_a units) more acidic than esters. To understand this difference, first recall that the standard free energy of ionization ΔG_a° and the ionization constant K_a are related by the equation $\Delta G_a^\circ = 2.3RT(pK_a)$ [p. 110]. If the free energy of an un-ionized carbonyl compound is lowered relative to that of its conjugate-base enolate ion, then ΔG_a° is

FIGURE 22.2 Resonance stabilization of an ester increases its standard free energy of ionization relative to that of a ketone and raises its pK_a. (Recall that $\Delta G° = 2.3RT(pK_a)$; Eq. 3.34, p. 110.) The free energies of the conjugate-base enolate ions have been placed at the same level for comparison purposes, and the resonance structures of the enolate ions (Eqs. 22.3 and 22.4) are not shown.

---

increased, and its pK_a is also increased; that is, its acidity is reduced (Fig. 22.2). The standard free energy of the ester is lowered relative to that of the ketone by the resonance interaction of the ester oxygen with the carbonyl group, which is shown in Fig. 22.2. This resonance effect overrides the polar effect of the ester oxygen, which, in the absence of resonance, would increase the acidity of esters relative to ketones. In the enolate ion, the analogous resonance structure is *much less important* because of the repulsion between negative charges.

$$\left[\text{RO}-\text{C}-\text{CH}_2 \longleftrightarrow \text{RO}=\text{C}-\text{CH}_2\right]$$

repulsion between negative charges

this resonance structure
is relatively unimportant

Therefore, the partial loss of "ester resonance" on ionization increases $\Delta G_a°$ for the ionization of the ester.

Amide N—H hydrogens are also α-hydrogens; that is, they are attached to an atom that is adjacent to a carbonyl group. The N—H hydrogens are the most acidic hydrogens in primary and secondary amides.

$$\text{B:}^- + \text{R}-\text{C}-\text{NH}_2 \rightleftharpoons \left[\text{R}-\text{C}=\text{NH} \longleftrightarrow \text{R}-\text{C}=\text{NH}\right] + \text{B}-\text{H} \quad (22.5)$$

$pK_a = 15-17$

Similarly, carboxylic acid OH hydrogens ($pK_a \approx 4-5$) are also α-hydrogens. We can think of amide conjugate-base anions as "nitrogen analogs" of enolate ions and carboxylate anions as "oxygen analogs" of enolate anions. Notice that the acidity order carboxylic acids > amides > (aldehydes, ketones) corresponds to the relative electronegativities of the atoms to which the acidic hydrogens are bound—oxygen, nitrogen, and carbon, respectively (element effect; Sec. 3.6A).

22.1 Give the structures of (a) diethyl malonate ($pK_a = 12.9$) and (b) ethyl acetoacetate (ethyl 3-oxobutanoate, $pK_a = 10.7$), identify the acidic hydrogen in each, and explain why these compounds are much more acidic than ordinary esters.

22.2 Which is more acidic: the diamide of succinic acid or the imide succinimide (p. 1049)? Why?

B. Introduction to the Reactions of Enolate Ions

The acidities of aldehydes, ketones, and esters are particularly important because enolate ions are key reactive intermediates in many important reactions of carbonyl compounds. Let's consider the types of reactivity we can expect to observe with enolate ions.

First, enolate ions are Brønsted bases, and they react with Brønsted acids. (The reaction of an enolate ion with an acid is the reverse of Eq. 22.3 or 22.4.) The formation of enolate ions and their reactions with Brønsted acids have two simple but important consequences. First, the α-hydrogens of an aldehyde or ketone—and no others—can be exchanged for deuterium by treating the carbonyl compound with a base in D_2O.

$$(22.6)$$

22.3 Write a mechanism involving an enolate ion intermediate for the reaction shown in Eq. 22.6. Explain why *only* the α-hydrogens are replaced by deuterium.

22.4 Sketch the proton NMR spectrum of 2-butanone, and explain how this spectrum would change if the compound were treated with D_2O and a base.

The second consequence of enolate-ion formation and protonation is that if an optically active aldehyde or ketone owes its chirality solely to an asymmetric α-carbon, and if this carbon bears a hydrogen, the compound will be racemized by base.

$$(22.7)$$

The reason racemization occurs is that the enolate ion, which forms in base, is *achiral* because of the sp^2 hybridization and trigonal planar geometry at its anionic carbon (Fig. 22.1). That is, the ionic α-carbon and its attached groups lie in one plane. The anion can be reprotonated at either face to give either enantiomer with equal probability. Although not very much enolate ion is present at any one time, the reactions involved in the ionization equilibrium are relatively fast, and racemization occurs relatively quickly if the carbonyl compound is left in contact with base.

sp^2-hybridized (trigonal planar) carbon

a chiral ketone

enolate ion: achiral

(22.8a)

(a)

enantiomers

(b)

(22.8b)

α-Hydrogen exchange and racemization reactions of aldehydes and ketones occur much more readily than those of esters. The reason is that aldehydes and ketones are more acidic than esters and therefore form enolate ions more rapidly and under milder conditions.

Enolate ions are not only Brønsted bases but Lewis bases as well. Hence, enolate ions react as *nucleophiles*. Like other nucleophiles, enolate ions react at the carbons of carbonyl groups:

enolate ion

carbonyl compound

further reactions (22.9)

This type of process is the first step in a variety of *carbonyl addition* reactions and *nucleophilic acyl substitution* reactions involving enolate ions as nucleophiles. Much of this chapter will be devoted to a study of such reactions.

Enolate ions, like other nucleophiles, are alkylated by nucleophilic substitution reactions with alkyl halides and sulfonate esters:

enolate ion

(22.10)

This type of reaction, too, is an important part of the chemistry discussed in this chapter.

PROBLEMS

22.5 Explain why the following compound does *not* undergo base-catalyzed exchange in D_2O even though it has an α-hydrogen. (*Hint:* See Secs. 7.6C and 15.6B.)

does not exchange in base/D_2O

continued

continued

22.6   Indicate which hydrogen(s) in each of the following molecules (if any) would be exchanged for deuterium following base treatment in D_2O.

(a)

$(CH_3)_3C—\overset{\displaystyle O}{\overset{\displaystyle \|}{C}}—CH_3$

(b)

(c)

22.2   ENOLIZATION OF CARBONYL COMPOUNDS

Carbonyl compounds with α-hydrogens are in equilibrium with small amounts of their enol isomers. The equilibrium constants shown in the following equations are typical.

$$H_3C—\overset{\displaystyle O}{\overset{\displaystyle \|}{C}}—H \quad \underset{}{\overset{K_{eq} = 5.9 \times 10^{-7}}{\rightleftharpoons}} \quad H_2C=\overset{\displaystyle OH}{\overset{\displaystyle |}{C}}—H \qquad (22.11)$$

acetaldehyde   **acetaldehyde enol**
                               (vinyl alcohol)

$$\text{cyclohexanone} \quad \underset{}{\overset{K_{eq} = 4.2 \times 10^{-7}}{\rightleftharpoons}} \quad \text{cyclohexanone enol} \qquad (22.12)$$

cyclohexanone   **cyclohexanone enol**

Unsymmetrical ketones are in equilibrium with more than one enol. (See Problem 22.8.) Esters contain even smaller amounts of enol isomers than aldehydes or ketones.

$$H_3C—\overset{\displaystyle O}{\overset{\displaystyle \|}{C}}—OEt \quad \underset{}{\overset{K_{eq} \approx 10^{-20}}{\rightleftharpoons}} \quad H_2C=\overset{\displaystyle OH}{\overset{\displaystyle |}{C}}—OEt \qquad (22.13)$$

ethyl acetate   **enol of ethyl acetate**

You may hear the word *tautomers* used to describe the relationship between enols and their corresponding carbonyl compounds. The term **tautomers** means "constitutional isomers that undergo such rapid interconversion that they cannot be independently isolated." Indeed, under most common circumstances, carbonyl compounds and their corresponding enols are in rapid equilibrium. However, chemists now know that the interconversion of enols and their corresponding carbonyl compounds is catalyzed by acids and bases (see following discussion). This reaction can be very slow in dilute solution in the *absence* of acid or base catalysts, and indeed, enols have actually been isolated under very carefully controlled conditions. Hence, the term *tautomers* is not very accurate and is of such limited utility that it is falling into disuse.

As the equilibrium constants in Eqs. 22.11–22.13 suggest, most carbonyl compounds are considerably more stable than their corresponding enols. Furthermore, these equations illustrate the fact that enolizations of esters and carboxylic acids are even less favorable than enolizations of most aldehydes and ketones. The major reason for the instability of enols is that the C=O double bond of a carbonyl group is a stronger bond than the C=C double bond of an enol. With esters and acids, the additional instability of enols results from loss of the stabilizing resonance interaction between the carboxylate oxygen and the carbonyl π electrons that is present in the carbonyl forms. (See Eq. 21.30, p. 1072).

A few enols are more stable than their corresponding carbonyl compounds. Notice that phenol is conceptually an enol—a "vinylic alcohol." However, it is more stable than its keto isomers because phenol is *aromatic*.

(22.14)

phenol
(a stable "enol")

unstable keto isomers of phenol

The enols of β-*dicarbonyl compounds* are also relatively stable. (β-**Dicarbonyl compounds** have two carbonyl groups separated by one carbon.)

(22.15)

2,4-pentanedione
(acetylacetone)
(a β-dicarbonyl compound)

enol form
92% in hexane solution

There are two reasons for the stability of these enols. First, they are conjugated, whereas their parent carbonyl compounds are not. The resonance stabilization (π-electron overlap) associated with conjugation provides additional bonding that stabilizes the enol.

(22.16)

The second stabilizing effect is the intramolecular hydrogen bond present in each of these enols. This provides another source of increased bonding and, hence, increased stabilization.

an intramolecular
hydrogen bond

PROBLEMS

22.7 Draw all enol isomers of the following compounds. If there are none, explain why.

(a) 2-methylcyclohexanone   (b) 2-methylpentanoic acid

(c) benzaldehyde   (d) *N,N*-dimethylacetamide   (e) *N,N*-dimethylformamide

continued

continued ───

22.8 Draw all of the enol forms of 2-butanone. Which is the least stable? Explain why. (*Hint:* Apply what you know about alkene stability.)

22.9 Draw the "enol" isomers of the following compounds. (The "enol" of a nitro compound is called an *aci*-nitro compound, and the "enol" of an amide is called an *imidic acid*.)

(a)

$$H_3C-\overset{+}{\underset{\underset{\ddot{\overset{\cdot\cdot}{O}}:^-}{\parallel}}{\overset{\overset{\ddot{\overset{\cdot\cdot}{O}}:}{\parallel}}{N}}}$$

nitromethane

(b)

$$Ph-\overset{\overset{\ddot{O}:}{\parallel}}{C}-\ddot{N}H_2$$

benzamide

22.10 (a) Explain why 2,4-pentanedione (Eq. 22.15) contains much less enol form in water (15%) than it does in hexane (92%).

(b) Explain why the same compound has a strong UV absorption in hexane solvent (λ_{max} = 272 nm, ϵ = 12,000), but a weaker absorption in water (λ_{max} = 274 nm, ϵ = 2050).

───

**STUDY GUIDE
LINK 22.2**
Kinetic versus
Thermodynamic
Stability of Enols

The formation of enols and the reverse reaction, the conversion of enols into carbonyl compounds, are catalyzed by both acids and bases. Although enols have been isolated and observed under carefully controlled conditions, their rapid conversion into carbonyl compounds under most ordinary circumstances accounts for the fact that enols are difficult to isolate as pure compounds. (See Study Guide Link 22.2.)

The conversion of a carbonyl compound into its enol is called **enolization**. *Base-catalyzed enolization* involves the intermediacy of an *enolate ion*, and is thus a consequence of the acidity of the α-hydrogen.

$$\text{(22.17a)}$$

aldehyde or
ketone

enolate ion
(conjugate base of both the
carbonyl compound and the enol)

enol

Protonation of the enolate anion by water on the α-carbon gives back the carbonyl compound; protonation on oxygen gives the enol. Notice that *the enolate ion is the conjugate base of both the carbonyl compound and the enol.*

Acid-catalyzed enolization involves the conjugate acid of the carbonyl compound. Recall that this ion is also a carbocation (Sec. 19.6). Loss of the proton from oxygen gives back the starting carbonyl compound; loss of the proton from the α-carbon gives the enol. *An enol and its carbonyl isomer have the same conjugate acid.*

$$\text{(22.17b)}$$

aldehyde or
ketone

(conjugate acid of
both the carbonyl compound and the enol)

enol

Exchange of α-hydrogens for deuterium, as well as racemization at the α-carbon, are catalyzed not only by bases (Sec. 22.1B) but also by acids.

(22.18)

optically active

racemate

(22.19)

Both acid-catalyzed processes can be explained by the intermediacy of enols. As you can see by following Eq. 22.17b in the reverse direction, formation of a carbonyl compound from an enol introduces hydrogen from solvent at the α-carbon; this fact accounts for the observed isotope exchange. This carbon of an enol, like that of an enolate ion, has planar geometry and is therefore not an asymmetric carbon. The absence of chirality in the enol accounts for the racemization observed in acid.

PROBLEM

22.11 Give a curved-arrow mechanism for (a) the racemization shown in Eq. 22.19; (b) the deuterium exchange shown in Eq. 22.18.

22.3   α-HALOGENATION OF CARBONYL COMPOUNDS

A. Acid-Catalyzed α-Halogenation

This section begins a survey of reactions that involve enols and enolate ions as reactive intermediates. Halogenation of an aldehyde or ketone in *acidic* solution usually results in the replacement of *one* α-hydrogen by halogen.

p-bromoacetophenone

1-(4-bromophenyl)-2-bromoethanone
(69–72% yield)

(22.20)

cyclohexanone

2-chlorocyclohexanone
(61–66% yield)

(22.21)

Enols are reactive intermediates in these reactions.

aldehyde or
ketone

enol

(22.22a)

Like other "alkenes," enols react with halogens; but unlike ordinary alkenes, enols add only one halogen atom. After addition of the first halogen to the double bond, the resulting carbocation intermediate loses a proton instead of adding the second halogen. (Addition of the second halogen would form a tetrahedral addition intermediate which, in this case, is relatively unstable.)

$$\text{(22.22b)}$$

Acid-catalyzed halogenation provides a particularly instructive case study that shows the importance of the rate law in determining the mechanism of a reaction. Under the usual reaction conditions, the rate law for acid-catalyzed halogenation is

$$\text{rate} = k[\text{ketone}][\text{H}_3\text{O}^+] \qquad (22.23)$$

where k is the rate constant. This rate law implies that, even though the reaction is a halogenation, *the rate is independent of the halogen concentration.* Thus, halogens *cannot* be involved in the transition state for the rate-limiting step of the reaction (Sec. 9.3B). From this observation and others, it was deduced that *enol formation* (Eq. 22.22a) *is the rate-limiting process in the acid-catalyzed halogenation of aldehydes and ketones.* Because the halogen is not involved in enol formation, it does not appear in the rate law at the concentrations of halogen ordinarily used.

$$\text{(22.24)}$$

Enol formation is described in this equation as the rate-limiting *process.* This process consists of two elementary steps, as shown in Eq. 22.17b (p. 1112). The rate-limiting *step* of acid-catalyzed enolization is the second step, removal of the α-proton. The same step, therefore, is also the rate-limiting *step* of α-halogenation.

Because only one halogen is introduced at a given α-carbon in acidic solution, it follows that introduction of a second halogen is much slower than introduction of the first. The slower halogenation is probably a consequence of the decreased stability of the carbocation intermediate that is formed by reaction of the halogen with the halogenated enol. This carbocation is destabilized by the electron-withdrawing polar effect of *two* halogens:

$$\text{(22.25)}$$

halogenated enol

carbocation intermediate
is destabilized by the polar effect
of two bromines

If the rate-limiting transition state resembles this carbocation, then the transition state should have very high energy and the rate should be small.

PROBLEMS

22.12 (a) Sketch a reaction free-energy diagram for acid-catalyzed enol formation using the mechanism in Eq. 22.17b as your guide. Assume that the second step, proton removal from the α-carbon, is rate-limiting.

(b) Incorporating the results of part (a), sketch a reaction free-energy diagram for the acid-catalyzed halogenation of an aldehyde or ketone.

22.13 Explain why:

(a) the rate of iodination of optically active 1-phenyl-2-methyl-1-butanone in acetic acid/HNO_3 is identical to its rate of racemization under the same conditions.

(b) the rates of bromination and iodination of acetophenone are identical at a given acid concentration.

B. Halogenation of Aldehydes and Ketones in Base: The Haloform Reaction

Halogenation of aldehydes and ketones with α-hydrogens also occurs in base. In this reaction, *all* α-hydrogens are substituted by halogen.

$$3NaOH + (CH_3)_3C\overset{O}{\overset{\|}{C}}CH_3 + 3Br_2 \xrightarrow[\substack{H_2O/dioxane \\ 0\,°C}]{NaOH} (CH_3)_3C\overset{O}{\overset{\|}{C}}CBr_3 + 3Na^+Br^- + 3H_2O \quad (22.26a)$$

α-hydrogens (pointing to the CH_3)

no α-hydrogens (pointing to the $(CH_3)_3C$)

When the aldehyde or ketone starting material is either acetaldehyde or a methyl ketone (as in Eq. 22.26a), the product of halogenation is a trihalo carbonyl compound, which is not stable under the reaction conditions. This compound reacts further to give, after acidification of the reaction mixture, a carboxylic acid and a haloform. (Recall from Sec. 8.2A that a *haloform* is a trihalomethane—that is, a compound of the form HCX_3, where X = halogen.)

$$(CH_3)_3C\overset{O}{\overset{\|}{C}}CBr_3 \xrightarrow{^-OH} \xrightarrow{H_3O^+} (CH_3)_3C\overset{O}{\overset{\|}{C}}OH + HCBr_3 \quad (22.26b)$$

(71–74% yield)   **bromoform**

The conversion of acetaldehyde or a methyl ketone into a carboxylic acid and a haloform by halogen in base, followed by acidification, as exemplified by Eq. 22.26a–b, is called the **haloform reaction**. Notice that, in a haloform reaction, a carbon–carbon bond is broken.

The mechanism of the haloform reaction involves the formation of an *enolate ion* as a reactive intermediate.

$$R\overset{O}{\overset{\|}{C}}CH_3 + OH^- \rightleftharpoons R\overset{O}{\overset{\|}{C}}\ddot{C}H_2 + H_2O \quad (22.27a)$$

enolate ion

The enolate ion reacts as a nucleophile with halogen to give an α-halo carbonyl compound.

$$R\overset{O}{\overset{\|}{C}}\ddot{C}H_2 + :\ddot{B}r\overset{\frown}{—}\ddot{B}r: \longrightarrow R\overset{O}{\overset{\|}{C}}CH_2\ddot{B}r: + :\ddot{B}r:^- \quad (22.27b)$$

However, halogenation does not stop here, because the enolate ion of the α-halo ketone is formed even more rapidly than the enolate ion of the starting ketone. The reason is that the polar effect of the halogen stabilizes the enolate ion and, by Hammond's postulate, the transition state for enolate-ion formation. Consequently, a second bromination occurs.

$$R-\overset{\overset{O}{\|}}{C}-\overset{\overset{H}{|}}{\underset{\underset{H}{|}}{C}}-Br \longrightarrow R-\overset{\overset{O}{\|}}{C}-\overset{\overset{H}{|}}{\underset{\cdot\cdot}{C}}-Br \xrightarrow{Br-Br} R-\overset{\overset{O}{\|}}{C}-CHBr_2 + Br^- \qquad (22.27c)$$

The dihalo carbonyl compound brominates again, even more rapidly. (Why?)

$$R-\overset{\overset{O}{\|}}{C}-CHBr_2 \xrightarrow{Br_2, OH^-} R-\overset{\overset{O}{\|}}{C}-CBr_3 + Br^- + H_2O \qquad (22.27d)$$

A carbon–carbon bond is broken when the trihalo carbonyl compound undergoes a *nucleophilic acyl substitution reaction.*

$$R-\overset{\overset{:O:}{\|}}{C}-CBr_3 \rightleftharpoons R-\overset{\overset{:\overset{\cdot\cdot}{O}:}{|}}{\underset{\underset{:\overset{\cdot\cdot}{O}H}{|}}{C}}-CBr_3 \rightleftharpoons R-\overset{\overset{:O:}{\|}}{C}-\overset{\cdot\cdot}{O}-H + {}^-:CBr_3 \longrightarrow R-\overset{\overset{:O:}{\|}}{C}-\overset{\cdot\cdot}{O}:^- + H-CBr_3 \qquad (22.27e)$$
a trihalomethyl
anion

The leaving group in this reaction is a trihalomethyl anion. Usually, carbanions are too basic to serve as leaving groups; but trihalomethyl anions are much less basic than ordinary carbanions. (Why?) However, the basicity of trihalomethyl anions, although low enough for them to act as leaving groups, is high enough for them to react irreversibly with the carboxylic acid by-product, as shown in the last part of Eq. 22.27e. This acid–base reaction drives the overall haloform reaction to completion. (This situation is analogous to saponification, which is also driven to completion by ionization of the carboxylic acid product; Sec. 21.8A.) The carboxylic acid itself can be isolated by acidifying the reaction mixture, as shown in Eq. 22.26b.

Occasionally, the haloform reaction can be used to prepare carboxylic acids from readily available methyl ketones. This reaction was also once used as a qualitative test for methyl ketones, called the **iodoform test**. In the iodoform test, a compound of unknown structure is mixed with alkaline I_2. A yellow precipitate of iodoform (HCI_3) is taken as evidence for a methyl ketone (or acetaldehyde, the "methyl aldehyde"). The iodoform test is specific for methyl ketones because only by replacement of *three* hydrogens with halogen does the carbon become a good enough leaving group for the nucleophilic acyl substitution reaction shown in Eq. 22.27e to occur.

Alcohols of the form shown in Eq. 22.28 also undergo the iodoform reaction because they are oxidized to methyl ketones (or to acetaldehyde, in the case of ethanol) by the basic iodine solution.

$$R-\overset{\overset{OH}{|}}{C}H-CH_3 \xrightarrow{I_2, \text{ base}} R-\overset{\overset{O}{\|}}{C}-CH_3 \qquad (22.28)$$
undergoes the
iodoform reaction

PROBLEMS

22.14   Give the products expected (if any) when each of the following compounds reacts with Br_2 in NaOH.

(a)

(b)

(c) OH
|
Ph—CH—CH$_3$

22.15 Give the structure of a compound $C_6H_{10}O_2$ that gives succinic acid and iodoform on treatment with a solution of I_2 in aqueous NaOH, followed by acidification.

C. α-Bromination of Carboxylic Acids

Carboxylic acids can be brominated at their α-carbons. A bromine is substituted for an α-hydrogen when a carboxylic acid is treated with Br_2 and a catalytic amount of red phosphorus or PBr_3. (The actual catalyst is PBr_3; phosphorus can be used because it reacts with Br_2 to give PBr_3.)

$$CH_3CH_2CH_2CH_2CH_2CO_2H + Br_2 \xrightarrow{\text{P or } PBr_3} CH_3CH_2CH_2CH_2\underset{\underset{Br}{|}}{C}HCO_2H + HBr \quad (22.29)$$

hexanoic acid

2-bromohexanoic acid
(83–89% yield)

This reaction, called the **Hell–Volhard–Zelinsky reaction** after its discoverers, is sometimes nicknamed the **HVZ reaction**.

The first stage in the mechanism of the HVZ reaction is the conversion of a small amount of the carboxylic acid into the acid bromide by the catalyst PBr_3 (Sec. 20.9A).

$$3\ R_2CH\overset{O}{\overset{||}{-}}C\!-\!OH + PBr_3 \longrightarrow 3\ R_2CH\overset{O}{\overset{||}{-}}C\!-\!Br + P(OH)_3 \quad (22.30a)$$

carboxylic acid
with α-hydrogens
(R = H, alkyl, or aryl)

an acid
bromide

From this point, the mechanism closely resembles that for the acid-catalyzed bromination of ketones (Eqs. 22.22a and 22.22b). The *enol* of the acid bromide is the species that actually brominates.

$$R_2CH\overset{O}{\overset{||}{-}}C\!-\!Br \;\rightleftarrows\; R_2C\overset{OH}{\overset{|}{=}}C\!-\!Br \xrightarrow{Br_2} R_2\underset{\underset{Br}{|}}{C}\overset{O}{\overset{||}{-}}C\!-\!Br + HBr \quad (22.30b)$$

acid bromide   enol form

α-bromo acid
bromide

When *a small amount* of PBr_3 catalyst is used, the α-bromo acid bromide reacts with the carboxylic acid to form more acid bromide, which is then brominated as shown in Eq. 22.30b.

$$R_2CH\overset{O}{\overset{||}{-}}C\!-\!OH + R_2\underset{\underset{Br}{|}}{C}\overset{O}{\overset{||}{-}}C\!-\!Br \;\rightleftarrows\; R_2CH\overset{O}{\overset{||}{-}}C\!-\!Br + R_2\underset{\underset{Br}{|}}{C}\overset{O}{\overset{||}{-}}C\!-\!OH \quad (22.31)$$

enters the bromination
sequence at Eq. 22.30b

α-bromo acid

Thus, when a catalytic amount of PBr_3 is used, the reaction product is the α-bromo acid.

If one full equivalent of PBr_3 is used, the α-bromo acid bromide is the reaction product; this can be used in many of the reactions of acid halides discussed in Sec. 21.8A. For example, the reaction mixture can be treated with an alcohol to give an α-bromo ester:

$$CH_3CH_2\overset{O}{\overset{||}{-}}C\!-\!OH \xrightarrow[\text{P (1 equiv.)}]{Br_2} CH_3\underset{\underset{Br}{|}}{C}H\overset{O}{\overset{||}{-}}C\!-\!Br \xrightarrow[\substack{(CH_3)_2\ddot{N}\!-\!Ph \\ \text{(a base)}}]{(CH_3)_3COH} CH_3\underset{\underset{Br}{|}}{C}H\overset{O}{\overset{||}{-}}C\!-\!OC(CH_3)_3 + (CH_3)_2\overset{+}{\underset{\underset{Ph}{|}}{N}}H\ Br^- \quad (22.32)$$

propanoic acid

***tert*-butyl 2-bromopropanoate**

D. Reactions of α-Halo Carbonyl Compounds

Most α-halo carbonyl compounds are very reactive in S_N2 reactions and can be used to prepare other α-substituted carbonyl compounds.

$$(CH_3)_2\ddot{S}: \quad :\ddot{Br}-CH_2-\overset{\overset{O}{\parallel}}{C}-Ph \xrightarrow[25\ °C,\ 30\ min]{acetone/H_2O} H_3C-\overset{CH_3}{\underset{+}{\overset{|}{S}}}-CH_2-\overset{\overset{O}{\parallel}}{C}-Ph \quad :\ddot{Br}:^- \quad (22.33)$$
$$(85\%\ yield)$$

In the case of α-halo ketones, nucleophiles used in these reactions must not be too basic. For example, dimethyl sulfide, used in Eq. 22.33, is a very weak base but a fairly good nucleophile. (Stronger bases promote enolate-ion formation; and the enolate ions of α-halo ketones undergo other reactions.) More basic nucleophiles can be used with α-halo acids because, under basic conditions, α-halo acids are ionized to form their carboxylate conjugate-base anions; a second ionization to give an enolate ion, which would introduce a second negative charge into the molecule, does not occur.

$$\underset{\substack{\text{2,4-dichlorophenol}\\ \text{(ionized by NaOH)}}}{Cl-\!\!\!\!\!\bigcirc\!\!\!\!\!-OH} + \underset{\text{chloroacetic acid}}{Cl-CH_2-CO_2H} \xrightarrow[2)\ H_3O^+]{1)\ NaOH\ (2\ equiv.)} \underset{\substack{\text{2,4-dichlorophenoxyacetic acid}\\ \text{(2,4-D, a selective herbicide;}\\ \text{87\% yield)}}}{Cl-\!\!\!\!\!\bigcirc\!\!\!\!\!-O-CH_2-CO_2H} + Cl^- \quad (22.34)$$

$$:N\!\!\equiv\!\!C:^- + \underset{\substack{\text{chloroacetate}\\ \text{anion}}}{Cl-CH_2-CO_2^-} \longrightarrow \xrightarrow{H_3O^+} \underset{\substack{\text{cyanoacetic acid}\\ \text{(77–80\% yield)}}}{:N\!\!\equiv\!\!C-CH_2-CO_2H} + Cl^- \quad (22.35)$$

The following comparison gives a quantitative measure of the S_N2 reactivity of α-halo carbonyl compounds:

$$Cl-CH_2-\overset{\overset{O}{\parallel}}{C}-CH_3 + KI \xrightarrow{acetone} I-CH_2-\overset{\overset{O}{\parallel}}{C}-CH_3 + KCl \qquad \begin{array}{c}\textit{relative rate:}\\ 35{,}000\end{array} \quad (22.36a)$$

$$Cl-CH_2CH_2CH_3 + KI \xrightarrow{acetone} I-CH_2CH_2CH_3 + KCl \qquad 1 \quad (22.36b)$$

The explanation for the enhanced reactivity is probably similar to that for the increased reactivity of allylic alkyl halides in S_N2 displacements (Fig. 17.3, p. 850).

In contrast, α-halo carbonyl compounds react so slowly by the S_N1 mechanism that this reaction is not useful.

$$H_3C-\overset{CH_3}{\underset{\underset{:\ddot{Cl}:}{|}}{\overset{|}{C}}}-\overset{\overset{O}{\parallel}}{C}-CH_3 \xrightarrow{very\ slow} \overset{H_3C}{\underset{H_3C}{>}}\overset{+}{C}-\overset{\overset{O}{\parallel}}{C}-CH_3 + :\ddot{Cl}:^- \quad (22.37)$$

In fact, *reactions that require the formation of carbocations alpha to carbonyl groups generally do not occur.* Although it might seem that an α-carbonyl carbocation should be resonance-stabilized like an allylic cation, its resonance structure is not important because it involves an electron-deficient oxygen.

$$
\left[\begin{array}{c} H_3C \\ \underset{H_3C}{\overset{+}{C}} - \overset{:O:}{\overset{\parallel}{C}} - CH_3 \end{array} \right] \longleftrightarrow\kern-1.5em\times\kern0.5em \left[\begin{array}{c} H_3C \\ C = \overset{:O:^+}{C} - CH_3 \\ H_3C \end{array} \right] \tag{22.38}
$$

electron-deficient oxygen

not an important
resonance structure

Moreover, the carbocation is destabilized by its unfavorable electrostatic interaction with the bond dipole of the carbonyl group—that is, with the partial positive charge on the carbonyl carbon atom.

$$
(CH_3)_2\overset{+}{C} \underset{}{\overset{\overset{\textstyle :O:^{\delta-}}{\parallel}}{\underset{\delta+}{C}}} CH_3
$$

destabilizing
electrostatic interaction

PROBLEMS

22.16 What product is formed when
(a) phenylacetic acid is treated first with Br_2 and one equivalent of PBr_3, then with a large excess of ethanol?
(b) propionic acid is treated first with Br_2 and one equivalent of PBr_3, then with a large excess of ammonia?

22.17 Give the structure of the product expected in each of the following reactions.

(a)

$$
\underset{\textbf{1-bromo-2-butanone}}{CH_3CH_2\overset{\overset{\textstyle O}{\parallel}}{C}CH_2Br} + \underset{\textbf{pyridine}}{\overset{}{\underset{\ddot N}{\bigcirc}}} \longrightarrow
$$

(b)

$$
\underset{\boldsymbol{\alpha}\textbf{-bromoacetophenone}}{BrCH_2\overset{\overset{\textstyle O}{\parallel}}{C}-Ph} + \underset{\textbf{sodium acetate}}{CH_3\overset{\overset{\textstyle O}{\parallel}}{C}-O^- \ Na^+} \longrightarrow
$$

22.18 Give a curved-arrow mechanism for the reaction in Eq. 22.34. Your mechanism should show why *two equivalents* of NaOH must be used.

22.4 ALDOL ADDITION AND ALDOL CONDENSATION

A. Base-Catalyzed Aldol Reactions

In aqueous base, acetaldehyde undergoes a reaction called the *aldol addition*.

$$
\underset{\textbf{acetaldehyde}}{2 \ H_3C-\overset{\overset{\textstyle O}{\parallel}}{C}H} \xrightarrow[\substack{H_2O \\ 4\text{–}5\,°C}]{NaOH} \underset{\substack{\textbf{3-hydroxybutanal} \\ \textbf{(aldol)} \\ \text{(50\% yield)}}}{H_3C-\overset{\overset{\textstyle OH}{|}}{C}H-CH_2-\overset{\overset{\textstyle O}{\parallel}}{C}H} \tag{22.39}
$$

The term **aldol** is both a traditional name for 3-hydroxybutanal and a generic name for β-hydroxy carbonyl compounds. An **aldol addition** is a reaction of two aldehyde molecules to form a β-hydroxy aldehyde. The aldol addition is a very important and general reaction of aldehydes and ketones that have α-hydrogens. Notice that this reaction provides another method of forming carbon–carbon bonds.

The base-catalyzed aldol addition involves an *enolate ion* as an intermediate. In this reaction, an enolate ion, formed by the reaction of acetaldehyde with aqueous NaOH, adds to a second molecule of acetaldehyde.

(22.40a)

(22.40b)

The aldol addition is another nucleophilic addition to a carbonyl group. In this reaction, the nucleophile is an enolate ion. The reaction may *look* more complicated than some additions because of the number of carbon atoms in the product. However, it is not conceptually different from other nucleophilic additions, such as cyanohydrin formation.

(22.41)

PROBLEM

22.19   Use the reaction mechanism to deduce the product of the aldol addition reaction of (a) PhCH$_2$CH=O (phenylacetaldehyde); (b) propionaldehyde.

The aldol addition is reversible. Like many other carbonyl addition reactions (Sec. 19.7B), the equilibrium for the aldol addition is more favorable for aldehydes than for ketones.

$$2\,H_3C-\overset{\displaystyle O}{\overset{\displaystyle \|}{C}}-CH_3 \quad \underset{\text{(equilibrium lies to the left)}}{\overset{Ba(OH)_2}{\rightleftharpoons}} \quad H_3C-\overset{\overset{\displaystyle OH}{|}}{\underset{\underset{\displaystyle CH_3}{|}}{C}}-CH_2-\overset{\displaystyle O}{\overset{\displaystyle \|}{C}}-CH_3 \qquad (22.42)$$

acetone

4-hydroxy-4-methyl-2-pentanone
(diacetone alcohol)

In this aldol addition reaction of acetone, the equilibrium favors the ketone reactant rather than the addition product, diacetone alcohol. This product can be isolated in good yield only if an apparatus is used that allows the product to be removed from the base catalyst as it is formed. This strategy drives the reaction toward formation of product by Le Châtelier's principle.

Under more severe conditions (higher base concentration, or heat, or both), the product of aldol addition undergoes a dehydration reaction.

$$2\,CH_3CH_2CH_2CH\!=\!O \quad \underset{80\,°C}{\overset{1\,M\,NaOH}{\rightleftharpoons}} \quad CH_3CH_2CH_2\overset{\overset{\displaystyle OH}{|}}{CH}-\underset{\underset{\displaystyle CH_2CH_3}{|}}{CH}CH\!=\!O \quad \longrightarrow$$

butanal

2-ethyl-3-hydroxyhexanal
(the aldol addition product)

$$CH_3CH_2CH_2CH\!=\!\underset{\underset{\displaystyle CH_2CH_3}{|}}{C}CH\!=\!O \;+\; H_2O \qquad (22.43)$$

2-ethyl-2-hexenal
(86% yield)

The sequence of reactions consisting of the aldol addition followed by dehydration, as in Eq. 22.43, is called the **aldol condensation**. (A **condensation** is a reaction in which two molecules combine to form a larger molecule with the elimination of a small molecule, in many cases water.)

> The term *aldol condensation* has been used historically to refer to the aldol addition reaction as well as to the addition and dehydration reactions together. To eliminate ambiguity, *aldol condensation* is used in this text only for the addition–dehydration sequence. The term *aldol reactions* is used to refer generically to both addition and condensation reactions.

The dehydration part of the aldol condensation is a β-elimination reaction catalyzed by base, and it occurs in two distinct steps through an *enolate-ion intermediate*.

$$(22.44)$$

aldol addition
product

carbanion
intermediate

α,β-unsaturated
carbonyl compound

This is not a concerted β-elimination. In this respect, it differs from the E2 reaction. (See Problem 9.17, p. 407.) A β-elimination mechanism that involves the unimolecular decomposition of the conjugate base is sometimes called an **E1cB mechanism**.

elimination ———→ E1cB ←——— conjugate base

↑

unimolecular

A base-catalyzed dehydration reaction of simple alcohols is unknown; ordinary alcohols do *not* dehydrate in base. However, β-hydroxy aldehydes and β-hydroxy ketones do for two reasons. First, their α-hydrogens are relatively acidic. Recall that base-promoted β-eliminations are particularly rapid when acidic hydrogens are involved (Secs. 17.3B and 9.5B). Second, the product is conjugated and therefore is particularly stable. To the extent that the transition state of the dehydration reaction resembles the α,β-unsaturated ketone, it too is stabilized by conjugation, and the elimination reaction is accelerated (Hammond's postulate).

The product of the aldol condensation is an α,β-unsaturated carbonyl compound. The aldol condensation is an important method for the preparation of certain α,β-unsaturated carbonyl compounds. Whether the aldol addition product or the condensation product is formed depends on reaction conditions, which must be worked out on a case-by-case basis. You can assume for purposes of problem-solving, unless stated otherwise, that either the addition product or the condensation product can be prepared.

Musical History of the Aldol Condensation

III. Notturno

Discovery of the aldol condensation in 1872 is usually attributed solely to Charles-Adolphe Wurtz, a French chemist who trained Friedel and Crafts. However, the reaction was first investigated during the period 1864–1873 by Aleksandr Borodin (1833–1887), a Russian chemist who was also a self-taught and proficient musician and composer. (Borodin was one of "The Five," a circle of self-trained St. Petersburg composers who sought to reshape Russian music along nationalistic lines.) Borodin found it difficult to compete with Wurtz's large, modern, well-funded laboratory. Borodin also lamented that his professional duties so burdened him with "examinations and commissions" that he could only compose when he was at home ill. Knowing this, his musical friends used to greet him, "Aleksandr, I hope you are ill today!"

B. Acid-Catalyzed Aldol Condensation

Aldol condensations are also catalyzed by acid.

mesityl oxide
(79% yield)

$$2\,H_3C\!-\!\overset{\overset{\displaystyle O}{\|}}{C}\!-\!CH_3 \xrightarrow{\text{acid}} \overset{\displaystyle H_3C}{\underset{\displaystyle H_3C}{>}}C\!=\!CH\!-\!\overset{\overset{\displaystyle O}{\|}}{C}\!-\!CH_3 + H_2O \qquad (22.45)$$

Acid-catalyzed aldol condensations, as in this example, generally give α,β-unsaturated carbonyl compounds as products; addition products cannot be isolated.

In acid-catalyzed aldol condensations, the conjugate acid of the aldehyde or ketone is a key reactive intermediate.

$$H_3C\!-\!\overset{\overset{\displaystyle :O:}{\|}}{C}\!-\!CH_3 \;\; \overset{\longleftarrow}{\rightleftharpoons} \;\; H_3C\!-\!\overset{\overset{\displaystyle \overset{+}{:}O\!\diagup H}{\|}}{C}\!-\!CH_3 + \overset{..}{O}H_2 \qquad (22.46a)$$

This protonated ketone plays two roles. First, it serves as a source of the *enol*, as shown in Eq. 22.17b on p. 1112. Second, the protonated ketone is the electrophilic species in the reaction. It reacts as an electrophile with the π electrons of the enol to give the conjugate acid of the addition product:

(22.46b)

As the second part of Eq. 22.46b shows, the loss of a proton gives the β-hydroxy ketone product. Under the acidic conditions, this material spontaneously undergoes acid-catalyzed dehydration to give an α,β-unsaturated carbonyl compound:

(22.46c)

STUDY GUIDE LINK 22.3
Dehydration of β-Hydroxy Carbonyl Compounds

This dehydration drives the aldol condensation to completion. (Recall that without this dehydration, the aldol condensation of ketones is unfavorable; Eq. 22.42).

Let's contrast the species involved in the acid- and base-catalyzed aldol reactions. An *enol*, not an enolate ion, is the nucleophilic species in an acid-catalyzed aldol condensation. *Enolate ions are too basic to exist in acidic solution.* Although an enol is much less nucleophilic than an enolate ion, it reacts at a useful rate because the protonated carbonyl compound (an α-hydroxy carbocation) with which it reacts is a potent electrophile. In a base-catalyzed aldol reaction, an *enolate ion* is the nucleophile. A protonated carbonyl compound is *not* an intermediate because it is too acidic to exist in basic solution. The electrophile that reacts with the enolate ion is a *neutral* carbonyl compound. To summarize:

Reaction	*Nucleophile*	*Electrophile*
Base-catalyzed aldol reaction	enolate ion	neutral carbonyl compound
Acid-catalyzed aldol condensation	enol	protonated carbonyl compound

C. Special Types of Aldol Reactions

Crossed Aldol Reactions The preceding discussion considered only aldol reactions between two molecules of the same aldehyde or ketone. When two *different* carbonyl compounds are used, the reaction is called a **crossed aldol reaction**. In many cases, the result of a crossed aldol reaction is a difficult-to-separate mixture, as Study Problem 22.1 illustrates.

STUDY PROBLEM 22.1

Give the structures of the aldol addition products expected from the base-catalyzed reaction of acetaldehyde and propionaldehyde.

SOLUTION Such a reaction involves four different species: acetaldehyde (A) and its enolate ion (A'), as well as propionaldehyde (P) and its enolate ion (P'):

$$
\underset{A}{H_3C-\overset{\overset{\displaystyle O}{\|}}{C}-H} \qquad \underset{A'}{H_2\overset{..}{C}-\overset{\overset{\displaystyle O}{\|}}{C}-H} \qquad \underset{P}{CH_3CH_2-\overset{\overset{\displaystyle O}{\|}}{C}-H} \qquad \underset{P'}{CH_3\overset{..}{C}H-\overset{\overset{\displaystyle O}{\|}}{C}-H}
$$

Four possible addition products can arise from the reaction of each enolate ion with each aldehyde:

$$
\underset{A+A'}{\overset{\overset{\displaystyle OH}{|}}{CH_3CHCH_2CH}=O} \qquad \underset{P+A'}{\overset{\overset{\displaystyle OH}{|}}{CH_3CH_2CHCH_2CH}=O} \qquad \underset{\underset{A+P'}{CH_3}}{\overset{\overset{\displaystyle OH}{|}}{CH_3CHCHCH}=O} \qquad \underset{\underset{P+P'}{CH_3}}{\overset{\overset{\displaystyle OH}{|}}{CH_3CH_2CHCHCH}=O}
$$

(Be sure you see how each product is formed; write a mechanism for each, if necessary.) To complicate the situation even further, diastereomers are possible for the last two products because each has two asymmetric carbons.

Crossed aldol reactions that provide complex mixtures, such as the one in Study Problem 22.1, are not very useful because the product of interest is not formed in very high yield, and because isolation of one product from a complex mixture is in most cases extremely tedious. Although conditions that favor one product or another in crossed aldol reactions have been worked out in specific cases, as a practical matter, under the usual conditions (aqueous or alcoholic acid or base), useful crossed aldol reactions are limited to situations in which *a ketone with α-hydrogens is condensed with an aldehyde that has no α-hydrogens*. An important example of this type is the **Claisen–Schmidt condensation**. In a Claisen–Schmidt condensation, a ketone with α-hydrogens—acetone in the following example—is condensed with an aromatic aldehyde that has no α-hydrogens—benzaldehyde in this case.

$$
\underset{\substack{\textbf{benzaldehyde}}}{PhCH=O} + \underset{\substack{\textbf{acetone} \\ \textbf{(excess)}}}{H_3C-\overset{\overset{\displaystyle O}{\|}}{C}-CH_3} \xrightarrow{\text{aqueous NaOH}} \underset{\substack{\textbf{benzalacetone} \\ \textbf{(4-phenyl-3-buten-2-one)} \\ \textbf{(65–78\% yield)}}}{\overset{\displaystyle Ph \diagup \overset{\displaystyle H}{} \diagdown}{\underset{\displaystyle H}{} \overset{\displaystyle}{C}=C \diagup \overset{\overset{\overset{\displaystyle O}{\|}}{C}-CH_3}{}}} + H_2O \qquad (22.47)
$$

Notice that the addition product is not isolated in this reaction; only the condensation product is isolated because it is highly conjugated and thus very stable. Additionally, only the trans stereoisomer is isolated because it is much more stable than the cis stereoisomer.

In view of the complex mixture obtained in the example used in Study Problem 22.1, it is reasonable to ask why only one product is obtained from the crossed aldol condensation in Eq. 22.47. The analysis of this case highlights several important principles of carbonyl-compound reactivity. First, *because the aldehyde in the Claisen–Schmidt reaction has no α-hydrogens, it cannot act as the enolate component of the aldol condensation*; consequently, two of the four possible crossed aldol products cannot form. The other possible side reaction is the aldol addition reaction of the ketone with itself, as in Eq. 22.42; why doesn't this reaction occur? The enolate ion from acetone can react either with another molecule of acetone or with benzaldehyde. Recall that *addition to a ketone occurs more slowly than* *addition to an aldehyde* (Sec. 19.7C). Furthermore, even if addition to acetone does occur, *the aldol addition reaction of two ketones is reversible* (Eq. 22.42) and *addition to an aldehyde has a more favorable equilibrium constant than addition to a ketone* (Sec. 19.7B). Thus, in Eq. 22.47, both the rate and equilibrium for addition to benzaldehyde are more favorable than they are for addition to a second molecule of acetone. Thus, the product shown in Eq. 22.47 is the only one formed.

The Claisen–Schmidt condensation, like other aldol condensations, can also be catalyzed by acid.

$$Ph—CH \!=\! O + H_3C—\overset{\overset{\displaystyle O}{\|}}{C}—Ph \xrightarrow[\text{CH}_3\text{CO}_2\text{H}]{\text{H}_2\text{SO}_4} \quad \underset{\substack{Ph \qquad H \\ \text{(95\% yield)}}}{\overset{\substack{H \qquad \overset{\overset{\displaystyle O}{\|}}{C}—Ph}}{C\!=\!C}} \qquad (22.48)$$

**STUDY GUIDE
LINK 22.4**
Understanding
Condensation
Reactions

Directed Aldol Reactions   There have been many approaches to the crossed-aldol problem. A well-established approach is to *pre-form* an enolate ion, or a compound that behaves like an enolate ion, from one carbonyl compound. This enolate component is then allowed to react with the carbonyl group of a second carbonyl compound. If we can dictate successfully, by a suitable choice of reagents and conditions, which component is to be the enolate component and which is to be the carbonyl component, we have solved the crossed-aldol problem. An aldol reaction that meets these criteria is usually called a **directed aldol reaction**.

To illustrate a directed aldol reaction, we'll use the reaction of an enolate ion derived from a ketone with the carbonyl group of an aldehyde. Consider, for example, the reaction of 2-pentanone with propanal.

2-pentanone   **propanal**

The solution to this problem involves formation of an enolate of the ketone, 2-pentanone, and then allowing it to react with the propanal, the aldehyde.

In the case of 2-pentanone, there are two possible enolate ions (or three, if we count stereoisomers). We show these with a lithium counter-ion:

A (*E* and *Z*)   *B*

Enolate ions *A* are more stable than enolate ion *B* because the double bond in *A* has more alkyl substituents. The question is whether any of these enolates can be formed selectively.

It was discovered in the early 1970s that a family of very strong, highly branched nitrogen bases, such as the following two examples, can be used to form stable enolate ions rapidly and irreversibly at –78 °C from ketones (and esters, as we shall see in Sec. 22.8B).

**lithium
diisopropylamide
(LDA)**

**lithium
cyclohexylisopropylamide
(LCHIA)**

pK_a of conjugate acids ≈ 35

(Do not confuse the term *amide* in the names of these bases with the carboxylic acid derivative. This term has a double usage. As used here, an *amide* is the conjugate-base anion of an amine.) The conjugate acids of these bases are amines, which have pK_a values near 35. These amide bases themselves are generated from the corresponding amines and butyllithium (a commercially available organolithium reagent) at –78 °C in tetrahydrofuran (THF) solvent.

$$\underset{\substack{\textbf{diisopropylamine}\\ \text{p}K_a \approx 35}}{\text{N—H}} + \underset{\textbf{butyllithium}}{\text{CH}_3\text{CH}_2\text{CH}_2\text{CH}_2\text{—Li}} \xrightarrow[-78\,°C]{\text{THF}} \quad \text{N:}^- \;\; \text{Li}^+ + \underset{\substack{\textbf{butane}\\ \text{p}K_a \approx 55}}{\text{CH}_3\text{CH}_2\text{CH}_2\text{CH}_3} \qquad (22.49)$$

Because ketones have pK_a values near 25, these amide bases are strong enough to convert ketones completely into their conjugate-base enolate ions. When LDA reacts with 2-pentanone, the α-proton of the methyl group is removed *very rapidly* to give the conjugate-base lithium enolate *B*:

$$\underset{\substack{\textbf{2-pentanone}\\ (\text{p}K_a \approx 19)}}{} + \underset{\textbf{LDA}}{(i\text{-Pr})_2\ddot{\text{N}}{:}^- \;\text{Li}^+} \longrightarrow \left[\underset{B}{} \right] + \underset{\substack{\textbf{diisopropylamine}\\ (\text{p}K_a \approx 35)}}{(i\text{-Pr})_2\ddot{\text{N}}\text{H}} \qquad (22.50)$$

Notice that this is the *less stable* enolate ion of 2-pentanone.

Both the enolate ion and LDA are drawn for simplicity as simple ionic compounds, but in reality they are more complex. They can have several different structures that depend on concentration and solvent, but they typically are multi-molecular aggregates that include several solvent molecules. (For the aggregated structure of the amide base, see Further Exploration 23.2 in the *Study Guide*, Chapter 23.) The enolate structure shown in Fig. 22.3 is typical. What you should notice about this structure is that the oxygens of the enolate are heavily coordinated to the lithium atoms with partial covalent bonds, and the α-carbons are exposed along the periphery. The aggregated character of these reagents is important to their reactivity, as we'll soon see.

Why is the *less stable* enolate ion formed selectively in Eq. 22.50? The reason is that *the less stable enolate is formed more rapidly.* This is another example of kinetic versus thermodynamic control (Sec. 15.4C). The isopropyl groups of the amide base play a key role in this selectivity. The large, highly branched base reacts with a methyl proton—that is, a proton at the carbon with fewer alkyl substituents—more rapidly for steric reasons. Van der Waals repulsions occur in the vicinity of the other α-carbon to retard the reaction of LDA with

FIGURE 22.3 Typical aggregated structure of a lithium enolate in THF solvent. Solvent molecules are shown in blue. Notice that the each enolate oxygen is coordinated to three lithiums. (This bonding, indicated by dashed lines, is similar to a strong hydrogen bond with lithium instead of hydrogen.) The α-carbons of the enolate (*red*) are accessible along the periphery of the aggregate.

the proton at that carbon. Steric repulsion, then, is used to *retard* proton removal and more branched (and therefore more congested) carbons. The aggregated character of the amide base probably intensifies this steric effect.

Why doesn't the amide base react as a nucleophile at the carbonyl carbon? Again, steric repulsion is important. A reaction at a carbonyl group with two alkyl substituents is much slower than a reaction at a tiny, unhindered proton. The aggregated character of the amide base undoubtedly intensifies this steric effect as well. For an amide base to react at the carbonyl is analogous to trying to put a dinner plate in a vending machine slot intended for a quarter (or a credit card).

Why couldn't the enolate *B*, once formed, react with un-ionized ketone to form a mixture of *A* and *B*?

$$(22.51)$$

The answer lies in the rate at which the proton is removed from the ketone. The base LDA is so strong that *it deprotonates the ketone instantly*, whereas the enolate that is formed is a much weaker base and reacts more slowly with the ketone. Moreover, the ketone is *added to* the base. This experimental protocol ensures that very little of the ketone is present in solution simultaneously with the enolate.

Once the enolate is formed, the aldehyde is added to it, and a rapid addition occurs. This addition is assisted by coordination of both oxygens—the oxygen of the enolate and the oxygen of the aldehyde—to the lithium. (The aldehyde oxygen probably replaces a solvent molecule in the aggregate structure of Fig. 22.3.) This coordination brings the exposed enolate α-carbon (*red*) into proximity with the aldehyde carbonyl carbon (*blue*) through a six-membered cyclic transition state:

$$(22.52a)$$

This equation shows why the α-carbon rather than the oxygen is the nucleophilic site of the enolate. Protonation of the lithium alkoxide affords the aldol addition product when dilute aqueous acid is added.

$$(22.52b)$$

6-hydroxy-4-octanone
(62% yield)

We could go on to ask why the addition product of Eq. 22.52a does not react with another equivalent of ketone enolate. Remember that ketone carbonyl groups are much less reactive than aldehyde carbonyl groups; therefore, the reaction of the ketone enolate with the aldehyde is the faster addition reaction.

This discussion shows that all of the potential side reactions are largely avoided by the character of the reagents and the experimental protocol (addition of aldehyde to pre-formed enolate at low temperature).

We can expect the directed aldol addition to be useful when the enolate component is derived from a methyl ketone or another ketone in which a single enolate constitutional isomer can be formed, and the carbonyl component is derived from an aldehyde or another unusually reactive carbonyl compound.

PROBLEM

22.20 (a) Give the simplified (nonaggregated) ionic structure of the enolate formed by treatment of 2,2-dimethylcyclohexanone with LDA; then, give the product of its directed aldol reaction with acetaldehyde following acidic workup.

2,2-dimethylcyclohexanone

(b) Two diastereomers of the product in part (a) are formed. Explain. Are they formed in equal amounts or different amounts?

Intramolecular Aldol Condensations   When a molecule contains more than one aldehyde or ketone group, an *intramolecular* reaction (a reaction within the same molecule) is possible. In such a case, the aldol condensation results in formation of a ring. Intramolecular aldol condensations are particularly favorable when five- and six-membered rings can be formed because of the proximity effect (Sec. 11.8).

$$+ \text{H}_2\text{O} \qquad (22.53)$$

PROBLEM

22.21 Predict the product(s) in each of the following aldol condensations.

(a)

(equal molar amounts)

(b) acetophenone + hexanal $\xrightarrow{\text{NaOH}}$

(c)

(d)

D. Synthesis with the Aldol Condensation

The aldol condensation can be applied to the synthesis of a wide variety of α,β-unsaturated aldehydes and ketones, and it is also another method for the formation of carbon–carbon bonds.

(See the complete list in Appendix VI.) If you want to prepare a particular α,β-unsaturated aldehyde or ketone by the aldol condensation, you must ask two questions: (1) What starting materials are required in the aldol condensation? (2) With these starting materials, is the aldol condensation of these compounds a feasible one?

The starting materials for an aldol condensation can be determined by mentally "splitting" the α,β-unsaturated carbonyl compound at the double bond:

That is, work backward from the desired synthetic objective by replacing the double bond on the carbonyl side by two hydrogens and on the other side by a carbonyl oxygen ($=O$) to obtain the structures of the starting materials in the aldol condensation.

Knowing the potential starting materials for an aldol condensation is not enough; you must also know whether the condensation is one that works, or whether instead it is one that is likely to give troublesome mixtures. (This issue was introduced in Study Problem 4.9 on p. 158.) In other words, you can't make every conceivable α,β-unsaturated aldehyde or ketone by the aldol condensation—only certain ones. This point is illustrated in Study Problem 22.2.

STUDY PROBLEM 22.2

Determine whether the following α,β-unsaturated ketone can be prepared by an aldol reaction.

SOLUTION   Following the procedure in Eq. 22.54, analyze the desired product as follows:

The desired product requires a crossed aldol condensation between an aldehyde and a ketone: acetaldehyde and 2-butanone. The question, then, is whether the desired product is the only one that could form, or whether other competing aldol reactions would occur.

First, either acetaldehyde and 2-butanone could serve as the enolate component of an aldol addition. Although the aldehyde should be more reactive toward addition of the enolate than the ketone, its reaction with the enolate ions from both 2-butanone and another acetaldehyde molecule would give a mixture. To complicate matters even more, 2-butanone has two nonequivalent α-carbons at which enolate ions (or enols) could form. This opens yet other possibilities for aldol reactions and thus for complex product mixtures. To summarize all of these possibilities:

desired:
$$\overset{O}{\underset{\|}{H_3C-CH}} \quad \overset{O}{\underset{\|}{H_2\ddot{C}-C-CH_2CH_3}}$$

but also possible:
$$\overset{O}{\underset{\|}{H_3C-CH}} \quad \overset{O}{\underset{\|}{H_3C-C-\ddot{C}HCH_3}} \quad or \quad \overset{O}{\underset{\|}{H_3C-CH}} \quad \overset{O}{\underset{\|}{H_2\ddot{C}-CH}}$$

Hence, the base-catalyzed or acid-catalyzed aldol reaction of acetaldehyde and 2-butanone would *not* be useful for preparing the desired ketone because a large number of other products would be expected.

However, because 2-butanone is a methyl ketone, and because the carbonyl component is an aldehyde, a directed aldol addition followed by dehydration should work. Therefore, we could form the less stable (kinetic) lithium enolate of 2-butanone with LDA in THF at low temperature and allow it to react with acetaldehyde. Acid- or base-catalyzed dehydration of the addition product should give the desired product in the more stable *E* configuration.

$$\overset{O}{\underset{\|}{H_3C-C-CH_2CH_3}} \xrightarrow[\substack{THF \\ -78\,°C}]{LDA} \overset{Li^+\;O}{\underset{\|}{H_2\bar{C}-C-CH_2CH_3}} \xrightarrow{CH_3CH=O} \xrightarrow{H_3O^+}$$

$$\overset{OH}{\underset{|}{H_3C-CH}}-CH_2-\overset{O}{\underset{\|}{C}}-CH_2CH_3 \xrightarrow[heat]{H_3O^+} \overset{H}{\underset{H_3C}{}}C=C\overset{\overset{O}{\underset{\|}{C}-CH_2CH_3}}{\underset{H}{}} + H_2O$$

PROBLEMS

22.22 Some of the following molecules can be synthesized in good yield using an aldol condensation (or an aldol addition followed by a dehydration). Identify these and give the structures of the required starting materials. Others cannot be synthesized in good yield by an aldol condensation. Identify these, and explain why the required aldol condensation would not be likely to succeed.

(a)

$$CH_3O-\underset{}{\bigcirc}-CH=\underset{\underset{CH_3}{|}}{C}-\overset{O}{\underset{\|}{C}}-CH_2CH_3$$

(b)

$$\bigcirc-CH=\underset{\underset{CH_3}{|}}{C}-\overset{O}{\underset{\|}{C}}-CH_2CH_2CH_3$$

(c)

$$\underset{}{\bigcirc}\overset{CH=O}{\underset{CH_3}{}}$$

(d)

(e)

$$\underset{Ph}{\overset{O}{\bigcirc}}\overset{Ph}{\underset{Ph}{}}$$

(f)

$$Ph\underset{}{\nearrow}\overset{O}{\underset{}{\diagdown}}\nearrow Ph$$

(g)

$$(CH_3)_3C\overset{O}{\underset{}{\diagdown}}\nearrow CH_3$$

(h)

22.23 Analyze the aldol condensation in Eq. 22.53 on p. 1128 using the method given in Eq. 22.54. Show that four possible aldol condensation products might in principle result from the starting material. Explain why the observed product is the most reasonable one.

22.5   ALDOL REACTIONS IN BIOLOGY

The aldol addition, or variations of it, appear in a number of biochemical pathways. In this section we'll examine one of the best-known examples, an aldol addition catalyzed by an enzyme called, appropriately enough, *aldolase*. There are two major variations of this enzyme, and we'll discuss the reaction catalyzed by the enzyme (Class I aldolase) found in animals (including humans) and plants. (A different aldolase, Class II aldolase, is found in fungi and bacteria.)

This enzyme catalyzes the formation of fructose-1,6-bisphosphate from a ketone, dihydroxyacetone phosphate (DHAP) and an aldehyde, glyceraldehyde-3-phosphate (G3P).

(R)-glyceraldehyde-3-phosphate (G3P)   **1,3-dihydroxyacetone phosphate (DHAP)**   **fructose-1,6-bisphosphate**   (22.55)

Don't be distracted by the stereochemical details in this reaction; notice primarily that *a new carbon–carbon bond is formed at the α-carbon of the ketone*. We want to focus on how this bond is formed.

The aldolase active site contains a lysine residue, which has a primary amine in its side chain:

structure of a lysine residue in a protein
(conjugate-base form)

This lysine forms an *imine*, or *Schiff base*, derivative with DHAP (Sec. 19.11A). This imine can isomerize to an *enamine* (Sec. 19.11B). Recall that enamines have the same relationship to imines as enols have to aldehydes and ketones.

DHAP   imine (Schiff base)   enamine   (22.56a)
+ H$_2$O

In the enamine, the unshared electron pair on the nitrogen is delocalized by resonance onto the α-carbon.

(22.56b)

In other words, this enamine has some *carbanion character* and can be thought of as a *disguised carbanion*. The negative charge at the α-carbon is further stabilized by the polar effect of the —OH group. When the enamine is brought together in proximity to the carbonyl group of glyceraldehyde-3-phosphate in the enzyme active site, the aldol addition occurs. In this addition, the carbanion reacts as a nucleophile at the carbonyl carbon of glyceraldehyde:

$$(22.56c)$$

The oxygen of a tyrosine residue in the active site facilitates the proton transfers that complete the addition. Notice that an imine derivative of the product is formed in the active site.

$$(22.56d)$$

imine derivative of fructose-1,6-bisphosphate

Hydrolysis of the imine gives fructose-1,6-bisphosphate and regenerates the enzyme for another catalytic cycle.

imine derivative of fructose-1,6-bisphosphate   **fructose-1,6-bisphosphate**   free enzyme

$$(22.56e)$$

This aldol addition reaction has a very favorable $\Delta G°$ in the addition direction (-23.8 kJ mol^{-1} at pH $= 7$); however, at the concentrations of the reactants and products present in the cell, the actual ΔG is only slightly negative. Therefore, the reaction can run in either direction under cellular conditions. It is utilized in the cell in two pathways. In the reverse-aldol direction, it is part of the *glycolysis* pathway, which breaks down glucose (by way of fructose-1,6-bisphosphate); and in the aldol addition direction shown above, it is part of the *gluconeogenesis* pathway, which operates in periods of glucose deprivation to form glucose (also by way of fructose-1,6-bisphosphate) from three-carbon compounds.

Fructose-1,6-bisphosphate spontaneously forms cyclic hemiacetals (Sec. 19.10A), which are the major forms in solution.

noncyclic form
(~2% at equilibrium)

cyclic hemiacetal form
(~98% at equilibrium;
a mixture of both
C-2 diastereomers)

$$(22.57)$$

The formation of the cyclic hemiacetals contributes to the favorable $\Delta G°$ for the addition. In the reverse reaction, the enzyme specifically binds the noncyclic form of the sugar.

The utilization of an enamine as a "disguised enolate ion" (Eq. 22.56b) is well precedented in laboratory chemistry (Problem 22.25).

PROBLEMS

22.24 (a) The enzyme "KDPG aldolase" catalyzes the aldol addition reaction between pyruvate and glyceraldehyde-3-phosphate.

(R)-glyceraldehyde-3-phosphate (G3P)

pyruvate

3-keto-2-deoxy-6-phosphogluconate (KDPG)

The reaction is known to involve the formation of an imine (Schiff base) between a lysine residue of the enzyme and pyruvate. Give the product of the reaction (KDPG) and outline a curved-arrow mechanism, assuming acids and bases are present as needed. (The stereochemistry of the asymmetric carbons is the same as in Eq. 22.57.)

(b) The product of this reaction exists as two diastereomeric cyclic acetals. Give their structures.

22.25 The following reaction is known to involve enamine formation between the amine catalyst and the carbonyl marked with an asterisk (*).

(an enamine)

Draw a curved-arrow mechanism for this reaction, using the following steps as your guide.

(a) Begin your mechanism by showing the formation of enamine X.

(b) Complete the mechanism by showing the acid-catalyzed aldol condensation of the enamine that forms the ring.

(c) Draw the hydrolysis mechanism for the reaction that gives the ketone and regenerates the catalyst.

22.6 CONDENSATION REACTIONS INVOLVING ESTER ENOLATE IONS

A. Claisen Condensation

The base-catalyzed aldol reactions discussed in Sec. 22.4A involve enolate ions derived from *aldehydes* and *ketones*. This section discusses condensation reactions that involve the enolate ions of *esters*.

Ethyl acetate undergoes a condensation reaction in the presence of one equivalent of sodium ethoxide in ethanol to give ethyl 3-oxobutanoate, which is known commonly as ethyl acetoacetate.

ethyl acetate

ethyl acetoacetate
(75–76% yield)

This is the best-known example of the *Claisen condensation*, which is named for Ludwig Claisen (1851–1930), who was a professor at the University of Kiel. (Don't confuse this reaction with the Claisen–Schmidt condensation in the previous section—same Claisen, different reaction.) The product of this reaction, ethyl acetoacetate, is an example of a **β-keto ester**: a compound with a ketone carbonyl group β to an ester carbonyl group.

a ketone group β to an ester group

$$H_3C-\underset{\beta}{\overset{O}{\overset{\|}{C}}}-\underset{\alpha}{CH_2}-\overset{O}{\overset{\|}{C}}-OEt$$

Thus, a **Claisen condensation** is the base-promoted condensation of two ester molecules to give a β-keto ester.

The first step in the mechanism of the Claisen condensation is formation of an *enolate ion* by the reaction of the ester with the ethoxide base.

$$EtO\!:^-$$

$$H_2C-\overset{O}{\overset{\|}{C}}-OEt \;\rightleftarrows\; \left[H_2\ddot{C}-\overset{:O:}{\overset{\|}{C}}-OEt \;\longleftrightarrow\; H_2C=\overset{:\ddot{O}:^-}{\overset{|}{C}}-OEt \right] + Et\ddot{O}H \qquad (22.59a)$$

$$pK_a \approx 25 \qquad\qquad\qquad\qquad\qquad\qquad\text{enolate ion}$$

Because ethoxide ion is a nucleophile, we might ask whether it can also react at the carbonyl group of the ester to give the usual nucleophilic acyl substitution reaction. This reaction undoubtedly takes place, but the products are the same as the reactants! This is why ethoxide ion is used as a base with ethyl esters in the Claisen condensation (see Study Guide Link 22.1 and Problem 22.27).

Although the ester enolate ion is formed in very low concentration, it is a strong base and good nucleophile, and it undergoes a *nucleophilic acyl substitution reaction* with a second molecule of ester (Eq. 22.59b). The usual two-step substitution mechanism is observed—that is, formation of a tetrahedral addition intermediate followed by loss of a leaving group:

$$H_3C-\overset{:O:}{\overset{\|}{C}}-OEt \quad H_2\ddot{C}-\overset{O}{\overset{\|}{C}}-OEt \;\rightleftarrows\; H_3C-\underset{\overset{|}{:OEt}}{\overset{:\ddot{O}:^-}{\overset{|}{C}}}-CH_2-\overset{O}{\overset{\|}{C}}-OEt \;\rightleftarrows\;$$

tetrahedral addition intermediate

$$H_3C-\overset{:O:}{\overset{\|}{C}}-CH_2-\overset{O}{\overset{\|}{C}}-OEt + Et\ddot{O}:^- \qquad (22.59b)$$

The overall equilibrium as written in Eqs. 22.59a–b lies far on the side of the reactants; that is, *all β-keto esters are less stable than the esters from which they are derived.* For this reason, the Claisen condensation must be driven to completion by applying Le Châtelier's principle. The most common technique is to use one full equivalent of ethoxide catalyst. In the β-keto ester product, the hydrogens on the carbon adjacent to both carbonyl groups (red in Eq. 22.59c) are especially acidic (why?), and the ethoxide removes one of these protons to form quantitatively the conjugate base of the product.

$$H_3C-\overset{O}{\overset{\|}{C}}-CH_2-\overset{O}{\overset{\|}{C}}-OEt + Na^+\,EtO^- \;\rightleftarrows\; H_3C-\overset{O}{\overset{\|}{C}}-\underset{Na^+}{\overset{\overset{..}{}}{CH}}-\overset{O}{\overset{\|}{C}}-OEt + EtOH \qquad (22.59c)$$

$$pK_a = 10.7 \qquad\qquad\qquad\qquad\qquad\qquad pK_a = 15\text{--}16$$

The un-ionized β-keto ester product in Eq. 22.58 is formed when acid is added subsequently to the reaction mixture.

Notice that ethoxide ion is a *catalyst* for the reactions in Eqs. 22.59a–b, but it is consumed in Eq. 22.59c. Thus, ethoxide is a reactant rather than a catalyst in the overall reaction, and for this reason *one full equivalent* of ethoxide must be used in the Claisen condensation.

The removal of a product by ionization is the same strategy employed to drive ester saponification to completion (Sec. 21.8A). The importance of this strategy in the success of the Claisen condensation is evident if the condensation is attempted with an ester that has only one α-hydrogen: *No condensation product is formed.* In this case, the desired condensation product has a quaternary α-carbon, and therefore it has no α-hydrogens acidic enough to react completely with ethoxide.

$$2\,(CH_3)_2CH-\overset{\overset{\textstyle O}{\|}}{C}-OEt \underset{EtOH}{\overset{^-OEt}{\longleftrightarrow}} (CH_3)_2CH-\overset{\overset{\textstyle O}{\|}}{C}-\underset{\underset{\textstyle CH_3}{|}}{\overset{\overset{\textstyle CH_3}{|}}{C}}-CO_2Et \qquad (22.60)$$

no acidic hydrogen here

(no product observed)

Furthermore, if the product of Eq. 22.60 (prepared by another method) is subjected to the conditions of the Claisen condensation, it readily decomposes back to starting materials because of the reversibility of the Claisen condensation.

The Claisen condensation is another example of *nucleophilic acyl substitution*. In this reaction, the nucleophile is an enolate ion derived from an ester. Although the reaction may seem complex because of the number of carbon atoms in the product, it is not conceptually different from other nucleophilic acyl substitutions, such as ester saponification:

	Saponification:	*Claisen condensation:*
nucleophile + ester	$\overset{^-:\ddot{O}H}{}$ $H_3C-C\!\!=\!\!\ddot{O}\!:$ $\overset{|}{:\ddot{O}Et}$	$H_2\bar{\ddot{C}}-CO_2Et$ $H_3C-C\!\!=\!\!\ddot{O}\!:$ $\overset{|}{:\ddot{O}Et}$
tetrahedral addition intermediate	$:\ddot{O}H$ $H_3C-\overset{|}{C}-\ddot{O}\!:^-$ $\overset{|}{(:\ddot{O}Et}$	H_2C-CO_2Et $H_3C-\overset{|}{C}-\ddot{O}\!:^-$ $\overset{|}{(:\ddot{O}Et}$
substitution product	$:\ddot{O}H$ $H_3C-C\!\!=\!\!\ddot{O}\!:\; + EtÖ\!:^-$	H_2C-CO_2Et $H_3C-C\!\!=\!\!\ddot{O}\!:\; + EtÖ\!:^-$
acid–base reaction	$:\ddot{O}\!:^-$ $H_3C-C\!\!=\!\!\ddot{O}\!:\; + Et\ddot{O}H$	$H\bar{\ddot{C}}-CO_2Et$ $H_3C-C\!\!=\!\!\ddot{O}\!:\; + Et\ddot{O}H$ (22.61)

You have now studied two types of condensation reactions: the aldol condensation and the Claisen condensation. These condensations are quite different and should not be confused. To compare:

1. The aldol condensation is an *addition* reaction of an enolate ion or an enol with an aldehyde or ketone followed by a *dehydration*. The Claisen condensation is a *nucleophilic acyl substitution* reaction of an enolate ion with an ester group.

2. The aldol condensation is catalyzed by both base and acid. The Claisen condensation requires a full equivalent of base and is *not* catalyzed by acid.

3. The aldol addition requires only one α-hydrogen. A second α-hydrogen is required, however, for the dehydration step of the aldol condensation. In the Claisen condensation, the ester starting material must have at least *two* α-hydrogens, one for each of the ionizations shown in Eqs. 22.59a and 22.59c.

PROBLEMS

22.26 Give the Claisen condensation product formed in the reaction of each of the following esters with one equivalent of NaOEt, followed by neutralization with acid.

(a) ethyl phenylacetate   (b) ethyl butyrate

22.27 Hydroxide ion is about as basic as ethoxide ion. Would NaOH be a suitable base for the Claisen condensation of ethyl acetate? Explain by writing suitable equations. (*Hint:* See Study Guide Link 22.1.)

B. Dieckmann Condensation

Intramolecular Claisen condensations, like intramolecular aldol condensations, take place readily when five- or six-membered rings can be formed. The intramolecular Claisen condensation reaction is called the **Dieckmann condensation**.

diethyl adipate

$$+ \text{ HOEt}$$

ethyl 2-oxocyclopentane-carboxylate
(74–81% yield)

(22.62)

Like the Claisen condensation, the Dieckmann condensation requires one full equivalent of base to form the enolate ion of the product and thus to drive the reaction to completion.

PROBLEM

22.28 (a) Explain why compound *A*, when treated with one equivalent of NaOEt, followed by acidification, is completely converted into compound *B*.

(b) Write a curved-arrow mechanism for this conversion.

(c) Give the structure of the only product formed when diethyl α-methyladipate (compound *C*) reacts in the Dieckmann condensation. Explain your reasoning.

A   *B*   *C*

C. Crossed Claisen Condensation

The Claisen condensation of two *different* esters is called a **crossed Claisen condensation**. The crossed Claisen condensation of two esters that both have α-hydrogens gives a mixture of four compounds that are typically difficult to separate. Such reactions in most cases are not synthetically useful.

$$H_3C-CO_2Et + CH_3CH_2-CO_2Et \xrightarrow{NaOEt} \xrightarrow{H_3O^+} H_3C-\overset{O}{\overset{\|}{C}}-CH_2-\overset{O}{\overset{\|}{C}}-OEt +$$

$$CH_3CH_2-\overset{O}{\overset{\|}{C}}-CH_2-\overset{O}{\overset{\|}{C}}-OEt + H_3C-\overset{O}{\overset{\|}{C}}-\underset{\underset{CH_3}{|}}{CH}-\overset{O}{\overset{\|}{C}}-OEt + CH_3CH_2-\overset{O}{\overset{\|}{C}}-\underset{\underset{CH_3}{|}}{CH}-\overset{O}{\overset{\|}{C}}-OEt \quad (22.63)$$

This problem is conceptually similar to the problem with crossed aldol reactions, discussed in Study Problem 22.1 on p. 1124.

Crossed Claisen condensations are useful, however, if one ester is especially reactive or has no α-hydrogens. For example, formyl groups (—CH=O) are readily introduced with esters of formic acid such as ethyl formate:

$$EtO_2C\diagdown\diagup^{CO_2Et} + H-\overset{O}{\overset{\|}{C}}-OEt \xrightarrow[\substack{toluene \\ (solvent)}]{\substack{Na \\ (1\ equiv.) \\ EtOH \\ (trace)}} \xrightarrow{H_3O^+} EtO_2C\diagdown\underset{CO_2Et}{\overset{\overset{\displaystyle O}{\overset{\|}{HC}}\ H}{\diagup}} + EtOH \quad (22.64)$$

diethyl succinate   **ethyl formate**

diethyl formylsuccinate
(60–70% yield)

Formate esters fulfill both of the criteria for a crossed Claisen condensation. First, they have no α-hydrogens; second, their carbonyl reactivity is considerably greater than that of other esters. The reason for their higher reactivity is that the carbonyl group in a formate ester is "part aldehyde," and aldehydes are particularly reactive toward nucleophiles (Eq. 21.62, p. 1085).

A less reactive ester without α-hydrogens can be used if it is present in excess. For example, an ethoxycarbonyl group can be introduced with diethyl carbonate.

$$PhCH_2-\overset{O}{\overset{\|}{C}}-OEt + EtO-\overset{O}{\overset{\|}{C}}-OEt \xrightarrow[heat]{\substack{NaOEt \\ (1\ equiv.)}} \xrightarrow{H_3O^+} Ph-\underset{\overset{|}{\overset{\displaystyle O=C-OEt}{}}}{CH}-\overset{O}{\overset{\|}{C}}-OEt + EtOH \quad (22.65)$$

ethyl phenylacetate   **diethyl carbonate**
(excess)

ethoxycarbonyl group

diethyl phenylmalonate
(86% yield)

In this example, the enolate ion of ethyl phenylacetate condenses preferentially with diethyl carbonate rather than with another molecule of itself because of the much higher concentration of diethyl carbonate. The excess diethyl carbonate must then be separated from the product.

Another type of crossed Claisen condensation is the reaction of ketones with esters. In this type of reaction, the enolate ion of a ketone reacts at the carbonyl group of an ester.

$$EtO-\overset{O}{\overset{\|}{C}}-H + \text{[cyclohexanone]} \xrightarrow[ether]{\substack{NaOEt \\ (1\ equiv.)}} \xrightarrow{H_3O^+} \text{[2-oxocyclohexanecarbaldehyde]} + EtOH \quad (22.66)$$

ethyl formate

cyclohexanone

2-oxocyclohexanecarbaldehyde
(70–74% yield)

$$Ph\overset{O}{\underset{\|}{-C}}-CH_3 + EtO\overset{O}{\underset{\|}{-C}}-CH_3 \xrightarrow[\text{xylene}]{\text{NaOEt} \atop \text{(1 equiv.)}} \xrightarrow{H_3O^+} Ph\overset{O}{\underset{\|}{-C}}-CH_2\overset{O}{\underset{\|}{-C}}-CH_3 + EtOH \quad (22.67)$$

acetophenone   **ethyl acetate**   **1-phenyl-1,3-butanedione**
                   (large excess)   (a β-diketone)
                                                               (64–70% yield)

In Eq. 22.66, the enolate ion derived from the ketone cyclohexanone is acylated by the ester ethyl formate. In Eq. 22.67, the enolate ion of the ketone acetophenone is acylated by the ester ethyl acetate. In these reactions, several side reactions are possible in principle but in fact do not interfere. The analysis of these cases again highlights important principles of carbonyl-compound reactivity.

In Eq. 22.66, a possible side reaction is the aldol addition of cyclohexanone with itself. However, *the equilibrium for the aldol addition of two ketones favors the reactants*, whereas *the Claisen condensation is irreversible* because one equivalent of base is used to form the enolate ion of the product. *Because the ester has no α-hydrogens, it cannot condense with itself.*

The ester in Eq. 22.67, however, does have α-hydrogens and is known to condense with itself (Eq. 22.58, p. 1133). Why is such a condensation not an interfering side reaction? The answer is that ketones are far more acidic than esters (by about 5–7 pK_a units; see Eqs. 22.3–22.4 on p. 1104). Thus, *the enolate ion of the ketone is formed in much greater concentration than the enolate ion of the ester.* The ketone enolate ion can react with another molecule of ketone—an unfavorable equilibrium—or it can be intercepted by the excess of ethyl acetate to give the observed product, which is a β-diketone. Even though esters are less reactive than ketones, a β-diketone is especially acidic (like a β-keto ester) and is ionized completely by the one equivalent of NaOEt. (Be sure to identify the acidic hydrogens of the product in Eq. 22.67.) Hence, *β-diketone formation is observed because ionization makes this an irreversible reaction.*

These examples illustrate that the crossed Claisen condensation can be used for the synthesis of a wide variety of β-dicarbonyl compounds.

PROBLEM

22.29   Complete the following reactions. Assume that one equivalent of NaOEt is present in each case.

(a)
$$H_3C\overset{O}{\underset{\|}{-C}}-CMe_3 + EtO\overset{O}{\underset{\|}{-C}}-OEt \xrightarrow{\text{NaOEt}} \xrightarrow{H_3O^+}$$
                                    (excess)

(b)
$$Ph\overset{O}{\underset{\|}{-C}}-CH_3 + Ph\overset{O}{\underset{\|}{-C}}-OEt \xrightarrow{\text{NaOEt}} \xrightarrow{H_3O^+}$$
                                 (excess)

(c)

$\xrightarrow{\text{NaH}} \xrightarrow{H_3O^+} (C_{11}H_{16}O_4)$

D. Synthesis with the Claisen Condensation

As the examples in the previous sections have shown, the Claisen condensation and related reactions can be used for the synthesis of β-dicarbonyl compounds: β-keto esters, β-diketones, and the like. Compare these types of compounds with those prepared by the aldol condensation, and notice carefully the differences.

In planning the synthesis of a β-dicarbonyl compound, we adopt the usual two-step strategy: examine the target molecule and work backward to reasonable starting materials.

Then we must analyze the reaction of these starting materials to see whether the desired reaction is reasonable or whether other reactions will occur instead.

To determine the starting materials for a Claisen condensation, mentally reverse the condensation by adding the elements of ethanol (or another alcohol) across either of the carbon–carbon bonds *between* the carbonyl groups. Because there are two such bonds, we will generally find two possible "disconnections" (labeled (a) and (b) in the following equation) and two corresponding sets of starting materials by this procedure.

(22.68)

A β-diketone can be similarly analyzed in two different ways:

(22.69)

Having determined the possible starting materials required in a Claisen condensation, we then ask whether the Claisen condensation of the required materials will give mostly the desired product or a complex mixture. Such an analysis of a target β-keto ester is illustrated in Study Problem 22.3.

STUDY PROBLEM 22.3

Determine whether the following compound can be prepared by a Claisen condensation or one of its variations; if so, give the possible starting materials.

SOLUTION   This is a β-diketone, a type of compound for which a Claisen or Dieckmann condensation might be appropriate. To determine the possible starting materials, follow the foregoing procedure: Add EtOH in turn across each of the bonds indicated:

Addition across bond (a) gives the following possible starting material:

$$
\text{(3)} \quad O \overset{B}{=} C \overset{OEt}{\diagdown} \quad O \atop CH_2-\overset{\|}{C}-CH_3 \atop \underset{(1)}{} \quad \underset{A}{} \quad (2)
$$

Now let's think about *all* possible Dieckmann condensation reactions that can occur with this compound. Three possible sets of α-hydrogens could ionize to give enolate ions. Hydrogens (1) and (2), because they are adjacent to a ketone carbonyl, are more acidic than hydrogens (3), which are adjacent to an ester carbonyl. Formation of an enolate ion at (1) and reaction of this enolate at carbonyl *B* give the desired product, and this reaction is driven to completion by using one equivalent of NaOEt. Formation of an enolate ion at (2) and reaction of this enolate at carbonyl *B* would give a β-diketone product containing a seven-membered ring:

Because five-membered rings typically form much more rapidly than seven-membered rings (Sec. 11.8), the desired product should be the major one, although formation of the seven-membered ring is a potential complication.

Breaking bond (b) in the target gives the following starting materials:

cyclopentanone   **ethyl acetate**

In this case, the ketone, cyclopentanone, is more acidic than the ester, ethyl acetate. Because of its symmetry, cyclopentanone can give only one enolate ion. Aldol addition of this enolate ion to another molecule of cyclopentanone is an unfavorable equilibrium; recall that the equilibria for aldol additions of ketones are unfavorable. If an excess of ethyl acetate is used, this potential side reaction can be further suppressed, if it occurs at all. The desired Claisen condensation can be made irreversible by use of one equivalent of NaOEt to ionize the products. Consequently, this set of starting materials—cyclopentanone and ethyl acetate—should give the desired reaction.

Evidently, *both* sets of potential starting materials would work, and in fact are acceptable answers. Which would be best in practice? Cyclopentanone and ethyl acetate are inexpensive and readily available. The other starting material would probably have to be prepared in a multistep synthesis. Consequently, cyclopentanone and ethyl acetate are the starting materials of choice. (This synthesis is conceptually the same as the one in Eq. 22.67.)

**STUDY GUIDE
LINK 22.5**
Variants of the
Aldol and Claisen
Condensations

PROBLEMS

22.30 Analyze each of the following compounds and determine what starting materials would be required for its synthesis by a Claisen condensation. Then decide which if any of the possible Claisen condensations would be a reasonable route to the desired product.

(a)

(b) $Et-\overset{O}{\overset{\|}{C}}-\underset{\underset{CH_3}{|}}{CH}-\overset{O}{\overset{\|}{C}}-OEt$

(c)

(d)

22.31 Give the starting material required for the synthesis of each of the following compounds by a Dieckmann condensation.

(a)

(b)

22.7   THE CLAISEN CONDENSATION IN BIOLOGY: BIOSYNTHESIS OF FATTY ACIDS

The utility of the Claisen condensation and the aldol reactions is not confined to the laboratory; these reactions are also important in the biological world. The biosynthesis of *fatty acids* (Sec. 20.5) illustrates how nature uses the Claisen condensation to build long carbon chains.

The carbons of fatty acids are all derived from **acetyl-coenzyme A**, abbreviated *acetyl-CoA*, a thiol ester of acetic acid.

$$H_3C - \overset{\overset{\displaystyle O}{\|}}{C} - S - CoA$$

acetyl-CoA

The full structure of acetyl-CoA is given in Fig. 25.1 on p. 1284. (We'll learn more about the role of acetyl-CoA and other thioesters in biology in Chapter 25.) The complexity of the CoA part of the molecule is important for its binding interactions with proteins, but can be ignored in the chemical transformations we're going to learn about here.

The overall chemical transformation of acetyl-CoA to the 16-carbon fatty acid, palmitic acid, is as follows:

$$8\ H_3C \overset{\overset{\displaystyle O}{\|}}{\underset{}{C}} SCoA\ +\ 14\ NADPH\ +\ 13\ H^+\ \longrightarrow$$

acetyl-CoA   (Sec. 19.8B)

palmitate $\overset{\overset{\displaystyle O}{\|}}{C}{-}O^-\ +\ 14\ NADP^+\ +\ 6\ H_2O\ +\ 8\ HSCoA$ (22.70)

palmitate (the conjugate base of palmitic acid)
(all carbons are derived from acetyl-CoA)

As a prelude to the actual condensation that forms the new carbon–carbon bonds of fatty acids, a molecule of acetyl-CoA is activated by carboxylation to form malonyl-CoA.

$$H_3C - \overset{\overset{\displaystyle O}{\|}}{C} - S - CoA\ \longrightarrow\ ^-O - \overset{\overset{\displaystyle O}{\|}}{C} - CH_2 - \overset{\overset{\displaystyle O}{\|}}{C} - S - CoA$$ (22.71)

acetyl-CoA   **malonyl-CoA**

Carbon dioxide itself is present in too low a concentration to be useful in this conversion; rather, it is present as an *N*-carboxy derivative of *biotin*. The structures of biotin (sometimes called vitamin H or vitamin B$_7$) and *N*-carboxybiotin are shown in Fig. 22.4 on p. 1142. Biotin is nature's "carbon dioxide carrier." (The biosynthesis of *N*-carboxybiotin is discussed in Sec. 25.7B.)

FIGURE 22.4 The structure of biotin, nature's CO_2 carrier, and *N*-carboxybiotin, its carboxylated derivative.

biotin is typically attached covalently to enzymes here by amide formation with an amino group of lysine

biotin

abbreviated structure for biotin used in the text

***N*-carboxybiotin**

The enzyme that catalyzes the carboxylation of acetyl-CoA is *acetyl-CoA carboxylase*. In the active site of this enzyme, *N*-carboxybiotin decomposes to CO_2 and the biotin anion; this transformation is assisted by hydrogen-bond donation (*blue*) from the enzyme to the carbonyl oxygen, which reduces the basicity of the biotin anion and makes it a better leaving group. The biotin anion abstracts a proton from acetyl-CoA to give the conjugate-base enolate ion. This enolate ion reacts with the just-liberated CO_2 in a carbonyl addition to give malonyl-CoA.

(22.72)

Malonyl-CoA and acetyl-CoA are then transferred to two different thiol groups on the next enzyme by *transthioesterification* reactions. (Transthioesterification is a transesterification reaction of a thioester with a thiol.) This enzyme, fatty acid synthase, is a large protein that, in animals, consists of several tightly associated component proteins, each of which catalyzes one of the subsequent reactions. In one transthioesterification reaction, the malonyl-CoA is transferred to a thiol-containing protein subunit of the enzyme called *acyl carrier protein* (HS—ACP) to give malonyl-ACP. The acetyl-CoA is transferred in another transthioesterification reaction to a different thiol of the protein that we'll abbreviate simply as "SYNTH—SH." Each transfer liberates a molecule of CoA—SH as a by-product. (E = the overall enzyme.)

(22.73a)

The purpose of these transthioesterifications is to position the α-carbon of the malonyl thioester and the carbonyl group of the acetyl thioester in proximity for the next reaction, which is the Claisen condensation.

Loss of CO_2 from the malonyl group gives a transient enolate ion, which immediately reacts at the carbonyl of the nearby acetyl thioester to displace the thiolate anion and form the condensation product, acetoacetyl-ACP.

+ CO$_2$

(22.73b)

It may seem odd that nature would install a CO_2 group on acetyl-CoA (Eq. 22.72) only to lose it again in the Claisen condensation. Recall, however, that the Claisen condensation in the laboratory is thermodynamically unfavorable, and it is driven to completion by ionization of an α-hydrogen of the acetoacetate ester product (Eq. 22.59c). In the enzyme-catalyzed Claisen condensation, no group in the active site of the enzyme or in solution is basic enough to ionize this hydrogen completely. Rather, the loss of CO_2, which ultimately escapes as a gas, drives the reaction to completion.

In subsequent steps of fatty-acid biosynthesis, the ACP serves as a remarkable "swinging arm" that rotates the acetoacetyl group successively into the other active sites on the fatty acid synthase complex to undergo carbonyl reduction (what do you think is the reducing agent?), alcohol dehydration, and double-bond reduction.

(22.74)

Notice that the overall result is the conversion of a two-carbon thioester (acetyl-CoA) into a four-carbon thioester (butyryl-ACP).

The butyryl-ACP is finally transthioesterified onto the thiol of the SYNTH—SH protein, the site previously occupied by the acetyl group. The ACP—SH is "re-charged" with another malonyl group from malonyl-CoA, and an analogous series of reactions takes place again that converts the four-carbon thioester into a six-carbon thioester.

(22.75)

This process continues until the growing chain contains 16 carbons. The addition of carbons to the growing chain two carbons at a time accounts for the fact that *ordinary fatty acids have even numbers of carbons*. Hydrolysis of the 16-carbon thioester gives palmitic acid, the 16-carbon fatty acid.

palmitate
(conjugate-base anion of **palmitic acid**)

Longer fatty acids and unsaturated fatty acids are produced by other reactions.

PROBLEMS

22.32 Using abbreviated structures like the ones used in this section, outline the steps that convert hexanoyl-ACP into octanoyl-ACP during fatty-acid biosynthesis.

22.33 Suppose a sample of acetyl-CoA labeled at the carbonyl carbon with the radioisotope ^{14}C is introduced into the fatty-acid synthase system, and palmitic acid (the 16-carbon fatty acid; Eq. 22.76) is isolated. Which carbons of palmitic acid should be radiolabeled?

acetyl-CoA (carbonyl ^{14}C-labeled)

22.34 Fatty acids are degraded to acetyl-CoA in fatty-acid metabolism. The enzyme that catalyzes this conversion, *acyl-CoA acetyl transferase*, contains a nucleophilic thiol group in its active site and catalyzes the following reactions (enzyme = E):

(a) What is species *X*?

(b) Assuming that acids and bases are provided in the enzyme active site as needed, outline curved-arrow mechanisms for these transformations.

(c) What is the relationship of the first reaction to the Claisen condensation?

22.8 ALKYLATION AND ALDOL REACTIONS OF ESTER ENOLATE IONS

Sections 22.6 and 22.7 described reactions in which the enolate ions of esters react as nucleophiles in carbonyl substitution reactions. This section considers reactions in which ester enolates are used as nucleophiles in S_N2 reactions and aldol additions.

A. Malonic Ester Synthesis

Diethyl malonate (malonic ester), like many other β-dicarbonyl compounds, has unusually acidic α-hydrogens. (Why?) Consequently, its conjugate-base enolate ion can be formed nearly completely with alkoxide bases such as sodium ethoxide.

$$EtÖ:^- + EtO-\overset{\overset{O}{\|}}{C}-CH_2-\overset{\overset{O}{\|}}{C}-OEt \rightleftharpoons EtÖ-H + EtO-\overset{\overset{O}{\|}}{C}-\overset{..}{\overset{-}{C}}H-\overset{\overset{O}{\|}}{C}-OEt \qquad (22.77a)$$

diethyl malonate
$pK_a = 12.9$
           enolate ion of diethyl malonate

The conjugate-base anion of diethyl malonate is nucleophilic, and it reacts with alkyl halides and sulfonate esters in typical S_N2 reactions. Such reactions can be used to introduce alkyl groups at the α-position of malonic ester.

$$\underset{}{\overset{CO_2Et}{\underset{CO_2Et}{|}}}\;Br\quad :CH\;Na^+ \xrightarrow{EtOH} \underset{CO_2Et}{\overset{CO_2Et}{CH}} + Na^+\;Br^- \qquad (22.77b)$$

(83% yield)

FURTHER EXPLORATION 22.1
Malonic Ester Alkylation

As this example shows, even secondary halides can be used in this reaction. (See Further Exploration 22.1.)

The importance of this reaction is that it can be extended to the preparation of carboxylic acids. Saponification (Sec. 21.8A) of the diester and acidification of the resulting solution gives a substituted malonic acid derivative. Recall that heating any malonic acid derivative causes it to *decarboxylate* (Sec. 20.11). The result of the alkylation, saponification, and decarboxylation sequence is a carboxylic acid that conceptually is a substituted acetic acid—an acetic acid molecule with an alkyl group on its α-carbon.

$$\underset{CO_2Et}{\overset{CO_2Et}{CH}} \xrightarrow[H_2O]{NaOH} \underset{CO_2^-\;Na^+}{\overset{CO_2^-\;Na^+}{CH}} \xrightarrow{H_3O^+} \underset{CO_2H}{\overset{CO_2H}{CH}} \xrightarrow[heat]{\boxed{\text{decarboxylation (Sec. 20.11A)}}} \overset{\overbrace{}}{CH_2-CO_2H} \qquad (22.77c)$$

with labels: ester saponification (Sec. 21.8A); protonation; a "substituted acetic acid"; $+ CO_2$

The overall sequence of ionization, alkylation, saponification, and decarboxylation starting from diethyl malonate (Eqs. 22.77a–c) is called the **malonic ester synthesis**. Notice that the alkylation step of the malonic ester synthesis (Eq. 22.77b) results in the formation of a new carbon–carbon bond.

The anion of malonic ester can be alkylated twice in two successive reactions with different alkyl halides (if desired) to give, after hydrolysis and decarboxylation, a *disubstituted* acetic acid. This possibility allows us to think of any disubstituted acetic acid in terms of diethyl malonate and two alkyl halides, as follows (X = halogen):

$$\underset{R'}{\overset{\overbrace{\text{"substituted acetic acid"}}}{R-CH-CO_2H}} \Rightarrow \underset{R'}{\overset{CO_2Et}{R-C-CO_2Et}} \Rightarrow \overset{CO_2Et}{H_2C-CO_2Et},\; R-X,\; R'-X \qquad (22.78)$$

If the alkyl halides R—X and R′—X are among those that will undergo the S_N2 reaction, then the target carboxylic acid can in principle be prepared by the malonic ester synthesis. This analysis is illustrated in Study Problem 22.4.

STUDY PROBLEM 22.4

Outline a malonic ester synthesis of the following carboxylic acid:

CH₃

2-methylheptanoic acid

SOLUTION Using the analysis in the text, identify the "acetic acid" unit in the carboxylic acid. The two alkyl groups—in this case, a methyl group and a pentyl group—are derived from alkyl halides.

CH₃ ◄— derived from CH₃I

derived from CH₃(CH₂)₄Br ➔

CH—CO₂H

"substituted acetic acid"

This analysis leads to the following synthesis:

formation of the enolate ion

formation of the enolate ion | introduction of the second alkyl group

$$CH_2(CO_2Et)_2 \xrightarrow[\text{EtOH}]{\text{NaOEt}} \qquad \xrightarrow{\text{Br}} \qquad CH(CO_2Et)_2 \xrightarrow[\text{EtOH}]{\text{NaOEt}} \xrightarrow{H_3C-I} \qquad (22.79)$$

diethyl malonate

introduction of the first alkyl group

CH₃

C(CO₂Et)₂ + NaI

(80% yield)

Ester saponification, acidification, and decarboxylation, as in Eq. 22.77c, give the desired product.

The two enolate-forming and alkylation reactions must be performed as *separate steps*. Adding two different alkyl halides and two equivalents of NaOEt to malonic ester at the same time would not give the desired product. (Why?)

PROBLEMS

22.35 Indicate whether each of the following compounds could be prepared by a malonic ester synthesis. If so, outline a preparation from diethyl malonate and any other reagents. If not, explain why.

(a) 3-phenylpropanoic acid   (b) 2-ethylbutanoic acid   (c) 3,3-dimethylbutanoic acid

22.36 Give the product of the following reaction sequence and explain your answer.

$$CH_2(CO_2Et)_2 + BrCH_2CH_2CH_2Cl \xrightarrow[\text{EtOH}]{2\,\text{NaOEt}} \xrightarrow{\text{NaOH}} \xrightarrow[\text{heat}]{\text{HCl}} (C_5H_8O_2)$$

22.37 (a) When the conjugate-base enolate of diethyl malonate is treated with bromobenzene, no diethyl phenylmalonate is formed. Explain why bromobenzene is inert.

⁻:CH(CO₂Et)₂ + ⟨ ⟩—Br ✕➔ ⟨ ⟩—CH(CO₂Et)₂ + Br⁻

diethyl phenylmalonate

(b) When the same enolate ion is treated with bromobenzene and a catalytic amount of Pd[P(*t*-Bu)₃]₄, diethyl phenylmalonate is formed in excellent yield. Explain the role of the catalyst with a mechanism. (*Hint:* See Secs. 18.5 and 18.6.)

B. Direct Alkylation of Enolate Ions Derived from Monoesters

In the synthesis of carboxylic acids by malonic ester alkylation, a —CO_2Et group is "wasted" because it is later removed. Why not avoid this altogether and alkylate directly the enolate ion of an acetic acid ester?

$$B:^- + H_3C-\overset{\overset{\displaystyle O}{\|}}{C}-OR \longrightarrow H_2\overset{\overset{\displaystyle ..}{..}}{C}-\overset{\overset{\displaystyle O}{\|}}{C}-OR \xrightarrow{R'-I} R'-CH_2-\overset{\overset{\displaystyle O}{\|}}{C}-OR + I^- \quad (22.80)$$
$$\text{(a base)} \qquad\qquad\qquad + B-H$$

At one time this idea could not be used in practice because enolate ions derived from esters, once formed, undergo another, faster reaction: Claisen condensation with the parent ester (Sec. 22.6A). However, the development of the strong amide bases introduced in the discussion of the directed aldol addition (Sec. 22.4C, p. 1123) made it possible to form the lithium enolates of esters. These enolates can be alkylated with alkyl halides as shown conceptually in Eq. 22.80. The following equation shows a specific example.

ethyl 2-methyl-propanoate + lithium cyclohexylisopropylamide (LICHA) $\xrightarrow[\substack{THF \\ < 15\ min}]{-78\ ^\circ C}$ enolate $\xrightarrow[DMSO]{H_3C-I}$ ethyl 2,2-dimethylpropanoate (ethyl pivalate) (87% yield) + $Li^+ I^-$ (22.81)

This method of ester alkylation is considerably more expensive than the malonic ester synthesis. It also requires special inert-atmosphere techniques because the strong bases that are used react vigorously with both oxygen and water. For these reasons, the malonic ester synthesis remains very useful, particularly for large-scale syntheses. However, for the preparation of laboratory samples, or for the preparation of compounds that are unavailable from the malonic ester synthesis, the preparation and alkylation of enolate ions with amide bases is particularly valuable.

As with the directed aldol condensation, we'll consider the possible side reactions that might occur with this method and why they are avoided. The possibility of the Claisen condensation as a side reaction was noted in the discussion of Eq. 22.80. The use of a very strong amide base avoids the Claisen condensation because the reaction is run by *adding the ester to the base*. When a molecule of ester enters the solution, it can react either with the strong base to form an enolate ion or with a molecule of already formed enolate ion in the Claisen condensation. The reaction of esters with strong amide bases is so much faster at –78 °C than the Claisen condensation that the enolate ion is formed instantly and never has a chance to undergo the Claisen condensation. In other words, the Claisen condensation is avoided because the ester and its enolate ion are never present simultaneously (except for an instant) in the reaction flask.

Another potential side reaction is the nucleophilic reaction of the amide base (or even its conjugate acid amine, which is, after all, also a base) at the ester carbonyl group. Because amines react with esters to give products of aminolysis (Sec. 21.8C), it might be reasonable to expect the *conjugate bases* of amines—very strong bases, indeed—to react even more rapidly as nucleophiles with esters. That this reaction does not happen is once again the result of a competition. When an amide base reacts with the ester, it can either remove a proton or react at the carbonyl carbon. A reaction at the carbonyl carbon is retarded by van der Waals repulsions between groups on the carbonyl compound and the large branched groups on the bases, as in the directed aldol reaction. If the amide base could be in contact with the ester long enough, it would eventually react at the carbonyl carbon; but the base instead reacts more rapidly in a different way: it abstracts an α-proton. Reaction with a tiny hydrogen does not involve the van der Waals repulsions that would occur if the base were to react at the carbonyl carbon.

PROBLEMS

22.38 Outline a synthesis of each of the following compounds from either diethyl malonate or ethyl acetate. Because the branched amide bases are relatively expensive, you may use them in only one reaction.

(a)

$$CH-CO_2H$$
$$|$$
$$CH_3$$

(b)

$$-CO_2H$$

valproic acid
(used in treatment of epilepsy)

(c)

$$CH_2CH_3$$
$$|$$
$$CH_3CH_2-C-CO_2Et$$

22.39 Predict the product formed when the conjugate-base enolate ion of ethyl 2-methylpropanoate (shown in Eq. 22.81) is treated with bromobenzene and a catalytic amount of Pd[P(t-Bu)$_3$]$_4$, and explain the role of the catalyst.

C. Acetoacetic Ester Synthesis

Recall that β-keto esters, like malonic esters, are substantially more acidic than ordinary esters (Eq. 22.59c, p. 1134) and are completely ionized by alkoxide bases.

$$EtO^- + H_3C-\overset{O}{\overset{||}{C}}-CH_2-\overset{O}{\overset{||}{C}}-OEt \longrightarrow EtO-H + H_3C-\overset{O}{\overset{||}{C}}-\overset{-}{C}H-\overset{O}{\overset{||}{C}}-OEt \quad (22.82)$$

ethyl acetoacetate
pK_a = 10.7

ethanol
pK_a = 16

The enolate ions derived from β-keto esters, like those from malonate ester derivatives, can be alkylated by primary or unbranched secondary alkyl halides or sulfonate esters.

$$H_3C-\overset{O}{\overset{||}{C}}-\overset{Na^+}{\overset{-}{C}H}-\overset{O}{\overset{||}{C}}-OEt \longrightarrow H_3C-\overset{O}{\overset{||}{C}}-CH-\overset{O}{\overset{||}{C}}-OEt + Na^+ \ :\ddot{Br}:^- \quad (22.83)$$

1-bromobutane

ethyl 2-acetylhexanoate
(70% yield)

Dialkylation of β-keto esters is also possible.

$$2 H_3C-\overset{O}{\overset{||}{C}}-OEt \xrightarrow[\text{(1 equiv.)}]{\text{NaOEt}} H_3C-\overset{O}{\overset{||}{C}}-\overset{-}{C}H-\overset{O}{\overset{||}{C}}-OEt \xrightarrow{I}$$

Claisen condensation

first alkylation

$$H_3C-\overset{O}{\overset{||}{C}}-CH-\overset{O}{\overset{||}{C}}-OEt \xrightarrow{\text{NaOEt}} \xrightarrow{H_3C-I} H_3C-\overset{O}{\overset{||}{C}}-\overset{CH_3}{\underset{|}{C}}-\overset{O}{\overset{||}{C}}-OEt \quad (22.84)$$

second alkylation

Alkylation of a Dieckmann condensation product is the same type of reaction:

(from a Dieckmann condensation)

ethyl 2-oxo-1-propyl-cyclopentanecarboxylate
(85% yield)

(22.85)

Like esters of substituted malonic acids, the alkylated derivatives of ethyl acetoacetate can be hydrolyzed and decarboxylated to give ketones. Ester saponification and protonation gives a substituted β-keto acid; and β-keto acids spontaneously decarboxylate at room temperature (Sec. 20.11). This series of reactions is illustrated as carried out on the product of Eq. 22.83:

(22.86a)

If the β-keto ester alkylation product has no especially acidic α-hydrogen at carbon-2, as in Eq. 22.84, the basic conditions of saponification will bring about a reverse Claisen condensation (see Eq. 22.60, p. 1135, and the accompanying discussion). In cases like this, *acid-catalyzed* ester hydrolysis should be used. The resulting β-keto acid decarboxylates spontaneously after it forms under the acidic conditions.

(22.86b)

The alkylation of ethyl acetoacetate followed by saponification, protonation, and decarboxylation to give a ketone is called the **acetoacetic ester synthesis**. The alkylation part of this sequence, like the alkylation of diethyl malonate, involves the construction of new carbon–carbon bonds.

Whether a target ketone can be prepared by the acetoacetic ester synthesis can be determined by mentally reversing the synthesis.

(22.87)

This analysis involves replacing an α-hydrogen of the target ketone with a —CO_2Et group. This process unveils the β-keto ester required for the synthesis. The β-keto ester, in turn, can either be prepared directly by a Claisen condensation or can be prepared from other β-keto esters by alkylation or dialkylation with appropriate alkyl halides, as indicated by the possibilities in Eq. 22.87.

STUDY GUIDE LINK 22.6
Further Analysis of the Claisen Condensation

STUDY PROBLEM   22.5

Outline a preparation of 2-methyl-3-pentanone by a reaction sequence that involves at least one Claisen condensation.

SOLUTION   The discussion in the text leads to the following analysis:

$$\underset{\textbf{2-methyl-3-pentanone}}{CH_3CH_2\overset{\overset{\displaystyle O}{\|}}{C}-\underset{\underset{\displaystyle CH_3}{|}}{\overset{\overset{\displaystyle H}{|}}{C}}-CH_3} \Rightarrow \underset{A}{CH_3CH_2\overset{\overset{\displaystyle O}{\|}}{C}-\underset{\underset{\displaystyle CH_3}{|}}{\overset{\overset{\displaystyle CO_2Et}{|}}{C}}-CH_3}$$

The symbol $\Rightarrow$, as usual, means "implies as a starting material." The β-keto ester A cannot be prepared directly by a Claisen condensation because it would require a crossed Claisen condensation (see Eq. 22.69, p. 1139), and because the reaction could not be made irreversible by deprotonation. A second option is to provide one of the methyl groups by alkylation of the enolate ion derived from β-keto ester B:

$$\underset{A}{CH_3CH_2\overset{\overset{\displaystyle O}{\|}}{C}-\underset{\underset{\displaystyle CH_3}{|}}{\overset{\overset{\displaystyle CO_2Et}{|}}{C}}-CH_3} \Rightarrow \underset{B}{CH_3CH_2\overset{\overset{\displaystyle O}{\|}}{C}-\overset{\overset{\displaystyle CO_2Et}{|}}{CH}-CH_3},\ H_3C-I$$

The enolate ion of compound B, in turn, can be prepared directly by the Claisen condensation of ethyl propionate. (This follows from the analysis shown in Eq. 22.69, p. 1139.)

$$\underset{\textbf{ethyl propionate}}{2\,CH_3CH_2CO_2Et} \xrightarrow[\text{EtOH}]{\substack{\text{NaOEt}\\ \text{(1 equiv.)}}} \underset{\text{enolate ion of }B}{CH_3CH_2\overset{\overset{\displaystyle O}{\|}}{C}-\underset{\underset{\displaystyle CH_3}{|}}{\overset{\overset{\displaystyle CO_2Et}{|}}{\underset{\cdot\cdot}{C}}}-CH_3} \xrightarrow{H_3C-I} A$$

Saponifying A and acidifying the solution will give the β-keto acid, which will decarboxylate spontaneously under the acidic reaction conditions to give the desired ketone.

$$\underset{A}{CH_3CH_2\overset{\overset{\displaystyle O}{\|}}{C}-\underset{\underset{\displaystyle CH_3}{|}}{\overset{\overset{\displaystyle CO_2Et}{|}}{C}}-CH_3} + H_2O \xrightarrow[\text{heat}]{H_3O^+} \underset{\text{target molecule}}{CH_3CH_2\overset{\overset{\displaystyle O}{\|}}{C}-\underset{\underset{\displaystyle CH_3}{|}}{\overset{\overset{\displaystyle H}{|}}{C}}-CH_3} + CO_2 + EtOH$$

This section has discussed the reactions of ester enolates with alkylating agents such as alkyl halides. Conceptually, we might ask whether the enolate of an aldehyde or a ketone might also be alkylated. They can be alkylated, but such alkylations are less useful because they occur at both the α-carbon and the oxygen of the enolate.

an aldehyde or ketone enolate   *C*-alkylation   *O*-alkylation   (22.88)

O-alkylation is not a problem with ester enolates. (Further Exploration 22.2 explains this aspect of enolate chemistry further.)

Therefore, alkylation at the α-carbon of a ketone requires further activation of the α-carbon by an ester group, which is, after alkylation, removed by saponification and decarboxylation.

**FURTHER
EXPLORATION 22.2**
Alkylation of Enolate Ions
Derived from Ketones

PROBLEMS

22.40   Outline a synthesis of each of the following compounds from ethyl acetoacetate and any other reagents.

(a) 5-methyl-2-hexanone   (b) 4-phenyl-2-butanone

22.41   Outline a synthesis of each of the following compounds from a β-keto ester; then show how the β-keto ester itself can be prepared.

(a)   PhCH₂CH—C—CH₂CH₃
            | ‖
           CH₃   O

(b)   PhCH—C—CH₂Ph
           | ‖
          CH₃   O

22.42   Predict the outcome of the following reaction by identifying *A*, then *B*, then the final product. (*Hint:* How do nucleophiles react with epoxides under basic conditions?)

$$\text{diethyl malonate} \xrightarrow[\text{EtOH}]{\text{NaOEt}} A$$

$$A + \underset{\substack{H_3C}}{\overset{\substack{H_3C \quad O}}{\diagdown C{-}CH_2}} \xrightarrow{\text{EtOH}} B \longrightarrow (C_9H_{14}O_4)$$

D. Aldol Reactions of Ester Enolates

Ester enolates react as nucleophiles with aldehyde and ketone carbonyl groups and give useful aldol addition or aldol condensation reactions. For example, the lithium enolate formed from an acetate ester gives an aldol addition with aldehydes and ketones.

$$H_3C{-}CO_2tBu \xrightarrow[\substack{THF \\ -78\,°C}]{LDA} Li^+ \; H_2\ddot{C}{-}CO_2tBu \xrightarrow[\textbf{acetone}]{(CH_3)_2C=O} \underset{\substack{| \\ CH_3}}{H_3C{-}\overset{\substack{O^- \; Li^+ \\ |}}{C}{-}CH_2{-}CO_2tBu} \xrightarrow[\substack{H_3O^+}]{\text{dilute}}$$

$$\underset{\substack{| \\ CH_3}}{H_3C{-}\overset{\substack{OH \\ |}}{C}{-}CH_2{-}CO_2tBu} + Li^+ \quad (22.89)$$

***tert*-butyl 3-hydroxy-
3-methylbutanoate**
(>90% yield)

One of the oldest but nevertheless widely used examples of an aldol addition is called the **Reformatsky reaction** after its discoverer, Sergei Nikolaevich Reformatsky (1860–1934), a Russian chemist who worked at the University of Kiev in Ukraine.

ethyl (1-hydroxycyclo-pentyl)acetate
(72% yield)   (22.90a)

In this reaction, an enolate is formed by the reaction of powdered zinc with an α-bromo ester. (See Sec. 22.3C for the preparation of α-bromo esters.) The zinc metal undergoes an insertion reaction with an α-bromo ester to form the zinc analog of a Grignard reagent (Eq. 9.62, p. 430). Unlike Grignard reagents, which have to be formed in a separate step, Reformatsky reagents are formed in the presence of the aldehyde or ketone. Although the actual structure of the reagent is more complex, we can think of it as a zinc enolate of the ester.

conceptual enolate structure
of the Reformatsky reagent

Organozinc compounds are much less reactive than Grignard reagents; consequently, the Reformatsky reagent reacts with aldehydes and ketones, but not with esters. Therefore, the ester group of the reagent is unaffected. In Eq. 22.90a, the zinc enolate undergoes an addition reaction with the carbonyl group of the ketone to give a zinc alkoxide. Addition of aqueous acid gives the aldol addition product.

Recall that the aldol addition reactions of ketones (and some aldehydes) are reversible. The aldol additions of the ester enolates derived from LDA and the Reformatsky reaction are *not* reversible because the nucleophile is a much stronger base than the product alkoxide. From an acid–base perspective, we can think of these reactions in the following way (M^+ = metal):

The practical significance of this irreversibility is that aldol addition products can be isolated in the reactions of ester enolates of simple esters following addition of dilute acid (as in Eqs. 22.89 and 22.90a).

The enolates derived from malonic esters or acetoacetic esters can also be used in aldol condensations. An aldol condensation reaction of malonic ester, acetoacetic ester, and other relatively acidic carbonyl compounds is called a **Knoevenagel reaction** (pronounced approximately kuh-NOER-vuh-NAH-gul), after the German chemist Emil Knoevenagel (1865–1921), who developed this reaction. For example, in the following reaction, the enolate ion of malonic ester undergoes an aldol condensation with the imine formed between benzaldehyde and piperidine.

$$Ph-\overset{\overset{O}{\|}}{CH} \;+\; \underset{\text{diethyl malonate}}{H_2C-CO_2Et} \xrightarrow[\substack{\text{benzene, heat}}]{\substack{\text{piperidine}\\ \text{(catalysts)}}} Ph-CH=C\overset{CO_2Et}{\underset{CO_2Et}{}} \qquad (22.92)$$

benzaldehyde   diethyl malonate
(malonic ester)

and PhCO₂H

piperidine

(86–91% yield)

As in many aldol reactions, the addition step is reversible because the nucleophile, the enolate conjugate base of diethyl malonate, is not very basic. For that reason, the addition product is not isolated; the reaction is driven to completion by dehydration of the addition product to give an aldol condensation product. (B: = the amine catalyst.)

$$Ph-\underset{\underset{H\;\;:B}{|}}{\overset{\overset{:\ddot{O}H}{|}}{CH}}-\underset{|}{\overset{CO_2Et}{C}}-CO_2Et \rightleftharpoons Ph-\underset{+\;^+BH}{\overset{:\ddot{O}H}{CH}}-\underset{\ddot{}}{\overset{CO_2Et}{C}}-CO_2Et \longrightarrow Ph-CH=C\overset{CO_2Et}{\underset{CO_2Et}{}} \qquad (22.93a)$$

pK$_a \approx 12$ pK$_a = 11.2$ $+\;:\ddot{O}H$

aldol addition product

Reaction of the hydroxide by-product with the conjugate acid of the catalyst is a favorable reaction that regenerates the catalyst.

$$\underset{\underset{H \quad H}{\overset{+}{N}}}{\bigcirc} \;+\; :\ddot{O}H \;\rightleftharpoons\; \underset{\underset{H}{\overset{\cdot\cdot}{N}}}{\bigcirc} \;+\; H_2\ddot{O}: \qquad (22.93b)$$

$^+BH\;(pK_a = 11.2)$ B :

pK$_a = 15.7$

The diester products of Knoevenagel condensations can be subjected to hydrolysis and decarboxylation to give carboxylic acids.

Aldol Addition of Ester Enolates in Biology: HMG-CoA Biosynthesis The aldol addition reaction of ester enolates serves as a model for similar reactions in biology. An important example is the reaction of acetoacetyl-CoA with acetyl-CoA to give (S)-3-hydroxy-3-methylglutaryl-CoA, known in biology as *HMG-CoA*.

$$\underset{\text{acetyl-CoA}}{CoAS\overset{O}{\frown}} \;+\; \underset{\text{acetoacetyl-CoA}}{\overset{O\quad O}{\frown\frown}SCoA} \xrightarrow{\text{HMG-CoA synthase}} \underset{\substack{\textbf{(S)-3-hydroxy-3-methylglutaryl-CoA}\\ \textbf{(HMG-CoA)}}}{^-O\overset{O\;\;HO\;\;CH_3\;\;O}{\frown\frown\frown}SCoA} \;+\; CoASH \qquad (22.94)$$

We can conceptualize this reaction mechanistically as the aldol addition reaction of an enolate derived from acetyl-CoA with the ketone carbonyl group of acetoacetyl-CoA.

$$CoAS\overset{\overset{O}{\|}}{\underset{\ddot{}}{C}}\overset{-}{\underset{}{CH_2}} \qquad H_3C\overset{\overset{:\ddot{O}:}{\|}}{\underset{}{C}}\overset{}{\underset{}{CH_2}}\overset{\overset{O}{\|}}{\underset{}{C}}SCoA$$

(The mechanism of this reaction is explored in Problem 22.45.) The reduction of HMG-CoA is the rate-limiting step in isoprenoid and cholesterol biosynthesis, as we'll see in Sec 25.5C.

In this reduction, HMG-CoA is reduced to mevalonate, which is converted, in turn, into isopentenyl pyrophosphate (Problem 20.55, p. 1041), the key starting material in isoprenoid and steroid biosynthesis (Sec. 17.6B,C). Acetoacetyl-CoA is itself generated from fatty-acid metabolism (Problem 22.34), and fatty acids ultimately originate from acetyl-CoA (Sec. 22.7).

$$(22.95)$$

Therefore, a remarkably diverse array of materials—fatty acids, isoprenoids, and steroids (and many other compounds we haven't discussed)—all originate from the two-carbon compound acetyl-CoA. As we have seen in this chapter, aldol and Claisen condensations are very important reactions in the biosynthesis of these compounds.

In this text we've presented a number of examples of how chemistry is carried out in living systems, and we have shown that *all of these processes have close analogies in laboratory chemistry*. With the benefit of hindsight, it might seem obvious that natural chemistry and laboratory chemistry should be closely related. However, this point was far from obvious to early chemists. The serendipitous synthesis of urea by Friedrich Wöhler in 1828 (p. 3) signaled the beginning of an age in which the chemistry of living systems and laboratory chemistry are regarded as branches of the same basic science.

The "traditional" way of learning biochemistry is to memorize the many pathways and to try to understand the relationships between them. The better way to learn biochemical pathways is to see them as logical sequences of transformations that make sense in terms of the organic chemistry involved. (The problem and section references in Eq. 22.95 show how we have taken this approach to this pathway.) Students who bring an understanding of the fundamental mechanisms of organic chemistry to their study of biochemistry are empowered to take this more logical, and certainly less tedious, approach.

PROBLEMS

22.43 The following aldol addition gives two diastereomeric addition products, *A* and *B*, in different amounts. Compounds *A* and *B* are both racemates. Give their structures and the mechanisms for their formation.

22.44 Give the addition or condensation products of each of the following reactions indicated by letter, and explain your reasoning.

(a)

continued

continued

(b)

22.45 In HMG-CoA biosynthesis (Eq. 22.94), the first step is transfer of the acetyl group from acetyl-CoA to a thiol group of HMG-CoA synthase, the catalyzing enzyme (= E in the equations below). The resulting thioester undergoes aldol addition to aceto-acetyl-CoA followed by hydrolysis of the enzyme thioester:

Assuming that acids and bases are provided by groups on the enzyme as needed, give a curved-arrow mechanism for each of these reactions.

22.9 CONJUGATE-ADDITION REACTIONS

A. Conjugate Addition to α,β-Unsaturated Carbonyl Compounds

The conjugated arrangement of C=C and C=O bonds endows α,β-unsaturated carbonyl compounds with unique reactivity, which is illustrated by the reaction of an α,β-unsaturated ketone with HCN.

In this reaction, the elements of HCN appear to have added across the C=C bond. Yet this is not a reaction of ordinary double bonds:

Nucleophilic addition to the double bond in an α,β-unsaturated carbonyl compound occurs because it gives a resonance-stabilized enolate ion intermediate:

(Nucleophilic addition to the alkene in Eq. 22.97, in contrast, would give a very unstable alkyl anion.) The enolate ion can be protonated on either oxygen or carbon. In either case, a carbonyl group is eventually regenerated because enols spontaneously form carbonyl compounds (Sec. 22.2). The *overall* result of the reaction is net addition to the double bond.

(22.98b)

observed product   enol form of product

Nucleophilic addition to the carbon–carbon double bonds of α,β-unsaturated alde-hydes, ketones, esters, and nitriles is a rather general reaction that can be observed with a variety of nucleophiles. Some additional examples follow; try to write the mechanisms of these reactions.

Conjugate additions to α,β-unsaturated esters:

(22.99)

(22.100)

2-propanethiol   **methyl acrylate**   **methyl 3-(isopropylthio)propanoate**
(97% yield)

Conjugate addition to an α,β-unsaturated ketone:

(22.101)

(85% yield)

Conjugate addition to an α,β-unsaturated nitrile:

$$CH_3SH \;+\; H_2C{=}CH{-}CN \xrightarrow[\text{MeOH}]{\text{NaOMe}} CH_3S{-}CH_2{-}CH_2{-}CN \qquad (22.102)$$

methanethiol   **acrylonitrile**   **3-(methylthio)propanenitrile**
(91% yield)

Notice that the addition of cyanide in Eq. 22.99 forms a new carbon–carbon bond, and that the nitrile group can then be converted into a carboxylic acid group by hydrolysis. The addition of a nucleophile to acrylonitrile (as in Eq. 22.102) is a useful reaction called **cyanoethylation**.

Because quinones (Sec. 18.8A) are α,β-unsaturated carbonyl compounds, they also undergo similar conjugate-addition reactions.

$$+ \ CH_3SH \ \xrightarrow[\text{0 °C}]{\text{MeOH}} \ \text{intermediate} \ \longrightarrow \ \text{product} \tag{22.103}$$

In this example, the reaction is driven to completion by enolization of the ketone in brackets to the phenol, which is aromatic. (See Eq. 22.14, p. 1111.)

A Conjugate Addition Involved in Drug Toxicity

Acetaminophen (sometimes called paracetamol) is a well-known drug that is used in a number of commercial analgesic medications for the relief of headaches and other minor aches and pains. Although it is generally regarded as safe at the recommended dosages, an overdose of acetaminophen can be very toxic. The basis of this toxicity involves a conjugate-addition reaction to a metabolite of acetaminophen.

In Sec. 17.5B, you learned that the liver enzyme *cytochrome P450* (CyP450) is involved in phase I drug metabolism. CyP450 brings about the hydroxylation of acetaminophen at the nitrogen to give the *N*-hydroxy derivative.

This derivative readily undergoes an elimination of water to form a quinone-like intermediate called an *iminoquinone.* Although aromaticity is lost in this reaction, the reaction is driven by loss of the very weak N—O bond.

The resulting iminoquinone is usually scavenged by a conjugate-addition reaction with the thiol group of a small peptide, *glutathione.* The resulting product is converted into a water-soluble derivative that is carried into the urine and excreted. (A reaction with, and removal of, foreign electrophiles is a very important biological role of glutathione.)

γ-glutamylcysteinylglycine (glutathione)

glutathione adduct of
acetaminophen iminoquinine
(converted into a derivative that
is excreted in the urine)

This conjugate addition is highly favorable because it regenerates the aromatic ring system.

If too much iminoquinone forms because of a drug overdose, the glutathione supply in the liver is exhausted, and the iminoquinone can then react with the thiol groups of other proteins. These reactions of the iminoquinone are the source of the liver toxicity.

A conjugate-addition reaction of the iminoquinone is also used in the treatment of acetaminophen overdose. Treatments of acetaminophen overdose typically involve the administration of another thiol, the amino acid derivative *N*-acetylcysteine.

N-acetylcysteine

The large excess of this thiol competes effectively for the iminoquinone and prevents the undesired reactions with protein thiols.

For a related biological reaction of quinones, see the sidebar, "Poison Ivy and Itchy Quinones," on p. 923.

The preceding examples occur under basic or neutral conditions, but acid-catalyzed additions to the carbon–carbon double bonds of α,β-unsaturated carbonyl compounds are also known.

$$H_2C{=}CH{-}CO_2Me + HBr \xrightarrow{Et_2O} Br{-}CH_2CH_2{-}CO_2Me \qquad (22.104)$$

methyl acrylate   **methyl β-bromopropionate**
(80–84% yield)

$$H_2C{=}CH{-}CH{=}O + HCl \xrightarrow{-15\,°C} Cl{-}CH_2CH_2{-}CH{=}O \qquad (22.105)$$

Although such reactions appear to be nothing more than simple additions to the carbon–carbon double bond, this is not the case. The more basic site of an α,β-unsaturated carbonyl compounds is not the double bond, but rather the carbonyl oxygen. Protonation on the carbonyl oxygen is followed by a reaction with the halide ion. The electrophilic oxygen can accept electrons as a result of a nucleophilic reaction of the halide ion either at the carbonyl carbon or, because of the conjugated arrangement of π bonds, at the β-carbon:

(22.106)

A reaction of Br^- at the carbonyl carbon yields a relatively unstable tetrahedral addition intermediate, which loses the Br^- leaving group and reverts back to the protonated ketone; a reaction at the β-carbon yields an enol, which rapidly reverts to the observed carbonyl product.

An addition to the double bond of an α,β-unsaturated carbonyl compound is an example of *conjugate addition*. The mechanism of the conjugate addition of HBr shown in Eq. 22.106 is similar to the conjugate addition of HBr to 1,3-butadiene (Sec. 15.4A); both involve carbocation intermediates. However, the *nucleophilic conjugate addition*, such as the addition of cyanide in Eq. 22.99, has no parallel in the reactions of simple conjugated dienes.

B. Conjugate-Addition Reactions versus Carbonyl-Group Reactions

Any conjugate-addition reaction *competes* with a carbonyl-group reaction. In the case of aldehydes and ketones, conjugate addition competes with addition to the carbonyl group. (Nuc = nucleophile; for example, in cyanide addition, H—Nuc = H—CN.)

(22.107)

In the case of esters, conjugate addition competes with *nucleophilic acyl substitution*.

$$R-CH=CH-\overset{\overset{\displaystyle O}{\|}}{C}-OEt + H-Nuc$$

$$R-\underset{\underset{\displaystyle Nuc}{|}}{CH}-CH_2-\overset{\overset{\displaystyle O}{\|}}{C}-OEt$$
(conjugate addition)

$$R-CH=CH-\overset{\overset{\displaystyle O}{\|}}{C}-Nuc + EtOH$$
(nucleophilic acyl substitution)

(22.108)

When can we expect to observe conjugate addition, and when can we expect reactions at the carbonyl carbon?

Consider first the reactions of aldehydes and ketones. Relatively weak bases that give *reversible* carbonyl-addition reactions with ordinary aldehydes and ketones tend to give conjugate addition with α,β-unsaturated aldehydes and ketones. Among the relatively weak bases in this category are cyanide ion, amines, thiolate ions, and enolate ions derived from β-dicarbonyl compounds. Conjugate addition is observed with these nucleophiles because *the conjugate-addition products are more stable than the carbonyl-addition products.* If carbonyl addition is reversible—even if it occurs more rapidly—then conjugate addition can drain the carbonyl compound from the addition equilibrium, and the conjugate-addition product is formed ultimately.

$$R-CH=CH-\overset{\overset{\displaystyle O}{\|}}{C}-R + H-CN$$

faster but reversible

$$R-CH=CH-\underset{\underset{\displaystyle CN}{|}}{\overset{\overset{\displaystyle OH}{|}}{C}}-R$$
carbonyl-addition (kinetic) product
(less stable)

slower but irreversible

$$R-\underset{\underset{\displaystyle CN}{|}}{CH}-CH_2-\overset{\overset{\displaystyle O}{\|}}{C}-R$$
conjugate-addition (thermodynamic) product
(more stable)

(22.109)

This, then, is another case of *kinetic versus thermodynamic control of a reaction* (Sec. 15.4C). The conjugate-addition product is the thermodynamic (more stable) product of the reaction.

The greater stability of the conjugate-addition product can be understood with a bond energy argument. Conjugate addition retains a carbonyl group at the expense of a carbon–carbon double bond. Carbonyl addition retains a carbon–carbon double bond at the expense of a carbonyl group. Because a C=O bond is considerably stronger than a C=C bond (Table 5.3, p. 216), conjugate addition gives a more stable product. (Other bonds are broken and formed as well, but the major effect is the relative strengths of the two kinds of double bonds.) These same factors are reflected in the relative heats of formation of the isomers allyl alcohol and propionaldehyde:

$$H_2C=CH-CH_2-OH \qquad H_3C-CH_2-CH=O \qquad (22.110)$$

	allyl alcohol	**propionaldehyde**
ΔH_f°	-124 kJ mol^{-1}	-189 kJ mol^{-1}
	(-29.6 kcal mol^{-1})	(-45.2 kcal mol^{-1})

As Eq. 22.109 suggests, carbonyl addition is in many cases the kinetically favored process; that is, it is faster than conjugate addition. When nucleophiles are used that undergo *irreversible* carbonyl additions, then the carbonyl-addition product is observed rather than the conjugate-addition product. This is exactly what happens with very powerful nucleophiles such as $LiAlH_4$ and organolithium reagents: These species add irreversibly to carbonyl groups and form carbonyl-addition products whether the reactant carbonyl compound is α,β-unsaturated or not. (These reactions are discussed further in Secs. 22.10 and 22.11A.)

Many of the same nucleophiles that undergo conjugate addition with aldehydes and ketones also undergo conjugate addition with esters. In contrast, stronger bases that react irreversibly at the carbonyl carbon react with esters to give nucleophilic acyl substitution products. Thus, hydroxide ion reacts with an α,β-unsaturated ester to give products of saponification, a nucleophilic acyl substitution reaction, because saponification is not reversible. Likewise, $LiAlH_4$ reduces α,β-unsaturated esters at the carbonyl group because the reaction of hydride ion at the carbonyl group is irreversible.

To summarize: Conjugate addition usually occurs with nucleophiles that are relatively weak bases. Stronger bases give irreversible carbonyl addition or nucleophilic acyl substitution reactions.

PROBLEMS

22.46 Give the product expected when methyl methacrylate (methyl 2-methylpropenoate) reacts with each of the following reagents.

(a) ^-CN and HCN in MeOH

(b) C_2H_5SH and NaOMe catalyst in MeOH

(c) HBr

(d) NaOH

22.47 Give a curved-arrow mechanism for each of the following reactions.

(a) $Me\ddot{N}H_2 + 2\,H_2C{=}CH{-}CO_2Me \longrightarrow MeO_2C{-}CH_2CH_2{-}\underset{\underset{Me}{|}}{\ddot{N}}{-}CH_2CH_2{-}CO_2Me$

(b)

(mixture of stereoisomers; why?)

(c)

C. Conjugate Addition of Enolate Ions

Enolate ions, especially those derived from malonic ester derivatives, β-keto esters, and the like, undergo conjugate-addition reactions with α,β-unsaturated carbonyl compounds, as in the following example:

$$\text{(22.111)}$$

(65–71% yield)

The mechanism of this reaction follows exactly the same pattern established for other nucleophilic conjugate additions; the nucleophile is the enolate ion formed in the reaction of ethoxide with diethyl malonate (Eq. 22.77a, p. 1146). Notice that, in contrast to the

Claisen ester condensation (Sec. 22.6A), this reaction requires only a catalytic amount of base. The reaction does *not* rely on ionization of the product to drive it to completion. It goes to completion because a carbon–carbon π bond in the starting α,β-unsaturated carbonyl compound is replaced by a stronger carbon–carbon σ bond.

enolate ion from
diethyl malonate
(see Eq. 22.77a)

$$^-{:}CH(CO_2Et)_2 \ + \qquad\qquad\qquad\qquad\qquad (22.112)$$

Conjugate additions of carbanions to α,β-unsaturated carbonyl compounds are called **Michael additions**, after Arthur Michael (1853–1942), a Harvard professor who investigated these reactions extensively.

Proper planning is needed to use a Michael addition in a synthesis. The product of a given Michael addition might originate from two different pairs of reactants. For example, in the reaction shown in Eq. 22.112, the same product (in principle) might be obtained by the Michael addition reaction of either of the following pairs of reactants (convince yourself of this point):

(a)

(b)
(This is the pair used in Eq. 22.112.)

Which pair of reactants should be used? To answer this question, use the result in Sec. 22.9B: Weaker bases tend to give conjugate addition, and stronger bases tend to give carbonyl-group reactions. Hence, to maximize conjugate addition, *choose the pair of reactants with the less basic enolate ion*—pair (b) in the foregoing example.

In one useful variation of the Michael addition, called the **Robinson annulation**, the immediate product of the addition can be subjected to an aldol condensation that closes a ring. (An annulation is a ring-forming reaction, from the Latin *annulus*, meaning "ring.")

(63–65% yield)

The Michael addition involves the enolate ion formed by ionization of the acidic proton of the β-diketone. (Write the curved-arrow mechanism of this addition.) The mechanism of the aldol condensation that follows is explored in Problem 22.25 on p. 1133. This type of reaction sequence was named for Sir Robert Robinson (1886–1975), a British chemist at Oxford University who pioneered its use. (Robinson received the 1947 Nobel Prize in Chemistry for his work in alkaloids, which are discussed in Sec. 23.12B.)

STUDY PROBLEM   22.6

Outline a synthesis of tricarballylic acid from diethyl fumarate and any other reagents.

diethyl fumarate   **tricarballylic acid**

SOLUTION   Two of the carboxylic acid groups required in the target are already in place as the ester units in diethyl fumarate. A Michael addition of some species that could be converted into a —CH_2CO_2H group is required. Notice that the desired product is conceptually a substituted acetic acid:

**STUDY GUIDE
LINK 22.7**
Synthetic
Equivalents
in Conjugate
Addition

Recall that one way of preparing substituted acetic acids is the malonic ester synthesis (Sec. 22.8A). A variation of the malonic ester synthesis can be employed here in which alkylation of the conjugate-base anion of diethyl malonate is carried out by a Michael addition with diethyl fumarate instead of an S_N2 reaction with an alkyl halide.

Saponification of all four ester groups, protonation, and decarboxylation yields the desired tricarboxylic acid:

PROBLEMS

22.48   Provide structures for the missing nucleophiles that could be used in the following transformations.

(a)

(b)

$$Y + H_2C{=}CH{-}CN \xrightarrow[\text{EtOH}]{\text{NaOEt}} \xrightarrow[\text{heat}]{\text{H}_3\text{O}^+} H_3C{-}\overset{\displaystyle O}{\overset{\displaystyle \|}{C}}{-}\underset{\displaystyle CH_2CH_2CO_2H}{\overset{\displaystyle CH_2CH_2CO_2H}{CH}}$$
(excess)

22.49 Give a curved-arrow mechanism for each of the following reactions. In each reaction identify the intermediate indicated by A or B.

(a)

(b)

D. Conjugate Addition in Biology: Fumarase

Conjugate additions occur in a number of biological pathways. An example that we've already studied is the conversion of fumaric acid (fumarate) into malic acid (malate), a reaction that occurs in the Krebs cycle.

$$(22.114)$$

fumarate   **(S)-malate**

In Sec. 4.9C, this reaction was introduced as our first example of enzyme catalysis. In Sec. 7.7A, we considered the stereochemistry of this reaction.

This reaction is a conjugate addition of water to the double bond of fumarate. Let's consider first this reaction at physiological pH and why enzyme catalysis is necessary. If water is the nucleophile, the first step of the mechanism would be the nucleophilic reaction of water at the β-carbon of the double bond:

$$(22.115)$$

fumarate

carbanion intermediate

Water is a very weak base (conjugate acid $pK_a = -1.7$) and therefore is a poor nucleophile. Hydroxide would be a much better nucleophile, but the concentration of hydroxide ion at neutral pH is minuscule. Therefore, catalysis has to overcome the "poor nucleophile" problem.

FIGURE 22.5 The mechanism of fumarate hydration, a conjugate addition, catalyzed by the enzyme fumarase. Enzyme active-site residues are shown in blue. The water shown in red is an immobilized, permanent part of the active site. The pointers indicate important aspects of fumarase catalysis discussed in the text.

Notice in Eq. 22.115 that the carbanion formed on addition of the nucleophile is destabilized by repulsion with the negative charge of the carboxylate ion. In fact, carboxylates normally do not undergo conjugate additions under normal conditions for exactly this reason. We have seen (Eqs. 22.99 and 22.100) that esters are good conjugate-addition electrophiles; so, we might imagine that *un-ionized* carboxylic acids would also be good conjugate-addition electrophiles. However, under the usual basic conditions of the nucleophilic conjugate addition carboxylic acids are fully ionized and cannot take on another negative charge without excessive charge repulsion. Therefore, catalysis has to overcome the "poor electrophile" problem.

The mechanism of fumarase catalysis is shown in Fig. 22.5. The electron pair from the basic nitrogen of a histidine in the enzyme active site is "relayed" to a proton of the nucleophilic water (*red*) by way of a second, immobilized, water molecule, thus partially converting the water into more nucleophilic hydroxide. The acceptor carboxylate ion interacts strongly by both hydrogen bonding and electrostatic attraction with the protonated amino group of a lysine. This interaction not only helps to bind the fumarate ion in the active site but also neutralizes the negative charge on one carboxylate group; this carboxylate, then, is more like a carboxylic acid. Furthermore, delivery of a proton to the double bond by an acidic group in the enzyme active site avoids the formation of a highly basic carbanion and further reduces charge repulsion. Finally, the alignment of all of these elements within the active site provides a very large rate acceleration by the proximity effect (Sec. 11.8). All of the proton transfers are believed to occur in a concerted manner. The enzyme active site is therefore constructed to convert what would be a very unfavorable conjugate-addition reaction into a very rapid one.

PROBLEM

22.50 The reducing agent in the double-bond reduction step of fatty-acid biosynthesis (the last step of Eq. 22.74, repeated below) is NADPH (Sec. 19.8B). (ACP = acyl carrier protein.)

$$H_3O^+ + NADPH + \text{trans-crotonyl-ACP} \longrightarrow \text{butyryl-ACP} + NADP^+ + H_2O$$

(a) Using an abbreviated structure for NADPH, and assuming that acids and bases are available as needed in the enzyme active site, draw a curved-arrow mechanism for this reaction.

(b) Explain why this reaction is an example of a conjugate addition.

(c) When NADPD (a deuterium-labeled NADPH, below) and D$_2$O are used in the enzyme-catalyzed reduction of *trans*-crotonyl-ACP, the product is the (2S,3R) stereoisomer of the doubly deuterated butyryl-ACP. Is this conjugate-addition reaction a syn- or an anti-addition? Give your reasoning.

22.10 REDUCTION OF α,β-UNSATURATED CARBONYL COMPOUNDS

The carbonyl group of an α,β-unsaturated aldehyde or ketone, like that of an ordinary aldehyde or ketone (Sec. 19.8), is reduced to an alcohol with lithium aluminum hydride.

This reaction, like other LiAlH$_4$ reductions, involves the nucleophilic reaction of hydride at the carbonyl carbon and is therefore a carbonyl addition.

The reason that carbonyl addition, rather than conjugate addition, is observed follows from the discussion in Sec. 22.9B. Carbonyl addition is not only *faster* than conjugate addition but, in this case, is also *irreversible*. It is irreversible because hydride is a very poor leaving group. Because carbonyl addition of LiAlH$_4$ is irreversible, conjugate addition never has a chance to occur and is therefore not observed.

In other words, reduction of the carbonyl group with LiAlH$_4$ is a *kinetically controlled* reaction.

Many α,β-unsaturated aldehydes and ketones are reduced by NaBH$_4$ to give mixtures of both carbonyl-addition products and conjugate-addition products. Because mixtures are obtained, NaBH$_4$ reductions of α,β-unsaturated ketones are not useful. Why conjugate addition is observed with NaBH$_4$ is not well understood. Although some cases of conjugate addition with LiAlH$_4$ are known, this reagent usually reduces carbonyl groups, including the carbonyl groups of esters, without affecting double bonds.

The carbon–carbon double bond of an α,β-unsaturated carbonyl compound can in most cases be reduced selectively by catalytic hydrogenation. (See also Eq. 19.36, p. 975.)

$$\text{Ph—CH=CH—}\overset{\displaystyle O}{\overset{\displaystyle \|}{\text{C}}}\text{—Ph} + H_2 \xrightarrow[\substack{\text{3 atm}\\ \text{ethyl acetate}\\ \text{(solvent)}}]{\text{Pt}} \text{Ph—CH}_2\text{CH}_2\text{—}\overset{\displaystyle O}{\overset{\displaystyle \|}{\text{C}}}\text{—Ph} \qquad (22.118)$$

1,3-diphenyl-2-propen-1-one   **1,3-diphenyl-1-propanone**
                                                                     (81–95% yield)

PROBLEM

22.51   Show how ethyl 2-butenoate can be used as a starting material to prepare (a) ethyl butanoate and (b) 2-buten-1-ol.

The reduction of double bonds in α,β-unsaturated carbonyl compounds occurs in biological pathways. (An example of such a reduction by NADPH is explored in Problem 22.50, p. 1166.)

22.11   REACTIONS OF α,β-UNSATURATED CARBONYL COMPOUNDS WITH ORGANOMETALLIC REAGENTS

A. Addition of Organolithium Reagents to the Carbonyl Group

Organolithium reagents react with α,β-unsaturated carbonyl compounds to yield products of carbonyl addition.

$$\substack{\text{H}_3\text{C}\\ \\ \text{H}_3\text{C}}\text{C=CH—}\overset{\displaystyle O}{\overset{\displaystyle \|}{\text{C}}}\text{—CH}_3 + \text{PhLi} \longrightarrow \xrightarrow{\text{H}_2\text{O}} \substack{\text{H}_3\text{C}\\ \\ \text{H}_3\text{C}}\text{C=CH—}\underset{\underset{\displaystyle \text{Ph}}{|}}{\overset{\overset{\displaystyle \text{OH}}{|}}{\text{C}}}\text{—CH}_3 \qquad (22.119)$$

4-methyl-3-penten-2-one   **4-methyl-2-phenyl-3-penten-2-ol**
(mesityl oxide)   (67% yield)

$$\text{H}_2\text{C=}\underset{\underset{\displaystyle \text{CH}_3}{|}}{\text{C}}\text{—}\overset{\displaystyle O}{\overset{\displaystyle \|}{\text{C}}}\text{—OCH}_3 + 2\,\text{Bu—Li} \xrightarrow{\text{THF}} \xrightarrow{\text{H}_3\text{O}^+} \text{H}_2\text{C=}\underset{\underset{\displaystyle \text{CH}_3}{|}}{\text{C}}\text{—}\underset{\underset{\displaystyle \text{Bu}}{|}}{\overset{\overset{\displaystyle \text{OH}}{|}}{\text{C}}}\text{—Bu} \qquad (22.120)$$

methyl   **3-butyl-2-methyl-1-hepten-3-ol**
2-methylpropenoate   (89% yield)

The reason carbonyl addition occurs rather than conjugate addition is the same as in the case of LiAlH₄ reduction (Sec. 22.10): carbonyl addition is more rapid than conjugate addition and it is also irreversible.

Because Grignard and organolithium reagents undergo many of the same types of reactions, it is reasonable to ask whether Grignard reagents also undergo carbonyl addition. Grignard reagents in many cases give mixtures of conjugate addition and carbonyl addition. (The reason is discussed in Sec. 22.11B.) Because both types of addition occur with Grignard reagents, organolithium reagents are used with α,β-unsaturated carbonyl compounds when only carbonyl addition is the desired reaction.

B. Conjugate Addition of Lithium Dialkylcuprate Reagents

Lithium dialkylcuprate reagents (Secs. 11.5C and 21.10B) give exclusively products of *conjugate addition* when they react with α,β-unsaturated esters and ketones.

$$\text{2-cyclohexenone} \xrightarrow[\text{ether, } -78\,°C]{(CH_3)_2Cu^-\ Li^+} \xrightarrow{H_2O} \text{3-methylcyclohexanone} \quad (22.121)$$

2-cyclohexenone

3-methylcyclohexanone
(97% yield)

Even α,β-unsaturated aldehydes, which are normally very reactive at the carbonyl group, give all or mostly products of conjugate addition, especially at low temperature.

$$Et_2C{=}CH{-}CH{=}O \xrightarrow[\substack{\text{ether} \\ -50\,°C}]{(CH_3)_2CuLi} \xrightarrow{H_3O^+} \underset{\substack{CH_3 \\ (95\% \text{ of product})}}{Et_2C{-}CH_2{-}CH{=}O} + \underset{\substack{CH_3 \\ (5\% \text{ of product})}}{Et_2C{=}CH{-}\overset{OH}{\underset{|}{CH}}} \quad (22.122)$$

70% total yield

The fact that lithium dialkylcuprate reagents undergo conjugate addition might seem to contradict the notion that strong bases undergo carbonyl addition. However, there is good evidence that conjugate addition of lithium dialkylcuprate reagents proceeds by a special mechanism promoted by the presence of copper, and that this mechanism is particularly favorable for conjugate addition. (See Further Exploration 22.3.) For our purposes, however, the reaction can be envisioned mechanistically to be similar to other conjugate additions. The nucleophilic reaction of an anion—in this case the "alkyl anion" of the dialkylcuprate reagent—at the double bond gives a resonance-stabilized enolate ion.

FURTHER EXPLORATION 22.3
Conjugate Addition of Organocuprate Reagents

$$ \quad (22.123)$$

enolate ion of the product

When water is added to the reaction mixture, protonation of the enolate ion gives the conjugate-addition product.

We noted in Sec. 22.10 that Grignard reagents react with α,β-unsaturated carbonyl compounds to give mixtures of carbonyl-addition and conjugate-addition products. Some chemists have theorized that the conjugate-addition products are due to small amounts of transition metals known to be present in commercial magnesium. Indeed, certain transition metals are known to promote conjugate addition of Grignard reagents. In fact, if a Grignard reagent is treated with CuCl, magnesium organocuprate reagents are formed, and these give exclusively conjugate addition like their lithium organocuprate counterparts.

To summarize: To carry out a *carbonyl-addition* reaction with an organometallic reagent, use an organolithium reagent. To carry out a *conjugate-addition* reaction, use a lithium organocuprate (or a Grignard reagent with added CuCl).

PROBLEMS

22.52 Outline a synthesis of each of the following compounds from mesityl oxide (4-methyl-3-penten-2-one). Use an organometallic reagent in at least one step of each synthesis.

(a)

$$(CH_3)_3CCH_2{-}\overset{O}{\underset{\|}{C}}{-}CH_3$$

(b)

$$(CH_3)_2C{=}CH{-}\overset{OH}{\underset{|}{C}}(CH_3)_2$$

(c)

$$CH_3CH_2{-}\overset{CH_3}{\underset{\underset{CH_3}{|}}{\overset{|}{C}}}{-}CH{=}C(CH_3)_2$$

continued

continued _____

22.53   Complete the following reactions and explain your reasoning.

(a)

$+ \ Me_2CuLi \ \longrightarrow \ \xrightarrow{H_3O^+}$

(b) $H_3C-C\equiv C-CO_2Me \ + \ Me_2CuLi \ \longrightarrow \ \xrightarrow{H_3O^+}$
(1 equiv.)

22.12   ORGANIC SYNTHESIS WITH CONJUGATE-ADDITION REACTIONS

When is a conjugate-addition reaction useful in an organic synthesis? One way to think of this problem is that any group at the β-position of a carbonyl compound (or nitrile) can *in principle* be delivered as a nucleophile in a conjugate addition. Hence, a conjugate addition can be mentally reversed by subtracting a nucleophilic group from the β-position of the target molecule, and a positive fragment (usually a proton) from the α-position:

$$R-CH_2-\underset{\underset{H}{|}}{CH}-\overset{\overset{O}{\|}}{C}-R' \ \Rrightarrow \ \text{``}R^-,\ H^+\text{''} \ + \ H_2C=CH-\overset{\overset{O}{\|}}{C}-R' \qquad (22.124)$$

This approach is explored in the Study Problem 22.7.

STUDY PROBLEM 22.7

Outline a preparation of 2-octanone by a conjugate-addition reaction.

SOLUTION   Two groups are attached to the β-carbon of 2-octanone: a hydrogen (*blue*) and a butyl group (*red*).

$$\overset{\beta\text{-carbon}}{CH_3CH_2CH_2CH_2-\overset{\overset{H}{|}}{CH}-CH_2-\overset{\overset{O}{\|}}{C}-CH_3}$$

2-octanone

One choice for the "R^-" group in Eq. 22.124 is a butyl group, which can be introduced as a "butyl anion" by the reaction of lithium dibutylcuprate with methyl vinyl ketone; the α-proton is provided in the subsequent protonolysis step:

$$Li^+ \ (CH_3CH_2CH_2CH_2)_2Cu^- \ + \ H_2C=CH-\overset{\overset{O}{\|}}{C}-CH_3 \ \longrightarrow \ \xrightarrow{H_3O^+} \ \text{2-octanone} \qquad (22.125)$$

lithium dibutylcuprate   **methyl vinyl ketone**

Another choice for "R^-" is the hydrogen. Although we've not considered any ways for adding "H^-" in a conjugate addition (there are some!), a process with the same outcome is the hydrogenation of an α,β-unsaturated ketone:

$$CH_3CH_2CH_2CH_2CH=CH-\overset{\overset{O}{\|}}{C}-CH_3 \ \xrightarrow{H_2/\text{catalyst}} \ \text{2-octanone} \qquad (22.126)$$

The type of analysis illustrated here will be even more useful if you keep in mind the notion of "synthetic equivalents" in Study Guide Link 22.7.

PROBLEM

22.54   Show how a conjugate addition can be used to prepare each of the following compounds.

(a), (b) 3,4-dimethyl-2-hexanone (2 ways)

(c)

$$H_3C-\overset{\overset{\displaystyle O}{\|}}{C}-CH_2CH_2CH_2-\overset{\overset{\displaystyle O}{\|}}{C}-OH$$

(d)

$$H_3C-\overset{\overset{\displaystyle O}{\|}}{C}-CH_2CH_2-\overset{\overset{\displaystyle O}{\|}}{C}-OH$$

levulinic acid

A great many of the reactions discussed in this chapter can be used to form carbon–carbon bonds.

1. aldol addition and condensation reactions (Sec. 22.4)
2. Claisen and Dieckmann condensations (Sec. 22.6)
3. malonic ester synthesis (Sec. 22.8A)
4. alkylation of ester enolates with amide bases and alkyl halides or tosylates (Sec. 22.8B)
5. acetoacetic ester synthesis (Sec. 22.8C)
6. Aldol addition of ester enolates (Sec. 22.8D)
7. conjugate addition of cyanide ions (Sec. 22.9A) and enolate ions (Sec. 22.9C) to α,β-unsaturated carbonyl compounds
8. reaction of lithium dialkylcuprates with α,β-unsaturated carbonyl compounds (Sec. 22.11B)

(A complete list of methods for forming carbon–carbon bonds is given in Appendix VI.) Their utility for carbon–carbon bond formation accounts in large measure for the importance of these reactions in organic chemistry.

KEY IDEAS IN CHAPTER 22

- Hydrogens on carbon atoms α to carbonyl groups and cyano groups are acidic. Ionization of these hydrogens gives enolate ions.

- Most carbonyl compounds with α-hydrogens are in equilibrium with small amounts of enol (vinylic alcohol) isomers. Generally, carbonyl–enol equilibria favor the carbonyl compounds, but there are important exceptions, such as phenols and β-diketones, in which enol isomers are the major forms.

- The equilibria between carbonyl compounds and their enol isomers are catalyzed by acids and bases.

- Enolate ions can act as nucleophiles in a number of reactions. Enolate ions can

  1. undergo α-halogenation (haloform reaction)
  2. add to carbonyl groups (aldol addition and condensation, Reformatsky reaction, and Knoevenagel condensation)

  3. act as nucleophiles in carbonyl-substitution reactions (Claisen and Dieckmann condensations)
  4. react with alkyl halides (enolate alkylation, as in the malonic ester synthesis and the acetoacetic ester synthesis)
  5. add to α,β-unsaturated carbonyl compounds (Michael addition)

- Enols of aldehydes and ketones undergo α-halogenation and aldol condensation reactions in acidic solution.

- The aldol addition and the Claisen condensation are reversible reactions. The aldol addition is generally favorable for aldehydes, but unfavorable for most ketones. It can be driven to completion by dehydration of the addition product, a β-hydroxy carbonyl compound (aldol condensation). The Claisen condensation is driven to completion by ionization of the product.

- Aldol additions to aldehydes and ketones can be carried out with lithium and zinc enolates (Reformatsky reagent) of esters and with lithium enolates of ketones (followed by protonation) because these reactions are irreversible.

- Two types of addition to α,β-unsaturated carbonyl compounds are possible: addition to the carbonyl group and addition to the double bond (conjugate addition,

or 1,4-addition). When carbonyl addition is reversible, conjugate addition is observed because it gives the more stable product. When carbonyl addition is irreversible, it is usually observed instead because it is faster.

- Reactions closely resembling aldol additions and Claisen condensations are observed in living systems. Many such reactions employ acetyl-CoA as a starting material.

 REACTION REVIEW *For a summary of reactions discussed in this chapter, see the* **Reaction Review** *section of Chapter 22 in the* Study Guide and Solutions Manual.

ADDITIONAL PROBLEMS

22.55 Give the principal organic product expected when 3-buten-2-one (methyl vinyl ketone) reacts with each of the following reagents.
  - (a) HBr
  - (b) H_2, Pt (cat.)
  - (c) $LiAlH_4$, then H_2O
  - (d) HCN in water, pH 10
  - (e) Et_2CuLi, then H_3O^+
  - (f) diethyl malonate and NaOEt, then H_3O^+
  - (g) ethylene glycol, HCl (cat.)
  - (h) 1,3-butadiene

22.56 Give the principal organic products expected when ethyl *trans*-2-butenoate (ethyl crotonate) reacts with each of the following reagents.
  - (a) $^-$CN in ethanol, then H_2O/H_3O^+, heat
  - (b) Me_2NH, room temperature
  - (c) NaOH, H_2O, heat
  - (d) CH_3Li (excess), then H_3O^+
  - (e) H_2, catalyst
  - (f) 1,3-cyclopentadiene

22.57 Give the principal organic products expected when isobutyraldehyde (2-methylpropanal) reacts with each of the following reagents:
  - (a) the lithium enolate of acetone followed by H_2O/H_3O^+
  - (b) the lithium enolate of ethyl 2-methylpropanoate followed by dilute H_3O^+
  - (c) ethyl α-bromoacetate + Zn, then H_3O^+
  - (d) diethyl malonate + a secondary amine ($R_2\overset{..}{N}H$) catalyst
  - (e) ethyl acetoacetate + a secondary amine catalyst

22.58 Give the structure of a compound that meets each criterion.
  - (a) an optically active compound $C_6H_{12}O$ that racemizes in base

  - (b) an achiral compound $C_6H_{12}O$ that does not give a positive Tollens test (Sec. 19.14)
  - (c) an optically active compound $C_6H_{12}O$ that neither racemizes in base nor gives a positive Tollens test
  - (d) an optically active compound $C_6H_{12}O$ that gives a positive Tollens test and does not racemize in base

22.59 Give the structure of a compound that meets each criterion.
  - (a) A carbonyl compound C_4H_8O that gives a precipitate of iodoform with I_2 in base.
  - (b) A carbonyl compound C_4H_8O that does not give a precipitate of iodoform with I_2 in base.
  - (c) An alcohol C_4H_8O that gives a precipitate of iodoform with I_2 in base.

22.60 Each of the following compounds is unstable and either exists as an isomer or spontaneously decomposes to other compounds. In each case, give the more stable isomer or decomposition product and explain.
  - (a) $H_3C-C\equiv C-OH$
  - (b)
  - (c)
$$H_3C-CH-O-CH=CH-CH_3$$
    with OH on the CH

22.61 (a) Draw the structure of the conjugate base of each of the following compounds. What is the relationship between the two conjugate bases?

A   *B*

(b) Which compound is more acidic? Use an energy diagram to explain your reasoning.

22.62 (a) Show that the two following compounds have the same conjugate base.

A   *B*

(b) Which compound is more acidic? Explain your reasoning using an argument invoking the relative free energies of *A* and *B*.

22.63 (a) Give the products that result from the ester hydrolysis of 1-cyclohexenyl acetate (compound *A*) below.

A
$\Delta G°$ (hydrolysis) = –67.3 kJ mol^{-1}

B
$\Delta G°$ (hydrolysis) = –30.9 kJ mol^{-1}

(b) As the above data show, the $\Delta G°$ for hydrolysis of *A* is *much* more negative than the $\Delta G°$ for hydrolysis of phenyl acetate (*B*). Explain why the equilibrium for the hydrolysis of *A* is much more favorable than the equilibrium for the hydrolysis of *B*.

22.64 Which compound in each of the sets shown in Fig. P22.64 is most acidic? Explain.

22.65 Arrange the following compounds in order of increasing acidity and explain.

(1) isobutyramide   (2) octanoic acid

(3) toluene   (4) ethyl acetate

(5) phenylacetylene   (6) phenol

22.66 (a) Give the structures of the three separable monobromo derivatives that could form when 2-methylcyclohexanone is treated with Br$_2$ in the presence of HBr.

(b) In fact, only *one* of these derivatives is formed. Assuming that this derivative results from bromination of the *more stable enol*, predict which of the three isomers in part (a) is formed, and explain your choice.

22.67 When 1,3-diphenyl-2-propanone is treated with Br$_2$ in acid, 1,3-dibromo-1,3-diphenyl-2-propanone is obtained in good yield. On further characterization, however, this product proves to have a very broad melting point (79–87 °C), a fact suggesting a mixture of compounds. Account for this observation.

22.68 When acetoacetic acid is decarboxylated in the presence of bromine, bromoacetone is isolated (see Fig. P22.68). The rate of appearance of bromoacetone is described by the following rate law:

$$\text{rate} = k[\text{acetoacetic acid}]$$

(The reaction rate is zero order in bromine.) Suggest a mechanism for the reaction that is consistent with this rate law.

22.69 Account for the fact that treatment of 1,3-diphenyl-1,3-propanedione with I$_2$ and NaOH gives a precipitate of iodoform even though it is not a methyl ketone. Besides iodoform, the other product of the reaction, after acidification, is two equivalents of benzoic acid.

1,3-diphenyl-1,3-propanedione

Figure P22.64

acetoacetic acid   **bromoacetone**

Figure P22.68

22.70 Indicate which hydrogens are replaced by deuterium when each of the following compounds is treated with dilute NaOD in a large excess of CH_3OD.

(a)

(b)

22.71 When compound *A* in Fig. P22.71 is treated with $NaOCH_3$ in CH_3OH, isomerization to compound *B* occurs.

(a) Give a mechanism for the reaction, and explain why the equilibrium favors compound *B*.

(b) Explain why, when compound *C* is subjected to the same conditions, no isomerization occurs.

22.72 Although one enantiomer of the drug thalidomide (sidebar, p. 243) is a sedative, the other is teratogenic (causes birth defects). Unfortunately, thalidomide is racemized in the body and, for that reason, both enantiomers are teratogenic. (See Fig. P22.72.) Assuming the availability of acids and bases as needed, give a curved-arrow mechanism for the base-catalyzed racemization of thalidomide in water.

22.73 In either acid or base, 3-cyclohexenone comes to equilibrium with 2-cyclohexenone:

(a) Explain why the equilibrium favors the α,β-unsaturated ketone over its β,γ-unsaturated isomer.

(b) Give a mechanism for this reaction in aqueous NaOH.

(c) Give a mechanism for the same reaction in dilute aqueous H_2SO_4. (*Hint:* The enol 1,3-cyclohexadienol is an intermediate in the acid-catalyzed reaction.)

(d) Is the equilibrium constant for the analogous reaction of 4-methyl-3-cyclohexenone expected to be greater or smaller? Explain.

22.74 In 3-methyl-2-cyclohexenone the eight hydrogens H^a, H^b, H^c, and H^d can be exchanged for deuterium in CH_3O^-/CH_3OD.

(a) Write curved-arrow mechanisms for the base-catalyzed exchange of hydrogens H^a, H^c, and H^d.

(b) Explain why hydrogen H^b is much less acidic than hydrogens H^a, H^c, and H^d, even though it is an α-hydrogen.

(c) Although hydrogen H^b is not unusually acidic, it nevertheless exchanges readily in base. Write a mechanism for the exchange of H^b. (*Hint:* Consider the equilibrium in Problem 22.73.)

(d) Which hydrogens of the sex hormone testosterone would be exchanged for deuterium in CH_3O^-/CH_3OD?

testosterone

A   *B*   *C*

Figure P22.71

(S)-thalidomide   **(R)-thalidomide**

Figure P22.72

22.75 (a) Compound A, γ-pyrone, is usually basic with a conjugate-acid pK_a of -0.4. The pK_a of the conjugate acid of compound B, in contrast, is about -3.

A   *B*
γ-pyrone

Draw the structures of the conjugate acids of both molecules, and explain why A is more basic than B.

(b) Tropone reacts with one equivalent of HBr to give a stable crystalline conjugate acid salt with a pK_a of -0.6, which is considerably greater (that is, less negative) than the pK_a values of most protonated α,β-unsaturated ketones. Give the structure of the conjugate acid of tropone, and explain why tropone is unusually basic.

tropone

22.76 (a) The resonance structures shown in part (a) of Fig. P22.76 can be written for an α,β-unsaturated carboxylic acid. Would this type of resonance interaction increase or diminish the acidity of an α,β-unsaturated carboxylic acid relative to that of an ordinary carboxylic acid? Explain.

(b) Consider the pK_a data given in part (b) of Fig. P22.76. Show how these data are consistent with both the polar effect of the double bond and the resonance effect in part (a).

22.77 (a) The pK_a of 2-nitropropane is 10. Give the structure of its conjugate base, and suggest reason(s) why 2-nitropropane has a particularly acidic C—H bond.

2-nitropropane

(b) When the conjugate base of 2-nitropropane is protonated, an isomer of 2-nitropropane is formed, which, on standing, is slowly converted into 2-nitropropane itself. Give the structure of this isomer.

(c) What product forms when 2-nitropropane reacts with ethyl acrylate ($H_2C{=}CH{-}CO_2Et$) in the presence of NaOEt in EtOH?

22.78 Identify the intermediates A and B in the transformation shown in Fig. P22.78, and show how they are formed.

22.79 Explain the following findings.

(a) One full equivalent of base must be used in the Claisen or Dieckmann condensation.

(b) Ethyl acetate readily undergoes a Claisen condensation in the presence of one equivalent of sodium ethoxide, but phenyl acetate does *not* undergo a Claisen condensation in the presence of one equivalent of sodium phenoxide.

(c) Although the aldol condensation can be catalyzed by acid, the Claisen condensation cannot.

(a)

$$\left[R{-}CH{=}CH{-}\overset{:\overset{..}{O}:}{\overset{\|}{C}}{-}OH \longleftrightarrow R{-}\overset{+}{C}H{-}CH{=}\overset{:\overset{..}{O}{:}^{-}}{C}{-}OH \right]$$

(b)

$$H_2C{=}CH{-}CH_2{-}CO_2H$$
$$pK_a = 4.37$$

$$pK_a = 4.70$$

$$CH_3CH_2CH_2{-}CO_2H$$
$$pK_a = 4.87$$

Figure P22.76

Figure P22.78

22.80 A student, Cringe Labrack, has suggested each of the faulty synthetic procedures shown in Fig. P22.80. Explain why each one cannot be expected to work as shown.

22.81 When 2,4-pentanedione in ether is treated with one equivalent of sodium hydride (NaH), a gas is evolved and an ionic compound *A* is formed.

(a) Give the structure of *A*. Which atoms of *A* should be nucleophilic? Explain.

(b) When *A* reacts with CH_3I, three isomeric compounds, *B*, *C*, and *D* ($C_6H_{10}O_2$), are formed. Suggest structures for these compounds.

22.82 Propose syntheses of each of the following compounds from the indicated starting materials and any other reagents.

(a) 3-ethylcyclopentanol from 2-cyclopentenone

(b) 1,3,3-trimethylcyclohexanol from 3-methyl-2-cyclohexenone

(c) 2-benzylcyclohexanone from $EtO_2C(CH_2)_5CO_2Et$

(d) 2,2-dimethyl-1,3-propanediol from diethyl malonate

(e) $H_2NCH_2CH_2CH_2CH_2OH$ from ethyl acrylate (ethyl 2-propenoate)

(f)

$$Et\!-\!\!\!\overset{\displaystyle}{\underset{\displaystyle Et}{N}}\!\!\!-CH_2CH_2CH_2NH_2 \quad \text{from acrylonitrile (propenenitrile)}$$

(g) 2-phenylbutanoic acid from phenylacetic acid (Do not use branched amide bases; see Eq. 22.65 on p. 1137.)

(h) 1,3-diphenyl-1-butanone from acetophenone

(i)

$$Ph\!-\!\!\overset{\displaystyle OH}{\overset{\displaystyle |}{CH}}\!\!-CD_2\!-\!Ph \quad \text{from phenylacetic acid}$$

(j)

from cyclohexanone

(k)

from acetyl chloride

(l)

$$CH_3CH_2CH_2CH_2\overset{\displaystyle O}{\overset{\displaystyle \|}{C}}\overset{\displaystyle}{\underset{\displaystyle CH_3}{\overset{\displaystyle |}{CH}}}CH_2CH_2CH_3 \quad \text{from diethyl malonate}$$

(m)

from

22.83 A useful diketone, *dimedone*, can be prepared in high yield by the synthesis shown in Fig. P22.83. Provide structures for both the intermediate *A* (a Michael-addition product) and dimedone, and give a curved-arrow mechanism for each step up to compound *B*.

22.84 When the diethyl ester of a substituted malonic acid is treated with sodium ethoxide and urea, Veronal, a *barbiturate*, is formed (see Fig. P22.84). (Barbiturates are hypnotic drugs; some are actively used in modern anes-

---

(a) $CH_3CH_2CO_2Et \xrightarrow[\text{EtOH}]{\text{NaOEt}} \xrightarrow{CH_3I} (CH_3)_2CHCO_2Et$

(b) $CH_2(CO_2Et)_2 \xrightarrow[\text{EtOH}]{\text{NaOEt}} \xrightarrow{\text{PhBr}} Ph\!-\!CH(CO_2Et)_2$

(c)

$$CH_3\overset{\displaystyle OH}{\overset{\displaystyle |}{CH}}CO_2Et \xrightarrow[\text{heat}]{H_3O^+} H_2C\!=\!CH\!-\!CO_2Et$$

(d)

$$CH_3CH_2CO_2H \xrightarrow[\text{Br}_2]{\text{PBr}_3\text{ (cat.)}} \xrightarrow[\text{ether}]{\text{Mg}} \xrightarrow{\begin{array}{l}\text{1) } CH_3CH\!=\!O \\ \text{2) } H_3O^+\end{array}} H_3C\!-\!\overset{\displaystyle OH}{\overset{\displaystyle |}{CH}}\!-\!\overset{\displaystyle CH_3}{\overset{\displaystyle |}{CH}}\!-\!CO_2H$$

(e)

$$CH_3CH_2\!-\!\overset{\displaystyle O}{\overset{\displaystyle \|}{C}}\!-\!CH_3 + H_3C\!-\!\overset{\displaystyle O}{\overset{\displaystyle \|}{CH}} \xrightarrow{^-OH} H_3C\!-\!CH\!=\!CH\!-\!\overset{\displaystyle O}{\overset{\displaystyle \|}{C}}\!-\!CH_2CH_3$$

(f)

Figure P22.80

thesia.) Using the curved-arrow notation, give a mechanism for the Veronal synthesis.

22.85 Using a reaction similar to the one in Problem 22.84, outline a synthesis of pentothal from diethyl malonate and any other reagents. (The sodium salt of pentothal is a widely used injectable anesthetic.)

pentothal

22.86 When the epoxide 2-vinyloxirane reacts with lithium dibutylcuprate, followed by protonolysis, a compound *A* is the major product formed. Oxidation of *A* with PCC yields *B*, a compound that gives a positive Tollens test and has an intense UV absorption around 215 nm. Treatment of *B* with Ag_2O, followed by catalytic hydrogenation, gives octanoic acid. Identify *A* and *B*, and outline a mechanism for the formation of *A*.

2-vinyloxirane

22.87 Using the curved-arrow notation, provide mechanisms for each of the reactions given in Fig. P22.87 (p. 1178).

22.88 Complete the reactions given in Fig. P22.88 on p. 1179 by giving the major organic products. Explain your reasoning.

22.89 A biochemist, Sal Monella, has come to you to ask your assistance in testing a promising biosynthetic hypothesis. She wishes to have two samples of methylsuccinic acid specifically labeled with ^{14}C as shown in the following structures. The source of the isotope, for financial reasons, is to be the salt $Na^{14}CN$. Outline syntheses that will accomplish the desired objective. (*$C = ^{14}C$)

(a)

(b)

22.90 The reversibility of the aldol addition reaction is a major factor in each of the following problems. Provide a plausible curved-arrow mechanism for each transformation.

(a) When the terpene *pulegone* is heated with aqueous NaOH, acetone and 3-methylcyclohexanone are formed.

pulegone

Figure P22.83

Figure P22.84

(b)

be used to form three-membered rings. (Part (a) is an example of the *Darzens glycidic ester condensation*.)

22.91 Using the curved-arrow notation, provide mechanisms for each of the reactions given in Fig. P22.91 on p. 1180.

22.92 Outline curved-arrow mechanisms for each of the known transformations shown in Fig. P22.92 on p. 1180 that can

22.93 Treatment of (S)-(+)-5-methyl-2-cyclohexenone with lithium dimethylcuprate gives, after protonolysis, a good yield of a mixture containing mostly a dextrorotatory ketone *A* and a trace of an optically inactive isomer *B*. Treatment of *A* with zinc amalgam and HCl affords an optically active, dextrorotatory hydrocarbon *C*. Identify *A*, *B*, and *C*, and give the absolute stereochemical configurations of *A* and *C*.

(a)

(b)

(c)

(d)

(e)

(f)

(g)

chelidonic acid

Figure P22.87

(a)

$$H_3C-\overset{O}{\underset{\|}{C}}-CH_2-\overset{O}{\underset{\|}{C}}-OEt \;+\; Br(CH_2)_4Br \xrightarrow[\text{EtOH}]{\text{NaOEt (excess)}} \xrightarrow[\text{heat}]{H_3O^+} \;(C_7H_{12}O)$$

(b)

γ-butyrolactone

$$\xrightarrow{Li^+\;[(CH_3)_2CH]_2\ddot{N}{:}^-} \xrightarrow{CH_3I}$$

(c)

$$+ \; Na^+ \;^-{:}CH(CO_2Et)_2 \longrightarrow \xrightarrow[\substack{H_2O \\ \text{heat}}]{H_3O^+}$$

(d)

$$\xrightarrow[\text{2) }H_3O^+]{\text{1) LiAlH}_4}$$

(e)

$$+ \; CH_3CH(CO_2Et)_2 \xrightarrow[\text{EtOH}]{\text{NaOEt}}$$

(f)

$$CH_2(CO_2Et)_2 \;+\; Ph-CH{=}O \xrightarrow[\text{(catalyst)}]{\overset{\text{NH}}{\bigcirc}} \xrightarrow[\text{EtOH}]{K^+\,^-CN} \xrightarrow[\text{heat}]{H_3\overset{+}{O},\,H_2O} \;\text{(a dicarboxylic acid }C_{10}H_{10}O_4)$$

(g)

$$CH_2(CO_2Et)_2 \xrightarrow[\text{EtOH}]{Na^+\,EtO^-} \xrightarrow{} \xrightarrow[\text{heat}]{H_3O^+,\,H_2O}$$

(h)

$$H_3C-\overset{O}{\underset{\|}{C}}-CH_2-CO_2Et \;+\; H_2C{=}CH-CO_2Et \xrightarrow[\text{EtOH}]{EtO^-} \xrightarrow[\text{H}_2\text{O}]{H_3O^+}$$

(i)

$$Cl(CH_2)_3CH\overset{O}{\underset{O}{\big\langle}} \xrightarrow{\text{Mg, CuBr}} \xrightarrow[\text{benzene}]{H_3O^+} \;(C_{10}H_{14}O) \;+\; H_2O$$

(*Hint:* Grignard reagents treated with CuBr or CuCl react like lithium dialkylcuprate reagents; see Sec. 22.11B.)

(j)

$$\xrightarrow{\text{CuCl}} \xrightarrow{H_2O}$$ [Give the stereochemistry of the product and explain; see the hint for part (i).]

(k)

$$(CH_3)_2C{=}\underset{CO_2Et}{\overset{CO_2Et}{C}} \;+\; CH_3MgI \;+\; CuCl \xrightarrow[\text{ether}]{} \xrightarrow[\text{heat}]{H_3\overset{+}{O},\,H_2O}$$

(l)

$$+ \; Ph-CH{=}CH-\overset{O}{\underset{\|}{C}}-Ph \xrightarrow{\text{AlCl}_3} \xrightarrow{H_3O^+} \;\text{(a ketone }C_{21}H_{18}O)$$

(m)

$$MeO_2C\!-\!\!\left\langle\!\!\bigcirc\!\!\right\rangle\!\!-Br \;+\; CH_2(CO_2Et)_2 \;+\; NaH \xrightarrow[\text{THF}]{\text{Pd[P(}t\text{-Bu)}_3]_4 \text{ catalyst}} \xrightarrow[\text{H}_2\text{O, heat}]{H_3O^+}$$

Figure P22.88

(a)

$$H_3C-\overset{\overset{\displaystyle O}{\|}}{C}-CH_2-CO_2Et \;+\; PhCO_2Et \;\xrightarrow{\text{NaOEt}}\; \xrightarrow{\text{H}_3\text{O}^+}$$

$$Ph\overset{\overset{\displaystyle O}{\|}}{C}-CH_2-CO_2Et \;+\; CH_3CO_2Et \quad \text{(removed as it is formed in the first reaction)}$$

(b)

$$CH_2(CO_2Et)_2 \;+\; \underset{H}{PhC}\overset{\overset{\displaystyle O}{\diagdown}}{-}CH_2 \;\xrightarrow[\text{EtOH}]{\text{NaOEt}}\; \xrightarrow[\text{H}_2\text{O}]{\text{NaOH}}\; \xrightarrow[\text{heat}]{\text{H}_2\text{O, H}_3\text{O}^+}$$

(c)

$$PhO-CH=CH-\overset{\overset{\displaystyle O}{\|}}{C}-CH_3 \;\xrightarrow{\text{H}_2\text{O, H}_3\text{O}^+}\; O=CH-CH_2-\overset{\overset{\displaystyle O}{\|}}{C}-CH_3 \;+\; PhOH$$

(d)

$$PhO-CH=CH-\overset{\overset{\displaystyle O}{\|}}{C}-CH_3 \;\xrightarrow[\text{2) H}_2\text{O, H}_3\text{O}^+]{\text{1) H}_2\text{O, }^-\text{OH}}\; O=CH-CH_2-\overset{\overset{\displaystyle O}{\|}}{C}-CH_3 \;+\; PhOH$$

(e)

$$Ph-\overset{\overset{\displaystyle OH}{|}}{CH}-CH=CH-CH_3 \;\xrightarrow[\text{EtOH/H}_2\text{O}]{\text{KOH}}\; Ph-\overset{\overset{\displaystyle O}{\|}}{C}-CH_2CH_2CH_3$$

(f)

$$\underset{H_3C}{\overset{H}{\diagdown}}C=C\underset{H}{\overset{\overset{\displaystyle O}{\|}}{-C-OCH_3}} \;+\; H_3C-NH-OH \;\longrightarrow\; A \;\longrightarrow\; \text{(Identify A.)}$$

(g)

$$Ph(CH_2)_3CH=O \;+\; N\equiv C-CH_2-CO_2Et \;\xrightarrow[\text{2) KCN/EtOH}]{\text{1) }^-\text{OH (1 equiv.)}}\; \xrightarrow[\text{4) heat}]{\text{3) H}_3\text{O}^+, \text{H}_2\text{O}}\; \underset{CH_2C\equiv N}{Ph(CH_2)_3CHC\equiv N}$$

(h)

$$\xrightarrow{^-\text{CN}}\; \xrightarrow[\text{heat}]{\text{H}_2\text{O, H}_3\text{O}^+}$$

(i)

$$+ \; Ph_3P\colon \;\longrightarrow\; A \;\xrightarrow{\text{CH}_3\text{CH}_2\text{CH}_2\text{CH}=\text{O}}\; \xrightarrow[\text{heat}]{\text{H}_2\text{O, H}_3\text{O}^+}$$

$$CH_3CH_2CH_2CH=\underset{CO_2H}{\overset{\overset{\displaystyle CH_2CO_2H}{|}}{C}} \;+\; Ph_3\overset{+}{P}-\overset{..}{\underset{..}{O}}\colon^-$$

Figure P22.91

(a)

$$Ph-CH=O \;+\; Br-CH_2-\overset{\overset{\displaystyle O}{\|}}{C}-OEt \;\xrightarrow[\text{t-BuOH}]{\text{K}^+ \text{ t-BuO}^-}\; Ph-CH\overset{\overset{\displaystyle O}{\diagdown}}{-}CH-\overset{\overset{\displaystyle O}{\|}}{C}-OEt \;+\; KBr$$
(mixture of stereoisomers)

(b)

$$H_3C-\overset{\overset{\displaystyle O}{\|}}{C}-CH_2CH_2CH_2-Cl \;+\; KOH \;\xrightarrow{\text{H}_2\text{O}}\; H_3C-\overset{\overset{\displaystyle O}{\|}}{C}-\overset{\overset{\displaystyle CH_2}{\diagup}}{CH}-CH_2$$

Figure P22.92

22.94 Ethyl vinyl ether, EtO—CH=CH$_2$, hydrolyzes in weakly acidic water to acetaldehyde and ethanol. Under the same conditions, diethyl ether does not hydrolyze. Quantitative comparisons of the hydrolysis rates of the two ethers under comparable conditions show that ethyl vinyl ether hydrolyzes about 10^{13} times faster than diethyl ether. The rapid hydrolysis of ethyl vinyl ether suggests an unusual mechanism for this reaction. The acetaldehyde formed when the hydrolysis of ethyl vinyl ether is carried out in D$_2$O/D$_3$O$^+$ contains one deuterium in its methyl group (that is, DCH$_2$—CH=O). Suggest a hydrolysis mechanism for ethyl vinyl ether consistent with these facts. (*Hint:* Vinylic ethers are also called *enol ethers*. Where do enols protonate? See Eq. 22.17b on p. 1112.)

22.95 Bearing in mind the hydrolysis reaction discussed in Problem 22.94, predict the final product of the reaction sequence shown in Fig. P22.95.

22.96 In the Krebs cycle (tricarboxylic acid, or citric acid, cycle), the enzyme *citrate synthase* catalyzes the synthesis of citrate from oxaloacetate and acetyl-CoA as shown in Fig. P22.96. Assuming that acids and bases are provided as needed in the enzyme active site, propose a curved-arrow mechanism for this reaction.

22.97 The enzyme *dehydroquinate synthase* catalyzes reactions (1) and (3) in the sequence shown in Fig. P22.97 on p. 1182, which is part of the pathway for the biosynthesis

of aromatic amino acids such as tryptophan; reactions (2), (4), and (5) occur spontaneously.

(a) Assuming that acids and bases are present as needed, propose a curved-arrow mechanism for reaction (2). To what laboratory reaction is this reaction analogous? [*Hint:* Phosphate is a leaving group in reaction (2).]

(b) The two enzyme-catalyzed steps (1) and (3) are, respectively, an oxidation and a reduction that consume and then regenerate an NAD$^+$ and result in no *net* change at the carbon involved. What is the chemical rationale for these two reactions? [*Hint:* Why is reaction (1) necessary for the success of reaction (2)?]

(c) Assuming that acids and bases are present as needed, propose curved-arrow mechanisms for reactions (4) and (5). To what laboratory reactions are these similar?

22.98 *Clavulanic acid* (Fig. P22.98, p. 1182) is an inhibitor of β-lactamase enzymes, which cause penicillin resistance. (See Sec. 21.8D.) Clavulanic acid inhibits β-lactamases by first reacting with the side-chain hydroxy group of an active-site serine residue (abbreviated as shown in Fig. P22.98). The resulting derivative *A* then reacts with the side-chain amino group of a lysine residue in the same active site to give the product *B*, which permanently blocks the active site, along with a by-product *C*, which is released into solution. Assuming acids and bases are present as needed, provide curved-arrow mechanisms for these reactions, and draw a structure of by-product *C*.

CH$_3$OCH$_2$Cl + Ph$_3$P $\longrightarrow$ $\xrightarrow{\text{Bu—Li}}$ $\xrightarrow{\hspace{1cm}}$ $\xrightarrow{\text{H}_3\text{O}^+,\ \text{H}_2\text{O}}$

Figure P22.95

Figure P22.96

**3-deoxy-D-arabinoheptulosonate-
7-phosphate (DAHP)**

3-dehydroquinate

Figure P22.97

from a serine
residue of the
enzyme

clavulanic acid
(conjugate base)

A

from a lysine
residue of the
enzyme

B

Figure P22.98

The Chemistry of Amines

Amines are organic derivatives of ammonia in which the ammonia hydrogens are replaced by alkyl or aryl groups. Amines are classified by the number of alkyl or aryl substituents (R groups) on the amine nitrogen. A **primary amine** has one substituent; a **secondary amine** has two; and a **tertiary amine** has three.

$$\ddot{N}H_3 \qquad R—\ddot{N}H_2 \qquad R—\ddot{N}H—R \qquad \overset{\overset{\textstyle R}{\textstyle |}}{R—N—R}$$

ammonia　　primary amine　　secondary amine　　tertiary amine

Examples:

$$CH_3CH_2—\ddot{N}H_2 \qquad CH_3CH_2—\ddot{N}H—CH_2CH_3 \qquad \overset{\overset{\textstyle CH_3CH_2}{\textstyle |}}{CH_3CH_2—N—CH_2CH_3}$$

ethylamine　　　　**diethylamine**　　　　**triethylamine**
(a primary amine)　　(a secondary amine)　　(a tertiary amine)

(This classification is like that of amides; see Sec. 21.1E.) It is important to distinguish between the classifications of alcohols and amines; alcohols are classified according to the number of alkyl or aryl groups on the α-carbon, but amines (like amides) are classified according to the number of alkyl or aryl groups *on nitrogen*.

$$\overset{\overset{\textstyle CH_3}{\textstyle |}}{\underset{\underset{\textstyle CH_3}{\textstyle |}}{H_3C—C—OH}}$$

the α-carbon has three alkyl branches

a tertiary alcohol

$$\overset{\overset{\textstyle CH_3}{\textstyle |}}{\underset{\underset{\textstyle CH_3}{\textstyle |}}{H_3C—C—NH_2}}$$

one alkyl group on the nitrogen

a primary amine
(even though the α-carbon is tertiary)

The reason for the difference in the classification of amines and alcohols is that the substitution of the OH hydrogen of an alcohol by an alkyl or aryl group results in a different functional group—an ether. In contrast, substitutions of the hydrogens of amines give other amines, and a way is needed to classify amines by their degree of nitrogen substitution.

Besides amines, this chapter also considers briefly some other nitrogen-containing compounds that are formed from, or converted into, amines: quaternary ammonium salts (Sec. 23.6), azobenzenes (Sec. 23.10B), diazonium salts (Sec. 23.10A), acyl azides (Sec. 23.11D), and nitro compounds (Sec. 23.11B).

23.1 NOMENCLATURE OF AMINES

A. Common Nomenclature

In common nomenclature, an amine is named by appending the suffix *amine* to the name of the alkyl group; the name of the amine is written as one word.

$$CH_3CH_2NH_2 \qquad (CH_3)_3N$$
ethylamine   **trimethylamine**

When two or more alkyl groups in a secondary or tertiary amine are different, the compound is named as an *N*-substituted derivative of the larger group.

$$(CH_3)_2N—CH_2CH_2CH_2CH_3 \qquad CH_3CH_2—NH—\bigcirc$$
N,N-dimethylbutylamine

                N-ethylcyclohexylamine

This type of notation is required to indicate that the substituents are on the amine nitrogen and not on an alkyl group carbon.

Aromatic amines are named as derivatives of aniline.

$$\bigcirc—NH_2 \qquad O_2N—\bigcirc—NH_2 \qquad \bigcirc—NH—CH_2CH_3$$

   aniline   **3-nitroaniline**   **N-ethylaniline**
              (*m*-nitroaniline)

B. Substitutive Nomenclature

Because the IUPAC system for amine nomenclature is not logically consistent with IUPAC nomenclature of other organic compounds, the most widely used system of substitutive amine nomenclature is that of *Chemical Abstracts* (p. 67). In this system, an amine is named in much the same way as the analogous alcohol, except that the suffix *amine* is used.

$$\overset{OH}{\underset{|}{CH_3CHCH_2CH_2CH_3}} \qquad \overset{NH_2}{\underset{|}{CH_3CHCH_2CH_2CH_3}} \qquad \bigcirc—NH—CH_3$$

   2-pentanol   **2-pentanamine**

                          N-methylcyclohexanamine

$$H_2N—CH_2CH_2CH_2CH_2CH_2—NH_2 \qquad H_3C—NH—CH_2CH_2CH_2—NH—CH_2CH_3$$
1,5-pentanediamine   **N-ethyl-N'-methyl-1,3-propanediamine**

In diamine nomenclature, as in diol nomenclature, the final *e* of the hydrocarbon name is retained. In the last of the preceding examples, the prime is used to indicate that the ethyl and methyl groups are on different nitrogens.

The priority of citation of amine groups as principal groups is just below that of alcohols:

$$-\overset{\overset{\displaystyle O}{\|}}{C}-OH \text{ (carboxylic acid} > -\overset{\overset{\displaystyle O}{\|}}{C}-H \text{ (aldehyde)} \quad -\overset{\overset{\displaystyle O}{\|}}{C}- \text{ (ketone)} \ > \ -OH \ > \ -NR_2 \qquad (23.1)$$
and derivatives)

$(R = H, alkyl,$
or aryl$)$

(A complete list of group priorities is given in Appendix I.)

When cited as a substituent, the $-NH_2$ group is called the **amino** group.

$$H_2N-CH_2CH_2-OH \qquad H_2N-\overset{1}{C}H_2-\overset{2}{C}H=\overset{3}{C}H_2$$

2-aminoethanol   **2-propen-1-amine**
(OH has priority)   (NH₂ has priority)

$$CH_3CH_2NCH_2CH_2\overset{\overset{\displaystyle CH_3}{|}}{C}HCH_2CH_2Cl \qquad\qquad (CH_3)_2N-CH_2CH_2CH_2-OH$$

5-chloro-*N*-ethyl-*N*,3-dimethyl-1-pentanamine   **3-(dimethylamino)-1-propanol**

An *N* designation in the last example is unnecessary because the position of the methyl groups is clear from the parentheses.

Although *Chemical Abstracts* calls aniline *benzenamine*, the usual practice is to use the common name *aniline* in substitutive nomenclature.

The nomenclature of *heterocyclic compounds* was introduced in Sec. 8.2C in the discussion of ether nomenclature. Many important nitrogen-containing heterocyclic compounds are known by specific names that should be learned. Some important saturated heterocyclic amines are the following:

piperidine   **morpholine**   **pyrrolidine**   **aziridine**

As in the oxygen heterocyclics, numbering generally begins with the heteroatom. The following are examples of substituted derivatives:

2-methylaziridine   ***N*-ethylmorpholine**
(**4-ethylmorpholine**)   **3-pyrrolidinecarboxylic acid**

To a useful approximation, much of the chemistry of the saturated heterocyclic amines parallels the chemistry of the corresponding acyclic amines. There are also a number of unsaturated aromatic heterocyclic amines. Among these are pyridine and pyrrole, which were considered briefly in Sec. 15.7D.

pyridine   **pyrrole**

The chemistry of the aromatic heterocycles, which is quite different from that of their saturated counterparts, is considered in Chapter 26.

PROBLEMS

23.1   Draw the structure of each of the following compounds.
  (a) *N*-isopropylaniline   (b) *tert*-butylamine
  (c) 3-methoxypiperidine   (d) 2,2-dimethyl-3-hexanamine
  (e) ethyl 2-(diethylamino)pentanoate   (f) *N,N*-diethyl-3-heptanamine

23.2   Give an acceptable name for each of the following compounds.

(a) [structure: (CH₃)₂CHCH₂—N(CH₃)—CH₂CH₃]

(b) [structure: O₂N—C₆H₄—N(CH₃)CH₃]

(c) [structure: cyclohexyl—NH—cyclohexyl]

(d) CH₃NHCHCH₂CH₂OH
        |
        CH₂CH₃

(e) [structure: pyrrolidine with CH₂CH₂CH₃ substituent and N-CH₂CH₂Cl]

23.2   STRUCTURE OF AMINES

The C—N bonds of aliphatic amines are longer than the C—O bonds of alcohols, but shorter than the C—C bonds of alkanes, as expected from the effect of atomic size on bond length (Sec. 1.3B).

bond:	C—C	C—N	C—O	C–F
typical length:	1.54 Å	1.47 Å	1.43 Å	1.39 Å

Aliphatic amines have a pyramidal shape (or approximately tetrahedral shape, if the electron pair is considered to be a "group"). Most amines undergo rapid *inversion* at nitrogen, which occurs through a planar transition state and converts a chiral amine into its mirror image. (See Fig. 6.18, p. 261.)

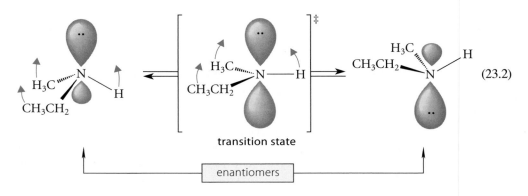

(23.2)

transition state

enantiomers

Because of this rapid inversion, amines in which the only asymmetric atom is the amine nitrogen cannot be resolved into enantiomers.

The C—N bond in aniline, with a length of 1.40 Å, is shorter than the C—N bond in aliphatic amines. This reflects both the sp^2 hybridization of the adjacent carbon and the overlap of the unshared electrons on nitrogen with the π-electron system of the ring. This overlap, shown by the following resonance structures, gives some double-bond character to the C—N bond.

(23.3)

PROBLEM

23.3 Within each set, arrange the compounds in order of increasing C—N bond length. Explain your answers.

(a) *p*-nitroaniline, aniline, cyclohexylamine

(b) HN=CH₂   H₃C—NH₂   H₂C=CH—NH₂

  $\quad\quad\quad$ A $\quad\quad\quad\quad\quad$ B $\quad\quad\quad\quad\quad$ C

23.3 PHYSICAL PROPERTIES OF AMINES

Most amines are somewhat polar liquids with unpleasant odors that range from fishy to putrid. Primary and secondary amines, which can both donate and accept hydrogen bonds, have higher boiling points than isomeric tertiary amines, which cannot donate hydrogen bonds.

	CH_3 $\vert$ $CH_3CH_2CHCH_3$	$CH_3CH_2N(CH_3)_2$	$(CH_3CH_2)_2NH$	$CH_3CH_2CH_2CH_2NH_2$
	isopentane	**ethyldimethylamine**	**diethylamine**	**butylamine**
boiling point:	27.8 °C	37.5 °C	56.3 °C	77.8 °C
dipole moment:	0 D	0.6 D	1.2–1.3 D	1.4 D

Increasing hydrogen bonding and polarity $\longrightarrow$

The fact that primary and secondary amines can both donate and accept hydrogen bonds also accounts for the fact that they have higher boiling points than ethers. On the other hand, alcohols are better hydrogen-bond donors than amines because alcohols are more acidic than amines. (See Sec. 23.5D.) Therefore, alcohols have higher boiling points than amines.

	$(CH_3CH_2)_2NH$	$(CH_3CH_2)_2O$	$(CH_3CH_2)_2CH_2$	$CH_3CH_2CH_2CH_2OH$	$CH_3CH_2CH_2CH_2NH_2$
	diethylamine	**diethyl ether**	**pentane**	**1-butanol**	**1-butanamine**
boiling point:	56.3 °C	37.5 °C	36.1 °C	117.3 °C	77.8 °C

The water miscibility of most primary and secondary amines with four or fewer carbons, as well as trimethylamine, is consistent with their hydrogen-bonding abilities. Amines with large carbon groups have little or no water solubility.

23.4   **SPECTROSCOPY OF AMINES**

A. IR Spectroscopy

The most important absorptions in the infrared spectra of primary amines are the N—H stretching absorptions, which usually occur as two peaks at 3200–3375 cm^{-1} corresponding to the NH_2 symmetric and asymmetric stretching vibrations. Also characteristic of primary amines is an NH_2 scissoring absorption (Fig. 12.8, p. 582) near 1600 cm^{-1}. These absorptions are illustrated in the IR spectrum of butylamine (Fig. 23.1). Most secondary amines show a single N—H stretching absorption rather than the two peaks observed for primary amines, and the absorptions associated with the various NH_2 bending vibrations of primary amines are not present. For example, diethylamine lacks the NH_2 scissoring absorption present in the butylamine spectrum. Tertiary amines show no absorptions associated with N—H vibrations. The C—N stretching absorptions of amines, which occur in the same general part of the spectrum as C—O stretching absorptions (1050–1225 cm^{-1}), are not very useful.

B. NMR Spectroscopy

The characteristic resonances in the proton NMR spectra of amines are those of the protons adjacent to the nitrogen (the α-protons) and the N—H protons. In alkylamines, the α-protons are observed in the δ 2.5–3.0 region of the spectrum. In aromatic amines, the α-protons of *N*-alkyl groups have somewhat greater chemical shifts, near δ 3. (Why? See Sec 13.3C.) The following chemical shifts are typical:

δ 0.9 (s) δ 1.0 (t) δ 3.0 (q)

CH_3CH_2—NH—CH_2—CH_3 ⟨phenyl⟩—NH—CH_2—CH_3

δ 2.6 (q) δ 3.2 (s) δ 1.1 (t)

The chemical shift of the N—H proton, like that of the O—H proton in an alcohol, depends on the concentration of the amine and on other conditions of the NMR experiment. In alkylamines, this resonance occurs at rather small chemical shift—typically around δ 1. In aromatic amines, this resonance is at greater chemical shift, as in the second of the preceding examples.

Like the OH protons of alcohols, phenols, and carboxylic acids, the NH protons of amines under most conditions undergo rapid exchange (Secs. 13.6 and 13.7D). For this reason, splitting between the amine N—H and adjacent C—H groups is usually not observed. Thus, in the NMR spectrum of diethylamine, the N—H resonance is a singlet rather than the triplet expected from splitting by the adjacent —CH_2— protons. In some amine samples, the N—H

FIGURE 23.1 The IR spectrum of butylamine, a typical primary amine. The N—H stretching and NH_2 bending (scissoring) absorptions are the most important amine absorptions.

resonance is broadened and, like the O—H proton of alcohols, it can be obliterated from the spectrum by exchange with D_2O (the "D_2O shake," p. 653).

The characteristic ^{13}C NMR absorptions of amines are those of the α-carbons—the carbons attached directly to the nitrogen. These absorptions occur in the δ 30–50 chemical-shift range. As expected from the relative electronegativities of oxygen and nitrogen, these shifts are somewhat less than the α-carbon shifts of ethers.

PROBLEMS

23.4 Identify the compound that has an M + 1 ion at m/z = 136 in its CI mass spectrum, an IR absorption at 3279 cm^{-1}, and the following NMR spectrum: δ 0.91 (1H, s), δ 1.07 (3H, t, J = 7 Hz), δ 2.60 (2H, q, J = 7 Hz), δ 3.70 (2H, s), δ 7.18 (5H, apparent s).

23.5 A compound has IR absorptions at 3400–3500 cm^{-1} and the following NMR spectrum: δ 2.07 (6H, s), δ 2.16 (3H, s), δ 3.19 (broad, exchanges with D_2O), δ 6.63 (2H, s). To which one of the following compounds do these spectra belong? Explain.

(1) 2,4-dimethylbenzylamine (2) 2,4,6-trimethylaniline
(3) *N,N*-dimethyl-*p*-methylaniline (4) 3,5-dimethyl-*N*-methylaniline
(5) 4-ethyl-2,6-dimethylaniline

23.6 Explain how you could distinguish between the two compounds in each of the following sets using *only* ^{13}C NMR spectroscopy.

(a) 2,2-dimethyl-1-propanamine and 2-methyl-2-butanamine
(b) *trans*-1,2-cyclohexanediamine and *trans*-1,4-cyclohexanediamine

---

23.5   BASICITY AND ACIDITY OF AMINES

A. Basicity of Amines

Amines, like ammonia, are strong enough bases that they are completely protonated in dilute acid solutions.

$$H_3C—\overset{..}{N}H_2 + H—Cl \;\rightleftharpoons\; H_3C—\overset{+}{N}H_3 \; Cl^- + H_2O \qquad (23.4)$$

methylamine   **methylammonium chloride**

The salts of protonated amines are called **ammonium salts**. The ammonium salts of simple alkylamines are named as substituted derivatives of the ammonium ion. Other ammonium salts are named by replacing the final *e* in the name of the amine with the suffix *ium*.

$$(CH_3)_2\overset{+}{N}H_2 \; Cl^-$$

dimethylammonium chloride

anilinium benzoate
(aniline + *ium*)

Always remember that ammonium salts are fully ionic compounds. Although ammonium chloride is often written as NH_4Cl, the structure is more properly represented as $^+NH_4 \; Cl^-$. Although the N—H bonds are covalent, there is no covalent bond between the nitrogen and the chlorine. (A covalent bond would violate the octet rule.)

Many pharmaceuticals contain one or more amine functional groups and are marketed as salts. The basic amine nitrogen is protonated and positively charged, and the drug is accompanied by the pharmacologically inert anion. As noted in Sec. 8.6, salts are much more soluble in water than many neutral organic compounds. The choice of counter ion is important because its identity affects the desired physical properties of the compound, including its melting point and rate of solubilization. In addition, protonating the nitrogen makes it resistant to oxidation, which can lead to degradation of the compound and a decrease in its efficacy.

(S)-propranolol hydrochloride (Inderal®)
(used to treat hypertension and anxiety)

$CH_3SO_3^-$

imatinib (Gleevec®)
(leukemia treatment)

sildenafil citrate (Viagra®)
(a phosphodiesterase type 5 inhibitor)

Recall that the basicity of any compound, including an amine, is expressed in terms of the pK_a of its conjugate acid (Sec. 3.4D). The higher the pK_a of an ammonium ion, the more basic is its conjugate-base amine. (A discussion of the relationship between basicity constants K_b and dissociation constants K_a is given in Sec 3.4D, p. 103.)

B. Substituent Effects on Amine Basicity

The pK_a values for the conjugate acids of some representative amines are given in Table 23.1. As this table shows, the exact basicity of an amine depends on its structure. Four factors influ-

TABLE 23.1 Basicities of Some Amines

(Each pK_a value is for the dissociation of the corresponding conjugate-acid ammonium ion.)

Amine	pK_a	Amine	pK_a	Amine	pK_a
CH_3NH_2	10.62	$(CH_3)_2NH$	10.64	$(CH_3)_3N$	9.76
$CH_3CH_2NH_2$	10.63	$(CH_3CH_2)_2NH$	10.98	$(CH_3CH_2)_3N$	10.65
$PhCH_2NH_2$	9.34				
$PhNH_2$	4.62	$PhNHCH_3$	4.85	$PhN(CH_3)_2$	5.06
O_2N—⬡—NH_2	≈1.0	(3-nitroaniline)	2.45	(2-nitroaniline)	−0.26
Cl—⬡—NH_2	3.81	(3-chloroaniline)	3.32	(2-chloroaniline)	2.62
H_3C—⬡—NH_2	5.07	(3-methylaniline)	4.67	(2-methylaniline)	4.38

ence the basicity of amines; these are the same effects that influence the acid–base properties of other compounds. They are:

1. the effect of alkyl substitution
2. the polar effect
3. the resonance effect
4. the effect of charge

Recall that the pK_a of an ammonium ion, like that of any other acid, is directly related to the standard free-energy difference ΔG_a° between it and its conjugate base by the following equation (Eq. 3.34, p. 110).

$$\Delta G_a^\circ = 2.3RT(pK_a) \tag{23.5}$$

The effect of a substituent group on pK_a can be analyzed in terms of how it affects the energy of either an ammonium ion or its conjugate-base amine, as shown in Fig. 23.2. For example, if a substituent stabilizes an amine more than it stabilizes the conjugate-acid ammonium ion (Fig. 23.2a), the standard free energy of the amine is lowered, ΔG_a° is decreased, and the pK_a of the ammonium ion is reduced; that is, the amine is less basic than the amine without the substituent. If a substituent stabilizes the ammonium ion more than its conjugate-base amine (Fig. 23.2b), the opposite effect is observed: the pK_a is increased, and the amine basicity is also increased.

Consider first the effect of alkyl substitution. Most common alkylamines are somewhat more basic than ammonia in aqueous solution:

pK_a *in aqueous solution:*

$$\overset{+}{N}H_4 \qquad Me\overset{+}{N}H_3 \qquad Me_2\overset{+}{N}H_2 \qquad Me_3\overset{+}{N}H \qquad\qquad (23.6)$$
$$9.21 \qquad\quad 10.62 \qquad\quad 10.64 \qquad\quad 9.76$$

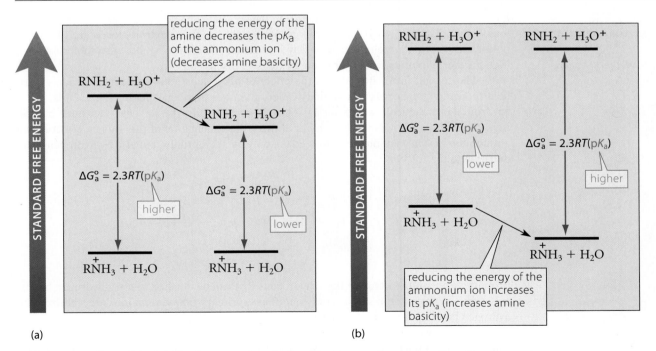

(a)   (b)

FIGURE 23.2 Effect of the relative free energies of ammonium ions and amines on the pK_a values of the ammonium ions. (a) Reducing the energy of an amine decreases the pK_a of its conjugate-acid ammonium ion and thus reduces the basicity of the amine. (b) Reducing the energy of an ammonium ion increases its pK_a and thus increases the basicity of its conjugate-base amine.

However, the increase in basicity that results from substitution of one hydrogen of ammonia by a methyl group is reversed as the number of alkyl substituents is increased to three. How can we explain this "turnaround" in amine basicity?

Two opposing factors are actually at work here. The first is the tendency of alkyl groups to stabilize charge through a *polarization* effect. The electron clouds of the alkyl groups distort so as to create a net attraction between them and the positive charge of the ammonium ion:

Because the ammonium ion is stabilized by this effect, its pK_a is increased (Fig. 23.2b). This effect is evident in the *gas-phase* basicities of amines. In the gas phase, the acidity of ammonium ions decreases regularly with increasing alkyl substitution:

Gas-phase acidity:

$$\overset{+}{N}H_4 > Me\overset{+}{N}H_3 > Me_2\overset{+}{N}H_2 > Me_3\overset{+}{N}H \tag{23.7}$$

FURTHER EXPLORATION 23.1
Alkyl Group Polarization in Ionization Reactions

The same polarization effect also operates in the gas-phase acidity of alcohols (Sec. 10.1D), except in the opposite direction. In other words, the polarization of alkyl groups can act to stabilize *either* positive or negative charge. In the presence of a positive charge, as in ammonium ions, electrons polarize *toward* the charge; in the presence of a negative charge, as in alkoxide ions, they polarize *away from* the charge. We might say that electron clouds are like some politicians: they polarize in whatever way is necessary to create the most favorable situation. (See Further Exploration 23.1 for a more extensive discussion.)

The second factor involved in the effect of alkyl substitution on amine basicity must be a *solvent effect*, because the basicity order of amines in the gas phase (Eq. 23.7) is different from that in aqueous solution (Eq. 23.6). In other words, the solvent water must play an important role in the solution basicity of amines. An explanation of this solvent effect is that ammonium ions in solution are stabilized not only by alkyl groups, but also by hydrogen-bond donation to the solvent:

Primary ammonium salts have three hydrogens that can be donated to form hydrogen bonds, but a tertiary ammonium salt has only one. Thus, primary ammonium ions are stabilized by hydrogen bonding more than tertiary ones.

The pK_a values of alkylammonium salts reflect the operation of both hydrogen bonding and alkyl-group polarization. Because these effects work in opposite directions, the basicity in Eq. 23.6 maximizes at the secondary amine.

Ammonium-ion pK_a values, like the pK_a values of other acids, are also sensitive to the *polar effects* of substituents.

$$\underset{4.55}{\overset{+}{Et_2NHCH_2C\equiv N}} \qquad \underset{7.65}{\overset{+}{Et_2NHCH_2CH_2C\equiv N}} \qquad \underset{10.08}{\overset{+}{Et_2NH(CH_2)_4C\equiv N}} \qquad \underset{10.65}{\overset{+}{Et_3NH}} \qquad (23.8)$$

pK$_a$:

An electronegative (electron-withdrawing) group such as halogen or cyano destabilizes an ammonium ion because of a repulsive electrostatic interaction between the positive charge on the ammonium ion and the positive end of the substituent bond dipole.

Notice that the polar effects of substituent groups operate largely on the *conjugate acid* of the amine—the alkylammonium ion—because this cation is the charged species in the acid–base equilibrium. (Recall from Sec. 3.6C that polar effects on stability are greatest on the charged species in a chemical equilibrium.) In the case of a carboxylic acid, alcohol, or phenol, the *conjugate-base anion* is the charged species; consequently, the polar effects of substituents are reflected primarily in their effects on the stabilities of these anions.

The data in Eq. 23.8 show that the base-weakening effect of electron-withdrawing substituents, like all polar effects, decreases significantly with distance between the substituent and the charged atom.

Resonance effects on amine basicity are illustrated by the difference between the pK$_a$ values of the conjugate acids of aniline and cyclohexylamine, two primary amines of almost the same shape and molecular mass.

	conjugate acid of aniline	conjugate acid of cyclohexylamine	
pK$_a$:	4.62	10.64	(23.9)

(Notice that the pK$_a$ values of the substituted anilinium ions in Table 23.1 are considerably lower than the pK$_a$ values of the alkylammonium ions.) Aniline is stabilized by a resonance interaction of the unshared electron pair on nitrogen with the aromatic ring. (This resonance interaction is shown in Eq. 23.3, p. 1187.) When aniline is protonated, this resonance stabilization is no longer present, because the unshared pair is bound to a proton and is "out of circulation." The stabilization of aniline relative to its conjugate acid reduces its basicity (Eq. 23.5 and Fig. 23.2). In other words, the resonance stabilization of aniline lowers the energy required for its formation from its conjugate acid, and thus lowers its basicity relative to that of cyclohexylamine, in which the resonance effect is absent.

The electron-withdrawing polar effect of the aromatic ring also contributes significantly to the reduced basicity of aromatic amines. Recall that a similar polar effect is responsible for the increased acidities of phenols relative to alcohols (Sec. 18.7A).

STUDY PROBLEM   23.1

Arrange the following three amines in order of increasing basicity.

p-chloroaniline   **4-chlorocyclohexanamine**   **aniline**

SOLUTION Because the chloro substituent is an electron-withdrawing group, it reduces the basicity of an aniline. (The slight electron-donating resonance effect of a chloro substituent is much less important than its electron-withdrawing polar effect.) Hence, *p*-chloroaniline is less basic than aniline. Because of the resonance interaction of the amine nitrogen with the ring π-electron system, *p*-chloroaniline is also less basic than 4-chlorocyclohexanamine. But what about the relative basicity of aniline and 4-chlorocyclohexanamine? The resonance and polar effects of the aromatic ring and the polar effect of the chlorine are all base-weakening effects. The problem is to decide whether the effect of the aromatic ring or the effect of the chlorine is more important in reducing basicity. To make this decision, reason *by analogy*. Examine the effect of an electron-withdrawing group on the basicity of an amine in which the polar effect is the only effect that can operate. For example, the series in Eq. 23.8 shows that an electronegative group four carbons away from the amine nitrogen has a very modest effect. On the other hand, the comparison in Eq. 23.9 shows that the resonance and polar effects of an aromatic ring change the pK_a of an amine by about six units. Consequently, the base-weakening effect of "changing" a cyclohexane ring to a phenyl ring is much more important by many orders of magnitude than the base-weakening effect of "replacing" a hydrogen with a 4-chloro group in cyclohexanamine. Hence, the basicity order is

p-chloroaniline < aniline << 4-chlorocyclohexanamine

PROBLEMS

23.7 Arrange the amines within each set in order of increasing basicity in aqueous solution, least basic first.

(a) propylamine, ammonia, dipropylamine

(b) methyl 3-aminopropanoate, *sec*-butylamine, $H_3\overset{+}{N}CH_2CH_2NH_2$

(c) aniline, methyl *m*-aminobenzoate, methyl *p*-aminobenzoate

(d) benzylamine, *p*-nitrobenzylamine, cyclohexylamine, aniline

23.8 Explain the basicity order of the following three amines: *p*-nitroaniline (*A*), *m*-nitroaniline (*B*), and aniline (*C*). The structures and pK_a data are shown in Table 23.1.

C. Separations Using Amine Basicity

Because ammonium salts are ionic compounds, many have appreciable water solubilities. Hence, when a water-insoluble amine is treated with dilute aqueous acid, such as 5% HCl solution, *the amine dissolves as its ammonium salt.* Upon treatment with base, the ammonium salt is converted back into the amine. These observations can be used to design separations of amines from other compounds, as Study Problem 23.2 illustrates.

STUDY PROBLEM 23.2

A chemist has treated *p*-chloroaniline with acetic anhydride and wants to separate the amide product from any unreacted amine. Design a separation based on the basicities of the two compounds.

SOLUTION If the mixture is treated with 5% aqueous HCl, the amine will form the hydrochloride salt and dissolve in the aqueous solution. The amide, however, is not basic enough to be protonated in 5% HCl and therefore does not dissolve.

(23.10)

The water-insoluble amide can be filtered or extracted away from the aqueous solution of the ammonium salt. Then the aqueous solution of the ammonium salt can be treated with NaOH to liberate the free amine.

Amine basicities can play a key role in the design of enantiomeric resolutions. The enantiomeric resolution discussed in Sec. 6.8B should be reviewed with this idea in mind.

PROBLEMS

23.9   Using their solubilities in acidic or basic solution, design a separation of *p*-chlorobenzoic acid, *p*-chloroaniline, and *p*-chlorotoluene from a mixture containing all three compounds.

23.10   Design an enantiomeric resolution of racemic 2-phenylpropanoic acid using a pure enantiomer of 1-phenylethanamine as resolving agent. (Assume that a solvent can be found in which the diasteromeric salts have different solubilities.) Discuss the importance of amine basicity to the success of this scheme. (See Sec. 6.8B.)

D. Acidity of Amines

Although amines are normally considered to be bases, primary and secondary amines are also very weakly acidic. In other words, amines are *amphoteric compounds* (p. 99). The conjugate base of an amine is called an **amide** (not to be confused with amide derivatives of carboxylic acids).

The amide conjugate base of ammonia itself is usually prepared by dissolving an alkali metal such as sodium in liquid ammonia in the presence of a trace of ferric ion. When sodium is used, the resulting base is called *sodium amide* (or *sodamide*).

$$2\,Na + 2\,\ddot{N}H_3 \xrightarrow{Fe^{3+}} 2\,(Na^+\ ^-\!:\ddot{N}H_2) + H_2 \qquad (23.11)$$

**sodium amide
(sodamide)**

The conjugate bases of alkylamines are prepared by treating the amine with butyllithium in an ether solvent such as THF.

diisopropylamine **butyllithium** **butane**

**lithium
diisopropylamide
(LDA)**

$$(23.12)$$

Although the structures of these amide bases are conventionally drawn as in Eq. 23.12, they are actually more complex. (See Further Exploration 23.2 for a more extensive discussion.)

The pK_a of a typical amine is about 35. Thus, amide anions are *very* strong bases. This is why they can be used to form acetylide ions or enolate ions (Secs. 14.7A, 22.4C, and 22.8B). The difference in the pK_a values of ammonium ions and amines illustrates the effect of charge on pK_a. A positive charge on the nitrogen increases the acidity of the attached hydrogen by more than 20 pK_a units.

**FURTHER
EXPLORATION 23.2**
Structures of
Amide Bases

$$R_2\ddot{N}\!-\!H \qquad\qquad R_3\overset{+}{N}\!-\!H$$

amines   ammonium ions
$pK_a \approx 32\text{–}35$ $pK_a \approx 9\text{–}11$

E. Summary of Acidity and Basicity

We have now surveyed the acidity and basicity of the most important organic functional groups. This information is summarized in the tables in Appendix VII. Although the values given in these tables are typical values, remember that acidity and basicity are affected by alkyl substitution, polar effects, and resonance effects. The acid–base properties of organic compounds are important not only in predicting many of their chemical properties, but also in their industrial and medicinal applications.

23.6   QUATERNARY AMMONIUM AND PHOSPHONIUM SALTS

Closely related to ammonium salts are compounds in which all four hydrogens of $\overset{+}{N}H_4$ are replaced by alkyl or aryl groups. Such compounds are called **quaternary ammonium salts**. The following compounds are examples:

$$(CH_3)_4N^+ \ Cl^-$$

tetramethylammonium chloride

$$Ph\overset{+}{N}Et_3 \ Br^-$$

N,N,N-triethylanilinium bromide

$$PhCH_2\overset{+}{N}(CH_3)_3 \ {}^-OH$$

benzyltrimethylammonium hydroxide (Triton B)

$$CH_3(CH_2)_{15}\overset{\overset{\displaystyle CH_3}{|}}{\underset{\underset{\displaystyle CH_3}{|}}{N^+}}-CH_2-Ph \ Cl^-$$

"benzalkonium chloride"
a cationic surfactant (Sec. 20.5)

Like the corresponding ammonium ions, quaternary ammonium salts are fully ionic compounds. Many quaternary ammonium salts containing large alkyl or aryl groups are soluble in nonaqueous solvents. Triton B, for example, is used as a source of hydroxide ion that is soluble in organic solvents. Benzalkonium chloride, a common antiseptic, acts as a surfactant in water (Sec. 20.5) and is also soluble in several organic solvents. The quaternary ammonium ions in such compounds can be conceptualized as "positive charges surrounded by greasy groups." (The term "greasy" is used to refer to materials soluble in hydrocarbon solvents.)

Quaternary phosphonium salts are the phosphorus analogs of quaternary ammonium salts.

$$CH_3(CH_2)_{14}CH_2-\overset{\overset{\displaystyle CH_2CH_2CH_2CH_3}{|}}{\underset{\underset{\displaystyle CH_2CH_2CH_2CH_3}{|}}{P^+}}-CH_2CH_2CH_2CH_3 \ Br^-$$

cetyltributylphosphonium bromide
(a quaternary phosphonium salt)

Recall (Eq. 19.79a, p. 992) that phosphonium salts are the starting materials for the preparation of Wittig reagents.

An interesting and important use of both ammonium and phosphonium salts with large alkyl substituents is to catalyze organic reactions between an ionic reactant that is soluble in water and an organic reactant that is water insoluble. Such reactions typically involve two mutually insoluble layers—an aqueous phase and an organic phase. For example, if 1-bromooctane (alone or dissolved in a water-insoluble solvent) is treated with aqueous sodium cyanide, no reaction takes place, even with rapid stirring, because sodium cyanide, an ionic compound, is soluble only in the aqueous layer and cannot come into contact with the alkyl halide, which is insoluble in water. But when only a few percent of the quaternary phosphonium salt shown

FIGURE 23.3 Phase-transfer catalysis by a quaternary phosphonium salt. (a) At the beginning of the reaction, the ionic nucleophile (*red*) is soluble in the aqueous layer. (b) Rapid equilibration of the nucleophile with the counterion of the quaternary salt brings the nucleophile into the organic phase. (c) The nucleophile, now in the organic phase, can come into contact with the organic reactant, and a reaction occurs, forming the product and regenerating the phase-transfer catalyst.

above is added to the reaction mixture and the solution is stirred vigorously, the S_N2 reaction between the cyanide ion and the alkyl halide takes place readily.

$$CH_3(CH_2)_6CH_2—Br \ + \ Na^+ \ ^-CN \ \xrightarrow[\text{H}_2\text{O}]{\substack{\text{quaternary} \\ \text{phosphonium salt}}} \ CH_3(CH_2)_6CH_2—CN \ + \ Na^+ Br^- \quad (23.13)$$

1-bromooctane **nonanenitrile**

When used in this way, the quaternary salt is called a **phase-transfer catalyst**. The name describes the mechanism of action (Fig. 23.3). Because of its large alkyl groups, the quaternary salt is soluble in the organic phase, which consists of a water-insoluble solvent and a mixture of the organic reactant and, ultimately, the product. The bromide ion in the organic phase readily exchanges with the cyanide ion from the aqueous phase, thus bringing the cyanide ion nucleophile into the organic phase, where it can react with the alkyl halide.

PROBLEMS

23.11 Draw the structure of each of the following quaternary ammonium salts.

(a) tetraethylammonium fluoride (b) dibenzyldimethylammonium bromide

23.12 Explain why the quaternary ammonium salt *A* can be isolated in optically active form, but the trialkylammonium salt *B* cannot.

$$\underset{A}{\overset{\overset{\displaystyle CH_3}{|}}{PhCH_2—\overset{+}{\underset{\underset{\displaystyle CH_2CH_2CH_3}{|}}{N}}—CH_2CH_3}} \ Cl^- \qquad \underset{B}{\overset{\overset{\displaystyle CH_3}{|}}{PhCH_2—\overset{+}{\underset{\underset{\displaystyle H}{|}}{N}}—CH_2CH_3}} \ Cl^-$$

23.7 ALKYLATION AND ACYLATION REACTIONS OF AMINES

The previous section showed that amines are *Brønsted bases*. Amines, like many other Brønsted bases, are also *nucleophiles* (Lewis bases). Three reactions of nucleophiles are

1. S_N2 reaction with alkyl halides, sulfonate esters, or epoxides (Secs. 9.1, 9.4, 10.4, and 11.5)

2. addition to aldehydes, ketones, and α,β-unsaturated carbonyl compounds (Secs. 19.7, 19.11, and 22.9A)

3. nucleophilic acyl substitution at the carbonyl groups of carboxylic acid derivatives (Sec. 21.8)

This section covers or reviews reactions of amines that fit into each of these categories.

A. Direct Alkylation of Amines

Treatment of ammonia or an amine with an alkyl halide or other alkylating agent results in alkylation of the nitrogen.

$$R_3\ddot{N}: + H_3C—I \longrightarrow R_3\overset{+}{N}—CH_3 \ I^- \tag{23.14}$$

This process is an example of an S_N2 reaction in which the amine acts as the nucleophile.

The product of the reaction shown in Eq. 23.14 is an alkylammonium ion. If this ammonium ion has N—H bonds, further alkylations can take place to give a complex product mixture, as in the following example:

$$\ddot{N}H_3 + CH_3I \longrightarrow CH_3\overset{+}{N}H_3 \ I^- + (CH_3)_2\overset{+}{N}H_2 \ I^- + (CH_3)_3\overset{+}{N}H \ I^- + (CH_3)_4N^+ \ I^- \tag{23.15}$$

A mixture of products is formed because the methylammonium ion produced initially is partially deprotonated by the ammonia starting material. Because the resulting methylamine is also a good nucleophile, it too reacts with methyl iodide.

$$\ddot{N}H_3 + H_3C—I \longrightarrow H_3\overset{+}{N}—CH_3 \ I^- \tag{23.16a}$$

$$\ddot{N}H_3 + H—\overset{+}{N}H_2—CH_3 \ I^- \rightleftarrows \overset{+}{N}H_4 \ I^- + H_2\ddot{N}—CH_3 \tag{23.16b}$$

$$H_3C—\ddot{N}H_2 + H_3C—I \longrightarrow (CH_3)_2\overset{+}{N}H_2 \ I^- \tag{23.16c}$$

Analogous deprotonation–alkylation reactions give the other products of the mixture shown in Eq. 23.15 (see Problem 23.21).

Epoxides, as well as α,β-unsaturated carbonyl compounds and α,β-unsaturated nitriles, also react with amines and ammonia. As the following results show, multiple alkylations can occur with these alkylating agents as well.

$$(CH_3)_3CNH_2 + H_2C\overset{O}{\overset{\diagup \diagdown}{—}}CH_2 \xrightarrow{H_2O} (CH_3)_3CNH—CH_2CH_2—OH + (CH_3)_3CN(CH_2CH_2OH)_2 \tag{23.17}$$

$$NH_3 \text{ (excess)} + H_2C{=}CH—CN \longrightarrow H_2N—CH_2CH_2CN + HN(CH_2CH_2CN)_2 \tag{23.18}$$
$$\text{(32\% yield)} \qquad \text{(57\% yield)}$$

In an alkylation reaction, the exact amount of each product obtained depends on the precise reaction conditions and on the relative amounts of starting amine and alkyl halide. Because a mixture of products results, the utility of alkylation as a preparative method for amines is limited, although, in specific cases, conditions have been worked out to favor particular products. Section 23.11 discusses other methods that are more useful for the preparation of amines.

PROBLEM

23.13 Provide reaction mechanisms for the following equations.
   (a) Eq. 23.17 (*Hint:* Refer to Sec. 11.5A, p. 524.)
   (b) Eq. 23.18 (*Hint:* Refer to Eqs. 22.101 and 22.102, p. 1157.)
   (c) Suggest a reason why the reaction in Eq. 23.18 stops after two additions, and a third doesn't occur in high yield. How might you change the reaction conditions to get a third addition?

Quaternization of Amines   Amines can be converted into quaternary ammonium salts with excess alkyl halide. This process, called **quaternization**, is one of the most important synthetic applications of amine alkylation. The reaction is particularly useful when especially reactive alkyl halides, such as methyl iodide or benzylic halides, are used.

$$PhCH_2\ddot{N}Me_2 \quad + \quad MeI \quad \xrightarrow[EtOH]{} \quad PhCH_2\overset{+}{N}Me_3 \ I^- \qquad (23.19)$$

benzyldimethylamine   **benzyltrimethylammonium
iodide**
(94–99% yield)

$$CH_3(CH_2)_{15}NMe_2 \quad + \quad PhCH_2{-}Cl \quad \xrightarrow[acetone]{} \quad CH_3(CH_2)_{15}\overset{+}{N}Me_2 \ Cl^- \qquad (23.20)$$
                                                                                                    |
N,N-dimethyl-1-hexadecanamine   **benzyl chloride** CH_2Ph

**benzylhexadecyldimethylammonium
chloride**

$$CH_3CHNHMe \quad + \quad MeI \text{ (excess)} \quad \xrightarrow[ether]{heat} \quad CH_3CHN\overset{+}{N}Me_3 \ I^- \quad + \quad HI \quad (23.21)$$
     | |
     CH_2CH_3 CH_2CH_3

***sec*-butylmethylamine**   ***sec*-butyltrimethylammonium
iodide**

Conversion of an amine into a quaternary ammonium salt with excess methyl iodide (as in Eqs. 23.19 and 23.21) is called **exhaustive methylation**.

B. Reductive Amination

When primary and secondary amines react with either aldehydes or ketones, they form imines and enamines, respectively (Sec. 19.11). In the presence of a reducing agent, imines and enamines are reduced to amines.

$$EtNH_2 + H_3C\overset{\overset{O}{\|}}{-}C{-}CH_3 \xrightarrow{-H_2O} \left[H_3C\overset{\overset{NEt}{\|}}{-}C{-}CH_3 \right] \xrightarrow[EtOH]{H_2, Pt \atop 30 \ psi} H_3C\overset{\overset{NHEt}{|}}{-}CH{-}CH_3 \quad (23.22)$$

acetone   an imine   **ethylisopropylamine**
                                 (not isolated)

Reduction of the C=N double bond is analogous to reduction of the C=O double bond (Sec. 19.8). Notice that the imine or enamine does not have to be isolated, but is reduced within the reaction mixture as it forms. *Because imines and enamines are reduced more rapidly than carbonyl compounds*, reduction of the carbonyl compound is not a competing reaction.

PROBLEM

23.14   Provide a reaction mechanism for step 1, formation of the imine, in Eq. 23.22 (*Hint:* Refer to Sec. 19.11A).

The formation of an amine from the reaction of an aldehyde or ketone with another amine and a reducing agent is called **reductive amination**. Two hydride reducing agents, sodium triacetoxyborohydride, $NaBH(OAc)_3$, and sodium cyanoborohydride, $NaBH_3CN$, find frequent use in reductive amination.

$$Ph-\overset{\overset{\displaystyle O}{\|}}{C}-H + H_2N-C(CH_3)_3 \xrightarrow[\substack{\text{1,2-dichloroethane}\\ \text{(solvent)}}]{\substack{NaBH(OAc)_3 \\ HOAc}} \xrightarrow{NaOH} Ph-\overset{\overset{\displaystyle NHC(CH_3)_3}{|}}{CH}-H \quad + H_2O \quad (23.23)$$

benzaldehyde   *tert*-butylamine   *N-tert*-butylaniline
                                                                                  (95% yield)

$$\text{cyclohexanone} + Me_2NH \xrightarrow[MeOH]{\substack{NaBH_3CN \\ HCl\ (1\ equiv.)}} \xrightarrow{KOH} \text{N,N-dimethylcyclohexanamine} \quad (23.24)$$

dimethylamine

cyclohexanone   ***N,N*-dimethylcyclohexanamine**
                                                                        (71% yield)

Both sodium triacetoxyborohydride and sodium cyanoborohydride are commercially available, easily handled solids, and sodium cyanoborohydride can even be used in aqueous solutions above pH > 3. Reductive amination with $NaBH_3CN$ is known as the **Borch reaction**, after Richard F. Borch, who discovered and developed the reaction while he was a professor of chemistry at the University of Minnesota in 1971. Like $NaBH_4$ reductions, the Borch reduction requires a protic solvent or one equivalent of acid. A proton source is also required for reduction with sodium triacetoxyborohydride. In some cases, the water generated in the reaction is adequate for this purpose, and in other situations, a weak acid can be added. (Acetic acid serves this role in Eq. 23.23.)

PROBLEM

23.15 Why is base (NaOH or KOH) added as a second step in each of the reactions shown in Eqs. 23.23 and 23.24? (*Hint:* What is the form of the product after the reduction step? Refer to Sec. 23.5A.)

Reductive amination, like catalytic hydrogenation, typically involves the imines or enamines and their conjugate acids as intermediates.

$$R-\ddot{N}H_2 + \overset{\overset{\displaystyle O}{\|}}{C} \longrightarrow \underset{\text{imine}}{\overset{\overset{\displaystyle R\diagdown N:}{\|}}{C}} \underset{+ H_2O}{\overset{}{}} \underset{H_3O^+}{\rightleftharpoons} \overset{\overset{\displaystyle R\diagdown \overset{+}{N}\diagup H}{\|}}{C} \xrightarrow{H-\bar{B}H_2CN\ Na^+} \overset{\overset{\displaystyle R\diagdown \ddot{N}H}{|}}{CH} \quad (23.25)$$

The success of reductive amination depends on the discrimination by the reducing agents between the imine intermediate and the carbonyl group of the aldehyde or ketone starting material. Each reagent is a sodium borohydride ($NaBH_4$) derivative in which one or more of the hydrides have been substituted with electron-withdrawing groups (—OAc or —CN). The polar effect of these groups reduces the effective negative charge on the hydride and, as a result, each reagent is less reactive than sodium borohydride itself. Each reagent is effectively "tuned" to be just reactive enough to reduce imines, but not reactive enough to reduce aldehydes or ketones. When $NaCNBH_3$ is used in protic solvents, hydrogen-bond donation by the solvent to the imine nitrogen (which is more basic than a carbonyl oxygen) catalyzes the reduction.

Formaldehyde can be reductively aminated with primary and secondary amines using the Borch reaction. This provides a way to introduce methyl groups to the level of a tertiary amine:

$$(23.26)$$

(84% yield)

$$(23.27)$$

benzylethylamine **formaldehyde** **benzylethylmethylamine**
(80% yield)

PROBLEM

23.16 Quaternization (Sec. 23.6) does not occur in the reactions shown in Eqs. 23.26 and 23.27. Explain.

Neither an imine nor an enamine can be an intermediate in the reaction of a secondary amine with formaldehyde (Eq. 23.27). (Why?) In this case a small amount of a cationic intermediate, an *imminium ion*, is formed in solution by protonation of a carbinolamine intermediate and loss of water. The imminium ion, which is also a carbocation, is rapidly and irreversibly reduced by its reaction with hydride.

$$(23.28)$$

Suppose you want to prepare a given amine and want to determine whether reductive amination would be a suitable preparative method. How do you determine the required starting materials? Adopt the usual strategy for analyzing a synthesis: Start with the target molecule and mentally reverse the reductive amination process. Mentally break one of the C—N bonds and replace it on the nitrogen side with an N—H bond. On the carbon side, drop a hydrogen from the carbon and add a carbonyl oxygen.

$$(23.29)$$

H added   O added

As this analysis shows, the target amine must have a hydrogen on the "disconnected" carbon. This process is applied in Study Problem 23.3.

Outline a preparation of *N*-ethyl-*N*-methylaniline from suitable starting materials using a reductive amination sequence.

$$\langle\!\!\bigcirc\!\!\rangle\!\!-\!\!N\!\!-\!\!CH_3$$
$$\qquad\qquad|$$
$$\qquad\qquad CH_2CH_3$$

N-ethyl-N-methylaniline

SOLUTION   Either the *N*-methyl or *N*-ethyl bond can be used for analysis. (The *N*-phenyl bond cannot be used because the carbon in the C—N bond has no hydrogen.) We arbitrarily choose the N—CH$_3$ bond and make the appropriate replacements as indicated in Eq. 23.29 to reveal the following starting materials:

$$\langle\!\!\bigcirc\!\!\rangle\!\!-\!\!N\!\!-\!\!CH_2\!\!-\!\!H \implies \langle\!\!\bigcirc\!\!\rangle\!\!-\!\!N\!\!-\!\!H \ + \ O\!\!=\!\!CH_2$$
$$\qquad\quad|\qquad\qquad\qquad\qquad\qquad|$$
$$\qquad CH_2CH_3 \qquad\qquad\qquad CH_2CH_3 \quad \textbf{formaldehyde}$$

N-ethylaniline

Thus, treatment of *N*-ethylaniline with formaldehyde and NaBH$_3$CN should give the desired amine. (See Problem 23.18.)

C. Acylation of Amines

Amines can be converted into amides by reaction with acid chlorides, anhydrides, or esters. These reactions are covered in Sec. 21.8.

$$R\overset{O}{\overset{\|}{-C}}-Cl + 2\,R_2NH \longrightarrow R\overset{O}{\overset{\|}{-C}}-NR_2 + R_2\overset{+}{N}H_2\ Cl^- \qquad (23.30)$$

$$R\overset{O}{\overset{\|}{-C}}-O\overset{O}{\overset{\|}{-C}}-R + 2\,R_2NH \longrightarrow R\overset{O}{\overset{\|}{-C}}-NR_2 + R\overset{O}{\overset{\|}{-C}}-O^- + R_2\overset{+}{N}H_2 \qquad (23.31)$$

$$R\overset{O}{\overset{\|}{-C}}-OR + R_2NH \longrightarrow R\overset{O}{\overset{\|}{-C}}-NR_2 + R\,OH \qquad (23.32)$$

In this type of reaction, a bond is formed between the amine and a carbonyl carbon. These are all examples of *acylation*: a reaction involving the transfer of an *acyl group*.

Recall that the reaction of an amine with an acid chloride or an anhydride requires either *two equivalents* of the amine or one equivalent of the amine and an additional equivalent of another base such as a tertiary amine or hydroxide ion. These and other aspects of amine acylation can be reviewed in Sec. 21.8.

PROBLEMS

23.17   Suggest two syntheses of *N*-ethylcyclohexanamine by reductive amination.

23.18   Outline a second synthesis of *N*-ethyl-*N*-methylaniline (the target molecule in Study Problem 23.3) by reductive amination.

23.19   Outline a synthesis of the quaternary ammonium salt (CH$_3$)$_3$$\overset{+}{N}CH_2$Ph Br$^-$ from each of the following combinations of starting materials.

(a) dimethylamine and any other reagents   (b) benzylamine and any other reagents

23.20   (a) A chemist Caleb J. Cookbook heated ammonia with bromobenzene expecting to form tetraphenylammonium bromide. Can Caleb expect this reaction to succeed? Explain.

(b) What type of catalyst might be used to bring about this reaction under relatively mild conditions? (See Sec. 23.11C.)

23.21   Continue the sequence of reactions in Eqs. 23.16a–c to show how trimethylammonium iodide is formed as one of the products in Eq. 23.15.

23.22   Outline a preparation of each of the following from an amine and an acid chloride.

(a) *N*-phenylbenzamide   (b) *N*-benzyl-*N*-ethylpropanamide

23.8   HOFMANN ELIMINATION OF QUATERNARY AMMONIUM HYDROXIDES

The previous section discussed ways to *make* carbon–nitrogen bonds. In these reactions, amines react as *nucleophiles*. The subject of this section is an elimination reaction used to *break* carbon–nitrogen bonds. In this reaction, which involves *quaternary ammonium hydroxides* (R_4N^+ ^-OH) as starting materials, amines act as *leaving groups*.

When a quaternary ammonium hydroxide is heated, a β-elimination reaction takes place to give an alkene, which distills from the reaction mixture.

$$\text{a quaternary ammonium hydroxide} \xrightarrow{\text{heat}} \text{methylenecyclohexane (74\% yield)} + H—OH + \ddot{N}Me_3 \quad (23.33)$$

This type of elimination reaction is called a **Hofmann elimination**, after August Wilhelm Hofmann (1818–1892), a German chemist who became professor at the Royal College of Chemistry in London and later, at the University of Berlin. Hofmann was particularly noted for his work on amines.

A quaternary ammonium hydroxide used as the starting material in Hofmann eliminations is formed by treating a quaternary ammonium salt with silver hydroxide (AgOH), which is essentially a hydrated form of silver(I) oxide (Ag_2O).

$$\text{—}CH_2\overset{+}{N}Me_3 \; I^- + Ag_2O \cdot H_2O \longrightarrow \text{—}CH_2\overset{+}{N}Me_3 \; ^-OH + AgI \quad (23.34)$$

Alkenes, then, can be formed from amines by a three-step process: exhaustive methylation (see Eq. 23.21, p. 1199), conversion of the ammonium salt to the hydroxide (Eq. 23.34), and Hofmann elimination (Eq. 23.33).

The Hofmann elimination is conceptually analogous to the E2 reaction of alkyl halides (Sec. 9.5), in which a proton and a halide ion are eliminated; in the Hofmann elimination, a proton and a tertiary amine are eliminated. Because the amine leaving group is very basic, and therefore a relatively poor leaving group, the conditions of the Hofmann elimination are typically harsh.

Like the analogous E2 reaction of alkyl halides, the Hofmann elimination generally occurs as an anti-elimination (Sec. 9.5E).

$$ (23.35) $$

PROBLEMS

23.23 What product (including its stereochemistry) is expected from the Hofmann elimination of each of the following stereoisomers?

(a)

$$\overset{+}{N}(CH_3)_3 \ \ \bar{O}H$$
$$|$$
$$(2R,3R)\text{-Ph}-CH-CH-Ph$$
$$|$$
$$CH_3$$

(b)

$$\overset{+}{N}(CH_3)_3 \ \ \bar{O}H$$
$$|$$
$$(2R,3S)\text{-Ph}-CH-CH-Ph$$
$$|$$
$$CH_3$$

23.24 Give the major product formed when each of the following amines is treated exhaustively with methyl iodide and then heated with Ag_2O. Explain your reasoning.

(a)

(b)

23.25 (+)-*Coniine* is the toxic component of the poison hemlock, the plant believed to have killed Socrates. Coniine has the molecular formula $C_8H_{17}N$. When coniine is exhaustively methylated, and the resulting product is then heated with Ag_2O, the mixture of compounds *A–C* is formed. [Compounds *A* and *B* are the (*E*) stereoisomers.]

$$Me_2NCH_2CH_2CH_2CH_2CH=CHCH_2CH_3 \qquad Me_2NCH_2CH_2CH_2CH=CHCH_2CH_2CH_3$$
$$\qquad\qquad A \qquad\qquad\qquad\qquad\qquad\qquad\qquad\qquad B$$

$$H_2C=CHCH_2CH_2CHCH_2CH_2CH_3$$
$$|$$
$$NMe_2$$
$$C$$

Propose a structure for coniine. (The absolute configuration of coniine cannot be determined from the data.)

23.9 AROMATIC SUBSTITUTION REACTIONS OF ANILINE DERIVATIVES

Aromatic amines can undergo *electrophilic aromatic substitution* reactions on the ring (Sec. 16.4). The amino group is one of the most powerful ortho, para-directing groups in electrophilic substitution. If the conditions of the reaction are not too acidic, aniline and its derivatives undergo rapid ring substitution. For example, aniline, like phenol, brominates three times under mild conditions.

$$\text{(aniline)} + 3\,Br_2 \longrightarrow \text{(2,4,6-tribromoaniline)} + 3\,HBr \qquad (23.36)$$

PROBLEM

23.26 Provide a reaction mechanism for the reaction shown in Eq. 23.36. (*Hint:* Refer to Sec. 16.4.)

When aniline is nitrated with a mixture of nitric and sulfuric acid, a mixture of meta and para nitration products is obtained.

$$(+2\% \text{ ortho isomer}) \qquad (23.37)$$

(51%) (47%)

(95% yield)

The formation of *m*-nitroaniline as one of the products is understandable because the major form of the starting material in the strongly acidic solution is the conjugate-acid ammonium ion. Because the ammonium ion has no unshared electron pair on nitrogen, it cannot donate electrons by a resonance effect; consequently, its electron-withdrawing polar effect makes it a meta-directing group. It is likely that the *p*-nitroaniline product arises by the rapid nitration of the minuscule amount of highly reactive unprotonated aniline in the reaction mixture. As the unprotonated aniline reacts, it is replenished by Le Châtelier's principle.

Aniline can be nitrated regioselectively at the para position if the nitrogen is first *protected* from protonation. (The general idea of a *protecting group* was introduced in Sec. 19.10B.) This strategy is used in the solution to Study Problem 23.4.

**STUDY GUIDE
LINK 23.1**
Nitration of Aniline

STUDY PROBLEM 23.4

Outline a preparation of *p*-nitroaniline from aniline and any other reagents.

SOLUTION An amide group is much less basic than an amino group. Hence, acylation of the amino group of aniline with acetyl chloride to give *N*-phenylacetamide (acetanilide) will protect the nitrogen from protonation. The acetamido group, although much less activating than a free amino group, is nevertheless an activating, ortho, para-directing group in aromatic substitution (Table 16.2 on p. 812). Following nitration of acetanilide, the acetyl group is removed to give *p*-nitroaniline, the target compound.

Because the acetamido group is considerably less basic than an amino group, it is only partially protonated under the acidic reaction conditions of nitration. Because the acetamido group is less activating than a free amino group (why?), nitration occurs only once.

23.27 Outline a preparation of sulfanilamide, a sulfa drug, from aniline and any other reagents. (*Hint:* See Eq. 20.27, p. 1025; also note that sulfonyl chlorides react with amines to form amides in much the same manner as carboxylic acid chlorides.)

$$H_2N \text{—} \underset{\underset{O}{\parallel}}{\overset{\overset{O}{\parallel}}{S}} \text{—} NH_2$$

sulfanilamide

23.28 Outline a preparation of each of the following compounds from aniline and any other reagents.
(a) 2,4-dinitroaniline
(b) sulfathiazole, a sulfa drug (*Hint:* 2-Aminothiazole is a readily available amine.)

sulfathiazole   **2-aminothiazole**

23.10   DIAZOTIZATION; REACTIONS OF DIAZONIUM IONS

A. Formation and Substitution Reactions of Diazonium Salts

The reactions considered in previous sections show that the chemistry of the amino group vaguely resembles that of the hydroxy group. Amino groups, like hydroxy groups, can both donate and accept hydrogen bonds. Amino groups, like hydroxy groups, are basic and nucleophilic (only more so). Amino groups, like hydroxy groups (with suitable activation), can serve as leaving groups. And amino groups, like hydroxy groups, activate aromatic rings toward electrophilic aromatic substitution.

In contrast, when it comes to oxidation reactions, there is no parallel between amines and alcohols or phenols. Oxidation of amines generally occurs at the amino nitrogen, whereas oxidation of alcohols and phenols occurs at the α-carbon. An important oxidation reaction of amines that illustrates this point is called **diazotization**: the reaction of primary amines with nitrous acid (HNO_2) to form diazonium salts. A **diazonium salt** is a compound of the form $R\text{—}^+N\equiv N: X^-$, in which X^- is a typical anion (chloride, bromide, sulfate, and so on). Because nitrous acid is unstable, it is usually generated as needed by the reaction of sodium nitrite ($NaNO_2$) with a strong acid such as HCl or H_2SO_4. Both aliphatic and aromatic primary amines are readily diazotized:

FURTHER EXPLORATION 23.3
Mechanism of Diazotization

$$\underset{\underset{NH_2}{|}}{CH_3CHCH_2CH_3} \xrightarrow[\text{H}_2\text{O}]{\text{NaNO}_2,\ \text{HCl}} \left[\underset{\underset{+}{\overset{|}{N\equiv N:}}\ Cl^-}{CH_3CHCH_2CH_3} \right] \tag{23.38}$$

2-butanamine
(***sec*-butylamine**)

2-butanediazonium chloride
(an aliphatic diazonium salt)
unstable; cannot be isolated

$$Ph\text{—}NH_2 \xrightarrow[\text{H}_2\text{O}]{\text{NaNO}_2,\ \text{HCl}} Ph\text{—}\overset{+}{N}\equiv N: Cl^- \tag{23.39}$$

aniline

benzenediazonium chloride
(an aromatic diazonium salt)
can be isolated

Notice that diazonium salts incorporate one of the very best leaving groups—molecular nitrogen (*red* in Eq. 23.38). For this reason, *aliphatic* diazonium salts react immediately as they are formed by S_N1, E1, and/or S_N2 mechanisms to give substitution and elimination products along with dinitrogen, a gas.

$$CH_3CHCH_2CH_3 \xrightarrow[H_2O]{NaNO_2, H_2SO_4} \left[\begin{array}{c} CH_3CHCH_2CH_3 \\ | \\ N\equiv N: \\ + \end{array} \right] \longrightarrow \left[CH_3\overset{+}{C}HCH_2CH_3 \right] + :N\equiv N:\uparrow$$

dinitrogen (gas)

$$H_3O^+ + CH_3CHCH_2CH_3 + H_2C=CHCH_2CH_3 + CH_3CH=CHCH_3 \qquad (23.40)$$

(9% of product)   (31% of product; 10% cis and 21% trans)

(60% of product)

(The rapid liberation of dinitrogen on treatment with nitrous acid is a qualitative test for primary alkylamines.) Because of the complex mixture of solvolysis and elimination products that results, the reactions of aliphatic diazonium salts are not generally useful in organic synthesis.

Recall that benzene rings bearing good leaving groups do not readily undergo S_N1 or S_N2 reactions (Secs. 18.1 and 18.3). For this reason, *aryl*diazonium salts can be isolated and used in a variety of reactions. In practice, though, they are usually prepared in solution at 0–5 °C and used without isolation, because they lose nitrogen on heating and because they are explosive in the dry state.

Among the most important reactions of aryldiazonium salts are substitution reactions with cuprous halides; in these reactions, the diazonium group is replaced by a halogen.

p-methylaniline (p-toluidine)   **p-methylbenzenediazonium chloride**   **p-chlorotoluene** (70–71% yield)

(23.41)

2-methoxyaniline   **1-bromo-2-methoxybenzene (o-bromoanisole)** (88–93% yield)

(23.42)

An analogous reaction occurs with cuprous cyanide, CuCN.

4-methylbenzonitrile (67% yield)

(23.43)

This reaction is another way of forming a carbon–carbon bond, in this case to an aromatic ring (see Appendix VI). The resulting nitrile can be converted by hydrolysis into a carboxylic acid, which can, in turn, serve as the starting material for a variety of other types of compounds. The reaction of an aryldiazonium ion with a cuprous salt is called the **Sandmeyer reaction**. This reaction is an important method for the synthesis of aryl halides and nitriles.

Aryl iodides can also be made by the reaction of diazonium salts with the potassium salt KI.

(23.44)

(74–76% yield)

The analogous reactions with KBr and KCl do not work; cuprous salts are required.

Aryldiazonium salts can be hydrolyzed to phenols by heating them in water. A variation of this reaction reminiscent of the Sandmeyer reaction is the use of cuprous oxide (Cu_2O) and an excess of aqueous cupric nitrate [$Cu(NO_3)_2$] at room temperature.

(23.45)

p-bromoaniline

p-bromophenol
(95% yield)

Finally, the diazonium group is replaced by hydrogen when the diazonium salt is treated with hypophosphorous acid, H_3PO_2.

(23.46)

2,4,6-tribromoaniline   **2,4,6-tribromobenzenediazonium chloride**   **1,3,5-tribromobenzene**
(70% yield)

The product of Eq. 23.46, 1,3,5-tribromobenzene, cannot be prepared by the bromination of benzene itself. (Why?) Recall that the starting material, 2,4,6-tribromoaniline, is prepared by the bromination of aniline (Eq. 23.36). In this bromination reaction, the positions of the bromines are determined by the powerful directing effect of the amino nitrogen. Once the amino group has fulfilled its role as an activating and directing group, it can be removed using the reaction in Eq. 23.46.

The diazonium salt reactions shown in Eqs. 23.41–23.46 are all substitution reactions, but none are S_N2 or S_N1 reactions because aromatic rings do *not* undergo substitution by these mechanisms (Secs. 18.1 and 18.3). It turns out that the Sandmeyer and related reactions occur by radical mechanisms involving the copper. The reaction of diazonium salts with KI (Eq. 23.44), although not involving copper, probably occurs by a similar mechanism. The reaction of diazonium salts with H_3PO_2 (Eq. 23.46) has been shown definitively to be a free-radical chain reaction.

PROBLEMS

23.29 Outline a synthesis for each of the following compounds from the indicated starting materials using a reaction sequence involving a diazonium salt.

(a) 2-bromobenzoic acid from *o*-toluidine (*o*-methylaniline)

(b) 2,4,6-tribromobenzoic acid from aniline

23.30 As shown in the following equation, when (*R*)-1-deuterio-1-butanamine is diazotized with nitrous acid in water, the alcohol product formed has the *S* configuration (D = ^{2}H).

(*R*)-1-deuterio-1-butanamine

(a) Give the stereochemical configuration of the diazonium ion formed as an intermediate in this reaction. Draw its structure.

(b) What mechanism for reaction of the diazonium ion with water is consistent with the stereochemical result in the preceding equation?

B. Aromatic Substitution with Diazonium Ions

Aryldiazonium ions react with aromatic compounds containing strongly activating substituent groups, such as amines and phenols, to give substituted *azobenzenes*. (Azobenzene itself is Ph—N=N—Ph.)

butter yellow
(an azobenzene)

This is an electrophilic aromatic substitution reaction in which the terminal nitrogen of the diazonium ion is the electrophile. The mechanism follows the usual pattern of electrophilic aromatic substitution (Sec. 16.4B). First, the π electrons of the aromatic compound are donated to the electrophilic nitrogen to give a resonance-stabilized carbocation:

This carbocation then loses a proton to give the substitution product.

(Why does substitution occur at the para position? See Sec. 16.5A.)

The azobenzene derivatives formed in these reactions have extensive conjugated π-electron systems, and most of them are colored (Sec. 15.2C). Some of these compounds are used as dyes and indicators; as a class, they are known as **azo dyes**. (An azo dye is a colored derivative of azobenzene.) For example, the azo dye methyl orange is a well-known acid–base indicator.

$$\bar{O}_3S-\!\!\langle\ \rangle\!\!-\ddot{N}=\!\ddot{N}-\!\!\langle\ \rangle\!\!-\ddot{N}Me_2 \quad\underset{\longleftarrow}{\overset{H_3O^+}{\longrightarrow}}\quad \left[\ \bar{O}_3S-\!\!\langle\ \rangle\!\!-\overset{+}{\underset{H}{N}}=\!\ddot{N}-\!\!\langle\ \rangle\!\!-\ddot{N}Me_2 \quad\updownarrow\quad \bar{O}_3S-\!\!\langle\ \rangle\!\!-\ddot{N}H-\!\ddot{N}=\!\!\langle\ \rangle\!\!=\!\overset{+}{N}Me_2\ \right] + H_2O \quad (23.49)$$

methyl orange (yellow)
an azo dye

protonated methyl orange (red)
$pK_a = 3.5$

Because methyl orange changes color when it is protonated, it can be used as an acid–base indicator at pH values near its pK_a of 3.5. Some azo dyes are used in dyeing fabrics, foodstuffs, and cosmetics. For example, FD & C Yellow No. 6 (FD & C = food, drug, and cosmetic) is a compound used to color gelatin desserts, ice cream, beverages, candy, and so on.

$$Na^+\ \ ^-O_3S$$

$$OH$$

$$N=N-\!\!\langle\ \rangle\!\!-SO_3^-\ \ Na^+$$

FD & C Yellow No. 6 ("Sunset Yellow")

C. Reactions of Secondary and Tertiary Amines with Nitrous Acid

Secondary amines react with nitrous acid to give *N*-nitrosoamines, compounds of the form $R_2N-\!N=\!O$, usually called simply **nitrosamines**.

$$Me_2\ddot{N}H\ +\ HNO_2\ \longrightarrow\ Me_2\ddot{N}-\!\!\overset{O}{\underset{\ddot{N}}{\parallel}}\ +\ H_2O \quad (23.50)$$

***N,N*-dimethylnitrosamine**
(89–90% yield)

$$Ph-NH-CH_3\ \xrightarrow{NaNO_2,\ HCl}\ Ph-\underset{\underset{N=O}{|}}{N}-CH_3 \quad (23.51)$$

***N*-methyl-*N*-nitrosoaniline**

Nitrosamines, Cancer, and Breakfast Bacon

The fact that many nitrosamines are known to be potent carcinogens created a debate over the use of sodium nitrite ($NaNO_2$) as a meat preservative. The meat-packing industry argued that sodium nitrite is important in preventing the botulism that results from meat spoilage. But because sodium nitrite is, in combination with acid, a diazotizing reagent, it has the capacity for producing nitrosamines in the acidic environment of the stomach. For example, the frying of bacon generates nitrosamines that concentrate in the fat. [Well-drained bacon contains fewer nitrosamines, and nitrosamines are destroyed by ascorbic acid (vitamin C), which is present in fruit and vegetable juices. Perhaps this is a good reason for drinking orange juice when having bacon for breakfast!] The potential hazards of sodium nitrite led to a long campaign by consumer groups to have it banned as a meat preservative; the campaign was fought by the meat-packing industry. Then researchers found that nitrite is produced by the bacteria in the normal human intestine. It became questionable whether the risk from nitrite in meat is any greater than the risk faced all along from normal intestinal flora. These findings caused the Food and Drug Administration in 1980 to back away from banning sodium nitrite as a preservative, recommending only that it be kept to a minimum.

A tertiary amine cannot form a nitrosamine. However, N,N-disubstituted aromatic amines undergo electrophilic aromatic substitution on the benzene ring. The electrophile is the nitrosyl cation, $^+\ddot{N}{=}\ddot{O}$, which is generated from nitrous acid under acidic conditions.

N,N-dimethylaniline

N,N-dimethyl-4-nitrosoanilinium chloride
(89–90% yield)

$$(23.52)$$

PROBLEMS

23.31 (a) Write a Lewis structure for HNO_2 in Eq. 23.50.

(b) Write a mechanism for the reaction shown in Eq. 23.50.

23.32 Design a synthesis of methyl orange (Eq. 23.49) using aniline as the only aromatic starting material.

23.33 What two compounds would react in a diazo coupling reaction to form FD & C Yellow No. 6?

23.34 (a) Using the curved-arrow notation, show how the nitrosyl cation, $^+\ddot{N}{=}\ddot{O}$, is generated from HNO_2 under acidic conditions.

(b) Draw a curved-arrow mechanism for the electrophilic aromatic substitution reaction shown in Eq. 23.52.

23.11 SYNTHESIS OF AMINES

Several reactions discussed in previous sections can be used for the synthesis of amines. In this section, five additional methods will be presented, and, in Sec. 23.11E, all of the methods for preparing amines are summarized.

A. Synthesis of Primary Amines: The Gabriel Synthesis and the Staudinger Reaction

Recall that direct alkylation of ammonia is generally not a good synthetic method for the preparation of amines because multiple alkylation takes place (Sec. 23.7A). This problem can be avoided by protecting the amine nitrogen so that it can react only once with alkylating reagents. One approach of this sort begins with the imide *phthalimide*. Because the pK_a of phthalimide is 8.3, its conjugate-base anion is easily formed with KOH or NaOH. This anion is a good nucleophile, and is alkylated by alkyl halides or sulfonate esters in S_N2 reactions.

phthalimide
pK_a = 8.3

$$(23.53a)$$

N-butylphthalimide

The alkyl halides and sulfonates used in this reaction are primary or unbranched secondary. Because the *N*-alkylated phthalimide formed in this reaction is really a double amide, it can be converted into the free amine by amide hydrolysis in either strong acid or base.

$$\text{(23.53b)}$$

butylammonium bromide

N-butylphthalimide

In this example, acidic hydrolysis gives the ammonium salt, which can be converted into the free amine by neutralization with base.

The alkylation of phthalimide anion followed by hydrolysis of the alkylated derivative to the primary amine is called the **Gabriel synthesis**, after Siegmund Gabriel (1851–1924), a professor at the University of Berlin, who developed the reaction in 1887. Because the nitrogen in phthalimide has only one acidic hydrogen, it can be alkylated only once. Although *N*-alkylphthalimides also have a pair of unshared electrons on nitrogen, they do not alkylate further, because neutral imides are *much* less basic (why?), and therefore less nucleophilic, than the phthalimide anion. Hence, multiple alkylation, which occurs in the direct alkylation of ammonia, is avoided in the Gabriel synthesis.

alkyl halide

$$\text{(23.54)}$$

Another method for preparing primary amines while avoiding the multiple alkylation problem utilizes alkyl azides. Alkyl azides can be prepared from an alkyl halide (or a sulfonate ester, such as a tosylate) and a source of azide ion, or N_3^-, such as sodium azide (NaN_3). (See Table 9.1, p. 385.) Like other S_N2 reactions, the reaction takes place most readily with relatively unhindered alkyl halides.

$$CH_3CH_2CH_2\text{—}Br \ + \ Na^+\,N_3^- \ \xrightarrow[\text{H}_2\text{O, THF}]{} \ CH_3CH_2CH_2\text{—}N_3 \ + \ Na^+\,Br^- \quad \text{(23.55)}$$

propyl azide
(88% yield)

Like the phthalimide in the Gabriel synthesis, the alkyl azide can be thought of as a protected amine nitrogen. The alkyl azide reacts with triphenylphosphine to form a phosphazide intermediate.

triphenyl-phosphine   an alkyl azide   a phosphazide

$$\text{(23.56a)}$$

The phosphazide can undergo a rearrangement to yield a second intermediate called an iminophosphorane, producing dinitrogen as a gaseous by-product. (Recall that phosphorous, a third-row element, can exceed an octet of valence electrons, as discussed in Sec. 10.10A.)

phosphazide   an iminophosphorane   **nitrogen gas**

$$\text{(23.56b)}$$

Notice the similarity between the rearrangement of the phosphazide in Eq. 23.56b and the oxaphosphetane in the Wittig alkene synthesis (Eq. 19.78a, p. 991).

In the presence of water, the iminophosphorane is hydrolyzed to yield triphenylphosphine oxide and a primary amine. This reaction is analogous to imine hydrolysis discussed in Sec. 19.11A.

$$Ph_3P{=}\ddot{N}{-}\bigcirc \quad + \ H_2O \quad \longrightarrow \quad Ph_3P{=}O \quad + \quad H_2\ddot{N}{-}\bigcirc \qquad (23.57)$$

cyclohexyl-   **triphenylphosphine**   **cyclohexylamine**
iminophosphorane   **oxide**   a primary amine
                                               (95% yield)

This reaction is called the **Staudinger reaction** after another German chemist, Hermann Staudinger (1881–1965), who developed the method in 1919 as a professor at the Swiss Federal Institute of Technology in Zürich, Switzerland. In 1953, he received the Nobel Prize in Chemistry for other pioneering work in the field of polymers, work that ultimately led to everyday products such as nylon and polyesters.

PROBLEM

23.35 (a) Which one of the following three amines can be prepared by either the Gabriel synthesis or the Staudinger reaction: 2,2-dimethyl-1-propanamine, 3-methyl-1-pentanamine, or *N*-butylaniline?

(b) Starting with an alkyl halide, propose a synthesis for the compound you chose in part (a), using the Gabriel synthesis.

(c) Propose an alternative synthesis for the same compound using the Staudinger reaction and the same alkyl halide starting material.

(d) Explain why the other two amines cannot be prepared by either method.

B. Reduction of Nitro Compounds

Nitro compounds can be reduced easily to amines by catalytic hydrogenation:

$$(23.58)$$

1,2-dimethoxy-4-nitrobenzene   **3,4-dimethoxyaniline**
                                                 (97% yield)

In an older, but nevertheless effective, method, finely divided tin or iron powders and HCl can be used to convert aromatic nitro compounds into aniline derivatives.

$${-}NO_2 \xrightarrow[\text{Fe/HCl}]{\text{Sn/HCl or}} \xrightarrow{^{-}OH} \cdots {-}NH_2 + \ Sn^{2+} \ or \ Fe^{3+} \ salts \qquad (23.59)$$

1-bromo-3-nitrobenzene   ***m*-bromoaniline**
                                            (80% yield)

In this reaction, the nitro compound is reduced *at nitrogen*, and the metal, which is oxidized to a metal salt, is the reducing agent. Although the methods shown in both Eqs. 23.58 and 23.59 also work with aliphatic nitro compounds, they are particularly important with aromatic nitro compounds as methods for introducing an amino group into an aromatic ring.

In view of the utility of lithium aluminum hydride ($LiAlH_4$) and sodium borohydride ($NaBH_4$) as reducing agents for other compounds, what happens when nitro compounds are treated with these reagents? Aromatic nitro compounds do react with LiAlH$_4$, but the reduction products are azobenzenes (Sec. 23.10B), not amines:

(23.60)

Nitro groups do not react at all with sodium borohydride under the usual conditions.

(23.61)

Hence, LiAlH$_4$ and NaBH$_4$ are *not* useful in forming aromatic amines from nitro compounds.

PROBLEM

23.36 Outline syntheses of the following compounds from the indicated starting materials.
   (a) *p*-iodoanisole from phenol and any other reagents
   (b) *m*-bromoiodobenzene from nitrobenzene

C. Amination of Aryl Halides and Aryl Triflates

Arylamines can be prepared by the direct amination of aryl chlorides and aryl bromides in the presence of a base and a Pd(0) catalyst.

(23.62)

N-(2,5-dimethylphenyl)pyrrolidine
(98% yield)

The direct amination of aryl halides is sometimes called **Buchwald–Hartwig amination** to recognize the two chemistry professors who led the research groups that developed these reactions: Stephen L. Buchwald of MIT, and John F. Hartwig of the University of California, Berkeley.

A number of different catalysts have been explored for direct amination. These are typically of the form PdL$_2$, where L is a sterically demanding ligand such as the following:

(Cy = cyclohexyl)

These catalysts are formed by mixing palladium(II) acetate or other Pd precursors and two equivalents of the ligands.

These amination reactions have been shown to operate by more than one mechanism. All of the mechanisms, however, like the mechanisms of the Heck, Suzuki, and Stille reactions (Secs. 18.6A, 18.6B, and 18.10B), involve the key steps of oxidative addition and reductive elimination (Sec. 18.5E). The following scheme summarizes these features.

$$PdL_2 \rightleftharpoons PdL + L$$

PdL + Ar—Cl $\xrightarrow[\text{addition}]{\text{oxidative}}$ L→Pd—Ar(Cl) $\xrightarrow[\substack{\text{ligand}\\\text{substitution}}]{\text{Base:}^-\ R_2NH}$ L→Pd—Ar(NR_2) $\xrightarrow[\text{elimination}]{\text{reductive}}$ PdL + Ar—NR_2   (23.63)

a 12e^- complex
Pd(0)

a 14e^- complex
Pd(II)

+ Base—H
+ Cl$^-$

The amine used as the starting material in the amination reaction must lose a hydrogen in the reaction. Consequently, when a tertiary amine is the amination product, it cannot react further. However, when a primary amine is used as the starting material, the product is a secondary amine. It can in principle serve as the starting material in a competing second amination.

Ar—Cl $\xrightarrow[\text{catalyst}]{RNH_2,\ \text{base,}}$ Ar—NHR $\xrightarrow[\text{catalyst}]{Ar-Cl,\ \text{base,}}$ Ar—NR(Ar)   (23.64)

The product of this second amination becomes an unwanted by-product. Nevertheless, amination with primary amines is practical if the primary amine is itself an arylamine, or if it has a large or highly branched alkyl group. In such cases, steric hindrance is used to advantage. The catalyst complex leading to the tertiary amine has significant steric repulsions; as a result, the undesired second amination is relatively slow and does not occur to a significant extent. This is one reason that the catalysts involve sterically demanding ligands.

[*p*-chlorotoluene structure] + H_2N—$CH_2(CH_2)_4CH_3$ + $Na^+\ ^-Ot\text{-}Bu$ $\xrightarrow[\text{toluene}]{\text{Pd(0) catalyst}}$

hexylamine

***p*-chlorotoluene**

[*N*-hexyl-4-methylaniline structure] $NHCH_2(CH_2)_4CH_3$ + $Na^+\ Cl^-$ + $t\text{-BuOH}$ (23.65)

***N*-hexyl-4-methylaniline**
(85% yield)

As you have learned, reductive amination is another way to prepare tertiary arylamines. (See Eq. 23.23 on p. 1200.) Some tertiary arylamines, however, such as those containing nitrogen heterocycles (Eq. 23.62), would be difficult to prepare by reductive amination. Direct amination provides a straightforward route to these amines. Another attractive aspect of direct amination is that it, like other Pd-catalyzed coupling reactions, tolerates a wide variety of other functional groups, as Study Problem 23.5 illustrates.

STUDY PROBLEM 23.5

Outline a synthesis of *p*-dipropylaminoacetophenone from chlorobenzene.

SOLUTION   Considering the problem retrosynthetically gives the following synthetic pathway, starting with the target molecule:

p-dipropylaminoacetophenone

Direct amination of *p*-chloroacetophenone with dipropylamine and an appropriate Pd(0) catalyst gives the target:

Reductive amination would not have worked, because the acetyl group would have been reduced under conditions of reductive amination.

The starting material for the amination, *p*-chloroacetophenone, can in turn be prepared by a Friedel–Crafts acylation reaction.

Aryl triflates can also be used as starting materials in amination. Because aryl triflates can be prepared from the corresponding phenols (Sec. 18.10B), this reaction provides a synthetic path from phenols to arylamines.

p-tert-butylphenyl triflate
(prepared from a phenol)

dibutylamine

N,N-dibutyl-4-tert-butylaniline
(73% yield)

$$+ \ Na^+ \ ^-OTf \ + \ t\text{-BuOH} \qquad (23.66)$$

PROBLEM

23.37 Outline a synthesis of each of the following compounds from the indicated starting material and any other reagents.

(a) *N*-(*sec*-butyl)-*N*-ethylaniline from chlorobenzene

(b)

from phenol

(c)

from chlorobenzene

D. Curtius and Hofmann Rearrangements

A very useful synthesis of amines starts with a class of compounds called *acyl azides*. An **acyl azide** has the following general structure:

$$R-\overset{\overset{\displaystyle O}{\|}}{C}-\overset{..}{\underset{..}{N}}-\overset{+}{N}\equiv N: \quad \text{or} \quad R-\overset{\overset{\displaystyle O}{\|}}{C}-N_3 \qquad (23.67)$$

(The synthesis of acyl azides is discussed below.) When an acyl azide is heated in an inert solvent such as benzene or toluene, it is transformed with loss of nitrogen into an **isocyanate**, a compound of the general structure R—N=C=O.

$$CH_3(CH_2)_{10}-\overset{\overset{\displaystyle O}{\|}}{C}-N_3 \xrightarrow[\text{benzene}]{\text{heat}} CH_3(CH_2)_{10}-N=C=O + N_2 \qquad (23.68)$$

dodecanoyl azide   **undecyl isocyanate**
                                        (81–86% yield)

This reaction, called the **Curtius rearrangement**, is a concerted reaction that can be represented as follows:

$$R-\overset{\overset{\displaystyle :O:}{\|}}{\underset{\overset{|}{\underset{..}{N}}}{C}}\overset{+}{N}\equiv N: \longrightarrow \overset{..}{\underset{..}{O}}=C=\overset{..}{N}-R + :N\equiv N: \qquad (23.69)$$

The rearrangement is named for its discoverer, Theodor Curtius (1857–1928), who was professor of chemistry at Heidelberg University.

The isocyanate product of a Curtius rearrangement can be transformed into an amine by hydration in either acid or base. Hydration involves, first, addition of water across the C=N bond to give a carbamic acid:

$$H-OH + R-N=C=O \xrightarrow{H_3O^+} R-NH-\overset{\overset{\displaystyle O}{\|}}{C}-OH \qquad (23.70)$$

        an isocyanate   a carbamic acid

Carbamic acids are among those types of carboxylic acids that spontaneously decarboxylate (see Eq. 20.43b, p. 1031). Decarboxylation gives the amine, which is protonated under the acidic conditions of the reaction. The free amine is obtained by neutralization:

$$R-NH-\overset{\overset{\displaystyle O}{\|}}{C}-OH \xrightarrow{H_3O^+} R-\overset{+}{N}H_3 \xrightarrow{{}^-OH} R-NH_2 + H_2O \qquad (23.71)$$
$$+ CO_2$$

The overall transformation that occurs as a result of the Curtius rearrangement followed by hydration is the loss of the carbonyl carbon of the acyl azide as CO_2.

$$R-\overset{\overset{\displaystyle O}{\|}}{C}-N_3 \xrightarrow[\text{heat}]{-N_2} R-N=C=O \xrightarrow{H_2O, H_3O^+}$$

acyl azide   isocyanate

$$\left[R-NH-\overset{\overset{\displaystyle O}{\|}}{C}-OH \right] \longrightarrow R\overset{+}{N}H_3 \xrightarrow{{}^-OH} RNH_2 + H_2O \qquad (23.72)$$

        carbamic acid $+ CO_2\uparrow$ amine
        (unstable)

**STUDY GUIDE
LINK 23.2**
Mechanism of
the Curtius
Rearrangement

**STUDY GUIDE
LINK 23.3**
Formation and
Decarboxylation of
Carbamic Acids

An important use of the Curtius rearrangement is for the preparation of carbamic acid derivatives (see Sec. 21.1G). Such derivatives are produced by allowing the isocyanate products to react with nucleophiles other than water. The reaction of isocyanates with alcohols or phenols yields carbamate esters, whereas the reaction with amines yields ureas.

$$
R-N=C=O \xrightarrow{
\begin{array}{c} R\,OH \\ \text{(alcohol or} \\ \text{phenol)} \end{array}}
\underset{\text{a carbamate ester}}{R-NH-\overset{\overset{\displaystyle O}{\|}}{C}-OR}
$$

(from the Curtius rearrangement)

$$
\xrightarrow{H_2O} \underset{\text{an amine}}{CO_2 + R-NH_2}
$$

$$
\xrightarrow{\begin{array}{c} R\,NH_2 \\ \text{(amine)} \end{array}} \underset{\text{a urea}}{R-NH-\overset{\overset{\displaystyle O}{\|}}{C}-NH-R}
$$

(23.73)

The key to the preparation of acyl azides used in the Curtius rearrangement is to recognize that these compounds are carboxylic acid derivatives. The most straightforward preparation is the reaction of an acid chloride with sodium azide.

$$
\underset{\textbf{phenylacetyl chloride}}{Ph-CH_2-\overset{\overset{\displaystyle O}{\|}}{C}-Cl} + NaN_3 \longrightarrow \underset{\substack{\textbf{phenylacetyl azide} \\ \text{(an acyl azide)}}}{Ph-CH_2-\overset{\overset{\displaystyle O}{\|}}{C}-N_3} + NaCl
$$

(23.74)

Another widely used method is to convert an ethyl ester into an acyl derivative of hydrazine (H_2N-NH_2) by aminolysis (Sec. 21.8C). The resulting amide, an *acyl hydrazide*, is then diazotized with nitrous acid to give the acyl azide.

$$
\underset{\textbf{ethyl phenylacetate}}{Ph-CH_2-\overset{\overset{\displaystyle O}{\|}}{C}-OEt} + \underset{\textbf{hydrazine}}{NH_2NH_2} \xrightarrow{-EtOH} \underset{\substack{\textbf{phenylacetyl hydrazide} \\ \text{(an acyl hydrazide)} \\ \text{(80–100\% yield)}}}{Ph-CH_2-\overset{\overset{\displaystyle O}{\|}}{C}-NHNH_2} \xrightarrow[\substack{HCl \\ -10\,°C}]{NaNO_2,} \underset{\textbf{phenylacetyl azide}}{PhCH_2-\overset{\overset{\displaystyle O}{\|}}{C}-N_3}
$$

(23.75)

Notice the similarity of this diazotization to the diazotization of alkylamines:

Compare:

$$
R-CH_2-NH_2 + HONO \longrightarrow R-CH_2-\overset{+}{N}\equiv N:
$$

$$
R-\overset{\overset{\displaystyle O}{\|}}{C}-NH-NH_2 + HONO \longrightarrow R-\overset{\overset{\displaystyle O}{\|}}{C}-\overset{\overset{\displaystyle H}{|}}{N}-\overset{+}{N}\equiv N: \xrightarrow{H_2O} R-\overset{\overset{\displaystyle O}{\|}}{C}-\overset{..}{\overset{-}{N}}-\overset{+}{N}\equiv N: + H_3O^+
$$

(23.76)

conjugate acid of the acyl azide

Because the conjugate acid of the acyl azide is quite acidic (why?), it loses a proton from the adjacent nitrogen to the give the neutral acyl azide.

A reaction closely related to the Curtius rearrangement is the **Hofmann rearrangement** or **Hofmann hypobromite reaction**. The starting material for this reaction is a primary amide rather than an acyl azide. Treatment of an amide with bromine in base gives rise to a rearrangement.

$$Br_2 + 2\ NaOH + (CH_3)_3CCH_2\overset{\overset{\displaystyle O}{\|}}{-C}-NH_2 \longrightarrow (CH_3)_3CCH_2-NH_2 + O{=}C{=}O + 2\ NaBr + H_2O \quad (23.77)$$

3,3-dimethylbutanamide   **2,2-dimethyl-1-propanamine**
(neopentylamine)

The first step in the mechanism of the Hofmann rearrangement is ionization of the amide N—H (Sec. 22.5); the resulting anion is then brominated.

$$\qquad (23.78a)$$

an *N*-bromoamide

$$\qquad (23.78b)$$

(This reaction is analogous to the α-bromination of a ketone in base; Sec. 22.3B.) The *N*-bromoamide product is even more acidic than the amide starting material, and it too ionizes.

$$\qquad (23.78c)$$

The *N*-bromo anion then rearranges to an isocyanate.

an isocyanate

$$\qquad (23.78d)$$

Notice that the rearrangement steps of the Hofmann and Curtius reactions are conceptually identical; the only difference is the leaving group.

Hofmann:

$$R-N{=}C{=}O \quad (23.79)$$

Curtius:

Because the Hofmann rearrangement is carried out in aqueous base, the isocyanate cannot be isolated as it is in the Curtius rearrangement. It spontaneously hydrates to form a carbamate ion, which then decarboxylates to the amine product under the strongly basic reaction conditions. (See Study Guide Link 23.3.)

$$R-N{=}C{=}O + {}^-OH \longrightarrow RNH\overset{\overset{\displaystyle O}{\|}}{-C}-O^- \rightleftharpoons RNH_2 + CO_2 \xrightarrow{\ ^-OH\ } HCO_3^- \quad (23.80)$$

isocyanate   carbamate ion   amine   bicarbonate ion

Although the reaction of amines with CO_2 is reversible, formation of the amine in the Hofmann rearrangement is driven to completion by the reaction of hydroxide ion with CO_2 to form bicarbonate ion (or carbonate ion) under the strongly basic conditions of the reaction.

An interesting and very useful aspect of both the Hofmann and Curtius rearrangements is that they take place with complete *retention of stereochemical configuration* in the migrating alkyl group:

(23.81)

Hence, optically active carboxylic acid derivatives can be used to prepare optically active amines of known stereochemical configuration. (Eq. 23.81 is a further illustration of the fact that there is no simple correlation between a compound's absolute configuration and the sign of its optical rotation.)

The advantage of the Curtius rearrangement over the Hofmann rearrangement is that the Curtius reaction can be run under mild, neutral conditions, and the isocyanate can be isolated if desired. The disadvantage is that some acyl azides in the pure state can detonate without warning, and extreme caution is required in handling them. Amides, in contrast, are stable and easily handled organic compounds.

PROBLEMS

23.38   (a) Could *tert*-butylamine be prepared by the Gabriel synthesis? If so, write out the synthesis. If not, explain why.

(b) Propose a synthesis of *tert*-butylamine by another route.

23.39   Write a curved-arrow mechanism for each of the following reactions.

(a) ethyl isocyanate ($CH_3CH_2—N{=}C{=}O$) with ethanol to yield ethyl *N*-ethylcarbamate

(b) ethyl isocyanate with ethylamine to yield *N*,*N*'-diethylurea

23.40   What product is formed when 2-methylpropanamide is subjected to the conditions of the Hofmann rearrangement (a) in ethanol solvent? (b) in aqueous NaOH?

23.41   When hexanamide is subjected to the conditions of the Hofmann rearrangement, pentanamine (*A*) is obtained as expected. However, a significant by-product is *N*,*N*'-dipentylurea (*B*). Explain the origin of *B*. (*Hint:* Neither pentyl isocyanate nor pentanamine has appreciable solubility in aqueous base.)

E. Synthesis of Amines: Summary

The following amine syntheses have been covered in this and previous sections:

1. reduction of amides and nitriles with $LiAlH_4$ (Secs. 21.9B and 21.9C)
2. direct alkylation of amines (Sec. 23.7A)
3. reductive amination (Sec. 23.7B)
4. aromatic substitution reactions of anilines (Sec. 23.9)
5. direct amination of aryl halides (Sec. 23.11C)

6. Gabriel synthesis of primary amines (Sec. 23.11A)

7. Staudinger reaction (Sec. 23.11A)

8. reduction of nitro compounds (Sec. 23.11B)

9. Hofmann and Curtius rearrangements (Sec. 23.11D)

Methods 2, 3, 4, and 5 are used to prepare amines from other amines, and method 2 is really only useful for preparing quaternary ammonium salts. When an amide used in method 1 is prepared from an amine, this method, too, is a method for obtaining one amine from another. Methods 6–9, as well as nitrile reduction in method 1, are limited to the preparation of primary amines, and methods 1, 3, 8, and 9 can be used for obtaining amines from other functional groups.

PROBLEM

23.42 Show how 2-cyclopentyl-*N*,*N*-dimethylethanamine could be synthesized from each of the following starting materials.

(a) $\bigcirc$—CH_2—CO_2H (b) $\bigcirc$—CH_2—CN

(c) $\bigcirc$—CH_2CH_2—CO_2H (d) $\bigcirc$—CH_2—$CH{=}O$ (two ways)

23.12   USE AND OCCURRENCE OF AMINES

A. Industrial Use of Amines and Ammonia

Among the relatively few industrially important amines is hexamethylenediamine, $H_2N(CH_2)_6NH_2$, used in the synthesis of nylon-6,6 (Sec. 21.12A). Ammonia is also an important "amine" and is a key source of nitrogen in a number of manufacturing processes. In agricultural chemistry, for example, liquid ammonia itself and urea, which is made from ammonia and CO_2, are important nitrogen fertilizers. Ammonia is manufactured by the hydrogenation of N_2. Although it might not seem that the industrial synthesis of ammonia has anything to do with organic chemistry, the hydrogen used in its manufacture in fact comes from the cracking of alkanes (p. 219). Thus, the availability of ammonia is presently tied to the availability of hydrocarbons. However, there is significant interest in the development of methods for utilizing solar energy for *water splitting*—the conversion of water into H_2 and O_2. Should water splitting become practical, the production of ammonia would be completely uncoupled from the availability of petroleum.

B. Naturally Occurring Amines

Alkaloids   Among the many types of naturally occurring amines are the **alkaloids**: nitrogen-containing bases that occur naturally in plants. This simple definition encompasses a highly diverse group of compounds; the structures of a few alkaloids are shown in Fig. 23.4 on p. 1222. Because amines are the most common organic bases, it is not surprising that most alkaloids are amines, including heterocyclic amines. It is believed that the first alkaloid ever isolated and studied was morphine, discovered in 1805. Many alkaloids have biological activity (Fig. 23.4); others have no known activity, and their functions within the plants from which they come are, in many cases, obscure. Investigations dealing with the isolation, structure, and medicinal properties of alkaloids continue to be major research activities in organic chemistry.

PROBLEM

23.43   Illustrate the Brønsted basicity of (a) morphine and (b) mescaline (Fig. 23.4) by giving the structures of their conjugate acids.

FIGURE 23.4 Structures of some alkaloids. Each compound has at least one basic amine group.

quinine
(an antimalarial drug)

cocaine
(a stimulant of the central nervous system; induces euphoria; widely abused)

morphine (R = H)
codeine (R = CH$_3$)
(medically important analgesics)

nicotine
(the principal alkaloid from tobacco)

mescaline
(a hallucinogen from peyote cactus)

Hormones and Neurotransmitters   Epinephrine (adrenaline) is an amine secreted by both the adrenal medulla and sympathetic nerve endings; it is an example of a **hormone**—a compound that regulates the biochemistry of multicellular organisms, particularly vertebrates.

epinephrine

Epinephrine, for example, is associated with the "fight-or-flight" response to external stimuli; you might feel the effects of epinephrine secretion when you walk unprepared into your organic chemistry class and your instructor says, "Pop quiz today." The mechanisms by which hormones exert their effects are important research areas in contemporary biochemistry.

Norepinephrine, another amine, and acetylcholine, a quaternary ammonium ion, are examples of *neurotransmitters*.

norepinephrine

acetylcholine

Neurotransmitters are molecules that are involved in the communication between nerve cells or between nerve cells and their target organs. This communication occurs at cellular junctions called *synapses*. A nerve impulse is transmitted when a neurotransmitter is released from a nerve cell on one side of the synapse, moves by diffusion across the synapse, and binds to a protein receptor molecule of another nerve cell or a target organ on the other side. (The involvement of pi–cation interactions in the binding of acetylcholine to its receptor protein is discussed in Sec. 15.8C, p. 776.) This binding triggers either the transmission of the impulse down the nerve cell to the next synapse or a response by the target organ. Different neurotransmitters are involved in different parts of the nervous system.

Significant advances have occurred in understanding the chemistry that takes place in the human brain (*neurochemistry*). These advances are being made by teams of molecular biologists, biochemists, and organic chemists. It is conceivable that a deeper understanding of neurochemistry will lead to treatments for such widespread and tragic afflictions as Parkinson's disease and Alzheimer's disease. Sigmund Freud perhaps anticipated these developments when he wrote in 1930, "The hope of the future lies in organic chemistry. . . ."

KEY IDEAS IN CHAPTER 23

- Amines are classified as primary, secondary, or tertiary. Quaternary ammonium salts are compounds in which all four hydrogens of the ammonium ion are replaced by alkyl or aryl groups.

- Most amines undergo rapid inversion at nitrogen. This inversion interconverts a chiral amine and its mirror image.

- Simple amines are liquids with unpleasant odors. Amines of low molecular mass are miscible with water.

- The N—H stretching absorptions are the most important infrared absorptions of primary and secondary amines. In the NMR spectra of amines, the α-protons have chemical shifts in the δ 2.5–3.0 range. The ^{13}C NMR chemical shifts of amine α-carbons are in the δ 30–50 range.

- Basicity is one of the most important chemical properties of amines. The basicity of an amine is expressed by the pK_a of its conjugate-acid ammonium ion. Ammonium-ion pK_a values are affected by alkyl substitution on the nitrogen, the polar effects of nearby substituent groups, and resonance interaction of the amine's unshared electrons with an adjacent aromatic ring.

- Because amines are basic, they are also nucleophilic. The nucleophilicity of amines is important in many of their reactions, such as alkylation, imine formation, and acylation.

- Amines can serve as leaving groups. Quaternary ammonium hydroxides, when heated, undergo the Hofmann elimination, in which an amine is lost from the α-carbon and a proton is lost from a β-carbon atom. However, because amines are basic, they are relatively poor leaving groups.

- Amines are weak acids with pK_a values in the 32–40 range.

- The amino group is ortho, para directing in electrophilic aromatic substitution reactions. However, in many reactions of this type, such as nitration and sulfonation, protection of the amine nitrogen as an amide is required to prevent its protonation under the acidic reaction conditions.

- Treatment of primary amines with nitrous acid gives diazonium ions. Aliphatic diazonium ions decompose under the diazotization conditions to give complex mixtures of products. Aryldiazonium ions, however, can be used in a number of substitution reactions: the Sandmeyer reaction (substitution of N_2 by halide or cyanide groups) or hydrolysis (substitution with —OH). Because aryldiazonium ions are electrophiles, they react with activated aromatic compounds, such as aromatic amines and phenols, to give substituted azobenzenes, some of which are dyes. Secondary amines react with nitrous acid to give nitrosamines. Under acidic conditions, tertiary amines do not react, although aromatic tertiary amines give ring-nitrosated products.

• Amines can be synthesized from amides, nitriles, nitro compounds, and other amines, as summarized in

Sec. 23.11E. The Curtius and Hofmann rearrangements yield amines with one less carbon atom. The Curtius rearrangement can also be used to prepare isocyanates, which react with nucleophiles by addition across the C=N bond to give carbamic acid derivatives.

 REACTION REVIEW *For a summary of reactions discussed in this chapter, see the* Reaction Review *section of Chapter 23 in the* Study Guide and Solutions Manual.

ADDITIONAL PROBLEMS

23.44 Give the principal organic product(s) expected when *p*-chloroaniline or other compound indicated reacts with each of the following reagents.

(a) dilute HBr

(b) CH₃CH₂MgBr in ether

(c) NaNO₂, HCl, 0 °C

(d) *p*-toluenesulfonyl chloride

(e) product of part (c) with H₂O, Cu₂O, and excess Cu(NO₃)₂

(f) product of part (c) with CuBr

(g) product of part (c) with H₃PO₂

(h) product of part (c) with CuCN

(i) product of part (d) + NaOH, 25 °C

23.45 Give the principal organic product(s) expected when *N*-methylaniline reacts with each of the following reagents.

(a) Br₂

(b) benzoyl chloride

(c) benzyl chloride (excess), then dilute ⁻OH

(d) *p*-toluenesulfonic acid

(e) NaNO₂, HCl

(f) excess CH₃I, heat, then Ag₂O

(g) CH₃CH=O, NaBH(OAc)₃, and HOAc in ClCH₂CH₂Cl, then KOH

(h) chlorobenzene, K⁺ *t*-BuO⁻, and a Pd(0) catalyst

23.46 Give the principal organic product(s), if any, expected when isopropylamine or other compound indicated reacts with each of the following reagents.

(a) dilute H₂SO₄

(b) dilute NaOH solution

(c) butyllithium in THF, −78 °C

(d) acetyl chloride, pyridine

(e) NaNO₂, aqueous HBr, 0 °C

(f) acetone, H₂, Pd/C

(g) excess CH₃I, heat

(h) benzoic acid, 25 °C

(i) formaldehyde, NaBH₃CN, EtOH

(j) 2,4-dimethylchlorobenzene, K⁺ *t*-BuO⁻, and a Pd(0) catalyst

(k) product of part (g) + Ag₂O, then heat

(l) product of part (d) with LiAlH₄, then H₃O⁺, then ⁻OH

23.47 Give the structure of a compound that fits each description. (There may be more than one correct answer for each.)

(a) a chiral primary amine C₄H₇N with no triple bonds

(b) a chiral primary amine C₄H₁₁N

(c) two secondary amines, which, when treated with CH₃I, then Ag₂O and heat, give propene and *N,N*-dimethylaniline

(d) a compound C₄H₉N that reacts with NaBH(OAc)₃ and 1 equivalent of HOAc, then KOH, to give *N*-methyl-2-propanamine

23.48 Explain how you would distinguish the compounds within each set by a simple chemical test with readily observable results, such as solubility in acid or base, evolution of a gas, and so forth.

(a) *N*-methylhexanamide; 1-octanamine; *N,N*-dimethyl-1-hexanamine

(b) *p*-methylaniline, benzylamine, *p*-cresol, anisole

23.49 On the package insert for the drug *labetalol*, used in the control of blood pressure and hypertension, is given the following structure:

labetalol hydrochloride

(a) Labetalol is claimed to be a salt. Explain by giving a more detailed structure.

(b) What happens to labetalol·HCl when it is treated with one equivalent of NaOH at room temperature?

(c) What happens to labetalol when it is treated with an excess of aqueous NaOH and heat?

(d) What are the products formed when labetalol is treated with 6 *M* aqueous HCl and heat?

23.50 (a) Give the structure of cocaine (Fig. 23.4) as it would exist in 1 *M* aqueous HCl solution.

(b) What products would form if cocaine were treated with an excess of aqueous NaOH and heat?

(c) What products would form if cocaine were treated with an excess of concentrated aqueous HCl and heat?

23.51 Design a separation of a mixture containing the following four compounds into its pure components. Describe exactly what you would do and what you would expect to observe.

nitrobenzene, aniline, *p*-chlorophenol, and *p*-nitrobenzoic acid

23.52 How would the basicity of trifluralin, a widely used herbicide, compare with that of *N,N*-diethylaniline: much greater, about the same, or much less? Explain.

trifluralin

23.53 (a) When anthranilic acid is treated with NaNO$_2$ in aqueous HCl solution, and the resulting solution is treated

with *N,N*-dimethylaniline, a dye called *methyl red* is formed. Give the structure of methyl red.

anthranilic acid

(b) When an acidic solution of methyl red is titrated with base, the dye behaves as a diprotic acid with pK_a values of 2.3 and 5.0. The color of the methyl red solution changes very little as the pH is raised past 2.3, but, as the pH is raised past 5.0, the color of the solution changes dramatically from red to yellow. Explain.

23.54 Alizarin yellow R is an azo dye that changes color from yellow to red between pH 10.2 and 12.2.

alizarin yellow R

(a) Outline a synthesis of alizarin yellow R from aniline, salicylic acid (*o*-hydroxybenzoic acid), and any other reagents.

(b) Draw the structure of alizarin yellow R as it exists in its yellow form at pH = 9. Note that the conjugate acid of a diazo group has a pK_a near 5.

(c) Draw the structure of alizarin yellow R as it exists at pH > 12. Why does it change color?

23.55 Amanda Amine, an organic chemistry student, has proposed the reactions given in Fig. P23.55. Indicate in each case why the reaction would not succeed as written.

Figure P23.55

23.56 Outline a sequence of reactions that would bring about the conversion of aniline into each of the following compounds.

(a) benzylamine

(b) benzyl alcohol

(c) 2-phenylethanamine

(d) *N*-phenyl-2-butanamine

(e) *p*-chlorobenzoic acid

(f) diphenylamine

23.57 When *p*-aminophenol reacts with one molar equivalent of acetic anhydride, a compound acetaminophen (*A*, C$_8$H$_9$NO$_2$) is formed that dissolves in dilute NaOH. When *A* is treated with one equivalent of NaOH followed by ethyl iodide, an ethyl ether *B* is formed. What is the structure of acetaminophen? Explain your reasoning.

23.58 When 1,5-dibromopentane reacts with ammonia, among several products isolated is a water-soluble compound *A* that rapidly gives a precipitate of AgBr with acidic AgNO$_3$ solution. Compound *A* is unchanged when treated with dilute base, but treatment of *A* with concentrated NaOH and heat gives a new compound *B* (C$_{10}$H$_{19}$N) that decolorizes Br$_2$ in CCl$_4$. Compound *B* is identical to the product obtained from the reaction sequence shown in Fig. P23.58. Identify *A* and *B* and explain your reasoning.

23.59 Give an explanation for each of the following facts.

(a) The barrier to internal rotation about the *N*-phenyl bond in *N*-methyl-*p*-nitroaniline is considerably higher (42–46 kJ mol^{-1}, or 10–11 kcal mol^{-1}) than that in *N*-methylaniline itself (about 25 kJ mol^{-1}, or 6 kcal mol^{-1}).

(b) *Cis*- and *trans*-1,3-dimethylpyrrolidine rapidly interconvert.

(c) CH$_3$NH—CH$_2$—NHCH$_3$ is unstable in aqueous solution.

(d) The following compound exists as the enamine isomer shown rather than as an imine:

$$\text{H}_3\text{C}-\text{NH} \qquad \text{O}$$
$$\text{H}_3\text{C}-\text{C}=\text{CH}-\text{C}-\text{OC}_2\text{H}_5$$

(e) Diazotization of 2,4-cyclopentadien-1-amine gives a diazonium salt, which, unlike most aliphatic diazonium ions, is relatively stable and does not decompose to a carbocation.

23.60 Imagine that you have been given a sample of racemic 2-phenylbutanoic acid. Outline steps that would allow you to obtain pure samples of each of the following compounds from this starting material and any other reagents. (Note that enantiomeric resolutions are time-consuming. One resolution that would serve all five syntheses would be most efficient.)

(a)
$$\text{(R)-Ph}-\overset{\overset{\text{Et}}{|}}{\text{CH}}-\text{NH}-\overset{\overset{\text{O}}{||}}{\text{C}}-\text{OMe}$$

(b)
$$\text{(S)-Ph}-\overset{\overset{\text{Et}}{|}}{\text{CH}}-\overset{\overset{\text{O}}{||}}{\text{C}}-\text{OEt}$$

(c)
$$\text{(R,R)-Ph}-\overset{\overset{\text{Et}}{|}}{\text{CH}}-\text{NH}-\overset{\overset{\text{O}}{||}}{\text{C}}-\text{NH}-\overset{\overset{\text{Et}}{|}}{\text{CH}}-\text{Ph}$$

(d)
$$\text{meso-Ph}-\overset{\overset{\text{Et}}{|}}{\text{CH}}-\text{NH}-\overset{\overset{\text{O}}{||}}{\text{C}}-\text{NH}-\overset{\overset{\text{Et}}{|}}{\text{CH}}-\text{Ph}$$

(e)
$$\text{(R)-Ph}-\overset{\overset{\text{Et}}{|}}{\text{CH}}-\text{NH}-\langle \text{ring} \rangle-\text{CO}_2\text{H}$$

23.61 Show how the insecticide *carbaryl* can be prepared from methyl isocyanate, H$_3$C—N=C=O.

$$\langle \text{naphthalene} \rangle - \text{O}-\overset{\overset{\text{O}}{||}}{\text{C}}-\text{NHCH}_3$$

carbaryl

23.62 Offer an explanation for each of the following observations, including the structure of each product and the role of the quaternary ammonium salt.

(a) When sodium benzenethiolate, Na$^+$ PhS$^-$, is mixed with 1-bromooctane in water, no reaction takes place. However, when 1–2 mole percent tetrabutylammonium bromide, (CH$_3$CH$_2$CH$_2$CH$_2$)$_4$N$^+$ Br$^-$, is included in the reaction mixture, a product is formed readily.

(b) When phenylacetonitrile, Ph—CH$_2$—C≡N, is mixed with aqueous sodium hydroxide and 1,4-dibromobutane, a separate organic layer forms and no reaction takes place. However, when the three components were rapidly stirred in dichloromethane solvent with a few mole percent of tetrabutylammonium bromide [structure in part (a)], a compound with the formula C$_{12}$H$_{13}$N was formed in high yield.

(c) When morpholine (p. 1185) and bromobenzene are allowed to react in toluene solvent in the pres-

---

4-pentenoic acid $\xrightarrow{\text{SOCl}_2}$ piperidine $\xrightarrow{\quad}$ $\xrightarrow[\substack{\text{1) LiAlH}_4 \\ \text{2) H}_3\text{O}^+ \\ \text{3) dilute }^-\text{OH}}]{}$ *B*

Figure P23.58

ence of the catalyst Pd[P(*t*-Bu)$_3$]$_2$, a reaction takes place when aqueous NaOH is used as the base and 1 mole percent of cetyltrimethylammonium bromide, $CH_3(CH_2)_{14}CH_2\overset{+}{N}(CH_3)_3$ Br$^-$, is added to the reaction mixture.

23.63 A compound *A* (C$_{22}$H$_{27}$NO) is insoluble in acid and base but reacts with concentrated aqueous HCl and heat to give a clear aqueous solution from which, on cooling, benzoic acid precipitates. When the supernatant solution is made basic, a liquid *B* separates. Compound *B* is achiral. Treatment of *B* with benzoyl chloride in pyridine gives back *A*. Evolution of gas is not observed when *B* is treated with an aqueous solution of NaNO$_2$ and HCl. Treatment of *B* with excess CH$_3$I, then Ag$_2$O and heat, gives a compound *C*, C$_9$H$_{19}$N, plus styrene, Ph—CH=CH$_2$. Compound *C*, when treated with excess CH$_3$I, then Ag$_2$O and heat, gives a *single* alkene *D* that is identical to the compound obtained when cyclohexanone is treated with the ylid :$\overset{-}{C}H_2$—$\overset{+}{P}Ph_3$. Give the structure of *A* and explain your reasoning.

23.64 Three bottles *A*, *B*, and *C* have been found, each of which contains a liquid and is labeled "amine C$_8$H$_{11}$N." As an expert in amine chemistry, you have been hired as a consultant and asked to identify each compound. Compounds *A* and *B* give off a gas when they react with NaNO$_2$ and HCl at 0 °C; *C* does not. However, when the aqueous reaction mixture from the diazotization of *C* is warmed, a gas is evolved. Compound *A* is optically inactive, but when it reacts with (+)-tartaric acid, two isomeric salts with different physical properties are obtained. Titration of *C* with aqueous HCl reveals that its conjugate acid has a pK_a = 5.1. Oxidation of *C* with H$_2$O$_2$ (a reagent known to oxidize amino groups to nitro groups), followed by vigorous oxidation with KMnO$_4$, gives *p*-nitrobenzoic acid. Oxidation of *B* in a similar manner yields 1,4-benzenedicarboxylic acid (terephthalic acid), and oxidation of *A* yields benzoic acid. Identify compounds *A*, *B*, and *C*.

23.65 Complete the reactions given in Fig. P23.65 by giving the structure(s) of the major product(s). Explain how you arrived at your answers.

(a) NH$_3$ + HNO$_2$ $\longrightarrow$

(b)

(c)

(d)

(e)

(f)

CH$_3$CH$_2$CH$_2$CH$_2$—NH$_2$ + H$_2$C—CH$_2$ (excess) $\xrightarrow{\text{H}_2\text{O}}$

(g)

Et$_2$NH + (CH$_3$)$_2$C—CH$_2$ $\longrightarrow$

(h)

(i)

+ HN⟨⟩ $\longrightarrow$ (Give the stereochemistry as well as the structure of the product.)

(j) H$_2$C=C—CH$_2$Br + EtNH$_2$ (excess) $\longrightarrow$ (a compound with 5 carbons)
　　　|
　　　Br

(k)

—NO$_2$ + 2 CH$_3$CH$_2$CH$_2$CH=O + H$_2$ $\xrightarrow{\text{Pd/C}}$

Figure P23.65

23.66 Outline a synthesis for each of the following compounds from the indicated starting materials and any other reagents. The starting material for the compounds in parts (a) through (e) is pentanoic acid.

(a) *N*-methyl-1-hexanamine

(b) pentylamine

(c) *N,N*-dimethyl-1-pentanamine

(d) butylamine

(e) hexylamine

(f)

$$PhCH_2\overset{\overset{\displaystyle CH_3}{|}}{\underset{\underset{\displaystyle CH_3}{|}}{N^+}}CH_2CH_2CH_2CH_3 \quad Br^- \text{ from}$$
butyraldehyde

(g) *N*-ethyl-3-phenyl-1-propanamine from toluene

(h) 2-pentanamine from diethyl malonate

(i) isobutylamine from acetone

(j) isopentylamine from acetone

(k) *m*-chlorobromobenzene from nitrobenzene

(l) *p*-chlorobromobenzene from nitrobenzene

(m) *p*-methoxybenzonitrile from phenol

(n) (*S*)-CH₃CHCH₂NH₂ from (*R*)-CH₃CHOH
        | |
        D   D

(o)

structure from H₃C—C(=O)—CH₂CH₂—CO₂H   **levulinic acid**

23.67 (a) An amine *A* has an EI mass spectrum with a base peak at $m/z = 72$. An amine *B* has an EI mass spectrum with a base peak at $m/z = 58$. One of the amines is 2-methyl-2-heptanamine, and the other is *N*-ethyl-4-methyl-2-pentanamine. Which is which? (*Hint:* A major fragmentation mechanism of amines in EI and CI mass spectra is α-cleavage; Eq. 12.29, p. 600.)

(b) Tributylamine has a CI mass spectrum with a strong M + 1 peak and one other major peak resulting from α-cleavage. At what m/z value does this peak occur?

23.68 In the NMR spectrum of a concentrated (4.5 *M*) aqueous solution of methylamine, the methyl group appears as a quartet when the solution pH = 1. At intermediate pH, the methyl group appears as a broad line. At pH = 9 the methyl group is observed as a single sharp line. Explain these observations.

23.69 Aniline has a UV spectrum with peaks at $\lambda_{max} = 230$ nm ($\epsilon = 8600$) and 280 nm ($\epsilon = 1430$). In the presence of dilute HCl, the spectrum of aniline changes dramatically: $\lambda_{max} = 203$ ($\epsilon = 7500$) and 254 ($\epsilon = 160$). This spectrum is nearly identical to the UV spectrum of benzene. Account for the effect of acid on the UV spectrum of aniline.

23.70 Imagine that you have samples of the following four isomeric amines, but you don't know which is which. Explain how you could use proton NMR to distinguish among them.

NH₂
|
PhCH₂CHCH₃   PhCH₂N(CH₃)₂
      A   *B*

NH₂
|
PhCHCH₂CH₃   PhCH₂CH₂NHCH₃
      C   *D*

23.71 In the warehouse of the company Tumany Amines, Inc., three unidentified compounds have been found. The president of the company, Wotta Stench, has hired you to identify them from their spectra.

(a) Compound *A* (C₉H₁₃NO): IR spectrum: 3360, 3280 cm⁻¹ (doublet); 1611 cm⁻¹; no carbonyl absorption. NMR spectrum shown in Fig. P23.71a.

(b) Compound *B* (C₆H₁₆N₂): IR spectrum: 3281 cm⁻¹. NMR spectrum: δ 1.1 (8*H*, t, *J* = 7 Hz), δ 2.66 (4*H*, q, *J* = 7 Hz), δ 2.83 (4*H*, s). (*Hint:* The triplet at δ 1.1 conceals another broad resonance that contributes to the integral.)

(c) Compound *C* (C₆H₁₃N): IR spectrum: 3280, 1653, 898 cm⁻¹. NMR spectrum in Fig. 23.71b.

23.72 Propose a structure for the compound *A* (C₆H₁₅O₂N) that is unstable in aqueous acid and has the following NMR spectra:

Proton NMR: δ 2.30 (6*H*, s); δ 2.45 (2*H*, d, *J* = 6 Hz);
                      δ 3.27 (6*H*, s); δ 4.50 (1*H*, t, *J* = 6 Hz)
¹³C NMR: δ 46.3, δ 53.2, δ 68.8, δ 102.4

23.73 (a) Propose a structure for an amine *A* (C₄H₉N), which liberates a gas when treated with NaNO₂ and HCl. The ¹³C NMR spectrum of *A* is as follows, with attached protons in parentheses: δ 14(2), δ 34.3(2), δ 50.0(1).

(b) Propose a structure for an amine *B* (C₄H₉N), which does *not* liberate a gas when treated with NaNO₂ and HCl, and has IR absorptions at 917 cm⁻¹, 990 cm⁻¹, and 1640 cm⁻¹, as well as N—H absorption at 3300 cm⁻¹. The ¹³C NMR spectrum of *B* is as follows: δ 36.0, δ 54.4, δ 115.8, δ 136.7.

23.74 Draw a curved-arrow mechanism for each of the rearrangement reactions given in Fig. P23.74.

23.75 Provide a curved-arrow mechanism for the example of the *Bayliss–Hilman reaction* shown in Fig. P23.75 on p. 1230. Be sure that the role of the triethylamine catalyst is clearly indicated. (*Hint:* The role of the catalyst is *not* to remove the α-proton of the ester; this proton is not acidic. Why?)

23.76 A chemist, Mada Meens, treated ammonia with pentanal in the presence of hydrogen gas and a catalyst in the expectation of obtaining 1-pentanamine by reductive amination. In addition to 1-pentanamine, however, she also obtained dipentylamine and tripentylamine (see Fig. P23.76, p. 1230). Explain how the by-products are formed.

Figure P23.71 (a) The NMR spectrum for compound *A*, Problem 23.71(a). (b) The NMR spectrum for compound *C*, Problem 23.71(c). Integrals are shown in red above their respective resonances.

Figure P23.74

23.77 Around 1912, Swiss chemist Richard Willstätter (who subsequently was awarded the 1915 Nobel Prize in Chemistry) treated diamine *A* with methyl iodide and then with Ag$_2$O and heat, whereupon a hydrocarbon *B*, C$_8$H$_8$, distilled from the reaction mixture. Compound *B* reacted rapidly with Br$_2$ under mild conditions. Treatment of compound *C* in the same way gave a hydrocarbon *D*, C$_6$H$_6$, which did not react with Br$_2$.

Identify the two hydrocarbons *B* and *D*, and explain their very different behavior toward Br$_2$. (Willstätter concluded from these observations that compound *D* could not be an alkene.)

23.78 Explain the transformations shown in Fig. P23.78 by showing relevant intermediates, providing analogies to known reactions, and, where appropriate, giving curved-arrow mechanisms.

23.79 Explain the fact that the amines shown in Fig. P23.79, despite their similarities in structure, have considerably different basicities. (*Hint:* Make a model of compound *B*. Look at the relationship of the nitrogen unshared electron pair to the *p* orbitals of the benzene ring.)

23.80 (See Fig. P23.80.) Amide *A*, δ-valerolactam, is a typical amide with a conjugate-acid pK_a of 0.8. The two cyclic

tertiary amines *B* and *C* also have typical conjugate-acid pK_a values. In contrast, the conjugate-acid pK_a of amide *D* is unusually high for an amide, and it hydrolyzes much more rapidly than other amides. Draw the structure of the conjugate acid of amide *D*, and suggest a reason for both its unusual pK_a and its rapid hydrolysis.

23.81 The Staudinger *ligation* is a method that has found increasing use in the growing field of chemical biology for labeling biomolecules, such as proteins, with fluorescent probes. Proteins possessing an azide functional group can be *ligated* onto a phosphine bearing the probe. (See Fig. P23.81.)

(a) Using your knowledge of the Staudinger *reaction* (Sec. 23.11A), propose a mechanism that accounts for the formation of a key intermediate in the Staudinger *ligation*, shown below.

(b) Instead of reacting with water to release the free amine, as in the Staudinger reaction (Eq. 23.57), the probe-labeled protein product is formed. Draw a curved-arrow mechanism for the conversion of the intermediate in part (a) to the probe-labeled protein product in Fig. P23.81.

(c) In the absence of the methyl ester functional group, what would be the product of the reaction in the presence of water?

Figure P23.75

Figure P23.76

Figure P23.78 (*continues*)

(d)

(e)

Figure P23.78 (*continued*)

	A	B	C
conjugate-acid pK_a:	5.20	7.79	10.95

Figure P23.79

	A	B	C	D
conjugate-acid pK_a:	0.8	10.65	10.95	5.33

Figure P23.80

Figure P23.81

Carbohydrates

Because of their abundance in the natural world and their importance to living things, sugars have been the subject of intense investigation since the earliest days of scientific inquiry. Scientists refer to sugars and their derivatives as *carbohydrates*. As this name implies, most of the common sugars have molecular formulas that fit a "hydrate-of-carbon" pattern—that is, a formula of the form $C_n(H_2O)_m$. For example, sucrose (table sugar) has the formula $C_{12}(H_2O)_{11}$ or $C_{12}H_{22}O_{11}$, and both glucose and fructose (sugars prevalent in honey) have the formula $C_6(H_2O)_6$ or $C_6H_{12}O_6$. This hydrate-of-carbon pattern is more than an apparent relationship. Anyone familiar with the conversion of table sugar into carbon by concentrated H_2SO_4 (or anyone who has made caramel sauce, a less extreme example of the same phenomenon) has witnessed in practice the dehydration of carbohydrates:

> A quarter pound of nice white lump sugar put into a breakfast cup with the smallest possible dash of boiling water and then the addition of plenty of oil of vitriol [H_2SO_4] is a truly wonderful spectacle, and more instructive than much reading, to see the white sugar turn black, then boil spontaneously, and now, rising out of the cup in solemn black, it heaves and throbs as the oil of vitriol continues its work in the lower part of the cup, emitting volumes of steam. . . . [J.W. Pepper, *Scientific Amusements for Young People,* 1863]

As the result of a more modern understanding of their structures, **carbohydrates** are now defined as aldehydes and ketones containing a number of hydroxy groups on an unbranched carbon chain, as well as their chemical derivatives.

Two common carbohydrate structures:

$$O{=}CH-\underset{\underset{OH}{|}}{CH}-\underset{\underset{OH}{|}}{CH}-\underset{\underset{OH}{|}}{CH}-\underset{\underset{OH}{|}}{CH}-CH_2OH \qquad HOCH_2-\underset{\underset{O}{\|}}{C}-\underset{\underset{OH}{|}}{CH}-\underset{\underset{OH}{|}}{CH}-\underset{\underset{OH}{|}}{CH}-CH_2OH$$

Less precisely, but more descriptively, carbohydrate chemistry can be regarded as the chemistry of sugars and their derivatives.

Carbohydrates are among the most abundant organic compounds on Earth. In polymerized form as cellulose, carbohydrates account for 50–80% of the dry weight of plants. Carbohydrates are a major source of food; sucrose (table sugar) and lactose (milk sugar) are examples. Even the shells of arthropods such as lobsters consist largely of carbohydrate.

The study of carbohydrates relies heavily on the principles of stereochemistry (Chapter 6) and on the conformational aspects of cyclohexane rings (Chapter 7). Therefore, molecular models should be very helpful as you study the material in this chapter.

24.1 CLASSIFICATION AND PROPERTIES OF CARBOHYDRATES

Carbohydrates can be classified in several ways based on structure. One type of classification is based on the type of carbonyl group in the carbohydrate. A carbohydrate with an aldehyde carbonyl group is called an **aldose**; a carbohydrate with a ketone carbonyl group is called a **ketose**. Carbohydrates can also be classified by the number of carbon atoms they contain. A six-carbon carbohydrate is called a **hexose**, and a five-carbon carbohydrate is called a **pentose**. These two classifications can be combined: an **aldohexose** is an aldose containing six carbon atoms, and a **ketopentose** is a ketose containing five carbon atoms. A ketose can also be indicated with the suffix *ulose*; thus, a five-carbon ketose is also called a **pentulose**. These classifications are illustrated by the following examples.

$$HOCH_2-\underset{\underset{OH}{|}}{CH}-\underset{\underset{OH}{|}}{CH}-\underset{\underset{OH}{|}}{CH}-\underset{\underset{OH}{|}}{CH}-CH{=}O \qquad HOCH_2-\underset{\underset{OH}{|}}{CH}-\underset{\underset{OH}{|}}{CH}-\underset{\underset{O}{||}}{C}-CH_2OH$$

an *aldose* (aldehyde carbonyl group)
a *hexose* (six carbon atoms)
an *aldohexose* (combination of
the above classifications)

a *ketose* (ketone carbonyl group)
a *pentose* (five carbon atoms)
a *ketopentose* or *pentulose* (combination
of the above classifications)

Another type of classification scheme is based on the hydrolysis of certain carbohydrates to simpler carbohydrates. **Monosaccharides** cannot be converted into simpler carbohydrates by hydrolysis. Glucose and fructose are examples of monosaccharides. Sucrose, however, is a **disaccharide**—a compound that can be converted by hydrolysis into two monosaccharides.

$$\text{sucrose } (C_{12}H_{22}O_{11}) + H_2O \xrightarrow{\text{acid or enzymes}} \underbrace{\text{glucose } (C_6H_{12}O_6) + \text{fructose } (C_6H_{12}O_6)}_{\text{monosaccharides}} \qquad (24.1)$$

a disaccharide

Likewise, **trisaccharides** can be hydrolyzed to three monosaccharides, **oligosaccharides** to a "few" monosaccharides, and **polysaccharides** to a very large number of monosaccharides.

Because of their many hydroxy groups, carbohydrates are very soluble in water. The ease with which a large amount of table sugar dissolves in water to make syrup is an example from common experience of carbohydrate solubility. Carbohydrates are virtually insoluble in apolar aprotic solvents.

24.2 FISCHER PROJECTIONS

Almost all of the carbohydrates are chiral molecules, and most have more than one asymmetric carbon. Many carbohydrates have several contiguous asymmetric carbons in an unbranched chain. For example, the aldoses have four such carbons, which are indicated by asterisks in the following structure:

$$HOCH_2-\overset{*}{\underset{\underset{OH}{|}}{CH}}-\overset{*}{\underset{\underset{OH}{|}}{CH}}-\overset{*}{\underset{\underset{OH}{|}}{CH}}-\overset{*}{\underset{\underset{OH}{|}}{CH}}-CH{=}O$$

aldohexoses
four asymmetric carbons (*)

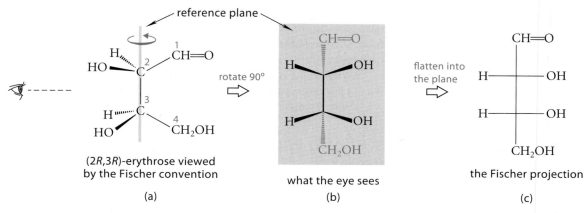

FIGURE 24.1 How to derive a Fischer projection for an aldotetrose. (a) The eclipsed conformation used to derive the projection, with the reference plane perpendicular to the page. (b) The view of the conformation in (a) as seen by the eye. The reference plane is now the plane of the page. The groups behind the plane are shown in gray. (c) The Fischer projection. The asymmetric carbons are located at the intersection of vertical and horizontal lines.

To show the stereochemistry of such molecules, we could use line-and-wedge structures. However, a simpler system of showing stereochemistry was developed by the German chemist Emil Fischer, whose landmark work on the structure of glucose we'll take up in Sec. 24.10. Fischer developed a way to represent three-dimensional structures on a two-dimensional surface (paper or blackboard) that does not require the use of wedges and dashed wedges. Such structures are called **Fischer projections**. We'll use Fischer projections extensively in this chapter. In this section, you'll learn how to draw and manipulate Fischer projections.

To illustrate the process of drawing a Fischer projection, we'll use the 2*R*,3*R* enantiomer of erythrose, an aldotetrose with two asymmetric carbons (Fig. 24.1). *You should follow this discussion with a molecular model.* To represent this molecule in a Fischer projection, arrange the molecule in an *all-eclipsed conformation* about the C2–C3 bond—the bond connecting the two asymmetric carbons. View the molecule as shown by the eye in Fig. 24.1a. Next, impose a reference plane containing the C2–C3 bond on the molecule. (This plane will ultimately be the plane of the page.) The plane should be oriented so that the other two carbon–carbon bonds are *receding behind this plane*, and the bonds to the OH and H groups *are emerging in front of this plane*. The view seen by the eye is shown in Fig. 24.1b. Finally, we project this structure onto the plane—that is, flatten it into the page—to give the Fischer projection. *The asymmetric carbons themselves are not drawn*, but are assumed to be located at the intersections of vertical and horizontal bonds (Fig. 24.1c). (As one student pointed out, the Fischer projection is the way that the molecule would look if we were to put it on the floor and step on it!)

The following five rules summarize the conventions used in the construction of Fischer projections.

1. A Fischer projection is based on an eclipsed molecular conformation.
2. The bonds connecting the asymmetric carbons are arranged in a vertical line.
3. The asymmetric carbons are located at the intersections of vertical and horizontal bonds and are not drawn explicitly.
4. Vertical bonds to the asymmetric carbons recede behind the page, away from the observer in the three-dimensional model.
5. Horizontal bonds to the asymmetric carbons emerge from the page, toward the observer in the three-dimensional model.

In other words, a flat Fischer projection can convey three-dimensional information if it requires a specific viewing mode (rules 1 and 2) and a strict adherence to a convention about the relationship of the horizontal and vertical bonds to the plane of the page (rules 4 and 5).

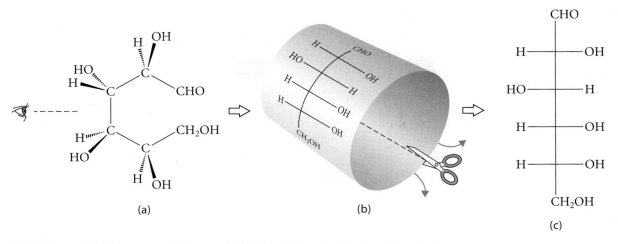

FIGURE 24.2 How to derive a Fischer projection for d-glucose, the naturally occurring enantiomer. (a) The eclipsed conformation is viewed with the chain of carbons oriented vertically and curving away from the observer, and the horizontal bonds projecting toward the observer. (b) This view is projected onto an imaginary curved cylinder. (c) Mentally cutting the cylinder and flattening it gives the Fischer projection.

Rule 3 alerts us to the fact that we are (or might be) dealing with a Fischer projection and not an "ordinary" Lewis structure.

To derive the Fischer projection of a molecule with more than two asymmetric carbons, a molecule is first placed (or imagined) in an eclipsed conformation in which the chain of asymmetric carbons is vertical and curving away from the observer, as if this chain were drawn on a convex surface such as a paper cylinder. This conformation is illustrated for the naturally occurring enantiomer of glucose, an aldohexose, in Fig. 24.2a. The horizontal bonds project toward the observer from this surface. All bonds are then projected onto this surface (Fig. 24.2b). Mentally cutting the cylinder and flattening it gives the Fischer projection (Fig. 24.2c).

The use of an eclipsed conformation to derive a Fischer projection does *not* mean that the molecule actually has such a conformation. As you've learned, most molecules actually exist in staggered conformations (Sec. 2.3). Fischer projections convey *no* information about molecular conformations. Their only purpose is to show the *absolute configuration* of each asymmetric carbon.

To derive a three-dimensional model of a molecule from its Fischer projection, reverse the process just described. Always remember that the vertical bonds in the Fischer projection extend *away from* the observer, and the horizontal bonds extend *toward* the observer.

$$
\begin{array}{c}
\text{Fischer projection} \qquad \text{the corresponding line-and-wedge structures}
\end{array}
\qquad (24.2)
$$

For any given molecule, *several valid Fischer projections can be drawn*. It is useful to be able to draw the different Fischer projections of a molecule without going back and forth to a three-dimensional model. For this purpose, some rules for manipulation of Fischer projections are helpful. Be sure to use models to convince yourself that these rules are valid.

1. *A Fischer projection may be turned 180° in the plane of the paper.*
    By this rule, the following two Fischer projections represent the same stereoisomer.

$$\text{(24.3)}$$

This manipulation is allowed because it leaves horizontal bonds horizontal and vertical bonds vertical; therefore, it does not alter the meaning of the Fischer projection.

2. *A Fischer projection may* not *be turned 90° in the plane of the page.*

$$\text{(24.4)}$$

This manipulation violates the Fischer convention that all asymmetric carbons should be aligned vertically. When we attempt this operation on a Fischer projection containing a *single* asymmetric carbon, a further problem becomes evident:

$$\text{(24.5)}$$

The 90° rotation exchanges horizontal and vertical groups and, in the process, *interconverts the original structure into its enantiomer.* This is disastrous, because the whole idea of Fischer projections is to convey stereochemical information.

3. *A Fischer projection may not be lifted from the plane of the paper and turned over.*

$$\text{(24.6)}$$

This rule has a similar rationale to rule 2. By flipping the structure over as shown, the stereochemistry of each asymmetric carbon is reversed.

4. *The three groups at either end of a Fischer projection may be interchanged in a cyclic permutation. That is, all three groups can be moved at the same time in a closed loop so that each occupies an adjacent position.*

$$\underset{\substack{\text{CH}=\text{O} \\ \text{H}\!-\!\!\!-\!\text{OH} \\ \text{CH}_2\text{OH}}}{} \xrightarrow{\text{ALLOWED}} \underset{\substack{\text{CH}=\text{O} \\ \text{HOCH}_2\!-\!\!\!-\!\text{H} \\ \text{OH}}}{} \qquad (24.7)$$

$$\underset{\substack{\text{CH}_2\text{OH} \\ \text{H}\!-\!\!\!-\!\text{OH} \\ \text{HO}\!-\!\!\!-\!\text{H} \\ \text{CH}_2\text{OH}}}{} \xrightarrow{\text{ALLOWED}} \underset{\substack{\text{CH}_2\text{OH} \\ \text{H}\!-\!\!\!-\!\text{OH} \\ \text{HOCH}_2\!-\!\!\!-\!\text{OH} \\ \text{H}}}{} \xrightarrow{\text{ALLOWED}} \underset{\substack{\text{OH} \\ \text{HOCH}_2\!-\!\!\!-\!\text{H} \\ \text{HOCH}_2\!-\!\!\!-\!\text{OH} \\ \text{H}}}{} \qquad (24.8)$$

Fischer projections of the same molecule

This operation is equivalent to an internal rotation. This point should become clear if you convert any one of the structures in Eq. 24.8 into a model. Leaving the model in an eclipsed conformation, carry out an internal rotation of 120° about the central carbon–carbon bond as shown by the colored arrows in Eq. 24.8, and then form a new Fischer projection from the resulting structure. Each 120° internal rotation is equivalent to one cyclic permutation described by rule 4. A different Fischer projection of the same molecule results from each different eclipsed conformation.

5. *An interchange of any two of the groups bound to an asymmetric carbon changes the configuration of that carbon.*
   (Verify this rule with models.) This rule applies not only to Fischer projections, but also to three-dimensional models as well. It follows that a *pair* of interchanges leaves the configuration of the carbon unaffected; the first interchange changes the configuration, and the second interchange changes the configuration back to the original. In fact, the cyclic permutation in rule 4 is equivalent to a pair of interchanges.

It is particularly easy to recognize enantiomers and meso compounds from the appropriate Fischer projections, because planes of symmetry in the actual molecules reduce to lines of symmetry in their projections.

mirror line

$$\underset{R}{\underset{\substack{\text{CH}=\text{O} \\ \text{H}\!-\!\!\!-\!\text{OH} \\ \text{CH}_2\text{OH}}}{}} \qquad \underset{S}{\underset{\substack{\text{CH}=\text{O} \\ \text{HO}\!-\!\!\!-\!\text{H} \\ \text{CH}_2\text{OH}}}{}} \qquad \qquad \underset{\substack{\text{CO}_2\text{H} \\ \text{H}\!-\!\!\!-\!\text{OH} \\ \text{H}\!-\!\!\!-\!\text{OH} \\ \text{CO}_2\text{H}}}{} \text{line of symmetry} \qquad (24.9)$$

enantiomers

a meso compound

The *R,S* system can be applied to a Fischer projection without using a model. If the group of lowest priority is in either of the two vertical positions, simply apply the *R,S* priority rules to the remaining three groups.

$$\underset{\substack{4 \\ \boxed{\text{H}} \\ {}_1\boxed{\text{HO}}\!\!-\!\!\oplus\!\!-\!\boxed{\text{CH}_2\text{OH}}{}_3 \\ \boxed{\text{HO}\!-\!\!\!-\!\text{H}} \\ \text{CH}_2\text{OH} \\ 2}}{} \quad \Longrightarrow \quad \underset{\substack{\boxed{4} \\ \boxed{1}\!-\!\!\!-\!\boxed{3} \\ \boxed{2}}}{} \qquad (24.10)$$

asymmetric carbon being examined

group of lowest priority is in a vertical position

counterclockwise order of descending priorities; therefore, the configuration is *S*

This method works because, if the lowest-priority group is in a vertical position in the Fischer projection, it is oriented away from the observer as required for application of the priority rules.

If the lowest-priority group is in a horizontal position, proceed in the same manner; but, since the molecule is being viewed incorrectly for assigning configuration, reverse the assignment. (This is a rare situation in which two wrongs make a right!)

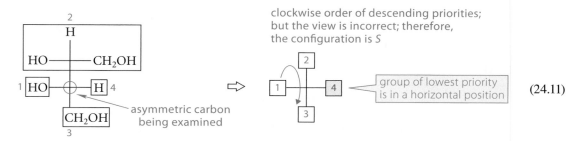

(24.11)

PROBLEMS

24.1   Draw at least two Fischer projections for each of the following molecules.

(a)
$$HO_2C \overset{R}{-}CH \overset{S}{-}CH \overset{S}{-}CH -CH_2OH$$
         | | |
         OH   OH   OH

(b) (S)-2-butanol

24.2   Indicate whether the structures in each of the following pairs are enantiomers, diastereomers, or identical molecules.

(a)

(b)

24.3   Which of the following are Fischer projections of a meso compound?

| A | B | C | D |

24.3   STRUCTURES OF THE MONOSACCHARIDES

A. Stereochemistry and Configuration

We'll consider the stereochemistry of carbohydrates by focusing largely on the aldoses with six or fewer carbons. The aldohexoses have four asymmetric carbons and therefore exist as 2^4 or 16 possible stereoisomers. These can be divided into two enantiomeric sets of eight diastereomers.

$$HOCH_2-CH-CH-CH-CH-CH=O$$
$$| \quad | \quad | \quad |$$
$$OH \quad OH \quad OH \quad OH$$

aldohexoses
$2^4 = 16$ stereoisomers

Similarly, there are two enantiomeric sets of four diastereomers (eight stereoisomers total) in the aldopentose series. Each diastereomer is a *different carbohydrate* with *different properties*, known by a *different name*. The aldoses with six or fewer carbons are given in Fig. 24.3 on p. 1240 as Fischer projections.

Each of the monosaccharides in Fig. 24.3 has an enantiomer. For example, the two enantiomers of glucose have the following structures:

enantiomers of glucose

It is important to specify the enantiomers of carbohydrates in a simple way. Suppose you have a model of one of these glucose enantiomers in your hand; how would you explain to someone who cannot see the model (for example, over the telephone) which enantiomer you are holding? You could use the R,S system to describe the configuration of one or more of the asymmetric carbon atoms. A different system, however, was in use long before the R,S system was established. The **D,L system**, which came from proposals made in 1906 by a New York University chemist, M. A. Rosanoff, is still used today for this purpose. As this system is applied to carbohydrates, the configuration of a carbohydrate enantiomer is specified by applying the following conventions:

1. The configuration of the naturally occurring triose (+)-glyceraldehyde is designated as D, and the configuration of its enantiomer, (−)-glyceraldehyde, is designated as L. The OH group in the D stereoisomer is on the right when the CH=O group is in the upper vertical position and the CH_2OH group in the lower vertical position.

D-(+)-glyceraldehyde   L-(−)-glyceraldehyde

The basis for the use of the letters D and L was simply the fact that the D stereoisomer of glyceraldehyde is dextrorotatory and the L stereoisomer is levorotatory. As with the R,S system, however, there is no *general* correlation between configuration and the sign of optical rotation.

2. The other aldoses or ketoses are written in a Fischer projection with their carbon atoms in a straight vertical line, and the carbons are numbered consecutively as they would be in systematic nomenclature, so the carbonyl carbon receives the lowest possible number.

3. The *asymmetric carbon of highest number* is designated as a *reference carbon*. If this carbon has the H, OH, and CH_2OH groups in the same relative configuration

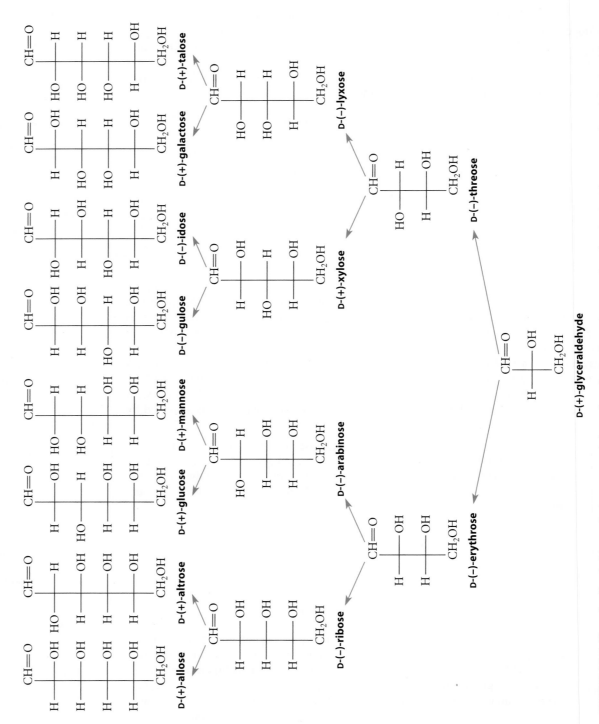

FIGURE 24.3 The D family of aldoses. Each compound shown here has an enantiomer in the L family. The blue arrows show how the aldoses are related by the Kiliani–Fischer synthesis (Sec. 24.9).

as the same three groups of D-glyceraldehyde, the carbohydrate is said to have the D configuration. If this carbon has the same configuration as L-glyceraldehyde, then the carbohydrate is said to have the L configuration.

The application of these conventions is illustrated in Study Problem 24.1.

STUDY PROBLEM 24.1

Determine whether the following carbohydrate derivative, shown in Fischer projection, has the D or L configuration.

$$
\begin{array}{c}
\text{OH}\\
\text{HOCH}_2\!\!-\!\!\!\overset{\displaystyle|}{\underset{\displaystyle|}{\rule{0pt}{0pt}}}\!\!-\!\!\text{H}\\
\text{H}\!\!-\!\!\!\overset{\displaystyle|}{\underset{\displaystyle|}{\rule{0pt}{0pt}}}\!\!-\!\!\text{OH}\\
\text{HO}\!\!-\!\!\!\overset{\displaystyle|}{\underset{\displaystyle|}{\rule{0pt}{0pt}}}\!\!-\!\!\text{H}\\
\text{CO}_2\text{H}
\end{array}
$$

SOLUTION   First redraw the structure so that the carbon with the lowest number in substitutive nomenclature—the carbon of the carboxylic acid group—is at the top. This can be done by rotating the structure 180° in the plane of the page. Then carry out a cyclic permutation of the three groups at the bottom so that all carbons lie in a vertical line. Recall from Sec. 24.2 that these are allowed manipulations of Fischer projections.

Finally, compare the configuration of the highest-numbered asymmetric carbon with that of D-glyceraldehyde. Because the configuration is different, the molecule has the L configuration.

D-glyceraldehyde   therefore L configuration

The monosaccharides shown in Fig. 24.3 constitute the D family of enantiomers. Each of the compounds in this figure has an enantiomer with the L configuration. This figure illustrates the point that *there is no general correspondence between configuration and the sign of the optical rotation* (see Sec. 6.3C). For example, some D-aldoses have positive rotations, but others have negative rotations. Also, *there is no simple relationship between the D,L system and the R,S system.* The R,S system is used to specify the configuration of *each* asymmetric carbon atom in a molecule, but *the D,L system specifies a particular enantiomer of a molecule that might contain many asymmetric carbons.*

An annoying aspect of the D,L system is that each diastereomer is given a different nonsystematic name. This is one reason that the D,L system has been generally replaced with the R,S system, which can be used with systematic nomenclature. Nevertheless, the common names of many carbohydrates are so well entrenched that they remain important. Although use of the D,L system is fairly straightforward for carbohydrates and amino acids, it has been virtually abandoned for other compounds.

A few of the aldoses in Fig. 24.3 are particularly important, and you will find it helpful to learn their structures. D-Glucose, D-mannose, and D-galactose are the most important aldohexoses because of their wide natural occurrence. The structures of the latter two aldoses are easy to remember once the structure of glucose is learned, because their configurations differ in a simple way from the configuration of glucose. D-Glucose and D-mannose differ in configuration only at carbon-2; D-glucose and D-galactose differ only at carbon-4. Diastereomers that differ in configuration at only one of several asymmetric carbons are called **epimers**. Hence, D-glucose and D-mannose are *epimeric* at carbon-2, whereas D-glucose and D-galactose are *epimeric* at carbon-4.

D-Ribose is a particularly important aldopentose; its structure is easy to remember because all of its —OH groups are on the right in the standard Fischer projection.

D-Fructose is an important naturally occurring ketose:

$$
\begin{array}{c}
CH_2OH \\
| \\
C{=}O \\
HO{-\!\!\!-}H \\
H{-\!\!\!-}OH \\
H{-\!\!\!-}OH \\
CH_2OH
\end{array}
$$

D-fructose

Notice that carbons 3, 4, and 5 of D-fructose have the same stereochemical configuration as carbons 3, 4, and 5 of D-glucose.

PROBLEMS

24.4   Classify each of the following aldoses as D or L.

(a)

$$
\begin{array}{c}
CH{=}O \\
H{-\!\!\!-}OH \\
HO{-\!\!\!-}H \\
H{-\!\!\!-}OH \\
H_3C{-\!\!\!-}H \\
OH
\end{array}
$$

6-deoxyglucose

(b) the glucose enantiomer with the *R* configuration at carbon-3.

24.5   Which pair of the following aldoses are epimers and which pair are enantiomers?

$$
\begin{array}{ccc}
\begin{array}{c}
CH{=}O \\
H{-}OH \\
HO{-}H \\
H{-}OH \\
H{-}CH_2OH \\
OH
\end{array}
&
\begin{array}{c}
OH \\
O{=}CH{-}H \\
HO{-}H \\
H{-}OH \\
HO{-}CH_2OH \\
H
\end{array}
&
\begin{array}{c}
OH \\
H{-}CH{=}O \\
H{-}OH \\
HO{-}H \\
HO{-}H \\
CH_2OH
\end{array} \\
A & B & C
\end{array}
$$

B. Cyclic Structures of the Monosaccharides

Recall from Sec. 19.10A that γ- or δ-hydroxy aldehydes exist predominantly as *cyclic hemiacetals.*

$$\text{(24.12a)}$$

$$\text{(24.12b)}$$

The same is true of aldoses and ketoses. Although monosaccharides are often written by convention as acylic carbonyl compounds, they exist predominantly as cyclic acetals. For example, in aqueous solution, glucose consists of about 0.003% aldehyde and a trace of the hydrate; the rest—more than 99.99%—is cyclic hemiacetals.

PROBLEM

24.6   Provide reaction mechanisms for the following equations.
   (a) Eq. 24.12a   (b) Eq. 24.12b

In many carbohydrates, both five- and six-membered cyclic hemiacetals are possible, depending on which hydroxy group undergoes cyclization.

$$\text{(24.13)}$$

A five-membered cyclic acetal form of a carbohydrate is called a **furanose** (after furan, a five-membered oxygen heterocycle); a six-membered cyclic acetal form of a carbohydrate is called a **pyranose** (after pyran, a six-membered oxygen heterocycle).

furan   pyran

The aldohexoses and aldopentoses exist predominantly as pyranoses in aqueous solution, but the furanose forms of some carbohydrates are important.

A name such as *glucose* is used when referring to any or all forms of the carbohydrate. To name a cyclic hemiacetal form of a carbohydrate, start with a prefix derived from the name of the carbohydrate (for example, *gluco* for glucose or *manno* for mannose) followed by a suffix that indicates the type of hemiacetal ring (*pyranose* for a six-membered ring; *furanose* for a five-membered ring). Thus, a six-membered cyclic hemiacetal form of D-glucose is called D-glucopyranose, and a five-membered cyclic hemiacetal form of D-mannose is called D-mannofuranose.

Although the cyclic structures of aldoses were originally proved by chemical degradations, these cyclic structures are readily apparent today from NMR spectroscopy. For example, the aldehydic proton resonance of glucose in the δ 9–10 region of its proton NMR spectrum is too weak to detect under ordinary circumstances, yet there is a doublet at δ 5.2 corresponding to a proton α to two oxygens: the proton at carbon-1 of the pyranose structure. Similarly, in the ^{13}C NMR spectra of aldoses, the resonance of carbon-1 occurs at about δ 92, which is very similar to the analogous carbons in acetals. There is no evidence of a carbonyl absorption near δ 200.

Anomers *The furanose or pyranose form of a carbohydrate has one more asymmetric carbon than the open-chain form—carbon-1, in the case of the aldoses. Thus, there are two possible diastereomers of D-glucopyranose.*

anomers of D-glucopyranose

(24.14)

(The rings in the Fischer projections of these cyclic compounds are closed with a rather strange-looking long bond. We'll learn shortly how to draw more conventional representations of these cyclic structures.) Both of these compounds are forms of D-glucopyranose, and in fact, glucose in solution exists as a mixture of both. They are diastereomers and are therefore separable compounds with different properties. When two cyclic forms of a carbohydrate differ in configuration only at their hemiacetal carbons, they are said to be **anomers**. In other words, anomers are cyclic forms of carbohydrates that are epimeric at the hemiacetal carbon. Thus, the two forms of D-glucopyranose are anomers of glucose. The hemiacetal carbon (carbon-1 of an aldose) is sometimes called the **anomeric carbon**.

As the preceding structures illustrate, anomers are named with the Greek letters α and β. This nomenclature refers to the Fischer projection of the cyclic form of a carbohydrate, written with all carbon atoms in a straight vertical line. *In the α-anomer, the hemiacetal —OH group is on the same side of the Fischer projection as the oxygen at the configurational carbon.* (The configurational carbon is the one used for specifying the D,L designation; for example, carbon-5 for the aldohexoses.) Conversely, in the β-anomer, the hemiacetal —OH group is on the side of the Fischer projection opposite the oxygen at the configurational carbon. The application of these definitions to the nomenclature of the D-glucopyranose anomers is as follows:

FURTHER EXPLORATION 24.1
Nomenclature of Anomers

(24.15)

α-anomer

α-D-glucopyranose

β-anomer

β-D-glucopyranose

Haworth Projections, Line-and-Wedge Structures, and Conformational Representations of Pyranoses and Furanoses   The goal of this section is to convert the Fischer projections of the pyranoses and furanoses into other representations of these cyclic carbohydrate structures. Study Problem 24.2 shows how to make these conversions in a systematic manner for the pyranoses.

STUDY PROBLEM   24.2

Convert the Fischer projection of *β*-D-glucopyranose into a Haworth projection, a line-and-wedge structure, and a chair conformation. (A Haworth projection is defined below.)

SOLUTION   First redraw the Fischer projection for *β*-D-glucopyranose in an equivalent Fischer projection in which the ring oxygen is in a down position. This projection is derived by using a cyclic permutation of the groups on carbon-5, an allowed manipulation of Fischer projections (Eq. 24.8, p. 1237).

Recall that the carbon backbone of such a Fischer projection is imagined to be folded around a barrel or drum (Fig. 24.2b, p. 1235). Such an interpretation of the Fischer projection of *β*-D-glucopyranose yields the following structure, in which the ring lies in a plane that emerges from the page. (The ring hydrogens are not shown.)

Haworth projection

When the plane of the ring is turned 90° so that the *anomeric carbon is on the right and the ring oxygen is in the rear*, the groups in *up* positions are those that are on the *left* in the Fischer projection, whereas the groups in *down* positions are those that are on the *right* in the Fischer projection. A planar structure of this sort is called a **Haworth projection**. In a Haworth projection, the ring is drawn in a plane perpendicular to the page and the positions of the substituents are indicated with up or down bonds. The wedged bonds are in front of the page, and the others are in back.

The corresponding line-and-wedge structure follows directly from the Haworth projection by viewing it from above. Substituents that are "up" in the Haworth projection become solid wedges, and substituents that are "down" in the Haworth projection become dashed wedges.

Neither a Haworth projection nor a line-and-wedge structure indicates the conformation of the ring. Six-membered carbohydrate rings resemble substituted cyclohexanes, and, like substituted cyclohexanes, they exist in chair conformations. Thus, to complete the conformational representation of β-D-glucopyranose, we apply the process given in Sec 7.4A (p. 286) for substituted cyclohexanes. Draw either one of the two chair conformations in which the anomeric carbon and the ring oxygen are in the same relative positions as they are in the preceding Haworth projection or line-and-wedge structure. Then place *up* and *down* groups in axial or equatorial positions, as appropriate.

Remember: Although the chair interconversion changes equatorial groups to axial, and vice versa, it does not change whether a group is up or down. Consequently, it doesn't matter which of the two possible chair conformations you draw first. Once you have drawn one chair conformation and have converted it into the other, you can then decide which is the more stable conformation. For β-D-glucopyranose, the more stable conformation is the one on the left, because all of the OH groups are equatorial.

To summarize the conclusions of Study Problem 24.2: When a carbohydrate ring is drawn with the anomeric carbon on the right and the ring oxygen in the rear, substituents that are on the left in the Fischer projection are *up* in all of the other representations, whereas groups that are on the right in the Fischer projection are *down* in the other representations.

Although the five-membered rings of furanoses are nonplanar, they are close enough to planarity that Haworth projections are reasonable approximations to their actual structures. Haworth projections are frequently used for furanoses for this reason. Thus, a Haworth projection of β-D-ribofuranose is derived as follows:

(24.16)

β-D-ribofuranose
(Haworth projection)

The corresponding line-and-wedge structure follows from the Haworth projection.

Haworth projection

line-and-wedge structure

β-D-ribofuranose

The Haworth projection is named for Sir Walter Norman Haworth (1883–1950), a noted British carbohydrate chemist who carried out important research on the cyclic structures of carbohydrates. Haworth received the Nobel Prize in Chemistry in 1937 and was knighted in 1947.

Although the procedure in Study Problem 24.2 can be used for any carbohydrate, in some cases it is simpler to derive a cyclic structure from its relationship to another cyclic structure. First, notice that the structure of β-D-glucopyranose is easy to remember because, in the more stable chair conformation, all ring substituents are equatorial (Study Problem 24.2). Suppose, now, that we want to draw the conformation of β-D-galactopyranose. Because D-galactose and D-glucose are epimers at carbon-4, the conformational representation of β-D-galactopyranose can be quickly derived by interchanging the —H and —OH groups at carbon-4 of β-D-glucopyranose.

(24.17)

β-D-glucopyranose   **β-D-galactopyranose**

Likewise, because mannose and glucose are epimeric at carbon-2, the structure of a D-mannopyranose can be simply derived by interchanging the —H and —OH groups at carbon-2 of the corresponding D-glucopyranose structure.

Sometimes it becomes necessary to draw the conformation of a carbohydrate that either is a mixture of anomers or is of uncertain anomeric composition. In such cases, the configuration at the anomeric carbon is represented by a "squiggly bond."

the squiggly bond indicates mixed or uncertain anomeric composition

24.7   Draw a Fischer projection, a Haworth projection, and a line-and wedge structure for each of the following compounds. For the pyranoses, draw the two possible chain conformations.

(a) α-D-glucopyranose

(b) β-D-mannopyranose

(c) β-D-xylofuranose

(d) α-D-fructopyranose (The structure of D-fructose, a ketose, is given on p. 1242.)

(e) α-L-glucopyranose

(f) a mixture of the α- and β-anomers of L-glucopyranose

24.8   Name each of the following aldoses. In part (a), work back to the Fischer projection and consult Fig. 24.3. In part (b), decide which carbons have configurations different from those of glucose, and which have the same configurations; then use Fig. 24.3.

24.4   MUTAROTATION OF CARBOHYDRATES

When pure α-D-glucopyranose is dissolved in water, its specific rotation is found to be +112 degrees mL g^{-1} dm^{-1}. With time, however, the specific rotation of the solution decreases, ultimately reaching a stable value of +52.7 degrees mL g^{-1} dm^{-1}. When pure β-D-glucopyranose is dissolved in water, it has a specific rotation of +18.7 degrees mL g^{-1} dm^{-1}. The specific rotation of this solution increases with time, also to +52.7 degrees mL g^{-1} dm^{-1}. This change of optical rotation with time is called **mutarotation** (*muta*, meaning "change"). Mutarotation also occurs when pure anomers of other carbohydrates are dissolved in aqueous solution.

The mutarotation of glucose is caused by the conversion of the α- and β-glucopyranose anomers into an equilibrium mixture of both. The same equilibrium mixture is formed, as it must be, from either pure α-D-glucopyranose or β-D-glucopyranose. Mutarotation is catalyzed by both acid and base, but it also occurs slowly in pure water.

(24.18)

a-anomer
[*a*]$_D$ = +112 degrees mL g^{-1} dm^{-1}

β-anomer
[*a*]$_D$ = +18.7 degrees mL g^{-1} dm^{-1}

equilibrium mixture: [*a*]$_D$ = +52.7 degree mL g^{-1} dm^{-1}

Mutarotation is characteristic of the *cyclic hemiacetal* forms of glucose; an aldehyde cannot undergo mutarotation, because an aldehyde carbon is not an asymmetric carbon. Mutarotation was one of the phenomena that suggested to early carbohydrate chemists that aldoses might exist as cyclic hemiacetals.

Mutarotation occurs, first, by opening of the pyranose ring to the free aldehyde form. This is nothing more than the reverse of hemiacetal formation (Sec. 19.10A). Then a 180° rotation about the carbon–carbon bond to the carbonyl group permits reclosure of the hemiacetal ring by the reaction of the hydroxy group at the opposite face of the carbonyl carbon.

opening of the hemiacetal

H_3O^+ or ^-OH

α-anomer

180° internal rotation

closing of the hemiacetal

H_3O^+ or ^-OH

β-anomer

(24.19)

The mutarotation of glucose is due almost entirely to the interconversion of its two pyranose forms. Other carbohydrates undergo more complex mutarotations. An example of this behavior is provided by D-fructose, a 2-ketohexose. The structures of the cyclic hemiacetal forms of D-fructose can be derived from its carbonyl (ketone) form using the methods described in Sec. 24.3B (see also Problem 24.7d):

D-fructose — this oxygen is involved in pyranose formation

D-fructose — this oxygen is involved in furanose formation

It happens that the crystalline form of D-fructose is *β*-D-fructopyranose. When crystals of this form are dissolved in water, it equilibrates to both pyranose and furanose forms.

β-D-fructopyranose
(57%)

α-D-fructopyranose
(3%)

β-D-fructofuranose
(31%)

α-D-fructofuranose
(9%)

(24.20)

TABLE 24.1 Compositions of Monosaccharides at Equilibrium in Aqueous Solution at 40 °C

| | Percent at equilibrium | | | | |
| | pyranose | | furanose | | |
Sugar	α	β	α	β	aldehyde or ketone
D-glucose	36	64	trace*		0.003
D-galactose[†]	27–36	64–73	trace		trace
D-mannose	68	32	trace	0	trace
D-allose	18	70	5	7	
D-altrose	27	40	20	13	
D-idose[‡]	39	36	11	14	
D-talose	40	29	20	11	
D-arabinose[§]	63	34	3		
D-xylose	37	63			
D-ribose	20	56	6	18	0.02
D-fructose	0–3	57–75	4–9	21–31	0.25

* In 10% aqueous dioxane, glucose contains 0.1–0.2% of each furanose.

[†] At 25 °C, galactose contains 29% α-pyranose, 64% β-pyranose, 3% α-furanose, and 4% β-furanose.

[‡] 25 °C

[§] At 25 °C, arabinose contains 60% α-pyranose, 35% β-pyranose, 3% α-furanose, and 2% β-furanose.

Glucose in solution also contains furanose forms, but these are present in very small amounts—about 0.2% each.

The foregoing discussion shows that a single hexose can exist in at least five forms: the acyclic aldehyde or ketone form, the α- and β-pyranose forms, and the α- and β-furanose forms. We've also learned that these forms are in equilibrium in aqueous solution. The percentage of each form for some monosaccharides are summarized in Table 24.1.

Some general conclusions from this table are

1. Most aldohexoses and aldopentoses exist primarily as pyranoses, although a few have substantial amounts of furanose forms.
2. Most monosaccharides contain relatively small amounts of their noncyclic carbonyl forms.
3. Mixtures of α- and β-anomers are usually found, although the exact amounts of each vary from case to case.

NMR spectroscopy has been one of the primary techniques that chemists have used to determine the ratios of the various forms of monosaccharides in solution. Using D-glucose as an example, Table 24.1 shows that at equilibrium there is nearly two times as much of the β-pyranose form (64%) as there is of the α-pyranose form (36%). Since the two pyranose forms are diastereomers (specifically, anomers), we would expect them to have different proton NMR spectra. However, since they are in equilibrium with each other, an NMR sample of D-glucose will contain both forms, and peaks from each will be seen.

Let's consider the proton attached to carbon-1 (the anomeric carbon) in each pyranose form. In the β-form, this proton has a chemical shift of 4.63 ppm, and in the α-form it has a shift of 5.21 ppm. But how do we know which peak corresponds to which anomer? Both signals will be split into a doublet by the neighboring proton on carbon-2. However, the coupling constants between that proton and the carbon-1 protons will be different, depending on whether the anomeric proton is axial or equatorial (recall the discussion of the Karplus curve in Sec. 13.5A). Since the two protons are approximately anti to one another in the β-form, we expect a larger coupling constant ($J_{1,2} = 7.9$ Hz) than when the two protons are nearly gauche, as in the α-form ($J_{1,2} = 3.8$ Hz).

$$\delta\ 5.21\ ppm,\ J_{1,2}= 3.8\ Hz \qquad \delta\ 4.63\ ppm,\ J_{1,2}= 7.9\ Hz \qquad (24.21)$$

Since these two peaks integrate in a 64:36 ratio, the amounts of each form present at equilibrium can be extrapolated. (The other peaks for protons attached to other carbons also integrate at the same ratio, but the anomeric protons are usually the most easily identified and are thus most often used for this purpose of determining ratios of anomers.)

The fraction of any form in solution at equilibrium is determined by its stability relative to that of all other forms. To predict the data in Table 24.1 for a given monosaccharide would require an understanding of all the factors that contribute to the stability or instability of *every* one of its isomeric forms in aqueous solution. In some cases, though, the principles of cyclohexane conformational analysis (Secs. 7.3 and 7.4) can be applied, as Problem 24.11 illustrates.

PROBLEMS

24.9   Using the curved-arrow notation, fill in the details for acid-catalyzed mutarotation of glucopyranose shown in Eq. 24.19. Begin by protonating the ring oxygen.

24.10   Using the curved-arrow notation, fill in the details for base-catalyzed mutarotation of glucopyranose. Begin by removing a proton from the hydroxy group at carbon-1.

24.11   Consider the β-D-pyranose forms of glucose and talose (Fig. 24.3, p. 1240). Suggest one reason why talose contains a smaller fraction of β-pyranose form than glucose.

24.12   Draw a conformational representation of
(a) β-D-allopyranose   (b) α-D-idofuranose

24.13   From the specific rotations shown in Eq. 24.18, calculate the percentages of α- and β-D-glucopyranose present at equilibrium. (Assume that the amounts of aldehyde and furanose forms are negligible.) Compare your answer to the data given in Table 24.1.

24.5   BASE-CATALYZED ISOMERIZATION OF ALDOSES AND KETOSES

In base, aldoses and ketoses rapidly equilibrate to mixtures of other aldoses and ketoses.

This transformation is an example of the **Lobry de Bruyn–Alberda van Ekenstein reaction**, named for two Dutch chemists, Cornelius Adriaan van Troostenbery Lobry de Bruyn (1857–1904) and Willem Alberda van Ekenstein (1858–1907). Despite its rather formidable name, this reaction is a relatively simple one that is closely related to processes you have already studied.

Although glucose in solution exists mostly in its cyclic hemiacetal forms, it is also in equilibrium with a small amount of its acyclic aldehyde form. This aldehyde, like other carbonyl compounds with α-hydrogens, ionizes to give small amounts of its enolate ion in base. Protonation of this enolate ion at one face of the double bond gives back glucose; protonation at the other face gives mannose. This is much like the process shown in Eq. 22.8a–b, p. 1109.

$$(24.23)$$

The enolate ion can also be protonated on oxygen to give a new enol, called an **enediol**. An enediol contains a hydroxy group at each end of a double bond. The enediol derived from glucose is simultaneously the enol of not only the aldoses glucose and mannose but also the ketose fructose.

$$(24.24a)$$

$$(24.24b)$$

Such base-catalyzed epimerizations and aldose–ketose equilibria need not stop at carbon-2. For example, D-fructose epimerizes at carbon-3 on prolonged treatment with base. (Why?)

Several transformations of this type are important in metabolism. One such reaction, the conversion of D-glucose-6-phosphate into D-fructose-6-phosphate, occurs in the breakdown of D-glucose (glycolysis), the series of reactions by which D-glucose is utilized as a food source. Because biochemical reactions occur near pH 7, too little hydroxide ion is present to catalyze the reaction. Instead, the reaction is catalyzed by an enzyme, D-glucose-6-phosphate isomerase, and the enzyme-catalyzed reaction also involves an enediol intermediate.

D-glucose-6-phosphate   **D-fructose-6-phosphate**

(24.25)

The conversion of D-glucose derived from corn into D-fructose is also an enzyme-catalyzed process that undoubtedly involves reactions that are similar if not identical. This conversion is central to the commercial production of high-fructose corn syrup, a widely used sweetener.

PROBLEM

24.14 Into what other aldose and 2-ketose would each of the following aldoses be transformed on treatment with base? Give the structure and name of the aldose, and the structure of the 2-ketose.

(a) D-galactose   (b) D-allose

24.6 GLYCOSIDES

Most monosaccharides react with alcohols under acidic conditions to yield cyclic acetals.

D-glucose

methyl α-D-glucopyranoside   **methyl β-D-glucopyranoside**
(83–85% yield; separated by fractional crystallization)   (24.26)

Such compounds are called **glycosides**. They are special types of acetals in which one of the oxygens of the acetal group is the ring oxygen of the pyranose or furanose.

acetal group

Contrast the reaction of a cyclic hemiacetal (such as glucopyranose) with the corresponding reaction of an ordinary aldehyde under the same conditions:

Glycoside formation:

one —OR group is incorporated

$$\text{+ ROH} \quad \xrightleftharpoons{\text{acid}} \quad \text{+ } H_2O \qquad (24.27a)$$

Formation of an aldehyde acetal:

two —OR groups are incorporated

$$-CH{=}O + 2\,ROH \quad \xrightleftharpoons{\text{acid}} \quad \overset{\displaystyle OR}{\underset{|}{-}CH{-}OR} + H_2O \qquad (24.27b)$$

As Eqs. 24.27a and 24.27b show, a cyclic hemiacetal such as glucopyranose incorporates one alcohol —OR group, whereas an ordinary aldehyde incorporates two —OR groups. This difference between aldoses and ordinary aldehydes is another reason that early carbohydrate chemists suspected that aldoses exist as cyclic hemiacetals.

As illustrated in Eq. 24.26, glycosides are named as derivatives of the parent carbohydrate. The term *pyranoside* indicates that the glycoside ring is six-membered. The term *furanoside* is used for a five-membered ring.

Glycoside formation, like acetal formation, is catalyzed by acid and involves an α-alkoxy carbocation intermediate (Sec. 19.6).

STUDY GUIDE
LINK 24.1
Acid Catalysis of
Carbohydrate
Reactions

an α-alkoxy carbocation   (24.28a)

There are then two stereochemically different fates for α-alkoxycarbocation:

methyl β-D-glycopyranoside

$$+ \ H_2\overset{\cdot\cdot}{\overset{+}{O}}CH_3 \qquad (24.28b)$$

carbocation
from Eq. 24.28a

reaction at the
upper face

reaction at the
lower face

methyl α-D-glucopyranoside

$$+ \ H_2\overset{\cdot\cdot}{\overset{+}{O}}CH_3 \qquad (24.28c)$$

Like other acetals, glycosides are *stable to base*, but, in dilute aqueous acid, they are hydrolyzed back to their parent carbohydrates.

$$+ \ H_2O \xrightarrow{\ H_3O^+\ } \qquad + \ HOCH_3 \qquad (24.29)$$

Many compounds occur naturally as glycosides; two examples are shown in Fig. 24.4. In addition, glycoside formation plays an important role in the removal of some chemicals

---

salicin
(from extract of willow bark)

doxorubicin
(an antitumor drug)

FIGURE 24.4 Two naturally occurring glycosides of medicinal interest. The carbohydrate part of each glycoside is shown in red.

from the body in phase II metabolism. In this process, a carbohydrate is joined to an —OH group of the substance to be removed. The added carbohydrate group makes the substance more soluble in water and, hence, more easily excreted. (See Eq. 8.11, p. 358.)

Like simple methyl glycosides, the glycoside of a natural product can be hydrolyzed to its component alcohol or phenol and carbohydrate.

$$\text{(24.30)}$$

converted into an alcohol or phenol

PROBLEMS

24.15 (a) Name the following glycoside.

(b) Into what products will this glycoside be hydrolyzed in aqueous acid?

24.16 Vanillin (the natural vanilla flavoring) occurs in nature as a β-glycoside of glucose. Suggest a structure for this glycoside.

vanillin

24.17 Draw structures for

(a) methyl β-D-fructofuranoside   (b) isopropyl α-D-galactopyranoside

---

24.7   ETHER AND ESTER DERIVATIVES OF CARBOHYDRATES

Because carbohydrates contain many —OH groups, carbohydrates undergo many of the reactions of alcohols. One such reaction is ether formation. In the presence of concentrated base, carbohydrates are converted into ethers by reactive alkylating agents such as dimethyl sulfate, methyl iodide, or benzyl chloride.

dimethyl sulfate
(see Eq. 10.31, p. 469)

$$\text{(24.31)}$$

methyl 2,3,4,6-tetra-*O*-methyl-D-glucopyranoside

(Notice that the ethers are named as *O*-alkyl derivatives of the carbohydrates.) These reactions are examples of the Williamson ether synthesis (Sec. 11.2A). The Williamson synthesis with most alcohols requires a base stronger than ⁻OH to form the conjugate-base alkoxide. The hydroxy groups of carbohydrates, however, are more acidic ($pK_a \approx 12$) than those of ordinary alcohols. (The higher acidity of carbohydrate hydroxy groups is attributable to the polar effect of the many neighboring oxygens in the molecule.) Consequently, substantial concentrations of their conjugate-base alkoxide ions are formed in concentrated NaOH. A large excess of the alkylating reagent is used because hydroxide itself, present in large excess, also reacts with alkylating agents. Little or no base-catalyzed epimerization (Sec. 24.5) is observed in this reaction despite the strongly basic conditions used. Evidently, alkylation of the hydroxy group at the anomeric carbon is much faster than epimerization. Once this oxygen is alkylated, epimerization can no longer occur. (Why?)

Other reagents used to form methyl ethers of carbohydrates include CH_3I/Ag_2O, and the strongly basic $NaNH_2$ (sodium amide) in liquid NH_3 followed by CH_3I.

Remember that the alkoxy group at the anomeric carbon is different from the other alkoxy groups in an alkylated carbohydrate because it is part of the glycosidic linkage. Because it is an acetal, it can be hydrolyzed in aqueous acid under mild conditions:

$$\text{(24.32)}$$

The other alkoxy groups are ordinary ethers and do not hydrolyze under these conditions. They require *much* harsher conditions for cleavage (Sec. 11.4).

Another reaction of alcohols is esterification; indeed, the hydroxy groups of carbohydrates, like those of other alcohols, can be esterified.

$$\text{(24.33)}$$

D-glucopyranose   **1,2,3,4,6-penta-*O*-acetyl-D-glucopyranose**
(83% yield)

Ester derivatives of carbohydrates can be saponified in base or removed by transesterification with an alkoxide such as methoxide:

$$\text{(24.34)}$$

Ethers and esters are used as protecting groups in reactions involving carbohydrates. For example, acetate esters are used as protecting groups in the synthesis of ^{18}F-fluoro-D-glucose (FDG), an important imaging agent in positron emission tomography. (See the sidebar on pp. 401–403.) Because ethers and esters of carbohydrates have broader solubility characteristics and greater volatility than the carbohydrates themselves, they also find use in the characterization of carbohydrates by chromatography and mass spectrometry.

STUDY PROBLEM   24.3

Outline a sequence of reactions by which D-glucose can be converted into methyl 2,3,4,6-tetra-*O*-acetyl-D-glucopyranoside.

methyl 2,3,4,6-tetra-*O*-acetyl-D-glucopyranoside

SOLUTION   In solving problems of this sort, in which apparently similar hydroxy groups are converted into different derivatives, the key is to recognize that hydroxy groups and alkoxy groups at the anomeric position (carbon-1 in aldoses) behave quite differently from these groups at other positions. As the earlier text discussion shows, the hydroxy group at carbon-1 of glucose is part of a *hemiacetal* group, and alkoxy groups at the same carbon are part of an *acetal* group. Acetals are formed and hydrolyzed under much milder conditions than ordinary ethers. Consequently, the methyl "ether" (actually, a methyl acetal) at carbon-1 can be formed by treating D-glucose with methanol and acid. The remaining hydroxy groups can then be esterified with excess acetic anhydride (as in Eq. 24.33) to give the desired product:

(24.35)

PROBLEMS

24.18   Explain why acetals hydrolyze more rapidly than ordinary ethers. (*Hint:* Consider hydrolysis by a carbocation mechanism.)

24.19   Outline a sequence of reactions that will bring about each of the following conversions.

(a) D-galactopyranose to ethyl 2,3,4,6-tetra-*O*-methyl-D-galactopyranoside

(b) D-glucopyranose to 2,3,4,6-tetra-*O*-benzyl-D-glucopyranose

24.8   OXIDATION AND REDUCTION REACTIONS OF CARBOHYDRATES

Like simpler aldehydes, the aldehyde group of aldoses can be both oxidized and reduced. It is also possible to selectively oxidize the primary alcohol and aldehyde groups of an aldose without oxidizing the secondary alcohols. The structures and names of the common oxidation and reduction products of aldoses are summarized in Table 24.2.

TABLE 24.2 Structures of Common Oxidation and Reduction Products of Aldoses

General structure: $X-\left(\underset{\underset{OH}{|}}{CH}\right)_n-Y$

Derivative structure			
X— =	**—Y =**	**General name**	**Example derived from glucose**
$HOCH_2—$	$—CH{=}O$	aldose	glucose
$HOCH_2—$	$—CO_2H$	aldonic acid	gluconic acid
$HO_2C—$	$—CO_2H$	aldaric acid	glucaric acid
$HOCH_2—$	$—CH_2OH$	alditol	glucitol
$HO_2C—$	$—CH{=}O$	uronic acid	glucuronic acid

PROBLEM

24.20 Using Table 24.2 to assist you, draw a Fischer projection for the structure of (a) galacturonic acid, the uronic acid derived from galactose; (b) ribitol, the alditol derived from ribose.

A. Oxidation to Aldonic Acids

Treatment of an aldose with bromine water oxidizes the aldehyde group to a carboxylic acid. The oxidation product is an *aldonic acid* (see Table 24.2).

$$\text{(24.36)}$$

D-glucose

D-gluconic acid
(an aldonic acid;
77–96% yield as Ca^{2+} salt)

Although it is customary to represent aldonic acids in the free carboxylic acid form, they, like other γ- and δ-hydroxy acids (Sec. 20.8A), exist in acidic solution as lactones called *aldono-lactones*. The lactones with five-membered rings are somewhat more stable than those with six-membered rings.

$$\text{(24.37)}$$

D-gluconic acid

D-γ-gluconolactone
(an aldonolactone)

Oxidation with bromine water is a useful test for aldoses. Aldoses can also be oxidized with other reagents, such as the Tollens reagent [$Ag^+(NH_3)_2$; Sec. 19.14]. However, because the Tollens reagent is alkaline and causes base-catalyzed epimerization of aldoses (Sec. 24.5), it is less useful synthetically. Because the alkaline conditions of the Tollens test also promote the equilibration of aldoses and ketoses, ketoses also give positive Tollens tests. Glycosides are *not* oxidized by bromine water, because the aldehyde carbonyl group is protected as an acetal.

B. Oxidation to Aldaric Acids

Dilute nitric acid oxidizes aldehydes and primary alcohols to carboxylic acids *without affecting secondary alcohols*. Consequently, this is a very useful reagent for converting aldoses (or aldonic acids) into aldaric acids (Table 24.2).

Like aldonic acids, aldaric acids in acidic solution form lactones. Two different five-membered lactones are possible, depending on which carboxylic acid group undergoes lactonization. Furthermore, under certain conditions, some aldaric acids can be isolated as dilactones, in which both carboxylic acid groups are lactonized.

STUDY GUIDE LINK 24.2
Configurations of Aldaric Acids

PROBLEMS

24.21 Give Fischer projections for the aldaric acids derived from both D-glucose and L-gulose. What is the relationship between these structures?

24.22 Draw a Fischer projection for the aldaric acid, and a structure of the 1,4-lactone, derived from the oxidation of (a) D-galactose; (b) D-mannose.

24.23 Give the product formed when each of the following alcohols is oxidized by dilute HNO_3.

(a) (b) $HOCH_2CH_2CH_2CH_2OH$

C. Periodate Oxidation

Many carbohydrates contain vicinal glycol units and, like other 1,2-glycols, are oxidized by periodic acid (Sec. 11.6B). A complication arises when, as in many carbohydrates, more than two adjacent carbons bear hydroxy groups. When one of the oxidation products is an α-hydroxy aldehyde, as in the following example, it is oxidized further to formic acid and another aldehyde.

By analogy, an α-hydroxy ketone is oxidized to an aldehyde and a carboxylic acid.

Because it is possible to determine accurately both the amount of periodate consumed and the amount of formic acid produced, periodate oxidation can be used to differentiate between pyranose and furanose structures of saccharide derivatives. For example, periodate oxidation of methyl α-D-glucopyranoside liberates one equivalent of formic acid:

A furanose form of this glycoside, however, gives formaldehyde:

The periodate oxidation of carbohydrates was developed by Claude S. Hudson (1881–1952), a noted American carbohydrate chemist at the National Institutes of Health. It was used extensively to relate the anomeric configurations of many carbohydrate derivatives. How this was done is suggested by Problems 24.24 and 24.25.

PROBLEMS

24.24 Explain why the methyl α-D-pyranosides of *all* D-aldohexoses give, in addition to formic acid, the same compound when oxidized by periodate.

24.25 Assuming you knew the properties of the compound obtained in Problem 24.24, including its optical rotation, show how you could use periodate oxidation to distinguish methyl α-D-galactopyranoside from methyl β-D-galactopyranoside.

D. Reduction to Alditols

Aldohexoses, like ordinary aldehydes, undergo many of the usual carbonyl reductions. For example, sodium borohydride ($NaBH_4$, Sec. 19.8A) reduces aldoses to alditols (Table 24.2).

$$
\begin{array}{ccc}
\text{CH} =\!\!=\! \text{O} & & \text{CH}_2\text{OH} \\
\text{H} \!-\!\!\!-\! \text{OH} & & \text{H} \!-\!\!\!-\! \text{OH} \\
\text{HO} \!-\!\!\!-\! \text{H} & \xrightarrow[\text{H}_2\text{O}]{\text{NaBH}_4} & \text{HO} \!-\!\!\!-\! \text{H} \\
\text{HO} \!-\!\!\!-\! \text{H} & & \text{HO} \!-\!\!\!-\! \text{H} \\
\text{H} \!-\!\!\!-\! \text{OH} & & \text{H} \!-\!\!\!-\! \text{OH} \\
\text{CH}_2\text{OH} & & \text{CH}_2\text{OH}
\end{array}
$$

(24.42)

D-galactose   **galactitol (dulcitol)**
an alditol
(90% yield; neither D nor L
because it is meso)

Catalytic hydrogenation (for example, H_2 with a Raney nickel catalyst in aqueous ethanol) can also be used for the same transformation.

In the oxidation and reduction reactions discussed in this section, aldoses have been depicted in their carbonyl forms rather than in their cyclic hemiacetal forms. Do not lose sight of the fact that all forms are present at equilibrium, and the aldehyde form can react even though it is present in a very small amount. Once it reacts, it is immediately replenished (Le Châtelier's principle). Thus, when the aldehyde group reacts with $NaBH_4$ to give alditol, the equilibrium provides more of the aldehyde form:

$$\text{cyclic (hemiacetal) forms} \;\rightleftharpoons\; \text{aldehyde form} \xrightarrow[\text{H}_2\text{O}]{\text{NaBH}_4} \text{alditol} \qquad (24.43)$$

Furthermore, the acidic or basic conditions of many aldehyde reactions (basic conditions in the case of $NaBH_4$ reduction) catalyze this equilibrium. Not only is more aldehyde formed, but it is also formed *rapidly*.

24.9   KILIANI–FISCHER SYNTHESIS

Aldoses, like other aldehydes, undergo addition of hydrogen cyanide to give cyanohydrins (Sec. 19.7). This reaction, like several others that have been discussed, involves the aldehyde form of the sugar.

$$
\begin{array}{ccc}
\text{CH} =\!\!=\! \text{O} & & \\
\text{HO} \!-\!\!\!-\! \text{H} & & \\
\text{H} \!-\!\!\!-\! \text{OH} \;+\; \text{HCN} & \xrightarrow[\text{H}_2\text{O}]{\substack{\text{NaCN}\\ \text{pH 8}}} & \\
\text{H} \!-\!\!\!-\! \text{OH} & & \\
\text{CH}_2\text{OH} & &
\end{array}
$$

additional asymmetric carbon

$$
\begin{array}{ccccc}
\text{C}\!\equiv\!\text{N} & & & & \text{C}\!\equiv\!\text{N} \\
\text{H} \!-\!\!\!-\! \text{OH} & & & & \text{HO} \!-\!\!\!-\! \text{H} \\
\text{HO} \!-\!\!\!-\! \text{H} & & + & & \text{HO} \!-\!\!\!-\! \text{H} \\
\text{H} \!-\!\!\!-\! \text{OH} & & & & \text{H} \!-\!\!\!-\! \text{OH} \\
\text{H} \!-\!\!\!-\! \text{OH} & & & & \text{H} \!-\!\!\!-\! \text{OH} \\
\text{CH}_2\text{OH} & & & & \text{CH}_2\text{OH}
\end{array}
$$

(24.44)

D-arabinose   **D-gluconitrile**   **D-mannonitrile**
(29% yield)   (51% yield)

Because the cyanohydrin product has an additional asymmetric carbon, it is formed as a mixture of two epimers. Because these epimers are diastereomers, they are typically formed in different amounts (Sec. 7.7B), as in Eq. 24.44. Although the exact amount of each is not easily predicted, in most cases significant amounts of both are obtained.

The mixture of cyanohydrins can be converted into a mixture of aldoses by catalytic hydrogenation, and these aldoses can be separated.

$$(24.45)$$

As this equation shows, the hydrogenation reaction involves reduction of the nitrile to an imine (or a cyclic carbinolamine derivative of the imine). Under the reaction conditions, the imine hydrolyzes readily to the aldose and ammonium ion.

This example shows that cyanohydrin formation followed by reduction converts an aldose into two epimeric aldoses with one additional carbon. That is, two aldohexoses, epimeric at carbon-2, are formed from an aldopentose. Notice particularly that this synthesis does not affect the stereochemistry of carbons 2, 3, and 4 in the starting material.

The formation of cyanohydrins from aldoses was developed by Heinrich Kiliani (1855–1945), head of the medicinal chemistry laboratory at the University of Freiburg. Kiliani also showed that the cyanohydrins could be hydrolyzed to aldonic acids. Emil Fischer, whose remarkable accomplishments in carbohydrate chemistry are described in Sec 24.10, developed a method to reduce the aldonic acids (as their lactones) to aldoses. The three processes—cyanohydrin formation, hydrolysis, and reduction—provided a way to convert an aldose into two other aldoses with one additional carbon. The overall transformation came to be known as the **Kiliani–Fischer synthesis**. The chemistry shown in Eqs. 24.44 and 24.45, which was developed in the 1970s, is a modern variation of the Kiliani–Fischer synthesis.

Kiliani, a noted authority on carbohydrates, also proved the structures of several monosaccharides, including the 2-ketose structure of fructose.

PROBLEM

24.26 Assuming the D configuration, identify *A* and *B*.

24.10   THE PROOF OF GLUCOSE STEREOCHEMISTRY

The aldohexose structure of (+)-glucose (that is, the structure without any stereochemical details) was established around 1870. The van't Hoff–LeBel theory of the tetrahedral carbon atom, published in 1874 (Sec. 6.10), suggested the possibility that glucose and the other aldohexoses could be stereoisomers. Which one of the 2^4 possible stereoisomers, then, is (+)-glucose? This problem was solved in two stages.

A. Which Diastereomer? The Fischer Proof

The first (and major) part of the solution to the problem of glucose stereochemistry was published in 1891 by Emil Fischer. (See Fischer's biography on p. 1267.) It would be reason enough to study Fischer's proof as one of the most brilliant pieces of reasoning in the history of chemistry. However, it also will serve to sharpen your understanding of stereochemical relationships.

It is important to understand that in Fischer's day, the tools for determining the absolute stereochemical configurations of chemical compounds had not yet been developed. Consequently, Fischer adopted the *arbitrary convention* that, at carbon-5 of (+)-glucose (the configurational carbon in the D,L system), the —OH is on the right in the standard Fischer projection; that is, Fischer assumed that (+)-glucose has what we now call the D configuration. Fischer, then, proved the stereochemistry of (+)-glucose *relative* to an assumed configuration at carbon-5. In other words, Fischer, of necessity, reduced the problem of 2^4 stereoisomers—two enantiomeric sets of 8 diastereomers—to a problem of eight diastereomers.

Fischer had, as a starting point, pure samples of the naturally occurring aldopentose (−)-arabinose and the two aldohexoses, (+)-glucose and (+)-mannose. (+)-Glucose and (+)-mannose were known to be stereoisomers. The remarkable thing about Fischer's proof is that it allowed him to assign relative configurations in space for each asymmetric carbon in these compounds using only chemical reactions and optical activity. The logic involved is direct, simple, and elegant, and it can be summarized in four steps:

Step 1. **(−)-Arabinose, an aldopentose, is converted into both (+)-glucose and (+)-mannose by a Kiliani–Fischer synthesis.** From this observation (see Sec. 24.9), Fischer deduced that (+)-glucose and (+)-mannose are epimeric at carbon-2, and that the configuration of (−)-arabinose at carbons 2, 3, and 4 is the same as that of (+)-glucose and (+)-mannose at carbons 3, 4, and 5, respectively.

(24.46)

Step 2. **(−)-Arabinose can be oxidized by dilute HNO_3 (Sec. 24.8B) to an optically active aldaric acid.** From this observation, Fischer concluded that the —OH group at carbon-2 of arabinose must be on the left. If this —OH group were on the right, then the aldaric acid of arabinose would have to be meso, and thus optically inactive, *regardless of the configuration of the —OH group at carbon-3.* (Be sure you see why this is so; if necessary, draw both possible structures for (−)-arabinose to verify this deduction.)

(24.47)

The relationships among arabinose, glucose, and mannose established in steps 1 and 2 require the following partial structures for (+)-glucose and (+)-mannose.

(24.48)

(−)-arabinose   one is **(+)-glucose**; one is **(+)-mannose**

Step 3.   **Oxidations of both (+)-glucose and (+)-mannose with HNO₃ give optically active aldaric acids.** From this observation, Fischer deduced that the —OH group at carbon-4 is on the right in both (+)-glucose and (+)-mannose. Recall that, whatever the configuration at carbon-4 in these two aldohexoses, it must be the same in both. Only if the —OH is on the right will *both* structures yield, on oxidation, optically active aldaric acids. If the —OH were on the left, *one* of the two aldohexoses would have given a meso (and hence, an optically inactive) aldaric acid.

(24.49)

Because the configuration at carbon-4 of (+)-glucose and (+)-mannose is the same as that at carbon-3 of (−)-arabinose (step 1), at this point Fischer could deduce the complete structure of (−)-arabinose.

(24.50)

structure *A*   structure *B*

(−)-arabinose

one is **(+)-mannose**; one is **(+)-glucose**

The previous steps had established that (+)-glucose had one of the two structures in Eq. 24.50 and (+)-mannose had the other, but Fischer did not yet know which structure goes with which sugar. This point is confusing to some students. Fischer's situation was like that of a young woman who has just met two brothers, but she doesn't know their names. So she asks a friend, "What are their names?" The friend says, "Oh, they are Mannose and Glucose; only I don't know which is which!" Just because the woman knows *both* names doesn't mean that she can associate *each* name with *each* face. Similarly, although Fischer knew the structures associated with both (+)-glucose and (+)-mannose, he did not yet know how to correlate *each* aldose with *each* structure. This last problem was solved in Step 4.

Step 4.   **Another aldose, (+)-gulose, can be oxidized with HNO_3 to the same aldaric acid as (+)-glucose.** How does this fact differentiate between (+)-glucose and (+)-mannose? Two *different* aldoses can give the same aldaric acid only if their —CH=O and —CH_2OH groups are *at opposite ends of an otherwise identical molecule* (Problem 24.21). Of the two structures in Eq. 24.50, only in structure *A* does an interchange of the —CH_2OH and —CH=O groups result in a different aldohexose. Fischer actually interconverted these two groups chemically on (+)-glucose (by a series of reactions discussed in Further Exploration 24.2 in the Study Guide) and obtained a different sugar, which he named (+)-gulose.

FURTHER EXPLORATION 24.2
More on the Fischer Proof

(24.51)

(+)-glucose   **(+)-gulose**

different aldohexoses

oxidation of either aldohexose gives the same product

To be sure that the CH_2OH and CH=O interconversions had gone as expected, Fischer verified that both (+)-glucose and (+) gulose were oxidized to the same aldaric acid, as shown in Eq. 24.51. The inescapable conclusion, then, is that the structure of (+)-glucose in Eq. 19.50 is *A*.

Completion of the proof requires that the same interconversions, when carried out on structure *B*, give the *same* aldohexose. (Verify this point by rotating either structure 180° in the plane of the page.)

CH=O
HO——H
HO——H
H——OH
H——OH
CH₂OH

interconvert
CH₂OH
and CH=O
→

CH₂OH
HO——H
HO——H
H——OH
H——OH
CH=O

(24.52)

the same aldohexose,
and therefore,
(+)-mannose

Therefore, structure *B* cannot be (+)-glucose. Because the only other possibility for *B* was (+)-mannose, the structure of this aldohexose was proved as well.

Emil Fischer

Emil Fischer (1852–1919) studied with Adolph von Baeyer and ultimately became Professor at Berlin University in 1892. Fischer carried out important research on sugars, proteins, and heterocycles. Fischer was a technical advisor to Kaiser Wilhelm II. The following story gives some indication of the authority that Fischer commanded in Germany. It is said that one day he and the Kaiser were arguing questions of science policy, and the Kaiser sought to end debate by pounding his fist on the table, shouting, "Ich bin der Kaiser!" (I am the Emperor!) Fischer, not to be silenced, responded in kind: "Ich bin Fischer!" Another story, perhaps apocryphal, attributes an important laboratory function to Fischer's beard. It was said that when a student had difficulty crystallizing a sugar derivative (some of which are notoriously difficult to crystallize), Fischer would shake his beard over the flask containing the recalcitrant compound. The accumulated seed crystals in his beard would fall into the flask and bring about the desired crystallization. Fischer was awarded the Nobel Prize in Chemistry in 1902.

PROBLEMS

24.27 An aldopentose *A* can be oxidized with dilute HNO₃ to an optically active aldaric acid. A Kiliani–Fischer synthesis starting with *A* gives two new aldoses: *B* and *C*. Aldose *B* can be oxidized to an achiral, and therefore optically inactive, aldaric acid, but aldose *C* is oxidized to an optically active aldaric acid. Assuming the D configuration, give the structures of *A*, *B*, and *C*.

24.28 An aldohexose *A* is either D-idose or D-gulose (see Fig. 24.3). It is found that a different aldohexose, L-(−)-glucose, gives the same aldaric acid as *A*. What is the identity of *A*?

B. Which Enantiomer? The Absolute Configuration of D-(+)-Glucose

Fischer never learned whether his arbitrary assignment of the absolute configuration of (+)-glucose was correct—that is, whether the —OH at carbon-5 of (+)-glucose was really on the right in its Fischer projection (as assumed) or on the left. The groundwork for solving this problem was laid when the configuration of (+)-glucose was correlated to that of (−)-tartaric acid. (Stereochemical correlation was introduced in Sec. 6.5.) This correlation was carried out in the following way.

(+)-Glucose was converted into (−)-arabinose by a reaction called the **Ruff degradation**. In this reaction sequence, an aldose is oxidized to its aldonic acid (Sec. 24.8), and the calcium salt of the aldonic acid is treated with ferric ion and hydrogen peroxide. This treatment decarboxylates the calcium salt and simultaneously oxidizes carbon-2 to an aldehyde.

$$
2 \;
\begin{array}{c}
\text{CH}=\text{O} \\
\text{H} \!-\!\!-\! \text{OH} \\
\text{HO} \!-\!\!-\! \text{H} \\
\text{H} \!-\!\!-\! \text{OH} \\
\text{H} \!-\!\!-\! \text{OH} \\
\text{CH}_2\text{OH}
\end{array}
\quad
\xrightarrow[\text{2) Ca(OH)}_2]{\text{1) Br}_2/\text{H}_2\text{O}}
\quad
\left(
\begin{array}{c}
\text{CO}_2^- \\
\text{H} \!-\!\!-\! \text{OH} \\
\text{HO} \!-\!\!-\! \text{H} \\
\text{H} \!-\!\!-\! \text{OH} \\
\text{H} \!-\!\!-\! \text{OH} \\
\text{CH}_2\text{OH}
\end{array}
\right)_2 \!\!\! \text{Ca}^{2+}
\quad
\xrightarrow[\text{30\% H}_2\text{O}_2]{\text{Fe(OAc)}_3}
\quad
2 \;
\begin{array}{c}
\text{CH}=\text{O} \\
\text{HO} \!-\!\!-\! \text{H} \\
\text{H} \!-\!\!-\! \text{OH} \\
\text{H} \!-\!\!-\! \text{OH} \\
\text{CH}_2\text{OH}
\end{array}
\;+\; 2\,\text{CO}_2 \qquad (24.53)
$$

(+)-glucose   calcium gluconate   (−)-arabinose
                                                (41% yield)

In other words, an aldose is degraded to another aldose with one fewer carbon atom, *its stereochemistry otherwise remaining the same*. Because the relationship between (+)-glucose and (−)-arabinose was already known from the Kiliani–Fischer synthesis (step 1 of the Fischer proof in the previous section), this reaction served to establish the course of the Ruff degradation. Next, (−)-arabinose was converted into (−)-erythrose by another cycle of the Ruff degradation.

$$
(-)\text{-arabinose} \quad \xrightarrow{\text{Ruff degradation}} \quad (-)\text{-erythrose} \qquad (24.54)
$$

D-(+)-Glyceraldehyde, in turn, was related to D-(−)-erythrose by a Kiliani–Fischer synthesis:

$$
\begin{array}{c}
\text{CH}=\text{O} \\
\text{H} \!-\!\!-\! \text{OH} \\
\text{CH}_2\text{OH}
\end{array}
\quad
\xrightarrow{\text{Kiliani–Fischer}}
\quad
\begin{array}{c}
\text{CH}=\text{O} \\
\text{H} \!-\!\!-\! \text{OH} \\
\text{H} \!-\!\!-\! \text{OH} \\
\text{CH}_2\text{OH}
\end{array}
\;+\;
\begin{array}{c}
\text{CH}=\text{O} \\
\text{HO} \!-\!\!-\! \text{H} \\
\text{H} \!-\!\!-\! \text{OH} \\
\text{CH}_2\text{OH}
\end{array}
\qquad (24.55)
$$

D-(+)-glyceraldehyde   D-(−)-erythrose   D-(−)-threose
(absolute configuration   (configurations at carbon–3
assumed by convention)   assumed by convention)

This sequence of reactions showed that (+)-glucose, (−)-erythrose, (−)-threose, and (+)-glyceraldehyde were all of the same stereochemical series: the D series. Oxidation of D-(−)-threose with dilute HNO₃ gave D-(−)-tartaric acid.

In 1950, the absolute configuration of naturally occurring (+)-tartaric acid (as its potassium rubidium double salt) was determined by a special technique of X-ray crystallography called *anomalous dispersion*. This determination was made by J. M. Bijvoet, A. F. Peerdeman, and A. J. van Bommel, Dutch chemists who worked, appropriately enough, at the van't Hoff laboratory in Utrecht. If Fischer had made the right choice for the configuration at carbon-5 of (+)-glucose—what we now call the D configuration—the assumed structure for D-(−)-tartaric acid and the experimentally determined structure of (+)-tartaric acid determined by the Dutch crystallographers would be enantiomers. If Fischer had guessed incorrectly, the assumed structure for (−)-tartaric acid would be the same as the experimentally determined structure of (+)-tartaric acid, and would have to be reversed. To quote Bijvoet and his colleagues: *"The result is that Emil Fischer's convention* [for the D configuration] *appears to answer to reality."*

$$(24.56)$$

D-(−)-threose → **D-(−)-tartaric acid**   **L-(+)-tartaric acid**
(by X-ray crystallography)

enantiomers

PROBLEMS

24.29 Given the structure of D-glyceraldehyde, how would you assign a structure to each of the two aldoses obtained from it by Eq. 24.55, assuming that these compounds were previously unknown?

24.30 Imagine that a scientist reexamines the crystallographic work that established the absolute configuration of (+)-tartaric acid and finds that the structure of this compound is the mirror image of the one given in Eq. 24.56. What changes would have to be made in Fischer's structure of D-(+)-glucose?

24.11   DISACCHARIDES AND POLYSACCHARIDES

A. Disaccharides

Disaccharides consist of two monosaccharides connected by a glycosidic linkage. **(+)-Lactose** is an example of a disaccharide. [(+)-Lactose is present to the extent of about 4.5% in cow's milk and 6–7% in human milk.]

galactose unit   glucose unit

(+)-lactose
or **4-O-(β-D-galactopyranosyl)-D-glucopyranose**

In (+)-lactose, a D-glucopyranose molecule is linked by its oxygen at carbon-4 to carbon-1 of D-galactopyranose. In effect, (+)-lactose is a glycoside in which galactose is the carbohydrate and glucose is the "alcohol." Recall that the glycosidic linkage is an acetal, and acetals hydrolyze under acidic conditions (Sec. 24.6). Therefore, (+)-lactose can be hydrolyzed in acidic solution to give one equivalent each of D-glucose and D-galactose, in the same sense that a methyl glycoside can be hydrolyzed to give methanol and a carbohydrate.

$$\text{D-galactose} \qquad \text{D-glucose} \qquad (24.57a)$$

Compare:

$$(24.57b)$$

Equation 24.57a demonstrates the structural basis for the definition of disaccharides presented in Sec. 24.1: A disaccharide is a carbohydrate that can be hydrolyzed to two monosaccharides. Hydrolysis occurs at the glycosidic bond between the two monosaccharide residues.

The stereochemistry of the glycosidic bond in (+)-lactose is β. That is, the stereochemistry of the oxygen linking the two monosaccharide residues in the glycosidic bond corresponds to that in the β-anomer of D-galactopyranose. This stereochemistry is very important in biology, because higher animals possess an enzyme, β-galactosidase, that catalyzes the hydrolysis of this β-glycosidic linkage near neutral pH; this hydrolysis allows lactose to act as a source of glucose. α-Glycosides of galactose are inert to the action of this enzyme. (People who suffer from lactose intolerance must take a form of this enzyme orally to digest lactose-containing foods.)

Because carbon-1 of the galactose residue in (+)-lactose is involved in a glycosidic linkage, it cannot be oxidized. However, carbon-1 of the glucose residue is part of a hemiacetal group, which, like the hemiacetal group of monosaccharides, is in equilibrium with the free aldehyde and can undergo characteristic aldehyde reactions. Thus, treatment of (+)-lactose with bromine water (Sec. 24.7A) effects oxidation of the glucose residue:

(+)-lactose

lactobionic acid

$$(24.58)$$

Carbohydrates such as (+)-lactose that can be oxidized in this way are called **reducing sugars** because, in being oxidized, they reduce the oxidizing agents. The glucose residue is said to be at the *reducing end* of the disaccharide, and the galactose residue at the *nonreducing end*. Because of its hemiacetal group, (+)-lactose also undergoes many other reactions of aldose hemiacetals, such as mutarotation.

(+)-Sucrose, or table sugar, is another important disaccharide. More than 120 million tons of sucrose is produced annually in the world. Sucrose consists of a D-glucopyranose residue and a D-fructofuranose residue connected by glycosidic bonds (*color*) at the anomeric carbons of *both* monosaccharides.

α glycosidic bond at glucose

β glycosidic bond at fructose

(+)-sucrose
or α-D-glucopyranosyl-β-D-fructofuranoside

The glycosidic bond in (+)-sucrose is different from the one in lactose. Only one of the residues of lactose—the galactose residue—contains an acetal (glycosidic) carbon. In contrast, *both* residues of (+)-sucrose have an acetal carbon. The glycosidic bond in (+)-sucrose bridges carbon-2 of the fructofuranose residue and carbon-1 of the glucopyranose residue. These are the carbonyl carbons in the noncyclic forms of the individual monosaccharides; remember that the carbonyl carbons become the acetal or hemiacetal carbons in the cyclic forms.

double glycoside linkage

C1 of glucose   C2 of fructose

Thus, neither the fructose nor the glucose part of sucrose has a free hemiacetal group. Hence, (+)-sucrose cannot be oxidized by bromine water, nor does it undergo mutarotation. Carbohydrates such as (+)-sucrose that cannot be oxidized by bromine water are classified as **nonreducing sugars**.

Like other glycosides, (+)-sucrose can be hydrolyzed to its component monosaccharides. Sucrose is hydrolyzed by aqueous acid or by enzymes (called *invertases*) to an equimolar mixture of D-glucose and D-fructose. This mixture is sometimes called *invert sugar* because, as hydrolysis of sucrose proceeds, the positive rotation of the solution changes to a negative rotation characteristic of the glucose–fructose mixture. This rotation is negative because the strongly negative rotation (-92 degrees mL g^{-1} dm^{-1}) of fructose (sometimes called *levulose*) has a greater magnitude than the positive rotation ($+52.7$ degrees mL g^{-1} dm^{-1}) of glucose (sometimes called *dextrose*). Fructose, which is the sweetest of the common sugars (about twice as sweet as sucrose), accounts for the intense sweetness of honey, which is mostly invert sugar.

The biosynthesis of disaccharides is discussed briefly in Sec. 25.7D.

PROBLEMS

24.31 What products are expected from each of the following reactions?
  (a) lactobionic acid (Eq. 24.58) + 1 *M* aqueous HCl $\longrightarrow$
  (b) (+)-lactose + dimethyl sulfate, NaOH $\longrightarrow$
  (c) product of part (b) + 1 *M* aqueous H$_2$SO$_4$ $\longrightarrow$

continued

continued

24.32   Consider the structure of cellobiose, a disaccharide obtained from the hydrolysis of the polysaccharide cellulose. Into what monosaccharide(s) is cellobiose hydrolyzed by aqueous HCl?

cellobiose

Tales of Serindipitous Sweet Discovery: Artificial Sweeteners

Artificial sweeteners are synthetic substitutes for sucrose and other natural sweeteners. Such compounds are in high demand because they allow consumers to enjoy a sweet taste without the calories of natural sweeteners. The annual worldwide market in artificial sweeteners is more than $5 billion. It has been estimated that well over 100 million Americans use artificial sweeteners. Artificial sweeteners have been sought since the ancient Romans, who used lead acetate ("lead sugar") as an alternative to sugar, and many Romans suffered severe lead toxicity as a result.

To qualify as an artificial sweetener, a compound must have sweetness many times that of sugar so that very little sweetener has to be used to achieve the desired effect. Because so little is used, almost no calories are consumed. However, finding sweet compounds is not the major hurdle to the development of an artificial sweetener. Rather, a candidate compound must undergo rigorous testing to be sure that it is not toxic and does not have undesired side effects. Almost every sweetener that has been introduced has attracted its share of public suspicion despite the large amount of biological testing involved in getting it to market. The most widely used sweeteners in modern times have been sodium saccharin, sodium cyclamate, aspartame, and, most recently, sucralose.

sodium saccharin (1879)
300 times as sweet as sucrose

sodium cyclamate (1937)
30–50 times as sweet as sucrose

aspartame (1965)
180 times as sweet as sucrose

sucralose (1989)
600 times as sweet as sucrose

The discoveries of all of these sweeteners were serendipitous, and all resulted from the tasting of laboratory samples by the researchers involved. Tasting new compounds was actually once a common laboratory practice, and the tastes of new compounds were routinely reported in the chemical literature. However, tasting new compounds is no longer condoned as a safe laboratory practice.

Sodium saccharin was discovered accidentally in 1879 by Constantin Fahlberg, a student in the laboratory of Prof. Ira Remsen at The Johns Hopkins University. At dinner, Fahlberg noticed a sweet taste on his fingers and concluded that it must have come from a compound he was working with. A "taste test" of many of his compounds led to sodium saccharin as the culprit. Fahlberg became wealthy from commercialization of the discovery, and when Prof. Remsen did not receive any financial benefit, he became quite bitter toward his former student.

Aspartame was discovered accidentally in 1965 by Jim Schlatter, a chemist at G. D. Searle and Company, while he was working on the discovery of drugs to treat stomach ulcers. He noticed a sweet taste on his fingers after handling the compound. His supervisor convinced Searle that the compound was worth development, and the result was the NutraSweet brand of aspartame.

The most intriguing story surrounds the discovery of sucralose. Scientists from Tate & Lyle, a British sugar company, were working in 1989 with researcher Leslie Hough and his student, Shashikant Phadnis, at Queen Elizabeth College (now part of King's College) in London on a project that involved testing chlorinated sugars as chemical intermediates for the synthesis of other compounds. Hough asked Phadnis to "test" sucralose, but Phadnis misunderstood the word "test," thinking that he had been asked to "taste" the compound. The rest is history.

B. Polysaccharides

In principle, any number of monosaccharide residues can be linked together with glycosidic bonds to form chains. When such chains are long, the sugars are called **polysaccharides**. This section surveys a few important polysaccharides.

Cellulose   Cellulose, the principal structural component of plants, is the most abundant organic compound on Earth. Cotton is almost pure cellulose; wood is cellulose combined with a polymer called *lignin*. About 5×10^{14} kg of cellulose is biosynthesized and degraded annually on Earth.

Cellulose is a regular polymer of D-glucopyranose units connected by β-1,4-glycosidic linkages.

cellulose

general structure

Like disaccharides, polysaccharides can be hydrolyzed to their constituent monosaccharides. Thus, cellulose can be hydrolyzed to D-glucose residues. Mammals lack the enzymes that catalyze the hydrolysis of the β-glycosidic linkages of cellulose; this is why humans cannot digest grasses, which are principally cellulose. Cattle, though, can derive nourishment from grasses, but this is because the bacteria in their rumens provide the appropriate enzymes that break down plant cellulose to glucose.

Processed cellulose (cellulose that has been specially treated) has many other uses. It can be spun into fibers (rayon) or made into wraps (cellophane). The paper on which this book is printed is largely processed cellulose. Nitration of the cellulose hydroxy groups gives nitrocellulose, a powerful explosive. Cellulose acetate, in which the hydroxy groups of cellulose are esterified with acetic acid, is known by the trade names Celanese, Arnel, and so on, and is used in knitting yarn and decorative household articles.

cellulose acetate

Cellulose is potentially important as an alternative energy source. Biomass is largely cellulose, and cellulose is merely polymerized glucose. The glucose derived from the hydrolysis of cellulose can be fermented to ethanol, which can be used as a fuel (as in gasohol); and plants obtain the energy to manufacture cellulose from the Sun. Thus, the cellulose in plants—the most abundant source of carbon on Earth—can be regarded as a storehouse of solar energy. An important research problem is how to convert the more abundant sources of cellulose, such as wild grasses, into glucose without expending large amount of energy. A solution to this problem would reduce or eliminate the need to use cultivated crops, such as corn, as a source of ethanol.

Starch   Starch, like cellulose, is also a polymer of glucose. In fact, starch is a mixture of two different types of glucose polymer. In one, *amylose*, the glucose units are connected by α-1,4-glycosidic linkages. Conceptually, the only chemical difference between amylose and cellulose is the stereochemistry of the glycosidic bond.

α-1,4-glycosidic linkage

amylose
($n \approx 400$)

The other constituent of starch is *amylopectin*, a branched polysaccharide. Amylopectin contains relatively short chains of glucose units in α-1,4-linkages. In addition, it contains branches that involve α-1,6-glycosidic linkages. Part of a typical amylopectin molecule might look as follows:

glucose unit

an amylopectin branch

Starch is the important storage polysaccharide in corn, potatoes, and other starchy vegetables. Humans have enzymes that catalyze the hydrolysis of the α-glycosidic bonds in starch and can therefore use starch as a source of glucose.

Chitin   *Chitin* is a polysaccharide that also occurs widely in nature—notably, in the shells of arthropods (for example, lobsters and crabs). Crab shell is an excellent source of nearly pure chitin.

β-1,4-glycosidic linkage

chitin

Chitin is a polymer of *N*-acetyl-D-glucosamine (or, as it is named systematically, 2-acetamido-2-deoxy-D-glucose). Residues of this carbohydrate are connected by β-1,4-glycosidic linkages within the chitin polymer. *N*-Acetyl-D-glucosamine is liberated when chitin is hydrolyzed in aqueous acid. Stronger acid brings about hydrolysis of the amide bond to give D-glucosamine hydrochloride and acetic acid.

$$\text{chitin} \xrightarrow[\text{H}_2\text{O}]{2\,M\,\text{HCl}} \quad \xrightarrow[\text{H}_2\text{O, heat}]{6\,M\,\text{HCl}} \quad + \text{CH}_3\text{CO}_2\text{H} \tag{24.59}$$

N-acetyl-D-glucosamine
(2-acetamido-2-deoxy-D-glucose)

D-glucosamine HCl salt
(2-amino-2-deoxy-D-glucose)

Glucosamine and *N*-acetylglucosamine are the best-known examples of the **amino sugars**. A number of amino sugars occur widely in nature. Amino sugars linked to proteins (glycoproteins) are found at the outer surfaces of cell membranes, and some of these are responsible for blood-group specificity.

Discovery of D-Glucosamine

In 1876 Georg Ledderhose was a premedical student working in the laboratory of his uncle, Friedrich Wöhler (the same chemist who first synthesized urea; p. 3). One day, Wöhler had lobster for lunch, and returned to the laboratory carrying the lobster shell. "Find out what this is," he told his nephew. History does not record Ledderhose's thoughts on receiving the refuse from his uncle's lunch, but he proceeded to do what all chemists did with unknown material—he boiled it in concentrated HCl. After hydrolysis of the shell, crystals of the previously unknown D-glucosamine hydrochloride precipitated from the cooled solution (see Eq. 24.59).

Principles of Polysaccharide Structure   Studies of many polysaccharides have revealed the following generalizations about polysaccharide structure:

1. Polysaccharides are mostly long chains with some branches; there are no highly cross-linked, three-dimensional networks. Some cyclic oligosaccharides are known.

2. The linkages between monosaccharide units are in every case glycosidic linkages; thus, monosaccharides can be liberated from all polysaccharides by acid hydrolysis.

3. A given polysaccharide contains only one stereochemical type of glycoside linkage. Thus, the glycoside linkages in cellulose are all β; those in starch are all α.

PROBLEM

24.33 What product(s) would be obtained when cellulose is treated first exhaustively with dimethyl sulfate/NaOH, then with 1 *M* aqueous HCl?

KEY IDEAS IN CHAPTER 24

• Carbohydrates are aldehydes and ketones that contain a number of hydroxy groups on an unbranched carbon chain, as well as their chemical derivatives.

• The structures of chiral compounds can be drawn in planar representations called Fischer projections, which are especially useful for depicting molecules that contain contiguous asymmetric carbons in an unbranched chain. A Fischer projection is derived from an eclipsed conformation in which all asymmetric carbons are aligned vertically. Asymmetric carbons are represented as the intersection points of vertical and horizontal lines. All vertical bonds at each asymmetric carbon are assumed to be oriented away from the observer and all horizontal bonds toward the observer. Several valid Fischer projections can be drawn for any chiral molecule. These are derived by the rules given in Sec. 24.2.

• A Fischer projection shows the configuration of each asymmetric carbon in a chiral molecule but implies nothing about its conformation.

• The D,L system is an older but widely used method for specifying carbohydrate enantiomers. The D enantiomer is the one in which the asymmetric carbon of highest number has the same configuration as (*R*)-glyceraldehyde (D-glyceraldehyde).

• Monosaccharides exist in cyclic furanose or pyranose forms in which a hydroxy group and the carbonyl group of the aldehyde or ketone have reacted to form a cyclic hemiacetal.

• The cyclic forms of monosaccharides are in equilibrium with small amounts of their respective aldehydes or

ketones and can therefore undergo a number of aldehyde and ketone reactions. These include oxidation (bromine water or dilute nitric acid); reduction with sodium borohydride; cyanohydrin formation (the first step in the Kiliani–Fischer synthesis); and base-catalyzed enolization and enolate-ion formation (the Lobry de Bruyn–Alberda van Eckenstein reaction).

• The —OH groups of carbohydrates undergo many typical reactions of alcohols and glycols, such as ether formation, ester formation, and glycol cleavage with periodate.

• Because the hemiacetal carbons of monosaccharides are asymmetric, the cyclic forms of monosaccharides exist as diastereomers called anomers. The equilibration of anomers is the reason that carbohydrates undergo mutarotation.

• In a glycoside, the —OH group at the anomeric carbon of a carbohydrate is substituted with an ether (—OR) group. In disaccharides or polysaccharides, the —OR group is derived from another saccharide residue. The —OR group of glycosides can be replaced with an —OH group by hydrolysis. Thus, higher saccharides can be hydrolyzed to their component monosaccharides in aqueous acid.

• Disaccharides, trisaccharides, and so on, can be classified as reducing or nonreducing sugars. Reducing sugars have at least one free hemiacetal group. In nonreducing sugars, all anomeric carbons are involved in glycosidic linkages.

 REACTION REVIEW *For a summary of reactions discussed in this chapter, see the* Reaction Review *section of Chapter 24 in the* Study Guide and Solutions Manual.

ADDITIONAL PROBLEMS

24.34 Give the product(s) expected when D-mannose (or other compound indicated) reacts with each of the following reagents. (Assume that cyclic mannose derivatives are pyranoses.)

(a) $Ag^+(NH_3)_2$

(b) dilute HCl

(c) dilute NaOH

(d) Br_2/H_2O, then H_3O^+

(e) CH_3OH, HCl

(f) acetic anhydride/pyridine

(g) product of part (d) + $Ca(OH)_2$, then $Fe(OAc)_3$, H_2O_2

(h) product of part (e) + $PhCH_2Cl$ (excess) and NaOH

24.35 Give the products expected when D-ribose (or other compound indicated) reacts with each of the following reagents.

(a) dilute HNO_3

(b) ^-CN, H_2O

(c) product of part (b) + $H_2/Pd/BaSO_4$ + H_3O^+/H_2O

(d) CH_3OH, HCl (four isomeric compounds; two pyranosides and two furanosides)

(e) products of part (d) + $(CH_3)_2SO_4$ (excess) and NaOH

24.36 Draw the indicated type of structure for each of the following compounds.

(a) α-D-talopyranose (chair)

(b) propyl β-L-arabinopyranoside (chair)

24.37 Name the specific form of each aldose shown here.

(a) HOCH₂ ... OH OH OH (pyranose ring structure)

(b) HOCH₂ HO ... O OH OH OH (pyranose ring structure)

(c) HOCH₂ C H OH OH O OH OH (furanose ring structure)

24.38 Draw the structure(s) of

(a) all the 2-ketohexoses

(b) an achiral ketopentose $C_5H_{10}O_5$

(c) α-D-galactofuranose

(d) β-D-idofuranose

24.39 Specify the relationship(s) of the compounds in each of the following sets. Choose among the following terms: identical compounds, epimers, anomers, enantiomers, diastereomers, constitutional isomers, none of the above. (More than one answer may be correct.)

(a) α-D-glucopyranose and β-D-glucopyranose

(b) α-D-glucopyranose and α-D-mannopyranose

(c) β-D-mannopyranose and β-L-mannopyranose

(d) α-D-ribofuranose and α-D-ribopyranose

(e) aldehyde form of D-glucose and α-D-glucopyranose

(f) methyl α-D-fructofuranoside and 2-O-methyl-α-D-fructofuranose

24.40 Tell whether each structure or term is a correct description of the L-sorbose structure shown here or a form with which it is in equilibrium.

$$\begin{array}{c} CH_2OH \\ | \\ C{=}O \\ HO{-}\!\!-{-}H \\ H{-}\!\!-{-}OH \\ HO{-}\!\!-{-}H \\ | \\ CH_2OH \end{array}$$

L-sorbose

(a) a hexose

(b) a ketohexose

(c) a glycoside

(d) an aldohexose

(e) $$\begin{array}{c} CH_2OH \\ | \\ C{=}O \\ HO{-}\!\!-{-}H \\ H{-}\!\!-{-}OH \\ HOCH_2{-}\!\!-{-}OH \\ | \\ H \end{array}$$

(f) $$\begin{array}{c} H \\ HOCH_2{-}\!\!-{-}OH \\ HO{-}\!\!-{-}H \\ H{-}\!\!-{-}OH \\ C{=}O \\ | \\ CH_2OH \end{array}$$

(g) HOCH₂ O CH₂OH HO HO OH (furanose ring structure)

(h) HOCH₂ O CH₂OH OH OH OH (furanose ring structure)

(i)

(j)

(k)

24.41 Consider the structure of *raffinose*, a trisaccharide found in sugar beets and a number of higher plants.

raffinose

(a) Classify raffinose as a reducing or nonreducing sugar, and tell how you know.

(b) Identify the glycoside linkages in raffinose, and classify each as either α or β.

(c) Name the monosaccharides formed when raffinose is hydrolyzed in aqueous acid.

(d) What products are formed when raffinose is treated with dimethyl sulfate in NaOH, and then with aqueous acid and heat?

24.42 Draw the structure of 3-*O*-β-D-glucopyranosyl-α-D-arabinofuranose, a disaccharide that is the β-glycoside formed between D-glucopyranose at the nonreducing end and the —OH group at carbon-3 of α-D-arabinofuranose at the reducing end.

24.43 Fucose, a carbohydrate with the following structure, has been identified as a component of the cell-surface antigens of certain tumor cells.

fucose

(a) Is this the D- or L-enantiomer of fucose? Explain.

(b) Is this the α- or β-anomer?

(c) Is fucose an aldose, a ketose, or neither? Explain.

(d) Draw a Fischer projection of the carbonyl form of this carbohydrate.

24.44 An important reaction used by Emil Fischer in his research on carbohydrate chemistry was the reaction of aldoses and ketoses with phenylhydrazine to give *osazones*, shown in Fig. P24.44. Osazones, unlike many carbohydrates, form crystalline solids that are useful in characterizing carbohydrates.

(a) Glucose and mannose give the same osazone. Given that these two compounds are aldohexoses, what could a scientist who knows nothing about the stereochemistry of these carbohydrates deduce about their stereochemical relationship from this fact?

(b) What aldopentose gives the same osazone as D-arabinose?

24.45 Complete the reactions shown in Fig. P24.45 by giving the major organic product(s).

24.46 A biologist, Simone Spore, needs the following isotopically labeled aldoses for some feeding experiments. Realizing your expertise in the saccharide field, she has come to you to ask whether you will synthesize these compounds for her. She has agreed to provide an adequate

Figure P24.44

supply of D-(−)-arabinose as a starting material. (* = ^{14}C, T = ^{3}H = tritium)

(a)

$$*CH{=}O$$

H	—	OH
HO	—	H
H	—	OH
H	—	OH

$$CH_2OH$$

(b)

$$CT{=}O$$

H	—	OH
HO	—	H
H	—	OH
H	—	OH

$$CH_2OH$$

(c)

$$CH{=}O$$

HO	—*C—	H
HO	—	H
H	—	OH
H	—	OH

$$CH_2OH$$

Available commercial sources of isotopes include Na_2*CO_3, $Na*CN$, $^3H_2$, and ^3H_2O. Outline a synthesis of each isotopically labeled compound.

24.47 Compound *A*, known to be a monomethyl ether of D-glucose, can be oxidized to a carboxylic acid *B* with bromine water. When the calcium salt of *B* is subjected to

the Ruff degradation, another aldose monomethyl ether is obtained that can also be oxidized with bromine water. When *A* is subjected to the Kiliani–Fischer synthesis, two new methyl ethers are obtained. Both are optically active, and one of them can be oxidized with dilute nitric acid to an optically inactive compound. Suggest a structure for *A*, including its stereochemistry.

24.48 Chlorotris(triphenylphosphine)rhodium brings about the decarbonylation of aldehydes:

$$RCH{=}O + (Ph_3P)_3RhCl \longrightarrow$$

chlorotris(triphenyl-phosphine)rhodium

$$R{-}H + (Ph_3P)_3Rh(CO)$$

carbonyltris(triphenyl-phosphine)rhodium

(a) What product is obtained when this reaction is applied to D-galactose?

(b) Suggest a reason why the reaction of an aldose requires a much higher temperature (130 °C) than the same reaction of an ordinary aldehyde (70 °C).

(c) Outline a mechanism for this reaction by showing the elementary steps involved and the important organo-metallic intermediates. (*Hint:* See Sec. 18.5E.)

24.49 When an optically active aldopentose *A* was subjected to the decarbonylation reaction in Problem 24.48, an opti-

---

(a) phenyl β-D-glucopyranoside + CH_3OH (solvent) $\xrightarrow{H_2SO_4}$

(b) $HOCH_2CH_2CH_2CH_2CH{=}O + CH_3OH$ (solvent) $\xrightarrow{HCl}$

(c) [cyclohexene structure] $\xrightarrow[\text{2) H}_2\text{O, NaHSO}_3]{\text{1) OsO}_4}$ $\xrightarrow{H_5IO_6}$ $\xrightarrow[\text{CH}_3\text{OH}]{\text{NaBH}_4}$

(d) [cyclohexene structure] $\xrightarrow[\text{2) H}_2\text{O, NaHSO}_3]{\text{1) OsO}_4}$ $\xrightarrow[\text{acetone}]{\text{dil. HCl}}$

(e) (+)-sucrose + CH_3I (excess) $\xrightarrow{Ag_2O}$

(f) (+)-lactose + C_2H_5OH $\xrightarrow[\text{heat}]{H_2SO_4}$

(g)

$$\begin{array}{c} CH_3O \quad OCH_3 \\ CH \end{array}$$

H	—	OH
HO	—	H
H	—	OH
H	—	OH

$$CH_2OH$$

$\xrightarrow{CH_3OH, HCl}$

D-glucose dimethyl acetal

Figure P24.45

cally inactive product *B* was obtained. When aldopentose *A* was subjected to the Kiliani–Fischer synthesis, two aldoses *C* and *D* were obtained. Treatment of *C* and *D* with HNO_3 gave optically active aldaric acids *E* and *F*, respectively. Identify compounds *A–E*.

24.50 When D-ribose-5-phosphate was treated with an extract of mouse spleen, an optically *inactive* compound *X*, $C_5H_{10}O_5$, was produced. Treatment of *X* with $NaBH_4$ gave a mixture of the alditols ribitol and xylitol. (See Table 24.2, p. 1258.) Treatment of *X* with periodic acid produced two molar equivalents of formaldehyde. Suggest a structure for *X*.

24.51 The *Wohl degradation*, shown in Fig. P24.51, can be used to convert an aldose into another aldose with one fewer carbon. Give the structure of the missing compounds as well as the curved-arrow mechanisms for the conversion of *B* to *C* and *C* to arabinose.

24.52 The sequence of reactions shown in Fig. P24.52, called the *Weerman degradation*, can be used to degrade an aldose to another aldose with one less carbon atom. Using glucose as the aldose, explain what is happening in each step of the sequence. Your explanation should include the identity of compounds *A* and *B*. (*Hint:* Compound *A* is a lactone, and a lactone is a type of ester.)

24.53 L-Rhamnose is a 6-deoxyaldose with the following structure. When a methyl glycoside of L-rhamnose, methyl α-L-rhamnopyranoside, was treated with periodic acid, compound *A*, $C_6H_{12}O_5$, was obtained that showed no evi-

dence of a carbonyl group in its IR spectrum. Treatment of *A* with CH_3I/Ag_2O gave a derivative *B*, $C_8H_{16}O_5$. Treatment of *A* with H_2/Ni or $NaBH_4$ gave compound *C*, shown here in Fischer projection. Give the structure of *A*. Explain why *A* gives no detectable carbonyl absorption in its IR spectrum, yet reacts with $NaBH_4$.

L-rhamnose
(Haworth projection)

C

24.54 Oligosaccharides of the type shown in Fig. P24.54 are obtained from the partial hydrolysis of starch amylopectin. What ratio of erythritol to glycerol would be obtained from successive treatment of a 12-unit oligosaccharide of the type shown with periodic acid, then $NaBH_4$, and then hydrolysis in aqueous acid?

erythritol   **glycerol**

Figure P24.51

Figure P24.52

24.55 Maltose is a disaccharide obtained from the hydrolysis of starch. Maltose can be hydrolyzed to two equivalents of glucose and can be oxidized to an acid, maltobionic acid, with bromine water. Treatment of maltose with dimethyl sulfate and sodium hydroxide, followed by hydrolysis of the product in aqueous acid, yields one equivalent each of 2,3,4,6-tetra-O-methyl-D-glucose and 2,3,6-tri-O-methyl-D-glucose. Hydrolysis of maltose is catalyzed by α-amylase, an enzyme known to affect only α-glycosidic linkages. Give *two* structures of maltose consistent with these data, and explain your answers.

Treatment of maltobionic acid with dimethyl sulfate and sodium hydroxide followed by hydrolysis of the product in aqueous acid gives 2,3,4,6-tetra-O-methyl-D-glucose and 2,3,5,6-tetra-O-methyl-D-gluconic acid. (See Eq. 24.36, p. 1259, for the structure of D-gluconic acid.) Give the structure of maltose.

24.56 Planteose, a carbohydrate isolated from tobacco seeds, can be hydrolyzed in dilute acid to yield one equivalent each of D-fructose, D-glucose, and D-galactose. Almond emulsin (an enzyme preparation that hydrolyzes α-galactosides) catalyzes the hydrolysis of planteose to D-galactose and sucrose. Planteose does not react with bromine water. Treatment of planteose with $(CH_3)_2SO_4$/NaOH, followed by dilute acid hydrolysis, yields, among other compounds, 1,3,4-tri-O-methyl-D-fructose. Suggest a structure for planteose.

24.57 A process called *sizing* chemically modifies the cellulose in paper. As a result, the paper resists wetting (and thus prevent inks from running). In addition, sizing leaves the paper in a slightly alkaline state. (Acid-free paper lasts much longer than paper that is not acid-free.) One sizing process involves treatment of cellulose with 2-alkylsuccinic anhydrides (where R and R′ are short alkyl groups—for example, ethyl or propyl groups):

(a) What general reaction occurs when this sizing agent reacts with cellulose at pH 7?

(b) Why should this treatment cause the cellulose to become more resistant to wetting? (In answering this question, think of wetting as a solvation phenomenon.)

(c) Why does this treatment cause the paper to be slightly alkaline? That is, what basic group does this treatment introduce?

24.58 Outline a mechanism for the reaction shown in Fig. P24.58, which is an example of the *Maillard reaction* followed by the *Amadori rearrangement*.

24.59 Explain with a curved-arrow mechanism why treatment of the 2-deoxy-2-amino derivative of D-glucose (D-glucosamine) with aqueous NaOH liberates ammonia.

D-glucosamine

Figure P24.54

Figure P24.58

24.60 L-Ascorbic acid (vitamin C) has the following structure:

(a) Ascorbic acid has $pK_a = 4.21$, and is thus about as acidic as a typical carboxylic acid. Identify the acidic hydrogen and explain.

(b) Thousands of tons annually of ascorbic acid are made commercially from D-glucose. In the synthesis shown in Fig. P24.60, give the structures of the compounds A–C.

24.61 At 100 °C, D-idose exists mostly (about 86%) as a 1,6-anhydropyranose:

1,6-anhydro-D-idopyranose

(a) Draw the chair conformation of this compound.

(b) Explain why D-idose has more of the anhydro form than D-glucose. (Under the same conditions, glucose contains only 0.2% of the 1,6-anhydro form.)

24.62 The proton NMR of the C1 proton region of D-glucopyranose shows that both anomers are present. (The large peak in the middle is residual HDO in the H$_2$O solvent.) The integrals are given in arbitrary units.

(a) Which is the resonance of the α-anomer, and which is the resonance of the β-anomer? How do you know?

(b) How much of each anomer is present in the mixture?

Figure P24.60

The Chemistry of Thioesters, Phosphate Esters, and Phosphate Anhydrides

This chapter covers two classes of compounds. The first class is thioesters, which are the sulfur analogs of carboxylate esters. The second class is phosphate esters and phosphate anhydrides, which are derivatives of phosphoric acid. Unlike previous chapters, which focus on laboratory chemistry, the emphasis of this chapter is on reactions that are important in biology, because both of these compound classes are very important in biological chemistry.

One goal of this chapter is to understand the strong analogy of thioester, phosphate ester, and phosphate anhydride chemistry to the chemistry of the analogous carboxylic acid derivatives that we considered in Chapters 20 and 21. A mastery of the principles in these chapters will serve you well in understanding the material in this chapter. Despite this connection, there are important differences. Our second goal is to understand the unique aspects of thioester, phosphate ester, and phosphate anhydride chemistry that led to the evolution of these functional groups to their position of importance in biology.

25.1 THIOESTERS

Thioesters are the sulfur analogs of esters. There are actually two types of thioesters. The more common type, the *S*-alkyl (or aryl) thioester, has a C=O and a sulfur-linked alkyl or aryl group. The much less common type, the *O*-alkyl (or aryl) thioester, has a C=S and an oxygen-linked alkyl or aryl group.

$$H_3C-\overset{\overset{\textstyle O}{\|}}{C}-S-CH_2CH_3 \qquad\qquad H_3C-\overset{\overset{\textstyle S}{\|}}{C}-O-CH_2CH_3$$

S-ethyl thioacetate
(an *S*-alkyl thioester)

O-ethyl thioacetate
(an *O*-alkyl thioester)

The *S*-alkyl thioesters are the only ones we'll be concerned with, and we'll use the term *thioester* in this text to refer to them. If you consult older literature, you will also see them referred to as *thiol esters* ("esters of thiols").

The common nomenclature of thioesters uses the prefix *thio-* before the parent name of the acid and an *S* prefix before the name of the alkyl group to indicate that it is bonded to sulfur. This nomenclature is illustrated by the previous example. *In common usage, the S prefix is understood.* Thus, *S*-ethyl thioacetate is commonly called ethyl thioacetate.

In systematic IUPAC nomenclature, *thio* is inserted before the *ate* suffix in the name of the corresponding oxygen ester. A final *e* must be added to the carboxylic acid stem (*pentan* in this case, which becomes *pentane* because the stem name precedes a consonant).

methyl pentanoate
(an oxygen ester)

***S*-methyl pentan**ethioate
(a thioester)

A few thioesters are biologically important; of these, the most important is **acetyl-CoA**, which is the thioester of acetic acid and a thiol called **coenzyme A**. The structure of acetyl-CoA and a few of its common abbreviations are shown in Fig. 25.1. As with many other biologically important compounds, a very small part of the molecule (the thioester in this case) is involved in its chemistry; the functionally rich remainder of the molecule is involved in noncovalent interactions when it binds to enzymes. We have considered some of the chemistry involving acetyl-CoA in Sections 22.7 and 22.8D.

PROBLEMS

25.1   Draw a structure for each of the following thioesters:

(a) cyclohexyl thiobenzoate

(b) *S*-isopropyl butanethioate

(c) *S*-phenyl cyclohexanecarbothioate

25.2   Provide both common and IUPAC systematic names for the following thioester:

FIGURE 25.1 The structure of acetyl-coenzyme A (acetyl-CoA) and two abbreviated structures. Coenzyme A itself is the thiol HSCoA, shown at the lower right.

25.2   PHOSPHORIC ACID DERIVATIVES

The organic derivatives of phosphoric acid are conceptually similar to the corresponding carboxylic acid derivatives. Thus, esters (phosphate esters), acid chlorides (phosphoryl chlorides), and amides (phosphoramides) are well known.

phosphoric acid
(un-ionized form)

trimethyl phosphate
a phosphate ester

phosphoryl trichloride
(commonly known as
phosphorus oxychloride)
a phosphoric acid chloride

N,N,N´,N´,N″,N″-**hexamethyl-**
phosphorotriamide
(HMPT)
a phosphoramide

We'll concern ourselves primarily with the two types of derivatives that are most important in biology: phosphate esters and anhydrides.

A. Phosphate Esters

One, two, or three of the acidic —OH groups of phosphoric acid can be esterified to give monoesters, diesters, or triesters, respectively.

methyl phosphate
(monomethyl phosphate)
a phosphate monoester

dimethyl phosphate
a phosphate diester

trimethyl phosphate
a phosphate triester

One of the most important differences between phosphate esters and carboxylate esters is that both phosphate diesters and phosphate monoesters have acidic hydrogens. Their pK_a values are very similar to the corresponding pK_a values of phosphoric acid.

$pK_a = 2.2, 7.2, 12.3$ $pK_a = 2.3, 6.7$ $pK_a = 2.3$

At physiological pH (7.4), the major form of phosphate monoesters is the conjugate-base di-anion, and the major form of phosphate diesters is the conjugate-base anion:

major forms at pH = 7.4

Therefore, at physiological pH, phosphate monoesters and phosphate diesters are *ionic compounds*. Because they are ionic, they have *significant water solubility*.

Phosphate diesters occur throughout the biological world. Deoxyribonucleic acid (DNA, the carrier of genetic information), and ribonucleic acid (RNA) are polymeric phosphate diesters. The following structure shows two repeating units of a DNA polymer chain. A DNA polymer can have thousands of repeating units.

DNA covalent structure
(two units)

The B groups shown in red are heterocyclic nitrogen-containing groups, called *DNA bases*, or *nucleobases*, which vary from unit to unit. These can have any of four different structures. (A detailed discussion of DNA structure is given in Sec. 26.5B.) What you should notice for now is the phosphate diester part of the molecule. Notice also that DNA at physiological pH is ionized; each diester unit has one negative charge. The negative charges of DNA are a very important aspect of the DNA three-dimensional structure (Sec. 26.5B).

An interesting and important phosphate diester that isn't a polymer is *cyclic AMP*, a very strained cyclic phosphate diester, which serves as an intracellular signaling molecule.

cyclic adenosine monophosphate
(cyclic AMP)

In this molecule, adenine (*blue*) is one of the nucleobases that occur in DNA and RNA. (Don't be concerned with the detailed structure of the nucleobase at this point. We'll revisit the structures of the bases in Sec. 26.5A.)

Phosphate monoesters occur widely in biology. For example, glucose-6-phosphate is a metabolic intermediate in *glycolysis*, the biological pathway by which glucose is converted into pyruvate.

glucose-6-phosphate
a phosphate monoester

A particularly interesting process involving phosphate monoesters is the biological conversion of tyrosine or serine residues of proteins, which are uncharged at physiological pH, into their phosphate monoesters. This conversion changes the charge of the tyrosine or serine from 0 to −2. As we'll learn in Sec. 27.8E, this charge alteration can result in a profound change in the three-dimensional structure of a protein in which this conversion occurs.

tyrosine residue in a protein → phosphotyrosine residue in a protein

change in charge = −2

(25.1)

B. Phosphate Anhydrides

A phosphate anhydride contains two (or more) linked phosphate groups. The simplest anhydride of phosphoric acid is pyrophosphoric acid, which exists at physiological pH as its conjugate-base tri-anion. Though not an organic compound, it plays an important role as a biological leaving group (Secs. 10.4E and 17.6B) when it is linked to an organic group as a pyrophosphate monoester.

pyrophosphoric acid
(un-ionized form)
pK_a values = 0.9, 2.0, 6.6, 9.4

pyrophosphate
(ionized form at pH = 7.4)
often abbreviated ⁻OPP

γ,γ-dimethylallyl pyrophosphate
a pyrophosphate monoester
(Sec. 17.6B)

The linkage of more than two phosphate groups is also possible. Such polyphosphates serve as phosphate reservoirs in biological systems.

polyphosphate

Among the most important phosphate anhydrides are the **nucleoside triphosphates** and **nucleoside diphosphates**. These contain three structural units: a heterocyclic nucleobase, a sugar (ribose), and the phosphate anhydride group. In a nucleoside triphosphate, the phosphate anhydride actually contains two linked phosphate anhydrides.

general structure of a nucleoside triphosphate
(shown in its fully ionized form)

The most widely occurring nucleoside triphosphate is **adenosine triphosphate** (abbreviated **ATP**). The ATP molecule is the principal chemical storage unit in living cells for the energy derived from glucose metabolism, as we shall see. The corresponding nucleoside diphosphate is **adenosine diphosphate** (abbreviated **ADP**).

adenosine triphosphate (ATP)

adenosine diphosphate (ADP)

The heterocyclic nucleobase in ATP and ADP is adenine (*blue*), the same nucleobase that occurs in the structure of cyclic AMP.

Under physiological conditions, both ATP and ADP (and other nucleoside triphosphates and diphosphates) exist primarily as their complexes with Mg^{2+}, in which the Mg^{2+} is held, or chelated, by interaction with oxygen anions on the two terminal phosphate groups.

Mg^{2+} complex with ATP

Mg^{2+} complex with ADP

A few mixed anhydrides of phosphoric and carboxylic acids are important in biology. This type of compound is called an **acyl phosphate**. An example is carbamoyl phosphate, a mixed anhydride of carbamic acid (p. 1031) and phosphoric acid.

an acyl phosphate

carbamoyl phosphate

PROBLEMS

25.3 Given the pK_a values of methyl phosphate shown in this section, calculate the percentage of the un-ionized form, the mono-anion form, and the di-anion form at pH 7.4.

25.4 The side chain —R of the amino acid serine is —CH_2OH. Draw the structure of the phosphate monoester of serine.

25.5 In the structure of acetyl-CoA (Fig. 25.1), point out and identify the phosphorus-containing functional groups.

25.3 STRUCTURES OF THIOESTERS AND PHOSPHATE ESTERS

A. Structures of Thioesters

The structure of a simple thioester, methyl thioacetate, is shown in Fig. 25.2 along with the structure of an oxygen ester, dimethyl carbonate. The longer carbon–sulfur bonds in comparison to the carbon–oxygen bonds are expected from the greater size of sulfur. The most important comparative aspect of the structures, however, is the relative lengths of the two

FIGURE 25.2 A comparison of the structures of (a) a thioester, (b) a carboxylic acid ester, and (c) a phosphate ester. In comparing the thioester and the oxygen ester, notice that the lengths of the two carbon–sulfur bonds of the thioester are nearly the same, whereas the lengths of the two carbon–oxygen bonds of the oxygen ester are significantly different. In the phosphate ester, notice the tetrahedral structure and the relative bond lengths of the phosphorus–oxygen double and single bonds.

types of carbon–oxygen and carbon–sulfur bonds. In the oxygen ester, the carbonyl–oxygen bond (1.34 Å) is about 6% shorter than the other C—O single bond (1.42 Å). This shortening reflects the partial double-bond character associated with the resonance overlap of the oxygen unshared electron pair with the π electrons of the carbonyl group.

resonance stabilization
of an ester

(25.2a)

In the thioester, however, the carbonyl–sulfur bond (1.78 Å) is only 1% shorter than the other C—S single bond (1.81 Å). Resonance overlap of a sulfur unshared pair with the π electrons of the carbonyl group is less important than it is in an oxygen ester; consequently, there is less double-bond character in the carbonyl–sulfur bond.

this resonance structure
is not very important

(25.2b)

The reason for the weaker resonance interaction in a thioester is that the sulfur resonance interaction involves $3p$ orbitals, which do not overlap well with the $2p$ orbitals of the carbonyl group. (See Fig. 16.7 for a similar situation.) *In other words, resonance does not stabilize a thioester as much as it stabilizes an oxygen ester.* This point will prove to be very important in understanding both the rates and equilibria for the reactions of thioesters.

B. Structures of Phosphate Esters

Because the phosphorus in phosphate derivatives is surrounded by four groups, it has approximately tetrahedral geometry. This geometry stands in contrast to the trigonal planar geometry at carbonyl carbon atoms. The structure of trimethyl phosphate is compared with the structure of dimethyl carbonate in Fig. 25.2. As expected from the greater size of phosphorus, bond lengths to phosphorus are greater than those to carbon.

Phosphorus in phosphate esters and anhydrides is in the phosphorus(V) oxidation state. As usually shown, the phosphorus atom in the structures of these compounds has 10 shared electrons. This is another case of *octet expansion* (Sec. 10.10A). The phosphorus in these derivatives can also be shown with an electronic octet if the P=O double bond is shown as a dipolar single bond.

$$\left[\begin{array}{ccc} & :\overset{..}{O}: & \\ & \parallel & \\ CH_3O\text{····} & P & \diagup OCH_3 \\ & \diagup & \\ & CH_3O & \end{array} \longleftrightarrow \begin{array}{ccc} & :\overset{..}{\underset{..}{O}}:^- & \\ & | & \\ CH_3O\text{····} & \overset{+}{P} & \diagup OCH_3 \\ & \diagup & \\ & CH_3O & \end{array} \right]$$

Comparisons of bond lengths, as well as theoretical calculations, have shown that the P=O bond is a true double bond. In Fig. 25.2, for example, notice that the P=O bond is shorter than the P—O single bonds. In addition, the P=O double bond has a large bond dissociation energy of about 540 kJ mol^{-1} (130 kcal mol^{-1}); the bond energy of a P—O single bond is about 380 kJ mol^{-1} (90 kcal mol^{-1}). As indicated in Sec. 10.10A, octet expansion requires the involvement of phosphorus 3*d* orbitals in forming the double bond. This type of bonding is shown for sulfur in Fig. 10.4 (p. 496); the bonding picture for phosphorus is similar.

Because of its tetrahedral geometry, the phosphorus of a phosphate ester can be an asymmetric atom when the four groups attached to phosphorus are different.

PROBLEMS

25.6   In Fig. 25.2, the C—O—P bond angle (118°) suggests that the oxygen is sp^2-hybridized. Use resonance structures to show why this hybridization and geometry is reasonable.

25.7   The following chiral phosphate ester cannot be isolated in optically active form. Explain.

$$\begin{array}{c} O \\ \parallel \\ CH_3O-P-OH \\ | \\ OCH_2CH_3 \end{array}$$

25.4   PROTON AND CARBON NMR SPECTROSCOPY OF PHOSPHORUS-CONTAINING MOLECULES

Phosphorus (^{31}P) has a nuclear spin of $\pm\frac{1}{2}$. Therefore, when protons or ^{13}C nuclei are being observed, a nearby phosphorus can cause splitting. This is analogous to the splitting of protons caused by ^{19}F (Sec. 13.7C). The splitting of the ^{13}C and ^{1}H resonances in trimethyl phosphate is illustrative.

Carbon NMR:
(proton coupling eliminated)
δ 54, doublet, J_{C-P} = 6 Hz

Proton NMR:
δ 3.8, doublet, J_{H-P} = 11 Hz

$$\begin{array}{c} O \\ \parallel \\ CH_3O-P-OCH_3 \\ | \\ OCH_3 \end{array}$$

Comparison of the chemical shifts in this example with the chemical shifts in carboxylic esters given in Chapter 21 (p. 1055) shows that the chemical shifts of both carbons and protons are very similar in phosphate esters and carboxylic esters.

The ^{31}P NMR of the phosphorus atoms can be observed at an operating frequency different from the frequencies used to observe ^{13}C or ^1H (Sec. 13.9, Table 13.4). In the ^{31}P NMR of trimethyl phosphate, for example, the phosphorus resonance is split into a 10-line pattern with $J = 11$ Hz by the nine methyl hydrogens. As with proton spectra, no splitting of the phosphorus resonance by ^{13}C is observed because of the low natural abundance of the ^{13}C isotope. Additional splitting by ^{13}C would be observed in a sample that is enriched with ^{13}C.

PROBLEM

25.8   Describe the splitting expected in the proton resonance of the —CH$_2$— group in triethyl phosphate. (The coupling constants for H–H splitting and P–H splitting happen to be identical in this case.)

25.5   REACTIONS OF THIOESTERS WITH NUCLEOPHILES

Thioesters, like esters, undergo nucleophilic substitution reactions. In these reactions, a thiol serves as the leaving group. In this section, we'll focus on the hydrolysis of thioesters and a few other reactions that have biological importance.

A. Hydrolysis of Thioesters

Thioesters, like esters, undergo saponification.

Like the saponification of esters, the saponification of thioesters is driven toward products by ionization of the carboxylic acid product. Thioesters have about the same reactivity in saponification as the corresponding oxygen esters. Let's consider why this similar reactivity is reasonable.

Recall (Sec. 21.7E) that relative reactivity in ester hydrolysis is governed by both the stabilization of the ester, which reduces reactivity, and the stabilization of the tetrahedral intermediate, which increases reactivity. As we learned in Sec. 25.3A, the resonance stabilization of thioesters is much weaker than the resonance stabilization of esters.

This fact alone suggests that thioesters should be *more reactive* than esters. However, sulfur is considerably less electronegative than oxygen and chlorine. Therefore, the polar stabilization of the thioester transition state by sulfur is less than the polar stabilization of the ester transition state by oxygen. This fact alone suggests that thioesters should be *less reactive* than esters. Therefore, the two effects work in opposite directions and tend to cancel; in other words, the reactivities of oxygen esters and thioesters are about the same.

The acid-catalyzed hydrolysis of thioesters, like that of esters, requires fairly strong acid.

$$
\underset{\textbf{ethyl thioacetate}}{H_3C-\overset{\displaystyle O}{\overset{\|}{C}}-SCH_2CH_3} + H_2O \xrightarrow[\substack{\text{acetone–water} \\ 40\,°C}]{0.1\ M\ \text{HCl}} H_3C-\overset{\displaystyle O}{\overset{\|}{C}}-OH + HSCH_2CH_3 \quad (25.4)
$$

Thioesters are about 2% as reactive as the corresponding esters in acid-catalyzed hydrolysis, but this difference is not large in the overall reactivity spectrum of carboxylic acid derivatives. In summary, then, thioesters, depending on conditions, have comparable or somewhat lower reactivity toward hydrolysis than oxygen esters.

At pH 7.4 and 37 °C—physiological conditions—the hydrolysis rates of thioesters are negligible; their hydrolysis takes *years*. This means that thioesters such as acetyl-CoA can survive under cellular conditions until they are needed for enzyme-catalyzed reactions in biological processes.

Why did thioesters such as acetyl-CoA evolve in biology rather than the corresponding oxygen esters? We can see that their relative reactivity is *not* the reason, because esters and thioesters have similar reactivities toward hydrolysis. We'll consider this question in Sec. 25.8C.

B. Reactions of Thioesters with Other Nucleophiles

Thioesters, like oxygen esters, can react with a variety of nucleophiles. Here we'll focus on the types of reactions that are most important in biology.

One of the most common biological reactions of thioesters is the reaction with another thiol to form another thioester, called **transthioesterification**. A relatively common example of transthioesterification is the reaction of an ester of CoA—SH, such as malonyl-CoA, to a thiol group of a protein. This thiol group is invariably provided by the side chain of the amino acid residue cysteine.

cysteine residue
in a protein
(such as acyl carrier
protein in fatty-acid
biosynthesis)

malonyl-CoA

a malonylated protein

$$(25.5)$$

This reaction was illustrated in our discussion of fatty-acid biosynthesis (Eq. 22.73a, p. 1143).

We learned about the reactions of thioesters with nucleophilic enolate ions in the Claisen condensation steps of fatty-acid biosynthesis (Eq. 22.73b, p. 1143).

The reaction of a thioester with an alcohol is illustrated by the biosynthesis of acetylcholine, an important neurotransmitter in the brain. This reaction involves the enzyme-catalyzed displacement of the thiol CoA—SH by the oxygen of choline.

acetyl-CoA   **choline**

acetylcholine
a neurotransmitter

$$(25.6)$$

The role of the enzyme is, first, to bind the acetyl-CoA and choline molecules in proximity. The charged nitrogen of choline and the many functional groups in the CoA part of acetyl-CoA (Fig. 25.1) are molecular "handles" that provide sites for significant noncovalent attractions. Second, the enzyme provides a basic histidine residue that enhances the nucleophilicity of the choline hydroxy group by partial proton removal. The enzyme also provides a hydrogen-bond donor—the OH group of a serine residue—which enhances the electrophilicity of the carbonyl group of acetyl-CoA by forming a hydrogen bond to the carbonyl oxygen.

(25.7)

A very important aspect of this reaction is its favorable equilibrium constant ($K_{eq} \approx 13$). A favorable equilibrium constant (that is, a negative $\Delta G°$) is a characteristic of all reactions of acetyl-CoA and other thioesters with alcohols and water. The basis of this observation is considered in Sec. 25.8C.

PROBLEM

25.9   (a) Complete the following reaction.

(b) Although the reaction of the thioester and the amine in part (a) is thermodynamically favorable under neutral conditions, the reaction is not observed in the presence of acid. Explain.

C. Reduction of Thioesters: HMG-CoA Reductase

Thioesters undergo reduction with hydride reducing agents such as sodium borohydride ($NaBH_4$) and lithium aluminum hydride ($LiAlH_4$). (Recall that esters are also reduced by $LiAlH_4$; Sec. 21.9A.) The hydride reduction of esters has a medically important biological counterpart in the reduction of a thioester, (S)-3-hydroxy-3-methylglutaryl-CoA (HMG-CoA), by NADPH to the corresponding alcohol.

(25.8)

The reduction product, (R)-mevalonate, is the biological precursor of isopentenyl pyrophosphate, which is the starting material for the biosynthesis of isoprenoids and steroids, particularly cholesterol (Sec. 17.6C). This reaction, catalyzed by the enzyme HMG-CoA reductase, is the rate-limiting step in the biosynthesis of cholesterol. Inhibiting this enzyme has proven to be a highly effective strategy for lowering cholesterol (see sidebar).

We learned in Sec. 19.8B that one of the most common hydride donors for carbonyl-group reduction in biology is NADH or its phosphorylated variant, NADPH. In the reduction of HMG-CoA in humans, NADPH is the hydride donor. As in other enzyme-catalyzed hydride reductions, a hydrogen-bond donor (in this case a protonated amino group of a lysine residue) activates the carbonyl group by partial protonation. As the CoA—S⁻ leaving group is lost, it is protonated by another group on the enzyme, the conjugate acid of an imidazole group of a histidine residue. The first intermediate, as in laboratory ester reductions, is an aldehyde called *mevaldehyde*:

(25.9a)

Mevaldehyde is then reduced by a second molecule of NADPH to mevalonate. The mechanism of this second step is analogous to the one introduced in Sec. 19.8B. (See Problem 25.10, p. 1296.)

(25.9b)

The Statins: Blockbuster Drugs That Inhibit HMG-CoA Reductase

Coronary heart disease is caused by the formation of cholesterol-containing plaques in the coronary arteries, the blood vessels that supply blood to the heart. Therefore, inhibiting the biosynthesis of cholesterol should be a good strategy for eliminating or reducing these occlusive plaques. Key steps in the biosynthetic pathway that produces cholesterol can be summarized as follows:

The reduction of HMG-CoA is the rate-limiting step in the biosynthesis of cholesterol. Therefore, inhibiting (blocking) the enzyme that catalyzes this step offered the promise of shutting down this pathway. As with many other scientific discoveries, the development of a drug for this purpose was somewhat serendipitous.

In the 1970s, Japanese microbiologist Akira Endo of the Sankyo Company in Tokyo was screening fermentation broths of *Penicillium citrinum* for new compounds that might be used as antibiotics. Fortunately, Endo was also interested in cholesterol metabolism, and he also screened the compounds he discovered for their ability to inhibit HMG-CoA reductase. He found a compound, called *compactin*, that strongly inhibited this enzyme. This turned out to be the first of many drugs, called as a group *statins*, that were effective in inhibiting HMG-CoA reductase. In 1978, scientists at Merck Research Laboratories found another fermentation product, subsequently called *lovastatin*, which also inhibited HMG-CoA reductase. The development of a number of statin analogs followed. The most widely used drug in this class in recent years has been *atorvastatin* (model above), marketed under the trade name Lipitor® by the pharmaceutical company Pfizer.

The sales of Lipitor reached $12 *billion* annually before patent protection for the drug expired in 2011. High-revenue drugs such as Lipitor that are taken indefinitely for chronic conditions are sometimes referred to as "blockbuster drugs."

The mode of action of the statins was further clarified by Michael S. Brown (b. 1941) and Joseph L. Goldstein (b. 1940), physician–scientists at the University of Texas Health Sciences Center in Dallas. Cholesterol is deposited into arterial plaques from transporter proteins called LDL (low-density lipoproteins). Brown and Goldstein showed that when cholesterol biosynthesis is inhibited, ordinary cells, starved for needed cholesterol by the inhibitory action of statins, internalize LDL-cholesterol by expressing LDL receptors on the cell surface. These receptors bind the LDL and transport it into the cell. This takes cholesterol (as the offending LDL-cholesterol) out of circulation. For this work, Brown and Goldstein received the 1985 Nobel Prize for Physiology or Medicine.

PROBLEM

25.10 Using abbreviated structures like the ones in Eq. 25.9a, give a curved-arrow mechanism for the reduction of mevaldehyde shown in Eq. 25.9b.

25.6 HYDROLYSIS OF PHOSPHATE ESTERS AND ANHYDRIDES

Next, we turn to the reactions of phosphate esters and anhydrides with nucleophiles. In this section, we consider the hydrolysis reactions of phosphate esters and anhydrides with the goal of understanding the biologically important examples of these reactions. In Sec. 25.7 we'll consider the reactions of these compounds with other nucleophiles.

A. Hydrolysis of Phosphate Esters

Hydrolysis of Phosphate Triesters Superficially, carboxylate ester hydrolysis and phosphate ester hydrolysis are similar reactions. For example, trialkyl phosphates undergo base-promoted hydrolysis (saponification) reactions just as carboxylic esters do.

$$EtO-\underset{\underset{OEt}{|}}{\overset{\overset{O}{\|}}{P}}-OEt \ + \ ^-OH \ \xrightarrow[40\ °C]{EtOH} \ EtO-\underset{\underset{OEt}{|}}{\overset{\overset{O}{\|}}{P}}-O^- \ + \ EtOH \tag{25.11}$$

Notice that one ester group can be hydrolyzed without affecting the others, a significant point that we'll return to below. Triethyl phosphate saponification is about 3% as fast as ethyl acetate saponification; in other words, phosphate triesters are somewhat less reactive than carboxylate esters.

There are two possible pathways for the saponification reaction: C—O cleavage and P—O cleavage. These can be distinguished by using ⁻OH containing the heavy oxygen isotope ^{18}O (*red*) and determining which product (diethyl phosphate or ethanol) contains the isotope.

$$ \tag{25.12}$$

In one pathway, ⁻OH reacts as a nucleophile at the α-carbon of the alkyl group in an S_N2 reaction and the nucleophilic oxygen is incorporated into the alcohol product. This pathway is analogous to the S_N2 reaction of sulfonate esters (Sec. 10.4A). In the other, ⁻OH reacts at the phosphorus and the nucleophilic oxygen is incorporated into the phosphate diester product. This pathway is analogous to the reaction of ⁻OH at the carbonyl group in ester saponification (Eq. 21.9a, p. 1061). For trimethyl phosphate, C—O cleavage accounts for about 90% of the product. However, ethyl derivatives are less reactive than methyl derivatives in S_N2 reactions (Table 9.3, p. 396). Therefore, C—O cleavage should be considerably slower and the P—O cleavage pathway should be observed for ethyl esters.

The observation of both C—O and P—O cleavage shows that the rates of these two processes are not very different. The P—O cleavage pathway is the most commonly observed one in biology with one exception, the cleavage of alkyl pyrophosphates, which occurs with C—O cleavage. (We'll consider that process in Sec. 25.7C.) For now, we'll focus on the mechanisms of P—O cleavage.

By analogy to the mechanism for carboxylate ester saponification, we can write phosphate ester saponification as an addition–elimination process:

FURTHER EXPLORATION 25.1
Pentacovalent Intermediates in Phosphate Ester Hydrolysis

The addition intermediate in phosphate ester saponification is *pentacovalent*, whereas the addition intermediate in carboxylic ester saponification is tetrahedral. However, phosphorus, because of octet expansion, can be pentacovalent. (For example, PF_5 is a well-known compound.) Although there is no direct evidence for such an intermediate in the saponification of simple phosphate esters, in special cases (see Further Exploration 25.1) there is definitive evidence for pentacovalent addition intermediates.

An alternative mechanism for phosphate ester saponification is a concerted (single-step) substitution of hydroxide on phosphorus, which is analogous to an S_N2 reaction at carbon:

In this case, the pentacovalent species is a transition state rather than an intermediate.

We'll use the concerted substitution mechanism, but you may encounter either mechanism if you study phosphate esters in biochemistry, and either is acceptable.

Hydrolysis of Phosphate Diesters   As shown in Eq. 25.11, a phosphate triester can be hydrolyzed in base to a phosphate diester product, which can be isolated in high yield. The isolation of the diester shows that *phosphate diesters must hydrolyze much more slowly than phosphate triesters*. In fact, the saponification of phosphate diesters by the same P—O cleavage mechanism occurs at about 10^{-6} to 10^{-7} (that is, *one ten-millionth*) the rate of phosphate triester saponification. How do we account for this drastic difference in rates?

In the P—O mechanism for saponification, the transition state for triester saponification contains one negative charge, whereas the transition state for diester saponification contains two negative charges:

$$
\left[
\begin{array}{c}
\overset{\delta-}{:}\overset{..}{O}: \\
\parallel \\
H\overset{..}{\underset{..}{O}}\text{-----}P\text{-----}\overset{..}{\underset{..}{O}}Et \\
EtO^{\diagup}\quad OEt
\end{array}
\right]^{\ddagger}
\qquad
\left[
\begin{array}{c}
\overset{\delta-}{:}\overset{..}{O}: \\
\parallel \\
H\overset{..}{\underset{..}{O}}\text{-----}P\text{-----}\overset{..}{\underset{..}{O}}Et \\
EtO^{\diagup}\quad O^{-}
\end{array}
\right]^{\ddagger}
$$

triester saponification:
one negative charge
in the transition state

diester saponification:
two negative charges
in the transition state

(Three $\delta-$ = one delocalized – charge)

$\qquad\qquad\qquad$ (25.15)

The repulsion between negative charges raises the energy of the transition state. This repulsion accounts for most of the difference in hydrolysis rates. To see that charge repulsion can have an effect of this magnitude, consider the first and second pK_a values of phosphoric acid:

$$
H_2O \;+\; HO-\overset{\displaystyle O}{\overset{\parallel}{\underset{\underset{\displaystyle OH}{|}}{P}}}-OH \;\rightleftharpoons\; HO-\overset{\displaystyle O}{\overset{\parallel}{\underset{\underset{\displaystyle OH}{|}}{P}}}-O^- \;+\; H_3O^+
$$

$pK_a = 2.3$

$$
H_2O \;+\; HO-\overset{\displaystyle O}{\overset{\parallel}{\underset{\underset{\displaystyle OH}{|}}{P}}}-O^- \;\longleftarrow\; HO-\overset{\displaystyle O}{\overset{\parallel}{\underset{\underset{\displaystyle O^-}{|}}{P}}}-O^- \;+\; H_3O^+
$$
$\qquad\qquad$ (25.16)

$pK_a = 7.2$

repulsion accounts for
pK_a difference

The introduction of one negative charge by the first ionization reduces the K_a for the second ionization by a factor of about 10^5. Charge repulsion has a similar effect on rate.

Now we are in a position to appreciate why phosphate diesters are ideally suited as the connections between the nucleic acid monomer units in DNA and RNA. First, these compounds are *polyanions* at pH 7.4. Anions do not readily cross the hydrophobic interior of membranes. This means that DNA and RNA molecules, once formed, do not "leak" from the cell. Secondly, because of the minuscule rate of phosphodiester hydrolysis, DNA and RNA do not undergo spontaneous hydrolysis. The integrity of the DNA and RNA polymers is crucial to the transmission of genetic information. It has been estimated that the spontaneous hydrolysis of a single phosphate ester bond in DNA under physiological conditions occurs with a half-life of greater than 100,000 years!

> Half-life is a way of expressing a rate in terms of the lifetime of the reacting species rather than as a rate constant. The **half-life** is the time required for 50% of a compound to react. Therefore, in one half-life, 50% reacts; in two half-lives, 75%; in three half-lives, 87.5%; and so on. It takes about seven half-lives for 99% of a compound to react.

Although phosphate diesters are very unreactive in solution, the hydrolysis reactions of specific phosphate diesters are important in biology and are catalyzed by enzymes. Let's examine one such reaction: the hydrolysis of DNA by *staphylococcal nuclease*, which we'll abbreviate as *StaphNuc*. A **nuclease** is an enzyme that breaks a phosphate diester bond in DNA or RNA. Staphylococcal nuclease is an enzyme secreted from the bacterium *Staphylococcus aureus*, variations of which are responsible for "staph" infections. Other than the fact that StaphNuc is somehow involved in bacterial infection, its biological role is not well characterized. The following reaction is catalyzed by StaphNuc:

the DNA polymer chain is broken   (25.17)

(Don't worry about the detailed structure of the DNA sugars; focus on the phosphate diester.)

Our goals in examining the catalytic mechanism of StaphNuc are (1) to see how enzymes overcome the intrinsic lack of reactivity of phosphate diesters, and (2) to gain mechanistic insights that we might apply to the reactions of other phosphate esters. The mechanism of StaphNuc catalysis is shown in Fig. 25.3, along with the corresponding curved-arrow notation. Recall that the main reason for the low hydrolytic reactivity of phosphate diesters is their negative charge, and the repulsion between negative charges in the transition state of the hydrolysis reaction. In StaphNuc, two arginine residues in their positively charged, conjugate-acid form—the form present at physiological pH—neutralize the two negative charges in the transition state for hydrolysis and, at the same time, form hydrogen bonds to the phosphate oxygens that contribute to phosphate binding to the active site. A calcium ion also interacts with an oxygen of the phosphate ester to neutralize charge. The carboxylate anion of a glutamic acid residue acts as a base to remove a proton from the water nucleophile as it reacts at the phosphorus. This water, then, is made more nucleophilic by its partial conversion into

FIGURE 25.3 (a) The transition state for DNA (phosphate diester) cleavage catalyzed by the enzyme staphylococcal nuclease. The groups in blue are the metal ion and the catalytically important groups (derived from two arginine residues and a glutamic acid residue) in the active site of the enzyme. The dashed red lines indicate hydrogen bonds that are important for binding and/or catalysis. (b) The structures of the amino acid residues arginine and glutamic acid as they occur in proteins.

a stronger base, hydroxide. The protonated arginines also reduce the basicity of the leaving oxygen by donating hydrogen bonds. In other words, the basicity of the leaving group is reduced by partial protonation. Finally, this reaction is strongly accelerated by the *proximity effect* (Sec. 11.8), because all reactants, acids, and bases are pre-positioned in the active site and do not have to find each other by random diffusion as they would in an ordinary reaction. It has been estimated that the rate acceleration provided by enzyme catalysis in this reaction is a factor of more than 10^{17}.

An important aspect of this mechanism is the stereochemistry of the reaction. The opposite-side substitution mechanism shown in Fig. 25.3 predicts that if the phosphorus were an asymmetric center, the reaction would proceed with inversion. This stereochemical outcome has been verified by examining the StaphNuc-catalyzed conversion of a synthetic chiral diester substrate in which the two enantiotopic oxygens are differentiated by two isotopes of oxygen, ^{17}O and ^{18}O. Reaction of the optically pure diester of known absolute configuration with ordinary water (which contains ^{16}O) gives a chiral product:

(Recall from the Cahn–Ingold–Prelog priority rules, discussed in Sec. 4.2B, that atomic mass can be used to rank priorities for atoms of the same atomic number.) Some ingenious methods have been devised for determining the absolute configuration of phosphate diesters that are chiral by virtue of isotopic substitution. We won't concern ourselves with these methods; we only need to recognize that absolute configurations of isotopically chiral phosphate derivatives can be determined. Analysis of the absolute configuration shows whether the reaction has occurred with inversion, retention, or loss of configuration. The result is (as shown in Eq. 25.18) that *the reaction proceeds with inversion of stereochemistry*. This stereochemistry, as well as the active-site structure shown in Fig. 25.3, fully supports the opposite-side substitution pathway for phosphate ester hydrolysis. In other words, *the stereochemistry of nucleophilic substitution at phosphorus is like the stereochemistry of the S_N2 reaction at carbon.* The term *in-line displacement* is sometimes used by chemists to describe this stereochemistry at phosphorus. This term is completely equivalent to the term *opposite-side substitution*.

Hydrolysis of Phosphate Monoesters Phosphate monoesters contain *two* negative charges, and for that reason we might expect that these compounds would hydrolyze in base by P—O cleavage even more slowly than phosphate diesters. In fact, the nucleophilic substitution mechanism observed for phosphate triesters and phosphate diesters is so unfavorable that it does not occur; however, another P—O cleavage mechanism intervenes. In this mechanism, the cleavage is preceded by a very unfavorable acid–base reaction in which the proton is transferred from its normal position (at physiological pH) to the leaving group. Following this proton transfer, the monoester then undergoes a *unimolecular dissociation* to a very unstable species called **metaphosphate** (PO_3^-), which is the phosphorus analog of nitrate (NO_3^-).

metaphosphate

overlap of nitrogen and oxygen 2*p* orbitals
in the nitrate ion
(a)

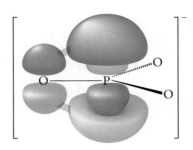

overlap of oxygen 2*p* orbitals
and phosphorus 3*p* orbitals
in the metaphosphate ion
(b)

FIGURE 25.4 The overlap of *p* orbitals in (a) nitrate and (b) metaphosphate. In nitrate, the overlap between oxygen and nitrogen 2*p* orbitals, and therefore the corresponding resonance interaction, is strong. In metaphosphate, the overlap of the oxygen 2*p* orbitals with the phosphorus 3*p* orbitals, and therefore the resonance interaction, is relatively inefficient. The weakness of the resonance interaction accounts for the instability of metaphosphate.

This mechanism occurs because there are *two* negatively charged oxygens to provide the "electronic push" to expel the leaving group, and because the leaving group, once protonated, is a weak base. Although this is the favored mechanism, a reaction that occurs by this mechanism is *very slow* because the initial proton transfer is very unfavorable and the reactive species is present in very low concentration. This reaction is so slow that *phosphate monoesters are the least reactive phosphate esters.* An estimate of the hydrolysis rate at pH 7 and 37 °C suggests that the hydrolysis of an ordinary phosphate monoester di-anion has a half-life of 100 *billion* years!—a time greater than the lifetime of our universe. Therefore, phosphate monoesters can survive indefinitely under physiological conditions until their hydrolysis reactions are catalyzed by specific enzymes.

This dissociative mechanism is somewhat analogous to the S_N1 reaction at carbon (Sec. 9.6), in which dissociation of a leaving group from carbon gives a carbocation. Metaphosphate, unlike nitrate, is a *very unstable* species. Its trigonal-planar geometry requires that the phosphorus is sp^2-hybridized. This means that oxygen 2*p* orbitals must overlap with phosphorus 3*p* orbitals, which have an additional node (Fig. 25.4; compare with a similar situation in Fig. 16.7, p. 819). Hence, resonance structures contribute much less to the stability of metaphosphate than they do to the stability of nitrate, in which overlap occurs among 2*p* orbitals.

Metaphosphate is instantly consumed by nucleophiles such as water to form phosphate:

$$\text{H}_2\overset{..}{\text{O}}: \quad \overset{:\text{O}:}{\underset{^-\overset{..}{\text{O}}}{\overset{\parallel}{\text{P}}}} \overset{\frown}{\text{O}}: \longrightarrow \text{H}_2\overset{+}{\overset{..}{\text{O}}} \overset{\text{O}}{\underset{:\overset{..}{\text{O}}:}{\overset{\parallel}{\text{P}}}} \overset{..}{\text{O}}: \xrightarrow{\text{H}_2\text{O}} \text{H}\overset{..}{\text{O}} \overset{\text{O}}{\underset{:\overset{..}{\text{O}}:}{\overset{\parallel}{\text{P}}}} \overset{..}{\text{O}}^- \; + \; \text{H}_3\text{O}^+ \qquad (25.19b)$$

metaphosphate   **phosphate**

Enzymes that hydrolyze phosphate monoesters are called **phosphatases**. One example of a phosphatase is *fructose 1,6-bisphosphatase*, which we'll abbreviate F16BP. This enzyme catalyzes the hydrolysis of the 1-phosphate ester of fructose-1,6-diphosphate to fructose-6-phosphate.

$$\text{H}_2\text{O} + {}^-\text{O}\overset{\text{O}}{\underset{\text{O}^-}{\overset{\parallel}{\text{P}}}}\text{O} \xrightarrow[\text{bisphosphatase}]{\text{fructose-1,6-}} \text{HO}\overset{\text{O}}{\underset{\text{O}^-}{\overset{\parallel}{\text{P}}}}\text{O}^- + \text{HO} \qquad (25.20)$$

fructose-1,6-bisphosphate   **phosphate**   **fructose-6-phosphate**

FIGURE 25.5 The transition state for hydrolysis of the 1-phosphate monoester in fructose-1,6-bisphosphate catalyzed by the enzyme fructose-1,6-bisphosphatase (F16BP). The carboxylate group in blue is part of the protein structure. The blue coordination arrows indicate that the magnesium ions are also coordinated to groups on the protein (typically carboxylate groups of glutamic and/or aspartic acid residues). Compare the catalytic features of this active site with those in Fig. 25.3 (p. 1299).

The structure of this enzyme has been well characterized, and the transition state of the reaction hypothesized from the structure is shown in Fig. 25.5. The enzyme active site has evolved to overcome all the difficulties of phosphate monoester hydrolysis in solution. Recall that the major problem is the large amount of negative charge on the phosphate oxygens, which makes introduction of a third negative charge in the transition state very unfavorable. The active site contains possibly as many as *three* Mg^{2+} ions that are strongly associated with the phosphate oxygens of the ester that is hydrolyzed. This association in effect neutralizes the negative charge. The nucleophilic water molecule is deprotonated by a second water molecule, which in turn is deprotonated by a carboxylate ion within the active site. This deprotonation enhances the basicity of the nucleophilic water. The leaving group basicity is reduced by its interaction with a Mg^{2+} ion, which, for purposes of the mechanism, we can regard as a "fat proton." Notice that the geometry at phosphorus is trigonal planar, which is consistent with either a trapped metaphosphate, a pentacovalent species, or a concerted substitution. The stereochemistry implied by this mechanism is *inversion*, as in the stereochemistry of phosphate diester hydrolysis. (This stereochemistry has been verified independently with an analog of fructose-1,6-bisphosphate; see Problem 25.11).

The catalysis of phosphate monoester hydrolysis by some phosphatases, unlike F16BP catalysis, shows overall *retention of stereochemistry.* For example, catalysis of phosphate ester hydrolysis by the well-known enzyme *alkaline phosphatase* from the bacterium *Escherichia coli* is in this class.

$$\text{phenyl phosphate (isotopically substituted)} + \text{HO}-\text{R} \xrightarrow{\text{alkaline phosphatase}} \text{phosphate (R = H) a phosphate monoester (R} \neq \text{H)} + \text{HO}-\text{C}_6\text{H}_5 \quad (25.21)$$

phenyl phosphate
(isotopically substituted)

water (R = H)
an alcohol (R ≠ H)

phosphate (R = H)
a phosphate monoester (R ≠ H)

retention of configuration
when R ≠ H

(Although the proton is shown on a specific oxygen, the proton can be on any of the oxygens because proton exchange is very fast.) Because there are only three isotopes of oxygen, we can't study the stereochemistry of the hydrolysis reaction itself. (Notice that transfer to water—hydrolysis—gives an achiral product.) However, the enzyme is fairly promiscuous in the types of molecules that can be used as nucleophiles, and we make the assumption that using a nucleophile other than water will not affect the stereochemistry of the reaction. To study the stereochemistry of this reaction, chemists used an alcohol as a nucleophile, as shown in Eq. 25.21.

Phosphatases that give overall retention have, in all cases so far, been shown to involve a **phosphoenzyme intermediate**. That is, the *overall* substitution reaction occurs as a sequence of *two* substitution processes that involve a nucleophilic group in the enzyme active site. For example, the nucleophilic group in the active site of alkaline phosphatase is the alcohol group in the side chain of a serine residue:

(25.22)

When two successive substitutions with inversion of configuration take place at the same asymmetric atom, overall retention of configuration is observed. This same type of result is obtained in double displacements at an asymmetric carbon, the first of which is intramolecular because of the proximity effect (Sec. 11.8D). In other words, each substitution at phosphorus occurs with inversion.

> We won't concern ourselves with why catalysis of phosphate monoester hydrolysis by some enzymes involves phosphoenzyme intermediates and why catalysis by others does not. We also won't discuss the biological purpose of phosphoenzyme mechanisms. We only need to recognize that when phosphoenzyme intermediates are involved, phosphate ester hydrolysis must involve two substitutions: formation of the phosphoenzyme followed by hydrolysis of the phosphoenzyme.

Although differing in detail from enzyme to enzyme, the mechanisms of both the formation and hydrolysis reactions of phosphoenzymes have the same general mechanistic features as other enzyme-catalyzed phosphate ester hydrolysis reactions, which we now summarize:

1. Enzyme-catalyzed substitution reactions at phosphorus in phosphate ester derivatives occur with inversion of stereochemistry. When retention of stereochemistry is observed in a hydrolysis reaction, it is because the mechanism involves two successive inversion steps.

2. Enzymes overcome the low hydrolytic reactivity of both phosphate diesters and phosphate monoesters in several ways:

   a. removal of a proton from the water nucleophile, thus making it more basic and therefore more nucleophilic;

   b. donation of a proton or metal ion to the oxygen of the leaving group, thus making it less basic and thus a better leaving group;

c. neutralization of charge on the phosphate oxygens either by hydrogen-bond donation from positively charged amino acid residues or by interactions with positively charged metal ions such as Mg^{2+} and Ca^{2+} within the active site, or both; and

d. the proximity effect: all reacting species—acids, bases, substrates, nucleophiles, and metal ions—are bound within the enzyme–substrate complex in ideal, or nearly ideal, arrangements for reaction.

These principles will appear repeatedly as we study the enzyme-catalyzed reactions of other phosphate-containing molecules.

PROBLEMS

25.11 The stereochemistry of substitutions in phosphate esters is sometimes studied with compounds in which sulfur is used in place of one oxygen. For example, the F16BP-catalyzed hydrolysis of fructose-1,6-bisphosphate in $H_2^{17}O$ was studied with the following sulfur-substituted analog:

a. Why is the substitution of sulfur for oxygen important when studying the stereochemistry of this reaction?

b. Assuming that the substitution of sulfur for oxygen does not change the mechanism of the reaction, what is the product of this reaction and what is its stereochemistry?

25.12 The hydrolysis of phosphotyrosine esters in proteins is catalyzed by a family of enzymes called *protein phosphotyrosine phosphatases.*

These hydrolyses in many cases involve phosphoenzyme intermediates. The nucleophilic group of the phosphatase (the enzyme) is the thiol group of a cysteine residue in the active site.

Using abbreviated structures like the ones in Eq. 25.22, write a curved-arrow mechanism for the hydrolysis that includes the structure of the phosphoenzyme intermediate. Assume that acids and bases are present in the enzyme active site as needed.

B. Hydrolysis of Phosphate Anhydrides

Phosphate anhydrides, like carboxylic anhydrides, are much more reactive than the corresponding monoesters, but their hydrolysis rates are small enough that their spontaneous hydrolysis reactions do not compromise their survival under biological conditions. The reason for their low reactivity is the same as for the phosphate monoesters: the large amount of negative charge that causes repulsion when the phosphorus reacts with a nucleophile (see

Eq. 25.15, p. 1298). For example, ATP at physiological pH bears four negative charges and ADP bears three.

adenosine triphosphate (ATP)

adenosine triphosphate (ADP)

It has been estimated that the spontaneous hydrolysis of ATP at physiological pH and 25 °C has a half-life of 6.3 years. ATP forms a complex with divalent ions, such as Mg^{2+} (Sec. 25.2B); as we learned from studying the enzyme-catalyzed hydrolysis of phosphate esters, such ions accelerate hydrolysis, but only by a factor of about 10, because ATP forms a complex with only one Mg^{2+}. The half-lives of ADP and its Mg^{2+} complex are similar. These half-lives, although much smaller than phosphate ester half-lives, are nevertheless long enough that their spontaneous hydrolysis does not compromise their biological function.

The biologically important reactions of ATP involve its reactions with nucleophiles other than water. These reactions are considered in the discussion of the reactions of anhydrides with nucleophiles in Sec. 25.7A (p. 1307).

An anhydride that undergoes a biologically important hydrolysis reaction is the inorganic ion *pyrophosphate*.

pyrophosphate

pyrophosphate
(complex with magnesium ion)

Inorganic pyrophosphate bears three negative charges and is therefore resistant to reaction with nucleophiles. Its spontaneous hydrolysis at 37 °C has a half-life of about 35 years. Divalent ions such as Mg^{2+} increase the hydrolysis rate by a factor of about 4, but a half-life of 9 to 10 years is still very long in comparison to the lifetime of pyrophosphate in biological systems, in which it undergoes rapid enzyme-catalyzed hydrolysis.

The enzyme-catalyzed hydrolysis of magnesium pyrophosphate by water is a biologically significant process.

magnesium pyrophosphate

phosphate

$$\text{(25.23)}$$

In biology, this reaction is catalyzed by a family of enzymes called *inorganic pyrophosphatases*, which accelerate the pyrophosphate hydrolysis by a factor of about 10^{12}. Why is pyrophosphate hydrolysis important? Pyrophosphate serves as a leaving group in a number of biological substitution reactions. Two examples we have discussed are farnesylation (Sec. 10.4E, Eq. 10.38) and isoprenoid biosynthesis (Sec. 17.6B, Eq. 17.44). The enzyme-catalyzed hydrolysis of pyrophosphate, by removal of pyrophosphate as a reaction product, ensures that the substitution reactions are completely irreversible (Le Châtelier's principle).

The active sites of pyrophosphatases show the same features we learned about in phosphate ester hydrolysis. You can examine the active site of one pyrophosphatase for yourself in Problem 25.13 and Fig. 25.6.

PROBLEM

25.13   The transition state for enzyme-catalyzed pyrophosphate hydrolysis, deduced from the structure of the enzyme inorganic pyrophosphatase, is shown in Fig. 25.6.

(a) Indicate the catalytic role of each numbered feature—that is, how each feature enhances catalysis.

(b) If the stereochemistry of the reaction could be determined by isotopic and sulfur substitution of the oxygens, would the result be retention or inversion of configuration at phosphorus?

FIGURE 25.6 The active site for pyrophosphate hydrolysis catalyzed by yeast inorganic pyrophosphatase. (See Problem 25.13.) The groups in blue are from amino acid residues (identified in parentheses) of the protein in the active site. The numbers refer to aspects of catalysis to be identified when working the problem.

25.7 REACTIONS OF PHOSPHATE ANHYDRIDES WITH OTHER NUCLEOPHILES

A. Reactions of ATP with Nucleophiles

Phosphate anhydrides, like carboxylic acid anhydrides, react not only with water, but also with other nucleophiles, such as alcohols. Just as the hydrolysis reactions of phosphate anhydrides are very slow at pH 7, their reactions with alcohols are also very slow. However, many such reactions are catalyzed by enzymes. Adenosine triphosphate (ATP) reacts with a number of biologically important nucleophiles, many of which are alcohols or phenols.

ATP can in principle react with nucleophiles at any of the three phosphorus atoms, and all reactions are known biologically. Reaction at the terminal phosphorus (γ-phosphate) gives a phosphorylated nucleophile and ADP:

(25.25a)

In this reaction, the proton transfer steps are not shown explicitly. Moreover, the association of Mg^{2+} with ATP and ADP, which is important biologically, is not shown but is understood. Enzymes that catalyze this type of reaction are called **kinases**. (Contrast this reaction with the hydrolysis of phosphate monoesters, catalyzed by *phosphatases*, as in Eq. 25.20.)

Reaction at the middle phosphorus (β-phosphate) gives a nucleophile pyrophosphate and AMP:

(25.25b)

In principle, there are two possible phosphate leaving groups in this reaction, but only the loss of AMP is biologically important. This leaving group is selectively activated by the catalyzing enzyme. Reaction at the β-phosphate is the least common of the three modes of reaction of ATP in biology.

Reaction at the phosphorus closest to the sugar (the α-phosphorus) could also in principle result in the loss of two possible leaving groups, but, of the two, the pyrophosphate leaving group is a much weaker base and a better leaving group. Loss of pyrophosphate results in *adenylation* of the nucleophile and the formation of pyrophosphate as a by-product. The pyrophosphate is hydrolyzed to phosphate in a reaction catalyzed by inorganic pyrophosphatase (Sec. 25.6B).

(25.25c)

pyrophosphate
(hydrolyzed to
phosphate by
pyrophosphatase)

the nucleophile is
adenylated

Of the three pathways just shown, the most widely occurring reaction of ATP with nucleophiles in biology is the reaction of a nucleophile at the γ-phosphate to give a phosphorylated nucleophile and ADP (Eq. 25.25a). An example of this reaction is the phosphorylation of glucose in the first step of *glycolysis*, the reaction pathway by which glucose is broken down into smaller fragments and chemical energy is produced. The formation of glucose-6-phosphate is catalyzed by an enzyme called *hexokinase*.

glucose   **glucose-6-phosphate**   (25.26)

This enzyme-catalyzed reaction follows the now-familiar pattern for substitutions at phosphorus. It occurs with *inversion of configuration* at phosphorus, which was demonstrated with isotopically chiral ATP:

(25.27)

The active site of hexokinase contains features that we might expect from our consideration of phosphate-processing enzymes in Sec. 25.6A—namely, a base that removes a proton from the nucleophilic hydroxy group of glucose, a Mg^{2+} ion that bridges between the oxygens on the β- and γ-phosphate groups, and several groups that donate hydrogen bonds to the oxygens of the ADP leaving group.

Although we won't illustrate in the text the reactions of nucleophiles at the β- and γ-phosphates, these reactions fit the general pattern of nucleophilic substitutions at phosphorus—namely, opposite-side substitution, activation of the nucleophile by base catalysis within the enzyme active site, activation of the leaving group by proton donation or metal-ion activation within the enzyme active site, and neutralization of negative charge in the substrate. The mode of reactivity observed in each case—substitution at the α-, β-, or γ-phosphorus—is governed by the way that the nucleophile and ATP are bound in the enzyme active site. For example, if the β-phosphate reacts with a nucleophile, the nucleophile and the β-phosphorus are close together in the enzyme–substrate complex. Therefore, *proximity* determines which mode of reactivity is accelerated, and therefore observed, in a specific case. Each catalyzing enzyme has evolved not only for catalytic efficiency, but also for *selectivity*. (For examples of α-phosphate reactivity, see Problems 25.14 and 25.40; for an example of β-phosphate reactivity, see Problem 25.39.)

PROBLEM

25.14 The conversion of a fatty acid into a fatty acyl-CoA is one step in the utilization of fats:

This process occurs in two steps. In the first step, the fatty acid is adenylated by a nucleophilic reaction of the carboxylate with the α-phosphate group of ATP. In the second step, the fatty acyl adenylate reacts with coenzyme A (CoA—SH).

(a) Show each step of this process.

(b) What additional reaction ensures that the reaction goes fully to completion? (See Eq. 25.23.)

B. Reactions of Acyl Phosphates with Nucleophiles

The reactions of a few mixed carboxylic phosphate anhydrides (acyl phosphates; p. 1288) occur biologically. In one example, the side-chain carboxylic acid group of the amino acid glutamate is phosphorylated by ATP. The resulting acyl phosphate reacts with ammonia to give the amide form of glutamate, called *glutamine*. Both reactions are catalyzed by the enzyme *glutamine synthetase*.

(25.28a)

(25.28b)

Although this reaction is enzyme-catalyzed, the preferential reaction of the nucleophile ammonia at the carbonyl carbon rather than the phosphorus is consistent with the greater reactivity of carboxylic anhydrides.

Another interesting occurrence of an acyl phosphate is in the carboxylation of biotin to form *N*-carboxybiotin. (See Fig. 22.4, p. 1142, for the structure of biotin and *N*-carboxybiotin.) Recall that biotin is nature's "carbon dioxide carrier," and that the purpose of *N*-carboxybiotin is to deliver a carboxy group to the α-carbon of acetyl-CoA to form malonyl-CoA in fatty-acid biosynthesis (Eq. 22.71, p. 1141). The carboxylation of biotin is catalyzed by the enzyme *biotin carboxylase*. The source of the carboxyl group in *N*-carboxybiotin is *not* cellular CO_2, which is present in far too low a concentration to be useful. Rather the source is the bicarbonate ion, which is present in much higher cellular concentration than CO_2. Bicarbonate itself is not reactive enough to carboxylate biotin; so, bicarbonate is activated by an enzyme-catalyzed reaction with ATP to form carboxyphosphate, an acyl phosphate. This reaction is catalyzed by the enzyme biotin carboxylase.

bicarbonate + ATP $\xrightarrow{\text{biotin carboxylase}}$ **carboxyphosphate** (an acyl phosphate) + ADP   (25.29a)

The phosphate group of the carboxyphosphate provides an ionic group that is used to bind this molecule to the enzyme. This acyl phosphate is decomposed in a β-elimination reaction to CO_2. This reaction occurs very close to the biotin that is bound in the active site of the same enzyme.

β-elimination   (25.29b)

biotin   **biotin**

The eliminated phosphate serves as a base to abstract a proton from the nearby biotin and convert it into its conjugate-base anion. Because of its proximity to the biotin anion, CO_2 reacts in a carbonyl addition to give *N*-carboxybiotin before it can escape from the active site into solution.

biotin   a resonance-stabilized anion   *N*-carboxybiotin   (25.29c)

In other words, the source of the carboxy group is not *free* CO_2 from solution, but rather a CO_2 molecule that is produced locally, exactly where it is needed. Notice that one ATP $\rightarrow$ ADP conversion is the price of generating a CO_2 molecule in proximity to biotin.

C. Reactions of Alkyl Pyrophosphates at Carbon

The reactions of alkyl pyrophosphates illustrate a completely different mode of reactivity of phosphate anhydrides. We've been considering the reactions of nucleophiles at phosphorus. However, nucleophiles can also react with alkyl pyrophosphates at *carbon* in S_N2 or S_N1 reactions to expel pyrophosphate as a leaving group (Nuc:$^-$ = a nucleophile):

$$H_2O + R—CH_2—O—P—O—P—O^- \longrightarrow R—CH_2—Nuc + HO—P—O—P—O^- + {}^-OH \qquad (25.30)$$

an alkyl pyrophosphate   **pyrophosphate**

We might expect a nucleophilic reaction at carbon to be particularly favorable when the alkyl group has high S_N2 or S_N1 reactivity. In the two examples we considered in previous chapters, the alkyl groups are both allylic (farnesyl pyrophosphate, Sec. 10.4E; and dimethylallyl pyrophosphate, Sec. 17.6B). Recall (Secs. 17.1 and 17.4) that allylic systems are particularly reactive in S_N1 and S_N2 reactions. This reaction is conceptually similar to the S_N2 and S_N1 reactions of sulfonate esters (Sec. 10.4A). It appears that nature has evolved pyrophosphate as the preferred leaving group for nucleophilic reactions at carbon, and the reason is that these reactions can then be driven to completion by pyrophosphatase-catalyzed hydrolysis of pyrophosphate (Eq. 25.23, p. 1305).

D. Reactions of Other Phosphate Anhydrides at Carbon

Diphosphates other than pyrophosphate serve as leaving groups in the biosynthesis of disaccharides and polysaccharides, and in the biosynthesis of *glycoproteins*—biomolecules in which sugars are linked to proteins. In the biosynthesis of the disaccharide lactose, for example, UDP-galactose, a conjugate of the sugar galactose in its pyranose form and the nucleotide diphosphate UDP, reacts with glucose, also in its pyranose form:

$$+ H_3O^+ \qquad (25.31)$$

This reaction is believed to be an S_N1-like reaction involving a trapped carbocation intermediate. The key features of enzyme catalysis are ones that we have encountered in previous examples—namely, a metal ion (in this case Mn^{2+}) that neutralizes charge on the diphosphate and makes it a better leaving group; a basic group on the enzyme (a carboxylate group of an ionized aspartic acid residue (*blue*) that removes a proton from the nucleophilic oxygen (*red*) of glucose, therefore making this oxygen more basic and more nucleophilic; and a binding site that holds all reactants in proximity, thus making all the reactions effectively intramolecular.

Mn²⁺–UDP leaving group (25.32)

This reaction shows inversion of stereochemical configuration at the anomeric carbon of galactose, but other reactions that form disaccharides with retained (α) configuration at the anomeric carbon are well known. These reactions require either a double-displacement mechanism, or a carbocation mechanism in which the face of the carbocation opposite to the leaving group is shielded by the active site (Problem 25.16).

The glycosylation of proteins is discussed in Sec. 27.8E.

PROBLEMS

25.15 In the biosynthesis of *S*-adenosylmethionine (SAM; Sec. 11.7B), the amino acid methionine undergoes an enzyme-catalyzed reaction with ATP to give SAM and triphosphate. (This is one of only two known biochemical reactions in which triphosphate serves as a leaving group.)

(a) Give a curved-arrow mechanism for this reaction, assuming it is an S_N2 process.

(b) The enzyme active site of the catalyzing enzyme, SAM synthase, contains two Mg^{2+} ions. What role might these ions have in the catalytic process?

25.16 Sucrose synthase is a plant enzyme that catalyzes the biosynthesis of the disaccharide sucrose (structure on p. 1271) from UDP-glucose and β-D-fructofuranose.

(a) Draw the structure of UDP-glucose, assuming it has the same stereochemistry at the anomeric carbon as UDP-galactose.

(b) Draw out the reaction for sucrose biosynthesis. Does this reaction occur with retention or inversion of stereochemistry at the anomeric carbon of UDP-glucose?

(c) Sucrose synthase contains *no* metal cations in its active site. However, it does contain arginine and lysine residues in their cationic, conjugate-acid form (shown below) near the UDP-glucose binding site. What role are these residues likely to play in the synthase mechanism? (*Hint:* See Fig. 25.3, p. 1299.)

structure of an arginine (Arg) residue in a protein   structure of a lysine (Lys) residue in a protein

25.8   "HIGH-ENERGY" COMPOUNDS

A. The Concept of a "High-Energy" Compound

One of the important aspects of ATP and acetyl-CoA in biology is not their reactivities, but rather their utilization as a source of chemical energy. To understand this idea, we must first understand the convention that is commonly used for expressing free energy in biochemical systems. Recall (Sec. 3.5) that the definition of a standard free-energy change for a reaction, $\Delta G°$, is the free energy required to convert an ideal 1 M solution of reactants into an ideal 1 M solution of products at 298 K (25 °C). If protons are involved in the reaction, their concentration is also taken as 1 M (that is, pH = 0). However, in biology, the standard state of the proton is taken as 10^{-7} M (that is, pH = 7), because this is the whole-number pH closest to the physiological pH of 7.4. If the equilibrium for a reaction does not involve the uptake or release of protons, or the acid dissociation of a reactant or product, the $\Delta G°$ for a reaction in the two conventions is the same. However, as we now appreciate, many important biomolecules have dissociable groups with pK_a values within a few units of 7. In these cases, the $\Delta G°$ values in the two conventions are different. To be sure that the convention being used is clear, standard free-energy changes in the pH = 7 convention are labeled with a prime symbol (´) and notated as $\Delta G°´$.

To illustrate a common situation in which the two conventions give different $\Delta G°$ values, consider the $\Delta G°$ for the hydrolysis of a simple ester, such as ethyl acetate.

$$CH_3CO_2Et + H_2O \rightleftharpoons EtOH + CH_3CO_2H \qquad G° = -6.95 \text{ kJ mol}^{-1} (-1.66 \text{ kcal mol}^{-1}) \qquad (25.33a)$$

ethyl acetate   **ethanol**   **acetic acid**

The $\Delta G°$ for hydrolysis of ethyl acetate has been determined experimentally to be -6.95 kJ mol^{-1} (-1.66 kcal mol^{-1}). However, at pH 7, the hydrolysis product acetic acid ($pK_a = 4.76$) is ionized, and this ionization pulls the equilibrium to the right, as we learned in the discussion of ester saponification (Sec. 21.7A):

$$CH_3CO_2Et + H_2O \rightleftharpoons EtOH + CH_3CO_2H \xrightarrow[\text{pH} = 7]{H_2O} CH_3CO_2^- + H_3O^+ \qquad (25.33b)$$

$$pK_a = 4.76$$

At pH = 7, therefore, the free energy of hydrolysis is more favorable than it is at pH = 0—that is, $\Delta G°´$ is more negative than $\Delta G°$. The value of $\Delta G°´$ can be calculated from $\Delta G°$ by the general formula for free energy:

$$\Delta G°´ = \Delta G° + 2.3RT \log \frac{(r_{CH_3CO_2H})(r_{EtOH})}{(r_{CH_3CO_2Et})(r_{H_2O})} \qquad (25.34a)$$

where the r values are the ratios of concentrations of the various species in the equilibrium at pH $= 7$ to their concentrations under standard conditions (pH $= 0$). The only species in this reaction affected significantly by pH is acetic acid. Therefore, Eq. 25.34a becomes

$$\Delta G^{\circ\prime} = \Delta G^{\circ} + 2.3RT \log(r_{CH_3CO_2H}) \qquad (25.34b)$$

The ratio of CH_3CO_2H at pH $= 7$ to its concentration at pH $= 0$ can be calculated from the fractional dissociation formulas at each pH (Eqs. 3.31a and b, p. 107). The fraction of un-ionized acetic acid at pH $= 0$ is 1.0, and the fraction of un-ionized acetic acid at pH $= 7$ is 0.0057. Therefore, $r = 0.0057$. Substituting this into Eq. 25.34b gives

$$\begin{aligned}
\Delta G^{\circ\prime} &= \Delta G^{\circ} + 2.3RT \log (0.0057) \\
&= -6.95 + 5.71(-2.24) \\
&= -19.8 \text{ kJ mol}^{-1} \; (-4.73 \text{ kcal mol}^{-1})
\end{aligned} \qquad (25.34c)$$

As Le Châtelier's principle suggests, the standard free energy in the pH $= 7$ convention is indeed more favorable (more negative).

Now that we've established the pH $= 7$ convention for free energies, we're ready to consider "high-energy" compounds. Let's begin with a simple laboratory example. We learned that we can prepare ethyl acetate by letting acetic acid come to equilibrium with an excess of ethanol in the presence of an acid catalyst. However, we also learned that ethyl acetate can be prepared more rapidly by the reaction of ethanol with acetic anhydride or acetyl chloride. When we studied these reactions, our focus was on the *reactivity* of acetic anhydride and acetyl chloride. However, another reason that acetic anhydride or acetyl chloride is used is that *their esterification reactions are essentially irreversible because the equilibrium constants for these reactions are very large.* A large equilibrium constant means that the standard free-energy change for the reaction is very favorable (negative). The source of this large, negative standard free-energy change is *both* the high standard free energy of the reactants (acetic anhydride or acetyl chloride) and the low standard free energy of reaction products (an ester and either acetate ion or chloride ion). To take the reaction of an anhydride as an example, we showed in Eqs. 21.26–21.27 (p. 1069) that the resonance stabilization of acetic anhydride is less effective than the resonance stabilization of both an ester and the acetate ion. The gain in resonance stabilization releases significant energy (Fig. 25.7). Therefore, the relative free energies of *both* reactants and products contribute to the large negative $\Delta G^{\circ\prime}$ for ester formation. However, biologists tend to focus on the high-energy side of the equilibrium, and they would say that acetic anhydride and acetyl chloride are "high-energy" compounds.

In biology, a **high-energy compound** is one that has a large negative $\Delta G^{\circ\prime}$ for hydrolysis. (Typically this concept is applied to carboxylic and phosphoric acid derivatives, all of which undergo hydrolysis to the corresponding acids.) *A compound is considered to be a high-energy compound if its $\Delta G^{\circ\prime}$ for hydrolysis is more negative than about –30 kJ mol⁻¹ (–7.2 kcal mol⁻¹).* Let's see how this idea can be applied to the esterification of acetic acid. The $\Delta G^{\circ\prime}$ for the esterification of acetic acid is the negative of the $\Delta G^{\circ\prime}$ for ester hydrolysis (+19.8 kJ mol⁻¹ [+4.73 kcal mol⁻¹]; Eq. 25.34c). Remember, this $\Delta G^{\circ\prime}$ value is actually the standard free energy for the esterification of acetic acid as it exists at pH $= 7$—in its conjugate-base acetate form—and any protons present are at a concentration of 10^{-7} M (pH $= 7$). The $\Delta G^{\circ\prime}$ for acetic anhydride hydrolysis has been experimentally determined to be –91.2 kJ mol⁻¹ (–21.8 kcal mol⁻¹). This value puts acetic anhydride into the category of a "high-energy compound." Remember, now, that *we can calculate the energy change for any reaction as the sum of the energy changes for other reactions.* (We used this idea with heats of formation in Sec. 4.5A.) If we add the $\Delta G^{\circ\prime}$ for the esterification of acetic acid to the $\Delta G^{\circ\prime}$ for the hydrolysis of acetic anhydride, we can cancel one acetate ion, one water molecule, and one proton, and the result is the $\Delta G^{\circ\prime}$ for the formation of ethyl acetate from acetic anhydride.

FIGURE 25.7 The reaction of ethanol with acetic anhydride releases considerable free energy because the resonance stabilization of both products is more effective than the resonance stabilization of the anhydride.

$$H^+ + CH_3CO_2^- + EtOH \rightleftharpoons CH_3CO_2Et + H_2O \qquad G^{\circ\prime} = +19.8 \text{ kJ mol}^{-1} (+4.7 \text{ kcal mol}^{-1})$$

$$\underset{\substack{O \quad O \\ \| \quad \|}}{CH_3COCCH_3} + H_2O \rightleftharpoons 2\,CH_3CO_2^- + 2\,H^+ \qquad G^{\circ\prime} = -91.2 \text{ kJ mol}^{-1} (-21.8 \text{ kcal mol}^{-1})$$

Sum: $\underset{\substack{O \quad O \\ \| \quad \|}}{CH_3COCCH_3} + EtOH \rightleftharpoons CH_3CO_2Et + CH_3CO_2^- + H^+ \qquad G^{\circ\prime} = -71.4 \text{ kJ mol}^{-1} (-17.1 \text{ kcal mol}^{-1})$

$$(25.35)$$

The $\Delta G^{\circ\prime}$ for the formation of ethyl acetate from acetic acid at pH = 7 and ethanol is actually unfavorable because of the ionization of acetic acid. The relatively large chemical free energy of acetic anhydride provides the free-energy "push" that makes esterification with the anhydride so favorable, and the $\Delta G^{\circ\prime}$ for the hydrolysis of acetic anhydride is our standard measure of this free energy (Fig. 25.8, p. 1316).

B. ATP as a "High-Energy" Compound

ATP is the most ubiquitous "high-energy" compound in biology. For example, Eq. 25.26 (p. 1308) shows an example of the use of ATP as a phosphorylating reagent to form phosphate monoesters. The $\Delta G^{\circ\prime}$ for the hydrolysis of ATP (to ADP and phosphate) at pH = 7 in the presence of 10^{-3} *M* Mg^{2+} is -30.5 kJ mol^{-1} (-7.3 kcal mol^{-1}). The $\Delta G^{\circ\prime}$ for the formation of glucose-6-phosphate from phosphate and glucose is $+13.8$ kJ mol^{-1} ($+3.3$ kcal mol^{-1}). The positive $\Delta G^{\circ\prime}$ value shows that the formation of glucose-6-phosphate from phosphate and glucose is unfavorable. However, utilization of the phosphate anhydride ATP as the phosphorylating reagent makes phosphate monoester formation favorable:

FIGURE 25.8 Diagram of Eq. 25.35. The relatively high free energy of acetic anhydride (see Fig. 25.7) makes its reaction with nucleophiles such as ethanol a very favorable process. The $\Delta G°'$ for its hydrolysis is the standard way of expressing the "high-energy" character of acetic anhydride. Notice that ester formation from acetic acid and ethanol is an unfavorable process because acetic acid is ionized to acetate at pH = 7, the standard state for the energy calculation.

$$\text{ATP} + \text{H}_2\text{O} \xrightleftharpoons[]{10\ m\text{M Mg}^{2+}} \text{ADP} + \text{phosphate} \qquad G°' = -30.5 \text{ kJ mol}^{-1}\ (-7.3 \text{ kcal mol}^{-1})$$

$$\text{glucose} + \text{phosphate} \rightleftharpoons \text{glucose-6-phosphate} + \text{H}_2\text{O} \qquad G°' = +13.8 \text{ kJ mol}^{-1}\ (+3.3 \text{ kcal mol}^{-1})$$

Sum: $$\text{glucose} + \text{ATP} \xrightleftharpoons[]{10\ m\text{M Mg}^{2+}} \text{glucose-6-phosphate} + \text{ADP} \qquad G°' = -16.7 \text{ kJ mol}^{-1}\ (-4.0 \text{ kcal mol}^{-1})$$

$$(25.36)$$

In reactions like this, it is sometimes said, "ATP hydrolysis is coupled to phosphate ester formation." As you can see, ATP is not really hydrolyzed. But a product of the reaction, ADP, is the same one that would form if ATP *were* hydrolyzed. This terminology is making use of the idea that ATP hydrolysis is a way of measuring the fact that it is a high-energy compound (Fig. 25.9).

ATP hydrolysis releases a lot of free energy for the same reason that hydrolysis of carboxylic anhydrides does: because there is a greater resonance stabilization in the product (phosphate, in this case) than there is in the anhydride. However, the resonance interaction in phosphate is weaker than in a carboxylate because phosphorus is a third-period atom, and so we expect this effect to be less important for ATP. Indeed, comparing the $\Delta G°'$ for the hydrolysis of ATP (Eq. 25.36) with the $\Delta G°'$ for the hydrolysis of acetic anhydride (Eq. 25.35), we find that substantially less energy is released from ATP hydrolysis.

Another reason that ATP hydrolysis is so favorable is that the unfavorable electrostatic repulsion between four negative charges in ATP is relieved. The $\Delta G°'$ for the hydrolysis of ATP cited in Eq. 25.36 is for its complex with Mg^{2+}, which is the form of ATP present under physiological conditions. (The product ADP is also present as a Mg^{2+} complex.) The presence of Mg^{2+} partially offsets the charge repulsion in ATP and ADP. Indeed, when Mg^{2+} is not present, the $\Delta G°'$ for the hydrolysis of ATP is more negative by about 10 kJ mol^{-1} (2.4 kcal mol^{-1}).

C. Thioesters as "High-Energy" Compounds

Thioesters, and specifically acetyl-CoA, are examples of high-energy compounds. The $\Delta G°'$ for the hydrolysis of acetyl-CoA is –31.4 kJ mol^{-1} (–7.5 kcal mol^{-1}). (The $\Delta G°'$ for the hydrolysis of ethyl acetate, an oxygen ester, is –19.8 kJ mol^{-1} [–4.73 kcal mol^{-1}].)

FIGURE 25.9 Diagram of Eq. 25.36. The $\Delta G°'$ for the phosphorylation of glucose by ATP is equal to the $\Delta G°'$ for the phosphorylation of glucose by phosphate plus the $\Delta G°'$ for ATP hydrolysis. The $\Delta G°'$ for ATP hydrolysis is the standard way of showing the "high-energy" character of ATP.

The reason that thioesters are high-energy compounds can be seen from an analysis of their resonance structures. Resonance overlap between the sulfur unshared electron pair and the carbonyl π orbital is weak, because the sulfur $3p$ orbitals have nodes that weaken this overlap. (See Figs. 16.7 and 25.4 for similar situations.) In the carboxylate hydrolysis product, resonance stabilization is very strong. It is the gain of resonance stabilization in the hydrolysis product that accounts for the highly negative $\Delta G°'$ for the hydrolysis of thioesters. Another way to express the same idea is that the high energy of thioesters is a reflection of their weak resonance stabilization. Resonance stabilization in oxygen esters is more effective. Therefore, less energy is released (the $\Delta G°'$ is "less negative") when oxygen esters undergo hydrolysis (Fig. 25.10, p. 1318).

The observations in this section suggest a reason that thioesters evolved to become important in biology. Thioesters are not particularly reactive; under physiological conditions, acetyl-CoA and other thioesters are stable toward hydrolysis and other nucleophilic substitution reactions. Therefore, they will survive in the cell until they are needed for metabolically important enzyme-catalyzed reactions. However, their favorable $\Delta G°'$ values for reactions with water and alcohols show that, when they do undergo enzyme-catalyzed reactions with these nucleophiles, their reactions will go essentially to completion.

PROBLEMS

25.17 The $\Delta G°'$ for ester hydrolysis of the neurotransmitter acetylcholine is –25.1 kJ mol^{-1} (–6.0 kcal mol^{-1}), and the $\Delta G°'$ for the hydrolysis of acetyl-CoA is –31.4 kJ mol^{-1} (–7.5 kcal mol^{-1}).

$$H_3C-\overset{\overset{\displaystyle O}{\|}}{C}\diagdown O\diagup\overset{+}{N}(CH_3)_3 \qquad HO\diagup\overset{+}{N}(CH_3)_3$$

acetylcholine
a neurotransmitter

choline

(a) Draw a free-energy diagram that shows how the $\Delta G°'$ for the biosynthesis of acetylcholine from acetyl-CoA and choline can be derived from the $\Delta G°'$ for these two hydrolysis reactions.

(b) What is the $\Delta G°'$ for this biosynthetic reaction?

(c) What is the equilibrium constant for this reaction at 37 °C? ($2.3RT$ at 310K = 5.92 kJ mol^{-1}.)

continued

continued

25.18 (a) Phosphocreatine (creatine phosphate) is a compound in muscle tissue that serves as a reservoir of high-energy phosphate. It reacts with ADP to replenish ATP stores in muscle:

phosphocreatine   **creatine**

The $\Delta G^{\circ\prime}$ for this reaction is -12.5 kJ mol^{-1} (-3.0 kcal mol^{-1}). Calculate the $\Delta G^{\circ\prime}$ for the hydrolysis of phosphocreatine. Is phosphocreatine a "high-energy" compound?

(b) Using resonance structures of creatine and phosphocreatine, explain why the hydrolysis of phosphocreatine releases a lot of energy. (*Hint:* Use an octet structure for the phosphorus.)

D. Free Energy in Living Systems

A final point about biochemical free energy: Most reactions in biochemical systems are *not* at equilibrium. (If they were, we would all be dead!) The actual free-energy change in biochemical reactions is calculated from the general formula for free energy. For a reaction A + B $\rightleftharpoons$ C + D, the actual $\Delta G'$ under cellular conditions is

$$\Delta G' = \Delta G^{\circ\prime} + 2.3RT \log \frac{[C][D]}{[A][B]} \qquad (25.37)$$

FIGURE 25.10 The weak resonance interaction between sulfur and the carbonyl π electrons results in a relatively high standard free energy for thioesters such as acetyl-CoA. Hydrolysis of the thioester or any other nucleophilic substitution resulting in replacement of the sulfur with nitrogen or oxygen releases significant free energy because a derivative is formed in which the resonance interaction is much stronger.

where $\Delta G°'$ is the standard free-energy change at pH = 7 (that is, $-2.3RT \log K_{eq}$) and the concentrations [A], [B], [C], and [D] are the *actual* concentrations in the cell. (If water is a reactant or product, its concentration is taken into account as part of the $\Delta G°'$. This means that the water concentration can be ignored or, equivalently, taken as 1.0, as we do with acid dissociation constants.) Even though $\Delta G'$ is the *actual* chemical free-energy change under specific conditions, the $\Delta G°'$ is a useful quantity because it provides the baseline, or standard, from which the actual $\Delta G'$ is calculated.

STUDY PROBLEM 25.1

The $\Delta G°'$ for ATP hydrolysis is -30.5 kJ mol^{-1} (-7.3 kcal mol^{-1}). In the red blood cell at 37 °C (310 K) the concentrations of ATP, ADP, and phosphate are 2.25, 0.25, and 1.65 mM, respectively. Show that these are not the equilibrium concentrations, and calculate the actual $\Delta G'$ under these conditions. ($2.3RT$ at 37 °C is 5.93 kJ mol^{-1} [1.42 kcal mol^{-1}].)

SOLUTION   The ATP hydrolysis reaction is

$$\text{ATP} + \text{H}_2\text{O} \underset{\xrightarrow{\text{10 m}M\text{ Mg}^{2+}}}{\rightleftharpoons} \text{ADP} + \text{phosphate}$$

First, we calculate the equilibrium constant from the relation $\Delta G°' = -2.3RT \log K_{eq}$, from which we find that $\log K_{eq} = -30.5/-5.93 = 5.14$. Therefore $K_{eq} = 1.39 \times 10^5$. This large equilibrium constant means that ATP hydrolysis is very favorable, as we expect for the hydrolysis of an anhydride. The equilibrium constant is equal to the ratio [ADP][phosphate]/[ATP] *at equilibrium* at pH = 7. At the concentrations given in the problem, this ratio is actually 1.83×10^{-4}. Hence, under biological conditions, there is much more ATP than there would be at equilibrium. This is typically true in living systems, because ATP must be available to drive the many energy-requiring biological processes that depend on it.

To calculate the actual $\Delta G'$, we apply Eq. 25.37 and use the actual cellular concentrations of the three components of the reaction. The result is

$$\Delta G' = \Delta G°' + 2.3RT \log \frac{\text{[ADP] [phosphate]}}{\text{[ATP]}}$$

$$= -30.5 + (5.93)\log(1.83 \times 10^{-4})$$

$$= -30.5 + (5.93)(-3.74) = -52.7 \text{ kJ mol}^{-1}$$

This calculation shows that the chemical driving force $\Delta G'$ for ATP hydrolysis is greater (that is, more negative) than the $\Delta G°'$. However, as this calculation shows, the $\Delta G°'$ is a significant component of this total free-energy change.

PROBLEMS

25.19   What would the $\Delta G'$ be for ATP hydrolysis in a cell in which the concentrations of ATP, ADP, and phosphate were all 1 mM?

25.20   This question refers to the biosynthesis of *S*-adenosylmethionine (SAM) introduced in Problem 25.15 (p. 1312). The hydrolysis of the triphosphate leaving group in this reaction is catalyzed by the enzyme SAM synthase, and the hydrolysis of the pyrophosphate product of this hydrolysis is hydrolyzed by pyrophosphatase.

ADP rather than ATP could have evolved as an adenosyl donor, in which case pyrophosphate rather than triphosphate would have been the leaving group. What is the advantage of producing triphosphate as a leaving group rather than pyrophosphate? (*Hint:* What if methionine were present in very small concentration? Think about $\Delta G'$.)

KEY IDEAS IN CHAPTER 25

- Thioesters are conceptually similar to esters. However, thioesters have much less resonance stabilization than esters because of the relatively poor orbital overlap between sulfur $3p$ orbitals and the $2p$ orbitals of the carbonyl group. The weaker resonance stabilization of thioesters accounts for their status as "high-energy" compounds.

- In spite of their weaker resonance stabilization, thiols have about the same reactivity as esters in hydrolysis reactions because the sulfur, being much less electronegative than oxygen, does not stabilize the transition state for hydrolysis as well as oxygen.

- Acetyl-CoA is the most important and widely occurring biological thioester. Thioesters formed between acyl groups and the thiol groups of cysteine residues of proteins are also important intermediates in biology.

- Phosphoric acid derivatives have tetrahedral geometry. One, two, or all three of the OH groups can be derivatized. Thus, phosphoric acid can have monoester, diester, and triester derivatives. Monoester and diester derivatives have acidic protons.

- The NMR spectra of phosphoric acid derivatives are similar to those of carboxylic acid derivatives except that ^{31}P (nuclear spin $= \pm\frac{1}{2}$) can cause splitting of nearby ^{13}C and ^{1}H nuclei. Phosphorus NMR can be used to detect ^{31}P.

- Phosphate esters undergo substitution reactions at phosphorus that are conceptually similar to those of carboxylate esters. The substitution mechanism most often observed in biology is an enzyme-catalyzed S_N2 mechanism at phosphorus.

- Pyrophosphate monoesters in biology typically react by S_N1 or S_N2 substitution at the carbon of the ester, and the resulting pyrophosphate leaving group is hydrolyzed to two molecules of phosphate in a reaction catalyzed by the enzyme pyrophosphatase. The biosynthesis reactions of di- and polysaccharides involve conceptually similar nucleophilic substitutions by sugar nucleophiles at an anomeric carbon in which a nucleotide diphosphate such as UDP acts as a leaving group.

- Phosphate triesters have roughly the same reactivity in hydrolysis as carboxylic acid esters. Phosphate diesters are much less reactive in base-promoted hydrolysis because of repulsion between negative charges in the transition state. Phosphate monoester di-anions are even less reactive; they react very slowly by an acid-

catalyzed dissociative mechanism to give a very unstable metaphosphate intermediate, which rapidly reacts with nucleophiles.

- The catalysis of hydrolysis reactions of phosphate diesters and monoesters by enzymes typically involves

  a. reaction of the nucleophile at phosphorus, and cleavage of the bond between phosphorus and the leaving group;

  b. catalysis by proximity: all of the reactive groups are bound in an ideal arrangement for reaction in the enzyme active site;

  c. base catalysis: a proton is removed from a nucleophilic water or alcohol molecule by a basic group of the enzyme;

  d. acid catalysis: the leaving-group basicity is reduced by partial protonation by an acidic group of the enzyme or by interaction with a metal ion (typically Mg^{2+}); and

  e. electrophilic catalysis: the negative charge in the transition state for hydrolysis is neutralized by divalent metal ions, by hydrogen-bond donation, and/or by positively charged groups in the enzyme active site.

- Phosphate monoester hydrolysis is catalyzed by phosphatases. Some phosphatase mechanisms involve net retention of configuration at phosphorus. This stereochemical outcome is the result of two successive opposite-side displacements involving a phosphoenzyme intermediate.

- Phosphate anhydrides, although much more reactive than phosphate esters, are stable under physiological conditions. Pyrophosphate hydrolysis is catalyzed by inorganic pyrophosphatase. ATP can react with nucleophiles at any of its three phosphorus atoms; the particular reaction observed depends on the enzyme catalyst. Reaction at the γ-phosphorus, catalyzed by kinases, results in phosphorylation of the nucleophile and loss of ADP as a leaving group. Reaction at the β-phosphorus results in loss of AMP and pyrophosphorylation of the nucleophile. Reaction at the α-phosphorus results in adenylation of the nucleophile and loss of pyrophosphate as a leaving group. The mechanistic and stereochemical aspects of these reactions are much like those observed in the enzyme-catalyzed hydrolysis of phosphate esters.

- The $\Delta G°$ most commonly used in biology is $\Delta G°'$, which is the $\Delta G°$ value adjusted to pH $= 7$.

- "High-energy" compounds release a substantial amount of free energy (as measured by $\Delta G°'$) when they are hydrolyzed. Examples in laboratory chemistry are carboxylic anhydrides and carboxylic acid chlorides. Examples in biology are ATP and other nucleoside triphosphates, thioesters, and acyl phosphates. The reactions of these compounds with ordinary nucleophiles such as alcohols are typically irreversible.

• Most biochemical reactions are not at equilibrium in living systems. The driving force for biochemical reactions at pH = 7 is $\Delta G'$, the actual free-energy change of the reaction under nonequilibrium conditions (Eq. 25.37, p. 1318).

 REACTION REVIEW *For a summary of reactions discussed in this chapter, see the* Reaction Review *section of Chapter 25 in the* Study Guide and Solutions Manual.

ADDITIONAL PROBLEMS

25.21 Identify the phosphorus-containing functional groups in each of the biologically occurring compounds shown in Fig. P25.21. Choose between phosphate monoester, phosphate diester, phosphate anhydride, and pyrophosphate monoester.

25.22 The anomeric proton (*red*) in the α-anomer of glucose-1-phosphate in the proton NMR spectrum appears at δ 5.45. It is split into a doublet of doublets—four lines of equal size—with coupling constants of 3.5 Hz and 7.3 Hz. Explain the origin of each splitting.

glucose-1-phosphate
(dipotassium salt)

25.23 In each of the following sets, arrange the three compounds in order of increasing reactivity toward base-promoted hydrolysis, least reactive first, and explain your reasoning. (No enzymes are involved.)

(a)

$$H_3C - \overset{O}{\overset{\|}{C}} - OCH_2CH_3 \qquad EtO - \overset{O}{\overset{\|}{\underset{O^-}{P}}} - O^- \qquad EtO - \overset{O}{\overset{\|}{\underset{O^-}{P}}} - OEt$$

A   B   C

(b)

$$EtO - \overset{O}{\overset{\|}{\underset{OEt}{P}}} - O - \overset{O}{\overset{\|}{\underset{OEt}{P}}} - OEt \qquad EtO - \overset{O}{\overset{\|}{\underset{OEt}{P}}} - OEt$$

A   B

$$Me_2N - \overset{O}{\overset{\|}{\underset{NMe_2}{P}}} - NMe_2 \qquad Cl - \overset{O}{\overset{\|}{\underset{Cl}{P}}} - Cl$$

C   D

5-phosphoribosy-1-pyrophosphate

cyclic GMP

glycerol-1,3-diphosphate

Figure P25.21

25.24 Complete the reactions shown in Fig. P25.24 by drawing the structures of the products, and explain your reasoning.

25.25 When *S*-ethyl thiobenzoate (*A*) is allowed to react with 2-aminoethanol (*B*), an amide *C* is formed if one equivalent of triethylamine (Et$_3$N:) is included in the reaction mixture, but an ester *D* is formed if one equivalent of a strong acid such as *p*-toluenesulfonic acid is added as a catalyst. (See Fig. P25.25.) Give the structures of *C* and *D*, and explain why different products are formed under different conditions.

25.26 When *S*-cyclohexyl thioacetate is reduced by LiAlH$_4$ in ether, followed by protonolysis, cyclohexanethiol is formed. However, when a large excess of the Lewis acid BF$_3$ is added to the reaction mixture before the reduction, cyclohexyl ethyl sulfide is formed (Fig. P25.26). Account mechanistically for the effect of BF$_3$ in changing the outcome of the reaction.

25.27 Sarin is an acid fluoride and ester of methylphosphonic acid.

sarin
(propan-2-yl methylphosphonofluoridate)

Sarin is a deadly nerve gas that was outlawed by the Chemical Weapons Convention of 1993. However, it has been used at various times as a weapon of mass destruction by terrorists, including a 1995 attack in the Tokyo metro and a 2013 attack by the Syrian government on rebel forces during the Syrian civil war.

(a) Sarin is chiral; the *S* enantiomer is more active than the *R* enantiomer. Draw a line-and-wedge structure of the more active enantiomer.

Figure P25.24

Figure P25.25

Figure P25.26

(b) Sarin's biological effects are due to its reaction with a serine residue in the active site of acetylcholinesterase, an enzyme that is very important in nerve transmission.

a serine residue in
acetylcholinesterase

Give the product of the reaction of the more active enantiomer of sarin with serine. (You can represent the serine as R—CH₂OH.)

25.28 (a) Give the products that result from the hydrolysis of the phosphate ester group of phosphoenolpyruvate (PEP), an important compound in the glycolysis pathway.

phosphoenolpyruvate (PEP)

(b) The $\Delta G°'$ for the hydrolysis of PEP is -61.9 kJ mol^{-1} (14.8 kcal mol^{-1}), which puts it well into the category of a "high-energy" compound. Most phosphate esters have a $\Delta G°'$ for hydrolysis of about -13.8 kJ mol^{-1}

(3.3 kcal mol^{-1}). Explain why the hydrolysis of PEP releases much more energy than the hydrolysis of an ordinary phosphate ester.

25.29 Methanol containing the oxygen isotope ^{18}O is allowed to react, in separate reactions, with each of the acid chlorides shown in Fig. P25.29a–c, and each of the resulting compounds A, C, and E is treated with one equivalent of sodium hydroxide. This treatment regenerates methanol in all three cases, along with hydrolysis products B, D, and F, respectively. Identify the compounds A–F and account for the distribution of the isotope in all three cases. Specifically, why is isotopically substituted methanol formed in reaction (b) but not in reactions (a) and (c)?

25.30 Account with a mechanism for the fact that the hydrolysis of trimethyl phosphate to dimethyl phosphate in acidic solution containing ^{18}O-labeled water gives methanol containing ^{18}O and dimethyl phosphate containing no isotope, as shown in Fig. P25.30.

25.31 The tetrabutylammonium salt of isotopically chiral phenyl phosphate was heated in the polar aprotic solvent acetonitrile containing excess tert-butyl alcohol, and isotopically substituted tert-butyl phosphate was isolated. An analysis of its stereochemistry showed that it was completely racemic. (See Fig. P25.31, p. 1324.) Explain this result with a mechanism. (*Hint:* Think about the reactive intermediate, and also consider the possible role of the solvent.)

Figure P25.29

Figure P25.30

25.32 The reaction of water with metaphosphate ion is shown in Eq. 25.19b on p. 1301. Could nitrate ion undergo an analogous reaction with aqueous NaOH? If so, draw the structure of the product. If not, explain why.

25.33 The conjugation of alcohols containing nonpolar groups (for example, phenol) to glucuronic acids is a key process by which these alcohols can be transformed into, and excreted as, water-soluble derivatives called *glucuronides* in phase II metabolism. (See Eq. 8.11, p. 358.) These conjugation reactions involve the reaction of UDP-glucuronic

acid and the alcohol catalyzed by a family of enzymes called *UDP-glucuronyl transferases*. (See Fig. P25.33.) What would you expect to find as key catalytic features in the active sites of these enzymes?

25.34 Ribonuclease A, a well-characterized enzyme from beef pancreas, catalyzes the hydrolysis of RNA into its component ribonucleic acids. When this reaction is studied with an isotopically chiral sulfur analog, as shown in Fig. P25.34, the reaction occurs with overall *retention* of stereochemistry. No phosphoenzyme intermediate is

Figure P25.31

Figure P25.33

a thio-substituted ribodinucleotide

Figure P25.34

formed. Give a curved-arrow mechanism that explains this stereochemical result. (Assume that acidic and basic groups are present in the enzyme active site as needed.)

25.35 Phosphoglycerate mutase catalyzes the interconversion of (R)-3-phosphoglycerate and (R)-2-phosphoglycerate, as shown in reaction (1) of Fig. P25.35. The enzyme contains a histidine residue in its active site, and the reaction involves a phosphohistidine intermediate, as shown in reaction (2) of Fig. P25.35.

(a) Assuming the presence of acidic and basic groups in the active site of the enzyme as needed, draw a curved-arrow mechanism for the transformation of 3-phosphoglycerate to 2-phosphoglycerate that is consistent with the information given.

(b) Suppose that an enantiomerically pure 2-phosphoglycerate derivative containing an isotopically chiral phosphate group is subjected to this reaction. Would the isotopically chiral 3-phosphoglycerate formed have the *same* configuration at phosphorus, the *opposite configuration* at phosphorus, or equal amounts of the two configurations? Explain your reasoning.

25.36 The enzyme-catalyzed formation of fatty acyl-carnitine from fatty acyl-CoA (Fig. P25.36) is an early step in the metabolism of fatty acids. The $\Delta G^{\circ\prime}$ for the hydro-

lysis of an acyl-CoA is -31.4 kJ mol^{-1} (-7.5 kcal mol^{-1}), and the $\Delta G^{\circ\prime}$ for the hydrolysis of an acyl-carnitine is -17.4 kJ mol^{-1} (-4.2 kcal mol^{-1}).

(a) Calculate the $\Delta G^{\circ\prime}$ for the reaction shown in Fig. P25.36.

(b) Calculate the equilibrium constant (at pH = 7) for this reaction at 37 °C (310 K).

(c) Give a structural reason that the reaction is so favorable.

25.37 Calculate the $\Delta G^{\circ\prime}$ for the reaction of acetyl phosphate with CoA—SH, given that the $\Delta G^{\circ\prime}$ for the hydrolysis of acetyl phosphate is -43.1 kJ mol^{-1} (-10.3 kcal mol^{-1}) and the $\Delta G^{\circ\prime}$ for the hydrolysis of acetyl-CoA is -31.4 kJ mol^{-1} (-7.52 kcal mol^{-1}). As part of your calculation, draw a free-energy diagram relating the $\Delta G^{\circ\prime}$ of the two hydrolysis reactions to the $\Delta G^{\circ\prime}$ of the reaction between acetyl phosphate and acetyl-CoA.

25.38 The pK_a of the thiol group of acetyl-CoA is 9.6. How would the ΔG° for the hydrolysis of acetyl-CoA change (relative to its value at pH = 7) if the hydrolysis were carried out at pH = 11? Explain.

25.39 The enzyme *phosphoribosyl pyrophosphate synthetase* (PRPP synthetase) catalyzes the conversion of ribose-5-

Figure P25.35

Figure P25.36

phosphate into its 1-pyrophosphate derivative, as shown in Fig. P25.39. The pyrophosphate group in the product is provided by ATP, not shown in the figure.

(a) At which phosphate of ATP does the nucleophilic hydroxy group of ribose-5-phosphate react? What is the by-product of the reaction?

PRPP is an important intermediate in the biosynthesis of nucleic acids and amino acids. It reacts with nucleophiles at carbon-1 in typical S_N2-like processes.

(b) To illustrate this reactivity, show the reaction of ammonia with PRPP. Be sure to show the stereochemistry.

(c) A by-product of the reaction in part (b) is pyrophosphate. What other biochemical reaction ensures that the reaction you drew in part (b) runs to completion?

25.40 Aspartic acid and glutamic acid are the two naturally occurring α-amino acids that have carboxylate groups in their side chains. Asparagine and glutamine are the corresponding α-amino acids in which the side chains are primary amides. (All of these amino acids have the *S* configuration.)

Asparagine synthetase catalyzes the biosynthesis of asparagine from glutamine according to the equation shown in Fig. P25.40. Note that the amide —NH_2 group in asparagine comes from the amide —NH_2 group of glutamine.

(a) Suggest a sequence of steps for this reaction that involves the adenylation of aspartic acid.

(b) Glutamine and asparagine could in principle be equilibrated without ATP. Why is ATP utilized in this way?

(c) Calculate the $\Delta G°´$ for the biosynthesis of asparagine using the following $\Delta G°´$ values for hydrolysis (in kJ mol^{-1}): ATP to AMP + pyrophosphate, −45.6; asparagine to aspartic acid, −15.1; glutamine to glutamic acid, −14.2. (The corresponding values in kcal mol^{-1} are −10.9, −3.61, and −3.39.)

(d) It is sometimes said that asparagine biosynthesis is coupled to ATP hydrolysis (to AMP). What is meant by this statement? Is ATP actually hydrolyzed?

(e) What other reaction (catalyzed by a different enzyme) ensures that asparagine biosynthesis goes to completion?

aspartic acid

asparagine

glutamic acid

glutamine

ribose-5-phosphate

5-phosphoribosyl-1-pyrophosphate (PRPP)

Figure P25.39

$$H_2O + glutamine + aspartic\ acid + ATP \xrightarrow{\text{asparagine synthetase}} glutamic\ acid + asparagine + AMP + pyrophosphate$$

Figure P25.40

The Chemistry of the Aromatic Heterocycles and Nucleic Acids

Heterocyclic compounds are compounds with rings that contain more than one element. The heterocyclic compounds of greatest interest to organic chemists have rings containing carbon and one or more **heteroatoms**—atoms other than carbon. Although the chemistry of many *saturated* heterocyclic compounds is analogous to that of their noncyclic counterparts, a significant number of *unsaturated* heterocyclic compounds exhibit aromatic behavior. This chapter focuses primarily on the unique chemistry of a few of these aromatic heterocycles. The principles that emerge should enable you to understand the chemistry and properties of other heterocyclic compounds that you may encounter.

In addition, we'll consider the structure of DNA and examine some of the chemistry of pyridoxal phosphate, an important coenzyme in biology that is one of the forms of vitamin B_6. The medical importance of heterocyclic compounds is not restricted to their role in biochemistry; about 95% of small-molecule drugs contain heterocyclic groups.

26.1 NOMENCLATURE AND STRUCTURE OF THE AROMATIC HETEROCYCLES

A. Nomenclature

The names and structures of some important aromatic heterocyclic compounds are given in Fig. 26.1 on p. 1328. This figure also shows how the rings are numbered in substitutive nomenclature. In all but a few cases, a heteroatom is given the number 1. (Isoquinoline is an exception.) As illustrated by thiazole and oxazole, the convention is that oxygen and sulfur are given a lower number than nitrogen. Substituent groups are given the lowest number con-

FIGURE 26.1 Common aromatic heterocyclic compounds. The numbers (*red*) are used in substitutive nomenclature.

pyridine   pyrimidine   quinoline   isoquinoline   purine

pyrrole   thiazole   oxazole   imidazole   indole

furan   thiophene   benzofuran   benzothiophene   benzimidazole

sistent with the ring numbering. (These are the same rules used in numbering and naming saturated heterocyclic compounds; see Secs. 8.1C and 23.1B.)

3-ethylpyrrole   **5-methoxyindole**   **2-nitrofuran**   **4-oxazolecarboxylic acid**

PROBLEMS

26.1 Draw the structure of
(a) 4-(dimethylamino)pyridine   (b) 4-ethyl-2-nitroimidazole

26.2 Name the following compounds.
(a)   (b)   (c)   (d)

B. Structure and Aromaticity

The aromatic heterocyclic compounds furan, thiophene, and pyrrole can be written as resonance hybrids, illustrated here for furan.

$$ \qquad\qquad\qquad\qquad\qquad\qquad\qquad\qquad\qquad\qquad\qquad (26.1) $$

Because separation of charge is present in all but the first structure, the first structure is considerably more important than the others. Nevertheless, the importance of the other structures is evident in a comparison of the dipole moments of furan and tetrahydrofuran, a saturated heterocyclic ether.

	tetrahydrofuran	furan
dipole moment:	1.7 D	0.7 D
boiling point:	67 °C	31.4 °C

The dipole moment of tetrahydrofuran is attributable mostly to the bond dipoles of its polar C—O single bonds. That is, electrons in the σ bonds are pulled toward the oxygen because of its electronegativity. This same effect is present in furan, but in addition there is a second effect: the resonance delocalization of the oxygen unshared electrons into the ring shown in Eq. 26.1. This tends to push electrons away from oxygen into the π-electron system of the ring.

(26.2)

| dipole moment contribution of the C—O σ bonds | + | dipole moment contribution of π-electron delocalization | = | net dipole moment of furan |

Because these two effects in furan nearly cancel, furan has a very small dipole moment. The relative boiling points of tetrahydrofuran and furan reflect the difference in their dipole moments (Sec. 8.5B).

Pyridine, like benzene, can be represented by two equivalent neutral resonance structures. Three additional pyridine structures of less importance reflect the relative electronegativities of nitrogen and carbon.

(26.3)

minor resonance contributors

The aromaticity of some heterocyclic compounds was considered in the discussion of the Hückel $4n + 2$ rule (Sec. 15.7D). It is important to understand which unshared electron pairs in a heterocyclic compound are part of the $4n + 2$ aromatic π-electron system, and which are not (Fig. 26.2, p. 1330). Vinylic heteroatoms, such as the nitrogen of pyridine, contribute one π electron to the six π-electron aromatic system, just like each of the carbon atoms in the π system. The orbital containing the unshared electron pair of the pyridine nitrogen is perpendicular to the $2p$ orbitals of the ring and is therefore not involved in π bonding. In contrast, allylic heteroatoms, such as the nitrogen of pyrrole, contribute two electrons (an unshared pair) to the aromatic π-electron system. This nitrogen adopts sp^2 hybridization and

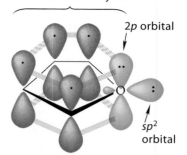

(a) **pyridine**

The unshared pair is vinylic and thus is *not* part of the $4n + 2$ π-electron system.

(b) **pyrrole**

The unshared pair is allylic and thus *is* part of the $4n + 2$ π-electron system.

(c) **furan**

One unshared pair is allylic and is part of the $4n + 2$ π-electron system; the other unshared pair is vinylic and is not.

FIGURE 26.2 The orbital configurations in pyridine, pyrrole, and furan. (a) In pyridine, the nitrogen 2p orbital, like the carbon 2p orbitals, contributes one electron to the $4n + 2$ π-electron system, and the nitrogen unshared electron pair occupies an sp^2 orbital (*blue*), which is not part of the π system. (b) In pyrrole, the unshared electron pair occupies a 2p orbital on nitrogen (*blue*), and it contributes two electrons to the $4n + 2$ π-electron system. (c) Furan has two unshared electron pairs (*red*). One unshared pair occupies a 2p orbital and contributes two electrons to the $4n + 2$ π-electron system, as in pyrrole. The other unshared pair occupies an sp^2 orbital. Like the unshared pair in pyridine, it does not contribute to the $4n + 2$ π-electron system.

trigonal planar geometry so that its unshared electron pair can occupy a *2p* orbital, which has the optimum shape and orientation to overlap with the carbon *2p* orbitals and thus to be part of the aromatic π-electron system. Consequently, the hydrogen of pyrrole lies in the plane of the ring. The oxygen of furan contributes one unshared electron pair to the aromatic π-electron system, and the other unshared electron pair occupies a position analogous to the carbon–hydrogen bond of pyrrole—in the ring plane, perpendicular to the *2p* orbitals of the ring.

The *empirical resonance energy* can be used to estimate the additional stability of a heterocyclic compound due to its aromaticity. (This concept was introduced in the discussion of the aromaticity of benzene in Sec. 15.7C.) The empirical resonance energies of benzene and some heterocyclic compounds are given in Table 26.1. To the extent that resonance energy is a measure of aromatic character, furan has the least aromatic character of the heterocyclic compounds in the table. The modest resonance energy of furan has significant consequences for its reactivity, as we'll learn in Sec. 26.3B.

PROBLEMS

26.3   Draw the important resonance structures for pyrrole.

26.4   (a) The dipole moments of pyrrole and pyrrolidine are similar in magnitude but have opposite directions. Explain, indicating the direction of the dipole moment in each compound. (*Hint:* Use the result in Problem 26.3.)

$\mu = 1.80$ D $\mu = 1.57$ D

(b) Explain why the dipole moments of furan and pyrrole have opposite directions.

(c) Should the dipole moment of 3,4-dichloropyrrole be greater than or less than that of pyrrole? Explain.

26.5   Each of the following NMR chemical shifts goes with a proton at carbon-2 of either pyridine, pyrrolidine, or pyrrole. Match each chemical shift with the appropriate heterocyclic compound, and explain your answer: δ 8.51; δ 6.41; and δ 2.82.

TABLE 26.1 Empirical Resonance Energies of Some Aromatic Compounds

Compound	Resonance energy		Compound	Resonance energy	
	kJ mol^{-1}	kcal mol^{-1}		kJ mol^{-1}	kcal mol^{-1}
benzene	138–151	33–36	thiophene	121	29
pyridine	96–117	23–28	pyrrole	89–92	21–22
			furan	67	16

26.2 BASICITY AND ACIDITY OF THE NITROGEN HETEROCYCLES

A. Basicity of the Nitrogen Heterocycles

Pyridine and quinoline act as ordinary amine bases.

$$\text{(26.4)}$$

pyridine

$pK_a = 5.2$

$$\text{(26.5)}$$

quinoline

$pK_a = 4.9$

Pyridine and quinoline are much less basic than aliphatic tertiary amines (Sec. 23.5A) because of the sp^2 hybridization of their nitrogen unshared electron pairs. (Recall from Sec. 14.7A that the basicity of an unshared electron pair decreases with increasing s character.)

Because pyrrole and indole look like amines, it may come as a surprise that neither of these two heterocycles has appreciable basicity. These compounds are protonated only in strong acid, and protonation occurs on carbon, not nitrogen.

$$\text{(26.6)}$$

pyrrole

$pK_a \approx -4$

The marked contrast between the basicities of pyridine and pyrrole can be understood by considering the role of the nitrogen unshared electron pair in the aromaticity of each compound (Fig. 26.2). Protonation of the pyrrole nitrogen would disrupt the aromatic system of six π electrons by taking the nitrogen's unshared pair "out of circulation."

$$\text{(26.7)}$$

not aromatic;
is not formed

Although protonation of the carbon of pyrrole (Eq. 26.6) also disrupts the aromatic π-electron system, at least the resulting cation is resonance-stabilized. On the other hand, protonation of the pyridine unshared electron pair occurs easily because this electron pair is *not* part of the π-electron system. Hence, protonation of this electron pair does not disrupt pyridine's aromaticity.

The basicity of the nitrogen heterocycle imidazole, explored in Study Problem 26.1, is particularly relevant to biology because the amino acid histidine in proteins has an imidazole group in its side chain.

imidazole

a histine residue
in a protein

This group in many cases serves as an acid–base catalyst in enzyme-catalyzed reactions.

STUDY PROBLEM 26.1

Imidazole is a base; the pK_a of its conjugate acid is 6.95. On which nitrogen does imidazole protonate?

SOLUTION Imidazole has two nitrogens: one has the electronic configuration of the nitrogen in pyridine—that is, it is vinylic—but the other is like the nitrogen of pyrrole—it is allylic. Consequently, protonation occurs on the pyridine-like nitrogen—the nitrogen whose unshared electron pair is *not* part of the aromatic sextet.

$$pK_a = 6.95$$

(26.8)

does not form

According to the resonance structures in Eq. 26.8, the two nitrogens of protonated imidazole are equivalent; consequently, deprotonation to give imidazole can occur at either nitrogen. Imidazole is more basic than pyridine because of the resonance stabilization of its conjugate-acid cation.

B. Acidity of Pyrrole and Indole

Pyrrole and indole are weak acids.

(26.9)

With pK_a values of about 17.5, pyrrole and indole are about as acidic as alcohols and about 15–17 pK_a units more acidic than primary and secondary amines (Sec. 23.5D). The greater acidity of pyrroles and indoles is a consequence of the resonance stabilization of their conjugate-base anions (Eq. 26.9; draw the three missing resonance structures in this equation). Pyrrole and indole are rapidly deprotonated by Grignard and organolithium reagents.

FURTHER EXPLORATION 26.1
Relative Acidities of 1,3-Cyclopentadiene and Pyrrole

$$\text{pyrrole} + CH_3CH_2{-}MgBr \longrightarrow \text{pyrrole-MgBr} + CH_3CH_3 \qquad (26.10)$$

PROBLEMS

26.6 (a) Suggest a reason why pyridine is miscible with water, whereas pyrrole has little water solubility.

(b) Indicate whether you would expect imidazole to have high or low water solubility, and why.

26.7 (a) The compound 4-(dimethylamino)pyridine protonates to give a conjugate acid with a pK_a value of 9.9. This compound is thus 4.7 pK_a units more basic than pyridine itself. Draw the structure of the conjugate acid of 4-(dimethylamino)pyridine, and, using resonance structures, explain why 4-(dimethylamino)pyridine is much more basic than pyridine.

(b) What product is expected when 4-(dimethylamino)pyridine reacts with CH_3I?

26.8 Protonation of aniline causes a dramatic shift of its UV spectrum to lower wavelengths, but protonation of pyridine has almost no effect on its UV spectrum. Explain the difference.

26.3 THE CHEMISTRY OF FURAN, PYRROLE, AND THIOPHENE

A. Electrophilic Aromatic Substitution

Furan, thiophene, and pyrrole, like benzene, undergo electrophilic aromatic substitution reactions. Let's try to predict the ring carbon at which substitution occurs in these compounds by examining the carbocation intermediates involved in the substitution reactions at the two different positions and applying Hammond's postulate.

STUDY PROBLEM 26.2

Using the nitration of pyrrole as an example, predict whether electrophilic aromatic substitution occurs predominantly at carbon-2 or carbon-3.

SOLUTION Recall (p. 803) that the electrophile in nitration is the nitronium ion, $^+NO_2$. Substitution at the two different positions of pyrrole by the nitronium ion gives different carbocation intermediates:

Substitution at carbon-2:

$$(26.11a)$$

Substitution at carbon-3:

$$(26.11b)$$

The carbocation resulting from substitution at carbon-2 has more important resonance structures and is therefore more stable than the carbocation resulting from substitution at carbon-3. Applying Hammond's postulate, we predict that the reaction involving the more stable intermediate should be the faster reaction. Consequently, nitration should occur at carbon-2. The experimental facts are as follows:

$$\text{(26.12)}$$

2-Nitropyrrole is the major nitration product of pyrrole, as predicted. Nothing is really wrong with substitution at carbon-3; substitution at carbon-2 is simply more favorable. As Eq. 26.12 shows, some 3-nitropyrrole is obtained in the reaction.

As Study Problem 26.2 suggests, electrophilic substitution of pyrrole occurs predominantly at the 2-position. Similar results are observed with furan and thiophene:

(a Friedel–Crafts reaction)   (75–92% yield) $$\text{(26.13)}$$

(70% yield)   (5% yield) $$\text{(26.14)}$$

Pyrrole, furan, and thiophene are all much more reactive than benzene in electrophilic aromatic substitution. Although precise reactivity ratios depend on the particular reaction, the relative rates of bromination are typical:

$$\text{pyrrole} \quad > \quad \text{furan} \quad > \quad \text{thiophene} \quad > \quad \text{benzene} \qquad \text{(26.15)}$$
$$3 \times 10^{18} \qquad 6 \times 10^{11} \qquad 5 \times 10^{9} \qquad 1$$

Milder reaction conditions must be used with more reactive compounds. (Reaction conditions that are too vigorous in many cases bring about so many side reactions that polymerization and tar formation occur.) For example, a less reactive acylating reagent is used in the acylation of furan than in the acylation of benzene. (Recall that anhydrides are less reactive than acid chlorides; Eq. 21.24, p. 1068.)

$$\text{(26.16a)}$$

(97% yield)

(75–92% yield) $$\text{(26.16b)}$$

The reactivity order of the heterocycles (Eq. 26.15) is a consequence of the relative abilities of the heteroatoms to stabilize positive charge in the intermediate carbocations (as in

Eq. 26.11a, for example). Both pyrrole and furan have heteroatoms from the second period of the periodic table. Because nitrogen is better than oxygen at delocalizing positive charge— nitrogen is less electronegative—pyrrole is more reactive than furan. The sulfur of thiophene is a third-period element and, although it is less electronegative than oxygen, its $3p$ orbitals overlap less efficiently with the $2p$ orbitals of the aromatic π-electron system (see Fig. 16.7, p. 819). In fact, the reactivity *order* of the heterocycles in aromatic substitution parallels the reactivity order of the correspondingly substituted benzene derivatives:

Relative reactivities:

$$(26.17)$$

N,N-dimethylaniline   **anisole**   **thioanisole**

When we consider the activating and directing effects of substituents in furan, pyrrole, and thiophene rings, the usual activating and directing effects of substituents in aromatic substitution apply (see Table 16.2, p. 812). Superimposed on these effects is the normal effect of the heterocyclic atom in directing substitution to the 2-position. The following example illustrates these effects:

$$(26.18)$$

3-thiophenecarboxylic acid   **5-bromo-3-thiophenecarboxylic acid**
                                        (69% yield)

The position of substitution follows from an analysis of the resonance and polar effects in the possible arenium ion intermediates. This process, which is the same one used to understand the directing effects of substituents in benzene chemistry (Sec. 16.5), is illustrated in Study Problem 26.3.

STUDY PROBLEM 26.3

Explain the position of substitution observed in the bromination of thiophene-3-carboxylic acid shown in Eq. 26.18. Draw the arenium ion intermediates for all possible positions of substitution and show that the intermediate in the observed substitution is the most stable one.

SOLUTION   The arenium ion intermediates involved in the three possible positions of substitution are the following.

A	*B*	*C*
(substitution at C-5)	(substitution at C-4)	(substitution at C-2)

Arenium ion *A* has three resonance structures:

resonance structures for arenium ion *A*

Arenium ion *B* has only two resonance structures:

resonance structures for arenium ion *B*

Arenium ion *C*, like ion *A*, has three resonance structures. However, in one of the structures (*red*), the electron deficiency and positive charge are located on the carbon adjacent to an electron-withdrawing group, the carboxylic acid.

resonance structures for arenium ion *C*

Therefore, arenium ion *A* is the most stable carbocation intermediate because it has the most resonance structures without placing the positive charge next to the electron-withdrawing substituent. As a result, the most favorable substitution involves this intermediate—that is, substitution at C-5.

If we count around the carbon framework from the carboxylic acid substituent, we see that the substitution in Eq. 26.18 occurs at the carbon that is in a 1,3-relationship to the substituted carbon. In other words, substitution has occurred meta to the carboxylic acid group. Recall (Table 16.2, p. 812) that the carboxylic acid group is a "meta-directing" substituent.

Thiophene itself substitutes at the carbon next to the sulfur. Therefore, the observed product satisfies the directing effects of both the heteroatom and the substitutions. The directing effects of substituents are exactly as they are in benzene substitution, provided that we view the ortho, para, or meta relationship *through the carbon framework and not through the heteroatom.*

In the following example, the chloro group is an ortho, para-directing group. Because the position "para" to the chloro group is also a 2-position, both the sulfur of the ring and the chloro group direct the incoming nitro group to the same position.

2-chlorothiophene   **2-chloro-5-nitrothiophene**
                                            (57% yield)

$$(26.19)$$

When the directing effects of substituents and the ring compete, it is not unusual to observe mixtures of products.

the NO₂ group directs here

the sulfur directs here

2-nitrothiophene

$\xrightarrow[0\,°C]{HNO_3}$ O₂N—[S]—NO₂ + [O₂N thiophene]—NO₂   (26.20)

2,5-dinitrothiophene (44%)   **2,4-dinitrothiophene** (56%)

(60% yield)

Finally, if both 2-positions are occupied, 3-substitution takes place.

$$H_3C\text{—furan—}CH_3 + H_3C-\overset{O}{\underset{}{C}}-O-\overset{O}{\underset{}{C}}-CH_3 \xrightarrow[\text{acetic acid}]{BF_3} H_3C\text{—furan(}\overset{O}{\overset{||}{C}-CH_3\text{)}}-CH_3 + CH_3\overset{O}{\overset{||}{C}}OH \quad (26.21)$$

2,5-dimethylfuran   **acetic anhydride**   (65% yield)

B. Addition Reactions of Furan

The previous sections focused on the aromatic character of furan, pyrrole, and thiophene. A furan, pyrrole, or thiophene could, however, be viewed as a 1,3-butadiene with its terminal carbons "tied down" by a heteroatom bridge.

"butadiene" unit within furan

Do the heterocycles ever behave chemically as if they are conjugated dienes? Of the three heterocyclic compounds furan, pyrrole, and thiophene, furan has the least resonance energy (Table 26.1) and, by implication, the least aromatic character. Consequently, of the three compounds, furan has the greatest tendency to behave like a conjugated diene.

One characteristic reaction of conjugated dienes is *conjugate addition* (Sec. 15.4A). Indeed, furan does undergo some conjugate addition reactions. One example of such a reaction occurs in bromination. For example, furan undergoes conjugate addition of bromine and methanol in methanol solvent; the conjugate-addition product then undergoes an S_N1 reaction with the methanol. (Write mechanisms for both parts of this reaction; refer to Sec. 15.4A, if necessary.)

a conjugate addition   an S_N1 reaction

$$\text{[furan]} + Br_2 + CH_3OH \longrightarrow CH_3O\text{—[}\underset{H\ \ O\ \ H}{}\text{]—}Br \xrightarrow{CH_3OH} CH_3O\text{—[}\underset{H\ \ O\ \ H}{}\text{]—}OCH_3 \quad (26.22)$$

+ HBr   + HBr
mixture of stereoisomers
(72–76% yield)

Another manifestation of the conjugated-diene character of furan is that it undergoes Diels–Alder reactions (Sec. 15.3) with reactive dienophiles such as maleic anhydride.

furan   **maleic anhydride**   (>90% yield)   (26.23)

C. Side-Chain Reactions

Many reactions occur at the side chains of heterocyclic compounds without affecting the rings, just as some reactions occur at the side chain of a substituted benzene (Secs. 17.1–17.5).

$$\text{3-thiophenecarbaldehyde} \xrightarrow[\text{(Sec. 19.14)}]{\text{Ag}_2\text{O}} \text{3-thiophenecarboxylic acid} \qquad (26.24)$$

3-thiophenecarbaldehyde

3-thiophenecarboxylic acid
(95–97% yield)

A particularly useful example of a side-chain reaction is removal of a carboxy group directly attached to the ring (*decarboxylation*). This reaction is effected by strong heating, in some cases with catalysts.

$$\xrightarrow[200\ °C]{\text{heat}} \text{furan} + \text{CO}_2 \qquad (26.25a)$$

2-furancarboxylic acid   furan

The conditions of the uncatalyzed reaction are much harsher than the conditions of the decarboxylations discussed in Sec. 20.11 because there is no "electron sink" to accept the electrons from the carboxylic acid group. It may be that this reaction proceeds by a very unfavorable internal protonation (or protonation by a second carboxylic acid molecule) to give a carbocation, which is actually the species that decarboxylates.

$$(26.25b)$$

$pK_a = 3.1$ $pK_a \approx -5$
a resonance-stabilized
carbocation

Heat is required for the first, highly unfavorable, reaction to proceed at a reasonable rate. The carbocation serves as the "electron sink" for the decarboxylation reaction.

PROBLEMS

26.9 Complete each of the following reactions by giving the principal organic product(s). For (b), write a curved-arrow mechanism that shows the arenium ion intermediate and its resonance structures.

(a)

$$+ \text{HNO}_3 \xrightarrow{\text{acetic acid}}$$

(b)

$$+ \text{H}_3\text{C}-\overset{O}{\underset{||}{C}}-O-\overset{O}{\underset{||}{C}}-\text{CH}_3 \xrightarrow[\text{acetic acid}]{\text{BF}_3}$$

(c)

$$+ \text{H}_3\text{C}-\overset{O}{\underset{||}{C}}-\text{Ph} \xrightarrow{\text{NaOH}}$$

(d)

$$\xrightarrow[\text{CCl}_4]{\text{light}}$$

26.10   Write a curved-arrow mechanism for the following reaction.

Erlich's reagent
(used for detecting pyrroles
and indoles)

 26.4   THE CHEMISTRY OF PYRIDINE

A. Electrophilic Aromatic Substitution

In general, it is difficult to prepare monosubstituted pyridines by electrophilic aromatic substitution because pyridine has a very low reactivity; it is much less reactive than benzene. An important reason for this low reactivity is that the nitrogen of pyridine is protonated under the very acidic conditions of most electrophilic aromatic substitution reactions (Eq. 26.4, p. 1331). The resulting positive charge on nitrogen makes it difficult to form a carbocation intermediate, which would place a second positive charge within the same ring.

Fortunately, a number of monosubstituted pyridines are available from natural sources. Among these are the methylpyridines, or *picolines*:

α-picoline   *β*-picoline   *γ*-picoline

The picolines (and other methylated pyridines) are obtained from *coal tar* (Sec. 16.7). Another very useful monosubstituted derivative of pyridine is *nicotinic acid* (pyridine-3-carboxylic acid), which is conveniently prepared in a number of ways, one of which is side-chain oxidation of nicotine, an alkaloid present in tobacco (Fig. 23.4, p. 1222).

nicotine   **nicotinic acid**
(70% yield)

(26.26)

(Nitric acid in this reaction is used as an oxidizing agent.)

Although electrophilic aromatic substitution reactions are not very useful for introducing substituents into pyridine itself, pyridine rings substituted with activating groups such as methyl groups do undergo such reactions.

2,6-dimethylpyridine   **2,6-dimethyl-3-nitropyridine**
(2,6-lutidine)   (81% yield)

(26.27)

Reactions such as this, which occur in acidic solution, undoubtedly take place on the very small amount of the unprotonated pyridine that is in rapid equilibrium with the much larger amount of its conjugate acid. Because the reactive species—the unprotonated heterocycle—is present in very small concentration, the methyl-substituted pyridines are not very reactive, despite the presence of the activating methyl substituents.

As the example in Eq. 26.27 illustrates, substitution in pyridine generally takes place in the 3-position. Although the methyl groups in Eq. 26.27 also direct substitution to the 3-position, the tendency of pyridine to undergo 3-substitution is general even in the absence of such directing groups. As with other electrophilic substitutions, an understanding of this directing effect comes from an examination of the carbocation intermediates formed in substitution at different positions. Substitution in the 3-position gives a carbocation with three different resonance structures:

3-Substitution:

$$(26.28a)$$

Substitution at the 4-position also involves a carbocation intermediate with three resonance structures, but the one shown in red is particularly unstable and thus unimportant because *the nitrogen, an electronegative atom, is electron-deficient.* (See the discussion of guideline 3 for drawing resonance structures, on p. 753).

4-Substitution:

$$(26.28b)$$

not an important
resonance structure

Be sure to understand that the nitrogen in the red structure is *very* different from the nitrogen in pyrrole during electrophilic aromatic substitution (Eq. 26.11a, p. 1333). The pyrrole nitrogen is also positively charged, but it is not electron-deficient because it has a complete octet. In contrast, an *electron-deficient* electronegative atom such as the one in Eq. 26.28b is very unfavorable energetically. Consequently, the carbocation intermediate in 4-substitution is less stable than the intermediate in 3-substitution. By Hammond's postulate, 3-substitution is therefore the faster reaction.

If electrophilic substitution in pyridine occurs at the 3-position, how can we obtain pyridine derivatives substituted at other positions? One compound used to obtain 4-substituted pyridines is pyridine-*N*-oxide, formed by the oxidation of pyridine with 30% hydrogen peroxide.

$$(26.29)$$

pyridine   **pyridine-*N*-oxide**
                                 (90% yield)

An analogy to pyridine-*N*-oxide from benzene chemistry is phenoxide, the conjugate base of phenol. Just as phenol or phenoxide is much more reactive in electrophilic aromatic substitution than benzene (Sec. 18.9), pyridine-*N*-oxide is much more reactive than pyridine. Because the nitrogen of pyridine-*N*-oxide has a positive charge, this compound is *much* less reactive than phenol or phenoxide. Nevertheless, pyridine-*N*-oxide undergoes useful aromatic substitution reactions, and substitution occurs in the 4-position.

$$(26.30)$$

Once the *N*-oxide function is no longer needed, it can be removed by catalytic hydrogenation; this procedure also reduces the nitro group. A reaction with trivalent phosphorus compounds, such as PCl$_3$, removes the *N*-oxide function without reducing the nitro group.

$$(26.31)$$

PROBLEMS

26.11 Which should be more reactive in nitration: β-picoline or α-picoline? Explain using resonance structures, and give the major nitration product(s) in each case.

26.12 By drawing resonance structures for the carbocation intermediates, show why aromatic substitution in pyridine-*N*-oxide occurs at the 4-position rather than at the 3-position.

26.13 (a) Draw a curved-arrow mechanism for the reduction of pyridine-*N*-oxide by :PCl$_3$. (*Hint:* In the first step, the oxygen of the *N*-oxide reacts as a nucleophilic center with the phosphorus as an electrophilic center. Remember that phosphorus can expand its octet.)

(b) This is an oxidation–reduction reaction. What is oxidized and what is reduced? How many electrons are involved in this redox reaction?

B. Nucleophilic Aromatic Substitution

In contrast to its low reactivity in *electrophilic* aromatic substitution, the pyridine ring readily undergoes *nucleophilic* aromatic substitution. A rather unusual reaction of this type can be used to prepare 2-aminopyridine. In this reaction, called the **Chichibabin reaction**, treatment of a pyridine derivative with the strong base sodium amide (Na^+ $^-NH_2$; Eq. 14.21, p. 699) brings about the direct substitution of an amino group for a ring hydrogen.

pyridine   **2-aminopyridine**
                                (66–76% yield)

(26.32)

In the first step of the mechanism, the amide ion reacts as a nucleophile at the 2-position of the ring to form a *tetrahedral addition intermediate*.

tetrahedral addition intermediate

(26.33a)

This step of the mechanism can be understood by recognizing that *the* C=N *linkage of the pyridine ring is somewhat analogous to a carbonyl group*; that is, carbon at the 2-position has some of the character of a carbonyl carbon and can react with nucleophiles. The C=N group of pyridine, though, is *much* less reactive than a carbonyl group because a nitrogen anion is considerably more basic than an oxygen anion and because it is part of an aromatic system.

Compare:

($^-$:Nuc = nucleophile)

(26.33b)

In the second step of the mechanism, the leaving group, a *hydride ion*, is lost.

(26.33c)

Hydride ion is a very poor, and thus very unusual, leaving group because it is very basic. This reaction occurs for two reasons. First, the aromatic pyridine ring is reformed; aromaticity lost in the formation of the tetrahedral addition intermediate is regained when the leaving group departs. Second, the basic hydride produced in the reaction reacts with the —NH_2 group irreversibly to form dihydrogen (a gas) and the resonance-stabilized conjugate-base anion of 2-aminopyridine.

(26.33d)

The neutral 2-aminopyridine is formed when water is added in a separate step.

(26.33e)

A reaction similar to the Chichibabin reaction occurs with organolithium reagents.

$$\text{pyridine} + \text{PhLi} \xrightarrow[\text{toluene}]{\text{heat}} \xrightarrow{\text{H}_2\text{O}} \text{2-phenylpyridine} + \text{LiH} \qquad (26.34)$$

pyridine

2-phenylpyridine
(40–49% yield)

When pyridine is substituted with a much better leaving group than hydride at the 2-position, it reacts more rapidly with nucleophiles. The 2-halopyridines, for example, readily undergo substitution of the halogen by other nucleophiles under conditions that are much milder than those used in the Chichibabin reaction.

$$\text{2-chloropyridine} + \text{Na}^+\ ^-\text{OMe} \xrightarrow{\text{MeOH}} \text{2-methoxypyridine} + \text{Na}^+\text{Cl}^- \qquad (26.35)$$

2-chloropyridine

2-methoxypyridine
(95% yield)

This nucleophilic substitution can also be related to the analogous reaction of a carbonyl compound. This reaction of a 2-chloropyridine resembles the nucleophilic acyl substitution reaction of an acid chloride—except that acid chlorides are *much* more reactive than 2-halopyridines.

Compare:

$$(26.36)$$

The nucleophilic substitution reactions of pyridines can be classified as *nucleophilic aromatic substitution* reactions. Recall that aryl halides undergo nucleophilic aromatic substitution when the benzene ring is substituted with electron-withdrawing groups (Sec. 18.4). The "electron-withdrawing group" in the reactions of pyridines is the pyridine nitrogen itself. The tetrahedral addition intermediate (Eq. 26.33a) is analogous to the Meisenheimer complex of nucleophilic aromatic substitution (Eqs. 18.15a–c, p. 886). Thus, there is a mechanistic parallel between three types of reactions: (1) nucleophilic acyl substitution, a typical reaction of carboxylic acid derivatives; (2) nucleophilic aromatic substitution; and (3) nucleophilic substitution on the pyridine ring.

The 2-aminopyridines formed in the Chichibabin reaction serve as starting materials for a variety of other 2-substituted pyridines. For example, diazotization of 2-aminopyridine gives a diazonium ion that can undergo substitution reactions (see Secs. 23.10A–B).

$$(26.37)$$

2-aminopyridine

2-pyridinediazonium bromide

2-bromopyridine
(86–92% yield)

When the diazonium salt reacts with water, it is hydrolyzed to 2-hydroxypyridine, which in most solvents exists in its carbonyl form, 2-pyridone.

$$(26.38)$$

2-pyridinediazonium ion

2-hydroxypyridine

2-pyridone

Let us consider briefly the equilibrium between 2-hydroxypyridine and 2-pyridone. This is analogous to a keto–enol equilibrium, except that the "keto" form is an amide in this case. In this equilibrium, the ratio of the hydroxy form to the carbonyl form is 1 : 910 in water, but the ratio varies with concentration and with solvent; in the vapor phase, the ratio is 1 : 0.4. The important points about this equilibrium, however, are (1) enough of each form is present so that either form can be involved in chemical reactions, and (2) *much* more carbonyl isomer is present than there is in phenol (Eq. 22.14, p. 1111). Why should this be so? A major factor that determines whether an aromatic hydroxy compound exists as a carbonyl or hydroxy ("enol") form is whether the energetic advantage of aromaticity—that is, the *resonance stabilization* of the aromatic hydroxy isomer—outweighs the large carbonyl C=O bond energy. In the case of phenol itself, the resonance stabilization of the benzene ring is large enough that the phenol isomer is strongly preferred. As Table 26.1 (p. 1331) shows, the resonance energy, and thus the resonance stabilization, of pyridine is considerably smaller than that of benzene. Moreover, the resonance interaction of the amide nitrogen with the carbonyl group further stabilizes 2-pyridone. As the following structure shows, the resulting resonance structure is aromatic. Consequently, 2-pyridone itself has a significant amount of aromatic character. The keto isomer of phenol has no stabilizing contribution of this sort.

$$\tag{26.39}$$

an aromatic
resonance structure

2-Pyridone undergoes some reactions that are similar to the reactions of hydroxy compounds that we have studied. For example, treatment of 2-pyridone with PCl_5 gives 2-chloropyridine.

$$+ \; PCl_5 \;\xrightarrow{\text{heat}}\; + \; POCl_3 \; + \; HCl \tag{26.40}$$

2-pyridone   **2-chloropyridine**

If we think of 2-pyridone in terms of its 2-hydroxypyridine isomer, this reaction is similar to the preparation of acid chlorides from carboxylic acids.

is similar to

$$\tag{26.41}$$

Notice again the analogy between pyridine chemistry and carbonyl chemistry.

Pyridines with leaving groups in the 4-position also undergo nucleophilic substitution reactions.

$$+ \; PhNH_2 \;\xrightarrow{\text{heat}}\; + \; HCl \;\rightleftharpoons\; Cl^- \tag{26.42}$$

4-chloropyridine   **4-(phenylamino)pyridine**

As the examples in this section suggest, nucleophilic substitution reactions at the 2- and 4-positions of a pyridine ring are particularly common. The reason follows from the mechanism of this type of reaction: Negative charge in the addition intermediate is delocalized onto the electronegative pyridine nitrogen.

Substitution at carbon-2: (Y = leaving group, ⁻:Nuc = nucleophile)

$$(26.43a)$$

Substitution at carbon-4:

$$(26.43b)$$

What about substitution at carbon-3? 3-Substituted pyridines are *not* reactive in nucleophilic substitution because negative charge in the addition intermediate *cannot* be delocalized onto the electronegative nitrogen:

Substitution at carbon-3:

$$(26.43c)$$

PROBLEMS

26.14   Give the structure of the product and a curved-arrow mechanism for its formation in the reaction of 4-chloropyridine with sodium methoxide. Draw all important resonance structures for the addition intermediate.

26.15   Which compound should readily undergo substitution of the bromine by phenolate anion: 4-bromopyridine or 3-bromopyridine? Explain, and give the structure of the product.

C. *N*-Alkylpyridinium Salts and Their Reactions

Pyridine, like many other bases, is a nucleophile. When pyridines react in S_N2 reactions with alkyl halides or sulfonate esters, *N-alkylpyridinium salts* are formed.

$$(26.44)$$

1-methylpyridinium iodide
(an *N*-alkylpyridinium salt)

N-Alkylpyridinium salts are activated toward nucleophilic reactions at the 2- and 4-positions of the ring much more than pyridines themselves because the positively charged nitrogen is more electronegative, and is therefore a better electron acceptor, than the neutral nitrogen of a pyridine. When the nucleophiles in such displacement reactions are anions, charge is neutralized. In the following reaction, for example, the pyridinium salt reacts as an electrophile at its 2-position with the hydroxide ion nucleophile; the resulting hydroxy compound is then oxidized by potassium ferricyanide [$K_3Fe(CN)_6$] present in the reaction mixture.

$$ (26.45) $$

A biological example of nucleophilic addition to the 4-position of a pyridinium ring is found in biological oxidations with NAD^+ (Eq. 10.58a, p. 487).

Pyridine-*N*-oxides are in one sense pyridinium ions, and they react with nucleophiles in much the same way as *N*-alkylpyridinium salts:

$$ (26.46) $$

pyridine N-oxide

2-phenylpyridine

D. Side-Chain Reactions of Pyridine Derivatives

The "benzylic" hydrogens of an alkyl group at the 2- or 4-position of a pyridine ring are about 7 pK_a units more acidic than ordinary benzylic hydrogens because the electron pair (and charge) in the conjugate-base anion is delocalized onto the electronegative pyridine nitrogen.

$$ (26.47) $$

relatively acidic
p$K_a \approx 34$

(Draw the resonance structures of this ion and verify that charge is delocalized onto the pyridine nitrogen.) As the example in Eq. 26.47 illustrates, strongly basic reagents such as organolithium reagents or $NaNH_2$ abstract a "benzylic" hydrogen from 2- or 4-alkylpyridines. The anion formed in this way has a reactivity much like that of other organolithium reagents. In Eq. 26.48, for example, it adds to the carbonyl group of an aldehyde to give an alcohol (Sec. 19.9).

1) $CH_3(CH_2)_3Li$
2)
3) H_2O

$$ (26.48) $$

This reaction is another example of the analogy between pyridine chemistry and carbonyl chemistry. If the C=N linkage of a pyridine ring is analogous to a carbonyl group, then the "benzylic" anion is analogous to an enolate anion.

is analogous to

$$ (26.49) $$

On the basis of this analogy, then, it is reasonable that these anions should undergo some of the reactions of enolate anions, such as the aldol-like addition in Eq. 26.48.

The "benzylic" hydrogens of 2- or 4-alkylpyridinium salts are much more acidic than those of the analogous pyridines because the conjugate-base "anion" is actually a neutral compound, as the following resonance structures show:

(26.50)

The "benzylic" hydrogens of 2- or 4-alkylpyridinium salts are acidic enough that the conjugate-base "anions" can be formed in useful concentrations by aqueous NaOH or amines. In the following reaction, which exploits this acidity, the conjugate base of a pyridinium salt is used as the "enolate" component in a variation of the Claisen–Schmidt condensation (Sec. 22.4C).

(26.51)

(85% yield)

Many side-chain reactions of pyridines are analogous to those of the corresponding benzene derivatives. For example, side-chain oxidation (Sec. 17.5C) is a useful reaction of both alkylbenzenes and alkylpyridines. The oxidation of nicotine to nicotinic acid (Eq. 26.26, p. 1339) is an example of such a reaction.

PROBLEMS

26.16 Give the principal organic product in the reaction of quinoline with each of the following reagents. (*Hint:* Consider the similar reactions of pyridine.)

(a) 30% H_2O_2 (b) $NaNH_2$, heat; then H_2O (c) product of part (a), then HNO_3, H_2SO_4

26.17 Outline a synthesis for each of the following compounds from the indicated starting material and any other reagents.

(a) 3-methyl-4-nitropyridine from β-picoline (p. 1339)

(b) 4-methyl-3-nitropyridine from γ-picoline

(c)

 from α-picoline (*Hint:* See Sec. 20.6.)

(d) 3-aminopyridine from β-picoline

26.18 Predict the predominant product in each of the following reactions. Explain your answer.

(a) 3,4-dimethylpyridine + butyllithium (1 equiv.), then $CH_3I \longrightarrow (C_8H_{11}N)$

(b) 3,4-dibromopyridine + NH_3, heat $\longrightarrow (C_5H_5BrN_2)$

E. Pyridinium Ions in Biology: Pyridoxal Phosphate

The chemistry in the previous three sections has as its basis the fact that the nitrogen of the pyridine ring can serve as an acceptor of electrons and that this electron-acceptor tendency is

pyridoxal phosphate   **pyridoxal**   abbreviated structure of pyridoxal and pyridoxal phosphate used in the text

pyridoxamine   **pyridoxol (pyridoxine)**

FIGURE 26.3 Various forms of vitamin B_6. Pyridoxol was the first form to be isolated, but any of the compounds shown can serve as a source of the vitamin. (For example, pyridoxol can be oxidized and phosphorylated to give pyridoxal phosphate.) Pyridoxal phosphate is the form of the vitamin involved in most biochemical transformations; pyridoxamine phosphate is an intermediate in some transformations. All compounds are shown in the ionization states in which they exist at pH 7.4 (physiological pH). An abbreviated structure of pyridoxal and pyridoxal phosphate, also shown, is used in the text.

particularly enhanced in pyridinium ions. Review this idea by noticing in Eq. 26.45 that the pyridinium ion is strongly activated toward reactions with nucleophiles; *notice particularly the electron flow onto the positively charged nitrogen.* Notice also in Eq. 26.50 that the positively charged nitrogen of the pyridinium ion serves to stabilize the attached carbanion by resonance. This chemistry has some close parallels in the biological world. For example, reviewing Sec. 10.8 will show how the pyridinium ion of NAD$^+$ serves as an electron acceptor in biochemical reductions. (Notice particularly Eq. 10.58a on p. 487.) Another biologically important pyridine derivative, *pyridoxal phosphate*, fulfills a similar mechanistic role in other reactions.

As shown in Fig. 26.3, pyridoxal phosphate is one of several forms of *vitamin B_6*. Pyridoxol was the first form of the vitamin discovered as a nutritional factor in 1934, but, in 1944, Esmond Snell (1914–2003), of the University of Texas, noticed that metabolites of pyridoxol secreted in the urine are more active. These metabolites turned out to be pyridoxal and pyridoxamine. Through the next decade, Snell and his co-workers elucidated the chemical role of these compounds.

Pyridoxal phosphate is an essential *coenzyme* (p. 486) in several important biochemical transformations. Here are only three of many:

Decarboxylation of α-amino acids:

$$H_3O^+ + RCH_2CH\underset{\overset{|}{\underset{^+NH_3}{}}}{-}\overset{\overset{O}{\parallel}}{C}-O^- \longrightarrow RCH_2CH_2\overset{+}{N}H_3 + O=C=O + H_2O \qquad (26.52a)$$

This transformation is utilized for the production of biologically important amines, such as the neurotransmitters serotonin and dopamine in the brain, and the vasoconstrictor histamine. The biosynthesis of serotonin, for example, employs the α-amino acid L-tryptophan as a starting material. L-Tryptophan is oxidized and the resulting α-amino acid, 5-hydroxy-L-tryptophan, is then decarboxylated.

L-tryptophan → *oxidation* → **5-hydroxy-L-tryptophan** → *decarboxylation* $-CO_2$ →

**serotonin
(5-hydroxy-L-tryptophan)
(conjugate acid)**   (26.52b)

The α-amino acid L-tyrosine is oxidized and decarboxylated to give dopamine, and the α-amino acid L-histidine is decarboxylated to give histamine.

L-tyrosine → *oxidation* → **3,4-dihydroxy-L-phenylalanine
(L-dopa)** → *decarboxylation* $-CO_2$ →

L-dopamine
(conjugate acid)   (26.52c)

L-histidine → *decarboxylation* $-CO_2$ → **histamine**
(conjugate acid)   (26.52d)

Interconversion of α-amino acids and α-keto acids:

an α-amino acid   an α-keto acid   (26.53)

This process is an important one in the biological synthesis and degradation of amino acids.

Loss of formaldehyde from serine:

serine
(an α-amino acid)   **formaldehyde**
(trapped by
tetrahydrofolate,
another vitamin)   **glycine**
(an α-amino acid)   (26.54)

This conversion is important as a source of single-carbon units for biological processes that involve single-carbon transfer. (The carbon of formaldehyde ends up as the methyl group in S-adenosylmethionine; Sec. 11.7B.)

In biological systems, each of these reactions is catalyzed by pyridoxal phosphate and an appropriate enzyme. These reactions can also be catalyzed by pyridoxal alone in the laboratory at elevated temperatures in the presence of certain metal ions.

Let's examine the first of these reactions to illustrate the essentials of pyridoxal phosphate catalysis. In the discussion that follows, keep your eye on the protonated pyridine ring of pyridoxal phosphate and relate the various transformations to the reactions of the previous sections.

In the biological world, pyridoxal phosphate typically is covalently attached as an imine (Schiff base) to various enzymes, usually through the side-chain amino group of the amino acid lysine within the protein structure:

a lysine residue in an enzyme

In the attachment reaction, the amino group of lysine acts as a nucleophile to form an imine (Schiff base) with the aldehyde of pyridoxal phosphate (Sec. 19.11A). (The abbreviated structure shown in Fig. 26.3 for pyridoxal phosphate is used in this and subsequent equations.)

$$(26.55)$$

pyridoxal phosphate
(abbreviated structure)

pyridoxal phosphate
attached to the enzyme
as an imine (Schiff base)

In the first step of all of the reactions of α-amino acids in Eqs. 26.51–26.54, the amino group of the α-amino acid, acting as a nucleophile in its unprotonated form, reacts with the imine product of Eq. 26.55 to form a new imine. This is exactly like imine formation from an amine and an aldehyde (Sec. 19.11A), except that the reaction of the amino group is with the C=N bond of an imine rather than with the C=O bond of an aldehyde.

$$(26.56)$$

α-amino acid
(unprotonated form)

enzyme-attached
pyridoxal phosphate

imine derivative of
the α-amino acid and
pyridoxal phosphate

(Because imines are sometimes called *Schiff bases*, this reaction is sometimes whimsically called "trans-Schiffization.") Decarboxylation forms what appears to be a *carbanion intermediate*.

imine derivative of the *a*-amino acid   a carbanion intermediate
and pyridoxal phosphate

(26.57a)

(See also Eq. 22.73b, p. 1143, for a similar reaction in fatty-acid biosynthesis.) This, however, is no ordinary carbanion. Most carbanions are such strong bases that they cannot exist under physiological conditions; however, this carbanion is a much weaker base because *it is stabilized by resonance*:

three of the many resonance structures for the carbanion intermediate

(26.57b)

(Only three of the many possible resonance structures are shown; you should draw others.) The curved arrows in the middle structure, which result in the structure on the right, show how *the pyridinium ion stabilizes negative charge by accepting electrons*. In fact, the red part of the structure on the right shows that the "carbanion" is really not a carbanion at all—it is a neutral molecule. (Compare with Eq. 26.50 on p. 1347.) The same type of "carbanion" is involved in all of the pyridoxal-catalyzed transformations discussed in this section. (See Problem 26.19.)

Recall from Sec. 20.11A that decarboxylation reactions typically require an "electron sink" to accept the electrons from the carboxy group that is lost. Eqs. 26.57a–b show that the pyridinium ring of pyridoxal phosphate serves this role effectively.

Protonation of this anion and hydrolysis of the resulting imine gives pyridoxal phosphate and the product of Eq. 26.52a. ($^+$B—H and B: are acidic and basic groups in the enzyme active site.)

pyridoxal phosphate

(26.57c)

Given how important the pyridinium ring is for delocalizing charge in the reactions of pyridoxal phosphate, a pertinent question is whether pyridoxal phosphate actually exists in the pyridinium-ion form. A typical pK_a of pyridinium ions is about 5 (Eq. 26.4, p. 1331). Yet the reactions promoted by pyridoxal phosphate take place at physiological pH values (about 7.4). If the pyridinium ion in pyridoxal phosphate had a pK_a near 5, most of it would exist as the conjugate-base pyridine form at pH 7.4; less that 1% of it would exist in the conjugate-acid pyridinium-ion form. (Verify this conclusion.) It turns out that the molecular architecture of pyridoxal phosphate ensures a much higher concentration of the crucial pyridinium-ion form. The key element in the structure is the —OH group in the 3-position and its ortho relationship to the aldehyde (see Eq. 26.58). This ortho relationship makes the phenolic —OH group of pyridoxal phosphate *unusually acidic*. (Why? See Problem 18.30(c), p. 916.) Ionization of the phenolic —OH group, in turn, raises the pK_a of the pyridinium ion because the negative charge of the phenolate stabilizes the positive charge of the pyridinium ion (and vice versa). In addition, the negatively charged phosphate also stabilizes the positive charge on the pyridinium nitrogen. As a result, the predominant form of pyridoxal phosphate at physiological pH is the form in which the phenol is ionized and the pyridine is protonated:

$$(26.58)$$

two forms of pyridoxal phosphate

But that's not all. When pyridoxal phosphate is bound to the enzymes that catalyze its reactions, the pyridinium form is further stabilized. In one well-studied case, this stabilization is the result of an ionized carboxylate group that interacts directly with the positively charged nitrogen:

As you can see, everything conspires to ensure that the pyridinium nitrogen stays protonated!

PROBLEMS

26.19 (a) Pyridoxal catalysis of Eq. 26.53 involves the following transformations. (Running these reactions in the reverse direction with a different α-keto acid completes Eq. 26.53.)

Using bases (B:) and acids (⁺BH) as needed, provide curved-arrow mechanisms for these reactions. Your mechanism should show the important intermediates. As part of your mechanism, explain the significance of the pyridinium ion in the catalysis of this reaction sequence.

(b) Using bases (B:) and acids (⁺BH) as needed, provide a curved-arrow mechanism for the conversion of the α-amino acid serine into formaldehyde and glycine (Eq. 26.54, p. 1349).

26.20 Isoniazid is an antituberculosis drug that operates by reacting with pyridoxal phosphate in the causative *Mycobacterium*. Show how isoniazid reacts with pyridoxal phosphate. (*Hint:* See Table 19.3, p. 985.)

isoniazid

| 26.5 | # NUCLEOSIDES, NUCLEOTIDES, AND NUCLEIC ACIDS |

Heterocyclic compounds occur widely in living systems. Heterocyclic compounds, specifically derivatives of purine and pyrimidine (Fig. 26.1, p. 1328), play a very important role in the structures of the nucleic acids DNA and RNA, polymers that are responsible for the storing and transmission of genetic information. This section introduces nucleic acids, and Sec. 26.6 introduces some of the other heterocyclic compounds that are important in living systems.

A. Nucleosides and Nucleotides

A **ribonucleoside** is a compound formed between the furanose form of D-ribose and a heterocyclic compound. The heterocyclic group is commonly referred to as the **base**, and the ribose as the **sugar**. The stereochemistry of the bond between the base and the ribose is most commonly β (pp. 1244–1245). A **deoxyribonucleoside** is a similar derivative of D-2-deoxyribose and a heterocyclic base. The prefix *deoxy* means "without oxygen"; thus, 2-deoxyribose is a ribose that has a second —H instead of an —OH group at carbon-2.

Notice in these structures that the sugar ring and the heterocyclic ring are numbered separately. To differentiate the two sets of numbers, primes (′) are used in referring to the atoms of the sugar. For example, the 2′ (pronounced *two-prime*) carbon of adenosine is carbon-2 of the sugar ring.

The sugar (ribose or deoxyribose) is often represented, especially in biochemistry texts, in a *Haworth projection* (see Sec. 24.3B) in which the C2′–C3′ bond is shown as a heavy line, implying that it is in the front of the page. Previously in the text we've represented cyclic compounds with planar, line-and-wedge structures. The relationship between the two is as follows:

adenosine
line-and-wedge structure
for the sugar

adenosine
Haworth projection
for the sugar

Neither representation conveys the fact that the five-membered ring actually exists as an equilibrium mixture of several rapidly interconverting puckered conformations (see Sec. 7.5A). However, we won't need to be concerned with conformations at this point.

The bases that occur most frequently in nucleosides are derived from two heterocyclic ring systems: *pyrimidine* and *purine*. (The numbering of these rings is shown in red.) Three bases of the pyrimidine type and two of the purine type occur most commonly.

pyrimidine

purine

cytosine
(C)

uracil
(U; occurs in RNA)

thymine
(T; occurs in DNA)

guanine
(G)

adenine
(A)

The base is attached to the sugar at N-9 of the purines and N-1 of the pyrimidines, as in the preceding examples.

In a nucleoside or nucleotide, the base is conventionally shown in either of two ways that differ by a 180° angle of internal rotation about the glycosidic bond—that is, the bond between the base and the sugar (C1′–N9 in adenosine).

adenosine
(anti conformation
about C1′–N9)

180° rotation

adenosine
(syn conformation
about C1′–N9)

As you can see from these structures, drawing the syn conformation for a purine base requires drawing an unrealistically long C1′–N9 bond. This long bond is simply an artifact of the Haworth projection. We'll use the anti conformation in most cases, but be aware that you may see nucleosides and nucleotides drawn in the syn conformation in other places.

In living systems, the 5′ —OH group of the ribose in a nucleoside is usually found esterified to a phosphate group. A 5′-phosphorylated nucleoside is called a **nucleotide**. A **ribonucleotide** is derived from the monosaccharide ribose; a **deoxyribonucleotide** is derived from 2′-deoxyribose. Some nucleotides contain a single phosphate group; others contain two or three phosphate groups condensed in phosphate anhydride linkages (See Sec. 25.2B).

uridylic acid (UMP)   **adenosine triphosphate (ATP)**

Although the ionization state of the phosphate groups depends on pH, these groups are written conventionally in the ionized form.

The nomenclature of the five common bases and their corresponding nucleosides and nucleotides is summarized in Table 26.2. This table gives the names of the ribonucleosides and ribonucleotides. To name the corresponding 2′-deoxy derivatives, the prefix *2′-deoxy* (or simply *deoxy*) is appended to the names of the corresponding ribose derivatives. For example, the 2′-deoxy analog of adenosine is called 2′-deoxyadenosine or simply deoxyadenosine. In addition, the names of the mono-, di-, and triphosphonucleotide derivatives are often abbreviated. Thus, adenylic acid is abbreviated AMP (for adenosine monophosphate); the di- and tri-phosphorylated derivatives are called ADP and ATP, respectively (see preceding example). The abbreviations for the corresponding deoxy derivatives contain a *d* prefix. Thus, 2′-deoxythymidylic acid can be abbreviated *d*TMP.

In addition to their role as the monomeric units of RNA, ribonucleotides also have other important biochemical functions, some of which have already been presented. NAD^+, one of nature's important oxidizing agents and its phosphorylated analog $NADP^+$ (Fig. 10.1, p. 486), *S*-adenosylmethionine (Fig. 11.1, p. 537), and coenzyme A (Fig. 25.1, p. 1284) are all ribonucleotides. One of the most ubiquitous ribonucleotides is adenosine triphosphate (ATP), whose biological roles were discussed in Sec. 25.7. The role of the ribonucleotide uridine diphosphate (UDP) in disaccharide biosynthesis was discussed in Sec. 25.7C.

TABLE 26.2 Nomenclature of Nucleic Acid Bases, Nucleosides, and Nucleotides*

Base	Nucleoside	Nucleotide (5′-monophosphate)	Abbreviation for the monophosphate
adenine (A)	adenosine	adenylic acid	AMP
uracil (U)	uridine	uridylic acid	UMP
thymine (T)	thymidine	thymidylic acid	TMP
cytosine (C)	cytidine	cytidylic acid	CMP
guanine (G)	guanosine	guanylic acid	GMP

* The deoxyribonucleosides and deoxyribonucleotides are named by appending the prefix *deoxy*, for example:

        deoxyadenosine   deoxyadenylic acid   *d*AMP

The prefix *deoxy* means 2′-deoxy unless stated otherwise.

26.21   Draw the structure of (a) deoxythymidine monophosphate (*d*TMP); (b) GDP.

26.22   Draw and label the syn and anti conformations of the nucleoside guanosine about the C1′–N9 bond.

B. The Structures of DNA and RNA

Nucleic acids are polymers of nucleotides. **Deoxyribonucleic acid (DNA)** is a polymer of deoxyribonucleotides and is the storehouse of genetic information throughout all of nature (with the exception of certain viruses). The monomeric units of the DNA polymer are called **residues**. A three-residue segment of DNA is shown in Fig. 26.4. This figure shows that the nucleotide residues in DNA are interconnected by phosphate groups that are esterified both to the 3′ —OH group of one ribose and the 5′ —OH of another. The DNA polymer incorporates adenine, thymine, guanine, and cytosine as the nucleotide bases. Although only three residues are shown in Fig. 26.4, a typical strand of DNA might be thousands or even millions of nucleotides long. *Each residue in a polynucleotide is differentiated by the identity of its base, and the sequence of bases encodes the genetic information in DNA.* The DNA polymer is thus a backbone of alternating phosphates and 2′-deoxyribose groups to which are connected bases that differ from residue to residue. The ends of the DNA polymer are labeled 3′ or 5′, corresponding to the deoxyribose carbon to which the terminal hydroxy group is attached.

Ribonucleic acid (RNA) polymers are conceptually much like DNA polymers, except that ribose, rather than 2′-deoxyribose, is the sugar. Three of the four bases in RNA are the same as in DNA; the fourth base, uracil, occurs in RNA instead of thymine (p. 1354), and some rare bases (not considered here) are found in certain types of RNA.

It was known for many years before the detailed structure of DNA was determined that DNA carries genetic information. It was also known that DNA is *replicated*, or copied, during cell reproduction. In 1950, Erwin Chargaff (1905–2002) of Columbia University showed that the ratios of adenine to thymine, and guanosine to cytosine, in DNA are

FIGURE 26.4 General structure of a single strand of DNA (base = A, T, G, or C; Table 26.2). Only three residues are shown here; a typical strand of DNA contains thousands or even millions of residues.

both 1.0; this observation has come to be known as *Chargaff's first parity rule*. How this rule relates to the storage and transmission of genetic information, however, remained a mystery. It became clear to a number of scientists that a knowledge of the three-dimensional structure of DNA would be essential to understand how DNA functions as it does. The importance of this problem was sufficiently obvious that several scientists worked feverishly to be the first to determine the three-dimensional structure of DNA. In 1953, James D. Watson (b. 1928) and Francis C. Crick (1916–2004), then at Cambridge University, proposed a structure for DNA. Their proposal was based on X-ray diffraction patterns of DNA fibers obtained by their colleagues at the Medical Research Council laboratory in England, Rosalind Franklin (1920–1958) and Maurice Wilkins (1916–2004). For their work on the structure of DNA, Watson, Wilkins, and Crick were awarded the Nobel Prize in Medicine or Physiology in 1962. Rosalind Franklin did not share the prize posthumously because the terms of Nobel's bequest stipulated that the prize should go only to living scientists.

The Watson–Crick structure of DNA is shown in Fig. 26.5 on p. 1358. The structure has the following important features:

1. The structure of DNA contains *two* right-handed helical polynucleotide chains coiled around a common axis; the structure is therefore that of a *double helix*. The helix makes a complete turn every 10 nucleotide residues. (Other helical conformations of DNA also occur.) The two polynucleotide chains run in opposite directions; that is, one chain runs in the $3' \to 5'$ direction, and the other in the $5' \to 3'$ direction.

2. The sugars and phosphates, which are rich in —OH groups and charges, are on the outside of the helix, where they can interact with solvent water or other hydrophilic compounds; the bases, which are hydrophobic, are buried in the interior of the double helix, away from water.

3. The chains are held together by hydrogen bonds between bases. *Each adenine (A) in one chain forms hydrogen bonds to a thymine (T) in the other, and each guanosine (G) in one chain forms hydrogen bonds to a cytosine (C) in the other.* Thus, every purine in one chain is hydrogen-bonded to a pyrimidine in the other. For this reason, A is said to be *complementary* to T, and G is *complementary* to C. The hydrogen-bonded A–T and G–C pairs are often referred to as **Waston–Crick base pairs**. Figure 26.6 (p. 1359) provides a closer look at these Watson–Crick base pairs. Notice that the A–T pair has about the same spatial dimensions as the G–C pair.

4. The planes of successive complementary base pairs are stacked, one on top of the other, and are perpendicular to the axis of the helix. The distance between each successive base-pair plane is 3.4 Å. Because the helix makes a complete turn every 10 residues, there is a distance of $10 \times 3.4 = 34$ Å along the helix per complete turn.

5. The double-helical structure of DNA results in two grooves that wrap around the double helix along its periphery. The larger groove is called the *major groove*, and the smaller is called the *minor groove*. These are shown in Fig. 26.5a. These grooves, particularly the major groove, are sites at which other macromolecules such as proteins are found to interact with DNA.

6. There is no intrinsic restriction on the sequence of bases in a polynucleotide; however, because of the base pairing described in point 3, the sequence of one polynucleotide strand (the "Watson" strand) in the double helix is complementary to that in the other strand (the "Crick" strand). Thus, everywhere there is an A in one strand, there is a T in the other; everywhere there is a G in one strand, there is a C in the other.

Hydrogen-bonding complementarity in DNA accounts nicely for the Chargaff parity rule: if A always hydrogen bonds to T and G always hydrogen bonds to C, then the number of A residues must equal the number of T residues, and the number of G residues must equal the number of C residues. This structure also suggests a reasonable mechanism for the duplication

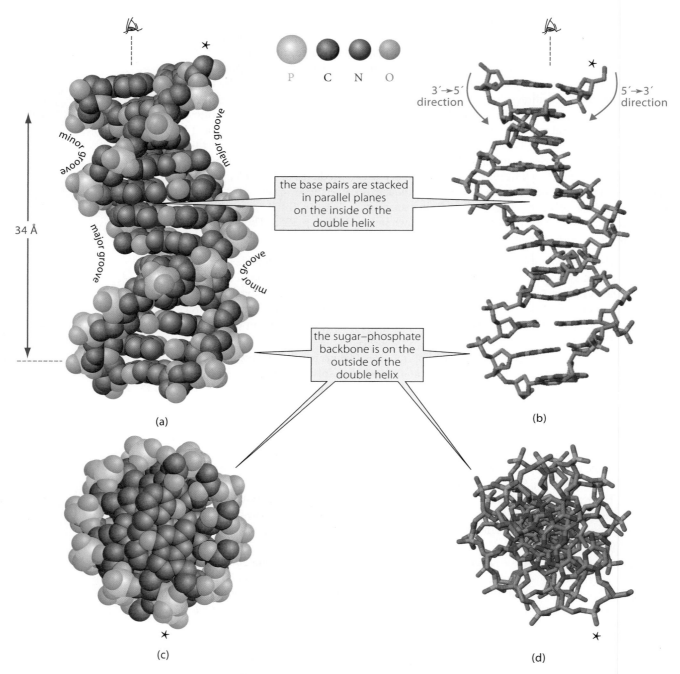

FIGURE 26.5 Molecular models of DNA from different perspectives. (a) A space-filling model in a side-on view, perpendicular to the helical axis. The major and minor grooves encircle the helix throughout its length. Many proteins that interact with DNA bind along the major groove. (b) A stick model in the same view as (a). This model shows more clearly the stacking of the bases in parallel planes. The parallel-offset stacking of the rings provides the major stabilization of the double-helical structure. (c) and (d) Views of the same models from the top (along the helical axis; the direction is indicated by the eyeball, with a common point indicated by the star, *). Notice that the sugars and phosphates are on the outside of the double helix, where they can readily interact with water, and the base pairs are on the inside of the helix, where they can form complementary hydrogen bonds with each other.

of DNA during cell division: the two strands come apart, and a new strand is grown as a complement of each original strand. In other words, *the proper sequence of each new DNA strand during cellular reproduction is ensured by hydrogen-bonding complementarity* (Fig. 26.7, p. 1360).

FIGURE 26.6 A closer look at the complementary base pairing in DNA. (a) A cytosine–guanine (C–G) base pair involves three hydrogen bonds. (b) A thymine–adenine (T–A) base pair involves two hydrogen bonds. (c) Superposition of the C–G (*white*) and T–A (*blue*) base pairs shows that the two occupy about the same space. (Regions of overlap are shown in gray.)

Although the complementary hydrogen bonds within the base pairs contribute to the stability of the double helix and account for the sequence fidelity of DNA replication, they are not the primary reason for the stability of double-helical DNA. There is good evidence that the stacking interactions between the rings on successive planes provide the major part of the free-energy driving force for the formation of the double helix. These stacking interactions can be classified as pi offset-stacking interactions (Sec. 15.8A).

One of the most significant achievements in the history of biochemical analysis has been the development of methods for rapidly determining the sequential arrangement of individual bases in DNA. This type of analysis is called *DNA sequencing.* (We won't consider these methods in detail here.) In 1990, the National Institutes of Health and the Department of Energy co-sponsored the *Human Genome Project,* the primary goal of which was to identify the 20,000–25,000 human genes and to sequence all of the DNA in the human genome. Preliminary sequences were published in 2001, and high-quality sequences of almost all of the 24 human chromosomes were nearly complete in 2008. The magnitude of this project can be appreciated from the fact that the human genome contains about three billion base pairs! The genome sequences of thousands of other organisms are either complete or nearing completion. These "sequenced organisms" come from every corner of the biological world—viruses (such as influenza A and human immunodeficiency virus (HIV)), bacteria (such as anthrax), plants (such as wheat and rice), insects (such as fruit flies), and higher animals (such as cows). Some viruses, called *retroviruses,* carry their genetic information in RNA rather than DNA. When these viruses infect a cell, a viral enzyme, *reverse transcriptase,* translates the RNA code of the virus to viral DNA, forming an RNA–DNA hybrid. The RNA is then hydrolyzed back to individual ribonucleotides, and the remaining viral DNA is incorporated into the reproductive

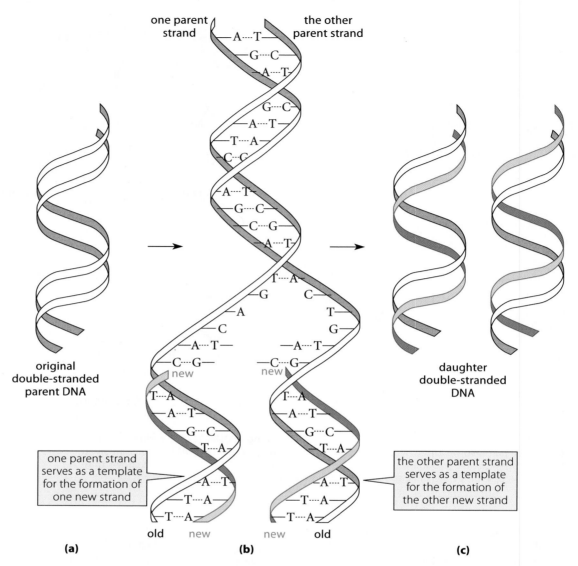

FIGURE 26.7 Complementary base pairing in DNA is crucial to its faithful replication. (a) A typical DNA double helix. (b) In the replicating DNA, a new strand grows on each of the original parent strands. (The synthesis of new DNA on a parent DNA template is catalyzed by several enzymes, which are not shown.) (c) Two new molecules of DNA, each containing one parent strand and one new strand.

machinery of the cell, where translation leads to the formation of many new virus particles and destruction of the host cell. Viral DNA can be produced in the laboratory from viral RNA with reverse transcriptase. Sequencing of this DNA provides the viral genome.

The central importance of DNA sequences is that they provide a linear code for biosynthesis of *messenger RNAs* (mRNAs), whose sequences, in turn, provide the code for the biosynthesis of proteins:

DNA sequence ⇨ mRNA sequence ⇨ linear sequence of amino acids in proteins

In other words, the linear sequence of amino acids in every protein is determined by a DNA code. (This process is discussed in more detail in Sec. 27.6B.) It is now possible to read the DNA "code" for a protein and thereby determine—without isolating the protein—the sequence of amino acids in the protein. These sequence data are unlocking the genetic basis of diseases as well as the significant genetic variations that occur among individuals. It is not unreasonable to imagine that, in the future, we will walk into our doctor's office or pharmacy with a small magnetic card containing the complete sequence of our individual genome, which will be used to assess our individual risk of disease and to prescribe just the right medication and to determine its proper dose.

PROBLEMS

26.23 Draw in detail the structure of a section of RNA four residues long that, from the 5′-end, has the following sequence of bases: A, U, C, G. Label the 3′ and 5′ ends.

26.24 Would you expect Chargaff's first parity rule to apply within an individual strand of DNA? Explain.

C. DNA Modification and Chemical Carcinogenesis

We've shown that the double-helical structure of DNA and DNA replication involve very specific Waston–Crick base-pairing complementarity. Other important processes, such as RNA biosynthesis and protein biosynthesis, also involve this type of complementarity. The molecular basis of this complementarity, as we've seen, is the specific hydrogen bonding between a pyrimidine and a purine base. If this hydrogen bonding is disrupted, the Waston–Crick base-pairing complementarity can also disrupted, and with it, some or all of the biological processes that rely on this phenomenon. There is strong circumstantial evidence that chemical damage to DNA can interfere with this hydrogen-bonding complementarity and can in some cases trigger the state of uncontrolled cell division known as cancer.

One type of chemical damage to DNA is caused by *alkylating agents* (Sec. 10.4B). Certain types of alkylating agents react with DNA by alkylating one or more of the nucleotide bases. These same alkylating agents are also *carcinogens* (cancer-causing compounds). A few such compounds are the following:

methyl methanesulfonate
(a weak carcinogen)

dimethyl sulfate
(a weak carcinogen)

N-methyl-N-nitrosourea
(a potent carcinogen)

methyldiazonium ion
(the actual alkylating agent)

When such alkylating agents (abbreviated $H_3C—X$ in the following equations) react with DNA, alkylated guanosines are among the products. The major product is alkylated on N-7 of the guanine base, but an important minor product is alkylated on the oxygen at C-6 (called the O-6 position).

deoxyribose

a G residue of DNA

N-7 alkylation product
(major)

O-6 alkylation product
(minor)

(An analogous alkylation occurs at O-4 of thymine; see Problem 26.25) Notice that the alkylation at O-6 prevents the N-1 nitrogen from acting as a hydrogen-bond donor in a Watson–Crick base pair (Fig. 26.6) because the hydrogen is lost from this nitrogen as a result of alkylation. (B: = a base.)

(26.61)

The N-7 alkylation, in contrast, does not directly affect any of the atoms involved in the hydrogen-bonding complementarity. It has been found that the alkylating agents that are the most potent carcinogens also yield the greatest amount of the guanines alkylated at O-6 and thymines alkylated at O-4. Although this correlation does not prove that these alkylations are primary events in carcinogenesis, it provides strong circumstantial evidence in this direction.

The nitrogen mustards used as antitumor drugs were described in the sidebar of Sec. 11.8B (pp. 545–546). These drugs, as dialkylating agents, react at the N-7 positions on guanine bases in opposite strands of the DNA double helix. This reaction forms a crosslink between the two strands and prevents the strand separation that must accompany cell division (Fig. 26.7).

a nitrogen mustard

G residues on opposite strands of DNA

dialkylated DNA
(crosslinked DNA strands)

(26.62)

Because tumor cells with crosslinked DNA cannot divide and reproduce, the tumor cannot grow.

The conversion of aromatic hydrocarbons into carcinogenic diol epoxides by enzymes in living systems was discussed in Sec. 16.7. These epoxides react with G residues of DNA. Although the reaction can occur at N-7 (as with the mustards), the reaction that leads to carcinogenicity is the reaction with the amino group at N-2.

a G residue in the
DNA double helix

diol epoxide of
benzo[*a*]pyrene
(Eq. 16.50, p. 828)

(26.63)

This nitrogen is also involved in the hydrogen-bonding interaction of G with C (Fig. 26.6). Thus, it may be that alkylation by aromatic hydrocarbon epoxides also triggers the onset of cancer by interfering with the base-pairing complementarity.

DNA damage can also be caused by ultraviolet radiation. Ultraviolet light promotes a cycloaddition of two pyrimidines when they occur in adjacent positions on a strand of DNA. (This type of cycloaddition is discussed in Sec. 28.3.) In the following example, a thymine dimer is formed from two adjacent thymines.

two thymines in DNA at
adjacent positions on same strand

thymine dimer

(26.64)

Most people have a biological repair system that effects the removal of the modified pyrimidines and repairs the DNA. People with a rare skin disease, *xeroderma pigmentosum*, have a genetic deficiency in the enzyme that initiates this repair. Most of these people contract skin cancer and die at an early age. Here, then, is a situation in which the chemical modification of DNA has been clearly associated with the onset of cancer.

PROBLEM

26.25 There is evidence that alkylation at O-4 of thymine, like alkylation at O-6 of guanine, is another mutagenic event that can lead to cancer.
   (a) Draw the structure of a thymine residue as it would exist after O-4 methylation.
   (b) Explain why O-4 alkylation at thymine would disrupt Watson–Crick base pairing.

26.6   OTHER BIOLOGICALLY IMPORTANT HETEROCYCLIC COMPOUNDS

Nitrogen heterocycles occur widely in nature. Sec. 23.12B introduced the *alkaloids* (Fig. 23.4, p. 1222), many of which contain heterocyclic ring systems. The naturally occurring amino acids proline, histidine, and tryptophan, which are covered in Chapter 27, contain, respectively, a pyrrolidine, imidazole, and an indole ring (Fig. 26.8, p. 1364). A number of vitamins are heterocyclic compounds; without these compounds, many important metabolic processes could not take place. For example, we have already discussed the importance of the pyridinium group in the vitamins NAD^+ (Sec. 10.8) and pyridoxal phosphate (Sec. 26.4E). Some other heterocycle-containing vitamins are shown in Fig. 26.8.

Heterocyclic compounds are involved in some of the colors of nature that have intrigued humankind from the earliest times. Why is blood red? Why is grass green? The color of blood is due to an iron complex of heme, a heterocycle composed of pyrrole units. This type of heterocycle is called a **porphyrin** (*red* in the following structure).

(S)-proline   **(S)-histidine**   **(S)-tryptophan**

folic acid   **thiamin** (vitamin B$_1$)

riboflavin (vitamin B$_2$)   **chlorophyll *a***

FIGURE 26.8 A few of the many naturally occurring heterocyclic compounds. The *S* enantiomers of proline, histidine, and tryptophan are α-amino acids. Folic acid, thiamin, and riboflavin are vitamins. The chlorophylls are the pigments responsible for the green color of plants. The C$_{20}$H$_{39}$ group is an isoprenoid side chain; see Sec. 17.6A.) NAD$^+$ (Fig. 10.1, p. 486) and pyridoxal phosphate (Fig. 26.3, p. 1348) are examples of important naturally occurring pyridine derivatives.

protein (globin)

coordinated O$_2$

plane of the heme molecule

imidazole from a globin histidine

heme
(occurs in the protein complex hemoglobin)

a schematic view of oxygenated heme in hemoblobin

Heme is the Fe(II) complex of an aromatic heterocycle that is found in red blood cells tightly bound to a protein called *globin*; the complex is called *hemoglobin*. The iron, held in position by coordination with the nitrogens of heme and an imidazole of globin, reversibly forms a complex with oxygen. Thus, hemoglobin is the oxygen carrier of blood, and the red color of blood is due to oxygenated hemoglobin. Carbon monoxide and cyanide, two well-known respiratory poisons, also form complexes with the iron in hemoglobin, as well as with iron in the heme groups of other respiratory proteins.

The green color of plants is caused by *chlorophyll*, a class of compounds closely related to the porphyrins (Fig. 26.8). The absorption of sunlight by chlorophylls is the first step in the conversion of sunlight into usable energy by plants. Thus, the chlorophyll molecules are nature's "solar energy collectors."

KEY IDEAS IN CHAPTER 26

- Heterocyclic organic compounds contain rings consisting of both carbon atoms and atoms of other elements.

- In aromatic heterocycles, a vinylic nitrogen contributes one electron to the aromatic π-electron system, and its unshared electron pair is in the plane of the ring. A nitrogen of this type (for example, in pyridine) is basic with $pK_a \approx 5$. An allylic nitrogen contributes its unshared electron pair to the aromatic π-electron system. A nitrogen of this type (for example, in pyrrole), is not basic, because protonation would disrupt the $4n + 2$ aromatic π-electron system. Furan has two unshared electron pairs, one of each type.

- Pyrrole, furan, and thiophene are all more reactive than benzene in electrophilic aromatic substitution and undergo substitution predominantly at the 2-position. The reactivity order is pyrrole > furan > thiophene.

- Because furan has a relatively small empirical resonance energy, it undergoes some diene conjugate-addition reactions, such as the Diels–Alder reaction.

- Pyridine reacts very slowly in electrophilic aromatic substitution. Pyridine and its derivatives undergo electrophilic substitution at the 3-position. Electrophilic substitution reactions of pyridine-*N*-oxides, however, occur at the 4-position.

- Many side-chain reactions of heterocyclic compounds proceed normally without disrupting the heterocyclic rings.

- Pyridine derivatives undergo nucleophilic aromatic substitution reactions at the 2- and 4-positions. Thus, pyridines react in the Chichibabin and related reactions;

2- and 4-chloropyridines undergo nucleophilic aromatic substitution reactions. Pyridinium salts are even more reactive than pyridines in these reactions. The chemistry of the pyridine C=N linkage has some similarity to the chemistry of the carbonyl group.

- The "benzylic" hydrogens of 2-alkyl- and 4-alkyl-pyridines and especially pyridinium salts are acidic enough to be removed by bases. In biology, the reactions of NAD^+ and pyridoxal phosphate are due to the electron-accepting ability of the pyridinium ions in these molecules.

- Ribonucleotides and deoxyribonucleotides, which are phosphorylated derivatives of ribonucleosides and deoxyribonucleosides, are the building blocks of RNA and DNA, respectively. These compounds are phosphorylated derivatives of either ribose or 2'-deoxyribose, respectively, and a purine or pyrimidine base, in which the base is connected to C-1 of the ribose with β stereo-chemistry. Adenine, guanine, and cytosine are bases in both DNA and RNA, whereas thymine is unique to DNA, and uracil is unique to RNA.

- An important conformation of DNA is the double helix, in which two right-handed helical strands of DNA running in opposite directions wrap around a common axis. The sugars and phosphate groups lie on the outside of the helix, and the bases are stacked in parallel planes on the inside. The two strands of the double helix are held together by Watson–Crick base pairs—that is, by hydrogen bonds between a purine base (A or G) and its complementary pyrimidine (T or C). A number of known carcinogens apparently modify DNA in such a way that this complementary hydrogen bonding is disrupted.

 REACTION REVIEW *For a summary of reactions discussed in this chapter, see the* Reaction Review *section of Chapter 26 in the* Study Guide and Solutions Manual.

ADDITIONAL PROBLEMS

26.26 Give the principal organic product(s) expected when 2-methylthiophene or other compound indicated reacts with each of the following reagents.

(a) acetic anhydride, BF_3, acid

(b) HNO_3

(c) *N*-bromosuccinimide, CCl_4, light

(d) dilute aqueous HCl

(e) dilute aqueous NaOH

(f) product of part (c) + Mg/ether, then CO_2, then H_3O^+

(g) product of part (a) + Ph—CH=O and NaOH

26.27 Give the principal organic product(s) expected when 2-methylpyridine or other compound indicated reacts with each of the following reagents.

(a) dilute aqueous HCl

(b) dilute aqueous NaOH

(c) $CH_3CH_2CH_2CH_2$—Li

(d) HNO_3, H_2SO_4, heat; then $^-$OH

(e) 30% H_2O_2

(f) CH_3I

(g) product of part (c) + PhCH=O, then H_2O

(h) product of part (e) + H_2, catalyst

26.28 Rank the following compounds in order of increasing reactivity toward nitration with HNO_3, and explain your choices: thiophene, benzene, 3-methylthiophene, and 2-methylfuran.

26.29 Rank each of the following compounds in order of increasing S_N1 solvolysis reactivity in ethanol, and explain your choices by drawing suitable structures.

(a)

A   *B*

C   *D*

(b)

E   *F*

26.30 Think of the compounds in the following sets as enols. Then draw the carbonyl isomers of the following com-

pounds. Which compound within each set contains the greatest percentage of carbonyl isomer? Explain.

(a) 2-hydroxyfuran or 2-hydroxypyrrole

(b) phenol or 4-hydroxypyridine

26.31 Rank the compounds within each of the following sets in order of increasing basicity, and explain your reasoning.

(a) pyridine, 4-methoxypyridine, 5-methoxyindole, 3-methoxypyridine

(b) pyridine, 3-nitropyridine, 3-chloropyridine

(c)

(d) imidazole and oxazole

(e) imidazole and thiazole

26.32 The following compound is a very strong base; its conjugate acid has a pK_a of about 13.5. Give the structure of its conjugate acid and show that it is stabilized by resonance.

26.33 Draw the structure of the major form of each of the following compounds present in an aqueous solution containing initially one molar equivalent of 1 *M* HCl. Explain your reasoning.

(a) quinine (Fig. 23.4, p. 1222)

(b) nicotine (Eq. 26.26, p. 1339 or Fig. 23.4, p. 1222)

(c)

tryptamine

(d) 3,4-diaminopyridine

(e)

1,4-diazaindene

(f)

1-methyl-1,2,3-benzotriazole

(g) See Fig. P26.33.

26.34 Complete the following reactions by giving the major organic product(s).

(a)

+ D$_2$O (excess) → $\xrightarrow{\text{NaOD (catalyst)}}$

(b)

+ PhLi →

(c)

+ Br$_2$ $\xrightarrow[\text{CCl}_4]{\text{dark}}$

Note: The starting material of (c) has no double bond between positions 2 and 3; that is, it is not an indole.)

(d)

+ H$_2$ $\xrightarrow[\text{25 °C}]{\text{Pt/C}}$ (C$_7$H$_9$N)

(e)

$\xrightarrow[\text{H}_2\text{SO}_4]{\text{fuming HNO}_3}$

(f)

$\xrightarrow[\text{−5 °C}]{\text{HNO}_3 \atop \text{Ac}_2\text{O}}$ (nitration occurs at a 5-position— but in which ring?)

(g)

$\xrightarrow{\text{1) fuming HNO}_3, \text{H}_2\text{SO}_4 \atop \text{2) neutralize with NaOH}}$

(h)

$\xrightarrow[\text{heat}]{\text{NH}_2\text{NH}_2, \text{ }^-\text{OH}}$

26.35 Doreen Dimwhistle has proposed the following variations on the Chichibabin reaction:

(a) indole + NaNH$_2$ ⟶ 2-aminoindole

(b) 2-chloropyridine + NaNH$_2$ ⟶ 2-amino-6-chloropyridine

She is shocked to find that neither of these reactions works as planned and has come to you for an explanation. Explain what reaction, if any, occurs instead in each case.

26.36 The following compound is isolated as a by-product in the Chichibabin reaction of pyridine and sodium amide. Give a curved-arrow mechanism for its formation.

26.37 When pyrrole is treated with 5.5 *M* HCl at 0 °C for 30 s, a crystalline product *B* is obtained (see Fig. P26.37). A likely intermediate in this reaction is compound *A*.

(a) Draw a curved-arrow mechanism for the formation of *A*.

(b) Draw a curved-arrow mechanism for the formation of *B* from *A*, pyrrole, and HCl.

26.38 Indole (Fig. 26.1, p. 1328) can undergo electrophilic substitution reactions. The following reaction is an example:

H$_2$N— —NO$_2$ $\xrightarrow[\text{H}_2\text{SO}_4]{\text{NaNO}_2,}$ $\xrightarrow{\text{indole}}$

The electrophilic species in this reaction that reacts with indole is the following diazonium ion.

:N≡N⁺— —NO$_2$

Draw a curved-arrow mechanism for the electrophilic aromatic substitution reaction of this ion with indole and use it to derive the structure of the product, an azo dye.

(g)

imatinib (Gleevec®)
(an anticancer drug; see Sec. 1.1B)

Figure P26.33

Figure P26.37

26.39 Outline a synthesis for each of the following compounds from the indicated starting material and any other reagents:

(a) ![2-(aminomethyl)pyridine structure with CH₂NH₂] from pyridine

(b) SEt ![4-(ethylthio)pyridine structure] from pyridine

(c) ![difurylpropenone structure] from furfural (furan-2-carbaldehyde) as the only source of furan rings

(d) ![propyl furan-2-carboxylate structure] from furfural (furan-2-carbaldehyde)

(e) ![N,N'-di(3-pyridyl)urea structure] from 3-methylpyridine

(f) $CH_3C(CH_2CH_2CN)_2$![N-methylpyridinium iodide structure with I⁻ and CH₃] from 4-ethylpyridine

(g) $CH_3CH_2CH_2CHCO_2H$![2-substituted pyridine structure] from 2-methylpyridine

26.40 Many furan derivatives are unstable in strong acid. Hydrolysis of 2,5-dimethylfuran in aqueous acid gives a compound A, $C_6H_{10}O_2$, that has a proton NMR spectrum consisting entirely of two singlets at δ 2.1 and δ 2.6 in the ratio 3 : 2, respectively. On treatment of compound A with very dilute NaOD in D_2O, both NMR signals disappear. Treatment of A with zinc amalgam and HCl gives hexane. Propose a structure for A, and then give a curved-arrow mechanism for its formation from 2,5-dimethylfuran.

26.41 Compound A, $C_8H_{11}NO$, smells as if it might have been isolated from an extract of dirty socks. This compound can be resolved into enantiomers and it dissolves in 5% aqueous HCl. Oxidation of A with concentrated HNO_3 and heat gives nicotinic acid (3-pyridinecarboxylic acid; see Eq. 26.26.) When A reacts with CrO_3 in pyridine, compound B (C_8H_9NO) is obtained. Compound B, when treated with dilute NaOD in D_2O, incorporates five deuterium atoms per molecule. Identify A, and explain your reasoning.

26.42 Aromatic sulfonation can be reversed by heating an arylsulfonic acid in steam:

![benzene-SO₃H + H₂O heat → benzene-H + H₂SO₄]

(a) Draw a curved-arrow mechanism for this transformation.

(b) Identify A, B, and C in the scheme shown in Fig. P26.42.

(c) Explain why compound C cannot be synthesized in one step from thiophene.

26.43 Provide curved-arrow mechanisms for each of the reactions given in Fig. P26.43. Give the structure for the intermediates A and B in parts (d) and (f).

26.44 Rank the three compounds in Fig. P26.44 in order of their reactivity toward amide hydrolysis in aqueous NaOH, least reactive first, and explain your reasoning.

26.45 Decarboxylation of the amino acid L-histidine in *Lactobacillus* species involves an enzyme-attached amide (*blue*) of pyruvic acid, as shown in Fig. P26.45 on p. 1370. (E = enzyme).

(a) Assuming that bases (B:) and acids ($^+$BH) are available as needed, suggest a curved-arrow mechanism for this transformation. (*Hint:* An imine intermediate is involved.)

(b) In your mechanism, what group serves as the "electron sink" for the decarboxylation?

26.46 Pyridoxal phosphate and an enzyme, *tryptophan synthetase*, catalyze the last step in the biosynthesis of the amino acid *tryptophan* (see Fig. P26.46, p. 1370).

(a) The first part of this reaction involves the reaction of pyridoxal phosphate with serine to form species A, an imine of the unstable amino acid *dehydroserine*. [The red carbon is for part (b).]

![structure A, dehydroserine imine of pyridoxal phosphate]

A
dehydroserine imine
of pyridoxal phosphate

Assuming that bases (B:) and acids ($^+$BH) are available as needed, give a curved-arrow mechanism for the formation of A.

(b) Show with appropriate resonance structures that the carbon shown in red in the structure A has carbocation character.

(c) Recognizing that indole derivatives readily undergo electrophilic aromatic substitution at carbon-3,

(Problem continues on p. 1370)

Figure P26.42

(a)

(*Hint:* Form the iminium ion $H_2C\overset{+}{=}NEt_2$ first.)

(b)

(c)

(d)

(e)

(f)

(g)

Figure P26.43

1-acetylimidazole

A

1-acetyl-3-methyl-imidazolium ion

B

N-acetylpiperazine

C

Figure P26.44

complete a curved-arrow mechanism for the biosynthesis of tryptophan. Assume that bases (B:) and acids (⁺BH) are available as needed.

26.47 The *racemization* of amino acids is an important reaction in a number of bacteria.

This is a pyridoxal-phosphate-catalyzed reaction. Outline a curved-arrow mechanism for this reaction showing clearly the role of pyridoxal phosphate. Assume that bases (B:) and acids (⁺BH) are available as needed.

26.48 Explain each of the following facts.

(a) In the following compound, the hydrogens of the methyl group shown in red, as well as the imide proton, are readily exchanged for deuterium by dilute NaOD in D_2O, but those of the other methyl group are not.

(b) In the following ion, the hydrogens of the methyl group shown in red are most acidic, even though the other methyl group is directly attached to the positively charged nitrogen.

(c) The reaction given in Fig. P26.48 takes place in aqueous base.

(d) The compound 2-pyridone does not hydrolyze in aqueous NaOH using conditions that bring about the rapid hydrolysis of δ-butyrolactam.

L-histidine → **histamine** (conjugate acid) + CO_2 + H_2O

Figure P26.45

indole glycerol phosphate + **L-serine** (an α-amino acid) → **L-tryptophan** + **D-glyceraldehyde-3-phosphate**

Figure P26.46

2-pyridone   **δ-butyrolactam**

(e) Treatment of 4-chloropyridine with ammonia gives 4-aminopyridine, but treatment of 3-chloropyridine under the same conditions gives no reaction.

26.49 You work for a pharmaceutical company whose management has decided to produce synthetic vitamin B$_6$. The company is in possession of some fragmentary notes from Strong E. Nuff, one of their early chemists, that outline the synthesis shown in Fig. P26.49 of pyridoxine (a form of vitamin B$_6$). Unfortunately, reagents for each of the numbered steps have been omitted. They have hired you as a consultant; suggest the reagents that would accomplish each step.

26.50 Although the synthesis of heterocyclic rings was not discussed in the text, many such syntheses employ reactions that are similar or identical to reactions in other parts of the text. Give curved-arrow mechanisms for the reactions involved in each of the heterocyclic syntheses in

Fig. P26.50 on p. 1372. Begin, as always, by analyzing the relationship between the atoms in the reactants and products.

26.51 One theory of genetic mutation postulates that some mutations arise as the result of mispairing of bases in DNA caused by the existence of relatively rare isomeric forms of the bases. Show the hydrogen-bonding complementarity that can result from (a) the pairing of an imine isomer of C with A; (b) the pairing of an enol isomer of T with G.

26.52 When RNA is treated with periodic acid, and the product of that reaction treated with base, only the nucleotide residue at the 3′-end is removed.

(a) Explain this transformation by showing its chemistry.

(b) Would the same reactions occur with DNA? Explain.

26.53 When DNA is treated with 0.5 *M* NaOH at 26 °C, no reaction takes place, but when RNA is subjected to the same conditions, it is rapidly cleaved into mononucleotide 2- and 3-phosphates. Explain. (*Hint:* What is the only structural difference between RNA and DNA? How can this difference promote the observed behavior? See Sec. 11.8.)

Figure P26.48

Figure P26.49

(a)

(b) Hinsberg thiophene synthesis:

$$MeO_2C-CH_2-S-CH_2-CO_2Me + Ph-\overset{O}{\underset{||}{C}}-\overset{O}{\underset{||}{C}}-Ph \xrightarrow[\text{3) HCl}]{\text{1) NaOMe, MeOH}\atop\text{2) H}_2\text{O, heat}}$$

(c) Friedlander quinoline synthesis:

(d) Combes quinoline synthesis:

(e) Hantzsch dihydropyridine synthesis:

$$2\,H_3C-\overset{O}{\underset{||}{C}}-CH_2-\overset{O}{\underset{||}{C}}-OEt + NH_3 + H_2C=O \xrightarrow{\text{Et}_2\text{NH}} \quad + 3\,H_2O$$

(f) Reissert indole synthesis. Identify compounds *A* and *B*, and give the mechanisms for the formation of compounds *A* and *C*.

o-nitrotoluene

diethyl oxalate

A ($C_{11}H_{11}O_5N$) $\xrightarrow[\text{HOAc}]{\text{H}_2,\text{ Pt}}$

+ EtOH

B (not isolated) $\xrightarrow{\text{HOAc}}$

+ H_2O

C

ethyl 2-indolecarboxylate

(g) Larsen–Chen indole synthesis. Show the different catalytic intermediates. (Assume that appropriate ligands are available for the catalyst; abbreviate these as "L.")

Figure P26.50

Amino Acids, Peptides, and Proteins

Amino acids, as the name implies, are compounds that contain both an amino group and a carboxylic acid group.

alanine
(an α-amino acid)

***p*-aminobenzoic acid**
(PABA, a component
of folic acid, a vitamin)

As these structures show, a neutral amino acid—an amino acid with an *overall* charge of zero—can contain within the same molecule two groups of opposite charge. Molecules containing oppositely charged groups are known as **zwitterions** (German, meaning "hybrid ion"). A zwitterionic structure is possible because the basic amino group can accept a proton and the acidic carboxylic acid group can lose a proton. Each of the **α-amino acids**, of which alanine is an example, has an amino group on the α-carbon—the carbon adjacent to the carboxylic acid group.

 Peptides are biologically important polymers in which α-amino acids are joined into chains through amide bonds, called **peptide bonds**. A peptide bond is derived from the amino group of one amino acid and the carboxylic acid group of another. Peptides are heteropolymers because the R-groups in the repeating units can differ.

general peptide structure

Proteins are very large peptides, and some proteins are aggregates of more than one peptide. The name *protein* (from a Greek word meaning "of first rank") is particularly apt because peptides and proteins serve many important roles in biology. For example, almost all enzymes (biological catalysts) and some hormones are peptides or proteins.

27.1   NOMENCLATURE OF AMINO ACIDS AND PEPTIDES

A. Nomenclature of Amino Acids

Some amino acids are named substitutively as carboxylic acids with amino substituents.

$$H_2N—CH_2CH_2CH_2—CO_2H \rightleftharpoons \overset{+}{H_3N}—CH_2CH_2CH_2—CO_2^-$$

**4-aminobutanoic acid
(γ-aminobutyric acid)**

**2-aminobenzoic acid
(o-aminobenzoic acid,
anthranilic acid)**

$$(CH_3)_2\overset{+}{N}H—CH_2CH_2—\overset{\displaystyle O}{\overset{\|}{C}}—O^-$$

3-(dimethylamino)propanoic acid

Even if they exist as zwitterions, amino acids are named as uncharged compounds.

Twenty α-amino acids are known by widely accepted traditional names. These are the amino acids that occur commonly as constituents of proteins. The names and structures of these amino acids are given in Table 27.1 on pp. 1376–1377.

Two points about the structures of the α-amino acids will help you to remember them. First, with the exception of proline, all α-amino acids have the same general structure, differing only in the identity of the side chain R.

general structure

phenylalanine
(R = CH₂Ph)

serine
(R = CH₂OH)

Proline is the only naturally occurring amino acid with a secondary amino group. In proline the —NH— and the side chain are "tied together" in a ring.

proline

Second, as Table 27.1 shows, the amino acids can be organized into six groups according to the nature of their side chains.

- amino acids with —H or aliphatic hydrocarbon side chains
- amino acids with side chains containing aromatic groups
- amino acids with aliphatic side chains containing —SH, —SCH_3, or —OH groups
- amino acids with side chains containing carboxylic acid or amide groups
- amino acids with basic side chains
- proline

The α-amino acids are often designated by either three-letter or single-letter abbreviations, which are given in Table 27.1.

B. Nomenclature of Peptides

The terminology and nomenclature associated with peptides are best illustrated by an example. Consider the following peptide formed from the three amino acids alanine, valine, and lysine.

abbreviated **Ala-Val-Lys** or **A-V-K**

The **peptide backbone** is the repeating sequence of nitrogen, α-carbon, and carbonyl groups shown in blue in the foregoing structure. The characteristic amino acid side chains are attached to the peptide backbone at the respective α-carbon atoms. Each amino acid unit in the peptide is called a **residue**. For example, the part of the peptide derived from valine, the *valine residue*, is outlined in red. The ends of a peptide are labeled as the **amino end** or **amino terminus** and the **carboxy end** or **carboxy terminus**. A peptide can be characterized by the number of residues it contains. For example, the preceding peptide is a **tripeptide** because it contains three amino acid residues. A peptide containing two, three, or five amino acids would be called a **dipeptide**, **tripeptide**, or **pentapeptide**, respectively. A relatively short peptide of unspecified length containing a few amino acids is sometimes referred to as an **oligopeptide** (from a Greek root meaning "scant" or "few").

A peptide is conventionally named by giving successively the names of the amino acid residues, *starting at the amino end*. The names of all but the carboxy-terminal residue are formed by dropping the final ending (*ine*, *ic*, or *an*) and replacing it with *yl*. Thus, the foregoing peptide is named alanylvalyllysine. In practice, this type of nomenclature is cumbersome for all but the smallest peptides. A simpler way of naming peptides is to connect with hyphens the three-letter (or one-letter) abbreviations of the component amino acid residues beginning with the amino-terminal residue. Thus, the preceding peptide is also written as Ala-Val-Lys or A-V-K.

Large peptides of biological importance are known by their common names. Thus, *insulin* is an important peptide hormone that contains 51 amino acid residues; *ribonuclease*, an enzyme, is a protein containing 124 amino acid residues (and a rather small protein at that!).

(text continues on p. 1378)

TABLE 27.1 Names, Structures, Abbreviations, and Properties of the Twenty Common Naturally Occurring Amino Acids

General structure:

$$H_3\overset{+}{N}-\underset{\underset{R}{|}}{CH}-\overset{\overset{O}{\|}}{C}-O^-$$

Name and abbreviations	R*	Optical rotation of L enantiomer in H₂O (sign of $[\alpha]_D$)	pK_{a1}	pK_{a2}	pK_{a3}	Isoelectric point, pI
Amino acids with simple aliphatic side chains						
glycine, Gly, G	—H		2.34	9.60	—	5.97
alanine, Ala, A	—CH₃	(+)	2.35	9.69	—	6.02
valine, Val, V	—CH(CH₃)₂	(+)	2.32	9.62	—	5.97
leucine, Leu, L	—CH₂CH(CH₃)₂	(−)	2.36	9.60	—	5.98
isoleucine, Ile, I	—CH—CH₂CH₃, CH₃ [S configuration]	(+)	2.36	9.68	—	6.02
Amino acids with aromatic side chains						
phenylalanine, Phe, F	—CH₂Ph	(−)	1.83	9.13	—	5.48
tryptophan, Trp, W	—CH₂[indole]	(−)	2.38	9.39	—	5.88
tyrosine, Tyr, Y	—CH₂[phenol, OH]	(−)	2.20	9.11	10.07	5.65
histidine, His, H†	—CH₂[imidazole]	(−)	1.82	6.00	9.17	7.58

Amino acids with aliphatic side chains containing —OH, —SH, and —SCH₃ groups

Amino acid	Side chain		pK_{a1}	pK_{a2}	pK_{a3}	pI
serine, Ser, S	$-CH_2-OH$	(–)	2.21	9.15	—	5.68
threonine, Thr, T	$-CH-OH$, CH_3 [R configuration]	(–)	2.71	9.62	—	5.16
methionine, Met, M	$-CH_2CH_2-SCH_3$	(–)	2.28	9.21	—	5.75
cysteine, Cys, C‡	$-CH_2-SH$	(–)	1.71	8.18	10.28	5.02

Amino acids with side chains containing carboxylic acid or amide groups

Amino acid	Side chain		pK_{a1}	pK_{a2}	pK_{a3}	pI
aspartic acid, Asp, D	$-CH_2-\overset{O}{\underset{\parallel}{C}}-OH$	(+)	1.88	3.65	9.60	2.76
glutamic acid, Glu, E	$-CH_2CH_2-\overset{O}{\underset{\parallel}{C}}-OH$	(+)	2.16	4.32	9.67	3.24
asparagine, Asn, N	$-CH_2-\overset{O}{\underset{\parallel}{C}}-NH_2$	(–)	2.02	8.80	—	5.41
glutamine, Gln, Q	$-CH_2CH_2-\overset{O}{\underset{\parallel}{C}}-NH_2$	(+)	2.17	9.13	—	5.65

Amino acids with side chains containing strongly basic groups

Amino acid	Side chain		pK_{a1}	pK_{a2}	pK_{a3}	pI
lysine, Lys, K	$-(CH_2)_4-NH_2$	(+)	2.18	9.12	10.53	9.82
arginine, Arg, R	$-(CH_2)_3NH-\overset{NH}{\underset{\parallel}{C}}-NH_2$	(+)	2.17	9.04	12.48	10.76

Cyclic (secondary) amino acid

Amino acid	Side chain		pK_{a1}	pK_{a2}	pK_{a3}	pI
proline, Pro, P	(pyrrolidine ring with CO_2H, N—H)	(–)	1.99	10.60	—	6.10

* Side chains are shown in their uncharged form.

† Histidine is a weakly basic amino acid.

‡ Cysteine often occurs in proteins as a disulfide dimer, called cystine: $H_2N-CH-CH_2-S-S-CH_2-CH-NH_2$, with CO_2H on each CH. For this reason, cysteine is sometimes called **half-cystine** and abbreviated Cys/2.

Sometimes we want to focus on a single residue in a large protein. For example, suppose we wanted to focus on the role of a lysine residue in a protein such as an enzyme. We can employ the following condensed notations, which we have used throughout this text:

($\boxed{E}$ = enzyme)

The "E" indicates that the protein is an enzyme.

PROBLEMS

27.1 Draw the structures of the following peptides.

(a) tryptophylglycylisoleucylaspartic acid   (b) Glu-Gln-Phe-Arg (or E-Q-F-R)

27.2 Using three-letter abbreviations for the amino acid residues, name the following peptide.

27.3 Draw condensed structures for each of the following:

(a) a phenylalanine residue in a large protein

(b) a serine residue in an enzyme

27.2 STEREOCHEMISTRY OF THE α-AMINO ACIDS

With the exception of glycine, all common naturally occurring α-amino acids have an asymmetric α-carbon atom and are chiral molecules. The chiral amino acids in Table 27.1 are found within naturally occurring proteins in only one enantiomeric form, which has the following configuration:

stereochemical configuration of the
naturally occurring α-amino acids
(*S* for all chiral α-amino acids except cysteine)

(two of the many possible line-and-wedge projections)

(Remember that we can draw many different valid line-and-wedge projections corresponding to views from different perspectives.) The same *configuration in space* for cysteine, unlike the other chiral α-amino acids, is *R* in the *R,S* system because of the presence of sulfur in the side chain, which raises the side-chain priority (Problem 27.6).

The stereochemistry of α-amino acids is often specified with an older system, the **D,L system**. The D,L system relates the configuration of the α-carbon of an amino acid to that of the 3-carbon aldose *glyceraldehyde*. (You may have learned about the D,L system for carbohydrates in Sec. 24.3A.) Serine is the reference α-amino acid that makes the comparison to glyceraldehyde easiest to see. Corresponding groups are shown in the same color.

D-glyceraldehyde   **D-serine**   **L-glyceraldehyde**   **L-serine**
                             [(R)-serine]   **[(S)-serine]**

In D-serine and D-glyceraldehyde, the relative positions of corresponding groups are the same. By extension, all of the α-amino acids with the same configuration as L-serine are L-amino acids.

L-serine   general structure of
                          an L-amino acid

Notice that both L-serine and L-glyceraldehyde have the *S* configuration in the *R,S* system; both D-serine and D-glyceraldehyde have the *R* configuration. However, *this correspondence is not general*. For example, L-cysteine has the *R* configuration. (Why? See Problem 27.6.) All of the naturally occurring amino acids have the L configuration, with a few rare exceptions. (For example, D-alanine occurs in bacterial cell walls; and certain antibiotics contain D-amino acids.)

As with the carbohydrates (Sec. 24.3A), the D and L designations specify the configuration of a *single reference carbon*. For α-amino acids, the reference carbon is the α-carbon. When an α-amino acid contains more than one asymmetric carbon, as in threonine and isoleucine, the diastereomers are given different names. For example, L-threonine has the 2*S*,3*R* configuration; its diastereomer with the 2*S*,3*S* configuration is L-allothreonine. (The prefix *allo* comes from the Greek root *allos*, meaning "other"; L-allothreonine is the "other" threonine.)

L-threonine   **L-allothreonine**
(2*S*, 3*R*)   (2*S*, 3*S*)

Although the use of different names for diastereomers undoubtedly makes organic nomenclature more colorful, this aspect of the D,L system is a major annoyance, especially for the beginner, because it is not systematic. Except for carbohydrates and amino acids, where it is widely used, the D,L system has been largely (and mercifully) abandoned.

PROBLEMS

27.4   Draw three valid line-and-wedge general structures for naturally occurring α-amino acids in addition to the ones shown in this section.

27.5   (a) L-Isoleucine has two asymmetric carbons and has the 2S,3S configuration. Complete the following line-and-wedge structure for L-isoleucine.

(b) Alloisoleucine is the diastereomer of isoleucine. Complete the line-and-wedge structure in part (a) for D-alloisoleucine.

27.6   (a) What is the α-carbon configuration of L-cysteine in the R,S system?

(b) Explain why L-cysteine and L-serine have different configurations in the R,S system.

27.3   ACID–BASE PROPERTIES OF AMINO ACIDS AND PEPTIDES

A. Zwitterionic Structures of Amino Acids and Peptides

As suggested in the introduction to this chapter, the neutral forms of the α-amino acids are *zwitterions*. Some of the evidence for zwitterionic structures is as follows:

1. Amino acids are insoluble in apolar aprotic solvents such as ether. On the other hand, most unprotonated amines and un-ionized carboxylic acids dissolve in ether.

2. Amino acids have very high melting points. For example, glycine melts at 262 °C (with decomposition), and tyrosine melts at 310 °C (also with decomposition). Hippuric acid, a much larger molecule than glycine, and glycinamide, the amide of glycine, have much lower melting points. The former compound lacks the amino group, the latter lacks the carboxylic acid group, and neither can exist as a zwitterion.

N-benzoylglycine (hippuric acid)
mp 190 °C

glycinamide
mp 67–68 °C

glycine
mp 262 °C (d)

The high melting points and greater solubilities in water than in ether are characteristics expected of salts, not uncharged organic compounds. These salt-like characteristics are, however, what *would* be expected of a zwitterionic compound. The strong forces in the solid states of the amino acids that result from the attractions between full positive and negative charges on *different molecules* are much like those between the ions in a salt. These attractions stabilize the solid state and resist conversion of the solid into a liquid—whether a pure liquid melt or a solution. Water is the best solvent for most amino acids because it solvates ionic groups much as it solvates the ions of a salt (Sec. 8.6F).

3. The dipole moments of the amino acids are very large—much larger than those of similar-sized molecules with only one amine or carboxylic acid group.

glycine
μ = 14 D

propanoic acid
μ = 1.7 D

butylamine
μ = 1.4 D

A large dipole moment is expected for molecules that contain a great deal of separated charge (Sec. 1.2D and Further Exploration 1.1).

4. The pK_a values for amino acids are what would be expected for the zwitterionic forms of the neutral molecules.

Suppose a neutral amino acid is titrated with acid. When one equivalent of acid is added, the *basic* group of the amino acid will have been protonated. When this experiment is carried out with glycine, the pK_a of the basic group is found to be 2.3. If glycine is indeed a zwitterion, this basic group can only be the carboxylate ion. If glycine is not a zwitterion, this basic group has to be the amine.

the basic group of the zwitterionic form

$$H_3O^+ + H_3\overset{+}{N}-CH_2-\overset{O}{\underset{||}{C}}-\overset{..}{\underset{..}{O}}:^- \rightleftharpoons H_3\overset{+}{N}-CH_2-\overset{O}{\underset{||}{C}}-\overset{..}{\underset{..}{O}}H + H_2O \quad (27.1a)$$

the basic group of the nonzwitterionic form

$$H_3O^+ + H_2\overset{..}{N}-CH_2-\overset{O}{\underset{||}{C}}-OH \rightleftharpoons H_3\overset{+}{N}-CH_2-\overset{O}{\underset{||}{C}}-OH + H_2O \quad (27.1b)$$

Which is the correct description of the titration? The pK_a of 2.3 is that expected of a carboxylic acid in a molecule containing a nearby electron-withdrawing group (in this case, the H_3N^+— group). In contrast, the conjugate acids of amines have pK_a values in the 8–10 range. This analysis suggests that the zwitterion, not the uncharged form, is being titrated.

Along the same line, if NaOH is added to neutral glycine, a group is titrated with $pK_a = 9.6$. This is a reasonable pK_a value for an alkylammonium ion, but would be very unusual for a carboxylic acid. This comparison also suggests that the neutral form of glycine is a zwitterion.

The acid–base equilibria for glycine can be summarized as follows:

principal form in
neutral aqueous solution

$$H_3\overset{+}{N}-CH_2-CO_2^-$$

$pK_a = 9.6$ $pK_a = 2.3$

H_3O^+ H_3O^+

principal form in aqueous base $H_2N-CH_2-CO_2^-$ $H_3\overset{+}{N}-CH_2-CO_2H$ principal form in aqueous acid   (27.2)

H_3O^+ H_3O^+

$$H_2N-CH_2-CO_2H$$

minor neutral form
(about 1 part in 10^5)

Thus, the major neutral form of any α-amino acid is the zwitterion. In fact, it can be estimated that the ratio of the uncharged form of an α-amino acid to the zwitterion form is about 1 part in 10^5, as shown for glycine in Eq. 27.2.

Peptides also exist as zwitterions; that is, at pH values near 7, amino groups are protonated and carboxylic acid groups are ionized.

B. Acid–Base Equilibria of Amino Acids

The goal of this section is to understand how the various forms of the amino acids vary with pH. Consider the acid–base equilibria of the amino acid alanine.

$$
\overset{+}{H_3}N-\underset{\underset{CH_3}{|}}{CH}-CO_2H
\;\underset{H_3O^+}{\overset{H_2O}{\rightleftharpoons}}\;
\overset{+}{H_3}N-\underset{\underset{CH_3}{|}}{CH}-CO_2^-
\;\underset{H_3O^+}{\overset{H_2O}{\rightleftharpoons}}\;
H_2N-\underset{\underset{CH_3}{|}}{CH}-CO_2^-
\quad (27.3)
$$

with $pK_{a1} = 2.3$ and $pK_{a2} = 9.7$, labeled A, N, and B.

In very acidic solution alanine is fully protonated, and A is the only species present. As the pH is raised, the most acidic proton dissociates; this is the carboxylic acid proton, which has the lower pK_a. In this first dissociation equilibrium, the acid A and its conjugate base N are the only species involved. *When a molecule can undergo two (or more) proton dissociations, we can treat the two equilibria independently if the pK_a values are separated by 2 or more units.* Therefore, we can apply Eqs. 3.31a–b (p. 107) with A as the acid and N as the base. When we plot these equations, we get a plot exactly like Fig. 3.1 (p. 108), in which the fraction of A decreases and the fraction of N increases with increasing pH; the fractions of A and N are equal when pH $= pK_{a1} = 2.3$. This plot is shown on the left half of Fig. 27.1.

As the pH is raised further, the second dissociation becomes important. In this case, N is now the acid and B is its conjugate base. When we plot Eqs. 3.31a–b for this acid–base pair, we obtain the plot on the right half of Fig. 27.1. The fractions of N and B are equal with pH $= pK_{a2} = 9.7$. As the pH is raised further, B becomes the only species present.

The process we have just outlined can be applied to any two-dissociation system (not just amino acids), and it is readily extended to any number of dissociations. If two of the pK_a values involved differ by less than 2 units, the process is conceptually the same, except that the intermediate form N never becomes 100%.

PROBLEM

27.7 Obtain the three pK_a values of the α-amino acid *histidine* from Table 27.1. Because there are three pK_a values, the dissociation equilibria contain four species.

(a) With the species present at low pH on the left, and the fully dissociated species on the right, draw and label by letter all of the species involved in the acid–base dissociation equilibria of histidine. (*Hint:* The side-chain group protonates on the vinylic nitrogen of the imidazole ring.)

(b) Sketch a graph similar to Fig. 27.1 in which the fraction of all of the species are shown as a function of pH. Label each curve with a letter corresponding to one of the forms you drew in part (b). Indicate the pH values at which the fractions of two species are equal.

C. Isoelectric Points of Amino Acids and Peptides

An important measure of the acidity or basicity of an amino acid is its **isoelectric point** or **isoelectric pH**. (The two terms mean the same thing.) This is the pH of a dilute aqueous solution of the amino acid at which the total charge on all molecules of the amino acid is zero. Let's use the acid–base equilibria of alanine (Eq. 27.3) to illustrate the isoelectric point. Two conditions are met at the isoelectric point. First, the concentration of conjugate-acid molecules A equals the concentration of conjugate-base molecules B. Because the conjugate acid A is positively charged and the conjugate base B is negatively charged, the equality of the two concentrations means that the total charge on all molecules of the amino acid is zero. (The amount of N present doesn't matter because its net charge is zero.) Second, at the isoelectric point, the relative concentration of the zwitterion form N is greater than at any other pH. The

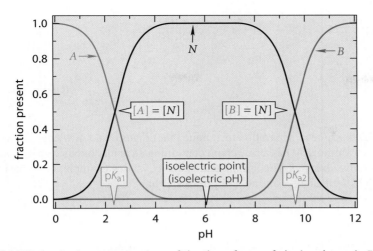

FIGURE 27.1 Variation in the concentrations of the three forms of alanine shown in Eq. 27.3 as a function of pH. The blue line represents the concentration of the conjugate acid A, the red line the concentration of the conjugate base B, and the black line the concentration of the neutral zwitterionic form N. The dissociation constant of A is K_{a1}, and that of N is K_{a2}. The concentration of the neutral form N is a maximum at a pH that is halfway between pK_{a1} and pK_{a2}; this pH is the isoelectric point.

first part of the definition—the equality of the B and A concentrations—is sufficient to calculate the isoelectric pH.

If the concentrations of A and B are equal at the isoelectric pH, their fractions are also equal. We use the following definitions: isoelectric pH = pH_i, $\Delta_1 = pH_i - pK_{a1}$, and $\Delta_2 = pH_i - pK_{a2}$. Applying Eqs. 3.31a–b for the fractions of each species, and setting these fractions equal, we have

$$f_A = \frac{1}{1 + 10^{\Delta_1}} = f_B = \frac{1}{1 + 10^{-\Delta_2}} \tag{27.4}$$

Simplifying,

$$1 + 10^{\Delta_1} = 1 + 10^{-\Delta_2}$$

or

$$10^{\Delta_1} = 10^{-\Delta_2}$$

Taking logs and applying the definitions of Δ_1 and Δ_2, we have

$$pH_i - pK_{a1} = -(pH_i - pK_{a2})$$

or

$$pH_i = \frac{(pK_{a1} + pK_{a2})}{2} \tag{27.5}$$

This result shows that *the isoelectric pH occurs at the average of the two pK_a values.* Therefore, alanine has an isoelectric pH (or isoelectric point) of $(2.3 + 9.7)/2 = 6.0$. This point is marked on the pH axis of Fig. 27.1.

The isoelectric point is significant because it indicates not only the pH value at which a solution of the amino acid contains the greatest amount of zwitterion form N, but also the sign of the net charge on the amino acid at *any* pH. For example, at a pH value lower (more acidic) than the isoelectric point, more molecules of an amino acid are in form A than in form B; in this situation, the amino acid has a net *positive charge*. At a pH value greater (more basic) than the isoelectric point, more molecules of an amino acid are in form B than in form A. In this situation, the amino acid has a net *negative charge*. To summarize:

$$\overset{+}{H_3N}-\underset{\underset{CH_3}{|}}{CH}-CO_2H \underset{H_3O^+}{\overset{H_2O}{\rightleftharpoons}} \overset{+}{H_3N}-\underset{\underset{CH_3}{|}}{CH}-CO_2^- \underset{H_3O^+}{\overset{H_2O}{\rightleftharpoons}} H_2N-\underset{\underset{CH_3}{|}}{CH}-CO_2^- \qquad (27.6)$$

A   *N*   *B*
predominates at pH values much   predominates at pH values much
lower than the isoelectric point   *higher* than the isoelectric point

Section 27.3D discusses a separation technique that hinges on a knowledge of the net charge.

When an amino acid has a side chain containing an acidic or basic group, the isoelectric point is markedly changed. The amino acid lysine (Lys), for example, has a basic side-chain amino group as well as its α-amino and carboxy groups.

$$\boxed{pK_{a2} = 9.1}$$

$$\overset{+}{H_3N}-\underset{\underset{(CH_2)_4}{|}}{CH}-CO_2^- \underset{H_3O^+}{\overset{H_2O}{\rightleftharpoons}} H_2N-\underset{\underset{(CH_2)_4}{|}}{CH}-CO_2^- \underset{H_3O^+}{\overset{H_2O}{\rightleftharpoons}} H_2N-\underset{\underset{(CH_2)_4}{|}}{CH}-CO_2^- \qquad (27.7)$$

$$\underset{+NH_3}{} \qquad \underset{+NH_3 \boxed{pK_{a3}=10.5}}{} \qquad \underset{NH_2}{}$$

A   *N*   *B*

The isoelectric point of lysine is 9.82, which is the average of its two *highest* pK_a values of 9.12 and 10.53. At the isoelectric point of lysine, equal amounts of forms *A* and *B* of lysine are present, and form *N* has its maximum concentration. The lowest pK_a of lysine—the pK_a of the carboxylic acid group, 2.2—doesn't enter the picture because neither of the equilibria involving the neutral form *N* involves ionization of the carboxylic acid group.

Let's compare the charge state of alanine and lysine at pH 6. Because pH 6 is the isoelectric point of alanine, its charge is zero. Because pH 6 is much lower than the isoelectric point of lysine, the net charge on lysine molecules is positive at this pH. Amino acids with high isoelectric points are classified as *basic amino acids*. Lysine and arginine are the two most basic of the common naturally occurring amino acids (Table 27.1). As indicated by its isoelectric point, arginine is the more basic of the two. Its side chain carries the basic guanidino group, the conjugate acid of which has a pK_a of 12.5. The basicity of this group results from the resonance stabilization of its conjugate acid.

$$\underset{\text{guanidino group}}{H_2\ddot{N}-\underset{\underset{:NH}{|}}{C}=NH} + H_3O^+ \rightleftharpoons \left[H_2\ddot{N}-\underset{\underset{:NH}{|}}{C}-\overset{+}{N}H_2 \leftrightarrow H_2\overset{+}{N}=\underset{\underset{:NH}{|}}{C}-\ddot{N}H_2 \leftrightarrow H_2\ddot{N}-\underset{\underset{+NH}{||}}{C}-\ddot{N}H_2 \right] \qquad (27.8)$$

$$+ H_2O$$

The amino acids aspartic acid (Asp) and glutamic acid (Glu) have carboxylic acid groups on their side chains and have low isoelectric points. The isoelectric point of aspartic acid, for example, is 2.76, the average of its two *lowest* pK_a values. (You should show why this is reasonable.) Amino acids with low isoelectric points are classified as *acidic amino acids*. Molecules of an acidic amino acid carry a net negative charge at pH 6. Aspartic acid and glutamic acid are the two most acidic of the common naturally occurring amino acids.

Amino acids with isoelectric points near 6, such as glycine or alanine, are classified as *neutral amino acids*. A pH of 6 is not *exactly* neutral; "neutral" amino acids are actually slightly acidic because the carboxylic acid group is somewhat more acidic than the amino group is basic. However, as Fig. 27.1 shows, even at pH 7 (neutral pH), the neutral amino acids are almost completely in form *N*.

Let's summarize the charge situation in basic, neutral, and acidic amino acids at pH 6:

Principal forms at pH 6:

net +1 charge   net 0 charge   net –1 charge

$$H_3\overset{+}{N}—CH—CO_2^-$$
$$|$$
$$(CH_2)_4$$
$$|$$
$$^+NH_3$$

$$H_3\overset{+}{N}—CH—CO_2^-$$
$$|$$
$$CH_3$$

$$H_3\overset{+}{N}—CH—CO_2^-$$
$$|$$
$$CH_2$$
$$|$$
$$CO_2^-$$

lysine   **alanine**   **aspartic acid**
(a *basic* amino acid)   (a *neutral* amino acid)   (an *acidic* amino acid)

Peptides with both acidic and basic groups also have isoelectric points. We can tell by inspection whether a peptide is acidic, basic, or neutral by examining the number of acidic and basic groups that it contains. A peptide with more amino and guanidino groups than carboxylic acid groups, for example, will have a high isoelectric point. Conversely, a peptide with more carboxylic acid groups than amino or guanidino groups will have a low isoelectric point.

PROBLEMS

27.8 (a) Point out the ionizable groups of the amino acid *tyrosine* (Table 27.1).

(b) What is the net charge on tyrosine at pH 6? How do you know?

(c) Draw the structure of the major form(s) of tyrosine present at this pH.

27.9 (a) Estimate the isoelectric point of each of the following peptides.

A-K-V-I-M   G-D-G-L-F

(b) Draw the structures of these peptides, indicating the predominant ionization state of each at its isoelectric point.

27.10 Classify the following peptides as acidic, basic, or neutral. What is the net charge on each peptide at pH = 6?

(a) Gly-Leu-Val

(b) Leu-Trp-Lys-Gly-Lys

(c) *N*-acetyl-Asp-Val-Ser-Arg-Arg (*N*-acetyl means that the terminal amino group of the peptide is acetylated.)

(d) Glu-Lys-Asp-Ala-Phe-Ile

D. Separations of Amino Acids and Peptides Using Acid–Base Properties

Isoelectric points are often used to design separations of amino acids and peptides. Consider, for example, the water solubilities of amino acids and peptides. Most peptides and amino acids, like carboxylic acids and amines, are most soluble when they carry a net charge and are least soluble in their neutral forms. Thus, some peptides, proteins, and amino acids precipitate from water when the pH is adjusted to their isoelectric points. These same compounds are more soluble in water at pH values far from their isoelectric points, because they carry a net charge at these pH values.

A separation technique used a great deal in amino acid, peptide, and protein chemistry is *ion-exchange chromatography*. This method, too, depends on the isoelectric points of amino acids and peptides.

Recall from Sec. 6.8A that *chromatography* is a separation technique based on the relative adsorptions of compounds to a material called a *stationary phase*. If the stationary phase is contained in a column, compounds that are adsorbed weakly move through the column most rapidly and are eluted early; compounds that are more strongly adsorbed move through the column more slowly and emerge later. Thus, compounds are separated by their *differential adsorption* to the stationary phase. (Review Fig, 6.15, p. 254.)

The type of chromatography that is used depends on the mechanism used to effect differential adsorption to the stationary phase. Recall from Sec. 6.8A that a chiral stationary phase is used to separate enantiomers. In **ion-exchange chromatography**, the column is filled with a buffer solution, and the stationary phase is a polymer called an *ion-exchange resin*. This resin bears charged groups. One popular resin, for example, is a sulfonated polystyrene—a polystyrene in which the phenyl rings contain strongly acidic sulfonic acid groups. If the pH of the buffer is such that the sulfonic acid groups are ionized, the resin bears a negative charge. This charge is the key to ion-exchange separations.

$$\cdots \text{—CH}_2\text{—CH—CH}_2\text{—CH—CH}_2\text{—CH—} \cdots \tag{27.9}$$

structure of sulfonated polystyrene with ionized sulfonic acid groups

The way ion-exchange chromatography works is illustrated in Fig. 27.2. Suppose the buffer in the column has a pH of 6. Because the pK_a of the sulfonic acid groups on the resin is about 1, these groups are ionized; therefore, at pH = 6, the resin is *anionic*. A solution con-

FIGURE 27.2 (a) The cation-exchange separation of valine (Val) and lysine (Lys). (b) The amino acid concentration in the eluent (the buffer emerging from the column) as a function of time. Lysine, which carries a positive charge at the pH of the buffer, is attracted to the negatively charged resin and moves through the column more slowly than valine, which carries zero charge. (The colors are for emphasis; Val and Lys are colorless.)

taining a mixture of the two amino acids Val and Lys in the same buffer is added to the top of the column. Buffer is then allowed to flow through the column; a frit (a porous glass plate) keeps the resin from washing out. Because valine has zero charge at this pH, it is not attracted by the ionic groups on the column and is washed through the column with a relatively small volume of buffer. Lysine, on the other hand, has an isoelectric point of 9.8 and therefore bears a net positive charge at pH 6. Hence, lysine is strongly attracted to the negatively charged resin. Because of this attraction, lysine is retained on the column and emerges only after a considerably larger amount of buffer has passed through the column. The two amino acids are thus separated. Thus, whether an amino acid or peptide is adsorbed by the column depends on its charge—which, in turn, depends on the relationship of its isoelectric point to the pH of the buffer.

In the experiment shown in Fig. 27.2, the ion-exchange resin is negatively charged and adsorbs cations; it is therefore called a *cation-exchange resin*. Resins that bear positively charged pendant groups adsorb anions, and are called *anion-exchange resins*.

Ion Exchange and Water Softeners

Ion exchange has very important commercial applications, such as water treatment. Commercial water softeners contain cation-exchange resins much like the one used in this example, which adsorb the more highly charged calcium and magnesium ions in hard water and replace them with sodium ions with which the column is supplied. When the supply of sodium ions is exhausted, the column has to be flushed extensively, or regenerated, with concentrated NaCl solution to replace the adsorbed calcium and magnesium ions with sodium ions.

PROBLEM

27.11   (a) How might the structure of the resin in Eq. 27.9 be altered to make the resin an anion exchanger (that is, an anion-binding resin)? (*Hint:* What type of organic functional group carries a positive charge at neutral pH?)

(b) Predict the order of elution of the following peptides from an anion-exchange resin at pH 6: A-V-G, D-E-E-G, D-N-N-G. Explain your reasoning.

27.4   SYNTHESIS AND ENANTIOMERIC RESOLUTION OF α-AMINO ACIDS

A. Alkylation of Ammonia

Some α-amino acids can be prepared by the alkylation of ammonia with α-bromo carboxylic acids.

$$\underset{\substack{|\\ CH_3}}{CH_3CH_2CH}\overset{Br}{\underset{|}{-}}CH-CO_2^- + \underset{\substack{(large\\ excess)}}{NH_3} \longrightarrow \underset{\substack{|\\ CH_3}}{CH_3CH_2CH}\overset{+NH_3}{\underset{|}{-}}CH-CO_2^- + Br^- \quad (27.10)$$

isoleucine
(49% yield)

This is an S_N2 reaction in which ammonia acts as the nucleophile. (Recall from Sec. 22.3D that α-halo carbonyl compounds are very reactive in S_N2 reactions.) Alkylation of ammonia is usually not a good method for preparing primary amines because ammonia can be alkylated more than once to give complex mixtures (Sec. 23.7A). However, the use of a large excess of ammonia in this synthesis favors monoalkylation. Furthermore, amino acids are less reactive toward alkylating agents than simple alkylamines for two reasons: (1) the amino groups of amino acids are less basic, and therefore less nucleophilic, than ammonia and simple alkyl-amines; and (2) branching in amino acids provides a steric impediment to further alkylation.

B. Alkylation of Aminomalonate Derivatives

One of the most widely used methods for preparing α-amino acids is a variation of the malonic ester synthesis (Sec. 22.8A). The malonic ester derivative used is one in which a protected amino group is already in place: diethyl α-acetamidomalonate. This derivative is treated with sodium ethoxide in ethanol to form the enolate ion, which is then alkylated with an alkyl halide (benzyl chloride in the following example):

diethyl acetamidomalonate

enolate ion

(27.11a)

The resulting compound is then treated with hot aqueous HCl or HBr. This acid treatment accomplishes three things: First, the ester groups are hydrolyzed to carboxylic acids (Sec. 21.7A), yielding a disubstituted malonic acid. Second, the malonic acid derivative decarboxylates under the reaction conditions (Sec. 20.11A). Third, the acetamido group, an amide, is also hydrolyzed (Sec. 21.7B). Neutralization gives the α-amino acid.

phenylalanine hydrobromide
(65% yield)

C. Strecker Synthesis

An important method for synthesizing carboxylic acids is the hydrolysis of nitriles (Secs. 21.7C and 21.11). Thus, α-amino nitriles can be hydrolyzed to give α-amino acids. α-Amino nitriles, in turn, are prepared by the treatment of aldehydes with ammonia in the presence of cyanide ion.

acetaldehyde

2-aminopropanenitrile
(an α-amino nitrile)

(27.12)

alanine
(52–60% yield)

This preparation of α-amino acids is called the **Strecker synthesis**.

The mechanism of α-amino nitrile formation probably involves an imine intermediate.

$$H_3C\!-\!CH\!=\!O + \overset{\cdot\cdot}{N}H_3 \longrightarrow H_2O + H_3C\!-\!CH\!=\!\overset{\cdot\cdot}{N}H \underset{\longleftarrow}{\overset{+NH_4}{\longrightarrow}} H_3C\!-\!CH\!=\!\overset{+}{N}H_2 + NH_3 \quad (27.13a)$$

an imine

The conjugate acid of the imine reacts with cyanide under the conditions of the reaction to give the α-amino nitrile.

$$H_3C\!-\!CH\!=\!\overset{+}{N}H_2 + :\bar{C}N \longrightarrow H_3C\!-\!\underset{\underset{CN}{|}}{CH}\!-\!\overset{\cdot\cdot}{N}H_2 \quad (27.13b)$$

The addition of cyanide to an imine is analogous to the formation of a cyanohydrin from an aldehyde or ketone (Sec. 19.7A).

$$HCN + H_3C\!-\!CH\!=\!O \longrightarrow H_3C\!-\!\underset{\underset{CN}{|}}{CH}\!-\!OH \quad (27.14)$$

Recall that the trapping of an imine intermediate by a nucleophile also occurs in *reductive amination* (Sec. 23.7B). In reductive amination, the nucleophile is the hydride ion derived from $Na^+\,{}^-BH_3CN$ or $Na^+\,{}^-BH(OAc)_3$. In the Strecker synthesis, the nucleophile is cyanide ion.

$$R\!-\!CH\!=\!\overset{+}{N}HR'$$

conjugate acid of an imine

"H:⁻" ↓   ↓ ⁻CN

$$R\!-\!\underset{\underset{H}{|}}{CH}\!-\!NHR' \qquad\qquad R\!-\!\underset{\underset{CN}{|}}{CH}\!-\!NHR' \xrightarrow[\text{heat}]{H_2O,\,H_3O^+} R\!-\!\underset{\underset{CO_2H}{|}}{CH}\!-\!\overset{+}{N}H_2R' \quad (27.15)$$

reductive amination   Strecker synthesis

PROBLEM

27.12 Indicate which of the methods in this section can be used to prepare each of the following amino acids. For each method that can be used, give an equation. For each case in which a method would not work, give a reason.

(a) α-phenylglycine   (b) leucine

D. Enantiomeric Resolution of α-Amino Acids

Amino acids synthesized by common laboratory methods, such as the ones discussed in Secs. 27.4A–C, are *racemic*. Because enantiomerically pure amino acids are often needed, the racemic mixtures must be resolved. As useful as the diastereomeric salt method is (Sec. 6.8B), it can be tedious and time-consuming. An alternative approach to the preparation of enantiomerically pure amino acids, and one that is used industrially, is the synthesis of amino acids by microbiological fermentation. Some cultures of microorganisms can be used to produce industrial quantities of certain amino acids in the natural L form.

Certain enzymes can be used to resolve racemic amino acids into enantiomers by catalyzing the enantioselective hydrolysis of an *N*-amidated amino acid derivative. For example, a preparation of the enzyme *acylase* from hog kidney selectively catalyzes the hydrolysis of

N-acetyl-L-amino acids and leaves the corresponding D isomers unaffected. Thus, treatment of the *N*-acetylated racemate with this enzyme affords the free L-amino acid only:

$$H_2O + H_3C-\overset{\overset{O}{\|}}{C}-NH-\underset{\underset{CH_3}{|}}{CH}-CO_2H \xrightarrow[\text{(an enzyme)}]{\text{hog-kidney acylase}}$$

N-acetyl-D,L-alanine
(racemate)

$$H_3\overset{+}{N}\cdots\underset{\underset{H}{|}}{\overset{\overset{CO_2^-}{|}}{C}}CH_3 \quad + \quad H_3C-\underset{\underset{O}{\|}}{C}-NH\underset{H}{\overset{\overset{CO_2^-}{|}}{C}}CH_3 \quad + \quad CH_3CO_2^- \qquad (27.16)$$

L-(+)-alanine   **N-acetyl-D-alanine**
(insoluble in EtOH)   (soluble in EtOH)

In this example, the liberated L-alanine is precipitated from ethanol; the *N*-acetyl-D-alanine remains in solution, from which it can be recovered and hydrolyzed in aqueous acid to D-alanine.

The enzyme differentiates between the two enantiomers of *N*-acetylalanine because it is an *enantiomerically pure chiral catalyst*. Recall that enantiomers have different reactivities with chiral reagents (Sec. 7.7A). The use of enzymes in this way is a practical example of the *principle of enantioneric differentiation* (Sec. 6.8).

27.5 ACYLATION AND ESTERIFICATION REACTIONS OF AMINO ACIDS

Amino acids undergo many of the reactions characteristic of both amines and carboxylic acids. *Acylation* is an amine reaction that is very important in amino acid chemistry. Acylation by acetic anhydride is shown in Eq. 27.17.

$$H_3\overset{+}{N}-\underset{\underset{CH(CH_3)_2}{\overset{|}{\underset{|}{CH_2}}}}{CH}-CO_2^- + H_3C-\overset{\overset{O}{\|}}{C}-O-\overset{\overset{O}{\|}}{C}-CH_3 \xrightarrow[CH_3CO_2H]{} \xrightarrow{H_2O} H_3C-\overset{\overset{O}{\|}}{C}-NH-\underset{\underset{CH(CH_3)_2}{\overset{|}{\underset{|}{CH_2}}}}{CH}-CO_2H \qquad (27.17)$$

leucine   **acetic anhydride**   **N-acetylleucine**
(85–95% yield)

Acylation by acid chlorides is also a useful reaction (Sec. 21.8A).

In Eq. 27.17, the amino group is protonated; yet the *neutral* form of the amine is required to serve as a nucleophile in the acylation reaction. Even in acidic solution, a *very small* amount of neutral amine is present. When this form reacts, the acid–base equilibrium shifts rapidly to replenish this form. More generally, a very minor component of an equilibrium can serve as a reactant in a reaction provided that (a) this component is sufficiently reactive and (b) the equilibrium can shift quickly enough to replenish the minor form once it reacts.

Amino acids, like ordinary carboxylic acids, are easily esterified by heating with an alcohol and a strong acid catalyst (acid-catalyzed esterification; Sec. 20.8A).

$$H_2N-\underset{}{\bigcirc}-\overset{\overset{O}{\|}}{C}-OH \ + \ EtOH \ \xrightarrow[heat]{H_2SO_4} \ \xrightarrow{NaHCO_3} \ H_2N-\underset{}{\bigcirc}-\overset{\overset{O}{\|}}{C}-OEt \ + \ H_2O \quad (27.18)$$

p-aminobenzoic acid (PABA)   **ethyl p-aminobenzoate**
                                                                    (**benzocaine,** a local anesthetic)

PROBLEMS

27.13   Draw the structure of the major product expected when

(a) leucine is treated with *p*-toluenesulfonyl chloride (tosyl chloride).

(b) alanine is heated in methanol solvent with HCl catalyst.

27.14   If the hydrochloride salt of glycine methyl ester is neutralized and allowed to stand in solution, a polymer forms. If the hydrochloride itself is allowed to stand, the polymerization reaction does not occur. Explain these observations by writing the reaction that occurs.

27.6   PEPTIDE AND PROTEIN SYNTHESIS

This section addresses the synthesis of peptides and proteins from individual α-amino acids. In Sec. 27.6A, we'll discuss the laboratory synthesis of peptides. In Sec. 27.6B, we'll consider the biosynthesis of proteins: how living systems utilize the genetic code to assemble proteins from α-amino acids.

A. Solid-Phase Peptide Synthesis

A number of methods have been developed for peptide synthesis, but the most widely used are variations of an ingenious method called **solid-phase peptide synthesis**. In this method, the carboxy-terminal amino acid is covalently anchored to an *insoluble* polymer, and the peptide is "grown" by adding one amino acid residue at a time to this polymer. Solutions containing the appropriate reagents are shaken with the polymer. At the conclusion of each step, the polymer containing the peptide is simply filtered away from the solution, which contains soluble by-products and impurities. The completed peptide is removed from the polymer by a reaction that breaks its bond to the resin, just as a plant is harvested by cutting it away from the ground. The advantage of this method is the ease with which the peptide is separated from soluble by-products of the reaction. The reactions used in solid-phase peptide synthesis also illustrate some important amino acid and peptide chemistry.

Solid-Phase Peptide Synthesis

Solid-phase peptide synthesis was devised by R. Bruce Merrifield (1921–2006) of The Rockefeller University, and first reported in the early 1960s. A particularly impressive achievement of the method was the synthesis of an active enzyme by Merrifield's research group in 1969 using a homemade machine in which the various steps of the method were preprogrammed. (Modern instruments for automated solid-phase peptide synthesis are available commercially.) The enzyme that was synthesized, ribonuclease, contains 124 amino acid residues; the synthesis required 369 separate reactions and 11,931 individual operations, yet was carried out in 17% overall yield. (Several other proteins have since been prepared by solid-phase peptide synthesis.) For his invention and development of the solid-phase method, Merrifield was awarded the 1984 Nobel Prize in Chemistry.

Before considering an actual solid-phase synthesis of a peptide, we need to understand two particularly important aspects of peptide synthesis: the use of *protecting groups* and the use of *active esters*.

Protecting groups were introduced in Sec. 19.10B. Peptide synthesis can employ a number of protecting groups for amino groups, carboxy groups, and side-chain functional groups, but we'll consider only the use of an amino protecting group in our synthesis. The purpose of an amino protecting group is to block an amine from reacting as a nucleophile at some point in the synthesis. The amino protecting group that we'll use is the (9-fluorenyl)methyl-oxycarbonyl group, known mercifully throughout the peptide-chemistry world as the **Fmoc group** (pronounced "eff-mock"). For example, an Fmoc-protected alanine has the following structure:

Fmoc-Ala

Fmoc group

The rationale behind the design of this group, which was developed in 1972 by Prof. Louis A. Carpino (b. 1927) of the University of Massachusetts, becomes evident in Eq. 27.22.

Active esters are more reactive toward nucleophiles than ordinary alkyl esters. Two widely used active esters are based on two unusual "alcohols," *N*-hydroxysuccinimide and 1-hydroxy-1,2,3-benzotriazole. Both compounds have N—O bonds. These "alcohols" have very low pK_a values that are closer to the pK_a values of carboxylic acids than ordinary alcohols.

***N*-hydroxysuccinimide**
(abbreviated **NHS**)
pK_a = 6.0

an NHS ester

1-hydroxy-1,2,3-benzotriazole
(abbreviated **HOBt**)
pK_a = 4.6

an HOBt ester

abbreviation for
an HOBt ester

Weak bases generally are good leaving groups in carbonyl substitution reactions, just as they are in S_N2 reactions (Sec. 9.4F), because the same property that makes them weak bases—their electron-withdrawing character—stabilizes the transition state for substitution (Sec. 21.7E). NHS and HOBt esters are much more reactive in nucleophilic substitution reactions than alkyl esters; HOBt esters verge on anhydride-like reactivity. Other rationales for use of these unusual leaving groups will become apparent in the discussion of Eq. 27.27.

We'll illustrate solid-phase peptide synthesis with the preparation of a tripeptide, Phe-Gly-Ala (F-G-A). The synthesis begins with the preparation of Fmoc-Ala, the protected alanine derivative shown above.

$$\text{(27.19)}$$

Fmoc-Ala
(85% yield)
(a carbamate ester)

(NHS)

This reaction illustrates the superior leaving-group properties of the NHS group. Fmoc-NHS is a carbonate ester; potentially, two alcohols could be leaving groups. Because of its low pK_a, NHS is a much better leaving group than (9-fluorenyl)methanol. Consequently, the NHS group is displaced and the (9-fluorenyl)methyl group remains intact as a carbamate-ester protecting group.

The next step of the synthesis introduces the solid phase for which the method is named. The Fmoc-Ala formed in Eq. 27.19 is anchored onto an insoluble solid polymeric support, called a *resin*, using the reactivity of its free carboxylate group. A variety of such resins are available commercially, but a popular one is the following:

$$\text{(27.20)}$$

This is, in effect, an "insoluble *p*-alkoxybenzyl chloride," and it has the enhanced reactivity generally associated with benzylic halides (Sec. 17.4). (The reason for the para —OCH_2— group becomes apparent in Eq. 27.29.) An S_N2 reaction between the cesium salt of Fmoc-Ala and the chloromethyl group of the resin results in the formation of an ester linkage to the resin by alkylation of the carboxylate ion. (See Sec. 20.8B for related chemistry.)

FURTHER EXPLORATION 27.1
Solid-Phase Peptide Synthesis

Fmoc-Ala linked to the resin

$$\text{(27.21)}$$

The role of the Fmoc protecting group in this reaction is to keep the amino group from competing with the carboxylate group as a nucleophile for the benzylic halide group on the resin.

The resin is supplied as a powder consisting of tiny spherical beads. Although the preceding equations show only one peptide on the resin, many peptide chains are anchored to each polymer bead, and many polymer beads are used in each synthesis.

Once the Fmoc-amino acid is anchored to the resin, the Fmoc protecting group is removed by treatment with piperidine, an amine base.

$$(27.22)$$

(reacts further
with piperidine)

This is an E2 reaction. Recall from Sec. 17.3B that E2 reactions are particularly fast when the β-hydrogen is particularly acidic. Here we can appreciate the ingenious design of the Fmoc protecting group. The β-hydrogen of this group is particularly acidic because the anion that would be formed by removal of this hydrogen as a proton is *aromatic* and therefore particularly stable. (Note the "imbedded" cyclopentadienyl anion in red; see Sec. 15.7D and Eq. 15.47, p. 768.)

an aromatic anion

Because the product of the β-elimination in Eq. 27.22 is a carbamate anion, it decarboxylates under the reaction conditions. (See Eq. 20.43b, p. 1031.)

$$(27.23)$$

resin-bound Ala

This reaction exposes the amino group of the resin-bound amino acid. This amino group serves as a nucleophile in the next reaction.

Next comes the formation of the first peptide bond. Coupling of Fmoc-glycine to the free amino group of the resin-bound Ala is effected by the reagent 1,3-diisopropylcarbodiimide (DIC) in the presence of HOBt. (In this equation, iPr = isopropyl.)

$$\text{Fmoc}-\text{NH}-\text{CH}_2-\overset{\overset{\displaystyle O}{\|}}{\text{C}}-\text{OH} + i\text{PrN}=\text{C}=\text{N}i\text{Pr} + \text{H}_2\text{N}-\underset{\underset{\displaystyle \text{CH}_3}{|}}{\text{CH}}-\overset{\overset{\displaystyle O}{\|}}{\text{C}}-\text{O}-\text{CH}_2-\bigcirc\text{-}\xi \quad \xrightarrow{\begin{array}{c}\text{HOBt}\\\text{DMF}\end{array}}$$

Fmoc-Gly   **DIC**   **Ala**-resin

$$\text{Fmoc}-\text{NH}-\text{CH}_2-\overset{\overset{\displaystyle O}{\|}}{\text{C}}-\text{NH}-\underset{\underset{\displaystyle \text{CH}_3}{|}}{\text{CH}}-\overset{\overset{\displaystyle O}{\|}}{\text{C}}-\text{O}-\text{CH}_2-\bigcirc\text{-}\xi \quad + \quad i\text{PrNH}-\overset{\overset{\displaystyle O}{\|}}{\text{C}}-\text{NH}i\text{Pr}$$

Fmoc-Gly-Ala-resin   ***N,N′*-diisopropylurea (DIU)**

$$(27.24)$$

In Eq. 27.24, addition of the carboxylic acid group to a double bond of DIC gives a derivative called an *O*-acyl isourea.

$$\text{R}-\overset{\overset{\displaystyle O}{\|}}{\text{C}}-\text{OH} + i\text{PrN}=\text{C}=\text{N}i\text{Pr} \longrightarrow \text{R}-\overset{\overset{\displaystyle O}{\|}}{\text{C}}-\text{O}-\underset{\underset{\displaystyle \text{N}i\text{Pr}}{\|}}{\overset{\overset{\displaystyle \text{NH}i\text{Pr}}{|}}{\text{C}}} \quad \longleftarrow \boxed{\text{a very good leaving group}} \quad (27.25a)$$

the protected   **DIC**
amino acid

an *O*-acyl isourea

This derivative behaves somewhat like an anhydride and is an excellent acylating agent. It reacts rapidly with the HOBt to form an HOBt ester of the acylating amino acid.

$$\text{FmocNH}-\text{CH}_2-\overset{\overset{\displaystyle O}{\|}}{\text{C}}-\text{O}-\underset{\underset{\displaystyle \text{N}i\text{Pr}}{\|}}{\overset{\overset{\displaystyle \text{NH}i\text{Pr}}{|}}{\text{C}}} + \text{HOBt} \longrightarrow \text{FmocNH}-\text{CH}_2-\overset{\overset{\displaystyle O}{\|}}{\text{C}}-\text{OBt} + \text{O}=\underset{\underset{\displaystyle \text{NH}i\text{Pr}}{|}}{\overset{\overset{\displaystyle \text{NH}i\text{Pr}}{|}}{\text{C}}} \quad (27.25b)$$

an *O*-acyl isourea   HOBt ester of the   **DIU**
protected amino acid

We have already seen (Eq. 27.19) that NHS esters readily undergo aminolysis. The same reaction happens here with the HOBt ester. The HOBt ester reacts with the amino group of the resin-bound Ala to form the peptide bond—that is, an amide linkage.

$$\text{FmocNH}-\text{CH}_2-\overset{\overset{\displaystyle O}{\|}}{\text{C}}-\text{OBt} + \text{H}_2\text{N}-\underset{\underset{\displaystyle \text{CH}_3}{|}}{\text{CH}}-\overset{\overset{\displaystyle O}{\|}}{\text{C}}-\text{O}-\text{CH}_2-\bigcirc\text{-}\xi \longrightarrow$$

HOBt ester of the protected amino acid   **Ala**-resin

$$\text{FmocNH}-\text{CH}_2-\overset{\overset{\displaystyle O}{\|}}{\text{C}}-\text{HN}-\underset{\underset{\displaystyle \text{CH}_3}{|}}{\text{CH}}-\overset{\overset{\displaystyle O}{\|}}{\text{C}}-\text{O}-\text{CH}_2-\bigcirc\text{-}\xi + \text{HOBt} \quad (27.26)$$

the new
peptide bond

Fmoc-Gly-Ala-resin

Notice that HOBt is used as a nucleophile, and, immediately afterward, as a leaving group. Why not let the amine nucleophile of the alanine on the resin react with the *O*-acylisourea, which, as we have already observed, is an excellent leaving group? Why bother "wasting" a molecule of HOBt?

In fact, the *O*-acylisourea *does* react with the resin-bound amino acid. Unfortunately, it also undergoes an undesirable intramolecular side reaction, called an O → N acyl shift.

$$\text{(27.27)}$$

an *O*-acyl isourea

an *N*-acyl urea

The *N*-acyl urea is an undesirable "dead-end" by-product because it cannot react further to give a peptide. In addition, the *O*-acyl isoureas, in some cases, undergo other side reactions. This O → N acyl shift, because it occurs in solution, is faster than the heterogeneous coupling reaction with the resin. The HOBt ester, because of its structure, cannot form such a by-product, and it is formed rapidly in the solution phase—evidently, so rapidly that the O → N acyl shift does not occur. Therefore, the HOBt is used to "trap" the *O*-acylisourea and form the HOBt "active ester" for which the reaction with the amino group of the Ala-resin is the only possible reaction.

Completion of the peptide synthesis requires deprotection of the dipeptide-resin as in Eqs. 27.22 and 27.23 and a final coupling step with Fmoc-Phe, DIC, and HOBt:

$$\text{(27.28)}$$

Phe-Gly-Ala-resin

Once all the peptide bonds in the desired tripeptide are assembled, the completed peptide must be removed from the resin. The ester linkage that connects the peptide to the resin, like most esters, is more easily cleaved than the peptide (amide) bonds (Sec. 21.7E). The particular ester linkage used in this case is broken by a carbocation mechanism using 50–60% trifluoroacetic acid (TFA) in dichloromethane.

linker-resin;
indicated by LR in subsequent structures

(27.29)

The acidic conditions promote breaking of the ester linkage by an S_N1 mechanism. Protonation of the peptide carbonyl converts this group into a good leaving group because it is the conjugate acid of a very weak base. The S_N1 cleavage yields a benzylic carbocation that is resonance-stabilized, not only by the benzene ring but also by the para oxygen. (Draw resonance structures that show this stabilization.) The reason for inclusion of the para —OCH_2— group in the design of the resin is that it accelerates ester cleavage, and thus release of the peptide, by acid. As a result of this reaction, the peptide is liberated into solution, from which it can be readily isolated.

Notice that the conditions of peptide synthesis and deprotection do *not* affect the ester group by which the peptide is linked to the resin. Benzylic esters undergo aminolysis very sluggishly with secondary amines such as piperidine because of steric hindrance between the phenyl hydrogens and the amine. Furthermore, the piperidine treatment required for removal of the Fmoc group takes only one minute. This is too brief a time for aminolysis of the ester to occur. However, the ester is cleaved by acidic conditions because of the ease with which it forms a relatively stable carbocation.

The method of solid-phase peptide synthesis just discussed, which employs the Fmoc group as the amino-terminal protecting group, is one of two major methods in common use today. The other important method involves a conceptually similar stepwise approach but employs a different protecting group, the *tert*-butoxycarbonyl (Boc) group, which can be removed by anhydrous acid.

$$(27.30)$$

This deprotection scheme relies on the formation of a relatively stable *tert*-butyl cation by an S_N1 mechanism that is very similar to the mechanism in Eq. 27.29. Because an *acid* (TFA) is used for deprotection, the linkage of the peptide to the resin must be stable to treatment with TFA. Hence, a different type of resin linkage is used when Boc protection is employed:

$$(27.31)$$

Compare this with the resin used in Fmoc chemistry (Eq. 27.20); although both are benzylic esters, there is no para oxygen in this case. This peptide linkage is therefore much more stable to acid because the carbocation S_N1 cleavage mechanism is less favorable. (Why?) In fact, liquid HF is required to cleave the peptide from the resin. (Boc protection and HF cleavage were used in much of the original work on solid-phase peptide synthesis.) Although the HF cleavage procedure is relatively simple with the proper apparatus, liquid HF is extremely hazardous. Avoiding liquid HF provided much of the impetus for the development of the Fmoc scheme.

The same reagents used for solid-phase peptide synthesis can also be used for peptide synthesis in solution, but removal of the DIU by-product from the product peptide (Eq. 27.25b) is sometimes difficult. The advantage of the solid-phase method, then, is the ease with which dissolved impurities and by-products are removed from the resin-bound peptide by simple filtration.

Despite its advantages, solid-phase peptide synthesis has one unique problem. Suppose, for example, that a coupling reaction is incomplete, or that other side reactions take place to give impurities that remain covalently bound to the resin. These are then carried along to the end of the synthesis, when they are also removed from the resin and must be separated (in some cases tediously) from the desired peptide product. To avoid impurities, then, each step in the solid-phase synthesis must occur with virtually 100% yield. Remarkably, this ideal is often approached closely in practice. (See Problem 27.15.)

PROBLEMS

27.15 Calculate the average yield of each of the 369 steps in the synthesis of ribonuclease by the solid-phase method discussed in the sidebar on p. 1391, assuming the reported overall yield of 17%.

27.16   What average yield per amino acid would be required to synthesize a protein containing 100 amino acids in 50% overall yield?

27.17   (a) Arrange the following esters in order of increasing reactivity toward glycinamide (the amide of the amino acid glycine), least reactive first, and explain your choice.

A   *B*   *C*   *D*

(b) Give the product of the reaction of glycinamide with *B*.

27.18   (a) An aspiring peptide chemist, Mo Bonds, has decided to attempt the synthesis of the peptide Gly-Lys-Ala using the solid-phase method. To the Ala-resin he couples the following derivative of lysine:

$$Fmoc-NH-CH-CO_2H$$
$$|$$
$$(CH_2)_4$$
$$|$$
$$NH-Fmoc$$

a,ε-diFmoc-lysine

Why are *two* protecting groups necessary for lysine?

(b) After the coupling, he deprotects his resin-bound peptide with 20% piperidine in DMF, and then completes the synthesis in the usual way by coupling Fmoc-Gly, deprotecting the peptide, and removing it from the resin. He is shocked to find a mixture of several peptide products. Two of them contain one residue of each of the amino acids Ala, Gly, and Lys, and one contains two residues of Gly, one residue of Ala, and one residue of Lys. Suggest a structure for each product and explain how each is formed.

27.19   Consider the following solid-phase peptide synthesis:

(a) Give the structure of each compound *A–E* and *P*.

(b) Explain the reason for the Boc group on the side chain of the Lys group in the reaction *B* ⟶ *C*.

(c) Explain why Boc-Val rather than Fmoc-Val is used in the *D* ⟶ *E* step of the synthesis.

Problem 27.18 shows that certain amino acid side chains can also react under the conditions of peptide synthesis. Special *protecting groups* (Sec. 19.10B) must be introduced on these side chains. (Problem 27.19 illustrates this point.) These protecting groups must survive the entire synthesis, including the removal of the amino-protecting group at each stage, yet they must be removable at the end of the synthesis. A variety of side-chain protecting groups has been developed. The choice of protecting groups that can meet these exacting requirements is an important aspect of any peptide synthesis.

B. The Biosynthesis of Proteins

Proteins are biosynthesized in nature by a system that translates the sequence of nucleotides in DNA into a sequence of amino acids in proteins. First, we'll outline the translation process, and then we'll consider the chemical aspects of the synthesis itself.

As discussed in Sec. 26.5B, the sequence of nucleotides in DNA forms a linear code for every protein and RNA molecule. To understand this point, let's see how the following strand of DNA, which might be imagined as part of a gene, could be used biologically to direct the synthesis of a specific protein. If this were the DNA from a cell, it would be one of the two strands of the double helix; each letter identifies a residue of DNA by its particular base.

◄——— 3′ end   5′ end ———►
··· A–A–A–G–A–T–T–C–A–C–C–C–C–T–C–A–T–C ···

First, a strand of DNA directs the biosynthesis of a complementary strand of RNA. This process is called **transcription**. The RNA product of transcription is called **messenger RNA (mRNA)**. The sequence of the mRNA transcript is complementary to the DNA strand from which it was transcribed; that is, each base of the DNA has hydrogen-bonding complementarity with a base of RNA (see Fig. 26.6, p. 1359). For example, everywhere there is a G (guanine) in DNA, there is a C (cytosine) in mRNA, and everywhere there is an A (adenine) in DNA, there is a U (uracil) in mRNA. (Messenger RNA contains U instead of T; U is simply a T without the methyl group.) Thus, the foregoing gene fragment would be transcribed as follows:

DNA template

◄——— 3′ end   5′ end ———►
··· A–A–A–G–A–T–T–C–A–C–C–C–C–T–C–A–T–C ···
··· U–U–U–C–U–A–A–G–U–G–G–G–G–A–G–U–A–G ···
◄——— 5′ end   3′ end ———►

mRNA transcript

Notice that the complementary sequence of mRNA runs in the direction opposite to that of its parent DNA—the 5′-end of RNA matches the 3′-end of DNA, and vice versa.

Once the mRNA synthesis is complete, the mRNA sequence is used by the cell to direct the synthesis of a specific protein from its component amino acids. This process is called **translation**. Each successive three-residue triplet, called a **codon**, in the sequence of mRNA is translated as a specific amino acid in the sequence of a protein according to the **genetic code** given in Table 27.2. Thus, the particular stretch of mRNA shown above would be translated into a protein sequence as follows:

DNA contains regions that do not code for proteins; these regions, when transcribed, are called "introns." They are excised before the RNA is translated.

Just as a sequence of dots and dashes in Morse code can be used to form words, *the precise sequence of bases in DNA (by way of its complementary mRNA transcription product) codes for the successive amino acids of a protein.* Morse code has two coding units—the dot and the dash. In DNA or mRNA, there are four: A, T (U in mRNA), G, and C, the four nucleotide bases. Notice that the sequences of DNA and mRNA contain no "commas." The protein-synthesizing system of the cell knows where one amino acid code ends and another starts because, as Table 27.2 shows, there is a specific "start" signal—either of the nucleotide sequences AUG or GUG—at the appropriate point in the mRNA. Because mRNA also contains "stop" signals (UAA, UGA, or UAG), protein synthesis is also terminated at the right place.

Some amino acids have multiple codes. For example, Table 27.2 shows that glycine, the most abundant amino acid in proteins, is coded by GGU, GGC, GGA, and GGG.

It is possible for the change of only one base in the DNA (and consequently in the mRNA) of an organism to cause the change of an amino acid in the corresponding protein. A dramatic example of such a change is the genetic disease *sickle-cell anemia*. In this painful disease, the red blood cells take on a peculiar sickle shape that causes them to clog capillaries. The molecular basis for this disease is a single amino acid substitution in hemoglobin, the protein that transports oxygen in the blood (Fig. 27.17, p. 1431). In sickle-cell hemoglobin,

TABLE 27.2 The Genetic Code

5′-Terminal base of mRNA	Middle base of mRNA				3′-Terminal base of mRNA
	U	C	A	G	
U	Phe	Ser	Tyr	Cys	U
	Phe	Ser	Tyr	Cys	C
	Leu	Ser	(Stop)	(Stop)	A
	Leu	Ser	(Stop)	Trp	G
C	Leu	Pro	His	Arg	U
	Leu	Pro	His	Arg	C
	Leu	Pro	Gln	Arg	A
	Leu	Pro	Gln	Arg	G
A	Ile	Thr	Asn	Ser	U
	Ile	Thr	Asn	Ser	C
	Ile	Thr	Lys	Arg	A
	Met*	Thr	Lys	Arg	G
G	Val	Ala	Asp	Gly	U
	Val	Ala	Asp	Gly	C
	Val	Ala	Glu	Gly	A
	Val*	Ala	Glu	Gly	G

* Sometimes used as "start" codons.

glutamic acid at position 6 in one of the protein chains of normal hemoglobin is changed to valine. That is, sickle-cell disease results from a change in but one of the 141 amino acids in this hemoglobin chain! The mRNA genetic code for Glu is GAA and GAG, whereas the code for Val is GUA and GUG (among others). In other words, a change of only one nucleotide (A → U) of the (3 × 141) = 423 nucleotides that code for this chain of hemoglobin is responsible for the disease.

Now let's consider the chemical aspects of peptide-bond formation in the biosynthesis of proteins. A different type of RNA, called **transfer RNA (tRNA)**, serves as the bridge, or adaptor, between the mRNA code and each amino acid. Each tRNA molecule contains between 73 and 94 nucleotides, depending on the amino acid to be encoded. Each amino acid has its own tRNA. The structure of a typical tRNA and the abbreviation that we'll use for tRNA in chemical equations are shown in Fig. 27.3. At one end of each tRNA molecule is a three-residue segment, called the **anticodon**, which forms a complementary hydrogen-bonding interaction with the codon of mRNA. *The codon–anticodon complementarity is the basis for the faithful translation of an mRNA sequence to a protein sequence.* For example, one of the codons for the amino acid alanine in mRNA (Table 27.2) is GCU. The anticodon in

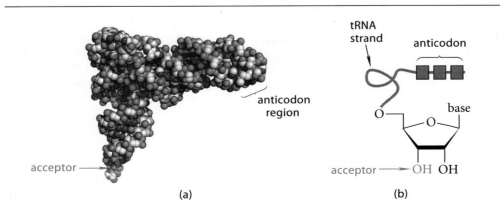

FIGURE 27.3 Transfer RNA (tRNA). (a) A space-filling model of a phenylalanine tRNA from yeast. (b) The abbreviated structure of tRNA used in the text.

(a)   (b)

alanine-tRNA that would "read" this code is CGA. (There is more than one alanine-tRNA; why?) At the other end of alanine-tRNA is the **acceptor stem**; this is where the amino acid is covalently attached.

The amino acid must be activated to react chemically with the acceptor stem of tRNA. The amino acid is converted into an *amino acid adenylate* by a reaction with ATP. (Adenylation was discussed in Sec. 25.7A, Eq. 25.25c.) In this reaction, the α-amino acid is converted into an *acyl phosphate*.

an α-amino acid ATP

(27.32a)

the adenylated α-amino acid
(an acyl phosphate)

pyrophosphate
(hydrolyzed to phosphate by
pyrophosphatase)

In this reaction, pyrophosphate is lost as a leaving group; the reaction is driven to the right by the hydrolysis reaction of pyrophosphate with water, catalyzed by pyrophosphatase, to give two molecules of phosphate (Eq. 25.23, p. 1305). Recall that acyl phosphates are excellent acylating agents (Sec. 25.7B). The aminoacyl phosphate reacts with the 3'-hydroxy group of the ribose at the 3'-end of the tRNA to give the aminoacyl-tRNA.

the adenylated α-amino acid
(an acyl phosphate)

the rest
of the tRNA
molecule

the acceptor end of the
tRNA molecule

an aminoacyl-tRNA

AMP

(27.32b)

(In some cases, the 2'-hydroxy group reacts, but in those cases the resulting 2'-aminoacyl-tRNA derivative is converted intramolecularly into the 3'-aminoacyl-tRNA.) Both reactions

(Eqs. 27.32a and 27.32b) are catalyzed by the same enzyme, an aminoacyl-tRNA synthetase that is specific for the particular amino acid.

Proteins are synthesized from the amino end to the carboxy end. A large protein–RNA complex called the **ribosome** controls the reading of the mRNA and catalyzes the peptide bond formation. The RNA component of the ribosome (rRNA) actually catalyzes the peptide bond formation; it is therefore an unusual example of a nonprotein enzyme (called colloquially a "ribozyme"). The formation of the peptide bond is shown in Fig. 27.4 on p. 1404. At the first codon of mRNA (which codes for the first amino acid), the aminoacyl-tRNA with the complementary anticodon forms a complex within the ribosome at a site called the "P-site." At an adjacent site on the ribosome, called the "A-site," the second aminoacyl-tRNA binds; it has an anticodon that is complementary to the second codon of mRNA. The binding of the two aminoacyl-tRNAs places the amino group of the second aminoacyl group in proximity to the carbonyl group of the first, and an ester aminolysis (Sec. 21.8C) occurs to produce the first peptide bond (Figs. 27.4a–b).

The empty tRNA formed exits the "P-site." The ribosome moves down the mRNA so that the dipeptidyl-tRNA now occupies the "P-site," and a new aminoacyl-tRNA enters the "A-site," and the process repeats (Fig. 27.4c).

To stress the chemical aspects of protein biosynthesis, we have omitted any discussion of other protein factors, called *elongation factors* and *release factors*, involved in this process. Hydrolysis of two guanosine triphosphate (GTP) molecules is involved in each elongation step, and part of the elongation process is a "proofreading" mechanism that ensures fidelity of translation. This hydrolysis, along with the two high-energy phosphates of ATP utilized during the formation of aminoacyl-tRNA molecules, results in an "energy cost" per peptide bond of four high-energy phosphate bonds—roughly 122 kJ mol^{-1} (29.2 kcal mol^{-1}) for the formation of each peptide bond. This high energetic cost ensures not only that the equilibrium for peptide-bond formation favors the product, but also that the peptide bond to the *correct* amino acid is formed. When we think about an enzyme active site as a highly organized arrangement of acidic and/or basic groups, along with residues in exactly the correct place to bind substrates noncovalently (Sec. 11.8C), we can now understand the high energetic price that must be paid for all of this organization.

PROBLEMS

27.20 (a) Give all of the mRNA codes for the peptide Phe-Arg-Gly-His-Trp.

(b) What are the DNA codes for the same peptide?

(c) What is the anticodon sequence in the tRNA for the Trp residue? (*Hint:* Be sure to specify the direction.)

27.21 In some tRNAs the anticodon contains an *inosine*. The heterocyclic base in inosine is *hypoxanthine*.

hypoxanthine

Inosine can form hydrogen-bonded base pairs with A, U, or C. This means that the inosine in tRNA can pair any of these bases in mRNA. Show the hydrogen-bonded base pair of hypoxanthine (a) with adenine; (b) with uracil.

27.7 HYDROLYSIS OF PEPTIDES

A. Complete Hydrolysis and Amino Acid Analysis

One reaction that all peptides have in common is amide hydrolysis. When peptides are heated with moderately concentrated aqueous acid or base, they are hydrolyzed to their component amino acids. This hydrolysis is typically carried to completion in 6 *M* aqueous HCl at 110 °C for 20–24 hours.

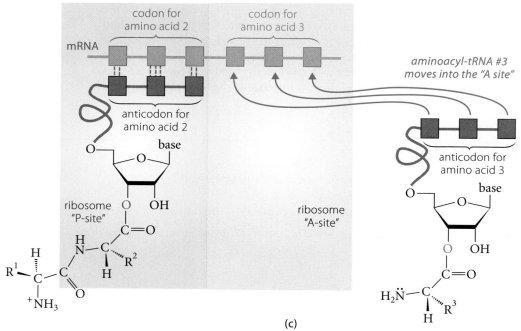

FIGURE 27.4 Peptide-bond formation on the ribosome in the biosynthesis of proteins. (a) The N-terminal aminoacyl-tRNA binds at the "P-site" of the ribosome. Its anticodon forms Watson–Crick hydrogen bonds with the mRNA codon. The aminoacyl-tRNA of the second amino acid binds at the ribosome "A-site"; its α-amino group acts as a nucleophile in an ester aminolysis that forms the first peptide bond. (Tetrahedral intermediates and proton transfers are not shown explicitly.) (b) The aminoacyl-tRNA of the dipeptide now occupies the "A-site." The ribosome moves to the next codon on the mRNA. As a result, the aminoacyl-tRNA of the dipeptide is shifted to the "P-site." (c) The ribosomal "A-site" is now available to the aminoacyl-tRNA of the third amino acid, which binds there, and the process repeats.

$$\text{Ala-Val} \xrightarrow[110\,°C,\ 22\,h]{6\,M\,HCl} \text{Ala} + \text{Val} \qquad (27.33)$$

An important reason for hydrolyzing peptides of unknown structure is that the amino acid products that result from this hydrolysis can be separated and quantified. The determination of the identities and relative amounts of amino acids in a peptide is called **amino acid analysis**.

Several reliable techniques are available for carrying out amino acid analysis. The methods in most common use today involve the conversion of the mixture of amino acids formed in the hydrolysis of a peptide into derivatives that are readily detected by spectroscopy. For example, in one method, the mixture of amino acids resulting from hydrolysis is allowed to react with 1-{[(6-quinolylamino)carbonyl]oxy}-2,5-pyrrolidinedione, a compound whose name in common usage is mercifully shortened to the acronym "AQC-NHS."

This is the same type of reaction that is used in peptide synthesis (Eq. 27.19, p. 1393). It results in the "tagging" of each amino acid in a hydrolysis mixture with the AQC group, which absorbs strongly at 254 nm in UV spectroscopy. This group is also *fluorescent*, emitting fluorescence at 395 nm, in the blue region of the visible spectrum. (Fluorescence was discussed in Sec. 15.2D.) After the various AQC-amino acids are separated, they can be quantified by measuring either UV absorption at 254 nm or fluorescence at 395 nm, because both techniques depend on the concentration of the absorbing or fluorescing species. (Fluorescence is more sensitive; that is, one can detect smaller quantities with fluorescence.)

Before the relative amounts of AQC-amino acids in a hydrolysis mixture can be determined, they must be separated. The separation of nearly 20 compounds of rather closely related structure might seem to be a daunting task, but conditions have been carefully worked out so that this separation is a routine matter. Again, liquid chromatography (Sec. 27.3C) is used; this type of liquid chromatography is called *C18 high-performance liquid chromatography*, or C18–HPLC.

Recall that, in chromatography, compounds are separated by their differential adsorptions on a stationary phase (Secs. 6.8A and 27.3D). In C18–HPLC, the stationary phase is a powder that consists of microscopic glass beads to which 18-carbon unbranched alkyl groups (that is, octadecyl groups) have been covalently bonded. We can represent the stationary phase schematically as follows:

$$\text{(27.35)}$$

[R = some other group, e.g., $(CH_3)_3Si-$]

We can think of this stationary phase as glass with a hydrocarbon coat, and we can regard adsorption simply as a solubility phenomenon, with adsorption governed by the same non-covalent interactions that govern solubility. Compounds that are more soluble in hydrocarbons are adsorbed more strongly by the column. If we consider the structures of the various AQC-amino acids in this light, we would expect that the derivatives of amino acids with hydrocarbon side chains, such as leucine, isoleucine, and phenylalanine, would be adsorbed more strongly on the stationary phase. We would expect the AQC derivatives of amino acids with polar side chains, such as serine and aspartic acid, to be adsorbed less strongly for the same reasons that alcohols and carboxylic acids are not very soluble in hydrocarbons. This is exactly what happens. The C18–HPLC separation of a mixture of AQC-amino acids is shown in Fig. 27.5. The C18 column is first eluted with water. The AQC-amino acids with polar

FIGURE 27.5 Separation of a mixture containing 50 pico-moles (pmol) $(50 \times 10^{-12}$ mole) of each AQC-amino acid by C18–HPLC in an aqueous buffer at pH 5.0 containing increasing percentages of acetonitrile. The percentage of acetonitrile in the eluting solvent is plotted in the blue overlay. Detection of the AQC-amino acids is by fluorescence. AQC-tryptophan (Trp) is not shown because tryptophan is destroyed by the strongly acidic conditions of peptide hydrolysis, but special base-hydrolysis methods can be used to detect Trp. AQC-glutamine (Gln) and AQC-asparagine (Asn) are also not shown because the side-chain amide groups of Asn and Gln are hydrolyzed under the conditions of amide hydrolysis; see Problem 27.22. Cysteine (Cys) and lysine (Lys) are present at half the concentration of the other amino acids. The compound that elutes first, AMQ, is an ester-hydrolysis product of AQC-NHS that is formed in a side reaction during derivatization. Notice that the AQC-amino acids with the greatest hydrocarbon character are eluted last. Fluorescence intensity grows to the right because it is solvent-dependent and is greater in the solvents with a higher percentage of acetonitrile.

side chains are more soluble in the solvent and are less strongly attracted to the column, so they elute first. The AQC-amino acids with less polar, more hydrocarbonlike side chains are adsorbed by the column. They are eluted by changing the solvent composition gradually to about 20% acetonitrile; the adsorbed compounds are more soluble in acetonitrile than they are in water and are removed from the column by acetonitrile. As they emerge from the column, the various AQC-amino acids are detected by their fluorescence.

Once a standard mixture of AQC-amino acids has been through the C18 column and the relative fluorescences of the different compounds have been determined, the hydrolysate of a peptide of unknown structure can be "tagged" with AQC and treated in exactly the same way. The relative amounts of each amino acid are then calculated from the data.

A second and older method of amino acid analysis involves separation of the amino acids themselves by ion-exchange chromatography under carefully defined conditions. As the amino acids emerge from the chromatography column, they are mixed with a compound called *ninhydrin*, a reagent that reacts with primary amines and α-amino acids to give a dye called *Ruhemann's purple*, which has an intense blue-violet color.

ninhydrin

FURTHER EXPLORATION 27.2
Reaction of α-Amino Acids with Ninhydrin

Ruhemann's purple
$\lambda_{max} = 570$ nm

$$\text{(27.36)}$$

The intensity of the resulting color is proportional to the amount of the amino acid present.

As an example of amino acid analysis, imagine that a hypothetical peptide *P* has been hydrolyzed, tagged with AQC, and subjected to C18–HPLC, and that the results are as follows:

$P:$ (Asp or Asn),Gly_2,His,NH_3,Arg,Ala_3,Pro,Tyr,Val,Met,Lys,Ile,Leu,Phe,Trp

According to this analysis, the peptide contains three times as much Ala and twice as much Gly as Arg, His, Lys, or the other amino acids present. The absolute number of each amino acid residue is unknown unless the molecular mass of the peptide is known. The relative order of the amino acid residues within the peptide is also unknown. In this sense, amino acid analysis is to the amino acid composition of a peptide as elemental analysis is to the molecular formula of an organic compound.

PROBLEMS

27.22   (a) Notice in peptide *P* (see previous discussion) that Asn and Asp are not distinguished by amino acid analysis. Explain. (*Hint:* Why is ammonia present in the amino acid analysis of peptide *P*?)

(b) What other pair of amino acids are not differentiated by amino acid analysis?

27.23   AQC-tryptophan is not shown in Fig. 27.5 because the indole ring does not survive the acid hydrolysis. In what general region of the chromatogram would you expect to find AQC-Trp if it were present? Explain.

continued

continued ——

27.24 The amino acids Lys and Cys, after "tagging" with AQC, are each found to contain *two* AQC groups. Explain; your explanation should involve the structures of the AQC-amino acids.

B. Enzyme-Catalyzed Peptide Hydrolysis

Peptides can be hydrolyzed at specific amino acid residues by treating them with certain enzymes, called **proteases**, **peptidases**, or **proteolytic enzymes**. These methods are very useful in determining the structures of peptides, as we'll find in Sec. 27.8C.

One of the most widely used proteases is the enzyme *trypsin*. This is a digestive enzyme (obtained commercially from cattle), the biological role of which is to catalyze the hydrolytic breakdown of dietary proteins in the intestinal tract. Trypsin catalyzes the hydrolysis of peptides or proteins at the carbonyl group of arginine or lysine residues, provided that these residues are (a) not at the amino end of the protein, and (b) not followed by a proline residue. (In the following equations, Pep^N stands for the amino-terminal part of the peptide—the part attached by an amide bond to the α-nitrogen of arginine or lysine, in this case—and Pep^C stands for the carboxy-terminal part of the peptide.)

$$(27.37a)$$

$$(27.37b)$$

(The mechanism of trypsin-catalyzed hydrolysis, and the reason for its lysine and arginine specificity, are discussed in Sec. 27.10.) Because trypsin catalyzes the hydrolysis of peptides at internal rather than terminal residues, it is called an **endopeptidase**. (Enzymes that cleave peptides only at terminal residues are called **exopeptidases**. The prefixes *endo* and *exo* come from Greek roots meaning "inside" and "outside," respectively.)

Biochemists have developed an arsenal of different proteolytic enzymes that are used for the hydrolysis of peptides at specific sites. For example, chymotrypsin, another mammalian digestive protein related to trypsin, is used to catalyze the hydrolysis of peptides at amino acid residues with aromatic side chains and, to a lesser extent, at residues with large hydrocarbon side chains. Thus, chymotrypsin cleaves peptides at Phe, Trp, Tyr, and occasionally, at Leu and Ile residues. An important endopeptidase from a microorganism, *Staphylococcus aureus*, catalyzes the hydrolysis of peptides at glutamic acid residues.

PROBLEMS

27.25 *A peptide P has the sequence of amino acids E-R-G-A-N-I-K-K-H-E-M. What products would be formed if this peptide were subjected to trypsin-catalyzed hydrolysis?*

27.26   When a peptide Q with the amino acid analysis (A,F,G$_2$,I,K,N,P,R,S,Y) is treated with trypsin, three new peptides are formed (amino acid analysis in parentheses): $T1$(A,F,R,S); $T2$(G,I,K); and $T3$(G,N,P,Y). When peptide Q is treated with chymotrypsin, four peptides are formed: $C1$(A,N,R,S,Y); $C2$(F,K); $C3$(G,I); and $C4$(G,P). What can you deduce about the order in which the amino acids in P are connected? Explain. What are the points of uncertainty?

27.8   PRIMARY STRUCTURE OF PEPTIDES AND PROTEINS

A. The Elements of Primary Structure

The structures of molecules as large as peptides and proteins can be described at different levels of complexity. The simplest description of a peptide or protein structure is its covalent structure, or **primary structure**. The most important aspect of any primary structure is the **amino acid sequence**, which is the order in which the amino acid residues are connected.

Peptide bonds are not the only covalent bonds that can connect amino acid residues. **Disulfide bonds** (Sec. 10.10B) link cysteine residues in different parts of a sequence.

(In some proteins, all Cys residues are involved in disulfide bond formation; in others, some Cys residues are not.) Disulfide bonds thus serve as crosslinks between different parts of a peptide chain. A number of proteins contain several peptide chains; disulfide bonds help to hold these chains together. The primary structure of a peptide or protein, then, includes its amino acid sequence and its disulfide bonds. The primary structure of *lysozyme*, a small enzyme that is abundant in hen egg white, is shown in Fig. 27.6 on p. 1410. Lysozyme is a single polypeptide chain of 129 amino acids that includes eight cysteine residues incorporated into four disulfide bonds.

The disulfide bonds of a protein are readily reduced to free cysteine thiols by other thiols. Two commonly used thiol reagents are 2-mercaptoethanol (HSCH$_2$CH$_2$OH), and dithiothreitol (known to biochemists as DTT, or Cleland's reagent).

(27.38)

This reaction is a biological example of the thiol–sulfide equilibrium shown in Eqs. 10.71a and b, p. 498. Typically, when the extraneous thiols are removed, the thiols of the protein spontaneously re-oxidize in air back to disulfides.

The reaction in Eq. 27.38 is a very clever application of the proximity effect (Sec. 11.8). The equilibrium constant between one thiol–disulfide pair and another

FIGURE 27.6 The primary structure of the enzyme lysozyme from hen egg white. Physiologically, lysozyme catalyzes the hydrolysis of bacterial cell walls. Different variants of this enzyme are found in tears, nasal mucus, and even viruses—anywhere antibacterial action is important. Lysozyme is one of the smallest known enzymes. Individual amino acid residues, connected by peptide bonds, are numbered from the amino terminus. The cysteine residues involved in the disulfide bonds are shown in orange.

would typically be close to 1.0. However, the formation of six-membered rings is favored entropically. As a result, the equilibrium strongly favors the right side of the equation.

A Practical Example of Disulfide-Bond Reduction

An interesting example of the biological effects of disulfide-bond reduction occurs in the use of permanent-wave preparations to curl the hair. Hair (protein) is treated with a thiol solution; this solution is responsible for the unpleasant smell of permanents. The thiol solution reduces the disulfide bonds in the hair. With the hair in curlers, the disulfides are allowed to re-oxidize. The hair is thus set by disulfide-bond reformation into the shape dictated by the curlers. Only after a long time do the disulfide bonds rescramble to their normal configuration, when another permanent becomes necessary.

An industrial example of the use of disulfide bonds is the process of vulcanization (Sec. 15.5), which introduces disulfide bonds into both natural rubber and synthetic polymers. Vulcanization provides a polymer with greater rigidity.

Disulfide bonds are the most common type of crosslink between peptide chains, but other types of crosslinks are possible. For example, the side-chain amino group of a Lys residue, or the side-chain carboxy groups of Asp or Glu residues, could form amide bonds, thus creating branches in peptide chains. One of the best known examples of peptide crosslinking occurs in the bacterial cell wall. A rigid, two-dimensional network is formed by an all-glycine pentapeptide connected by amide bonds to the side-chain amino group of a Lys residue in one peptide chain and the carboxy group of a D-alanine residue in another.

Crosslinking of this sort is generally found in structural proteins, and even in these cases it is relatively uncommon. Most proteins consist of unbranched peptide chains crosslinked by disulfide bonds.

The determination of the primary structure of a peptide or protein would appear to be a complex problem, because a given amino acid composition can correspond to a very large number of amino-acid arrangements, or *sequences*, for even a small peptide. For example, 120 unique sequences are possible for a pentapeptide containing five different amino acids, and more than 3.6 *million* sequences are possible for a decapeptide containing 10 different amino acids! Despite this apparent complexity, there are well-established methods for determining the primary structures of peptides. The determination of a primary structure is called **peptide sequencing**. In Sec. 27.8B, we'll discuss the use of mass spectrometry (Sec. 12.6) for peptide sequencing, and in Sec. 27.8C, we'll discuss a well-established chemical method. Finally, in Sec. 27.8D, we'll learn how the primary amino acid sequences of large proteins are determined.

B. Peptide Sequencing by Mass Spectrometry

A typical peptide can be sequenced in the mass spectrometer using electrospray ionization (ESI), which we can think of as a chemical ionization (CI) technique (Sec. 12.6D). The peptide in the gas phase is protonated to give an M + 1 ion, from which the molecular mass M of the peptide is determined. In a special type of mass spectrometer, the M + 1 ion is subjected to a technique called **tandem mass spectrometry**, or simply **MS–MS**. In this technique, the ion is first energized in some way. Increasing the energy of this ion causes it to undergo fragmentation. For example, in one method, the M + 1 ion is routed into a collision cell into which a noble gas such as argon or xenon is admitted. The collision of the ion with the gas molecules causes it to gain energy and undergo fragmentation. (This is something like what happens to a window if you throw baseballs at it.) Recall that, in fragmentation, a cationic fragment and a neutral fragment are produced, and the mass spectrometer detects the cation. Peptide fragmentation can in principle occur at every bond along the peptide backbone. When fragmentation at a particular bond takes place, the charge can remain with the amino-terminal fragment, or it can remain with the carboxy-terminal fragment. Using a four-residue peptide as an example, the possible fragments are categorized as follows:

fragmentations in which charge is carried by the C-terminal fragment

$$(27.39)$$

fragmentations in which charge is carried by the N-terminal fragment

It is important to understand that each fragment shown above arises from a *single* fragmentation of the M + 1 ion. In other words, some M + 1 ions undergo b_2 fragmentation, others undergo y_2 fragmentation, others undergo c_3 fragmentation, and so on.

In many cases, the b-type and y-type fragmentations occur most frequently. That is, *fragmentation often occurs at the peptide bond.* To illustrate the type of data obtained from a peptide mass spectrum, we'll imagine, for simplicity, a case in which only b-type fragmentation is significant. An M + 1 ion can be formed by protonation of any one of the carbonyl groups of the peptide as well as protonation of basic side-chain groups. In other words, the M + 1 ion is actually a mixture of ions that differ in their sites of protonation. Let's consider the b_2 fragmentation of the peptide in Eq. 27.39. Most of the protonation probably occurs at the carbonyl oxygen (Sec. 21.5). The carbonyl-protonated peptide can undergo a cyclization in which the carbonyl oxygen of the previous residue acts as a nucleophile. The resulting product is a protonated *oxazolone*, which is the ion detected by the mass spectrometer.

$$(27.40a)$$

a protonated oxazolone

Another possible mechanism for b-type cleavage involves protonation of an amide nitrogen. Amide nitrogens are *much* less basic than amide carbonyl oxygens because all amide resonance is obliterated in the nitrogen-protonated species. However, once protonated, this nitrogen can serve as a good leaving group in the formation of an *acylium ion* (Eq. 16.23, p. 808). The acylium ion has the same molecular mass as the protonated oxazolone.

$$(27.40b)$$

an *N*-protonated amide   an acylium ion

Now imagine that b-type fragments are produced from other M + 1 ions that are protonated at other amide carbonyls or nitrogens, and further imagine that each of these fragmentations occurs in a measurable amount. As a result, fragmentation of the M + 1 ions results

FIGURE 27.7 The MS–MS of an *N*-acylated tripeptide, V-F-M. The M + 1 ion was observed before it was subjected to fragmentation and found to have a mass of 729. The mass differences between the major peaks correspond to the residues between successive b-type fragmentation points. The fragmentation points are marked with dashed lines. The mass differences correspond to the residue masses in Table 27.3.

TABLE 27.3 Amino Acid Residue Masses

the masses of $-NHCHC-$ (or isomeric structures),

where R = H for Gly, R = CH_3 for Ala, etc.

Residue	Mass	Residue	Mass	Residue	Mass
Ala (A)	71.0	Glu (E)	129.0	Pro (P)	97.1
Arg (R)	156.1	His (H)	137.1	Ser (S)	87.0
Asn (N)	114.0	Ile (I)	113.1	Thr (T)	101.0
Asp (D)	115.0	Leu (L)	113.1	Trp (W)	186.1
Cys (C)	103.0	Lys (K)	128.1	Tyr (Y)	163.1
Gln (Q)	128.1	Met (M)	131.0	Val (V)	99.1
Gly (G)	57.0	Phe (F)	147.1		

in a family of b-type ions—b_1, b_2, and b_3—the masses of which *differ by the atomic masses of the intervening residues*:

mass difference between fragments = the mass of residue 3

fragment from b_2 cleavage

fragment from b_3 cleavage

$$H_2N-CH-C-NH-CH-C-NH-CH-C-NH-CH-C-OH \quad (27.41)$$

$R^1 \quad R^2 \quad R^3 \quad R^4$

$b_1 \quad b_2 \quad b_3$

The sequence from the C-terminus can then be read directly from the mass spectrum, right-to-left, using the mass differences between the peaks. Such a sequence determination is illustrated in Fig. 27.7 for a synthetic *N*-acylated tripeptide with the structure shown in the figure. The masses of the amino acid residues used to make the identification are given in Table 27.3.

When several types of fragmentation occur, each type of fragmentation gives a family of peaks, but each peak in a given family differs from the next by a residue mass. If the mass spectra are complex because of the occurrence of multiple fragmentation modes, computer programs are available that can assist in the analysis.

In MS–MS, all of the sequence information is obtained in one experiment. But MS–MS has another, even more powerful, capability. Suppose a large peptide is subjected to trypsin-catalyzed hydrolysis, and four or five smaller peptides are obtained. Traditionally, the sequencing of peptides required pure samples, and purification of a complex peptide mixture typically required laborious chromatography. However, with MS–MS, each peptide in the mixture can be sequenced directly from the mixture—no purification required—provided that these peptides differ in mass. The MS–MS analyzer can sequester the M + 1 ions of each peptide in turn, energize them, and thus produce separate mass spectra for all of them in the same experiment. In effect, the mass spectrometer performs the separation on the basis of the differing masses of the M + 1 ions. This high-throughput capability has revolutionized peptide sequence analysis.

The number of residues that can be sequenced by MS–MS depends on the specific case. Sequencing 7–10 residues is often possible, and 20 residues can be sequenced in favorable cases. Longer peptides, however, can be hydrolyzed with trypsin or other proteolytic enzymes to produce shorter peptides. The order of the shorter peptides in the overall sequence can be established by comparing the results of two or more digests resulting from the use of proteolytic enzymes with differing specificities. This process is called the *method of overlapping peptides*. This strategy is illustrated by Study Problem 27.1.

STUDY PROBLEM 27.1

A peptide *P* with the amino acid composition (A,E,F,G$_2$,H,K,L,M,P,R,V,Y) did not yield to MS–MS sequencing, so it was hydrolyzed with trypsin to three peptides, *T1*, *T2*, and *T3*, which were found by MS–MS to have the following structures:

| *T1:* A-H-K | *T2:* E-M-V | *T3:* (L,P)-F-G-G-Y-R |

(The parentheses in *T3* mean that the order of L and P could not be determined.) Hydrolysis of *P* catalyzed by chymotrypsin yielded three peptides, *C1*, *C2*, and *C3*, which were sequenced by MS–MS and found to have the following structures:

| *C1:* A-H-K-L-P-F | *C2:* G-G-Y | *C3:* R-E-M-V |

What is the amino acid sequence of peptide *P*?

SOLUTION   Because trypsin breaks peptides at the C-terminal side of K and R, and because only peptide *T2* does not have one of these residues at its C-terminus, the sequence of peptide *T2* must have been at the C-terminus of peptide *P*. Hence, the two possible sequences of *P* are *T1-T3-T2* and *T3-T1-T2*. The sequence of the chymotryptic peptide *C3* overlaps parts of the sequences of *T3* and *T2*. This overlap shows that, in the sequence of peptide *P*, an R residue precedes the E residue and that the sequence of *T3* precedes the sequence of *T2* in peptide *P*. The chymotryptic peptide *C1* resolves the ambiguity in the positions of L and P in peptide *T3*. Because the sequence of *C1* overlaps parts of the sequences of *T1* and *T3*, this sequence also confirms the sequence order *T1-T3*. Therefore, the sequence of peptide *P* is *T1-T3-T2*, or

P: A-H-K-L-P-F-G-G-Y-R-E-M-V

PROBLEMS

27.27   (a) Give the *m/z* values of the fragment ions expected from b-type fragmentation of an M + 1 ion of the peptide N-F-E-S-G-K.

   (b) Give the *m/z* values of the fragment ions expected from y-type fragmentation of an M + 1 ion of the peptide in part (a). All y-type fragments contain a protonated terminal amino group; that is, H$_3$N$^+$—.

27.28   Give the curved-arrow mechanism for the formation of each of the following fragment ions in Fig. 27.7 from an M + 1 ion.

   (a) the fragment at *m/z* = 551.94. [*Hint:* This fragment results from an a-type cleavage (Eq. 27.39).]

   (b) the fragment ion at *m/z* = 710.97

   (c) the fragment ion at 727.09. Show how this mechanism might be tested with a deuterium-labeled peptide.

C. Peptide Sequencing by the Edman Degradation

Prior to the advent of peptide sequencing by MS–MS, the standard sequencing method was a chemical process called the **Edman degradation**, named after Pehr Victor Edman (1916–1977), a Swedish biochemist who devised the method in 1952. In an Edman degradation, the peptide is treated with *phenyl isothiocyanate* (often called the **Edman reagent**). The peptide reacts with the Edman reagent at its amino groups to give thiourea derivatives. Although reaction with the Edman reagent also occurs at the side-chain amino groups of lysine residues (see Problem 27.49, p. 1442), only the reaction at the terminal amino group is relevant to the degradation. (As before, the abbreviation Pep^C is used for the carboxy-terminal part of a peptide.)

$$\text{Ph—N}{=}\text{C}{=}\text{S} + \text{H}_2\text{N—CH—C—NH—}\boxed{\text{Pep}^C} \xrightarrow[\text{pyridine/H}_2\text{O}]{\text{Me}_2\text{NPh}}$$

phenyl isothiocyanate
(Edman reagent)

a thiourea   (27.42a)

This reaction is exactly analogous to the reaction of amines with *isocyanates*, the oxygen analogs of *isothiocyanates* (Eq. 23.73, p. 1218). Any remaining phenyl isothiocyanate is removed, and the modified peptide is then treated with anhydrous trifluoroacetic acid. As a result of this treatment, the sulfur of the thiourea, which is nucleophilic, displaces the amino group of the adjacent residue to yield a five-membered heterocycle called a *thiazolinone*; the other product of the reaction is *a peptide that is one residue shorter.*

a thiourea   a thiazolinone   a new peptide, one residue shorter   (27.42b)

When treated subsequently with aqueous acid, the thiazolinone derivative forms an isomer called a **phenylthiohydantoin**, or **PTH**. This probably occurs by reopening of the thiazolinone to the thiourea, followed by ring formation involving the thiourea nitrogen. Notice in this and the previous equation the intramolecular formation of five-membered rings.

thiazolinone   a thiourea   phenylthiohydantoin (PTH) derivative of the amino-terminal residue   (27.42c)

side chain of the amino-terminal residue

Because the PTH derivative carries the characteristic side chain of the amino-terminal residue, identification of the PTH identifies the amino acid residue that was removed. Methods for identifying PTH derivatives by liquid chromatography are well established. The peptide liberated in Eq. 27.42b can be subjected in turn to the Edman degradation again to yield the PTH derivative of the next amino acid and a new peptide that is shorter by yet another residue.

In principle, the Edman degradation can be continued indefinitely for as many residues as necessary to define completely the sequence of a peptide. In practice, because the yields at each step are not perfectly quantitative, an increasingly complex mixture of peptides is formed with each successive step in the cleavage, and, after a number of such steps, the results become ambiguous. Hence, the number of residues in a sequence that can be determined by the Edman method is limited. Nevertheless, instruments are now in use that can apply Edman chemistry to the structure determination of peptides in a highly standardized, automated, and reproducible form. In such instruments, the sequential degradation of 20 residues is common, and the degradation of as many as 60 or 70 amino acid residues is sometimes possible.

When using the Edman degradation, a researcher has to wait for the completion of one cycle before initiating the next cycle. Furthermore, a peptide or protein must be purified before subjecting it to the Edman degradation. Unless multiple sequencing instruments are available in the laboratory, only one peptide can be sequenced at a time. Because of its chemistry, the Edman method works progressively from the amino end of a peptide, and, if a carboxy-terminal sequence is needed, the Edman method is of no use. (Another limitation is the subject of Problem 27.30.) Sequencing by MS–MS has none of these limitations. However, the Edman method is extremely reliable, and generally more residues can be sequenced with the Edman technique than with MS–MS. Instruments for Edman sequencing are also much less expensive than MS–MS instruments. Hence, the Edman method is still used, but it is used much less than it once was.

PROBLEMS

27.29 Using the curved-arrow notation, write in detail the mechanisms for the reactions in
(a) Eq. 27.42a   (b) Eq. 27.42b   (c) Eq. 27.42c

27.30 Some peptides found in nature have an amino-terminal acetyl group (*red*):

$$\text{H}_3\text{C}-\overset{\overset{\text{O}}{\|}}{\text{C}}-\text{NH}-\text{CH}-\overset{\overset{\text{O}}{\|}}{\text{C}}-\text{NH}-\cdots$$
$$|$$
$$\text{R}$$

(a) Can these peptides undergo the Edman degradation? Explain.
(b) Does *N*-acylation have any adverse effect on sequencing by MS–MS? Explain.

D. Protein Sequencing

In the early history of structural biology, the complete primary sequence of a protein was determined by isolation of the individual peptide chains followed by tryptic and chymotryptic hydrolysis to afford overlapping peptides of manageable size. These peptides were then sequenced by the Edman degradation. Determining the complete sequence of a protein took several years of work. Nevertheless, many protein sequences were determined in this way, and these have been proven to be quite reliable.

The development, in the late 1970s, of methods for rapidly sequencing the nucleotides in DNA made it possible to read the complete sequences of proteins directly from DNA sequences. The biosynthesis of proteins is coded within DNA by contiguous sequences of nucleotides in which each amino acid is represented by a three-base code. (This process was explained in Sec. 27.6B.) In principle, then, identification of the gene for a protein in a DNA sequence results automatically in the knowledge of the protein sequence. The one fly in the ointment, however, is that DNA contains noncoding sequences that serve regulatory functions, code for certain types of RNA, or, in some cases, have no known function. However, biochemists have learned how to produce DNA, called **complementary DNA**, or **cDNA**, in which the noncoding sequences have been excised. (We leave the details of this process for your study of biochemistry.) The sequence of a cDNA, then, contains all of the information necessary to read the sequence of a protein for which it codes. Almost all protein sequences are now determined by "translating" their cDNA codes.

As part of the effort to sequence the DNA of the complete genomes of many different species (including humans; see Sec. 26.5B), databases have been developed that show the protein-coding regions. When a researcher isolates an unknown protein from one of these species, a partial sequence is in many cases sufficient to obtain a unique match to a DNA coding region, from which the entire sequence of the protein can then be read. Standard computer software for matching partial sequences to genomic DNA sequences is available. This is one of the many reasons for the ascendancy of peptide sequencing by MS–MS. Tryptic digestion of a protein followed by the *simultaneous* partial sequencing, without purification, of the peptide products is often sufficient to establish a cDNA match, from which the sequence of the entire protein can then be determined.

E. Posttranslational Modification of Proteins

Once proteins are produced within a living cell, the structures of many of them are modified by subsequent reactions. These reactions are called collectively **posttranslational modifications** because they occur after *translation*—the biosynthesis of the protein itself. More than 200 posttranslational modifications have been documented. The DNA-coding sequence for a protein carries no information about any posttranslational modifications that might occur. Currently, the only way to elucidate the posttranslational modifications of a protein is by studying the protein itself.

Most posttranslational modification reactions involve the chemical alteration of side-chain functional groups. These types of modifications serve many roles. They can provide unique structures so that the protein can be recognized by another molecule. They can serve as "molecular switches," turning on or off enzyme activity. They can control the lifetime of a protein within the cell, and they can be involved in *protein trafficking*—that is, in directing a protein to the proper destination within a cell. In short, posttranslational modifications increase the diversity of amino acid side chains available in living systems. MS–MS is widely used in studying these alterations. We'll illustrate posttranslational modification with two of the most common types: *protein phosphorylation* and *protein glycosylation*.

Protein Phosphorylation   Proteins are phosphorylated by the transfer of a γ-phosphate group from ATP to a serine, tyrosine, cysteine or (more rarely) a histidine residue of proteins. These reactions are catalyzed by *protein kinases*. (The mechanism of phosphate transfer and the action of kinases was discussed in Sec. 25.7A; see Eq. 25.25a.) An analysis of the human genome predicts that there are more that 500 different protein kinases. Phosphorylation of serine is illustrated in the following equation.

(27.43)

This reaction has an equilibrium constant $K'_{eq} \approx 100$ at pH = 7, the favorable K_{eq} resulting from the anhydride character of ATP (see Sec. 25.8B).

Dephosphorylations of phosphoserine, phosphotyrosine, or phosphothreonine occur by an enzyme-catalyzed reaction of the phosphorylated amino acid residue with water. The enzymes that catalyze this reaction are called *phosphatases*; over 100 phosphatases occur in humans. Phosphatases were discussed on pp. 1301–1303.

a phosphoserine residue in a protein

a serine residue in a protein

inorganic phosphate

(27.44)

Phosphorylation in biology has several roles. One role is a regulatory function. Phosphorylation converts a neutral Ser, Tyr, Cys, or His residue into a residue with two negative charges. This charged state can result in a conformational change in the protein driven by the electrostatic and/or hydrogen-bonding interaction of the phosphate with a positively charged residue, such as a lysine or an arginine residue elsewhere in the protein, as shown schematically in Fig. 27.8.

a Ser residue

HOCH₂—

a protein with an unphosphorylated serine residue

an Arg residue

protein–serine kinase
ATP

conformational change

conformational change

protein–serine phosphatase
H₂O

favorable hydrogen-bonding and electrostatic interactions

a phosphorylated Ser residue

a protein with a phosphorylated serine residue

FIGURE 27.8 A diagram showing how phosphorylation of a serine residue can induce a conformational change in a protein. The gray strand represents the protein chain, and the green arrows show how the chain moves in the conformational change. Phosphorylation causes the change, and dephosphorylation reverses it.

An impressive example of a phosphorylation-induced conformational change in a protein occurs in the enzyme *glycogen phosphorylase*. The biological role of this enzyme is to initiate the cleavage of glycogen, a polymer of glucose, by inorganic phosphate to give glucose-1-phosphate, which is then processed by the glycolysis pathway and the citric acid cycle to provide the products of glucose metabolism along with chemical energy. The activity of this enzyme is regulated by phosphorylation. One serine residue of the enzyme and a few of its neighboring residues cover the active site like a lid on a bucket and completely prevent access to the site by the substrates. Surrounding this serine are a number of aspartic acid and glutamic acid residues, which have a negative charge. Fig. 27.9a shows a stick model of glycogen phosphorylase with the serine shown as a space-filling model for emphasis. This serine becomes phosphorylated by ATP when metabolic conditions require more glucose. When this serine is phosphorylated, the negative charge on the resulting serine phosphate is repelled by its negative environment; as a result, a conformational change of the protein occurs in which the phosphoserine moves 34 Å to form a favorable electrostatic attraction with two arginine residues. Fig. 27.9b shows the complex of the serine phosphate and one of the arginines as space-filling models. This conformational change removes the "lid" from the active site, and the enzyme can then bind both of its substrates, glycogen and phosphate.

PROBLEMS

27.31   (a) Draw the structure of a phosphotyrosine residue.

(b) Would the equilibrium constant for formation of a phosphotyrosine residue from ATP (by a reaction analogous to the one shown in Eq. 27.43) be greater than, less than, or about the same as K_{eq} for the phosphorylation of a serine residue? Explain.

27.32   Draw the structure of a phosphocysteine residue that shows the configuration on the asymmetric carbons with lines and wedges.

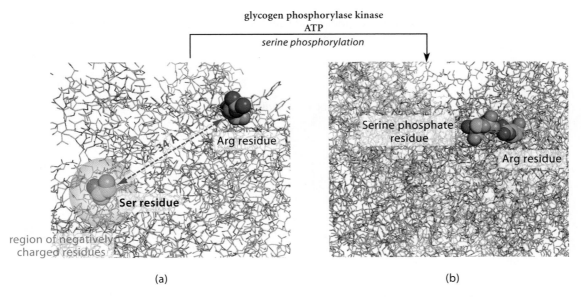

(a)   (b)

FIGURE 27.9 Part of the structure of the enzyme glycogen phosphorylase showing a phosphorylation-induced conformational change. The bonds of the enzyme are shown as green stick models except for one serine and one arginine residue, which are shown as space-filling models. (a) The enzyme in the unphosphorylated, inactive, state. The pink halo indicates a region of negatively charged residues surrounding the serine. When the serine is phosphorylated, the resulting serine phosphate is repelled by the surrounding negative charge. A conformational change occurs in the enzyme that brings the serine phosphate close to the arginine, with which it has a favorable electrostatic interaction (see Fig. 27.8). (A second arginine in the same region is not shown.) This conformational change unveils the active site and thus allows the enzyme to become active.

Protein Glycosylation   The most prevalent posttranslational modification of proteins is the attachment of oligosaccharides (Sec. 24.11B) to give **glycoproteins**. Glycoproteins serve a variety of functions. Perhaps the most common glycoproteins in common experience are the mucins, proteins that are constituents of mucus. Glycosylation is responsible for the slimy feel of mucus. Glycoproteins are important in cell–cell recognition, in the immune response, and in connective tissue. Your blood type depends on the oligosaccharides attached to proteins and lipids of red-blood-cell membranes. A few enzymes and peptide hormones are glycoproteins. The determination of the structure of a glycoprotein generally involves application of a combination of specific glycoside-hydrolyzing enzymes and mass spectrometry.

The attachment of a saccharide to a protein is called **glycosylation.** There are two broad types of glycosylation, **N-glycosylation** and **O-glycosylation**. In *N*-glycosylation, an oligosaccharide is attached by the reducing end of an *N*-acetylglucosamine residue to the side-chain amide nitrogen of an asparagine (Asn) residue that is part of a sequence Asn-X-Ser/Thr; X is any amino acid, and the residue on the carboxy-terminal side of residue X must be either serine or threonine. Notice that these residues contain —OH groups in their side chains.

local structure of an *N*-linked glycoprotein

The oligosaccharide, prior to attachment to the protein, is built up, residue-by-residue, as a *dolichol diphosphate* derivative.

abbreviated structure

Dolichol is a long terpene hydrocarbon (typically 75–95 carbons). As we might expect from its hydrocarbon character, dolichol is anchored within the lipid bilayer of a membrane-like cellular structure called the *endoplasmic reticulum* (ER). Proteins are synthesized at the external surface of the ER, and the attachment of the oligosaccharide to the protein takes place as the protein chain is being elongated. The hydrocarbon chain of the dolichol ensures that the oligosaccharide-transfer process is localized at the surface of the ER, where the polypeptide chain is being formed—another example of a proximity effect (Sec. 11.8).

An apparent chemical problem in the process of *N*-glycosylation is that the amide —NH_2 group of the asparagine is normally not nucleophilic, because the nitrogen electron pair is delocalized into the carbonyl group.

Also, an amide —NH_2 group is not acidic enough ($pK_a \approx 15$) to form measurable amounts of its conjugate base at physiological pH. The catalyzing enzyme, a *glycosyl transferase*, overcomes this problem by forcing a 90° rotation of the amide nitrogen out of conjugation with the carbonyl group (shown below as a sawhorse projection):

$$ \text{(27.45)} $$

The —NH_2 group, after rotation, behaves more like an ordinary amino group because its electron pair is no longer delocalized. (See the sidebar, "An Amide with a Twist," p. 1059.) This rotation requires energy, and the energetic cost of this "de-conjugation" is paid by the formation of two hydrogen bonds, one from each of the N—H hydrogens to a carboxylate group in the enzyme active site. These hydrogen bonds also place more negative charge on the nitrogen, thus making it more basic and more nucleophilic. This nitrogen then acts as a nucleophile toward the anomeric carbon of the sugar-phosphate, displacing dolichol diphosphate and forming the *N*-glycoside. This substitution occurs with inversion of stereochemistry.

the *N*-glycosylated protein (27.46)

(The proton transfers are not shown.) This reaction is another example of the C—O cleavage of a pyrophosphate, a pattern that we have observed in the biosynthesis of terpenes (Secs. 17.6B and 25.7C).

In most cases of *O*-glycosylation, one sugar residue is attached to a serine or threonine residue of the protein acceptor, and subsequent sugars are added, one at a time, to complete the oligosaccharide. (Contrast this process with *N*-glycosylation, in which the pre-formed oligosaccharide is transferred to the protein.) The formation of the *O*-glycosyl bond between *N*-acetylgalactosamine and the oxygen of Ser or Thr in the proteins of mucin is typical. The sugar is transferred as an α-UDP-*N*-acetylgalactosamine derivative. The UDP serves as a leaving group.

UDP-*N*-acetylgalactosamine

a serine (R = H) or threonine (R = CH$_3$) residue in an acceptor protein

a transferase enzyme
Mn^{2+}

the *O*-glycosylated protein (27.47)

Once again we have an example of a pyrophosphate as a leaving group in a C—O cleavage. (The enzyme-catalyzed transfers of most saccharide groups, whether to proteins or to other sugars, always involve a nucleoside-diphosphate derivative of the sugar, and UDP-sugars are very common.)

Stereochemically, some *O*-glycosylations involve inversion of configuration, and some involve retention of configuration. The particular substitution considered here occurs with *retention of configuration*, as shown in Eq. 27.47. If the mechanism of substitution is strictly S_N2, then retention implies that two substitutions must occur at carbon-1 of the sugar (see Sec. 11.8D). However, another possibility is that the substitution mechanism resembles an S_N1 reaction with a carbocation intermediate.

(27.48)

One stereochemical hallmark of S_N1 mechanisms in solution is that both retention and inversion take place (see Figs. 9.12 and 9.13, pp. 425 and 426). For retention to occur exclusively, the face of the sugar carbocation opposite to the leaving group must be blocked by groups in the enzyme active site so that the leaving group departs and the nucleophile enters from the same side. Although the mechanism is not known with certainty, the carbocation mechanism is supported by the fact that the sugar binds to the enzyme in a flattened conformation that very much resembles the conformation of a carbocation.

PROBLEMS

27.33   Using abbreviated structures, draw the structure of an *N*-acetylgalactosamine conjugate with a threonine residue of an acceptor protein formed with *inversion* of configuration at the anomeric carbon.

27.34   (a) Draw a resonance structure for the carbocation intermediate in Eq. 27.48.
(b) All of the enzyme-catalyzed glycosylations require Mn^{2+}, a divalent cation. Suggest a role for the metal ion in the glycosylation mechanism.

Glycosylation in Clinical Diagnostics: The Hemoglobin A1c Test

Glucose is present in the blood of all humans, and chronically elevated blood glucose is an indicator of diabetes. Physicians traditionally order "fasting blood glucose" tests to detect such elevated levels. This test determines the blood glucose level after a 12-hour fast (so that temporary elevations in glucose resulting from a recent meal don't skew the results). In the mid-1970s, Dr. Anthony Cerami (b. 1940), Professor of Medical Biochemistry at the Rockefeller University, and his students found in blood an unusual variant of hemoglobin, the iron-containing protein of red blood cells, and they observed that the level of this unusual protein correlated with fasting blood glucose levels. They called this variant *hemoglobin A1c*, abbreviated HbA1c. They subsequently established that HbA1c results from the *nonenzymatic* glycosylation by glucose of the amino-terminal amino group in the β-chain of hemoglobin (*photo*). In this reaction, the aldehyde form of glucose reacts with the amino group of hemoglobin to form an imine (Schiff base), which can re-close to the *N*-glycoside.

$$\text{D-glucopyranose (either anomer)} \quad \longleftrightarrow \quad \text{aldehyde form of glucose} \quad \xrightarrow{\underset{\textbf{hemoglobin}}{H_2\ddot{N}-\boxed{Hb}}}$$

$$H_2O + \text{glucose–Hb imine (Schiff base)} \quad \longleftarrow \quad \text{Hb } N\text{-glycoside of D-glucopyranose} \qquad (27.49)$$

The imine undergoes a reaction called an *Amadori rearrangement* to give a new set of products. In this reaction, the imine carbon is converted into a methylene group and carbon-2 is converted into a ketone. (See Problem 27.35.)

$$\boxed{\text{Amadori rearrangement}}$$

glucose–Hb imine (Schiff base)

internal rotation

Amadori product   (27.50)

(The Amadori rearrangement is the nitrogen analog of the *Lobry de Bruyn–Alberda van Eckenstein* rearrangement; Sec. 24.5.) The products of the Amadori rearrangement constitute the "unusual" HbA1c found by Cerami. Over time, these products can also react further to give colored pigments.

All of these reactions are in equilibrium, and they are relatively slow, in part because the concentration of glucose in the blood is relatively low ($4–8 \times 10^{-3}$ mol L^{-1}). It takes about 60–90 days for the equilibrium to be established. Therefore, unlike "fasting blood glucose," which gives a "snapshot" of blood glucose level at the time the blood sample is taken, the HbA1c test gives a long-term average. Because the reactions are reversible, a reduction in blood glucose level will, over time, reduce the HbA1c level as well. It took over 20 years for the test to be accepted, but now it is a staple in the battery of tests for diabetes and pre-diabetes. In a normal person, less than 6% of the hemoglobin is present as Hb1Ac. An HbA1c level of greater than 6.5% is an indicator of diabetes.

Why does this one amino group of hemoglobin react, and not the many amino groups in the side chains of lysine residues? The reason is that the terminal amino group has a lower conjugate-acid pK_a than the lysine amino groups, and it is the *conjugate-base* form of the amine that reacts as a nucleophile with the aldehyde group of glucose.

If glucose can react with hemoglobin, then it can also react with other proteins. For example, glucose can react with the lens proteins of the eye, and the glycosylated proteins and their by-products can be one cause of cataracts. This is the reason that cataracts can result from uncontrolled diabetes.

PROBLEMS

27.35 Draw a curved-arrow mechanism for the Amadori rearrangement in Eq. 27.50. Let H_3O^+ and H_2O be the acid–base pair, present as necessary.

27.36 It is possible to envision an Amadori rearrangement of an *N*-glycoside formed at an asparagine amide-NH_2 group. In fact, this does *not* occur because the imine required for the rearrangement does not form. Use the mechanism of imine formation to explain why. (The inability of asparagine *N*-glycosides to undergo this reaction is probably why asparagine rather than lysine evolved as the amino acid residue involved in *N*-glycoside formation.)

27.37 When the formation of HbA1c reverses, not only glucose, but also mannose, is formed. Explain the origin of the mannose.

27.9   HIGHER-ORDER STRUCTURES OF PROTEINS

Just as the simple Lewis structure of a small molecule contains no information about its conformation, the primary structure of a protein does not indicate how the molecule actually looks in three dimensions. The three-dimensional aspects of protein structure are described at three levels, which are called *secondary structure*, *tertiary structure*, and *quaternary structure*.

A. Secondary Structure

With few exceptions, all of the amide units in most peptides and proteins are planar. Recall that rotation about the carbonyl–nitrogen bond of most amides is relatively slow, and that the preferred conformation about this bond is *Z* (Sec. 21.2); the same is true for the amide bonds in a peptide or protein.

The **secondary structure** of a peptide or protein describes the relative orientations of the planes of its peptide bonds. The possibilities are described by the two internal rotations shown in Fig. 27.10; these are the internal rotations about the single bonds to the α-carbon.

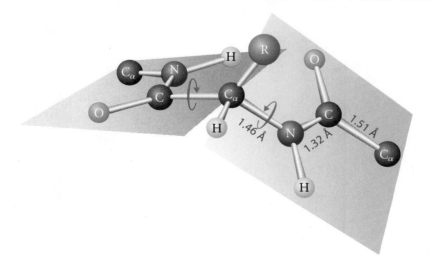

FIGURE 27.10 Typical dimensions of a peptide bond. The two planes are those of the adjacent amide groups, and the amino acid side chain is represented by R. In principle, rotations about the bonds to the α-carbons (marked with *green* arrows) are possible.

Because a protein contains many such bonds, it might seem that a very large number of conformations could occur in a protein or large peptide. However, studies in the late 1940s and early 1950s by Linus Pauling (1901–1994) and his co-workers (of the California Institute of Technology) showed that protein conformations are governed by the hydrogen-bonding interactions of their backbone amide groups and the avoidance of van der Waals repulsions between the residue side chains. For this work (and earlier work on the nature of the chemical bond), Pauling received the 1954 Nobel Prize in Chemistry. (Pauling also received the 1962 Nobel Peace Prize.) Two major conformations, the α-*helix* and the β-*sheet*, are very common in proteins; as we shall see, hydrogen bonding plays a key role in maintaining these conformations. Within proteins are also found some regions of disorder, called *random coil*.

In the **right-handed α-helix**, shown in Fig. 27.11, the peptide chain adopts a conformation in which it turns in a clockwise manner along a helical axis. In this conformation, the side-chain groups are positioned on the outside of the helix, and the helix is stabilized by *hydrogen bonds* between the carbonyl oxygen of one residue and the amide N—H four residues further along the helix. The alpha (α) terminology refers to a characteristic X-ray diffraction pattern that was observed for certain proteins before chemists fully understood their structures. The α type of pattern was eventually shown to be associated with the right-handed helix—thus the name α-helix.

Another commonly occurring X-ray diffraction pattern, called a β pattern, was eventually found to be characteristic of a second peptide conformation, called **β-structure** or **pleated sheet**. In this type of structure, a peptide chain adopts an open, zigzag conformation, and is engaged in hydrogen bonding with another peptide chain (or a different part of the same chain) in a similar conformation. The successive hydrogen-bonded chains can run (in the amino-terminal to carboxy-terminal sense) in the same direction (**parallel pleated sheet**) or, more commonly, in opposite directions (**antiparallel pleated sheet**). The antiparallel pleated sheet structure is shown in Fig. 27.12. The name "pleated sheet" is derived from the pleated

FIGURE 27.11 A peptide α-helix. (a) Hydrogen atoms are shown, and the side chains R are represented by *green* spheres. The side chains extend away from the helix on the outside. (b) Backbone atoms only, which form the α-helix itself, are shown. The typical α-helix has a pitch of 26°, a distance of 5.1 Å between turns, and 3.6 residues per turn.

carbonyl group at the *i*th residue

N—H at the *i* + 4th residue

hydrogen bonds→

helical axis

5.1 Å

26°

3.6 residues

(a)   (b)

interchain hydrogen bonds

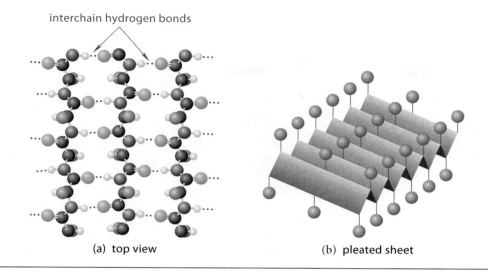

(a) top view

(b) pleated sheet

FIGURE 27.12 The β-antiparallel pleated-sheet structure of proteins. The amino acid side chains are shown as green spheres. (a) A top view of the antiparallel peptide chains. Notice the hydrogen bonds between chains (*gray dotted lines*). (b) The imaginary pleated-sheet surface formed by the backbone atoms.

surface described by the aggregate of several hydrogen-bonded chains (Fig. 27.12b). Notice that the side-chain R-groups alternate between positions above and below the sheet. When two parts of the same peptide chain interact in a β-sheet, they are typically connected by a very short turn, called a *β-turn*.

Peptides and proteins can contain regions of local disorder, which are called **random coil**. As the name implies, peptides and proteins that adopt a random coil show no discernible pattern in their conformations. An apt analogy for the random coil is the appearance of a tangled ball of yarn after an hour's encounter with a playful house cat.

Although other conformations are known in peptides, the α-helix and β-sheet are the major ones. Some peptides and proteins exist entirely in one conformation. For example, the α-keratins, major proteins of hair and wool, exist in the α-helical conformation. In these proteins, several α-helices are coiled about one another to form "molecular ropes." These structures have considerable physical strength. In contrast, silk fibroin, the fiber secreted by the silkworm, adopts the β-antiparallel pleated sheet conformation. Despite these examples, proteins that contain a single type of conformation are relatively rare. Rather, most proteins consist of regions of α-helix and β-sheet separated by short regions of random coil.

B. Tertiary Structure

The complete three-dimensional description of protein structure at the atomic level is called **tertiary structure**. The tertiary structures of proteins are determined by X-ray crystallography; each crystallographic structure analysis requires significant effort. Nevertheless, since the first protein crystallographic structure was determined in 1960, hundreds of protein structures have been elucidated, and more structures are continually appearing. High-field NMR is also being used with greater frequency for determining protein structures.

The tertiary structure of any given protein is an aggregate of α-helix, β-sheet, random coil, and other structural elements. In recent years it has become evident that, in many proteins, certain higher-order structural motifs are common. For example, a common motif is a bundle of four helices (called a *four-helix bundle*), each running approximately antiparallel to the next and separated by short turns in the peptide chain. Another common structural motif is the *β-barrel*, literally a bag consisting of β-sheets connected by short turns. Several such motifs can occur within a given protein, so a protein might consist of several smaller, relatively ordered structures connected by short turns. These ordered substructures are sometimes called **domains**.

A useful way to portray secondary structure as part of an overall protein structure is a *ribbon structure*, illustrated for hen-egg lysozyme in Fig. 27.13 on p. 1428. In a **ribbon structure**, the *peptide backbone* (Sec. 27.1B, p. 1375) is portrayed as a ribbon. Recall from Sec. 27.9A that the amide bond is planar; that is, the carbonyl group, its attached N—H, and the

FIGURE 27.13 The three-dimensional structure of lysozyme from hens' eggs shown as a ribbon structure. The face of the ribbon shows the relative orientations of the planes of the peptide bond. Lysozyme contains regions of α-helix, β-sheet, and random coil. The two domains of lysozyme occur on either side of the plane indicated by the dashed line. (The primary structure of lysozyme is shown in Fig. 27.6.)

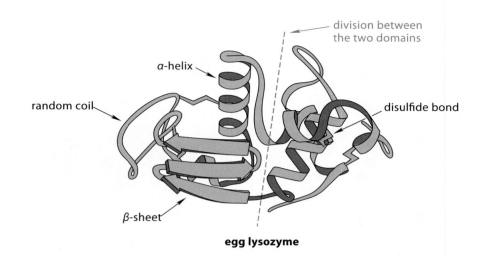

egg lysozyme

two α-carbons attached to these groups lie in a common plane. The face of the ribbon defines the orientation of the planes of the peptide bonds. Twists and turns in the ribbon are defined by the two angles shown by arrows in Fig. 27.10. The ribbon structure of lysozyme clearly shows regions of α-helix and β-sheet. In this structure, the two domains of the protein are also clearly discernible.

Richardson Ribbons: The Origin of Ribbon Structures

Ribbon structures were devised in 1980 by Jane Richardson (b. 1941), a crystallographer and biochemistry professor at Duke University, as a way to see the secondary structural elements within proteins at a glance. (One reviewer has said, "[These dia-grams provide] protein structure with a sense of substance and design.") These diagrams (sometimes called *Richardson diagrams*) are now universally used to show secondary structure. Prof. Richardson's first ribbon structures were hand-drawn. Now, all computer programs that render protein structures have a "ribbon structure" option that provide such structures automatically.

In general, the tertiary structures of proteins are determined by the preferred geometries of the peptide bonds, by the presence and location of disulfide bonds, and by *noncovalent interactions*. The effect of noncovalent interactions is a *balance* between interactions of groups in the protein with each other and interactions with the solvent water. Let's review the noncovalent interactions we have learned about and consider how these might affect protein structure. (These interactions are shown diagrammatically in Fig. 27.14.)

1. *Hydrogen bonds.* We have already seen the key role of hydrogen bonds within the protein backbone in the protein α-helix and β-sheet. In addition, the carbonyls and N—H groups of the backbone can serve as acceptors and donors, respectively, toward amino acid side chains that have hydrogen-bonding capability: Asp, Glu, His, Lys, Arg, Ser, Thr, Asn, Gln, Tyr, and the N—H proton in the indole ring of Trp, and in some cases these amino acids can form hydrogen bonds with each other. A number of cases are known in which one or more water molecules are immobilized within the structure of a protein. These waters can serve as hydrogen-bonded anchors for three-dimensional structure. Likewise, the amino acids just listed can form hydrogen bonds with the external solvent water.

2. *Interactions between hydrocarbon groups.* This type of interaction includes van der Waals interactions (Sec. 8.5A), pi-offset stacking (Sec. 15.8A), and associations caused by hydrophobic bonding—that is, the association of hydrocarbon

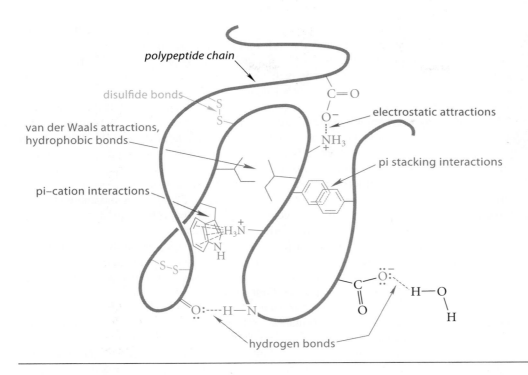

polypeptide chain

disulfide bonds

van der Waals attractions,
hydrophobic bonds

pi–cation interactions

electrostatic attractions

pi stacking interactions

hydrogen bonds

FIGURE 27.14 A diagram showing the various interactions responsible for the three-dimensional structures of proteins. The noncovalent interactions on the interior of the protein are balanced against the noncovalent interactions with the surrounding milieu.

groups driven by the unfavorable entropy of solvation by water (Sec. 8.6D). The hydrocarbonlike amino acids are Ala, Leu, Ile, Val, Phe, Trp, and the aromatic ring of Tyr.

3. *Electrostatic interactions.* These interactions include attractions between oppositely charged groups, pi–cation interactions, ion–dipole attractions, and dipole–dipole attractions. The effect of an attraction between oppositely charged groups was illustrated in Fig. 27.9 with the conformational change caused by the attraction between a phosphorylated serine and the conjugate base of an arginine. Pi–cation interactions were discussed in Sec. 15.8B. An important example of dipole–dipole attractions is illustrated by the antiparallel arrangement of α-helices when they are adjacent in a protein structure. The α-helix has a net dipole moment that is aligned along the helical axis from the amino end to the carboxyl end (Fig. 27.15). This dipole moment results from the dipole moments of individual residues, which are oriented nearly parallel to the helical axis and oriented in the N-to-C direction. When α-helices are oriented in opposite directions, oppositely charged ends of the helices are adjacent and have attractive interactions.

helix dipole moments

N to C direction

N to C direction

FIGURE 27.15 The antiparallel alignment of α-helices in proteins is favored because of the favorable interactions of the helical dipoles.

The overall conformation of a protein is the balance of all these forces acting within the protein against the attractions with the surrounding aqueous solvent or other environment, such as the interior of a membrane. In water-soluble proteins, enough charged residues and residues that can form hydrogen bonds or favorable ion–dipole interactions with water are located at the protein surface to provide adequate solvation for the protein. However, even water-soluble proteins have a significant number of hydrocarbon groups on their surfaces. As a result, many water-soluble proteins are *roughly* globular in shape so as to minimize the water-exposed hydrocarbon surface area of hydrocarbon-containing residues. A number of proteins, such as ion channels (Sec. 8.7B) and many receptors, span the cell membrane or other membrane-like structures within the cell. Many of these proteins contain bundled anti-parallel α-helices as a significant part of their structures. In the α-helix, the amino acid side chains point outward along the periphery of the helix (Fig. 27.11). In membrane-embedded proteins, the surfaces of the helices exposed to the phospholipid bilayer, predictably, contain a high percentage of the hydrocarbonlike amino acid residues.

Suppose we were to synthesize a protein. Would the finished protein automatically "know" what conformation to assume, or is some external agent required to direct the protein into its naturally occurring conformation? This question was first answered by two elegant experiments with the enzyme ribonuclease. First of all, synthetic ribonuclease was prepared by the solid-phase method (Sec. 27.6A) and found to be an active enzyme. Because the enzyme must have its natural, or native, conformation to be active, it follows that ribonuclease, once synthesized, spontaneously folds into this conformation.

The second type of experiment involved the denaturation and renaturation of ribonuclease. **Denaturation** is illustrated schematically in Fig. 27.16. When a protein is denatured, it is converted entirely into a random-coil structure. (A common example of *irreversible* protein denaturation occurs when an egg is fried; the denaturation and precipitation of the proteins in egg white are responsible for the change in appearance of the white as it is cooked.) Some proteins, including ribonuclease, can be denatured *reversibly* by chemical agents. Typically, a protein is denatured by breaking its disulfide bonds with thiols, such as DTT or 2-mercaptoethanol (Eq. 27.38, p. 1409), and then by treating it with 8 *M* urea, detergents, or heat. Ribonuclease was denatured by treatment with 2-mercaptoethanol and 8 *M* urea. After the urea was removed, and the cysteine —SH groups were allowed to re-oxidize back to disulfides, the protein spontaneously reassumed its original, or *native*, conformation. This process is called **renaturation**. This experiment showed that *the amino acid sequence of ribonuclease specifies its conformation; that is, the native structure is the most stable structure.* If this were not so, another, more stable structure would have formed when the protein was allowed to refold after the urea was removed.

This renaturation experiment [for which Christian B. Anfinsen (1916–1995) of the U.S. National Institutes of Health shared the 1972 Nobel Prize in Chemistry] indicates that proteins spontaneously assume their native conformations at the time of their biosynthesis. In other words, *primary structure dictates tertiary structure.*

FIGURE 27.16 When a protein is denatured, its disulfide bonds are broken and it is converted entirely into a random coil. When the denaturing agents are removed, the protein renatures.

FIGURE 27.17 The tertiary and quaternary structure of hemoglobin. The hydrogen atoms are not shown. The two α chains are shown in red and orange, and the two β chains are shown in blue and turquoise. The heme groups are shown in magenta as framework models. The subunits are located at the corners of a regular tetrahedron.

In some cases, protein folding is assisted by other proteins called *chaperones*. It seems likely that these molecules speed the folding process by helping proteins to avoid conformations other than the most stable ones. This is an active area of current research.

Protein and peptide misfolding in living organisms can be pathogenic. For example, Alzheimer's disease is known to result from the aggregation in the brain of abnormally folded peptides called β-amyloids into plaques. (See the sidebar on p. 552.) The misfolding of a neural protein, α-synuclein, occurs in Parkinson's disease. Mad cow disease (bovine spongiform encephalopathy) and the related Creuzfeldt–Jakob disease in humans also result from the formation of protein plaques in the brain caused by infectious misfolded proteins called *prions* that somehow gain entry into the brain.

C. Quaternary Structure

Many proteins are aggregates of other proteins. The best-known example of such proteins is *hemoglobin*, which transports oxygen in the bloodstream. Hemoglobin (Fig. 27.17) is an aggregate of four smaller proteins, or *subunits*, two of one type (called α subunits) and two of another (called β subunits). (This terminology has nothing to do with α- and β-structure.) The α and β subunits are similar, but differ somewhat in their primary structures. These subunits are held together solely by noncovalent forces. Notice in Fig. 27.17 that the individual subunits lie more or less at the vertices of a regular tetrahedron. This shape is the most compact arrangement that can be assumed by four objects. Many important proteins are aggregates of individual polypeptide subunits. In some proteins, the subunits are identical; in other cases, they are different. The description of the subunit arrangement in a protein is called **quaternary structure**.

PROBLEM

27.38   What would you expect to happen when hemoglobin is treated with a denaturant such as 8 M urea? Explain.

27.10   ENZYMES: BIOLOGICAL CATALYSTS

A. The Catalytic Action of Enzymes

Enzymes are the catalysts for biological reactions (Sec. 4.9C). Except for a few instances of biological catalysis by ribonucleic acids (ribosomal RNA, "ribozymes"), all enzymes are proteins. We have discussed specific aspects of enzyme catalysis throughout this text. Let's now summarize the key points about enzymes.

1. *Enzymes are catalysts.* They substantially increase the rates of biological reactions, but they do not affect equilibrium constants. That is, *they lower the standard free energy of activation* ($\Delta G°^{‡}$) *for a reaction, but they do not affect its overall standard free energy* ($\Delta G°$). The rate accelerations associated with enzyme catalysis depend on the reaction, but factors of 10^{12} or more are common.

2. *The binding of substrates at the enzyme active site is an obligatory part of enzyme catalysis.* The compounds on which enzymes act are called **substrates**. A substrate is bound into a specific region of the enzyme called the **active site** prior to the actual reaction that converts bound substrate to bound product.

$$\underset{\substack{\text{enzyme} \quad \text{substrate}}}{E \;+\; S} \;\rightleftharpoons\; \underset{\substack{\text{noncovalent} \\ \text{enzyme–substrate} \\ \text{complex}}}{E \cdot S} \;\rightleftharpoons\; \underset{\substack{\text{noncovalent} \\ \text{enzyme–product} \\ \text{complex}}}{E \cdot P} \;\rightleftharpoons\; \underset{\substack{\text{enzyme} \quad \text{product}}}{E \;+\; P} \qquad (27.51)$$

3. *The binding of substrates to enzymes is noncovalent.* Noncovalent attractions, such as van der Waals attractions, hydrogen bonding, and electrostatic attractions of various types are used to pay the entropic price of substrate binding. In addition, solvating water can be "stripped off" of a substrate as part of substrate binding; the return of low-entropy solvation water to higher-entropy solvent water can contribute a favorable entropic component to binding. Because substrate binding involves noncovalent forces, binding is typically very fast; in many cases it occurs at the rate that the enzyme and substrate can diffuse together (10^{8}–10^{9} $M^{-1}s^{-1}$).

4. *The rate accelerations observed in enzyme catalysis are largely due to the fact that the reactions within the active site are intramolecular* (Sec. 11.8D). Because the enzyme structure and the active site gather several reacting groups into proximity, concerted reaction mechanisms are observed that would be entropically impossible in the absence of enzymes. In addition, there may be a "solvent effect" associated with the active-site environment in some cases. When the active site is free of water, a protein active site may have an effective dielectric constant (Eq. 8.10 and discussion, p. 353) that closely resembles that of a dipolar aprotic solvent ($\epsilon = 15$–20). Recall that substitution reactions are strongly accelerated in such solvents (Sec. 9.4E). Moreover, there is evidence in some cases that the noncovalent binding forces are capable of distorting a bound substrate toward its transition state by bending or stretching some of the bonds involved in the enzyme-catalyzed reaction.

5. *Enzymes are structurally specific for their substrates.* Each biological reaction has its own unique enzyme. In most cases, enzymes have very little tolerance for variations in substrate structure. For cases in which variations are tolerated, certain parts of the structure cannot be varied. The reason for this structural specificity is that each enzyme has evolved so that its specific substrate "fits" its active site. This idea, first enunciated by Emil Fischer in 1894, came to be called the **lock-and-key hypothesis**. This hypothesis implies that the substrate (the "key") fits into a rigid enzyme active site (the "lock"). However, enzymes are flexible and can undergo conformational changes, and such changes have been observed in many cases. In many situations an enzyme changes its conformation during the binding event to match the structure of its substrate. This idea, called **induced fit**, was proposed in 1958 by Daniel E. Koshland, Jr. (1920–2007), then a scientist at Brookhaven National Laboratory and subsequently a professor of biochemistry at the University of California, Berkeley.

6. *Enzymes are stereochemically specific for their substrates.* As we have learned, enzymes are linear polymers of L-amino acids. They are therefore *chiral catalysts*. With very few exceptions, each enzyme is specific for a single stereoisomer of its

substrate (Sec. 7.7A). Enzymes provide a powerful illustration of the *principle of enantiomeric differentiation.*

If an enzyme consisting of all L-amino acid residues catalyzes a reaction of a chiral substrate, then the enantiomeric enzyme consisting of all D-amino acid residues should catalyze the same reaction of the *enantiomeric* substrate. Nature does not afford enantiomeric enzymes that can be used to test this prediction. However, an all-D enzyme was prepared in 1993 by chemists at The Scripps Research Institute in La Jolla, California, using solid-phase peptide synthesis. The natural enzyme that they prepared, HIV protease, is a small (and thus synthetically accessible) peptide-hydrolyzing enzyme produced by the viral agent for the disease AIDS. (This enzyme, as a drug target, is discussed in Sec. 27.10B.) This enzyme catalyzes the hydrolysis of certain peptides consisting of amino acids with the L configuration. The synthetic all-D enzyme was found to be inactive with the natural substrates, but—as predicted—it hydrolyzes the *enantiomers* of the natural substrates with exactly the same catalytic efficiency as the natural enzyme hydrolyzes its all-L substrates.

7. *Enzymes are large molecules.* The smallest enzymes have molecular masses around 9000 g mol^{-1}, but many enzymes are much larger, and some have molecular masses exceeding 100,000 g mol^{-1}. The biosynthesis of an enzyme represents a substantial investment of biochemical energy (Sec. 27.6B). For that reason, we assume that the very specific organization of an active site that provides the necessary noncovalent interactions for substrate binding and the precise placement of catalytic groups would not be possible in a much smaller protein. In addition, a large protein cannot escape from a cell or cellular compartment by passive diffusion as some small molecules can. Some enzymes consist of single polypeptide chains, such as lysozyme (Fig. 27.6); others consist of several polypeptide chains held together by disulfide bonds, such as trypsin (see the discussion of trypsin that follows). Other enzymes consist of several identical subunits, each containing the same active site. Still others merge several different types of subunits that catalyze sequential reactions in a pathway; fatty acid synthase (Sec. 22.7) is such an enzyme.

8. *Coenzymes and other cofactors are involved in some enzyme-catalyzed reactions.* A **coenzyme** is a nonprotein biomolecule that is involved covalently in an enzyme-catalyzed reaction, but is itself either unchanged or later "recycled." Examples of coenzymes are NAD$^+$/NADH (Sec. 10.8, 19.8B), *S*-adenosylmethionine (Sec. 11.7), thiamin pyrophosphate (Sec. 20.11B), biotin (Sec. 22.7), and pyridoxal phosphate (Sec. 26.4E). Metal ions (such as Mg^{2+}) are obligatory cofactors in some enzyme-catalyzed reactions. As we have learned, they are sometimes part of the catalytic mechanism.

Let's consider the mechanism by which the enzyme *trypsin* catalyzes the hydrolysis of peptide bonds to see each of these points in context. Trypsin is the mammalian digestive enzyme used in the sequencing of proteins (Sec. 27.7B). With a molecular weight of about 24,000, trypsin is an enzyme of modest size. It is a soluble, globular protein containing three polypeptide chains held together by disulfide bonds. The following comparison provides some idea of the catalytic effectiveness of trypsin. Peptides in the presence of trypsin are rapidly hydrolyzed at 37 °C and pH = 8. In the absence of trypsin, peptide hydrolysis under the same conditions requires hundreds of years. As we learned in Sec. 27.7A, to hydrolyze proteins at a reasonable rate requires boiling them in 6 *M* HCl for several hours. Moreover, trypsin, in contrast to hot HCl solution, does not catalyze hydrolysis at just *any* peptide bonds. It is specific for the hydrolysis of the peptide bonds at lysine and arginine residues (Eq. 27.37, Sec. 27.7B). Moreover, it is specific only for substrates with the L stereochemical configuration.

The active site of trypsin consists of a cavity, or "pocket," that just accommodates the amino acid side chain of a lysine or arginine residue from the substrate. Several hydrophobic residues line this cavity. At the bottom of the cavity is the side-chain carboxylic acid group of an aspartic acid residue (Asp-189 in the trypsin sequence). This group is ionized, and therefore *negatively charged*, at neutral pH. The amino group of a lysine side chain and the guanidino group of an arginine side chain are both protonated, and therefore *positively charged*, at neutral pH. The favorable electrostatic attraction between the ionized Asp-189 side chain of the enzyme and the positively charged side chain of the substrate helps stabilize the enzyme–substrate complex. This complex is also stabilized by the van der Waals interactions between the —CH_2— groups of the substrate side chain and the hydrophobic residues that line the cavity. These, then, are some of the reasons for the *specificity* of trypsin. The active site just "fits" the substrate (and vice versa), and it contains groups that are noncovalently attracted to groups on the substrate.

Near the mouth of the active site are two amino acid residues, Ser-195 and His-57, that serve a critical catalytic function. The way that these residues act to catalyze peptide bond hydrolysis at an Arg residue is shown in Fig. 27.18. The —OH group of the serine side chain acts as a nucleophile to displace the peptide leaving group from the carbonyl group of the substrate. The resulting product is an *acyl-enzyme*; in this transient covalent complex, the residual peptide substrate is actually esterified to the enzyme. The imidazole group of His-57 serves as a base catalyst to remove the proton from the nucleophilic serine hydroxy group. When water enters the active site, it too is deprotonated by the histidine as it reacts as a nucleophile at the carbonyl carbon of the acyl-enzyme, to give the free carboxy group of the substrate and thus regenerate the enzyme. After the product leaves the active site, the enzyme is ready for a new substrate molecule.

The *catalytic efficiency* of trypsin, as well as that of other enzymes, is attributable mostly to the *proximity effect* (Sec. 11.8D). That is, all of the necessary reactive groups—the substrate carbonyl, a nucleophile (the serine —OH group), and an acid–base catalyst (the imidazole of the histidine)—are positioned in proximity within the enzyme–substrate complex. These groups do not have to "find" one another by random collision, as they would if they were all free in solution.

If scientists understand the details of enzyme catalysis, they should be able to design and synthesize artificial enzymes that bind specific compounds and act on them catalytically and stereospecifically. Success in this endeavor would yield an arsenal of rationally designed molecules that could catalyze industrially important transformations under mild, environmentally friendly conditions. General success in this sort of activity is yet to be realized, and research in rational catalyst design is pursued by a number of chemists. However, there is some progress. The asymmetric epoxidation catalyst (Sec. 11.11) and the transition-metal catalysts for aryl and vinylic substitution reactions (Secs. 18.6, 18.10B, and 23.11C) bring about the assembly of reactants on metal "templates" that results in reactions that are impossible in the absence of the catalysts. Along the same lines, chemists and biochemists are beginning to learn how to custom-design protein enzymes by altering known enzymes so as to change their specificities in predictable ways. Such "designer enzymes" can then be produced by synthesizing their genes and inserting them into the genomes of bacteria, which then become miniature "protein factories" when they are grown in the laboratory.

B. Enzymes as Drug Targets: Enzyme Inhibition

Suppose an enzyme is known to be an essential factor in the development of a certain disease state. By "knocking out" the enzyme (that is, by preventing it from catalyzing a reaction), we could prevent the disease. This strategy lies at the heart of drug development. One way to inactivate an enzyme is to treat it with an *inhibitor*. An **inhibitor** is a compound that prevents an enzyme from fulfilling its catalytic role. The most common inhibitors are *competitive inhibitors*. A compound is a **competitive inhibitor** when it binds to the enzyme's active site so tightly that the enzyme can no longer bind its usual substrate. Because substrate binding must precede catalysis (Eq. 27.51), prevention of binding results in the loss of catalysis.

Consider two impressive examples of enzyme inhibition. The human immunodeficiency virus (HIV, the virus responsible for AIDS) requires several unique enzymes for its replication and cellular infection; these enzymes have been used as drug targets. The anti-AIDS drug AZT and the HIV-protease inhibitors are two classes of anti-AIDS drug whose effectiveness is based on enzyme inhibition. (We'll examine an HIV-protease inhibitor in more detail subsequently.) Another example is the family of the modern cholesterol-lowering drugs. These

FIGURE 27.18 A stylized representation of the trypsin active site showing the mechanism of the trypsin-catalyzed hydrolysis of an arginyl–peptide bond, beginning with the enzyme–substrate complex and ending with the enzyme–product complex. The imidazole in the side chain of the His-57 residue acts alternately as a base and, in its protonated form, as an acid to bring about the necessary proton transfers. Pep^N and Pep^C denote the amino-terminal and carboxy-terminal parts of the peptide substrate, respectively. Groups of the enzyme are colored blue.

drugs act by inhibiting an important enzyme (HMG-CoA reductase) in the biochemical pathway by which cholesterol is synthesized in the body (Sec. 25.5C). [Not all drug targets are enzymes; other proteins, RNA, and even DNA can serve as drug targets; see, for example, the story of the mustards (sidebar, p. 545). Nevertheless, enzyme drug targets are fairly common.]

Let's consider enzyme inhibition by starting with a simple case: the enzyme trypsin. Trypsin isn't involved in a disease state, but its inhibition provides an example in which the molecular basis for inhibition is particularly clear. Recall from Sec. 27.10A that trypsin specificity is based on a *hydrophobic pocket* containing an *ionized carboxy group* (Asp-189). Trypsin acts mostly on peptide bonds adjacent to Arg and Lys residues; these residues both have positively charged groups at the end of hydrocarbon side chains. It seems likely that a compound that has some or all of these same features—namely, a hydrocarbon group of appropriate size and a positively charged group attached to it, might be an inhibitor of trypsin. The idea is that if the inhibitor binds tightly enough, it should clog the active site and make the active site inaccessible to substrate.

Many such inhibitors for trypsin are known. One is the benzamidinium ion.

$$H_3O^+ + \text{benzamidine} \xrightleftharpoons{pH\ 7.4} \text{benzamidinium ion} + H_2O \qquad (27.52)$$

benzamidine   **benzamidinium ion**
$pK_a = 11.6$

(This cation, the conjugate acid of benzamidine, has a pK_a value of 11.6, and is thus fully protonated at the physiological pH of 7.4.) The design of this inhibitor utilizes its resemblance to the arginine residue at the trypsin cleavage site of peptides (Eq. 27.37b, p. 1408).

Arg residue

hydrocarbon group

amidinium group

benzamidinium ion

In the presence of 10^{-3} M benzamidine at pH 7.4, trypsin is inactive as a catalyst because the benzamidinium ion binds to the active site of trypsin, thus blocking the access of substrates to the active site. A model of trypsin containing a bound benzamidinium ion is shown in Fig. 27.19a. (This model comes from X-ray crystallography.) The benzamidinium ion is "stuck" in the active site like a counterfeit coin stuck in the slot of a vending machine. Fig. 27.19b shows that the benzamidinium ion binds just as we would expect. The positively charged group of the cation is very near the ionized, and therefore anionic, side-chain carboxylic acid group of Asp-189. The benzamidinium ion is also probably involved in significant hydrogen bonding with the carboxylate ion as well.

The modern design of enzyme inhibitors as drugs can be illustrated with inhibitors for HIV protease, one of the enzymes of human immunodeficiency virus (HIV) involved in the virus's replication and infectious activity. Before the crystal structure of this enzyme was known, chemists had recognized that this enzyme resembled the digestive enzyme pepsin in its mechanism of action. Both pepsin and HIV protease catalyze peptide hydrolysis by a mechanism involving the side-chain carboxylic acid groups of two active-site Asp residues and a tightly bound water molecule (see Problem 27.40). (The enzyme groups are shown in blue.)

(a)   (b)

FIGURE 27.19 The trypsin–benzamidinium ion complex. (a) A space-filling model of trypsin with the bound benzamidinium ion shown in *magenta*. (Hydrogen atoms are not shown.) (b) A detailed diagram of the active site, showing the crucial Asp-189 and its spatial relationship to the positively charged group of the benzamidinium ion.

(27.53)

(The enzymes in this protease family are called *aspartyl proteases* for this reason.) A number of pepsin inhibitors were known, and these compounds were also found to be inhibitors of HIV protease. These would not make good drugs, however, because, ideally, a suitable drug should discriminate between the enzyme we want to inhibit—HIV protease—and enzymes such as pepsin, whose functions we do not want to compromise.

   The determination of the X-ray crystal structure of HIV protease was a very important development in the design of inhibitors for this enzyme. Fig. 27.20a on p. 1439 shows the HIV protease structure.

   The active site of HIV protease was easily identified by the presence of the two active-site Asp residues and the bound water. (The active site is exactly where it was expected—within the large "hole" in the enzyme structure.) Inhibitors for HIV protease were designed by *molecular modeling.* Molecular modeling is conceptually like building models with model sets, with two important exceptions. First, it is done on a computer, so that *really large* molecules such as proteins can be handled easily and sophisticated graphics tools can be employed; and second, the interaction energies between molecules (or between groups within molecules) can be calculated. These capabilities enable chemists to determine which inhibitors are likely to bind strongly to the enzyme. Inhibitors bind to HIV protease principally by an extensive network of hydrogen-bonding interactions, as well as by van der Waals attractions between nonpolar groups. (An electrostatic interaction like the one in trypsin is not present.) The best inhibitors have hydrogen-bonding sites that "mate" well with the corresponding sites on the protease and have nonpolar groups of appropriate size that interact well with similar groups on the enzyme. Molecular modeling studies suggested some possible unique structures for inhibitors. Organic chemists then prepared compounds with those and related structures, and they were tested as inhibitors. Crystallographers in some cases examined the complexes of these inhibitors with the protease to see whether the predictions of molecular modeling were correct; as a result of this work, refined inhibitor structures were proposed. Eventually, through the collaboration of crystallographers, molecular modelers, organic chemists, and biologists, satisfactory inhibitors were developed. Ritonavir (Norvir) and Indinavir (Crixivan) are HIV-protease inhibitors that evolved from such studies. Today these compounds are actively used as anti-HIV drugs. Darunavir was subsequently developed as a second-generation drug for the treatment of HIV infections that had become resistant to the other drugs. (Viruses, like bacteria, mutate into drug-resistant strains; see sidebar, Sec. 21.8D.)

Norvir

Crixivan

Darunavir

These compounds form very strong noncovalent complexes with HIV protease. For example, the dissociation constant for the complex of Norvir with the protease is $15 \times 10^{-12}\ M$, and the dissociation constant for the protease–Darunavir complex is $4 \times 10^{-12}\ M$. These are *very* strong bindings! The complex of Norvir with HIV protease is shown in Fig. 27.20b. In comparing HIV protease with and without inhibitor, notice how the "arms" of the protease at the base of the structure come together when the inhibitor is bound. (This has been called a "fireman's grip.") This change in the enzyme on binding the inhibitor illustrates the *induced fit* concept (point 5, Sec. 27.10A).

The strong binding of an inhibitor is a necessary but insufficient condition for it to be an effective drug. A good drug must also be nontoxic. It must not be metabolized (destroyed by the body) at too great a rate. It must have just the right water solubility. It must penetrate the appropriate tissues (that is, it must be *bioavailable*). Considerations such as these were very important in developing the final drugs, and a number of candidate compounds with excellent binding properties were tested before the final candidates were chosen. These and other *ribbon structures:*

FIGURE 27.20 (a) The HIV protease. (Hydrogen atoms are not shown.) (b) The HIV protease containing an inhibitor (Norvir) bound to the active site. The carbons of the inhibitor are shown in *magenta*. Compare the ribbon structures in parts (a) and (b). Notice in the ribbon structures how the "arms" at the base of the protease come together around the inhibitor. This motion illustrates the concept of *induced fit*.

space-filling models:

the inhibitor is bound in the active site

(a)   (b)

HIV-protease drugs have led to a dramatic improvement in the life expectancy of patients with HIV infections. These cases illustrate how organic chemistry, when teamed with areas of the life sciences, can be used for the improvement of human health.

PROBLEMS

27.39   Aeruginosin-B is a recently discovered natural product that inhibits trypsin.

aeruginosin-B

Postulate one structural reason that aeruginosin-B binds to trypsin. Explain.

27.40   Eq. 27.53 shows the first step in the curved-arrow mechanism of peptide hydrolysis catalyzed by HIV protease. Complete the mechanism, using the two aspartic acid residues as catalytic groups.

27.41   The scientists who developed Norvir stated that one of the significant interactions of the inhibitor with the enzyme is hydrogen bonding of a thiazole nitrogen (the thiazole on the right side of the structure on p. 1438) with a backbone N—H of a nearby peptide bond on the enzyme. Draw a thiazole such as the one in Norvir and, using it, show such a hydrogen-bonding interaction.

27.42   Show all of the potential hydrogen-bonding sites of Darunavir; explain whether they are acceptor or donor sites.

KEY IDEAS IN CHAPTER 27

- Amino acids are compounds that contain both an amino group and a carboxylic acid group. Peptides and proteins are polymers of the α-amino acids.

- The common naturally occurring chiral amino acids have the L configuration, which is the same as the *S* configuration for all amino acids except cysteine.

- Amino acids, as well as peptides that contain both acidic and basic side-chain groups, exist at neutral pH as zwitterions.

- The charge on a peptide can be deduced from a knowledge of its isoelectric point relative to the pH of the solution. For example, a peptide with an isoelectric point $\gg 7$ is positively charged at pH 7.

- Common methods for the synthesis of α-amino acids include the alkylation of ammonia with α-halo acids, the alkylation of acetamidomalonate esters followed by hydrolysis and decarboxylation, and the Strecker synthesis.

- Amino acids react both as amines and carboxylic acids. Thus, the amino group can be acylated, and the carboxylic acid group can be esterified.

- The amide bonds of proteins and peptides can be hydrolyzed by heating them for several hours in 6 *N* HCl or 6 *N* NaOH solution. The constituent α-amino acids are formed. In amino acid analysis, the amino acids (or their derivatives) are separated by chromatography and quantified. In this way, the relative amount of each amino acid in the peptide or protein is determined.

- Peptide synthesis strategically involves the attachment of amino acids one at a time to a peptide chain beginning at the carboxy terminus. Amino-protecting groups such as the Fmoc group must be used to prevent competing reactions. In solid-phase peptide synthesis, an amino-protected amino acid is covalently attached to an insoluble resin; the protecting group is removed; an amino-protected amino acid is attached using DIC along with an agent used to form an active ester, such as HOBt; the new peptide is deprotected; and the cycle is continued. At the conclusion of the synthesis, the peptide is removed from the resin with trifluoroacetic acid.

- Proteins are synthetized biologically from the amino terminus. The genetic code from DNA is translated into messenger RNA (mRNA). Each three-base triplet in mRNA codes for one amino acid of a protein. Amino acids are introduced as aminoacyl-transfer RNA (tRNA) derivatives, which are carboxylate esters of the 3′-hydroxy group of a tRNA. Adjacent mRNA triplets dictate the binding of aminoacyl-tRNAs with complementary anticodons to adjacent sites on the ribosome, and the peptide bond is formed by an ester aminolysis reaction.

- Specific hydrolysis reactions catalyzed by enzymes are important in determining the primary structures of proteins. Peptides and proteins undergo trypsin-catalyzed hydrolysis at the carbonyl group of arginine and lysine residues. Chymotrypsin catalyzes the hydrolysis of peptides at hydrophobic residues such as Phe, Tyr, Trp, and occasionally Leu and Ile.

- The primary structure of a peptide or protein includes its amino-acid sequence and the location of its disulfide bonds. Peptides and proteins can be sequenced by MS–MS and by the Edman degradation.

- In MS–MS sequencing, the M + 1 ions of a peptide undergo fragmentation at the peptide bonds. In many cases, the sequence can be read directly from the mass differences of the fragment peaks.

- In the Edman degradation, a peptide is treated with phenyl isothiocyanate followed by acid. Each cycle of this degradation yields a phenylthiohydantoin (PTH) derivative of the amino-terminal amino acid plus a peptide that is one residue shorter, which is then subjected to the same procedure.

- Posttranslational modifications of proteins are chemical alterations that occur following its biosynthesis. Two examples are protein phosphorylation, which occurs at Ser, Tyr, or Cys residues; and protein glycosylation, which occurs at the amide nitrogen of Asn or the oxygen of Ser or Thr residues.

- The higher-order structure of proteins includes secondary, tertiary, and quaternary structure. The secondary structure of a protein is a description of the relative orientations of the planes of its peptide bonds. Common types of secondary structure are the α-helix and the β-sheet. Proteins are aggregates of localized regions of these primary structures connected by regions of random coil. The secondary structure of a protein can be conveyed with a ribbon structure.

- The tertiary structure of a protein is a complete description of its three-dimensional structure. The quaternary structure of a protein is the manner in which the subunits of the protein aggregate to form larger structures. In addition to the peptide and the disulfide bonds, noncovalent forces determine the three-dimensional structures of proteins. The noncovalent attractions within the protein are balanced against the noncovalent attractions of the protein with the solvent or surrounding milieu.

- The quaternary structure of a protein is the manner in which the subunits of the protein aggregate to form larger structures.

- Protein secondary, tertiary, and quaternary structures are disrupted when the protein is denatured by heat or by treatment with thiols, such as DTT, and 8 *M* aqueous urea.

- Enzymes are highly specific and efficient biological catalysts. A substrate is bound noncovalently at the enzyme active site before it is converted into products.

- A compound that binds tightly to an enzyme active site is called a competitive inhibitor. Competitive inhibitors block the access of substrates to the enzyme active site. The development of competitive inhibitors is an important basis of drug design. The anti-HIV drugs constitute one example of drugs based on the concept of competitive inhibition.

 REACTION REVIEW *For a summary of reactions discussed in this chapter, see the* Reaction Review *section of Chapter 27 in the* Study Guide and Solutions Manual.

ADDITIONAL PROBLEMS

27.43 Give the structures of the products expected when (1) valine and (2) proline (or other compounds indicated) react with each of the following reagents:

(a) ethanol (solvent), H_2SO_4 catalyst

(b) benzoyl chloride, Et_3N

(c) aqueous HCl solution

(d) aqueous NaOH solution

(e) benzaldehyde, heat, NaCN

(f) Fmoc-NHS ester, Na_2CO_3, aqueous 1,2-dimethoxyethane, then neutralize with H_3O^+

(g) product of part (f) + DIC/HOBt + glycine *tert*-butyl ester

(h) product of part (g) + anhydrous CF_3CO_2H

(i) product of part (h) + 20% piperidine in DMF

(j) product of part (i) + 6 *M* aqueous HCl, heat

27.44 Referring to Table 27.1, pp. 1376–1377, identify the amino acid(s) that satisfy each of the following criteria.

(a) the most acidic amino acid

(b) the most basic amino acid

(c) the amino acids that can exist as diastereomers

(d) the amino acid that has zero optical rotation under all conditions

(e) the amino acids that are converted into other amino acids on treatment with concentrated hot aqueous NaOH solution followed by neutralization.

27.45 In repeated attempts to synthesize the dipeptide Val-Leu, aspiring peptide chemist Polly Styreen performs each of the following operations. Explain what, if anything, is wrong with each procedure.

(a) The cesium salt of leucine is allowed to react with chloromethyl resin (as in Eq. 27.21, p. 1393). The

resulting derivative is treated with Fmoc-Val and DIC/HOBt, then with trifluoroacetic acid.

(b) The cesium salt of Fmoc-Leu is allowed to react with the chloromethyl resin. The resulting derivative is then treated with Fmoc-Val and DIC/HOBt, then with trifluoroacetic acid.

27.46 According to its amino acid composition (Fig. 27.6, p. 1410), lysozyme has an isoelectric point that is (choose one and explain):

(1) <<6   (2) about 6   (3) >>6

27.47 Consider the following two peptides, which may have been isolated from the floor of a Big Ten basketball arena circa 2000:

A: K-E-A-D-Y   *B:* K-N-I-G-H-T

(a) Which peptide is more basic (has the higher isoelectric point)? Explain.

(b) Which peptide will emerge first from an anion-exchange column at pH = 6.0?

(c) Which one of the following nucleotide sequences (3′-end on the left) could be the DNA sequence that codes for the biosynthesis of peptide *A*? Explain your reasoning.

Sequence 1: CGUGAAGCCGACUACUAA

Sequence 2: TTCCTTCGGCTGATA

Sequence 3: ATTATGGTGCGGGATGCA

Sequence 4: GCACUUCGGCUGAUGAUU

(d) Using abbreviated structures like those used in Fig. 27.4, give the structure of the aminoacyl-tRNA that codes for the first amino acid in the biosynthesis of peptide *A*. Include the sequence of residues in the anticodon region and the structure of the 3′ (acceptor) end.

27.48 Which of the following statements would correctly describe the isoelectric point of *cysteic acid*, an oxidation product of cysteine? Explain your answer.

$$H_3\overset{+}{N}-CH-\overset{O}{\overset{\|}{C}}-O^-$$
$$|$$
$$CH_2$$
$$|$$
$$O=S=O$$
$$|$$
$$O^-$$

cysteic acid

(1) lower than that of aspartic acid
(2) about the same as that of aspartic acid
(3) about the same as that of cysteine
(4) about the same as that of lysine
(5) higher than that of lysine

27.49 A peptide was subjected to one cycle of the Edman degradation, and the following compound was obtained. What is the amino-terminal residue of the peptide?

$$\begin{array}{c}Ph\\ \diagdown N-\overset{O}{\overset{\|}{C}}\\ | \quad\quad CH(CH_2)_4NH-\overset{S}{\overset{\|}{C}}-NH-Ph\\ S=\overset{C}{\underset{|}{}}\diagup N\\ \quad\quad H\end{array}$$

27.50 *Dansyl chloride* (5-dimethylamino-1-naphthalenesulfonyl chloride) reacts with amino groups to give a fluorescent derivative. After a peptide *P* with the composition (Arg,Asp,Gly,Leu$_2$,Thr,Val) reacts with dansyl chloride at pH 9, it is hydrolyzed in 6 *M* aqueous HCl. The derivative shown in the equation given in Fig. P27.50, detected by its fluorescence, is isolated after neutralization, along with the free amino acids Arg, Asp, Gly, Leu, and Thr. What conclusion can be drawn about the structure of the peptide from this result?

27.51 A peptide Q has the following composition by amino acid analysis:

$$Q: \quad Ala,Arg,Asp,Gly_2,Glu,Leu,Val_2,NH_3$$

Treatment of *Q* once with the Edman reagent followed by anhydrous acid gives a new peptide *R* with the following composition by amino acid analysis:

$$R: \quad Ala,Arg,Asp,Gly_2,Glu,Val_2,NH_3$$

Treatment of *Q* and *R* with the enzyme *dipeptidylaminopeptidase* (DPAP) yields a mixture of the following peptides:

$$Q \xrightarrow{DPAP} \text{Arg-Gly, Gln-Ala, Leu-Val, Val-Asp, Gly}$$

$$R \xrightarrow{DPAP} \text{Ala-Gly, Asp-Gln, Gly-Val, Val-Arg}$$

What is the amino acid sequence of *Q*?

27.52 The peptide hormone glucagon has the following amino acid sequence:

His-Ser-Gln-Gly-Thr-Phe-Thr-Ser-Asp-Tyr-Ser-Lys-Tyr-Leu-Asp-Ser-Arg-Arg-Ala-Gln-Asp-Phe-Val-Gln-Trp-Leu-Met-Asn-Thr

Give the products that would be obtained when this protein is treated with

(a) trypsin at pH 8
(b) Ph—N≡C≡S, then CF$_3$CO$_2$H, then aqueous acid

27.53 A peptide *C* was found to have a molecular mass of about 1000. Amino acid analysis of *C* revealed its composition to be (Ala$_2$,Arg,Gly,Ile). The peptide was unchanged on treatment with the Edman reagent, then CF$_3$CO$_2$H. Treatment of *C* with trypsin gave a single peptide *D* with an amino acid analysis identical to that of *C*. Three cycles of the Edman degradation applied to *D* revealed the partial sequence Ala-Ile-Gly.

(a) Suggest a structure for peptide *C* and explain how you arrived at that structure.

(b) Describe what you would expect to see for the b-type fragmentation of the M + 1 ion of both peptides *C* and *D* in MS–MS.

27.54 When bovine insulin is treated with the Edman reagent followed by anhydrous CF$_3$CO$_2$H, then by aqueous acid, the PTH derivatives of *both* glycine and phenylalanine are

---

dansyl chloride + peptide *P* $\xrightarrow[\text{pH 9}]{}$ $\xrightarrow[\text{2) neutralize}]{\text{1) 6 }M\text{ HCl, H}_2\text{O, heat}}$ (fluorescent) + Arg, Asp, Gly, Leu, Thr (free amino acids)

Figure P27.50

obtained in nearly equal amounts. What can be deduced about the structure of insulin from this information?

27.55 An amino acid *A*, isolated from the acid-catalyzed hydrolysis of a peptide antibiotic, gave a positive test with ninhydrin and had a specific optical rotation (HCl solution) of $+37.5°$ mL g^{-1} dm^{-1}. Compound *A* was not identical to any of the amino acids in Table 27.1, pp. 1376–1377. The isoelectric point of compound *A* was found to be 9.4. Compound *A* could be prepared by the reaction of L-glutamine with Br$_2$ in NaOH, followed by neutralization. (See Sec. 23.11D.) Suggest a structure for *A*.

27.56 A previously unknown amino acid, γ-carboxyglutamic acid (Gla), was discovered to be a posttranslational modification in the amino acid sequence of the blood-clotting protein prothrombin.

$$H_3\overset{+}{N}\!-\!CH\!-\!CO_2^-$$
$$|$$
$$CH_2$$
$$|$$
$$CH$$
$$^-O_2C\diagup\quad\diagdown CO_2^-$$

γ-carboxyglutamic acid (Gla)

This amino acid escaped detection for many years because, on acid hydrolysis, it is converted into another common amino acid. Explain.

27.57 (a) What reagent would be used to convert the corresponding chloromethyl polystyrene resin into the following resin?

$$-\!(CH_2\!-\!CH)_n\!-$$

$$CH_2\!-\!\overset{+}{N}(CH_3)_3 \quad Cl^-$$

(b) To a column containing this resin suspended in a pH 6 buffer is added a mixture of the amino acids Arg, Glu, and Leu, and the column is eluted with the same buffer. In what order will the amino acids emerge from the column? Explain.

27.58 In *paper electrophoresis*, amino acids and peptides can be separated by their differential migration in an electric field. To the center of a strip of paper is applied a mixture of the following three peptides in a single small spot: Gly-Lys, Gly-Asp, and Gly-Ala. The paper is soaked in a pH = 6 buffer, a positively charged electrode (anode) is attached to the left side of the paper, and a negatively charged electrode (cathode) is attached to the right side. A voltage is applied across the ends of the paper for a time, after which the peptides have separated into three spots: one near the cathode, one near the anode, and one in the center, at the location of the original spot. Which peptide is in each spot? Explain.

27.59 When a mixture of the amino acids Phe and Gly is subjected to chromatography in a pH 6 buffer on the ion-exchange resin shown in Eq. 27.9 on p. 1386, the Phe emerges from the column much later than the Gly, even though the two amino acids have the same isoelectric point. Explain.

27.60 Suppose a mixture of AQC-amino acids is subjected to HPLC on a stationary phase that consists of C8-silica rather than C18-silica; that is, the glass stationary phase (Eq. 27.35, p. 1406) contains covalently attached octyl groups rather than octadecyl groups. Assuming all other conditions are the same, how would this change affect the separation of the AQC-amino acids? Explain.

27.61 Explain each of the following observations.
   (a) The optical rotations of alanine are different in water, 1 *M* HCl, and 1 *M* NaOH.
   (b) Two mono-*N*-acetyl derivatives of lysine are known.
   (c) The peptide Gly-Ala-Arg-Ala-Glu is readily hydrolyzed by trypsin in water at pH = 8, but it is inert to trypsin in 8 *M* urea at the same pH.
   (d) After peptides containing cysteine are treated with HSCH$_2$CH$_2$OH, then with aziridine, they can be cleaved by trypsin at their (modified) cysteine residues.

$$\overset{\displaystyle NH}{\overset{\diagup\diagdown}{H_2C\!-\!CH_2}}$$

aziridine

   (e) When L-methionine is oxidized with H$_2$O$_2$, *two* separable methionine sulfoxides with the following structure are formed:

$$H_3\overset{+}{N}\!-\!CH\!-\!CO_2^-$$
$$|$$
$$CH_2CH_2\!-\!\overset{..}{\underset{\parallel}{S}}\!-\!CH_3$$
$$O$$

27.62 (a) When proteins are prepared for sequencing, they are treated with DTT (Eq. 27.38, p. 1409) and then with an excess of iodoacetic acid, I—CH$_2$—CO$_2$H, at pH = 8–9. Explain how iodoacetic acid reacts with the side-chain thiol group of a cysteine residue, and why this reaction is a necessary prelude to sequencing.
   (b) Another reaction that accomplishes the same objective is oxidation of the disulfide bonds with H$_2$O$_2$. What is the product of this oxidation?

27.63 One posttranslational side-chain modification of proteins is the methylation of aspartic acid residues to give a side-chain Asp-methyl ester. Draw the structure of this residue, and indicate what coenzyme is involved in this reaction. (*Hint:* See Sec. 11.7B.)

27.64 Sometimes preparations of chymotrypsin are contaminated with small amounts of trypsin. This can be a problem if the specific hydrolysis of peptides with *only* chymotrypsin is desired. How could trypsin-catalyzed hydrolysis with this chymotrypsin–trypsin mixture be avoided without having to separate the two enzymes?

27.65 Sometimes it is necessary in solid-phase peptide synthesis to use a resin linker that is more sensitive (that is, more reactive) to acid than the linker shown in Eq. 27.20 on p. 1393. The following group (*blue*) is one such linker. Explain why the peptide can be removed from this linker with much more dilute acid than is required for the linker in Eq. 27.20. (*Hint:* Consider the mechanism in Eq. 27.29, p. 1397.)

$$Pep^N—NH—CH—\overset{\overset{\displaystyle O}{\|}}{C}—O—CH_2$$

with OCH₃ and OCH₂C—O substituents on the aromatic ring

27.66 When either Norvir or Crixivan bind to the active site of HIV protease, the —OH group in the middle of each molecule is found by X-ray crystallography to displace the tightly bound water present in the free enzyme. (See Eq. 27.53, p. 1437.) Show how the two aspartic acid residues of the enzyme could interact with this hydroxy group in such a way that binding of the inhibitor is enhanced.

27.67 Poly-L-lysine (a peptide containing only lysine residues) exists entirely in an α-helical conformation at pH > 11. Below pH 10, however, the peptide becomes a random coil. Poly-L-glutamic acid, on the other hand, exists in the α-helical conformation at pH < 4, but above pH 5 it becomes a random coil. Explain the effect of pH on the secondary structure of both polymers. That is, explain why low pH destroys the helical conformation of one peptide while high pH destroys the helical conformation of the other. (*Hint:* Look carefully at the location of the amino acid side chains in Fig. 27.11, p. 1426.)

27.68 (a) For many years it was difficult to determine the X-ray structures of proteins that are imbedded in membranes because, when they are extracted into an aqueous buffer, they denature. Explain why this denaturation occurs.

(b) In some cases, this denaturation can be prevented by extraction of the protein from the membrane with detergents (Sec. 20.5). Explain.

27.69 Following is a ribbon diagram for one type of *opioid receptor*, a family of proteins that bind morphine and other opioids and initiate their physiological effects. The opioid receptor consists mostly of α-helices; the receptor spans the cell membrane. Parts of the helix (R-groups in the following diagram) are adjacent to the lipid bilayer of the membrane.

Which group of amino acid residues are likely to be found in greatest proportion near the phospholipid bilayer (choose one)? Explain why.

A: Asp and Glu   *B:* Lys, Arg, and His

C: Gly   *D:* Pro

E: Phe, Leu. Ile, and Val   *F:* Ser, Thr, Gln, and Asn

27.70 Complete the reactions given in Fig. P27.70 assuming the amino acid residue is part of a peptide in aqueous solution and is at neither the amino nor the carboxy terminus.

27.71 Outline a synthesis of each of the following compounds from the indicated starting material and any other reagents.

(a)
$$Ph—\overset{\overset{\displaystyle O}{\|}}{C}—NH—\text{⟨benzene ring⟩}—\overset{\overset{\displaystyle O}{\|}}{C}—OEt \quad \text{from } p\text{-aminobenzoic acid}$$

(b) $Ph—CD—CO_2^-$ from benzoic acid
with $^+NH_3$ substituent

(c) $CD_3—CH—CO_2^-$ from $CD_3—CH=O$
with $^+NH_3$ substituent

(d) L-Lys-L-Ala-L-Pro from L-proline, L-alanine, and the following compound using a solid-phase peptide synthesis. Use Fmoc protection for the α-amino groups in each coupling step but the last.

$$tBuO—\overset{\overset{\displaystyle O}{\|}}{C}—NH—CH—\overset{\overset{\displaystyle O}{\|}}{C}—OH$$

with side chain $(CH_2)_4$ then $NH—\overset{\overset{\displaystyle O}{\|}}{C}—OtBu$

a,ε-diBoc-L-lysine

(e)

$$EtO-\overset{\overset{\displaystyle O}{\|}}{C}-\underset{\underset{\displaystyle C\equiv N}{|}}{CH}\overset{\displaystyle \overset{O}{\|}}{\underset{\displaystyle}{NH-C-CH_3}}$$

from $EtO-\overset{\overset{\displaystyle O}{\|}}{C}-CH=O$

(f)

$$\underset{\underset{\displaystyle +NH_3}{|}}{CH}-\overset{\overset{\displaystyle O}{\|}}{C}-O^-$$

(cyclopentenyl ring attached)

from the product of part (e) and cyclopentene. (Cyclopentenyl amino acids are produced by certain plants. *Hint:* Hydrogens α to a cyano group are about as acidic as those α to an ester group.)

(g) The polymer *p*-aramid from terephthalic acid (1,4-benzenedicarboxylic acid). (This polymer is used in tire cord and other applications that require rigidity and strength.)

$$\left(HN-\!\!\!\bigcirc\!\!\!-NH-\overset{\overset{\displaystyle O}{\|}}{C}-\!\!\!\bigcirc\!\!\!-\overset{\overset{\displaystyle O}{\|}}{C}\right)_n$$

***p*-aramid**

27.72 Show how the acetamidomalonate method can be used to prepare the following unusual amino acids from the indicated starting material and any other reagents.

(a) $(CH_3)_2CDCH_2-\underset{\underset{\displaystyle +NH_3}{|}}{CH}-CO_2^-$ from isobutylene (2-methylpropene)

(b) $Ph-CHD-\underset{\underset{\displaystyle +NH_3}{|}}{CH}-CO_2^-$ from benzaldehyde

(c)

$$HO-\!\!\!\bigcirc\!\!\!-\overset{\overset{\displaystyle O}{\|}}{C}-CH_2-\underset{\underset{\displaystyle +NH_3}{|}}{CH}-CO_2^-$$ from anisole (methoxybenzene)

γ-oxohomotyrosine

27.73 When peptides containing a 2,3-diaminopropanoic acid (DAPA) residue are treated with the Edman reagent and then with acid, a peptide cleavage occurs in addition to degradation of the amino-terminal residue (see Fig. P27.73, p. 1446). Using the curved-arrow notation to rationalize your answer, propose a structure for *X*.

27.74 The artificial sweetener *aspartame* (sidebar, p. 1272) was withheld from the market for several years because, on storage for extended periods of time in aqueous solution, it

---

(a) lysine residue + $H_2C=O$ + $NaCNBH_3$ $\xrightarrow{\text{pH 9}}$
(excess of both)

(b) lysine residue + [succinic anhydride structure] $\xrightarrow{\text{pH 8}}$

succinic anhydride

(c) cysteine residue + [maleimide structure] $\xrightarrow{\text{pH > 5}}$ (a conjugate-addition product)

maleimide

(d) aspartic acid residue + [cyclohexyl]$-N=C=N-CH_2CH_2-\overset{+}{N}$[morpholine with H_3C] Cl^- + $H_2N-CH_2-CO_2CH_3$ $\longrightarrow$

a water-soluble carbodiimide that
reacts much like DIC

(e) tyrosine residue + $Ph\overset{+}{N}\equiv N$ Cl^- $\xrightarrow{\text{pH 9}}$

Figure P27.70

forms a *diketopiperazine* (see Fig. P27.74). (Extensive biological testing was required to show that this by-product was safe for consumers.) Give a curved-arrow mechanism for the formation of the diketopiperazine.

27.75 Complete the reactions given in Fig. P27.75 by giving the structure of the major organic product(s).

27.76 Identify each of the compounds *A–D* in the reaction scheme shown in Fig. P27.76. Explain your answers.

27.77 Draw a curved-arrow mechanism for each of the reactions given in Fig. P27.77.

a DAPA residue

Figure P27.73

aspartame

a diketopiperazine

Figure P27.74

(a) ethylamine + Ph—N=C=S ⟶

(b)

PhCH=O + KCN + CH_3NH_2 ⟶ $\xrightarrow[\text{heat}]{H_3O^+/H_2O}$ $\xrightarrow[\text{(neutralize)}]{\text{dil. NaOH}}$

(c)

$H_3\overset{+}{N}$—CH—CO_2^- $\xrightarrow[\text{(1 equiv.)}]{\text{NaOH}}$ $\xrightarrow[H_2O]{(CH_3)_2CH—CH=O}$ $\xrightarrow[H_2O]{NaCNBH_3}$
  |
  CH_3

(d)

$(CH_3)_3C$—O—$\overset{O}{\overset{\|}{C}}$—NH—CH—$\overset{O}{\overset{\|}{C}}$—$OCH_3$ + NaOH (1 equiv.) ⟶
        |
        CH_3

(e)

Ph—$\overset{O}{\overset{\|}{C}}$—NH—$NH_2$ + $NaNO_2$ $\xrightarrow{HCl}$

(f) product of part (e) $\xrightarrow{\text{heat in benzene}}$

(g) product of part (f) + valine methyl ester ⟶

(h)

H—$\overset{O}{\overset{\|}{C}}$—NH—$CH(CO_2Et)_2$ $\xrightarrow[\text{EtOH}]{\text{NaOEt}}$ $\xrightarrow{\underset{H_3C}{\overset{H_2C}{\diagup}}C—CH_2Cl}$ $\xrightarrow[\text{heat}]{H_2O, H_3O^+}$

(i)

N≡C—CH—$\overset{O}{\overset{\|}{C}}$—OEt + H_2N—NH_2 ⟶ $\xrightarrow[\text{EtOH, heat}]{NaNO_2, HCl}$ $\xrightarrow[\text{heat}]{HCl/H_2O}$
     |
     $CH(CH_3)_2$

Figure P27.75

27.78 When peptides containing the Asn-Gly sequence, such as *H* in the equation given in Fig. P27.78 are stored in aqueous solution at neutral or slightly basic solution, ammonia is liberated and a derivative *I* is formed. On continued storage, species *I* reacts to give two new peptides: *J* and *K*. Peptide *J* is the same as peptide *H* except that Asn is replaced by Asp, and peptide *K* is an isomer of peptide *J*. Propose structures for peptides *J* and *K*, and rationalize their formation using the curved-arrow notation. (These reactions are believed to be a major source of deterioration associated with aging in naturally occurring peptides and proteins.)

Figure P27.76

(a)

o-phthalaldehyde

fluorescent; used to detect amino acids

(b)

(c)

(d)

Figure P27.77

Figure P27.78

27.79 In 2007, scientists in New Zealand isolated a peptide *P* from enzymatic digests of the pili of gram-positive bacteria. (A *pilus* is a fibrous appendage that a bacterium uses, among other things, to bind to cells, beginning the process of infection.) This peptide was sequenced by MS–MS, and the following two partial sequences were reconstructed from the mass spectrum:

A: L-T-V-T-K-N-L   *B:* N-S-L

The mass M of *P*, deduced from the *m/z* of its M + 1 ion, equaled the mass of (*A* + *B*) – 17 mass units. Two other fragment ions, *C* and *D*, were also observed. The mass of *C* equaled the mass of (T-K-N-L + N-S-L) – 17 mass units, and the mass of *D* equaled the mass of (V-T-K-N-L + N-S-L) – 17 mass units.

(a) The scientists used these data to propose a primary structure for *P* that contains an unusual peptide bond. What is this structure? Show how it is consistent with the 17 mass-unit differences.

(b) The scientists also suggested that the unusual peptide bond forms spontaneously by a reaction between two peptide chains within the pilus. Show this reaction and its mechanism.

(c) No such reaction is observed when the peptides *A* and *B* are mixed at physiological pH, and, evidently, no enzyme is involved. Why might this reaction nevertheless proceed at a rapid rate within the pilus structure? (*Hint:* See Sec. 11.8.)

27.80 (a) In most peptides, the amide bonds have the *Z* conformation; explain why.

(b) One particular amino acid residue in the Pep^C position adopts the *E* conformation in some cases. Which amino acid residue should be most likely to assume an *E* conformation, and why?

27.81 (a) Explain why two monomethyl esters of *N*-acetyl-L-aspartic acid are known. Draw their structures.

(b) Explain why a mixture of these two compounds can be separated by cation-exchange chromatography at pH = 3.0, but not at pH = 7. (*Hint:* Use the pK_a values of aspartic acid in Table 27.1, pp. 1376–1377.) Your explanation should indicate which of the two compounds would emerge first from a cation-exchange column at pH = 3.0. Explain.

27.82 When *N*-acetyl-L-aspartic acid is treated with acetic anhydride, an optically active compound *A*, $C_6H_7NO_4$, is formed. Treatment of *A* with the amino acid L-alanine yields two separable, isomeric peptides, *B* and *C*, that are both converted into a mixture of L-alanine and L-aspartic acid by acid hydrolysis. Suggest structures for *A*, *B*, and *C*.

27.83 Lysozyme (Fig. 27.6, p. 1410) is an antibacterial enzyme that hydrolyzes polysaccharides in bacterial cell walls. It also catalyzes the hydrolysis of a β-1,4-linked hexasaccharide oligomer of *N*-acetylglucosamine into a tetrasaccharide and a disaccharide with *retention* of stereochemistry, as shown in Fig. P27.83. The active site of lysozyme runs through the enzyme at the junction of the two domains (Fig. 27.13, p. 1428). Near one end of the active site are two aspartic acid residues, Glu-35 and Asp-52, which are believed to be essential residues involved in the catalysis of hydrolysis. The pK_a values of the carboxylic acid groups in the side chains of these residues are about 6.0 and 3.5, respectively. The enzyme functions optimally at pH = 5.

(a) Draw a curved-arrow mechanism for the lysozyme-catalyzed oligosaccharide hydrolysis at pH = 5 shown in Fig. P27.83. Your mechanism should show the roles of the Glu-35 and Asp-52 carboxylic acid groups, and it should account for the stereochemistry of the reaction.

(b) Treatment of lysozyme with triethyloxonium tetrafluoroborate (pp. 536–537) results in a reaction of the carboxylic acid group of Asp-52 that completely eliminates enzyme activity. What is this reaction? In the light of the mechanism you proposed in (a), account for the effect of this reaction on enzyme activity.

Figure P27.83   Ac = acetyl = $H_3C-\overset{\displaystyle O}{\overset{\|}{C}}-$

Pericyclic Reactions

Pericyclic reactions occur by a *concerted cyclic shift of electrons*. This definition states two key elements of the reaction. First, a pericyclic reaction is *concerted*. In a *concerted reaction*, reactant bonds are broken and product bonds are formed at the same time, without intermediates. Second, a pericyclic reaction involves a *cyclic shift of electrons*. (The word *pericyclic* means "around the circle.") The Diels–Alder reaction (Sec. 15.3) and the S_N2 reaction (Sec. 9.4) are both concerted reactions, but only the Diels–Alder reaction occurs by a *cyclic electron shift*. Hence, the Diels–Alder reaction is a pericyclic reaction, but the S_N2 reaction is not.

This chapter is concerned with three major types of pericyclic reactions, although there are others. The first type is the **electrocyclic reaction**: an intramolecular reaction of an acyclic π-electron system in which a ring is formed with a new σ bond, and the product has one less π bond than the reactants.

$$\text{(28.1)}$$

one less π bond · a new σ bond closes a ring

The second type of reaction is the **cycloaddition**: a reaction of two separate π-electron systems in which a ring is formed with two new σ bonds, and the product has two fewer π bonds than the reactants.

$$\text{(28.2)}$$

separate π systems · new σ bonds close a ring · two fewer π bonds

The third type of reaction is the **sigmatropic reaction**: a reaction in which an allylic σ bond at one end of a π-electron system appears to migrate to the other end of the π-electron system. The π bonds change positions in the process, but their total number is unchanged.

$$(28.3)$$

$$(28.4)$$

Three features of any given type of pericyclic reaction are intimately related:

1. the way the reaction is activated (heat or light)
2. the number of electrons involved in the reaction
3. the stereochemistry of the reaction

Before illustrating these points, let's clarify the first two terms in this list. Point 1 refers to the fact that many pericyclic reactions require no catalysts or reagents other than the reacting partners. Such reactions take place either on heating or on irradiation with ultraviolet light. Reactions that are activated by heat are not activated by light, and vice versa. Recall, for example, that Diels–Alder reactions occur merely on heating the diene and dienophile together (Sec. 15.3). These reactions are not activated by light.

The number of electrons involved in a pericyclic reaction (point 2), as in any heterolytic process, is twice the number of curved arrows required to write the reaction mechanism in the curved-arrow notation. For example:

three curved arrows;
six electrons

$$(28.5)$$

The direction of "electron flow" in many pericyclic reactions indicated by the curved arrows is arbitrary. Although it is clockwise in Eq. 28.5, it could be written counterclockwise and be equally correct.

Specifying any two of the features in the foregoing list for a particular type of reaction specifies the third. To illustrate, consider the following electrocyclic reactions:

$$(28.6a)$$

$$(28.6b)$$

$$\text{(28.6c)}$$

First compare Eqs. 28.6a and 28.6b. Both are activated by heat; however, the former reaction, involving four electrons, gives only the trans-disubstituted isomer of the cyclic product, whereas the latter reaction, involving six electrons, gives only the cis-disubstituted isomer.

Next compare Eqs. 28.6b and 28.6c. Both reactions involve six electrons. When the starting material is heated, only the cis-disubstituted isomer of the cyclic product is obtained. When the starting material is irradiated with ultraviolet light, the only product obtained is the trans-disubstituted isomer.

Correlations such as these had been observed for many years, but the reasons for them were not understood. In 1965, a theory that clearly explained these observations and successfully predicted many new ones was proposed by Robert B. Woodward (1917–1979), then a professor of chemistry at Harvard University, and Roald Hoffmann (b. 1937), at the time a junior fellow at Harvard and presently Professor Emeritus at Cornell University. For this theory, called *conservation of orbital symmetry*, Hoffmann received the 1981 Nobel Prize in Chemistry. He shared the prize with Kenichi Fukui (1918–1998), a professor of chemistry at Kyoto University in Japan, who had advanced a related theory, called *frontier orbital theory*. (The two theories make the same predictions; they are alternative ways of looking at the same reactions.) Woodward undoubtedly would have also shared the Nobel Prize had he not died prior to its announcement. (The terms of Nobel's bequest require that the prize be awarded only to living scientists.) Woodward had, however, received an earlier Nobel Prize (1965) for his work in organic synthesis. This chapter presents elements of the Woodward–Hoffmann–Fukui theory that will enable you to understand and predict the outcome of pericyclic reactions.

PROBLEM

28.1 Classify each of the following pericyclic reactions as an electrocyclic, cycloaddition, or sigmatropic reaction. Give the curved-arrow notation for each reaction, and tell how many electrons are involved.

(a)

(b)

(c)

(d)

(e)

28.1   MOLECULAR ORBITALS OF CONJUGATED
π-ELECTRON SYSTEMS

Understanding the theory of pericyclic reactions requires an understanding of some basics of *molecular orbital theory*, particularly as it applies to molecules containing π electrons. Molecular orbital theory was introduced in Secs. 1.8, 4.1B, and 15.1A; these sections should be reviewed carefully.

A. Molecular Orbitals of Conjugated Alkenes

When *p* orbitals can overlap, pi (π) molecular orbitals can form. The overlap of *p* orbitals to give π molecular orbitals is described by the mathematics of quantum theory. However, the mathematical aspects of this theory are not required to appreciate the results. This section considers the molecular orbital theory of ethylene and conjugated alkenes. The π molecular orbitals for such molecules can be constructed according to the following generalizations, which are applied to ethylene and 1,3-butadiene in Figs. 28.1 and 28.2. In these and subsequent figures, the carbons are flattened into the plane of the page and the hydrogens are not shown so that the nodes and symmetry relationships within the orbitals can be seen clearly. (Perspective illustrations of these MOs can be found in Fig. 4.6, p. 129, for ethylene and in Fig. 15.1, p. 715, for 1,3-butadiene.) Using Figs. 28.1 and 28.2, convince yourself that each generalization applies to each example.

1. When a number (say *j*) of atomic *p* orbitals interact, the resulting π-electron system contains the same number *j* of molecular orbitals (MOs), all with different energies.

Because two 2*p* orbitals contribute to the π-electron system of ethylene, this molecule has the same number—two—of π MOs, which are designated as π_1 and π_2. Similarly, the four 2*p* orbitals of 1,3-butadiene combine to form four MOs, π_1, π_2, π_3, and π_4.

2. Half of the molecular orbitals have lower energies than the isolated *p* orbitals. These are called **bonding molecular orbitals**. The other half have higher energies than the isolated *p* orbitals. These are called **antibonding molecular orbitals**.

To emphasize this distinction, antibonding MOs are indicated with asterisks. Thus, ethylene has one bonding MO (π_1) and one antibonding MO (π_2^*); 1,3-butadiene has two bonding MOs (π_1 and π_2) and two antibonding MOs (π_3^* and π_4^*).

FIGURE 28.1 The π molecular orbitals (MOs) of ethylene. The carbons are shown as black dots, and the hydrogens are not shown. Wave peaks are shown in blue and wave troughs in green. The symmetry classification is described in Fig. 28.3 and the associated discussion. The node in π_2^*, shown with a heavy gray line, is perpendicular to the plane of the page. The nodal plane of the 2p orbitals is common to all π MOs. Only the nodes in addition to this one ("new nodes") are shown in this and subsequent figures.

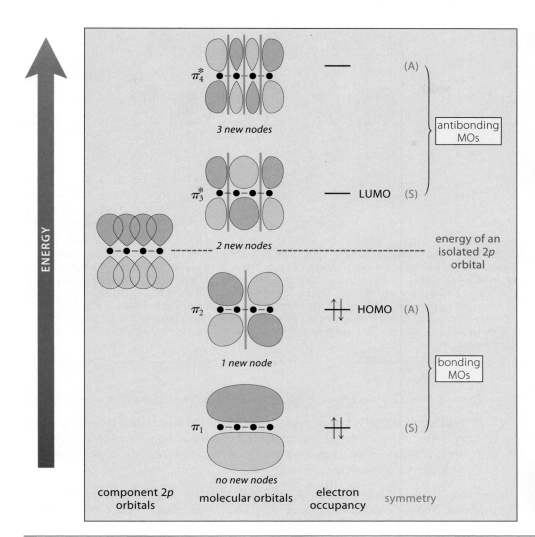

FIGURE 28.2 The π molecular orbitals of 1,3-butadiene. The carbons are flattened into the plane of the page and the hydrogens are not shown so that the nodes and symmetry relationships within the orbitals can be seen clearly. The symmetry classification is described in Fig. 28.3 on p. 1454 and the associated discussion.

3. The bonding molecular orbital of lowest energy, π_1, has no new nodes. (It does retain a node in the plane of the molecule, which is common to all $2p$ orbitals and to all π MOs). Each MO of increasingly higher energy has one additional node.

Recall from Sec. 1.6B that a *node* is a surface, in this case a plane, at which an electron wave (orbital) is zero; that is, when an electron is in a given MO, there is zero *probability of finding the electron*, or zero *electron density*, at the node. A particularly important feature of the node for understanding pericyclic reactions is that the electron wave has a *peak* on one side of a node (*blue*) and a *trough* on the other side (*green*). Because of the symmetry relationships within the MOs (discussed in point 5, below), it doesn't matter whether a particular lobe of an MO is treated as a peak or a trough, *provided that a peak changes to a trough and a trough changes to a peak each time a node is crossed*.

Thus, π_1 of ethylene has no new nodes, and π_2^* has one new node. In 1,3-butadiene, π_1 has no new nodes, π_2 has one new node, π_3^* has two, and π_4^* has three.

4. The nodes occur *between* atoms and are arranged symmetrically with respect to the center of the π-electron system.

The node in π_2^* of ethylene is between the two carbon atoms, in the center of the π system. The node in π_2 of 1,3-butadiene is also symmetrically placed in the center of the π system. The two nodes in π_3^* are placed between carbons 1 and 2, and between carbons 3 and 4, respectively—equidistant from the center of the π system. Each of the three nodes in π_4^*, the MO of highest energy, must occur between carbon atoms.

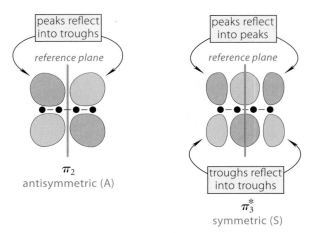

FIGURE 28.3 The antisymmetric and symmetric MO symmetry classifications illustrated for the π_2 and the π_3^* molecular orbitals of 1,3-butadiene. These classifications refer to the symmetry of an MO with respect to a reference plane (*gray*) through the center of the molecule and perpendicular to the plane of the page. (This plane is not the same thing as a node, although it may coincide with a node, as in the π_2 MO.) The symmetry classifications of the MOs in Figs. 28.1 and 28.2 are indicated by the abbreviations (A) for *antisymmetric* and (S) for *symmetric*. Although the molecules are flattened, the same symmetry relationships apply to their s-cis and s-trans conformations.

The next generalization relates to the *symmetry* of the molecular orbitals.

5. Odd-numbered MOs (π_1, π_3, π_5, . . .) are symmetric with respect to an imaginary *reference plane* at the center of the π-electron system and perpendicular to the plane of the molecule. Even-numbered MOs (π_2, π_4, π_6, . . .) are antisymmetric with respect to this plane.

The *reference plane* in this generalization is shown for the 1,3-butadiene molecule in Fig. 28.3. A **symmetric MO** is an MO in which peaks reflect across the reference plane into peaks and troughs reflect into troughs, as shown for π_3^* of 1,3-butadiene in Fig. 28.3. An **antisymmetric MO** is an MO in which peaks reflect into troughs, as shown for π_2 of 1,3-butadiene in Fig. 28.3. Of particular importance for the analysis of pericyclic reactions is the *relative phase* of each MO at its *terminal carbons*. The **relative phase** of an MO refers to the relative orientation of peaks and troughs at two different points. Within any symmetric MO, such as π_3^* of 1,3-butadiene, the relative phase at the two terminal carbons is *the same*. That is, the peaks on the two carbons are on the same side of the carbon chain (the upper side in Fig. 28.3), and the two troughs are on the same side. Within any antisymmetric MO, such as π_2 of 1,3-butadiene, the relative phase at the two terminal carbons is *different*. Be sure to verify for yourself that the other MOs in Figs. 28.1 and 28.2 fit this pattern.

The last generalization deals with the distribution of the available π electrons within the MOs.

6. Electrons are placed pairwise into each molecular orbital, beginning with the orbital of lowest energy (aufbau principle).

This point is illustrated in Figs. 28.1 and 28.2 in the column labeled "electron occupancy." An alkene has the same number of π electrons as it has $2p$ orbitals. Thus, ethylene, with two $2p$ atomic orbitals, has two π electrons. These are both placed (with opposite spin) into π_1 (Fig. 28.1). 1,3-Butadiene, with four $2p$ atomic orbitals, has four π electrons. Two are placed in π_1 and two in π_2 (Fig. 28.2). These examples show that the bonding MOs are fully filled in both simple and conjugated alkenes and that the antibonding MOs are empty.

The presence of unconjugated substituents (for example, alkyl groups), to a useful approximation, does not alter the MO nodal properties of a conjugated alkene. For example, the π molecular orbital nodes in 1,3-butadiene and 1,3-pentadiene are essentially the same.

Recall from Sec. 15.1B that the π-electron contribution to the energy of a molecule is determined by the energies of its *occupied* MOs. Because bonding MOs have lower energies than isolated $2p$ orbitals, there is an energetic advantage to π molecular orbital formation; this is why π bonds exist.

Two MOs are of particular importance in understanding pericyclic reactions. One is the occupied molecular orbital of highest energy, called the **highest occupied molecular orbital (HOMO)**. The other is the unoccupied molecular orbital of lowest energy, called the **lowest unoccupied molecular orbital (LUMO)**. These are labeled in Figs. 28.1 and 28.2. In ethylene, π_1 is the HOMO and π_2^* the LUMO; in 1,3-butadiene, π_2 is the HOMO and π_3^* the LUMO. *The HOMO and LUMO of a conjugated alkene have opposite symmetries.* Also, *the HOMO has lower energy than the LUMO.*

The HOMO and LUMO are sometimes collectively referred to as the **frontier orbitals** because they are the molecular orbitals at the energy extremes: the HOMO is the occupied molecular orbital of *highest* energy, and the LUMO is the unoccupied molecular orbital of *lowest* energy. *The analysis of pericyclic reactions focuses heavily on the symmetries of frontier orbitals.*

PROBLEMS

28.2 Answer the following questions for 1,3,5-hexatriene, the conjugated triene containing six carbons.

(a) How many π MOs are there?

(b) Classify each MO as symmetric or antisymmetric. (See Fig. 28.3.)

(c) Which MOs are bonding? Which are antibonding?

(d) Which MOs are the frontier molecular orbitals?

(e) Within the HOMO, is the phase at the terminal carbons the same or different?

(f) Within the LUMO, is the phase at the terminal carbons the same or different?

28.3 Without drawing the MOs, state whether the π molecular orbital π_6 in 1,3,5,7,9-decapentaene (a 10-carbon conjugated alkene) is symmetric or antisymmetric with respect to the reference plane; is bonding or antibonding; is a frontier MO; and, if so, is a HOMO or a LUMO.

B. Molecular Orbitals of Conjugated Ions and Radicals

Conjugated unbranched ions and radicals have an odd number of carbon atoms. For example, the allyl cation has three carbon atoms and three $2p$ orbitals, and hence, three MOs.

$$\left[H_2C\!=\!CH\!-\!\overset{+}{C}H_2 \quad \longleftrightarrow \quad H_2\overset{+}{C}\!-\!CH\!=\!CH_2 \right]$$

allyl cation

The MOs of such species follow many but not all of the same patterns as those of conjugated alkenes. The MOs for the allyl and 2,4-pentadienyl systems are shown in Figs. 28.4 and 28.5, respectively, on pp. 1456 and 1457. (A perspective illustration of the allyl MOs is given in Fig. 15.14, p. 744.) These figures show two important differences between these MOs and those of conjugated alkenes. First, in each case, one MO is neither bonding nor antibonding but has the same energy as the isolated $2p$ orbitals; this MO is called a **nonbonding molecular orbital**. The nonbonding MO in the allyl system is π_2. The remaining orbitals are either bonding or antibonding, and there are an equal number of each type. Second, in some of the MOs, nodes pass through carbon atoms. For example, in the allyl system, there is a node on the central carbon of π_2. This means that electrons in π_2 have no electron density on

FIGURE 28.4 The π molecular orbitals of the allyl system. The nodal properties of the MOs are the same for the cation, the radical, and the anion. (The electron occupancy does affect the orbital energies somewhat, but this effect can be ignored.) In a radical, the molecular orbital containing the unpaired electron is called the SOMO (singly occupied molecular orbital). The designation of the HOMO and the LUMO depends on the electron occupancy.

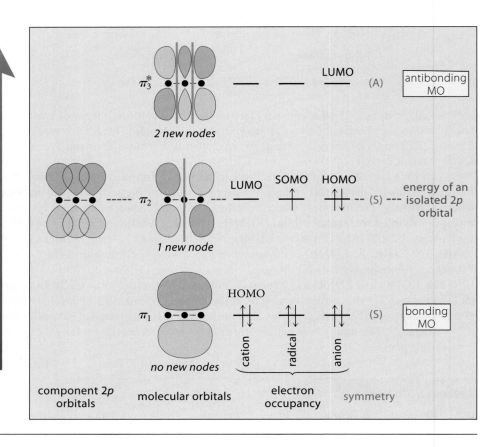

the central carbon. This is why, for example, the charge in the allyl anion resides only on the terminal carbons:

allyl anion

Just as the charge in an atomic anion is associated with an excess of valence electrons, the charge in a conjugated carbanion can be associated with the electrons in its HOMO. (See the discussion on p. 745, and Problem 15.25, p. 745.)

The MOs of cations, radicals, and anions involving the same π system have the same nodal properties. For example, the MOs of the allyl system apply equally well to the allyl cation, allyl radical, and allyl anion because all three species contain the same arrangement of $2p$ orbitals. These species differ only in the *number* of π electrons, as shown in the "electron occupancy" column of Figs. 28.4 and 28.5. (The relative energies of the MOs differ somewhat, but we can ignore these differences.) Thus, the HOMO of the allyl cation is π_1 and the LUMO is π_2^*. In contrast, the HOMO of the allyl anion is π_2, and the LUMO is π_3^*. The HOMO of the allyl radical contains a single electron. Because this molecular orbital is "half-occupied," it is sometimes referred to as a **singly occupied molecular orbital (SOMO)**.

PROBLEMS

28.4 Answer the following questions for the 2,4,6-heptatrienyl cation.

$$H_2C=CH-CH=CH-CH=CH-\overset{+}{C}H_2$$

2,4,6-heptatrienyl cation

(a) Which MO is nonbonding?

(b) Classify each MO as symmetric or antisymmetric.

(c) To which carbon atoms in this cation is the positive charge delocalized? Explain with both resonance structures and molecular orbital arguments.

28.5   Explain using (a) resonance arguments and (b) molecular orbital arguments why the unpaired electron in the allyl radical is delocalized to carbon-1 and carbon-3 but not to carbon-2.

C. Excited States

The molecules and ions we have been discussing can absorb energy from light of certain wavelengths. This process, which is also responsible for the UV spectra of these species (Sec. 15.2B), is shown schematically in Fig. 28.6 on p. 1458 for 1,3-butadiene. The normal electronic configuration of any molecule is called the **ground state**. Energy from absorbed

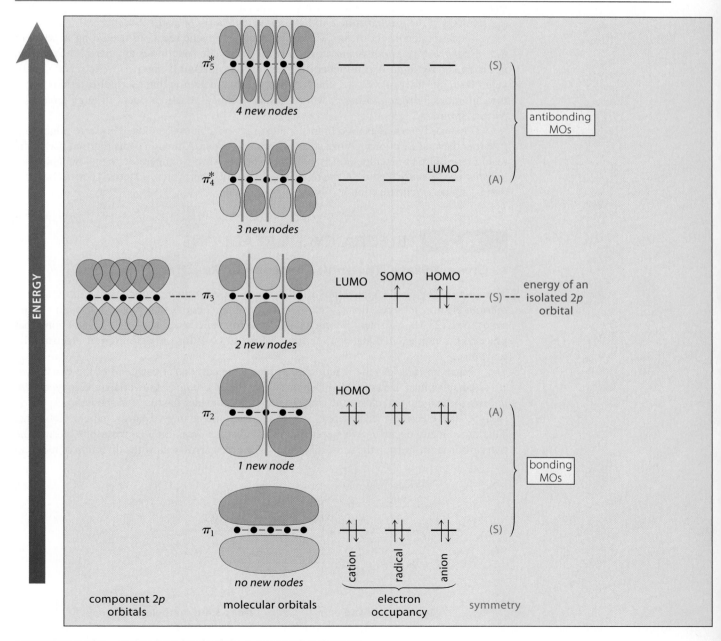

FIGURE 28.5 The π molecular orbitals of the 2,4-pentadienyl system.

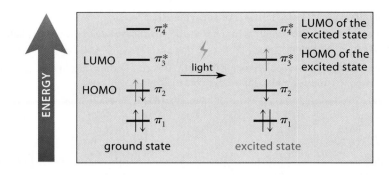

FIGURE 28.6 Light absorption by a conjugated species such as (in this example) 1,3-butadiene promotes an electron from the HOMO to the LUMO and produces an excited state.

light is used to promote an electron from the HOMO of ground-state 1,3-butadiene (π_2) into the LUMO (π_3^*). A species with a promoted electron is called an **excited state**. In the excited state of 1,3-butadiene, π_3^* is the HOMO, even though it contains only one electron. Notice that *the HOMOs of the ground state and the excited state have opposite symmetries.*

In subsequent sections, we will differentiate pericyclic reactions according to whether they are *thermal* or *photochemical*. A **thermal pericyclic reaction** is any pericyclic reaction *not* activated by light. A **photochemical pericyclic reaction** is any pericyclic reaction activated by light. As you will see, the fundamental distinction is that thermal reactions occur through molecular *ground states*, whereas photochemical reactions occur through molecular *excited states*.

The word *thermal* as used in this context may be a bit misleading. This term might suggest that thermal reactions require strong heating to occur. Although some thermal pericyclic reactions require high temperatures, others can occur at room temperature or below. The word *thermal* in this sense means "not activated by light." In other words, a thermal pericyclic reaction is any pericyclic reaction involving a ground state.

28.2 ELECTROCYCLIC REACTIONS

A. Ground-State (Thermal) Electrocyclic Reactions

This section begins the application of MO theory to pericyclic reactions with a discussion of *thermal* electrocyclic reactions—that is, electrocyclic reactions that proceed through ground states. Sec. 28.2B considers *photochemical* electrocyclic reactions—that is, reactions that proceed through excited states. (See Eq. 28.1 on p. 1449 for the definition of electrocyclic reactions.)

When an electrocyclic reaction takes place, the carbons at each end of the conjugated π system must turn in a concerted fashion so that the 2*p* orbitals can overlap (and rehybridize) to form the σ bond that closes the ring. To illustrate, consider the reaction shown in Eq. 28.6a (p. 1450), the electrocyclic closure of (2*E*,4*E*)-2,4-hexadiene to give 3,4-dimethylcyclobutene. This turning can occur in two stereochemically distinct ways. In a **conrotatory** closure the two carbon atoms turn in the same direction. (The green arrows show the direction of motion.)

(28.8)

(2E,4E)-2,4-hexadiene   ***trans*-3,4-dimethylcyclobutene**

a conrotatory reaction (observed)

the carbons rotate in the same direction

(Of the two conrotatory modes, clockwise and counterclockwise, the clockwise mode is shown, but the counterclockwise mode in this case is equally probable.) In the second mode of ring closure, called a **disrotatory** mode, the carbon atoms turn in opposite directions.

(2E,4E)-2,4-hexadiene   **cis-3,4-dimethylcyclobutene**

a disrotatory reaction (does not occur)   (28.9)

The two modes of ring closure can be distinguished by the stereochemistry of the product. As noted in Eq. 28.8, *trans*-, not *cis*-3,4-dimethylcyclobutene is the observed product. Hence, *the mode of ring closure is conrotatory.*

Molecular orbital theory explains this result. A simple way to look at the reaction is to focus on the *HOMO of the diene.* This molecular orbital contains the π electrons of highest energy. *These π electrons are to a molecule as valence electrons are to an atom.* Just as the atomic valence electrons are involved in most chemical reactions, the electrons in the HOMO govern the course of pericyclic reactions. When the ring closure takes place, the two $2p$ orbitals on the ends of the π system must overlap. But simple overlap is not enough: they must overlap *in phase.* That is, the wave peak on one carbon must overlap with the wave peak on the other, or a wave trough must overlap with a wave trough. If a peak were to overlap with a trough, the electron waves would cancel and no bond would form.

Let's see what it takes to provide the required bonding overlap. First, the diene must assume the *s*-cis conformation; only in this conformation are the terminal carbons of the π-electron system close enough to each other that their $2p$ orbitals can overlap. (The *s*-cis and *s*-trans conformations of dienes are discussed in Sec. 15.1B.) Next, recall from Sec. 28.1A that alkyl substituents, to a useful approximation, do not affect the nodal properties of the MOs of a conjugated alkene. Consequently, the nodes of the π molecular orbitals of 2,4-hexadiene are the same as those of 1,3-butadiene (Fig. 28.2). In other words, the methyl groups at each end of the molecule can be largely ignored when considering the MOs of the system. An examination of the HOMO of a conjugated diene (π_2 in Fig. 28.2) reveals that, because of the antisymmetric nature of π_2, conrotatory ring closure is required for in-phase, or bonding, overlap, as observed:

π_2 (HOMO)   bonding overlap   (28.10)

In contrast, disrotatory ring closure gives out-of-phase overlap, an antibonding (and hence unstable) situation:

π_2 (HOMO)   antibonding overlap   (28.11)

Thus, it is the relative orbital phase at the terminal carbon atoms of the HOMO—the *orbital symmetry*—that determines whether the reaction is conrotatory or disrotatory. This observation suggests that *all* conjugated polyenes with *antisymmetric* HOMOs should undergo conrotatory ring closure, and indeed, such is the case. The electrocyclic reactions of conjugated alkenes with symmetric HOMOs can be predicted by a similar analysis, as Study Problem 28.1 illustrates.

STUDY PROBLEM   28.1

Predict the stereochemistry of the thermal electrocyclic ring closure of (2*E*,4*Z*,6*E*)-2,4,6-octatriene to 5,6-dimethyl-1,3-cyclohexadiene.

(2*E*,4*Z*,6*E*)-2,4,6-octatriene   **5,6-dimethyl-1,3-cyclohexadiene**

SOLUTION   First, ignore the substituent groups and examine the HOMO of the simpler triene, 1,3,5-hexatriene (Problem 28.2). Because the HOMO of this triene (π_3) is *symmetric*, the HOMO has the *same phase* at each end of the π system. Hence, bonding overlap can occur only if the ring closure is *disrotatory*.

$$\text{(28.12)}$$

π_3 (HOMO)   *cis*-5,6-dimethyl-1,3-cyclohexadiene

The disrotatory motion, as Eq. 28.12 shows, requires that the methyl groups have a cis relationship in the product. As Eq. 28.6b shows, this is indeed the observed stereochemistry of the reaction.

To summarize: Electrocyclic closure of a conjugated diene is conrotatory, and that of a conjugated triene is disrotatory. The reason for the difference is the phase relationships within the HOMO at the terminal carbons of these π systems. In the diene, the HOMO has opposite phase at these two carbons; in the triene, the HOMO has the same phase. A different type of rotation is thus required in each case for bonding overlap.

This result can be generalized. Conjugated alkenes with $4n$ π electrons (n = any integer) have antisymmetric HOMOs and undergo conrotatory ring closure; those with $4n + 2$ π electrons have symmetric HOMOs and undergo disrotatory ring closure. That is, conrotatory ring closure is *allowed* for systems with $4n$ π electrons; it is *forbidden* for systems with $4n + 2$ π electrons. Conversely, disrotatory ring closure is *allowed* for systems with $4n + 2$ π electrons; it is *forbidden* for systems with $4n$ π electrons.

B. Excited-State (Photochemical) Electrocyclic Reactions

When a molecule absorbs light, it reacts through its *excited state* (Sec. 28.1C). The HOMO of the excited state is different from the HOMO of the ground state, and has different

TABLE 28.1 Selection Rules for Electrocyclic Reactions

Number of electrons*	Mode of activation	Allowed stereochemistry
$4n$	thermal	conrotatory
	photochemical	disrotatory
$4n + 2$	thermal	disrotatory
	photochemical	conrotatory

* n = an integer

symmetry. For example, as Eq. 28.6c (p. 1451) shows, the *photochemical* ring closure of (2E,4Z,6E)-2,4,6-octatriene is *conrotatory*. This is understandable in terms of the symmetry of π_4^*, the HOMO of the excited state.

$$\pi_4^* \text{(HOMO of the excited state)}$$

trans-5,6-dimethyl-1,3-cyclohexadiene

(28.13)

Contrast the stereochemistry of the product with that observed in the ground-state reaction of the same triene in Eq. 28.12. *The stereochemical result is different because the symmetry of the HOMO is different.*

To generalize this result: the mode of ring closure in *photochemical* electrocyclic reactions—reactions that occur through electronically excited states—differs from that of thermal electrocyclic reactions, which occur through electronic ground states. These results can be summarized with a series of *selection rules* for electrocyclic reactions, given in Table 28.1.

C. Selection Rules and Microscopic Reversibility

The selection rules in Table 28.1 are based on the orbital symmetry of the open-chain (conjugated alkene) reactant. However, these rules (as well as others to be considered) refer to the *rates* of pericyclic reactions but have nothing to say about the *positions of the equilibria* involved. Thus, the electrocyclic reaction of the diene in Eq. 28.6a on p. 1450 to give a cyclobutene favors the diene at equilibrium because of the strain in the cyclobutene, but the electrocyclic reaction of the conjugated triene in Eq. 28.6b favors the cyclic compound because σ bonds are stronger than π bonds, and because six-membered rings are relatively stable.

It is also common for a photochemical reaction to favor the less stable isomer of an equilibrium because the energy of light is harnessed to drive the equilibrium energetically "uphill." For example, in the following reaction, the conjugated alkene absorbs UV light, but the bicyclic compound does not; hence, the photochemical reaction favors the latter.

$$\xrightarrow[\text{light}]{\text{disrotatory}}$$

(28.14)

(42% yield)

In summary, the selection rules do not indicate which component of an equilibrium will be favored—only whether the equilibrium will be established at a reasonable rate.

The *principle of microscopic reversibility* (p. 175) assures us that selection rules apply equally well to the forward and reverse of any pericyclic reaction, because the reaction in both directions must proceed through the same transition state. Hence, an electrocyclic ring *opening* must follow the same selection rules as its reverse, an electrocyclic ring *closure*. Thus, the thermal ring-opening reaction of the cyclobutene in Eq. 28.15, like the reverse ring-closure reaction, must be a conrotatory process (Table 28.1).

$$\xrightarrow[\text{heat}]{\textit{conrotatory}}$$

(28.15)

In the following electrocyclic ring-opening reaction, the allowed thermal conrotatory process would give a highly strained molecule containing a trans double bond within a small ring.

$$\xrightarrow{\textit{conrotatory}}$$

a trans double bond

(28.16)

Although the selection rules suggest that the reaction could occur, it does not because of the strain in the product (Sec. 7.6C). In other words, "allowed" reactions are sometimes prevented from occurring for reasons having nothing to do with the selection rules. Furthermore, concerted ring opening to the relatively unstrained all-cis diene also does not occur, because this would be a disrotatory process—a process "forbidden" by the selection rules in Table 28.1 for a concerted 4*n*-electron reaction.

$$\xrightarrow[\text{forbidden}]{\textit{disrotatory}}$$

(28.17)

Hence, the bicyclic compound is effectively "trapped into existence"; that is, there is no concerted thermal pathway by which it can reopen. Recall from Eq. 28.14 (p. 1461) that it is formed *photochemically* from the all-cis diene.

Two rather spectacular examples of the effect of the selection rules for pericyclic reactions are the benzene isomers *prismane* (p. 759) and *Dewar benzene*:

prismane   Dewar benzene

Despite the tremendous amount of strain in both molecules and the aromatic stability of benzene, neither prismane nor Dewar benzene is readily transformed into benzene because, in each case, such a transformation violates pericyclic selection rules (see Problem 28.54, p. 1485). Because no low-energy pathway exists for its conversion into the much more stable isomer benzene, prismane has been referred to in the chemical literature as a "caged tiger."

PROBLEMS

28.6   Which one of the following electrocyclic reactions should occur readily by a concerted mechanism?

Reaction 1:

Reaction 2:

28.7   Show both conrotatory processes for the thermal electrocyclic conversion of (2E,4E)-2,4-hexadiene into 3,4-dimethylcyclobutene (Eq. 28.8). Explain why the two processes are equally likely.

28.8   In the thermal ring opening of *trans*-3,4-dimethylcyclobutene, *two* products could be formed by a conrotatory mechanism, but only one is observed. Give the two possible products. Which one is observed and why?

28.9   After heating to 200 °C, the following compound is converted in 95% yield into an isomer A that can be hydrogenated to cyclodecane. Give the structure of A, including its stereochemistry.

28.3   CYCLOADDITION REACTIONS

A *cycloaddition reaction* (Eq. 28.2, p. 1449) is classified, first, by the number of electrons involved in the reaction. The reaction in Eq. 28.18a is a [4 + 2] cycloaddition because the reaction involves four electrons from one reacting component and two electrons from the other. The reaction in Eq. 28.18b is a [2 + 2] cycloaddition.

$$[4 + 2] \tag{28.18a}$$

$$[2 + 2] \tag{28.18b}$$

As in electrocyclic reactions, the number of electrons involved is determined by writing the reaction mechanism in the curved-arrow notation. The number of electrons contributed by a given reactant is equal to twice the number of curved arrows originating from that component (two electrons per arrow).

   A cycloaddition reaction is also classified by its stereochemistry with respect to *the plane of each reacting molecule.* (Recall that the carbons and their attached atoms in π-electron systems are coplanar.) This classification is shown for a [4 + 2] cycloaddition in Fig. 28.7 on p. 1464. A cycloaddition may in principle occur either across the same face, or across opposite faces, of the planes in each reacting component. If the reaction occurs across the same face of a π system, the reaction is said to be **suprafacial** with respect to that π system. A suprafacial addition is the same thing as a syn-addition (Sec. 7.8A). If the reaction occurs at opposite faces

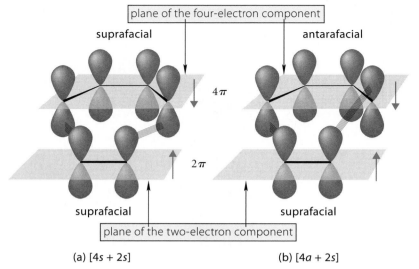

(a) [4s + 2s]   (b) [4a + 2s]

FIGURE 28.7 Classification of cycloaddition reactions, illustrated for (a) a [4s + 2s] cycloaddition and (b) a [4a + 2s] cycloaddition. The s and a designations refer to the stereochemistry of the cycloaddition (suprafacial or antarafacial) with respect to the planes of the reacting components. The two components approach each other in parallel planes (*green arrows*). For example, in part (b), the addition (*blue lines*) occurs below the plane of the 4π-electron component at one end and above the plane at the other and is therefore classified as 4a, or antarafacial on the 4π-electron component. The addition occurs on the 2π-electron component on the same side of its plane at both ends and is therefore classified as 2s, or suprafacial on the 2π-electron component. This reaction is therefore a [4a + 2s] cycloaddition.

**STUDY GUIDE
LINK 28.1**
Frontier Orbitals

of a π system, it is said to be **antarafacial**. An antarafacial addition is an anti-addition. Thus, a [4s + 2s] cycloaddition is one that occurs suprafacially (or syn) on both the 4π component and the 2π component. A [4a + 2s] cycloaddition occurs antarafacially (or anti) on the 4π component, but suprafacially (or syn) on the 2π component.

For a cycloaddition to occur, bonding overlap must take place between the $2p$ orbitals at the terminal carbons of each π-electron system, because these are the carbons at which new bonds are formed. This bonding overlap begins when the HOMO of one component interacts with the LUMO of the other. The electrons in the HOMO of one component are analogous to the valence electrons in an atom: They are the reacting electrons. The LUMO of the other component is the empty orbital of lowest energy into which the electrons from the HOMO must flow. It doesn't matter whether we consider the HOMO from the 4π-electron component and the LUMO from the 2π-electron component, or vice versa. The important point is that the two frontier MOs involved in the interaction must have *matching phases* if bonding overlap is to be achieved.

This phase match is achieved when a [4 + 2] cycloaddition occurs *suprafacially* on each component—that is, when the cycloaddition is a [4s + 2s] process. (A [4a + 2a] process is also theoretically allowed but is geometrically impossible.) Using the HOMO from the 4π-electron component and the LUMO from the 2π-electron component, this overlap can be represented as follows:

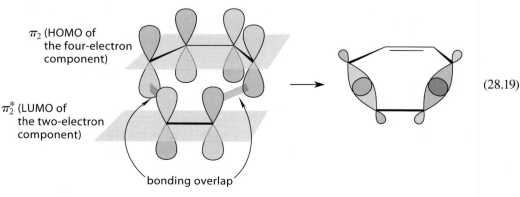

π_2 (HOMO of the four-electron component)

π_2^* (LUMO of the two-electron component)

bonding overlap

(28.19)

Recall from Sec. 15.3C that the Diels–Alder reaction, the most important example of a $[4s + 2s]$ cycloaddition, occurs suprafacially on each component. This can be seen from the retention of stereochemistry observed in both the diene and dienophile in the following example:

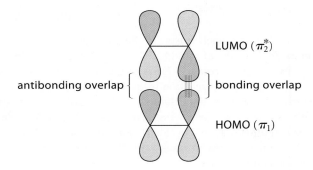

(28.20)

You should also convince yourself that the $[4s + 2a]$ and $[4a + 2s]$ modes of cycloaddition do *not* provide bonding overlap at both ends of the π-electron systems. You should also convince yourself that it doesn't matter which component provides the HOMO and which provides the LUMO (Problem 28.10, p. 1466).

The situation is different in a $[2 + 2]$ cycloaddition. Again, we use the HOMO of one component and the LUMO of the other. Notice that the orbital symmetries do *not* accommodate a cycloaddition that is suprafacial on both components:

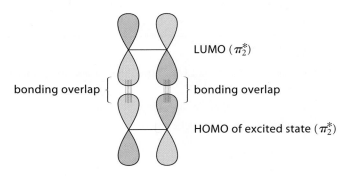

However, an addition that is suprafacial on one component but antarafacial on the other is allowed by orbital symmetry but is geometrically impossible. For this reason, the thermal $[2 + 2]$ cycloaddition is a much less common reaction than the Diels–Alder reaction. All of the known thermal $[2 + 2]$ additions occur by nonconcerted mechanisms and therefore do not fall under the purview of the rules for pericyclic reactions.

Although the $[2s + 2s]$ cycloaddition is forbidden by orbital symmetry under *thermal* conditions, it is allowed under *photochemical* conditions. Under these conditions, the *excited state* of one alkene reacts with the other alkene, which is not photochemically excited. (Only a small fraction of the alkenes exist in an excited state under photochemical conditions.) The HOMO of the excited state has the proper symmetry to interact in a bonding way with the LUMO of the reacting partner:

Indeed, many examples of photochemical $[2s + 2s]$ cycloadditions are known. Such processes are widely used for making cyclobutanes.

TABLE 28.2 Selection Rules for Cycloaddition Reactions

Number of electrons*	Mode of activation	Allowed stereochemistry†
4n	thermal	supra–antara antara–supra
	photochemical	supra–supra antara–antara
4n + 2	thermal	supra–supra antara–antara
	photochemical	supra–antara antara–supra

* n = an integer
† supra = suprafacial; antara = antarafacial

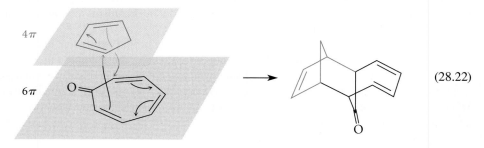

$$(28.21)$$

(excess)

(95% yield)

Notice the retention of alkene stereochemistry in this reaction.

An important example of the allowed [2s + 2s] cycloaddition in biology is the photochemical reaction of two thymine bases in DNA to give a thymine dimer. (This reaction is shown in Eq. 26.64 on p. 1363.) This reaction is known to cause certain types of cancer, and is probably a contributor to the onset of melanoma, a particularly virulent type of skin cancer.

The results of this section can be generalized to the cycloaddition selection rules given in Table 28.2. Notice that all-suprafacial cycloadditions are allowed thermally for systems in which the total number of reacting electrons is $4n + 2$, and they are allowed photochemically for systems in which the number is $4n$.

The following reaction is an example of a suprafacial cycloaddition involving more than six π electrons. This all-suprafacial cycloaddition is a [6s + 4s] process involving 10 electrons (five curved arrows). Notice that $4n + 2$ equals 10 for $n = 2$.

4π

6π

$$(28.22)$$

PROBLEMS

28.10 Show that using the HOMO from the 2π-electron component and the LUMO from the 4π-electron component also gives bonding overlap in a [4s + 2s] cycloaddition.

28.11 Show by a frontier orbital analysis that the [4a + 2s] and [4s + 2a] modes of cycloaddition are not allowed.

28.12 Give the product of the following reaction, which involves an [8s + 2s] cycloaddition:

$$+ \ EtO_2C—C{\equiv}C—CO_2Et \longrightarrow$$

28.13   The photochemical cycloaddition of two molecules of *cis*-2-butene gives a mixture of two products: *A* and *B*. The analogous photochemical cycloaddition of *trans*-2-butene also gives a mixture of two products: *B* and *C*. The photochemical reaction of a *mixture* of *cis*- and *trans*-2-butene gives a mixture of *A*, *B*, and *C*, along with a fourth product, *D*. Propose structures for all four compounds.

28.4   THERMAL SIGMATROPIC REACTIONS

A. Classification and Stereochemistry

Sigmatropic reactions were defined in Eqs. 28.3 and 28.4 on p. 1450. In this section, we consider only thermal sigmatropic reactions—sigmatropic reactions that occur through ground states—because photochemical sigmatropic reactions are not very common.

Sigmatropic reactions are classified by using bracketed numbers to indicate the number of atoms over which a σ bond appears to migrate. In some reactions, both ends of a σ bond migrate. In the following reaction, for example, each end of a σ bond migrates over three atoms. (Count the point of original attachment as atom #1.) The following reaction is therefore a [3,3] sigmatropic reaction.

transition state

(28.23)

In other reactions, one end of a σ bond remains fixed to the same group and the other end migrates. For example, the following reaction is a [1,5] sigmatropic reaction because one end of the bond "moves" from atom #1 to atom #1 (that is, it doesn't move), and the other end moves over five atoms.

transition state

(28.24)

Sigmatropic reactions, like other pericyclic reactions, are also classified by their stereochemistry. This classification is based on whether the migrating bond moves over the same face, or between opposite faces, of the π-electron system. If the migrating bond moves across one face of the π system, the reaction is said to be *suprafacial*. For example, if the [1,5] sigmatropic reaction of Eq. 28.24 were suprafacial, it would occur in the following manner:

suprafacial

(28.25)

If the reaction were *antarafacial*, it would occur instead as shown in Eq. 28.26:

antarafacial

(28.26)

When both ends of a σ bond migrate, the reaction can be suprafacial or antarafacial with respect to either π system. For example, if the [3,3] sigmatropic reaction in Eq. 28.23 were suprafacial on both π systems, it could occur as follows:

suprafacial

suprafacial

(28.27)

The stereochemistry of a sigmatropic reaction is revealed experimentally only if the molecules involved have stereocenters at the appropriate carbons. This point is illustrated in Study Problem 28.2.

STUDY PROBLEM   28.2

Classify the following sigmatropic reaction by giving its bracketed-number designation and its stereochemistry with respect to the plane of the π-electron system.

SOLUTION First identify the bond that is migrating. Because the hydrogen atom migrates, one end of the migrating bond remains fixed; in other words, this is a [1,?] sigmatropic reaction in which we have to determine "?" by counting the carbons over which the migration takes place:

The original point of attachment is counted as carbon-1. Consequently, this is a [1,7] sigmatropic reaction. To determine the stereochemistry, imagine that carbons 1 through 7 all lie in the same plane. As the molecule is depicted, the migrating hydrogen is *below* that plane (*dashed wedge*). Because rotation about the C6–C7 bond cannot occur until after the reaction is over (it is a double bond), depict the T and D in the same relative orientations in the starting material and product (as they are depicted in the problem). This reveals that the hydrogen is *above* the plane of the π-electron system in the product. Consequently, the hydrogen has migrated from the lower to the upper face of the π-electron system. Therefore, the reaction is a [1,7] *antarafacial* sigmatropic reaction.

PROBLEMS

28.14 (a) Refer to Study Problem 28.2 and, assuming an antarafacial migration, give the structure of a starting material that would give a stereoisomer of the product with the *R* configuration at the isotopically substituted carbon.

(b) Give the structure of *another* starting material that would give the same stereoisomer as in part (a).

28.15 Classify the following sigmatropic reactions with bracketed numbers.

Orbital symmetry governs the connection between the type of sigmatropic reaction and its stereochemistry. Consider, for example, a [1,5] sigmatropic migration of hydrogen across a π-electron system. Think of this reaction as the migration of a proton from one end of the 2,4-pentadienyl anion (Fig. 28.5) to the other:

For H₂C⟩   CH₂ think of the transition state as: [H₂C:⁻   CH₂ ⟷ H₂C   ⁻:CH₂]‡

The interaction of the proton LUMO—an empty 1*s* orbital—with the HOMO of the π system controls the stereochemistry of the reaction. For the 2,4-pentadienyl anion (Fig. 28.5), the HOMO is symmetric. This means that bonding overlap can occur only if the migration occurs suprafacially:

STUDY GUIDE LINK 28.2
Orbital Analysis of Sigmatropic Reactions

HOMO (π₃) of the 2,4-pentadienyl anion

H⁺

suprafacial hydrogen migration

An ingenious experiment published in 1970 by Wolfgang Roth and his collaborators at the University of Cologne revealed the stereochemistry of the thermal [1,5] sigmatropic hydrogen shift. In the isotopically labeled, optically active alkene shown in Eq. 28.28, a suprafacial [1,5] hydrogen shift is possible from each of the conformations shown:

(28.28)

It follows from these equations that if the migration is suprafacial, the 3*E* isomer of the product must have the *R* configuration and the 3*Z* isomer of the product must have the *S* configuration. These were exactly the stereochemical results observed.

The [1,3] hydrogen shift involves the HOMO of the allyl anion, an antisymmetric orbital:

HOMO (π_2) of the allyl anion

antarafacial hydrogen migration

For a [1,3] hydrogen shift to occur, the migrating hydrogen must pass from one face of the allyl π system to the other. Despite the fact that this reaction is "allowed" by orbital symmetry, it requires that the migrating proton bridge too great a distance for adequate bonding. Alternatively, the terminal lobes of the allyl π system could twist; but then a new problem would arise: these lobes would not overlap with the $2p$ orbital of the central carbon. The resulting loss of orbital overlap would raise the energy of the transition state. As these arguments suggest, the concerted thermal sigmatropic [1,3] hydrogen shift is nonexistent in organic chemistry. (See Problem 28.16, p. 1473.)

The interconversion of enols and their isomeric carbonyl compounds are [1,3] hydrogen shifts, which are disallowed as a concerted process by orbital symmetry.

an enol   ⇌   the isomeric carbonyl compound

Yet, from Sec. 22.2, enols are rapidly converted into their corresponding carbonyl compounds. This is not a violation of orbital symmetry because all of the reactions by which enols and carbonyl compounds are interconverted involve *nonconcerted* pathways. (See, for example,

Eqs. 22.17a and 22.17b, p. 1112.) In fact, chemists have succeeded in preparing enols in the absence of catalysts that promote their conversion into carbonyl compounds, and these enols have proven to be quite stable thermally, as the foregoing discussion of [1,3] hydrogen shifts suggests.

Several interesting experiments have been conducted in which a molecule could in principle undergo both [1,5] and [1,3] hydrogen shifts. In one particularly elegant experiment carried out in 1964, also by Roth, 1,3,5-cyclooctatriene was labeled at carbons 7 and 8 with deuterium and then allowed to undergo many hydrogen shifts for a long time. When the molecule undergoes [1,5] hydrogen shifts, the D should migrate part of the time, and the H part of the time. However, after a long time, the D should eventually scramble to all positions that have a 1,5-relationship. In such a case, only carbons 3, 4, 7, and 8 would be partially deuterated:

(28.29a)

(You should write a series of steps for this transformation to convince yourself that it is the predicted result.) On the other hand, if the molecule undergoes successive [1,3] hydrogen (or deuterium) shifts, the deuterium should be scrambled eventually to all positions.

(28.29b)

The experimental result was that, even after very long reaction times, deuterium appeared only in the positions predicted by the [1,5] shift.

Although the suprafacial [1,3] shift of a hydrogen is *not* allowed, the corresponding shift of a carbon atom *is* allowed, provided that a stringent stereochemical condition is met. Suppose that an alkyl group (suitably substituted so that its stereochemistry can be traced) were to undergo a suprafacial [1,3] sigmatropic shift. This shift could occur in two stereochemically distinct ways. In the first way, the carbon migrates with *retention* of configuration.

(28.30a)

(In this and the following equation, one carbon of the allyl group is marked with an asterisk so that its fate can be traced.) In the second way, the carbon migrates with *inversion* of configuration.

(28.30b)

Consider the orbital symmetry relationships in these two modes of reaction. Think of the migrating group as an alkyl cation migrating between the ends of an allyl anion. The

LUMO of the alkyl cation—an empty $2p$ orbital—interacts with the HOMO of an allyl anion (π_2 in Fig. 28.4, p. 1456). In the case of migration with *retention*, the phase relationships between the orbitals involved lead to antibonding overlap:

Retention:

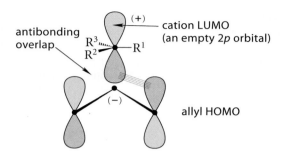

Hence, suprafacial carbon migration with retention is forbidden by orbital symmetry, in the same sense that hydrogen migration is forbidden. If migration occurs with *inversion*, however, bonding overlap can occur in the transition state.

Inversion:

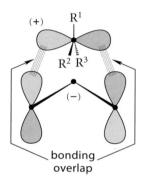

Thus, carbon migration with *inversion* is allowed by orbital symmetry.

This analysis shows that it is the node in the $2p$ orbital of the migrating carbon that makes the [1,3] suprafacial migration of this carbon possible; each of the two lobes of the $2p$ orbital, which have opposite phase, can overlap with each end of the allyl π system. Because a bond is broken at one side of the migrating carbon and formed at the other side, inversion of configuration is observed. In the migration of a hydrogen, the orbital involved is a $1s$ orbital, which has no nodes. Hence, [1,3] suprafacial migration of hydrogen is not allowed.

Orbital symmetry, then, makes a very straightforward prediction: The suprafacial [1,3] sigmatropic migration of carbon must occur with inversion of configuration. The following result confirms this prediction:

$$\text{(28.31)}$$

(See Problem 28.39, p. 1482, for another example.) Migration with retention of configuration might have been expected to be the most straightforward, least contorted pathway that the rearrangement could take; yet the theory of orbital symmetry predicts otherwise. One of the remarkable things about the theory is that it correctly predicts so many reactions that otherwise would have appeared unlikely.

As might be expected, orbital symmetry dictates the opposite stereochemistry for [1,5] migrations. Carbon, like hydrogen, undergoes suprafacial [1,5] migrations with retention of configuration (Problem 28.18).

PROBLEMS

28.16 Explain why the hydrogen migration shown in reaction (1) occurs readily and why the very similar migration shown in (2) does *not* take place even under forcing conditions. (The asterisked carbons indicate a carbon isotope present so that the rearrangement can be detected.)

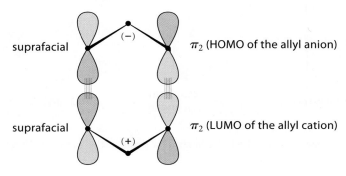

28.17 Predict the result that would have been expected in the experiment described by Eq. 28.28 (p. 1470) for an antarafacial migration.

28.18 (a) Carry out an orbital symmetry analysis to show that suprafacial [1,5] carbon migrations should occur with retention of configuration in the migrating group.

   (b) Indicate what type of sigmatropic reactions are involved in the following transformations. Is the stereochemistry of the first step in accord with the predictions of orbital symmetry?

B. Thermal [3,3] Sigmatropic Reactions

Let's now examine a sigmatropic reaction in which both ends of a σ bond change positions. One of the most common and useful examples of this type of reaction is the [3,3] sigmatropic rearrangement. Using the same logic as before, the transition state for this rearrangement can be visualized as the interaction of two allylic systems, one a cation and one an anion.

For ⬡ think of the transition state as: [⬡ ⟷ ⬡]‡

The frontier orbitals involved are the HOMO of the anion and the LUMO of the cation, which are the same orbital (π_2) of the allyl system (Fig. 28.4, p. 1456).

suprafacial π_2 (HOMO of the allyl anion)

suprafacial π_2 (LUMO of the allyl cation)

The two MOs involved achieve bonding overlap when the [3,3] sigmatropic rearrangement occurs suprafacially on both components. (You should convince yourself that a reaction that is antarafacial on both π systems is also allowed by orbital symmetry, but one that is suprafacial on one component and antarafacial on the other is forbidden.)

One of the best known types of [3,3] sigmatropic rearrangement is the *Cope rearrangement*, which was extensively investigated by Professor Arthur C. Cope (1909–1966) of the Massachusetts Institute of Technology long before the principles of orbital symmetry were known. The **Cope rearrangement** is simply a 1,5-diene isomerization:

$$\text{(28.32)}$$

An interesting variation of the Cope rearrangement is the "anionic oxyCope" reaction. In this reaction, the alkoxide ion derived from an alcohol undergoes a thermal [3,3] sigmatropic reaction. The alkoxide is generated with potassium hydride. The alkoxide is then heated in boiling THF. An enolate ion is formed initially; when it reacts with acid, the corresponding ketone is formed.

$$\text{(28.33)}$$

An oxyCope reaction can also take place without alkoxide formation, but the reaction then requires very strong heating. Alkoxide formation accelerates the reaction by a factor of at least 10^{10}. The reaction of the alkoxide calls to mind the breakdown of a tetrahedral addition intermediate in a carbonyl addition reaction. The extra unshared electron pair on oxygen helps initiate the anionic oxyCope reaction by "pushing out" the bond to the neighboring carbon.

In the **Claisen rearrangement**, an ether that is both allylic and vinylic or an allylic aryl ether undergoes a [3,3] sigmatropic rearrangement. (The asterisk shows the fate of a carbon atom.)

$$\text{(28.34)}$$

allyl phenyl ether   keto form of a phenol   **2-allylphenol**

If both ortho positions are blocked by substituent groups, the para-substituted derivative is obtained:

(28.35)

(95% yield)

This reaction occurs by a sequence of two Claisen rearrangements, followed by isomerization of the product to the phenol:

the same compound redrawn

(28.36a)

(28.36b)

Problem 28.20 considers the Claisen rearrangement of an aliphatic ether.

C. Summary: Selection Rules for Thermal Sigmatropic Reactions

The stereochemistry of sigmatropic reactions is a function of the number of electrons involved. (As with other pericyclic reactions, the number of electrons involved is determined from the curved-arrow notation: Count the curved arrows and multiply by 2.) All-suprafacial thermal sigmatropic reactions occur when $4n + 2$ electrons are involved in the reaction—that is, an odd number of electron pairs or curved arrows. In contrast, a thermal sigmatropic reaction must be antarafacial on one component and suprafacial on the other when $4n$ electrons (an even number of electron pairs or curved arrows) are involved. When a single carbon migrates, the term *suprafacial* is taken to mean "retention of configuration," and the term *antarafacial* is taken to mean "inversion of configuration."

These generalizations are summarized in Table 28.3 as selection rules for thermal sigmatropic reactions.

TABLE 28.3 Selection Rules for Thermal Sigmatropic Reactions

Number of electrons[†]	Allowed stereochemistry*	
	Generalized stereochemistry	Stereochemistry of single-atom migrations
$4n$	supra–antara	supra–inversion
	antara–supra	antara–retention
$4n + 2$	supra–supra	supra–retention
	antara–antara	antara–inversion

* supra = suprafacial; antara = antarafacial
[†] n = an integer

PROBLEMS

28.19 (a) What allowed and reasonable sigmatropic reaction(s) can account for the following transformation?

(b) What product(s) are expected from a similar reaction of 2,3-dimethyl-1,3-cyclopentadiene?

28.20 (a) Aliphatic allylic vinylic ethers undergo the Claisen rearrangement. Complete the following reaction:

$$(CH_3)_2C = CH - CH_2 - O - CH = CH_2 \xrightarrow{\text{heat}}$$

(b) What starting material would give the following compound in an aliphatic Claisen rearrangement?

28.21 Show how the transition state for a [3,3] sigmatropic reaction can be analyzed as the interaction of two allylic radicals, and that the same stereochemical outcome is predicted. (See Study Guide Link 28.2.)

28.22 (a) Give the curved-arrow mechanism for the anionic oxyCope reaction of compound *A*, and explain why the stereoisomer *B* does not react under the same conditions.

(b) Give the structure of the product, including its stereochemistry, expected from the anion oxyCope reaction of the following compound.

28.23 Show by an orbital symmetry analysis that a [3,3] sigmatropic reaction that is antarafacial on both components is allowed. Would you expect such a reaction to be very common? Why?

28.5   **FLUXIONAL MOLECULES**

A number of compounds continually undergo rapid sigmatropic rearrangements at room temperature. One such compound is *bullvalene*, which was first prepared in 1963.

bullvalene

The [3,3] sigmatropic rearrangements in bullvalene rapidly interconvert *identical forms of the molecule*:

If the carbons could be individually labeled, there would be 1,209,600 equivalent structures of bullvalene in equilibrium! Each one of these forms is converted into another at a rate of about 2000 times per second at room temperature. (These reactions can be observed by the dynamic NMR methods discussed in Sec. 13.8.) At high temperature (above 100 °C), both the proton NMR spectrum and the ^{13}C NMR spectrum of bullvalene consist of a single line; the rapid fluxional isomerization averages the NMR absorptions so that all of the protons, as well as all of the carbons, become equivalent over time.

Molecules such as bullvalene that undergo rapid bond shifts are called **fluxional molecules**. Their atoms are in a continual state of motion associated with the rapid changes in bonding. Fluxional molecules are *not* resonance structures because the nuclei actually move during their interconversion.

"The Bull" and Bullvalene

The fluxional nature of bullvalene was predicted by William von E. Doering (1917–2011) during his tenure as a professor at Yale University. The intriguing name of this compound seems to have originated from a seminar of his research group in late 1961 in which Doering first disclosed his prediction. To understand the origin of the name, one must know that Doering's research students had nicknamed him "The Bull," and also that Doering had been interested in a class of compounds known as the fulvalenes. When Doering wrote the structure on the board, one graduate student in the back of the room blurted out, "Bullvalene!" The name stuck. Bullvalene was first prepared serendipitously in 1963 by Gerhard Schröder, a chemist at Union Carbide Corporation in Brussels.

PROBLEM

28.24 Each of the following compounds exists as a fluxional molecule that is interconverted into one or more identical forms by the sigmatropic process indicated. Draw one structure in each case that demonstrates the process involved, and explain why each process is an allowed pericyclic reaction.

(a) [3,3]

(b) [1,2]

28.6 BIOLOGICAL PERICYCLIC REACTIONS. THE FORMATION OF VITAMIN D

It has long been known that, in areas of the world where winters are long and there is little sunlight, children suffer from a disease called *rickets* (from Old English, *wrickken*, meaning "to twist"). This disease is characterized by inadequate calcification of the bones. A similar disease in adults, *osteomalacia*, is particularly prominent among Bedouin Arab women who must remain completely covered when they are outdoors.

Rickets can be prevented by administration of any one of the forms of vitamin D, a hormone that controls calcium deposition in bone. The human body manufactures a chemical precursor to vitamin D called 7-dehydrocholesterol (structure in Eq. 28.38). This is converted into vitamin D_3, or *cholecalciferol* (structure in Eq. 28.39), only when the skin receives adequate ultraviolet radiation from the Sun or other source.

The reaction by which 7-dehydrocholesterol is converted into vitamin D_3 is a sequence of two pericyclic reactions. The first is a photochemical electrocyclic reaction:

7-dehydrocholesterol   **previtamin D₃** (28.38)

$$R = \overset{H_3C}{\underset{H}{\diagdown}} C \diagdown CH_2CH_2CH_2CH(CH_3)_2$$

This reaction is a *conrotatory* process. (Be sure you understand why this is so; examine the reverse reaction if necessary.) This is precisely the stereochemistry required for a *photochemically allowed* electrocyclic reaction involving $4n + 2$ electrons. Sunlight ordinarily provides the UV radiation necessary for this reaction to occur in humans.

The final step in the formation of vitamin D_3 is a thermal [1,7] sigmatropic hydrogen shift:

$$\frac{[1,7]}{37\,°C}$$

cholecalciferol (vitamin D₃) (28.39)

(The stereochemistry of this process is not defined by the structures of the reactant and product; see Problem 28.28.) Vitamin D_3 exists in the more stable *s*-trans conformation, attained by internal rotation (*green arrow*) about the bond shown in *red*:

s-cis   180° internal rotation   *s*-trans (28.40)

Vitamin D_2 (ergocalciferol or calciferol), a compound closely related to vitamin D_3, is produced commercially by irradiation of a steroid called *ergosterol*. Ergosterol is identical to 7-dehydrocholesterol (Eq. 28.38) except for the side-chain R:

7-dehydrocholesterol

$$R = \quad H_3C_{\cdots\cdots}C\!\!-\!\!H \quad CH_2CH_2CH_2CH(CH_3)_2$$

$$R = \quad ergosterol$$

ergosterol

Irradiation of ergosterol gives successively previtamin D_2 and vitamin D_2, which are identical to the products of Eqs. 28.38 and 28.39, respectively, except for the R-group. Vitamin D_2, sometimes called "irradiated ergosterol," is the form of vitamin D that is commonly added to milk and other foods as a dietary supplement.

PROBLEMS

28.25 When previtamin D_2 (which is identical to previtamin D_3, p. 1478, except for the R-group) is isolated and irradiated, ergosterol is obtained along with a stereoisomer, *lumisterol*. Explain mechanistically the origin of lumisterol.

lumisterol

28.26 When previtamin D_2 is *heated*, two compounds, *A* and *B*, are obtained that are stereoisomers of both ergosterol and lumisterol. Suggest structures for these compounds, and explain mechanistically how they are formed.

28.27 When the compounds *A* and *B* in the previous problem are *irradiated*, two stereoisomeric compounds, *C* and *D*, respectively, are obtained, each of which contains a cyclobutene ring. Suggest structures for *C* and *D*, and explain mechanistically how they are formed. Explain why irradiation of either *A* or *B* does *not* give back previtamin D_2.

28.28 Although the stereochemistry of Eq. 28.39 cannot be determined from the reaction, what stereochemistry is expected from orbital symmetry considerations?

KEY IDEAS IN CHAPTER 28

- Pericyclic reactions are concerted reactions that occur by cyclic electron shifts. Electrocyclic reactions, cycloadditions, and sigmatropic rearrangements are important pericyclic reactions.

- Electrocyclic reactions are stereochemically classified as conrotatory or disrotatory; cycloadditions and sigmatropic rearrangements are classified as suprafacial or antarafacial.

- The stereochemical course of a pericyclic reaction is governed largely by the symmetry of the reactant HOMO (highest occupied molecular orbital), or, if there are two reacting components, by the relative symmetries of the HOMO of one component and the LUMO (lowest unoccupied molecular orbital) of the other.

- Considerations of orbital symmetry lead to selection rules for pericyclic reactions. Whether a pericyclic reaction is allowed or forbidden depends on the number of electrons involved, the mode of activation (thermal or photochemical), and the stereochemical course of

the reaction. The selection rules are summarized in Tables 28.1–28.3.

 REACTION REVIEW *For a summary of reactions discussed in this chapter, see the* Reaction Review *section of Chapter 28 in the* Study Guide and Solutions Manual.

ADDITIONAL PROBLEMS

28.29 Without consulting tables or figures, answer the following questions:

(a) Is a thermal disrotatory electrocyclic reaction involving 12 electrons allowed or forbidden?

(b) Is a [8s + 4s] photochemical cycloaddition allowed?

(c) Is the HOMO of (Z)-3,4-dimethyl-1,3,5-hexatriene symmetric or antisymmetric?

28.30 What do the pericyclic selection rules have to say about the *position of equilibrium* in each of the reactions given in Fig. P28.30? Which side of each equilibrium is favored and why?

28.31 (a) Predict the stereochemistry of compounds *B* and *C* in Fig. P28.31.

Figure P28.30

Figure P28.31

(b) What stereoisomer of *A* also gives compound *C* on heating?

28.32 (a) Classify the following pericyclic reaction. (More than one classification is possible.)

(80% yield)

(b) Suppose the migrating methyl group in part (a) were labeled with the hydrogen isotopes deuterium (D) and tritium (T) so that it is a —CHDT group with the *S* configuration. What would be the configuration of this group in the product? Explain your reasoning.

28.33 Complete the following reactions by giving the major organic product(s), including stereochemistry.

(a)

$$Ph—C{\equiv}C—Ph + \underset{H_3C}{\overset{H_3C}{}}C{=}C\underset{CH_3}{\overset{CH_3}{}} \xrightarrow{\text{light}} (C_{20}H_{22})$$

(b)

(c)

(d)

28.34 When compound *A* is irradiated with ultraviolet light for 115 hours in pentane, an isomeric compound *B* is obtained that decolorizes bromine in CH_2Cl_2 and reacts with ozone to give, after the usual workup, a compound *C*.

(a) Give the structure of *B* and the stereochemistry of both *B* and *C*.

(b) On heating to 90 °C, compound *D*, a stereoisomer of *B*, is converted into *A*, but compound *B* is virtually inert under the same conditions. Identify compound *D* and account for these observations.

28.35 When 1,3-cyclopentadiene and *p*-benzoquinone are allowed to react at room temperature, a compound *X* is obtained. Irradiation of compound *X* gives compound *Y* (see Fig. P28.35). Give the structure and stereochemistry of *X* and explain its conversion into *Y*.

28.36 *Heptafulvalene* undergoes a thermal reaction with tetracyanoethylene (TCNE) to give the adduct shown in Fig. P28.36. What is the stereochemistry of this adduct? Explain.

1,3-cyclopentadiene   *p*-benzoquinone   *Y*

Figure P28.35

heptafulvalene   **TCNE**

Figure P28.36

28.37 (a) Explain why the following equilibrium lies far to the right.

A   **toluene** *(B)*

(b) Chemists had always assumed that this reaction would be so fast that compound *A* could never be isolated. However, this compound was prepared in 1962 and shown to be stable in the gas phase at 70 °C, despite the favorable equilibrium constant for its transformation to *B*. Show why the conversion of *A* into *B* above would *not* be expected to occur as a concerted reaction.

(c) Would you expect a concerted mechanism for the following reaction to be equally slow? Why?

C   D

28.38 Suggest a mechanism for each of the following transformations. Some involve pericyclic reactions only; others involve pericyclic reactions as well as other steps. Invoke the appropriate selection rules to explain any stereochemical features observed.

(a)

(b)

(c)

28.39 Classify the following sigmatropic reaction, give the curved-arrow notation, and show that the stereochemistry is that expected for a thermal concerted reaction. (This reaction, discovered by Prof. J. A. Berson at Yale University, was the first example of this type of pericyclic reaction.)

28.40 When 1,3,5-cyclooctatriene, *A*, is heated to 80–100 °C, it comes to equilibrium with an isomeric compound *B*. Treatment of the mixture of *A* and *B* with $CH_3O_2C-C\equiv C-CO_2CH_3$ gives a compound *C*, which, when heated to 200 °C for 20 minutes, gives dimethyl phthalate and cyclobutene. Identify compounds *B* and *C*, and explain what reactions have occurred.

dimethyl phthalate

28.41 The reaction in Fig. P28.41 occurs as a sequence of two pericyclic reactions. Identify the intermediate *A*, and describe the two reactions.

28.42 Certain black bugs of the order Hemiptera, generally observed in the tropical regions of India immediately after the rainy season, give off a characteristic nauseating smell whenever they are disturbed or crushed. Substance *A*, the compound causing the odor, can be obtained either by extracting the bugs with petroleum ether (which no doubt disturbs them greatly), or it can be prepared by heating the compound below at 170–180 °C for a short time. Give the structure of compound *A*.

28.43 When each of the compounds shown in Fig. P28.43 is heated in the presence of maleic anhydride, an intermediate is trapped as a Diels–Alder adduct. What is the intermediate formed in each reaction, and how is it formed from the starting material?

α-phellandrene

Figure P28.41

28.44 When 2-methyl-2-propenal is treated with allylmagnesium chloride (H₂C=CH—CH₂—MgCl) in ether, then with dilute aqueous acid, a compound *A* is obtained, which, when heated strongly, yields an aldehyde *B*. Give the structures of compounds *A* and *B*.

28.45 A compound *A* ($C_{11}H_{14}O_3$) is insoluble in base and gives an isomeric compound *B* when heated strongly. Compound *B* gives a sodium salt when treated with NaOH. Treatment of the sodium salt of *B* with dimethyl sulfate gives a new compound *C* ($C_{12}H_{16}O_3$) that is identical in all respects to a natural product *elemicin*. Ozonolysis of elemicin followed by oxidation gives the carboxylic acid *D*. Propose structures for compounds *A*, *B*, and *C*.

$$OCH_3$$

H₃CO⟍ ⟍OCH₃

CH₂CO₂H

D

28.46 Using phenol and any other reagents as starting materials, outline a synthesis of each of the following compounds.

(a) 1-ethoxy-2-propylbenzene

(b)   OCH₃

CH₂CH₂CO₂H

28.47 Each of the following reactions involves a sequence of two pericyclic reactions. Identify the intermediate *X* or *Y*

involved in each reaction, and describe the pericyclic reactions involved.

(a)

$$\xrightarrow[140\ °C]{heat} X \rightleftharpoons$$

(b)

$$\xrightarrow{light} Y \xrightarrow{light} \boxed{\ } + CO_2$$

28.48 An all-suprafacial [3,3] sigmatropic rearrangement could in principle take place through either a chair-like or a boat-like transition state:

chair-like
transition state

boat-like
transition state

(a) According to the following results, which of these two transition states is preferred?

Me H   H   Me

$$\xrightarrow{180\ °C}$$ Me

Me H   H   H

(mostly)

Me H   H   H

$$\xrightarrow{180\ °C}$$ Me   Me

Me H   H   H

(mostly)

(a)

$$\xrightarrow[heat]{maleic\ anhydride}$$

(b)

$$\xrightarrow[heat]{maleic\ anhydride}$$

(mixture of stereoisomers)

(c)

$$\xrightarrow[250\ °C]{maleic\ anhydride}$$

Figure P28.43

(b) When the terpene germacrone is distilled under reduced pressure at 165 °C, it is transformed to β-elemenone by a Cope rearrangement. Deduce the structure of germacrone, including its stereochemistry.

germacrone $\xrightarrow{165\ °C}$

β-elemenone

28.49 Ions as well as neutral molecules undergo pericyclic reactions. Classify the pericyclic reactions of the cation involved in the transformation shown in Fig. P28.49. Tell whether the methyl groups are cis or trans and why.

28.50 (a) The transformation shown in Fig. P28.50, which involves a sequence of two pericyclic reactions, was used as a key step in a synthesis of the sex hormone estrone. Identify the unstable intermediate A, and give the mechanism for both its formation and its subsequent reaction.

(b) Show how the product of part (a) can be converted into estrone.

estrone

28.51 In 1985, two researchers at the University of California, Riverside, carried out the reaction given in Fig. P28.51. The equilibrium mixture contained compound A (22%), a *single* stereoisomer of B (47%), and a *single* stereoisomer of C (31%). Predict the stereochemistry of compounds B and C at the carbon marked with the asterisk (*). Explain your prediction.

28.52 An interesting heterocyclic compound C was prepared and trapped by the sequence of reactions given in Fig. P28.52. Give the structure of all missing compounds, and explain what happens in each reaction.

Figure P28.49

Figure P28.50

Figure P28.51

28.53 Anticipating the isolation of the potentially aromatic hydrocarbon *B*, a group of chemists irradiated compound *A* with ultraviolet light. Compound *C* was obtained as a product instead of *B*.

(a) Explain why compound *B* might be expected to be unstable in spite of its cyclic array of $4n + 2$ π electrons.

(b) Explain why the formation of compound *B* is allowed by the pericyclic selection rules.

(c) Account for the formation of the observed product *C*.

28.54 (a) What type of pericyclic reaction is required to form benzene from Dewar benzene?

Dewar benzene → **benzene**

(b) Explain why Dewar benzene, although a very unstable molecule, is not readily transformed to benzene. (Although Dewar benzene forms benzene when heated, this reaction requires a surprisingly high temperature and is believed not to be concerted.)

28.55 (a) Identify the hydrocarbon *B* and the intermediate *A* (both with the formulas $C_{11}H_{10}$) in the following reaction sequence. Compound *B* is formed spontaneously from *A* in a pericyclic reaction.

(b) The proton NMR spectrum of *B* consists of a complex absorption at δ 7.1 (8*H*) and a singlet at δ (−0.5) (2*H*). Account for the absorption at a negative chemical shift; that is, to which protons does this absorption correspond, and why do they absorb at a negative chemical shift?

Figure P28.52

Appendices

APPENDIX I. SUBSTITUTIVE NOMENCLATURE OF ORGANIC COMPOUNDS

The substitutive name of an organic compound is based on its *principal group* and *principal chain*.

The *principal group* is *assigned* according to the following priorities:

$$-\overset{\overset{O}{\|}}{C}-OH \text{ (carboxylic acid)} > -\overset{\overset{O}{\|}}{C}-O-\overset{\overset{O}{\|}}{C}- \text{ (anhydride)} > -\overset{\overset{O}{\|}}{C}-OR \text{ (ester)} >$$

$$-\overset{\overset{O}{\|}}{C}-X \text{ (acid halide)} > -\overset{\overset{O}{\|}}{C}-NR_2 \text{ (amide)} > -C \equiv N \text{ (nitrile)} > -\overset{\overset{O}{\|}}{C}-H \text{ (aldehyde)} >$$

$$-\overset{\overset{O}{\|}}{C}- \text{ (ketone)} > -OH \text{ (alcohol, phenol)} > -SH \text{ (thiol)} > -NR_2 \text{ (amine)}$$

The *principal chain* is *identified* by applying the following criteria in order until a decision can be made.

1. Maximum number of substituents corresponding to the principal group
2. Maximum number of double and triple bonds considered together
3. Maximum length
4. Maximum number of substituents cited as prefixes

A *principal chain* is *numbered* by applying the following criteria in order until there is no ambiguity. Where multiple numbers are possible, comparisons are made at the first point of difference.

1. Lowest number for the principal group cited as a suffix—that is, the group on which the name is based
2. Lowest numbers for multiple bonds, with double bonds having priority over triple bonds in case of ambiguity
3. Lowest numbers for other substituents, taking into account the "first point of difference" rule (p. 64, rule 8)
4. Lowest number for the substituent named as a prefix that is cited first in the name

The *name* is *constructed* starting with the hydrocarbon corresponding to the principal chain.

1. Cite the principal group by its suffix and number; its number is the last one cited in the name.
2. If there is no principal group, name the compound as a substituted hydrocarbon.
3. Cite the names and numbers of the other substituents in alphabetical order at the beginning of the name.

These lists cover most of the cases cited in the text. (See Study Problems 8.1–8.3, pp. 330–333, for illustrations.) For a more complete discussion of nomenclature, see *Nomenclature of Organic Chemistry* (1979 Edition), by the International Union of Pure and Applied Chemistry, published by Pergamon Press.

In 1993, the IUPAC released *A Guide to IUPAC Nomenclature of Organic Compounds Recommendations 1993,* by R. Panico, W. H. Powell, and Jean-Claude Richer (senior editor), Blackwell Science. This publication advocated one major change that affects the nomenclature of relatively simple compounds. This change involves the way that principal groups are cited. The *1993 Recommendations* cite the principal group or multiple bond position with a number preceding the suffix itself, whereas the *1979 Recommendations* (followed in this text) cite the principal group or multiple bond position with a number preceding the hydrocarbon name. These differences are best illustrated by example.

	$H_2C\!\!=\!\!CHCH_2CH_2CH_3$	$HOCH_2CH_2CH_2CH_2CH_3$	$HOCH_2CH_2CH_2CH\!\!=\!\!CH_2$
1979 Recommendations:	**1-pentene**	**1-pentanol**	**4-penten-1-ol**
1993 Recommendations:	**pent-1-ene**	**pentan-1-ol**	**pent-4-en-1-ol**

The *1993 Recommendations* have not yet been generally adopted. Thus, names that adhere to either set of recommendations are acceptable.

APPENDIX II. INFRARED ABSORPTIONS OF ORGANIC COMPOUNDS

This table presents a summary of the important infrared absorptions discussed in this text. For more detailed tables, the reader may wish to consult more specialized texts, such as *Infrared Absorption Spectroscopy,* by Koji Nakanishi and Philippa H. Solomon, San Francisco: Holden-Day, 1977; or *Organic Structure Analysis,* by Philip Crews, Jaime Rodríguez, and Marcel Jaspars, 1998, Oxford University Press, Chapter 8.

Type of absorption	Frequency, cm^{-1} (Intensity)*	Comment
Alkanes		
C—H stretch	2850–3000 (m)	occurs in all compounds with aliphatic C—H bonds
Alkenes		
C=C stretch —CH=CH$_2$	1640 (m)	
\C=CH$_2$/	1655 (m)	
others	1660–1675 (w)	not observed if alkene is symmetrical
=C—H stretch	3000–3100 (m)	
=C—H bend		
—CH=CH$_2$	910–990 (s)	
\C=CH$_2$/	890 (s)	
H\C=C/ /H (cis)	960–980 (s)	
H\C=C/H	675–730 (s)	position is highly variable
H\C=C/	800–840 (s)	
Alcohols and Phenols		
O—H stretch	3200–3400 (s)	
C—O stretch	1050–1250 (s)	also present in other compounds with C—O bonds: ethers, esters, etc.
Alkynes		
C≡C stretch	2100–2200 (m)	not present or weak in many internal alkynes
≡C—H stretch	3300 (s)	present in 1-alkynes only
Aromatic Compounds		
C=C stretch	1500, 1600 (s)	two absorptions
C—H bend	650–750 (s)	
overtone	1660–2000 (w)	

* (s) = strong; (m) = medium; (w) = weak.

(Table continues)

Type of absorption	Frequency, cm^{-1} (Intensity)*	Comment
Aldehydes		
C=O stretch		
ordinary	1720–1725 (s)	
α,β-unsaturated	1680–1690 (s)	
benzaldehydes	1700 (s)	
C—H stretch	2720 (m)	
Ketones		
C=O stretch		
ordinary	1710–1715 (s)	increases with decreasing ring size (Table 21.3, p. 1054)
α,β-unsaturated	1670–1680 (s)	
aryl ketones	1680–1690 (s)	
Carboxylic Acids		
C=O stretch		
ordinary	1710 (s)	
benzoic acids	1680–1690 (s)	
O—H stretch	2400–3000 (s)	very broad
Esters and Lactones		
C=O stretch	1735–1745 (s)	increases with decreasing ring size (Table 21.3, p. 1054)
Acid Chlorides		
C=O stretch	1800 (s)	a second weaker band sometimes observed at 1700–1750
Anhydrides		
C=O stretch	1760, 1820 (s)	two bands; increases with decreasing ring size in cyclic anhydrides
Amides and Lactams		
C=O stretch	1650–1655 (s)	increases with decreasing ring size (Table 21.3, p. 1054)
N—H bend	1640 (s)	
N—H stretch	3200–3400 (m)	doublet absorption observed for some primary amides
Nitriles		
C≡N stretch	2200–2250 (m)	
Amines		
N—H stretch	3200–3375 (m)	several absorptions sometimes observed, especially for primary amines

* (s) = strong; (m) = medium; (w) = weak.

APPENDIX III. PROTON NMR CHEMICAL SHIFTS IN ORGANIC COMPOUNDS

This appendix is subdivided into a table of chemical shifts for protons that are *part of* functional groups and a table of chemical shifts for protons that are *adjacent to* functional groups.

A. Protons within Functional Groups

Group	Chemical shift, ppm	Group	Chemical shift, ppm
—C—C—H	0.7–1.5	O ‖ —C—H	9–11
C=C (with H)	4.6–5.7	O ‖ —C—N—H	7.5–9.5
—O—H	varies with solvent and with acidity of O—H	—C—NH—	0.5–1.5
—C≡C—H	1.7–2.5		
(benzene ring with H)	6.5–8.5	(benzene ring)—NH—	2.5–3.5

B. Protons Adjacent to Functional Groups

In this table, a range of chemical shifts is given for protons in the general environment

$$H-\overset{|}{\underset{|}{C}}-G$$

in which G is a group listed in column 1, and the two other bonds are to carbon or hydrogen. The remaining columns give the approximate chemical shifts for methyl protons (H_3C—G),

methylene protons (—CH_2—G), and methine protons (—$\overset{|}{C}H$—G), respectively. The shifts in the following table are typical; some variation with structure of a few tenths of a ppm can be expected. The chemical shifts of methine protons are usually further downfield than those of methylene protons, which are further downfield than methyl protons. Each additional carbon substitution increases the chemical shift by 0.3–1.0 ppm.

Group, G	Chemical shift of H_3C—G, ppm	Chemical shift of —CH_2—G, ppm	Chemical shift of —CH—G, ppm
—H	0.2		
—CR_3	0.9	1.2	1.4
—F	4.3	4.5	4.8
—Cl	3.0	3.4	4.0
—Br	2.7	3.4	4.1
—I	2.2	3.2	4.2
—CR=CR_2 (R = H, alkyl)	1.8	2.0	2.3
—C≡CR (R = alkyl, H)	1.8	2.2	2.8
(phenyl)	2.3	2.6	2.8
RO— (R = alkyl, H)	3.3 (R = alkyl) 3.5 (R = H)	3.4	3.6
RO— (R = aryl)	3.7	4.0	4.6
RS— (R = alkyl, H)	2.4	2.6	3.0
R—C(=O)—	2.1 (R = alkyl) 2.6 (R = aryl)	2.4 (R = alkyl) 2.7 (R = aryl)	2.6 (R = alkyl) 3.4 (R = aryl)
RO—C(=O)— (R = alkyl, H)	2.1	2.2	2.5
R—C(=O)—O— (R = alkyl, H)	3.6 (R = alkyl) 3.8 (R = aryl)	4.1 (R = alkyl, aryl)	5.0 (R = alkyl, aryl)
R_2N—C(=O)— (R = alkyl, H)	2.0	2.2	2.4
R—C(=O)—N—R (R = alkyl, H)	2.8	3.4	3.8
—NR_2 (R = alkyl, H)	2.2	2.4	2.8
—N(R)(phenyl) (R = alkyl, H)	2.6	3.1	3.6
N≡C—	2.0	2.4	2.9

APPENDIX IV. ^{13}C NMR CHEMICAL SHIFTS IN ORGANIC COMPOUNDS

This section is divided into a table of chemical shifts for carbons within functional groups and a table of chemical shifts for alkyl carbons adjacent to functional groups. A typical range of shifts is given for each case.

A. Chemical Shifts of Carbons within Functional Groups

Group	Chemical shift range, ppm
—CH$_3$	8–23
—CH$_2$—	20–30
—CH—	21–33
—C—	17–29
C=C	105–150*
—C≡C—	66–93*
benzene ring —R	125–150*
ketone (C=O)	200–220
—C(=O)—O—R R = H, alkyl	170–180
—C(=O)—N(R)—R R = H, alkyl	165–175
—C≡N	110–120

* Alkyl substitution typically increases chemical shift.

B. Chemical Shifts of Carbons Adjacent to Functional Groups

In most cases, alkyl substitution on the carbon increases chemical shift. Methyl carbons will have shifts in the low end of the range; tertiary and quaternary carbons will have shifts in the upper end of the range.

Group G	Chemical shift of carbon in G—C—
R₂C=CR—	14–40
HC≡C—	18–28
⬡—	29–45
F—	83–91
Cl—	44–68
Br—	32–65
I—	5–42
HO—	62–70
RO— R = alkyl, H	70–79
R—C(O)— R = alkyl, H	43–50
RO—C(O)— R = alkyl, H	33–44
R₂N— R = alkyl, H	41–51 (R = H) 53–60 (R = alkyl)
N≡C—	16–28

APPENDIX V. SUMMARY OF SYNTHETIC METHODS

The following methods are listed in order of their occurrence in the text; the section reference follows each reaction in parentheses. Thus, a review at any point in the text is possible by considering the methods listed for earlier sections.

Don't forget that in many cases, a method can be applied to compounds containing more than one functional group. Thus, catalytic hydrogenation can be used to convert phenols into alcohols, but it is listed under "Synthesis of Alkanes and Aromatic Hydrocarbons" because the actual transformation is the formation of —CH₂—CH₂— groups from —CH=CH— groups; the presence of the —OH group is incidental.

Reaction summaries for each chapter are found in the Study Guide.

A. Synthesis of Alkanes and Aromatic Hydrocarbons

1. Catalytic hydrogenation of alkenes (4.9A)
2. Protonolysis of Grignard or related reagents (9.8C)
3. Cyclopropane formation by the addition of carbenoids to alkenes (Simmons–Smith reaction; 9.9B)
4. Catalytic hydrogenation of alkynes (14.6A)
5. Friedel–Crafts alkylation of aromatic compounds (16.4E)
6. Catalytic hydrogenation of aromatic compounds (16.6)
7. Stille reaction of aryl triflates and aryl- or alkylstannanes to form substituted aromatic hydrocarbons (18.10B)
8. Wolff–Kishner or Clemmensen reductions of aldehydes or ketones (19.12)
9. Reaction of aryldiazonium salts with hypophosphorous acid (23.10A)

B. Synthesis of Alkenes

1. β-Elimination reactions of alkyl halides or sulfonates (9.5, 10.4A, 17.3B)
2. Acid-catalyzed dehydration of alcohols (10.2)
3. Catalytic hydrogenation of alkynes (gives cis-alkenes when used with internal alkynes; 14.6A)
4. Reduction of alkynes with alkali metals in liquid ammonia (gives trans-alkenes when used with internal alkenes; 14.6B)
5. Diels–Alder reactions of dienes and alkenes to give cyclic alkenes (15.3, 28.3)
6. Heck reaction of aryl halides and alkenes to give aryl-substituted alkenes (18.6A)
7. Suzuki coupling of aryl or vinylic halides with aryl or vinylic boronic acids (18.6B)
8. Alkene metathesis (18.6C)
9. Wittig reaction of aldehydes and ketones (19.13)
10. Aldol condensation reactions of aldehydes or ketones to give α,β-unsaturated aldehydes or ketones (22.4)
11. Hofmann elimination of quaternary ammonium hydroxides (23.8)

C. Synthesis of Alkynes

1. Alkylation of acetylenic anions with alkyl halides or sulfonates (14.7B)
2. β-Elimination reactions of alkyl dihalides or vinylic halides (18.2)

D. Synthesis of Alkyl, Aryl, and Vinylic Halides

1. Addition of hydrogen halides to alkenes (4.7, 15.4A)
2. Addition of halogens to alkenes to give vicinal dihalides (5.2)
3. Peroxide-promoted addition of HBr to alkenes (5.6)
4. Addition of halogens or HBr to alkynes (14.4)
5. Synthesis of dihalocyclopropanes by the addition of dihalomethylene to alkenes (9.9A)
6. Reaction of alcohols with HBr, thionyl chloride, or triphenylphosphine dibromide (10.3, 10.4D)
7. Reaction of sulfonate esters or other alkyl halides with halide ions (10.4A, 17.4)
8. Halogenation of aromatic compounds (16.4A)
9. Allylic and benzylic bromination of alkenes or aromatic hydrocarbons (17.2)
10. α-Halogenation of aldehydes, ketones, or carboxylic acids (22.3A,C)
11. Synthesis of aryl halides by the reaction of cuprous chloride, cuprous bromide, or potassium iodide with aryldiazonium salts (Sandmeyer and related reactions; 23.10A)

E. Synthesis of Grignard Reagents and Related Organometallic Compounds

1. Reaction of alkyl or aryl halides with metals (9.8A)
2. Preparation of lithium dialkylcuprates by the reaction of alkyllithium reagents with cuprous halides (9.8B)
3. Preparation of acetylenic Grignard reagents by the metal–hydrogen exchange (14.7A)
4. Preparation of alkyl- and arylstannanes by the reaction of Grignard reagents with trialkylstannyl chlorides (18.10B)

F. Synthesis of Alcohols and Phenols

(Syntheses apply only to alcohols unless noted otherwise.)

1. Acid-catalyzed hydration of alkenes (used industrially, but generally not a good laboratory method; 4.9B)
2. Synthesis of halohydrins from alkenes (5.2B)
3. Oxymercuration–reduction of alkenes (5.4A)
4. Hydroboration–oxidation of alkenes (5.4B)
5. Ring-opening reactions of epoxides (11.5A,B)
6. Reaction of ethylene oxide with Grignard reagents (11.5C)
7. Reaction of epoxides with lithium organocuprates (11.5C)
8. Reduction of aldehydes or ketones (19.8, 22.10, 24.8D)
9. Reaction of aldehydes or ketones with Grignard or related reagents (19.9, 22.11A)
10. Reduction of carboxylic acids to primary alcohols (20.10)
11. Reduction of esters to primary alcohols (21.9A)
12. Reaction of esters with Grignard or related reagents (21.10A)
13. Aldol addition reactions of aldehydes or some ketones to give β-hydroxy aldehydes or ketones (22.4)
14. Reaction of diazonium salts with water to give phenols (23.10A)
15. Synthesis of phenols by the Claisen rearrangement of allylic aryl ethers (28.4B)

G. Synthesis of Glycols

1. Acid-catalyzed hydrolysis of epoxides (11.5B)
2. Reaction of alkenes with osmium tetroxide or alkaline potassium permanganate (11.6A)

H. Synthesis of Ethers, Acetals, and Sulfides

1. Alkylation of alkoxides, phenoxides, or thiolates with alkyl halides or alkyl sulfonates (Williamson synthesis; 11.2A, 18.8B, 24.7)
2. Alkoxymercuration–reduction of alkenes (11.2B)
3. Acid-catalyzed dehydration of alcohols (11.2C)
4. Acid-catalyzed addition of alcohols to alkenes (11.2C)
5. Reaction of epoxides with alkoxides and alcohols (11.5A,B)
6. Reaction of 2- and 4-nitroaryl halides with alkoxides (18.4)
7. Acetal formation by the acid-catalyzed reaction of alcohols with aldehydes or ketones (19.10A, 24.6)
8. Conjugate addition of thiols to α,β-unsaturated carbonyl compounds (22.9A)

I. Synthesis of Epoxides

1. Oxidation of alkenes with peroxycarboxylic acids (11.3A)
2. Cyclization of halohydrins (11.3B)
3. Asymmetric epoxidation of allylic alcohols (11.11)

J. Synthesis of Disulfides

1. Oxidation of thiols (10.10, 27.8)

K. Synthesis of Aldehydes

1. Ozonolysis of alkenes (of limited utility because carbon–carbon bonds are broken; 5.5)
2. Oxidation of primary alcohols (10.7A)
3. Oxidative cleavage of glycols (of limited utility because carbon–carbon bonds are broken; 11.65B, 24.8C)
4. Hydroboration–oxidation of alkynes (14.5B)
5. Oxidation of allylic and benzylic alcohols with MnO_2 (Sec. 17.5A)
6. Reduction of acid chlorides (21.9D)
7. Aldol addition reactions of aldehydes to give β-hydroxy aldehydes (22.4)
8. Aldol condensation reactions of aldehydes to give α,β-unsaturated aldehydes (22.4)
9. Synthesis of aldoses from other aldoses by the Kiliani–Fischer synthesis (24.9) and the Ruff degradation (24.10B)

L. Synthesis of Ketones

1. Ozonolysis of alkenes (of limited utility because carbon–carbon bonds are broken; 5.5)
2. Oxidation of secondary alcohols (10.7A)
3. Oxidative cleavage of glycols (of limited utility because carbon–carbon bonds are broken; 11.6B)
4. Mercuric-ion catalyzed hydration of alkynes (14.5A)
5. Friedel–Crafts acylation of aromatic compounds (16.4F)
6. Oxidation of phenols to quinones (18.8A)
7. Reaction of acid chlorides with lithium dialkylcuprates (21.10B)
8. Aldol condensation reactions of ketones to give α,β-unsaturated ketones (22.4)
9. Claisen and Dieckmann condensation reactions of esters to give β-keto esters (22.6A, B)
10. Crossed Claisen condensation reactions of esters to give β-diketones (22.6C)
11. Acetoacetic ester synthesis (22.8C)
12. Conjugate addition reactions of α,β-unsaturated ketones (22.9), including the addition of lithium dialkylcuprate reagents (22.11B)

M. Synthesis of Sulfoxides and Sulfones

1. Oxidation of sulfides (11.9)

N. Synthesis of Carboxylic and Sulfonic Acids

(Syntheses apply only to carboxylic acids unless noted otherwise.)

1. Ozonolysis of alkenes (of limited utility because carbon–carbon bonds are broken; 5.5)
2. Oxidation of primary alcohols (10.7A)
3. Oxidation of thiols to sulfonic acids (10.10)
4. Sulfonation of aromatic compounds to give arylsulfonic acids (16.4D)
5. Side-chain oxidation of alkylbenzenes (17.5C)
6. Oxidation of aldehydes (19.14)

7. Reaction of Grignard or related reagents with carbon dioxide (20.6)

8. Hydrolysis of carboxylic acid derivatives, especially nitriles (21.7, 21.11, 25.5A, 27.4C)

9. Haloform reaction of methyl ketones (of limited utility because carbon–carbon bonds are broken; 22.3B)

10. Malonic ester synthesis (22.8A, 27.4B)

11. Strecker synthesis of α-amino acids (27.4C)

O. Synthesis of Esters

1. Reaction of alcohols and phenols with sulfonyl chlorides (for sulfonate esters; 10.4A, 18.10B)

2. Acid-catalyzed esterification of carboxylic acids with primary or secondary alcohols (20.8A, 24.7, 27.5)

3. Alkylation of carboxylic acids with diazomethane (20.8B)

4. Alkylation of carboxylate salts with alkyl halides (20.8B)

5. Reaction of acid chlorides, anhydrides, or esters with alcohols and phenols (21.8, 24.7)

6. Claisen and Dieckmann condensation reactions of esters to give β-keto esters (22.6A,B)

7. Alkylation of ester enolate ions; includes malonic ester synthesis, acetoacetic ester synthesis, and direct alkylation (22.8)

8. Conjugate addition reactions of α,β-unsaturated esters (22.9, 22.11B)

P. Synthesis of Anhydrides

1. Reaction of carboxylic acids with dehydrating agents (20.9B)

2. Reaction of acid chlorides with carboxylate salts (21.8A)

Q. Synthesis of Acid Chlorides

1. Reaction of carboxylic or sulfonic acids with thionyl chloride, phosphorus pentachloride, or related reagents (20.9A)

2. Synthesis of sulfonyl halides by chlorosulfonation of aromatic compounds (20.9A)

R. Synthesis of Amides

1. Reaction of acid chlorides, anhydrides, or esters with amines (21.8, 23.7C, 27.5)

2. Condensation of amines and carboxylic acids with diisopropylcarbodiimide (27.6)

S. Synthesis of Nitriles

1. Formation of cyanohydrins from aldehydes and some ketones (19.7A,B, 24.9)

2. Reaction of alkyl halides or alkyl sulfonates with cyanide ion (21.11)

3. Conjugate addition of cyanide ion to α,β-unsaturated carbonyl compounds (22.9A)

4. Reaction of cuprous cyanide with aryldiazonium salts (23.10A)

T. Synthesis of Amines

1. Reduction of amides (21.9B)

2. Reduction of nitriles to primary amines (21.9C)

3. Direct alkylation of ammonia or amines (of limited utility because of the possibility of over-alkylation; 23.7A, 27.4A)

4. Reductive amination of aldehydes and ketones (23.7B)

   5. Aromatic substitution reactions of aniline derivatives (23.9)
   6. Gabriel and Staudinger syntheses of primary amines (23.11A)
   7. Reduction of nitro compounds (23.11B)
   8. Pd-catalyzed amination of aryl halides and triflates (23.11C)
   9. Curtius and Hofmann rearrangements (23.11D)
   10. Strecker synthesis of α-amino acids (27.4C)

U. Synthesis of Nitro Compounds

   1. Nitration of aromatic compounds (16.4C, 18.9)

APPENDIX VI. REACTIONS USED TO FORM CARBON–CARBON BONDS

Reactions that form carbon–carbon bonds have central importance in organic chemistry, because these reactions can be used to form carbon chains or rings. These reactions are listed in the order in which they are discussed in the text. The section reference follows each reaction in parentheses.

   1. Cyclopropane formation by addition of carbenes or carbenoids to alkenes (9.9)
   2. Reaction of Grignard reagents with ethylene oxide (11.5C)
   3. Reaction of epoxides with lithium organocuprates (11.5C)
   4. Reaction of acetylenic anions with alkyl halides or sulfonates (14.7B)
   5. Diels–Alder reactions (15.3, 28.3)
   6. Friedel–Crafts alkylation (16.4E) and acylation reactions (16.4F)
   7. The Heck reaction of alkenes with aryl halides (18.6A)
   8. Suzuki coupling of aryl or vinylic halides with aryl or vinylic boronic acids (18.6B)
   9. Alkene metathesis (18.6C)
   10. The Stille reaction of organostannanes with aryl triflates (18.10B)
   11. Cyanohydrin formation (19.7, 24.9, 27.4C)
   12. Reaction of Grignard and related reagents with aldehydes and ketones (19.9)
   13. Wittig alkene synthesis (19.13)
   14. Reaction of Grignard and related reagents with carbon dioxide (20.6)
   15. Reaction of Grignard and related reagents with esters (21.10A)
   16. Reaction of lithium dialkylcuprates with acid chlorides (21.10B)
   17. Reaction of cyanide ion with alkyl halides or sulfonates (21.11)
   18. Aldol addition and condensation reactions (22.4)
   19. Claisen and related condensation reactions (22.6)
   20. Malonic ester synthesis (22.8A, 27.4B)
   21. Alkylation of ester enolate ions with alkyl halides or tosylates (22.8B)
   22. Acetoacetic ester synthesis (22.8C)
   23. Conjugate-addition reactions of cyanide ion (22.9A) or enolate ions (22.9C) to α,β-unsaturated carbonyl compounds
   24. Conjugate addition of lithium dialkylcuprate reagents to α,β-unsaturated carbonyl compounds (22.11B)
   25. Reaction of aryldiazonium salts with cuprous cyanide (23.10A)
   26. Formation of rings by electrocyclic reactions (28.2)
   27. Claisen rearrangement (28.4B)

APPENDIX VII. TYPICAL ACIDITIES AND BASICITIES OF ORGANIC FUNCTIONAL GROUPS

A. Acidities of Groups That Ionize to Give Anionic Conjugate Bases

Functional group	Structure*	Structure of conjugate base	Typical pK_a
sulfonic acid	$R-\overset{\overset{O}{\|\|}}{\underset{\underset{O}{\|\|}}{S}}-O-H$	$R-\overset{\overset{O}{\|\|}}{\underset{\underset{O}{\|\|}}{S}}-O^-$	−2.8 (strong acid)
carboxylic acid	$R-\overset{\overset{O}{\|\|}}{C}-O-H$	$R-\overset{\overset{O}{\|\|}}{C}-O^-$	4–5
phenol	X—⟨ring⟩—O—H†	X—⟨ring⟩—O⁻†	9–11
thiol	$R-S-H$	$R-S^-$	10–12
sulfonamide	$R-\overset{\overset{O}{\|\|}}{\underset{\underset{O}{\|\|}}{S}}-\overset{\overset{H}{\|}}{N}-R$	$R-\overset{\overset{O}{\|\|}}{\underset{\underset{O}{\|\|}}{S}}-\bar{N}-R$	10
amide	$R-\overset{\overset{O}{\|\|}}{C}-\overset{\overset{H}{\|}}{N}-R$	$R-\overset{\overset{O}{\|\|}}{C}-\bar{N}-R$	15–17
alcohol	$R-O-H$	$R-O^-$	15–19
aldehyde, ketone	$R-\overset{\overset{O}{\|\|}}{C}-\overset{\overset{H}{\|}}{C}R_2$	$R-\overset{\overset{O}{\|\|}}{C}-\bar{C}R_2$	17–20
ester	$R_2\overset{\overset{H}{\|}}{C}-\overset{\overset{O}{\|\|}}{C}-OR$	$R_2\bar{C}-\overset{\overset{O}{\|\|}}{C}-OR$	25
alkyne	$R-C\equiv C-H$	$R-C\equiv C^-$	25
nitrile	$R_2\overset{\overset{H}{\|}}{C}-C\equiv N$	$R_2\bar{C}-C\equiv N$	25
amine	R_2N-H	R_2N^-	32
alkene	$\underset{R}{\overset{R}{C}}=\underset{R}{\overset{H}{C}}$	$\underset{R}{\overset{R}{C}}=\underset{R}{\overset{}{C^-}}$	42
benzylic alkyl group	$Ar-\overset{\overset{H}{\|}}{C}R_2$	$Ar-\bar{C}R_2$	42
alkane	R_3C-H	R_3C^-	55–60

* In the structures, R = alkyl or H. The acidic hydrogen is shown in red.

† X = a general ring substituent group.

B. Basicities of Groups That Protonate to Give Cationic Conjugate Acids

One should be careful to distinguish between the behavior of a particular functional group as an acid and the same functional group as a base. For example, when an alcohol acid acts as an acid, it loses the RO—H proton to form an alkoxide (see the table in part A of this section.) When it acts as a base, it *gains* a proton to form $R\overset{+}{O}H_2$. These are very different processes with different pK_a values. When we discuss the *acidity* of an alcohol, the relevant pK_a is that for the alcohol itself (see the table in part A). This same pK_a describes the *basicity* of the alkoxide, RO^-, which is the conjugate base of the alcohol. When we discuss the *basicity* of the alcohol itself, the relevant pK_a is the value for the acidity of $R\overset{+}{O}H_2$, given in the following table.

Functional group	Structure*,‡	Structure of conjugate acid*,‡	Typical conjugate-acid pK_a
alkylamine	R_3N	$R3\overset{+}{N}$—H	9–11
pyridine			5
aromatic amine			4–5
amide			–1
alcohol, ether	R—O—R		–2 to –3
ester, carboxylic acid			–6
phenol, aromatic ether†			–6 to –7
thiol, sulfide	R—S—R		–6 to –7
aldehyde, ketone			–7

* In the structures, R = alkyl or H. In the conjugate acid, the acidic hydrogen is shown in red.

‡ X = a general ring substituent group.

† A phenol or aromatic ether can be protonated on a ring carbon if the resulting carbocation can be strongly stabilized by the substituent groups.

(Table continues)

Functional group	Structure*	Structure of conjugate acid	Typical conjugate-acid pK_a
alkene			−8 to −10
nitrile	$R—C≡N$		−10

* In the structures, R = alkyl or H.

Credits

Page 600 Courtesy of Dr. Karl Wood, Purdue Mass Spectrometry Center

Page 612 (Figure 13.1), page 628 (Figure 13.6), page 631 (Figure 13.7), page 642 (Figure 13.13), page 644 (Figure 13.14), page 648 (Figure 13.16), page 649 (Figure 13.18), page 652 (Figure 13.19), page 659 (Figure 13.23), page 673 (Figure P13.40), page 674 (Figure P13.45), page 675 (Figure P13.46), page 676 (Figure P13.51), and page 678 (Figure P13.53) Purdue Magnetic Resonance Laboratory

Page 616 © Monkey Business Images Ltd./ Dreamstime.com

Page 655 (Figure 13.21) Reprinted with permission from F. A. Bovey, *Chemical and Engineering News*, August 30, 1965. Copyright © by The American Chemical Society

Page 660 (Figure 13.24), page 720 (Figure 15.5), page 798 (Figure 16.4), page 937 (Figure P18.59), page 956 (Figure 19.5), and page 957 (Figure 19.6) © Bio-Rad Laboratories, Inc., Informatics Division, Sadtler Software & Databases. All Rights Reserved.

Page 661 (Figure 13.25) DEPT-NMR spectrum of camphor, courtesy of John Kozlowski, Purdue University

Page 666 Photo © Marc Loudon

Page 670 (a) Courtesy, Philips Corporation; (b) Karen Hellerhof

Page 680 (Figure P13.66) Adapted with permission from "Conformational preferences in monosubstituted cyclohexanes determined by nuclear resonance spectroscopy," F. R. Jensen, C. H. Bushweller, and B. H. Beck, *J. Amer. Chem. Soc.* (1969), **91**(2), 344–351, Fig. 1c.

Page 705 © Marc Loudon

Page 724 (Figure 15.8) © Marc Loudon

Page 726 © Blend Images/SuperStock, Inc.

Page 731 (Figure 15.11c) Douglas Gaston IV/DG Media; (Figure 15.12a) Sierra Blakely/Creative Commons; (Figure 15.12b) Riccardo Cassiani-Ingoni/Science Source; (Figure 15.12c) AP Photo/Mayo Clinic/Rex Features

Page 746 © Marius G/Shutterstock.com

Page 777 © Dean Mitchell/iStockphoto.com

Page 778 (Figure 15.21) Based on Coordinates from the Protein Data Bank (PDB). (See p. 1401 credits for full source.)

Page 810 David del Rio/ NOTIMEX/Newscom.com

Page 858 Louvre, Paris/Bridgeman International Art Library

Page 859 © Marc Loudon

Page 923 © Chris Hill/Shutterstock.com

Page 973 Wilco Oelen

Page 993 © Editorial/Alamy

Page 1059 © Oksama Kuzmina/Shutterstock.com

Page 1079 © molekuul.be/Shutterstock.com

Page 1122 © Pictorial Press Ltd./Alamy

Page 1158 © Sherry Yates Young/Shutterstock.com

Page 1210 © Lauri Patterson/iStockphoto.com

Page 1267 Image Asset Management Ltd./ SuperStock, Inc.

Page 1272 © Steve Hamblin/Alamy

Page 1275 © ex0rzist/Shutterstock.com

Page 1295 (Atorvastatin model) Ben Mills

Page 1318 (Figure 25.10) Ben Mills

Page 1360 (Figure 267) WATSON, MOLECULAR BIO GENE, 3rd © 1976. Printed and electronically reproduced by permission of Pearson Education, Inc., New York, New York

Page 1387 © BanksPhotos/iStockphoto.com

Page 1391 © Editorial/Alamy

Page 1401 (Figure 27.3), page 1419 (Figure 27.9), page 1428 (Figure 27.13), page 1431 (Figure 27.17), page 1437 (Figure 27.19), and page 1439 (Figure 27.20) are all based on coordinates obtained from the Protein Data Bank (PDB), operated by the Research Collaboratory for Structural Biology (RCSB), supported by the National Science Foundation (NSF), the National Institute of General Medical Sciences (NIGMS), the Office of Science, Department of Energy (DOE), the National Library of Medicine (NLM), the National Cancer Institute (NCI), the National Center for Research Resources (NCRR), the National Institute of Biomedical Imaging and Bioengi¬neering (NIBIB), the National Institute of Neurological Disorders and Stroke (NINDS), and the National Institute of Diabetes and Digestive and Kidney Diseases (NIDDK). We gratefully acknowledge the assistance of Professor Markus Lill, Purdue University, in obtaining the images in Figures 27.17 and 27.19, and the assistance of Prof. Carol Post in obtaining the images in Figure 27.20.

Page 1410 (bottom) © Radu Razvan Gheorghe/ Dreamstime.com

Page 1413 (Figure 27.7), courtesy of Dr. Karl Wood (Purdue Mass Spectrometry Laboratory), Prof. Animesh Aditya, and the late Prof. Richard Gibbs.

Page 1424 Kallistra Images/SuperStock, Inc.

Page 1428 Courtesy, Duke Photography

Page 1477 © Odua Images/Shutterstock.com

Index

INDEX GUIDE

Leading Greek letters, numbers, stereochemical designations (e.g., *S*, D), and other locators (e.g., *N*-) are ignored in alphabetizing except for cases in which they are the only elements that differentiate entries. Spaces and hyphens (except hyphens associated with nomenclature prefixes and numbers) take alphabetical precedence over letters and numbers. For example, Acetyl-CoA comes before Acetyl group, which comes before Acetylacetone. **Boldfaced** page numbers give the locations of boldfaced definitions in the text. The *italic* letters *f*, *p*, *sp*, *t*, and *v* following page numbers refer to entries found in figures, problems, study problems, tables, and vignettes (sidebars), respectively.

Abietic acid, 787
Absolute configuration, **244**
    determination, 244–246
Absolute stereochemistry. *See* Absolute configuration
Absorbance, 575, **720**
    Beer-Lambert law, **721**
Absorption spectroscopy, **572**
Absorption, in UV spectroscopy
    $n \rightarrow \pi^*$, 956
        forbidden in aldehydes and ketones, 957
    $\pi \rightarrow \pi^*$
        in aldehydes and ketones, 956
        in benzene derivatives, 797–798
        in conjugated alkenes, 722
Abstract, 67*s*
Acceptor. *See* Hydrogen-bond acceptor
Acceptor stem, of tRNA, **1402**
Acebutolol, 86*p*
Acetaldehyde
    aldol addition, 1119–1120
    enolization K_{eq}, 1110
    formation of paraldehyde, 981*v*
    from decarboxylation of pyruvate, 1032–1035
    from ethanol in biological oxidation, 485–488
    MOs of enolate ion, 1105
    reduction to ethanol by NADH in yeast, 975*p*
    structure, 947*f*, 1009*f*
Acetals, **979**
    cyclic
        from reaction of ethylene glycol with aldehydes or ketones, 979

from 4- and 5-hydroxy aldehydes, 981
    glycosides, 1253
    hydrolysis, **980**
    preparation from aldehydes and ketones, 975–981
    protecting groups, 983*sp*
Acetamido group, 1050*t*
    protecting group for amines, 1205*p*
α-Acetamidomalonate. *See* Diethyl α-acetamidomalonate
Acetaminophen, toxicity, 1158–1159*v*
Acetanilide, nitration, 1205*sp*
Acetate, abbreviation, 1045
Acetato, ligand in transition-metal complexes, 891*t*
Acetic acid, 1006*t*
    dimerization, 1010
    esterification with ethanol, standard free energy, 1313–1314
    pK_a, 115, 1013
    structure, 1009*f*
Acetic anhydride
    acetylation of amino acids, 1390
    acetylation of carbohydrates, 1257
    acylating agent for furan, 1334
    as a high-energy compound, 1314–1315
    dehydrating agent for *N*-hydroxydehydropyridines, 1346
    reagent for cyclic anhydride formation, 1027
    standard free energy of hydrolysis, 1315
Acetoacetic acid, decarboxylation, 1030

Acetoacetic ester synthesis, 1149–1152, **1150**
    synthetic analysis, 1150
Acetoacetyl-ACP, intermediate in fatty-acid biosynthesis, 1142–1143
Acetoacetyl-CoA, in biosynthesis of HMG-CoA, 1154
Acetone, 948
    acid-catalyzed aldol condensation, 1122–1123
    aldol addition, 1121
    cyanohydrin in manufacture of poly(methyl methacrylate), 996
    EPM, 952
    industrial preparation, 932–933, 996
    industrial use, 996
    structure, 947*f*, 1009*f*
Acetophenone
    carbonyl stretching absorption, 953
    pK_a, 1104
    preparation by Friedel–Crafts acylation of benzene, 808
Acetoxy group, 1050*t*
    abbreviation, 190
Acetyl-CoA, 1284
    in biosynthesis
        of acetylcholine, 1317p
        of fatty acids, 1141–1145
        of HMG-CoA, 1154–1155
    pK_a of thiol, 1325*p*
    reaction with oxaloacetate, 1181*p*
    standard free energy of hydrolysis, 1316
    structure, 1284*f*
Acetyl group, 948
Acetyl-coenzyme A. *See* Acetyl-CoA

A Periodic Table of the Elements

—— Group numbers recommended by the IUPAC are in gray.

—— Older group numbers. (These are used in the text for calculating formal charge for the A-group elements.)

(Atomic weights in parentheses are for the isotope of longest life.)

The shaded elements will be encountered most frequently in the text.

Transition elements

Period	1 / 1A	2 / 2A	3 / 3B	4 / 4B	5 / 5B	6 / 6B	7 / 7B	8 / 8B	9 / 8B	10 / 8B	11 / 1B	12 / 2B	13 / 3A	14 / 4A	15 / 5A	16 / 6A	17 / 7A	18 / 8A
1	1 H 1.008																	2 He 4.003
2	3 Li 6.941	4 Be 9.012											5 B 10.811	6 C 12.011	7 N 14.007	8 O 15.999	9 F 18.998	10 Ne 20.18
3	11 Na 22.99	12 Mg 24.305											13 Al 26.982	14 Si 28.086	15 P 30.974	16 S 32.065	17 Cl 35.453	18 Ar 39.948
4	19 K 39.098	20 Ca 40.078	21 Sc 44.956	22 Ti 47.867	23 V 50.942	24 Cr 51.996	25 Mn 54.938	26 Fe 55.845	27 Co 58.933	28 Ni 58.693	29 Cu 63.546	30 Zn 65.38	31 Ga 69.723	32 Ge 72.64	33 As 74.922	34 Se 78.96	35 Br 79.904	36 Kr 83.798
5	37 Rb 85.468	38 Sr 87.62	39 Y 88.906	40 Zr 91.224	41 Nb 92.906	42 Mo 95.96	43 Tc (98)	44 Ru 101.07	45 Rh 102.906	46 Pd 106.42	47 Ag 107.868	48 Cd 112.411	49 In 114.818	50 Sn 118.71	51 Sb 121.76	52 Te 127.6	53 I 126.904	54 Xe 131.293
6	55 Cs 132.905	56 Ba 137.327	57 La 138.905	72 Hf 178.49	73 Ta 180.948	74 W 183.84	75 Re 186.21	76 Os 190.23	77 Ir 192.217	78 Pt 195.084	79 Au 196.967	80 Hg 200.59	81 Tl 204.383	82 Pb 207.2	83 Bi 208.98	84 Po (209)	85 At (210)	86 Rn (222)
7	87 Fr (223)	88 Ra (226)	89 Ac (227)	104 Rf (267)	105 Db (268)	106 Sg (271)	107 Bh (272)	108 Hs (277)	109 Mt (276)	110 Ds (281)	111 Rg (280)	112 Cn (285)	113 (284)	114 (287)	115 (288)	116 (293)	117 (293)	118 (294)

Lanthanides

58 Ce 140.116	59 Pr 140.908	60 Nd 144.242	61 Pm (145)	62 Sm 150.36	63 Eu 151.964	64 Gd 157.25	65 Tb 158.925	66 Dy 162.5	67 Ho 164.93	68 Er 167.259	69 Tm 168.934	70 Yb 173.054	71 Lu 174.967

Actinides

90 Th 232.038	91 Pa 231.036	92 U 238.029	93 Np (237)	94 Pu (244)	95 Am (243)	96 Cm (247)	97 Bk (247)	98 Cf (251)	99 Es (252)	100 Fm (257)	101 Md (258)	102 No (259)	103 Lr (262)